Contents—1997 FUNDAMENTALS

Contents—1996 HVAC SYSTEMS AND EQUIPMENT

1999 ASHRAE® HANDBOOK

Heating, Ventilating, and Air-Conditioning APPLICATIONS

Inch-Pound Edition

American Society of Heating, Refrigerating and Air-Conditioning Engineers, Inc.

1791 Tullie Circle, N.E., Atlanta, GA 30329

(404) 636-8400

http://www.ashrae.org

DEDICATED

TO THE ADVANCEMENT OF

THE PROFESSION

AND ITS ALLIED INDUSTRIES

ISBN 1-883413-71-0

CONTENTS

BUILDING OPERATIONS AND MANAGEMENT

GENERAL APPLICATIONS

ADDITIONS AND CORRECTIONS

INDEX

CONTRIBUTORS

In addition to the Technical Committees, the following individuals contributed significantly to this volume. The appropriate chapter numbers follow each contributor's name.

Hugh I. Henderson, Jr. (1)
CDH Energy Corporation

John L. Harrod (2)
C.H. Guernsey & Company

John C. Mentzer (2)
Giffels Associates, Inc.

Lynn F. Werman (2, 5, 6)
HDR Architecture, Inc.

E. Douglas Fitts (3, 5)
St. Louis County Department
of Public Works

Itzhak H. Maor (3, 5)
Fresh Air Solutions

Joseph F. Scolaro (3)
Scolaro Engineering
Consultants

Kenneth W. Cooper (4)
Poolpak Inc.

Ralph Kittler (4)
Dectron Inc. (USA)

Reinhold Kittler (4)
Dectron Inc.

Mark S. Lentz (4, 16, 50)
Lentz Engineering Associates

Richard Hermans (6)
Ellerbe Becket

Robert G. Baker (7)
BBJ Chemical Compounds, Inc.

Paul T. Ninomura (7)
PHS/Indian Health Service

Chris P. Rousseau (7)
Newcomb & Boyd

David C. Allen (8)
D C Allen Inc.

James J. Bushnell (8, 10)
HVAC Consulting Services

Manuel C. Gameiro da Silva (8)
Universidade de Coimbra

Kenneth R. Hesser (8)
LTK Engineering Services

S. Lim Kwon (8)
Thermo King Corporation

Gary J. Prusak (8)
Adtranz, ABB Daimler-Benz
Transportation

Nicholas G. Zupp (8)
General Motors Corporation

David R. Space (9)
Boeing Commercial Airplane
Group

Kenneth L. Waters (9)
Boeing Commercial Airplane
Group

Duane L. Willse (9)
Hamilton Standard

Norm Maxwell (11, 17, 18,
19, 26, 27)
Gayle King Carr & Lynch Inc.

Arthur G. Bendelius (12)
Parsons Brinckerhoff

Kelly A. Giblin (12)
New Jersey Transit

Joseph J. Grella (12)
National Railroad Passenger
Corp. (Amtrak)

Gary W. Kile (12)
Sverdrup Corporation

Sam S. Levy (12)
Parsons Brinckerhoff

Roger A. Lichtenwald (12)
American Warming and
Ventilating

John A. Murphy (12)
Jogram, Inc.

Dharam Pal (12)
The Port Authority of New York
& New Jersey

Joseph D. Rago (12)
National Railroad Passenger
Corp. (Amtrak)

Louis Hartman (13)
Harley Ellington Design

Leonard H. Schwartz (13)
Strategic Planning and
Engineering

Richard A. Evans (14, 16, 23, 25)

Mel J. Crichton (15)
Eli Lilly & Company

Larry J. Hughes (15)
Alpha Engineering

Robert E. Swezey (15)
TG Associates

Derald G. Welles (15, 44)
Tech-HVAC

William A. Kumpf (16)
Campos Engineering

S. Louis Kelter (20)
Kelter & Gilligo, P.C.

William P. Lull (20)
Garrison/Lull Inc.

William B. Rose (20)
University of Illinois

Stefan Michalski (20)
Canadian Conservation Institute

Alexander M. Zhivov (20, 28, 29)
University of Illinois

Albert J. Heber (21)
Purdue University

Farhad Memarzadeh (21)
National Institutes of Health

Gerald L. Riskowski (21)
University of Illinois

Yuanhui Zhang (21)
University of Illinois

Roger C. Brook (22)
Michigan State University

Charlie C. Shieh (23, 24)
Goldman Copeland, PC

Deep Ghosh (24)
Southern Company Services

John B. Riley (24)
Black & Veatch

Ravisankar Ganta (25)
Bechtel

Hassan M. Bagheri (28)
P2S Engineering, Inc.

Eugene O. Shilkrot (28)
Central Research Institute
for Industrial Buildings
(TsNIIpromzdanii)

Todd A. Talbott (28, 29)
United McGill Corporation

Vladimir N. Posokhin (29)
Kazan Civil Engineering
Academy

Andrey S. Strongin (29)
Central Research Institute
for Industrial Buildings
(TsNIIpromzdanii)

Thomas E. Carter (30)
Garland Commercial
Ranges, Ltd.

Philip Morton (30)
Gaylord Industries Inc.

Vernon A. Smith (30)
Architectural Energy
Corporation

David W. Wolbrink (30)
Broan Mfg Co., Inc.

Steven P. Kavanaugh (31)
University of Alabama

William E. Murphy (31)
Paducah Community College

Kevin D. Rafferty (31)
Oregon Institute of Technology

Henry M. Healey (32)
Healey & Associates

Earl E. Rush (32)
Sandia National Laboratories

Jim Deichert (33)
Steffes ETS, Inc.

Chad B. Dorgan (33)
Dorgan Associates, Inc.

James S. Elleson (33)
University of Wisconsin
HVAC&R Center

Clifford C. Federspiel (33)
University of California—
Berkeley

Charles W. Frazell (33)
TU Electric

Budd W. Lee (33)
Dow Chemical

Russell E. Lindemann (33)
Baltimore Aircoil Company

Mark M. MacCracken (33)
Calmac Mfg Corp.

Kirby P. Nelson (33)
Paul Mueller Company

Sidney A. Parsons (33)
Parsons & Lumsden CC

David C.J. Peters (33)
Southland Industries

Douglas T. Reindl (33)
University of Wisconsin—
Madison

Peter Simmonds (33)
Flack & Kurtz Consulting
Engineers, LLP

Sandra L. Steinmann (33)
CBI Technical Services Co.

Charles G. Tharp (33)
Pelham-Phillips-Hagerman

Steven M. Tredinnick (33)
Kattner/FVB

Richard P. Mazzucchi (34, 39)
Resource Performance
Management

Lawrence G. Spielvogel (34)
Lawrence G. Spielvogel, Inc.

Terry L. Cornell (35)
GARD Analytics Inc.

CONTRIBUTORS (*Concluded*)

David G. Guckelberger (35)
The Trane Company

Wayne K. Robertson (35)
Heery International Inc.

Alexander H. Sleiman (35)
District Energy St. Paul Inc.

William J. Thomaston (35)
Alabama Gas Corporation

Joseph J. Watson (35)
Tozour Energy Systems

Gerald J. Kettler (36)
Air Engineering and
Testing, Inc.

William K. Thomas (36)
Thomas-Young Associates, Inc.

Dennis H. Tuttle (36)
Wessels Company

Richard A. Greco (37)
Mazzetti & Associates

Duane A. Barrett (38)
Bentley Systems, Inc.

David J. Branson (38)
Compliance Services
Group, Inc.

Brian K. Kammers (38)
Johnson Controls, Inc.

Richard T. Linton (38)
University of Wisconsin

Ron M. Nelson (38)
Iowa State University

Anil Saigal (38)
Honeywell, Inc.

Michael C.A. Schwedler (38)
The Trane Company

Hashem Akbari (39)
Lawrence Berkeley National
Laboratory

Jeff S. Haberl (39)
Texas A&M University

Kristin H. Heinemeier (39)
Honeywell Technology Center

J. Michael MacDonald (39)
Oak Ridge National Laboratory

Danny S. Parker (39)
Florida Solar Energy Center

John Phelan (39)
Architectural Energy Corp.

T. Agami Reddy (39)
Drexel University

James E. Braun (40)
Purdue University

David W. Bevirt (41)
National Environmental
Balancing Bureau

Wayne A. Dunn (41)
Sunbelt Engineering, Inc.

William C. Brown (42)
National Research Council
Canada

Jeffrey E. Christian (42)
Oak Ridge National Laboratory

Adrian N. Tuluca (42)
Steven Winter Associates

Michael A. Ratcliff (43)
RWDI, Inc.

David J. Wilson (43)
University of Alberta

Gemma Kerr (44)
Inair Environmental

Eugene L. Valerio (44)
Technical Marketing Services

Douglas W. Vanosdell (44)
Research Triangle Institute

Frank W. Mayhew (45)
ECTC Consulting

John R. Sosoka (45)
P2S Engineering, Inc.

Lawrence Uebele (45)
Siemens Building
Technologies, Inc.

Ted N. Carnes (46)
Pelton Marsh Kinsella

Brian Guenther (46)
Vibro-Acoustics

Richard J. Peppin (46)
Scantek

Alfred J. Warnock (46)
National Research Council
Canada

Mark Hodgson (47)
Clayton Environmental
Consultants

William R. Leizear, Sr. (47)
Claude Laval Corp.

Mark Leonardelli (47)
Nalco Chemical

Robert Petterson (47)
Marley Cooling Tower Co.

Charles H. Sanderson (47)
Superior Manufacturing
Division

Anne M. Wilson (47)
Nalco Chemical

John W. Calhoun, Jr. (48)
Alabama Power Company

Fredric S. Goldner (48)
Energy Management &
Research Associates

Wilbur L. Haag, Jr. (48)
A.O. Smith Water Products

Robert J. Hemphill (48)
Gas Research Institute

Daryl L. Hosler (48)
Southern California Gas
Company

Lawrence H. Chenault (49)
Hume Snow Melting
Systems, Inc.

James W. Ramsey (49)
University of Minnesota

Leon E. Shapiro (50)
ADA Systems

Patricia T. Thomas (50)
Munters Corporation

John A. Clark (51)
BKBM Engineers

John H. Klote (51)
John H. Klote, Inc.

Gary D. Lougheed (51)
National Research Council
Canada

Gary B. Hayden (52)
Burnham Radiant Heating
Corporation

James M. Porter (52)

J. Marx Ayres (53)
Ayres & Ezer Associates, Inc.

James A. Carlson (53)
Leo A. Daly & Associates

Patrick J. Lama (53)
Mason Industries, Inc.

Terry E. Townsend (53)
Townsend Engineering, Inc.

ASHRAE HANDBOOK COMMITTEE

Evans J. Lizardos, Chair

1999 Applications Volume Subcommittee: **Eugene L. Valerio,** Chair

Richard S. Armstrong   **Kenneth W. Cooper**   **Patrick J. Lama**   **Hsien-Sheng J. Pei**   **James M. Porter**

ASHRAE HANDBOOK STAFF

Robert A. Parsons, Editor   **Christina D. Tate,** Associate Editor

Scott A. Zeh, Nancy F. Thysell, and **Jayne E. Jackson,** Publishing Services

W. Stephen Comstock,
Director, Communications and Publications
Publisher

ASHRAE TECHNICAL COMMITTEES AND TASK GROUPS

SECTION 1.0—FUNDAMENTALS AND GENERAL
1.1 Thermodynamics and Psychrometrics
1.2 Instruments and Measurements
1.3 Heat Transfer and Fluid Flow
1.4 Control Theory and Application
1.5 Computer Applications
1.6 Terminology
1.7 Operation and Maintenance Management
1.8 Owning and Operating Costs
1.9 Electrical Systems
1.10 Energy Resources

SECTION 2.0—ENVIRONMENTAL QUALITY
2.1 Physiology and Human Environment
2.2 Plant and Animal Environment
2.3 Gaseous Air Contaminants and Gas Contaminant Removal Equipment
2.4 Particulate Air Contaminants and Particulate Contaminant Removal Equipment
2.6 Sound and Vibration Control
2.7 Seismic Restraint Design
TG Buildings' Impacts on the Environment
TG Global Climate Change

SECTION 3.0—MATERIALS AND PROCESSES
3.1 Refrigerants and Secondary Coolants
3.2 Refrigerant System Chemistry
3.3 Refrigerant Contaminant Control
3.4 Lubrication
3.5 Desiccant and Sorption Technology
3.6 Corrosion and Water Treatment
3.8 Refrigerant Containment

SECTION 4.0—LOAD CALCULATIONS AND ENERGY REQUIREMENTS
4.1 Load Calculation Data and Procedures
4.2 Weather Information
4.3 Ventilation Requirements and Infiltration
4.4 Building Materials and Building Envelope Performance
4.5 Fenestration
4.6 Building Operation Dynamics
4.7 Energy Calculations
4.8 Indoor Environmental Modeling
4.10 Smart Building Systems
TG Integrated Building Design

SECTION 5.0—VENTILATION AND AIR DISTRIBUTION
5.1 Fans
5.2 Duct Design
5.3 Room Air Distribution
5.4 Industrial Process Air Cleaning (Air Pollution Control)
5.5 Air-to-Air Energy Recovery
5.6 Control of Fire and Smoke
5.7 Evaporative Cooling
5.8 Industrial Ventilation
5.9 Enclosed Vehicular Facilities
5.10 Kitchen Ventilation

SECTION 6.0—HEATING EQUIPMENT, HEATING AND COOLING SYSTEMS AND APPLICATIONS
6.1 Hydronic and Steam Equipment and Systems
6.2 District Heating and Cooling
6.3 Central Forced Air Heating and Cooling Systems
6.4 In Space Convection Heating
6.5 Radiant Space Heating and Cooling
6.6 Service Water Heating
6.7 Solar Energy Utilization
6.8 Geothermal Energy Utilization
6.9 Thermal Storage
6.10 Fuels and Combustion

SECTION 7.0—PACKAGED AIR-CONDITIONING AND REFRIGERATION EQUIPMENT
7.1 Residential Refrigerators and Food Freezers
7.4 Combustion Engine Driven Heating and Cooling Equipment
7.5 Mechanical Dehumidification Equipment and Heat Pipes
7.6 Unitary and Room Air Conditioners and Heat Pumps

SECTION 8.0—AIR-CONDITIONING AND REFRIGERATION SYSTEM COMPONENTS
8.1 Positive Displacement Compressors
8.2 Centrifugal Machines
8.3 Absorption and Heat Operated Machines
8.4 Air-to-Refrigerant Heat Transfer Equipment
8.5 Liquid-to-Refrigerant Heat Exchangers
8.6 Cooling Towers and Evaporative Condensers
8.7 Humidifying Equipment
8.8 Refrigerant System Controls and Accessories
8.10 Pumps and Hydronic Piping
8.11 Electric Motors and Motor Control

SECTION 9.0—AIR-CONDITIONING SYSTEMS AND APPLICATIONS
9.1 Large Building Air-Conditioning Systems
9.2 Industrial Air Conditioning
9.3 Transportation Air Conditioning
9.4 Applied Heat Pump/Heat Recovery Systems
9.5 Cogeneration Systems
9.6 Systems Energy Utilization
9.7 Testing and Balancing
9.8 Large Building Air-Conditioning Applications
9.9 Building Commissioning
9.10 Laboratory Systems
9.11 Clean Spaces
TG Combustion Gas Turbine Inlet Air Cooling Systems
TG Tall Buildings

SECTION 10.0—REFRIGERATION SYSTEMS
10.1 Custom Engineered Refrigeration Systems
10.2 Automatic Icemaking Plants and Skating Rinks
10.3 Refrigerant Piping, Controls, and Accessories
10.4 Ultra-Low Temperature Systems and Cryogenics
10.5 Refrigerated Distribution and Storage Facilities
10.6 Transport Refrigeration
10.7 Commercial Food and Beverage Cooling Display and Storage
10.8 Refrigeration Load Calculations
10.9 Refrigeration Application for Foods and Beverages
10.MOC Mineral Oil Circulation

PREFACE

This handbook describes heating, ventilating, and air conditioning for a broad range of applications. Most of the chapters from the 1995 *ASHRAE Handbook* have been revised for this volume to reflect current requirements and design approaches. New chapters on HVAC for museums and power plants, information on air quality in aircraft, additional information on maintaining a proper environment for indoor swimming pools and new information on sound control and building operation make this a particularly useful reference. Because this book focuses on specific applications for HVAC, it provides background information to designers new to the application as well as to those needing a refresher on the topic. In addition, many chapters include valuable data for design. Some of the revisions that have been made are as follows.

- Chapter 4, Places of Assembly, provides more comprehensive design information on natatoriums.
- Chapter 5, Hotels, Motels, and Dormitories, includes more information on hotels and motels, which is reflected in a change in the title of the chapter.
- Chapter 8, Surface Transportation, has been substantially revised. It now includes information about European bus air conditioning and the state of the art in railcar air conditioning.
- Chapter 9, Aircraft, has been completely rewritten. It describes the environmental control systems used in commercial aircraft today and their operation during a typical flight. Applicable regulations are summarized and the section on air quality has been expanded. Information on air-cycle equipment has been deleted.
- Chapter 12, Enclosed Vehicular Facilities, includes more ventilation design information. Research from a tunnel fire test has provided new ventilation design criteria for tunnel ventilation. The chapter now covers ventilation for toll booths, railroad tunnels, and areas with vehicles that use alternative fuels.
- Chapter 13, Laboratories, has additional information on scale-up laboratories and compressed gas storage. The sections on internal heat load and Biosafety Level 3 have been expanded.
- Chapter 15, Clean Spaces, greatly expands on pharmaceutical and biomanufacturing cleanrooms. A new section covers high bay cleanrooms.
- Chapter 20, Museums, Libraries, and Archives, is a new chapter. It discusses in detail the importance of relative humidity on collections and suggests the degree of environmental control for various types of collections and historic buildings.
- Chapter 21, Environmental Control for Animals and Plants, provides new information on levels of contaminants in livestock buildings and suggests several methods of control. It includes new findings on ventilation for laboratory animals.
- Chapter 24, Power Plants, is a new chapter that introduces HVAC design criteria for the various facilities in electrical generating stations and in facilities that produce process heat and power.
- Chapter 29, Industrial Local Exhaust Systems, has been greatly expanded with much more information about specific types of hoods and their design and application.
- Chapter 30, Kitchen Ventilation, is updated to reflect code changes including changed terminology for nonlisted hoods and new information on exhaust system effluent control.
- Chapter 33, Thermal Storage, has an expanded discussion of control strategies.
- Chapter 34, Energy Management, has new and more comprehensive energy consumption data for commercial and residential buildings in the United States.
- Chapter 35, Owning and Operating Costs, has updated information of the impact of refrigerant phaseouts. New information is included on financing alternatives, on district energy service and on-site electric generation in view of deregulation, and on computer analysis.

- Chapter 39, Building Energy Monitoring, reflects the new direction in this field and focuses on monitoring designed to answer specific questions rather than broad-based research-oriented programs. The section on accuracy and uncertainty is rewritten.
- Chapter 40, Supervisory Control Strategies and Optimization, has a new title to reflect its reorganization and the significant amount of new material. The first section defines systems and control variables. The second section, which is intended for practitioners, presents computerized control strategies. The third section presents basic optimization methods and is for researchers and developers of advanced control strategies.
- Chapter 41, Building Commissioning, has expanded the information on the various phases of commissioning.
- Chapter 42, Building Envelopes, is moved from the 1997 *ASHRAE Handbook—Fundamentals* with minor editing.
- Chapter 43, Building Air Intake and Exhaust Design, is a revision of the last half of Chapter 15 in the 1997 *ASHRAE Handbook*.
- Chapter 44, Control of Gaseous Indoor Air Contaminants, has a greatly expanded section on air cleaning. It includes more information on the equipment used and on its design, energy use, startup procedures, operation, maintenance, and testing procedures.
- Chapter 45, Design and Application of Controls, is slightly reorganized because the information on control fundamentals was moved to the 1997 *ASHRAE Handbook*.
- Chapter 46, Sound and Vibration Control, describes all currently recognized criteria methods—dBA, NC, RC, RC Mark II, and NCB. New sections include information on: (1) uncertainties that can reasonably be expected from the data in the chapter, (2) chiller and air-cooled condenser noise, and (3) data for estimating ceiling plenum insertion loss.
- Chapter 47, Water Treatment, is reorganized and has added information on biological growth control. A new section covers startup and shutdown procedures.
- Chapter 49, Snow Melting, includes expanded equations for heating requirements and new load data including maps. Information on piping materials for hydronic systems has been updated.
- Chapter 51, Fire and Smoke Management, includes new sections on fire management (i.e., through-penetration fire stopping) and on smoke management in large spaces.
- Chapter 53, Seismic and Wind Restraint Design, introduces the proposed International Building Code seismic design equations and describes several new seismic snubbers. The chapter also includes a new section on wind restraint design.

Each Handbook is published in two editions. One edition contains inch-pound (I-P) units of measurement, and the other contains the International System of Units (SI).

Look for corrections to the 1996, 1997, and 1998 volumes of the Handbook on the Internet at http://www.ashrae.org. Any changes to this volume will be reported in the 2000 *ASHRAE Handbook* and on the Internet.

If you have suggestions for improving a chapter or you would like more information on how you can help revise a chapter, e-mail bparsons@ashrae.org; write to Handbook Editor, ASHRAE, 1791 Tullie Circle, Atlanta, GA 30329; or fax (404) 321-5478.

Robert A. Parsons
ASHRAE Handbook Editor

RESIDENCES

SPACE-CONDITIONING systems for residential use vary with both local and application factors. Local factors include energy source availability (both present and projected) and price; climate; socioeconomic circumstances; and the availability of installation and maintenance skills. Application factors include housing type, construction characteristics, and building codes. As a result, many different systems are selected to provide combinations of heating, cooling, humidification, dehumidification, and air filtering. This chapter emphasizes the more common systems for space conditioning of both single-family (i.e., traditional site-built and modular or manufactured homes) and multifamily residences. Low-rise multifamily buildings generally follow single-family practice because constraints favor compact designs. Retrofit and remodeling construction also adopt the same systems as those for new construction, but site-specific circumstances may call for unique designs.

Systems

The common residential heating systems are listed in Table 1. Three generally recognized groups are central forced air, central hydronic, and zoned systems. System selection and design involve such key decisions as (1) source(s) of energy, (2) means of distribution and delivery, and (3) terminal device(s).

Climate determines the services needed. Heating and cooling are generally required. Air cleaning (by filtration or electrostatic devices) can be added to most systems. Humidification, which can also be added to most systems, is generally provided in heating systems only when psychrometric conditions make it necessary for comfort and health (as defined in ASHRAE *Standard* 55). Cooling systems dehumidify as well. Typical residential installations are shown in Figures 1 and 2.

Figure 1 shows a gas furnace, a split-system air conditioner, a humidifier, and an air filter. The system functions as follows: Air returns to the equipment through a return air duct (1). It passes initially through the air filter (2). The circulating blower (3) is an integral part of the furnace (4), which supplies heat during winter.

An optional humidifier (10) adds moisture to the heated air, which is distributed throughout the home from the supply duct (9). When cooling is required, the circulating air passes across the evaporator coil (5), which removes heat and moisture from the air. Refrigerant lines (6) connect the evaporator coil to a remote condensing unit (7) located outdoors. Condensate from the evaporator is removed through a drainline with a trap (8).

Figure 2 shows a split-system heat pump, supplemental electric resistance heaters, a humidifier, and an air filter. The system functions as follows: Air returns to the equipment through the return air duct (1) and passes through the air filter (2). The circulating blower (3) is an integral part of the indoor unit (or air handler) of the heat pump (4), which supplies heat via the indoor coil (6) during the heating season. Optional electric heaters (5) supplement heat from the heat pump during periods of low ambient temperature and counteract airstream cooling during the defrost cycle. An optional humidifier (10) adds moisture to the heated air, which is distributed throughout the home from the supply duct (9). When cooling is required, the circulating air passes across the indoor coil (6), which removes heat and moisture from the air. Refrigerant lines (11) connect the indoor coil to the outdoor unit (7). Condensate from the indoor coil drains away through a drainline with a trap (8).

Table 1 Residential Heating and Cooling Systems

	Forced Air	Hydronic	Zoned
Most common energy sources	Gas Oil Electricity Resistance Heat pump	Gas Oil Electricity Resistance Heat pump	Gas Electricity Resistance Heat pump
Heat distribution medium	Air	Water Steam	Air Water Refrigerant
Heat distribution system	Ducting	Piping	Ducting Piping or None
Terminal devices	Diffusers Registers Grilles	Radiators Radiant panels Fan-coil units	Included with product

The preparation of this chapter is assigned to TC 7.6, Unitary Air Conditioners and heat pumps

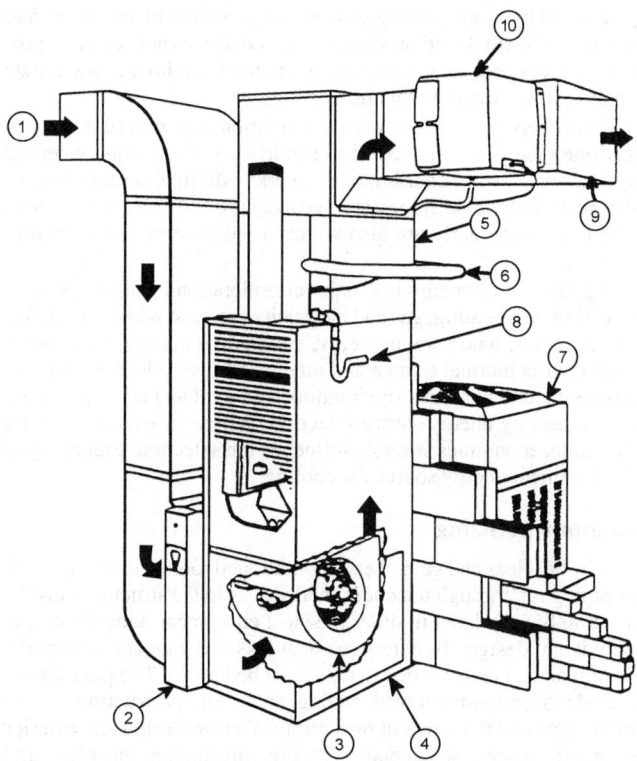

Fig. 1   Typical Residential Installation of Heating, Cooling, Humidifying, and Air Filtering System

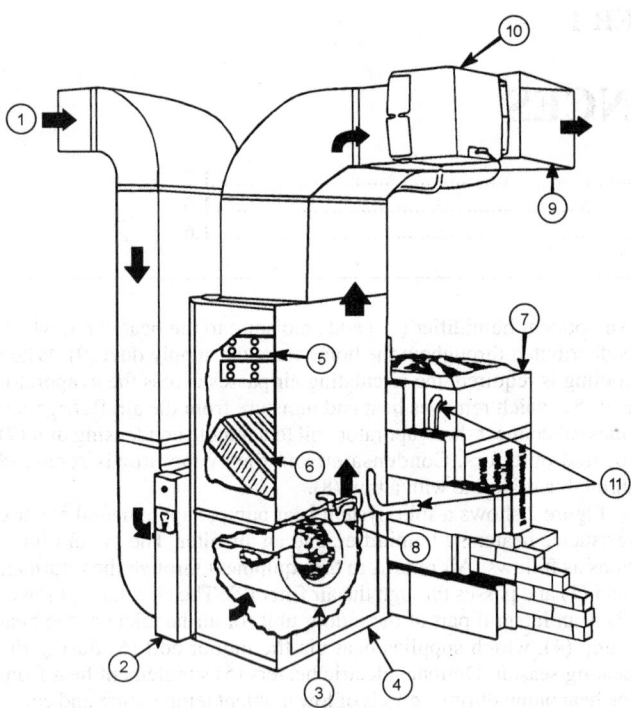

Fig. 2   Typical Residential Installation of Heat Pump

Single-package systems, where all equipment is contained in one cabinet, are also popular in the United States. They are used extensively in areas where residences have duct systems in crawlspaces beneath the main floor and in areas such as the Southwest, where they are typically rooftop-mounted and connected to an attic duct system.

Central hydronic heating systems are popular both in Europe and in parts of North America where central cooling is not normally provided. If desired, central cooling is often added through a separate cooling-only system with attic ducting.

Zoned systems are designed to condition only part of a home at any one time. They may consist of individual room units or central systems with zoned distribution networks. Multiple central systems that serve individual floors or serve sleeping and common portions of a home separately are also widely used in large single-family houses.

The source of energy is a major consideration in heating system selection. For heating, gas and electricity are most widely used, followed by oil, wood, solar energy, geothermal energy, waste heat, coal, district thermal energy, and others. Relative prices, safety, and environmental concerns (both indoor and outdoor) are further factors in heating energy source selection. Where various sources are available, economics strongly influence the selection. Electricity is the dominant energy source for cooling.

Equipment Sizing

The heat loss and gain of each conditioned room and of ductwork or piping run through unconditioned spaces in the structure must be accurately calculated in order to select equipment with the proper output and design. To determine heat loss and gain accurately, the floor plan and construction details must be known. The plan should include information on wall, ceiling, and floor construction as well as the type and thickness of insulation. Window design and exterior door details are also needed. With this information, heat loss and gain can be calculated using the Air-Conditioning Contractors of America (ACCA) *Manual* J or similar calculation procedures. To conserve energy, many jurisdictions require that the building be

designed to meet or exceed the requirements of ASHRAE *Standard* 90.2, Energy-Efficient Design of New Low-Rise Residential Buildings, or similar requirements.

Proper matching of equipment capacity to the design heat loss and gain is essential. The heating capacity of air-source heat pumps is usually supplemented by auxiliary heaters, most often of the electric resistance type; in some cases, however, fossil fuel furnaces or solar systems are used.

The use of undersized equipment results in an inability to maintain indoor design temperatures at outdoor design conditions and slow recovery from setback or set-up conditions. Grossly oversized equipment can cause discomfort due to short on-times, wide indoor temperature swings, and inadequate dehumidification when cooling. Gross oversizing may also contribute to higher energy use due to an increase in cyclic thermal losses and off-cycle losses. Variable capacity equipment (heat pumps, air conditioners, and furnaces) can more closely match building loads over specific ambient temperature ranges, usually reducing these losses and improving comfort levels; in the case of heat pumps, supplemental heat needs may also be reduced.

Recent trends toward tightly constructed buildings with improved vapor retarders and low infiltration may cause high indoor humidity conditions and the buildup of indoor air contaminants in the space. Air-to-air heat-recovery equipment may be used to provide tempered ventilation air to tightly constructed houses. Outdoor air intakes connected to the return duct of central systems may also be used when lower installed costs are the most important factor. Simple exhaust systems with passive air intakes are also becoming popular. In all cases, minimum ventilation rates, as outlined in ASHRAE *Standard* 62 should be maintained.

SINGLE-FAMILY RESIDENCES

Heat Pumps

Heat pumps for single-family houses are normally unitary systems; i.e., they consist of single-package units or two or more factory-built modules as illustrated in Figure 2. These differ from applied or built-up heat pumps, which require field engineering to select compatible components for complete systems.

Most commercially available heat pumps (particularly in North America) are electrically powered air-source systems. Supplemental heat is generally required at low outdoor temperatures or during defrost. In most cases, supplemental or backup heat is provided by electric resistance heaters.

Heat pumps may be classified by thermal source and distribution medium in the heating mode as well as the type of fuel used. The most commonly used classes of heat pump equipment are air-to-air and water-to-air. Air-to-water and water-to-water types are also used.

Heat pump systems, as contrasted to the actual heat pump equipment, are generally described as air-source or ground-source. The thermal sink for cooling is generally assumed to be the same as the thermal source for heating. Air-source systems using ambient air as the heat source/sink are generally the least costly to install and thus the most commonly used. Ground-source systems usually employ water-to-air heat pumps to extract heat from the ground via groundwater or a buried heat exchanger.

Ground-Source (Geothermal) Systems. As a heat source/sink, groundwater (from individual wells or supplied as a utility from community wells) offers the following advantages over ambient air: (1) heat pump capacity is independent of ambient air temperature, reducing supplementary heating requirements; (2) no defrost cycle is required; (3) for equal equipment rating point efficiency, the seasonal efficiency is usually higher for heating and for cooling; and (4) peak heating energy consumption is usually lower. Ground-coupled or surface-water-coupled systems offer the same advantages. However, they circulate brine or water in a buried or submerged heat exchanger to transfer heat from the ground. Direct expansion ground-source systems, with evaporators buried in the ground, are occasionally used. The number of ground-source systems is growing rapidly, particularly of the ground-coupled type. Water-source systems that extract heat from surface water (e.g., lakes or rivers) or city (tap) water are also used where local conditions permit.

Water supply, quality, and disposal must be considered for groundwater systems. Bose et al. (1985) provides detailed information on these subjects. Secondary coolants for ground-coupled systems are discussed in this manual and in Chapter 20 of the 1997 *ASHRAE Handbook—Fundamentals*. Buried heat exchanger configurations may be horizontal or vertical, with the vertical including both multiple-shallow- and single-deep-well configurations. Ground-coupled systems avoid water quality, quantity, and disposal concerns but are sometimes more expensive than groundwater systems. However, ground-coupled systems are usually more efficient, especially when pumping power for the groundwater system is considered.

Add-On Heat Pumps. In add-on systems, a heat pump is added—often as a retrofit—to an existing furnace or boiler system. The heat pump and combustion device are operated in one of two ways: (1) alternately, depending on which is most cost-effective, or (2) in parallel. In unitary bivalent heat pumps, the heat pump and combustion device are grouped in a common chassis and cabinets to provide similar benefits at lower installation costs.

Fuel-Fired Heat Pumps. Extensive research and development has been conducted to develop fuel-fired heat pumps. They are beginning to be marketed in North America.

Water-Heating Options. Heat pumps may be equipped with desuperheaters (either integral or field-installed) to reclaim heat for domestic water heating. Integrated space-conditioning and water-heating heat pumps with an additional full-size condenser for water heating are also available.

Furnaces

Furnaces are fueled by gas (natural or propane), electricity, oil, wood, or other combustibles. Gas, oil, and wood furnaces may draw combustion air from the house or from outdoors. If the furnace space is located such that combustion air is drawn from the outdoors, the arrangement is called an isolated combustion system (ICS). Furnaces are generally rated on an ICS basis. When outdoor air is ducted to the combustion chamber, the arrangement is called a direct vent system. This latter method is used for manufactured home applications and some mid- and high-efficiency equipment designs. Using outside air for combustion eliminates both the infiltration losses associated with the use of indoor air for combustion and the stack losses associated with atmospherically induced draft hood-equipped furnaces.

Two available types of high-efficiency gas furnaces are noncondensing and condensing. Both increase efficiency by adding or improving heat exchanger surface area and reducing heat loss during furnace off-times. The higher efficiency condensing type also recovers more energy by condensing water vapor from the combustion products. The condensate is developed in a high-grade stainless steel heat exchanger and is disposed of through a drain line. Condensing furnaces generally use PVC for vent pipes and condensate drains.

Wood-fueled furnaces are used in some areas. A recent advance in wood furnaces is the addition of catalytic converters to enhance the combustion process, increasing furnace efficiency and producing cleaner exhaust.

Chapters 28 and 29 of the 1996 *ASHRAE Handbook—Systems and Equipment* include more detailed information on furnaces and furnace efficiency.

Hydronic Heating Systems—Boilers

With the growth of demand for central cooling systems, hydronic systems have declined in popularity in new construction, but still account for a significant portion of existing systems in northern climates. The fluid is heated in a central boiler and distributed by piping to terminal units (fan coils, radiators, radiant panels, or baseboard convectors) in each room. Most recently installed residential systems use a forced circulation, multiple zone hot water system with a series-loop piping arrangement. Chapters 12, 27, and 32 of the 1996 *ASHRAE Handbook—Systems and Equipment* have more information on hydronics and hydronic.

Design water temperature is based on economic and comfort considerations. Generally, higher temperatures result in lower first costs because smaller terminal units are needed. However, losses tend to be greater, resulting in higher operating costs and reduced comfort due to the concentrated heat source. Typical design temperatures range from 180 to 200°F. For radiant panel systems, design temperatures range from 110 to 170°F. The preferred control method allows the water temperature to decrease as outdoor temperatures rise. Provisions for the expansion and contraction of the piping and heat distributing units and for the elimination of air from the hydronic system are essential for quiet, leaktight operation.

Fossil fuel systems that condense water vapor from the flue gases must be designed for return water temperatures in the range of 120 to 130°F for most of the heating season. Noncondensing systems must maintain high enough water temperatures in the boiler to prevent this condensation. If rapid heating is required, both terminal unit and boiler size must be increased, although gross oversizing should be avoided.

Zoned Heating Systems

Zoned systems offer the potential for lower operating costs, because unoccupied areas can be kept at lower temperatures in the winter and at higher temperatures in the summer. Common areas can be maintained at lower temperatures at night and sleeping areas at lower temperatures during the day.

One form of this system consists of individual heaters located in each room. These heaters are usually electric or gas-fired. Electric heaters are available in the following types: baseboard free-convection, wall insert (free-convection or forced-fan), radiant panels for walls and ceilings, and radiant cables for walls, ceilings, and floors. Matching equipment capacity to heating requirements is critical for individual room systems. Heating delivery cannot be adjusted by adjusting air or water flow, so greater precision in room-by-room sizing is needed.

Individual heat pumps for each room or group of rooms (zone) are another form of zoned electric heating. For example, two or more small unitary heat pumps can be installed in two-story or large one-story homes.

The multisplit heat pump consists of a central compressor and an outdoor heat exchanger to service up to eight indoor zones. Each zone uses one or more fan coils, with separate thermostatic control for each zone. Such systems are used in both new and retrofit construction.

A method for zoned heating in central ducted systems is the zone-damper system. This consists of individual zone dampers and thermostats combined with a zone control system. Both variable-air-volume (damper position proportional to zone demand) and on-off (damper fully open or fully closed in response to thermostat) types are available. Such systems sometimes include a provision to modulate to lower capacities when only a few zones require heating.

Solar Heating

Both active and passive solar energy systems are sometimes used to heat residences. In typical active systems, flat plate collectors heat air or water. Air systems distribute heated air either to the living space for immediate use or to a thermal storage medium (i.e., a rock pile). Water systems pass heated water through a secondary heat exchanger and store extra heat in a water tank. Due to low delivered water temperatures, radiant floor panels requiring moderate temperatures are generally used.

Trombe walls and sunspaces are two common passive systems. Glazing facing south (with overhangs to reduce solar gains in the summer) and movable insulated panels can reduce heating requirements.

Backup heating ability is generally needed with solar energy systems. Chapter 32 has information on sizing solar heating equipment.

Unitary Air Conditioners

In forced-air systems, the same air distribution duct system can be used for both heating and cooling. Split-system central cooling, as illustrated in Figure 1, is the most widely used forced-air system. Upflow, downflow, and horizontal airflow units are available. Condensing units are installed on a noncombustible pad outside and contain a motor- or engine-driven compressor, condenser, condenser fan and fan motor, and controls. The condensing unit and evaporator coil are connected by refrigerant tubing that is normally field-supplied. However, precharged, factory-supplied tubing with quick-connect couplings is also common where the distance between components is not excessive.

A distinct advantage of split-system central cooling is that it can readily be added to existing forced-air heating systems. Airflow rates are generally set by the cooling requirements to achieve good performance, but most existing heating duct systems are adaptable to cooling. Airflow rates of 350 to 450 cfm per nominal ton of refrigeration are normally recommended for good cooling performance. As with heat pumps, these systems may be fitted with desuperheaters for domestic water heating.

Some cooling equipment includes forced-air heating as an integral part of the product. Year-round heating and cooling packages with a gas, oil, or electric furnace for heating and a vapor-compression system for cooling are available. Air-to-air and water-source heat pumps provide cooling and heating by reversing the flow of refrigerant.

Distribution. Duct systems for cooling (and heating) should be designed and installed in accordance with accepted practice. Useful information is found in ACCA *Manuals* D and G. Chapter 9 of the 1996 *ASHRAE Handbook—Systems and Equipment* also discusses air distribution design for small heating and cooling systems.

Because weather is the primary influence on the load, the cooling load in each room changes from hour to hour. Therefore, the owner or occupant should be able to make seasonal or more frequent adjustments to the air distribution system to obtain improved comfort. Such adjustments may involve opening additional outlets in second-floor rooms during the summer and throttling or closing heating outlets in some rooms during the winter. Manually adjustable balancing dampers may be provided to facilitate these adjustments. Other possible refinements are the installation of a heating and cooling system sized to meet heating requirements, with additional self-contained cooling units serving rooms with high summer loads, or of separate central systems for the upper and lower floors of a house. On deluxe applications, zone-damper systems can be used.

Operating characteristics of both heating and cooling equipment must be considered when zoning is used. For example, a reduction in the air quantity to one or more rooms may reduce the airflow across the evaporator to such a degree that frost forms on the fins. Reduced airflow on heat pumps during the heating season can cause overloading if airflow across the indoor coil is not maintained at above 350 cfm per ton. Reduced air volume to a given room would reduce the air velocity from the supply outlet and could cause unsat-

isfactory air distribution in the room. Manufacturers of zoned systems normally provide guidelines for avoiding such situations.

Special Considerations. In split-level houses, cooling and heating are complicated by air circulation between various levels. In many such houses, the upper level tends to overheat in winter and undercool in summer. Multiple outlets, some near the floor and others near the ceiling, have been used with some success on all levels. To control airflow, the homeowner opens some outlets and closes others from season to season. Free circulation between floors can be reduced by locating returns high in each room and keeping doors closed.

In existing homes, the cooling that can be added is limited by the air-handling capacity of the existing duct system. While the existing duct system is usually satisfactory for normal occupancy, it may be inadequate during large gatherings. In all cases where new cooling (or heating) equipment is installed in existing homes, supply-air ducts and outlets must be checked for acceptable air-handling capacity and air distribution. Maintaining upward airflow at an effective velocity is important when converting existing heating systems with floor or baseboard outlets to both heat and cool. It is not necessary to change the deflection from summer to winter for registers located at the perimeter of a residence. Registers located near the floor on the inside walls of rooms may operate unsatisfactorily if the deflection is not changed from summer to winter.

Occupants of air-conditioned spaces usually prefer minimum perceptible air motion. Perimeter baseboard outlets with multiple slots or orifices directing air upwards effectively meet this requirement. Ceiling outlets with multidirectional vanes are also satisfactory.

A residence without a forced-air heating system may be cooled by one or more central systems with separate duct systems, by individual room air conditioners (window-mounted or through-the-wall), or by mini-split-room air conditioners.

Cooling equipment must be located carefully. Because cooling systems require higher indoor airflow rates than most heating systems, the sound levels generated indoors are usually higher. Thus, indoor air-handling units located near sleeping areas may require sound attenuation. Outdoor noise levels should also be considered when locating the equipment. Many communities have ordinances regulating the sound level of mechanical devices, including cooling equipment. Manufacturers of unitary air conditioners often certify the sound level of their products in an ARI program (ARI *Standard* 270). ARI *Standard* 275 gives information on how to predict the dBA sound level when the ARI sound rating number, the equipment location relative to reflective surfaces, and the distance to the property line are known.

An effective and inexpensive way to reduce noise is to put distance and natural barriers between sound source and listener. However, airflow to and from air-cooled condensing units must not be obstructed. Most manufacturers provide recommendations regarding acceptable distances between condensing units and natural barriers. Outdoor units should be placed as far as is practical from porches and patios, which may be used while the house is being cooled. Locations near bedroom windows and neighboring homes should also be avoided.

Evaporative Coolers

In dry climates, evaporative coolers can be used to cool residences. Further details on evaporative coolers can be found in Chapter 19 of the 1996 *ASHRAE Handbook—Systems and Equipment* and in Chapter 50 of this volume.

Humidifiers

For improved winter comfort, equipment that increases indoor relative humidity may be needed. In a ducted heating system, a central humidifier can be attached to or installed within a supply plenum or main supply duct, or installed between the supply and return duct systems. When applying supply-to-return duct humidifiers on heat pump systems, care should be taken to maintain proper airflow across the indoor coil. Self-contained humidifiers can be used in any residence. Even though this type of humidifier introduces all the moisture to one area of the home, moisture will migrate and raise humidity levels in other rooms. Overhumidification, which can cause condensate to form on the coldest surfaces in the living space (usually the windows), should be avoided.

Central humidifiers may be rated in accordance with ARI *Standard* 610. This rating is expressed in the number of gallons per day evaporated by 140°F entering air. Some manufacturers certify the performance of their product to the ARI standard. Selecting the proper size humidifier is important and is outlined in ARI *Guideline* F.

Since moisture migrates through all structural materials, vapor retarders should be installed near the warmer inside surface of insulated walls, ceilings, and floors in most temperature climates. Improper attention to this construction detail allows moisture to migrate from inside to outside, causing damp insulation, possible structural damage, and exterior paint blistering. Humidifier cleaning and maintenance schedules should be followed to maintain efficient operation and prevent bacteria buildup. Chapter 20 of the 1996 *ASHRAE Handbook—Systems and Equipment* contains more information on residential humidifiers.

Dehumidifiers

Many homes also use dehumidifiers to remove moisture and control indoor humidity levels. In cold climates, dehumidification is sometimes required during the summer in basement areas to control mold and mildew growth and to reduce zone humidity levels. Traditionally, portable dehumidifiers have been used to control humidity in this application. While these portable units are not always as efficient as central systems, their low first cost and the ability to serve a single zone make them appropriate in many circumstances.

In hot and humid climates, the importance of providing sufficient dehumidification with sensible cooling is increasingly recognized. While conventional air conditioning units provide some dehumidification as a consequence of sensible cooling, in some cases space humidity levels can still exceed the upper limit of 60% relative humidity specified in ASHRAE *Standard* 55.

Several dehumidification enhancements to conventional air conditioning systems are possible to improve moisture removal characteristics and lower the space humidity level. Some simple improvements include lowering the supply air flow rate and eliminating off-cycle fan operation. Additional equipment options such as condenser/reheat coils, sensible-heat-exchanger-assisted evaporators (e.g., heat pipes), and subcooling/reheat coils can further improve dehumidification performance. Desiccants—applied as either thermally-activated units or heat recovery systems (i.e., enthalpy wheels)—can also increase dehumidification capacity and lower the indoor humidity level. Some dehumidification options add heat to the conditioned zone that, in some cases, increases the load on the sensible cooling equipment.

Air Filters

Most comfort conditioning systems that circulate air incorporate some form of air filter. Usually they are disposable or cleanable filters that have relatively low air-cleaning efficiency. Higher efficiency alternatives include pleated media filters and electronic air filters. These high-efficiency filters may have high static pressure drops. The air distribution system should be carefully evaluated before the installation of such filters.

Air filters are mounted in the return air duct or plenum and operate whenever air circulates through the duct system. Air filters are rated in accordance with ARI *Standard* 680, which is based on

ASHRAE *Standard* 52.1. Atmospheric dust spot efficiency levels are generally less than 20% for disposable filters and vary from 60 to 90% for electronic air filters.

To maintain optimum performance, the collector cells of electronic air filters must be cleaned periodically. Automatic indicators are often used to signal the need for cleaning. Electronic air filters have higher initial costs than disposable or pleated filters, but generally last the life of the air-conditioning system. Chapter 24 of the 1996 *ASHRAE Handbook—Systems and Equipment* covers the design of residential air filters in more detail.

Controls

Historically, residential heating and cooling equipment has been controlled by a wall thermostat. Today, simple wall thermostats with bimetallic strips are often replaced by microelectronic models that can set heating and cooling equipment at different temperature levels, depending on the time of day. This has led to night setback control to reduce energy demand and operating costs. For heat pump equipment, electronic thermostats can incorporate night setback with an appropriate scheme to limit use of resistance heat during recovery. Chapter 45 contains more details about automatic control systems.

MULTIFAMILY RESIDENCES

Attached homes and low-rise multifamily apartments generally use heating and cooling equipment comparable to that used in single-family dwellings. Separate systems for each unit allow individual control to suit the occupant and facilitate individual metering of energy use.

Central Forced-Air Systems

High-rise multifamily structures may also use unitary heating and cooling equipment comparable to that used in single-family dwellings. Equipment may be installed in a separate mechanical equipment room in the apartment, or it may be placed in a soffit or above a drop ceiling over a hallway or closet.

Small residential warm-air furnaces may also be used, but a means of providing combustion air and venting combustion products from gas- or oil-fired furnaces is required. It may be necessary to use a multiple-vent chimney or a manifold-type vent system. Local codes should be consulted. Direct vent furnaces that are placed near or on an outside wall are also available for apartments.

Another concept for multifamily residences (also applicable to single-family dwellings) is a combined water heating/space heating system that uses water from the domestic hot water storage tank to provide space heating. Water circulates from the storage tank to a hydronic coil in the system air handler. Space heating is provided by circulating indoor air across the coil. A split-system central air conditioner with the evaporator located in the system air handler can be included to provide space cooling.

Hydronic Central Systems

Individual heating and cooling units are not always possible or practical in high-rise structures. In this case, applied central systems are used. Two- or four-pipe hydronic central systems are widely used in high-rise apartments. Each dwelling unit has either individual room units located at the perimeter or interior, or ducted fan-coil units.

The most flexible hydronic system with the lowest operating costs is the four-pipe type, which provides heating or cooling for each apartment dweller. The two-pipe system is less flexible in that it cannot provide heating and cooling simultaneously. This limitation causes problems during the spring and fall when some apartments in a complex require heating while others require cooling due to solar or internal loads. This spring/fall problem may be overcome by operating the two-pipe system in a cooling mode and providing the relatively low amount of heating that may be required by means of individual electric resistance heaters. Chapter 12 of the 1996 *ASHRAE Handbook—Systems and Equipment* discusses hydronic design in more detail.

Through-the-Wall Units

Through-the-wall room air conditioners, packaged terminal air conditioners (PTACs), and packaged terminal heat pumps (PTHPs) give the highest flexibility for conditioning single rooms. Each room with an outside wall may have such a unit. These units are used extensively in the renovation of old buildings because they are self-contained and do not require complex piping or ductwork renovation.

Room air conditioners have integral controls and may include resistance or heat pump heating. PTACs and PTHPs have special indoor and outdoor appearance treatments, making them adaptable to a wider range of architectural needs. PTACs can include gas, electric resistance, hot water, or steam heat. Integral or remote wall-mounted controls are used for both PTACs and PTHPs. Further information may be found in Chapter 45 of the 1996 *ASHRAE Handbook—Systems and Equipment* and in ARI *Standard* 310/380.

Water-Loop Heat Pumps

Any mid- or high-rise structure having interior zones with high internal heat gains that require year-round cooling can efficiently use a water-loop heat pump. Such systems have the flexibility and control of a four-pipe system while using only two pipes. Water-source heat pumps allow for the individual metering of each apartment. The building owner pays only the utility cost for the circulating pump, cooling tower, and supplemental boiler heat. Existing buildings can be retrofitted with heat flow meters and timers on fan motors for individual metering. Economics permitting, solar or ground heat energy can provide the supplementary heat in lieu of a boiler. The ground can also provide a heat sink, which in some cases can eliminate the cooling tower.

Special Concerns for Apartment Buildings

Many ventilation systems are used in apartment buildings. Local building codes generally govern air quantities. ASHRAE *Standard* 62 requires minimum outdoor air values of 50 cfm intermittent or 20 cfm continuous or operable windows for baths and toilets, and 100 cfm intermittent or 25 cfm continuous or operable windows for kitchens.

In some buildings with centrally controlled exhaust and supply systems, the systems are operated on time clocks for certain periods of the day. In other cases, the outside air is reduced or shut off during extremely cold periods. If known, these factors should be considered when estimating heating load.

Buildings using exhaust and supply air systems 24 h a day may benefit from air-to-air heat recovery devices (see Chapter 42 of the 1996 *ASHRAE Handbook—Systems and Equipment*). Such recovery devices can reduce energy consumption by transferring 40 to 80% of the sensible and latent heat between the exhaust air and supply air streams.

Infiltration loads in high-rise buildings without ventilation openings for perimeter units are not controllable on a year-round basis by general building pressurization. When outer walls are pierced to supply outdoor air to unitary or fan-coil equipment, combined wind and thermal stack effects create other infiltration problems.

Interior public corridors in apartment buildings need positive ventilation with at least two air exchanges per hour. Conditioned supply air is preferable. Some designs transfer air into the apartments through acoustically lined louvers to provide kitchen and toilet makeup air, if necessary. Supplying air to, instead of exhausting air from, corridors minimizes odor migration from apartments into corridors.

Air-conditioning equipment must be isolated to reduce noise generation or transmission. The design and location of cooling towers must be chosen to avoid disturbing occupants within the building and neighbors in adjacent buildings. An important load, frequently overlooked, is heat gain from piping for hot water services.

In large apartment houses, a central panel may allow individual apartment air-conditioning systems or units to be monitored for maintenance and operating purposes.

MANUFACTURED HOMES

Manufactured homes are constructed at a factory and constitute over 7% of all housing units and about 25% of all new single-family homes sold each year. In the United States, heating and cooling systems in manufactured homes, as well as other facets of construction such as insulation levels, are regulated by HUD Manufactured Home Construction and Safety Standards. Each complete home or home section is assembled on a transportation frame—a chassis with wheels and axles—for transport. Manufactured homes vary in size from small, single-floor section units starting at 400 ft^2 to large, multiple sections, which when joined together provide over 2500 ft^2 and have the same appearance as site-constructed homes.

Heating systems are factory-installed and are primarily forced air downflow units feeding main supply ducts built into the subfloor, with floor registers located throughout the home. A small percentage of homes in the far South and in the Southwest use upflow units feeding overhead ducts in the attic space. Typically there is no return duct system. Air returns to the air handler from each room through hallways. The complete heating system is a reduced clearance type (0 in.) with the air-handling unit installed in a small closet or alcove usually located in a hallway. Sound control measures may

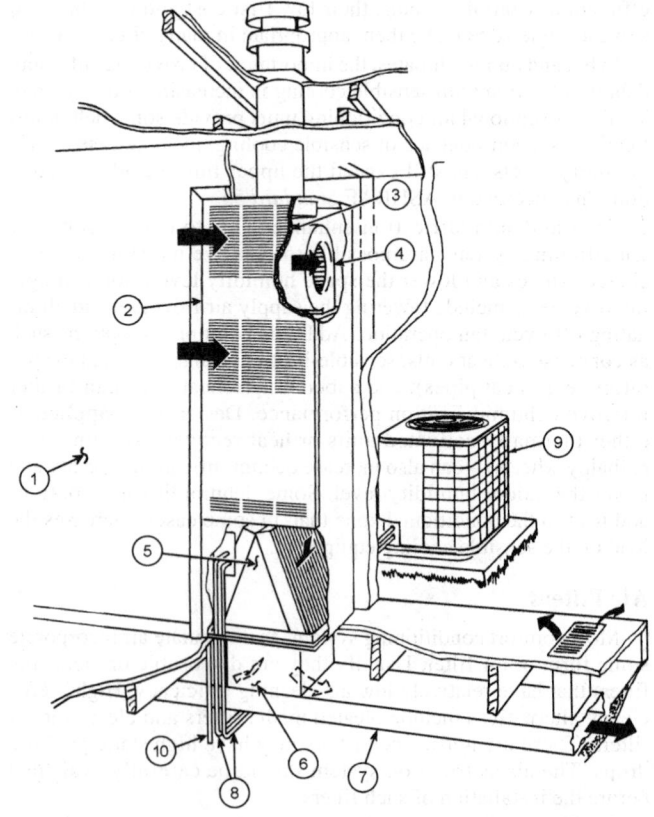

Fig. 3 Typical Installation of Heating and Cooling Equipment for a Manufactured Home

be required if large forced-air systems are installed close to sleeping areas. Gas, oil and electric furnaces or heat pumps may be installed by the home manufacturer to satisfy market requirements.

Gas and oil furnaces are compact direct vent types that have been approved for installation in a manufactured home. The special venting arrangement used is a vertical through-the-roof concentric pipe-in-pipe system that draws all air for combustion directly from the outdoors and discharges the combustion products through a windproof vent terminal. Gas furnaces must be easily convertible from liquefied petroleum to natural gas and back as required at the final site.

Manufactured homes may be cooled with add-on split or single-package air-conditioning systems when the supply ducts are adequately sized and rated for that purpose according to HUD requirements. The split-system evaporator coil may be installed in the integral coil cavity provided with the furnace. A high static pressure blower is used to overcome resistance through the furnace, the evaporator coil and the compact air duct distribution system. Supply air from a single-package air conditioner is connected with flexible air ducts to feed existing factory in-floor or overhead ducts. Dampers or other means are required to prevent the cooled, conditioned air from backflowing through a furnace cabinet.

A typical installation of a downflow gas or oil furnace with a split-system air conditioner is illustrated in Figure 3. Air returns to the furnace directly through the hallway (1), passing through a louvered door (2) on the front of the furnace. The air then passes through air filters (3) and is drawn into the top-mounted blower (4), which during the winter months forces air down over the heat exchanger, where it picks up heat. For summer cooling, the blower forces air through the split-system evaporator coil (5), which removes heat and moisture from the passing air. During heating and cooling, the conditioned air then passes through a combustible floor base and duct connector (6), before flowing into the floor air distribution duct (7). The evaporator coil is connected via quick-connect refrigerant lines (8) to a remote air-cooled condensing unit (9). The condensate collected at the evaporator is drained by a flexible hose (10), which is routed to the outdoors through the floor construction and connected to a suitable drain.

REFERENCES

ACCA. 1970. Selection of distribution systems. *Manual G.* Air-Conditioning Contractors of America, Washington, DC.

ACCA. 1995. Duct design for residential winter and summer air conditioning and equipment selection. *Manual* D, 3rd ed. Air-Conditioning Contractors of America, Washington, DC.

ACCA. 1986. Load calculation for residential winter and summer air conditioning. *Manual* J, 7th ed. Air-Conditioning Contractors of America, Washington, DC.

ARI. 1984. Application of sound rated outdoor unitary equipment. *Standard* 275-84. Air Conditioning and Refrigeration Institute, Arlington, VA.

ARI. 1995. Sound rating of outdoor unitary equipment. *Standard* 270-95. Air Conditioning and Refrigeration Institute, Arlington, VA.

ARI. 1988. Selection, installation and servicing of residential humidifiers. *Guideline* F-1988. Air Conditioning and Refrigeration Institute, Arlington, VA.

ARI. 1989. Central system humidifiers for residential applications. *Standard* 610-89. Air Conditioning and Refrigeration Institute, Arlington, VA.

ARI. 1993. Packaged terminal air-conditioners and heat pumps. *Standard* 310/380-93. Air Conditioning and Refrigeration Institute, Arlington, VA.

ARI. 1993. Residential air filter equipment. *Standard* 680-93. Air Conditioning and Refrigeration Institute, Arlington, VA.

ASHRAE. 1989. Ventilation for acceptable indoor air quality. *Standard* 62-1989.

ASHRAE. 1992. Gravimetric and dust spot procedures for testing air-cleaning devices used in general ventilation for removing particulate matter. *Standard* 52.1-1992.

ASHRAE. 1992. Thermal environmental conditions for human occupancy. *Standard* 55-1992.

ASHRAE. 1993. Energy-efficient design of new low-rise residential buildings. *Standard* 90.2-1993.

Bose, J.E., J.D. Parker, and F.C. McQuiston. 1985. Design/Data manual for closed-loop ground-coupled heat pump systems. ASHRAE.

CHAPTER 2

RETAIL FACILITIES

THIS chapter covers the design and application of air-conditioning and heating systems for various retail merchandising facilities. Load calculations, systems, and equipment are covered elsewhere in the Handbook series.

GENERAL CRITERIA

To apply equipment properly, it is necessary to know the construction of the space to be conditioned, its use and occupancy, the time of day in which greatest occupancy occurs, the physical building characteristics, and the lighting layout.

The following must also be considered:

- Electric power—size of service
- Heating—availability of steam, hot water, gas, oil, or electricity
- Cooling—availability of chilled water, well water, city water, and water conservation equipment
- Internal heat gains
- Rigging and delivery of equipment
- Structural considerations
- Obstructions
- Ventilation—opening through roof or wall for outdoor air duct, number of doors to sales area, and exposures
- Orientation of store
- Code requirements
- Utility rates and regulations
- Building standards

Specific design requirements, such as the increase in outdoor air required for exhaust where lunch counters exist, must be considered. The requirements of ASHRAE ventilation standards must be followed. Heavy smoking and objectionable odors may necessitate special filtering in conjunction with outdoor air intake and exhaust. Load calculations should be made using the procedure outlined in Chapter 28 of the 1997 *ASHRAE Handbook—Fundamentals*.

In almost all localities there is some form of energy code in effect that establishes strict requirements for insulation, equipment efficiencies, system designs, and so forth, and places strict limits on fenestration and lighting. The requirements of ASHRAE *Standard 90* should be met as a minimum guideline for retail facilities.

The selection and design of the HVAC for retail facilities are normally determined by economics. First cost is usually the determining factor for small stores; for large retail facilities, operating and maintenance costs are also considered. Generally, decisions about mechanical systems for retail facilities are based on a cash flow analysis rather than on a full life-cycle analysis.

SMALL STORES

The large glass areas found at the front of many small stores may cause high peak solar heat gain unless they have northern exposures. High heat loss may be experienced on cold, cloudy days. The HVAC system for this portion of the small store should be designed

The preparation of this chapter is assigned to TC 9.8, Large Building Air-Conditioning Applications.

to offset the greater cooling and heating requirements. Entrance vestibules and heaters may be needed in cold climates.

Many new small stores are part of a shopping center. While exterior loads will differ between stores, the internal loads will be similar; the need for proper design is important.

Design Considerations

System Design. Single-zone unitary rooftop equipment is common in store air conditioning. The use of multiple units to condition the store involves less ductwork and can maintain comfort in the event of partial equipment failure. Prefabricated and matching curbs simplify installation and ensure compatibility with roof materials.

The heat pump, offered as packaged equipment, readily adapts to small-store applications and has a low first cost. Winter design conditions, utility rates, and operating cost should be compared to those for conventional heating systems before this type of equipment is chosen.

Water-cooled unitary equipment is available for small-store air conditioning, but many communities in the United States have restrictions on the use of city and ground water for condensing purposes and require the installation of a cooling tower. Water-cooled equipment generally operates efficiently and economically.

Retail facilities often have a high sensible heat gain relative to the total heat gain. Unitary HVAC equipment should be designed and selected to provide the necessary sensible heat removal.

Air Distribution. The external static pressures available in small-store air-conditioning units are limited, and ducts should be designed to keep duct resistances low. Duct velocities should not exceed 1200 fpm and pressure drop should not exceed 0.10 in. of water per 100 ft. Average air quantities range from 350 to 450 cfm per ton of cooling in accordance with the calculated internal sensible heat load.

Attention should be paid to suspended obstacles, such as lights and displays, that interfere with proper air distribution.

The duct system should contain enough dampers for air balancing. Dampers should be installed in the return and outdoor air duct for proper outdoor air/return air balance. Volume dampers should be installed in takeoffs from the main supply duct to balance air to the branch ducts.

Control. Controls for small stores should be kept as simple as possible while still able to perform the required functions. Unitary equipment is typically available with manufacturer-supplied controls for ease of installation and operation.

Automatic dampers should be placed in the outdoor air intake to prevent outdoor air from entering when the fan is turned off.

Heating controls vary with the nature of the heating medium. Duct heaters are generally furnished with manufacturer-installed safety controls. Steam or hot water heating coils require a motorized valve for heating control.

Time clock control can limit unnecessary HVAC operation. Unoccupied reset controls should be provided in conjunction with timed control.

Maintenance. To protect the initial investment and ensure maximum efficiency, the maintenance of air-conditioning units in small stores should be contracted out to a reliable service company on a yearly basis. The contract should clearly specify responsibility for filter replacements, lubrication, belts, coil cleaning, adjustment of controls, compressor maintenance, replacement of refrigerant, pump repairs, electrical maintenance, winterizing, system startup, and extra labor required for repairs.

Improving Operating Cost. Outdoor air economizers can reduce the operating cost of cooling in most climates. They are generally available as factory options or accessories with roof-mounted units. Increased exterior insulation generally reduces operating energy requirements and may in some cases allow the size of installed equipment to be reduced. Many codes now include minimum requirements for insulation and fenestration materials.

DISCOUNT AND BIG BOX STORES

Large warehouse or big box stores attract customers with discount prices when large quantities are purchased. These stores typically have high bay fixture displays and usually store merchandise in the sales area. They feature a wide range of merchandise and may include such diverse areas as a lunch counter, an auto service area, a supermarket area, a pharmacy, and a garden shop. Some stores sell pets, including fish and birds. This variety of merchandise must be considered in designing air conditioning. The design and application suggestions for small stores also apply to discount stores.

Another type of big box facility provides both dry good and grocery areas. The grocery area is typically treated as a traditional stand-alone grocery. Conditioning of outside air into the dry goods areas must be considered to limit the introduction of excess moisture that will migrate to the freezer aisles.

Hardware, lumber and furniture, etc. is also retailed from big box facilities. An particular concern in this type of facility is ventilation for material handling equipment, such as forklift trucks.

In addition to the sales area in any of these facilities such areas as stockrooms, rest rooms, offices, and special storage rooms for perishable merchandise may require air conditioning or refrigeration.

Load Determination

Operating economics and the spaces served often dictate the indoor design conditions. Some stores may base summer load calculations on a higher inside temperature (e.g., 80°F db) but then set the thermostats to control at 72 to 75°F db. This reduces the installed equipment size while providing the desired inside temperature most of the time.

Special rooms for storage of perishable goods are usually designed with separate unitary air conditioners.

The heat gain from lighting will not be uniform throughout the entire area. For example, jewelry and other specialty displays have lighting heat gains as high as 6 to 8 W per square foot of floor area, while the typical sales area has an average value of 2 to 4 W/ft^2. For stockrooms and receiving, marking, toilet, and rest room areas, a value of 2 W/ft^2 may be used. When available, actual lighting layouts rather than average values should be used for load computation.

The store owner usually determines the population density for a store based on its location, size, and past experience.

Food preparation and service areas in discount and outlet stores range from small lunch counters having heat-producing equipment (ranges, griddles, ovens, coffee urns, toasters) in the conditioned space to large deluxe installations with kitchens separate from the conditioned space. Chapter 30 has specific information on HVAC systems for kitchen and eating spaces.

Data on the heat released by special merchandising equipment, such as amusement rides for children or equipment used for preparing speciality food items (e.g., popcorn, pizza, frankfurters, hamburgers, doughnuts, roasted chickens, cooked nuts, etc.), should be obtained from the equipment manufacturers.

Ventilation and outdoor air must be provided as required in ASHRAE standards and local codes.

Design Considerations

Heat released by the installed lighting is usually sufficient to offset the design roof heat loss. Therefore, the interior areas of these stores need cooling during business hours throughout the year. The perimeter areas, especially the storefront and entrance areas, may have highly variable heating and cooling requirements. Proper zone control and HVAC design are essential. The location of checkout lanes in this area makes proper environmental control even more important.

System Design. The important factors in selecting discount and outlet store air-conditioning systems are (1) installation costs, (2) floor space required for equipment, (3) maintenance requirements and equipment reliability, and (4) simplicity of control. Roof-mounted units are the most commonly used.

Air Distribution. The air supply for large sales areas should generally be designed to satisfy the primary cooling requirement. In designing air distribution for the perimeter areas, the variable heating and cooling requirements must be considered.

Because of the store requirement to maintain high, clear areas for display and restocking, air is generally distributed from heights of 14 ft and greater. Air distribution at these heights requires high velocities in the heating season to overcome the buoyancy of the hot air. The velocity of this discharge air creates turbulence in the space, and induces air from the ceiling area complete mixing of the air.

The designer should take advantage of stratification to reduce the equipment load. By introducing air near the customers at low velocity, the air will stratify during the cooling season. With this method of air distribution, the set-point temperature can be maintained in the occupant zone and the temperature in the upper space can be allowed to rise. However, this strategy requires equipment to destratify the air during the heating season. Space-mounted fans, and radiant heating at the perimeter, entrance, and sales areas may be required.

Control. Because the controls are usually operated by personnel who have little knowledge of air-conditioning, it should be simple, dependable, and fully automatic systems—it should be as simple to operate as a residential system. Most unitary equipment has automatic electronic controls for ease of operation.

Maintenance. Most stores do not employ trained maintenance personnel; they rely instead on service contracts with either the installer or a local service company. For suggestions on lowering operating costs, see the previous section on Small Stores.

SUPERMARKETS

Load Determination

Heating and cooling loads should be calculated using the methods outlined in Chapter 28 of the 1997 *ASHRAE Handbook—Fundamentals*. Data for calculating the loads due to people, lights, motors, and heat-producing equipment should be obtained from the store owner or manager or from the equipment manufacturer. In supermarkets, space conditioning is required both for human comfort and for proper operation of refrigerated display cases. The air-conditioning unit should introduce a minimum quantity of outdoor air. This quantity is either the volume required for ventilation based on ASHRAE *Standard* 62 or the volume required to maintain slightly positive pressure in the space, whichever is larger.

Many supermarkets are units of a large chain owned or operated by a single company. The standardized construction, layout, and equipment used in designing many similar stores simplify load calculations.

It is important that the final air-conditioning load be correctly determined. Refer to manufacturers' data for information on the total heat extraction, sensible heat, latent heat, and percentage of latent to total load for display cases. Engineers report considerable fixture heat removal (case load) variation as the relative humidity and temperature vary in comparatively small increments. Relative humidity above 55% substantially increases the load, while reduced relative humidity substantially decreases the load, as shown in Figure 1. Trends in store design, which include more food refrigeration and more efficient lighting, reduce the sensible component of the load even further.

To calculate the total load and percentage of latent and sensible heat that the air conditioning must handle, the refrigerating effect imposed by the display fixtures must be subtracted from the building's gross air-conditioning requirements.

Modern supermarket designs have a high percentage of closed refrigerated display fixtures. These vertical cases have large glass display doors and greatly reduce the problem of latent and sensible heat removal from the occupied space. The doors do, however, require heaters to prevent condensation and fogging. These heaters should be cycled based on some kind of automatic control.

Design Considerations

Store owners and operators frequently complain about cold aisles, heaters that operate even when the outdoor temperature is above 70°F, and air-conditioners that operate infrequently. These problems are usually attributed to spillover of cold air from open refrigerated display equipment.

Although refrigerated display equipment may be the cause of cold stores, the problem is not due to excessive spillover or improperly operating equipment. Heating and air-conditioning systems must compensate for the effects of open refrigerated display equipment. Design considerations include the following:

1. Increased heating requirement due to removal of large quantities of heat, even in summer.
2. Net air-conditioning load after deducting the latent and sensible refrigeration effect (Item 1). The load reduction and change in sensible-latent load ratio have a major effect on equipment selection.
3. Need for special air circulation and distribution to offset the heat removed by open refrigerating equipment.
4. Need for independent temperature and humidity control.

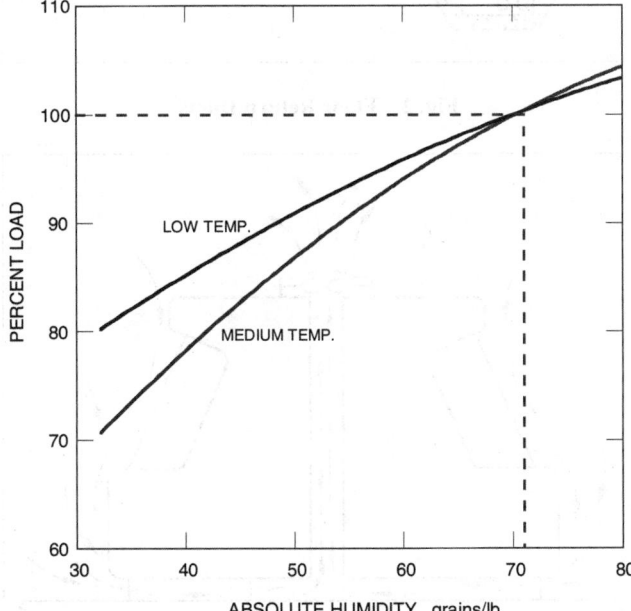

**Fig. 1   Refrigerated Case Load Variation with Store
Air Humidity**

Each of these problems is present to some degree in every supermarket, although situations vary with climate and store layout. Methods of overcoming these problems are discussed in the following sections. Extremely high energy cost may ensue if the year-round air-conditioning system has not been designed to compensate for the effects of refrigerated display equipment.

Heat Removed by Refrigerated Displays. The display refrigerator not only cools a displayed product but envelops it in a blanket of cold air that absorbs heat from the room air in contact with it. Approximately 80 to 90% of the heat removed from the room by vertical refrigerators is absorbed through the display opening. Thus, the open refrigerator acts as a large air cooler, absorbing heat from the room and rejecting it via the condensers outside the building. Occasionally, this conditioning effect can be greater than the design air-conditioning capacity of the store. The heat removed by the refrigeration equipment *must* be considered in the design of the air-conditioning and heating systems because this heat is being removed constantly, day and night, summer and winter, regardless of the store temperature.

The display cases increase the heating requirement of the building such that heat is often required at unexpected times. The following example is an indication of the extent of this cooling effect. The desired store temperature is 75°F. Store heat loss or gain is assumed to be 15,000 Btu/h per °F of temperature difference between outdoor and store temperature. (This value varies with store size, location, and exposure.) The heat removed by refrigeration equipment is 190,000 Btu/h. (This value varies with the number of refrigerators.) The latent heat removed is assumed to be 19% of the total, leaving 81% or 154,000 Btu/h sensible heat removed, which will cool the store 154,000/15,000 = 10°F. By constantly removing sensible heat from its environment, the refrigeration equipment in this store will cool the store 10°F below outdoor temperature in winter and in summer. Thus, in mild climates, heat must be added to the store to maintain comfort conditions.

The designer has the choice of discarding or reclaiming the heat removed by refrigeration. If economics and store heat data indicate that the heat should be discarded, heat extraction from the space must be included in the heating load calculation. If this internal heat loss is not included, the heating system may not have sufficient capacity to maintain the design temperature under peak conditions.

The additional sensible heat removed by the cases may change the air-conditioning latent load ratio from 32% to as much as 50% of the net heat load. Removal of a 50% latent load by means of refrigeration alone is very difficult. Normally, it requires specially designed equipment with reheat or chemical adsorption.

Multishelf refrigerated display equipment requires 55% rh or less. In the dry-bulb temperature ranges of average stores, humidity in excess of 55% can cause heavy coil frosting, product zone frosting in low-temperature cases, fixture sweating, and substantially increased refrigeration power consumption.

A humidistat can be used during summer cooling to control humidity by transferring heat from the condenser to a heating coil in the airstream. The store thermostat maintains proper summer temperature conditions. Override controls prevent conflict between the humidistat and the thermostat.

The equivalent result can be accomplished with a conventional air-conditioning system by using three- or four-way valves and reheat condensers in the ducts. This system borrows heat from the standard condenser and is controlled by a humidistat. For higher energy efficiency, specially designed equipment should be considered. Desiccant dehumidifiers and heat pipes have also been used.

Humidity. Cooling from the refrigeration equipment does not preclude the need for air conditioning. On the contrary, it increases the need for humidity control.

With increases in store humidity, heavier loads are imposed on the refrigeration equipment, operating costs rise, more defrost periods are required, and the display life of products is shortened. The dew point rises with the relative humidity, and sweating can become profuse to the extent that even nonrefrigerated items such as shelving superstructures, canned products, mirrors, and walls may sweat.

There are two commonly used devices for achieving humidity control. One is an electric vapor compression air conditioner, which overcools the air to condense the moisture and then reheats it to a comfortable temperature. Condenser waste heat is often used. Vapor compression holds humidity levels at 50 to 55% rh.

The second dehumidification device is a desiccant dehumidifier. A desiccant absorbs or adsorbs moisture directly from the air to its surface. The desiccant material is reactivated by passing hot air at 180 to 230°F through the desiccant base. Condenser waste heat can provide as much as 40% of the heat required. Desiccant systems in supermarkets can maintain humidity in the lower ranges of the comfort index. Lower humidity results in lower operating costs for the refrigerated cases.

System Design. The same air-handling equipment and distribution system are generally used for both cooling and heating. The entrance area is the most difficult section to heat. Many supermarkets in the northern United States are built with vestibules provided with separate heating equipment to temper the cold air entering from the outdoors. Auxiliary heat may also be provided at the checkout area, which is usually close to the front entrance. Methods of heating entrance areas include the use of (1) air curtains, (2) gas-fired or electric infrared radiant heaters, and (3) waste heat from the refrigeration condensers.

Air-cooled condensing units are the most commonly used in supermarkets. Typically, a central air handler conditions the entire sales area. Specialty areas like bakeries, computer rooms, or warehouses are better served with a separate air handler because the loads in these areas vary and require different control than the sales area.

Most installations are made on the roof of the supermarket. If air-cooled condensers are located on the ground outside the store, they must be protected against vandalism as well as truck and customer traffic. If water-cooled condensers are used on the air-conditioning equipment and a cooling tower is required, provisions should be made to prevent freezing during winter operation.

Air Distribution. Designers overcome the concentrated load at the front of a supermarket by discharging a large portion of the total air supply into the front third of the sales area.

The air supply to the space with the vapor compression system has typically been 1 cfm per square foot of sales area. This value should be calculated based on the sensible and latent internal loads. The desiccant system typically requires less air supply due to its high moisture removal rate. In most cases, only about 40% of the circulation rate passes through the dehumidifier.

Being more dense, the air cooled by the refrigerators settles to the floor and becomes increasingly colder, especially in the first 36 in. above the floor. If this cold air remains still, it causes discomfort and serves no purpose, even when other areas of the store need more cooling. Cold floors or areas in the store cannot be eliminated by the simple addition of heat. Any reduction of air-conditioning capacity without circulation of the localized cold air would be analogous to installing an air conditioner without a fan. To take advantage of the cooling effect of the refrigerators and provide an even temperature in the store, the cold air must be mixed with the general store air.

To accomplish the necessary mixing, air returns should be located at floor level; they should also be strategically placed to remove the cold air near concentrations of refrigerated fixtures. The returns should be designed and located to avoid the creation of drafts. There are two general solutions to this problem:

1. **Return Ducts in Floor.** This is the preferred method and can be accomplished in two ways. The floor area in front of the refrigerated display cases is the coolest area. Refrigerant lines are run to all of these cases, usually in tubes or trenches. If the trenches or tubes are enlarged and made to open under the cases for air return, air can be drawn in from the cold area (see Figure 2). The air is returned to the air-handling unit through a tee connection to the trench before it enters the back room area. The opening through which the refrigerant lines enter the back room should be sealed.

   If refrigerant line conduits are not used, the air can be returned through inexpensive underfloor ducts. If refrigerators have insufficient undercase air passage, the manufacturer should be consulted. Often they can be raised off the floor approximately 1.5 in. Floor trenches can also be used as ducts for tubing, electrical supply, and so forth.

   Floor-level return relieves the problem of localized cold areas and cold aisles and uses the cooling effect for store cooling, or increases the heating efficiency by distributing the air to the areas that need it most.

2. **Fans Behind Cases.** If ducts cannot be placed in the floor, circulating fans can draw air from the floor and discharge it above the cases (see Figure 3). While this approach prevents objectionable cold aisles in front of the refrigerated display cases, it does not prevent the area with a concentration of refrigerated fixtures from remaining colder than the rest of the store.

Control. Store personnel should only be required to change the position of a selector switch to start or stop the system or to change

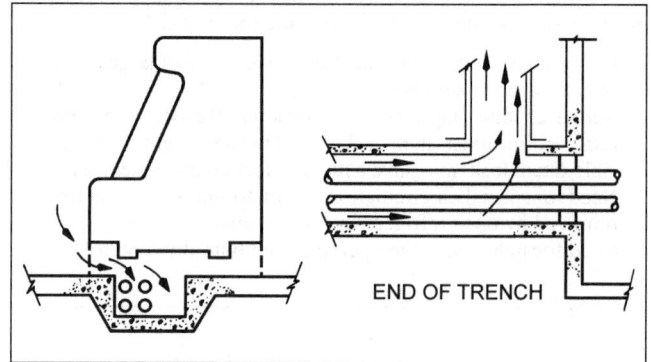

Fig. 2 Floor Return Ducts

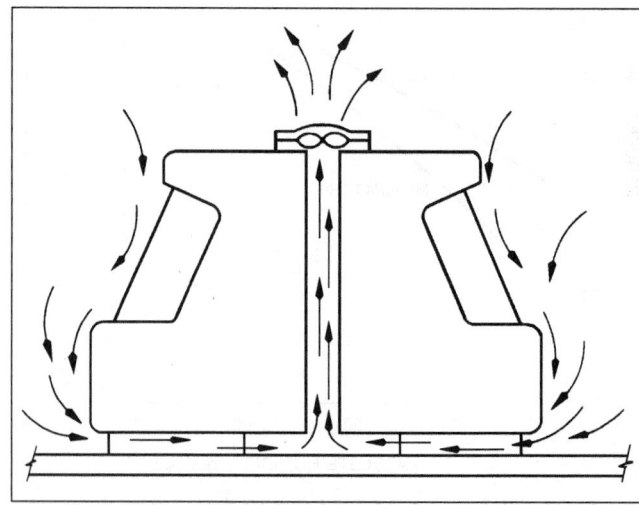

Fig. 3 Air Mixing Using Fans Behind Cases

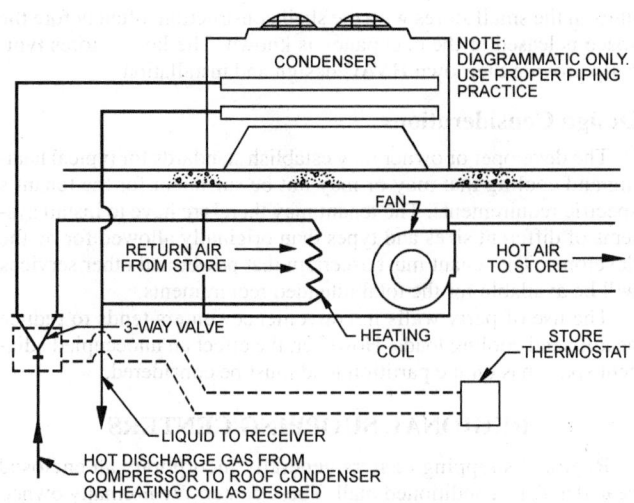

Fig. 4   Heat Reclaiming Systems

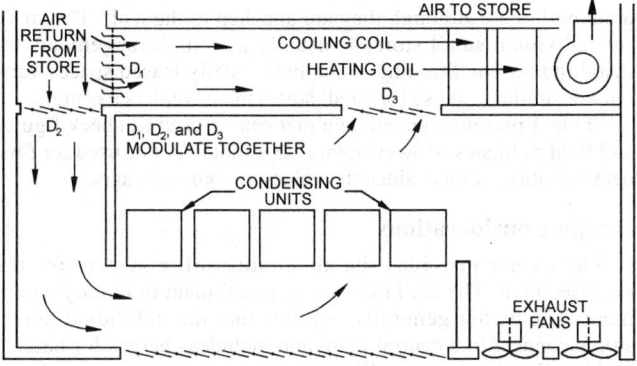

Fig. 5   Machine Room with Automatic Temperature Control Interlocked with Store Temperature Control

from heating to cooling or from cooling to heating. Control systems for heat recovery applications are more complex and should be coordinated with the equipment manufacturer.

Maintenance and Heat Reclamation. Most supermarkets, except large chains, do not employ trained maintenance personnel, but rather rely on service contracts with either the installer or a local service company. This relieves the store management of the responsibility of keeping the air conditioning operating properly.

The heat extracted from the store and the heat of compression may be reclaimed for heating cost saving. One method of reclaiming the rejected heat is to use a separate condenser coil located in the air conditioner's air handler, either alternately or in conjunction with the main refrigeration condensers, to provide heat as required (see Figure 4). Another system uses water-cooled condensers and delivers its rejected heat to a water coil in the air handler.

The heat rejected by conventional machines using air-cooled condensers may be reclaimed by proper duct and damper design (see Figure 5). Automatic controls can either reject this heat to the outdoors or recirculate it through the store.

DEPARTMENT STORES

Department stores vary in size, type, and location, so air-conditioning design should be specific to each store. An ample minimum quantity of outdoor air reduces or eliminates odor problems. Essential features of a quality system include (1) an automatic control system properly designed to compensate for load fluctuations, (2) zoned air distribution to maintain uniform conditions under shifting loads, and

(3) use of outdoor air for cooling during intermediate seasons and peak sales periods. It is also desirable to adjust indoor temperature for variations in the outdoor temperature. While the close control of humidity is not necessary, a properly designed system should operate to maintain relative humidity at 50% or below with a corresponding dry-bulb temperature of 78°F. This humidity limit eliminates musty odors and retards perspiration, particularly in fitting rooms.

Load Determination

Because the occupancy (except store personnel) is transient, indoor conditions are commonly set not to exceed 78°F db and 50% rh at outdoor summer design conditions, and 70°F db at outdoor winter design conditions. Winter humidification is seldom used in store air conditioning.

The number of customers and store personnel normally found on each conditioned floor must be ascertained, particularly in specialty departments or other areas having a greater-than-average concentration of occupants. Lights should be checked for wattage and type. Table 1 gives approximate values for lighting in various areas; Table 2 gives approximate occupancies.

Other loads, such as those from motors, beauty parlor and restaurant equipment, and any special display or merchandising equipment, should be determined.

The minimum outdoor air requirement should be as defined in the ASHRAE *Standard* 62, which are generally acceptable and adequate for removing odors and keeping the building atmosphere fresh. However, local ventilation ordinances may require greater quantities of outdoor air.

Paint shops, alteration rooms, rest rooms, eating places, and locker rooms should be provided with positive exhaust ventilation, and their requirements must be checked against local codes.

Design Considerations

Before performing load calculations, the designer should examine the store arrangement to determine what will affect the load and the system design. For existing buildings, a survey can be made of actual construction, floor arrangement, and load sources. For new buildings, an examination of the drawings and a discussion with the architect or owner will be required.

Larger stores may contain beauty parlors, restaurants, lunch counters, or auditoriums. These special areas may operate during all store hours. If one of these areas has a load that is small in proportion to the total load, the load may be met by the portion of the air conditioning serving that floor. If present or future operation could for any reason be compromised by such a strategy, this space should be served by separate air conditioning. Because of the concentrated load in the beauty parlor, separate air distribution should be provided for this area.

Table 1   Approximate Lighting Load for Department Stores

Area	W/ft^2
Basement	3 to 5
First floor	4 to 7
Upper floors, women's wear	3 to 5
Upper floors, house furnishings	2 to 3

Table 2   Approximate Occupancy for Department Stores

Area	ft^2 per person
Basement, metropolitan area	25 to 100
Basement, other with occasional peak	25 to 100
First floor, metropolitan area	25 to 75
First floor, suburban	25 to 75
Upper floors, women's wear	50 to 100
Upper floors, house furnishings	100 or more

The restaurant, because of its required service facilities, is generally centrally located. It is often used only during the noon hours. For control of odors, a separate air-handling system should be considered. Future plans for the store must be ascertained because they can have a great effect on the type of air conditioning and refrigeration to be used.

System Design. Air conditioning for department stores may be of the unitary or central-station type. Selection should be based on owning and operating costs as well as any other special considerations for the particular store, such as store hours, load variations, and size of load.

Large department stores often use central-station systems consisting of air-handling units having chilled water cooling coils, hot water heating coils, fans, and filters. Air systems must have adequate zoning for varying loads, occupancy, and usage. Wide variations in people loads may justify consideration of variable volume air distribution systems. The water chilling and heating plants distribute water to the various air handlers and zones and may take advantage of some load diversity throughout the building.

Air-conditioning equipment should not be placed in the sales area; instead, it should be located in ceiling, roof, and mechanical equipment room areas whenever practicable. Maintenance and operation after installation should be considered in the location of equipment.

Air Distribution. All buildings must be studied for orientation, wind exposure, construction, and floor arrangement. These factors affect not only load calculations, but also zone arrangements and duct locations. In addition to entrances, wall areas with significant glass, roof areas, and population densities, the expected locations of various departments (e.g., the lamp department) should be considered. Flexibility must be left in the duct design to allow for future movement of departments. The preliminary duct layout should also be checked in regard to winter heating to determine any special situations. It is usually necessary to design separate air systems for entrances, particularly in northern areas. This is also true for storage areas where cooling is not contemplated.

Air curtains may be installed at entrance doorways to limit or prevent infiltration of unconditioned air, at the same time providing greater ease of entry.

Control. The necessary extent of automatic control depends on the type of installation and the extent to which it is zoned. Control must be such that correctly conditioned air is delivered to each zone. Outdoor air intake should be automatically controlled to operate at minimum cost. Partial or full automatic control should be provided for cooling to compensate for load fluctuations. Completely automatic refrigeration plants should be considered.

Maintenance. Most department stores employ personnel for routine operation and maintenance, but rely on service and preventive maintenance contracts for refrigeration cycles, chemical treatment, central plant systems, and major repairs.

Improving Operating Cost. An outdoor air economizer can reduce the operating cost of cooling in most climates. These are generally available as factory options or accessories with the air-handling units or control systems. Heat recovery and thermal storage should also be analyzed.

CONVENIENCE CENTERS

Many small stores, discount stores, supermarkets, drugstores, theaters, and even department stores are located in convenience centers. The space for an individual store is usually leased. Arrangements for installing air conditioning in leased space vary. In a typical arrangement, the developer builds a shell structure and provides the tenant with an allowance for typical heating and cooling and other minimum interior finish work. The tenant must then install an HVAC system. In another arrangement, developers install HVAC units in the small stores with the shell construction, often before the space is leased or the occupancy is known. The larger stores typically provide their own HVAC design and installation.

Design Considerations

The developer or owner may establish standards for typical heating and cooling that may or may not be sufficient for the tenant's specific requirements. The tenant may therefore have to install systems of different sizes and types than originally allowed for by the developer. The tenant must ascertain that power and other services will be available for the total intended requirements.

The use of party walls in convenience centers tends to reduce heating and cooling loads. However, the effect an unoccupied adjacent space has on the partition load must be considered.

REGIONAL SHOPPING CENTERS

Regional shopping centers generally incorporate an enclosed heated and air-conditioned mall. These centers are normally owned by a developer, who may be an independent party, a financial institution, or one of the major tenants in the center.

The major department stores are typically considered to be separate buildings, although they are attached to the mall. The space for individual small stores is usually leased. Arrangements for installing air conditioning in the individually leased spaces vary, but are similar to those for small stores in convenience centers.

Table 3 presents typical data that can be used as check figures and field estimates. However, this table should not be used for final determination of load, since the values are only averages.

Design Considerations

The owner provides the air-conditioning system for the enclosed mall. The mall may use a central plant or unitary equipment. The owner generally requires that the individual tenant stores connect to a central plant and includes charges for heating and cooling in the rent. Where unitary systems are used, the owner generally requires that the individual tenant install a unitary system of similar design.

The owner may establish standards for typical heating and cooling systems that may or may not be sufficient for the tenant's specific requirements. Therefore, the tenant may have to install systems of different sizes than originally allowed for by the developer.

Leasing arrangements may include provisions that have a detrimental effect on conservation (such as allowing excessive lighting and outdoor air or deleting requirements for economizer systems). The designer of HVAC for tenants in a shopping center must be well aware of the lease requirements and work closely with leasing agents to guide these systems toward better energy efficiency.

Many regional shopping centers now contain specialty food court areas that require special considerations for odor control, outdoor air requirements, kitchen exhaust, heat removal, and refrigeration equipment.

System Design. Regional shopping centers vary widely in physical arrangement and architectural design. Single-level and smaller centers usually use unitary systems for mall and tenant air conditioning; multilevel and larger centers usually use a central plant. The owner sets the design of the mall and generally requires that similar systems be installed for tenant stores.

A typical central plant may distribute chilled air to the individual tenant stores and to the mall air-conditioning system and employ variable volume control and electric heating at the local use point. Some plants distribute both hot and chilled water. All-air systems have also been used that distribute chilled or heated air to the individual tenant stores and to the mall air conditioning and use variable volume control at the local use point. The central plant provides improved efficiency and better overall economics of operation. They also provide the basic components required for smoke control.

Table 3 Typical Installed Capacity and Energy Use in Enclosed Mall Centers[a]

Based on 1979 Data—Midwestern United States

Type of Space	Installed Cooling Btu/h · ft^2	Annual Consumption, kWh/ft^2			
		Lighting[b]	Cooling[c]	Heating[d]	Miscellaneous
Candy store	44.8 to 78.0	23.9 to 29.6	7.8 to 23.5	9.6 to 5.7	2.4 to 70.1
Clothing store	37.3 to 45.5	14.1 to 24.9	8.0 to 9.6	8.9 to 5.1	1.2 to 6.7
Fast food	48.2 to 78.0	16.7 to 32.4	9.8 to 23.5	9.6 to 3.8	38.1 to 70.0
Game room	33.8 to 44.5	6.8 to 12.5	7.2 to 7.3	13.8 to 4.7	0.2 to 13.8
Gen. merchandise	32.9 to 43.6	13.2 to 23.3	6.3 to 9.7	6.3 to 5.6	1.4 to 9.3
Gen. service	39.3 to 50.3	15.7 to 17.4	7.8 to 9.2	12.0 to 5.9	7.6 to 8.2
Gift store	36.4 to 51.1	12.8 to 22.9	6.3 to 10.0	7.1 to 4.9	0.4 to 2.2
Grocery	69.5 to 86.2	10.9 to 19.3	5.3 to 8.9	6.0 to 7.0	18.6 to 21.0
Jewelry	53.5 to 66.1	37.2 to 44.9	10.6 to 12.3	5.6 to 3.6	7.6 to 9.3
Mall	30.0 to 48.0	5.4 to 12.0	8.3 to 13.7	11.5 to 5.6	0.2 to 1.5
Restaurant	40.0 to 53.0	5.0 to 22.0	7.8 to 12.3	19.0 to 16.7	17.2 to 21.0
Shoe store	36.9 to 50.6	20.4 to 32.1	7.4 to 11.4	7.2 to 4.6	1.0 to 2.3
Center average	30.0 to 48.0	18.0 to 28.0	8.0 to 15.0	6.0 to 3.0	1.0 to 3.0

[a]Operation of center assumed to be 12 h/day and 6.5 days/week.
[b]Lighting includes miscellaneous and receptacle loads.
[c]Cooling includes blower motor and is for unitary-type system.
[d]Heating includes blower and is for electric resistance heating.

Air Distribution. Air distribution for individual stores should be designed for a particular space occupancy. Some tenant stores maintain a negative pressure relative to the mall for odor control.

The air distribution should maintain a slight positive pressure relative to atmospheric pressure and a neutral pressure relative to most of the individual tenant stores. Exterior entrances should have vestibules with independent heating systems.

Smoke management is required by many building codes, so the air distribution should be designed to easily accommodate smoke control requirements.

Maintenance. Methods for ensuring the operation and maintenance of HVAC systems in regional shopping centers are similar to those used in department stores. Individual tenant stores may have to provide their own maintenance.

Improving Operating Cost. Methods for lowering operating costs in shopping centers are similar to those used in department stores. Some shopping centers have successfully employed cooling tower heat exchanger economizers.

Central plant systems for regional shopping centers typically have much lower operating costs than unitary systems. However, the initial cost of the central plant system is typically higher.

MULTIPLE-USE COMPLEXES

Multiple-use complexes are being developed in most metropolitan areas. These complexes generally combine retail facilities with other facilities such as offices, hotels, residences, or other commercial space into a single site. This consolidation of facilities into a single site or structure provides benefits such as improved land use; structural savings; more efficient parking; utility savings; and opportunities for more efficient electrical, fire protection, and mechanical systems.

Load Determination

The various occupancies may have peak HVAC demands that occur at different times of the day or even of the year. Therefore, the HVAC loads of the various occupancies should be determined independently. Where a combined central plant is under consideration, a block load should also be determined.

Design Considerations

Retail facilities are generally located on the lower levels of multiple-use complexes, and other commercial facilities are on upper levels. Generally, the perimeter loads of the retail portion differ from those of the other commercial spaces. Greater lighting and population densities also make the HVAC demands for the retail space different from those for the other commercial space.

The differences in HVAC characteristics for the various occupancies within a multiple-use complex indicate that separate air handling and distribution should be used for the separate spaces. However, combining the heating and cooling requirements of the various facilities into a central plant can achieve a substantial saving. A combined central heating and cooling plant for a multiple-use complex also provides good opportunities for heat recovery, thermal storage, and other similar functions that may not be economical in a single-use facility.

Many multiple-use complexes have atriums. The stack effect created by atriums requires special design considerations for tenants and space on the main floor. Areas near entrances require special considerations to prevent drafts and accommodate extra heating requirements.

System Design. Individual air-handling and distribution systems should be designed for the various occupancies. The central heating and cooling plant may be sized for the block load requirements, which may be less than the sum of each occupancy's demand.

Control. Multiple-use complexes typically require centralized control. It may be dictated by requirements for fire and smoke control, security, remote monitoring, billing for central facilities use, maintenance control, building operations control, and energy management.

COMMERCIAL AND PUBLIC BUILDINGS

THIS chapter summarizes the load characteristics and both the general and specific design criteria that apply to commercial buildings. Design criteria include such factors as comfort level; cost; and fire, smoke, and odor control. Specific information is included on dining and entertainment centers, office buildings, bowling centers, communication centers, transportation centers, and warehouses are covered. Museums, libraries, and archives are covered in Chapter 20.

GENERAL CRITERIA

In theory, most systems, if properly applied, can be successful in any building. However, in practice, such factors as initial and operating costs, space allocation, architectural design, location, and the engineer's evaluation and experience limit the proper choices for a given building type.

Heating and air-conditioning systems that are simple in design and of the proper size for a given building generally have fairly low maintenance and operating costs. For optimum results, as much inherent thermal control as is economically possible should be built into the basic structure. The relationship between the shape, orientation, and air-conditioning capacity of a building should also be considered. Because the exterior load may vary from 30 to 60% of the total air-conditioning load when the fenestration area ranges from 25 to 75% of the exterior envelope surface area, it may be desirable to minimize the perimeter area. For example, a rectangular building with a four-to-one aspect ratio requires substantially more refrigeration than a square building with the same floor area.

Building size, shape, and component selection is normally determined by the building architect and/or owner. Bringing about changes in any of these areas in an attempt to reach optimum results requires the cooperation of these individuals or groups.

Proper design also considers controlling noise and minimizing pollution of the atmosphere and water systems around the building.

Retrofitting existing buildings is also an important part of the construction industry because of increased costs of construction and the necessity of reducing energy consumption. Table 1 lists factors to consider before selecting a system for any building. The selection is often made by the owner and may not be based on an engineering study. To a great degree, system selection is based on the engineer's ability to relate those factors involving higher first cost or lower life-cycle cost and benefits that have no calculable monetary value.

Some buildings are constructed with only heating and ventilating systems. For these buildings, greater design emphasis should be placed on natural or forced ventilation systems to minimize occupant discomfort during hot weather. If such buildings are to provide for future cooling, humidification, or both, the design principles are the same as those for a fully air-conditioned building.

Load Characteristics

Load characteristics for the building must be understood to ensure that systems respond adequately at part load as well as at full load. Systems must be capable of responding to load fluctuations based on the combination of occupancy variations, process load shifts, solar load variations, and atmospheric weather conditions (e.g., temperature and humidity changes). Some building loads may be brief and infrequent such as an annual meeting of a large group in a conference room, while others may be more constant and long lasting such as the operation of data processing equipment.

Analysis of any building for heat recovery or total energy systems requires sufficient load profile and load duration information on all forms of building input to (1) properly evaluate the instantaneous effect of one on the other when no energy storage is contemplated and (2) evaluate short-term effects (up to 48 h) when energy storage is used.

Load profile curves consist of appropriate energy loads plotted against the time of day. Load duration curves indicate the accumulated number of hours at each load condition, from the highest to the lowest load for a day, a month, or a year. The area under load profile and load duration curves for corresponding periods is equivalent to the load multiplied by the time. These calculations must consider the type of air and water distribution systems in the building.

Load profiles for two or more energy forms during the same operating period may be compared to determine load-matching characteristics under diverse operating conditions. For example, when thermal energy is recovered from a diesel-electric generator at a rate equal to or less than the thermal energy demand, the energy can be used instantaneously, avoiding waste. But it may be worthwhile to store thermal energy when it is generated at a greater rate than demanded. A load profile study helps determine the economics of thermal storage.

Similarly, with internal source heat recovery, load matching must be integrated over the operating season with the aid of load duration curves for overall feasibility studies. These curves are useful in energy consumption analysis calculations as a basis for hourly input values in computer programs (see Chapter 30 of the 1997 *ASHRAE Handbook—Fundamentals*).

Aside from environmental considerations, the economic feasibility of district heating and cooling is influenced by load density and diversity factors for branch feeds to buildings along distribution mains. For example, the load density or energy per unit length of distribution main can be small enough in a complex of low-rise, lightly loaded buildings located at a considerable distance from one another, to make a central heating, cooling, or heating and cooling plant uneconomical.

Concentrations of internal loads peculiar to each application are covered later in this chapter and in Chapters 27, 28, and 29 of the 1997 *ASHRAE Handbook—Fundamentals*.

The preparation of this chapter is assigned to TC 9.8, Large Building Air-Conditioning Applications.

Table 1   General Design Criteria[a, b]

General Category	Specific Category	Inside Design Conditions		Air Movement	Circulation, air changes per hour
		Winter	Summer		
Dining and Entertainment Centers	Cafeterias and Luncheonettes	70 to 74°F 20 to 30% rh	78°F[d] 50% rh	50 fpm at 6 ft above floor	12 to 15
	Restaurants	70 to 74°F 20 to 30% rh	74 to 78°F 55 to 60% rh	25 to 30 fpm	8 to 12
	Bars	70 to 74°F 20 to 30% rh	74 to 78°F 50 to 60% rh	30 fpm at 6 ft above floor	15 to 20
	Nightclubs and Casinos	70 to 74°F 20 to 30% rh	74 to 78°F 50 to 60% rh	below 25 fpm at 5 ft above floor	20 to 30
	Kitchens	70 to 74°F	85 to 88°F	30 to 50 fpm	12 to 15[g]
Office Buildings		70 to 74°F 20 to 30% rh	74 to 78°F 50 to 60% rh	25 to 45 fpm 0.75 to 2 cfm/ft^2	4 to 10
Museums, Libraries, and Archives (Also see Chapter 20.)	Average	68 to 72°F 40 to 55% rh		below 25 fpm	8 to 12
	Archival	See Chapter 20, Museums, Libraries, and Archives		below 25 fpm	8 to 12
Bowling Centers		70 to 74°F 20 to 30% rh	75 to 78°F 50 to 55% rh	50 fpm at 6 ft above floor	10 to 15
Communication Centers	Telephone Terminal Rooms	72 to 78°F 40 to 50% rh	72 to 78°F 40 to 50% rh	25 to 30 fpm	8 to 20
	Radio and Television Studios	74 to 78°F 30 to 40% rh	74 to 78°F 40 to 55% rh	below 25 fpm at 12 ft above floor	15 to 40
Transportation Centers (Also see Chapter 12, Enclosed Vehicular Facilities.)	Airport Terminals	70 to 74°F 20 to 30% rh	74 to 78°F 50 to 60% rh	25 to 30 fpm at 6 ft above floor	8 to 12
	Ship Docks	70 to 74°F 20 to 30% rh	74 to 78°F 50 to 60% rh	25 to 30 fpm at 6 ft above floor	8 to 12
	Bus Terminals	70 to 74°F 20 to 30% rh	74 to 78°F 50 to 60% rh	25 to 30 fpm at 6 ft above floor	8 to 12
	Garages[j]	40 to 55°F	80 to 100°F	30 to 75 fpm	4 to 6
Warehouses		Inside design temperatures for warehouses often depend on the materials stored.			1 to 4

Table 1 General Design Criteria[a, b] (Concluded)

Noise[c]	Filtering Efficiencies (ASHRAE *Standard* 52.1)	Load Profile	Comments
NC 40 to 50[e]	35% or better	Peak at 1 to 2 P.M.	Prevent draft discomfort for patrons waiting in serving lines
NC 35 to 40	35% or better	Peak at 1 to 2 P.M.	
NC 35 to 50	Use charcoal for odor control with manual purge control for 100% outside air to exhaust ±35% prefilters	Peak at 5 to 7 P.M.	
NC 35 to 45[f]	Use charcoal for odor control with manual purge control for 100% outside air to exhaust ±35% prefilters	Nightclubs peak at 8 P.M. to 2 A.M. Casinos peak at 4 P.M. to 2 A.M. Equipment, 24 h/day	Provide good air movement but prevent cold draft discomfort for patrons
NC 40 to 50	10 to 15% or better	[h]	Negative air pressure required for odor control. (See also Chapter 30, Kitchen Ventilation.)
NC 30 to 45	35 to 60% or better	Peak at 4 P.M.	
NC 35 to 40	35 to 60% or better	Peak at 3 P.M.	
NC 35	35% prefilters plus charcoal filters 85 to 95% final[i]	Peak at 3 P.M.	
NC 40 to 50	10 to 15%	Peak at 6 to 8 P.M.	
to NC 60	85% or better	Varies with location and use	Constant temperature and humidity required
NC 15 to 25	35% or better	Varies widely due to changes in lighting and people	Constant temperature and humidity required
NC 35 to 50	35% or better and charcoal filters	Peak at 10 A.M. to 9 P.M.	Positive air pressure required in terminal
NC 35 to 50	10 to 15%	Peak at 10 A.M. to 5 P.M.	Positive air pressure required in waiting area
NC 35 to 50	35% with exfiltration	Peak at 10 A.M. to 5 P.M.	Positive air pressure required in terminal
NC 35 to 50	10 to 15%	Peak at 10 A.M. to 5 P.M.	Negative air pressure required to remove fumes; positive air in pressure adjacent occupied spaces
to NC 75	10 to 35%	Peak at 10 A.M. to 3 P.M.	

Notes to Table 1, General Design Criteria

[a]This table shows design criteria differences between various commercial and public buildings. It should not be used as the sole source for design criteria. Each type of data contained here can be determined from the *ASHRAE Handbooks* and *Standards*.

[b]Consult governing codes to determine minimum allowable requirements. Outdoor air requirements may be reduced if high-efficiency adsorption equipment or other odor- or gas-removal equipment is used. See ASHRAE *Standard* 62 for calculation procedures. Also see Chapter 44 in this volume and Chapter 13 of the 1997 *ASHRAE Handbook—Fundamentals*.

[c]Refer to Chapter 46.

[d]Food in these areas is often eaten more quickly than in a restaurant, therefore, the turnover of diners is much faster. Because diners seldom remain for long periods, they do not require the degree of comfort necessary in restaurants. Thus, it may be possible to lower design criteria standards and still provide reasonably comfortable conditions. Although space conditions of 80°F and 50% rh may be satisfactory for patrons when it is 95°F and 50% rh outside, indoor conditions of 78°F and 40% rh are better.

[e]Cafeterias and luncheonettes usually have some or all of the food preparation equipment and trays in the same room with the diners. These eating establishments are generally noisier than restaurants, so that noise transmission from the air-conditioning equipment is not as critical.

[f]In some nightclubs, the noise from the air-conditioning system must be kept low so that all patrons can hear the entertainment.

[g]Usually determined by kitchen hood requirements.

[h]Peak kitchen heat load does not generally occur at peak dining load, although in luncheonettes and some cafeterias where cooking is done in the dining areas, peaks may be simultaneous.

[i]Methods for removal of chemical pollutants must also be considered.

[j]Also includes service stations.

Design Concepts

If a structure is characterized by several exposures and multipurpose use, especially with wide load swings and noncoincident energy use in certain areas, multiunit or unitary systems may be considered for such areas, but not necessarily for the entire building. The benefits of transferring heat absorbed by cooling from one area to other areas, processes, or services that require heat may enhance the selection of such systems. Systems such as incremental closed-loop heat pumps may be cost-effective.

When the cost of energy is included in the rent with no means for permanent or checkmetering, tenants tend to consume excess energy. This energy waste raises operating costs for the owner, decreases profitability, and has a detrimental effect on the environment. While design features can minimize excess energy penalties, they seldom eliminate waste. For example, U.S. Department of Housing and Urban Development nationwide field records for total-electric housing show that rent-included dwellings use approximately 20% more energy than those directly metered by a public utility company.

Diversity factor benefits for central heating and cooling in rent-included buildings may result in lower building demand and connected loads. However, energy waste may easily result in load factors and annual energy consumption exceeding that of buildings where the individual has a direct economic incentive to reduce energy consumption. Heat flow (Btu) meters should be considered for charging for energy consumption.

Design Criteria

In many applications, design criteria are fairly evident, but in all cases, the engineer should understand the owner's and user's intent because any single factor may influence system selection. The engineer's personal experience and judgment in the projection of future needs may be a better criterion for system design than any other single factor.

Comfort Level. Comfort, as measured by temperature, humidity, air motion, air quality, noise, and vibration, is not identical for all buildings, occupant activities, or uses of space. The control of static electricity may be a consideration in humidity control.

Costs. Owning and operating costs can affect system selection and seriously conflict with other criteria. Therefore, the engineer must help the owner resolve these conflicts by considering factors such as the cost and availability of different fuels, the ease of equipment access, and the maintenance requirements.

Local Conditions. Local, state, and national codes, regulations, and environmental concerns must be considered in the design. Chapters 25 and 26 of the 1997 *ASHRAE Handbook—Fundamentals* give information on calculating the effects of weather in specific areas.

Automatic Temperature Control. Proper automatic temperature control maintains occupant comfort during varying internal and external loads. Improper temperature control may mean a loss of customers in restaurants and other public buildings. An energy management control system can be combined with a building automation system to allow the owner to manage energy, lighting, security, fire protection, and other similar systems from one central control point. Chapters 34 and 45 include more details.

Fire, Smoke, and Odor Control. Fire and smoke can easily spread through elevator shafts, stairwells, ducts, and other routes. Although an air-conditioning system can spread fire and smoke by (1) fan operation, (2) penetrations required in walls or floors, or (3) the stack effect without fan circulation, a properly designed and installed system can be a positive means of fire and smoke control.

Chapter 51 has information on techniques for positive control after a fire starts. Effective attention to fire and smoke control also helps prevent odor migration into unventilated areas (see Chapter 44). The design of the ventilation system should consider applicable National Fire Protection Association standards, especially NFPA *Standards* 90A and 96.

DINING AND ENTERTAINMENT CENTERS

Load Characteristics

Air conditioning of restaurants, cafeterias, bars, and nightclubs presents common load problems encountered in comfort conditioning, with additional factors pertinent to dining and entertainment applications. Such factors include

- Extremely variable loads with high peaks, in many cases, occurring twice daily
- High sensible and latent heat gains because of gas, steam, electric appliances, people, and food
- Sensible and latent loads that are not always coincident
- Large quantities of makeup air normally required
- Localized high sensible and latent heat gains in dancing areas
- Unbalanced conditions in restaurant areas adjacent to kitchens which, although not part of the conditioned space, still require special attention
- Heavy infiltration of outdoor air through doors during rush hours
- Smoking versus nonsmoking areas

Internal heat and moisture loads come from occupants, motors, lights, appliances, and infiltration. Separate calculations should be made for patrons and employees. The sensible and latent heat load must be proportioned in accordance with the design temperature selected for both sitting and working people, because the latent to sensible heat ratio for each category decreases as the room temperature decreases.

Hoods required to remove heat from appliances may also substantially reduce the space latent loads.

Infiltration is a considerable factor in many restaurant applications because of short occupancy and frequent door use. It is increased by the need for large quantities of makeup air, which should be provided by mechanical means to replace air exhausted through hoods and for smoke removal. Systems for hood exhaust makeup should concentrate on exhausting nonconditioned and minimally heated makeup air. Wherever possible, vestibules or revolving doors should be installed to reduce infiltration.

Design Concepts

The following factors influence system design and equipment selection:

- High concentrations of food, body, and tobacco-smoke odors require adequate ventilation with proper exhaust facilities.
- Step control of refrigeration plants gives satisfactory and economical operation under reduced loads.
- Exhausting air at the ceiling removes smoke and odor.
- Building design and space limitations often favor one equipment type over another. For example, in a restaurant having a vestibule with available space above it, air conditioning with condensers and evaporators remotely located above the vestibule may be satisfactory. Such an arrangement saves valuable space, even though self-contained units located within the conditioned space may be somewhat lower in initial cost and maintenance cost. In general, small cafeterias, bars, and the like, with loads up to 10 tons, can be most economically conditioned with packaged units; larger and more elaborate establishments require central plants.
- Smaller restaurants with isolated plants usually use direct-expansion systems.
- Mechanical humidification is typically not provided due to high internal latent loads.
- Some air-to-air heat recovery equipment can reduce the energy required for heating and cooling ventilation air. Chapter 42 of the 1996 *ASHRAE Handbook—Systems and Equipment* includes details. The potential for grease condensation on heat recovery surfaces must also be considered.
- A vapor compression or desiccant-based dehumidifier should be considered for makeup air handling and enhanced humidity control.

Because eating and entertainment centers generally have low sensible heat factors and require high ventilation rates, fan-coil and induction systems are usually not applicable. All-air systems are more suitable. Space must be established for ducts, except for small systems with no ductwork. Large establishments are often served by central chilled water systems.

In cafeterias and luncheonettes, the air distribution system must keep food odors at the serving counters away from areas where patrons are eating. This usually means heavy exhaust air requirements at the serving counters, with air supplied into, and induced from, eating areas. Exhaust air must also remove the heat from hot trays, coffee urns, and ovens to minimize patron and employee discomfort and to reduce air-conditioning loads. These factors often create greater air-conditioning loads for cafeterias and luncheonettes than for restaurants.

Odor Removal. Transferring air from dining areas into the kitchen keeps odors and heat out of dining areas and cools the kitchen. Outdoor air intake and kitchen exhaust louvers should be located so that exhaust air is neither drawn back into the system nor allowed to cause discomfort to passersby.

Where odors can be drawn back into dining areas, activated charcoal filters, air washers, or ozonators are used to remove odors. Kitchen, locker room, toilet, or other malodorous air should not be recirculated unless air purifiers are used.

Kitchen Air Conditioning. If planned in the initial design phases, kitchens can often be air conditioned effectively without excessive cost. It is not necessary to meet the same design criteria as for dining areas, but kitchen temperatures can be reduced significantly. The relatively large number of people and food loads in dining and kitchen areas produce a high latent load. Additional cooling required to eliminate excess moisture increases refrigeration plant, cooling coil, and air-handling equipment size.

Advantageously located self-contained units with air distribution designed so as not to produce drafts off hoods and other equipment, can be used to spot cool intermittently. High-velocity air distribution may be effective. The costs are not excessive, and kitchen personnel efficiency can be improved greatly.

Even in climates with high wet-bulb temperatures, direct or indirect evaporative cooling may be a good compromise between the expense of air conditioning and the lack of comfort in ventilated kitchens. For more information, see Chapter 30, Kitchen Ventilation.

Special Considerations

In establishing design conditions, the duration of individual patron occupancy should be considered. Patrons entering from outdoors are more comfortable in a room with a high temperature than those who remain long enough to become acclimated. Nightclubs and deluxe restaurants are usually operated at a lower effective temperature than cafeterias and luncheonettes.

Often, the ideal design condition must be rejected for an acceptable condition because of equipment cost or performance limitations. Restaurants are frequently affected in this way because ratios of latent to sensible heat may result in uneconomical or oversized equipment selection, unless an enhanced dehumidification system or a combination of lower design dry-bulb temperature and higher relative humidity (which gives an equal effective temperature) is selected.

In severe climates, entrances and exits in any dining establishment should be completely shielded from diners to prevent drafts. Vestibules provide a measure of protection. However, both vestibule doors are often open simultaneously. Revolving doors or local means for heating or cooling infiltration air may be provided to offset drafts.

Uniform employee comfort is difficult to maintain because of (1) temperature differences between the kitchen and dining room and (2) the constant motion of employees. Since customer satisfaction is essential to a dining establishment's success, patron comfort is the primary consideration. However, maintenance of satisfactory temperature and atmospheric conditions for customers also helps alleviate employee discomfort.

One problem in dining establishments is the use of partitions to separate areas into modular units. Partitions create such varied load conditions that individual modular unit control is generally necessary.

Baseboard radiation or convectors, if required, should be located so as not to overheat patrons. This is difficult to achieve in some layouts because of movable chairs and tables. For these reasons, it is desirable to enclose all dining room and bar heating elements in insulated cabinets with top outlet grilles and baseboard inlets. With heating elements located under windows, this practice has the additional advantage of directing the heat stream to combat window downdraft and air infiltration. Separate smoking and nonsmoking areas may be required. The smoking area should be exhausted or served by separate air-handling equipment. Air diffusion device selection and placement should minimize smoke migration toward nonsmoking areas. Smoking areas must have a negative air pressure relationship with adjacent occupied areas.

Restaurants. In restaurants, people are seated and served at tables, while food is generally prepared in remote areas. This type of dining is usually enjoyed in a leisurely and quiet manner, so the ambient atmosphere should be such that the air conditioning is not noticed. Where specialized or open cooking is a feature, provisions should be made for the control and handling of cooking odors and smoke.

Bars. Bars are often a part of a restaurant or nightclub. If they are establishments on their own, they often serve food as well as drinks, and they should be classified as restaurants with food preparation in remote areas. Alcoholic beverages produce pungent vapors, which must be drawn off. In addition, smoking at bars is generally considerably heavier than in restaurants. Therefore, outdoor air requirements are relatively high by comparison.

Nightclubs and Casinos. Both nightclubs and casinos may include a restaurant, bar, stage, and dancing area. The bar should be treated as a separately zoned area, with its own supply and exhaust system. People in the restaurant area who dine and dance may require twice the air changes and cooling required by patrons who dine and then watch a show. The length of stay generally exceeds that encountered in most eating places. In addition, eating in nightclubs and casinos is usually secondary to drinking and smoking. Patron density usually exceeds that of conventional eating establishments.

Kitchens. The kitchen has the greatest concentration of noise, heat load, smoke, and odors; ventilation is the chief means of removing these objectionable elements and preventing them from entering dining areas. To ensure odor control, kitchen air pressure should be kept negative relative to other areas. Maintenance of reasonably comfortable working conditions is important. For more information, see Chapter 30, Kitchen Ventilation.

OFFICE BUILDINGS

Load Characteristics

Office buildings usually include both peripheral and interior zone spaces. The peripheral zone extends from 10 to 12 ft inward from the outer wall toward the interior of the building and frequently has a large window area. These zones may be extensively subdivided. Peripheral zones have variable loads because of changing sun position and weather. These zone areas typically require heating in winter. During intermediate seasons, one side of the building may require cooling, while another side requires heating. However, the interior zone spaces usually require a fairly uniform cooling rate throughout the year because their thermal loads are derived almost entirely from lights, office equipment, and people. Interior space conditioning is often by systems that have variable air volume control for low- or no-load conditions.

Most office buildings are occupied from approximately 8:00 A.M. to 6:00 P.M.; many are occupied by some personnel from as early as 5:30 A.M. to as late as 7:00 P.M. Some tenants' operations may require night work schedules, usually not to extend beyond 10:00 P.M. Office buildings may contain printing plants, communications operations, broadcasting studios, and computing centers, which could operate 24 h per day. Therefore, for economical air-conditioning design, the intended uses of an office building must be well established before design development.

Occupancy varies considerably. In accounting or other sections where clerical work is done, the maximum density is approximately one person per 75 ft^2 of floor area. Where there are private offices, the density may be as little as one person per 200 ft^2. The most serious cases, however, are the occasional waiting rooms, conference rooms, or directors' rooms where occupancy may be as high as one person per 20 ft^2.

The lighting load in an office building constitutes a significant part of the total heat load. Lighting and normal equipment electrical loads average from 1 to 5 W/ft^2 but may be considerably higher, depending on the type of lighting and the amount of equipment. Buildings with computer systems and other electronic equipment can have electrical loads as high as 5 to 10 W/ft^2. An accurate appraisal should be made of the amount, size, and type of computer equipment anticipated for the life of the building to size the air-handling equipment properly and provide for future installation of air-conditioning apparatus.

About 30% of the total lighting heat output from recessed fixtures can be withdrawn by exhaust or return air and, therefore, will not enter into space-conditioning supply air requirements. By connecting a duct to each fixture, the most balanced air system can be provided. However, this method is expensive, so the suspended ceiling is often used as a return air plenum with the air drawn from the space to above the suspended ceiling.

Miscellaneous allowances (for fan heat, duct heat pickup, duct leakage, and safety factors) should not exceed 12% of the total load.

Building shape and orientation are often determined by the building site, but certain variations in these factors can produce increases of 10 to 15% in the refrigeration load. Shape and orientation should therefore be carefully analyzed in the early design stages.

Design Concepts

The variety of functions and range of design criteria applicable to office buildings have allowed the use of almost every available air-conditioning system. While multistory structures are discussed here, the principles and criteria are similar for all sizes and shapes of office buildings.

Attention to detail is extremely important, especially in modular buildings. Each piece of equipment, duct and pipe connections, and the like may be duplicated hundreds of times. Thus, seemingly minor design variations may substantially affect construction and operating costs. In initial design, each component must be analyzed not only as an entity, but also as part of an integrated system. This systems design approach is essential for achieving optimum results.

There are several classes of office buildings, determined by the type of financing required and the tenants who will occupy the building. Design evaluation may vary considerably based on specific tenant requirements; it is not enough to consider typical floor patterns only. Included in many larger office buildings are stores, restaurants, recreational facilities, data centers, telecommunication centers, radio and television studios, and observation decks.

Built-in system flexibility is essential for office building design. Business office procedures are constantly being revised, and basic building services should be able to meet changing tenant needs.

The type of occupancy may have an important bearing on the selection of the air distribution system. For buildings with one owner or lessee, operations may be defined clearly enough that a system can be designed without the degree of flexibility needed for a less well-defined operation. However, owner-occupied buildings may require considerable design flexibility because the owner will pay for all alterations. The speculative builder can generally charge alterations to tenants. When different tenants occupy different floors, or even parts of the same floor, the degree of design and operation complexity increases to ensure proper environmental comfort conditions to any tenant, group of tenants, or all tenants at once. This problem is more acute if tenants have seasonal and variable overtime schedules.

Stores, banks, restaurants, and entertainment facilities may have hours of occupancy or design criteria that differ substantially from those of office buildings; therefore, they should have their own air distribution systems and, in some cases, their own heating and/or refrigeration equipment.

Main entrances and lobbies are sometimes served by a separate system because they buffer the outside atmosphere and the building interior. Some engineers prefer to have a lobby summer temperature 4 to 6°F above office temperature to reduce operating cost and the temperature shock to people entering or leaving the building.

The unique temperature and humidity requirements of data processing installations and the fact that they often run 24 h per day for extended periods generally warrant separate refrigeration and air distribution systems. Separate backup systems may be required for data processing areas in case the main building HVAC system fails. Chapter 16 has further information.

The degree of air filtration required should be determined. The service cost and the effect of air resistance on energy costs should be analyzed for various types of filters. Initial filter cost and air pollution characteristics also need to be considered. Activated charcoal filters for odor control and reduction of outdoor air requirements are another option to consider.

Providing office buildings with continuous 100% outdoor air is seldom justified; therefore, most office buildings are designed to minimize outdoor air use, except during economizer operation. However, attention to indoor air quality may dictate higher levels of ventilation air. In addition, the minimum volume of outdoor air should be maintained in variable volume air-handling systems. Dry-bulb or enthalpy controlled economizer cycles should be considered for reducing energy costs.

When an economizer cycle is used, systems should be zoned so that energy waste will not occur due to heating of outside air. This is often accomplished by a separate air distribution system for the interior and each major exterior zone.

High-rise office buildings have traditionally used perimeter dual-duct, induction, or fan-coil systems. Where fan-coil or induction systems have been installed at the perimeter, separate all-air systems have generally been used for the interior. More recently, variable air volume systems, including modulated air diffusers and self-contained perimeter unit systems, have also been used. If variable air volume systems serve the interior, perimeters are usually served by variable volume or dual-duct systems supplemented with fan-powered terminals, terminals with reheat coils, or radiation (ceiling panels or baseboard). The perimeter systems can be hydronic or electric.

Many office buildings without an economizer cycle have a bypass multizone unit installed on each floor or several floors with a heating coil in each exterior zone duct. Variable air volume variations of the bypass multizone and other floor-by-floor, all-air systems are also being used. These systems are popular due to low fan power, low initial cost, and energy savings resulting from independent operating schedules, which are possible between floors occupied by tenants with different operating hours.

Perimeter radiation or infrared systems with conventional, single-duct, low-velocity air conditioning that furnishes air from packaged air-conditioning units or multizone units may be more economical for small office buildings. The need for a perimeter system, which is a function of exterior glass percentage, external wall thermal value, and climate severity, should be carefully analyzed.

A perimeter heating system separate from the cooling system is preferable, since air distribution devices can then be selected for a specific duty rather than as a compromise between heating and cooling performance. The higher cost of additional air-handling or fan-coil units and ductwork may lead the designer to a less expensive option, such as fan-powered terminal units with heating coils serving perimeter zones in lieu of a separate heating system. Radiant ceiling panels for the perimeter zones are another option.

Interior space usage usually requires that interior air-conditioning systems permit modification to handle all load situations. Variable air volume systems are often used. When using these systems, a careful evaluation of low-load conditions should be made to determine whether adequate air movement and outdoor air can be provided at the proposed supply air temperature without overcooling. Increases in supply air temperature tend to nullify energy savings in fan power, which are characteristic of variable air volume systems. Low-temperature air distribution for additional savings in transport energy is seeing increased use, especially when coupled with an ice storage system.

In small to medium-sized office buildings, air-source heat pumps may be chosen. In larger buildings, internal source heat pump systems (water-to-water) are feasible with most types of air-conditioning systems. Heat removed from core areas is rejected to either a cooling tower or perimeter circuits. The internal source heat pump

can be supplemented by a central heating system or electrical coils on extremely cold days or over extended periods of limited occupancy. Removed excess heat may also be stored in hot water tanks.

Many heat recovery or internal source heat pump systems exhaust air from conditioned spaces through lighting fixtures. Approximately 30% of lighting heat can be removed in this manner. One design advantage is a reduction in required air quantities. In addition, lamp life is extended by operation in a much cooler ambient environment.

Suspended ceiling return air plenums eliminate sheet metal return air ductwork to reduce floor-to-floor height requirements. However, suspended ceiling plenums may increase the difficulty of proper air balancing throughout the building. Problems often connected with suspended ceiling return plenums are as follows:

- Air leakage through cracks, with resulting smudges
- Tendency of return air openings nearest to a shaft opening or collector duct to pull too much air, thus creating uneven air motion and possible noise
- Noise transmission between office spaces

Air leakage can be minimized by proper workmanship. To overcome drawing too much air, return air ducts can be run in the suspended ceiling pathway from the shaft, often in a simple radial pattern. The ends of the ducts can be left open or dampered. Generous sizing of return air grilles and passages lowers the percentage of circuit resistance attributable to the return air path. This bolsters effectiveness of supply air balancing devices and reduces the significance of air leakage and drawing too much air. Structural blockage can be solved by locating openings in beams or partitions with fire dampers, where required.

Spatial Requirements

Total office building electromechanical space requirements vary tremendously based on types of systems planned; however, the average is approximately 8 to 10% of the gross area. Clear height required for fan rooms varies from approximately 10 to 18 ft, depending on the distribution system and equipment complexity. On office floors, perimeter fan-coil or induction units require approximately 1 to 3% of the floor area. Interior air shafts and pipe chases require approximately 3 to 5% of the floor area. Therefore, ducts, pipes, and equipment require approximately 4 to 8% of each floor's gross area.

Where large central units supply multiple floors, shaft space requirements depend on the number of fan rooms. In such cases, one mechanical equipment room usually furnishes air requirements for 8 to 20 floors (above and below for intermediate levels), with an average of 12 floors. The more floors served, the larger the duct shafts and equipment required. This results in higher fan room heights and greater equipment size and mass.

The fewer floors served by an equipment room, the more equipment rooms will be required to serve the building. This axiom allows greater flexibility in serving changing floor or tenant requirements. Often, one mechanical equipment room per floor and complete elimination of vertical shafts requires no more total floor area than a few larger mechanical equipment rooms, especially when there are many small rooms and they are the same height as typical floors. Equipment can also be smaller, although maintenance costs will be higher. Energy costs may be reduced, with more equipment rooms serving fewer areas, because the equipment can be shut off in unoccupied areas, and high-pressure ductwork will not be required. Equipment rooms on upper levels generally cost more to install because of rigging and transportation logistics.

In all cases, mechanical equipment rooms must be thermally and acoustically isolated from office areas.

Cooling Towers. Cooling towers are the largest single piece of equipment required for air-conditioning systems. Cooling towers require approximately 1 ft^2 of floor area per 400 ft^2 of total building area and are from 13 to 40 ft high. When towers are located on the roof, the building structure must be capable of supporting the cooling tower and dunnage, full water load (approximately 120 to 150 lb/ft^2), and seismic and wind load stresses.

Where cooling tower noise may affect neighboring buildings, towers should be designed to include sound traps or other suitable noise baffles. This may affect tower space, mass of the units, and motor power. Slightly oversizing cooling towers can reduce noise and power consumption due to lower speeds, but this may increase the initial cost.

Cooling towers are sometimes enclosed in a decorative screen for aesthetic reasons; therefore, calculations should ascertain that the screen has sufficient free area for the tower to obtain its required air quantity and to prevent recirculation.

If the tower is placed in a rooftop well or near a wall, or split into several towers at various locations, design becomes more complicated, and initial and operating costs increase substantially. Also, towers should not be split and placed on different levels because hydraulic problems increase. Finally, the cooling tower should be built high enough above the roof so that the bottom of the tower and the roof can be maintained properly.

Special Considerations

Office building areas with special ventilation and cooling requirements include elevator machine rooms, electrical and telephone closets, electrical switchgear, plumbing rooms, refrigeration rooms, and mechanical equipment rooms. The high heat loads in some of these rooms may require air-conditioning units for spot cooling.

In larger buildings having intermediate elevator, mechanical, and electrical machine rooms, it is desirable to have these rooms on the same level or possibly on two levels. This may simplify the horizontal ductwork, piping, and conduit distribution systems and permit more effective ventilation and maintenance of these equipment rooms.

An air-conditioning system cannot prevent occupants at the perimeter from feeling direct sunlight. Venetian blinds and drapes are often provided but seldom used. External shading devices (screens, overhangs, etc.) or reflective glass are preferable.

Tall buildings in cold climates experience severe stack effect. The extra amount of heat provided by the air-conditioning system in attempts to overcome this problem can be substantial. The following features help combat infiltration due to the stack effect:

- Revolving doors or vestibules at exterior entrances
- Pressurized lobbies or lower floors
- Tight gaskets on stairwell doors leading to the roof
- Automatic dampers on elevator shaft vents
- Tight construction of the exterior skin
- Tight closure and seals on all dampers opening to the exterior

BOWLING CENTERS

Bowling centers may also contain a bar, a restaurant, a children's play area, offices, locker rooms, and other types of facilities. Such auxiliary areas are not discussed in this section, except as they may affect design for the bowling area, which consists of alleys and a spectator area.

Load Characteristics

Bowling alleys usually have their greatest period of use in the evenings, but weekend daytime use may also be heavy. Thus, when designing for the peak air-conditioning load on the building, it is necessary to compare the day load and its high outside solar load and off-peak people load with the evening peak people load and zero solar load. Because bowling areas generally have little fenestration, the solar load may not be important.

If the building contains auxiliary areas, these areas may be included in the refrigeration, heating, and air distribution systems for the bowling alleys, with suitable provisions for zoning the different areas as dictated by load analysis. Alternatively, separate systems may be established for each area having different load operation characteristics.

Heat buildup due to lights, external transmission load, and pin-setting machinery in front of the foul line can be reduced by exhausting some air above the alleys or from the area containing the pin-setting machines; however, this gain should be compared against the cost of conditioning additional makeup air. In the calculation of the air-conditioning load, a portion of the unoccupied alley space load is included. Because this consists mainly of lights and some transmission load, about 15 to 30% of this heat load may have to be taken into account. The higher figure may apply when the roof is poorly insulated, no exhaust air is taken from this area, or no vertical baffle is used at the foul line. One estimate is 5 to 10 Btu/h per square foot of vertical surface at the foul line, depending mostly on the type and intensity of the lighting.

The heat load from bowlers and spectators may be found in Table 3 in Chapter 28 of the 1997 *ASHRAE Handbook—Fundamentals*. The proper heat gain should be applied for each person to avoid too large a design heat load.

Design Concepts

As with other building types having high occupancy loads, heavy smoke and odor concentration, and low sensible heat factors, all-air systems are generally the most suitable for bowling alley areas. Since most bowling alleys are almost windowless structures except for such areas as entrances, exterior restaurants, and bars, it is uneconomical to use terminal unit systems because of the small number required. Where required, radiation in the form of baseboard or radiant ceiling panels is generally placed at perimeter walls and entrances.

It is not necessary to maintain normal indoor temperatures down the length of the alleys; temperatures may be graded down to the pin area. Unit heaters are often used at this location.

Air Pressurization. Spectator and bowling areas must be well shielded from entrances so that no cold drafts are created in these areas. To minimize infiltration of outdoor air into the alleys, the exhaust and return air system should handle only 85 to 90% of the total supply air, thus maintaining a positive pressure within the space when outside air pressurization is taken in consideration.

Air Distribution. Packaged units without ductwork produce uneven space temperatures, and unless they are carefully located and installed, the units may cause objectionable drafts. Central ductwork is recommended for all but the smallest buildings, even where packaged refrigeration units are used. Because only the areas behind the foul line are air conditioned, the ductwork should provide comfortable conditions within this area.

The return and exhaust air systems should have a large number of small registers uniformly located at high points, or pockets, to draw off the hot, smoky, and odorous air. In some parts of the country and for larger bowling alleys, it may be desirable to use all outdoor air to cool during intermediate seasons.

Special Considerations

People in sports and amusement centers engage in a high degree of physical activity, which makes them feel warmer and increases their rate of evaporation. In these places, odor and smoke control are important environmental considerations.

Bowling centers are characterized by the following:

- A large number of people concentrated in a relatively small area of a very large room. A major portion of the floor area is unoccupied.
- Heavy smoking, high physical activity, and high latent heat load.
- Greatest use from about 6:00 P.M. to midnight.

The first two items make furnishing large amounts of outdoor air mandatory to minimize odors and smoke in the atmosphere.

The area between the foul line and the bowling pins need not be air conditioned or ventilated. Transparent or opaque vertical partitions are sometimes installed to separate the upper portions of the occupied and unoccupied areas so that air distribution is better contained within the occupied area.

COMMUNICATION CENTERS

Communication centers include telephone terminal buildings, radio stations, television studios, and transmitter and receiver stations. Most telephone terminal rooms are air conditioned because constant temperature and relative humidity help prevent breakdowns and increase equipment life. In addition, air conditioning permits the use of a lower number of air changes, which, for a given filter efficiency, decreases the chances of damage to relay contacts and other delicate equipment.

Radio and television studios require critical analysis for the elimination of heat buildup and the control of noise. Television studios have the added problem of air movement, lighting, and occupancy load variations. This section deals with television studios because they present most of the problems also found in radio studios.

Load Characteristics

Human occupancy is limited, so the air-conditioning load for telephone terminal rooms is primarily equipment heat load.

Television studios have very high lighting capacities, and the lighting load may fluctuate considerably in intensity over short periods. The operating hours may vary every day. In addition, there may be from one to several dozen people onstage for short times. The air-conditioning system must be extremely flexible and capable of handling wide load variations quickly, accurately, and efficiently, similar to the conditions of a theater stage. The studio may also have an assembly area with a large number of spectator seats. Generally, studios are located so that they are completely shielded from external noise and thermal environments.

Design Concepts

The critical areas of a television studio are the performance studio and control rooms. The audience area may be treated much like a place of assembly. Each area should have its own air distribution system or at least its own zone control separate from the studio system. The heat generated in the studio area should not be allowed to permeate the audience atmosphere.

The air distribution system selected must have the capabilities of a dual-duct, single-duct system with cooling and heating booster coils, a variable air volume systems, or a multizone system to satisfy design criteria. The air distribution system should be designed so that simultaneous heating and cooling cannot occur unless heating is achieved solely by heat recovery.

Studio loads seldom exceed 100 tons of refrigeration. Even if the studio is part of a large communications center or building, the studio should have its own refrigeration system in case of emergencies. The refrigeration equipment in this size range may be reciprocating units, which require a remote location so that machine noise is isolated from the studio.

Special Considerations

On-Camera Studios. This is the stage of the television studio and requires the same general considerations as a concert hall stage. Air movement must be uniform, and, because scenery, cameras, and equipment may be moved during the performance, ductwork must be planned carefully to avoid interference with proper studio operation.

Control Rooms. Each studio may have one or more control rooms serving different functions. The video control room, which

is occupied by the program and technical directors, contains monitors and picture-effect controls. The room may require up to 30 air changes per hour to maintain proper conditions. The large number of necessary air changes and the low sound level that must be maintained require special analysis of the air distribution system.

If a separate control room is furnished for the announcer, the heat load and air distribution problems will not be as critical as those for control rooms for the program, technical, and audio directors.

Thermostatic control should be furnished in each control room, and provisions should be made to enable occupants to turn the air conditioning on and off.

Noise Control. Studio microphones are moved throughout the studio during a performance, and they may be moved past or set near air outlets or returns. These microphones are considerably more sensitive than the human ear; therefore, air outlets or returns should be located away from areas where microphones are likely to be used. Even a leaky pneumatic thermostat can be a problem.

Air Movement. It is essential that air movement within the stage area, which often contains scenery and people, be kept below 25 fpm within 12 ft of the floor. The scenery is often fragile and will move in air velocities above 25 fpm; also, actors' hair and clothing may be disturbed.

Air Distribution. Ductwork must be fabricated and installed so that there are no rough edges, poor turns, or improperly installed dampers to cause turbulence and eddy currents within the ducts. Ductwork should contain no holes or openings that might create whistles. Air outlet locations and the distribution pattern must be carefully analyzed to eliminate turbulence and eddy currents within the studio that might cause noise that could be picked up by studio microphones.

At least some portions of supply, return, and exhaust ductwork will require acoustical material to maintain noise criterion (NC) levels from 20 to 25. Any duct serving more than one room should acoustically separate each room by means of a sound trap. All ductwork should be suspended by means of neoprene or rubber in shear-type vibration mountings. Where ductwork goes through wall or floor slabs, the openings should be sealed with acoustically deadening material. The supply fan discharge and the return and exhaust fan inlets should have sound traps; all ductwork connections to fans should be made with nonmetallic, flexible material. Air outlet locations should be coordinated with ceiling-mounted tracks and equipment. Air distribution for control rooms may require a perforated ceiling outlet or return air plenum system.

Piping Distribution. All piping within the studio, as well as in adjacent areas that might transmit noise to the studio, should be supported by suitable vibration isolation hangers. To prevent transmission of vibration, piping should be supported from rigid structural elements to maximize absorption.

Mechanical Equipment Rooms. These rooms should be located as remotely from the studio as possible. All equipment should be selected for very quiet operation and should be mounted on suitable vibration-eliminating supports. Structural separation of these rooms from the studio is generally required.

Offices and Dressing Rooms. The functions of these rooms are quite different from each other and from the studio areas. It is recommended that such rooms be treated as separate zones, with their own controls.

Air Return. Whenever practicable, the largest portion of studio air should be returned over the banks of lights. This is similar to theater stage practice. Sufficient air should also be removed from studio high points to prevent heat buildup.

TRANSPORTATION CENTERS

The major transportation facilities are airports, ship docks, bus terminals, and passenger car garages. Airplane hangars and freight and mail buildings are also among the types of buildings to be considered. Freight and mail buildings are usually handled as standard warehouses.

Load Characteristics

Airports, ship docks, and bus terminals operate on a 24 h basis, with a reduced schedule during late night and early morning hours.

Airports. Terminal buildings consist of large, open circulating areas, one or more floors high, often with high ceilings, ticketing counters, and various types of stores, concessions, and convenience facilities. Lighting and equipment loads are generally average, but occupancy varies substantially. Exterior loads are, of course, a function of architectural design. The largest single problem often results from thermal drafts created by large entranceways, high ceilings, and long passageways, which have many openings to the outdoors.

Ship Docks. Freight and passenger docks consist of large, high-ceilinged structures with separate areas for administration, visitors, passengers, cargo storage, and work. The floor of the dock is usually exposed to the outdoors just above the water level. Portions of the side walls are often open while ships are in port. In addition, the large ceiling (roof) area presents a large heating and cooling load. Load characteristics of passenger dock terminals generally require the roof and floors to be well insulated. Occasional heavy occupancy loads in visitor and passenger areas must be considered.

Bus Terminals. This building type consists of two general areas: the terminal building, which contains passenger circulation, ticket booths, and stores or concessions, and the bus loading area. Waiting rooms and passenger concourse areas are subject to a highly variable people load. Occupancy density may reach 10 ft^2 per person and, at extreme periods, 3 to 5 ft^2 per person. Chapter 12, Enclosed Vehicular Facilities, has further information on bus terminals.

Design Concepts

Heating and cooling is generally centralized or provided for each building or group in a complex. In large, open circulation areas of transportation centers, any all-air system with zone control can be used. Where ceilings are high, air distribution is often along the side wall to concentrate the air conditioning where desired and avoid disturbing stratified air. Perimeter areas may require heating by radiation, a fan-coil system, or hot air blown up from the sill or floor grilles, particularly in colder climates. Hydronic perimeter radiant ceiling panels may be especially suited to these high-load areas.

Airports. Airports generally consist of one or more central terminal buildings connected by long passageways or trains to rotundas containing departure lounges for airplane loading. Most terminals have portable telescoping-type loading bridges connecting departure lounges to the airplanes. These passageways eliminate the heating and cooling problems associated with traditional permanent structure passenger loading.

Because of difficulties in controlling the air balance due to the many outside openings, high ceilings, and long, low passageways (which often are not air conditioned), the terminal building (usually air conditioned) should be designed to maintain a substantial positive pressure. Zoning is generally required in passenger waiting areas, in departure lounges, and at ticket counters to take care of the widely variable occupancy loads.

Main entrances may be designed with vestibules and windbreaker partitions to minimize undesirable air currents within the building.

Hangars must be heated in cold weather, and ventilation may be required to eliminate possible fumes (although fueling is seldom permitted in hangars). Gas-fired, electric, and low- and high-intensity radiant heaters are used extensively in hangars because they provide comfort for employees at relatively low operating costs.

Hangars may also be heated by large air blast heaters or floor-buried heated liquid coils. Local exhaust air systems may be used to evacuate fumes and odors that result in smaller ducted systems.

Under some conditions, exhaust systems may be portable and may possibly include odor-absorbing devices.

Ship Docks. In severe climates, occupied floor areas may contain heated floor panels. The roof should be well insulated, and, in appropriate climates, evaporative spray cooling substantially reduces the summer load. Freight docks are usually heated and well ventilated but seldom cooled.

High ceilings and openings to the outdoors may present serious draft problems unless the systems are designed properly. Vestibule entrances or air curtains help minimize cross drafts. Air door blast heaters at cargo opening areas may be quite effective.

Ventilation of the dock terminal should prevent noxious fumes and odors from reaching occupied areas. Therefore, occupied areas should be under positive pressure, and the cargo and storage areas exhausted to maintain negative air pressure. Occupied areas should be enclosed to simplify any local air conditioning.

In many respects, these are among the most difficult buildings to heat and cool because of their large open areas. If each function is properly enclosed, any commonly used all-air or large fan-coil system is suitable. If areas are left largely open, the best approach is to concentrate on proper building design and heating and cooling of the openings. High-intensity infrared spot heating is often advantageous (see Chapter 15 of the 1996 *ASHRAE Handbook—Systems and Equipment*). Exhaust ventilation from tow truck and cargo areas should be exhausted through the roof of the dock terminal.

Bus Terminals. Conditions are similar to those for airport terminals, except that all-air systems are more practical because ceiling heights are often lower, and perimeters are usually flanked by stores or office areas. The same systems are applicable as for airport terminals, but ceiling air distribution is generally feasible.

Properly designed radiant hydronic or electric ceiling systems may be used if high-occupancy latent loads are fully considered. This may result in smaller duct sizes than are required for all-air systems and may be advantageous where bus loading areas are above the terminal and require structural beams. This heating and cooling system reduces the volume of the building that must be conditioned. In areas where latent load is a concern, heating-only panels may be used at the perimeter, with a cooling-only interior system.

The terminal area air supply system should be under high positive pressure to ensure that no fumes and odors infiltrate from bus areas. Positive exhaust from bus loading areas is essential for a properly operating total system (see Chapter 12).

Special Considerations

Airports. Filtering outdoor air with activated charcoal filters should be considered for areas subject to excessive noxious fumes from jet engine exhausts. However, locating outside air intakes as remotely as possible from airplanes is a less expensive and more positive approach.

Where ionization filtration enhancers are used, outdoor air quantities are sometimes reduced due to cleaner air. However, care must be taken to maintain sufficient amounts of outside air for space pressurization.

Ship Docks. Ventilation design must ensure that fumes and odors from forklifts and cargo in work areas do not penetrate occupied and administrative areas.

Bus Terminals. The primary concerns with enclosed bus loading areas are health and safety problems, which must be handled by proper ventilation (see Chapter 12). Although diesel engine fumes are generally not as noxious as gasoline fumes, bus terminals often have many buses loading and unloading at the same time, and the total amount of fumes and odors may be quite disturbing.

Enclosed Garages. In terms of health and safety, enclosed bus loading areas and automobile parking garages present the most serious problems in these buildings. Three major problems are encountered. The first and most serious is emission of carbon monoxide (CO) by cars and oxides of nitrogen by buses, which can cause serious illness and possibly death. The second problem is oil and gasoline fumes, which may cause nausea and headaches and can also create a fire hazard. The third is lack of air movement and the resulting stale atmosphere that develops because of the increased carbon dioxide content in the air. This condition may cause headaches or grogginess. Most codes require a minimum ventilation rate to ensure that the CO concentration does not exceed safe limits. Chapter 12 covers the ventilation requirements and calculation procedures for enclosed vehicular facilities in detail.

All underground garages should have facilities for testing the CO concentration or should have the garage checked periodically. Clogged duct systems; improperly operating fans, motors, or dampers; clogged air intake or exhaust louvers, etc., may not allow proper air circulation. Proper maintenance is required to minimize any operational defects.

WAREHOUSES

Warehouses are used to store merchandise and may be open to the public at times. They are also used to store equipment and material inventory at industrial facilities. The buildings are generally not air conditioned, but often have sufficient heat and ventilation to provide a tolerable working environment. Facilities such as shipping, receiving, and inventory control offices associated with warehouses and occupied by office workers are generally air conditioned.

Load Characteristics

Internal loads from lighting, people, and miscellaneous sources are generally low. Most of the load is thermal transmission and infiltration. An air-conditioning load profile tends to flatten where materials stored are massive enough to cause the peak load to lag.

Design Concepts

Most warehouses are only heated and ventilated. Forced flow unit heaters may be located near heat entrances and work areas. Large central heating and ventilating units are widely used. Even though comfort for warehouse workers may not be considered, it may be necessary to keep the temperature above 40°F to protect sprinkler piping or stored materials from freezing.

A building designed for the addition of air conditioning at a later date will require less heating and be more comfortable. For maximum summer comfort without air conditioning, excellent ventilation with noticeable air movement in work areas is necessary. Even greater comfort can be achieved in appropriate climates by adding roof spray cooling. This can reduce the roof's surface temperature by 40 to 60°F, thereby reducing ceiling radiation inside. Low- and high-intensity radiant heaters can be used to maintain the minimum ambient temperature throughout a facility above freezing. Radiant heat may also be used for occupant comfort in areas permanently or frequently open to the outside.

If the stored product requires defined indoor conditions, an air-conditioning system must be added. Using only ventilation may help in maintaining lower space temperature, but caution should be exercised not to damage the stored product with uncontrolled humidity. Direct or indirect evaporative cooling may also be an option.

Special Considerations

Forklifts and trucks powered by gasoline, propane, and other fuels are often used inside warehouses. Proper ventilation is necessary to alleviate the buildup of CO and other noxious fumes. Proper ventilation of battery-charging rooms for electrically powered forklifts and trucks is also required.

REFERENCES

ASHRAE. 1992. Gravimetric and dust-spot procedures for testing air-cleaning devices used in general ventilation for removing particulate matter. ANSI/ASHRAE *Standard* 52.1-1992.

ASHRAE. 1989. Ventilation for acceptable indoor air quality. ANSI/ASHRAE *Standard* 62-1989.

NFPA. 1994. Ventilation control and fire protection of commercial cooking operations. *Standard* 96-94. National Fire Protection Agency, Quincy, MA.

NFPA. 1996. Installation of warm air heating and air conditioning systems. ANSI/NFPA *Standard* 90A-96.

PLACES OF ASSEMBLY

ASSEMBLY rooms are generally large, have relatively high ceilings, and are few in number for any given facility. They usually have a periodically high density of occupancy per unit floor area, as compared to other buildings, and thus have a relatively low design sensible heat ratio.

This chapter summarizes some of the design concerns for enclosed assembly buildings. (Chapter 3, which covers general criteria for commercial and public buildings, also includes information that applies to public assembly buildings.)

GENERAL CRITERIA

Energy conservation codes and standards must be considered because they have a major impact on design and performance.

Assembly buildings may have relatively few hours of use per week and may not be in full use when maximum outdoor temperatures or solar loading occur. Often they are fully occupied for as little as 1 to 2 h, and the load may be materially reduced by precooling. The designer needs to obtain as much information as possible regarding the anticipated hours of use, particularly the times of full seating, so that simultaneous loads may be considered to obtain optimum performance and operating economy. Dehumidification requirements should be considered before reducing equipment size. The intermittent or infrequent nature of the cooling loads may allow these buildings to benefit from thermal storage systems.

The occupants usually generate the major room cooling and ventilation load. The number of occupants is best determined from the seat count, but when this is not available, it can be estimated at 7.5 to 10 ft^2 per person for the entire seating area, including exit aisles but not the stage, performance areas, or entrance lobbies.

Outdoor Air

Outdoor air ventilation rates as prescribed by ASHRAE *Standard* 62 can be a major portion of the total load. The latent load (dehumidification and humidification) and energy used to maintain the relative humidity within prescribed limits is also a concern. Humidity must be maintained at proper levels to prevent mold and mildew growth and to maintain acceptable indoor air quality and comfort.

Lighting Loads

Lighting loads are one of the few major loads that vary from one type of assembly building to another. Lighting may be at the level of 150 footcandles in convention halls where color television cameras are expected to be used, or lighting may be virtually absent, as in a movie theater. In many assembly buildings, lights are controlled by dimmers or other means to present a suitably low level of light during performances, with much higher lighting levels during cleanup, when the house is nearly empty. The designer should ascertain what light levels will be associated with maximum occupancies, not only in the interest of economy but also to determine the proper room sensible heat ratio.

The preparation of this chapter is assigned to TC 9.8, Large Building Air-Conditioning Applications.

Indoor Air Conditions

Indoor air temperature and humidity should follow the ASHRAE comfort recommendations (see Chapter 8 of the 1997 *ASHRAE Handbook—Fundamentals* and ASHRAE *Standard* 55). In addition, the following should be considered:

1. In arenas, stadiums, gymnasiums, and movie theaters, people generally dress informally. Summer indoor conditions may favor the warmer end of the thermal comfort scale while the winter indoor temperature may favor the cooler end of the scale.
2. In churches, concert halls, and theaters, most men wear jackets and ties and women often wear suits. The temperature should favor the middle range of design, and there should be little summer-to-winter variation.
3. In convention and exhibition centers, the public is continually walking. Here the indoor temperature should favor the lower range of comfort conditions both in summer and in winter.
4. In spaces with a high population density or with a sensible heat factor of 0.75 or less, a lower dry-bulb temperature reduces the latent heat load from people, thus saving energy by reducing the need for reheat.
5. Energy conservation codes must be considered in both the design and during operation.

Assembly areas generally require some reheat to maintain the relative humidity at a suitably low level during periods of maximum occupancy. Refrigerant hot gas or condenser water is well suited for this purpose. Face and bypass control of low-temperature cooling coils is also effective. In colder climates, it may also be desirable to provide humidification. High rates of internal gain may make evaporative humidification attractive during economizer cooling.

Filtration

Most places of assembly are minimally filtered with filters rated at 30 to 35% efficiency, as tested in accordance with ASHRAE *Standard* 52.1. Where smoking is permitted, however, filters with a minimum rating of 80% are required to remove tobacco smoke effectively. Filters with 80% or higher efficiency are also recommended for those facilities having particularly expensive interior decor. Because of the few operating hours of these facilities, the added expense of higher efficiency filters can be justified by their longer life. Low-efficiency prefilters are generally used with high-efficiency filters to extend their useful life. Ionization and chemically reactive filters should be considered where high concentrations of smoke or odors are present.

Noise and Vibration Control

The desired noise criteria (NC) vary with the type and quality of the facility. The need for noise control may be minimal in a gymnasium or natatorium, but it is important in a concert hall. Facilities that are used for varied functions require noise control evaluation over the entire spectrum of use.

In most cases, sound and vibration control is required for both equipment and duct systems, as well as in the selection of diffusers and grilles. When designing for a theater or concert hall, an experienced acoustics engineer should be consulted. In these projects, the quantity and quality or characteristic of the noise is very important.

Transmission of vibration and noise can be decreased by mounting pipes, ducts, and equipment on a separate structure independent of the music hall. If the mechanical equipment space is close to the music hall, the entire mechanical equipment room may need to be floated on isolators, including the floor slab, structural floor members, and other structural elements such as supporting pipes or similar materials that can carry vibrations. Properly designed inertia pads are often used under each piece of equipment. The equipment is then mounted on vibration isolators.

Manufacturers of vibration isolating equipment have devised methods to float large rooms and entire buildings on isolators. Where subway and street noise may be carried into the structure of a music hall, it is necessary to float the entire music hall on isolators. If the music hall is isolated from outside noise and vibration, it also must be isolated from mechanical equipment and other internal noise and vibrations.

External noise from mechanical equipment such as cooling towers should not enter the building. Care should be taken to avoid designs that permit noises to enter the space through air intakes or reliefs and carelessly designed duct systems.

Ancillary Facilities

Ancillary facilities are generally a part of any assembly building; almost all have some office space. Convention centers and many auditoriums, arenas, and stadiums have restaurants and cocktail lounges. Churches may have apartments for the clergy or a school. Many facilities have parking structures. These varied ancillary facilities are discussed in other chapters of this volume. However, for reasonable operating economy, these facilities should be served by separate systems when their hours of use differ from those of the main assembly areas.

Air Conditioning

Because of their characteristic large size and need for considerable ventilation air, assembly buildings are frequently served by single-zone or variable-volume systems providing 100% outdoor air. Separate air-handling units usually serve each zone, although multizone, dual-duct, or reheat types can also be applied with lower operating efficiency. In larger facilities, separate zones are generally provided for the entrance lobbies and arterial corridors that surround the seating space. Low intensity radiant heating is often an efficient alternative. In some assembly rooms, folding or rolling partitions divide the space for different functions, so that a separate zone of control for each resultant space is best. In extremely large facilities, several air-handling systems may serve a single space, due to the limits of equipment size and also for energy and demand considerations.

Precooling

Cooling the building mass several degrees below the desired indoor temperature several hours before it is occupied allows it to absorb a portion of the peak heat load. This cooling reduces the equipment size needed to meet short-term loads. The effect can be used if cooling time of at least one hour is available prior to occupancy, and then only when the period of peak load is relatively short (2 h or less).

The designer must advise the owner that the space temperature will be cold to most people as occupancy begins, but that it will warm up as the performance progresses. This may be satisfactory, but it should be understood by all concerned before proceeding with a precooling concept. Precooling is best applied when the space is used only occasionally during the hotter part of the day and when provision of full capacity for an occasional purpose is not economically justifiable.

Stratification

Because most assembly buildings relatively high ceilings, some heat may stratify above the occupied zone, thereby reducing the load on the equipment. Heat from lights can be stratified, except for the radiant portion (about 50% for fluorescent and 65% for incandescent or mercury-vapor fixtures). Similarly, only the radiant effect of the upper wall and roof load (about 33%) reaches the occupied space. Stratification only occurs when air is admitted and returned at a sufficiently low elevation so that it does not mix with the upper air. Conversely, stratification may increase heating loads during periods of minimal occupancy in winter months. In these cases, ceiling fans, air-handling systems, or high/low air distribution may be desirable to reduce stratification. Balconies may also be affected by stratification and should be well ventilated.

Air Distribution

In assembly buildings, people generally remain in one place throughout a performance, so they cannot avoid drafts. Therefore, good air distribution is essential.

Heating is seldom a major problem, except at entrances or during warm-up prior to occupancy. Generally, the seating area is isolated from the exterior by lobbies, corridors, and other ancillary spaces. For cooling, air can be supplied from the overhead space, where it mixes with heat from the lights and occupants. Return air openings can also aid air distribution. Air returns located below seating or at a low level around the seating can effectively distribute air with minimum drafts. For returns located below the seats, register velocities in excess of 275 fpm may cause objectionable drafts and noise.

Because of the configuration of these spaces, supply jet nozzles with long throws of 50 to 150 ft may need to be installed on sidewalls. For ceiling distribution, downward throw is not critical provided returns are low. This approach has been successful in applications that are not particularly noise-sensitive, but the designer needs to select air distribution nozzles carefully.

The air-conditioning systems must be quiet. This is difficult to achieve if the supply air is expected to travel 30 ft or more from sidewall outlets to condition the center of the seating area. Due to the large size of most houses of worship, theaters, and halls, high air discharge velocities from the wall outlets are required. These high velocities can produce objectionable noise levels for people sitting near the outlets. This can be avoided if the return air system does some of the work. The supply air must be discharged from the air outlet (preferably at the ceiling) at the highest velocity consistent with an acceptable noise level. Although this velocity does not allow the conditioned air to reach all seats, the return air registers, which are located near seats not reached by the conditioned air, pull the air to cool or heat the audience, as required. In this way, the supply air blankets the seating area and is pulled down uniformly by the return air registers under or beside the seats.

A certain amount of exhaust air should be taken from the ceiling of the seating area, preferably over the balcony (if there is one) to prevent pockets of hot air, which can produce a radiant effect and cause discomfort, as well as increase the cost of air conditioning. Where the ceiling is close to the audience (e.g., below balconies and mezzanines), specially designed plaques or air-distributing ceilings should be provided to absorb noise.

Regular ceiling diffusers placed more than 30 ft apart normally give acceptable results if careful engineering is applied in the selection of the diffusers. Because large air quantities are generally involved and because the building is large, fairly large capacity diffusers, which tend to be noisy, are frequently selected. Linear diffusers are more acceptable architecturally and perform well if selected properly. Integral dampers in diffusers should not be used as the only means of balancing because they generate intolerable amounts of noise, particularly in larger diffusers.

Mechanical Equipment Rooms

The location of mechanical and electrical equipment rooms affects the degree of sound attenuation treatment required. Mechanical equipment rooms located near the seating area are more critical because of the normal attenuation of sound through space. Mechanical equipment rooms located near the stage area are critical because the stage is designed to project sound to the audience. If possible, mechanical equipment rooms should be located in an area separated from the main seating or stage area by buffers such as lobbies or service areas. The economies of the structure, attenuation, equipment logistics, and site must be considered in the selection of locations for mechanical equipment rooms.

At least one mechanical equipment room is placed near the roof to house the toilet exhaust, general exhaust, cooling tower, kitchen, and emergency stage exhaust fans, if any. Individual roof-mounted exhaust fans may be used, thus eliminating the need for a mechanical equipment room. However, to reduce sound problems, mechanical equipment should not be located on the roof over the music hall or stage but rather over offices, storerooms, or auxiliary areas.

HOUSES OF WORSHIP

Houses of worship seldom have full or near-full occupancy more than once a week, but they have considerable use for smaller functions (meetings, weddings, funerals, christenings, or daycare) throughout the balance of the week. It is important to determine how and when the building will be used. When thermal storage is used, longer operation of equipment prior to occupancy may be required due to the high thermal mass of the structure. The seating capacity of houses of worship is usually well defined. Some houses of worship have a movable partition to form a single large auditorium for special holiday services. It is important to know how often this maximum use is expected.

Houses of worship test a designer's ingenuity in locating equipment and air diffusion devices in architecturally acceptable places. Because occupants are often seated, drafts and cold floors should be avoided. Many houses of worship have a high vaulted ceiling, which creates thermal stratification. Where stained glass is used, a shade coefficient equal to solar glass (SC = 0.70) is assumed.

Houses of worship may also have auxiliary rooms that should be air conditioned. To ensure privacy, sound transmission between adjacent areas should be considered in the air distribution scheme. Diversity in the total air-conditioning load requirements should be evaluated to take full advantage of the characteristics of each area.

In houses of worship, it is desirable to provide some degree of individual control for the platform, sacristy, and bema or choir area.

AUDITORIUMS

The types of auditoriums considered are movie theaters, playhouses, and concert halls. Auditoriums in schools and the large auditoriums in some convention centers may follow the same principles, with varying degrees of complexity.

Movie Theaters

Motion picture theaters are the simplest of the auditorium structures mentioned here. They run continuously for periods of 4 to 8 h or more and, thus, are not a good choice for precooling techniques, except for the first matinee peak. They operate frequently at low occupancy levels, and low-load performance must be considered.

Motion picture sound systems make noise control less important than it is in other kinds of theaters. The lobby and exit passageways in a motion picture theater are seldom densely occupied, although some light to moderate congestion can be expected for short times in the lobby area. A reasonable design for the lobby space is one person per 20 to 30 ft^2.

The lights are usually dimmed when the house is occupied; full lighting intensity is used only during cleaning. A reasonable value for lamps above the seating area during a performance is 5 to 10% of the installed wattage. Designated smoking areas should be handled with separate exhaust or air-handling systems to avoid contamination of the entire facility.

Projection Booths. The projection booth represents the major problem in motion picture theater design. For large theaters using high-intensity lamps, projection room design must follow applicable building codes. If no building code applies, the projection equipment manufacturer usually has specific requirements. The projection room may be air conditioned, but it is normally exhausted or operated at negative pressure. Exhaust is normally taken through the housing of the projectors. Additional exhaust is required for the projectionist's sanitary facilities. Other heat sources include the sound and dimming equipment, which require a continuously controlled environment and necessitate a separate system.

Smaller theaters have fewer requirements for projection booths. It is a good idea to condition the projection room with filtered supply air to avoid soiling lenses. In addition to the projector light, heat sources in the projection room include the sound equipment, as well as the dimming equipment.

Legitimate Theaters

The legitimate theater differs from the motion picture theater in the following ways:

- Performances are seldom continuous. Where more than one performance occurs in a day, the performances are usually separated by a period of 2 to 4 h. Accordingly, precooling techniques are applicable, particularly for afternoon performances.
- Legitimate theaters seldom play to houses that are not full or near full.
- Legitimate theaters usually have intermissions, and the lobby areas are used for drinking and socializing. The intermissions are usually relatively short, seldom exceeding 15 to 20 min; however, the load may be as dense as one person per 5 ft^2.
- Because sound amplification is less used than that in a motion picture theater, background noise control is more important.
- Stage lighting contributes considerably to the total cooling load in the legitimate theater. Lighting loads can vary from performance to performance.

Stages. The stage presents the most complex problem. It consists of the following loads:

- A heavy, mobile lighting load
- Intricate or delicate stage scenery, which varies from scene to scene and presents difficult air distribution requirements
- Actors, who may perform tasks that require exertion

Approximately 40 to 60% of the lighting load can be eliminated by exhausting air around the lights. This procedure works for lights around the proscenium. However, it is more difficult to place exhaust air ducts directly above lights over the stage because of the scenery and light drops. Careful coordination is required to achieve an effective and flexible layout.

Conditioned air should be introduced from the low side and back stages and returned or exhausted around the lights. Some exhaust air must be taken from the top of the tower directly over the stage containing lights and equipment (i.e., the fly). The air distribution design is further complicated because pieces of scenery may consist of light materials that flutter in the slightest air current. Even the vertical stack effect created by the heat from lights may cause this motion. Therefore, low air velocities are essential and air must be distributed over a wide area with numerous supply and return registers.

With multiple scenery changes, low supply or return registers from the floor of the stage are almost impossible to provide. However, some return air at the footlights and for the prompter should be considered. Air conditioning should also be provided for the stage manager and the control board areas.

One phenomenon encountered in many theaters with overhead flies is the billowing of the stage curtain when it is down. This situation is primarily due to the stack effect created by the height of the main stage tower, the heat from the lights, and the temperature difference between the stage and seating areas. Proper air distribution and balancing can minimize this phenomenon. Bypass damper arrangements with suitable fire protection devices may be feasible.

Loading docks adjacent to stages located in cold climates should be heated. The doors to these areas may be open for long periods, for example, while scenery is being loaded or unloaded for a performance.

On the stage, local code requirements must be followed for emergency exhaust ductwork or skylight (or blow-out hatch) requirements. These openings are often sizable and should be incorporated in the early design concepts.

Concert Halls

Concert halls and music halls are similar to legitimate theaters. They normally have a full stage, complete with fly gallery, and dressing areas for performers. Generally, the only differences between the two are in size and decor, with the concert hall usually being larger and more elaborately decorated.

Air-conditioning design must consider that the concert hall is used frequently for special charity and civic events, which may be preceded by or followed by parties (and may include dancing) in the lobby area. Concert halls often have cocktail lounge areas that become very crowded, with heavy smoking during intermissions. These areas should be equipped with flexible exhaust-recirculation systems. Concert halls may also have full restaurant facilities.

As in theatres, noise control is important. The design must avoid characterized or narrow-band noises in the level of audibility. Much of this noise is structure-borne, resulting from inadequate equipment and piping vibration isolation. An experienced acoustical engineer is essential for help in the design of these applications.

ARENAS AND STADIUMS

Functions at arenas and stadiums may be quite varied, so the air-conditioning loads will vary. Arenas and stadiums are not only used for sporting events such as basketball, ice hockey, boxing, and track meets but may also house circuses; rodeos; convocations; social affairs; meetings; rock concerts; car, cycle, and truck events; and special exhibitions such as home, industrial, animal, or sports shows. For multipurpose operations, the designer must provide systems having a high degree of flexibility. High-volume ventilation may be satisfactory in many instances, depending on load characteristics and outside air conditions.

Load Characteristics

Depending on the range of use, the load may vary from a very low sensible heat ratio for events such as boxing to a relatively high sensible heat ratio for industrial exhibitions. Multispeed fans often improve the performance at these two extremes and can aid in sound control for special events such as concerts or convocations. When using multispeed fans, the designer should consider the performance of the air distribution devices and cooling coils when the fan is operating at lower speeds.

Because total comfort cannot be ensured in an all-purpose facility, the designer must determine the level of discomfort that can be tolerated, or at least the type of performances for which the facility is primarily intended.

As with other assembly buildings, seating and lighting combinations are the most important load considerations. Boxing events, for example, may have the most seating, because the arena area is very small. For the same reason, however, the area that needs to be intensely illuminated is also small. Thus, boxing matches may represent the largest latent load situation. Other events that present large latent loads are rock concerts and large-scale dinner dances, although the audience at a rock concert is generally less concerned with thermal comfort. Ventilation is also essential in removing smoke or fumes at car, cycle, and truck events. Circuses, basketball, and hockey have a much larger arena area and less seating. The sensible load from lighting the arena area does improve the sensible heat ratio. The large expanse of ice in hockey games represents a considerable reduction in both latent and sensible loads. High latent loads caused by occupancy or ventilation can create severe problems in ice arenas such as condensation on interior surfaces and fog. Special attention should be paid to the ventilation system, air distribution, humidity control, and construction materials.

Enclosed Stadiums

An enclosed stadium may have either a retractable or a fixed roof. When the roof is closed, ventilation is needed, so ductwork must be run in the permanent sections of the stadium. The large air volumes required and the long air throws make proper air distribution difficult to achieve; thus, the distribution system must be very flexible and adjustable.

Some open stadiums have radiant heating coils in the floor slabs of the seating areas. Gas-fired or electric high- or low-intensity radiant heating located above the occupants is also used.

Open racetrack stadiums may present a ventilation problem if the grandstand is enclosed. The grandstand area may have multiple levels and be in the range of 1300 ft long and 200 ft deep. The interior (ancillary) areas must be ventilated to control odors from toilet facilities, concessions, and the high population density. General practice provides about four air changes per hour for the stand seating area and exhausts the air through the rear of the service areas. More efficient ventilation systems may be selected if architectural considerations permit. Fogging of windows is a winter concern with glass-enclosed grandstands. This can be minimized by double glazing, humidity control, moving dry air across the glass, or a radiant heating system for perimeter glass areas.

Air-supported structures require the continuous operation of a fan to maintain a properly inflated condition. The possibility of condensation on the underside of the air bubble should be considered. The U-factor of the roof should be sufficient to prevent condensation at the lowest expected ambient temperature. Heating and air-conditioning functions can be either incorporated into the inflating system or furnished separately. Solar and radiation control is also possible through the structure's skin. Applications, though increasing rapidly, still require working closely with the enclosure manufacturer to achieve proper and integrated results.

Ancillary Spaces

The concourse areas of arenas and stadiums are heavily populated during entrance, exit, and intermission periods. Considerable odor is generated in these areas by food, drink, and smoke, requiring considerable ventilation. If energy conservation is an important factor, carbon filters and controllable recirculation rates should be considered. Concourse area air systems should be considered for their flexibility of returning or exhausting air. The economics of this type of flexibility should be evaluated with regard to the associated problem of air balance and freeze-up in cold climates.

Ticket offices, restaurants, and similar facilities are often expected to be open during hours that the main arena is closed; and, therefore, separate systems should be considered for these areas.

Locker rooms require little treatment other than excellent ventilation, usually not less than 2 or 3 cfm per square foot. To reduce the outdoor air load, excess air from the main arena or stadium may be transferred into the locker rooms. However, reheat or recooling by water or primary air should be considered to maintain the locker room temperature. To maintain proper air balance under all conditions, locker rooms should have separate supply and exhaust systems.

When an **ice rink** is designed into the facility, the concerns of groundwater conditions, site drainage, structural foundations, insulation, and waterproofing become even more important, with the potential of freezing of soil or fill under the floor and subsequent expansion. The rink floor may have to be strong enough to support heavy trucks. The floor insulation also must be strong enough to take this load. Ice-melting pits of sufficient size with steam pipes may have to be furnished. If the arena is to be air conditioned, the possibility of combining the air-conditioning system with the ice rink system may be analyzed. The designer should be aware that both systems operate at vastly different temperatures and may operate at different capacity levels at any given time. The radiant effects of the ice on the people and of the heat from the roof and lights on the ice must be considered in the design and operation of the system. Also, low air velocity at the floor is related to minimizing the refrigeration load. High air velocities will cause moisture to be drawn from the air by the ice sheet. Fog is caused by the uncontrolled introduction of airborne moisture through ventilation with warm, moist outside air and generally develops within the boarded area (playing area). Fog can be controlled by reducing outdoor air ventilation rates, using appropriate air velocities to bring the air in contact with the ice, and using a dehumidification system. Air-conditioning systems have limited impact on reducing the dew-point temperature sufficiently to prevent fog.

The type of lighting used over ice rinks must be carefully considered when precooling is used prior to hockey games and between periods. Main lights should be capable of being turned off, if feasible. Incandescent lights require no warm-up time and are more applicable than types requiring warm-up. Low emissivity ceilings with reflective characteristics successfully reduce condensation on roof structures; they also reduce lighting requirements.

Gymnasiums

Smaller gymnasiums, such as those found in school buildings, are miniature versions of arenas and often incorporate multipurpose features. For further information, see Chapter 6, Educational Facilities.

Many school gymnasiums are not air conditioned. Low-intensity perimeter radiant heaters with central ventilation supplying four to six air changes per hour are effective and energy efficient. Unit heaters located in the ceiling area are also effective. Ventilation must be provided due to the high activity levels and the resulting odors.

Most gymnasiums are located in schools. However, public and private organizations and health centers may also have gymnasiums. During the day, gymnasiums are usually used for physical activities, but in the evening and on weekends, they may be used for sports events, social affairs, or meetings. Thus, their activities fall within the scope of those of a civic center. More gymnasiums are being considered for air conditioning to make them more suitable for civic center activities.

The design criteria are similar to arenas and civic centers when used for nonstudent training activities. However, for schooltime use, space temperatures are often kept between 65 and 68°F during the heating season. Occupancy and the degree of activity during daytime use does not usually require high quantities of outdoor air, but if used for other functions, system flexibility is required.

CONVENTION AND EXHIBITION CENTERS

Convention-exhibition centers schedule diverse functions similar to those at arenas and stadiums and present a unique challenge to the designer. The center generally is a high-bay, long-span space. These centers can be changed weekly, for example, from an enormous computer room into a gigantic kitchen, large machine shop, department store, automobile showroom, or miniature zoo. They can also be the site of gala banquets or used as major convention meeting rooms.

The income earned by these facilities is directly affected by the time it takes to change from one activity to the next, so highly flexible utility distribution and air conditioning equipment are needed.

Ancillary facilities include restaurants, bars, concession stands, parking garages, offices, television broadcasting rooms, and multiple meeting rooms varying in capacity from small (10 to 20 people) to large (hundreds or thousands of people). Often, an appropriately sized full-scale auditorium or arena is also incorporated.

By their nature, these facilities are much too large and diverse in their use to be served by a single air-handling system. Multiple air handlers with several chillers can be economical.

Load Characteristics

The main exhibition room is subject to a variety of loads, depending on the type of activity in progress. Industrial shows provide the highest sensible loads, which may have a connected capacity of 20 W/ft^2 along with one person per 40 to 50 ft^2. Loads of this magnitude are seldom considered because large power-consuming equipment is seldom in continuous operation at full load. An adequate design accommodates (in addition to lighting load) about 10 W/ft^2 and one person per 40 to 50 ft^2 as a maximum continuous load.

Alternative loads that are very different in character may be encountered. When the main hall is used as a meeting room, the load will be much more latent in character. Thus, multispeed fans or variable volume systems may provide a better balance of load during these high latent, low sensible periods of use. The determination of accurate occupancy and usage information is critical in any plan to design and operate such a facility efficiently and effectively.

System Applicability

The main exhibition hall is normally handled by one or more all-air systems. This equipment should be capable of operating on all outdoor air, because during set-up time, the hall may contain a number of highway-size trucks bringing in or removing exhibit materials. There are also occasions when the space is used for equipment that produces an unusual amount of fumes or odors, such as restaurant or printing industry displays. It is helpful to build some flues into the structure to duct fumes directly to the outside. Perimeter radiant ceiling heaters have been successfully applied to exhibition halls with large expanses of glass.

Smaller meeting rooms are best conditioned with either individual room air handlers, or with variable volume central systems, because these rooms have high individual peak loads but are not used frequently. Constant volume systems of the dual- or single-duct reheat type waste considerable energy when serving empty rooms, unless special design features are incorporated.

Offices and restaurants often operate for many more hours than the meeting areas or exhibition areas and should be served separately. Storage areas can generally be conditioned by exhausting excess air from the main exhibit hall through these spaces.

NATATORIUMS

Environmental Control

A natatorium requires year-round humidity levels between 40 and 60% for comfort, energy consumption, and building protection. The designer must address the following concerns: humidity control, ventilation requirements for air quality (outdoor and exhaust air), air distribution, duct design, pool water chemistry, and evaporation rates. A humidity control system will not provide satisfactory results if any of these items are overlooked.

Humidity Control

Humans are very sensitive to relative humidity. Fluctuations in relative humidity outside the 40 to 60% range can increase levels of bacteria, viruses, fungi and other factors that reduce air quality. For swimmers, 50 to 60% rh is most comfortable. High relative humidity levels are destructive to building components. Mold and mildew can attack wall, floor, and ceiling coverings; and condensation can degrade many building materials. In the worst case, the roof could collapse due to corrosion from water condensing on the structure.

Load Estimation

Loads for a natatorium include building heat gains and losses from outdoor air, lighting, walls, roof, and glass. Internal latent loads are generally from people and evaporation. Evaporation loads in pools and spas are significant relative to other load elements and may vary widely depending on pool features, the areas of water and wet deck, water temperature, and activity level in the pool.

Evaporation. The rate of evaporation (Carrier 1918) can be estimated from empirical Equation (1). This equation is valid for pools at normal activity levels, allowing for splashing and a limited area of wetted deck. Other pool uses may have more or less evaporation (Smith et al. 1993).

$$w_p = \frac{A}{Y}(p_w - p_a)(95 + 0.425V) \tag{1}$$

where

w_p = evaporation of water, lb/h
A = area of pool surface, ft^2
V = air velocity over water surface, fpm
Y = latent heat required to change water to vapor at surface water temperature, Btu/lb
p_a = saturation pressure at room air dew point, in. Hg
p_w = saturation vapor pressure taken at surface water temperature, in. Hg

The units for the constant 95 are Btu/(h·ft^2·in. Hg). The units for the constant 0.425 are Btu·min/(h·ft^3·in. Hg).

Equation (1) may be modified by multiplying it by an activity factor F_a to alter the estimate of evaporation rate based on the level of activity supported. For Y values of about 1000 Btu/lb and V values ranging from 10 to 30 fpm, Equation (1) can be reduced to

$$w_p = 0.1A(p_w - p_a)F_a \tag{2}$$

The following activity factors should be applied to the areas of specific features, and not to the entire wetted area:

Type of Pool	Typical Activity Factor (F_a)
Residential pool	0.5
Condominium	0.65
Therapy	0.65
Hotel	0.8
Public, schools	1.0
Whirlpools, spas	1.0
Wavepools, water slides	1.5 (minimum)

The effectiveness of controlling the natatorium environment depends on the correct estimation of water evaporation rates. Applying the correct activity factors is extremely important in determining water evaporation rates. The difference in peak evaporation rates between private pools and active public pools of comparable size may be more than 100%.

Table 1 Typical Natatorium Design Conditions

Type of Pool	Air Temperature, °F	Water Temperature, °F	Relative Humidity, %
Recreational	75 to 85	75 to 85	50 to 60
Therapeutic	80 to 85	85 to 95	50 to 60
Competition	78 to 85	76 to 82	50 to 60
Diving	80 to 85	80 to 90	50 to 60
Whirlpool/spa	80 to 85	97 to 104	50 to 60

Actual operating temperatures and relative humidity conditions should be established prior to design. How the area will be used usually dictates design. The elderly prefer significantly warmer operating temperatures than those listed Table 1.

Air temperatures in public and institutional pools should be maintained 2 to 4°F above the water temperature (but not above the comfort threshold of 86°F) to reduce the evaporation rate and avoid chill effects on swimmers.

Ventilation Requirements

Air Quality. Outdoor air ventilation rates prescribed by ASHRAE *Standard* 62 are intended to provide acceptable air quality conditions for the average pool using chlorine for its primary disinfection process. The ventilation requirement may be excessive for private pools and installations with low use. They may also prove inadequate for high occupancy public installations.

Air quality problems in pools and spas are caused by water quality problems, so simply increasing ventilation rates may prove both expensive and ineffective. Water quality conditions are a direct function of pool use and the type and effectiveness of the water disinfection process used.

Because an indoor pool usually has a high ceiling, temperature stratification can have a detrimental effect on indoor air quality. Care must be taken with duct layout to ensure that the space receives the proper air changes and homogeneous air quality throughout. Some air movement at the deck and pool water level is essential to ensure acceptable air quality. Complaints from swimmers tend to indicate that the greatest chloramine (see the section on Pool Water Chemistry) concentrations occur at the water surface. Children are especially vulnerable to chloramine poisoning.

Exhaust air from pool is rich in moisture and may contain high levels of chloramine compounds. While most codes permit pool air to be used as makeup for showers, toilets, and locker rooms, these spaces should be provided with separate ventilation and maintained at a positive pressure with respect to the pool.

Pool and spa areas should be maintained at a negative pressure of 0.05 to 0.15 in. of water relative to adjacent areas of the building to prevent moisture and chloramine odor migration. Active methods of pressure control may prove more effective than static balancing and may be necessary where outdoor air is used as a part of an active humidity control strategy. Openings from the pool to other areas should be minimized and controlled. Passageways should be equipped with doors with automatic closers to inhibit migration of moisture and air.

Air Delivery Rates. Total airflow should be determined by a psychrometric analysis. Most codes require a minimum of six (6) air changes per hour, except where mechanical cooling is used. This rate may prove inadequate for some anticipated conditions of occupancy and use.

Where mechanical cooling is provided, air delivery rates should be established to maintain appropriate conditions of temperature and humidity. The following rates are typically desired:

Pools with no spectator areas	4 to 6 air changes per hour
Spectator areas	6 to 8 air changes per hour
Therapeutic pools	4 to 6 air changes per hour

Outdoor air delivery rates may be constant or variable, depending on the design. Minimum rates, however, must provide adequate dilution of contaminants generated by pool water and must maintain acceptable ventilation for occupancy.

Where a minimum outdoor air ventilation rate is established to protect against condensation in a building's structural elements, the rates are typically used for 100% outdoor air systems. These rates usually result in excessive humidity levels under most operating conditions and are generally not adequate to produce acceptable indoor air quality, especially in public facilities subject to heavy use.

Duct Design

As with any installation, proper duct design and installation is necessary for proper equipment performance. Poorly installed return duct connections, for example, can significantly reduce the performance of a dehumidifier. The following duct construction practices apply to natatoriums:

- Fiberglass duct liner should not be used. Where condensation may occur the insulation must be applied to the exterior of the duct.
- Duct materials and hardware must be resistant to chemical corrosion from the pool atmosphere. The 400 series stainless steels are readily attacked by chlorides in moist environments. The 300 series stainless steel, painted galvanized, or aluminum sheet metal may be used for exposed duct systems. Buried ductwork should be constructed from non-metallic fiberglass reinforced or PVC materials due to the difficulty of replacing damaged materials.
- Grilles, registers and diffusers should be constructed from aluminum. They should be selected for low static pressure loss and for appropriate throws for proper air distribution.
- Supply air should be directed against interior envelope surfaces prone to condensation (walls, glass, and doors). A portion of the supply air should be directed over the water surface to move contaminated air toward an exhaust point and control chloramines released at the water surface.
- Return air inlets should be located to recover the warm humid air and return it to the ventilation system for treatment, to prevent the supply air from short-circuiting, and minimize recirculation of chloramines.
- Exhaust air inlets should be located to maximize capture effectiveness and minimize the recirculation of chloramines. Exhausting from directly above whirlpools is also desirable. Exhaust air should be taken directly to the outside, through heat recovery devices where provided.
- Filtration should be selected to provide 45 to 65% efficiencies (as defined in ASHRAE *Standard* 52.1) and be installed in locations selected to prevent condensation in the filter bank. Filter media and support materials should be resistant to moisture degradation.
- Air systems may be designed for noise levels of NC 45-50, however, wall, floor, and ceiling surfaces should be evaluated for their attenuation effect.

Envelope Design

Glazing in exterior walls becomes susceptible to condensation when the outdoor temperature drops below the pool room dew point. The design goal is to maintain the surface temperature of the glass and the window frames a minimum of 5°F above the pool room dew point. Windows must allow unobstructed air movement on inside surfaces. Thermal break frames should be used. Recessed windows and protruding window frames should be avoided. Skylights are especially vulnerable and attention should be given to control condensation on them. Wall and roof vapor retarder designs should be carefully reviewed, especially at wall-to-wall and wall-to-roof junctures and at window, door, and duct penetrations. The pool enclosure must be suitable for year round operation at 50 to 60% relative humidity. A vapor barrier analysis (as in Figure 11 in Chapter 22 of the 1997 *ASHRAE Handbook—Fundamentals*) should be prepared.

Failure to install an effective vapor retarder will result in condensation forming in the structure and potentially serious damage.

Pool Water Chemistry

Failure to maintain proper chemistry in the pool water causes serious air quality problems and deterioration of mechanical and building systems. Water treatment equipment should be installed in a separate, dedicated, well-ventilated space that is under negative pressure. Pool water treatment consists of primary disinfection, pH control, water filtration and purging, and water heating. For further information, refer to Kowalsky (1990).

Air quality problems are usually caused by the reaction of chlorine with biological wastes, and particularly with ammonia, which is a by-product of the breakdown of urine and perspiration. Chlorine reacts with these wastes, creating chloramines (monochloramine, dichloramine, and nitrogen trichloride) that are commonly measured as combined chlorine. The addition of chemicals to pool water increases total contaminant levels. In high occupancy pools, water contaminant levels can double in a single day of operation.

The reduction of ammonia by chlorine is affected by several factors including water temperature, water pH, total chlorine concentration, and the level of dissolved solids in the water. Because of their higher operating temperature and higher ratio of occupancy per unit water volume, spas produce greater quantities of air contaminants than pools.

The following measures have demonstrated a potential to reduce chloramine concentrations in the air and water:

- **Ozonation.** In low concentrations, ozone has substantially reduced the concentration of combined chlorine in the water. In high concentrations, ozone can replace chlorine as the primary disinfection process, however, ozone is unable to maintain sufficient residual levels in the water to maintain a latent biocidal effect. This necessitates the maintenance of chlorine as a residual process at concentrations of 0.5 to 1.5 ppm.
- **Water Exchange Rates.** High concentrations of dissolved solids in water have been shown to directly contribute to high combined chlorine (chloramine) levels. Adequate water exchange rates are necessary to prevent the buildup of biological wastes and their oxidized components in pool and spa water. Conductivity measurement is an effective method to control the exchange rate of water in pools and spas to effectively maintain water quality and minimize water use. In high-occupancy pools, heat recovery may prove useful in reducing water heating energy requirements.

Energy Considerations

Natatoriums can be major energy burden on facilities, so they represent a significant opportunity for energy conservation. Several design solutions are possible using both dehumidification and ventilation strategies. When evaluating a system, the energy consumed by all elements should be considered, including primary heating and cooling systems, fan motors, water heaters, and pumps.

Natatoriums with fixed outdoor air ventilation rates without dehumidification generally have seasonally fluctuating space temperature and humidity levels. Systems designed to provide minimum ventilation rates without dehumidification are unable to maintain relative humidity conditions within prescribed limits. These systems may facilitate mold and mildew growth and may be unable to provide acceptable indoor air quality. Peak dehumidification loads vary with activity levels and during the cooling season when ventilation air becomes an additional dehumidification load to the space.

FAIRS AND OTHER TEMPORARY EXHIBITS

Occasionally, large-scale exhibits are constructed to stimulate business, present new ideas, and provide cultural exchanges. Fairs of this type take years to construct, are open from several months to several years, and are sometimes designed considering future use of some of the buildings. Fairs, carnivals, or exhibits, which may consist of prefabricated shelters and tents that are moved from place to place and remain in a given location for only a few days or weeks, are not covered here because they seldom require the involvement of architects and engineers.

Design Concepts

One consultant or agency should be responsible for setting uniform utility service regulations and practices to ensure proper organization and operation of all exhibits. Exhibits that are open only during the intermediate spring or fall months require a much smaller heating or cooling plant than those designed for for peak summer or winter design conditions. This information is required in the earliest planning stages so that system and space requirements can be properly analyzed.

Occupancy

Fair buildings have heavy occupancy during visiting hours, but patrons seldom stay in any one building for a long period. The length of time that patrons stay in a building determines the air-conditioning design. The shorter the anticipated stay, the greater the leeway in designing for less-than-optimum comfort, equipment, and duct layout. Also, whether patrons wear coats and jackets while in the building influences operating design conditions.

Equipment and Maintenance

Heating and cooling equipment used solely for maintaining comfort and not for exhibit purposes may be secondhand or leased, if available and of the proper capacity. Another possibility is to rent the air-conditioning equipment to reduce the capital investment and eliminate disposal problems when the fair is over.

Depending on the size of the fair, the length of time it will operate, the types of exhibitors, and the policies of the fair sponsors, it may be desirable to analyze the potential for a centralized heating and cooling plant versus individual plants for each exhibit. The proportionate cost of a central plant to each exhibitor, including utility and maintenance costs, may be considerably less than having to furnish space and plant utility and maintenance costs. The larger the fair, the more savings may result. It may be practical to make the plant a showcase, suitable for exhibit and possibly added revenue. A central plant may also form the nucleus for the commercial or industrial development of the area after the fair is over.

If exhibitors furnish their own air-conditioning plants, it is advisable to analyze shortcuts that may be taken to reduce equipment space and maintenance aids. For a 6-month to 2-year maximum operating period, for example, tube pull or equipment removal space is not needed or may be drastically reduced. Higher fan and pump motor power and smaller equipment is permissible to save on initial costs. Ductwork and piping costs should be kept as low as possible because these are usually the most difficult items to salvage; cheaper materials may be substituted wherever possible. The job must be thoroughly analyzed to eliminate all unnecessary items and reduce all others to bare essentials.

The central plant may be designed for short-term use as well. However, if the plant is to be used after the fair closes, the central plant should be designed in accordance with the best practice for long-life plants. It is difficult to determine how much of the piping distribution system can be used effectively for permanent installations. For that reason, piping should be simply designed initially, preferably in a grid, loop, or modular layout, so that future additions can be made easily and economically.

Air Cleanliness

The efficiency of the filters needed for each exhibit is determined by the nature of the area served. Because the life of an exhibit is very short, it is desirable to furnish the least expensive filtering system. If possible, one set of filters should be selected to last for the life of the exhibit. In general, the filtering efficiencies do not have to exceed 30% (see ASHRAE *Standard* 52.1).

System Applicability

If a central air-conditioning plant is not built, the equipment installed in each building should be the least costly to install and operate for the life of the exhibit. These units and systems should be designed and installed to occupy the minimum usable space.

Whenever feasible, heating and cooling should be performed by one medium, preferably air, to avoid running a separate piping and radiation system for heating and a duct system for cooling. Air curtains used on an extensive scale may, on analysis, simplify the building structure and lower total costs.

Another possibility when both heating and cooling are required is a heat pump system, which may be less costly than separate heating and cooling plants. Economical operation may be possible, depending on the building characteristics, lighting load, and occupant load. If well or other water is available, it may produce a more economical installation than an air-source heat pump.

ATRIUMS

Atriums have diverse functions and occupancies. An atrium may (1) connect buildings; (2) serve as an architectural feature, leisure space, greenhouse, and/or smoke reservoir; and (3) afford energy and lighting conservation. The temperature, humidity, and hours of usage of an atrium are directly related to those of the adjacent buildings. Glass window walls and skylights are common. Atriums are generally large in volume with relatively small floor areas. The temperature and humidity conditions, air distribution, impact from adjacent buildings, and fenestration loads to the space must be considered in the design of an atrium.

Perimeter radiant heating (e.g., overhead radiant type, wall finned-tube or radiant type, floor radiant type, or combinations thereof) is commonly used for the expansive glass windows and skylights. Air-conditioning systems can heat, cool, and control smoke. The distribution of air across windows and skylights can also control heat transfer and condensation. Low supply and high return air distribution can control heat stratification, as well as wind and stack effects. Some atrium designs include a combination of high/low supply and high/low return air distribution to control heat transfer, condensation, stratification, and wind/stack effects.

The energy use of an atrium can be reduced by installing double- and triple-panel glass and mullions with thermal breaks, as well as shading devices such as external, internal, and interior screens, shades, and louvers.

Extensive landscaping is common in atriums. Humidity levels are generally maintained between 10 and 35%. Hot and cold air should not be distributed directly onto plants and trees.

BIBLIOGRAPHY

ASHRAE. 1989. Ventilation for acceptable indoor air quality. ANSI/ASHRAE *Standard* 62-1989.

ASHRAE. 1992. Gravimetric and dust-spot procedures for testing air-cleaning devices used in general ventilation for removing particulate matter. ANSI/ASHRAE *Standard* 52.1-1992.

Carrier, W.H. 1918. The temperature of evaporation. *ASHVE Transactions* 24:25-50.

Kittler, R. 1983. Indoor natatorium design and energy recycling. *ASHRAE Transactions* 95(1):521-26.

Smith, C.C., R.W. Jones, and G.O.G. Löf. 1993. Energy requirements and potential savings for heated indoor swimming pools. *ASHRAE Transactions* 99(2):864-74

Kowalsky, L., ed. 1990. *Pool/spa operators handbook*. National Swimming Pool Foundation, Merrick, NY.

HOTELS, MOTELS, AND DORMITORIES

HOTELS, motels, and dormitories may be single-room or multiroom, long- or short-term dwelling (or residence) units; they may be stacked sideways and/or vertically. Information in the first three sections of the chapter applies generally; the last three sections are devoted to the individual types of facilities.

LOAD CHARACTERISTICS

1. Ideally each room served by an HVAC unit should be able to be ventilated, cooled, heated, or dehumidified independently of any other room. If not, air conditioning for each room will be compromised.
2. Typically, the space is not occupied at all times. For adequate flexibility, each unit's ventilation and cooling should be able to be shut off (except when humidity control is required), and its heating to be shut off or turned down.
3. Concentrations of lighting and occupancy are typically low; activity is generally sedentary or light. Occupancy is transient in nature, with greater use of bedrooms at night.
4. Kitchens, whether integrated with or separate from residential quarters, have the potential for high appliance loads, odor generation, and large exhaust requirements.
5. Rooms generally have an exterior exposure, kitchens, toilets, and dressing rooms may not. The building as a whole usually has multiple exposures, as may many individual dwelling units.
6. Toilet, washing, and bathing facilities are almost always incorporated in the dwelling units. Exhaust air is usually incorporated in each toilet area.
7. The building has a relatively high hot water demand, generally for periods of an hour or two, several times a day. This demand can vary from a fairly moderate and consistent daily load profile in a senior citizens building to sharp, unusually high peaks at about 6:00 P.M. in dormitories. Chapter 48 includes details on service water heating.
8. Load characteristics of rooms, dwelling units, and buildings can be well defined with little need to anticipate future changes to the design loads, other than the addition of a service such as cooling that may not have been incorporated originally.
9. The prevalence of shifting, transient interior loads and exterior exposures with glass results in high diversity factors; the long-hour usage results in fairly high load factors.

DESIGN CONCEPTS AND CRITERIA

Wide load swings and diversity within and between rooms require the design of a flexible system for 24-hour comfort. Besides opening windows, the only method of providing flexible temperature control is having individual room components under individual room control that can cool, heat, and ventilate independently of the equipment in other rooms.

The preparation of this chapter is assigned to TC 9.8, Large Building Air-Conditioning Applications.

In some climates, summer humidity becomes objectionable because of the low internal sensible loads that result when cooling is on-off controlled. Modulated cooling and/or reheat may be required to achieve comfort. Reheat should be avoided unless some sort of heat recovery is involved.

Dehumidification can be achieved by lowering cooling coil temperatures and reducing airflow. Another means of dehumidification is desiccant dehumidifiers.

Some people have a noise threshold low enough that certain types of equipment disturb their sleep. Higher noise levels may be acceptable in areas where there is little need for air conditioning. Medium and better quality equipment is available with noise criteria (NC) 35 levels at 10 to 14 ft in medium to soft rooms and little sound change when the compressor cycles.

Perimeter fan coils are usually quieter than unitary systems, but unitary systems provide more redundancy in case of failure.

SYSTEMS

Energy-Efficient Systems

The most efficient systems generally include water-source and air-source heat pumps. In areas with ample solar radiation, water-source heat pumps may be solar assisted. Energy-efficient equipment generally has the lowest operating cost and is relatively simple, an important factor where skilled operating personnel are unlikely to be available. Most systems allow individual operation and thermostatic control. The typical system allows individual metering so that most, if not all, of the cooling and heating costs can be metered directly to the occupant (McClelland 1983). Existing buildings can be retrofitted with heat flow meters and timers on fan motors for individual metering.

The water-loop heat pump has a lower operating cost than air-cooled unitary equipment, especially where electric heat is used. The lower installed cost encourages its use in mid- and high-rise buildings where individual dwelling units have floor areas of 800 ft² or larger. Some systems incorporate sprinkler piping as the water loop.

Except for the central circulating pump, heat rejector fans, and supplementary heater, the water-loop heat pump is predominantly decentralized; individual metering allows most of the operating cost to be paid by the occupant. Its life should be longer than for other unitary systems because most of the mechanical equipment is in the building and not exposed to outdoor conditions. Also, the load on the refrigeration circuit is not as severe because the water temperature is controlled for optimum operation. Operating costs are low due to the energy conservation inherent in the system. Excess heat may be stored during the day for the following night, and heat may be transferred from one part of the building to another.

While heating is required in many areas during cool weather, cooling may be needed in rooms having high solar or internal loads. On a mild day, surplus heat throughout the building is frequently transferred into the hot water loop by water-cooled condensers on cooling cycle so that water temperature rises. The heat remains stored in water from which it can be extracted at night; a water heater is not needed. This heat storage is improved by the presence of a greater mass of water in the pipe loop; some systems include a storage tank for this reason. Because the system is designed to operate during the heating season with water supplied at a temperature as low as 60°F, the water-loop heat pump lends itself to solar assist; relatively high solar collector efficiencies result from the low water temperature.

The installed cost of the water-loop heat pump is higher in very small buildings. In severe cold climates with prolonged heating seasons, even where natural gas or fossil fuels are available at reasonable cost, the operating cost advantages of this system may diminish unless heat can be recovered from some another source, such as solar collectors, geothermal, or internal heat from a commercial area served by the same system.

Energy-Neutral Systems

Energy-neutral systems do not allow simultaneous cooling and heating. Some examples are (1) packaged terminal air conditioners (PTACs) (through-the-wall units), (2) window units or radiant ceiling panels for cooling combined with finned or baseboard radiation for heating, (3) unitary air conditioners with an integrated heating system, (4) fan coils with remote condensing units, and (5) variable air volume (VAV) systems with either perimeter radiant panel heating or baseboard heating. To qualify as energy-neutral, a system must have controls that prevent simultaneous operation of the cooling and heating cycles. For unitary equipment, control may be as simple as a heat-cool switch. For other types, dead-band thermostatic control may be required.

PTACs are frequently installed to serve one or two rooms in buildings with mostly small, individual units. In a common two-room arrangement, a supply plenum diverts a portion of the conditioned air serving one room into the second, usually smaller, room. Multiple PTAC units allow additional zoning in dwellings having more rooms. Additional radiation heat is sometimes needed around the perimeter in cold climates.

Heat for a PTAC may be supplied either by electric resistance heaters or by hot water or steam heating coils. Initial costs are lower for a decentralized system using electric resistance heat. Operating costs are lower for coils heated by combustion fuels. Despite having relatively inefficient refrigeration circuits, the operating cost of a PTAC is quite reasonable, mostly because individual thermostatic control each machine, which eliminates the use of reheat while preventing the space from being overheated or overcooled. Also, because the equipment is located in the space being served, little power is devoted to circulating the room air. Servicing is simple—a defective machine is replaced by a spare chassis and forwarded to a service organization for repair. Thus, building maintenance requires relatively unskilled personnel.

Noise levels are generally no higher than NC 40, but some units are noisier than others. Installations near a seacoast should be specially constructed (usually with stainless steel or special coatings) to avoid the accelerated corrosion to aluminum and steel components caused by salt. In high-rise buildings of more than 12 stories, special care is required, both in the design and construction of outside partitions and in the installation of air conditioners, to avoid operating problems associated with leakage (caused by stack effect) around and through the machines.

Frequently, the least expensive installation is finned or baseboard radiation for heating and window-type room air conditioners for cooling. The window units are often purchased individually by the building occupants. This choice offers a reasonable operating cost and is relatively simple to maintain. However, window units have the shortest equipment life, the highest operating noise level, and the poorest distribution of conditioned air of any of the systems discussed in this section.

Fan coils with remote condensing units are used in smaller buildings. Fan coil units are located in closets, and the ductwork distributes air to the rooms in the dwelling. Condensing units may be located on roofs, at ground level, or on balconies.

Low capacity residential warm air furnaces may be used for heating, but with gas- or oil-fired units, the products of combustion must be vented. In a one- or two-story structure, it is possible to use individual chimneys or flue pipes, but in a high-rise structure requires a multiple-vent chimney or a manifold vent. Local codes should be consulted.

Sealed combustion furnaces draw all the combustion air from outside and discharge the flue products through a windproof vent to the outdoors. The unit must be located near an outside wall, and exhaust gases must be directed away from windows and intakes. In one- or two-story structures, outdoor units mounted on the roof or on a pad at ground level may also be used. All of these heating units can be obtained with cooling coils, either built-in or add-on. Evaporative-type cooling units are popular in motels, low-rise apartments, and residences in mild climates.

Desiccant dehumidification should be considered when independent control of temperature and humidity is required to avoid reheat.

Energy-Inefficient Systems

Energy-inefficient systems allow simultaneous cooling and heating. Examples include two-, three-, and four-pipe fan coil units, terminal reheat systems, and induction systems. Some units, such as the four-pipe fan coil, can be controlled so that they are energy-neutral. They are primarily used for humidity control.

Four-pipe systems and two-pipe systems with electric heaters can be designed for complete temperature and humidity flexibility during summer and intermediate season weather, although none provides winter humidity control. Both systems provide full dehumidification and cooling with chilled water, reserving the other two pipes or an electric coil for space heating or reheat. The equipment and necessary controls are expensive, and only the four-pipe system, if equipped with an internal-source heat-recovery design for the warm coil energy, can operate at low cost. When year-round comfort is essential, four-pipe systems or two-pipe systems with electric heat should be considered.

Total Energy Systems

A total energy system is an option for any multiple or large housing facility with large year-round service water heating requirements. Total energy systems are a form of cogeneration in which all or most electrical and thermal energy needs are met by on-site systems as described in Chapter 7 of the 1996 *ASHRAE Handbook—Systems and Equipment*. A detailed load profile must be analyzed to determine the merits of using a total energy system. The reliability and safety of the heat-recovery system must also be considered.

Any of the previously described systems can perform the HVAC function of a total energy system. The major considerations as they apply to total energy in choosing a HVAC system are as follows:

- Optimum use must be made of the thermal energy recoverable from the prime mover during all or most operating modes, not just during conditions of peak HVAC demand.
- Heat recoverable via the heat pump may become less useful because the heat required during many of its potential operating hours will be recovered from the prime mover. The additional investment for heat pump or heat recovery cycles may be more difficult to justify because operating savings are lower.
- The best application for recovered waste heat is for those services that use only heat (i.e., service hot water, laundry facilities, and space heating).

Special Considerations

Local building codes govern ventilation air quantities for most buildings. Where they do not, ASHRAE *Standard* 62 should be followed. The quantity of outdoor air introduced into the rooms or corridors is usually slightly in excess of the exhaust quantities to pressurize the building. To avoid adding any load to the individual systems, outdoor air should be treated to conform to indoor air temperature and humidity conditions. In humid climates, special attention must be given to controlling the humidity from outdoor air. Otherwise, the outdoor air may reach corridor temperature while still retaining a significant amount of moisture.

In buildings having a centrally controlled exhaust and supply, the system is regulated by a time clock or a central management system for certain periods of the day. In other cases, the outside air may be reduced or shut off during extremely cold periods, although this practice is not recommended and may be prohibited by local codes. These factors should be considered when estimating heating load.

For buildings using exhaust and supply air on a 24 hour basis, air-to-air heat recovery devices may be merited (see Chapter 42 of the 1996 *ASHRAE Handbook—Systems and Equipment)*. Such recovery devices can reduce energy consumption by capturing 60 to 80% of the sensible and latent heat extracted from the air source.

Infiltration loads in high-rise buildings without ventilation openings for perimeter units are not controllable year-round by general building pressurization. When outer walls are pierced to supply outdoor air to unitary or fan-coil equipment, combined wind and thermal stack-effect forces create equipment operating problems. These factors must be considered for high-rise buildings (see Chapter 25 of the 1997 *ASHRAE Handbook—Fundamentals*).

Interior public corridors should have tempered supply air with transfer into individual area units, if necessary, to provide kitchen and toilet makeup air requirements. The transfer louvers need to be acoustically lined. Corridors, stairwells, and elevators should be pressurized for fire and smoke control (see Chapter 51).

Kitchen air can be recirculated through hoods with activated charcoal filters rather than exhausted. Toilet exhaust can be VAV with a damper operated by the light switch. A controlled source of supplementary heat in each bathroom is recommended to ensure comfort while bathing.

Air-conditioning equipment must be isolated to reduce noise generation or transmission. The cooling tower or condensing unit must be designed and located to avoid disturbing occupants of the building or of adjacent buildings.

An important but frequently overlooked load is the heat gain from piping for hot water services. Insulation thickness should conform to the latest local energy codes and standards at a minimum. In large, luxury-type buildings, a central energy or building management system allows supervision of individual air-conditioning units for operation and maintenance.

Some facilities achieve energy conservation by reducing the indoor temperature during the heating season. Such a strategy should be pursued with caution because it could affect the competitiveness of the hotel/motel, for example.

HOTELS AND MOTELS

Hotel and motel accommodations are usually single guest rooms with a toilet and bath adjacent to a corridor, and flanked on both sides by other guest rooms. The building may be single-story, low-rise, or high-rise. Multipurpose subsidiary facilities range from stores and offices to ballrooms, dining rooms, kitchens, lounges, auditoriums, and meeting halls. Luxury motels may be built with similar facilities. Occasional variations are seen, such as kitchenettes, multiroom suites, and outside doors to patios and balconies. Hotel classes range from the deluxe hotel to the economy hotel/motel as outlined in Table 1.

A hotel can be divided into three main areas:

1. Guest rooms
2. Public areas
   - Lobby, atrium, and lounges
   - Ballrooms
   - Meeting rooms
   - Restaurants and dining rooms
   - Stores
   - Swimming pools
   - Health clubs
3. Back-of-the-house (BOTH) areas
   - Kitchens
   - Storage areas
   - Laundry
   - Offices
   - Service areas and equipment rooms

The two main areas of use are the guest rooms and the public areas. Maximum comfort in these areas is critical to the success of any hotel. Normally the BOTH spaces are less critical than the remainder of the hotel with the exception of a few spaces where a controlled environment is required or recommended.

Guest Rooms

Air conditioning in hotel rooms should be quiet, easily adjustable, and draft free. It must also provide ample outside air. Because the hotel business is so competitive and space is at a premium, systems that require little space and have low total owning and operating costs should be selected.

Design Concepts and Criteria. Table 2 lists design criteria for hotel guest rooms. In addition, the design criteria for hotel room HVAC services must consider the following factors:

Table 1 Hotel Classes

Type of Facility	Typical Occupancy, Persons per Room	Characteristics
Deluxe hotel	1.2	Large rooms, suites, specialty restaurants
Luxury/first class, full-service hotel	1.2 to 1.3	Large rooms, large public areas, business center, pool and health club, several restaurants
Mid-scale, full-service hotel	1.2 to 1.3	Large public areas, business center, several restaurants
Convention hotel	1.4 to 1.6	Large number of rooms, very large public areas, extensive special areas, rapid shifting of peak loads
Limited-service hotel	1.1	Limited public areas, few restaurants, may have no laundry
Upscale, all-suites hotel	2.0	Rooms are two construction bays, in-room pantries, limited public areas, few restaurants
Economy, all-suites hotel	2.0 to 2.2	Smaller suites, limited public areas and restaurants
Resort hotel	1.9 to 2.4	Extensive public areas, numerous special and sport areas, several restaurants
Conference center	1.3 to 1.4	Numerous special meeting spaces, limited dining options
Casino hotel	1.5 to 1.6	Larger rooms, large gaming spaces, extensive entertainment facilities, numerous restaurants
Economy hotel/motel	1.6 to 1.8	No public areas, little or no dining, usually no laundry

Table 2 Hotel Design Criteria[a,b]

Category	Inside Design Conditions				Ventilation[d]	Exhaust[d,e]	Filter Efficiency[f]	Noise, RC Level
	Winter		Summer					
	Temperature	Relative Humidity[c]	Temperature	Relative Humidity				
Guest rooms	74 to 76°F	30 to 35%	74 to 78°F	50 to 60%	30 to 60 cfm per room	20 to 50 cfm per room	10 to 15%	25 to 35
Lobbies	68 to 74°F	30 to 35%	74 to 78°F	40 to 60%	15 cfm per person	–	35% or better	35 to 45
Conference/ Meeting rooms	68 to 74°F	30 to 35%	74 to 78°F	40 to 60%	20 cfm per person	–	35% or better	25 to 35
Assembly rooms	68 to 74°F	30 to 35%	74 to 78°F	40 to 60%	15 cfm per person	–	35% or better	25 to 35

[a] This table should not be used as the only source for design criteria. The data contained here can be determined from ASHRAE handbooks, standards, and governing local codes.

[b] Design criteria for stores, restaurants, and swimming pools are in Chapters 2, 3, and 4, respectively.

[c] Minimum recommended humidity.

[d] As per ASHRAE *Standard* 62, Ventilation for Acceptable Indoor Air Quality.

[e] Air exhaust from bath and toilet area.

[f] As per ASHRAE *Standard* 52.1, Gravimetric and Dust-Spot Procedures for Testing Air-Cleaning Devices Used in General Ventilation for Removing Particulate Matter.

- Individual and quick responding temperature control
- Draft-free air distribution
- Toilet room exhaust
- Ventilation (makeup) air supply
- Humidity control
- Acceptable noise level
- Simple controls
- Reliability
- Ease of maintenance
- Operating efficiency
- Use of space

Load Characteristics. The great diversity in the design, purpose, and use of hotels and motels makes analysis and load studies very important. The fact that guest rooms have a transient occupancy and the diversity associated with the operation of the support facilities provides an opportunity to take advantage of load diversification.

The envelope cooling and heating load is dominant because the guest rooms normally have exterior exposures. Other load sources such as people, lights, appliances, etc. are a relatively small portion of the space sensible and latent loads. The ventilation load can represent up to 15% of the total cooling load.

Because of the nature of the changing envelope sensible load and the transient occupancy of the guest room, large fluctuations in the space sensible load in a one-day cycle are common. The ventilation sensible cooling load can vary from 0 to 100% in a single day, whereas the ventilation latent load can remain almost constant for the entire day. A low sensible heat ratio is common in moderate to very humid climates. Usually the HVAC equipment must only handle part or low loads and peak loads rarely occur. For example, in humid climates, introducing untreated outside air directly into the guest room or into the return air plenum of the HVAC unit operating at part or low load creates a severe high-humidity problem, which is one of the causes of mold and mildew. The situation is further aggravated when the HVAC unit operates in on-off cycle during part- or low-load conditions.

Applicable Systems. Most hotels use all-water or unitary refrigerant-based equipment for guest rooms. All-water systems include

- Two-pipe fan coils
- Two-pipe fan coil with electric heat
- Four-pipe fan coils

Unitary refrigerant-based systems include

- Packaged terminal air conditioner or packaged terminal heat pump (with electric heat)
- Air to air heat pump (ductless, split)
- Water-source heat pump

With the exception of the two-pipe fan coil, all of these systems cool, heat, or dehumidify independently of any other room and regardless of the season. A two-pipe fan coil system should be selected only when economics and design objectives dictate a compromise to performance. Selection of a particular system should be based on

- First cost,
- Economical operation, especially at part load, and
- Maintainability.

Compared to unitary refrigerant-based units, all-water systems offer the following advantages:

- Reduced total installed cooling capacity due to load diversity
- Lower operating cost due to a more efficient central cooling plant
- Lower noise level (compared to PTAC and water-source heat pump)
- Longer service life
- Less equipment to be maintained in the occupied space
- Less water in circulation (compared to water-source heat pump)
- Smaller pipes and pumps (compared to water-source heat pump)

The unitary refrigerant-based system on the other hand offer the following advantages:

- Lower first cost
- Immediate all year availability of heating and cooling
- No seasonal changeover required
- Cooling available without operating a central refrigeration plant
- Can transfer energy from spaces being cooled to spaces being heated (with water-source heat pump)
- Range of circulated water temperature requires no pipe insulation (for water-source heat pump)
- Less dependence on a central plant for heating and cooling
- Simplicity, which results in lower operating and maintenance staff costs

The type of facility, sophistication, and quality desired by the owner/operator, as well as possible code requirements typically influence the selection. An economic analysis (life-cycle cost) is particularly important when selecting the most cost-effective system. Chapter 35 has further information on economic analysis techniques. Computer software like the NIST Building Life-Cycle Cost Program (BLCC) performs life-cycle cost analyses quickly and accurately (NIST 1994).

Chapter 4 and Chapter 5 of the 1996 *ASHRAE Handbook—Systems and Equipment* provide additional information about all-water systems and unitary refrigerant-based systems.

Room fan coils and room unitary refrigerant-based units are available in many configurations including horizontal, vertical, exposed, and concealed. The unit should be located in the guest room so that it provides excellent air diffusion without creating

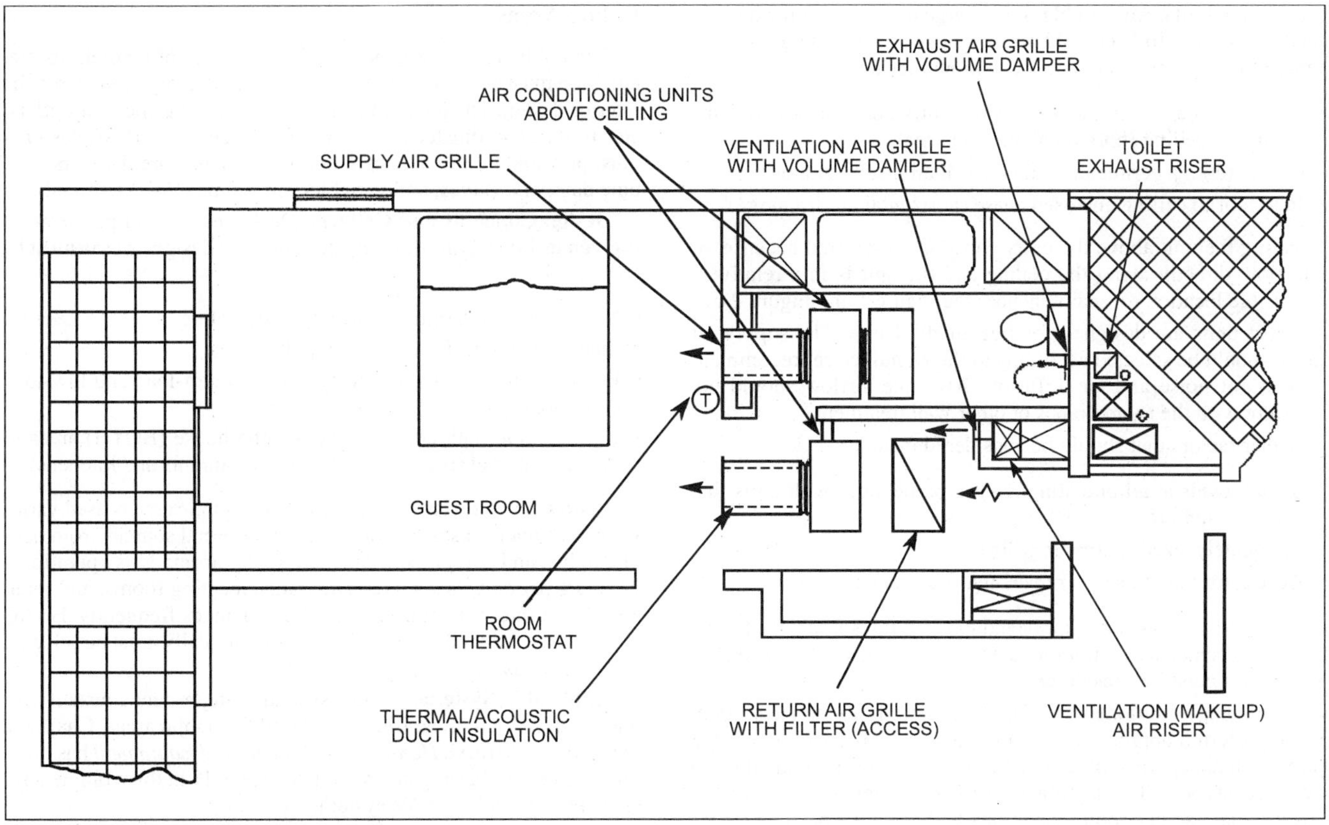

Fig. 1   Alternate Location for Hotel Guest Room Air Conditioning Unit above Hung Ceiling

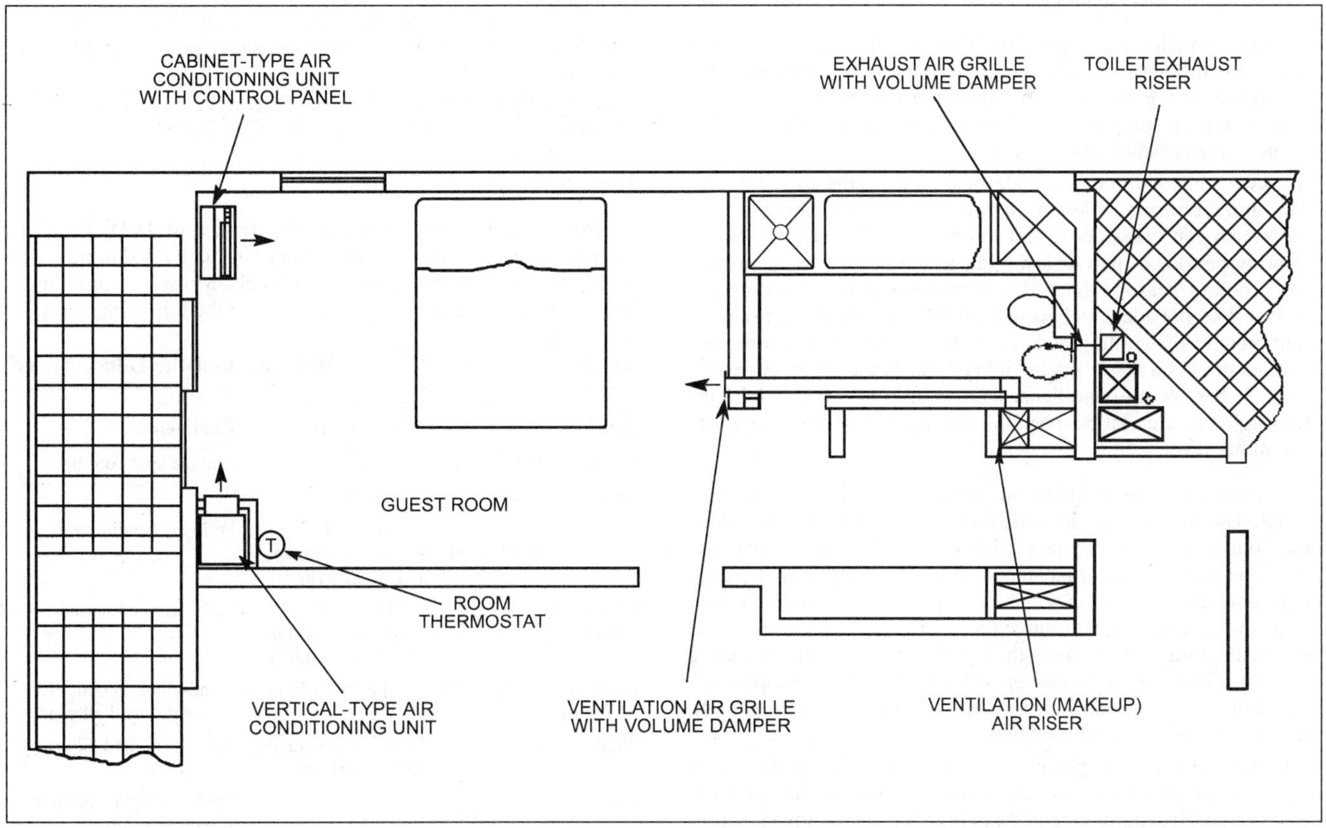

Fig. 2   Alternate Location for Hotel Guest Room Air Conditioning Unit on Room Perimeter and Chase-Enclosed

unpleasant drafts. Air should not discharge directly over the head of the bed in order to keep cold air away from a sleeping guest. The fan-coil/heat pump unit is most commonly located

- Above the ceiling in the guest room entry corridor or above the bathroom ceiling (horizontal air discharge),
- On the room's perimeter wall (vertical air discharge), or
- In a floor to ceiling enclosed chase (horizontal air discharge).

Locating the unit above the entry corridor is preferred because air can flow directly along the ceiling and the unit is in a relatively accessible location for maintenance (see Figure 1 and Figure 2).

Most units are designed for free-air discharge. The supply air grille should be selected according to the manufacturer recommendations for noise and air diffusion. Also, the airflow should not interfere with the room drapes or other wall treatment.

Other factors that should be considered include

- Sound levels at all operating modes, particularly with units that cycle on and off
- Adequately sized return air grille
- Access for maintenance, repair, and filter replacement

Ventilation (makeup) supply and exhaust rates must meet local code requirements. Ventilation rates vary and the load imposed by ventilation must be considered.

Providing conditioned ventilation air directly to the guest room is the preferred approach. Normally outside air is conditioned in a primary makeup air unit and distributed via a primary air duct to every guest room. This approach controls the supply air conditions, ensures satisfactory room conditions and room air balance (room pressurization) even during part or no load conditions, and controls mold and mildew.

Other ventilation techniques are to

- Transfer conditioned ventilation air from the corridor to each guest room. This approach controls ventilation air conditions better; however, the air balance (makeup versus exhaust) in the guest room may be compromised. This approach is prohibited under many code jurisdictions.
- Introduce unconditioned outside air directly to the air-conditioning unit's return air plenum (perimeter wall installations). This approach can cause mold and mildew and should be avoided. During periods of part or low load, which occur during most of the cooling season, the thermostatically-controlled air conditioner does not adequately condition the constant flow of outside air because the cooling coil valve closes and/or the compressor cycles off. As a result the humidity level in the room increases. Also, when the air conditioner's fan is off, outside air infiltrates through the ventilation opening and again elevates the room humidity level.

Guest-room HVAC units are normally controlled by a room thermostat. Thermostats for fan coils normally control valves in two-pipe, four-pipe, and two-pipe chilled water with electric heat systems. The control should include dead-band operation to separate the heating and cooling set points. Two-pipe system control valves are normally equipped with automatic changeover, which senses the water temperature and changes the mode of operation from heating to cooling. The thermostat may provide modulation or two-position control of the water control valve. The fan can be adjusted to high, medium, or low speed on most units.

Typical unitary refrigerant-based units have a push button off/fan/heat/cool selector switch, adjustable thermostat, and fan cycle switch. Heat pumps include a defrost cycle to remove ice from the outdoor coil. Chapter 45 has more information on control for fan coils.

Public Areas

The public areas are generally the showcase of a hotel. Special attention must be paid to incorporating a satisfactory system into the interior design. The location of supply diffusers, grilles. air outlets. etc. must be coordinated to satisfy the architect. The HVAC designer must pay attention to access doors for servicing fire dampers, volume dampers. valves, and variable-air-volume (VAV) terminals.

Design Concepts and Criteria. Design criteria for public areas is given in Table 2. In addition, the following design criteria must be considered:

- Year round availability of heating and cooling
- Independent unit for each main public area
- Economical and satisfactory operation at part-load and low-load conditions
- Coordination with adjacent back-of-the-house (BOTH) areas to insure proper air pressurization (i.e., restaurants and kitchens)

Load Characteristics. The hours each public area is used varies widely. In many cases the load is due to internal sources from people, lights, and equipment. The main lobby normally is operational 24 hours per day. Areas like restaurants, meeting rooms, and retail areas have intermittent use, so the load changes frequently. HVAC systems that respond effectively and economically must be selected for these areas.

Applicable Systems. All-air systems, single-duct constant-volume, and VAV are most frequently used for public areas. Chapter 2 of the 1996 *ASHRAE Handbook—Systems and Equipment* has more information on these systems while Chapter 45 in this volume covers control for all-air VAV systems.

Back-of-the-House (BOTH) Areas

The BOTH area is normally considered a service or support area. The climatic conditions in these areas are typically less critical than in the remainder of the hotel. However, a few spaces require special attention.

Design Concepts and Criteria. Recommended design criteria for several areas in the BOTH are shown in Table 3.

Special Concerns

Humidity, Mildew, Moisture Control, and IAQ. Humidity control is critical to ensure satisfactory air quality and to minimize costly mold and mildew problem in hotels. Moisture can be introduced and infiltrate into the guest rooms in the following ways:

Table 3　Design Criteria for Hotel Back-of-the-House Areas[a]

Category	Inside Design Conditions	Comments
Kitchen (general)[b]	82°F	Provide spot cooling
Kitchen (pastry)[b]	76°F	
Kitchen (chef's office)[b]	74 to 78°F 50 to 60% rh (summer) 30 to 35% rh (winter)	Fully air conditioned
Housekeeper's office	74 to 78°F 50 to 60% rh (summer) 30 to 35% rh (winter)	Fully air conditioned
Telephone equipment room	Per equipment criteria	Stand-alone air conditioner; air conditioned all year
Wine storage	Per food and beverage manager criteria	Air conditioned all year
Laundry		Spot cooling as required at workstations

[a] Governing local codes must be followed for design of the HVAC.
[b] Consult Chapter 30 for details on kitchen ventilation.

1. Unconditioned ventilation air is delivered directly into the guest room through the HVAC unit. At part or low sensible loads or in situations where the unit cycles on and off, the air-conditioning unit will not dehumidify the air adequately to remove the excess moisture.
2. Outdoor humid air infiltrates through openings, cracks, gaps. shafts, etc. due to insufficient space pressurization.
3. Moisture migrates through external walls and building elements due to a vapor pressure differential.
4. An internal latent load or moisture is generated.

Removing water vapor from the air is the most feasible way to control mold and mildew, particularly when the problem spreads to walls and carpeting. Good moisture control can be achieved by applying the following techniques:

1. Introduce adequately dried ventilation (makeup) air [i.e., with a dew point of 53°F (60 grains/lb of dry air) or less] directly to the guest room.
2. Maintain slightly positive pressure in the guest room to minimize infiltration of hot and humid air into the room. Before a new HVAC system is accepted by the owner, a certified air balance contractor should be engaged to demonstrate that the volume of dry makeup air exceeds the volume of exhaust air. As the building ages, it is important to maintain this slight positive pressure; otherwise humid air that infiltrates into the building cavities will be absorbed regardless of how dry the room in maintained (Banks 1992).
3. Provide additional dehumidification capability to the ventilation (makeup air) by dehumidifying the air to a lower level than the desired space humidity ratio. For example, introducing 60 cfm of makeup air at 55 gr/lb can provide approximately 400 Btu/h of internal latent cooling (assuming 65 gr/lb is a desirable space humidity ratio).
4. Allow air conditioning to operate in unoccupied rooms instead of turning the units off, especially in humid areas.
5. Improve the room envelope by increasing its vapor and infiltration resistance.

Method 3 allows the ventilation air to handle part of the internal latent load (people, internal moisture generation, and moisture migration from external walls and building elements). In addition, this method can separate the internal sensible cooling, internal latent cooling, and ventilation loads. Independent ventilation/dehumidification allows room pressurization and space humidity control regardless of the mode of operation or magnitude of the air-conditioning load. Desiccant dehumidifiers can be retrofitted to solve existing moisture problems.

Ventilation (Makeup) Air Units. Makeup air units are designed to condition ventilation air introduced into a space and to replace air exhausted from the building. The geographic location and class of the hotel dictate the functions of the makeup air units, which may filter, heat, cool, humidify, and/or dehumidify the ventilation air. Makeup air may be treated directly or by air-to-air heat recovery (sensible or combined sensible and latent) and other heat recovery techniques. Equipment to condition the air by air-to-air heat recovery and final heating, cooling, humidification, and/or dehumidification is also available.

Chapter 26 of the 1997 *ASHRAE Handbook—Fundamentals* provides design weather data for ventilation. Analyzing and selecting the proper makeup unit for the full range of entering conditions are critical for an efficient and sufficient all-year operation. Air-to-air heat recovery helps stabilize the entering conditions, which helps provide an efficient and stable operation. However, heat recovery may not always be feasible. Often, exhaust air comes from many individual stacks. In this case, the cost of combining many exhausts for heat recovery may not be warranted.

Table 4 Design Criteria for Hotel Guest Room Makeup Air Units

Supply Air Conditions				Filter Efficiency (ASHRAE *Std.* 52.1)
Winter		Summer		
Temperature	Relative Humidity	Temperature	Relative Humidity	
68 to 76°F	30 to 45%	74 to 78°F	40 to 50%	30%

Notes:
1. Follow local codes when applicable.
2. The building location may dictate the optimum supply condition in the recommended range.

Typical design criteria for ventilation (makeup) air units are listed in Table 4.

Makeup air units can be stand alone packaged (unitary) or integrated in an air handler. A typical makeup air unit usually has the following features:

- Heating, cooling, and dehumidification
  - Chilled/hot water or steam coils in the air handling unit
  - Unitary refrigerant-based unit (direct-expansion cooling and gas furnace or electric heat)
  - Air-to-air energy recovery combined with mechanical cooling (DX or chilled water) and heating
  - Desiccant-based dehumidifier combined with air-to-air energy recovery, indirect/direct evaporative cooling and supplementary mechanical cooling and heating

- Heating only
  - Hot water or steam coils in the air handling unit
  - Stand alone gas-fired or electric makeup units
  - Air-to-air energy recovery with supplement heat

Humidification should be considered for all cold climates. The HVAC designer must also consider avoiding coil freeze up in water based systems. Chapter 31 and Chapter 42 of the 1996 *ASHRAE Handbook—Systems and Equipment* provide information about makeup air units and air-to-air energy recovery, respectively.

Hotel location, environmental quality desired by the owner, and design sophistication determines the system selected. For example, in locations with cool summers, dehumidification with mechanical cooling only is satisfactory. For humid locations or where enhanced dehumidification is required, a desiccant-based unit can provide lower supply air humidity. Such a unit can help prevent mold and mildew and provide internal latent cooling.

Central Mechanical Plant. Designing a reliable and energy efficient mechanical plant is essential element toward ensuring a profitable hotel. The chiller plant must operate efficiently at part load conditions. Some redundancy should be considered in case of equipment failure. Designs often include spare critical equipment where spare parts and qualified service are not readily available. Chillers with multistage compressors should be considered because they provide partial cooling during failures and enhance part-load operation. When using two chillers, each should provide at least 60% of the total load. Combinations of three chillers providing 40% each or four chillers providing 30% each are better for tracking part-load conditions. Cooling towers, pumps, etc. can be sized in a similar manner.

The heating plant should be designed to accommodate the heating load during the winter end could provide domestic hot water, swimming pool heating, and service to kitchens and laundries as well. The type of fuel used depends on location, availability, use, and cost.

Multipurpose boiler design for the kitchen and laundry should offer redundancy, effective part-load handling, and efficient operation during the summer when the HVAC heating load does not exist.

In areas with mild winters, a two-pipe system or an air-to-water heat pump chiller/heater can be considered. In any event, the HVAC designer must understand the need for all-year cooling and heating availability in the public areas. In this case a combination of air-to-water heat pump, chiller/heater for the guest rooms and independent heat pumps for the public areas can be installed.

Acoustics and Noise Control. The sound level in guest room and public areas is a major design element. Both the level and constancy of the noise generated by the HVAC unit are of concern. Normally, packaged-terminal air conditioners/heat pumps and water-source heat pumps are noisier due to the compressor. Some equipment, however, has extra sound insulation, which reduces the noise significantly.

Lowering the fan speed, which is usually acceptable, can reduce fan noise levels. On/off cycling of the fan and the compressor can be objectionable even if the generated noise is low. Temperature control by cycling the fan only (no flow control valve) should not be used.

Another source of noise is sound that transfers between guest rooms through the toilet exhaust duct. Internal duct lining and sound attenuators are commonly used to minimize this problem.

Noise from equipment located on the roof or in a mechanical room located next to a guest room should be avoided. Proper selection of vibration isolators should prevent vibration transmission. In critical cases, an acoustician must be consulted.

New Technology in Hotels. Modern hotels are implementing new techniques to enhance comfort and convenience. For example, the telephone, radio, TV, communications, lighting, and air-conditioning unit are being integrated into one control system. Occupancy sensors are being used to conserve energy by resetting the temperature control when the room is occupied or when the guests leave. As soon as a new guest checks in at the front desk, the room temperature is automatically reset. But even with this improved technology, it is important to remember that temperature reset may create humidity problems.

DORMITORIES

Dormitory buildings frequently have large commercial dining and kitchen facilities, laundering facilities, and common areas for indoor recreation and bathing. Such ancillary loads may make heat pump or total energy systems appropriate, economical alternatives, especially on campuses with year-round activity.

When dormitories are shut down during cold weather, the heating system must supply sufficient heat to prevent freeze-up. If the dormitory contains non-dwelling areas such as administrative offices or eating facilities, these facilities should be designed as a separate zone or with a separate system for flexibility, economy, and odor control.

Subsidiary facilities should be controlled separately for flexibility and shutoff capability, but they may share common refrigeration and heating plants. With internal-source heat pumps, this

interdependence of unitary systems permits the reclamation of all internal heat usable for building heating, domestic water preheating, and snow melting. It is easier and less expensive to place heat reclaim coils in the building's exhaust than to use air-to-air heat recovery devices. Heat reclaim can easily be sequence controlled to add heat to the building's chilled water system when required.

MULTIPLE-USE COMPLEXES

Multiple-use complexes combine retail facilities, office space, hotel space, residential space, and/or other commercial space into a single site. The peak HVAC demands of the various facilities may occur at different times of the day and year. Loads should be determined independently for each occupancy. Where a central plant is considered, a block load should also be determined.

Separate air handling and distribution should serve the separate facilities. However, heating and cooling units can be combined economically into a central plant. A central plant provides good opportunities for heat recovery, thermal storage, and other techniques that may not be economical in a single-use facility. A multiple-use complex is a good candidate for central fire and smoke control, security, remote monitoring, billing for central facility use, maintenance control, building operations control, and energy management.

REFERENCES

Banks, N.J. 1992. Field test of a desiccant-based HVAC system for hotels. *ASHRAE Transactions* 98(1):1303-10.

Kokayko, M.J. 1997. Dormitory renovation project reduces energy use by 69%. *ASHRAE Journal* 39(6).

McClelland, L. 1983. Tenant paid energy costs in multi-family rental housing. DOE, University of Colorado, Boulder.

NIST. 1994. Building life cycle cost (BLCC) computer program, Version 4.2-95. National Institute of Standards and Technology, Gaithersburg, MD.

BIBLIOGRAPHY

ASHRAE. 1993. *Air-conditioning system design manual*, pp. 8-21 and 8-22.

Haines, R.W. and D.C. Hittle. 1983. *Control systems for heating, ventilation and air conditioning*, 5th ed., pp. 178-83. Chapman & Hall, New York.

Harriman, L.G., D. Plager, and D. Kosar. 1997. Dehumidification and cooling load from ventilation air. *ASHRAE Journal* 39(11).

Kimbrough, J. 1990. The essential requirements for mold and mildew. Plant Pathology Department, University of Florida, Gainesville, FL.

Lehr, V.A. 1995. Current trends in hotel HVAC design. Heating/Piping/Air Conditioning, February.

Peart, V. 1989. Mildew and moisture problems in hotels and motels in Florida. Institute of Food and Agricultural Sciences, University of Florida, Gainesville, FL.

Wong, S.P. and S.K. Wang. 1990. Fundamentals of simultaneous heat and mass transfer between the building envelope and the conditioned space. *ASHRAE Transactions* 96(2).

CHAPTER 6

EDUCATIONAL FACILITIES

THIS chapter contains technical and environmental factors and considerations that will assist the design engineer in the proper application of heating, ventilating, and air-conditioning systems and equipment for educational facilities. For further information on the HVAC systems mentioned in this chapter, see the 1996 *ASHRAE Handbook—Systems and Equipment.*

GENERAL CONSIDERATIONS

All forms of educational facilities require an efficiently controlled atmosphere for a proper learning environment. This involves the selection of heating, ventilating, and air-conditioning systems, equipment, and controls to provide adequate ventilation, comfort, and a quiet atmosphere. The system must also be easily maintained by the facility's maintenance staff.

The selection of the HVAC equipment and systems depends upon whether the facility is new or existing and whether it is to be totally or partially renovated. For minor renovations, existing HVAC systems are often expanded in compliance with current codes and standards with equipment that matches the existing types. For major renovations or new construction, new HVAC systems and equipment can be installed if the budget allows. The remaining useful life of existing equipment and distribution systems should be considered.

HVAC systems and equipment energy use and associated life-cycle costs should be evaluated. Energy analysis may justify new HVAC equipment and systems when a good return on investments can be shown. The engineer must take care to review all assumptions in the energy analysis with the school administration. Assumptions, especially as they relate to hard-to-measure items such as infiltration and part-load factors, can have a significant influence on the energy use calculated.

Other considerations for existing facilities are (1) whether the central plant is of adequate capacity to handle the additional loads of new or renovated facilities, (2) the age and condition of the existing equipment, pipes, and controls, and (3) the capital and operating costs of new equipment. Educational facilities usually have very limited budgets. Any savings in capital expenditures and energy costs may be available for the maintenance and upkeep of the HVAC systems and equipment and for other facility needs.

When new facilities are built or major renovations to existing facilities are made, seismic bracing of the HVAC equipment should be considered. Refer to Chapter 53, Seismic and Wind Restraint Design, for further information. Seismic codes may also apply in areas where tornados and hurricanes necessitate additional bracing. This consideration is especially important if there is an agreement with local officials to let the facility be used as a disaster relief shelter.

The type of HVAC equipment selected also depends upon the climate and the months of operation. In hot, dry climates, for instance, evaporative cooling may be the primary type of cooling; some school districts may choose not to provide air conditioning. In

The preparation of this chapter is assigned to TC 9.8, Large Building Air-Conditioning Applications.

hot, humid climates, it is recommended that air conditioning or dehumidification be operated year-round to prevent the growth of mold and mildew.

ENERGY CONSIDERATIONS

Most new buildings and renovated portions of existing buildings must comply with the energy codes and standards enforced by local code officials. Use of energy-efficient equipment, systems, and building envelope materials should always be considered and evaluated. Methods of reducing energy usage should be discussed by the design team.

Outdoor air economizer cycles use cool outdoor air to accommodate high internal loads for many hours of the year, as dry- and wet-bulb temperatures permit. A relief air system sized to remove up to 100% of the outdoor air that could be supplied is required. Some packaged rooftop air-conditioning units do not have adequate relief air outlets, which could lead to overpressurization of the space. Additional relief vents or an exhaust system may be necessary.

Night setback, activated by a time clock controller and controlled by night setback thermostats, conserves energy by resetting the heating and cooling space temperatures during unoccupied hours. Intermittent use of educational facilities sometimes requires that individual systems have their own night setback control times and settings.

Coupled with night setback is **morning warm-up and cool-down**. This allows the space temperatures to be reset for normal occupancy at a predetermined time. This can be achieved either by setting a time clock for approximately one hour before occupancy or through the optimization program of a central energy management system. Such a program determines the warm-up or cool-down time based on the mass of the building, the outdoor air temperature, and a historical log of building occupancy.

Hot water temperature reset can save energy by resetting the building hot water temperature based upon outdoor air temperature. Care must be taken to not lower the water temperature below that recommended by the boiler manufacturer.

During cool weather, **cooling-tower free cooling** uses the cooling towers and either chiller refrigerant migration or a heat exchanger to cool the chilled water system without using the chillers. This system is useful for facilities with high internal heat loads.

Heat recovery systems can save energy when coupled to systems that generate a substantial amount of exhaust air. A study must be performed to determine whether a heat recovery system is cost-effective. Various heat recovery systems include runaround loops, heat wheels, and heat exchangers.

Boiler economizers can save energy by recovering the heat lost in the stacks of large boilers or in central plants. The recovered heat can be used to preheat the boiler makeup water or combustion air. Care should be exercised not to reduce the stack temperature enough to cause condensation. Special stack materials are required for systems in which condensation may occur.

Variable-speed drives for large fans and pumps can save energy, especially if coupled with energy-efficient motors. The

HVAC systems should be evaluated to determine whether there is enough variation in the airflows or water flows to justify the application of these devices. A cost analysis should also be performed.

Passive solar techniques, such as the use of external shading devices at fenestrations, can be incorporated into the building design. Coordination with the owner and design team is essential for this strategy. A life-cycle cost analysis may be necessary to evaluate the cost-effectiveness of this method.

Solar energy systems can either supplement other energy sources or be the sole energy source for an HVAC system. A life-cycle cost analysis should be performed to evaluate the cost-effectiveness of such a system.

DESIGN CONSIDERATIONS

Climate control systems for educational facilities may be similar to those of other types of buildings. To suit the mechanical systems to an educational facility, the designer must understand not only the operation of the various systems but also the functions of the school. Year-round school programs, adult education, night classes, and community functions put ever-increasing demands on these facilities; the environmental control system must be designed carefully to meet these needs. When occupied, classrooms, lecture rooms, and assembly rooms have uniformly dense occupancies. Gymnasiums, cafeterias, laboratories, shops, and similar functional spaces have variable occupancies and varying ventilation and temperature requirements. Many schools use gymnasiums as auditoriums and community meeting rooms. Wide variation in use and occupancy makes flexibility important in these multiuse rooms.

Design considerations for the indoor environment of educational facilities include heat loss, heat gain, humidity, air movement, ventilation methods, ventilation rates, noise, and vibration. Winter and summer design dry-bulb temperatures for the various spaces in schools are given in Table 1. These values should be evaluated for each application to avoid energy waste while ensuring comfort.

Sound levels are also an important consideration in the design of educational facilities. Typical ranges of sound design levels are given in Table 2. Whenever possible and as the budget allows, the design sound level should be at the lower end of these ranges. For additional information, see Chapter 46, Sound and Vibration Control.

Proper ventilation can control odors and inhibit the spread of respiratory diseases. Refer to ASHRAE *Standard* 62 for the ventilation rates required for various spaces within educational facilities.

Equipment for school facilities should be reliable and require minimal maintenance. Equipment location is also an important consideration. The easier equipment is to access and maintain, the more promptly it will be serviced. Roof hatches, ladders, and walkways should be provided for access to rooftop equipment.

Controls. Controls for most educational facilities should be kept as simple as possible because a wide variety of people need to use, understand, and maintain them. The different demands of the various spaces indicate the need for individual controls within each space. However, unless each space is served by its own self-contained unit, this approach is usually not practical. Many times, a group of similar spaces will be served by a single thermostat controlling one heating or cooling unit. Rooms grouped together should have similar occupancy schedules and the same outside wall orientation so that their cooling loads peak at approximately the same time.

Because most facilities are unoccupied at night and on weekends, night setback, boiler optimization, and hot water temperature reset controls are usually incorporated. When setbacks are used, morning warm-up and cool-down controls should be considered. After morning warm-up, automatic controls should change systems from heating to cooling as spaces heat up due to light and occupancy loads.

Table 1 Recommended Winter and Summer Design Dry-Bulb Temperatures for Various Spaces Common in Schools[a]

Space	Winter Design, °F	Summer Design, °F
Laboratories	72	78[d]
Auditoriums, libraries administrative areas, etc.	72	78
Classrooms		
Pre-K through 3rd	75	78
4th through 12th	72	78
Shops	72	78[b]
Locker, shower rooms	75[d]	c,d
Toilets	72	c
Storage	65	c,d
Mechanical rooms	60	c
Corridors	68	80[b]

[a]For spaces of high population density and where sensible heat factors are 0.75 or less, lower dry-bulb temperatures will result in generation of less latent heat, which may reduce the need for reheat and thus save energy. Therefore, optimum dry-bulb temperatures should be the subject of detailed design analysis.
[b]Frequently not air conditioned.
[c]Usually not air conditioned.
[d]Provide ventilation for odor control.

Table 2 Typical Ranges of Sound Design Levels

Space	A-Sound Levels, dB	Desired NC (Noise Criteria)
Libraries, classrooms	35-45	30-40
Laboratories, shops	40-50	35-45
Gyms, multipurpose corridors	40-55	35-50
Kitchens	45-55	40-55

Computerized systems are desirable. They can do much more than start and stop equipment; capabilities include resetting water temperatures, optimizing chillers and boilers, and keeping maintenance logs. See Chapter 38, Computer Applications, for more information about computerized management systems. For further information on controls, see Chapter 45.

Preschools

Commercially operated preschools are generally provided with a standard architectural layout based on owner-furnished designs. This may include the HVAC systems and equipment. All preschool facilities require quiet and economical systems. The equipment should be easy to operate and maintain. The design should provide for warm floors and no drafts. Radiant floor heating systems can be highly effective. Consideration should be given to the fact that the children are smaller and will play on the floor, while the teacher is taller and will be walking around. The teacher also requires a place for desk work; the designer should consider treating this area as a separate zone.

Preschool facilities generally operate on weekdays from early in the morning to 6:00 or 7:00 P.M. This schedule usually coincides with the normal working hours of the children's parents. The HVAC systems will therefore operate 12 to 14 hours per work day and be either off or on night setback at night and on weekends.

Supply air outlets should be located so that the floor area is maintained at about 75°F without the introduction of drafts. Both supply and return air outlets should be placed where they will not be blocked by furniture positioned along the walls. Coordination with the architect about the location of these outlets is essential. Proper ventilation is essential for controlling odors and helping prevent the spread of diseases among the children.

The use of floor-mounted heating equipment, such as electric baseboard heaters, should be avoided, as children must be prevented from coming in contact with hot surfaces or electrical devices.

Elementary Schools

Elementary schools are similar to preschools except that they have more facilities, such as gymnasiums and cafeterias. Such facilities are generally occupied from about 7:00 A.M to about 3:00 P.M. Peak cooling loads usually occur at the end of the school day. Peak heating is usually early in the day, when classrooms begin to be occupied and outdoor air is introduced into the facility. The following are facilities found in elementary schools:

Classrooms. Each classroom should at a minimum be heated and ventilated. Air conditioning should be seriously considered for school districts that have year-round classes in warm, humid climates. In humid climates, serious consideration should be given to providing dehumidification during the summer months, when the school is unoccupied, to prevent the growth of mold and mildew. Economizer systems during the winter months, to provide cooling and ventilation after morning warm-up cycles, should also be considered.

Gymnasiums. Gyms may be used after regular school hours for evening classes, meetings, and other functions. There may also be weekend use for group activities. The loads for these occasional uses should be considered when selecting and sizing the systems and equipment. Independent gymnasium HVAC systems with control capability allow for flexibility with smaller part load conditions.

Administrative Areas. The school's office area should be set up for individual control because it is usually occupied after school hours. As the offices are also occupied before school starts in the fall, air conditioning for the area should be considered or provisions should be allowed for the future.

Science Classrooms. Science rooms are now being provided for elementary schools. Although the children do not usually perform experiments, odors may be generated if the teacher demonstrates an experiment or if animals are kept in the classroom. Under these conditions, adequate ventilation for science classrooms is essential along with an exhaust fan with a local on-off switch for the occasional removal of excessive odors.

Libraries. If possible, libraries should be air conditioned to preserve the books and materials stored in them. See Chapter 3, Commercial and Public Buildings, for additional information.

Middle and Secondary Schools

Middle and secondary schools are usually occupied for longer periods of time than elementary schools, and they have additional special facilities. The following are typical special facilities for these schools:

Auditoriums. These facilities require a quiet atmosphere as well as heating, ventilating, and in some cases, air conditioning. Auditoriums are not often used with exception for special assemblies, practice for special programs, and special events. For other considerations see Chapter 4, Places of Assembly.

Computer Classrooms. These facilities have a high sensible heat load due to the computer equipment. Computer rooms may require additional cooling equipment such as small spot-cooling units to offset the additional load. Humidification may also be required. See Chapter 16, Data Processing and Electronic Office Areas, for additional information.

Science Classrooms. Science facilities in middle and secondary schools may require fume hoods with special exhaust systems. A makeup air system may be required if there are several fume hoods within a room. If there are no fume hoods, a room exhaust system is recommended for odor removal, depending on the type of experiments conducted in the room and whether animals are kept within the room. Any associated storage and preparation rooms are generally exhausted continuously to remove odors and vapors emanating from stored materials. The amount of exhaust and the location of exhaust grilles may be dictated by local or National Fire Protection Association (NFPA) codes. See Chapter 13, Laboratories, for further information.

Auto Repair Shops. These facilities require outdoor air ventilation to remove odors and fumes and to provide makeup air for exhaust systems. The shop is usually heated and ventilated but not air conditioned. To contain odors and fumes, return air should not be supplied to other spaces, and the shop should be kept at a negative pressure relative to the surrounding spaces. Special exhaust systems such as welding exhaust or direct-connected carbon monoxide exhaust systems may be required. See Chapter 29, Industrial Local Exhaust Systems, for further information.

Ice Rinks. These facilities require special air-conditioning and dehumidification systems to keep the spectators comfortable, maintain the ice surface at freezing conditions, and prevent the formation of fog at the surface. See Chapter 4, Places of Assembly, and Chapter 34 of the 1998 *ASHRAE Handbook—Refrigeration* for further discussion of these systems.

School Stores. These facilities contain school supplies and paraphernalia and are usually open for short periods of time. The heating and air-conditioning systems serving these areas should be able to be shut off when the store is closed to save energy.

Natatoriums. These facilities, like ice rinks, require special humidity control systems. In addition, special materials of construction are required. See Chapter 4, Places of Assembly, for further information.

Industrial Shops. These facilities are similar to auto repair shops and have special exhaust requirements for welding, soldering, and paint booths. In addition, a dust collection system is sometimes provided and the collected air is returned to the space. Industrial shops have a high sensible load due to the operation of the shop equipment. When calculating loads, the design engineer should consult the shop teacher about the shop operation, and where possible, diversity factors should be applied. See Chapter 29, Industrial Local Exhaust Systems, for more information.

Locker Rooms. Model building codes in the United States require that these facilities be exhausted directly to the outside when they contain toilets and/or showers. They are usually heated and ventilated only. Makeup air is required; the exhaust and makeup air systems should be coordinated and should operate only when required.

Home Economics Rooms. These rooms usually have a high sensible heat load due to appliances such as washing machines, dryers, stoves, ovens, and sewing machines. Different options should be considered for the exhaust of the stoves and dryers. If local codes allow, residential-style range hoods may be installed over the stoves. A central exhaust system could be applied to the dryers as well as to the stoves. If enough appliances are located within the room, a makeup air system may be required. These areas should be maintained at negative pressure in relationship to adjacent classrooms and administrative areas. See Chapter 30, Kitchen Ventilation, for more information.

Colleges and Universities

College and university facilities are similar to those of middle and secondary schools, except that there are more of them, and they may be located in several buildings on the campus. Some colleges and universities have satellite campuses scattered throughout a city or a state. The design criteria for each building are established by the requirements of the users. The following is a list of major facilities commonly found on college and university campuses but not in middle and secondary schools:

Administrative Buildings. These buildings should be treated as office buildings. They usually have a constant interior load and additional part-time loads associated with such spaces as conference rooms.

Animal Facilities. These spaces are commonly associated with laboratories, but usually have their own separate areas. Animal facilities need close temperature control and require a significant amount of outdoor air ventilation to control odors and prevent the spread of disease among the animals. Discussion of animal facilities is found in Chapter 13, Laboratories.

Dormitory Buildings. A discussion of dormitories can be found in Chapter 5, Hotels, Motels, and Dormitories.

Storage Buildings. These buildings are usually heated and ventilated only; however, a portion of the facility may require temperature and humidity control, depending upon the materials stored there.

Student and Faculty Apartment Buildings. A discussion of these facilities can be found in Chapter 5.

Large Gymnasiums. These facilities are discussed in Chapter 4, Places of Assembly.

Laboratory Buildings. These buildings house research facilities and may contain fume hoods, machinery, lasers, animal facilities, areas with controlled environments, and departmental offices. The HVAC systems and controls must be able to accommodate the diverse functions of the facility, which has the potential for 24-hour, year-round operation, and yet be easy to service and quick to repair. Variable air volume (VAV) systems can be used in laboratory buildings, but a proper control system should be applied to introduce the required quantities of outdoor air. Significant energy savings can be achieved by recovering the energy from the exhaust air and tempering the outdoor makeup air. Some other energy-saving systems employed for laboratory buildings are (1) ice storage, (2) economizer cycles, (3) heat reclaim chillers to produce domestic hot water or hot water for booster coils in the summer, and (4) cooling-tower free cooling.

The design engineer should discuss expected contaminants and concentrations with the owner to determine which materials of construction should be used for fume hoods and fume exhaust systems. Backup or standby systems for emergency use should be considered, as should alarms on critical systems. The maintenance staff should be thoroughly trained in the upkeep and repair of all systems, components, and controls. See Chapter 13, Laboratories, for additional information.

Museums. These facilities are discussed in Chapter 20, Museums, Libraries, and Archives.

Central Plants. These facilities contain the generating equipment and other supporting machinery for campus heating and air conditioning. Central plants are generally heated and ventilated only. Central plants are covered in Chapter 11 of the 1996 *ASHRAE Handbook—Systems and Equipment.*

Television and Radio Studios. A discussion of these facilities can be found in Chapter 3, Commercial and Public Buildings.

Hospitals. Hospitals are discussed in Chapter 7, Health Care Facilities.

Chapels. A discussion of these facilities can be found in Chapter 4, Places of Assembly.

Student Unions. These facilities may contain a dining room, a kitchen, meeting rooms, lounges, and game rooms. Student unions are generally open every day from early in the morning to late at night. An HVAC system such as VAV cooling with perimeter heating may be suitable. Extra ventilation air is required as makeup air for the kitchen systems and due to the high occupancy of the building. See Chapter 3, Commercial and Public Buildings, and Chapter 30, Kitchen Ventilation, for additional information.

Lecture Halls. These spaces house large numbers of students for a few hours several times a day and are vacant the rest of the time, so heat loads vary throughout the day. The systems that serve lecture halls must respond to these load variations. In addition, because the rooms are configured such that students at the rear of the room are

seated near the ceiling, the systems must be quiet and must not produce any drafts. Coordination with the design team is essential for determining the optimal location of diffusers and return grilles and registers. Placement of the supply diffusers at the front of the room and the return air grilles and registers near the front and rear of the room or under the seats should be considered. An analysis of the system noise levels expected in the hall is recommended. See Chapter 46, Sound and Vibration Control, for further information.

EQUIPMENT AND SYSTEM SELECTION

The architectural configuration of an educational building has a considerable influence on the building's heating and cooling loads and therefore on the selection of its HVAC systems. Other influences on system selection are the budget and whether the project is to expand an existing system or to integrate an HVAC system into the design of a new building. ASHRAE *Standard* 100 has recommendations for energy conservation in existing institutional buildings. ASHRAE *Standard* 90.1 should be followed for the energy-efficient design of new buildings. ASHRAE *Standard* 62 addresses the criteria necessary to meet indoor air quality requirements.

The only trend evident in HVAC design for educational facilities is that air-conditioning systems are being installed whenever possible. In smaller single buildings, systems such as unit ventilators, heat pumps, small rooftop units, and VAV systems are being applied. In larger facilities, VAV systems, large rooftop units with perimeter heating systems, and heat pump systems coupled with central systems are utilized. When rooftop units are used, stairs to the roof or penthouses make maintenance easier.

Single-Duct with Reheat

Reheat systems are restricted from use by ASHRAE *Standard* 90.1 unless recovered energy is used as the reheat medium. Booster heating coils should be considered if the heat is to be applied after mechanical cooling has been turned off or locked out (e.g., during an economizer cycle). The booster heating coil system can use recovered energy to further reduce energy costs.

Multizone

The original multizone design, which simply heated or cooled by individual zones, is seldom used because mixing hot and cold airstreams is not energy-efficient. Multizone system units can be equipped with reset controllers that receive signals from zone thermostats, allowing them to maintain optimum hot and cold temperatures and minimize the mixing of airstreams.

Another energy-conserving technique for a multizone design is the incorporation of a neutral zone that mixes heated or cooled air with return air, but does not mix heated and cooled airstreams. This technique minimizes the energy-wasting practice of simultaneously producing both heating and cooling.

Multizone systems are also available that have individual heating and cooling within each zone, therefore avoiding simultaneous heating and cooling. The zone thermostats energize either the heating system or the cooling system as required.

Replacement multizone units with the energy-saving measures (see the section of Energy Considerations) are also available for existing systems. The replacement of multizone units with VAV systems described in the next section has been successful.

Variable Air Volume

Types of variable air volume systems include single-duct, single-duct with perimeter heating, and dual-duct systems. The supply air volumes from the primary air system are adjusted according to the space load requirements. VAV systems are most often zoned by exposure and occupancy schedules. The primary system usually provides supply air to the zones at a constant temperature either for cooling only or, in the case of dual-duct systems, for heating or cooling. To conserve energy, heating and cooling should not be provided simultaneously to the same zone. Heating should be provided only during economizer cooling or when mechanical cooling has been turned off or locked out.

Perimeter heating can be provided by duct-mounted heating coils, fan terminal units with duct- or unit-mounted heating coils, perimeter fin tubes, baseboard heaters, cabinet unit heaters, or convectors. Radiant ceiling or floor heating systems can also be applied around the perimeter. Each zone or exposure should be controlled to prevent overheating of the space. Perimeter air-distribution systems should be installed so that drafts are not created.

Fan terminal units are utilized when constant air motion is required. These units operate either constantly or only during heating periods. Noisy units should be avoided. A system of fan terminal units with heating coils allows the primary heating system to be shut off during night setback or unoccupied periods.

Packaged Units

Unit ventilators or through-the-wall packaged units with either split or integral direct-expansion cooling may serve one or two rooms with a common exposure. Heat pumps can also be used. When a single packaged unit covers more than one space, zoning is critical. As is the case for VAV systems, spaces with similar occupancy schedules are grouped together. If exterior spaces are to be served, exterior wall orientation must also match.

For water-source heat pumps, a constant flow of water is circulated to the units. This water is tempered by boilers for heating in the winter or by a fluid cooler provided for heat rejection (cooling). Heat rejection may be necessary for much of the school year, depending upon the internal loads. Certain operating conditions balance the water loop for water-source heat pumps, which could result in periods in which neither primary heating nor cooling is required. There may be simultaneous heating and cooling by different units throughout the facility, depending upon exposure and room loads. A water storage tank may be incorporated into the water loop to store heated water during the day for use at night.

Air-to-air heat pumps are also used in schools, as are small rooftop and self-contained unit ventilators featuring heat pump cycles.

REFERENCES

ASHRAE. 1989. Energy efficient design of new buildings except low-rise residential buildings. ANSI/ASHRAE *Standard* 90.1-1989.

ASHRAE. 1989. Ventilation for acceptable indoor air quality. ANSI/ASHRAE *Standard* 62-1989.

ASHRAE. 1995. Energy conservation in existing buildings. ANSI/ASHRAE *Standard* 100-1995.

HEALTH CARE FACILITIES

CONTINUAL advances in medicine and technology necessitate constant reevaluation of the air-conditioning needs of hospitals and medical facilities. While medical evidence has shown that proper air conditioning is helpful in the prevention and treatment of many conditions, the relatively high cost of air conditioning demands efficient design and operation to ensure economical energy management (Woods et al. 1986, Murray et al. 1988, Demling and Maly 1989, and Fitzgerald 1989).

Health care occupancy classification, based on the latest occupancy guidelines from the National Fire Protection Association (NFPA), should be considered early in the design phase of the project. Health care occupancy is important for fire protection (smoke zones, smoke control) and for the future adaptability of the HVAC system for a more restrictive occupancy.

Health care facilities are becoming increasingly diversified in response to a trend toward outpatient services. The term **clinic** may refer to any building from the ubiquitous residential doctor's office to a specialized cancer treatment center. Prepaid health maintenance provided by integrated regional health care organizations is becoming the model for medical care delivery. These organizations, as well as long-established hospitals, are constructing buildings that look less like hospitals and more like office buildings.

For the purpose of this chapter, health care facilities are divided into the following categories:

- Hospital facilities
- Outpatient health care facilities
- Nursing home facilities
- Dental care facilities

The specific environmental conditions required by a particular medical facility may vary from those in this chapter, depending on the agency responsible for the environmental standard. Among the agencies that may have standards for medical facilities are state and local health agencies, the U.S. Department of Health and Human Services, Indian Health Service, Public Health Service, Medicare/Medicaid, the U.S. Department of Defense, the U.S. Department of Veterans Affairs, and the Joint Commission on Accreditation of Healthcare Organizations. It is advisable to discuss infection control objectives with the hospital's infection control committee.

The general hospital was selected as the basis for the fundamentals outlined in the first section, Hospital Facilities, because of the variety of services it provides. Environmental conditions and design criteria apply to comparable areas in other health facilities.

The general acute care hospital has a core of critical care spaces, including operating rooms, labor rooms, delivery rooms, and a nursery. Usually the functions of radiology, laboratory, central sterile, and the pharmacy are located close to the critical care space. Inpatient nursing, including intensive care nursing, is in the complex.

The facility also incorporates an emergency room, a kitchen, dining and food service, a morgue, and central housekeeping support.

Criteria for outpatient facilities are given in the second section, Outpatient Health Care Facilities. Outpatient surgery is performed with the anticipation that the patient will not stay overnight. An outpatient facility may be part of an acute care facility, a freestanding unit, or part of another medical facility such as a medical office building.

Nursing homes are addressed separately in the third section, Nursing Home Facilities, because their fundamental requirements differ greatly from those of other medical facilities.

Dental facilities are briefly discussed in the fourth section, Dental Facilities. Requirements for these facilities differ from those of other health care facilities because many procedures generate aerosols, dusts, and particulates.

AIR CONDITIONING IN THE PREVENTION AND TREATMENT OF DISEASE

Hospital air conditioning assumes a more important role than just the promotion of comfort. In many cases, proper air conditioning is a factor in patient therapy; in some instances, it is the major treatment.

Studies show that patients in controlled environments generally have more rapid physical improvement than do those in uncontrolled environments. Patients with thyrotoxicosis do not tolerate hot, humid conditions or heat waves very well. A cool, dry environment favors the loss of heat by radiation and evaporation from the skin and may save the life of the patient.

Cardiac patients may be unable to maintain the circulation necessary to ensure normal heat loss. Therefore, air conditioning cardiac wards and rooms of cardiac patients, particularly those with congestive heart failure, is necessary and considered therapeutic (Burch and Pasquale 1962). Individuals with head injuries, those subjected to brain operations, and those with barbiturate poisoning may have hyperthermia, especially in a hot environment, due to a disturbance in the heat regulatory center of the brain. Obviously, an important factor in recovery is an environment in which the patient can lose heat by radiation and evaporation—namely, a cool room with dehumidified air.

A hot, dry environment of 90°F dry bulb and 35% rh has been successfully used in treating patients with rheumatoid arthritis.

Dry conditions may constitute a hazard to the ill and debilitated by contributing to secondary infection or infection totally unrelated to the clinical condition causing hospitalization. Clinical areas devoted to upper respiratory disease treatment and acute care, as well as the general clinical areas of the entire hospital, should be maintained at a relative humidity of 30 to 60%.

Patients with chronic pulmonary disease often have viscous respiratory tract secretions. As these secretions accumulate and increase in viscosity, the patient's exchange of heat and water dwindles. Under these circumstances, the inspiration of warm,

The preparation of this chapter is assigned to TC 9.8, Large Building Air-Conditioning Applications.

humidified air is essential to prevent dehydration (Walker and Wells 1961).

Patients needing oxygen therapy and those with tracheotomies require special attention to ensure warm, humid supplies of inspired air. Cold, dry oxygen or the bypassing of the nasopharyngeal mucosa presents an extreme situation. Rebreathing techniques for anesthesia and enclosure in an incubator are special means of addressing impaired heat loss in therapeutic environments.

Burn patients need a hot environment and high relative humidity. A ward for severe burn victims should have temperature controls that permit adjusting the room temperature up to 90°F dry bulb and the relative humidity up to 95%.

HOSPITAL FACILITIES

Although proper air conditioning is helpful in the prevention and treatment of disease, the application of air conditioning to health facilities presents many problems not encountered in the usual comfort conditioning design.

The basic differences between air conditioning for hospitals (and related health facilities) and that for other building types stem from (1) the need to restrict air movement in and between the various departments; (2) the specific requirements for ventilation and filtration to dilute and remove contamination in the form of odor, airborne microorganisms and viruses, and hazardous chemical and radioactive substances; (3) the different temperature and humidity requirements for various areas; and (4) the design sophistication needed to permit accurate control of environmental conditions.

Infection Sources and Control Measures

Bacterial Infection. Examples of bacteria that are highly infectious and transported within air or air and water mixtures are *Mycobacterium tuberculosis* and *Legionella pneumophila* (Legionnaire's disease). Wells (1934) showed that droplets or infectious agents of 5 μm or less in size can remain airborne indefinitely. Isoard et al. (1980) and Luciano (1984) have shown that 99.9% of all bacteria present in a hospital are removed by 90 to 95% efficient filters (ASHRAE *Standard* 52.1). This is because bacteria are typically present in colony-forming units that are larger than 1 μm. Some authorities recommend the use of high efficiency particulate air (HEPA) filters having dioctyl phthalate (DOP) test filtering efficiencies of 99.97% in certain areas.

Viral Infection. Examples of viruses that are transported by and virulent within air are *Varicella* (chicken pox/shingles), *Rubella* (German measles), and *Rubeola* (regular measles). Epidemiological evidence and other studies indicate that many of the airborne viruses that transmit infection are submicron in size; thus, there is no known method to effectively eliminate 100% of the viable particles. HEPA and/or ultra low penetration (ULPA) filters provide the greatest efficiency currently available. Attempts to deactivate viruses with ultraviolet light and chemical sprays have not proven reliable or effective enough to be recommended by most codes as a primary infection control measure. Therefore, isolation rooms and isolation anterooms with appropriate ventilation-pressure relationships are the primary means used to prevent the spread of airborne viruses in the health care environment.

Molds. Evidence indicates that some molds such as *Aspergillis* can be fatal to advanced leukemia, bone marrow transplant, and other immunocompromised patients.

Outdoor Air Ventilation. If outdoor air intakes are properly located, and areas adjacent to outdoor air intakes are properly maintained, outdoor air, in comparison to room air, is virtually free of bacteria and viruses. Infection control problems frequently involve a bacterial or viral source within the hospital. Ventilation air dilutes the viral and bacterial contamination within a hospital. If ventilation systems are properly designed, constructed, and maintained to preserve the correct pressure relations between

functional areas, they remove airborne infectious agents from the hospital environment.

Temperature and Humidity. These conditions can inhibit or promote the growth of bacteria and activate or deactivate viruses. Some bacteria such as *Legionella pneumophila* are basically waterborne and survive more readily in a humid environment. Codes and guidelines specify temperature and humidity range criteria in some hospital areas as a measure for infection control as well as comfort.

AIR QUALITY

Systems must also provide air virtually free of dust, dirt, odor, and chemical and radioactive pollutants. In some cases, outside air is hazardous to patients suffering from cardiopulmonary, respiratory, or pulmonary conditions. In such instances, systems that intermittently provide maximum allowable recirculated air should be considered.

Outdoor Intakes. These intakes should be located as far as practical (on directionally different exposures whenever possible), but not less than 30 ft, from combustion equipment stack exhaust outlets, ventilation exhaust outlets from the hospital or adjoining buildings, medical-surgical vacuum systems, cooling towers, plumbing vent stacks, smoke control exhaust outlets, and areas that may collect vehicular exhaust and other noxious fumes. The bottom of outdoor air intakes serving central systems should be located as high as practical (12 ft recommended) but not less than 6 ft above ground level or, if installed above the roof, 3 ft above the roof level.

Exhaust Outlets. These exhausts should be located a minimum of 10 ft above ground level and away from doors, occupied areas, and operable windows. Preferred location for exhaust outlets is at roof level projecting upward or horizontally away from outdoor intakes. Care must be taken in locating highly contaminated exhausts (e.g., from engines, fume hoods, biological safety cabinets, kitchen hoods, and paint booths). Prevailing winds, adjacent buildings, and discharge velocities must be taken into account (see Chapter 15 of the 1997 *ASHRAE Handbook—Fundamentals*). In critical or complicated applications, wind tunnel studies or computer modeling may be appropriate.

Air Filters. A number of methods are available for determining the efficiency of filters in removing particulates from an airstream (see Chapter 24 of the 1996 *ASHRAE Handbook—Systems and Equipment*). All central ventilation or air-conditioning systems should be equipped with filters having efficiencies no lower than those indicated in Table 1. Where two filter beds are indicated, Filter Bed No. 1 should be located upstream of the air-conditioning equipment, and Filter Bed No. 2 should be downstream of the supply fan, any recirculating spray water systems, and water-reservoir type humidifiers. Appropriate precautions should be observed to prevent wetting of the filter media by free moisture from humidifiers. Where only one filter bed is indicated, it should be located upstream of the air-conditioning equipment. All filter efficiencies are based on ASHRAE *Standard* 52.1.

The following are guidelines for filter installations:

1. HEPA filters having DOP test efficiencies of 99.97% should be used on air supplies serving rooms for clinical treatment of patients with a high susceptibility to infection due to leukemia, burns, bone marrow transplant, organ transplant, or human immunodeficiency virus (HIV). HEPA filters should also be used on the discharge air from fume hoods or safety cabinets in which infectious or highly radioactive materials are processed. The filter system should be designed and equipped to permit safe removal, disposal, and replacement of contaminated filters.

2. All filters should be installed to prevent leakage between the filter segments and between the filter bed and its supporting frame. A small leak that permits any contaminated air to escape through the filter can destroy the usefulness of the best air cleaner.

Table 1 Filter Efficiencies for Central Ventilation and Air-Conditioning Systems in General Hospitals

Minimum Number of Filter Beds	Area Designation	Filter Efficiencies, %		
		Filter Bed		
		No. 1[a]	No. 2[a]	No. 3[b]
3	Orthopedic operating room Bone marrow transplant operating room Organ transplant operating room	25	90	99.97[c]
2	General procedure operating rooms Delivery rooms Nurseries Intensive care units Patient care rooms Treatment rooms Diagnostic and related areas	25	90	
1	Laboratories Sterile storage	80		
1	Food preparation areas Laundries Administrative areas Bulk storage Soiled holding areas	25		

[a]Based on ASHRAE *Standard* 52.1.
[b]Based on DOP test.
[c]HEPA filters at air outlets.

3. A manometer should be installed in the filter system to measure the pressure drop across each filter bank. Visual observation is not an accurate method for determining filter loading.
4. High-efficiency filters should be installed in the system, with adequate facilities provided for maintenance without introducing contamination into the delivery system or the area served.
5. Because high-efficiency filters are expensive, the hospital should project the filter bed life and replacement costs and incorporate these into their operating budget.
6. During construction, openings in ductwork and diffusers should be sealed to prevent intrusion of dust, dirt, and hazardous materials. Such contamination is often permanent and provides a medium for the growth of infectious agents. Existing or new filters may rapidly become contaminated by construction dust.

Air Movement

The data given in Table 2 illustrate the degree to which contamination can be dispersed into the air of the hospital environment by one of the many routine activities for normal patient care. The bacterial counts in the hallway clearly indicate the spread of this contamination.

Because of the dispersal of bacteria resulting from such necessary activities, air-handling systems should provide air movement patterns that minimize the spread of contamination. Undesirable airflow between rooms and floors is often difficult to control because of open doors, movement of staff and patients, temperature differentials, and stack effect, which is accentuated by the vertical openings such as chutes, elevator shafts, stairwells, and mechanical shafts common to hospitals. While some of these factors are beyond practical control, the effect of others may be minimized by terminating shaft openings in enclosed rooms and by designing and balancing air systems to create positive or negative air pressure within certain rooms and areas.

Systems serving highly contaminated areas, such as autopsy and infectious isolation rooms, should maintain a negative air pressure within these rooms relative to adjoining rooms or the corridor (Murray et al. 1988). The negative pressure difference is obtained by supplying less air to the area than is exhausted from it (CDC 1994). This

Table 2 Influence of Bedmaking on Airborne Bacterial Count in Hospitals

Item	Count per Cubic Foot	
	Inside Patient Room	Hallway near Patient Room
Background	34	30
During bedmaking	140	64
10 min after	60	40
30 min after	36	27
Background	16	
Normal bedmaking	100	
Vigorous bedmaking	172	

Source: Greene et al. (1960).

pressure differential causes air to flow into the room through various leakage areas (e.g., the perimeter of doors and windows, utility/fixture penetrations, cracks, etc.) and prevents outward airflow. Protective isolation rooms exemplify positive pressure conditions or the opposite conditions. Exceptions to normally established negative and positive pressure conditions include operating rooms where highly infectious patients may be treated (e.g., operating rooms in which bronchoscopy or lung surgery is performed) and infectious isolation rooms that house immunosuppressed patients with airborne infectious diseases such as tuberculosis (TB). These areas should include an anteroom between the operating or isolation room and the corridor or other contiguous space. The anteroom should either be positive to both the room and contiguous space or negative to both the room and the contiguous space, depending on local fire and smoke management regulations. Either technique minimizes cross contamination between the patient area and surrounding areas.

Differential air pressure can be maintained only in an entirely sealed room. Therefore, it is important to obtain a reasonably close fit of all doors and seal all wall and floor penetrations between pressurized areas. This is best accomplished by using weather stripping and drop bottoms on doors. The opening of a door between two areas instantaneously reduces any existing pressure differential between them to such a degree that its effectiveness is nullified. When such openings occur, a natural interchange of air takes place between the two rooms due to turbulence created by the door opening and closing combined with personnel ingress/egress.

For critical areas requiring both the maintenance of pressure differentials to adjacent spaces and personnel movement between the critical area and adjacent spaces, the use of appropriate air locks or anterooms is indicated.

Figure 1 shows the bacterial count in a surgery room and its adjoining rooms during a normal surgical procedure. These bacterial counts were taken simultaneously. The relatively low bacterial counts in the surgery room, compared with those of the adjoining rooms, are attributable to the lower level of activity and higher air pressure within operating rooms.

In general, outlets supplying air to sensitive ultraclean areas should be located on the ceiling, and perimeter or several exhaust outlets should be near the floor. This arrangement provides a downward movement of clean air through the breathing and working zones to the floor area for exhaust. Infectious isolation rooms should have supply air above and near the doorway and exhaust air from near the floor, behind the patient's bed. This arrangement is such that clean air first flows to parts of the room where workers or visitors are likely to be, and then flows across the infectious source and into the exhaust. Thus, noninfected persons are not positioned between the infectious source and the exhaust location (CDC 1994). The bottoms of the return or exhaust openings should be at least 3 in. above the floor.

The laminar airflow concept developed for industrial clean room use has attracted the interest of some medical authorities. There are advocates of both vertical and horizontal laminar airflow systems,

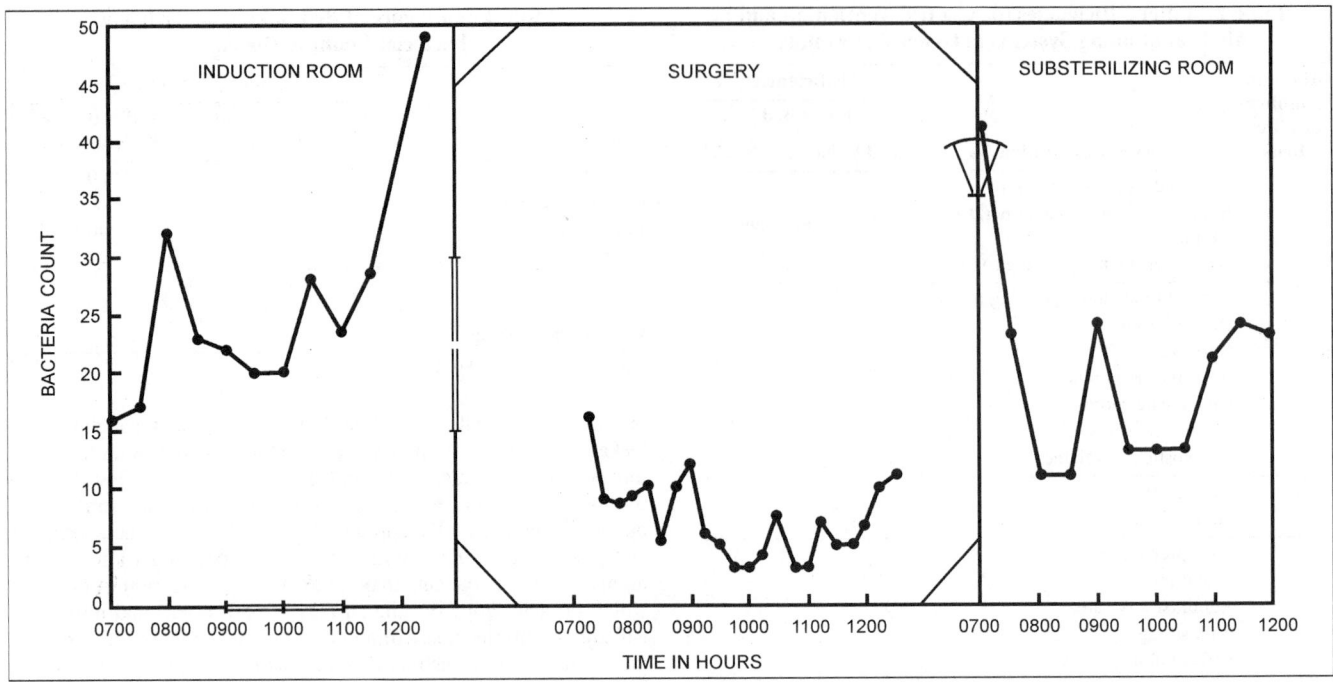

Fig. 1 Typical Airborne Contamination in Surgery and Adjacent Areas

with and without fixed or movable walls around the surgical team (Pfost 1981). Some medical authorities do not advocate laminar airflow for surgeries but encourage air systems similar to those described in this chapter.

Laminar airflow in surgical operating rooms is airflow that is predominantly unidirectional when not obstructed. The unidirectional laminar airflow pattern is commonly attained at a velocity of 90 ± 20 fpm.

Laminar airflow has shown promise in rooms used for the treatment of patients who are highly susceptible to infection (Michaelson et al. 1966). Among such patients would be the badly burned and those undergoing radiation therapy, concentrated chemotherapy, organ transplants, amputations, and joint replacement.

Temperature and Humidity

Specific recommendations for design temperatures and humidities are given in the next section, Specific Design Criteria. Temperature and humidity for other inpatient areas not covered should be 75°F or less and 30% to 60% rh.

Pressure Relationships and Ventilation

Table 3 covers ventilation recommendations for comfort, asepsis, and odor control in areas of acute care hospitals that directly affect patient care. Table 3 does not necessarily reflect the criteria of the American Institute of Architects (AIA) or any other group. If specific organizational criteria must be met, refer to that organization's literature. Ventilation in accordance with ASHRAE *Standard* 62, Ventilation for Acceptable Indoor Air Quality, should be used for areas where specific standards are not given. Where a higher outdoor air requirement is called for in ASHRAE *Standard* 62 than in Table 3, the higher value should be used. Specialized patient care areas, including organ transplant and burn units, should have additional ventilation provisions for air quality control as may be appropriate.

Design of the ventilation system must as much as possible provide air movement from clean to less clean areas. In critical care areas, constant volume systems should be employed to assure proper pressure relationships and ventilation, except in unoccupied rooms. In noncritical patient care areas and staff rooms, variable air

volume (VAV) systems may be considered for energy conservation. When using VAV systems within the hospital, special care should be taken to ensure that minimum ventilation rates (as required by codes) are maintained and that pressure relationships between various spaces are maintained. With VAV systems, a method such as air volume tracking between supply, return, and exhaust could be used to control pressure relationships (Lewis 1988).

The number of air changes may be reduced to 25% of the indicated value, when the room is unoccupied, if provisions are made to ensure that (1) the number of air changes indicated is reestablished whenever the space is occupied, and (2) the pressure relationship with the surrounding rooms is maintained when the air changes are reduced.

In areas requiring no continuous directional control (±), ventilation systems may be shut down when the space is unoccupied and ventilation is not otherwise needed.

Because of the cleaning difficulty and potential for buildup of contamination, recirculating room heating and/or cooling units must not be used in areas marked "No." Note that the standard recirculating room unit may also be impractical for primary control where exhaust to the outside is required.

In rooms having hoods, extra air must be supplied for hood exhaust so that the designated pressure relationship is maintained. Refer to Chapter 13, Laboratories, for further discussion of laboratory ventilation.

For maximum energy conservation, use of recirculated air is preferred. If all-outdoor air is used, an efficient heat recovery method should be considered.

Smoke Control

As the ventilation design is developed, a proper smoke control strategy must be considered. Passive systems rely on fan shutdown, smoke and fire partitions, and operable windows. Proper treatment of duct penetrations must be observed.

Active smoke control systems use the ventilation system to create areas of positive and negative pressures that, along with fire and smoke partitions, limit the spread of smoke. The ventilation system may be used in a smoke removal mode in which the products of

Table 3   General Pressure Relationships and Ventilation of Certain Hospital Areas

Function Space		Pressure Relationship to Adjacent Areas[a]	Minimum Air Changes of Outdoor Air per Hour[b]	Minimum Total Air Changes per Hour[c]	All Air Exhausted Directly to Outdoors	Air Recirculated Within Room Units[d]
SURGERY AND CRITICAL CARE						
Operating room	(all outdoor air system)	P	15[e]	15	Yes	No
	(recirculating air system)	P	5	25	Optional	No
Delivery room	(all outdoor air system)	P	15	15	Optional	No
	(recirculating air system)	P	5	25	Optional	No
Recovery room		E	2	6	Optional	No
Nursery suite		P	5	12	Optional	No
Trauma room[f]		P	5	12	Optional	No
Anesthesia storage (see code requirements)		±	Optional	8	Yes	No
NURSING						
Patient room		±	2	4	Optional	Optional
Toilet room[g]		N	Optional	10	Yes	No
Intensive care		P	2	6	Optional	No
Protective isolation[i]		P	2	15	Yes	Optional
Infectious isolation[h]		±	2	6	Yes	No
Isolation alcove or anteroom		±	2	10	Yes	No
Labor/delivery/recovery/postpartum (LDRP)		E	2	4	Optional	Optional
Patient corridor		E	2	4	Optional	Optional
ANCILLARY						
Radiology	X-ray (surgery and critical care)	P	3	15	Optional	No
	X-ray (diagnostic and treatment)	±	2	6	Optional	Optional
	Darkroom	N	2	10	Yes[j]	No
Laboratory, general		N	2	6	Yes	No
Laboratory, bacteriology		N	2	6	Yes	No
Laboratory, biochemistry		P	2	6	Optional	No
Laboratory, cytology		N	2	6	Yes	No
Laboratory, glasswashing		N	Optional	10	Yes	Optional
Laboratory, histology		N	2	6	Yes	No
Laboratory, nuclear medicine		N	2	6	Yes	No
Laboratory, pathology		N	2	6	Yes	No
Laboratory, serology		P	2	6	Optional	No
Laboratory, sterilizing		N	Optional	10	Yes	No
Laboratory, media transfer		P	2	4	Optional	No
Autopsy		N	2	12	Yes	No
Nonrefrigerated body-holding room[k]		N	Optional	10	Yes	No
Pharmacy		P	2	4	Optional	Optional
ADMINISTRATION						
Admitting and Waiting Rooms		N	2	6	Yes	Optional
DIAGNOSTIC AND TREATMENT						
Bronchoscopy, sputum collection, and pentamidine administration		N	2	10	Yes	Optional
Examination room		±	2	6	Optional	Optional
Medication room		P	2	4	Optional	Optional
Treatment room		±	2	6	Optional	Optional
Physical therapy and hydrotherapy		N	2	6	Optional	Optional
Soiled workroom or soiled holding		N	2	10	Yes	No
Clean workroom or clean holding		P	2	4	Optional	Optional
STERILIZING AND SUPPLY						
Sterilizer equipment room		N	Optional	10	Yes	No
Soiled or decontamination room		N	2	6	Yes	No
Clean workroom and sterile storage		P	2	4	Optional	Optional
Equipment storage		±	2 (Optional)	2	Optional	Optional
SERVICE						
Food preparation center[l]		±	2	10	Yes	No
Warewashing		N	Optional	10	Yes	No
Dietary day storage		±	Optional	2	Optional	No
Laundry, general		N	2	10	Yes	No
Soiled linen sorting and storage		N	Optional	10	Yes	No
Clean linen storage		P	2 (Optional)	2	Optional	Optional
Linen and trash chute room		N	Optional	10	Yes	No
Bedpan room		N	Optional	10	Yes	No
Bathroom		N	Optional	10	Optional[f]	No
Janitor's closet		N	Optional	10	Optional	No

P = Positive   N = Negative   E = Equal   ± = Continuous directional control not required

[a]Where continuous directional control is not required, variations should be minimized, and in no case should a lack of directional control allow the spread of infection from one area to another. Boundaries between functional areas (wards or departments) should have directional control. Lewis (1988) describes methods for maintaining directional control by applying air-tracking controls.

[b]Ventilation in accordance with ASHRAE *Standard* 62, Ventilation for Acceptable Indoor Air Quality, should be used for areas for which specific ventilation rates are not given. Where a higher outdoor air requirement is called for in *Standard* 62 than in Table 3, the higher value should be used.

[c]Total air changes indicated should be either supplied or, where required, exhausted.

[d]Recirculating HEPA filter units used for infection control (without heating or cooling coils) are acceptable.

[e]For operating rooms, 100% outside air should be used only when codes require it and only if heat recovery devices are used.

[f]The term "trauma room" as used here is the first aid room and/or emergency room used for general initial treatment of accident victims. The operating room within the trauma center that is routinely used for emergency surgery should be treated as an operating room.

[g]See section on Patient Rooms for discussion on design of central toilet exhaust systems.

[h]The infectious isolation rooms described in this table are those that might be used for infectious patients in the average community hospital. The rooms are negatively pressurized. Some isolation rooms may have a separate anteroom. Refer to the discussion in the chapter for more detailed information. Where highly infectious respirable diseases such as tuberculosis are to be isolated, increased air change rates should be considered.

[i]Protective isolation rooms are those used for immunosuppressed patients. The room is positively pressurized to protect the patient. Anterooms are generally required and should be negatively pressurized with respect to the patient room.

[j]All air need not be exhausted if darkroom equipment has scavenging exhaust duct attached and meets ventilation standards of NIOSH, OSHA, and local employee exposure limits.

[k] The nonrefrigerated body-holding room is only for facilities that do not perform autopsies on-site and use the space for short periods while waiting for the body to be transferred.

[l]Food preparation centers should have an excess of air supply for positive pressure when hoods are not in operation. The number of air changes may be reduced or varied for odor control when the space is not in use. Minimum total air changes per hour should be that required to provide proper makeup air to kitchen exhaust systems. See Chapter 30, Kitchen Ventilation.

combustion are exhausted by mechanical means. As design of active smoke control systems continues to evolve, the engineer and code authority should carefully plan system operation and configuration. Refer to Chapter 51 and NFPA *Standards* 90A, 92A, 99, and 101.

SPECIFIC DESIGN CRITERIA

There are seven principal divisions of an acute care general hospital: (1) surgery and critical care, (2) nursing, (3) ancillary, (4) administration, (5) diagnostic and treatment, (6) sterilizing and supply, and (7) service. The environmental requirements of each of the departments/spaces within these divisions differ to some degree according to their function and the procedures carried out in them. This section describes the functions of these departments/spaces and covers details of design requirements. Close coordination with health care planners and medical equipment specialists in the mechanical design and construction of health facilities is essential to achieve the desired conditions.

Surgery and Critical Care

No area of the hospital requires more careful control of the aseptic condition of the environment than does the surgical suite. The systems serving the operating rooms, including cystoscopic and fracture rooms, require careful design to reduce to a minimum the concentration of airborne organisms.

The greatest amount of the bacteria found in the operating room comes from the surgical team and is a result of their activities during surgery. During an operation, most members of the surgical team are in the vicinity of the operating table, creating the undesirable situation of concentrating contamination in this highly sensitive area.

Operating Rooms. Studies of operating-room air distribution devices and observation of installations in industrial clean rooms indicate that delivery of the air from the ceiling, with a downward movement to several exhaust inlets located on opposite walls, is probably the most effective air movement pattern for maintaining the concentration of contamination at an acceptable level. Completely perforated ceilings, partially perforated ceilings, and ceiling-mounted diffusers have been applied successfully (Pfost 1981).

Operating room suites are typically in use no more than 8 to 12 h per day (excepting trauma centers and emergency departments). For energy conservation, the air-conditioning system should allow a reduction in the air supplied to some or all of the operating rooms when possible. Positive space pressure must be maintained at reduced air volumes to ensure sterile conditions. The time required for an inactive room to become usable again must be considered. Consultation with the hospital surgical staff will determine the feasibility of this feature.

A separate air exhaust system or special vacuum system should be provided for the removal of anesthetic trace gases (NIOSH 1975). Medical vacuum systems have been used for removal of non-flammable anesthetic gases (NFPA *Standard* 99). One or more outlets may be located in each operating room to permit connection of the anesthetic machine scavenger hose.

Although good results have been reported from air disinfection of operating rooms by irradiation, this method is seldom used. The reluctance to use irradiation may be attributed to the need for special designs for installation, protective measures for patients and personnel, constant monitoring of lamp efficiency, and maintenance.

The following conditions are recommended for operating, catheterization, cystoscopic, and fracture rooms:

1. The temperature set point should be adjustable by surgical staff over a range of 62 to 80°F.
2. Relative humidity should be kept between 45 and 55%.
3. Air pressure should be maintained positive with respect to any adjoining rooms by supplying 15% excess air.
4. Differential pressure indicating device should be installed to permit air pressure readings in the rooms. Thorough sealing of

all wall, ceiling, and floor penetrations and tight-fitting doors is essential to maintaining readable pressure.
5. Humidity indicator and thermometers should be located for easy observation.
6. Filter efficiencies should be in accordance with Table 1.
7. Entire installation should conform to the requirements of NFPA *Standard* 99, Health Care Facilities.
8. All air should be supplied at the ceiling and exhausted or returned from at least two locations near the floor (see Table 3 for minimum ventilating rates). Bottom of exhaust outlets should be at least 3 in. above the floor. Supply diffusers should be of the unidirectional type. High-induction ceiling or side-wall diffusers should be avoided.
9. Acoustical materials should not be used as duct linings unless 90% efficient minimum terminal filters are installed downstream of the linings. Internal insulation of terminal units may be encapsulated with approved materials. Duct-mounted sound traps should be of the packless type or have polyester film linings over acoustical fill.
10. Any spray-applied insulation and fireproofing should be treated with fungi growth inhibitor.
11. Sufficient lengths of watertight, drained stainless steel duct should be installed downstream of humidification equipment to assure complete evaporation of water vapor before air is discharged into the room.

Control centers that monitor and permit adjustment of temperature, humidity, and air pressure may be located at the surgical supervisor's desk.

Obstetrical Areas. The pressure in the obstetrical department should be positive or equal to that in other areas.

Delivery Rooms. The design for the delivery room should conform to the requirements of operating rooms.

Recovery Rooms. Postoperative recovery rooms used in conjunction with the operating rooms should be maintained at a temperature of 75°F and a relative humidity between 45 and 55%. Because the smell of residual anesthesia sometimes creates odor problems in recovery rooms, ventilation is important, and a balanced air pressure relative to the air pressure of adjoining areas should be provided.

Nursery Suites. Air conditioning in nurseries provides the constant temperature and humidity conditions essential to care of the newborn in a hospital environment. Air movement patterns in nurseries should be carefully designed to reduce the possibility of drafts.

All air supplied to nurseries should enter at or near the ceiling and be removed near the floor with the bottom of exhaust openings located at least 3 in. above the floor. Air system filter efficiencies should conform to Table 1. Finned tube radiation and other forms of convection heating should not be used in nurseries.

Full-Term Nurseries. A temperature of 75°F and a relative humidity from 30 to 60% are recommended for full-term nurseries, examination rooms, and work spaces. The maternity nursing section should be controlled similarly to protect the infant during visits with the mother. The nursery should have a positive air pressure relative to the work space and examination room, and any rooms located between the nurseries and the corridor should be similarly pressurized relative to the corridor. This prevents the infiltration of contaminated air from outside areas.

Special Care Nurseries. These nurseries require a variable range temperature capability of 75 to 80°F and a relative humidity from 30 to 60%. This type of nursery is usually equipped with individual incubators to regulate temperature and humidity. It is desirable to maintain these same conditions within the nursery proper to accommodate both infants removed from the incubators and those not placed in incubators. The pressurization of special care nurseries should correspond to that of full-term nurseries.

Observation Nurseries. Temperature and humidity requirements for observation nurseries are similar to those for full-term nurseries. Because infants in these nurseries have unusual clinical symptoms, the air from this area should not enter other nurseries. A negative air pressure relative to the air pressure of the workroom should be maintained in the nursery. The workroom, usually located between the nursery and the corridor, should be pressurized relative to the corridor.

Emergency Rooms. Emergency rooms are typically the most highly contaminated areas in the hospital as a result of the soiled condition of many arriving patients and the relatively large number of persons accompanying them. Temperatures and humidities of offices and waiting spaces should be within the normal comfort range.

Trauma Rooms. Trauma rooms should be ventilated in accordance with requirements in Table 3. Emergency operating rooms located near the emergency department should have the same temperature, humidity, and ventilation requirements as those of operating rooms.

Anesthesia Storage Rooms. Anesthesia storage rooms must be ventilated in conformance with NFPA *Standard* 99. However, mechanical ventilation only is recommended.

Nursing

Patient Rooms. When central systems are used to air condition patients' rooms, the recommendations in Tables 1 and 3 for air filtration and air change rates should be followed to reduce cross-infection and to control odor. Rooms used for isolation of infected patients should have all air exhausted directly outdoors. A winter design temperature of 75°F with 30% rh is recommended; 75°F with 50% rh is recommended for summer. Each patient room should have individual temperature control. Air pressure in patient suites should be neutral in relation to other areas.

Most governmental design criteria and codes require that all air from toilet rooms be exhausted directly outdoors. The requirement appears to be based on odor control. Chaddock (1986) analyzed odor from central (patient) toilet exhaust systems of a hospital and found that large central exhaust systems generally have sufficient dilution to render the toilet exhaust practically odorless.

Where room unit systems are used, it is common practice to exhaust through the adjoining toilet room an amount of air equal to the amount of outdoor air brought into the room for ventilation. The ventilation of toilets, bedpan closets, bathrooms, and all interior rooms should conform to applicable codes.

Intensive Care Units. These units serve seriously ill patients, from postoperative to coronary patients. A variable range temperature capability of 75 to 80°F, a relative humidity of 30% minimum and 60% maximum, and positive air pressure are recommended.

Protective Isolation Units. Immunosuppressed patients (including bone marrow or organ transplant, leukemia, burn, and AIDS patients) are highly susceptible to diseases. Some physicians prefer an isolated laminar airflow unit to protect the patient; others are of the opinion that the conditions of the laminar cell have a psychologically harmful effect on the patient and prefer flushing out the room and reducing spores in the air. An air distribution of 15 air changes per hour supplied through a nonaspirating diffuser is often recommended. The sterile air is drawn across the patient and returned near the floor, at or near the door to the room.

In cases where the patient is immunosuppressed but not contagious, a positive pressure should be maintained between the patient room and adjacent area. Some jurisdictions may require an anteroom, which maintains a negative pressure relationship with respect to the adjacent isolation room and an equal pressure relationship with respect to the corridor, nurses' station, or common area. Exam and treatment rooms should be controlled in the same manner. A positive pressure should also be maintained between the entire unit and the adjacent areas to preserve sterile conditions.

When a patient is both immunosuppressed and contagious, isolation rooms within the unit may be designed and balanced to provide a permanent equal or negative pressure relationship with respect to the adjacent area or anteroom. Alternatively, when it is permitted by the jurisdictional authority, such isolation rooms may be equipped with controls that enable the room to be positive, equal, or negative in relation to the adjacent area. However, in such instances, controls in the adjacent area or anteroom must maintain the correct pressure relationship with respect to the other adjacent room(s).

A separate, dedicated air-handling system to serve the protective isolation unit simplifies pressure control and quality (Murray et al. 1988).

Infectious Isolation Unit. The infectious isolation room is used to protect the remainder of the hospital from the patients' infectious diseases. Recent multidrug-resistant strains of tuberculosis have increased the importance of pressurization, air change rates, filtration, and air distribution design in these rooms (Rousseau and Rhodes 1993). Temperatures and humidities should correspond to those specified for patient rooms.

The designer should work closely with health care planners and the code authority to determine the appropriate isolation room design. It may be desirable to provide more complete control, with a separate anteroom used as an air lock to minimize the potential that airborne particulates from the patients' area reach adjacent areas.

Switchable isolation rooms (rooms that can be set to function with either positive or negative pressure) have been installed in many facilities. AIA (1996) and CDC (1994) have, respectively, prohibited and recommend against this approach. The two difficulties associated with this approach are (1) maintaining the mechanical dampers and controls required to accurately provide the required pressures, and (2) that it provides a false sense of security on the part of staff who think that this provision is all that is required to change a room between protective isolation and infectious isolation, to the exclusion of other sanitizing procedures.

Floor Pantry. Ventilation requirements for this area depend on the type of food service adopted by the hospital. Where bulk food is dispensed and dishwashing facilities are provided in the pantry, the use of hoods above equipment, with exhaust to the outdoors, is recommended. Small pantries used for between-meal feedings require no special ventilation. The air pressure of the pantry should be in balance with that of adjoining areas to reduce the movement of air into or out of it.

Labor/Delivery/Recovery/Postpartum (LDRP). The procedures for normal childbirth are considered noninvasive, and rooms are controlled similarly to patient rooms. Some jurisdictions may require higher air change rates than in a typical patient room. It is expected that invasive procedures such as cesarean section are performed in a nearby delivery or operating room.

Ancillary

Radiology Department. Among the factors that affect the design of ventilation systems in these areas are the odorous characteristics of certain clinical treatments and the special construction designed to prevent radiation leakage. The fluoroscopic, radiographic, therapy, and darkroom areas require special attention.

Fluoroscopic, Radiographic, and Deep Therapy Rooms. These rooms require a temperature from 75 to 80°F and a relative humidity from 40 to 50%. Depending on the location of air supply outlets and exhaust intakes, lead lining may be required in supply and return ducts at the points of entry to the various clinical areas to prevent radiation leakage to other occupied areas.

The darkroom is normally in use for longer periods than the X-ray rooms, and it should have an independent system to exhaust the air to the outdoors. The exhaust from the film processor may be connected into the darkroom exhaust.

Laboratories. Air conditioning is necessary in laboratories for the comfort and safety of the technicians (Degenhardt and Pfost 1983). Chemical fumes, odors, vapors, heat from equipment, and the undesirability of open windows all contribute to this need.

Particular attention should be given to the size and type of equipment heat gain used in the various laboratories, as equipment heat gain usually constitutes the major portion of the cooling load.

The general air distribution and exhaust systems should be constructed of conventional materials following standard designs for the type of systems used. Exhaust systems serving hoods in which radioactive materials, volatile solvents, and strong oxidizing agents such as perchloric acid are used should be fabricated of stainless steel. Washdown facilities should be provided for hoods and ducts handling perchloric acid. Perchloric acid hoods should have dedicated exhaust fans.

Hood use may dictate other duct materials. Hoods in which radioactive or infectious materials are to be used must be equipped with ultrahigh efficiency filters at the exhaust outlet and have a procedure and equipment for the safe removal and replacement of contaminated filters. Exhaust duct routing should be as short as possible with a minimum of horizontal offsets. This applies especially to perchloric acid hoods because of the extremely hazardous, explosive nature of this material.

Determining the most effective, economical, and safe system of laboratory ventilation requires considerable study. Where the laboratory space ventilation air quantities approximate the air quantities required for ventilation of the hoods, the hood exhaust system may be used to exhaust all ventilation air from the laboratory areas. In situations where hood exhaust exceeds air supplied, a supplementary air supply may be used for hood makeup. The use of VAV supply/ exhaust systems in the laboratory has gained acceptance but requires special care in design and installation.

The supplementary air supply, which need not be completely conditioned, should be provided by a system that is independent of the normal ventilating system. The individual hood exhaust system should be interlocked with the supplementary air system. However, the hood exhaust system should not shut off if the supplementary air system fails. Chemical storage rooms must have a constantly operating exhaust air system with a terminal fan.

Exhaust fans serving hoods should be located at the discharge end of the duct system to prevent any possibility of exhaust products entering the building. For further information on laboratory air conditioning and hood exhaust systems, see Chapter 13; NFPA *Standard 99*; and Control of Hazardous Gases and Vapors in Selected Hospital Laboratories (Hagopian and Doyle 1984).

The exhaust air from the hoods in the biochemistry, histology, cytology, pathology, glass washing/sterilizing, and serology-bacteriology units should be discharged to the outdoors with no recirculation. Typically, exhaust fans discharge vertically at a minimum of 7 ft above the roof at velocities up to 4000 fpm. The serology-bacteriology unit should be pressurized relative to the adjoining areas to reduce the possibility of infiltration of aerosols that could contaminate the specimens being processed. The entire laboratory area should be under slight negative pressure to reduce the spread of odors or contamination to other hospital areas. Temperatures and humidities should be within the comfort range.

Bacteriology Laboratories. These units should not have undue air movement, so care should be exercised to limit air velocities to a minimum. The sterile transfer room, which may be within or adjoining the bacteriology laboratory, is a room where sterile media are distributed and where specimens are transferred to culture media. To maintain a sterile environment, an ultrahigh efficiency HEPA filter should be installed in the supply air duct near the point of entry to the room. The media room, essentially a kitchen, should be ventilated to remove odors and steam.

Infectious Disease and Virus Laboratories. These laboratories, found only in large hospitals, require special treatment. A minimum ventilation rate of 6 air changes per hour or makeup equal to hood exhaust volume is recommended for these laboratories, which should have a negative air pressure relative to any other area in the vicinity to prevent the exfiltration of any airborne contaminants. The exhaust air from fume hoods or safety cabinets must be sterilized before being exhausted to the outdoors. This may be accomplished by the use of electric or gas-fired heaters placed in series in the exhaust systems and designed to heat the exhaust air to 600°F. A more common and less expensive method of sterilizing the exhaust is to use HEPA filters in the system.

Nuclear Medicine Laboratories. Such laboratories administer radioisotopes to patients orally, intravenously, or by inhalation to facilitate diagnosis and treatment of disease. There is little opportunity in most cases for airborne contamination of the internal environment, but exceptions warrant special consideration.

One important exception involves the use of iodine 131 solution in capsules or vials to diagnose disorders of the thyroid gland. Another involves use of xenon 133 gas via inhalation to study patients with reduced lung function.

Capsules of iodine 131 occasionally leak part of their contents prior to use. Vials emit airborne contaminants when opened for preparation of a dose. It is common practice for vials to be opened and handled in a standard laboratory fume hood. A minimum face velocity of 100 fpm should be adequate for this purpose. This recommendation applies only where small quantities are handled in simple operations. Other circumstances may warrant provision of a glove box or similar confinement.

Use of xenon 133 for patient study involves a special instrument that permits the patient to inhale the gas and to exhale back into the instrument. The exhaled gas is passed through a charcoal trap mounted in lead and is often vented outdoors. The process suggests some potential for escape of the gas into the internal environment.

Due to the uniqueness of this operation and the specialized equipment involved, it is recommended that system designers determine the specific instrument to be used and contact the manufacturer for guidance. Other guidance is available in U.S. Nuclear Regulatory Commission Regulatory Guide 10.8 (NRC 1980). In particular, emergency procedures to be followed in case of accidental release of xenon 133 should include temporary evacuation of the area and/or increasing the ventilation rate of the area.

Recommendations concerning pressure relationships, supply air filtration, supply air volume, recirculation, and other attributes of supply and discharge systems for histology, pathology, and cytology laboratories are also relevant to nuclear medicine laboratories. There are, however, some special ventilation system requirements imposed by the NRC where radioactive materials are used. For example, NRC (1980) provides a computational procedure to estimate the airflow necessary to maintain xenon 133 gas concentration at or below specified levels. It also contains specific requirements as to the amount of radioactivity that may be vented to the atmosphere; the disposal method of choice is adsorption onto charcoal traps.

Autopsy Rooms. Susceptible to heavy bacterial contamination and odor, autopsy rooms, which are part of the hospital's pathology department, require special attention. Exhaust intakes should be located both at the ceiling and in the low sidewall. The exhaust system should discharge the air above the roof of the hospital. A negative air pressure relative to adjoining areas should be provided in the autopsy room to prevent the spread of contamination. Where large quantities of formaldehyde are used, special exhaust hoods may be needed to keep concentration below legal maximums.

In smaller hospitals where the autopsy room is used infrequently, local control of the ventilation system and an odor control system with either activated charcoal or potassium permanganate-impregnated activated alumina may be desirable.

Animal Quarters. Principally due to odor, animal quarters (found only in larger hospitals) require a mechanical exhaust system that discharges the contaminated air above the hospital roof. To prevent the spread of odor or other contaminants from the animal quarters to other areas, a negative air pressure of at least 0.1 in. of water relative to adjoining areas must be maintained. Chapter 13 has further information on animal room air conditioning.

Pharmacies. Local ventilation may be required for chemotherapy hoods and chemical storage. Room air distribution and filtration must be coordinated with any laminar airflow benches that may be needed. See Chapter 13, Laboratories, for more information.

Administration

This department includes the main lobby and admitting, medical records, and business offices. Admissions and waiting rooms are areas where there are potential risks of the transmission of undiagnosed airborne infectious diseases. The use of local exhaust systems that move air toward the admitting patient should be considered. A separate air-handling system is considered desirable to segregate this area from the hospital proper because it is usually unoccupied at night.

Diagnostic and Treatment

Bronchoscopy, Sputum Collection, and Pentamidine Administration Areas. These spaces are remarkable due to the high potential for large discharges of possibly infectious water droplet nuclei into the room air. Although the procedures performed may indicate the use of a patient hood, the general room ventilation should be increased under the assumption that higher than normal levels of airborne infectious contaminants will be generated.

Magnetic Resonance Imaging (MRI) Rooms. These rooms should be treated as exam rooms in terms of temperature, humidity, and ventilation. However, special attention is required in the control room due to the high heat release of computer equipment; in the exam room, due to the cryogens used to cool the magnet.

Treatment Rooms. Patients are brought to these rooms for special treatments that cannot be conveniently administered in the patients' rooms. To accommodate the patient, who may be brought from bed, the rooms should have individual temperature and humidity control. Temperatures and humidities should correspond to those specified for patients' rooms.

Physical Therapy Department. The cooling load of the electrotherapy section is affected by the shortwave diathermy, infrared, and ultraviolet equipment used in this area.

Hydrotherapy Section. This section, with its various water treatment baths, is generally maintained at temperatures up to 80°F. The potential latent heat buildup in this area should not be overlooked. The exercise section requires no special treatment, and temperatures and humidities should be within the comfort zone. The air may be recirculated within the areas, and an odor control system is suggested.

Occupational Therapy Department. In this department, spaces for activities such as weaving, braiding, artwork, and sewing require no special ventilation treatment. Recirculation of the air in these areas using medium-grade filters in the system is permissible.

Larger hospitals and those specializing in rehabilitation offer patients a greater diversity of skills to learn and craft activities, including carpentry, metalwork, plastics, photography, ceramics, and painting. The air-conditioning and ventilation requirements of the various sections should conform to normal practice for such areas and to the codes relating to them. Temperatures and humidities should be maintained within the comfort zone.

Inhalation Therapy Department. This department treats pulmonary and other respiratory disorders. The air must be very clean, and the area should have a positive air pressure relative to adjacent areas.

Workrooms. Clean workrooms serve as storage and distribution centers for clean supplies and should be maintained at a positive air pressure relative to the corridor.

Soiled workrooms serve primarily as collection points for soiled utensils and materials. They are considered contaminated rooms and should have a negative air pressure relative to adjoining areas. Temperatures and humidities should be within the comfort range.

Sterilizing and Supply

Used and contaminated utensils, instruments, and equipment are brought to this unit for cleaning and sterilization prior to reuse. The unit usually consists of a cleaning area, a sterilizing area, and a storage area where supplies are kept until requisitioned. If these areas are in one large room, air should flow from the clean storage and sterilizing areas toward the contaminated cleaning area. The air pressure relationships should conform to those indicated in Table 3. Temperature and humidity should be within the comfort range.

The following guidelines are important in the central sterilizing and supply unit:

1. Insulate sterilizers to reduce heat load.
2. Amply ventilate sterilizer equipment closets to remove excess heat.
3. Where ethylene oxide (ETO) gas sterilizers are used, provide a separate exhaust system with terminal fan (Samuals and Eastin 1980). Provide adequate exhaust capture velocity in the vicinity of sources of ETO leakage. Install an exhaust at sterilizer doors and over the sterilizer drain. Exhaust aerator and service rooms. ETO concentration sensors, exhaust flow sensors, and alarms should also be provided. ETO sterilizers should be located in dedicated unoccupied rooms that have a highly negative pressure relationship to adjacent spaces and 10 air changes per hour. Many jurisdictions require that ETO exhaust systems have equipment to remove ETO from exhaust air. See OSHA 29 CFR, Part 1910.
4. Maintain storage areas for sterile supplies at a relative humidity of no more than 50%.

Service

Service areas include dietary, housekeeping, mechanical, and employee facilities. Whether these areas are air conditioned or not, adequate ventilation is important to provide sanitation and a wholesome environment. Ventilation of these areas cannot be limited to exhaust systems only; provision for supply air must be incorporated into the design. Such air must be filtered and delivered at controlled temperatures. The best-designed exhaust system may prove ineffective without an adequate air supply. Experience has shown that reliance on open windows results only in dissatisfaction, particularly during the heating season. The use of air-to-air heat exchangers in the general ventilation system offers possibilities for economical operation in these areas.

Dietary Facilities. These areas usually include the main kitchen, bakery, dietitian's office, dishwashing room, and dining space. Because of the various conditions encountered (i.e., high heat and moisture production and cooking odors), special attention in design is needed to provide an acceptable environment. Refer to Chapter 30 for information on kitchen facilities.

The dietitian's office is often located within the main kitchen or immediately adjacent to it. It is usually completely enclosed to ensure privacy and noise reduction. Air conditioning is recommended for the maintenance of normal comfort conditions.

The dishwashing room should be enclosed and minimally ventilated to equal the dishwasher hood exhaust. It is not uncommon for the dishwashing area to be divided into a soiled area and a clean area. In such cases, the soiled area should be kept at a negative pressure relative to the clean area.

Ventilation of the dining space should conform to local codes. The reuse of dining space air for ventilation and cooling of food preparation areas in the hospital is suggested, provided the reused air is passed through 80% efficient filters. Where cafeteria service is provided, serving areas and steam tables are usually hooded. The air-handling capacities of these hoods should be at least 75 cfm per square foot of perimeter area.

Kitchen Compressor/Condenser Spaces. Ventilation of these spaces should conform to all codes, with the following additional considerations: (1) 350 cfm of ventilating air per compressor horsepower should be used for units located within the kitchen; (2) condensing units should operate optimally at 90°F maximum ambient temperature; and (3) where air temperature or air circulation is marginal, combination air- and water-cooled condensing units should be specified. It is often worthwhile to use condenser water coolers or remote condensers. Heat recovery from water-cooled condensers should be considered.

Laundry and Linen Facilities. Of these facilities, only the soiled linen storage room, the soiled linen sorting room, the soiled utility room, and the laundry processing area require special attention.

The room provided for storage of soiled linen prior to pickup by commercial laundry is odorous and contaminated and should be well ventilated and maintained at a negative air pressure.

The soiled utility room is provided for inpatient services and is normally contaminated with noxious odors. This room should be exhausted directly outside by mechanical means.

In the laundry processing area, equipment such as washers, flatwork ironers, and tumblers should have direct overhead exhaust to reduce humidity. Such equipment should be insulated or shielded whenever possible to reduce the high radiant heat effects. A canopy over the flatwork ironer and exhaust air outlets near other heat-producing equipment capture and remove heat best. The air supply inlets should be located to move air through the processing area toward the heat-producing equipment. The exhaust system from flatwork ironers and tumblers should be independent of the general exhaust system and equipped with lint filters. Air should exhaust above the roof or where it will not be obnoxious to occupants of other areas. Heat reclamation from the laundry exhaust air may be desirable and practicable.

Where air conditioning is contemplated, a separate supplementary air supply, similar to that recommended for kitchen hoods, may be located in the vicinity of the exhaust canopy over the ironer. Alternatively, spot cooling for the relief of personnel confined to specific areas may be considered.

Mechanical Facilities. The air supply to boiler rooms should provide both comfortable working conditions and the air quantities required for maximum rates of combustion of the particular fuel used. Boiler and burner ratings establish maximum combustion rates, so the air quantities can be computed according to the type of fuel. Sufficient air must be supplied to the boiler room to supply the exhaust fans as well as the boilers.

At workstations, the ventilation system should limit temperatures to 90°F effective temperature. When ambient outside air temperature is higher, indoor temperature may be that of the outside air up to a maximum of 97°F to protect motors from excessive heat.

Maintenance Shops. Carpentry, machine, electrical, and plumbing shops present no unusual ventilation requirements. Proper ventilation of paint shops and paint storage areas is important because of fire hazard and should conform to all applicable codes. Maintenance shops where welding occurs should have exhaust ventilation.

CONTINUITY OF SERVICE AND ENERGY CONCEPTS

Zoning

Zoning—using separate air systems for different departments—may be indicated to (1) compensate for exposures due to orientation

or for other conditions imposed by a particular building configuration, (2) minimize recirculation between departments, (3) provide flexibility of operation, (4) simplify provisions for operation on emergency power, and (5) conserve energy.

By ducting the air supply from several air-handling units into a manifold, central systems can achieve a measure of standby capacity. When one unit is shut down, air is diverted from noncritical or intermittently operated areas to accommodate critical areas, which must operate continuously. This or other means of standby protection is essential if the air supply is not to be interrupted by routine maintenance or component failure.

Separation of supply, return, and exhaust systems by department is often desirable, particularly for surgical, obstetrical, pathological, and laboratory departments. The desired relative balance within critical areas should be maintained by interlocking the supply and exhaust fans. Thus, exhaust should cease when the supply airflow is stopped in areas otherwise maintained at positive or neutral pressure relative to adjacent spaces. Likewise, the supply air should be deactivated when exhaust airflow is stopped in spaces maintained at a negative pressure.

Heating and Hot Water Standby Service

The number and arrangement of boilers should be such that when one boiler breaks down or is temporarily taken out of service for routine maintenance, the capacity of the remaining boilers is sufficient to provide hot water service for clinical, dietary, and patient use; steam for sterilization and dietary purposes; and heating for operating, delivery, birthing, labor, recovery, intensive care, nursery, and general patient rooms. However, reserve capacity is not required in climates where a design dry-bulb temperature of 25°F is equaled or exceeded for 99.6% of the total hours in any one heating period as noted in the tables in Chapter 26 of the 1997 *ASHRAE Handbook—Fundamentals.*

Boiler feed pumps, heat circulation pumps, condensate return pumps, and fuel oil pumps should be connected and installed to provide both normal and standby service. Supply and return mains and risers for cooling, heating, and process steam systems should be valved to isolate the various sections. Each piece of equipment should be valved at the supply and return ends.

Some supply and exhaust systems for delivery and operating room suites should be designed to be independent of other fan systems and to operate from the hospital emergency power system in the event of power failure. The operating and delivery room suites should be ventilated such that the hospital facility retains some surgical and delivery capability in cases of ventilating system failure.

Boiler steam is often treated with chemicals that cannot be released in the air-handling units serving critical areas. In this case, a clean steam system should be considered for humidification.

Mechanical Cooling

The source of mechanical cooling for clinical and patient areas in a hospital should be carefully considered. The preferred method is to use an indirect refrigerating system using chilled water or antifreeze solutions. When using direct refrigerating systems, consult codes for specific limitations and prohibitions. Refer to ASHRAE *Standard* 15, Safety Code for Mechanical Refrigeration.

Insulation

All exposed hot piping, ducts, and equipment should be insulated to maintain the energy efficiency of all systems and protect building occupants. To prevent condensation, ducts, casings, piping, and equipment with outside surface temperature below ambient dew point should be covered with insulation having an external vapor barrier. Insulation, including finishes and adhesives on the exterior surfaces of ducts, pipes, and equipment, should have a flame spread rating of 25 or less and a smoke-developed rating of 50 or less, as determined by an independent testing laboratory in accordance with NFPA *Standard* 255, as required by NFPA 90A. The smoke-developed rating for pipe insulation should not exceed 150 (DHHS 1984a).

Linings in air ducts and equipment should meet the erosion test method described in Underwriters Laboratories *Standard* 181. These linings, including coatings, adhesives, and insulation on exterior surfaces of pipes and ducts in building spaces used as air supply plenums, should have a flame spread rating of 25 or less and a smoke developed rating of 50 or less, as determined by an independent testing laboratory in accordance with ASTM *Standard* E 84.

Duct linings should not be used in systems supplying operating rooms, delivery rooms, recovery rooms, nurseries, burn care units, or intensive care units, unless terminal filters of at least 90% efficiency are installed downstream of linings. Duct lining should be used only for acoustical improvement; for thermal purposes, external insulation should be used.

When existing systems are modified, asbestos materials should be handled and disposed of in accordance with applicable regulations.

Energy

Health care is an energy-intensive, energy-dependent enterprise. Hospital facilities are different from other structures in that they operate 24 h a day year-round, require sophisticated backup systems in case of utility shutdowns, use large quantities of outside air to combat odors and to dilute microorganisms, and must deal with problems of infection and solid waste disposal. Similarly, large quantities of energy are required to power diagnostic, therapeutic, and monitoring equipment; and support services such as food storage, preparation, and service and laundry facilities.

Hospitals conserve energy in various ways, such as by using larger energy storage tanks and by using energy conversion devices that transfer energy from hot or cold building exhaust air to heat or cool incoming air. Heat pipes, runaround loops, and other forms of heat recovery are receiving increased attention. Solid waste incinerators, which generate exhaust heat to develop steam for laundries and hot water for patient care, are becoming increasingly common. Large health care campuses use central plant systems, which may include thermal storage, hydronic economizers, primary/secondary pumping, cogeneration, heat recovery boilers, and heat recovery incinerators.

The construction design of new facilities, including alterations of and additions to existing buildings, has a major influence on the amount of energy required to provide such services as heating, cooling, and lighting. The selection of building and system components for effective energy use requires careful planning and design. Integration of building waste heat into systems and use of renewable energy sources (e.g., solar under some climatic conditions) will provide substantial savings (Setty 1976).

OUTPATIENT HEALTH CARE FACILITIES

An outpatient health care facility may be a free-standing unit, part of an acute care facility, or part of a medical facility such as a medical office building (clinic). Any surgery is performed without anticipation of overnight stay by patients (i.e., the facility operates 8 to 10 h per day).

If physically connected to a hospital and served by the hospital's HVAC systems, spaces within the outpatient health care facility should conform to requirements in the section on Hospital Facilities. Outpatient health care facilities that are totally detached and have their own HVAC systems may be categorized as diagnostic clinics, treatment clinics, or both.

Table 4 Filter Efficiencies for Central Ventilation and Air-Conditioning Systems in Nursing Homes[a]

Area Designation	Minimum Number of Filter Beds	Filter Efficiency of Main Filter Bed, %
Patient care, treatment, diagnostic, and related areas	1	80
Food preparation areas and laundries	1	80
Administrative, bulk storage, and soiled holding areas	1	30

[a]Ratings based on ASHRAE *Standard* 52.1-92.

DIAGNOSTIC CLINICS

A diagnostic clinic is a facility where patients are regularly seen on an ambulatory basis for diagnostic services or minor treatment, but where major treatment requiring general anesthesia or surgery is not performed. Diagnostic clinic facilities have design criteria as shown in Tables 4 and 5 (see the section on Nursing Home Facilities).

TREATMENT CLINICS

A treatment clinic is a facility where major or minor procedures are performed on an outpatient basis. These procedures may render patients incapable of taking action for self-preservation under emergency conditions without assistance from others (NFPA *Standard* 101).

Design Criteria

The system designer should refer to the following paragraphs from the section on Hospital Facilities:

- Infection Sources and Control Measures
- Air Quality
- Air Movement
- Temperature and Humidity
- Pressure Relationships and Ventilation
- Smoke Control

Air-cleaning requirements correspond to those in Table 1 for operating rooms. A recovery area need not be considered a sensitive area. Infection control concerns are the same as in an acute care hospital. The minimum ventilation rates, desired pressure relationships, desired relative humidity, and design temperature ranges are similar to the requirements for hospitals shown in Table 3 except for operating rooms, which may meet the criteria for trauma rooms.

The following departments in a treatment clinic have design criteria similar to those in hospitals:

- Surgical—operating rooms, recovery rooms, and anesthesia storage rooms
- Ancillary
- Diagnostic and Treatment
- Sterilizing and Supply
- Service—soiled workrooms, mechanical facilities, and locker rooms

Continuity of Service and Energy Concepts

Some owners may desire that the heating, air-conditioning, and service hot water systems have standby or emergency service capability and that these systems be able to function after a natural disaster.

To reduce utility costs, facilities should include energy-conserving measures such as recovery devices, variable air volume, load shedding, or devices to shut down or reduce the ventilation of certain areas when unoccupied. Mechanical ventilation should take advantage of outside air by using an economizer cycle, when appropriate, to reduce heating and cooling loads.

Table 5   Pressure Relationships and Ventilation of Certain Areas of Nursing Homes

Function Area	Pressure Relationship to Adjacent Areas	Minimum Air Changes of Outdoor Air per Hour Supplied to Room	Minimum Total Air Changes per Hour Supplied to Room	All Air Exhausted Directly to Outdoors	Air Recirculated Within Room Units
PATIENT CARE					
Patient room	±	2	2	Optional	Optional
Patient area corridor	±	Optional	2	Optional	Optional
Toilet room	N	Optional	10	Yes	No
DIAGNOSTIC AND TREATMENT					
Examination room	±	2	6	Optional	Optional
Physical therapy	N	2	6	Optional	Optional
Occupational therapy	N	2	6	Optional	Optional
Soiled workroom or soiled holding	N	2	10	Yes	No
Clean workroom or clean holding	P	2	4	Optional	Optional
STERILIZING AND SUPPLY					
Sterilizer exhaust room	N	Optional	10	Yes	No
Linen and trash chute room	N	Optional	10	Yes	No
Laundry, general	±	2	10	Yes	No
Soiled linen sorting and storage	N	Optional	10	Yes	No
Clean linen storage	P	Optional	2	Yes	No
SERVICE					
Food preparation center	±	2	10	Yes	Yes
Warewashing room	N	Optional	10	Yes	Yes
Dietary day storage	±	Optional	2	Yes	No
Janitor closet	N	Optional	10	Yes	No
Bathroom	N	Optional	10	Yes	No

P = Positive   N = Negative   ± = Continuous directional control not required

The subsection on Continuity of Service and Energy Concepts in the section on Hospital Facilities includes information on zoning and insulation that applies to outpatient facilities as well.

NURSING HOME FACILITIES

Nursing homes may be classified as follows:

Extended care facilities are for the recuperation of hospital patients who no longer require hospital facilities but do require the therapeutic and rehabilitative services of skilled nurses. This type of facility is either a direct hospital adjunct or a separate facility having close ties with the hospital. Clientele may be of any age, usually stay from 35 to 40 days, and usually have only one diagnostic problem.

Skilled nursing homes are for the care of people who require assistance in daily activities; many of them are incontinent and non-ambulatory, and some are disoriented. Clientele come directly from home or residential care homes, are generally elderly (with an average age of 80), stay an average of 47 months, and frequently have multiple diagnostic problems.

Residential care homes are generally for elderly people who are unable to cope with regular housekeeping chores but have no acute ailments and are able to care for all their personal needs, lead normal lives, and move freely in and out of the home and the community. These homes may or may not offer skilled nursing care. The average length of stay is four years or more.

Functionally, these buildings have five types of areas that are of concern to the HVAC designer: (1) administrative and supportive areas, inhabited by the staff, (2) patient areas that provide direct normal daily services, (3) treatment areas that provide special medical services, (4) clean workrooms for storage and distribution of clean supplies, and (5) soiled workrooms for collection of soiled and contaminated supplies and for sanitization of nonlaundry items.

DESIGN CONCEPTS AND CRITERIA

Controlling bacteria levels in nursing homes is not as critical as it is in acute care hospitals. Nevertheless, the designer should be aware of the necessity for odor control, filtration, and airflow control between certain areas.

Table 4 lists recommended filter efficiencies for air systems serving specific nursing home areas. Table 5 lists recommended minimum ventilation rates and desired pressure relationships for certain areas in nursing homes.

Recommended interior winter design temperature is 75°F for areas occupied by patients and 70°F for nonpatient areas. Provisions for maintenance of minimum humidity levels in winter depend on the severity of the climate and are best left to the judgment of the designer. Where air conditioning is provided, the recommended interior summer design temperature and humidity is 75°F and 50% rh.

The general design criteria in the sections on Heating and Hot Water Standby Service, Insulation, and Energy for hospital facilities apply to nursing home facilities as well.

APPLICABILITY OF SYSTEMS

Nursing homes occupants are usually frail, and many are incontinent. Though some occupants are ambulatory, others are bedridden, suffering from the advanced stages of illnesses. The selected HVAC system must dilute and control odors and should not cause drafts. Local climatic conditions, costs, and designer judgment determine the extent and degree of air conditioning and humidification. Odor may be controlled with large volumes of outside air and some form of heat recovery. To conserve energy, odor may be controlled with activated carbon or potassium permanganate-impregnated activated alumina filters instead.

Temperature control should be on an individual room basis. In geographical areas with severe climates, patients' rooms should have supplementary heat along exposed walls. In moderate climates, i.e., where outside winter design conditions are 30°F or above, heating from overhead may be used.

DENTAL CARE FACILITIES

Institutional dental facilities include reception and waiting areas, treatment rooms (called operatories), and workrooms where supplies are stored and instruments are cleaned and sterilized; they may include laboratories where restorations are fabricated or repaired.

Many common dental procedures generate aerosols, dusts, and particulates (Ninomura and Byrns 1998). The aerosols/dusts may contain microorganisms (both pathogenic and nonpathogenic), metals (such as mercury fumes), and other substances (e.g., silicone dusts, latex allergens, etc.). Some measurements indicate that levels of bioaerosols during and immediately following a procedure can be extremely high (Earnest and Loesche 1991). Lab procedures have been shown to generate dusts and aerosols containing metals. At this time, only limited information and research is available regarding the level, nature, or persistence of bioaerosol and particulate contamination in dental facilities.

Nitrous oxide is used as an analgesic/anesthetic gas in many facilities. The design for the control of nitrous oxide should consider (1) that nitrous oxide is heavier than air and may accumulate near the floor if air mixing is inefficient, and (2) that nitrous oxide be exhausted directly outside. NIOSH (1996) includes recommendations for the ventilation/exhaust system.

REFERENCES

AIA. 1996. Guidelines for design and construction of hospital and health care facilities. The American Institute of Architects, Washington, D.C.

ASHRAE. 1989. Ventilation for acceptable indoor air quality. ANSI/ASHRAE *Standard* 62-1989.

ASHRAE. 1992. Gravimetric and dust-spot procedures for testing air-cleaning devices used in general ventilation for removing particulate matter. ANSI/ASHRAE *Standard* 52.1-1992.

ASHRAE. 1994. Safety code for mechanical refrigeration. ANSI/ASHRAE *Standard* 15-1994.

ASTM. 1998. Standard test method for surface burning characteristics of building materials. ANSI/ASTM *Standard* E 84. American Society for Testing and Materials, West Conshohocken, PA.

Burch, G.E. and N.P. Pasquale. 1962. *Hot climates, man and his heart.* C.C. Thomas, Springfield, IL.

CDC. 1994. Guidelines for preventing the transmission of *Mycobacterium tuberculosis* in health-care facilities, 1994. U.S. Dept. of Health and Human Services, Public Health Service, Centers for Disease Control and Prevention, Atlanta.

Chaddock, J.B. 1986. Ventilation and exhaust requirements for hospitals. *ASHRAE Transactions* 92(2A):350-95.

Degenhardt, R.A. and J.F. Pfost. 1983. Fume hood design and application for medical facilities. *ASHRAE Transactions* 89(2B):558-70.

Demling, R.H. and J. Maly. 1989. The treatment of burn patients in a laminar flow environment. Annals of the New York Academy of Sciences 353: 294-259.

DHHS. 1984. Guidelines for construction and equipment of hospital and medical facilities. *Publication* No. HRS-M-HF, 84-1. United States Department of Health and Human Services, Washington, D.C.

Earnest, R. and W. Loesche. 1991. Measuring harmful levels of bacteria in dental aerosols. *The Journal of the American Dental Association.* 122:55-57.

Fitzgerald, R.H. 1989. Reduction of deep sepsis following total hip arthroplasty. *Annals of the New York Academy of Sciences* 353:262-69.

Greene, V.W., R.G. Bond, and M.S. Michaelsen. 1960. Air handling systems must be planned to reduce the spread of infection. *Modern Hospital* (August).

Hagopian, J.H. and E.R. Hoyle. 1984. Control of hazardous gases and vapors in selected hospital laboratories. *ASHRAE Transactions* 90(2A):341-53.

Isoard, P., L. Giacomoni, and M. Payronnet. 1980. Proceedings of the 5th International Symposium on Contamination Control, Munich (September).

Lewis, J.R. 1988. Application of VAV, DDC, and smoke management to hospital nursing wards. *ASHRAE Transactions* 94(1):1193-1208.

Luciano, J.R. 1984. New concept in French hospital operating room HVAC systems. *ASHRAE Journal* 26(2):30-34.

Michaelson, G.S., D. Vesley, and M.M. Halbert. 1966. The laminar air flow concept for the care of low resistance hospital patients. Paper presented at the annual meeting of American Public Health Association, San Francisco (November).

Murray, W.A., A.J. Streifel, T.J. O'Dea, and F.S. Rhame. 1988. Ventilation protection of immune compromised patients. *ASHRAE Transactions* 94(1):1185-92.

NFPA. 1996. Standard method of test of surface burning characteristics of building materials. ANSI/NFPA *Standard* 255-96. National Fire Protection Agency, Quincy, MA.

NFPA. 1996. Standard for health care facilities. ANSI/NFPA *Standard* 99-96.

NFPA. 1996. Standard for the installation of air conditioning and ventilation systems. ANSI/NFPA *Standard* 90A-96.

NFPA. 1996. Recommended practice for smoke-control systems. ANSI/NFPA *Standard* 92A-96.

NFPA. 1997. Life safety code. ANSI/NFPA *Code* 101-97.

Ninomura, P.T. and G. Byrns. 1998. Dental ventilation theory and applications. *ASHRAE Journal* 40(2):48-32.

NIOSH. 1975. Elimination of waste anesthetic gases and vapors in hospitals, *Publication* No. NIOSH 75-137 (May). United States Department of Health, Education, and Welfare, Washington, D.C.

NIOSH. 1996. Controls of nitrous oxide in dental operatories. *Publication* No. NIOSH 96-107 (January). National Institute for Occupational Safety and Health, Cincinnati, OH.

NRC. 1980. *Regulatory Guide* 10.8. Nuclear Regulatory Commission.

OSHA. Occupational exposure to ethylene oxide. OSHA 29 CFR, Part 1910. United States Department of Labor, Washington, D.C.

Pfost, J.F. 1981. A re-evaluation of laminar air flow in hospital operating rooms. *ASHRAE Transactions* 87(1):729-39.

Rousseau, C.P. and W.W. Rhodes. 1993. HVAC system provisions to minimize the spread of tuberculosis bacteria. *ASHRAE Transactions* 99(2):1201-04.

Samuals, T.M. and M. Eastin. 1980. ETO exposure can be reduced by air systems. *Hospitals* (July).

Setty, B.V.G. 1976. Solar heat pump integrated heat recovery. *Heating, Piping and Air Conditioning* (July).

UL. 1996. Factory-made air ducts and connectors, 9th ed. *Standard* 181. Underwriters Laboratories, Northbrook, IL.

Walker, J.E.C. and R.E. Wells. 1961. Heat and water exchange in the respiratory tract. *American Journal of Medicine* (February):259.

Wells, W.F. 1934. On airborne infection. Study II: Droplets and droplet nuclei. *American Journal of Hygiene* 20:611.

Woods, J.E., D.T. Braymen, R.W. Rasussen, G.L. Reynolds, and G.M. Montag. 1986. Ventilation requirement in hospital operating rooms—Part I: Control of airborne particles. *ASHRAE Transactions* 92(2A): 396-426.

BIBLIOGRAPHY

DHHS. 1984. Energy considerations for hospital construction and equipment. *Publication* No. HRS-M-HF, 84-1A. United States Department of Health and Human Services, Washington, D.C.

Gustofson, T.L. et al. 1982. An outbreak of airborne nosocomial Varicella. *Pediatrics* 70(4):550-56.

Rhodes, W.W. 1988. Control of microbioaerosol contamination in critical areas in the hospital environment. *ASHRAE Transactions* 94(1):1171-84.

CHAPTER 8

SURFACE TRANSPORTATION

AUTOMOBILE AIR CONDITIONING

ENVIRONMENTAL control in modern automobiles consists of one or more of the following systems: (1) heater-defroster, (2) ventilation, and (3) cooling and dehumidifying (air-conditioning). All passenger cars sold in the United States must meet federal defroster requirements, so ventilation systems and heaters are included in the basic vehicle design. The integration of the heater-defroster and ventilation systems is common. Air conditioning remains an extra-cost option on many vehicles.

Heating

Outdoor air passes through a heater core, using engine coolant as a heat source. To avoid visibility-reducing condensation on the glass due to raised air dew point from occupant respiration and interior moisture gains, interior air should not recirculate through the heater.

Temperature control is achieved by either water flow regulation or heater air bypass and subsequent mixing. A combination of ram effect from forward movement of the car and the electrically driven blower provides the airflow.

Heater air is generally distributed into the lower forward compartment, under the front seat, and up into the rear compartment. Heater air exhausts through body leakage points. At higher vehicle speeds, the increased heater air quantity (ram assist through the ventilation system) partly compensates for the infiltration increase. Air exhausters are sometimes installed to increase airflow and reduce the noise of air escaping from the car.

The heater air distribution system is usually adjustable between the diffusers along the floor and on the dashboard. Supplementary ducts are sometimes required when consoles, panel-mounted air conditioners, or rear seat heaters are installed. Supplementary heaters are frequently available for third-seat passengers in station wagons and for the rear seats in limousines and luxury sedans.

Defrosting

Some heated outdoor air is ducted from the heater core to defroster outlets at the base of the windshield. This air absorbs moisture from the interior surface of the windshield and raises the glass temperature above the interior dew point. Induced outdoor air has a lower dew point than the air inside the vehicle, which absorbs moisture from the occupants and car interior. Heated air provides the energy necessary to melt or sublime ice and snow from the glass exterior. The defroster air distribution pattern on the windshield is developed by test for conformity with federal standards, satisfactory distribution, and rapid defrost.

Most automobiles operate the air-conditioning compressor to dry the induced outdoor air and/or to prevent a wet evaporator from increasing the dew point when the compressor is disengaged. Some vehicles are equipped with side window demisters that direct a small amount of heated air and/or air with lowered dew point to the front side windows. Rear windows are defrosted primarily by heating wires embedded in the glass.

Ventilation

Fresh air is introduced either by (1) ram air or (2) forced air. In both systems, air enters the vehicle through a screened opening in the cowl just forward of the base of the windshield. The cowl plenum is usually an integral part of the vehicle structure. Air entering this plenum can also supply the heater and evaporator cores.

In the ram air system, ventilation air flows back and up toward the front seat occupants' laps and then over the remainder of their bodies. Additional ventilation occurs by turbulence and air exchange through open windows. Directional control of ventilation air is frequently unavailable. Airflow rate varies with relative wind-vehicle velocity but may be adjusted with windows or vents.

Forced air ventilation is available in many automobiles. The cowl inlet plenum and heater/air-conditioning blower are used together with instrument panel outlets for directional control. Positive air pressure from the ventilation fan or blower helps reduce the amount of exterior pollutants entering the passenger compartment. In air-conditioned vehicles, the forced air ventilation system uses the air-conditioning outlets. Body air exhausts and vent windows exhaust air from the vehicle. With the increased popularity of air conditioning and forced ventilation, most late model vehicles are not equipped with vent windows.

Air Conditioning

Air conditioners are installed either with a combination evaporator-heater or as an add-on system. The combination evaporator-heater in conjunction with the ventilation system is the prevalent type of factory-installed air conditioning. This system is popular because (1) it permits dual use of components such as blower motors, outdoor air ducts, and structure; (2) it permits compromise standards where space considerations dictate (ventilation reduction on air-conditioned cars); (3) it generally reduces the number and complexity of driver controls; and (4) it typically features capacity control innovations such as automatic reheat.

Outlets in the instrument panel distribute air to the car interior. These are individually adjustable, and some have individual shut-offs. The dashboard end outlets are for the driver and front seat passenger; center outlets are primarily for rear seat passengers.

The dealer-installed add-on air conditioner is normally available only as a service or after-market installation. In recent designs, the air outlets, blower, and controls built into the automobile are used. Evaporator cases are styled to look like factory-installed units. These units are integrated with the heater as much as possible to provide outdoor air and to take advantage of existing air-mixing

---

The preparation of this chapter is assigned to TC 9.3, Transportation Air Conditioning.

dampers. Where it is not possible to use existing air ducts, custom ducts distribute the air in a manner similar to those in factory-installed units. On occasion, rather than installing a second duct, the dealer will create a slot in the existing duct to accept an evaporator.

As most of the air for the rear seat occupants flows through the center outlets of the front evaporator unit, a passenger in the center of the front seat impairs rear seat cooling. Supplemental trunk units for luxury sedans and roof units for station wagons improve the cooling of rear seat passengers.

The trunk unit is a blower-evaporator, complete with expansion valve, installed in the trunk of limousines and premium line vehicles. The unit uses the same high-side components as the front evaporator unit. Most auxiliary or rear units recirculate air from the passenger compartment.

DESIGN FACTORS

General considerations for design include ambient temperatures and contaminants, vehicle and engine concessions, flexibility, physical parameters, durability, electrical power, refrigeration capacity, occupants, infiltration, insulation, solar effect, and noise.

Ambient and Vehicle Criteria

Ambient Temperature. Heaters are evaluated for performance at temperatures from −40 to 70°F. Air-conditioning systems with reheat are evaluated from 40 to 110°F; add-on units are evaluated from 50 to 100°F, although ambient temperatures above 125°F are occasionally encountered. Because the system is an integral part of the vehicle detail, the effects of vehicle heat and local heating must be considered.

Ambient Contaminants. Airborne bacteria, pollutants, and corrosion agents must also be considered when selecting materials for seals and heat exchangers. Electronic air cleaners and filters have been installed in premium automobiles.

Vehicle Performance. Proper engine coolant temperature, freedom from gasoline vapor lock, adequate electrical charging system, acceptable vehicle ride, minimum surge due to compressor clutch cycling, and good handling must be maintained.

Flexibility. Engine coolant pressure at the heater core inlet ranges up to 40 psig in cars and 55 psig on trucks. The engine coolant thermostat remains closed until coolant temperature reaches 160 to 205°F. Coolant flow is a function of pressure differential and system restriction but ranges from 0.6 gpm at idle to 10 gpm at 60 mph (lower for water valve regulated systems because of the added restriction).

Present-day antifreeze coolant solutions have specific heats from 0.65 to 1.0 Btu/lb·°F and boiling points from 250 to 272°F (depending on concentration) when a 15 psi radiator pressure cap is used.

Multiple-speed (usually four-speed) blowers supplement the ram air effect through the ventilation system and produce the necessary velocities for distribution. Heater air quantities range from 125 to 190 cfm. Defroster air quantities range from 90 to 145 cfm. Other considerations are (1) compressor/engine speed, from 500 to 5500 rpm; (2) drive ratio, from 0.89:1 to 1.41:1; (3) condenser air, from 50 to 125°F and from 325 to 3000 cfm (corrected for restriction and distribution factors); and (4) evaporator air, from 100 to 400 cfm (limits established by design but selective at operator's discretion) and from 35 to 150°F.

Physical Parameters

Parameters include engine rock, proximity to adverse environments, and durability.

Engine Rock. Relative to the rest of the car, the engine moves both fore and aft because of inertia and in rotation because of torque. Fore and aft movement may be as much as 0.25 in.; rotational movements at the compressor may be as much as 0.75 in. due to acceleration and 0.5 in. due to deceleration.

Proximity to Adverse Environments. Wiring, refrigerant lines, hoses, vacuum lines, and so forth, must be protected from exhaust manifold heat and sharp edges of sheet metal. Normal service items such as oil filler caps, power steering filler caps, and transmission dipsticks must be accessible. Air-conditioning components should not have to be removed for servicing other components.

Durability. Hours of operation are short compared to commercial systems (160,000 miles ÷ 40 mph = 4000 h), but the shock and vibration the vehicle receives or produces must not cause a malfunction or failure.

Equipment is designed to meet the recommendations of SAE *Recommended Practice* J 639, which states that the burst strength of those components subjected to high-side refrigerant pressure must be at least 2.5 times the venting pressure (or pressure equivalent to venting temperature) of the relief device. SAE J 639 also requires electrical cut-out of the clutch coil prior to pressure relief to prevent unnecessary refrigerant discharge. Components for the low-pressure side frequently have burst strengths in excess of 300 psi. The relief device should be located as close as possible to the discharge gas side of the compressor, preferably in the compressor itself.

Power and Capacity

Fan size is kept to a minimum not only to save space and reduce weight, but also to reduce energy consumption. If a standard vehicle has a heater that draws 10 A, and the air conditioner requires 20 A, the alternator and wiring must be redesigned to supply this additional 10 A, with obvious cost and weight penalties.

The refrigeration capacity must be adequate to reduce the vehicle interior to a comfortable temperature quickly and then to maintain the selected temperature at reasonable humidity during all operating conditions and environments. A design may be established by mathematical modeling or empirical evaluation of all the known and predicted factors. A design tradeoff in capacity is sought relative to the vehicle's weight, component size, and fuel economy needs.

Occupancy per unit volume is high in automotive applications. The air conditioner (and auxiliary evaporators and systems) must be matched to the intended vehicle occupancy.

Other Concerns

Infiltration varies with relative wind-vehicle velocity. It also varies with assembly quality. Body sealing is part of air-conditioning design for automobiles. Occasionally, sealing beyond that required for dust, noise, and draft control is required.

Due to cost, insulation is seldom added to reduce thermal load. Insulation for sound control is generally considered adequate. Roof insulation is of questionable benefit, as it retards heat loss during the nonoperating, or soak, periods. Additional dashboard and floor insulation helps reduce cooling load. Typical maximum ambient temperatures are 200°F above mufflers and catalytic converters, 120°F for other floor areas, 145°F for dash and toe board, and 110°F for sides and top. The following three solar effects add to the cooling load:

Vertical. Maximum intensity occurs at or near noon. Solar heat gain through all glass surface area normal to the incident light is a substantial fraction of the cooling load.

Horizontal and Reflected Radiation. The intensity is significantly less, but the glass area is large enough to merit consideration.

Surface Heating. The temperature of the surface is a function of the solar energy absorbed, the interior and ambient temperatures, and the automobile's velocity.

The temperature control should not produce objectionable sounds. During maximum heating or cooling operation, a slightly higher noise level is acceptable. Thereafter, it should be possible to maintain comfort at a lower blower speed with an acceptable noise level. Compressor-induced vibrations, gas pulsations, and noise must be kept to a minimum. Suction and discharge mufflers are often used to reduce noise. Belt-induced noises, engine torsional vibration, and compressor mounting all require particular attention.

COMPONENTS

Compressors

Scroll and piston compressors are used in mobile applications. They have the following characteristics:

Displacement. Fixed displacement compressors have displacements of 6.1 to 12.6 in^3/rev. Variable displacement compressors have a minimum displacement of about 1 in^3/rev, about 10% of their maximum displacement. A typical variable capacity scroll compressor has a maximum displacement of 7.3 in^3/rev and a minimum displacement of just 3% of the maximum.

Physical Size. Fuel economy, lower hood lines, and more engine accessories all decrease compressor installation space. These features, along with the fact that smaller engines have less accessory power available, promote the use of smaller compressors.

Speed Range. Because compressors are belt driven directly from the engine, they must withstand speeds of over 6000 rpm and remain smooth and quiet down to 500 rpm. In the absence of a variable drive ratio, there are instances when the maximum compressor speed may need to be higher in order to achieve sufficient pumping capacity at idle.

Torque Requirements. Because torque pulsations cause or aggravate vibration problems, it is best to minimize them. Minimizing peak torque benefits the compressor drive and mount systems. Multicylinder reciprocating and rotary compressors aid in reducing vibration. An economical single-cylinder compressor reduces cost; however, any design must reduce peak torques and belt loads, which would normally be at a maximum in a single-cylinder design.

Compressor Drives. A magnetic clutch, energized by power from the car engine electrical system, drives the compressor. The clutch is always disengaged when air conditioning is not required. The clutch can also be used to control evaporator temperature (see the section on Controls).

Variable Displacement Compressors. Wobble plate compressors are used for automobile air conditioning. The angle of the wobble plate changes in response to the evaporator and discharge pressure to achieve a constant suction pressure just above freezing, regardless of load. A bellows valve or electronic sensor-controlled valve routes internal gas flow to control the wobble plate.

Refrigerants and Lubricants

New Vehicles. The phaseout of R-12 has led to the use of non-CFC alternatives in vehicles. The popular choice is R-134a because its physical and thermodynamic properties closely match those of R-12. Because its miscibility with mineral oil is poor, R-134a requires the use of synthetic lubricants such as polyalkylene glycol (PAG) or polyol ester (POE). PAGs are currently preferred. These lubricants—especially PAGs—are more hygroscopic than mineral oil; production and service procedures must emphasize the need for system dehydration and limited exposure of the lubricants to atmospheric moisture.

Older Vehicles. Vehicles originally equipped with R-12 do not require retrofit to another refrigerant as long as an affordable supply of new or reclaimed R-12 exists. When circumstances require it, most air conditioners can be modified to use R-134a and POE. When PAG oil is used in a retrofit, the remaining mineral oil should be removed to levels established by the original equipment manufacturer to prevent residual chlorides from breaking down the PAG.

R-22 is not suitable for topping off or retrofit in R-12 systems, as it will produce unacceptable discharge pressures (up to 640 psig at idle) and temperatures. R-12 idle discharge pressures can reach 435 psig; in addition, it may be incompatible with the elastomeric seals and hose materials.

Several blends of two or more refrigerants are being sold as "drop-in" replacements for R-12. They should not be used without a thorough investigation of their effect on performance and reliability under all operating conditions. Other concerns include (1) flammability, (2) fractionation, (3) evaporator temperature glide, (4) increased probability of adding the wrong refrigerant or oil in future service, and (5) compounded complexity of refrigerant reclaim.

Condensers

Condensers must be properly sized. High discharge pressures reduce the compressor capacity and increase power requirements. When the condenser is in series with the radiator, the air restriction must be compatible with the engine cooling fan and engine cooling requirements. Generally, the most critical condition occurs at engine idle under high load conditions. An undersized condenser can raise head pressures sufficiently to stall small displacement engines.

Automotive condensers must meet internal and external corrosion, pressure cycle, burst, and vibration requirements. They may be of the following designs: (1) tube-and-fin with mechanically bonded fins, (2) serpentine tube with brazed, multilouvered fins, and (3) header extruded tube brazed to multilouvered fins. Aluminum is popular for its low cost and weight.

An oversized condenser may produce condensing temperatures significantly below the engine compartment temperature. This can result in evaporation of refrigerant in the liquid line where the liquid line passes through the engine compartment (the condenser is ahead of the engine and the evaporator is behind it). Engine compartment air has not only been heated by the condenser but by the engine and radiator as well. Typically, this establishes a minimum condensing temperature of 30°F above ambient.

Liquid flashing occurs more often at reduced load, when the liquid line velocity decreases, allowing the liquid to be heated above saturation temperature before reaching the expansion valve. This is more apparent on cycling systems than on systems that have a continuous liquid flow. Liquid flashing is audibly detected as gas enters the expansion valve. This problem can be reduced by adding a subcooler or additional fan power to the condenser.

Condensers generally cover the entire radiator surface to prevent air bypass. Accessory systems designed to fit several different cars are occasionally over-designed. Internal pressure drop should be minimized to reduce power requirements. Condenser-to-radiator clearances as low as 0.25 in. have been used, but 0.5 in. is preferable. Primary-to-secondary surface area ratio varies from 8:1 to 16:1. Condensers are normally painted black so that they are not visible through the vehicle's grille.

A condenser ahead of the engine-cooling radiator not only restricts air but also heats the air entering the radiator. The addition of air conditioning requires supplementing the engine cooling system. Radiator capacity is increased by adding fins, depth, or face area or by raising pump speed to increase coolant flow. Pump cavitation at high speeds is a limiting factor. Also, increased coolant velocity may cause excessive tube erosion. In the case of direct fan drive, increasing the speed of the water pump increases the engine cooling fan speed, which not only supplements the engine cooling system but also provides more air for the condenser.

When air conditioning is installed in an automobile, fan size, number of blades, and blade width and pitch are frequently increased, and fan shrouds are often added. Increases in fan speed, diameter, and pitch raise the noise level and power consumption. Temperature- and torque-sensitive drives (viscous drives or couplings) or flex-fans reduce these increases in noise and power. They rely on the airflow produced by the forward motion of the car to reduce the amount of air the radiator fan must move to maintain adequate coolant temperatures. As vehicle speed increases, fan requirements drop.

Front-wheel-drive vehicles typically have electric motor-driven cooling fans. Some vehicles also have a side-by-side condenser and radiator, each with its own motor-driven fan.

Evaporators

Current automotive evaporator materials and construction include (1) copper or aluminum tube and aluminum fin, (2) brazed aluminum plate and fin, and (3) brazed serpentine tube and fin. Design parameters include air pressure drop, capacity, and condensate carryover. Fin spacing must permit adequate condensate drainage to the drain pan below the evaporator.

Condensate must drain outside the vehicle. At road speeds, the vehicle exterior is generally at a higher pressure than the interior by 1 to 2 in. of water. Drains are usually on the high-pressure side of the blower; they sometimes incorporate a trap and are as small as possible. Drains can become plugged not only by contaminants but also by road splash. Vehicle attitude (slope of the road and inclines), acceleration, and deceleration must be considered when designing condensate systems.

High refrigerant pressure loss in the evaporator requires externally equalized expansion valves. A bulbless expansion valve, which provides external pressure equalization without the added expense of an external equalizer, is available. The evaporator must provide stable refrigerant flow under all operating conditions and have sufficient capacity to ensure rapid cool-down of the vehicle after it has been standing in the sun.

The conditions affecting evaporator size and design are different from those in residential and commercial installations in that the average operating time, from a hot-soaked condition, is less than 20 min. Inlet air temperature at the start of the operation can be as high as 150°F, and it decreases as the duct system is ventilated. In a recirculating system, the temperature of inlet air decreases as the car interior temperature decreases; in a system using outdoor air, inlet air temperature decreases to a few degrees above ambient (perpetual heating by the duct system). During longer periods of operation, the system is expected to cool the entire vehicle interior rather than just produce a flow of cool air.

During sustained operation, vehicle occupants want less air noise and velocity, so the air quantity must be reduced; however, sufficient capacity must be preserved to maintain satisfactory interior temperatures. Ducts must be kept as short as possible and should be insulated from engine compartment and solar-ambient heat loads. Thermal lag resulting from the added heat sink of ducts and housings increases cool-down time.

Filters and Hoses

Air filters are not common. Coarse screening prevents such objects as facial tissues, insects, and leaves from entering fresh-air ducts. Studies show that wet evaporator surfaces reduce the pollen count appreciably. In one test, an ambient of 23 to 96 mg/mm^3 showed 53 mg/mm^3 in a non-air-conditioned car and less than 3 mg/mm^3 in an air-conditioned car. Rubber hose assemblies are installed where flexible refrigerant transmission connections are needed due to relative motion between components or because stiffer connections cause installation difficulties and noise transmission. Refrigerant effusion through the hose wall is a design concern. Effusion occurs at a reasonably slow and predictable rate that increases as pressure and temperature increase. Hose with a nylon core is less flexible (pulsation dampening), has a smaller OD, is generally cleaner, and allows practically no effusion. It is recommended for Refrigerant 134a.

Heater Cores

The heat transfer surface in an automotive heater is generally either copper/brass cellular, aluminum tube and fin, or aluminum brazed tube and center. Each of these designs can currently be found

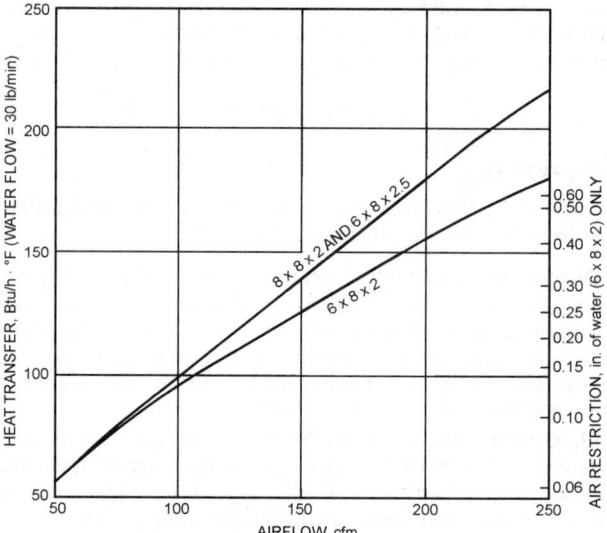

Fig. 1 Typical Copper-Brass Cellular Heater Core Capacity

in production in either straight-through or U-flow designs. The basics of each of the designs are outlined below.

The copper/brass cellular design (Figure 1) uses brass tube assemblies (0.006 to 0.016 in.) as the water course and convoluted copper fins (0.003 to 0.008 in.) held together with a lead-tin solder. The tanks and connecting pipes are usually brass (0.026 to 0.034 in.) and again are attached to the core by a lead-tin solder. Capacity is adjusted by varying the face area of the core to increase or decrease the heat transfer surface area.

The aluminum tube and fin design generally uses round copper or aluminum tubes mechanically joined to aluminum fins. U-tubes can take the place of a conventional return tank. The inlet/outlet tank and connecting pipes are generally plastic and clinched onto the core with a rubber gasket. Capacity can be adjusted by varying face area, adding coolant-side turbulators, or varying air-side surface geometry for turbulence and air restriction.

The aluminum brazed tube and center design uses flat aluminum tubes and convoluted fins or centers as the heat transfer surface. Tanks can be either plastic and clinched onto the core or aluminum and brazed to the core. Connecting pipes can be constructed of various materials and attached to the tanks a number of ways, including brazing, clinching with an o-ring, fastening with a gasket, and so forth. Capacity can be adjusted by varying face area, core depth, or air-side surface geometry.

Receiver-Drier Assembly

The receiver-drier assembly accommodates charge fluctuations from changes in system load (refrigerant flow and density). It accommodates an overcharge of refrigerant (8 to 16 oz) to compensate for system leaks and hose effusion. The assembly houses the high-side filter and desiccant. Several types of desiccant are used, the most common of which is spherical molecular sieves; silica gel is occasionally used. Mechanical integrity (freedom from powdering) is important because of the vibration to which the assembly is exposed. For this reason, molded desiccants have not obtained wide acceptance.

Moisture retention at elevated temperatures is also important. The rate of release with temperature increase and the reaction while accumulating high concentration should be considered. Design temperatures of at least 140°F should be used.

The receiver-drier often houses a sight glass that allows visual inspection of the system charge level. It houses safety devices such as fusible plugs, rupture disks, or high-pressure relief valves. High-pressure relief valves are gaining increasing acceptance because they do not vent the entire charge. Location of the relief devices is important. Vented refrigerant should be directed so as not to endanger personnel.

Receivers are usually (though not always) mounted on or near the condenser. They should be located so that they are ventilated by ambient air. Pressure drops should be minimal.

Expansion Valves

Thermostatic expansion valves (TXVs) control the flow of refrigerant through the evaporator. These are applied as shown in Figures 4, 5, and 6. Both liquid- and gas-charged power elements are used. Internally and externally equalized valves are used as dictated by system design. Externally equalized valves are necessary where high evaporator pressure drops exist. A bulbless expansion valve, usually block-style, that senses evaporator outlet pressure without the need for an external equalizer, is now widely used. There is a trend toward variable compressor pumping rate expansion valves.

Orifice Tubes

An orifice tube instead of an expansion valve has come into widespread use to control refrigerant flow through the evaporator, primarily due to its lower cost. Components must be matched to obtain proper performance. Even so, under some conditions liquid refrigerant floods back to the compressor with this device. Chapter 45 of the 1998 *ASHRAE Handbook—Refrigeration* covers the design of orifice tubes.

Suction Line Accumulators

A suction line accumulator is required with an orifice tube to ensure uniform return of refrigerant and oil to the compressor to prevent slugging and to cool the compressor. It also stores excess refrigerant. A typical suction line accumulator is shown in Figure 2. A bleed hole at the bottom of the standpipe meters oil and liquid refrigerant back to the compressor. The filter and desiccant are contained in the accumulator because no receiver-drier is used with this system. The amount of refrigerant charge is more critical when a suction line accumulator is used than it is with a receiver-drier.

Refrigerant Flow Control

The cycling clutch designs shown in Figures 3 and 4 are common for both factory- and dealer-installed units. The clutch is cycled by a thermostat that senses evaporator temperature or by a pressure switch that senses evaporator pressure. Some dealer-installed units use an adjustable thermostat, which controls car temperature by

controlling evaporator temperature. The thermostat also prevents evaporator icing. Most units use a fixed thermostat or pressure switch set to prevent evaporator icing. Temperature is then controlled by blending the air with warm air coming through the heater.

Cycling the clutch sometimes causes noticeable surges as the engine is loaded and unloaded by the compressor. This is more evident in cars with smaller engines. Reevaporation of condensate from the evaporator during the off-cycle may cause objectionable temperature fluctuation or odor. This system cools faster and at lower cost than a continuously running system.

In orifice tube-accumulator systems, the clutch cycling switch disengages at about 25 psig and cuts in at about 45 psig. Thus, the evaporator defrosts on each off-cycle. The flooded evaporator has enough thermal inertia to prevent rapid clutch cycling. It is desirable to limit clutch cycling to a maximum of 4 cycles per minute because heat is generated by the clutch at engagement. The pressure switch can be used with a thermostatic expansion valve in a dry evaporator if the pressure switch is damped to prevent rapid cycling of the clutch.

Continuously running systems, once widely used, are rarely seen today because they require more power and, consequently, more fuel to operate. In a continuously running system, an evaporator pressure regulator (EPR) keeps the evaporator pressure above the condensate freezing level. Temperature is controlled by reheat or by blending the air with warm air from the heater core.

The continuously running system possesses neither of the previously mentioned disadvantages of the cycling clutch system, but it does increase the suction line pressure drop, which reduces performance slightly at maximum load. A solenoid version of this valve,

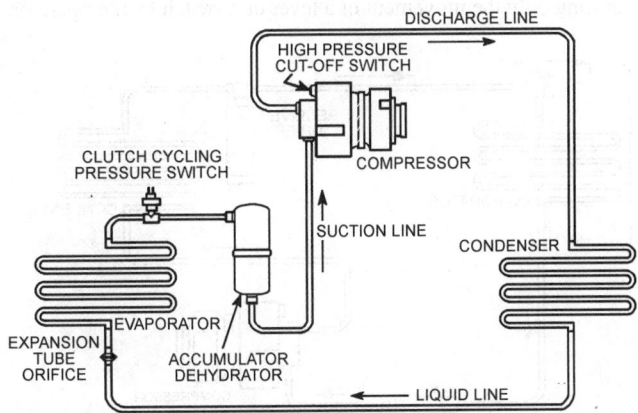

**Fig. 3   Clutch Cycling Orifice Tube
Air-Conditioning Schematic**

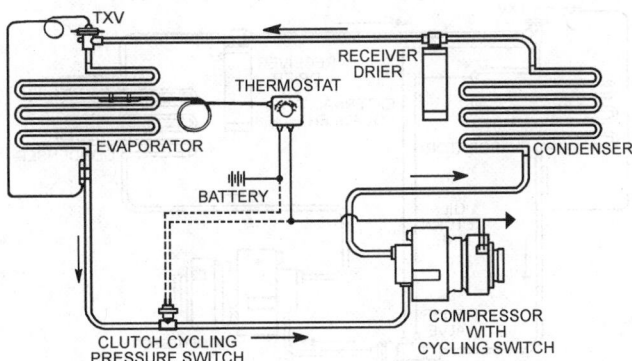

**Fig. 4   Clutch Cycling System with Thermostatic
Expansion Valve (TXV)**

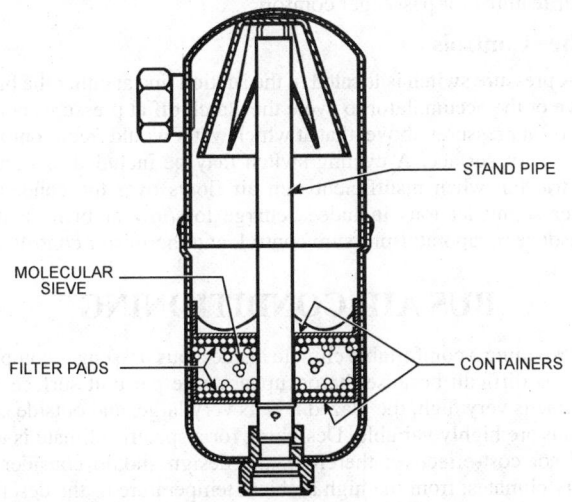

Fig. 2   Typical Suction Line Accumulator

which is controlled by an antifreeze switch that senses evaporator fin temperature, has also been used.

Two refrigeration circuits use the EPR. Figure 5 shows a conventional dry circuit. Figure 6 shows a flooded evaporator circuit, which uses a plate-and-separator type of evaporator with a tank at the top and bottom. It also has a unique piping arrangement. The TXV external equalizer line is connected downstream from the EPR valve. Also, a small oil return line containing an internal pressure relief valve is connected downstream from the EPR. These two lines may be connected to the housing of the EPR valve downstream of its valve mechanism. This piping arrangement causes the TXV to open wide when the EPR throttles, thus flooding the evaporator and causing the EPR to act as an expansion valve. The system allows the air conditioner to run at ambients as low as 35°F to defog windows while maintaining adequate refrigerant flow and ensuring oil return to the compressor.

CONTROLS

Manual

The fundamental control mechanism is a flexible control cable that transmits linear motion with reasonable efficiency and little hysteresis. Rotary motion is obtained by a crank mechanism. Common applications are (1) movement of damper doors that blend air to control the discharge air temperature, (2) regulation of the amount of defrost bleed, and (3) regulation of valves to control the flow of engine coolant through the heater core.

Vacuum

Vacuum provides a silent, powerful method of manual control, requiring only the movement of a lever or a switch by the operator.

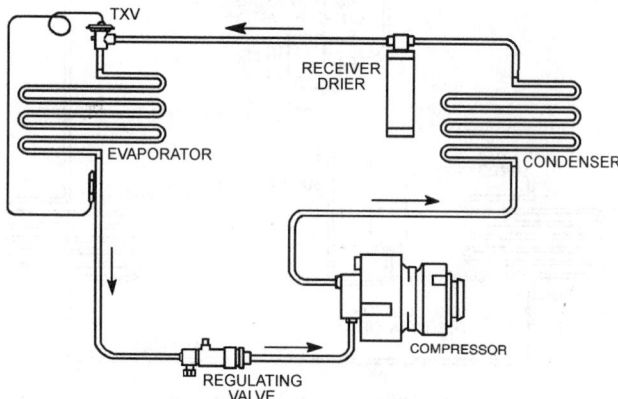

Fig. 5   Suction Line Regulation System

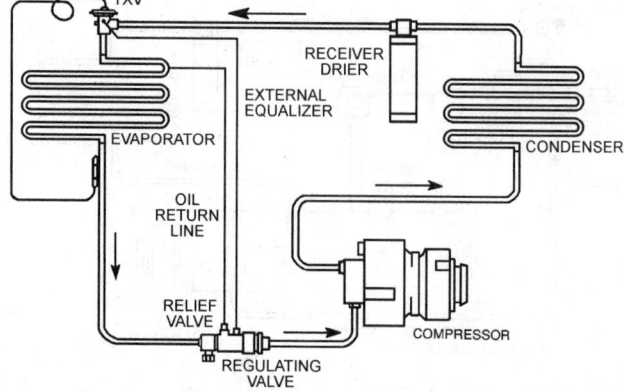

Fig. 6   Flooded Evaporator Suction Line Regulation System

Vacuum is obtained from the engine intake manifold. A vacuum reservoir (25 to 250 in³) may be used to ensure an adequate source. Most systems are designed to function at a minimum vacuum of 5 to 6 in. Hg, although as much as 26 in. Hg may be available at times.

Linear or rotary slide valves select functions. Vacuum-modulating, temperature-compensating (bimetal) valves control thermostatic coolant valves. Occasionally, solenoid valves are used, but they are generally avoided because of their cost and associated wiring.

Electrical

Electrical controls regulate blower motors. Blowers have three or four speeds. The electrically operated compressor clutch frequently sees service as a secondary system control operated by function (mode) selection integration or evaporator temperature or pressure sensors.

Temperature Control

Air-conditioning capacity control to regulate car temperature is achieved in one of two ways—the clutch can be cycled in response to an adjustable thermostat sensing evaporator discharge air, or the evaporator discharge air can be blended with or reheated by air flowing through the heater core. The amount of reheat or blend is usually controlled by driver-adjusted damper doors.

Solid-state logic interprets system requirements and automatically adjusts to heating or air conditioning, depending on the operator's selection of temperature and on ambient temperature. Manual override enables the occupant to select the defrost function. The system not only regulates this function but also controls capacity and air quantity. An in-car thermistor measures the temperature of the air in the passenger compartment and compares it to the setting at the temperature selector. An ambient sensor, sometimes a thermistor, senses ambient temperature to prevent offset or droop. These elements, along with a vacuum supply line from an engine vacuum reservoir, are coupled to the control package, which consists of an electronic amplifier and a transducer. The output is a vacuum or electrical signal regulated by the temperature inputs.

This regulated signal is supplied to servo units that control the quantity of coolant flowing in the heater core, the position of the heater blending door, or both; the air-conditioning evaporator pressure; and the speed of the blower. The same regulated signal controls sequencing of damper doors, either resulting in the discharge of air at the occupant's feet or at waist level or causing the system to work on the recirculation of interior air. Several interlocking devices (1) prevent blower operation before the engine coolant is up to temperature, (2) prevent the air-conditioning compressor clutch from being energized at low ambient temperatures, and (3) provide other features for passenger comfort.

Other Controls

A pressure switch is located in the suction line at either the block valve or the accumulator to cycle the clutch off at pressures below, and on at pressures above, that at which water would freeze onto the evaporator surface. A cycling switch may be included to start an electric fan when insufficient ram air flows over the condenser. Other sophistications include a charge loss/low ambient switch, transducer evaporator pressure control, and thermistor control.

BUS AIR CONDITIONING

Providing a comfortable climate inside a bus passenger compartment is difficult because the occupancy rate per unit surface and volume is very high, the glazed area is very large, and outside conditions are highly variable. Designing for a specific climate is usually not cost-effective; therefore, the design should consider all likely climates, from the high ambient temperature of the desert to the high humidity of the tropics. Units should operate satisfactorily in ambient conditions up to 120°F. A unit designed for both extremes has a greater sensible cooling capacity in hot, dry climates than in humid climates.

Ambient air quality must also be considered. Frequently, intakes are subjected to thermal contamination either from road surfaces, condenser air recirculation, or radiator air discharge. Vehicle motion also introduces pressure variables that affect condenser fan performance. In addition, engine speed affects compressor capacity. Bus air conditioners are initially performance-tested as units in small climate-controlled test cells. Larger test cells that can hold the whole bus are commonly used to verify as-installed performance.

Heat Load

The main parameters that must be considered in the design of a bus air conditioning system include:

- Occupancy data (number of passengers, distance traveled, distance traveled between stops, typical permanence time)
- Dimensions and optical properties of glass
- Outside weather conditions (temperature, relative humidity, solar radiation)
- Dimensions and thermal properties of materials in the bodies of the bus and indoor design conditions (temperature, humidity, and air velocity)

The heating or cooling load in a passenger bus may be estimated by summing the following loads:

- Heat flux from solid walls (side panels, roof, floor)
- Heat flux from glass (side, front and rear windows)
- Heat flux from passengers
- Heat flux from the engine, passengers, and ventilation (difference in enthalpy between outside and inside air)
- Heat flux from the air conditioner

The extreme loads for both summer and winter should be calculated. The cooling load is the most difficult load to handle; the heating load is normally handled by heat recovered from the engine. An exception is that an idling engine provides marginal heat in very cold climates. James and He (1993) and Andre et al. (1994) describe computational models for calculating the heat load in vehicles, as well as for simulating the thermal behavior of the passenger compartment.

The following conditions can be assumed for calculating the summer heat load in an interurban vehicle similar to that shown in Figure 7:

- Capacity of 50 passengers
- Insulation thickness of 1 to 1.5 in.
- Double-pane tinted windows
- Outdoor air intake of 400 cfm
- Road speed of 65 mph
- Inside design temperatures of 80°F dry bulb and 67°F wet bulb, or 20°F lower than ambient

Loads from 3.5 to 10 tons are calculated, depending on the outside weather conditions and on the geographic location of the bus. The typical distribution of the different heat loads during a summer day at 40° North latitude is shown in Figure 8.

Inlets and Outlets

Correct positioning of external air inlets and outlets to the passenger compartment is important on interurban buses that operate mostly at a high, constant speed. Figure 9 shows the pressure coefficient distribution around a typical bus. The main features, resulting from the analysis of the figure are

- On the front surface, most of the pressure is positive, with the stagnation point located at 1/3 of the height.
- At the frontal leading edge, the pressure is strongly negative.

Fig. 7   Distribution of Heat Load (Summer)

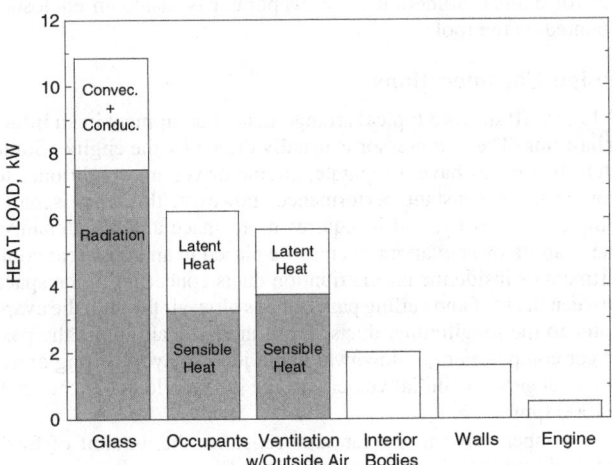

Fig. 8   Main Heat Fluxes in a Bus

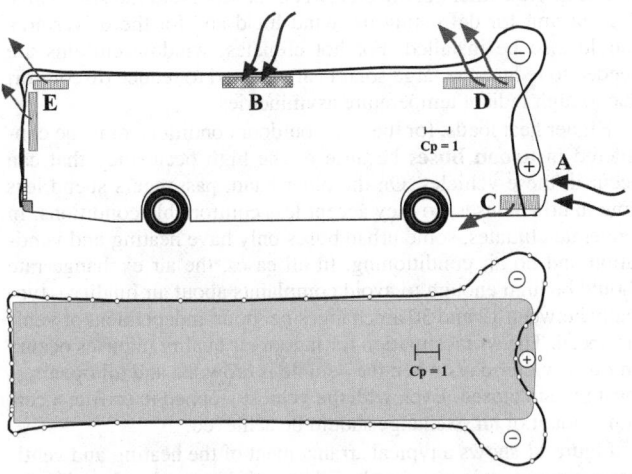

Fig. 9   Pressure Distribution Around a Moving Bus

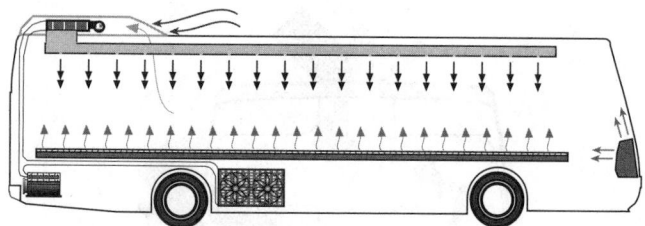

Fig. 10   Typical Arrangement of Air Conditioning in an Interurban Bus

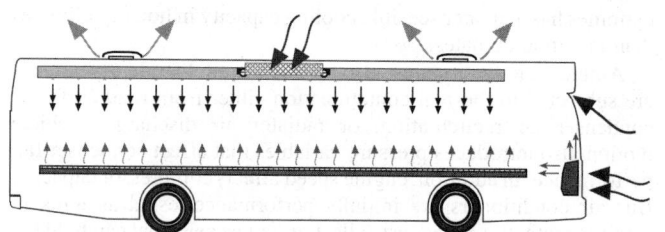

Fig. 11   Typical Arrangement of Air Conditioning in an Urban Bus

• Behind the recirculation bubble in the front, the pressure on the roof is nearly atmospheric.
• In the rear, the pressure coefficient is always slightly negative.

Thus, the best location for inlets is the lower part of the front surface (A). An alternative is near the top and center of the side (B), which is also out of any rain. The areas with strong negative pressure coefficients (C, D) in the side panels just behind the front, vertical pillars are the best locations for the outlets because vehicle movement drives the flow. Slightly negative pressure at the rear and near the rear lateral panels (E) also makes those areas a good location for outlets. Sometimes, the evaporator is inside an enclosure mounted on the roof.

Design Considerations

Figure 10 shows a typical arrangement of equipment in an **interurban bus**. The compressor is usually driven by the engine. Some interurban buses have a separate, engine-driven air conditioner to provide more constant performance; however, this unit is more complex and costly, and it requires more space and maintenance. The evaporator or evaporators may be placed in an upper rear compartment or inside the air distribution ducts concealed in the space between the roof and ceiling panels. Fans blow air through the evaporator to the longitudinal ducts. From there the air enters the passenger compartment in downward cold jets. To avoid strong drafts on passengers, the initial velocity of the jets should not exceed 800 to 1000 fpm.

A damper valve in the fan area regulates the amount of fresh and recirculated air. To maintain its efficiency, the condenser should not be placed in an area with a negative pressure coefficient or high temperatures.

A heater for cold weather is also important to offset downdrafts. Hot coolant from the engine is usually used to heat convectors placed in the corners defined between the floor and the side walls. A front unit for defrosting the windshield and for the driver area should also be installed. For hot climates, window curtains are needed to reduce the large solar heat gain and to reduce discomfort due to high radiant temperature asymmetries.

Higher heat loads, for the same outdoor conditions must be considered in **urban buses** because of the high occupancy that can occur in these vehicles. On the other hand, passengers spend less time in urban buses, so they accept less comfortable conditions. In moderate climates, some urban buses only have heating and ventilation and no air conditioning. In all cases, the air exchange rate should be high enough to avoid complaints about air quality—typically between 12 and 30 air changes per hour, independent of vehicle speed. The worst situation for indoor air quality in buses occurs on rainy winter days when the vehicle is crowded and all openings must remain closed. Even with the vehicle stopped in traffic, a certain amount of air exchange should be achieved.

Figure 11 shows a typical arrangement of the heating and ventilation equipment in an urban bus. The air inlets can be placed in the front, just beneath the destination announcement box and under the windshield, or in the upper part of the lateral panels near the center

of the bus. This location is preferred if some independence to vehicle speed is desirable. Roof ventilators can be used to extract air from the passenger compartment. Their rotation speed can not be very high to avoid noise complaints, which occur when noise exceeds 65 dB(A) at the passengers' head height when the engine is idling. Increasing the number of ventilators and decreasing their rotation speed can be a solution to the problem

RAILROAD AIR CONDITIONING

Passenger railcar air conditioners are generally electromechanical units with direct-expansion evaporators, usually using R-22 (an HCFC). As HCFCs are phased out, the most likely replacements are HFC-407C or HFC-410A. Because HFC-410A operates at a higher pressure, it is not a viable replacement refrigerant for existing R-22 systems. HFC-134a, a medium-pressure refrigerant, is being used in a few new applications and as a replacement refrigerant for existing R-12 equipment. But for the same reason that R-12 was limited (i.e., its lower density in the gas phase), R-134a is not usually chosen because it requires larger suction piping, heat exchangers, and compressor.

Electronic, automatic controls are common, with a trend toward microprocessor control with fault monitoring, logging, and feedback to a central car information system. Electric heating elements installed in the air-conditioning unit or supply duct temper outdoor air brought in for ventilation. They are often used to control humidity by reheating the conditioned supply air during partial cooling load conditions.

Passenger Car Construction

Passenger car design has emphasized lighter car construction to lower operating and maintenance costs. This trend has caused a reduction in the size and weight of air-conditioning equipment and other auxiliaries.

Vehicle Types

Mainline intercity passenger rail service generally operates single and bilevel cars hauled by a locomotive. Locomotive-driven alternators or static inverters distribute power via an intercar cable power bus to the air conditioners. A typical railcar has a control package and two air-conditioning systems. The units are usually either split, with the compressor/condenser units located in the car undercarriage area and the evaporator blower portion mounted in the ceiling area, or self contained packages mounted in interior equipment rooms. Underfloor and roof-mounted package units are less common.

Commuter cars operating around large cities are similar in size to mainline cars and generally have two evaporator-heater fan units mounted above the ceiling with a common or two separate under floor compressor-condenser unit(s) and a control package. These cars may be locomotive hauled and have air conditioning arrangements similar to mainline intercity cars; but they are often self propelled by high-voltage direct current (dc) or alternating current (ac) power supplied from an overhead catenary or from a dc-supplied third rail system. On such cars, the air conditioning may operate on ac or dc power. Self-propelled diesel-driven vehicles still operate in a few areas that use onboard generated power for the air conditioners.

Subway and **elevated rapid-transit cars** usually operate on a third rail dc power supply. The car air conditioning operates on the normal dc supply voltage or on three-phase ac supply from an alternator or inverter mounted under the car. Split air-conditioning systems are common, with evaporators in the interior ceiling area and underfloor-mounted condensing sections.

Streetcars, light rail vehicles, and downtown **people movers** usually run on the city ac or dc power supply and have air-conditioning equipment similar to rapid transit cars. Roof-mounted packages are used more often than under car or split systems. This is partly because of the lack of undercar space availability and in cases, where a sufficient clearance envelope will allow, the roof mounted location offers the most convenient location for add-on systems and for modifying rail vehicles designed without HVAC.

Equipment Selection

The source and type of power dictate the type of air conditioning installed on a passenger railcar; weight, the type of vehicle, and the service parameters of the vehicle to which the system is applied are other major concerns. Thus, ac-powered, semi-hermetic reciprocating, vane, or scroll hermetic compressors, which are lighter than open machines with dc motor drives, are a common choice. However, each car design must be examined in this respect because dc/ac inverters may increase not only the total weight, but also the total power draw, due to conversion losses.

Other concerns in equipment selection include the space required, location, accessibility, reliability, and maintainability. Interior and exterior equipment noise levels must be considered both during the early stages of design and later, when the equipment is coordinated with the car builder's ductwork and grilles.

Design Limitations

Space underneath and inside a railroad car is at a premium, especially on self-propelled light rail vehicles and rapid transit and commuter cars; this generally rules out unitary interior or underfloor-mounted systems. Components are usually built to fit the configuration of the available space. Overall car height, roof profile, ceiling cavity, and undercar clearance restrictions determine the shape and size of equipment.

Because a mainline passenger railcar must operate in various parts of the country, the air conditioning must be designed to handle the national seasonal extreme design days. Commuter cars and rapid transit cars operate in a small geographical area, so only the local design temperatures and humidities need be considered.

Dirt and corrosion constitute an important design factor, especially if the equipment is beneath the car floor, where it is subject to extremes of weather and severe dirt conditions. For this reason, corrosion-resistant materials and coatings must be selected. Aluminum has not proved durable in exterior exposed applications; the sandblasting effect tends to degrade any surface treatment on it. Because dirt pickup cannot be avoided, the equipment must be designed for quick and easy cleaning; access doors are provided and evaporator and condenser fin spacing is usually limited to 8 to 10 fins per inch. Closer spacing causes more rapid dirt buildup and higher cleaning costs. Dirt and severe environmental conditions must also be considered in selecting motors and controls.

Railroad HVAC equipment requires more maintenance and servicing than stationary units. A passenger railcar, having sealed windows and a well-insulated structure, becomes almost unusable if the air conditioning fails. The equipment therefore has many additional components to permit quick diagnosis and correction of a failure. Motors, compressors, control compartments, diagnostic test ports and other maintenance points must be easily accessible for inspection or repair. However, changes in maintenance strategies and environmental regulations have caused some users to move away from fully on-car serviceable air conditioners, and toward modular, self-contained units with hermetically sealed refrigerant systems. These units are designed for rapid removal and replacement and off-car repair in a dedicated air-conditioning service area. Microprocessor-based controls are designed to identify, log, and indicate system faults in a manner that facilitates expeditious diagnosis and repair.

Security of the air-conditioning equipment attachment to the vehicle must be considered, especially on equipment located beneath the car. Vibration isolators and supports should be designed to safely retain the equipment on the vehicle, even if the vibration isolators or fasteners fail completely. A piece of equipment that dangles or drops off could cause a train derailment. All belt drives and other revolving items must be safety guarded. High-voltage controls and equipment must be labeled by approved warning signs. Pressure vessels and coils must meet ASME test specifications for protection of the passengers and maintenance personnel. Materials selection criteria include low flammability, low toxicity, and low smoke emission.

The design, location, and installation of air-cooled condenser sections must allow for considerable discharge air recirculation and/or temporary extreme inlet air temperatures from adjacent equipment that may occur at passenger loading platforms or in tunnels. To prevent a total system shutdown due to high discharge pressure, a capacity reduction control device is typically used to reduce the cooling capacity, thus temporarily reducing discharge pressure.

Interior Comfort

Air-conditioning and heating comfort and ventilation parameters may be selected in accordance with ASHRAE *Standards* 55 and 62. The duration of exposure and the effect of prior passenger activity should also be considered. For example, a person running for a ride and/or waiting in a hot or cold climate prior to boarding for a 10 minute ride in clothing fit for outside is not a typical building occupant. Jones et al. (1994) evaluated the heat load imposed by people under transient weather and activity conditions as opposed to steady state metabolic rates traditionally used. An application program TRANMOD was developed that allows a designer to predict the thermal loads imposed by passengers (Jones and He 1993).

Door openings, ventilation air sources (ambient contamination), and loading profiles need to be considered when locating inlets and calculating the amount of makeup air. Evaporator sectional staging, variable compressor and evaporator fan speed control, and/or compressor cylinder unloading, coupled with electric reheat, are used to provide part-load humidity control. In winter, humidity control is usually not provided.

The dominant summer cooling load is due to passengers, followed by ventilation, car body transmission, solar gain, and internal heat. Heating loads are due to car body losses and ventilation. The heating load calculation does not credit heat from passengers and internal sources. Comfortable internal conditions in ventilated non-air-conditioned cars can be maintained only when ambient conditions permit. Interior conditions are difficult to maintain because the passenger and solar loads in mass transit cars vary continuously.

Air conditioners in North American cars are selected to maintain temperatures of 73 to 76°F, with a maximum relative humidity of 55 to 60%. In Europe and elsewhere, the conditions are usually set at a dry-bulb temperature approximately 10°F below ambient, with a coincidental relative humidity of 50 to 66%. In the heating mode, the car interior is kept in the 65 to 70°F range.

Other Requirements

Most cars are equipped with both overhead and floor heat. The overhead heat raises the temperature of the recirculated and ventilation air mixture to slightly above the car design temperature. A duct limit thermostat is often used to prevent the supply air from overheating, which can cause passenger discomfort due to temperature stratification. The floor heat offsets heat loss through the car body. The times of maximum occupancy, outdoor ambient, and solar gain must be ascertained. The peak cooling load on urban transit cars usually coincides with the evening rush hour, while the peak load on intercity railcars occurs in the mid afternoon.

Heating capacity for the car depends on body construction, car size, and the design area-averaged relative wind-vehicle velocity. In some instances, minimum car warm-up time may be the governing factor. In long-distance trains, the toilets, galley, and lounges often have exhaust fans. Ventilation airflow must exceed forced exhaust air rates sufficiently to maintain positive car pressure. Ventilation air pressurizes the car and reduces infiltration.

Air Distribution

The most common air distribution system is a centerline supply duct running the length of the car in the space between the ceiling and the roof. The air outlets are usually ceiling-mounted linear slot air diffusers. Louvered or egg crate recirculation grilles are positioned in the ceiling beneath the evaporator units. The main supply duct must be insulated from the ceiling cavity to prevent thermal gain/loss and condensation. Taking ventilation air from both sides of the roof line helps overcome the effect of wind. Adequate snow and rain louvers, and in some cases internal baffles, must be installed on the outdoor air intakes. Separate outdoor air filters are usually paired with a return air filter. Disposable media or permanent, cleanable air filters are used and are usually serviced every month. Some long-haul cars, such as sleeper cars, require a network of delivered-air and return ducts. Duct design should consider noise and static pressure losses.

Piping Design

Standard refrigerant piping practice is followed. Pipe joints should be accessible for inspection and, on split systems, not concealed in car walls. Evacuation, leak testing, and dehydration must be completed successfully after installation and prior to charging. Piping should be supported adequately and installed without traps that could retard the flow of lubricant back to the compressor. Pipe sizing and arrangement should be in accordance with Chapter 2 of the 1998 *ASHRAE Handbook—Refrigeration*. Evacuation, dehydration and charging, should be performed as described in Chapter 46 of the 1998 *ASHRAE Handbook*. Piping on packaged units should also conform to these recommendations.

Control Requirements

Car HVAC systems are automatically controlled for year-round comfort. In split systems cooling load variations are usually handled by two-stage direct-expansion coils and compressors equipped with suction pressure unloaders, electric unloaders, or speed control. Unitary systems may have hot gas bypass to keep the compressor from cycling and reduce the cooling effect. Under low load conditions, cooling is provided by outdoor air supplied by ventilation. During part-load cooling, electric reheat maintains humidity. Under low loads not requiring humidity control, the system assumes the ventilation mode. A pump-down cycle and low ambient lockout are recommended on split systems to protect the compressor from damage caused by liquid slugging. In addition, the compressor may be fitted with a crankcase heater that is energized during the compressor off-cycle. During the heating mode, floor and overhead heaters are staged to maintain the car interior temperature. Today's controls use thermistors and solid-state electronics instead of electromechanical devices. The control circuits are normally powered by low-voltage DC or, occasionally, by single-phase AC.

Future Trends

Air cycle technology is being introduced in passenger railcar air conditioning in Germany (Giles et al. 1997); however, issues of greater weight, higher cost, and low efficiency need to be addressed before air-cycle equipment will be widely accepted. Other trends include the proliferation of hermetic scroll compressors, significantly increased self-diagnostic capabilities included in microprocessor-based controls, and continued efforts to identify the best refrigerant for the long term.

FIXED GUIDEWAY VEHICLE AIR CONDITIONING

Fixed guideway systems, commonly called people movers, can be monorails or rubber-tired cars running on an elevated or grade level guideway, as seen at airports and in urban areas such as Miami. The guideway directs and steers the vehicle and provides the electrical power to operate the car's traction motors, lighting, electronics, air conditioner, and heater. People movers are usually unmanned and computer controlled from a central point. Operations control determines vehicle speed and headway and the length of time doors stay open based on telemetry from the individual cars or trains. Therefore, reliable and effective environmental control is essential.

People movers are smaller than most other mass transit vehicles, generally having spaces for 20 to 40 seated passengers and generous floor space for standing passengers. Under some conditions of passenger loading, a 40-ft car can accommodate 100 passengers. The wide range of passenger loading and the continual movement of the car from full sunlight to deep shade make it essential that the car's air conditioner be especially responsive to the amount of cooling required at a given moment.

System Types

The HVAC for a people mover is usually one of three types:

1. Conventional undercar condensing unit connected with refrigerant piping to an evaporator/blower unit mounted above the car ceiling
2. Packaged, roof-mounted unit having all components in one enclosure and mated to an air distribution system built into the car ceiling
3. Packaged, undercar-mounted unit mated to supply and return air ducts built into the car body

Two systems are usually installed in each car, one at each end; each system provides one-half of the maximum cooling requirement. The systems, whether unitary or split, operate on the guideway's power supply, which, in the United States, is usually 60 Hz, 460 to 600 V (ac).

Refrigeration Components

Because commercial electrical power is available, standard semihermetic motor-compressors and commercially available fan motors and other components can be used. Compressors generally have one or two stages of unloaders, and hot gas bypass is used to maintain cooling at low loads. Condenser and cooling coils are copper tube, copper, or aluminum fin units. Generally, flat fins are preferred for undercar systems to make it simpler to clean the coils. Evaporator/blower sections must often be designed for the specific vehicle and fitted to its ceiling contours. The condensing units must also be arranged to fit in the limited space available and still ensure good airflow across the condenser coil. R-22 is used in almost all of these units.

Heating

Where heating must be provided, electric resistance heaters that operate on the guideway power supply are installed at the evaporator unit discharge. One or two stages of heat control are used, depending on the size of the heaters.

Controls

A solid-state control is usually used to maintain interior conditions. The cooling set point is typically between 74 to 76°F. For heating, the set point is 68°F or lower. Some controls provide humidity control by using the electric heat. Between the cooling and heating set points, blowers continue to operate on a ventilation cycle. Often, two-speed blower motors are used, switching to low speed for the heating cycle. Some controls have internal diagnostic capability; they are able to signal the operations center when a cooling or heating malfunction occurs.

Fresh Air

With overhead air-handling equipment, fresh air is introduced into the return airstream at the evaporator entrance. Fresh air is usually taken from a grilled or louvered opening in the end or side of the car and, depending on the configuration of components, filtered separately or directed so that the return air filter can handle both airstreams. For undercar systems, a similar procedure is used, except the air is introduced into the system through an intake in the undercar enclosure. In some cases, a separate fan is used to induce fresh air into the system.

The amount of mechanical outdoor air ventilation is usually expressed as cfm per passenger on a full load continuous basis. Passenger loading is not continuous at full load in this application, with the net result that more outside air is provided than indicated. The passengers may load and unload in groups, which causes additional air exchange with the outside. Frequent opening of doors, sometimes on both sides at once, allows additional natural ventilation. The effective outside air ventilation per passenger is a summation of all these factors. Some vehicles currently in service have no mechanical outside air supply. Others have up to 10 cfm per passenger. Lower values of mechanical ventilation, typically 5 cfm or less per passenger, are associated with travel times of less than 2 min and large passenger turnover. Longer rides justify higher rates of mechanical ventilation.

Air Distribution

With overhead equipment, air is distributed through linear ceiling diffusers that are often constructed as a part of the overhead lighting fixtures. Undercar equipment usually makes use of the void spaces in the sidewalls and below fixed seating. In all cases, the spaces used for air supply must be adequately insulated to prevent condensation on surfaces and, in the case of voids below seating, to avoid cold seating surfaces. The supply-air discharge from undercar systems is typically through a windowsill diffuser. Recirculation air from overhead equipment flows through ceiling-mounted grilles. For undercar systems, return air grilles are usually found in the door wells or beneath seats.

Because of the small size of the vehicle and its low ceilings, extreme care must be taken to design the air supply so that it does not blow directly on passengers' heads or shoulders. High rates of diffusion are needed, and diffuser placement and arrangement should cause the discharge to hug the ceiling and walls of the car. Total air quantity and discharge temperature must be carefully balanced to minimize cold drafts and air currents.

REFERENCES

Andre, J.C.S., E.Z.E. Conceição, M.C.G. Silva, and D.X. Viegas. 1994. Integral simulation of air conditioning in passenger buses. Fourth International Conference on Air Distribution in Rooms (ROOMVENT 94).

ASHRAE. 1989. Ventilation for acceptable indoor air quality. ANSI/ASHRAE *Standard* 62-1989.

ASHRAE. 1992. Thermal environmental conditions for human occupancy. ANSI/ASHRAE *Standard* 55-1992.

Giles, G.R., R.G. Hunt, and G.F. Stevenson. 1997. Air as a refrigerant for the 21st century. *Proceedings* ASHRAE/NIST Refrigerants Conference, Refrigerants for the 21st Century.

Jones, B.W., Q. He, J.M. Sipes, and E.A. McCullough. 1994. The transient nature of thermal loads generated by people. *ASHRAE Transactions* 100(2):432-38.

Jones, B.W. and Q. He. 1993. User manual. Transient human heat transfer model (includes application TRANMOD). Institute of Environmental Research, Kansas State University, Manhattan.

SAE. 1994. Safety and containment of refrigerant for mechanical vapor compression systems used for mobile air-conditioning systems. *Recommended Practice* J 639 1994. SAE International, Warrendale, PA.

BIBLIOGRAPHY

Conceição, E.Z.E., M.C.G. Silva, and D.X. Viegas. 1997. Airflow around a passenger seated in bus. *International Journal of HVAC&R Research* 3(4):311-23.

Conceição, E.Z.E., M.C.G. Silva, and D.X. Viegas. 1997. Air quality inside the passenger compartment of a bus. *Journal of Exposure Analysis & Environmental Epidemiology* 7:521-34.

Silva, M.C.G. and D.X. Viegas. 1994. External flow field around an intercity bus. Second International Conference on Experimental Fluid Mechanics.

CHAPTER 9

AIRCRAFT

ENVIRONMENTAL control system (ECS) is a generic term used in the aircraft industry for the systems and equipment associated with the ventilation, heating, cooling, humidity/contamination control, and pressurization in the occupied compartments, cargo compartments, and electronic equipment bays. The term ECS often encompasses other functions such as windshield defog, airfoil anti-ice, oxygen systems, and other pneumatic demands. The regulatory or design requirements of these related functions are not covered in this chapter.

Environmental control systems of various types and complexity are used in military and civil aircraft, helicopter, and spacecraft applications. This chapter applies to commercial transport aircraft that predominately use air-cycle air conditioning for the ECS.

Commercial users categorize their ECS equipment in accordance with the Air Transport Association of America *Specification No.* 100, Specification for Manufacturers Technical Data. The following ATAA 100 chapters define ECS functions and components:

- **Chapter 21**, Air Conditioning, includes heating, cooling, moisture/contaminant control, temperature control, distribution, and cabin pressure control. Common system names are the air-conditioning system (ACS) and the cabin pressure control system (CPCS).
- **Chapter 30**, Ice and Rain Protection, includes airfoil ice protection; engine cowl ice protection; and windshield ice, frost, or rain protection.
- **Chapter 35**, Oxygen, includes components that store, regulate, and deliver oxygen to the passengers and crew.
- **Chapter 36**, Pneumatic, covers ducts and components that deliver compressed air (bleed air) from a power source (main engine or auxiliary power unit) to connecting points for the using systems (Chapters 21, 30, and Chapter 80, Starting). The pneumatic system is also commonly called the engine bleed air system (EBAS).

REGULATIONS

The Federal Aviation Administration (FAA) regulates the design of transport category aircraft for operation in the United States under Federal Aviation Regulation (FAR) Part 25. ECS equipment and systems must meet these requirements, which are primarily related to health and safety of the occupants. Certification and operation of these aircraft in the U.S. is regulated by the FAA in FAR Part 121. Similar regulations are applied to European nations by the European Joint Aviation Authorities (JAA), which represents the combined requirements of the airworthiness authorities of the participating nations; the equivalent design regulation is JAR Part 25. Operating rules based on FAA or JAA regulations are applied individually by the nation of registry. Regulatory agencies may impose special conditions on the design, and compliance is mandatory.

The preparation of this chapter is assigned to TC 9.3, Transportation Air Conditioning.

Several FAR and JAR Part 25 paragraphs apply directly to transport category aircraft ECSs; those most germane to the ECS design requirements are as follows:

FAR/JAR 25.831	Ventilation
FAR 25.832	Cabin ozone concentration
FAR/JAR 25.841	Pressurized cabins
FAR/JAR 25.1309	Equipment, systems, and installations
FAR/JAR 25.1438	Pressurization and pneumatic systems
FAR/JAR 25.1461	Equipment containing high energy rotors

These regulatory requirements are summarized in the following sections; however, the applicable FAR and JAR paragraphs, amendments and advisory circulars should be consulted for the latest revisions and full extent of the rules.

Ventilation

FAR/JAR *Paragraph* 25.831

- Each passenger and crew compartment must be ventilated.
- Each crew member must have enough fresh air to perform their duties without undue fatigue or discomfort (minimum of 10 cfm).
- Crew and passenger compartment air must be free from hazardous concentration of gases and vapors:

    - Carbon monoxide limit is 1 part in 20,000 parts of air
    - Carbon dioxide limit is 3% by volume, sea level equivalent (An FAA amendment reduces the carbon dioxide limit to 0.5%)
    - Conditions must be met after reasonably probable failures

- Smoke evacuation from the cockpit must be readily accomplished without depressurization.
- The occupants of the flight deck, crew rest area, and other areas must be able to control the temperature and quantity of ventilating air to their compartments independently.

FAR *Amendment* No. 25-87:

- Under normal operating conditions and in the event of any probable failure, the ventilation system must be designed to provide each occupant with an airflow containing at least 0.55 lb of fresh air per minute (or about 10 cfm at 8000 ft).
- The maximum exposure an any given temperature is specified as a function of the temperature exposure.

JAR ACJ (*Advisory Circular-Joint*) 25.831:

- The supply of fresh air in the event of loss of one source, should not be less than 0.4 lb/min per person for any period exceeding 5 min. However, reductions below this flow rate may be accepted provided that the compartment environment can be maintained at a level that is not hazardous to the occupant.
- Where the air supply is supplemented by a recirculating system, it should be possible to stop the recirculating system.

Cabin Ozone Concentration

FAR 25.832 specifies the cabin ozone concentration during flight must be shown not to exceed:

- 0.25 ppm by volume, sea level equivalent, at any time above flight level 320 (32,000 ft)
- 0.10 ppm by volume, sea level equivalent, time weighted average during any 3 hour interval above flight level 270 (27,000 ft)

At present, JAR 25 has no requirement for cabin ozone concentration.

Pressurized Cabins

FAR/JAR 25.841:

- Limits the maximum cabin pressure altitude to 8000 ft at the maximum aircraft operating altitude under normal operating conditions.
- For operation above 25,000 ft, a cabin pressure altitude of not more than 15,000 ft must be maintained in the event of any reasonably probable failure or malfunction in the pressurization system.
- The makeup of the cabin pressure control components, instruments, and warning indication is specified to ensure the necessary redundancy and flight crew information.

FAR *Amendment* No. 25-87 imposes additional rules for high altitude operation.

Equipment, Systems, and Installations

FAR/JAR 25.1309:

- Systems and associated components must be designed such that occurrence of any failure that would prevent continued safe flight and landing is extremely improbable.
- The occurrence of any other failure that reduces the capability of the aircraft or the ability of the crew to cope with adverse operating conditions is improbable.
- Warning information must be provided to alert the crew to unsafe system operating conditions to enable them to take corrective action.
- Analysis in compliance with these requirements must consider possible failure modes, probability of multiple failures, undetected failures, current operating condition, crew warning, and fault detection.

FAR *Advisory Circular* AC 25.1309-1A and JAR ACJ *No.* 1 to JAR 25.1309 define the required failure probabilities for the various failure classifications: probable, improbable, and extremely improbable for the FAR requirements; and frequent, reasonably probable, remote, and extremely remote for the JAR requirements.

Pressurization and Pneumatic Systems

FAR 25.1438 specifies the proof and burst pressure factors for pressurization and pneumatic systems as follows:

- Pressurization system elements (air conditioning)
  - Burst pressure: 2.0 times maximum normal pressure
  - Proof pressure: 1.5 times max normal pressure
- Pneumatic system elements (bleed)
  - Burst pressure: 3.0 times maximum normal pressure
  - Proof pressure: 1.5 times maximum normal pressure

JAR 25.1438 and ACJ 25.1438 specify the proof and burst pressure factors for pressurization and pneumatic systems as follows:

- Proof pressure
  - 1.5 times worst normal operation
  - 1.33 times worst reasonable probable failure
  - 1.0 times worst remote failure
- Burst pressure
  - 3.0 times worst normal operation
  - 2.66 times worst reasonably probable failure

- 2.0 times worst remote failure
- 1.0 times worst extremely remote failure

Equipment Containing High-Energy Rotors

FAR/JAR 25.1461:

Equipment must comply with at least one of the following three requirements:

1. High-energy rotors contained in equipment must be able to withstand damage caused by malfunctions, vibration, and abnormal temperatures.

   - Auxiliary rotor cases must be able to contain damage caused by high-energy rotor blades.
   - Equipment control devices must reasonably ensure that no operating limitations affecting the integrity of high-energy rotors will be exceeded in service.

2. Testing must show that equipment containing high-energy rotors can contain any failure that occurs at the highest speed attainable with normal speed control devices inoperative.

3. Equipment containing high-energy rotors must be located where rotor failure will neither endanger the occupants nor adversely affect continued safe flight.

DESIGN CONDITIONS

Design conditions for aircraft applications differ in several ways from other HVAC applications. Commercial transport aircraft operate in a physical environment that is not survivable by unprotected humans. This requires a complex ECS to provide passengers and crew with safety and comfort without health risks.

Aircraft ECSs operate under unique conditions. Outside air at altitude is extremely cold, dry, and can contain high levels of ozone. On the ground outside air can be hot, humid, and contain many pollutants such as particulate matter, aerosols, and hydrocarbons. These ambient conditions change quickly from ground operations to flight. The hot-day, high-humidity ground condition usually dictates the size of air-conditioning equipment, and high altitude flight conditions determine the impedance necessary to provide adequate ventilation and pressurization bleed airflow. Maximum heating requirements can be determined by either cold day ground or flight operations.

In addition to essential safety requirements, the ECS should provide a comfortable environment for the passengers and crew. This presents a unique challenge because of the high density seating of the passengers and the changes in cabin pressure and the outside environment during flight. Also aircraft systems must be lightweight, accessible for quick inspection and servicing, highly reliable, tolerant of a wide range of environmental conditions, able to withstand aircraft vibratory and maneuver loads, and able to accommodate failures occurring during flight.

Ambient Temperature, Humidity, and Pressure

Figure 1 shows typical design ambient temperature profiles for hot, standard, and cold days. The ambient temperatures used for the design of a particular aircraft may be higher or lower than those shown in Figure 1, depending on the regions in which the aircraft is to be operated. The design ambient moisture content at various altitudes that is recommended for commercial aircraft is shown in Figure 2. However, operation at moisture levels exceeding 200 grains per pound of dry air is possible in some regions. The variation in ambient pressure with altitude is shown in Figure 3.

ECS Performance

The ECS is designed to provide a comfortable cabin temperature, acceptable ventilation, good airflow distribution across the cabin width, and sufficient bleed airflow to maintain cabin pressurization

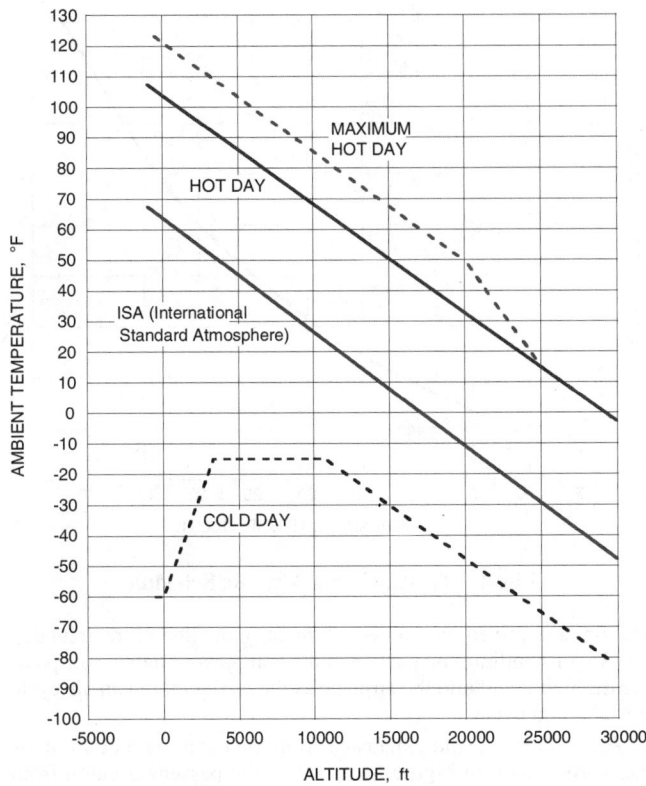

Fig. 1 Typical Ambient Temperature Profiles

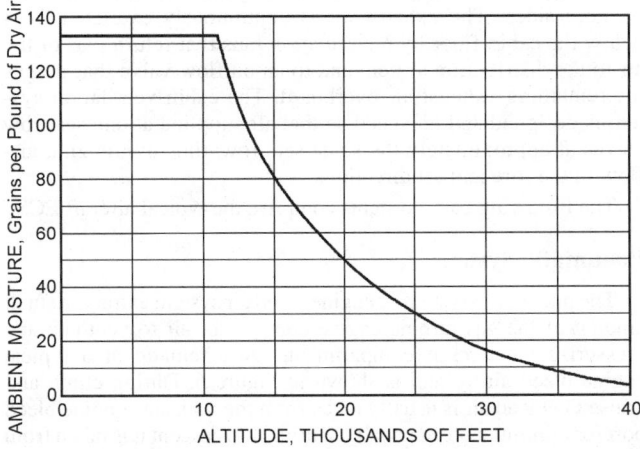

Fig. 2 Design Moisture for Equipment Performance

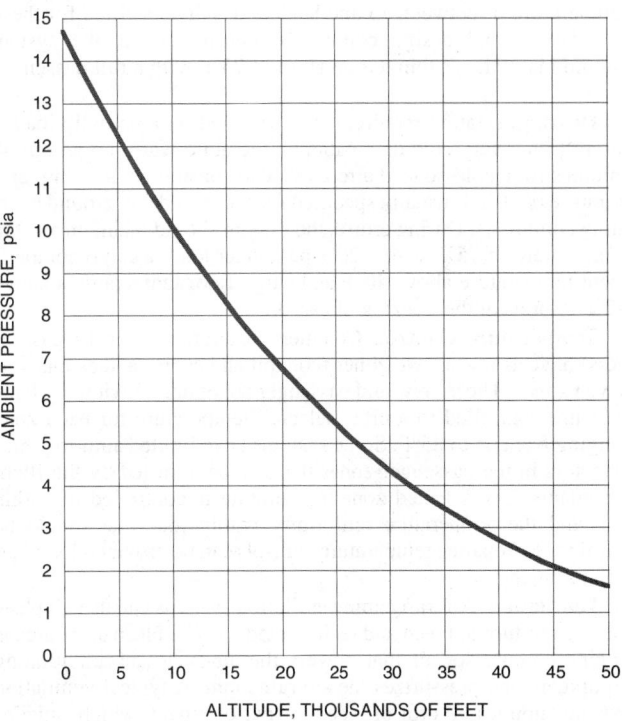

Fig. 3 Variation of Ambient Pressure with Altitude

• Convection and radiation from internal sources of heat such as electrical equipment

The heat transfer analysis should include all possible flow paths through the complex aircraft structure. Air film coefficients vary with altitude and should be considered. During flight, the increase in air temperature and pressure due to ram effects is appreciable and may be calculated from the following equations:

$$\Delta t_r = 0.2 M^2 T_a F_r$$

$$\Delta p_r = (1 + 0.2 M^2)^{3.5} p_a - p_a$$

where

F_r = recovery factor, dimensionless
Δt_r = increase in temperature due to ram effect, °F
M = Mach number, dimensionless
T_a = absolute static temperature of ambient air, °R
Δp_r = increase in pressure due to ram effect, psi
p_a = absolute static pressure of ambient air, psi

The average increase in aircraft skin temperature for subsonic flight is generally based on a recovery factor F_r of 0.9.

Ground and flight requirements may be quite different. For an aircraft sitting on the ground in bright sunlight, surfaces that have high absorptivity and are perpendicular to the sun's rays may have much higher temperatures than ambient. This skin temperature is reduced considerably if a breeze blows across the aircraft. Other considerations for ground operations include cool-down or warm-up requirements, and the time that doors are open for passenger and galley servicing.

Cooling. The sizing criteria for the air conditioning is usually ground operation on a hot, humid day with the aircraft fully loaded and the doors closed. A second consideration is cool-down of a empty, heat-soaked aircraft prior to passenger loading; a cool-down time of less than 30 minutes is usually specified. A cabin

and repressurization during descent, while accounting for certain failure conditions. The ECS also includes provisions to dehumidify the cabin supply during cooling operations.

Load Determination. The steady-state cooling and heating loads for a particular aircraft model are determined by a heat transfer study of the several elements that comprise the air-conditioning load. The heat transfer involves the following factors:

• Convection between the boundary layer and the outer aircraft skin
• Radiation between the outer aircraft skin and the external environment
• Solar radiation through transparent areas
• Conduction through cabin walls and the aircraft structure
• Convection between the interior cabin surface and the cabin air
• Convection between the cabin air and occupants or equipment

temperature of between 75 and 80°F is usually specified for these hot day ground design conditions. During cruise, the system should maintain a cabin temperature of 75°F with a full passenger load.

Heating. Heating requirements are based on a partially loaded aircraft on a very cold day; cabin temperature warm-up within 30 minutes for a cold-soaked aircraft is also considered. A cabin temperature of 70°F is usually specified for these cold day ground operating conditions. During cruise, the system should maintain a cabin temperature of 75°F with a 20% passenger load, a cargo compartment temperature above 40°F and cargo floor temperatures above 32°F to prevent the freezing of cargo.

Temperature Control. Commercial aircraft (over 19 passengers) have as few as two zones (cockpit and cabin) and as many as seven zones. These crew and passenger zones are individually temperature controlled to a crew selected temperature for each zone ranging from 65 to 85°F. Some systems have limited authority bias selectors in the passenger zones that can be adjusted by the flight attendants. The selected zone temperature is controlled to within 2°F and the temperature uniformity within the zone should be within 5°F. Separate temperature controls can be provided for cargo compartments.

Ventilation. Aircraft cabin ventilation systems can use all bleed air or a mixture of bleed and recirculated air. The bleed air source is engine compressor air that powers the air-cycle air-conditioning equipment and pressurizes the aircraft cabin. A typical ventilation system supplies 10 cfm of bleed air per occupant, which satisfies regulatory (FAA/JAA) requirements for ventilation and concentrations of carbon monoxide and carbon dioxide.

An equal amount of filtered, recirculated cabin air can be provided for improved distribution and circulation. The use of filtered, recirculated air reduces the bleed air penalty and increases cabin humidity. Most systems have no filtration of the engine bleed air supply, although bleed air centrifugal cleaners are sometimes offered as optional equipment. Some aircraft that fly where high ambient ozone levels can be expected require catalytic ozone converters.

Particulate filters are used to filter the recirculated cabin air. High-efficiency particulate air (HEPA) filters are preferred. Charcoal filters are sometimes offered as optional equipment, and are installed in series with the particulate filters. The demand for charcoal filters is low for domestic carriers because of the smoking ban on domestic flights. Their use worldwide is also limited due to cost, frequency of maintenance (replacement), and disposal requirements.

Pressurization. Cabin pressurization achieves the required partial pressures of oxygen for the crew and passengers during high altitude flight. While the aircraft operates at altitudes above 40,000 ft, the occupied cabin must be pressurized to an equivalent altitude of 8000 ft or less to permit normal physiological functions without supplemental oxygen. The maximum pressure difference between the cabin and the outside environment is limited by aircraft structural design limits. The differential pressure control provides a cabin pressure based on the flight altitude of the aircraft. A typical cabin altitude schedule is shown in Figure 4. The cabin pressure control must also limit the normal maximum rates of change of cabin pressure; the recommended limits are 500 fpm for increasing altitude (ascent) and 300 fpm for decreasing altitude (descent). Provisions separate from the normal cabin pressure controls must be provided for positive and negative pressure relief to protect the aircraft structure.

SYSTEM DESCRIPTION

The outside air supplied to the airplane cabin is provided by the engine compressors (or auxiliary power unit in the aircraft tailcone), cooled by **air-conditioning packs** located under the wing center

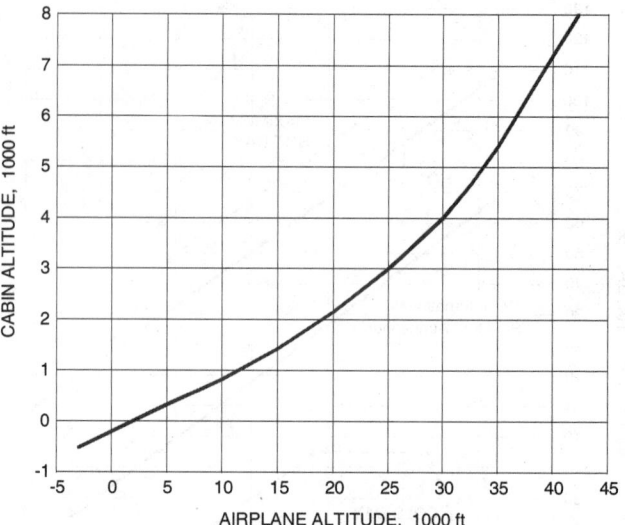

Fig. 4   Typical Cabin Altitude Schedule

section, and mixed with an equal quantity of filtered, recirculated air. An air-conditioning pack uses the compressed outside air passing through it and into the airplane as the refrigerant in an air-cycle refrigeration cycle.

Air is supplied and exhausted from the cabin on a continuous basis. As shown in Figure 5, air enters the passenger cabin from overhead distribution outlets that run the length of the cabin. The exhaust air leaves the cabin through return air grilles located in the sidewalls near the floor, and running the length of the cabin on both sides. The exhaust air is continuously extracted from below the cabin floor by recirculation fans that return part of the air to the distribution system and to an outflow valve that purges the remaining exhaust air overboard. The cabin ventilation system is designed and balanced so that air supplied at one seat row leaves at approximately the same seat row, thus minimizing airflow in the fore and aft directions.

The following basic systems comprise the typical aircraft ECS.

Pneumatic System

The pneumatic system or engine bleed air system extracts a small amount of the gas turbine engine compressor air to ventilate and pressurize the aircraft compartments. A schematic of a typical engine bleed air system is shown in Figure 6. During climb and cruise the bleed air is usually taken from the midstage engine bleed port for minimum bleed penalty. During idle descent it is taken from the high-stage engine bleed port where maximum available pressure is required to maintain cabin pressure and ventilation. The bleed air is pressure controlled to meet the requirements of the using system; and it is usually cooled to limit bleed manifold temperatures to be compatible with fuel safety requirements. In fan jets, the fan air is used as the heat sink for the bleed air heat exchanger (precooler); for turboprop engines, ram air is used, which usually requires an ejector or fan for static operation. Other components include bleed-shutoff and modulating valves, fan-air modulating valve, sensors, controllers, and ozone converters. The pneumatic system is also used intermittently for airfoil and engine cowl anti-icing, engine start, and several other pneumatic functions.

Each engine has an identical bleed air system for redundancy and to equalize the compressor air bled from the engines. The equipment is sized to provide the necessary temperature and airflow for airfoil and cowl anti-icing, or cabin pressurization and air conditioning with one system or engine inoperative. The bleed air used for airfoil anti-icing is controlled by valves feeding piccolo tubes

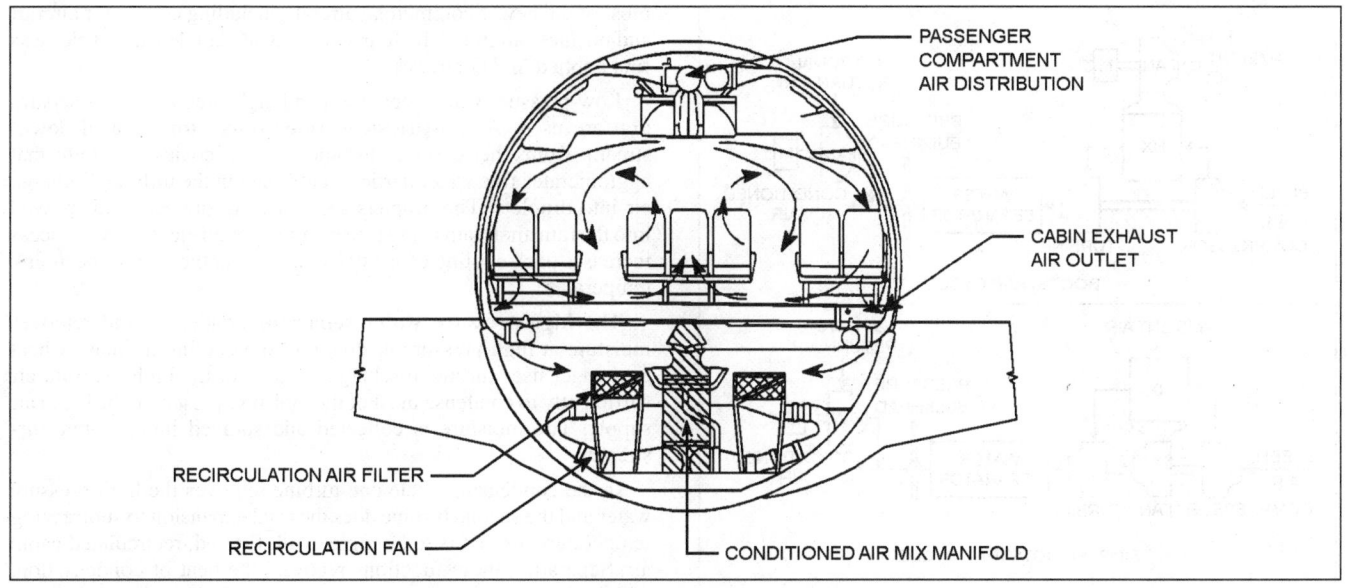

Fig. 5   Cabin Airflow Patterns
(Hunt et al. 1995)

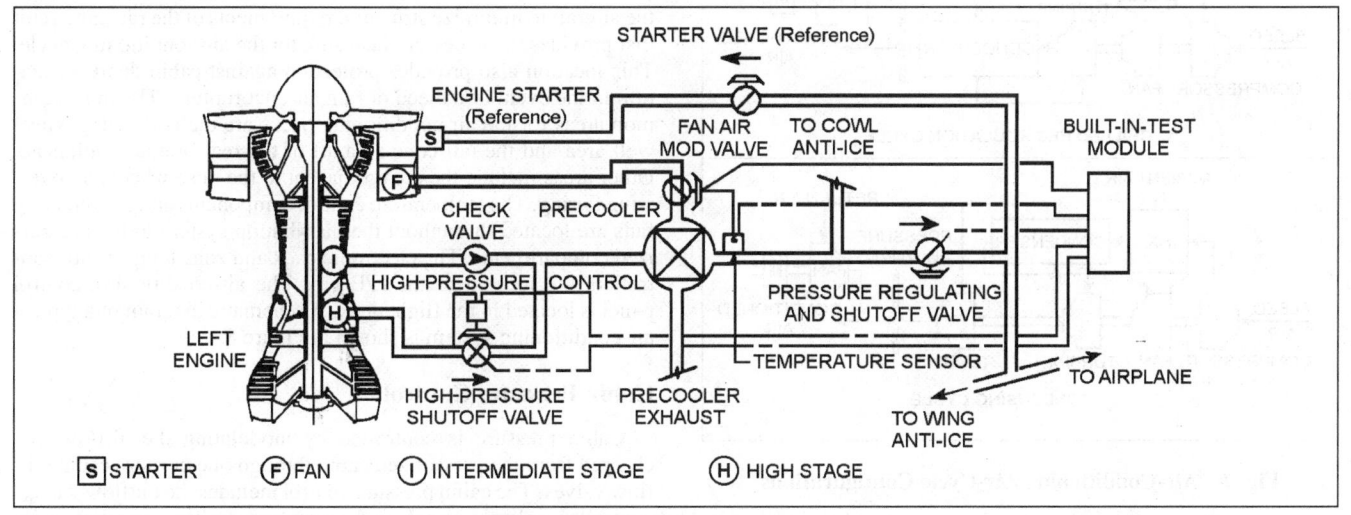

Fig. 6   Typical Engine Bleed Air System Schematic
(Hunt et al. 1995)

extending along the wing leading edge. Similar arrangements are used for anti-icing the engine cowl and tail section.

Air Conditioning

Air-cycle refrigeration is the predominant means of air conditioning for commercial and military aircraft. The reverse-Brayton cycle or Brayton refrigeration cycle is used as opposed to the Brayton power cycle that is used in gas turbine engines. The difference between the two cycles is that in the power cycle fuel in a combustion chamber adds heat, and in the refrigeration cycle a ram-air heat exchanger removes heat. The familiar Rankine vapor cycle, which is used in building and automotive air conditioning and in domestic and commercial refrigeration, has limited use for aircraft air conditioning. The Rankine cycle is used mostly in unpressurized helicopters and general aviation applications.

In an air cycle, compression of the ambient air by the gas turbine engine compressor provides the power input. The heat of compression is removed in a heat exchanger using ambient air as the heat sink. This cooled air is refrigerated by expansion across a turbine powered by the compressed bleed air. The turbine energy resulting from the isentropic expansion is absorbed by a second rotor, which is either a ram air fan, bleed air compressor, or both. This assembly is called an **air-cycle machine** (ACM). Moisture condensed during the refrigeration process is removed by a water separator.

The compartment supply temperature is controlled by mixing hot bleed air with the refrigerated air to satisfy the range of heating and cooling. Other more sophisticated means of temperature control are often used; these include ram air modulation, various bypass schemes in the air-conditioning pack, and downstream controls that add heat for individual zone temperature control.

The bleed airflow is controlled by a flow control or pressure regulating valve at the inlet of the air-conditioning pack. This valve controls bleed air used to minimize engine bleed penalties while satisfying the aircraft outside air ventilation requirements.

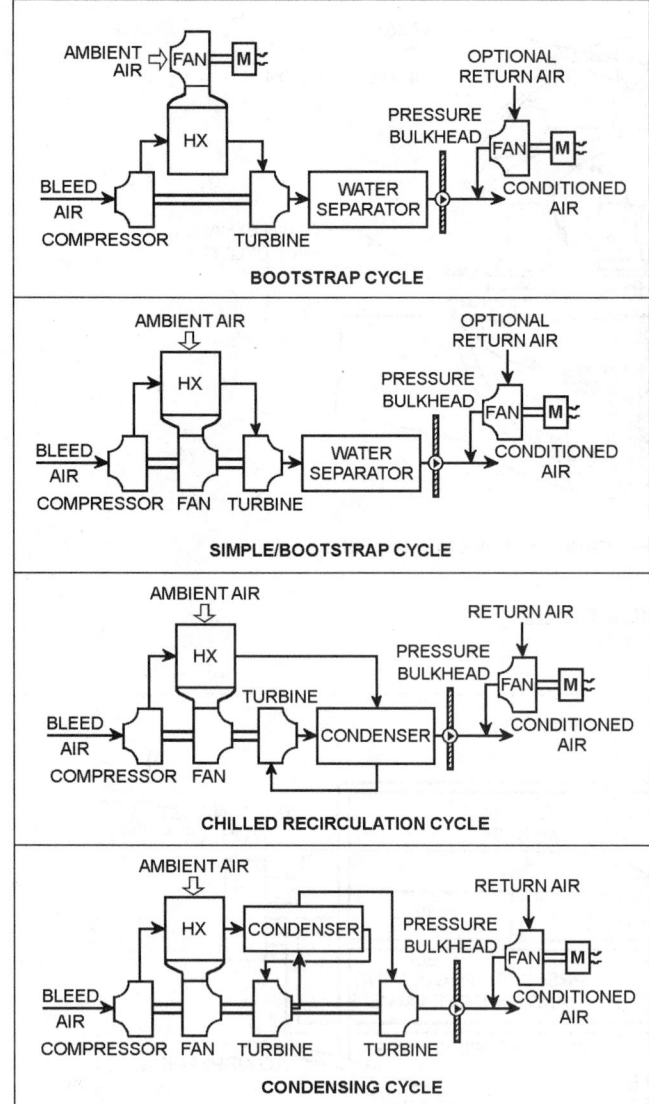

Fig. 7 Air-Conditioning Air-Cycle Configurations

Most aircraft use two or three air-cycle packs operating in parallel to compensate for failures during flight and to allow the aircraft to be dispatched with certain failures. However, many business and commuter aircraft use a single-pack air conditioner; high-altitude aircraft that have single pack also have emergency pressurization equipment that uses precooled bleed air.

The most common types of air-conditioning cycles in use on commercial transport aircraft are shown in Figure 7. All equipment in common use on commercial and military aircraft are open loop, although many commercial aircraft systems include various means of recirculating cabin air to minimize engine bleed air use without sacrificing cabin comfort. The basic differences between the systems are the type of air-cycle machine used and its means of water separation.

The most common air-cycle machines in use are the bootstrap ACM consisting of a turbine and compressor; the three-wheel ACM consisting of a turbine, compressor, and fan; and the four wheel ACM consisting of a two turbines, a compressor, and a fan. The bootstrap ACM is most commonly used for military applications, although many older commercial aircraft models use the bootstrap cycle. The three-wheel ACM (simple bootstrap cycle) is used on

most of the newer commercial aircraft, including commuter aircraft and business aircraft. The four-wheel ACM (condensing cycle) was first applied in 777 aircraft.

Low-pressure water separation and high-pressure water separation are used. A **low-pressure water separator**, located downstream from the cooling turbine, has a coalescer cloth that agglomerates fine water particles entrained in the turbine discharge air into droplets. The droplets are collected, drained, and sprayed into the ram airstream using a bleed air powered ejector; this process increases pack cooling capacity by depressing the ram-air heat-sink temperature.

The **high-pressure water separator** condenses and removes moisture at high pressure upstream of the cooling turbine. A heat exchanger uses turbine discharge air to cool the high-pressure air sufficiently to condense most of the moisture present in the bleed air supply. The moisture is collected and sprayed into the ram airstream.

In the condensing cycle one turbine removes the high-pressure water and the second turbine does the final expansion to subfreezing temperature air that is to be mixed with filtered, recirculated cabin air. Separating these functions recovers the heat of condensation, which results in a higher cycle efficiency. It also eliminates condenser freezing problems because the condensing heat exchanger is operated above freezing conditions.

The air-conditioning packs are located in unpressurized areas of the aircraft to minimize structural requirements of the ram air circuit that provides the necessary heat sink for the air-conditioning cycle. This location also provides protection against cabin depressurization in the event of a bleed or ram air duct rupture. The most common areas for the air-conditioning packs are the underwing/wheel well area and the tail cone area aft of the rear pressure bulkhead. Other areas include the areas adjacent to the nose wheel and overwing fairing. The temperature control components and recirculating fans are located throughout the distribution system in the pressurized compartments. The electronic pack and zone temperature controllers are located in the E/E bay. The air-conditioning control panel is located in the flight deck. A schematic diagram of a typical air-conditioning system is shown in Figure 8.

Cabin Pressure Control

Cabin pressure is controlled by modulating the airflow discharged from the pressurized cabin through one or more cabin outflow valves. The cabin pressure control includes the outflow valves, controller, selector panel, and redundant positive pressure relief valves. Provisions for negative pressure relief are incorporated in the relief valves and/or included in the aircraft structure (door). The system controls the cabin ascent and descent rates to acceptable comfort levels, and maintains cabin pressure altitude in accordance with cabin-to-ambient differential pressure schedules. Modern controls usually set landing field altitude if not available from the flight management system (FMS), and monitor aircraft flight via the FMS and the air data computer (ADC) to minimize the cabin pressure altitude and rate of change.

The cabin pressure modulating valves and safety valves (positive pressure relief valves) are located either on the aircraft skin in the case of large commercial aircraft, or on the fuselage pressure bulkhead in the case of commuter, business and military aircraft. Locating the outflow valves on the aircraft skin precludes the handling of large airflows in the unpressurized tailcone or nose areas and provides some thrust recovery; however, these double gate valves are more complex than the butterfly valves or poppet-type valves used for bulkhead installations. The safety valves are poppet-type valves for either installation. Most commercial aircraft have electronic controllers located in the E/E bay. The cabin pressure selector panel is located in the flight deck.

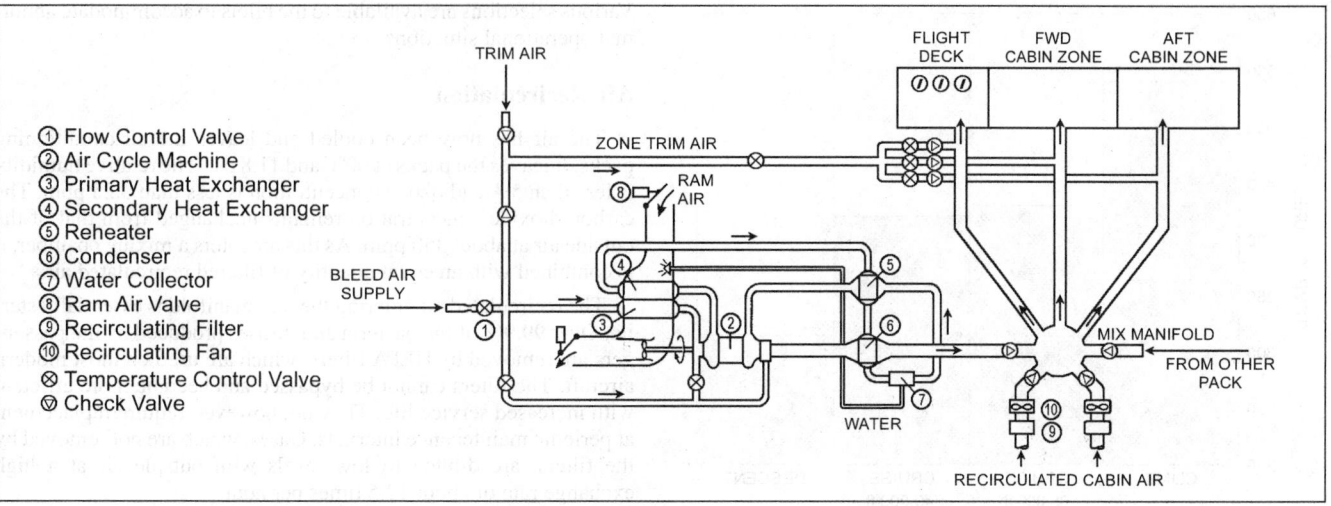

Fig. 8 Typical Aircraft Air-Conditioning Schematic

① Flow Control Valve
② Air Cycle Machine
③ Primary Heat Exchanger
④ Secondary Heat Exchanger
⑤ Reheater
⑥ Condenser
⑦ Water Collector
⑧ Ram Air Valve
⑨ Recirculating Filter
⑩ Recirculating Fan
⊗ Temperature Control Valve
⊘ Check Valve

TYPICAL FLIGHT

A typical flight scenario from London's Heathrow Airport to Los Angeles International Airport would be as follows:

While the aircraft is at the gate, the ECS can be powered by bleed air supplied by the auxiliary power unit (APU), bleed air from a ground cart, or bleed air from the main engines after engine start. The APU or ground-cart bleed air is ducted directly to the bleed air manifold upstream of the air-conditioning packs. Engine bleed air is preconditioned by the engine bleed air system prior to being ducted to the air-conditioning packs. The air-conditioning packs are normally switched to operation using engine bleed air after the engines are started.

Taxiing from the gate at Heathrow, the outside air temperature is 59°F with an atmospheric pressure of 14.7 psia. The aircraft engines are at low thrust, pushing the aircraft slowly along the taxiway.

Engine Bleed Air Control

As outside air enters the compressor stages of the engine, it is compressed to 32 psia and a temperature of 330°F. Some of this air is then extracted from the engine core through one of two bleed port openings in the side of the engine. Which bleed port extracts the air depends on the positioning of valves that control the ports. One bleed port is at the engine's fifteenth compressor stage, commonly called high stage. The second is at the eighth compressor stage, commonly called low stage or intermediate stage. The exact stage varies depending on engine type. The high stage is the highest air pressure available from the engine compressor. At low engine power, the high stage is the only source of air at sufficient pressure to meet the needs of the bleed system. The bleed system is totally automatic, except for a shutoff selection available to the pilots on the overhead panel in the flight deck.

As the aircraft turns onto the runway, the pilots advance the engine thrust to takeoff power. The engine's high stage compresses the air to 1200°F and 430 psia. This energy level exceeds the requirements for the air-conditioning packs and other pneumatic services—approximately 50% of the total energy available at the high stage port cannot be used. However, the bleed system automatically switches to the low-stage port, which conserves energy.

Because the engine must cope with widely varying conditions from ground level to flight at an altitude of up to 43,100 ft, during all seasons and throughout the world, the air at the high or low stage of the engine compressor will seldom exactly match the needs of the pneumatic systems. Excess energy must be discarded as waste heat.

The bleed system constantly monitors engine conditions and selects the least wasteful port. Even so, bleed temperatures often exceed safe levels at which fuel will not auto-ignite. The precooler automatically discharges excess energy to the atmosphere to ensure that the temperature of the pneumatic manifold is well below that which could ignite fuel. The precooler is particularly important in the event of a fuel leak.

The aircraft climbs to a cruise altitude of 39,000 ft where the outside air temperature is −70°F at an atmospheric pressure of 2.9 psia, and the partial pressure of oxygen in 0.5 psi. Until the start of descent to Los Angeles, the low compressor stage is able to compress the low-pressure cold outside air to a pressure of more than 30 psia and temperature above 400°F. This conditioning of the air is all accomplished through the heat of compression—fuel is added only after the air has passed through the compressor stages of the engine core.

Figure 9 shows the temperature of the air being extracted from the engine compressor to the bleed system from the time of departure at London to the time of arrival at Los Angeles. The temperature of the air supplied to the bleed system far exceeds that required to destroy any microorganisms present in the outside air during any point of the trip. At cruising altitudes, outside air contains very few biological particles (a few dark fungal spores per cubic metre of air), although there are higher concentrations at lower altitudes. Because of the high air temperatures from the engine compressor, the air supplied to the air-conditioning packs is sterile.

Air leaving the bleed system while in cruise enters the pneumatic manifold at a temperature of 400°F and a pressure of 30 psia. The air then passes through an ozone converter on its way to the air-conditioning packs located under the wing at the center of the aircraft.

Ozone Protection

While flying at 39,000 ft, several ozone plumes are encountered. Some have ozone concentrations as high as 0.8 parts per million (ppm) or 0.62 ppm sea level equivalent (SLE). This assumes a worst case flight during the month of April, when ozone concentrations are highest. If this concentration of ozone were introduced into the cabin, passengers and crew could experience some chest pain, coughing, shortness of breath, fatigue, headache, nasal congestion, and eye irritation.

Atmospheric ozone dissociation occurs when the ozone goes through the compressor stages of the engine, the ozone catalytic converter (which is on aircraft with a route structure that can encounter high ozone concentrations), and the air-conditioning packs. The ozone further dissociates when contacting airplane

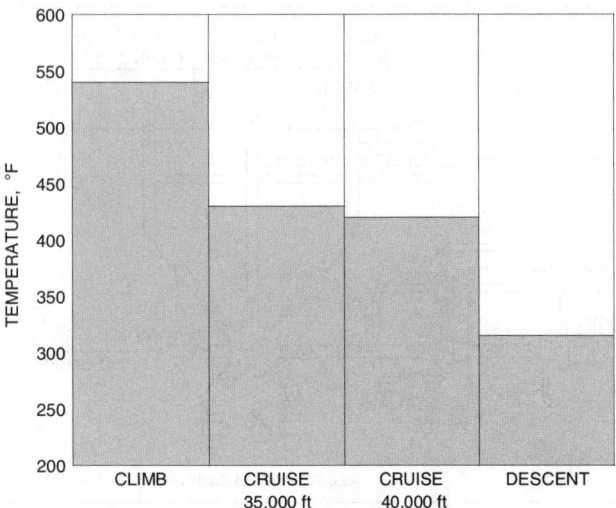

Fig. 9 Typical Air Temperature Supplied to Bleed System During Flight (Hunt et al. 1995)

ducts, interior surfaces, and the airplane recirculation system. The ozone converter dissociates ozone to oxygen molecules by the catalyzing action of a noble catalyst such as palladium. A new converter dissociates approximately 95% of the ozone entering the converter to oxygen. It has a useful life of about 12,000 flight hours.

As the air leaves the ozone converter, it is still at 400°F and a pressure of 30 psia. Assuming a worst case when the converter is approaching the end of its useful life with an ozone conversion efficiency of 60%, the ozone concentration leaving the converter is about 0.25 ppm SLE. This air goes through the air-conditioning packs and enters the cabin. The ozone concentration in the cabin is about 0.09 ppm. As mentioned in the section on Regulations, the FAA sets a three-hour time-weighted average ozone concentration limit in the cabin of 0.1 ppm and a peak ozone concentration limit of 0.25 ppm.

Air Conditioning and Temperature Control

The air next enters the air-conditioning packs. The air-conditioning pack provides essentially dry, sterile, and dust free conditioned air to the airplane cabin at the proper temperature, flow rate, and pressure to satisfy pressurization and temperature control requirements. For most aircraft, this is approximately 5 cfm per passenger. To ensure redundancy, two air-conditioning packs (two are typical, some aircraft have more) provide a total of about 10 cfm of conditioned air per passenger. An equal quantity of filtered, recirculated air is mixed with the air from the air-conditioning packs for a total of approximately 20 cfm per passenger. This quantity of supply air results in a complete cabin air exchange about every 2.5 min, or about 25 air changes per hour. The high air exchange rate is necessary to control temperature gradients, prevent stagnant cold areas, maintain air quality, and dissipate smoke and odors in the cabin. Temperature control is the predominant driver of outside airflow requirements.

The automatic control for the air-conditioning packs constantly monitors airplane flight parameters, the flight crew's selection for the temperature zones, the cabin zone temperature, and the mixed distribution air temperature. The control automatically adjusts the various valves for a comfortable environment under normal conditions. The pilot's controls are located on the overhead panel in the flight deck along with the bleed system controls. Normally, pilots are required only to periodically monitor the compartment temperatures from the overhead panel. Temperatures can be adjusted based on flight attendant reports of passengers being too hot or too cold.

Various selections are available to the pilots to accommodate abnormal operational situations.

Air Recirculation

The air has now been cooled and leaves the air-conditioning packs. It leaves the packs at 60°F and 11.8 psi. The relative humidity is less than 5% and ozone concentration is less than 0.25 ppm. The carbon dioxide concentration remains unchanged from that of the outside air at about 350 ppm. As this air enters a mixing chamber, it is combined with an equal quantity of filtered recirculated air.

The recirculated air entering the mix manifold is essentially sterile. Over 99.9% of the bacteria and viruses produced by the passengers are removed by HEPA filters, which are used on most modern aircraft. The filters cannot be bypassed and become more efficient with increased service life. They do, however, require replacement at periodic maintenance intervals. Gases, which are not removed by the filters, are diluted to low levels with outside air at a high exchange rate of about 12.5 times per hour.

Air Distribution

The air flows from the mix manifold into duct risers dedicated to each seating zone. The risers direct the air from below the floor to the overhead cabin ventilation system. Trim air (hot bleed air from the pneumatic manifold) is added in the risers to increase the air temperature, if needed. The supply air temperature per seating zone can vary due to differences in seating densities between seating zones.

The overhead air distribution network runs the length of the cabin. The air is dust free and sterile with a relative humidity of 10% to 20%. The temperature is 65 to 85°F, depending upon the seating zone the air is being supplied to, and the carbon dioxide concentration is about 1050 ppm. The carbon dioxide is generated by passenger respiration.

Due to the large quantity of air entering the relatively small volume of the cabin, as compared to a building, control of the airflow patterns is required to give comfort without draftiness. Air enters the passenger cabin from overhead distribution outlets that run the length of the cabin. These outlets create circular airflow patterns in the cabin (Figure 5). Air leaves the outlets at a velocity of more than 500 fpm, becomes entrained with cabin air, and maintains sufficient momentum to sweep the cabin walls and floor and to wash out any cold pockets of air in the cabin. The air direction is oriented to avoid exposed portions of a seated passenger, such as the arms, hand, face, neck, and legs; yet it is of sufficient velocity to avoid the sensation of stagnant air. This requires seated passenger impingement velocities between 20 and 70 fpm.

The air volume circulates in the cabin while continuously mixing with cabin air for 2 to 3 min before it enters the return air grilles that are located in the sidewalls near the floor and run the length of the cabin along both sides. While this air is in the cabin, about 0.33% of the oxygen is consumed by human metabolism. The oxygen is replaced by an equal quantity of carbon dioxide from passenger respiration. In addition, the return air entrains microorganisms or other contaminants from passengers or the cabin itself. Approximately one-half of the return air is exhausted overboard and the other half recirculated to sterile conditions through HEPA filters.

In the aft section, exhaust air is extracted by the cabin pressure outflow valve and exhausted overboard. In the forward section, it is continuously extracted from below the floor by recirculation fans filtered, and then mixed with the outside air being supplied by the air-conditioning packs.

The cabin ventilation is balanced so that air supplied at one seat row leaves at approximately the same seat row. This minimizes airflow in the fore and aft directions, to minimize the spread of passenger-generated contaminants.

Cabin Pressure Control

The cabin pressure control continuously monitors ground and flight modes, altitude, climb, cruise or descent modes, as well as the airplane's holding patterns at various altitudes. It uses this information to position the cabin pressure outflow valve to maintain cabin pressure as close to sea level as practical, without exceeding a cabin-to-outside pressure differential of 8.60 psi. At a 39,000 ft cruise altitude, the cabin pressure is equivalent to 6900 ft or a pressure of 11.5 psia. In addition, the outflow valve repositions itself to allow more or less air to escape as the airplane changes altitude. The resulting cabin altitude is consistent with airplane altitude within the constraints of keeping pressure changes comfortable for passengers. Normal pressure change rates are 0.26 psi per minute ascending and 0.16 psi per minute descending.

The cabin pressure control system panel is located in the pilot's overhead panel near the other air-conditioning controls. Normally, the cabin pressure control system is totally automatic, requiring no attention from the pilots.

AIR QUALITY

Aircraft cabin air quality is a complex function of many variables including ambient air quality, the design of the cabin volume, the design of the ventilation and pressurization systems, the way the systems are operated and maintained, the presence of sources of contaminants, and the strength of such sources. The following factors can individually or collectively affect aircraft cabin air quality.

Airflow

The total volume of air is exchanged approximately every 2.5 to 3 min in a wide-body aircraft, and every 2 to 3 min on a standard-body aircraft. The airflow per unit length of the airplane is the same for all sections. However, economy class has a lower airflow per passenger because of its greater seating density as compared to first class and business class. A high air exchange rate and sufficient quantity of outside air is supplied to each cabin zone to maintain air quality, control temperature gradients, prevent stagnant cold areas, and dissipate smoke and odors in the cabin.

The flight deck is provided with a higher airflow per person than the cabin in order to maintain a positive pressure in the cockpit (1) to prevent smoke ingress from adjacent areas (abnormal condition), (2) to provide cooling for electrical equipment, (3) to account for increased solar loads and night heat loss through the airplane skin and windows, and (4) to minimize temperature gradients.

Typical commercial transport aircraft provide approximately 50% conditioned (outside) and 50% filtered recirculated air to the passenger cabin on a continuous basis. Increasing the quantity of outside air beyond 50% to the cabin would lower the cabin CO_2 concentration slightly, but it would also increase the potential cabin ozone concentration and lower the cabin relative humidity.

The recirculated air is cleaned by drawing it through HEPA filters; the filters cannot be bypassed. The air distribution system is designed to provide approximately 10 cfm outside air and 10 cfm filtered, recirculated air per passenger. A fully-loaded, all-tourist-class passenger aircraft, which has the maximum seating density throughout the airplane, provides an outside airflow per passenger of 6.5 cfm (standard-body) to 8 cfm (wide-body).

The outside air quantity supplied on some aircraft models can be lowered by shutting off one air-conditioning pack. The flight crew has control of these packs to provide flexibility in case of a system failure or for special use of the aircraft. Packs should be in full operation whenever passengers are on board.

Environmental Tobacco Smoke (ETS)

Currently, there are no governmental, occupational, or ambient standards for environmental tobacco smoke (ETS). However, in 1986, the National Academy of Sciences recommended banning smoking on U.S. domestic flights to eliminate the possibility of fires caused by cigarettes, to lessen irritation and discomfort to passengers and crew, and to reduce potential health hazards.

On flights where smoking is allowed, ETS in the cabin is controlled indirectly by controlling the concentration of carbon monoxide (CO) and respirable suspended particulates (RSP). CO and RSP are tracer constituents of ETS and several standards do list acceptable maximum levels for these constituents. (Table 1 and Table 2 in Chapter 9 of the 1997 *ASHRAE Handbook—Fundamentals* summarize these standards.) This method of control does not consider other constituents present in ETS.

Measured CO levels in the smoking section(s) of aircraft during peak smoking are within acceptable limits. RSP concentrations in the smoking sections can exceed recommended levels during peak smoking, which is also true of most heavy smoking areas (e.g. restaurants, bowling alleys, etc.). Of 92 airplanes tested in a DOT sponsored study, average RSP values of 40 µg/m^3 and 175 µg/m^3 were measured in the nonsmoking and smoking sections, respectively.

Ozone

Ozone is present in the atmosphere as a consequence of the photochemical conversion of oxygen by solar ultraviolet radiation. Ozone levels vary with season, altitude, latitude, and weather systems. A marked and progressive increase in ozone concentration occurs in the flight altitude of commercial aircraft. The mean ambient ozone concentration increases with increasing latitude, is maximal during the spring (fall season for southern latitudes), and often varies when weather causes high ozone plumes to descend.

Residual cabin ozone concentration is a function of the ambient concentration, the design of the air distribution system and how it is operated and maintained, and whether or not catalytic ozone converters are installed.

Cabin ozone limits are set by FAR 121.578 and FAR 25.832. The use of catalytic ozone converters is generally required on airplanes flying mission profiles where the cabin ozone levels are predicted to exceed these FAR limits (refer to the FAA Code of Federal Regulations for other compliance methods).

Microbial Aerosols

Biologically derived particles that become airborne include viruses, bacteria, actinomycetes, fungal spores and hyphae, arthropod fragments and droppings, and animal and human dander. Only one study has documented the occurrence of an outbreak of infectious disease related to airplane use. In 1977, because of an engine malfunction, an airliner with 54 persons onboard was delayed on the ground for 3 h, during which the airplane ventilation system was reportedly turned off. Within 3 days of the incident, 72% of the passengers became ill with influenza. One passenger (the index case) was ill while the airplane was delayed.

With the ventilation system shut off, no fresh air was introduced into the cabin to displace microbial aerosols and CO_2 or to control cabin temperatures. It is believed that had the ventilation system been operating during the delay, the possibility of other passengers becoming ill would have been minimal.

The airplane ventilation system should never be shut off when passengers are on board; an exception to this recommendation is during no-pack takeoffs when the air packs (but not the recirculation fans) are shut off for the short time only on takeoff.

To remove particulates and biological particles from the recirculated air, filter assemblies that contain a HEPA filter that has a minimum efficiency of 94 to 99.97% DOP as measured by MIL-STD-282 should be used. A HEPA filter is rated using 0.3 µm size particles. A filter's efficiency increases over time as particulates become trapped by the filter. Due to the overlap of capture mechanisms in a filter, the efficiency also increases for particles smaller and larger than the most penetrating particle size (MPPS). For an airplane filter, the MPPS is about 0.1 to 0.2 μm.

The efficiency of the filter to remove 0.003 μm particles from the air is in excess of 99.9+%. Most bacteria (99%) are larger than 1 μm. Viruses are approximately 0.003 to 0.05 μm in size. Test results in a DOT study conducted on 92 randomly selected flights showed that bacteria and fungi levels measured in the airplane cabin are similar to or lower than those found in the home. These low microbial contaminant levels are due to the large quantity of outside airflow and high filtration of the recirculation system.

Volatile Organic Compounds

Volatile organic compounds (VOCs) can be emitted by material used in furnishings, pesticides, disinfectants, cleaning fluids, and food and beverages. In-flight air quality testing on revenue flights sponsored by The Boeing Company and the Air Transport Association of America (ATAA 1994) detected trace quantities of VOCs, which were considered well below levels that could result in adverse health effects.

Carbon Dioxide

Carbon dioxide is the product of normal human metabolism, which is the predominant source in aircraft cabins. The CO_2 concentration in the cabin varies with outside air rate, the number of people present, and their individual rates of CO_2 production that vary with activity and (to a smaller degree) with diet and health. CO_2 has been widely used as an indicator of indoor air quality, typically serving the function of a surrogate. Per a DOT-sponsored study, measured cabin CO_2 values of 92 randomly selected smoking and nonsmoking flights average 1500 ppm.

The Environmental Exposure Limit adopted by the American Conference of Governmental Industrial Hygienists (ACGIH) is 5000 ppm as the time-weighted average (TWA) limit for CO_2; this value corresponds to a fresh air ventilation rate of 2.3 cfm per person. The TWA is the concentration, for a normal 8 h workday and a 40 hour workweek, to which nearly all workers can be repeatedly exposed, day after day, without adverse effects. As mentioned in the Regulations section under Ventilation, an FAA amendment also limits CO_2 to 5000 ppm (0.5%). Aircraft cabin CO_2 concentrations are below this limit.

ASHRAE *Standard* 62, which does not apply to aircraft per se, states, "Comfort (odor) criteria are likely to be satisfied if the ventilation rate is set so that 1000 ppm CO_2 is not exceeded." It further states, "This level is not considered a health risk but is a surrogate for human comfort (odor)." An interpretation of this standard noted that 1000 ppm CO_2 is not a requirement of the standard, but it can be considered a target concentration level.

Humidity

The relative humidity in airplanes tested in a DOT-sponsored study ranged from approximately 5% to 35% with an average of 15% to 20%. The humidity is made up mainly of moisture from passengers and will increase with more passengers and decrease with increased outside airflow. A major benefit of filtered, recirculated air supplied to the passenger cabin is an increase in cabin humidity compared to airplanes with only outside supply air.

After three or four hours of exposure to relative humidity in the 5 to 10% range, some passengers may experience such discomfort as dryness of the eyes, nose, and throat. However, no serious adverse health effects of low relative humidity on the flying population have been documented.

Cabin Pressure/Oxygen

At a normal airplane cruise altitude, aircraft cabins are pressurized to a maximum cabin altitude of 8000 ft (to compress the ambient air to a form that is physiologically acceptable). A DOT-sponsored National Academy of Sciences study concluded that current pressurization criteria and regulations are generally adequate to protect the traveling public. The Academy also noted that the normal maximum rates of change of cabin pressure (approximately 500 ft/min in increasing altitude and 300 ft/min in decreasing altitude) are such that they do not pose a problem for the typical passenger.

However, pressurization of the cabin to equivalent altitudes of up to 8000 ft, as well as changes in the normal rates of pressure during climb and descent, may create discomfort for some people such as those suffering from upper respiratory or sinus infections, obstructive pulmonary diseases, anemias, or certain cardiovascular conditions. In those cases, supplemental oxygen may be recommended. Children and infants sometimes experience discomfort or pain because of pressure changes during climb and descent. Injury to the middle ear has occurred to susceptible people, but is rare.

Some articles and reports state that substandard conditions exist in airplane cabins due to a lack of oxygen. Some reports suggest that this condition is exacerbated by reduced fresh air ventilation rates or through the use of recirculated air. These arguments imply that the oxygen content of cabin air is depleted through the consumption by occupants. Humans at rest breathe at a rate of approximately 0.32 cfm while consuming oxygen at a rate of 0.015 cfm. The percent oxygen makeup of the supply air remains at approximately 21% at cruise altitude. A person receiving 10 cfm of outside air and 10 cfm of recirculation air would therefore receive approximately 4.2 cfm of oxygen. Consequently, the content of oxygen in cabin air is little affected by breathing as it is replaced in sufficient quantities compared to the human consumption rate.

Although the percentage of oxygen in cabin air remains virtually unchanged (21%) at all normal flight altitudes, the partial pressure of oxygen decreases with increasing altitude, which decreases the amount of oxygen held by the blood's hemoglobin. The increase in cabin altitude may cause low grade hypoxia (reduced tissue oxygen levels) to some people. Low grade hypoxia in combination with other stresses is the main cause of passenger fainting and fatigue. However, the National Academy of Sciences concluded that pressurization of the cabin to an equivalent altitude of 5000 to 8000 ft is physiologically safe—no supplemental oxygen is needed to maintain sufficient arterial oxygen saturation.

REFERENCES

ATAA. Specification for manufacturers technical data. *Specification No. 100.* Air Transport Association of America, Washington, DC.

ATAA. 1994. *Airline cabin air quality study.* Air Transport Association of America, Washington, DC.

ASHRAE. 1991. Air quality, ventilation, temperature and humidity in aircraft. *ASHRAE Journal* (4).

ASHRAE 1981. Ventilation for acceptable indoor air quality. *Standard* 62-1981.

ASHRAE 1989. Ventilation for acceptable indoor air quality. ANSI/ASHRAE *Standard* 62-1989.

DOT. 1989. Airliner cabin environment: Contaminant measurements, health risks, and mitigation options. U.S. Department of Transportation, Washington, DC.

FAA. Airworthiness standards: Transport category airplanes. *Federal Aviation Regulations*, Part 25.

FAA. Certification and operations: Domestic, flag and supplemental air carriers and commercial operators of large aircraft. *Federal Aviation Regulations*, Part 121.

Hunt, E.H. and D.R. Space. 1995. The airplane cabin environment: Issues pertaining to flight attendant comfort. The Boeing Company, Seattle.

Hunt, E.H., D.H. Reid, D.R. Space, and F.E. Tilton. 1995. Commercial airliner environmental control system: Engineering aspects of cabin air quality. Presented at the Aerospace Medical Association annual meeting, Anaheim, CA.

JAA. Joint airworthiness requirements: Part 25: Large aeroplanes. Airworthiness Authorities Steering Committee. Publisher: Civil Aviation Authority, Cheltenham, England.

NAS. 1986. *The airliner cabin environment: Air quality and safety.* National Academy of Sciences, National Academy Press, Washington, DC.

SHIPS

THIS chapter covers air conditioning for oceangoing surface vessels, including luxury liners, tramp steamers, and naval vessels. Although the general principles of air conditioning that apply to land installations also apply to marine installations, some types are not suitable for ships due to their inability to meet shock and vibration requirements. The chapter focuses on load calculations and air distribution for these ships.

BASIC CRITERIA

Air conditioning in ships provides an environment in which personnel can live and work without heat stress. It also increases crew efficiency, improves the reliability of electronic and similar critical equipment, and prevents rapid deterioration of special weapons equipment aboard naval ships.

The following factors should be considered in the design of air conditioning for shipboard use:

1. It should function properly under conditions of roll and pitch.
2. The construction materials should withstand the corrosive effects of salt air and seawater.
3. It should be designed for uninterrupted operation during the voyage and continuous year-round operation. Because ships en route cannot be easily serviced, some standby capacity, spare parts for all essential items, and extra refrigerant charges should be carried.
4. It should have no objectionable noise or vibration, and must meet the noise criteria required by shipbuilding specifications.
5. It should meet the special requirements for operation given in Section 4 of ASHRAE Standard 26.
6. The equipment should occupy a minimum of space commensurate with its cost and reliability. Its weight should be kept to a minimum.
7. Because a ship may pass through one or more complete cycles of seasons on a single voyage and may experience a change from winter to summer operation in a matter of hours, the system should be flexible enough to compensate for climatic changes with minimal attention of the ship's operating personnel.
8. Infiltration through weather doors is generally disregarded. However, specifications for merchant ships occasionally require an assumed infiltration load for heating steering gear rooms and the pilothouse.
9. Sun load must be considered on all exposed surfaces above the waterline. If a compartment has more than one exposed surface, the surface with the greatest sun load is used, and the other exposed boundary is calculated at outside ambient temperature.

10. Cooling load inside design conditions are given as a dry-bulb temperature with a maximum relative humidity. For merchant ships, the cooling coil leaving air temperature is assumed to be 49°F dry bulb. For naval ships, it is assumed to be 51.5°F dry bulb. For both naval and merchant ships, the wet bulb is consistent with 95% rh. This off-coil air temperature is changed only when humidity control is required in the cooling season.
11. When calculating winter heating loads, heat transmission through boundaries of machinery spaces in either direction is not considered.

Calculations for Merchant Ship Heating, Ventilation, and Air Conditioning Design, a bulletin available from the Society of Naval Architects and Marine Engineers (SNAME), gives sample calculation methods and estimated values.

MERCHANT SHIPS

DESIGN CRITERIA

Outdoor Ambient Temperature

The service and type of vessel determines the proper outdoor design temperature. Some luxury liners make frequent off-season cruises where more severe heating and cooling loads may be encountered. The selection of the ambient design should be based on the temperatures prevalent during the voyage. In general, for the cooling cycle, outdoor design conditions for North Atlantic runs are 95°F dry bulb and 78°F wet bulb; for semitropical runs, 95°F dry bulb and 80°F wet bulb; and for tropical runs, 95°F dry bulb and 82°F wet bulb. For the heating cycle, 0°F is usually selected as the design temperature, unless the vessel will always operate in higher temperature climates. The design temperatures for seawater are 85°F in summer and 28°F in winter.

Indoor Temperature

Effective temperatures (ETs) from 71 to 74°F are generally selected as inside design conditions for commercial oceangoing surface ships.

Inside design temperature ranges from 76 to 80°F dry bulb and approximately 50% rh for summer and from 65 to 75°F dry bulb for winter.

Comfortable room conditions during intermediate outside ambient conditions should be considered in the design. Quality systems are designed to (1) provide optimum comfort when outdoor ambient conditions of 65 to 75°F dry bulb and 90 to 100% rh exist, and (2) ensure that proper humidity and temperature are maintained during periods when sensible loads are light.

Ventilation Requirements

Ventilation must meet the requirements given in ASHRAE Standard 62, except when special consideration is required for ships.

The preparation of this chapter is assigned to TC 9.3, Transportation Air Conditioning.

Air-Conditioned Spaces. In public spaces (e.g., mess rooms, dining rooms, and lounges), 20 to 30 cfm of outdoor air per person or 3 air changes per hour (ACH) are required. In all other spaces, at least 15 cfm of outside air per person or 2 ACH must be provided. However, the maximum outside air for any air-conditioned space is 50 cfm per person.

Ventilated Spaces. The fresh air to be supplied to a space is determined by the required rate of change or the limiting temperature rise. The minimum quantity of air to a space is 30 cfm per occupant or 35 cfm per terminal. In addition to these requirements, exhaust requirements must be balanced.

Load Determination. The cooling load estimate for air conditioning considers factors discussed in Chapter 28 of the 1997 *ASHRAE Handbook—Fundamentals*, including the following:

- Solar radiation
- Heat transmission through hull, decks, and bulkheads
- Heat (latent and sensible) dissipation from occupants
- Heat gain due to lights
- Heat (latent and sensible) gain due to ventilation air
- Heat gain due to motors or other electrical equipment
- Heat gain from piping, machinery, and equipment

The cooling effect of adjacent spaces is not considered unless temperatures are maintained with refrigeration or air-conditioning equipment. The latent heat from the scullery, galley, laundry, washrooms, and similar ventilated spaces is assumed to be fully exhausted overboard.

The heating load estimate for air conditioning should consist of the following:

- Heat losses through decks and bulkheads
- Ventilation air
- Infiltration (when specified)

No allowances are made for heat gain from warmer adjacent spaces.

Heat Transmission Coefficients

The overall heat transmission coefficient *U* for the composite structures common to shipboard construction do not lend themselves to theoretical derivation; they are most commonly obtained from full-scale panel tests. SNAME *Bulletin* 4-7 gives a method for determining these coefficients when tested data are unavailable.

Heat Dissipation from People

The rate at which heat and moisture dissipate from people depends on their activity levels and the ambient dry-bulb temperature. Values that can be used at 80°F room dry bulb are listed in Table 1.

Heat Gain from Sources Within a Space

Information regarding the heat gain from motors, appliances, lights, and other equipment should be obtained from the manufacturers. Data in Chapter 28 of the 1997 *ASHRAE Handbook—Fundamentals* may be used when manufacturers' data are unavailable.

Table 1   Heat Gain from Occupants

Activity at 80°F	Heat Rate, Btu/h		
	Sensible	Latent	Total
Dancing	245	605	850
Eating (mess rooms and dining rooms)	220	330	550
Waiters	300	700	1000
Moderate activity (lounge, ship's office, chart rooms)	200	250	450
Light activity (staterooms, crew's berthing)	195	205	400
Workshops	250	510	760

EQUIPMENT SELECTION

The principal equipment required for an air-conditioning system can be divided into four categories:

1. Central station air-handling, consisting of fans, filters, central heating and cooling coils, and sound treatment
2. Distribution network, including air ductwork, water piping, and steam piping
3. Required terminal treatment, consisting of heating and cooling coils, terminal mixing units, and diffusing outlets
4. Refrigeration equipment

When selecting air-conditioning equipment, the following factors should be considered:

- Installed initial cost
- Space available in fan rooms, passageways, machinery rooms, and staterooms
- Operation cost, including maintenance
- Noise levels
- Weight

High-velocity air distribution offers many advantages. The use of unitary (factory-assembled) central air-handling equipment and prefabricated piping, clamps, and fittings facilitates installation for both new construction and conversions. Substantial space-saving is possible as compared to the conventional low-velocity sheet metal ducts. Maintenance is also reduced.

Fans must be selected for stable performance over their full range of operation and should have adequate isolation to prevent transmission of vibration to the deck. Because fan rooms are often located adjacent to or near living quarters, effective sound treatment is essential.

In general, equipment used for ships is considerably more rugged than equipment for land applications, and it must withstand the corrosive environment of the salt air and sea. Materials such as stainless steel, nickel-copper, copper-nickel, bronze alloys, and hot-dipped galvanized steel are used extensively. Sections 6 through 10 of ASHRAE *Standard* 26 list equipment requirements.

Fans

The U.S. Maritime Administration (USMA) specifies a family of standard vaneaxial, tubeaxial, and centrifugal fans. Selection curves found on the standard drawings are used in selecting fans (*Standard Plans* S38-1-101, S38-1-102, S38-1-103). Belt-driven centrifugal fans must conform to these requirements, except for speed and drive.

Cooling Coils

Cooling coils are based on an inlet water temperature of 42°F with a temperature rise of about 10°F and must meet the following requirements:

- Maximum face velocity of 500 fpm
- At least six rows of tubes and 25% more rows than required by the manufacturer's published ratings
- Construction and materials as specified in USMA (1965)

Heating Coils

Heating coils must meet the following requirements:

- Maximum face velocity of 1000 fpm
- Supply preheaters with a final design temperature between 55 and 60°F with outside air at 0°F
- Preheaters for air-conditioning systems suit design requirements but have a discharge temperature not less than 45°F with 100% outside air at 0°F
- Tempering heaters for supply systems have a final design temperature between 50 and 70°F, with outside air at 0°F

- Pressure drop for preheater and reheater in series, with full fan volume, does not exceed 0.5 in. of water
- Capacity of steam heaters is based on a steam pressure of 35 psi gage, less an allowed 5 psi line pressure drop and the design pressure drop through the control valve

Filters

Filters must meet the following requirements:

- All supply systems fitted with cooling and/or heating coils have manual roll, renewable marine-type air filters
- Maximum face velocity of 500 fpm
- Weather protection
- Located so that they are not bypassed when the fan room door is left open
- Clean, medium rolls are either fully enclosed or arranged so that the outside surface of the clean roll becomes the air-entering side as the medium passes through the airstream
- Dirty medium is wound with the dirty side inward

Medium rolls are 65 ft long and are readily available as standard factory-stocked items in nominal widths of 2, 3, 4, or 5 ft. A permanently installed, dry-type filter gage, graduated to read from 0 to 1 in. of water, is installed on each air filter unit.

Air-Mixing Boxes

If available, air-mixing boxes should fit between the deck beams. Other requirements include

- Volume regulation over the complete mixing range within +5% of the design volume, with static pressure in the hot and cold ducts equal to at least the fan design static pressure
- Leakage rate of the hot and cold air valve(s) less than 2% when closed
- Calculations for space air quantities allow for leakage through the hot air damper

Box sizes are based on manufacturer's data, as modified by the above requirements.

Air Diffusers

Air diffusers ventilate the space without creating drafts. In addition, diffusers must meet the following requirements:

- Diffusers serving air-conditioned spaces should be a high-induction type and constructed so that moisture will not form on the cones when a temperature difference of 30°F is used with the space dew point at least 9°F above that corresponding to summer inside design conditions. Compliance with this requirement should be demonstrated in a mock-up test, unless previously tested and approved.
- Volume handled by each diffuser should not, in general, exceed 500 cfm.
- Diffusers used in supply ventilation are of the same design, except have adjustable blast skirts.
- Manufacturer's published data are used in selecting diffusers.

Air-Conditioning Compressors

Compressors should be the same types used for ship's service and cargo refrigeration, except for the number of cylinders or the speed (see Chapter 30, Marine Refrigeration, in the 1998 *ASHRAE Handbook—Refrigeration*).

TYPICAL SYSTEMS

Comfort air-conditioning systems installed on merchant ships are classified as those serving (1) passenger staterooms, (2) crew's quarters and similar small spaces, and (3) public spaces.

Single-Zone Central Air Conditioning

Public spaces are treated as one zone and are effectively handled by single-zone central air conditioning. Exceptionally large spaces may require two systems. Generally, either built-up or factory-assembled fan-coil central station air conditioning is selected for large public spaces. A schematic of a typical air conditioning system, also known as the **Type A** system, is shown in Figure 1. Under certain conditions, it is desirable to avoid the use of return ducts and to use all outdoor air, which requires more refrigeration. Outdoor air and return air are both filtered before being preheated, cooled, and reheated as required at the central station unit. A room thermostat maintains the desired temperature by modulating the valve regulating the flow of steam to the reheat coil. Humidity control is often provided. During mild weather, the dampers frequently are opened to admit 100% outdoor air automatically.

Multizone Central Air Conditioning

Multizone central air conditioning, also known as the **Type C** system, is usually confined to the crew's and officers' quarters (Figure 2). Spaces are divided into zones in accordance with similarity of loads and exposures. Each zone has a reheat coil to supply air at a temperature adequate for all the spaces served. These coils, usually steam-heated, are controlled thermostatically. Manual control of the air volume is the only means for occupant control of conditions. Filters, cooling coils, and dampers are essentially the same as for Type D systems.

Each internal sensible heat load component, particularly solar and lights, varies greatly. The thermostatic control methods cannot

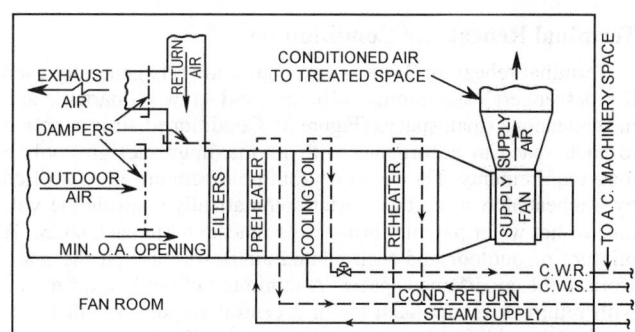

Fig. 1 Single-Zone Central (Type A) System

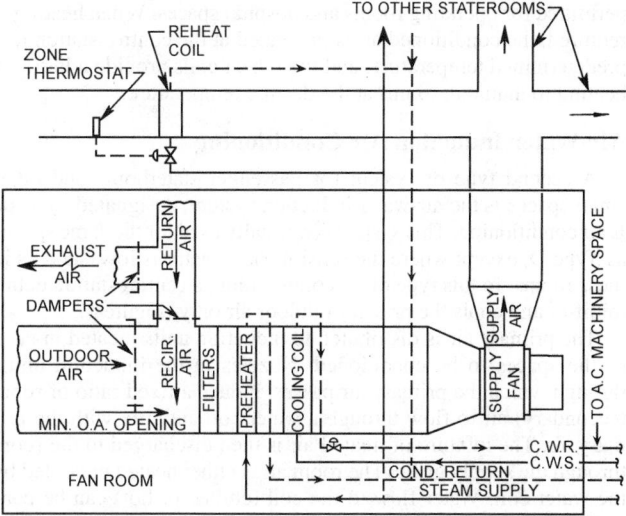

Fig. 2 Multizone Central (Type C) System

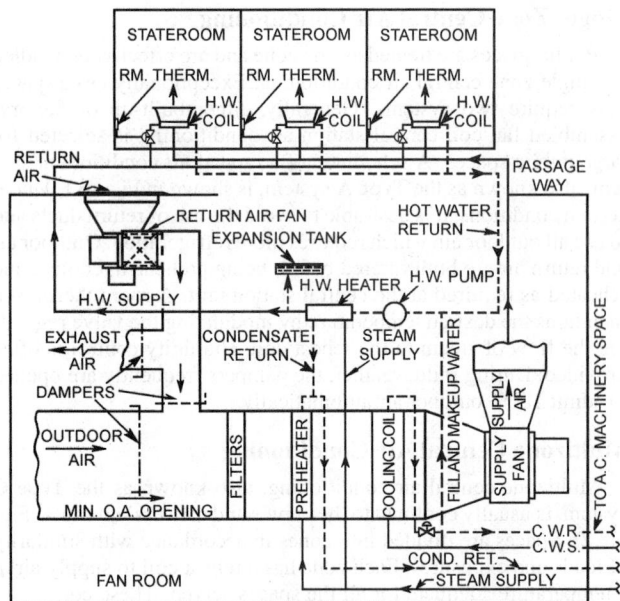

Fig. 3   Terminal Reheat (Type D) System

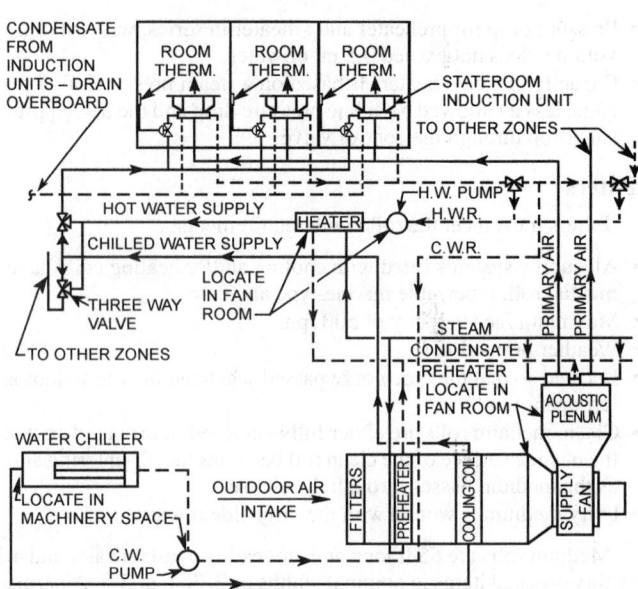

Fig. 4   Air-Water Induction (Type E) System

compensate for these large variations; therefore, this system cannot always satisfy individual space requirements. Volume control is conducive to noise, drafts, and odors.

Terminal Reheat Air Conditioning

Terminal reheat, or **Type D** air conditioning, is generally used for passengers' staterooms, officers' and crew's quarters, and miscellaneous small spaces (Figure 3). Conditioned air is supplied to each space in accordance with its maximum design cooling load requirements. The room dry-bulb temperature is controlled by a reheater. A room thermostat automatically controls the volume of hot water passing through the reheat coil in each space. A mixture of outdoor and recirculated air flows through the duct-work to the conditioned spaces. A minimum of outdoor air mixed with return or recirculated air in a central station system is filtered, dehumidified, and cooled by the chilled water cooling coils, and then distributed by the supply fan through conventional ductwork to the spaces.

Dampers control the volume of outdoor air. No recirculated air is permitted for operating rooms and hospital spaces. When heating is required, the conditioned air is preheated at the control station to a predetermined temperature, and the reheat coils provide additional heating to maintain rooms at the desired temperature.

Air-Water Induction Air Conditioning

A second type of system for passenger staterooms and other small spaces is the air-water induction system, designated as **Type E** air conditioning. This system is normally used for the same spaces as Type D, except where the sensible heat factor is low, such as in mess rooms. In this type of air conditioning, a central station dehumidifies and cools the primary outdoor air only (Figure 4).

The primary air is distributed to induction units located in each of the spaces to be conditioned. Nozzles in the induction units, through which the primary air passes, induce a fixed ratio of room (secondary) air to flow through a water coil and mix with the primary air. The mixture of treated air is then discharged to the room through the supply grille. The room air is either heated or cooled by the water coil. Water flow to the coil (chilled or hot) can be controlled either manually or automatically to maintain the desired room conditions.

This system requires no return or recirculated air ducts because only a fixed amount of outdoor (primary) air needs to be conditioned by the central station equipment. This relatively small amount of conditioned air must be cooled to a sufficiently low dew point to take care of the entire latent load (outdoor air plus room air). It is distributed at high velocity and pressure and thus requires relatively little space for air distribution ducts. However, this saving of space is offset by the additional space required for water piping, secondary water pumps, induction cabinets in staterooms, and drain piping.

Space design temperature is maintained during intermediate conditions (i.e., the outdoor temperature is above the changeover point and chilled water is at the induction units), with primary air heated at the central station unit according to a predetermined temperature schedule. Unit capacity in spaces requiring cooling must be sufficient to satisfy the room sensible heat load plus the load of the primary air.

When outdoor temperature is below the changeover point, the chiller is secured, and space design temperature is maintained by circulating hot water to the induction units. Preheated primary air provides cooling for spaces that have a cooling load. Unit capacity in spaces requiring heating must be sufficient to satisfy the room heat load plus the load of the primary air. Detailed analysis is required to determine the changeover point and temperature schedules for water and air.

High-Velocity Dual-Duct System

High-velocity dual-duct air conditioning, also known as the **Type G** system, is normally used for the same kinds of spaces as Types D and E. In Type G air conditioning, all air is filtered, cooled, and dehumidified in the central units (Figure 5). Blow-through coil arrangements are essential to ensure efficient design. A high-pressure fan circulates air at high velocity approaching 6000 fpm through two ducts or pipes, one carrying cold air and the other warm air. A steam reheater in the central unit heats the air as required.

In each space served, the warm and cold air flow to an air-mixing unit that has a control valve to proportion hot and cold air to obtain the desired temperature. Within the capacity limits of the equipment, any temperature can quickly be obtained and maintained, regardless of load variations in adjacent spaces. The air-mixing unit also incorporates self-contained regulators that maintain a constant volume of total air delivery to the various spaces, regardless of adjustments in the air supplied to rooms down the line.

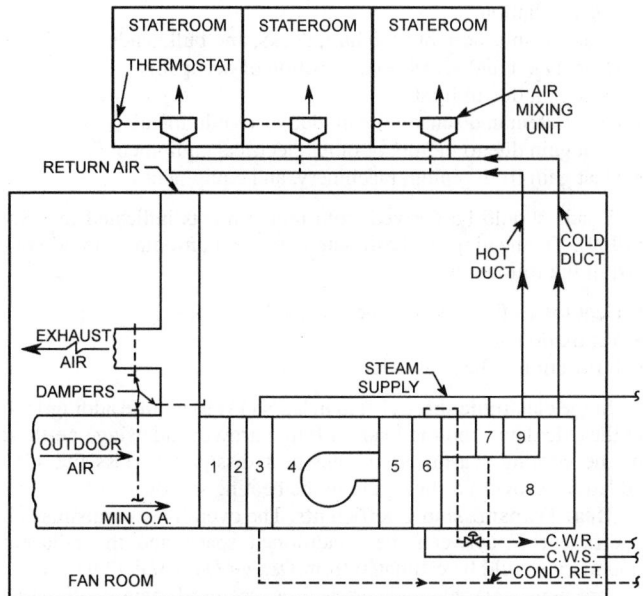

Fig. 5   Dual-Duct (Type G) System

The following are some advantages of this system:

- All conditioning equipment is centrally located, simplifying maintenance and operation
- The system can heat and cool adjacent spaces simultaneously without cycle changeover and with a minimum of automatic controls
- Because only air is distributed from fan rooms, no water or steam piping, electrical equipment, or wiring are in the conditioned spaces
- With all conditioning equipment centrally located, direct-expansion cooling using halocarbon refrigerants is possible, eliminating all intermediary water-chilling equipment

AIR DISTRIBUTION METHODS

Good air distribution in staterooms and public spaces is difficult to achieve due to low ceiling heights and compact space arrangements. The design should consider room dimensions, ceiling height, volume of air handled, air temperature difference between supply and room air, location of berths, and allowable noise. For major installations, mock-up tests are often used to establish the exacting design criteria required for satisfactory performance.

Air usually returns from individual small spaces either by a sight-tight louver mounted in the door or by an undercut in the door leading to the passageway. An undercut door can only be used with small air quantities of 75 cfm or less. Louvers are most commonly sized for a velocity of 400 fpm based on net area.

Ductwork

Ductwork on merchant ships is constructed of steel. Ducts, other than those requiring heavier construction because of susceptibility to damage or corrosion, are usually made with riveted seams sealed with hot solder or fire-resistant duct sealer, welded seams, or hooked seams and laps. They are fabricated of hot-dipped, galvanized, copper-bearing sheet steel, suitably stiffened externally. The minimum thickness of material is determined by the diameter of round ducts or by the largest dimension of rectangular ducts, as listed in Table 2.

The increased application of high-velocity, high-pressure systems has resulted in a greater use of prefabricated round pipe and fittings, including spiral formed sheet metal ducts. It is important that the field fabrication of ducts and fittings ensure that all are airtight.

Table 2   Minimum Thickness of Steel Ducts

All vertical exposed ducts	16 USSG	0.0598 in.
Horizontal or concealed vertical ducts less than 6 in.	24 USSG	0.0239 in.
Horizontal or concealed vertical ducts 6.5 to 12 in.	22 USSG	0.0299 in.
Horizontal or concealed vertical ducts ducts 12.5 to 18 in.	20 USSG	0.0359 in.
Horizontal or concealed vertical ducts 18.5 to 30 in.	18 USSG	0.0476 in.
Horizontal or concealed vertical ducts over 30 in.	16 USSG	0.0598 in.

The use of factory-fabricated fittings, clamps, and joints effectively minimizes air leakage for these high-pressure ducts.

In addition to the space advantage, small ductwork saves weight, another important consideration for this application.

CONTROL

The conditioning load, even on a single voyage, varies over a wide range in a short period. Not only must the refrigeration plant meet these variations in load, but the controls must readily adjust the system to sudden climatic changes. Accordingly, it is general practice to equip the plant with automatic controls. Since comfort is a matter of individual taste, adjustable room thermostats are placed in living spaces.

Manual volume control has also been used to regulate temperatures in cabins and staterooms. For low-velocity air distribution, however, manual volume control tends to disturb the air balance of the remainder of the spaces served. Manual controls are also applied on high-velocity single-duct systems if a constant volume regulator is installed in the terminal box.

Design conditions in staterooms and public spaces can be controlled by one or a combination of the following:

Volume Control. (Used in Type C systems.) This is the least expensive and simplest control. Its disadvantages are an inability to meet simultaneous heating and cooling demands in adjacent spaces, unsatisfactory air distribution, objectionable noise, and inadequate ventilation because of reduction in air delivery.

Reheater Control. Zone reheaters of Type C systems are controlled by regulating the amount of steam to the zone coils by one of two methods. The first method uses a room thermostat in a *representative* space with other spaces in the zone using manual dampers. Spaces served by this type of control do not always develop the desired comfort conditions. Master-submaster control, used in the second method, adjusts the reheater discharge temperature according to a predetermined schedule so that full heat is applied at the design outside heating temperature and less heat is applied as the temperature rises. This control does not adjust to meet variations in individual room loads; however, it is superior to the first method because it is more dependable.

Preheater Control. A duct thermostat, which modulates steam through the valve, controls the preheaters. To prevent bucking, the set point usually is a few degrees below the design cooling off-coil setting.

Coil Control. Except for Type A air conditioning with humidity control and Type E air conditioning primary air coils, cooling coils (water) are controlled by a dew-point thermostat to give a constant off-coil temperature during the entire cooling cycle. No control is provided for coils of Type E air conditioning because they are in series with the flow-through induction units, and maximum dehumidification must be accomplished by the primary coil to maintain dry-coil operation in the room units.

Damper Control. Outdoor, return, and exhaust dampers are either manually (as a group) or automatically controlled. In automatic control, one of two methods is used: (1) controlling the damper settings by thermostats exposed to weather air, or (2) using a duct thermostat that restricts the outdoor airflow only when the design temperature leaving the cooling coil cannot be achieved with full flow through the coil.

REGULATORY AGENCIES

Merchant vessels that operate under the United States flag come under the jurisdiction of the Coast Guard. Accordingly, the installation and components must conform to the Marine Engineering Rules and Marine Standards of the Coast Guard. Equipment design and installation must also comply with the requirements of both the U.S. Public Health Service and the U.S. Department of Agriculture. Principally, this involves rat-proofing.

Comfort air-conditioning installations do not primarily come under the American Bureau of Shipping. However, equipment should be manufactured, wherever possible, to comply with the American Bureau of Shipping Rules and Regulations. This is important when the vessels are equipped for carrying cargo refrigeration, because the air-conditioning compressors may serve as standby units in the event of a cargo compressor failure. This compliance eliminates the necessity of a separate spare cargo compressor.

NAVAL SURFACE SHIPS

DESIGN CRITERIA

Outdoor Ambient Temperature

Design conditions for naval vessels have been established as a compromise, considering the large cooling plants required for internal heat loads generated by machinery, weapons, electronics, and personnel. Temperatures of 90°F dry bulb and 81°F wet bulb are used as design requirements for worldwide applications, together with 85°F seawater temperatures. Heating season temperatures are assumed to be 10°F for outdoor air and 28°F for seawater.

Indoor Temperature

Naval ships are generally designed for space temperatures of 80°F dry bulb with a maximum of 55% rh for most areas requiring air conditioning. The *Air Conditioning, Ventilation and Heating Design Criteria Manual for Surface Ships of the United States Navy* (USN 1969) gives design conditions established for specific areas. *Standard Specification for Cargo Ship Construction* (USMA 1965) gives temperatures for ventilated spaces.

Ventilation Requirements

Ventilation must meet the requirements given in ASHRAE *Standard* 62, except when special consideration is required for ships.

Air-Conditioned Spaces. Naval ship design requires that air-conditioning systems serving living and berthing areas on surface ships replenish air in accordance with damage control classifications, as specified in USN (1969):

1. Class Z systems: 5 cfm per person.
2. Class W systems for troop berthing areas: 5 cfm per person.
3. All other Class W systems: 10 cfm per person. The flow rate is increased only to meet either a 75 cfm minimum branch requirement or to balance exhaust requirements. Outdoor air should be kept at a minimum to prevent the air-conditioning plant from becoming excessively large.

Load Determination

The cooling load estimate consists of coefficients from *Design Data Sheet* DDS511-2 of USN *General Specification for Building Naval Shipss* or USN (1969) and includes allowances for the following:

- Solar radiation
- Heat transmission through hull, decks, and bulkheads
- Heat (latent and sensible) dissipation of occupants
- Heat gain due to lights
- Heat (latent and sensible) gain due to ventilation air
- Heat gain due to motors or other electrical equipment
- Heat gain from piping, machinery, and equipment

Loads should be derived from requirements indicated in USN (1969). The heating load estimate for air conditioning should consist of the following:

- Heat losses through hull, decks, and bulkheads
- Ventilation air
- Infiltration (when specified)

Some electronic spaces listed in USN (1969) require adding 15% to the calculated cooling load for future growth and using one-third of the cooling season equipment heat dissipation (less the 15% added for growth) as heat gain in the heating season.

Heat Transmission Coefficients. The overall heat transmission coefficient U between the conditioned space and the adjacent boundary should be estimated from *Design Data Sheet* DDS511-2. Where new materials or constructions are used, new coefficients may be used from SNAME or calculated using methods found in DDS511-2 and SNAME.

Heat Dissipation from People. USN (1969) gives heat dissipation values for people in various activities and room conditions.

Heat Gain from Sources Within the Space. USN (1969) gives heat gain from lights and motors driving ventilation equipment. Heat gain and use factors for other motors and electrical and electronic equipment may be obtained from the manufacturer or from Chapter 28 of the 1997 *ASHRAE Handbook—Fundamentals*.

EQUIPMENT SELECTION

The equipment described for merchant ships also applies for U.S. naval vessels, except as follows:

Fans

A family of standard fans is used by the navy, including vane-axial, tubeaxial, and centrifugal fans. Selection curves used for system design are found on NAVSEA *Standard Drawings* 810-921984, 810-925368, and 803-5001058. Manufacturers are required to furnish fans dimensionally identical to the standard plan and within 5% of the delivery. No belt-driven fans are included in the fan standards.

Cooling Coils

The navy uses eight standard sizes of direct-expansion and chilled water cooling coils. All coils have eight rows in the direction of airflow, with a range in face area of 0.6 to 10.0 ft^2.

The coils are selected for a face velocity of 500 fpm maximum; however, sizes 54 DW to 58 DW may have a face velocity up to 620 fpm if the bottom of the duct on the discharge is sloped up at 15° for a distance equal to the height of the coil.

Chilled water coils are most commonly used and are selected based on 45°F inlet water with approximately a 6.7°F rise in water temperature through the coil. This is equivalent to 3.6 gpm per ton of cooling.

Construction and materials are specified in MIL-C-2939.

Heating Coils

The standard naval steam and electric duct heaters have specifications as follows:

Steam Duct Heaters

- Maximum face velocity is 1800 fpm
- Preheater leaving air temperature is 42 to 50°F

- Steam heaters are served from a 50 psig steam system

Electric Duct Heaters

- Maximum face velocity is 1400 fpm.
- Temperature rise through the heater is per MIL-H-22594A, but is in no case more than 48°F.
- Power supply for the smallest heaters is 120 V, 3 phase, 60 Hz. All remaining power supplies are 440 V, 3 phase, 60 Hz.
- Pressure drop through the heater must not exceed 0.35 in. of water at 1000 fpm. Manufacturers' tested data should be used in system design

Filters

Characteristics of the seven standard filter sizes the navy uses are as follows:

- Filters are available in steel or aluminum
- Filter face velocity is between 375 and 900 fpm
- A filter-cleaning station on board ship includes facilities to wash, oil, and drain filters

Air Diffusers

Although it also uses standard diffusers for air-conditioning, the navy generally uses a commercial type similar to those used for merchant ships.

Air-Conditioning Compressors

The navy uses reciprocal compressors up to approximately 150 tons. For larger capacities, open, direct-drive centrifugal compressors are used. Seawater is used for condenser cooling at the rate of 5 gpm per ton for reciprocal compressors and 4 gpm per ton for centrifugal compressors.

Table 3 Minimum Thickness of Materials for Ducts

Diameter or Longer Side	Sheet for Fabricated Ductwork			
	Nonwatertight		Watertight	
	Galvanized Steel	Aluminum	Galvanized Steel	Aluminum
Up to 6	0.018	0.025	0.075	0.106
6.5 to 12	0.030	0.040	0.100	0.140
12.5 to 18	0.036	0.050	0.118	0.160
18.5 to 30	0.048	0.060	0.118	0.160
Above 30	0.060	0.088	0.118	0.160

	Welded or Seamless Tubing	
Tubing Size	Nonwatertight Aluminum	Watertight Aluminum
2 to 6	0.035	0.106
6.5 to 12	0.050	0.140

	Spirally Wound Duct (Nonwatertight)	
Diameter	Steel	Aluminum
Up to 8	0.018	0.025
Over 8	0.030	0.032

Note: All dimensions in inches.

Typical Air Systems

On naval ships, zone reheat is used for most applications. Some ships with sufficient electric power have used low-velocity terminal reheat systems with electric heaters in the space. Some newer ships have used a fan coil unit with fan, chilled water cooling coil, and electric heating coil in spaces with low to medium sensible heat per unit area of space requirements. The unit is supplemented by conventional systems serving spaces with high sensible or latent loads.

Air Distribution Methods

Methods used on naval ships are similar to those discussed in the section for merchant ships. The minimum thickness of materials for ducts is listed in Table 3.

Control

The navy's principal air-conditioning control uses a two-position dual thermostat that controls a cooling coil and an electric or steam reheater. This thermostat can be set for summer operation and does not require resetting for winter operation.

Steam preheaters use a regulating valve with (1) a weather bulb controlling approximately 25% of the valve's capacity to prevent freeze-up, and (2) a line bulb in the duct downstream of the heater to control the temperature between 42 and 50°F.

Other controls are used to suit special needs. For example, pneumatic/electric controls can be used when close tolerances in temperature and humidity control are required, as in operating rooms. Thyristor controls are sometimes used on electric reheaters in ventilation systems.

REFERENCES

ASHRAE. 1989. Ventilation for acceptable indoor air quality. ANSI/ASHRAE *Standard* 62-1989.

ASHRAE. 1996. Mechanical refrigeration and air-conditioning installations aboard ship. *Standard* 26-1996.

SNAME. 1963. Thermal insulation report. *Technical and Research Bulletin No.* 4-7. Society of Naval Architects and Marine Engineers, Jersey City, NJ.

SNAME. 1980. Calculations for merchant ship heating, ventilation and air conditioning design. *Technical and Research Bulletin No.* 4-16. Society of Naval Architects and Marine Engineers, Jersey City, NJ.

SNAME. 1992. *Marine engineering.* R. Harrington, ed. Society of Naval Architects and Marine Engineers, Jersey City, NJ.

USMA. 1965. Standard specification for cargo ship construction. U.S. Maritime Administration, Washington, D.C.

USMA. *Standard Plan* S38-1-101, *Standard Plan* S38-1-102, and *Standard Plan* S38-1-103. U.S. Maritime Administration, Washington, D.C.

USN. 1969. The air conditioning, ventilation and heating design criteria manual for surface ships of the United States Navy. *Document* No. 0938-018-0010. Washington, D.C.

USN. NAVSEA *Drawing* No. 810-921984, NAVSEA *Drawing* No. 810-925368, and NAVSEA *Drawing* No. 803-5001058. Naval Sea Systems Command, Dept. of the Navy, Washington, D.C.

USN. *General specifications for building naval ships.* Naval Sea Systems Command, Dept. of the Navy, Washington, D.C.

Note: MIL specifications are available from Commanding Officer, Naval Publications and Forms Center, ATTN: NPFC 105, 5801 Tabor Ave., Philadelphia, PA 19120.

CHAPTER 11

INDUSTRIAL AIR CONDITIONING

INDUSTRIAL plants, warehouses, laboratories, nuclear power plants and facilities, and data processing rooms are designed for specific processes and environmental conditions that include proper temperature, humidity, air motion, air quality, and cleanliness. Airborne contaminants generated must be collected and treated before being discharged from the building or returned to the area.

Many industrial buildings require large quantities of energy, both in manufacturing and in the maintenance of building environmental conditions. Energy can be saved by the proper use of insulation, ventilation, and solar energy and by the recovery of waste heat and cooling.

For worker efficiency, the environment should be comfortable, minimize fatigue, facilitate communication, and not be harmful to health. Equipment should (1) control temperature and humidity or provide spot cooling to prevent heat stress, (2) have low noise levels, and (3) control health-threatening fumes.

GENERAL REQUIREMENTS

Typical temperatures, relative humidities, and specific filtration requirements for the storage, manufacture, and processing of various commodities are listed in Table 1. Requirements for a specific application may differ from those in the table. Improvements in processes and increased knowledge may cause further variation; thus, systems should be flexible to meet future requirements.

Inside temperature, humidity, filtration levels, and allowable variations should be established by agreement with the owner. A compromise between the requirements for product or process conditions and those for comfort may optimize quality and production costs.

A work environment that allows a worker to perform assigned duties without fatigue caused by temperatures that are too high or too low and without exposure to harmful airborne contaminants results in better, continued performance. It may also improve worker morale and reduce absenteeism.

PROCESS AND PRODUCT REQUIREMENTS

A process or product may require control of one or more of the following: (1) moisture regain; (2) rates of chemical reactions; (3) rates of biochemical reactions; (4) rate of crystallization; (5) product accuracy and uniformity; (6) corrosion, rust, and abrasion; (7) static electricity; (8) air cleanliness; and (9) product formability. Discussion of each of these factors follows.

Moisture Regain

In the manufacture or processing of hygroscopic materials such as textiles, paper, wood, leather, and tobacco, the air temperature and relative humidity have a marked influence on the production rate and on product mass, strength, appearance, and quality.

Moisture in vegetable or animal materials (and some minerals) reaches equilibrium with the moisture of the surrounding air by

regain. Regain is defined as the percentage of absorbed moisture in a material compared to its bone-dry mass. If a material sample with a mass of 110 g has a mass of 100 g after a thorough drying under standard conditions of 220 to 230°F, the mass of absorbed moisture is 10 g—10% of the sample's bone-dry mass. The regain, therefore, is 10%.

Table 2 lists typical values of regain for materials at 75°F in equilibrium at various relative humidities. Temperature change affects the rate of absorption or drying, which generally varies with the nature of the material, its thickness, and its density. Sudden temperature changes cause a slight regain change, even with fixed relative humidity; but the effect of temperature on regain is small compared to the effect of relative humidity.

Hygroscopic Materials. In absorbing moisture from the air, hygroscopic materials deliver sensible heat to the air in an amount equal to that of the latent heat of the absorbed moisture. Moisture gains or losses by materials in processes are usually quite small, but if they are significant, the amount of heat liberated should be included in the load estimate. Actual values of regain should be obtained for a particular application. Manufacturing economy requires regain to be maintained at a level suitable for rapid and satisfactory manipulation. Uniform humidity allows high-speed machinery to operate efficiently.

Conditioning and Drying. Materials may be exposed to the required humidity during manufacturing or processing, or they may be treated separately after conditioning and drying. Conditioning removes or adds hygroscopic moisture. Drying removes both hygroscopic moisture and free moisture in excess of that in equilibrium. Free moisture may be removed by evaporation, physically blowing it off, or other means.

Drying and conditioning may be combined to remove moisture and accurately regulate final moisture content in, for example, tobacco and some textile products. Conditioning or drying is frequently a continuous process in which the material is conveyed through a tunnel and subjected to controlled atmospheric conditions. Chapter 22 of the 1996 *ASHRAE Handbook—Systems and Equipment* describes dehumidification and pressure-drying equipment.

Rates of Chemical Reactions

Some processes require temperature and humidity control to regulate chemical reactions. For example, in rayon manufacture, pulp sheets are conditioned, cut to size, and passed through a mercerizing process. The temperature directly controls the rate of the reaction, while the relative humidity maintains a solution of constant strength and a constant rate of surface evaporation.

The oxidizing process in drying varnish depends on temperature. Desirable temperatures vary with the type of varnish. High relative humidity retards surface oxidation and allows internal gases to escape as chemical oxidizers cure the varnish from within. Thus, a bubble-free surface is maintained with a homogeneous film throughout.

The preparation of this chapter is assigned to TC 9.2, Industrial Air Conditioning.

Table 1   Temperatures and Humidities for Industrial Air Conditioning

Process	Dry Bulb, °F	rh, %
ABRASIVE		
Manufacture	79	50
CERAMICS		
Refractory	110 to 150	50 to 90
Molding room	80	60 to 70
Clay storage	60 to 80	35 to 65
Decalcomania production	75 to 80	48
Decorating room	75 to 80	48

Use high-efficiency filtration in decorating room. To minimize the danger of silicosis in other areas, a dust-collecting system or medium-efficiency particulate air filtration may be required.

Process	Dry Bulb, °F	rh, %
DISTILLING		
General manufacturing	60 to 75	45 to 60
Aging	65 to 72	50 to 60

Low humidity and dust control are important where grains are ground. Use high-efficiency filtration for all areas to prevent mold spore and bacteria growth. Use ultrahigh-efficiency filtration where bulk flash pasteurization is performed.

Process	Dry Bulb, °F	rh, %
ELECTRICAL PRODUCTS		
Electronics and X-ray		
Coil and transformer winding	72	15
Semiconductor assembly	68	40 to 50
Electrical instruments		
Manufacture and laboratory	70	50 to 55
Thermostat assembly and calibration	75	50 to 55
Humidistat assembly and calibration	75	50 to 55
Small mechanisms		
Close tolerance assembly	72*	40 to 45
Meter assembly and test	75	60 to 63
Switchgear		
Fuse and cutout assembly	73	50
Capacitor winding	73	50
Paper storage	73	50
Conductor wrapping with yarn	75	65 to 70
Lightning arrester assembly	68	20 to 40
Thermal circuit breakers assembly and test	75	30 to 60
High-voltage transformer repair	79	5
Water wheel generators		
Thrust runner lapping	70	30 to 50
Rectifiers		
Processing selenium and copper oxide plates	73	30 to 40

*Temperature to be held constant.

Dust control is essential in these processes. Minimum control requires medium-efficiency filters. Degree of filtration depends on the type of function in the area. Smaller tolerances and miniature components suggest high-efficiency particulate air (HEPA) filters.

Process	Dry Bulb, °F	rh, %
FLOOR COVERING		
Linoleum		
Mechanical oxidizing of linseed oil*	90 to 100	
Printing	80	
Stoving process	160 to 250	

*Precise temperature control required.

Medium-efficiency particulate air filtration is recommended for the stoving process.

Process	Dry Bulb, °F	rh, %
FOUNDRIES*		
Core making	60 to 70	
Mold making		
Bench work	60 to 70	
Floor work	55 to 65	
Pouring	40	
Shakeout	40 to 50	
Cleaning room	55 to 65	

*Winter dressing room temperatures. Spot coolers are sometimes used in larger installations.

In mold making, provide exhaust hoods at transfer points with wet-collector dust removal system. Use 600 to 800 cfm per hood.

In shakeout room, provide exhaust hoods with wet-collector dust removal system. Exhaust 400 to 500 cfm in grate area. Room ventilators are generally not effective.

In cleaning room, provide exhaust hoods for grinders and cleaning equipment with dry cyclones or bag-type collectors. In core making, oven and adjacent cooling areas require fume exhaust hoods. Pouring rooms require two-speed powered roof ventilators. Design for minimum of 2 cfm per square foot of floor area at low speed. Shielding is required to control radiation from hot surfaces. Proper introduction of air minimizes preheat requirements.

Process	Dry Bulb, °F	rh, %
FUR		
Drying	110	
Shock treatment	18 to 20	
Storage	40 to 50	55 to 65

Shock treatment or eradication of any insect infestations requires lowering the temperature to 18 to 20°F for 3 to 4 days, then raising it to 60 to 70°F for 2 days, then lowering it again for 2 days and raising it to the storage temperature.

Furs remain pliable, oxidation is reduced, and color and luster are preserved when stored at 40 to 50°F.

Humidity control is required to prevent mold growth (which is prevalent with humidities above 80%) and hair splitting (which is common with humidities lower than 55%).

Process	Dry Bulb, °F	rh, %
GUM		
Manufacturing	77	33
Rolling	68	63
Stripping	72	53
Breaking	73	47
Wrapping	73	58

Process	Dry Bulb, °F	rh, %
LEATHER		
Drying	68 to 125	75
Storage, winter room temperature	50 to 60	40 to 60

After leather is moistened in preparation for rolling and stretching, it is placed in an atmosphere of room temperature and 95% relative humidity.

Leather is usually stored in warehouses without temperature and humidity control. However, it is necessary to keep humidity sufficiently low to prevent mildew. Medium-efficiency particulate air filtration is recommended for fine finish.

Process	Dry Bulb, °F	rh, %
LENSES (OPTICAL)		
Fusing	75	45
Grinding	80	80

Table 1 Temperatures and Humidities for Industrial Air Conditioning (*Concluded*)

Process	Dry Bulb, °F	rh, %		Process	Dry Bulb, °F	rh, %
MATCHES				**PLASTICS**		
Manufacture	72 to 73	50		Manufacturing areas		
Drying	70 to 75	60		Thermosetting molding compounds	80	25 to 30
Storage	60 to 63	50		Cellophane wrapping	75 to 80	45 to 65

Water evaporates with the setting of the glue. The amount of water evaporated is 18 to 20 lb per million matches. The match machine turns out about 750,000 matches per hour.

In manufacturing areas where plastic is exposed in the liquid state or molded, high-efficiency particulate air filters may be required. Dust collection and fume control are essential.

Process	Dry Bulb, °F	rh, %
PAINT APPLICATION		
Lacquers: Baking	300 to 360	
Oil paints: Paint spraying	60 to 90	80

Process	Dry Bulb, °F	rh, %
PLYWOOD		
Hot pressing (resin)	90	60
Cold pressing	90	15 to 25
RUBBER-DIPPED GOODS		
Manufacture	90	
Cementing	80	25 to 30*
Dipping surgical articles	75 to 80	25 to 30*
Storage prior to manufacture	60 to 75	40 to 50*
Testing laboratory	73	50*

*Dew point of air must be below evaporation temperature of solvent.

The required air filtration efficiency depends on the painting process. On fine finishes, such as car bodies, high-efficiency particulate air filters are required for the outdoor air supply. Other products may require only low- or medium-efficiency filters.

Makeup air must be preheated. Spray booths must have 100 fpm face velocity if spraying is performed by humans; lower air quantities can be used if robots perform spraying. Ovens must have air exhausted to maintain fumes below explosive concentration. Equipment must be explosion-proof. Exhaust must be cleaned by filtration and solvents reclaimed or scrubbed.

Solvents used in manufacturing processes are often explosive and toxic, requiring positive ventilation. Volume manufacturers usually install a solvent-recovery system for area exhaust systems.

Process	Dry Bulb, °F	rh, %
PHOTO STUDIO		
Dressing room	72 to 74	40 to 50
Studio (camera room)	72 to 74	40 to 50
Film darkroom	70 to 72	45 to 55
Print darkroom	70 to 72	45 to 55
Drying room	90 to 100	35 to 45
Finishing room	72 to 75	40 to 55
Storage room (b/w film and paper)	72 to 75	40 to 60
Storage room (color film and paper)	40 to 50	40 to 50
Motion picture studio	72	40 to 55

Process	Dry Bulb, °F	rh, %
TEA		
Packaging	65	65

Ideal moisture content is 5 to 6% for quality and mass. Low-limit moisture content for quality is 4%.

The above data pertain to average conditions. In some color processes, elevated temperatures as high as 105°F are used, and a higher room temperature is required.

Conversely, ideal storage conditions for color materials necessitate refrigerated or deep-freeze temperatures to ensure quality and color balance when long storage times are anticipated.

Heat liberated during printing, enlarging, and drying processes is removed through an independent exhaust system, which also serves the lamp houses and dryer hoods. All areas except finished film storage require a minimum of medium-efficiency particulate air filters.

Process	Dry Bulb, °F	rh, %
TOBACCO		
Cigar and cigarette making	70 to 75	55 to 65*
Softening	90	85 to 88
Stemming and stripping	75 to 85	70 to 75
Packing and shipping	73 to 75	65
Filler tobacco casing and conditioning	75	75
Filter tobacco storage and preparation	77	70
Wrapper tobacco storage and conditioning	75	75

*Relative humidity fairly constant with range as set by cigarette machine.
Before stripping, tobacco undergoes a softening operation.

Rates of Biochemical Reactions

Fermentation requires temperature and humidity control to regulate the rate of biochemical reactions.

Rate of Crystallization

The cooling rate determines the size of crystals formed from a saturated solution. Both temperature and relative humidity affect the cooling rate and change the solution density by evaporation.

In the coating pans for pills, a heavy sugar solution is added to the tumbling mass. As the water evaporates, sugar crystals cover each pill. Blowing the proper quantity of air at the correct dry- and wet-bulb temperatures forms a smooth opaque coating. If cooling and drying are too slow, the coating is rough, translucent, and unsatisfactory in appearance; if cooling and drying are too fast, the coating chips through to the interior.

Product Accuracy and Uniformity

In the manufacture of precision instruments, tools, and lenses, air temperature and cleanliness affect the quality of work. If manufacturing tolerances are within 0.0002 in., close temperature control

prevents expansion and contraction of the material. Constant temperature is more important than the temperature level; thus conditions are usually selected for personnel comfort and to prevent a film of moisture on the surface. High- or ultrahigh-efficiency particulate air filtration may be required.

Corrosion, Rust, and Abrasion

In the manufacture of metal articles, the temperature and relative humidity are kept sufficiently low to prevent hands from sweating, thus protecting the finished article from fingerprints, tarnish, and/or etching. The salt and acid in body perspiration can cause corrosion and rust within a few hours. The manufacture of polished surfaces usually requires medium- to high-efficiency particulate air filtering to prevent surface abrasion. This is also true for steel-belted radial tire manufacturing.

Static Electricity

In processing light materials such as textile fibers and paper, and where explosive atmospheres or materials are present, humidity can be used to reduce static electricity, which is often detrimental

Table 2 Regain of Hygroscopic Materials[a]

Classification	Material	Description	10	20	30	40	50	60	70	80	90
Natural	Cotton	Sea island—roving	2.5	3.7	4.6	5.5	6.6	7.9	9.5	11.5	14.1
textile	Cotton	American—cloth	2.6	3.7	4.4	5.2	5.9	6.8	8.1	10.0	14.3
fibers	Cotton	Absorbent	4.8	9.0	12.5	15.7	18.5	20.8	22.8	24.3	25.8
	Wool	Australian merino—skein	4.7	7.0	8.9	10.8	12.8	14.9	17.2	19.9	23.4
	Silk	Raw chevennes—skein	3.2	5.5	6.9	8.0	8.9	10.2	11.9	14.3	18.3
	Linen	Table cloth	1.9	2.9	3.6	4.3	5.1	6.1	7.0	8.4	10.2
	Linen	Dry spun—yarn	3.6	5.4	6.5	7.3	8.1	8.9	9.8	11.2	13.8
	Jute	Average of several grades	3.1	5.2	6.9	8.5	10.2	12.2	14.4	17.1	20.2
	Hemp	Manila and sisal rope	2.7	4.7	6.0	7.2	8.5	9.9	11.6	13.6	15.7
Rayons	Viscose nitrocellulose	Average skein	4.0	5.7	6.8	7.9	9.2	10.8	12.4	14.2	16.0
	Cuprammonium cellulose acetate		0.8	1.1	1.4	1.9	2.4	3.0	3.6	4.3	5.3
Paper	M.F. newsprint	Wood pulp—24% ash	2.1	3.2	4.0	4.7	5.3	6.1	7.2	8.7	10.6
	H.M.F. writing	Wood pulp—3% ash	3.0	4.2	5.2	6.2	7.2	8.3	9.9	11.9	14.2
	White bond	Rag—1% ash	2.4	3.7	4.7	5.5	6.5	7.5	8.8	10.8	13.2
	Comm. ledger	75% rag—1% ash	3.2	4.2	5.0	5.6	6.2	6.9	8.1	10.3	13.9
	Kraft wrapping	Coniferous	3.2	4.6	5.7	6.6	7.6	8.9	10.5	12.6	14.9
Miscellaneous	Leather	Sole oak—tanned	5.0	8.5	11.2	13.6	16.0	18.3	20.6	24.0	29.2
organic	Catgut	Racquet strings	4.6	7.2	8.6	10.2	12.0	14.3	17.3	19.8	21.7
materials	Glue	Hide	3.4	4.8	5.8	6.6	7.6	9.0	10.7	11.8	12.5
	Rubber	Solid tires	0.11	0.21	0.32	0.44	0.54	0.66	0.76	0.88	0.99
	Wood	Timber (average)	3.0	4.4	5.9	7.6	9.3	11.3	14.0	17.5	22.0
	Soap	White	1.9	3.8	5.7	7.6	10.0	12.9	16.1	19.8	23.8
	Tobacco	Cigarette	5.4	8.6	11.0	13.3	16.0	19.5	25.0	33.5	50.0
Miscellaneous	Asbestos fiber	Finely divided	0.16	0.24	0.26	0.32	0.41	0.51	0.62	0.73	0.84
inorganic	Silica gel		5.7	9.8	12.7	15.2	17.2	18.8	20.2	21.5	22.6
materials	Domestic coke		0.20	0.40	0.61	0.81	1.03	1.24	1.46	1.67	1.89
	Activated charcoal	Steam activated	7.1	14.3	22.8	26.2	28.3	29.2	30.0	31.1	32.7
	Sulfuric acid		33.0	41.0	47.5	52.5	57.0	61.5	67.0	73.5	82.5

[a]Moisture content expressed in percent of dry mass of the substance at various relative humidities, temperature 75°F.

to processing and extremely dangerous in explosive atmospheres. Static electricity charges are minimized when the air has a relative humidity of 35% or higher. The power driving the processing machines is converted into heat that raises the temperature of the machines above that of the adjacent air, where humidity is normally measured. Room relative humidity may need to be 65% or higher to maintain the high humidity required in the machines.

Air Cleanliness

Each application must be evaluated to determine the filtration needed to counter the adverse effects on the product or process of (1) minute dust particles, (2) airborne bacteria, and (3) other air contaminants such as smoke, radioactive particles, spores, and pollen. These effects include chemically altering production material, spoiling perishable goods, and clogging small openings in precision machinery. See Chapter 24 of the 1996 *ASHRAE Handbook—Systems and Equipment*.

Product Formability

The manufacture of pharmaceutical tablets requires close control of humidity for optimum tablet formation.

EMPLOYEE REQUIREMENTS

Space conditions required by health and safety standards to avoid excess exposure to high temperatures and airborne contaminants are often established by the American Conference of Governmental Industrial Hygienists (ACGIH). In the United States, the National Institute of Occupational Safety and Health (NIOSH) does research and recommends guidelines for workspace environmental

control. The Occupational Safety and Health Administration (OSHA) sets standards from these environmental control guidelines and enforces them through compliance inspection at industrial facilities. Enforcement may be delegated to a corresponding state agency.

Standards for safe levels of contaminants in the work environment or in air exhausted from facilities do not cover all contaminants encountered. Minimum safety standards and facility design criteria are available from various U.S. Department of Health, Education, and Welfare (DHEW) agencies such as the National Cancer Institute, the National Institutes of Health, and the Public Health Service (Centers for Disease Control and Prevention). For radioactive substances, standards established by the U.S. Nuclear Regulatory Commission (NRC) should be followed.

Thermal Control Levels

In industrial plants that need no specific control for products or processes, conditions range from 68 to 100°F and 25 to 60% rh. For a more detailed analysis, the work rate, air velocity, quantity of rest, and effects of radiant heat must be considered (see Chapter 8 of the 1997 *ASHRAE Handbook—Fundamentals*). To avoid stress to workers exposed to high work rates and hot temperatures, the ACGIH established guidelines to evaluate high-temperature air velocity and humidity levels in terms of heat stress (Dukes-Dobos and Henschel 1971).

If a comfortable environment is of concern rather than the avoidance of heat stress, the thermal control range becomes more specific (McNall et al. 1967). In still air, nearly sedentary workers (410 Btu/h metabolism) prefer 72 to 75°F dry-bulb temperature with 20 to 60% rh, and they can detect a 2°F change per hour. Workers at a high rate of activity (1020 Btu/h metabolism) prefer 63 to 66°F dry bulb with 20 to 50% rh, but they are less sensitive to temperature change (ASHRAE *Standard* 55). Workers at a high activity level can also be cooled by increasing the air velocity (ASHRAE *Standard* 55).

Contaminant Control Levels

Toxic materials are present in many industrial plants and laboratories. In such plants, the air-conditioning and ventilation systems must minimize human exposure to toxic materials. When these materials become airborne, their range expands greatly, thus exposing more people. Chapter 12 of the 1997 *ASHRAE Handbook—Fundamentals*, current OSHA regulations, and ACGIH (1998) give guides to evaluating the health impact of contaminants.

In addition to being a health concern, gaseous flammable substances must also be kept below explosive concentrations. Acceptable concentration limits are 20 to 25% of the lower explosive limit of the substance. Chapter 12 of the 1997 *ASHRAE Handbook—Fundamentals* includes information on flammable limits and means of control.

Instruments are available to measure concentrations of common gases and vapors. For less common gases or vapors, the air is sampled by drawing it through an impinger bottle or by inertial impaction-type air samplers, which support biological growth on a nutrient gel for subsequent identification after incubation.

Gases and vapors are found near acid baths and tanks holding process chemicals. Machine processes, plating operations, spraying, mixing, and abrasive cleaning operations generate dusts, fumes, and mists. Many laboratory procedures, including grinding, blending, homogenizing, sonication, weighing, dumping of animal bedding, and animal inoculation or intubation, generate aerosols.

DESIGN CONSIDERATIONS

To select equipment, the required environmental conditions, both for the product and personnel comfort, must be known. Consultation with the owner establishes design criteria such as temperature and humidity levels, energy availability and opportunities to recover it, cleanliness, process exhaust details, location and size of heat-producing equipment, lighting levels, frequency of equipment use, load factors, frequency of truck or car loadings, and sound levels. Consideration must be given to separating dirty processes from manufacturing areas that require relatively clean air. If uncontrolled by physical barriers, hard-to-control contaminants—such as mists from presses and machining, or fumes and gases from welding—can migrate to areas of final assembly, metal cleaning, or printing and can cause serious problems.

Because of changing mean radiant temperature, insulation should be evaluated for initial and operating cost, saving of heating and cooling, elimination of condensation (on roofs in particular), and maintenance of comfort. When high levels of moisture are required within buildings, the structure and air-conditioning system must prevent condensation damage to the structure and ensure a quality product. Condensation can be prevented by (1) proper selection of insulation type and thickness, (2) proper placing of vapor retarders, and (3) proper selection and assembly of construction components to avoid thermal short circuits. Chapters 22 and 23 of the 1997 *ASHRAE Handbook—Fundamentals* have further details.

Personnel engaged in some industrial processes may be subject to a wide range of activity levels for which a broad range of indoor temperature and humidity levels are desirable. Chapter 8 of the 1997 *ASHRAE Handbook—Fundamentals* addresses recommended indoor conditions at various activity levels.

If layout and construction drawings are not available, a complete survey of existing premises and a checklist for proposed facilities is necessary (Table 3).

New industrial buildings are commonly single-story with flat roofs and ample height to distribute utilities without interfering with process operation. Fluorescent fixtures are commonly mounted at heights up to 12 ft, high-output fluorescent fixtures up to 20 ft, and high-pressure sodium or metal halide fixtures above 20 ft.

Lighting design considers light quality, degree of diffusion and direction, room size, mounting height, and economics. Illumination levels should conform to the recommended levels of the Illuminating Engineering Society of North America (IESNA 1993). Air-conditioning systems can be located in the top of the building. However, the designs require coordination because air-handling equipment and ductwork compete for space with sprinkler systems, piping, structural elements, cranes, material-handling systems, electric wiring, and lights.

Operations within the building must also be considered. Production materials may be moved through outside doors and large amounts of outdoor air be allowed to enter. Some operations require close control of temperature, humidity, and contaminants.

Table 3 Facilities Checklist

Construction
1. Single or multistory
2. Type and location of doors, windows, crack lengths
3. Structural design live loads
4. Floor construction
5. Exposed wall materials
6. Roof materials and color
7. Insulation type and thicknesses
8. Location of existing exhaust equipment
9. Building orientation

Use of Building
1. Product needs
2. Surface cleanliness; acceptable airborne contamination level
3. Process equipment: type, location, and exhaust requirements
4. Personnel needs, temperature levels, required activity levels, and special workplace requirements
5. Floor area occupied by machines and materials
6. Clearance above floor required for material-handling equipment, piping, lights, or air distribution systems
7. Unusual occurrences and their frequency, such as large cold or hot masses of material moved inside
8. Frequency and length of time doors open for loading or unloading
9. Lighting, location, type, and capacity
10. Acoustical levels
11. Machinery loads, such as electric motors (size, diversity), large latent loads, or radiant loads from furnaces and ovens
12. Potential for temperature stratification

Design Conditions
1. Design temperatures—indoor and outdoor dry and wet bulb
2. Altitude
3. Wind velocity
4. Makeup air required
5. Indoor temperature and allowable variance
6. Indoor relative humidity and allowable variance
7. Indoor air quality definition and allowable variance
8. Outdoor temperature occurrence frequencies
9. Operational periods: one, two, or three
10. Waste heat availability and energy conservation
11. Pressurization required
12. Mass loads from the energy release of productive materials

Code and Insurance Requirements
1. State and local code requirements for ventilation rates, etc.
2. Occupational health and safety requirements
3. Insuring agency requirements

Utilities Available and Required
1. Gas, oil, compressed air (pressure), electricity (characteristics), steam (pressure), water (pressure), wastewater, interior and site drainage
2. Rate structures for each utility
3. Potable water

A time schedule of operation helps in estimating the heating and cooling load.

LOAD CALCULATIONS

Table 1 and specific product chapters discuss product requirements. Chapter 28 of the 1997 *ASHRAE Handbook—Fundamentals* covers load calculations for heating and cooling.

Solar and Transmission

The solar load on the roof is generally the largest perimeter load and is usually a significant part of the overall load. Roofs should be light colored to minimize solar heat gain. Wall loads are often insignificant, and most new plants have no windows in the manufacturing area, so a solar load through glass is not present. Large windows in old plants may be closed in.

Internal Heat Generation

The process, product, facility utilities, and employees generate internal heat. People, power or process loads, and lights are internal sensible loads. Of these, production machinery often creates the largest sensible load. The design should consider anticipated brake power rather than connected motor loads. The lighting load is generally significant. Heat gain from people is usually negligible in process areas.

Heat from operating equipment is difficult to estimate. Approximate values can be determined by studying the load readings of the electrical substations that serve the area.

In most industrial facilities, the latent heat load is minimal, with people and outdoor air being the major contributors. In these areas, the sensible heat factor approaches 1.0. Some processes, such as papermaking, release large amounts of moisture. This moisture and its condensation on cold surfaces must be managed.

Stratification Effect

The cooling load may be dramatically reduced in a work space that takes advantage of temperature stratification, which establishes a stagnant blanket of air directly under the roof by keeping air circulation to a minimum. The convective component of high energy sources such as the roof, upper walls, and high-level lights has little impact on the cooling load in the lower occupied zones. A portion of the heat generated by high-energy sources (such as process equipment) in the occupied zone is not part of the cooling load. Due to radiation and a buoyancy effect, 20-60% of the energy rises to the upper strata. The amount that rises depends on the building construction, air movement, and temperatures of the source, other surfaces, and air.

Supply- and return-air ducts should be as low as possible to avoid mixing the warm boundary layers. The location of supply-air diffusers generally establishes the boundary of the warmer stratified air. For areas with supply air quantities greater than 2 cfm per square foot, the return air temperature is approximately that of the air entrained by the supply airstream and only slightly higher than that at the end of the throw. With lower air quantities, the return air is much warmer than the supply air at the end of the throw. The amount depends on the placement of return inlets relative to internal heat sources. The average temperature of the space is higher, thus reducing the effect of outside conditions on heat gain. Spaces with a high area-to-employee ratio adapt well to low quantities of supply air and to spot cooling. For design specifics, refer to Chapter 28, the section on General Comfort and Dilution Ventilation.

Makeup Air

Makeup air that has been filtered, heated, cooled, humidified, and/or dehumidified is introduced to replace exhaust air, provide ventilation, and pressurize the building. For exhaust systems to function, air must enter the building by infiltration or the air-conditioning equipment. The space air-conditioning system must be large enough to heat or cool the outside air required to replace the exhaust. Cooling or heat recovery from exhaust air to makeup air can substantially reduce the outdoor air load.

Exhaust from air-conditioned buildings should be kept to a minimum by proper hooding or by relocating exhausted processes to areas not requiring air conditioning. Frequently, excess makeup air is provided to pressurize the building slightly, thus reducing infiltration, flue downdrafts, and ineffective exhaust under certain wind conditions. Makeup air and exhaust systems can be interlocked so that outdoor makeup air can be reduced as needed, or makeup air units may be changed from outdoor air heating to recirculated air heating when process exhaust is off.

In some facilities, outdoor air is required for ventilation because of the function of the space. Recirculation of air is avoided to reduce concentrations of health-threatening fumes, airborne bacteria, and radioactive substances. Ventilation rates for human occupancy should be determined from ASHRAE *Standard* 62 and applicable codes.

Outdoor air dampers of industrial air-conditioning units should handle 100% supply air, so modulating the dampers satisfies the room temperature under certain weather conditions. For 100% outside air, a modulating exhaust damper or fan must be used. The infiltration of outdoor air often creates considerable load in industrial buildings due to poorly sealed walls and roofs. Infiltration should be minimized to conserve energy.

Fan Heat

The heat from air-conditioning return-air fans or supply fans goes into the refrigeration load. This energy is not part of the room sensible heat with the exception of supply fans downstream of the conditioning apparatus.

SYSTEM AND EQUIPMENT SELECTION

Industrial air-conditioning equipment includes heating and cooling sources, air-handling and air-conditioning apparatus, filters, and an air distribution system. To provide low life-cycle cost, components should be selected and the system designed for long life with low maintenance and operating costs.

Systems may consist of (1) heating only in cool climates, where ventilation air provides comfort for workers; (2) air washer systems, where high humidities are desired and where the climate requires cooling; or (3) heating and mechanical cooling, where temperature and/or humidity control are required by the process and where the activity level is too high to be satisfied by other means of cooling. All systems include air filtration appropriate to the contaminant control required.

A careful evaluation will determine the zones that require control, especially in large, high-bay areas where the occupied zone is a small portion of the space volume. ASHRAE *Standard* 55 defines the occupied zone as 3 to 72 in. high and more than 24 in. from the walls.

Air-Handling Units

Air-handling units heat, cool, humidify, and dehumidify air that is distributed to a workspace. They can supply all or any portion of outdoor air so that in-plant contaminants do not become too concentrated. Units may be factory assembled or field constructed.

HEATING SYSTEMS

Floor Heating

Floor heating is often desirable in industrial buildings, particularly in large high-bay buildings, garages, and assembly areas where workers must be near the floor, or where large or fluctuating outdoor air loads make maintenance of ambient temperature difficult.

As an auxiliary to the main heating system, floors may be tempered to 65 or 70°F by embedded hydronic systems, electrical resistance cables, or warm-air ducts. The heating elements may be buried deep (6 to 18 in.) in the floor to permit slab warm-up at off-peak times, thus using the floor mass for heat storage to save energy during periods of high use.

Floor heating may be the primary or sole heating means, but floor temperatures above 85°F are uncomfortable, so such use is limited to small, well-insulated spaces.

Unit Heaters

Gas, oil, electricity, hot water, or steam unit heaters with propeller fans or blowers are used for spot heating areas or are arranged in multiples for heating an entire building. Temperatures can be varied by individual thermostatic control. Unit heaters are located so that the discharge (or throw) will reach the floor and flow adjacent to and parallel with the outside wall. They are spaced so that the discharge of one heater is just short of the next heater, thus producing a ring of warm air moving peripherally around the building. In industrial buildings with heat-producing processes, much heat stratifies in high-bay areas. In large buildings, additional heaters should be placed in the interior so that their discharge reaches the floor to reduce stratification. Downblow unit heaters in high bays and large areas may have a revolving discharge.

Gas- and oil-fired unit heaters should not be used where corrosive vapors are present. Unit heaters function well with regular maintenance and periodic cleaning in dusty or dirty conditions. Propeller fans generally require less maintenance than centrifugal fans. Centrifugal fans or blowers usually are required if heat is to be distributed to several areas. Gas- and oil-fired unit heaters require proper venting.

Ducted Heaters

Ducted heaters include large direct- or indirect-fired heaters, door heaters, and heating and ventilating units. They generally have centrifugal fans.

Code changes and improved burners have led to increased use of direct-fired gas heaters (the gas burns in the air supplied to the space) for makeup air heating. With correct interlock safety precautions for supply air and exhaust, no harmful effect from direct firing occurs. The high efficiency, high turndown ratio, and simplicity of maintenance make these units suitable for makeup air heating.

For industrial applications of ducted heaters, the following is a list of problems commonly encountered and their solutions:

- **Steam coil freeze-up.** Use steam-distributing-type (sometimes called nonfreeze) coils, face-and-bypass control with the steam valve wide open for entering air below 35°F, and a thermostat in the exit to stop the airflow when it falls below 40°F. Ensure free condensate drainage.
- **Hot water coil freeze-up.** Ensure adequate circulation through drainable coil at all times; a thermostat in the air and in the water leaving the coil opens the water valve wide and stops the airflow under freezing conditions.
- **Temperature override.** Because of wiping action on coil with face-and-bypass control, zoning control is poor. Locate face damper carefully and use room thermostat to reset the discharge air temperature controller.
- **Bearing failure.** Follow bearing manufacturer's recommendations for application and lubrication.
- **Insufficient air quantity.** Require capacity data based on testing; keep forward-curved blade fans clean or use backward-inclined blade fans; inspect and clean coils; install low- to medium-efficiency particulate air filters and change when required.

Door Heating

Unit heaters and makeup air heaters commonly temper outdoor air that enters the building at open doors. These door heaters may have directional outlets to offset the incoming draft or may resemble a vestibule where air is recirculated.

Unit heaters successfully heat air at small doors open for short periods. They temper the incoming outdoor air through mixing and quickly bring the space to the desired temperature after the door is closed. The makeup air heater should be applied as a door heater in buildings where the doors are large (those that allow railroad cars or large trucks to enter) and open for long periods. Unit heaters are also needed in facilities with a sizable negative pressure or that are not tightly constructed. These units help pressurize the door area, mix the incoming cold air and temper it, and bring the area quickly back to the normal temperature after the door is closed.

Often, door heater nozzles direct heated air at the top or down the sides of a door. Doors that create large, cold drafts when open can best be handled by introducing air in a trench at the bottom of the door. When not tempering cold drafts through open doors, some door heaters direct heated air to nearby spots. When the door is closed, they can switch to a lower output temperature.

The door heating units that resemble a vestibule operate with air flowing down across the opening and recirculating from the bottom, which helps reduce cold drafts across the floor. This type of unit is effective on doors routinely open and no higher than 10 ft.

Infrared

High-intensity infrared heaters (gas, oil, or electric) transmit heat energy directly to warm the occupants, floor, machines, or other building contents, without appreciably warming the air. Some air heating occurs by convection from objects warmed by the infrared. These units are classed as near- or far-infrared heaters, depending on the closeness of the wavelengths to visible light. Near-infrared heaters emit a substantial amount of visible light.

Both vented and unvented gas-fired infrared heaters are available as individual radiant panels, or as a continuous radiant pipe with burners 15 to 30 ft apart and an exhaust vent fan at the end of the pipe. Unvented heaters require exhaust ventilation to remove flue products from the building, or moisture will condense on the roof and walls. Insulation reduces the exhaust required. Additional information on both electric and gas infrared is given in Chapter 15 of the 1996 *ASHRAE Handbook—Systems and Equipment*.

Infrared heaters are used in the following areas:

1. High-bay buildings, where the heaters are usually mounted 10 to 30 ft above the floor, along outside walls, and tilted to direct maximum radiation to the floor. If the building is poorly insulated, the controlling thermostat should be shielded to avoid influence from the radiant effect of the heaters and the cold walls.
2. Semiopen and outdoor areas, where people can comfortably be heated directly and objects can be heated to avoid condensation.
3. Loading docks, where snow and ice can be controlled by strategic placement of near-infrared heaters.

COOLING SYSTEMS

Common cooling systems include refrigeration equipment, evaporative coolers, and high-velocity ventilating air.

For manufacturing operations, particularly heavy industry where mechanical cooling cannot be economically justified, evaporative cooling systems often provide good working conditions. If the operation requires heavy physical work, spot cooling by ventilation, evaporative cooling, or refrigerated air can be used. To minimize summer discomfort, high outdoor air ventilation rates may be adequate in some hot process areas. In all these operations, a mechanical air supply with good distribution is needed.

Refrigerated Cooling Systems

The refrigeration cooling source may be located in a central equipment area, or smaller packaged cooling systems may be placed near or combined with each air-handling unit. Central mechanical equipment uses positive-displacement, centrifugal, or absorption refrigeration to chill water. Pumped through unit cooling coils, the chilled water absorbs heat, then returns to the cooling equipment.

Central system condenser water rejects heat through a cooling tower. The heat may be transferred to other sections of the building where heating or reheating is required, and the cooling unit becomes a heat recovery heat pump. Refrigerated heat recovery is particularly advantageous in buildings with a simultaneous need for heating exterior sections and cooling interior sections.

When interior spaces are cooled with a combination of outdoor air and chilled water obtained from reciprocating or centrifugal chillers with heat recovery condensers, hot water at temperatures up to 110°F is readily available. Large quantities of air at room temperature, which must be exhausted because of contaminants, can first be passed through a chilled water coil to recover heat. Heating or reheating obtained by refrigerated heat recovery occurs at a COP approaching 4, regardless of outdoor temperature, and can save considerable energy.

Mechanical cooling equipment should be selected in multiple units. This enables the equipment to match its response to fluctuations in the load and allows maintenance during nonpeak operation. Small packaged refrigeration equipment commonly uses positive-displacement (reciprocating or screw) compressors with air-cooled condensers. These units usually provide up to 207 tons of cooling. Because equipment is often on the roof, the condensing temperature may be affected by warm ambient air, which is often 10 to 20°F higher than the design outdoor air temperature. In this type of system, the cooling coil may receive refrigerant directly.

ASHRAE *Standard* 15 limits the type and quantity of refrigerant in direct air-to-refrigerant exchangers. Fins on the air side improve heat transfer, but they increase the pressure drop through the coil, particularly as they get dirty.

Evaporative Cooling Systems

Evaporative cooling systems may be evaporative coolers or air washers. Evaporative coolers have water sprayed directly on a wetted surface through which air moves. An air washer recirculates water, and the air flows through a heavily misted area. Water atomized in the airstream evaporates, and the water cools the air. Refrigerated water simultaneously cools and dehumidifies the air.

Evaporative cooling offers energy conservation opportunities, particularly in intermediate seasons. In many industrial facilities, evaporative cooling controls both temperature and humidity. In these systems, the sprayed water is normally refrigerated, and a reheat coil is often used. Temperature and humidity of the exit airstream may be controlled by varying the temperature of the chilled water and the reheat coil and by varying the quantity of air passing through the reheat coil with a dewpoint thermostat.

Care must be taken that accumulation of dust or lint does not clog the nozzles or evaporating pads of the evaporative cooling systems. It may be necessary to filter the air before the evaporative cooler. Fan heat, air leakage through closed dampers, and the entrainment of room air into the supply airstream all affect design. Chemical treatment of the water may be necessary to prevent mineral buildup or biological growth on the pads or in the pans.

AIR FILTRATION SYSTEMS

Air filtration systems remove contaminants from air supplied to, or exhausted from, building spaces. Supply air filtration (most frequently on the intake side of the air-conditioning apparatus) removes particulate contamination that may foul heat transfer surfaces, contaminate products, or present a health hazard to people, animals, or plants. Gaseous contaminants must sometimes be removed to prevent exposing personnel to health-threatening fumes or odors. The supply airstream may consist of air recirculated from building spaces and/or outdoor air for ventilation or exhaust air makeup. Return air with a significant potential for carrying contaminants should be recirculated only if it can be filtered enough to minimize personnel exposure. If monitoring and contaminant control cannot be ensured, the return air should be exhausted.

The supply air filtration system usually includes a collection medium or filter, a medium-retaining device or filter frame, and a filter house or plenum. The filter medium is the most important component of the system; a mat of randomly distributed small-diameter fibers is commonly used.

Depending on fiber material, size, density, and arrangement, fibrous filters have a wide range of performance. Low-density filter media with relatively large-diameter fibers remove large particles, such as lint. These roughing filters collect a large percentage by mass of the particulates, but are ineffective in reducing the total particle concentration. The fibers are sometimes coated with an adhesive to reduce particle reentrainment.

Small-fiber, high-density filter media effectively collect essentially all particulates. Ultrahigh-efficiency particulate air filters reduce total particle concentration by more than 99.9%. As filter efficiency increases, so does the resistance to airflow, with typical pressure drops ranging from 0.05 to 1 in. of water. Conversely, dust-holding capacity decreases with increasing filter efficiency, and fibrous filters should not be used for dust loading greater than 100 μg/ft^3 of air. For more discussion of particulate filtration systems, refer to Chapter 24 of the 1996 *ASHRAE Handbook—Systems and Equipment*.

Exhaust Air Filtration Systems

Exhaust air systems are either (1) general systems that remove air from large spaces or (2) local systems that capture aerosols, heat, or gases at specific locations within a room and transport them so they can be collected (filtered), inactivated, or safely discharged to the atmosphere. The air in a general system usually requires minimal treatment before discharging to the atmosphere. The air in local exhaust systems can sometimes be safely dispersed into the atmosphere, but sometimes contaminants must be removed so that the emitted air meets air quality standards. Chapters 28 amd 29 have more information on industrial ventilation and exhaust systems.

Many types of contamination collection or inactivation systems are applied in exhaust air emission control. Fabric bag filters, glass-fiber filters, venturi scrubbers, and electrostatic precipitators all collect particulates. Packed bed or sieve towers can absorb toxic gases. Activated carbon columns or beds, often with oxidizing agents, are frequently used to absorb toxic or odorous organics and radioactive gases.

Outdoor air intakes should be carefully located to avoid recirculation of contaminated exhaust air. Because wind direction, building shape, and the location of the effluent source strongly influence concentration patterns, exact patterns are not predictable.

Air patterns resulting from wind flow over buildings are discussed in Chapter 15 of the 1997 *ASHRAE Handbook—Fundamentals*. The leading edge of a roof interrupts smooth airflow, resulting in reduced air pressure at the roof and on the lee side. To prevent fume damage to the roof and roof-mounted equipment, and to keep fumes from the building air intakes, fumes must be discharged through either (1) vertical stacks terminating above the turbulent air boundary or (2) short stacks with a velocity high enough to project the effluent through the boundary into the undisturbed air passing over the building. A high vertical stack is the safest and simplest solution to fume dispersal.

Contaminant Control

In addition to maintaining thermal conditions, air-conditioning systems should control contaminant levels to provide (1) a safe and healthy environment, (2) good housekeeping, and (3) quality control for the processes. Contaminants may be gases, fumes, mists, or airborne particulate matter. They may be produced by a process within the building or contained in the outside air.

Contamination can be controlled by (1) preventing the release of aerosols or gases into the room environment and (2) diluting room air contaminants. If the process cannot be enclosed, it is best to capture aerosols or gases near their source of generation with local exhaust systems that include an enclosure or hood, ductwork, a fan, a motor, and an exhaust stack.

Dilution controls contamination in many applications but may not provide uniform safety for personnel within a space (West 1977). High local concentrations of contaminants can exist within a room, even though the overall dilution rate is quite high. Further, if tempering of outdoor air is required, high energy costs can result from the increased airflow required for dilution.

EXHAUST SYSTEMS

An exhaust system draws the contaminant away from its source and removes it from the space. An exhaust hood that surrounds the point of generation contains the contaminant as much as is practical. The contaminants are transported through ductwork from the space, cleaned as required, and exhausted to the atmosphere. The suction air quantity in the hood is established by the velocities required to contain the contaminant. Chapter 29 has more information on local exhaust systems.

Design values for average and minimum face velocities are a function of the characteristics of the most dangerous material that the hood is expected to handle. Minimum values may be prescribed in codes for exhaust systems. Contaminants with greater mass may require higher face velocities for their control.

Properly sized ductwork keeps the contaminant flowing. This requires very high velocities for heavy materials. The selection of materials and the construction of exhaust ductwork and fans depend on the nature of the contaminant, the ambient temperature, the lengths and arrangement of duct runs, the method of hood fan operation, and the flame and smoke spread rating.

Exhaust systems remove chemical gases, vapors, or smokes from acids, alkalis, solvents, and oils. Care must be taken to minimize the following:

1. **Corrosion**, which is the destruction of metal by chemical or electrochemical action; commonly used reagents in laboratories are hydrochloric, sulfuric, and nitric acids, singly or in combination, and ammonium hydroxide. Common organic chemicals include acetone, benzene, ether, petroleum, chloroform, carbon tetrachloride, and acetic acid.
2. **Dissolution**, which is a dissolving action. Coatings and plastics are subject to this action, particularly by solvent and oil fumes.
3. **Melting**, which can occur in certain plastics and coatings at elevated hood operating temperatures.

Low temperatures that cause condensation in ferrous metal ducts increase chemical destruction. Ductwork is less subject to attack when the runs are short and direct to the terminal discharge point. The longer the runs, the longer the period of exposure to fumes and the greater the degree of condensation. Horizontal runs allow moisture to remain longer than it can on vertical surfaces. Intermittent fan operation can contribute to longer periods of wetness (because of condensation) than continuous operation. High loading of condensables in exhaust systems should be avoided by installing condensers or scrubbers as close to the source as possible.

OPERATION AND MAINTENANCE OF COMPONENTS

All designs should allow ample room to clean, service, and replace any component quickly so that design conditions are affected

as little as possible. Maintenance of refrigeration and heat-rejection equipment is essential for proper performance without energy waste.

For system dependability, water treatment is essential. Air washers and cooling towers should not be operated unless the water is properly treated.

Maintenance of heating and cooling systems includes changing or cleaning system filters on a regular basis. Industrial applications are usually dirty, so proper selection of filters, careful installation to avoid bypassing, and prudent changing of filters to prevent overloading and blowout are required. Dirt that lodges in ductwork and on forward-curved fan blades reduces air-handling capacity appreciably.

Fan and motor bearings require lubrication, and fan belts need periodic inspection. Infrared and panel systems usually require less maintenance than equipment with filters and fans, although gas-fired units with many burners require more attention than electric heaters.

The direct-fired makeup heater has a relatively simple burner that requires less maintenance than a comparable indirect-fired heater. The indirect oil-fired heater requires more maintenance than a comparable indirect gas-fired heater. With either type, the many safety devices and controls require periodic maintenance to ensure constant operation without nuisance cutout. Direct- and indirect-fired heaters should be inspected at least once per year.

Steam and hot water heaters have fewer maintenance requirements than comparable equipment having gas and oil burners. When used for makeup air in below-freezing conditions, however, the heaters must be correctly applied and controlled to prevent frozen coils.

REFERENCES

ACGIH. 1998. *Industrial ventilation: A manual of recommended practice*, 23rd ed. American Conference of Governmental Industrial Hygienists, Cincinnati, OH.

ASHRAE. 1989. Ventilation for acceptable indoor air quality. ANSI/ASHRAE *Standard* 62-1989.

ASHRAE. 1992. Thermal environmental conditions for human occupancy. ANSI/ASHRAE *Standard* 55-1992.

ASHRAE. 1994. Safety code for mechanical refrigeration. ANSI/ASHRAE *Standard* 15-1994.

Dukes-Dobos, F. and A. Henschel. 1971. The modification of the WNGT Index for establishing permissible heat exposure limits in occupational work. U.S. DHEW, U.S. PHS, ROSH joint *Publication* TR-69.

IESNA. 1993. *Lighting handbook*, 8th ed. Illuminating Engineering Society of North America, New York.

McNall, P.E., J. Juax, F.H. Rohles, R.G. Nevins, and W. Springer. 1967. Thermal comfort (thermally neutral) conditions for three levels of activity. *ASHRAE Transactions* 73(1):I.3.1-14.

West, D.L. 1977. Contaminant dispersion and dilution in a ventilated space. *ASHRAE Transactions* 83(1):125-40.

BIBLIOGRAPHY

Azer, N.Z. 1982a. Design guidelines for spot cooling systems. Part 1—Assessing the acceptability of the environment. *ASHRAE Transactions* 88(2):81-95.

Azer, N.Z. 1982b. Design guidelines for spot cooling systems. Part 2—Cooling jet model and design procedure. *ASHRAE Transactions* 88(2): 97-116.

Gorton, R.L. and Bagheri. 1987a. Verification of stratified air-conditioning design. *ASHRAE Transactions* 93(2):211-27.

Gorton, R.L. and Bagheri. 1987b. Performance characteristics of a system designed for stratified cooling operating during the heating season. *ASHRAE Transactions* 93(2):367-81.

Harstad, J., et al. 1967. Air filtration of submicron virus aerosols. *American Journal of Public Health* 57:2186-93.

O'Connell, W.L. 1976. How to attack air-pollution control problems. *Chemical Engineering* (October), desktop issue.

Whitby, K.T. and D.A. Lundgren. 1965. Mechanics of air cleaning. *ASAE Transactions* 8:3, 342. American Society of Agricultural Engineers, St. Joseph, MI.

Yamazaki, K. 1982. Factorial analysis on conditions affecting the sense of comfort of workers in the air-conditioned working environment. *ASHRAE Transactions* 88(1):241-54.

ENCLOSED VEHICULAR FACILITIES

THIS chapter deals with the ventilation requirements for cooling, contaminant control, and emergency smoke and temperature control for road tunnels, rapid transit tunnels and stations, enclosed parking structures, bus garages, and bus terminals. Also included are the design approach and type of equipment applied to these ventilation systems.

ROAD TUNNELS

A road tunnel is defined as any enclosed facility through which an operating roadway carrying motor vehicles passes. Road tunnels may be of the underwater, mountain, or urban type. Tunnels may also be created by development of air-right structures over a roadway and overbuilds of the roadway.

All road tunnels require ventilation to remove the contaminants produced by vehicle engines during normal operation. Ventilation may be provided by natural means, by the traffic-induced piston effect, or by mechanical equipment. The method selected should be the most economical in terms of construction and operating costs. Natural and traffic-induced ventilation is adequate for relatively short tunnels and tunnels with low traffic volume (or density). Long and heavily traveled tunnels should have mechanical ventilation.

The exhaust gas constituent of greatest concern from spark-ignition engines is carbon monoxide (CO), because of its asphyxiate nature. From compression-ignition (diesel) engines the predominant contaminant is nitrogen dioxide (NO_2). Tests and operating experience indicate that when the level of carbon monoxide is properly diluted, the other dangerous and objectionable exhaust byproducts are also diluted to acceptable levels. An exception to this condition is the large amount of unburned hydrocarbons from vehicles with diesel engines. When the diesel-engine vehicle portion of the traffic mix exceeds 15%, visibility in the tunnel can become a serious concern. The section on Bus Terminals includes further information on diesel engine contaminants and their dilution.

VENTILATION

Three types of mechanical ventilation are considered for road tunnels. **Normal ventilation** is used during normal traffic operations to maintain acceptable levels of contaminants in the tunnel.

The preparation of this chapter is assigned to TC 5.9, Enclosed Vehicular Facilities.

Emergency ventilation is used during a fire emergency to remove and control smoke and hot gases. The primary objective of emergency ventilation is to provide an environment sufficiently clear of smoke and hot gases and at a sufficiently low temperature to permit safe evacuation of motorists, and to allow relatively safe access for firefighters.

Temporary ventilation (usually removed after construction is complete) is needed by workers during construction of the facility or while working in the finished tunnel. These ventilation requirements, which are not covered in this chapter, are specified by state or local mining laws, industrial codes, or in the standards set by the U.S. Occupational Safety and Health Administration (OSHA).

VENTILATION SYSTEMS

Any ventilation must dilute contaminants during normal tunnel operations and control smoke during emergency operations. Factors that determine the system selected include tunnel length, cross section, and grade; surrounding environment; traffic volume; and construction cost.

Natural Ventilation

Naturally-ventilated tunnels rely primarily on atmospheric conditions to maintain airflow and a satisfactory environment in the tunnel. The piston effect of traffic provides additional airflow when the traffic is moving. The chief factor affecting the tunnel environment is the pressure differential that is created by differences in elevation, ambient temperature, or wind between the tunnels two portals. Unfortunately, none of these factors can be relied on for continued, consistent results. A change in wind direction or speed can negate all these natural effects, including the piston effect. The pressure must be large enough to overcome the tunnel resistance, which is influenced by tunnel length, cross-sectional geometry, wall roughness, number of vehicles in the tunnel, and air density.

Air can flow through a naturally ventilated tunnel from portal-to-portal (Figure 1A) or portal-to-shaft (Figure 1B). Portal-to-portal flow functions best with unidirectional traffic, which produces a consistent, positive airflow. The air speed in the roadway area is relatively uniform, and the contaminant concentration increases to a maximum at the exit portal. Under adverse atmospheric conditions, the air speed decreases and the contaminant concentration increases, as shown by the dashed line on Figure 1A. The introduction of

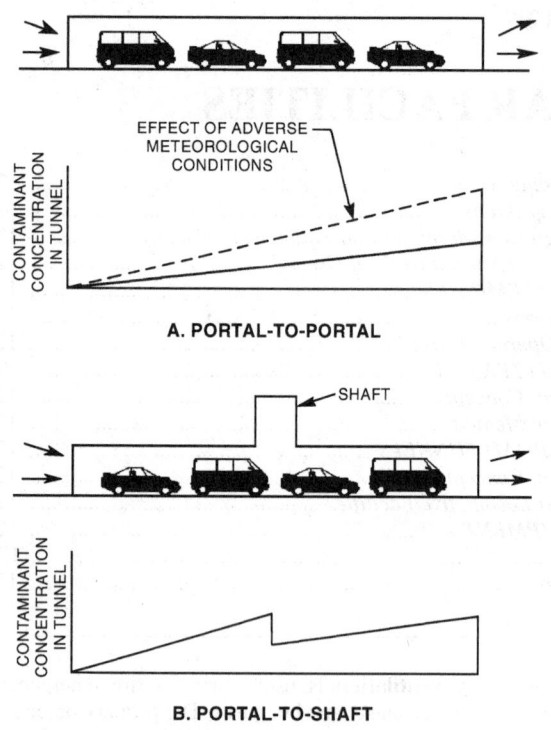

A. PORTAL-TO-PORTAL

B. PORTAL-TO-SHAFT

Fig. 1   Natural Ventilation

Table 1   Smoke Movement During Natural Ventilation Tests
(Memorial Tunnel Fire Ventilation Test Program)

Test No.	Fire Heat Release Rate, 10^6 Btu/h		Smoke Layer Begins Descent, min	Smoke Fills Tunnel Road-way, min	Peak Smoke Velocity, fpm
	Nominal	Peak			
501	68	99	3+	5	1200
502	170	194	1+	3	1600

Note: Tunnel grade is 3.2%.

bidirectional traffic into such a tunnel will further reduce the tunnel airflow and increases the contaminant concentration.

The naturally ventilated tunnel with an intermediate shaft (Figure 1B) is best suited for bidirectional traffic. However, airflow through such a shafted tunnel is also at the mercy of the elements. The benefit of the stack effect of the shaft depends on air and rock temperatures, wind, and shaft height. The addition of more than one shaft to a naturally ventilated tunnel may be more of a disadvantage than an advantage because a pocket of contaminated air can be trapped between the shafts.

Naturally ventilated urban tunnels over 800 ft long require emergency mechanical ventilation to extract smoke and hot gases generated during a fire. This emergency system may also be used to remove stagnate contaminants during adverse atmospheric conditions. Because of the uncertainties of natural ventilation, especially the effect of adverse meteorological and operating conditions, reliance on natural ventilation for tunnels over 800 ft long should be thoroughly evaluated. This is particularly important for a tunnel with an anticipated heavy or congested traffic flow. If the natural ventilation is inadequate, the installation of a mechanical system with fans should be considered for normal operations.

Smoke from a fire in a tunnel with only natural ventilation moves up the grade driven primarily by the buoyant effect of the hot smoke and gases. The steeper the grade the faster the smoke will move thus restricting the ability of motorists trapped between the incident and the portal at the higher elevation to evacuate the tunnel safely. As shown in Table 1, MHD/FHWA (1995) demonstrates how smoke moves in a naturally ventilated tunnel with a 3.2% grade.

Mechanical Ventilation

A tunnel that is sufficiently long, has heavy traffic flow, or experiences adverse atmospheric conditions requires mechanical ventilation with fans. Mechanical ventilation layouts in road tunnels include longitudinal ventilation, semitransverse ventilation, and full transverse ventilation.

Longitudinal Ventilation. This type of ventilation introduces or removes air from the tunnel at a limited number of points, thus creating a longitudinal flow of air along the roadway. Ventilation is either by injection, by jet fans, or by a combination of injection or extraction at intermediate points in the tunnel.

Injection longitudinal ventilation is frequently used in rail tunnels and is also found in road tunnels. Air injected at one end of the tunnel mixes with air brought in by the piston effect of the incoming traffic (Figure 2A). This type of ventilation is most effective where traffic is unidirectional. The air speed remains uniform throughout the tunnel, and the concentration of contaminants increases from zero at the entrance to a maximum at the exit. Adverse external atmospheric conditions can reduce the effectiveness of this system. The contaminant level at the exit portal increases as the airflow decreases or the tunnel length increases.

Injection longitudinal ventilation with the supply at a limited number of locations in the tunnel is economical because it requires the least number of fans, places the least operating burden on these fans, and requires no distribution air ducts. As the length of the tunnel increases, however, disadvantages such as excessive air velocities in the roadway and smoke being drawn the entire length of the roadway during an emergency become apparent. Uniform air distribution would alleviate these problems, however the system then becomes a semitransverse ventilation system.

A longitudinal system with a fan shaft (Figure 2B) is similar to the naturally ventilated system with a shaft, except that it provides a positive stack effect. Bidirectional traffic in a tunnel ventilated in this manner causes a peak contaminant concentration at the shaft location. For unidirectional tunnels, the contaminant levels become unbalanced.

Another form of longitudinal system has two shafts near the center of the tunnel: one for exhaust and one for supply (Figure 2C). This arrangement reduces contaminant concentration in the second half of the tunnel. A portion of the air flowing in the roadway is replaced in the interaction at the shafts. Adverse wind conditions can reduce the airflow, causing the contaminant concentration to rise in the second half of the tunnel, and short-circuit the flow from fan to exhaust.

Jet fan longitudinal ventilation has been installed in a number of tunnels worldwide. Longitudinal ventilation is achieved with specially designed axial fans (jet fans) mounted at the tunnel ceiling (Figure 2D). Such a system eliminates the space needed to house ventilation fans in a separate structure or ventilation building; however, it may require a tunnel of greater height or width to accommodate jet fans so that they are out of the tunnel's **dynamic clearance envelope**. This envelope, formed by the vertical and horizontal planes surrounding the roadway pavement in a tunnel, define the maximum limits of predicted vertical and lateral movement of vehicles traveling on the road at design speed. As the length of the tunnel increases, however, the disadvantages of longitudinal systems, such as excessive air speed in the roadway and smoke being drawn the entire length of the roadway during an emergency, become apparent.

The longitudinal form of ventilation is the most effective method of smoke control in a highway tunnel with unidirectional traffic. A ventilation system must generate sufficient longitudinal air velocity to prevent the backlayering of smoke. Backlayering is the movement of smoke and hot gases contrary to the direction of the ventilation airflow in the tunnel roadway. The air velocity necessary to prevent backlayering of smoke over the stalled

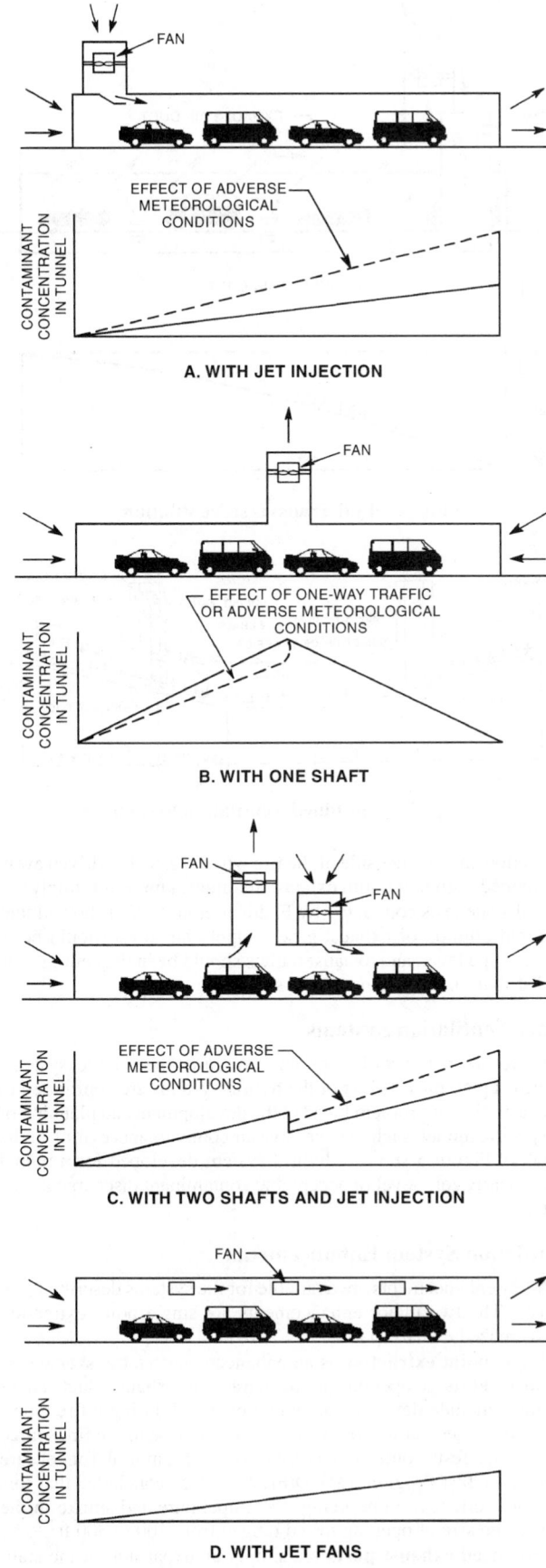

Fig. 2 Longitudinal Ventilation

motor vehicles is the minimum velocity needed for smoke control in a longitudinal ventilation system and is known as the critical velocity.

Semitransverse Ventilation. Semitransverse ventilation can be configured as either a supply system or an exhaust system. This type of ventilation incorporates the distribution of (supply system) or collection of (exhaust) air uniformly throughout the length of a road tunnel. Semitransverse ventilation is normally used in tunnels up to about 7000 ft; beyond that length the tunnel air speed near the portals become excessive.

Supply semitransverse ventilation applied to a tunnel with bidirectional traffic produces a uniform level of contaminants throughout the tunnel because the air and the vehicle exhaust gases enter the roadway area at the same uniform rate. In a tunnel with unidirectional traffic, additional airflow is generated in the roadway by the movement of the vehicles, thus reducing the contaminant level in portions of the tunnel, as shown in Figure 3A.

Because the tunnel airflow is fan-generated, this type of ventilation is not adversely affected by atmospheric conditions. The air flows the length of the tunnel in a duct fitted with supply outlets spaced at predetermined distances. Fresh air is best introduced at vehicle exhaust pipe level to dilute the exhaust gases immediately. An adequate pressure differential between the duct and the roadway must be generated and maintained to counteract piston effect and adverse atmospheric winds.

If a fire occurs in the tunnel, the supply air initially dilutes the smoke. Supply semitransverse ventilation should be operated in a reversed mode for the emergency so that fresh air enters the tunnel through the portals to create a respirable environment for fire-fighting efforts and emergency egress. Therefore, the ventilation configuration for a supply semitransverse system should preferably have a ceiling supply and reversible fans so that smoke can be drawn up to the ceiling in an emergency.

Exhaust semitransverse ventilation (Figure 3B) installed in a unidirectional tunnel produces a maximum contaminant concentration at the exit portal. In a bidirectional tunnel, the maximum level of contaminants is located near the center of the tunnel. A combination supply and exhaust system (Figure 3C) should be applied only in a unidirectional tunnel where air entering with the traffic stream is exhausted in the first half, and air supplied in the second half is exhausted through the exit portal.

In a fire emergency both the exhaust semitransverse ventilation system and the reversed semitransverse supply system create a longitudinal air velocity in the tunnel roadway, which extracts smoke and hot gases at uniform intervals.

Full Transverse Ventilation. Full transverse ventilation is used in extremely long tunnels and in tunnels with heavy traffic volume. Full transverse ventilation includes both a supply duct and an exhaust duct to achieve uniform distribution of supply air and uniform collection of vitiated air (Figure 4) throughout the tunnel length. With this arrangement, pressure along the roadway is uniform and there is no longitudinal airflow except that generated by the traffic piston effect, which tends to reduce contaminant levels. An adequate pressure differential between the ducts and the roadway is required to assure proper air distribution under all ventilation conditions.

During a fire emergency the exhaust system should be operated at the highest available capacity while the supply is operated at a somewhat lower capacity. This operation allows smoke that has stratified towards the ceiling to remain at that higher elevation and to be extracted by the exhaust without mixing. This then maintains a respirable environment at the roadway and allows fresh air to enter through the portals and create a respirable environment for fire fighting and emergency egress.

In longer tunnels a means should be provided to control the individual sections or zones so that the section with traffic trapped behind a fire is provided with maximum supply and no exhaust and

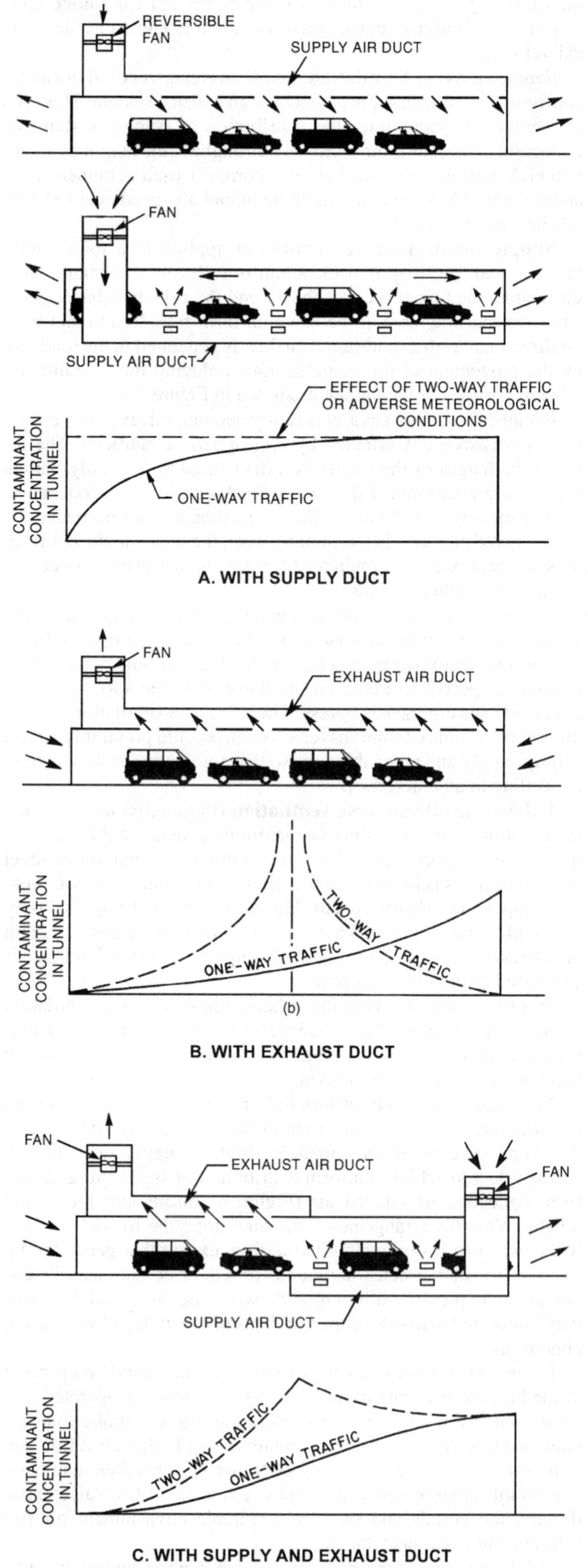

REVERSIBLE FAN — SUPPLY AIR DUCT

FAN

SUPPLY AIR DUCT

— EFFECT OF TWO-WAY TRAFFIC OR ADVERSE METEOROLOGICAL CONDITIONS

ONE-WAY TRAFFIC

CONTAMINANT CONCENTRATION IN TUNNEL

A. WITH SUPPLY DUCT

FAN — EXHAUST AIR DUCT

TWO-WAY TRAFFIC
ONE-WAY TRAFFIC
TWO-WAY TRAFFIC

CONTAMINANT CONCENTRATION IN TUNNEL

(b)

B. WITH EXHAUST DUCT

FAN — EXHAUST AIR DUCT — FAN

SUPPLY AIR DUCT

TWO-WAY TRAFFIC
ONE-WAY TRAFFIC

CONTAMINANT CONCENTRATION IN TUNNEL

C. WITH SUPPLY AND EXHAUST DUCT

Fig. 3   Semitransverse Ventilation

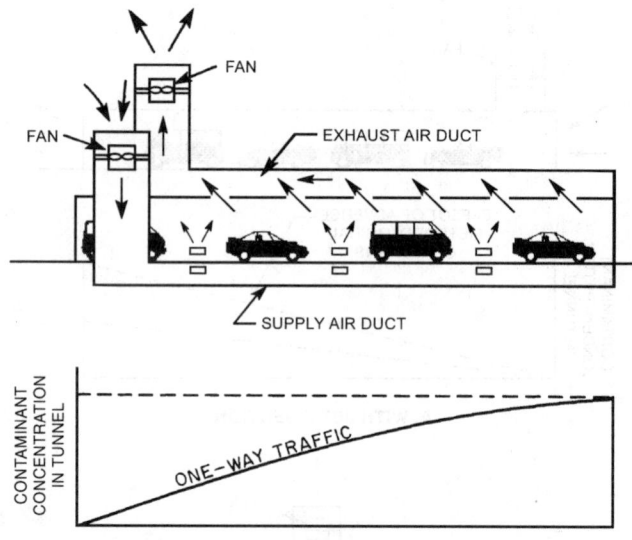

FAN
FAN — EXHAUST AIR DUCT

SUPPLY AIR DUCT

ONE-WAY TRAFFIC

CONTAMINANT CONCENTRATION IN TUNNEL

Fig. 4   Full Transverse Ventilation

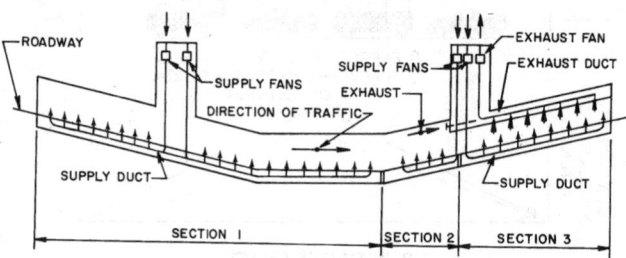

ROADWAY
SUPPLY FANS
EXHAUST FAN
SUPPLY FANS
EXHAUST DUCT
EXHAUST
DIRECTION OF TRAFFIC
SUPPLY DUCT
SUPPLY DUCT
SECTION 1   SECTION 2   SECTION 3

Fig. 5   Combined Ventilation System

the section on the other side of the fire where traffic has driven away is provided with maximum exhaust and minimum or no supply.

Full-scale tests conducted by Fieldner et al. (1921) showed that for rapid dilution of exhaust gases, supply air inlets should be at exhaust pipe level, and exhaust outlets should be in the ceiling. The air distribution can be one or two sided.

Other Ventilation Systems

There are many variations and combinations of the systems described previously. Most of the hybrid systems are configured to solve a particular problem faced in the development and planning of the specific tunnel, such as excessive air contaminants exiting at the portal(s). Figure 5 shows a hybrid system developed for a tunnel with a nearly zero level of acceptable contaminant discharge at one portal.

Ventilation System Enhancements

A few enhancements are available for the systems described previously. The two major enhancements are single point extraction and oversized exhaust ports.

Single point extraction is an enhancement to a transverse system that adds large openings to the extraction (exhaust) duct. These openings include devices that can be operated during a fire emergency to extract a large volume of smoke as close to the fire source as possible. Tests conducted as a part of the Memorial Tunnel Fire Ventilation Test Program (MHD/FHWA 1995) concluded that this concept is effective in reducing the temperature and smoke in the tunnel. The size of opening tested ranged from 100 to 300 ft^2.

Oversized exhaust ports are simply an expansion of the standard exhaust port installed in the exhaust duct of a transverse or semitransverse ventilation system. Two methods are used to create such a configuration. One is to install on each port expansion a damper with a fusible link; the other uses a material that when heated to a specific temperature melts and opens the airway. Several tests of such meltable material were conducted as part of the Memorial Tunnel Fire Ventilation Test Program but with limited success.

NORMAL VENTILATION AIR QUANTITIES

Contaminant Emission Rates

The vehicle emissions of carbon monoxide (CO), oxides of nitrogen (NO_x), and hydrocarbons for any given calendar year can be predicted in the United States by using the model developed and maintained by the EPA (1992).

All suggested ventilation rates given in this chapter are derived using uncontrolled emission rates, but the results can be corrected as indicated to reflect the impact of emission control devices.

Allowable Carbon Monoxide

In 1975, the EPA issued a supplement to its *Guidelines for Review of Environmental Statements for Highway Projects* that evolved into a design approach based on keeping a CO concentration of 125 ppm or below for a maximum of one hour exposure time for tunnels located at or below an altitude of 3280 ft. In 1988, the EPA revised its recommendations for maximum CO levels in tunnels located at or below an altitude of 5000 ft to the following:

- Maximum 120 ppm for 15 min exposure time
- Maximum 65 ppm for 30 min exposure time
- Maximum 45 ppm for 45 min exposure time
- Maximum 35 ppm for 60 min exposure time

These guidelines do not apply to tunnels in operation prior to the adoption date. Outdoor air standards and regulations such as those from the Occupational Safety and Health Standards (OSHA) and the American Conference of Governmental Industrial Hygienists (ACGIH) are discussed in the section on Bus Terminals.

At higher elevations the CO emission of vehicles is greatly increased, and human tolerance to CO exposure is reduced. For tunnels above 5000 ft, the engineer should consult with medical authorities to establish a proper design value for CO concentrations. Unless specified otherwise, the material in this chapter refers to tunnels at or below an altitude of 5000 ft.

Computer Programs Available

SES. The Subway Environment Simulation (SES) computer program can be used to examine longitudinal airflow in a road tunnel (DOT 1997). The SES is a one dimensional network model is the predominant worldwide tool used to evaluate longitudinal airflow in tunnels.

TUNVEN. This program solves the coupled one-dimensional steady-state tunnel aerodynamic and advection equations. It can predict quasi-steady-state longitudinal air velocities and the concentrations of CO, oxides of nitrogen (NO_x), and total hydrocarbons along a highway tunnel for a wide range of tunnel designs, traffic loads, and external ambient conditions. The program can model all common ventilation system: natural, longitudinal, semitransverse, and transverse. The program is available from NTIS (1980).

EMERGENCY VENTILATION AIR QUANTITIES

The ventilation system must have the capacity to protect the traveling public during the most adverse and dangerous conditions, as well as during normal conditions. Establishing air requirements is difficult because many uncontrollable variables, such as the many possible vehicle combinations and traffic situations, could occur during the lifetime of the facility. Longitudinal flow and extraction and dilution are two primary methods of controlling smoke from fire in a tunnel.

Table 2 Typical Fire Size Data for Road Vehicles

Cause of Fire	Equivalent Size of Gasoline Pool, ft^2	Fire Heat Release Rate, 10^6 Btu/h	Smoke Generation Rate, 1000 cfm	Maximum Temp.,[a] °F
Passenger car	22	≈17	42	750
Bus/Truck	86	≈68	127	1290
Gasoline tanker	300–1100	≈340	200–400	1830

Source: PIARC (1995).
[a]Temperature 30 ft downwind of fire with the minimum air velocity necessary to prevent back-layering.

Table 3 Maximum Temperature Experienced at Ventilation Fans (Memorial Tunnel Fire Ventilation Test Program)

Nominal FHRR, 10^6 Btu/h	Temperature at Central Fans[a], °F	Temperature at Jet Fan[b], °F
340	325	1250
170	255	700
68	225	450

[a]Central fans were located about 700 ft from fire site.
[b]Jet fans were located 170 ft downstream of fire site.

Fire Size. The design fire size selected has a significant effect on the magnitude of the critical air velocity necessary to prevent backlayering. The data in Table 2 provide a guideline to the typical fire size for a selection of road vehicles. The method for calculating the required critical velocity is shown below.

Temperature. A fire in a tunnel significantly increases the air temperature in the tunnel roadway or duct. This means that the structure and equipment are exposed to the high gas and smoke temperature. The temperatures shown in Table 3 provide guidance in selecting the design temperature for structures and equipment.

Critical Velocity. The simultaneous solution of Equations (1) and (2), by iteration, determines critical velocity (Kennedy et al. 1996), which is the minimum steady-state velocity of the ventilation air moving toward the fire that is needed to prevent backlayering:

$$V_C = K_1 K_g \left(\frac{gHq}{\rho c_p A T_f} \right)^{1/3} \tag{1}$$

$$T_f = \left(\frac{q}{\rho c_p A V_C} \right)^{1/3} + T \tag{2}$$

where

A = area perpendicular to the flow, ft^2
c_p = specific heat of air, Btu/lb·°R
g = acceleration caused by gravity, ft/s^2
H = height of duct or tunnel at the fire site, ft
K_1 = 0.606
K_g = grade factor (see Figure 6).
q = heat that fire adds directly to air at the fire site, Btu/s
T = temperature of approach air, °R
T_f = average temperature of fire site gases, °R
V_C = critical velocity, ft/s
ρ = average density of approach (upstream) air, lb/ft^3

Memorial Tunnel Fire Ventilation Test Program. This full-scale test program was conducted to evaluate the effectiveness of various tunnel ventilation systems and ventilation airflow rates to control smoke from a fire (MHS/FHWA 1995). The results are useful in the development of emergency tunnel ventilation design and emergency operational procedures.

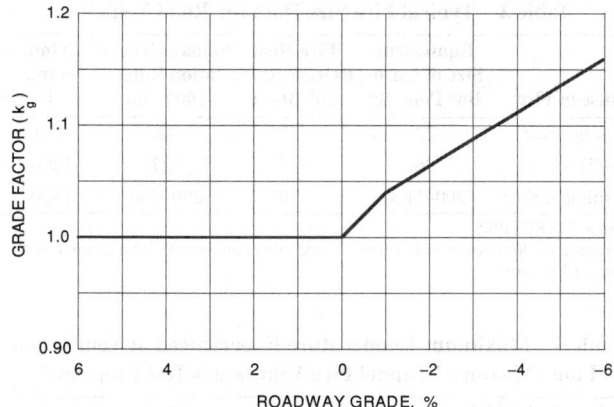

Fig. 6   Roadway Grade Factor

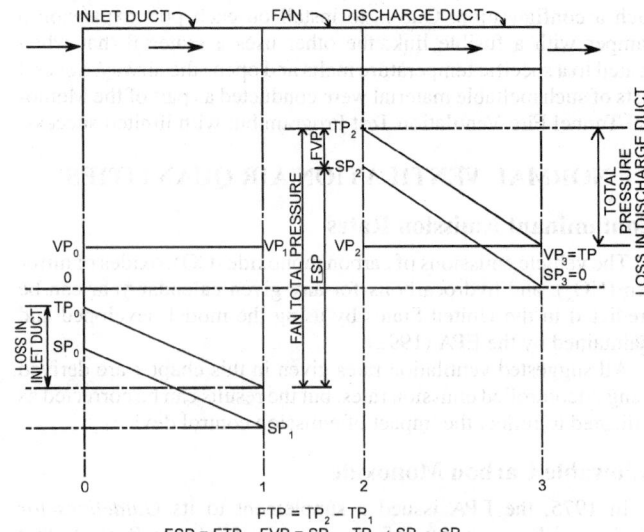

$$FTP = TP_2 - TP_1$$
$$FSP = FTP - FVP = SP_2 - TP_1 \neq SP_2 - SP_1$$

Fig. 7   Fan Total Pressure

Computer Programs Available

SES. This program contains a fire model that can be used to determine how much longitudinal airflow is required to overcome backlayering and control smoke movement in a tunnel.

CFD. Computational fluid dynamics software has become the design tool of choice of engineers from different disciplines to obtain an optimum design due to the high cost, complexity, and limited information obtained from experimental methods. CFD can analyze different tunnel configurations without having to perform actual tests.

PRESSURE EVALUATION

Air pressure losses in the tunnel ducts must be evaluated to compute fan pressure and drive requirements. Fan selection should be based on total pressure across the fans, not on static pressure only.

Fan total pressure (FTP) is defined by ASHRAE *Standard* 51/AMCA *Standard* 210 as the algebraic difference between the total pressures at the fan discharge (TP_2) and at the fan inlet (TP_1), as shown in Figure 7. The fan velocity pressure (FVP) is defined as the pressure (VP_2) corresponding to the bulk air velocity and air density at the fan discharge.

$$FVP = VP_2 \tag{3}$$

The fan static pressure (FSP) equals the difference between the fan total pressure and the fan velocity pressure.

$$FSP = FTP - FVP \tag{4}$$

The total pressure at the fan discharge (TP_2) must equal the total pressure losses ($\Delta TP_{2\text{-}3}$) in the discharge duct and the exit velocity pressure (VP_3).

$$TP_2 = \Delta TP_{2\text{-}3} + VP_3 \tag{5}$$

Likewise, the total pressure at the fan inlet (TP_1) must equal the total pressure losses in the inlet duct and the inlet pressure.

$$TP_1 = TP_0 + \Delta TP_{0\text{-}1} \tag{6}$$

Straight Ducts

Straight ducts in tunnel ventilation systems can be classified as (1) those that transport air and (2) those that uniformly distribute (supply) or uniformly collect (exhaust) air. Several methods have been developed to predict pressure losses in a duct of constant cross-sectional area that uniformly distributes or collects air. The most widely used method was developed for the Holland Tunnel in New York (Singstad 1929). The following relationships based on the Singstad's work, give pressure losses at any point in the duct.

Total pressure for a **supply duct**,

$$p_T = p_1 + 12 \frac{\rho_a V_0^2}{2g\rho_w} \left[\frac{aLZ^3}{3H} - (1-K)\frac{Z^2}{2} \right] \tag{7}$$

Static pressure loss for an **exhaust duct**,

$$p_s = p_1 + 12 \frac{\rho_a V_0^2}{2g\rho_w} \left[\frac{aLZ^3}{(3+c)H} - \frac{3Z^2}{(2+c)} \right] \tag{8}$$

where

p_T = total pressure loss at any point in duct, in. of water
p_s = static pressure loss at any point in duct, in. of water
p_1 = pressure at last outlet, in. of water
ρ_a = density of air, lb/ft^3
ρ_w = density of water, lb/ft^3
V_0 = velocity of air entering duct, ft/s
g = gravity constant = 32.2 ft/s^2
a = constant related to coefficient of friction for concrete = 0.0035
L = total length of duct, ft
Z = $(L - X)/L$
X = distance from duct entrance to any location, ft
H = hydraulic radius, ft
K = constant accounting for turbulence = 0.615
c = constant relating to turbulence of exhaust port = 0.25

The geometry of the exhaust air slot connection to the main duct is a concern in the derivation of the exhaust duct equation. The derivation is based on a 45° angle between the slot discharge and the main airstream axes. Variations in this angle can greatly affect the energy losses at the convergence from each exhaust slot, with total pressure losses for a 90° connection increasing by 50 to 100% over the 45° values (Haerter 1963).

For **distribution ducts** with sections that differ along their length, these equations may also be solved sequentially for each constant area section, with transition losses considered at each change in section area. For a **transport duct** having constant cross-sectional area and constant velocity, the pressure losses are due to friction alone and can be computed using the standard expressions for losses in ducts and fittings (see Chapter 32 of the 1997 *ASHRAE Handbook—Fundamentals*).

CARBON MONOXIDE ANALYZERS AND RECORDERS

The air quality in a tunnel should be monitored continuously at several key points. Carbon monoxide is the contaminant usually selected as the prime indicator of tunnel air quality. CO-analyzing instruments base their measurements on one of three processes: catalytic oxidation, infrared absorption, or electrochemical oxidation.

The **catalytic oxidation (metal oxide) instrument**, the most widely used in road tunnels, offers reliability and stability at a moderate initial cost. Maintenance requirements are low, and the instruments can be calibrated and serviced by maintenance personnel after only brief instruction.

The **infrared analyzer** has the advantage of sensitivity and response but the disadvantage of high initial cost. A precise and complex instrument, it requires a highly trained technician for maintenance and servicing.

The **electrochemical analyzer** is precise, compact, and light. The units are of moderate cost; but have a limited life, usually not exceeding two years, thus requiring periodic replacement.

No matter which type of CO analyzer is selected, each air sampling point must be located where significant readings can be obtained. For example, the area near the entrance of a unidirectional tunnel usually has very low CO concentrations, and sampling points there yield little information for ventilation control. The length of piping between the sampling point and the CO analyzer should be as short as possible to maintain a reasonable air sample transport time.

Where intermittent analyzers are used, provisions should be made to prevent the loss of more than one sampling point during an air pump outage. Each analyzer should be provided with a strip chart recorder to keep a permanent record of tunnel air conditions. Usually, recorders are mounted on the central control board.

Haze or smoke detectors have been used on a limited scale, but most of these instruments are optical devices and require frequent or constant cleaning with a compressed air jet. If traffic is predominantly diesel powered, smoke haze and oxides of nitrogen require monitoring in addition to CO.

CONTROLS

To reduce the number of operating personnel at a tunnel, all ventilating equipment should be controlled at a central location. At many older tunnel facilities, fan operation is manually controlled by an operator at the central control board. Many new tunnels, however, have partial or total automatic control for fan operation.

CO Analyzer Control

In this system, adjustable contacts in CO analyzers turn on additional fans or increase fan speeds as CO readings increase. The reverse occurs as CO levels decrease. The system requires fairly complex wiring because fan operation must normally respond to the single highest level being recorded at several analyzers. A manual override and delay devices are required to prevent the ventilation system from responding to short-lived high or low levels.

Time Clock Control

This automatic fan control is best suited for installations that experience heavy rush-hour traffic. A time clock is set to increase the ventilation level in preset increments before the anticipated traffic increase. The system is simple and easily revised to suit changing traffic patterns. Because it anticipates an increased air requirement, the ventilation system can be made to respond slowly and thus avoid expensive demand charges by the power company. As with the CO system, a manual override is needed to cope with unanticipated conditions.

Traffic-Actuated Control

Several automatic fan controls based on recorded traffic flow have been devised. Most of them require the installation of computers and other electronic equipment needing maintenance expertise.

Local Fan Control

In addition to a central control board, each fan should have local control in sight of the unit. It should be interlocked to permit positive isolation of the fan from remote or automatic operation during maintenance and servicing.

TOLLBOOTHS

Toll plazas for vehicular tunnels, bridges, and toll roads generally include a series of individual tollbooths. An overhead weather canopy and a utility tunnel located below the roadway surface are frequently provided. The canopy facilitates the installation of roadway signs and lighting. The utility tunnel is used for installation of electrical and mechanical systems; also it provides access to each tollbooth. An administration building is usually situated nearby.

Toll collectors and supervisors are exposed to adverse environmental conditions similar to those in bus terminals and underground parking garages. The normal highway automotive emission levels are increased considerably by vehicles decelerating, idling, and accelerating at the toll facility. The presence of the increased levels of CO, NO_x, diesel and gasoline fumes, and other automotive emissions has a potential detrimental effect on health.

The toll collectors can not rely on physical barriers to isolate them totally from the automotive emissions because toll windows are necessary for collecting tolls. The frequent opening and closing of the window makes the heating and cooling loads of each booth fluctuate independently. The heat loss or gain is extremely high as all four sides (and frequently the ceiling) of the relatively small tollbooth are exposed to the outdoor ambient temperature.

Workplace air quality standards are mandated by local, state, and federal agencies. The adverse ambient conditions to which the toll collecting personnel are exposed and the need to maintain an acceptable environment inside the tollbooths require a careful evaluation of air distribution when designing the HVAC system.

AIR QUALITY CRITERIA

ACGIH (1998) recommends a threshold limit of 25 ppm of CO for an 8 hour exposure. The OSHA standard for repeated (day after day) 8 hour exposure to CO in the ambient air is 35 ppm. The deceleration, idling, and acceleration of varying number of vehicles make it difficult to estimate the CO levels in the vicinity of the toll collecting facilities. However, the resultant CO levels must be estimated using computer programs now available.

Longitudinal tunnel ventilation systems that use jet fans or Saccardo nozzles are being used increasingly for vehicular tunnels with unidirectional traffic flow. These longitudinal ventilation systems discharge air contaminants from the tunnel towards the exit portal. The resultant CO levels in the vicinity of the tunnel toll collecting facilities may be higher than other toll facilities.

Any contaminants that enter booths would remain in the ventilation air if recirculation is used. The ventilation system should be designed to distribute 100% outside air to each booth in order to prevent the intrusion of air contaminants.

DESIGN CONSIDERATIONS

The ventilation system for toll plazas should pressurize the booths to keep out contaminants exhausted by vehicular traffic. The opening of the window during toll collection varies with booth design and individual toll collector habits. The amount of ventilation air required for pressurization also varies.

A variable air volume (VAV) system that varies the air supply rate based on the pressure differential between the booth and outdoor or toll window position is achievable with controls now available. A constant volume maximum-minimum arrangement may also be used at booths with a VAV central unit.

Because open window area varies with individual toll collectors and booth architecture, a design window opening that is based on an estimated average opening may be used for determining the air supply rate. The minimum air supply when the booth window is closed should be based on the amount of air required to meet the heating/cooling requirement and to prevent infiltration of air contaminants through door and window cracks. Where the minimum air supply exceeds the air exfiltration rate, provisions to relieve the excess air should be made to prevent overpressurization.

The space between the booth roof and the overhead canopy may be used for installation of individual HVAC units, fan coil units, or VAV boxes. Air ducts and piping may be installed on top of the plaza canopy or in the utility tunnel.

The amount of ventilation air is high compared to the size of the booth. The resultant rate of air change is extremely high. Supply air outlets should be sized and arranged to deliver air at low velocity. Air reheating should be considered where the supply air temperature is considered too low.

In summer, the ideal air supply is from the booth ceiling to allow the cold air to descend through the booth. In winter, the ideal air supply is from the bottom of the booth or at floor level. Distribution that is ideal for both cooling and heating is not always possible. When air is supplied from the ceiling, other means of providing heat at the floor level, such as electric forced air heaters, electric radiant heating, or heating coils in the floor should be considered.

The supply air intake should be located such that the air drawn into the system is as free as practicable of vehicle exhaust fumes. The prevailing wind should be considered when locating the intake, and it should be placed as far off the side of the roadway as is practicable to ensure better quality ventilation air.

EQUIPMENT SELECTION

Two types of HVAC systems for toll plazas are commonly used: individual HVAC units and central HVAC systems. Individual HVAC units allow each toll collector the freedom of choosing either heating, cooling, or ventilation mode. Maintenance of individual units can be performed without affecting other lanes. In contrast, a central HVAC system should have redundancy to avoid a shut down of the system for the entire toll plaza during maintenance.

The design emphasis on booth pressurization requires the use of 100% outside air. Where a VAV system is used to reduce the operating cost, the varying supply rate of 100% outside air requires a complex temperature control that is normally not available for individual HVAC units. The individual HVAC units should be considered only where the toll plaza is small or tollbooths are so dispersed that a central HVAC system is not economically justifiable.

Where hot water, chilled water, or secondary water service are available from an adjacent administration building, individual fan coils for each booth and a central air handler for supplying ventilation air may be economical. When operating hours for booths and the administration building are significantly different, separate heating and cooling for the toll collecting facility should be considered. The selection of a central air distribution system should be based on the number of maximum open traffic lanes during peak hours and minimum open traffic lanes during moderate traffic.

The HVAC for toll plazas is generally required to operate continuously. A minimum amount of air may be supplied to tollbooths that are not occupied to prevent infiltration of exhaust fumes into tollbooths. Otherwise consideration should be given to remotely flushing the closed booths with ventilation air prior to occupancy.

PARKING GARAGES

Automobile parking garages are either fully enclosed or partially open. Fully enclosed parking areas are usually underground and require mechanical ventilation. Partially open parking levels are generally above-grade structural decks having open sides (except for barricades), with a complete deck above. Natural ventilation, mechanical ventilation, or a combination of both can be used for partially open garages.

The operation of automobiles presents two concerns. The most serious is the emission of carbon monoxide, with its known risks. The second concern is the presence of oil and gasoline fumes, which may cause nausea and headaches as well as present a fire hazard. Additional concerns regarding oxides of nitrogen and smoke haze from diesel engines may also require consideration. However, the ventilation required to dilute carbon monoxide to acceptable levels also controls the other contaminants satisfactorily, provided the percentage of diesel vehicles does not exceed 20%.

For many years, the model codes and ASHRAE *Standard* 62 and its predecessor standards have recommended a flat exhaust rate of either 1.5 cfm/ft^2 or 6 air changes per hour (ACH) for enclosed parking garages. But because vehicle emissions have been reduced over the years, ASHRAE sponsored a study to determine ventilation rates required to control contaminant levels in enclosed parking facilities (Krarti et al. 1998). The results of that study, which are summarized here, found that in some cases much less ventilation than 1.5 cfm/ft^2 is satisfactory. However, local codes may still require 1.5 cfm/ft^2 or 6 ACH, so the engineer may be required to request a variation or waiver from the authorities having jurisdiction to implement the recommendations given here.

VENTILATION REQUIREMENTS

The design ventilation rate required for an enclosed parking facility depends on four factors:

• Contaminant level acceptable in the parking facility
• Number of cars in operation during peak conditions
• Length of travel and operation time of cars in the parking garage
• Emission rate of a typical car under various conditions

Contaminant Level Criteria

ACGIH (1998) recommends a threshold limit of CO of 25 ppm for an 8-hour exposure, and the EPA has determined that at or near sea level, exposure to a CO concentration of 35 ppm for up to 1 hour is acceptable. For installations above 3500 ft, more stringent limits would be required. European countries maintain an average of 35 ppm and a maximum level of 200 ppm.

Opinions of the various agencies and countries differ on acceptable CO levels. A reasonable solution at this time would be a ventilation rate designed to maintain a CO level of 35 ppm for 1 hour exposure with a maximum of 120 ppm or 25 ppm for an 8 hour exposure. Because the time involved in parking or driving out of the garage is in the order of minutes, the 35 ppm level should be an acceptable level of exposure.

Table 4 Average Entrance and Exit Times for Vehicles

Level	Average Entrance Time, s	Average Exit Time, s
1	35	45
3[a]	40	50
5	70	100

Source: Stankunas et al. (1980).
[a]Average pass-through time = 30 s.

Table 5   Predicted CO Emissions in Parking Garages

Season	Hot Emission, (Stabilized), g/min		Cold Emission, g/min	
	1991	1996	1991	1996
Summer (90°F)	2.54	1.89	4.27	3.66
Winter (32°F)	3.61	3.38	20.74	18.96

Results from EPA Mobile 3, version NYC-2.2; sea level location.
Note: Assumed vehicle speed is 5 mph.

Number of Cars in Operation

The number of cars operating at any one time depends on the type of facility served by the parking garage. For a distributed, continuous use such as an apartment house or shopping area, the variation is generally from 3 to 5% of the total vehicle capacity. It could reach 15 to 20% for peak use, such as in a sports stadium or short-haul airport.

Length of Time of Operation

The length of time that a car remains in operation in a parking garage is a function of the size and layout of the garage and the number of cars attempting to enter or exit at a given time. This time could vary from 60 to 600 s, but on the average it usually ranges from 60 to 180 s. Table 4 lists approximate data for vehicle movements. These data should be adjusted to suit the specific physical configuration of the facility.

Car Emission Rate

The operation of a car engine in a parking garage differs considerably from normal vehicle operation, including normal operation in a road tunnel. On entry the car travels slowly. As the car proceeds from the garage, the engine is cold and operates with a rich fuel mixture and in low gear. Emissions for a cold start are considerably higher, so the distinction between hot and cold emission plays a critical role in determining the ventilation rate. Motor vehicle emission factors for hot and cold start operation are presented in Table 5. An accurate analysis requires correlation of CO readings with the survey data on car movements (Hama et al. 1974). These data should be adjusted to suit the specific physical configuration of the facility and design year.

Design Method

To determine the design flow rate required to ventilate an enclosed parking garage, the following procedure can be followed:

Step 1. Collect the following data:

- Number of cars in operation during hour of peak use (N)
- Average CO emission rate for typical car (E), g/h
- Average length of operation and travel time for typical car (θ), s
- CO concentration acceptable in garage (CO_{max}), ppm
- Total floor area of parking area (A_f), ft^2

Step 2. (a) Determine peak generation rate (G) in g/h·ft^2, for parking garage per unit floor area using Equation (9):

$$G = NE/A_f \qquad (9)$$

(b) Normalize value of generation rate using reference value $G_0 = 2.48$ g/h·ft^2. This reference value is based on an actual enclosed parking facility (Krarti et al. 1998):

$$f = 100 G/G_0 \qquad (10)$$

Step 3. Determine the minimum required ventilation rate per unit floor area (Q) using Figure 8 or the correlation presented by Equation (11) depending on the maximum level of CO concentration CO_{max}:

$$Q = Cf\theta \qquad (11)$$

where

C = 2.370×10^{-4} cfm/ft^2·s for CO_{max} = 15 ppm
 = 1.363×10^{-4} cfm/ft^2·s for CO_{max} = 25 ppm
 = 0.468×10^{-4} cfm/ft^2·s for CO_{max} = 35 ppm

Example 1. Consider a two level enclosed parking garage with a total capacity of 450 cars, a total floor area of 90,000 ft^2, and an average height of 9 ft. The total length of time for a typical car operation is 2 min (120 s). Determine the required ventilation rate for the enclosed parking garage in cfm/ft^2 and in ACH so that the CO level never exceeds 25 ppm. Assume that the number of cars in operation during peak use is 40% of the total vehicle capacity.

Step 1. Garage data:

N = 450×0.4 = 180 cars;
E = 11.66 g/min = 700 g/h (average of all values of the
        emission rate for a winter day obtained from Table 5)
CO_{max} = 25 ppm, θ = 120 s.

Step 2. Calculate CO generation rate:

$$G = (180 \times 700 \text{ g/h})/90,000 = 1.40 \text{ g/h·ft}^2$$

$$f = 100 \times 1.40/2.48 = 56$$

Step 3. Determine the ventilation requirement:

Using Figure 8 or the correlation of Equation (11) for CO_{max} = 25 ppm

$$Q = 1.363 \times 10^{-4} \text{ cfm/s·ft}^2 \times 56 \times 120 \text{ s} = 0.92 \text{ cfm/ft}^2$$

Or, in air changes per hour:

$$\text{ACH} = (0.92 \text{ cfm/ft}^2 \times 60 \text{ min/h})/9 \text{ ft} = 6.1$$

Notes:

1. If the emission was E = 6.6 g/min (due, for instance, to better emission standards or better maintained cars), the required minimum ventilation rate would be 3.5 ACH (i.e., 0.52 cfm/ft^2).
2. Once the calculations are made and a decision is reached to use CO demand ventilation control, increasing the airflow by a safety margin does not increase operating cost as the larger fans work for shorter periods to sweep the garage and maintain satisfactory conditions.

CO Demand Ventilation Control

The ventilation system, whether mechanical, natural, or both, should meet applicable codes and maintain an acceptable contaminant level. To conserve energy, fans should be controlled by a CO monitor to vary the amount of air supplied, if permitted by local codes. For example, ventilation could consist of multiple fans with single- or variable-speed motors or variable-pitch blades. In multilevel parking garages or single-level structures of extensive area, independent fan systems, each under individual control are preferred. Figure 9 shows the maximum CO level in a tested garage (Krarti et al. 1998) for three car movement profiles (as illustrated in Figure 10) and the following ventilation control strategies:

- Constant volume (CV) where ventilation is on during the entire occupancy period
- On-Off control, with fans stopped and started based on input from CO sensors
- Variable air volume (VAV) control, with fan speeds adjusted depending on the CO level in the garage, based on input from CO sensors

Figure 9 also indicates typical fan energy saving achieved by the On-Off and VAV systems relative to the fan energy used by the constant-volume systems. As illustrated in the figure, a significant saving in fan energy can be obtained when a CO-based demand ventilation control strategy is used to operate the ventilation system

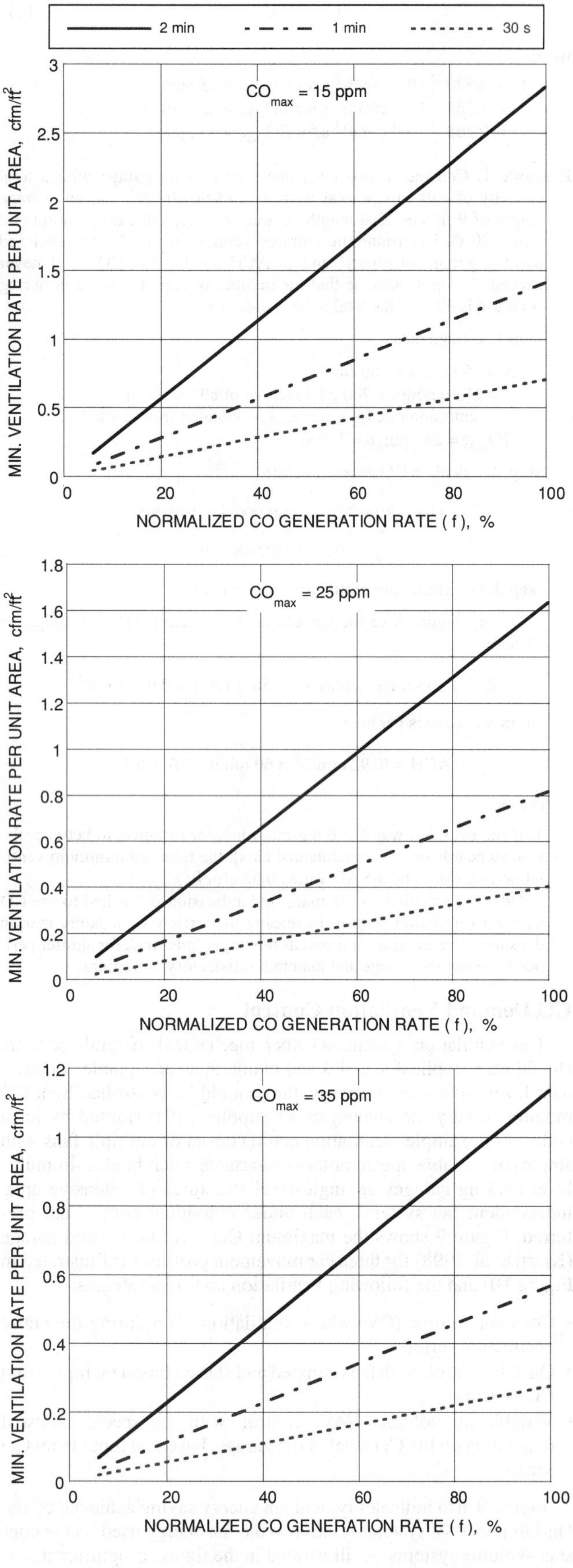

Fig. 8 Ventilation Requirement for Enclosed Parking Garage

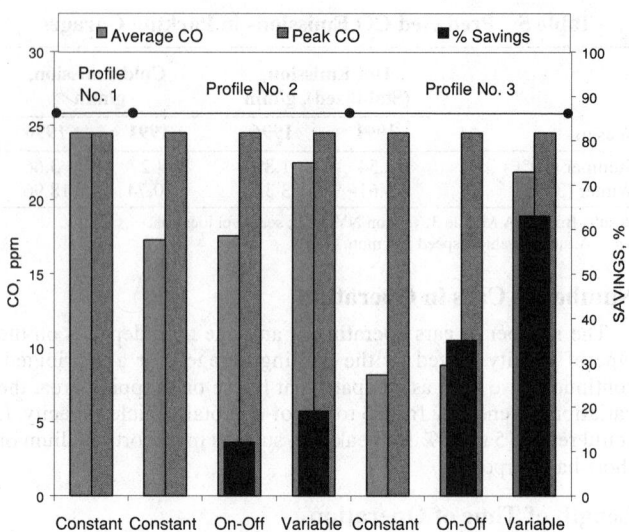

Fig. 9 Typical Energy Savings and Maximum CO Level Obtained for Demand CO-Ventilation Controls

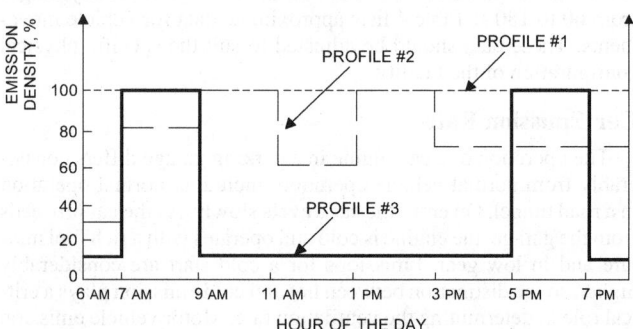

Fig. 10 Three Car Movement Profiles

while maintaining CO levels below 25 ppm. Also, wear and tear and maintenance on the mechanical and electrical equipment is reduced.

In cold climates, the additional cost of heating the makeup air is also reduced. Even if only outside air openings are used or infiltration is permitted, energy is lost to the cold incoming air as energy stored in the mass of the structure helps maintain the garage temperature at an acceptable level.

Ventilation System Configuration

Systems can be classified as supply-only, exhaust-only, or combined. Regardless of which design is chosen, the following should be considered:

- Contaminant level of outside air drawn in for ventilation
- Avoiding short circuiting of supply air
- Avoiding long flow fields that permit the contaminant levels to build up above an acceptable level at the end of the flow field
- Providing short flow fields in areas of high contaminant emission, thereby limiting the extent of mixing
- Providing an efficient, adequate flow throughout the parking structure
- Stratification of engine exhaust gases

Other Considerations

Access tunnels or long, fully enclosed ramps should be designed the same way as road tunnels. When natural ventilation is used, the wall opening free area should be as large as possible. A portion of the free area should be at floor level. For parking levels with large interior floor areas, a central emergency smoke exhaust system should be considered for smoke removal or fume removal during calm weather.

Noise

In general, parking garage ventilation systems move large quantities of air through large openings without extensive ductwork. These conditions, in addition to the highly reverberant nature of the space, contribute to high noise levels. For this reason, sound attenuation should be considered. This is a safety concern as well, since high noise levels may mask the sound of an approaching car.

Ambient Standards and Contaminant Control

The air exhausted from the garage should meet state and local air pollution control requirements.

BUS GARAGES

Bus garages generally include a maintenance and repair area, a service lane where buses are fueled and cleaned, a storage area where buses are parked, and support areas such as offices, a stock room, a lunch room, and locker rooms. The location and layout of these spaces can depend on such factors as climate, bus fleet size, and type of fuel used by the buses. Servicing and storage functions may be located outdoors in temperate regions, but are often found indoors in colder climates. While large fleets cannot always be stored indoors, maintenance areas may double as storage space for small fleets. Local building and/or fire codes may prohibit the dispensing of certain types of fuel indoors.

In general, for areas where buses are maintained or serviced, ventilation should be accomplished using 100% outside air with no recirculation. Therefore, the use of heat recovery devices should be considered in colder climates. Tailpipe emissions should be exhausted directly in maintenance and repair areas. Offices and similar support areas should be kept under positive pressure to prevent the infiltration of vehicle emissions.

MAINTENANCE AND REPAIR AREAS

ASHRAE *Standard* 62 and most model codes require a minimum outdoor air ventilation rate of 1.5 cfm per square foot of floor area in vehicle repair garages, with no recirculation recommended. However, because the interior ceiling height may vary greatly from garage to garage, the designer should consider making a volumetric analysis of contaminant generation and air exchange rates. The section on Bus Terminals has information on diesel engine emissions and the ventilation rates necessary to control their concentrations in areas where buses are operated.

Maintenance and repair areas often include below-grade inspection and repair pits for working underneath buses. Because the vapors produced by conventional bus fuels are heavier than air, they tend to settle in these pit areas, so a separate exhaust system should be provided to prevent their accumulation. NFPA *Standard* 88B recommends a minimum of 12 air changes per hour in pit areas and the location of exhaust registers near the pit floor.

Fixed repair stations, such as inspection and repair pits or hydraulic lifts, should include a direct exhaust for tailpipe emissions. Such systems have a flexible hose and coupling that is attached to the bus tailpipe; emissions are discharged to the outdoors through an exhaust fan. The system may be of the overhead reel type, overhead tube type, or underfloor duct type, depending on the location of the tailpipe. For heavy diesel engines, a minimum exhaust rate of 600 cfm per station is recommended to capture emissions without creating excessive back pressure in the vehicle. Fans, ductwork, and hoses should be able to receive vehicle exhaust at temperatures exceeding 500°F without degradation.

Garages often include areas for battery charging, a process that can produce potentially explosive concentrations of corrosive, toxic gases. There are no published code requirements for the ventilation of battery-charging areas, but DuCharme (1991) has suggested using a combination of floor and ceiling exhaust registers to remove gaseous by-products. Recommended exhaust rates are 2.25 cfm per square foot of room area at floor level to remove acid vapors and 0.75 cfm per square foot of room area at ceiling level to remove hydrogen gases. Supply air volume should be 10 to 20% less than exhaust air volume and directed to provide 100 fpm terminal velocity at floor level. If the battery-charging space is located in the general maintenance area rather than in a dedicated room, an exhaust hood should be provided to capture gaseous by-products. Chapter 29 contains specific information on exhaust hood design. Makeup air should be provided to replace that removed by the exhaust hood.

Garages may also contain spray booths or rooms for painting buses. Most model codes reference NFPA *Standard* 33 for spray booth requirements, and that document should be consulted when designing heating and ventilating systems for such areas.

SERVICING AREAS

For indoor service lanes, ASHRAE *Standard* 62 and several model codes specify a minimum of 1.5 cfm/ft² of ventilation air, while NFPA *Standard* 30A requires only 1 cfm/ft². The designer should determine which minimum requirement is applicable for the project location; additional ventilation may be necessary due to the nature of servicing operations. Buses often queue up in the service lane with their engines running during the service cycle. Depending on the length of the queue and the time to service each bus, the contaminant levels may exceed maximum allowable concentrations. A volumetric analysis examining contaminant generation should thus be considered.

Because of the increased potential for concentrations of flammable or combustible vapor, service lane HVAC systems should not be interconnected with systems serving other parts of the garage. Service lane systems should be interlocked with the dispensing equipment to prevent operation of the latter if the former is shut off or fails. Exhaust inlets should be located at ceiling level and between 3 and 12 in. above the finished floor, with supply and exhaust diffusers/registers arranged to provide air movement across all planes of the dispensing area.

Another feature common to modern service lanes is the cyclone cleaning system, which is used to vacuum out the interior of a bus. These devices have a dynamic connection to the front door(s) of the bus, through which a large-volume fan vacuums dirt and debris from inside the bus. A large cyclone assembly then removes the dirt and debris from the airstream and deposits it into a large hopper for disposal. Because of the large volume of air involved, the designer should consider the discharge and makeup air systems required to complete the cycle. Recirculation or energy recovery should be considered, especially in the winter months. To aid in contaminant and heat removal during the summer months, some systems currently in use discharge the cyclone air to the outdoors and provide untempered makeup air through relief hoods above the service lane.

STORAGE AREAS

Where buses are stored indoors, the minimum ventilation standard of 1.5 cfm/ft² should be provided, subject to volumetric considerations. The designer should also consider the increased contaminant levels present during peak traffic periods. An example is the morning pull-out, when the majority of the bus fleet is dispatched for rush-hour commute. It is common practice to start and idle a large number of buses during this period to warm up the engines and check for defects. As a result, the concentration of emissions in the storage area rises, and additional ventilation may be required to maintain contaminant levels in acceptable limits. The use of supplemental purge fans is a common solution to this problem. These fans can be (1) interlocked with a timing device to operate during peak traffic periods, (2) started manually on an as-needed basis, or (3) connected to an air quality monitoring system that activates them when contaminant levels exceed some preset limit.

DESIGN CONSIDERATIONS AND EQUIPMENT SELECTION

Most model codes require that open flame heating equipment such as unit heaters be located at least 8 ft above the finished floor or, where located in active trafficways, 2 ft above the tallest vehicle. Fuel-burning equipment located outside of the garage area, such as boilers in a mechanical room, should be installed with the combustion chamber at least 18 in. above the floor. Combustion air should be taken from outside the building. Exhaust fans should be of the nonsparking type, with the motor located outside the airstream.

Infrared heating systems and air curtains are often considered for repair garages because of the size of the building and the high amount of infiltration through the large overhead doors needed to move buses in and out of the garage. However, caution must be exercised in applying infrared heating in areas where buses are parked or stored for extended periods; the buses may absorb most of the heat, which is then lost when they leave the garage. This is especially true during the morning pull-out. Infrared heating is more successfully applied in the service lane or at fixed repair positions. Air curtains should be considered for high-traffic doorways to limit heat loss and infiltration of cold air.

Where air quality monitoring systems are considered for controlling ventilation equipment, maintainability is a key factor in determining the success of the application. The high concentration of particulate matter in bus emissions can adversely affect the performance of the monitoring equipment, which often has filtering media at sampling ports to protect the sensors and instrumentation. Location of sampling ports, effects of emissions fouling, and calibration requirements should be considered when selecting the monitoring equipment to control the ventilation systems and air quality of the garage. Nitrogen dioxide and CO exposure limits published by OSHA and the EPA should be consulted to determine the contaminant levels at which exhaust fans should be activated.

EFFECTS OF ALTERNATIVE FUEL USE

Because of legislation limiting contaminant concentrations in diesel bus engine emissions, the transit industry has begun using buses that operate on alternative fuels. Such fuels include methanol, ethanol, hydrogen, compressed natural gas (CNG), liquefied natural gas (LNG), and liquefied petroleum gas (LPG). The flammability, emission, and vapor dispersion characteristics of these fuels differ from the conventional fuels for which current code requirements and design standards were developed. For this reason the requirements may not be valid for garage facilities in which alternative-fuel vehicles are maintained, serviced, and stored. The designer should consult current literature regarding the design of HVAC systems for these facilities rather than relying on conventional practice. One source is the Alternative Fuels Data Center at the U.S. Department of Energy in Washington, DC.

For facilities housing CNG buses, NFPA *Standard* 52 provides requirements for indoor fueling and gas processing/storage areas. It requires a separate mechanical ventilation system providing at least 5 air changes per hour for these areas. The ventilation system should operate continuously, or be activated by a continuously monitoring natural gas detector when a gas concentration of not more than one-fifth the lower flammability limit is present. The fueling or compression equipment should be interlocked to shut down in the event of a failure in the mechanical ventilation system. Supply inlets should be located near floor level, and exhaust outlets at high points in the roof or exterior wall structure.

DOT (1996) guidelines for CNG facilities address bus storage and maintenance areas as well as fueling areas. Recommendations include minimizing the potential for "dead air zones" and gas pockets (which may necessitate coordination with architectural and structural designers); using a normal ventilation rate of 6 air changes per hour, and increasing that rate by an additional 6 air changes per hour in the event of a gas release; using nonsparking exhaust fans rated for use in Class 1, Division 2 areas (as defined by NFPA *Standard* 70); and, increasing the minimum ventilation rate in smaller facilities to maintain similar dilution levels as would be found in larger facilities. Open-flame heating equipment should not be used, and the surface temperature of heating units should not exceed 800°F. Consideration should also be given to de-energizing supply fans that discharge near the ceiling level in the event of a gas release to avoid spreading the gas plume. The DOT guideline has more information concerning these recommendations.

For facilities housing LNG-fueled buses, the only published requirements are found in NFPA *Standard* 57, which are limited to fueling areas. It also requires a separate mechanical ventilation system providing at least 5 air changes per hour for indoor fueling areas. Again, the ventilation system should operate continuously, or be activated by a continuously monitoring natural gas detection system when a gas concentration of not more than one-fifth the lower flammability limit is present; and the fueling equipment should be interlocked to shut down in the event of a failure in the mechanical ventilation system.

NFPA *Standard* 58 is the only document that contains provisions relating specifically to LPG-fueled buses. This standard prohibits indoor fueling of all LPG vehicles, allowing only an "adequately ventilated" weather shelter or canopy, although the term adequately ventilated is not defined by any prescriptive rate. However, vehicles are allowed to be stored and serviced indoors provided they are not parked near sources of heat, open flames or similar sources of ignition, or "inadequately ventilated" pits. The standard does not specify any required ventilation rates for these repair and storage facilities, although it does require a minimum ventilation rate of 1 cfm/ft^2 in building and structures housing LPG distribution facilities.

No U.S. standards relate specifically to hydrogen powered vehicles. NFPA *Standard* 50A provides general requirements for point-of-use gaseous hydrogen systems, and provides limited guidance.

BUS TERMINALS

Bus terminals vary considerably in physical configuration. Most terminals consist of a fully enclosed space containing passenger waiting areas, ticket counters, and some retail spaces. Buses load and unload outside the building, generally under a canopy for weather protection. In larger cities, where space is at a premium and bus service is extensive or integrated with subway service, comprehensive customer services and multiple levels with attendant busway tunnels and/or ramps may be required.

Waiting rooms and consumer spaces should have a controlled environment in accordance with normal practice for public terminal occupancy. The space should be pressurized against intrusion of the busway environment. Pressurized vestibules should be installed at each doorway to further reduce contaminant migration. Waiting rooms and passenger concourse areas are subject to a highly variable people load. The occupant density may reach 10 ft^2 per person; during periods of extreme congestion, 3 to 5 ft^2 per person.

Basically, there are two types of bus service: urban-suburban and long distance. Urban-suburban service is characterized by frequent bus movements and the requirement of rapid loading and unloading. Therefore, ideal passenger platforms are long and narrow, and buses can drive through and do not have to backup. Long-distance operations and those with greater headways generally use sawtooth gate configurations.

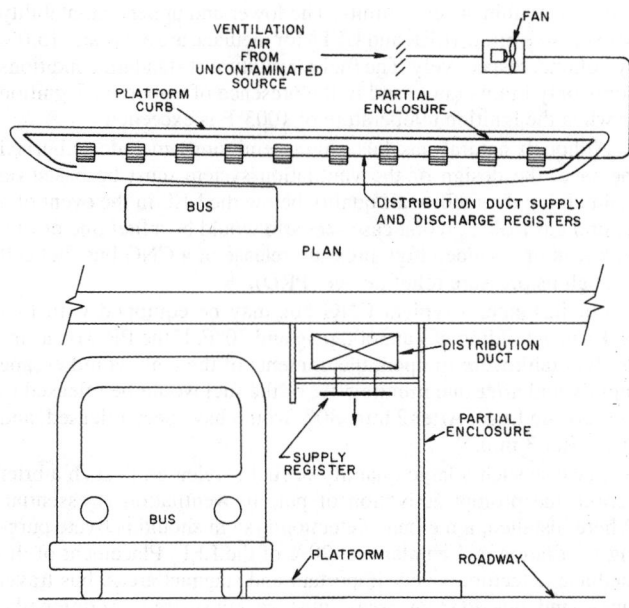

Fig. 11   Partially Enclosed Platform, Drive-Through Type

Either natural or forced ventilation can be used. When natural ventilation is selected, the bus levels should be open on all sides, and the slab-to-ceiling dimension should be sufficiently high, or the space contoured, to permit free air circulation. Jet fans improve the natural airflow with relatively low energy. Mechanical systems that ventilate open platforms or gate positions should be configured to serve the bus operating areas, as shown in Figures 11 and 12.

PLATFORMS

Naturally ventilated drive-through platforms may expose passengers to inclement weather and strong winds. Enclosed platforms (except for an open front) with appropriate mechanical equipment should be considered. Even partially enclosed platforms may trap contaminants and require mechanical ventilation.

Multilevel bus terminals have limited headroom, thus restricting natural ventilation. For such terminals, mechanical ventilation should be selected, and all platforms should be partially or fully enclosed. Ventilation should induce little contaminated air from the busway. Figure 11 shows a partially enclosed drive-through platform with an air distribution system. Supply air velocity should be limited to 250 fpm to avoid drafty conditions on the platform. Partially enclosed platforms require large amounts of outside air to hinder fume penetration; experience indicates that a minimum of 17 cfm per square foot of platform area is required during rush hours and about half of this quantity during the remaining time.

Platform air quality remains essentially the same as that of the ventilation air introduced. Because of the piston effect, however, some momentary higher concentrations of contaminants occur on the platform. Separate ventilation systems with two-speed fans for each platform permit operational flexibility. Fans should be controlled automatically to conform to bus operating schedules. In cold climates, mechanical ventilation may need to be reduced during extreme winter weather.

Fully enclosed platforms are strongly recommended for large terminals with heavy bus traffic. They can be pressurized adequately and ventilated with the normal heating and cooling air quantities, depending on the tightness of construction and the number of boarding doors and other openings. Conventional air distribution can be used; air should not be recirculated. Openings around doors and in the enclosure walls are usually adequate to relieve air pressure unless platform construction is extraordinarily tight.

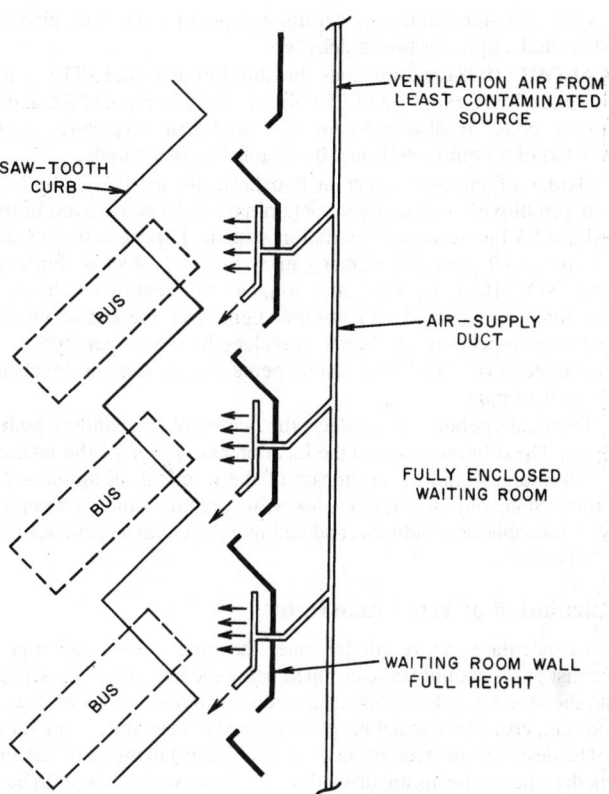

Fig. 12   Fully Enclosed Waiting Room with Sawtooth Gates

BUS OPERATION AREAS

Most buses are powered by diesel engines. Certain models have small auxiliary gasoline engines to drive the air-conditioner. Tests performed on the volume and composition of exhaust gases emitted from diesel engines in various traffic conditions indicate large variations depending on temperature and humidity; manufacturer, size, and adjustment of the engine; and the fuel burned.

Contaminants

The components of diesel exhaust gases that affect the ventilation design are oxides of nitrogen, hydrocarbons, formaldehyde, odor constituents, aldehydes, smoke particulates, and a relatively small amount of carbon monoxide. Diesel engines operating in enclosed spaces reduce visibility and create odors and contaminants. Table 6 lists major health-threatening contaminants typically found in diesel engine exhaust gas. The nature of bus engines should be determined for each project, however.

The U.S. Code of Federal Regulations (CFR) Occupational Safety and Health Standards, Subpart Z, sets the contaminant levels for an 8-hour time-weighted average (TWA) exposure as follows: carbon monoxide (CO), 50 ppm, nitric oxide (NO), 25 ppm, and formaldehyde (HCHO), 0.75 ppm. Subpart Z also

Table 6   Approximate Diesel Bus Engine Emissions (ppm)

	Idling 117 scfm	Accelerating 476 scfm	Cruising 345 scfm	Decelerating 302 scfm
Carbon monoxide	215	500	230	130
Hydrocarbons	390	210	90	330
Oxides of nitrogen (NO_x)	60	850	235	30
Formaldehydes (HCHO)	9	17	11	30

Note: For more data on diesel bus and truck engine emissions, see Watson et al. (1988).

sets the short-term exposure ceiling as 5 ppm for nitrogen dioxide (NO_2) and 2 ppm for formaldehyde.

ACGIH (1998) recommends threshold limit values (TLVs) for an 8-hour TWA exposure of 25 ppm for CO, 25 ppm for NO, and 3 ppm for NO_2. It also recommends short-term exposure limits (STELs) of 5 ppm for NO_2 and 0.3 ppm for formaldehyde.

Oxides of nitrogen occur in two basic forms: NO_2 and NO. Nitrogen dioxide is the major contaminant to be considered in the design of a bus terminal ventilation system. Exposure to concentrations of 10 ppm and higher causes health problems. Furthermore, NO_2 affects light transmission, causing visibility reduction. It is intensely colored and absorbs light over the entire visible spectrum, especially at shorter wavelengths. Odor perception is immediate at 0.42 ppm and can be perceived by some at levels as low as 0.12 ppm.

Terminal operation also affects the quality of surrounding ambient air. The dilution rate and the location and design of the intakes and discharges control the impact of the terminal on ambient air quality. State and local regulations, which require consideration of local atmospheric conditions and ambient contaminant levels, must be followed.

Calculation of Ventilation Rate

To calculate the ventilation rate, the total amount of engine exhaust gases should be determined using the bus operating schedule, the amount of time that buses are in different modes of operation (i.e., cruising, decelerating, idling, and accelerating) and Table 6. The designer must ascertain the grade (if any) in the terminal and whether the platforms are drive-through, drive-through with bypass lanes, or sawtooth. Bus headways, speed, and operating modes must also be evaluated.

For instance, with sawtooth platforms, the departing bus must accelerate backward, brake, and then accelerate forward. The drive-through platform requires a different mode of operation. Certain codes prescribe a maximum idling time for engines, usually 3 to 5 min. Normally, 1 to 2 min of engine operation is required to build up brake air pressure.

The discharged contaminant quantities should be diluted by natural and/or forced ventilation to acceptable, legally prescribed levels. To maintain odor control and visibility, the exhaust gas contaminants should be diluted with outside air in the proportion of 75 to 1. Where urban-suburban operations are involved, the ventilation rate varies considerably throughout the day and between weekdays and weekends. Control of fan speed or blade pitch should be used to conserve energy. It follows that the required airflow may be reduced by removing contaminant emissions as quickly as possible. This can be achieved by mounting exhaust capture hoods on the ceiling above each of the bus exhaust stacks. The exhaust air collected by the capture hoods is then discharged outside of the facility through a dedicated exhaust system.

Effects of Alternative Fuel Use

As indicated in the section on Bus Garages, alternative fuels are being more widely used in lieu of conventional diesel. Alternative fuel buses are more likely to be operated on urban-suburban routes than on long distance bus service.

Current codes and design standards developed for conventional fuels may not be valid for alternative fuel buses. Special attention should be given to the design of the HVAC and electrical systems for these facilities in case of a fuel tank or fuel line leak on an alternative fuel bus. Research continues in this application and further information may be available from U.S. DOT Federal Transit Administration and NFPA.

Natural Gas (NG) Buses. Fuel burned in LNG and CNG buses has a composition of up to 98% methane (CH_4). Methane burns in a self-sustained reaction only when the volume percentage of fuel and air is within specific limits. The lower and upper flammability or explosive limits (LEL and UEL) for methane are 5.3% and 15.0% by volume, respectively. The fuel-air mixture at standard conditions burns only in this range and in the presence of a source of ignition or when the ignition temperature of 1003°F is exceeded.

Although natural gas bus engine emissions include unburned methane, the design of the ventilation system must be based on maintaining the facility air quality below the LEL in the event of a natural gas leak. A worst case scenario would be a fuel line or fuel tank leak or a sudden high pressure release of a CNG bus fuel cell through its pressure relief device (PRD).

For instance, a typical CNG bus may be equipped with fuel tanks each holding gas at 3600 psig and 70°F. If the PRD on a single fuel tank were to open, the contents of the tank would escape rapidly and after one minute 50% of the fuel would be released to the surroundings. After 2 min, 80% would have been released, and 90% after 3 min.

Because such a large quantity of fuel is released in such a brief period, the prompt activation of purging ventilation is essential. Where installed, a methane detection system should activate purging ventilation and an alarm at 20% of the LEL. Placement of the methane detectors is very important and stagnant areas, bus travel lanes, and bus loading areas must be considered. Additionally, although methane is lighter than air (relative density = 0.55), some research has indicated that it may not rise immediately after a leak. In the case of a CNG bus PRD release, the rapid throttle-like flow through the small diameter orifice of the device may actually cool the fuel, making it denser than air. Under these conditions, the fuel may migrate towards the floor until thermal equilibrium with the surrounding environment is reached, at which time buoyancy drives the fuel-air mixture to the ceiling. Thus, the designer may consider locating methane detectors at both ceiling and floor levels.

Although no specific ventilation criteria has been published for natural gas vehicles in bus terminals, NFPA *Standard* 52 recommends a blanket rate of 5 air changes per hour (ACH) in fueling areas. Also, the Federal Transit Administration guidelines for CNG transit facility design recommend a slightly more conservative 6 ACH normal ventilation rate for bus storage areas, with a 12 ACH purge ventilation rate upon methane sensor activation. The designer can also calculate a purging ventilation rate based on the volumetric flow rate of methane released, the duration of the release, and the volume of the facility. The volume of the facility has a significant effect on the amount of ventilation required to maintain maximum average concentrations of methane below 10% of the LEL; e.g., the larger the facility, the lower the number of air changes that are required. However, concentrations of methane exceeding the LEL can be expected in the immediate area of the leak regardless of the ventilation rate used. Plume size and location and duration of unsafe concentrations may be determined using a comprehensive modeling analysis such as computational fluid dynamics.

Source of Ventilation Air

Because dilution is the primary means of contaminant level control, the source of ventilation air is extremely important. The cleanest available ambient air, which in an urban area is generally above the roof, should be used. Surveys of ambient air contaminant levels should be conducted and the most favorable source located. The possibility of short circuiting of exhaust air due to prevailing winds and building airflow patterns should also be evaluated.

Control by Contaminant Level Monitoring

Time clocks or tapes coordinated with bus movement schedules and smoke monitors (obscurity meters) provide the most useful means of controlling the ventilation system.

Instrumentation is available for monitoring various contaminants, namely electrochemical cells for NO_2 and CO. Control by instrumentation can be simplified by monitoring carbon dioxide (CO_2) levels, as studies have shown a relationship between the levels of various diesel engine contaminants and CO_2. However, the mix and quantity of contaminants varies with the rate of operation and the condition of the bus engines. Therefore, if CO_2 is monitored, actual conditions under specific bus traffic conditions should be determined in order to verify the selected CO_2 settings.

Dispatcher's Booth

The dispatcher's booth should be kept under positive pressure to prevent the intrusion of engine fumes. Because the booth is occupied for sustained periods, normal interior comfort conditions and OSHA contaminant levels must be maintained.

RAPID TRANSIT

Today's high-performance, air-conditioned subway vehicles consume most of the energy required to operate rapid transit and are the greatest source of heat in underground areas of the system. Therefore, minimizing the traction power consumed has a strong effect on the operating cost and comfort. The environmental control system (ECS) maintains a reasonably comfortable environment during normal operations and assists in keeping patrons safe during a fire emergency. The large amount of heat produced by the rolling stock, if not properly controlled, can cause patron discomfort, shorten equipment life, and increase maintenance. Tropical climates present an additional stress that makes environment control more critical.

Temperature, humidity, air velocity, air pressure change, and rate of air pressure change are among the conditions needed to determine ECS performance. These conditions are affected by the time of day (morning, evening, or off-peak), the circumstance (normal, congested, or emergency operations), and the location in the system (tunnels, station platforms, entrances, and stairways). The *Subway Environmental Design Handbook* (DOT 1976), based on unproven experience but validated by field and model tests, provides comprehensive and authoritative design aids.

Operations are normal when trains are moving through the system according to schedule and passengers are traveling smoothly through stations to and from transit vehicles. Since this is the predominant category of operations, considerable effort is made to optimize performance of the ECS during normal operations.

Congested operations result from delays or operational problems that prevent the free flow of trains. Trains may wait in stations, or stop at predetermined locations in tunnels. Delays usually range from 30 s to 20 min, although longer delays may occasionally be experienced. Passenger evacuation or exposure of passengers to danger does not occur during congested operations. The analyses focus on the ventilation required to support the continued operation of train air-conditioning units and thus maintain patron comfort during congested operations.

Emergency operations generally result from a malfunction of the transit vehicle. The most serious emergency is a train on fire stopped in a tunnel, disrupting traffic, and requiring passenger evacuation. The analyses focus on determining the ventilation required to maintain a single evacuation path from the train clear of smoke and hot gases.

Patron comfort is a function of ambient temperature and humidity, air velocity, and the time of exposure to an environment. For example, a person that enters a 84°F station from a 90°F outdoors will momentarily feel more comfortable. However, in a short time, usually in about 6 min, the person's metabolism will adjust to the new environment and he will feel nearly as uncomfortable as before. If a train were to arrive during this period, the relatively high temperature of the station would be acceptable. The Relative Warmth Index (RWI) provides a means quantifying this transient effect thus allowing the designer to select an appropriate design temperature based on the transient sensation of comfort rather than the steady-state sensation of comfort. This temperature is higher (often 5 to 9°F) than that which would be selected by the steady-state approach and hence results in reduced cooling or air-conditioning requirements).

DESIGN CONCEPTS

The factors to be considered, although interrelated, may be divided into four categories: natural ventilation, mechanical ventilation, emergency ventilation, and station air conditioning.

Natural Ventilation

Natural ventilation in subway systems (infiltration and exfiltration) is primarily the result of train operation in tightly fitting trainways, where air generally moves in the direction of train travel. The positive pressure in front of a train expels air through portals and station entrances; the negative pressure in the wake of the train induces airflow through these same openings.

Considerable short-circuiting occurs in subway structures when two trains traveling in opposite directions pass each other, especially in stations or in tunnels with porous walls (walls with intermittent openings to allow air passage between tracks) or nonporous walls. Such short circuiting reduces the net ventilation rate and causes increased air velocities on station platforms and in station entrances. During the time of peak operation and peak ambient temperatures, it can cause an undesirable heat buildup.

To help counter these negative effects, ventilation shafts are customarily placed near the interfaces between tunnels and stations. Shafts in the approach tunnel are often called blast shafts; part of the air pushed ahead of the train is expelled through them. Relief shafts in the departure tunnel relieve the negative pressure created during the departure of the train and induce outside air through the shaft rather than through station entrances. Additional ventilation shafts may be provided between stations (or between portals, for underwater crossings), as dictated by tunnel length. The high cost of such ventilation structures necessitates a design for optimum effectiveness. Internal resistance due to offsets and bends should be kept to a minimum, and shaft cross-sectional areas should be approximately equal to the cross-sectional area of a single-track tunnel (DOT 1976).

Mechanical Ventilation

Mechanical ventilation in subways (1) supplements the ventilation effect created by train piston action, (2) expels heated air from the system, (3) introduces cool outside air, (4) supplies makeup air for exhaust, (5) restores the cooling potential of the heat sink through extraction of heat stored during off-hours or system shutdown, (6) reduces the flow of air between the tunnel and the station, (7) provides outside air for passengers in stations or tunnels in an emergency or during other unscheduled interruptions of traffic, and (8) purges smoke from the system in case of fire.

The most cost-effective design for mechanical ventilation is one that serves two or more purposes. For example, a vent shaft provided for natural ventilation may also be used for emergencies if a fan is installed in parallel with a bypass (Figure 13).

Several vent shafts working together may be capable of meeting many, if not all, of the eight objectives. Depending on shaft location and the situation, a shaft with the bypass damper open and the fan damper closed may serve as a blast or relief shaft. With the fan in operation and the bypass damper closed, air can be supplied or exhausted by mechanical ventilation, depending on the direction of fan rotation.

Except for emergencies, fan rotation is usually predetermined based on the overall ventilation concept. If subway stations are not air conditioned, the hot air in the subway should be exchanged at a maximum rate with cooler outside air. If stations are air conditioned below the ambient temperature, the inflow of warmer outside air should be limited and controlled.

Figure 14 illustrates a typical tunnel ventilation system between two subway stations. Here the flow of heated tunnel air into a cooler station is kept to a minimum. The dividing wall separating Tracks No. 1 and 2 is discontinued in the vicinity of the emergency fans. As shown in Figure 14A, air pushed ahead of the train on Track No. 2 diverts partially to the emergency fan bypasses and partially into the wake of a train on Track No. 1 as a result of pressure differences. Figure 14B shows an alternative operation with the same ventilation system. When outdoor temperatures are favorable, the mid-tunnel fans operate as exhaust fans, with makeup air introduced through the emergency fan

bypasses. This system can also provide or supplement station ventilation. To achieve this goal, emergency fan bypasses would be closed, and the makeup air for mid-tunnel exhaust fans would enter through station entrances.

A more direct ventilation system removes station heat at its primary source, the underside of the train. Figure 15 illustrates trackway ventilation. Tests have shown that such ventilation not only reduces the upwelling of heated air onto platform areas, but also removes significant portions of the heat generated by dynamic braking resistor grids and air-conditioning condensers underneath the train (DOT 1976). Ideally, makeup air for the exhaust should be introduced at track level to provide a positive control over the direction of airflow (Figure 15A).

An underplatform exhaust without makeup supply air, as illustrated in Figure 15B, is the least effective and, under certain conditions, could cause heated tunnel air to flow into the station. Figure 15C shows a cost-effective compromise in which makeup air is introduced at the ceiling above the platform. Although the heat removal effectiveness may not be as good as that of the system illustrated in Figure 15A, the inflow of hot tunnel air that might occur without supply air makeup is negated.

Emergency Ventilation

During a subway tunnel fire, mechanical ventilation is a major part of the control strategy. An increase in air supply over that required for combustion reduces the progression of the fire by lowering the flame temperature. Further, ventilation can control the direction of smoke migration to permit safe evacuation of passengers and facilitate access by fire fighters (see NFPA *Standard* 130 and the section on Road Tunnels).

Emergency ventilation must allow for the unpredictable location of a disabled train or the source of fire and smoke. Therefore,

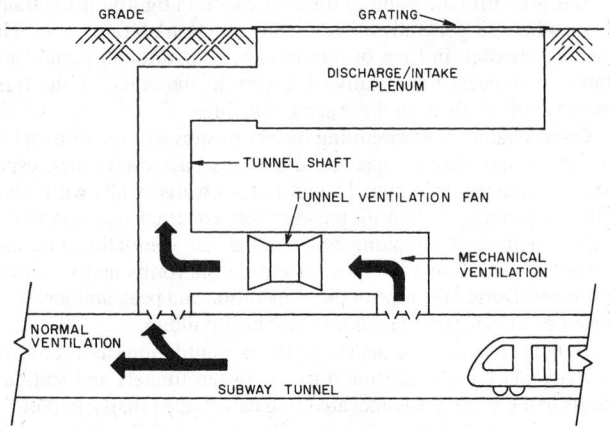

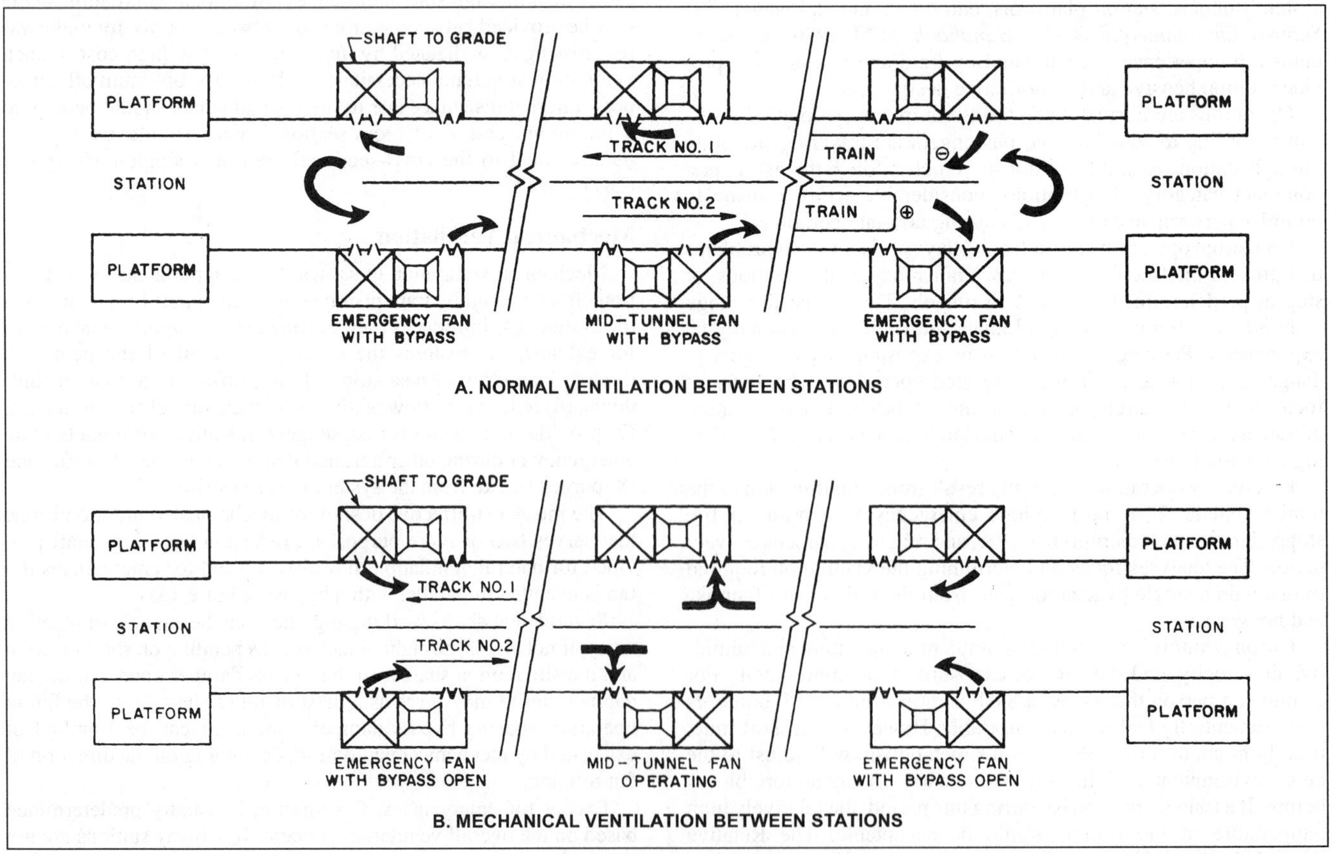

Fig. 13   Tunnel Ventilation Shaft

Fig. 14   Tunnel Ventilation Concept

emergency ventilation fans should have nearly full reverse flow capability so that fans on either side of a stalled train can operate together to control the direction of airflow and counteract the migration of smoke. When a train is stalled between two stations and smoke is present, outside air is supplied from the nearest station and contaminated air is exhausted at the opposite end of the train (unless the location of the fire dictates otherwise). Passengers can then be evacuated along walkways in the tunnel via the shortest route (Figure 16).

Provisions must be made to (1) quickly assess any emergency situation; (2) communicate the situation to central control; (3) establish the location of the train; and (4) quickly start, stop, and reverse emergency ventilation fans from the central console to establish smoke control. Mid-tunnel and station trackway ventilation fans may be used to enhance the emergency ventilation; therefore, these fans must withstand elevated temperatures for a long period and have reverse flow capacity.

Station Air Conditioning

Higher approach speeds and closer headways, made possible by computerized train control, have increased the amount of heat gains in a subway. The net internal sensible heat gain in a typical double-track subway station, with 40 trains per hour per track traveling at top speeds of 50 mph, may reach 5×10^6 Btu/h even after heat is removed by the heat sink, by station underplatform exhausts, and by tunnel ventilation. To remove such a quantity of heat by station ventilation with outside air at a temperature rise of 3°F, for example, requires roughly 1.4×10^6 cfm.

Not only would such a system be costly, but the air velocities on station platforms would be objectionable to passengers. The same amount of sensible heat gain, plus latent heat, plus outside air load with a station design temperature 7°F lower than ambient, could be handled by about 630 tons of refrigeration. Even if station air conditioning is initially more expensive, long-term benefits include (1) reduced design airflow rates, (2) improved environment for passengers, (3) increased equipment life, (4) reduced maintenance of equipment and structures, and (5) increased acceptance of the subway as a viable means of public transportation.

Air conditioning should be considered for other ancillary station areas such as concourses, concession areas, and transfer levels. However, unless these walk-through areas are designed to attract patronage to concessions, the cost of air conditioning is not usually warranted.

The physical configuration of the platform level usually determines the cooling distribution pattern. Areas with high ceilings, local hot spots due to train location, high-density passenger accumulation, or high-level lighting may need spot cooling. Conversely, where train length equals platform length and ceiling height above the platform is limited to 10 to 11.5 ft, isolation of heat sources and application of spot cooling is not usually feasible.

The use of available space in the station structure for air distribution systems should be of prime concern because of the high cost of underground construction. Overhead distribution ductwork, which adds to building height in commercial construction, could add to the depth of excavation in subway construction. The space beneath a subway platform is normally an excellent area for low-cost distribution of supply, return, and/or exhaust air.

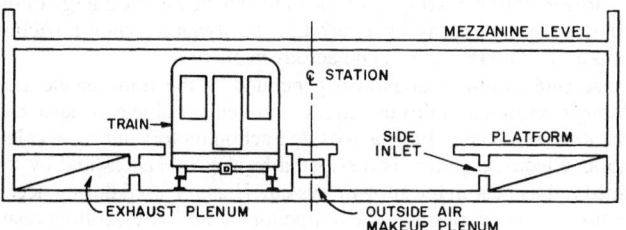

A. OUTSIDE AIR MAKEUP AT TRACK LEVEL

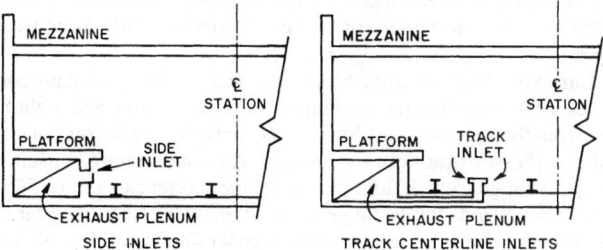

B. NO MAKEUP AIR SUPPLY

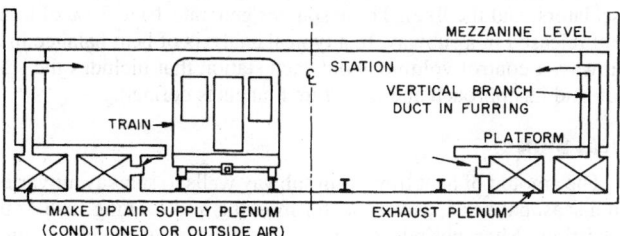

C. MAKEUP AIR SUPPLY AT CEILING LEVEL

Fig. 15   Trackway Ventilation Concepts (Cross-Sections)

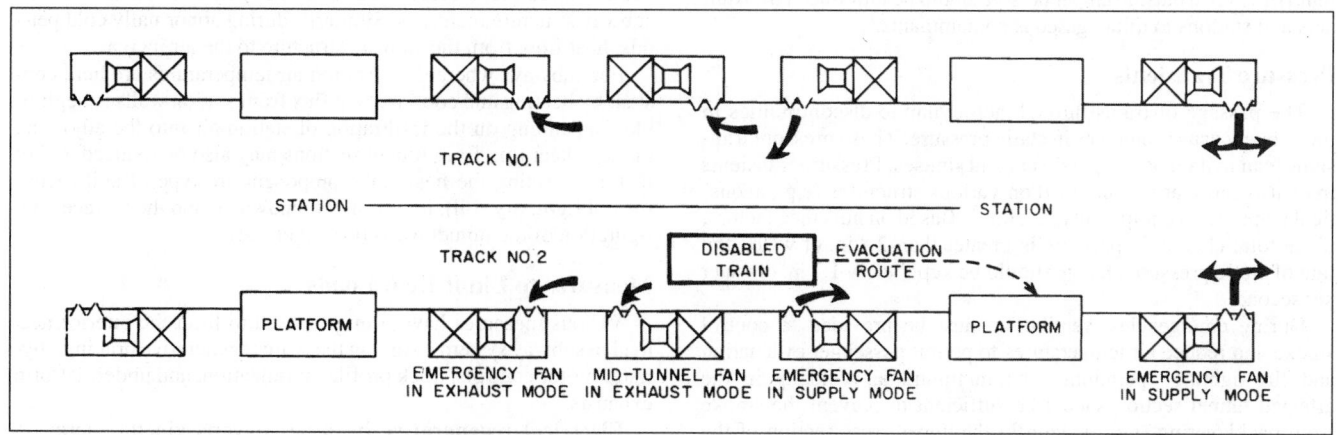

Fig. 16   Emergency Ventilation Concept

DESIGN METHOD

A subway may have to satisfy two separate sets of environmental criteria: one for normal operations and one for emergencies. Criteria for normal operations generally include limits on temperature and humidity for various times of the year, a minimum ventilation rate to dilute contaminants generated in the subway, and limits on air velocity and the rate of air pressure change to which commuters may be exposed. Some of these criteria are subjective and may vary based on demographics. Criteria for emergencies generally include a minimum purge time for sections of the subway in which smoke or fire may occur and minimum and maximum fan-induced tunnel air velocities.

Given a set of criteria, a set of outdoor design conditions, and appropriate tools for estimating interior heat loads, earth heat sink, ventilation, air velocity, and air pressure changes, design engineers select the elements of the environmental control system. Controls for air temperature, air velocity, air quality, and air pressure should be considered. The system selected generally includes a combination of unpowered ventilation shafts, powered ventilation shafts, underplatform exhaust, and air conditioning.

The train propulsion/braking systems and the configurations of tunnels and stations greatly affect the subway environment, which must often be considered during the early stages of design. The factors affecting a subway environmental control system are discussed in this section. The Subway Environmental Design Handbook (DOT 1976) and NFPA *Standard* 130, Standard for Fixed Guideway Transit Systems, have additional information.

Comfort Criteria

Because of the transient nature of a person's exposure to the subway environment, comfort criteria are not as strict as for continuous occupancy. As a general principle, the environment in a subway station should provide a smooth transition between outside conditions and those in the transit vehicles. Based on nuisance concerns, peak air velocities in public areas should be limited to 1000 fpm.

Air Quality

Air quality in a subway system is influenced by many factors, some of which are not under the direct control of the HVAC engineer. Some particulates, gaseous contaminants, and odorants existing on the outside can be prevented from entering the subway system by the judicious location of ventilation shafts. Particulate matter, including iron and graphite dust generated by train operations, is best controlled by a regular cleaning of the subway. However, the only viable way to control gaseous contaminants such as ozone (from electrical equipment) and CO_2 (generated by human respiration) is through adequate ventilation from outside. A minimum of 7.5 cfm outside air per person should be introduced into tunnels and stations to dilute gaseous contaminants.

Pressure Transients

The passage of trains through aerodynamic discontinuities in the subway causes changes in static pressure. These pressure transients can irritate passengers' ears and sinuses. Pressure transients may also cause additional load on various structures (e.g., acoustical panels) and equipment (e.g., fans). Based on nuisance factors, if the total change in pressure is greater than 2.8 in. of water, the rate of static pressure change should be kept below 1.7 in. of water per second.

During emergencies, ventilation must be provided to control smoke and reduce air temperatures to permit passenger evacuation and fire-fighting operations. The minimum air velocity in the affected tunnel section should be sufficient to prevent the smoke from backlayering (i.e., flowing in the upper cross-section of the tunnel in the direction opposite to the forced outside ventilation air). The method for ascertaining this minimum velocity is provided in

DOT (1997). The maximum air velocity experienced by evacuating passengers should be 2200 fpm.

Interior Heat Loads

Heat in a subway system is generated mostly by (1) train braking; (2) train acceleration; (3) car air conditioning and miscellaneous equipment; (4) station lights, people, and equipment; and (5) ventilation (outside) air.

Deceleration. From 40 to 60% of the heat generated in a subway arises from the braking of trains. Many rapid transit vehicles use nonregenerative braking systems, in which the kinetic energy of the train is dissipated as heat through the dynamic and/or friction brakes, rolling resistance, and aerodynamic drag.

Acceleration. Heat is also generated as the train accelerates. Many operational trains use cam-controlled variable resistance elements to regulate voltage across dc traction motors during acceleration. Electrical power is dissipated by these resistors and by the third rail as heat in the subway system. Heat released during acceleration also includes that due to traction motor losses, rolling resistance, and aerodynamic drag and generally amounts to 10 to 20% of the total heat released in a subway.

For closely spaced stations, more heat is generated because trains frequently only accelerate or brake, and operate little at constant speed.

Car Air Conditioning. Most new cars are fully climate controlled. Air-conditioning equipment removes patron and lighting heat from the cars and transfers it, along with the condenser fan heat and compressor heat, into the subway. Air-conditioning capacities generally range from 10 tons per car for the shorter cars (about 50 ft) up to about 20 tons for the longer cars (about 70 ft). Heat from car air conditioners and other accessories is generally 25 to 30% of total heat generated in a subway.

Other Sources. Heat also comes from people, lighting, induced outside air, and miscellaneous equipment (fare vending machines, escalators, and the like). These sources generate 10 to 30% of total heat released in a subway. In a typical analysis of heat balance in a subway, a control volume about each station that includes the station and its approach and departure tunnels is defined.

Heat Sink

The amount of heat flow from subway walls to subway air varies on a seasonal basis, as well as for morning and evening rush-hour operations. Short periods of abnormally high or low outside air temperature may cause a temporary departure from the heat sink effect in portions of a subway that are not heated or air conditioned. This departure causes a change in the subway air temperature. However, the change from normal is diminished by the thermal inertia of the subway structure. During abnormally hot periods, heat flow to the subway structure increases. Similarly, during abnormally cold periods, heat flow from the subway structure to the air increases.

For subways where daily station air temperatures are held constant by heating and cooling, heat flux from station walls is negligible. Depending on the infiltration of station air into the adjoining tunnels, heat flux from tunnel sections may also be reduced. Other factors affecting the heat sink component are type of soil (dense rock or light, dry soil), migrating groundwater, and the surface configuration of the tunnel walls (ribbed or flat).

Measures to Limit Heat Loads

Various measures have been proposed to limit the interior heat load in subway systems. Among these are regenerative braking, thyristor motor controls, track profile optimization, and underplatform exhausts.

Electrical regenerative braking converts kinetic energy to electrical energy for use by other trains. Flywheel energy storage, an alternative form of regenerative braking, stores part of the braking energy in high-speed flywheels for subsequent use in vehicle acceleration. Using these methods, the reduction in heat generated by braking is limited by present technology to approximately 25%.

Conventional cam-controlled propulsion applies a set of resistance elements to regulate traction motor current during acceleration. Electrical energy dissipated by these resistors appears as waste heat in a subway. Thyristor motor controls replace the acceleration resistors with solid-state controls, which reduce acceleration heat loss by about 10% on high-speed subways and about 25% on low-speed subways.

Track profile optimization refers to a trackway that is lower between stations. With this type of track, less power is used for acceleration, because some of the potential energy of the standing train is converted to kinetic energy as the train accelerates toward the tunnel low point. Conversely, some of the kinetic energy of the train at maximum speed is converted to potential energy as it approaches the next station. This profile reduces the maximum vehicle heat loss from acceleration and braking by about 10%.

An **underplatform exhaust** uses a hood to remove some of the heat generated by vehicle underfloor equipment (e.g., resistors and air-conditioning condensers) from the station environment. Exhaust ports beneath the edge of the station platform withdraw heated air under the car.

For preliminary calculations, it may be assumed that (1) train heat release (due to braking and train air conditioning) in the station box is about two-thirds of the control volume heat load and (2) the underplatform exhaust is about 50% effective.

A quantity of air equal to that withdrawn by the underplatform exhaust enters the control volume from outside. Thus, when the ambient temperature is higher than the station design temperature, an underplatform exhaust reduces the subway heat load by removing undercar heat, but it increases the heat load by drawing in outside air. A proposed technique to reduce the uncontrolled infiltration of outside air is to provide a complementary supply of outside air on the opposite side of the underplatform exhaust ports. While in principle the underplatform exhaust with complementary supply tends to reduce the mixing of outside air with air in public areas of the subway, test results on such systems are unavailable.

RAILROAD TUNNELS

Railroad tunnels serving diesel locomotive operation require ventilation to remove residual diesel exhaust so that the succeeding train can be exposed to a relatively clean environment. Ventilation is also required to prevent the locomotives from overheating. For short tunnels, the ventilation generated by the piston effect of the train followed by natural ventilation is usually sufficient to purge the tunnel of diesel exhaust in a reasonable time period. Mechanical ventilation for locomotive cooling is also usually not required because the time a train is in the tunnel is typically less than the time it would take for a locomotive to overheat. Under certain conditions, however, such as excessively slow trains or during hot weather, overheating can become a problem. For long tunnels, mechanical ventilation is required to purge the tunnel following the passage of a train. Mechanical ventilation may also be required for locomotive cooling, depending on the speed of the train and the number and arrangement of locomotives used.

The diesel locomotive is essentially an electrically driven vehicle. The diesel engine drives a generator which in turn supplies electrical power to the traction motors. The power of these engines ranges from about 1000 to 6000 horsepower.

Since the overall efficiency of the locomotive is generally under 30%, most of energy generated by the combustion process must be dissipated as heat to the surrounding environment. Most of this heat is released above the locomotive via the exhaust stack and the radiator (Figure 17). When operating in a tunnel, this heat

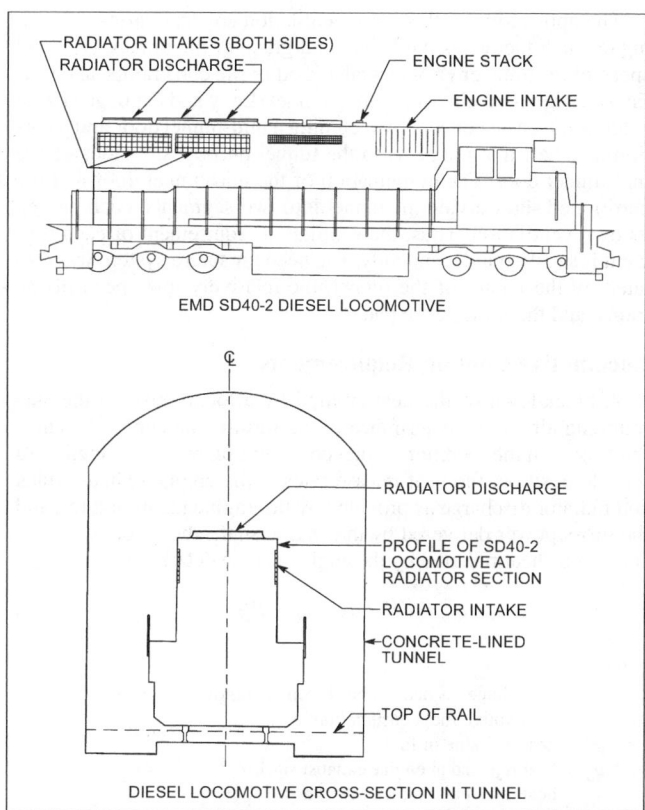

Fig. 17 Typical Diesel Locomotive Arrangement

is confined to the region surrounding the train. Most commercial trains are powered by more than one locomotive, so the last unit is subjected to the heat released from the preceding units. If sufficient ventilation is not provided, the air temperature entering the radiator of the last unit will exceed its allowable limit. Depending on the engine protection system, the locomotive will either shut down or drop to a lower throttle position. When this occurs, the train slows down. But, as discussed in the next section, the train relies on its speed to generate sufficient ventilation for cooling. As a result a domino effect takes place, which may cause the train to stall in the tunnel.

DESIGN CONCEPTS

Most long railroad tunnels (over five miles) in the western hemisphere that serve diesel operation use a ventilation concept that includes a combination of a tunnel door and a system of fans and dampers, all located at one end of the tunnel. When a train moves through the tunnel, ventilating air required for locomotive cooling is generated by the piston effect of the train moving toward or away from the closed portal door. This effect often creates a sufficient flow of air past the train for self-cooling. Under certain conditions when the piston effect cannot provide the required airflow, a fan supplements the effect. This mode of fan operation is commonly referred to as "cooling." When the train leaves the tunnel, the tunnel is purged of residual smoke and diesel contaminants by running the fans with the door closed to move fresh air from one end of the tunnel to the other. This mode of fan operation is referred to as "purge." Because the airflow and pressure required for the cooling and purge modes may be substantially different, separate fans or variable-volume fans may be required to perform the two operations. Also, dampers are provided to relieve the pressure across the door to facilitate its operation while the train is in the tunnel.

The application of this basic ventilation concept varies depending on such factors as the length and grade of the tunnel, type and speed of the train, environmental and other site constraints, and traffic. A design for a nine-mile- long tunnel (Levy and Danziger 1985) extends the basic concept by including a mid-tunnel door and a partitioned shaft that connects to the tunnel on opposite sides of the mid-tunnel door. The combination of the mid-tunnel door and the partitioned shaft divide the tunnel into two segments, each having its own ventilation. Thus, the ventilation requirement of each segment is satisfied independently. The need for such a system was dictated by the length of the tunnel, the relatively low speed of the trains, and the train traffic pattern.

Locomotive Cooling Requirements

A breakdown of the heat emitted by a locomotive to the surrounding air can be determined by performing an energy balance. Starting with the locomotive fuel consumption rate as a function of throttle position, the heat release rates at the engine exhaust stack and radiator discharge as provided by the engine manufacturer, and the gross power delivered by the engine shaft, the amount of miscellaneous heat radiated by the engine can be determined:

$$q_m = FH - q_S - q_R - P_G \qquad (12)$$

where

q_m = miscellaneous heat radiated from locomotive engine
F = locomotive fuel consumption
H = heating value of fuel
q_S = heat rejected at engine exhaust stack
q_R heat rejected at radiator discharge
P_G = gross power at engine shaft

Because the locomotive auxiliaries are driven off the engine shaft with the remaining power used for traction power via the main engine generator, the heat released by the main engine generator is as follows:

$$q_G = (P_G - L_A)(1 - \varepsilon_G) \qquad (13)$$

where

q_G = main generator heat loss
L_A = power driving locomotive auxiliaries
ε_G = main generator efficiency

Heat loss from the traction motors and gear trains is as follows:

$$q_{TM} = P_G - L_A - q_G - P_{TE} \qquad (14)$$

where

q_{TM} = heat loss from traction motors and gear trains
P_{TE} = locomotive tractive effort power

The total locomotive heat release rate q_T is

$$q_T = q_S + q_R + q_m + L_A + q_G + q_{TM} \qquad (15)$$

For a locomotive consist containing N locomotives, the average air temperature approaching the last locomotive is

$$t_{AN} = t_{AT} + \frac{q_T(N-1)}{\rho c_p Q_R} \qquad (16)$$

where

t_{AN} = average tunnel air temperature approaching the Nth unit
t_{AT} = average tunnel air temperature approaching locomotive consist
ρ = density of tunnel air approaching locomotive consist
c_p = specific heat of air
Q_R = tunnel airflow rate relative to the train

The inlet air temperature to the locomotive radiators is used to judge the adequacy of the ventilation. For most locomotive units running at maximum throttle position, the maximum allowable inlet air temperature recommended by locomotive manufacturers is about 115°F. Field tests conducted in operating tunnels (Aisiks and Danziger 1969, Levy and Elpidorou 1991) have shown, however, that some units can operate continuously with radiator inlet temperatures as high as 135°F. The allowable inlet temperature for each locomotive type should be obtained from the manufacturer when contemplating a design.

To determine the airflow rate required to prevent locomotive overheating, the relationship between the average tunnel air temperature approaching the last unit and the radiator inlet air temperature must be known or conservatively estimated. This relationship, however, depends on such variables as the number of locomotives in the consist, the air velocity relative to the train, tunnel cross-section, type of tunnel lining, and locomotive orientation (i.e., facing forward or backward). For trains traveling under 20 mph, Levy and Elpidorou (1991) showed that a reasonable estimate would be to assume the radiator inlet air temperature to be about 10°F higher than the average temperature approaching the unit. For trains moving at 30 mph or more, a reasonable estimate is to assume that the radiator inlet air temperature is equal to the average temperature approaching the unit. When the last unit of the consist is facing forward, thereby putting the exhaust stack ahead of its own radiators, the stack heat release rate must be included when evaluating the radiator inlet air temperature.

Tunnel Aerodynamics

When designing a ventilation system for a railroad tunnel, the airflow and pressure distribution throughout the tunnel as a function of train type, train speed, and mechanical ventilation system operations must be determined. This information is required to determine: (1) whether sufficient ventilation is provided for locomotive cooling; (2) the pressures the fans are required to deliver; and (3) the pressures the structural and ventilation elements of the tunnel must be designed to withstand.

The following equation (DOT 1997) relates the piston effect of the train, the steady-state airflow delivered by the fans to the tunnel, and the pressure across the tunnel door. The expression assumes the leakage across the tunnel door is negligible. Figure 18 shows the dimensional variables on a schematic of a typical tunnel.

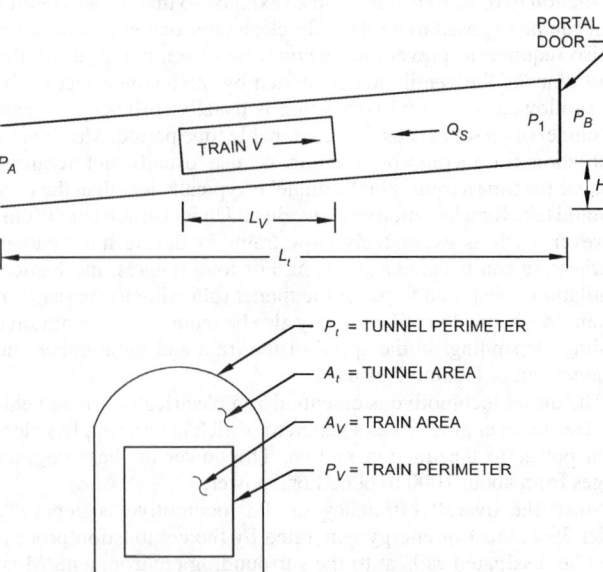

Fig. 18 Railroad Tunnel Aerodynamic Related Variables

$$\frac{\Delta p}{\rho} = \frac{(p_A - p_B)}{\rho} - Hg$$

$$+ \left[\frac{(A_V^2 + A_V A_t C_{DVB})}{(A_t - A_V)^2} + \frac{A_V C_{DVF}}{A_t} \right] \frac{(A_t V + Q_s)^2}{2A_t^2}$$

$$+ \frac{f_t L_V P_t (A_V V + Q_s)^2}{8(A_t - A_V)^3} + \frac{\lambda_V L_V P_V (A_t V + Q_s)^2}{8(A_t - A_V)^3}$$

$$+ \frac{f_t (L_t - L_V) P_t Q_s^2}{8 A_t^3} + \frac{K Q_s^2}{2 A_t^2} \qquad (17)$$

where

Δp = static pressure across tunnel door, lb_f/ft^2
p_A = barometric pressure at Portal A, lb_f/ft^2
p_B = barometric pressure at Portal B, lb_f/ft^2
H = difference in elevation between portals, ft
ρ = density of air, $slug/ft^3$
g = acceleration of gravity = 32.2 ft/s^2
A_V = train cross-sectional area, ft^2
A_t = tunnel cross-sectional area, ft^2
V = velocity of train, ft/s
Q_S = airflow delivered by fan, ft^3/s
C_{DVB} = drag coefficient at back end of train
C_{DVF} = drag coefficient at front end of train
L_t = tunnel length, ft
L_V = train length, ft
f_t = tunnel wall friction factor
λ_V = train skin friction factor
P_t = tunnel perimeter, ft
P_V = train perimeter, ft
K = miscellaneous tunnel loss coefficient

The pressure across the tunnel door generated by train piston action only is evaluated by setting Q_S equal to zero. The airflow rate relative to the train required to evaluate locomotive cooling requirements is

$$Q_{rel} = A_t V + Q_S \qquad (18)$$

where

Q_{rel} = airflow rate relative to train, ft^3/s

Typical values for C_{DVB} and C_{DVF} are about 0.5 and 0.8, respectively. Because trains passing through a tunnel are often more than a mile long, the parameter that most affects the magnitude of the pressure generated is the train skin friction coefficient. For dedicated coal or grain trains, which essentially use uniform cars throughout, a value of 0.09 for the skin friction coefficient results in pressure predictions which closely conform to those observed at various tunnels (Flathead, Cascade, Moffat, Mount Macdonald). For trains with a nonuniform makeup of cars, the skin friction coefficient may be as high as 1.5 times that of a uniform distribution.

The wall surface friction factor corresponds to the coefficient used in the Darcy-Weisbach equation for friction losses in pipe flow. Typical effective values for tunnels constructed with a formed concrete lining having a ballasted track range from 0.015 to 0.017.

Tunnel Purge

Train operations through a tunnel require that the leading end of a train be exposed to an environment that is relatively free of smoke and diesel contaminants emitted by preceding trains. The tunnel is usually purged by displacing the contaminated tunnel air with fresh air by mechanical means after a train has left the tunnel. With the tunnel door closed, air is either supplied to or exhausted from the tunnel, thereby moving fresh air from one end to the other. Observations at the downstream ends of tunnels have found that an effective purge time is usually based on displacing 1.25 times the tunnel volume with fresh air.

The time for purging is primarily determined by operating schedule needs. Too long a time limits traffic, while too short a time may require very high ventilation rates with resulting high electrical energy demand and consumption. Consequently, multiple factors must be considered, including the overall ventilation concept, when establishing purge time.

DIESEL LOCOMOTIVE FACILITIES

Diesel locomotive facilities include shops where locomotives are maintained and repaired, enclosed servicing areas where supplies are replenished, and overbuilds where locomotives routinely operate inside an enclosed space. In general, these areas should be kept under slightly negative pressure to aid in the removal of fumes and contaminants. Ventilation should be accomplished using 100% outside air with no recirculation. However, recirculation may be acceptable to maintain space temperature during unoccupied periods when engines are not running. Heat recovery devices should be considered in colder climates when heating is needed.

Maintenance and Repair Areas

ASHRAE *Standard* 62 and most model codes require a minimum outdoor air ventilation rate of 1.5 cfm/ft² in vehicle repair garages, with no recirculation recommended. Because the interior ceiling height is usually large in locomotive shops, the designer should consider making a volumetric analysis of contaminant generation and air exchange rates rather than using the above value as a blanket standard. The section on contaminants at the end of this section has more information on diesel engine exhaust emissions.

Information in the section on Bus Garages applies to locomotive shops, especially for below-grade pits, battery charging areas, and paint spray booths. However, diesel locomotives generally have much larger engines (ranging up to 6035 hp). Ventilation is needed to reduce crew and worker exposure to exhaust gas contaminants and to remove rejected heat from the engine radiators. Diesel engines should not be operated in shops where possible. The shop practices should reflect this and policies established to define engine speeds and time limits. Some shops require locomotives to be load tested at higher engine speeds. One dedicated area where engines are operated should be identified and hoods used to capture engine exhaust. If hoods are impractical because of physical obstructions, then dilution ventilation must be used.

In designing hoods, the physical location of each exhaust on each type of locomotive must be identified so that each hood can be centered and as close as possible to each exhaust point. Local and state railroad clearance regulations must be followed along with occupational safety requirements. Hoods should not increase the backpressure on the locomotive exhaust by designing a throat velocity that is at least twice the exhaust discharge velocity. Duct design should include access doors and provisions for cleaning the oily residue that increases the risk of fire. Fans and other equipment in the airstream should be selected with regard to the elevated temperatures and the effects of the emission's residue.

Sometimes high ceilings or overhead cranes limit the use of hoods. The *Manual for Railway Engineering* (AREA 1996) notes that six (6) air changes per hour is usually sufficient to provide adequate dilution for idling locomotives and for short runs at a higher speed. The design should take advantage of the thermal buoyancy of the hot gases and exhaust them high before they cool and drop to the floor. Locomotive radiator cooling fans recirculate cooled gases making them more difficult to remove. Makeup air should be introduced into the shop at floor level and tempered as needed.

Heating of shops is necessary in colder climates for the comfort of workers and to prevent freezing of equipment and building piping. Heating may consist of a combination of perimeter convectors to offset building transmission losses, underfloor slab or infrared radiation for workers, and makeup air units for ventilation. Where natural gas is available and codes allow, direct-fired gas heaters can be an economical compromise for obtaining a high degree of worker comfort. Usually air curtains or door heaters are not needed because of infrequent opening.

Enclosed Servicing Areas

While most locomotive servicing is done outside, some railroads use enclosed servicing areas for protection from weather and extreme cold. Operations include refilling fuel tanks, replenishing sand (used to aid traction), draining toilet holding tanks, checking lubrication oil and radiator coolant levels, and performing minor repairs. Generally, the time a locomotive spends in the servicing area is less than one hour. Ventilation is needed to reduce crew and worker exposure to exhaust gas contaminants and to remove rejected heat from engine radiators. The designer should also consider the presence of vapors from fuel oil dispensing and silica dust from sanding operations. Heating may be included depending on the need for worker comfort and the operations performed.

Ventilation of servicing areas should be similar to that for maintenance and repair areas. Hoods rather than dilution ventilation should be used where possible. However, the locating of exhaust points may be difficult due to the coupling of different locomotives together in consists of two or more. Also, elevated sanding towers and distribution piping may interfere. The designer should be aware that contaminant levels may be higher in servicing areas than in shops due to constantly idling locomotives and their occasional higher speed movement in and out of the facility. For dilution ventilation, the designer should ascertain the operation and make a volumetric analysis of contaminant generation and air exchange rates.

For heating, the designer should consider the use of infrared radiation for workers. As with maintenance and repair areas, direct-fired gas heaters may be economical. Air curtains or door heaters may be justified due to the frequent opening of doors or lack of doors.

Overbuilds

With increasing real estate costs, the space above tracks and stations platforms are more commonly built over to enclose the locomotive operation area. Ventilation is needed to reduce crew and passenger exposure to exhaust and to remove rejected heat from engine radiators and train air conditioners. Overbuilds are generally not heated.

Diesel locomotives are more appropriately termed diesel-electric locomotives because the transmission is actually an electric generator (alternator) driven by the diesel engine supplying power to electric traction motors. The throttle, which usually has eight position (notches), controls engine speed and power. Modern passenger trains use auxiliary power from the locomotive, termed head end power (HEP) to provide electricity for lighting, heating, ventilating, air conditioning, receptacles, and other use throughout the train.

Exhaust emissions from passenger diesel locomotives in overbuilds are higher than idling locomotives because of the HEP power requirements. The designer should determine the types of locomotives and operating practices in the proposed overbuild. As with shops and servicing areas, hoods are recommended. According to *Ventilation Requirements for Overbuild Construction* (Amtrak 1991) the temperature at the source will be between 350 and 705°F. A typical design could have hoods approximately 18 to 23 ft above the top of the rail with throat velocities between 30 and 36.7 ft/s. For dilution ventilation, the designer should make a volumetric analysis of contaminant generation and air exchange rates.

Contaminants

The components of diesel exhaust gases that affect the ventilation system design are oxides of nitrogen (NO_x) hydrocarbons (HC), formaldehyde, sulfur dioxide (SO_2), odor constituents, aldehydes, smoke particulate, and carbon monoxide (CO). Operation of diesel engines in enclosed spaces causes visibility obstruction and odors, as well as contaminants. A diesel locomotive that has been operating for several minutes in a lightly loaded mode often produces large amounts of visible smoke and emissions from accumulated, unburned fuel.

Information in the section on Bus Operation Areas, except for engine emissions' data, applies to diesel locomotives. The *Tunnel Engineering Handbook* describes that all exhaust gas contaminants will be maintained with acceptable limits as long as the oxides of nitrogen are maintained in specified acceptable limits. Table 7 provides approximate data on the contaminants found in a typical locomotive's engine exhaust gases and other design information. The designer should determine the nature of the applicable locomotive engines for each project, and consult the manufacturers.

Table 7 Typical Diesel Locomotive Emission Data

Throttle Pos. (Notch)	Engine Speed, rpm	Engine Power, hp	Exhaust Volume, cfm	Exhaust Temp., °F	NO_x, g/min	CO, g/min	HC, g/min	SO_2, g/min
Engine Nominal Power = 3200 horsepower, with Head End Power (HEP)								
8	892	3359	19,545	695	816	29	5.6	43.0
7	892	3002	18,160	690	690	28	5.0	38.5
6	892	2247	14,890	635	410	17	4.5	29.9
5	892	1681	13,835	525	320	19	4.8	23.8
4	893	1213	12,700	425	233	16	6.2	19.2
3	893	857	12,100	370	172	17	6.5	15.7
2	894	584	11,560	325	134	17	6.5	13.0
1	895	312	11,010	280	103	17	6.8	10.7
Idle	895	241	10,865	270	145	26	10.0	8.3
Engine Nominal Power = 3200 horsepower, with No Head End Power (HEP)								
8	906	3222	18,500	662	708	26	10.2	33.8
7	826	2845	16,550	665	630	27	9.0	30.0
6	727	2081	12,300	688	398	43	7.0	22.5
5	651	1656	10,500	630	329	24	5.8	17.9
4	565	1123	8,250	525	235	9.9	4.3	12.3
3	491	708	6,500	424	148	7.9	3.7	8.2
2	343	398	4,850	356	93	4.8	2.7	4.7
1	271	93	3,400	239	36	3.5	2.0	1.7
Idle	199	9	3,300	99	17	1.7	1.3	0.7

EQUIPMENT

The ability of an enclosed vehicular facility to function depends mostly on the effectiveness and reliability of its ventilation system. The ventilation system must be capable of operating effectively under the most adverse environmental and traffic conditions and during periods when not all equipment is operational. For a tunnel, the ventilation system should also have more than one dependable source of power to prevent an interruption of service.

FANS

The fan manufacturer should be prequalified and should be responsible under one contract for furnishing and installing the fans, bearings, drives (including variable-speed components), motors, vibration devices, sound attenuators, discharge and inlet dampers, and limit switches.

The prime concerns in selecting the type, size and number of fans include the total theoretical ventilating air capacity required and a reasonable safety factor. Selection is also influenced by the manner in which reserve ventilation capacity is provided when a fan is inoperative and during repair of equipment or of the power supply.

Selection of fans (number and size) to meet normal and reserve capacity requirements is based on the principle of parallel fan operation. Actual capacities can be determined by plotting fan performance and system curves on the same pressure-volume diagram.

It is important that fans selected for parallel operation operate in the region of their performance curves so that the transfer of capacity back and forth between fans does not occur. This is accomplished by selecting a fan size and speed such that the duty point, no matter how many fans are operating, falls well below the unstable range of fan performance. Fans operating in parallel should be of equal size and have identical performance curves. If flow is regulated by speed control, all fans should operate at the same speed. If flow is regulated by dampers or inlet vane controls, all dampers or inlet vanes should be at the same angle. For axial flow fans, blades on all fans should be at the same pitch or stagger angle.

Jet fans can be used for longitudinal ventilation to provide a positive means of smoke and temperature management in tunnels. This ventilation concept has been proven by the Memorial Tunnel Fire Ventilation Test Program (MHD/FHWA 1995). Jet fans deliver relatively small air quantities at high velocity. The energy, which is transmitted in the form of thrust, induces airflow through the tunnel in the desired direction. Jet fans can be unidirectional or reversible.

Number and Size of Fans

The number and size of fans should be selected by comparing several fan arrangements. Comparison should be based on feasibility, efficiency, and overall economy of the arrangement and the duty required. Several factors should be studied, including (1) annual power cost for operation, (2) annual capital cost of the equipment (usually capitalized over an assumed equipment life of 30 years for rapid transit tunnel fans and 50 years for road tunnel fans), and (3) annual capital cost of the structure required to house equipment (usually capitalized over an arbitrary structure life of 50 years).

Two views are held regarding the proper number and size of fans. The first advocates a few large-capacity fans; the second prefers numerous small units. In most cases, a compromise arrangement produces the greatest efficiency. Regardless of the design philosophy, the number and size of fans should be selected to build sufficient flexibility into the system to meet the varying ventilation demands created by daily and seasonal traffic fluctuations and emergency conditions.

The size of jet fans is usually limited by the space available for their installation in the tunnel. They are typically mounted on the tunnel ceiling, over the traffic lanes or on the walls. Jet fans are sometimes placed in niches to minimize the vertical height of the tunnel boundary. Such a mounting reduces the thrust of the fan.

For longitudinal ventilation using jet fans, the required number of fans (once the fan size and tunnel airflow requirements have been determined) is defined by the total thrust required to overcome the tunnel resistance (pressure loss) divided by the individual jet fan thrust (which is a function of the mean velocity in the tunnel). Jet fans installed longitudinally should be kept at least 100 fan diameters apart so that the jet velocity does not affect the performance of the downstream fan(s). Fans installed side by side should be kept at least two (2) fan diameters between centers.

Type of Fan

Normally, the ventilation system of a vehicular facility requires large air volumes at relatively low pressure. Some fans have low efficiencies under these conditions, so the choice of suitable fan type is often limited to either centrifugal, vane axial or jet fan.

Special Considerations. Special attention must be given to any fan in areas with **flow and pressure transients** caused by train passage. If the transient tends to increase flow to the fan (i.e., the positive flow in front of a train to an exhaust fan or negative flow behind the train to a supply fan), it is important that blade loading does not become high enough to produce long-term fatigue failures.

If the disturbance tends to decrease flow to the fan (i.e., the negative flow behind the train to an exhaust fan or the positive flow in front of the train to a supply fan), the fan performance characteristic must have adequate margins to prevent an aerodynamic stall.

The ability to **rapidly reverse** the rotation of tunnel supply, exhaust, and emergency fans is usually important in an emergency. This requirement must be considered in the selection and design of the fan and drive system.

Fan Design and Operation

Fans and components (e.g., blade-positioning mechanisms, drives, bearings, motors, controls, etc.) that are required to operate in the exhaust airstream during a fire or a smoke emergency should be capable of operating at maximum speed under the temperature conditions specified by the following standards or calculations:

- Rapid transit tunnels, NFPA *Standard* 130
- Road tunnels, NFPA *Standard* 502
- Rail tunnels, at maximum expected temperatures based on computer simulations or other calculations

Fans and dampers that are operated infrequently or for emergency service only should be operated at least once every three months to ensure that all rotating elements are in good condition and are properly lubricated. The period of operation should be long enough to achieve stabilized temperatures in the bearings and motor windings. Multispeed motors should be operated at all speeds.

Inlet boxes can be used to protect the bearings and drives of centrifugal fans from the high temperatures, corrosive gases, and particles present in the exhaust air during the emergency operating mode. This arrangement requires that special attention be given to the design of the fan shaft due to the overhung drive loads as discussed in the following section on Fan Shafts.

Axial flow fans that must be reversible should be able to be rapidly reversed from maximum design speed in one direction to maximum design speed in the opposite direction. The time from rotation at full speed in one direction to full speed in the opposite direction should be less than 60 s. The design should include the effects of any temperature changes associated with the reversal of the flow direction. All components of these fans should be designed for a minimum of four reversal cycles per year for 50 years without damage.

Housings for variable-pitch axial flow fans should be furnished with flow-measuring instruments designed to measure the airflow in both directions. Capped connections should be provided for measuring the pressure developed across the fan. The fan should be protected from operating in a stall region.

To minimize blade failure in axial flow fans, the following precautions should be taken:

- Blades should be secured to the hub with positive locking devices.
- The fan inlet (and discharge if reversible) should be protected to prevent the entry of foreign objects which could damage the rotating assembly.
- The natural frequency (both static and rotating) of the blade and the maximum stress on the surface of the blade during operation at all points on the fan characteristic curve should be measured during the factory tests.
- For rapid transit systems, fans that are subjected to flow and pressure reversals caused by train passage should be designed (and tested to verify) to withstand 4,000,000 cycles of flow reversals.

When the fan includes a variable frequency drive (VFD), a factory test using a production version of the VFD should be conducted to ensure adequate operation and compatibility.

Jet fan blades should be made of such material that their strength in a fire emergency is sufficient to withstand the temperature specified. The design calculations for the fans should consider the fan or fans at the fire location to be destroyed by the fire.

Fan Shafts. Fan shafts should be designed so that the maximum deflection of the assembled fan components, including the forces associated with the fan drive, does not exceed 0.005 in. per foot of shaft length between centers of bearings. For centrifugal fans where the shaft overhangs the bearing, the maximum deflection at the centerline of the fan drive pulley should not exceed 0.005 in. per foot of shaft length between the center of the bearing and the center of the fan drive pulley.

Good practice suggests that the fundamental bending mode frequency of the assembled shaft, wheel, or rotor be more than 50% higher than the highest fan speed. The first resonant speed of all rotational components should be at least 125% above the maximum speed. The fan assembly should be designed to withstand for at least 3 min all stresses and loads resulting from an overspeed test at 110% of the maximum design fan speed.

Bearings. Fan and motor bearings should have a minimum equivalent L10 rated life of 10,000 hours as defined by the Anti-Friction Bearing Manufacturers Association. Special attention must be given to belt-driven fans as the improper tensioning (over tension) of the belts can drastically reduce the life of the bearings that support the drive, the belts and possibly the life of the shafts as well.

Axial flow and jet fans. Each fan motor bearing and fan bearing should be equipped with a monitoring system that senses individual bearing vibrations and temperatures and provides a warning alarm if either rises above the normal range.

Centrifugal fans. Due to their low speed (generally less than 450 rpm), centrifugal fans are not always provided with bearing vibration sensors, but they do require temperature sensors with warning alarm and automatic fan shutdown. Bearing pedestals for centrifugal fans should provide rigid support for the bearing with negligible impediments to airflow. The static and dynamic loading of the shaft and impeller and the maximum force due to the tension in the belts should be considered.

Corrosion-Resistant Materials. Choosing a particular material or coating to protect a fan from corrosive gas is a matter of economics. The selection of the material and/or coating should be based on the site environment, the fan duty, and an expected service life of 50 years.

Sound. For the normal ventilation mode, the construction documents should specify the following:

- Speed and direction of airflow and number of fans operating

- Maximum dBA rating or NC curve or curves acceptable under installed conditions and the locations at the fan inlet and exhaust stack outlet where these requirements apply
- That jet fans meet OSHA or local requirements, which will probably require silencers 1 or 2 diameters in length
- That the dBA rating measured at locations such as intake louvers, discharge louvers, or discharge stacks may not exceed OSHA or local requirements
- That if the measured sound values exceed the specified maximum values, the fan manufacturer must furnish and install the acoustic treatment needed to bring the sound level to an acceptable value

DAMPERS

Dampers play a major role in overall tunnel safety and the successful operation of a tunnel ventilation system. A damper's day-to-day function is to regulate fresh air to maintain comfortable temperatures in the tunnel area through natural or fan forced ventilation. Dampers also perform a pressure relief function as they allow air to be pushed out of openings in front of trains and sucked into areas behind trains as they travel through the tunnel. Additionally, they are used, in conjunction with fans, to dilute or remove carbon monoxide, flammable gases, or other toxic fumes from the tunnel. However, their most important function is to control heat and smoke during a fire emergency. In this function, dampers and fans are used to exhaust the smoke caused by the fire, and control the smoke so people can safely evacuate the tunnel.

Damper Design

Tunnel damper design requires a thorough understanding of the design criteria, installation methods, environmental surroundings, life expectancy, and maintainability and a general understanding of the operating system. Damper construction varies from one tunnel project to another, but the general construction is based on the following design criteria:

- Maximum fan operating pressure
- Day-to-day and rogue train pressures
- Maximum temperature
- Maximum velocity
- Corrosion protection desired
- Maintainability and life expectancy of the equipment
- Maximum damper module size limitations

Fan Pressure. The maximum operating pressure that the damper will withstand during normal or emergency ventilation is typically the maximum pressure that the fan can generate at shutoff. This pressure is generally between 4 and 50 in. of water.

Day-to-Day and Rogue Train Pressures. Some dampers in the track area of a train tunnel see much higher positive and negative pressure pulses, caused by the piston action of trains moving through the tunnel, than the maximum pressure generated by the fan. A closed damper is subjected to a positive pressure in front of a train and a negative pressure behind it as the train travels through the tunnel. This pressure reversal subjects the damper blades and components to reverse bending loads that must be evaluated to prevent premature fatigue failure. The magnitude of the pulsating pressure depends on such factors as maximum train speed, length of the tunnel, clearance between the train and the tunnel walls, and amount of air that can be pushed through the open dampers and/or the station entrances.

This pulsating pressure is part of normal day-to-day operation. However, a rogue train condition could occur once or twice during the life of a tunnel ventilation system. This condition occurs when a train operates at high speed during an emergency or during a runaway. Dampers must be designed both for day-to-day fatigue and for maximum train-speed conditions.

Design specifications should require that the damper and its components meet reverse bending load criteria for anywhere from 1 to 6 million reverse bend cycles for the normal day-to-day train operation. This number of reverse bend cycles equates to a train passing by a damper once every 5 to 20 min for 30 to 50 years. This number can be adjusted for each application. Additionally, the specifying engineer should indicate what pressure could result from a once or twice in a lifetime rogue train condition.

Typically, actuators for tunnel dampers must be sized to operate against the maximum fan pressure. Because reversing pressure only occurs for a short duration, and because trains will not normally be traveling down the track during an emergency ventilation situation, the actuators are not expected to operate under such conditions.

Temperature. The maximum temperature can vary for each tunnel project. Some specifying engineers use the temperature limits set by NFPA. One typical tunnel specification states that the dampers, actuators, and accessories shall meet the operational requirements of the emergency ventilation fans as described in NFPA *Standard* 130, which states, "Emergency ventilation fans, their motors, and all related components exposed to the exhaust airflow shall designed to operate in an ambient atmosphere of 482°F for a minimum of one hour with actual values to be determined by design analysis."

Some tunnel design engineers have specified higher temperature criteria based on additional design considerations. A few automobile tunnels have been designed for the possibility of two truck tankers with flammable liquids exploding from an accident in the tunnel, which could subject the tunnel dampers to much higher temperatures. Projects of this type or others with special design considerations have been designed for maximum temperatures up to 800°F. In any event, the specifying engineer must evaluate the design conditions for each tunnel project and determine what the maximum temperature could be.

Dampers, and especially damper actuators, must be specially constructed to operate reliably in high temperature conditions for extended periods. Verifying that the proposed equipment can provide this required safety function is important. Because standard test procedures have not been developed, a custom, high-temperature test of a sample damper and actuator should be considered for inclusion in the equipment specifications.

Air Velocity. The maximum velocity for a tunnel damper design is determined from the maximum flow that can be expected to go through the damper during any one of its operating conditions. That flow can occur during normal ventilation or during an emergency set of conditions. Additionally, the flow could come from more than one fan depending on design. Actuators for tunnel dampers are typically selected to operate against the maximum flow that the dampers will see in their worst case scenario. Thus, the specifying engineer must advise what the maximum flow will be.

Corrosion Protection. Materials of construction for tunnel jobs vary considerably, and their selection is usually determined from one or more of the following reasons:

- Initial project cost
- Environmental conditions
- Life expectancy of the equipment
- Success or failure of previous materials used on similar projects
- Engineer's knowledge and/or experience of the materials required to provide corrosion protection
- Design criteria; specifically, pressure, temperature, and velocity

The corrosion resistant characteristics of a damper should be determined by the environment in which it will operate. A damper installation near saltwater or heavy industrial areas may need superior corrosion protection than one in a rural, nonindustrialized city. Dampers exposed to underground indoor environments may need less corrosion protection. However, many underground dampers are also exposed to rain, snow, and sleet. These and other factors must be evaluated by the engineer before a proper specification can be written.

Tunnel dampers have been made from commercial-quality galvanized and hot-dipped galvanized steel, anodized aluminum, aluminum with a duranodic finish, carbon steel with various types of finishes, 304, 304L, 316, 316L, and 317 stainless steel.

Maintainability and Life Expectancy of Equipment. These issues are of great concern when specifying dampers may be difficult to access for servicing, inspection, or maintenance on a regular basis. Additionally, the equipment may be difficult to replace, should it fail prematurely. Reasons could be that the damper was marginally designed for the pressures, temperatures, corrosion resistance, etc. required for the application.

Thus, some clients and specifying engineers purposely design the dampers of a more robust construction. They specify heavier material and/or more corrosion resistant materials than may be required for the application in hopes of reducing operational problems and cost while extending the life expectancy of the product. Typical methods used by specifying engineers to design dampers of a more robust construction are

- Limiting blade, frame, and linkage deflections to a maximum of $L/360$
- Selecting actuators for 200 to 300% of actual damper torque required
- Using large safety factors for stresses and deflections of high stress components.
- Specifying heavier material sizes and gages than necessary
- Using more corrosion resistant materials and finishes than required

Numerous specifications include a Quality Assurance (QA) or System Assurance Program (SAP) to assure that the required levels of performance are met. Others include an experience criteria that requires manufacturers to have five installations with five or more years of operating experience. This specification requires a list of projects and the names of contacts that the customer can communicate with regarding the product performance. These techniques help insure that reliable products are supplied.

Maximum Damper Module Size. The maximum damper module size is one of the most important initial cost factors. Many dampers can be fabricated as a single module assembly or in several sections that can be field assembled into a single module damper. However, some damper openings are very large and it may not be practical to manufacture the damper in a one-piece frame construction due to shipping, handling, and/or installation problems.

Generally speaking, the initial damper cost is less with fewer modules because they have fewer blades, frames, jackshafts, actuators, and mullion supports. However, other factors such as job site access, lifting capabilities, and installation labor costs must also be included in the initial cost analysis. These factors change for each project and the specifying engineer must evaluate each application.

Damper Applications and Types

Dampers allow or restrict flow in the tunnel. **Fan isolation dampers** (FID) can be installed in multiple-fan systems (1) to isolate any parallel, nonoperating fan from those operating to prevent short circuiting and consequent pressure and flow loss through the inoperative fan; (2) to prevent serious windmilling of an inoperative fan; and (3) to provide a safe environment for maintenance and repair work on each fan. Single-fan installations also may have a FID to prevent serious windmilling due to natural or piston-effect drafts and to facilitate fan maintenance.

Ventilation dampers (VD) control the amount of fresh air supplied to and exhausted from the tunnel and station areas. In some cases, these dampers may also serve as the **smoke exhaust dampers** (SED) and/or **fire dampers** (FD) depending on their location and the design. Two types of ventilation dampers are generally used: (1) the trapdoor type, which is installed in a vertical duct such that the door lies horizontal when closed; and (2) the multiple-blade louver type with parallel operating blades. Both types are usually driven by an electric or pneumatic actuator, which is operated by the fan controller. During normal operation, the damper usually closes when the fan is shut off and opens when the fan is turned on.

The trapdoor damper is simple and works satisfactorily where a vertical duct enters a plenum-type fan room through an opening in the floor. This damper is usually constructed of steel plate, reinforced with welded angle iron, is hinged on one side and closes by gravity against the embedded angle frame of the opening. The opening mechanism is usually a shaft sprocket-and-chain device. The drive motor and gear drive mechanism or actuator must develop sufficient force to open the damper door against the maximum static pressure differential that the fan can develop. This pressure can be obtained from fan performance curves. Limit switches start and stop the gear-motor drive or actuator at the proper position.

Fan isolation and ventilation dampers placed in other than vertical ducts should have multiblade louvers. These dampers usually consist of a rugged channel frame, the flanges of which are bolted to the flanges of the fan, duct, wall, or floor opening. Damper blades are assembled with shafts that turn in bearings mounted on the outside of the channel frame. This arrangement requires access outside the duct for bearing and shaft lubrication, maintenance, and space for linkages to operate. Multiple-blade dampers should have blade edge and/or end seals to meet the air leakage requirements for the application.

The trapdoor-type damper, if properly fabricated, is inherently a low leakage design due to its weight and the overlap at its edges. Multiblade dampers can also be of a low leakage design, but they must be carefully constructed to ensure tightness on closing. The pressure drop across a fully opened damper and the leakage rate across a fully closed damper should be verified by the appropriate test procedure in AMCA *Standard* 500. A damper that leaks excessively under pressure can cause the fan to rotate counter to its power rotation, thus making restarting dangerous and possibly damaging the fans' drive motor.

Actuators and Accessory Selections

Tunnel damper specifications typically call for dampers, actuators and accessories to meet the operational requirements of the **emergency ventilation fans** described by NFPA *Standard* 130. Damper actuators are also normally specified to be electric or pneumatic. This requirement is determined by the designer or the customer and is, in most cases, decided by the available power or the initial and/or long term operating cost.

Pneumatic Actuators. Pneumatic actuators are available in many designs and sizes. Rack and pinion, air cylinder, or Scotch yoke configurations are common. They can all be supplied in a double-acting (air supplied to operate the damper in both directions) or a spring-return construction. A spring-return design uses air to power it in one direction and a spring to drive it in the opposite direction. The spring return is selected when it is desirable to have the damper fail to the open or closed position upon a loss of air supply. Many manufacturers make pneumatic actuators, and several manufacture both double-acting and spring-return designs capable of operating at 482°F for one hour.

Electric Actuators. Electric actuators are also available in a variety of designs and sizes. They can be supplied to be powered in both directions to open and close the damper. In this case, the actuator usually fails in its last position upon loss of power. Electric actuators are also available that are powered in one direction and spring driven in the opposite direction. As with the pneumatic design, the spring return is selected when it is desirable to have the damper fail to the open or closed position upon a loss of power. There are fewer electric than pneumatic actuator manufacturers and most of them do not market a spring-return design, especially

in the larger torque models. Also, very few electric actuators are manufactured that are capable of operating at 482°F for one hour—particularly for the spring-return designs.

Actuator Selection. Actuators for tunnel dampers are typically sized to operate against the maximum flow or velocity and pressure conditions that will occur in the worst case scenario. The maximum air velocity corresponds to the maximum flow that is expected to go through the damper during any of its operating conditions. That flow can occur during normal operation or during an emergency. Additionally, this flow could come from more than one fan depending on the design. The maximum pressure that the damper will see during normal or emergency ventilation is typically the maximum pressure that the fan can generate at shutoff.

Additionally, actuators are sized (1) to overcome the frictional resistance of blade bearings, linkage pivots, jackshafting assemblies, etc. and (2) to compress the blade and jamb seals to meet the specified leakage requirements. Thus, in order to properly select the actuator, the specifying engineer must determine what the maximum flow or velocity and pressure conditions will be, as well as the leakage criteria.

Safety factors, regarding actuator selection, are not always addressed in tunnel damper specifications. This could result in potential operational problems if a manufacturer were to select his actuators close to the operating torque required. These dampers are expected to function for many years and in some cases are not maintained as frequently as they should be. Additionally, damper manufacturers determine their torque requirements based on square, plumb, and true installations. All of these factors, plus the fact that dirt and debris buildup can increase damper torque, suggests that a minimum safety factor of at least 50% should be specified. Greater safety factors can be specified for some applications; however, larger actuators require larger drive shafts, which results in a higher initial cost.

Supply Air Intake

Supply air intakes require careful design to ensure that the quality of air drawn into the system is the best available. Such factors as recirculation of exhaust air or intake of contaminants from nearby sources should be considered.

Louvers or grilles are usually installed over air intakes for aesthetic, security, or safety reasons. Bird screens are also necessary if openings between louver blades or grilles are large enough to allow birds to enter.

Due to the large air volume required in some ventilation systems, it may not be possible for intake louvers to have sufficiently low face velocities to be weatherproof. Therefore, intake plenums, shafts, fan rooms, and fan housings need water drains. Blowing snow can also fill the fan room or plenum, but this snow usually does not stop the ventilation system from operating satisfactorily if additional floor drains are located in the area of the louvers.

Noise elimination devices may have to be installed in fresh air intakes to keep fan and air noise from disturbing the outside environment. If sound reduction is required, the total system—fan, fan plenum, building, fan housing, and air intake (location and size)—should be investigated. Fan selection should also be based on a total system, including the pressure drop that results from sound attenuation devices.

Exhaust Outlets

Exhaust air should be discharged away from the street level and away from areas with human occupancy. Contaminant concentrations in the exhaust air are not of concern if the system is working effectively. However, odors and entrained particulate matter make this air undesirable in occupied areas. Exhaust stack discharge velocity should be high enough to disperse contaminants into the atmosphere. A minimum of 2000 fpm is usually necessary.

In the past, evasé (flared) outlets were used to regain some static pressure and thereby reduce the energy consumption of the exhaust fan. Unless the fan discharge velocity is in excess of 2000 fpm, however, the energy saving may not offset the cost of the evasé.

In a vertical or near-vertical exhaust fan discharge connection to an exhaust duct or shaft, rainwater will run down the inside of the stack into the fan. This water will dissolve material deposited from vehicle exhausts on the inner surface of the stack and become extremely corrosive. Therefore, fan housing should be made of a corrosion-resistant material or specially coated to protect the metal from corrosion.

Discharge louvers and gratings should be sized and located so that their discharge is not objectionable to pedestrians or contaminate intake air louvers; at the same time, the air resistance across the louver or grating should be minimized. Discharge velocities through sidewalk gratings are usually limited to 500 fpm. Bird screens should be provided if the exhaust air is not continuous (24 h, 7 days a week) and the openings between the louver blades are large enough to allow birds to enter.

The corrosion-resistant characteristics of the louver or grating should be determined by the corrosiveness of the exhaust air and the installation environment. The pressure drop across the louvers should be verified by the appropriate test procedure in AMCA *Standard* 500.

REFERENCES

ACGIH. 1998. *Industrial ventilation: A manual of recommended practice*, 23rd ed., Appendix A. American Conference of Governmental Industrial Hygienists, Cincinnati, OH.

Aisiks, E.G. and N.H. Danziger. 1969. Ventilation research program at Cascade Tunnel, Great Northern Railway. American Railway Engineering Association, pp. 114.

AMCA. 1989. Test methods for louvers, dampers and shutters. *Standard* 500. Air Movement and Control Association, Arlington Heights, IL.

Amtrak. 1991. *Ventilation requirements for overbuild construction*. Report issued by Office of Engineering Design, National Railroad Passenger Corporation, Philadelphia, PA.

AREA. 1996. Buildings and support facilities. Chapter 6 of Part 4 of Section 4.7 in *Manual for Railway Engineering*. American Railway Engineering Association, Washington, DC.

ASHRAE. 1985. Laboratory methods of testing fans for rating. *Standard* 51-1985 (AMCA *Standard* 210-85).

ASHRAE. 1989. Ventilation for acceptable indoor air quality. ANSI/ASHRAE *Standard* 62-1989.

Bendelius, A.G. 1982. Tunnel ventilation. Chapter 19 in *Tunnel Engineering Handbook* edited by J.O. Bickel and T.R. Kuesel. Van Nostrand Reinhold, New York.

Code of Federal Regulations. 29 CFR 1910. Occupational Safety and Health Standards, Subpart Z.

DOT. 1997. Subway environment simulation (SES) computer program Version 4: User's manual and programmer's manual. Issued as Volume II of *Subway Environmental Design Handbook*. Pub. No. FTA-MA-26-7022-97-1. US Department of Transportation, Washington, DC. Also available from Volpe Transportation Center, Cambridge, MA.

DOT. 1996. Design guidelines for bus transit systems using compressed natural gas as an alternative fuel. Federal Transit Administration, U.S. Government Printing Office, Washington, DC.

DOT. 1976. *Subway environmental design handbook*. Urban Mass Transportation Administration, U.S. Government Printing Office, Washington, DC.

DuCharme, G.N. 1991. Ventilation for battery charging. *Heating/Piping/Air Conditioning* (February).

EPA. 1992. Mobile 5a mobile emissions factor model. EPA-AA-TEB-92. Environmental Protection Agency, Research Triangle Park, NC.

Fieldner, A.C. et al. 1921. Ventilation of vehicular tunnels. Report of the U.S. Bureau of Mines to New York State Bridge and Tunnel Commission and New Jersey Interstate Bridge and Tunnel Commission. American Society of Heating and Ventilating Engineers (ASHVE).

Hama, G.M., W.G. Frederick, and H.G. Monteith. 1974. How to design ventilation systems for underground garages. Air Engineering. Study by the Detroit Bureau of Industrial Hygiene, Detroit (April).

Haerter, A. 1963. Flow distribution and pressure change along slotted or branched ducts. *ASHVE Transactions* 69:124-37.

Kennedy, W.D., J.A. Gonzales, and J.G. Sanchez. 1996. Derivation and application of the SES critical velocity equations. *ASHRAE Transactions* 102(2):40-44.

Krarti, M. and A. Ayari. 1998. Evaluation of fixed and variable rate ventilation system requirement for enclosed parking facilities. *ASHRAE Transactions* 99(1).

Levy, S.S. and N.H Danziger. 1985. Ventilation of the Mount Macdonald Tunnel. BHRA Fluid Engineering, Fifth International Symposium on Aerodynamics and Ventilation of Vehicle Tunnels, Lille, France.

Levy, S.S. and D.P. Elpidorou. 1991. Ventilation of Mount Shaughnessy Tunnel. Seventh International Symposium on Aerodynamics and Ventilation of Vehicle Tunnels, Brighton, UK.

MHD/FHWA. 1995. Memorial tunnel fire ventilation test program, comprehensive test report. Massachusetts Highway Dept., Boston, and Federal Highway Administration, Washington, DC.

NFPA 1994. Standard for gaseous hydrogen systems at consumer sites. *Standard* 50A. National Fire Protection Association, Quincy, MA.

NFPA. 1995. Standard for parking structures. *Standard* 88A.

NFPA. 1995. Standard for spray application using flammable and combustible materials. *Standard* 33-95.

NFPA. 1995. Standard for compressed natural gas (CNG) vehicular fuel systems. *Standard* 52.

NFPA. 1996. Automotive and marine service station code. *Standard* 30A.

NFPA. 1996. Standard for Liquefied Natural Gas (LNG) vehicular fuel systems. *Standard* 57.

NFPA. 1996. National electrical code. *Standard* 70.

NFPA. 1996. Recommended practice on fire protection for limited access highways, tunnels, bridges, elevated roadways, and air right structures. *Standard* 502.

NFPA. 1997. Standard for repair garages. *Standard* 88B-97.

NFPA. 1997. Standard for fixed guideway transit systems. *Standard* 130-97.

NFPA. 1998. Liquefied petroleum gas code. *Standard* 58.

NTIS. 1980. User's guide for the TUNVEN and DUCT programs. *Publication* PB80141575. National Technical Information Service, Springfield, VA.

PIARC. 1995. Road tunnels. XXth World Road Congress, Montreal.

Singstad, O. 1929. Ventilation of vehicular tunnels. World Engineering Congress, Tokyo, Japan.

Stankunas, A.R., P.T. Bartlett, and K.C. Tower. 1980. Contaminant level control in parking garages. *ASHRAE Transactions* 86(2):584-605.

Watson, A.Y., R.R. Bates, and D. Kennedy. 1988. Air pollution, the automobile, and public health. Sponsored by the Health Effects Institute. National Academy Press, Washington, DC.

BIBLIOGRAPHY

Ball, D. and J. Campbell. 1973. Lighting, heating and ventilation in multistory and underground car park. Paper presented at the Institution for Structural Engineers and The Institution of Highway Engineers' Joint Conference on Multi-story and Underground Car Parks (May).

Federal Register. 1974. 39(125), June.

Ricker, E.R. 1948. *The traffic design of parking garages*. ENO Foundation for Highway Traffic Control, Saugatuck, CT.

Round, F.G. and H.W. Pearall. Diesel exhaust odor: Its evaluation and relation to exhaust gas composition. Research Laboratories, General Motors Corporation, *Society of Automotive Engineers Technical Progress Series* 6.

Turk, A. 1963. Measurements of odorous vapors in test chambers: Theoretical. *ASHRAE Journal* 5(10):55-58.

Wendell, R.E., J.E. Norco, and K.G. Croke. 1973. Emission prediction and control strategy: Evaluation and pollution from transportation systems. *Air Pollution Control Association Journal* (February).

LABORATORIES

MODERN laboratories require regulated temperature, humidity, relative static pressure, air motion, air cleanliness, sound, and exhaust. This chapter addresses biological, chemical, animal, and physical laboratories. Within these generic categories, some laboratories have unique requirements. This chapter provides an overview of the heating, ventilating, and air-conditioning (HVAC) characteristics and design criteria for laboratories, including a brief overview of architectural and utility concerns. This chapter does not cover pilot plants, which are essentially small manufacturing units.

The function of a laboratory is important in determining the appropriate HVAC system selection and design. Air-handling, hydronic, control, life safety, and heating and cooling systems must function as a unit and not as independent systems. HVAC systems must conform to applicable safety and environmental regulations.

Providing a safe environment for all personnel is a primary objective in the design of HVAC systems for laboratories. A vast amount of information is available, and HVAC engineers must study the subject thoroughly to understand all the factors that relate to proper and optimum design. This chapter serves only as an introduction to the topic of laboratory HVAC design.

HVAC systems must integrate with architectural planning and design, electrical systems, structural systems, other utility systems, and the functional requirements of the laboratory. The HVAC engineer, then, is a member of a team that includes other facility designers, users, industrial hygienists, safety officers, operators, and maintenance staff. Decisions or recommendations by the HVAC engineer may significantly affect construction, operation, and maintenance costs.

Laboratories frequently use 100% outside air, which broadens the range of conditions to which the systems must respond. They seldom operate at maximum design conditions, so the HVAC engineer must pay particular attention to partial load operations that are continually changing due to variations in internal space loads, exhaust requirements, external conditions, and day-night variances.

Most laboratories will be modified at some time. Consequently, the HVAC engineer must consider to what extent laboratory systems should be adaptable for other needs. Both economics and integration of the systems with the rest of the facility must be considered.

The preparation of this chapter is assigned to TC 9.10, Laboratory Systems.

LABORATORY TYPES

Laboratories can be divided into the following generic types:

- **Biological laboratories** are those that contain biologically active materials or involve the chemical manipulation of these materials. This includes laboratories that support such disciplines as biochemistry, microbiology, cell biology, biotechnology, immunology, botany, pharmacology, and toxicology. Both chemical fume hoods and biological safety cabinets are commonly installed in biological laboratories.
- **Chemical laboratories** support both organic and inorganic synthesis and analytical functions. They may also include laboratories in the material and electronic sciences. Chemical laboratories commonly contain a number of fume hoods.
- **Animal laboratories** are areas for manipulation, surgical modification, and pharmacological observation of laboratory animals. They also include animal holding rooms, which are similar to laboratories in many of the performance requirements but have an additional subset of requirements.
- **Physical laboratories** are spaces associated with physics; they commonly incorporate lasers, optics, nuclear material, high- and low-temperature material, electronics, and analytical instruments.

HAZARD ASSESSMENT

Laboratory operations potentially involve some hazard; nearly all laboratories contain some type of hazardous materials. A comprehensive hazard assessment, which must be completed before the laboratory can be designed, should be performed by the owner's designated **safety officers**. They include, but are not limited to, the chemical hygiene officer, radiation safety officer, biological safety officer, and fire and loss prevention official. The hazard assessment should be incorporated into the chemical hygiene plan, radiation safety plan, and biological safety protocols.

Hazard study methods such as hazard and operability analysis (HAZOP) can be used to evaluate design concepts and certify that the HVAC design conforms to the applicable safety plans. The nature and quantity of the contaminant, types of operations, and degree of hazard dictate the types of containment and local exhaust devices. For functional convenience, operations posing less hazard potential are conducted in devices that use directional airflows for personnel protection (e.g., laboratory fume hoods and biological safety cabinets). However, these devices do not provide absolute containment. Operations having a significant hazard potential are conducted in devices that provide greater protection but are more restrictive (e.g., sealed glove boxes).

The design team should visit similar laboratories to assess successful design approaches and safe operating practices. Each laboratory is somewhat different. Its design must be evaluated using appropriate, current standards and practices rather than duplicating existing and possibly outmoded facilities.

Laboratory Safety Resource Materials

ACGIH. *Industrial Ventilation: A Manual of Recommended Practice*, 23rd ed. 1998. American Conference of Governmental Industrial Hygienists, Cincinnati, OH.

AIA. *Guidelines for Design and Construction of Hospital and Health Care Facilities*, 1996-97 ed. American Institute of Architects, Washington, D.C.

AIHA. Laboratory Ventilation. ANSI/AIHA *Standard* Z9.5-93. American Industrial Hygiene Association, Fairfax, VA.

Americans with Disabilities Act (ADA).

BOCA. Building, Mechanical, and Fire Prevention Model Codes. Building Officials and Code Administrators International, Country Club Hills, IL.

CAP. *Medical Laboratory Planning and Design*. College of American Pathologists, Northfield, IL.

DHHS. *Biosafety in Microbiological and Biomedical Laboratories*, 3rd ed. 1993. U.S. Department of Health and Human Services *Publication* (CDC) 93-8395.

ICBO. Uniform Building, Mechanical, and Fire Prevention Model Codes. International Conference of Building Officials, Whittier, CA.

NFPA. Fire Protection for Laboratories Using Chemicals. ANSI/NFPA *Standard* 45-96. National Fire Protection Association, Quincy, MA.

NFPA. Hazardous Chemicals Data. ANSI/NFPA *Standard* 49-94.

NFPA. Health Care Facilities. ANSI/NFPA *Standard* 99-96.

NRC. *Biosafety in the Laboratory: Prudent Practices for Handling and Disposal of Infectious Materials*. 1989. National Research Council, National Academy Press, Washington, DC.

NRC. *Prudent Practices in the Laboratory: Handling and Disposal of Chemicals*. 1995. National Research Council, National Academy Press, Washington, D.C.

OSHA. Occupational Exposure to Chemicals in Laboratories. Appendix VII, 29 CFR 1910.1450. Available from U.S. Government Printing Office, Washington, D.C.

SEFA. Laboratory Fume Hoods Recommended Practices. SEFA 1.2-1996. Scientific Equipment and Furniture Association, Hilton Head, SC.

Other regulations and guidelines may apply to laboratory design. All applicable institutional, local, state, and federal requirements should be identified prior to the start of design.

DESIGN PARAMETERS

The following design parameters must be established for a laboratory space:

- Temperature and humidity, both indoor and outdoor
- Air quality from both process and safety perspectives, including the need for air filtration and special treatment (e.g., charcoal, HEPA, or other filtration of supply or exhaust air)
- Equipment and process heat gains, both sensible and latent
- Minimum ventilation rates
- Equipment and process exhaust quantities
- Exhaust and air intake locations
- Style of the exhaust device, capture velocities, and usage factors
- Need for standby equipment and emergency power
- Alarm requirements
- Potential changes in the size and number of fume hoods
- Anticipated increases in internal loads
- Room pressurization requirements

It is important to (1) review design parameters with the safety officers and scientific staff, (2) determine limits that should not be exceeded, and (3) establish the desirable operating conditions. For areas requiring variable temperature or humidity, these parameters must be carefully reviewed with the users to establish a clear understanding of expected operating conditions and system performance.

Because laboratory HVAC systems often incorporate 100% outside air systems, the selection of design parameters has a substantial effect on capacity, first cost, and operating costs. The selection of proper and prudent design conditions is very important.

Internal Thermal Considerations

In addition to the heat gain from people and lighting, laboratories frequently have significant sensible and latent loads from equipment and processes. Often, data for equipment used in laboratories is unavailable or the equipment has been custom built. Heat release from animals that may be housed in the space can be found in Chapter 10 of the 1997 *ASHRAE Handbook—Fundamentals* and in Alereza and Breen (1984).

Careful review of the equipment to be used, a detailed understanding of how the laboratory will be used, and prudent judgment are required to obtain good estimates of the heat gains in a laboratory. The convective portion of heat released from equipment located within exhaust devices can be discounted. Heat from equipment that is directly vented or heat from water-cooled equipment should not be considered part of the heat released to the room. Any unconditioned makeup air that is not directly captured by an exhaust device must be included in the load calculation for the room. In many cases, additional equipment will be obtained by the time a laboratory facility has been designed and constructed. The design should allow for this additional equipment.

Internal load as measured in watts per square foot is the average continuous internal thermal load discharged into the space. It is not a tabulation of the connected electrical load because it is rare for all equipment to operate simultaneously, and most devices operate with a duty cycle that keeps the average electrical draw below the nameplate information. When tabulating the internal sensible heat load in a laboratory, the duty cycle of the equipment should be obtained from the manufacturer. This information, combined with the nameplate data for the item, may provide a more accurate assessment of the average thermal load.

The HVAC system designer should evaluate equipment nameplate ratings, applicable use and usage factors, and overall diversity. Much laboratory equipment includes computers, automation, sample changing, or robotics; this can result in high levels of use even during unoccupied periods. The HVAC designer must evaluate internal heat loads under all anticipated laboratory operating modes. Due to highly variable equipment heat gain, individual laboratories should have dedicated temperature controls.

Two cases encountered frequently are (1) building programs based on generic laboratory modules and (2) laboratory spaces that are to be highly flexible and adaptive. Both situations require the design team to establish heat gain on an area basis. The values for area-based heat gain vary substantially for different types of laboratories. Heat gains of 5 to 25 W/ft^2 or more are common for laboratories with high concentrations of equipment.

Architectural Considerations

The integration of utility systems into the architectural planning, design, and detailing is essential to providing successful research facilities. The architect and the HVAC system engineer must seek an early understanding of each other's requirements and develop integrated solutions. HVAC systems may fail to perform properly if the architectural requirements are not addressed correctly. Quality assurance of the installation is just as important as proper specifications. The following play key roles in the design of research facilities:

Modular Planning. Most laboratory programming and planning is based on developing a module that becomes the base building block for the floor plan. Laboratory planning modules are frequently 10 to 12 ft wide and 20 to 30 ft deep. The laboratory modules may be developed as single work areas or combined to form multiple-station work areas. Utility systems should be arranged to reflect the architectural planning module, with services provided for each module or pair of modules, as appropriate.

Development of Laboratory Units. National Fire Protection Association (NFPA) *Standard* 45 requires that laboratory units be designated. Similarly, the International, Uniform, and Building Officials and Code Administrators International (BOCA) model codes require the development of control areas. Laboratory units or control areas should be developed, and the appropriate hazard levels should be determined early in the design process. The HVAC designer should review the requirements for maintaining separations between laboratories and note requirements for exhaust ductwork to serve only a single laboratory unit.

Additionally, NFPA *Standard* 45 requires that no fire dampers be installed in laboratory exhaust ductwork. Building codes offer no relief from maintaining required floor-to-floor fire separations. These criteria and the proposed solutions should be reviewed early in the design process with the appropriate building code officials. The combination of the two requirements commonly necessitates the construction of dedicated fire-rated shafts from each occupied floor to the penthouse or building roof.

Provisions for Adaptability and Flexibility. Research objectives frequently require changes in laboratory operations and programs. Thus, laboratories must be flexible and adaptable, able to accommodate these changes without significant modifications to the infrastructure. For example, the utility system design can be flexible enough to supply ample cooling to support the addition of heat-producing analytical equipment without requiring modifications to the HVAC system. Adaptable designs should allow programmatic research changes that require modifications to the laboratory's infrastructure within the limits of the individual laboratory area and/or interstitial and utility corridors. For example, an adaptable design would allow the addition of a fume hood without requiring work outside that laboratory space. The degree of flexibility and adaptability for which the laboratory HVAC system is designed should be determined from discussion with the researchers, laboratory programmer, and laboratory planner. The HVAC designer should have a clear understanding of these requirements and their financial impact.

Early Understanding of Utility Space Requirements. The amount and location of utility space are significantly more important in the design of research facilities than in that of most other buildings. The available ceiling space and the frequency of vertical distribution shafts are interdependent and can significantly affect the architectural planning. The HVAC designer must establish these parameters early, and the design must reflect these constraints. The designer should review alternate utility distribution schemes, weighing their advantages and disadvantages.

High-Quality Envelope Integrity. Laboratories that have stringent requirements for the control of temperature, humidity, relative static pressure, and background particle count generally require architectural features to allow the HVAC systems to perform properly. The building envelope may need to be designed to handle relatively high levels of humidification and slightly negative building pressure without moisture condensation in the winter or excessive infiltration. Some of the architectural features that the HVAC designer should evaluate include

- Vapor barriers—position, location, and kind
- Insulation—location, thermal resistance, and kind
- Window frames and glazing
- Caulking
- Internal partitions—their integrity in relation to air pressure, vapor barriers, and insulation value
- Finishes—vapor permeability and potential to release particles into the space
- Doors
- Air locks

Air Intakes and Exhaust Locations. Mechanical equipment rooms and their air intakes and exhaust stacks must be located to avoid intake of fumes into the building. As with other buildings, air intake locations must be chosen to minimize fumes from loading docks, cooling tower discharge, vehicular traffic, etc.

LABORATORY EXHAUST AND CONTAINMENT DEVICES

FUME HOODS

The Scientific Equipment and Furniture Association (SEFA 1996) defines a laboratory fume hood as "a ventilated enclosed work space intended to capture, contain, and exhaust fumes, vapors, and particulate matter generated inside the enclosure. It consists basically of side, back and top enclosure panels, a floor or counter top, an access opening called the face, a sash(es), and an exhaust plenum equipped with a baffle system for airflow distribution." Figure 1 shows the basic elements of a general-purpose benchtop fume hood.

Fume hoods may be equipped with a variety of accessories, including internal lights, service outlets, sinks, air bypass openings, airfoil entry devices, flow alarms, special linings, ventilated base storage units, and exhaust filters. Undercounter cabinets for storage of flammable materials require special attention to ensure safe installation. NFPA *Standard* 30, *Flammable and Combustible Liquids Code*, does not recommend venting these cabinets; however, ventilation is often required to avoid accumulation of toxic or hazardous vapors. Ventilation of these cabinets by a separately ducted supply and exhaust that will maintain the temperature rise of the cabinet interior within the limits defined by NFPA *Standard* 30 should be considered.

Types of Fume Hoods

The following are the primary types of fume hoods and their applications:

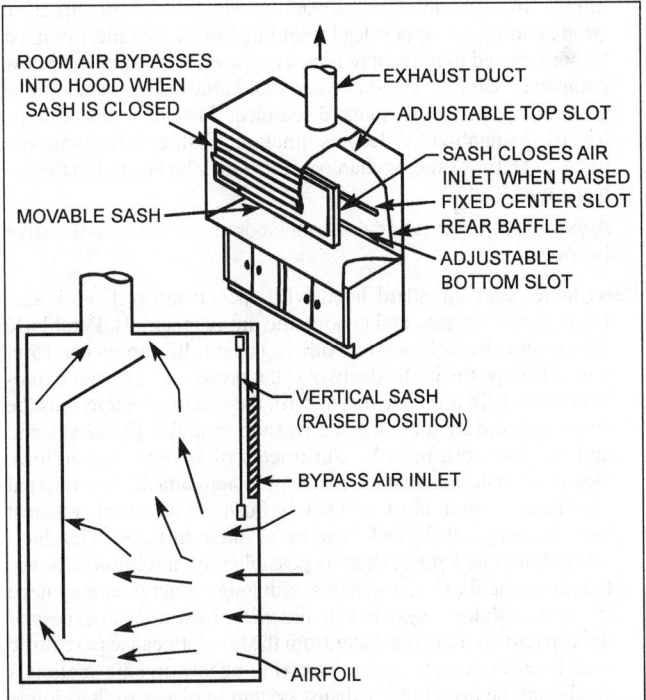

Fig. 1 Bypass Fume Hood with Vertical Sash and Bypass Air Inlet

Standard (approximately constant volume airflow with variable face velocity). Hood that meets basic SEFA definition. Sash may be vertical, horizontal, or combination type.

Application: Research laboratories—frequent or continuous use. Moderate to highly hazardous processes; varying procedures.

Bypass (approximately constant volume airflow with approximately constant face velocity). Standard vertical sash hood modified with openings above and below the sash. The openings are sized to minimize the change in the face velocity, which is generally to 3 or 4 times the full-open velocity, as the sash is lowered.

Application: Research laboratories—frequent or continuous use. Moderate to highly hazardous processes; varying procedures.

Variable Volume (constant face velocity). Hood has an opening or bypass designed to provide a prescribed minimum air intake when the sash is closed and an exhaust system designed to vary airflow in accordance with sash opening. Sash may be vertical, horizontal, or a combination of both.

Application: Research laboratories—frequent or continuous use. Moderate to highly hazardous processes; varying procedures.

Auxiliary Air (approximately constant volume airflow with approximately constant face velocity). A plenum above the face receives air from a secondary air supply that provides partially conditioned or unconditioned outside air.

Application: Research laboratories—frequent or continuous use. Moderate to highly hazardous processes; varying procedures.

Note: Many organizations restrict the use of this type of hood.

Process (approximately constant volume airflow with approximately constant face velocity). Standard hood without a sash. By some definitions, this is not a fume hood. Considered a ventilated enclosure.

Application: Process laboratories—intermittent use. Low-hazard processes; known procedures.

Radioisotope. Standard hood with special integral work surface, linings impermeable to radioactive materials, and structure strong enough to support lead shielding bricks. The interior must be constructed to prevent radioactive material buildup and allow complete cleaning. The ductwork should have flanged neoprene gasketed joints with quick disconnect fasteners that can be readily dismantled for decontamination. High-efficiency particulate air (HEPA) and/or charcoal filters may be needed in the exhaust duct.

Application: Process and research laboratories using radioactive isotopes.

Perchloric Acid. Standard hood with special integral work surfaces, coved corners, and nonorganic lining materials. Perchloric acid is an extremely active oxidizing agent. Its vapors can form unstable deposits in the ductwork that present a potential explosion hazard. To alleviate this hazard, the exhaust system must be equipped with an internal water washdown and drainage system, and the ductwork must be constructed of smooth, impervious, cleanable materials that are resistant to acid attack. The internal washdown system must completely flush the ductwork, exhaust fan, discharge stack, and fume hood inner surfaces. The ductwork should be kept as short as possible with minimum elbows. Perchloric acid exhaust systems with longer duct runs may need a zoned washdown system to avoid water flow rates in excess of the capacity to drain the water from the hood. Because perchloric acid is an extremely active oxidizing agent, organic materials should not be used in the exhaust system in places such as joints and gaskets. Ducts should be constructed of a stainless steel material, with a chromium and nickel content not less than that of 316 stainless steel, or of a suitable nonmetallic material. Joints should be welded and ground smooth. A perchloric acid exhaust system should only be used for work involving perchloric acid.

Application: Process and research laboratories using perchloric acid. Mandatory use because of explosion hazard.

California. Special hood with sash openings on multiple sides (usually horizontal).

Application: For enclosing large and complex research apparatus that require access from two or more sides.

Walk-In. Standard hood with sash openings to the floor. Sash can be either horizontal or vertical.

Application: For enclosing large or complex research apparatus. Not designed for personnel to enter while operations are in progress.

Distillation. Standard fume hood with extra depth and 1/3- to 1/2-height benches.

Application: Research laboratory. For enclosing tall distillation apparatus.

Canopy. An open hood with an overhead capture structure.

Application: Not a true fume hood. Useful for heat or water vapor removal from some work areas. Not to be substituted for a fume hood. Not recommended when workers must bend over the source of heat or water vapor.

Fume Hood Sash Configurations

The work opening has operable glass sash(es) for observation and shielding. A sash may be vertically operable, horizontally operable, or a combination of both. A vertically operable sash can incorporate single or multiple vertical panels. A horizontally operable sash incorporates multiple panels that slide in multiple tracks, allowing the open area to be positioned across the face of the hood. The combination of a horizontally operable sash mounted within a single vertically operable sash section allows the entire hood face to be opened for setup. Then the opening area can be limited by closing the vertical panel, with only the horizontally sliding sash sections used during experimentation. Either multiple vertical sash sections or the combination sash arrangement allow the use of larger fume hoods with limited opening areas, resulting in reduced exhaust airflow requirements. Fume hoods with vertically rising sash sections should include provisions around the sash to prevent the bypass of ceiling plenum air into the fume hood.

Fume Hood Performance

Containment of hazards in a fume hood is based on the principle that a flow of air entering at the face of the fume hood, passing through the enclosure, and exiting at the exhaust port prevents the escape of airborne contaminants from the hood into the room.

The following variables affect the performance of the fume hood:

- Face velocity
- Size of face opening
- Sash position
- Shape and configuration of entrance
- Shape of any intermediate posts
- Inside dimensions and location of work area relative to face area
- Location of service fittings inside the fume hood
- Size and number of exhaust ports
- Back baffle and exhaust plenum arrangement
- Bypass arrangement, if applicable
- Auxiliary air supply, if applicable
- Arrangement and type of replacement supply air outlets
- Air velocities near the hood
- Distance from openings to spaces outside the laboratory
- Movements of the researcher within the hood opening
- Location, size, and type of research apparatus placed in the hood
- Distance from the apparatus to the researcher's breathing zone

Air Currents. Air currents external to the fume hood can jeopardize the hood's effectiveness and expose the researcher to materials used in the hood. Detrimental air currents can be produced by

- Air supply distribution patterns in the laboratory
- Movements of the researcher
- People walking past the fume hood
- Thermal convection
- Opening of doors and windows

Caplan and Knutson (1977, 1978) conducted tests to determine the interactions between room air motion and fume hood capture velocities with respect to the spillage of contaminants into the room. Their tests indicated that the effect of room air currents is significant and of the same order of magnitude as the effect of the hood face velocity. Consequently, improper design and/or installation of the replacement air supply lowers the performance of the fume hood.

Disturbance velocities at the face of the hood should be no more than one-half and preferably one-fifth the face velocity of the hood. This is an especially critical factor in designs that use low face velocities. For example, a fume hood with a face velocity of 100 fpm could tolerate a maximum disturbance velocity of 50 fpm. If the design face velocity were 60 fpm, the maximum disturbance velocity would be 30 fpm.

To the extent possible, the fume hood should be located so that traffic flow past the hood is minimal. Also, the fume hood should be placed to avoid any air currents generated from the opening of windows and doors. To ensure the optimum placement of the fume hoods, the HVAC system designer must take an active role early in the design process.

Use of Auxiliary Air Fume Hoods. AIHA *Standard* Z9.5 discourages the use of auxiliary air fume hoods. These hoods incorporate an air supply at the fume hood to reduce the amount of room air exhausted. The following difficulties and installation criteria are associated with auxiliary air fume hoods:

- The auxiliary air supply must be introduced outside the fume hood to maintain appropriate velocities past the researcher.
- The flow pattern of the auxiliary air must not degrade the containment performance of the fume hood.
- Auxiliary air must be conditioned to avoid blowing cold air on the researcher; often the air must be cooled to maintain the required temperature and humidity within the hood.
- Auxiliary air may introduce additional heating and cooling loads in the laboratory.
- Only vertical sash may be used in the hood.
- Controls for the exhaust, auxiliary, and supply airstreams must be coordinated.
- Additional coordination of utilities during installation is required to avoid spatial conflicts caused by the additional duct system.
- Humidity control can be difficult.

Fume Hood Performance Criteria. ASHRAE *Standard* 110, Method of Testing Performance of Laboratory Fume Hoods, describes a quantitative method of determining the containment performance of a fume hood. The method requires the use of a tracer gas and instruments to measure the amount of tracer gas that enters the breathing zone of a mannequin; this simulates the containment capability of the fume hood as a researcher conducts operations in the hood.

The following tests are commonly used to judge the performance of the fume hood: (1) face velocity test, (2) flow visualization test, (3) large volume flow visualization, (4) tracer gas test, and (5) sash movement test. These tests should be performed under the following conditions:

- Usual amount of research equipment in the hood; the room air balance set
- Doors and windows in their normal positions

- Fume hood sash set in varying positions to simulate both static and dynamic performance

All fume hoods should be tested annually and their performance certified. The following descriptions partially summarize the test procedures. ASHRAE *Standard* 110 provides specific requirements and procedures.

Face Velocity Test

The desired face velocity should be determined by the safety officer and the researcher. The velocity is a balance between safe operation of the fume hood, airflow needed for the hood operation, and energy cost. Face velocity measurements are taken on a vertical/horizontal grid, with each measurement point representing not more than 1 ft². The measurements should be taken with a device that is accurate in the intended operating range, and an instrument holder should be used to improve accuracy. Computerized multipoint grid measurement devices provide the greatest accuracy.

Flow Visualization

1. Swab a strip of titanium tetrachloride along both walls and the hood deck in a line parallel to the hood face and 6 in. back into the hood. *Caution*: Titanium tetrachloride forms smoke and is corrosive to the skin and extremely irritating to the eyes and respiratory system.
2. Swab an 8 in. circle on the back of the hood. Define air movement toward the face of the hood as reverse airflow and lack of movement as dead airspace.
3. Swab the work surface of the hood, being sure to swab lines around all equipment in the hood. All smoke should be carried to the back of the hood and out.
4. Test the operation of the deck airfoil bypass by running the cotton swab under the airfoil.
5. Before going to the next test, move the cotton swab around the face of the hood; if there is any outfall, the exhaust capacity test (large capacity flow visualization) should not be made.

Large Volume Flow Visualization

Appropriate measures should be taken prior to undertaking a smoke test to avoid accidental activation of the building's smoke detection system.

1. Ignite and place a smoke generator near the center of the work surface 6 in. behind the sash. Some smoke sources generate a jet of smoke that produces an unacceptably high challenge to the hood. Care is required to ensure that the generator does not disrupt the hood performance, leading to erroneous conclusions.
2. After the smoke bomb is ignited, pick it up with tongs and move it around the hood. The smoke should not be seen or smelled outside the hood.

Tracer Gas Test

1. Place the sulfur hexafluoride gas ejector in the required test locations (i.e., the center and near each side). Similarly position a mannequin with a detector in its breathing zone in the corresponding location at the hood.
2. Release the tracer gas and record measurements over a 5 min time span.
3. After testing with the mannequin is complete, remove it, traverse the hood opening with the detector probe, and record the highest measurement.

Sash Movement Test

Verify containment performance of the fume during operation of the fume hood sash as described in ASHRAE *Standard* 110.

BIOLOGICAL SAFETY CABINETS

A biological safety cabinet protects the researcher and, in some configurations, the research materials as well. Biological safety

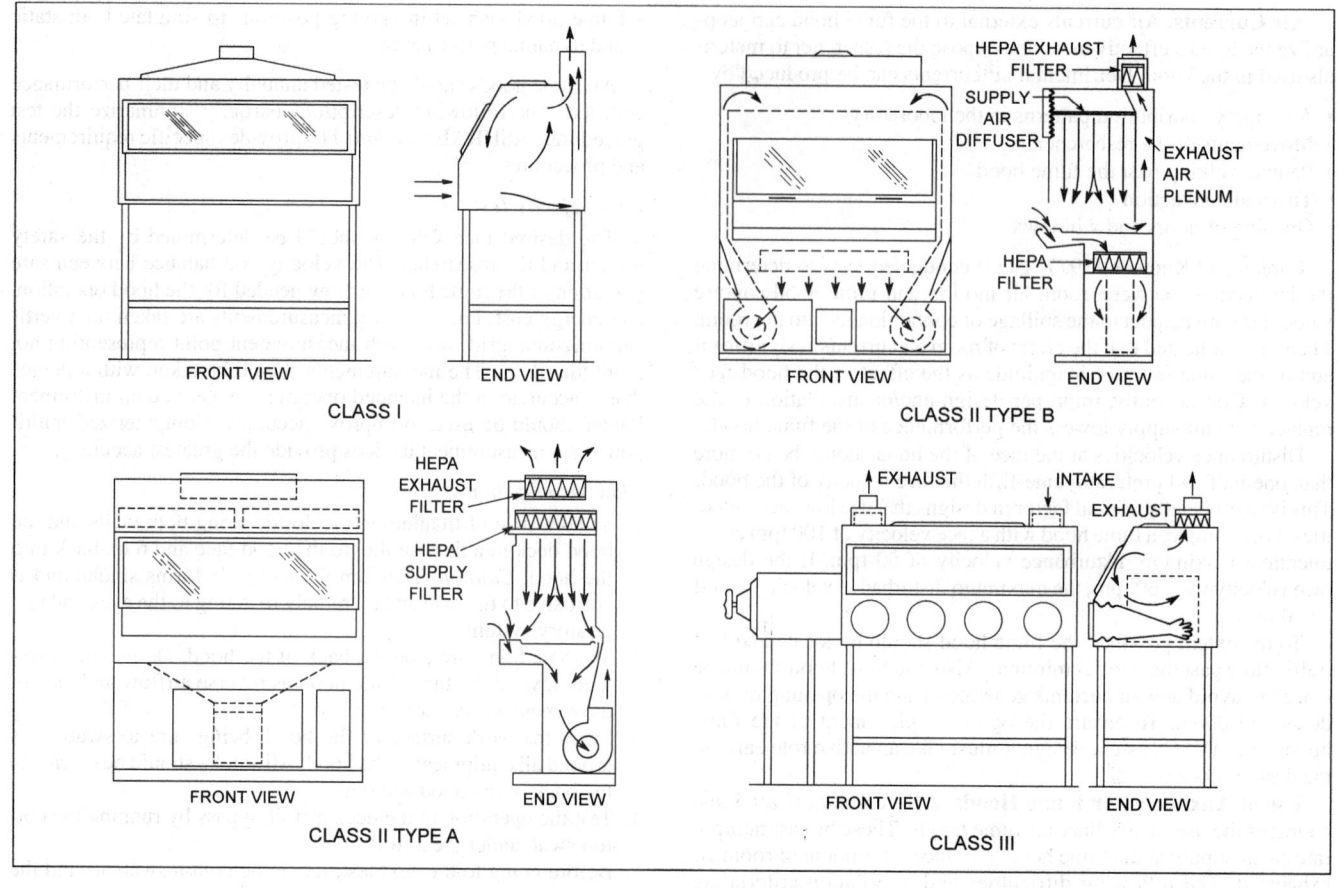

Fig. 2   Types of Biological Safety Cabinets

cabinets are sometimes called safety cabinets, ventilated safety cabinets, laminar flow cabinets, and glove boxes. Biological safety cabinets are categorized into six groups (several are shown in Figure 2):

Class I	Similar to chemical fume hood, no research material protection, 100% exhaust through a HEPA filter
Class II	
Type A	70% recirculation within the cabinet; 30% exhaust through a HEPA filter; common plenum configuration; can be recirculated into the laboratory
Type B1	30% recirculation within the cabinet; 70% exhaust through a HEPA filter; separate plenum configuration, must be exhausted to the outside
Type B2	100% exhaust through a HEPA filter to the outside
Type B3	70% recirculation within the cabinet; 30% exhaust through a HEPA filter; common plenum configuration; must be exhausted to the outside
Class III	Special applications; 100% exhaust through a HEPA filter to the outside; researcher manipulates material within cabinet through physical barriers (gloves)

Several key decisions must be made by the researcher prior to the selection of a biological safety cabinet (Eagleston 1984). An important difference in biological safety cabinets is their ability to handle chemical vapors properly (Stuart et al. 1983). Of special concern to the HVAC engineer are the proper placement of the biological safety cabinet in the laboratory and the room's air distribution. Rake (1978) concluded the following:

"A general rule of thumb should be that, if the crossdraft or other disruptive room airflow exceeds the velocity of the air curtain at the unit's face, then problems do exist. Unfortunately, in most laboratories such disruptive room airflows are present to various extent. Drafts from open windows and doors are the most hazardous sources because they can be far in excess of 200 fpm and accompanied by substantial turbulence. Heating and air-conditioning vents perhaps pose the greatest threat to the safety cabinet because they are much less obvious and therefore seldom considered.... It is imperative then that all room airflow sources and patterns be considered before laboratory installation of a safety cabinet."

Class II biological safety cabinets should only be placed in the laboratory in compliance with NSF International *Standard* 49, Class II (Laminar Flow) Biohazard Cabinetry. Assistance in procuring, testing, and evaluating performance parameters of Class II biological safety cabinets is available from NSF as part of the standard. The cabinets should be located away from drafts, active walkways, and doors. The air distribution system should be designed to avoid air patterns that impinge on the cabinet.

The different biological safety cabinets have varying static pressure resistance requirements. Generally, Class II Type A cabinets have pressure drops ranging between 0.005 and 0.1 in. of water. Class II Type B1 cabinets have pressure drops in the range of 0.6 to 1.2 in. of water, and Class II Type B2 cabinets have pressure drops ranging from 1.5 to 2.3 in. of water. The manufacturer must be consulted to verify specific requirements.

The pressure requirements also vary based on filter loadings and the intermittent operation of individual biological safety cabinets. Exhaust systems for biological safety cabinets must be designed with these considerations in mind. Care must be exercised when manifolding biological safety cabinet exhausts to ensure that the varying pressure requirements are met.

The manufacturer of the biological safety cabinet may be able to supply the transition to the duct system. The transition should include an access port for testing and balancing and an airtight damper for decontamination. As with any containment ductwork, high-integrity duct fabrication and joining systems are necessary. The responsible safety officer should be consulted to determine the need for and placement of isolation dampers to facilitate decontamination operations.

Class I Cabinets

The Class I cabinet is a partial containment device designed for research operations with low- and moderate-risk etiologic agents. It does not provide protection for the materials used in the cabinet. Room air flows through a fixed opening and prevents aerosols that may be generated within the cabinet enclosure from escaping into the room. Depending on cabinet usage, air exhausted through the cabinet may be HEPA filtered prior to being discharged into the exhaust system. The fixed opening through which the researcher works is usually 8 in. high. To provide adequate personnel protection, the air velocity through the fixed opening is usually at least 75 fpm.

If approved by the appropriate safety officer, it is possible to modify the Class I cabinet to contain chemical carcinogens by adding appropriate exhaust air treatment and increasing the velocity through the opening to 100 fpm. Large pieces of research equipment can be placed in the cabinet if adequate shielding is provided.

The Class I cabinet is not appropriate for containing systems that are vulnerable to airborne contamination because the air flowing into the cabinet is untreated. Also, the Class I cabinet is not recommended for use with highly infectious agents because an interruption of the inward airflow may allow aerosolized particles to escape.

Class II Cabinets

Class II cabinets provide protection to personnel, product, and the environment. The cabinets feature an open front with inward airflow and HEPA-filtered recirculated and exhaust air.

The Class II Type A cabinet has a fixed opening with a minimum inward airflow velocity of 75 fpm. The average minimum downward velocity of the internal airflow is 75 fpm. The Class II Type A cabinet is suitable for use with agents meeting Biosafety Level 2 criteria (DHHS 1993), and, if properly certified, can meet Biosafety Level 3. However, because approximately 70% of the airflow is recirculated, the cabinet is not suitable for use with flammable, toxic, or radioactive agents.

The Class II Type B1 cabinet has a vertical sliding sash and maintains an inward airflow of 100 fpm at a sash opening of 8 in. The average downward velocity of the internal airflow is 100 fpm. The Class II Type B1 cabinet is suitable for use with agents meeting Biosafety Level 3. Approximately 70% of the internal airflow is exhausted through HEPA filters; this allows the use of biological agents treated with limited quantities of toxic chemicals and trace amounts of radionuclides, provided the work is performed in the direct exhaust area of the cabinet.

The Class II Type B2 cabinet maintains an inward airflow velocity of 100 fpm through the work opening. The cabinet is 100% exhausted through HEPA filters to the outdoors; all downward velocity air is drawn from the laboratory or other supply source and is HEPA filtered prior to being introduced into the work space. The Class II Type B2 cabinet may be used for the same level of work as the Class II Type B1 cabinet. In addition, the design permits use of toxic chemicals and radionuclides in microbiological studies.

The Class II Type B3 cabinet maintains an inward airflow velocity of 100 fpm and is similar in performance to the Class II Type A cabinet.

In Class II Type A and Type B3 cabinets, exhaust air delivered to the outlet of the cabinet by internal blowers must be handled by the laboratory exhaust system. This arrangement requires a delicate balance between the cabinet and the laboratory's exhaust system, and it may incorporate a thimble-type connection between the cabinet and the laboratory exhaust ductwork. **Thimble (or canopy) connections** incorporate an air gap between the biological safety cabinet and the exhaust duct. The exhaust system must pull more air than is exhausted by the biological safety cabinet to make air flow in through the gap. The designer should confirm the amount of air to be drawn through the air gap. A minimum flow is required to provide the specified level of containment, and a maximum flow cannot be exceeded without causing an imbalance through aspiration.

Class II Type B1 and Type B2 cabinets rely on the building exhaust system to pull the air from the cabinet's work space and through the exhaust HEPA filters. The pressure resistance that must be overcome by the building exhaust system can be obtained from the cabinet manufacturer. Because containment in this type of cabinet depends on the building's exhaust system, the exhaust fan(s) should have redundant backups.

Class III Cabinets

The Class III cabinet is a gastight, negative pressure containment system that physically separates the agent from the worker. These cabinets provide the highest degree of personnel protection. Work is performed through arm-length rubber gloves attached to a sealed front panel. Room air is drawn into the cabinet through HEPA filters. Particulate material entrained in the exhaust air is removed by HEPA filtration or incineration before discharge to the atmosphere. A Class III system may be designed to enclose and isolate incubators, refrigerators, freezers, centrifuges, and other research equipment. Double-door autoclaves, liquid disinfectant dunk tanks, and pass boxes are used to transfer materials into and out of the cabinet.

Class III systems can contain highly infectious materials and radioactive contaminants. Although there are operational inconveniences with these cabinets, they are the equipment of choice when a high degree of personnel protection is required. It should be noted that explosions have occurred in Class III cabinets used for research involving volatile substances.

MISCELLANEOUS EXHAUST DEVICES

Snorkels are used in laboratories to remove heat or nontoxic particulates that may be generated from benchtop research equipment. Snorkels usually have funnel-shaped inlet cones connected to 3 to 6 in. diameter flexible or semiflexible ductwork extending from the ceiling to above the benchtop level.

Typically, **canopy hoods** are used to remove heat or moisture generated by a specific piece of research apparatus (e.g., steam sterilizer) or process. Canopy hoods cannot contain hazardous fumes adequately to protect the researcher.

The **laboratory**, if maintained at negative relative static pressure, provides a second level of containment, protecting occupied spaces outside of the laboratory from operations and processes undertaken therein.

LAMINAR FLOW CLEAN BENCHES

Laminar flow clean benches are available in two configurations—horizontal (crossflow) and vertical (downflow). Both configurations filter the supply air and usually discharge the air out the front opening into the room. Clean benches protect the experiment or product but do not protect the researcher; therefore, they should not be used with any potentially hazardous or allergenic substances. Clean benches are not recommended for any work involving hazardous biological, chemical, or radionuclide materials.

COMPRESSED GAS STORAGE AND VENTILATION

Gas Cylinder Closets

Most laboratory buildings require storage closets for cylinders of compressed gases, which may be inert, flammable, toxic, corrosive, or poisonous. The requirements for storage and ventilation are covered in building codes and NFPA standards and codes (NFPA 15, 30, 45, 50A, 55, 58, 70, and 90A). Water sprinklers are usually required, but other types of fire suppression may be needed based on the gases stored. Explosion containment requires a separate structural study, and closets generally require an outside wall for venting. One design used by a large chemical manufacturer to house gases with explosion potential specifies a completely welded 0.25 in. steel inner liner for the closet, heavy-duty door latches designed to hold under the force of an internal explosion, and venting out the top of the closet.

The closet temperature should not exceed 125°F per NFPA *Standard* 55. Ventilation for cylinder storage is established in NFPA *Standard* 55 at a minimum of 1 cfm/ft^2. Ventilation rates can be calculated by determining both the amount of gas that could be released by complete failure of the cylinder outlet piping connection and the time the release would take, and then finding the dilution airflow required to reduce any hazard below the maximum allowable limit. Design principles for biohazardous materials may be different than for chemical hazards. An investigation for biohazard containment can start with NFPA *Standard* 99, Health Care Facilities.

Ventilation air is usually exhausted from the closet; makeup air comes from the surrounding space through openings in and around the door or through a transfer duct. That makeup air must be added into the building air balance. Ventilation for a closet to contain materials with explosion potential must be carefully designed, with safety considerations taken into account. NFPA *Standard* 68 is a reference on explosion venting.

Cylinder closet exhausts should be connected through a separate duct system to a dedicated exhaust fan or to a manifold system in which constant volume can be maintained under any possible manifold condition. A standby source of emergency power should be considered for the exhaust system fan(s).

Gas Cylinder Cabinets

Compressed gases that present a physical or health hazard are often placed in premanufactured gas cylinder cabinets. Gas cylinder cabinets are available for single-, dual-, or triple-cylinder configurations and are commonly equipped with valve manifolds, fire sprinklers, exhaust connections, access openings, and operational and safety controls. The engineer must fully understand safety, material, and purity requirements associated with specific compressed gases when designing and selecting cylinder cabinets and the components that make up the compressed gas handling system.

Exhaust from the gas cylinder cabinets is provided at a high rate. Air is drawn into the gas cylinder cabinet from the surrounding space through a filtered opening, usually on the lower front of the cylinder cabinet. Depending on the specific gas in the cabinet, the exhaust system may require emission control equipment and a source of emergency power.

LABORATORY VENTILATION SYSTEMS

The total airflow rate for a laboratory is dictated by one of the following:

1. Total amount of exhaust from containment and exhaust devices
2. Cooling required to offset internal heat gains
3. Minimum ventilation rate requirements

Fume hood exhaust requirements (including evaluation of alternate sash configurations as described in the section on Fume Hoods) must be determined in consultation with the safety officers. The HVAC engineer must determine the expected heat gains from the research equipment after consulting with the research staff (see the section on Internal Thermal Considerations).

Minimum airflow rates are generally in the range of 6 to 10 air changes per hour when the space is occupied; however, some spaces (e.g., animal holding areas) may have minimum airflow rates established by specific standards or by internal facility policies. For example, the National Institutes of Health (NIH 1996a, 1996b) recommend a minimum of 6 air changes per hour for occupied laboratories but a minimum of 15 air changes per hour for animal housing and treatment areas. The maximum airflow rate for the laboratory should be reviewed to ensure that appropriate supply air delivery methods are chosen such that supply airflows do not impede the performance of the exhaust devices.

Laboratory ventilation systems can be arranged for either constant volume or variable volume airflow. The specific type should be selected with the research staff, safety officers, and maintenance personnel. Special attention should be given to unique areas such as glass washing areas, hot and cold environmental rooms and labs, fermentation rooms, and cage washing rooms. Emergency power systems to operate the laboratory ventilation equipment should be considered based on hazard assessment or other specific requirements. Care should be taken to ensure that an adequate amount of makeup air is available whenever exhaust fans are operated on emergency power. Additional selection criteria are described in the sections on Hazard Assessment and Operation and Maintenance.

Usage Factor

In many laboratories, all hoods and safety cabinets are seldom needed at the same time. A system **usage factor** represents the maximum number of exhaust devices with sashes open or in use simultaneously. The system usage factor depends on

- Type and size of facility
- Total number of fume hoods
- Number of fume hoods per researcher
- Type of fume hood controls
- Type of laboratory ventilation systems
- Number of devices that must operate continuously due to chemical storage requirements or contamination prevention
- Number of current and projected research programs

Usage factors should be applied carefully when sizing equipment. For example, teaching laboratories may have a usage factor of 100% when occupied by students. If too low a usage factor is selected, design airflow and containment performance cannot be maintained. It is usually expensive and disruptive to add capacity to an operating laboratory's supply or exhaust system. Detailed discussions with research staff are required to ascertain maximum usage rates of exhaust devices.

Noise

Noise level in the laboratory should be considered at the beginning of the design so that noise criterion (NC) levels suitable for scientific work can be achieved. For example, at the NIH, sound levels of NC 45 (including fume hoods) are required in regularly occupied laboratories. The requirement is relaxed to NC 55 for instrument rooms. If noise criteria are not addressed as part of the design, NC levels can be 65 or greater, which is unacceptable to most occupants. Sound generated by the building HVAC equipment should be evaluated to ensure that excessive levels do not escape to the outdoors. Remedial correction of excessive sound levels can be difficult and expensive.

SUPPLY AIR SYSTEMS

Supply air systems for laboratories provide the following:

- Thermal comfort for occupants
- Minimum and maximum airflow rates
- Replacement for air exhausted through fume hoods, biological safety cabinets, or other exhaust devices
- Space pressurization control
- Environmental control to meet process or experimental criteria

The design parameters must be well defined for selection, sizing, and layout of the supply air system. Installation and setup should be verified as part of the commissioning process. Design parameters are covered in the section on Design Parameters, and commissioning is covered in the section on Commissioning.

Laboratories in which chemicals and compressed gases are used generally require nonrecirculating or 100% outside air supply systems. The selection of 100% outside air supply systems versus return air systems should be made as part of the hazard assessment process, which is discussed in the section on Hazard Assessment. A 100% outside air system must have a very wide range of heating and cooling capacity, which requires special design and control.

Supply air systems for laboratories include both constant volume and variable volume systems that incorporate either single-duct reheat or dual-duct configurations, with distribution through low-, medium-, or high-pressure ductwork.

Filtration

The filtration for the air supply depends on the requirements of the laboratory. Conventional chemistry and physics laboratories commonly use 85% dust spot efficient filters (ASHRAE *Standard* 52.1). Biological and biomedical laboratories usually require 85 to 95% dust spot efficient filtration. HEPA filters should be provided for spaces where research materials or animals are particularly susceptible to contamination from external sources. HEPA filtration of the supply air is necessary for such applications as environmental studies, studies involving specific pathogen-free research animals or nude mice, dust-sensitive work, and electronic assemblies. In many instances, biological safety cabinets or laminar flow clean benches (which are HEPA filtered) may be used rather than HEPA filtration for the entire laboratory.

Air Distribution

Air supplied to a laboratory must be distributed to keep temperature gradients and air currents to minimum. Air outlets (preferably nonaspirating diffusers) must not discharge into the face of a fume hood, a biological safety cabinet, or an exhaust device. Acceptable room air velocities are covered in the sections on Fume Hoods and Biological Safety Cabinets. Special techniques and diffusers are often needed to introduce the large air quantities required for a laboratory without creating disturbances at exhaust devices.

EXHAUST SYSTEMS

Laboratory exhaust systems remove air from containment devices and from the laboratory itself. The exhaust system must be controlled and coordinated with the supply air system to maintain correct pressurization. Additional information on the control of exhaust systems is included in the section on Control.

Design parameters must be well defined for selection, sizing, and layout of the exhaust air system. Installation and setup should be verified as part of the commissioning process. See the sections on Design Parameters and Commissioning.

Laboratory exhaust systems should be designed for high reliability and ease of maintenance. One method of achieving this is to provide redundant exhaust fans and to sectionalize equipment so that maintenance work may be performed on an individual exhaust fan

while the system is operating. Another option is to use predictive maintenance procedures to detect problems prior to failure and to allow for scheduling shutdowns for maintenance. To the extent possible, components of exhaust systems should allow maintenance without exposing maintenance personnel to the exhaust airstream. Access to filters and the need for bag-in, bag-out filter housings should be considered during the design process.

Depending on the effluent of the processes being conducted, the exhaust airstream may require filtration, scrubbing, or other emission control to remove environmentally hazardous materials. Any need for emission control devices must be determined early in the design so that adequate space can be provided and cost implications can be recognized.

Types of Exhaust Systems

Laboratory exhaust systems can be constant volume, variable volume, or high-low volume systems with low-, medium-, or high-pressure ductwork. Each fume hood may have its own exhaust fan, or fume hoods may be manifolded and connected to central exhaust fans. Maintenance, functional requirements, and safety must be considered when selecting an exhaust system. Part of the hazard assessment analysis is to determine the appropriateness of variable volume systems and the need for individually ducted exhaust systems. Laboratories with a high hazard potential should be analyzed carefully before variable volume airflow is selected. Airflow monitoring and pressure-independent control may be required even with constant volume systems. In addition, fume hoods or other devices in which extremely hazardous or radioactive materials are used should receive special review to determine whether they should be connected to a manifolded exhaust system.

All exhaust devices installed in a laboratory are seldom used simultaneously at full capacity. This allows the HVAC engineer to conserve energy and, potentially, to reduce equipment capacities by installing a variable volume system that includes an overall system usage factor. The selection of an appropriate usage factor is discussed in the section on Usage Factor.

Manifolded Exhaust Systems. These can be classified as pressure-dependent or pressure-independent. **Pressure-dependent systems** are constant volume only and incorporate manually adjusted balancing dampers for each exhaust device. If an additional fume hood is added to a pressure-dependent exhaust system, the entire system must be rebalanced, and the speed of the exhaust fans may need to be adjusted. Because pressure-independent systems are more flexible, pressure-dependent systems are not common in current designs.

A **pressure-independent system** can be constant volume, variable volume, or a mix of the two. It incorporates pressure-independent volume regulators with each device. The system offers two advantages: (1) the flexibility to add exhaust devices without having to rebalance the entire system and (2) variable volume control.

The volume regulators can incorporate either direct measurement of the exhaust airflow rate or positioning of a calibrated pressure-independent air valve. The input to the volume regulator can be (1) a manual or timed switch to index the fume hood airflow from minimum to operational airflow, (2) sash position sensors, or (3) fume hood cabinet pressure sensors. The section on Control covers this topic in greater detail. Running many exhaust devices into the manifold of a common exhaust system offers the following benefits:

- Lower ductwork cost
- Fewer pieces of equipment to operate and maintain
- Fewer roof penetrations and exhaust stacks
- Opportunity for energy recovery
- Centralized locations for exhaust discharge
- Ability to take advantage of exhaust system diversity
- Ability to provide a redundant exhaust system by adding one spare fan per manifold

Individually Ducted Exhaust Systems. These comprise a separate duct, exhaust fan, and discharge stack for each exhaust device or laboratory. The exhaust fan can be single-speed, multiple-speed, or variable-speed and can be configured for constant volume, variable volume, or a combination of the two. An individually ducted exhaust system has the following potential benefits:

• Provision for installation of special exhaust filtration or treatment systems
• Customized ductwork and exhaust fan corrosion control for specific applications
• Provision for selected emergency power backup
• Simpler initial balancing

Maintaining correct flow at each exhaust fan requires (1) periodic maintenance and balancing and (2) consideration of the flow rates with the fume hood sash in different positions. One problem encountered with individually ducted exhaust systems occurs when an exhaust fan is shut down. In this case, air is drawn in reverse flow through the exhaust ductwork into the laboratory because the laboratory is maintained at a negative pressure.

A challenge in designing independently ducted exhaust systems for multistory buildings is to provide extra vertical ductwork, extra space, and other provisions for the future installation of additional exhaust devices. In multistory buildings, dedicated fire-rated shafts are required from each floor to the penthouse or roof level. As a result, individually ducted exhaust systems (or vertically manifolded systems) consume greater floor space than horizontally manifolded systems. However, less height between floors may be required.

Ductwork Leakage

Ductwork should have low leakage rates and should be tested to confirm that the specified leakage rates have been attained. Leaks from positive pressure exhaust ductwork can contaminate the building, so they must be kept to a minimum. Designs that minimize the amount of positive-pressure ductwork are desirable. All positive-pressure ductwork should be of the highest possible integrity. The fan discharge should connect directly to the vertical discharge stack. Careful selection and proper installation of airtight flexible connectors at the exhaust fans are essential, including providing flexible connectors on the exhaust fan inlet only. Flexible connectors are not used on the discharge side of the exhaust fan because a connector failure could result in the leakage of hazardous fumes into the equipment room. Machine rooms that house exhaust fans should be ventilated to minimize exposure to exhaust effluent (e.g., leakage from the shaft openings of exhaust fans).

The leakage of the containment devices themselves must also be considered. For example, in vertical sash fume hoods, the clearance to allow sash movement creates an opening from the top of the fume hood into the ceiling space or area above. The air introduced through this leakage path also contributes to the exhaust airstream. The amount that such leakage sources contribute to the exhaust airflow depends on the fume hood design. Edge seals can be placed around sash tracks to minimize leaks. Although the volumetric flow of air exhausted through a fume hood is based on the actual face opening, appropriate allowances for air introduced through paths other than the face opening must be included.

Materials and Construction

The selection of materials and the construction of exhaust ductwork and fans depend on the following:

• Nature of the effluents
• Ambient temperature
• Effluent temperature
• Length and arrangement of duct runs
• Constant or intermittent flow
• Flame spread and smoke developed ratings
• Duct velocities and pressures

Effluents may be classified generically as organic or inorganic chemical gases, vapors, fumes, or smoke; and qualitatively as acids, alkalis (bases), solvents, or oils. Exhaust system ducts, fans, dampers, flow sensors, and coatings are subject to (1) corrosion, which destroys metal by chemical or electrochemical action; (2) dissolution, which destroys materials such as coatings and plastics; and (3) melting, which can occur in certain plastics and coatings at elevated temperatures.

Common reagents used in laboratories include acids and bases. Common organic chemicals include acetone, ether, petroleum ether, chloroform, and acetic acid. The HVAC engineer should consult with the safety officer and scientists because the specific research to be conducted determines the chemicals used and therefore the necessary duct material and construction.

The ambient temperature in the space housing the ductwork and fans affects the condensation of vapors in the exhaust system. Condensation contributes to the corrosion of metals, and the chemicals used in the laboratory may further accelerate corrosion.

Ducts are less subject to corrosion when runs are short and direct, the flow is maintained at reasonable velocities, and condensation is avoided. Horizontal ductwork may be more susceptible to corrosion if condensate accumulates in the bottom of the duct. Applications with moist airstreams (cage washers, sterilizers, etc.) may require condensate drains that are connected to process sewers.

If flow through the ductwork is intermittent, condensate may remain for longer periods because it will not be able to reevaporate into the airstream. Moisture can also condense on the outside of ductwork exhausting cold environmental rooms.

Flame spread and smoke developed ratings, which are specified by codes or insurance underwriters, must also be considered when selecting duct materials.

In determining the appropriate duct material and construction the HVAC engineer should

• Determine the types of effluents (and possibly combinations) handled by the exhaust system
• Classify effluents as either organic or inorganic, and determine whether they occur in the gaseous, vapor, or liquid state
• Classify decontamination materials
• Determine the concentration of the reagents used and the temperature of the effluents at the hood exhaust port (this may be impossible in research laboratories)
• Estimate the highest possible dew point of the effluent
• Determine the ambient temperature of the space housing the exhaust system
• Estimate the degree to which condensation may occur
• Determine whether flow will be constant or intermittent (intermittent flow conditions may be improved by adding time delays to run the exhaust system long enough to dry the duct interior prior to shutdown)
• Determine whether insulation, watertight construction, or sloped and drained ductwork are required
• Select materials and construction most suited for the application

Considerations in selecting materials include resistance to chemical attack and corrosion, reaction to condensation, flame and smoke ratings, ease of installation, ease of repair or replacement, and maintenance costs.

Appropriate materials can be selected from standard references and by consulting with manufacturers of specific materials. Materials for chemical fume exhaust systems and their characteristics include the following:

Galvanized steel. Subject to acid and alkali attack, particularly at cut edges and under wet conditions; cannot be field welded without destroying galvanization; easily formed; low in cost.

Stainless steel. Subject to acid and chloride compound attack depending on the nickel and chromium content of the alloy;

relatively high in cost. The most common stainless steel alloys used for laboratory exhaust systems are 304 and 316. Cost increases with increasing chromium and nickel content.

Asphaltum-coated steel. Resistant to acids; subject to solvent and oil attack; high flame and smoke rating; base metal vulnerable when exposed by coating imperfections and cut edges; cannot be field welded without destroying galvanization; moderate cost.

Epoxy-coated steel. Epoxy phenolic resin coatings on mild black steel can be selected for particular characteristics and applications; they have been successfully applied for both specific and general use, but no one compound is inert or resistive to all effluents. Requires sand blasting to prepare the surface for a shop-applied coating, which should be specified as pinhole-free, and field touch-up of coating imperfections or damage caused by shipment and installation; cannot be field welded without destroying coating; cost is moderate.

Polyvinyl-coated galvanized steel. Subject to corrosion at cut edges; cannot be field welded; easily formed; moderate in cost.

Fiberglass. When additional glaze coats are used, this is particularly good for acid applications, including hydrofluoric acid. May require special fire suppression provisions. Special attention to hanger types and spacing is needed to prevent damage.

Plastic materials. Have particular resistance to specific corrosive effluents; limitations include physical strength, flame spread and smoke developed rating, heat distortion, and high cost of fabrication. Special attention to hanger types and spacing is needed to prevent damage.

Borosilicate glass. For specialized systems with high exposure to certain chemicals such as chlorine.

FIRE SAFETY FOR VENTILATION SYSTEMS

Most local authorities have laws that incorporate NFPA *Standard* 45, Fire Protection for Laboratories Using Chemicals. Laboratories located in patient care buildings require fire standards based on NFPA *Standard* 99, Health Care Facilities. NFPA *Standard* 45 design criteria include the following:

Air balance. "The air pressure in the laboratory work areas shall be negative with respect to adjacent corridors and non-laboratory areas." (para 6-3.3)

Controls. "Controls and dampers…shall be of a type that, in the event of failure, will fail in an open position to assure a continuous draft." (para. 6-5.7)

Diffuser locations. "The location of air supply diffusion devices shall be chosen to avoid air currents that would adversely affect performance of laboratory hoods…" (para. 6-3.4)

Fire dampers. "Automatic fire dampers shall not be used in laboratory hood exhaust systems. Fire detection and alarm systems shall not be interlocked to automatically shut down laboratory hood exhaust fans…" (para. 6-10.3)

Hood alarms. "A flow monitor shall be installed on each new laboratory hood." (para.6-8.7.1)

"A flow monitor shall also be installed on existing hoods whenever any modifications or changes are made…" (para. 6-8.7.2)

Hood placement. "For new installations, laboratory hoods shall not be located adjacent to a single means of access or high traffic areas." (para. 6-9.2)

Recirculation. "Air exhausted from laboratory hoods or other special local exhaust systems shall not be recirculated." (para. 6-4.1)

"Air exhausted from laboratory work areas shall not pass unducted through other areas." (para. 6-4.3)

The designer should review the entire NFPA *Standard* 45 and local building codes to determine applicable requirements. Then the designer should inform the other members of the design team of their responsibilities (such as proper fume hood placement).

Incorrect placement of exhaust devices is a frequent design error and a common cause of costly redesign work.

CONTROL

Laboratory controls must regulate temperature and humidity, control and monitor laboratory safety devices that protect personnel, and control and monitor secondary safety barriers used to protect the environment outside the laboratory from laboratory operations (West 1978). Reliability, redundancy, accuracy, and monitoring are important factors in controlling the lab environment. Many laboratories require precise control of temperature, humidity, and airflows; components of the control system must provide the necessary accuracy and corrosion resistance if they are exposed to corrosive environments.

Laboratory controls should provide **fail-safe operation**, which should be defined jointly with the safety officer. A fault tree can be developed to evaluate the impact of the failure of any control system component and to ensure that safe conditions are maintained.

Thermal Control

Temperature in laboratories with a constant volume air supply is generally regulated with a thermostat that controls the position of a control valve on a reheat coil in the supply air. In laboratories with a variable volume ventilation system, room exhaust device(s) are generally regulated as well. The room exhaust device(s) are modulated to handle greater airflow in the laboratory when additional cooling is needed. The exhaust device(s) may determine the total supply air quantity for the laboratory.

Most microprocessor-based laboratory control systems are able to use proportional-integral-derivative (PID) algorithms to eliminate the error between the measured temperature and the temperature set point. Anticipatory control strategies increase accuracy in temperature regulation by recognizing the increased reheat requirements associated with changes in the ventilation flow rates and adjusting the position of reheat control valves before the thermostat measures space temperature changes (Marsh 1988).

Constant Air Volume (CAV) Versus Variable Air Volume (VAV) Room Airflow Control

In the past, the only option for airflow in a laboratory setting was fixed airflow. Many laboratories used chemical fume hoods controlled by on-off switches located at the hood that significantly affected the actual air balance and airflow rate in the laboratory. Now, true CAV or VAV control can be successfully achieved. The question is which system is most appropriate for a contemporary laboratory.

Many laboratories that were considered CAV systems in the past were not truly constant. Even when the fume hoods operated continuously and were of the bypass type, considerable variations in airflow could occur. The variations in airflow resulted from

- Static pressure changes due to filter loading
- Wet or dry cooling coils
- Wear of fan belts that change fan speed
- Position of chemical fume hood sash or sashes
- Outside wind speed and direction
- Position of doors and windows

Current controls can achieve good conformance to the requirements of a CAV system, subject to normal deviations in control performance (i.e., the dead band characteristics of the controller and the hysteresis present in the control system). The same is true for VAV systems, although they are more complex.

Systems may be either uncontrolled or controlled. An uncontrolled CAV system can be designed with no automatic controls associated with airflow other than two-speed fan motors to reduce flow during unoccupied periods. These systems are balanced by means of manual dampers and adjustable drive pulleys. They provide reasonable airflow rates relating to design values but do not provide true CAV under varying conditions, maintain constant fume hood face velocity, or maintain relative static pressures in the spaces. For laboratories that are not considered hazardous and do not have stringent safety requirements, uncontrolled CAV may be satisfactory.

For laboratories housing potentially hazardous operations (i.e., involving toxic chemicals or biological hazards), a true CAV or VAV system ensures that proper airflow and room pressure relationships are maintained at all times. A true CAV system requires volume controls on the supply and exhaust systems.

The principal advantages of a VAV system are its ability to (1) ensure that the face velocities of chemical fume hoods are maintained within a set range and (2) reduce energy use by reducing laboratory airflow. The appropriate safety officer and the users should concur with the choice of a VAV system or a CAV system with reduced airflow during unoccupied periods. Consideration should be given to providing laboratory users with the ability to reset VAV systems to full airflow volume in the event of a chemical spill. Education of the laboratory occupants in proper use of the system is essential. The engineer should recognize that the use of variable volume exhaust systems may result in higher concentrations of contaminants in the exhaust airstream, which may increase corrosion, which influences the selection of materials.

Room Pressure Control

For the laboratory to act as a secondary confinement barrier, the air pressure in the laboratory must be maintained slightly negative with respect to adjoining areas. Exceptions are sterile facilities or clean spaces that may need to be maintained at a positive pressure with respect to adjoining spaces. See Chapter 25, Nuclear Facilities, for examples of secondary containment for negative pressure control.

The common methods of room pressure control include manual balancing, direct pressure, volumetric flow tracking, and cascade control. All methods modulate the same control variable—supply airflow rate; however, each method measures a different variable.

Direct Pressure Control. This method measures the pressure differential across the room envelope and adjusts the amount of supply air into the laboratory to maintain the required differential pressure. Challenges encountered include (1) maintaining the pressure differential when the laboratory door is open, (2) finding suitable sensor locations, (3) maintaining a well-sealed laboratory envelope, and (4) obtaining and maintaining accurate pressure sensing devices. The direct pressure control arrangement requires tightly constructed and compartmentalized facilities and may require a vestibule on entry/exit doors. Engineering parameters pertinent to envelope integrity and associated flow rates are difficult to predict.

Because direct pressure control works to maintain the pressure differential, the control system automatically reacts to transient disturbances. Entry/exit doors may need a switch to disable the control system when they are open. Pressure controls recognize and compensate for unquantified disturbances such as stack effects, infiltration, and influences of other systems in the building. Expensive, complex controls are not required, but the controls must be sensitive and reliable. In noncorrosive environments, controls can support a combination of exhaust applications, and they are insensitive to minimum duct velocity conditions. Successful pressure control provides the desired directional airflow but cannot guarantee a specific volumetric flow differential.

Volumetric Flow Tracking Control. This method measures both the exhaust and supply airflows and controls the amount of supply air to maintain the desired pressure differential. Volumetric control requires that the air at each supply and exhaust point be controlled. It does not recognize or compensate for unquantified disturbances such as stack effects, infiltration, and influences of other systems in the building. Flow tracking is essentially independent of

room door operation. Engineering parameters are easy to predict, and extremely tight construction is not required. Balancing is critical and must be addressed across the full operating range.

Controls may be located in corrosive and contaminated environments; however, the controls may be subject to fouling, corrosive attack, and/or loss of calibration. Flow measurement controls are sensitive to minimum duct velocity conditions. Volumetric control may not guarantee directional airflow.

Cascade Control. This method measures the pressure differential across the room envelope to reset the flow tracking differential set point. Cascade control includes the merits and problems of both direct pressure control and flow tracking control; however, first cost is greater and the control system is more complex to operate and maintain.

Fume Hood Control

Criteria for fume hood control differ depending on the type of hood. The exhaust volumetric flow is kept constant for standard, auxiliary air, and air-bypass fume hoods. In variable volume fume hoods, the exhaust flow is varied to maintain a constant face velocity. Selection of the fume hood control method should be made in consultation with the safety officer.

Constant volume fume hoods can further be split into either pressure-dependent or pressure-independent systems. Although simple in configuration, the pressure-dependent system is unable to adjust the damper position in response to any fluctuation in system pressure across the exhaust damper.

Variable volume fume hood control strategies can be grouped into two categories. The first either measures the air velocity entering a small sensor in the wall of the fume hood or determines face velocity by other techniques. The measured variable is used to infer the average face velocity based on an initial calibration. This calculated face velocity is then used to modulate the exhaust flow rate to maintain the desired face velocity.

The second category of variable volume fume hood control measures the fume hood sash opening and computes the exhaust flow requirement by multiplying the sash opening by the face velocity set point. The controller then adjusts the exhaust device (e.g., by a variable-frequency drive on the exhaust fan or a damper) to maintain the desired exhaust flow rate. The control system may measure the exhaust flow for closed-loop control, or it may use linear calibrated flow control dampers.

STACK HEIGHTS AND AIR INTAKES

Laboratory exhaust stacks should release effluent to the atmosphere without producing undesirable high concentrations at fresh air intakes, operable doors and windows, and locations on or near the building where access is uncontrolled.

Three primary factors that influence the proper disposal of effluent gases are stack/intake separation, stack height, and stack height plus momentum. Chapter 15 of the 1997 *ASHRAE Handbook—Fundamentals* covers the criteria and formulas to calculate the effects of these physical relationships. For complex buildings or buildings with unique terrain or other obstacles to the airflow around the building, either scale model wind tunnel testing or computational fluid dynamics should be considered. However, standard k-ε computational fluid dynamics methods as applied to airflow around buildings need further development (Murakami et al. 1996; Zhou and Stathopoulos 1996).

Stack/Intake Separation

Separation of the stack discharge and air intake locations allows the atmosphere to dilute the effluent. Separation is simple to calculate with the use of short to medium-height stacks; however, to achieve adequate atmospheric dilution of the effluent, greater separation than is physically possible may be required, and the building roof near the stack will be exposed to higher concentrations of the effluent.

Stack Height

Chapter 15 of the 1997 *ASHRAE Handbook—Fundamentals* describes a geometric method to determine the stack discharge height high enough above the turbulent zone around the building that little or no effluent gas impinges on air intakes of the emitting building. The technique is conservative and generally requires tall stacks that may be visually unacceptable or fail to meet building code or zoning requirements. Also, the technique does not ensure acceptable concentrations of effluents at air intakes (e.g., if there are large releases of hazardous materials or elevated intake locations on nearby buildings). A minimum stack height of 10 ft is required by AIHA *Standard* Z9.5 and is recommended by Appendix A of NFPA *Standard* 45.

Stack Height plus Momentum

To increase the effective height of the exhaust stacks, both the volumetric flow and the discharge velocity can be increased to increase the discharge momentum (Momentum Flow = Density × Volumetric Flow × Velocity). The momentum of the large vertical flow in the emergent jet lifts the plume a substantial distance above the stack top, thereby reducing the physical height of the stack and making it easier to screen from view. This technique is particularly suitable when (1) many small exhaust streams can be clustered together or manifolded prior to the exhaust fan to provide the large volumetric flow and (2) outside air can be added through automatically controlled dampers to provide constant exhaust velocity under variable load. The drawbacks to the second arrangement are the amount of energy consumed to achieve the constant high velocity and the added complexity of the controls to maintain constant flow rates. Dilution equations presented in Chapter 15 of the 1997 *ASHRAE Handbook—Fundamentals* or mathematical plume analysis (e.g., Halitsky 1989) can be used to predict the performance of this arrangement, or performance can be validated through wind tunnel testing. Current mathematical procedures tend to have a high degree of uncertainty, and the results should be judged accordingly.

Architectural Screens

Rooftop architectural screens around exhaust stacks are known to adversely affect exhaust dispersion. In general, air intakes should not be placed within the same screen enclosure as laboratory exhausts. Petersen et al. (1997) describe a method of adjusting dilution predictions of Chapter 15 of the 1997 *ASHRAE Handbook—Fundamentals* using a stack height adjustment factor, which is essentially a function of screen porosity.

Criteria for Suitable Dilution

An example criterion based on Halitsky (1988) is that the release of 15 cfm of pure gas through any stack in a moderate wind (3 to 18 mph) from any direction with a near-neutral atmospheric stability (Pasquill Gifford Class C or D) must not produce concentrations exceeding 3 ppm at any air intake. This criterion is meant to simulate an accidental release such as would occur in a spill of an evaporating liquid or after the fracture of the neck of a small lecture bottle of gas in a fume hood.

The intent of this criterion is to limit the concentration of exhausted gases at the air intake locations to levels below the odor thresholds of gases released in fume hoods, excluding highly odorous gases such as mercaptans. Laboratories that use extremely hazardous substances should conduct a chemical-specific analysis based on published health limits. A more lenient limit may be justified for laboratories with low levels of chemical usage. Project-specific requirements must be developed in consultation with the safety officer.

The equations in Chapter 15 of the 1997 *ASHRAE Handbook—Fundamentals* are presented in terms of dilution, defined as the ratio of stack exit concentration to receptor concentration. The exit concentration, and therefore the dilution required to meet the criterion, varies with the total volumetric flow rate of the exhaust stack. For the above criterion with the emission of 15 cfm of a pure gas, a small stack with a total flow rate of 1000 cfm will have an exit concentration of 15/1000 or 15,000 ppm. A dilution of 1:5000 is needed to achieve an intake concentration of 3 ppm. A larger stack with a flow rate of 10,000 cfm will have a lower exit concentration of 15/10,000 or 1500 ppm and would need a dilution of only 1:500 to achieve the 3 ppm intake concentration.

The above criterion is preferred over a simple dilution standard because a defined release scenario (15 cfm) is related to a defined intake concentration (3 ppm) based on odor thresholds or health limits. A simple dilution requirement may not yield safe intake concentrations for a stack with a low flow rate.

APPLICATIONS

LABORATORY ANIMAL FACILITIES

Laboratory animals must be housed in comfortable, clean, temperature- and humidity-controlled rooms. Animal welfare must be considered in the design; the air-conditioning system must provide the microenvironment in the animal's primary enclosure or cage specified by the facility's veterinarian (Woods 1980, Besch 1975, ILAR 1996). Early detailed discussions with the veterinarian concerning airflow patterns, cage layout, and risk assessment help ensure a successful animal room HVAC design. The elimination of research variables (fluctuating temperature and humidity, drafts, and spread of airborne diseases) is another reason for a high-quality air-conditioning system. See Chapter 21 for additional information on environments for laboratory animals.

Temperature and Humidity

Due to the nature of research programs, air-conditioning design temperature and humidity control points may be required. Research animal facilities require more precise environmental control than farm animal or production facilities because variations affect the experimental results. A totally flexible system permits control of the temperature of individual rooms to within ±2°F for any set point in a range of 64 to 85°F. This flexibility requires significant capital expenditure, which can be mitigated by designing the facility for selected species and their specific requirements.

Table 1 lists dry-bulb temperatures recommended by ILAR (1996) for several common species. In the case of animals in confined spaces, the range of daily temperature fluctuations should be kept to a minimum. Relative humidity should also be controlled. ASHRAE *Standard* 62 recommends that the relative humidity in habitable spaces be maintained between 30 and 60% to minimize growh of pathogenic organisms. ILAR (1996) suggests the acceptable range of relative humidity is 30 to 70%.

Table 1 Recommended Dry-Bulb Temperatures for Common Laboratory Animals (ILAR 1996)

Animal	Temperature, °F
Mouse, rat, hamster, gerbil, guinea pig	64 to 79
Rabbit	61 to 72
Cat, dog, nonhuman primate	64 to 84
Farm animals and poultry	61 to 81

Note: The above ranges permit the scientific personnel who will use the facility to select optimum conditions (set points). The ranges do not represent acceptable fluctuation ranges.

Table 2   Heat Generated by Laboratory Animals

Species	Weight, lb	Heat Generation, Btu/h per Normally Active Animal		
		Sensible	Latent	Total
Mouse	0.046	1.11	0.54	1.65
Hamster	0.260	4.02	1.98	6.00
Rat	0.62	7.77	3.83	11.6
Guinea pig	0.90	10.2	5.03	15.2
Rabbit	5.41	39.2	19.3	58.5
Cat	6.61	45.6	22.5	68.1
Nonhuman primate	12.0	71.3	35.1	106
Dog	22.7	105	56.4	161
Dog	50.0	231	124	355

Ventilation

A guideline of 10 to 15 fresh-air changes per hour has been used for secondary enclosures for many years. Although it is effective in many settings, the guideline does not consider the range of possible heat loads; the species, size, and number of animals involved; the type of bedding or frequency of cage changing; the room dimensions; or the efficiency of air distribution from the secondary to the primary enclosure. In some situations, such a flow rate might overventilate a secondary enclosure that contains few animals and waste energy or underventilate a secondary enclosure that contains many animals and allow heat and odor to accumulate.

The air-conditioning load and flow rate for an animal room should be determined by the following factors:

- Desired animal microenvironment (Besch 1975, 1980; ILAR 1996)
- Species of animal(s)
- Animal population
- Recommended ambient temperature (Table 1)
- Heat produced by motors on specialized animal housing units (e.g., laminar flow racks or HEPA-filtered air supply units for ventilated racks)
- Heat generated by the animals (Table 2)

Additional design factors include method of animal cage ventilation; operational use of a fume hood or a biological safety cabinet during procedures such as animal cage cleaning and animal examination; airborne contaminants (generated by animals, bedding, cage cleaning, and room cleaning); and institutional animal care standards (Besch 1980, ILAR 1996). It should be noted that the ambient conditions of the animal room might not reflect the actual conditions within a specific animal cage.

Animal Heat Production

Air-conditioning systems must remove the sensible and latent heat produced by laboratory animals. The literature concerning the metabolic heat production appears to be divergent, but new data are consistent. Current recommended values are given in Table 2. These values are based on experimental results and the following formula:

$$ATHG = 2.5M$$

where

ATHG = average total heat gain, Btu/h per animal
M = metabolic rate of animal, Btu/h per animal = $6.6W^{0.75}$
W = weight of animal, lb

Conditions in animal rooms must be maintained constant. This may require year-round availability of refrigeration and, in some cases, dual/standby chillers and emergency electrical power for motors and control instrumentation. Storage of critical spare parts is one alternative to the installation of a standby refrigeration system.

Design Considerations

If the entire animal facility or extensive portions of it are permanently planned for species with similar requirements, the range of individual adjustments should be reduced. Each animal room or group of rooms serving a common purpose should have separate temperature and humidity controls.

The animal facility and human occupancy areas should be conditioned separately. The human areas may use a return air HVAC system and may be shut down on weekends for energy conservation. Separation prevents exposure of personnel to biological agents, allergens, and odors present in animal rooms.

Control of air pressure in animal housing and service areas is important to ensure directional airflow. For example, quarantine, isolation, soiled equipment, and biohazard areas should be kept under negative pressure, whereas clean equipment and pathogen-free animal housing areas and research animal laboratories should be kept under positive pressure (ILAR 1996).

Supply air outlets should not cause drafts on research animals. Efficient air distribution for animal rooms is essential; this may be accomplished effectively by supplying air through ceiling outlets and exhausting air at floor level (Hessler and Moreland 1984). Supply and exhaust systems should be sized to minimize noise.

A study by Neil and Larsen (1982) showed that predesign evaluation of a full-size mock-up of the animal room and its HVAC system was a cost-effective way to select a system that distributes air to all areas of the animal-holding room. Wier (1983) describes many typical design problems and their resolutions. Room air distribution should be evaluated using ASHRAE *Standard* 113 procedures to evaluate drafts and temperature gradients.

HVAC ductwork and utility penetrations must present a minimum number of cracks in animal rooms so that all wall and ceiling surfaces can be easily cleaned. Exposed ductwork is not generally recommended; however, if constructed of 316 stainless steel in a fashion to facilitate removal for cleaning, it can provide a cost-effective alternative. Joints around diffusers, grilles, and the like should be sealed. Exhaust air grilles with 1 in. washable or disposable filters are normally used to prevent animal hair and dander from entering the ductwork.

Noise from the HVAC system and sound transmission from nearby spaces should be evaluated. Sound control methods should be used required.

Multiple cubicles animal rooms enhance the operational flexibility of the animal room (i.e., housing multiple species in the same room, quarantine, and isolation). Each cubicle should be treated as if it were a separate animal room, with air exchange/balance, temperature, and humidity control.

CONTAINMENT LABORATORIES

With the initiation of biomedical research involving recombinant DNA technology, federal guidelines on laboratory safety were published that influence design teams, researchers, and others. Containment describes safe methods for managing hazardous chemicals and infectious agents in laboratories. The three elements of containment are laboratory operational practices and procedures, safety equipment, and facility design. Thus, the HVAC design engineer helps decide two of the three containment elements during the design phase.

In the United States, the U.S. Department of Health and Human Services (DHHS), Centers for Disease Control and Prevention (CDC), and National Institutes of Health (NIH) classify biological laboratories into four levels—Biosafety Levels 1 through 4—listed in DHHS (1993).

Biosafety Level 1

Biosafety Level 1 is suitable for work involving agents of no known hazard or of minimal potential hazard to laboratory personnel and the environment. The laboratory is not required to be separated from the general traffic patterns in the building. Work may be conducted either on an open benchtop or in a chemical fume hood. Special containment equipment is neither required nor generally used. The laboratory can be cleaned easily and contains a sink for washing hands. The federal guidelines for these laboratories contain no specific HVAC requirements, and typical college laboratories are usually acceptable. Many colleges and research institutions require directional airflow from the corridor into the laboratory, chemical fume hoods, and approximately 3 to 4 air changes per hour of outside air. Directional airflow from the corridor into the laboratory helps to control odors.

Biosafety Level 2

Biosafety Level 2 is suitable for work involving agents of moderate potential hazard to personnel and the environment. DHHS (1993) contains lists that explain the levels of containment needed for various hazardous agents. Laboratory access is limited when certain work is in progress. The laboratory can be cleaned easily and contains a sink for washing hands. Biological safety cabinets (Class I or II) are used whenever

- Procedures with a high potential for creating infectious aerosols are conducted. These include centrifuging, grinding, blending, vigorous shaking or mixing, sonic disruption, opening containers of infectious materials, inoculating animals intranasally, and harvesting infected tissues or fluids from animals or eggs.
- High concentrations or large volumes of infectious agents are used. Federal guidelines for these laboratories contain minimum facility standards.

At this level of biohazard, most research institutions have a full-time safety officer (or safety committee) who establishes facility standards. The federal guidelines for Biosafety Level 2 contain no specific HVAC requirements; however, typical HVAC design criteria can include the following:

- 100% outside air systems
- 6 to 15 air changes per hour
- Directional airflow into the laboratory rooms
- Site-specified hood face velocity at fume hoods (many institutions specify 80 to 100 fpm)
- An assessment of research equipment heat load in a room
- Inclusion of biological safety cabinets

Most biomedical research laboratories are designed for Biosafety Level 2. However, the laboratory director must evaluate the risks and determine the correct containment level before design begins.

Biosafety Level 3

Biosafety Level 3 applies to facilities in which work is done with indigenous or exotic agents that may cause serious or potentially lethal disease as a result of exposure by inhalation. The Biosafety Level 3 laboratory uses a physical barrier of two sets of self-closing doors to separate the laboratory work area from areas with unrestricted personnel access. This barrier enhances biological containment to within the laboratory work area.

The ventilation system must be single-pass, nonrecirculating and configured to maintain the laboratory at a negative pressure relative to surrounding areas. Both audible alarms and visual gages are required to notify personnel if the laboratory pressure relationship changes from a negative condition. The designer may want to have these alarms reported to a constantly monitored location.

Gastight dampers are required in the supply and exhaust ductwork to allow decontamination of the laboratory. The ductwork between these dampers and the laboratory must also be gastight. All penetrations of the Biosafety Level 3 laboratory envelope must be

sealed for containment and to facilitate gaseous decontamination of the work area.

All procedures involving the manipulation of infectious materials are conducted inside biological safety cabinets. The engineer must ensure that the connection of the cabinets to the exhaust system does not adversely affect the performance of the biological safety cabinets or the exhaust system. Refer to the section on Biological Safety Cabinets for further discussion.

The exhaust air from biological safety cabinets and/or the laboratory work area may require HEPA filtration. The need for filtration should be reviewed with the appropriate safety officers. If required, HEPA filters should be equipped with provisions for bag-in, bag-out filter handling systems and gastight isolation dampers for biological decontamination of the filters.

The engineer should review with the safety officer the need for special exhaust or filtration of exhaust from any scientific equipment located in the Biosafety Level 3 laboratory.

Biosafety Level 4

Biosafety Level 4 is required for work with dangerous and exotic agents that pose a high risk of aerosol-transmitted laboratory infections and life-threatening disease. The design of HVAC systems for these areas will have stringent requirements that must be determined by the biological safety officer.

SCALE-UP LABORATORIES

Scale-up laboratories are defined differently depending on the nature and volume of work being conducted. For laboratories performing recombinant DNA research, large-scale experiments involve 2.5 gal or more. A chemical or biological laboratory is defined as scale-up when the principal holding vessels are glass or ceramic. When the vessels are constructed primarily of metals, the laboratory is considered a pilot plant, which this chapter does not currently address.

The amount of experimental materials present in scale-up laboratories is generally significantly greater than the amount found in the small-scale laboratory. The experimental equipment is also larger and therefore requires more space. The result is larger chemical fume hoods or reaction cubicles that may be of the walk-in type. Significantly higher laboratory airflow rates are needed to maintain the face velocity of the chemical fume hoods or reaction cubicles, although their size frequently presents problems of airflow uniformity over the entire face area. Walk-in hoods are sometimes entered during an experimental run, so provisions for breathing-quality air stations and other forms of personnel protection should be considered. Environmental containment or the ability to decontaminate the laboratory, the laboratory exhaust airstream, or other effluent may be needed in the event of an upset. Scale-up laboratories are frequently airflow intensive and at times may be in operation for sustained periods.

TEACHING LABORATORIES

Laboratories in academic settings can generally be classified as those used for instruction and those used for research. Research laboratories vary significantly depending on the work being performed; they generally fit into one of the categories of laboratories described previously.

The design requirements for teaching laboratories also vary based on their function. The designer should become familiar with the specific teaching program, so that a suitable hazard assessment can be made. For example, the requirements for the number and size of fume hoods vary greatly between undergraduate inorganic and graduate organic chemistry teaching laboratories.

Unique aspects of teaching laboratories include the need of the instructor to be in visual contact with the students at their work stations and to have ready access to the controls for the fume hood operations and any safety shutoff devices and alarms. Frequently, students have not received extensive safety instruction, so easily understood controls and labeling are necessary. Because the teaching environment depends on verbal communication, sound from the building ventilation system is an important concern.

CLINICAL LABORATORIES

Clinical laboratories are found in hospitals and as stand-alone operations. The work in these laboratories generally consists of handling human specimens (blood, urine, etc.) and using chemical reagents for analysis. Some samples may be infectious; because it is impossible to know which samples may be contaminated, good work practices require that all be handled as biohazardous materials. The primary protection of the staff at clinical laboratories depends on the techniques and laboratory equipment (e.g., biological safety cabinets) used to control aerosols, spills, or other inadvertent releases of samples and reagents. People outside the laboratory must also be protected.

Additional protection can be provided by the building HVAC system with suitable exhaust, ventilation, and filtration. The HVAC engineer is responsible for providing an HVAC system that meets the biological and chemical safety requirements. The engineer should consult with appropriate senior staff and safety professionals to ascertain what potentially hazardous chemical or biohazardous conditions will be in the facility and then provide suitable engineering controls to minimize risks to the staff and the community. The use of biological safety cabinets, chemical fume hoods, and other specific exhaust systems should be considered by appropriate laboratory staff and the design engineer.

RADIOCHEMISTRY LABORATORIES

In the United States, laboratories located in Department of Energy (DOE) facilities are governed by DOE regulations. All other laboratories using radioactive materials are governed by the Nuclear Regulatory Commission (NRC), state, and local regulations. Other agencies may be responsible for the regulation of other toxic and carcinogenic materials present in the facility.

Laboratory containment equipment for nuclear processing facilities are treated as primary, secondary, or tertiary containment zones, depending on the level of radioactivity anticipated for the area and the materials to be handled. Chapter 25 has additional information on nuclear laboratories.

OPERATION AND MAINTENANCE

Due to long-term research studies, laboratories may need to maintain design performance conditions with no interruptions for long periods. Even when research needs are not so demanding, systems that maintain air balance, temperature, and humidity in laboratories must be highly reliable, with a minimal amount of downtime. The designer should work with operation and maintenance personnel as well as users early in the design of systems to gain their input and agreement.

System components must be of adequate quality to achieve reliable HVAC operation, and they should be reasonably accessible for maintenance. Laboratory work surfaces should be protected from possible leakage of coils, pipes, and humidifiers. Changeout of supply and exhaust filters should require minimum downtime.

Centralized monitoring of laboratory variables (e.g., pressure differentials, face velocity of fume hoods, supply flows, and exhaust flows) is useful for predictive maintenance of equipment and for ensuring safe conditions.

For their safety, laboratory users should be instructed in the proper use of laboratory fume hoods, safety cabinets, ventilated enclosures, and local ventilation devices. They should be trained to understand the operation of the devices and the indicators and alarms that show whether they are safe to operate. Users should request periodic testing of the devices to ensure that they and the connected ventilation systems are operating properly.

Personnel that know the nature of the contaminants in a particular laboratory should be responsible for decontamination of equipment and ductwork before they are turned over to maintenance personnel for work.

Maintenance personnel should be trained to keep laboratory systems in good operating order and should understand the critical safety requirements of those systems. Preventive maintenance of equipment and periodic checks of air balance should be scheduled. High-maintenance items should be placed outside the actual laboratory (in service corridors or interstitial space) to reduce disruption of laboratory operations and exposure of the maintenance staff to laboratory hazards. Maintenance personnel must be aware of and trained in procedures for maintaining good indoor air quality (IAQ) in laboratories. Many IAQ problems have been traced to poor maintenance due to poor accessibility (Woods et al. 1987).

ENERGY

Due to the nature of the functions they support, laboratory HVAC systems consume large amounts of energy. Efforts to reduce energy use must not compromise the safety standards established by safety officers. Typically, HVAC systems supporting laboratories and animal areas use 100% outside air and operate continuously. All HVAC systems serving laboratories can benefit from energy reduction techniques that are either an integral part of the original design or added later. Energy reduction techniques should be analyzed in terms of both appropriateness to the facility and economic payback.

The HVAC engineer must understand and respond to the scientific requirements of the facility. Research requirements typically include continuous control of temperature, humidity, relative static pressure, and air quality. Energy reduction systems must maintain the required environmental conditions during both occupied and unoccupied modes.

Energy Conservation

Energy can be conserved in laboratories by reducing the exhaust air requirements. For example, the exhaust air requirements for fume hoods can be reduced by closing part of the hood opening during operation, thereby reducing the airflow needed to obtain the desired capture velocities (an exception is bypass hoods, which require similar quantities of exhaust air whether open or fully closed). The sash styles that may be adjusted are described in the section on Laboratory Exhaust and Containment Devices.

Another way to reduce exhaust airflow is to use variable volume control of exhaust air through the fume hoods to reduce exhaust airflow when the fume hood sash is not fully open. A variation of this arrangement incorporates user-initiated selection of the fume hood airflow from a minimum flow rate to a maximum flow rate when the hood is in use. Any airflow control must be integrated with the laboratory control system, described in the section on Control, and must not jeopardize the safety and function of the laboratory.

A third energy conservation method uses night setback controls when the laboratory is unoccupied to reduce the exhaust volume to one-quarter to one-half the minimum required when the laboratory is occupied. Timing devices, sensors, manual override, or a combination of these can be used to set back the controls at night. If this strategy is a possibility, the safety and function of the laboratory must be considered, and appropriate safety officers should be consulted.

Energy Recovery

Energy can often be recovered economically from the exhaust airstream in laboratory buildings with large quantities of exhaust air. Many energy recovery systems are available, including rotary air-to-air energy exchangers or heat wheels, coil energy recovery loops (runaround cycle), twin tower enthalpy recovery loops, heat pipe heat exchangers, fixed-plate heat exchangers, thermosiphon heat exchangers, and direct evaporative cooling. Some of these technologies can be combined with indirect evaporative cooling for further energy recovery. See Chapter 42 of the 1996 *ASHRAE Handbook—Systems and Equipment* for more information.

Concerns about the use of energy recovery devices in laboratory HVAC systems include (1) the potential for cross-contamination of chemical and biological materials from the exhaust air to the intake airstream and (2) the potential for corrosion and fouling of the devices located in the exhaust airstream. NFPA *Standard* 45 specifically prohibits the use of latent heat recovery devices in fume hood exhaust systems.

Energy recovery is also possible for hydronic systems associated with HVAC. Rejected heat from centrifugal chillers can be used to produce low-temperature reheat water. Potential also exists in plumbing systems, where waste heat from washing operations can be recovered to heat makeup water.

COMMISSIONING

In addition to HVAC systems, electrical systems and chemical handling and storage areas must be commissioned. Training of technicians, scientists, and maintenance personnel is a critical aspect of the commissioning process. Users must understand the systems and their operation.

It should be determined early in the design process whether any laboratory systems must comply with Food and Drug Administration (FDA) regulations because these systems have additional design and commissioning requirements.

Commissioning is defined in Chapter 41, and the process is outlined in ASHRAE *Guideline* 1. Laboratory commissioning is more demanding than that described in ASHRAE *Guideline* 1 because areas must be considered that are not associated with the normal office complex. Requirements for commissioning should be clearly understood by all participants, including the contractors and the owner's personnel. Roles and responsibilities should be defined, and responsibilities for documenting results should be established.

Laboratory commissioning starts with the intended use of the laboratory and should include development of a commissioning plan, as outlined in ASHRAE *Guideline* 1. The validation of individual components should come first; after individual components are successfully validated, the entire system should be evaluated. This requires verification and documentation that the design meets applicable codes and standards and that it has been constructed in accordance with the design intent. Before general commissioning begins, the following data must be obtained:

- Complete set of the laboratory utility drawings
- Definition of the use of the laboratory and an understanding of the work being performed
- Equipment requirements
- All test results
- Understanding of the intent of the system operation

For HVAC system commissioning, the following should be verified and documented:

- Fume hood design face velocities have been met.
- Manufacturer's requirements for airflows for biological safety cabinets and laminar flow clean benches have been met.

- Exhaust system configuration, damper locations, and performance characteristics, including any required emission equipment, are correct.
- Control system operates as specified. Controls include fume hood alarm; miscellaneous safety alarm systems; fume hood and other exhaust airflow regulation; laboratory pressurization control system; laboratory temperature control system; and main ventilation unit controls for supply, exhaust, and heat recovery systems. Control system performance verification should include speed of response, accuracy, repeatability, turndown, and stability.
- Desired laboratory pressurization relationships are maintained throughout the laboratory, including entrances, adjoining areas, air locks, interior rooms, and hallways. Balancing terminal devices within 10% of design requirements will not provide adequate results. Additionally, internal pressure relationships can be affected by airflow around the building itself. See Chapter 15 of the 1997 *ASHRAE Handbook—Fundamentals* for more information.
- Fume hood containment performance is within specification. ASHRAE *Standard* 110 provides criteria for this evaluation.
- Dynamic response of the laboratory's control system is satisfactory. One method of testing the control system is to open and shut laboratory doors during fume hood performance testing.
- System fault tree and failure modes are as specified.
- Standby electrical power systems function properly.
- Design noise criterion) (NC levels of occupied spaces have been met.

ECONOMICS

In laboratories, HVAC systems make up a significant part (often 30 to 50%) of the overall construction budget. The design criteria and system requirements must be reconciled with the budget allotment for HVAC early in the planning stages and continually throughout the design stages to ensure that the project remains within budget.

Every project must be evaluated on both its technical features and its economics. The following common economic terms are discussed in Chapter 35 and defined here as follows:

Initial cost. Cost to design, install, and test an HVAC system such that it is fully operational and suitable for use.

Operating cost. Cost to operate a system (including energy, maintenance, and component replacements) such that the total system can reach the end of its normal useful life.

Life-cycle cost. Cost related to the total cost over the life of the HVAC system, including capital cost, considering the time value of money.

Mechanical and electrical costs related to HVAC systems are commonly assigned a depreciation life based on current tax policies. This depreciation life may be different from the projected functional life of the equipment, which is influenced by the quality of the system components and of the maintenance they receive. Some portions of the system, such as ductwork, could last the full life of the building. Other components, such as air-handling units, may have a useful life of 15 to 30 years, depending on their original quality and ongoing maintenance efforts. Estimated service life of equipment is listed in Chapter 35.

Engineering economics can be used to evaluate life-cycle costs of configuration (utility corridor versus interstitial space), systems, and major equipment. The user or owner makes a business decision concerning quality of the system and its ongoing operating costs. The HVAC engineer may be asked to provide an objective analysis of energy, maintenance, and construction costs, so that an appropriate life-cycle cost analysis can be made. Other considerations that may be appropriate include economic influences related to the long-term use of energy and governmental laws and regulations.

Many technical considerations and the great variety of equipment available influence the design of HVAC systems. Factors affecting design must be well understood to ensure appropriate comparisons between various systems and to determine the impact on either first or operating costs.

REFERENCES

AIHA. 1993. Laboratory ventilation. ANSI/AIHA *Standard* Z9.5-93. American Industrial Hygiene Association, Fairfax, VA.

Alereza, T. and J. Breen, III. 1984. Estimates of recommended heat gains due to commercial appliances and equipment. *ASHRAE Transactions* 90(2A):25-58.

ASHRAE. 1990. Method of testing for room air diffusion. ANSI/ASHRA *Standard* 113-1990.

ASHRAE. 1992. Gravimetric and dust-spot procedures for testing air-cleaning devices used in general ventilation for removing particulate matter. ANSI/ASHRAE *Standard* 52.1-1992.

ASHRAE. 1995. Method of testing performance of laboratory fume hoods. ANSI/ASHRAE *Standard* 110-1995.

ASHRAE. 1996. Guideline for the HVAC commissioning process. *Guideline* 1-1996.

Besch, E. 1975. Animal cage room dry bulb and dew point temperature differentials. *ASHRAE Transactions* 81(2):459-58.

Besch, E. 1980. Environmental quality within animal facilities. *Laboratory Animal Science* 30(2II):385-406.

Caplan, K. and G. Knutson. 1977. The effect of room air challenge on the efficiency of laboratory fume hoods. *ASHRAE Transactions* 83(1):141-56.

Caplan, K. and G. Knutson. 1978. Laboratory fume hoods: Influence of room air supply. *ASHRAE Transactions* 84(1):511-37.

DHHS. 1993. Biosafety in microbiological and biomedical laboratories, 3rd ed. Publication No. (CDC) 93-8395. U.S. Department of Health and Human Services, NIH, Bethesda, MD.

Eagleston, J., Jr. 1984. Aerosol contamination at work. In *The international hospital federation yearbook*. Sabrecrown Publishing, London.

Halitsky, J. 1988. Dispersion of laboratory exhaust gas by large jets, 81st Annual Meeting of the Air Pollution Control Association, June, Dallas.

Halitsky, J. 1989. A jet plume model for short stacks. *APCA Journal* 39(6).

Hessler, J. and A. Moreland. 1984. Design and management of animal facilities. In *Laboratory animal medicine*, J. Fox, B. Cohen, F. Loew, eds. Academic Press, San Diego, CA.

ILAR. 1996. *Guide for the care and use of laboratory animals.* Institute of Laboratory Animal Resources, National Academy of Sciences, National Academy Press, Washington, D.C.

Marsh, C.W. 1988. DDC systems for pressurization, fume hood face velocity and temperature control in variable air volume laboratories. *ASHRAE Transactions* 94(2):1947-68.

Murakami, S., A. Mochida, R. Ooka, S. Kato, and S. Iizuka. 1996. Numerical prediction of flow around buildings with various turbulence models: Comparison of k-ϵ, EVM, ASM, DSM, and LES with wind tunnel tests. *ASHRAE Transactions* 102(1):741-53.

NFPA. 1994. Gaseous hydrogen systems at consumer sites. ANSI/NFPA *Standard* 50A-94. National Fire Protection Association, Quincy, MA.

NFPA. 1994. Guide for venting of deflagrations. ANSI/NFPA *Standard* 68-94.

NFPA. 1996. Fire protection for laboratories using chemicals. ANSI/NFPA *Standard* 45-96.

NFPA. 1996. *Flammable and combustible liquids code.* ANSI/NFPA *Standard* 30-96.

NFPA. 1996. Health care facilities. ANSI/NFPA *Standard* 99-96.

NFPA. 1996. Installation of air conditioning and ventilating systems. ANSI/NFPA *Standard* 90A-96.

NFPA. 1996. *National electrical code.* ANSI/NFPA *Standard* 70-96.

NFPA. 1996. Water spray fixed systems for fire protection. ANSI/NFPA *Standard* 15-96

NFPA. 1998. *Liquefied petroleum gas code. Standard* 58-98.

NFPA. 1998. Storage, use, and handling of compressed and liquefied gases in portable cylinders. *Standard* 55-98.

NIH. 1996a. Research laboratory design policy and guidelines. Office of Research Services, National Institutes of Health, Bethesda, MD.

NIH. 1996b. Vivarium design policy and guidelines. Office of Research Services, National Institutes of Health, Bethesda, MD.

NSF. 1992. Class II (laminar flow) biohazard cabinetry. *Standard* 49-92. NSF International, Ann Arbor, MI.

Neil, D. and R. Larsen. 1982. How to develop cost-effective animal room ventilation: Build a mock-up. *Laboratory Animal Science* (Jan-Feb): 32-37.

Petersen, R.L., J. Carter, and M. Ratcliff. 1997. The influence of architectural screens on exhaust dilution. ASHRAE *Research Project* RP-805. Draft Report approved by Technical Committee June 1997.

Rake, B. 1978. Influence of crossdrafts on the performance of a biological safety cabinet. *Applied and Environmental Microbiology* (August): 278-83.

SEFA. 1996. Laboratory fume hoods recommended practices. SEFA 1-96. Scientific Equipment and Furniture Association, Hilton Head, SC.

Stuart, D., M. First, R. Rones, and J. Eagleston. 1983. Comparison of chemical vapor handling by three types of Class II biological safety cabinets. *Particulate & Microbial Control* (March/April).

West, D.L. 1978. Assessment of risk in the research laboratory: A basis for facility design. *ASHRAE Transactions* 84(1):547-57.

Wier, R.C. 1983. Toxicology and animal facilities for research and development. *ASHRAE Transactions* 89(2B):533-41.

Woods, J. 1980. The animal enclosure—A microenvironment. *Laboratory Animal Science* 30(2II):407-13.

Woods, J., J. Janssen, P. Morey, and D. Rask. 1987. Resolution of the "sick" building syndrome. *Proceedings* of ASHRAE Conference: Practical Control of Indoor Air Problems, pp. 338-48.

Zhou, Y. and T. Stathopoulos. 1996. Application of two-layer methods for the evaluation of wind effects on a cubic building. *ASHRAE Transactions* 102(1):754-64.

BIBLIOGRAPHY

Abramson, B. and T. Tucker. 1988. Recapturing lost energy. *ASHRAE Journal* 30(6):50-52.

Adams, J.B., Jr. 1989. Safety in the chemical laboratory: Synthesis—Laboratory fume hoods. *Journal of Chemical Education* 66(12).

Ahmed, O. and S.A Bradley. 1990. An approach to determining the required response time for a VAV fume hood control system. *ASHRAE Transactions* 96(2):337-42.

Ahmed, O., J.W. Mitchell, and S.A. Klein. 1993. Dynamics of laboratory pressurization. *ASHRAE Transactions* 99(2):223-29.

Albern, W., F. Darling, and L. Farmer. 1988. Laboratory fume hood operation. *ASHRAE Journal* 30(3):26-30.

Anderson, S. 1987. Control techniques for zoned pressurization. *ASHRAE Transactions* 93(2B):1123-39.

Anderson, C.P. and K.M. Cunningham. 1988. HVAC controls in laboratories—A systems approach. *ASHRAE Transactions* 94(1):1514-20.

ASHRAE. 1989. Ventilation for acceptable indoor air quality. *Standard* 62-1989.

Barker, K.A., O. Ahmed, and J.A. Parker. 1993. A methodology to determine laboratory energy consumption and conservation characteristics using an integrated building automation system. *ASHRAE Transactions* 99(2):1155-67.

Baylie, C.L. and S.H. Schultz. 1994. Manage change: Planning for the validation of HVAC Systems for a clinical trials production facility. *ASHRAE Transactions* 100(1):1660-68.

Bertoni, M. 1987. Risk management considerations in design of laboratory exhaust stacks. *ASHRAE Transactions* 93(2B):2149-64.

Bossert, K.A. and S.M. McGinley. 1994. Design characteristics of clinical supply laboratories relating to HVAC systems. *ASHRAE Transactions* 94(100):1655-59.

Brow, K. 1989. AIDS research laboratories—HVAC criteria. In *Building Systems: Room Air and Air Contaminant Distribution*, L.L. Christianson, ed. ASHRAE cosponsored *Symposium*, pp. 223-25.

Brown, W.K. 1993. An integrated approach to laboratory energy efficiency. *ASHRAE Transactions* 99(2):1143-54.

Carnes, L. 1984. Air-to-air heat recovery systems for research laboratories. *ASHRAE Transactions* 90(2A):327.

Code of Federal Regulations. (Latest edition). Good laboratory practices. CFR 21 Part 58. U.S. Government Printing Office, Washington, D.C.

Coogan, J.J. 1994. Experience with commissioning VAV laboratories. *ASHRAE Transactions* 100(1):1635-40.

Crane, J. 1994. Biological laboratory ventilation and architectural and mechanical implications of biological safety cabinet selection, location, and venting. *ASHRAE Transactions* 100(1):1257-65.

CRC. 1989. *CRC handbook of laboratory safety*, 3rd ed. CRC Press, Boca Raton, FL.

Dahan, F. 1986. HVAC systems for chemical and biochemical laboratories. *Heating, Piping and Air Conditioning* (May):125-30.

Davis, S. and R. Benjamin. 1987. VAV with fume hood exhaust systems. *Heating, Piping and Air Conditioning* (August):75-78.

Degenhardt, R. and J. Pfost. 1983. Fume hood system design and application for medical facilities. *ASHRAE Transactions* 89(2B):558-70.

DHHS. 1981. NIH guidelines for the laboratory use of chemical carcinogens. NIH *Publication No.* 81-2385. Department of Health and Human Services, National Institutes of Health, Bethesda, MD.

DiBeradinis, L., J. Baum, M. First, G. Gatwood, E. Groden, and A. Seth. 1992. *Guidelines for laboratory design: Health and safety considerations.* John Wiley and Sons, Boston.

Doyle, D.L., R.D. Benzuly, and J.M. O'Brien. 1993. Variable volume retrofit of an industrial research laboratory. *ASHRAE Transactions* 99(2): 1168-80.

Flanherty, R.J. and R. Gracilieri. 1994. Documentation required for the validation of HVAC systems. *ASHRAE Transactions* 100(1):1629-34.

Ghidoni, D.A. and R.L. Jones, Jr. 1994. Methods of exhausting a BSC to an exhaust system containing a VAV component. *ASHRAE Transactions* 100(1):1275-81.

Hayter, R.B. and R.L. Gorton. 1988. Radiant cooling in laboratory animal caging. *ASHRAE Transactions* 94(1):1834-47.

Hitchings, D.T. and R.S. Shull 1993. Measuring and calculating laboratory exhaust diversity—Three case studies. *ASHRAE Transactions* 99(2): 1059-71.

ILAR. 1977. Laboratory animal management—Rodents. *ILAR News* 20(3).

ILAR. 1978. Laboratory animal management—Cats. *ILAR News* 21(3).

ILAR. 1980. Laboratory animal management—Nonhuman primates. *ILAR News* 23(2-3).

Knutson, G. 1984. Effect of slot position on laboratory fume hood performance. *Heating, Piping and Air Conditioning* (February):93-96.

Knutson, G. 1987. Testing containment laboratory hoods: A field study. *ASHRAE Transactions* 93(2B):1801-12.

Koenigsberg, J. and H. Schaal. 1987. Upgrading existing fume hood installations. *Heating, Piping and Air Conditioning* (October):77-82.

Koenigsberg, J. and E. Seipp. 1988. Laboratory fume hood—An analysis of this special exhaust system in the post "Knutson-Caplan" era. *ASHRAE Journal* 30(2):43-46.

Laboratory & health facilities programming & planning. Planning *Monograph* 1.0. Planning Collaborative Ltd., Reston, VA.

Lacey, D.R. 1994. HVAC for a low temperature biohazard facility. *ASHRAE Transactions* 100(1):1282-86.

Lentz, M.S. and A.K. Seth. 1989. A procedure for modeling diversity in laboratory VAV systems. *ASHRAE Transactions* 95(1):114-20.

Maghirang, R.G., G.L. Riskowski, P.C. Harrison, H.W. Gonyou, L. Sebek, and J. McKee. 1994. An individually ventilated caging system for laboratory rats. *ASHRAE Transactions* 100(1):913-20.

Maust, J. and R. Rundquist. 1987. Laboratory fume hood systems—Their use and energy conservation. *ASHRAE Transactions* 93(2B):1813-21.

McDiarmid, M.D. 1988. A quantitative evaluation of air distribution in full scale mock-ups of animal holding rooms. *ASHRAE Transactions* 94(1B):685-93.

Mikell, W. and F. Fuller. 1988. Safety in the chemical laboratory: Good hood practices for safe hood operation. *Journal of Chemical Education* 65(2).

Moyer, R.C. 1983. Fume hood diversity for reduced energy consumption. *ASHRAE Transactions* 89(2B):552-57.

Moyer, R. and J. Dungan. 1987. Turning fume hood diversity into energy savings. *ASHRAE Transactions* 93(2B):1822-34.

Murray, W., A. Streifel, T. O'Dea, and F. Rhame. 1988. Ventilation for protection of immune compromised patients. *ASHRAE Transactions* 94(1):1185-92.

NIH. 1996. Clinical center design policy and guidelines. Office of Research Services, National Institutes of Health, Bethesda, MD.

NIH. 1996. Reference material for the design policy and guidelines. Office of Research Services, National Institutes of Health, Bethesda, MD.

Neuman, V. 1989. Design considerations for laboratory HVAC system dynamics. *ASHRAE Transactions* 95(1):121-24.

Neuman, V. 1989. Disadvantages of auxiliary air fume hoods. *ASHRAE Transactions* 95(1):70-75.

Neuman, V. 1989. Health and safety in laboratory plumbing. *Plumbing Engineering* (March):21-24.

Neuman, V. and H. Guven. 1988. Laboratory building HVAC systems optimization. *ASHRAE Transactions* 94(2):432-51.

Neuman, V. and W. Rousseau. 1986. VAV for laboratory hoods—Design and costs. *ASHRAE Transactions* 92(1A):330-46.

Neuman, V., F. Sajed, and H. Guven. 1988. A comparison of cooling thermal storage and gas air conditioning for a lab building. *ASHRAE Transactions* 94(2):452-68.

Parker, J.A., O. Ahmed, and K.A. Barker. 1993. Application of building automation system (BAS) in evaluating diversity and other characteristics of a VAV laboratory. *ASHRAE Transactions* 99(2):1081-89.

Peterson, R. 1987. Designing building exhausts to achieve acceptable concentrations of toxic effluents. *ASHRAE Transactions* 93(2):2165-85.

Peterson, R.L., E.L. Schofer, and D.W. Martin. 1983. Laboratory air systems—Further testing. *ASHRAE Transactions* 89(2B):571-96.

Pike, R. 1976. Laboratory-associated infections: Summary and analysis of 3921 cases. *Health Laboratory Science* 13(2):105-14.

Rabiah, T.M. and J.W. Wellenbach. 1993. Determining fume hood diversity factors. *ASHRAE Transactions* 99(2):1090-96.

Rhodes, W.W. 1988. Control of microbioaerosol contamination in critical areas in the hospital environment. *ASHRAE Transactions* 94(1):1171-82.

Richardson, G. 1994. Commissioning of VAV laboratories and the problems encountered. *ASHRAE Transactions* 100(1):1641-45.

Rizzo, S. 1994. Commissioning of laboratories: A case study. *ASHRAE Transactions* 100(1):1646-52.

Sandru, E. 1996. Evaluation of the laboratory equipment Component of cooling loads. *ASHRAE Transactions* 102(1):732-37

Schuyler, G. and W. Waechter. 1987. Performance of fume hoods in simulated laboratory conditions. Report No. 487-1605 by Rowan Williams Davies & Irwin, Inc., under contract for Health and Welfare Canada.

Schwartz, Leonard. 1994. Heating, ventilating and air conditioning considerations for pharmaceutical companies. *Pharmaceutical Engineering* 14(4).

Sessler, S. and R. Hoover. 1983. Laboratory fume hood noise. *Heating, Piping, and Air Conditioning* (September):124-37.

Simons, C.G. 1991. Specifying the correct biological safety cabinet. *ASHRAE Journal* 33(8).

Simons, C.G. and R. Davoodpour. 1994. Design considerations for laboratory facilities using molecular biology techniques. *ASHRAE Transactions* 100(1):1266-74.

Smith, W. 1994. Validating the direct digital control (DDC) system in a clinical supply laboratory. *ASHRAE Transactions* 100(1):1669-75.

Streets, R.A. and B.S.V. Setty. 1983. Energy conservation in institutional laboratory and fume hood systems. *ASHRAE Transactions* 89(2B): 542-51.

Stuart, D., R. Greenier, R. Rumery, and J. Eagleson. 1982. Survey, use, and performance of biological safety cabinets. *American Industrial Hygiene Association Journal* 43:265-70.

Varley, J.O. 1993. The measurement of fume hood use diversity in an industrial laboratory. *ASHRAE Transactions* 99(2):1072-80.

Wilson, D.J. 1983. A design procedure for estimating air intake contamination from nearby exhaust vents. *ASHRAE Transactions* 89(2A):136.

Yoshida, K., H. Hachisu, J.A. Yoshida, and S. Shumiya. 1994. Evaluation of the environmental conditions in a filter-capped cage using a one-way airflow system. *ASHRAE Transactions* 100(1):901-905.

ENGINE TEST FACILITIES

INDUSTRIAL testing of turbine and internal combustion engines is performed in enclosed test spaces to control noise and isolate the test for safety or security. These spaces are ventilated or conditioned to control the facility environment and fumes. Isolated engines are tested in test cells; engines situated inside automobiles are tested on chassis dynamometers. The ventilation and safety principals for test cells also apply when large open areas in the plant are used for production testing and emissions measurements.

Enclosed test cells are normally found in research or emissions test facilities. Test cells may require instruments to measure cooling system water flow and temperature; exhaust gas flow, temperature, and emission concentrations; fuel flow; power output; and combustion air volume and temperature. Changes in the temperature and humidity of the test cell affect these measurements. Accurate control of the testing environment is becoming more critical. For example, the U.S. Environmental Protection Agency requires tests to demonstrate control of automobile contaminants in both hot and cold environments.

Air conditioning and ventilation of test cells must (1) supply and exhaust proper quantities of air to remove heat and control temperature; (2) exhaust sufficient air at proper locations to prevent buildup of combustible vapors; (3) supply and modulate large quantities of air to meet changing conditions; (4) remove exhaust fumes; (5) supply combustion air; (6) prevent noise transmission through the system; and (7) provide for human comfort and safety during setup, testing, and tear-down procedures. Supply and exhaust systems for test cells may be unitary, central, or a combination of the two. Mechanical exhaust is necessary in all cases.

ENGINE HEAT RELEASE

The special air-conditioning requirements of an engine test facility stem from the burning of the fuel used to run the engine. The heat released by an engine divides into work done, exhaust, cooling water, and radiation. For internal combustion engines at full load, 10% of the total heat content of the fuel is radiated and convected into the room or test cell atmosphere and 90% is fairly evenly divided between the shaft output (work), exhaust gas heating, and heating of the jacket cooling water.

Air-cooled engines create a forced convection load on the test space equal to the jacket water heat that it replaces. For turbine engines, the exhaust gas carries double the heat of the internal combustion engine exhaust and there is no jacket water to heat. The engine manufacturer can provide a more precise analysis of the heat release characteristics at various speeds and power outputs.

Test facilities use dynamometers to determine the power being supplied to the engine shaft. The dynamometer converts the shaft work into heat that must be accounted for by a cooling system or as heat load into the space. Often the shaft work is converted into electricity through a generator and the electric power is dissipated by a resistance load bank. Inefficiencies of the various pieces of equipment add to the load of the space in which they are located.

The preparation of this chapter is assigned to TC 9.2, Industrial Air Conditioning.

Heat released into the jacket water must also be removed. If a closely connected radiator is used, the heat load is added to the room load. Many test facilities include a heat exchanger and a secondary cooling circuit transfers the heat to a cooling tower. Some engines require an oil cooler separate from the jacket water. Whichever system is used, the cooling water flow, temperature, and pressure is usually monitored as part of the test operation and heat from these sources need to be accommodated by the facility's air conditioning.

Exhaust systems present several challenges to the design of an engine test cell. Exhaust gases can exit the engine at 1500°F or higher. Commonly the exhaust gas is augmented by inserting the exhaust pipe into a larger bore exhaust system (laboratory fixed system). The exhaust gas draws room air into the exhaust to both cool the gas and ventilate the test cell. Both the exhausted room air and combustion air must be supplied to the room from the HVAC or from outside.

Radiation and convection from the exhaust pipes, catalytic converter, muffler, etc. also add to the load. In most cases the test cell HVAC system should account for an engine that can fully load the dynamometer. It should have capacity control for operation at partial load and no load conditions.

Large gas turbine engines have unique noise and airflow requirements, and as such they usually are provided with dedicated test cells. Small gas turbines can often be tested in a regular engine test cell with minor modifications.

ENGINE EXHAUST

Engine exhaust systems remove combustible products, unburned fuel vapors, and water vapor. Flow loads and the operating pressure need to be established for the design of the supporting HVAC.

Flow loads are calculated based on the number of engines, the engine sizes and loads, and the use factors or diversity.

Operating pressure is the engine discharge pressure at the connection to the exhaust. Systems may operate at positive pressure using available engine tail-pipe pressure to force the flow of gas, or at negative pressure with mechanically induced flow.

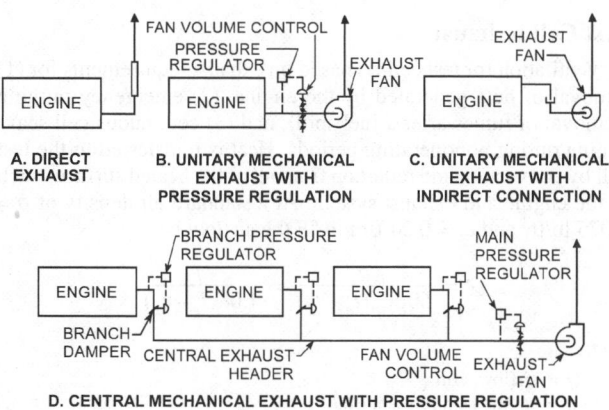

Fig. 1   Engine Exhaust Systems

The simplest way to induce engine exhaust from a test cell is to size the exhaust pipe to minimize variations in pressure on the engine and to connect it directly to the outside (Figure 1A). Exhausts directly connected to the outside are subject to wind currents and air pressure and can be hazardous due to the positive pressure in the system.

Mechanical engine exhausts are either unitary or central. A **unitary exhaust** (Figure 1B) serves only one test cell. It can be closely regulated to match the operation of the engine. A central exhaust (Figure 1D) serves multiple test cells with one or more exhaust fans and a duct system with branch connections to the individual test cells. Relief of a possible explosion in the ductwork should be considered.

Engine exhaust pressures fluctuate with changes in engine load and speed. Central exhausts should be designed so that load variations in individual test cells have a minimum effect on the system. Engine characteristics and diversity of operation determine the maximum airflow to be handled. Dampers and pressure regulators may be required to keep the pressures within test tolerances.

An indirect connection between the engine exhaust pipe and mechanical exhaust gas removal (Figure 1C) eliminate variation in back pressure and augment the exhaust gas flow by inducing room air into the exhaust stream. In this system the engine exhaust pipe terminates by being centered and inserted about 3 in. into the augmentation pipe, which is at least 1 in. larger in diameter. The induced room air is mixed with the exhaust gases yielding a much cooler exhaust flow. However, the potential for increased corrosion in a cooler exhaust must be considered when selecting the construction materials. The engine muffler should be located upstream of the augmentation connection to control noise. The indirect connection should be considered a potential point of ignition if the exhaust is fuel rich and the tail pipe reaches temperatures above 700°F.

Exhaust pipes and mufflers run very hot. A ventilated heat shield or a water jacketed pipe reduces this cell heat load and some exhausts are equipped with direct water injection. Thermal expansion, stress, and pressure fluctuations must also be considered in the design of the exhaust fan and ducting. The equipment must be adequately supported and anchored to relieve the thermal expansion.

Exhaust systems for chassis dynamometer installations must capture the high velocity exhaust from the tailpipe to prevent fume buildup in the room. An exhaust flow rate of 700 cfm has been used effectively for automobiles at a simulated speed of 65 mph.

Engine exhaust should discharge through a stack extending above the roof to an elevation sufficient to allow the fumes to clear the building. Chapter 43, Building Air Intake and Exhaust Design, has further details about exhaust stacks. Codes or air emission standards may require that exhaust gases be cleaned prior to being discharged to atmosphere.

INTERNAL COMBUSTION ENGINE TEST CELLS

Test Cell Exhaust

Ventilation for test cells is based on exhaust requirements for (1) removal of heat generated by the engine, (2) emergency purging (removal of fumes after a fuel spill), and (3) continuous cell scavenging during nonoperating periods. Heat is transferred to the test cell by convection and radiation from all of the heated surfaces such as the engine and exhaust system. At a standard air density of $\rho = 0.075$ lb/ft^3 and $c_p = 0.24$ Btu/lb·°F the airflow is:

$$Q = \frac{q}{60\rho c_p (t_e - t_s)} = \frac{q}{1.08(t_e - t_s)}$$

where

Q = airflow, cfm
q = engine heat release, Btu/h
t_e = temperature of exhaust air, °F
t_s = temperature of supply air, °F

The constant (1.08) should be corrected for other temperatures and pressures.

Heat radiated from the engine, dynamometer, and exhaust piping warms the surrounding surfaces, which release the heat to the air by convection. The value for $(t_e - t_s)$ in the equation cannot be arbitrarily set when a portion of q is radiated heat. The previous section on Engine Heat Release discusses other factors required to determine the overall q.

Vapor Removal. The exhaust should remove vapors as quickly as possible. Emergency purging, often 10 cfm per square foot of floor area, should be controlled by a manual overriding switch for each test cell. In case of fire, provisions need to be made to shut down all equipment, close fire dampers at all openings, and shut off the fuel-flow solenoid valves.

Cell Scavenging. Exhaust air is the minimum amount of air required to keep combustible vapors from fuel leaks from accumulating. In general, 2 cfm per square foot of floor area is sufficient. Because gasoline vapors are heavier than air, exhaust grilles should be low even when an overhead duct is used. Exhausting close to the engine minimizes the convective heat that escapes into the cell.

In some installations, all air is exhausted through a floor grating surrounding the engine bed plate and into a cubicle or duct below. In this arrangement, slots in the ceiling over the engine supply a curtain of air to remove the heat. This scheme is particularly suitable for a central exhaust (Figure 2). Water sprays in the under-floor exhaust lessen the danger of fire or explosion in case of fuel spills.

Trenches and pits should be avoided in test cells. If they exist, as in most chassis dynamometer rooms, they should be mechanically exhausted at all times. Long trenches may require multiple exhaust takeoffs. The exhaust should sweep the entire area, leaving no dead air spaces. Because of fuel spills and vapor accumulation, suspended ceilings or basements should not be located directly below the engine test cell. If such spaces exist, they should be ventilated continuously and have no fuel lines running through them.

Table 1 lists exhaust quantities used in current practice; although the exhaust should be calculated for each test cell on the basis of heat to be removed, evaporation of possible fuel spills, and the minimum ventilation needed during downtime.

TEST CELL SUPPLY

The air supply to a test cell should be balanced to yield a slightly negative pressure. Test cell air should not be recirculated. Air taken from non-test areas can be used, provided good ventilation practices

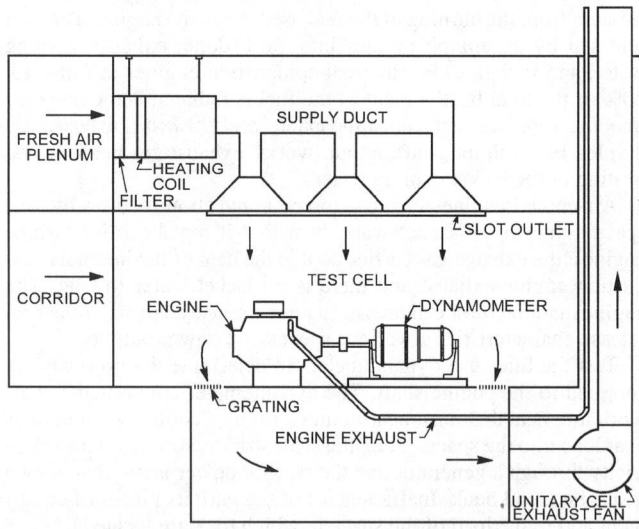

**Fig. 2   Engine Test Cell Showing Direct Engine
Exhaust—Unitary Ventilation System**

Table 1 Exhaust Quantities for Test Cells

	Minimum Exhaust Rates per Square Foot of Floor Area	
	cfm	Air changes per h[a]
Engine testing—cell operating	10	60[b]
Cell idle	2	12
Trenches[c] and pits	10	—
Accessory testing	4	24
Control rooms and corridors	1	6

[a] Based on a cell height of 10 ft.
[b] For chassis dynamometer rooms, this quantity is usually set by test requirements.
[c] For large trenches, use 100 fpm across the cross-sectional area of the trench.

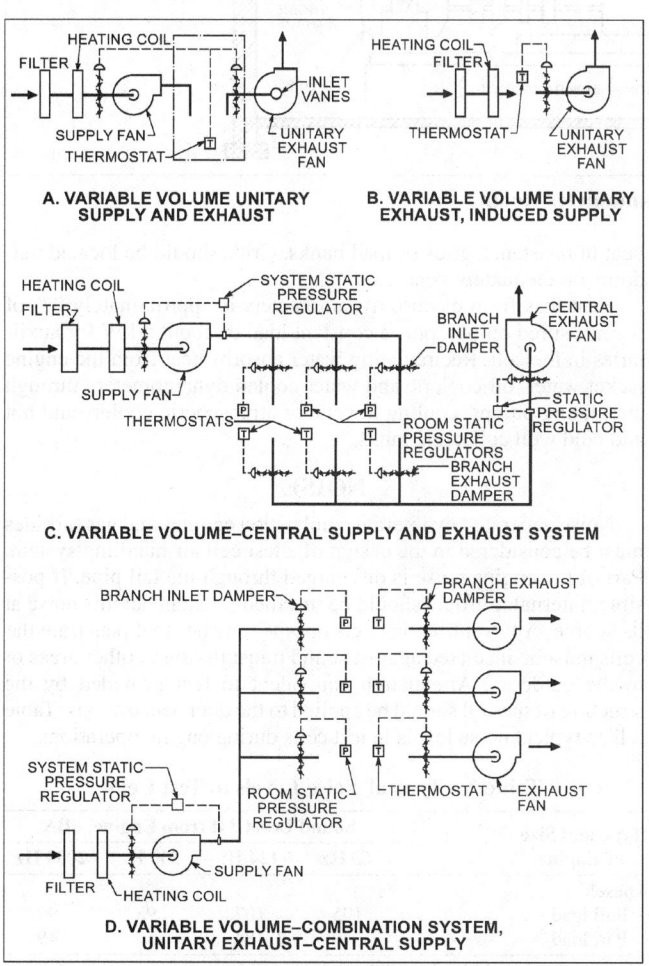

A. VARIABLE VOLUME UNITARY SUPPLY AND EXHAUST

B. VARIABLE VOLUME UNITARY EXHAUST, INDUCED SUPPLY

C. VARIABLE VOLUME–CENTRAL SUPPLY AND EXHAUST SYSTEM

D. VARIABLE VOLUME–COMBINATION SYSTEM, UNITARY EXHAUST–CENTRAL SUPPLY

Fig. 3 Heat Removal Ventilation Systems

are followed; such as using air that is free of unacceptable contaminants and is sufficient for temperature control and can maintain the proper test cell pressure.

Ventilation air should keep the released heat from the engine away from the cell occupants. Slot outlets with automatic dampers to maintain a constant discharge velocity have been used with variable volume systems.

A variation of systems C and D in Figure 3 includes a separate air supply sized for the minimum (downtime) ventilation rate and for a cooling coil with room thermostat to regulate the coil to control the temperature in the cell. This system is useful in installations where much time is devoted to the setup and preparation of tests, or where constant temperature is required for complicated or sensitive instrumentation. Except for production and endurance testing, the actual engine operating time in test cells may be surprisingly low. The average test cell is used approximately 15 to 20% of the time.

Air should be filtered to remove particulates and insects. The degree of filtration is determined by the type of tests. Facilities in clean environments sometimes use unfiltered outdoor air.

Heating coils are needed to temper supply air if there is danger of freezing equipment or if low temperatures adversely affect tests.

GAS-TURBINE TEST CELLS

Large gas-turbine test cells must handle large quantities of air required by the turbine, attenuate the noise generated, and operate safely with respect to the large flow of fuel. Such cells are unitary and use the turbine to draw in the untreated air and exhaust it through noise attenuators.

Small gas turbine engines can generally be tested in a conventional test cell with relatively minor modifications. The test-cell ventilation air supply and exhausts are sized for the heat generated by the turbine as for a conventional engine. The combustion air supply for the turbine is considerable; it may be drawn from the cell, from outdoors, or through separate conditioning units that handle only combustion air.

Exhaust quantities are higher than from internal combustion engines and are usually ducted directly to the outdoors through muffling devices that provide little restriction to airflow. The exhaust air may be water cooled, as temperature may exceed 1300°F.

CHASSIS DYNAMOMETER ROOMS

A chassis dynamometer (Figure 4) simulates road driving and acceleration conditions. The drive wheels of the vehicle rest on a large roll which drives the dynamometer. Air quantities, which are calibrated to correspond to the air velocity at a particular road speed, flow across the front of the vehicle for radiator cooling and to approximate the effects air speed on the body of the vehicle. Additional refinements may vary the air temperature within prescribed limits from 59 to 130°F, control relative humidity, and/or add shakers to simulate road conditions. Air is usually introduced through an area approximating the frontal area of the vehicle. A duct with a return grille at the rear of the vehicle may be lowered so that the air remains near the floor rather than cycling through a ceiling return air grille. The air is recirculated to the air-handling equipment above the ceiling.

Chassis dynamometers are also installed in:

- Cold rooms, where temperatures may be as low as −100°F
- Altitude chambers, where elevations up to 12,000 ft can be simulated
- Noise chambers for sound evaluation
- Electromagnetic cells for evaluation of electrical components
- Environmental chambers
- Full-sized wind tunnels with throat areas much larger than the cross sectional area of the vehicle. Combustion air is drawn directly from the room, but the engine exhaust must be installed in the wall that will preserve the low temperature and humidity.

A temperature soak space is often placed near chassis dynamometer rooms having a controlled temperature. This space is used to cool or heat those automobiles scheduled to enter the room. Generally, 18 to 24 h is required before the temperature of the vehicle stabilizes to the temperature of the room. The soak space and the temperature controlled room are often isolated from the rest of the facility, with entry and egress through an air lock.

VENTILATION

Constant-volume systems with variable supply temperatures can be used; however, variable-volume, variable-temperature systems are usually selected. Ventilation is generally controlled on the exhaust side (Figure 3). Unitary variable-volume systems (Figure

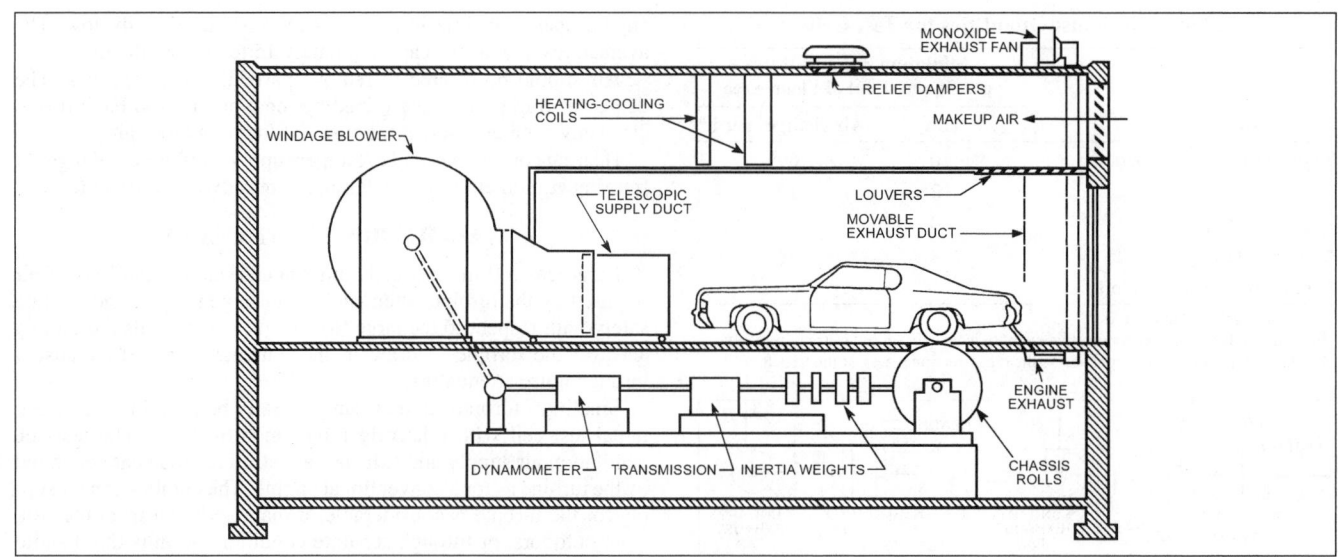

Fig. 4   Chassis Dynamometer Room

3A) use an individual exhaust fan and makeup air supply for each cell. Supply and exhaust fans are interlocked, and their operation is coordinated with the engine, usually by sensing the temperature of the cell. Some systems have exhaust only, with supply induced directly from outside (Figure 3B). The volume is varied by changing fan speed or damper position.

Ventilation with central supply fans, central exhaust fans, or both (Figure 3C) regulate air quantities by test cell temperature control of individual dampers or by two-position switches actuated by dynamometer operations. Air balance is maintained by static pressure regulation in the cell. Constant pressure in the supply duct is obtained by controlling the supply-fan inlet vanes, modulating dampers, or varying fan speed.

In systems with individual exhaust fans and central supply air, the exhaust is controlled by cell temperature or a two-position switch actuated by dynamometer operation. The central supply system is controlled by a static pressure device located in the cell to maintain room pressure (Figure 3D). Variable-volume exhaust airflow should not drop below minimum requirements. Exhaust requirements should override cell temperature requirements; thus reheat may be needed.

Ventilation should be interlocked with fire protection to shut down the supply to and exhaust from the cell in case of fire. Exhaust fans should be nonsparking, and makeup air should be tempered.

COMBUSTION AIR SUPPLY

Combustion air is usually drawn from the test cell or introduced directly from the outdoors. Separate units that only condition the combustion air can be used if combustion air must be closely regulated and conditioning of the entire test cell is impractical. These units filter, heat, and cool the supply air and regulate its humidity and pressure; they usually provide air directly to the engine air intake. Combustion air systems may be central units or portable packaged units.

COOLING WATER SYSTEMS

Dynamometers absorb and measure the useful output of an engine or its components. Two classes of dynamometers are the water-cooled induction type and the electrical type. In the water-cooled dynamometer the engine work is converted to heat, which is absorbed by circulating water. Electrical dynamometers convert engine work to electrical energy, which can be used or dissipated as

heat in resistance grids or load banks. Grids should be located outdoors or adequately ventilated.

Heat loss from electric dynamometers is approximately 8% of the measured output, plus a constant load of about 5 kW for auxiliaries in the cell. Recirculating water absorbs heat from the engine jacket water, oil coolers, and water cooled dynamometers through circulating pumps, cooling towers, or atmospheric coolers and hot and cold well collecting tanks.

NOISE

Noise generated by internal combustion engines and gas turbines must be considered in the design of a test cell air-handling system. Part of the engine noise is discharged through the tail pipe. If possible, internal mufflers should be installed to attenuate this noise at its source. Any ventilation ducts or pipe trenches that penetrate the cells must be insulated against sound transmission to other areas or to the outdoors. Attenuation equivalent to that provided by the structure of the cell should be applied to the duct penetrations. Table 2 lists typical noise levels in test cells during engine operations.

Table 2   Typical Noise Levels in Test Cells

Type and Size of Engine	Sound Level 3 ft from Engine, dBA			
	63 Hz	124 Hz	500 Hz	2000 Hz
Diesel				
Full load	105	107	98	99
Part load	70	84	56	49
Gasoline engine, 440 in³ at 5000 rpm				
Full load	107	108	104	104
Part load	75	—	—	—
Rotary engine, 100 hp				
Full load	90	90	83.5	86
Part load	79	78	75	72

BIBLIOGRAPHY

Bannasch, L.T. and G.W. Walker. 1993. Design factors for air-conditioning systems serving climatic automobile emission test facilities. *ASHRAE Transactions* 99(2):614-23.

Computer controls engine test cells. *Control Engineering* 16(75):69.

Paulsell, C.D. 1990. Description and specification for a cold weather emissions testing facility. U.S. Environmental Protection Agency, Washington, D.C.

Schuett, J.A. and T.J. Peckham. 1986. Advancements in test cell design. *SAE Transactions*, Paper no. 861215. Society of Automotive Engineers, Warrendale, PA.

CLEAN SPACES

DESIGN of clean spaces or cleanrooms encompasses much more than the traditional control of temperature and humidity. In clean space design, other factors may include control of particulate, microbial, electrostatic discharge (ESD), and gaseous contamination, airflow pattern control, pressurization, sound and vibration control, industrial engineering aspects, and manufacturing equipment layout. The objective of a good cleanroom design is to control these variables while maintaining reasonable installation and operating costs.

TERMINOLOGY

Acceptance criteria. The upper and lower limits of the room environment (critical parameters) if these limits are exceeded, the pharmaceutical product may be considered adulterated.

Air lock. A small room between two rooms of different air pressure, with interlocked doors to prevent loss of pressure in the higher pressure room.

As-built cleanroom. A cleanroom that is complete and ready for operation, with all services connected and functional, but without production equipment or personnel in the room.

Aseptic space. A space controlled such that bacterial growth is contained within acceptable limits. This is not a sterile space, in which absolutely no life exists.

At-rest cleanroom. A cleanroom that is complete with production equipment installed and operating, but without personnel in the room.

CFU (colony forming unit). A measure of bacteria present in a pharmaceutical processing room measured via sampling as part of performance qualification.

Challenge. An airborne dispersion of particles of known sizes and concentration used to test filter integrity and efficiency.

Cleanroom. A specially constructed enclosed area environmentally controlled with respect to particulate, temperature, humidity, air pressure, air pressure flow patterns, air motion, vibration, noise, viable organisms, and lighting.

Clean space. A defined area in which the concentration of particles and environmental conditions are controlled at or below specified limits.

Contamination. Any unwanted material, substance, or energy.

Conventional-flow cleanroom. A cleanroom with nonunidirectional or mixed airflow patterns and velocities.

Critical parameter. A room variable (such as temperature, humidity, air changes, room pressure, particulates, viable organisms, etc.) that, by law or by determination from pharmaceutical product development data, affects product strength, identity, safety, purity, or quality (SISPQ).

Critical surface. The surface of the work part to be protected from particulate contamination.

Design conditions. The environmental conditions for which the clean space is designed.

DOP. Dioctyl phthalate, an aerosol formerly used for testing efficiency and integrity of HEPA filters.

E.C. (or E.E.C. or EU) European Economic Community guidelines for GMP pharmaceutical manufacturing.

ESD. Electrostatic discharge.

Exfiltration. Leakage of air from a room through cracks in doors and pass throughs or through material transfer openings, etc., due to differences in space pressures.

First air. The air that issues directly from the HEPA filter before it passes over any work location.

FS 209. A United States Federal Standard that specifies airborne particulate cleanliness classes in cleanrooms and clean zones. **ISO Standard 14644-1** is a comparable international standard for cleanrooms. Table 1 and Figure 1 compares and summarizes the classes of both the Federal and the ISO standard. Typical FS 209 classes are as follows:

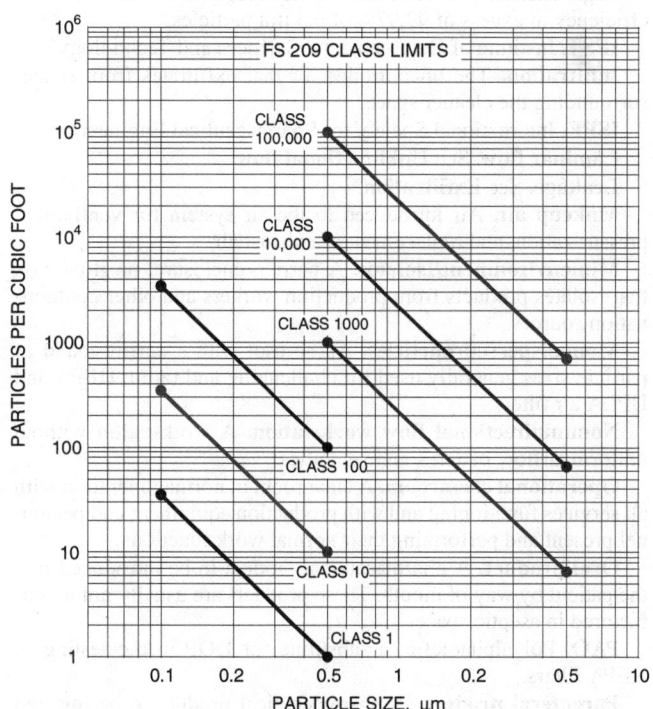

Fig. 1   Air Cleanliness Class Limits as Defined by U.S. *Federal Standard* 209

The preparation of this chapter is assigned to TC 9.11, Clean Spaces.

15.1

Table 1   Comparison of Airborne Particle Concentration Limits from FS 209 and ISO/FDIS 14644-1

FS 209 Class	ISO Class	0.1 μm			0.5 μm			5.0 μm		
		Federal *Standard* 209		ISO	Federal *Standard* 209		ISO	Federal *Standard* 209		ISO
		Particles/ft³	Particles/m³	Particles/m³	Particles/ft³	Particles/m³	Particles/m³	Particles/ft³	Particles/m³	Particles/m³
	1			10						
	2			100			4			
1	3	35	1230	1000	1	35	35			
10	4	345	12,200	10,000	10	353	352			
100	5	3,450	122,000	100,000	100	3,530	3,520			29
1000	6	34,500	1,220,000	1,000,000	1000	35,300	35,200	7	247	293
10,000	7	345,000	1.22×10^7		10,000	353,000	352,000	65	2300	2930
100,000	8	3,450,000	1.22×10^8		100,000	3,530,000	3,520,000	700	24,700	29,300
	9	3.45×10^7	1.22×10^9				35,200,000			293,000

Note: Values shown are the concentration limits for particles equal to and larger than the sizes shown.

$$C_n = N(0.5/D)^{2.2} \text{ where } C_n = \text{concentration limits in particles/ft}^3, N = \text{FS 209 class, and } D = \text{particle diameter in μm}$$

$$C_n = 10N(0.1/D)^{2.08} \text{ where } C_n = \text{concentration limits in particles/m}^3, N = \text{ISO class, and } D = \text{particle diameter in μm}$$

- **FS 209 Class 1.** Particle count not to exceed 1 particle per cubic foot of a size 0.5 μm (or appropriate other size, in accordance with Figure 1) and larger. This criterion should be based on sampling of counts.
- **FS 209 Class 10.** Particle count not to exceed 10 particles per cubic foot (350 particles/m³) of a size 0.5 μm and larger, with no particle exceeding 5.0 mm.
- **FS 209 Class 100.** Particle count not to exceed 100 particles per cubic foot of a size 0.5 μm and larger.
- **FS 209 Class 10,000.** Particle count not to exceed 10,000 particles per cubic foot of a size 0.5 μm and larger or 65 particles per cubic foot of a size 0.5 μm and larger.
- **FS 209 Class 100,000.** Particle count not to exceed 100,000 particles per cubic foot of a size 0.5 μm and larger or 700 particles per cubic foot of a size 5 μm and larger.

GMP. Good manufacturing practice. As defined by CFR 21 (also, cGMP = current GMP).

High-efficiency particulate air (HEPA) filter. A filter with an efficiency in excess of 99.97% of 0.3 μm particles.

IEST. Institute of Environmental Sciences and Technology.

Infiltration. The uncontrolled air that exfiltrates from spaces surrounding the cleaner space.

ISPE. International Society for Pharmaceutical Engineering.

Laminar flow. See **Unidirectional flow**.

Leakage. See **Exfiltration**.

Makeup air. Air introduced to the air system for ventilation, pressurization, and replacement of exhaust air.

Minienvironment/Isolator. A barrier, enclosure, or glove box that isolates products from production workers and other contamination sources.

Monodispersed particles. An aerosol with a narrow band of particle sizes generally used for challenging and rating HEPA and UPLA air filters.

Nonunidirectional flow workstation. A workstation without uniform airflow patterns and velocities.

Operational cleanroom. A cleanroom in normal operation with all services functioning and with production equipment and personnel present and performing their normal work functions.

Oral product. A pharmaceutical product to be introduced into the patient by way of mouth. These products are usually not manufactured in aseptic spaces.

PAO. Polyalphaolefin, a substitute for DOP in the testing of HEPA filters.

Parenteral product. A pharmaceutical product to be injected into the patient. Parenterals are manufactured under aseptic conditions or are terminally sterilized to destroy bacteria and meet aseptic requirements.

Particle concentration. The number of individual particles per unit volume of air.

Particle size. The apparent maximum linear dimension of a particle in the plane of observation.

Polydispersed particles. An aerosol with a broad band of particle sizes generally used to leak test filters and filter framing systems.

Qualification. Formal commissioning and operating of a system through established installation, operational, and performance qualification procedures (with approvals).

Qualification protocol. A written description of activities necessary to qualify a pharmaceutical facility, with required approval signatures.

Room classification. Room air quality class (see Figure 1 and Table 1).

SISPQ. A pharmaceutical product's strength, identify, safety, purity, and quality.

SOP. Standard operating procedure.

Topical product. A pharmaceutical product to be applied to the skin or soft tissue in forms of liquid, cream, or ointment, which therefore needs not be aseptic. Sterile ophthalmic products, though, are usually manufactured aseptically.

ULPA (ultralow penetration air) filter. A filter with a minimum of 99.999% efficiency on 0.12 μm particles.

Unidirectional flow. Formerly called laminar flow. Air flowing at constant and uniform velocity in the same direction.

Workstation. An open or enclosed work surface with direct air supply.

CLEAN SPACES AND CLEANROOM APPLICATIONS

The use of clean space environments in manufacturing, packaging, and research continues to grow as technology advances and the need for cleaner work environments increases. The following major industries use clean spaces for their products:

Pharmaceuticals. Preparation of pharmaceutical, biological, and medical products requires clean spaces to control viable (living) particles that would produce undesirable bacteria growth and other contaminants.

Electronics. The advances in semiconductor microelectronics continue to drive cleanroom design. Semiconductor facilities account for a significant percentage of all cleanrooms in operation in the United States, with most of the newer semiconductor cleanrooms being FS 209 Class 100 or cleaner.

Aerospace. Cleanrooms were first developed for aerospace applications to manufacture and assemble satellites, missiles, and aerospace electronics. Most applications involve clean spaces of large volumes with cleanliness levels of Class 100,000 or cleaner.

Miscellaneous Applications. Cleanrooms are also used in aseptic food processing and packaging; manufacture of artificial limbs and joints; automotive paint booths; crystal; laser/optic industries; and advanced materials research.

Hospital operating rooms may be classified as cleanrooms, yet their primary function is to limit particular types of contamination rather than the quantity of particles present. Cleanrooms are used in patient isolation and surgery where risks of infection exist. For more information, see Chapter 7, Health Care Facilities.

AIRBORNE PARTICLES AND PARTICLE CONTROL

Airborne particles occur in nature as pollen, bacteria, miscellaneous living and dead organisms, and windblown dust and sea spray. Industry generates particles from combustion processes, chemical vapors, and friction in manufacturing equipment. People in the workspace are a prime source of particles in the form of skin flakes, hair, clothing lint, cosmetics, respiratory emissions, and bacteria from perspiration. These airborne particles vary in size from 0.001 μm to several hundred micrometres. Particles larger than 5 μm tend to settle quickly. With many manufacturing processes, these airborne particles are viewed as a source of contamination.

Particle Sources in Clean Spaces

In general, particle sources, with respect to the clean space, are grouped into two categories: external and internal.

External Sources. External sources are those particles that enter the clean space from the outside, normally via infiltration through doors, windows, and wall penetrations for pipes, ducts, etc. However, the largest external source is usually outside makeup air entering through the air conditioning.

In an operating cleanroom, external particle sources normally have little effect on overall cleanroom particle concentration because HEPA filters clean the supply air. However, the particle concentration in clean spaces at rest relates directly to ambient particle concentrations. External sources are controlled primarily by air filtration, room pressurization, and sealing of space penetrations.

Internal Sources. Particles in the clean space are generated by people, cleanroom surface shedding, process equipment, and the manufacturing process itself. Cleanroom personnel can be the largest source of internal particles. Workers may generate several thousand to several million particles per minute in a cleanroom. Personnel-generated particles are controlled with new cleanroom garments, proper gowning procedures, and airflow designed to continually shower the workers with clean air. As personnel work in the cleanroom, their movements may reentrain airborne particles from other sources. Other activities, such as writing, may also cause higher particle concentrations.

Particle concentrations in the cleanroom may be used to define cleanroom class, but actual particle deposition on the product is of greater concern. The sciences of aerosols, filter theory, and fluid motions are the primary sources of understanding of contamination control. Cleanroom designers may not be able to control or prevent internal particle generation completely, but they may anticipate internal sources and design control mechanisms and airflow patterns to limit their effect on the product.

Fibrous Air Filters

Proper air filtration prevents most externally generated particles from entering the cleanroom. Present technology for high-efficiency air filters centers around two types: high-efficiency particulate air (HEPA) filters and ultralow penetration air (ULPA) filters. HEPA and ULPA filters use glass fiber paper technology. Laminates and nonglass media for special applications have been developed. HEPA and ULPA filters are usually constructed in a deep pleated form with either aluminum, coated string, filter

paper, or hot-melt adhesives as pleating separators. Filters may vary from 1 to 12 in. in depth; available media area increases with deeper filters and closer pleat spacing.

Theories and models verified by empirical data indicate that interception and diffusion are the dominant capture mechanisms for HEPA filters. Fibrous filters have their lowest removal efficiency at the most penetrating particle size (MPPS), which is determined by filter fiber diameter, volume fraction or packing density, and air velocity. For most HEPA filters, the MPPS is between 0.1 and 0.3 μm. Thus, HEPA and ULPA filters have rated efficiencies based on 0.3 μm and 0.12 μm particle sizes, respectively.

AIR PATTERN CONTROL

Air turbulence in the clean space is strongly influenced by air supply and return configurations, foot traffic, and process equipment layout. Selection of the air pattern configurations is the first step of good cleanroom design. User requirements for cleanliness level, process equipment layout, available space for installation of air pattern control equipment (i.e., air handlers, clean workstations, environmental control components, etc.), and project financial considerations all influence the final air pattern design selection. Financial aspects govern air pattern control design where operating and capital costs may limit the type and size of air-handling equipment that can be used.

Numerous air pattern configurations are in use, but in general they fall into two categories: unidirectional airflow (often mistakenly referred to as laminar flow) and nonunidirectional airflow (commonly called turbulent or mixed airflow).

Unidirectional airflow, though not truly laminar airflow, is characterized as air flowing in a single pass in a single direction through a cleanroom or clean zone with generally parallel streamlines. Ideally, the flow streamlines would be uninterrupted, and although personnel and equipment in the airstream do distort the streamlines, a state of constant velocity is approximated. Most particles that encounter an obstruction in a unidirectional airflow strike the obstruction and continue around it as the airstream reestablishes itself downstream of the obstruction.

Nonunidirectional airflow does not meet the definition of unidirectional airflow because it has either multiple-pass circulating characteristics or nonparallel flow.

Nonunidirectional Airflow

Variations of nonunidirectional airflow are based primarily on the location of supply air inlets and outlets and air filter locations. Some examples of unidirectional and nonunidirectional airflow systems are shown in Figures 2 and 3. Airflow is typically supplied to the space through supply diffusers containing HEPA filters (Figure 2) or through supply diffusers with HEPA filters located in the ductwork or air handler (Figure 3). In a mixed-flow room, air is prefiltered in the supply and then HEPA-filtered at the workstations located in the clean space (see the left side of Figure 3).

Nonunidirectional airflow may provide satisfactory contamination control results for cleanliness levels of Class 1000 through Class 100,000. Attainment of desired cleanliness classes with designs similar to Figures 2 and 3 presupposes that the major space contamination is from makeup air and that contamination is removed in air-handler or ductwork filter housings or through HEPA filter supply devices. When internally generated particles are of primary concern, clean workstations are provided in the clean space.

Air turbulence is harmful for flow control, but air turbulence is needed to enhance the mixing of low and high particle room concentrations and produce a homogeneous particle concentration level acceptable to the process.

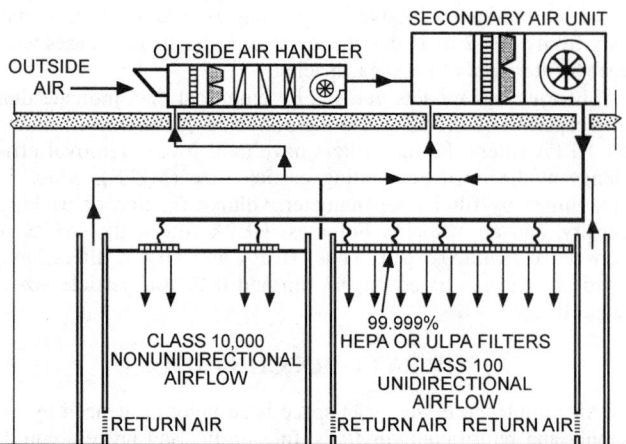

Fig. 2   Class 10,000 Nonunidirectional Cleanroom with Ducted HEPA Filter Supply Elements and Class 100 Unidirectional Cleanroom with Ducted HEPA or ULPA Filter Ceiling

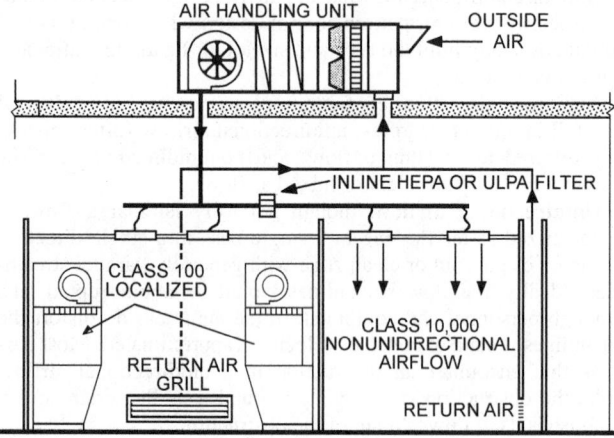

Fig. 3   Class 10,000 Nonunidirectional Cleanroom with HEPA Filters Located in Supply Duct and Class 100 Local Workstations

Unidirectional Airflow

Air patterns are optimized, and air turbulence is minimized in unidirectional airflow. In a **vertical laminar flow (VLF) room**, air is typically introduced through the ceiling HEPA or ULPA filters and returned through a raised access floor or at the base of sidewalls. Because air enters from the entire ceiling area, this configuration produces nominally parallel airflow. In a **horizontal flow cleanroom**, air enters one wall and returns on the opposite wall.

A **downflow cleanroom** has a ceiling with HEPA filters, but air flows from only a portion of the ceiling. In a cleanroom with a low class number the greater part of the ceiling requires HEPA filters. For an FS 209 Class 100 or better room, the entire ceiling usually requires HEPA filtration. Ideally, a grated or perforated floor serves as the air exhaust. This type of floor is inappropriate in pharmaceutical cleanrooms, which typically have solid floors and low-level wall returns.

In a downflow cleanroom, a uniform shower of air bathes the entire room in a downward flow of ultraclean air. Contamination generated in the space will not move laterally against the downward flow of air (it is swept down and out through the floor) or contribute to a buildup of contamination in the room.

Care must be taken in design, selection, and installation to seal a HEPA or ULPA filter ceiling. Properly sealed HEPA or ULPA filters installed in the ceiling can provide the cleanest air presently available in a workroom environment.

In a **horizontal flow cleanroom**, the supply wall consists entirely of HEPA or ULPA filters supplying air at approximately 90 fpm across the entire section of the room. The air then exits through the return wall at the opposite end of the room and recirculates. As with the downflow room, this design removes contamination generated in the space and minimizes cross-contamination perpendicular to the airflow. However, a major limitation to this design is that downstream air becomes contaminated. The air leaving the filter wall is the cleanest; it then becomes contaminated by the process as it flows past the first workstation. The process activities can be oriented to have the most critical operations at the clean end of the room, with the progressively less critical operations located toward the return air, or dirty end of the room.

U.S. *Federal Standard* 209E does not specify velocity requirements, so the actual velocity is as specified by the owner or the owner's agent. The 90 fpm velocity specified in an earlier standard (FS 209B) is still widely accepted in the cleanroom industry. Current research suggests lower velocities may be possible. Careful testing should be performed to ensure that required cleanliness levels are maintained. Other reduced air volume designs may use a mixture of high and low pressure drop HEPA filters, reduced coverage in high traffic areas, or lower velocities in personnel corridor areas.

Unidirectional airflow systems have a predictable airflow path that airborne particles tend to follow. Without good filtration practices, unidirectional airflow only indicates a predictable path for particles. However, superior cleanroom performance may be obtained with a good understanding of unidirectional airflow. This airflow remains parallel (or within 18° of parallel) to below the normal work surface height of 30 to 36 in. However, this flow deteriorates when the air encounters obstacles such as process equipment and work benches, or if air has traveled excessive distances. Personnel movement also degrades the flow. The result is a cleanroom with areas of good unidirectional airflow and areas of turbulent airflow.

Turbulent zones have countercurrents of air with high velocities, reverse flow, or no flow at all (stagnancy). Countercurrents can produce stagnant zones where small particles may cluster and finally settle onto surfaces or product; countercurrents may also lift particles from contaminated surfaces. Once lifted, these particles may deposit on product surfaces.

Cleanroom mockups may help the designer avoid turbulent airflow zones and countercurrents. Smoke, neutral-buoyancy helium-filled soap bubbles, and nitrogen vapor fogs can make air streamlines visible within the cleanroom mockup.

Computer-Aided Flow Modeling

Computer models of particle trajectories transport mechanisms, and contamination propagation are commercially available. Flow analysis with computer models may compare flow fields associated with different process equipment, work benches, robots, and building structural design. Airflow analysis of flow patterns and air streamlines is performed by computational fluid dynamics for laminar and turbulent flow where incompressibility and uniform thermophysical properties are assumed. Design parameters may be modified to determine the effect of airflow on particle transport and flow streamlines; thus avoiding the cost of mock-ups.

Major features and benefits associated with most computer flow models are

1. Two- or three-dimensional modeling of simple cleanroom configurations, including people and equipment
2. Modeling of both laminar and turbulent airflows
3. Multiple air inlets and outlets of varying sizes and velocities

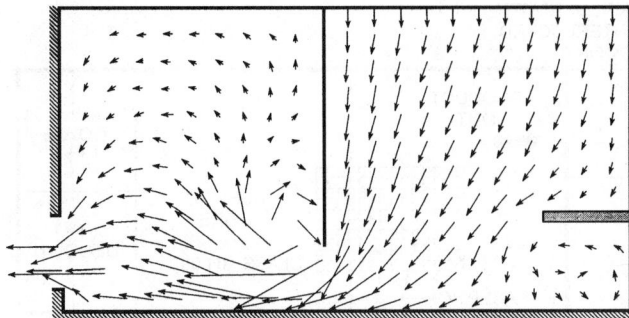

**Fig. 4   Cleanroom Airflow Velocity Vectors
Generated by Computer Simulation**

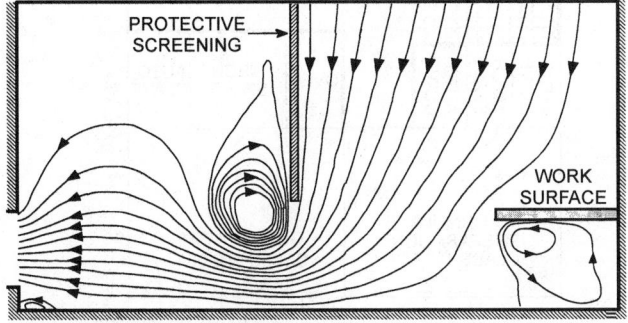

**Fig. 5   Computer Modeling of Cleanroom
Airflow Streamlines**

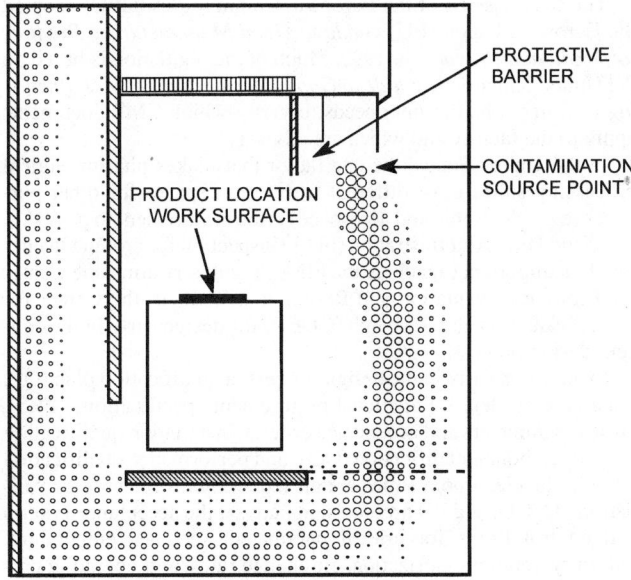

**Fig. 6   Computer Simulation of Particle
Propagation in a Cleanroom**

4. Allowances for varying boundary conditions associated with walls, floors, and ceilings
5. Aerodynamic effects of process equipment, work benches, and people
6. Prediction of specific airflow patterns, velocities, and temperature gradients of all or part of a cleanroom
7. Reduced cost associated with new cleanroom design verification
8. Graphical representation of flow streamlines and velocity vectors to assist in flow analysis (Figures 4 and 5)
9. Graphical representation of simulated particle trajectories and propagation (Figure 6)

Research has shown good correlation between flow modeling by computer and that done in simple mockups. However, computer flow modeling software should not be considered a panacea for cleanroom design.

TESTING CLEAN AIR AND CLEAN SPACES

Because early cleanrooms were largely for governmental use, testing procedures have been set by government standards. U.S. *Federal Standard* 209 is widely accepted, as it defines air cleanliness levels for clean spaces around the world. Standardized testing methods and practices have been developed and published by the Institute of Environmental Sciences and Technology (IEST), the American Society for Testing and Materials (ASTM), and others.

Three basic test modes for cleanrooms are used to evaluate a facility properly: (1) as built, (2) at rest, and (3) operational. A cleanroom cannot be fully evaluated until it has performed under full occupancy, and the process to be performed in it is operational. Thus, the techniques for conducting initial performance tests and operational monitoring must be similar.

Sources of contamination, as previously described, are both external and internal for both laminar and nonlaminar flow cleanrooms. The primary air loop is the major source of external

contamination. Discrete particle counters using laser or light-scattering principles may be used to detect particles of 0.01 to 5 μm. For particles 5 μm and larger, microscopic counting can be used, with the particles collected on a membrane filter through which a sample of air has been drawn.

HEPA filters in unidirectional flow and Class 100 ceilings should be tested for pinhole leaks at the following places: the filter media, the sealant between the media and the filter frame, the filter frame gasket, and the filter bank supporting frames. The area between the wall or ceiling and the frame should also be tested. A pinhole leak at the filter bank can be extremely critical, since the leakage rate varies inversely as the square of the pressure drop across the hole. (The industry term "pinhole" used to describe the leak site is a misnomer. The size is almost never that of a hole formed by a pin, but is actually many times smaller.)

The testing procedure of the IEST (1993) describes 12 tests for cleanrooms. Which tests are applicable to each specific cleanroom project must be determined.

PHARMACEUTICAL AND BIOMANUFACTURING CLEAN SPACES

Facilities for the manufacture of pharmaceutical products require careful assessment of many entities including HVAC, controls, room finishes, process equipment, room operations, and utilities. Flow of equipment, personnel, and product must also be considered. It is important to involve the designers, operators, commissioning staff quality control, maintenance, constructors, and the production representative during the conceptual stage of design. Critical variables for room environment and types of controls vary greatly with the clean space's intended purpose. It is particularly important to determine critical parameters with quality assurance to set limits for temperature, humidity, room pressure, and other control requirements.

In the United States, regulatory requirements and specification documents such as the *Code of Federal Regulations* 210 and 211, ISPE Guides, and National Fire Protection Association (NFPA) standards are available. These documents describe good manufacturing practice (GMP). The goal of GMP is to achieve a proper and repeatable method of producing sterile products free from microbial and particle contaminants.

The Commission of the European Communities (CEC) adopted the European Union (EU) *Guide to Good Manufacturing Practice for Medicinal Products* in 1992. Much of the regulation is based on the United Kingdom's *Guide to Good Pharmaceutical Manufacturing Practice*. The designer needs to verify which GMP documents apply to the facility and which are advisory.

In the United States, the one factor that makes pharmaceutical processing suites most different from clean spaces for other purposes (e.g., electronic and aerospace) is the requirement to pass U.S. Food and Drug Administration (FDA) inspection for product licensing. It is important to include the FDA's regulatory arms, the Center for Biologics Evaluation and Research (CBER) or the Center for Drug Evaluation and Research (CDER) for design early in the concept design process.

In addition, early in the design process, a qualification plan (QP) must be considered. Functional requirement specifications (FRS), critical parameters and acceptance criteria, installation qualification (IQ), operational qualification (OQ), and performance qualification (PQ) in the cleanroom suites are all requirements for process validation. lQ, OQ, and PQ protocols, in part, set the acceptance criteria and control limits for critical parameters such as temperature, humidity, room pressurization, air change rates, and operating particle counts (or air classifications). These protocols must receive defined approvals in compliance with the owner's guidelines. The qualification plan must also address standard operating procedures (SOPs), preventive maintenance (PM), and operator and maintenance personnel training.

Biomanufacturing and pharmaceutical aseptic clean spaces are typically arranged in suites, with clearly defined operations in each suite. For example, common convention positions an aseptic core (FS 209 Class 100) filling area in the innermost room, which is at the highest pressure surrounded by areas of descending pressure and increasing particulate classes and bacterial levels (see Figure 7).

In aseptic processing facilities, the highest quality area is intentionally placed within the lower quality areas and separated by room air classification and air pressure differences via air locks. A commonly used pressure difference is 0.05 to 0.06 in. of water between air classifications, with the higher quality room having the higher pressure. Lower pressure differences may be acceptable if they are proven effective. A pressure differential is generally accepted as good manufacturing practice to inhibit particles from entering a clean suite.

Where product containment is an issue (e.g., pathogens or toxic materials), the suite requires a lower pressure than the adjacent rooms, but it may still require a higher air quality. In this case, the air lock may be designed to maintain one pressure level higher than the adjacent room, and also higher than the contained suite (Figure 8). Biological containment requirements are addressed by the U.S. National Institutes of Health where potentially hazardous organisms are grouped in biosafety levels BL-1 to BL-4.

Design Concerns for Pharmaceutical Cleanrooms

The owner and designer must define the maximum range of variable value (**acceptance criterion**) for each critical parameter. In that range, the product's safety, identity, strength, purity, and quality will not be affected. The owner should define **action alarm** points at the limits of acceptance criteria, such that product exposed to conditions outside these action limits may have been adulterated. The designer should select tighter (but achievable) target design values for critical parameters (in the range of acceptance criteria) and for non-critical parameters.

Facilities manufacturing penicillin or similar antibiotics (such as cephalosporins) must be physically isolated from other manufacturing areas and served by their own HVAC system. Certain other products may require similar separation.

Facilities manufacturing aseptic/sterile products derived from chemical synthesis may have different requirements than those

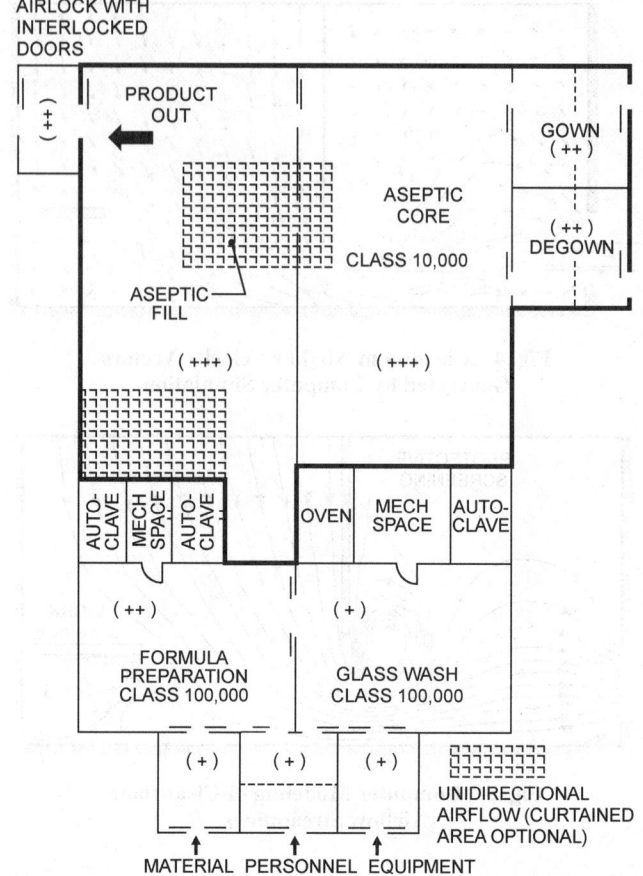

(+) = ROOM PRESSURIZATION LEVEL ABOVE ZERO REFERENCE

Fig. 7   Typical Aseptic Suite

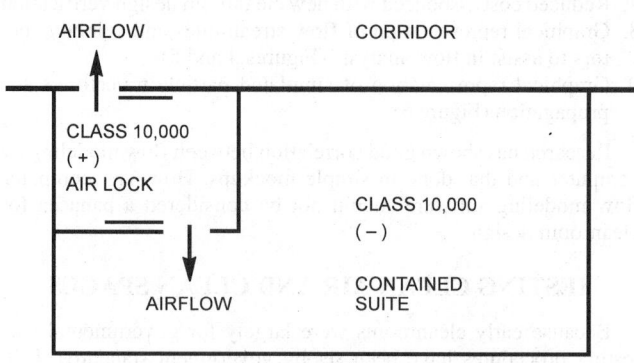

(−) = ROOM PRESSURIZATION LEVEL ABOVE ZERO REFERENCE

Fig. 8   Contained Suite Arrangement

manufacturing biological or biotechnological products. The owner must define the inspecting agency's requirements.

The United States Pharmacopoeia (USP) limits the temperatures to which finished pharmaceutical products may be exposed to 59 to 86°F. The production facility may need tighter limits than these, based on the owner's observed product data. Personnel comfort will be a factor in design. Personnel can perspire in their protective overgarments and cause particulate and microbial counts to increase, so lower temperatures and tighter temperature control may be necessary.

Relative humidity may be critical to the product's integrity. Some products are processed or packaged cold and need low room dew point to prevent condensation on equipment and vials. Certain products are hygroscopic and require lower humidity than a condensing coil can provide; desiccant dehumidification should be considered. Humidification is usually needed for personnel comfort but not usually for product needs. It may also be needed where dust might present an explosion hazard or where low humidity may hinder the handling of dry materials. Clean steam (free of chemicals and other additives) is preferred for humidification because it is free of bacteria, but the humidification system should be free of amines or other contaminants if room air might contact the product.

Although airborne particles and viable organisms may be minimized by dilution (high air changes) and by supplying high-quality HVAC air, the most effective control is to minimize the release of these contaminants within the room. Personnel and machinery are the most common sources of contamination, and can be isolated from the product by gowning, masks, and isolation barriers. A careful study of the manner in which each room will operate should reveal the most probable sources of contaminants and will help the HVAC designer determine dilution air quantities and locate air supply outlets and returns. Avoid duct liners and silencers in supply air ductwork where contaminants can collect and bacteria can grow. Ensure special attention is paid to cleaning and degreasing of metal sheeting and air ductwork before installation.

Airborne particulate and microbe levels in aseptic processing areas are limited by government regulations, with lower limits for the more critical operations rooms. European and U.S. FDA particulate limits are for the room in operation, but may also define limits for the room at rest.

Facilities meeting U.S. GMPs must meet particulate levels with manufacturing underway (the exception being aseptic powder processing, where airborne particulate levels at powder filling heads will exceed limits. These rooms are tested with production running but no product present). GMPs suggest a minimum of 20 air changes per hour for rooms with an air particulate classification, with exposed sterile product under a unidirectional flow hood or inside a FS 209 Class 100 barrier enclosure. The U.S. GMPs currently require only two classes: FS 209 Class 100 for sterile product exposure, and FS 209 Class 100,000 for adjoining spaces. However, common practice places exposed products in a FS 209 Class 100 unidirectional flow zone inside a FS 209 Class 10,000 room. Facilities designed for European Community regulations require a minimum of 20 air changes per hour in every room for which particulate levels must be controlled (i.e., where aseptic product or equipment is handled). Where isolation barriers and containment or spot exhaust are not practical, airborne contaminants may be minimized by increased room air changes. There are no minimum air change requirements for facilities manufacturing nonaseptic products.

Note that once product is stoppered and capped in containers, the need for particulate control and minimal air changes is reduced or eliminated, depending on the degree of protection provided by product packaging. The owner should determine the necessary critical environmental parameters and acceptance criteria for each room and processing step.

Return openings for room HVAC should be located at low levels on the walls, to promote a downward flow of air from supply to return, sweeping contaminants to the floor and away from the product. In larger rooms, internal return air columns may be necessary. Perforated floors are discouraged due to cleanability problems.

Aseptic facilities usually require pinhole-scanned (integrity-tested) HEPA filters on supply air (not ULPA). Many operations install HEPA filters in the supply air to nonaseptic production facilities to minimize cross-contamination from other manufacturing areas served by the HVAC. To increase the life of terminal HEPA filters in aseptic facilities, and to minimize the need to rebalance the supply system due to differential loading of terminal HEPA filters, many designers install a high-capacity HEPA bank downstream of the supply air fan, with constant volume control to compensate for primary filter pressure changes and any dehumidifier airflow. The final HEPA filter is usually in a sealed gel frame or composed of a one-piece lay-in design that can be caulked to the ceiling frame, maintaining the integrity of the room envelope. Standard supply air diffusers are not recommended for a classified room since they are difficult to clean.

Aseptic product must be protected by pressurizing the room in which it is exposed, to either 0.05 in. of water in the U.S. or 15 Pa in Europe above the next lower cleanliness room classification. To keep pressure differential from dropping to zero when a door is opened, air locks are recommended between rooms of different air pressure, especially at the entrance to the aseptic production room itself. Room pressure is a function of airflow resistance through cracks, openings, and permeable surfaces in the room shell. Consider all potential openings, slots, and door leakage that can affect the amount of air needed to pressurize the space. Because net room airflow and room pressure are closely related, outdoor or makeup air requirements are often dictated by room pressures rather than by the number of occupants. The HVAC should be able to handle more makeup air than needed for commissioning, because door seals can deteriorate over time.

FS 209 Class 100 Unidirectional hoods are basically banks of HEPA filters, integrity tested to be pinhole-free. Since it is difficult to maintain unidirectional flow for long distances or over large areas, the hood should be located as closely as possible to product exposure sites (the work surface). Hood-face velocity is usually 90 fpm ±20%; velocity requirements differ somewhat from the FDA's. The user should specify velocity and uniformity requirements. A unidirectional hood usually has clear sidewalls (curtains) to promote downward airflow and prevent entrainment of room particles into the hood's zone of protection. Curtains should extend below the product exposure site and be designed to prevent accidental disruption of airflow patterns by personnel. Many production facilities prefer rigid curtains to facilitate cleaning and sanitization.

Hood fan heat may become a problem, forcing the designer to overcool the room from which the hood draws its air or to provide sensible cooling air directly into the hood's circulating system.

Barrier Technology

Cleanrooms designed to meet FS 209 Class 100 or better require considerable equipment, space, and maintenance. Operating this equipment is an expensive commitment. Furthermore, cleanrooms typically require gowned operators to be inside to manipulate the product and adjust the machinery. Since the operator generates a large portion of particulate and contamination, it would be better to place the operator outside the controlled environment. With the operator removed from the controlled space, its volume can be reduced leading to substantially reduced capital and operating cost, hence the increasing popularity of barrier technology or isolators.

Each isolator is designed to fit a specific application and is customized for its tasks. Applications vary widely depending on product, process equipment, and throughput volume. Sterile barriers are typically positive pressure envelopes around the filling equipment with a multiple of glove ports for operator access, constructed of polished stainless steel with clear rigid view ports.

"Mouse holes" permit passage of vials in and out of the unit. Special transfer ports are used for stoppers and tools. Important design concerns include accessibility, ergonomics, interference with mating equipment, decontamination or sterilization procedures, access to service equipment, filter change, filter certification, process validation, and environmental control.

Extra attention must be paid to product filling, vial, and stopper protection; access to the barrier for sterilized stoppers; interface to the vial sterilization (depyrogenation) device; sterilizing of product path including pumps and tubing; and airflow patterns inside the barrier, especially at critical points. If hydrogen peroxide is to be used as a surface sterilant, care must be taken to assure good circulation and adequate concentration inside the barrier, as well as removal of residual vapor in the required time frame.

Other barrier applications offer operator protection from potent compounds while maintaining a sterile internal environment. These tend to be total containment isolators with totally contained product transfer ports. This system maintains all internal surfaces sealed from the external environment or operator exposure. Because of potential chamber leaks, its internal pressure may be kept negative with respect to the ambient space via exhaust fans, posing an additional potential risk to the product that must be addressed with the owner.

Nonsterile powder control may incorporate more passive barrier designs such as a downflow sample weighing hood. This arrangement takes advantage of unidirectional airflow to wash particles down and away from the operator's breathing zone. Low-wall air returns positioned at the back of the cubicle capture the dust. An arrangement of roughing and final filters permits the air to return to the air handlers and back to the work zone through the ceiling. Product compounds involving noxious or solvent vapors require a once-through air design. Barrier technology, by the nature of its design, affords the opportunity for installation in environments that might require no special control or particulate classification. Since isolators and containment chambers are still relatively new to the pharmaceutical industry, installations for sterile product must be in a controlled ambient room condition of FS 209 Class 100,000.

Maintainability

A facility that considers maintainability in its design will be much more reliable and should have fewer operational and regulatory concerns. Such maintainability concerns as accessibility, frequency of maintenance, and spare parts should be considered. Many pharmaceutical facilities have been designed such that routine maintenance can be performed from outside the facility, the exception being unidirectional and terminal HEPA filters, which must be tested twice a year. Quality of materials is important to reliability, especially where failure can compromise a critical parameter. Consider how much exposure and risk to product and personnel is required during maintenance activities (i.e., how to clean the inside of a glove box contaminated by a toxic product). Beside cleanable room surfaces that must be sanitized, consider if and how HVAC equipment may be sanitized using the owner's procedures. Determine if ductwork must be internally cleaned, and how.

Controls, Monitors, Alarms

Room pressure may be maintained by a passive (statically balanced) HVAC if there are few airflow variables. For example, an HVAC of a few pressurized rooms may be statically balanced if there is a method of maintaining supply airflow volume to compensate for filter loading to assure minimum room air changes. More complex designs may require active room pressure control, usually by controlling exhaust or return air volume in a control loop that senses and controls room pressure, not airflow differentials. Pressure controls should not overreact to doors opening and closing, since it is virtually impossible to pressurize a room to 0.05 in. of water with a room door standing open. Some time delay is usually advisable in pressure controls and pressure alarms to allow time for doors to close. Duration of the time delay can be determined by observing actual processes, but should be as short as possible.

If room humidity must be maintained to tolerances tighter than the broad range that normal comfort cooling can maintain, active relative humidity control should be considered. If a desiccant dehumidifier is needed, operation of the unit over its range of flow must not adversely affect the ability of the HVAC to deliver a constant air supply volume to the facility.

Monitor and alarm critical parameters to prove they are under control. Log alarm data and parameter values during excursions. Logging may range from a local recorder to DDC disk data storage with controlled access. Software source code should be traceable, with changes to software under the owner's control after qualification is complete. Commercial HVAC graphic programming software is usually acceptable, but verify this with regulatory agencies before detailed design begins. Also, keep complete calibration records for sensors, alarms, and recorders of critical parameter data.

Noise Concerns

Noise from the HVAC can be a common problem because fans must overcome the pressure drop of additional air filtration. The amount of noise generated must be reduced rather than adding duct silencers, which may harbor bacteria and are difficult to clean. Separate supply and return fans running at lower tip speeds instead of a single-fan air handler can reduce the noise. HVAC noise may not be an issue if production equipment is considerably noisier.

Nonaseptic Products

Nonaseptic pharmaceutical facilities (such as topical and oral products) are similar in design to those described for aseptic product manufacturing, but with fewer critical components to be qualified. However, critical parameters such as room humidity may become more important while airborne particulate counts do not apply in the United States. If the product is potent, barrier isolation may still be advisable. Room differential pressures or airflow directions, and air changes are usually critical in that they are needed to control cross-contamination of products, but no regulatory minimum pressure or air change values apply.

START-UP AND QUALIFICATION OF PHARMACEUTICAL CLEANROOMS

Qualification of HVAC for Aseptic Pharmaceutical Manufacturing

Qualification of the pharmaceutical cleanroom HVAC is part of the overall commissioning of the building and its equipment, except that documentation is more rigorous. The qualification covers the equipment affecting critical parameters and their control. Other groups in the manufacturing company, such as the safety or environmental groups may also require similar commissioning documentation for their areas of concern. The most important objective in meeting the requirements of the approving agency is to (1) state what procedures will be followed and then verify that it was done, and (2) show that product is protected and room acceptance criteria are met.

Qualification Plan and Acceptance Criteria

Early in the design, the owner and designer should discuss who will be responsible for as-built drawings, setting up maintenance files, and training. They should create a qualification plan for the HVAC, which includes (1) a functional description of what the systems will do; (2) maps of room pressures, airflow diagrams, and cleanliness zones served by each air handler; (3) a list of critical components to be qualified including the computer controlling the HVAC; (4) a list of owner's procedures that must be followed for the qualification of equipment and systems that affect critical variables; (5) a list of qualification procedures (IQ/OQ/PQ protocols) that must be written especially for the project; and (6) a list of needed commissioning equipment.

The approval procedure should also be defined in the QP. It is important to be able to measure and document the critical variables of a system (such as room pressure), but it is also important to document the performance of components that affect that critical variables (i.e., room pressure sensors, temperature sensors, airflow volume monitor, etc.) for GMP as well as business records. Documentation helps ensure that replacement parts (such as motors) can be specified, purchased, and installed to satisfy critical variables.

It is integral to determine what the critical components are (including instruments), the performance of which could affect critical parameters, and the undetected failure of which could lead to adulteration of the product. This may be accomplished by a joint effort between the HVAC engineer, owner, and a qualified protocol writer. If performance data are in the qualification records, replacement parts of different manufacture can be installed without major change control approvals, as long as they meet performance requirements. Owner approvals for the qualification plan should be obtained while proceeding with detailed design.

Qualification is the successful completion of the following activities for critical components and systems. The designer should understand the requirements for owner's approval of each protocol (usually, the owner approves the blank protocol form and the subsequently executed protocol).

The installation qualification (IQ) protocol is a record of construction inspection to verify compliance with contract documents, including completion of punchlist work, for critical components. It may include material test reports, receipt verification forms, shop inspection reports, motor rotation tests, and contractor-furnished testing and balancing. This record also includes calibration records for instrumentation used in commissioning and for installed instrumentation (such as sensors, recorders, transmitters, controllers, and actuators) traceable to National Institute of Standards and Technology (NIST) instruments.

Control software should be bench tested, and preliminary (starting) tuning parameters should be entered. Control loops should be dry-loop checked to verify that installation is correct. Equipment and instruments should be tagged and wiring labeled. Commissioning documentation must attest the completion of these activities. This includes as-built drawings and installation-operation-maintenance (IOM) manuals from the contractors and vendors.

The operational qualification (OQ) protocol documents the start-up that includes critical components. This includes individual performance testing of control loops under full operating pressure performed in a logical order (i.e., fan control before room pressure control). The commissioning agent must produce verification that operating parameters are within acceptance criteria.

The HVAC may be challenged under extremes of design load (where possible) to verify operation of alarms and recorders, to determine (and correct, if significant) weak points, and to verify control and door interlocks. Based on observations, informal alert values of critical parameters, which might signify abnormal operation, may be considered. Although product would not be adulterated at these parameter values, the owner's staff could assess an alarm and react to it before further deviations from normal operation were experienced.

Documented smoke tests will verify space pressure and airflow in critical rooms or inside containment hoods, and show airflow patterns and directions around critical parts of production equipment. Many smoke tests have been documented on videotape, especially when room pressure differentials are lower than acceptance criteria require and cannot be corrected.

Files should include an updated description of the HVAC, which describes how it operates, schematics, airflow diagrams, and room pressure maps that accompany it. Copies should be readily accessible and properly filed. Operating personnel should be familiar with the data shown in these GMP records and be able to explain it to an agency inspector.

Other Documents. GMP documents should also include test reports for HEPA filters (efficiency or pinhole-scan integrity tests) at final operating velocities. If the filter installer performed the tests, the data should have been part of the IQ package.

Documents should verify that instruments display, track, and store critical parameters and action alarms. (Consider recording data by exception and routine documentation of data at minimal regular frequency.)

Systems and equipment should be entered into the owner's maintenance program, with rough drafts of new maintenance procedures (final drafts should reflect commissioning experience).

Records should attest to the completion of these activities, including final as-built and air and water balance reports.

Performance qualification (PQ) is proof that the entire HVAC performs as intended under actual production conditions. PQ is the beginning of the ongoing verification that the system meets the acceptance criteria of the product (often called validation). This includes documentation of

- Maintenance record keeping and final operating and maintenance procedures in place, with recommended frequency of maintenance. The owner may also want a procedure for periodic challenge of the controls and alarms
- Logs of critical parameters that prove the system maintains acceptance criteria over a prescribed time
- Training records of operators and maintenance personnel
- Final loop tuning parameters

After acceptance of PQ, the owner's change control procedure should limit further modifications to critical components (as shown on IQ and OQ forms) that affect the product. Although much of the building's HVAC equipment should not need qualification, records for the entire facility should be kept and problems corrected before they become significant. Records of the corrections should also be maintained.

Once the system is operational, pharmaceutical product trial lots are run in the facility (process validation) and the owner should regularly monitor viable (microbial) and nonviable particles in the room.

SEMICONDUCTOR CLEANROOMS

Since the middle 1990s, most microelectronic facilities manufacturing semiconductors have required a cleanroom providing FS 209 Class 100 and cleaner. State-of-the-art cleanroom technology has been driven by the decreasing size of microelectronics circuitry and larger wafer sizes. A deposited particle having a diameter of 10% of the circuit width may cause a circuit to fail. With circuit line widths approaching 0.1 μm, particles of 0.01 μm are of concern. Many of today's facilities are designed to meet as-built air cleanliness of less than 1 particle 0.1 μm and larger per cubic foot of air.

Semiconductor Cleanroom Configuration

Semiconductor cleanrooms today are of two major configurations: the clean tunnel or the open-bay design. The **clean tunnel** is composed of narrow modular cleanrooms that may be completely isolated from each other. Fully HEPA or ULPA filtered pressurized plenums, ducted HEPA or ULPA filters, or individual fan modules are used in clean tunnel installations. Production equipment may be located within the tunnel or installed through the wall where a lower cleanliness level (nominally Class 10,000 or cleaner) service chase is adjacent to the clean tunnel. The service chase is used in conjunction with either sidewall return or a raised floor. A basement return may also be used with a raised floor.

The primary advantage of the tunnel is reduced HEPA or ULPA filter coverage and ease of expanding additional tunnel modules into unfiltered areas. The tunnel is typically between 8 and 14 ft wide. If the tunnel is narrower, production equipment cannot be placed on both sides. If it is wider, the flow becomes too turbulent and tends to break toward the walls before it leaves the work plane. Figure 9 is a clean tunnel element.

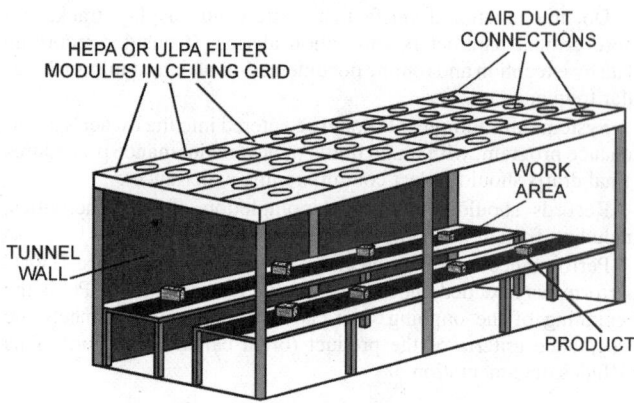

Fig. 9 Elements of a Clean Tunnel

The tunnel design also has the drawback of restricting new equipment layouts. Cleanroom flexibility is very valuable to semiconductor manufacturing logistics. As processes change and new equipment is installed, the clean tunnel may restrict equipment location to the point that a new module must be added. The tunnel approach may complicate the movement of product from one type of equipment to another.

The **open-bay design** involves large (up to 100,000 ft^2) open-construction cleanroom layouts. Interior walls may be placed wherever manufacturing logistics dictate, thus providing maximum equipment layout flexibility. Replacement of process equipment with newer equipment is an ongoing process for most wafer fabrication facilities where support services must be designed to handle different process equipment layouts and even changes in process function. Many times a manufacturer may completely redo the equipment layout if a new product is being made.

In the open-bay design, pressurized plenum or ducted filter modules are used, but pressurized plenums are becoming more common. When the pressurized plenum is used, either one large plenum with multiple supply fans or small adjacent plenums may be configured. Small plenums provide the ability to shut down areas of the cleanroom without disturbing other clean areas. Small plenums may also include one or more supply fans.

Major semiconductor facilities, with total manufacturing areas of 30,000 ft^2 and larger, may incorporate both open-bay and tunnel design configurations. Flexibility to allow equipment layout revisions warrants the open-bay design. Process equipment suitable for through-the-wall installation, such as diffusion furnaces, may use the tunnel design or open bay, whereas process equipment such as lithographic steppers and coaters must be located entirely under laminar flow conditions; thus, open-bay designs are more suitable. The decisions on which method to use, tunnel or open bay, should be discussed among the cleanroom designer, production personnel, and the contamination control specialist.

Many semiconductor facilities contain separate cleanrooms for process equipment ingress into the main factory These ingress areas are staged levels of cleanliness. For instance, the equipment receiving area may be Class 100,000 for equipment uncrating while the next stage may be Class 10,000 for preliminary equipment setup and inspection. The final stage, where equipment is cleaned and final installation preparations are made prior to the fabrication entrance, is Class 1000. In some cases, these staged cleanrooms must have adequate clear heights to allow forklifting of equipment subassemblies.

Airflow in Semiconductor Cleanrooms

Current semiconductor industry cleanrooms use **vertical unidirectional airflow**, which produces a uniform shower of clean air throughout the entire cleanroom. Particles are swept from personnel and process equipment, with contaminated air leaving at the floor level; this produces clean air for all space above the work surface. In contrast, horizontal airflow produces decreasing cleanliness downstream from the air inlet.

In vertical unidirectional airflow with cleanliness FS 209 Classes 1, 10, and 100, the cleanroom ceiling area consists of HEPA or ULPA filters set in a nominal grid size of 2 ft by 4 ft HEPA or ULPA filters are set into a T-bar-style grid with gasketed or caulked seals for many FS 209 Class 100 systems; Class 1 and Class 10 systems often use either low-vapor pressure petrolatum fluid or silicone dielectric gel to seal the HEPA or ULPA filters into a channel-shaped ceiling grid. Whether T-bar or channel-shaped grids are used, the HEPA or ULPA filters normally cover 85 to 95% of the ceiling area, with the remainder of the ceiling area composed of gridwork, lighting, and fire protection sprinkler panels.

HEPA or ULPA filters in vertical unidirectional airflow designs are installed (1) with a pressurized plenum above the filters, (2) through individually ducted filters, or (3) with individually fan-powered filter modules. A system with a plenum must provide even pressurization to maintain uniform airflow through each filter. Ducted HVAC typically has higher static pressure loss from the ducting and balance dampers. Higher maintenance costs may also be incurred due to the balance method involved with ducted HVAC.

Individual fan-powered filter modules use fractional horsepower fans (usually forward-curved fans) that provide airflow through one filter assembly. This method allows airflow to be varied throughout the cleanroom and requires less space for mechanical components. The disadvantages of this method are the large number of fans involved, low fan and motor efficiencies due to the small sizes, potentially higher fan noises, and higher maintenance costs.

When through-the-floor return grating is used, a basement return is normally included to provide a more uniform return as well as floor space for "dirty" production support equipment.

Sidewall returns are an alternative to through-the-floor returns; however, airflow may not be uniform throughout the work area. As previously stated, these returns are most applicable for cleanrooms with double sidewall returns and widths less than 14 ft.

Adequate prefiltration is an economical way to increase ULPA filter life. Prefilters are located in the recirculation airflow, either in the return basement or in the air handler. The prefilters should be located in the recirculated air to allow replacement without disrupting production.

Air Ionization. In addition to cleanroom particle control with fiber filters, air ionization is sometimes used to control particle attraction to product surfaces by eliminating electrostatic discharge and static charge buildup. However, the emitter tip material must be selected to prevent deposit of particles on the product.

HIGH-BAY CLEANROOMS

High-bay cleanrooms are those facilities with ceiling heights between 40 and 120 ft. They are used primarily in the aerospace industry for the production and testing of missiles, launch vehicles, rocket engineers, communication and observation satellites, and jet aircraft assembly, painting, and cleaning operations. Crystal-pulling areas in semiconductor labs also use high-bay designs.

Most high-bay cleanrooms are designed to meet Class 10,000, Class 100,000 or an unofficial Class 300,000 called out in some U.S. Air Force and U.S. Navy specifications. If Class 1000 or 100 is specified, the end user must be made to understand the tremendous economic impact this will have on the project costs.

Downflow and Horizontal Flow Designs

Downflow. In this design, air is delivered in a unidirectional (or simulated unidirectional) flow pattern from the ceiling and returned through floor return openings or low sidewall returns. The objective is to shower the object from above so that all particulate matter is flushed to the returns. The supply air terminals may be HEPA filter diffusers or high-volume air diffusers. Downflow rooms allow space flexibility because more than one device may be worked on in the room at the same air cleanliness level.

The disadvantage of downflow high-bay cleanrooms is the relative difficulty of balancing the airflow. High-bay cleanrooms typically have concrete floors that may include some trenches to return some of the air that is not taken in at low sidewall returns. Special care must be taken to ensure clean air at the object because the laminar flow pattern of the air disintegrates. At the low velocities associated with unidirectional design, pathways may be created toward the returns, causing the clean air to miss the object. Any activity in the cleanroom that generates even a small amount of heat produce updrafts in the downward-flowing supply air.

Horizontal Flow. Horizontal flow designs are always unidirectional with the cleanest air always available to wash the object in the room. Properly designed horizontal rooms are easier to balance than vertical flow rooms because the volumes of supply and return air may be controlled at different horizontal levels in the room.

The principal disadvantage of horizontal flow high-bay rooms is that they provide clean air for only one object, or at best, several objects, as long as they are in the same plane exposed to the supply air. Once past the object, the air cleanliness degrades to the extent the process generates that particles.

Downflow designs are most widely used, but certain projects such as the Space Telescope use horizontal high-bay cleanrooms.

Air-Handling

Due to the large volume of air in a high-bay cleanroom, a central recirculating fan system is commonly used with minimum heating and cooling capability. A separate injection air handler provides heating, cooling, and makeup air. The injection system must include volumetric controls to ensure proper building pressure.

Equipment and Filter Access

Air-handling equipment and prefilters should be accessible from outside the cleanroom. Adequate provision must be made for filter change-out if air is distributed to the cleanroom with HEPA filters at the room entry. In horizontal cleanrooms, access should be from the upstream (pressure) side, and service scaffolds should be incorporated at least every 8 ft in height of the filter bank. Downflow ceiling filters in T-bar ceilings or gel-seal ceilings must be accessed from below using an approved gantry crane with full mobility across the ceiling. Prefilters in the main air supply should be placed in built-up frames with both upstream and downstream access. A HEPA filter bank remote from the room air distribution system should be installed in a built-up bank with a gel or clamp seal. Access doors must then be installed upstream and downstream for certification, scanning, and qualification testing.

Prefilter Selection

In any high-bay cleanroom class (except Class 300,000), air will pass through a final HEPA filter before entering the room, and these final filters must be protected by prefilters. HEPA filters in recirculating air should be protected with 85% rated bag or rigid media filters with no other prefilters present than required. Makeup air should include minimum 85% ASHRAE filters on the fan inlet and minimum 95% DOP filters on the fan discharge. This design cleans the makeup air sufficiency so that it is no longer a variable, regardless of the particle count of the outside air.

ENVIRONMENTAL SYSTEMS

Cooling Loads and Cooling Methods

Two major internal heat load components in cleanroom facilities are process equipment and fans. Because most cleanrooms are located entirely within conditioned space, traditional heat sources of infiltration, fenestration, and heat conductance from adjoining spaces are typically less than 2 to 3% of the total load. Some cleanrooms have been built with windows to the outside, usually for daylight awareness, and a corridor separating the cleanroom window from the exterior window.

The major cooling sources designed to remove cleanroom heat and maintain environmental conditions are makeup air units, primary and secondary air units, and the process equipment cooling system. Some process heat, typically heat from electronic sources in process equipment computers and controllers, may be removed by the process exhaust.

Fan energy is a very large heat source in FS 209 Class 10 or better cleanrooms. Recirculated room airflow rates of 90 fpm, which equals 500 plus air changes per hour, are typical for FS 209 Class 10 or better cleanrooms.

Latent loads are primarily associated with makeup air dehumidification. A low dry-bulb leaving air temperature (35 to 45°F), associated with dehumidified makeup air, supplements sensible cooling. Supplemental cooling by makeup air may account for as much as 300 Btu/h per square foot of cleanroom.

Process cooling water (PCW) is used in process equipment heat exchangers, performing either simple heat transfer to cool internal heat sources, or process specific heat transfer, in which the PCW contributes to the process reaction.

The diversity of the manufacturing heat sources, that is, the portion of the total heat that is transferred to each cooling medium, should be well understood. When bulkhead or through-the wall equipment is used, the equipment heat loss to support chases versus to the production area will affect the cooling design when the support chase is served by a different cooling system than the production area.

Makeup Air

The control of makeup air and cleanroom exhaust affects cleanroom pressurization, humidity, and room cleanliness. The flow requirements of makeup air are dictated by the amount required to replace process exhaust and by air volumes for pressurization. Makeup air volumes can be much greater than the total process exhaust volume in order to provide adequate pressurization and safe ventilation.

Makeup air is frequently introduced into the primary air path on the suction side of the primary fan(s). Makeup volumes are adjusted with zone dampers and makeup fan controls using speed controllers, inlet vanes, etc. Opposed-blade dampers should have low leak characteristics and minimum hysteresis.

Makeup air should be filtered prior to injection into the cleanroom because it is the primary source of external particles in the cleanroom. If the makeup air is injected upstream of the cleanroom ceiling ULPA or HEPA filters, minimum 95% efficient filters (ASHRAE *Standard* 52.1) should be used to avoid high dust loading and reduced life of HEPA filters.

In addition, 30% efficient prefilters followed by 85% filters may be used to prolong the life of the 95% filter. When makeup air is injected downstream of the main HEPA filter, additional HEPA filtering of the makeup air should be added in with the previously mentioned prefilters. In addition to particle filtering, many makeup air handlers require filters to remove chemical contaminants present in the outside air. These contaminants include salts and pollutants from industries and automobiles. Chemical filtration may be accomplished with absorbers such as activated carbon or potassium permanganate impregnated with activated alumina or zeolite.

Process Exhaust

Process exhausts for semiconductor facilities handle acid, solvent, toxic, pyrophoric (self-igniting) fumes, and process heat exhaust. Process exhaust should be dedicated for each fume category, dedicated by process area, or based on the chemical nature of the fume and its compatibility with exhaust duct material. Typically, process exhausts are segregated into corrosive fumes, which are ducted through plastic or fiberglass reinforced plastic (FRP) ducts, and flammable (normally from solvents) gases and heat exhaust, which are ducted in metal ducts. Care must be taken to ensure that gases cannot combine into hazardous compounds that can ignite or explode within the ductwork. Segregated heat exhausts are sometimes installed to recover heat, or hot uncontaminated air that may be exhausted into the suction side of the primary air path.

Required process exhaust volumes vary from 1 cfm per square foot of cleanroom for photolithographic process areas, to 10 cfm/ft^2 for wet etch, diffusion, and implant process areas. When specific process layouts have not been designated prior to exhaust design, an average of 5 cfm/ft^2 is normally acceptable for fan and abatement equipment sizing. Fume exhaust ductwork should be sized at low velocities (1000 fpm) to allow for future needs.

For many airborne substances, the American Conference of Governmental Industrial Hygienists (ACGIH) has established requirements to avoid excessive worker exposure. The U.S. Occupational Safety and Health Administration (OSHA) has set specific standards for the allowable concentration of airborne substances. These limits are based on working experience, laboratory research, and medical data; and they are subject to constant revision. *Industrial Ventilation: A Manual of Recommended Practices* (ACGIH 1998) may be referred to when limits are to be determined.

Fire Safety for the Exhaust

The Uniform Building Code (UBC) designates semiconductor fabrication facilities as Group H, Division 6, occupancies. The H-6 occupancy class should be reviewed even if the local jurisdiction does not use the UBC because it is currently the only major code in the United States specially written for the semiconductor industry and, hence, can be considered usual practice. This review is particularly helpful if the local jurisdiction has few semiconductor facilities.

Article 51 in the *Uniform Fire Code* (UFC) addresses the specific requirements for process exhaust relating to fire safety and minimum exhaust standards. UFC Article 80, Hazardous Materials, is relevant to many semiconductor cleanroom projects due to the large quantities of hazardous materials stored in these areas. Areas covered include ventilation and exhaust standards for production and storage areas, control requirements, use of gas detectors, redundancy and emergency power, and duct fire protection.

Temperature and Humidity

Precise temperature control is required in most semiconductor cleanrooms. Specific chemical processes may change under different temperatures, or masking alignment errors may occur due to product dimensional changes as a result of the coefficient of expansion. Temperature tolerances of ±1°F are common, and precision of 0.1 to 0.5°F is likely in wafer or mask writing process areas. Wafer reticle writing by electron beam technology requires ±0.1°F while photolithographic projection printers require ±0.5°F tolerance. Specific process temperature control zones must be small enough to counteract the large air volume inertia in vertical laminar flow cleanrooms. Internal environmental controls, which allow for room tolerances of ±1°F and larger temperature control zones, are used in many process areas.

Within temperature zones of the typical semiconductor factory, latent heat loads are normally small enough to be offset by incoming makeup air. Sensible temperature is controlled with either (1) cool-ing coils in the primary air stream, or (2) unitary sensible cooling units that bypass primary air through the sensible air handler and blend conditioned air with unconditioned primary air.

In most cleanrooms of FS 209 Class 1000 or better, production personnel wear full coverage protective smocks that require cleanroom temperatures of 68°F or less. If full-coverage smocks are not used, lower temperature set points are recommended for comfort. Process temperature set points may be higher as long as product tolerances are maintained.

Semiconductor humidity levels vary from 30 to 50% rh. Humidity control and precision are functions of process requirements, prevention of condensation on cold surfaces within the cleanroom, and control of static electric forces. Humidity tolerances vary from 0.5 to 5% rh, primarily dictated by process requirements. Photolithographic areas have the more precise standards and lower set points. Photoresists are chemicals used in photolithography, and their exposure timing can be affected by varying relative humidity. Negative resists typically require low (35 to 45%) relative humidity. Positive resists tend to be more stable, so the relative humidity can go up to 50% rh where there is less of a static electricity problem.

Independent makeup units should control the dew point in places where direct-expansion refrigeration, chilled water/glycol cooling coils, or chemical dehumidification are used. Chemical dehumidification is rarely used in semiconductor facilities due to the high maintenance cost and the potential for chemical contamination in the cleanroom. While an operating cleanroom generally does not require reheat, systems are typically designed to provide heat to the space when new cleanrooms are being built and no production equipment has been installed.

Makeup air is humidified by steam humidifiers or atomizing equipment. Steam humidifiers are most commonly used. Care should be taken to avoid the potential release of water treatment chemicals. Stainless-steel unitary packaged boilers with high-purity water and stainless-steel piping have also been used. Water sprayers, located in the cleanroom return, use air-operated water jet sprayers. Evaporative coolers have been used, taking advantage of the sensible cooling effect in dry climates.

Pressurization

Pressurization of semiconductor cleanrooms is another method of contamination control, providing resistance to infiltration of external sources of contaminants. Outside particulate contaminates enter the cleanroom by infiltration through doors, cracks, passthroughs, and other process-related penetrations for pipes, ducts, etc. Positive pressure in the cleanroom (as referenced to any less clean space) ensures that air flows from the cleanest space to the less clean space. Positive differential pressure in the cleanroom inhibits the entrance of unfiltered external particulate contamination.

A differential pressure of 0.05 in. of water is a widely used standard. The cleanest cleanroom should maintain the highest pressure, with pressure decreasing corresponding to decreasing cleanliness.

Pressure in the cleanroom is principally established by the balance between process exhaust, leakage, makeup air volumes, and the supply and return air volumes. Process equipment vendors and industrial hygienists dictate process exhaust requirements, and they cannot be changed without safety risks. Cleanroom supply air volumes are set by contamination control specialists. Control of makeup air and return air volumes are the primary means for pressure control. Pressure differences should be kept as low as possible while still creating the proper flow direction. Large pressure differences can create eddy currents at wall openings and cause vibration problems.

Static or active control methods are normally used in cleanroom pressure control. One method is used in lieu of the other based upon pressure control tolerance. Pressure control precision is typically ±0.01 to ±0.03 in. of water; the owner's contamination control specialist specifies the degree of precision required. Many semiconductor processes where cleanroom pressure affects the process itself (e.g., glass deposition with saline gas) require process chamber pressure precision of ±0.0025 in. of water.

Static pressure control methods are suited for unchanging cleanroom environments, where the primary pressure control parameters (process exhaust and supply air volumes) either do not change or change slowly over weeks or months at a time. Static controls provide initial room pressure, and monthly or quarterly maintenance adjusts makeup and return volumes if the pressure level has changed. Static systems may include differential pressure gages for visual monitoring by maintenance personnel.

Active systems provide closed-loop control where pressure control is critical. Standard controls normally cannot maintain the differential pressure when doors without an air lock are opened. Active systems should be evaluated as to their need, however.

Air locks may be used to segregate pressures in the factory, but typically, air locks are used between uncontrolled personnel corridors, entrance foyers, and the protective-clothing gowning area. Air locks may also be used between the gowning room and the main wafer fabrication area and for process equipment staging areas prior to ingress into the wafer fabrication area. Within the main portion of the factory, air locks are rarely used because they restrict personnel access, evacuation routes, and traffic control.

Sizing and Redundancy

The design of environmental HVAC must consider future requirements of the factory. Semiconductor products can become obsolete in as little as two years, and process equipment may be replaced as new product designs dictate. As new processes are added or old ones deleted (e.g., wet etch versus dry etch) the function of one cleanroom may change from high-humidity requirements to low humidity, or the heat load many increase or decrease. Thus, the cleanroom designer must design for flexibility and growth. Unless specific process equipment layouts are available, maximum cooling capability should be provided in all process areas at the time of installation, or space should be provided for future installations.

Because cleanroom space relative humidity must be held to close tolerances, and humidity excursions cannot be tolerated, the latent load removal should be based on high ambient dew points and not on the high mean coincident dry-bulb/wet-bulb data.

In addition to proper equipment sizing, redundancy is also desirable when economics dictate it. Many semiconductor wafer facilities operate 24 hours per day, seven days per week, and shut down only during holidays and scheduled nonwork times. Mechanical and electrical redundancy is required if the loss of such equipment would shut down critical and expensive manufacturing processes. For example, process exhaust fans must operate continuously for safety reasons, while particularly hazardous exhaust should have two fans, both running. The majority of the process equipment is computer-controlled with interlocks to provide safety for personnel and products. Electrical redundancy or uninterrupted power supplies may be necessary to prevent costly downtime during power outages. Decisions regarding redundancy should be based on life-cycle economics.

ENERGY CONSERVATION IN CLEANROOMS

The major operating costs associated with a cleanroom include conditioning of makeup air, air movement in the cleanroom, and process exhaust. Environmental control, contamination control, and process equipment electrical loads can be as much as 300 W/ft². Besides process equipment electrical loads, most energy is used for cooling, air movement, and process liquid transport (i.e., deionized water and process cooling water pumping). A life-cycle cost analysis should determine the extent of energy reductions.

Fan Energy. Because flow rates in typical semiconductor facilities are 90 to 100 times greater than in conventional HVAC, and still very high in other cleanrooms, the fan system should be closely examined for ways to conserve energy. Static pressures and total airflow requirements should be designed to reduce operating costs. The fan energy required to move recirculation air may be decreased by reducing the air volume or static pressure. Energy conservation operating modes should be verified during qualification of the system. If these modes are not part of the original design, the control procedure must be changed and the operational change validated.

Air volumes may be lowered by decreasing HEPA or ULPA filter coverage or by reducing the cleanroom average air velocity. When air volumes are reduced, each square foot of reduced HEPA coverage saves 25 to 50 W/ft² in fan energy and the same amount in cooling load. Reducing room average velocity from 90 to 80 fpm saves 5 W/ft² in fan energy and the same amount in cooling energy. If the amount of air supplied to the cleanroom cannot be lowered, reductions in static pressure can produce significant savings. With good fan selection and transport design, up to 15 W/ft² can be saved per 1 in. of water reduction in static pressure. Installing low pressure drop HEPA filters, pressurized plenums in lieu of ducted filters, and proper fan inlets and outlets, may reduce static pressure. Many cleanrooms operate for only one shift. Air volume may be reduced during nonworking hours by using two-speed motors, variable-frequency drives, inverters, inlet vanes, and variable-pitch fans or, in multifan systems, by using only some of the fans.

Additional energy may be saved by installing high-efficiency motors on fans instead of standard-efficiency motors. Good fan selection also influences energy cost. The choice of forward-curved centrifugal fans versus backward-inclined, airfoil, or vaneaxial fans affects efficiency. The number of fans used in a pressurized plenum design influences redundancy as well as total energy use. The fan size changes power requirements as well. Different options should be investigated.

Makeup and Exhaust Energy. As stated previously, process exhaust requirements in the typical semiconductor facility vary from 1 to 10 cfm per square foot. Makeup air requirements vary correspondingly, with an added amount for leakage and pressurization. The energy required to supply the conditioned makeup air can be quite large. The type of equipment installed normally determines the quantity of exhaust in a given facility. Heat recovery has been used effectively in process exhaust. When heat recovery is used, the heat exchanger material must be selected carefully due to the potentially corrosive atmosphere. Also, heat recovery equipment has the potential to cross-contaminating products in pharmaceutical facilities.

Makeup air cannot normally be reduced without decreasing the process exhaust. Due to safety and contamination control requirements, process exhaust may be difficult to reduce. Therefore, the costs of conditioning the makeup air should be investigated. Conventional HVAC methods, such as use of high-efficiency chillers, good equipment selection, and precise control design can also save energy. One energy-saving method uses multiple-temperature chillers to bring outdoor air temperature for large facilities to a desired dew point in steps.

NOISE AND VIBRATION CONTROL

Noise is difficult to control. Noise generated by contamination control equipment requires particular attention, although production equipment noise may be more significant than HVAC noise. Prior to the start of the design, the noise and vibration criteria should be established. Chapter 46 provides more complete information on sound control.

In normal applications of microelectronics contamination control, equipment vibration displacement levels need not be dampened below 0.5 μm in the 1 to 50 Hz range. However, electron microscopes and other ultrasensitive microelectronics cleanroom instruments may require smaller deflections in different frequency ranges. Photolithographic areas may prohibit floor deflections greater than 0.075 μm. As a general rule, the displacement should not exceed one-tenth the line width.

For highly critical areas, vaneaxial fans may be considered. These fans generate less noise in the lower frequencies, and they may be dynamically balanced to displacements of less than 4 μm, which decreases the likelihood of transmitting vibration to sensitive areas in electronics cleanrooms.

ROOM CONSTRUCTION AND OPERATION

Control of particulate contamination from sources other than the supply air depends on the classification of the space, the type of system, and the operation involved. Typical details that may vary with the room class include the following:

Construction Finishes

- **General.** Smooth, monolithic, cleanable, and chip-resistant, with minimum seams, joints, and no crevices or moldings
- **Floors.** Sheet vinyl, epoxy, or polyester coating with wall base carried-up, or raised floor (where approved) with and without perforations using the above materials
- **Walls.** Plastic, epoxy-coated drywall, baked enamel, polyester, or porcelain with minimum projections
- **Ceilings.** Plaster covered with plastic, epoxy, or polyester coating or with plastic-finished clipped acoustical tiles (no tiles in pharmaceutical cleanrooms) when entire ceiling is not fully HEPA or ULPA filtered
- **Lights.** Teardrop-shaped single lamp fixtures mounted between filters, or flush-mounted and sealed
- **Service penetrations.** All penetrations for pipes, ducts, conduit runs, etc. fully sealed or gasketed
- **Appurtenances.** All doors, vision panels, switches, clocks, etc. either flush-mounted or sloped tops
- **Windows.** All windows flush with wall; no ledges on cleanest side

Personnel and Garments

- Hands and face cleaned before entering area
- Lotions and soap containing lanolin to lessen the emission of skin particles
- No cosmetics and skin medications
- No smoking and eating
- Lint-free smocks, coveralls, gloves, head covers, and shoe covers

Materials and Equipment

- Equipment and materials are cleaned before entry.
- Nonshedding paper and ballpoint pens are used. Pencils and erasers are not permitted.
- Work parts are handled with gloved hands, finger cots, tweezers, and other methods to avoid transfer of skin oils and particles.
- Sterile pharmaceutical product containers must be handled with sterilized tools only.

Particulate Producing Operations

- Electronics grinding, welding, and soldering operations are shielded and exhausted.
- Nonshedding containers and pallets are used for transfer and storage of materials.

Entries

- Air locks and pass-throughs maintain pressure differentials and reduce contamination.

BIBLIOGRAPHY

ACGIH. 1989. Guidelines for the assessment of bioaerosols in the indoor environment.

ACGIH. 1998. *Industrial ventilation: A manual of recommended practice*, 23rd ed. American Conference of Governmental Industrial Hygienists, Cincinnati.

British Standard. 1989. Environmental cleanliness in enclosed spaces. BS 5295.

CFR. 1992. Boiler water additives. Chapter 1, Section 173.310, *Code of Federal Regulations* 21.

European Union. 1997. Manufacture of sterile medical products. Revision of Annex I to the *EU guide to good manufacturing practice*.

FDA. 1991. Current good manufacturing practice for finished pharmaceuticals. CFR 21, Parts 210,211.

FDA. 1987. Guidelines for sterile drug products producted by septic processing. Republished 1991.

Federal Standard. 1992. Airborne particulate cleanliness classes in cleanrooms and clean zones. FS 209E.

The following publications are available from the Institute of Environmental Sciences and Technology, Mount Prospect, Illinois.

IEST-RP-CC001.3	HEPA and ULPA Filters
IEST-RP-CC002	Laminar-Flow Clean-Air Devices
IEST-RP-CC003.2	Garment System Considerations for Cleanrooms and Other Controlled Environments
IEST-RP-CC004.2	Evaluating Wiping Materials Used in Cleanrooms and Other Controlled Environments.
IEST-RP-CC005	Cleanroom Gloves and Finger Cots
IEST-RP-C-006.2	Testing Cleanrooms
IEST-RP-CC007.1	Testing ULPA Filters
IEST-RP CC008	Gas-Phase Adsorber Cells
IEST-RP-CC009.2	Compendium of Standards, Practices, Methods, and Similar Documents Relating to Contamination Control
IEST-RP-CC011.2	A Glossary of Terms and Definitions Relating to Contamination Control
IEST-RP-CC012.1	Cleanroom Design Considerations
IEST-RP-CC013	Procedures for the Calibration or Validation of Equipment
IEST-RP-CC014	Calibrating Particle Counters
IEST-RP-CC015	Installation of Cleanroom Production Equipment
IESR-RP-CC016	The Rate of Deposition of Nonvolatile Residue in Cleanrooms
IEST-RP-CC017	Ultrapure Water: Contamination Analysis and Control
IEST-RP-CC018	Cleanroom Housekeeping---Operating and Monitoring Procedures
IEST-RP-CC019	Qualifications for Agencies and Personnel Engaged in the Testing and Certification of Cleanrooms and Clean Air Devices
IEST-RP-CC020	Substrates and Forms for Documentation in Cleanrooms
IEST-RP-CC021	Testing HEPA and ULPA Filter Media
IEST-RP-CC022.1	Electrostatic Charge in Cleanrooms and Other Controlled Environments
IEST-RP-CC023.1	Microorganisms in Cleanrooms
IEST-RP-CC024	Measuring and Reporting Vibration in Microelectronics Facilities
IEST-RP-CC025	Evaluation of Swabs Used in Cleanrooms
IEST-RP-CC026	Cleanroom Operations
IEST-RP-CC027	Personnel in Cleanrooms
IEST-RP-CC028	Minienvironments
IEST-RP-CC029	Automotive Paint Spray Applications
IEST-RP-CC100	Federal Standard 209
IEST-RP-CC1246	Military Standard 1246

ISO/FDIS. Cleanrooms and associated controlled environments—Part 1: Classification of airborne particulates. *Standard* 14644-1.

ISPE/FDA. 1997. Baseline guide for oral solid dosage facilities. International Society for Pharmaceutical Engineering, Tampa, FL.

ISPE/FDA. 1999. Baseline guide for sterile manufacturing facilities. International Society for Pharmaceutical Engineering, Tampa, FL.

NFPA. 1995. Standard for the protection of cleanrooms. *Standard* 318. National Fire Protection Association, Quincy, MA.

U.K. Dept. of Health. 1997. Guide to good pharmaceutical manufacturing practice.

Whyte, W. 1991. *Cleanroom design*. John Wiley, New York.

CHAPTER 16

DATA PROCESSING AND
ELECTRONIC OFFICE AREAS

DATA PROCESSING and electronic office areas contain computers, electronic equipment, and peripheral equipment needed for data processing functions. Computer equipment has expanded beyond the traditional computer room to become an integral part of the entire office environment.

Computers and electronic equipment generate heat and gaseous contaminants; such equipment usually contains components vulnerable to temperature and humidity variations, dust and other air impurities, and static electrical discharge. Exposure to conditions deviating from prescribed limits can cause improper operation or complete shutdown of equipment and may substantially reduce equipment useful life expectancy (Weschler and Shields 1991). Loss of equipment function or data can cause significant, sometimes irreparable, damage to organizations that depend on it.

Ancillary spaces for computer-related activities or for storage of computer components and materials may require environmental conditions comparable to those where the computers are housed, although tolerances may be wider and the degree of criticality is usually much lower. Other ancillary areas housing such auxiliary equipment as engine generators, motor generators, uninterruptible power supplies (UPSs), and transformers have less stringent air-conditioning and ventilating requirements than do computer areas, but the continuing satisfactory operation of this equipment is vital to the proper functioning of the computer system.

DESIGN CRITERIA

Computer Rooms

Data processing spaces that house mainframe computers, computer personnel, and associated equipment require air conditioning

Table 1 Typical Computer Room Design Conditions

Condition	Recommended Level
Temperature control range[a]	72°F ± 2°F
Relative humidity control range[a]	50 ± 5%
Filtration quality[b]	45%, min. 20%

[a]These conditions are typical of those recommended by most computer equipment manufacturers.
[b]From ASHRAE *Standard* 52.1 dust-spot efficiency test.

Table 2 Design Conditions for Air Supply Direct to Computer Equipment

Condition	Recommended Level
Temperature	As required for heat dissipation
Relative humidity[a]	Maximum 65%
Filtration quality[b]	45% minimum

[a] Some manufacturers permit up to 80% rh.
[b] As per dust spot efficiency test in ASHRAE *Standard* 52.1.

The preparation of this chapter is assigned to TC 9.2, Industrial Air Conditioning.

to maintain proper environmental conditions for both the equipment and personnel. Computer room environments should be maintained within established limits and not be subjected to rapid changes.

The environmental conditions required by computer equipment vary, depending on the manufacturer and the type of equipment supported. Table 1 lists conditions recommended for most computer rooms. Most computer manufacturers recommend that the computer equipment draw conditioned air from the room. Some equipment may require direct cooling using conditioned air, chilled water, or refrigerant. Criteria for air-cooled computers may differ from those cooled by room air in that this air must remove computer equipment heat adequately and preclude the possibility of condensation of moisture within the equipment (see Table 2).

Because of the large amount of heat released in data processing areas, designs should minimize new energy use for cooling, humidification, and dehumidification.

Room Environmental Requirements

The distribution of air in the room and/or the response of controls may not match the nonuniform heat distribution of the equipment. Modern computer equipment is controlled to minimize energy input, which results in a much wider variation in power consumption and heat release than for older equipment.

High relative humidity may cause improper feeding of paper, corrosion, and, in extreme cases, condensation on cold surfaces of direct-cooled equipment. Low relative humidity, combined with other factors, may result in static discharge, which can destroy or adversely affect the operation of data processing and electronic equipment. Volatile organic compounds can plate onto critical surfaces like disk drives, causing the head to crash, and onto printed circuit boards where dust and static electricity can induce stray currents in very low voltage applications.

Before it is introduced into the computer room, outdoor air should be treated and preconditioned to remove dust, salts, and corrosive gases. Dust can adversely affect the operation of data processing equipment, so high-quality filtration and proper filter maintenance are essential. Corrosive gases can quickly destroy the thin metal films and conductors used in printed circuit boards, and corrosion can cause high resistance at terminal connection points.

Computer room air conditioning must provide adequate outdoor air (1) to dilute internally generated contaminants and (2) to maintain the room under positive pressure relative to surrounding spaces. The need to maintain positive pressure to keep contaminants out of the room is usually a controlling design criterion. An outdoor air quantity of 6 to 8 air changes per day usually satisfies contaminant dilution requirements (Weschler and Shields 1991). Although most computer rooms have few occupants, ventilation for human occupancy should be provided in accordance with ASHRAE *Standard* 62. Where outdoor air is diluted with recirculated air from the computer room or adjacent areas, higher ventilation rates are required.

The air-conditioning equipment for computer spaces should be served by electrically isolated power sources to prevent electrical noise from adversely affecting computer equipment operation and reliability. Equipment located in computer spaces should be independently supported and isolated to prevent vibration transmission to the computer equipment. Computer equipment manufacturers should be consulted regarding computer equipment sound tolerance and specific requirements for vibration isolation.

Temperature, humidity, and filtration requirements for the equipment are within the comfort range for room occupants, but drafts and cold surfaces must be minimized in occupied areas. Some manufacturers have established criteria for allowable rates of environmental change to prevent shock to the computer equipment. These can usually be satisfied by high-quality, commercially available controls with control ranges of ± 1°F and ± 5% rh. The manufacturer's requirements should be reviewed and fulfilled to ensure that the system will function properly during normal operation and during periods of start-up and shutdown. Computer equipment usually tolerates a somewhat wider range of environmental conditions when idle, but it may be desirable to operate the air conditioning to keep the room within those limits and minimize thermal shock to the equipment.

Because computer technology is continually changing, computer equipment in a given space will be changed and/or rearranged during the life of an air-conditioning system. The system must be sufficiently flexible to permit rearrangement of components and expansion without requiring rebuilding. In the typical installation, it should be possible to modify the system without extensive air-conditioning shutdowns. In critical applications, it should be possible to modify the system without shutdown.

Isolation of Computer Spaces

Computer equipment spaces are usually isolated for security and environmental control. To maintain proper relative humidity in computer rooms in otherwise unhumidified spaces, vapor retarders sufficient to restrain moisture migration during the maximum expected vapor pressure differences between the computer room and surrounding areas should be installed around the entire envelope. Cable and pipe entrances should be sealed and caulked with a vapor-retarding material. Door jambs should fit tightly. In exterior walls in colder climates, windows should be double- or triple-glazed and door seals are required.

Ancillary Spaces Environment

Storage spaces for products such as paper and tapes generally require conditions similar to those in the computer room.

Electrical power supply and conditioning equipment can tolerate more variation in temperature and humidity than computer equipment. Equipment in this category includes motor generators, UPSs, batteries, voltage regulators, and transformers. Ventilation to remove heat from the equipment is normally sufficient. Manufacturers' data should be checked to determine the amount of heat release and the design conditions for satisfactory operation.

Battery rooms for UPSs require ventilation to remove hydrogen and to control the space temperature. The optimum space temperature is 77°F. Temperatures maintained higher or lower reduce the ability of batteries to hold a charge. Hydrogen accumulation may be no greater than 3% by volume, and the ventilation should be designed to prevent pockets of concentration, particularly at the ceiling.

Engine generators used for primary or emergency power require large amounts of ventilation when running. This equipment is easier to start if a low ambient temperature is avoided.

COOLING LOADS

The major heat source in a computer room is the equipment. This heat tends to be highly concentrated, nonuniformly distributed, and increasingly variable. Computer components that generate large

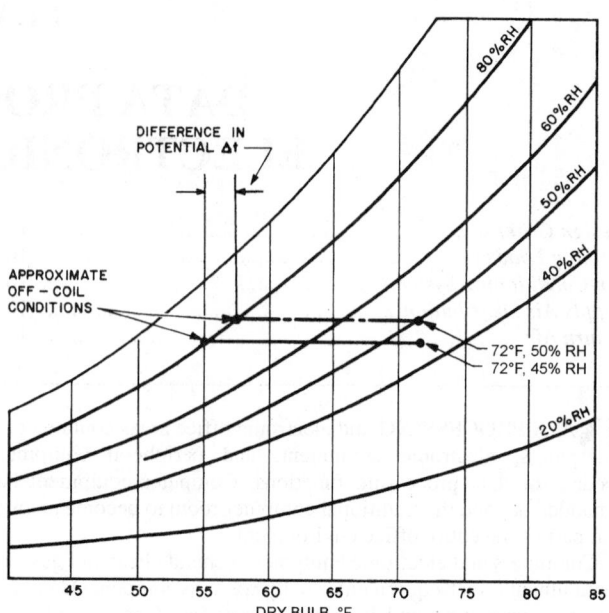

Fig. 1 Potential Effect of Room Design Conditions on Design Supply Air Quantity

quantities of heat are normally constructed with internal fans and passages to convey cooling air, usually drawn from the space, through the machine. Heat gain from lights should be no greater than that in good-quality office space; occupancy loads will be low to moderate. Heat gains through the structure depend on the location and construction of the room. Transmission heat gain to the space should be carefully evaluated and provided for in the design. Vapor retarder analyses should be performed where humidity-controlled spaces contain windows on outside walls.

Information on computer equipment heat release should be obtained from the specific computer manufacturers. In general, systems should be sized without a reduction for diversity, unless the computer manufacturer or experience with a similar installation recommends it.

Heat generated in computer rooms is almost entirely sensible. For this reason and because of the low room design temperatures, air supply quantity per unit of cooling load will be greater than for most comfort applications. Figure 1 shows that choosing room design at 72°F, 45% rh can reduce air supply quantity by approximately 15% compared to a slightly more humid room design of 72°F, 50% rh. A sensible heat ratio between 0.9 and 1.0 is common for computer room applications.

AIR-CONDITIONING SYSTEMS

It may be desirable for air-handling systems for data processing areas to be independent of other systems in the building, although cross-connection with other systems may be desirable for backup. Redundant air-handling equipment is frequently used, normally with automatic operation. The air-handling facilities should provide ventilation air, air filtration, cooling and dehumidification, humidification, and heating.

The refrigeration systems should be independent of other systems and may be required year-round, depending on design approach.

Computer rooms are being successfully conditioned with a wide variety of systems, including packaged precision air-conditioning units and central station air-handling systems. Air-handling and refrigeration equipment may be located either inside or outside computer rooms.

Precision Air-Conditioning Units

Precision air-conditioning units should be specifically designed for computer room applications and built to and tested in accordance with the requirements of ASHRAE *Standard* 127. Precision air-conditioning units are available with chilled water or multiple refrigerant compressors with separate refrigeration circuits; air filters; humidifiers; reheat; and integrated control systems with remote monitoring panels and interfaces. These units may also be equipped with dry coolers and propylene glycol precooling coils to permit water-side economizer operation where weather conditions make this strategy economical.

Self-contained air conditioners are usually located within the computer room but may also be remotely located and ducted to the conditioned space. If they are remote, their temperature and humidity controls should be in the conditioned space. Locating the air conditioner close to the load improves system flexibility to accommodate the changing load patterns common in computer rooms. Location in the computer room is practical because remote units may not have security protection beyond standard building maintenance. Where security is an issue, special attention should be paid to protecting the refrigerant condensing equipment for remote units.

In systems using precision air-conditioning units, it may be advantageous to introduce outdoor air through a dedicated system serving all data processing areas. Redundancy can be achieved with multiple units so that the loss of one or more units has less impact on overall system performance. Expansion of an existing data processing facility is often easier with precision air-conditioning units than with central station systems.

A set point of 72°F with constant volume precision air-conditioning units generally permits equipment to remain at an acceptable temperature within the established range for satisfactory operation. This low control set-point temperature provides a cushion for short-term peak load temperature rise without adversely affecting computer operation. Compared to constant volume units, variable volume equipment can be sized to provide excess capacity but will operate at discharge temperatures appropriate for optimum humidity control, minimize operational fan horsepower requirements, provide superior control over space temperature, and reduce the need for reheat.

When chilled water precision air-conditioning units are used, the reliability of the remote refrigeration system must be considered. This system generally must be capable of operating 24 h per day, and provisions must be made for a year-round supply of chilled water. Chilled water precision air-conditioning units do not contain refrigeration equipment; and they generally require less servicing, can be more efficient, and more readily support heat recovery strategies than direct-expansion equipment.

Central Station Air-Handling Units

Central station supply systems must be designed to accommodate expanding loads in the computer areas. These systems must include filters, humidification, dehumidification, and controls. Central station supply systems have a larger capacity than precision air-conditioning units and offer significantly greater opportunities for energy conservation. Systems that use direct evaporative humidification with mechanical cooling and variable volume control allow the use of outdoor air for free cooling without a humidification energy penalty (see Chapter 50, Evaporative Cooling Applications). Waste heat from computer equipment can be used to provide the heat of vaporization.

By using discharge air temperature as a method of dew-point control, variable volume ventilation from air-handling equipment, using evaporative equipment with a saturation effectiveness of 90% or higher, can provide effective humidity control without humidity sensing or active control. Figure 2 illustrates a central station unit

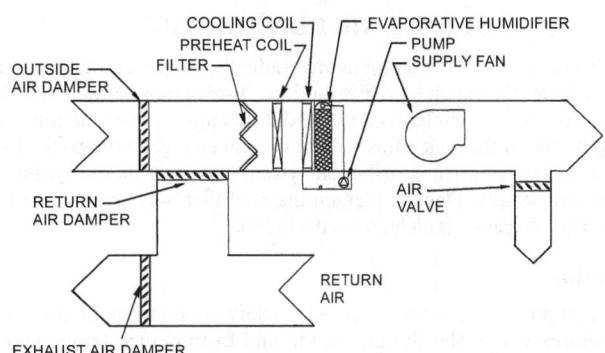

Fig. 2   Constant Temperature/Dew-Point Air-Handling Unit

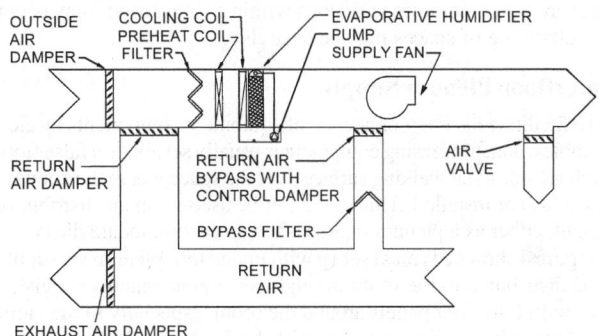

Fig. 3   Air-Handling Unit with Return Air Bypass

configuration that provides air at a constant temperature and dew point. As this cool air is provided at rates to maintain space temperature design conditions, space relative humidity becomes a direct function of space temperature, eliminating the need for expensive and unreliable humidity controls. Efficiency is increased by the evaporative process, which can use waste heat from the computers to provide the necessary heat of vaporization for humidification.

Because data processing areas have an extremely high sensible heat ratio, variable volume becomes a viable strategy, further reducing energy use. The evaporative process eliminates the need for refrigeration at temperatures below the supply air temperature of the system and reduces the need for refrigeration for significant periods of time when outside air conditions permit an evaporative cooling benefit. Stable, less-than-saturated air conditions can be achieved by bypassing system return air around the evaporative process to warm and dehumidify the supply air (see Figure 3).

Central station air-handling equipment should be arranged for convenient servicing and maintenance. Flexibility and redundancy can be achieved by using variable volume distribution, oversizing, cross-connecting multiple systems, or providing standby equipment. No floor space in the computer room is required, and sources of water and refrigeration may be eliminated from the computer room. Virtually all servicing and maintenance operations are performed in areas devoted specifically to air-conditioning equipment. System security issues, however, must be addressed.

Central systems serving computer rooms may serve other building functions if designed to provide continuous, year-round operation and to meet the cooling and humidification requirements. Central systems are superior to precision air-conditioning units for the electronic office area where overall environmental requirements may approach those of the computer room. Care should be taken to exercise redundant equipment frequently to prevent conditions that enhance the growth of mold and mildew in filters, insulated unit enclosures, and outdoor air pathways where spores and food sources for microbial growth may accumulate.

SUPPLY AIR DISTRIBUTION

To minimize room temperature gradients, supply air distribution should closely match load distribution. Distribution systems should be sufficiently flexible to accommodate changes in the location and magnitude of the heat gains with minimum change in the basic distribution system. Air distribution system materials should ensure a clean air supply. Duct or plenum material that may erode must be avoided. Access for cleaning is desirable.

Zoning

Computer rooms should be adequately zoned to maintain temperatures within the design criteria and to minimize temperature variations due to load fluctuations. Individual control for each major zone is desirable. The controlling thermostat must be located where it will sense the average conditions in the area it serves. In larger areas, temperature variations within a single room may occur, and subzoning of spaces may be desirable.

Underfloor Plenum Supply

To facilitate the interconnection of equipment components by electric cables, data processing equipment is usually set above a false floor, which affords a flat walking surface over the space where the connecting cables are installed. This space can be used as an air distribution channel, either as a plenum or, less often, to accommodate ducts.

Figure 4 shows a typical set up with underfloor plenum air supply. Air is distributed to the room through perforated panels or registers set or built into floor panels around the room, especially in the vicinity of computer equipment with high heat release. Airflow can be controlled by the size, location, and quantity of floor registers and perforated floor panels. These systems have the flexibility to accommodate relocation of computer equipment and future additional heat loads. All air should flow through openings in the underfloor cavity unless direct flow to a computer unit is desired. The potential adverse effect of direct air supply on the computer equipment (i.e., condensation within the machine) is serious enough to discourage its use unless the manufacturer requires it. Most openings between the underfloor space and the equipment usually exist to accommodate cables; collars are available that fit the cable and seal the opening.

Floor panels can be similar in appearance to, and interchangeable with, conventional computer room floor panels. The free area of floor panels may vary significantly between manufacturers. Because they are completely flush with the floor, perforated floor panels are suitable for installation in normal traffic aisles. With moderate airflows, panels located fairly close to equipment can produce a high degree of mixing near the point of discharge and will better prevent drafts and injection of unmixed conditioned air into computers than floor registers. Figure 5 shows typical floor panel air performance.

Floor-mounted registers allow volume adjustment and are capable of longer throws and better directional control than perforated floor outlets. Floor registers, however, should be located outside traffic areas. Some are not flush-mounted, and almost all tend to be draftier for nearby personnel than perforated floor panels with lower induction ratios.

Sufficient clearance to permit airflow must exist beneath a raised plenum floor: 12 in. of clearance is desirable, and 10 in. is the usual minimum. In applications where cabling is extensive and/or air quantities are especially high, additional floor clearance may be required. The supply air connections into an underfloor cavity should be designed to minimize air turbulence. Where possible, the supply to the cavity should be central to the area served, and abrupt changes in direction should be avoided. Piping, cable, and conduit should not be permitted to interfere with supply airflow.

Where multiple zones are served from an underfloor plenum, dividing baffles may be omitted because air will follow the path of least resistance. Dividing baffles required to meet fire codes can impede the modification of computer systems having cabling

interconnections between zones because any cable change may entail penetration of a zone baffle.

Plenums should be airtight, thoroughly cleaned, and smoothly finished to prevent entrainment of foreign materials in the airstream. The use and method of construction of plenums may be restricted by local codes or fire underwriter regulations. Surfaces in underfloor plenums (e.g., water, chilled water, brine, or refrigerant piping) may require insulation and a vapor retarder if surface temperatures are low enough to cause condensation. If this is the case, water detectors should be installed to protect computers and underfloor wiring.

Ceiling and Ceiling Plenum Supply

Overhead supply through ceiling diffusers, shown in Figure 6, may be suitable for computer rooms. Ceiling and ceiling plenum supply systems can satisfy equipment and personnel comfort requirements, but they are generally not as flexible as underfloor plenum supply systems. This arrangement is compatible with both central station and packaged unitary equipment.

Distribution of air can be regulated by selective placement of acoustical pads on the perforated panels of a metal panel ceiling or by placement of active perforated sections in a lay-in acoustical tile plenum ceiling. Where precise distribution is essential, active ceiling diffusers or air supply zones may be equipped with air valves. Ceiling plenums, if properly constructed and cleaned, are more likely to remain clean than underfloor plenums.

Ceiling plenums must be sufficiently deep to permit airflow without turbulence; the required depth depends on air quantities. Best conditions may be achieved by using distribution ductwork,

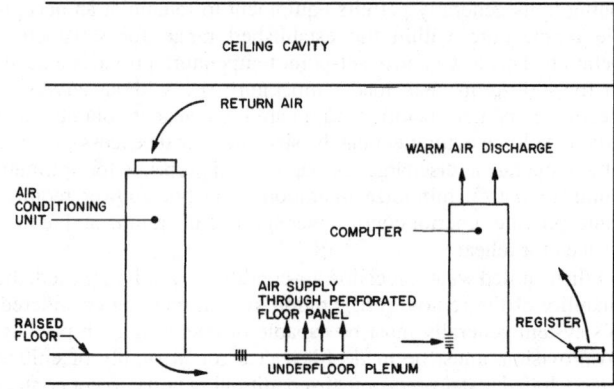

Fig. 4   Typical Underfloor Distribution

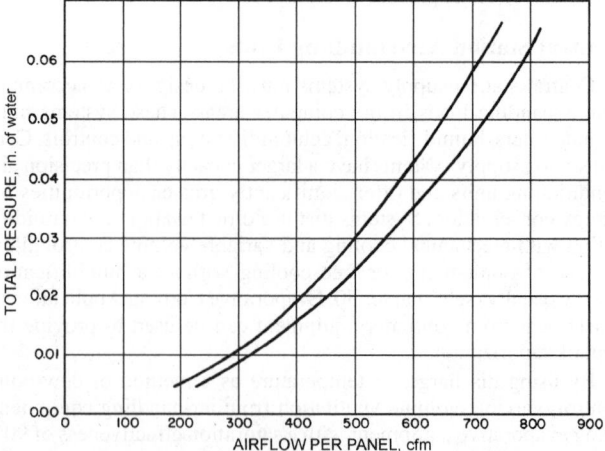

Fig. 5   Floor Panel Air Performance

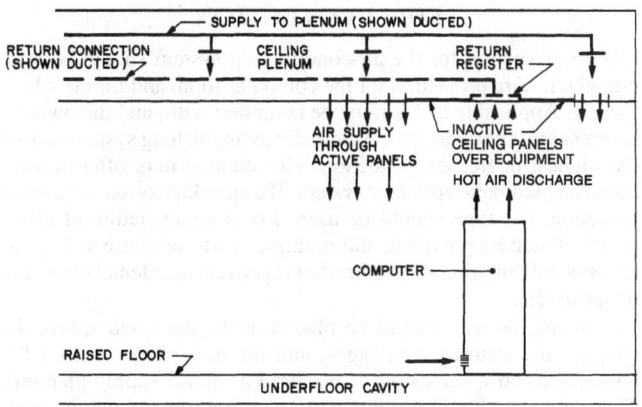

Fig. 6 Typical Ceiling Plenum Distribution

with the air discharged into the space through air valves or adjustable outlets above the ceiling. Overhead supply systems should be limited to applications where air supply concentrations are low or the need for flexibility is small. Where loads are high, aspirating diffusers may cause drafts, especially if the ceilings are low. It can be difficult to relocate rigidly ducted outlets in a facility that must remain in continuous operation.

RETURN AIR

The use of ceiling plenum returns is a common and effective strategy. Inlets should be located above equipment having high heat dissipation to take advantage of the thermal plume created above the equipment. Ceiling plenum returns can capture a portion of the heat from the computers and the lights directly in the return airstream, allowing a reduced air circulation rate. Ceiling plenum returns enhance the flexibility of the space to support future modifications of the computer installation.

DIRECT-COOLED COMPUTER EQUIPMENT

The majority of cooling in direct-cooled equipment is still accomplished with space air. Some computer equipment requires direct cooling to maintain the equipment environment within the limits established by the manufacturer. Heat released to the space from this equipment is usually equal to or greater than that for similar air-cooled computer systems. Generally, a closed system circulates distilled water or refrigerant through passages in the computer to cool it. Manufacturers normally supply the cooling system as part of the computer equipment.

Chilled water may be provided by either a small chiller matched in capacity to the computer equipment or a branch of the chilled water system serving the air-handling units. Design and installation of chilled water or refrigerant piping, and selection of the operating temperatures, should minimize the potential for leaks and condensation, especially within the computer room, while satisfying the requirements of the systems served. Chilled water systems for water-cooled computer equipment must be designed (1) to provide water at a temperature within the manufacturer's tolerances and (2) to be capable of operating year-round, 24 h per day.

AIR-CONDITIONING COMPONENTS

Controls

A well-planned control system must coordinate the performance of the temperature and humidity equipment. Controls for central station air-handling equipment may be included in a project-specific control system. Space temperature and humidity sensors should be carefully located to sample room conditions accurately.

Where multiple packaged units are provided, integration of control systems and regular calibration of controls is necessary to prevent individual units from working against each other. Errors in control system calibration, differences in unit set points, and sensor drift can cause multiple-unit installations to simultaneously heat and cool, and/or humidify and dehumidify, wasting a significant amount of energy.

Refrigeration

Refrigeration systems should be designed to match the anticipated cooling load, should be capable of expansion, and may be required to provide year-round, continuous operation. A separate refrigeration facility for the data processing area may be desirable where system requirements differ from those provided for other building and process systems and/or where emergency power requirements preclude combined systems. The system must provide the reliability and redundancy required to match the facility's needs. System operation, servicing, and maintenance should not interfere with facility operation. If required, the refrigeration system should be operable with emergency power. Fulfillment of these requirements may necessitate multiple units or cross-connections with other reliable systems. If the installation is especially critical, it may be necessary to install up to 100% standby capacity to meet the minimum requirements in the event of equipment failure.

Humidification

Many types of humidifiers may be used to serve data processing areas, including steam-generating (remote or local), pan types (with immersion elements or infrared lamps), and evaporative types (wetted pad and ultrasonic). Ultrasonic devices should use deionized water to prevent the formation of abrasive dusts from the crystallization of dissolved solids in the water. The humidifier must be responsive to control, maintainable, and free of moisture carryover. See Chapter 20 of the 1996 *ASHRAE Handbook—Systems and Equipment* for more information on humidifiers.

When an air-side economizer is used in the cooling of computer areas, evaporative humidification should be considered due to its ability to use waste heat energy to meet humidification requirements. Water-side economizers are economically feasible in some climates but do not have the energy efficiency potential of air-side approaches and may not provide adequate dilution ventilation.

Chilled Water Distribution Systems

Chilled water distribution systems should be designed to the same standards of quality, reliability, and flexibility as other computer room support systems. Where growth is likely, the chilled water system should be designed for expansion or the addition of new equipment without extensive shutdown. Figure 7 illustrates a looped chilled water system with sectional valves and multiple valved branch connections. The branches could serve air handlers or water-cooled computer equipment. The valves permit modifications or repairs without complete shutdown.

Where chilled water serves packaged equipment within the computer room, chilled water temperatures should be selected to satisfy the space sensible cooling loads while minimizing the risk of excessive condensation. Because computer room loads are primarily sensible, chilled water should be relatively warm. Water temperatures as high as 48°F are still slightly below the dew point of a 72°F, 45% rh room and several degrees below that of a 72°F, 50% rh room.

Chilled water and glycol piping must be pressure-tested, fully insulated, and protected with an effective vapor retarder. The test pressure should be applied in increments to all sections of pipe in the computer area. Drip pans piped to an effective drain should be placed below any valves or other components within the computer room that cannot be satisfactorily insulated. A good-quality strainer

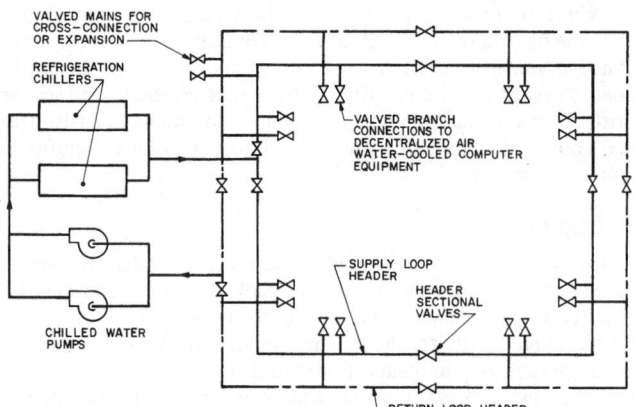

Fig. 7 Chilled Water Loop Distribution

should be installed in the inlet to local cooling equipment to prevent control valve and heat exchanger passages from clogging.

If cross-connections with other systems are made, possible effects on the computer room system of the introduction of dirt, scale, or other impurities must be addressed.

Redundancy

System reliability is so vital that the potential cost of system failure may justify redundant systems, capacity, and/or components. The designer should identify potential points of failure that could cause the system to interrupt critical data processing applications and should provide redundant or backup systems.

It may be desirable to cross-connect refrigeration equipment for backup, as suggested for air-handling equipment. Redundant refrigeration may be required; the extent of the redundancy depends on the importance of the computer installation. In many cases, standby power for the computer room air-conditioning system is justified.

Any complication to the basic conditioning system that might impair its reliability or otherwise adversely affect performance should be carefully considered before incorporation of heat recovery into the design. Potential monetary loss from system malfunction or unscheduled shutdown may outweigh savings from increased operating efficiency.

INSTRUMENTATION

Because computer equipment malfunctions may be caused by or attributed to improper regulation of the computer room thermal environment, it may be desirable to keep permanent records of the space temperature and humidity. If air is supplied directly to computer equipment, these records can be correlated with equipment function.

Alarms should be provided to signal when temperature or humidity limits are violated. With water-cooled computers, records should be kept of the entering and exiting water temperature and pressure. Indicating thermometers and pressure gages should be placed throughout the system so that operators can tell at a glance when unusual conditions prevail. Properly maintained and accurate gages for air-handling equipment filters can help prevent loss of system capacity and maintain correct computer room conditions. Sensing devices to indicate leaks or the presence of water in the computer room underfloor cavity are desirable in spaces served by precision air-conditioning units, especially if brine or chilled water distribution piping is installed within a computer room. Low points in drip pans installed beneath piping should also be monitored.

All monitoring and alarm devices should provide local indication. Monitoring devices many be interfaced with an integrated control or remote monitoring system that permits indications of system malfunctions to be transmitted to a remote location to initiate critical and maintenance alarms.

FIRE PROTECTION

Fire protection for the air-conditioning system should be integrated with fire protection for the computer room and for the whole facility. Applicable codes must be complied with, and the owner's insurers must be consulted. Automatic extinguishing systems afford the highest degree of protection. Fire underwriters often recommend an automatic sprinkler system. If a sprinkler system is used, a preaction, dry type should be used. Local annunciation of alarm points should be provided, and multiple stages of alarm and sprinkler system initiation should be used to prevent accidental discharge of sprinklers.

Sensing devices should be placed in the occupied space, the supply- and return-air passages, and the underfloor cavity (if it contains electric cables but is not used as an air supply plenum). These devices should provide early warning of combustion products, even if no smoke is visible and temperatures are at or near normal levels, to permit active measures to be taken to prevent damage or unscheduled shutdown of computer systems. All fire protection systems should provide for either manual or automatic shutdown of computer power, depending on the importance of the system and on the potential effect of an unwarranted shutdown (NFPA *Standard* 70).

Exhaust systems may be provided to ventilate computer rooms in the event of a fire suppression system discharge. Location of the exhaust pickup point below a supply plenum floor promotes quick purging of the space.

ENERGY CONSERVATION AND HEAT RECOVERY

Energy Conservation

Dramatic reductions in energy use can be achieved with conservation strategies. Where secondary uses for recovered heat do not exist or justify the use of heat recovery, central station air-conditioning systems using outdoor air for free cooling, variable volume ventilation, and evaporative cooling/humidification strategies offer significant opportunities for reducing energy use and improving air quality over precision air-conditioning systems. A dew-point control strategy can eliminate the need for troublesome humidity sensing devices and provide precise humidity control.

These same strategies may be employed to provide efficient and cost-effective solutions on a larger scale within the electronic office and other support environments.

Heat Recovery

If a use for recovered energy exists, computer rooms are good candidates for heat recovery because of their large year-round loads. Heat rejected during condensing can be used for space heating, domestic water heating, or other process heat. If the heat removed by the conditioning system can be efficiently transferred and applied elsewhere in the facility, some operational cost saving may be realized.

BIBLIOGRAPHY

Krzyzanowski, M.E. and B.T. Reagor. 1991. Measurement of potential contaminants in data processing environments. *ASHRAE Transactions* 97(1):464-76.

Lentz, M.S. 1991. Adiabatic saturation and VAV: A prescription for economy and close environmental control. *ASHRAE Transactions* 97(1):477-85.

Longberg, J.C 1991. Using a central air-handling unit system for environmental control of electronic data processing centers. *ASHRAE Transactions* 97(1):486-93.

Weschler, C.J. and H.C. Shields. 1991. The impact of ventilation and indoor air quality on electronic equipment. *ASHRAE Transactions* 97(1):455-63.

PRINTING PLANTS

THIS chapter outlines air-conditioning requirements for key printing operations. Air conditioning of printing plants can provide controlled, uniform air moisture content and temperature in working spaces. Paper, the principal material used in printing, is hygroscopic and very sensitive to variations in the humidity of the surrounding air. Printing problems caused by paper expansion and contraction can be avoided by controlling the moisture content throughout the manufacture and printing of the paper.

DESIGN CRITERIA

The following are three basic printing methods:

- **Relief printing (letterpress).** Ink is applied to a raised surface.
- **Lithography.** The inked surface is neither in relief nor recessed.
- **Gravure (intaglio printing).** The inked areas are recessed below the surface.

Figure 1 shows the general work flow through a printing plant. The operation begins at the publisher and ends with the finished printed product and paper waste. Paper waste, which may be as much as 20% of the total paper used, affects the profitability of a printing operation. Proper air conditioning can help reduce the amount of paper wasted.

In sheetfed printing, individual sheets are fed through a press from a stack or load of sheets and collected after printing. In webfed rotary printing, a continuous web of paper is fed through the press from a roll. The printed material is cut, folded, and delivered from the press as signatures, which form the sections of a book.

The preparation of this chapter is assigned to TC 9.2, Industrial Air Conditioning.

Sheetfed printing is a slow process in which the ink is essentially dry as the sheets are delivered from the press. **Offsetting**, the transference of an image from one sheet to another, is prevented by applying a powder or starch to separate each sheet as it is delivered from the press. Starches present a housekeeping problem: the particles (30 to 40 µm in size) tend to fly off, eventually settling on any horizontal surface.

If both temperature and relative humidity are maintained within normal human comfort limits, they have little to do with web breaks or the runnability of paper in a webfed press. At extremely low humidity, static electricity causes the paper to cling to the rollers, creating undue stress on the web, particularly with high-speed presses. Static electricity is also a hazard when flammable solvent inks are used.

Special Considerations

Various areas in printing plants require special attention to processing and heat loads.

Engraving and platemaking departments must have very clean air—not as clean as that for industrial clean rooms, but cleaner than that for offices. Engraving and photographic areas may also have special ventilation needs because of the chemicals used. Nitric acid fumes from powderless etching require careful duct material selection. Composing rooms, which contain computer equipment, can be treated the same as similar office areas. The excessive dust from cutting in the stitching and binding operations must be controlled. Stereotype departments have very high heat loads.

In pressrooms, air distribution must not cause the web to flutter or force contaminants or heat (which normally would be removed by roof vents) down to the occupied level. Air should be introduced immediately above the occupied zone wherever possible to

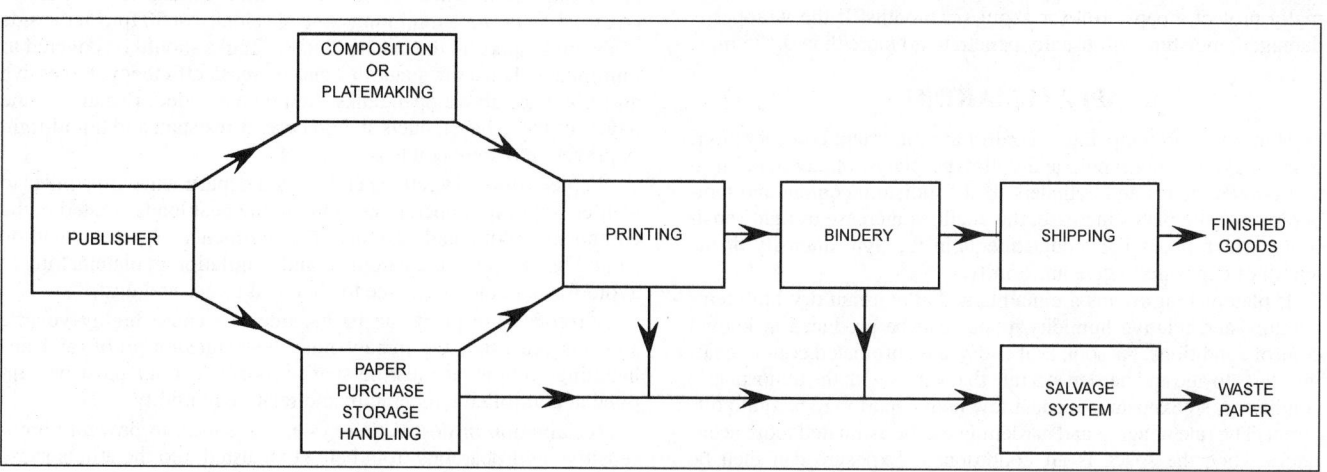

Fig. 1   Work Flow Through a Printing Plant

minimize total flow and encourage stratification. High air exchange rates may be required where solvent- or oil-based inks are used, due to the large quantity of organic solvent vapors that may be released from nonpoint sources. Exhaust emissions from dryer systems may contain substantial concentrations of solvent vapors, which must be captured and recovered or incinerated to satisfy local air pollution requirements. Where these measures are required, efforts should be made to maximize point source capture of vapors to minimize the size, cost, and energy requirements for vapor recovery/incineration equipment. Such efforts also minimize the impact of these requirements on general ventilation systems.

Conventional air-conditioning and air-handling equipment, particularly rooftop equipment, may be unable to handle the high outside air requirements of pressroom applications effectively. Stratified ventilation may be used in high-bay installations to reduce total system airflow and air-conditioning requirements. Pressrooms using oil- or solvent-based inks should be provided with a minimum of 0.5 cfm/ft^2 of outside air to ensure adequate dilution of internally generated volatile organic compounds. Ventilation of storage areas should be about 0.5 air changes per hour; bindery ventilation should be about 1 air change per hour. Storage areas with materials piled high may need roof-mounted smoke- and heat-venting devices.

In a bindery, loads of loose signatures are stacked near equipment, which makes it difficult to supply air to occupants without scattering the signatures. One solution is to run the main ducts at the ceiling with many supply branches dropped to within 8 to 10 ft of the floor. Conventional adjustable blow diffusers, often the linear type, are used.

CONTROL OF PAPER MOISTURE CONTENT

Controlling the moisture content and temperature of paper is important in all printing, particularly multicolor lithography. Paper should be received at the printing plant in moisture-proof wrappers, which are not broken or removed until the paper is brought to the pressroom temperature. When exposed to room temperature, paper at temperatures substantially below the room temperature rapidly absorbs moisture from the air, causing distortion. Figure 2 shows the time required to temperature-condition wrapped paper. Printers usually order paper with a moisture content approximately in equilibrium with the relative humidity maintained in their pressrooms. Papermakers find it difficult to supply paper in equilibrium with a relative humidity higher than 50%.

Digital hygrometers can be used to check the hygroscopic condition of paper relative to the surrounding air. The probes contain a moisture-sensitive element that measures the electrical conductivity of the paper. Intact mill wrappings and the tightness of the roll normally protect a paper roll for about six months. If the wrapper is damaged, moisture will usually penetrate no more than 0.125 in.

PLATEMAKING

Humidity and temperature control are important considerations when making lithographic and collotype plates, photoengravings, and gravure plates and cylinders. If the moisture content and temperature of the plates increase, the coatings increase in light sensitivity, which necessitates adjustments in the light intensity or the length of exposure to give uniformity.

If platemaking rooms are maintained at constant dry-bulb temperature and relative humidity, plates can be produced at known control conditions. As soon as it is dry, a bichromated colloid coating starts to age and harden at a rate that varies with the atmospheric conditions, so exposures made a few hours apart may be quite different. The rate of aging and hardening can be estimated more accurately when the space is air conditioned. Exposure can then be reduced progressively to maintain uniformity. An optimum relative humidity of 45% or less substantially increases the useful life of

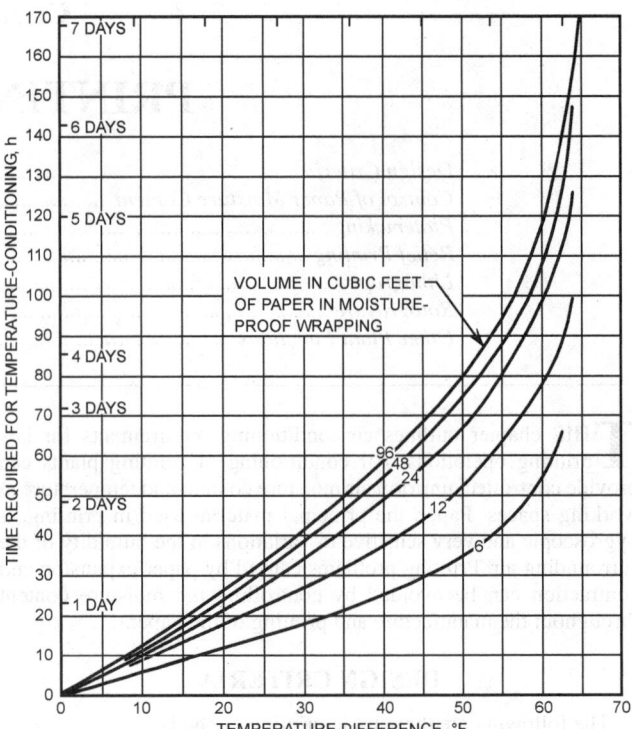

Fig. 2   Temperature-Conditioning Chart for Paper

bichromated colloid coatings; the relative humidity control should be within 2%. A dry-bulb temperature of 75 to 80°F maintained within 2°F is good practice. The ventilation air requirements of the plate room should be investigated. A plant with a large production of deep-etch plates should consider locating this operation outside the conditioned area.

Exhausts for platemaking operations consist primarily of lateral or downdraft systems at each operation. Because of their bulkiness or weight, plates or cylinders are generally conveyed by overhead rail to the workstation, where they are lowered into the tank for plating, etching, or grinding. Exhaust ducts must be below or to one side of the working area, so lateral exhausts are generally used for open-surface tanks.

Exhaust quantities vary, depending on the nature of the solution and shape of the tank, but they should provide exhaust in accordance with the recommendations of *Industrial Ventilation* by the American Conference of Governmental Industrial Hygienists (ACGIH 1998) for a minimum control velocity of 50 fpm at the side of the tank opposite the exhaust intake. Tanks should be covered to minimize exhaust air quantities and increase efficiency. Excessive air turbulence above open tanks should be avoided. Because of the nature of the exhaust, ducts should be acid-resistant and liquid-tight to prevent moisture condensation.

Webfed offset operations and related departments are similar to webfed letterpress operations, without the heat loads created in the composing room and stereotype departments. Special attention should be given to air cleanliness and ventilation in platemaking to avoid flaws in the plates due to chemical fumes and dust.

A rotogravure plant can be hazardous because highly volatile solvents are used. Equipment must be explosion-proof, and air-handling equipment must be spark-proof. Clean air must be supplied at controlled temperature and relative humidity.

Reclamation or destruction systems are used to prevent photosensitive hydrocarbons from being exhausted into the atmosphere. Some reclamation systems use activated carbon for continuous processing. Incineration or catalytic converters may be used to produce rapid oxidation to eliminate pollutants. The amount of solvents reclaimed may exceed that added to the ink.

RELIEF PRINTING

In relief printing (letterpress), rollers apply ink only to the raised surface of a printing plate. Pressure is then applied to transfer the ink from the raised surface directly to the paper. Only the raised surface touches the paper to transfer the desired image.

Air conditioning in newspaper pressrooms and other webfed letterpress printing areas minimizes problems caused by static electricity, ink mist, and expansion or contraction of the paper during printing. A wide range of operating conditions is satisfactory. The temperature should be selected for operator comfort.

At web speeds of 1000 to 2000 fpm, it is not necessary to control the relative humidity because inks are dried with heat. In some types of printing, moisture is applied to the web, and the web is passed over chill rolls to further set the ink.

Webfed letterpress ink is heat-set, made with high-boiling, slow-evaporating synthetic resins and petroleum oils dissolved or dispersed in a hydrocarbon solvent. The solvent must have a narrow boiling range with a low volatility at room temperatures and a fast evaporating rate at elevated temperatures. The solvent is vaporized in the printing press dryers at temperatures from 250 to 400°F, leaving the resins and oils on the paper. Webfed letterpress inks are dried after all colors are applied to the web.

The inks are dried by passing the web through dryers at speeds of 1000 to 2000 fpm. There are several types of dryers: open-flame gas cup, flame impingement, high-velocity hot air, and steam drum.

Exhaust quantities through a press dryer vary from about 7000 to 15,000 cfm at standard conditions, depending on the type of dryer used and the speed of the press. Exhaust temperatures range from 250 to 400°F.

Solvent-containing exhaust is heated to 1300°F in an air pollution control device to incinerate the effluent. A catalyst can be used to reduce the temperature required for combustion to 1000°F, but it requires periodic inspection and rejuvenation. Heat recovery reduces the fuel required for incineration and can be used to heat pressroom makeup air.

LITHOGRAPHY

Lithography uses a grease-treated printing image receptive to ink, on a surface that is neither raised nor depressed. Both grease and ink repel water. Water is applied to all areas of the plate, except the printing image. Ink is then applied only to the printing image and transferred to the paper in the printing process. In multicolor printing operations, the image may be printed up to four times on the same sheet of paper in different colors. Registration of images is critical to final color quality.

Offset printing transfers the image first to a rubber blanket and then to the paper. Sheetfed and web offset printing are similar to letterpress printing. The inks used are similar to those used in letterpress printing but contain water-resistant vehicles and pigment. In web offset and gravure printing, the relative humidity in the pressroom should be maintained constant, and the temperature should be selected for comfort or, at least, to avoid heat stress. It is important to maintain steady conditions to ensure the dimensional stability of the paper onto which the images are printed.

The pressroom for sheet multicolor offset printing has more exacting humidity requirements than other printing processes. The paper must remain flat with constant dimensions during multicolor printing, in which the paper may make six or more passes through the press over a period of a week or more. If the paper does not have the right moisture content at the start, or if there are significant changes in atmospheric humidity during the process, the paper will not retain its dimensions and flatness, and misregistering will result. In many cases of color printing, a register accuracy of 0.005 in. is

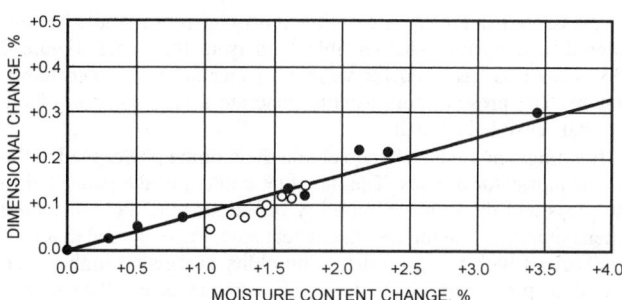

Fig. 3   Effects of Variation in Moisture Content on Dimensions of Printing Papers
(Weber and Snyder 1934)

required. Figure 3 shows the close control of the air relative humidity that is necessary to achieve this register accuracy. The data shown in this figure are for composite lithographic paper.

Maintaining constant moisture content of the paper is complicated because paper picks up moisture from the moist offset blanket during printing—0.1 to 0.3% for each impression. When two or more printings are made in close register work, the paper at the start of the printing process should have a moisture content in equilibrium with air at 5 to 8% rh above the pressroom air. At this condition, the moisture evaporated from the paper into the air nearly balances the moisture added by the press. In obtaining register, it is important to keep the sheet flat and free from wavy or tight edges. To do this, the relative humidity balance of the paper should be slightly above that of the pressroom atmosphere. This balance is not as critical in four-color roll-feed presses because the press moisture does not penetrate the paper quickly enough between colors to affect sheet dimensions or cause sheet distortion.

Recommended Environment

The Graphic Arts Technical Foundation recommends ideal conditions in a lithographic pressroom of 76 to 80°F dry-bulb temperature and 43 to 47% rh, controlled to ±2°F dry-bulb temperature and ±2% rh (Reed 1970). Comfort and economy of operation influence the choice of temperature. The effect of relative humidity variations on register can be estimated for offset paper from Figure 3. Closer relative humidity control of the pressroom air is required for multicolor printing of 76 in. sheets than for 22 in. sheets with the same register accuracy. Closer control is needed for multicolor printing, where the sheet makes two or more trips through the press, than for one-color printing.

Ink drying is affected by temperature and humidity, so uniform results are difficult to obtain without controlling the atmospheric conditions. Printing inks must dry rapidly to prevent offsetting and smearing. High relative humidity and high moisture content in paper tend to prevent ink penetration, so more ink remains on the surface than can be quickly oxidized. This affects drying time, intensity of color, and uniformity of ink on the surface. Relative humidity below 60% is favorable for drying at a comfortable temperature. Higher relative humidity may cause severe paper distortion and significant damage to the final product.

The air conditioning for the pressroom of a lithographic plant should control air temperature and relative humidity, filter the air, supply ventilation air, and distribute the air without pronounced drafts around the presses. The use of antioffset sprays to set the ink creates an additional air-filtering load from the pressroom. Drafts and high airflow over the presses lead to excessive drying of the ink and water, which causes scumming or other problems.

The operating procedures of the pressroom should be analyzed to determine the heat removal load. The lighting load is high and constant throughout the day. The temperature of the paper brought into the pressroom and the length of time it is in the room should be considered to determine the sensible load from the paper. Figure 2 shows the time required for wrapped paper to reach room temperature. The press motors usually generate a large portion of the internal sensible heat gain.

Readings should be taken to obtain the running power load of the larger multicolor presses. The moisture content of the paper fed to the press and the relative humidity of the air must be considered when computing the internal latent heat gain. Paper is used that is in equilibrium with air at a relative humidity somewhat higher than that of the pressroom, so the paper gives up moisture to the space as it absorbs moisture during printing. If the moisture transfer is in balance, the water used in the printing process should be included in the internal moisture load. It is preferable to determine the water evaporation from the presses by testing.

Air Conditioning

Precise multicolor offset lithography printing requires either refrigeration with provision for separate humidity control, or sorption dehumidifying equipment for independent humidity control with provision for cooling. The need for humidity control in the pressroom may be determined by calculating the dimensional change of the paper for each percent of change in relative humidity and checking this with the required register for the printing process.

Air conditioning of the photographic department is usually considered next in importance to that of the pressroom. Most of the work in offset lithography is done on film. Air conditioning controls cleanliness and comfort and maintains the size of the film for register work.

Air conditioning is important in the stripping department, both for comfort and for maintaining size and register. Curling of the film and flats, as well as shrinkage or stretch of materials, can be minimized by maintaining constant relative humidity. This is particularly important for close-register color work. The photographic area, stripping room, and platemaking area usually are maintained at the same conditions as the pressroom.

Dryers used for web offset printing are the same type as for webfed letterpress. Drying is not as complex because less ink is applied and presses run at lower speeds—800 to 1800 fpm.

ROTOGRAVURE

Rotogravure printing uses a cylinder with minute inkwells etched in the surface to form the printing image. Ink is applied to the cylinder, filling the wells. Excess ink is then removed from the cylinder surface by doctor blades, leaving only the ink in the wells. The image is then transferred to the paper as it passes between the printing cylinder and an impression cylinder.

In sheetfed gravure printing (as in offset printing), expansion, contraction, and distortion should be prevented to obtain correct register. The paper need not be in equilibrium with air at a relative humidity higher than that of the pressroom, because no moisture is added to the paper in the printing process. Humidity and temperature control should be exacting, like in offset printing. The relative humidity should be 45 to 50%, controlled to within ±2%, with a comfort temperature controlled to within ±2°F.

Gravure printing ink dries principally by evaporating the solvent in the ink, leaving a solid film of pigment and resin. The solvent is a low-boiling hydrocarbon, and evaporation takes place rapidly, even without the use of heat. The solvents have closed-cup flash points from 22 to 80°F and are classified as Group I or special hazard liquids by local code and insurance company standards. As a result, in areas adjacent to gravure press equipment and solvent and ink storage areas, electrical equipment must be Class I, Division 1 or 2, as described by the *National Electrical Code* (NFPA *Standard* 70), and ventilation requirements (both supply and exhaust) are stringent. Ventilation should be designed for high reliability, with sensors to

detect unsafe pollutant concentrations and then to initiate alarm or safety shutdown when necessary.

Rotogravure printing units operate in tandem, each superimposing print over that from the preceding unit. Press speeds range from 1200 to 2400 fpm. Each unit is equipped with its own dryer to prevent subsequent smearing or smudging.

A typical drying system consists of four dryers connected to an exhaust fan. Each dryer is equipped with fans to recirculate 5000 to 8000 cfm (at standard conditions) through a steam or hot water coil and then through jet nozzles. The hot air (130°F) impinges on the web and drives off the solvent-laden vapors from the ink. It is normal to exhaust half of this air. The system should be designed and adjusted to prevent solvent vapor concentration from exceeding 25% of its lower flammable limit (Marsailes 1970). If this is not possible, constant lower-flammable-limit (LFL) monitoring, concentration control, and safety shutdown capability should be included.

In exhaust design for a particular process, solvent vapor should be captured from the printing unit where paper enters and exits the dryer, from the fountain and sump area, and from the printed paper, which continues to release solvent vapor as it passes from one printing unit to another. Details of the process, such as ink and paper characteristics and rate of use, are required to determine exhaust quantities.

When dilution-type ventilation is used, exhaust of 1000 to 1500 cfm (at standard conditions) at the floor is often provided between each unit. The makeup air units are adjusted to supply slightly less air to the pressroom than that exhausted in order to keep the pressroom negative with respect to the surrounding areas.

OTHER PLANT FUNCTIONS

Flexography

Flexography uses rubber raised printing plates and functions much like a letterpress. Flexography is used principally in the packaging industry to print labels and also to print on smooth surfaces, such as plastics and glass.

Collotype Printing

Collotype or photogelatin printing is a sheetfed printing process related to lithography. The printing surface is bichromated gelatin with varying affinity for ink and moisture, depending on the degree of light exposure received. There is no mechanical dampening as in lithography, and the necessary moisture in the gelatin printing surface is maintained by operating the press in an atmosphere of high relative humidity, usually about 85%. Since the tonal values printed are very sensitive to changes in the moisture content of the gelatin, the relative humidity should be maintained within ±2%.

Because tonal values are also very sensitive to changes in ink viscosity, temperature must be closely maintained; 80 ± 3°F is recommended. Collotype presses are usually partitioned off from the main plant, which is kept at a lower relative humidity, and the paper is exposed to high relative humidity only while it is being printed.

Salvage

Salvage systems remove paper trim and shredded paper waste from production areas, and carry airborne shavings to a cyclone or baghouse collector, where they are baled for recycling. Air quantities required are 40 to 45 ft^3 per pound of paper trim, and the transport velocity in the ductwork is 4500 to 5000 fpm (Marsailes 1970). Humidification may be provided to prevent the buildup of a static charge and consequent system blockage.

Air Filtration

Ventilation and air-conditioning systems for printing plants commonly use automatic moving-curtain dry-media filters with renewable media having a weight arrestance of 80 to 90% (ASHRAE *Standard* 52.1).

In sheetfed pressrooms, a high-performance final filter is used to filter starch particles, which require about 85% ASHRAE dust-spot efficiency. In film processing areas, which require relatively dust-free conditions, high-efficiency air filters are installed, with 90 to 95% ASHRAE dust-spot efficiency.

A different type of filtration problem in printing is **ink mist** or **ink fly**, which is common in newspaper pressrooms and in heatset letterpress or offset pressrooms. Minute droplets of ink (5 to 10 μm) are dispersed by ink rollers rotating in opposite directions. The cloud of ink droplets is electrostatically charged. Suppressors, charged to repel the ink back to the ink roller, are used to control ink mist. Additional control is provided by automatic moving curtain filters.

Binding and Shipping

Some printed materials must be bound. Two methods of binding are perfect binding and stitching. In **perfect binding**, sections of a book (signatures) are gathered, ruffed, glued, and trimmed. The glued edge is flat. Large books are easily bound by this type of binding. Low-pressure compressed air and a vacuum are usually required to operate a perfect binder, and paper shavings are removed by a trimmer. The use of heated glue necessitates an exhaust system if the fumes are toxic.

In **stitching**, sections of a book are collected and stitched (stapled) together. Each signature is opened individually and laid over a moving chain. Careful handling of the paper is important. This has the same basic air requirements as perfect binding.

Mailing areas of a printing plant wrap, label, and ship the manufactured goods. Operation of the wrapper machine can be affected by low humidity. In winter, humidification of the bindery and mailing area to about 40 to 50% rh may be necessary to prevent static buildup.

REFERENCES

ACGIH. 1998. *Industrial ventilation: A manual of recommended practice*, 23rd ed. American Conference of Governmental Industrial Hygienists, Cincinnati, OH.

ASHRAE. 1992. Gravimetric and dust-spot procedures for testing air-cleaning devices used in general ventilation for removing particulate matter. ANSI/ASHRAE *Standard* 52.1-1992.

Marsailes, T.P. 1970. Ventilation, filtration and exhaust techniques applied to printing plant operation. *ASHRAE Journal* (December):27.

NFPA. 1996. *National electrical code.* ANSI/NFPA *Standard* 70-96. National Fire Protection Association, Quincy, MA.

Reed, R.F. 1970. What the printer should know about paper. Graphic Arts Technical Foundation, Pittsburgh, PA.

Weber, C.G. and L.W. Snyder. 1934. Reactions of lithographic papers to variations in humidity and temperature. *Journal of Research* 12(January). National Bureau of Standards. Available from NIST, Gaithersburg, MD.

TEXTILE PROCESSING PLANTS

THIS chapter covers (1) basic processes for making fiber, yarn, and fabric; (2) various types of air-conditioning systems used in textile manufacturing plants; (3) relevant health considerations; and (4) energy conservation procedures.

Most textile manufacturing processes may be put into one of three general classifications: synthetic fiber making, yarn making, or fabric making. Synthetic fiber manufacturing is divided into staple processing, tow-to-top conversion, and continuous fiber processing; yarn making is divided into spinning and twisting; and fabric making is divided into weaving and knitting. Although these processes vary, their descriptions reveal the principles on which air-conditioning design is based.

FIBER MAKING

Processes preceding fiber extrusion have diverse ventilating and air-conditioning requirements based on principles similar to those that apply to chemical plants.

Synthetic fibers are extruded from metallic spinnerets and solidified as continuous parallel filaments. This process, called **continuous spinning**, differs from the mechanical spinning of fibers or tow into yarn, which is generally referred to as **spinning**.

Synthetic fibers may be formed by melt-spinning, dry-spinning, or wet-spinning. Melt-spun fibers are solidified by cooling the molten polymer; dry-spun fibers by evaporating a solvent, leaving the polymer in fiber form; and wet-spun fibers by hardening the extruded filaments in a liquid bath. The selection of a spinning method is affected by economic and chemical considerations. Generally, nylons, polyesters, and glass fibers are melt-spun, acetates dry-spun, rayons and aramids wet-spun, and acrylics dry- or wet-spun.

For melt-spun and dry-spun fibers, the filaments of each spinneret are usually drawn through a long vertical tube called a **chimney** or **quench stack**, within which solidification occurs. For wet-spun fibers, the spinneret is suspended in a chemical bath where coagulation of the fibers takes place. Wet-spinning is followed by washing, applying a finish, and drying.

Synthetic continuous fibers are extruded as a heavy denier tow for cutting into short lengths called staple or somewhat longer lengths for tow-to-top conversion, or they are extruded as light denier filaments for processing as continuous fibers. An oil is then applied to lubricate, give antistatic properties, and control fiber cohesion. The extruded filaments are usually drawn (stretched) both to align the molecules along the axis of the fiber and to improve the crystalline structure of the molecules, thereby increasing the fiber's strength and resistance to stretching.

Heat applied to the fiber when drawing heavy denier or high-strength synthetics releases a troublesome oil mist. In addition, the mechanical work of drawing generates a high localized heat load. If the draw is accompanied by twist, it is called **draw-twist**; if not, it is called **draw-wind**. After draw-twisting, continuous fibers may be given additional twist or may be sent directly to warping.

The preparation of this chapter is assigned to TC 9.2, Industrial Air Conditioning.

When tow is cut to make staple, the short fibers are allowed to assume random orientation. The staple, alone or in a blend, is then usually processed as described in the section on Cotton System. However, tow-to-top conversion, a more efficient process, has become more popular. The longer tow is broken or cut to maintain parallel orientation. Most of the steps of the cotton system are bypassed; the parallel fibers are ready for blending and mechanical spinning into yarn.

In the manufacture of glass fiber yarn, light denier multifilaments are formed by attenuating molten glass through platinum bushings at high temperatures and speeds. The filaments are then drawn together while being cooled with a water spray, and a chemical size is applied to protect the fiber. This is all accomplished in a single process prior to winding the fiber for further processing.

YARN MAKING

The fiber length determines whether spinning or twisting must be used. Spun yarns are produced by loosely gathering synthetic staple, natural fibers, or blends into rope-like form; drawing them out to increase fiber parallelism, if required; and then twisting. Twisted (continuous filament) yarns are made by twisting mile-long monofilaments or multifilaments. Ply yarns are made in a similar manner from spun or twisted yarns.

The principles of mechanical spinning are applied in three different systems: cotton, woolen, and worsted. The cotton system is used for all cotton, most synthetic staple, and many blends. Woolen and worsted systems are used to spin most wool yarns, some wool blends, and synthetic fibers such as acrylics.

Cotton System

The cotton system was originally developed for spinning cotton yarn, but now its basic machinery is used to spin all varieties of staple, including wool, polyester, and blends. Most of the steps from raw materials to fabrics, along with the ranges of frequently used humidities, are outlined in Figure 1.

Opening, Blending, and Picking. The compressed tufts are partly opened, most foreign matter and some short fibers are removed, and the mass is put in an organized form. Some blending is desired to average the irregularities between bales or to mix different kinds of fiber. Synthetic staple, which is cleaner and more uniform, usually requires less preparation. The product of the picker is pneumatically conveyed to the feed rolls of the card.

Carding. This process lengthens the lap into a thin web, which is gathered into a rope-like form called a **sliver**. Further opening and fiber separation follows, as well as partial removal of short fiber and trash. The sliver is laid in an ascending spiral in cans of various diameters.

For heavy, low-count (length per unit of mass) yarns of average or lower quality, the card sliver goes directly to drawing. For lighter, high-count yarns requiring fineness, smoothness, and strength, the card sliver must first be combed.

Lapping. In sliver lapping, several slivers are placed side by side and drafted. In ribbon lapping, the resulting ribbons are laid one on

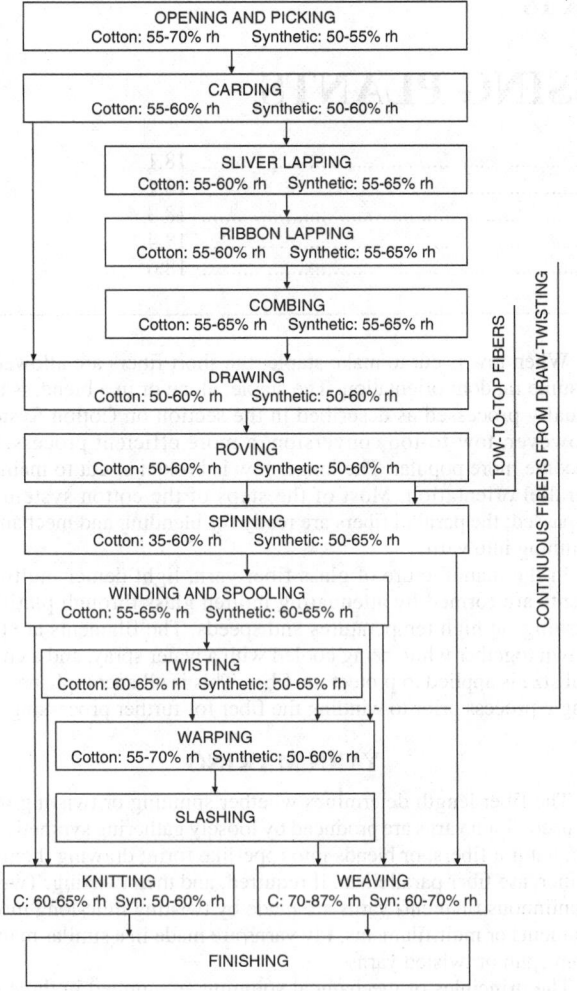

Fig. 1   Textile Process Flowchart and Ranges of Humidity

roll feed, spindle speed, and drag, which is related to the traveler weight.

The space between the nip or bite of the rolls is adjustable and must be slightly greater than the longest fiber. The speeds of front and back rolls are independently adjustable. Cotton spindles normally run at 8000 to 9000 rpm but may exceed 14,000 rpm. In ring twisting, drawing rolls are omitted, and a few spindles run as high as 18,000 rpm.

Open-end or turbine spinning combines drawing, roving, lapping, and spinning. Staple fibers are fragmented as they are drawn from a sliver and fed into a small, fast-spinning centrifugal device. In this device, the fibers are oriented and discharged as yarn; twist is imparted by the rotating turbine. This system is faster, quieter, and less dusty than ring spinning.

Spinning is the final step in the cotton system; the feature that distinguishes it from twisting is the application of draft. The amount and point of draft application accounts for many of the subtle differences that require different humidities for apparently identical processes.

Atmospheric Conditions. From carding to roving, the loosely bound fibers are vulnerable to static electricity. In most instances, static can be adequately suppressed with humidity, which should not be so high as to cause other problems. In other instances, it is necessary to suppress electrostatic properties with antistatic agents. Wherever draft is applied, constant humidity is needed to maintain optimum frictional uniformity between adjacent fibers and, hence, cross-sectional uniformity.

Woolen and Worsted Systems

The woolen system generally makes coarser yarns, while the worsted system makes finer ones of a somewhat harder twist. Both may be used for lighter blends of wool, as well as for synthetic fibers with the characteristics of wool. The machinery used in both systems applies the same principles of draft and twist but differs greatly in detail and is more complex than that used for cotton.

Compared to cotton, wool fibers are dirtier, greasier, and more irregular. They are scoured to remove grease and are then usually reimpregnated with controlled amounts of oil to make them less hydrophilic and to provide better interfiber behavior. Wool fibers are scaly and curly, so they are more cohesive and require different treatment. Wool, in contrast to cotton and synthetic fibers, requires higher humidities in the processes prior to and including spinning than it does in the processes that follow. Approximate humidities are presented in Table 1.

another and drafted again. The doubling and redoubling averages out sliver irregularities; drafting improves fiber parallelism. Some recent processes lap only once before combing.

Combing. After lapping, the fibers are combed with fine metal teeth to substantially remove all fibers below a predetermined length, to remove any remaining foreign matter, and to improve fiber arrangement. The combed lap is then attenuated by drawing rolls and again condensed into a single sliver.

Drawing. Drawing follows either carding or combing and improves uniformity and fiber parallelism by doubling and drafting several individual slivers into a single composite strand. Doubling averages the thick and thin portions; drafting further attenuates the mass and improves parallelism.

Roving. Roving continues the processes of drafting and paralleling until the strand is a size suitable for spinning. A slight twist is inserted, and the strand is wound on large bobbins used for the next roving step or for spinning.

Spinning. Mechanical spinning simultaneously applies draft and twist. The packages (any form into or on which one or more ends can be wound) of roving are creeled at the top of the frame. The unwinding strand passes progressively through gear-driven drafting rolls, a yarn guide, the C-shaped traveler, and then to the bobbin. The vertical traverse of the ring causes the yarn to be placed in predetermined layers.

The difference in peripheral speed between the back and front rolls determines the draft. Twist is determined by the rate of front

Table 1   Recommended Humidities for Wool Processing at 75 to 80°F

Departments	Humidity, %
Raw wool storage	50 to 55
Mixing and blending	65 to 70
Carding —worsted	60 to 70
—woolen	60 to 75
Combing, worsted	65 to 75
Drawing, worsted —Bradford system	50 to 60
—French system	65 to 70
Spinning —Bradford worsted	50 to 55
—French (mule)	75 to 85
—woolen (mule)	65 to 75
Winding and spooling	55 to 60
Warping, worsted	50 to 55
Weaving, woolen, and worsted	50 to 60
Perching or clothroom	55 to 60

Reprinted by permission of Interscience Division of John Wiley and Sons, New York.

Twisting Filaments and Yarns

Twisting was originally applied to silk filaments; several filaments were doubled and then twisted to improve strength, uniformity, and elasticity. Essentially the same process is used today, but it is now extended to spun yarns, as well as to single or multiple filaments of synthetic fibers. Twisting is widely used in the manufacture of sewing thread, twine, tire cord, tufting yarn, rug yarn, ply yarn, some knitting yarns, etc.

Twisting and doubling is done on a **down-** or **ring-twister**, which draws in two or more ends from packages on an elevated creel, twists them together, and winds them into a package. Except for the omission of drafting, down-twisters are similar to conventional ring-spinning frames.

When yarns are to be twisted without doubling, an **up-twister** is used. Up-twisters are primarily used for throwing synthetic monofilaments and multifilaments to add to or vary elasticity, light reflection, and abrasion resistance. As with spinning, yarn characteristics are controlled by making the twist hard or soft, right (S) or left (Z). Quality is determined largely by the uniformity of twist, which, in turn, depends primarily on the tension and stability of the atmospheric conditions (Figure 1 and Table 1). As the frame may be double- or triple-decked, twisting requires concentrations of power. The frames are otherwise similar to those used in spinning, and they present the same air distribution problems. In twisting, lint is not a serious problem.

FABRIC MAKING

Preparatory Processes

When spinning or twisting is complete, the yarn may be prepared for weaving or knitting by processes that include winding, spooling, creeling, beaming, slashing, sizing, and dyeing. These processes have two purposes: (1) to transfer the yarn from the type of package dictated by the preceding process to a type suitable for the next and (2) to impregnate some of the yarn with sizes, gums, or other chemicals that may not be left in the final product.

Filling Yarn. Filling yarn is wound on quills for use in a loom shuttle. It is sometimes predyed and must be put into a form suitable for package or skein dyeing before it is quilled. If the filling is of relatively hard twist, it may be put through a twist-setting or conditioning operation in which internal stresses are relieved by applying heat, moisture, or both.

Warp Yarn. Warp yarn is impregnated with a transient coating of size or starch that strengthens the yarn's resistance to the chafing it will receive in the loom. The yarn is first rewound onto a cone or other large package from which it will unwind speedily and smoothly. The second step is warping, which rewinds a multiplicity of ends in parallel arrangement on large spools, called **warp or section beams**. In the third step, slashing, the threads pass progressively through the sizing solution, through squeeze rolls, and then around cans, around steam-heated drying cylinders, or through an air-drying chamber. As much as several thousand pounds may be wound on a single loom beam.

Knitting Yarn. If hard-spun, knitting yarn must be twist-set to minimize kinking. Filament yarns must be sized to reduce stripbacks and improve other running qualities. Both must be put in the form of cones or other suitable packages.

Uniform tension is of great importance in maintaining uniform package density. Yarns tend to hang up when unwound from a hard package or slough off from a soft one, and both tendencies are aggravated by spottiness. The processes that require air conditioning, along with recommended relative humidities, are presented in Figure 1 and Table 1.

Weaving

In the simplest form of weaving, harnesses raise or depress alternate warp threads to form an opening called a **shed**. A shuttle containing a quill is kicked through the opening, trailing a thread of filling behind it. The lay and the reed then beat the thread firmly into one apex of the shed and up to the fell of the previously woven cloth. Each shuttle passage forms a pick. These actions are repeated at frequencies up to five per second.

Each warp thread usually passes through a drop-wire that is released by a thread break and automatically stops the loom. Another automatic mechanism inserts a new quill in the shuttle as the previous one is emptied, without stopping the loom. Other mechanisms are actuated by filling breaks, improper shuttle boxing, and the like, which stop the loom until it is manually restarted. Each cycle may leave a stop mark sufficient to cause an imperfection that may not be apparent until the fabric is dyed.

Beyond this basic machine and pattern are many complex variations in harness and shuttle control, which result in intricate and novel weaving effects. The most complex loom is the **jacquard**, with which individual warp threads may be separately controlled. Other variations appear in looms for such products as narrow fabrics, carpets, and pile fabrics. In the **Sulzer weaving machine**, a special filling carrier replaces the conventional shuttle. In the rapier, a flat, spring-like tape uncoils from each side and meets in the middle to transfer the grasp on the filling. In the **water jet loom**, a tiny jet of high-pressure water carries the filling through the shed of the warp. Other looms transport the filling with compressed air.

High humidity increases the abrasion resistance of the warp. Weave rooms require 80 to 85% humidity or higher for cotton and up to 70% humidity for synthetic fibers. Many looms run faster when room humidity and temperature are precisely controlled.

In the weave room, power distribution is uniform, with an average concentration somewhat lower than in spinning. The rough treatment of fibers liberates many minute particles of both fiber and size, thereby creating considerable amounts of airborne dust. Due to high humidity, air changes average from four to eight per hour. Special provisions must be made for maintaining conditions during production shutdown periods, usually at a lower relative humidity.

Knitting

Typical products are seamless articles knitted on circular machines (e.g., undershirts, socks, and hosiery) and those knitted flat (e.g., full-fashioned hosiery, tricot, milanese, and warp fabrics).

Knitted fabric is generated by forming millions of interlocking loops. In its simplest form, a single end is fed to needles that are actuated in sequence. In more complex constructions, hundreds of ends may be fed to groups of elements that function more or less in parallel.

Knitting yarns may be either single strand or multi-filament and must be of uniform high quality and free from neps or knots. These yarns, particularly the multifilament type, are usually treated with special sizes to provide lubrication and to keep broken filaments from stripping back.

The need for precise control of yarn tension, through controlled temperature and relative humidity, increases with the fineness of the product. For example, in finer gages of full-fashioned hosiery, a 2°F change in temperature is the limit, and a 10% change in humidity may change in the length of a stocking by 3 in. For knitting, desirable room conditions are approximately 76°F dry-bulb and 45 to 65% rh.

Dyeing and Finishing

Finishing, which is the final readying of a mill product for its particular market, ranges from cleaning to imparting special characteristics. The specific operations involved vary considerably, depending on the type of fiber, yarn, or fabric, and the end product usage. Operations are usually done in separate plants.

Inspection is the only finishing operation to which air conditioning is regularly applied, although most of the others require ventilation. Finishing operations that use wet processes usually keep their solutions at high temperatures and require special ventilation to prevent destructive condensation and fog. Spot cooling of workers may be necessary for large releases of sensible, latent, or radiant heat.

AIR-CONDITIONING DESIGN

Air washers are especially important in textile manufacturing and may be either conventional low-velocity or high-velocity units in built-up systems. Unitary high-velocity equipment using rotating eliminators, although no longer common, is still found in some plants.

Contamination of air washers by airborne oils often dictates the separation of air washers and process chillers by heat exchangers, usually of the plate or frame type.

Integrated Systems

Many mills use a refined air washer system that combines the air-conditioning system and the collector system (see the section on Collector Systems) into an integrated unit. The air handled by the collector system fans and any air required to make up total return air are delivered back to the air-conditioning apparatus through a central duct. The quantity of air returned by individual yarn-processing machine cleaning systems must not exceed the air-conditioning supply air quantity. The air discharged by these individual suction systems is carried by return air ducts directly to the air-conditioning system. Before entering the duct, some of the cleaning system air passes over the yarn-processing machine drive motor and through a special enclosure to capture heat losses from the motor.

When integrated systems occasionally exceed the supply air requirements of the area served, the surplus air must be reintroduced after filtering.

Individual suction cleaning systems that can be integrated with air conditioning are available for cards, drawing frames, lap winders, combers, roving frames, spinning frames, spoolers, and warpers. The following advantages result from this integration:

1. With a constant air supply, the best uniform air distribution can be maintained year-round.
2. Downward airflow can be controlled; crosscurrents in the room are minimized or eliminated; drift or fly from one process to another is minimized or eliminated. Room partitioning between systems serving different types of manufacturing processes further enhances the value of this integration by controlling room air pattern year-round.
3. Heat losses of the yarn-processing frame motor and any portion of the processing frame heat captured in the duct, as well as the heat of the collector system equipment, cannot affect room conditions; hot spots in motor alleys are eliminated, and although this heat goes into the refrigeration load, it does not enter the room. As a result, the supply air quantity can be reduced.
4. Uniform conditions in the room improve production; conditioned air is drawn directly to the work areas on the machines, minimizing or eliminating wet or dry spots.
5. Maximum cleaning use is made of the air being moved. A guide for cleaning air requirements follows:

Pickers	2500 to 4000 cfm per picker
Cards	700 to 1500 cfm per card
Spinning	4 to 8 cfm per spindle
Spooling	40 cfm per spool

Collector Systems

A collector system is a waste-capturing device that uses many orifices operating at high suction pressures. Each piece of production machinery is equipped with suction orifices at all points of major lint generation. The captured waste is generally collected in a fan and filter unit located either on each machine or centrally to accept waste from a group of machines.

A collector in the production area may discharge waste-filtered air either back into the production area or into a return duct to the air-conditioning system. It then enters the air washer or is relieved through dampers to the outdoors.

Figure 2 shows a mechanical spinning room with air-conditioning and collector systems combined into an integrated unit. In this case, the collector system returns all of its air to the air-conditioning system. If supply air from the air-conditioning system exceeds the maximum that can be handled by the collector system, additional air should be returned by other means.

Figure 2 also shows return air entering the air-conditioning system through damper T, passing through air washer H, and being delivered by fan J to the supply duct, which distributes it to maintain conditions within the spinning room. At the other end of each spinning frame are unitary-type filter-collectors consisting of enclosure N, collector unit screen O, and collector unit fan P.

Collector fan P draws air through the intake orifices spaced along the spinning frame. This air passes through the duct that runs lengthwise to the spinning frame, passes through the screen O, and is then discharged into the enclosure base (beneath the fan and screen). The air quantity is not constant; it drops slightly as material builds up on the filter screen.

Because the return air quantity must remain constant, and the air quantity discharged by fan P is slightly reduced at times, relief openings are necessary. Relief openings also may be required when the return air volume is greater than the amount of air the collector suction system can handle.

The discharge of fan P is split, so part of the air cools the spinning frame drive motor before rejoining the rest of the air in the return air tunnel. Regardless of whether the total return air quantity enters the return air tunnel through collector units, or through a combination of collector units and floor openings beneath spinning frames, return air fan R delivers it into the apparatus, ahead of return air damper T. Consideration should be given to filtering the return air prior to its delivery into the air-conditioning apparatus.

Mild-season operation causes more outdoor air to be introduced through damper U. This air is relieved through motorized damper S, which opens gradually as outdoor damper U opens, while return damper T closes in proportion. All other components perform as typical central station air-washer systems.

A system having the general configuration shown in Figure 2 may also be used for carding; the collector system portion of this arrangement is shown in Figure 3. A central collector filters the lint-laden air taken from multiple points on each card. This air is discharged to return air duct A and is then either returned to the air-conditioning system, exhausted outside, or returned directly to the room. A central collector filter may also be used with the spinning room system of Figure 2.

Air Distribution

Textile plants served by generally uniform air distribution may still require special handling for areas of load concentration.

Continuous Spinning Area. Methods of distribution are diverse and generally not critical. However, spot cooling or localized heat removal may be required. This area may be cooled by air conditioning, evaporative cooling, or ventilation.

Chimney (Quench Stack). Carefully controlled and filtered air or other gas is delivered to the chimneys; it is returned for conditioning and recovery of any valuable solvents present. Distribution of the air is of the utmost importance. Nonuniform temperature, humidity, or airflow disturbs the yarn, causing variations in fiber diameter, crystalline structure, and orientation. A fabric made of such fibers streaks when dyed.

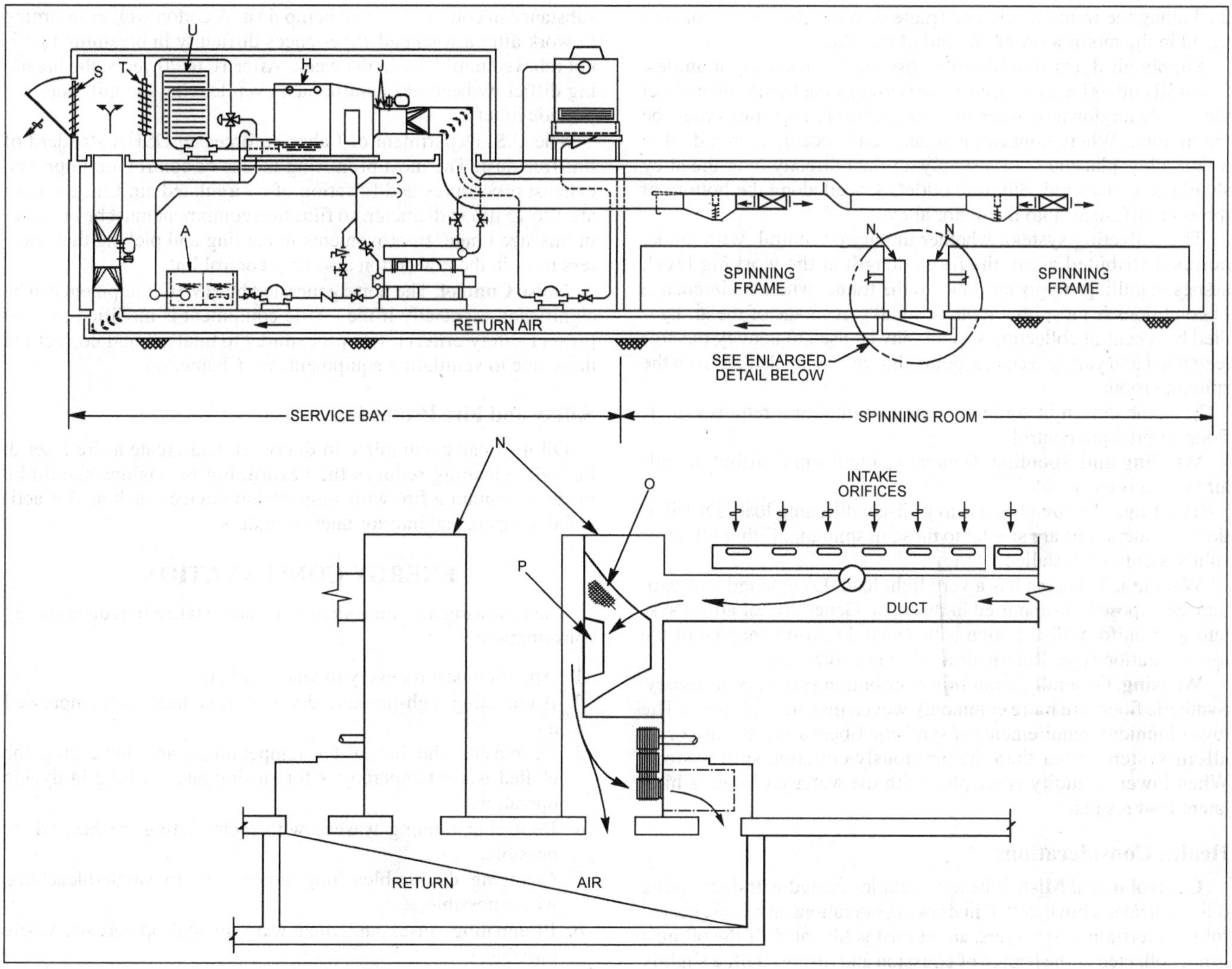

Fig. 2   Mechanical Spinning Room with Combined Air-Conditioning and Collector System

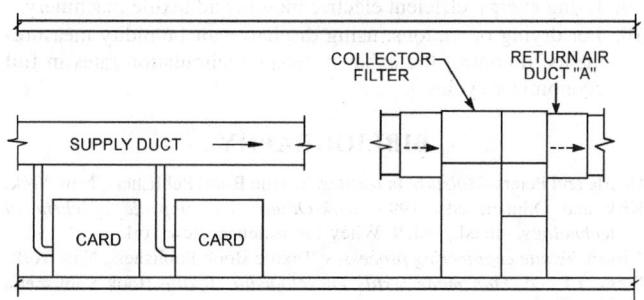

Fig. 3   Central Collector for Carding Machine

In melt spinning, the solvent concentration in the chimney air must be maintained below its explosive limit. Care is still required to prevent vapors from being ignited by a spark or flame. The air-conditioning system must be reliable, because interruption of the spinning causes the solution to solidify in the spinnerets.

Wind-Up or Take-Up Areas of Continuous Spinning. A heavy air-conditioning load is developed. Air is often delivered through branch ducts alongside each spinning machine. Diffusers are of the low-velocity, low-aspiration type, sized not to agitate delicate fibers.

Draw-Twist or Draw-Wind Areas of Fiber Manufacture. A heavy air-conditioning load is developed. Distribution, diffusion, and return systems are similar to those for the continuous spinning take-up area.

Opening and Picking. Usually, opening and picking require only a uniform distribution system. The area is subject to shutdown of machinery during portions of the day. Generally, an all-air system with independent zoning is installed.

Carding. A uniform distribution system is generally installed. There should be little air movement around the web in cotton carding. Central lint collecting systems are available but must be incorporated into the system design. An all-air system is often selected for cotton carding.

In wool carding, there should be less air movement than in cotton carding, not only to avoid disturbing the web, but also to reduce cross-contamination between adjacent cards. This is because different colors of predyed wool may be run side by side on adjacent cards. A split system may be considered for wool carding to reduce air movement. The method of returning air is also critical for achieving uniform conditions.

Drawing and Roving. Generally, a uniform distribution all-air system works well.

Mechanical Spinning Areas. A heavy air-conditioning load is generated, consisting of spinning frame power uniformly distributed along the frame length and frame driver motor losses concentrated in the motor alley at one end of the frame.

Supply air ducts should run across the frames at right angles. Sidewall outlets between each of the two adjacent frames then direct the supply air down between the frames, where conditions must be maintained. Where concentrated heat loads occur, as in a double motor alley, placement of a supply air duct directly over the alley should be considered. Sidewall outlets spaced along the bottom of the duct diffuse air into the motor alley.

The collecting system, whether unitary or central, with intake points distributed along the frame length at the working level, assists in pulling supply air down to the frame, where maintenance of conditions is most important. A small percentage of the air handled by a central collecting system may be used to convey the collected lint and yarn to a central point, thus removing that air from the spinning room.

Machine design in spinning systems sometimes requires interfloor air pressure control.

Winding and Spooling. Generally, a uniform distribution, all-air system is used.

Twisting. This area has a heavy air-conditioning load. Distribution considerations are similar to those in spinning. Either all-air or split systems are installed.

Warping. This area has a very light load. Long lengths of yarn may be exposed unsupported in this area. Generally, an all-air system with uniform distribution is installed. Diffusers may be of the low-aspiration type. Return air is often near the floor.

Weaving. Generally, a uniform distribution system is necessary. Synthetic fibers are more commonly woven than natural fibers. The lower humidity requirements of synthetic fibers allow the use of an all-air system rather than the previously common split system. When lower humidity is coupled with the water jet loom, a high latent load results.

Health Considerations

Control of Oil Mist. Whenever textiles coated with lubricating oils are heated above 200°F in drawing operations in ovens, heated rolls, tenterframes, or dryers, an oil mist is liberated. If the oil mist is not collected at the source of emission and disposed of, a slightly odorous haze results.

Various devices have been proposed to separate oil mist from the exhaust air, such as fume incinerators, electrostatic precipitators, high-energy scrubbers, absorption devices, high-velocity filters, and condensers.

Spinning operations that generate oil mist must be provided with a high percentage (30 to 75%) of outside air. In high-speed spinning, 100% outside air is commonly used.

Operations such as drum cooling and air texturizing, which could contaminate the air with oil, require local exhausts.

Control of Monomer Fumes. Separate exhaust systems for monomers are required, with either wet- or dry-type collectors, depending on the fiber being spun. For example, caprolactam nylon spinning requires wet exhaust scrubbers.

Control of Hazardous Solvents. Provisions must be made for the containment, capture, and disposal of hazardous solvents.

Control of Cotton Dust. Byssinosis, also known as brown or white lung disease, is believed to be caused by a histamine-releasing substance in cotton, flax, and hemp dust. A cotton worker returning to work after a weekend experiences difficulty in breathing that is not relieved until later in the week. After 10 to 20 years, the breathing difficulty becomes continuous; even leaving the mill does not provide relief.

The U.S. Department of Labor enforces an OSHA standard of lint-free dust. The most promising means of control are improved exhaust procedures and filtration of recirculated air. Lint particles are 1 to 15 μm in diameter, so filtration equipment must be effective in this size range. Improvements in carding and picking that leave less trash in the raw cotton also help control lint.

Noise Control. The noise generated by HVAC equipment can be significant, especially if the textile equipment is modified to meet present safety criteria. For procedures to analyze and correct the noise due to ventilating equipment, see Chapter 46.

Safety and Fire Protection

Oil mist can accumulate in ductwork and create a fire hazard. Periodic cleaning reduces the hazard, but provisions should be made to contain a fire with suppression devices such as fire-activated dampers and interior duct sprinklers.

ENERGY CONSERVATION

The following are some steps that can be taken to reduce energy consumption:

1. Applying heat recovery to water and air.
2. Automating high-pressure dryers to save heat and compressed air.
3. Decreasing the hot water temperatures and increasing the chilled water temperatures for rinsing and washing in dyeing operations.
4. Replacing running washes with recirculating washes, where possible.
5. Changing double-bleaching procedures to single-bleaching, where possible.
6. Eliminating rinses and final wash in dye operations, where possible.
7. Drying by means of the "bump and run" process.
8. Modifying the drying or curing oven air-circulation systems to provide counterflow.
9. Using energy-efficient electric motors and textile machinery.
10. For drying operations, using discharge air humidity measurements to control the exhaust versus recirculation rates in full economizer cycles.

BIBLIOGRAPHY

Hearle and Peters. *Moisture in textiles*. Textile Book Publishers, New York.

Kirk and Othmer, eds. 1993. *Kirk-Othmer encyclopedia of chemical technology*, 4th ed., Vol. 9. Wiley-Interscience, New York.

Nissan. *Textile engineering processes*. Textile Book Publishers, New York.

Press, J.J., ed. *Man made textile encyclopedia*. Textile Book Publishers, New York.

Sachs, A. 1987. Role of process zone air conditioning. *Textile Month* (October):42.

Schicht, H.H. 1987. Trends in textile air engineering. *Textile Month* (May): 41.

PHOTOGRAPHIC MATERIALS

THE processing and storage of sensitized photographic products requires temperature, humidity, and air quality control. Manufacturers of photographic products and processing equipment provide specific recommendations for facility design that should always be consulted. This chapter contains general information that can be used in conjunction with these recommendations.

STORING UNPROCESSED PHOTOGRAPHIC MATERIALS

Virtually all photosensitive materials deteriorate with age; the rate of photosensitivity deterioration depends largely on the storage conditions. Photosensitivity deterioration increases both at high temperature and at high relative humidity and usually decreases at lower temperature and humidity.

High humidity can accelerate loss of sensitivity and contrast, increase shrinkage, produce mottle, cause softening of the emulsion (which can lead to scratches), and promote fungal growth. Low relative humidity can increase the susceptibility of the film or paper to static markings, abrasions, brittleness, and curl.

Because different photographic products require different handling, product manufacturers should be consulted regarding proper temperature and humidity conditions for storage. Refrigerated storage may be necessary for some products in some climates.

Products not packaged in sealed vaportight containers are vulnerable to contaminants. Such products must be protected from solvent, cleanser, and formaldehyde vapors (emitted by particleboard and some insulation, plastics, and glues); industrial gases; and engine exhaust. In hospitals, industrial plants, and laboratories, all photosensitive products, regardless of their packaging, must be protected from x-rays, radium, and radioactive sources. For example, films stored 25 ft away from 100 mg of radium require the protection of 3.5 in. of lead.

PROCESSING AND PRINTING PHOTOGRAPHIC MATERIALS

Ventilation with clean, fresh air maintains a comfortable working environment and prevents fume-related health problems. It is also necessary for the high-quality processing, safe handling, and safe storage of photographic materials.

Processing produces odors, fumes, high humidity, and heat (from lamps, electric motors, dryers, mounting presses, and high-temperature processing solutions). Thus, it is important to supply plentiful clean, fresh air at the optimum temperature and relative humidity to all processing rooms. Modern ventilation systems filter, supply, and distribute air, keep it moving, control its temperature and humidity, and adjust its pressure.

Air Conditioning for Preparatory Operations

During receiving operations, exposed film is removed from its protective packaging for presplicing and processing. **Presplicing** combines many individual rolls of film into a long roll to be processed. At high relative humidity, photographic emulsions become

soft and can be scratched. At excessively low relative humidity, the film base is prone to static, sparking, and curl deformation. The presplice work area should be maintained at 50 to 55% rh and 70 to 75°F dry bulb.

Air Conditioning for Processing Operations

Processing exposed films or paper involves the use of a series of tempered chemical and wash tanks that emit heat, humidity, and fumes. Room exhaust must be provided, along with local exhaust at noxious tanks. To conserve energy, air from the pressurized presplice rooms can be used as makeup for processing room exhaust. Further supply air should maintain the processing space at a maximum of 80°F dry bulb and 50% rh.

The processed film or paper proceeds from the final wash to the dryer, which controls the moisture remaining in the product. Too little drying will cause the film to stick when wound, while too much drying will cause undesirable curl. Drying can be regulated by controlling drying time, humidity, and temperature.

The volume of incoming air should be sufficient to change the air in a processing room completely in about 8 min. Airflow should be diffused or distributed to avoid objectionable drafts. Apart from causing personnel discomfort, drafts can cause dust problems and disturb the surface temperature uniformity of drying drums and other heated equipment. For automated processing equipment, tempered fresh air should be supplied from the ceiling above the feed or head end of the machine at a minimum rate of 150 cfm per machine (see Figure 1). If the machine extends through a wall into another room, both rooms should be ventilated.

An exhaust system should be installed to remove humid or heated air and chemical vapors directly to the outside of the building. The room air from an open machine or tank area should be exhausted to the outdoors at a minimum rate of 170 cfm per machine. An exhaust rate higher than the supply rate produces a negative pressure and makes the escape of vapors or gases to

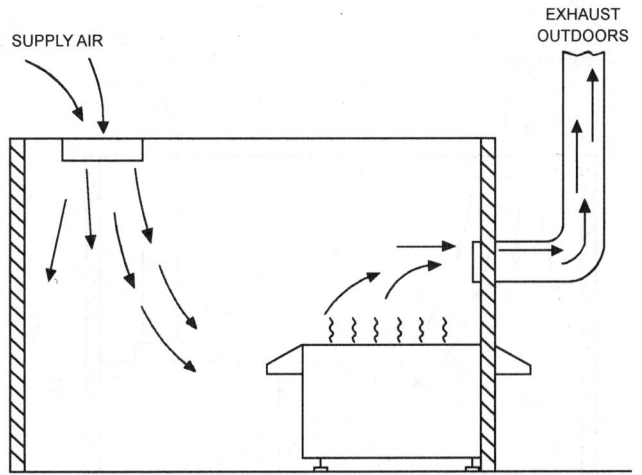

Fig. 1   Open Machine Ventilation

The preparation of this chapter is assigned to TC 9.2, Industrial Air Conditioning.

adjoining rooms less likely. The exhaust opening should be positioned so that the flow of exhausted air is away from the operator, as illustrated in Figure 2. This air should not be recirculated. The exhaust opening should always be as close as possible to the source of the contaminant for efficient removal. For a processing tank, an exhaust hood having a narrow opening at the back of and level with the top edge of the tank should be used.

If the processing tanks are enclosed and equipped with an exhaust connection, the minimum room air supply rate can be reduced to 90 cfm and the air removal rate to 100 cfm per machine (see Figure 3).

Air distribution to the drying area must provide a tolerable environment for operators. The exposed sides of the dryer should be insulated as much as is practical to reduce the large radiant and convected heat losses to the space. Return or exhaust grilles above the dryer can directly remove much of its rejected heat and moisture. The supply air should be directed to offset the remaining radiant losses.

The use of processor dryer heat to preheat cold incoming air during winter conditions can save energy. An economic evaluation is necessary to determine whether the energy savings justify the additional cost of the heat recovery equipment.

A canopy exhaust hood over the drying drum of continuous paper processors extracts heat and moisture. The dryers should be exhausted to the outside of the building to avoid undesirable humidity buildup. A similar hood over a sulfide-toning sink can be used to vent hydrogen sulfide. However, the exhaust duct must be placed on the side opposite the operator so that the vapor is not drawn toward the operator's face.

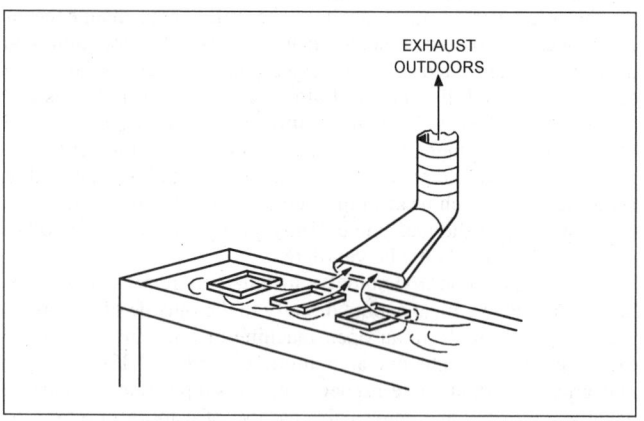

Fig. 2 Open-Tray Exhaust Ventilation from a Processing Sink

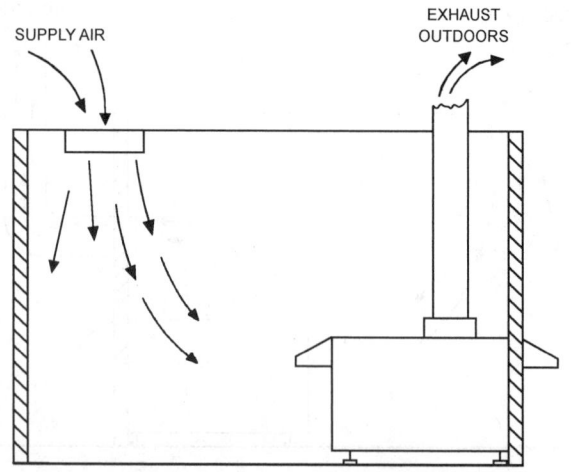

Fig. 3 Enclosed Machine Ventilation

Exhaust should draw off the vapor from the solvent and wax mixture that is normally applied to lubricate the motion picture film as it leaves the dryer.

Air Conditioning for the Printing/Finishing Operation

In printing, where a second sensitized product is exposed through the processed original, the amount of environmental control needed depends on the size and type of operation. For small-scale printing, close control of the environment is not necessary, except to minimize dust. In photofinishing plants, printers for colored products emit substantial heat. The effect on the room can be reduced by removing the lamphouse heat directly. Computer-controlled electronic printers transport the original film and raw film or paper at high speed. Proper temperature and humidity are especially important because, in some cases, two or three images from many separate films may be superimposed in register onto one film. For best results, the printing room should be maintained at between 70 and 75°F and at 50 to 60% rh to prevent curl, deformation, and static. Curl and film deformation affect the register and sharpness of the images produced. Static charge should be eliminated because it leaves static marks and may also attract dust to the final product.

Mounting of reversal film into slides is a critical finishing operation requiring a 70 to 75°F dry-bulb temperature with 50 to 55% rh.

Particulates in Air

Air conditioning for most photographic operations requires 85% efficient disposable bag-type filters with lower efficiency prefilters to extend filter life. In critical applications (such as high-altitude aerial films) and for microminiature images, filtering of foreign matter is extremely important. These products are handled in a laminar airflow room or workbench with HEPA filters plus prefilters.

Other Exhaust Requirements

A well-ventilated room should be provided for mixing the chemicals used in color processing and high-volume black-and-white work. The room should contain one or more movable exhaust hoods that provide a capture velocity of 100 fpm. Figure 4 shows a suitable mixing tank arrangement.

If prints are lacquered regularly, a spray booth is needed. A concentration of lacquer spray is both hazardous and very objectionable to personnel; spray booth exhaust must be discharged outside.

Processing Temperature Control

The density of a developed image on photographic material depends on characteristics of the emulsion, the exposure it has received, and the degree of development. With a particular emulsion, the degree of development depends on the time of development, the developer temperature, the degree of agitation, and the developer

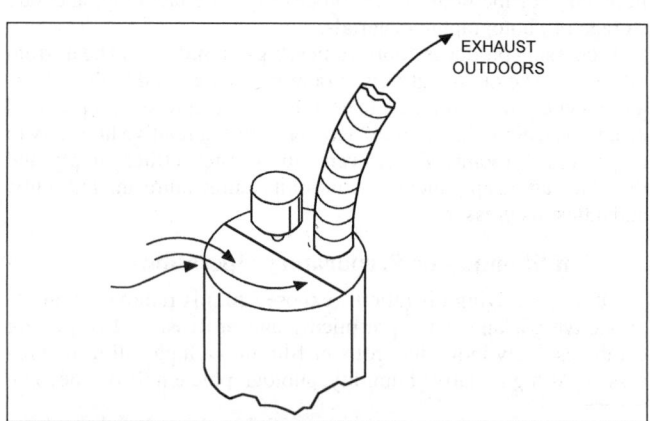

Fig. 4 Chemical Mixing Tank Exhaust

activity. Therefore, developer solution temperature control is critical. Once a developer temperature is determined, it should be maintained within 1°F for black-and-white products and within 0.5°F or better for color materials. Other solutions in the processor have a greater temperature tolerance.

Low processing volumes are typically handled in minilabs, which are often installed in retail locations. Minilabs are usually self-contained and equipped with temperature controls, heaters, and pumps. Typically, the owner only has to hook the lab up to water, electricity, exhaust (thimble connection), and a drain.

Higher volume processing is handled with processors that come from the manufacturer complete with controls, heat exchangers, pumps, and control valves designed for the process that the owner has specified. Electricity, hot water, cold water, drainage, and steam may be required, depending on the manufacturer, who typically provides the specifications for these utilities.

STORING PROCESSED FILM AND PAPER

Storage of developed film and paper differs from storage of the raw stock because the developed materials are no longer photosensitive, are seldom sealed against moisture, and are generally stored for much longer periods. Required storage conditions depend on (1) the value of the records, (2) the length of storage time, (3) whether the films are on nitrate or safety base, (4) whether the paper base is resin coated, and (5) the type of photographic image.

Photographic materials must be protected against fire, water, mold, chemical or physical damage, extreme relative humidity, and high temperature. Relative humidity is much more important than temperature. High relative humidity can cause films to stick together—particularly roll films, but also sheet films. High humidity also damages gelatin, encourages the growth of mold, increases dimensional changes, accelerates the decomposition of nitrate support, and accelerates the deterioration of both black-and-white and color images. Low relative humidity causes a temporary increase in curl and decrease in flexibility, but when the humidity rises again, these conditions are usually reversed. An exception occurs when motion picture film is stored for a long time in loosely wound rolls at very low humidities. The curl causes the film roll to resemble a polygon rather than a circle when viewed from the side. This **spokiness** occurs because a highly curled roll of film resists being bent in the length direction when it is already bent in the width direction. When a spoky roll is stored for a long time, the film flows permanently into the spoky condition, resulting in film distortion. Very low relative humidity in storage may also cause the film or paper to crack or break if handled carelessly.

Low temperature is desirable for film and paper storage provided that (1) the relative humidity of the cold air is controlled, and (2) the material is sufficiently warmed before opening to prevent moisture condensation. High temperature can accelerate film shrinkage, which may produce physical distortions and the fading of dye images. High temperature is also detrimental to the stability of any nitrate film still in storage.

Film Longevity

The American National Standards Institute (ANSI *Standard* IT9.11) defines longevities of films with a life expectancy (LE) rating. The **LE rating** is the minimum number of years information can be retrieved if the subject film is stored under long-term storage conditions. In order to achieve the maximum LE rating, a product must be stored under long-term storage conditions. Polyester black-and-white silver gelatin films have an LE rating of 500, while acetate black-and-white silver gelatin films have an LE rating of 100. No LE ratings have been assigned to color films or black-and-white silver papers. Medium-term storage conditions have been defined for materials that are to retain their information for at least 10 years.

Medium-Term Storage

Rooms for the medium-term storage of safety base film should be protected from accidental water damage by rain, flood, or pipe leaks. Air conditioning with controlled relative humidity is desirable but not always essential in moderate climates. Extremes of relative humidity are detrimental to film.

The most desirable storage relative humidity for processed film is about 50%, although the range from 30 to 60% is satisfactory. Air conditioning is required where the relative humidity of the storage area exceeds 60% for any appreciable period. For a small room, a dehumidifier may be used if air conditioning cannot be installed. The walls should be coated with a vapor retarder, and the controlling humidistat should be set at about 40% rh. If the prevailing relative humidity is under 25% for long periods, and problems from curl or brittleness are encountered, humidity should be controlled by a mechanical humidifier with a controlling humidistat set at 40%.

For medium-term storage, a room temperature between 68 and 77°F is recommended. Higher temperatures may cause shrinkage, distortion, and dye fading. Occasional peak temperatures of 95°F should not have a serious effect. Color films should be stored at temperatures below 50°F to reduce dye fading. Films stored below the ambient dew point should be allowed to warm up before being opened to prevent moisture condensation.

An oxidizing or reducing atmosphere may deteriorate the film base and gradually fade the photographic image. Oxidizing agents may also cause microscopically small colored spots on fine grain film such as microfilm (Adelstein et al. 1970). Typical gaseous contaminants include hydrogen sulfide, sulfur dioxide, peroxides, ozone, nitrogen oxides, and paint fumes. If such fumes are present in the intended storage space, they must be eliminated, or the film must be protected from contact with the atmosphere. Chapter 29 and Chapter 44 have further information on this subject.

Long-Term Storage

For films or records that are to be preserved indefinitely, long-term storage conditions should be maintained. The recommended space relative humidity ranges from 20 to 50% rh, depending on the film type. When several film types are stored within the same area, 30% rh is a good compromise. The recommended storage temperature is below 70°F. Low temperature aids preservation, but if the storage temperature is below the dew point of the outdoor air, the records must be allowed to warm up in a closed container before they are used, to prevent moisture condensation. Temperature and humidity conditions must be maintained year-round and should be continuously monitored.

Requirements of a particular storage application can be met by any one of several air-conditioning equipment combinations. Standby equipment should be considered. Sufficient outdoor air should be provided to keep the room under a slight positive pressure for ventilation and to retard the entrance of untreated air. The air-conditioning unit should be located outside the vault for ease of maintenance, with precautions taken to prevent water leakage into the vault. The conditioner casing and all ductwork must be well insulated. Room conditions should be controlled by a dry-bulb thermostat and either a wet-bulb thermostat, a humidistat, or a dew-point controller. For more information, see Chapter 45.

Air-conditioning installations and fire dampers in ducts carrying air to or from the storage vault should be constructed and maintained according to National Fire Protection Association (NFPA) recommendations for air conditioning (NFPA *Standard* 90A) and for fire-resistant file rooms (NFPA *Standard* 232).

All supply air should be filtered with noncombustible HEPA filters to remove dust, which may abrade the film or react with the photographic image. As is the case with medium-term storage, gaseous contaminants such as paint fumes, hydrogen sulfide, sulfur dioxide, peroxides, ozone, and nitrogen oxides may cause slow deterioration of the film base and gradual fading of the photographic image. When these substances cannot be avoided, an air scrubber, activated carbon adsorber, or other purification method is required.

Films should be stored in metal cabinets with adjustable shelves or drawers and with louvers or openings located to facilitate circulation of conditioned air through them. The cabinets should be arranged in the room to permit free circulation of air around them.

All films should be protected from water damage due to leaks, fire sprinkler discharge, or flooding. Drains should have sufficient capacity to keep the water from sprinkler discharge from reaching a depth of 3 in. The lowest cabinet, shelf, or drawer should be at least 6 in. off the floor and constructed so that water cannot splash through the ventilating louvers onto the records.

When fire-protected storage is required, the film should be kept in either fire-resistant vaults or insulated record containers (Class 150). Fire-resistant vaults should be constructed in accordance with NFPA *Standard* 232. Although the NFPA advises against air conditioning in valuable-paper record rooms because of the possible fire hazard from outside, properly controlled air conditioning is essential for long-term preservation of archival films. The fire hazard introduced by the openings in the room for air-conditioning ducts may be reduced by fire and smoke dampers activated by smoke detectors in the supply and return ducts.

Storage of Nitrate Base Film

Although photographic film has not been manufactured on cellulose nitrate film base for several decades, many archives, libraries, and museums still have valuable records on this material. The preservation of the nitrate film will be of considerable importance until the records have been printed on safety base.

Cellulose nitrate film base is chemically unstable and highly flammable. It decomposes slowly but continuously even under normal room conditions. The decomposition produces small amounts of nitric oxide, nitrogen dioxide, and other gases. Unless the nitrogen dioxide can escape readily, it reacts with the film base, accelerating the decomposition (Carrol and Calhoun 1955). The rate of decomposition is further accelerated by moisture and is approximately doubled with every 10°F increase in temperature.

All nitrate film must be stored in an approved vented cabinet or vault. Nitrate films should never be stored in the same vault with safety base films because any decomposition of the nitrate film will cause decomposition of the safety film. Cans in which nitrate film is stored should never be sealed, since this traps the nitrogen dioxide gas. Standards for the storage of nitrate film have been established (NFPA *Standard* 40). The National Archives and the National Institute of Standards and Technology have also investigated the effect of a number of factors on fires in nitrate film vaults (Ryan et al. 1956).

The storage temperature should be kept as low as economically possible. The film should be kept at below 50% rh. The temperature and humidity recommendations for the cold storage of color film in the following section also apply to nitrate film.

Storage of Color Film and Prints

All dyes fade in time. ANSI *Standard* IT9.11 does not define an LE for color films or black-and-white images on paper. However, many valuable color films and prints exist, and it is important to preserve them for as long as possible.

Light, heat, moisture, and atmospheric pollution contribute to the fading of color photographic images. Storage temperature should be as low as possible for the preservation of dyes. For maximum permanence of images, the materials should be stored in light-tight sealed containers or in moisture-proof wrapping materials at a temperature below freezing and at a relative humidity of 20 to 50%. The

containers should be warmed to room temperature prior to opening to avoid moisture condensation on the surface. Photographic films can be brought to the recommended humidity by passing them through a conditioning cabinet with circulating air at about 20% rh for about 15 min.

An alternative is the use of a storage room or cabinet controlled at a steady (noncycled) low temperature and maintained at the recommended relative humidity. This eliminates the necessity of sealed containers, but involves an expensive installation. The dye-fading rate decreases rapidly with decreasing storage temperature.

Storage of Black-and-White Prints

The recommended storage conditions for processed black-and-white paper prints are given by ANSI (1982). The optimum limits for relative humidity of the ambient air are 30 to 50%, but daily cycling between these limits should be avoided.

A variation in temperature can drive relative humidity beyond the acceptable range. A temperature between 59 and 77°F is acceptable, but daily variations of more than 7°F should be avoided. Prolonged exposure to temperatures above 86°F should also be avoided. The degradative processes in black-and-white prints can be slowed considerably by low storage temperature. Exposure to airborne particles and oxidizing or reducing atmospheres should also be avoided, as mentioned for films.

REFERENCES

Adelstein, P.Z., C.L. Graham, and L.E. West. 1970. Preservation of motion picture color films having permanent value. *Journal of the Society of Motion Picture and Television Engineers* 79(November):1011.

ANSI. 1982. Photography (film and slides)—Practice for storage of black-and-white photographic paper prints. *Standard* PH1.48-82. American National Standards Institute, New York.

ANSI. 1993. Imaging media—Processed safety photographic films—Storage. *Standard* IT9.11-93.

Carrol, J.F. and J.M. Calhoun. 1955. Effect of nitrogen oxide gases on processed acetate film. *Journal of the Society of Motion Picture and Television Engineers* 64(September):601.

NFPA. 1995. Protection of records. ANSI/NFPA *Standard* 232-95. National Fire Protection Association, Quincy, MA.

NFPA. 1996. Installation of air conditioning and ventilating systems. ANSI/NFPA *Standard* 90A-96.

NFPA. 1997. Storage and handling of cellulose nitrate motion picture film. ANSI/NFPA *Standard* 40-97.

Ryan, J.V., J.W. Cummings, and A.C. Hutton. 1956. Fire effects and fire control in nitro-cellulose photographic-film storage. Building Materials and Structures Report, No. 145. U.S. Department of Commerce, Washington, D.C. (April).

BIBLIOGRAPHY

ACGIH. 1998. *Industrial ventilation: A manual of recommended practice.* American Conference of Governmental Industrial Hygienists, Cincinnati, OH.

ANSI. 1991. Photography (photographic film)—Specifications for safety film. *Standard* IT9.6-91. American National Standards Institute, New York.

ANSI. 1996. Imaging materials—Ammonia-processed diazo photographic films—Specifications for stability. *Standard* IT9.5-96.

Carver, E.K., R.H. Talbot, and H.A. Loomis. 1943. Film distortions and their effect upon projection quality. *Journal of the Society of Motion Picture and Television Engineers* 41(July):88.

Kodak. 1987. Current information summary—General guidelines for ventilating photographic process areas. CIS-58. Eastman Kodak Company, Rochester, NY.

Kodak. 1990. Photolab design for professionals. *Publication* K-13. Eastman Kodak Company, Rochester, NY.

UL. 1995. Tests for fire resistance of record protection equipment, 14th ed. *Standard* 72-95. Underwriters Laboratories, Northbrook, IL.

MUSEUMS, LIBRARIES, AND ARCHIVES

APPRECIATION for the arts and history dictates a need for special temperature and humidity conditions in museums, libraries, and archives. Proper environments reduce the deterioration rate of stored materials and provide comfort for the occupants. Providing stability in the building environment is often as critical as establishing correct design criteria to protect the building contents. Facilities often have open traffic patterns and require different conditions between changing exhibits. Maintaining diverse environmental conditions that are so close to each other merits special consideration. The designer must also understand the effect environmental conditions will have on the building envelope, which is of serious concern in existing buildings.

In theory, a variety of systems can be successful in a building that requires environmental control, if properly applied. In practice, the available choices are limited by factors such as the material to be housed, collection value, length of time on exhibit (or in storage), operating costs, space allocation, architectural design, or project location. The selection process is most often driven by the past experience of the designers and available project funding. The operating benefits and constraints of the proposed system must be reviewed in detail with the building operators during conceptual stages to minimize the potential for misunderstandings at subsequent phases.

Heating and air conditioning that is simple in design and of the proper size for a given zone and building generally is economical to operate and maintain. For optimum results, the basic structure should include as much thermal and humidity control as is economically possible. Features might include materials with high thermal or impervious properties, insulation, multiple or special glazing, and shading devices.

The designer's input is essential in the early planning of the building. A good working relationship between the mechanical engineer, architect, interior designer, and owner is critical. All limitations must be defined at the beginning of the design. Some of the factors the designer must consider include

- The relationship between building shape, orientation, and environmental control.
- The amount of fenestration. When fenestration is a high percentage of the exterior envelope surface area the exterior load can be a significant part of the total air-conditioning load. Also, natural light may be detrimental to the buildings contents.
- Noise control, both inside and outside.
- Quality and quantity of indoor and outdoor air. Select appropriate filters to support the needs of the structure and consider the effects that off-gassing might have on its contents.
- Minimizing the building's influence on the environment, including reduction of pollution into the atmosphere.

Retrofitting existing buildings has become important because of the increased cost of new structures as well as the desire to preserve

older buildings of historical significance. In some cases the effect of the environmental systems may be of more concern relative to the structure than the contents. Older buildings with minimal insulation or without vapor retarders must be protected from the potential damage that could be created by an environmental control system that affords year round, interior stability.

Selection of the HVAC system should be a team effort. Often, the most likely solutions will be compromises that do not meet all the criteria for all spaces. To a great extent, the selection will be based on interior design conditions, which are driven by the primary use of the structure. There are several opinions relative to the inside environment. All are based on the idea that materials deteriorate because of vapor, heat, gases, and radiation found naturally in the environment. Deterioration can be reduced by controlling temperature, humidity, sunlight, and particulates. The disagreement is over the acceptable limits of control and where to compromise. Buildings that house exhibitions are usually quite open, to promote ease in circulation of the occupants. However different exhibits may require contrasting conditions. Some exhibits may off-gas materials that are detrimental to others. The acceptable limits, budget, and architectural constraints must be established early in the design.

DESIGN TARGETS FOR MECHANICAL SYSTEMS

Museums, archives, and libraries have two groups of indoor air requirements. The first falls into the general category of health and safety as listed in ASHRAE *Standard* 55 and ASHRAE *Standard* 62. These standards consist of requirements such as concentrations of hazardous substances, air temperature, velocity, and relative humidity.

The second group of requirements pertains to the collections. Collection requirements are not simple, they are not yet clearly understood, and they often conflict across collection types. Besides, the building envelope often is incapable of maintaining the chosen conditions. For these reasons, the question of what is the risk of compromising on the values of relative humidity and temperature is asked. The following sections summarize the best information on these issues.

In terms of both health and safety and collection requirements, it is useful to label building spaces three ways, as shown in Table 1. First, collection versus noncollection; second, public versus nonpublic; and third, "dirty" versus "clean". These subdivisions distinguish between areas that have very different thermal and indoor air quality requirements, outdoor ventilation rates and air supply strategies, etc. Typically these areas require separate HVAC systems. See Chapters 13, 28, and 29 for more information on "dirty" rooms. Noncollection rooms are not considered in this chapter because their HVAC requirements are similar to those in other public buildings, which are discussed in Chapter 3.

Derivation of Collection Requirements

The hierarchy in the following paragraphs (Michalski 1996a) provides a logic for developing appropriate climate and indoor air

The preparation of this chapter is assigned to TC 9.8, Large Building Air-Conditioning Applications.

Table 1   Classification of Rooms for Museums and Libraries

		High Internal Source of Contaminants (Dirty)	Low Internal Source of Contaminants (Clean)
Collection	Non-public access	Conservation laboratories, museum workshops (VOCs, fumes, dusts) "Wet" collections (alcohol evaporation from poorly sealed jars in natural history collections) Photographic collections ("vinegar syndrome" produces acetic acid vapors)	Most storage areas, vaults, library stacks
	Public access	Displays of conservation work in progress (unusual and temporary)	Galleries, reading rooms
Noncollection	Non-public access	Smoking offices (unusual)	Offices (nonsmoking)
	Public access	Cafeterias, rest rooms, spaces where smoking permitted	Public spaces without food preparation or smoking

quality (IAQ) parameters for different museums, libraries, and archives. Further, it demonstrates that a single target is a compromise between a large number of different, often contradictory, requirements. Fortunately, many collections have sufficient uniformity that useful generalizations can be made about their requirements. Throughout this hierarchy, the term "effective" includes institutional value judgements as well as the science of deterioration.

Museum, library, archive. A set of collections. The target for a single climate and the indoor air quality (IAQ) for an entire institution is that which provides the most effective environment for all collection needs. Sometimes, but not always, a museum has a single type of collection. If the museum is split into zones with different targets, then specific collections can be given specific targets.

Collection. A set of artifacts. A single climate and IAQ target for an entire collection is that which provides the most effective environment for the collection. Sometimes, but not always, a collection consists of artifacts with identical climate and IAQ needs, e.g., a coin collection, photo collection, a rare book library, etc.

Artifact. Assembly of materials. A single microclimate and contaminant deposition target for an entire artifact is that which provides the most effective common environment for preservation of both the individual constituent materials and of the assembly itself. Some artifacts have an elementary assembly; i.e., they are a single piece of material such as a blank sheet of paper, or a pure mineral specimen, and hence have a simpler target to specify.

Material. A single microclimate and pollutant deposition target for a single piece of material is that which provides the most effective common environment for preservation of the material.

ROLE OF RELATIVE HUMIDITY AND TEMPERATURE IN DETERIORATION

Role of High Relative Humidity

Dampness causes rapid mold on most surfaces and rapid corrosion of base metals. It is the single most important factor that causes damage in museums and archives.

The most comprehensive mold data exists in the feed and food literature. Fortunately, this provides a conservative outer limit to dangerous conditions, because mold in museum objects occurs first on surfaces contaminated with sugars, starch, oils, etc.; or on objects made of grass, skin, bone, and other feed- or food-like materials. Water activity is identical to and always measured as the equilibrium relative humidity of air adjacent to the material. The equilibrium relative humidity, rather than equilibrium moisture content (EMC), provides a better measure for mold germination and growth on a wide variety of materials (Beuchat 1987). In Figure 1, the combined role of temperature and relative humidity is shown. The study of the most vulnerable book material by Groom and Panisset (1933) is consistent with the general trend of culture studies from Ayerst (1968). Ohtsuki (1990) reported microscopic mold occurring on clean metal surfaces at 60% rh. The DNA helix is known to collapse near 55% rh (Beuchat 1987), so a conservative limit for no mold

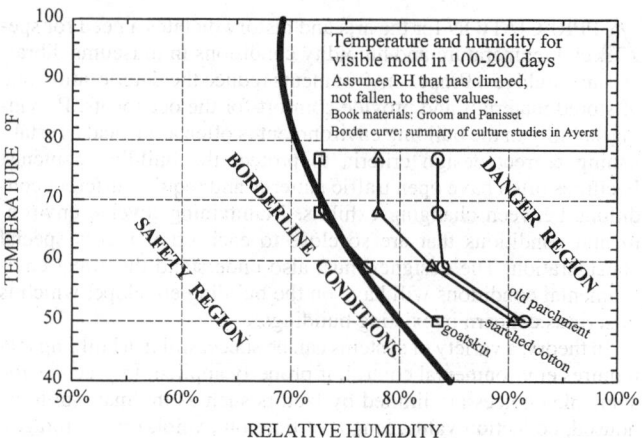

Fig. 1   Temperature and Humidity for Visible Mold in 100 to 200 Days

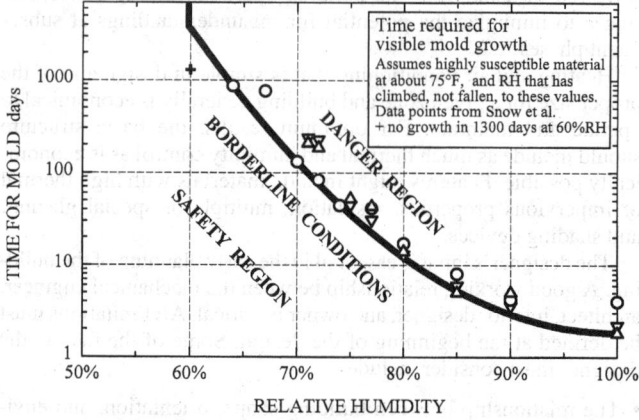

Fig. 2   Time Required for Visible Mold Growth

ever, on anything, at any temperature, is below 60% rh. Chapter 23, Figure 1, suggests a similar lower boundary for mold: 62% rh.

Snow et al. (1944) looked for visible mold growth on materials inoculated with a mixture of mold species. These are plotted in Figure 2, and follow the same trend as reported in the European building industry for wall mold (Hens 1993) and in Chapter 22, Figure 1.

Therefore, Figure 1 and Figure 2 show practical dangers, e.g., that growth in less than a summer season requires over 70% rh, in less than a week requires over 85% rh. The precise point between 32 and 23°F where mold growth ceases is not well characterized.

Conditions for rapid drying of already wet collections following some disaster is a separate issue. The boundaries for rapid mold growth shown in Figure 2 are still reasonably correct, but partially-dried porous objects are far from equilibrium relative humidity conditions.

Figure 1 and Figure 2 are conservative also for the onset of production of allergenic spores and human toxins in molds, since both are known to require slightly higher relative humidity than spore germination (Beuchat 1987). For both collection and occupant health, and given the need in museums to humidify air more than many applications in wintertime, special care must be taken to avoid cold spots along ductwork.

Rapid corrosion above 75% rh occurs for two reasons: increased surface adsorption of water, and contamination by salts. The adsorption of water on clean metal surfaces climbs rapidly from 3 molecules or less below 75% rh to bulk liquid layers above 75% rh (Graedel 1994). This phenomenon is aggravated by most surface contaminants, as shown in studies of the role of dust on clean steel corrosion. The most common contaminant of museum metals, sodium chloride, dissolves and becomes liquid (deliquesces) above 76% rh.

Role of Critical Relative Humidity

At some critical relative humidity, some minerals hydrate, dehydrate, or deliquesce. When such minerals are part of a salt-contaminated porous stone, a stone containing salt, a corroded metal, or a natural history specimen, they disintegrate the object. Dozens of minerals with distinct critical relative humidity values are known in natural history collections (Waller 1992). Pyrites that contaminate most fossils disintegrate if held above 60% rh (Howie 1992). One of the most important archaeological metals, bronze, has a complex chemistry of corrosion, with several critical relative humidity values (Scott 1990). This variety leads away from a generalized safe relative humidity, toward an emphasis on particular conditions for particular artifacts. Often this is achieved by local cabinets or small relative humidity controlled packages (Waller 1992). The only simple generalization for these materials is that relative humidity over 75% is almost always dangerous.

Role of Warmth and Relative Humidity above 0%

Warmth and moderate amounts of absorbed moisture lead to rapid decay in chemically unstable artifacts, especially some archival records. In archives, the most important deterioration mechanism of modern records is acid hydrolysis, which affects papers, photographic negatives, and magnetic media (both analog and digital). Michalski (1999) developed equations for relative lifetime as a function of relative humidity and temperature. Fortunately, for design purposes the dependence on relative humidity and temperature in all records is very similar. Although the precise quantification and meaning of record lifetime is debatable, all authorities agree that the most rapidly decaying records, such as videotape and acidic negatives, can become unusable within a few decades at normal room conditions, and much faster in hot, humid conditions. The relative increase in record lifetime due to cold, dry conditions is shown in Figure 3. The range in numbers on each line reflect the spread in the available data. The extension of the plots below 5% rh is uncertain; the rates of chemical decay may or may not approach zero, depending on slow, non-moisture-controlled mechanisms such as oxidation.

Lifetime improvement predictions specifically from photographic data has been derived by the Image Permanence Institute (Reilly 1993; Nishimura 1993) and used for a lifetime prediction wheel. Relative lifetime estimates from these wheels are less optimistic about improvement with low relative humidity, but the general trend is the same as the Michalski plots.

The rate of light fading of some colorants depends on relative humidity, although the rate does not drop to zero at 0% rh. Some lake pigments, some color photographs on display, can last twice as long at 10% rh as at 50% rh (Bailie et al. 1988; Bard et al. 1986).

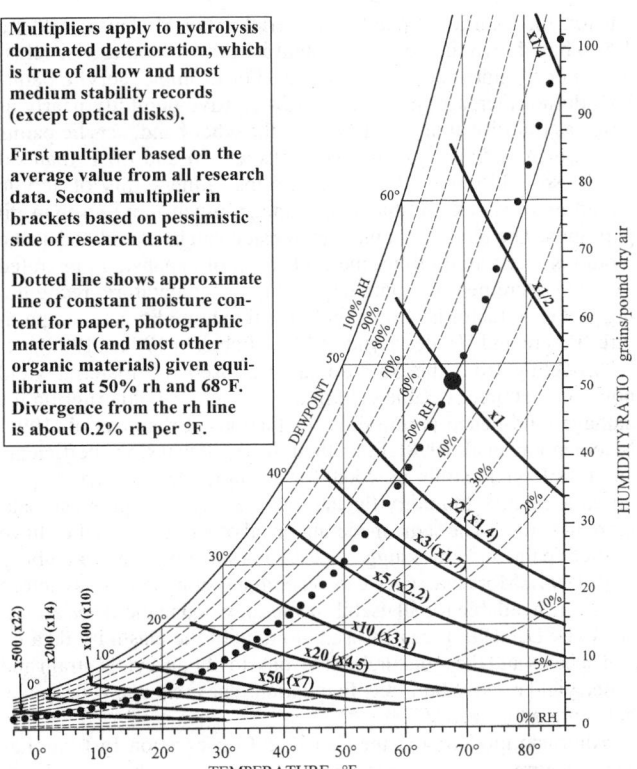

Multipliers apply to hydrolysis dominated deterioration, which is true of all low and most medium stability records (except optical disks).

First multiplier based on the average value from all research data. Second multiplier in brackets based on the pessimistic side of research data.

Dotted line shows approximate line of constant moisture content for paper, photographic materials (and most other organic materials) given equilibrium at 50% rh and 68°F. Divergence from the rh line is about 0.2% rh per °F.

Fig. 3   Lifetime Multipliers Relative to 68°F and 50% rh

Role of Fluctuations and Low Temperature

Fluctuations in relative humidity, fluctuations in temperature, and cold by itself, can each lead to mechanical damage in artifacts. The fundamental cause is expansion and contraction of materials, combined with some form of internal or external restraint. Very low humidity or temperature has the added effect of increasing the stiffness of many organic materials, making them more vulnerable to fracture. Traditionally, these concerns led to extremely narrow specifications, such as 50 ± 3% rh and 70 ± 2°F (Lafontaine 1979), which still form the basis of many museum and archive guidelines. These specifications were derived from extrapolation of a common observation, that very large fluctuations fractured some objects, but with no experimental or theoretical basis for precise extrapolation to smaller fluctuations.

Research (Erhardt et al. 1994, 1996; Mecklenburg et al. 1991, 1998; Michalski 1991a, 1993) on objects, experience in historic building museums, and a comparison of historic buildings with and without air conditioning (Oreszczyn et al. 1994) have led to reappraisal of these fluctuation specifications. Despite some small disagreements over the interpretation of these studies, the practical conclusions are the same; that is, the traditional extrapolation to very narrow tolerances such as ±3% rh and ±2°F was an exaggeration of artifact needs. Somewhere between fluctuations of about ±10% rh, ±18°F and ±20% rh, ±35°F, the risk of fracture or deformation does climb from insignificant to significant for a mixed historic collection. Beyond this range, risk climbs even more quickly for increasingly large fluctuations. It is the range of ±20% rh to ±40% rh that has yielded the common observations of cracked cabinetry and paintings.

All research studies begin with the model of a restrained sample of organic material, typically paint, glue, or wood, subject to lowered relative humidity or temperature. The material both shrinks and stiffens. Mecklenburg (1991), Daly and Michalski (1987), and Hedley (1988) have collected consistent data on the increase in tension for traditional painting materials; and Michalski (1991a, 1999) has shown this to be consistent in turn with other viscoelastic data on paints and their polymers. For example, acrylic paintings do not increase in tension at low relative humidity nearly as much as traditional oil paintings. On the other hand, acrylic paints increase in stiffness much more between 70 and 40°F than oil paints, so cold temperatures increase the vulnerability of acrylic paintings to shock and handling damage much more so than oil paintings. The increase in material tension can be calculated as the modulus of elasticity times the coefficient of expansion integrated over the decrement in relative humidity or temperature, recognizing that each factor is a function of relative humidity and temperature (Perera and Vanden Eynde 1987; Michalski 1991a, 1998).

Modeling from the Smithsonian has used the material yield point, i.e., nonrecoverable deformation, as a threshold criterion for damage, and hence permissible fluctuations (Erhardt et al. 1994; Mecklenburg et al. 1998). From their data on expansion coefficients and tensile yield strain in wood, for example, they estimate a permissible relative humidity fluctuation of ±10% rh in pine and oak, more in spruce. The same modeling on all other materials they have studied (paint and glue) suggest ±15% as a safe range (Mecklenburg et al. 1994). More extensive data on compression yield stress across the grain, available for all useful species of wood in the *Wood Handbook* (USDA 1987), can be used along with the elasticity data on that species and the moisture content data to obtain a yield strain and hence yield relative humidity fluctuation. These show a wide range, but centered near ±15% rh.

Alternate modeling at the Canadian Conservation Institute has used fracture as the criterion for damage, and used the general pattern of fatigue fracture in wood and polymers to extrapolate the effect of smaller multiple cycles (Michalski 1991a). Assuming a benchmark of high probability of single-cycle fracture at ±40% rh (e.g., a drop from 50% rh to 10% rh) known from common observation of many museum artifacts, one can extrapolate to the fatigue threshold stress (10^7 cycles or more) by an approximate factor of 0.5 in wood (±20% rh) and 0.25 in brittle polymers (±10% rh) such as old paint. Fortunately, this is consistent with the yield criterion, inasmuch as the yield stress in these materials corresponds to stresses that cause very small or negligible crack growth per cycle. Both groups of researchers also noted that the coefficient of expansion in wood and other materials is at a minimum at moderate relative humidity due to their sigmoidal absorption isotherms, so fluctuations at lower and higher relative humidity set points tend to be even more risky.

Both models assume uniformly restrained materials. As a first approximation, many laminar objects, such as paintings on stretchers and many photographic records fall in this class. Many other artifacts, however, are complex assemblies of materials. Some are less vulnerable due to lack of restraint, such as floating wood panels, or books and photographic papers with components that dilate in reasonable harmony; while other assemblies contain sites of severe stress concentration that initiate early fracture. A table of vulnerabilities for wooden objects has been developed (Michalski 1996b) in which assembly vulnerability is classified as Very High, High, Medium (uniformly restrained components), and Low. Each category differs from the next lower category by a factor of two, i.e. needs half the relative humidity fluctuation for the same risk of damage.

Response Times of Artifacts

Very brief relative humidity fluctuations may not affect artifacts, for example, very few museum objects respond significantly to fluctuations under an hour in duration, so a 15 min cycle in HVAC output does not affect most artifacts (unless it is so large as to cause sudden damp conditions). Many objects take days to respond. Figure 4 shows calculated humidity response times of wooden artifacts. Figure 5 shows the interaction of air leakage, wood coating, and

textile buffering on the response of a chest of drawers. It shows that museums cause greater risk than ordinary households if they display the piece empty and open, rather than closed and full, because response time falls from months to days.

Very long relative humidity and temperature fluctuations, such as seasonal changes, are slow enough to take advantage of stress relaxation in artifact components. Data on the effective modulus of elasticity of many oil and acrylic paints as a function of time, temperature, and relative humidity (Michalski 1991a) and direct stress relaxation data for both paint (Michalski 1995) and wood imply that the stress caused by a given strain applied over one day falls to about half or less if that strain is applied over 4 months at moderate room temperatures. Thus a four month seasonal ramp of ±20% rh

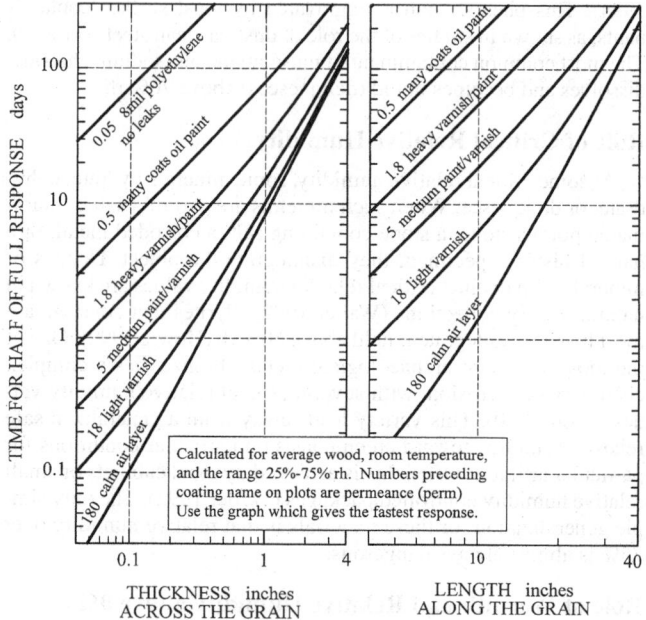

Fig. 4 Calculated Humidity Response Times of Wooden Artifacts

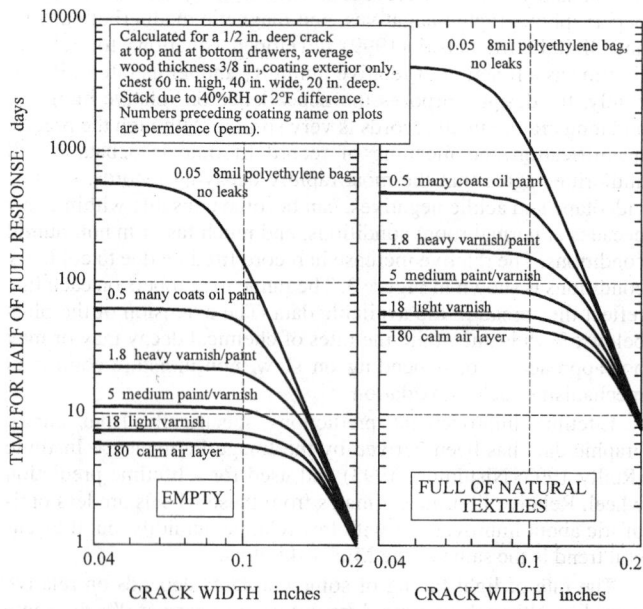

Fig. 5 Interaction of Air Leakage, Wood Coating, and Textile Buffering on Response of Chest of Drawers

(e.g. from 50% rh to 30% rh should cause less stress in most artifacts than a one week fluctuation of ±10% rh. Thus the most stressful relative humidity fluctuations are those shorter in rise-time than any stress relaxation mechanisms, but long enough for artifacts to respond.

Humidity Response Time of Cases, Cabinets, and Packages.
Leakage of museum display cases has been studied using either moisture (Thompson 1977; Michalski 1994) or contaminants (Cass 1988) as tracer gases. Michalski (1994) developed equations to model all the mechanisms that contribute to the leakage of humidity between enclosures and the room. Some results are given in Figure 6, and show the transition between diffusion-dominated and infiltration-dominated leakage in a 3 ft case, given a stack pressure in the case of 160 μin. of water due to a temperature difference of 2°F or a humidity difference of 40% rh. Most museum cases leak in the range of 10 to 100 air changes per day (ACD) or 0.4 to 4 air changes per hour (ACH), but careful design can limit this to 0.1 ACD (0.004 ACH).

Any material that sorbs and desorbs moisture (thereby tending to moderate relative humidity change in an enclosure) is known in museums as a **humidity buffer**. The hygrometric halftime of an enclosure with natural buffers like wood, textile, paper, etc., equals (Thomson 1977):

$$\theta_{0.5} = \frac{0.693 \alpha \rho f_p}{N C_{ws}}$$

where

$\theta_{0.5}$ = time for enclosure to reach half of an external step in relative humidity, days

α = moisture absorption coefficient, typically 0.1 for natural organic materials, 0.3 for regular silica gel

ρ = bulk density of the buffer, conservatively 31 lb/ft^3

f_p = packing factor, fraction of case volume filled with buffer, dimensionless

N = leakage or exchange rate of enclosure, air changes per day (ACD)

C_{ws} = absolute humidity of air at saturation, 0.00108 lb/ft^3 at 68°F

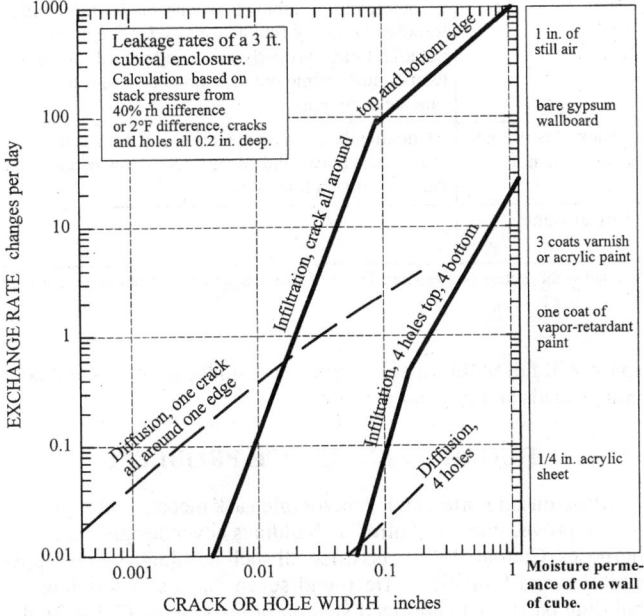

Fig. 6   Typical Contribution of Infiltration, Diffusion, and Permeation to Leakage of Museum Display Case or Cabinet

For wood, paper, textiles, leather, and similar natural organic materials, this reduces to approximately

$$\theta_{0.5} = \frac{2000 f_p}{N}$$

For ordinary cases and cabinets leaking at 10 to 100 ACD and f_p = 0.1, halftime would be 2 to 20 days. For a tight case, leaking 0.1 ACD, this improves to 2000 days, well beyond seasonal fluctuations. For an ordinary bookcase cabinet filled with books, f_p approaches 1, so halftimes would approach 20 to 200 days. These halftimes mean the artifacts ignore short fluctuations in relative humidity. On the other hand, there is a risk that after sustained periods of dampness or water entry these halftimes show a closed cabinet at 90% rh would take weeks to reach 70% rh, given a room at 50% rh. Such microenvironments can eliminate external humidity fluctuations; but the enclosure will settle at the mean relative humidity, which may be an undesirable value (e.g., most artifacts in humid climates and many artifacts in climates with winter heating and no humidification). Two solutions exist: periodic reconditioning of removable buffer, such as silica gel (Lafontaine 1984) or the active supply of correct humidity air to the exhibit case, as discussed in the section on Design Issues under System Selection.

SELECTION OF DESIGN TARGETS

The target specifications are summarized in Table 2, based on a balanced concern of all the deterioration processes previously described. The difficult issue of permissible fluctuations has been reduced to five classes: AA, A, B, C, and D. Gradients are conservatively considered to add to short-term fluctuations because artifacts can be moved from one part of a space to another, thereby adding a space gradient fluctuation to the dynamic fluctuations of the HVAC.

Class A is the optimum for most museums and galleries. Two possibilities with equivalent risks are given for Class A, either a larger gradient and short-term fluctuation, or a larger seasonal swing. Stress relaxation is used to equate an extra ±10% rh seasonal swing to a short-term ±5% rh.

A major institution with the mandate and resources to prevent even tiny risks legitimately demands the narrower fluctuations of Class AA. However, design for very long term reliability must take precedence over narrow fluctuations.

Classes B, C, and D should not be dismissed as ineffective. All of the material science described earlier points to a rapidly increasing risk with extreme conditions. Thus Class B and C control are useful and feasible for many medium to small institutions, and the best that can be done in most historic buildings. Class D recognizes that control of dampness is the single most important museum climate issue.

Library and Archive Dilemma

The whole issue of fluctuation control is secondary for libraries and archives. Except for rare and archaic records, their preservation problem is chemical decay of nineteenth and twentieth century records, and its acceleration by warmth and humidity. During retrieval and handling, however, the issues become mechanical, so here the conditions are identical to those required by museums and galleries. A detailed overview of all these issues is given by Michalski (1999) and in several standards on paper, photographic, and electronic records (ANSI *Standards* IT9.11, IT9.18, and IT9.25).

Specifications Apply to Entire Usable Space

The specifications in Table 2 are not simply HVAC targets for a single optimum point in the space or system. Museum and

Table 2 Temperature and Relative Humidity Specifications for Museum, Gallery, Library, and Archival Collections

Type	Set Point or Annual Average	Maximum Fluctuations and Gradients in Controlled Spaces			Collection Risks and Benefits
		Class of Control	Short Fluctuations plus Space Gradients	Seasonal Adjustments in System Set Point	
General Museums, Art Galleries, Libraries, and Archives All reading and retrieval rooms, rooms for storage of chemically stable collections, especially if mechanically medium to high vulnerability.	50% rh (or historic annual average for permanent collections) Temperature set between 59 and 77°F *Note:* Rooms intended for loan exhibitions must handle set point specified in loan agreement, typically 50% rh, 70°F, but sometimes 55% or 60% rh).	**AA** Precision control, no seasonal changes	±5% rh ±4°F	Relative humidity no change Up 9°F; down 9°F	No risk of mechanical damage to most artifacts and paintings. Some metals and minerals may degrade if 50% rh exceeds a critical relative humidity. Chemically unstable objects unusable within decades.
		A Precision control, some gradients or seasonal changes, not both	±5% rh ±4°F	Up 10% rh, down 10% rh Up 9°F; down 18°F	Small risk of mechanical damage to high vulnerability artifacts, no mechanical risk to most artifacts, paintings, photographs, and books. Chemically unstable objects unusable within decades.
			±10% rh ±4°F	RH no change Up 9°F; down 18°F	
		B Precision control, some gradients plus winter temperature setback	±10% rh ±9°F	Up 10%, down 10% rh Up 18°F, but not above 86°F Down as low as necessary to maintain RH control	Moderate risk of mechanical damage to high vulnerability artifacts, tiny risk to most paintings, most photographs, some artifacts, some books and no risk to many artifacts and most books. Chemically unstable objects unusable within decades, less if routinely at 86°F, but cold winter periods will double life.
		C Prevent all high risk extremes	Within 25 to 75% rh year-round Temperature rarely over 86°F, usually below 77°F		High risk of mechanical damage to high vulnerability artifacts, moderate risk to most paintings, most photographs, some artifacts, some books and tiny risk to many artifacts and most books. Chemically unstable objects unusable within decades, less if routinely at 86°F, but cold winter periods will double life.
		D Prevent dampness	Reliably below 75% rh		High risk of sudden or cumulative mechanical damage to most artifacts and paintings due to low humidity fracture; but high-humidity delamination and deformations, especially in veneers, paintings, paper, and photographs, is avoided. Mold growth and rapid corrosion avoided. Chemically unstable objects unusable within decades, less if routinely at 86°F, but cold winter periods will double life.
Archives, Libraries Storage of chemically unstable collections	Cold Store: −4°F, 40% rh	±10% rh ±4°F			Chemically unstable objects usable for millennia. Relative humidity fluctuations under one month do not affect most properly packaged records at these temperatures (time out of storage becomes the lifetime determinant).
	Cool Store: 50°F 30 to 50% rh	(Even if achieved only during winter setback, this is a net advantage to such collections, as long as damp is not incurred)			Chemically unstable objects usable for a century or more. Such books and papers tend to low mechanical vulnerability to fluctuations.
Special Metal Collections	Dry room: 0 to 30% rh	Relative humidity not to exceed some critical value, typically 30% rh			

Note: Short fluctuations means any fluctuation less than the seasonal adjustment. However, as noted in the section on Response Times of Artifacts, some fluctuations are too short to affect some artifacts or enclosed artifacts.

archive clients assume that stated specifications are performance targets for anywhere in the usable space, because they are defined by artifacts that can occupy any point in that space. These users usually monitor relative humidity and temperature independently of the HVAC sensors with thermohygrographs, data loggers, or handheld meters. These monitors often sit in shelves, in corners, against walls, or on floors, where temperature and moisture gradients are worst. Thus to meet a specification of ±10% rh in a space where the temperature gradient in the room may reach ±4°F, the system needs to maintain ±5% rh at the humidistat. Typical design implications for reduction of such gradients are better than average air distribution, and higher than average air change rates. It is important for client and engineer to agree on when and where the

space will fall within specifications, and under what conditions of indoor loads and outdoor extremes.

BUILDING ENVELOPE PROBLEM

Museums, libraries, and archives often ask mechanical engineers for "improved climate control" in buildings never designed for such purposes. Conrad (1995) classified all such buildings by their possibilities and limitations. He found seven classes of building (or building part). In an abridged version of his scheme (Table 3), the possible classes of fluctuation control possible are listed for each class of building. Local climate determines which possibility is most likely.

Table 3   Classification of the Climatic Control Potential in Buildings

Category of Control	Building Class	Typical Building Construction	Typical Type of Building	Typical Building Use	System Used	Practical Limit of Climate Control	Class of Control Possible
Uncontrolled	I	Open structure	Privy, stocks, bridge, sawmill, well	No occupancy, open to viewers all year	No system.	None	D (if benign climate)
	II	Sheathed post and beam	Cabins, barns, sheds, silos, icehouse	No occupancy. Special event access.	Exhaust fans, open windows, supply fans, attic venting. No heat.	Ventilation	C (if benign climate) D (unless damp climate)
Partial control	III	Uninsulated masonry, framed and sided walls, single-glazed windows	Boat, train, lighthouse, rough frame house, forge	Summer tour use. Closed to public in winter. No occupancy.	Low level heat, summer exhaust ventilation, humidistatic heating for winter control.	Heating, ventilating	C (if benign climate) D (unless hot damp climate)
	IV	Heavy masonry or composite walls with plaster. Tight construction; storm windows	Finished house, church, meeting house, store, inn, some office buildings	Staff in isolated rooms, gift shop. Walk-through visitors only. Limited occupancy. No winter use.	Ducted low level heat. Summer cooling, on/off control, DX cooling, some humidification. Reheat capability.	Basic HVAC	B (if benign climate) C (if mild winter) D
Climate controlled	V	Insulated structures, double glazing, vapor retardant, double doors	Purpose built museums, research libraries, galleries, exhibits, storage rooms	Education groups. Good open public facility. Unlimited occupancy.	Ducted heat, cooling, reheat, and humidification with control dead band.	Climate control, often with seasonal drift	AA (if mild winters) A B
	VI	Metal wall construction, interior rooms with sealed walls and controlled occupancy	Vaults, storage rooms, cases	No occupancy. Access by appointment.	Special heating, cooling, and humidity control with precision constant stability control.	Special constant environments	AA A Cool Cold Dry

Source: Adapted from Conrad (1995).

Historic Buildings

The New Orleans Charter, adopted by the American Institute for Conservation and the Association for Preservation Technology, calls for care in deciding on indoor environments in historic buildings. In particular, it warns against control of the indoor environment in ways that might be harmful to the building fabric. Generally, an indoor environment that varies seasonally and with changes in the weather is less detrimental to the building envelope than one that is designed for fixed set points of humidity and temperature. A common approach is to begin design with specified indoor conditions of temperature and humidity, then to expect the building envelope design (or redesign) to accommodate the indoor environment specification. This typically imposes thermal insulation, vapor retarders, and infiltration control. The New Orleans Charter suggests the following opposite approach for historic buildings:

1. Determine what tempering of the outdoor climate is already achieved by the building (including what fresh air requirements are provided by existing infiltration).
2. Determine what extremes of natural indoor climate swings are unacceptable.
3. Design mechanical services to prevent the undesirable extremes.

The advantages of this approach are (1) lowered stresses on the building envelope and (2) generally smaller equipment (beneficial in historic buildings, which rarely have generous service spaces). Of course, the mechanical equipment in historic buildings may be of historic value itself, and should not be removed or discarded without documentation.

Envelope Design

For detailed guidance on the thermal and moisture performance of building envelopes, see Chapter 22 and Chapter 23 of the 1997 *ASHRAE Handbook—Fundamentals*.

The often stringent requirements for humidity and filtration put special stresses on the building envelopes of museums and libraries. Museum buildings usually operate under positive pressure to help maintain good indoor air quality by ensuring that outdoor air enters the building at the air handler, thus allowing filtration. However, during cold weather, positive pressure in a humidified building may be detrimental. Humidified air contacts cold surfaces as it is forced outward through cracks and openings by the positive building pressure.

The building may be provided with either mechanical or natural relief vents to provide a path for exfiltrating air. With or without relief vents, however, a roughly quantifiable amount of moisture-laden air still passes through the building envelope. The pathways for the exfiltrating air should be elements of the building envelope design. In particular, the design should ensure that the exfiltration pathway will not wash potentially cold surfaces with humidified air. In analyzing the likelihood of condensation on potentially cold surfaces, the profile method described in Chapter 22 of the 1997 *ASHRAE Handbook—Fundamentals* should be followed. This analysis is important in order to determine the quantity of water that is likely to accumulate; and this quantity should be compared to the storage capacity of the critical building material. In many cases, even where the profile method indicates the possibility of condensation, the amount is small compared to the storage capacity of the material. It may be desirable in the design to include the capability to vary the pressurization of the building, so that it can be reduced during particularly cold weather.

Foundations. Water should not flow uncontrolled into any building through the foundation. This precaution is particularly true because the foundation areas are often used for storage where humidity control may be more critical than in public spaces. Chapter 23 of the 1997 *ASHRAE Handbook—Fundamentals* provides more information on this topic.

Windows. The classic and traditional means of determining excess humidity inside a building is to observe condensation on windows. The early indication of condensation conditions is a slight film of dew or frost on the pane (or occasionally on the frame if it is of metal and more thermally conductive than the glass). Often such a film evaporates with rising temperatures with no detrimental effect. If the rate of water accumulation is so great that the beads collect into droplets and run onto the window frame, the conditions for damage are in place. Water may damage window finishes, and may collect and puddle and lead to mold growth in members that contain the collected water.

A useful rule, which may not always be true, says that condensation occurs when the temperature of the inside surface of the glass is at or below the dew point of the indoor air. However, a clean, dry surface, as it is chilled, may go 9°F or more below the dew-point temperature of the air and still remain dry. Usually window condensation occurs at temperatures below the dew point of the air, but once it has occurred, the temperature of the condensation itself is an excellent indicator of the dew-point temperature of the air.

In short, window condensation is a problem only to the extent that running condensation can damage finishes on the window and below. Window condensation is not a reliable early indicator of damage in other parts of the building envelope. Chapter 29 of the 1997 *ASHRAE Handbook—Fundamentals* provides more information about condensation on windows.

Walls. The classic concern for moisture in wall assemblies in humidified buildings is for interstitial condensation—that is, moisture damage to cold surfaces that is invisible from indoors. For most walls, vapor retarders are used against moisture diffusion. However, each building requires a different approach. In general, the common solution involves preventing moisture from entering the wall by diverting rain or adjusting the interior humidity. All of these concerns are the same as those for attic spaces.

Attic Spaces. A classic concern for an attic space is the formation of frost on the underside of the roof deck. The classic solution to this and wall problems is ventilation. Several elements of these problems and their solution deserve clarification.

Most water problems in attics are due not to vapor transport but rather to roof leaks, snow entry through vent devices, condensate on cold exposed pipes and ducts, leaking mechanical equipment, etc.

Interstitial condensation is a problem only in assemblies with cavities. Most low-slope roofs are constructed with dense cellular foam insulation directly beneath the roof membrane. Such systems are called compact roof systems, to distinguish them from cavity roof systems. Where there is no cavity, interstitial condensation is not possible.

The materials contained within attic cavities can be porous or hygroscopic, and their moisture content will vary with changes in the ambient air next to them. The moisture performance of cavities can be estimated and predicted only using transient rather than steady-state methods.

The moisture content of the materials strongly depend on the humidity characteristics of the air. The concentration of air in the attic is a function of the concentrations indoors and out, and the relative contributions of indoor air and outdoor air. Simply placing a vent to the outdoors does not guarantee a high concentration of outdoor air and a low concentration of indoor air—both of which is desirable. The relative concentrations of indoor and outdoor air is most affected by the aerodynamics of the roof and by any mechanical fans.

To design attics for good performance when the building is humidified requires preventing humidified indoor air from being transported into the attic. The first step in achieving this with cavity construction is to provide an airtight ceiling plane. Chapter 23 of the 1997 *ASHRAE Handbook—Fundamentals* discusses roof venting in detail.

COCOONED ZONES

Amdur (1965) suggested the use of a room-within-a-room or "cocoon" for humidity-controlled museum spaces. This concept has been applied both as a specially built box in an existing room, and as a zone of well-controlled rooms surrounded by a zone of less well-controlled perimeter rooms. In terms of Table 3, it is the creation of a class VI building environment in a building that may be as low as class II. It has been especially useful for creating display and storage zones in an historic building. The intent in all cases is to avoid direct contact between humidity controlled air and the building envelope. Such cocoons also reduce the risk of thermal and relative humidity gradients near the walls of the controlled zone.

Assuming the outer unhumidified zone is at least one room wide, has normal levels of air change, and is not subject to unusually high vapor and air pressure differences, the inner wall need not be extraordinarily airtight, because exfiltration between zones is much lower than exfiltration through the outer envelope. All doors between zones, however, must be closed most of the time.

If the gap between the cocoon and the building envelope becomes very small, the cocoon simply becomes a good air retarder installed on the inner face of the old wall. Such a cocoon must be extraordinarily airtight. In practice, when cocooning near the envelope of an historic, leaky building, the risk of humidity damage to the historic envelope is much less if the space between cocoon and envelope is heated with unhumidified air and is wide enough for human access. This access allows easy installation of services through this space and inspection for airtightness, water leaks, pest entry, and fire risks. Cocoons also permit the conversion of historic rooms to modern display areas without damage to the original wall finish.

SYSTEM SELECTION

A typical project consists of many types of spaces, such as galleries, reading rooms, search rooms, laboratories, conference rooms, stacks, storage rooms, restaurants or cafeterias, auditoriums, rare book vaults, offices, lounges, and study rooms. Only a few of these spaces will hold environmentally sensitive objects. One of the primary prerequisites for most buildings of this type is a high expectation for performance with low or limited budgets. This generally shifts the focus to capital cost and minimizes features that may reduce operating costs and problems.

Design Issues

Functional Organization. While the HVAC system is the main determinant for actively maintaining the preservation environment, an effective preservation environment involves more than just the HVAC system. The basic architectural design (such as windows and vapor retarders) and building operation (24 h operation and availability of tempering sources) must complement these purposes and not pose impractical challenges to the HVAC design.

This suggests involvement by the HVAC engineer in the early planning stages to ensure that the space layout does not present unnecessary problems. In the best case, collections are housed separately from visitors, users, staff and all other functions. Where separation is not possible, processes and activities that pose a threat to collections should be physically and mechanically separated from the collections. Separate systems for collection and noncollection areas allow isolation of environments and can reduce project costs for noncollection areas.

A typical issue, particularly in fine art museums, is whether to treat executive offices as collection spaces. This issue should be considered carefully, not only for the added capital and operating cost, but for the risks to the collection if the offices are not treated and are nonetheless used for collections display.

Frequently used entrances, such as the lobby and loading dock, are one of the most environmentally disruptive elements in a museum or library. The designer should ensure that the loads from these spaces are managed and isolated from the primary collection areas.

Film, particularly a substantial amount, should be housed separately from other collections in order to minimize exposure to unfiltered acid gas emitted by many paper products. These gaseous pollutants are thought to contribute to vinegar syndrome and the formation of redox blemishes on film collections.

Reliability. While some collections cannot tolerate lost environmental conditions for more than a short period, most collections can tolerate several hours of lost conditions without major damage. The engineer should evaluate equipment failure scenarios against collection sensitivities and likely maintenance efforts to see what reasonable precautions can be taken to minimize down time and damage to the collection. In some cases spare equipment may need to be kept at the project site to allow timely repairs.

Loads. In most cases the HVAC system should operate 24 hours a day. Galleries and reading rooms generally only have a high occupancy at certain times. Some gallery occupancies are as high as 10 ft^2 per person, in contrast to stack or storage areas where the density may be 1000 ft^2 per person or less. Areas in the building may be used for receptions or parties with higher than normal occupancy density. Not only should the system be designed to handle this load, but it should be able to deal efficiently with the more common part-load condition.

Lighting loads vary widely from space to space and at different times of the day. The most common driver of sizing for cooling in a museum is the display lighting. Lighting typically varies from 2 to 8 W/ft^2 for display areas. Figures as high as 15 W/ft^2 are sometimes requested by lighting designers, but rarely needed in fact. With the trend to more concern for collection damage by light, display areas for light-sensitive objects should have low illumination levels, and associated low lighting power densities.

Humidistatically Controlled Heating. In this approach, the heating system is controlled by a humidistat rather than a thermostat (Lafontaine and Michalski 1984). Where interior temperatures drop consistently below 50°F, it solves the problem of humidity in a building that does not have an adequate envelope. Cold, damp air is heated until the relative humidity drops to 50%. Obviously, humidistatically controlled heating does not provide human comfort in winter, but many small museums, historic village buildings, and reserve collection buildings are essentially unoccupied in winter. A high limit thermostat is necessary to stop overheating during warm weather, and a low limit thermostat is optional if freezing of water pipes is a concern. This approach has been applied in Canada (Lafontaine 1982; Marcon 1987) in the United States, such as Shelbourne historic village (Kerschner 1992), and in many historic buildings in Britain (Staniforth 1984).

Some cautions apply. Foundations in a previously heated building may heave if the ground is waterlogged prior to freeze-up. Improved drainage, insulation of the ground near the footings, and heating of the basement reduce this risk. Problems have been noted (Padfield and Jensen 1990) for a building with dense object storage and very low infiltration rate, such as a specially sealed storage space. In this situation, a very slow supply of dehumidified air to the space has been successful (Padfield 1996).

Overall, the approach is cost effective in seasonal museums in colder climates like the northern United States and Canada, especially for low-mass wood frame buildings and in maritime regions. Humidistat control does not work in warm, humid weather; in fact, without a high-limit thermostat it will heat hot, humid conditions to dangerous temperatures. In this case, humidistat control has been supplemented by domestic dehumidifiers.

Exhibit Cases. Exhibit cases should be designed to protect the collection from environmental extremes and excesses. Both sealed cases and vented cases are used. Sealed cases rely on isolation from the ambient environment in the exhibition room, and they usually require passive or special conditioning systems independent of the regular room air. Because sealed cases are subject to a buildup of gaseous contaminants, they should be made of inert materials. Cases that are inside, built-in, or back up to a properly conditioned space can be vented to the conditioned space.

HVAC supply air should not be blown into exhibit cases, nor return air drawn through the cases. The temperature and humidity of supply air varies as it corrects space conditions and can cause extremes in temperature and humidity if blown into a case. Even if temperature and humidity conditions are sensed inside the case as part of the control system, the typical high ratio of supply air to the case volume has made virtually all such treatments problematic in the past. Problems are usually caused by poor design, or the drastic effects of an HVAC system component failure; acute temperature or humidity conditions from the supply air are undiluted and immediately impact on what are usually the most sensitive objects in a collection. While return air is more stable, most return air accumulates particles, which then accumulate in the display case.

In spite of the recommendation against doing so, exhibit cases have been actively conditioned successfully—but it has been done by using special equipment designed specifically for the purpose. Such a system is not configured like a typical HVAC system. It includes various features, such as desiccant beds to stabilize the supply air humidity, that are part of a successful application. Its use has been primarily in cold climates where conditioning large, historic galleries has been problematic, and case conditioning is often the only solution. Moreover, the stability of conditions achieved with such systems has been less than ideal, and certainly less stable than sealed exhibit cases with passive conditioning agents. The primary value is where many cases need to be conditioned where rigorous sealing and reconditioning passive agents are impractical.

Two approaches to such local mechanical control of display cases have been used: one machine per case, and one machine for many cases. The Royal Ontario Museum in Toronto has taken a one machine per case approach. The equipment circulates air in a closed loop between case and machine via small ducts (2 in. diameter). A prototype and design plans for a machine to supply many cases was developed by the Canadian Conservation Institute (Michalski 1982). The unit supplies filtered and humidity controlled air via small tubes to each case (typically 0.25 in. diameter) without return air, relying instead on compensating leakage from the case. It uses multistage centrifugal blowers at about 10 in. of water gage and a silica gel column, which controls humidity.

Such slow air exchange units take full advantage of the case buffering capacity, so that the relative humidity takes many days to rise or fall after mechanical failure, thereby making repair response time less critical. Sease (1990, 1991) described several units retrofitted in the Chicago Field Museum. Beale (1996) compared the cost and effectiveness of three approaches to relative humidity control in an existing museum building at the Boston Museum of Fine Arts: the CCI machine for many cases, HVAC retrofits for gallery zones, and silica gel in cases. Although case controlling machines are very cost effective compared to whole zone control, they currently suffer from inadequate support infrastructure, such as repair manuals, service parts, installation guidelines, and troubleshooting guides.

Cold Storage Vaults. Projects often require cold storage vaults to extend the life of materials particularly sensitive to thermal deterioration, such as acetate film and color photographic materials. These vaults usually require special equipment and most such successful systems are provided by experienced turnkey or design-build vendors (Wilhelm 1993).

Primary Elements and Features

The following primary functional elements and basic features of an HVAC system provide a good preservation environment for a museum or library, as suggested in Figure 7 (Lull 1990).

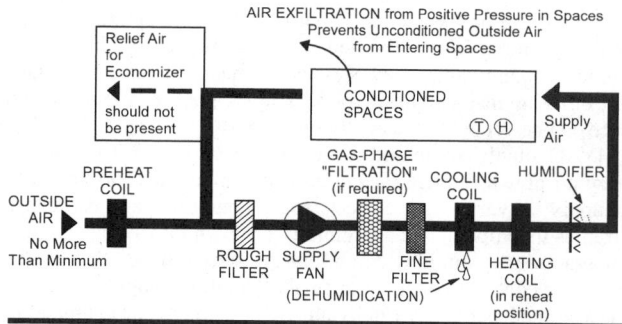

Fig. 7 Primary Elements of Preservation Environment HVAC System
(Lull 1990)

Constant Air Volume. Libraries and museums rely on mechanically assisted airflow to filter the air, ensure even temperature and humidity, suppress mold growth, and provide cooling or heating. Air should be constantly circulated at sufficient volume, regardless of space-tempering needs, to ensure good circulation throughout the collections space. In general, perimeter radiation and other sensible-only heating or cooling elements should be avoided, because they can create local humidity extremes near collections.

Cooling and Heating. The system should provide the necessary cooling and heating, but must subordinate temperature control, particularly excessive or capricious heating, to the need for stable relative humidity.

Humidification. Humidification should be provided by steam or water introduced in the air system. The humidification moisture source should be checked for potential contamination. In many cases the heating steam is treated with compounds that can pose a risk to the collection (Volent and Baer 1985). Systems should be selected and designed to prevent standing pools of water, and should follow good humidification design as described in Chapters 1 and 20 of the 1996 *ASHRAE Handbook—Systems and Equipment.*

While virtually all HVAC treatments are based on temperature control zones, for this type of building, humidity control is of more concern than temperature control. In spite of the averaging effect of a common mixed-return air and common humidifier on a central system, in several instances the retrofit of zone humidifiers have been required to recover from poor conditions even when the same humidity level is desired in each zone on the same system. However, maintaining widely different conditions in zones off the same air handler can be wasteful of energy and may ultimately be difficult to achieve. If possible, different zone conditions should have the same absolute moisture content, using zone reheat to modify space humidity for the different relative humidity conditions.

Dehumidification. The most common problem in museums and libraries is inadequate or ineffective dehumidification. Modest dehumidification can be achieved with most cooling systems, limited by the apparatus dew point at the cooling coil, and requiring adequate reheat. Most problems can be traced to compromises in the temperature of the cooling medium or the lack of reheat. Some chilled water systems may not be able to reliably deliver chilled water that is sufficiently cold, the chilled water temperature may be reset, or may have cooling coils too shallow to provide adequate dehumidification. Some reheat coils use a heating source that is unavailable in the summer or too expensive to operate.

Sebor (1995) notes the following typical approaches to more aggressive dehumidification, with increasing capability of reaching lower moisture content in the conditioned air:

- **Low-temperature chilled water**, which is usually based on a glycol solution. This method has the advantages of familiar operation and stable control and the disadvantage of managing glycol.

- **Direct-expansion (DX) refrigeration** is usually better for small systems and has lower capital costs; it generally is less reliable, requires more energy, and may require a defrost cycle.
- **Desiccant dehumidifiers** are not as familiar as the other approaches; desiccants can be quite effective if properly designed, installed, and maintained. Economy of operation is very sensitive to the cost of the regeneration heat source.

Desiccant and other dehumidification equipment needs to be applied with a cooling system; it cannot maintain comfort conditions alone. For library or archive projects requiring cool, dry conditions, a dehumidification system may be required. Chapter 22 of the 1996 *ASHRAE Handbook—Systems and Equipment* has further information on desiccants.

Outside Air. Economizer cooling is usually the primary cause of humidity fluctuations in museums and libraries. Outside air is seldom at a desirable temperature and humidity, and it can introduce particulate and gaseous contaminants as well. Air-side economizers should not be used unless (1) a bin analysis or other appropriate study shows outside air to have a favorable moisture content for an economical number of hours, and (2) favorable outside air can be reliably selected by the control system. Usually only the minimum amount of outside air should be provided for occupants as required by ASHRAE *Standard* 62, to pressurize the collection spaces, and to compensate for air evacuated by local exhausts. Relief air (see Figure 7) usually should not be used, unless required to support high outside air requirements for high occupancy conditions. Since peak occupancies are rare, outside air might be controlled by monitoring levels of carbon dioxide or other appropriate gases indicative of occupancy.

The added cost of preconditioning outside air, usually with a cooling coil, often benefits humidity control and economical operation. This cooling can reduce the amount of reheat needed for dehumidification, and may reduce the static pressure on the primary fan by keeping the cooling coil dry.

Outdoor air should be supplied by a separate duct with a fan connected to the inlet side of the HVAC system. The outdoor air supply rate for space pressurization depends on the building envelope and should be sufficient to create a small positive pressure of approximately 0.004 in. of water. A typical outdoor air supply rate should create 0.2 ACH in the space (calculated using the volume of the empty space).

Where the major museum concern is external urban pollution, Cass et al. (1988) observed that the outside air supply rate (or natural infiltration such as open windows) determines the concentration of ozone found in the building. The authors recommended that the outside air supply be adjusted as necessary and not be fixed for maximum occupancy levels.

Particulate Filtration. To avoid particle buildup and cleaning problems, air filters should be effective down to the contamination particle size, typically < 1 µm. This will likely require 90 to 95% atmospheric dust spot media filtration (as required in ASHRAE *Standard* 52.1) in an economical particulate filtration system.

High-voltage electrostatic air cleaners may not be suitable for this type of building because they may generate ozone. Any ozone runs the risk of oxidizing precious objects. An alternative filter is a permanently-polarized electret-plastic media, which cannot generate ozone.

Gas-Phase Filtration. Where new construction materials, interior contamination, collection off-gassing, or outdoor pollution present a gaseous contamination threat to a sensitive collection, active control of gaseous contaminants may be required. This is common practice for sensitive, valuable holdings, such as film media and rare books. Chapter 44 discusses filtration of gaseous contaminants.

Air Distribution. High, monumental spaces are prone to thermal stratification; if this condition places collections at risk then appropriate return and supply air may be required to ensure air motion across the entire space. Diffusers typically create a uniform temperature distribution along the room height. Displacement ventilation may have an application in some buildings. Temperature differences between the occupied zone air and the return air of 5 to 9°F can be expected in this method of ventilation.

Galleries and display areas may have changing use and loads from varied lighting and visitors. While this can sometimes be addressed through temperature control zones, an adjustable supply air may be more economical and ultimately more effective.

Supply air should not blow on collections. In other types of buildings the supply air may be diffused along a wall. This airflow pattern can be a major problem in a gallery because collections are displayed on the walls, and a wall jet would produce high temperature and humidity gradients. A floor supply should also be avoided because particulates at the dirtiest level will be entrained in the air.

Factors affecting uniformity of air temperature in the occupied zone (usually within 7 ft from the floor level) include the method of air supply, type of air diffuser used, the temperature difference between the supply and room air, the room air temperature, and the level of space obstruction (e.g. by partitions). However exhibits located above the normal height must be considered as well. Zhivov et al. (1996) define the maximum supply air temperature differential that does not cause greater than desired temperature differences in the occupied zone. The supply air temperature should be based on space loads moderated by the requirement for good air diffusion. Chapter 31 of the 1997 *ASHRAE Handbook—Fundamentals* has further information on air distribution design.

Controls. Sensors, thermostats, and humidistats must be located in the collection space itself, not in the return air stream. A variation in temperature is usually preferable to a prolonged swing in humidity. This method of control has strong implications for design, since conventional control approaches treat temperature as the primary goal and humidity as supplementary. In cases where comfort conditions are not required, humidity-controlled heating (Lafontaine 1982; Marcon 1987; Staniforth 1984) might be used. This modulates heating within a very broad temperature dead band to seek stable or moderated humidity conditions.

Types of Systems

Proper airflow is important to filter the air, control humidity, and suppress mold growth, as well as to heat and cool the collections space. Minimum airflow criteria vary from 6 to 8 air changes per hour (NBS 1983; Chapter 3 of this volume). This requirement is usually best met with a constant-volume system.

The problems most often overlooked in this type of building are maintenance access and risk to the collection from disruptions and leaks from overhead or decentralized equipment. Pipes carrying water or steam over or in collection areas always present the possibility of leaks. While some systems can provide full control without running any pipes to the zones, other systems require two to six pipes to be run to each zone. To reach collection spaces these pipes usually must be run over or in collection areas and are, unfortunately, the pipes most likely to leak. Leaks and maintenance can prevent effective use of spaces and result in lost space efficiency. For this reason, all-air systems are usually preferred.

Central air-handling stations keep filtration, dehumidification, humidification, maintenance, and monitoring equipment away from the collection. The investment in this added space and the expense of the more elaborate duct system provides major returns in reduced disruption to the collection spaces and a dramatically extended service life of the distribution system. Unlike most commercial projects, space turnover in museums and the like is low and rezoning is rarely needed. In some museums, 60-year old multizone duct systems have been reused; renovating the old system is quite economical, with the majority of the renovations confined entirely to the mechanical rooms. This is in comparison to common duct-distribution systems (such as terminal reheat, dual-duct, and variable-air-volume systems), where renovations often require a new duct system and terminal equipment, which involves a major expense from demolishing the old duct system, installing the new duct system, and demolishing and re-installing architectural finishes.

Constant-Volume Reheat. In many institutions, terminal reheat with steam or hot water coils located near or over collection spaces have caused chronic problems from steam and water leaks. Efficient zone-level humidification often suggests placing the humidifier downstream from the reheat coil; if the reheat coil is located near or over collection spaces, preventive maintenance on humidifiers further complicates the maintenance problems. Constant-volume reheat systems are quite effective when the reheat coils and humidifiers are installed entirely within the mechanical space, instead of at the terminal, feeding through what is effectively a "multizone" distribution system.

Multizone System. A multizone air handler with zone reheat and zone humidification can provide a stable and relatively energy-efficient solution. However, multizone systems without individual zone reheat and individual zone humidification have proved problematic in many institutions, requiring the retrofit of zone equipment for stable humidity control. With the proper layout and equipment compliment, a multizone system can reduce the amount of reheat and prove to be among the most energy efficient.

Bovill (1988, Figure 10) shows the preference for constant volume and multizone for this type of building. Without the requirement for many temperature control zones, but with the requirement for high air quality, the choices are limited to constant volume, multizone with bypass, dual-duct, and four-pipe induction. When air handlers are kept outside the collection areas (as recommended), the best choices are constant volume and multizone with bypass. When other systems are used, the client must be made fully aware of the compromises in performance, cost, and serviceability that may come with that system.

Dehumidification Coil. An important feature to multizone and dual-duct air handlers is a separate dehumidification coil upstream of both the hot and cold decks. This separate cooling coil, distinct from the cooling coil in the cold deck, is used when there is a dehumidification demand. In this way air might be cooled to dew point even if it eventually flows through the hot deck. Without this feature moist return air can be warmed in the hot deck and delivered back to the room without ever being dehumidified. An alternative is to locate a single cooling coil upstream of both decks, where the cold deck simply bypasses the hot deck, although this configuration can increase energy use.

Fan-Coil Units. Fan-coil units have been problematic when placed in and above collection areas. Fan-coil units expand and decentralize maintenance, causing much maintenance in the collection areas and a net increase in overall maintenance effort for the facility. Since they cool locally they need condensate drains. These drains have a history of leaking or backing up over time. As all-water systems, they require four pipes with pressurized water to run to each unit, increasing the chance of piping leaks in collection areas. On top of these problems, fan-coil units rarely, if ever, can provide the filtration and humidity control features required for collection spaces.

Variable Air Volume (VAV). VAV, while appropriate for other types of buildings, is generally inappropriate for museums, libraries, and archives. VAV systems used in this type of building have a reputation for poor humidity control, inadequate airflow, maintenance disruption, leaks in the collection spaces, and inflexibility to meet environmental needs. Many times the VAV system is chosen due to space and budget constraints. In virtually every case, the cost and space required for the properly designed VAV system (full filtration, local humidification, local dehumidification, minimum air-volume settings, well-planned piping, well-planned maintenance access, and well-documented operating instructions) gives no advantages over the more conservative, less problematic, and easier to maintain constant volume systems. While the VAV system can be made to save energy compared to the constant-volume system, it usually does so at the expense of the client's collection.

If used, a VAV system should look much like a constant-volume reheat system, which has minimum airflow to prevent mold growth, contamination buildup, and uneven conditions in the conditioned space. Terminal equipment should include reheat for each zone, with the terminal equipment located in mechanical rooms or other spaces where access and service will not place a collection at risk. If the performance of a VAV system becomes a problem, it can be easily converted to a constant-volume reheat system with only an adjustment of controls.

Fan-Powered Mixing Boxes. These are generally inappropriate for this type of building. While fan-powered mixing boxes can help ensure air circulation to suppress mold growth, they do not provide the opportunity to effectively filter the air for particulates and gases. These fans also increase local maintenance requirements and present an added fire risk. If they include reheat, there is an added risk from leaks if the reheat is water or steam or from fire if the reheat is electric.

Energy and Operating Costs

In many cases the operation of a preservation environment is a significant expense, but for the long-term protection of a valuable collection, the annual cost of a good environment are a necessary expenditure. For the institution with a small or limited operating budget that cannot afford a major increase in the annual energy cost, some compromises might be warranted or some initial capital investments might be made to reduce the recurring annual costs.

Scope of Special Environments. One of the best ways to reduce operating costs is to treat as little of the building as possible with the special environments. Spaces not requiring the higher preservation criteria should be on separate air systems so they can be operated only when occupied.

Energy Efficiency. The recovery of condenser heat is an efficient way to provide reheat for dehumidification; it also substantially reduces the energy cost for dehumidification. Although an air-side economizer can cause problems, a water-side economizer can allow efficient winter cooling when condenser water is used. Due to the load diversity between day and night operations, particularly in museums, night cooling loads can sometimes best be met with a smaller off-hours chiller. Similarly, primary-secondary pumping can be of value as the loads vary throughout the day and from area to area.

Daylighting. The use of natural light is often proposed for this type of building. Ayres et al. (1990) observed that this feature is always a net energy penalty. Where used, the daylighting aperture should be kept to a minimum, and avoided as much as possible in and over collection areas. For lower risk of leaks and better-managed illumination, clerestory daylighting is preferred over skylights.

Maintenance and Ease of Operation. A common failing in many designs is little concern for on-going operation and maintenance. Most designs will work if properly adjusted and maintained, but many institutions find they do not have the staff, budget or expertise to give the system the attention it needs. The regular maintenance required should be matched against the institution's staff capabilities. For large projects in larger cities, code-required staffing for the plant should be considered; in some situations smaller reciprocating chillers can be used at night to preclude the licensed engineer required to operate larger chillers. Small projects without HVAC maintenance staff may need package equipment that does not require daily attention.

REFERENCES

Amdur, E.J. 1965. Humidity control—Isolated area plan. *Museum News, Technical Supplement* No.5.

ANSI. 1993. Imaging media—Processed safety photographic films—Storage. *Standard* IT9.11-1993. American National Standards Institute, New York.

ANSI. 1996. Imaging materials—Processed photographic plates—Storage practices. *Standard* IT9.18-1996. American National Standards Institute, New York.

ANSI. 1998. Imaging materials—Optical disc media—Storage. *Standard* IT9.25-1998. American National Standards Institute, New York.

Ayerst, G. 1968. Prevention of biodeterioration by control of environmental conditions. *Biodeterioration of Materials*, ed. A.H. Walters and J.J. Elphick. Elsevier, Amsterdam: 223-41.

Ayres, J.M., H. Lau, and J.C. Haiad. 1990. Energy impact of various inside air temperatures and humidities in a museum when located in five U.S. cities. *ASHRAE Transactions* 96(2):100-11.

Note: Although the energy simulations in this research showed a net energy savings with skylights, this was only for the smallest skylight. Further, the only reason the smallest skylight showed a net energy savings is that it was assumed, unlike the larger skylights, to cause no increase in fan sizing and fan energy use. Instead, fan size and energy for the smallest skylight is increased proportionally on a par with the other skylights, it is a net energy loss. The only energy savings from the skylight in New York City is (apparently) from avoided reheat energy, and this does not offset its energy costs.

Bailie, C.W., R.M. Johnston-Feller, and R.L. Feller. 1988. The fading of some traditional pigments as a function of relative humidity. *Material issues in art and archaeology*, ed. E.V. Sayre et al. Materials Research Society, Pittsburgh: 287-92.

Beale, A. 1996. Environmental control options: Evaluating macro, micro, active and passive methods. Preservation of Collections, Workshop Notes. American Institute for Conservation, Washington DC: 56-62.

Beuchat, L.R. 1987. Influence of water activity on sporulation, germination, outgrowth, and toxin production. *Water activity: Theory and applications to food*, ed. L.B. Rockland and L.R. Beuchat. Marcel Dekker, New York: 137-52.

Bovill, C. 1988. Qualitative engineering. *ASHRAE Journal* 30(4):29-34.

Note: This article compared typical HVAC distribution systems based on various performance factors. The factors applicable to this building type differ from the emphasis of the article, but the final system analysis diagram (Figure 10) shows that without the requirement for many temperature control zones but with the requirement for high air quality the choices are limited to constant volume, multizone with bypass, dual-duct, and 4-pipe induction.

Cass, G.R., J.R. Druzik, D. Grosjean, W.W. Nazaroff, P.M. Whitmore, and C.L. Wittman. 1988. *Protection of works of art from photochemical smog, final report.* Environmental Quality Laboratory, California Institute of Technology, Pasadena.

Conrad, E. 1995. A table for classification of climatic control potential in buildings. Landmark Facilities Group, Inc., Norwalk, CT.

Daly, D. and S. Michalski. 1987. Methodology and status of the lining project, CCI. *ICOM Conservation Committee 8th Triennial Meeting Sydney.* The Getty Conservation Institute, Los Angeles:145-52.

Erhardt, D. and M. Mecklenburg. 1994. Relative humidity re-examined. *Preventive conservation practice, theory and research*, ed. A. Roy and P. Smith. International Institute for Conservation of Historic and Artistic Works, London: 32-38.

Erhardt, D., M. Mecklenburg, and C.S. Tumosa, 1996. New versus old wood: Differences and similarities in physical, mechanical, and chemical properties. *ICOM conservation committee 11th triennial meeting Edinburgh.* James and James, London:903-10.

Graedel, T.E. 1994. Mechanisms of chemical change in metals exposed to the atmosphere. *Durability and change: The science, responsibility, and cost of sustaining cultural heritage*, ed. W.E. Krumbein, P. Brimblecombe, D.E. Cosgrove, and S. Staniforth. John Wiley and Sons, London: 95-105.

Groom, P. and T. Panisset. 1933. Studies in Penicillium Chrysogenum Thom in relation to temperature and relative humidity of the air. *Annals of Applied Biology* 20:633-60.

Hedley, G. 1988. Relative humidity and the stress/strain response of canvas paintings: Uniaxial measurements of naturally aged samples. *Studies in Conservation* 33:133-48.

Hens, H.L.S.C. 1993. Mold risk: Guidelines and practice, commenting the results of the international energy agency, EXCO on energy conservation in buildings and community systems, Annex 14 Condensation and energy. *Bugs, Mold and Rot III: Moisture Specification and Control in Buildings*, ed. W. Rose and A. Tenwolde. National Institute of Building Sciences. Washington, DC:19-28.

Howie, F.M.P., ed. 1992. Pyrite and marcasite. *The care and conservation of geological materials*. Gutterworth-Heinemann, London: 70-84.

Kerschner, R.L. 1992. A practical approach to environmental requirements for collections in historic buildings. *J. American Institute for Conservation* 31:65-76.

Lafontaine, R.H. 1979. *Environmental norms for Canadian museums, art galleries, and archives. CCI Technical Bulletin #5*. Canadian Conservation Institute, Ottawa.

Lafontaine, R.H. 1982. Humidistatically-controlled heating: A new approach to relative humidity control in museums closed for the winter season. *J. International Institute for Conservation, Canadian Group* 7 (1 and 2):35-41.

Lafontaine, R.H. 1984. *Silica gel. CCI Technical Bulletin #10*. Canadian Conservation Institute, Ottawa.

Lafontaine, R.H. and S. Michalski. 1984. The control of relative humidity—Recent developments. *ICOM conservation committee 7th triennial meeting Copenhagen*. ICOM-CC, Paris: 84.17.33-37.

Lord, G.D. and B. Lord, eds. Zoning as a museum planning tool. *The manual of museum planning*. HMSO, London: 241-45.

Lull, W.P. 1990. *Conservation environment guidelines for libraries and archives*. New York State Library, New York.

Marcon, P.J. 1987. Controlling the environment within a new storage and display facility for the governor general's carriage. *J. International Institute for Conservation, Canadian Group* 12:37-42.

Mecklenburg, M.F. 1991. Some mechanical and physical properties of gilding gesso. *Gilded Wood: Conservation and History*, ed. D. Bigelow et al. Sound View Press, Madison, CT: 163-70.

Mecklenburg, M.F. and C.S. Tumosa. 1991. Mechanical behaviour of paintings subjected to changes in temperature and relative humidity. *Art in Transit*, ed. M.F. Mecklenburg. National Gallery of Art, Washington, DC: 173-216.

Mecklenburg, M.F., C.S. Tumosa, and D. Erhardt. 1998. Structural response of wood panel paintings to changes in ambient relative humidity. *Painted wood: History and conservation*. Getty Conservation Institute, Los Angeles: 464-83.

Michalski, S. 1982. A control module for relative humidity in display cases. *Science and Technology in the Service of Conservation*. IIC, London: 28-31.

Michalski, S. 1991a. Paintings, their response to temperature, relative humidity, shock and vibration. *Works of Art in Transit*, ed. M.F. Mecklenburg. National Gallery, Washington, DC:223-48.

Michalski, S. 1991b. Crack mechanisms in gilding. *Gilded Wood: Conservation and History*. Sound View Press, Madison, CT:171-81.

Michalski, S. 1993. Relative humidity in museums, galleries and archives: Specification and control. *Bugs, Mold and Rot III: Moisture Specification and Control in Buildings*, ed. W. Rose and A. Tenwolde. National Institute of Building Science, Washington, DC:51-62.

Michalski, S. 1994. Leakage prediction for buildings, cases, bags, and bottles. *Studies in Conservation* 39:169-86.

Michalski, S. 1995. *Wooden artifacts and humidity fluctuations: Different construction and different history mean different vulnerabilities*. Chart, version 3.0, Canadian Conservation Institute, Ottawa.

Michalski, S. 1996a. *Environmental guidelines: Defining norms for large and varied collections*. AIC presession workshop notes. American Institute for Conservation, Washington, D.C.

Michalski, S. 1996b. Quantified risk reduction in the humidity dilemma. *APT Bulletin* 37:25-30.

Michalski, S. 1999. *Relative humidity and temperature guidelines for Canadian archives*. Canadian Council of Archives and Canadian Conservation Institute, Ottawa.

NBS. 1983. *Air quality criteria for storage of paper-based archival records*. NBSIR 83-2795. National Institute of Standards and Technology, Gaithersburg, MD:21 and 25.

Ohtsuki, T. 1990. Studies on Eurotium tonophilum Ohtsuki. *Scientific Papers on Japanese Antiques and Art Crafts* No 35:28-34.

Oreszczyn, T., M. Cassar, and K. Fernandez. 1994. Comparative studies of air-conditioned and non air-conditioned museums. *Preventive conservation practice, theory and research*, ed. A. Roy and P. Smith. International Institute for Conservation of Historic and Artistic Works, London: 144-48.

Padfield, T. 1996. Low energy climate control in stores: A postscript. *ICOM Conservation Committee 11th Triennial Meeting Edinburgh*. James and James, London: 68-71.

Padfield, T. and P. Jensen. 1990. Low energy climate control in museum stores. *ICOM Conservation Committee 9th Triennial Meeting Dresden*. ICOM Conservation Committee, Los Angeles: 596-601.

Perera, D.Y. and D. Vanden Eynde. 1987. Moisture and temperature induced stresses (hygrothermal stresses) in organic coatings. *Journal of Coatings Technology* 59(5):55-63.

Reilly, J.M. 1993. *IPI storage guide for acetate film*. Image Permanence Institute, Rochester, NY.

Scott, D.A. 1990. Bronze disease: A review of some chemical problems and the role of relative humidity. *J. American Institute of Conservation* 29:193-206.

Sease, C. 1990. A new means of controlling relative humidity in exhibit cases. *Collection Forum* 6(1):12-20.

Sease, C. 1991. The development of the humidity control module at the Field Museum. *J. American Institute for Conservation* 30(2):187-96.

Sebor, A.J. 1995. Heating, ventilating, and air-conditioning systems. *Storage of natural history collections: A preventive conservation approach*, ed. C.L. Rose et. al. Society of the Preservation of Natural History Collections, Pittsburgh.

Snow, D., M.H.G. Crichton, and N.C. Wright. 1944. Mould deterioration of feeding stuffs in relation to humidity of storage. *Annals of Applied Biology* 31:102-10.

Staniforth, S. 1984. Environmental conservation. *Manual of Curatorship*, ed. J.M.A. Thompson. Butterworths, London: 192-202.

Thomson, G. 1977. Stabilization of RH in exhibition cases: Hygrometric half-time. *Studies in Conservation* 22:85-102.

Thomson, G. 1986. Appendix I. *The museum environment*, 2nd ed. Butterworths, London.

USDA. 1987. *Wood Handbook*. U.S. Department of Agriculture, Washington, DC.

Volent, P. and N.S. Baer. 1985. Volatile amines used as corrosion inhibitors in museum humidification systems. *The International Journal of Museum Management and Curatorship* 6 (July).

Waller, R. 1992. Temperature- and humidity-sensitive mineralogical and petrological specimens. *The Care and Conservation of Geological Material*, ed. F. Howie. Butterworth Heinemann, London:25-50.

Wilhelm, H. 1993. *The Permanence and Care of Color Photographs: Traditional and Digital Color Prints, Color Negatives, Slides, and Motion Pictures,* with C. Brower. Preservation Publishing Company, Grinnell, IA.

Williams, S.L. 1991. Investigation of the causes of structural damage to teeth in natural history collections. *Collection Forum* 7(1):13-25.

Zhivov, A.M., J.B. Priest, and L.L. Christianson. 1996. Air distribution design for realistic rooms. *Proceedings* ROOMVENT'96.

ENVIRONMENTAL CONTROL FOR ANIMALS AND PLANTS

THE design of plant and animal housing is complicated by the many environmental factors affecting the growth and production of living organisms. The financial constraint that equipment must repay costs through improved economic productivity must be considered by the designer. The engineer must balance the economic costs of modifying the environment against the economic losses of a plant or animal in a less-than-ideal environment.

Thus, the design of plant and animal housing is affected by (1) economics, (2) concern for both workers and the care and well-being of animals, and (3) regulations on pollution, sanitation, and health assurance.

DESIGN FOR ANIMAL ENVIRONMENTS

Typical animal production plants modify the environment, to some degree, by housing or sheltering animals year-round or for parts of a year. The degree of modification is generally based on the expected increase in production. Animal sensible heat and moisture production data, combined with information on the effects of environment on growth, productivity, and reproduction, help designers select optimal equipment (Chapter 10 of the 1997 ASHRAE Handbook—Fundamentals). Detailed information is available in a series of handbooks published by the MidWest Plan Service. These include Mechanical Ventilating Systems for Livestock Housing (MWPS 1990), Natural Ventilating Systems for Livestock Housing and Heating (MWPS 1989), and Cooling and Tempering Air for Livestock Housing (MWPS 1990). ASAE Monograph No. 6, Ventilation of Agricultural Structures (Hellickson and Walter 1983), also gives more detailed information.

Design Approach

Environmental control systems are typically designed to maintain thermal and air quality conditions within an acceptable range and as near the ideal for optimal animal performance as is practicable. Equipment is usually sized assuming steady-state energy and mass conservation equations. Experimental measurements confirm that heat and moisture production by animals is not constant and that there may be important thermal capacitance effects in livestock buildings. Nevertheless, for most design situations, the steady-state equations are acceptable.

Achieving the appropriate fresh air exchange rate and establishing the proper distribution within the room are generally the two most important design considerations. The optimal ventilation rate is selected according to the ventilation rate logic curve (Figure 1).

During the coldest weather, the ideal ventilation rate is that required to maintain indoor relative humidity at or below the

The preparation of this chapter is assigned to TC 2.2, Plant and Animal Environment.

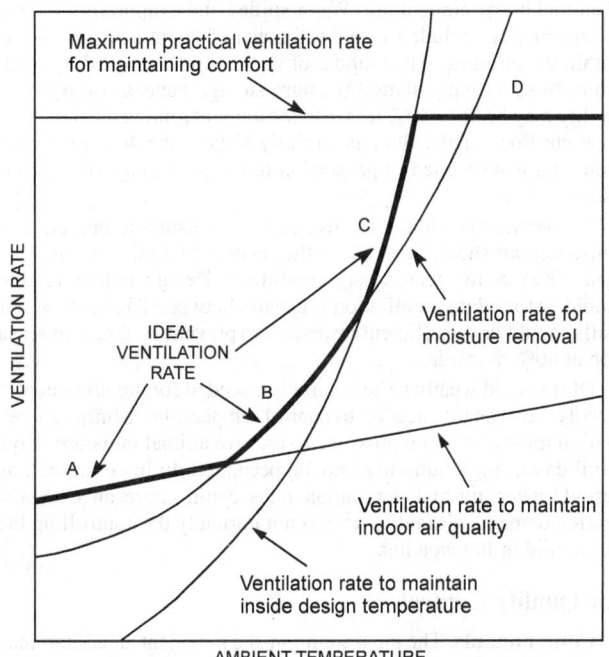

Fig. 1 Logic for Selecting the Appropriate Ventilation Rate in Livestock Buildings
(Adapted from Christianson and Fehr 1983)

maximum desired, and air contaminant concentrations within acceptable ranges (Rates A and B in Figure 1). Supplemental heating is often required to prevent the temperature from dropping below optimal levels.

In milder weather, the ventilation rate required for maintaining optimal room air temperature is greater than that required for moisture and air quality control (Rates C and D in Figure 1). In hot weather, the ventilation rate is chosen to minimize the temperature rise above ambient and to provide optimal air movement over animals. Cooling is sometimes used in hot weather. The maximum rate (D) is often set at 60 air changes per hour as a practical maximum.

Temperature Control

The temperature within an animal structure is computed from the sensible heat balance of the system, usually disregarding transient effects. Nonstandard buildings with low airflow rates and/or large thermal mass may require transient analysis. Steady-state heat transfer through walls, ceiling or roof, and ground is calculated as presented in Chapter 24 of the 1997 ASHRAE Handbook—Fundamentals.

Mature animals typically produce more heat per of unit floor area than do young stock. Chapter 10 of the 1997 *ASHRAE Handbook—Fundamentals* presents estimates of animal heat loads. Lighting and equipment heat loads are estimated from power ratings and operating times. Typically, the designer selects indoor and outdoor design temperatures and calculates the ventilation rate to maintain the temperature difference. Outdoor design temperatures are given in Chapter 26 of the 1997 *ASHRAE Handbook—Fundamentals*. The section on Recommended Practices by Species in this chapter presents indoor design temperature values for various livestock.

Moisture Control

Moisture loads produced in an animal building may be calculated from data in the 1997 *ASHRAE Handbook—Fundamentals*. The mass of water vapor produced is estimated by dividing the animal latent heat production by the latent heat of vaporization of water at animal body temperature. Water spilled and evaporation of fecal water must be included in the estimates of latent heat production within the building. The amount of water vapor removed by ventilation from a totally slatted (manure storage beneath floor) swine facility may be up to 40% less than the amount removed from a solid concrete floor. If the floor is partially slatted, the 40% maximum reduction is decreased in proportion to the percentage of the floor that is slatted.

The ventilation should remove enough moisture to prevent condensation but should not reduce the relative humidity so low (less than 40%) as to create dusty conditions. Design indoor relative humidity for winter ventilation is usually between 70 and 80%. The walls should have sufficient insulation to prevent surface condensation at 80% rh inside.

During cold weather, the ventilation needed for moisture control usually exceeds that needed to control temperature. Minimum ventilation must always be provided to remove animal moisture. Up to a full day of high humidity may be permitted during extreme cold periods when normal ventilation rates could cause an excessive heating demand. Humidity level is not normally the controlling factor in mild or hot weather.

Air Quality Control

Contaminants. The most common and prevalent air contaminant in animal buildings is particulates. In animal buildings, particulates originate mainly from the feed, litter, fecal materials, and animals. Particulates include solid particles, liquid droplets, microorganisms, and moisture, and can be deposited deep within the respiratory system. Particulates carry the allergens which cause discomfort and health problems for workers in laboratory rodent facilities. They also carry much of the odors in animal facilities, and can carry the odors for long distances from the facilities. Consequently, particulates pose major problems for animals, workers, and neighbors. Particulate levels in swine buildings have been measured to range from 0.028 to 0.43 mg/ft[3]. Dust has not been a major problem in dairy buildings; one two-year study found an average of only 0.014 mg/ft[3] in a naturally ventilated dairy barn. Poultry building dust levels average around 0.057 to 0.20 mg/ft[3], but levels up to 0.51 to 0.82 mg/ft[3] have been measured during high activity periods.

The most common gas contaminants are ammonia, hydrogen sulfide, other odorous compounds, carbon dioxide, and carbon monoxide. High moisture levels can also aggravate other contaminant problems. Ammonia, which results from the decomposition of manure, is the most important chronically present contaminant gas. Typical ammonia levels measured have been 10 to 50 ppm in poultry units, 0 to 20 ppm in cattle buildings, 5 to 30 ppm in swine units with liquid manure systems, and 10 to 50 ppm in swine units with solid floors (Ni et al. 1998a). Up to 200 ppm have been measured in swine units in winter. Ammonia should be maintained below 25 ppm and, ideally, below 10 ppm.

Zhang et al. (1992) and Maghirang et al. (1995) found ammonia levels in laboratory animal rooms to be negligible, but concentrations could reach 60 ppm in cages. Weiss et al. (1991) found ammonia levels in rat cages of up to 350 ppm with four male rats per cage and 68 ppm with four female rats per cage. Hasenau et al. (1993) found that ammonia levels varied widely among various mouse microisolation cages; ammonia ranged from negligible to 520 ppm nine days after cleaning the cage.

Hydrogen sulfide, a by-product of the microbial decomposition of stored manure, is the most important acute gas contaminant. During normal operation, hydrogen sulfide concentration is usually insignificant (i.e., below 1 ppm). A typical level of hydrogen sulfide in swine buildings is around 150 to 350 ppb (Ni et al. 1998b). However, levels can reach 200 to 330 ppm, and possibly up to 1000 to 8000 ppm during in-building manure agitation.

Odors from animal facilities are becoming an increasing concern, both in the facilities and in the surrounding areas. Odors result from both gases and particulates; particulates are of primary concern since odorous gases can be quickly diluted below odor threshold concentrations in typical weather conditions, while particulates can retain odor for long periods. Methods that control particulate and odorous gas concentrations in the air will reduce odors, but controlling odor generation at the source appears to be the most promising method of odor control.

Barber et al. (1993), reporting on 173 pig buildings, found that carbon dioxide concentrations were below 3000 ppm in nearly all instances when the external temperature was above 32°F but almost always above 3000 ppm when the temperature was below 32°F. The report indicated that there was a very high penalty in heating cost in cold climates if the maximum-allowed carbon dioxide concentration was less than 5000 ppm. Air quality control based on carbon dioxide concentrations was suggested by Donham et al. (1989). They suggested a carbon dioxide concentration of 1540 ppm as a threshold level, above which symptoms of respiratory disorders occurred in a population of swine building workers. For other industries, a carbon dioxide concentration of 5000 ppm is suggested as the time-weighted threshold limit value for 8 h of exposure (ACGIH 1998).

Other gas contaminants can also be important. Carbon monoxide from improperly operating unvented space heaters sometimes reaches problem levels. Methane is another occasional concern.

Control Methods. Three standard methods used to control air contaminant levels in animal facilities are

1. Reduce contaminant production at the source.
2. Remove contaminants from the air.
3. Reduce gas contaminant concentration by dilution (ventilation).

The first line of defense is to reduce release of contaminants from the source, or to at least intercept and remove them before they reach the workers and animals. Animal feces and urine are the largest sources of contaminants; but feed, litter, and the animals themselves are also a major source of contaminants, especially particulates. Successful operations effectively collect and remove all manure from the building within three days, before it decomposes sufficiently to produce large quantities of contaminants. Removing ventilation air uniformly from manure storage or collection areas helps remove contaminants before they reach animal or worker areas.

Ammonia production can be minimized by removing wastes from the room and keeping floor surfaces or bedding dry. Immediately covering manure solids in gutters and pits with water also reduces ammonia, which is highly soluble in water. Since adverse effects of hydrogen sulfide on production begin to occur at 20 ppm, ventilation systems should be designed to maintain hydrogen sulfide levels below 20 ppm during agitation. When manure is agitated and removed from the storage, the building should be well ventilated and all animals and occupants evacuated due to potentially fatal concentrations of gases.

For laboratory animals, changing the bedding frequently and keeping the bedding dry with lower relative humidities and appropriate cage ventilation can reduce ammonia release. Individually ventilated laboratory animal cages or the placement of cages in mass air displacement units reduce contaminant production by keeping litter drier. Using localized contaminant containment work stations for dust-producing tasks such as cage-changing may also help. For poultry or laboratory animals, the relative humidity of air surrounding the litter should be kept between 50% and 75% to reduce particulate and gas contaminant release. Relative humidities between 40% and 75% also reduce the viability of pathogens in the air. A moisture content of 25 to 30% (wet basis) in the litter or bedding, keeps dust to a minimum. Adding 0.5 to 2% of edible oil or fat can significantly reduce dust emission from the feed. Respirable dust (smaller than 10 μm), which is most harmful to the health and comfort of personnel and animals, is primarily from feces, animal skins, and dead microorganisms. Respirable dust concentration should be kept below 0.0065 mg/ft^3. Some dust control technologies are available. For example, sprinkling oil at 0.12 gal per 1000 ft^2 of floor area per day can reduce dust concentration by more that 80%. High animal activity levels release large quantities of particulates into the air, so management strategies to reduce agitation of animals are helpful.

Methods of removing contaminants from the air are essentially limited to particulate removal since gas removal methods are often too costly for animal facilities. Some animal workers wear personal protection devices (appropriate masks) to reduce inhaled particulates. Room air filters reduce animal disease problems, but they have not proven practical for large animal facilities due to the large quantity of particulates and the difficulty in drawing the particulates from the room and through a filter. Air scrubbers have the capability to remove gases and particulates, but the initial cost and maintenance make them impractical. Aerodynamic centrifugation is showing promise for removing the small particulates found in animal buildings.

Ventilation is the most prevalent method used to control gas contaminant levels in animal facilities. It is reasonably effective in removing gases, but not as effective in removing particulates. Pockets in a room with high concentrations of gas contaminants are common. These polluted pockets occur in dead air spots or near large contaminant sources. Providing high levels of ventilation can be costly in winter, can create drafts on the animals, and can increase the release of gas contaminants by increasing air velocity across the source.

Disease Control

Airborne microbes can transfer disease-causing organisms among animals. For some situations, typically with young animals where there are low-level infections, it is important to minimize air mixing among animal groups. It is especially important to minimize air exchange between different animal rooms, so buildings need to be fairly airtight.

Poor thermal environments and air contaminants can increase stress on the animals, which can make them more susceptible to disease. Therefore, a good environmental control system is important for disease control.

Air Distribution

Air speed should be maintained below 50 fpm for most animal species in both cold and mild weather. Animal sensitivities to draft are comparable to those of humans, although some animals are more sensitive at different stages. Riskowski and Bundy (1988) documented that air velocities for optimal rates of gain and feed efficiencies can be below 25 fpm for young pigs at thermoneutral conditions.

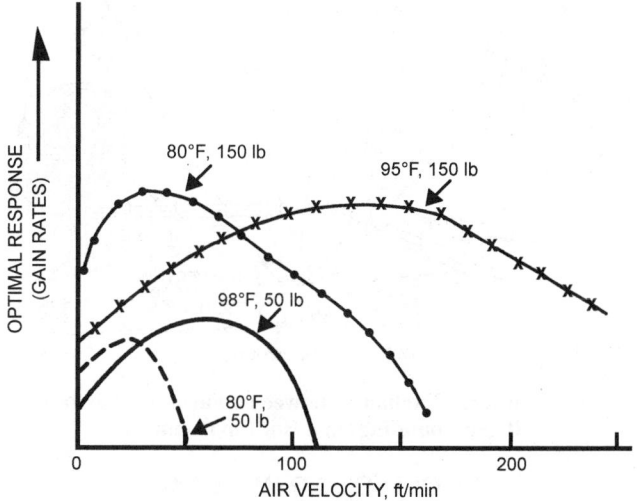

Fig. 2   Response of Swine to Air Velocity

Increased air movement during hot weather increases growth rates and improves heat tolerance. There are conflicting and limited data defining optimal air velocity in hot weather. Bond et al. (1965) and Riskowski and Bundy (1988) determined that both young and mature swine perform best when air speed is less than 200 fpm (Figure 2). Mount and Start (1980) did not observe performance penalties at air speeds increased to a maximum of 150 fpm.

Degree of Shelter

Livestock, especially young animals, need some protection from adverse climates. On the open range, mature cattle and sheep need protection during severe winter conditions. In winter, dairy cattle and swine may be protected from precipitation and wind with a three-sided, roofed shelter open on the leeward side. The windward side should also have approximately 10% of the wall surface area open to prevent a negative pressure inside the shelter; this pressure could cause rain and snow to be drawn into the building on the leeward side. Such shelters do not protect against extreme temperature or high humidity.

In warmer climates, shades often provide adequate shelter, especially for large, mature animals such as dairy cows. Shades are commonly used in Arizona; research in Florida has shown an approximate 10% increase in milk production and a 75% increase in conception efficiency for shaded versus unshaded cows. The benefit of shades has not been documented for areas with less severe summer temperatures. Although shades for beef cattle are also common practice in the southwest, beef cattle are somewhat less susceptible to heat stress, and extensive comparisons of various shade types in Florida have detected little or no differences in daily weight gain or feed conversion.

The energy exchange between an animal and various areas of the environment is illustrated in Figure 3. A well-designed shade makes maximum use of radiant heat sinks, such as the cold sky, and gives maximum protection from direct solar radiation and high surface temperature under the shade. Good design considers geometric orientation and material selection, including roof surface treatment and insulation material on the lower surface.

An ideal shade has a top surface that is highly reflective to solar energy and a lower surface that is highly absorptive to solar radiation reflected from the ground. A white-painted upper surface reflects solar radiation, yet emits infrared energy better than aluminum. The undersurface should be painted a dark color to prevent multiple reflection of shortwave energy onto animals under the shade.

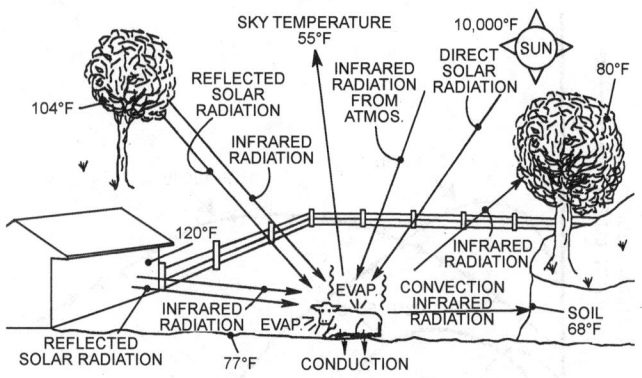

Fig. 3 Energy Exchange Between a Farm Animal and Its Surroundings in a Hot Environment

COOLING AND HEATING

Air Velocity

Increasing air velocity helps to facilitate the cooling of mature animals. It is especially beneficial when combined with skin wetting evaporative cooling. Mature swine benefit most with air velocities up to 200 fpm; cattle around 300 fpm; and poultry around 600 fpm. Air velocity can be increased with air circulation fans that blows air horizontally in circular patterns around the room, paddle fans that blow air downward, or tunnel cooling that moves air horizontally along the length of the building.

Evaporative Cooling

Supplemental cooling of animals in intensive housing conditions may be necessary during heat waves to prevent heat prostration, mortality, or serious losses in production and reproduction. Evaporative cooling, which may reduce ventilation air to 80°F or lower in most of the United States, is popular for poultry houses, and is sometimes used for swine and dairy housing.

Evaporative cooling is well suited to animal housing because the high air exchange rates effectively remove odors and ammonia, and increase air movement for convective heat relief. Initial cost, operating expense, and maintenance problems are all relatively low compared to other types of cooling systems. Evaporative cooling works best in areas with low relative humidity, but significant benefits can be obtained even in the humid southeastern United States.

Design. The pad area should be sized to maintain air velocities between 200 and 275 fpm through the pads. For most pad systems, these velocities produce evaporative efficiencies between 75 and 85%; they also increase pressures against the ventilating fans from 0.04 to 0.12 in. of water, depending on the pad design.

The building and pad system must be airtight because air leaks due to the negative pressure ventilation will reduce the airflow through the pads, and hence reduce the cooling effectiveness.

The most serious problem encountered with evaporative pads for agricultural applications is clogging by dust and other airborne particles. Whenever possible, fans should exhaust away from pads on adjacent buildings. Regular preventive maintenance is essential. Water bleed-off and the addition of algaecides to the water are recommended. When pads are not used in cool weather, they should be sealed to prevent dusty inside air from exhausting through them.

High-pressure fogging with water pressure of 500 psi is preferred to pad coolers for cooling the air in broiler houses with built-up litter. The high pressure creates a fine aerosol, causing minimal litter wetting. Timers and/or thermostats control the cooling. Evaporative efficiency and installation cost are about one-half those of a well-designed evaporative pad. Foggers can also be

used with naturally ventilated, open-sided housing. Low-pressure systems are not recommended for poultry, but may be used during emergencies.

Nozzles that produce water mist or spray droplets to wet animals directly are used extensively during hot weather in swine confinement facilities with solid concrete or slatted floors. Currently, misting or sprinkling systems with larger droplets that directly wet the skin surface of the animals (not merely the outer portion of the hair coat) are preferred. Timers that operate periodically, (e.g., 2 to 3 min on a 15 to 20 min cycle) help to conserve water.

Mechanical Refrigeration

Mechanical refrigeration can be designed for effective animal cooling, but it is considered uneconomical for most production animals. Air-conditioning loads for dairy housing may require 2.5 kW or more per cow. Recirculation of refrigeration air is usually not feasible due to high contaminant loads in the air in the animal housing. Sometimes, zone cooling of individual animals is used instead of whole-room cooling, particularly in swine farrowing houses where a lower air temperature is needed for sows than for unweaned piglets. It is also beneficial for swine boars and gestating sows. Refrigerated air, 18 to 36°F below ambient temperature, is supplied through insulated ducts directly to the head and face of the animal. Air delivery rates are typically 20 to 40 cfm per animal for snout cooling, and 60 to 80 cfm per sow for zone cooling.

Earth Tubes

Some livestock facilities obtain cooling in summer and heating in winter by drawing ventilation air through tubing buried 6 to 13 ft below grade. These systems are most practical in the north central United States for animals that benefit from both cooling in summer and heating in winter.

Cooling and Tempering Air for Livestock Housing (MWPS 1990) details design procedures for this method. A typical design uses 50 to 150 ft of 8 in. diameter pipe to provide 300 cfm of tempered air. Soil type and moisture, pipe depth, airflow, climate, and other factors affect the efficiency of buried pipe heat exchangers. The pipes must slope to drain condensation, and must not have dips that could plug with condensation.

Heat Exchangers

Ventilation accounts for 70 to 90% of the heat losses in typical livestock facilities during winter. Heat exchangers can reclaim some of the heat lost with the exhaust ventilating air. However, predicting fuel savings based on savings obtained during the coldest periods will overestimate yearly savings from a heat exchanger. Estimates of energy savings based on air enthalpy can improve the accuracy of the predictions.

Heat exchanger design must address the problems of condensate freezing and/or dust accumulation on the heat exchanging surfaces. If unresolved, these problems result in either reduced efficiency and/or the inconvenience of frequent cleaning.

Supplemental Heating

For poultry weighing 3.3 lb or more, for pigs heavier than 50 lb, and for other large animals such as dairy cows, the body heat of the animals at recommended space allocations is usually sufficient to maintain moderate temperatures (i.e., above 50°F) in a well-insulated structure. Combustion-type heaters are used to supplement heat for baby chicks and pigs. Supplemental heating also increases the moisture-holding capacity of the air, which reduces the quantity of air required for removal of moisture. Various types of heating equipment may be included in ventilation, but they need to perform well in dusty and corrosive atmospheres.

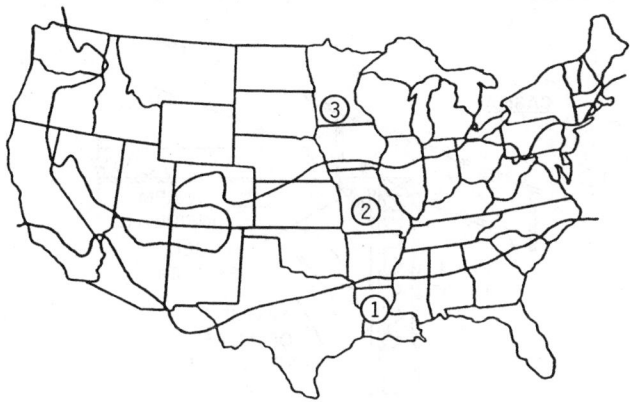

Fig. 4 Climatic Zones
(Reprinted with permission from ASAE *Standard* S401.2)

Table 1 Minimum Recommended Overall Coefficients of Heat Transmission *U* for Insulated Assemblies[a,b]

| Climatic Zone[d] | Recommended Minimum U, Btu/h·ft²·°F[c] | | | | | |
| | Cold | | Modified Environment | | Supplementally Heated | |
	Walls	Ceiling	Walls	Ceiling	Walls	Ceiling
1	—	0.17[e]	0.17[e]	0.071	0.071	0.045
2	—	0.17	0.17	0.059	0.071	0.040
3	—	0.17	0.083	0.040	0.050	0.030

[a]Use assembly U-factors that include framing effects, air spaces, air films, linings, and sidings. Determine assembly U-factors by testing the full assembly in accordance with ASTM C 236 or C 976 or calculate by the procedures presented in the 1997 *ASHRAE Handbook—Fundamentals*.
[b]Values shown are not the values necessary to provide a heat balance between heat produced by products or animals and heat transferred through the building.
[c]Current practice for poultry grow-out buildings uses a U of 0.11 to 0.14 Btu/h·ft²·°F in the roof and walls.
[d]Refer to Figure 4.
[e]Where ambient temperature and radiant heat load are severe, $U = 0.83$ Btu/h·ft²·°F.

Insulation Requirements

The amount of building insulation required depends on climate, animal space allocations, and animal heat and moisture production. Refer to Figure 4 and Table 1 for selecting insulation levels. In warm weather, ventilation between the roof and insulation helps reduce the radiant heat load from the ceiling. Insulation in warm climates can be more important for reducing radiant heat loads in summer than reducing building heat loss in winter.

Cold buildings have indoor conditions about the same as outside conditions. Examples are free-stall barns and open-front livestock buildings. Minimum insulation is frequently recommended in the roofs of these buildings to reduce solar heat gain in summer and to reduce condensation in winter.

Modified environment buildings rely on insulation, natural ventilation, and animal heat to remove moisture and to maintain the inside within a specified temperature range. Examples are warm free-stall barns, poultry production buildings, and swine finishing units.

Supplementary heated buildings require insulation, ventilation, and extra heat to maintain the desired inside temperature and humidity. Examples are swine farrowing and nursery buildings.

VENTILATION

Mechanical Ventilation

Mechanical ventilation uses fans to create a static pressure difference between the inside and outside of a building. Farm buildings use either positive pressure, with fans forcing air into a building, or negative pressure, with exhaust fans. Some ventilation systems use a combination of positive pressure to introduce air into a building and separate fans to remove air. These zero-pressure systems are particularly appropriate for heat exchangers.

Positive Pressure Ventilation. Fans blow outside air into the ventilated space, forcing humid air out through any planned outlets and through leaks in walls and ceilings. If vapor barriers are not complete, moisture condensation will occur within the walls and ceiling during cold weather. Condensation causes deterioration of building materials and reduces insulation effectiveness. The energy used by fan motors and rejected as heat is added to the building—an advantage in winter but a disadvantage in summer.

Negative Pressure Ventilation. Fans exhaust air from the ventilated space while drawing outside air in through planned inlets and leaks in walls, in ceilings, and around doors and windows. Air distribution in negative pressure ventilation is often less complex and costly than positive or neutral pressure systems. Simple openings and baffled slots in walls control and distribute air in the building.

However, at low airflow rates, negative pressure ventilation may not distribute air uniformly due to air leaks and wind pressure effects. Supplemental air mixing may be necessary.

Allowances should be made for reduced fan performance due to dust, guards, and corrosion of louver joints (Person et al. 1979). Totally enclosed fan motors are protected from exhaust air contaminants and humidity. Periodic cleaning helps prevent overheating. Negative pressure ventilation is more commonly used than positive pressure ventilation.

Ventilation should always be designed so that manure gases are not drawn into the building from manure storages connected to the building by underground pipes or channels.

Neutral Pressure Ventilation. Neutral pressure (push-pull) ventilation typically use supply fans to distribute air down a distribution duct to room inlets and exhaust fans to remove air from the room. Supply and exhaust fan capacities should be matched.

Neutral pressure systems are often more expensive, but they achieve better control of the air. They are less susceptible to wind effects and to building leakage than positive or negative pressure systems. Neutral pressure systems are most frequently used for young stock and for animals most sensitive to environmental conditions, primarily where cold weather is a concern.

Natural Ventilation

Either natural or mechanical ventilation is used to modify environments in livestock shelters. Natural ventilation is most common for mature animal housing, such as free-stall dairy, poultry growing, and swine finishing houses. Natural ventilation depends on pressure differences caused by wind and temperature differences. Well-designed natural ventilation keeps temperatures reasonably stable, if automatic controls regulate ventilation openings. Usually, a design includes an open ridge (with or without a rain cover) and openable sidewalls, which should cover at least 50% of the wall for summer operation. Ridge openings are about 2 in. wide for each 10 ft of house width, with a minimum ridge width of 6 in. to avoid freezing problems in cold climates. Upstand baffles on each side of the ridge opening greatly increase airflow (Riskowski et al. 1998). Small screens and square edges around sidewall openings can significantly reduce airflow through vents.

Openings can be adjusted automatically, with control based on air temperature. Some designs, referred to as flex housing, include a combination of mechanical and natural ventilation usually dictated by outside air temperature and/or the amount of ventilation required.

VENTILATION MANAGEMENT

Air Distribution

Pressure differences across walls and inlet or fan openings are usually maintained between 0.04 and 0.06 in. of water. (The exhaust fans are usually sized to provide proper ventilation at pressures up to 0.12 in. to compensate for wind effects.) This pressure difference creates inlet velocities of 600 to 1000 fpm, sufficient for effective air mixing, but low enough to cause only a small reduction in fan capacity. A properly planned inlet system distributes fresh air equally throughout the building. Negative pressure ventilation that relies on cracks around doors and windows does not distribute fresh air effectively. Inlets require adjustment, since winter airflow rates are typically less than 10% of summer rates. Automatic controllers and inlets are available to regulate inlet areas.

Positive pressure ventilation, with fans connected directly to perforated air distribution tubes, may combine heating, circulation, and ventilation in one system. Air distribution tubes or ducts connected to circulating fans are sometimes used to mix the air in negative pressure ventilation. Detailed design procedures for perforated ventilation tubes are described by Zhang (1994). However, dust in the ducts is of concern when air is recirculated, particularly when cold incoming air condenses moisture in the tubes.

Inlet Design. Inlet location and size most critically affect air distribution within a building. Continuous or intermittent inlets can be placed along the entire length of one or both outside walls. Building widths narrower than 20 ft may need only a single inlet along one wall. The total inlet area may be calculated by the system characteristic technique, which follows. Because the distribution of the inlet area is based on the geometry and size of the building, specific recommendations are difficult.

System Characteristic Technique. This technique determines the operating points for the ventilation rate and pressure difference across inlets. Fan airflow rate as a function of pressure difference across the fan should be available from the manufacturer. Allowances must be made for additional pressure losses from fan shutters or other devices such as light restriction systems or cooling pads.

Inlet flow characteristics are available for hinged baffle and center-ceiling flat baffle slotted inlets (Figure 5). Airflow rates can be calculated for the baffles in Figure 5 by the following:

For Case A:

$$Q = 285Wp^{0.5} \qquad (1)$$

For Case B:

$$Q = 183Wp^{0.5} \qquad (2)$$

For Case C (Total airflow from sum of both sides):

$$Q = 320Wp^{0.5}(D/T)^{0.08}e^{(-0.867\,W/T)} \qquad (3)$$

where

Q = airflow rate, cfm per foot length of slot opening
W = slot width, in.
p = pressure difference across the inlet, in. water gage
D = baffle width, in.
T = width of slot in ceiling, in.

Zhang and Barber (1995) measured infiltration rates of five rooms in a newly built swine building at 0.12 cfm/ft^2 of surface area at 0.08 in. water gage. Surface area included the area of walls and ceiling enclosing the room. It is important to include this infiltration rate into the ventilation design and management. For example, at 0.12 cfm/ft^2 of surface area, the infiltration represents 1.4 air changes per hour. In the heating season, the minimum ventilation is

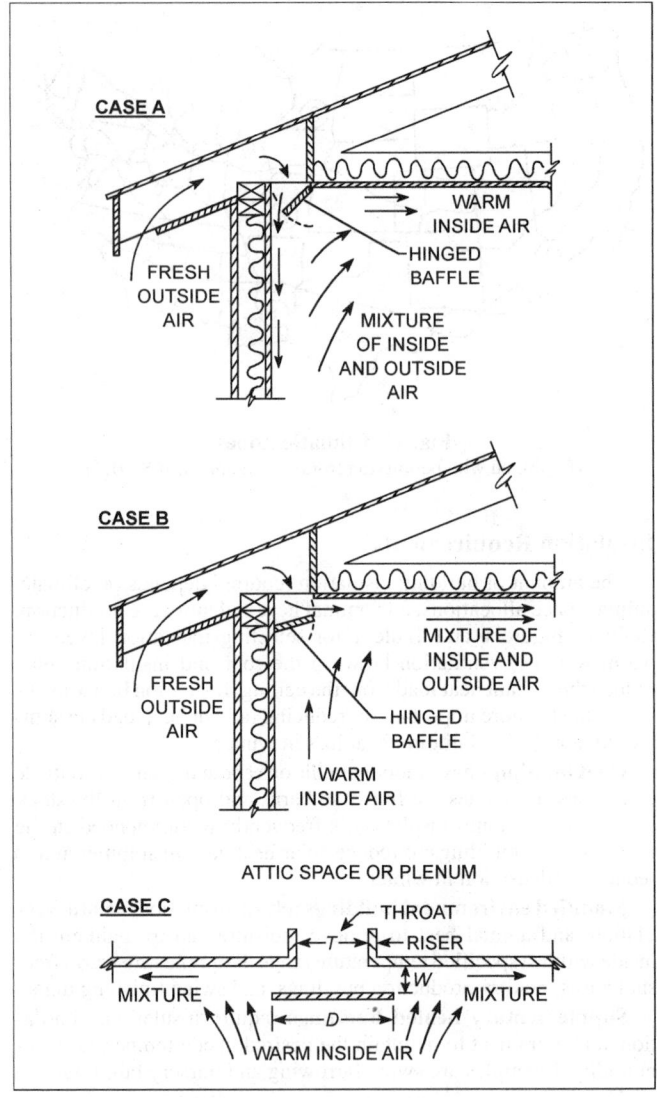

Fig. 5   Typical Livestock Building Inlet Configurations

usually about 3 air changes per hour. Thus, large infiltration rates greatly reduce the airflow from the controlled inlet and adversely affect the air distribution.

Room Air Velocity. The average air velocity inside a slot ventilated structure relates to the inlet air velocity, inlet slot width (or equivalent continuous length for boxed inlets), building width, and ceiling height. Estimates of air velocity within a barn, based on air exchange rates, may be very low due to the effects of jet velocity and recirculation. Conditions are usually partially turbulent, and there is no reliable way to predict room air velocity at animal level. General design guidelines keep the throw distance less than 20 ft from slots and less than 10 ft from perforated tubes.

Fans

Fans should not exhaust against prevailing winds, especially for cold-weather ventilation. If structural or other factors require installing fans on the windward side, fans rated to deliver the required capacity against at least 0.12 in. of water static pressure and with a relatively flat power curve should be selected. The fan motor should withstand a wind velocity of 30 mph, equivalent to a static pressure of 0.4 in. of water, without overloading beyond its service factor. Wind hoods on the fans or windbreak fences reduce the effects of wind.

Third-party test data should be used to obtain fan performance and energy efficiencies for fan selection (BESS Lab 1997). Fans should be tested with all accessories (such as louvers, guards, and hoods) in place, just as they will be installed in the building. The accessories have a major effect on fan performance.

Flow Control. Since the numbers and size of livestock and climatic conditions vary, means to modulate ventilation rates are often required beyond the conventional off/on thermostat switch. The minimum ventilation rate to remove moisture, reduce air contaminant concentrations, and keep water from freezing should always be provided. Methods of modulating ventilation rates include: (1) intermittent fan operation—fans operate for a percentage of the time controlled by a percentage timer with a 10 min cycle; (2) staging of fans using multiple units or fans with high/ low-exhaust capability; (3) the use of multispeed fans—larger fans (1/2 hp and up) with two flow rates, the lower being about 60% of the maximum rate; and (4) the use of variable-speed fans—split-capacitor motors designed to modulate fan speed smoothly from maximum down to 10 to 20% of the maximum rate (the controller is usually thermostatically adjusted).

Generally, fans are spaced uniformly along the winter leeward side of a building. Maximum distance between fans is 115 to 165 ft. Fans may be grouped in a bank if this range is not exceeded. In housing with side curtains, exhaust fans that can be reversed or removed and placed inside the building in the summer are sometimes installed to increase air movement in combination with doors, walls, or windows being opened for natural ventilation.

Thermostats

Thermostats should be placed where they respond to a representative temperature as sensed by the animals. Thermostats need protection and should be placed to prevent potential physical or moisture damage (i.e., away from animals, ventilation inlets, water pipes, lights, heater exhausts, outside walls, or any other objects that will unduly affect performance). Thermostats also require periodic adjustment based on accurate thermometer readings taken in the immediate proximity of the animal.

Emergency Warning

Animals housed in a high-density, mechanically controlled environment are subject to considerable risk of heat prostration if a failure of power or ventilation equipment occurs. To reduce this danger, an alarm and an automatic standby electric generator are highly recommended. Many alarms will detect failure of the ventilation. These alarms range from inexpensive power-off alarms to alarms that sense temperature extremes and certain gases. Automatic telephone-dialing systems are effective as alarms and are relatively inexpensive. Building designs that allow some side wall panels (e.g., 25% of wall area) to be removed for emergency situations are also recommended.

RECOMMENDED PRACTICES BY SPECIES

Mature animals readily adapt to a broad range of temperatures, but efficiency of production varies. Younger animals are more temperature sensitive. Figure 6 illustrates animal production response to temperature.

Relative humidity has not been shown to influence animal performance, except when accompanied by thermal stress. Relative humidity consistently below 40% may contribute to excessive dustiness; above 80%, it may increase building and equipment deterioration. Disease pathogens also appear to be more viable at either low or high humidity. Relative humidity has a major influence on the effectiveness of skin-wetting cooling methods.

Dairy Cattle

Dairy cattle shelters include confinement stall barns, free stalls, and loose housing. In a stall barn, cattle are usually confined to stalls approximately 4 ft wide, where all chores, including milking and feeding, are conducted. Such a structure requires environmental modification, primarily through ventilation. Total space requirements are 50 to 75 ft^2 per cow. In free-stall housing, cattle are not confined to stalls but can move freely. Space requirements per cow are 75 to 100 ft^2. In loose housing, cattle are free to move within a fenced lot containing resting and feeding areas. Space required in sheltered loose housing is similar to that in free-stall housing. The shelters for resting and feeding areas are generally open-sided and require no air conditioning or mechanical ventilation, but supplemental air mixing is often beneficial during warm weather. The milking area is in a separate area or facility and may be fully or partially enclosed, thus requiring some ventilation.

For dairy cattle, climate requirements for minimal economic loss are broad, and range from 35 to 75°F with 40 to 80% rh. Below 35°F, production efficiency declines and management problems increase. However, the effect of low temperature on milk production is not as extreme as are high temperatures, where evaporative coolers or other cooling methods may be warranted.

Ventilation Rates for Each 1100 lb Cow

Winter	Spring/Fall	Summer
36 to 47 cfm	142 to 190 cfm	230 to 470 cfm

Required ventilation rates depend on specific thermal characteristics of individual buildings and internal heating load. The relative humidity should be maintained between 50 and 80%.

Both loose housing and stall barns require an additional milk room to cool and hold the milk. Sanitation codes for milk production contain minimum ventilation requirements. The market being supplied should be consulted for all applicable codes. Some state codes require positive pressure ventilation of milk rooms. Milk rooms are usually ventilated with fans at rates of 4 to 10 air changes per hour to satisfy requirements of local milk codes and to remove heat from milk coolers. Most milk codes require ventilation in the passageway (if any) between the milking area and the milk room.

Beef Cattle

Beef cattle ventilation requirements are similar to those of dairy cattle on a unit weight basis. Beef production facilities often provide only shade and wind breaks.

Swine

Swine housing can be grouped into four general classifications:

1. Farrowing pigs, from birth to 30 lb, and sows
2. Nursery pigs, from 30 to 75 lb
3. Growing/finishing pigs, from 75 lb to market weight
4. Breeding and gestation

In farrowing barns, two environments must be provided: one for sows and one for piglets. Because each requires a different temperature, zone heating and/or cooling is used. The environment within the nursery is similar to that within the farrowing barn for piglets. The requirements for growing barns and breeding stock housing are similar.

Currently recommended practices for **farrowing houses**:

- Temperature: 50 to 68°F, with small areas for piglets warmed to 82 to 90°F by means of brooders, heat lamps, or floor heat. Avoid cold drafts and extreme temperatures. Hovers are sometimes used. Provide supplemental cooling for sows (usually drippers or zone cooling) in extreme heat.

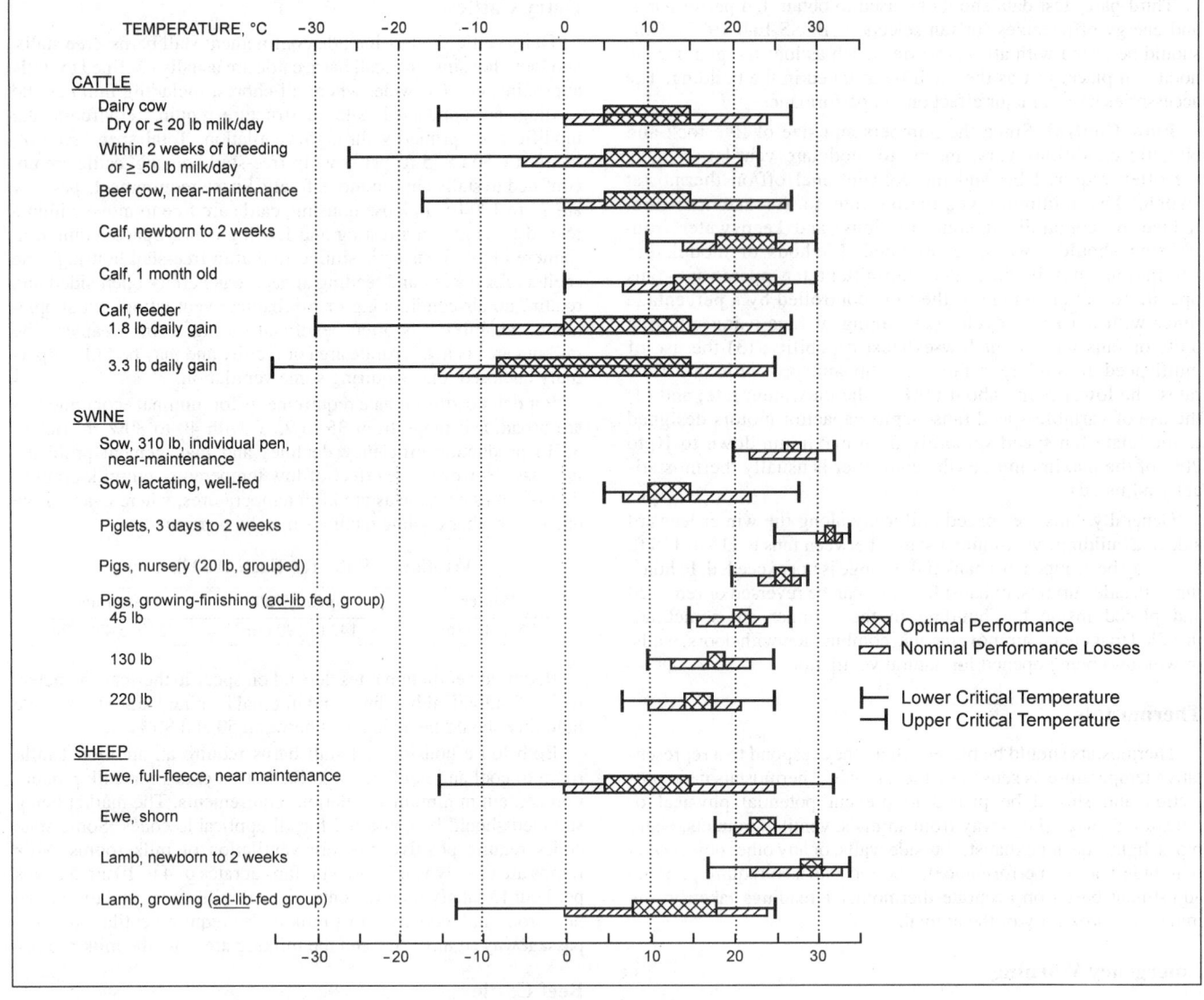

**Fig. 6   Critical Ambient Temperatures and Temperature Zone for Optimum Performance and
Nominal Performance Loss in Farm Animals**

(Hahn 1985)

- Relative humidity: Up to 70% maximum
- Ventilation rate: 20 to 500 cfm per sow and litter (about 400 lb total weight). The low rate is for winter; the high rate is for summer temperature control.
- Space: 35 ft^2 per sow and litter (stall); 65 ft^2 per sow and litter (pens)

Recommendations for **nursery barns**:

- Temperature:
  80°F for first week after weaning. Lower room temperature 3°F per week to 72°F. Provide warm, draft-free floors. Provide supplemental cooling for extreme heat (temperatures 85°F and above).
- Ventilation rate:
  2 to 2.5 cfm per pig, 12 to 30 lb each
  3 to 35 cfm per pig, 30 to 75 lb each
- Space:
  2 to 2.5 ft^2 per pig, 12 to 30 lb each
  3 to 4 ft^2 per pig, 30 to 75 lb each

Recommendations for **growing** and **gestation barns**:

- Temperature:
  55 to 72°F preferred. Provide supplemental cooling (sprinklers or evaporative coolers) for extreme heat.
- Relative humidity:
  75% maximum in winter; no established limit in summer
- Ventilation rate:
  Growing pig (75 to 150 lb), 7 to 75 cfm
  Finishing pig (150 to 220 lb), 10 to 120 cfm
  Gestating sow (325 lb), 12 to 150 cfm
  Boar/breeding sow (400 lb), 14 to 300 cfm
- Space:
  6 ft^2 per pig, 75 to 150 lb each
  8 ft^2 per pig, 150 to 220 lb each
  14 to 24 ft^2 per sow, 240 to 500 lb each

Poultry

In broiler and brooder houses, growing chicks require changing environmental conditions, and heat and moisture dissipation rates increase as the chicks grow older. Supplemental heat, usually from brooders, is used until the sensible heat produced by the birds is adequate to maintain an acceptable air temperature. At early stages of growth, moisture dissipation per bird is low. Consequently, low ventilation rates are recommended to prevent excessive heat loss. Litter is allowed to accumulate over 3 to 5 flock placements. Lack of low-cost litter material may justify the use of concrete floors. After each flock, caked litter is removed and fresh litter is added.

Housing for poultry may be open, curtain-sided or totally enclosed. Mechanical ventilation depends on the type of housing used. For open-sided housing, ventilation is generally natural airflow in warm weather, supplemented with stirring fans, and by fans with closed curtains in cold weather or during the brooding period. Mechanical ventilation is used in totally enclosed housing. Newer houses have smaller curtains and well-insulated construction to accommodate both natural and mechanical ventilation operation.

Recommendations for **broiler houses**:

- Room temperature: 60 to 80°F
- Temperature under brooder hover: 86 to 91°F, reducing 5°F per week until room temperature is reached
- Relative humidity: 50 to 80%
- Ventilation rate: Sufficient to maintain house within 2 to 4°F of outside air conditions during summer. Generally, rates are about 0.1 cfm per lb live weight during winter and 1 to 2 cfm per lb for summer conditions.
- Space: 0.6 to 1.0 ft^2 per bird (for the first 21 days of brooding, only 50% of floor space is used)
- Light: Minimum of 10 lx or 1 footcandle to 28 days of age; 1 to 20 lx or 0.1 to 2 footcandles for growout (in enclosed housing).

Recommendations for **breeder houses** with birds on litter and slatted floors:

- Temperature: 50 to 86°F maximum; consider evaporative cooling if higher temperatures are expected.
- Relative humidity: 50 to 75%
- Ventilation rate: Same as for broilers on live weight basis.
- Space: 2 to 3 ft^2 per bird
  Recommendations for **laying houses** with birds in cages:
- Temperature, relative humidity, and ventilation rate: Same as for breeders.
- Space: 50 to 65 ft^2 per hen minimum
- Light: Controlled day length using light-controlled housing is generally practiced (January through June).

Laboratory Animals

The well-being and experimental response of laboratory animals depends greatly on the design of the facilities. Cage type, noise levels, light levels, air quality, and thermal environment can affect animal well-being and, in many cases, affect how the animal responds to experimental treatments (Moreland 1975, Lindsey et al. 1978, Clough 1982, McPherson 1975). If any of these factors vary across treatments or even within treatments, it can affect the validity of experimental results, or at least increase experimental error. Consequently, laboratory animal facilities must be designed and maintained to expose the animals to appropriate levels of these environmental conditions and to ensure that all animals in an experiment are in a uniform environment. See Chapter 13 for additional information on laboratory animal facilities.

In the United States, recommended environmental conditions within laboratory animal facilities are usually dictated by the *Guide for the Care and Use of Laboratory Animals* (ILAR 1996). Temperature recommendations vary from 61 to 84°F depending on the species being housed. The acceptable range for relative humidity is 30 to 70%. For animals in confined spaces, daily temperature

fluctuations should be kept to a minimum. Relative humidity must also be controlled, but not as precisely as temperature.

Ventilation recommendations are based on room air changes; however, cage ventilation rates may be inadequate in some cages and excessive in other cages depending on cage and facility design. ILAR (1996) recommendations for room ventilation rates of 10 to 15 air changes per hour are an attempt to provide adequate ventilation for the room and the cages. This recommendation is based on the assumption that adequate ventilation in the macroenvironment (room) provides sufficient ventilation to the microenvironment (cage). This may be a reasonable assumption when cages have a top of wire rods or mesh. However, several studies have shown that covering cages with filter tops provides a protective barrier for rodents and reduces airborne infections and diseases, especially neonatal diarrhea; but the filter can create significant differences in microenvironmental conditions.

Maghirang et al. (1995) and Riskowski et al. (1996) surveyed room and cage environmental conditions in several laboratory animal facilities and found that the animal's environmental needs may not be met even though the facilities were designed and operated according to ILAR (1996). The microenvironments were often considerably poorer than the room conditions, especially in microisolator cages. For example, ammonia levels in cages were up to 60 ppm even though no ammonia was detected in a room. Cage temperatures were up to 7°F higher than room temperature and relative humidities up to 41% higher.

Furthermore, cage microenvironments within the same room were found to have significant variation (Riskowski et al. 1996). Within the same room, cage ammonia levels varied from 0 to 60 ppm; cage air temperature varied from 1 to 7°F higher than room temperature; cage relative humidity varied from 1 to 30% higher than room humidity, and average light levels varied from 2 to 337 lx. This survey found three identical rooms that had room ventilation rates from 4.4 to 12.5 air changes per hour (ACH) but had no differences in room or cage environmental parameters.

A survey of laboratory animal environmental conditions in seven laboratory rat rooms was conducted by Zhang et al. (1992). They found that room air ammonia levels were under 0.5 ppm for all rooms, even though room airflow varied from 11 to 24 ACH. Air exchange rates in the cages varied from less than 0.1 cfm to 2.5 cfm per rat, and ammonia levels ranged from negligible to 60 ppm. Riskowski et al. (1996) measured several environmental parameters in rat shoebox cages in full-scale room mockups with various room and ventilation configurations. Significant variations in cage temperature and ventilation rates within a room were also found. Varying room ventilation rate from 5 to 15 ACH did not have large effects on the cage environmental conditions. These studies verify that designs based only on room air changes do not guarantee desired conditions in the animal cages.

In order to analyze the ventilation performance of different laboratory animal research facilities, Memarzadeh (1998) used **computational fluid dynamics** (CFD) to undertake computer simulation of over 100 different room configurations. CFD is a three-dimensional mathematical technique used to compute the motion of air, water, or any other gas or liquid. However, all conditions must be correctly specified in the simulation to produce accurate results. Empirical work defined inputs for such parameters as heat dissipation and surface temperature as well as the moisture, CO_2, and NH_3 mass generation rates for mice.

This approach compared favorably with experimentally measured temperatures and gas concentrations in a typical animal research facility. To investigate the relationships between room configuration parameters and the room and cage environments in laboratory animal research facilities, the following parameters were varied:

- Supply air diffuser type and orientation, air temperature, and air moisture content

- Room ventilation rate
- Exhaust location and number
- Room pressurization
- Rack layout and cage density
- Change station location, design, and status
- Leakage between the cage lower and upper moldings
- Room width

Room pressurization, change station design, and room width had little effect on ventilation performance. However, other factors found to affect either the macroenvironment or microenvironment or both led to the following observations:

- Ammonia production depends on relative humidity. Ten days after the last change of bedding, a high-humidity environment produced ammonia at about three times the rate of cages in a low-humidity environment.
- Acceptable room and cage ammonia concentrations after 5 days without changing cage bedding are produced by room supply airflow rates of around 0.85 cfm per 100 g of body mass of mice. This is equivalent to 5 ACH for the room with single-density racks considered in this study, and 10 ACH for the room with double-density racks. The temperature of the supply air must be set appropriately for the heat load in the room. The room with single-density racks contained 1050 mice with a total mass of 21 kg and the room with double-density racks contained 2100 mice with a total mass of 42 kg.
- Increasing the room ventilation rate has a continuous beneficial effect on the room breathing zone ventilation (as measured by CO_2 and NH_3 concentrations). For single-density racks parallel to the walls, increasing the airflow from 5 ACH to 20 ACH reduces the breathing zone CO_2 concentration from 140 ppm to 63 ppm, a decrease of 55%. For double-density racks perpendicular to the walls, the reduction is even more dramatic—from over 300 ppm to 93 ppm (over 70% reduction) when the airflow is increased from 5 ACH to 20 ACH.
- Increasing the room ventilation rate does not have a large effect on the cage ventilation. Increasing the supply airflow from 5 ACH to 20 ACH around single-density racks parallel to the walls reduces the CO_2 concentration from 1764 ppm to 1667 ppm, a reduction of only 6%. For the double-density racks perpendicular to the walls, the reduction is larger, but still only from about 2300 ppm to 1800 ppm (around 20%)
- Both the cage and the room ammonia concentrations can be reduced by increasing the supply air temperatures. This reduces the relative humidity for a given constant moisture content in the air and the lower relative humidity leads to lower ammonia generation. Raising the supply discharge temperature from 66°F to 72°F at 15 ACH raises the room temperature by 5°F to around 73°F and the cages by 4°F to around 77°F. This can reduce ammonia concentrations by up to 50%.
- Using 72°F as the supply discharge temperature at 5 ACH (the lowest flow rate considered) for double-density racks produces a room temperature around 79°F with cage temperatures only slightly higher. Although this higher temperature provides a more comfortable environment for the mice (Gordon et al. 1997), the high room temperature may be unacceptable to the scientists working in the room.
- Ceiling or high-level exhausts tend to produce lower room temperatures (for a given supply air temperature, all CFD models were designed to have 72°F at the room exhaust) when compared to low-level exhausts. This indicates that low-level exhausts are less efficient at cooling the room.
- Low-level exhausts appear to ventilate the cages slightly better (up to 27% for the radial diffuser; much less for the slot diffuser) than ceiling or high-level exhausts when the cages are placed parallel to the walls, near the exhausts. Ammonia concentration in the cages decreased even further, although this is due to the higher

temperatures in the low-level exhaust cases when compared to the ceiling and high-level exhausts. The room concentrations of CO_2 and ammonia do not show that any type of supply or exhaust is significantly better or worse than the other type.

DESIGN FOR PLANT FACILITIES

Greenhouses, plant growth chambers, and other facilities for indoor crop production overcome adverse outdoor environments and provide conditions conducive to economical crop production. The basic requirements of indoor crop production are (1) adequate light; (2) favorable temperatures; (3) favorable air or gas content; (4) protection from insects and disease; and (5) suitable growing media, substrate, and moisture. Because of their lower cost per unit of usable space, greenhouses are preferred over plant growth chambers for protected crop production.

This section covers greenhouses and plant growth facilities, and Chapter 10 of the 1997 *ASHRAE Handbook—Fundamentals* describes the environmental requirements in these facilities. Figure 7 shows the structural shapes of typical commercial greenhouses. Other greenhouses may have Gothic arches, curved glazing, or simple lean-to shapes. Glazing, in addition to traditional glass, now includes both film and rigid plastics. High light transmission by the glazing is usually important; good location and orientation of the house are important in providing desired light conditions. Location also affects heating and labor costs, exposure to plant disease and air pollution, and material handling requirements. As a general rule in the northern hemisphere, a greenhouse should be placed at a distance of at least 2.5 times the height of the object closest to it in the eastern, western, and southern directions.

GREENHOUSES

Site Selection

Sunlight. Sunlight provides energy for plant growth and is often the limiting growth factor in greenhouses of the central and northern areas of North America during the winter. When planning greenhouses that are to be operated year-round, a designer should design for the greatest sunlight exposure during the short days of midwinter. The building site should have an open southern exposure, and if the land slopes, it should slope to the south.

Soil and Drainage. When plants are to be grown in the soil covered by the greenhouse, a growing site with deep, well-drained, fertile soil, preferably sandy loam or silt loam, should be chosen. Even though organic soil amendments can be added to poor soil, fewer problems occur with good natural soil. However, when good soil is not available, growing in artificial media should be considered. The greenhouse should be level, but the site can and often should be sloped and well-drained to reduce salt buildup and insufficient soil aeration. A high water table or a hardpan may produce water-saturated soil, increase greenhouse humidity, promote diseases, and prevent effective use of the greenhouse. If present, these problems can be alleviated by tile drains under and around the

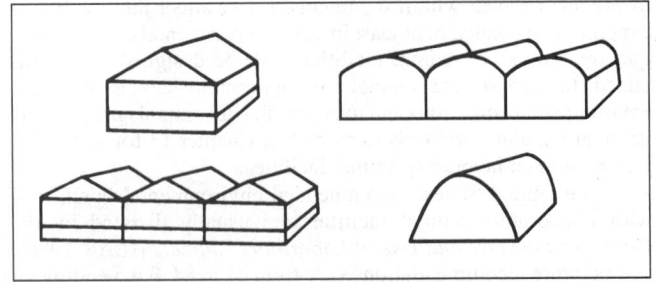

Fig. 7 Structural Shapes of Commercial Greenhouses

greenhouse. Ground beds should be level to prevent water from concentrating in low areas. Slopes within greenhouses also increase temperature and humidity stratification and create additional environmental problems.

Sheltered Areas. Provided they do not shade the greenhouse, surrounding trees act as wind barriers and help prevent winter heat loss. Deciduous trees are less effective than coniferous trees in midwinter, when the heat loss potential is greatest. In areas where snowdrifts occur, windbreaks and snowbreaks should be 100 ft or more from the greenhouse to prevent damage.

Orientation. Generally, in the northern hemisphere, for single-span greenhouses located north of 35° latitude, maximum transmission during winter is attained by an east-west orientation. South of 35° latitude, orientation is not important, provided headhouse structures do not shade the greenhouse. North-south orientation provides more light on an annual basis.

Gutter-connected or ridge-and-furrow greenhouses are oriented preferably with the ridge line north-south regardless of latitude. This orientation permits the shadow pattern caused by the gutter superstructure to move from the west to the east side of the gutter during the day. With an east-west orientation, the shadow pattern would remain north of the gutter, and the shadow would be widest and create the most shade during winter when light levels are already low. Also, the north-south orientation allows rows of tall crops, such as roses and staked tomatoes, to align with the long dimension of the house—an alignment that is generally more suitable to long rows and the plant support methods preferred by many growers.

The slope of the greenhouse roof is a critical part of greenhouse design. If the slope is too flat, a greater percentage of sunlight is reflected from the roof surface (Figure 8). A slope with a 1:2 rise-to-run ratio is the usual inclination for a gable roof.

HEATING

Structural Heat Loss

Estimates for heating and cooling a greenhouse consider conduction, infiltration, and ventilation energy exchange. In addition, the calculations must consider solar energy load and electrical input, such as light sources, which are usually much greater for greenhouses than for conventional buildings. Generally, conduction q_c plus infiltration q_i are used to determine the peak requirements q_t for heating.

$$q_t = q_c + q_i \qquad (4)$$

$$q_c = UA(t_i - t_o) \qquad (5)$$

$$q_i = 0.018 VN(t_i - t_o) \qquad (6)$$

where

U = overall heat loss coefficient, Btu/h·ft²·°F (Table 2 and Table 3)
A = exposed surface area, ft²
t_i = inside temperature, °F
t_o = outside temperature, °F
V = greenhouse internal volume, ft³
N = number of air exchanges per hour (Table 4)

Type of Framing

The type of framing should be considered in determining overall heat loss. Aluminum framing and glazing systems may have the metal exposed to the exterior to a greater or lesser degree, and the heat transmission of this metal is higher than that of the glazing material. To allow for such a condition, the U-factor of the glazing material should be multiplied by the factors shown in Table 3.

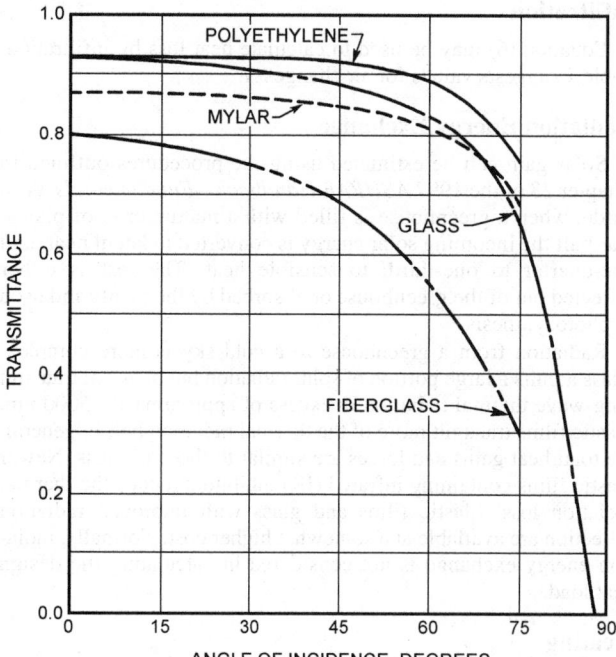

Fig. 8 Transmittance of Solar Radiation Through Glazing Materials for Various Angles of Incidence

Table 2 Suggested Heat Transmission Coefficients

		U, Btu/h·ft²·°F
Glass		
	Single glazing	1.13
	Double glazing	0.70
	Insulating	Manufacturers' data
Plastic film		
	Single film[a]	1.20
	Double film, inflated	0.70
	Single film over glass	0.85
	Double film over glass	0.60
Corrugated glass fiber		
	Reinforced panels	1.20
Plastic structured sheet[b]		
	16 mm thick	0.58
	8 mm thick	0.65
	6 mm thick	0.72

[a]Infrared barrier polyethylene films reduce heat loss; however, use this coefficient when designing heating systems because the structure could occasionally be covered with non-IR materials.
[b]Plastic structured sheets are double-walled, rigid plastic panels.

Table 3 Construction U-Factor Multipliers

Metal frame and glazing system, 16 to 24 in. spacing	1.08
Metal frame and glazing system, 48 in. spacing	1.05
Fiberglass on metal frame	1.03
Film plastic on metal frame	1.02
Film or fiberglass on wood	1.00

Table 4 Suggested Design Air Changes (N)

New Construction	
Single glass lapped (unsealed)	1.25
Single glass lapped (laps sealed)	1.0
Plastic film covered	0.6 to 1.0
Structured sheet	1.0
Film plastic over glass	0.9
Old Construction	
Good maintenance	1.5
Poor maintenance	2 to 4

Infiltration

Equation (6) may be used to calculate heat loss by infiltration. Table 4 suggests values for air changes N.

Radiation Energy Exchange

Solar gain can be estimated using the procedures outlined in Chapter 28 of the 1997 *ASHRAE Handbook—Fundamentals*. As a guide, when a greenhouse is filled with a mature crop of plants, one-half the incoming solar energy is converted to latent heat; and one-quarter to one-third, to sensible heat. The rest is either reflected out of the greenhouse or absorbed by the plants and used in photosynthesis.

Radiation from a greenhouse to a cold sky is more complex. Glass admits a large portion of solar radiation but does not transmit long-wave thermal radiation in excess of approximately 5000 nm. Plastic films transmit more of the thermal radiation but, in general, the total heat gains and losses are similar to those of glass. Newer plastic films containing infrared (IR) inhibitors reduce the thermal radiation loss. Plastic films and glass with improved radiation reflection are available at a somewhat higher cost. Normally, radiation energy exchange is not considered in calculating the design heat load.

Heating

Greenhouses may have a variety of heaters. One is a convection heater that circulates hot water or steam through plain or finned pipe. The pipe is most commonly placed along walls and occasionally beneath plant benches to create desirable convection currents. A typical temperature distribution pattern created by perimeter heating is shown in Figure 9. More uniform temperatures can be achieved when about one-third the total heat comes from pipes spaced uniformly across the house. These pipes can be placed above or below the crop, but temperature stratification and shading are avoided when they are placed below. Outdoor weather conditions affect temperature distribution, especially on windy days in loosely constructed greenhouses. Manual or automatic overhead pipes are also used for supplemental heating to prevent snow buildup on the roof. In a gutter-connected greenhouse in a cold climate, a heat pipe should be placed under each gutter to prevent snow accumulation.

An overhead tube heater consists of a unit heater that discharges into 12 to 30 in. diameter plastic film tubing perforated to provide uniform air distribution. The tube is suspended at 6 to 10 ft intervals and extends the length of the greenhouse. Variations include a tube and fan receiving the discharge of several unit heaters. The fan and tube system is used without heat to recirculate the air and, during cold weather, to introduce ventilation air. However, tubes sized for heat distribution may not be large enough for effective ventilation during warm weather.

Perforated tubing, 6 to 10 in. in diameter, placed at ground level heaters (underbench) can also improve heat distribution. Ideally, the ground-level tubing should draw air from the top of the greenhouse for recirculation or heating. Tubes on or near the floor have the disadvantage of being obstacles to workers and reducing usable floor space.

Underfloor heating can supply up to 25% or more of the peak heating requirements in cold climates. A typical underfloor system uses 0.75 in. plastic pipe spaced 12 to 16 in. on center, and covered with 4 in. of gravel or porous concrete. Hot water, not exceeding 104°F, circulates at a rate of 2 to 2.5 gpm per loop. Pipe loops should generally not exceed 400 ft in length. This can provide 16 to 20 Btu/h·ft^2 from a bare floor, and about 75% as much when potted plants or seedling flats cover most of the floor.

Similar systems can heat soil directly, but root temperature must not exceed 77°F. When used with water from solar collectors or other heat sources, the underfloor area can store heat. This storage

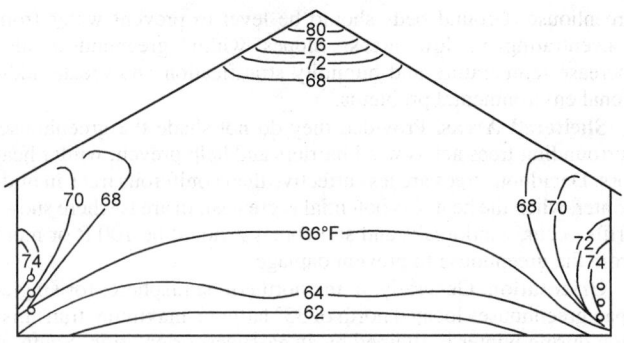

Fig. 9 Temperature Profiles in a Greenhouse Heated with Radiation Piping along the Sidewalls

consists of a vinyl swimming pool liner placed on top of insulation and a moisture barrier at a depth of 8 to 12 in. below grade, and filled with 50% void gravel. Hot water from solar collectors or other clean sources enters and is pumped out on demand. Some heat sources, such as cooling water from power plants, cannot be used directly but require closed-loop heat transfer to avoid fouling the storage and the power plant cooling water.

Greenhouses can also be bottom heated with 0.25 in. diameter EPDM tubing (or variations of that method) in a closed loop. The tubes can be placed directly in the growing medium of ground beds or under plant containers on raised benches. The best temperature uniformity is obtained by flow in alternate tubes in opposite directions. This method can supply all the greenhouse heat needed in mild climates.

Bottom heat, underfloor heating, and under-bench heating are, because of the location of the heat source, more effective than overhead or peripheral heating, and can reduce energy loss by 20 to 30%.

Unless properly located and aimed, overhead unit heaters, whether hydronic or direct fired, do not give uniform temperature at the plant level and throughout the greenhouse. Horizontal blow heaters positioned so that they establish a horizontal airflow around the outside of the greenhouse offer the best distribution. The airflow pattern can be supplemented with the use of horizontal blow fans or circulators.

When direct combustion heaters are used in the greenhouse, combustion gases must be adequately vented to the outside to minimize danger to plants and humans from products of combustion. One manufacturer recommends that combustion air must have access to the space through a minimum of two permanent openings in the enclosure, one near the bottom. A minimum of 1 in^2 of free area per 1000 Btu/h input rating of the unit, with a minimum of 100 in^2 for each opening, whichever is greater, is recommended. Unvented direct combustion units should not be used inside the greenhouse.

In many greenhouses, a combination of overhead and perimeter heating is used. Regardless of the type of heating, it is common practice to calculate the overall heat loss first, and then to calculate the individual elements such as the roof, sidewalls, and gables. It is then simple to allocate the overhead portion to the roof loss and the perimeter portions to the sides and gables, respectively.

The annual heat loss can be approximated by calculating the design heat loss and then, in combination with the annual degree-day tables using the 65°F base, estimating an annual heat loss and computing fuel usage on the basis of the rating of the particular fuel used. If a 50°F base is used, it can be prorated.

Heat curtains for energy conservation are becoming more important in greenhouse construction. Although this energy savings may be considered in the annual energy use, it should not be used when calculating design heat load; the practice is to open the heat curtains during snowstorms to facilitate the melting of snow, thereby nullifying its contribution to the design heat loss value.

Air-to-air and water-to-air heat pumps have been used experimentally on small-scale installations. Their usefulness is especially sensitive to the availability of a low-cost heat source.

Radiant (Infrared) Heating

Radiant heating is used in some limited applications for greenhouse heating. Steel pipes spaced at intervals and heated to a relatively high temperature by special gas heaters serve as the source of radiation. Because the energy is transmitted by radiation from a source of limited size, proper spacing is important to completely cover the heated area. Further, heavy foliage crops can shade the lower parts of the plants and the soil, thus restricting the radiation from warming the root zone, which is important to plant growth.

Cogenerated Sources of Heat

Greenhouses have been built near or adjacent to power plants to use the heat and electricity generated by the facility. While this energy may cost very little, an adequate standby energy source must be provided, unless the power supplier can assure that it will supply a reliable, continuous source of energy.

COOLING

Solar radiation is a considerable source of sensible heat gain; even though some of this energy is reflected from the greenhouse, some of it is converted into latent heat as the plants transpire moisture, and some is converted to plant material by photosynthesis. Natural ventilation, mechanical ventilation, shading, and evaporative cooling are common methods used to remove this heat. Mechanical refrigeration is seldom used to air condition greenhouses because the cooling load and resulting cost is so high.

Natural Ventilation

Most older greenhouses and many new ones rely on natural ventilation with continuous roof sashes on each side of the ridge and continuous sashes in the sidewalls. The roof sashes are hinged at the ridge, and the wall sashes are hinged at the top of the sash. During much of the year, vents admit enough ventilating air for cooling without the added cost of running fans.

The principles of natural ventilation are explained in Chapter 25 of the 1997 *ASHRAE Handbook—Fundamentals*. Ventilation air is driven by wind and thermal buoyancy forces. Proper vent openings take advantage of pressure differences created by wind. Thermal buoyancy caused by the temperature difference between the inside and the outside of the greenhouse is enhanced by the area of the vent opening and the stack height (vertical distance between the center of the lower and upper opening). Within the limits of typical construction, the larger the vents, the greater the ventilating air exchanged. For a single greenhouse, the combined area of the sidewall vents should equal that of the roof vents. In ranges of several gutter-connected greenhouses, the sidewall area cannot equal the roof vent area.

Mechanical (Forced) Ventilation

Exhaust fans provide positive ventilation without depending on wind or thermal buoyancy forces. The fans are installed in the side or end walls of the greenhouse and draw air through vents on the opposite side or end walls. The air velocity through the inlets should not exceed 400 fpm.

Air exchange rates between 0.75 and 1 change per minute effectively control the temperature rise in a greenhouse. As shown in Figure 10, the temperature inside the greenhouse rises rapidly at lower airflow rates. At higher airflow rates the reduction of the temperature rise is small, fan power requirements are increased, and plants may be damaged by the high air speed.

Shading

Shading compounds can be applied in varying amounts to the exterior of the roof of the greenhouse to achieve up to 50% shading. Durability of these compounds varies—ideally the compound will wear away during the summer and leave the glazing clean in the fall when shading is no longer needed. In practice, some physical cleaning is needed. Compounds used formerly usually contained lime, which corrodes aluminum and attacks some caulking. Most compounds used currently are formulated to avoid this problem.

Mechanically operated shade cloth systems with a wide range of shade levels are also available. They are mounted inside the greenhouse to protect them from the weather. Not all shading compounds or shade cloths are compatible with all plastic glazings, so the manufacturers' instructions and precautions should be followed.

Evaporative Cooling

Fan-and-Pad Systems. Fans for fan-and-pad evaporative cooling are installed in the same manner as fans used for mechanical ventilation. Pads of cellulose material in a honeycomb form are installed on the inlet side. The pads are kept wet continuously when evaporative cooling is needed. As air is drawn through the pads, the water evaporates and cools the air. New pads cool the air by about 80% of the difference between the outdoor dry-bulb and wet-bulb temperature, or to 3 to 4°F above the wet-bulb temperature.

The empirical base rate of airflow is 8 cfm per square foot of floor area. This flow rate is modified by multiplying it by factors for elevation (F_e), maximum interior light intensity (F_l), and the allowable temperature rise between the pad and the fans (F_t). These factors are listed in Table 5. The overall factor for the house is given by the following equation:

$$F_h = F_e F_l F_t \qquad (7)$$

The maximum fan-to-pad distance should be kept to 175 ft, although some greenhouses with distances of 225 ft have shown no

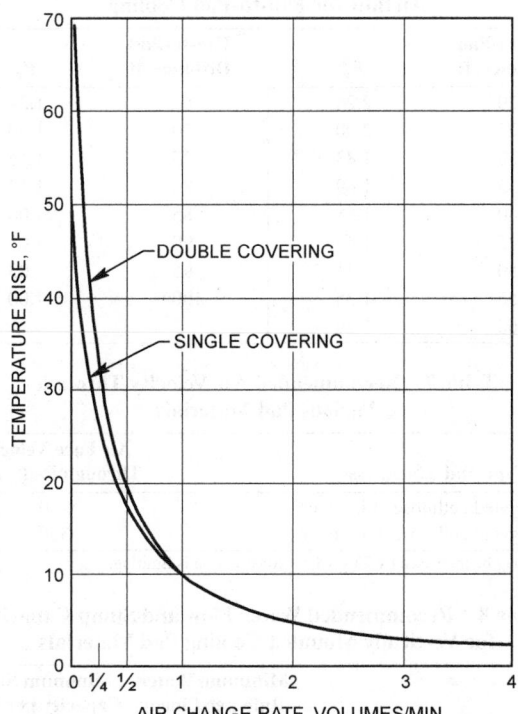

Fig. 10 Influence of Air Exchange Rate on Temperature Rise in Single- and Double-Covered Greenhouses

serious reduction in effectiveness. With short distances, the air velocity becomes so low that the air feels clammy and stuffy, even though the airflow is sufficient for cooling. Therefore, a velocity factor F_v listed in Table 6 is used for distances less than 100 ft. For distance less than 100 ft, F_v is compared to F_h. The factor that gives the greatest airflow is used to modify the empirical base rate. For fan-to-pad distances greater than 100 ft, F_v can be ignored.

For best performance, pads should be installed on the windward side, and fans spaced within 25 ft of each other. Fans should not blow toward pads of an adjacent house unless it is at least 50 ft away. Fans in adjacent houses should be offset if they blow toward each other and are within 15 ft of each other.

Recommended air velocities through commonly used pads are listed in Table 7. Water flow and sump capacities are shown in Table 8. The system should also include a small, continuous bleed-off of water to reduce the buildup of dirt and other impurities.

Unit Evaporative Coolers. This equipment contains the pads, water pump, sump, and fan in one unit. Unit coolers are primarily used for small compartments. They are mounted 15 to 20 ft apart on

Table 5   Multipliers for Calculating Airflow for Fan-and-Pad Cooling

Elevation (Above Sea Level)		Max. Interior Light Intensity		Fan-to-Pad Temp. Difference	
ft	F_e	footcandles	F_l	°F	F_t
<1000	1.00	4000	0.80	10	0.70
1000	1.04	4500	0.90	9	0.78
2000	1.08	5000	1.00	8	0.88
3000	1.12	5500	1.10	7	1.00
4000	1.16	6000	1.20	6	1.17
5000	1.20	6500	1.30	5	1.40
6000	1.25	7000	1.40	4	1.75
7000	1.30	7500	1.50		
8000	1.36	8000	1.60		

Table 6   Velocity Factors for Calculating Airflow for Fan-to-Pad Cooling

Fan-to-Pad Distance, ft	F_v	Fan-to-Pad Distance, ft	F_v
20	2.24	65	1.24
25	2.00	70	1.20
30	1.83	75	1.15
35	1.69	80	1.12
40	1.58	85	1.08
45	1.49	90	1.05
50	1.41	95	1.03
55	1.35	100	1.00
60	1.29		

Table 7   Recommended Air Velocity Through Various Pad Materials

Pad Type and Thickness	Air Face Velocity Through Pad[a], fpm
Corrugated cellulose, 4 in. thick	250
Corrugated cellulose, 6 in. thick	350

[a]Speed may be increased by 25% where construction is limiting.

Table 8   Recommended Water Flow and Sump Capacity for Vertically Mounted Cooling Pad Materials

Pad Type and Thickness	Minimum Water Rate per Linear Foot of Pad, gpm	Minimum Sump Capacity per Unit Pad Area, gal/ft^2
Corrugated cellulose, 4 in. thick	0.5	0.8
Corrugated cellulose, 6 in. thick	0.8	1.0

the sidewall and blow directly into the greenhouse. They cool a distance of up to 50 ft from the unit. A side sash on the outside opposite wall is the best outlet, but roof vents may also work. The roof vent on the same side as the unit should be slightly open for better air distribution. If the roof vent on the opposite side is opened instead, air may flow directly out the vent and not cool the opposite side of the greenhouse.

Fog. In a direct-pressure atomizer, a high-pressure pump forces water at 800 to 1000 psi through a special fog nozzle. Fog is considered to be a water droplet smaller than 40 μm in diameter. The direct-pressure atomizer generates droplets of 35 μm or less. This requires a superior filter to minimize clogging of the very small nozzle orifices.

A line of nozzles placed along the top of the vent opening can cool the entering air nearly to its wet-bulb temperature. Additional lines in the greenhouse continue to cool the air as it absorbs heat in the space.

Fogging will cool satisfactorily with less airflow than fan-and-pad systems, but the fan capacity must still be based on one air change per minute to ventilate the greenhouse when the cooler will be used without fog.

OTHER ENVIRONMENTAL CONTROLS

Humidity Control

At various times during the year, humidity may need to be controlled in the greenhouse. When the humidity is too high at night, it can be reduced by adding heat and ventilating simultaneously. When the humidity is too low during the day, it can be increased by turning on a fog or mist nozzle.

Winter Ventilation

During the winter, houses are normally closed tightly to conserve heat, but photosynthesis by the plants may lower the carbon dioxide level to such a point that it slows plant growth. Some ventilation helps maintain inside carbon dioxide levels. A normal rate of airflow for winter ventilation is 2 to 3 cfm per square foot of floor area.

Air Circulation

Continuous air circulation within the greenhouse reduces still-air conditions that favor plant diseases. Recirculating fans, heaters that blow air horizontally, and fans attached to polyethylene tubes are used to circulate air. The amount of recirculation has not been well defined, except that some studies have shown high air velocities (greater than 200 fpm) can harm plants or reduce growth.

Insect Screening

Insect screening is being used to cover vent inlets and outlets. These fine-mesh screens increase the resistance to airflow, which must be considered when selecting ventilation fans. The screen manufacturer should provide static pressure data for its screens. The pressure drop through the screen can be reduced by framing out from the vent opening to increase the area of the screen.

Carbon Dioxide Enrichment

Carbon dioxide is added in some greenhouse operations to increase growth and enhance yields. However, CO_2 enrichment is practical only when little or no ventilation is required for temperature control. Carbon dioxide can be generated from solid CO_2 (dry ice), bottled CO_2, and misting carbonated water. Bulk or bottled CO_2 gas is usually distributed through perforated tubing placed near the plant canopy. Carbon dioxide from dry ice is distributed by passing greenhouse air through an enclosure containing dry ice. Air movement around the plant leaf increases the efficiency with which the plant absorbs whatever CO_2 is available. One study found an air

Table 9 Constants to Convert to W/m²

Light Source	klx	$\mu mol/(s \cdot m^2)$
400 to 700 nm		
Incandescent (INC)	3.99	0.20
Fluorescent cool white (FCW)	2.93	0.22
Fluorescent warm white (FWW)	2.81	0.21
Discharge clear mercury (HG)	2.62	0.22
Metal halide (MH)	3.05	0.22
High-pressure sodium (HPS)	2.45	0.20
Low-pressure sodium (LPS)	1.92	0.20
Daylight	4.02	0.22

speed of 100 fpm to be equivalent to a 50% enrichment in CO_2 without forced air movement.

Radiant Energy

Light is normally the limiting factor in greenhouse crop production during the winter. North of the 35th parallel, light levels are especially inadequate or marginal in fall, winter, and early spring. Artificial light sources, usually high-intensity discharge (HID) lamps, may be added to greenhouses to supplement low natural light levels. High-pressure sodium (HPS), metal halide (MH), low-pressure sodium (LPS), and occasionally, mercury lamps coated with a color-improving phosphor, are currently used. Since differing irradiance or illuminance ratios are emitted by the various lamp types, the incident radiation is best described as radiant flux density (W/ft²) between 400 and 850 nm, or as photon flux density between 400 and 700 nm, rather than in photometric terms of lux or footcandles.

To assist in relating irradiance to more familiar illuminance values, Table 9 shows constants for converting illuminance (lux) and photon flux density [$\mu mol/(s \cdot m^2)$] of HPS, MH, LPS, and other lamps to the irradiance (W/m²). One footcandle is approximately 10 lux.

Table 10 gives values of suggested irradiance at the top of the plant canopy, duration, and time of day for supplementing natural light levels for specific plants.

HID lamps in luminaires developed specifically for greenhouse use are often placed in a horizontal position, which may decrease both the light output and the life of the lamp. These drawbacks may be balanced by improved horizontal and vertical uniformity as compared to industrial parabolic reflectors.

Photoperiod Control

Artificial light sources are also used to lengthen the photoperiod during the short days of winter. Photoperiod control requires much lower light levels than those needed for photosynthesis and growth. Photoperiod illuminance needs to be only 0.6 to 1.1 W/ft². The incandescent lamp is the most effective light source for this purpose due to its higher far-red component. Lamps such as 150 W (PS-30) silverneck lamps spaced 10 to 13 ft on centers and 13 ft above the plants provide a cost-effective system. Where a 13 ft height is not practical, 60 W extended service lamps on 6.5 ft centers are satisfactory. One method of photoperiod control is to interrupt the dark period by turning the lamps on at 2200 and off at 0200. The 4 h interruption, initially based on chrysanthemum response, induces a satisfactory long-day response in all photoperiodically sensitive species. Many species, however, respond to interruptions of 1 h or less. Demand charges can be reduced in large installations by operating some sections from 2000 to 2400 and others from 2400 to 0400. The biological response to these schedules, however, is much weaker than with the 2200 to 0200 schedule, so some varieties may flower prematurely. If the 4 h interruption period is used, it is not necessary to keep the light on throughout the interruption period. Photoperiod control of most plants can be accomplished by operating the lamps on light and

Table 10 Suggested Radiant Energy, Duration, and Time of Day for Supplemental Lighting in Greenhouses

Plant and Stage of Growth	W/ft²	Duration Hours	Duration Time
African violets	1 to 2	12 to 16	0600-1800
early-flowering			0600-2200
Ageratum	1 to 4.5	24	
early-flowering			
Begonias—fibrous rooted	1 to 2	24	
branching and early-flowering			
Carnation	1 to 2	16	0800-2400
branching and early-flowering			
Chrysanthemums	1 to 2	16	0800-2400
vegetable growth branching			
and multiflowering	1 to 2	8	0800-1600
Cineraria	0.6 to 1	24	
seedling growth (four weeks)			
Cucumber	1 to 2	24	
rapid growth and early-flowering			
Eggplant	1 to 4.5	24	
early-fruiting			
Foliage plants	0.6 to 1	24	
(Philodendron, Schefflera) rapid growth			
Geranium	1 to 4.5	24	
branching and early-flowering			
Gloxinia	1 to 4.5	16	0800-2400
early-flowering	0.6 to 1	24	
Lettuce	1 to 4.5	24	
rapid growth			
Marigold	1 to 4.5	24	
early-flowering			
Impatiens—New Guinea	1	16	0800-2400
branching and early-flowering			
Impatiens—Sultana	1 to 2	24	
branching and early-flowering			
Juniper	1 to 4.5	24	
vegetative growth			
Pepper	1 to 2	24	
early-fruiting, compact growth			
Petunia	1 to 4.5	24	
branching and early-flowering			
Poinsettia—vegetative growth	1	24	
branching and multiflowering	1 to 2	8	0800-1600
Rhododendron	1	16	0800-2400
vegetative growth (shearing tips)			
Roses (hybrid teas, miniatures)	1 to 4.5	24	
early-flowering and rapid regrowth			
Salvia	1 to 4.5	24	
early-flowering			
Snapdragon	1 to 4.5	24	
early-flowering			
Streptocarpus	1	16	0800-2400
early-flowering			
Tomato	1 to 2	16	0800-2400
rapid growth and early-flowering			
Trees (deciduous)	0.6	16	1600-0800
vegetative growth			
Zinnia	1 to 4.5	24	
early-flowering			

dark cycles with 20% "on" times; for example, 12 s/min. The length of the dark period in the cycle is critical, and the system may fail if the dark period exceeds about 30 min. Demand charges can be reduced by alternate scheduling of the "on" times between houses or benches without reducing the biological effectiveness of the interruption.

Plant displays in places such as showrooms or shopping malls require enough light for plant maintenance and a spectral distribution that best shows the plants. Metal halide lamps, with or without incandescent highlighting, are often used for this purpose. Fluorescent lamps, frequently of the special phosphor plant-growth type, enhance color rendition, but are more difficult to install in aesthetically pleasing designs.

Design Conditions

Plant requirements vary from season to season and during different stages of growth. Even different varieties of the same species of plant may vary in their requirements. State and local cooperative extension offices are a good source of specific information on design conditions affecting plants. These offices also provide current, area-specific information on greenhouse operations.

Alternate Energy Sources and Energy Conservation

Limited progress has been achieved in heating commercial greenhouses with solar energy. Collecting and storing the heat requires a volume at least one-half the volume of the entire greenhouse. Passive solar units work at certain times of the year and, in a few localities, year-round.

If available, reject heat is a possible source of winter heat. Winter energy and solar (photovoltaic) sources are possible future energy sources for greenhouses, but the development of such systems is still in the research stage.

Energy Conservation. A number of energy-saving measures (e.g., thermal curtains, double glazing, and perimeter insulation) have been retrofitted to existing greenhouses and incorporated into new construction. Sound maintenance is necessary to keep heating system efficiency at a maximum level.

Automatic controls, such as thermostats, should be calibrated and cleaned at regular intervals, and heating-ventilation controls should interlock to avoid simultaneous operation. Boilers that can burn more than one type of fuel permit use of the most inexpensive fuel available.

Modifications to Reduce Heat Loss

Film covers that reduce heat loss are used widely in commercial greenhouses, particularly for growing foliage plants and other species that grow under low light levels. Irradiance (intensity) is reduced 10 to 15% per layer of plastic film.

One or two layers of transparent 4 or 6 mil continuous-sheet plastic is stretched over the entire greenhouse (leaving some vents uncovered), or from the ridge to the sidewall ventilation opening. When two layers are used, (outdoor) air at a pressure of 0.2 to 0.25 in. of water is introduced continuously between the layers of film to maintain the air space between them. When a single layer is used, an air space can be established by stretching the plastic over the glazing bars and fastening it around the edges, or a length of polyethylene tubing can be placed between the glass and the plastic and inflated (using outside air) to stretch the plastic sheet.

Double-Glazing Rigid Plastic. Double-wall panels are manufactured from acrylic and polycarbonate plastics, with walls separated by about 0.4 in. Panels are usually 48 in. wide and 96 in. or longer. Nearly all types of plastic panels have a high thermal expansion coefficient and require about 1% expansion space (0.12 in/ft). When a panel is new, light reduction is roughly 10 to 20%. Moisture accumulation between the walls of the panels must be avoided.

Double-Glazing Glass. The framing of most older greenhouses must be modified or replaced to accept double glazing with glass.

Light reduction is 10% more than with single glazing. Moisture and dust accumulation between glazings increases light loss. As with all types of double glazing, snow on the roof melts slowly and increases light loss. Snow may even accumulate sufficiently to cause structural damage, especially in gutter-connected greenhouses.

Silicone Sealants. Transparent silicone sealant in the glass overlaps of conventional greenhouses reduces infiltration and may produce heat savings of 5 to 10% in older structures. There is little change in light transmission.

Precautions. The various methods described above reduce heat loss by reducing conduction and infiltration. They may also cause more condensation, higher relative humidity, lower carbon dioxide concentration, and an increase in ethylene and other pollutants. Combined with the reduced light levels, these factors may cause delayed crop production, elongated plants, soft plants, and various deformities and diseases, all of which reduce the marketable crop.

Thermal blankets are any flexible material that is pulled from gutter to gutter and end to end in a greenhouse, or around and over each bench, at night. Materials ranging from plastic film to heavy cloth, or laminated combinations, have successfully reduced heat losses by 25 to 35% overall. Tightness of fit around edges and other obstructions is more important than the kind of material used. Some films are vaportight and retain moisture and gases. Others are porous and permit some gas exchange between the plants and the air outside the blanket. Opaque materials can control crop day length when short days are part of the requirement for that crop. Condensation may drip onto and collect on the upper sides of some blanket materials to such an extent that they collapse.

Multiple-layer blankets, with two or more layers separated by air spaces, have been developed. One such design combines a porous-material blanket and a transparent film blanket; the latter is used for summer shading. Another design has four layers of porous, aluminum foil-covered cloths, with the layers separated by air.

Thermal blankets may be opened and closed manually as well as automatically. The decision to open or close should be based on the irradiance level and whether it is snowing, rather than on the time of day. Two difficulties with thermal blankets are the physical problems of installation and use in greenhouses with interior supporting columns, and the loss of space due to shading by the blanket when it is not in use during the day.

Other Recommendations. While the foundation can be insulated, the insulating materials must be protected from moisture, and the foundation wall should be protected from freezing. All or most of the north wall can be insulated with opaque or reflective-surface materials. The insulation reduces the amount of diffuse light entering the greenhouse and, in cloudy climates, causes reduced crop growth near the north wall.

Ventilation fan cabinets should be insulated, and fans not needed during the winter should be sealed against air leaks. Efficient management and operation of existing facilities are the most cost-effective ways to reduce energy use.

PLANT GROWTH ENVIRONMENTAL FACILITIES

Controlled-environment rooms (CERs), also called plant growth chambers, include all controlled or partially controlled environmental facilities for growing plants, except greenhouses. CERs are indoor facilities. Units with floor areas less than 50 ft^2 may be moveable with self-contained or attached refrigeration units. CERs usually have artificial light sources, provide control of temperature and, in some cases, control relative humidity and CO_2 level.

CERs are used to study all aspects of botany. Some growers use growing rooms to increase seedling growth rate, produce more uniform seedlings, and grow specialized, high-value crops. The main components of the CER are (1) an insulated room or an insulated box with an access door; (2) a heating and cooling mechanism with associated air-moving devices and controls; and (3) a lamp module at the top of the insulated box or room. CERs are similar to walk-in cold storage rooms, except for the lighting and larger refrigeration system needed to handle heat produced by the lighting.

Location

The location for a CER must have space for the outside dimensions of the chamber, refrigeration equipment, ballast rack, and control panels. Additional space around the unit is necessary for servicing the various components of the system and, in some cases, for substrate, pots, nutrient solutions, and other paraphernalia associated with plant research. The location also requires electricity (on the order of 28 W/ft² of controlled environment space), water, and compressed air.

Construction and Materials

Wall insulation should have a thermal conductance of less than 0.026 Btu/h·ft²·°F. Materials should resist corrosion and moisture. The interior wall covering should be metal, with a high-reflectance white paint, or specular aluminum with a reflectivity of at least 80%. Reflective films or similar materials can be used, but will require periodic replacement.

Floors and Drains

Floors that are part of the CER should be corrosion-resistant. Tar or asphalt waterproofing materials and volatile caulking compounds should not be used because they will likely release phytotoxic gases into the chamber atmosphere. The floor must have a drain to remove spilled water and nutrient solutions. The drains should be trapped and equipped with screens to catch plant and substrate debris.

Plant Benches

Three bench styles for supporting the pots and other plant containers are normally encountered in plant growth chambers: (1) stationary benches; (2) benches or shelves built in sections that are adjustable in height; and (3) plant trucks, carts, or dollies on casters, which are used to move plants between chambers, greenhouses, and darkrooms. The bench supports containers filled with moist sand, soil, or other substrate, and is usually rated for loads of at least 50 lb/ft². The bench or truck top should be constructed of nonferrous, perforated metal or metal mesh to allow free passage of air around the plants and to let excess water drain from the containers to the floor and subsequently to the floor drain.

Normally, benches, shelves, or truck tops are adjustable in height so that small plants can be placed close to the lamps and thus receive a greater amount of light. As the plants grow, the shelf or bench is lowered so that the tops of the plants continue to receive the original radiant flux density.

Control

Environmental chambers require complex controls to provide the following:

- Automatic transfer from heating to cooling with 2°F or less dead zone and adjustable time delay.
- Automatic daily switching of the temperature set point for different day and night temperatures (setback may be as much as 10°F).
- Protection of sensors from radiation. Ideally, the sensors are located in a shielded, aspirated housing, but satisfactory performance can be attained by placing them in the return air duct.

- Control of the daily duration of light and dark periods. Ideally, this control should be programmable to change the light period each day to simulate the natural progression of day length. Photoperiod control, however, is normally accomplished with mechanical time clocks, which must have a control interval of 5 min or less for satisfactory timing.
- Protective control to prevent the chamber temperature from going more than a few degrees above or below the set point. Control should also prevent short cycling of the refrigeration system, especially when the condensers are remotely located.
- Audible and visual alarms to alert personnel of malfunctions.
- Maintenance of relative humidity to prescribed limits.

Data loggers, recorders, or recording controllers are recommended to aid in monitoring daily operation. Solid-state, microprocessor-based control is not yet widely used. However, programming flexibility and control performance are expected to improve as microprocessor control is developed for CER use.

Heating, Air Conditioning, and Airflow

When the lights are on, cooling will normally be required, and the heater will rarely be called on to operate. When the lights are off, however, both heating and cooling may be needed. Conventional refrigeration is generally used with some modification. Direct expansion units usually operate with a hot-gas bypass to prevent numerous on-off cycles, and secondary coolant may use aqueous ethylene glycol rather than chilled water. Heat is usually provided by electric heaters, but other energy sources can be used, including hot gas from the refrigeration.

The plant compartment is the heart of the growth chamber. The primary design objective, therefore, is to provide the most uniform, consistent, and regulated environmental conditions possible. Thus, airflow must be adequate to meet specified psychrometric conditions, but it is limited by the effects of high air speed on plant growth. As a rule, the average air speed in CERs is restricted to about 100 fpm.

To meet the uniform conditions required by a CER, conditioned air is normally moved through the space from bottom to top, although an increasing number of CERs use top-to-bottom airflow. There is no apparent difference in plant growth between horizontal, upward, or downward airflow when the speed is less than 175 fpm. Regardless of the method, a temperature gradient is certain to exist, and the design should keep the gradient as small as possible. Uniform airflow is more important than the direction of flow; thus, selection of properly designed diffusers or plenums with perforations is essential for achieving it.

The ducts or false sidewalls that direct air from the evaporator to the growing area should be small, but not so small that the noise increases appreciably more than acceptable building air duct noise. CER design should include some provision for cleaning the interior of the air ducts.

Air-conditioning equipment for relatively standard chambers provides temperatures that range from 45 to 90°F. Specialized CERs that require temperatures as low as −5°F need low-temperature refrigeration equipment and devices to defrost the evaporator without increasing the growing area temperature. Other chambers that require temperatures as high as 115°F need high-temperature components. The air temperature in the growing area must be controlled with the least possible variation about the set point. Temperature variation about the set point can be held to 0.5°F using solid-state controls, but in most existing facilities, the variation is 1 to 2°F.

The relative humidity in many CERs is simply an indicator of the existing psychrometric conditions and is usually between 50 and 80%, depending on the temperature. Relative humidity in the chamber can be increased by steam injection, misting, hot-water evaporators, and other conventional humidification methods. Steam injection causes the least temperature disturbance, and sprays or

Table 11 Input Power Conversion of Light Sources

Lamp Identification		Total Input Power, W	Radiation (400-700 nm), %	Radiation (400-850 nm), %	Other Radiation, %	Conduction and Convection, %	Ballast Loss, %
Incandescent	INC, 100A	100	7	15	75	10	0
Fluorescent							
Cool white	FCW	46	21	21	32	34	13
Cool white	FCW	225	19	19	34	35	12
Warm white	FWW	46	20	20	32	35	13
Plant growth A	PGA	46	13	13	35	39	13
Plant growth B	PGB	46	15	16	34	37	13
Infrared	FIR	46	2	9	39	39	13
Discharge							
Clear mercury	HG	440	12	13	61	17	9
Mercury deluxe	HG/DX	440	13	14	59	18	9
Metal halide	MH	460	27	30	42	15	13
High-pressure sodium	HPS	470	26	36	36	13	15
Low-pressure sodium	LPS	230	27	31	25	22	22

Note: Conversion efficiency is for lamps without luminaire. Values compiled from manufacturers' data, published information, and unpublished test data by R.W. Thimijan.

misting cause the greatest disturbance. Complete control of relative humidity requires dehumidification as well as humidification.

A typical humidity control includes a cold evaporator or steam injection to adjust the chamber air dew point. The air is then conditioned to the desired dry-bulb temperature by electric heaters, a hot-gas bypass evaporator, or a temperature-controlled evaporator. A dew point lower than about 40°F cannot be obtained with a cold plate dehumidifier because of icing. Dew points lower than 40°F usually require a chemical dehumidifier in addition to the cold evaporator.

Lighting Environmental Chambers

The type of light source and the number of lamps used in CERs are determined by the desired plant response. Traditionally, cool-white fluorescent plus incandescent lamps that produce 10% of the fluorescent illuminance are used. Nearly all illumination data are based on either cool-white or warm-white fluorescent, plus incandescent lamps. A number of fluorescent lamps have special phosphors hypothesized to be the spectral requirements of the plant. Some of these lamps are used in CERs, but there is little data to suggest that they are superior to cool-white and warm-white lamps. In recent years, high-intensity discharge lamps have been installed in CERs, either to obtain very high radiant flux densities, or to reduce the electrical load while maintaining a light level equal to that produced by the less efficient fluorescent-incandescent systems.

One method to design lighting for biological environments is to base light source output recommendations on photon flux density $\mu mol/(s \cdot m^2)$ between 400 and 700 nm, or, less frequently, as radiant flux density between 400 and 700 nm, or 400 and 850 nm. Rather than basing illuminance measurements on human vision, this enables comparisons between light sources as a function of plant photosynthetic potential. Table 9 shows constants for converting various measurement units to W/m^2. However, instruments that measure the 400 to 850 nm spectral range are generally not available, and some controversy exists about the effectiveness of 400 to 850 nm as compared to the 400 to 700 nm range in photosynthesis. The power conversion of various light sources is listed in Table 11.

The design requirements for plant growth lighting differ greatly from those for vision lighting. Plant growth lighting requires a greater degree of horizontal uniformity and, usually, higher light levels than vision lighting. In addition, plant growth lighting should have as much vertical uniformity as possible—a factor rarely important in vision lighting. Horizontal and vertical uniformity are much easier to attain with linear or broad sources, such as fluorescent lamps, than with point sources, such as HID lamps. Tables 12 and 13 show the type and number of lamps, mounting height, and spacing required to obtain several levels of incident energy. Since

Table 12 Approximate Mounting Height and Spacing of Luminaires in Greenhouses

Lamp and Wattage	Irradiation, W/ft²			
	0.6	1.1	2.2	4.4
	Height and Spacing, in.			
HPS (400 W)	118	90	63	39
LPS (180 W)	94	67	47	31
MH (400 W)	106	79	55	35

the data were taken directly under lamps with no reflecting wall surfaces nearby, the incident energy is perhaps one-half of what the plants would receive if the lamps had been placed in a small chamber with highly reflective walls.

Extended-life incandescents or traffic signal lamps, which have a much longer life, will lower lamp replacement requirements. These lamps have lower lumen output, but are nearly equivalent in the red portion of the spectrum. For safety, porcelain lamp holders and heat-resistant lamp wiring should be used. Lamps used for CER lighting include fluorescent lamps (usually 1500 mA), 250, 400, and occasionally 1000 W HPS and MH lamps, 180 W LPS lamps, and various sizes of incandescent lamps. In many installations, the abnormally short life of incandescent lamps is due to vibration from the lamp loft ventilation or from cooling fans. Increased incandescent lamp life under these conditions can be attained by using lamps constructed with a C9 filament.

Energy-saving lamps have approximately equal or slightly lower irradiance per input watt. Since the irradiance per lamp is lower, there is no advantage to using these lamps, except in tasks that can be accomplished with low light levels. Light output of all lamps declines with use, except perhaps for low-pressure sodium (LPS) lamps, which appear to maintain approximately constant output but require an increase in input power during use.

Fluorescent and metal halide designs should be based on 80% of the initial light level. Most CER lighting systems have difficulty maintaining a relatively constant light level over considerable periods of time. Combinations of MH and HPS lamps compound the problem, because the lumen depreciation of the two light sources is significantly different. Thus, over time, the spectral energy distribution at plant level will shift toward the HPS. Lumen output can be maintained in two ways: (1) individual lamps, or a combination of lamps, can be switched off initially and activated as the lumen output decreases; and (2) the oldest 25 to 33% of the lamps can be replaced periodically. Solid-state dimmer systems are commercially available only for low-wattage fluorescent lamps and for mercury lamps.

Table 13   Height and Spacing of Luminaires

Light Source	Radiant Flux Density, W/ft²						
	0.03	0.08	0.28	0.84	1.67	2.5	4.6
Fluorescent—Cool White							
40 W single 4 ft lamp, 3.2 klm							
Radiant power, W/m², 400 to 700 nm	0.3	0.9	2.9	8.8			
Illumination, klx	0.10	0.30	1.0	3.0			
Lamps per 100 ft²	1.1	3.3	11	33			
Distance from plants, in.	114	67	36	21			
40 W 2-lamp fixtures (4 ft), 6.4 klm							
Radiant power, W/m², 400 to 700 nm	0.3	0.9	2.9	8.8			
Illumination, klx	0.10	0.30	1.0	3.0			
Fixtures per 100 ft²	0.6	1.7	5.5	16.7			
Distance from plants, in.	161	94	51	30			
215 W 2-8 ft lamps, 31.4 klm							
Radiant power, W/m², 400 to 700 nm	0.3	0.9	2.9	8.8	17.6	23.5	49.0
Illumination, klx	0.10	0.30	1.0	3.0	6.0	8.0	16.7
Lamps per 100 ft²	0.1+	0.4	1.2	3.6	7.1	9.3	20
Distance from plants, in.	346	201	110	63	43	39	28
High-Intensity Discharge							
Mercury-1 400 W parabolic reflector							
Radiant power, W/m², 400 to 700 nm	0.28	0.84	2.80	8.39	16.8	22.4	46.6
Illumination, klx	0.1	0.32	1.1	3.2	6.4	8.6	18.0
Lamps per 100 ft²	0.2	0.5	1.6	4.8	9.3	13.0	27
Distance from plants, in.	299	173	94	55	39	31	24
Metal halide-1 400 W							
Radiant power, W/m², 400 to 700 nm	0.77	0.80	2.68	8.03	16.1	21.4	44.6
Illumination, klx	0.09	0.26	0.88	2.6	5.3	7.0	15.0
Lamps per 100 ft²	0.09	0.2	0.7	2.2	4.4	5.8	12.0
Distance from plants, in.	445	256	142	83	59	51	34
High-pressure sodium 400 W							
Radiant power, W/m², 400 to 700 nm	0.22	0.65	2.18	6.52	13.0	17.4	36.2
Illumination, klx	0.09	0.27	0.89	2.7	5.3	7.1	15.0
Lamps per 100 ft²	0.05	0.14	0.5	1.4	2.8	3.6	7.6
Distance from plants, in.	559	323	177	102	71	63	43
Low-pressure sodium 180 W							
Radiant power, W/m², 400 to 700 nm	0.26	0.79	2.64	7.93	15.9	21.1	44.0
Illumination, klx	0.14	0.41	1.4	4.1	8.3	11.0	23.0
Lamps per 100 ft²	0.08	0.24	0.8	2.4	4.9	6.5	13.6
Distance from plants, in.	421	244	134	9	55	47	33
Incandescent							
Incandescent 100 W							
Radiant power, W/m², 400 to 700 nm	0.14	0.41	1.38	4.14	8.28	11.0	23.0
Illumination, klx	0.033	0.10	0.33	1.0	2.0	2.7	5.6
Lamps per 100 ft²	0.5	1.6	5.2	15.8	32	42	87
Distance from plants, in.	165	94	51	30	21	18	13
Incandescent 150 W flood							
Radiant power, W/m², 400 to 700 nm	0.14	0.41	1.38	4.14	8.28	11.0	23.0
Illumination, klx	0.033	0.098	0.33	1.0	2.0	2.6	5.5
Lamps per 100 ft²	0.3	0.9	3.3	9.3	19.5	26	54
Distance from plants, in.	212	122	67	39	28	24	16
Incandescent-Hg 160 W							
Radiant power, W/m², 400 to 700 nm	0.14	0.41	1.38	4.14	8.28	11.0	23.0
Illumination, klx	0.050	0.15	0.50	1.5	3.0	4.0	8.3
Lamps per 100 ft²	0.7	2.0	6.9	20.4	42	56	111
Distance from plants, in.	146	83	47	26	18	16	11
Sunlight							
Radiant power, W per 100 ft²	2.0	6.2	20.5	61.7	124	164	714
Illumination, klx	0.054	0.16	0.54	1.6	3.2	4.3	8.9

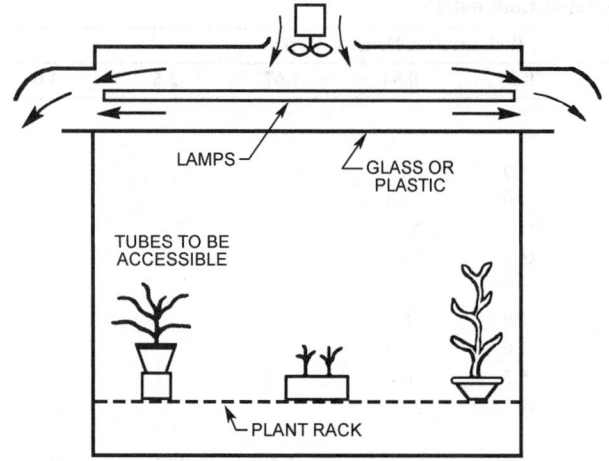

GROWTH CABINET – LIGHTS AIR-COOLED

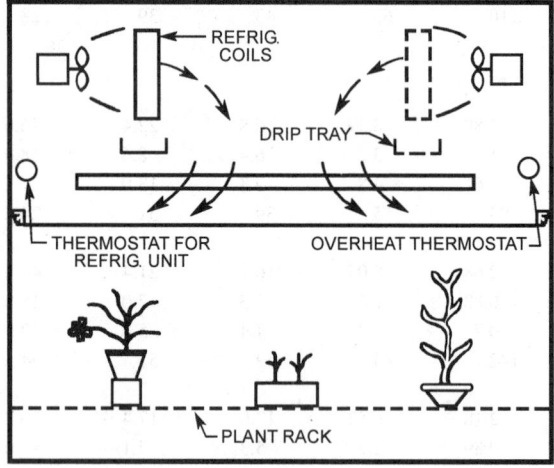

GROWTH CHAMBER – LIGHTS COOLED
BY REFRIGERATION

Fig. 11   Cooling Lamps in Growth Chambers

Large rooms, especially those constructed as an integral part of the building and retrofitted as CERs, rarely separate the lamps from the growing area with a transparent barrier. Rooms designed as CERs (at the time a building is constructed) and freestanding rooms or chambers usually separate the lamp from the growing area with a barrier of glass or rigid plastic. Light output from fluorescent lamps is a function of the temperature of the lamp. Thus, the barrier serves a two-fold purpose: (1) to maintain optimum lamp temperature when the growing area temperature is higher or lower than optimum, and (2) to reduce the thermal radiation entering the growing area. Fluorescent lamps should operate in an ambient temperature and airflow environment that will maintain the tube wall temperature at 104°F. Under most conditions, the light output of HID lamps is not affected by ambient temperature. The heat must be removed, however, to prevent high thermal radiation from causing adverse biological effects (see Figure 11).

Transparent glass barriers remove nearly all radiation from about 350 to 2500 nm. Rigid plastic is less effective than glass; however, the lighter weight and lower breakage risk of plastic makes it a popular barrier material. Ultraviolet is also screened by both glass and plastic (more by plastic). Special UV-transmitting plastic (which degrades rapidly) can be obtained if the biological process requires UV light. When irradiance is very high, especially from HID lamps

or large numbers of incandescent lamps or both, rigid plastic can soften from the heat and fall from the supports. Furthermore, very high irradiance and the resulting high temperatures can cause plastic to darken, which can increase the absorptivity and temperature enough to destroy it. Under these conditions, heat-resistant glass may be necessary. The lamp compartment and barrier absolutely require positive ventilation regardless of the light source, and the lamp loft should have limit switches that will shut down the lamps if the temperature rises to a critical level.

OTHER PLANT ENVIRONMENTAL FACILITIES

Plants may be held or processed in warehouse-type structures prior to sale or use in interior landscaping. Required temperatures range from slightly above freezing for cold storage of root stock and cut flowers, to 68 to 77°F for maintaining growing plants, usually in pots or containers. Provision must be made for venting fresh air to avoid CO_2 depletion.

Light duration must be controlled by a time clock. When they are in use, lamps and ballasts produce almost all the heat required in an insulated building. Ventilation and cooling may be required. Illumination levels depend on plant requirements. Table 14 shows approximate mounting heights for two levels of illumination. Luminaires mounted on chains permit lamp height to be adjusted to compensate for varying plant height.

The main concerns for interior landscape lighting are how it renders the color of plants, people, and furnishings, as well as how it meets the minimum irradiation requirements of plants. The temperature required for human occupancy is normally acceptable for plants. Light level and duration determine the types of plants that can be grown or maintained. Plants grow when exposed to higher levels, but do not survive below the suggested minimum. Plants may be grouped into three levels based on the following of irradiances:

Low (survival): A minimum light level of 0.07 W/ft^2 and a preferred level of 0.3 W/ft^2 irradiance for 8 to 12 h daily.

Medium (maintenance): A minimum of 0.3 W/ft^2 and a preferred level of 0.8 W/ft^2 irradiance for 8 to 12 h daily.

High (propagation): A minimum of 0.8 W/ft^2 and a preferred level of 2.2 W/ft^2 irradiance for 8 to 12 h daily.

Fluorescent (warm-white), metal halide, or incandescent lighting is usually chosen for public places. Table 13 lists the irradiance of various light sources.

Table 14   Mounting Height for Luminaires in Storage Areas

	Survival = 0.3 W/ft^2		Maintenance = 0.8 W/ft^2	
	Distance, ft	lux	Distance, ft	lux
Fluorescent (F)				
FCW two 40 W	3.0	1000	2.5	3000
FWW	3.0	1000	2.5	3000
FCW two 215 W	9.2	1000	5.2	3000
Discharge (HID)				
MH 400 W	10.8	800	6.6	2400
HPS 400 W	14.8	800	8.2	2400
LPS 180 W	11.2	1300	3.9	4000
Incandescent (INC)				
INC 160 W	4.3	350	1.0	1000
INC-HG 160 W	3.9	500	5.2	1500
DL	—	500	—	1500

REFERENCES

AGCIH. 1998. *Industrial ventilation: A manual of recommended practice*, 23rd ed. American Conference of Governmental Industrial Hygienists, Cincinnati, OH.

ASAE. 1997. Guidelines for use of thermal insulation in agricultural buildings. *Standard* S401.2. American Society of Agricultural Engineers, St. Joseph, MI.

Barber, E.M., J.A. Dosman, C.S. Rhodes, G.I. Christison, and T.S. Hurst. 1993. Carbon dioxide as an indicator of air quality in swine buildings. Proceedings of Third International Livestock Environment Symposium. American Society of Agricultural Engineers, St. Joseph, MI.

BESS Lab. 1997. Agricultural ventilation fans—Performance and efficiencies. Bioenvironmental and Structural Systems Laboratory, Dept. of Agricultural Engineering, Univ. of Illinois at Urbana-Champaign.

Bond, T.E., H.H. Heitman, Jr., and C.F. Kelly. 1965. Effect of increased air velocities on heat and moisture loss and growth of swine. *Transactions of the ASAE* 8(2):167-69; 174.

Christianson, L.L. and R.L. Fehr. 1983. Ventilation—Energy and economics. In *Ventilation of agricultural structures*, pp. 335-49. American Society of Agricultural Engineers, St. Joseph, MI.

Clough, G. 1982. Environmental effects on animals used in biomedical research. *Biol. Rev.* 57:487-523.

Donham, J.K., P. Haglind, Y. Peterson, R. Rylander, and L. Belin. 1989. Environmental and health studies of workers in Swedish swine confinement buildings. *British Journal of Industrial Medicine* 40:31-37.

Gordon, C.J., P. Becker, and J.S. Ali. 1997. Behavioural thermoregulatory responses of single- and group-housed mice. Neurotoxicology Division, National Health and Environmental Effects Research Laboratory, U.S. Environmental Protection Agency, Research Triangle Park, NC.

Hasenau, J.J., R.B. Baggs, and A.L. Kraus. 1993. Microenvironments in microisolator cages using BALB/c and CD-1 mice. *Contemporary Topics* 32(1):11-16.

Hellickson, M.A. and J.N. Walker, eds. 1983. Ventilation of agricultural structures. ASAE *Monograph* No. 6. American Society of Agricultural Engineers, St. Joseph, MI.

ILAR. 1996. *Guide for the care and use of laboratory animals.* National Institutes of Health, Bethesda, MD.

Lindsey, J.R., M.W. Conner, and H.J. Baker. 1978. Physical, chemical and microbial factors affecting biologic response. *Laboratory Animal Housing*, pp. 31-43, Institute of Laboratory Animal Resources, National Academy of Sciences, Washington, D.C.

Maghirang, R.G., G.L. Riskowski, L.L. Christianson, and P.C. Harrison. 1995. Development of ventilation rates and design information for laboratory animal facilities—Part 1 field study. *ASHRAE Transactions* 101(2):208-18.

McPherson, C. 1975. Why be concerned about the ventilation requirements of experimental animals. *ASHRAE Transactions* 81(2):539-41.

Memarzadeh, F. 1998. *Design handbook on animal research facilities using static microisolators*, Volumes I and II. National Institutes of Health. Bethesda, MD.

Moreland, A.F. 1975. Characteristics of the research animal bioenvironment. *ASHRAE Transactions* 81(2):542-48.

Mount, L.E. and I.B. Start. 1980. A note on the effects of forced air movement and environmental temperature on weight gain in the pig after weaning. *Animal Production* 30(2):295.

MWPS. 1989. Natural ventilating systems for livestock housing and heating. MidWest Plan Service, Ames, IA.

MWPS. 1990. Cooling and tempering air for livestock housing. MidWest Plan Service, Ames, IA.

MWPS. 1990. Mechanical ventilating systems for livestock housing. MidWest Plan Service, Ames, IA.

Ni, J., A.J. Heber, T.T. Lim, R.K. Duggirala, B.L. Haymore, and C.A. Diehl. 1998a. Ammonia emission from a tunnel-ventilated swine finishing building. ASAE *Paper* 984051. American Society of Agricultural Engineers, St. Joseph, MI.

Ni, J., A.J. Heber, T.T. Lim, R.K. Duggirala, B.L. Haymore, and C.A. Diehl. 1998b. Emissions of hydrogen sulfide from a mechanically-ventilated swine grow-finish unit. ASAE *Paper* 984050. American Society of Agricultural Engineers, St. Joseph, MI.

Person, H.L., L.D. Jacobson, and K.A. Jordan. 1979. Effect of dirt, louvers and other attachments on fan performance. *Transactions of ASAE* 22(3):612-16.

Riskowski, G.L. and D.S. Bundy. 1988. Effects of air velocity and temperature on weanling pigs. *Livestock environment* III. Proceedings of the Third International Livestock Environment Symposium. American Society of Agricultural Engineers, St. Joseph, MI.

Riskowski, G.L., S.E. Ford, and K.O. Mankell. 1998. Laboratory measurements of wind effects on ridge vent performance. *ASHRAE Transactions* 104(1).

Riskowski, G.L., R.G. Maghirang, and W. Wang. 1996. Development of ventilation rates and design information for laboratory animal facilities—Part 2 laboratory tests. *ASHRAE Transactions* 102(2):195-209.

Weiss, J., G.T. Taylor, and W. Nicklas. 1991. Ammonia concentrations in laboratory rat cages under various housing conditions. American Association for Laboratory Animal Science, Cordova, TN.

Zhang, Y. 1994. *Swine building ventilation.* Prairie Swine Centre, Saskatoon, Saskatchewan, Canada.

Zhang, Y. and E.M. Barber. 1995. Air leakage and ventilation effectiveness for confinement livestock housing. *Transactions of the ASAE* 38(5):1501-04.

Zhang, Y., L.L. Christianson, G.L. Riskowski, B. Zhang, G. Taylor, H.W. Gonyou, and P.C. Harrison. 1992. A survey on laboratory rat environments. *ASHRAE Transactions* 98(2):247-53.

BIBLIOGRAPHY

ANIMALS

Handbooks and Proceedings

Albright, L.D. 1990. *Environment control for animals and plants, with computer applications.* American Society of Agricultural Engineers, St. Joseph, MI.

ASAE. 1982. *Dairy housing* II. Second National Dairy Housing Conference Proceedings.

ASAE. 1982. *Livestock environment* II. Second International Livestock Environment Symposium.

ASAE. 1988. *Livestock environment* III. Proceedings of the Third International Livestock Environment Symposium.

ASAE. 1993. *Livestock environment* IV. Proceedings of the Fourth International Livestock Environment Symposium.

ASAE. 1993. Design of ventilation systems for livestock and poultry shelters. *Standard* EP270.5.

Curtis, S.E. 1983. *Environmental management in animal agriculture.* Iowa State University Press, Ames, IA.

Curtis, S.E., ed. 1988. Guide for the care and use of agricultural animals in agricultural research and teaching. Consortium for Developing a Guide for the Care and Use of Agricultural Animals in Agricultural Research and Teaching, 309 W. Clark Street, Champaign, IL 61820.

Hahn, G.L. 1985. Management and housing of farm animals in hot environments. In *Stress physiology in livestock*, Vol. II, pp. 151-76. CRC Press, Boca Raton, FL.

HEW. 1978. Guide for the care and use of laboratory animals. Publication No. (NIH)78-23. U.S. Department of Health, Education and Welfare, Washington, D.C.

Rechcigl, M., Jr., ed. 1982. *Handbook of agricultural productivity.* Vol. II, *Animal productivity.* CRC Press, Boca Raton, FL.

Straub, H.E. 1989. *Building systems: Room air and air contaminant distribution.* ASHRAE.

Air Cooling

Canton, G.H., D.E. Buffington, and R.J. Collier. 1982. Inspired-air cooling for dairy cows. *Transactions of ASAE* 25(3):730-34.

Hahn, G.L. and D.D. Osburn. 1969. Feasibility of summer environmental control for dairy cattle based on expected production losses. *Transactions of ASAE* 12(4):448-51.

Hahn, G.L. and D.D. Osburn. 1970. Feasibility of evaporative cooling for dairy cattle based on expected production losses. *Transactions of ASAE* 12(3):289-91.

Heard, L., D. Froelich, L. Christianson, R. Woerman, and R. Witmer. 1986. Snout cooling effects on sows and litters. *Transactions of ASAE* 29(4):1097-1101.

Morrison, S.R., M. Prokop, and G.P. Lofgreen. 1981. Sprinkling cattle for heat stress relief: Activation, temperature, duration of sprinkling, and pen area sprinkled. *Transactions of ASAE* 24(5):1299-1300.

Timmons, M.B. and G.R. Baughman. 1983. Experimental evaluation of poultry mist-fog systems. *Transactions of ASAE* 26(1):207-10.

Wilson, J.L., H.A. Hughes, and W.D. Weaver, Jr. 1983. Evaporative cooling with fogging nozzles in broiler houses. *Transactions of ASAE* 26(2): 557-61.

Air Pollution in Buildings

ACGIH. 1992. *1992-1993 Threshold limit values for chemical substances and physical agents and biological exposure indices.* American Conference of Governmental Industrial Hygienists, Cincinnati, OH.

Avery, G.L., G.E. Merva, and J.B. Gerrish. 1975. Hydrogen sulfide production in swine confinement units. *Transactions of ASAE* 18(1):149.

Bundy, D.S. and T.E. Hazen. 1975. Dust levels in swine confinement systems associated with different feeding methods. *Transactions of ASAE* 18(1): 137.

Deboer, S. and W.D. Morrison. 1988. *The effects of the quality of the environment in livestock buildings on the productivity of swine and safety of humans—A literature review.* Department of Animal and Poultry Science, University of Guelph, Ontario, Canada.

Grub, W., C.A. Rollo, and J.R. Howes. 1965. Dust problems in poultry environment. *Transactions of ASAE* 8(3):338.

Effects of Environment on Production and Growth of Animals

Cattle

Anderson, J.F., D.W. Bates, and K.A. Jordan. 1978. Medical and engineering factors relating to calf health as influenced by the environment. *Transactions of ASAE* 21(6):1169.

Garrett, W.N. 1980. Factors influencing energetic efficiency of beef production. *Journal of Animal Science* 51(6):1434.

Gebremedhin, K.G., C.O. Cramer, and W.P. Porter. 1981. Predictions and measurements of heat production and food and water requirements of Holstein calves in different environments. *Transactions of ASAE* 24(3): 715.

Holmes, C.W. and N.A. McLean. 1975. Effects of air temperature and air movement on the heat produced by young Friesian and Jersey calves, with some measurements of the effects of artificial rain. *New Zealand Journal of Agricultural Research* 18(3):277.

Morrison, S.R., G.P. Lofgreen, and R.L. Givens. 1976. Effect of ventilation rate on beef cattle performance. *Transactions of ASAE* 19(3):530.

General

Hahn, G.L. 1982. Compensatory performance in livestock: Influences on environmental criteria. Proceedings of the Second International Livestock Environment Symposium. American Society of Agricultural Engineers, St. Joseph, MI.

Hahn, G.L. 1981. Housing and management to reduce climatic impacts on livestock. *Journal of Animal Science* 52(1):175-86.

Pigs

Boon, C.R. 1982. The effect of air speed changes on the group postural behaviour of pigs. *Journal of Agricultural Engineering Research* 27(1):71-79.

Christianson, L.L., D.P. Bane, S.E. Curtis, W.F. Hall, A.J. Muehling, and G.L. Riskowski. 1989. *Swine care guidelines for pork producers using environmentally controlled housing.* National Pork Producers Council, Des Moines, IA.

Close, W.H., L.E. Mount, and I.B. Start. 1971. The influence of environmental temperature and plane of nutrition on heat losses from groups of growing pigs. *Animal Production* 13(2):285.

Driggers, L.B., C.M. Stanislaw, and C.R. Weathers. 1976. Breeding facility design to eliminate effects of high environmental temperatures. *Transactions of ASAE* 19(5):903.

McCracken, K.J. and R. Gray. 1984. Further studies on the heat production and affective lower critical temperature of early-weaned pigs under commercial conditions of feeding and management. *Animal Production* 39:283-90.

Nienaber, J.A. and G.L. Hahn. 1988. Environmental temperature influences on heat production of ad-lib-fed nursery and growing-finishing swine. *Livestock environment* III. Proceedings of the Third International Livestock Environment Symposium. American Society of Agricultural Engineers, St. Joseph, MI.

Phillips, P.A., B.A. Young, and J.B. McQuitty. 1982. Liveweight, protein deposition and digestibility responses in growing pigs exposed to low temperature. *Canadian Journal of Animal Science* 62:95-108.

Poultry

Buffington, D.E., K.A. Jordan, W.A. Junnila, and L.L. Boyd. 1974. Heat production of active, growing turkeys. *Transactions of ASAE* 17(3):542.

Carr, L.E., T.A. Carter, and K.E. Felton. 1976. Low temperature brooding of broilers. *Transactions of ASAE* 19(3):553.

Riskowski, G.L., J.A. DeShazer, and F.B. Mather. 1977. Heat losses of white leghorn laying hens as affected by intermittent lighting schedules. *Transactions of ASAE* 20(4):727-31.

Siopes, T.D., M.B. Timmons, G.R. Baughman, and C.R. Parkhurst. 1983. The effect of light intensity on the growth performance of male turkeys. *Poultry Science* 62:2336-42.

Sheep

Schanbacher, B.D., G.L. Hahn, and J.A. Nienaber. 1982. Photoperiodic influences on performance of market lambs. Proceedings of the Second International Livestock Environment Symposium. American Society of Agricultural Engineers, St. Joseph, MI.

Vesely, J.A. 1978. Application of light control to shorten the production cycle in two breeds of sheep. *Animal Production* 26(2):169.

Modeling and Analysis

Albright, L.D. and N.R. Scott. 1974a. An analysis of steady periodic building temperature variations in warm weather—Part I: A mathematical model. *Transactions of ASAE* 17(1):88-92, 98.

Albright, L.D. and N.R. Scott. 1974b. An analysis of steady periodic building temperature variations in warm weather—Part II: Experimental verification and simulation. *Transactions of ASAE* 17(1):93-98.

Albright, L.D. and N.R. Scott. 1977. Diurnal temperature fluctuations in multi-air spaced buildings. *Transactions of ASAE* 20(2):319-26.

Bruce, J.M. and J.J. Clark. 1979. Models of heat production and critical temperature for growing pigs. *Animal Production* 28:353-69.

Christianson, L.L. and H.A. Hellickson. 1977. Simulation and optimization of energy requirements for livestock housing. *Transactions of ASAE* 20(2):327-35.

Ewan, R.C. and J.A. DeShazer. 1988. Mathematical modeling the growth of swine. *Livestock environment* III. Proceedings of the Third International Livestock Environment Symposium. American Society of Agricultural Engineers, St. Joseph, MI.

Hellickson, M.L., K.A. Jordan, and R.D. Goodrich. 1978. Predicting beef animal performance with a mathematical model. *Transactions of ASAE* 21(5):938-43.

Teter, N.C., J.A. DeShazer, and T.L. Thompson. 1973. Operational characteristics of meat animals—Part I: Swine; Part II: Beef; Part III: Broilers. *Transactions of ASAE* 16:157-59; 740-42; 1165-67.

Timmons, M.B. 1984. Use of physical models to predict the fluid motion in slot-ventilated livestock structures. *Transactions of ASAE* 27(2):502-07.

Timmons, M.B., L.D. Albright, R.B. Furry, and K.E. Torrance. 1980. Experimental and numerical study of air movement in slot-ventilated enclosures. *ASHRAE Transactions* 86(1):221-40.

Shades for Livestock

Bedwell, R.L. and M.D. Shanklin. 1962. Influence of radiant heat sink on thermally-induced stress in dairy cattle. Missouri Agricultural Experiment Station *Research Bulletin* No. 808.

Bond, T.E., L.W. Neubauer, and R.L. Givens. 1976. The influence of slope and orientation of effectiveness of livestock shades. *Transactions of ASAE* 19(1):134-37.

Roman-Ponce, H., W.W. Thatcher, D.E. Buffington, C.J. Wilcox, and H.H. VanHorn. 1977. Physiological and production responses of dairy cattle to a shade structure in a subtropical environment. *Journal of Dairy Science* 60(3):424.

Transport of Animals

Ashby, B.H., D.G. Stevens, W.A. Bailey, K.E. Hoke, and W.G. Kindya. 1979. *Environmental conditions on air shipment of livestock.* USDA, SEA, Advances in Agricultural Technology, Northeastern Series No. 5.

Ashby, B.H., A.J. Sharp, T.H. Friend, W.A. Bailey, and M.R. Irwin. 1981. Experimental railcar for cattle transport. *Transactions of ASAE* 24(2): 452.

Ashby, B.H., H. Ota, W.A. Bailey, J.A. Whitehead, and W.G. Kindya. 1980. Heat and weight loss of rabbits during simulated air transport. *Transactions of ASAE* 23(1):162.

Grandin, R. 1988. *Livestock trucking guide.* Livestock Conservation Institute, Madison, WI.

Scher, S. 1980. Lab animal transportation receiving and quarantine. *Lab Animal* 9(3):53.

Stermer, R.A., T.H. Camp, and D.G. Stevens. 1982. Feeder cattle stress during handling and transportation. *Transactions of ASAE* 25(1):246-48.

Stevens, D.G., G.L. Hahn, T.E. Bond, and J.H. Langridge. 1974. *Environmental considerations for shipment of livestock by air freight.* USDA, APHIS (May).

Stevens, D.G. and G.L. Hahn. 1981. Minimum ventilation requirement for the air transportation of sheep. *Transactions of ASAE* 24(1):180.

Laboratory Animals

NIH. 1978. Laboratory animal housing. Proceedings of a symposium held at Hunt Valley, MD, September 1976. National Academy of Sciences, Washington, D.C.

McSheehy, T. 1976. *Laboratory animal handbook* 7—Control of the animal house environment. Laboratory Animals, Ltd.

Soave, O., W. Hoag, et al. 1980. The laboratory animal data bank. *Lab Animal* 9(5):46.

Ventilation Systems

Albright, L.D. 1976. Air flows through hinged-baffle, slotted inlets. *Transactions of ASAE* 19(4):728, 732, 735.

Albright, L.D. 1978. Air flow through baffled, center-ceiling, slotted inlets. *Transactions of ASAE* 21(5):944-47, 952.

Albright, L.D. 1979. Designing slotted inlet ventilation by the systems characteristic technique. *Transactions of ASAE* 22(1):158.

Pohl, S.H. and M.A. Hellickson. 1978. Model study of five types of manure pit ventilation systems. *Transactions of ASAE* 21(3):542.

Randall, J.M. 1980. Selection of piggery ventilation systems and penning layouts based on the cooling effects of air speed and temperature. *Journal of Agricultural Engineering Research* 25(2):169-87.

Randall, J.M. and V.A. Battams. 1979. Stability criteria for air flow patterns in livestock buildings. *Journal of Agricultural Engineering Research* 24(4):361-74.

Timmons, M.B. 1984. Internal air velocities as affected by the size and location of continuous inlet slots. *Transactions of ASAE* 27(5):1514-17.

Natural Ventilation

Bruce, J.M. 1982. Ventilation of a model livestock building by thermal buoyancy. *Transactions of ASAE* 25(6):1724-26.

Jedele, D.G. 1979. Cold weather natural ventilation of buildings for swine finishing and gestation. *Transactions of ASAE* 22(3):598-601.

Timmons, M.B., R.W. Bottcher, and G.R. Baughman. 1984. Nomographs for predicting ventilation by thermal buoyancy. *Transactions of ASAE* 27(6):1891-93.

PLANTS

Greenhouse and Plant Environment

Aldrich, R.A. and J.W. Bartok. 1984. *Greenhouse engineering.* Department of Agricultural Engineering, University of Connecticut, Storrs, CT.

ASAE. 1992. Guidelines for measuring and reporting environmental parameters for plant experiments in growth chambers. *Engineering Practice* EP411.2-92. American Society of Agricultural Engineers, St. Joseph, MI.

Clegg, P. and D. Watkins. 1978. *The complete greenhouse book.* Garden Way Publishing, Charlotte, VT.

Downs, R.J. 1975. *Controlled environments for plant research.* Columbia University Press, New York.

Langhans, R.W. 1985. *Greenhouse management.* Halcyon Press, Ithaca, NY.

Mastalerz, J.W. 1977. *The greenhouse environment.* John Wiley and Sons, New York.

Nelson, P.V. 1978. *Greenhouse operation and management.* Reston Publishing Co., Reston, VA.

Pierce, J.H. 1977. *Greenhouse grow how.* Plants Alive Books, Seattle, WA.

Riekels, J.W. 1977. *Hydroponics.* Ontario Ministry of Agriculture and Food, Fact Sheet No. 200-24, Toronto, Ontario, Canada.

Riekels, J.W. 1975. *Nutrient solutions for hydroponics.* Ontario Ministry of Agriculture and Food, Fact Sheet No. 200-532, Toronto, Ontario, Canada.

Sheldrake, R., Jr. and J.W. Boodley. *Commercial production of vegetable and flower plants.* Research Park, 1B-82, Cornell University, Ithaca, NY.

Tibbitts, T.W. and T.T. Kozlowski, eds. 1979. *Controlled environment guidelines for plant research.* Academic Press, New York.

Light and Radiation

Bickford, E.D. and S. Dunn. 1972. *Lighting for plant growth.* Kent State University Press, Kent, OH.

Carpenter, G.C. and L.J. Mousley. 1960. The artificial illumination of environmental control chambers for plant growth. *Journal of Agricultural Engineering Research* [England] 5:283.

Campbell, L.E., R.W. Thimijan, and H.M. Cathey. 1975. Special radiant power of lamps used in horticulture. *Transactions of ASAE* 18(5):952.

Cathey, H.M. and L.E. Campbell. 1974. Lamps and lighting: A horticultural view. *Lighting Design & Application* 4:41.

Cathey, H.M. and L.E. Campbell. 1979. Relative efficiency of high- and low-pressure sodium and incandescent filament lamps used to supplement natural winter light in greenhouses. *Journal of the American Society for Horticultural Science* 104(6):812.

Cathey, H.M. and L.E. Campbell. 1980. Light and lighting systems for horticultural plants. *Horticultural Reviews* 11:491. AVI Publishing Co., Westport, CT.

Cathey, H.M., L.E. Campbell, and R.W. Thimijan. 1978. Comparative development of 11 plants grown under various fluorescent lamps and different duration of irradiation with and without additional incandescent lighting. *Journal of the American Society for Horticultural Science* 103:781.

Hughes, J., M.J. Tsujita, and D.P. Ormrod. 1979. *Commercial applications of supplementary lighting in greenhouses.* Ontario Ministry of Agriculture and Food, Fact Sheet No. 290-717, Toronto, Ontario, Canada.

Kaufman, J.E., ed. 1981. IES *Lighting handbook*, Application Volume. IES, New York.

Kaufman, J.E., ed. 1981. IES *Lighting handbook*, Reference Volume. IES, New York.

Robbins, F.V. and C.K. Spillman. 1980. Solar energy transmission through two transparent covers. *Transactions of ASAE* 23(5).

Sager, J.C., J.L. Edwards, and W.H. Klein. 1982. Light energy utilization efficiency for photosynthesis. *Transactions of ASAE* 25(6):1737-46.

Photoperiod

Heins, R.D., W.H. Healy, and H.F. Wilkens. 1980. Influence of night lighting with red, far red, and incandescent light on rooting of chrysanthemum cuttings. *HortScience* 15:84.

Carbon Dioxide

Bailey, W.A., et al. 1970. CO_2 systems for growing plants. *Transactions of ASAE* 13(2):63.

Gates, D.M. 1968. Transpiration and leaf temperature. *Annual Review of Plant Physiology* 19:211.

Holley, W.D. 1970. CO_2 enrichment for flower production. *Transactions of ASAE* 13(3):257.

Kretchman, J. and F.S. Howlett. 1970. Enrichment for vegetable production. *Transactions of ASAE* 13(2):252.

Tibbitts, T.W., J.C. McFarlane, D.T. Krizek, W.L. Berry, P.A. Hammer, R.H. Hodgsen, and R.W. Langhans. 1977. Contaminants in plant growth chambers. *Horticulture Science* 12:310.

Wittwer, S.H. 1970. Aspects of CO_2 enrichment for crop production. *Transactions of ASAE* 13(2):249.

Heating, Cooling, and Ventilation

ASAE. 1990. Heating, ventilating and cooling greenhouses. *Engineering Practice* EP406.1-90. American Society of Agricultural Engineers, St. Joseph, MI.

Albright, L.D., I. Seginer, L.S. Marsh, and A. Oko. 1985. In situ thermal calibration of unventilated greenhouses. *Journal of Agricultural Engineering Research* 31(3):265-81.

Buffington, D.E. and T.C. Skinner. 1979. Maintenance guide for greenhouse ventilation, evaporative cooling, and heating systems. Publication No. AE-17. Department of Agricultural Engineering, University of Florida, Gainesville, FL.

Duncan, G.A. and J.N. Walker. 1979. Poly-tube heating ventilation systems and equipment. Publication No. AEN-7. Agricultural Engineering Department, University of Kentucky, Lexington, KY.

Elwell, D.L., M.Y. Hamdy, W.L. Roller, A.E. Ahmed, H.N. Shapiro, J.J. Parker, and S.E. Johnson. 1985. *Soil heating using subsurface pipes.* Department of Agricultural Engineering, Ohio State University, Columbus, OH.

Heins, R. and A. Rotz. 1980. Plant growth and energy savings with infrared heating. *Florists' Review* (October):20.

NGMA. 1989. Greenhouse heat loss. National Greenhouse Manufacturers' Association, Taylors, SC.

NGMA. 1989. Standards for ventilating and cooling greenhouses. National Greenhouse Manufacturers' Association, Taylors, SC.

NGMA. 1993. Recommendation for using insect screens in greenhouse structures. National Greenhouse Manufacturers' Association, Taylors, SC.

Roberts, W.J. and D. Mears. 1984. *Floor heating and bench heating extension bulletin for greenhouses.* Department of Agricultural and Biological Engineering, Cook College, Rutgers University, New Brunswick, NJ.

Roberts, W.J. and D. Mears. 1984. *Heating and ventilating greenhouses.* Department of Agricultural and Biological Engineering, Cook College, Rutgers University, New Brunswick, NJ.

Roberts, W.J. and D.R. Mears. 1979. *Floor heating of greenhouses.* Miscellaneous Publication, Rutgers University, New Brunswick, NJ.

Silverstein, S.D. 1976. Effect of infrared transparency on heat transfer through windows: A clarification of the greenhouse effect. *Science* 193:229 (see also Rogers).

Walker, J.N. and G.A. Duncan. 1975. Greenhouse heating systems. Publication No. AEN-31. Agricultural Engineering Department, University of Kentucky, Lexington, KY.

Walker, J.N. and G.A. Duncan. 1979. Greenhouse ventilation systems. Publication No. AEN-30. Agricultural Engineering Department, University of Kentucky, Lexington, KY.

Energy Conservation

Roberts, W.J., J.W. Bartok, Jr., E.E. Fabian, and J. Simpkins. 1985. *Energy conservation for commercial greenhouses.* NRAES-3. Department of Agricultural Engineering, Cornell University, Ithaca, NY.

Solar Energy Use

Albright, L.D., et al. 1980. Passive solar heating applied to commercial greenhouses. Publication No. 115, Energy in Protected Civilization. Acta Horticultura.

Cathey, H.M. 1980. Energy-efficient crop production in greenhouses. *ASHRAE Transactions* 86(2):455.

Duncan, G.A., J.N. Walker, and L.W. Turner. 1979. *Energy for greenhouses,* Part I: Energy Conservation. Publication No. AEES-16. College of Agriculture, University of Kentucky, Lexington, KY.

Duncan, G.A., J.N. Walker, and L.W. Turner. 1980. *Energy for greenhouses,* Part II: Alternative Sources of Energy. College of Agriculture, University of Kentucky, Lexington, KY.

Gray, H.E. 1980. Energy management and conservation in greenhouses: A manufacturer's view. *ASHRAE Transactions* 86(2):443.

Roberts, W.J. and D.R. Mears. 1980. Research conservation and solar energy utilization in greenhouses. *ASHRAE Transactions* 86(2):433.

Short, T.H., M.F. Brugger, and W.L. Bauerle. 1980. Energy conservation ideas for new and existing commercial greenhouses. *ASHRAE Transactions* 86(2):448.

DRYING AND STORING SELECTED FARM CROPS

CONTROL of moisture content and temperature during storage is critical to preserving the nutritional and economic value of farm crops as they move from the field to the market. Fungi (mold) and insects feed on poorly stored crops and reduce crop quality. Relative humidity and temperature affect mold and insect growth, which is reduced to a minimum if the crop is kept cooler than 50°F and if the relative humidity of the air in equilibrium with the stored crop is less than 60% (Figure 1).

Mold growth and spoilage are a function of elapsed storage time, temperature, and moisture content above critical values. The approximate allowable storage life for cereal grains is shown in Table 1. For example, corn at 60°F and 20% wet basis (w.b.) moisture has a storage life of about 25 days. If the corn is dried to 18% w.b. after 12 days, half of its storage life has elapsed. Thus, the remaining storage life at 60°F and 18% w.b. moisture content is 25 days, not 50 days.

Table 1 Approximate Allowable Storage Time (Days) for Cereal Grains

Moisture Content, % w.b.[a]	Temperature, °F					
	30	40	50	60	70	80
14	*	*	*	*	200	140
15	*	*	*	240	125	70
16	*	*	230	120	70	40
17	*	280	130	75	45	20
18	*	200	90	50	30	15
19	*	140	70	35	20	10
20	*	90	50	25	14	7
22	190	60	30	15	8	3
24	130	40	15	10	6	2
26	90	35	12	8	5	2
28	70	30	10	7	4	2
30	60	25	5	5	3	1

Table is based on composite of 0.5% maximum dry matter loss calculated on the basis of USDA research; *Transactions of ASAE* 333-337, 1972; and "Unheated Air Drying," Manitoba Agriculture Agdex 732-1, rev. 1986.

[a]Grain moisture content calculated as percent wet basis: (weight of water in a given amount of wet grain ÷ weight of the wet grain) × 100.

*Approximate allowable storage time exceeds 300 days.

Insects thrive in stored grain if the moisture content and temperature are not properly controlled. At low moisture contents and temperatures below 50°F, insects remain dormant or die.

Most farm crops must be dried to, and maintained at, a suitable moisture content. For most grains, a suitable moisture content is in the range of 12 to 15% w.b., depending on the specific crop, storage temperature, and length of storage. Oilseeds such as peanuts, sunflower seeds, and flaxseeds must be dried to a moisture content of 8 to 9% w.b. Grain stored for more than a year, grain that is damaged, and seed stock should be dried to a lower moisture content. Moisture levels above these critical values lead to the growth of fungi, which may produce toxic compounds such as aflatoxin.

The maximum yield of dry matter can be obtained by harvesting when the corn has dried in the field to an average moisture content of 26% w.b. However, for quality-conscious markets, the minimum damage occurs when corn is harvested at 21 to 22% w.b. Wheat can be harvested when it has dried to 20% w.b., but harvesting at these moisture contents requires expensive mechanical drying. Although field drying requires less expense than operating drying equipment, total cost may be greater because field losses generally increase as the moisture content decreases.

The price of grain to be sold through commercial market channels is based on a specified moisture content, with price discounts for moisture levels above the specified amount. These discounts compensate for the weight of excess water, cover the cost of water removal, and control the supply of wet grain delivered to market. Grain dried to below the base moisture content set by the market (15.0% w.b. for corn, 13.0% w.b. for soybeans, and 13.5% w.b. for wheat) is not generally sold at a premium; thus, the seller loses the opportunity to sell water for the price of grain.

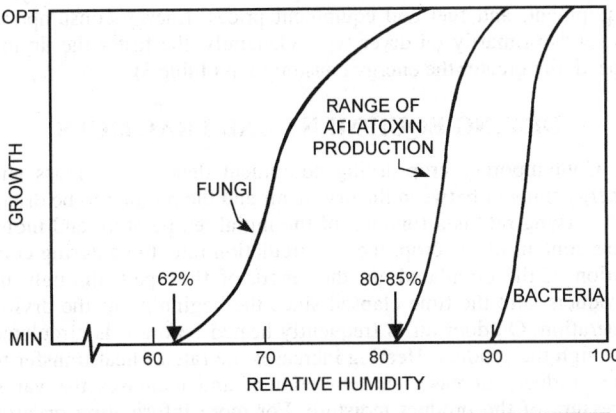

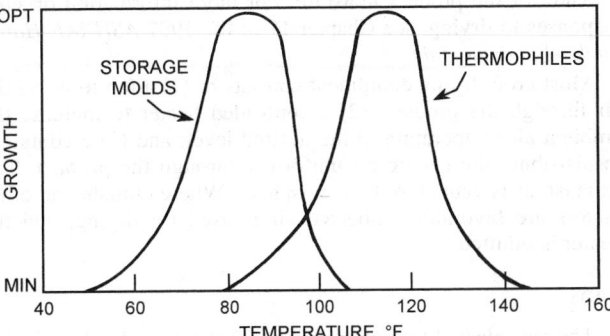

Fig. 1 Microbial Growth as Affected by Relative Humidity and Temperature
(University of Kentucky 1973)

The preparation of this chapter is assigned to TC 2.2, Plant and Animal Environment.

Table 2 Calculated Densities of Grains and Seeds Based on U.S. Department of Agriculture Data

	Bulk Density, lb/ft^3
Alfalfa	48.0
Barley	38.4
Beans, dry	48.0
Bluegrass	11.2 to 24.0
Canola	40.2 to 48.2
Clover	48.0
Corn[a]	
Ear, husked	28.0
Shelled	44.8
Cottonseed	25.6
Oats	25.6
Peanuts, unshelled	
Virginia type	13.6
Runner, Southeastern	16.8
Spanish	19.8
Rice, rough	36.0
Rye	44.8
Sorghum	40.0
Soybeans	48.0
Sudan grass	32.0
Sunflower	
Nonoil	19.3
Oilseed	25.7
Wheat	48.0

[a] 70 lb of husked ears of corn yield 1 bushel, or 56 lb of shelled corn. 70 lb of ears of corn occupy 2 volume bushels (2.5 ft^3).

Grain Quantity

The **bushel** is the common measure used for marketing grain in the United States. Most dryers are rated in bushels per hour for a specified moisture content reduction. The use of the bushel as a measure causes considerable confusion. A bushel is a volume measure equal to 1.244 ft^3. The bushel is used as a volume measure to estimate the holding capacity of bins, dryers, and other containers.

For buying and selling grain, for reporting production and consumption data, and for most other uses, the bushel weight is used. For example, the legal weight of a bushel is 56 lb for corn and 60 lb for wheat. When grain is marketed, bushels are computed as the load weight divided by the bushel weight. So, 56,000 lb of corn (regardless of moisture content) is 1000 bushels. Rice, grain sorghum, and sunflower are more commonly traded on the basis of the hundredweight (100 lb), a measure that does not connote volume. The relationship between bushel by volume and market bushel is the **bulk density** (listed for some crops in Table 2). For some crops, the market has defined a test weight parameter, lb/bu. Test weight is essentially the bulk density, with bushels and cubic feet related by the definition of 1 bushel = 1.244 cubic feet.

The terms **wet bushel** and **dry bushel** sometimes refer to the mass of grain before and after drying. For example, 56,000 lb of 25% moisture corn may be referred to as 1000 wet bushels or simply 1000 bushels. When the corn is dried to 15.5% moisture content (m.c.), only 49,704 lb or 49,704/56 = 888 bushels remain. Thus, a dryer rated on the basis of wet bushels (25% m.c.) shows a capacity 12.6% higher than if rated on the basis of dry bushels (15.5% m.c.).

The percent of weight lost due to water removed may be calculated by the following equation:

$$\text{Moisture shrink, \%} = \frac{M_o - M_f}{100 - M_f} \times 100$$

where

M_o = original or initial moisture content, wet basis

M_f = final moisture content, wet basis

Table 3 Estimated Corn Drying Energy Requirement

Dryer Type	Btu/lb of Water Removed
Unheated air	1000 to 1200
Low temperature	1200 to 1500
Batch-in-bin, continuous-flow in-bin	1500 to 2000
High temperature	
Air recirculating	1800 to 2200
Without air recirculating	2000 to 3000
Combination drying, dryeration	1400 to 1800

Note: Includes all energy requirements for fans and heat.

Applying the formula to drying a crop from 25% to 15%,

$$\text{Moisture shrink} = \frac{25 - 15}{100 - 15} \times 100 = 11.76\%$$

In this case, the moisture shrink is 11.76%, or an average 1.176% weight reduction for each percentage point of moisture reduction. The moisture shrink varies depending on the final moisture content. For example, the average shrink per point of moisture when drying from 20% to 10% is 1.111.

Economics

Producers generally have the choice of drying their grain on the farm before delivering it to market, or delivering wet grain with a price discount for excess moisture. The expense of drying on the farm includes both fixed and variable costs. Once a dryer is purchased, the costs of depreciation, interest, taxes, and repairs are fixed and minimally affected by volume of crops dried. The costs of labor, fuel, and electricity vary directly with the volume dried. Total drying costs vary widely, depending on the volume dried, the drying equipment, and fuel and equipment prices. Energy consumption depends primarily on dryer type. Generally, the faster the drying speed, the greater the energy consumption (Table 3).

DRYING EQUIPMENT AND PRACTICES

Contemporary crop-drying equipment depends on mass and energy transfer between the drying air and the product to be dried. The drying rate is a function of the initial temperature and moisture content of the crop, the air-circulation rate, the entering condition of the circulated air, the length of flow path through the products, and the time elapsed since the beginning of the drying operation. Outdoor air is frequently heated before it is circulated through the product. Heating increases the rate of heat transfer to the product, increases its temperature, and increases the vapor pressure of the product moisture. For more information on crop responses to drying, see Chapter 11 of the 1997 *ASHRAE Handbook—Fundamentals*.

Most crop-drying equipment consists of (1) a fan to move the air through the product, (2) a controlled heater to increase the ambient air temperature to the desired level, and (3) a container to distribute the drying air uniformly through the product. The exhaust air is vented to the atmosphere. Where climate and other factors are favorable, unheated air is used for drying, and the heater is omitted.

Fans

The fan selected for a given drying application should meet the same requirements important in any air-moving application. It must deliver the desired amount of air against the static resistance of the product in the bin or column, the resistance of the delivery system, and the resistance of the air inlet and outlet.

Foreign material in the grain can significantly change the required air pressure in the following ways:

- Foreign particles larger than the grain (straw, plant parts, and larger seeds) reduce airflow resistance. The airflow rate may be increased by 60% or more.
- Foreign particles smaller than the grain (broken grain, dust, and small seeds) increase the airflow resistance. The effect may be dramatic, decreasing the airflow rate by 50% or more.
- The method used to fill the dryer or the agitation or stirring of the grain after it is placed in the dryer can increase pressure requirements by up to 100%. In some grain, high moisture causes less pressure drop than does low moisture.

Vaneaxial fans are normally recommended when static pressures are less than 3 in. of water. Backward-curved centrifugal fans are commonly recommended when static pressures are higher than 4 in. of water column. Low-speed centrifugal fans operating at 1750 rpm perform well up to about 7 in. of water, and high-speed centrifugal fans operating at about 3500 rpm have the ability to develop static pressure up to about 10 in. of water. The in-line centrifugal fan consists of a centrifugal fan impeller mounted in the housing of an axial flow fan. A bell-shaped inlet funnels the air into the impeller. The in-line centrifugal fan operates at about 3450 rpm and has the ability to develop pressures up to 10 in. of water with 7.5 hp or larger fans.

After functional considerations are made, the initial cost of the dryer fan should be taken into account. Drying equipment has a low percentage of annual use in many applications, so the cost of dryer ownership per unit of material dried is sometimes greater than the energy cost of operation. The same considerations apply to other components of the dryer.

Heaters

Most crop dryer heaters are fueled by either natural gas, liquefied petroleum gas, or fuel oil, though some electric heaters are used. Dryers using coal, biomass (such as corn cobs, stubble, or wood), and solar energy have also been built.

Fuel combustion in crop dryers is similar to combustion in domestic and industrial furnaces. Heat is transferred to the drying air either indirectly, by means of a heat exchanger, or directly, by combining the combustion gases with the drying air. Direct combustion heating is generally limited to natural gas or liquefied petroleum (LP) gas heaters. Most grain dryers use direct combustion. Indirect heating is sometimes used in drying products such as hay because of its greater fire hazard.

Controls

In addition to the usual temperature controls for drying air, all heated air units must have safety controls similar to those found on space-heating equipment. These safety controls shut off the fuel in case of flame failure and stop the burner in case of overheating or excessive drying air temperatures. All controls should be set up to operate the machinery safely in the event of power failure.

SHALLOW-LAYER DRYING

Batch Dryers

The batch dryer cycles through the loading, drying, cooling, and unloading of the grain. Fans force hot air through columns (typically 12 in. wide) or layers (2 to 5 ft thick) of grain. Drying time depends on the type of grain and the amount of moisture to be removed. Some dryers circulate and mix the grain to prevent significant moisture content gradients from forming across the column. A circulation rate that is too fast or a poor selection of handling equipment may cause undue damage and loss of market quality. Batch dryers are suitable for farm operations and are often portable.

Continuous-Flow Dryers

This type of self-contained dryer passes a continuous stream of grain through the drying chamber. Some dryers use a second chamber to cool the hot, dry grain prior to storage. Handling and storage equipment must be available at all times to move grain to and from the dryers. These dryers have crossflow, concurrent flow, or counterflow designs.

Crossflow Dryers. A crossflow dryer is a column dryer that moves air perpendicular to the grain movement. These dryers commonly consist of two or more vertical columns surrounding the drying and cooling air plenums. The columns range in thickness from 8 to 16 in. Airflow rates range from 40 to 160 cfm per cubic foot of grain. The thermal efficiency of the drying process increases as column width increases and decreases as airflow rate increases. However, moisture uniformity and drying capacity increase as airflow rate increases and as column width decreases. Dryers are designed to obtain a desirable balance of airflow rate and column width for the expected moisture content levels and drying air temperatures. Performance is evaluated in terms of drying capacity, thermal efficiency, and dried product moisture uniformity.

As with the batch dryer, a moisture gradient forms across the column because the grain nearest the inside of the column is exposed to the driest air during the complete cycle. Several methods minimize the problem of uneven drying.

One method is to place in the columns turnflow devices that split the grain stream and move the inside half of the column to the outside and the outside half to the inside. Although effective, turnflow devices tend to plug if the grain is trashy. Under these conditions, a scalper/cleaner should be used to clean the grain before it enters the dryer.

Another method is to divide the drying chamber into sections and duct the hot air so that its direction through the grain is reversed in alternate sections. This method produces about the same effect as the turnflow method.

A third method is to divide the drying chamber into sections and reduce the drying air temperature in each section consecutively. This method is the least effective.

Rack-Type Dryers. In this special type of crossflow dryer, grain flows over alternating rows of heated air supply ducts and air exhaust ducts (Figure 2). This action mixes the grain and alternates exposure to relatively hot drying air and air cooled by previous contact with the grain, promoting moisture uniformity and equal exposure of the product to the drying air.

Concurrent-Flow Dryers. In the concurrent-flow dryer, grain and drying air move in the same direction in the drying chamber.

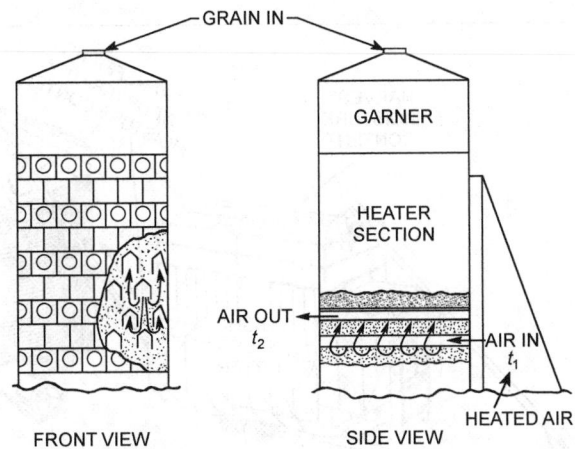

Fig. 2   Rack-Type Continuous-Flow Grain Dryer with Alternate Rows of Air Inlet and Outlet Ducts

The drying chamber is coupled to a counterflow cooling section. Thus, the hottest air is in contact with the wettest grain, allowing the use of higher drying air temperatures (up to 450°F). Rapid evaporative cooling in the wettest grain prevents the grain temperature from reaching excessive levels. Because higher drying air temperatures are used, the energy efficiency is better than that obtained with a conventional crossflow dryer. In the cooling section, the coolest air initially contacts the coolest grain. The combination of drying and cooling chambers results in lower thermal stresses in the grain kernels during drying and cooling and, thus, a higher-quality product.

Counterflow Dryers. The grain and drying air move in opposite directions in the drying chamber of this dryer. Counterflow is common for in-bin dryers. Drying air enters from the bottom of the bin and exits from the top. The wet grain is loaded from overhead, and floor sweep augers can be used to bring the hot, dry grain to a center sump, where it is removed by another auger. The travel of the sweep is normally controlled by moisture- or temperature-sensing elements.

A drying zone exists only in the lower layers of the grain mass and is truncated at its lower edge so that the grain being removed is not overdried. As a part of the counterflow process, the warm, saturated or near-saturated air leaving the drying zone passes through the cool incoming grain. Some energy is used to heat the cool grain, but some moisture may condense on the cool grain if the bed is deep and the initial grain temperature is low.

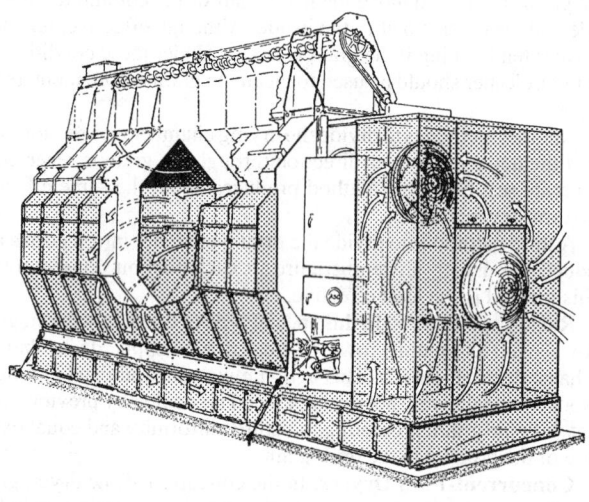

Fig. 3   Crop Dryer Recirculation Unit
(Courtesy Farm Fans, Inc.)

Reducing Energy Costs

Recirculation. In most commercially available continuous-flow dryers, optional ducting systems recycle some of the exhaust air from the drying and cooling chambers back to the inlet of the drying chamber (Figure 3). Systems vary, but most make it possible to recirculate all of the air from the cooling chamber and from the lower two-thirds of the drying chamber. The relative humidity of this recirculated air for most crossflow dryers is less than 50%. Energy savings of up to 30% can be obtained from a well-designed system.

Dryeration. This is another means of reducing energy consumption and improving grain quality. In this process, hot grain with a moisture content one or two percentage points above that desired for storage is removed from the dryer (Figure 4). The hot grain is placed in a dryeration bin, where it tempers without airflow for at least 4 to 6 h. After the first grain delivered to the bin has tempered, the cooling fan is turned on as additional hot grain is delivered to the bin. The air cools the grain and removes 1 to 2% of its moisture before the grain is moved to final storage. If the cooling rate equals the filling rate, cooling is normally completed about 6 h after the last hot grain is added. The crop cooling rate should equal the filling rate of the dryeration bin. A faster cooling rate cools the grain before it has tempered. A slower rate may result in spoilage, since the allowable storage time for hot, damp grain may be only a few days. The required airflow rate is based on dryer capacity and crop density. An airflow rate of 12 cfm for each bushel per hour (bu/h) of dryer capacity provides the cooling capacity to keep up with the dryer when it is drying corn that weighs 56 lb/bu. Recommended airflow rates for some crops are listed in Table 4.

Table 4   Recommended Airflow Rates for Dryeration

Crop	Weight, lb/bu	Recommended Dryeration Airflow Rate, cfm per bu/h
Barley	48	10
Corn	56	12
Durum	60	13
Edible beans	60	13
Flaxseeds	56	12
Millet	50	11
Oats	32	7
Rye	56	12
Sorghum	56	12
Soybeans	60	13
Nonoil sunflower seeds	24	5
Oil sunflower seeds	32	7
Hard red spring wheat	60	13

Note: Basic air volume is 12.9 ft³/lb.

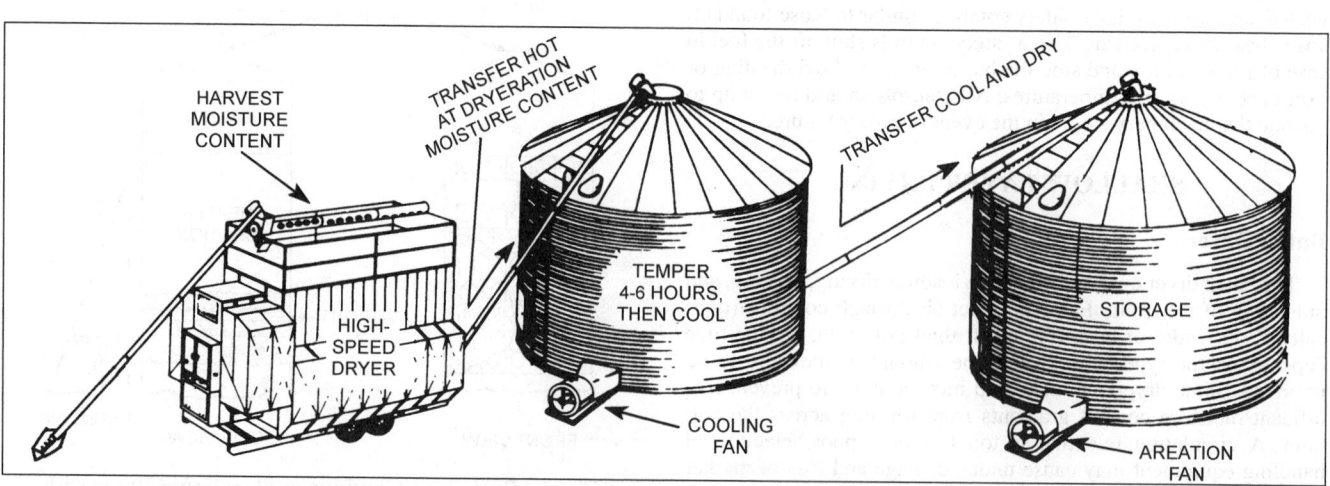

Fig. 4   Dryeration System Schematic

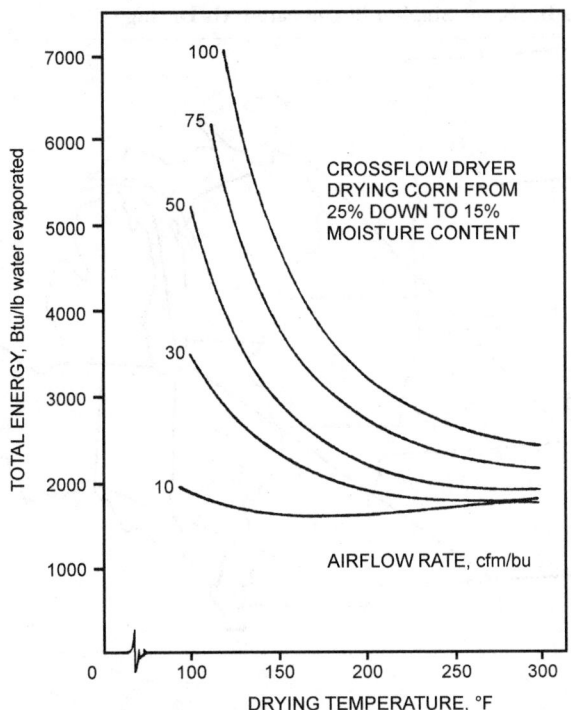

**Fig. 5 Energy Requirements of a Conventional Crossflow Dryer
as a Function of Drying Air Temperature and Airflow Rate**
(University of Nebraska)

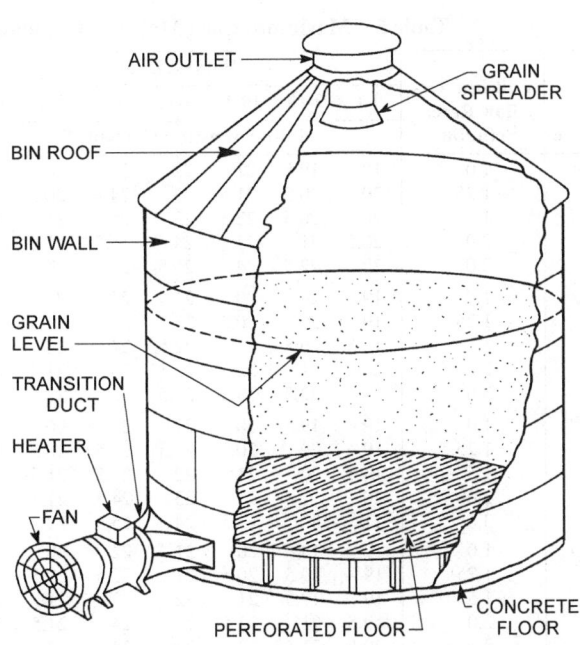

Fig. 6 Perforated Floor System for Bin Drying of Grain

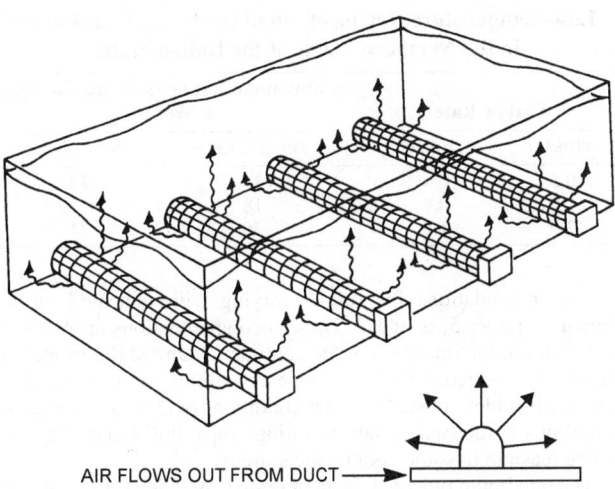

Fig. 7 Tunnel or Duct Air Distribution System

Combination Drying. This method was developed to improve drying thermal efficiency and corn quality. First, a high-temperature dryer dries the corn to 18 to 20% moisture content. Then it is transferred to a bin, where the full-bin drying system brings the moisture down to a safe storage level.

Dryer Temperature. For energy savings, operating temperatures of batch and continuous-flow dryers are usually set at the highest level that will not damage the product for its particular end use. The energy requirements of a conventional crossflow dryer as a function of drying air temperature and airflow rate are shown in Figure 5.

DEEP-BED DRYING

A deep-bed drying system can be installed in any structure that holds grain. Most grain storage structures can be designed or adapted for drying if a means of distributing the drying air uniformly through the grain is provided. A perforated floor (Figure 6) and duct systems placed on the floor of the bin (Figure 7) are the two most common means.

Perforations in the floor should have a total area of at least 10% of the floor area. A perforated floor distributes air more uniformly and offers less resistance to airflow than do ducts, but a duct system is less expensive for larger floor area systems. Ducts can be removed after the grain is removed, and the structure can be cleaned and used for other purposes. Ducts should not be spaced farther apart than one-half times the depth of the grain. The amount of perforated area or the duct length will affect airflow distribution uniformity.

Air ducts and tunnels that disperse air into the grain should be large enough to prevent the air velocity from exceeding 2000 fpm; slower speeds are desirable. Sharp turns, obstructions, or abrupt changes in duct size should be eliminated, as they cause pressure loss. Operating methods for drying grain in storage bins are (1) full-bin drying, (2) layer drying, (3) batch-in-bin drying, and (4) recirculating/continuous-flow bin drying.

Full-Bin Drying

Full-bin drying is generally performed with unheated air or air heated up to 10°F above ambient. A humidistat is frequently used to sense the humidity of the drying air and turn off the heater if the weather conditions are such that heated air would cause overdrying. A humidistat setting of 55% stops drying at approximately the 12% moisture level for most farm grains, assuming that the ambient relative humidity does not go below this point.

Airflow rate requirements for full-bin drying are generally calculated on the basis of cfm of air required per cubic foot or bushel of grain. The airflow rate recommendations depend on the weather conditions and on the type of grain and its moisture content. Airflow rate is important for successful drying. Because faster drying results from higher airflow rates, the highest economical airflow rate should be used. However, the cost of full-bin drying at high airflow rates may exceed the cost of using column dryers, or the electric power requirement may exceed the available capacity.

Table 5   Maximum Corn Moisture Contents, Wet Mass Basis, for Single-Fill Unheated Air Drying

Zone	Full-Bin Air-flow Rate, cfm/bu	Harvest Date 9-1	9-15	10-1	10-15	11-1	11-15	12-1
		Initial Moisture Content, %						
A	1.0	18	19.5	21	22	24	20	18
	1.25	20	20.5	21.5	23	24.5	20.5	18
	1.5	20	20.5	22.5	23	25	21	18
	2.0	20.5	21	23	24	25.5	21.5	18
	3.0	22	22.5	24	25.5	27	22	18
B	1.0	19	20	20	21	23	20	18
	1.25	19	20	20.5	21.5	24	20.5	18
	1.5	19.5	20.5	21	22.5	24	21	18
	2.0	20	21	22.5	23.5	25	21.5	18
	3.0	21	22.5	23.5	24.5	26	22	18
C	1.0	19	19.5	20	21	22	20	18
	1.25	19	20	20.5	21.5	22.5	20.5	18
	1.5	19.5	20	21	22	23.5	21.5	18
	2.0	20	21	22	23	24.5	21.5	18
	3.0	21	22	23.5	24.5	25.5	22	18
D	1.0	19	19.5	20	21	22	20	18
	1.25	19	19.5	20.5	21	22.5	20.5	18
	1.5	19	19.5	21	22	23	21	18
	2.0	19.5	21	21.5	23	24	21.5	18
	3.0	20.5	21.5	23	24	25	22	18

Midwest Plan Service, 1980.

Table 6   Minimum Airflow Rate for Unheated Air Low-Temperature Drying of Small Grains and Sunflower in the Northern Plains of the United States

Airflow Rate cfm/bu	cfm/ft³	Maximum Initial Moisture Content, % Wet Basis Small Grains	Sunflower
0.5	0.4	16	15
1.0	0.8	18	17
2.0	1.6	20	21

Recommendations for full-bin drying with unheated air are shown in Tables 5, 6, and 7. These recommendations apply to the principal production areas of the continental United States and are based on experience under average conditions; they may not be applicable under unusual weather conditions or even usual weather conditions in the case of late-maturing crops. Full-bin drying may not be feasible in some geographical areas.

The maximum practical depth of grain to be dried (distance of air travel) is limited by the cost of the fan, motor, air distribution system, and power required. This depth seems to be 20 ft for corn and soybeans, and about 15 ft for wheat.

To ensure satisfactory drying, heated air may be used during periods of prolonged fog or rain. Burners should be sized to raise the temperature of the drying air by no more than 10°F above ambient. The temperature should not exceed about 80°F after heating. Over-heating the drying air causes the grain to overdry and dry nonuniformly; heat is recommended only to counteract adverse weather conditions. Electric controllers are available to assist in controlling fan and heater operation to achieve the final desired grain moisture content.

Drying takes place in a drying zone, which advances upward through the grain (Figure 8). Grain above this drying zone remains at or slightly above the initial moisture content, while grain below the drying zone is at a moisture content in equilibrium with the drying air.

As the direction of air movement does not affect the rate of drying, other factors must be considered in choosing the direction. A pressure system moves the moisture-laden air up through the grain, and it is discharged under the roof. If there are insufficient

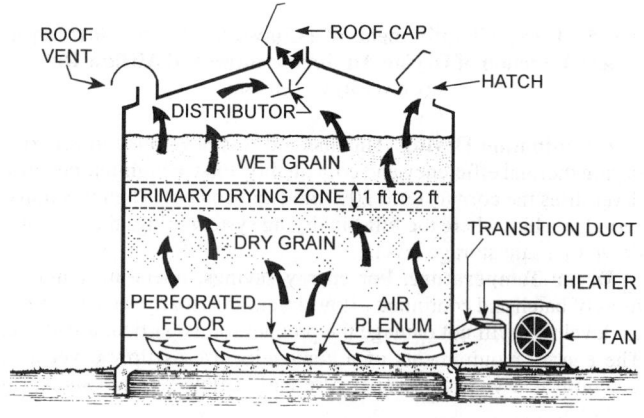

Fig. 8   Three Zones Within Grain During Full-Bin Drying

roof outlets, moisture may condense on the underside of metal roofs. During pressure system ventilation, the wettest grain is near the top surface and is easy to monitor. Fan and motor waste heat enter into the airstream and contribute to drying.

A negative-pressure system moves the air down through the grain. The moisture-laden air discharges from the fan into the outside atmosphere; thus, roof condensation is not a problem. Also, the air picks up some solar heat from the roof. However, the wettest grain is near the bottom of the mass and is difficult to sample. Of the two systems, the pressure system is recommended because it is easier to manage.

The following management practices must be observed to ensure the best performance of the dryer:

1. Minimize foreign material. A scalper-cleaner is recommended for cleaning the grain to reduce air pressure and energy requirements and to help provide uniform airflow for the elimination of wet spots.
2. Distribute the remaining foreign material uniformly by installing a grain distributor.
3. Place the grain in layers and keep it leveled.
4. Start the fan as soon as the floor or ducts are covered with grain.

Table 7 Recommended Unheated Air Airflow Rate for Different Grains and Moisture Contents in the Southern United States

Type of Grain	Grain Moisture Content, %	Recommended Airflow Rate, cfm per ft³ of grain	cfm/bu
Wheat	25	4.8	6.0
	22	4.0	5.0
	20	2.4	3.0
	18	1.6	2.0
	16	0.8	1.0
Oats	25	2.4	3.0
	20	1.6	2.0
	18	1.2	1.5
	16	0.8	1.0
Shelled Corn	25	4.0	5.0
	20	2.4	3.0
	18	1.6	2.0
	16	0.8	1.0
Ear Corn	25	6.4	8.0
	18	3.2	4.0
Grain Sorghum	25	4.8	6.0
	22	4.0	5.0
	18	2.4	3.0
	15	1.6	2.0
Soybeans	25	4.8	6.0
	22	4.0	5.0
	18	2.4	3.0
	15	1.6	2.0

Compiled from USDA *Leaflet* 332 (1952) and Univ. of Georgia *Bulletin* NS 33 (1958).

5. Operate the fan continuously with unheated air unless it is raining heavily or there is a dense ground fog. Once all the grain is within 1% of desired storage moisture content, run the fans only when the relative humidity is below 70%.

Layer Drying

In layer drying, successive layers of wet grain are placed on top of dry grain. When the top 6 in. has dried to within 1% of the desired moisture content, another layer is added (Figure 9). In comparison to full-bin drying, layering reduces the time that the top layers of grain remain wet. Because the effective airflow rate is greater for lower layers, allowable harvest moisture content of grain in these levels can be greater than that in the upper layers. Either unheated air or air heated 10 to 20°F above the ambient may be used, but the use of heated air controlled with a humidistat to prevent overdrying is most common. The first layer may be about 7 ft deep, with successive layers of about 3 ft.

Batch-in-Bin Drying

A storage bin adapted for drying may be used to dry several batches of grain during a harvest season, if the grain is kept to a shallow layer so that higher airflow rates and temperatures can be used. After the batch is dry, the bin is emptied, and the cycle is repeated. The drying capacity of the batch system (bu/yr) is greater than that of other in-storage drying systems. In a typical operation, batches of corn in 3 ft depths are dried from an initial moisture content of 25% with 130°F air at the rate of about 20 cfm per cubic foot. Considerable nonuniformity of moisture content may be present in the batch after drying is stopped; therefore, the grain should be well mixed as it is placed into storage. If the mixing is done well, grain that is too wet equalizes in moisture with grain that is too dry before spoilage can occur. Aeration of the grain in storage will facilitate the equalization of moisture.

Grain may be cooled in the dryer to ambient temperature before it is stored. Cooling is accomplished by operating the fan without the heater for about 1 h. Some additional drying occurs during the

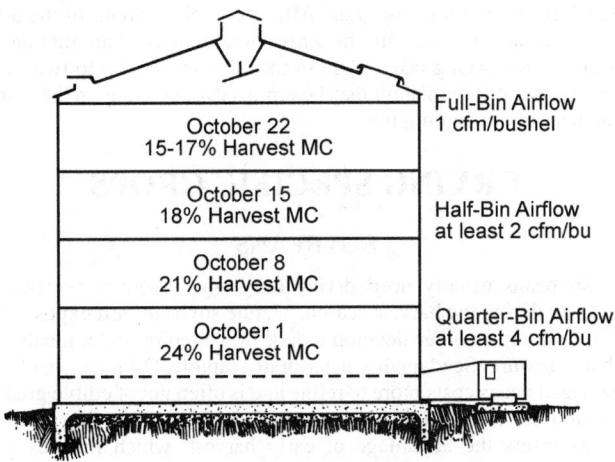

Fig. 9 Example of Layer Filling of Corn

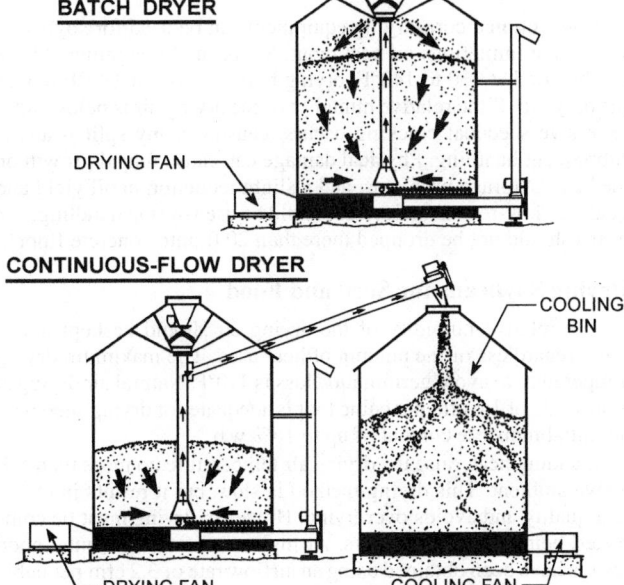

Fig. 10 Grain Recirculators Convert Bin Dryer to High-Speed Continuous-Flow Dryer

Grain stirring devices are used with both full-bin and batch-in-bin drying systems. Typically, these devices consist of one or more open, 2 in. diameter, standard pitch augers suspended from the bin roof and extending to near the bin floor. The augers rotate and simultaneously travel horizontally around the bin, mixing the drying grain to reduce moisture gradients and prevent overdrying of the bottom grain. The augers also loosen the grain, allowing a higher airflow rate for a given fan. Stirring equipment reduces bin capacity by about 10%. Furthermore, commercial stirring devices are available only for round storage enclosures.

Recirculating/Continuous-Flow Bin Drying

This type of drying incorporates a tapered sweep auger that removes uniform layers of grain from the bottom of the bin as it dries (Figure 10). The dry grain is then redistributed on top of the pile of grain or moved to a second bin for cooling. The sweep auger may be controlled by temperature or moisture sensors. When the desired condition is reached, the sensor starts the sweep auger, which removes a layer of grain. After a complete circuit of the bin, the sweep auger stops until the sensor determines that another layer is dry. Some drying takes place in the cooling bin. Up to two percentage points of moisture may be removed, depending on the management of the cooling bin.

DRYING SPECIFIC CROPS

SOYBEANS

Soybeans usually need drying only when there is inclement weather during the harvest season. Mature soybeans left exposed to rain or damp weather develop a dark brown color and a mealy or chalky texture. Seed quality deteriorates rapidly. Oil from weather-damaged beans costs more to refine and is often not of edible grade. In addition to preventing deterioration, the artificial drying of soybeans offers the advantage of early harvest, which reduces the chance of loss from bad weather and reduces natural and combine shatter loss. Soybeans harvested with a wet basis moisture content greater than 13.5% exhibit less damage.

Drying Soybeans for Commercial Use

Conventional corn-drying equipment can be used for soybeans, with some limitations on heat input. Soybeans for commercial use can be dried at 130 to 140°F; drying temperatures of 190°F reduce the oil yield. If the relative humidity of the drying air is below 40%, excessive seedcoat cracking occurs, causing many split beans in subsequent handling. Physical damage can cause fungal growth on the beans, storage problems, and a slight reduction in oil yield and quality. Flow-retarding devices should be used during handling, and beans should not be dropped more than 20 ft onto concrete floors.

Drying Soybeans for Seed and Food

The relative humidity of the drying air should be kept above 40%, regardless of the amount of heat used. The maximum drying temperature to avoid germination loss is 110°F. Natural air drying at a flow rate of 1.6 cfm per cubic foot is adequate for drying seed with an initial moisture content of up to 16% w.b.

If adding heat, raise the drying air temperature no more than 5°F above ambient. This drying method is slow, but it results in excellent quality and avoids overdrying. However, drying must be completed before spoilage occurs. At higher moisture contents, good results have been obtained using an airflow rate of 3.2 cfm per cubic foot with humidity control. Data on the allowable drying time for soybeans are unavailable. In the absence of better information, an estimate of storage life for oil crops can be made based on the values for corn, using an adjusted moisture content calculated by the following equation:

$$\text{Comparable moisture content} = \frac{\text{Oilseed moisture content}}{100 - \text{Seed oil content}} \times 100$$

A corn moisture content 2% greater than that of the soybeans should generally be used to estimate allowable drying time (e.g., 12% soybeans are comparable to 14% corn). Soybeans are dried from a lower initial moisture content than corn.

Dry high-moisture soybeans in a bin with the air temperature controlled to keep the relative humidity at 40% or higher. Airflow rates of 8.0 cfm per cubic foot are recommended, with the depth of the beans not to exceed 4 ft.

HAY

Hay normally contains 65 to 80% wet basis moisture at cutting. Field drying to 20% may result in a large loss of leaves. Alfalfa hay leaves average about 50% of the crop by weight, but

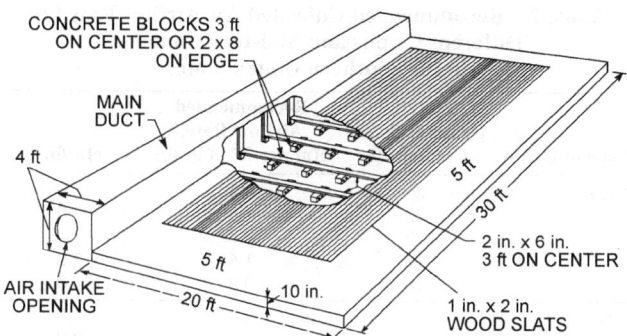

Fig. 11   Central Duct Hay-Drying System with Lateral Slatted Floor for Wide Mows

they contain 70% of the protein and 90% of the carotene. The quality of hay can be increased and the risk of loss due to bad weather reduced if the hay is put under shelter when partially field dried (35% moisture content) and then artificially dried to a safe storage moisture content. In good drying weather, hay conditioned by mechanical means can be dried sufficiently in one day and placed in the dryer. Hay may be long, chopped, or baled for this operation; unheated or heated air can be used.

In-Storage Drying

Unheated air is normally used for in-storage or mow drying. The hay is dried in the field to 30 to 40% moisture content before being placed in the dryer. For unheated air drying, the airflow should be at least 200 cfm per ton. The fan should be capable of delivering the required airflow against a static pressure of 1 to 2 in. of water.

Slotted floors, with at least 50% of the area open, are generally used for drying baled hay. For long or chopped hay in mows narrower than 36 ft wide, the center duct system is the most popular. A slotted floor should be placed on each side of the duct to within 5 ft of its ends and the outside walls (Figure 11). If the mow is wider than 36 ft, it should be divided crosswise into units of 28 ft or narrower. These should then be treated as individual dryers. If the storage depth exceeds about 13 ft, vertical flues and/or additional levels of ducts may be used. If tiered ducts are used, a vertical air chamber, about 75% of the probable hay depth, should be used. The supply ducts are then connected at 7 to 10 ft vertical intervals as the mow is filled. With either of these methods, hay in total depths up to 30 ft can be dried. The duct size should be such that the air velocity is less than 1000 fpm.

The maximum depth of wet hay that should be placed on a hay-drying system at any time depends on hay moisture content, weather conditions, the physical form of the hay, and the airflow rate. The maximum drying depth is about 16 ft for long hay, 13 ft for chopped hay, and 7 small rectangular bales deep for baled hay. Baled hay should have a density of about 8 lb/ft³. For best results, bales should be stacked tightly together on edge (parallel to the stems) to ensure that no openings exist between them.

For mow drying, the fan should run continuously during the first few days. Afterward, it should be operated only during low relative humidity weather. During prolonged wet periods, the fan should be operated only enough to keep the hay cool.

Batch Wagon Drying

Batch drying can be done on a slotted floor platform; however, because this method is labor-intensive, wagon dryers are more commonly used. With a wagon dryer system, hay is baled at about 45% moisture content to a density of about 11 lb/ft³. The hay is then stacked onto a wagon with tight, high sides and a slotted or expanded metal floor. Drying is accomplished most efficiently by forcing the heated air (up to 158°F) down the canvas duct of a plenum chamber secured to the top of the wagon. After 4 or 5 h of drying, the exhaust air is no longer saturated with moisture, and about 75% of it may be recirculated or passed through a second wagon of wet hay for greater drying efficiency.

In this method, the amount of hay harvested each day is limited by the capacity of the drying wagons. In this 24 h process, the hay cut one day is stored the following day; only enough hay to load the drying wagons should be harvested each day.

The airflow rate in this method is normally much higher than when unheated air is used. About 40 cfm per square foot of wagon floor space is required. As with mow drying, the duct size should be such that the air velocity is less than 1000 fpm.

COTTON

Producers normally allow cotton to dry naturally in the field to 12% moisture content or less before harvest. Cotton harvested in this manner can be stored in trailers, baskets, or compacted stacks for extended periods with little loss in fiber or seed quality. Thus, cotton is not normally aerated or artificially dried prior to ginning. Cotton harvested during inclement weather and stored cotton exposed to precipitation must be dried at the cotton gin within a few days to prevent self-heating and deterioration of the fiber and seed.

Though cotton may be safely stored at moisture contents as high as 12%, moisture levels near the upper limit are too high for efficient ginning and for obtaining optimum fiber grade. The cleaning efficiency of cotton is inversely proportional to its moisture content, with the most efficient level being 5% fiber moisture content. However, fiber quality is best preserved when the fiber is separated from the seed at moisture contents between 6.5 and 8%. Therefore, if cotton comes into the system below this level, it can be cleaned, but moisture should be added prior to separating the fiber from the seed to improve the ginning quality. Dryers in the cotton gins are capable of drying the cotton to the desired moisture level.

Although several types of dryers are available commercially, the tower dryer is the most commonly used. This device operates on a parallel flow principle: 14 to 24 cfm of drying air per pound of cotton also serves as the conveying medium. As it moves through the dryer's serpentine passages, cotton impacts on the walls. This action agitates the cotton for improved drying and lengthens its exposure time. The drying time depends on many variables, but total exposure seldom exceeds 15 s. For extremely wet cotton, two stages of drying are necessary for adequate moisture control.

Wide variations in initial moisture content dictate different drying amounts for each load of cotton. Rapid changes in drying requirements are accommodated by automatically controlling drying air temperature in response to moisture measurements taken before or after drying. These control systems prevent overdrying and reduce energy requirements. For safety and to preserve fiber quality, drying air temperature should not exceed 350°F in any portion of the drying system.

If the internal cottonseed temperature does not exceed 140°F, germination is unimpaired by drying. This temperature is not exceeded in a tower dryer; however, the moisture content of the seed after drying may be above the 12% level recommended for safe long-term storage. Wet cottonseed is normally processed immediately at a cottonseed oil mill. Cottonseed under the 12% level is frequently stored for several months prior to milling or prior to delinting and treatment at a seed processing plant. The aeration that cools deep beds of stored cottonseed effectively maintains viability and prevents an increase in free fatty acid content. For aeration, ambient air is normally drawn downward through the bed at a rate of at least 0.025 cfm per cubic foot of oil mill seed and 0.125 cfm per cubic foot of planting seed.

PEANUTS

Peanuts normally have a moisture content of about 50% at the time of digging. Allowing the peanuts to dry on the vines in the windrow for a few days removes much of this water. However, peanuts usually contain 20 to 30% moisture when removed from the vines, and some artificial drying is necessary. Drying should begin within 6 h after harvesting to keep the peanuts from self-heating. Both the maximum temperature and the rate of drying must be carefully controlled to maintain quality.

High temperatures result in an off flavor or bitterness. Drying too rapidly without high temperatures results in blandness or nuts that do not develop flavor during roasting. High temperatures, rapid drying, or excessive drying cause the skin to slip easily and the kernels to become brittle. These conditions result in high damage rates in the shelling operation but can be avoided if the moisture removal rate does not exceed 0.5% per hour. Because of these limitations, continuous-flow drying is not usually recommended for peanuts.

Peanuts can be dried in bulk bins using unheated air or air with supplemental heat. Under poor drying conditions, unheated air may cause spoilage, so supplemental heat is preferred. Air should be heated no more than 13 or 14°F to a maximum temperature of 95°F. An airflow rate of 10 to 25 cfm per cubic foot of peanuts should be used, depending on the initial moisture content.

The most common method of drying peanuts is bulk wagon drying. Peanuts are dried in depths of 5 to 6 ft, using airflow rates of 10 to 15 cfm per cubic foot of peanuts and air heated 11 to 14°F above ambient. This method retains quality and usually dries the peanuts in three to four days. Wagon drying reduces handling labor but may require additional investment in equipment.

RICE

Of all grains, rice is probably the most difficult to process without quality loss. Rice containing more than 13.5% moisture cannot be safely stored for long periods, yet the recommended harvest moisture content for best milling and germination ranges from 20 to 26%. When rice is harvested at this moisture content, drying must be started promptly to prevent souring. Normally, heated air is used in continuous-flow dryers, where large volumes of air are forced through 4 to 10 in. layers of rice. Temperatures as high as 130°F may be used, if (1) the temperature drop across the rice does not exceed 20 to 30°F, (2) the moisture reduction does not exceed two percentage points in a 0.5 h exposure, and (3) the rice temperature does not exceed 100°F. During the tempering period following drying, the rice should be aerated to ambient temperature prior to the next pass through the dryer. This removes additional moisture and eliminates one to two dryer passes. It is estimated that full use of aeration following dryer passes could increase the maximum daily drying capacity by about 14%.

Unheated air or air with a small amount of added heat (13°F above ambient, but not exceeding 95°F) should be used for deep-bed rice drying. Too much heat overdries the bottom, resulting in checking (cracking), reduced milling qualities, and possible spoilage in the top. Because unheated air drying requires less investment and attention than supplemental heat drying, it is preferred when conditions permit. In the more humid rice-growing areas, supplemental heat is desirable to ensure that the rice dries. The time required for drying varies with weather conditions, moisture content, and airflow rate. In California, the recommended airflow rate is 0.2 to 2.4 cfm per cubic foot. Because of less favorable drying conditions in Arkansas, Louisiana, and Texas, greater airflow rates are recommended (e.g., a minimum of 2.0 cfm per cubic foot is recommended in Texas). Whether unheated air or supplemental heat is used, the fan should be turned on as soon as rice uniformly covers the air distribution system. The fan should then run continuously until the moisture content in the top 1 ft of rice is reduced to about 15%. At this point, the supplemental heat should be turned off. The rice can then be dried to a safe storage level by operating the fan only when the relative humidity is below 75%.

STORAGE PROBLEMS AND PRACTICES

MOISTURE MIGRATION

Redistribution of moisture generally occurs in stored grain when grain temperature is not controlled (Figure 12). Localized spoilage can occur even when the grain is stored at a safe moisture level. Grain placed in storage in the fall at relatively high temperatures cools nonuniformly through contact with the outside surfaces of the storage bin as winter approaches. Thus, the grain near the outside walls and roof may be at cool outdoor temperatures while the grain nearer the center is still nearly the same temperature it was at harvest. These temperature differentials induce air convection currents that flow downward along the outside boundaries of the porous grain mass and upward through the center. When the cool air from the outer regions contacts the warm grain in the interior, the air is heated and its relative humidity is lowered, increasing its capacity to absorb moisture from the grain. When the warm, humid air reaches the cool grain near the top of the bin, it cools again and transfers vapor to the grain. Under extreme conditions, water condenses on the grain. The moisture concentration near the center of the grain surface causes significant spoilage if moisture migration is uncontrolled. During spring and summer, the temperature gradients are reversed. The grain moisture content increases most at depths of 2 to 4 ft below the surface. Daily variations in temperature do not cause significant moisture migration. Aside from seasonal temperature variations, the size of the grain mass is the most important factor in fall and winter moisture migration. In storages containing less

than 1200 ft³, there is less trouble with moisture migration. The problem becomes critical in large storages and is aggravated by incomplete cooling of artificially dried grain. Artificially dried grain should be cooled to near ambient temperature soon after drying.

GRAIN AERATION

Aeration by mechanically moving ambient air through the grain mass is the best way to control moisture migration. Aeration systems are also used to cool grain after harvest, particularly in warmer climates where grain may be placed in storage at temperatures exceeding 100°F. After the harvest heat is removed, aeration may be continued in cooler weather to bring the grain to a temperature within 20°F of the coldest average monthly temperature. The temperature must be maintained below 50°F.

Aeration systems are not a means of drying because airflow rates are too low. However, in areas where the climate is favorable, carefully controlled aeration may be used to remove small amounts of moisture. Commercial storages may have pockets of higher-moisture grain if, for example, some batches of grain are delivered after a rain shower or early in the morning. Aeration can control heating damage in the higher-moisture pockets.

Aeration Systems Design

Aeration systems include fans capable of delivering the required amount of air at the required static pressure, suitable ducts or floors to distribute the air into the grain, and controls to regulate the operation of the fan. The airflow rate determines how many hours are required to cool the crop (Table 8). Most aeration systems are designed with airflow rates between 0.05 and 0.2 cfm/bu.

Stored grain is aerated by forcing air up or down through the grain. Moving air up through the grain is more common because it is easier to observe when the cooling front has moved through the entire grain mass. In large, flat storages with long ducts, upward airflow results in more uniform air distribution than downdraft systems.

During aeration, a warming or cooling front moves through the crop (Figure 13); it is important to run the fan long enough to move the front completely through the crop.

Static pressure for an aeration system can be determined using the airflow resistance information in Chapter 11 of the 1997 *ASHRAE Handbook—Fundamentals*. All common types of fans are used in aeration systems. Attention should be given to noise levels with fans that are operated near residential areas or where people work for extended periods. The supply ducts connecting the fan to the distribution ducts in the grain should be designed and constructed according to the standards of good practice for

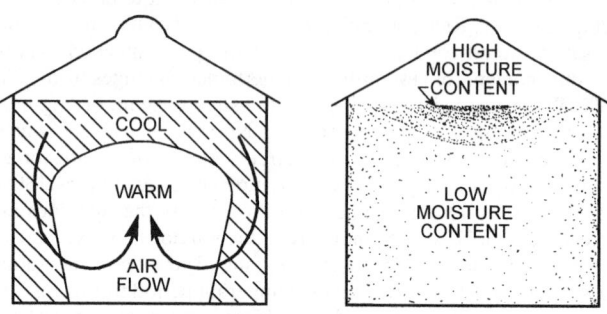

Fig. 12 Grain Storage Conditions Associated with Moisture Migration During Fall and Early Winter

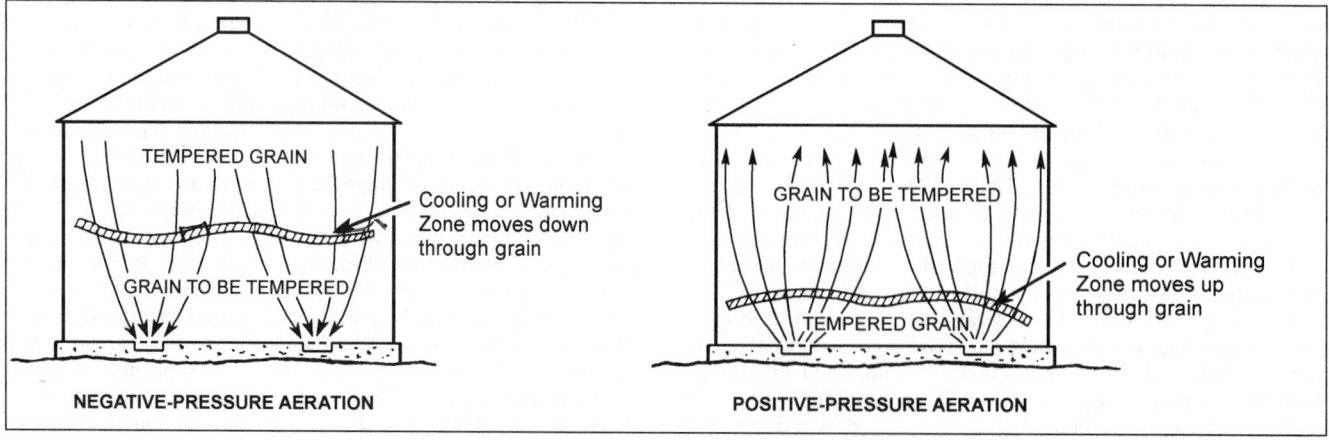

Fig. 13 Aerating to Change Grain Temperature

Table 8 Airflow Rates Corresponding to Approximate Grain Cooling Time

Airflow Rate, cfm/bu	Cooling Time, h
0.05	240
0.1	120
0.2	60
0.3	40
0.4	30
0.5	24
0.6	20
0.8	15
1.0	12

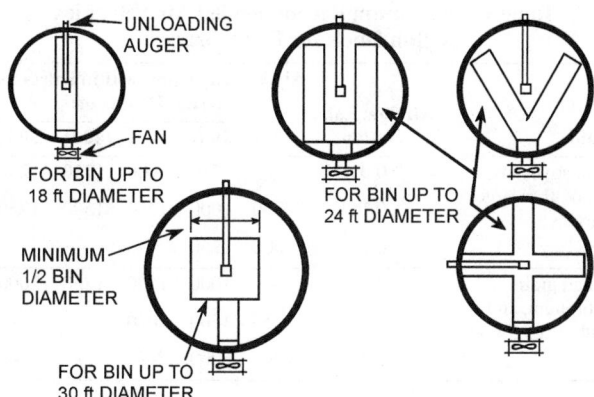

Fig. 14 Common Duct Patterns for Round Grain Bins

any air-moving application. A maximum air velocity of 2500 fpm may be used, but 1600 to 2000 fpm is preferred. In large systems, one large fan may be attached to a manifold duct leading to several distribution ducts in one or more storages, or smaller individual fans may serve individual distribution ducts. Where a manifold is used, valves or dampers should be installed at each takeoff to allow adjustment or closure of airflow when part of the aerator is not needed.

Distribution ducts are usually perforated sheet metal with a circular or inverted U-shaped cross section, although many functional arrangements are possible. The area of the perforations should be at least 10% of the total duct surface. The holes should be uniformly spaced and small enough to prevent the passage of the grain into the duct (e.g., 0.1 in. holes or 0.08 in. wide slots do not pass wheat).

Since most problems develop in the center of the storage, and the crop cools naturally near the wall, the aeration system must provide good airflow in the center. Flush floor systems work well in storages

with sweep augers and unloading equipment. Ducts should be easily removable for cleaning. Duct spacing should not exceed the depth of the crop; the distance between the duct and storage structure wall should not exceed one-half the depth of the crop for bins and flat storages. Common duct patterns for round bins are shown in Figure 14. Duct spacing for flat storages is shown in Figure 15.

When designing the distribution duct system for any type of storage, the following should be considered: (1) the cross-sectional area and length of the duct, which influences both the air velocity within the duct and the uniformity of air distribution; (2) the duct surface area, which affects the static pressure losses in the grain surrounding the duct; and (3) the distance between ducts, which influences the uniformity of airflow.

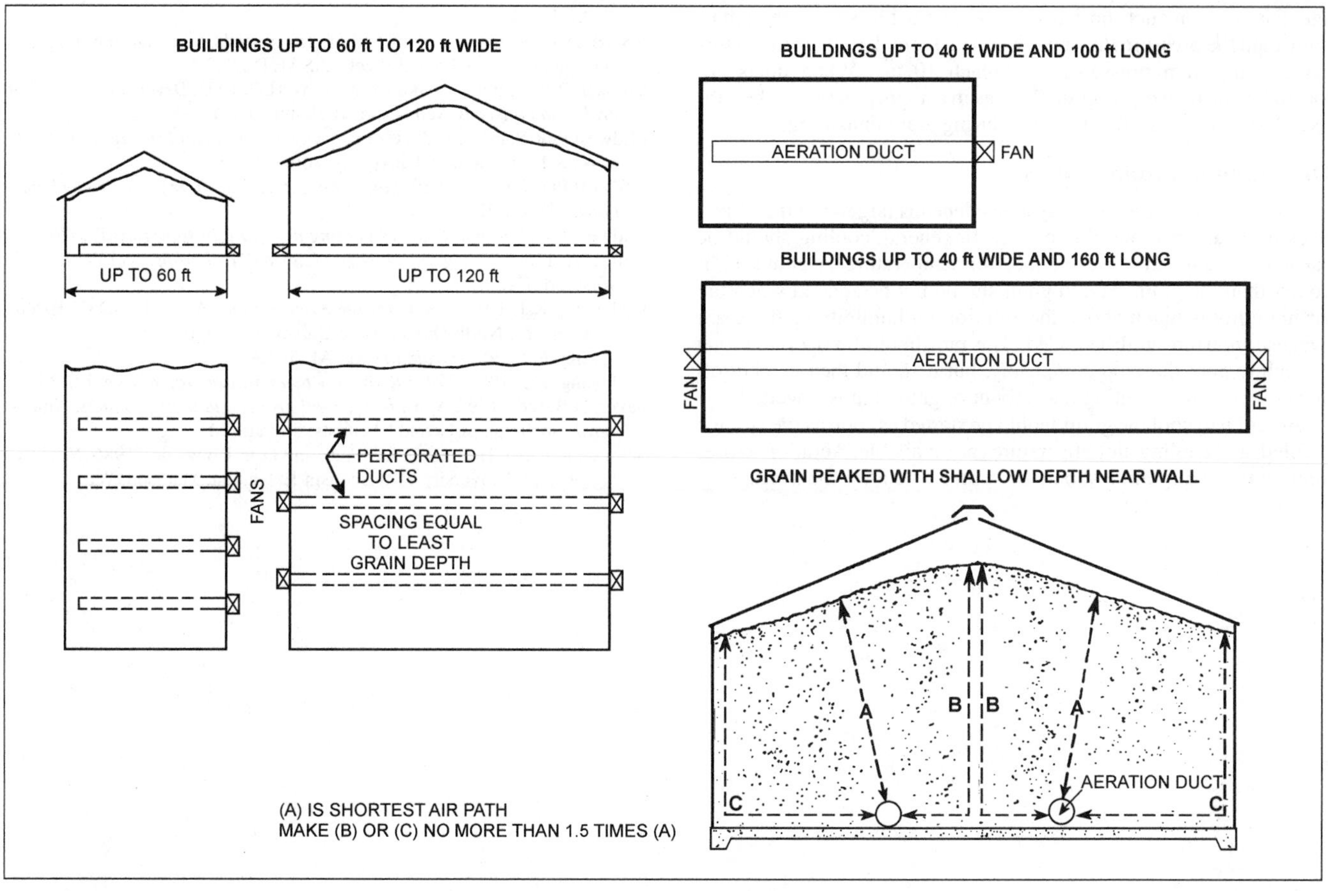

Fig. 15 Duct Arrangements for Large Flat Storages

Table 9 Maximum Recommended Air Velocities Within Ducts for Flat Storages

Grain	Airflow Rate, cfm/bu	Air Velocity (fpm) within Ducts for Grain Depths of:				
		10 ft	20 ft	30 ft	40 ft	50 ft
Corn, soybeans, and other large grains	0.05	—	750	1000	1250	1250
	0.1	750	1000	1250	1500	1750
	0.2	1000	1250	—	—	—
Wheat, grain sorghum, and other small grains	0.05	—	1000	1500	1750	2000
	0.1	750	1500	2000	—	—
	0.2	1000	2000	—	—	—

For upright storages where the distribution ducts are relatively short, distribution duct velocities up to 2000 fpm are permissible. Maximum recommended air velocities in ducts for flat storages are shown in Table 9. Furthermore, these velocities should not be exceeded in the air outlets from the storage; therefore, an air outlet area at least equal to the duct cross-sectional area should be provided.

The duct surface area that is perforated or otherwise open for air distribution must be great enough that the air velocity through the grain surrounding the duct is low enough to avoid excessive pressure loss. When a semicircular perforated duct is used, the entire surface area is effective; only 80% of the area of a circular duct resting on the floor is effective. For upright storages, the air velocity through the grain near the duct (duct face velocity) should be limited to 30 fpm or less; in flat storages, to 20 fpm or less.

Duct strength and anchoring are important. If ducts placed directly on the floor are to be held in place by the crop, the crop flow should be directly on top of the ducts to prevent movement and damage. Distribution ducts buried in the grain must be strong enough to withstand the pressure the grain exerts on them. In tall, upright storages, static grain pressures may reach 10 psi. When ducts are located in the path of the grain flow, as in a hopper, they may be subjected to many times this pressure during grain unloading.

Operating Aeration Systems

The operation of aeration systems depends largely on the objectives to be attained and the locality. In general, cooling should be carried out any time the outdoor air temperature is about 15°F cooler than the grain. Stored grain should not be aerated when the air humidity is much above the equilibrium humidity of the grain because moisture will be added. The fan should be operated long enough to cool the crop completely, but it should then be shut off and covered, thus limiting the amount of grain that is rewetted.

Aeration to cool the grain should be started as soon as the storage is filled, and cooling air temperatures are available. Aeration to prevent moisture migration should be started whenever the average air temperature is 10 to 15°F below the highest grain temperature. Aeration is usually continued as weather permits until the grain is uniformly cooled to within 20°F of the average temperature of the coldest month, or to 30 to 40°F.

Grain temperatures of about 32 to 50°F are desirable. In the northern corn belt, aeration may be resumed in the spring to equalize the grain temperature and raise it to between 40 and 50°F. This reduces the risk of localized heating from moisture migration. Storage problems are the only reason to aerate when air temperatures are above 60°F. Aeration fans and ducts should be covered when not in use.

In storages where the fans are operated daily in the fall and winter months, automatic controls work well when the air is not too warm or humid. One thermostat usually prevents fan operation when the air temperature is too high, and another prevents operation when the air is too cold. A humidistat allows operation when the air is not too humid. Fan controllers are available that determine the equilibrium moisture content of the crop based on existing air conditions and regulate the fan based on entered information.

SEED STORAGE

Seed must be stored in a cool, dry environment to maintain viability. Most seed storages have refrigeration equipment to maintain a storage environment of 45 to 55°F. Seed storage conditions must be achieved before mold and insect damage occur.

BIBLIOGRAPHY

ASAE. 1993. Density, specific gravity, and mass-moisture relationships of grain for storage. ANSI/ASAE D241.4. American Society of Agricultural Engineers, St. Joseph, MI.

ASAE. 1995. Moisture relationship of plant-based agricultural products. ASAE D245.5.

ASAE. 1996. Resistance of airflow of grains, seeds, other agricultural products, and perforated metal sheets. ASAE D272.2.

Brooker, D.B., F. Bakker-Arkema, and C.W. Hall. 1992. *Drying and storage of grains and oilseeds.* Van Nostrand, Reinhold, NY.

Midwest Plan Service. 1988. *Grain drying, handling and storage handbook.* MWPS-13. Iowa State University, Ames, IA.

Midwest Plan Service. 1980. *Low temperature and solar grain drying handbook.* MWPS-22.

Midwest Plan Service. 1980. *Managing dry grain in storage.* AED-20.

Hall, C.A. 1980. *Drying and storage of agricultural crops.* AVI Publishing, Westport, CT.

Hellevang, K.J. 1989. *Crop storage management.* AE-791. NDSU Extension Service, North Dakota State University, Fargo, ND.

Hellevang, K.J. 1987. *Grain drying.* AE-701.

Hellevang, K.J. 1983. *Natural air/Low temperature crop drying.* EB-35.

Saver, D.B. (ed.) 1992. *Storage of cereal grains and their products.* American Association of Cereal Chemists, St. Paul, MN.

Schuler, R.T, B.J. Holmes, R.J. Straub, and D.A. Rohweder. 1986. *Hay drying.* A3380. University of Wisconsin-Extension, Madison, WI.

CHAPTER 23

AIR CONDITIONING OF WOOD AND PAPER PRODUCT FACILITIES

THIS chapter covers some of the standard requirements for air conditioning of facilities that manufacture finished wood products as well as pulp and paper.

GENERAL WOOD PRODUCT OPERATIONS

Finished lumber products to be used in heated buildings should be stored in areas that are heated 10 to 20°F above ambient. This provides sufficient protection for furniture stock, interior trim, cabinet material, and stock for products such as ax handles and glue-laminated beams. Air should be circulated within the storage areas. Lumber that is kiln-dried to a moisture content of 12% or less can be kept within a given moisture content range through storage in a heated shed. The moisture content can be regulated either manually or automatically by altering the dry-bulb temperature (Figure 1).

Some special materials require close control of moisture content. For example, musical instrument stock must be dried to a given moisture level and maintained there because the moisture content of the wood affects the harmonics of most stringed wooden instruments. This control may require air conditioning with reheat and/or heating with humidification.

Process Area Air Conditioning

Temperature and humidity requirements in wood product process areas vary according to product, manufacturer, and governing code. For example, in match manufacturing, the match head must be cured (i.e., dried) after dipping. This requires careful control of the

The preparation of this chapter is assigned to TC 9.2, Industrial Air Conditioning.

humidity and temperature to avoid a temperature near the ignition point. Any process involving the application of flammable substances should follow the ventilation recommendations of the National Fire Protection Association, the National Fire Code, and the U.S. Occupational Safety and Health Act.

Finished Product Storage

Design engineers should be familiar with the client's entire operation and should be aware of the potential for moisture regain. Finished lumber products manufactured from predried stock (moisture content of 10% or less) regain moisture if exposed to a high relative humidity for an extended period. All storage areas housing finished products such as furniture and musical instruments should be conditioned to avoid moisture regain.

PULP AND PAPER OPERATIONS

The papermaking process comprises two basic steps: (1) wood is reduced to pulp, that is, wood fibers; and (2) the pulp is converted to paper. Wood can be pulped by either mechanical action (e.g., grinding in a groundwood mill), chemical action (e.g., kraft pulping), or a combination of both.

Many different types of paper can be produced from pulp, ranging from the finest glossy finish to newsprint to bleached board to fluff pulp for disposable diapers. To make newsprint, a mixture of mechanical and chemical pulps is fed into the paper machine. To make kraft paper (e.g., grocery bags and corrugated containers), however, only unbleached chemical pulp is used. Disposable diaper material and photographic paper require bleached chemical pulp with a very low moisture content of 6 to 9%.

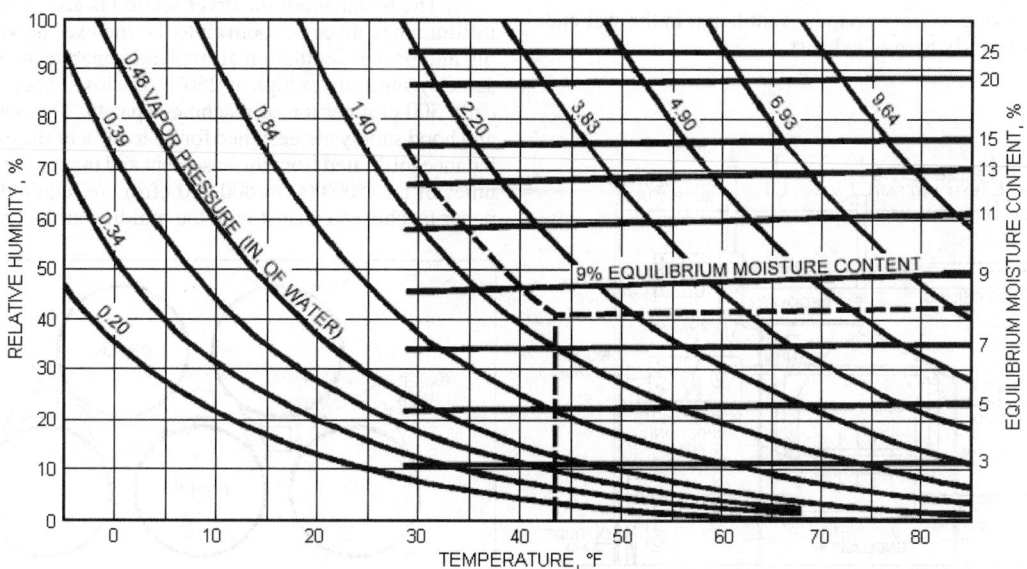

Fig. 1   Relationship of Temperature, Relative Humidity, and Vapor Pressure of Air and Equilibrium Moisture Content of Wood

Paper Machine Area

In papermaking, extensive air systems are required to support and enhance the process (e.g., by preventing condensation) and to provide reasonable comfort for operating personnel. Radiant heat from steam and hot water sources and mechanical energy dissipated as heat can result in summer temperatures in the machine room as high as 120°F. In addition, high paper machine operating speeds of 2000 to 4500 fpm and a stock temperature near 122°F produce warm vapor in the machine room.

Outside air makeup units and process exhausts absorb and remove room heat and water vapor released from the paper as it is dried (Figure 2). The makeup air is distributed to the working areas above and below the operating floor. Part of the air delivered to the basement migrates to the operating floor through hatches and stairwells. Motor cooling equipment distributes cooler basement air to the paper machine drive motors.

Wet and basement exhaust should be installed inside the room. Outside air intakes with insulated adjustable louvers should be installed on the outside wall to supplement the mechanical air supply. In facilities with no basement exterior wall, a sufficient mechanical air intake should be provided. The exhaust, adjustable louver, or mechanical air intake should be furnished with modulating control. When the ambient temperature drops to near freezing, outside airflow must be reduced to a minimum and the appropriate heater started to prevent freezing.

The most severe ventilation demand occurs in the area between the wet-end forming section and press section and the dryer section. In the forming section, the pulp slurry, which contains about 90% water, is deposited on a traveling screen. Gravity, rolls, foils, vacuum, steam boxes, and three or more press roll nips are sequentially used to remove up to 50% of the water in the forming section and press section. The wet end is very humid due to the evaporation of moisture and the mechanical generation of vapor by turning rolls and cleaning showers. Baffles and a custom-designed exhaust in the forming section help control the vapor. A drive-side exhaust in the wet end removes heat from the motor vent air and removes the process generated vapor.

To prevent condensation or accumulated fiber from falling on the traveling web, a false ceiling with ducts connected to roof exhausters remove humid air not captured at a lower point. At the wet end, heated inside air is usually circulated to scrub the underside of the roof to prevent condensation in cold weather. Additional roof exhaust may also remove accumulated heat from the dryer section and the dry end during warmer periods. Ventilation in the wet end should be predominantly by roof exhaust.

The large volume of moisture and vapor generated from the wet-end process rises and accumulates under the roof. To keep condensation from forming in winter, the roof is normally exhausted and hot air is distributed under the roof. Sufficient roof insulation should be installed to keep the inside surface temperature above the dewpoint. Heat transfer from the room to the interior surface is

$$\frac{t_r - t_{is}}{R_{r-is}} = \frac{t_{is} - t_o}{R_{is-o}} \tag{1}$$

where

t_r = room air temperature, °F
t_{is} = roof interior surface temperature, °F
t_o = outside air temperature, °F
R_{r-is} = heat transfer resistance from room air to roof interior surface. In winter, $R_{r-is} = 0.61$ ft^2·°F·h/Btu
R_{is-o} = required total R-value from roof interior surface to outside air, ft^2·°F·h/Btu

For a given project t_o and t_r have been determined and only t_{is} needs to be selected. For wet-end roof insulation and assuming 96% relative humidity, t_{is} can be shown on a psychrometric chart to be

$$t_{is} = t_r - 1.2°F \tag{2}$$

Then Equation (1) can be simplified to find the required roof R-value as

$$R_{is-o} = \frac{0.61}{1.2}(t_r - t_o - 1.2) \tag{3}$$

In the dryer section, the paper web is dried as it travels in a serpentine path around rotating steam-heated drums. Exhaust hoods remove heat from the dryers and moisture evaporated from the paper web. Most modern machines have enclosed hoods, which reduce the airflow required to less than 50% of that required for an open hood exhaust. The temperature inside an enclosed hood ranges from 130 to 140°F at the operating floor to 180 to 200°F in the hood exhaust plenum at 70 to 90% rh, with an exhaust rate generally ranging from 300,000 to 400,000 cfm.

Where possible, pocket ventilation air (see Figure 3) and hood supply air are drawn from the upper level of the machine room to take advantage of the preheating of makeup air by process heat as it rises. The basement of the dryer section is also enclosed to control infiltration of machine room air to the enclosed hood. The hood supply and pocket ventilation air typically operate at 200°F; however, some systems run as high as 250°F. Enclosed hood exhaust is typically 300 cfm per ton of machine capacity. The pocket ventilation and hood supply are designed for 75 to 80% of the exhaust, with the balance infiltrated from the basement and machine room. Large volumes of air (500,000 to 800,000 cfm) are required to balance the paper machine's exhaust with the building air balance.

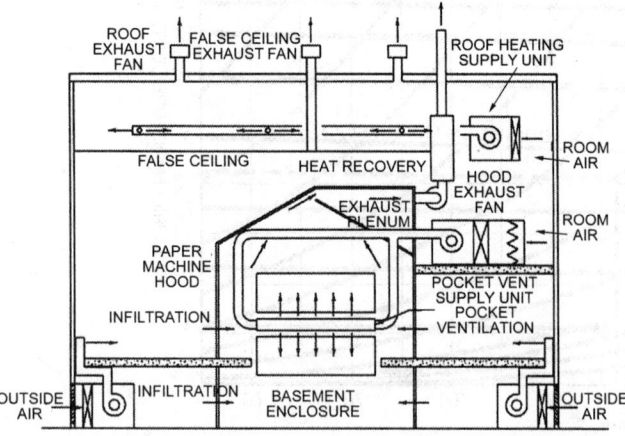

Fig. 2 Paper Machine Area

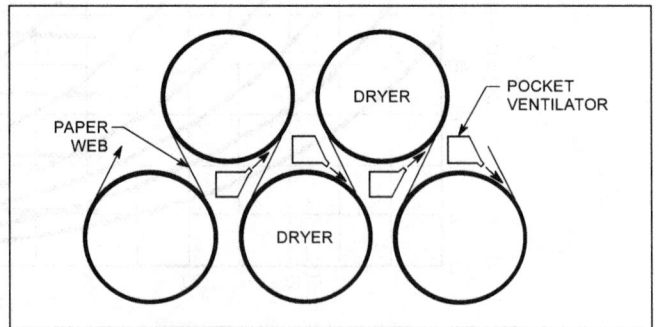

Fig. 3 Pocket Ventilation

The potential for heat recovery from the hood exhaust air should be evaluated. Most of the energy in the steam supplied to the paper dryers is converted to latent heat in the hood exhaust as water is evaporated from the paper web. Air-to-air heat exchangers are used where the air supply is located close to the exhaust. Air-to-liquid heat exchangers that recirculate water-glycol to heat remote makeup air units can also be used. Air-to-liquid systems provide more latent heat recovery, resulting in three to four times more total heat recovery than air-to-air units. Some machines use heat recovered from the exhaust air to heat process water. Ventilation in paper machine buildings in the United States ranges from 10 to 25 air changes per hour in northern mills to 20 to 50 in southern mills. In some plants, computers monitor the production rate and outside air temperature to optimize the operation and conserve energy.

After fine, bond, and cut papers have been bundled and/or packaged, they should be wrapped in a nonpermeable material. Most papers are produced with less than 10% moisture by weight, the average being 7%. Dry paper and pulp are hygroscopic and begin to swell noticeably and deform permanently when the relative humidity exceeds 38%. Therefore, finished products should be stored under controlled conditions to maintain their uniform moisture content.

Finishing Area

To produce a precisely cut paper that will stabilize at a desirable equilibrium moisture content, the finishing areas require temperature and humidity control. Further converting operations such as printing and die cutting require optimum sheet moisture content for efficient processing. Finishing room conditions range from 70 to 75°F db and from 40 to 45% rh. The rooms should be maintained within reasonably close limits of the selected conditions. Without precise environmental control, the paper equilibrium moisture content will vary, influencing dimensional stability, the tendency to curl, and further processing.

Process and Motor Control Rooms

In most pulp and paper applications, process control, motor control, and switchgear rooms are separate from the process environment. Air conditioning removes heat generated by equipment, lights, etc., and reduces the air-cleaning requirement. (See Chapter 16 for air conditioning in control rooms that include a computer, a computer terminal, or data processing equipment.) Ceiling grilles or diffusers should be located above access aisles to avoid the risk of condensation on control consoles or electrical equipment during start-up and recovery after an air-conditioning shutdown. Electrical rooms are usually maintained in the range of 75 to 80°F, with control rooms at 73°F; the humidity is maintained in the range of 45 to 55% in process control rooms and is not normally controlled in electrical equipment rooms.

Electrical control rooms for distributed control and process control contain electronic equipment that is susceptible to corrosion. The typical pulp and paper mill environment contains both particulate and vapor-phase contaminants with sulfur- and chloride-based compounds. To protect the equipment, multistage particulate and adsorbent filters should be used. They should have treated activated charcoal and potassium permanganate-impregnated alumina sections for vapor-phase contaminants, as well as fiberglass and cloth media for particulates.

Switchgear and motor control centers are not as heat-sensitive as control rooms, but the moisture-laden air carries chemical residues onto the contact surfaces. Arcing, corrosion, and general deterioration can result. A minimum amount of filtered, outside air and air conditioning is used to protect these areas.

In most projects, the electric distribution control system (DCS) is energized before the room air conditioning is installed and started. If a temporary air conditioner is used in the DCS room, a condensate drain pan and temporary drain pipe should be installed to keep condensate in the cable channel beneath the DCS panels.

Paper Testing Laboratories

Design conditions in paper mill laboratories must be followed rigidly. The most recognized standard for testing environments for paper and paper products (paperboard, fiberboard, and containers) is TAPPI (the Technical Association of the Pulp and Paper Industry) *Standard* T402, Standard Conditioning and Testing Atmospheres for Paper, Board, Pulp Handsheets, and Related Products. Other standards include ASTM E 171, Standard Specification for Standard Atmospheres for Conditioning and Testing Materials, and ISO/TC 125, Enclosures and Conditions for Testing.

Standard pulp and paper testing laboratories have three environments: preconditioning, conditioning, and testing. The physical properties of a sample are different if it is brought to the testing humidity from a high humidity than if it is brought from a lower humidity. Preconditioning at lower relative humidity tends to eliminate hysteresis. For a preconditioning atmosphere, TAPPI *Standard* T402 recommends 10 to 35% rh and 72 to 104°F db. Samples are usually conditioned in a controlled, conditioned cabinet.

Conditioning and testing atmospheres should be maintained at 50 ± 2.0% rh and 73 ± 2°F db. However, a change of 2°F db at 73°F without starting a humidifier causes the relative humidity to fluctuate as much as 3%. A dry-bulb temperature tolerance of ±1°F must be held to maintain a ±2% rh. A well-designed temperature and humidity control system should be provided.

Miscellaneous Areas

The pulp digester area contains many components that release heat and contribute to dusty conditions. For batch digesters, the chip feeders are a source of dust and need hooded exhaust and makeup air. The wash and screen areas have numerous components with hooded exhausts that require considerable makeup air. Good ventilation controls fumes and humidity. The lime kiln feed-end releases extremely large amounts of heat and requires high ventilation rates or air conditioning.

Recovery-boiler and power-boiler buildings have conditions similar to those of power plants; the ventilation rates are also similar. The control rooms are generally air conditioned. The grinding motor room, in which groundwood is made, contains many large motors that require ventilation to keep the humidity low.

System Selection

The system and equipment selected for air conditioning a pulp and paper mill depends on many factors, including the plant layout and atmosphere, geographic location, roof and ceiling heights (which can exceed 100 ft), and degree of control desired. Chilled water systems are economical and practical for most pulp and paper operations, because they have both the large cooling capacity needed by mills and the precision of control to maintain the proper temperature and humidity in laboratories and finishing areas. In the bleach plant, the manufacture of chlorine dioxide is enhanced by using water with a temperature of 45°F or lower; this water is often supplied by the chilled water system. If clean plant or process water is available, water-cooled chillers are satisfactory and may be supplemented by water-cooled direct-expansion package units for small, remote areas. However, if plant water is not clean enough, a separate cooling tower and condenser water system should be installed for the air conditioning.

Most manufacturers prefer water-cooled over air-cooled systems because of the gases and particulates present in most paper mills. The most prevalent contaminants are chlorine gas, caustic soda, borax, phosphates, and sulfur compounds. With efficient air cleaning, the air quality in and about most mills is adequate for properly placed air-cooled chillers or condensing units that have properly applied coil and housing coatings. Phosphor-free brazed coil joints are recommended in areas where sulfur compounds are present.

Heat is readily available from processing operations and should be recovered whenever possible. Most plants have quality hot water and steam, which can be used for unit heater, central station, or reheat quite easily. Evaporative cooling should be considered. Newer plant air-conditioning methods using energy conservation techniques, such as temperature destratification and stratified air conditioning, have application in large structures. Absorption systems should be considered for pulp and paper mills because they provide some degree of energy recovery from the high-temperature steam processes.

BIBLIOGRAPHY

ACGIH. 1998. *Industrial ventilation. A manual of recommended practice*, 23rd ed. American Conference of Governmental Industrial Hygienists, Cincinnati, OH.

ASTM. 1994. Standard specification for standard atmospheres for conditioning and testing flexible barrier materials. *Standard* E 171. American Society for Testing and Materials, West Conshohocken, PA.

ISO/TC 125. Enclosures and conditions for testing. International Organization for Standardization, Geneva, Switzerland.

TAPPI. 1988. Standard conditioning and testing atmospheres for paper, board, pulp, handsheets, and related products. *Test Method* T402. Technical Association of the Pulp and Paper Industry, Norcross, GA.

CHAPTER 24

POWER PLANTS

THIS chapter discusses heating, ventilating, and air-conditioning (HVAC) systems for industrial facilities for the production of process heat and power and for electrical generating stations. While not every type of power plant is specifically covered, the process areas addressed normally correspond to similar process areas in any plant. For example, wood-fired boilers are not specifically discussed, but the requirements for coal-fired boilers generally apply.

Aspects of HVAC system design unique to nuclear power plants are covered in Chapter 25.

GENERAL DESIGN CRITERIA

The space-conditioning systems in power plant buildings are designed to maintain an environment for the reliable operation of the power generation systems and equipment and for the convenience and safety of plant personnel. A balance is achieved between the cost of the process systems designed to operate in an environment and the cost of providing that environment.

Environmental criteria for personnel safety and comfort are governed by several sources. The U.S. Occupational Safety and Health Administration (OSHA) defines noise and air contaminant exposure limits. Chapter 11 and Chapter 28 of this volume and *Industrial Ventilation* by the American Conference of Governmental Industrial Hygienists (ACGIH 1998) also provide guidance for safety in work spaces. The degree of comfort for the worker is somewhat subjective and more difficult to quantify. The plant owner or operator ordinarily establishes the balance between cost and worker comfort.

Criteria should be clearly defined at the start of the design process because they document an understanding between the process designer and the HVAC system engineer that is fundamental to achieving the environment required for the various process areas. Typical criteria for a coal-fired power plant are outlined in Table 1. They should be reviewed for compliance with local codes, the plant operator's experience and preferences, and the overall financial objectives of the facility. Additional discussion of criteria may be found in the sections on specific areas.

Redundancy

Maintaining design operating temperatures within the power plant is essential for reliable operation. Operating electrical equipment in temperatures above its rated temperature reduces the life of the equipment. Sensitive electronic equipment, such as in the main control center, may not function reliably at high temperatures. Low temperatures also affect plant availability; for example, low temperatures in the batteries or freezing of pipes, instrument lines, or tanks could prevent normal plant operation.

HVAC systems or components essential for plant operation should be designed with redundancy to ensure plant availability. Automatic switchover to the backup system may be required in normally unoccupied areas. In areas where backup systems are

The preparation of this chapter is assigned to TC 9.2, Industrial Air Conditioning.

impractical, temperature monitoring and alarming systems should be considered to initiate temporary corrective measures.

HVAC systems that include multiple units (indicated as **multiplicity** in Table 1) also improve the reliability of the power plant. A space ventilated by multiple fans, such as four at 1/4 capacity, may retain sufficient ventilation even if one fan is out of service.

Filtration and Space Cleanliness

Filtration of ventilation air for process areas is usually not needed because some areas are generally dirtier than the outdoor surroundings, and the process equipment is designed to accommodate a dusty environment. However, the plant may be located in an area having sources of outdoor particulate contaminants that need to be managed to protect the process equipment. Power plants in dusty or sandy areas, or where there are seasonal nuisances such as cottonwood seeds, may require filtration of ventilation air. Plants at industrial sites such as refineries and paper mills may need to address gaseous contaminants and corrosive gases as well.

Indoor air cleanliness is a concern in control room HVAC system design. Even if the control center is in an independent building remote from the boiler-turbine building, coal transportation, coal crushing, fuel/ air distribution and combustion, ash handling, fume heat recovery, fume/smoke exhaust diffusion, and so forth, may contaminate the entire plant and its surroundings.

When potential outdoor contaminants are a factor, the quality of outdoor air may need to be evaluated. This may include collection of typical particulates and the use of corrosion coupons to quantify gaseous contaminants. The U.S. Environmental Protection Agency (EPA) is a source of data. Filtration requirements may include 30% dust-spot test efficiency prefilters, 65% to 90% efficiency final filters, and gas-phase filtration units.

Air-conditioned areas for people should meet ASHRAE *Standard* 62 requirements. Air-conditioned areas for control and electrical equipment should meet the requirements of the manufacturer. Guidelines for reliability of electrical equipment are found in Instrument Society of America (ISA) *Standard* S71.04.

Temperature and Humidity

Selection of outdoor design temperatures is based on the operating expectations of the plant. If the power production facility is critical and must operate during severe conditions, then the effect of local extreme high and low temperatures on the systems should be evaluated. Electrical power consumption is usually highest under extreme outdoor conditions, so the plant should be designed to operate when needed the most. Other temperature ranges, indicated in Table 1, may be more appropriate for less critical applications.

If the equipment is enclosed, the indoor temperatures must match the specified operating temperatures of the equipment. The electrical equipment, such as switchgear, motor control centers, and motors, typically determines the design temperature limits in the plant. Common ratings are 104 or 122°F. Elevator machine room equipment may include electronics with temperature restrictions.

Table 1 Design Criteria for Coal-Fired Power Plant

Building/Area	Design Outdoor Cooling/Heating Dry-Bulb[a]	Indoor Temperature, °F		Relative Humidity, %	Minimum Design Ventilation Rate, ACH	Filtration Efficiency, %	Pressurization	Redundancy[b]	Noise Criterion
		Maximum	Minimum						
Steam Turbine Area									
Suboperating level	0.4%/99.6%	Design outdoor + 10	45	None	30	None	None	Multiplicity	Background
Above operating floor	0.4%/99.6%	Design outdoor + 10	45	None	10	None	None	Multiplicity	Background
Combustion Turbine Area	0.4%/99.6%	Design outdoor + 15	45	None	20	None	None	Multiplicity	Background
Steam Generator Area									
Below burner elevation	0.4%/99.6%	Design outdoor + 10	45	None	30	None	None	Multiplicity	Background
Above operating floor	0.4%/99.6%	Design outdoor + 10	45	None	15	None	None	Multiplicity	Background
Air-Conditioned Areas									
Control rooms and control equipment rooms containing instruments and electronics	Extreme (see text)	75±5	75±5	30 to 65	ASHRAE Std. 62	85 to 90 (see text)	Positive	100%	NC-40[c]
Offices	1%/99%	78	72	30 to 65	ASHRAE Std. 62	ASHRAE Std. 62	Positive	None	See text
Laboratories	1%/99%	78	72	30 to 65	ASHRAE Std. 62	High	Positive	None	See text
Locker rooms and toilets	1%/99%	78	72	None	ASHRAE Std. 62	ASHRAE Std. 62	Negative	None	See text
Shops (not air-conditioned)	1%/99%	Design outdoor + 10	65	None	15	None	None	None	85 dBA
Mechanical Equipment									
Pumps, large power	0.4%/99.6%	Design outdoor + 10	45	None	30	None	None	Multiplicity	Background
Valve stations, miscellaneous	0.4%/99.6%	Design outdoor + 10	45	None	15	None	None	None	85 dBA
Elevator machine rooms	0.4%/99.6%	90	45	None	None	Low	Positive	None	85 dBA
Fire pump area	0.4%/99.6%	NFPA Std. 20	NFPA Std. 20	None	NFPA Std. 20	None	None	None	85 dBA
Diesel generator area	0.4%/99.6%	Design outdoor + 10	45	None	30	None	None	None	Background
Electrical Equipment									
Enclosed transformer equipment areas	0.4%/99.6%	Design outdoor + 10	45	None	60	Low	Positive	100%	85 dBA
Critical equipment	Extreme (see text)	Design outdoor + 10	45	None	30	None	None	100%	85 dBA
Miscellaneous electrical equipment	0.4%/99.6%	Design outdoor + 10	45	None	20	None	None	Multiplicity	85 dBA
Water Treatment									
Chlorine equipment rooms									
When occupied	0.4%/99.6%	Design outdoor + 10	None	None	60	None	Negative	None	85 dBA
When unoccupied	0.4%/99.6%	Design outdoor + 10	60	None	15	None	Negative	None	85 dBA
Chemical treatment	0.4%/99.6%	Design outdoor + 10	60	None	10	None	None	None	85 dBA
Battery Rooms	0.4%/99.6%	77	77	None	As required for hydrogen dilution	None	Negative or neutral	50%	85 dBA

[a]See Chapter 26 of the 1997 *ASHRAE Handbook—Fundamentals* for design dry-bulb temperature data corresponding to the given annual cumulative frequency of occurrence and the specific geographic location of the plant.

[b]Multiplicity indicates that the HVAC system should have multiple units.

[c]See Figure 4 in Chapter 7 of the 1997 *ASHRAE Handbook—Fundamentals* for noise criterion curves.

In plant areas where compressed gas containers are stored, the design temperature is according to the gas supplier. Typically, the minimum temperature should be high enough that the gas volume can be effectively released from the containers. If the gas is hazardous (e.g., chlorine), the minimum temperature does not apply during personnel occupancy periods, when high dilution ventilation rates are needed.

Practical ventilation rates for fuel-fired power plants provide indoor conditions 10 to 20°F above the outdoor ambient. Therefore, ventilation design criteria establish a temperature rise above the design outdoor temperature to produce an expected indoor temperature that matches the electrical equipment ratings. For example, an outdoor extreme design temperature of 112°F with a ventilation system designed for a 10°F rise would meet the requirements of 122°F-rated plant equipment unless a new record temperature occurred. However, the environment for workers should also be considered. Velocity (spot) cooling may be necessary in some areas to support work activities in such a building.

Low temperatures affect plant reliability due to the potential for freezing. The selection of the low design temperature should be balanced by the selection of the heating design margin. If the record low temperature is used in the design, indoor design temperatures of 35 to 40°F may be used. In the heating system design, credit is generally not taken for heat generated from operating equipment.

The selection of outdoor design humidity levels affects the selection of cooling towers and evaporative cooling processes and the sizing of air-conditioning coils for outdoor air loads. When values from Chapter 26 of the 1997 *ASHRAE Handbook—Fundamentals* are used for design, the mean coincident wet bulb is appropriate. If extreme dry-bulb temperatures are selected for the design basis, the use of extreme wet bulbs is too restrictive because the extremes are not coincident. It is prudent to use the wet bulb associated with the 1% dry bulb when extreme dry-bulb temperatures are used for the design.

Indoor design humidity is not a factor in ventilated areas unless the plant is in a harsh, corrosive environment. In this case, lower humidity reduces the potential for corrosion. In air-conditioned areas for personnel or electronic equipment, ASHRAE, ISA, and manufacturers' recommendations dictate the humidity criteria.

Ventilation Rates

Ventilation within plant structures provides heat removal and dilution of potentially hazardous gases. Ventilation rates for heat removal are calculated during HVAC system design to meet summer indoor design temperatures.

The numbers shown in Table 1 for air change rates are for estimating approximate ventilation needs. Actual heat emission rates should be obtained from equipment manufacturers or from the engineer's experience.

The ventilation rate for room heat removal is

$$Q = \frac{q}{(t_r - t_o)(60\rho c_p)} \tag{1}$$

where

Q = ventilation rate, cfm
q = room heat, Btu/h
t_r = allowed room temperature from Table 1, °F
t_o = outside air temperature, °F
ρ = air density, lb_m/ft^3
c_p = specific heat of air = 0.24 $Btu/lb_m \cdot °F$

Hazardous gases are mostly handled by the process system design functions. Natural gas and other combustible fuel gases are controlled by ignition safeties and may contain odorants for detection. Hydrogen and gases used for generator and bus cooling are monitored for leakage by pressure loss or makeup rates. Escaped gases are diluted by outside air infiltration. There are no specific ventilation rate criteria for these gases. For a building with very tight construction (i.e., very little natural infiltration), an analysis should be performed to verify that dilution rates are acceptable.

Flue gas is confined to the boiler and flue gas ductwork and generally poses no hazard. In a forced-draft boiler, however, flue gas is at a higher pressure than the surroundings and can leak into occupied areas. Also, special flue gas treatment gases such as ammonia can leak into the boiler building. In these cases, gas detection monitors should be used.

Ventilation for areas with hazardous gases (e.g., chlorine) should be designed by specific gas industry standards such as the *Chlorine Manual* or ACGIH (1998).

Infiltration

Infiltration of outside air into boiler and power generation structures is driven by the thermal buoyancy of the heated air. While this outside air is beneficial in that it dilutes fugitive fumes, it also adds to the cold weather load on the heating system.

Noise

Consideration should be given to noise levels produced by HVAC system equipment both inside and outside plant spaces. Indoor noise guidelines should be established for air-conditioned areas and ventilated areas with continuous occupancies. Outdoor noise levels are established by the environmental noise pollution concerns of adjacent areas.

Air-conditioned indoor spaces should meet the normal sound level guidelines for occupancies such as offices that are listed in Chapter 46. Special occupancies such as control rooms should follow the guidelines in Table 1.

Ventilated areas of the plant should be treated as other industrial areas following OSHA regulations. The sound levels indicated in Table 1 are suggested guidelines that may be appropriate in the absence of a specific engineered solution to meet the OSHA requirements.

HVAC system components contribute to the overall noise level outside the plant buildings either by generating noise or by having ventilation openings. HVAC designs for power plants situated in urban areas can be significantly influenced by outdoor noise level requirements. Equipment may have to include sound-absorbing materials or be located indoors in sound attenuation enclosures. Openings may require acoustical louvers.

VENTILATION APPROACH

Ventilation can be achieved by natural draft, forced mechanical supply and exhaust, or natural and mechanical combined systems. The HVAC engineer should select the approach that best suits the overall project objectives for performance and cost. **Natural-draft systems** use a combination of inlet louvers and open doors or windows and relieve warmed air through roof or high sidewall openings. **Mechanical systems** use fans, power roof ventilators (PRVs), or air-handling units to move air. A typical **combined** system would use mechanical supply fans with natural roof openings for relief. With any ventilation arrangement, consideration should be given to the physical separation of inlet and outlet openings to minimize recirculation, as discussed in Chapter 15 of the 1997 *ASHRAE Handbook—Fundamentals*.

Driving Forces

Natural ventilation uses the thermal buoyancy of the air as the motive force for air movement through a building. The equations for determining differential pressures for natural ventilation are found in Chapter 25 of the 1997 *ASHRAE Handbook—Fundamentals*. With natural ventilation, air enters the enclosed space and is heated by the plant equipment. The difference in density between the inside air and the outside air causes air to be drawn into the building at low elevations and relieved at high elevations.

Mechanical ventilation can provide the required ventilation rate regardless of the building configuration or temperature difference.

Air Distribution

With natural ventilation, small differential pressures drive air movement. Accordingly, air is drawn into the building at low velocities; it penetrates a short distance into the building and then disperses.

Mechanical ventilation supplied from the walls or from the roof can distribute air more effectively throughout the structure.

Inlet and Exhaust Areas

Due to the low differential pressure driving the air, natural ventilation requires large inlet louver and exhaust relief areas. This necessitates large wall spaces where the ventilation is needed and large roof areas for hoods, ridge vents, or gravity relief openings.

Mechanical ventilation requires fewer roof penetrations with the associated curbs, crickets, and flashing. The amount of wall louver can be reduced by a factor of four or more compared to natural ventilation.

Noise

Natural ventilation has the advantage of being noise-free. Noise can be an indirect concern, however, because the openings required for natural ventilation allow noise generated by other plant equipment to pass more easily to the outside.

Fans and PRVs generate noise directly, but the noise level can be managed by fan selection and sound treatment of PRV inlet hoods.

Impact on Plant Cleanliness

By its nature, natural ventilation creates negative pressure in the lower portions of the building. With the large openings low in the structure, the negative pressure may draw dust and fumes into the building from ground level.

Mechanical ventilation usually pressurizes the building and can draw air from relatively clean sources at higher elevations.

Economics

The primary advantage to natural ventilation is that there are no operating costs for fan power. Because natural ventilation is passive, it is more reliable and has lower maintenance costs than a mechanical system. However, natural ventilation may not always be the most economical selection. The cost of louvers and inlet openings, architectural features, and gravity relief openings to achieve an acceptable ventilation rate may be higher than the first cost for mechanical ventilation.

Another consideration is the average building temperature. Because internal heat is the driving force, the naturally ventilated building is normally warmer than the power ventilated building. This warmer average temperature may shorten the life of plant equipment such as expansion joints, seals, motors, electrical switchgear, and instrumentation. Warmer temperatures may also affect operator performance.

The large louver areas associated with natural ventilation may create greater infiltration, thereby increasing the winter heating load. This additional heating cost may offset some of the summer energy savings of natural ventilation.

STEAM GENERATOR BUILDINGS: INDUSTRIAL AND POWER FACILITIES

A steam generator is a device that uses heat energy to convert water to steam. The two basic subsystems of a steam generator are the heat energy system and the steam process system.

The heat energy system for a fueled (oil, gas, coal, etc.) steam generator includes fuel distribution piping or conveyors, preparation subsystems, and supply rate and ignition controls. Provision to supply and regulate combustion air is required at the combustion chamber; the flue gas is handled downstream of the combustion area. With ash-producing fuels, the bottom ash below the steam generator and the fly ash entrained in the flue gas must be processed. Figure 1 shows a steam generator building with typical components.

Steam process components typically found in the steam generator building include an enclosure for the fire and heat transfer surfaces and feedwater equipment such as pumps, piping, and controls. Steam lines for primary and reheat steam are typically routed from the steam drum and reheat sections of the steam generator to the steam turbine or process systems.

The heat energy and steam process systems impose requirements on the HVAC systems for specific areas of the steam generator building.

Burner Areas

Fuel (gas, oil, coal, etc.) is transferred to the furnace, mixed with combustion air, and ignited in the burner area of the steam generator. Instrumentation must modulate the fuel in response to combustion needs. View ports typically allow operators to monitor the combustion processes.

The burner area requires special attention for the steam generator building ventilation system. This area is often occupied by plant operators who monitor and inspect the controls and the combustion process. Heat is radiated and conducted to the adjacent spaces from inspection ports, penetrations, and the steam generator. Leakage of fumes and combustion gases is also possible.

Both the burner area operator and the controls require ventilation with outdoor air. Outdoor air also provides dilution for the fugitive fumes. Outdoor air can be ducted to burner areas and discharged by supply registers or blown directly into the area with wall-mounted fans, depending on the building arrangement. The flow rate is difficult to quantify; generally, 60 air changes per hour supplied to an area 15 to 20 ft around the steam generator provide adequate ventilation. Consideration should be given to providing velocity cooling of personnel workstations. In cold climates, outdoor air may need to be tempered with indoor air.

Steam Drum Instrumentation Area

A typical steam generator has a steam drum at the top of the boiler that provides the water-to-steam interface. The water level in the drum is monitored to regulate the flow of steam and feedwater. This is a critical steam generator control function, so accurate and reliable process flow measurement is important.

The steam drum instrumentation area may include sections of uninsulated furnace surface, which conducts and radiates heat to the surrounding area. The ventilation system should remove this heat to ensure that area temperatures are within instrumentation temperature limits. Instrumentation may need to be shielded from hot surfaces. Velocity cooling may be needed at operator workstations.

Wall-mounted panel fans in the outside walls are an option for providing ventilation air during warm weather. Heating is generally not a concern unless the steam generator is expected to be out of service during cold weather.

Local Control and Instrumentation Areas

In addition to the drum and burners, the steam generator building may house local control areas for such functions as fuel supply, draft fans, or ash handling. Because areas around a steam generator may be hot and dirty, the location and selection of the control equipment should be coordinated between the electrical system engineer and the HVAC system engineer.

The alternatives are (1) to locate the control equipment remotely from the steam generator, (2) to use electrical components that can withstand the environment, or (3) to provide a local environmentally controlled enclosure. The first alternative requires additional cable and raceway and perhaps additional signal boosters and conditioners. The second alternative requires increased cost for electrical equipment that can tolerate the extremely hot or dirty areas.

When the electrical and control system design dictates that the equipment be located near the steam generator, a dedicated enclosure with a supporting environmental control system may be necessary. A typical environmental control system may

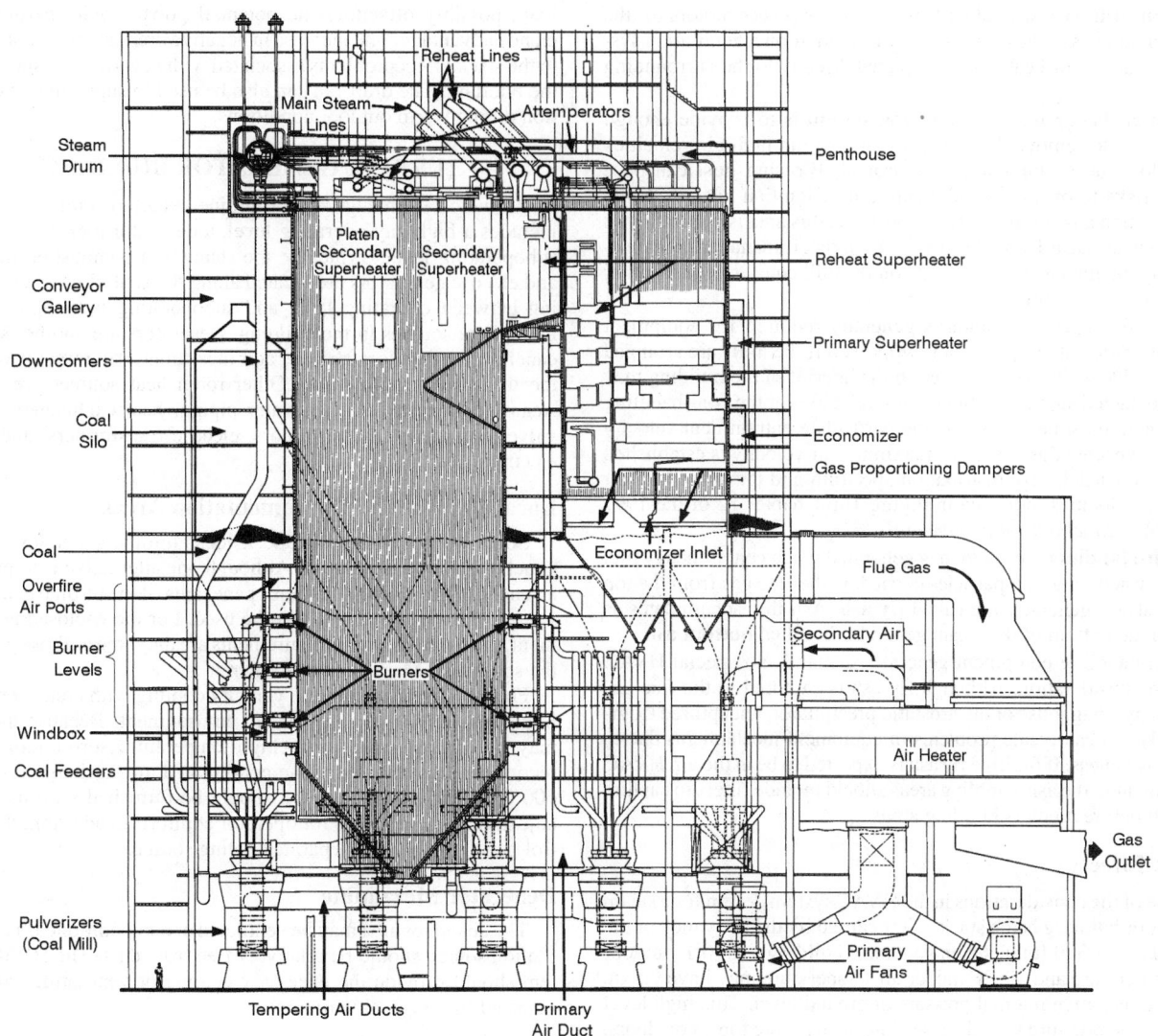

Fig. 1   Steam Generator Building
Courtesy Babcock and Wilcox

control an air-handling unit capable of providing adequate filtration, pressurization, and temperature control. The temperature control may be obtained with a chilled-water or direct-expansion coil with a remote condensing unit. An air-cooled condensing unit may be used if it is rated to match the surroundings.

Coal- and Ash-Handling Areas

Coal is typically stored on-site, either in piles or in storage structures. Material-handling equipment moves the coal to conveyors for transportation to preconditioning equipment (e.g., a crusher). The processed coal is conveyed to the steam generator building storage silos. Coal feed equipment regulates the supply of coal from the silos either to the burner or to final processing equipment such as pulverizing mills.

The coal-handling areas in the steam generator building that require special ventilation system consideration are the conveyor, silo, feeder, mill, and ash-handling areas.

Conveyor Areas. The primary concerns are dust control, outgassing from the coal, freezing of the coal and personnel access areas, and fire protection. Dust can be a concern due to the potential for environmental emission and also as a personnel and/or explosive hazard. Dust may be controlled by water-based spray systems or by

air induction pickups at the point of generation. Some types of coal may outgas small quantities of methane, which could accumulate in the conveyor and storage structures.

Natural or forced ventilation must remove heat from conveyor motors, other equipment, and envelope loads. Ventilation air can also remove outgassed fumes. Generally, the ventilation requirements are low, on the order of 2 to 5 air changes per hour. If air entrainment dust collection equipment is used, provisions for makeup should be included in the design. If natural openings are not sufficient for makeup air, supply ventilation fans may need to be electrically interlocked to operate with the dust suppression/collection equipment. The makeup air may have to be heated if freeze protection is a design criterion. Unit heaters are generally used for spot heating. The unit heater should be specified to the hazard classification for the area it serves. Because coal dust can produce acids when wet, consideration should be give to specifying noncorrosive materials and coatings.

Silo and Feeder Areas. Coal is generally fully contained by feeders and silos, so no special ventilation is needed. Occasionally, the coal systems include an inert gas purge system for fire prevention. Ventilation may be needed for life safety dilution ventilation of purge gases.

Coal Mill Areas. Coal mills require large power motors for the grinding process. These motors may have their own ventilation system, or the motor heat may be rejected directly to the surrounding space.

The challenge for the ventilation system is to provide enough ventilation to remove the heat without creating high air velocities that blow the accumulated dust around. Blowing dust can pose health risks to operators and create a dust ignition hazard. While dust ignition air-to-dust ratios are possible, this area is generally not classified as hazardous. The dust ignition risk is managed by housekeeping, maintenance of the seals on the mill equipment, and other dust-control measures.

Forced supply ventilation is generally required for equipment cooling. Sidewall propeller fans work well if the mills are arranged near outside walls. Mills located in the interior of the building may require ducted supply air. Supply air velocities at the coal-handling equipment must be lower than the particulate entrainment velocity for the expected dust size. The maximum air velocity is established using the particle size distribution spectrum and the associated air settling velocities indicated in Figure 1 in Chapter 12 of the 1997 *ASHRAE Handbook—Fundamentals*.

Ash-Handling Areas. Ash is generated when coal or heavy fuel oil is burned. Fine ash particles carried by the flue gas from the top of the steam generator are called **fly ash**. Ash that accumulates as slag in the bottom of the steam generator is called **bottom ash**.

Ash-handling equipment generally demands no special HVAC system consideration. Although fly ash is captured in the flue gas stream by a baghouse or electrostatic precipitator, uncaptured (fugitive) fly ash can create problems in equipment mechanisms due to its abrasiveness. If fugitive fly ash is expected to be in the air, HVAC equipment in the ash-handling areas should include filters to capture the ash before it enters building areas.

Stack Effect

One of the considerations in the HVAC system design for a steam generator building is the stack effect caused by the buoyancy of the heated air. A 300 ft tall steam generator building with 0°F outdoor air temperature and 100°F indoor air temperature may have 0.5 in. of water negative internal pressure at ground level. This high level of negative pressure would cause abnormally large forces on doors, creating a hazard for operators.

Sources of Combustion Air

Large-draft fans supply the combustion air for the steam generator. A positive pressure steam generator is supplied by forced-draft fans, and a negative pressure steam generator uses induced-draft fans. A balanced-draft steam generator, typical for a larger unit, uses both forced- and induced-draft fans. Because the air is heated to furnace temperatures in the combustion process, part of the fuel energy is used to heat the air. The forced-draft fans on a large steam generator can supply 100,000 cfm or more to the combustion process. A significant amount of energy is needed to preheat the combustion air.

Two prevailing methods of preheating combustion air are used. One method is to draw the air in from outdoors and heat it via steam or hot water coils using energy directly from the power cycle.

Another method is to use the heat rejected from the steam generator surfaces to the building space to heat combustion air; this method provides energy savings over heating outdoor air. Temperatures in the higher levels of the generator building can be 100°F or higher. The heat recovery is accomplished by locating the intake to the draft fan high in the building. Although the potential for savings is large in a cold climate, the total effect on the building heating and ventilation systems should be evaluated. One effect is that drawing the air from the building makes the building pressure more negative; this increases infiltration and adds to the building heating system

load, possibly offsetting the potential power cycle thermal efficiency advantage. The increase in negative pressure also contributes to the stack effect problems associated with negative pressure low in the building. The draft fan can also be used to supplement ventilation during warm outdoor conditions.

TURBINE GENERATOR BUILDING

As indicated in Figure 2, a turbine generator building usually includes a high-bay operating level, a deaerator mezzanine, and a suboperating level. Typically, the steam and combustion turbines and electric generators are located along the centerline of the building between operating level and suboperating level and are the major heat sources in the building. Deaerators are another significant heat contributor; the deaerator mezzanine is commonly open to the turbine operating level. Other room heat sources are steam, steam condensate and hot water piping, heat exchangers, steam valve stations and traps, motors, electric transformers, and other electrical equipment.

Local Control and Instrumentation Areas

Some power plants include a local turbine-generator control panel on the operating floor. Although one alternative is to provide an enclosure for the local control panel area of the turbine generator, an open arrangement may also be used. For the enclosed arrangement, the environmental requirements are the same as those given in the section on Main Control Center.

For an open arrangement, velocity cooling with conditioned air improves the operator's working environment. Because the area may be directly exposed to high-temperature surroundings, the recommended velocity of the conditioned air discharge is 300 to 600 fpm. The air distribution should be furnished with manually adjustable air deflectors for operator comfort. In addition, the control panel may need a separate cooling source.

Deaerator Mezzanine

The deaerator area may be ventilated or air conditioned. The ventilation source should be relatively free from dust. The HVAC system should provide the necessary cooling or ventilating capacity, adjusted for seasonal variations.

Bridge Crane Operating Rooms

Outside air entering the building is heated by the process heat, rises toward higher elevation, and is relieved through openings. The bridge crane operating room is as high as the roof beam and within the building exhaust airstream. If the outside air temperature is 95°F, the crane operating room may be surrounded by 105°F or hotter air. Hence, the bridge crane operating room is normally furnished with a cooled air supply.

Because the bridge crane operating room moves around in the building during its operation, through-the-wall mounted air conditioners are commonly used. To simplify the electrical work, an additional power plug in the crane for the air conditioner should be provided by the crane supplier. Provisions for the cooling coil condensate drain should be included in the design.

Suboperating Level

The turbine generator is located on the operating floor, which is a large deck surface that is open to the turbine building roof. The deck may be 70% or more of the turbine building area. Below the operating level are one or more suboperating levels. Ventilation supply air should be provided to the suboperating levels and at the lower elevations of the operating floor. The air rising through the operating levels brings room heat to the roof area, where it is relieved through high-elevation openings or exhausted by PRVs.

The major heat sources in the suboperating level are high-temperature mechanical and piping systems. Other heat contributors

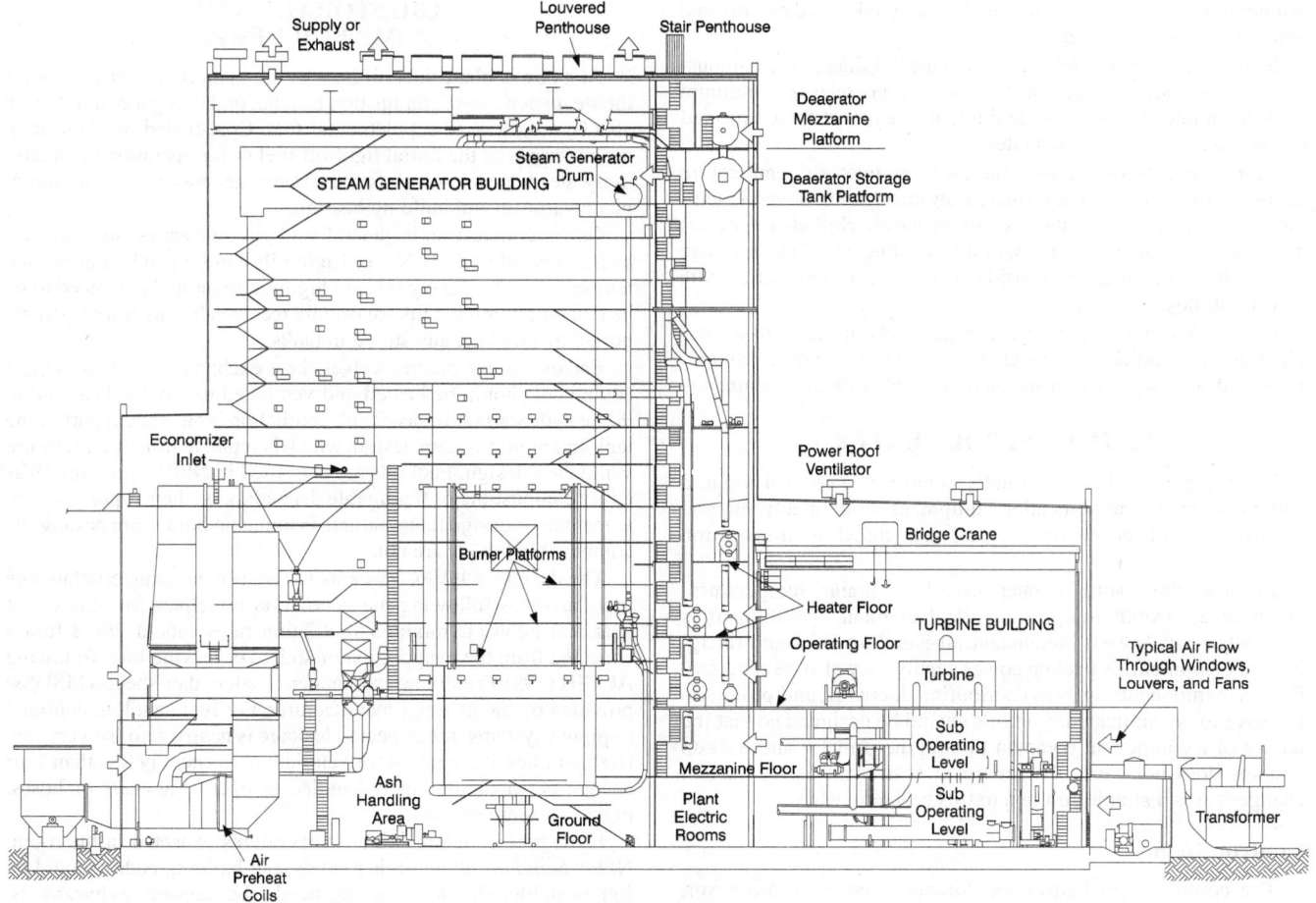

Fig. 2 Generation Building Arrangement
Courtesy Black & Veatch LLP

are electric transformer room exhaust, switchgear room exhaust, electric reactor room exhaust, electrical motor heat, etc. The electrical equipment exhaust heat in the turbine generator building is small compared to the heat emitted from mechanical and piping systems. Accordingly, ventilation air from the plant distribution electric room can generally be exhausted directly into the turbine building without ducting to the outside. Conditioned air from an air-conditioning system may be supplied to local instrumentation panel areas.

For plants in cold climates (temperatures below freezing), the engineer should consider spot heating and/or exterior door heated air curtains. Consideration should also be given to freeze protection of piping close to building walls because of the stack effect and wind-driven infiltration. This can increase local heating requirements at the building perimeter. This problem can be addressed by adding capacity to the installed heating systems or by providing mobile, temporary heating.

Electric Transformer Rooms

Transformer rooms are typically at a suboperating level between the turbine building and steam generator buildings. For the isolation of dust, the transformer rooms should not have inlet openings to the steam generator building.

The transformer room exhaust air temperature should not exceed the design limit of the transformers. Typically, air intake is from the turbine building suboperating floor level or from outdoors, and exhaust air discharges at the higher level of the transformer room toward the turbine building.

Plant Electrical Distribution Equipment and Switchgear/MCC Rooms

Air for the main station switchgear and motor control center (MCC) rooms should be relatively clean. Supply air from outdoors should be filtered with 30% efficiency air filters. Air can be relieved through louvers into the plant or to the outside.

A similar approach is used for the ventilation system for an electric reactor room. If the reactors have ducted connections for the exhausts, a removable section may be required so that the ducts can be disassembled when the reactor is lifted during maintenance.

COMBUSTION TURBINE AREAS

The heating and ventilation design issues unique to combustion turbines are combustion air intakes, ventilation of the equipment compartments, the high-temperature exhaust duct, fuel supply piping, inlet air cooling, and noise.

Airflows requirements associated with combustion air and separately ventilated compartments should be provided by the manufacturer and accommodated in the heating and ventilation system design.

High-temperature combustion turbine exhaust ductwork can create a significant heat load on the ventilation system. The duct should be insulated and arranged to allow air movement over its surfaces to minimize hot spots. Provisions should be made to isolate the hot duct wall and roof materials.

One method of addressing the high stack surface temperatures is the induction of room air into the stack to lower the exhaust temperature. The induced air must be included in the heating and ventilation system design.

Systems delivering fuel gas through the building to the combustion turbine should be designed to prevent gas release. Minimum ventilation rates should provide dilution of gases at their expected design leakage and detection rates.

Combustion turbine power increases as inlet air temperature decreases due to the increased mass flow throughput. Several methods are available to cool the inlet air, including chilled-water coils and both direct and indirect evaporative cooling coils. These power cycle enhancements can be considered alone or in conjunction with thermal storage.

Noise from combustion turbines is generally managed by a combination of sound-attenuating enclosures over the turbine components and acoustical treatments on building ventilation openings.

MAIN CONTROL CENTER

The main control center usually comprises a control room, a battery room, a communication equipment room, electronic and electric control panel rooms, and associated administration areas.

Because the control center usually contains temperature-sensitive electronic equipment critical to plant operation, it is generally provided with redundant air-handling units and refrigeration equipment. A backup power supply may also be required. Passive components such as distribution ductwork and piping do not have to be duplicated. Controls should be designed so that the failure of a component common to both the primary and backup systems does not cause the failure of both systems. A manual changeover is a simple solution to this problem.

Control Rooms

The control room houses the computerized microprocessor, printer, electronic and emergency response controls, fire protection controls, communication and security systems, regional system networks, accessories, and relevant wiring and tubing systems.

An air-conditioning system typical for office occupancy, with features to meet overall design requirements for reliability and the specific environmental needs of the control equipment, is generally appropriate.

Battery Rooms

The optimum temperature for battery rooms balances battery life against battery performance. Temperatures higher than the optimum reduce battery life, while lower temperatures reduce battery performance. The typical optimum temperature for wet cell battery rooms is approximately 77°F.

Another consideration in battery room design is the hydrogen gas generated during the charging process. The amount of hydrogen generated depends on the charging current and the duration of the charging process. Hydrogen generation rates vary among the different types of batteries. The battery manufacturer should be contacted for pertinent design-related information, and the guidelines of National Fire Protection Association (NFPA) *Standard* 70, the *National Electrical Code*, should be followed. The recommended hydrogen concentration in the battery room is 2% or less of the room volume. When no battery design information is available, a general ventilation exhaust rate of 2 to 4 air changes per hour may be adequate for preventing the hydrogen concentration in the battery room from reaching explosive limits. If dilution ventilation air is provided, the room does not have to be rated as hazardous. Standard HVAC materials should be used, and in the unlikely event of a spill, the room and the ventilation equipment should be cleaned.

OIL STORAGE AND PUMP BUILDINGS

At a power plant, fuel oil may be the main source of energy for the steam generator, combustion turbine, or diesel generator. It may also be a backup or supplemental fuel. Coal-fueled plants usually use oil or gas as the initial light-off fuel or for operation of an auxiliary steam generator. Auxiliary steam generators provide initial plant warm-up and building heating.

Oil for combustion is generally a light oil such as No. 2 fuel oil or a heavy oil such as No. 6. Light oils can be pumped at normal temperatures, but heavy oils are highly viscous and may need to be heated for pumping. Oils are usually received by rail or truck, transported by pipeline, and stored in tanks.

Enclosures for pumps, valves, heat exchangers, and associated equipment should be heated and ventilated to remove heat and to dilute hydrocarbon fumes. Tank ventilation is an integral part of the tank and piping system design, which is separate from the enclosure ventilation design. Fuel oils are classified in NFPA *Standard* 30 as either combustible or flammable depending on their vapor pressure at the indoor design temperature. Flammable liquids are hazardous; combustible liquids are not.

The design of HVAC systems for areas containing combustible fuels involves following the ventilation principles for removal of heat and good air mixing. Ventilation rates should dilute fumes expected from the evaporation of spilled or leaking fuel, following ACGIH (1998) guidelines and material safety data sheets (MSDSs) provided by the material manufacturer. For fuel handling confined to piping systems, the expected leakage is nearly zero, so very low fresh air rates are required for ventilation—generally less than 1 air change per hour. If fuel is handled in open containers or hoses, higher rates are prudent.

If the fuel is flammable at temperatures expected in the room, NFPA *Standard* 30 and other safety and building codes should be followed. Electrical systems may need to be classified as hazardous.

COAL CRUSHER AND COAL TRANSPORTATION SYSTEM BUILDINGS

The coal-handling facilities at a power plant receive and prepare coal and then transport it from the initial delivery point to the burners. Intermediate steps in the process may include long- or short-term storage, cleaning, and crushing. Receipt may include barge, railcar, or truck unloading. Storage may be in piles on the ground, underground, or in barns or silos. On-site, the coal is handled by mobile equipment or conveyor systems.

Regardless of the specifics, the following general HVAC issues apply for the structures involved.

Potential for Dust Ignition Explosion

Most types of coal readily break down into dust particles when handled or conveyed. The dust can become fine enough and occur in the right particle size distribution and concentration to create a dust explosion. The design engineer should review and apply the referenced NFPA standards and guidelines to determine the dust ignition potential for each ventilation system application.

Ventilation of Conveyor and Crusher Motors in a Coal Dust Environment

Heat from motors and process equipment should be removed through ventilation. The options are to use ducted, ventilated motors or to ventilate the building enclosures. Ventilation in enclosures containing coal should keep the velocity below the entrainment velocities of the expected particle sizes. Figure 1 in Chapter 12 of the 1997 *ASHRAE Handbook—Fundamentals* has information on settling velocity. Generally, air should be mechanically exhausted to allow ventilation air to enter the building through louvers at low velocities.

Cooling or Ventilation of Electrical and Control Equipment

Electrical and control equipment may be located near coal piles or other coal-handling facilities. Air-conditioned control rooms should be pressurized with filtered outdoor air. Ventilated motor control or switchgear areas should also be pressurized with filtered air. Due to the high dust concentrations in coal yards, ordinary filter media have a short life; a solution is the use of inertial filters. For air-conditioned areas, the inertial filters can be followed by higher efficiency media filters.

For electrical equipment rooms adjacent to an area with the potential for a dust ignition explosion, NFPA *Standard* 496 should be followed. This standard recommends the flow of clean air away from the electrical equipment into the dusty area.

Ventilation of Methane Fumes

Methane and other hydrocarbons are present in the coal both as free gas in the cracks and voids and as adsorbents within the coal. Although most of the methane is released from the interstitial coal structure during the mining and handling process, some methane or other potentially flammable gases may remain in the coal. Thus, flammable concentrations of methane can accumulate when large amounts of coal are stored. The design engineer should identify the potential for methane accumulation when designing for structures associated with silos or coal storage buildings. At the mine or mine mouth, methane gas emission rates as high as 5 ft^3/ton·day are possible. At locations removed from the mine mouth, the rate is usually less than 1 ft^3/ton·day Dust collection air exhaust or natural ventilation is often sufficient to prevent the methane level from reaching the 1% explosion limit. The design engineer should apply guidelines from NFPA *Standards* 120, 123, 850, 8503, 8504, and 8505.

Underground Tunnels and Conveyors

Enclosed conveyors are generally of loose construction and require no ventilation. Smoke or gases in underground conveyor tunnels, hoppers, or conveyor transfer points could cause a personnel safety hazard. Ventilation systems should be coordinated with escape route passages to move fresh air from the direction of the egress. Ventilation rates in the range of 2 to 5 air changes per hour are generally appropriate for normal system operation.

Makeup of Dust Collection Air

Coal dust can be controlled by high-velocity air pickup at locations where coal is transferred. The volumetric airflow associated with these pickup points may be sufficient to meet the ventilation requirements. Air inlets must be provided. If additional ventilation is needed, the ventilation fan must coordinate with the dust collection system. For heated structures, makeup air may need to be heated.

HEATING/COOLING SYSTEMS

Selection of the heating and cooling systems in a power plant depends on several variables, including the geographical location and orientation of the plant and the type of fuel used. Most plants are ventilated with outdoor air, but it is customary in hot climates to air condition many plant areas. Steam, hot water, gas, and electricity are alternative heating methods to be evaluated for the most economical choice. Electricity is the primary energy source for general-purpose cooling of the various areas of the plant. Areas such as the main control room, office areas, and electrical switchgear/MCC rooms are air conditioned for continuous human occupancy or for maintaining the operability of the electrical equipment and controls.

Cooling

The cooling source may be either a centrally located chilled-water system providing chilled water to various area coolers or individual direct-expansion area coolers with either air-cooled or water-cooled condensing units. Selection depends on the layout of the various areas to be cooled and the comparative costs of the two options. The condensing system of the water chiller may be either air cooled or water cooled with available water from the plant service water system. Air-cooled chillers are used when the air near the proposed chiller location is moderately clean and no fly ash or coal dust problem is anticipated. For a water-cooled system, a closed-loop cooling tower is sometimes used if service water is poor or unavailable (e.g., during start-up or plant outages). To protect the chillers from the fouling and corrosion effects of the service water, a heat exchanger is sometimes used between the chiller condenser and the service water source. In a power plant with several operating units and individual self-supporting chilled-water systems, the chilled-water systems of each unit are sometimes interconnected to provide backup and redundancy.

Heating

Heating in various areas of the power plant is usually provided by electric, steam, or hot water unit heaters or heating coils in air-handling units. In a hot water distribution system, glycol is usually added into the system for protection against freezing. Because the stack effect in the building induces large quantities of infiltration air, heating requirements in the lower levels of the steam generator building may increase when the steam generator is operating. Pressurization fans directing cooler, outside air into the warmer upper elevations of the plant can offset this infiltration. Also, the design engineer may evaluate redistribution of hotter air from higher to lower elevations.

An alternative to heating the open areas of the plant is to use pipeline heat tracing and spot heating at personnel workstations. For this approach, the design engineer should consider all components that may require heat tracing, such as instrument lines, small and large pipes, traps, pumps, tanks and other surfaces that may be subject to the freezing temperatures. Often the large number of components and surfaces to be heat traced and insulated makes this impractical.

REFERENCES

ACGIH. 1998. *Industrial ventilation: A manual of recommended practice*, 23rd ed. American Conference of Governmental Industrial Hygienists, Cincinnati, OH.

ASHRAE. 1989. Ventilation for acceptable indoor air quality. ANSI/ASHRAE *Standard* 62-1989.

CI. 1997. *The chlorine manual*, 6th ed. The Chlorine Institute, Washington, D.C.

IEEE. 1996. Recommended practice for installation design and installation of vented lead-acid batteries for stationary applications. ANSI/IEEE *Standard* 484-96. Institute of Electrical and Electronics Engineers, Piscataway, NJ.

ISA. 1986. Environmental conditions for process measurement and control systems: Airborne contaminants. ANSI/ISA *Standard* S71.04-86. International Society of Measurement and Control, Research Triangle Park, NC.

NFPA. 1992. Recommended practice for stoker operation. ANSI/NFPA *Standard* 8505-92. National Fire Protection Association, Quincy, MA.

NFPA. 1993. Purged and pressurized enclosures for electrical equipment. ANSI/NFPA *Standard* 496-93.

NFPA. 1994. Coal preparation plants. ANSI/NFPA *Standard* 120-94.

NFPA. 1995. Fire prevention and control in underground bituminous coal mines. ANSI/NFPA *Standard* 123-95.

NFPA. 1996. Atmospheric fluidized-bed boiler operation. ANSI/NFPA *Standard* 8504-96.

NFPA. 1996. *Flammable and combustible liquids code*. ANSI/NFPA *Standard* 30-96.

NFPA. 1996. Installation of centrifugal fire pumps. ANSI/NFPA *Standard* 20-96.

NFPA. 1996. Recommended practice for fire protection for electric generating plants and high voltage direct current converter stations. ANSI/NFPA *Standard* 850-96.

NFPA. 1997. Pulverized fuel systems. ANSI/NFPA *Standard* 8503-97.

NFPA. 1999. *National electrical code*. ANSI/NFPA *Standard* 70-96.

BIBLIOGRAPHY

China Ministry of Power. 1975 and 1994. Technical code for designing fossil fuel power plants. People's Republic of China *Standard* DL 5000-94.

Copelin, W. and R. Foiles. 1995. Explosion protection from low sulfur coal. *Power Engineering* (November).

NFPA. 1997. *Fire protection handbook*, 18th ed. Chapter 9. National Fire Protection Association, Quincy, MA.

Shieh, C. 1966. *HVAC handbook*. China Ministry of Power.

U.S. Bureau of Mines. 1978. Methane emissions from gassy coals in storage silos. *Report* of Investigation 8269.

NUCLEAR FACILITIES

THE HVAC requirements for facilities using radioactive materials are discussed in this chapter. Such facilities include nuclear power plants, fuel fabrication and processing plants, plutonium processing plants, hospitals, corporate and academic research facilities, and other facilities housing nuclear operations or materials. The information presented here should serve as a guide; however, careful and individual analysis of each facility is required.

BASIC TECHNOLOGY

Criticality, radiation fields, and regulation are three issues that are more important in the design of nuclear-related HVAC systems than in that of other special HVAC systems.

Criticality. Criticality considerations are unique to nuclear facilities. Criticality is the condition reached when the chain reaction of fissionable material, which produces extreme radiation and heat, becomes self-sustaining. Unexpected or uncontrolled conditions of criticality must be prevented at all cost. In the United States, only a limited number of facilities—including fuel plants, weapons facilities, and some national laboratories—handle **special nuclear material** (SNM) subject to criticality concerns.

Radiation Fields. Radiation fields are found in all facilities using nuclear materials. They pose problems of material degradation and personnel exposure. Although material degradation is usually addressed by regulation, it must be considered in all designs. The personnel exposure hazard is more difficult to measure than the amount of material degradation because a radiation field cannot be detected without special instruments. It is the responsibility of the designer and of the end user to monitor radiation fields and limit personnel exposure.

Regulation. In the United States, the Department of Energy (DOE) regulates weapons-related facilities and national laboratories; the Nuclear Regulatory Commission (NRC) controls civil, industrial, and power facilities. Further complicating the issue are other government regulations at the local, state, and federal levels. For example, meeting an NRC requirement does not relieve the designer or operator of the responsibility of meeting Occupational Safety and Health Administration (OSHA) requirements. The design of an HVAC system to be used near radioactive materials must follow all guidelines set by these agencies and by the local, state, and federal governments. For facilities outside the United States, the specific national and local regulations apply.

As Low as Reasonably Achievable

As low as reasonably achievable (ALARA) means that all aspects of a nuclear facility are designed to limit worker exposure to the minimum amount of radiation that is reasonably achievable. This refers not to meeting legal requirements, but rather to attaining the lowest cost-effective below-legal levels.

The preparation of this chapter is assigned to TC 9.2, Industrial Air Conditioning.

Design

HVAC requirements for a facility using or associated with radioactive materials depend on the type of facility and the specific service required. The following are design considerations:

- Physical layout of the HVAC system that minimizes the accumulation of material within piping and ductwork
- Control of the system so that portions can be safely shut down for maintenance and testing or in the case of any event, accident, or natural catastrophe that may cause radioactivity to be released
- Modular design for facilities that change operations regularly
- Preservation of confinement integrity to limit the spread of radioactive contamination in the physical plant and surrounding areas

The design basis in most nuclear facilities requires that safety-class systems and their components have active control during and after any event, accident, or natural catastrophe that causes radioactivity to be released.

Normal or Power Design Basis

The normal or power design basis for nuclear power plants covers normal plant operation, including normal operation mode and normal shutdown mode. This design basis imposes no requirements more stringent than those specified for standard indoor conditions.

Safety Design Basis

The safety design basis establishes special requirements necessary for a safe work environment and public protection from exposure to radiation. Any system designated **essential** or **safety-class** must mitigate the effect of any event, accident, or natural catastrophe that may cause the release of radioactivity into the surroundings or the plant atmosphere. These safety systems must be operative at all times. Safety analysis reports (SARs) determine which components must function during and after a **design basis accident** (DBA) or the simultaneous occurrence of such events as a safe shutdown earthquake (SSE), a tornado, a loss of coolant accident (LOCA), and loss of off-site electrical power (LOEP). Non-safety-related equipment should not adversely affect safety-related equipment.

System Redundancy. Systems important to safety must be redundant so that their function is performed even if a component of one system fails. Such a failure should not cause a failure in the backup system. For additional redundancy requirements, refer to the section on Nuclear Regulatory Commission Facilities.

Seismic Qualification. All safety-class components, including equipment, pipe, duct, and conduit, must be seismically qualified by testing or calculation to withstand and perform under the shock and vibration caused by an SSE or an operating basis earthquake (the largest earthquake postulated for the region). This qualification also covers any amplification by the building structure. In addition, any HVAC component, the failure of which could jeopardize the essential function of a safety-related component, must be seismically qualified or restrained to prevent such failure.

Environmental Qualification. Safety-class components must be environmentally qualified; that is, the useful life of the component in the environment in which it operates must be determined through a program of accelerated aging. Environmental factors such as temperature, humidity, pressure, acidity, and accumulated radioactivity must be considered.

Quality Assurance. All designs and components of safety-class systems must comply with the requirements of a quality assurance (QA) program for design control, inspection, documentation, and traceability of material. Refer to Appendix B of Title 10 of the *Code of Federal Regulations*, Part 50 (10 CFR 50) or ASME *Standard* NQA-1 for quality assurance program requirements.

Emergency Power. All safety-class systems must have a backup power source such as an emergency diesel generator.

Outdoor Conditions

Chapters 26 and 29 of the 1997 *ASHRAE Handbook—Fundamentals*, the National Oceanic and Atmospheric Administration, and site meteorology can provide information on outdoor conditions, temperature, humidity, solar load, altitude, and wind. DOE *Order* 6430.1A may specify outdoor conditions.

Nuclear facilities generally consist of heavy structures having high thermal inertia. Time lag should be considered in determining solar loads. For some applications, such as diesel generator buildings or safety-related pumphouses in nuclear power plants, 24 h averages suffice.

Indoor Conditions

Indoor temperatures are dictated by occupancy, equipment or process requirements, and personnel activities. HVAC system temperatures are dictated by the environmental qualification of the safety-class equipment located in the space and by ambient conditions during the different operating modes of the equipment.

Indoor Pressures

Where control of airflow pattern is required, a specific pressure relative to the outside atmosphere or to adjacent areas must be maintained. For process facilities having pressure zones, the pressure relationships are specified in the section on Confinement Systems.

In facilities where zoning is different from that in process facilities, and in cases where any airborne radioactivity must not spread to rooms within the same zone, this airborne radioactivity must be controlled by airflow.

Airborne Radioactivity

The level of airborne radioactivity within a facility and the amount released to the surroundings must be controlled to meet the requirements of DOE *Orders* 5400.5 and N 441.1, 10 CFR 20, 10 CFR 50, 10 CFR 61, and 10 CFR 100, or equivalent national regulations of the site country.

Tornado/Wind Protection

Protection from tornados and the objects or missiles launched by tornados or design basis wind is normally required to prevent the release of radioactive material to the atmosphere. A tornado passing over a facility causes a sharp drop in ambient pressure. If exposed to this transient pressure, ducts and filter housings could collapse because the pressure inside the structure would still be that of the environment prior to the pressure drop. Protection is usually provided by tornado dampers and missile barriers in all appropriate openings to the outside. Tornado dampers are heavy-duty, low-leakage dampers designed for pressure differences in excess of 3 psi. They are normally considered safety-class and are environmentally and seismically qualified.

Fire Protection

Fire protection for the HVAC and filtration systems must comply with the applicable requirements of Appendix R of 10 CFR 50 and NFPA, UL, and ANSI or equivalent standards. Design criteria should be developed for all building fire protection systems, including secondary sources, filter plenum protection, fire dampers, and detection/suppression systems. Fire protection systems may consist of a combination of building sprays, hoses and standpipes, and gaseous or foam suppression. The type of fire postulated in the SAR determines which kind of fire system is used.

Secondary sources may include one or more of the following: (1) a water tower with a diesel-powered or electric water pump, (2) a pressurized water tank, or (3) a second water main. These systems are not unique to nuclear facilities.

A requirement specific to nuclear facilities is protection of the filter plenums and the ventilation ductwork. Water sprays (window nozzles, fog nozzles, or standard dry pipe/wet pipe system spray heads) are usually used to protect the filters in filter plenums. In the case of metal filters, fire protection equipment may also be used for cleansing. Sprinkler location, system balancing, and spray patterns should be designed for maximum effect. Flow requirements should be consistent with the design criteria for the facility. Performance of the room ventilation system in response to fire scenarios must be analyzed to determine the operational requirements and optimum location of fire dampers. Various codes may limit alternatives.

Heat detectors and fire suppression systems should be considered for special equipment such as glove boxes. Application of the two systems in combination allows the shutdown of one system at a time for repairs, modifications, or maintenance. Smoke control criteria can be found in NFPA *Standards* 801 and 901.

Control Room Habitability Zone

The HVAC system in a control room is a safety-related system that must fulfill the following requirements during all normal and postulated accident conditions:

1. Maintain conditions comfortable to personnel, and ensure the continuous functioning of control room equipment.
2. Protect personnel from exposure to airborne radioactivity or toxic chemicals potentially present in the outside atmosphere or surrounding plant areas.
3. Protect personnel from the effects of breaks in high-energy lines in the surrounding plant areas.
4. Protect personnel from combustion products emitted from on-site fires.

Filtration

HVAC filtration systems can be designed to remove either radioactive particles or radioactive gaseous iodine from the airstream. They filter potentially contaminated exhaust air prior to discharge to the environment and may also filter potentially contaminated makeup air for power plant control rooms and technical support centers.

The composition of the filter train is dictated by the type and concentration of the contaminant, the process air conditions, and the filtration levels required by the applicable regulations (e.g., RG 1.52, RG 1.140, ASME AG-1, ASME N509, 10 CFR 20, and 10 CFR 100). Filter trains may consist of one or more of the following components: prefilters, high-efficiency particulate air (HEPA) filters, charcoal filters (adsorbers), sand filters, heaters, and demisters.

Dust Filters/Prefilters. Dust filters are selected for the efficiency required by the particular application. High-efficiency dust filters are often used as prefilters for the special filters listed below to prevent them from being loaded with atmospheric dust and to minimize replacement costs.

HEPA Filters. Nuclear HEPA filters are used where there is a risk of particulate airborne radioactivity. The construction and preuse quality assurance testing of HEPA filters is specified in DOE *Standards* NE F 3-43 and 3020. Filter performance requirements are based on penetration at a specified airflow and static pressure. For a 0.3 μm particle, the penetration at rated airflow must not exceed 0.03%. The construction and QA testing of HEPA filters for use in nuclear power plants is specified in ASME *Standards* N509, N510, and AG-1. HEPA filters are in-place tested and inspected when first installed, and then tested every 12 to 18 months thereafter. One or both of the following two methods may be used for in-place testing of a stage of filtration: (1) mass flow testing of the stage as a whole or (2) testing of the individual filters and frame that make up the stage. In both preuse and in-place tests, an approved challenge agent such as the aerosol dioctyl phthalate (DOP) or a similar agent must be used.

Sand Filters. Sand filters consist of multiple beds of sand and gravel through which air is drawn. The air enters an inlet tunnel that runs the entire length of the filter. Smaller cross-sectional laterals running perpendicular to the inlet tunnel distribute the air across the base of the sand. The air rises through several layers of various sizes of sand and gravel, typically at a rate of 5 fpm. It is then collected in the outlet tunnel for discharge to the atmosphere.

Charcoal Filters. Activated charcoal adsorbers are used mainly to remove radioactive iodine, which is a vapor or gas. Bed depths are typically 2 or 4 in. but may be deeper. These filters have an efficiency of 99.9% for elemental iodine and 95 to 99% for organic iodine. Charcoal filters lose efficiency rapidly as the relative humidity increases. They are preceded by a heating element to keep the relative humidity of the entering air below 70%.

To control the argon content in the primary coolant, argon is adsorbed on charcoal at extremely low (cryogenic) temperatures. In a high-temperature gas-cooled reactor or in the off-gas system of a boiling water reactor, helium is similarly adsorbed.

Heaters. Electric heating coils and/or demisters may be used to meet the relative humidity conditions requisite for charcoal filters. For safety-class systems, electric heating coils should be connected to the emergency power supply. Interlocks should be provided to prevent heater operation when the exhaust fan is deenergized.

Demisters (Mist Eliminators). Demisters are required to protect HEPA and charcoal filters if entrained moisture droplets are expected in the airstream. They should be fire resistant.

DEPARTMENT OF ENERGY FACILITIES

Nonreactor nuclear HVAC systems must be designed in accordance with DOE *Order* 6430.1A. Critical items and systems in plutonium processing facilities are designed to confine radioactive materials under both normal and DBA conditions, as required by 10 CFR 100.

CONFINEMENT SYSTEMS

Zoning

Typical process facility confinement systems are shown in Figure 1. Process facilities comprise several zones.

Primary Confinement Zone. This zone includes the interior of the hot cell, canyon, glove box, or other means of containing radioactive material. Containment must prevent the spread of radioactivity within or from the building under both normal conditions and upset conditions up to and including a facility DBA. Complete isolation from neighboring facilities is necessary. Multistage HEPA and/or sand filtration of the exhaust is required

Secondary Confinement Zone. This zone is bounded by the walls, floors, roofs, and associated ventilation exhaust systems of the cell or enclosure surrounding the primary confinement zone. Except for glove box operations, this zone is usually unoccupied.

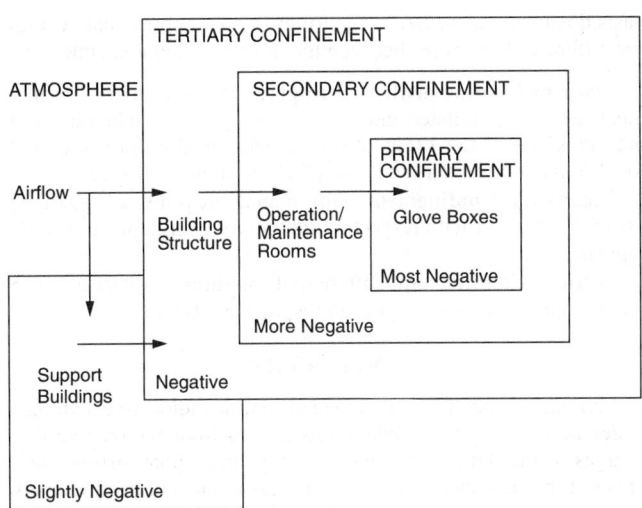

Fig. 1 Typical Process Facility Confinement Categories

Tertiary Confinement Zone. This zone is bounded by the walls, floors, roofs, and associated ventilation exhaust systems of the facility. They provide a final barrier against the release of hazardous material to the environment. Radiation monitoring may be required at exit points.

Uncontaminated Zone. This zone includes offices and cold shop areas.

Air Locks

Air locks in nuclear facilities are used as safety devices to maintain a negative differential pressure when a confinement zone is accessed. They are used for placing items in primary confinement areas and for personnel entry into secondary and tertiary confinement areas. Administrative controls ensure proper operation of the air lock doors.

There are three methods of ventilating personnel air locks (ventilated vestibules):

1. The clean conditioned supply (CCS) air method, where the air lock is at positive pressure with respect to the adjacent zones. For this method to be effective, the air lock must remain uncontaminated at all times.
2. The flow-through ventilation (FTV) air method, where no conditioned air is supplied to the air lock and the air lock stays at negative pressure with respect to the less contaminated zone.
3. The combined ventilation (CV) air method, which is a combination of the CCS and FTV methods. Testing has shown that this is the most effective method, when properly designed.

Zone Pressure Control

Negative static pressure increases (becomes more negative) from the uncontaminated zone to the primary confinement zone, causing any air leakage to be inward, toward areas of higher potential contamination. All zones should be maintained negative with respect to atmospheric pressure. Zone pressure control cannot be achieved through the ventilation system alone; confinement barrier construction must meet all applicable specifications.

Differential Pressures

Differential pressures help ensure that air flows in the proper direction in the event of a breach in a confinement zone barrier. The design engineer must incorporate the desired magnitudes of the differential pressures into the design early to avoid later operational problems. These magnitudes are normally specified in the design basis document of the SAR. The following are approximate values for differential pressures between the three confinement zones.

Primary Confinement. With respect to the secondary confinement area, air-ventilated glove boxes are typically maintained at pressures of −0.7 to −1.0 in. of water, inert gas glove boxes at −0.7 to −1.5 in., and canyons and cells at a minimum of −1.0 in.

Secondary Confinement. Differential pressures of −0.03 to −0.15 in. of water with respect to the tertiary confinement area are typical.

Tertiary Confinement. Differential pressures of −0.01 to −0.15 in. of water with respect to the atmosphere are typical.

VENTILATION

Ventilation systems are designed to confine radioactive materials under normal and DBA conditions and to limit radioactive discharges to the required minimum. They ensure that airflows are, under all normal conditions, toward areas (zones) of progressively higher potential radioactive contamination. Air-handling equipment should be sized conservatively so that upsets in the airflow balance do not cause the airflow to reverse direction. Examples of upsets include improper use of an air lock, a credible breach in the confinement barrier, or excessive loading of HEPA filters.

HEPA filters at the ventilation inlets in all primary confinement zone barriers prevent movement of contamination toward zones of lower potential contamination in the event of an airflow reversal. Ventilation system balancing helps ensure that the building air pressure is always negative with respect to the outside atmosphere.

Recirculating refers to the reuse of air in a particular zone or area. Room air recirculated from a space or zone may be returned to the primary air-handling unit for reconditioning and then, with the approval of health personnel, be returned to the same space (zone) or to a zone of greater potential contamination. All air recirculated from secondary and tertiary zones must be HEPA-filtered prior to reintroduction to the same space. Recirculating air is not permitted in primary confinement areas, except those with inert atmospheres.

A safety analysis is necessary to establish minimum acceptable response requirements for the ventilation system and its components, instruments, and controls under normal, abnormal, and accident conditions.

Analysis determines the number of exhaust filtration stages required in different areas of the facility to limit (in conformance with the applicable standards, policies, and guidelines) the amount of radioactive or toxic material released to the environment during normal and accident conditions. Consult DOE *Order* 6430.1A for air-cleaning system criteria.

Ventilation Requirements

A partial recirculating ventilation system may be considered for economic reasons. However, it must be designed to prevent contaminated exhaust from entering the room air-recirculating systems.

The exhaust system is designed to (1) clean radioactive contamination from the discharge air, (2) safely handle combustion products, and (3) maintain the building under negative pressure relative to the outside.

Provisions may be made for independent shutdown of ventilation systems or isolation of portions of the systems to facilitate operations, filter change, maintenance, or emergency procedures such as fire fighting. All possible effects of partial shutdown on the airflows in interfacing ventilation systems should be considered. Positive means must be provided to control the backflow of air that might transport contamination. A HEPA filter installed at the interface between the enclosure and the ventilation system minimizes contamination in the ductwork; a prefilter reduces HEPA filter loading. These HEPA filters should not be considered the first stage of an airborne contamination cleaning system.

Ventilation Systems

The following is a partial list of elements that may be included in the overall air filtration and air-conditioning system:

- Air-sampling devices
- Carbon bed adsorbers
- Prefilters and absorption, HEPA, sand, and glass fiber filters
- Scrubbers
- Demisters
- Process vessel vent systems
- Condensers
- Distribution baffles
- Fire suppression systems
- Fire and smoke dampers
- Exhaust stacks
- Fans
- Coils
- Heat removal systems
- Pressure- and flow-measuring devices
- Duct test ports
- Radiation-measuring devices
- Critically safe drain systems
- Tornado dampers

The ventilation system and associated fire suppression system are designed for fail-safe operation. The ventilation system is equipped with alarms and instruments that report and record its behavior through readouts in control areas and utility service areas.

Control Systems

Control systems for HVAC systems in nuclear facilities have some unique safety-related features. Because the exhaust system is to remain in operation during both normal and accident-related conditions, redundancy in the form of standby fans is often provided. These standby fans and their associated isolation dampers energize automatically upon a set reduction in either airflow rate or specific location pressure, as applicable. For DOE facilities, maintaining exhaust airflow is important, so fire dampers are excluded from all potentially contaminated exhaust ducts.

Pressure control in the facility interior maintains zones of increasing negative pressure in areas of increasing contamination potential. Care must be taken to prevent windy conditions from unduly effecting the atmospheric control reference. Pulsations can cause the pressure control system to oscillate strongly, resulting in potential reversal of relative pressures. One alternative is to use a variety of balancing and barometric dampers to establish an air balance at the desired differential pressures, lock the dampers in place, and then control the exhaust air to a constant flow rate.

Air and Gaseous Effluents Containing Radioactivity

Air and all other gaseous effluents are exhausted through a ventilation system designed to remove radioactive particulates. Exhaust ducts or stacks located downstream of final filtration that may contain radioactive contaminants should have two monitors, one a continuous air monitor (CAM) and the other a fixed sampler. These monitors may be a combination unit. Exhaust stacks from nuclear facilities are usually equipped with an isokinetic sampling system that relies on a relatively constant airflow rate. The isokinetic sensing probe is a symmetrically arranged series of pickup tubes connected through sweeping bends to a common tube, usually stainless steel, that leads to a nearby CAM. Typically, an exhaust system flow controller modulates the exhaust fan inlet dampers or motor speed to hold the exhaust airflow rate steady while the HEPA filters load.

Continuous air monitors can also be located in specific ducts where a potential for radiological contamination has been detected. These CAMs are generally placed beyond the final stage of HEPA filtration, as specified in ANSI *Standard* N13.1. Each monitoring system is connected to an emergency power supply.

The following are design considerations for CAM systems:

- Maintain fully developed turbulent flow near isokinetic sampler.
- For accurate CAM operation, heat tracing on the sampling air tubing may be required.
- Keep the ratio of the sample airflow rate to the total discharge airflow rate constant.

NUCLEAR REGULATORY COMMISSION FACILITIES

NUCLEAR POWER PLANTS

The two kinds of commercial light-water power reactors used in the United States today are the **pressurized water reactor** (PWR) and the **boiling water reactor** (BWR). For both types, the main objective of the HVAC systems, in addition to ensuring personnel comfort and reliable equipment operation, is protecting operating personnel and the general public from airborne radioactive contamination during all normal and emergency modes of plant operation. 10 CFR 20 sets forth the requirements for keeping radiation exposure as low as reasonably achievable (ALARA). The ALARA concept is the design objective of the HVAC system. In no case is the radiological dose allowed to exceed the limits as defined in 10 CFR 50 and 10 CFR 100.

The NRC has developed **regulatory guides** (RGs) that delineate techniques of evaluating specific problems and provide guidance to licensed applicants concerning information needed by the NRC for its review of the facility. Four regulatory guides that relate directly to HVAC system design are RG 1.52, RG 1.78, RG 1.95, and RG 1.140. Deviations from RG criteria must be justified by the owner and approved by the NRC.

The design of the HVAC systems for a nuclear power generating station must ultimately be approved by the NRC staff in accordance with Appendix A of 10 CFR 50. The NRC developed **standard review plans** (SRPs) as part of *Regulatory Report* NUREG-0800 to provide an orderly and thorough review. The SRP provides a good basis or checklist for the preparation of a **safety analysis report** (SAR). The **safety review plan** is based primarily on the information provided by an applicant in an SAR as required by Section 50.34 of 10 CFR 50. Technical specifications for nuclear power plant systems are developed by the owner and approved by the NRC as outlined in Section 50.36 of 10 CFR 50. Technical specifications define safety limits, limiting conditions for operation, and surveillance requirements for all systems important to plant safety.

Minimum requirements for the performance, design, construction, acceptance testing, and quality assurance of equipment used in safety-related air and gas treatment systems in nuclear facilities are found in ASME *Standard* AG-1.

PRESSURIZED WATER REACTORS

Reactor Containment Building

The containment building houses the reactor in a nuclear power plant. The temperature and humidity conditions are dictated by the **nuclear steam supply system** (NSSS). These conditions are generally specified for three modes of operation: normal operation, refueling operation, and loss of coolant accident (LOCA) condition. General design requirements are contained in ANS *Standard* 56.6.

Normal Operating Condition. Nuclear steam supply system temperature and humidity requirements are specified by the NSSS supplier. Some power plants require recirculation filtration trains in the containment building to control the level of airborne radioactivity. Cooling is provided by a reactor containment cooling system.

Refueling Condition. The maximum allowable temperature during refueling is determined by the refueling personnel. Because they work in protective clothing, their activities are slowed by discomfort, and the refueling outage is prolonged. Cooling can be by normal cooling units because the cooling load is low when the reactor is shut down. Ventilation with outdoor air is necessary.

Loss of Coolant Accident Condition. In the event of a LOCA (breakage of the primary cooling loop), circulating water at high pressure and temperature flashes and fills the containment building with radioactive steam. The major source of radioactivity is iodine in the water. The primary measures taken are directed at reducing the pressure in the containment building and lowering the amount of radioactive products in the containment atmosphere. Pressure is reduced by the reactor containment cooling units and/or sprays, which cool the atmosphere and condense the steam.

Containment Cooling. The following systems are typical for containment cooling:

Reactor containment coolers. These units remove most of the heat load. Distribution of the air supply depends on the containment layout and the location of the major heat sources.

Reactor cavity air-handling units or fans. These units are usually transfer fans without coils that provide cool air to the reactor cavity.

Control rod or control element drive mechanism (CRDM or CEDM) air-handling units. The CRDM and CEDM are usually cooled by induced-draft using exhaust fans. Because the flow rates, pressure drops, and heat loads are generally high, the air should be cooled before it is returned to the containment atmosphere.

Essential reactor containment cooling units. The containment air-cooling system, or a part of it, is normally designed to provide cooling after a postulated accident. The system must be able to perform at high temperature, pressure, humidity, and levels of radioactivity. Cooling coils are provided with essential service water.

System design must accommodate both normal and accident conditions. The ductwork must be able to endure the rapid pressure buildup associated with accident conditions, and fan motors must be sized to handle the high-density air.

Radioactivity Control. Airborne radioactivity is controlled by the following means:

Essential containment air filtration units. Some older power plants rely on redundant filter units powered by two Class 1E buses to reduce the amount of post-LOCA airborne radioactivity. The typical system consists of a demister, a heater, a HEPA filter bank, and a charcoal adsorber, possibly followed by a second filter. The electric heater is designed to reduce the relative humidity from 100% to less than 70% at the design inlet air temperature. All the components must be designed and manufactured to meet the requirements of a LOCA environment.

In the case of a LOCA and the subsequent operation of the filter train, the charcoal becomes loaded with radioactive iodine such that the decay heat could cause the charcoal to self-ignite if the airflow stops. If the primary fan stops, a secondary fan maintains a minimum airflow through the charcoal bed to remove the heat generated by the radioactive decay. The decay heat fan is powered by a Class 1E power supply. The filtration units are located inside the containment.

Containment power access purge or minipurge. Ventilation is needed during normal operation, when the reactor is under pressure, to control containment pressure or the level of airborne radioactivity within the containment. The maximum opening size allowed in the containment boundary during normal operation is 8 in.

The system consists of a supply fan, double containment isolation valves in each of the containment wall penetrations (supply and exhaust), and an exhaust filtration unit with a fan. The typical filtration unit contains a HEPA filter and a charcoal adsorber, possibly followed by a second filter.

This system should not be connected to any duct system inside the containment. It should include a debris screen within the containment over the inlet and outlet ducts, so that the containment isolation valves can close even if blocked by debris or collapsed ducts.

Containment refueling purge. Ventilation is required to control the level of airborne radioactivity during refueling. Because the reactor is not under pressure during refueling, there are no restrictions on the size of the penetrations through the containment boundary. Large openings of 42 to 48 in., each protected by double containment isolation valves, may be provided. The required ventilation rate is typically based on 1 air change per hour.

The system consists of a supply air-handling unit, double containment isolation valves at each supply and exhaust containment penetration, and an exhaust fan. Filters are recommended.

Containment combustible gas control. In the case of a LOCA, when a strong solution of sodium hydroxide or boric acid is sprayed into the containment, various metals react and produce hydrogen. Also, if some of the fuel rods are not covered with water, the fuel rod cladding can react with steam at elevated temperatures to release hydrogen into the containment. Therefore, redundant hydrogen recombiners are needed to remove the air from the containment atmosphere, recombine the hydrogen with the oxygen, and return the air to the containment. The recombiners may be backed up by special exhaust filtration trains.

BOILING WATER REACTORS

Primary Containment

The boiling water reactor (BWR) primary containment is a low-leakage, pressure-retaining structure that surrounds the reactor pressure vessel and related piping. Also known as the **drywell**, it is designed to withstand, with minimum leakage, the high temperature and pressure caused by a major break in the reactor coolant line. General design requirements are in ANS *Standard* 56.7.

The primary containment HVAC system consists of recirculating cooler units. It normally recirculates and cools the primary containment air to maintain the environmental conditions specified by the NSSS supplier. In an accident, the system performs the safety-related function of recirculating the air to prevent stratification of any hydrogen that may be generated. The cooling function may or may not be safety related, depending on the specific plant design.

Temperature problems have been experienced in many BWR primary containments due to temperature stratification and underestimation of heat loads. The ductwork should adequately mix the air to prevent stratification. Heat load calculations should include a safety factor sufficient to allow for deficiencies in insulation installation. In addition, a temperature monitoring system should be installed in the primary containment to ensure that bulk average temperature limits are not exceeded.

Reactor Building

The reactor building completely encloses the primary containment, auxiliary equipment, and refueling area. Under normal conditions, the reactor building HVAC system maintains the design space conditions and minimizes the release of radioactivity to the environment. The HVAC system consists of a 100% outside air cooling system. Outside air is filtered, heated, or cooled as required prior to being distributed throughout the various building areas. The exhaust air flows from areas with the least potential contamination to areas of most potential contamination. Prior to exhausting to the environment, potentially contaminated air is filtered with HEPA filters and charcoal adsorbers; all exhaust air is monitored for radioactivity. To ensure that no unmonitored exfiltration occurs during normal operations, the ventilation systems maintain the reactor building at a negative pressure relative to the atmosphere.

Upon detection of abnormal plant conditions, such as a line break, high radiation in the ventilation exhaust, or loss of negative pressure, the HVAC system's safety-related function is to isolate the reactor building. Once isolated via fast-closing, gastight isolation valves, the reactor building serves as a secondary containment boundary. This boundary is designed to contain any leakage from the primary containment or refueling area following an accident.

Once the secondary containment is isolated, pressure rises due to the loss of the normal ventilation system and the thermal expansion of the confined air. A safety-related exhaust system, the **standby gas treatment system** (SGTS), is started to reduce pressure and maintain the building's negative pressure. The SGTS exhausts air from the secondary containment to the environment through HEPA filters and charcoal adsorbers. The capacity of the SGTS is based on the amount of exhaust air needed to reduce the pressure in the secondary containment and maintain it at the design level, given the containment leakage rates and required drawdown times.

In addition to the SGTS, some designs include safety-related recirculating air systems within the secondary containment to mix, cool, and/or treat the air during accident conditions. These recirculation systems use portions of the normal ventilation system ductwork; therefore, the ductwork must be classified as safety related.

If the isolated secondary containment area is not to be cooled during accident conditions, it is necessary to determine the maximum temperature that could be reached during an accident. All safety-related components in the secondary containment must be environmentally qualified to operate at this temperature. In most plant designs, safety-related unit coolers handle the high heat release with **emergency core cooling system** (ECCS) pumps.

Turbine Building

Only a BWR supplies radioactive steam directly to the turbine, which could cause a release of airborne radioactivity to the surroundings. Therefore, areas of the BWR turbine building in which release of airborne radioactivity is possible should be enclosed. These areas must be ventilated and the exhaust filtered to ensure that no radioactivity is released to the surrounding atmosphere. Filtration trains typically consist of a prefilter, a HEPA filter, and a charcoal adsorber, possibly followed by a second filter. Filtration requirements are based on the plant and site configuration.

AREAS OUTSIDE PRIMARY CONTAINMENT

All areas located outside the primary containment are designed to the general requirements contained in ANS *Standard* 59.2. These areas are common to both PWRs and BWRs.

Auxiliary Building

The auxiliary building contains a large amount of support equipment, much of which handles potentially radioactive material. The building is air conditioned for equipment protection, and the exhaust is filtered to prevent the release of potential airborne radioactivity. The filtration trains typically consist of a prefilter, a HEPA filter, and a charcoal adsorber, possibly followed by a second filter.

The HVAC system is a once-through system, as needed for general cooling. Ventilation is augmented by local recirculation air-handling units in the individual equipment rooms requiring additional cooling due to localized heat loads. The building is maintained at negative pressure relative to the outside.

If the equipment in these rooms is not safety related, the area is cooled by normal air-conditioning units. If it is safety related, the area is cooled by safety-related or essential air-handling units powered from the same Class 1E (according to IEEE *Standard* 323) power supply as the equipment in the room.

The normal and essential functions may be performed by one unit having both a normal and an essential cooling coil and a safety-related fan served from a Class 1E bus. The normal coil is served with chilled water from a normal chilled water system, and the essential coil operates with chilled water from a safety-related chilled water system.

Control Room

The control room HVAC system serves the control room habitability zone—those spaces that must be habitable following a postulated accident to allow the orderly shutdown of the reactor—and performs the following functions:

- Control indoor environmental conditions
- Provide pressurization to prevent infiltration
- Reduce the radioactivity of the influent
- Protect the zone from hazardous chemical fume intrusion
- Protect the zone from fire
- Remove noxious fumes, such as smoke

The design requirements are described in detail in SRP 6.4 and SRP 9.4.1. Regulatory guides that directly affect control room design are RG 1.52, RG 1.78, and RG 1.95. NUREG-CR-3786 provides a summary of the documents affecting control room system design. ASME *Standards* N509 and AG-1 also provide guidance for the design of control room habitability systems and methods of analyzing pressure boundary leakage effects.

Control Cable Spreading Rooms

These rooms are located directly above and below the control room. They are usually served by the air-handling units that serve the electric switchgear room or the control room.

Diesel Generator Building

Nuclear power plants have auxiliary power plants to generate electric power for all essential and safety-related equipment in the event of loss of off-site electrical power. The auxiliary power plant consists of at least two independent diesel generators, each sized to meet the emergency power load. The heat released by the diesel generator and associated auxiliary systems is normally removed through outside air ventilation.

Emergency Electrical Switchgear Rooms

These rooms house the electrical switchgear that controls essential or safety-related equipment. The switchgear located in these rooms must be protected from excessive temperatures (1) to ensure that its useful life, as determined by environmental qualification, is not cut short and (2) to preserve power circuits required for proper operation of the plant, especially its safety-related equipment.

Battery Rooms

Battery rooms should be maintained at 77°F with a temperature gradient of not more than 5°F, according to IEEE *Standard* 484. The minimum room design temperature should be taken into account in determining battery size. Because batteries produce hydrogen gas during charging periods, the HVAC system must be designed to limit the hydrogen concentration to the lowest of the levels specified by IEEE *Standard* 484, OSHA, and the lower explosive limit (LEL). The minimum number of room air changes per hour is 5. Because hydrogen is lighter than air, the system exhaust duct inlet openings should be located on the top side of the duct to prevent hydrogen pockets from forming at the ceiling. If the ceiling is supported by structural beams, there should be an exhaust air opening in each beam pocket.

Fuel-Handling Building

New and spent fuel is stored in the fuel-handling building. The building is air conditioned for equipment protection and ventilated with a once-through air system to control potential airborne radioactivity. Normally, the level of airborne radioactivity is so low that the exhaust need not be filtered, although it should be monitored. If significant airborne radioactivity is detected, the building is sealed and kept under negative pressure by exhaust through filtration trains powered by Class 1E buses.

Personnel Facilities

For nuclear power plants, this area usually includes decontamination facilities, laboratories, and medical treatment rooms.

Pumphouses

Cooling water pumps are protected by houses that are often ventilated by fans to remove the heat from the pump motors. If the pumps are essential or safety related, the ventilation equipment must also be considered safety related.

Radioactive Waste Building

Radioactive waste other than spent fuel is stored, shredded, baled, or packaged for disposal in this building. The building is air conditioned for equipment protection and ventilated to control potential airborne radioactivity. The air may require filtration through HEPA filters and/or charcoal adsorbers prior to release to the atmosphere.

Technical Support Center

The technical support center (TSC) is an outside facility located close to the control room; it is used by plant management and technical support personnel to provide assistance to control room operators under accident conditions.

In case of an accident, the TSC HVAC system must provide the same comfort and radiological habitability conditions maintained in the control room. The system is generally designed to commercial HVAC standards. An outside air filtration system (HEPA-charcoal-HEPA) pressurizes the facility with filtered outside air during emergency conditions. The TSC HVAC system must be designed to safety-related standards.

NONPOWER MEDICAL AND RESEARCH REACTORS

The requirements for HVAC and filtration systems for nuclear nonpower medical and research reactors are set by the NRC. The criteria depend on the type of reactor (ranging from a nonpressurized swimming pool type to a 10 MW or more pressurized reactor), the type of fuel, the degree of enrichment, and the type of facility and environment. Many of the requirements discussed in the sections on various nuclear power plants apply to a certain degree to these reactors. It is therefore imperative for the designer to be familiar with the NRC requirements for the reactor under design.

LABORATORIES

Requirements for HVAC and filtration systems for laboratories using radioactive materials are set by the DOE and/or the NRC. Laboratories located at DOE facilities are governed by DOE regulations. All other laboratories using radioactive materials are regulated by the NRC. Other agencies may be responsible for regulating other toxic and carcinogenic material present in the facility.

Laboratory containment equipment for nuclear processing facilities is treated as a primary, secondary, or tertiary containment zone, depending on the level of radioactivity anticipated for the area and on the materials to be handled. For additional information see Chapter 13, Laboratories.

Glove Boxes

Glove boxes are windowed enclosures equipped with one or more flexible gloves for handling material inside the enclosure from the outside. The gloves are attached to a porthole in the enclosure and seal the enclosure from the surrounding environment. Glove boxes permit hazardous materials to be manipulated without being released to the environment.

Because the glove box is usually used to handle hazardous materials, the exhaust is HEPA filtered before leaving the box and prior to entering the main exhaust duct. In nuclear processing facilities, a glove box is considered primary confinement (Figure 1), and is therefore subject to the regulations governing those areas. For nonnuclear processing facilities, the designer should know the designated application of the glove box and design the system according to the regulations governing that particular application.

Laboratory Fume Hoods

Nuclear laboratory fume hoods are similar to those used in nonnuclear applications. Air velocity across the hood opening must be sufficient to capture and contain all contaminants in the hood. Excessive hood face velocities should be avoided because they cause contaminants to escape when an obstruction (e.g., an operator) is positioned at the hood face. For information on fume hood testing, refer to ASHRAE *Standard* 110.

Radiobenches

A radiobench has the same shape as a glove box except that in lieu of the panel for the gloves, there is an open area. Air velocity across the opening is generally the same as for laboratory hoods. The level of radioactive contamination handled in a radiobench is much lower than that handled in a glove box.

DECOMMISSIONING OF NUCLEAR FACILITIES

The exhaust air filtration system for decontamination and decommissioning (D&D) activities in nuclear facilities depends on the type and level of radioactive material expected to be found during the D&D operation. The exhaust system should be engineered to accommodate the increase in dust loading and more radioactive contamination than is generally anticipated because the D&D activities dislodge previously fixed materials, making them airborne. Good housekeeping measures include chemical fixing and vacuuming the D&D area as frequently as necessary.

The following are some design considerations for the ventilation systems required to protect the health and safety of the public and the D&D personnel:

- Maintain a higher negative pressure in the areas where D&D activities are being performed than in any of the adjacent areas.
- Provide an adequate capture velocity and transport velocity in the exhaust system from each D&D operation to capture and transport fine dust particles and gases to the exhaust filtration system.
- Exhaust system inlets should be as close to the D&D activity as possible to enhance the capture of contaminated materials and to minimize the amount of ductwork that is contaminated. Movable inlet capability is desirable.
- With portable enclosures, filtration of the enclosure inlet and exhaust air must maintain the correct negative internal pressure.

Low-Level Radioactive Waste (LLRW)

Requirements for the HVAC and filtration systems of LLRW facilities are governed by 10 CFR 61. Each facility must have a ventilation system to control airborne radioactivity. The exhaust air is drawn through a filtration system that typically includes a demister, heater, prefilter, HEPA filter, and charcoal adsorber, which may be followed by a second filter. Ventilation systems and their CAMs should be designed for the specific characteristics of the facility.

CODES AND STANDARDS

ANSI N13.1	Guide for Sampling Airborne Radioactive Materials in Nuclear Facilities
ANSI/ANS 56.6	Pressurized Water Reactor Containment Ventilation Systems
ANSI/ANS 56.7	Boiling Water Reactor Containment Ventilation Systems
ANSI/ANS 59.2	Safety Criteria for HVAC Systems Located Outside Primary Containment
ANSI/ASME AG-1	Code on Nuclear Air and Gas Treatment
ANSI/ASME N509	Nuclear Power Plant Air-Cleaning Units and Components
ANSI/ASME N510	Testing of Nuclear Air Treatment Systems
ANSI/ASME NQA-1	Quality Assurance Program Requirements for Nuclear Facility Applications
ANSI/ASHRAE 110	Method of Testing Performance of Laboratory Fume Hoods
10 CFR	Title 10 of the *Code of Federal Regulations*
Part 20	Standards for Protection Against Radiation (10 CFR 20)
Part 50	Domestic Licensing of Production and Utilization Facilities (10 CFR 50)
Part 61	Land Disposal of Radioactive Waste (10 CFR 61)
Part 100	Reactor Site Criteria (10 CFR 100)
DOE *Order* 5400.5	Radiation Protection of the Public and the Environment
DOE *Order* 6430.1A	General Design Criteria
DOE *Order* N 441.1	Radiological Protection for DOE Activities
DOE 3020	Specification for HEPA Filters Used by DOE Contractors
DOE NE F 3-43	Quality Assurance Testing of HEPA Filters
ANSI/IEEE 323	Standard for Qualifying Class 1E Equipment for Nuclear Power Generating Stations
ANSI/IEEE 484	Recommended Practice for Installation Design and Installation of Vented Lead-Acid Batteries for Stationary Applications
ANSI/NFPA 801	Standard for Fire Protection for Facilities Handling Radioactive Materials
ANSI/NFPA 901	Standard Classifications for Incident Reporting and Fire Protection Data
NUREG-0800	*Standard Review Plans*
SRP 6.4	Control Room Habitability Systems
SRP 9.4.1	Control Room Area Ventilation System
NUREG-CR-3786	A Review of Regulatory Requirements Governing Control Room Habitability
Regulatory Guides	Nuclear Regulatory Commission
RG 1.52	Design, Testing, and Maintenance Criteria for Engineered Safety Feature Atmospheric Cleanup System Air Filtration and Adsorption Units of LWR Nuclear Power Plants
RG 1.78	Assumptions for Evaluating the Habitability of Nuclear Power Plant Control Room During a Postulated Hazardous Chemical Release
RG 1.95	Protection of Nuclear Power Plant Control Room Operators Against Accidental Chlorine Release
RG 1.140	Design, Testing, and Maintenance Criteria for Normal Ventilation Exhaust System Air Filtration and Adsorption Units of LWR Nuclear Power Plants

MINE AIR CONDITIONING AND VENTILATION

IN underground mines, excess humidity, high temperature, and inadequate oxygen have always been points of concern because they lower worker efficiency and productivity and can cause illness and death. Air cooling and ventilation are needed in deep underground mines to minimize heat stress. As mines have become deeper, heat removal and ventilation problems have become more difficult to solve.

WORKER HEAT STRESS

Mine air must be conditioned to maintain a temperature and humidity that ensures the health and comfort of miners so they can work safely and productively. Chapter 8 of the 1997 *ASHRAE Handbook—Fundamentals* addresses human response to heat and humidity. The upper temperature limit for humans at rest in still, saturated air is about 90°F. If the air is moving at 200 fpm, the upper limit is 95°F. In a hot mine, a relative humidity of less than 80% is desirable.

Hot, humid environments are improved by providing air movement of 150 to 500 fpm. Although a greater air volume lowers the mine temperature, air velocity has a limited range in which it improves worker comfort.

Indices for defining acceptable temperature limits include the following:

- **Effective temperature scale.** An effective temperature of 80°F is the upper limit for ensuring worker comfort and productivity.
- **Wet-bulb globe temperature (WBGT) index.** A WBGT of 80°F is the permissible temperature exposure limit for moderate continuous work; a WBGT of 77°F is the limit for heavy continuous work.

See Figure 1 in Chapter 28 for recommended heat stress exposure limits.

SOURCES OF HEAT ENTERING MINE AIR

Adiabatic Compression

Air descending a shaft increases in pressure (due to the mass of air above it) and temperature. As air flows down a shaft, it is heated as if compressed in a compressor, even if there is no heat interchange with the shaft and no evaporation of moisture.

One Btu is added to each pound of air for every 778 ft decrease in elevation or is removed for the same elevation increase. For dry air, the specific heat is 0.24 Btu/lb·°F, and the dry-bulb temperature change is $1/(0.24 \times 778 \times 1) = 0.00535$°F per foot or 1°F per 187 ft of elevation. For constant air-vapor mixtures, the change in dry-bulb temperatures is $(1 + W)/(0.24 + 0.45W)$ per 778 ft of elevation, where W is the humidity ratio in pounds of water per pound of dry air.

Theoretically, when 100,000 cfm of standard air (density = 0.075 lb/ft³) is delivered underground via an inlet airway, the heat of autocompression for every 1000 ft of depth is calculated as follows:

The preparation of this chapter is assigned to TC 9.2, Industrial Air Conditioning.

$$100{,}000 \text{ cfm} \times 0.075 \frac{\text{lb}}{\text{ft}^3} \times \frac{1 \text{ Btu/lb}}{778 \text{ ft}} \times 1000 \text{ ft} = 9640 \frac{\text{Btu}}{\text{min}}$$
$$= 48.2 \text{ tons of refrigeration}$$

Autocompression of air may be masked by the presence of other heating or cooling sources, such as shaft wall rock, groundwater, air and water lines, or electrical facilities. The actual temperature increase for air descending a shaft does not usually match the theoretical adiabatic temperature increase, due to the following:

- Effect of night cool air temperature on the rock or shaft lining
- Temperature gradient of ground rock related to depth
- Evaporation of moisture within the shaft, which decreases the temperature while increasing the moisture content of the air

The seasonal variation in surface air temperature has a major effect on the temperature of air descending a shaft. When the surface air temperature is high, much of its heat is absorbed by the shaft walls; thus, the temperature rise for the descending air may not reach the adiabatic prediction. When the surface temperature is low, heat is absorbed from the shaft walls, and the temperature increases more than predicted adiabatically. Similar diurnal variations may occur. As air flows down a shaft and increases in temperature and density, its cooling ability and volume decrease. Additionally, the mine ventilation requirements increase with depth. Fan static pressures up to 10 in. of water gage are common in mine ventilation and raise the temperature of the air about 0.45°F per in. of water.

Electromechanical Equipment

Power-operated equipment transfers heat to the air. In mines, systems are commonly powered by electricity, diesel fuel, and compressed air.

For underground diesel equipment, about 90% of the heat value of the fuel consumed, or 125,000 Btu/gal, is dissipated to the air as heat. If exhaust gases are bubbled through a wet scrubber, the gases are cooled by adiabatic saturation, and both the sensible heat and the moisture content of the air are increased.

Vehicles with electric drives or electric-hydraulic systems release one-third to one-half the heat released by diesel equipment.

All energy used in a horizontal plane appears as heat added to the mine air. Energy required to elevate a load gives potential energy to the material and does not appear as heat.

Groundwater

Transport of heat by groundwater is the largest variable in mine ventilation. Groundwater usually has the same temperature as the virgin rock. If there is an uncovered ditch containing hot water, ventilation cooling air can pick up more heat from the ditch water than from the hot wall rock. Thus, hot drainage water should be contained in pipelines or in a covered ditch.

Heat release from open ditches becomes more significant as airways get older and the flow of heat from the surrounding rock decreases. In one Montana mine, water in open ditches was 40°F cooler than when it issued from the wall rock; the heat was transferred to the air. Evaporation of water from wall rock surfaces

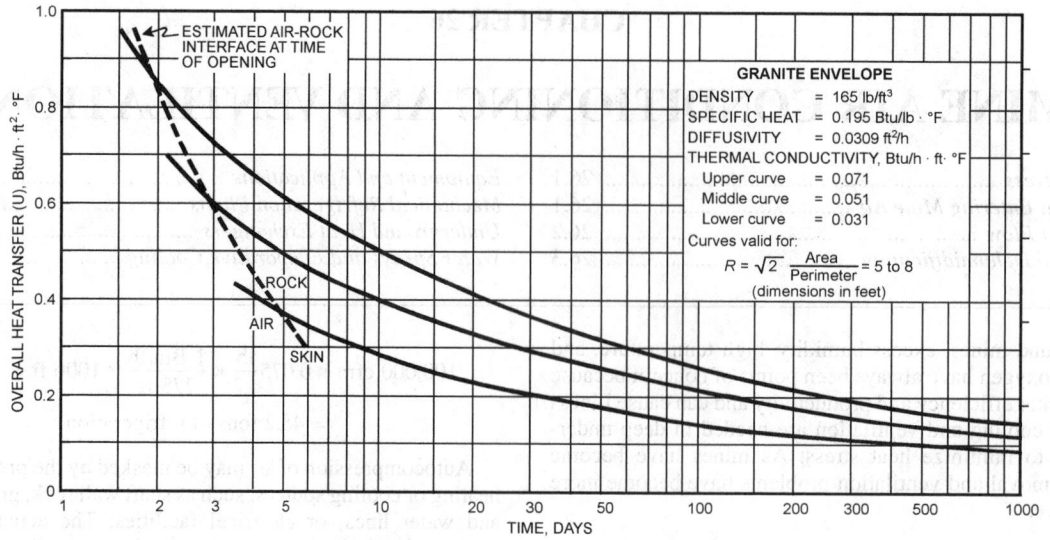

Fig. 1   Heat Flow for Salt Envelope Mine

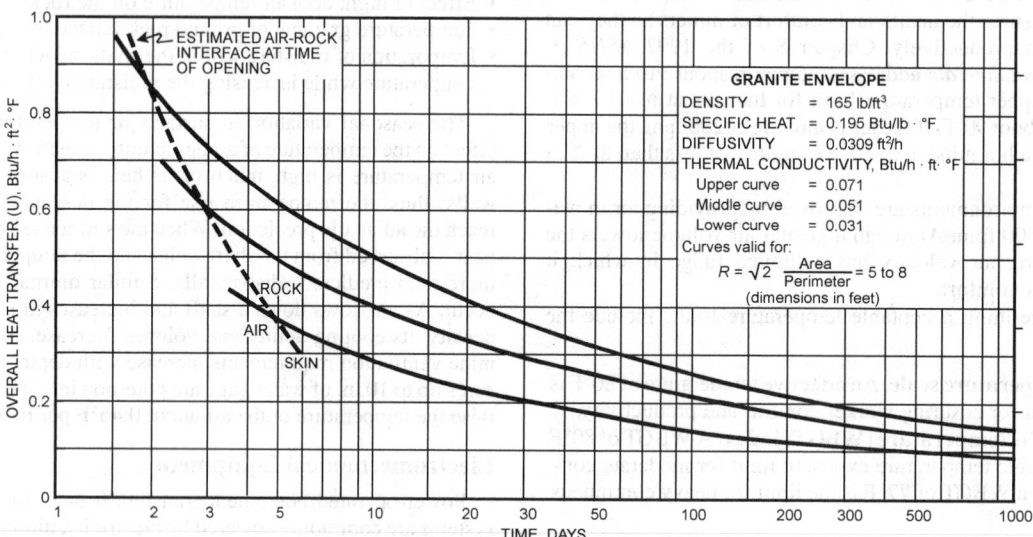

Fig. 2   Heat Flow for a Granite Envelope Mine

lowers the surface temperature of the rock, which increases the temperature gradient of the rock, depresses the dry-bulb temperature of the air, and allows more heat to flow from the rock. Most of this extra heat is expended in evaporation.

WALL ROCK HEAT FLOW

Wall rock is a major heat source in mines. The heat flow from wall rock with constant thermal conductivity is considerably higher when a mine is first excavated and transient heat flow applies than it is several years later, when steady-state conditions have developed. Heat flow from wall rock to air can be calculated using the steady-state heat transfer equation:

$$Q = UA\Delta t \qquad (1)$$

where

Q = heat flow, Btu/h
U = overall coefficient of heat transfer, Btu/h·ft^2·°F (from Figure 1 or 2)
A = area, ft^2
Δt = temperature difference between virgin rock and dry-bulb air temperature, °F

Equation (1) can be used with mine ventilation rates, virgin rock temperatures, and the age of airways to calculate rock heat flow throughout a mine.

In young mining areas, heat inflow must be computed by transient heat flow techniques. Figures 1 and 2 show heat flow from wall rock to air over 1000 days in two types of mines: a salt envelope mine and a granite envelope mine. The temperature difference between wall rock and air is not constant along the entire length of an airway, so the airway can be treated as a series of consecutive lengths. Figures 1 and 2 are based on the following:

- The cross-sectional areas of the mine openings range from 225 to 450 ft^2.
- The heat transfer coefficient decreases from the time the heading is excavated and ventilated.
- Steady-state heat transfer coefficients are approximately 0.10 to 0.15 Btu/h·ft^2·°F for salt or granite envelopes.

The Starfield and Goch-Patterson methods of calculating heat flow rates from underground exposed rock surfaces are often used. The Starfield method (Starfield 1966), which is theoretically more accurate, is frequently difficult to apply because it depends on air

Table 1 Maximum Virgin Rock Temperatures

Mining District	Depth, ft	Temperature, °F
Kolar Gold Field, India	11,000	152
South Africa	10,000	125 to 130
Morro Velho, Brazil	8,000	130
N. Broken Hill, Australia	3,530	112
Great Britain	4,000	114
Braloroe, BC, Canada	4,100	112.5
Kirkland Lake, Ontario	4,000 to 6,000	66 to 81
Falconbridge Mine, Ontario	4,000 to 6,000	70 to 84
Lockerby Mine, Ontario	3,000 to 4,000	67 to 96
Levac Borehold (Inco), Ontario	7,000 to 10,000	99 to 128
Garson Mine, Ontario	2,000 to 5,000	54 to 78
Lake Shore Mine, Ontario	6,000	73
Hollinger Mine, Ontario	4,000	58
Creighton Mine, Ontario	2,000 to 10,000	60 to 138
Superior, AZ	4,000	140
San Manuel, AZ	4,500	118
Butte, MT	5,200	145 to 150
Ambrosia Lake, NM	4,000	140
Brunswick, No. 12, New Brunswick, Canada	3,700	73
Belle Island Salt Mine, LA	1,400	88

Source: Fenton 1972.

velocity criteria. The errors in measuring or estimating air velocities and the length and cross-sectional dimensions of the rock surfaces are often greater than the error introduced by the Starfield method.

The Goch-Patterson method (Goch and Patterson 1940), which assumes that the rock face temperature is the same as that of the air, is widely used. Heat load calculated for airways is overestimated by 10 to 20%, so a contingency is normally not added to the heat load. When using the Goch-Patterson tables, the correct value of instantaneous versus average heat load data should be used used for shafts and older connecting airways. Time-average heat load data should be used for excavations that are continuous and active.

The rate of heat release from wall rock can be accurately estimated horizontally and vertically. The **virgin rock temperature (VRT)** should be measured at various elevations throughout the mine.

Before constructing a gradient to estimate the temperatures at various elevations, a base point for measurement must be selected. The VRT 50 ft below the earth's surface is often used; it is usually equal to the mean of surface air temperatures (dry-bulb) recorded over a number of years.

Table 1 lists maximum VRTs at the bottom of various mines.

Heat from Blasting

The heat produced by blasting can be appreciable. Virtually all the explosive energy is converted to heat. The typical heat potential in various types of explosives is similar to that of 60% dynamite, about 1800 Btu/lb.

AIR COOLING AND DEHUMIDIFICATION

Sources of heat in mines include wall rock, hot water issuing from underground sources, adiabatic compression of the inlet air down deep shafts, electromechanical equipment, hot compressed air lines and lines draining hot mine water, oxidation of timber and sulfide minerals, blasting, friction and shock losses from moving air, body metabolism, friction from rock movement, and internal combustion engines (Fenton 1972).

Thermal gradients of the world's major mining districts vary from approximately 0.5°F to over 4°F per 100 ft of elevation.

When air is cooled underground, the heat extracted must be removed. If enough cool, uncontaminated air is available, the warmed air can be mixed with it and removed from the mine. In a water heat exchanger, the water is warmed while the air is cooled. The warm water is returned in a closed circuit to the cooling plant (cooling tower, underground mechanical refrigeration unit, or heat exchanger), discharged to the mine drainage system, or sprayed into the mine exhaust air system.

The high heat capacity of water makes it a better heat transfer medium than air. For water, a relatively small pipeline can remove as much heat as would be removed by a large volume of air through a large shaft and inlet airways.

EQUIPMENT AND APPLICATIONS

Air-Conditioning Components and Practices

Figures 3 through 6 show components normally used in underground air-conditioning systems.

Cooling Surface Air

If cold air is available on the surface, as in far northern and far southern hemisphere winters, it can be forced into the mine. In deep mines with small cross-sectional airways, this supply of cold air may be insufficient for adequate cooling and may need to be supplemented by cooled air. In winter, it may even be necessary to heat the intake air.

Cooling air on the surface is the least expensive method of cooling. Factors favoring surface cooling are (1) low adiabatic compression (which applies to shafts that are not too deep) and (2) low heat gain in the shaft and intake airways. Cooling inlet air on the surface rather than in an underground cooling plant saves energy because the cooling water does not need to be returned to the surface.

In a surface plant installation, air is chilled using cold water or brine cooled by mechanical refrigeration units. The chilled air is then delivered to the intake shaft to be taken underground. Condenser water is usually cooled in towers or spray ponds. Such installations have been used in Morro Velho, Brazil; Robinson Deep Mine, Rand, South Africa; Kolar, India; and Rieu du Coeur, Belgium.

There are serious disadvantages to surface air cooling. By the time the chilled air reaches the working headings, it has been warmed considerably by autocompression and heat from the wall rock. Furthermore, a plant of sufficient size for peak load has an annual load factor as low as 60%. For these reasons, surface plant systems are no longer being installed. The trend is to install air-cooling plants underground with the cooling coils as close to the working areas of the mine as possible. If a large supply of cold water is available, it can be piped underground to the air-conditioning plants.

Cooling Surface Water

Surface-made ice or ice slurry in 32°F water can be delivered down a pipeline and used for activities such as drilling and wetting down muckpiles. The use of chilled water at 32°F reduces the required circulated flow rate to 50 to 70% of that in chilled water air-conditioning plants using warmer water. However, ice-making machines have a lower coefficient of performance and higher capital costs than do normal mechanical refrigeration units used to chill mine water.

In one Canadian mine, large, cold, mined-out areas are sprayed with water during the winter months, allowing ice to form in the openings. During the summer, intake air is cooled by drawing it over this ice.

Evaporative Cooling of Mine Chilled Water

Geographic areas with both low average winter temperatures and low relative humidity during the summer have a natural cooling capacity that is adaptable to mine air-cooling methods. A system in such an area cools water or brine on the surface, carries it through closed-circuit piping to an underground air-cooling plant near the working zone, and returns it to the cooling tower through a second pipeline. At the mine underground air-cooling plant, heat is absorbed

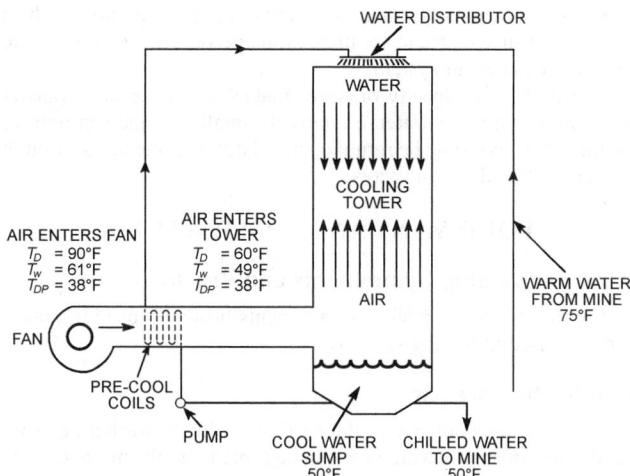

Fig. 3   Evaporative Cooling Tower System
(Richardson 1950)

by circulating water and is dissipated to the surface atmosphere in the surface cooling tower. The closed-circuit piping balances the hydrostatic pressure, so pumping power must only overcome frictional resistance.

An evaporative cooling tower was installed at a mine in the northwestern United States. This "dew-point" cooling system reduces the temperature of the cooling medium to below the wet-bulb temperature of the surface atmosphere (Figure 3). Precooling coils were installed between the fan and the cooling tower. Some cool water from the sump at the base of the cooling tower is pumped through the coils to the top of the cooling tower, where this heated flow joins the warm return water from the airflow; the moisture content in the airstream passing over the coils is unchanged, so that the dew-point temperature remains constant. The heat content of the air is reduced, and the equivalent heat is added to the water circulating in coil adsorbers. The dry- and wet-bulb temperature of the air entering the bottom of the cooling tower is less than the temperature of the air entering the fan. The temperature of the water leaving the tower approaches the wet-bulb temperature of the surface atmosphere.

Evaporative Cooling Plus Mechanical Refrigeration

On humid summer days, the wet-bulb temperature may increase over extended periods, severely hampering the effectiveness of evaporative cooling. This factor, plus warming of the air entering the mine, may necessitate the series installation of a mechanical refrigeration unit to chill the water delivered underground.

Performance characteristics at one northern United States mine on a spring day were as follows:

Evaporative cooling tower	1224 tons
Mechanical refrigeration unit (in series)	816 tons
Water volume circulated	1750 gpm
Water temperature entering mine	40°F
Water temperature leaving mine	68°F

Combination Systems

Components may be arranged in various ways for the greatest efficiency. For example, air-cooling towers may be used to cool water during the cool months of the year as well as to supplement a mechanical refrigerator during the warm months of the year.

Surface-installed mechanical refrigeration units provide the bulk of cooling in summer. In winter, much of the cooling comes from the precooling tower when the ambient wet-bulb temperature is usually lower than the temperature of water entering the tower.

The precooling tower is normally located above the return water storage reservoir. An evaporative cooling tower is more cost-effective (capital and operating costs) than mechanical refrigeration with comparable capacity.

Reducing Water Pressure

The use of underground refrigerated water chillers is increasing because they are efficient and can be located close to the work. Transfer of heat from the condensers is the major problem with these systems. If hot mine water is used to cool the condenser, efficiency is lost due to the high condensing temperature, the possibility of corrosion, and fouling. If surface water is used, it must be piped both in and out of the mine. If water is noncorrosive and nonfouling, fairly good chiller efficiencies can be obtained for entering condenser water with temperatures up to 125°F.

Surface water being delivered in a vertical pipe is usually allowed to flow into tanks located at different levels in the shaft to break the high water pressure that develops. In this process, energy is wasted, and the water temperature rises about 1°F for every 1000 ft of drop. The water pressure can be reduced for use at the mine level and after use, the water can be discharged to the drainage system. Although low-pressure mine cooling is convenient, the costs for pumping the water to the surface are high.

If a pipe 4000 ft high were filled with water (density = 62.4 lb/ft^3), the pressure would be $62.4 \times 4000/(144 \text{ in}^2/\text{ft}^2) = 1730$ psi. In an open piping system, the pressure at the bottom is further increased by the pressure necessary to raise the water up

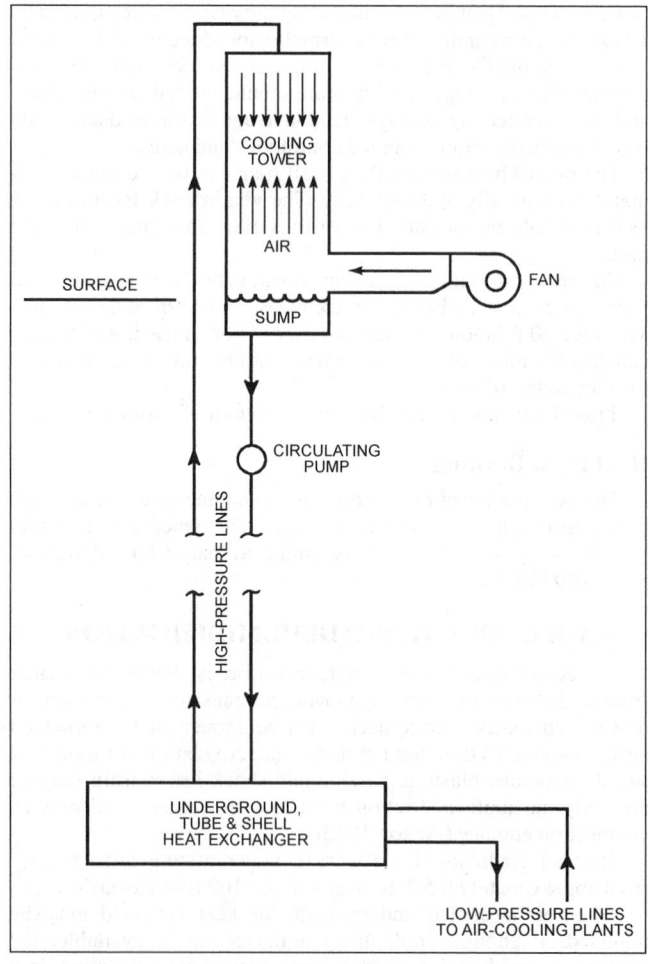

**Fig. 4   Underground Heat Exchanger,
Pressure Reduction System**

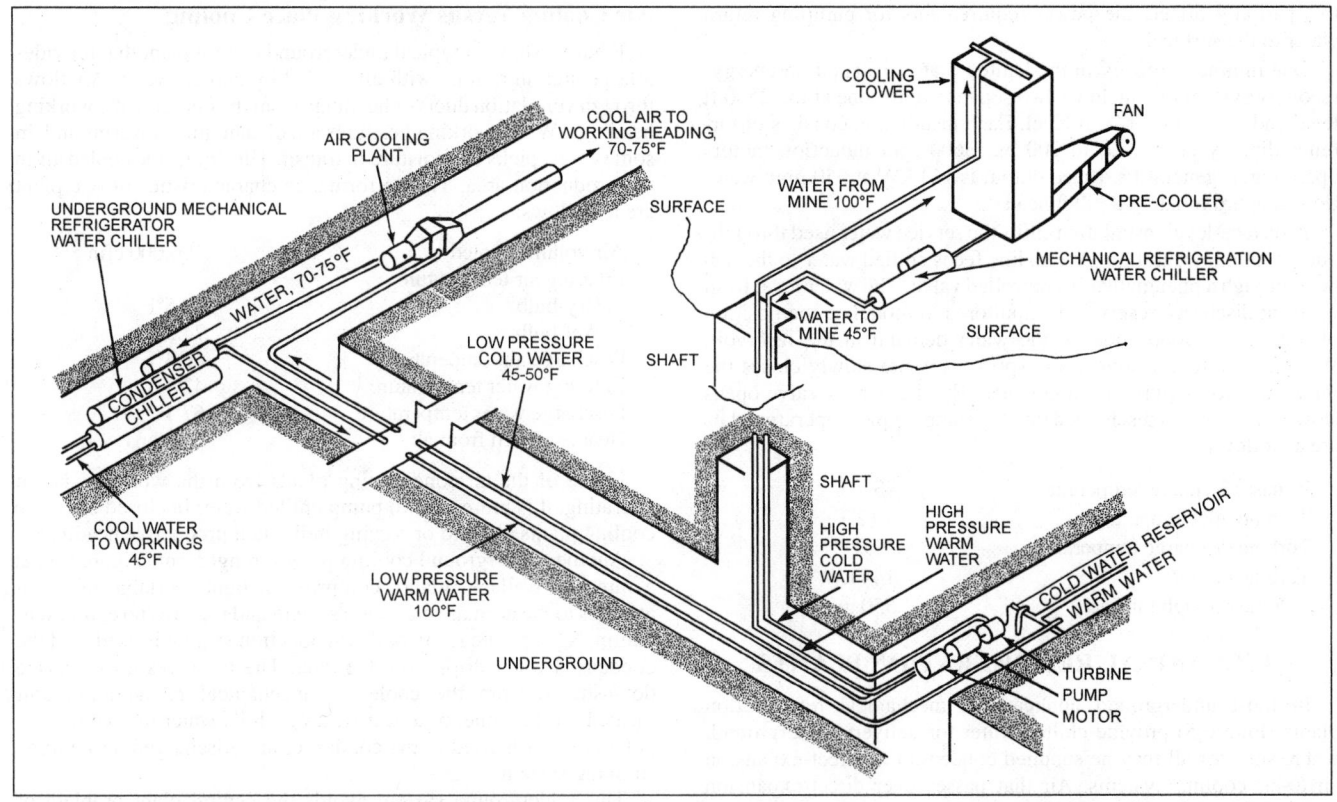

Fig. 5 Layout for Turbine-Pump-Motor Unit with Air-Cooling Plants and Mechanical Refrigeration

the pipe and out of the mine. Water pipes and coils in deep mines must be able to withstand this high pressure. Fittings and pipe specialties for high-pressure equipment are costly. Safety precautions and care must be taken when operating high-pressure equipment. Closed-circuit piping has the same static pressures, but pumps must only overcome pipe friction.

Frequent movement of surface cooling towers, a desired feature for shifting mining operations, results in high construction costs. Closed-circuit systems have been used in various mines in the United States to overcome the cost of pumping brine or cooling water out of the mine.

To take advantage of both low-pressure and closed-circuit systems, the Magma mine in Arizona installed heat exchangers underground at the mining horizon. Shell-and-tube heat exchangers convert surface chilled water in a high-pressure closed circuit to a low-pressure chilled water system on mine production levels. Air-cooling plants and chilled water lines can be constructed of standard materials, permitting frequent relocation (Figure 4). Although desirable, this system has not been widely used.

Energy-Recovery Systems

Pumping costs can be reduced by combining a water turbine with the pump. The energy of high-pressure water flowing to a lower pressure drives the pump needed for the low-pressure water circuit. Rotary-type water pumps have been developed to pump against a 5000 ft head, and water turbines are also available to operate under head. Figure 5 shows a turbine pump-motor combination. Only the shaft and the pipe and fittings to the unit on the working level need strong pipes. The system connects to underground refrigeration water-chilling units; the return chilled water is used for condensing before being pumped out.

Two types of turbines are suitable for mine use—the Pelton wheel and a pump in reverse. The Pelton wheel has a high-duty efficiency of about 80%, is simply constructed, and is readily con-

trolled. A pump in reverse is only 10 to 15% efficient, but mine maintenance and operating personnel are familiar with this equipment. Turbine energy recovery may encounter difficulties when operating on chilled mine service water because mine demands fluctuate widely, often outside the operating range of the turbine. Operating experience shows that coupling a Pelton wheel to an electric generator is the best approach.

South African mines have mechanical refrigeration units, surface heat-recovery systems, and turbine pumps incorporated into their air-cooling plants in a closed circuit.

A surface-sited plant has two disadvantages: (1) chilled water delivered underground at low operating pressure heats up at a rate of 1°F per 1000 ft of shaft, and (2) pumping this water back to the surface is expensive. Energy-recovery turbines underground and a heat-recovery system on the surface help offset pumping costs.

Descending chilled water is fed through a turbine mechanically linked to pumps operating in the return chilled water line. The energy recovery turbine reduces the rate of temperature increase in the descending chilled water column to about 0.3°F per 1000 ft of shaft.

Precooling towers on the surface reduce the water temperature a few degrees before it enters the refrigeration plant. Because of the unlimited supply of relatively cool ambient air for heat rejection, the operating cost of a surface refrigeration plant is about one-half that of a comparable underground plant.

In a uranium mine in South Africa, condensing water from surface refrigeration units is the heat source for a high condensing temperature heat pump, which discharges 130°F water. This water can be used as service hot water or as preheated feedwater for steam generation in the uranium plant. The total additional cost of a heat pump over a conventional refrigeration plant has a simple payback of about 18 months.

Pelton turbines are used by the South African gold-mining industry to recover energy from chilled water flowing down the shafts. About 1000 kW can be recovered at a typical installation;

this partially offsets the power requirements for pumping return water to the surface.

One mining company in the United States installed an energy-recovery system that includes two separate units, one at the 2500 ft level and one at the 5000 ft level. Each installation consists of turbines directly connected to 200 hp, 3600 rpm induction motors operating as generators. Rated output is 144 kW at 550 gpm water flow at approximately 2500 ft head.

A surface-level installation chills the service water used throughout the mining operation. A 6 in. line feeds chilled water to the turbine through a pneumatically controlled valve. The water level in an adjacent discharge reservoir is monitored to modulate the position of this valve proportional to the water demand in the reservoir. When the system is shut down, spring pressure slowly closes the inlet valve to the turbine. Simultaneously, the bypass valve opens slowly and water is discharged into the sump. Typical operating data are as follows:

Bypass discharge temperature	43.7°F
Turbine inlet temperature	40.2°F
Turbine discharge temperature	40.8°F
Turbine output	144 kW
Volume through turbine	550 gpm

MECHANICAL REFRIGERATION PLANTS

In most underground applications, mechanical refrigeration plants (Figure 5) provide chilled water for delivery underground, where some or all may be supplied condensed to direct-expansion air-to-air cooling systems. Air that passes over direct-expansion evaporator coils is cooled and delivered to working headings.

Underground mechanical refrigeration plants avoid the additional heat due to the autocompression of air coming down the shafts. The main disadvantage of these units is the disposal of heat from the condensers. The condenser discharge water can either be run into the mine drainage system, returned to the surface for cooling, or sprayed into the mine discharge air system.

Hot water discharged from the condensers of underground refrigeration units may be ejected into sumps of the mine water-pumping system. This procedure requires a constant resupply of condenser water from freshwater lines or mine drainage sources and increases mine pumping costs. Underground cooling towers are seldom used because they encounter air shortages, which result in higher condensing temperatures than those on the surface. It is difficult to predict the performance of underground cooling towers and the condensing temperatures of equipment over the life of a mine. High condenser temperatures cause excessive power consumption. In addition, water is often contaminated with dust and fumes, which cause fouling, scaling, and corrosion of the piping and condenser tubes. These problems make placing mechanical refrigeration units on the surface more favorable.

Spot Cooling

Spot cooling permits the driving of long headings for exploration or extended development prior to the installation of primary ventilation equipment. Using mechanical refrigeration systems with direct-expansion air coolers, spot cooling services a single heading or localized mining area. Spot coolers allow development headings to be advanced more rapidly and under more comfortable conditions. Development headings where rock temperatures exceed 100°F may also use spot cooling.

UNDERGROUND HEAT EXCHANGERS

Two types of heat exchangers are used underground: air-to-water and water-to-water. Air-to-water exchangers include those with (1) water sprays in the airstream and (2) finned tubes.

Air Cooling Versus Working Place Cooling

Figure 6 shows a typical underground central plant that provides a large area in a mine with air cooled by chilled water. Air flows through ventilation ducts or headings from the chiller to the working headings. When working headings are distant, the air warms and, in some cases, picks up moisture in transit. This limits the cooled air in the production area. The performance characteristics of one plant are as follows:

Air volume cooled	100,000 cfm
Entering air temperature	
Dry-bulb	80.5°F
Wet-bulb	74.3°F
Discharge air temperature	56.4°F
Entering water temperature	42°F
Discharge water temperature	67°F
Heat extracted from air	508 tons

If one of the air-conditioning plants from the workings is not operating, it is preferable to pump chilled water in closed circuit to cooling plants close to or serving individual productive headings.

Another underground cooling plant arrangement (Figure 6A) at a mine in Wallace, Idaho, comprises a light, portable unit that attaches to the normal ventilation system and requires no extra excavation. No separate spray or dust-collection system is used, and the condensed water drips from the duct. The fan gives good service downstream from the cooler. A mechanical refrigeration unit located on the same, or a nearby, level chills water to cool the air; potable water is used in the condenser and discharged to the mine sumping system.

This underground system avoids the central plant problem of cooled air being heated as it travels down a shaft. The cooling coils are normally located within 300 ft of the working place and reduce overall air-cooling plant requirements. Table 2 shows typical performance characteristics.

Cooling Coils and Fan Position

The fan may be upstream or downstream of the cooling coils in underground cooling plants. An upstream fan provides a discharge temperature that is a few degrees lower, but a downstream fan distributes air more efficiently over the coils.

WATER SPRAYS AND EVAPORATIVE COOLING

Finned-tube heat exchangers need periodic cleaning, especially when they are upstream from the fan. Downstream exchangers have spray water, which helps to wash away some of the dirt.

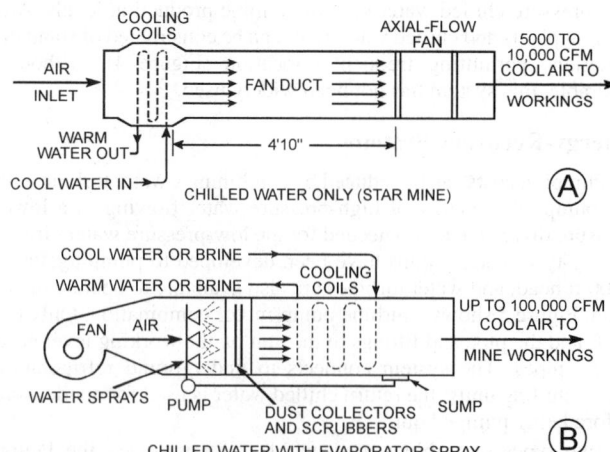

Fig. 6 Underground Air-Cooling Plants

Table 2 Typical Performance of Portable, Underground Cooling Units

Size Rating	30 by 48 in. (40 ton)		24 by 36 in. (20 ton)	
Location	Drift	Shaft	Stope	Stope
Entering air temperature	80°F sat.	80°F sat.	80°F sat.	80°F sat.
Discharge air temperature	70°F sat.	64°F sat.	68°F sat.	80°F sat.
Volume, cfm	12,000	12,000	6,000	6,000
Calculated tons	45	60	22.5	26

Another type of bulk-spraying cooling plant in South Africa consists of a spray chamber serving a section of isolated drift up to several hundred feet long. Chilled water is introduced through a manifold of spray nozzles. Warm, mild air flows countercurrent to the various stages of water sprays. Air cooled by this direct air-to-water cooling system is delivered to active mine workings by the primary and secondary ventilation system. These bulk-spray coolers are efficient and economical.

At one uranium mine, a portable bulk-spray cooling plant was developed that can be advanced with the working faces to overcome the high heat load between a stationary spray chamber and the production heading.

The 300 ton capacity plant has a stainless steel chamber 81 in. long, 87 in. wide, and 102 in. high that contains two stages of spray nozzles and a demister baffle. The skid-mounted unit weighs about 2200 lb and is divided into four components (spray chamber, demister assembly, and two sump halves).

Portable bulk-spray coolers have a wide range of cooling capacity and are cost-effective. A smaller portable spray cooling plant has been developed to cool mine air adjacent to the workplace. It cools and cleans the air through direct air-to-water contact. The cooler is tube-shaped and is normally mounted in a remote location. The mine inlet and discharge air ventilation ducts are connected to duct transitions from the unit. Chilled water is piped to an exposed manifold, and warm water is discharged from the unit into a sump drain.

Hot, humid air enters the cooler at the bottom; it then slows down and flows through egg crate flow straighteners. Initial heat exchange occurs as the air passes through plastic mesh and contacts suspended water droplets. Vertically sprayed water in a spray chamber then directly contacts the ascending warm air. The air passes through the mist eliminator, which removes suspended water droplets. Cool, dehumidified air exits from the cooler through the top outlet transition. The warmed spray water drops to the sump and is discharged through a drainpipe.

BIBLIOGRAPHY

Anonymous. 1980. Surface refrigeration proves energy efficient at Anglo mine. *Mine Engineering* (May).

Bell, A.R. 1970. Ventilation and refrigeration as practiced at Rhokana Corporation Ltd., Zambia. *Journal of the Mine Ventilation Society of South Africa* 23(3):29-35.

Beskine, J.M. 1949. Priorities in deep mine cooling. *Mine and Quarry Engineering* (December):379-84.

Bossard, F.C. 1983. *A manual of mine ventilation design practices.*

Bossard, F.C. and K.S. Stout. Underground mine air-cooling practices. USBM Sponsored Research Contract G0122137.

Bromilow, J.G. 1955. Ventilation of deep coal mines. *Iron and Coal Trades Review.* Part I, February 11:303-08; Part II, February 18:376; Part III, February 25:427-34.

Brown, U.E. 1945. Spot coolers increase comfort of mine workers. *Engineering and Mining Journal* 146(1):49-58.

Caw, J.M. 1953. Some problems raised by underground air cooling on the Kolar Gold Field. *Journal of the Mine Ventilation Society of South Africa* 2(2):83-137.

Caw, J.M. 1957. Air refrigeration. *Mine and Quarry Engineering* (March):111-17; (April):148-56.

Caw, J.M. 1958. Current ventilation practice in hot deep mines in India. *Journal of the Mine Ventilation Society of South Africa* 11(8):145-61.

Caw, J.M. 1959. Observations at an underground air conditioning plant. *Journal of the Mine Ventilation Society of South Africa* 12(11):270-74.

Cleland, R. 1933. Rock temperatures and some ventilation conditions in mines of Northern Ontario. C.I.M.M. *Bulletin Transactions Section* (August):370-407.

Fenton, J.L. 1972. *Survey of underground mine heat sources.* Masters Thesis, Montana College of Mineral Science and Technology.

Field, W.E. 1963. Combatting excessive heat underground at Bralorne. *Mining Engineering* (December):76-77.

Goch, D.C. and H.S. Patterson. 1940. The heat flow into tunnels. *Journal of the Chemical Metallurgical and Mining Society of South Africa* 41(3):117-28.

Hartman, H.L. 1961. *Mine ventilation and air conditioning.* The Ronald Press Company, New York.

Hill, M. 1961. Refrigeration applied to longwall stopes and longwall stope ventilation. *Journal of the Mine Ventilation Society of South Africa* 14(5):65-73.

Kock, H. 1967. Refrigeration in industry. *The South African Mechanical Engineer* (November):188-96.

Le Roux, W.L. 1959. Heat exchange between water and air at underground cooling plants. *Journal of the Mine Ventilation Society of South Africa* 12(5):106-19.

Marks, J. 1969. *Design of air cooler—Star Mine.* Hecla Mining Company, Wallace, ID.

Minich, G.S. 1962. The pressure recuperator and its application to mine cooling. *The South African Mechanical Engineer* (October):57-78.

Muller, F.T. and M. Hill. 1966. Ventilation and cooling as practiced on E.R.P.M. Ltd., South Africa. *Journal of the South African Institute of Mining and Metallurgy.*

Richardson, A.S. 1950. A review of progress in the ventilation of the mines of the Butte, Montana District. *Quarterly of the Colorado School of Mines* (April), Golden, CO.

Sandys, M.P.J. 1961. The use of underground refrigeration in stope ventilation. *Journal of the Mine Ventilation Society of South Africa* 14(6):93-95.

Schlosser, R.B. 1967. *The Crescent Mine cooling system.* Northwest Mining Association Convention (December).

Short, B. 1957. Ventilation and air conditioning at the Magma Mine. *Mining Engineering* (March):344-48.

Starfield, A.M. 1966. Tables for the flow of heat into a rock tunnel with different surface heat transfer coefficients. *Journal of the South African Institute of Mining and Metallurgy* 66(12):692-94.

Thimons, E., R. Vinson, and F. Kissel. 1980. Water spray vent tube cooler for hot stopes. USBM TPR 107.

Thompson, J.J. 1967. Recent developments at the Bralorne Mine. *Canadian Mining and Metallurgy Bulletin* (November):1301-05.

Torrance, B. and G.S. Minish. 1962. Heat exchanger data. *Journal of the Mine Ventilation Society of South Africa* 15(7):129-38.

Van der Walt, J., E. de Kock, and L. Smith. *Analyzing ventilation and cooling requirements for mines.* Engineering Management Services, Ltd., Johannesburg, Republic of South Africa.

Warren, J.W. 1958. *The science of mine ventilation.* Presented at the American Mining Congress, San Francisco (September).

Warren, J.W. 1965. Supplemental cooling for deep-level ventilation. *Mining Congress Journal* (April):34-37.

Whillier, A. 1972. Heat—A challenge in deep-level mining. *Journal of the Mine Ventilation Society of South Africa* 25(11):205-13.

INDUSTRIAL DRYING SYSTEMS

DRYING removes water and other liquids from gases, liquids, and solids. The term is most commonly used, however, to describe the removal of water or solvent from solids by thermal means. **Dehumidification** refers to the drying of a gas, usually by condensation or by absorption with a drying agent (see Chapter 21 of the 1997 *ASHRAE Handbook—Fundamentals*). **Distillation**, particularly **fractional distillation**, is used to dry liquids.

It is cost-effective to separate as much water as possible from a solid using mechanical methods *before* drying using thermal methods. Mechanical methods such as filtration, screening, pressing, centrifuging, or settling require less power and less capital outlay per unit mass of water removed.

This chapter describes systems used for industrial drying and their advantages, disadvantages, relative energy consumption, and applications.

MECHANISM OF DRYING

When a solid dries, two processes occur simultaneously: (1) the transfer of heat to evaporate the liquid and (2) the transfer of mass as vapor and internal liquid. Factors governing the rate of each process determine the drying rate.

The principal objective in commercial drying is to supply the required heat efficiently. Heat transfer can occur by convection, conduction, radiation, or a combination of these. Industrial dryers differ in their methods of transferring heat to the solid. In general, heat must flow first to the outer surface of the solid and then into the interior. An exception is drying with high-frequency electrical currents, where heat is generated within the solid, producing a higher temperature at the interior than at the surface and causing heat to flow from inside the solid to the outer surfaces.

APPLYING HYGROMETRY TO DRYING

In many applications, recirculating the drying medium improves thermal efficiency. The optimum proportion of recycled air balances the lower heat loss associated with more recirculation against the higher drying rate associated with less recirculation.

Because the humidity of drying air is affected by the recycle ratio, the air humidity throughout the dryer must be analyzed to determine whether the predicted moisture pickup of the air is physically attainable. The maximum ability of air to absorb moisture corresponds to the difference between saturation moisture content at wet-bulb (or adiabatic cooling) temperature and moisture content at supply air dew point. The actual moisture pickup of air is determined by heat and mass transfer rates and is always less than the maximum attainable.

ASHRAE psychrometric charts for normal and high temperatures (No. 1 and No. 3) can be used for most drying calculations. The process will not exactly follow the adiabatic cooling lines because some heat is transferred to the material by direct radiation or by conduction from the metal tray or conveyor.

The preparation of this chapter is assigned to TC 9.2, Industrial Air Conditioning.

Example 1. A dryer has a capacity of 90.5 lb of bone-dry gelatin per hour. Initial moisture content is 228% bone-dry basis, and final moisture content is 32% bone-dry basis. For optimum drying, the supply air is at 120°F dry bulb and 85°F wet bulb in sufficient quantity that the condition of exhaust air is 100°F dry bulb and 84.5°F wet bulb. Makeup air is available at 80°F dry bulb and 65°F wet bulb.

Find (1) the required amount of makeup and exhaust air and (2) the percentage of recirculated air.

Solution: In this example, the humidity in each of the three airstreams is fixed; hence, the recycle ratio is also determined. Refer to ASHRAE Psychrometric Chart No. 1 to obtain the humidity ratio of makeup air and exhaust air. To maintain a steady-state condition in the dryer, water evaporated from the material must be carried away by exhaust air. Therefore, the pickup (the difference in humidity ratio between exhaust air and makeup air) is equal to the rate at which water is evaporated from the material divided by the weight of dry air exhausted per hour.

Step 1. From ASHRAE Psychrometric Chart No. 1, the humidity ratios are as follows:

	Dry bulb, °F	Wet bulb, °F	Humidity ratio, lb/lb dry air
Supply air	120	85	0.018
Exhaust air	100	84.3	0.022
Makeup air	80	65.2	0.010

Moisture pickup is $0.022 - 0.010 = 0.012$ lb/lb dry air. The rate of evaporation in the dryer is

$$90.5 (228 - 32)/100 = 177 \text{ lb/h}$$

The dry air required to remove the evaporated water is $177/0.012 = 14{,}750$ lb/h.

Step 2. Assume $x =$ percentage of recirculated air and $(100 - x) =$ percentage of makeup air. Then

Humidity ratio of supply air =
 (Humidity ratio of exhaust and recirculated air) $(x/100)$
  + (Humidity ratio of makeup air)$(100 - x)/100$

Hence,

$$0.018 = 0.022(x/100) + 0.010(100 - x)/100$$

$x = 66.7\%$ recirculated air
$100 - x = 33.3\%$ makeup air

DETERMINING DRYING TIME

The following are three methods of finding drying time, listed in order of preference:

1. Conduct tests in a laboratory dryer simulating conditions for the commercial machine, or obtain performance data using the commercial machine.
2. If the specific material is not available, obtain drying data on similar material by either of the above methods. This is subject to the investigator's experience and judgment.
3. Estimate drying time from theoretical equations (see the section on Bibliography). Care should be taken in using the approximate values obtained by this method.

When designing commercial equipment, tests are conducted in a laboratory dryer that simulates commercial operating conditions. Sample materials used in the laboratory tests should be identical to the material found in the commercial operation. Results from several tested samples should be compared for consistency. Otherwise, the test results may not reflect the drying characteristics of the commercial material accurately.

When laboratory testing is impractical, commercial drying data can be based on the equipment manufacturer's experience.

Commercial Drying Time

When selecting a commercial dryer, the estimated drying time determines what size machine is needed for a given capacity. If the drying time has been derived from laboratory tests, the following should be considered:

- In a laboratory dryer, considerable drying may be the result of radiation and heat conduction. In a commercial dryer, these factors are usually negligible.
- In a commercial dryer, humidity conditions may be higher than in a laboratory dryer. In drying operations with controlled humidity, this factor can be eliminated by duplicating the commercial humidity condition in the laboratory dryer.
- Operating conditions are not as uniform in a commercial dryer as in a laboratory dryer.
- Because of the small sample used, the test material may not be representative of the commercial material.

Thus, the designer must use experience and judgment to modify the test drying time to suit the commercial conditions.

Dryer Calculations

To estimate preliminary cost for a commercial dryer, the circulating airflow rate, the makeup and exhaust airflow rate, and the heat balance must be determined.

Circulating Air. The required circulating or supply airflow rate is established by the optimum air velocity relative to the material. This can be obtained from laboratory tests or previous experience, keeping in mind that the air also has an optimum moisture pickup. (See the section on Applying Hygrometry to Drying.)

Makeup and Exhaust Air. The makeup and exhaust airflow rate required for steady-state conditions within the dryer is also discussed in the section on Applying Hygrometry to Drying. In a **continuously operating dryer**, the relationship between the moisture content of the material and the quantity of makeup air is given by

$$G_T(W_2 - W_1) = M(w_1 - w_2) \qquad (1)$$

where

G_T = dry air supplied as makeup air to the dryer, lb/h
M = stock dried in a continuous dryer, lb/h
W_1 = humidity ratio of entering air, lb of water vapor per lb of dry air
W_2 = humidity ratio of leaving air, lb of water vapor per lb of dry air (In a continuously operating dryer, W_2 is constant; in a batch dryer, W_2 varies during a portion of the cycle.)
w_1 = dry basis moisture content of entering material, lb of water per lb
w_2 = dry basis moisture content of leaving material, lb of water per lb

In **batch dryers**, the drying operation is given as

$$G_T(W_2 - W_1) = (M_1)\frac{dw}{d\theta} \qquad (2)$$

where

M_1 = mass of material charged in a discontinuous dryer, lb per batch
$dw/d\theta$ = instantaneous time rate of evaporation corresponding to w

The makeup air quantity is constant and is based on the average evaporation rate. Equation (2) then becomes identical to Equation (1), where $M = M_1/\theta$. Under this condition, the humidity in the batch

dryer varies from a maximum to a minimum during the drying cycle, whereas in the continuous dryer, the humidity is constant with constant load.

Heat Balance. To estimate the fuel requirements of a dryer, a heat balance consisting of the following is needed:

- Radiation and convection losses from the dryer
- Heating of the commercial dry material to the leaving temperature (usually estimated)
- Vaporization of the water being removed from the material (usually considered to take place at the wet-bulb temperature)
- Heating of the vapor from the wet-bulb temperature in the dryer to the exhaust temperature
- Heating of the total water in the material from the entering temperature to the wet-bulb temperature in the dryer
- Heating of the makeup air from its initial temperature to the exhaust temperature

The energy absorbed must be supplied by the fuel. The selection and design of the heating equipment is an essential part of the overall design of the dryer.

Example 2. Magnesium hydroxide is dried from 82% to 4% moisture content (wet basis) in a continuous conveyor dryer with a fin-drum feed (see Figure 7). The desired production rate is 3000 lb/h. The optimum circulating air temperature for drying is 160°F, which is not limited by the existing steam pressure of the dryer.

Step 1. Laboratory tests indicate the following:

Specific heats

air (c_a)	= 0.24 Btu/lb·°F
material (c_m)	= 0.3 Btu/lb·°F
water (c_w)	= 1.0 Btu/lb·°F
water vapor (c_v)	= 0.45 Btu/lb·°F
Temperature of material entering dryer	= 60°F
Temperature of makeup air	
dry bulb	= 70°F
wet bulb	= 60°F
Temperature of circulating air	
dry bulb	= 160°F
wet bulb	= 100°F
Air velocity through drying bed	= 250 fpm
Dryer bed loading	= 6.82 lb/ft²
Test drying time	= 25 min

Step 2. Previous experience indicates that the commercial drying time is 70% greater than the time obtained in the laboratory test. Therefore, the commercial drying time is estimated to be $1.7 \times 25 = 42.5$ min.

Step 3. The holding capacity of the dryer bed can be calculated as follows:

$$3000(42.5/60) = 2125 \text{ lb at 4\% (wet basis)}$$

The required conveyor area is 2125/6.82 = 312 ft². Assuming the conveyor is 8 ft wide, the length of the drying zone is 312/8 = 39 ft.

Step 4. The amount of water in the material entering the dryer is

$$3000[82/(100 + 4)] = 2370 \text{ lb/h}$$

The amount of water in the material leaving is

$$3000[4/(100 + 4)] = 115 \text{ lb/h}$$

Thus, the moisture removal rate is 2370 − 115 = 2255 lb/h.

Step 5. The air circulates perpendicular to the perforated plate conveyor, so the air volume is the face velocity times the conveyor area:

$$\text{Air volume} = 250 \times 312 = 78,000 \text{ cfm}$$

ASHRAE Psychrometric Charts 1 and 3 show the following air properties:

Supply air (160°F db, 100°F wb)

Humidity ratio	= 0.0285 lb per lb of dry air
Specific volume	= 16.33 ft³ per lb of dry air

Makeup air (70°F db, 60°F wb)

Humidity ratio W_1	= 0.0086 lb per lb of dry air

The mass flow rate of dry air is

$$(78,000 \times 60)/16.33 = 286,500 \text{ lb/h}$$

Step 6. The amount of moisture pickup is

$$2255/286,500 = 0.0079 \text{ lb per lb of dry air}$$

The humidity ratio of the exhaust air is

$$W_2 = 0.0285 + 0.0079 = 0.0364 \text{ lb per lb of dry air}$$

Substitute in Equation (1) and calculate G_T as follows:

$$G_T(0.0364 - 0.0086) = (3000/1.04)(82 - 4)/100$$

$$G_T = 81,000 \text{ lb dry air per hour}$$

Therefore,

Makeup air	$= 100 \times 81,000/286,500 = 28.2\%$
Recirculated air	$= 71.8\%$

Step 7. Heat Balance

Sensible heat of material	$=$	$M(t_{m2} - t_{m1})c_m$
	$=$	$(3000/1.04)(100 - 60)0.3$
	$=$	$34,600 \text{ Btu/h}$
Sensible heat of water	$=$	$M_{w1}(t_w - t_{m1})c_w$
	$=$	$2370(100 - 60)1.0$
	$=$	$94,800 \text{ Btu/h}$
Latent heat of evaporation	$=$	$M(w_1 - w_2)H$
	$=$	$2255 \text{ lb/h} \times 1037 \text{ Btu/lb}$
	$=$	$2,338,400 \text{ Btu/h}$
Sensible heat of vapor	$=$	$M(t_2 - t_w)c_v$
	$=$	$2255(160 - 100)0.45$
	$=$	$60,900 \text{ Btu/h}$
Required heat for material	$=$	$2,528,700 \text{ Btu/h}$

The temperature drop $(t_2 - t_3)$ through the bed is

$$\frac{\text{Required heat}}{\text{Supplied air, (lb/h)} \times c_a} = \frac{2,528,700}{286,500 \times 0.24} = 37°\text{F}$$

Therefore, the exhaust air temperature is $160 - 37 = 123°\text{F}$.

Required heat for makeup air	$=$	$G_T(t_3 - t_1)c_a$
	$=$	$81,000(123 - 70)0.24$
	$=$	$1,030,000 \text{ Btu/h}$

The total heat required for material and makeup air is

$$2,528,700 + 1,030,000 = 3,559,000 \text{ Btu/h}$$

Additional heat that must be provided to compensate for radiation and convection losses can be calculated from the known construction of the dryer surfaces.

DRYING SYSTEM SELECTION

A general procedure for selecting a drying system consists of the following:

1. Survey of suitable dryers.
2. Preliminary cost estimates of various types.
   (a) Initial investment
   (b) Operating cost
3. Drying tests conducted in prototype or laboratory units, preferably using the most promising equipment available. Sometimes a pilot plant is justified.
4. Summary of tests evaluating quality of samples of the dried products.

Factors that can overshadow the operating or investment cost include the following:

- Product quality, which should not be sacrificed
- Dusting, solvent, or other product losses
- Space limitation
- Bulk density of the product, which can affect packaging cost

Friedman (1951) and Parker (1963) discuss additional aids to dryer selection.

TYPES OF DRYING SYSTEMS

Radiant Infrared Drying

Thermal radiation may be applied by infrared lamps, gas-heated incandescent refractories, steam-heated sources, and, most often, electrically heated surfaces. Infrared heats only near the surface of a material, so it is best used to dry thin sheets.

The use of infrared heating to dry webs such as uncoated materials has been relatively unsuccessful because of process control problems. Thermal efficiency can be low; heat transfer depends on the emitter's characteristics and configuration, and on the properties of the material to be dried.

Radiant heating is used for drying ink and other coatings on paper, textile fabrics, paint films, and lacquers. Inks have been specifically formulated for curing with tuned or narrow wavelength infrared radiation.

Ultraviolet Radiation Drying

Ultraviolet (UV) drying uses electromagnetic radiation. Inks and other coatings based on monomers are cure dried when exposed to UV radiation. This method has superior properties (Chatterjee and Ramaswamy 1975): the print resists scuff, scratch, acid, alkali, and some solvents. Printing can also be done at higher speeds without damage to the web.

Major barriers to wider acceptance of UV drying include the high capital installation cost and the increased cost of inks. The cost and frequency of replacing UV lamps are greater than for infrared ovens.

Overexposure to radiation and ozone, which is formed by UV radiation's effect on atmospheric oxygen, can cause severe sunburn and possibly blood and eye damage. Safety measures include fitting the lamp housings with screens, shutters, and exhausts.

Conduction Drying

Drying rolls or drums (Figure 1), flat surfaces, open kettles, and immersion heaters are examples of direct-contact drying. The heating surface must have close contact with the material, and agitation may increase uniform heating or prevent overheating.

Conduction drying is used to manufacture and dry paper products. It (1) does not provide a high drying rate, (2) does not furnish uniform heat and mass transfer conditions, (3) usually results in a poor moisture profile across the web, (4) lacks proper control, (5) is costly to operate and install, and (6) usually creates undesirable working conditions in areas surrounding the machine. Despite these disadvantages, replacing existing systems with other forms of drying is expensive. For example, Joas and Chance (1975) report that RF (dielectric) drying of paper requires approximately four times the capital cost, six times the operating (heat) cost, and five times the maintenance cost of steam cylinder conduction drying.

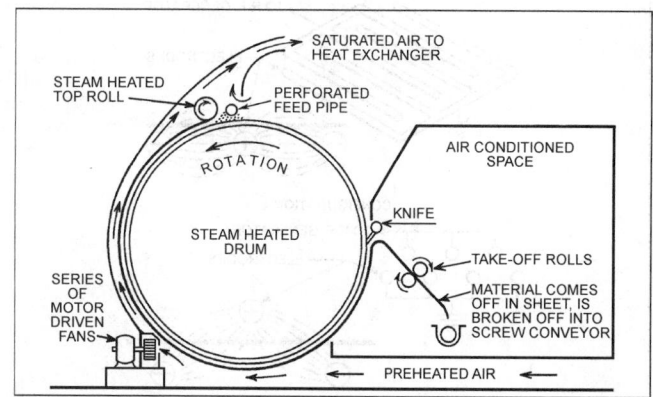

Fig. 1 Drum Dryer

However, augmenting conduction drying with dielectric drying sections offsets the high cost of RF drying and may produce savings and increased profits from greater production and higher final moisture content.

Further use of large conduction drying systems depends on reducing heat losses from the dryer, improving heat recovery, and incorporating other drying techniques to maintain quality.

Dielectric Drying

When wet material is placed in a strong, high-frequency (2 to 100 MHz) electrostatic field, heat is generated within the material. More heat is developed in the wetter areas than in the drier areas, resulting in automatic moisture profile correction. Water is evaporated without unduly heating the substrate. Therefore, in addition to its leveling properties, dielectric drying provides uniform heating throughout the web thickness.

Dielectric drying is controlled by varying field or frequency strength; varying field strength is easier and more effective. Response to this variation is quick, with neither time lag nor thermal lag in heating. The dielectric heater is a sensitive moisture meter.

Several electrode configurations are used. The platen type (Figure 2) is used for drying and baking foundry cores, heating plastic preforms, and drying glue lines in furniture. The rod or stray field types (Figure 3) are used for thin web materials such as paper and textile products. The double-rod types (over and under material) are used for thicker webs or flat stock, such as plywood.

Dielectric drying is popular in the textile industry. Because air is entrained between fibers, convection drying is slow and uneven. This can be overcome by dielectric drying after yarn drying. Because the yarn is usually transferred to large packages immediately after drying, even and correct moisture content can be

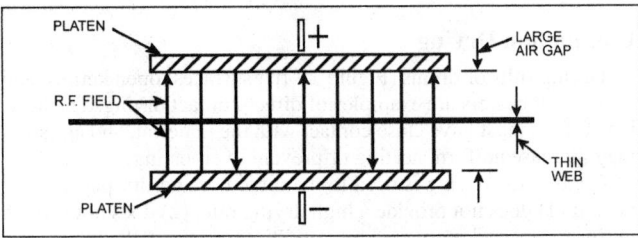

Fig. 2   Platen-Type Dielectric Dryer

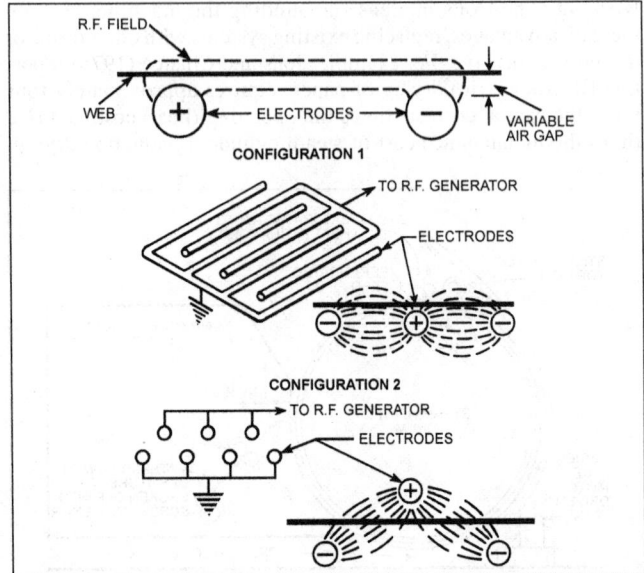

Fig. 3   Rod-Type Dielectric Dryers

obtained by dielectric drying. Knitting wool seems to benefit from internal steaming in hanks.

Warping caused by nonuniform drying is a serious problem for plywood and linerboard. Dielectric drying yields warp-free products.

Dielectric drying is not cost-effective for overall paper drying but has advantages when used at the dry end of a conventional steam drum dryer. It corrects moisture profile problems in the web without overdrying. This combination of conventional and dielectric drying is synergistic; the drying effect of the combination is greater than the sum of the two types of drying. This is more pronounced in thicker web materials, accounting for as much as a 16% line speed increase and a corresponding 2% energy input increase.

Microwave Drying

Microwave drying or heating uses ultrahigh-frequency (900 to 5000 MHz) radiation. It is a form of dielectric heating and is used for heating nonconductors. Because of its high frequency, microwave equipment is capable of generating extreme power densities.

Microwave drying is applied to thin materials in strip form by passing the strip through the gap of a split waveguide. Entry and exit shielding make continuous process applications difficult. Its many safety concerns make microwave drying more expensive than dielectric drying. Control is also difficult because microwave drying lacks the self-compensating properties of dielectrics.

Convection Drying (Direct Dryers)

Some convection drying occurs in almost all dryers. True convection dryers, however, use circulated hot air or other gases as the principal heat source. Each means of mechanically circulating air or gases has its advantages.

Rotary Dryers. These cylindrical drums cascade the material being dried through the airstream (Figure 4). The dryers are heated directly or indirectly, and air circulation is parallel or counterflow. A variation is the rotating-louver dryer, which introduces air beneath the flights to provide close contact.

Cabinet and Compartment Dryers. These batch dryers range from the heated loft (with only natural convection and usually poor and nonuniform drying) to self-contained units with forced draft and properly designed baffles. Several systems may be evacuated to dry delicate or hygroscopic materials at low temperatures. Material is usually spread in trays to increase the exposed surface. Figure 5 shows a dryer that can dry water-saturated products.

When designing dryers to process products saturated with solvents, special features must be included to prevent explosive gases from forming. Safe operation requires exhausting 100% of the air circulated during the initial drying period or during any part of the drying cycle when the solvent is evaporating at a high rate. At the end of the purge cycle, the air is recirculated and heat is gradually applied. To prevent explosions, laboratory dryers can be used to determine the amount of air circulated, the cycle lengths, and the rate that heat is applied for each product. In the drying cycle, dehumidified air, which is costly, should be recirculated as soon possible. The air *must not* be recirculated when cross-contamination of products is prohibited.

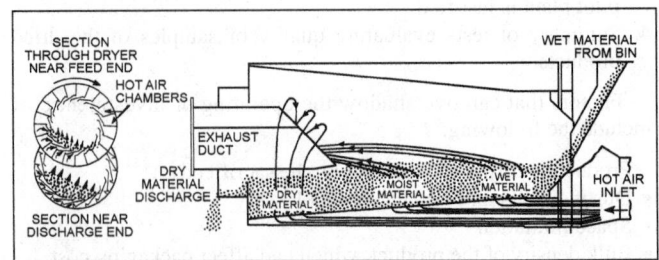

Fig. 4   Cross Section and Longitudinal Section of Rotary Dryer

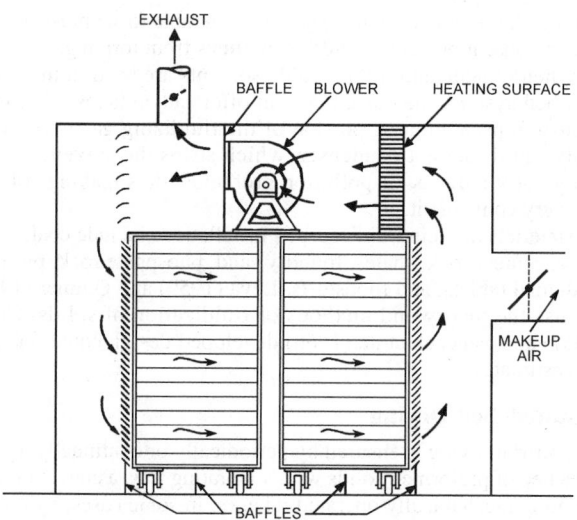

Fig. 5   Compartment Dryer Showing Trucks with Air Circulation

Dryers must have special safety features in case any part of the drying cycle fails. The following are some of the safety design features described in *Industrial Ovens and Driers* (FMEA 1990):

1. Each compartment must have separate supply and exhaust fans and an explosion-relief panel.
2. The exhaust fan blade tip speed should be 5000 fpm for a forward-inclined blade, 6800 fpm for a radial-tip blade, and 7500 fpm for a backward-inclined blade. These speeds produce high static pressures at the fan, ensuring constant air exhaust volumes under conditions such as negative pressures in the building or downdrafts in the exhaust stacks.
3. An airflow failure switch in the exhaust duct must shut off fans and the heating coil and must sound an alarm.
4. An airflow failure switch in the air supply system must shut off fans and the heating coil and must sound an alarm.
5. A high-temperature limit controller in the supply duct must shut off the heat to the heating coil and must sound an alarm.
6. An electric interlock on the dryer door must interrupt the drying cycle if the door is opened beyond a set point, such as that wide enough for a person to enter for product inspection.

Tunnel Dryers. Tunnel dryers are modified compartment dryers that operate continuously or semicontinuously. Heated air or combustion gas is circulated by fans. The material is handled on trays or racks on trucks and moves through the dryer either intermittently or continuously. The airflow may be parallel, counterflow, or a combination obtained by center exhaust (Figure 6). Air may also flow across the tray surface, vertically through the bed, or in any combination of directions. By reheating the air in the dryer or recirculating it, a high degree of saturation is reached before the air is exhausted, thus reducing the sensible heat loss.

The following problems with tunnel dryers have been experienced and should be considered in future designs:

- Operators may overload product trays to increase output, but this can overtax the system and increase the drying time.
- Sometimes air from the drying tunnel is discharged into the production area, increasing the humidity. Air from the drying tunnel should be discharged to the drying system return or outside.
- Overloaded product trays add pressure drop, which decreases flow through the tunnel dryer. The control panel should indicate validated flow through the tunnel. High and low flow and high moisture levels should trigger alarms.
- Cycle times can be reduced by designing dryers for cross-flow rather than end-to-end flow.

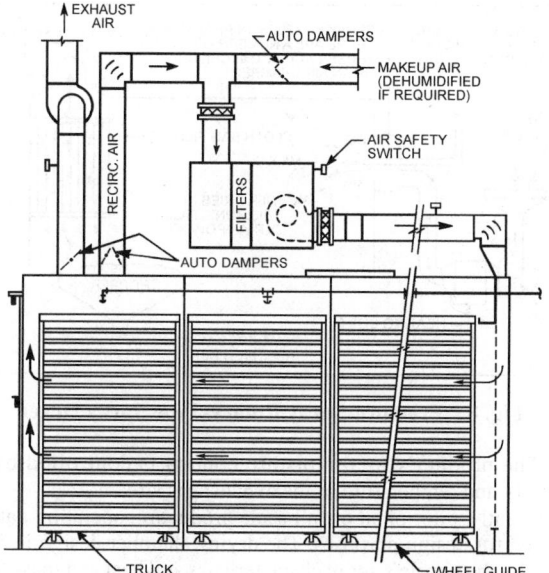

Fig. 6   Explosionproof Truck Dryer Showing Air Circulation and Safety Features

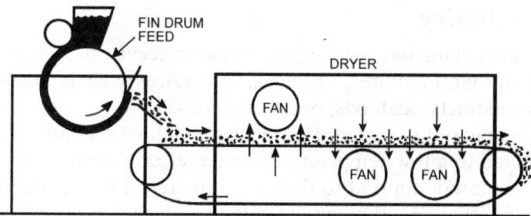

Fig. 7   Section of Blow-Through-Type Continuous Dryer

A variation of the tunnel dryer is the strictly continuous dryer, which has one or more mesh belts that carry the product through it, as shown in Figure 7. Many combinations of temperature, humidity, air direction, and velocity are possible. Hot air leaks at the entrance and exit can be minimized by baffles or inclined ends, with the material entering and leaving from the bottom.

High-Velocity Dryers. High-velocity hoods or dryers have been used as supplements to conventional cylinder dryers for drying paper. When used with conventional cylinder dryers, web instability and lack of process control result. Where internal diffusion is not the controlling factor in the drying rate, applications such as thin permeable webs offer more promise.

Spray Dryers. Spray dryers have been used in the production of dried milk, coffee, soaps, and detergents. Because the dried product (in the form of small beads) is uniform and the drying time is short (5 to 15 s), this drying method has become more important. When a liquid or slurry is dried, the spray dryer has high production rates.

Spray drying involves the atomization of a liquid feed in a hot-gas drying medium. The spray can be produced by a two-fluid nozzle, a high-pressure nozzle, or a rotating disk. Inlet gas temperatures range from 200 to 1400°F, with the high temperatures requiring special construction materials. Because thermal efficiency increases with the inlet gas temperature, high inlet temperatures are desirable. Even heat-sensitive products can be dried at higher temperatures because of the short drying time. Hot gas flow may be either concurrent or countercurrent to the falling droplets. Dried particles settle out by gravity. Fine material in the exhaust air is collected in cyclone separators or bag filters. Figure 8 shows a typical spray drying system.

The physical properties of the dried product (such as particle size, bulk density, and dustiness) are affected by atomization characteristics and the temperature and direction of flow of the drying

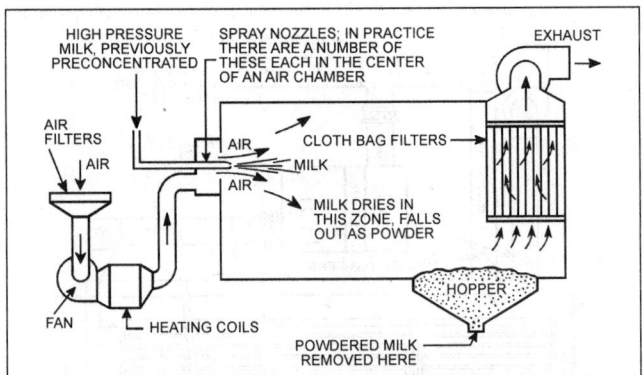

Fig. 8 Pressure-Spray Rotary-Type Spray Dryer

gas. The product's final moisture content is controlled by the humidity and temperature of the exhaust gas stream.

Currently, pilot-plant or full-scale production operating data are required for design purposes. The drying chamber design is determined by the nozzle's spray characteristics and heat and mass transfer rates. There are empirical expressions that approximate mean particle diameter, drying time, chamber volume, and inlet and outlet gas temperatures.

Freeze Drying

Freeze drying has been applied to pharmaceuticals, serums, bacterial and viral cultures, vaccines, fruit juices, vegetables, coffee and tea extracts, seafoods, meats, and milk.

The material is frozen, then placed in a high-vacuum chamber connected to a low-temperature condenser or chemical desiccant. Heat is slowly applied to the frozen material by conduction or infrared radiation, allowing the volatile constituent, usually water, to sublime and condense or be absorbed by the desiccant. Most freeze-drying operations occur between 14 and −40°F under minimal pressure. While this process is expensive and slow, it has advantages for heat-sensitive materials (see Chapter 15 of the 1998 *ASHRAE Handbook—Refrigeration*).

Vacuum Drying

Vacuum drying takes advantage of the decrease in the boiling point of water that occurs as the pressure is lowered. Vacuum drying of paper has been partially investigated. Serious complications arise if the paper breaks, and massive sections must be removed. Vacuum drying is used successfully for pulp drying, where lower speeds and higher weights make breakage relatively infrequent.

Fluidized-Bed Drying

A fluidized-bed system contains solid particles through which a gas flows with a velocity higher than the incipient fluidizing velocity but lower than the entrainment velocity. Heat transfer between the individual particles and the drying air is efficient because there is close contact between powdery or granular material and the fluidizing gas. This contact makes it possible to dry sensitive materials without danger of large temperature differences.

The dried material is free-flowing and, unlike that from convection-type dryers, is not encrusted on trays or other heat-exchanging surfaces. Automatic charging and discharging are possible, but the greatest advantage is reduced process time. Only simple controls are important (i.e., control over fluidizing air or gas temperatures and the drying time of the material).

All fluidized-bed dryers should have explosion relief flaps. Both the pressure and the flames of an explosion are dangerous. Also, when toxic materials are used, uncontrolled venting to the atmosphere is prohibited. Explosion suppression systems, such as pressure-actuated ammonium-phosphate extinguishers, have been used instead of relief

venting. An inert dryer atmosphere is preferable to suppression systems because it prevents explosive mixtures from forming.

When organic and inflammable solvents are used in the fluidized-bed system, the closed system offers advantages other than explosion protection. A portion of the fluidizing gas is continuously run through a condenser, which strips the solvent vapors and greatly reduces air pollution problems, thus making solvent recovery convenient.

Materials dried in fluidized-bed installations include coal, limestone, cement rock, shales, foundry sand, phosphate rock, plastics, medicinal tablets, and foodstuffs. Leva (1959) and Othmer (1956) discuss the theory and methods of fluidization of solids. Clark (1967) and Vanecek et al. (1966) developed design equations and cost estimates.

Agitated-Bed Drying

Uniform drying is ensured by periodically or continually agitating a bed of preformed solids with a vibrating tray, a conveyor, or a vibrating mechanically operated rake, or, in some cases, by partial fluidization of the bed on a perforated tray or conveyor through which recycled drying air is directed. Drying and toasting cereals is an important application.

Drying in Superheated Vapor Atmospheres

When drying solids with air or another gas, the vaporized solvent (water or organic liquid) must diffuse through a stagnant gas film to reach the bulk gas stream. Because this film is the main resistance to mass transfer, the drying rate depends on the solvent vapor diffusion rate. If the gas is replaced by solvent vapor, resistance to mass transfer in the vapor phase is eliminated, and the drying rate depends only on the heat transfer rate. Drying rates in solvent vapor, such as superheated steam, are greater than those in air for equal temperatures and mass flow of the drying media.

This method also has higher thermal efficiency, easier solvent recovery, and a lower tendency to overdry, and it eliminates oxidation or other chemical reactions that occur when air is present. In drying cloth, superheated steam reduces the migration tendency of resins and dyes. Superheated vapor drying cannot be applied to heat-sensitive materials because of the high temperatures.

Commercial drying equipment with recycled solvent vapor as the drying medium is available. Installations have been built to dry textile sheeting and organic chemicals.

Flash Drying

Finely divided solid particles that are dispersed in a hot gas stream can be dried by flash drying, which is rapid and uniform. Commercial applications include drying pigments, synthetic resins, food products, hydrated compounds, gypsum, clays, and wood pulp.

REFERENCES

Chatterjee, P.C. and R. Ramaswamy. 1975. Ultraviolet radiation drying of inks. *British Ink Maker* 17(2):76.

Clark, W.E. 1967. Fluid bed drying. *Chemical Engineering* 74(March 13): 177.

FMEA. 1990. Industrial ovens and driers. Data Sheet No. 6-9. Factory Mutual Engineering Association, Norcross, GA.

Friedman, S.J. 1951. Steps in the selection of drying equipment. *Heating and Ventilating* (February):95.

Joas, J.G. and J.L. Chance. 1975. Moisture leveling with dielectric, air impingement and steam drying—A comparison. *Tappi* 58(3):112.

Leva, M. 1959. *Fluidization*. McGraw-Hill, New York.

Othmer, D.F. 1956. *Fluidization*. Reinhold Publishing, New York.

Parker, N.H. 1963. Aids to drier selection. *Chemical Engineering* 70(June 24):115

Vanecek, Markvart, and Drbohlav. 1966. *Fluidized bed drying*. Chemical Rubber Company, Cleveland, OH.

VENTILATION OF THE INDUSTRIAL ENVIRONMENT

GENERAL ventilation controls heat, odors, and hazardous chemical contaminants that could affect the health and safety of industrial workers. For better control, heat and contaminants should be exhausted at their sources by local exhaust systems, which require lower airflows than general (dilution) ventilation (Goldfield 1980). Chapter 29, Industrial Local Exhaust Systems, supplements this chapter.

General ventilation can be provided by mechanical systems, by natural draft, or by a combination of the two. Examples of combination systems include (1) mechanical supply with air relief through louvers and/or other types of vents and (2) mechanical exhaust with air replacement inlet louvers and/or doors.

As a rule, mechanical supply systems provide the best control and the most comfortable environment. They consist of an inlet section, a filter, heating and/or cooling equipment, a fan, ducts, and air diffusers for distributing air within the building. When toxic gases and particles are not present, air that is cleaned in the general exhaust system or in free-hanging filter units can be recirculated via a return duct. Air recirculation can reduce heating costs in winter.

A general exhaust system, which removes air contaminated by gases or particles not captured by local exhausts, usually consists of inlets, ducts, an air cleaner, and a fan. After air passes through the filters, cleaned air is discharged outside or part is returned to the building. The cleaning efficiency of an air filter should conform to environmental regulations and depends on factors such as building location, background contaminant concentrations in the atmosphere, nature of the contaminants, and height and velocity of the discharge. In some cases, for example when the industrial zone is located away from residential areas, a general exhaust system may not have an air cleaner.

Many industrial ventilation systems must handle simultaneous exposures to heat and hazardous substances. In these cases, ventilation can be provided by a combination of local exhaust, general supply, and general exhaust systems. The ventilation engineer must carefully analyze supply and exhaust air requirements to determine the optimum balance between them. For example, air supply makeup for hood exhaust may be insufficient for control of heat exposure. It is also important to consider seasonal effects on the performance of ventilation systems.

In specifying acceptable design toxic chemical and heat exposure levels, the industrial hygienist or industrial hygiene engineer must consult the appropriate government standards and guidelines given either in this chapter or in reference materials. The standard levels for most chemical and heat exposures are time-weighted averages that allow excursions above the limit as long as they are balanced by equivalent excursions below the limit during the workday. However, exposure level standards for heat and contaminants are not lines of demarcation between safe and unsafe exposures. Rather, they represent conditions to which, it is believed, nearly all workers may be exposed day after day without adverse effects (ACGIH 1998b). Because a small percentage of workers may be overly stressed at exposure levels below the standards, it is prudent for the ventilation engineer to design for exposure levels below the limits.

The preparation of this chapter is assigned to TC 5.8, Industrial Ventilation.

In the case of exposure to toxic chemicals, the number of contaminant sources, their generation rates, and the effectiveness of exhaust hoods are rarely known. Consequently, the ventilation engineer must rely on common ventilation/industrial hygiene practice when designing toxic chemical controls. Close cooperation among the industrial hygienist, the process engineer, and the ventilation engineer is required (Schroy 1986).

This chapter describes principles of good ventilation practice and includes other information on hygiene in the industrial environment. Various publications from the U.S. National Institute for Occupational Safety and Health (NIOSH 1986), the British Occupational Hygiene Society (1987), the National Safety Council (1988), and the U.S. Department of Health and Human Services (1986) provide in-depth coverage of industrial hygiene principles and their application.

Ventilation control alone is frequently inadequate for meeting heat stress standards. Optimum solutions may involve additional controls, such as spot air cooling, changes in work-rest patterns, and radiation shielding. Goodfellow and Smith (1982) summarized the technical progress being made in the industrial ventilation field by different investigators throughout the world. Proceedings from international symposiums (e.g., Ventilation '85, '88, '91, '94, and '97) are also valuable sources of information on ventilation technology.

Supplemental information can be found in Chapters 16 through 21, 23, 24, and 25 of the 1996 *ASHRAE Handbook—Systems and Equipment*. Chapters 11 through 30 of this volume include ventilation requirements for specific applications, and Chapter 44 covers control of gaseous contaminants. Fundamentals of space air diffusion are covered in Chapter 31 of the 1997 *ASHRAE Handbook—Fundamentals*.

HEAT CONTROL IN INDUSTRIAL WORK AREAS

Ventilation for Heat Relief

Many industrial work situations involve processes that release large amounts of heat and moisture to the environment. In such environments, it may not be economically feasible to maintain comfort conditions (ASHRAE *Standard* 55), particularly during the hot summer months. Comfortable conditions are not physiologically necessary; the body must be in thermal balance with the environment, but this can occur at temperature and humidity conditions well above the comfort zone. In areas where heat and moisture gains from a process are low to moderate, comfort conditions may not be provided simply because personnel exposures are infrequent and of short duration. In such cases, ventilation is one of many controls that may be necessary to prevent excessive physiological strain from heat stress.

The engineer must distinguish between the control needs for hot-dry industrial areas and warm-moist conditions. In **hot-dry** areas, a process gives off only sensible and radiant heat without adding moisture to the air. This increases the heat load on exposed workers, but the rate of cooling by evaporation of sweat is not reduced. Heat balance may be maintained, but it may be at the expense of excessive sweating. Hot-dry work situations occur around furnaces, forges, metal-extruding and rolling mills, glass-forming machines, and so forth.

In **warm-moist** conditions, a wet process gives off mainly latent heat. The rise in the heat load on workers may be insignificant, but the increased moisture content of the air seriously reduces cooling by the evaporation of sweat. The warm-moist condition is potentially more hazardous than the hot-dry condition. Typical warm-moist operations are found in textile mills, laundries, dye houses, and deep mines where water is used extensively for dust control.

The industrial heat problem is affected by the local climate. Solar heat gain and elevated outdoor temperatures increase the heat load at the workplace, but these contributions may be insignificant compared to the process heat generated locally. The moisture content of the outdoor air is an important factor that can affect hot-dry work situations by seriously restricting an individual's evaporative cooling. For warm-moist conditions, solar heat gain and elevated outdoor temperatures are more important because the moisture contributed by the outdoor air is insignificant compared to that released by the process.

Both ASHRAE and the International Organization for Standardization (ISO) have standards for thermal comfort conditions for humans (ASHRAE *Standard* 55 and ISO *Standard* 7730). The research these standards are based on was performed mainly under environmental conditions similar to those in commercial and residential buildings, with relatively low activity levels (mainly sedentary, metabolic rate of 1.2 met), normal indoor clothing (insulation value of 0.5 to 1.0 clo), and a limited range of environmental parameters. One met is 18.4 Btu/h·ft^2; one clo is 0.88 ft^2·°F·h/Btu.

Analyses by Zhivov and Olesen (1993) and Olesen and Zhivov (1994) show that existing thermal comfort standards can be extended to workplaces with higher levels of activity.

Methods for evaluating the general thermal state of the body both in comfort conditions and under heat and cold stress are based on an analysis of the heat balance for the human body, which is discussed in Chapter 8 of the 1997 *ASHRAE Handbook—Fundamentals*. A person may find the thermal environment unacceptable or intolerable due to local effects on the body caused by asymmetric radiation, air velocity, vertical air temperature differences, or contact with hot or cold surfaces (floors, machinery, tools, etc.).

Moderate Thermal Environments

ISO *Standard* 7730 defines the **predicted mean vote and predicted percent dissatisfied (PMV and PPD) indices** (Fanger 1982) for evaluating moderate thermal environments. To quantify comfort, the PMV index gives a value on the seven-point ASHRAE thermal sensation scale:

    +3   hot
    +2   warm
    +1   slightly warm
      0   neutral
    −1   slightly cool
    −2   cool
    −3   cold

An equation in the standard for calculating the PMV index is based on six factors: clothing, activity, air temperature, mean radiant temperature, air speed, and humidity. Even if the PMV is 0, at least 5% of the occupants will be dissatisfied with the thermal environment. This method is also discussed in Chapter 8 of the 1997 *ASHRAE Handbook—Fundamentals*.

The PMV index is determined assuming that all evaporation from the skin is transported through the clothing to the environment; therefore, the PMV index is applicable only within −2 < PMV < +2, that is, for thermal environments where sweating is minimal. The PMV index is not applicable for hot environments.

Another method used to estimate combined effects in moderate environments is the **effective temperature ET***, which is described

in Chapter 8 of the 1997 *ASHRAE Handbook—Fundamentals*. In the comfort range, it gives similar results to the PMV index.

ASHRAE *Standard* 55 specifies ranges for operative temperatures that will be acceptable to at least 90% of the occupants. For example, for a sedentary activity level (1.2 met) and typical indoor clothing, the standard recommends the following operative temperature ranges: in winter (heating period, 0.9 to 1.0 clo), 68 to 75°F; in summer (cooling period, 0.5 clo), 73 to 79°F.

Operative temperatures for activities higher than 1.2 met (but less than 3 met) can be found from ISO *Standard* 7730 or can be calculated from the operative temperatures at sedentary conditions using the following equation (ASHRAE *Standard* 55):

$$t_o^{act} = t_o^{sed} - 5.4(1 + R_{cl})(M - 1.2) \tag{1}$$

where

t_o^{act} = operative temperature for activity, °F
t_o^{sed} = operative temperature for sedentary conditions, °F
R_{cl} = insulation value for garment ensemble, clo
M = metabolic rate, met

Heat Stress—Thermal Standards

Heat stress is the thermal condition of the environment that, in combination with metabolic heat generation of the body, causes the deep body temperature to exceed 100°F. The recommended heat stress index for evaluating an environment's heat stress potential is the **wet-bulb globe temperature** (WBGT), which is defined as follows:

Outdoors with solar load:

$$WBGT = 0.7 t_{nw} + 0.2 t_g + 0.1 t_{db} \tag{2}$$

Indoors or outdoors with no solar load:

$$WBGT = 0.7 t_{nw} + 0.3 t_g \tag{3}$$

where

t_{nw} = natural wet-bulb temperature (no defined range of air velocity; different from saturation temperature or psychrometric wet bulb), °F
t_{db} = dry-bulb temperature (shielded thermometer), °F
t_g = globe temperature (Vernon bulb thermometer, 6 in. diameter), °F

The **threshold limit value** (TLV) for heat stress is set for different levels of physical stress, as shown in Figure 1 (NIOSH 1986). This graph depicts the allowable work regime (in terms of rest periods and work periods each hour) for different levels of work over a range of WBGT. For applying Figure 1, it is assumed that the rest area has the same WBGT as the work area. If the rest area is at or below 75°F WBGT, the resting time is reduced by 25%. The curves are valid for workers acclimatized to heat. Refer to criteria of the National Institute for Occupational Safety and Health (NIOSH 1986) for recommended WBGT ceiling values and time-weighted average exposure limits for both acclimatized and unacclimatized workers.

The **WBGT index** is an international standard (ISO *Standard* 7243) for the evaluation of hot environments. The WBGT index and activity levels should be evaluated on 1 h mean values; that is, WBGT and activity are measured and estimated as time-weighted averages on a 1 h basis for continuous work, or on a 2 h basis when the exposure is intermittent. Although recommended by NIOSH, the WBGT has not been accepted as a legal standard by the Occupational Safety and Health Administration (OSHA). It is generally used in conjunction with other methods to determine heat stress.

Although Figure 1 is useful for evaluating heat stress, it is of limited use for control purposes or for the evaluation of comfort. Air velocity and psychrometric wet-bulb measurements are usually needed in order to specify proper controls, and neither is measured in WBGT determinations. However, Harris (1988) used the

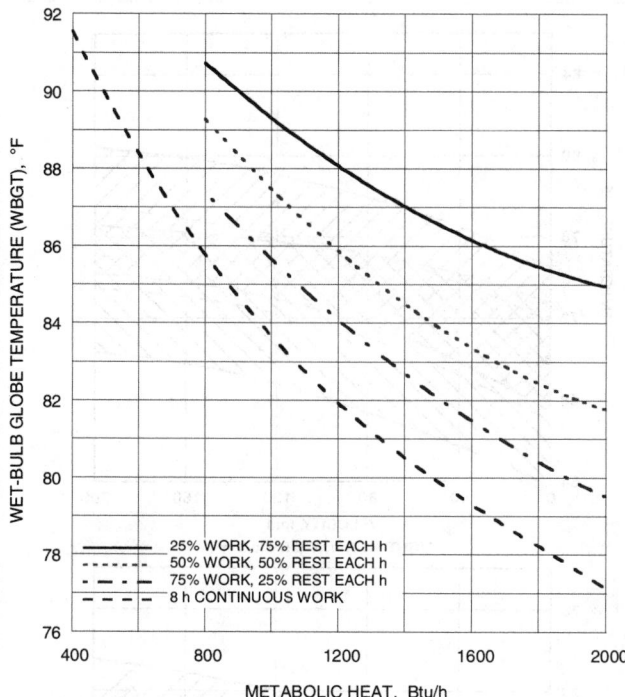

Fig. 1 Recommended Heat Stress Exposure Limits for Heat-Acclimatized Workers
[Adapted from NIOSH (1986)]

adiabatic wet-bulb temperature line on a psychrometric chart to represent natural wet bulb as a conservative substitute in heat stress situations. More useful tools, including the **heat stress index** (HSI), may be found in Chapter 8 of the 1997 *ASHRAE Handbook—Fundamentals* and in ISO *Standards* 7730 and 7933.

The thermal relationship between humans and their environment depends on four independent variables:

- Air temperature
- Radiant temperature
- Moisture content of the air
- Air velocity

Together with the rate of internal heat production (the metabolic rate), these factors may combine in various ways to create different degrees of heat stress. Heat stress index formulas include calculations of the relative contributions to stress resulting from metabolism, radiant heat gain (or loss), convective heat gain (or loss), and evaporative (sweat) heat gain (or loss). For supplemental information on the evaluation and control of heat stress using such methods as reduction of radiation, changes in work-rest pattern, spot cooling, and cooling vests and suits, refer to NIOSH (1986), ACGIH (1998b), Constance (1983), Caplan (1980), and Brief et al. (1983).

Local Discomfort and Individual Parameters

Even if the combined effect of comfort parameters provides a heat balance for the body as a whole, parts of the body may experience discomfort from drafts, radiant asymmetry, and/or vertical temperature differences.

Air Speed. Air at increased speeds is often used in industry to provide cooling in warm environments. Depending on the conditions, an air speed that is too high may cause a **draft**, which is defined as unwanted convective cooling of one or more parts of the body. Drafts are probably the largest source of complaints from occupants at low activity levels (sedentary, standing) in air-conditioned spaces.

The maximum value of the mean air speed in an occupied zone depends on the air temperature t_a and the turbulence intensity Tu. ASHRAE *Standard* 55 provides information on air speed and temperature limits for sedentary (less than 1.2 met), light (1.6 met), medium (2 met), and high (greater than 3 met) levels of human activity; however, this standard does not provide precise relationships. Air speeds as high as 300 fpm are acceptable, unless they cause problems unrelated to thermal comfort (e.g., blowing of papers or blowing of shield gas in semiautomatic welding).

Combinations of air temperature and speed related to comfort conditions in winter and summer periods for different levels of human activity (assuming typical indoor clothing) are presented in Figure 2, which is based on the data presented in ISO *Standard* 7730. Other research and normative data on acceptable and optimal thermal comfort environments are discussed by Olesen and Zhivov (1994).

The effect of turbulence intensity on the sensation of draft has been investigated by Fanger et al. (1988) for different methods of air distribution. **Turbulence intensity** is the ratio of the magnitude of the velocity fluctuation to the mean velocity:

$$Tu = 100 \frac{V'}{V} \qquad (4)$$

where

Tu = turbulence intensity, %
V' = standard deviation of air velocity fluctuation
V = mean air velocity

Based on studies of different air distribution methods, Fanger et al. (1988) found that turbulence intensity in ventilated spaces depends on the type of ventilation system and varies from 10 to 70%. For air velocities ranging from 10 to 80 fpm and air temperatures ranging from 68 to 79°F, tests were conducted at three levels of turbulence intensity: low (Tu less than 12%), medium (Tu between 20% and 35%), and high (Tu greater than 55%). For displacement systems having the same mean air velocity and temperature, Fanger et al. found that periodically fluctuating airflow is less comfortable than steady airflow. Very little data on turbulence intensity of different air distribution methods are available, and the effect of turbulence on draft sensation at higher activity levels has not been studied.

Radiant Temperature Asymmetry. Limits of 18°F and 9°F for the radiant asymmetry from cold walls and warm ceilings, respectively, are recommended in both ASHRAE *Standard* 55 and ISO *Standard* 7730. These data are based on testing with sedentary persons in 0.6 clo clothing (Fanger et al. 1985). Similar data are available for warm walls and cold ceilings. In industrial workplaces, radiant asymmetry from overhead radiant heaters or a hot roof is the main cause of problems. The values established for sedentary persons are too conservative for the higher activity levels and higher ceilings encountered in industry.

Langkilde et al. (1985) showed that significantly higher radiant asymmetry is acceptable. Based on criteria similar to those just described, and requiring that no more than 5% of the occupants are dissatisfied, they recommend an asymmetry limit of 18 to 25°F.

Vertical and Horizontal Air Temperature Differences. Another comfort parameter described in ASHRAE *Standard* 55 and ISO *Standard* 7730 is the temperature difference between the head and feet. The recommended limit of 5°F is based on studies with sedentary persons (Olesen et al. 1979) and cannot be used directly with higher activity levels. It is not the air temperature difference alone, but the combined effect of differences in air temperature, air velocity, and radiant temperature between the head and feet that causes discomfort. Data for industrial environments are available only in GOSSTROY (1992), in which the vertical temperature difference in occupied zones is limited to 7°F, and the horizontal temperature difference cannot exceed 7°F at light activity, 9°F at medium activity, and 11°F at high activity.

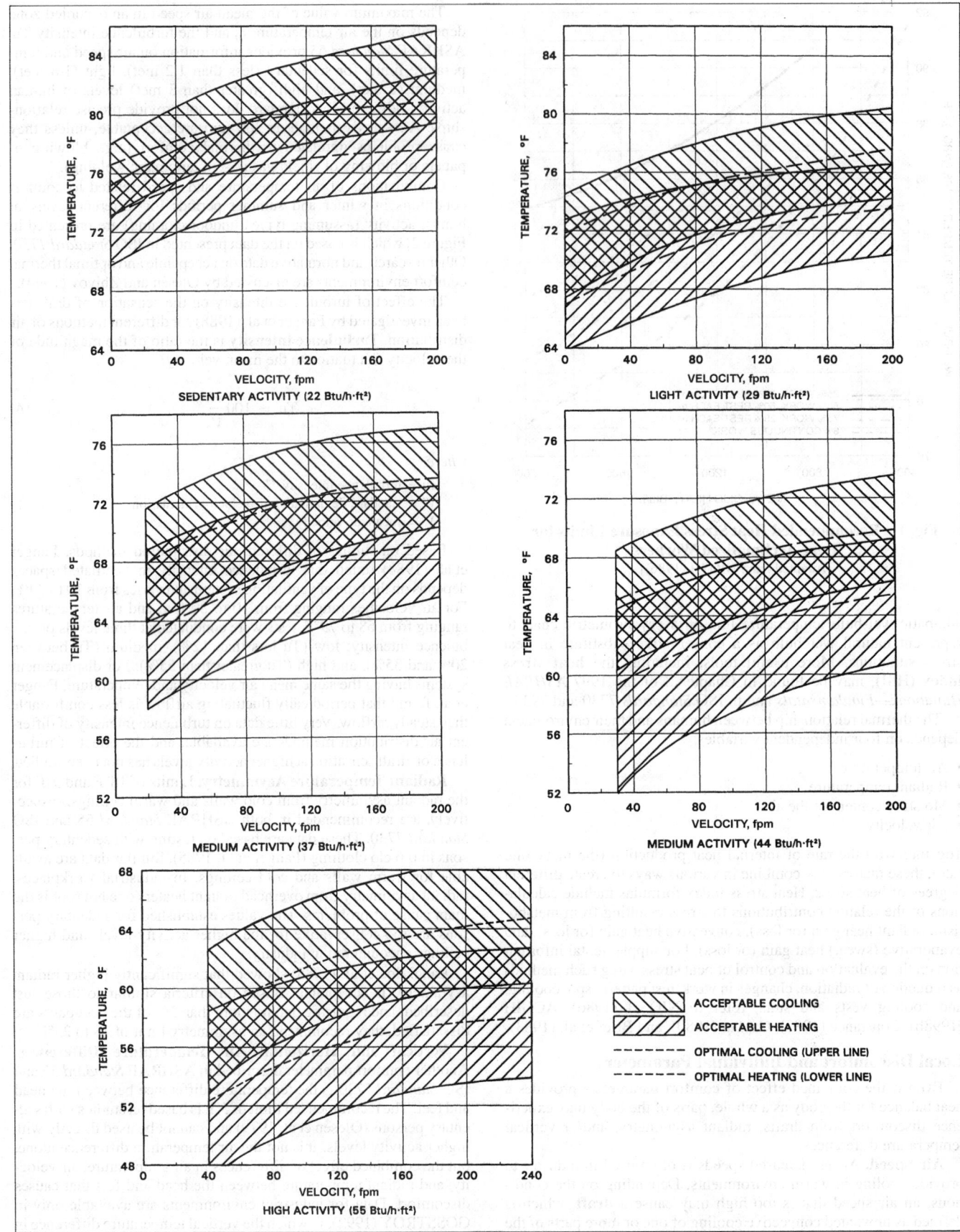

Fig. 2 Optimal and Acceptable Ranges of Air Temperature and Air Speed in Occupied Zone for Different Levels of Human Activity (ISO *Standard* 7730)

Table 1 Acceptable Air Speed in Workplace

Activity Level	Air Speed, fpm
Continuous exposure	
Air-conditioned space	50 to 75
Fixed workstation, general ventilation or spot cooling	
Sitting	75 to 125
Standing	100 to 200
Intermittent exposure, spot cooling, or relief stations	
Light heat loads and activity	1000 to 2000
Moderate heat loads and activity	2000 to 3000
High heat loads and activity	3000 to 4000

Table 2 Recommended Spot Cooling Air Speed and Temperature

Activity Level	Air Speed in Jet, fpm, Averaged on 1 ft² of Workplace	Average Air Temperature in Jet Cross Section, °F				
		Heat Flux Density, Btu/h·ft²				
		45-110	222	444	666	888
Light—I	200	82	75	70	61	—
	400	—	82	79	75	68
	600	—	—	82	79	75
	700	—	—	—	81	77
Moderate—II	200	81	72	—	—	—
	400	82	75	70	61	—
	600	—	81	75	70	64
	700	—	82	77	72	66
Heavy—III	400	77	66	61	—	—
	600	79	72	68	64	63
	700	—	73	72	68	66

Spot Cooling. If the workplace is located near a source of radiant heat that cannot be entirely controlled by radiation shielding, spot cooling is recommended (Olesen and Nielsen 1980, 1981, 1983; Azer 1982a, 1982b, 1984). The air temperature of the workplace is limited to 68 to 86°F, depending on the heat load and the air speed shown in Table 1.

Table 2 gives information about combinations of air temperature and speed in jets used for spot cooling. More detailed information is provided in GOSSTROY (1992). If the air temperature in the occupied zone is less than or greater than the temperatures listed in Table 2, the air temperature in the jet should be increased or decreased, respectively, by 0.4°F for each degree of temperature difference. The average temperature in the jet may not be lower than 61°F.

The temperature of the radiating surface should be averaged over the period of radiation. If the radiation lasts less than 15 min or more than 30 min, the jet air temperature can be respectively 4°F greater than or less than the values listed in Table 2.

HEAT EXPOSURE CONTROL

Control at Source

Heat exposure can be reduced by insulating hot equipment, locating such equipment in zones with good general ventilation or outdoors, covering steaming water tanks, providing covered drains for direct removal of hot water, and maintaining tight joints and valves where steam may escape.

Local Exhaust Ventilation

Local exhaust ventilation removes heated air generated by a hot process and/or nonbuoyant gases emitted by process equipment, while removing a minimum of air from the surrounding space.

The following criterion can be used to evaluate the economics of local exhaust versus general (dilution) ventilation:

$$\frac{C_{le} - C_o}{K_G(C_{oz} - C_o)} > (1.2 \text{ to } 1.5) \qquad (5)$$

where

C_{le} = concentration of gas, vapor, or particles evacuated by local exhausts, lb/lb

C_o = concentration of gas, vapor, or particles in air supplied, lb/lb

C_{oz} = concentration of gas, vapor, or particles in occupied zone, lb/lb

K_G = coefficient of air exchange efficiency for removing gas contaminants from occupied zone

If the value calculated is greater than 1.2 to 1.5, it is more economical to use local exhaust. Chapter 29 covers the design of exhaust hoods and duct systems for local exhaust.

Radiation Shielding

In some industries, the major environmental heat load is radiant heat from hot objects and surfaces, such as furnaces, ovens, furnace flues and stacks, boilers, molten metal, hot ingots, castings, and forgings. Because air temperature has no significant effect on radiant heat flow, ventilation is of little help in controlling such exposure. The only effective control is to reduce the amount of radiant heat impinging on the workers. Radiant heat exposure can be reduced by insulating or placing radiation shields around the source (Chapter 3 of the 1997 *ASHRAE Handbook—Fundamentals*).

Radiation shields are effective in the following forms:

- **Reflective shielding.** Sheets of reflective material or insulating board are semipermanently attached to the hot equipment or arranged in a semiportable floor stand.
- **Absorptive shielding (water-cooled).** These shields absorb and remove heat from hot equipment.
- **Transparent shields.** Heat-reflective tempered plate glass, reflective metal chain curtains, and close mesh wire screens moderate radiation without obstructing view of the hot equipment.
- **Flexible shielding.** Aluminum-treated fabrics give a high degree of radiation shielding.
- **Protective clothing.** Reflective garments such as aprons, gauntlet gloves, and face shields provide moderate radiation shielding. For extreme radiation exposures, complete suits with vortex tube cooling may be required.

If the shield is a good reflector, it will remain relatively cool in severe radiant heat. Bright or highly polished tinplate, stainless steel, and ordinary flat or corrugated aluminum sheets are efficient and durable. Foil-faced plasterboard, although less durable, gives reflects well on one side. To be efficient, however, the reflective shield must remain bright.

The best radiation shields are infrared reflectors, which should reflect the radiant heat back to the primary source, where it can be removed by local exhaust. However, unless the shield completely surrounds the primary source, some of the infrared energy will be reflected into the cooler surroundings and possibly into an occupied area. The direction of the reflected heat should be studied to ensure proper installation of the shielding.

VENTILATION DESIGN PRINCIPLES

General Ventilation

General ventilation supplies and/or exhausts air to provide heat relief, dilute contaminants to an acceptable level, and replace (make up) exhaust air. Ventilation can be provided by natural or mechanical supply and/or exhaust systems. Outdoor air is unacceptable for ventilation if it is known to contain any contaminant at a concentration above that given in ASHRAE *Standard* 62. If air is thought to contain any contaminant not listed in the standard, guidance on acceptable exposure levels should be obtained from OSHA standards

(ACGIH 1998a). General ventilation rates must be high enough to dilute the carbon dioxide produced by the occupants.

Ventilation design for the industrial environment can be established using the following techniques (Goodfellow 1985, 1986, 1987):

- Field testing
- Fluid dynamic modeling
- Computer modeling

For complex industrial ventilation problems, fluid dynamic modeling or computer modeling are often used in addition to field testing.

Field Testing

A field testing program must be organized and developed in detail prior to its trial in the field. Ventilation flow in large process buildings is usually complex, so an inexperienced sampling team could collect irrelevant or insufficient data for the subsequent analysis. Therefore, the field testing program should be reviewed and approved by plant operating personnel before the actual field testing begins. A good approach is to (1) perform preliminary calculations for all sources, (2) identify information gaps, and (3) determine what data must be collected.

Although each processing or manufacturing operation has unique ventilation flows, design parameters, and field testing protocol, many elements of a field testing program can be applied to all ventilation problems. The major engineering activities in a ventilation field testing program are listed in order in Table 3, which includes brief descriptions of the tasks and scope of the work required for each activity.

Step 1 in Table 3, gathering information, includes obtaining and studying all reports and drawings pertinent to the plant. Specific information about (1) current and future operating practices and (2) the nature of dust, chemical contaminants, and heat or cold stresses is required. A visit to the site should include a walk-through ventilation survey using a questionnaire or data sheet developed by the ventilation engineer and experienced process and operating personnel for the specific industry. The questionnaire is used to identify the operating practices (present and future) to be used as the design basis for the ventilation system. An industrial hygienist should work with the ventilation engineer to establish the scope and extent of the sampling program.

Step 4 in Table 3, developing details of the field testing program, includes preparation of the field data log sheets prior to the actual field testing.

Step 5 in Table 3, carrying out the field testing program, includes the following:

- Measurement of air velocities through all openings. Sufficient time must be allowed to determine representative velocities.
- Measurement of the temperature for each velocity measured. For air entering the building, the ambient temperature should be recorded hourly. The thermometer or temperature probe should not be exposed to sunlight or radiant heat from hot objects. Good temperature readings are important for heat balance calculations, and there must be sufficient temperature data to evaluate the air density distribution.
- Measurement of the mean surface temperatures of hot surfaces to determine the subsequent heat release and air movement.
- Recording weather data. Weather data also should be obtained from the nearest airport or meteorological station. The data, which should be recorded hourly, include ambient temperature, relative humidity, wind speed, and wind direction.
- Recording plant activities during the testing program. This includes the operational status of all major process and environmental equipment, as well as production levels. Plant records and

Table 3 Engineering Activities for Ventilation Field Testing Program

Activity	Specific Tasks
1. Gather information	Obtain drawings, reports, operating procedures Review existing data and studies Visit plant Define problem (summer, winter, heat/cold stress, chemical supply, exhaust)
2. Collect data on ventilation openings	Develop isometric drawings: plans, sections Develop schedule of openings (type, size, location)
3. Develop plant questionnaire	Develop process flow sheet and general layout Identify significant heat sources in building Identify typical operating practice Identify gaps in data to be filled in by field testing
4. Develop details of field testing program	Determine building ventilation flow rates Determine in-plant flows Develop field data log sheets for ventilation measurements, weather conditions, plant operating records, etc.
5. Carry out field testing program	Perform field measurements (velocity, temperature, pressure, etc.)
6. Analyze data and perform calculations	Determine plant ventilation flow balance Determine in-plant flow patterns Perform heat balance calculations (total plant/ventilation) Calculate air set-in-motion volumes
7. Perform computer simulation of ventilation flows (natural ventilation)	Calibrate using field test data Run program for different conditions (summer, winter, etc.)
8. Produce field testing report	Summarize test conditions and test results Make recommendations

charts from process operations should be obtained. It is helpful if the ventilation field testing team can work with plant personnel.

Step 8 in Table 3, the preparation of a report on the ventilation field testing program, includes reporting all field data, calculations, and test results. Using this report, an experienced ventilation engineer can recommend cost-effective solutions for any plant ventilation problem.

Fluid Dynamic Modeling

Fluid dynamic modeling (small-scale modeling) is a valuable tool for modeling problems and developing alternative cost-effective solutions. It has been used extensively for a wide variety of industrial ventilation applications (Baturin 1972; Goodfellow 1985, 1987). Applications of scale modeling include the following:

- Finalizing building ventilation flow rates and schemes
- Examining internal flow patterns and contaminant concentrations at any location
- Examining external flow patterns, including quantitative measurements of downwash and transport of contaminants to other buildings
- Establishing the effectiveness of source hoods

Fluid dynamic modeling can be used in solving ventilation problems in existing plants and in the design of ventilation systems for new plants. For an existing plant, fluid dynamic modeling can supplement the field testing program.

For the design of ventilation systems for new process buildings, the following steps are recommended: (1) develop the overall ventilation concepts and architectural constraints using computer models; (2) design the structural steel while the small-scale model is constructed and tested; and (3) use the results of the small-scale modeling to refine the ventilation design and to finalize all requirements.

Fluid dynamic modeling includes the following activities:

1. Define contaminant source characteristics.
2. Define environment flows, such as airflow rates, loads, and contaminant flows.
3. Develop details of the scope of the work.
4. Design the model system.
5. Construct the model system.
6. Test the program.
7. Produce the report.

In defining the contaminant source characteristics, the size of the industrial building and details about the source flux (e.g., heat and contaminant release rates) must be determined. Information about the major sources of heat is used to calculate heat balances and air volumes.

In defining environment flows, data are required on the external and internal flow conditions for the prototype. Information may be needed about site conditions such as wind speed and direction.

Design of the model system requires an examination of the scaling parameters and possible fluid media to select the best model for the specific ventilation problem. Although the most common media used in models are air and water, a variety of working or buoyancy-driven fluids are available. Fluid systems used include air and heated air, water and saltwater, water and carbon tetrachloride, and mercury and carbon tetrachloride. Usually, air models are the simplest and least expensive models to build and test, but they must be large to ensure fully turbulent flow. Because water has a smaller kinematic viscosity than air, a small model is required to ensure a high Reynolds number and turbulent flow. Flow visualization is easier with water-based models because velocities are lower than with air. Data measured in the model flow can be related quantitatively to the full-scale prototype flow by establishing dynamic similarity (geometric and kinematic) between the model and the prototype. If the model and prototype are to have similar ventilation, their Archimedes numbers must be equal (Baturin 1972):

$$\text{Ar}_m \quad = \quad \text{Ar}_p$$

$$\left(\frac{gL_o(t_o - t_s)}{V_o^2 T_s}\right)_m = \left(\frac{gL_o(t_o - t_s)}{V_o^2 T_s}\right)_p \tag{6}$$

where

g = gravitational acceleration rate, ft/s^2
L_o = length scale, ft
t_o = initial temperature of jet, °F
t_s = temperature of surrounding air, °F
V_o = initial air velocity of jet, fps
T_s = room air temperature (absolute), °R
p = subscript identifying prototype
m = subscript identifying model

An adequate simulation of the convective flows requires that the value of the Grashof-Prandtl number exceed 2×10^7:

$$\text{Gr} \times \text{Pr} = \frac{gL^3 \Delta t}{v^2 T_s} \times \frac{c_p v \rho}{k} = \frac{gL^3 \Delta t \, c_p \rho}{v T_s k} \geq 2 \times 10^7 \tag{7}$$

where

v = kinematic viscosity, ft^2/s
Δt = temperature difference between two points, °F
c_p = specific heat at constant pressure, Btu/lb·°F
k = thermal conductivity, Btu/s·ft·°F
ρ = density, lb/ft^3

The technique used to measure a contaminant must be evaluated for its impact on the cost of the testing program and on the degree of accuracy of the measurements.

For any model testing program, the use of photography and video cameras to record results is suggested. The photographs and videos are invaluable for analyzing test results and for presenting the proposed solutions to management.

Computer Modeling

Figure 3 shows the flowchart for a computer ventilation model based on design equations for flow and heat balances.

Once the observed ventilation flow rates are correctly modeled, the computer program can be used to study the effects of different weather conditions (e.g., summer/winter, wind direction and speed) on the ventilation. The computer program is useful also for evaluating and comparing proposed schemes to find the most cost-effective method of improving the ventilation.

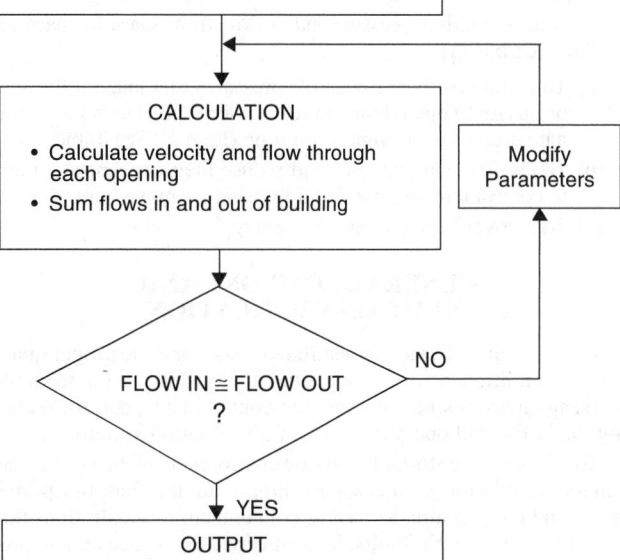

Fig. 3   Computer Ventilation Model Flow Diagram
(Goodfellow 1985)

Computer ventilation models can reliably predict the gross ventilation rates for complex process buildings. Computers enable the designer to examine the effect of architectural changes, wind conditions, or process changes on the performance of the proposed ventilation scheme. Problems such as contamination resulting from cross drafts or high temperatures in the work environment can be identified quickly and corrected. Other ventilation questions, such as internal flow patterns, intermittent flows, two- or three-dimensional flow, the location of fresh air, and contaminant concentrations in the breathing zone, can be answered using computer models.

Computer programs available for microcomputers can solve complex three-dimensional ventilation problems (Cawkwell and Goodfellow 1990). Proceedings for conferences such as Ventilation '85, '88, '91, '94, and '97 and RoomVent '90, '92, '94, and '96 include many technical papers on new computer programs for calculating contaminant levels as a function of the location and time spent in a workplace environment. These computer programs are used in the routine engineering design of ventilation systems.

Need for Makeup Air

For safe, effective operation, most industrial plants require makeup air to replace the large volumes of air exhausted to provide comfort and safety for personnel and good conditions for process operations. Makeup air consistently provided by good air distribution allows more effective cooling in the summer and more efficient and effective heating in the winter. The use of windows or other inlets that cannot function in stormy weather is discouraged. The most important functions of makeup air can be summarized as follows:

1. To replace air being exhausted through combustion processes and local and general exhaust systems (see Chapter 29).
2. To eliminate uncomfortable cross drafts by proper arrangement of supply air and to prevent infiltration (through doors, windows, and similar openings) that may make hoods unsafe or ineffective, defeat environmental control, bring in or stir up dust, or adversely affect processes by cooling or disturbances.
3. To obtain air from the cleanest source. Supply air can be filtered; infiltration air cannot.
4. To control building pressure and airflow from space to space for three reasons:
   (a) To avoid positive or negative pressures that make it difficult or unsafe to open doors, to replace air (Item 1), and to eliminate drafts and prevent infiltration (Item 2). See Table 4.
   (b) To confine contaminants and reduce their concentration and to control temperature, humidity, and air movement.
   (c) To recover heat and conserve energy.

GENERAL COMFORT AND DILUTION VENTILATION

Effective air diffusion in ventilated rooms and the proper quantity of conditioned air are essential for creating a comfortable working environment, for removing contaminants, and for reducing the initial and operating costs of a ventilation system.

To provide a comfortable and safe environment in the workplace, general ventilation systems must supply air that has the proper speed and temperature, as well as contaminant concentrations that are within permissible limits. In most cases, the objective is to provide tolerable (acceptable) working conditions rather than total comfort (optimal) conditions.

The design of a general ventilation system is based on the assumption that local exhaust ventilation, radiation shielding, and equipment insulation and encapsulation have been selected to minimize both the heat load and the contamination level in the workplace. In cold climates, infiltration and heat loss through the building shell must also be minimized.

Table 4 Negative Pressures That May Cause Unsatisfactory Conditions Within Buildings

Negative Pressure, in. of water	Adverse Conditions That May Result
0.01 to 0.02	Worker draft complaints—High-velocity drafts through doors and windows.
0.01 to 0.05	Natural draft stacks ineffective—Ventilation through roof exhaust ventilators and flow through stacks with natural draft greatly reduced.
0.02 to 0.05	Carbon monoxide hazard—Backdrafting will take place in hot water heaters, unit heaters, and other combustion equipment not provided with induced draft.
0.03 to 0.10	General mechanical ventilation reduced—Airflows reduced in propeller fans and low-pressure supply and exhaust systems.
0.05 to 0.10	Doors difficult to open—Serious injury may result from unchecked, slamming doors.
0.10 to 0.25	Local exhaust ventilation impaired—Centrifugal fan fume exhaust flow reduced.

Source: ACGIH (1998b).

Quantity of Supplied Air

Sufficient air must be supplied to replace air exhausted by process ventilation and local exhausts, to provide dilution of contaminants (gases, vapors, or airborne particles) not captured by local exhausts, and to provide the required thermal environment. The amount of supplied air should be the largest of the amounts needed for temperature control, dilution, and replacement.

The airflow rate Q_o to be supplied through diffusers for temperature control can be estimated from the following equation:

$$Q_o = Q_{exh} + \frac{q - \rho c_p Q_{exh}(t_{oz} - t_o)}{\rho c_p K_t (t_{oz} - t_o)} \qquad (8)$$

where

Q_o = required supply airflow rate, cfm
Q_{exh} = flow rate of air evacuated from occupied zone by general and/or local exhaust and process equipment, cfm
q = surplus heat gain or heat supplied to warm the air, Btu/min
c_p = specific heat of air at constant pressure, Btu/lb·°F
ρ = density of air, lb/ft^3
t_{oz} = air temperature in occupied zone, °F
t_o = temperature of supplied air, °F
t_{exh} = temperature of exhausted air, °F
K_t = coefficient of efficiency for removing heat from occupied zone (see Table 5)
  = $\dfrac{t_{exh} - t_o}{t_{oz} - t_o}$

The airflow rate Q_o required to remove water vapor released into the space at a rate M_w is given by the following equation:

$$Q_o = Q_{exh} + \frac{M_w - \rho Q_{exh}(W_{oz} - W_o)}{\rho K_w (W_{oz} - W_o)} \qquad (9)$$

where

M_w = rate of water vapor release, lb/min
W_o = humidity ratio, lb water per lb dry air supplied
W_{oz} = humidity ratio, lb water per lb dry air in occupied zone
W_{exh} = humidity ratio, lb water per lb exhausted air
K_w = coefficient of efficiency for removing water vapor from occupied zone (see Table 5)
  = $\dfrac{W_{exh} - W_o}{W_{oz} - W_o}$

The airflow rate Q_o required for dilution ventilation can be calculated as follows:

Table 5   Heat and Moisture Removal Efficiency Coefficients for Mechanically Ventilated Spaces with Insignificant Heat Load

| | Heat/Moisture Removal Efficiency Coefficients K_t/K_w | | | |
| | Air Change Rate, ACH | | | |
Air Supply Method	3	5	10	>15
Concentrated air jets	0.95/1.1	1.0/1.05	1.0/1.0	1.0/1.0
Concentrated air jets with vertical and/or horizontal directing jets	1.0/1.0	1.0/1.0	1.0/1.0	1.0/1.0
Inclined air jets from a height of				
Greater than 13 ft	1.15/1.4	1.1/1.2	1.0/1.1	1.0/1.0
Less than 13 ft	1.0/1.2	1.0/1.1	1.0/1.05	1.0/1.0
Trough ceiling-mounted air diffusers with				
Radial (linear) attached jets	0.95/1.1	1.0/1.05	1.0/1.0	1.0/1.0
Conical (compact) jets	1.05/1.1	1.0/1.05	1.0/1.0	1.0/1.0

Note: Insignificant heat load is defined as below 2.2 Btu/h·ft^3.

$$Q_o = Q_{exh} + \frac{G - \rho Q_{exh}(C_{oz} - C_o)}{\rho K_G} \qquad (10)$$

where

> G = rate of contaminant release into space, lb/min
>
> C_o = concentration of gas, vapor, or particles in air supplied, lb/lb
>
> C_{oz} = concentration of gas, vapor, or particles in occupied zone, lb/lb
>
> C_{exh} = concentration of gas, vapor, or particles in exhaust, lb/lb
>
> K_G = coefficient of efficiency for removing gaseous contaminants from occupied zone
>
> $\quad = \dfrac{C_{exh} - C_o}{C_{oz} - C_o}$

The values of the coefficients K_t, K_w, and K_G depend on the method of air distribution, characteristics of the ventilated space, and activities within the space. They can be determined by laboratory and field tests.

In some cases, the value of K_t can also be obtained analytically (Shilkrot 1993; Pozin 1993). For mixing-type air distribution, the values of K_t, K_w, and K_G are 1 ± 0.1. Displacement ventilation and air supply with inclined jets lower than 13 ft take advantage of thermal stratification. When the sources of heat and gaseous emissions are close to each other, both K_t and K_G can be greater than 1.2 and, in some cases, can reach 2.5 or more. Values of K_t and K_w for typical mixing-type air distribution are listed in Table 5. Values of K_t and K_w for displacement ventilation are close to those for natural ventilation and are listed in Table 10. When the sources of heat and gaseous emission are separated or are distributed uniformly in the occupied zone, the value of K_G for displacement ventilation does not differ much from 1, although it may be greater than 1.2 (Shilkrot and Zhivov 1992).

Local air supply does not provide the desired air quality for the entire occupied zone, but only for certain areas that are permanently occupied. This dramatically reduces the amount of supplied air.

The design value of the concentration C_{oz} of a contaminant in the occupied zone should not exceed an acceptable level of exposure such as a **permissible exposure limit** (PEL), a **threshold limit value** (TLV), or an in-house TLV. It is desirable to set design objectives below statutory PELs and TLVs because of variations in sensitivity to contaminants and because acceptable limits can be lowered. Determination of the generation rate G of the contaminant [in Equation (10)] can be based on production records, material balance, similar operations, and experience. However they are obtained, the acceptable exposure level and the contaminant generation rate are required for designing a dilution ventilation system properly. Designs based on the number of air changes per hour or other estimates are inadequate and could lead to unacceptably high exposure levels or to unnecessarily high installation costs and/or energy consumption.

Air Supply Methods

Air supply to industrial spaces can be by natural or mechanical ventilation systems. Although natural ventilation systems driven by gravity forces and/or wind effect are still widely used in industrial spaces (especially in hot premises in cold and moderate climates), they are inefficient in large buildings, may cause drafts, and cannot solve air pollution problems. Thus, most ventilation systems in industrial spaces are either mechanical or a combination of mechanical supply with natural exhaust. The most commonly used methods of air supply to industrial spaces are

- Mixing
- Displacement
- Unidirectional airflow (piston flow)
- Spiral vortex
- Localized

Mixing-Type Air Distribution. In mixing systems, air is normally supplied into the space at velocities much greater than those acceptable in the occupied zone. Supply air temperature can be above, below, or equal to the air temperature in the occupied zone, depending on the heating/cooling load. The supply air diffuser jet mixes with room air by entrainment, which reduces air velocities and equalizes the air temperature. The occupied zone is ventilated either directly by the air jet or by the reverse flow created by the jet. Properly selected and designed mixing air distribution creates relatively uniform air velocity, temperature, humidity, and air quality conditions in the occupied zone and over the room height.

In industrial spaces with mixing-type air distribution, air can be supplied with

- Horizontal air jets, attached or not attached to the ceiling, with the occupied zone ventilated by reverse airflow (Figure 4A and 4B)
- Horizontal concentrated air jets assisted by additional vertical and/or horizontal directing jets (Figure 4C)
- Inclined air jets through the grilles and nozzles installed on walls and/or columns at a height of 10 to 20 ft (Figure 5)
- Radial, conical, or compact air jets through ceiling-type air diffusers installed in or close to the ceiling (Figure 6A,B,C) or on the vertical duct drops (Figure 7)
- Horizontal compact or linear jets, attached to the ceiling, supplied through wall-mounted grilles (Figure 6D,E)
- Radial or linear jets through perforated surfaces of horizontal round or rectangular ducts (Figure 8)

Principles of mixing-type air distribution in industrial rooms are discussed by AIR-IX (1987), Cole (1995), Stroiizdat (1992b), Grimitlyn and Pozin (1993), Shepelev (1978), Tarnopolsky (1992), Zhivov (1992, 1993, 1994), and Zhivov et al. (1996).

Displacement Ventilation Systems. Conditioned air slightly cooler than the desired room air temperature in the occupied zone is supplied from air outlets at low air velocities—100 fpm or less

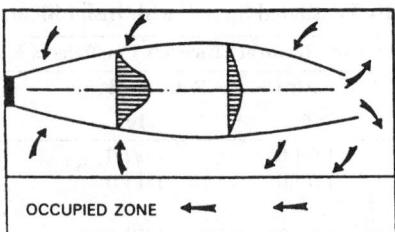

A. Air is supplied by nonattached horizontally projected jet, and occupied zone is ventilated by reverse flow.

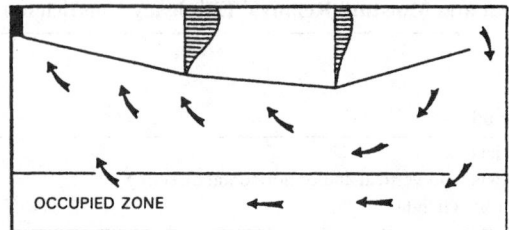

B. Air is supplied by horizontally projected jet attached to the ceiling, and occupied zone is ventilated by reverse flow.

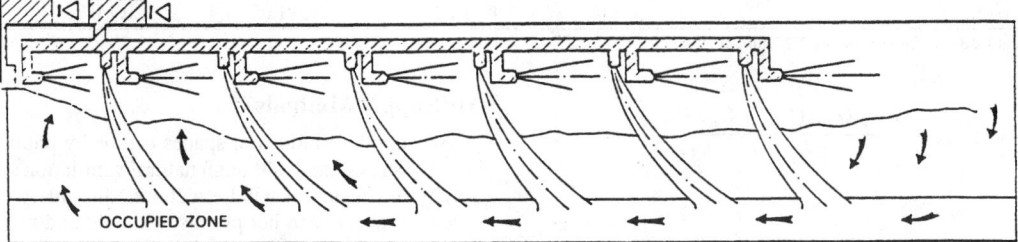

C. Air is supplied by horizontally projected concentrated air jets and vertical and/or horizontal directing jets, and occupied zone is ventilated by reverse flow and vertical directing jets.

Fig. 4   Concentrated Air Supply Methods

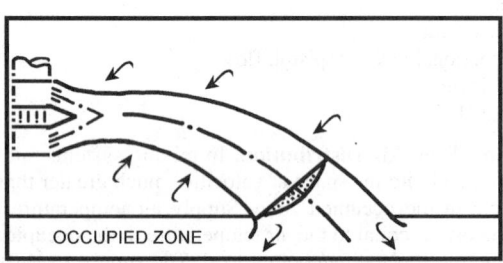

A. SCHEMATIC OF COOLED AIR SUPPLY

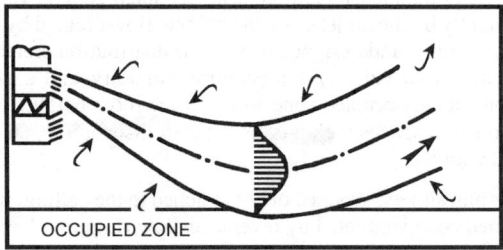

B. SCHEMATIC OF HEATED AIR SUPPLY

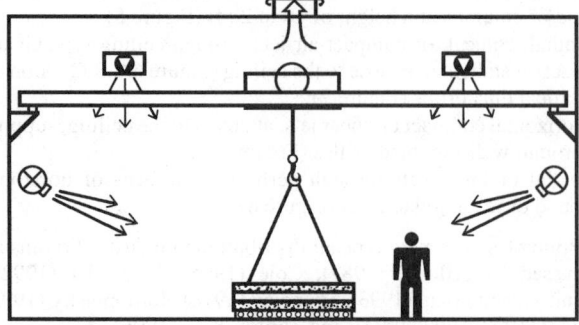

C. INCLINED AIR SUPPLY IN MECHANICAL SHOP

Fig. 5   Air Supply with Inclined Jets
(Zhivov 1992; AIR-IX 1987)

(Jackman 1991; Skåret 1985; Skistad 1994; Shilkrot and Zhivov 1996). Under the influence of buoyancy, cold air spreads along the floor and floods the lower zone of the room. The air close to the heat source is heated and rises upward as a convective airstream; in the upper zone, this stream spreads along the ceiling. The lower part of the convective airstream induces the cold air of the lower zone of the room, and the upper part of the convective airstream induces the heated air of the upper zone of the room. The height of the lower zone depends on the air volume discharged through the panels into the occupied zone (Figure 9A) and on the amount of convective heat discharged by the sources (Figure 9B).

Typically, the outlets are located at or near the floor, and the supply air is introduced directly into the occupied zone. In some applications of displacement ventilation (e.g., in computer rooms or hot industrial buildings), air may be supplied to the occupied zone through a false floor. In other applications, supply air outlets may be located above the occupied zone. Returns are located at or close to the ceiling/roof through which air is exhausted from the room.

Displacement ventilation has been common in Scandinavia in the past 20 years and is becoming popular in other European countries. Displacement ventilation is preferable when contaminants are released in combination with surplus heat, and contaminated air is warmer and/or lighter than the surrounding air. Displacement ventilation design guidelines and limitations are described by Skistad (1994), Laurikainen (1995) and Zhivov et al. (1997).

Unidirectional Airflow Ventilation. Airflow can be vertical [air supplied from the ceiling and exhausted through the floor or vice versa (Figure 10A)] or horizontal [air supplied through one wall and exhausted through returns located on the opposite wall (Figure 10B)]. The outlets are uniformly distributed over the ceiling and floor or walls to provide a low-turbulence plug-type flow across the entire room. This type of system is mainly used for ventilating clean rooms, in which the main objective is to remove contaminant particles within the room (see Chapter 15), or in halls with high heat and/or contaminant loads with double flooring or floor pedestals.

Spiral Vortex Flow. This method of air distribution can be used to localize air contaminants in certain room areas and evacuate the polluted air from those areas. Air is supplied through vertical supply ducts located along a closed contour (preferably the walls), thus generating a vertical spiral vortex (Figure 11). An exhaust outlet can

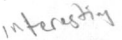

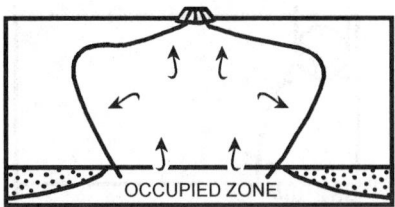

A. With radial jets attached to the ceiling

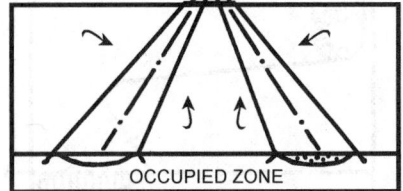

B. With downward projected conical jets

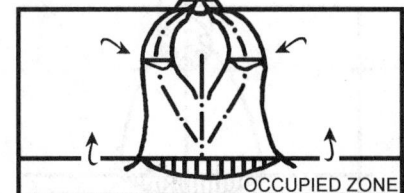

C. With downward projected compact jets

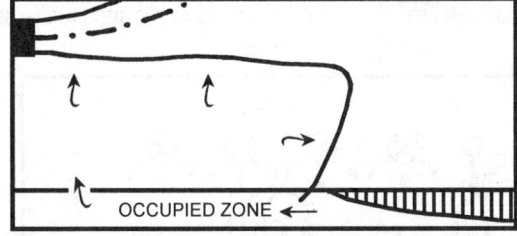

D. Through wall-mounted grille and linear diffusers; occupied zone ventilated by the jet directly

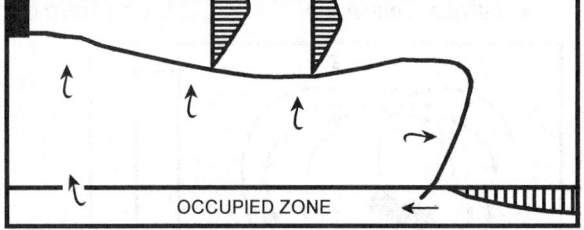

E. Through wall-mounted grille; occupied zone ventilated by jet and reverse flow

Fig. 6   Nonconcentrated Air Supply Methods

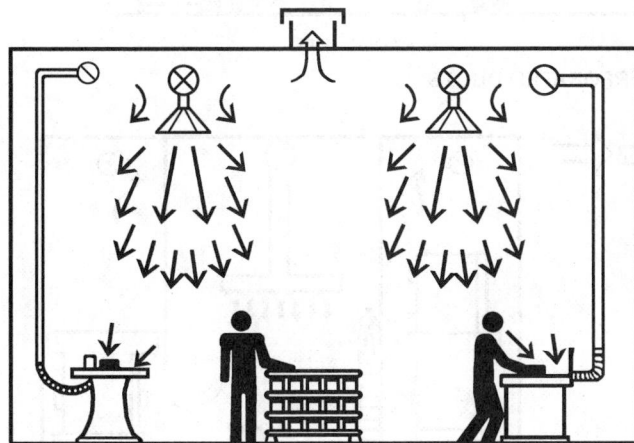

Fig. 7   Air Supply by Conical Air Jets Through Air Diffusers Installed on Vertical Duct Drops
(AIR-IX 1987)

be located in the ceiling near the center of the rotational flow. Low pressure in the vortex core concentrates contaminants in the core and prevents their diffusion to the clean space; contaminants are transported to the exhaust outlet along the core axis (Kuz'mina et al. 1986; Nagasawa et al. 1990)

Localized Ventilation. Air is supplied locally for occupied regions or a few permanent work areas (Figure 12). Conditioned air is supplied toward the breathing zone of the occupants to create comfortable conditions and/or to reduce the concentration of pollutants. These zones may have air 5 to 10 times cleaner than the surrounding air. In localized ventilation systems, air is supplied through one of the following devices:

- Nozzles or grilles (e.g., for spot cooling); specially designed low-velocity/low-turbulence devices (Kristensson and Lindqvist 1993)
- Perforated panels suspended on the vertical duct drops and positioned close to the work station
- Combination of vertical perforated supply air panels that creates an oasis around the work station and screens the work station

from the surrounding environment (Kristensson and Lindqvist 1993; AIR-IX 1987)

Guidelines for Selecting an Air Distribution Method

The economical and hygienic efficiencies of general ventilation systems depend mainly on the proper choice of air distribution method. Zhivov (1992) recommends the following steps for selecting a method of air distribution:

1. Select several possible methods of air distribution that are suitable for the given room, the comfort conditions required in the occupied zone, the size of the equipment, and the internal loads and operating modes of the HVAC or air-conditioning systems in a year-round cycle (variable/constant air volume system).
2. Determine the air volume requirements.
3. Determine the dimensions (length L and width B) of the area ventilated through one air diffuser, the type and size of air diffuser, the height of its installation, and the position of the controls for the design mode of system operation (e.g., highest heat loads in summer).
4. Check airflow pattern and air parameters in the occupied zone for other characteristic modes of system operation (e.g., heating regime). This is extremely important for VAV systems.
5. Compare the possible methods, and choose the most economical one.

Guidelines for the selection of an air distribution method for typical situations are presented in Table 6. Among the most important criteria used for selection are

- Room floor area and height
- Technological process and size of process equipment used; space obstruction of this equipment
- Number, location, and type (permanent or temporary) of work areas
- Type and amount of contaminants released into space, heating/cooling loads, air change rates
- Type of HVAC system used (variable air volume or constant air volume)
- Data on ventilation effectiveness for different air supply methods

If there are significant surplus heat gains in a room, the methods shown in Figure 5A, 6C, or 9 should be used with air diffusers

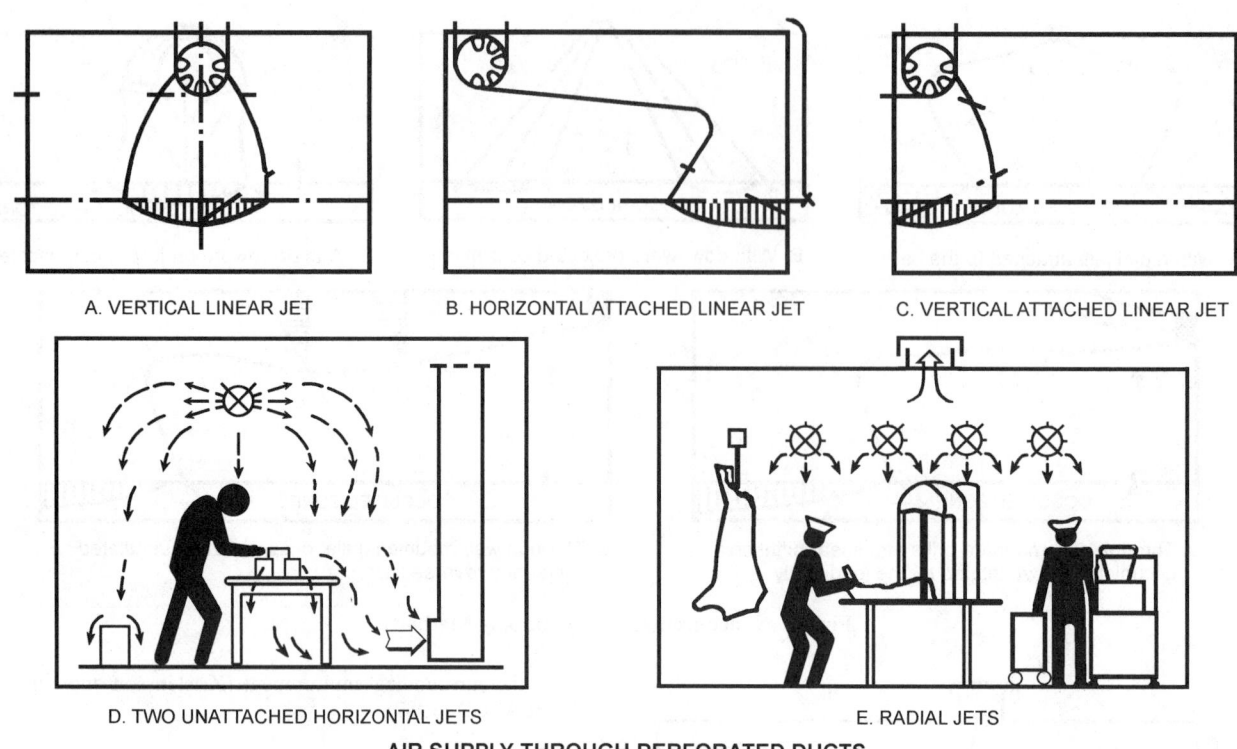

A. VERTICAL LINEAR JET   B. HORIZONTAL ATTACHED LINEAR JET   C. VERTICAL ATTACHED LINEAR JET

D. TWO UNATTACHED HORIZONTAL JETS   E. RADIAL JETS

AIR SUPPLY THROUGH PERFORATED DUCTS

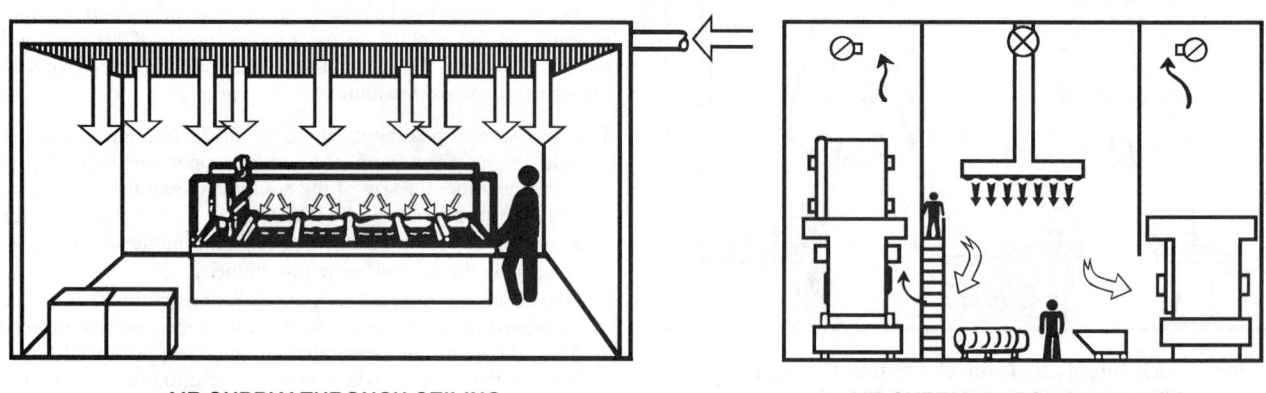

AIR SUPPLY THROUGH CEILING   **AIR SUPPLY THROUGH PANELS**

Fig. 8   Air Supply Through Perforated Ducts, Ceilings, and Panels
(AIR-IX 1987)

installed lower than 13 ft to supply conditioned air closer to the occupants and to take advantage of thermal stratification.

In variable air volume (VAV) systems, variations in the air volume supplied through diffusers and in the initial temperature difference over an annual cycle change the ratio between gravitational and inertial forces in the air jets. This affects the trajectory and throw of a jet, as well as the distance to the point where the jet separates from the ceiling. Undesirable airflow patterns or larger zones that are poorly ventilated may result, particularly in spaces with many obstructions. Zhivov (1990) discusses operation modes and air distribution design with VAV systems.

If the initial temperature differential $(t_o - t_{oz})$ of the supplied air is increased, the airflow supplied for temperature control can be reduced. Mixing-type air distribution methods provide high entrainment of the ambient air into the diffuser air jet, making it possible to increase the supply air temperature in the heating mode or decrease it in the cooling mode.

Displacement ventilation cannot be used to supply heated air and can be used to supply chilled air only if the temperature differential is less than 9 to 11°F.

When local exhaust ventilation is required for contaminant control, the exhaust rates are frequently greater than the required general ventilation for comfort. Under these conditions, the exhaust rate determines makeup air rates. A supply system must have air distribution arranged so that the makeup air is provided without disturbance at the hoods and process.

Caplan and Knutson (1978) and Peterson et al. (1983) established that the manner of air supply to a room with a hood has a major impact on the hood performance. The **hood performance factor** (or **hood index**) is defined as the logarithm (base 10) of the ratio of contaminant concentration within the hood to that just outside the hood. Adequate hood performance factors range from 4 to 6, with the higher number indicating a more effective hood installation (Fuller and Etchells 1979).

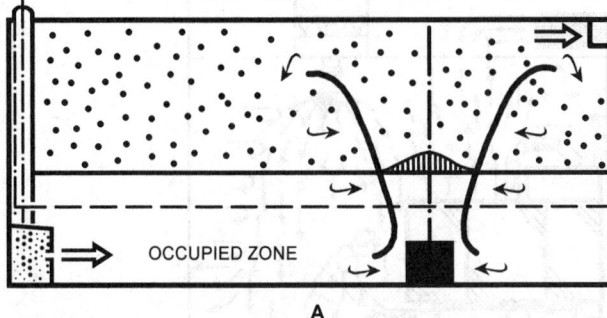

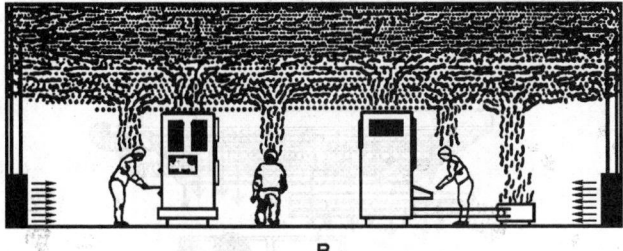

Fig. 9   Displacement Ventilation
(Kristensson and Lindqvist 1993)

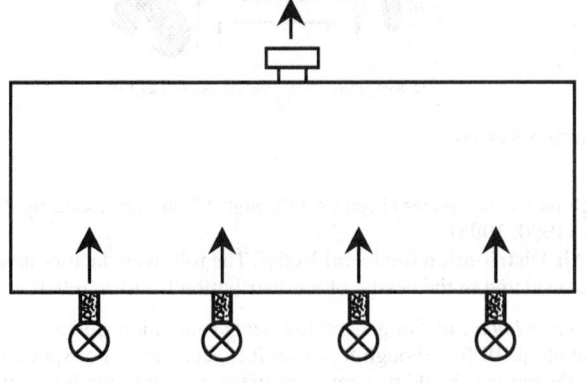

A. VERTICAL AIRFLOW

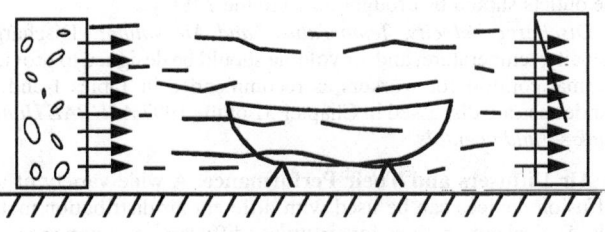

B. HORIZONTAL AIRFLOW

Fig. 10   Unidirectional Flow Systems
(AIR-IX 1987; LVIS 1996)

Local Area or Spot Cooling

In hot workplaces that have only a few work areas, it is impractical and wasteful to maintain a comfortable environment in the entire building. However, working conditions in areas occupied by workers can be improved by air-conditioned cabins, individual cooling, and spot cooling.

Air-conditioned cabins provide thermal comfort effectively, but they are expensive. Control rooms for monitoring production and manufacturing processes can use this technology.

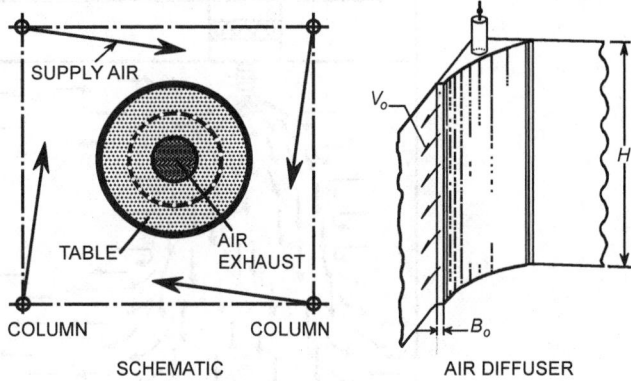

Fig. 11   Spiral Vortex Ventilation System
(Nagasawa et al. 1990; Kuz'mina et al. 1986)

Individual cooling can be provided by air- or water-cooled suits, vests, or helmets. **Air-cooled suits** are appropriate for moderately high temperatures and activity levels. The supply air can be cooled by a conventional heat exchanger or by a vortex tube. The suits are simple and self-regulating, provided that there is sufficient airflow. When the supply air is at skin temperature, heat is removed by increasing the evaporation of sweat. Lowering the supply air temperature improves the convection heat loss, thereby increasing the cooling capacity. **Water-cooled suits** have almost unlimited cooling capacity and lower pumping requirements, making them superior to air-cooled suits. A water-cooled garment needs a control system to protect the wearer from undercooling or overcooling. Various physiological measurements, including oxygen consumption and skin temperature measured at selected sites, have been used in the control of water-cooled garments.

Individual cooling with vests or helmets is appropriate if the thermal conditions are not extreme. Vests and helmets remove less heat than suits, but they cost less, are easier to use, and allow increased mobility.

Spot cooling is probably the most popular method of improving the thermal environment. Spot cooling can be provided by radiation (decreasing the mean radiant temperature), by convection (increasing the air velocity), or by a combination of the two methods. Spot cooling equipment is fixed at the workstation, whereas individual cooling has the worker wearing the cooling equipment.

Radiant spot cooling is provided by cooling panels installed near each workstation. Cooling panels decrease the mean radiant temperature, which allows greater radiant heat loss from the workers. The cooling power of radiant spot cooling can be changed either by altering the surface temperature of the cooling panels or by altering the angle factor between the cooling panels and the workers. (The angle factor is determined by the distance from the cooling panel to the worker and by the position of the cooling panel relative to the worker.) Radiant spot cooling is not very efficient. It may improve the thermal comfort of the workers, but it may also create local discomfort due to radiant asymmetry. In the design of radiant spot cooling, water condensation on the cooling panels should be considered. Condensation occurs when the surface temperature of the panel is equal to or less than the dew point temperature of the room air. If the surface temperature is below 32°F, the panel will be covered with ice. In practice, the poor efficiency, water condensation, and positioning of the panels may limit the application of radiant spot cooling.

Convective spot cooling by air jets is an efficient way of providing acceptable thermal comfort conditions. Convective spot cooling exposes workers to air with an increased velocity; the jet air may be the same temperature as the air in the workplace or cooler. Different combinations of jet outlet velocity and temperature can produce the same cooling effect. The optimal combination of jet velocity and temperature depends on the type of work performed, the clothing

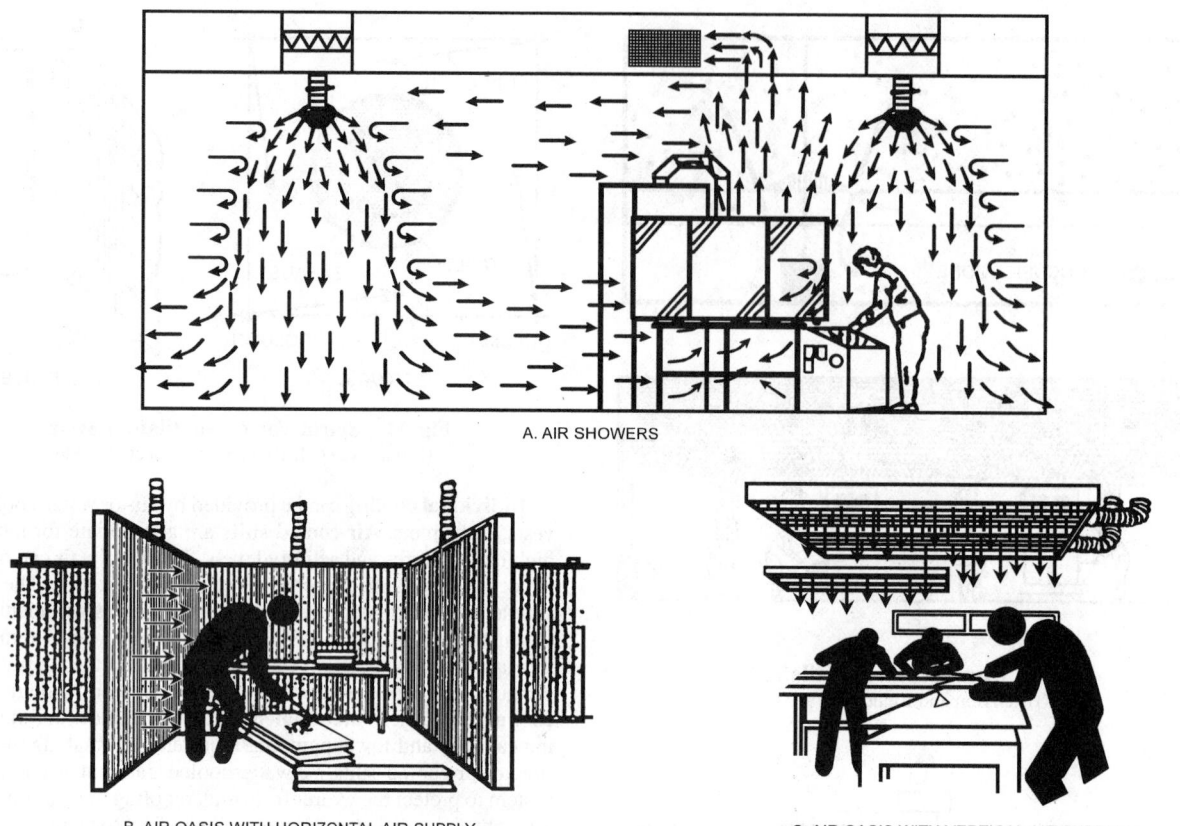

A. AIR SHOWERS

B. AIR OASIS WITH HORIZONTAL AIR SUPPLY   C. AIR OASIS WITH VERTICAL AIR SUPPLY

Fig. 12   Localized Ventilation Systems

worn, the surface area of exposed body parts, the size of the target area, and the direction of the jet.

A design procedure for spot cooling by air jets has been developed and optimized (Azer 1984, Hwang et al. 1984). The procedure is based on a semi-empirical model of the fully developed region of a cold jet projected downward in a hot environment. **Skin wettedness**, which is defined as the fraction of the subject's body surface area covered by evaporative moisture, is used as a physiological index. The cooling air jet is supposed to provide a skin wettedness of 0.5 as an acceptable physiological strain. For maximum cooling, studies (Robertson and Downie 1978; Melikov et al. 1994) show that the initial (potential core) region and the transition region of the jet should be used for spot cooling workers. The fully developed region of the jet does not cool as well because intensive mixing of cold jet air with warm room air increases the jet target temperature.

Subjective studies (Olesen and Nielsen 1983; Melikov et al. 1991) found significant differences in the target velocities preferred by individuals; therefore, individual control of the jet velocity and jet temperature are recommended. Spot cooling systems that also allow workers to adjust the distance to the jet outlet and the direction of the jet are preferable. Convective spot cooling, especially at high temperatures, reduces heat stress but may cause local discomfort due to draft. In hot environments, such as those in foundries and steel mills, velocities as high as 3000 to 4000 fpm are common (Table 1). When high-velocity air is used, it is important to avoid hot air convection, dust entrainment (which can be hazardous to eyes), and disturbance of local exhaust systems.

Air Distribution Design in Industrial Spaces

Chapter 31 of the 1997 *ASHRAE Handbook—Fundamentals* should be reviewed for basic principles of air distribution design and air diffuser selection. Design methods for air distribution in

large industrial spaces (Figures 4 through 12) are discussed by Zhivov (1990, 1993).

Air Distribution for Local Relief. The following factors should be considered in the design of air distribution for local relief:

Outlet Location. For general low-level ventilation, outlets should be at about 10 ft, although 8 to 11.5 ft is acceptable. For spot cooling, the outlets should be kept close to the worker to minimize mixing with warmer air in the space. In most spot cooling installations, the outlets should be brought down to the 7 ft level.

Discharge Velocity, Temperature, and Air Volume. Discharge velocity, temperature, and air volume should be designed to provide thermal comfort for workers as recommended in Tables 1 and 2. Guidelines are discussed in Chapter 31 of the 1997 *ASHRAE Handbook—Fundamentals.*

Air Diffusers and Their Performance. A wide variety of air diffusion devices can be used with different air distribution methods. Typical applications for air supply diffusers are summarized in Table 7.

Grilles are one of the most universal types of air diffusers. They can have one or two rows of vertical or horizontal vanes and different aspect and vane ratios. Vanes affect grille performance if their depth is at least equal to the distance between the vanes. A grille discharging air uniformly forward (with the vanes in a straight position) has a spread of 14 to 24°, depending on the duct approach, the discharge velocity, and the type of diffuser. Turning the vanes affects the direction and throw of the discharged airstream. Parallel, horizontal vanes direct the airstream vertically within 45°. If the **vane ratio** (vane depth divided by distance between vanes) is less than 2, the jet inclination will be smaller than the angle of the vanes.

Vertical vanes spread the air horizontally, and horizontal vanes spread the air vertically. A grille with diverging vanes (i.e., vertical

Table 6 Guidelines for Selection of Air Distribution Method

			Characteristics of Ventilated Space			
Height, ft	Air Change Rate, ACH	Human Activity	Requirement for Parameter Uniformity in Occupied Zone	Obstructions	HVAC System Type	Recommended Air Supply Method
> 20 to 26	< 5	Moderate and heavy	None	Insignificant	CAV	Concentrated air supply; air supply by horizontal concentrated jets and vertical and/or horizontal directing jets; air supply with inclined jets, air supply with vertical compact jets; displacement air supply; a spiral vortex air supply
					VAV	Air supply by horizontal concentrated jets and vertical and/or horizontal directing jets; air supply with inclined jets; downward air supply with compact jets
				> 10 ft	CAV	Air supply by horizontal concentrated jets and vertical and/or horizontal directing jets; air supply with inclined jets; displacement air supply
					VAV	Air supply by horizontal concentrated jets and vertical and/or horizontal directing jets; air supply with inclined jets
< 20 to 26	> 5	Sedentary and light	High degree of uniformity	Insignificant	CAV	Air supply with vertical radial, conical or compact jets, air supply with horizontal attached or non-attached jets; displacement air supply
					VAV	Air supply with vertical radial or compact jets; air supply with horizontal attached or nonattached jets
				> 7 ft	CAV	Air supply with vertical compact jets; air supply with horizontal attached or nonattached jets into the ails between process equipment displacement air supply
					VAV	Air supply with vertical compact jets; air supply with horizontal attached or nonattached jets into the ails between process equipment

Source: Gosstroy (1992).

Table 7 Typical Air Diffusers and Their Applications

Type of Air Diffuser	Air Diffuser Performance Characteristics		Method of Air Distribution	Application
	K_1*	K_2*		
Large grilles	2 to 6	1.8 to 5.1	Fig. 4A,B,C; Fig. 5	Large shops
Sidewall grilles	2 to 6	1.8 to 5.1	Fig. 6D,E	Low rooms (< 20 ft)
Grilles mounted on duct drops	2 to 4	1.8 to 3.5	Fig. 7	Large shops
Circular diffusers	1 to 3	0.9 to 3.2	Fig. 6A,B,C Fig. 7	Low rooms (< 20 ft) Large shops
Square diffusers	1 to 2.8	1.2 to 3.2	Fig. 6A,B,C Fig. 7	Low rooms (< 20 ft) Large shops
Linear diffusers	2.5	2	Fig. 6D	Low, small rooms
Perforated panels (round, half/quarter round, flat, mounted on or near floor)			Fig. 9 Fig. 10B Fig. 12B,C	Rooms higher than 12 ft with surplus heat or combined heat and contaminant emissions
Perforated panels mounted on duct drops	2.1	1.7	Fig. 8(3)	Shops with surplus heat and a few work areas or shops obstructed by process equipment
Perforated ducts	0.5	1.2	Fig. 8(1)	Low industrial rooms with surplus heat, high air exchange rate, and requirements for low velocities in the occupied zone
Perforated ceiling-mounted panels or perforated panels	2.1	1.7	Fig. 8(2)	Same as perforated ducts, plus special applications (e.g., clean rooms)
Nozzles Converging	6 to 6.8	4.2 to 4.8	Fig. 4, Fig. 5C	Large shops
Diverging or with swirl inserts	1 to 2.5	0.8 to 2.0	Fig. 5C, Fig. 7	Large shops; air supply into aisles

Note: Approximate ranges are given for air diffuser performance constants:
  K_1 = coefficient of velocity decay along the jet
  K_2 = coefficient of temperature decay along the jet
For actual values of these constants, consult manufacturers' guides.

vanes with uniformly increasing angular deflection from the centerline to a maximum at each end of 45°) has a spread of about 60° and reduces the throw considerably. With increasing divergence, the quantity of air discharged by the grille decreases for a given total upstream pressure.

A grille with converging vanes (i.e., vertical vanes with uniformly decreasing angular deflection from the centerline) has a slightly higher throw than a grille with straight vanes, but the spread is approximately the same for both. Compared to the airstream discharged from a grille with straight vanes, the airstream from a grille with converging vanes converges slightly for a short distance in front of the outlet and then spreads more rapidly.

Ceiling-mounted air diffusers can be round, rectangular, or linear and have outlets covered with grilles, perforations, flat plaques, or vanes forming a slot. They can be regulated or nonregulated. Depending on their design, they can form attached radial, concentrated, or linear jets as well as nonattached conical or concentrated jets. Rectangular air diffusers with triangular or four-sided grilles form nonuniform circular flow and can be considered as three or four separate jets.

For applications in industrial and commercial facilities with high ceilings, these diffusers can be mounted on duct drops (installation height 10 to 17 ft) supplying compact conical jets or nonattached radial jets. Nonattached radial and conical jets typically collapse into conical or compact jets under the influence of buoyant forces.

For VAV applications with a considerable air volume and initial temperature differential, air diffusers with a regulated outlet area and/or a regulated direction of air supply, as well as those with induction of room air, perform better within the year-round cycle of system operation.

Round, square, and rectangular nozzles with outlet sizes of 0.4 in. to 7 ft are commonly used for applications ranging from small rooms to large spaces in industrial buildings. They are also used to form directing jets for a mixing-type system.

Converging nozzles form air jets with considerably higher throw and lower noise levels than other air diffusers. Diverging nozzles supply compact jets with increased angle of divergence and reduced throw. Typically, the latter type of jet can be achieved either by placing two or more concentric cones at the supply side of the air diffuser or by placing a swirl insert inside the straight nozzle.

Round, half/quarter round, or flat perforated panels supply air directly into the occupied zone. They discharge air with low velocity (40 to 100 fpm) and low turbulence. These diffusers can be installed either near walls and columns or inside walls or other interior structures. Flat perforated plenums can be integrated into the ceiling construction to supply air with jets projected downward.

New features of perforated panels include

- Induction chambers, which allow supply air to mix with room air inside the air diffuser housing. This design allows supply air to have a greater air temperature differential without causing discomfort in the occupied zone.
- Internal deflectors to adjust the flow direction. These panels are capable of decreasing the restricted zone (zone with abnormal velocities) in front of the air diffuser.

Round or rectangular perforated ducts with partially or completely perforated or slotted walls are used primarily to supply air in spaces where a high air change rate is required and air velocity in the occupied zone is limited due to process restrictions (e.g., to prevent contaminant spillage from local exhausts). The supply surface may be created either by (1) by perforating the duct wall, (2) by cutting incomplete holes in the wall and bending metal peaks inside/outside the duct (to deflect air jets in the correct direction from the duct surface), or (3) by stamping converging nozzles in the desired areas of the sheet metal bend that is used to form the spiral duct.

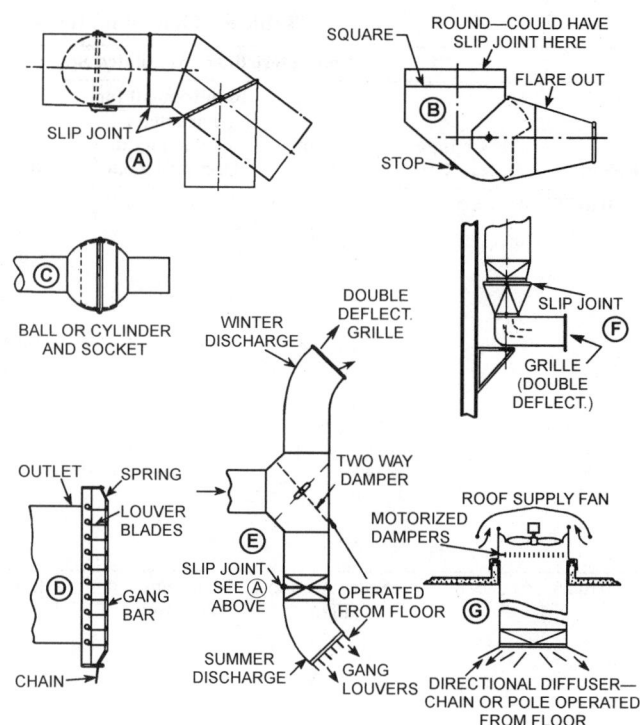

Fig. 13 Directional Outlets for Spot Cooling

Outlet dampers should always be provided for volume and directional control. Figure 13 shows some of the directional outlets used for low-level general ventilation and spot cooling. Outlet A (Navy Type E) has been used in various forms for many years in ship machinery spaces. Outlets B and C are excellent for local area or spot cooling. The adjustable louvers of D applied to directional outlets such as E or F provide excellent control; commercial directional grilles serve the same purpose. The two-way damper arrangement in E is used in local area or aisle ventilation; it directs the supply air upward in winter to mix discharge air with warm or hot air rising from internal sources. Outlet G is a commercial directional diffuser that can be adjusted to provide a variable downward airflow pattern that ranges from flat to vertical. The outlet is shown on a roof supply fan, which is a common application, but the outlet drop must extend through the layer of hot ceiling air to provide effective relief ventilation.

Locker Room, Toilet, and Shower Space Ventilation

The ventilation of locker rooms, toilets, and shower spaces is important in industrial facilities to remove odor and reduce humidity. In some industries, adequate control of workroom contamination requires prevention of ingestion, as well as inhalation, so adequate hygienic facilities, including appropriate ventilation, may be required in locker rooms, change rooms, showers, lunchrooms, and break rooms. State and local regulations should be consulted at early stages of design.

Supply air may be introduced through door or wall grilles. In some cases, plant air may be so contaminated that filtration or, preferably, mechanical ventilation, may be required. When control of workroom contaminants is inadequate or not feasible, total employee exposure can be reduced by minimizing the level of contamination in the locker rooms, lunchrooms, and break rooms by pressurizing these areas with excess supply air.

When mechanical ventilation is used, the supply system should have supply fixtures such as wall grilles, ceiling diffusers, or supply plenums to distribute the air adequately throughout the area.

Table 8   Ventilation for Locker Rooms, Toilets, and Shower Spaces

Description of Space	Ventilation
Locker Rooms	
Coat hanging or clean change room for nonlaboring shift employees with clean work clothes	1 cfm/ft^2
Change room for laboring employees with wet or sweaty clothes	2 cfm/ft^2; 7 cfm exhausted from each locker
Change room for laborers or worker assigned to heavy work; clothes will be wet or pick up odors	3 cfm/ft^2; 10 cfm exhausted from each locker
Toilet Spaces	2 cfm/ft^2; at least 25 cfm per toilet facility; 200 cfm minimum
Shower Spaces	2 cfm/ft^2; at least 50 cfm per shower head; 200 cfm minimum

Note: The source of this table is unknown. Nevertheless, the information has been used with apparent success for many years (*ASHRAE Guide and Data Book* 1970). Also refer to Article 16 of the BOCA *National Building Code* and ASHRAE *Standard 62*.

In the locker rooms, the exhaust should be taken primarily from the toilet and shower spaces, as needed, and the remainder from the lockers and the room ceiling. In the absence of specific codes, Table 8 provides a guide for ventilation of these spaces.

NATURAL VENTILATION

Natural ventilation is a controlled flow of air caused by thermal and wind pressures. It is commonly used in Canada and in European countries. Numerous studies of natural ventilation have been conducted, and methods of design have been developed (Baturin 1972; Shepelev 1978; Goodfellow 1985; Shilkrot 1993). Shilkrot in Stroiizdat (1992a) summarizes these design methods.

Natural ventilation can be used in spaces with significant heat release as long as contamination of the incoming air by gases and particles does not exceed 30% of the design exposure limits. Natural ventilation is not recommended when air filtration and treatment are required by the process or when the incoming outdoor air causes mist or condensation.

In the summer, air inlets for natural ventilation are located in exterior walls; the lower level of the openings is 1 to 6 ft above the floor. Air inlets can be arranged in one, two, or more rows in the longer exterior walls. Windows, doorways, other openings in the exterior walls, or apertures in floors over basements (with air transportation along special channels) also can be used as air inlets.

During seasons other than summer, air inlets must be located higher than 10 ft above floor level in rooms with ceilings lower than 20 ft. If the room height exceeds 20 ft, low-level inlets must be located higher than 13 ft. These inlets must be supplied with baffles to direct air upward. Schematics of and pressure loss coefficients for inlets are given in Table 9.

Interior ventilated halls can obtain outside air through ridge vents in adjacent "cold" halls. Cold halls must be separated from "hot" ones by screens dropped from the roof to create an airway above the floor that is 7 to 13 ft deep. Air supply via ridge vents on the roof is not recommended but can be used if necessary.

Air is evacuated from naturally ventilated spaces through wind-protected continuous ridge vents, skylights, or round roof ventilators. Many types of ridge vents and roof ventilators have been designed; their sizes and pressure loss coefficients are available from the manufacturers. All inlets and outlets for natural ventilation must be supplied with controls for easy opening and closing.

The air exchange rate G_o required for temperature control in the occupied zone can be calculated from the following room heat balance equation [similar to Equations (8) through (10)]:

Table 9   Pressure Loss Coefficients for Inlet Apertures

Schematic	Baffle Type	Aspect Ratio, h/b	Pressure Loss Coefficient ξ at Different Baffle Angles α				
			15°	30°	45°	60°	90°
	Single, top-hinged baffle	0	30.8	9.2	5.2	3.5	2.6
		0.5	20.6	6.9	4.0	3.2	2.6
		1	16.0	5.7	3.7	3.1	2.6
	Single, center-hinged baffle	0	59	13.6	6.6	3.2	2.7
		1	45.3	11.1	5.2	3.2	2.4
	Double, top-hinged baffle	0.5	30.8	9.8	5.2	3.5	2.4
		1.0	14.8	4.9	3.8	3.0	2.4
	Gate	—	—	—	—	—	2.4

Source: Stroiizdat (1992a).

Table 10   Coefficients K_t and K_w of Natural Ventilation Effectiveness for Industrial Spaces with Significant Heat Load

Room Category	K_t/K_w
Blacksmith and press shops, oven bays at foundries, rail rolling mills, bloomings, cooling bays, etc.	2.0/2.7
Heat treatment shops	1.9/2.6
Drying shops	1.8/2.5
Casting shops	1.7/2.3
Blast furnace shops	1.6/2.2
Rolling mills	1.5/2.1
Electrolysis shops, machine and compressor rooms	1.4/1.9
Shops for vulcanization and plastic production	1.3/1.8

Note: Significant heat load is defined as higher than 2.2 Btu/h·ft^3.

$$G_o = \frac{q}{c_p K_t (t_{uz} - t_{oz})} \qquad (11)$$

where

G_o = air exchange rate, lb/h
q = surplus heat released in the space, Btu/h
c_p = specific heat of air, Btu/lb·°F
t_{uz} = air temperature in upper zone, °F
K_t = coefficient of natural ventilation efficiency calculated from the air temperatures of the occupied zone t_{oz}, the air removed from the upper zone t_{exh}, and the outdoor air t_o

$$= \frac{t_{exh} - t_o}{t_{oz} - t_o}$$

The method of zone-by-zone heat balances (Shilkrot 1993) can be used to define K_t values. Computed values of K_t for different industrial applications are given in Table 10.

Both natural ventilation design and displacement ventilation design assume temperature stratification throughout the room height. Air close to the heat source is heated and rises as a convective stream. Part of this heated air is evacuated through outlets in the upper zone, and part of it remains in the upper zone, in the **heat cushion**. Separation level Z (Figure 14), which is the lower level of the upper zone, is defined in terms of the equality of G_{conv} and G_o, which are the airflow rate in the convective flows above the heat source and the airflow supplied to the occupied zone, respectively. It is assumed that the air temperature in the lower zone is equal to that in the occupied zone t_{oz} and that the air temperature t_{uz} in the upper zone is equal to that of the evacuated air t_{exh}.

The incoming air jet can destroy the air temperature stratification when the rarefaction Δp_{jet} created by this jet on the level of separation Z is strong enough to move the heated air from the upper zone to the lower zone:

$$\Delta p_{jet} \geq g(\rho_{oz} - \rho_{uz})(Z - h_o) \tag{12}$$

where

Δp_{jet} = rarefaction created by incoming air jet, lb_f/ft^2
g = gravitational acceleration, ft/s^2
ρ_{oz} = density of air in occupied zone, lb/ft^3
ρ_{uz} = density of air in upper zone, lb/ft^3
Z = separation level, ft
h_o = height of air supply opening, ft

Calculations by Shilkrot (1993) show that when the supply air velocity V_o in the inlet is 400 to 600 fpm and the separation level Z is more than 3 ft higher than the air supply opening h_o (i.e., $Z - h_o > 3$ ft), a temperature difference of 4 to 6°F between the air in the upper and lower zones ensures the stability of the stratification. Shilkrot has also shown that when the Richardson number R_N is greater than 5, the turbulent exchange between the upper and the lower zones can be neglected.

For single-bay shops, the following steps are recommended for natural ventilation design:

1. Determine the surplus heat load (i.e., heat gains minus heat losses) q (Btu/h) in the space.
2. Select K_t from Table 10.
3. Compute air exchange rate G_o (lb/h) for temperature control in the occupied zone:

$$G_o = \frac{q}{c_p K_t (t_{oz} - t_o)} \tag{13}$$

4. Compute the outgoing air temperature, °F:

$$t_{exh} = t_o + \frac{q}{c_p \rho_o Q} \tag{14}$$

5. Select the separation level Z:

$$Z = 0.4H \text{ to } 0.7H \tag{15}$$

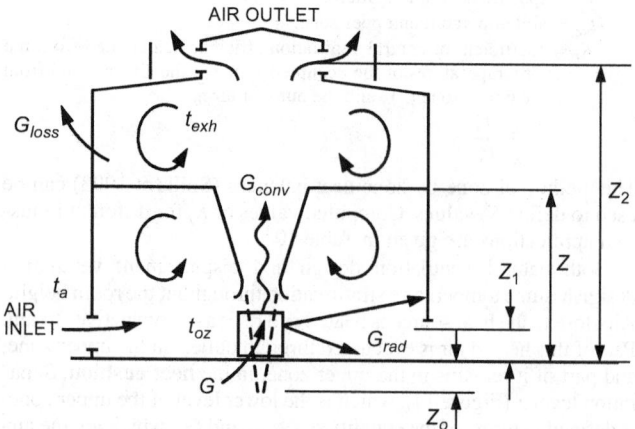

Fig. 14 Natural Ventilation of Single-Bay Building

where

H = ventilated space height, ft ($20 \leq H \leq 79$)
$Z = 0.7H$ when $H = 20$ ft; $Z = 0.4H$ when $H = 79$ ft

6. Determine the pressure difference available to transport air through the inlet and outlet apertures:

$$\Delta p = g(Z - Z_1)(\rho_o - \rho_{oz}) + g(Z_2 - Z)(\rho_o - \rho_{exh}) \tag{16}$$

where $\rho = \rho_{68°F} \times T_{68°F}/T = 0.075 \times 528/T = 40/T$, lb/ft^3.

7. Calculate the pressure loss in the air inlets, in. of water:

$$\Delta p_1 = \beta \Delta p \tag{17}$$

β is part of the pressure available for air transportation through inlets. To reduce the velocity of the incoming air, the inlet area can be increased to keep β in the range 0.1 to 0.4.

8. Calculate the area F_1 (ft^2) of the inlet apertures:

$$F_1 = \frac{G_o}{\sqrt{2\rho_{out}\Delta p_1/\xi_1}} \tag{18}$$

Values of the pressure loss coefficient ξ are listed in Table 9.

If the area of the inlets is given, the pressure drop Δp_1 can be calculated from the following equation:

$$\Delta p_1 = \frac{\xi_1}{2\rho_{out}}\left(\frac{G_o}{F_1}\right)^2 \tag{19}$$

9. Calculate the pressure drop available for the air outlets:

$$\Delta p_2 = \Delta p - \Delta p_1 \tag{20}$$

10. Determine the area F_2 of the air outlets (vent in roof ridge, roof ventilators, etc.):

$$F_2 = \frac{G_o}{\sqrt{2\rho_{exh}\Delta p_2/\xi_2}} \tag{21}$$

Stroiizdat (1992a) discusses a method for natural ventilation design applicable to multibay shops, multistory buildings, and combined natural and mechanical ventilation systems.

AIR CURTAINS

Air curtains are local ventilation devices that supply a high-velocity stream of air to reduce airflow through apertures in building shells (Asker 1970; Strongin 1993; Powlesland 1971, 1973; Stroiizdat 1992a) and in process equipment (Bintzer and Malehom 1976; Goodfellow 1985; Ivanitskaya et al. 1986; Strongin and Nikulin 1991). They are also used to localize gaseous and particulate emissions near their sources and to convey them toward local exhausts (Stoler and Savelyev 1977; Posokhin and Broida 1980; Posokhin 1985). Different applications of air curtains for jet assisted hoods are discussed in Chapter 29.

Air curtains provide better thermal environments for workstations located near doorways. They can also reduce the energy consumption of HVAC systems.

Shutter-type air curtains create a resistance to airflow through a door, thus reducing the incoming airflow (Figure 15). They direct air toward the incoming outside air at an angle ranging from 30 to 40°. Shutter-type air curtains may be single-sided or double-sided and projected upward or downward. Double-sided air curtains deter the incoming airflow better. Upward-projected air curtains are recommended when the gate width is greater than its height. They also provide better coverage of the lower area of the door opening.

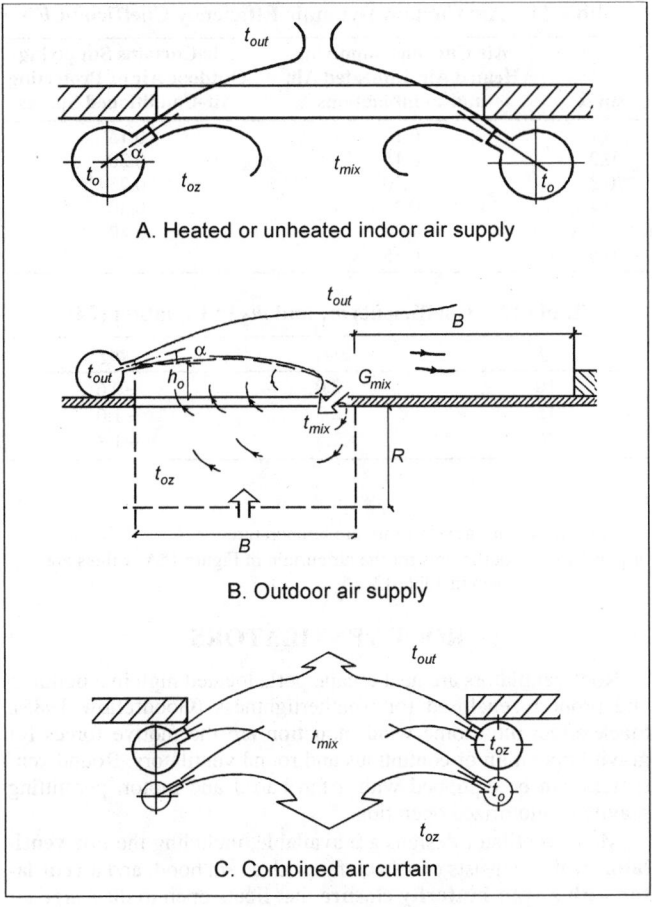

Fig. 15 Shutter-Type Air Curtains

A. Heated or unheated indoor air supply

B. Outdoor air supply

C. Combined air curtain

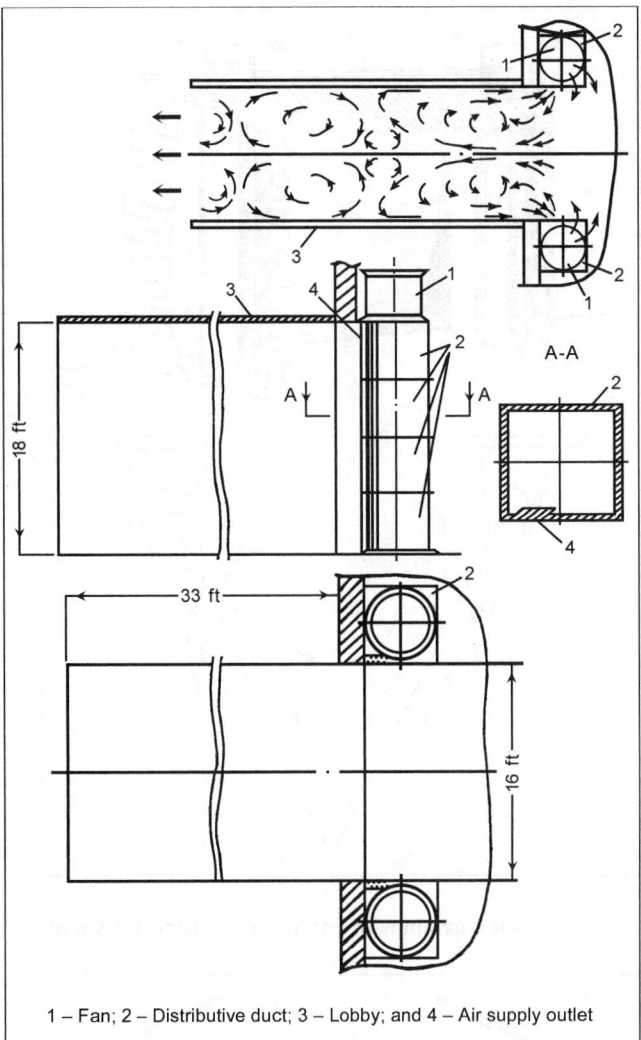

1 – Fan; 2 – Distributive duct; 3 – Lobby; and 4 – Air supply outlet

Fig. 16   Air Curtain for Medium-Sized Gate with Lobby

Air curtains can supply heated air, air at room temperature, or air at the outdoor temperature. Air curtains with heated air are recommended for doors smaller than 12 ft by 12 ft and for process apertures that are opened frequently (e.g., more than five times or for longer than 40 min during an 8 h shift) and that are located in regions with design outdoor winter temperatures of 5°F or lower. Air curtains provide desired air temperatures for workstations near apertures; however, the heat consumption is relatively high. Air curtains that supply unheated indoor air have application in spaces (1) with a heat surplus, (2) with temperature stratification over the room height, (3) with low air temperatures (less than 46°F) near the aperture area, and (4) in regions with a mild climate. Shutter-type air curtains also are recommended for use in cooled spaces.

Air curtains with a lobby are shown in Figure 16. Performance is based on transition of the supply air jet impulse into counterpressure, which prevents outdoor airflow into the room. Air is supplied counter to the outdoor airflow or at a small angle to it. A curtain jet propagates along the channel walls, slows down, and makes a U-turn, reversing along its axis. The length of the lobby is chosen to exceed 2.5 times its width (for double-sided air curtains) and so that air is not forced outside. To shorten the lobby, air is usually supplied by a jet with a coerced angle of divergence. Air curtains with a lobby for supplying outdoor air are shown in Figure 17.

Combined air curtains are used in very cold climates (winter temperatures as low as −85°F), for doors larger than 12 ft by 12 ft, and for spaces with several doors. A combined air curtain with a specially designed lobby is shown in Figure 18. The lobby has corrugated iron walls that reduce the wind pressure on the gate aperture. Air curtains in the lobby supply untreated outdoor air, while air curtains inside the building supply heated air. Combined air curtains without a lobby are shown in Figure 15C.

Air curtain controls should be designed to turn curtains on when the door is opened and turn them off when the door is closed or when the air temperature near the door reaches the target value.

Principles of Air Curtain Design

The velocity of the supplied air can be calculated from the following equation:

$$V_o = \sqrt{\frac{2\Delta p f g}{\beta_o \rho E}(60 \times 5.2)} = 1098 \sqrt{\frac{\Delta p f}{\beta_o \rho E}} \qquad (22)$$

where

V_o = velocity of supply air, fpm

Δp = average pressure difference between inside and outside air near the aperture with the air curtain turned on, in. of water

f = ratio of area A_{ap} of air supply slots to door area A_o. For air curtains with heated air or unheated indoor air and for combined air curtains, f = 10 to 20. For air curtains supplying outdoor air or protecting air-conditioned spaces, f = 20 to 40.

β_o = Boussinesq coefficient (describes uniformity of air velocity in opening cross section); for air curtain supply nozzle, β_o ranges from 1.05 to 1.1

ρ = density of air supplied by air curtain, lb/ft^3

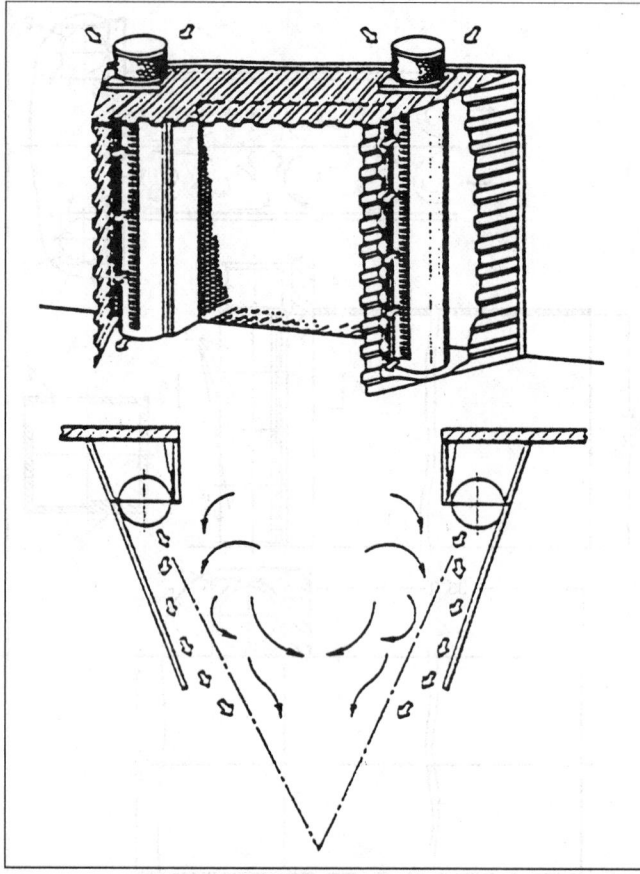

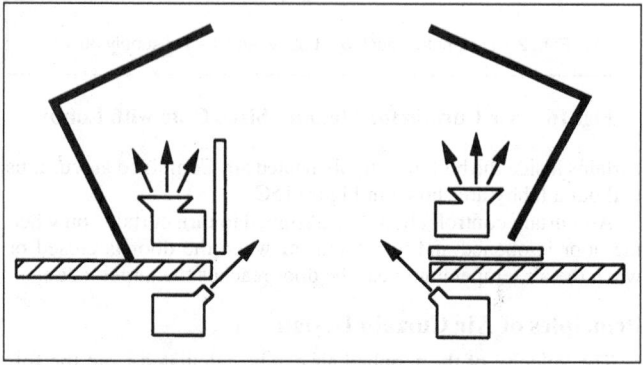

Fig. 17   Air Curtain with Lobby and Outside Air Supply

Fig. 18   Combined Air Curtain with Lobby
(Strongin and Nikulin 1987)

E = coefficient of air curtain dynamic efficiency (given in Table 11)
g = accelleration of gravity = 32.2 ft/s^2

Airflow supplied by the air curtains as shown in Figure 15 can be calculated from the following equation:

$$G_o = \rho V_o A_o \tag{23}$$

The temperature t_o of the supplied air can be estimated from the heat balance equation for the aperture under consideration:

$$t_o = t_{out} + m_1(t_{mix} - t_{out}) + m_2(t_{oz} - t_{out}) \tag{24}$$

Table 11   Air Curtain Dynamic Efficiency Coefficient E

sin α	Air Curtains Supplying Heated Air, Unheated Air, and Combinations	Air Curtains Supplying Outdoor Air or Protecting Air-Conditioned Spaces
0.1	0.10	0.15
0.2	0.15	0.20
0.3	0.20	0.25
0.4	0.25	0.30
0.5	0.30	0.40
0.6	0.35	—

Table 12   Coefficients m_1 and m_2 in Equation (24)

f	m_1	m_2
10	2.0	−0.6
15	2.3	−1.0
20	2.5	−1.3

where

t_{mix} = normative air mixture temperature
m_1 and m_2 = coefficients for the air curtain in Figure 15A; values are given in Table 12

ROOF VENTILATORS

Roof ventilators are heat escape ports located high in a building and properly enclosed for weathertightness (Goodfellow 1985). Stack effect plus some wind induction are the motive forces for gravity operation of continuous and round ventilators. Round ventilators can be equipped with a fan barrel and motor, permitting gravity or motorized operation.

Many ventilator designs are available, including the **low ventilator**, which consists of a stack fan with a rain hood, and a **ventilator with a split butterfly closure** that floats open to discharge air and closes by itself. Both use minimum enclosures and have little or no gravity capacity. Split butterfly dampers tend to make the fans noisy and are subject to damage from slamming during strong wind conditions. Because noise is frequently a problem in powered roof ventilators, the manufacturer's sound rating should be reviewed.

Roof ventilators can be listed in order of their heat removal capacity. The **continuous ventilation monitor** removes substantial, concentrated heat loads most effectively. One type, the streamlined continuous ventilator, is efficient, weathertight, and designed to prevent backdraft, and it usually has dampers that may be readily closed in winter to conserve building heat. Its capacity is limited only by the available roof area and the proper location and sizing of low-level air inlets. **Gravity ventilators** have low operating costs, do not generate noise, and are self-regulating (i.e., higher heat release results in higher airflow through the ventilators). Care must be taken to ensure that positive pressure exists at the ventilators; otherwise, outside air will enter the ventilators. This is of particular importance during the heating season.

Next according to their heat removal capacity are (1) the round gravity or windband ventilator, (2) the round gravity ventilator with fan and motor added, (3) the low-hood powered ventilator, and (4) the vertical upblast powered ventilator. The shroud for the vertical upblast design has a peripheral baffle to deflect the air upward instead of downward. Vertical discharge is highly desirable to reduce roof damage caused by the hot air if it contains condensable oil or solvent vapor. Ventilators with direct-connected motors are desirable because of the locations of the units and the belt maintenance required for units having short shaft centerline distances. Round gravity ventilators are applicable to warehouses with light heat loads and to manufacturing areas having high roofs and light loads.

Streamlined continuous ventilators must operate effectively without mechanical power. Efficient ventilator operation is generally obtained when the difference in elevation between the average air inlet level and the roof ventilation is at least 30 ft and the exit temperature is 25°F above the prevailing outdoor temperature. Chapter 25 of the 1997 *ASHRAE Handbook—Fundamentals* has further details. Under these conditions and with a wind velocity of 5 mph, the ventilator throat velocity will be about 375 fpm, and it will remove 0.24 Btu/lb·°F $\times$ 0.075 lb/ft^3 $\times$ 25°F $\times$ 375 fpm $\times$ 60 min/h = 10,000 Btu/h·ft^2.

To ensure this level of performance, sufficient low-level openings must be provided for the incoming air. Manufacturers recommend an inlet velocity of 250 to 500 fpm. Insufficient inlet area and significant air currents are the most common reasons gravity roof ventilators malfunction. A positive supply of air around the hot equipment may be necessary in large buildings where the external wall inlets are remote from the equipment.

The cost of electrical power for mechanical ventilation is offset by the advantage of constant airflow. Mechanical ventilation can also create the pressure differential necessary for good airflow, even with small inlets. Inlets should be sized correctly to avoid infiltration and other problems caused by high negative pressure in the building. Often, a mechanical system is justified to supply enough makeup air to maintain the work area under positive pressure.

Careful study of airflow around buildings is necessary to avoid reintroducing contaminants from the exhaust into the ventilation system. Discharging the exhaust at the roof level or from an area opposite the outside intake may not prevent the exhaust from reentering the building. Chapter 43 covers building exhaust and air intake design.

Fusible link dampers, which close in case of fire, may be required by building codes.

HEAT CONSERVATION AND RECOVERY

Because of the large air volumes required for ventilating industrial plants, heat conservation and recovery should be practiced and will provide substantial savings. Heat conservation and recovery should be incorporated into preliminary planning for an industrial plant.

In some cases, it is possible to provide unheated or partially heated makeup air to the building. Rotary, regenerative heat exchangers recover up to 80% of the heat available from the differential between the exhaust and the outdoor air temperatures. Reductions of 9 to 18°F in the heating requirement should be routine for most industrial systems. Although most of the heat conservation and recovery methods in this section apply to heating, the savings possible with air-conditioning systems are equally impressive. The following are some methods of heat conservation and recovery:

1. In the original design of the building, process, and equipment, provide insulation and heat shields to minimize heat loads. Vapor-proofing and reduction of glass area may be required. Changes in process design may be necessary to keep the building heat loads within reasonable bounds. Review the exhaust needs for hoods and process and keep those to a practical, safe minimum.
2. Design the supply and exhaust general ventilation systems for optimal operation throughout the year. Use VAV systems and efficient air distribution methods that provide minimal restrictions on the system operation. The air should be supplied as close to the occupied zone as possible (Zhivov 1990). Recirculated air should be used in the winter makeup, and unheated or partially heated air should be brought to the hoods or process (ACGIH 1998b; Holcomb and Radia 1986).
3. Design the system to achieve the highest efficiency and lowest residence time for contaminants. It is good practice to permit workers to adjust the air patterns to which they are exposed; people want direct personal control over their working environment.
4. When possible, recycle exhaust air. For example, office exhaust can be directed first to work areas, then to locker rooms or process areas, and finally to the outside. Air heated from cooling equipment in motor or generator rooms can be recycled. The cooling systems for many large motors and generators can be arranged to discharge into the building in the winter to provide heat and to the outside in the summer to avoid heat loads.
5. Supply air can be passed through air-to-air, liquid-to-air, or hot-gas-to-air heat exchangers to recover building or process heat. Rotary, regenerative, and air-to-air heat exchangers are discussed in Chapter 42 of the 1996 *ASHRAE Handbook—Systems and Equipment*.
6. Operate the system for economy. Shut the systems down at night or on weekends whenever possible, and operate the makeup air in balance with the needs of operating the process equipment and hoods. Keep heating supply air temperatures at the minimum, and cooling supply temperatures at the maximum, consistent with the needs of process and employee comfort. Keep the building in balance so that uncomfortable drafts do not necessitate excessive heating.

REFERENCES

ACGIH. 1993. Threshold limit values and biological exposure indices for 1993-94. American Conference of Governmental Industrial Hygienists, Cincinnati, OH.

ACGIH. 1998a. Guide to occupational exposure values—1998. Compiled by the ACGIH.

ACGIH. 1998b. *Industrial ventilation: A manual of recommended practice.* 23rd ed.

AIR-IX. 1987. Teollisuusilmanvaihdon suunnittelu. Kauppa-ja teollissus-ministerio, Helsinki (in Finnish).

ASHRAE. 1989. Ventilation for acceptable indoor air quality. ANSI/ASHRAE *Standard* 62-1989.

ASHRAE. 1992. Thermal environmental conditions for human occupancy. *Standard* 55-1992.

Asker, G.S.F. 1970. What, where and how of air curtain systems. *Heating, Piping and Air Conditioning* (June).

Azer, N.Z. 1982a. Design guidelines for spot cooling systems: Part 1—Assessing the acceptability of the environment. *ASHRAE Transactions* 88(1):81-95.

Azer, N.Z. 1982b. Design guidelines for spot cooling systems: Part 2—Cooling jet model and design procedure. *ASHRAE Transactions* 88(1):97-116.

Azer, N.Z. 1984. Design of spot cooling systems for hot industrial environments. *ASHRAE Transactions* 90(1B):460-75.

Baturin, V.V. 1972. *Fundamentals of industrial ventilation.* 3rd ed. Pergamon Press, Oxford, U.K.

Bintzer, W. and F.A. Malehom. 1976. Air curtains on electric arc furnaces at Lukens Steel Co. *Iron and Steel Engineer* (July):53-55.

BOCA. 1999. *National building code.* Article 16. BOCA International, Country Club Hills, IL.

Brief, R.S., S. Lipton, S. Amarmani, and R.W. Powell. 1983. Development of exposure control strategy for process equipment. *Annals American Conference of Governmental Industrial Hygienists* (5).

British Occupational Hygiene Society (BOHS). 1987. Controlling airborne contaminants in the workplace. Tech. Guide No. 7. Science Review Ltd. and H&H Sci. Consult., Leeds, U.K.

Caplan, K.J. 1980. Heat stress measurements. *Heating, Piping and Air Conditioning* (February):55-62.

Caplan, K.J. and G.W. Knutson. 1978. Laboratory fume hoods: Part 2—Influence of room air supply. *ASHRAE Transactions* 84(1):522-37.

Cawkwell, G.C. and H.D. Goodfellow. 1990. Multiple cell ventilation model with time-dependent emission sources. Proceedings of the 2nd International Conference on Engineering Aero- and Thermodynamics of Ventilated Rooms, A1-9 (Oslo, Norway, June 13-15).

Cole, J.P. 1995. Ventilation systems to accommodate the industrial process. *Heating, Piping, Air Conditioning* (May).

Constance, J.D. 1983. *Controlling in-plant airborne contaminants.* Marcel Dekker, Inc., New York.

Fanger, P.O. 1982. *Thermal comfort.* Robert E. Krieger Publishing Company, Malabar, FL.

Fanger, P.O., A.K. Melikov, H. Hanzawa, and J. Ring. 1988. Air turbulence and sensation of draft. *Energy and Buildings* 12(1).

Fanger, P.O., B.M. Ipsen, G. Langkilde, B.W. Olesen, M.K. Christiansen, and S. Tanabe. 1985. Comfort limits for asymmetric thermal radiation. *Energy and Buildings* 8:225-26.

Fuller, F.H. and A.W. Etchells. 1979. The rating of laboratory hood performance. *ASHRAE Journal* 21(10):49-53.

Goldfield, J. 1980. Contaminant concentration reduction general ventilation versus local exhaust ventilation. *American Industrial Hygienists Association Journal* 41 (November).

Goodfellow, H.D. 1985. *Advanced design of ventilation systems for contaminant control.* Elsevier Science Publishers, B.V., Amsterdam.

Goodfellow, H.D. 1986. Proceedings of Ventilation '85. Elsevier Science Publishers, B.V., Amsterdam.

Goodfellow, H.D. 1987. Ventilation, industrial. *Encyclopedia of physical science and technology* (14). Academic Press, San Diego, CA.

Goodfellow, H.D. and J.W. Smith. 1982. Industrial ventilation—A review and update. *American Industrial Hygiene Association Journal* 43 (March):175-84.

GOSSTROY. 1992. SNiP 2.04.05-92. Building norms and regulations. Heating, ventilation and air conditioning. Moscow (in Russian).

Grimitlyn, M.I. and G.M. Pozin. 1993. Fundamentals of optimizing air distribution in ventilated spaces. *ASHRAE Transactions* 99(1):1128-38.

Harris, R.L. 1988. Design of dilution ventilation for sensible and latent heat. *Applied Industrial Hygiene* 3(1).

Holcomb, M.L. and J.T. Radia. 1986. An engineering approach to feasibility assessment and design of recirculating exhaust systems. Proceedings of Ventilation '85. Elsevier Science Publishers, B.V., Amsterdam.

Hwang, C.L., F.A. Tillman, and M.J. Lin. 1984. Optimal design of an air jet for spot cooling. *ASHRAE Transactions* 90(1B):476-98.

ISO. 1989. Hot environments—An analytical determination and interpretation of thermal stress using calculation of required sweat rate. *Standard* 7933. International Organization for Standardization, Geneva.

ISO. 1989. Hot environments—Estimation of the heat stress on a working man, based on the WBGT-index (wet bulb globe temperature). *Standard* 7243.

ISO. 1994. Moderate thermal environments—Determination of the PMV and PPD indices and specifications of the conditions for thermal comfort. *Standard* 7730.

Ivanitskaya, M.Yu., A.S. Strongin, and E.A. Visotskaya. 1986. Studies of the air curtain application for the channel of the localizing ventilation. Proceedings of Heating and Ventilation. TsNIIpromzdanii (Moscow).

Jackman, P. 1991. Displacement ventilation. CIBSE National Conference (April). University of Kent, Canterbury.

Kristensson, J.A. and O.A. Lindqvist. 1993. Displacement ventilation systems in industrial buildings. *ASHRAE Transactions* 99(1):992-1006.

Kuz'mina, L.V., A.M. Kruglikova, and A.S. Gus'kov. 1986. Contaminant distribution in industrial hall with spiral vortex ventilation. Occupational Safety in Industry. *Transactions* of the All-Union Research Institutes for Labor Protection. PROFIZDAT, Moscow.

Langkilde, G., L. Gunnarsen, and N. Mortensen. 1985. Comfort limits during infrared radiant heating of industrial spaces. CLIMA 2000, Copenhagen.

Laurikainen, J. 1995. Displacement ventilation system design method. Seminar presentations, Part 2. INVENT *Report* 46. FIMET, Helsinki.

LVIS. 1996. LVIS 2000. Ilmastointi. Painopaikka: Kausalan Kirjapaino Oy (in Finnish).

Melikov, A.K., L. Halkjaer, R.S. Arakelian, and P.O. Fanger. 1991. Human response to cooling with air jets. ASHRAE Research Project 518-R.

Melikov, A.K., L. Halkjaer, R.S. Arakelian, and P.O. Fanger. 1994. Spot cooling—Parts 1 and 2. *ASHRAE Transactions* 100(2):476-510.

Nagasawa, Y., M. Nitadori, and S. Matsui. 1990. Characteristics of spiral vortex flow and its application to control indoor air quality. RoomVent '90. Proceedings of the International Conference on Air Distribution in Ventilated Rooms. Oslo, Norway.

National Safety Council. 1988. *Fundamentals of industrial hygiene*, 3rd ed. Chicago, IL.

NIOSH. 1986. Criteria for a recommended standard: Occupational exposure to hot environments. Revised Criteria. DHHS (NIOSH) *Publication* No. 86-113. National Institute for Occupational Safety and Health, Washington, D.C. Available from www.cdc.gov/NIOSH/86-113.html.

Olesen, B.W. and R. Nielsen. 1980. Spot cooling of workplaces in hot industries. Final Report to the European Coal and Steel Union. Research Programme Ergonomics—Rehabilitation III, No. 7245-35-004 (October):88.

Olesen, B.W. and R. Nielsen. 1981. Radiant spot cooling of hot working places. *ASHRAE Transactions* 87(1):593-608.

Olesen, B.W. and R. Nielsen. 1983. Convective spot-cooling of hot working environments. Proceedings of the XVIth International Congress of Refrigeration (September), Paris.

Olesen, B.W. and A.M. Zhivov. 1994. Evaluation of thermal environment in industrial work spaces. *ASHRAE Transactions* 100(2):623-35.

Olesen, B.W., M. Scholer, and P.O. Fanger. 1979. Discomfort caused by vertical air temperature differences. Indoor climate. Danish Research Institute, Copenhagen.

Peterson, R.L., E.L. Schofer, and D.W. Martin. 1983. Laboratory air systems—Further testing. *ASHRAE Transactions* 89(2B):571-98.

Posokhin, V.N. 1985. *Design of local ventilation systems for the process equipment with heat and gas release.* Mashinostroyeniye, Moscow (in Russian).

Posokhin, V.N. and V.A. Broida. 1980. Local exhausts incorporated with air curtain. Hydromechanics and heat transfer in sanitary technique equipment. KNTI, Kazan (in Russian).

Powlesland, J.W. 1971. Air curtains. *Canadian Mining Journal* (October): 84-93.

Powlesland, J.W. 1973. Air curtains in controlled energy flows. American Conference of Industrial Hygienists (February).

Pozin, G.M. 1993. Determination of the ventilating effectiveness in mechanically ventilated spaces. Proceedings of the 6th International Conference on Indoor Air Quality (IAQ '93), Helsinki.

Robertson, K.S. and Downie, R.H. 1978. Personal cooling in industry. Technical Report No. TR-18. Highett, Victoria.

RoomVent '90. 1990. Proceedings of the 2nd International Conference on Engineering Aero- and Thermodynamics of Ventilated Rooms (Oslo, Norway, June 13-15). ScanVac, Denmark.

RoomVent '92. 1992. Proceedings of the 3rd International Conference on Engineering Aero- and Thermodynamics of Ventilated Rooms (Aalborg, Denmark). ScanVac, Denmark.

RoomVent '94. 1994. Proceedings of the 4th International Conference on Engineering Aero- and Thermodynamics of Ventilated Rooms (Krakow, Poland, June 15-17). ScanVac, Denmark.

RoomVent '96. Proceedings of the 5th International Conference on Air Distribution in Rooms. Yokohama, Japan.

Schroy, J.M. 1986. A philosophy on engineering controls for workplace protection. *Annals of Occupational Hygiene* 30(2):231-36.

Shepelev, I.A. 1978. *Aerodynamics of air flows in rooms.* Stroiizdat, Moscow (in Russian).

Shilkrot, E.O. 1993. Determination of design loads on room heating and ventilation systems using the method of zone-by-zone balances. *ASHRAE Transactions* 99(1):987-91.

Shilkrot, E.O. and A.M. Zhivov. 1992. Room ventilation with designed temperature stratification. RoomVent '92. Proceedings of the 3rd International Conference on Engineering Aero- and Thermodynamics of Ventilated Rooms (Aalborg, Denmark).

Shilkrot, Eu.O. and A.M. Zhivov. 1996. Zonal model for displacement ventilation design. RoomVent '96. Proceedings of the 5th International Conference on Air Distribution in Rooms. Vol. 2. Yokohama, Japan.

Skåret, E. 1985. *Ventilation by displacement—Characterization and design applications.* Elsevier Science Publishers, B.V., Amsterdam.

Skistad, H. 1994. *Displacement ventilation.* Research Studies Press, John Wiley and Sons, West Sussex, UK.

SNiP 2.04.05-91. 1994. Building norms and regulations. Heating, Ventilation and Air Conditioning. Moscow: MINSTROY (in Russian).

Stoler, V.D. and Yu.L. Savelyev. 1977. Push-pull systems design for the etching tanks. *Heating, Ventilation, Water Supply, and Sewage Systems Design* 8(124). TsINIS, Moscow (in Russian).

Stroiizdat. 1992a. *Design manual: Ventilation and air conditioning*, 4th ed. Part 3(1). Moscow (in Russian).

Stroiizdat. 1992b. *Design manual: Ventilation and air conditioning.* Part 3(2). Moscow (in Russian).

Strongin, A.S. 1993. Aerodynamic protection of hangars against cold ingress through apertures. *Building Services Engineering Research and Technology* 14(1). CIBSE Ser. A, London, UK.

Strongin, A.S. and M.V. Nikulin. 1987. Air curtains with outdoors air supply. Heating and ventilation of industrial buildings. TsNIIpromzdanii, Moscow (in Russian).

Strongin, A.S. and M.V. Nikulin. 1991. A new approach to the air curtain design. *Construction and Architecture—Izvestiya VUZOV* (January):84-87 (in Russian).

Tarnopolsky, M. 1992. Design of air distribution systems in air conditioned spaces. Technion Research and Development Foundation, Haifa, Israel.

U.S. Department of Health and Human Services. 1986. Advanced industrial hygiene engineering. PB87-229621. Cincinnati, OH.

Ventilation '85. 1986. Proceedings of the 1st International Symposium on Ventilation for Contaminant Control. Elsevier Science Publishers, B.V., Amsterdam.

Ventilation '88. 1989. Proceedings of the 2nd International Symposium on Ventilation for Contaminant Control. Elsevier Science Publishers, B.V., Amsterdam.

Ventilation '91. 1993. Proceedings of the 3rd International Symposium on Ventilation for Contaminant Control. American Conference of Governmental Industrial Hygienists (ACGIH), Cincinnati, OH.

Ventilation '94. 1994. Proceedings of the 4th International Symposium on Ventilation for Contaminant Control. National Institute of Occupational Health. Stockholm, Sweden.

Ventilation '97. 1997. Proceedings of the 5th International Symposium on Ventilation for Contaminant Control. The Canadian Environment Industry Association. Ottawa, Canada.

Zhivov, A.M. 1990. Variable-air-volume ventilation systems for industrial buildings. *ASHRAE Transactions* 96(2):367-72.

Zhivov, A.M. 1992. Selection of general ventilation method for industrial spaces. Presented at 1992 ASHRAE Annual Meeting ("Supply Air Systems for Industrial Facilities" Seminar).

Zhivov, A.M. 1993. Theory and practice of air distribution with inclined jets. *ASHRAE Transactions* 99(1):1152-59.

Zhivov, A.M. 1994. Air supply with direction jets. Proceedings of the Fourth International Symposium on Ventilation for Contaminant Control Ventilator '94. Stockholm, Sweden.

Zhivov, A.M. and B.W. Olesen. 1993. Extending existing thermal comfort standards to work spaces. Proceedings of the 6th International Conference on Indoor Air Quality (IAQ'93), Helsinki.

Zhivov, A.M., J.B. Priest, and L.L. Christianson. 1996. Air distribution design for realistic rooms. 5th International Conference on Air Distribution in Rooms. RoomVent '96. Yokohama, Japan.

Zhivov, A.M., Eu.O. Shilkrot, P.V. Nielsen, and G.L. Riskowski. 1997. Displacement ventilation design. Proceedings of the 5th International Symposium on Ventilation for Contaminant Control (Ventilation '97). Vol. 1. Ottawa, Canada.

BIBLIOGRAPHY

Akinchev N. 1984. *General ventilation of hot shops*. Stroiizdat, Moscow (in Russian).

Alden, J.L. and J.M. Kane. 1982. *Design of industrial ventilation systems*. 5th ed. Industrial Press, New York.

Anderson, R. and M. Mehos. 1988. Evaluation of indoor air pollutant control techniques using scale experiments. ASHRAE Indoor Air Quality Conference.

Anufriev, L. et al. 1974. *Thermophysical calculations for agricultural industrial (production) buildings*. Stroiizdat, Moscow (in Russian).

Astelford, W. 1975. Engineering Control of Welding Fumes. DHEW (NIOSH) Pub. No. 75-115. NIOSH, Cincinnati, OH.

Balchin, N.C., ed. 1991. *Health and safety in welding and allied processes*. 4th ed. Abington Publishing, Cambridge.

Bartnecht, W. 1989. *Dust explosions: Course, prevention, protection*. Springer-Verlag, Berlin.

Baturin, V. and V. Elterman. 1963. *Aeration of industrial buildings*. Stroygiz, Moscow (in Russian).

BOCA. 1990. BOCA National building code. 11th ed. Building Officials and Code Administrators International, Country Club Hills, IL.

Boshnyakov, E. 1985. *Ventilation in the main productions of non-ferrous metallurgy*. Mettallurgiya, Moscow (in Russian).

Burgess, W.A., M.J. Ellenbecker, and R.D. Treitman. 1989. *Ventilation for control of the work environment*. John Wiley and Sons, New York.

Chamberlin, L.A. 1988. Use of controlled low velocity air patterns to improve operator environment at industrial work stations. Masters thesis, University of Massachusetts, September.

Cralley, L.V. and L.J. Cralley, eds. 1986. Industrial hygiene aspects of plant operations, Vol. 3 of *Patty's industrial hygiene and toxicology*. John Wiley and Sons, New York.

Crawford, M. 1976. *Air pollution control theory*. McGraw-Hill, New York.

Croome-Gale, D.J. and B.M. Roberts. 1981. *Air conditioning and ventilation of buildings*. 2nd ed. Pergamon Press, Oxford, U.K.

Curd, E.F. 1981. Possible applications of wall jets in controlling air contaminants. *Annals of Occupational Hygiene* 24(1):133-46.

Elterman V. 1985. *Pollution control at chemical and oil-chemical industry*. Chemia, Moscow (in Russian).

Eto, J.H. and C. Meyer. 1988. The HVAC costs of fresh air ventilation. *ASHRAE Journal* (9):31-35.

Flagan, R.C. and J.H. Seinfeld. 1988. *Fundamentals of air pollution engineering*. Prentice-Hall, Englewood Cliffs, NJ.

Godish, T. 1989. *Indoor air pollution control*. Lewis Publishers, Chelsea, MI.

Goodier, J.L., E. Boudreau, G. Coletta, and R. Lucas. 1975. Industrial health and safety criteria for abrasive blast cleaning operations. DHEW (NIOSH) Pub. No. 75-122. NIOSH, Cincinnati, OH.

Graedel, T.W. 1978. *Chemical compounds in the atmosphere*. Academic Press, New York.

Grimitlyn, M. 1983. *Ventilation and heating of plastics processing shops*. Chemia, Leningrad (in Russian).

Grimitlyn, M. 1993. *Ventilation and heating of shops of machinery building plants*. 2nd ed. Mashinostroyeniye, Moscow (in Russian).

Hagopian, J.H. and E.K. Bastress. 1976. Recommended industrial ventilation guidelines. DHEW (NIOSH) Pub. No. 76-162. NIOSH, Cincinnati, OH.

Hayashi, T., R.H. Howell, M. Shibata, and K. Tsuji. 1987. Industrial ventilation and air conditioning. CRC Press, Boca Raton, FL.

Heinsohn, R.J. 1991. *Industrial ventilation: Engineering principles*. John Wiley and Sons, New York.

Hughes, D. 1980. A literature survey and design study of fume cupboards and fume-dispersal systems. (Occup. Hyg. Manual 4). Science Review, Leeds, U.K.

Licht, W. 1988. *Air pollution control engineering*. 2nd ed. Marcel Dekker, Inc., New York.

McDermott, H.J. 1985. *Handbook of ventilation for contaminant control*. 2nd ed. Ann Arbor Science Publishers, Ann Arbor, MI.

Mehta, M.P., H.E. Ayer, B.E. Saltzman, and R. Ronk. 1988. Predicting concentration for indoor chemical spills. ASHRAE Indoor Air Quality Conference.

Murmann, H. 1980. *Lufttechnische Anlagen für gewerblichen Betriebe*, 2nd ed. (*Ventilation systems for industrial buildings*). C. Marhold, Berlin.

NFPA. 1973. National Fire Codes, Vol. 4 (NFPA 90-1973, NFPA 91-1973, and NFPA 96-1973). National Fire Protection Association, Quincy, MA.

NIOSH. 1973. The industrial environment—Its evaluation and control. National Institute for Occupational Safety and Health, Washington, D.C.

NIOSH. 1978. Engineering control technology assessment for the plastics and resins industry. DHEW (NIOSH) Pub. No. 78-159. NIOSH, Cincinnati, OH.

NIOSH. 1978. Proceedings of the Recirculation of Industrial Exhaust Air. DHEW (NIOSH) Pub. No.78-141. NIOSH, Cincinnati, OH.

NIOSH. 1989. Guidance for indoor air quality investigations. Cincinnati, OH.

O'Brien, D.M. and D.E. Hurley. 1981. An evaluation of engineering control technology for spray painting. DHEW (NIOSH) Pub. No. 81-121. NIOSH, Cincinnati, OH.

Partridge, L.J., Nayak, R.S. Stricoff, and J.H. Hagopian. 1978. A recommended approach to recirculation of exhaust air. DHEW (NIOSH) Pub. No. 78-124. NIOSH, Cincinnati, OH.

Perry, R. and P.W. Kirk, eds. 1988. Indoor and ambient air quality. Publication Division, Selper Ltd., London.

Pisarenko, V. and M. Roginsky. 1981. *Ventilation of working places in welding production*. Mashinostroyeniye, Moscow (in Russian).

RAPRA. 1979. *Ventilation handbook for the rubber and plastics industries*. Rubber and Plastics Research Association of Great Britain, Shrewsbury.

Rymkevich, A.A. 1990. *Optimization of general ventilation and air conditioning systems operation*. Stroiizdat, Moscow (in Russian).

Sandberg, M. 1983. Ventilation efficiency as a guide to design. *ASHRAE Transactions* 89(2B):455-79.

Sandberg, M. and C. Blomqvist. 1989. Displacement ventilation systems in office rooms. *ASHRAE Transactions* 95(2):1041-49.

Skåret, E. and H.M. Mathisen. 1989. Ventilation efficiency—A guide to efficient ventilation. *ASHRAE Transactions* 89(2B):480-95.

Speight, F. and H. Campbell, eds. 1979. Fumes and gases in the welding environment. American Welding Society, Miami, FL.

Stephanov, S.P. 1986. Investigation and optimization of air exchange in industrial halls ventilation. Proceedings of Ventilation '85. Elsevier Science Publishers, B.V., Amsterdam.

Teterevnikov, V.N., T.V. Kuksinskaya, and L.V. Pavlukhin. 1974. Industrial microclimate and air conditioning. TsNIIOT, Moscow (in Russian).

Theodore, L. and A. Buonicore, eds. 1982. *Air pollution control equipment: Selection, design, operation, and Maintenance.* Prentice-Hall, Englewood Cliffs, NJ.

Van Wagenen, H.D. 1979. Assessment of Selected Control Technology for Welding Fumes. DHEW (NIOSH) Pub. No. 79-125. NIOSH, Cincinnati, OH.

Vincent, J.H. 1989. *Aerosol sampling, science and practice.* John Wiley and Sons, Oxford, U.K.

Volkavein, J.C., M.R. Engle, and T.D. Raether. 1988. Dust control with clean air from an overhead air supply island (oasis). *Applied Industrial Hygiene* 3(August):8.

Wadden, R.A. and P.A. Scheff. 1982. *Indoor air pollution: Characterization, prediction, and control.* John Wiley and Sons, New York.

Wilson, D.J. 1982. A design procedure for estimating air intake contamination from nearby exhaust vents. *ASHRAE Transactions* 89(2A):136-52.

INDUSTRIAL LOCAL EXHAUST SYSTEMS

INDUSTRIAL exhaust ventilation systems collect and remove airborne contaminants consisting of particulates (dust, fumes, smokes, fibers), vapors, and gases that can create an unsafe, unhealthy, or undesirable atmosphere. Exhaust systems can also salvage usable material, improve plant housekeeping, or capture and remove excessive heat or moisture.

Local Exhaust Versus General Ventilation

Local exhaust ventilation systems are normally the most cost-effective method of controlling air pollutants and excessive heat. For many manual operations, capturing pollutants at or near their source is the only way to ensure compliance with threshold limit values (TLVs) in the worker's breathing zone. Especially where recirculation is not used, local exhaust ventilation optimizes ventilation airflow, thus optimizing system costs.

In some industrial ventilation designs, the main emphasis is on filtering the air captured by local exhausts prior to evacuating it to the outdoors or returning it to the production space (Chambers 1993). As a result, these systems are evaluated by the efficiency of their filters. However, if only a small percentage of the emission is captured, the degree of separation efficiency becomes almost irrelevant.

The pollutant capturing efficiency of local ventilation systems depends on the hood design, the hood's positioning near the source of contamination, and the exhaust airflow. The selection and layout of the hood has a significant influence on the initial and operating costs of both local and general ventilation systems. In addition, poorly designed and maintained local ventilation systems can cause deterioration of building structures and equipment, negative health effects, and lower working capacity.

No local exhaust ventilation system is 100% effective in capturing pollutants and/or surplus heat. In addition, the installation of local exhaust ventilation system may not be possible under some circumstances, due to the size or mobility of the process. In such situations, **general ventilation** is needed to dilute the pollutants and/or surplus heat. Air supplied by the general ventilation system is usually conditioned. Supply air replaces air extracted by the local and general exhaust systems and improves comfort conditions in the occupied zone.

Chapter 12, Air Contaminants, of the 1997 *ASHRAE Handbook—Fundamentals* covers definitions, particle sizes, allowable concentrations, and upper and lower explosive limits of various air contaminants. Chapter 28, Ventilation of the Industrial Environment, of this volume and Chapter 1 of *Industrial Ventilation: A Manual of Recommended Practice* (ACGIH 1998) detail steps to determine the air volumes necessary to dilute the contaminant concentration using general ventilation. Refer to Chapter 28 for further information on replacement and makeup air.

If insufficient replacement air is provided, the pressure of the building will be negative relative to local atmospheric pressure. Negative pressure allows air to infiltrate through open doors, window

cracks, and combustion equipment vents. As little as 0.05 in. of water of negative pressure can cause drafts and might cause backdrafts in combustion vents, thereby creating a potential health hazard. Negative plant pressure can also cause excessive energy use. If workers near the plant perimeter complain about cold drafts, unit heaters are often installed. Heat from these units is usually drawn into the plant interior because of the velocity of the infiltration air, leading to overheating. Too often, the solution is to exhaust more air from the interior, causing increased negative pressure and more infiltration. Negative plant pressure reduces the exhaust volumetric flow rate because of increased system resistance, which could decrease local exhaust efficiency. Wind effects on building balance are discussed in Chapter 15, Airflow Around Buildings, of the 1997 *ASHRAE Handbook—Fundamentals*.

Positive-pressure plants and balanced plants (those having equal exhaust and replacement air rates) use less energy. However, if there are clean and contaminated zones in the same building, surplus pressure in the contaminated zones could cause contaminants to move into the clean zones.

LOCAL EXHAUST FUNDAMENTALS

System Components

Local exhaust ventilation systems typically consist of the following basic elements:

- Hood or entry point of the system to capture pollutants and/or excessive heat
- Duct system to transport polluted air
- Air cleaning device to remove captured pollutants from the airstream for recycling or disposal
- Air-moving device (e.g., fan or high pressure air ejector), which provides the motive power to overcome system resistance
- Exhaust stack, which discharges system air to the atmosphere
- Return duct system to return cleaned air back to the plant

System Classification

By Contaminant Source Type. Knowledge of the process or operation is essential before a hood can be designed. The type and size of the hood depends on the type and geometry of the pollution source. There are three types of pollution sources, each of which creates different contaminant movement (Posokhin 1984). **Buoyant (heat) sources** cause contaminants to move in buoyant plumes over the heated surfaces. **Nonbuoyant (diffusion) sources** create contaminant diffusion in all directions due to the concentration gradient (e.g., in the case of emission from painted surface). The emission rate is significantly affected by the intensity of the ambient air turbulence and air velocity. **Dynamic sources** create contaminant movement with an air jet (e.g., linear jet over a tank with push-pull ventilation) or due to particle flow (e.g., from a grinding wheel). In some cases, the above factors influencing contaminant distribution in the room are combined.

The preparation of this chapter is assigned to TC 5.8, Industrial Exhaust.

The geometry of the contaminant source can be **compact** or **linear**. Hoods are round, rectangular, or slotted to accommodate the geometry of the source.

By Hood Type. Hoods are either enclosing or nonenclosing (Figure 1). **Enclosing hoods** provide better and more economical contaminant control because their exhaust rates and the effects of room air currents are minimal compared to those for nonenclosing hoods. Hood access openings for inspection and maintenance should be as small as possible and out of the natural path of the contaminant. Hood performance (i.e., how well it controls the contaminant) should be checked by an industrial hygienist.

A **nonenclosing hood** can be used if access requirements make it necessary to leave all or part of the process open. Careful attention must be paid to airflow patterns around the process and hood and to the characteristics of the process in order to make nonenclosing hoods functional. Nonenclosing hoods can be classified according to their location relative to the contaminant source (Posokhin 1984) as either **updraft coaxial**, **sidedraft (lateral)**, or **downdraft** (Figure 2).

By System Mobility. Local exhaust systems with nonenclosing hoods can be **stationary** (i.e., having a fixed hood position), **moveable**, **portable**, or **built-in** (into the process equipment). Moveable (turnable) hoods are used when the process equipment must be accessed for repair, loading, and unloading (e.g., in electric ovens for melting steel). Hoods attached to flexible extraction arms (Figure 3) are used when the source of contamination is not fixed, as in arc welding (Zhivov 1993; Zhivov and Ashe 1997). Flexible extraction arms usually have a hood connected to a duct 5.5 to 6.3 in. in

diameter and have higher efficiencies for lower airflow rates compared to stationary hoods. When the source of contamination is confined to a small, poorly ventilated space such as a tank, an additional flexible hose extension with a hood and a magnetic foot can be hooked on the fume extraction arm.

The portable extractor shown in Figure 4 is commonly used for the temporary extraction of fumes and solvents in confined spaces or during maintenance. It has a built-in fan and filter and a linear or round nozzle attached to a flexible hose about 1.8 in. in diameter. Built-in local exhausts, such as gun-mounted exhaust hoods and fume extractors built into stationary or turnover welding tables, are

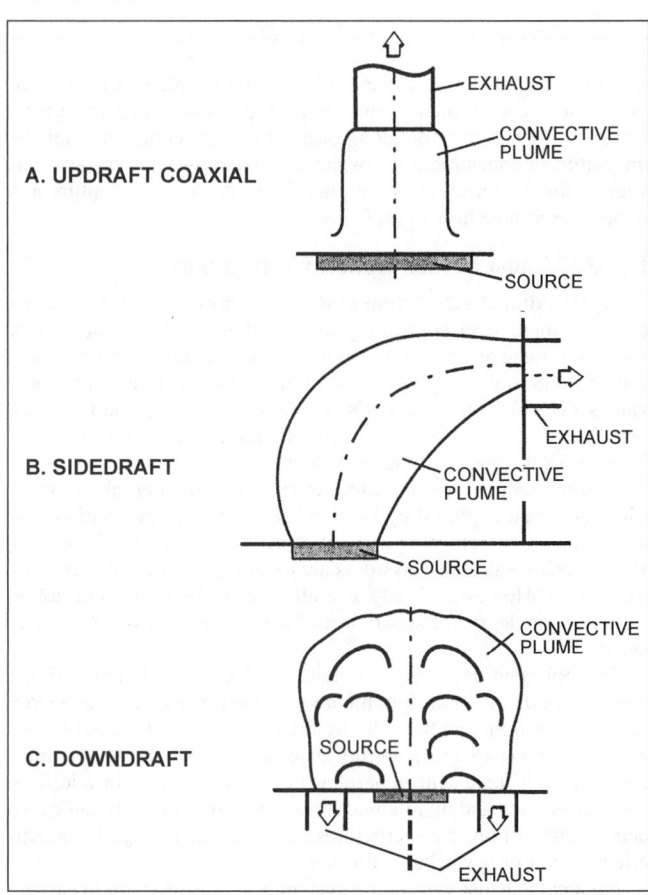

Fig. 2   Nonenclosing Hoods

Fig. 1   Enclosing and Nonenclosing Hoods

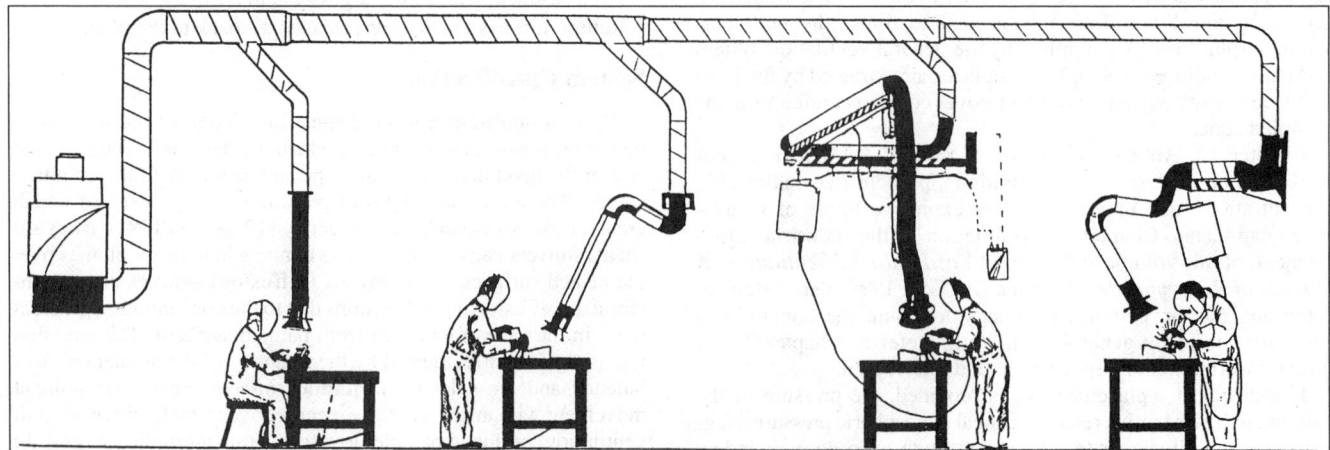

Fig. 3   Hoods Attached to Flexible Fume Extraction Arms

commonly used to evacuate welding fumes. Lateral exhaust hoods, which exhaust air through slots on the periphery of open vessels such as those used for galvanizing metals, are another example of built-in local exhausts.

Nonenclosing hoods should be located so that the contaminant is drawn away from the operator's breathing zone. **Canopy hoods** should not be used where the operator must bend over a tank or process (ACGIH 1998).

Effectiveness of Local Exhaust

The most effective hood uses the minimum exhaust airflow rate to provide maximum contaminant control. The **capturing effectiveness** should be high, but it would be difficult and costly to develop a hood that is 100% efficient. Makeup air supplied by general ventilation to replace exhausted air can dilute contaminants that are not captured by the hood (Posokhin 1984).

Capture Velocity. Capture velocity is the air velocity at the point of contaminant generation upstream of a hood. The contaminant enters the moving airstream at the point of generation and is conducted along with the air into the hood. Designers use capture velocity to select a volumetric flow rate Q to withdraw air through a hood. Table 1 shows ranges of capture velocities for several industrial operations. These figures are based on successful experience under ideal conditions. If velocities anywhere upstream of a hood are known $[V = f(Q_{x, y, z})]$, the capture velocity is set equal to V_c at point (x, y, z) where contaminants are to be captured, and Q is found. The transport equations between the source and the hood must be solved to ensure that contaminants enter an inlet.

Hood Volumetric Flow Rate. After the hood configuration and capture velocity are determined, the exhaust volumetric flow rate can be calculated. For **enclosing hoods**, the **target exhaust volumetric flow rate** (the airflow rate that allows contaminant capture) is

$$Q_o^* = V_o A_o \qquad (1)$$

where

- Q_o^* = target exhaust volumetric flow rate, cfm
- V_o = average air velocity in hood opening that ensures capture velocity at the point of contaminant release, fpm
- A_o = hood opening area, ft^2

The inflow velocity V_o is typically 100 fpm. However, research with laboratory hoods indicates that lower velocities can reduce the vortex downstream of the human body, thus lessening the reentrainment of contaminant into the operator's breathing zone (Caplan and Knutson 1977; Fuller and Etchells 1979). These lower face velocities require that the replacement air supply be distributed to mini-

mize the effects of room air currents. This is one reason replacement air systems must be designed with exhaust systems in mind. Because air must enter the hood uniformly, interior baffles are sometimes necessary (Figure 5).

For **nonenclosing hoods**, the target airflow rate is proportional to some characteristic flow rate Q_o that depends on the type of contaminant source (Posokhin and Zhivov 1997):

$$Q_o^* = KQ_o \qquad (2)$$

where

- K = dimensionless coefficient depending on hood design
- Q_o = characteristic airflow rate depending on contaminant source, ft^3/min (for example, for a buoyant source, Q_o can be equal to airflow in the convective plume; for a dynamic source, Q_o can be equal to airflow rate in the jet)

For a nonenclosing hood with a nonbuoyant contaminant source, the characteristic airflow can be calculated using the following equation:

$$Q_o = V_o A_o \qquad (3)$$

Table 1   Range of Capture (Control) Velocities

Condition of Contaminant Dispersion	Examples	Capture Velocity, fpm
Released with essentially no velocity into still air	Evaporation from tanks, degreasing, plating	50 to 100
Released at low velocity into moderately still air	Container filling, low-speed conveyor transfers, welding	100 to 200
Active generation into zone of rapid air motion	Barrel filling, chute loading of conveyors, crushing, cool shakeout	200 to 500
Released at high velocity into zone of very rapid air motion	Grinding, abrasive blasting, tumbling, hot shakeout	500 to 2000

Note: In each category above, a range of capture velocities is shown. The proper choice of values depends on several factors (Alden and Kane 1982):

Lower End of Range	Upper End of Range
1. Room air currents favorable to capture	1. Distributing room air currents
2. Contaminants of low toxicity or of nuisance value only	2. Contaminants of high toxicity
3. Intermittent, low production	3. High production, heavy use
4. Large hood; large air mass in motion	4. Small hood; local control only

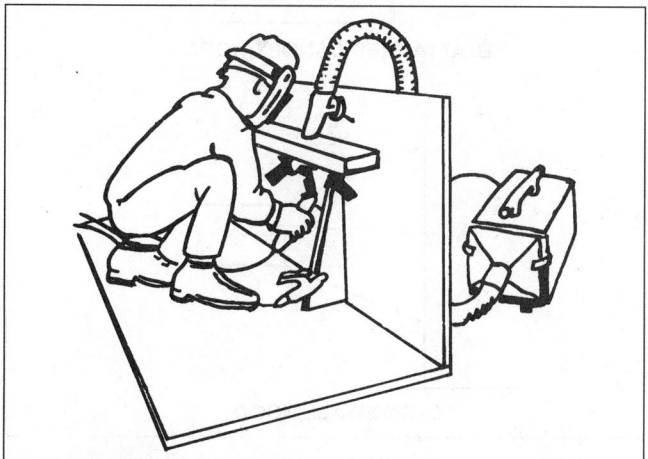

Fig. 4   Portable Fume Extractor with Built-in Fan and Filter

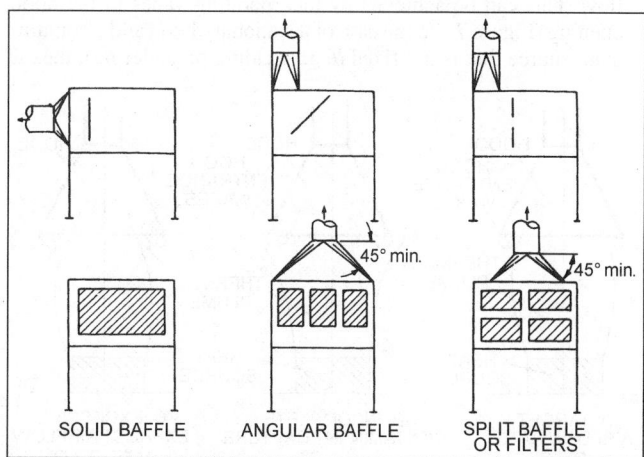

Fig. 5   Use of Interior Baffles to Ensure Good Air Distribution

An exhaust airflow rate lower than Q_o^* results in reduced contaminant capturing effectiveness. An exhaust airflow rate greater than Q_o^* results in excessive capturing effectiveness (Figure 6).

Airflow near the hood can be influenced by drafts from the supply air jets (spot cooling jets) or by turbulence of the ambient air caused by the jets, upward/downward convective flows, moving people, and drafts from doors and windows. Process equipment may be another source of air movement. For example, high-speed rotating machines such as pulverizers, high-speed belt material transfer systems, falling granular materials, and escaping compressed air from pneumatic tools all produce air currents.

These factors can significantly reduce the capturing effectiveness of local exhausts and should be accounted for in Equations (2) and (3) by the correction coefficient on room air movement. For example, Equation (2) will be replaced with the following:

$$Q_o^* = K_r K Q_o \qquad (4)$$

where K_r = coefficient on room air movement; $K_r > 1$.

The exhausted air may contain combustible pollutant-air mixtures. In this case, the exhaust airflow rate should be increased to dilute combustible mixture to less than 25% of the lower explosive limit of the pollutant (NFPA *Standard* 86). Thus,

$$Q_o > \frac{G}{0.25 C_{exp(min)}} \qquad (5)$$

where

G = amount of pollutant released by the source, lb/min
$C_{exp(min)}$ = lower explosive limit of pollutant, lb/ft^3

Principles of Hood Design Optimization

Numerous studies of local exhaust and common practices have led to the development of the following list of hood design principles (Posokhin 1984):

- The hood should be located as close as possible to the source of contamination.
- The hood opening should be positioned so that it causes the contaminant to deviate the least from its natural path.
- The hood should be located so that the contaminant is drawn away from the operator's breathing zone.
- The hood must be the same size as or larger than the cross section of the flow entering the hood. If the hood is smaller than the flow, a higher volumetric flow rate will be required.
- The velocity distribution in the hood opening cross section should be nonuniform, following the velocity profile of the incoming flow. This can be achieved by incorporating vanes in the hood opening (Figure 7). In the case of a stationary hood and a contaminant source that is not fixed (e.g., welding or soldering), the air

velocity along the hood must be uniform; this can be achieved using vanes or perforations.

AIR MOVEMENT IN VICINITY OF LOCAL EXHAUST

Theoretical Considerations

Airflow near the hood can be described using the incompressible, irrotational flow (i.e., potential flow) model. The total pressure p_{tot} in the area upstream of the hood remains constant and can be described with the following equation:

$$p_{tot} = p_{st} + p_d = \text{Constant} \qquad (6)$$

where

p_{st} = static pressure at any point of the flow, in. of water
p_d = $K\rho V^2/2g_c$ = dynamic pressure at any point of the flow, in. of water
K = units conversion factor = 0.0077
ρ = air density, lb/ft^3
V = air velocity, fpm
g_c = gravitational constant, 32.2 lb$_m$·ft/lb$_f$·s^2

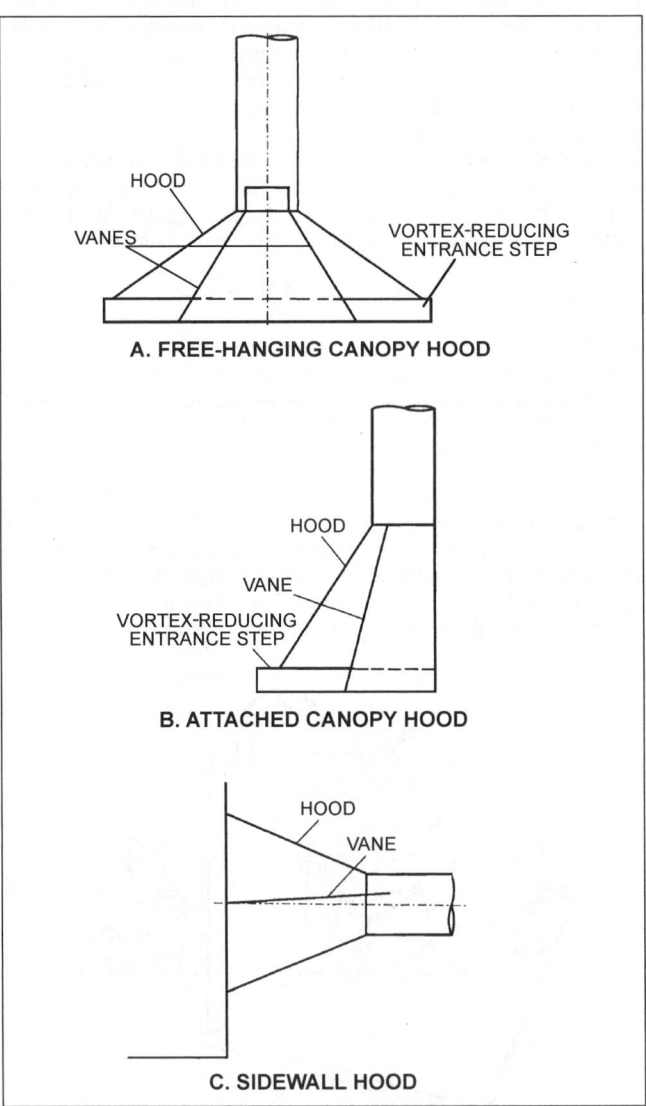

A. FREE-HANGING CANOPY HOOD

B. ATTACHED CANOPY HOOD

C. SIDEWALL HOOD

Fig. 7   Hoods with Nonuniform Velocities in Opening Cross Sections

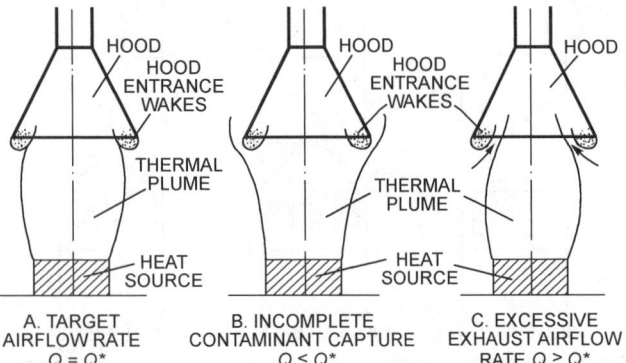

A. TARGET AIRFLOW RATE $Q = Q^*$

B. INCOMPLETE CONTAMINANT CAPTURE $Q < Q^*$

C. EXCESSIVE EXHAUST AIRFLOW RATE $Q > Q^*$

Fig. 6   Hood Performance at Different Exhaust Airflow Rates

At some distance from the hood, the total pressure p_{tot} in the airflow is equal to the ambient air pressure (i.e., $p_{tot} = 0$). Thus,

$$p_d = -p_{st} \tag{7}$$

The above discussion does not apply to wakes with vortex air movement (Figure 8).

Numerical simulation of hood performance is complex, and results depend on hood design, flow restriction by surrounding surfaces, source strength, and other boundary conditions. Thus, most currently used methods of hood design are based on analytical models and experimental studies.

According to these models, the exhaust airflow rate is calculated based on a desired capture velocity at a particular location in front of the hood. It is easier to understand the design process for a sink with vanishingly small dimensions—a point or a linear source of suction. The point source can approximate airflow near a round or square/rectangular hood, and the linear source approximates the airflow near a slot hood.

A point source will draw air equally from all directions. Given the exhaust airflow, the velocity at any distance can be calculated by the following equation:

$$V_x = Q/4\pi x^2 \tag{8}$$

where

> V_x = air velocity at distance x, fpm
> x = distance from hood, ft

A linear source will create a two-dimensional flow with the velocity V_x calculated as follows:

$$V_x = Q/2\pi x \tag{9}$$

Centerline velocities for different realistic hoods are presented in Table 2 (Posokhin and Zhivov 1997). Figure 9 compares relative velocity change for realistic hoods and a point source. At a distance greater than $x/R = 1$, velocities induced by realistic hoods are practically equal to those induced by a point source. This means that in

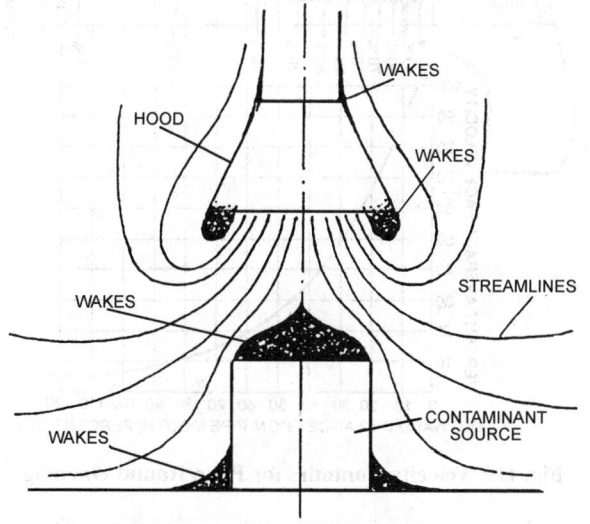

Fig. 8 Airflow in the Hood Vicinity

Table 2 Centerline Air Velocities Induced by Nonenclosing Hoods

Hoot Type	Schematic	Equation	Applicable Range	Reference
Round freestanding hood, unflanged		$\dfrac{V_x}{V_o} = (1 + 10x^2/A)^{-1}$	$x \le 1.7\sqrt{A}$; $\alpha \le 30°$	DallaValle (1952)
Round freestanding hood, flanged		$\dfrac{V_x}{V_o} = 1.1(0.07)^{x/D}$ $\dfrac{V_x}{V_o} = 0.1(x/D)^{-1.6}$	$0 \le \dfrac{x}{D} \le 0.5$; $C \ge D$ $0.5 \le \dfrac{x}{D} \le 1.5$; $C \ge D$	Garrison (1977)
Rectangular freestanding hood, unflanged		$\dfrac{V_x}{V_o} = (0.93 + 8.58\alpha_F^2)^{-1}$ $\alpha_F = (x/\sqrt{A})(a/b)^{\beta_F}$ $\beta_F = 0.2(x/\sqrt{A})^{-1.3}$	$1 \le \dfrac{a}{b} \le 16$; $0.05 \le \dfrac{x}{\sqrt{A}} \le 3$; $\alpha \le 30°$	Fletcher (1977)
Rectangular freestanding hood, flanged		$\left[\dfrac{V_x}{V_o} = 1 - \dfrac{2}{\pi}\text{atan}\left(\dfrac{2x\sqrt{x^2 + a^2 + b^2}}{ab}\right)\right]$	$1 \le \dfrac{a}{b} \le 16$; $0 \le \dfrac{x}{\sqrt{A}} \le 1.6$; $\dfrac{C}{\sqrt{A}} \ge 1$	Tyaglo and Shepelev (1970)
Slot in the pipe wall		$\dfrac{V_{cx}}{V_o} = \dfrac{V_{AD}}{V_o} = \dfrac{2R}{\pi x}\text{atan}\left[\dfrac{x + R}{x - R}\tan\left(\dfrac{\alpha}{2}\right)\right]$	$|x| \ge R$	Posokhin (1984)

C = flange width.

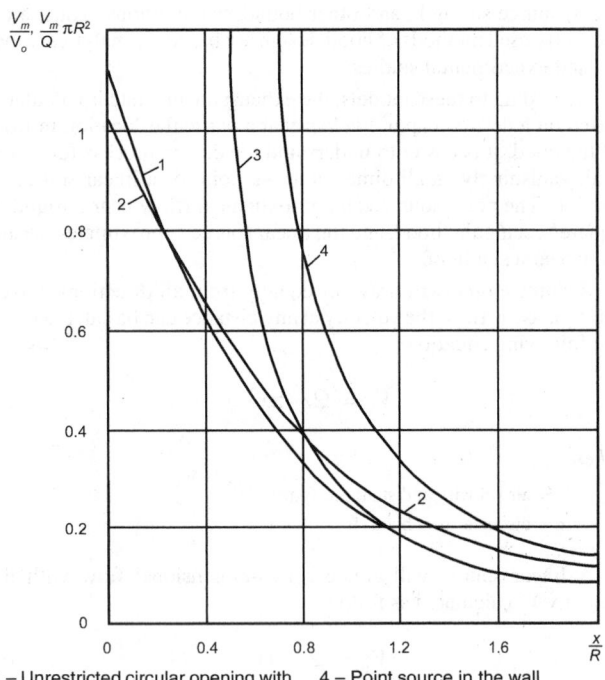

1 – Unrestricted circular opening with radius R
2 – Circular opening in the wall with radius R
3 – Unrestricted point source
4 – Point source in the wall
Q – Exhaust airflow rate
x – Distance from the source of suction to the point with air velocity V_m

Fig. 9   Relative Velocity Decay in Vicinity of a Point Source of Suction and of a Realistic Hood

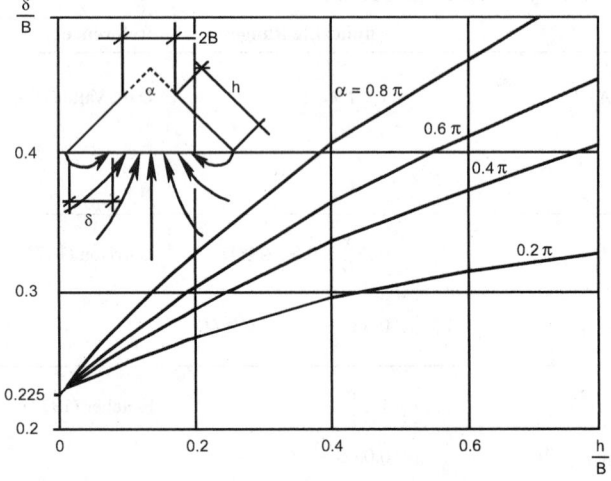

Fig. 10   Influence of Hood Configuration on Hood Entrance Wake Size δ

some cases, airflow in front of realistic hoods can be described using the simplified point source model.

Typically, velocity distribution in the hood face area is not uniform. Wakes formed close to the hood sides, or vena contracta, reduce the effective suction area of the hood. The size of these wakes and the level of velocity uniformity depend on hood design. Figure 10 shows the approximate relationship between wake size δ and the cone angle α of the hood (Posokhin 1984). Vanes, baffles, perforations, and other inserts can be used to control the size of the vena contracta and the velocity uniformity at the hood face area.

Air velocities in front of the hood suction opening depend on the exhaust airflow rate, the geometry of the hood, and the surfaces

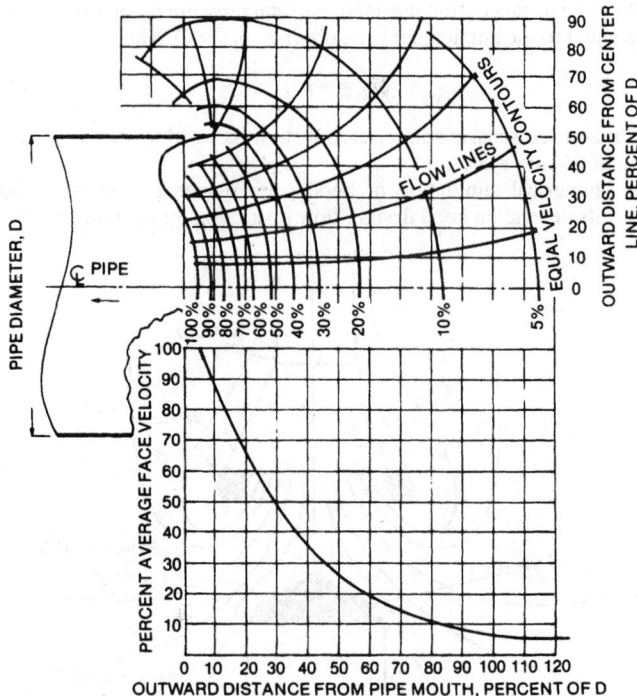

Fig. 11   Velocity Contours for Plain Round Opening

comprising the suction zone. Figure 11 shows lines of equal velocity (**velocity contours**) for a plain round opening. Studies have established the principle of similarity of velocity contours (expressed as a percentage of the hood face velocity) for zones with similar geometry (DallaValle 1952). Figure 12 (Alden and Kane 1972) shows velocity contours for a rectangular hood with an **aspect ratio** (width divided by length) of 0.333. The profiles are similar to those for the round hood but are more elongated. If the aspect ratio is lower than about 0.2 (0.15 for flanged openings), the shape of the flow pattern in front of the hood changes from approximately spherical to approximately cylindrical.

In the suction zone, the velocity decreases rapidly with distance from the hood. Velocity contours plotted in Figures 11 and 12 show that the velocity reaches 10% of the hood face velocity within the distance equal to the square root of the hood suction opening.

Air and Contaminant Distribution with Nonbuoyant Sources

Theories of hood performance with nonbuoyant pollution sources are based on the turbulent diffusion equation. The following equation allows the engineer to determine contaminant concentration decay in the uniform airflow upstream from the contaminant source:

$$C_x = C_o e^{-\frac{V}{D}x} \qquad (10)$$

where

x = distance from source, ft
C_o = contaminant concentration at source, lb/ft³
C_x = contaminant concentration at distance x from source, lb/ft³
V = air velocity in flow, fpm
D = coefficient of turbulent diffusion, ft²/min
e = 2.7182818 = base for natural logarithm

The value of D depends on the air change rate in the ventilated space and the method of air supply. Studies by Posokhin (1984) show that D for locations outside supply air jets is approximately

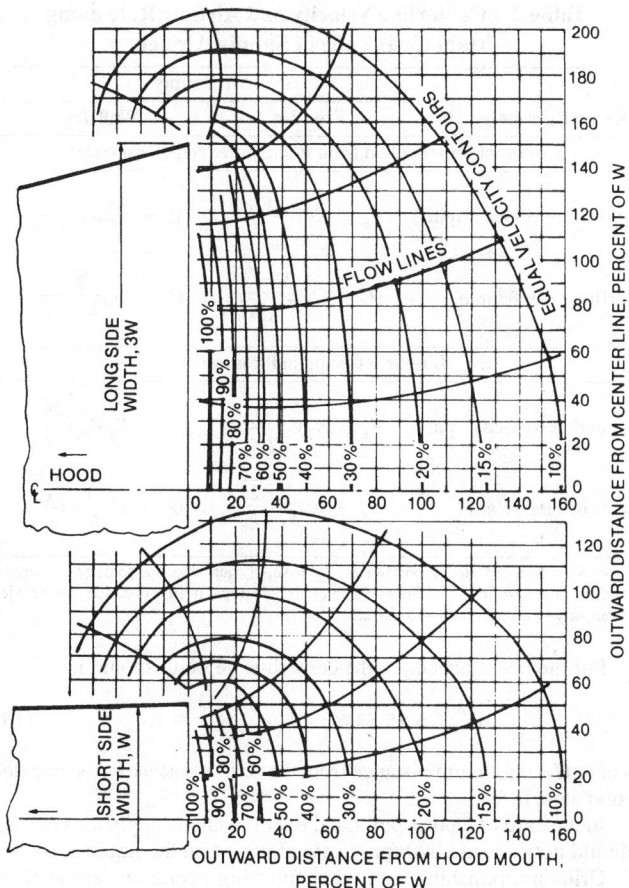

Fig. 12   Velocity Contours for Plain Rectangular Opening with Sides in a 1:3 Ratio

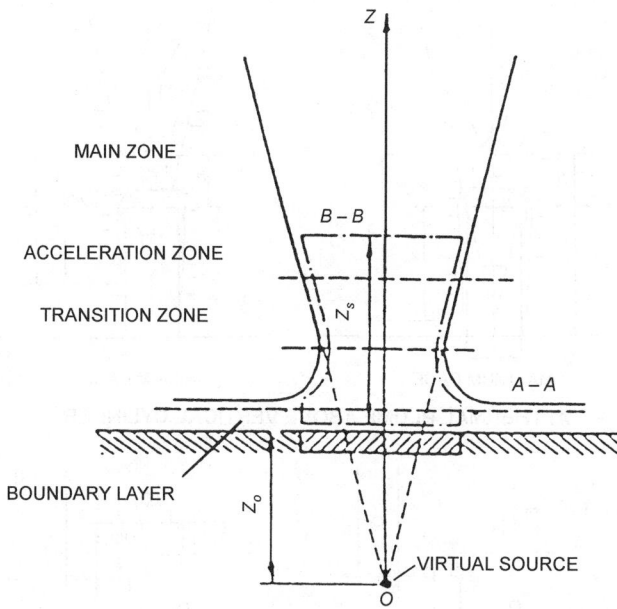

Z_o – distance from source surface to virtual source
Z_s – distance from source surface to thermal plume cross-section of interest

Fig. 13   Thermal Plume above Heat Source

16 ft²/min. Air disturbance caused by the operator or robot results in an increase of D by at least a factor of 2. Studies by Zhivov et al. (1997) showed that the value of D is affected by the velocity of cross drafts, their direction against the hood face, and the presence of an operator. For example, with a cross draft directed along the hood face with velocity $V = 100$ fpm, $D = 97$ ft²/min (with the presence of operator); an increase in cross draft velocity to $V = 200$ fpm results in $D = 194$ ft²/min.

Air and Contaminant Distribution with Buoyant Sources

The exhaust from hot processes requires special consideration because of the buoyant effect of heated air near the hot process. Determining the hood size and exhaust rate for a hot process requires an understanding of the convectional heat transfer rate (see the Chapter 3 of the 1997 *ASHRAE Handbook—Fundamentals*) and the physical size of the process. Convected heat and pollutants from the hot process are presumed to be contained in the thermal plume above the source, so the capture of the air transported with this plume will ensure the efficient capture of the contaminant (Burgess et al. 1989).

Analytical equations to calculate velocities, temperatures, airflow rates, and other parameters in thermal plumes over spot and linear heat sources with given heat loads were derived by Zeldovitch (1937), Schmidt (1941), Morton et al. (1956), and Shepelev (1961) based on the momentum and energy conservation equations and assuming Gaussian velocity and temperature difference (between plume and room air temperatures) distribution in thermal plume cross sections. These equations correspond to those received experimentally by other researchers (Popiolec 1981; Skäret 1986);

for example, the equation for the airflow rate in the thermal plume is as follows:

$$Q = C q_{conv}^{1/3} z^{5/3} \qquad (11)$$

where

$\quad Q$ = airflow rate, cfm
q_{conv} = convective component of the heat source, Btu/h
$\quad z$ = height above the source level, ft
$\quad C$ = experimental coefficient

Equation (11) was derived with the assumption that the heat source is very small; it does not account for the actual source dimensions.

Adjusting the point source model to realistic sources using the **virtual source method** (Figure 13) gives a reasonable estimate of the airflow rate in thermal plumes (Ivanitskaya et al. 1974; Elterman 1980; Holman 1989; Mundt 1992). The weak part of this method according to Skistad (1994) is estimating the location of the virtual point. The method of a "maximum case" and a "minimum case" (Skistad 1994) provides a tool for such estimation (Figure 14). According to the maximum case, the real source is replaced by the point source such that the border of the plume above the point source passes through the top edge of the real source (e.g., cylinder). The minimum case is when the diameter of the vena contracta of the plume is about 80% of the upper surface diameter and is located approximately 1/3 diameter above the source. For low-temperature sources, Skistad (1994) recommends the maximum case, whereas the minimum case best fits the measurements for larger, high-temperature sources.

Kofoed (1991) and Kofoed and Nielsen (1991) studied the interaction of the thermal plume with a wall and with another plume. In the case of the wall plume, the airflow rate should be decreased by a factor of 0.63; for interaction with another equal plume, it should be increased by a factor of 1.26.

Another approach to evaluating the thermal plume parameters (Nielsen 1993; Schaelin and Kofoed 1992; Davidson 1989; Aksenov and Gudzovskii 1994) is based on **computational fluid**

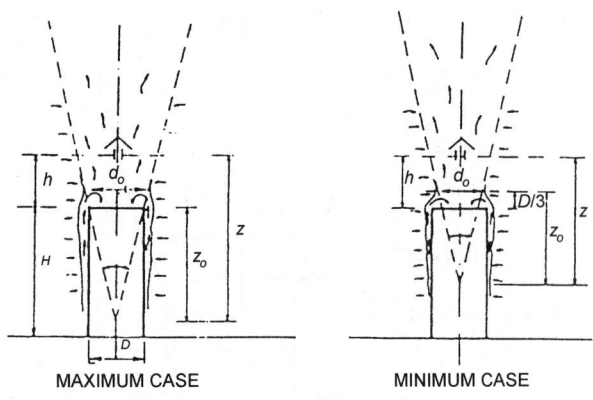

A. THERMAL PLUME ABOVE VERTICAL CYLINDER

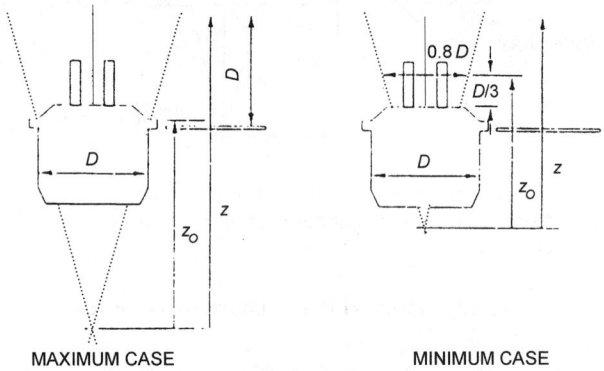

B. THERMAL PLUME ABOVE FURNACE

Fig. 14 "Minimum Case" and "Maximum Case" Approaches to Locating Virtual Source
[Reproduced from Skistad (1994)]

dynamics (CFD). With this approach, the airflow in the thermal plume is described by the Navier-Stokes equations and equations for energy and mass balances. Data from numerical and physical experiments on thermal plumes can be used to size overhead hoods above unusual sources or those with complex shapes.

Air Movement Created by Dynamic Sources

Push-pull systems (see the section on Jet-Assisted Hoods) supply air jets in the contaminated zone; they inject contaminated air and direct it toward the hood. Air jets in push-pull systems can be compact or linear. Table 3 presents velocity decay and airflow rates in free and attached compact and linear jets along the zone of practical interest.

To reduce the effect of room air movement on hood performance, the push air jet centerline velocity at the critical cross section (where the push air jet becomes weak, and the influence of the hood is not strong enough) should be from 200 to 400 fpm. According to Stroiizdat (1992), the centerline velocity V_x^{min} in the push jet attached to the heat source should be

$$V_x^{min} > 631 \sqrt{\frac{l(T_s - T_r)}{T_r}} \qquad (12)$$

where

l = distance between the supply nozzle and the exhaust hood, ft
T_s = temperature of the heat source surface, °R
T_r = room air temperature, °R

Table 3 Centerline Velocity and Airflow Rate along Main (3rd) Zone of Supply Air Jet

Design Parameter	Air Supply	
	Free Jet	Wall Jet
Compact jet (supplied from round or rectangular nozzle)		
Centerline velocity, fpm	$V_x = K_1 V_o \dfrac{\sqrt{A_o}}{x}$	$V_x = \sqrt{2} K_1 V_o \dfrac{\sqrt{A_o}}{x}$
Airflow rate, ft³/min	$Q_x = Q_o \dfrac{2}{K_1} \dfrac{x}{\sqrt{A_o}}$	$Q_x = Q_o \dfrac{\sqrt{2}}{K_1} \dfrac{x}{\sqrt{A_o}}$
Linear jet (supplied from slot)		
Centerline velocity, fpm	$V_x = K_1 V_o \sqrt{\dfrac{b_o}{x}}$	$V_x = K_1 V_o \sqrt{\dfrac{2b_o}{x}}$
Airflow rate, ft³/min	$Q_x = Q_o \dfrac{1}{K_1} \sqrt{\dfrac{2x}{b_o}}$	$Q_x = Q_o \dfrac{1}{K_1} \sqrt{\dfrac{x}{b_o}}$

Note: K_1 = velocity decay coefficient; V_o = supply air velocity, fpm; Q_o = supply airflow rate, ft³/min; A_o = effective area of air discharge, ft²; x = distance along air jet from supply nozzle, ft; b = slot width, ft.

For the nonattached jet, the centerline velocity should be

$$V_x^{min} > 45.53 (T_s - T_r)^{4/9} H^{1/3} \qquad (13)$$

where H = maximum distance from the source surface to the nozzle-hood axis, ft.

In the case of a push-pull hood over the tank, supply air velocity should not exceed 2000 fpm to avoid waves on the liquid surface.

Grinding, polishing, and other finishing operations are another type of dynamic contaminant source. These processes produce particles and impart them with some momentum. The receiving hood used for grinding operations is positioned and sized to catch the particles, which are thrown toward the hood (Burgess et al. 1989). The airflow required for the receiving hood can be calculated based on the capture velocity. It might seem that the distance x used in Equations (8) and (9) can be reduced to $x^* = x - S$, where S is the particle stopping distance (i.e., the distance a particle ejected into still air at the initial velocity will travel while decelerating to rest due to drag forces). However, the data from Hinds (1982) show the difficulty in throwing even fairly large particles an appreciable distance in still air. Any particles in the inhalable range (i.e., $d < 10\ \mu m$) should be considered immovable, so it should be always assumed that $x^* = x$ (Burgess et al. 1989). The hood used for grinding processes acts as a receiver for large particles ($d > 30\ \mu m$), but not for small and intermediate particles ($3 < d < 30\ \mu m$). Bastress et al. (1974) found that respirable particles were not captured efficiently by typical grinding wheel exhaust systems and escaped to the vicinity of the worker's breathing zone. Nonetheless, the standard hood designs recommended in ACGIH (1998), while not 100% efficient at capturing respirable particles, were sufficient to provide worker protection at or below the TLVs for total and respirable inert dust.

Pressure Losses in Local Exhausts

When air enters a hood, dynamic losses cause a loss of total pressure. This is called the **hood entry loss** and may have several components, each given by

$$\Delta p_e = C_o p_v \qquad (14)$$

where

Δp_e = hood entry loss, in. of water
C_o = loss factor depending on component geometry, dimensionless
p_v = appropriate velocity pressure, in. of water

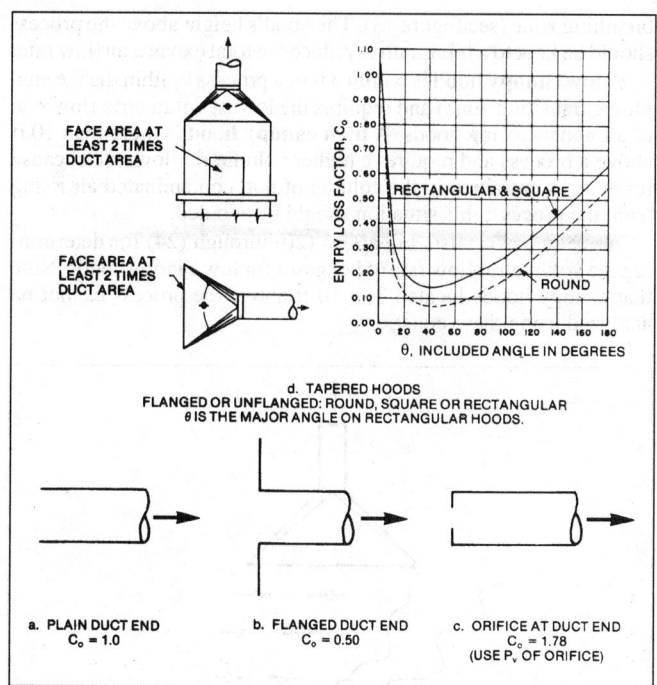

Fig. 15   Entry Losses for Typical Hoods

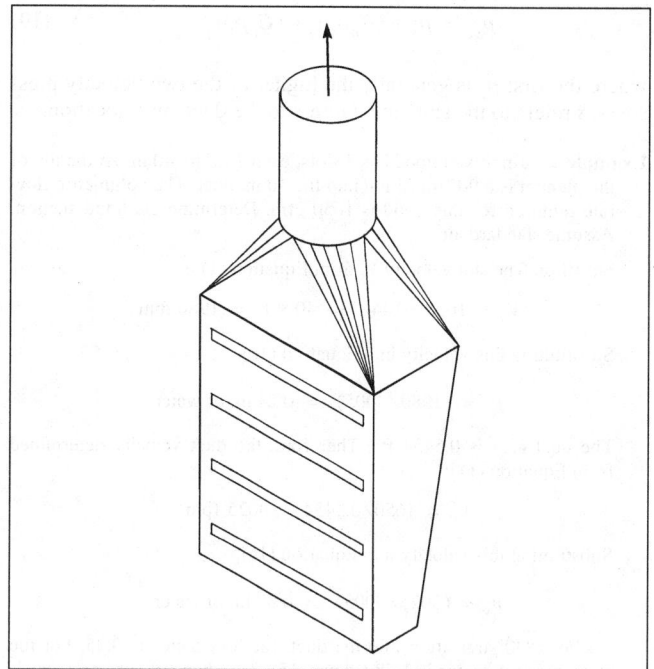

Fig. 16   Multislot Nonenclosing Hood

The following equation relates velocity to velocity pressure:

$$V = \sqrt{\frac{2p_v}{\rho}} \tag{15}$$

If the air temperature is 68°F ± 27°F, the ambient pressure is the standard 14.7 psia, the duct pressure is no more than 20 in. of water different from the ambient pressure, the dust loading is low (< 0.1 grains/ft^3), and moisture is not a consideration, then the density in Equation (15) is 0.075 lb/ft^3, and Equation (13) simplifies to

$$V = 4005\sqrt{p_v} \tag{16}$$

Loss factors C_o for various hood shapes are given in Figure 15. More information on loss factors can be found in Chapter 32, Duct Design, of the 1997 *ASHRAE Handbook—Fundamentals*, Idelchik et al. (1986), and ACGIH (1998). Figure 15 shows an optimum hood entry angle to minimize entry loss. However, this total included angle of 45° is impractical in many situations because of the required transition length. A 90° angle, with a corresponding loss factor of 0.25 (for rectangular openings), is standard for most tapered hoods.

The combination of several consecutive hood components may affect the values of their individual loss factors. Thus, a hood with multiple components should be treated as a single component with a pressure loss obtained in a laboratory or field test.

Total pressure is difficult to measure in a duct system because it varies from point to point across a duct, depending on the local velocity. On the other hand, static pressure remains constant across a straight duct. Therefore, measurement of static pressure in a straight duct at a single point downstream of the hood can monitor the volumetric flow rate. The absolute value of this static pressure, the **hood suction**, is given by

$$p_{st} = p_v + \Delta p_e \tag{17}$$

where p_{st} = hood suction, in. of water.

Hood suction is the negative static pressure measured about three duct diameters downstream of the hood. A larger distance is required for included angles of 180° or larger.

Simple Hoods. A simple hood has only one dynamic loss. The hood suction becomes

$$p_{st} = (1 + C_o)p_v \tag{18}$$

where p_v = duct velocity pressure, in. of water.

Example 1. A nonenclosing sidedraft flanged hood (Figure 20) with face dimensions of 1.5 ft by 4 ft rests on the bench. The required volumetric flow rate is 1560 cfm. The duct diameter is 9 in.; this gives a duct velocity of 3530 fpm. The hood is designed such that the largest angle of transition between the hood face and the duct is 90°. What is the suction for this hood? Assume standard air density.

Solution: The two transition angles cannot be equal. Whenever this is true, the larger angle is used to determine the loss factor from Figure 15. Because the transition piece originates from a rectangular opening, the curve marked "rectangular" must be used. This corresponds to a loss factor of 0.25. Equation (16), which assumes standard air density, can be used to determine the duct velocity pressure:

$$p_v = (3530/4005)^2 = 0.78 \text{ in. of water}$$

From Equation (18),

$$p_{st} = (1 + 0.25)(0.78) = 0.98 \text{ in. of water}$$

Compound Hoods. The losses for multislot hoods (see Figure 16) or single-slot hoods with a plenum (called compound hoods) must be analyzed somewhat differently. The slots distribute air over the hood face and do not influence capture efficiency. The slot velocity should be approximately 2000 fpm to provide the required distribution at the minimum energy cost. Higher velocities dissipate more energy.

Losses occur when air passes through the slot and when air enters the duct. Because the velocities, and therefore the velocity pressures, can be different at the slot and at the duct entry locations, the hood suction must reflect both losses and is given by

$$p_{st} = p_v + (C_o p_v)_s + (C_o p_v)_d \qquad (19)$$

where the first p_v is generally the higher of the two velocity pressures, s refers to the slot, and d refers to the duct entry location.

Example 2. A multislot hood has 3 slots, each 1 in. by 40 in. At the top of the plenum is a 90° transition into the 10 in. duct. The volumetric flow rate required for this hood is 1650 cfm. Determine the hood suction. Assume standard air.

Solution: The slot velocity V_s from Equation (1) is

$$V_s = 1650 \times 144/(3 \times 40 \times 1) = 1980 \text{ fpm}$$

Substituting this velocity into Equation (16),

$$p_v = (1980/4005)^2 = 0.24 \text{ in. of water}$$

The duct area is 0.5454 ft². Therefore, the duct velocity determined from Equation (1) is

$$V_d = 1650/0.5454 = 3025 \text{ fpm}$$

Substituting this velocity into Equation (16),

$$p_v = (3025/4005)^2 = 0.57 \text{ in. of water}$$

For a 90° transition into the duct, the loss factor is 0.25. For the slots, the loss factor is 1.78 (Figure 15). The duct velocity pressure is added to the sum of the two losses because it is larger than the slot velocity pressure. Using Equation (19),

$$p_{st} = 0.57 + (1.78 \times 0.24) + (0.25 \times 0.57) = 1.14 \text{ in. of water}$$

Exhaust volume requirements, minimum duct velocities, and entry loss factors for many specific operations are given in Chapter 10 of ACGIH (1998).

Influence of Air Movement on Local Exhaust Performance

Both air movement caused directly by supply air jets and turbulence of the ambient air resulting from general ventilation system operation, convective plumes, and moving people and process equipment are at least as important as hood face velocity in controlling contaminant spillage. Caplan and Knutson (1978) recommend that air movement caused by the above factors should be less than 1/2 to 2/3 times the hood face velocity.

In studies of hoods with a vertical face area, Zhivov et al. (1997) showed that the preferred orientation of the hood relative to the most likely direction of the cross draft is 135°. This orientation achieves both the lowest contaminant concentration in the operator's breathing zone and the highest capturing effectiveness. A moderate draft from behind the operator significantly increases the contaminant concentration in the operator's breathing zone. A cross draft has minimal effect on operator exposure, but the contaminant removal by the hood is low.

To reduce the influence of cross drafts greater than 80 fpm on the performance of a canopy hood above a buoyant source, Stroiizdat (1992) recommends attaching one-, two-, or three-sided removable shields to the hood that drop to a height above the source of 0.8 times the equivalent diameter of the design source.

Schematics in Figure 17 show how air jets can improve hood performance.

LOCAL EXHAUST FROM BUOYANT SOURCES

Overhead Hoods

If the process cannot be completely enclosed, the canopy hood should be placed above the process so that the contaminant moves toward the hood. Canopy hoods should be applied and designed with caution to avoid drawing contaminants across the operator's breathing zone (see Figure 18). The hood's height above the process should be kept to a minimum to reduce the total exhaust airflow rate.

A **low canopy hood** is within 3 ft of a process (within the thermal plume transition zone) and requires the lowest volumetric flow rate of all nonenclosing hoods. A **high canopy hood** is more than 10 ft above a process and requires a higher volumetric flow rate because room air is entrained in the column of hot, contaminated air rising from the process; this situation should be avoided.

Hemeon (1963) lists Equations (20) through (24) for determining the volumetric flow rate of hot gases for low canopy hoods. Note that canopy hoods located 3 to 10 ft above the process cannot be analyzed using these equations.

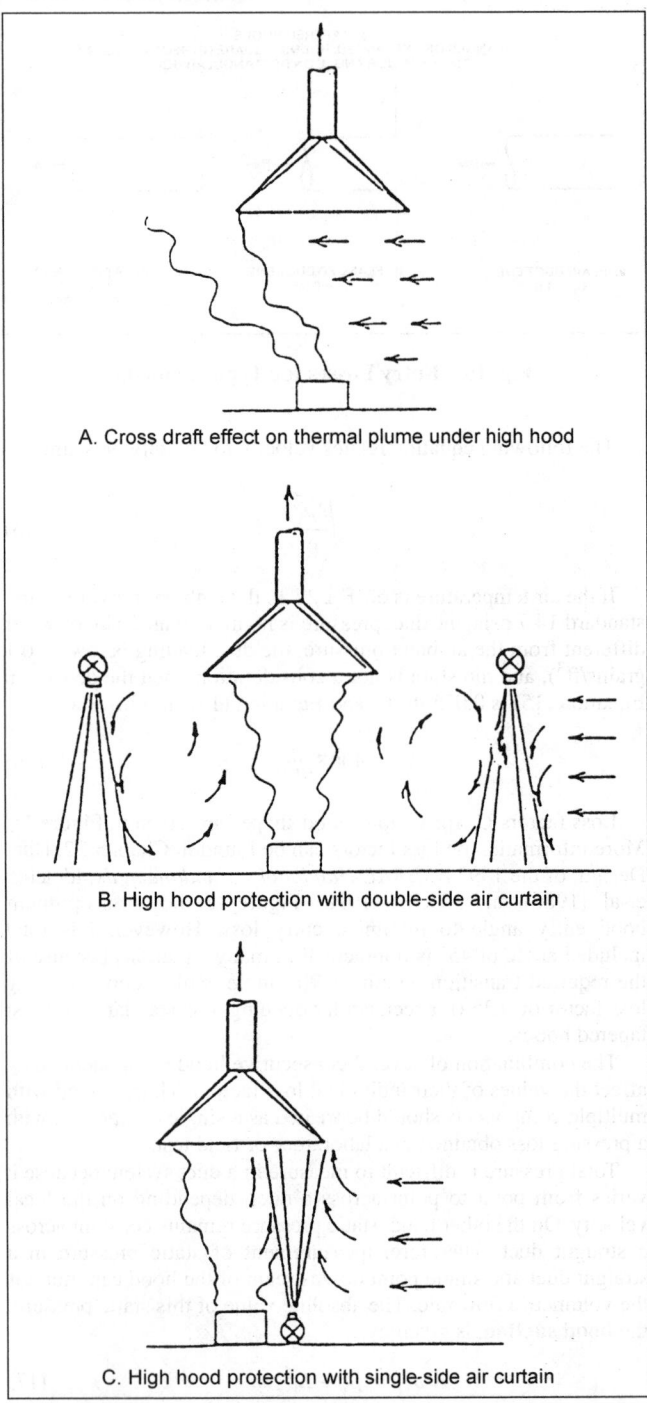

A. Cross draft effect on thermal plume under high hood

B. High hood protection with double-side air curtain

C. High hood protection with single-side air curtain

Fig. 17   Hood Performance Improvement with Air Jets

$$Q_o^* = \left[3600 \left(\frac{2gR}{pc_p} \right) \times q_{conv} LA_p^2 \right]^{1/3} \qquad (20)$$

where

Q_o^* = volumetric flow rate, cfm
g = gravitational acceleration, 32.2 ft/s^2
R = air gas constant, 53.352 ft·lb$_f$/lb$_m$·°R
p = local atmospheric pressure, lb$_f$/ft^2
c_p = constant pressure specific heat for air, 0.24 Btu/lb·°F
q_{conv} = convection heat transfer rate, Btu/min
L = vertical height of hot object, ft
A_p = cross-sectional area of airstream at upper limit of hot body, ft^2

For a standard atmospheric pressure of 2117 lb$_f$/ft^2, Equation (20) can be written as

Hermann p 122

$$Q_o^* = 29 \left(q_{conv} LA_p^2 \right)^{1/3} \qquad (21)$$

For three-dimensional bodies, the area A_p in Equations (20) and (21) is approximated by the plan view area of the hot body (Figure 19A). For horizontal cylinders, A_p is the product of the length and the diameter of the rod.

For vertical surfaces, the area A_p in Equations (20) and (21) is the area of the airstream (viewed from above) as the flow leaves the vertical surface (Figure 19B). As the airstream moves upward on a vertical surface, it appears to expand at an angle of approximately 4 to 5°. Thus, A_p is given by

$$A_p = wL\tan\theta \qquad (22)$$

where

w = width of vertical surface, ft
L = height of vertical surface, ft
θ = angle of air stream expansion, °

For horizontal heated surfaces, A_p is the surface area of the heated surface, and L is the longest length (conservative) of the horizontal surface or its diameter if it is round (Figure 19C).

If the heat transfer is caused by steam from a hot water tank,

$$q_{conv} = h_{fg} GA_p \qquad (23)$$

where

q_{conv} = convective heat transfer, Btu/min
h_{fg} = latent heat of vaporization, Btu/lb
G = steam generation rate, lb/min·ft^2
A_p = surface area of the tank, ft^2

At 212°F, the latent heat of vaporization is 970.3 Btu/lb. Using this value and Equation (23), Equation (20) simplifies to

$$Q_o^* = 287 A_p (GL)^{1/3} \qquad (24)$$

The exhaust volumetric flow rate determined by Equation (20) or (24) is the required exhaust flow rate when (1) a low canopy hood of the same dimensions as the hot object or surface is used and (2) side and back baffles are used to prevent room air currents from disturbing the rising air column. If side and back baffles cannot be used, the canopy hood size and the exhaust flow rate should be increased to reduce the possibility of contaminant escape around the hood. A good design provides a low canopy hood overhang equal to 40% of the distance from the hot process to the hood face on all sides (ACGIH 1998). The increased hood flow rate can be calculated using the following equation:

$$Q_t = Q_o^* + V_f (A_f - A_p) \qquad (25)$$

where

Q_t = total flow rate entering hood, cfm
Q_o^* = flow rate determined by Equation (20) or (24), cfm
V_f = desired indraft velocity through the perimeter area, fpm
A_f = hood face area, ft^2
A_p = plan view area of Equation (20) or (24)

A minimum indraft velocity of 100 fpm should be used for most design conditions. However, if room air currents are appreciable or if the contaminant discharge rate is high and the design exposure limit is low, higher values of V_f may be required.

The volumetric flow rate for a high canopy hood over a round, square, or rectangular (aspect ratio near 1) source can be predicted using Equation (11) with adjustments discussed in the section on Air and Contaminant Distribution with Buoyant Sources.

The diameter D_z of the plume at any elevation z above the virtual source can be determined by

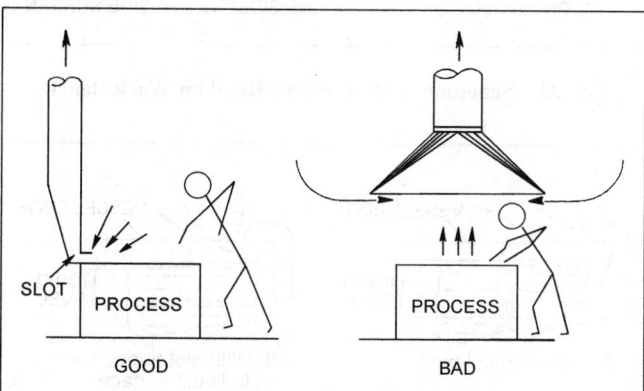

Fig. 18 Influence of Hood Location on Contamination of Air in the Operator's Breathing Zone

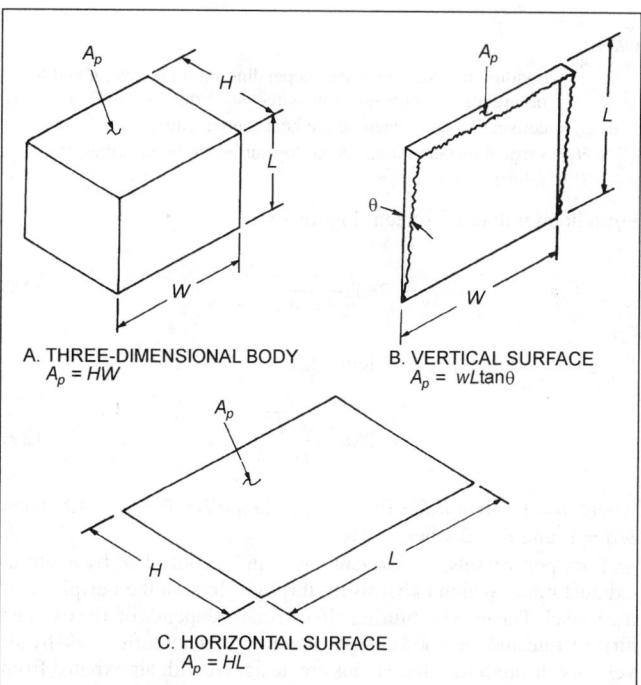

Fig. 19 A_p for Various Situations

$$D_z = 0.5z^{0.88} \tag{26}$$

High canopy hoods are extremely susceptible to room air currents. Therefore, they are typically much larger (often 100% larger) than indicated by Equation (26) and are used only if a low canopy hood cannot be used. The total flow rate exhausted from the hood can be evaluated using Equation (26) if Q_o is replaced by Q_z.

According to Posokhin (1984), the canopy hood is effective when

$$\frac{V_r(z + z_o)}{V_z b} \le 0.35$$

where

V_r = room air velocity, fpm
z_o = distance from virtual source to upper source level, ft
V_z = air velocity on thermal plume axis at hood face level, fpm
b = source width, ft

Sidedraft Hoods

Sidedraft hoods are typically used when the contaminant is drawn away from the operator's breathing zone (Figure 2B). With a buoyant source, a sidedraft hood requires a higher exhaust volumetric flow rate than a low canopy hood. If a low canopy hood restricts the operation, a sidedraft hood may be more cost-effective than a high canopy hood. Examples of sidedraft hoods include multislotted "pickling" hoods near welding benches (Figure 16), flanged hoods (Figure 20), and slot hoods on tanks (Figure 21).

Sidedraft hoods should be installed with the low edge of the suction area at the level of the top of the heat source. The distance b between the hood and the source may vary depending on the width of the source (Figure 22); maximum b is equal to the width B of the source. Based on studies by Kuz'mina (1959), the following airflow rate through the sidedraft hood is recommended (Stroiizdat 1992):

$$Q_o* = 194 c q_{conv}^{1/3}(H + B)^{5/3} \tag{27}$$

where

c = nondimensional coefficient depending on hood design and location relative to contaminant source [see Equations (28) and (29)]
q_{conv} = convective component of the heat source, Btu/h
H = vertical distance from source top surface to hood center, ft
B = source width, ft

For a hood without a screen (Figure 22A),

$$c = 280\left(\frac{I}{H + B}\right)^{2/3} \tag{28}$$

For a hood with a screen (Figure 22B),

$$c = 280m\sqrt{\frac{I}{H + B}} \tag{29}$$

where $m = 1$, when $b/B = 0$; $m = 1.5$, when $b/B = 0.3$; $m = 1.8$ when $b/B = 1$, and $m = 2$ when $b/B > 1$.

For open vessels, the contaminant can be controlled by a lateral exhaust hood, which exhausts air through slots on the periphery of the vessel. The hood capturing effectiveness depends on the exhaust airflow rate and the hood design; however, it is not influenced by air velocity through the slot. Hoods are designed with air exhaust from one side of the vessel or from two sides. Air exhaust from two sides requires a lower exhaust airflow rate. In most applications, a hood with a vertical face (Figure 23A) is used when the distance h_1

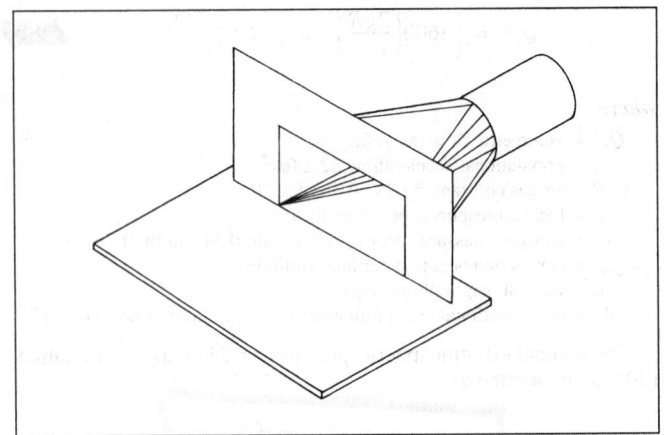

Fig. 20   Hood on Bench

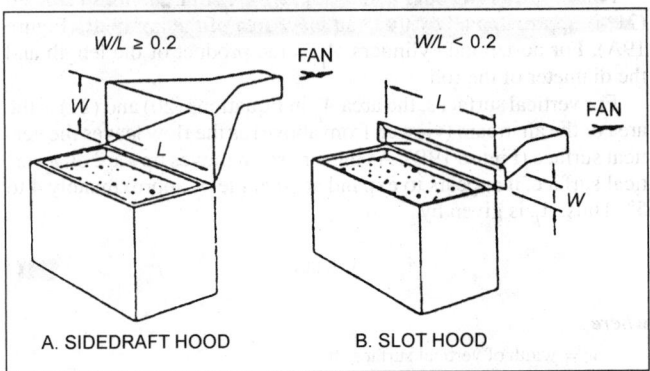

Fig. 21   Sidedraft Hood and Slot Hood on Tank

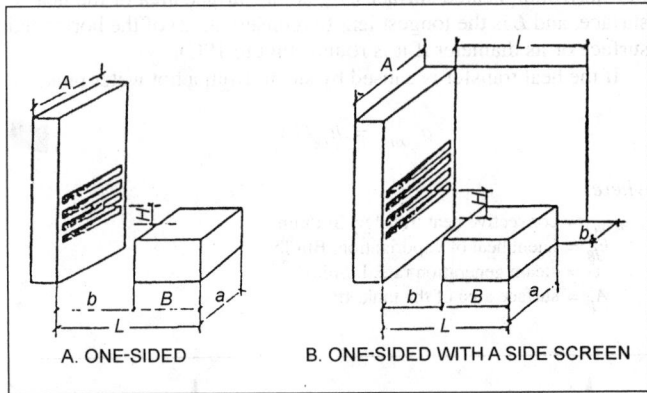

Fig. 22   Schematics of Sidedraft Hood on Work Bench

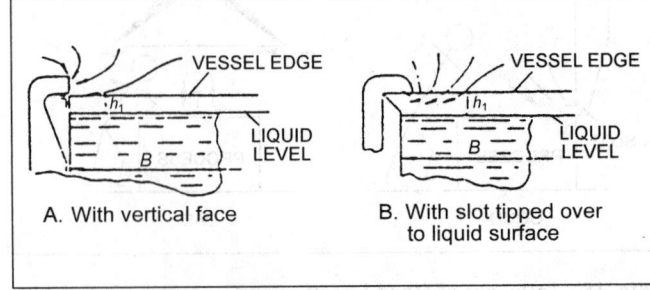

Fig. 23   Schematics of Sidedraft Slot Hood on Tank

Table 4 $K_{\Delta t}$ Coefficient Values

Liquid-to-Air Temperature Difference, °F								
32	50	68	86	104	122	140	158	176
1	1.16	1.31	1.47	1.63	1.79	1.94	2.1	2.26

(first column label $K_{\Delta t}$)

between the vessel edge and the liquid level is smaller than 4 in. (Stroiizdat 1992). When $h_l > 4$ in., hoods with the slot tipped over to the liquid surface (Figure 23B) are more effective.

Stroiizdat (1992) recommends the following exhaust airflow rate from one- and two-sided lateral slot hoods:

$$Q_o^* = 1.85 \times 10^5 \left(0.53\frac{Bl}{B+l} + h\right)^{1/3} BlK_1K_{\Delta t}K_t \qquad (30)$$

where

B = vessel width, ft
l = vessel length, ft
h = vertical distance between the liquid level and the hood face center, ft
K_1 = hood design coefficient: $K_1 = 1$ for two-sided hood; $K_1 = 1.8$ for one-sided hood
$K_{\Delta t}$ = coefficient reflecting liquid temperature (see Table 4)
K_t = coefficient reflecting process toxicity (from 1 to 2; e.g., for electroplating tanks, $K_t = 2$)

A more cost-effective alternative to a one- or two-sided lateral hood is a **push-pull hood**, described in the section on Jet-Assisted Hoods.

Downdraft Hoods

Downdraft hoods should be considered only when overhead or sidedraft hoods are impractical. Air can be exhausted through a slotted baffle (e.g., downdraft cutting table—see Figure 24) or through a circular slot with a round source (Figure 25A) or two linear slots along the long sides of a rectangular source (Figure 25B). To achieve higher capturing effectiveness, the exhaust should be located as close to the source as possible. Capturing effectiveness decreases with an increase in source height and increases when the top of the source is located below the hood face surface. With a buoyant source, the air velocity induced by the exhaust should be equal to or greater than the air velocity in the plume above the source (Posokhin 1984).

The target airflow rate for a circular downdraft hood is

$$Q_o^* = 6.1(q_{conv}d^5)^{1/3}\left(1 - 0.06\frac{q_{conv}^{vert}}{q_{conv}^{horiz}}\right)K_1K_v \qquad (31)$$

For a double linear slot downdraft hood,

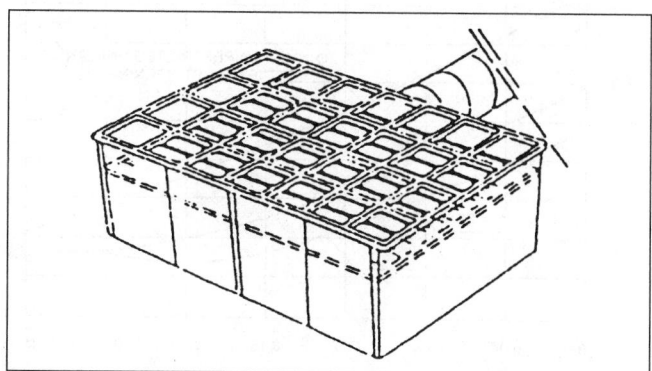

Fig. 24 Downdraft Welding Table

$$Q_o^* = 6.54q_{conv}^{1/3}lbK_1K_v \qquad (32)$$

where

d = source diameter, ft
l = source length, ft
b = source width, ft
q_{conv}^{vert} = convective heat component from the source vertical surfaces, Btu/h
q_{conv}^{horiz} = convective heat component from the source horizontal surface, Btu/h
K_1 = coefficient accounting for hood geometry that can be evaluated using graphs in Figure 25
K_v = coefficient accounting for room air movement V_r

$$= 1 + 0.0161\sqrt{V_r^3\frac{d}{q_{conv}}} \quad \text{for circular downdraft hood} \qquad (33)$$

$$= 1 + 0.0161\sqrt{V_r^3\frac{b}{q_{conv}}} \quad \text{for double slot downdraft hood} \qquad (34)$$

Example 3. A downdraft hood is to be designed to capture a contaminant from a rectangular source $l \times b \times h = 2$ ft $\times$ 1.6 ft $\times$ 0 ft. Convective heat component of the source $q_{conv} = 3400$ Btu/h. Room air movement $V_r = 80$ fpm. Two exhaust slots with a width $b = 4$ in. are located at the distance $B_1 = 2$ ft and $B_2 = 2.6$ ft. Determine the exhaust airflow rate.

Solution: Using the graph in Figure 25 for $B_2/B_1 = 2.6/2 = 1.3$, and $B_1/b = 2/1.6 = 1.25$, obtain $K_1 = 5$. Coefficient K_v accounting for room air movement [Equation (34)] is

$$K_v = 1 + 0.0161\sqrt{80^3\frac{1.6}{3400}} = 1.25$$

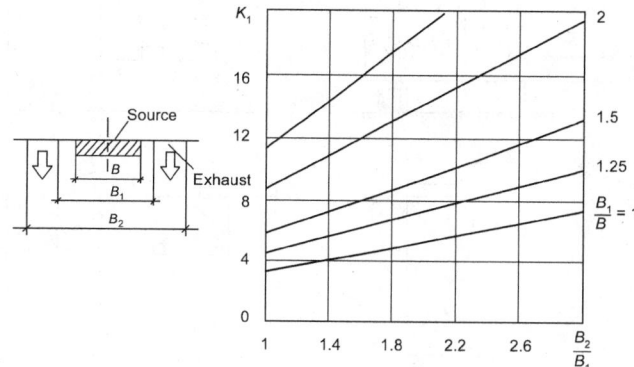

A. Circular slot with round source

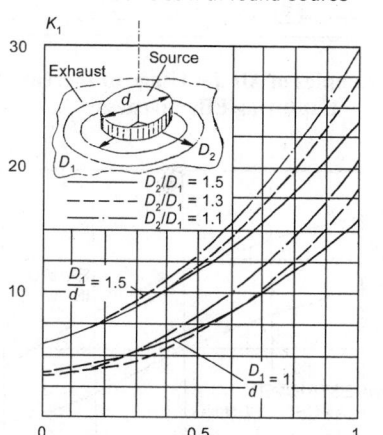

B. Linear slots along the long sides of rectangular source

Fig. 25 K_1 Coefficient Evaluation for Downdraft Hoods

The exhaust airflow rate [Equation (32)] is

$$Q_o{}^* = 6.54 \times 3400^{1/3} \times 2 \times 1.6 \times 5 \times 1.25 = 1967 \text{ cfm}$$

Airflow rate recommended for typical downdraft cutting table (Hagopian and Bastress 1976) can be calculated using Equation (1), where A_o is tabletop area, ft², and V_f = minimum face velocity, fpm.

JET-ASSISTED HOODS

Major Types and Application Recommendations

Jet-assisted hoods are nonenclosing hoods combined with compact, linear, or radial air jets. They are used to

- Increase capturing effectiveness (Figure 26) by transporting contaminants toward the face of the hood using supply jets (ACGIH 1998; Heinsohn 1991; Elinskii 1989; Stroiizdat 1992; Sciola 1993).
- Separate contaminated zones from relatively clean zones in working spaces (Boshnyakov 1975; Romeyko et al. 1976; Stoler and Savelyev 1977; Posokhin and Broida 1980; Anichkhin and Anichkhina 1984; Cesta 1988; Zhivov 1993). For example, jet-assisted hoods can be used over welding robots (Figure 29).

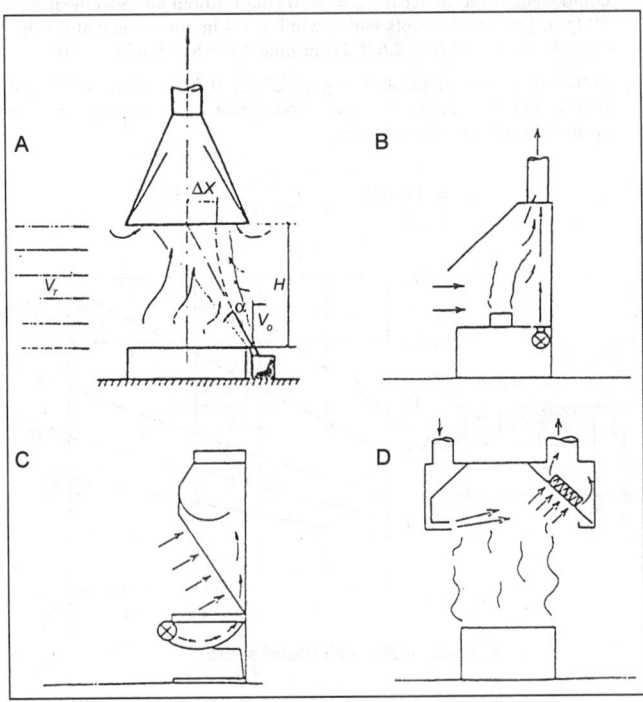

Fig. 26 Examples of Air Jet Usage to Increase Hood Capturing Effectiveness

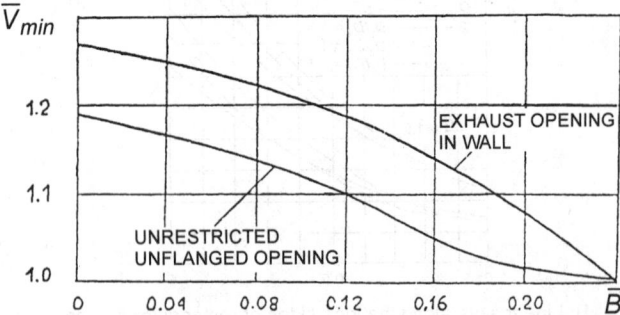

Fig. 27 Graph for Evaluation of $\overline{V}_{min}$

- Prevent contaminated air from moving into clean zones by creating positive static pressure (Strongin and Marder 1988); for example, in drying chambers and cooling tunnels of casting conveyors (Figure 30).
- Protect hoods from room air movement; for example, with some laboratory hoods and industrial ovens, hoods can be used in combination with air curtains (Figure 17).
- Create swirling airflows near the exhaust hood, increasing its capturing effectiveness (Ljungqvist and Waering 1988).

The capturing effectiveness of jet-assisted hoods is 15 to 20% higher than that of conventional hoods for the same operating costs (Strongin and Marder 1988). For the same capturing effectiveness, jet-assisted hoods are 25 to 30% more cost-effective due to lower airflow. When exhausted air must be cleaned, reduced cleaning costs make jet-assisted hoods 100 to 150% more cost-effective. If compensating-type jet-assisted hoods are used, operating costs for heating and cooling in general ventilation systems are also reduced.

Design Procedure

Push-Pull Hoods over Open Surface Tanks. A push-pull exhaust system is a more efficient alternative to lateral exhausts for open-surface vessels such as plating tanks. In a push-pull system, air is blown through one slot on the periphery of a vessel and exhausted from a slot on the opposite side. The inlet to the hood must be large enough to accommodate the jet and the room air entrained by the jet. Push-pull systems can be used on wider vessels than lateral exhaust hoods. Chapter 10 of ACGIH (1998) provides a design procedure for these systems.

Push-Pull Hood for Vertical Apertures. The design principle of a push-pull exhaust hood with a linear jet supplied upward toward the canopy hood above [e.g., the oven aperture (Figure 26A)] is based on maintaining the minimum velocity along the jet such that

$$V_{min} > 7811 \sqrt{\frac{\Delta p}{\rho_g}\left(\frac{\sqrt{1 + 142C^2} - 1}{89C^2}\right)} \qquad (35)$$

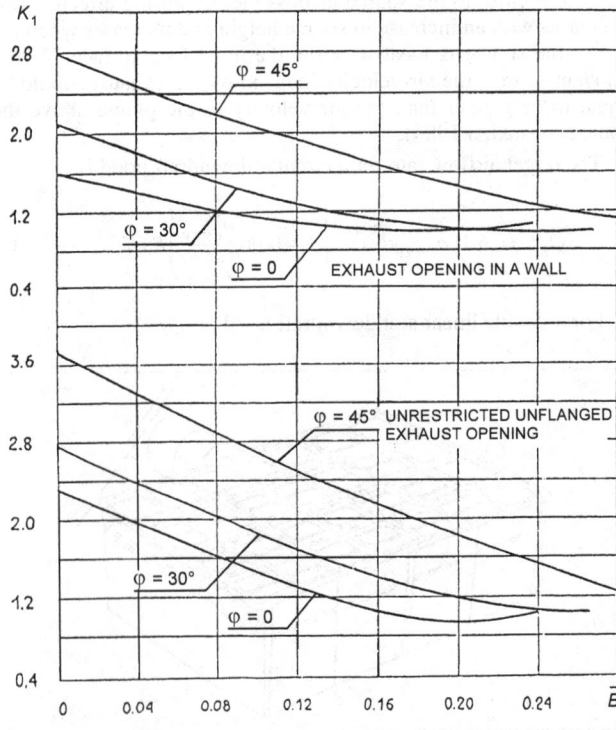

Fig. 28 Coefficient K_1 Evaluation for Pull-Push Hoods with Different Hood Configurations

where

$$C = \frac{1}{1 + 3.74(\rho_g/\rho_{air})} \qquad (36)$$

V_{min} = minimum velocity along jet, fpm
Δp = excessive pressure inside the process equipment, in. of water
ρ_{air} = density of room air, lb/ft^3
ρ_g = density of gas mixture releasing through the aperture in the process equipment, lb/ft^3

The supply and exhaust airflow rates Q_{sup} and Q_o, ft^3/min, can be determined as follows:

For a nonattached jet,

$$Q_{sup} = 0.43 \frac{V_{min}}{\overline{V}_{min}} a \sqrt{bl} \qquad (37)$$

$$Q_o = 0.201 \frac{V_{min}}{\overline{V}_{min}} al K_1 K_v \qquad (38)$$

For a wall jet,

$$Q_{sup} = 0.3 \frac{V_{min}}{\overline{V}_{min}} a \sqrt{bl} \qquad (39)$$

$$Q_o = 0.1 \frac{V_{min}}{\overline{V}_{min}} al K_1 K_v \qquad (40)$$

where

$\overline{V}_{min}$ = from graph in Figure 27
B = relative width of exhaust hood
  = $B/2l$ for a nonattached jet and B/l for a wall jet
B = width of exhaust hood, ft
a = length of exhaust hood, ft
b = width of supply slot, ft
K_1 = coefficient accounting for hood geometry can be evaluated using graphs in Figure 28
K_v = coefficient accounting for room air movement V_r

$$= 1 + \frac{V_r}{V_{min}} \qquad (41)$$

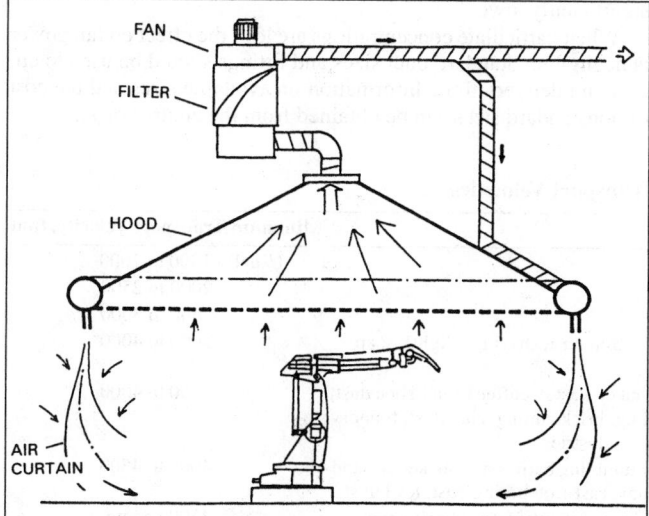

Fig. 29 Push-Pull Hood over Welding Robot

The following are some design considerations:

- Push-pull hoods are economically feasible if $l > 3$ ft.
- The jet should be considered a wall jet when the distance H between the supply nozzle and the vertical surface is smaller than $0.15l$. Otherwise, the jet is nonattached.
- When flange width $h > H + B$, the hood is treated as an opening in an infinite surface; when $h \le H + B$, the hood is treated as free-standing.
- The value of the minimum velocity V_{min} along the jet should be greater than 300 fpm.
- The width b of the supply air slot is typically chosen to be $0.01l$. However, it should be greater than 0.25 in. to prevent fouling. The length a of the supply slot should be equal to the length of the aperture.
- The supply air velocity V_o should not exceed 300 fpm. This can be achieved by selection of the appropriate slot width b.

Example 4. A push-pull hood is to capture a contaminant from an oven aperture. The surplus pressure in the oven $\Delta p = 0.008$ in. of water, and the temperature inside the oven $t_g = 1472°F$ ($\rho_g = 0.0206$ lb/ft^3). Canopy hood is installed at the height of $l = 48$ in. from the low edge of the oven aperture. The hood projection $B = 23$ in, and the hood width is equal to the aperture width $a = 6$ ft; the aperture height is 40 in. The room air velocity near the hood $V_r = 80$ fpm and the room air temperature $t_{air} = 68°F$ ($\rho_{air} = 0.075$ lb/ft^3). Determine the supply and exhaust airflow rates.

Solution: Using the graph in Figure 27 for $B = 23/(2 \times 48) = 0.24$, obtain $\overline{V}_{min} = 1$.
From Equations (35) and (36) obtain parameter C and velocity V_{min}:

$$C = \frac{1}{1 + 3.74(0.021/0.075)} = 0.494$$

$$V_{min} = 7811 \sqrt{\frac{0.008}{0.0206}\left(\frac{\sqrt{1 + 142 \times 0.494^2} - 1}{89 \times 0.494^2}\right)} = 1107 \text{ fpm}$$

Assuming $b = 1$ in., calculate supply airflow rate [Equation (37)]:

$$Q_{sup} = 0.43 \times \frac{1107}{1} \times (6)\sqrt{\frac{1}{12} \times 4} = 1650 \text{ cfm}$$

Coefficient K_v accounting for room air movement [Equation (41)]:

$$K_v = 1 + \frac{80}{1107} = 1.07$$

From the graph in Figure 28, $K_1 = 1$.
The exhaust airflow rate [Equation (38)]:

$$Q_{sup} = 0.201 \times \frac{1107}{1} \times 6 \times 4 \times 1 \times 1.07 = 5714 \text{ cfm}$$

Push-Pull Hood above Contaminated Area. A canopy hood with an incorporated slotted nozzle installed around the perimeter of the hood is used to prevent contaminant transfer from contaminated areas, for example, the operating zone of one or several welding robots (Figure 29), where enclosing hoods or other types of nonenclosing hoods are impractical (U.S. Patent). Air supplied through the nozzle creates steady air curtain protection along the contour. Due to the negative pressure created by the hood, the air curtain jet turns at or below the level of the contaminant source toward the center. To minimize the supply airflow rate, the nozzle is equipped with a honeycomb attachment that produces a low-turbulence jet. The width of the nozzle can be determined as follows:

$$b = \frac{A/P}{45\left(\frac{A}{PH}\right)^2\left(0.566\sqrt{\frac{\overline{H}}{b}} - 1\right)^2 - 0.25\left(0.566\sqrt{\frac{\overline{H}}{b}} + 1\right)^2} \qquad (42)$$

where

> b = nozzle width, ft
> A = hood cross-sectional area, ft^2
> P = hood perimeter, ft
> H = height of hood above contaminant source, ft

Push-Pull Protection System. These systems are used (Strongin et al. 1986; Strongin and Marder 1988) to prevent contaminant release from process equipment when the process requires that entering and/or exiting apertures remain open (e.g., conveyer painting chambers, cooling tunnels, etc.). The open aperture must be equipped with a tunnel and supply and exhaust air systems (Figure 30). The aperture is protected by the air jet(s) supplied through one or two slots installed along one side or two opposite sides of the tunnel and directed at angle α = 80 to 85° to the tunnel cross section. Air supplied through the slot(s) is thus directed toward the incoming room air. Moving along the tunnel, the jet(s) slow down, and their dynamic pressure is converted into static pressure, preventing room air from entering the chamber. After reaching the point with a zero centerline velocity, the jet(s) make a U-turn and redirect into the chamber. The air jet(s) can be supplied vertically (with supply air ducts installed along vertical walls) or horizontally (with supply air ducts installed along horizontal walls). The distance X (Figure 30) from the entrance of a tunnel (with cross-sectional area $B \times H$) to the supply slot location should be greater than or equal to $5B$ with a single vertical jet ($5H$ with a single horizontal jet) and $2.5B$ ($2.5H$) when air is supplied by two jets.

The air supply slot is equipped with diverging vanes (angle β between 30 to 90°) creating an air jet with an increased angle of divergence; the number n of these vanes should be greater than or equal to $\beta/10$. The increased angle of divergence of supply air jets allows a decrease in the distance X between the tunnel entrance and the slot.

Airflow rate supplied by the jet is determined as

$$Q_o^* = \sqrt{\frac{A_o b_o L_o \Delta p}{J}} \tag{43}$$

where

> A_o = cross-sectional area of the tunnel, ft^2
> b_o = supply slot width, ft
> L_o = supply slot length, ft
> J = supply jet parameter
> $= \sin\alpha + 2.5\dfrac{A_o}{A_c}\left[2.13(1+\psi)^2 + \left(\dfrac{\psi}{1+1/\psi}\right)^2 - \psi^2\right]$ (44)
>
> for $\psi = \dfrac{Q_{exh}}{Q_o}$
>
> Δp = chamber to room pressure difference, lb$_f$/ft^2
> $= 0.5gH(\rho_{room} - \rho_c)$ (45)

H = chamber height, ft
g = gravitational acceleration, 32.2 ft/s^2
ρ_{room} = room chamber air density, lb/ft^3
ρ_c = chamber air density, lb/ft^3

The minimum airflow rate to be exhausted outside from the chamber and the corresponding amount of outdoor air to be supplied through the slot should dilute the contaminants in the chamber to the desired concentration. In the case of prevention of contaminant release from a drying chamber, the solvent vapor concentration should not exceed 25% of the lower explosive limit $C_{exp(min)}$. In this case, the exhaust airflow rate can be determined as follows:

$$Q_{exh} = \frac{GK}{0.25C_{exp(min)}} \tag{46}$$

where

> G = amount of vapor release into the chamber, lb/min
> K = coefficient accounting for the nonuniformity of solvent evaporation and other irregularities; typically, $2 \leq K \leq 5$
> $C_{exp(min)}$ = lower explosive limit of pollutant, lb/ft^3

OTHER LOCAL EXHAUST SYSTEM COMPONENTS

Duct Design and Construction

Duct Considerations. The second component of a local exhaust ventilation system is the duct through which contaminated air is transported from the hood(s). Round ducts are preferred because they (1) offer a more uniform air velocity to resist settling of material and (2) can withstand the higher static pressures normally found in exhaust systems. When design limitations require rectangular ducts, the aspect ratio (height-to-width ratio) should be as close to unity as possible.

Minimum transport velocity is the velocity required to transport particulates without settling. Table 5 lists some generally accepted transport velocities as a function of the nature of the contaminants (ACGIH 1998). The values listed are typically higher than theoretical and experimental values to account for (1) damage to ducts, which would increase system resistance and reduce volumetric flow and duct velocity; (2) duct leakage, which tends to decrease velocity in the duct system upstream of the leak; (3) fan wheel corrosion or erosion and/or belt slippage, which could reduce fan volume; and (4) reentrainment of settled particulate caused by improper operation of the exhaust system. Design velocities can be higher than the minimum transport velocities but should never be significantly lower.

When particulate concentrations are low, the effect on fan power is negligible. Standard duct sizes and fittings should be used to cut cost and delivery time. Information on available sizes and the cost of nonstandard sizes can be obtained from the contractor(s).

Table 5   Contaminant Transport Velocities

Nature of Contaminant	Examples	Minimum Transport Velocity, fpm
Vapor, gases, smoke	All vapors, gases, smoke	Usually 1000 to 2000
Fumes	Welding	2000 to 2500
Very fine light dust	Cotton lint, wood flour, litho powder	2500 to 3000
Dry dusts and powders	Fine rubber dust, molding powder dust, jute lint, cotton dust, shavings (light), soap dust, leather shavings	3000 to 4000
Average industrial dust	Grinding dust, buffing lint (dry), wool jute dust (shaker waste), coffee beans, shoe dust, granite dust, silica flour, general material handling, brick cutting, clay dust, foundry (general), limestone dust, asbestos dust in textile industries	3500 to 4000
Heavy dust	Sawdust (heavy and wet), metal turnings, foundry tumbling barrels and shakeout, sandblast dust, wood blocks, hog waste, brass turnings, cast-iron boring dust, lead dust	4000 to 4500
Heavy and moist dust	Lead dust with small chips, moist cement dust, asbestos chunks from transite pipe cutting machines, buffing lint (sticky), quicklime dust	4500 and up

Source: Adapted from *Industrial Ventilation: A Manual of Recommended Practice* (ACGIH 1998).

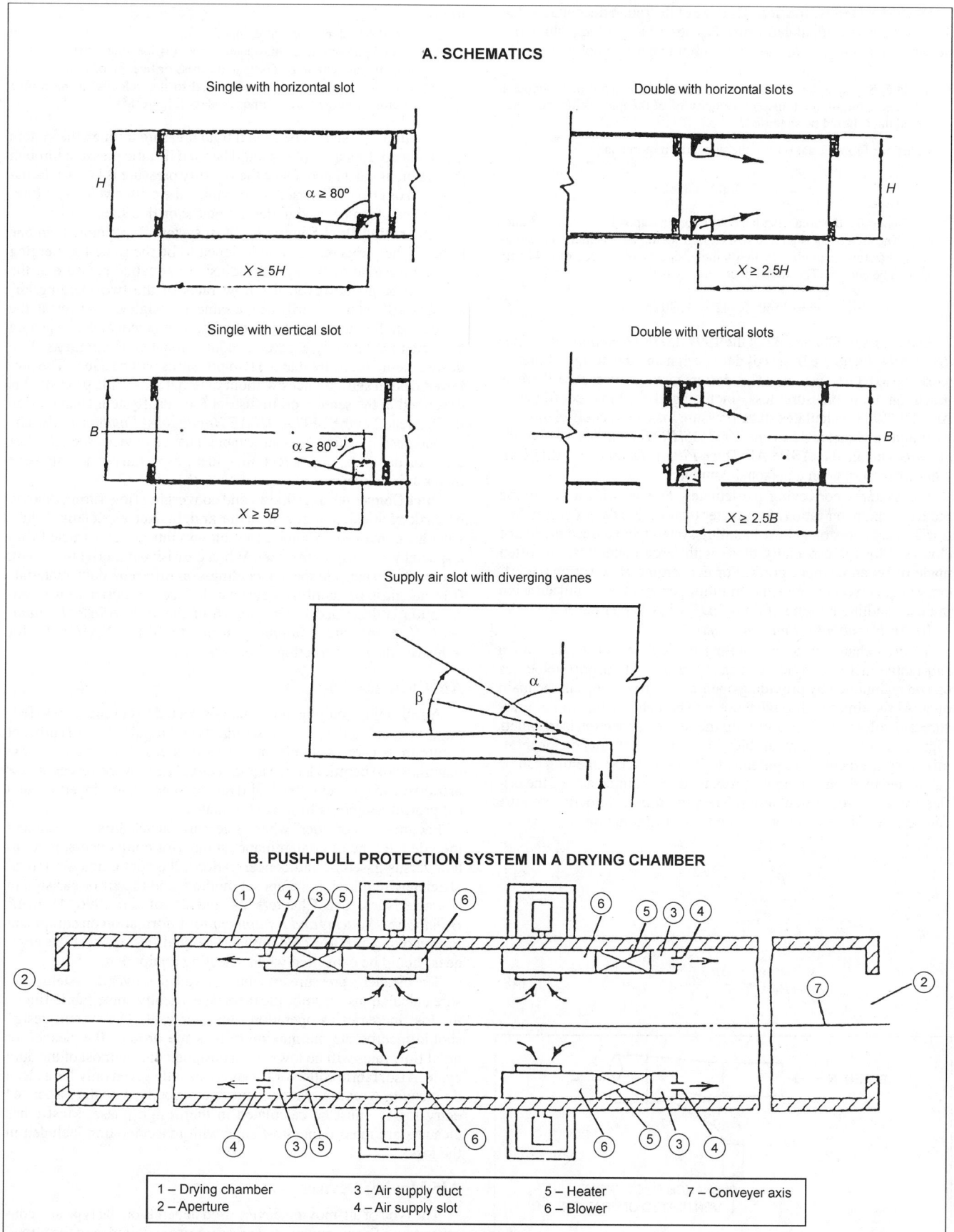

A. SCHEMATICS

Single with horizontal slot

Double with horizontal slots

$\alpha \geq 80°$

$X \geq 5H$

$X \geq 2.5H$

Single with vertical slot

Double with vertical slots

$\alpha \geq 80°$

$X \geq 5B$

$X \geq 2.5B$

Supply air slot with diverging vanes

β α

B. PUSH-PULL PROTECTION SYSTEM IN A DRYING CHAMBER

| 1 – Drying chamber | 3 – Air supply duct | 5 – Heater | 7 – Conveyer axis |
| 2 – Aperture | 4 – Air supply slot | 6 – Blower | |

Fig. 30   Push-Pull Protection System

Duct Size Determination. The size of the round duct attached to the hood can be calculated using Equation (1) for the volumetric flow rate and Table 5 for the minimum transport velocity.

Example 5. Suppose the contaminant captured by the hood in Example 1 requires a minimum transport velocity of 3000 fpm. What diameter round duct should be specified?

Solution: From Equation (1), the duct area required is

$$A = 1560/3000 = 0.52 \text{ ft}^2$$

Generally, the area calculated will not correspond to a standard duct size. The area of the standard size chosen should be less than that calculated. For this example, a 9 in. diameter duct with an area of 0.442 ft^2 should be chosen. The actual duct velocity is then

$$V = 1560/0.442 = 3530 \text{ fpm}$$

Duct Losses. Chapter 32 of the 1997 *ASHRAE Handbook—Fundamentals* covers the basics of duct design and the design of metalworking exhaust systems. The design method presented there is based on total pressure loss, including the fitting coefficients; ACGIH (1998) calculates static pressure loss. Loss coefficients can be found in Chapter 32 of the 1997 *ASHRAE Handbook—Fundamentals* and in the ASHRAE *Duct Fitting Database* (ASHRAE 1994), which runs on a personal computer.

For systems conveying particulates, elbows with a centerline radius-to-diameter ratio (*r/D*) greater than 1.5 are the most suitable. If *r/*D ≤ 1.5, abrasion in dust-handling systems can reduce the life of elbows. Elbows, especially those with large diameters, are often made of seven or more gores. For converging flow fittings, a 30° entry angle is recommended to minimize energy losses and abrasion in dust-handling systems (Fitting ED5-1 in Chapter 32 of the 1997 *ASHRAE Handbook—Fundamentals*).

Where exhaust systems handling particulates must allow for a substantial increase in future capacity, required transport velocities can be maintained by providing open-end stub branches in the main duct. Air is admitted through these stub branches at the proper pressure and volumetric flow rate until the future connection is installed. Figure 31 shows such an air bleed-in. The use of outside air minimizes replacement air requirements. The size of the opening can be calculated by determining the pressure drop required across the orifice from the duct calculations. Then the orifice velocity pressure can be determined from one of the following equations:

$$p_{v,o} = \frac{\Delta p_{t,o}}{C_o} \tag{47}$$

or

$$p_{v,o} = \frac{\Delta p_{s,o}}{C_o + 1} \tag{48}$$

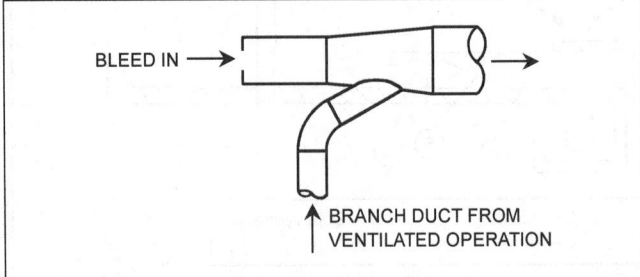

Fig. 31   Air Bleed-In

where

$p_{v,o}$ = orifice velocity pressure, in. of water
$\Delta p_{t,o}$ = total pressure to be dissipated across orifice, in. of water
$\Delta p_{s,o}$ = static pressure to be dissipated across orifice, in. of water
C_o = orifice loss coefficient referenced to the velocity at the orifice cross-sectional area, dimensionless (Figure 15)

Equation (47) should be used if total pressure through the system is calculated; Equation (48) should be used if static pressure through the system is calculated. Once the velocity pressure is known, Equation (15) or (16) can be used to determine the orifice velocity. Equation (1) can then be used to determine the orifice size.

Integrating Duct Segments. Most systems have more than one hood. If the pressures are not designed to be the same for merging parallel airstreams, the system adjusts to equalize pressure at the common point; however, the flow rates of the two merging airstreams will not necessarily be the same as designed. As a result, the hoods can fail to control the contaminant adequately, exposing workers to potentially hazardous contaminant concentrations. Two design methods ensure that the two pressures will be equal. The preferred design self-balances without external aids. This procedure is described in the section on Industrial Exhaust System Duct Design in Chapter 32 of the 1997 *ASHRAE Handbook—Fundamentals*. The second design, which uses adjustable balance devices such as blast gates or dampers, is not recommended, especially when abrasive material is conveyed.

Duct Construction. Elbows and converging flow fittings should be made of thicker material than the straight duct, especially if abrasives are conveyed. In some cases, elbows must be constructed with a special wear strip in the heel. When corrosive material is present, alternatives such as special coatings or different duct materials (fibrous glass or stainless steel) can be used. Industrial duct construction is described in Chapter 16 of the 1996 *ASHRAE Handbook—Systems and Equipment*. Refer to SMACNA (1990) for industrial duct construction standards.

Air Cleaners

Air-cleaning equipment is usually selected to (1) conform to federal, state, or local emissions standards and regulations; (2) prevent reentrainment of contaminants to work areas; (3) reclaim usable materials; (4) permit cleaned air to recirculate to work spaces and/or processes; (5) prevent physical damage to adjacent properties; and (6) protect neighbors from contaminants.

Factors to consider when selecting air-cleaning equipment include the type of contaminant (number of components, particulate versus gaseous, and concentration), the contaminant removal efficiency required, the disposal method, and the air or gas stream characteristics. See Chapters 24 and 25 of the 1996 *ASHRAE Handbook—Systems and Equipment* for information on equipment for removing airborne contaminants. A qualified applications engineer should be consulted when selecting equipment.

The cleaner's pressure loss must be added to overall system pressure calculations. In some cleaners, specifically some fabric filters, the loss increases as operation time increases. The system design should incorporate the maximum pressure drop of the cleaner, or hood flow rates will be lower than designed during most of the duty cycle. Also, fabric collector losses are usually given only for a clean air plenum. A reacceleration to the duct velocity, with the associated entry losses, must be calculated in the design phase. Most other cleaners are rated flange-to-flange with reacceleration included in the loss.

Air-Moving Devices

The type of air-moving device used depends on the type and concentration of contaminant, the pressure rise required, and the allowable noise levels. Fans are usually selected. Chapter 18 of the 1996 *ASHRAE Handbook—Systems and Equipment* describes available fans and refers the reader to Air Movement and Control Association (AMCA) *Publication* 201, *Fans and Systems,* for proper connection of the fan(s) to the system. The fan should be located downstream of the air cleaner whenever possible to (1) reduce possible abrasion of the fan wheel blades and (2) create negative pressure in the air cleaner so that air leaks into it and positive control of the contaminant is maintained.

In some instances, however, the fan is located upstream from the cleaner to help remove dust. This is especially true with cyclone collectors, for example, which are used in the woodworking industry. If explosive, corrosive, flammable, or sticky materials are handled, an injector can transport the material to the air-cleaning equipment. Injectors create a shear layer that induces airflow into the duct. Injectors should be the last choice because their efficiency seldom exceeds 10%.

Energy Recovery

The transfer of energy from exhausted air to replacement air may be economically feasible, depending on (1) the location of the exhaust and replacement air ducts, (2) the temperature of the exhausted gas, and (3) the nature of the contaminants being exhausted. The efficiency of heat transfer depends on the type of heat recovery system used. Rotary air-to-air exchangers have the best efficiency, 70-80%. Cross flow fixed-surface plate exchangers and energy recovery loops with liquid coupled coils have efficiencies of 50 and 60% (Aro and Kovula 1992).

If exhausted air contains particulate matter (e.g., dust, lint) or oil mist, the exhausted air should be filtered to prevent fouling the heat exchanger. If the exhausted air contains gaseous and vaporous contaminants such as hydrocarbons and water-soluble chemicals, their effect on the heat recovery device should be investigated (Aro and Kovula 1992).

Exhaust Stacks

The exhaust stack must be designed and located to prevent the reentrainment of discharged air into supply system inlets. The building's shape and surroundings determine the atmospheric airflow over it. Chapter 15 of the 1997 *ASHRAE Handbook—Fundamentals* and Chapter 43 of this volume cover exhaust stack design.

If rain protection is important, stackhead design is preferable to weathercaps. Weathercaps, which are not recommended, have three disadvantages:

1. They deflect air downward, increasing the chance that contaminants will recirculate into air inlets.
2. They have high friction losses.
3. They provide less rain protection than a properly designed stackhead.

Figure 32 contrasts the flow patterns of weathercaps and stackheads. Loss data for weathercaps and stackheads are presented in the ASHRAE *Duct Fitting Database* (ASHRAE 1994). Losses in the straight duct form of stackheads are balanced by the pressure regain at the expansion to the larger-diameter stackhead.

OPERATION

System Testing

After installation, an exhaust system should be tested to ensure that it operates properly with the required flow rates through each hood. If the actual installed flow rates are different from the design values, they should be corrected before the system is used. Testing is also necessary to obtain baseline data to determine (1) compliance with federal, state, and local codes; (2) by periodic inspections, whether maintenance on the system is needed to ensure design operation; (3) whether a system has sufficient capacity for additional airflow; and (4) whether system leakage is acceptable. AMCA *Pub-*

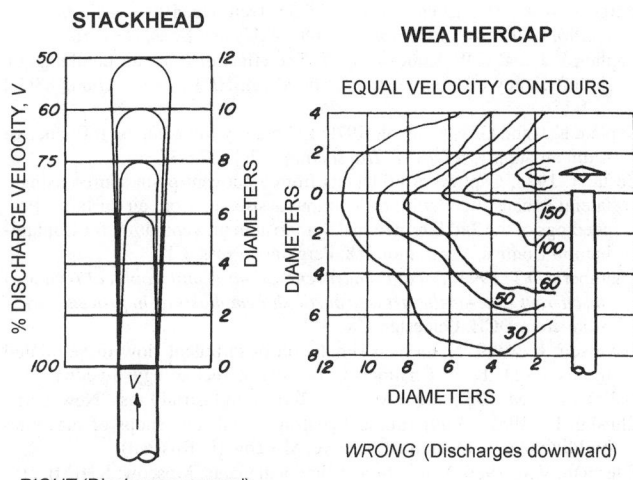

RIGHT (Discharges upward)

Fig. 32 Comparison of Flow Pattern for Stackheads and Weathercaps

lication 203 and Chapter 9 of ACGIH (1998) contain detailed information on the preferred methods for testing systems.

Operation and Maintenance

Periodic inspection and maintenance are required for the proper operation of exhaust systems. Systems are often changed or damaged after installation, resulting in low duct velocities and/or incorrect volumetric flow rates. Low duct velocities can cause the contaminant to settle and plug the duct, reducing flow rates at the affected hoods. Adding hoods to an existing system can change volumetric flow at the original hoods. In both cases, changed hood volumes can increase worker exposure and health risks. The maintenance program should include (1) inspecting ductwork for particulate accumulation and damage by erosion or physical abuse, (2) checking exhaust hoods for proper volumetric flow rates and physical condition, (3) checking fan drives, and (4) maintaining air-cleaning equipment according to manufacturers' guidelines.

REFERENCES

ACGIH. 1998. *Industrial ventilation: A manual of recommended practice,* 23rd edition. Committee on Industrial Ventilation, American Conference of Governmental Industrial Hygienists, Cincinnati, OH.

Aksenov, A.A. and A.V. Gudzovskii. 1994. Numerical simulation of turbulent thermal plumes in the stratified space. Proceedings of the First National Conference on Heat Transfer. Part 2—Free Convection. 21-25 November. Moscow (in Russian).

Alden, J.L. and J.M. Kane. 1982. Design of Industrial Ventilation Systems, 5th ed. Industrial Press, New York.

AMCA. 1995. Fans and systems. *Publication* 201-95. Air Movement and Control Association International, Arlington Heights, IL.

AMCA. 1995. Field performance measurement of fan systems. *Publication* 203-95.

Anichkhin, A.G. and G.N. Anichkhina. 1984. Ventilation of laboratories in Research Institutions. In "Energy efficiency improvement of mechanical systems". Nauka, Moscow.

Aro, T. and K. Kovula. 1992. Learning from experiences with Industrial Ventilation. Center for the Analysis and Dissemination of Demonstrated Energy Technologies. AIR-IX Consulting Engineers, Finland.

ASHRAE. 1994. *Duct fitting database.*

Bastress, E., J. Niedzwecki, and A. Nugent. 1974. Ventilation required for grinding, buffing, and polishing operations. U.S. Department of Health, Education, and Welfare. NIOSH. *Publication* No. 75-107. Washington, D.C.

Boshnyakov, E.N. 1975. Local exhaust with air curtains. *Water Supply and Sanitary Techniques,* #3. Moscow (in Russian).

Burgess, W.A., M.J. Ellenbecker, and R.D. Treitman. 1989. *Ventilation for control of the work environment.* John Wiley and Sons, New York.

Caplan, K.J. and G.W. Knutson. 1977. The effect of room air challenge on the efficiency of laboratory fume hoods. *ASHRAE Transactions* 83(1): 141-156.

Caplan, K.J. and G.W. Knutson. 1978. Laboratory fume hoods: Influence of room air supply. *ASHRAE Transactions* 82(1):522-37.

Cesta, T. 1988. Capture of pollutants from a buoyant point source using a lateral exhaust hood with and without assistance from air curtains. Proceedings of the 2nd International Symposium on Ventilation for Contamination Control, Ventilation '88. Pergamon Press, UK.

Chambers, D.T. 1993. *Local exhaust ventilation: A philosophical review of the current state-of-the-art with particular emphasis on improved worker protection.* DCE, Leicester, UK.

Davidson, L. 1989. Numerical simulation of turbulent flow in ventilated rooms. Ph.D. thesis, Chalmers University of Technology, Sweden.

DallaValle, J.M. 1952. *Exhaust hoods*, 2nd ed. Industrial Press, New York.

Elinskii, I.I. 1989. Ventilation and heating of galvanic shops of machine-building plants. Mashinostroyeniye, Moscow (in Russian).

Elterman, V.M. 1980. Ventilation of chemical plants. Moscow: KHIMIA (in Russian).

Fletcher, B. 1977. Center line velocity characteristics of rectangular unflanged slots and slots under suction. *Ann. Occup. Hyg.* 20:141-46.

Fuller, F.H. and A.W. Etchells. 1979. The rating of laboratory hood performance. *ASHRAE Journal* 21(10):49-53.

Garrison, R.P. 1977. Nozzle performance and design for high-velocity/low-volume exhaust ventilation. Ph.D. thesis. University of Michigan, Ann Arbor, MI.

Hagopian, J.H. and E.K. Bastress. 1976. Recommended industrial ventilation guidelines. U.S. Department of Health, Education, and Welfare, NIOSH. *Publication* No. 76-162. Washington, D.C.

Heinsohn, R.J. 1991. *Industrial ventilation: Engineering principles.* John Wiley and Sons, New York.

Hemeon, W.C.L. 1963. *Plant and process ventilation*, p. 77. Industrial Press, New York.

Hinds, W. 1982. Aerosol technology: Properties, behavior, and measurement of airborne particles. John Wiley and Sons, New York.

Holman, J.P. 1989. *Heat transfer.* McGraw Hill, Singapore.

Idelchik, I.E., G.R. Malyavskaya, O.G. Martynenko, and E. Fried. 1986. *Handbook of hydraulic resistance*, 2nd ed. Hemisphere Publishing Corporation, subsidiary of Harper and Row, New York.

Ivanitskaya, M.Yu., and V.I. Kunitsa. 1974. Experimental studies of thermal plumes above a round heat source. Proceedings of TsNIIPromzdanii. V.37. TsNIIPromzdanii, Moscow (in Russian).

Kofoed, P. 1991. Thermal plumes in ventilated rooms. Ph.D. thesis, Aalborg University, Denmark.

Kofoed, P. and P.V. Nielsen. 1991. Thermal plumes in ventilated rooms—Vertical volume flux influenced by enclosing walls. 12th AIVC Conference, Ottawa.

Kuz'mina, L.V. 1959. Sidedraft and cornerdraft hoods. Transactions of the Institutes for Labor Protection of the VTsSPS (All-Union Central Council of Trade Unions). No. 2. Moscow: PROFIZDAT, pp. 25-34 (in Russian).

Ljungqvist, B. and C. Waering. 1988. Some observations on "modern" design of fume cupboards. Proceedings of the 2nd International Symposium on Ventilation for Contaminant Control, Ventilation '88. Pergamon Press, UK.

Morton, B.R., G. Taylor, and J.S. Turner. 1956. Turbulent gravitational convection from maintained and instantaneous sources. Proceedings of Royal Society. Vol. 234A, p. 1.

Mundt, E. 1992. Convection flows in rooms with temperature gradients—Theory and measurements. RoomVent '92. Proceedings of the Third International Conference on Air Distribution in Rooms. Vol. 3. Aalborg.

Nielsen, P.V. 1993. Displacement ventilation—Theory and design. Department of Building Technology and Structural Engineering. Aalborg University, Denmark.

NFPA. 1995. Standard for ovens and furnaces. ANSI/NFPA *Standard* 86-95. National Fire Protection Association, Quincy, MA.

Popiolec, Z. 1981. Problems of testing and mathematical modeling of plumes above human body and other extensive heat sources. A4-seria. No. 54. KTH, Stockholm.

Posokhin, V.N. 1984. Design of local ventilation systems for process equipment with heat and gas release. Mashinostroyeniye, Moscow (in Russian).

Posokhin, V.N. and V.A. Broida. 1980. Local exhausts incorporated with air curtains. Hydromechanics and heat transfer in sanitary technique equipment. KHTI, Kazan (in Russian).

Posokhin, V.N. and A.M. Zhivov. 1997. Principles of local exhaust design. Proceedings of the 5th International Symposium on Ventilation for Contaminant Control. Vol.1. The Canadian Environment Industry Association (CEIA), Ottawa.

Romeyko, N.F., N.E. Siromyatnikova, and E.V. Schibraev. 1976. Design of air curtains near an oven opening supplied with a hood. Heating and Ventilation. Proceedings of the A.I. Mikoyan Institute of Civil Engineers (in Russian).

Schaelin, A. and P. Kofoed. 1992. Numerical simulation of thermal plumes in rooms. RoomVent '92. Proceedings of the Third International Conference on Air Distribution in Rooms. Vol. 1. Aalborg, Denmark.

Schmidt, W. 1941. Turbulente Ausbreitung eines Stromes erhitzter Luft. ZAMM. Bd. 21 # 5 (in German).

Sciola, V. 1993. The practical application of reduced flow push-pull plating tank exhaust systems. 3rd International Symposium on Ventilation for Contaminant Control, Ventilation '91 (Cincinnati, OH).

Shepelev, I.A. 1961. Turbulent convective stream above a heat source. Proceedings of Acad. Sci. USSR. Mechanics and Machinery Construction #4 (in Russian).

Skäret, E. 1986. Ventilasjonsteknikk. Textbook. Institute of Heating, Ventilation and Sanitary Techniques, NTH. Trondheim (in Norwegian).

Skistad, H. 1994. *Displacement ventilation.* Research Studies Press, John Wiley and Sons, West Sussex. UK.

SMACNA. 1977. Round industrial duct standards. Sheet Metal and Air Conditioning Contractors' National Association, Vienna, VA.

SMACNA. 1980. Rectangular industrial duct construction standards.

Stoler, V.D. and Yu. L. Savelyev. 1977. Push-pull systems design for etching tanks. *Heating, Ventilation, Water Supply, and Sewage Systems Design* 8(124). TsINIS, Moscow (in Russian).

Stroiizdat. 1992. *Designer's guide.* Ventilation and air conditioning, 4th ed. Part 3(1). Stroiizdat, Moscow (in Russian).

Strongin, A.S. and M.L. Marder. 1988. Complex solution of painting shops ventilation. Proceedings of the conference "Utilization of Natural Resources and New Ventilation and Dust Transportation Systems Design". Penza (in Russian).

Strongin, A.S., M.Yu. Ivanitskaya, and E.A. Visotskaya. 1986. Studies of the application of air curtains in tunnels for local ventilation. *Heating and Ventilation. Transactions* of TsNIIpromzdanii (in Russian).

Tyaglo, I.G. and I.A. Shepelev. 1970. Air flow near an exhaust opening. *Vodosnabzheniye i Sanitarnaya Tekhnika* #5, pp. 24-25 (in Russian).

U.S. Patent. Device for removal of deleterious impurities from room atmosphere. U.S. Patent # 5,716,268. February 1998.

Zeldovitch, Y.B. 1937. Fundamental principles for free convective plumes. *Journal of Experimental and Technical Physics* 7(12). Moscow (in Russian).

Zhivov, A.M. 1993. Principles of source capturing and general ventilation design for welding premises. *ASHRAE Transactions* 99(1):979-86.

Zhivov, A.M. and J.T. Ashe. 1997. Principles of welding fume control. Proceedings of the 5th International Symposium on Ventilation for Contaminant Control. Vol. 1. The Canadian Environment Industry Association (CEIA), Ottawa.

Zhivov, A.M., L.L. Christianson, and G.L. Riskowski. 1997. Influence of space air movement on hood performance. ASHRAE *Research Project* RP-744.

BIBLIOGRAPHY

AICVF. 1991. Principles of airflow applied to the HVAC field. Collection des guides de l'AICVF. Association d'Ingenierie de Chauffage, Ventilation et Froid, Paris (in French).

Balchin, N.C. (Ed.) 1991. Health and safety in welding and allied processes, 4th ed. Abington Publishing, Cambridge, UK.

Baturin, V.V. 1972. *Fundamentals of industrial ventilation*, 3rd English ed. Pergamon Press, New York.

Braconnier, R. 1988. Bibliographic review of velocity field in the vicinity of local exhaust hood openings. *American Industrial Hygiene Association Journal* 49(4):185-98.

Brandt, A.D., R.J. Steffy, and R.G. Huebscher. 1947. Nature of air flow at suction openings. *ASHVE Transactions* 53:55-76.

British Occupational Hygiene Society (BOHS). 1987. Controlling airborne contaminants in the workplace. *Technical Guide* No. 7. Science Review Ltd. and H&H Sci. Consult., Leeds, UK.

CIBSE. 1986. Guide B: Installation and equipment data. Chartered Institution of Building Services Engineers, London.

Elterman V. 1985. Pollution control at chemical and oil-chemical industry. Chemia, Moscow (in Russian).

Flynn, M.R. and M.J. Ellenbecker. 1985. The potential flow solution for air flow into a flanged circular hood. *American Industrial Hygiene Journal* 46(6):318-22.

Glinski, M. 1978. Influence of disturbing streams on efficiency of suction pipes in local ventilation installations. *Transactions of Central Institute for Labor Protection* 28:45-60. Warsaw.

Goodfellow, H.D. 1985. Advanced design of ventilation systems for contaminant control. *Chem Eng. Monograph* 231. Elsevier, Amsterdam.

Goodfellow, H.D. 1986. Ventilation '85 (Conference Proceedings). Elsevier, Amsterdam.

Grimitlyn, M. et al. 1983. Ventilation and heating of plastics processing shops. Chemia, Leningrad (in Russian).

Grimitlyn, M. et al. 1993. Ventilation and heating of shops of machinery building plants. 2nd ed. Mashinostroyeniye, Moscow (in Russian).

Heinsohn, R.J., K.C. Hsieh, and C.L. Merkle. 1985. Lateral ventilation systems for open vessels. *ASHRAE Transactions* 91(1B):361-82.

Huebener, D.J. and R.T. Hughes. 1985. Development of push-pull ventilation. *American Industrial Hygiene Association Journal* 46(5):262-67.

INRS. 1983. Ventilation of foundry knock-out workplace. *Guide for ventilation practice* #4. ED662. Institut National de Recherche et de Sécurité, Paris (in French).

INRS. 1985. Extraction and air cleaning from oil mist. *Guide for ventilation practice* #6. ED680 (in French).

INRS. 1986. General principles of ventilation. *Guide for ventilation practice* #0. ED695 (in French).

INRS. 1987. Painting of large and/or bulky equipment. *Guide for ventilation practice* #10. ED713 (in French).

INRS. 1988. Screen printing. *Guide for ventilation practice* #11. ED711 (in French).

INRS. 1988. Ventilation of arc welding operations. *Guide for ventilation practice* #8. ED668 (in French).

INRS. 1988. Ventilation of confined spaces. *Guide for ventilation practice* #9. ED703 (in French).

INRS. 1989. Production processes in laminated polyester workshops. *Guide for ventilation practice* #3. ED665 (in French).

INRS. 1989. Small articles gluing workshops. *Guide for ventilation practice* #5. ED672 (in French).

INRS. 1989. Workroom air cleaning. *Guide for ventilation practice* #1. ED657 (in French).

INRS. 1990. Ventilation of open surface tanks. *Guide for ventilation practice* #2. ED651 (in French).

INRS. 1991. Lead accumulator manufacturing. *Guide for ventilation practice* #13. ED746 (in French).

INRS. 1991. Ventilation of painting booths and workplace. *Guide for ventilation practice* #7. ED663 (in French).

INRS. 1992. Automobile radiator repairs. *Guide for ventilation practice* #15. ED752 (in French).

INRS. 1992. Denture manufacturing workshops. *Guide for ventilation practice* #16. ED760 (in French).

INRS. 1992. Woodworking. *Guide for ventilation practice* #12. ED750 (in French).

INRS. 1993. Use of powders. *Guide for ventilation practice* #17. ED767 (in French).

Pozin, G.M. 1977. Calculation of the effect of limitation planes on suction flows. *Transactions of the Central Institute for Labor Protection of the VCSPS* 105:8-13. Profizdat, Moscow (in Russian).

Qiang, Y.L. 1984. The effectiveness of hoods in windy conditions. Kungl Tekniska Hoggskolan. Stockholm, Sweden.

Safemazandarani, P. and H.D. Goodfellow. 1989. Analysis of remote receptor hoods under the influence of cross-drafts. *ASHRAE Transactions* 95(1):465-71.

Sepsy, C.F. and D.B. Pies. 1973. An experimental study of the pressure losses in converging flow fittings used in exhaust systems. *Document* PB 221 130. Prepared by Ohio State University for National Institute for Occupational Health.

Shibata, M., R.H. Howell, and T. Hayashi. 1982. Characteristics and design method for push-pull hoods: Part 1—Cooperation theory on air flow; Part 2—Streamline analysis of push-pull flows. *ASHRAE Transactions* 88(1):535-70.

Silverman, L. 1942. Velocity characteristics of narrow exhaust slots. *Journal of Industrial Hygiene and Toxicology* 24 (November):267.

Sutton, O.G. 1950. The dispersion of hot gases in the atmosphere. *Journal of Meteorology* 7(5):307.

Ventilation '85. 1986. Proceedings of the 1st International Symposium on Ventilation for Contaminant Control. Elsevier Science Publishers, Amsterdam.

Ventilation '88. 1989. Proceedings of the 2nd International Symposium on Ventilation for Contaminant Control. Elsevier Science Publishers, Amsterdam.

Ventilation '91. 1993. Proceedings of the 3nd International Symposium on Ventilation for Contaminant Control. American Conference of Governmental Industrial Hygienists (ACGIH), Cincinnati, OH.

Ventilation '94. 1994. Proceedings of the 4th International Symposium on Ventilation for Contaminant Control. Arbets Miljo Institutet (National Institute of Occupational Health), Stockholm.

Ventilation '97. 1997. Proceedings of the 5th International Symposium on Ventilation for Contaminant Control. The Canadian Environment Industry Association (CEIA), Ottawa.

Zarouri, M.D., R.J. Heinsohn, and C.L. Merkle. 1983. Computer-aided design of a grinding booth for large castings. *ASHRAE Transactions* 89 (2A):95-118.

Zarouri, M.D., R.J. Heinsohn, and C.L. Merkle. 1983. Numerical computation of trajectories and concentrations of particles in a grinding booth. *ASHRAE Transactions* 89(2A):119-35.

KITCHEN VENTILATION

K ITCHEN ventilation is a complex application of HVAC systems. System design includes aspects of air conditioning, fire safety, ventilation, building pressurization, refrigeration, air distribution, and food service equipment. Kitchens are in many buildings, including hotels, hospitals, retail malls, single- and multi-family dwellings, and correctional facilities. Each of these building types has special requirements for its kitchens, but many of the basic needs are common to all.

Kitchen ventilation has at least two purposes: (1) to provide a comfortable environment in the kitchen and (2) to enhance the safety of personnel working in the kitchen and of other building occupants. "Comfortable" in this context has different meanings because, depending on the local climate, some kitchens are not air conditioned. Obviously, the kitchen ventilation system can affect temperature and humidity in the kitchen. The ventilation system can also affect the acoustics of a kitchen.

The centerpiece of almost any kitchen ventilation system is an exhaust hood, which is used primarily to remove effluent from kitchens. **Effluent** includes the gaseous, liquid, and solid contaminants produced by the cooking process. These contaminants must be removed for both comfort and safety. Effluent can range from simply annoying to potentially life-threatening and, under certain conditions, flammable. The arrangement of the food service equipment and its coordination with the hood(s) greatly affect the operating costs of the kitchen.

HVAC system designers are most frequently involved in commercial kitchen applications, in which cooking effluent contains large amounts of grease or water vapor. Residential kitchens typically use a totally different type of hood. The amount of grease produced in residential applications is significantly less than in commercial applications, so the health and fire hazard is much lower.

COOKING EFFLUENT

Effluent Generation

Cooking is the process of creating chemical and physical changes in food by applying heat to the raw or precooked food. Cooking improves edibility, taste, or appearance or delays decay. As heat is applied to the food, effluent is released into the surrounding atmosphere. This effluent includes heat that has not transferred to the food, water vapor, and organic material released from the food. The heat source, especially if it involves combustion, may release other contaminants.

All cooking methods release some heat, some of which radiates from all hot surfaces; but most is dissipated by natural convection via a rising **plume** of heated air. Most of the effluent released from the food and the heat source is entrained in this plume, so primary contaminant control should be based on capturing and removing the air and effluent that constitute the plume. A quantitative analysis, or even a relative determination, of plume and combustion product volumetric flow rates is not available at present.

The preparation of this chapter is assigned to TC 5.10, Kitchen Ventilation.

Plume Behavior

The most common method of contaminant control is to install an air inlet device (a **hood**) where the plume can enter it and be conveyed away by an exhaust system. The hood is generally located above or behind the heated surface to intercept the normal upward flow path. Understanding the behavior of the plume is central to designing effective ventilation systems.

Effluent released from a noncooking cold process, such as metal grinding, is captured and removed by placing air inlets so that they catch forcibly ejected material, or by creating airstreams with sufficient velocity to induce the flow of effluent into an inlet. This technique has led to an empirical concept of **capture velocity** that is often misapplied to hot processes. Effluent released from a hot process and contained in a plume may be captured by locating an inlet hood so that the plume flows into it by buoyancy. The hood exhaust rate must equal or slightly exceed the plume flow rate, but the hood need not actively capture the effluent if the hood is large enough at its height above the cooking operation to encompass the plume as it expands during its rise. Additional exhaust airflow may be needed to resist crosscurrents that carry the plume away from the hood.

A plume, in the absence of crosscurrents or other interference, rises vertically. As it rises, it entrains additional air, which causes the plume to enlarge and its average velocity and temperature to decrease. In most cooking processes, the distance between the heated surface and the hood is so short that entrainment is negligible, and the plume loses very little of its velocity or temperature before it reaches the hood. If a surface parallel to the plume centerline (e.g., a back wall) is located nearby, the plume will attach to the surface by the **Coanda effect**. This tendency also directs the plume into the hood.

Appliance Types (Steam, Electric, Solid Fuel, Gas)

The heat source affects the type and quantity of effluent released. When steam is the heat source, it releases no contaminants because it is contained in a closed vessel. Electric heating sources similarly release no significant contaminants.

Solid (wood or charcoal) or gaseous (natural gas or liquefied petroleum gas) fuels are common sources of heat for cooking. Their combustion generates water vapor and carbon dioxide, and it may also generate carbon monoxide and other potentially harmful gases. These effluents must be controlled along with those released from the food. In some cases, the food or its container is directly exposed to the flame; as a result, the combustion effluent and the food effluent are mixed, and a single plume is generated. In other cases, such as ovens, the combustion products are ducted to an outlet adjacent to the plume, and the effluents still mix.

EXHAUST HOODS

The kitchen exhaust hood captures, contains, and evacuates heat, smoke, odor, steam, grease, vapor, and other contaminants generated from cooking in order to provide a safe, healthy, comfortable, and productive work environment for kitchen personnel.

This section discusses all aspects of kitchen hood design; it is based primarily on model codes and standards in the United States.

The design, engineering, construction, installation, and maintenance of commercial kitchen exhaust hoods are controlled by the major nationally recognized standards (e.g., NFPA *Standard* 96) and model codes. In some cases, local codes may prevail. Prior to designing a kitchen ventilation system, the designer should identify governing codes and consult the authority having jurisdiction. Local authorities having jurisdiction may have amendments or additions to these standards and codes.

Hood Types

Many types, categories, and styles of hoods are available, and hood selection depends on many factors. Hoods are classified based on whether they are designed to handle grease. Type I refers to hoods designed for removal of grease and smoke, and Type II refers to all other hoods. The model codes distinguish between grease-handling and non-grease-handling hoods, but not all model codes use Type I/Type II terminology. A Type I hood may be used where a Type II hood is required, but the reverse is not allowed. However, the characteristics of the cooking equipment under the hood, and not the hood type, determine the requirements for the entire exhaust system, including the hood.

A **Type I hood** is used for collection and removal of grease and smoke. It includes (1) listed grease filters, baffles, or extractors for removal of the grease and (2) fire suppression equipment. Type I hoods are required over restaurant equipment, such as ranges, fryers, griddles, broilers, ovens, and steam kettles, that produce smoke or grease-laden vapors.

A **Type II hood** is for collection and removal of steam, vapor, heat, and odors where grease is not present. It may or may not have grease filters or baffles and typically does not have a fire suppression system. It is typically used over dishwashers, steam tables, and so forth. The Type II hood is sometimes used over ovens, steamers, or kettles if they do not produce smoke or grease-laden vapor and if the authority having jurisdiction allows it.

Type I Hoods Categories

Type I hoods fall into two categories. One is the **conventional (nonlisted)** category, which meets the design, construction, and performance criteria of the applicable national and local codes. Conventional (nonlisted) hoods are not allowed to have fire-actuated exhaust dampers.

The second category comprises hoods that are **listed** in Underwriters Laboratories (UL) *Standard* 710. Listed hoods are not generally designed, constructed, or operated in accordance with requirements of the model codes, but are constructed in accordance with the terms of the hood manufacturer's listing. This is allowed because the model codes include exceptions for hoods listed to show equivalency with the safety criteria of the model code requirements.

The two basic subcategories of Type I listed hoods, as defined in UL *Standard* 710, are exhaust hoods without exhaust dampers and exhaust hoods with exhaust dampers. The UL listings do not distinguish between water-wash and dry hoods, except that water-wash hoods with fire-actuated water systems are identified in UL's product directory.

All listed hoods are subjected to electrical tests, temperature tests, and fire and cooking smoke capture tests. The listed exhaust hood with exhaust damper includes a fire-actuated damper, typically located at the exhaust duct collar (and at the replacement air duct collar, depending on the hood configuration). In the event of a fire, the damper closes to prevent fire from entering the duct. Fire-actuated dampers are permitted only as part of a hood listing.

Listed exhaust hoods with fire-actuated water systems are typically water-wash hoods in which the wash system also operates as

a fire-extinguishing system. In addition to meeting the requirements of UL *Standard* 710, these hoods are tested under UL *Standard* 300 and may be listed for plenum extinguishment, duct extinguishment, or both.

Type I Hoods—Grease Removal

Most grease removal devices in Type I hoods operate on the same general principle—the exhaust air passes through a series of baffles in which a centrifugal force that throws the grease particles out of the airstream is created as the exhaust air passes around the baffles. The amount of grease removed varies with the design of the baffles, the air velocity, the temperature, the type of cooking, and other factors. A recognized test protocol is not available at present. Mesh filters cannot meet the requirements of UL *Standard* 1046 and therefore cannot be used as primary grease filters. Grease removal devices generally fall into the following categories:

- **Baffle filter.** The baffle filter is a series of vertical baffles designed to capture grease and drain it into a container. The filters are arranged in a channel or bracket for easy insertion and easy removal from the hood for cleaning. Each hood usually has two or more baffle filters. The filters are typically constructed of aluminum, steel, or stainless steel, and they come in various standard sizes. Filters are cleaned by running them through a dishwasher or by soaking and rinsing. NFPA *Standard* 96 requires that grease filters be listed. Listed grease filters are tested and certified by a nationally recognized test laboratory under UL *Standard* 1046.

- **Removable extractor.** Removable extractors are an integral component of listed exhaust hoods designed to use them. They are typically constructed of stainless steel and contain a series of horizontal baffles designed to remove grease and drain it into a container. Removable extractors come in various sizes. They are cleaned by running them through a dishwasher or by soaking and rinsing.

- **Stationary extractor.** The stationary extractor (also called a **water-wash hood**) is an integral component of listed exhaust hoods that use them. They are typically constructed of stainless steel and contain a series of horizontal baffles that run the full length of the hood. The baffles are not removable for cleaning. The stationary extractor includes one or more water manifolds with spray nozzles that, upon activation, wash the grease extractor with hot, detergent-injected water, removing accumulated grease. The wash cycle is typically activated at the end of the day, after the cooking equipment and fans have been turned off; however, it can be activated more frequently. The cycle lasts for 5 to 10 min, depending on the hood manufacturer, the type of cooking, the duration of operation, and the water temperature and pressure. Most water-wash hood manufacturers recommend a water temperature of 130 to 180°F and water pressure of 30 to 80 psi. Average water consumption varies from 0.50 to 1.50 gpm per linear foot of hood, depending on the hood manufacturer. Most water-wash hood manufacturers provide a manual and/or an automatic means of activating the water-wash system in the event of a fire.

Some manufacturers of water-wash hoods provide continuous cold water as an option. The cold water runs continuously during cooking and may or may not be recirculated, depending on the manufacturer. Typical cold water usage is 1 gph per linear foot of hood. The advantage of continuous cold water is that it improves grease extraction and removal, partly through condensation of the grease. Many hood manufacturers recommend continuous cold water in hoods that are located over solid-fuel-burning equipment, as the water also extinguishes hot embers that may be drawn up into the hood and helps cool the exhaust stream.

UL *Standards* 1046 and 710 do not include grease extraction tests because no industry-accepted tests are available at present in the United States. Grease extraction rates published by filter and hood manufacturers are usually derived from tests conducted by independent test laboratories retained by the manufacturer. Test methods and results therefore vary greatly.

Type I Hoods—Styles

Figure 1 shows the six basic hood styles for Type I applications. These style names are not used universally in all standards and codes but are well accepted in the industry. The styles are as follows:

1. **Wall-mounted canopy.** Used for all types of cooking equipment located against a wall.
2. **Single-island canopy.** Used for all types of cooking equipment in a single-line island configuration.
3. **Double-island canopy.** Used for all types of cooking equipment mounted back-to-back in an island configuration.
4. **Back shelf.** Used for counter-height equipment typically located against a wall, but could be freestanding.
5. **Eyebrow.** Used for direct mounting to ovens and some dishwashers.
6. **Pass-over.** Used over counter-height equipment when pass-over configuration (from the cooking side to the serving side) is required.

Type I Hoods—Sizing

The size of the exhaust hood in relation to cooking appliances is an important aspect of hood performance. Usually the hood must extend beyond the cooking appliances—on all open sides on canopy-style hoods and over the ends on back shelf and pass-over hoods—to capture the expanding thermal currents rising from the appliances. This **overhang** varies with the style of the hood, the distance between the hood and the cooking appliance, and the characteristics of the cooking equipment. With back shelf and pass-over hoods, the front of the hood must be kept behind the front of the cooking equipment (**set back**) to allow head clearance for the cooks. These hoods may require a higher front inlet velocity to catch and contain the expanding thermal currents. All styles may have full or partial side panels to close the area between the appliances and the hood. This may eliminate the overhang requirement and generally reduces the exhaust flow rate requirement.

For conventional hoods, hood size is dictated by the prevailing model code, and for listed hoods, by the terms of the manufacturer's listing. Typically, the overhang requirements applied to listed hoods are similar to those for conventional hoods. General overhang requirements are shown in Table 1.

Type I Hoods—Exhaust Flow Rates

Exhaust flow rate requirements to capture, contain, and remove the effluent vary considerably depending on the hood style, the amount of overhang, the distance from the cooking surface to the hood, the presence and size of side panels, and the cooking equipment and product involved. The hot cooking surfaces and product vapors create thermal air currents that are received or captured by the hood and then exhausted. The velocity of these currents depends largely on the surface temperature and tends to vary from 15 fpm over steam equipment to 150 fpm over charcoal broilers. The actual required flow rate is determined by these thermal currents, a safety allowance to absorb crosscurrents and flare-ups, and a safety factor for the style of hood.

Overhangs, the distance from the cooking surface to the hood, and the presence or absence of side panels all help determine the safety factor for different hood styles. Use of gas-fired cooking equipment may require an additional allowance for the exhaust of combustion products and combustion air. Because it is not practical to place a separate hood over each piece of equipment, general practice is to categorize the equipment into four groups. While published lists vary, and accurate documentation does not yet exist, the following is a consensus opinion list (great variance in product or volume could shift an appliance into another category):

1. Light duty, such as ovens, steamers, and small kettles (up to 400°F)
2. Medium duty, such as large kettles, ranges, griddles, and fryers (up to 400°F)
3. Heavy duty, such as upright broilers, charbroilers, and woks (up to 600°F)
4. Extra heavy duty, such as solid-fuel-burning equipment (up to 700°F)

The exhaust volumetric flow rate requirement is based on the group of equipment under the hood. If there is more than one group, the flow rate is based on the heaviest duty group unless the hood design permits different rates over different sections of the hood.

For areas where model codes or other regional codes have been adopted, the exhaust flow rate requirement for conventional hoods is dictated by the codes; therefore, the manufacturers' calculation methods may not be used without consultation with the authority having jurisdiction. The model code required exhaust flow rates for conventional canopy hoods are typically calculated by multiplying the area A of the hood opening by a given air velocity. Table 2 indicates typical formulas, taken from the model codes, for determining the exhaust flow rate Q for conventional canopy hoods. Some jurisdictions may use the length of the open perimeter of the hood times the vertical height between the hood and the appliance instead of the horizontal hood area.

The *International Mechanical Code* (IMC) and some state codes have alternate formulas that allow lower flow rates for equipment that produces less heat and smoke; however, the IMC does require 200 cfm per square foot of hood area for hoods covering charbroilers. Back shelf and pass-over style nonlisted hoods are usually calculated at 300 cfm per linear foot of exhaust hood.

Listed hoods are allowed to operate at their listed exhaust flow rates by exceptions in the model codes. Most manufacturers of listed hoods verify their listed flow rates by conducting tests per UL *Standard* 710. Typically, the average flow rates are much lower than those dictated by the model codes. It should be noted that these listed values are established under draft-free laboratory conditions. The four categories of equipment groups mentioned are tested and marked according to cooking surface temperature: light and medium duty up to 400°F, heavy duty up to 600°F, and extra heavy duty up to 700°F. Each of these groups has an air quantity factor

Table 1 Typical Overhang Requirements for Both Listed and Conventional (Nonlisted) Type I Hoods

Type of Hood	End Overhang	Front Overhang	Rear Overhang
Wall-mounted canopy	6 in.	12 in.	—
Single-island canopy	12 in.	12 in.	12 in.
Double-island canopy	6 in.	12 in.	12 in.
Eyebrow	0 in.	12 in.	—
Back shelf/Pass-over	0 in.	6 to 12 in. front setback	

Note: The model codes typically require a 6 in. minimum overhang, but most manufacturers design for a 12 in. overhang.

Table 2 Typical Model Code Exhaust Flow Rates for Conventional Type I Hoods

Wall-mounted canopy	$Q = 100A$
Single-island canopy	$Q = 150A$
Double-island canopy	$Q = 100A$
Eyebrow	$Q = 100A$
Back shelf/Pass-over	$Q = 300 \times$ Length of hood

Note: Q = exhaust flow rate, cfm; A = area of hood exhaust aperture, ft^2

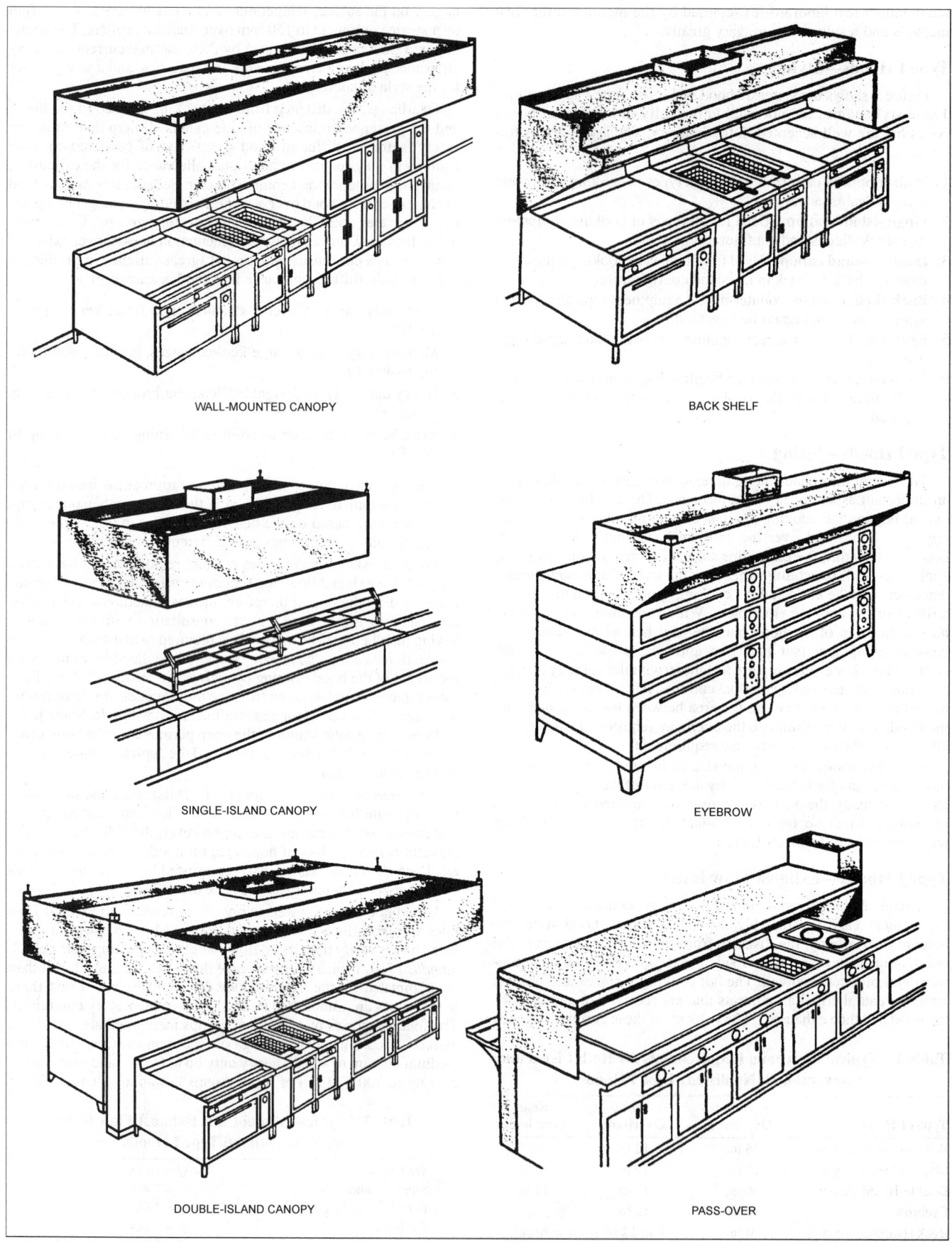

WALL-MOUNTED CANOPY

BACK SHELF

SINGLE-ISLAND CANOPY

EYEBROW

DOUBLE-ISLAND CANOPY

PASS-OVER

Fig. 1   Styles of Commercial Kitchen Exhaust Hoods

**Table 3 Typical Minimum Exhaust Flow Rates for
Listed Type I Hoods by Cooking Equipment Type**

| Type of Hood | Minimum Exhaust Flow Rate, cfm per linear foot of hood | | | |
	Light Duty	Medium Duty	Heavy Duty	Extra Heavy Duty
Wall-mounted canopy	150 to 200	200 to 300	200 to 400	350+
Single-island	250 to 300	300 to 400	300 to 600	550+
Double-island (per side)	150 to 200	200 to 300	250 to 400	500+
Eyebrow	150 to 250	150 to 250	—	—
Back shelf/ Pass-over	100 to 200	200 to 300	300 to 400	Not recommended

assigned for each style of hood, with the total exhaust flow rate typically calculated by multiplying this factor times the length of the hood.

Minimum exhaust flow rates for listed hoods serving single categories of equipment vary from manufacturer to manufacturer but are generally as shown in Table 3.

Actual exhaust flow rates for hoods with internal short-circuit replacement air are typically higher than those in Table 3, although the net exhaust (actual exhaust less replacement air quantity) may be similar. The specific hood manufacturer should be contacted for exact exhaust and replacement flow rates.

ASTM *Standard* F 1704 details a laboratory flow visualization procedure for determining the capture and containment threshold of an appliance/hood system. This procedure is consistent with the UL *Standard* 710 capture test and can be applied to all hood types and configurations operating over any cooking appliances.

Type I Hoods—Replacement (Makeup) Air Options

Air exhausted from the kitchen space must be replaced. Replacement air can be brought in through the traditional method of **ceiling registers**; however, they must be located so that the discharged air does not disrupt the pattern of air entering the hood. Air should be supplied either (1) as far from the hood as possible or (2) close to the hood and directed away from it or straight down at very low velocity. Exhaust and replacement air fans should be interlocked.

Another way of distributing replacement air is through systems built as an integral part of the hood. Figure 2 shows three available designs for **internal** replacement air. Combinations of these designs are also available. Because the actual flows and percentages vary with all hoods, the manufacturer should be consulted about specific applications. The following are typical descriptions:

Front Face Discharge. This method of introducing replacement air into the kitchen is flexible and has many advantages. Typical supply volume is 70 to 80% of the exhaust, depending on the air balance desired. Supply air temperature should range from 60 to 65°F but may be as low as 50°F, depending on flow rates, distribution, and internal heat load. This air should be directed away from the hood, but the closer the air outlet's lower edge is to the bottom of the hood, the lower the velocity must be to avoid drawing effluent out of the hood.

Down Discharge. This method of introducing replacement air to the kitchen area is typically used when spot cooling of the cooking staff is desired to help relieve the effects of severe radiant heat generated from such equipment as charbroilers. The air must be heated and/or cooled, depending on the climate. Discharge velocities must be carefully selected to avoid air turbulence at the cooking surface, discomfort to personnel, and cooling of food. The amount of supply air introduced may be up to 70% of the exhaust, depending on the cooking equipment involved. Air temperature should be between 50 and 65°F.

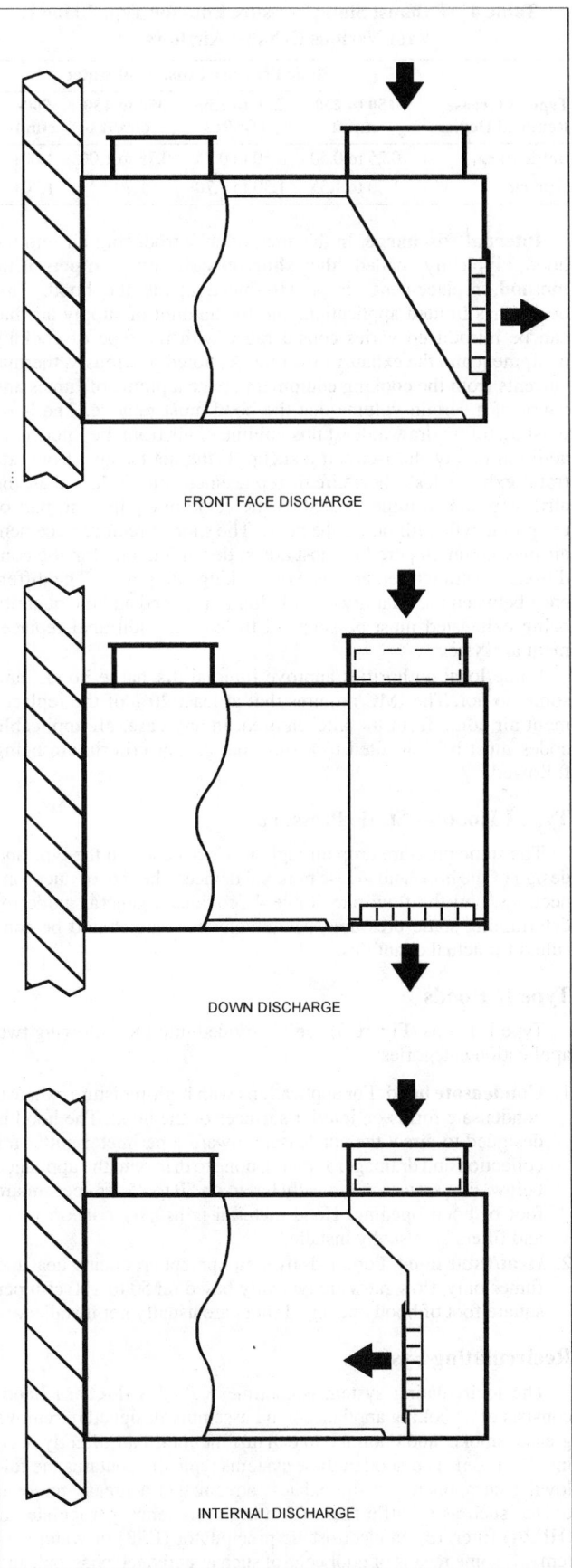

FRONT FACE DISCHARGE

DOWN DISCHARGE

INTERNAL DISCHARGE

Fig. 2 Internal Methods of Introducing Replacement Air

Table 4 Exhaust Static Pressure Loss for Type I Hoods for Various Exhaust Airflows

| Type of Grease Removal Device | Static Pressure Loss, in. of water | | | |
	150 to 250 cfm/ft	250 to 350 cfm/ft	350 to 450 cfm/ft	500+ cfm/ft
Baffle filter	0.25 to 0.50	0.50 to 0.75	0.75 to 1.00	1.00+
Extractor	1.00 to 1.35	1.30 to 1.70	1.70+	1.70+

Internal Discharge. In this method of introducing air into the hood, typically called the **short-circuit or compensating method**, replacement air is introduced inside the hood. This design has limited application, and the amount of supply air that can be introduced varies considerably with the type of cooking equipment and the exhaust flow rate. As noted previously, thermal currents from the cooking equipment create a plume of fumes and vapor of a certain volume that the hood must remove. The hood must therefore draw at least this volume of air from the kitchen, in addition to any short-circuit makeup. If the net exhaust flow rate (total exhaust less short-circuit replacement air) is less than the airflow plume volume created by the cooking equipment, part of the plume will spill out of the hood. The short-circuit replacement air may be untempered in most areas, depending on climatic conditions, manufacturer, and type of cooking equipment. The difference between the quantity of air being introduced and the quantity being exhausted must be supplied through a traditional replacement air system

Some local authorities approve internal discharge hoods, and some do not. The IMC requires that at least 20% of the replacement air come from the kitchen area. In any case, all applicable codes must be consulted to assure that proper criteria are being followed.

Type I Hoods—Static Pressure

The static pressure drop through hoods depends on the type and design of the hood and grease removal devices, the size of duct connections, and the flow rate. Table 4 provides a general guide for determining static pressures. Manufacturers' data should be consulted for actual quantities.

Type II Hoods

Type II hoods (Figure 3) can be divided into the following two application categories:

1. **Condensate hood.** For applications with high-moisture exhaust, condensate forms on interior surfaces of the hood. The hood is designed to direct the condensate toward a perimeter gutter for collection and drainage, allowing none to drip onto the appliance below. Flow rates are typically based on 50 to 75 cfm per square foot of hood opening. Hood material is usually noncorrosive, and filters are usually installed.

2. **Heat/fume hood.** For hoods over equipment producing heat and fumes only, flow rates are typically based on 50 to 100 cfm per square foot of hood opening. Filters are usually not installed.

Recirculating Systems

The recirculating system, sometimes called a **ductless hood**, consists of a cooking appliance/hood assembly designed to remove grease, smoke, and odor and to exhaust the return air directly back into the room. The hood in these systems typically contains the following components in the exhaust stream: (1) a grease removal device such as a baffle filter, (2) a high-efficiency particulate air (HEPA) filter, (3) an electrostatic precipitator (ESP) or water system, (4) some means of odor control such as activated charcoal, and (5) an exhaust fan. NFPA *Standard* 96, Chapter 10, is devoted entirely to recirculating systems and contains specific requirements

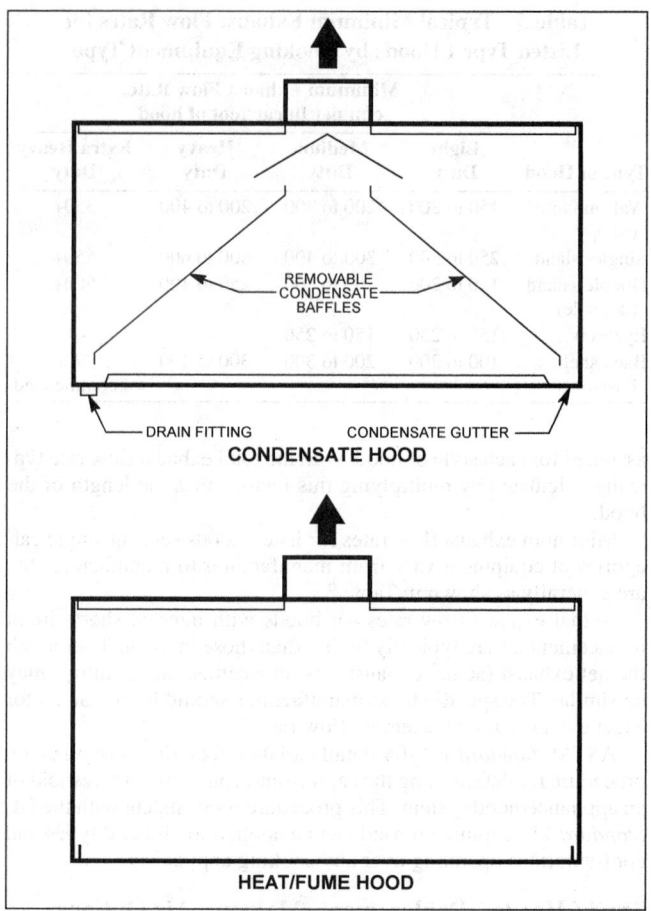

Fig. 3   Type II Hoods

such as (1) design, including interlocks of all critical components to prevent operation of the cooking appliance if any of the components are not operating; (2) fire extinguishment, including specific nozzle locations; (3) maintenance, including a specific schedule for cleaning filters, ESPs, hoods, and blowers; and (4) inspection and testing of the total operation and interlocks. In addition, NFPA *Standard* 96 requires that all recirculating systems be listed with a testing laboratory. The recognized test standard for a recirculating system is UL *Standard* 197. Ductless systems should not be used over gas-fired or solid-fuel-fired equipment, which must be exhausted directly to the outdoors to remove combustion products.

Designers should thoroughly review NFPA *Standard* 96 requirements and contact a manufacturer of recirculating systems to obtain specific information prior to incorporating this type of system into a food service design. It is important to note that recirculating systems discharge the total heat and moisture of the cooking operation back into the kitchen space, adding to the air-conditioning load.

EXHAUST SYSTEMS

Exhaust systems remove effluent from appliances and cooking processes to promote fire and health safety, comfort, and aesthetics. Typical exhaust systems simultaneously incorporate fire prevention designs and fire suppression equipment. In most cases, these functions complement each other, but in other cases they may seem to conflict. Designs must balance the various functions. For example, fire-actuated dampers may be installed to minimize the spread of fire to ducts, but maintaining an open duct might be better for removing the smoke of an appliance fire from the kitchen.

Effluent Control

Effluents generated by the cooking process include grease in the solid, liquid, and vapor states; smoke particles; and volatile organic compounds (VOCs or low-carbon aromatics, commonly referred to as odors). Effluent controls in the vast majority of today's kitchen ventilation systems are limited to the removal of solid and liquid grease particles by grease removal devices located in the hood. With currently available equipment, effluent control is typically a three-stage process: (1) grease removal, (2) smoke removal, and (3) VOC/odor removal.

Grease removal typically starts in the hood with baffle filters or grease removal devices. The more effective devices reduce grease buildup downstream of the hood, lowering the frequency of duct cleaning and reducing the fire hazard. Higher efficiency grease removal devices increase the efficiency of smoke and odor control equipment, if present.

The term grease extraction filters may be a misnomer. These filters are tested and listed not for their grease extraction ability, but for their ability to limit (not totally prevent) flame penetration into the hood plenum and duct. Additionally, research is beginning to indicate that grease particles are generally small, aerodynamic particles that are not easily removed by the centrifugal impingement principle used in most grease extraction devices (Kuehn et al. 1999).

If removal of these small particles is required, the next device is typically a particulate removal unit that removes a large percentage of the grease that was not removed by the grease removal device in the hood and a large percentage of smoke particles.

The following technologies are available today and applied to varying degrees for control of cooking effluent. Following the description of each technology are some qualifications and concerns about its use.

Electrostatic precipitators (ESPs). Particulate removal is by high-voltage ionization, then collection on flat plates.

- In a cool environment, collected grease can block airflow.
- As the ionizer section becomes dirty, efficiency drops because the effective plate surface area is reduced.

Water mist, waterfall, and water bath. Passage of the effluent stream through water mechanically entraps particulates.

- Airflow separates the bath and the waterfall, so they are less effective.
- Bath types have a very high static pressure loss.
- Spray nozzles need much attention. Water may need softening to minimize clogging.
- Drains tend to become blocked.

Pleated or bag filters of fine natural and synthetic fibers. Very fine particulate removal is by mechanical filtration. Some types have an activated carbon face coating for odor control.

- Filters become blocked quickly if too much grease enters.
- Static loss builds quickly with extraction, and airflow drops.
- Almost all filters are disposable and very expensive.

Activated carbon filters. VOC control is through adsorption by fine activated charcoal particles.

- Require a large volume and thick bed to be effective.
- Heavy and can be difficult to replace.
- Expensive to change and recharge. Many are disposable.
- Ruined quickly if they are grease-coated or subjected to water.
- Some concern that carbon is a source of fuel for a fire.

Oxidizing pellet bed filters. VOC and odor control is by oxidation of gaseous effluent into solid compounds.

- Require a large volume and long bed to be effective.

- Heavy to handle and can be difficult to replace.
- Expensive to change.
- Some concern about increased oxygen available in fire.

Incineration. Particulate, VOC, and odor control is by high-temperature oxidation (burning) into solid compounds.

- Must be at system terminus and clear of combustibles.
- Expensive to install with adequate clearances.
- Can be difficult to access for service.
- Very expensive to operate.

Catalytic conversion. A catalytic or assisting material, when exposed to relatively high-temperature air, provides additional heat adequate to decompose (oxidize) most particulates and VOCs.

- Requires high temperature (500°F minimum).
- Expensive to operate due to high temperature requirement.

Duct Systems

The exhaust ductwork conveys the exhaust air from the hood to the outdoors, along with any grease, smoke, VOCs, and odors that are not extracted from the airstream along the way. In addition, this ductwork may be used to exhaust smoke from a fire. To be effective, the ductwork must be greasetight; it must be clear of combustibles, or the combustible material must be protected so that it cannot be ignited by a fire in a duct; and ducts must be sized to convey the volumetric flow of air necessary to remove the effluent. Building codes set the minimum air velocity for exhaust ducts at 1500 fpm. Maximum velocities are limited by pressure drop and noise and should not exceed 2300 fpm. At the present time, 1800 fpm is considered the optimum design velocity.

The ductwork should have no traps that can hold grease, which would be an extra fuel source in the event of a fire, and ducts should pitch toward the hood for constant drainage of liquefied grease or condensates. On long duct runs, allowance must be made for possible thermal expansion due to a fire, and the slope back to the hood must be at least 1%.

Single-duct systems carry effluent from a single hood or section of a large hood to a single exhaust termination. In multiple-hood systems, several **branch ducts** carry effluent from several hoods to a single master duct that has a single termination.

For correct flow through the branch duct in multiple-hood systems, the static pressure loss of the branch must match the static pressure loss of the common duct upstream from the point of connection. Any exhaust points subsequently added or removed must be designed to comply with the minimum velocities required by code and to maintain the balance of the remaining system.

Ducts may be constructed of round or rectangular sections. The standards and model codes contain minimum specifications for duct materials, including gage, joining methods, and minimum clearances to combustible materials. UL-listed prefabricated duct systems may also be used. These systems typically allow reductions in the clearance to combustible materials.

Types of Exhaust Fans

Exhaust fans for kitchen ventilation must be capable of handling hot, grease-laden air. The fan should be designed to keep the motor out of the airstream and should be effectively cooled to prevent premature failure. To prevent roof damage, the fan should contain and properly drain all grease removed from the airstream.

The following types of exhaust fans are in common use (all have centrifugal wheels with backward-inclined blades):

- **Upblast.** These fans (Figure 4) are designed for roof mounting directly on top of the exhaust stack, and they discharge upward. Upblast fans are generally aluminum and must be listed for the service. They typically can provide static pressures only up to 1 in. of water gage but are available with higher pressures. They

Fig. 4   Upblast Exhaust Fan

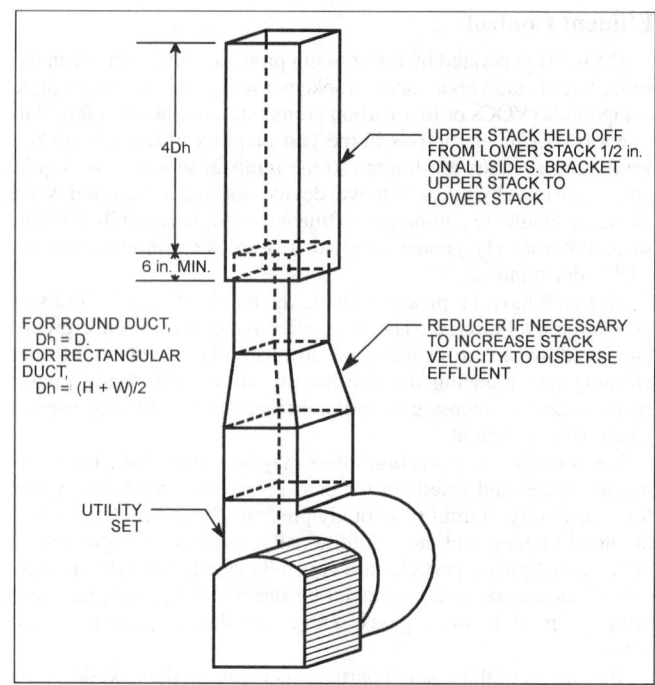

Fig. 5   Rooftop Utility Set with Vertical Discharge

may have an integral grease drainage path and collection container. These fans allow easy access for duct cleaning because they generally hinge back from the duct.

- **Utility set.** These fans are usually constructed of steel and roof-mounted. The inlet and outlet are at 90° to each other (single width, single inlet), and the outlet can usually be rotated to discharge at many different angles around a vertical circle. They can operate at medium to high and (with special provisions) very high static pressures. Care must be taken to drain the low part of the fan to a safe remote container.
- **Inline.** These fans are typically located in the duct run inside a building where exterior fan mounting is not practical for wall or roof exhaust. They are always constructed of steel. The gasketed flange mounting must be greasetight yet removable for service. A catch pan should be placed under the entire assembly in case of a grease leak at the flanges. If the fan housing is lower than the duct, there must be a drain to a separate collection container.

Exhaust Terminations

Rooftop. Rooftop terminations are preferred because the discharge can be directed away from the building, the fan is at the end of the system, and the fan is accessible. Common concerns with rooftop terminations are as follows:

- The discharge of the exhaust system should be arranged to minimize reentry of effluent into any fresh air intake or other opening to any building. This requires not only separation of the exhaust from intakes, but also knowledge of the direction of the prevailing winds. Some codes specify a minimum distance to air intakes.
- In the event of a fire, neither the flames, nor the radiant heat, nor the dripping grease should be able to ignite the roof or other nearby structures.
- All grease from the fan or the duct termination should be collected and drained to a remote closed container to preclude ignition.
- Rainwater should be kept out of the exhaust system, especially out of the grease container. If this is not possible, then the grease container should be designed to separate the water from the grease and drain the water back onto the roof. Figure 5 shows a rooftop utility set with a stackhead fitting, which directs the exhaust away from the roof and minimizes rain penetration. Discharge caps should not be used because they direct the exhaust back toward the roof and can become grease-fouled.

Outside Wall. Wall terminations are less common today but still exist. The fan may or may not be the terminus of the system, located on the outside of the wall. Common concerns with wall terminations are as follows:

- Discharge from the exhaust system should not be able to enter any fresh air intake or other opening to any building.
- Adequate clearance to combustibles must be maintained.
- To avoid having grease drain down the side of the building, duct sections should pitch back to the hood inside, or a grease drain should be provided to drain the grease back into a safe container inside the building.
- The discharge must not be directed downward or toward any pedestrian areas.
- Louvers should be designed to minimize their grease extraction effect and to prevent staining of the building facade.

Recirculating Systems. With these units, it is extremely critical to keep the components in good working order to maintain optimal performance. Otherwise, excessive grease and odor will accumulate in the premises.

As with other terminations, containing and removing the grease and keeping the discharge as far as possible from combustibles are the main concerns. Some units are fairly portable and can readily be set in an unsafe location. The operator should be made aware of the importance of safely locating the unit.

A major concern is system maintenance. If the hood is not properly maintained, it will allow grease, heat, and odors to accumulate in the space. These units are best for large, unconfined areas that have a separate outside exhaust to keep the environment comfortable.

REPLACEMENT (MAKEUP) AIR SYSTEMS

The kitchen air that is exhausted to remove the cooking effluent must be replaced with air from outside the building to avoid excess negative pressure in the space, which may degrade the exhaust system performance. The standards and model codes require 100% replacement (makeup) air. Some manufacturers offer internal discharge (short-circuit) hoods, which introduce replacement air directly into the hood capture cavity.

Replacement (Makeup) Air Through HVAC System

In many cases, the HVAC system is the ideal means of providing replacement air because the air is comfort conditioned and enhances the kitchen environment. In smaller stand-alone restaurants, rooftop units are typical; they can be obtained with outdoor air dampers that are controlled to automatically open for the proper amount of outside air when the hoods are operational and close when they are turned off. Even better is the use of an enthalpy or temperature control to provide not only the makeup air, but also a degree of free cooling, by keeping the compressors off when outdoor conditions warrant and adjusting the outside and return air dampers. Designers must exercise extreme caution when applying outdoor air economizer cycles to kitchen systems in multiple-occupancy applications because the economizer can upset the air balance between occupancies and cause odor migration.

The amount of air brought in through the HVAC units should be limited so that comfort conditions are maintained. Typically, dining room units should be limited to 25 to 30% outside air to avoid unacceptable variations in discharge temperature. Replacement air through these units must have a ready path to the exhaust hoods. Kitchen units, on the other hand, may be set to bring in up to 50% outside air because temperature swings are less critical. For commercial kitchens where there is no on-site dining (e.g., take-out stores, commissaries, etc.), the entire area can be treated under the kitchen design criteria. Rooftop units are available that deliver 100% outside air with discharge air temperature control on cooling and heating.

Non-HVAC System Replacement (Makeup) Air

In situations where the outside air from the HVAC system is insufficient to replace all the air being drawn from the kitchen, air must be brought directly to the kitchen with limited or no conditioning. In some climates, outdoor conditions may be temperate enough that outside air does not require heating or cooling, particularly if it is mixed with the conditioned air such that comfort is not unduly compromised. In colder climates, it is essential to heat the air at least to a minimal comfort level. This may be done using electric resistance or direct or indirect gas-fired equipment (local codes must be consulted before direct gas-fired equipment is selected). In more arid areas, evaporative cooling can be effective, and direct-expansion HVAC equipment may not be needed in either the kitchen or the dining areas.

Again, replacement air is used primarily for proper operation of the exhaust hoods and not for comfort of the kitchen personnel. Comfort can be achieved through adequate flow of air around the workers, and higher velocity air movement can compensate for higher temperature and humidity, up to a point. Still, a restaurant owner may desire greater comfort for the employees; in this case, comfort becomes an important design concern.

Attempts have been made to temper outside air using heat from the exhaust system itself. However, effective heat transfer from the exhaust stream to the incoming airstream is made difficult by the strict requirements of the exhaust duct design (i.e., cleanability, minimal grease entrapment, etc.). Careful selection of the exhaust hood system to minimize the amount of air exhausted helps minimize both the cost of conditioning the replacement air and the cost of operating the exhaust and replacement air fans.

Replacement (Makeup) Air for Large Buildings

In large buildings with commercial kitchens (e.g., malls, hospitals, supermarkets, schools, etc.), adequate outside air from other parts of the building is generally available to provide replacement air for the hoods. However, in designing such a facility, the kitchen needs should be factored into the outside air provisions even if the amount is a small fraction of the overall requirement. Also, the source of the replacement air must be considered, so that air contaminated with odors, germs, dust, and so forth, is not drawn to the kitchen area, where cleanliness is required. Most malls and multiple-occupancy buildings specify a minimum amount of air to be taken from their space to keep cooking odors in the kitchen, but they also specify a maximum amount that may be taken to hold down the cost of tempering the outside air.

Air Distribution

Outside replacement air should be distributed in the general vicinity of the hoods. It should be delivered so that high velocities, eddies, swirls, or stray currents do not degrade the performance of the hood. Replacement air from any source should be distributed throughout the kitchen with low-velocity diffusers. Some hood manufacturers place diffusers in the face or lower front edge of the canopy to attempt to introduce replacement air into the hood in an effective but nonintrusive manner. The important point is to have replacement air delivered to the hoods (1) with proper velocity and (2) uniformly from all directions to which the hood is open in order to minimize excessive crosscurrents that could cause spillage from the capture area.

Indoor Air Quality

ASHRAE *Standard* 62 requires that 20 cfm outdoor air per person, based on a maximum occupancy of 70 persons per 1000 square feet (100 persons per 1000 square feet for cafeterias and fast-food restaurants), be brought into the dining area. The requirement is 30 cfm per person in bars/cocktail lounges and 15 cfm per person in kitchens based on 20 persons per 1000 square feet. These requirements may be increased or decreased in certain areas by local code, and they sometimes affect the size of an HVAC system more than thermal loading or necessitate the use of another means of introducing outside air. A further requirement of *Standard* 62, that outside air be sufficient to provide for an exhaust rate of at least 1.5 cfm per square foot of kitchen space, is generally easily met.

SYSTEM INTEGRATION AND BALANCING

System integration and balancing bring the many ventilation components together to provide the most comfortable, efficient, and economical performance of each component and of the entire system. In commercial kitchen ventilation, the supply air system (typically referred to as the HVAC system) must integrate and balance with the exhaust system. Optimal performance is achieved more by effective use of components through controls and airflow adjustments than by the selection or design of the system and components. The following fundamentals for restaurants, and kitchens in particular, should be considered and applied within the constraints of the particular location and its equipment and systems.

Principles

Although there are exceptions that will be addressed, the following are the fundamental principles of integrating and balancing restaurant systems for both comfort control and economical operation:

1. In a freestanding restaurant, the overall building should be at all times slightly pressurized compared to the outdoors to minimize the infiltration of untempered air, as well as dirt, dust, and insects, when doors are opened. In multiple-occupancy buildings, a slight negative pressure in the restaurant is desirable to minimize odor migration from the restaurant to other occupancies.

2. Every kitchen should be at all times slightly negative compared to the rooms or areas immediately surrounding it for the following reasons:

- To better contain the unavoidable grease vapors in the kitchen area and limit the extent of cleanup necessary
- To keep cooking odors in the kitchen area
- To prevent the generally hotter and more humid kitchen air from diminishing the comfort level of the adjacent spaces, especially the dining areas

3. Cross-zoning of airflow should be minimal, especially in the temperate seasons, when adjacent zones may be in different modes (e.g., economizer versus air-conditioning or heating). Two situations to consider are the following:

- In the transitions from winter to spring and from summer to fall, the kitchen zone could be in the economizer mode, or even the mechanical cooling mode, while the dining areas are in the heating mode. Bringing heated dining area supply air into a kitchen that is in the cooling mode only adds to the cooling load. In some areas, this situation is present every day.
- When all zones are in the same mode, it is more acceptable, and even more economical, to bring dining area air into the kitchen. However, the controls to automatically effect this method of operation are more complex and costly.

4. Typically, no drafts should be noticeable, and temperatures should vary no more than 1°F in dining areas and 3°F in kitchen areas. These conditions can be achieved with even distribution and thorough circulation of air in each zone by an adequate number of registers sized to preclude high air velocities. If there are noticeable drafts or temperature differences, customers and restaurant personnel will be distracted and will not enjoy dining or working.

Both design concepts and operating principles for proper integration and balance are involved in bringing about the desired results under varying conditions. The same principles are important in almost every aspect of restaurant ventilation.

In designing restaurant ventilation, all the exhaust is assumed to be in operation at one time, and the design replacement air quantity is a maximum requirement because it is for these maximum design conditions that the heating and cooling equipment are sized.

In restaurants with a single large exhaust hood, balancing should be set for this one operation only. In restaurants with multiple exhaust hoods, some may be operated only during heavy business hours or for special menu items. In this case, replacement air must be controlled to maintain minimum building positive pressure and to maintain the kitchen at a negative pressure under all operating conditions. The more variable the exhaust, the more complex the design; the more numerous and smaller the zones involved, the more complex the design. But the overall pressure relationship principles must be maintained to provide optimum comfort, efficiency, and economy.

A different application is a kitchen having one side exposed to a larger building with common or remote dining. Examples are a food court in a mall or a small restaurant in a hospital, airport, or similar building. Pressurizing the kitchen at the front of its space might cause some of the cooking grease, vapor, and odors to spread into the common building space, which would be undesirable. In such a case, the kitchen area is held at a negative pressure relative to other common building areas as well as to its own back room storage or office space.

Air Balancing

Balancing is best performed when the manufacturers of all the various equipment are able to provide a certified reference method of measuring the airflows, rather than depending on generic measurements of duct flows or other forms of measurement in the field. These field measurements can be in error by 20% or more. The manufacturer of the equipment should be able to develop a reference method of measuring airflow in a portion of the equipment that is

dynamically stable in the laboratory as well as in the field. This method should relate directly to airflow by graph or formula.

The general steps for air balancing in restaurants are as follows:

1. The exhaust hoods should first be set to their proper flow rates. This should be done with the supply and exhaust fans on.

   Next, the supply air flow rate, whether part of combined HVAC units or separate replacement air units, should be set to the design values through the coils and the design supply flows from each outlet, with the approximate correct settings on the outside air flow rate. Then, the correct outside and return air flow rates should be set proportionally for each unit, as applicable. These settings should be made with the exhaust on, to ensure adequate relief for the outside air.

   Where the outside air and return air flows of a particular unit are expected to modulate, there should ideally be similar static losses through both airflow paths to preclude large changes in total supply air from the unit. Such changes, if large enough, could affect the efficiency of heat exchange and could also change the airflows within and between zones, thereby upsetting the air distribution and balance.

2. Next, outside air should be set with all fans (exhaust and supply) operating.

   The pressure difference between inside and outside should be checked to see that (1) the nonkitchen zones of the building are at a positive pressure compared to outside and (2) the kitchen zone pressure is negative compared to the surrounding zones and negative or neutral compared to outside.

   For applications with modulating exhaust, every step of exhaust and replacement should be shut off, one step at a time. Each combination of operation should be rechecked to be sure that the design pressures and flows are maintained within each zone and between zones. This requires that the replacement airflow rate compensate automatically with each increment of exhaust. It may require some adjustments in controls or in damper linkage settings to get the correct proportional response.

3. When the above steps are complete, the system is properly integrated and balanced. At this time, all fan speeds and damper settings (at all modes of operation) should be permanently marked on the equipment and in the test and balance report. The air balance records of exhaust, supply, return, fresh air, and individual register airflows must also be completed. These records should be kept by the food service facility for future reference.

4. For new facilities, after two or three days in operation (no longer than a week and usually before the facility opens), all belts in the system should be checked and readjusted because new belts wear in quickly and could begin slipping.

5. Once the facility is operational, the performance of the ventilation system should be checked to verify that the design is adequate for the actual operation, particularly at maximum cooking and at outdoor environmental extremes. Any necessary changes should be made, and all the records should be updated to show the changes.

   Rechecking the air balance should not be necessary more than once every 2 years unless basic changes are made in facility operation. If there are any changes, such as a new type of cooking equipment or added or deleted exhaust connections, the system should be modified accordingly.

Multiple-Hood Systems

Kitchen exhaust systems serving more than a single hood present several design challenges not encountered with single-hood systems. One of the main challenges of multiple-hood exhaust systems is air balancing. Because balancing dampers are not permitted in the exhaust ductwork, the system must be balanced by design. Most filters come in varying sizes to allow pressure loss equalization at varying airflows. Some hoods and grease filters have adjustable

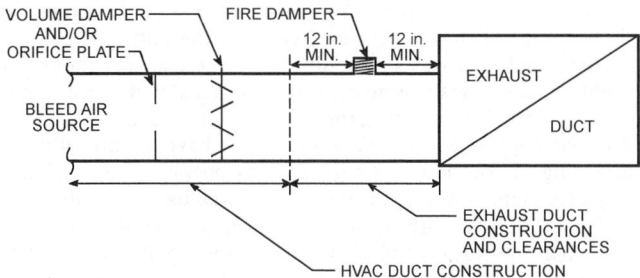

Fig. 6 Method of Introducing Replacement Air Directly into Exhaust Duct

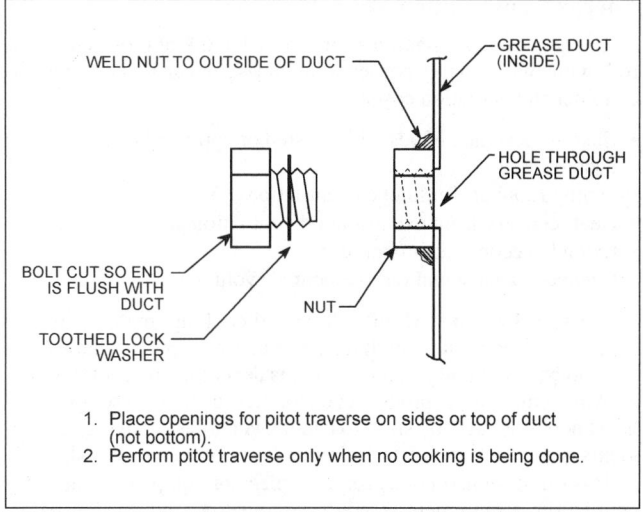

1. Place openings for pitot traverse on sides or top of duct (not bottom).
2. Perform pitot traverse only when no cooking is being done.

Fig. 7 Metal-to-Metal Seal for Pitot Tube Access Port in Grease Duct

baffles that allow airflow to be adjusted at the hood. These may be helpful for relatively fine balancing, but the system must provide most of the balancing. Adjustable filters should not be used when they can be interchanged between hoods or within the same hood because an interchange could disrupt the previously achieved balance. Balancing can also be accomplished by changing the number and/or the size of filters.

If one designer selects and lays out the entire exhaust system, such as for a large kitchen with several hoods, the ductwork can be designed for controlled pressure loss in all branches. In cases such as master kitchen exhaust systems, which are sometimes used in shopping center food courts, no single group is responsible for the entire design. The base building designer typically lays out the ductwork to (or through) each tenant space, and each tenant selects the hood and lays out the connecting ductwork. Often the base building designer has incomplete information regarding the tenants' exhaust requirements. Therefore, one engineer must be responsible for defining criteria for each tenant's design and for evaluating proposed tenant work to ensure that tenants' designs match the system's capacity as much as possible. The engineer should also evaluate any proposed changes to the system, such as changing tenancy. Rudimentary computer modeling of the exhaust system may be helpful (Elovitz 1992). Given the unpredictability and volatility of tenant requirements, it may not be possible to balance the entire system perfectly. However, without adequate supervision, chances are very good that at least parts of the system will be badly out of balance.

For greatest success with multiple-hood exhaust systems, pressure losses in the ductwork should be minimized by keeping velocities low, minimizing sharp transitions, and using hoods with relatively high pressure drops. When the pressure loss in the ductwork is low compared to the loss through the hood, changes in pressure loss in the ductwork due to field conditions or changes in design airflow will have a smaller effect on total pressure loss and thus on actual airflow.

The minimum, code-required air velocity must be maintained in all portions of the exhaust ductwork at all times. If fewer or smaller hoods are installed than the design anticipated, resulting in low velocity in portions of the ductwork, the velocity must be brought up to the minimum. One way is to introduce air, preferably untempered replacement air, directly into the exhaust duct where it is required (see Figure 6). The bypass duct should connect to the top or sides (at least 2 in. from the bottom) of the exhaust duct to prevent backflow of water or grease through the bypass duct when the fans are off. This arrangement is shown in NFPA *Standard* 96 and should be discussed with the authority having jurisdiction.

A fire damper should be provided in the bypass duct, located close to the exhaust duct. The bypass duct construction should be the same as the exhaust duct construction, including enclosure and clearance requirements, for at least several feet beyond the fire damper. Means to adjust the bypass airflow must be provided upstream of the fire damper. All dampers must be in the clean

bypass air duct so that they are not exposed to grease-laden exhaust air. The difference in pressure between the replacement air and the exhaust air duct may be great; the balancing device must be able to make a fine airflow adjustment against this pressure difference. It is best to provide two balancing devices in series, such as an orifice plate or blast gate for coarse adjustment followed by an opposed-blade damper for fine adjustment.

Direct measurement of air velocities in the exhaust ductwork to assess exhaust system performance may be desirable. **Velocity (pitot-tube) traverses** may be performed in kitchen exhaust systems, but the holes drilled for the pitot tube must be liquidtight to maintain the fire-safe integrity of the ductwork. Figure 7 shows a means of providing a metal-to-metal seal for these holes. Holes should never be drilled in the bottom of a duct, where they may collect grease. Velocity traverses should not be performed when cooking is in progress because grease collects on the instrumentation.

ENERGY CONSIDERATIONS

Energy conservation in restaurants depends on the following variables:

Climate. Average outdoor temperature is the principal predictor of restaurant energy use. Because climatic zones vary dramatically in temperature and humidity, kitchen ventilation conservation designs have widely varying economic recovery periods. In new facilities, the designer can select conservation measures suitable for the climatic zone and the HVAC system to maximize the economic benefits.

Restaurant Type. Claar et al. (1985) defined five restaurant types: fast-food, full-service, coffee shop, pizza, and cafeteria. Restaurant associations identify many additional restaurant types, and energy use varies significantly among the types. As a result, some energy conservation measures work in all restaurants while others may work only in specific types. For simplicity, this discussion classifies restaurants as cafeteria, fast-food, full-menu, and pizza.

Hood Type and Equipment Characteristics. The type of exhaust hood selected depends on such factors as restaurant type, restaurant menu, and food service equipment. Exhaust flow rates are largely determined by the food service equipment and the hood style. The effectiveness of conservation measures, whether in a new design or as a retrofit, is also affected by the type of hood and food service equipment.

Energy Conservation Measures

Energy costs in restaurants vary from 2 to 6% of gross sales. The following major energy conservation measures can reduce commercial kitchen ventilation costs:

- Custom-designed hoods, either listed or approved by the local code official
- Dining room air for kitchen ventilation
- Heat recovery from exhaust hood ventilation air
- Extended economizer operation
- Reduced exhaust and replacement air volumes

Custom-Designed Hoods. A typical cooking lineup combines food service equipment with different thermal updraft characteristics and exhaust flow rate requirements depending on cooking load, cooking temperature, product cooked, fuel source, and so forth. Exhaust hoods are usually sized to handle worst-case cooking requirements, so kitchen exhaust requirements are often overstated.

Custom-designed hoods for each piece of equipment can lower ventilation requirements and, consequently, reduce fan sizes, energy use, and energy costs for all types of restaurants. Hood manufacturers recommend exhaust flow rates based on either thermal current charts or empirical tests. The total flow rate and size of the hood exhaust are determined by summing the individual equipment airflow rate and size requirements for the entire cooking lineup. To operate at flow rates lower than required by code, custom-designed hoods must be either listed or approved by the local code official.

Dining Room Air for Kitchen Ventilation. Most fast-food and pizza restaurants do not physically segregate the kitchen from the dining room, so conditioned air moves from the dining area to the kitchen. Dining room air helps cool the kitchen to some extent and provides some replacement air for the kitchen exhaust.

Full-menu and cafeteria food service facilities are frequently designed so that minimal replacement air transfers between the dining room and the kitchen. This segregation means that the majority of kitchen exhaust must be made up within the kitchen by conditioned or unconditioned air.

In restaurants in which the kitchen and dining room are physically segregated, ducts between the two areas permit conditioned air to flow to the kitchen. This design can reduce replacement air requirements and enhance employee comfort, especially if the kitchen is not air conditioned. Of course, the dining room must have replacement air to replace the air transferred to the kitchen. Also, care must be taken to ensure that enough replacement air is introduced into the kitchen to meet the requirements of applicable codes and standards.

Heat Recovery from Exhaust Hood Ventilation Air. High-temperature effluent, often in excess of 400°F, heats the exhaust air to 100 to 200°F as it travels over the cooking surfaces and areas. It is frequently assumed that this heated exhaust air is suitable for heat recovery; however, smoke and grease in the exhaust air will, with time, cover any heat transfer surface. Under these conditions, the heat exchangers require constant maintenance (e.g., automatic wash-down) to maintain acceptable heat recovery.

Because heat recovery systems are very expensive, only food service facilities with large amounts of cooking equipment and large cooking loads are good candidates for this equipment. Heat recovery may work in full-menu and cafeteria restaurants and in institutional food service facilities such as prisons and colleges. An exhaust hood equipped with heat recovery is more likely to be cost-effective where the climate is extreme, such as the northern tier states or the southern states. A mild climate, such as in California, is not conducive to use of this conservation measure.

Extended Economizer Operation. In most restaurants, the mechanical cooling is called on as soon as someone is too warm in order to keep the staff and customers cool. By operating an economizer system instead, similar cooling can be achieved at less cost

because only fans are operated and not compressors. Economizers can be used in lieu of mechanical cooling when the outside temperature (and sometimes humidity) is lower than the return air condition. Economizer systems may be designed to increase ventilation by 25% or less over normal ventilation, or about 50 to 60% of system supply capacity. Yet economizers have the capability of increasing the outside air to 100%. This operation can dramatically extend the time that the mechanical cooling is kept off.

Use of full-capacity economizer operation requires variable rather than one-step control of the exhaust and replacement air. The savings can justify the initial cost, especially in moderate climates where temperatures range between about 77 and 83°F for most of the summer.

To allow economizer operation with canopy hoods, it may be adequate to increase the exhaust at each hood. With back shelf hoods, which are lower than the heads of the crew, it may be better to increase the flow of exhaust from the ceiling of the kitchen in order to lower the upper space temperature.

To take full advantage of maximum economizer operation, the entire food service area, including all cooking and dining zones, must be in the economizer mode. It may be uncomfortable or uneconomical to maximize economizer operation when some zones remain in the heating or mechanical cooling mode.

Reduced Exhaust and Replacement (Makeup) Airflow Rates. The tempering of outdoor replacement air can account for a large part of a food service facility's heating and cooling costs. By reducing the exhaust flow rates (and the corresponding replacement air quantity) when no product is being cooked, the cost of energy can be significantly reduced. Field evaluations by one large restaurant chain suggest that cooking appliances may be at zero load for 75% or more of an average business day (Spata and Turgeon 1995). During these periods when no smoke or grease-laden vapors are being produced, NFPA *Standard* 96 allows reduction of exhaust quantities. The only restriction on the reduced exhaust quantity is that it be "... sufficient to capture and remove flue gases and residual vapors..."

FIRE PROTECTION

The combination of flammable grease and particulates carried by kitchen ventilation systems and the potential of the cooking equipment to be an ignition source creates a higher hazard level than normally found in HVAC systems. The design of an exhaust system serving cooking equipment that may produce grease-laden vapors must provide, at a minimum, a reasonable level of protection for the safety of building occupants and fire fighters. The design can be enhanced to provide extra protection for property.

Replacement air systems, air-conditioning systems serving a kitchen, and exhaust systems serving only cooking equipment that does not produce grease-laden vapor have no specific fire protection requirements beyond those applicable to similar systems not located in kitchens. However, an exhaust system that serves any grease-producing cooking equipment must be considered a grease exhaust system even if it also serves non-grease-producing equipment.

Fire protection starts with proper operation and maintenance of the cooking equipment and the exhaust system. After that, the two primary aspects of fire protection in a grease exhaust system are (1) to extinguish a fire quickly once it has started and (2) to prevent the spread of fire from or to the grease exhaust system.

Fire Suppression

NFPA *Standard* 96 requires that exhaust systems serving grease-producing equipment must include a fire-extinguishing system, except where certain listed grease removal devices or systems are installed. The fire-extinguishing system must protect the cooking surfaces, the hood interior, the hood filters or grease extractors, the ductwork, and any additional grease removal devices in the system.

The most common fire-extinguishing systems are wet chemical and water spray systems.

Operation. Actuation of any fire-extinguishing system should not depend on normal building electricity. If actuation relies on electricity, it should be supplied with standby power.

Any extinguishing system must automatically shut off all supplies of fuel and heat to all equipment protected by that system. Any gas appliance not requiring protection but located under the same ventilating equipment must also be shut off. Upon operation of a wet chemical or water fire-extinguishing system, all electrical sources located under the ventilating equipment, if subject to exposure to discharge from the fire-extinguishing system, must be shut off. If the hood is in a building with a fire alarm system, actuation of the hood extinguishing system should send a signal to the fire alarm.

Dry and Wet Chemical Systems. Wet chemical fire-extinguishing systems are the most common in new construction for the protection of hoods and exhaust systems. Dry chemical systems were popular; however, most manufacturers have removed them from the market because they have not passed UL *Standard* 300. Dry chemical systems are covered in NFPA *Standard* 17, and wet chemical systems are covered in NFPA *Standard* 17A. Both standards provide detailed application information. These systems are tested for their ability to extinguish fires occurring in cooking operations in accordance with UL *Standard* 300. To date, only wet chemical systems are listed to UL *Standard* 300.

Both dry and wet chemicals extinguish a fire by reacting with fats and grease to **saponify**, or form a soapy layer of foam that prevents oxygen from reaching the hot surface. This suppresses the fire and prevents reignition. Saponification is particularly important with deep fat fryers, where the frying medium may be hotter than its autoignition temperature for some time after the fire is extinguished. Should the foam layer disappear or be disturbed before the frying medium has cooled below its autoignition temperature, it could reignite.

Frying media commonly used today have autoignition points of about 685 to 710°F when they are new. Contamination through normal use lowers the autoignition point. The chemical agent that extinguishes the fire also contaminates the frying medium, which can further reduce the autoignition point by 50 to 60°F. One advantage of wet chemical systems over dry chemical systems is that the wet chemical provides extra cooling to the frying medium, so that it falls below the autoignition point more quickly.

For a chemical system protecting the entire exhaust system, fire-extinguishing nozzles are located over the cooking equipment being protected, in the hood to protect the grease removal devices and hood plenum, and at the duct collar (downstream from any fire dampers) to protect the ductwork. The duct nozzle is rated to protect an unlimited length of ductwork, so additional nozzles are not required further downstream in the ductwork. Fire detection is required at the entrance to each duct (or ducts, in hoods with multiple duct takeoffs) and over each piece of cooking equipment that requires protection. The detector at the duct entrance may also cover the piece of cooking equipment directly below it.

Chemical fire-extinguishing systems are available as listed, pre-engineered (packaged) systems. Chemical systems typically consist of one or more tanks of chemical agent (dry or wet), a propellant gas, piping to the suppression nozzles, fire detectors, and auxiliary equipment. The fire detectors are typically fusible links that melt at a set temperature associated with a fire, although electronic devices are also available. Auxiliary equipment may include manual pull stations, gas shutoff valves (spring-loaded or solenoid-actuated), and auxiliary electric contacts.

Actuation of dry and wet chemical suppression systems is typically completely mechanical, requiring no electric power. The fire detectors are typically interconnected with the system actuator by steel cable in tension, so that melting of any of the fusible links releases the tension on the steel cable, causing the actuator to release

the propellant and suppressant. The total length of the steel cable and the number of pulley elbows permitted are limited. A manual pull station is typically connected to the system actuator by steel cable. If a mechanical gas valve is used, it is also connected to the system actuator by steel cable. Actuation of the system also switches auxiliary dry electrical contacts, which can be used to shut off electrical cooking equipment, operate an electric gas valve, shut off a replacement air fan, and/or send an alarm signal to the building fire alarm system.

Manual pull stations are generally required to be at least 10 ft from the cooking appliance and in a path of egress. Some authorities may prefer that the pull station be installed closer to the cooking equipment, for faster response. However, if the pull station is too close to the cooking equipment, it may not be possible to approach it once a fire has started. Refer to the applicable code requirements for each jurisdiction to determine specific requirements for location and mounting heights of pull stations.

Water Systems. Water can be used for protection of cooking appliances, hoods, and grease exhaust systems. Standard fire sprinklers may be used throughout the system, except over deep fat fryers, where special automatic spray nozzles specifically listed for the application must be used. These nozzles must be aimed properly and be supplied with the correct water pressure. Many hood manufacturers market a pre-engineered water spray system that typically includes a cabinet containing the necessary plumbing and electrical components to monitor the system and initiate fuel shutoff and building alarms. A water system for cleaning or grease removal in a hood can also protect the hood or duct in the case of fire, if the system is listed for this purpose.

Application of standard fire sprinklers for protection of cooking appliances, hoods, and grease exhaust systems is covered by NFPA *Standard* 13. NFPA *Standard* 96 covers maintenance of sprinkler systems serving an exhaust system. The sprinklers must connect to a wet-pipe building sprinkler system installed in compliance with NFPA *Standard* 13.

Water systems can be used to protect cooking equipment if the water spray is a fine mist. The water spray works in two ways to suppress the fire: the mist spray first absorbs heat from the fire, and then it becomes steam, which displaces air and suffocates the fire.

Although standard sprinklers may be used to protect cooking equipment other than deep fat fryers, care must be taken to ensure that sprinklers are properly selected for a fine mist discharge; if the pressure is too high or the spray too narrow, the water spray could push the flames off of the cooking equipment. Most hood manufacturers that use water to protect the cooking equipment use the sprinklers listed for deep fryers over all the cooking equipment.

One advantage of a sprinkler system is that it has virtually unlimited capacity, whereas chemical systems have limited chemical supplies. This can be an advantage in suppressing a serious fire, but all the water must be safely removed from the space. Where sprinklers are used in ductwork, the ductwork should be pitched to drain safely.

When sprinklers are used to protect ductwork, NFPA *Standard* 13 requires that they be installed every 10 ft on center in horizontal ductwork, at the top of every vertical riser, and in the middle of any vertical offset. Care must be taken to protect any sprinklers exposed to freezing temperatures.

Hoods that use water either for periodic cleaning (water-wash) or for grease removal (cold water mist) can use this feature as a fire-extinguishing system to protect the hood, grease removal device, and/or ductwork in the event of a fire, if the system has been tested and listed for the purpose. The water supply for these systems may be from the kitchen water supply if flow and pressure requirements are met. These hoods can also act as a fire stop because of their multipass configuration and the fact that the grease extractors are not removable.

Combination Systems. Different system types may protect different parts of the grease exhaust system, as long as the entire system is protected. Examples include (1) an approved water-wash or water mist system to protect the hood in combination with a dry or wet chemical system to protect the ductwork and the cooking surface or (2) a chemical system in the hood backed up by water sprinklers in the ductwork. Combination systems are more common in multiple-hood systems, where a separate extinguishing system may be used to protect a common duct. If combination systems will discharge their suppressing agents together in the same place, the agents must be compatible.

Multiple-Hood Systems. All hoods connected to a multiple-hood exhaust system must be on the same floor of the building to prevent the spread of fire through the duct from floor to floor. Preferably, there should be no walls requiring greater than a 1 h fire resistance rating between hoods.

The multiple-hood exhaust system must be designed (1) to prevent a fire in one hood or in the duct from spreading through the ductwork to another hood and (2) to protect against a fire starting in the common ductwork. Of course, the first line of protection for the ductwork is to keep it clean. Especially in a multiple-tenant system, a single entity must assume responsibility for cleaning the common ductwork frequently.

Each hood must have its own fire-extinguishing system to protect the hood and cooking surface. A single system could serve more than one hood, but in the event of fire under one hood, the system would discharge its suppressant under all of the hoods served, resulting in unnecessary cleanup expense and inconvenience. A water mist system could serve multiple hoods if the sprinkler heads were allowed to operate independently.

Because of the possibility of a fire spreading through the ductwork from one hood to another, the common ductwork must have its own fire protection system. The appendices of NFPA *Standards* 17 and 17A present detailed examples of how the common ductwork can be protected either by one system or by a combination of separate systems serving individual hoods. Different types of fire-extinguishing systems may be used to protect different portions of the exhaust system; however, in any case where two different types of system can discharge into the common duct at the same time, the agents must be compatible.

As always, actuation of the fire-extinguishing system protecting any hood must shut off fuel or power to all cooking equipment under that hood. When a common duct, or portion thereof, is protected by a chemical fire-extinguishing system, and it activates from a fire in a single hood, NFPA *Standards* 17 and 17A require shutoff of fuel or power to the cooking equipment under every hood served by that common duct, or portion of it protected by the activated system, even if there was no fire in the other hoods served by that duct.

From an operational standpoint, it is usually most sensible to provide one or more fire-extinguishing systems to detect and protect against fire in the common ductwork and a separate system to protect each hood and its connecting ductwork. This (1) prevents a fire in the common duct from causing discharge of fire suppressant under an unaffected hood and (2) allows unaffected hoods to continue operation in the event of a fire under one hood unless the fire spreads to the common duct.

Preventing Fire Spread

The exhaust system must be designed and installed both to prevent a fire starting in the grease exhaust system from spreading to other areas of the building or damaging the building and to prevent a fire in one area of the building from spreading to other parts of the building via the grease exhaust system. This protection has two main aspects: (1) maintaining a clearance from the duct to other portions of the building and (2) enclosing the duct in a fire resistance rated enclosure. Both aspects are sometimes addressed by a single action.

Clearance to Combustibles. A grease exhaust duct fire can generate gas temperatures of 2000°F or greater in the duct. In such a grease fire, heat radiating from the hot duct surface can ignite combustible materials in the vicinity of the duct. Most codes require a minimum clearance of 18 in. from the grease exhaust duct to any combustible material. However, even 18 in. may not be sufficient clearance to prevent ignition of combustibles in the case of a major grease fire, especially in larger ducts.

Several methods to protect combustible materials from the radiant heat of a grease duct fire and permit reduced clearance to combustibles are described in NFPA *Standard* 96. Previous editions required that these protections be applied to the combustible material rather than to the duct, on the theory that if the duct is covered with an insulating material, the duct itself will become hotter than it would if it were free to radiate its heat to the surroundings. The hotter temperatures could result in structural damage to the duct. NFPA *Standard* 96 allows materials to be applied to the actual duct. However, because this edition may not have been accepted by some jurisdictions, the authority having jurisdiction should be consulted before clearances to combustible materials are reduced.

Listed grease ducts, typically double-wall ducts with or without insulation between the walls, may be installed with reduced clearance to combustibles in accordance with manufacturers' installation instructions, which should include specific information regarding the listing. Listed grease ducts are tested under UL *Standard* 1978.

NFPA *Standard* 96 also requires a minimum clearance of 3 in. to "limited combustible" materials (e.g., gypsum wallboard on metal studs). At present, none of the standards and model codes requires any clearance to noncombustible materials except for enclosures.

Enclosures. Normally, when a duct penetrates a fire resistance rated wall or floor, a fire damper is used to maintain the integrity of the wall or floor. Because fire dampers may not be installed in a grease exhaust duct unless specifically approved for such use, there must be an alternate means of maintaining the integrity of rated walls or floors. Therefore, grease exhaust ducts that penetrate a fire resistance rated wall or floor-ceiling assembly must be continuously enclosed in a fire-rated enclosure from the point the duct penetrates the first fire barrier until the duct leaves the building. Listed grease ducts are also subject to these enclosure requirements. The requirements are similar to those for a vertical shaft (typically 1 h rating if the shaft penetrates fewer than three floors, 2 h rating if the shaft penetrates three or more floors), except that the shaft can be both vertical and horizontal. In essence, the enclosure extends the room containing the hood through all the other compartments of the building without creating any unprotected openings to those compartments.

Where a duct is enclosed in a rated enclosure, whether vertical or horizontal, clearance must be maintained between the duct and the shaft. NFPA *Standard* 96 and IMC require minimum 6 in. clearance and that the shaft be vented to the outdoors. IMC requires that each exhaust duct have its own dedicated enclosure.

Some available materials are listed to serve as a fire resistance rated enclosure for a grease duct when used to cover a duct directly with minimal clear space between the duct and the material. These listed materials must be applied in strict compliance with the manufacturer's installation instructions, which may limit the size of duct to be covered and specify required clearances for duct expansion and other installation details. When a duct is directly covered with an insulating material, there is a greater chance of structural damage to the duct due to the heat of a severe fire. The structural integrity of exhaust ducts should be assessed after any serious duct fire.

Insulation materials that have not been specifically tested and approved for use as fire protection for grease exhaust ducts should not be used in lieu of rated enclosures or to reduce clearance to combustibles. Even insulation that is approved for other fire protection applications, such as to protect structural steel, may not be appropriate for grease exhaust ducts because of the high temperatures that may be encountered in a grease fire.

Exhaust and Supply Fire-Actuated Dampers. Because of the risk that the damper may become coated with grease and become a source of fuel in a fire, balancing and fire-actuated dampers are not permitted at any point in a grease exhaust system except where specifically listed for each use or required as part of a listed device or system. Typically, fire dampers are found only at the hood collar and only if provided by the manufacturer as part of a listed hood.

Opinions differ regarding whether any fire-actuated dampers should be provided in the exhaust hood. On one hand, a fire-actuated damper at the exhaust collar may prevent a fire under the hood from spreading to the exhaust duct. However, like anything in the exhaust airstream, the fire-actuated damper and linkage may become coated with grease if not properly maintained, which may impede damper operation. On the other hand, without fire-actuated dampers, the exhaust fan will draw smoke and fire away from the hood. While this cannot be expected to remove all of the smoke from the kitchen in the event of a fire, it can help to contain the smoke in the kitchen and minimize migration of smoke to other areas of the building.

A fire-actuated damper will generally close only in the event of a severe fire; most kitchen fires are extinguished before enough heat is released to trigger the fire-actuated damper. Thus the hood fire-actuated damper remains open during relatively small fires, allowing the hood to remove smoke, but can close in the event of a severe fire, helping to contain the fire in the kitchen area.

Fan Operations. If it is over 2000 cfm, the replacement air supply to the kitchen is generally required to be shut down in case of fire to avoid feeding air to the fire. However, if the exhaust system is intended to operate during a fire to remove smoke from the kitchen (as opposed to just containing it in the kitchen), the replacement air system must operate as well. If the hood has an integral replacement air plenum, a fire-actuated damper must be installed in the replacement air collar to prevent a fire in the hood from entering the replacement air ductwork. NFPA *Standard* 96 details the instances where fire-actuated dampers are required in a hood replacement air assembly.

Regardless of whether fire-actuated dampers are installed in the exhaust system, NFPA *Standard* 96 calls for the exhaust fan to continue to run in the event of a fire unless fan shutdown is required by a listed component of the ventilating system or of the fire-extinguishing system. Dry and wet chemical fire-extinguishing systems protecting ductwork are tested both with and without airflow; exhaust airflow is not necessary for proper operation.

Special Filtration Systems

Concerns about air quality have emphasized the need for higher efficiency grease extraction from the exhaust airstream than can be provided by filters or grease extractors in the exhaust hoods. Cleaner exhaust discharge to outdoors may be required by increasingly stringent air quality regulations or where the exhaust discharge configuration is such that grease, smoke, or odors in the discharge would create a nuisance. In some cases, the exhaust air is cleaned so that it can be discharged indoors (e.g., through recirculating hoods). Several systems have been developed to clean the exhaust airstream, each of which presents special fire protection risks. These systems are as follows:

- Systems that use one or more stages of filtration to remove the grease from the airstream capture and collect the grease on the filters. In the event of a fire, grease-coated filters may become a source of fuel. Each bank of filters must be protected by an extinguishing system. If filters with deep pleats (e.g., HEPA filters) are used, the extinguishing agent must be able to penetrate deep into the pleats to extinguish the fire.
- Systems that use **electrostatic precipitators** (ESPs) to clean the exhaust stream must also have appropriate fire protection. ESPs use electricity, which could be a source of ignition. ESPs in grease

ducts should use self-limiting (cold spark) transformers to minimize the possibility of electrical fire. ESP plates must be cleaned frequently to prevent grease buildup and to maintain operating efficiency. ESPs are available with automatic washdown features that simplify the task of keeping the plates clean.

Where odor control is required in addition to grease removal, activated charcoal or other oxidizing bed filters are typically used downstream of the grease filters. Because much of the cooking odor is gaseous, and therefore not removed by air filtration, the filtration upstream of the charcoal filters must remove virtually all of the grease in the airstream in order to prevent grease buildup on the charcoal filters. However, charcoal filters themselves are flammable and must be protected from fire.

OPERATION AND MAINTENANCE

Operation and maintenance are closely related; operation creates the need for maintenance, and maintenance permits continued operation.

Operation

All components of the ventilation system are designed to operate in balance with each other, even under variable loads, to properly capture, contain, and remove the cooking effluent and heat and maintain proper temperature control in the spaces in the most efficient and economical manner. Deterioration in any of these components unbalances the system, affecting one or more of its design concepts. The system design intent should be fully understood by the operator so that any deviations in the operation can be noted and corrected. In addition to creating health and fire hazards, normal cooking effluent deposits can also unbalance the system, so they must be regularly removed.

All the various components of the exhaust and replacement air systems affect proper capture, containment, and removal of the cooking effluent. In the exhaust system, this includes the cooking equipment itself, the exhaust hood, all filtration devices, the ductwork, the exhaust fan, and any dampers in the exhaust system. In the replacement air system, this includes the air-handling unit(s) with the intake louvers, dampers, filters, fan wheels, heating and cooling coils, ductwork, and supply registers. In systems that obtain their replacement air from the general HVAC system for the space, this also includes the return air registers and ductwork.

When the system is first set up and balanced in new condition, these components are set to optimum efficiency. In time, all components become dirty; filtration devices, dampers, louvers, heating and cooling coils, and ducts become restricted; fan blades change shape as they accumulate dirt and grease; and fan belts loosen. In addition, dampers can come loose and change position, even closing, and ductwork can develop leaks or be shut off when internal insulation sheets fall down.

All these changes deteriorate the performance of the system. The operator should know how the system performed when it was new, in order to better recognize when it is no longer performing the same way. This knowledge allows problems to be found and corrected sooner and the peak efficiency and safety of the system operation to better be maintained.

Maintenance

Maintenance may be classified as preventive or emergency (breakdown). **Preventive maintenance** keeps the system operating at the smallest deviation from optimal performance. Preventive maintenance allows continued optimal performance of the system, including the maximum production and least shutdown. It is the most effective maintenance and is preferred.

Preventive maintenance is able to prevent most emergency shutdowns and emergency maintenance. Preventive maintenance has a modest ongoing cost and fewer unexpected costs. It is clearly the lowest cost maintenance in the long run because it keeps the system components in peak condition, which extends the operating life of all components.

Emergency maintenance must be applied when a breakdown occurs. Sufficient staffing and money must be applied to the situation to bring the system back on line in the shortest possible time. Such emergencies can be of almost any nature. They are impossible to predict or address in advance, except to presume the type of component failures that could shut the system down and keep spares of these components on hand, or readily accessible, so they can be quickly replaced. Preventive maintenance, which includes regular inspection of critical system components, is the most effective way to avoid emergency maintenance.

Following are brief descriptions of typical operations of various components of kitchen ventilation systems and the type of maintenance and cleaning that would be applied to the abnormally operating system to bring it back to normal. Many nontypical operations are not listed here.

Cooking Equipment

Normal Operation. Produces properly cooked product, of correct temperature, within expected time period. Minimum smoke during cooking cycles.

Abnormal Operation. Produces undercooked product, of lower temperature, with longer cooking times. Increased amount of smoke during cooking cycles.

Cleaning/Maintenance. Clean solid cooking surfaces between each cycle if possible, or at least once a day. Baked-on product insulates and retards heat transfer. Filter frying medium daily and change it on schedule recommended by supplier. Check that (1) fuel source is at correct rating, (2) thermostats are correctly calibrated, and (3) conditioned air is not blowing on cooking surface.

Exhaust Systems

Normal Operation. All cooking vapors are readily drawn into the exhaust hood, where they are captured and removed from the space. The environment immediately around the cooking operation is clear and fresh.

Abnormal Operation. Many cooking vapors do not enter the exhaust hood at all, and some that enter subsequently escape. The environment around the cooking operation, and likely in the entire kitchen, is contaminated with cooking vapors.

Cleaning/Maintenance. Clean all grease removal devices in the exhaust system. Hood filters should be cleaned at least daily. For other devices, follow the minimum recommendations of the manufacturer; even these may not be adequate at very high flow rates or with products producing large amounts of effluent. Check that (1) all dampers are in their original position, (2) fan belts are properly tensioned, (3) the exhaust fan is operating at the proper speed and turning in the proper direction, (4) the exhaust duct is not restricted, and (5) the fan blades are clear.

NFPA *Standard* 96 design requirements for access to the system should be followed to facilitate cleaning of the exhaust hood, ductwork, and fan. If the cleaner cannot get to parts of the system, these parts will not be cleaned. Cleaning should be done before the grease has built up to 0.25 in. in any part of the system. Cleaning should be by a method that cleans to bare metal, and the metal should be left clean. Cleaning agents should be thoroughly rinsed off, and all loose grease particles should be removed, as they can ignite more readily. Agents should not be added to the surface after cleaning, as their textured surfaces merely collect more grease more quickly. Fire-extinguishing systems may need to be disarmed prior to cleaning to prevent accidental discharge and then reset by authorized personnel after cleaning. All access panels removed must be reinstalled after

cleaning, with the proper gasketing in place to prevent grease leaks and escape of fire.

Supply, Replacement, and Return Air Systems

Normal Operation. The environment in the kitchen area is clear, fresh, comfortable, and free of drafts and excessive air noise.

Abnormal Operation. The kitchen is smoky, choking, hot, and humid, and perhaps very drafty with excessive air noise.

Cleaning/Maintenance. Check that the replacement air system is operating and that it is providing the correct amount of air to the space. If it is not, the exhaust system cannot operate properly. Check that the dampers are set correctly, the filters and exchangers are clean, the belts are tight, the fan is turning in the correct direction, and the supply and return ductwork and registers are open with the supply air discharging in the correct direction and pattern. If drafts persist, the system may need to be rebalanced. If noise persists in a balanced system, system changes may be required.

Filter cleaning or changing frequency varies widely depending on the quantity of airflow and the contamination of the local air. Once determined, the cleaning schedule must be maintained.

With replacement air systems, the air-handling unit, coils, and fan are usually cleaned in spring and fall, at the beginning of the seasonal change. More frequent cleaning or better quality filtering may be required in some contaminated environments. Duct cleaning for the system is on a much longer cycle, but local codes should be checked as stricter requirements are invoked. Ventilation systems should be cleaned by professionals to ensure that none of the expensive system components are damaged. Cleaning companies should be required to carry adequate liability insurance. The International Kitchen Exhaust Cleaning Association (IKECA) and the National Air Duct Cleaners Association (NADCA), both located in Washington, D.C., can provide descriptions of proper cleaning and inspection techniques and lists of their members.

Control Systems (Operation and Safety)

Normal Operation. Control systems should not permit the cooking equipment to operate unless both the exhaust and replacement air systems are operating and the fire suppression system is armed. With multiple exhaust and replacement air systems, the controls maintain the proper balance as cooking equipment is turned on and off. In the event that a fire suppression system operates, the energy source for the cooking equipment it serves is shut off. On ducted systems, the exhaust fan usually keeps running to remove fire and smoke from building. On ductless systems, the fan may or may not keep running, but a discharge damper closes to keep the flames away from the ceiling. The replacement air system may continue to run, or it may be shut off by a separate local area fire and smoke sensor. If the control system does not operate in this way, changing to this operation should be considered.

Abnormal Operation. The cooking equipment can operate when the exhaust and supply are turned off, perhaps because the fire suppression system is unarmed or has been bypassed. When extra cooking systems are turned on or off, the operator must remember to manually turn the exhaust and replacement air fans on or off as well.

When the exhaust and replacement air systems are not interlocked, the system can be out of balance. This can cause many of the kinds of abnormal operation described for the other systems. With gas-fired cooking equipment, the fire suppression system may have a false discharge if the exhaust system is not operating. If cooking is permitted when the fire suppression system inoperable, there is a great chance of a serious fire, and the operator is liable because insurance usually does not cover this situation.

Cleaning/Maintenance. Cleaning is usually restricted to the mechanical operators and electrical sensors in the fire suppression system and within the hood that are exposed to grease. If they become excessively coated with grease, mechanical operators cannot move and sensors cannot sense. The result is decreased control and safety. The mechanical operators should be cleaned as often as required to maintain free movement. Fire suppression system operators may need to be changed annually rather than cleaned. Check with a local fire suppression system dealer before attempting to clean any of these components.

Maintenance of control systems (presuming they were properly designed) is mostly a matter of checking the performance of the entire system regularly (minimum every three months) to be sure it is still performing as designed. Mechanical linkages on dampers and cleanliness of sensors should be checked regularly (minimum once a month). All electrical screw terminals in the components should be checked for tightness, relay contacts should be checked for cleanliness, and all exposed conducting surfaces should be checked for corrosion on an annual basis.

RESIDENTIAL KITCHEN VENTILATION

Although commercial and residential cooking processes are similar, ventilation requirements and procedures for the two are different. Differences include exhaust airflow rate and installation height. In addition, residential kitchen ventilation has no replacement air requirement, and energy consumption is insignificant due to lower airflow, smaller motors, and intermittent operation.

Residential Cooking Equipment and Processes

Although the physics of the cooking process and the resulting effluent are about the same, residential cooking is done more conservatively. Equipment such as upright broilers and solid-fuel-burning equipment, described in the section on Exhaust Hoods as Group 3 (heavy duty) and Group 4 (extra heavy duty), is not used. Therefore, the high ventilation rates and ventilation equipment associated with these rates are not found in residential kitchens. On the other hand, cooking effluent and by-products of open flame combustion must be more closely controlled in a residence than in a commercial kitchen because any escaping effluent is dispersed throughout a residence whereas a commercial kitchen is negative compared to surrounding spaces. A residence also has a much lower background ventilation rate, so escaping contaminant is more persistent. This situation presents a different challenge because it is not possible to overcome problems by simply increasing the ventilation rate at the cooking process.

Residential cooking always produces a convective plume that carries upward with it both cooking effluent and any by-products of combustion. Sometimes there is spatter as well, but those particles are so large that they are not removed by ventilation. Residential kitchen hoods depend almost entirely on buoyancy capture.

Hoods and Other Residential Kitchen Ventilation Equipment

Most residential kitchens are ventilated by wall-mounted, conventional range hoods. Overall, they do the best job at the lowest installed cost. A variation is the deep decorative canopy hood, which is usually chosen for aesthetic reasons. The canopy hood is significantly more costly, but somewhat more effective. **Downdraft range-top ventilators** are currently popular in higher priced homes. These are an exception because they are designed to capture the contaminants by producing velocities over the cooking surface greater than those of the convective plume. With sufficient velocity, their operation can be satisfactory; however, with gas cooking, velocity is limited to prevent an adverse effect on the flame.

Kitchen exhaust fans were more widely used in the past, but they are still used and probably will be for several years. Mounted in the wall or ceiling, they ventilate the kitchen generally rather than at the source of contaminant. For general ventilation in all possible kitchen arrangements, 15 air changes per hour are typical; for

ceiling-mounted fans, this is usually sufficient, but for wall-mounted fans, it may be marginal.

Continuous low-level ventilation throughout the house is used increasingly as houses are built with greater airtightness and public awareness of the importance of good indoor air quality grows. This continuous ventilation is for a distinctly different purpose than kitchen ventilation. ASHRAE *Standard* 62 recommends 0.35 air changes per hour but not less than 15 cfm per person. Some installations with equipment for continuous ventilation use that same equipment for kitchen ventilation by intermittently boosting the flow to a considerably higher level or by simply running longer to achieve the needed reduction in effluent concentration. These dual-purpose systems require good design for satisfactory operation and seldom achieve the results of a conventional range hood. If kitchen exhaust is introduced into a general ventilation system, bionutrients must be removed.

Differences Between Commercial and Residential Equipment

Residential hoods usually meet UL *Standard* 507 requirements. Fire-actuated dampers are never part of the hood and are almost never used. Grease filters in residential hoods are much simpler; grease collection through channels is undesirable because inadequate maintenance could allow grease to pool, creating a fire and health hazard.

Conventional residential wall hoods are usually produced with standard height (from 6 to 9 in. or 12 in.) and standard depth (17 to 22 in.). Width is usually the same as the cooking surface; 30 in. is almost a standard in the United States. The current HUD Manufactured Home Construction and Safety Standards call for 3 in. overhang per side.

Mounting height is usually 18 to 24 in., with some mounted even higher at a sacrifice in collection efficiency. The lower the hood, the more effective the capture; the velocity of the convective plume is not great, and the plume is easily disrupted by air currents and disturbances. Studies show that 18 in. is the minimum height for suitable access to the cooking surface. Some codes call for a minimum of 30 in. from the cooking surface to combustible cabinets. In that case, the bottom of a 6 in. hood can be 24 in. above the cooking surface.

A minimum flow rate (exhaust capacity) of 40 cfm per linear foot of hood width is recommended by the Home Ventilating Institute (HVI 1998). Informal tests were conducted in which hood users were told to keep adjusting the variable speed control on the hood until it was running slow enough to minimize the sound level while still maintaining good performance. After a trial period, the chosen setting was evaluated; most users had selected a speed providing flow within 15% of the recommended minimum 40 cfm/ft. Although these tests were too few to be statistically valid and were quite informal, the outcome was predicted. Additional capacity, with speed control, is desirable for handling unusually vigorous cooking and cooking mistakes. The airflow can be increased for a brief period to clear the air.

Exhaust Systems

Residential hoods offer little opportunity for custom design of an exhaust system. A duct connector is built into the hood, and the installation should use the same size duct. A hood includes either an axial or a centrifugal fan. The centrifugal fan has higher pressure capabilities, but the axial fan is usually adequate for low-volume hoods. Virtually all residential hoods on the market in the United States have HVI-certified airflow performance.

Replacement (Makeup) Air

The exhaust rate of residential hoods is generally low enough to avoid the need for replacement air systems, so hoods rely on natural infiltration. This may cause slight negative pressurization of the residence, but it is usually less than that caused by other equipment such as the clothes dryer. Still, the penalties for backdrafts through the flue of a combustion appliance are high. If the residence has a gas furnace and water heater, the flue should be checked for adequate flow with all exhaust devices turned on unless the furnace and water heater are of the sealed combustion type.

Sometimes commercial-style cooking equipment approved for residential use, with its associated higher ventilation requirements, is installed in residences. In such cases, designers should refer to the earlier sections of this chapter, possibly including the section on Replacement (Makeup) Air Systems.

Energy Conservation

The energy cost of residential hoods is quite low because of the few hours of running time and the low rate of exhaust. For example, it typically costs less than $10 per heating season in Chicago to run a hood and heat the replacement air, based on running at 150 cfm for an hour a day and using gas heat.

Fire Protection for Residential Hoods

Residential hoods must be installed with metal (preferably steel) duct, and the duct should be positioned to prevent pooling of grease. Residential hood exhaust ducts are almost never cleaned, and there is no evidence that this causes fires. Several attempts have been made to make fire extinguishers available in residential hoods, but none has met with broad acceptance. In contrast, residential grease fires on the cooking surface are still a problem. Although the fires always ignite at the cooking surface when cooking grease is left unattended and overheats, the hood is sometimes blamed. That problem is not solved with fire-extinguishing equipment.

Maintenance of Residential Kitchen Ventilation Equipment

All UL-listed hoods and kitchen exhaust fans are designed for simple cleaning, which should be done at intervals consistent with the cooking practices of the user. Although cleaning is often thought to be for fire prevention only, it also benefits health by removing nutrients available for the growth of organisms.

REFERENCES

ASHRAE. 1989. Ventilation for acceptable indoor air quality. ANSI/ASHRAE *Standard* 62-1989.

ASTM. 1996. Test method for performance of commercial kitchen ventilation systems. *Standard* F 1704-96. American Society for Testing and Materials, West Conshohocken, PA.

Claar, C.N., R.P. Mazzucchi, and J.A. Heidell. 1985. The project on restaurant energy performance (PREP)—End use monitoring and analysis. Office of Building Energy Research and Development. U.S. Department of Energy, Washington, DC.

Elovitz, G. 1992. Design considerations to master kitchen exhaust systems. *ASHRAE Transactions* 98(1):1199-1213.

HUD. Manufactured home construction and safety standards. *Code of federal regulations* Title 24, Part 3280. Department of Housing and Urban Development, Washington, DC.

HVI. 1998. Home ventilating guide. Home Ventilating Institute, Air Movement and Control Association International, Arlington Heights, IL.

ICC. 1996. *International mechanical code.* International Code Council, Falls Church, VA.

Kuehn, T.H., W.D. Gerstler, D.Y.H. Pui, and J.W. Ramsey. 1999. Comparison of emissions from selected commercial kitchen appliances and food products. *ASHRAE Transaction* 105(2).

NFPA. 1994. Dry chemical extinguishing systems. ANSI/NFPA *Standard* 17-94. National Fire Protection Association, Quincy, MA.

NFPA. 1994. Ventilation control and fire protection of commercial cooking operations. *Standard* 96-94. National Fire Protection Association, Quincy, MA.

NFPA. 1994. Wet chemical extinguishing systems. ANSI/NFPA *Standard* 17A-94.

NFPA. 1996. Installation of sprinkler systems. ANSI/NFPA *Standard* 13-96.

Spata, A.J. and S.M. Turgeon. 1995. Impact of reduced exhaust and ventilation rates at 'no-load' cooking conditions in a commercial kitchen during winter operation. *ASHRAE Transactions* 101(2):606-10.

UL. 1979. Grease filters for exhaust ducts, 2nd ed. *Standard* 1046-79. Underwriters Laboratories, Northbrook, IL.

UL. 1994. Electric fans, 8th ed. ANSI/UL *Standard* 507-94.

UL. 1995. Exhaust hoods for commercial cooking equipment, 5th ed. *Standard* 710-95.

UL. 1995. Grease ducts, 1st ed. *Standard* 1978-95.

UL. 1996. Fire testing of fire extinguishing systems for protection of restaurant cooking areas, 2nd ed. *Standard* 300-96.

BIBLIOGRAPHY

Abrams, S. 1989. Kitchen exhaust systems. *The Consultant* (FCSI) 22(4).

Alereza, T. and J.P. Breen. 1984. Estimates of recommended heat gains due to commercial appliances and equipment. *ASHRAE Transactions* 89(2A):25-58.

Annis, J.C. and P.J. Annis. 1989. Size distributions and mass concentrations of naturally generated cooking aerosols. *ASHRAE Transactions* 95(1):735-43.

Balogh, E.A. 1989. Tandem wet/dry chemical cooking area fire extinguishing concept. *The Consultant* (FCSI) 22(4).

Bevirt, W.D. 1994. What engineers need to know about testing and balancing. *ASHRAE Transactions* 100(1):705-14.

Black, D.K. 1989. Commercial kitchen ventilation—Efficient exhaust and heat recovery. *ASHRAE Transactions* 95(1):780-86.

Capalbo, F.J. 1991. Efficient kitchen ventilation: Grease filters are the key. *The Consultant.*

Chih-Shan, L., L. Wen-Hai, and J. Fu-Tien. 1993. Removal efficiency of particulate matter by a range exhaust fan. *Environmental International* 19:371-80.

Chih-Shan, L., L. Wen-Hai, and J. Fu-Tien. 1993. Size distributions of submicrometer aerosols from cooking. *Environmental International.*

Cini, J.C. 1989. Innovative kitchen exhaust systems. *The Consultant* (FCSI) 22(4).

Collison, R. 1979. Energy consumption during cooking. *Journal of Food Technology* 14:173-79.

Conover, D.R. 1992. Building construction regulation impacts on commercial kitchen ventilation and exhaust systems. *ASHRAE Transactions* 98(1):1227-35.

Crabtree, S. 1989. Ventilation: Where less is better. *The Consultant* (FCSI) 22(4).

Dallavalle, J. 1953. Design of kitchen range hoods. *Heating and Ventilating* (August).

de Albuquerue, A.J. 1972. Equipment loads in laboratories. *ASHRAE Journal* 14(9):59-62.

Department of Mechanical Engineering, University of Manitoba-BETT Program. 1984. Analysis of commercial kitchen exhaust systems design and energy conservation. Report prepared for Energy, Mines and Resources Canada.

Deppisch, J.R. and L.J. Irwin. 1974. Energy efficiencies of gas and electric range top sections. *Research Report* 1504. American Gas Association Laboratories, Cleveland, OH.

Donnelly, F. 1966. Preventing fires in kitchen ventilating ducts. *Air Conditioning, Heating and Ventilating* (December).

Drees, K.H., J.D. Wenger, and G. Janu. 1992. Ventilation airflow measurement for ASHRAE *Standard* 62-1989. *ASHRAE Journal* 34(10):40-44.

Electrification Council. 1975. New electric kitchen ventilation design criteria—Leader's guide. The Electrification Council, New York.

EPRI. 1989. Analysis of building codes for commercial kitchen ventilation systems. *Final Report* CU-6321 for National Conference of States on Building Codes and Standards. Electric Power Research Institute, Palo Alto, CA.

EPRI. 1989. Assessment of building codes, standards and regulations impacting commercial kitchen design. *Final Report* for National Conference of States on Building Codes and Standards. Electric Power Research Institute, Palo Alto, CA.

EPRI. 1991. Proposed testing protocols for commercial kitchen ventilation research. *Final Report* CU-7210 for Underwriters Laboratories. Electric Power Research Institute, Palo Alto, CA.

Esmen, N.A., D.A. Wegel, and F.P. McGuigan. 1986. Aerodynamic properties of exhaust hoods. *American Industrial Hygiene Association Journal* 47(8).

Farnsworth, C.A. and D.E. Fritzsche. Improving the ventilation of residential gas ranges. American Gas Association Laboratories, Cleveland, OH.

Farnsworth, C., A. Waters, R.M. Kelso, and D. Fritzsche. 1989. Development of a fully vented gas range. *ASHRAE Transactions* 95(1):759-68.

Fitzgerald, H. 1969. Code reduces kitchen exhaust fires. *Air Conditioning, Heating and Ventilating* (May).

Frey, D.J., C.N. Claar, and P.A. Oatman. 1989. The model electric restaurant. *Final Report* CU-6702. Electric Power Research Institute, Palo Alto, CA.

Frey, D.J., K.F. Johnson, and V.A. Smith. 1993. Computer modeling analysis of commercial kitchen equipment and engineered ventilation. *ASHRAE Transactions* 99(2):890-908.

Fritz, R.L. 1989. A realistic evaluation of kitchen ventilation hood designs. *ASHRAE Transactions* 95(1):769-79.

Fugler, D. 1989. Canadian research into the installed performance of kitchen exhaust fans. *ASHRAE Transactions* 95(1):753-58.

Garrett, R.T. 1984. Commercial/Institutional kitchen and make-up air analysis: Design and selection application guide. Muckler Industries, St. Louis, MO.

Garrett, R.T. 1987. Psychrometric limitations of exhaust and makeup air conditions associated with commercial kitchen ventilation. IAQ 87, *Practical control of indoor air problems*, pp. 359-71. ASHRAE.

Garrett, R.T. 1989. Are you willing to save money on heating costs just to blow your air conditioning out the exhaust? *The Consultant* (FCSI) 22(4).

Gordon, E.B. and N.D. Burk. 1993. A two-dimensional finite-element analysis of a simple commercial kitchen ventilation system. *ASHRAE Transactions* 99(2):90914.

Gordon, E.B., D.J. Horton, and F.A. Parvin. 1994. Development and application of a standard test method for the performance of exhaust hoods with commercial cooking appliances. *ASHRAE Transactions* 100(2):988-99.

Gordon, E.B., D.J. Horton, and F.A. Parvin. 1995. Description of a commercial kitchen ventilation (CKV) laboratory facility. *ASHRAE Transactions* 101(1):249-61.

Gotoh, N., N. Ohira, T. Fusegi, T. Sakurai, and T. Omori. 1993. A study of ventilation of a kitchen with a range hood fan. *Room air convection and ventilation effectiveness*, pp. 371-77. 1992 ISRACVE, Tokyo, Japan. ASHRAE.

Greenheck Fan Corporation. 1981. Cooking equipment ventilation—Application and design. CEV-5-82. Greenheck Fan Corporation, Schofield, WI.

Guffey, S.E. 1992. A computerized data acquisition and reduction system for velocity traverses in a ventilation laboratory. *ASHRAE Transactions* 98(1):98-106.

Hane, A.E. 1989. Discussion of ventilation and exhaust hoods. *The Consultant* (FCSI) 22(4).

Hicks, B. 1989. Who wants ventilation anyway? And why? *The Consultant* (FCSI) 22(4).

Himmel, R.L. 1983. Effects of ventilation on equipment performance. *Commercial Cooking Equipment Improvement*, Volume IV. GRI-80/0079.4 Final Report. American Gas Association Laboratories, Cleveland, OH.

Horton, D.J., J.N. Knapp, and E.J. Ladewski. 1993. Combined impact of ventilation rates and internal heat gains on HVAC operating costs in commercial kitchens. *ASHRAE Transactions* 99(2):877-83.

Hunt, C.M., D.R. Showalter, and S.J. Treado. 1974. Tests of a grease interceptor similar to those used in galleys. Center for Building Technology, National Bureau of Standards, Washington, D.C.

Jamboretz, J.S. 1982. Before you specify a "no-heat" hood. *The Consultant* (FCSI) 15(2):40.

Jin, Y. and J.R. Ogilvie. 1992. Isothermal airflow characteristics in a ventilated room with a slot inlet opening. *ASHRAE Transactions* 98(2):296-306.

Kelso, R.M. and C. Rousseau. 1995. Kitchen ventilation. *ASHRAE Journal* 38(9):32-36.

Kelso, R.M., A.J. Baker, and S. Roy. An efficient CFD algorithm for prediction of contaminant dispersion in room air motion. University of Tennessee, Knoxville.

Kelso, R.M., L.E. Wilkening, E.G. Schaub, and A.J. Baker. 1992. Computational simulation for kitchen airflows with commercial hoods. *ASHRAE Transactions* 98(1):1219-26.

Kim, I.G. and H. Homma. 1992. Possibility for increasing ventilation efficiency with upward ventilation. *ASHRAE Transactions* 98(1):723-29.

Knapp, J. 1989. Ventilation in the food service industry. *The Consultant* (FCSI) 22(4).

Knapp, J.N. and W.A. Cheney. 1993. Development of high-efficiency air cleaners for grilling and deep-frying operations. *ASHRAE Transactions* 99(2):884-89.

Kuehn, T.H., J. Ramsey, H. Han, M. Perkovich, and S. Youssef. 1989. A study of kitchen range exhaust systems. *ASHRAE Transactions* 95(1):744-52.

Laderoute, M. 1989. Under the hood (don't let fire put you out of business). *The Consultant* (FCSI) 22(4).

Leipman, M.A. 1989. Who do you trust? *The Consultant* (FCSI) 22(4).

Levenback, G. 1989. The kitchen ventilation troika: Underwriters Laboratories, National Fire Protection Association, and Council of American Building Officials. *The Consultant* (FCSI) 22(4).

Lockhart, C. 1989. Restaurant ventilation: Our environmental concerns. *The Consultant* (FCSI) 22(4).

Locklin, D.W., R.D. Giammer, and S.G. Talbert. 1971. Preliminary study of ventilation requirements for commercial kitchens. *ASHRAE Journal* 13(11):51.

Loving, W.H. 1964. Make-up air for commercial kitchens. *Air Engineering* (December).

Marn, W.L. 1962. Commercial gas kitchen ventilation studies. *Research Bulletin* 90. American Gas Association Laboratories, Cleveland, OH.

McGuire, A.B. 1993. Commercial and institutional kitchen exhaust systems. *Heating, Piping and Air Conditioning* (May).

Murakami, S., S. Kato, T. Tanaka, D.-H. Choi, and T. Kitazawa. 1992. The influence of supply and exhaust openings on ventilation efficiency in an air-conditioned room with a raised floor. *ASHRAE Transactions* 98(1):738-55.

Niemeier, R.E. 1989. Acoustical considerations for commercial kitchen ventilators. *The Consultant* (FCSI) 22(4).

Otenbaker, J. 1989. Parting the curtain on ventilation codes: A challenge for specifiers and manufacturers. *The Consultant* (FCSI) 22(4).

Parikh, J.S. 1989. UL's certification program for restaurant ventilation equipment. *The Consultant* (FCSI) 22(4).

Parikh, J.S. 1992. Testing and certification of fire and smoke dampers. *ASHRAE Journal* 34(11):30-33.

Pekkinen, J. and T.H. Takki-Halttunen. 1992. Ventilation efficiency and thermal comfort in commercial kitchens. *ASHRAE Transactions* 98(1):1214-18.

Pekkinen, J.S. 1993. Thermal comfort and ventilation effectiveness in commercial kitchens. *ASHRAE Journal* 35(7):35-38.

Riemenschneider, J. 1989. Ventilation and heat reclaim systems. *The Consultant* (FCSI) 22(4).

Romano, S. 1989. Kitchen cooking equipment ventilation is a lot of hot air. *The Consultant* (FCSI) 22(4).

Schaelin, D., J. van der Maas, and A. Moser. 1992. Simulation of airflow through large openings in buildings. *ASHRAE Transactions* 98(2):319-28.

Schmid, F.P., V.A. Smith, and R.T. Swierczyna. 1997. Schlieren flow visualization in commercial kitchen ventilation research. *ASHRAE Transactions* 103(2):937-42.

Seller, J. and B. Ward. 1991. The environment and you, Part 1. *Foodservice Equipment & Supplies Specialist*.

Shaub, E.G., A.J. Baker, N.D. Burk, E.B. Gordon, and P.G. Carswell. 1995. On development of a CFD platform for prediction of commercial kitchen ventilation flow fields. *ASHRAE Transactions* 101(2):581-93.

Shibata, M., R.H. Howell, and T. Hayashi. 1982. Characteristics and design method for push-pull hoods: Part 2—Streamline analyses of push-pull flows. *ASHRAE Transactions* 88(1):557-70.

Shute, R.W. 1992. Integrating access floor plenums for HVAC air distribution. *ASHRAE Journal* 34(10):46-51.

Snyder, O.P., D.R. Thompson, and J.F. Norwig. 1983. Comparative gas/electric food service equipment energy consumption ratio study. University of Minnesota, St. Paul.

Soling, S.P. and J. Knapp. 1985. Laboratory design of energy efficient exhaust hoods. *ASHRAE Transactions* 91(1B):383-92.

Smith, V.A., D.J. Frey, and C.V. Nicoulin. 1997. Minimum-energy kitchen ventilation for quick service restaurants. *ASHRAE Transactions* 103(2):950-61.

Smith, V.A., R.T. Swierczyna, and C.N. Claar. 1995. Application and enhancement of the standard test method for the performance of commercial kitchen ventilation systems. *ASHRAE Transactions* 101(2):594-605.

Swierczyna, R.T., V.A. Smith, and F.P. Schmid. 1997. New threshold exhaust flow rates for capture and containment of cooking effluent. *ASHRAE Transactions* 103(2):943-49.

Talbert, S.G., L.J. Flanigan, and J.A. Eibling. 1973. An experimental study of ventilation requirements of commercial electric kitchens. *ASHRAE Transactions* 79(1):34-47.

UL. 1989. Outline of investigation for power ventilators for restaurant exhaust appliances, Issue No. 1. *Subject* 762-89. Underwriters Laboratories, Northbrook, IL.

VDI Verlag. 1984. Raumlufttechnische Anlagen für Küchen (Ventilation equipment for kitchens). VDI 2052.

Wolbrink, D.W. and J.R. Sarnosky. 1992. Residential kitchen ventilation—A guide for the specifying engineer. *ASHRAE Transactions* 91(1):1187-98.

Wood, C.A. 1989. Hot air. *The Consultant* (FCSI) 22(4).

Zhang, J. S., G.J. Wu, and L.L. Christianson. 1992. Full-scale experimental results on the mean and turbulent behavior of room ventilation flows. *ASHRAE Transactions* 98(2):307-18.

Zimmerman, B.A. 1989. Fire dampers...Facts and fallacies. *The Consultant* (FCSI) 22(4).

GEOTHERMAL ENERGY

THE use of geothermal resources can be broken down into three general categories: high temperature (>300°F) electric power production, intermediate- and low-temperature direct-use applications (<300°F), and ground-source heat pump applications (generally <90°F). This chapter covers only the direct-use and ground-source heat pump categories. After an overview of resources, the chapter is divided into two sections. The section on Direct-Use Systems contains information on wells, equipment and applications. The section on Ground-Source Heat Pump Systems includes information only on the ground-source portion. Information on design within the building may be found in Chapter 8 of the 1996 ASHRAE Handbook—Systems and Equipment.

RESOURCES

Geothermal energy is the thermal energy within the earth's crust—the thermal energy in rock and fluid (water, steam, or water containing large amounts of dissolved solids) that fills the pores and fractures within the rock, and sand and gravel. Calculations show that the earth, originating from a completely molten state, would have cooled and become completely solid many thousands of years ago without an energy input, in addition to that of the sun. It is believed that the ultimate source of geothermal energy is radioactive decay within the earth (Bullard 1973).

Through plate motion and vulcanism, some of this energy is concentrated at high temperature near the surface of the earth. Energy is also transferred from the deeper parts of the crust to the earth's surface by conduction and by convection in regions where geological conditions and the presence of water permit.

Because of variation in volcanic activity, radioactive decay, rock conductivities, and fluid circulation, different regions have different heat flows (through the crust to the surface), as well as different temperatures at a particular depth. The normal increase of temperature with depth (i.e., the normal geothermal gradient) is about 13.7°F per 1000 ft of depth, with gradients of 5 to 27°F per 1000 ft being common. The areas that have the higher temperature gradients and/or higher-than-average heat flow rates constitute the most interesting and viable economic resources. However, areas with normal gradients may be valuable resources if certain geological features are present.

Geothermal resources of the United States are categorized into the following types:

- Igneous point sources
- Deep convective circulation in areas of high regional heat flow
- Geopressured resources
- Concentrated radiogenic heat sources
- Deep regional aquifers in areas of near normal gradient

Igneous point resources are associated with magma bodies, which result from volcanic activity. These bodies heat the surrounding and overlying rock by conduction and convection, as permitted by the rock permeability and fluid content in the rock pores.

Deep circulation of water in areas of high regional heat flow can result in hot fluids near the surface of the earth. Known as **hydrothermal convection systems**, these geothermal resources are widely used. The fluids near the surface have risen from natural convection between the hotter, deeper formation and the cooler formations near the surface. The passageway that provides for this deep convection must consist of adequate permeable fractures and faults.

The **geopressured resource**, present widely in the Gulf Coast of the United States, consists of regional occurrences of confined hot water in deep sedimentary strata, where pressures of greater than 10,000 psi are common. This resource also contains methane, which is dissolved in the geothermal fluid.

Radiogenic heat sources exist in various regions as granitic plutonic rocks that are relatively rich in uranium and thorium. These plutons have a higher heat flow than the surrounding rock; if the plutons are blanketed by sediments of low thermal conductivity, an elevated temperature at the base of the sedimentary section can result. This resource has been identified in the eastern United States.

Deep regional aquifers of commercial value can occur in deep sedimentary basins, even in areas of only normal temperature gradient. For deep aquifers to be of commercial value, (1) the basins must be deep enough to provide usable temperature levels at the prevailing gradient, and (2) the permeability within the aquifer must be adequate for flow.

The thermal energy in geothermal resources exists primarily in the rocks and only secondarily in the fluids that fill the pores and fractures within them. Thermal energy is usually extracted by bringing to the surface the hot water or steam that occurs naturally in the open spaces in the rock. Where rock permeability is low, the energy extraction rate is low. To extract the thermal energy from the rock itself, water must be recharged to the ground as the initial water is extracted. In permeable aquifers, the produced fluid may be injected back into the aquifer at some distance from the production well to pass through the aquifer again and recover some of the energy in the rock.

Temperature

The temperature of fluids produced in the earth's crust and used for their thermal energy content varies from below 40°F to 680°F. As indicated in Figure 1, local gradients also vary with geologic conditions. The lower value represents the fluids used as the low-temperature energy source for heat pumps, and the higher temperature represents an approximate value for the HGP-A well at Hilo, Hawaii.

The following classification by temperature is used in the geothermal industry:

High temperature	$t > 300°F$
Intermediate temperature	$195°F < t < 300°F$
Low temperature	$t < 195°F$

The preparation of this chapter is assigned to TC 6.8, Geothermal Energy Utilization.

Electric generation is generally not economical for resources with temperatures below about 300°F, which is the reason for the division between high- and intermediate-temperature. However, binary power plants, with the proper set of circumstances, have demonstrated that it is possible to generate electricity economically above 230°F. In 1988, there were 86 binary plants worldwide, generating a total of 126.3 MW (Di Pippo 1988).

The 195°F division between intermediate and low temperatures is common in resource inventories, but is somewhat arbitrary. At 195°F and above, applications such as district heating can be readily implemented with equipment used in conventional applications of the same type; at lower temperatures, these applications require redesign to take the greatest advantage of the geothermal resource.

Geothermal resources at the lower temperatures are more common. The frequency by reservoir temperature of identified convective systems above 195°F is shown in Figure 2.

Geothermal Fluids

Geothermal energy is currently extracted from the earth through the naturally occurring fluids in rock pores and fractures. In the future, an additional fluid may be introduced into the geothermal system and circulated through it to recover the energy. The fluids being produced are steam, hot water, or a two-phase mixture of both. These may contain various amounts of impurities, notably dissolved gases and dissolved solids.

Geothermal resources that produce essentially dry steam are **vapor-dominated**. While these are valuable resources, they are rare. Hot water (fluid-dominated) resources are much more common than vapor-dominated systems and can be produced either as hot water or as a two-phase mixture of steam and hot water, depending on the pressure maintained on the production plant. If the pressure in the production casing or in the formation around the casing is reduced below the saturation pressure at that temperature, some of the fluid will flash, and a two-phase fluid will result. If the pressure is maintained above the saturation pressure, the fluid will remain single-phase. In fluid-dominated resources, both dissolved gases and dissolved solids are significant. The quality of the fluid varies from site to site, from a few hundred parts per million (ppm) to over 300,000 ppm dissolved solids.

Table 1 presents the composition of fluids from three geothermal wells in the United States. The list illustrates the types of substances and the range of concentrations that can be expected in the fluids. The harshness of the fluid is not only site-dependent but also temperature-dependent; hardness increases with temperature.

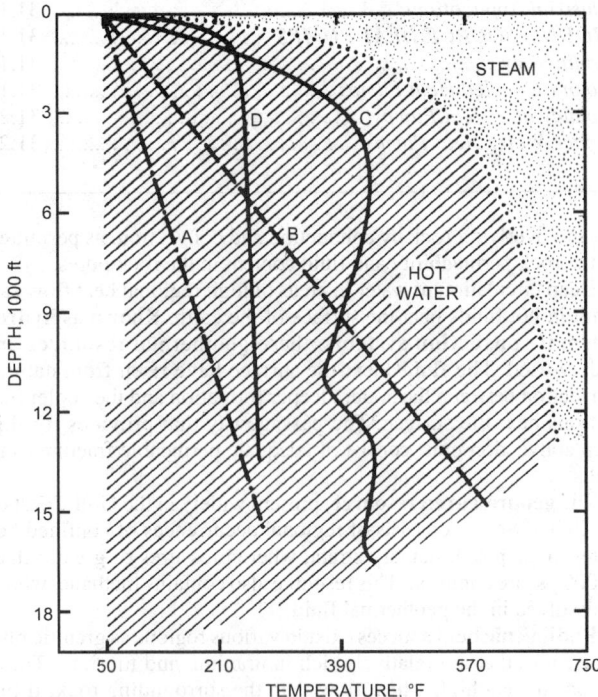

A Near-normal temperature gradient
B High conductive gradient
C and D Temperature resulting from convective flow

Fig. 1   Representative Temperature-Depth Relations in the Earth's Crust
(Combs et al. 1980)

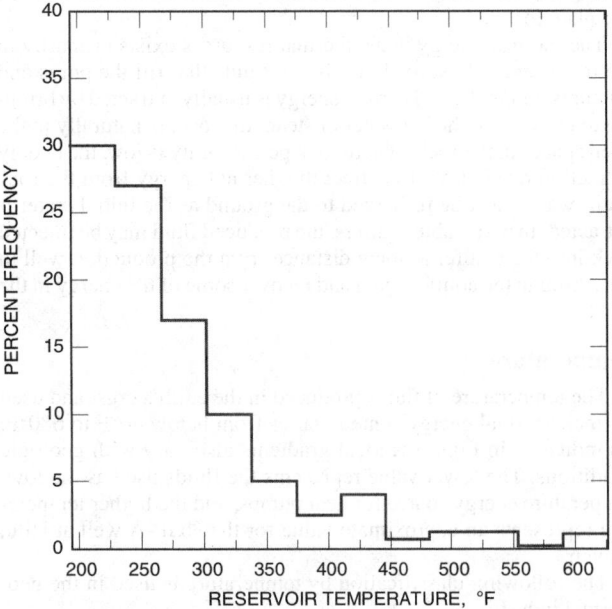

Fig. 2   Frequency of Identified Hydrothermal Convection Resources Versus Reservoir Temperature
(Muffler et al. 1979)

Table 1   Representative Fluid Composition from Geothermal Wells in Various Resource Areas

	Concentration, ppm		
Material	Boise, ID[a]	Klamath Falls, OR[b]	Salton Sea, CA
Total dissolved solids	290	795	220,000
SiO_2	160	48	350
Na^+	90	205	5,100
K^+	1.6	4.3	12,500
Ca^{2+}	1.7	26	23,000
Mg^{2+}	0.05	—	150
Cl^-	10	51	133,000
F^+	14	1.5	13
SO_4^{2-}	23	330	5
NO_3^-	—	4.9	—
NH_4^+	—	1.3	—
H_2S	trace	1.5	—
HCO_3^-	70	20	7,025
CO_3^{2-}	4	15	—
CO_2	0.2	—	—
B^{3+}	0.14	—	350
Fe^{2+}, Fe^{3+}	0.13	0.3	1,300
O_2	0.0029	0.2	—
Temperature, °F	176	192	482

Source: Boise and Salton Sea data from Cosner and Apps (1978);
   Klamath Falls data from Ellis and Conover (1981).
[a]Well name unknown, but near old penitentiary.   [b]Wendling Well (Lund et al. 1976).

This chapter concentrates on resources produced as a single-phase hot liquid.

Present Use

Discoveries of concentrated radiogenic heat sources and deep regional aquifers in areas of near normal temperature gradient indicate that 37 states in the United States have economically exploitable direct-use geothermal resources (Interagency Geothermal Coordinating Council 1980).

The Geysers, a resource area in northern California, is the largest single geothermal development in the world. The total electricity generated by geothermal development in the world was 5175 MW in 1988 (Di Pippo 1988). The direct application of geothermal energy for space heating and cooling, water heating, agricultural growth-related heating, and industrial processing represents about 2.9×10^{10} Btu/h worldwide in 1988 (Lienau et al. 1988 and Gundmundsson 1985). In the United States in 1988, direct-use installed capacity amounted to 5.7×10^9 Btu/h, providing 17×10^{12} Btu/yr (Lienau et al. 1988).

The major uses of geothermal energy in agricultural growth applications are for heating greenhouse and aquaculture facilities. The principal industrial uses of geothermal energy in the United States are for food processing (dehydration) and gold processing. Worldwide, the main applications include space and water heating, space cooling, agricultural growth, and food processing. Exceptions are diatomaceous earth processing in Iceland, and pulp and paper processing in New Zealand.

DIRECT-USE SYSTEMS

Figure 3 is a schematic of a typical direct-use system. Such a system may consist of five subsystems: (1) the production system, including the producing wellbore and associated wellhead equipment, (2) the transmission and distribution system that transports the geothermal energy from the resource site to the user site and then distributes it to the individual user loads, (3) the user system, (4) the disposal system, which can be either surface disposal or injection back into a formation, and (5) an optional peaking/backup system. None of the 14 major geothermal district heating systems in the United States includes a peaking/backup component as part of the main distribution system. Backup is most commonly included in the end user systems and peaking is rarely used.

In a typical direct-use system, the geothermal fluid is produced from the production borehole by a lineshaft multistage centrifugal pump. (For free-flowing wells with adequate quantities of fluid, a pump is not required. However, most commercial operations require pumping to provide the necessary flow.) When the geothermal fluid reaches the surface, it is delivered to the application site through the transmission and distribution system.

In the system shown in Figure 4, the geothermal production and disposal system are closely coupled, and they are both separated from the contact with the equipment by a heat exchanger. This sec-

ondary loop is especially desirable when the geothermal fluid is particularly corrosive and/or causes scaling. Then the geothermal fluid is pumped directly back into the ground without loss to the surrounding surface.

CHARACTERISTICS

The following characteristics influence the cost of energy delivered from geothermal resources:

- Depth of resource
- Distance between resource location and application site
- Well flow rate
- Resource temperature
- Temperature drop
- Load size
- Load factor
- Composition of fluid
- Ease of disposal
- Resource life

Many of these characteristics have a major influence because the cost of geothermal systems is primarily front-end capital cost; annual operating cost is relatively low.

Depth of the Resource

The cost of the well is usually one of the larger items in the overall cost of a geothermal system, and the cost increases with the depth of the resource. Compared to many other geothermal areas worldwide, well depth requirements in the western United States are relatively shallow; most larger geothermal systems operate with production wells of less than 2000 ft, and many at less than 1000 ft.

Distance Between Resource Location and Application Site

The direct use of geothermal energy must occur near the resource. The reason is primarily economic; although the geothermal fluid (or a secondary fluid) could be transmitted over moderately long distances (greater than 60 miles) without a great temperature loss, such transmission would not generally be economically feasible. Most existing geothermal projects are characterized by transmission distances of less than 1 mile.

Well Flow Rate

The energy output from a production well varies directly with the fluid flow rate. The energy cost at the wellhead varies inversely with the well flow rate. A typical good resource has a production rate of 400 to 800 gpm per production well; however, geothermal direct-use wells have been designed to produce up to 2000 gpm.

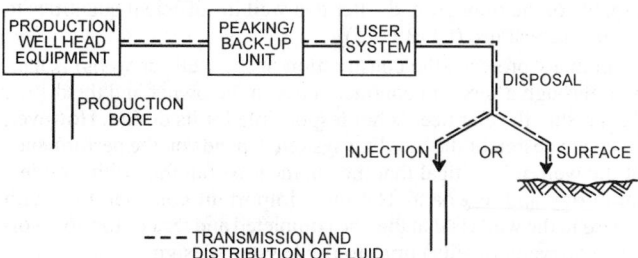

Fig. 3 Basic Geothermal Direct-Use System

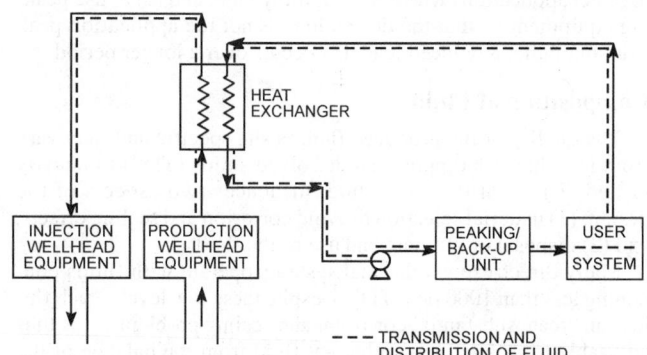

Fig. 4 Geothermal Direct-Use System with Wellhead Heat Exchanger and Injection Disposal

Resource Temperature

In geothermal resources, the available temperature is associated with the prevailing resource. This temperature is an nearly fixed value for a given resource. Although the temperature may or may not increase with deeper drilling. Natural convection in fluid-dominated resources keeps the temperature relatively uniform throughout the depth of the resource (see Figure 1); but if there are deeper, separate aquifers (producing zones) in the area, deeper drilling can recover energy at a higher temperature.

The temperature can restrict applications. It often requires a reevaluation of accepted application temperature, because they have been developed for uses served by conventional fuels for which the application temperature could be selected at any value within a relatively broad range. When geothermal energy is used directly, the application temperature must be lower than the temperature of the produced fluid; however, when heat pumps or other means of temperature boosting are used, the application temperature may be somewhat higher than the produced fluid temperature.

Temperature Drop

Because well flow is limited, the power output from the geothermal well is directly proportional to the temperature drop of the geothermal fluid. Consequently, a larger temperature drop results in lower energy cost at the wellhead. This concept is differs with many conventional and solar systems that circulate a heating fluid with a small temperature drop—so a different design philosophy and different equipment are required.

Cascading the geothermal fluid to uses with lower temperature requirements can be advantageous in achieving a large temperature difference (Δt). Most geothermal systems have been designed for a Δt of between 30 and 50°F; although, one system was designed for a Δt of 100°F with a 190°F resource temperature.

Load Size

Large-scale applications benefit from economy of scale, particularly in regard to reduced resource development and transmission system costs. For smaller developments, matching the size of the application with the production rate from the geothermal resource is important because the total output varies in increments of one well's output.

Load Factor

Defined as the ratio of the average load to the design capacity of the system, the load factor effectively reflects the fraction of time that the initial investment in the system is working. Again, because geothermal cost is primarily initial cost rather than operating cost, this factor significantly affects the viability of a geothermal system. As the load factor increases, so does the economy of using geothermal energy. The two main ways of increasing the load factor are (1) to select applications where it is naturally high, and (2) to use peaking equipment so that the design load is not the application peak load, but rather a reduced load that occurs over a longer period.

Composition of Fluid

The quality of the produced fluid is site specific and may vary from less than 1000 ppm total dissolved solids (TDS) to heavily brined. The quality of the fluid influences two aspects of the design: (1) material selection to avoid corrosion and scaling effects, and (2) disposal or ultimate end use of the fluid.

Many direct-use geothermal systems operate with fluids containing less than 1000 ppm TDS. Despite these low levels, such fluids can create substantial corrosion and scaling problems. It is thus advisable to isolate the geothermal fluid from the balance of the system. While it is more expensive than using the fluid directly in the process or heating equipment, isolation is preferred for minimizing long-term maintenance requirements.

Ease of Disposal

The costs associated with disposal, particularly when injection is involved, can substantially affect development costs. Most geothermal effluent is disposed of on the surface, including discharge to irrigation, rivers, and lakes. This method of disposal is considerably less expensive than the construction of injection wells. However, the magnitude of geothermal development in certain areas where surface disposal has historically been used (e.g., Klamath Falls, Oregon and Boise, Idaho) has caused the aquifer water level to decline. As a result, regulatory authorities in these and many other areas favor the use of injection in order to maintain reservoir fluid levels.

In addition, geothermal fluids sometimes contain chemical constituents that cause surface disposal to become a problem. Some of these constituents are listed in Table 2.

Table 2   Selected Chemical Species Affecting Fluid Disposal

Species	Reason for Control
Hydrogen sulfide (H_2S)	Odor
Boron (B^{3+})	Damage to agricultural crops
Fluoride (F^-)	Level limited in drinking water sources
Radioactive species	Levels limited in air, water, and soil

Source: Lunis (1989).

If injection is required, the depth at which the fluid can be injected affects well cost substantially. Some jurisdictions allow considerable latitude of injection level; others require the fluid be returned to the same or similar aquifers. In the latter case, it may be necessary to bore the injection well to the same depth as the production well.

Resource Life

The life of the resource has a direct bearing on the economic viability of a particular geothermal application. There is little experience on which to base projections of resource life for heavily developed geothermal resources. However, resources can readily be developed in a manner that will allow useful lives of 30 to 50 years and greater. In some heavily developed, direct-use areas, major systems have been in operation for many years. For example, the Boise Warm Springs Water District system (a district heating system serving some 240 residential users) has been in continuous operation since 1892.

WATER WELLS

Terminology

Although moisture exists to some extent at most subsurface locations, below a certain depth there exists a zone in which all of the pores and spaces between the rock are filled with water. This is called the **zone of saturation**. The object of constructing wells is to gain access to **groundwater**. Groundwater is the water that exists within the zone of saturation. An **aquifer** is a geologic unit that is capable of yielding groundwater to a well in sufficient quantities to be of practical use (UOP 1975).

In many projects, the construction of the well (or wells) is handled through a separate contract between the owner and the driller. As a result, the engineer is not responsible for its design. However, since the design of the building system depends on the performance of the well, it is critical that the engineer be familiar with well terminology and test data. The most important consideration with regard to the wells is that they be completed and tested (for flow volume and water quality) prior to final system design.

Figure 5 presents a summary of the more important terms relating to wells. Several references (Anderson 1984, EPA 1975, Roscoe

Moss Company 1985, Campbell and Lehr 1973) cover well drilling and well construction in detail.

Static water level (SWL) is the level that exists under static (non-pumping) conditions. In some cases, this level is much closer to the surface than that at which the driller encounters water during drilling. **Pumping water level** (PWL) is the level that exists under specific pumping conditions. Generally, this level is different for different pumping rates (higher pumping rates minus lower pumping levels). The difference between the SWL and the PWL is the **drawdown**. The **specific capacity** of a well is frequently quoted in gpm per foot of drawdown. For example, for a well with a static level of 50 ft that produces 150 gpm at a pumping level of 95 ft, drawdown = 95 − 50 = 45 ft; specific capacity = 150/450 = 3.33 gpm per foot.

For groundwater characterized by carbonate scaling potential, **water entrance velocity** (through the screen or perforated casing) is an important design consideration. Velocity should be limited to a minimum of 0.1 fps to avoid incrustation of the entrance area. A common course of high entrance velocity is **overpumping** (installation of a pump that is too large relative to the well's capacity and produces excessive drawdown).

The **pump bowl** assembly (impeller housings and impellers) is always placed sufficiently below the expected pumping level to prevent cavitation at the peak production rate. For the previous example, this pump should be placed at least 115 ft below the casing top (pump setting depth = 115 ft) to allow for adequate submergence at peak flow. The specific net positive suction head (NPSH) required for a pump varies with each application and should be carefully considered during design.

For the well pump, **total pump head** is composed of four primary components: lift, column friction, surface requirements and injection head. **Lift** is the vertical distance that water must be pumped to reach the surface. In the example, lift would be 95 ft. The additional 20 ft of submergence imposes no static pump head.

Column friction is calculated from pump manufacturer data in a similar manner to other pipe friction calculations (see Chapter 33 of the 1997 *ASHRAE Handbook—Fundamentals*). Surface pressure requirements account for friction losses through piping, heat exchangers, controls, and injection pressure (if any). Injection pressure requirements are a function of well design, aquifer conditions, and water quality. In theory, an injection well penetrating the same aquifer as the production well will experience a water level rise (assuming equal flows) that mirrors the drawdown in the production well. Using the earlier example, an injection well, with a 50 ft static level would experience a water level rise of 45 ft resulting in a surface injection pressure of 45 − 50 = − 5 ft or a water level which remains 5 ft below the ground surface. Thus no additional pump pressure is requires for injection.

In practice, injection pressure requirements usually exceed the theoretical value. With good (non-scaling) water, careful drilling, and little sand production, injection pressure should be near the theoretical value. For poor water quality, high sand production, or poor well construction, injection pressure may be 30% to 60% higher.

The well casing diameter depends on the diameter of the pump (bowl assembly) necessary to produce the required flow rate. Table 3 presents nominal casing sizes for a range of water flow rates.

In addition to the production well, most systems should include an injection well to dispose of the fluid after it has passed through the system. Injection stabilizes the aquifer from which the fluid is withdrawn and helps to assure long-term productivity. A brief discussion of injection wells is presented in Chapter 33.

Table 3 Nominal Well Surface Casing Sizes

Pump Bowl Dia., in.	Suggested Casing Size, in.	Minimum Casing Size, in.	Submersible Flow Range (3450 rpm), gpm	Lineshaft Flow Range (1750 rpm), gpm
4	6	5	<80	<50
6	10	8	80 − 350	50 − 175
7	12	10	250 − 600	150 − 275
8	12	10	360 − 800	250 − 500
9	14	12	475 − 850	275 − 550
10	14	12		500 − 1000
12	16	14		900 − 1300

Flow Testing

When possible, well testing should be completed prior to the mechanical design. Only with actual flow test data and water chemical analysis information can accurate design proceed.

Flow testing can be divided into three different types of tests: rig, short-term, and long-term (Stiger et al. 1989). Rig tests are generally shorter than 24 h and are accomplished while the drilling rig is on site. The primary purpose of this test is to purge the well of remaining drilling fluids and cuttings and to get a preliminary indication of yield. The length of the test is generally governed by the time required for the water to run clean. The rate is determined by the available pumping equipment. Frequently the well is blown or pumped with the drilling rig's air compressor. As a result, little can be learned about the production characteristics of the well from a rig test. If the well is air lifted, it may not be useful to collect water samples for chemical analysis because certain chemical constituents may be oxidized by the compressed air.

If properly conducted, short-term, single-well tests of 4 h to 3 days (with most in the 4 to 12 h range) duration yield information about the well flow rate, temperature, pressure, drawdown, and recovery. These data can provide initial estimates of reservoir parameters. The test is generally run with a temporary electric submersible

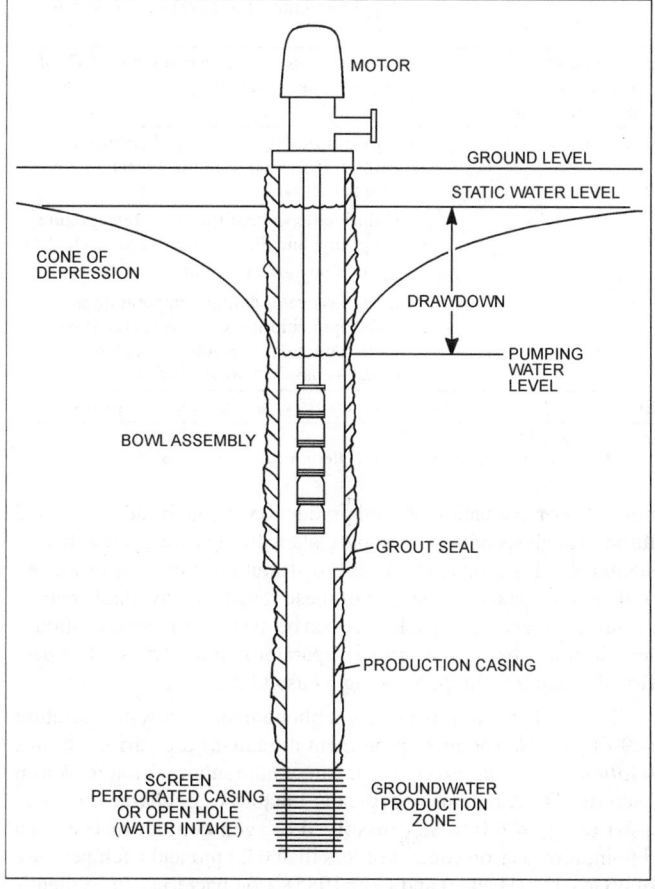

Fig. 5 Water Well Terminology

pump or lineshaft turbine pump driven by an internal combustion engine. The work may be performed by the drilling contractor or by a well pump contractor.

The test should involve at least three production rates, the largest being equal to the design flow rate for the system served. The three points are the minimum required to determine a productivity curve for the well that relates production to drawdown (Stiger et al. 1989). Water level and pumping rate should be stabilized at each point before the flow is increased. In most cases, water level is monitored with a "bubbler" or an electric sounder, and flow is measured using an orifice meter. This short-term test is generally used for small projects or in areas where reservoir parameters are already established.

Long-term tests of up to 30 days provide information on the reservoir. Normally these tests involve monitoring of nearby wells to evaluate interference effects. The data are useful in calculating average permeability thickness, storativity, reservoir boundaries, and recharge areas (Stiger et al. 1989).

It is also important to collect background information prior to the test and water level recovery data after pumping has ceased. Recovery data in particular can be used to evaluate skin effect, which is a type of well flow resistance caused by residual drilling fluids, insufficient screen or slotted liner area, or improper filter pack.

Water Quality Testing

Geothermal fluids commonly contain seven key chemical species that produce a significant corrosive effect (Ellis 1989). These include:

- Oxygen (generally from aeration)
- Hydrogen ion (pH)
- Chloride ion
- Sulfide species
- Carbon dioxide species
- Ammonia species
- Sulfate ion

The principal effects of these species are summarized in Table 4. Except as noted, the described effects are for carbon steel. Kindle and Woodruff (1981) present recommended procedures for complete chemical analysis of geothermal well water.

Two of these species are not reliably detected by standard water chemistry tests and deserve special mention. Dissolved oxygen does not occur naturally in low-temperature (120 to 220°F) geothermal fluids that contain traces of hydrogen sulfide. However, because of slow reaction kinetics, oxygen from air in-leakage may persist for some minutes. Once the geothermal fluid is produced, it is extremely difficult to prevent contamination, especially if pumps other than downhole submersible or lineshaft turbine pumps are used to move the fluid. Even if the fluid systems are maintained at positive pressure, air in-leakage at the pump seals is likely, particularly with the low level of maintenance in many installations.

Hydrogen sulfide is ubiquitous in extremely low concentrations in geothermal fluids above 120°F. This corrosive species also occurs naturally in many cooler ground waters. For alloys such as cupronickel, which are strongly affected by it, hydrogen sulfide concentrations in the low parts per billion (10^9) range may have a serious detrimental effect, especially if oxygen is also present. At these levels, the characteristic rotten egg odor of hydrogen sulfide may be absent, so field methods may be required for detection. Hydrogen sulfide levels down to 50 ppb can be detected using a simple field kit; however, absence of hydrogen sulfide at this low level may not preclude damage by this species.

Two other key species that should be measured in the field are pH and carbon dioxide concentrations. This is necessary because most geothermal fluids release carbon dioxide rapidly, causing a rise in pH. Production of suspended solids (sand) from a well should be evaluated during the well completion with gravel pack, screen or

Table 4   Principle Effects of Key Corrosive Species

Species	Principle Effects
Oxygen	• Extremely corrosive to carbon and low alloy steels; 30 ppb shown to cause fourfold increase in carbon steel corrosion rate.
	• Concentrations above 50 ppb cause serious pitting.
	• In conjunction with chloride and high temperature, <100 ppb dissolved oxygen can cause chloride-stress corrosion cracking (chloride-SCC) of some austenitic stainless steels.
Hydrogen ion (pH)	• Primary cathodic reaction of steel corrosion in air-free brine is hydrogen ion reduction. Corrosion rate decreases sharply above pH 8.
	• Low pH (5) promotes sulfide stress cracking (SSC) of high strength low alloy (HSLA) steels and some other alloys coupled to steel.
	• Acid attack on cements.
Carbon dioxide species (dissolved carbon dioxide, bicarbonate ion, carbonate ion)	• Dissolved carbon dioxide lowers pH, increasing carbon and HSLA steel corrosion.
	• Dissolved carbon dioxide provides alternative proton reduction pathway, further exacerbating carbon and HSLA steel corrosion.
	• May exacerbate SSC.
	• Strong link between total alkalinity and corrosion of steel in low-temperature geothermal wells.
Hydrogen sulfide species (hydrogen sulfide, bisulfide ion, sulfide ion)	• Potent cathodic poison, promoting SSC of HSLA steels and some other alloys coupled to steel.
	• Highly corrosive to alloys containing both copper and nickel or silver in any proportions.
Ammonia species (ammonia, ammonium ion)	• Causes stress corrosion cracking (SCC) of some copper-based alloys.
Chloride ion	• Strong promoter of localized corrosion of carbon, HSLA, and stainless steel, as well as of other alloys.
	• Chloride-dependent threshold temperature for pitting and SCC. Different for each alloy.
	• Little if any effect on SSC.
	• Steel passivates at high temperature in 6070 ppm chloride solution (pH = 5) with carbon dioxide. 133,500 ppm chloride destroys passivity above 300°F.
Sulfate ion	• Primary effect is corrosion of cements.

Source: Ellis (1989).
Note: Except as indicated, the described effects are for carbon steel.

both. Proper evaluation of sand production should be accomplished through analysis of water samples taken during flow testing. If substantial sand is produced, the size distribution should be evaluated with a sieve analysis. Only after these results are available can an accurate screen/gravel pack selection be made. Surface separation is less desirable because a surface separator requires the sand to pass first through the pump, reducing its useful life.

Biological fouling is largely a phenomena of low-temperature (<90°F) wells. The most prominent organisms are various strains (*Galionella*, *Crenothrix*) of what are commonly referred to as iron bacteria. These organisms typically inhabit water characterized by a pH range of 6.0 to 8.0, dissolved oxygen content of less than 5 ppm, ferrous iron content of less than 0.2 ppm and a temperature of 46 to 61°F (Hackett and Lehr 1985). Iron bacteria can be identified microscopically.

The most common treatment for iron bacteria infestation is chlorination, surging and flashing. Successful use of this treatment is dependent upon the maintenance of proper pH (less than 8.5), dosage, free residual chlorine content (200 to 500 gpm), contact time (24 h minimum), and agitation or surging. Hackett and Lehr (1985) provided additional detail on treatment.

EQUIPMENT AND MATERIALS

The primary equipment used in geothermal systems includes pumps, heat exchangers, and piping. While some aspects of these components are unique to geothermal applications, many of them are of routine design. However, the great variability and general aggressiveness of the geothermal fluid necessitate limiting corrosion and scale buildup rather than relying on system cleanup. Corrosion and scaling can be limited by (1) proper system and equipment design or (2) treatment of the geothermal fluid, which is generally precluded by cost and environmental regulations relating to disposal.

Performance of Materials

Carbon Steel. The Ryznar Index has traditionally been used to estimate the corrosivity and scaling tendencies of potable water supplies. However, one study found no significant correlation (at the 95% confidence level) between carbon steel corrosion and the Ryznar Index (Ellis and Smith 1983). Therefore, the Ryznar and other indices based on calcium carbonate saturation should not be used to predict corrosion in geothermal systems.

In Class Va geothermal fluids [as described by Ellis (1989) <5000 ppm total key species (TKS), total alkalinity 207 to 1329 ppm as $CaCO_3$, pH 6.7 to 7.6] corrosion rates of about 5 to 20 mils per year (mil/yr) can be expected, often with severe pitting.

In Class Vb geothermal fluids [as described by Ellis (1989) <5000 ppm TKS, total alkalinity <210 ppm as $CACO_3$, pH 7.8 to 9.85], carbon steel piping has given good service in a number of systems, provided the system design rigorously excluded oxygen. However, introduction of 30 ppb oxygen under turbulent flow conditions causes a fourfold increase in uniform corrosion. Saturation with air often increases the corrosion rate at least 15 times. Oxygen contamination at the 50 ppb level often causes severe pitting. Chronic oxygen contamination causes rapid failure.

In the case of buried steel pipe, the external surfaces must be protected from contact with groundwater. Groundwater is aerated and has caused pipe failures by external corrosion. Required external protection can be obtained by coatings, pipe-wrap, or preinsulated piping, provided the selected material resists the system operating temperature and thermal stress.

At temperatures above 135°F, galvanizing (zinc coating) does not reliably protect steel from either geothermal fluid or groundwater. Hydrogen blistering can be prevented by using of void free (killed) steels.

Low alloy steels (steels containing not more than 4% alloying elements) have corrosion resistance similar, in most respects, to carbon steels. As in the case of carbon steels, sulfide promotes entry of atomic hydrogen into the metal lattice. If the steel exceeds a hardness of Rockwell C22, sulfide stress cracking may occur.

Copper and Copper Alloys. Copper fan coil units and copper-tubed heat exchangers have a consistently poor performance due to traces of sulfide species found in geothermal fluids in the United States. Copper tubing rapidly becomes fouled with cuprous sulfide films more than 1 mm thick. Serious crevice corrosion occurs at cracks in the film, and uniform corrosion rates of 2 to 6 mil/yr appear typical, based on failure analyses.

Experience in Iceland also indicates that copper is unsatisfactory for heat exchange service and that most brasses (Cu-Zn) and bronzes (Cu-Sn) are even less suitable. Cupronickel often perform more poorly than copper does in low-temperature geothermal service because of trace sulfide.

Much less information is available regarding copper and copper alloys in non-heat transfer service. Copper pipe shows corrosion behavior similar to copper heat exchange tubes under conditions of moderate turbulence (Reynolds numbers of 40,000 to 70,000). An internal inspection of the few yellow brass valves showed no significant corrosion. However, silicon bronze CA 875 (12-16Cr, 3-5Si, <0.05Pb, <0.05P), an alloy normally resistant to dealloying, failed in less than three years when used as a pump impeller. Leaded red brass (CA 836 or 838) and leaded red bronze (SAE 67) appear viable as pump internal parts. Based on a few tests at Class Va sites, aluminum bronzes have shown potential for corrosion in heavy-walled components.

Solder is yet another problem area for copper equipment. Lead-tin solder (50Pb, 50Sn) was observed to fail by dealloying after a few years exposure. Silver solder (1Ag, 7P, Cu) was completely removed from joints in under two years. If the designer elects to accept this risk, solders containing at least 70% tin should be used.

Stainless Steel. Unlike copper and cupronickel, stainless steels are not affected by traces of hydrogen sulfide. Their most likely application is heat exchange surfaces. For economic reasons, most heat exchangers are probably of the plate-and-frame type, most of which will be fabricated with one of two standard alloys, Type 304 and Type 316 austenitic stainless steel. Some pump and valve trim also are fabricated from these or other stainless steels.

These alloys are subject to pitting and crevice corrosion above a threshold chloride level which depends on the chromium and molybdenum content of the alloy and on the temperature of the geothermal fluid. Above this temperature, the passivation film, which gives the stainless steel its corrosion resistance, is ruptured, and local pitting and crevice corrosion occur. Figure 6 shows the relationship between temperature, chloride level, and occurrence of localized corrosion of Type 304 and Type 316 stainless steel. This figure indicates, for example, that localized corrosion of Type 304 may occur in 80°F geothermal fluid if the chloride level exceeds approximately 210 ppm; while Type 316 is resistant at that temperature until the chloride level reaches approximately 510 ppm. Due to its 2 to 3% molybdenum content. Type 316 is always more resistant to chlorides than is Type 304.

Aluminum. Aluminum alloys are not acceptable in most cases because of catastrophic pitting.

Titanium. This material has extremely good corrosion resistance and could be used for heat exchanger plates in any low-temperature geothermal fluid, regardless of dissolved oxygen content.

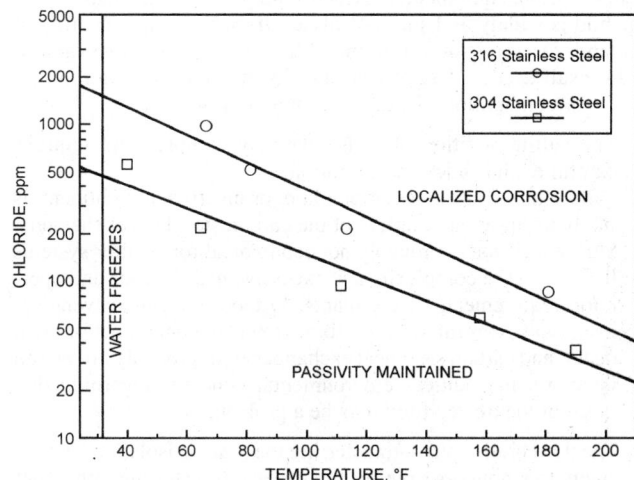

Fig. 6   Chloride Concentration Required to Produce Localized Corrosion of Stainless Steel as Function of Temperature
(Efrid and Moeller 1978)

Great care is required if acid cleaning is to be performed. The vendor's instructions must be followed. The titanium should not be scratched with iron or steel tools, as this can cause pitting.

Chlorinated Polyvinyl Chloride (CPVC) and Fiber Reinforced Plastic (FRP). These materials are easily fabricated and are not adversely affected by oxygen intrusion. External protection against groundwater is not required. The mechanical properties of these materials at higher temperatures may vary greatly from those at ambient temperature, and the mechanical limits of the materials should not be exceeded. The usual mode of failure is creep rupture—strength decays with time. Manufacturer's directions for joining should be followed to avoid premature failure of joints.

Elastomeric Seals. Tests on o-ring materials in a low-temperature system in Texas indicated that a fluoroelastomer is the best material for piping of this nature; Buna-N is also acceptable. Neoprene, which developed extreme compression set, was a failure. Natural rubber and Buna-S should also be avoided. Ethylene-propylene terpolymer (EPDM) has been used successfully in gasket, o-ring, and valve seats in many systems.

Corrosion Engineering and Design

The design of the geothermal plant is as critical to controlling corrosion as the selection of suitable materials. Furthermore, the design and material selection process is interactive in that certain design decisions force the use of certain materials; while, selection of material may dictate design. In all cases, the objective should be the same—to produce an adequately reliable system with the lowest possible lifetime cost.

Three philosophies are followed for corrosion design:

1. Use corrosion-resistant materials throughout.
2. Exclude or remove oxygen and use carbon steel throughout.
3. Transfer the heat via an isolation heat exchanger to a non-corrosive working medium so that the kind and number of components contacting the geothermal fluid are minimized. In addition, make those components of corrosion-resistant materials.

The first philosophy produces a reliable, low-maintenance system. However, the cost would be high, and many of the desired components are not available in the required alloys.

The second philosophy may be considered for district-sized heating projects with attendant surface storage and potential for oxygen intrusion; for such projects, it may be economical to inhibit the geothermal fluid by continuous addition of excess sulfite as an oxygen scavenger. When this is done, carbon steel can be used for heat exchange equipment, provided fluid pH is greater than 8. This method is widely and successfully used for municipal heating in Iceland. Design should minimize the introduction of oxygen to reduce sulfite costs. Use of vented tanks should also be minimized.

The second philosophy has three major drawbacks:

1. The sulfite addition plant is relatively complex and requires careful maintenance and operation.
2. Failure of the sulfite addition plant, or insufficient treatment, is likely to cause rapid failure of the carbon steel heat exchangers.
3. Sulfite addition is probably not economical for smaller systems because of the complexity and excessive maintenance and operation requirements of the plants. Without oxygen scavenging, even with careful design, some oxygen contamination will occur, and carbon steel heat exchangers will probably not be satisfactory. In addition, environmental concerns regarding disposal of the treated fluid may be a problem.

The third philosophy—transferring the heat via isolation heat exchangers to a noncorrosive secondary heat transfer medium—has several advantages for systems of all sizes. The system is much simpler to operate and therefore more reliable than those mentioned previously. Only a small number of corrosion-resistant components are required, and it is feasible to design systems in which oxygen

exclusion is not critical. Even with careful material selection and design, geothermal heating equipment requires more maintenance than conventional equipment. Minimizing the number and kinds of components in contact with geothermal fluid further reduces maintenance costs.

Pumps

Pumps are used for production, circulation, and disposal. For circulation and disposal, whether surface disposal or injection, standard hot water circulating pumps, almost exclusively of centrifugal design, are used. These are routine engineering design selections, the only special concern is the selection of appropriate materials. Small circulating in-line pumps of all-iron construction have shown acceptable performance in many applications. For larger pumps (base-mounted, vertical in-line, double suction), cast-iron impellers and volute with stainless steel shaft, screws, keys, and washers are typically used (Rafferty 1989a). In addition, mechanical seals are preferred over packing.

Production well pumps are among the most critical components in a geothermal system and have in the past been the source of much system downtime. Therefore, proper selection and design of the production well pump is extremely important. Well pumps are available for larger systems in two general configurations—lineshaft and submersible. The lineshaft type is most often used for direct-use systems (Rafferty 1989a).

Lineshaft Pumps. Lineshaft pumps are similar to those typically used in irrigation applications. An above-ground driver, typically an electric motor, rotates a vertical shaft extending down the well to the pump. The shaft rotates the pump impellers in the pump bowl assembly, which is positioned at such a depth in the wellbore that adequate net positive suction head (NPSH) is available when the unit is operating. Two designs for the shaft/bearing portion of the pump is available—open and enclosed.

In the **open lineshaft pump**, the shaft bearings are supported in "spiders," which are anchored to the pump column pipe at 5- to 10-ft intervals. The shaft and bearings are lubricated by the fluid flowing up the pump column. In geothermal applications, bearing materials for open lineshaft designs have consisted of both bronze and various elastomer compounds. The shaft material is typically stainless steel. Experience with this design in geothermal applications has been mixed. Two large district heating systems have successfully operated such pumps for approximately 10 years; many more systems, however, initiated operation with open-lineshaft pumps and subsequently changed to enclosed-lineshaft design after numerous failures. Open lineshaft pumps are generally less expensive than enclosed lineshaft pumps for the same application. In addition, the open lineshaft type is easier to apply to artesian wells.

In an **enclosed lineshaft pump**, an enclosing tube protects the shaft and bearings from exposure to the pumped fluid. A lubricating fluid is admitted to the enclosed tube at the wellhead. It flows down the tube, lubricates the bearings, and exits where the column attaches to the bowl assembly. The bowl shaft and bearings are lubricated by the pumped fluid. Oil-lubricated, enclosed lineshaft pumps have the longest service life in low-temperature, direct use applications.

These pumps typically include carbon or stainless steel shafts and bronze bearings in the lineshaft assembly, and stainless steel shafts and leaded red bronze bearings in the bowl assembly. Keyed-type impeller connections (to the pump shaft) are superior to collet-type connections (Rafferty 1989a).

Because of the lineshaft bearings, the reliability of lineshaft pumps decreases as the pump-setting depth increases. Nichols (1978) indicates that at depths greater than about 800 ft, reliability is questionable, even under good pumping conditions.

Submersible Pumps. The electrical submersible pump consists of three primary components located downhole—the pump, the drive motor, and the motor protector. The pump is a vertical multistage centrifugal type. The motor is usually a three-phase induction type that is oil filled for cooling and lubrication; it is cooled by heat transfer to the pumped fluid moving up the well. The motor protector is located between the pump and the motor and isolates the motor from the well fluid while allowing pressure equalization between the pump intake and the motor cavity.

The electrical submersible pump has several advantages over lineshaft pumps, particularly for wells requiring greater pump bowl setting depths. The deeper the well, the greater the economic advantage the submersible pump enjoys. Moreover, it is more versatile in that it adapts more easily to different depths. The break-over point is at a pump depth of 800 ft; the submersible pump is desirable at greater pump depths, and the lineshaft is preferred at lesser depths.

Submersible pumps have not demonstrated acceptable lifetimes in most geothermal applications. Although they are commonly used in high-temperature, downhole applications in the oil and gas industry, the acceptable overhaul interval in that industry is much shorter than in a geothermal application. In addition, most submersibles are 3600 rpm machines, which results in greater susceptibility to erosion in aquifers that produce moderate amounts of sand. They have, however, been applied in geothermal projects where an existing well that is of relatively small diameter must be used. Because submersibles operate at 3600 rpm, they provide greater flow capacity for a given bowl size than an equivalent 1750-rpm lineshaft pump.

Variable-Speed Drives

In many geothermal applications, well flow requirements vary over a substantial range. To avoid inefficient throttling of the pump to meet those requirements, variable-speed drives are frequently applied. In most cases, one of the two varieties of controls has been employed—fluid coupling or variable-frequency drive. One major geothermal system employs a two-speed induction motor on the production pump.

The fluid coupling was most often used prior to the general availability of electronic frequency controls. These units are typically installed on the lineshaft pumps between the electric motor and the pump shaft. They cannot be installed on submersible pumps. Advantages of the fluid coupling include relatively low cost and mechanical simplicity.

The efficiency of the fluid coupling is a function of the ratio of the output rpm to the input speed. As a result, in applications in which a large turndown is required (to less than about 70% of full speed), operating economy declines. In most geothermal applications, the minimum pump speed is controlled by the necessity to generate sufficient pressure to raise the fluid from the pumping level to the wellhead, resulting in only moderate turndown requirements.

The variable-frequency drive is commonly used on geothermal well pumps. It provides a more efficient overall drive than a fluid coupling. The savings from the use of a variable-speed drive in geothermal applications are subject to some unique pumping concerns. Circulation of fluid through a pipe where only friction is involved follows a cubic relationship between flow and power. Pumping from a well requires a considerable static pressure to lift the fluid. As a result, for an equal reduction in flow rate, the pump energy saving is less than that for a circulation pump.

Heat Exchangers

In the system in Figure 3, heat exchangers are located at the individual uses or process. For the system of Figure 4, one or more large heat exchangers are located near the wellhead. In both cases, the principle is to isolate the geothermal fluid from complicated systems or those which cannot be readily designed to be compatible with the geothermal fluid.

The principal types of heat exchangers used in transferring energy from the geothermal fluid are plate, shell-and-tube, and downhole.

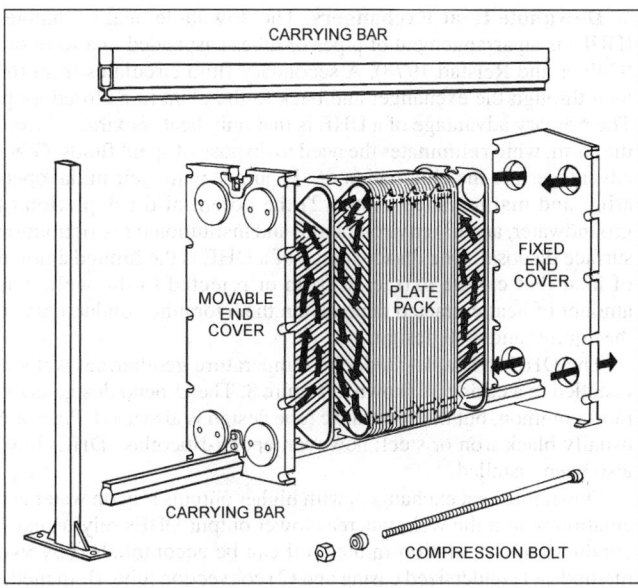

Fig. 7 Plate Heat Exchanger

Plate Heat Exchangers. In a plate heat exchanger (Figure 7), the geothermal fluid is routed along one side of each plate and the heated fluid is routed along the other side. The plates can be readily manifolded for various combinations of series and parallel flow. Plate heat exchangers have been widely used for many years in the food-processing industry and in marine applications. They have two main characteristics that make them desirable for many geothermal applications:

1. They are readily cleaned. By loosening the main bolts, the header plate and the individual heat exchanger plates can be removed and cleaned.
2. The stamped plates are very thin and may be made of a wide variety of materials. When expensive materials are required, the thinness of the plates keeps this type of heat exchanger much less expensive than other types.

Plate heat exchangers have additional characteristics that influence their selection in specific applications:

- Approach temperature differences are usually smaller than those for shell-and-tube heat exchangers; this is particularly important in low-temperature geothermal applications. Many geothermal applications have approach temperatures of 10°F and some as low as 5°F.
- Because overall heat transfer coefficients approach 1000 Btu/h·ft²·°F in water-to-water applications, heat transfer per unit volume is usually larger than for shell-and-tube heat exchangers.
- Increased capacity can be accommodated easily by adding plates.

In most low-temperature direct-use applications, plate heat exchangers are constructed of Type 316 stainless steel plates and Buna-N gaskets (Rafferty 1989a). Stainless steel plates have failed in at least two applications in which the geothermal fluid was used to heat swimming pool water. This is probably due to the high chlorine content in the pool water, which promotes pitting of the plates.

Shell-and-Tube Heat Exchangers. This type of heat exchanger is used in only a limited number of geothermal applications because plate heat exchangers are more economic when specialized materials are required to minimize corrosion. However, when mild steel shells and copper or silicon bronze tubes can be used, shell-and-tube heat exchangers can be more economical. With the geothermal fluid passing through the tube side of a shell-and-tube heat exchanger, the tubes should be in straight to facilitate mechanical cleaning.

Downhole Heat Exchangers. The downhole heat exchanger (DHE) is an arrangement of pipes or tubes suspended in a wellbore (Culver and Reistad 1978). A secondary fluid circulates from the load through the exchanger and back to the plant in a closed loop. The primary advantage of a DHE is that only heat is extracted from the earth, which eliminates the need to dispose of spent fluids. Other advantages are the elimination of (1) pumps with their initial operating and maintenance costs, (2) the potential for depletion of groundwater, and (3) environmental and institutional restrictions on surface disposal. One disadvantage of a DHE is the limited amount of heat that can be extracted from or rejected to the well. The amount of heat extracted depends on the hydraulic conductivity of the aquifer and well design.

The DHE in low-to-moderate temperature geothermal wells is installed in a casing as shown in Figure 8. The U-bend design is the most common, but the concentric tube design is also used. Pipes are usually black iron or steel; however, epoxy-fiberglass DHEs have also been installed.

Downhole heat exchangers with higher outputs rely on water circulation within the well, whereas lower output DHEs rely on earth conduction. Circulation in the well can be accomplished by two methods: (1) undersized casing and (2) convection tube. Both methods rely on the difference in density between the water surrounding the DHE and that in the aquifer.

Circulation provides the following advantages:

• Water circulates around the DHE at velocities that, at optimum conditions, can approach those in the shell of a shell-and-tube exchanger.
• Hot water moving up the annulus heats the upper rocks and the well becomes nearly isothermal.
• Some of the cool water, being more dense than the water in the aquifer, sinks into the aquifer and is replaced by hotter water, which flows up the annulus.

Figure 8 shows well construction in competent formation, i.e., where the wellbore will stand open without a casing. An undersized casing having perforations at the lowest producing zone (usually near the bottom) and just below the static water level is installed. A packer near the top of the competent formation permits the installation of cement between it and the surface. When the DHE is installed and heat extracted, thermosyphoning causes cooler water inside the casing to move to the bottom, and hotter water moves up the annulus outside the casing.

Because most DHEs are used for space heating (an intermittent operation), the heated rocks in the upper portion of the well store heat for the next cycle.

In areas where the well will not stand open without casing, the convection tube can be used. This is pipe one-half the diameter of the casing either hung with its lower end above the well bottom and its upper end below the surface or set on the bottom with perforations at the bottom and below the static water level. If a U-bend DHE is use, it can be either be outside the convection tube or have one leg in the convection tube. A concentric tube DHE should be outside the convection tube.

DHEs operates best in aquifers with a high hydraulic conductivity and that provide the water movement for heat and mass transfer.

Valves

In large (>2-1/2 in.) pipe sizes, resilient-lined butterfly valves have been the choice for geothermal applications. The lining material protects the valve body from exposure to the geothermal fluid. The rotary rather than reciprocating motion of the stem makes the valve less susceptible to leakage and the buildup of scale deposits. For many direct-use applications, these valves are composed of Buna-N or EPDM seats, stainless steel shafts, and bronze or stainless steel disks.

Gate valves have been used in some larger geothermal systems but have been subject to stem leakage and seizure. After several years of use, they are no longer capable of 100% shutoff.

Piping

Due to the aggressive nature of most geothermal fluids, nonmetallic piping has been widely applied to direct-use projects. Although steel has occasionally been used for distribution and transmission piping, such materials as fiberglass and polybutylene are much more common. Asbestos-cement (AC) piping has been widely applied, but is now being phased out.

Fiberglass reinforced plastic (FRP) piping—available with a wide variety of joining methods—has been used in a number of projects. The epoxy adhesive bell and spigot joint have seen the widest use.

In comparison to metallic piping, fiberglass piping is lighter and has smoother interior surfaces for lower pressure drop. Although it can be formulated for service at a temperature as high as 300°F, most cost-competitive products are limited to approximately 200 to 250°F. The rate of expansion for FRP is about twice that of steel. Due to the very low modulus, however, the forces exerted are only about 3 to 5% of those of steel. As a result, in buried application, expansion loops and joints are generally not required. Fitting costs, particularly in large sizes, can constitute a significant portion of the overall material costs. Fiberglass piping is generally available in sizes larger than 2 in.

Polyethylene (PE), another polyolefin has seen only limited use in geothermal direct-use applications. However, it can be used in direct buried applications without expansion joints or loops. It costs less than polybutylene, but has a lower maximum service temperature (140 to 150°F). It is much lower in cost than polybutylene. Cross-linked polyethylene (PEX) has been used extensively in European district heating applications. It carries higher temperature and pressure ratings (100 psi at 180°F) than standard polyethylene, along with much higher cost.

Because **steel** piping is susceptible to corrosion from both soil moisture (external) and geothermal fluid (internal), it has not been used widely in direct burial applications for transporting geothermal

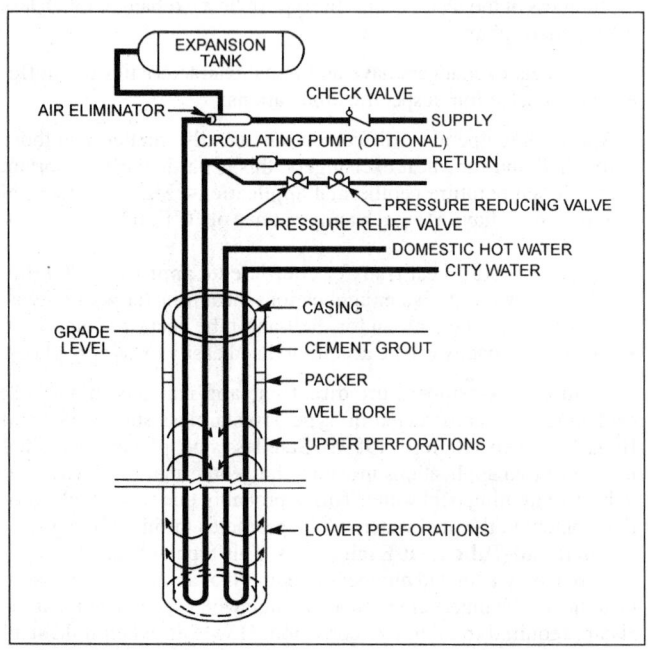

Fig. 8   Typical Connection of Downhole Heat Exchanger for Space and Domestic Hot Water Heating
(Reistad et al. 1979)

fluid. In two large projects in which steel piping was used for geothermal transmission lines were installed in a tunnel or in a building, so the piping was not exposed to soil moisture.

Steel piping is often used for mechanical room and in-building piping. Oxygen must be excluded from the system when steel pipe is used to transport geothermal fluid.

Preinsulated Pipe. Most large transmission and distribution piping for transporting geothermal fluids is preinsulated. These products consist of a carrier pipe through which the geothermal fluid flows, a layer (nominal 1 to 2 in.) of polyurethane insulation, and a jacket for protecting the insulation from physical damage and moisture penetration. The most common material combinations are FRP/FRP, FRP/PVC, steel/FRP, steel/PE, and PE/PE. Chapter 11 of the 1996 *ASHRAE Handbook—Systems and Equipment* has additional information on the design of insulated piping systems and thermal calculations.

RESIDENTIAL AND COMMERCIAL APPLICATIONS

The primary applications for the direct use of geothermal energy in the residential and commercial area are space heating, sanitary water heating, and space cooling (using the absorption process). While geothermal space and domestic hot water heating are widespread, space cooling is rare. Where space heating is accomplished, service water heating, or at least preheating, is almost universally accomplished as well.

Space Heating

Figure 9 illustrates a system that uses geothermal fluid at 170°F (Austin 1978). The geothermal fluid is used in two main equipment components for heating of the buildings: (1) a plate heat exchanger that supplies energy to a closed heating loop previously heated by a natural gas boiler (the boiler remains as a standby unit) and (2) a water-to-air coil used for preheating ventilation air. In this system, proper control is crucial for economical operation.

The average temperature of the discharged fluid is 120 to 130°F. The geothermal fluid is used directly in the preheat terminal equipment within the buildings (this would probably not be the case if the system were being designed today). Several corrosion problems have arisen in the direct use, mainly because of the action of hydrogen sulfide on copper-based equipment parts (Mitchell 1980). Even with these difficulties, the geothermal system appears to be highly cost-effective (Lienau 1979).

Figure 10 illustrates a geothermal district heating system. The geothermal fluid is produced from three wells. Depending on the load, one, two or three wells may be in operation. Each well is equipped with a single-speed electric motor and is capable of producing a constant flow rate. The main flowmeter controls which pumps are in operation. Flow is also controlled by a large throttling valve located in the main production line. This valve responds to a signal from a pressure transducer located in the downtown area on the supply side of the distribution system.

The open distribution system is composed of preinsulated pipe for supply fluid and uninsulated for disposal fluid. Temperature-sensitive, self-actuated valves are located at six strategic points around the distribution system. These valves open when necessary to maintain acceptable supply water temperature. Water meters are located at each valve to allow operators to monitor the amount of fluid required for temperature maintenance.

Figure 11 shows a geothermal district heating system that is unique in terms of its design based on a peak load Δt of 100°F using a 190°F resource. It is of the closed-loop design with central heat exchangers. The production well has an artesian shut-in pressure of 25 psi, so the system operates with no production pump for most of the year. During colder weather, a surface centrifugal pump located at the wellhead boosts the pressure.

Geothermal flow from the production well is initially controlled by a throttling valve on the supply line to the main heat exchanger, which responds to a temperature signal from the supply water on the closed-loop side of the heat exchanger. When the throttling valve has reached the full open position, the production booster pump is

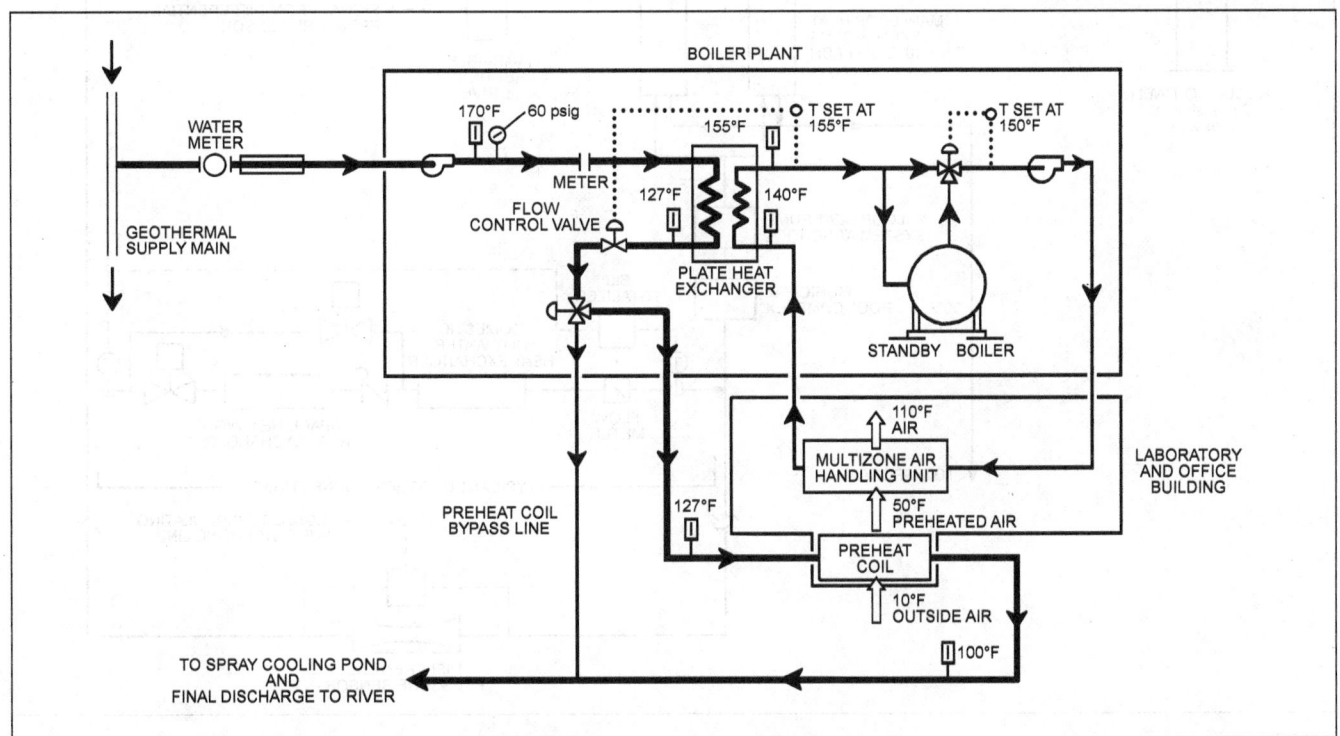

Fig. 9   Heating System Schematic

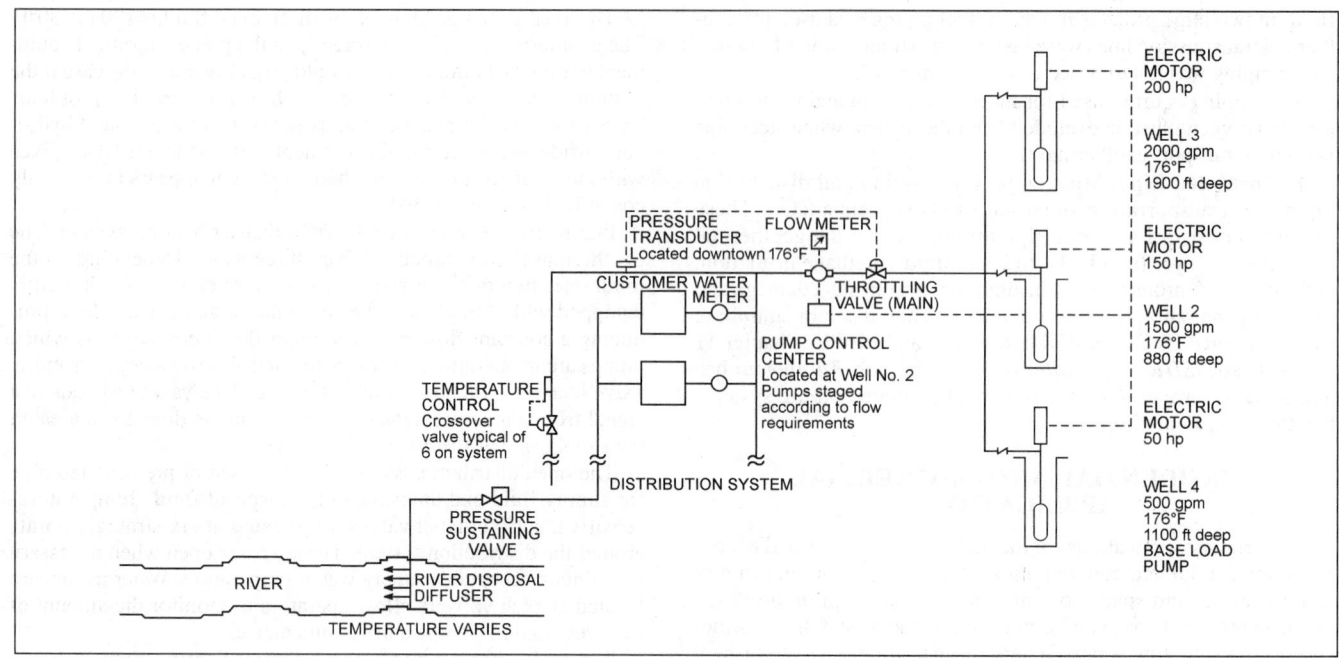

Fig. 10   Open-Type Geothermal District Heating System
(Rafferty 1989)

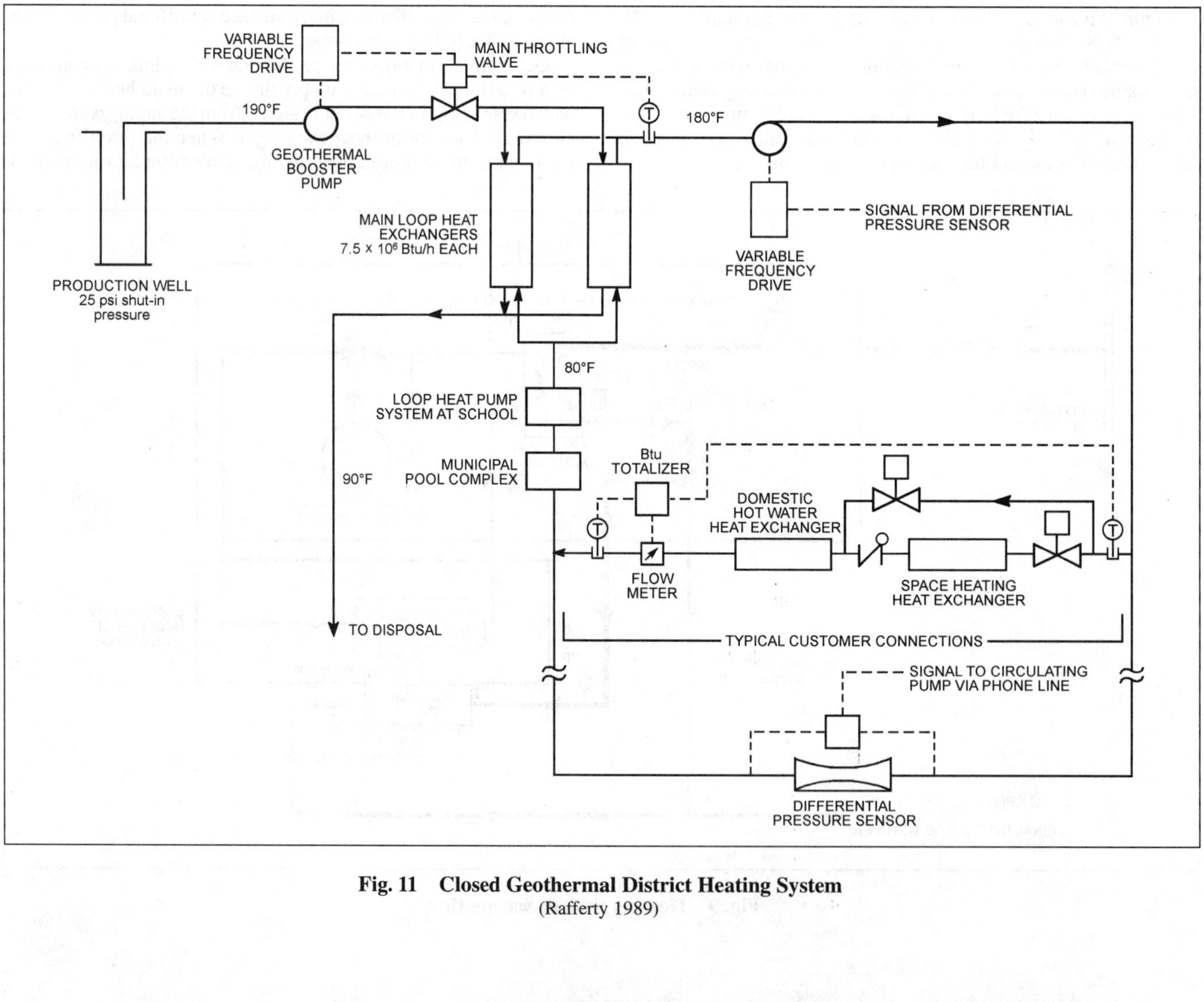

Fig. 11   Closed Geothermal District Heating System
(Rafferty 1989)

enabled. The pump is controlled through a variable-frequency drive that responds to the same supply water signal as the throttling valve. The booster pump is designed for a peak flow rate of 300 gpm of 190°F water.

Types of Terminal Heating Equipment. The terminal equipment used in geothermal systems is the same as that used in nongeothermal heating systems. However, certain types of equipment are better suited to geothermal design than others.

In many cases, buildings heated by geothermal sources operate their heating equipment at less than conventional temperatures due to the low temperature of the resource and the use of heat exchangers to isolate the fluids from the building loop. Because many geothermal sources are designed to take advantage of a large Δt, proper selection of equipment can enhance the feasibility of the project.

In many cases, the low temperature of the geothermal resource and the use of heat exchangers to isolate the fluid from the building loop require that heating equipment operate at less than conventional temperatures. In addition, many geothermal sources are designed to take advantage of a large temperature drop (Δt), so proper selection of equipment can enhance the feasibility of the project. Finned coil, forced air systems generally function best in this low-temperature/high Δt situation. One or two additional rows of coil depth compensate for the lower supply water temperature. While an increased Δt affects coil circuiting, it improves controllability. This type of system should be capable of using a supply water temperature as low as 120°F.

Radiant floor panels are well suited to the use of very low water temperature, particularly in industrial applications with little or no floor covering. In industrial settings, with a bare floor and a relatively low space temperature requirement, the average water temperature could be as low as 95°F. For a higher space temperature and/or thick floor coverings, a higher water temperature may be required.

Baseboard convectors and similar equipment are the least capable of operating at low supply water temperature. At 150°F average water temperatures, derating factors for this type of equipment are about 0.43 to 0.45. As a result, the quantity of equipment required to meet the design load is generally uneconomical. This type of equipment can be operated at low temperatures from the geothermal source to provide base-load heating. Peak load can be supplied by a conventional boiler.

Heat pumps take advantage of the lowest temperature geothermal resources. Loop heat-pump systems operate with a water temperature in the 25 to 100°F range, and central station heat pump plants (supplying four-pipe systems) in the 45°F range.

Domestic Water Heating

Domestic water heating in a district space heating system is beneficial because it increases the overall size of the energy load, the energy demand density, and the load factor. For those resources that cannot heat water to the required temperature, preheating is usually possible. Whenever possible, the domestic hot water load should be placed in series with the space heating load in order to reduce system flow rates and increase Δt.

Space Cooling

Geothermal energy has seldom been used for cooling, although emphasis on solar energy and waste heat has created interest in cooling with thermal energy. The absorption cycle is most often used and lithium-bromide/water absorption machines are commercially available in a wide range of capacities. Temperature and flow requirements for absorption chillers run counter to the general design philosophy for geothermal systems. They require high supply water temperatures and a small Δt on the hot water side. Figure

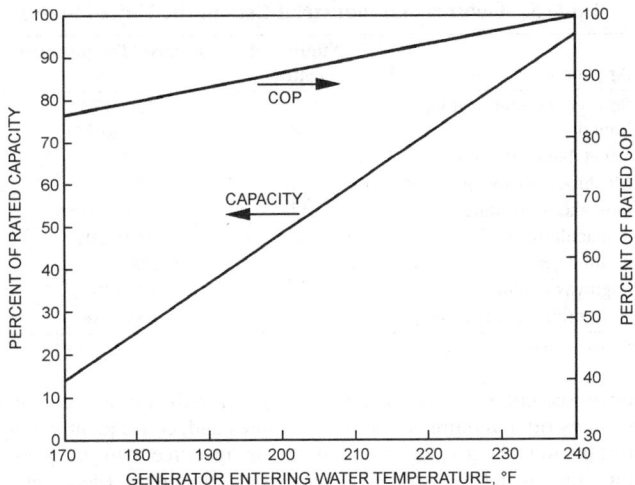

Fig. 12 Typical Lithium Bromide Absorption Chiller Performance Versus Temperature
(Christen 1977)

12 illustrates the effect of reduced supply water temperature on machine performance. The machine is rated at a 240°F input temperature, so derating factors must be applied if the machine is operated below this temperature. For example, operation at a 200°F supply water temperature would result in a 50% decrease in capacity, which seriously affects the economics of absorption cooling at a low resource temperature.

Coefficient of performance (COP) is less seriously affected by a reduction in supply water temperature. The nominal COP of a single-stage machine at 240°F is 0.65 to 0.70; that is, for each ton of cooling output, a heat input of 12,000 Btu/h divided by 0.65, or 18,460 Btu/h, is required.

Most absorption equipment is designed for steam input (an isothermal process) to the generator section. When this equipment is operated from a hot water source, a relatively small Δt must be used. This creates a mismatch between the building flow requirements for space heating and cooling. For example, assume a 200,000 ft² building is to use a geothermal resource for heating and cooling. At 25 Btu/h·ft² and a design Δt of 40°F, the flow requirement for heating is 250 gpm. At 30 Btu/h·ft², a Δt of 15°F, and a COP of 0.65, the flow requirement for cooling is 1230 gpm.

Some small-capacity (3- to 25-ton) absorption equipment has been optimized for low-temperature operation in conjunction with solar heat. While this equipment could also be applied to geothermal resources, the prospects for this are questionable. Small absorption equipment would generally compete with package direct-expansion units in this range; the absorption equipment requires a great deal more mechanical auxiliary equipment for a given capacity. The cost of the chilled water piping, pump and coil; cooling water piping, pump and tower; and hot water piping raises the capital cost of the absorption equipment substantially. Only in large sizes (>10 tons) and in areas with high electric rates and high cooling requirements (>2000 full load hours) would this type of equipment offer an attractive investment to the owner (Rafferty 1989b).

INDUSTRIAL APPLICATIONS

Design philosophy for the use of geothermal energy in industrial applications, including agricultural facilities, is similar to that for space conditioning. However, these applications have the potential for much more economical use of the geothermal resource, primarily because (1) they operate year-round, which gives them greater load factors than possible with space conditioning applications; (2) they do not require an extensive (and expensive) distribution to

Table 5   Geothermal Industrial Uses in the United States

Application	Number of Sites	Resource Temperature, °F
Sewage digester heating	2	130, 170
Laundry	4	104 to 181[a]
Vegetable dehydration	1	270
Mushroom growing	1	235
Greenhouse heating	37	95 to 210[a]
Aquaculture	8	61 to 205[a]
Grain drying	1	200
Highway deicing	3	47, 190
Gold mine heap leaching	2	238, 186

Source: Lienau et al. (1988).   [a]Varies with site.

dispersed energy consumers, as is common in district heating; and (3) they often require various temperatures and, consequently, may be able to make greater use of a particular resource than space conditioning, which is restricted to a specific temperature. Lienau et al. (1988) summarize the industrial applications of geothermal energy in the United States (Table 5).

GROUND-SOURCE HEAT PUMPS

Ground-source heat pumps were originally developed in the residential arena and are now being applied in the commercial sector. Many of the installation recommendations and design guides appropriate to residential design must be amended for large buildings. Kavanaugh and Rafferty (1997) and Caneta Research (1995) have a more complete overview of ground-source heat pumps. OSU (1988a,b) and Kavanaugh (1991) provide a more detailed treatment of the design and installation of ground-source heat pumps, but the focus of these two documents is primarily residential and light commercial applications. Comprehensive coverage of commercial and institutional design and construction of ground-source heat pump systems is provided in CSA (1993).

TERMINOLOGY

The term **ground-source heat pump** (GSHP) is applied to a variety of systems that use the ground, groundwater, or surface water as a heat source and sink. Included under the general term are **ground-coupled (GCHP), groundwater (GWHP),** and **surface water (SWHP) heat pumps**. Many parallel terms exist (e.g., **geothermal heat pumps (GHP)**, earth energy systems and ground-source (GS) systems) and are used to meet a variety of marketing or institutional needs (Kavanaugh 1992). Chapter 8 of the 1996 *ASHRAE Handbook—Systems and Equipment* should be consulted for a discussion of the merits of various other non-geothermal heat sources/sinks.

Ground-Coupled Heat Pumps

The GCHP is a subset of the GSHP and is often called a closed-loop ground-source heat pump. A GCHP refers to a system that consists of a reversible vapor compression cycle that is linked to a closed ground heat exchanger buried in soil (Figure 13). The most widely used unit is a water-to-air heat pump, which circulates a water or a water-antifreeze solution through a liquid-to-refrigerant heat exchanger and a buried thermoplastic piping network. A second type of GCHP is the direct expansion (DX) GCHP, which uses a buried copper piping network through which refrigerant is circulated. To distinguish them from DX GCHPs, systems using water-to-air and water-to-water heat pumps are often referred to as GCHPs with secondary solution loops.

The GCHP is further subdivided according to ground heat exchanger design: vertical and horizontal. The **vertical** GCHP (Figure 14) generally consist of two small-diameter high-density

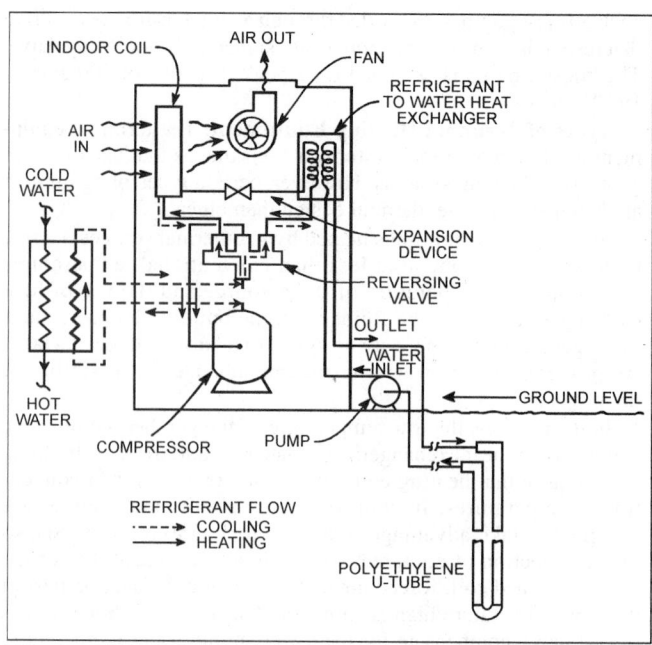

Fig. 13   Vertical Closed-Loop Ground-Coupled Heat Pump System
(Kavanaugh 1985)

PE tubes that have been placed in a vertical borehole that is subsequently filled with a solid medium. The tubes are thermally fused at the bottom of the bore to a close return U-bend. Vertical tubes range from 3/4- to 1-1/2-in. nominal diameter. Bore depths range from 50 to 600 ft depending on local drilling conditions and available equipment.

A minimum base separation distance of 20 ft is recommended when loops are placed in a grid pattern. This distance may be reduced when the bores are placed in a single row, the annual heating load is much greater than the annual cooling load, or vertical water movement mitigates the affect of heat build up in the loop field.

The advantages of vertical the GCHP is it (1) requires relatively small plots of ground, (2) is in contact with soil that varies very little in temperature and thermal properties, (3) requires the smallest amount of pipe and pumping energy, and (4) can yield the most efficient GCHP system performance. The disadvantage is it is typically higher in cost because of expensive equipment needed to drill the borehole and the limited availability of contractors to perform such work.

Hybrid systems are a variation of ground-coupled systems in which a smaller ground loop is used which is augmented, during the cooling mode, by a cooling tower. This approach can have merit in large cooling-dominated applications. The ground loop is sized to meet the heating requirements. The downsized loop is used in conjunction with the cooling tower (usually the closed circuit fluid cooler type) to meet the heat rejection load. Use of the tower reduces the capital cost of the ground loop in such applications, but somewhat increases maintenance requirements.

Horizontal GCHPs (Figure 15) can be divided into at least three subgroups: single-pipe, multiple-pipe, and spiral. Single-pipe horizontal GCHPs were initially placed in narrow trenches at least four deep. These designs require the greatest amount of ground area. Multiple pipes (usually two or four) placed in a single trench can reduce the amount of required ground area. With multiple pipe GCHPs, if trench length is reduced, total pipe length must be increased in order to overcome thermal interference from adjacent pipes. The spiral coil is reported to further reduce required ground area. These horizontal ground heat exchangers are made by

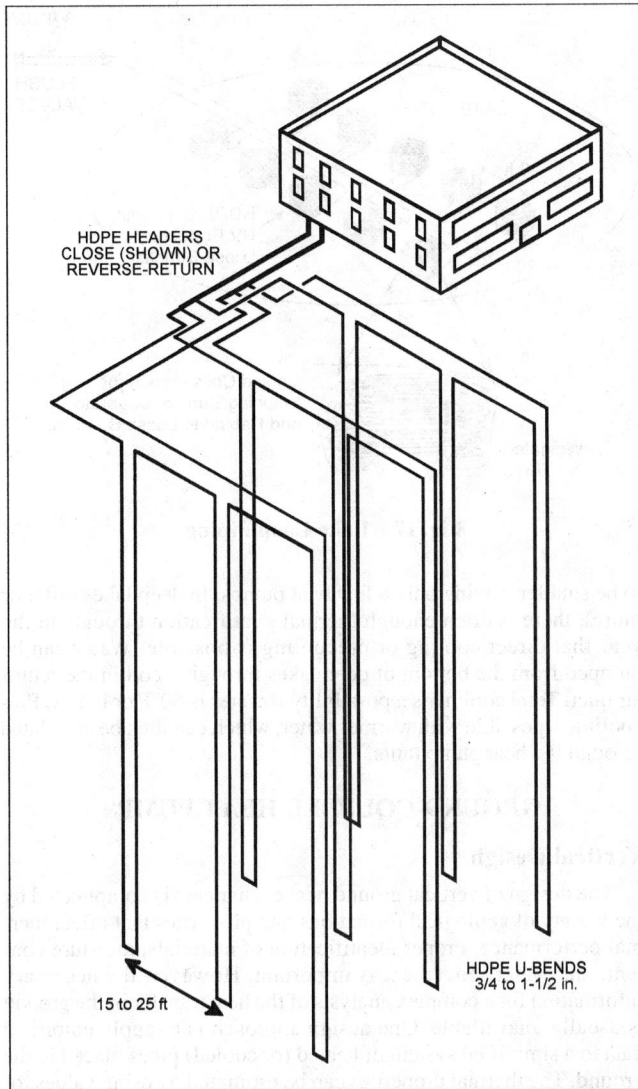

Fig. 14 Vertical Ground-Coupled Heat Pump Piping

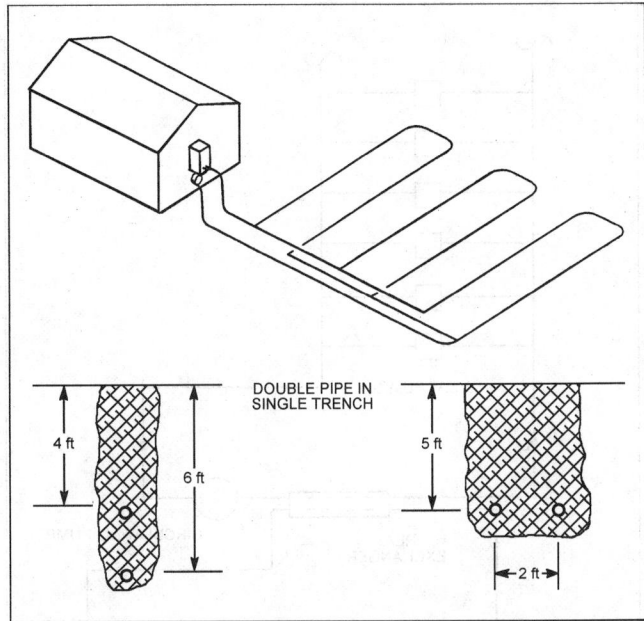

Fig. 15 Horizontal Ground-Coupled Heat Pump Piping

stretching small-diameter PE tubing from the tight coil in which it is shipped into an extended coil that can be placed vertically in a narrow trench or laid flat at the bottom of a wide trench. Recommended trench lengths are only 20 to 30% of single-pipe horizontal GCHPs, but trench lengths may need to be double this amount for equivalent thermal performance.

The advantages of horizontal GCHPs are that (1) they are typically less expensive than vertical GCHPs because relatively low cost installation equipment is widely available, (2) many residential applications have adequate ground area, and (3) trained equipment operators are more widely available. Disadvantages include, in addition to larger ground area requirement, greater adverse variations in performance because (1) ground temperatures and thermal properties fluctuate with season, rainfall, and burial depth, (2) slightly higher pumping energy requirements, and (3) lower system efficiencies. OSU (1988a,b) and Svec (1990) cover the design and installation of horizontal GCHPs.

Groundwater Heat Pumps

The second subset of GSHPs is groundwater heat pumps (Figure 16). Until the recent development of GCHPs, they were the most widely used type of GSHP. In the commercial sector,

GWHPs can be an attractive alternative because large quantities of water can be delivered from and returned to relatively inexpensive wells that require very little ground area. While the cost per unit capacity of the ground heat exchanger is relatively constant for GCHPs, the cost per unit capacity of a well water system is much lower for a large GWHP system. A single pair high-volume wells can serve an entire building. Properly designed groundwater loops with well-developed water wells require no more maintenance than conventional air and water central HVAC. When the groundwater is injected back into the aquifer by a second well, net water use is zero.

A widely used design places a central water-to-water heat exchanger between the groundwater and a closed water loop which is connected to water-to-air heat pumps located in the building. A second possibility is to circulate groundwater through a heat recovery chiller (isolated with a heat exchanger), and to heat and cool the building with a distributed hydronic loop.

Both types and other variations may be suited for direct preconditioning in much of the United States. Groundwater below 60°F can be circulated directly through hydronic coils in series or in parallel with heat pumps. The cool groundwater can displace a large amount of energy which would otherwise have to be generated by mechanical refrigeration.

The advantages of GWHPs under suitable conditions are (1) they cost less than GCHP equipment, (2) the space required for the water well is very compact, (3) water well contractors are widely available, and (4) the technology has been used for decades in some of the largest commercial systems.

Disadvantages are that (1) local environmental regulations may be restrictive, (2) water availability may be limited, (3) fouling precautions may be necessary if the wells are not properly developed or water quality is poor, and (4) pumping energy may be high if the system is poorly designed or draws from a deep aquifer.

Surface Water Heat Pumps

Surface water heat pumps have been included as a subset of GSHPs because of the similarities in applications and installation methods. SWHPs can be either closed-loop systems similar to GCHPs or open-loop systems similar to GWHPs. However, the thermal characteristics of surface water bodies are quite different

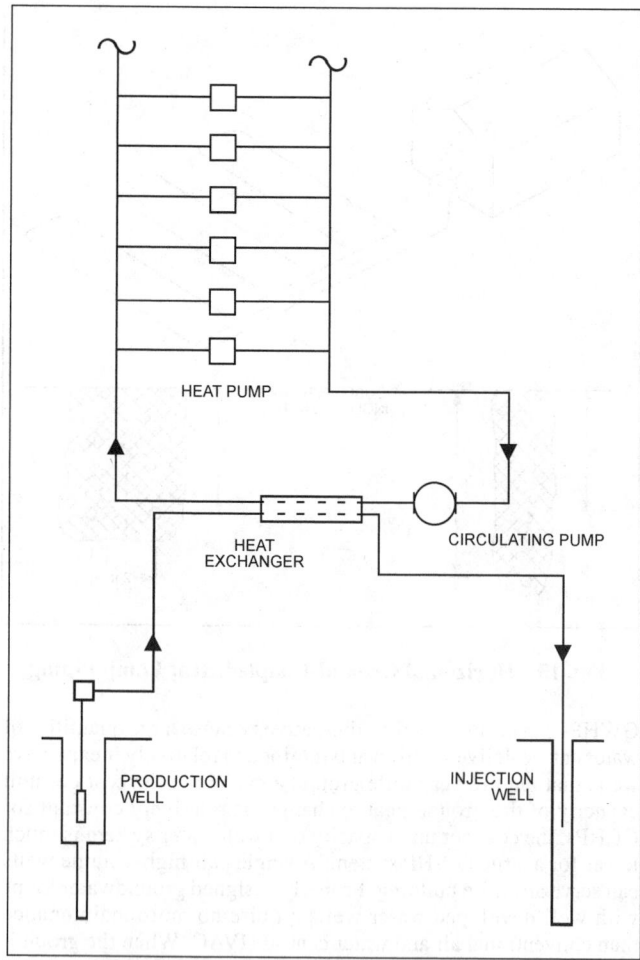

Fig. 16   Unitary Groundwater Heat Pump System

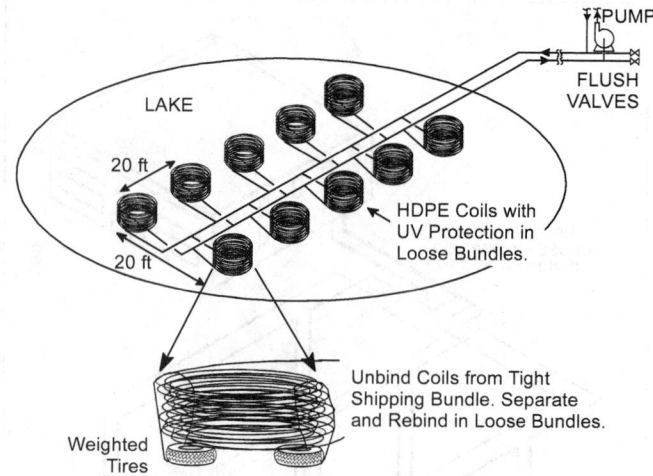

Fig. 17   Lake Loop Piping

than those of the ground or groundwater. Some unique applications are possible through special precautions may be warranted.

Closed-loop SWHPs (Figure 17) consist of water-to-air or water-to-water heat pumps connected to a piping network placed in a lake, river, or other open body of water. A pump circulates water or a water-antifreeze solution through the heat pump water-to-refrigerant heat exchanger and the submerged piping loop, which transfers heat to or from the body of water. The recommended piping material is thermally fused high-density PE tubing with ultra-violet (UV) radiation protection.

The advantages of closed-loop SWHPs are (1) relatively low cost (compared to GCHPs) due to reduced excavation costs, (2) low pumping energy requirements, (3) low maintenance requirements, and (4) low operating cost. Disadvantages are (1) the possibility of coil damage in public lakes and (2) wide variation in water temperature with outdoor conditions if lakes are small and/or shallow. Such variation in water temperature would cause undesirable variations in efficiency and capacity, though not as severe as with air-source heat pumps.

Open-loop SWHPs can use surface water bodies the way cooling towers are used, but without the need for fan energy or frequent maintenance. In warm climates, lakes can also serve as heat sources during the winter heating mode, but in colder climates closed-loop systems are the only viable option for heating in colder climates, where water temperatures drop below 45°F.

Lake water can be pumped directly to water-to-air or water-to-water heat pumps or through an intermediate heat exchanger that is connected to the units with a closed piping loop. Direct systems tend

to be smaller, having only a few heat pumps. In deep lakes (40 ft or more), there is often enough thermal stratification throughout the year that direct cooling or precooling is possible. Water can be pumped from the bottom of deep lakes through a coil in the return air duct. Total cooling is a possibility if water is 50°F or below. Precooling is possible with warmer water, which can then be circulated through the heat pump units.

GROUND-COUPLED HEAT PUMPS

Vertical Design

The design of vertical ground heat exchangers is complicated by the variety of geological formations and properties that affect thermal performance. Proper identification of materials, moisture content, and water movement is important. However, the necessary information for a complex analysis of the heat transfer in the ground is usually unavailable. One design approach is to apply empirical data to a simplified system of heated (or cooled) pipes placed in the ground. The thermal properties can be estimated by using values for soils in a particular group and a moisture content that is characteristic of local conditions.

This method has proven successful for residential and light commercial GCHPs. Some permanent change in the local ground temperature may be expected for systems with large annual differences between the amount of heat extracted in the heating mode and the amount rejected in the cooling mode. This problem is compounded in commercial systems because earth heat exchangers are more likely to be installed in close proximity due to limited ground area.

The method described here makes use of a limited amount of information from commercial systems. A major missing component is long-term, field-monitored data. These data are needed to further validate the design method so that the effects of water movement and long-term heat storage are more fully addressed. Currently, the conservative designer can assume no benefit from water movement, while the gambler can assume maximum benefit by ignoring annual imbalances in heat rejection and absorption.

One design method is based upon the solution of the equation for heat transfer from a cylinder buried in the earth. This equation was developed and evaluated by Carslaw and Jaeger (1947) and was suggested by Ingersoll and Zobel (1954) as an appropriate method of sizing ground heat exchangers. Kavanaugh (1985) adjusted the method to account for the U-bend arrangement and hourly heat rate variations.

Eskilson (1987) demonstrated that the thermal performance of a ground heat exchanger is a strong function of the amount of heat that has previously been extracted from or rejected to the ground. Minimum and maximum temperatures may take many years to occur, especially if multiple vertical bores are located in close proximity. However, short-term variations in heat transfer are damped by the large thermal mass of the ground between the bores. Therefore, an estimate of the annual net amount is sufficient for design purposes. This concept can also be extended to monthly variations with good accuracy. However, greater care must be exercised in handling daily or hourly variations in heat transfer to the ground.

The method of Ingersoll and Zobel (1954) can be used to handle these shorter term variations. It uses the following steady-state heat transfer equation:

$$q = \frac{L(t_g - t_w)}{R} \tag{1}$$

where

    q = heat transfer rate, Btu/h
    L = required bore length, ft
    t_g = ground temperature, °F
    t_w = liquid temperature, °F
    R = effective thermal resistance of the ground, h·ft·°F/Btu

The equation is rearranged to solve for the required bore length L. The steady-state equation is modified to represent the variable heat rate of a ground heat exchanger by using a series of constant heat rate "pulses." The thermal resistance of the ground per unit length is calculated as a function of time corresponding to the time span over which a particular heat pulse occurs. A term is also included to account for the thermal resistance of the pipe wall and interfaces between the pipe and fluid and the pipe and the ground. The resulting equation takes the form for cooling:

$$L_c = \frac{q_a R_{ga} + (q_{lc} - 3.41 W_c)(R_p + PLF_m R_{gm} + R_{gd} F_{sc})}{t_g - \dfrac{t_{wi} + t_{wo}}{2} - t_p} \tag{2}$$

The required length for heating is

$$L_h = \frac{q_a R_{ga} + (q_{lh} - 3.41 W_h)(R_p + PLF_m R_{gm} + R_{gd} F_{sc})}{t_g - \dfrac{t_{wi} + t_{wo}}{2} - t_p} \tag{3}$$

The terms used in the Equations (2) and (3) are

    F_{sc} = short circuit heat loss factor
    L_c = required bore length for cooling, ft
    L_h = required bore length for heating, ft
    PLF_m = part load factor during design month
    q_a = net annual average heat transfer to the ground, Btu/h
    q_{lc} = building design cooling block load, Btu/h
    q_{lh} = building design heating block load, Btu/h
    R_{ga} = effective thermal resistance of ground (annual pulse), h·ft·°F/Btu
    R_{gd} = effective thermal resistance of ground (daily pulse), h·ft·°F/Btu
    R_{gm} = effective thermal resistance of ground—monthly pulse, h·ft·°F/Btu
    R_p = thermal resistance of pipe, h·ft·°F/Btu
    t_g = undisturbed ground temperature, °F
    t_p = temperature penalty for interference of adjacent bores, °F
    t_{wi} = liquid temperature at heat pump inlet, °F
    t_{wo} = liquid temperature at heat pump outlet, °F
    W_c = power input at design cooling load, W
    W_h = power input at design heating load, W

Note: Heat transfer rate, building loads, and temperature penalties are positive for heating and negative for cooling.

Equations (2) and (3) consider three different pulses of heat to account for long-term heat imbalances (q_a), average monthly heat

rates during the design month, and maximum heat rates for a short-term period during a design day. This period could be as short as 1 h, but a 4-h block is recommended.

The required bore is the larger of the two lengths L_c and L_h found from Equations (2) and (3). If the L_c is larger than L_h, the benefits of an oversized coil could be enjoyed during the heating season. A second option is to install the smaller heating length along with a cooling tower to compensate for the undersized coil. If the L_h is larger, the designer should install this length, and during the cooling mode the efficiency benefits of an oversized ground coil could be used to compensate for the higher first cost.

The thermal resistance of the ground is calculated from ground properties, pipe dimensions, and operating periods of the representative heat rate pulses. Table 6 lists typical thermal properties for soils, and Table 7 gives equivalent thermal resistances and diameters for vertical U-bend heat exchangers.

The most difficult parameters to evaluate in Equations (2) and (3) are the equivalent thermal resistance of the ground. The solutions of Carslaw and Jaeger (1947) require that the time of operation, the outside pipe diameter, and the thermal diffusivity of the ground be related in the dimensionless Fourier Number (Fo):

$$Fo = \frac{4\alpha_g \tau}{d_b^2} \tag{4}$$

where

    α_g = thermal diffusivity of the ground
    τ = time of operation
    d = outside pipe diameter

The method may be modified to permit calculation of equivalent thermal resistances for varying heat pulses. A system can be modeled by three heat pulses, a 10-year (3650-day) pulse of q_a, a one-month (30-day) pulse of q_m, and a 6-hour (0.25-day) pulse of q_d. Three times are defined as:

$$\tau_1 = 3650 \text{ days}$$
$$\tau_2 = 3650 + 30 = 3680 \text{ days}$$
$$\tau_f = 3650 + 30 + 0.25 = 3680.25 \text{ days}$$

Table 6 Thermal Properties of Selected Soils, Rocks, and Backfills

	Dry Density, lb/ft³	Conductivity, Btu/h·ft·°F	Diffusivity, ft²/day
Soils			
Heavy clay (15% water)	120	0.8 – 1.1	0.45 – 0.65
Heavy clay (5% water)	120	0.6 – 0.8	0.5 – 0.65
Light clay (15% water)	80	0.4 – 0.6	0.35 – 0.5
Light clay (5% water)	80	0.3 – 0.5	0.35 – 0.6
Heavy sand (15% water)	120	1.6 – 2.2	0.9 – 1.2
Heavy sand (5% water)	120	1.2 – 1.9	1.0 – 1.5
Light sand (15% water)	80	0.6 – 1.2	0.5 – 1.0
Light sand (5% water)	80	0.5 – 1.1	0.6 – 1.3
Rocks			
Granite	165	1.3 – 2.1	0.9 – 1.4
Limestone	150 – 175	1.4 – 2.2	0.9 – 1.4
Sandstone	160 – 170	1.2 – 2.0	0.7 – 1.2
Wet shale		0.8 – 1.4	0.7 – 0.9
Dry shale		0.6 – 1.2	0.6 – 0.8
Grouts/Backfills			
Bentonite (20% soilds)		0.42 – 0.43	
Cement		0.40 – 0.45	
20% Bent.-40% SiO₂ sand		0.85	
Concrete (50% SiO₂ sand)		1.2 – 1.6	

Source: Kavanaugh and Rafferty (1997).

Table 7   Polyethylene U-Tube Pipe and Bore Hole Backfill Thermal Resistance

U-Tube Dimensions		Pipe Resistance (R_p), h·ft²·°F/Btu	Backfill Resistance, h·ft²·°F/Btu					
			4 in. Diameter Bore Hole k_{grout}, Btu/h·ft²·°F			6 in. Diameter Bore Hole k_{grout}, Btu/h·ft²·°F		
Dia.	SDR		0.5	1.0	2.0	0.5	1.0	2.0
3/4 in.	11	0.075	0.26	0.13	0.06	0.33	0.16	0.08
	9	0.093	0.22	0.13	0.06	0.33	0.16	0.08
1 in.	11	0.075	0.22	0.11	0.06	0.29	0.14	0.07
	9	0.093	0.22	0.11	0.06	0.29	0.14	0.07
1-1/4 in.	11	0.075	–	–	–	0.25	0.12	0.06
	9	0.093	–	–	–	0.25	0.12	0.06
1-1/2 in.	11	0.075	–	–	–	0.23	0.11	0.06
	9	0.093	–	–	–	0.23	0.11	0.06

Source: Kavanaugh and Rafferty (1997).
Based on Re = 10000+. Add 0.008 h·ft²·°F/Btu for Re = 4000; add 0.025 h·ft²·°F/Btu for Re = 1500.

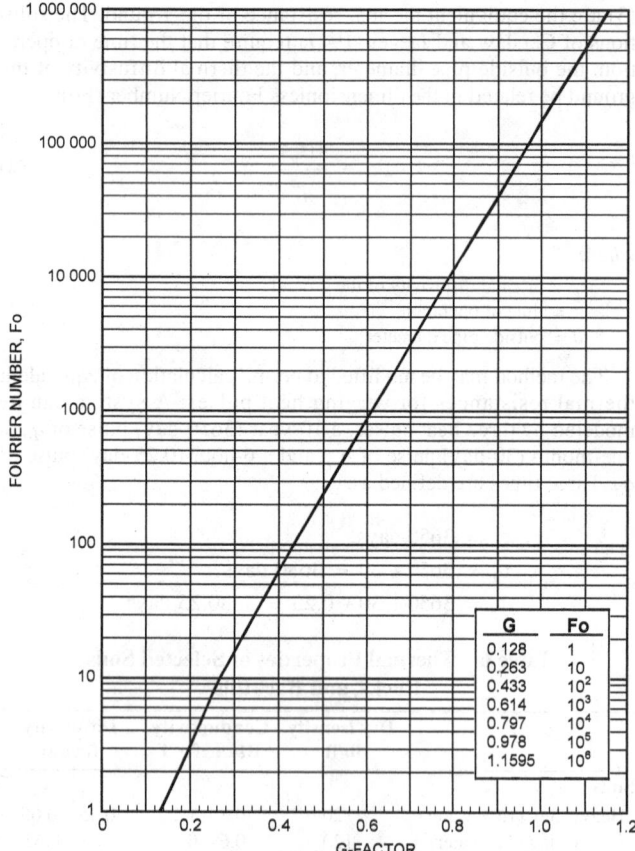

G	Fo
0.128	1
0.263	10
0.433	10^2
0.614	10^3
0.797	10^4
0.978	10^5
1.1595	10^6

Fig. 18   Fourier/G-Factor Graph for Ground Thermal Resistance
(Kavanaugh and Rafferty 1997)

The Fourier number is then computed with the following values:

$$Fo_f = 4\alpha\tau_f/d^2$$
$$Fo_1 = 4\alpha(\tau_f - \tau_1)/d^2$$
$$Fo_2 = 4\alpha(\tau_f - \tau_2)/d^2$$

An intermediate step in the computation of the thermal resistance of the ground using the methods of Ingersoll and Zobel (1954) is the determination of a G-factor, which is then determined from Figure 18 for each Fourier value.

$$R_{gA} = (G_f - G_1)/k_g \tag{5a}$$

$$R_{gB} = (G_1 - G_2)/k_g \tag{5b}$$

$$R_{gC} = G_2/k_g \tag{5c}$$

Ranges of the ground thermal conductivity k_g are given in Table 6. State geological surveys are a good source of soil and rock data. However, geotechnical site surveys are highly recommended to determine load soil and rock types, and drilling conditions. In-situ testing of a U-bend can be performed and thermal properties determined by inverse heat transfer methods.

Performance degrades somewhat due to short circuiting heat losses between the upward- and downward-flowing legs of a conventional U-bend loop. When flow is 3 gpm per ton, the degradation is approximately 4%. Losses can be accounted for by multiplying the equivalent thermal resistance for the daily pulse R_{gd} by 1.04 (Kavanaugh and Rafferty 1997). The loss is reduced considerably if there are multiple boreholes on a single parallel loop (see the following table).

Bores per Loop	F_{sc}	
	2 gpm/ton	3 gpm/ton
1	1.06	1.04
2	1.03	1.02
3	1.02	1.01

Temperature. The remaining terms in Equations (2) and (3) are temperatures. The local deep ground temperature t_g can best be obtained from local water well logs and geological surveys. A second, less accurate source is temperature contour maps similar to Figure 19 prepared by state geological surveys. A third source, which can yield ground temperatures within 4°F is a map with contours, such as Figure 20. Comparison of Figures 19 and 20 indicates the complex variations that would not be accounted for without detailed contour maps.

Selecting the temperature t_{wi} of the water entering the unit is a critical choice in the design process. Choosing a value close to the ground temperature results in higher system efficiency, but the required ground coil length will be very long and thus unreasonably expensive. Choosing a value far from t_g allows the selection of a small, inexpensive ground coil, but the system's heat pumps will have both greatly reduced capacity during heating and high demand when cooling. Selecting t_{wi} to be 20 to 30°F higher than t_g in cooling and 10 to 20°F lower than t_g in heating is a good compromise between first cost and efficiency in many regions of the United States.

A final temperature to consider is the temperature penalty t_p resulting from thermal interferences from adjacent bores. The designer must select a reasonable separation distance in order to minimize required land area without causing large increases in the required bore length (L_c, L_h). Table 8 presents the temperature penalty for a 10 by 10 vertical grid of bores for various operating

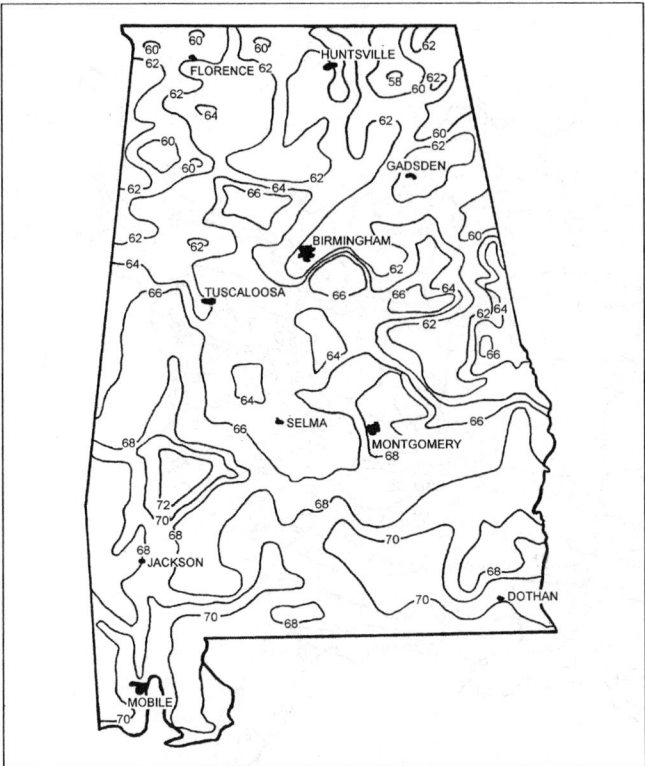

Fig. 19   Water and Ground Temperatures in Alabama at 50 to 150 ft Depth
(Chandler 1987)

Table 8   Long-Term Change in Ground Field Temperature for 10 by 10 Vertical Grid with a 100-Ton Block Load
(Kavanaugh and Rafferty 1997)

Equiv. Full Load Hours Heating/Cooling	Bore Separation, ft	Temperature Penalty, °F	Base Bore Length, ft/ton (refrig.)
1000/500	15	Negligible	180
1000/1000	15	4.7	225
	20	2.4	206
500/1000	15	7.6	260
	20	3.9	228
	15	12.8	345
500/1500	20	6.7	254
	25	3.5	224
	15	Not advisable	
0/2000	20	10.4	316
	25	5.5	252

Correction Factors for Other Grid Patterns			
1 × 10 grid $C_f = 0.36$	2 × 10 grid $C_f = 0.45$	5 × 5 grid $C_f = 0.75$	20 × 20 grid $C_f = 1.14$

conditions after 10 years of operation. Correction factors are included to find the permanent temperature change in four other grid patterns. Note the higher the number of internal bores, the larger the correction factor.

The table includes the length of bore per ton of peak block load to which the temperature change corresponds. Smaller bore lengths per ton of peak block load result in larger temperature changes. An inverse linear relationship for bore length and temperature change exists.

The values in this table represent worst-case scenarios and the temperature change will usually be mitigated by groundwater recharge (vertical flow), groundwater movement (horizontal flow), and evaporation (and condensation) of water in the soil.

Groundwater movement has a large impact upon the long-term temperature change in a densely packed ground loop field. Because the effect has not been thoroughly studied, the design engineer must establish a range of design lengths between one based on minimal groundwater movement, as in very tight clay soils with poor percolation rates, and a second based on the higher rates characteristic of porous aquifers.

Horizontal Design

The buried pipe of a closed-loop GSHP may theoretically produce a change in temperature in the ground up to 16 ft away. For all practical purposes, however, the ground temperature is essentially unchanged beyond about 3 ft from the pipe loop. For that reason, the pipe can be buried relatively near the ground surface and still benefit from the moderating temperatures that the earth provides. Because the ground temperature may fluctuate as much as ±10°F at a depth of 6 ft, an antifreeze solution must be used in most heating dominated regions. The critical design aspect of horizontal applications is to have enough buried pipe loop within the available land area to serve the equipment. The design guidelines for residential horizontal loop installations can be found in OSU (1988).

A horizontal loop design has several advantages over a vertical loop design for a closed-loop ground source system:

- Installation of the ground loop is usually less expensive than for vertical well designs because the capital cost of a backhoe or trencher is only a fraction of the cost of most drilling rigs.
- Most ground source contractors must have a backhoe for construction of the header pits and trenches to the building, so they can perform the entire job without the need to schedule a special contractor.
- Large drilling rigs may not be able to get to all locations due to their size and weight.
- There is usually no potential for aquifer contamination due to the shallow depth of the trench.
- There is minimal residual temperature effect from unbalanced annual loads on the ground loop because the heat transfer to or from the ground loop is small compared to the normal heat transfer occurring at the ground surface.

Some limitations on selecting a horizontal loop design include the following:

- The minimum land area needed for most non-spiral horizontal loop designs for an average house is about 0.5 acre. Horizontal systems are not feasible for most urban houses, which are commonly built on smaller lots.
- The larger length of pipe buried relatively near the surface is more susceptible to being cut during excavations for other utilities.
- Soil moisture content must be properly accounted for in computing the required ground loop length, especially in sandy soils or on hilltops that may dry out in summer.
- Rocks and other obstructions near the surface may make excavation with a backhoe or trencher impractical.

Multiple pipes are often placed in a single trench to reduce the land area needed for horizontal loop applications. Some common multiple pipe arrangements are shown in Figure 21. When pipes are placed at two depths, the bottom row is placed first, and then the trench is partially backfilled before the upper row is put in place. Rarely are more than two layers of pipe used in a single trench due to the extra time needed for the partial backfilling. Higher pipe densities in the trench provide diminishing returns because thermal interference between multiple pipes reduces the heat transfer effectiveness of each pipe. The most common multiple pipe applications

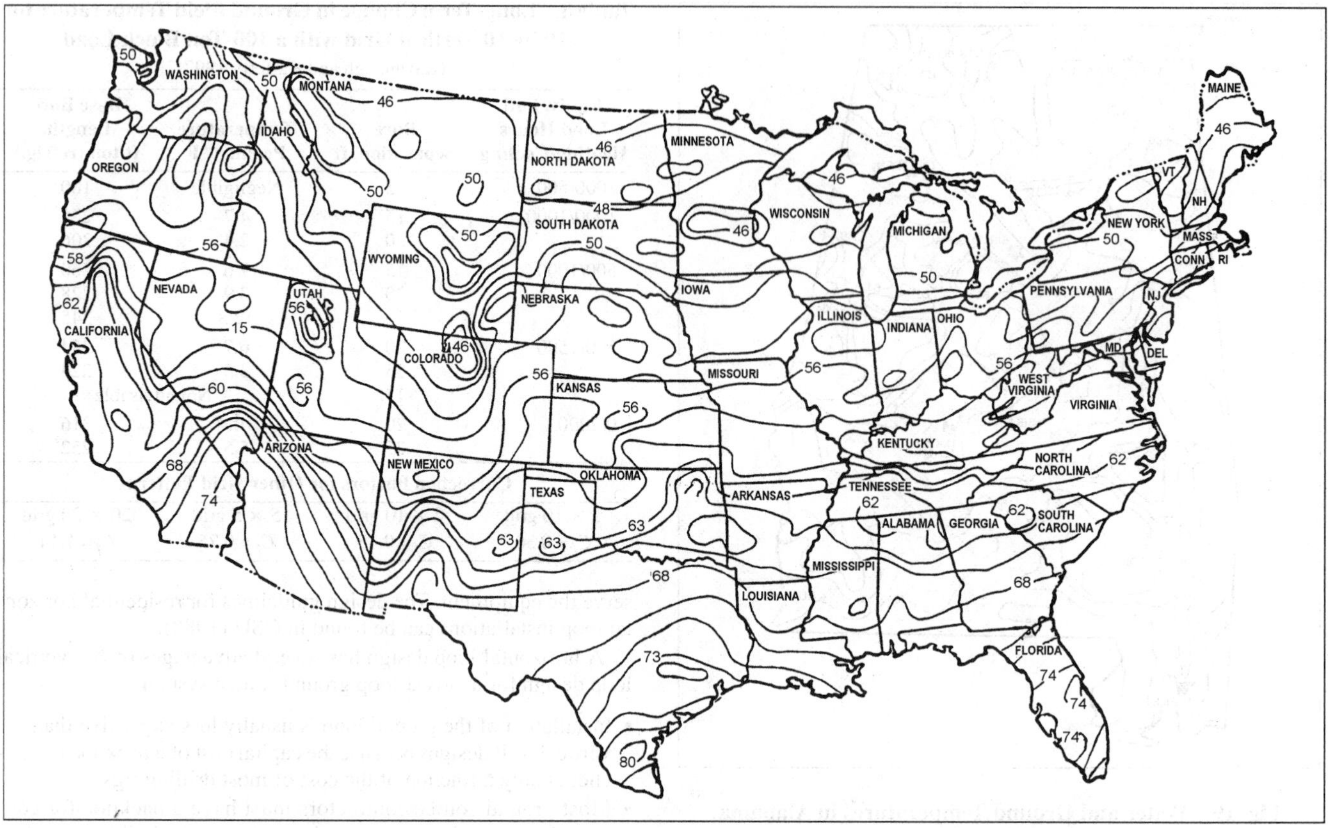
WASHINGTON
OREGON
IDAHO
MONTANA
NORTH DAKOTA
MINNESOTA
SOUTH DAKOTA
WYOMING
NEVADA
CALIFORNIA
UTAH
COLORADO
NEBRASKA
IOWA
KANSAS
MISSOURI
ARIZONA
NEW MEXICO
OKLAHOMA
TEXAS
ARKANSAS
LOUISIANA
MISSISSIPPI
ALABAMA
GEORGIA
TENNESSEE
KENTUCKY
WISCONSIN
MICHIGAN
ILLINOIS
INDIANA
OHIO
WEST VIRGINIA
VIRGINIA
NORTH CAROLINA
SOUTH CAROLINA
FLORIDA
PENNSYLVANIA
NEW YORK
MD
DEL
NJ
CONN
RI
MASS
VT
NH
MAINE
Fig. 20 Approximate Groundwater Temperatures (°F) in the United States

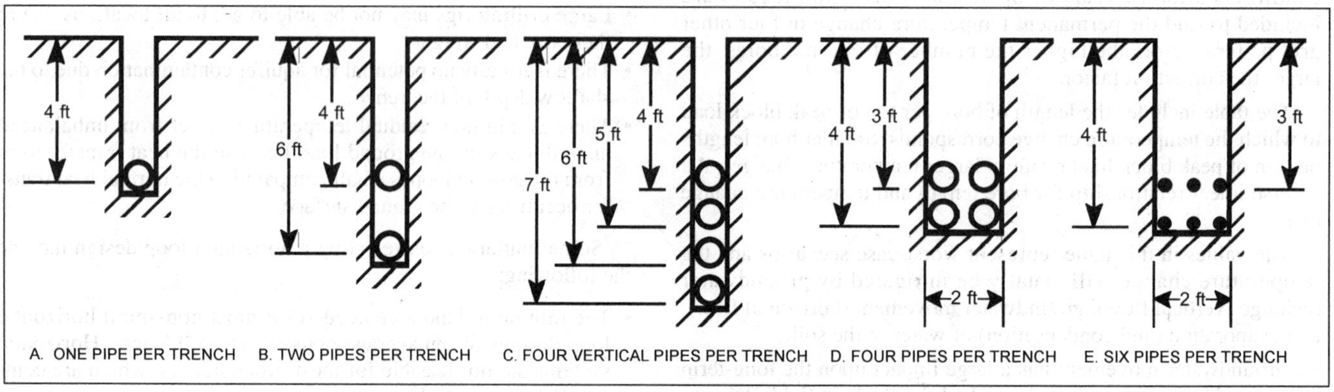
4 ft
6 ft
5 ft
7 ft
3 ft
2 ft
A. ONE PIPE PER TRENCH
B. TWO PIPES PER TRENCH
C. FOUR VERTICAL PIPES PER TRENCH
D. FOUR PIPES PER TRENCH
E. SIX PIPES PER TRENCH
Fig. 21 Horizontal Ground Loop Configurations

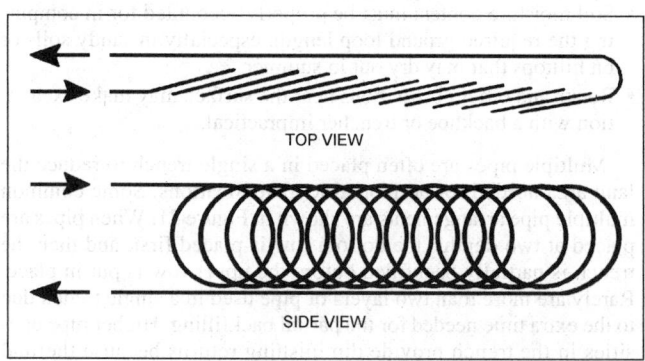

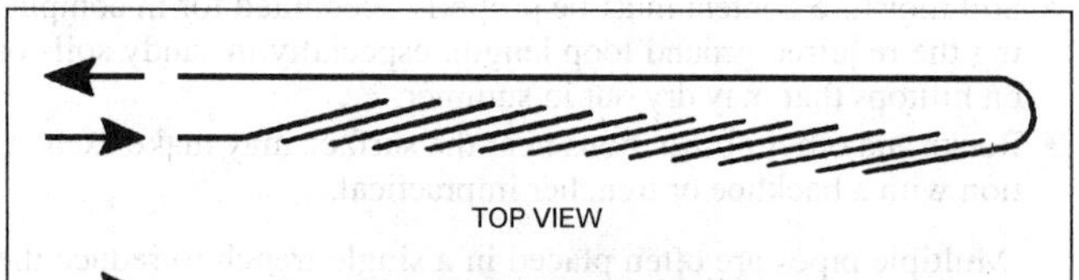
TOP VIEW
SIDE VIEW
Fig. 22 General Layout of a Spiral Earth Coil

are the two-pipe arrangement used with chain trenchers and the four- or six-pipe arrangements placed in trenches made with a wide backhoe bucket.

An overlapping spiral configuration shown in Figure 22, has also been used with some success. However, it requires special attention during the backfilling process to ensure soil fills all the pockets formed by the overlapping pipe. Large quantities of water must be added to compact the soil around the overlapping pipes. The backfilling must be performed in stages to guarantee complete filling around the pipes and good soil contact. The high pipe density (up to 10 ft of pipe per linear foot of trench) may cause problems in prolonged extreme weather conditions, either from soil dry out during cooling or from freezing during heating.

The extra time needed to backfill and the extra pipe length required make spiral configurations nearly as expensive to install as straight pipe configurations. However, the reduced land area needed for the more compact design may permit their use on smaller residential lots. The spiral pipe configuration laid flat in a horizontal pit arrangement is used commonly in the northern midwest part of the United States, where sandy soil causes trenches to collapse. A large open pit is excavated by a bulldozer, and then the overlapping pipes laid flat on the bottom of the pit. The bulldozer is also used to cover the pipe; the pipes should not be run over with the bulldozer tread.

Most horizontal loop installations place flow loops in a parallel rather than a single (series) loop to reduce pumping power (Figure 23). Parallel loops may require slightly more pipe, but may use smaller pipe and thus have smaller internal volumes requiring less antifreeze (if needed). Also, the smaller pipe is typically much cheaper for a given length, so total pipe cost is less for parallel loops. An added benefit is that parallel loops can be flushed out with a smaller purge pump than would be required for a larger single-pipe loop. A disadvantage of parallel loops is the potential for unequal flow in the loops and thus reduced heat exchange efficiency.

The time required to install a horizontal loop is not much different from that for a vertical system. For the arrangements described above, a two-person crew can typically install the ground loop for an average house in a single day.

While not restricted to single-family residential applications, horizontal loops are rarely used in larger commercial buildings due to the land area that is required. Even if the land adjacent to the building were initially available, installation of a horizontal loop could prevent any future construction above the loop field, tying up

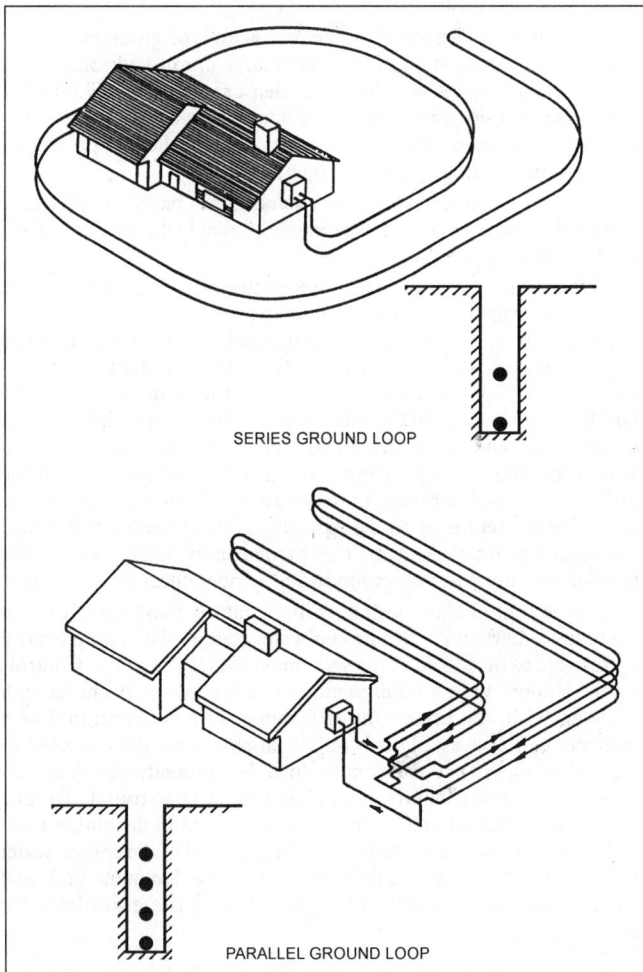

Fig. 23   Parallel and Series Ground Loop Configurations

a considerable investment in vacant land. Placement of horizontal loops under parking lots may have a negative impact on the effectiveness of the ground loop due to the greater surface heat exchange.

Soil characteristics are an important concern for any ground loop design. With horizontal loops, the soil type can be more easily determined because the excavated soil can be inspected and tested. EPRI et al. (1989) compiled a list of criteria and simple test procedures that can be used to classify soil and rock adequately enough for horizontal ground loop design.

Leaks in the heat-fused plastic pipe are rare when attention is paid to pipe cleanliness and proper fusion techniques. Should a leak occur, it is usually best to try to isolate the leaking parallel loop and abandon it in place. The time and effort required to find the source of the leak usually far outweighs the cost of replacing the defective loop. Because the loss of as little as 0.25 gal of water from the system will cause it to shut down, leaks cannot be located by looking for wet soil, as is commonly done with water lines.

GROUNDWATER HEAT PUMPS

A groundwater heat pump system (GWHP) removes groundwater from a well and delivers it to a heat pump (or an intermediate heat exchanger) to serve as a heat source or sink. Both unitary or central plant designs are used. In the unitary type, a large number of small water-to-air heat pumps are distributed throughout the building. The central plant design uses one or a small number of large-capacity chillers supplying hot and chilled water to a two- or four-pipe distribution system.

Regardless of the type of equipment installed in the building, the specific components for handling the groundwater are similar. The primary items include (1) the wells (supply and, if required, injection), (2) well pump, and (3) groundwater heat exchanger. The specifics of these items are discussed in the section on Direct-Use Systems. In addition to those comments, the following considerations apply.

Groundwater Flow Requirements

Generally, the greater the groundwater flow, the better the performance (COP or EER) of the heat pumps. However, increasing heat pump performance can be compromised quickly by well pump power at high groundwater flow rates. For this reason, optimum groundwater flow should be based on electrical power requirements of the well pump, heat pumps, and circulating pump. Optimum groundwater flow (for minimum system energy consumption) is a function of groundwater temperature, well pump head, heat exchanger design, loop pump power and heat pump performance.

For moderate-efficiency heat pumps (EER of 14.2), efficient loop pump design (7.5 hp/100 tons), and a heat exchanger approach of 3°F, Figure 24 provides curves for two different groundwater temperatures (70°F and 50°F) and two well pump heads (100 ft and 300 ft).

Although the four curves show a clear optimum flow, sometimes operating at a lower groundwater flow reduces the well/pump capital cost and reduces the problem of fluid disposal. These considerations are highly project specific, but do afford the designer some latitude in flow selection.

Well Pumps

Submersible pumps have not performed well in higher-temperature, direct-use projects. However, in a normal groundwater temperature as encountered in heat pump applications, the submersible pump is a cost-effective option. The low temperature eliminates the need to specify an industrial design for the motor/protector, thereby greatly reducing the first cost relative to direct-use. Caution should still be exercised for wells that are expected to produce moderate

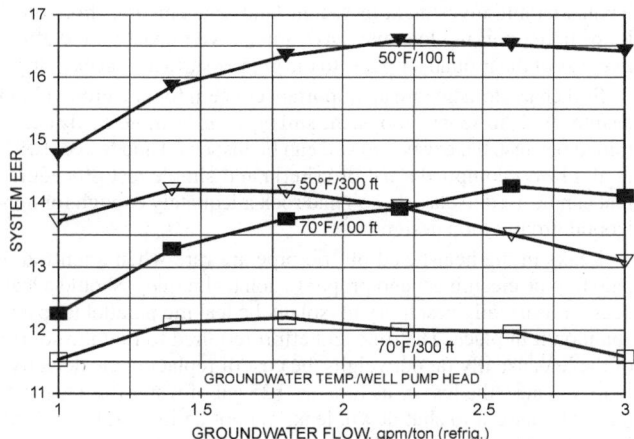

Fig. 24 Optimum Groundwater Flow for Maximum EER
(Kavanaugh and Rafferty 1997)

amounts of sand. The high speed (3500 rpm) of most submersibles makes them susceptible to erosion damage under these conditions.

Small groundwater systems have frequently been identified with excessive well pump energy consumption. In many cases this is true; however, the reasons for excessive pump energy consumption (high water flow rate, coupling to the domestic pressure tank, and low efficiency of small submersible pumps) are generally not present in large, commercial groundwater systems. In large systems, the groundwater flow per unit capacity is frequently less than half of residential systems. The pressure at the wellhead is not the 30 to 50 psi typical of domestic systems, but rather a function only of the losses through the groundwater loop. Finally, large lineshaft well pumps are characterized by efficiencies of up to 83% compared to the 35 to 40% range for small submersible pumps.

Heat Exchangers

Design of a plate-and-frame heat exchanger is largely a trade-off between pressure drop, which influences pumping (operating cost), and overall heat transfer coefficient, which influences surface area (capital cost). In general, exchangers in GWHP systems can be economically selected for a 3°F approach (between loop return and groundwater leaving temperatures) and a pressure drop of <10 psi. Excessive fouling factors (>0.0002 $h \cdot ft^2 \cdot °F/Btu$) should not be specified for plate heat exchangers because they can be easily disassembled and cleaned.

Heat exchanger cost may be reduced for groundwater applications by using Type 304 stainless steel plates rather than the Type 316 or titanium plates common in direct-use projects. The low temperature and generally low chloride content of heat pump fluids frequently make the less expensive Type 304 material acceptable.

Direct Systems

A direct system (in which in which the groundwater flows directly to the heat pump without an intermediate heat exchanger) is not recommended except on the very smallest installations. Although some systems with open designs have been successful, others have had serious difficulty even with groundwater of apparently benign chemistry. As a result, prudent design for commercial/industrial-scale projects isolates the groundwater from the building system with a heat exchanger. The increased capital cost arising from the installation of the heat exchanger amounts to a small percentage of the total cost. In view of the greatly reduced maintenance requirements of closed systems, the increased capital cost is quickly recovered.

Indirect Systems

Unitary System. In the unitary design, a groundwater heat exchanger is placed at the point occupied by the tower/boiler in a conventional water-loop system. If a tower/boiler is to be used for backup, the groundwater exchanger should be placed upstream of it. For most large building applications, the design of the heat exchanger will be governed by the cooling duty. The exception to this would be a building having little or no central core area in a northern climate with very low temperature groundwater.

The designer should carefully evaluate the need for insulation on the water-loop piping in the building. If a significant heating load exists and groundwater temperature is low, the loop temperature will likely be such that insulation will be required to avoid sweating.

Groundwater flow has been controlled by a several methods including cycling, multiple wells, two-speed well pumps, and well pumps equipped with variable-speed drives. The most common control strategy has been to operate the well pump with a dual set-point controller. The pump is operated above a loop temperature set point in the cooling mode and below a loop temperature set point in the heating mode. The loop temperature then floats between the two set points. Establishment of the temperature set points should be based on system performance (the well pump should only operate when it results in a net reduction in power requirements). Controller differential may need to consider thermal mass in order to prevent short cycling of the well pump. In some applications with a single well pump and a substantial difference between the heating and cooling flow requirements, pump capacity may need to be adjusted seasonally with a variable-speed drive.

Central Plant Systems. For central plant type groundwater systems, two heat exchangers are normally employed, one in the chilled water loop and one in the condenser water loop (Figure 25). The evaporator-loop exchanger provides a heat source for heating-dominated operation and the condenser-loop exchanger provides a heat sink for cooling-dominated operation.

Sizing of the **condenser loop exchanger** is based on providing sufficient capacity to reject the condenser load in the absence of any building heating requirement.

Sizing of the **chilled water loop exchanger** must consider two loads. The primary criteria is the load required during heating-dominant operation. The exchanger must transfer sufficient heat (when combined with compressor heat) from the groundwater to the chilled water loop to meet the space heating requirement of the building. Depending on the relative groundwater and chilled water temperatures and on the design rise, the exchangers may also provide some free cooling during cooling-dominant operation. If the groundwater temperature is lower than the temperature of the chilled water returning to the exchanger, some chilled water load can be met by the exchanger. This mode is most likely be available in northern climates with groundwater temperatures below 60°F.

For the central plant design, chiller controls must allow for the unique operation with a groundwater source. Well pump control options are as discussed in the section on Unitary Systems. Controls for the chiller can be similar to those on a heat recovery chiller with a tower, with one important difference. In a conventional heat recovery chiller, waste heat is only available when there is a building chilled water (or conditioning) load. In a groundwater system, a heat source (the groundwater) is available year-round. To take advantage of this source during the heating season the chiller must be loaded in response to the heating load instead of the chilled water load. That is, the control must include a heating-dominant mode and a cooling-dominant mode. Two general designs are available for this:

1. Chiller capacity remains controlled by chilled water (supply or return) temperature, and groundwater flow through the chilled water exchanger is varied in response to heating load, or

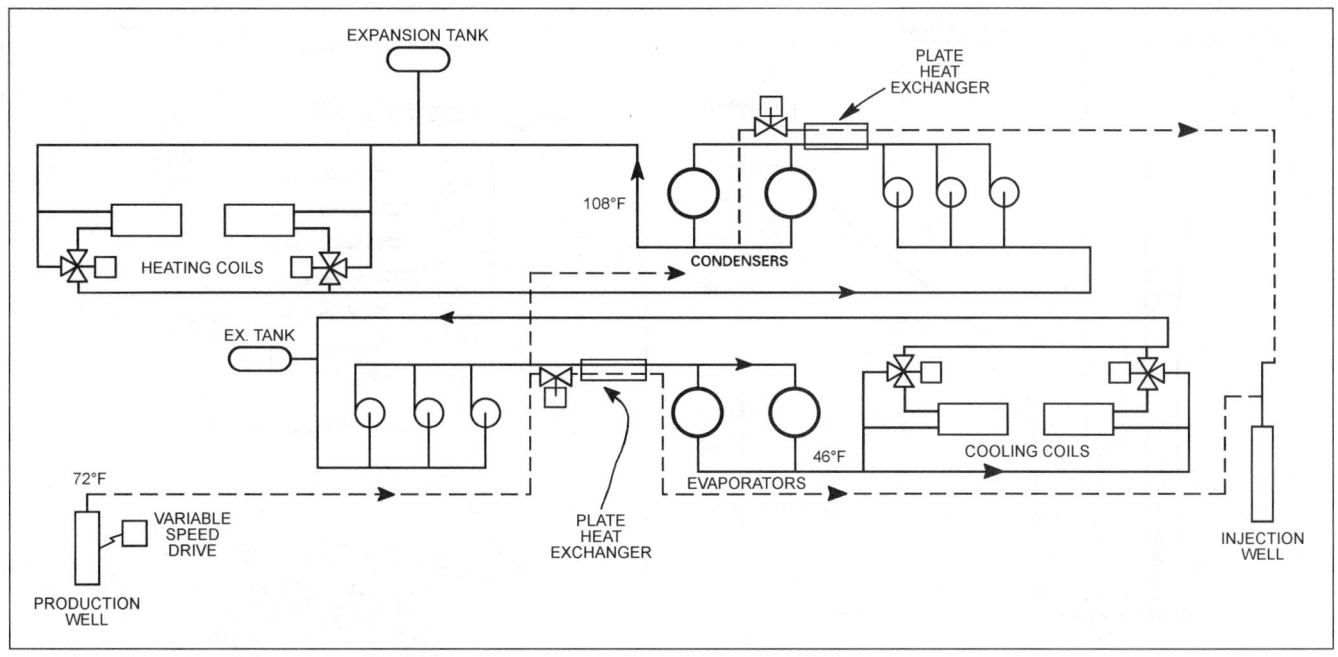

Fig. 25   Central Plant Groundwater Systems

2. Chiller capacity is controlled by the heating water (condenser) loop temperature, and the groundwater flow through the chilled water exchanger is controlled by chilled water temperature.

For buildings with a significant heating load, the former may be more attractive, while the latter may be appropriate for conventional building in moderate-to-warm climates.

SURFACE WATER HEAT PUMPS

Surface water bodies can be very good heat sources and sinks if properly used. In some cases, lakes can be the very best water supply for cooling. A variety of water circulation designs are possible and several of the more common are presented.

In a **closed-loop system**, a water-to-air heat pump is linked to a submerged coil. Heat is exchanged to (cooling mode) or from (heating mode) the lake by the fluid (usually a water-antifreeze mixture) circulating inside the coil. The heat pump transfers heat to or from the air in the building.

In an **open-loop system**, water is pumped from the lake through a heat exchanger and returned to the lake some distance from the point at which it was removed. The pump can be located either slightly above or submerged below the lake water level. For heat pump operation in the heating mode, this type is restricted to warmer climates; water temperature must remain above at least 42°F.

Thermal stratification of water often keeps large quantities of cold water undisturbed near the bottom of deep lakes. This water is cold enough to adequately cool buildings by simply being circulated through heat exchangers. A heat pump is not needed for cooling, and energy use is substantially reduced. Closed-loop coils may also be used in colder lakes. Heating can be provided by a separate source or with heat pumps in the heating mode. Precooling or supplemental total cooling are also permitted when water temperature is between 50 and 60°F.

Heat Transfer in Lakes

Heat is transferred to lakes by three primary modes: radiant energy from the sun, convective heat transfer from the surrounding air (when the air temperature is greater than the water temperature), and conduction from the ground. Solar radiation, which can exceed 300 Btu/h per square foot of lake area, is the dominant heating mechanism, but it occurs primarily in the upper portion of the lake unless the lake is very clear. About 40% of the solar radiation is absorbed at the surface (Pezent and Kavanaugh 1990). Approximately 93% of the remaining energy is absorbed at depths visible to the human eye.

Convection transfers heat to the lake when the lake surface temperature is lower than the air temperature. Wind speed increases the rate at which heat is transferred to the lake, but maximum heat gain by convection is usually only 10 to 20% of maximum solar heat gain. The conduction gain from the ground is even less than convection gain (Pezent and Kavanaugh 1990).

Cooling of lakes is accomplished primarily by evaporative heat transfer at the surface. Convective cooling or heating in warmer months will contribute only a small percentage of the total because of the relatively small temperature difference between the air and the lake surface temperature. Back radiation typically occurs at night when the sky is clear, and can account for significant amount of cooling. The relatively warm water surface will radiate heat to the cooler sky. For example, on a clear night, a cooling rate of up to 50 Btu/h·ft² from a lake 25°F warmer than the sky. The last major mode of heat transfer, conduction to the ground, does not play a major role in lake cooling (Pezent and Kavanaugh 1990).

To put these heat transfer rates in perspective, consider a 1-acre (43,560 ft²) lake that is used in connection with a 10-ton (120,000 Btu/h) heat pump. In the cooling mode, the unit will reject approximately 150,000 Btu/h to the lake. This is 3.4 Btu/h·ft², or approximately 1% of the maximum heat gain from solar radiation in the summer. In the winter, a 10-ton heat pump would absorb only about 90,000 Btu/h, or 2.1 Btu/h·ft², from the lake.

Thermal Patterns in Lakes

The maximum density of water occurs at 39.2°F, not at the freezing point of 32°F. This phenomenon, in combination with the normal modes of heat transfer to and from lakes, produces temperature profiles advantageous to efficient heat pump operation. In the winter, the coldest water is at the surface. It tends to remain at the surface and freeze. The bottom of a deep lake stays 5 to 10°F warmer than the surface. This condition is referred to as winter stagnation. The warmer water is a better heat source than the colder water at the surface.

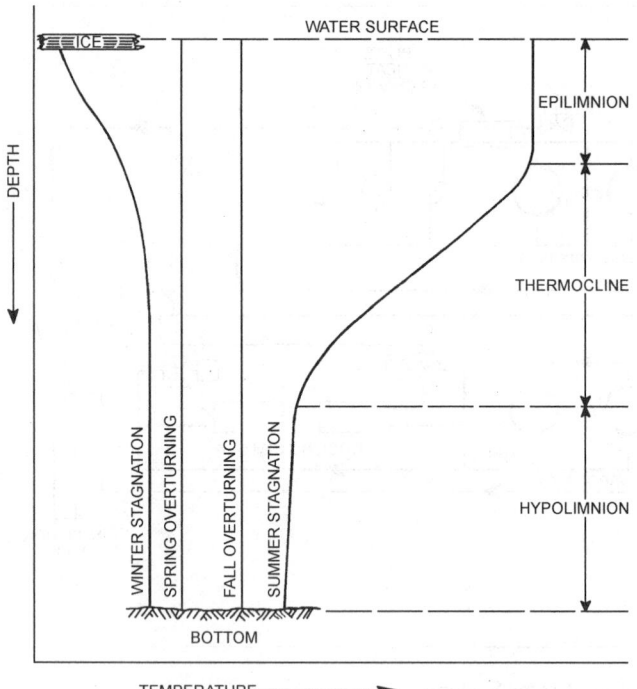

**Fig. 26 Idealized Diagram of Annual Cycle of
Thermal Stratification in Lakes**

As spring approaches, surface water warms until the temperature approaches the maximum density point of 39.2°F. The winter stratification becomes unstable and circulation loops begin to develop from top to bottom. This condition of spring overturn (Peirce 1964) causes the lake temperature to become fairly uniform.

Later in the spring as the water temperatures rise above 45°F, the circulation loops are in the upper portion of the lake. This pattern continues throughout the summer. The upper portion of the lake remains relatively warm, with evaporation cooling the lake and solar radiation warming it. The lower portion (hypolimnion) of the lake remains cold because most radiation is absorbed in the upper zone. Circulation loops do not penetrate to the lower zone and conduction to the ground is quite small. The result is that in deeper lakes with small or medium inflows, the upper zone is 70 to 90°F, the lower zone is 40 to 55°F, and the intermediate zone (thermocline) has a sharp change in temperature within a small change in depth. This condition is referred to as summer stagnation.

As fall begins, the water surface begins to cool by radiation and evaporation. With the approach of winter, the upper portion begins to cool towards the freezing point and the lower levels approach the maximum density temperature of 39.2°F. An ideal temperature versus depth chart is shown in Figure 26 for each of the four seasons (Peirce 1964).

Many lakes do exhibit near-ideal temperature profiles. However, a variety of circumstances can disrupt the profile. These characteristics include (1) high inflow/outflow rates, (2) insufficient depth for stratification, (3) level fluctuation, (4) wind, and (5) lack of enough cold weather to establish sufficient amounts of cold water necessary for summer stratification. Therefore, a thermal survey of the lake should be conducted or existing surveys of similar lakes in similar geographic locations should be consulted.

Closed-Loop Lake Water Heat Pump

The closed-loop lake water heat pump shown in Figure 17 has several advantages over the open-loop. One advantage is the reduced fouling resulting from the circulation of clean water (or

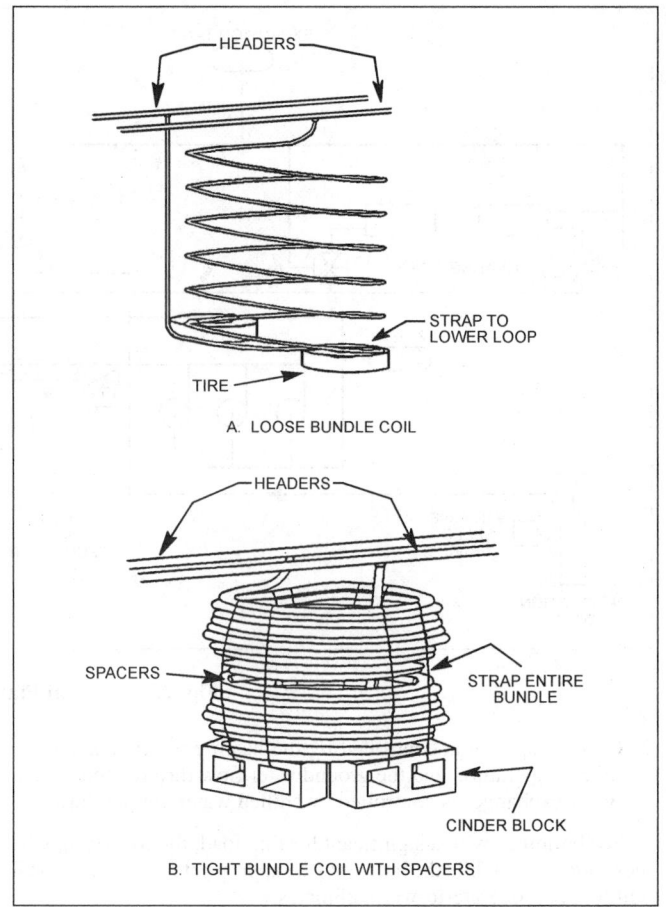

Fig. 27 Closed Loop Lake Coil in Bundles
(Kavanaugh 1991)

water-antifreeze solution) through the heat pump. A second advantage is the reduced pumping power requirement. This results from the absence of an elevation head from the lake surface to the heat pumps. A third advantage of a closed-loop is that it is the only type recommended if a lake temperature below 40°F is possible. The outlet temperature of the fluid will be about 6°F below that of the inlet at a flow of 3 gpm per ton. Frosting will occur on the heat exchanger surfaces when the bulk water temperature is in the 34 to 38°F range.

A closed-loop system has several disadvantages. Performance of the heat pump lowers slightly because the circulation fluid temperature drops 4 to 12°F below the lake temperature. A second disadvantage is the possibility of damage to coils located in public lakes. Thermally fused polyethylene loops are much more resistant to damage than copper, glued plastic (PVC), or tubing with bandclamped joints. The third possible disadvantage is fouling on the outside of the lake coil—particularly in murky lakes or where coils are located on or near the lake bottom.

Polyethylene (PE 3408) is recommended for all intake piping. All connections must be either thermally socket fused or butt fused. These plastic pipes should also have protection from UV radiation, especially when near the surface. Polyvinyl chloride (PVC) pipe and plastic pipe with band-clamped joints is not recommended.

The piping networks of closed-loop systems resemble those used in ground-coupled heat pump systems. Both a large-diameter header between the heat pump and lake coil and several parallel loops of piping in the lake are required. The loops are spread out to limit thermal interference, hot spots, and cold pockets. While this

Table 9   Suitability of Selected GCHP Antifreeze Solutions

Category	Methanol	Ethanol	Propylene Glycol	Potassium Acetate	CMA	Urea
Life cycle cost	***	***	**1	**1	**1	***
Corrosion	**2	**3	***	**	**4	*5
Leakage	***	**6	**6	*7	*8	*9
Health hazard risk	*10,11	**10,12	***10	***10	***10	***10
Fire risk	*13	*13	***14	***	***	***
Environmental risk	**15	**15	***	**15	**15	***
Risk of future use	*16	**17	***	**18	**19	**19

Key: *   Potential problems, caution in use required
　　　　**   Minor potential for problems
　　　　*** Little or no potential for problems

Category	Notes
Life cycle cost	1. Higher than average installation and energy costs.
Corrosion	2. High black iron and cast iron corrosion rates.
	3. High black iron and cast iron, copper and copper alloy corrosion rates.
	4. Medium black iron, copper and copper alloy corrosion rates.
	5. Medium black iron, high cast iron, and extremely high copper and copper alloy corrosion rates.
Leakage	6. Minor leakage observed.
	7. Moderate leakage observed. Extensive leakage reported in installed systems.
	8. Moderate leakage observed.
	9. Massive leakage observed.
Health risk	10. Protective measures required with use. See MSDS.
	11. Prolonged exposure can cause headaches, nausea, vomiting, dizziness, blindness, liver damage, and death. Use of proper equipment and procedures reduces risk significantly.
	12. Confirmed human carcinogen.
Fire Risk	13. Pure fluid only. Little risk when diluted with water in antifreeze.
	14. Very minor potential for pure fluid fire at elevated temperatures.
Environmental risk	15. Water pollution risk.
Risk of future use	16. Toxicity and fire concerns. Prohibited in some locations.
	17. Toxicity, fire and environmental concerns.
	18. Potential leakage concerns.
	19. Not currently used as GSHP antifreeze solution. May be difficult to obtain approval for use.

Source: Heinonen and Tapscott (1996)

layout is preferred in terms of performance, installation is more time consuming. Many contractors simply unbind plastic pipe coils and submerged them in a loose bundle. Some compensation for thermal interference is obtained by making the bundled coils longer than the spread coils. A diagram of this type of installation is shown in Figure 27.

Copper coils have also been used successfully. Copper tubes have a very high thermal conductivity, so coils only one-fourth to one-third the length of plastic coils are required. However, copper pipe does not have the durability of PE 3408 or polybutylene, and if the possibility of fouling exists, coils must be significantly longer.

Antifreeze Requirements

Closed loop horizontal and surface water heat exchanger systems will often require an antifreeze be added to the circulating water in locations with significant heating seasons. Antifreeze may not be needed in a comparable vertical borehole heat exchanger since the deep ground temperature will be essentially constant. At a depth of 6 ft, a typical value for horizontal heat exchangers, the ground temperature varies by approximately ±10°F. Even if the mean ground temperature were 60°F in late winter, the natural ground temperature would drop to 50°F. The heat extraction process would lower the temperature even further around the heat exchanger pipes, probably by an additional 10°F or more. Even with good heat transfer to the circulating water, the entering water temperature (leaving the ground heat exchanger) would be around 40°F. Lakes which freeze at the surface in the winter approach 39°F at the bottom, yielding nearly the same margin of safety against freezing of the circulating fluid. An additional 10°F temperature difference is usually needed in the heat pump's refrigerant-to-water heat exchanger to transfer the heat to the refrigerant. Having a refrigerant-to-water coil surface temperature below the freezing point of water risks the possibility of growing a layer of ice on the water side of the heat exchanger. In the best case, icing of the coil would restrict and may eventually block the flow of water and cause a shutdown. In the worst case, the ice could burst the tubing in the coil and require a major service expense.

Several factors must be considered when selecting an antifreeze for a ground loop heat exchanger. The most important considerations are: (1) impact on system life cycle cost, (2) corrosivity, (3) leakage, (4) health risks, (5) fire risks, (6) environmental risks from spills or disposal, and (7) risk of future use (will the antifreeze be acceptable over the life of the system). A study by Heinonen et al. (1997) of six antifreezes against these seven criteria is summarized in Table 9. No single material satisfies all criteria. Methanol and ethanol have good viscosity characteristics at low temperatures, yielding lower than average pumping power requirements. However, they both pose a significant fire hazard when in concentrated forms. Methanol is also toxic, eliminating it from consideration in areas that require non-toxic antifreeze to be used. Propylene glycol had no major concerns, with only leakage and pumping power requirements prompting minor concerns. Potassium acetate, calcium magnesium acetate (CMA), and urea have favorable environmental and safety performance; but they are all subject to significant leakage problems, which has limited their use in the past.

REFERENCES

Anderson, K.E. 1984. *Water well handbook*, Missouri Water Well and Pump Contractors Association, Belle, MD.

Austin, J.C. 1978. A low temperature geothermal space heating demonstration project. *Geothermal Resources Council Transactions* 2(2).

Bullard, E. 1973. Basic theories (Geothermal energy; Review of research and development). UNESCO, Paris.

Caneta Research. 1995. Commercial/institutional ground-source heat pump engineering manual. ASHRAE, Atlanta.

CSA. 1993. Design and construction of earth-energy heat pump systems for commercial and institutional buildings. *Standard* C447-93. Canadian Standards Association, Rexdale, ON.

Campbell, M.D. and J.H. Lehr. 1973. *Water well technology*. McGraw-Hill, New York.

Carslaw, H.S. and J.C. Jaeger. 1947. *Heat conduction in solids*. Claremore Press, Oxford.

Chandler, R.V. 1987. Alabama streams, lakes, springs and ground waters for use in heating and cooling. *Bulletin* 129. Geological Survey of Alabama, Tuscaloosa, AL.

Christen, J.E. 1977. *Central cooling—Absorption chillers*. Oak Ridge National Laboratories, Oak Ridge, TN.

Combs, J., J.K. Applegate, R.O. Fournier, C.A. Swanberg, and D. Nielson. 1980. Exploration, confirmation and evaluation of the resource. In *Special Report No. 7*, Direct utilization of geothermal energy: Technical handbook. Geothermal Resources Council.

Cosner, S.R. and J.A. Apps. 1978. A compilation of data on fluids from geothermal resources in the United States. DOE *Report* LBL-5936. Lawrence Berkeley Laboratory, Berkeley, CA.

Culver, G.G. and G.M. Reistad. 1978. Evaluation and design of downhole heat exchangers for direct applications. DOE *Report* No. RLO-2429-7.

Di Pippo, R. 1988. Industrial developments in geothermal power production. *Geothermal Resources Council Bulletin* 17(5).

Efrid, K.D. and G.E. Moeller. 1978 Electrochemical characteristics of 304 and 316 stainless steels in fresh water as functions of chloride concentration and temperature. *Paper 87*, Corrosion/78, Houston, TX.

EPRI. 1989. Soil and rock classification for the design of ground-coupled heat pump systems. International Ground Source Heat Pump Association, Stillwater, OK. Electric Power Research Institute, National Rural Electric Cooperative Association, Oklahoma State University.

Ellis, P. 1989. Materials selection guidelines. *Geothermal Direct Use Engineering and Design Guidebook* Ch. 8. Oregon Institute of Technology, Geo-Heat Center, Klamath Falls, OR.

Ellis, P. and C. Smith. 1983. Addendum to material selection guidelines for geothermal energy utilization systems. Radian Corporation, Austin, TX.

Ellis, P.F. and M.F. Conover. 1981. Material selection guidelines for geothermal energy utilization systems. DOE *Report* RA/27026-1, Radian Corporation, Austin, TX.

EPA. 1975. Manual of water well construction practices. EPA-570/9-75-001. U.S. Environmental Protection Agency, Washington, D.C.

Eskilson, P. 1987. *Thermal analysis of heat extraction boreholes*. University of Lund, Sweden.

Gudmundsson, J.S. 1985. Direct uses of geothermal energy in 1984. Geothermal Resources Council Proceedings, 1985 International Symposium on Geothermal Energy, International Volume, Davis, CA.

Hackett, G. and J. H. Lehr. 1985. *Iron bacteria occurrence problems and control methods in water wells*. National Water Well Association, Worthington, OH.

Heinonen, E.W. And R.E. Tapscott. 1996. Assessment of anti-freeze solutions for ground-source heat pump systems. New Mexico Engineering Research Institute for *ASHRAE RP-863*. ASHRAE.

Heinonen, E.W., R.E. Tapscott, M.W. Wildin, and A.N. Beall. 1997. Assessment of anti-freeze solutions for ground-source heat pump systems. *ASHRAE Research Report 90BRP.*

Ingersoll, L.R. and A.C. Zobel. 1954. *Heat conduction with engineering and geological application*, 2nd ed. McGraw-Hill, New York.

Interagency Geothermal Coordinating Council. Geothermal energy, research, development and demonstration program. DOE *Report* RA-0050, IGCC-5. U.S. Department of Energy, Washington, D.C.

Kavanaugh, S.P. 1985. Simulation and experimental verification of a vertical ground-coupled heat pump system. Ph.D. thesis. Oklahoma State University, Stillwater, OK.

Kavanaugh, S.P. 1991. *Ground and water source heat pumps*. Oklahoma State University, Stillwater, OK.

Kavanaugh, S.P. 1992. Ground-coupled heat pumps for commercial building. *ASHRAE Journal* 34(9):30-37.

Kavanaugh, S.P. and M.C. Pezent. 1990. Lake water applications of water-to-air heat pumps. *ASHRAE Transactions* 96(1):813-20.

Kavanaugh, S.P. and K. Rafferty. 1997. Ground-source heat pumps—Design of geothermal systems for commercial and institutional buildings. ASHRAE, Atlanta.

Kindle, C.H. and E.M. Woodruff. 1981. Techniques for geothermal liquid sampling and analysis. Battelle Pacific Northwest Laboratory, Richland, WA.

Lienau, P.J. 1979. Materials performance study of the OIT geothermal heating system. Geo-Heat Utilization Center *Quarterly Bulletin*, Oregon Institute of Technology, Klamath Falls, OR.

Lienau, P.J., G.G. Culver and J.W. Lund. 1988. Geothermal direct use developments in the United States. Oregon Institute of Technology, Geo-Heat Center, Klamath Falls, OR.

Lund, J.W., P.J. Lienau, G.G. Culver and C.H. Higbee, C.V. 1976. Klamath Falls geothermal heating district. *Geothermal Resources Council Transactions* 3.

Lunis, B. 1989. Environmental considerations. *Geothermal direct use engineering and design guidebook*, Ch. 20. Oregon Institute of Technology, Geo-Heat Center, Klamath Falls, OR.

Mitchell, D.A. 1980. Performance of typical HVAC materials in two geothermal heating systems. *ASHRAE Transactions* 86(1):763-68.

Muffler, L.J.P., ed. 1979. Assessment of geothermal Resources of the United States—1978. U.S. Geological Survey *Circular* No. 790.

Nichols, C.R. 1978. Direct utilization of geothermal energy: DOE's resource assessment program. Direct Utilization of Geothermal Energy: A Symposium. Geothermal Resources Council.

OSU. 1988a. *Closed-loop/ground-source heat pump systems installation guide*. International Ground Source Heat Pump Association, Oklahoma State University, Stillwater, OK.

OSU. 1988b. *Closed loop ground source heat pump systems*. Oklahoma State University, Stillwater, OK.

Peirce, L.B. 1964. Reservoir temperatures in north central alabama. Geological Survey of Alabama *Bulletin* 8. Tuscaloosa, AL.

Pezent, M.C. and S.P. Kavanaugh. 1990. Development and verification of a thermal model of lakes used with water-source heat pumps. *ASHRAE Transactions* 96(1).

Rafferty, K. 1989a. A materials and equipment review of selected U.S. geothermal district heating systems. Oregon Institute of Technology, Geo-Heat Center, Klamath Falls, OR.

Rafferty, K. 1989b. Absorption refrigeration. *Geothermal direct use engineering and design guidebook*, Ch. 14. Oregon Institute of Technology, Geo-Heat Center, Klamath Falls, OR.

Reistad, G.M., G.G. Culver, and M. Fukuda. 1979. Downhole heat exchangers for geothermal systems: Performance, economics and applicability. *ASHRAE Transactions* 85(1):929-39.

Roscoe Moss Company. 1985. *The engineers manual for water well design*. Roscoe Moss Company, Los Angeles, CA.

Stiger, S., J. Renner, and G. Culver. 1989. Well testing and reservoir evaluation. *Geothermal and direct use engineering and design guidebook*, Ch. 7. Oregon Institute of Technology, Geo-Heat Center, Klamath Falls, OR.

Svec, O. J. 1990. Spiral ground heat exchangers for heat pump applications. Proceedings of 3rd IEA Heat Pump Conference. Pergamon Press, Tokyo.

UOP. 1975. *Ground water and wells*. Johnson Division, UOP Inc., St. Paul, MN.

BIBLIOGRAPHY

Allen, E. 1980. Preliminary inventory of western U.S. cities with proximate hydrothermal potential. Eliot Allen and Associates, Salem, OR.

Anderson, D.A. and J.W. Lund, eds. 1980. Direct utilization of geothermal energy: Technical handbook. Geothermal Resources Council *Special Report* No. 7.

Caneta Research. 1995. Operating experiences with commercial ground-source heat pumps. *ASHRAE Research Project 863*.

SOLAR ENERGY USE

THE major obstacles encountered in solar heating and cooling are economic—the equipment needed to collect and store solar energy is high in cost. In some cases, the cost of the solar equipment is greater than the resulting savings in fuel costs. Some of the problems inherent in the nature of solar radiation include:

- It is relatively low in intensity, rarely exceeding 300 Btu/h·ft^2. Consequently, when large amounts of energy are needed, large collectors must be used.

- It is intermittent because of the variation in solar radiation intensity from zero at sunrise to a maximum at noon and back to zero at sunset. Some means of energy storage must be provided at night and during periods of low solar radiation.

- It is subject to unpredictable interruptions because of clouds, rain, snow, hail, or dust.

Systems should make maximum use of the solar energy input by effectively using the energy at the lowest temperatures possible.

QUALITY AND QUANTITY OF SOLAR ENERGY

Solar Constant

Solar energy approaches the earth as electromagnetic radiation, with wavelengths ranging from 0.1 μm (X rays) to 100 m (radio waves). The earth maintains a thermal equilibrium between the annual input of shortwave radiation (0.3 to 2.0 μm) from the sun and the outward flux of longwave radiation (3.0 to 30 μm). Only a limited band need be considered in terrestrial applications, because 99% of the sun's radiant energy has wavelengths between 0.28 and 4.96 μm. The current value of the solar constant (which is defined as the intensity of solar radiation on a surface normal to the sun's rays, just beyond the earth's atmosphere at the average earth-sun distance) is 433 Btu/h·ft^2. The section on Determining Incident Solar Flux in Chapter 29 of the 1997 *ASHRAE Handbook—Fundamentals* has further information on this topic.

Solar Angles

The axis about which the earth rotates is tilted at an angle of 23.45° to the plane of the earth's orbital plane and the sun's equator. The earth's tilted axis results in a day-by-day variation of the angle between the earth-sun line and the earth's equatorial plane, called the **solar declination** δ. This angle varies with the date, as shown in Table 1 for the year 1964 and in Table 2 for 1977. For other dates, the declination may be estimated by the following equation:

$$\delta = 23.45 \sin[360°(284 + N)/365] \quad (1)$$

The preparation of this chapter is assigned to TC 6.7, Solar Energy Utilization.

where N = year day, with January 1 = 1. For values of N, see Tables 1 and 2.

The relationship between δ and the date varies to an insignificant degree. The daily change in the declination is the primary reason for the changing seasons, with their variation in the distribution of solar radiation over the earth's surface and the varying number of hours of daylight and darkness.

The earth's rotation causes the sun's apparent motion (Figure 1). The position of the sun can be defined in terms of its altitude β above the horizon (angle HOQ) and its azimuth ϕ, measured as angle HOS in the horizontal plane.

At solar noon, the sun is exactly on the meridian, which contains the south-north line. Consequently, the solar azimuth ϕ is 0°. The **noon altitude** $β_N$ is given by the following equation as

$$β_N = 90° - \text{LAT} + δ \quad (2)$$

where LAT = latitude.

Because the earth's daily rotation and its annual orbit around the sun are regular and predictable, the solar altitude and azimuth may be readily calculated for any desired time of day when the latitude, longitude, and date (declination) are specified. Apparent solar time (AST) must be used, expressed in terms of the hour angle H, where

$$\begin{aligned} H &= (\text{number of hours from solar noon}) \times 15° \\ &= (\text{number of minutes from solar noon})/4 \end{aligned} \quad (3)$$

Solar Time

Apparent solar time (AST) generally differs from local standard time (LST) or daylight saving time (DST), and the difference can be significant, particularly when DST is in effect. Because the sun

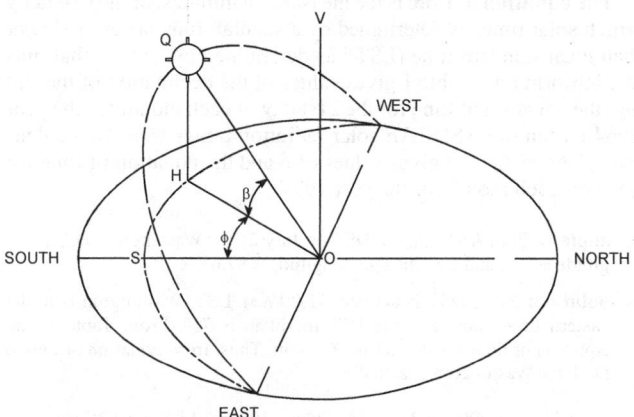

Fig. 1 Apparent Daily Path of the Sun Showing Solar Altitude (β) and Solar Azimuth (ϕ)

**Table 1   Date, Declination, and Equation of Time for the 21st Day of Each Month of 1964, with Data (A, B, C)
Used to Calculate Direct Normal Radiation Intensity at the Earth's Surface**

	Jan	Feb	Mar	Apr	May	June	July	Aug	Sept	Oct	Nov	Dec
Year Day	21	52	80	111	141	172	202	233	264	294	325	355
Declination δ, degrees	−19.9	−10.6	0.0	+11.9	+20.3	+23.45	+20.5	+12.1	0.0	−10.7	−19.9	−23.45
Equation of time, minutes	−11.2	−13.9	−7.5	+1.1	+3.3	−1.4	−6.2	−2.4	+7.5	+15.4	+13.8	+1.6
Solar noon		late			early			late			early	
A, Btu/h·ft^2	390	385	376	360	350	345	344	351	365	378	387	391
B, dimensionless	0.142	0.144	0.156	0.180	0.196	0.205	0.207	0.201	0.177	0.160	0.149	0.142
C, dimensionless	0.058	0.060	0.071	0.097	0.121	0.134	0.136	0.122	0.092	0.073	0.063	0.057

A = apparent solar irradiation at air mass zero for each month.
B = atmospheric extinction coefficient.
C = ratio of the diffuse radiation on a horizontal surface to the direct normal irradiation.

Table 2   Solar Position Data for 1977

Date		Jan	Feb	Mar	Apr	May	June	July	Aug	Sept	Oct	Nov	Dec
1	Year Day	1	32	60	91	121	152	182	213	244	274	305	335
	Declination δ	−23.0	−17.0	−7.4	+4.7	+15.2	+22.1	+23.1	+17.9	+8.2	−3.3	−14.6	−21.9
	Eq of Time	−3.6	−13.7	−12.5	−4.0	+2.9	+2.4	−3.6	−6.2	+0.0	+10.2	+16.3	+11.0
6	Year Day	6	37	65	96	126	157	187	218	249	279	310	340
	Declination δ	−22.4	−15.5	−5.5	+6.6	+16.6	+22.7	+22.7	+16.6	+6.7	−5.3	−16.1	−22.5
	Eq of Time	−5.9	−14.2	−11.4	−2.5	+3.5	+1.6	−4.5	−5.8	+1.6	+11.8	+16.3	+9.0
11	Year Day	11	42	70	101	131	162	192	223	254	284	315	345
	Declination δ	−21.7	−13.9	−3.5	−8.5	+17.9	+23.1	+22.1	+15.2	+4.4	−7.2	−17.5	−23.0
	Eq of Time	−8.0	−14.4	−10.2	−1.1	+3.7	+0.6	−5.3	−5.1	+3.3	+13.1	+15.9	+6.8
16	Year Day	16	47	75	106	136	167	197	228	259	289	320	350
	Declination δ	−20.8	−12.2	−1.6	+10.3	+19.2	+23.3	+21.3	+13.6	+2.5	−8.7	−18.8	−23.3
	Eq of Time	−9.8	−14.2	−8.8	+0.1	+3.8	−0.4	−5.9	−4.3	+5.0	+14.3	+15.2	+4.4
21	Year Day	21	52	80	111	141	172	202	233	264	294	325	355
	Declination δ	−19.6	−10.4	+0.4	+12.0	+20.3	+23.4	+20.6	+12.0	+0.5	−10.8	−20.0	−23.4
	Eq of Time	−11.4	−13.8	−7.4	+1.2	+3.6	−1.5	−6.2	−3.1	+6.8	+15.3	+14.1	+2.0
26	Year Day	26	57	85	116	146	177	207	238	269	299	330	360
	Declination δ	−18.6	−8.6	+2.4	+13.6	+21.2	+23.3	+19.3	+10.3	−1.4	−12.6	−21.0	−23.4
	Eq of Time	−12.6	−13.1	−5.8	+2.2	+3.2	−2.6	−6.4	−1.8	+8.6	+15.9	+12.7	−0.5

Source: ASHRAE *Standard* 93-1986 (Reaffirmed 1991).

Notes: Units for declination are angular degrees; units for equation of time are minutes. Values of declination and equation of time vary slightly for specific dates in other years.

appears to move at the rate of 360° in 24 h, its apparent rate of motion is 4 min per degree of longitude. The AST can be determined from the following equation:

$$AST = LST + \text{Equation of Time} + (4\text{ min})(\text{LST Meridian} - \text{Local Longitude}) \quad (4)$$

The longitudes of the seven standard time meridians that affect North America are Atlantic ST, 60°; Eastern ST, 75°; Central ST, 90°; Mountain ST, 105°; Pacific ST, 120°; Yukon ST, 135°; and Alaska-Hawaii ST, 150°.

The **equation of time** is the measure, in minutes, of the extent by which solar time, as determined by a sundial, runs faster or slower than local standard time (LST), as determined by a clock that runs at a uniform rate. Table 1 gives values of the declination of the sun and the equation of time for the 21st day of each month for the year 1964 (when the ASHRAE solar radiation tables were first calculated), while Table 2 gives values of δ and the equation of time for six days each month for the year 1977.

Example 1. Find AST at noon DST on July 21 for Washington, D.C., longitude = 77°, and for Chicago, longitude = 87.6°.

Solution: Noon DST is actually 11:00 A.M. LST. Washington is in the eastern time zone, and the LST meridian is 75°. From Table 1, the equation of time for July 21 is −6.2 min. Thus, from Equation (4), noon DST for Washington is actually

$$AST = 11:00 - 6.2 + 4(75 - 77) = 10:45.8 \quad AST = 10.76 \text{ h}$$

Chicago is in the central time zone, and the LST meridian is 90°. Thus, from Equation (4), noon central DST is

$$AST = 11:00 - 6.2 + 4(90 - 87.6) = 11:03.4 \quad AST = 11.06 \text{ h}$$

The hour angles H, for these two examples (see Figure 2) are

for Washington, $H = (12.00 - 10.76)15° = 18.6°$ east
for Chicago, $H = (12.00 - 11.06)15° = 14.10°$ east

To find the solar altitude β and the azimuth φ when the hour angle H, the latitude LAT, and the declination δ are known, the following equations may be used:

$$\sin \beta = \cos (LAT) \cos \delta \cos H + \sin (LAT) \sin \delta \quad (5)$$

$$\sin \phi = \cos \delta \sin H / \cos \beta \quad (6)$$

or

$$\cos \phi = \frac{\sin \beta \sin(LAT) - \sin \delta}{\cos \beta \cos(LAT)} \quad (7)$$

Tables 15 through 21 in Chapter 29 of the 1997 ASHRAE *Handbook—Fundamentals* give values for latitudes from 16 to 64° north. For any other date or latitude, interpolation between the tabulated values will give sufficiently accurate results. More precise values, with azimuths measured from the north, are given in the U.S. Hydrographic Office Bulletin No. 214 (1958).

Incident Angle

The angle between the line normal to the irradiated surface (OP′ in Figure 2) and the earth-sun line OQ is called the incident angle θ. It is important in solar technology because it affects the intensity of the direct component of the solar radiation striking the surface and the ability of the surface to absorb, transmit, or reflect the sun's rays.

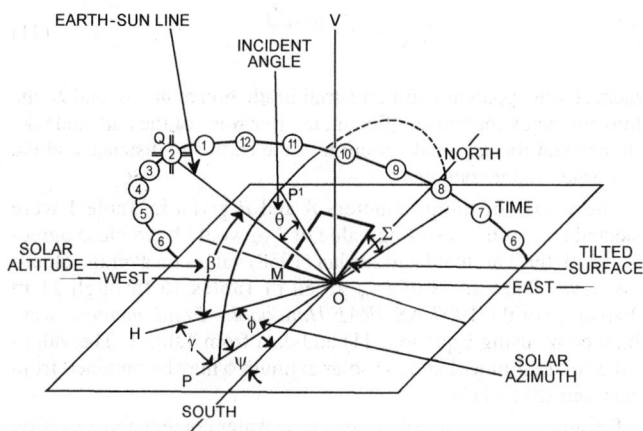

Fig. 2   Solar Angles with Respect to a Tilted Surface

To determine θ, the surface azimuth ψ and the surface-solar azimuth γ must be known. The surface azimuth (angle POS in Figure 2) is the angle between the south-north line SO and the normal PO to the intersection of the irradiated surface with the horizontal plane, shown as line OM. The surface-solar azimuth, angle HOP, is designated by γ and is the angular difference between the solar azimuth φ and the surface azimuth ψ. For surfaces facing *east* of south, γ = φ − ψ in the morning and γ = φ + ψ in the afternoon. For surfaces facing *west* of south, γ = φ + ψ in the morning and γ = φ − ψ in the afternoon. For south-facing surfaces, ψ = 0°, so γ = φ for all conditions. The angles δ, β, and φ are always positive.

For a surface with a tilt angle Σ (measured from the horizontal), the angle of incidence θ between the direct solar beam and the normal to the surface (angle QOP′ in Figure 2) is given by:

$$\cos \theta = \cos \beta \cos \gamma \sin \Sigma + \sin \beta \cos \Sigma \qquad (8)$$

For vertical surfaces, Σ = 90°, cos Σ = 0, and sin Σ = 1.0, so Equation (8) becomes

$$\cos \theta = \cos \beta \cos \gamma \qquad (9)$$

For horizontal surfaces, Σ = 0°, sin Σ = 0, and cos Σ = 1.0, so Equation (8) leads to

$$\theta_H = 90° - \beta \qquad (10)$$

Example 2. Find θ for a south-facing surface tilted upward 30° from the horizontal at 40° north latitude at 4:00 P.M., AST, on August 21.

Solution: From Equation (3), at 4:00 P.M. on August 21,

$$H = 4 \times 15° = 60°$$

From Table 1,

$$\delta = 12.1°$$

From Equation (5),

$$\sin \beta = \cos 40° \cos 12.1° \cos 60° + \sin 40° \sin 12.1°$$
$$\beta = 30.6°$$

From Equation (6),

$$\sin \phi = \cos 12.1° \sin 60° / \cos 30.6°$$
$$\phi = 79.7°$$

The surface faces south, so φ = γ. From Equation (8),

$$\cos \theta = \cos 30.6° \cos 79.7° \sin 30° + \sin 30.6° \cos 30°$$
$$\theta = 58.8°$$

ASHRAE *Standard* 93, Methods of Testing to Determine the Thermal Performance of Solar Collectors, provides tabulated values of q for horizontal and vertical surfaces and for south-facing surfaces tilted upward at angles equal to the latitude minus 10°, the latitude, the latitude plus 10°, and the latitude plus 20°. These tables cover the latitudes from 24° to 64° north, in 8° intervals.

Solar Spectrum

Beyond the earth's atmosphere, the effective black body temperature of the sun is 10,370°R. The maximum spectral intensity occurs at 0.48 μm in the green portion of the visible spectrum (Figure 3). Thekaekara (1973) presents tables and charts of the sun's extraterrestrial spectral irradiance from 0.120 to 100 μm, the range in which most of the sun's radiant energy is contained. The ultraviolet portion of the spectrum below 0.40 μm contains 8.73% of the total, another 38.15% is contained in the visible region between 0.40 and 0.70 μm, and the infrared region contains the remaining 53.12%.

Solar Radiation at the Earth's Surface

In passing through the earth's atmosphere, some of the sun's direct radiation I_D is scattered by nitrogen, oxygen, and other molecules, which are small compared to the wavelengths of the radiation; and by aerosols, water droplets, dust, and other particles with diameters comparable to the wavelengths (Gates 1966). This

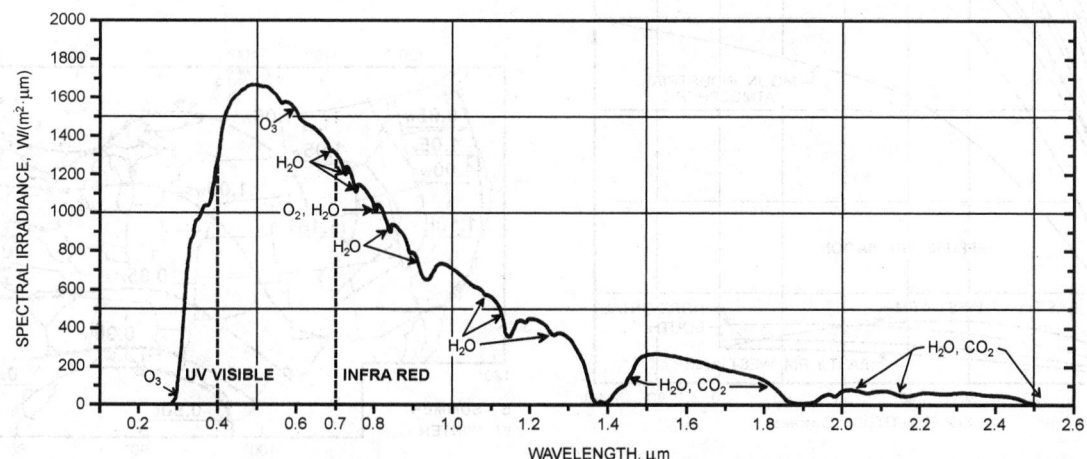

Fig. 3   Spectral Solar Irradiation at Sea Level for Air-Mass = 1.0

scattered radiation causes the sky to appear blue on clear days, and some of it reaches the earth as diffuse radiation I.

Attenuation of the solar rays is also caused by absorption, first by the ozone in the outer atmosphere, which causes a sharp cutoff at 0.29 μm of the ultraviolet radiation reaching the earth's surface. In the longer wavelengths, there are a series of absorption bands caused by water vapor, carbon dioxide, and ozone. The total amount of attenuation at any given location is determined by (1) the length of the atmospheric path through which the rays traverse and (2) the composition of the atmosphere. The path length is expressed in terms of the air mass m, which is the ratio of the mass of atmosphere in the actual earth-sun path to the mass that would exist if the sun were directly overhead at sea level ($m = 1.0$). For all practical purposes, at sea level, $m = 1.0/\sin \beta$. Beyond the earth's atmosphere, $m = 0$.

Prior to 1967, solar radiation data was based on an assumed solar constant of 419.7 Btu/h·ft² and on a standard sea level atmosphere containing the equivalent depth of 2.8 mm of ozone, 20 mm of precipitable moisture, and 300 dust particles per cubic centimeter. Threlkeld and Jordan (1958) considered the wide variation of water vapor in the atmosphere above the United States at any given time, and particularly the seasonal variation, which finds three times as much moisture in the atmosphere in midsummer as in December, January, and February. The basic atmosphere was assumed to be at sea level barometric pressure, with 2.5 mm of ozone, 200 dust particles per cm³, and an actual precipitable moisture content that varied throughout the year from 8 mm in midwinter to 28 mm in mid-July. Figure 4 shows the variation of the direct normal irradiation with solar altitude, as estimated for clear atmospheres and for an atmosphere with variable moisture content.

Stephenson (1967) showed that the intensity of the direct normal irradiation I_{DN} at the earth's surface on a clear day can be estimated by the following equation:

$$I_{DN} = Ae^{-B/\sin\beta} \tag{11}$$

where A, the apparent extraterrestrial irradiation at $m = 0$, and B, the atmospheric extinction coefficient, are functions of the date and take into account the seasonal variation of the earth-sun distance and the air's water vapor content.

The values of the parameters A and B given in Table 1 were selected so that the resulting value of I_{DN} would be in close agreement with the Threlkeld and Jordan (1958) values on average cloudless days. The values of I_{DN} given in Tables 15 through 21 in Chapter 29 of the 1997 AS*HRAE Handbook—Fundamentals*, were obtained by using Equation (11) and data from Table 1. The values of the solar altitude β and the solar azimuth ϕ may be obtained from Equations (5) and (6).

Because local values of atmospheric water content and elevation can vary markedly from the sea level average, the concept of **clearness number** was introduced to express the ratio between the actual clear-day direct irradiation intensity at a specific location and the intensity calculated for the standard atmosphere for the same location and date.

Figure 5 shows the Threlkeld-Jordan map of winter and summer clearness numbers for the continental United States. Irradiation values should be adjusted by the clearness numbers applicable to each particular location.

Design Values of Total Solar Irradiation

The total solar irradiation $I_{t\theta}$ of a terrestrial surface of any orientation and tilt with an incident angle θ is the sum of the direct component $I_{DN} \cos \theta$ plus the diffuse component $I_{d\theta}$ coming from the sky plus whatever amount of reflected shortwave radiation I_r may reach the surface from the earth or from adjacent surfaces:

$$I_{t\theta} = I_{DN} \cos \theta + I_{d\theta} + I_r \tag{12}$$

The diffuse component is difficult to estimate because of its non-directional nature and its wide variations. Figure 4 shows typical values of diffuse irradiation of horizontal and vertical surfaces. For clear days, Threlkeld (1963) has derived a dimensionless parameter (designated as C in Table 1), which depends on the dust and moisture content of the atmosphere and thus varies throughout the year:

$$C = I_{dH}/I_{DN} \tag{13}$$

where I_{dH} is the diffuse radiation falling on a horizontal surface under a cloudless sky.

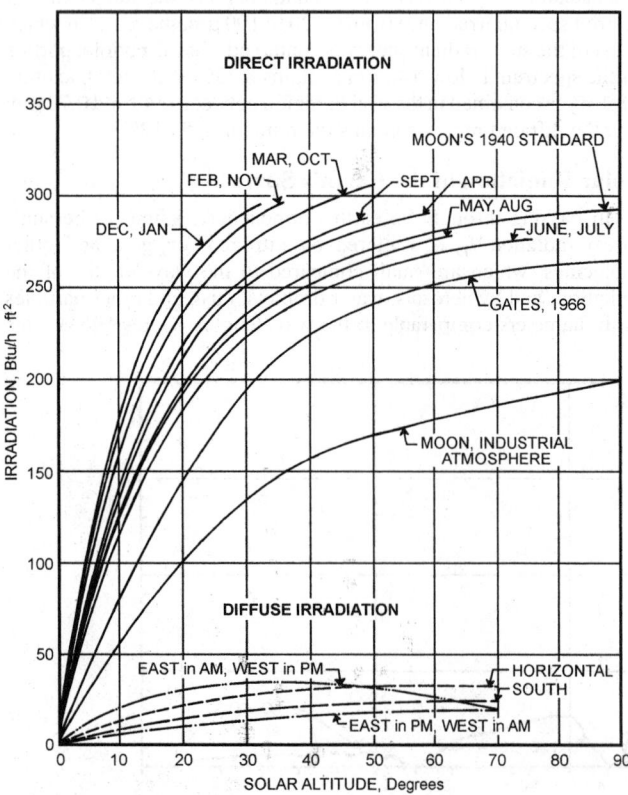

Fig. 4 Variation with Solar Altitude and Time of Year for Direct Normal Irradiation

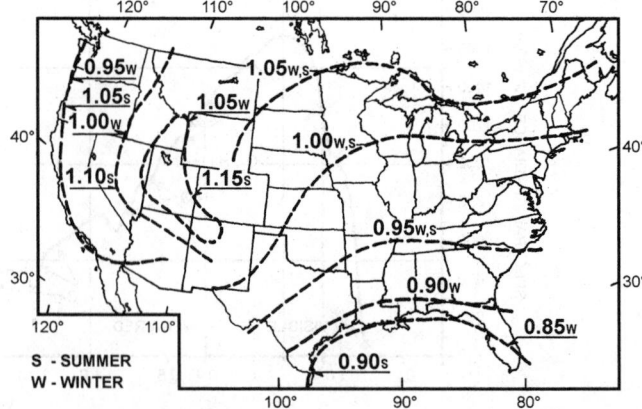

Fig. 5 Clearness Numbers for the United States

The following equation may be used to estimate the amount of diffuse radiation $I_{d\theta}$ that reaches a tilted or vertical surface:

$$I_{d\theta} = C\,I_{DN}F_{ss} \qquad (14)$$

where

$$F_{ss} = (1 + \cos \Sigma)/2 \qquad (15)$$
$$= \text{angle factor between the surface and the sky}$$

The reflected radiation I_r from the foreground is given by the following equation:

$$I_r = I_{tH}\,\rho_g F_{sg} \qquad (16)$$

where

ρ_g = reflectance of the foreground
I_{tH} = total horizontal irradiation

$$F_{sg} = (1 - \cos \Sigma)/2 \qquad (17)$$
$$= \text{angle factor between the surface and the earth}$$

The intensity of the reflected radiation that reaches any surface depends on the nature of the reflecting surface and on the incident angle between the sun's direct beam and the reflecting surface. Many measurements made of the reflection (albedo) of the earth under varying conditions show that clean, fresh snow has the highest reflectance (0.87) of any natural surface.

Threlkeld (1963) gives values of reflectance for commonly encountered surfaces at solar incident angles from 0 to 70°. Bituminous paving generally reflects less than 10% of the total incident solar irradiation; bituminous and gravel roofs reflect from 12 to 15%; concrete, depending on its age, reflects from 21 to 33%. Bright green grass reflects 20% at $\theta = 30°$ and 30% at $\theta = 65°$.

The maximum daily amount of solar irradiation that can be received at any given location is that which falls on a flat plate with its surface kept normal to the sun's rays so it receives both direct and diffuse radiation. For fixed flat-plate collectors, the total amount of clear day irradiation depends on the orientation and slope. As shown by Figure 6 for 40° north latitude, the total irradiation of horizontal surfaces reaches its maximum in midsummer, while vertical south-facing surfaces experience their maximum irradiation during the winter. These curves show the combined effects of the varying length of days and changing solar altitudes.

In general, flat-plate collectors are mounted at a fixed tilt angle Σ (above the horizontal) to give the optimum amount of irradiation for each purpose. Collectors intended for winter heating benefit from higher tilt angles than those used to operate cooling systems in summer. Solar water heaters, which should operate satisfactorily throughout the year, require an angle that is a compromise between the optimal values for summer and winter. Figure 6 shows the monthly variation of total day-long irradiation on the 21st day of each month at 40° north latitude for flat surfaces with various tilt angles.

Tables in ASHRAE *Standard* 93 give the total solar irradiation for the 21st day of each month at latitudes 24° to 64° north on surfaces with the following orientations: normal to the sun's rays (direct normal data *do not* include diffuse irradiation); horizontal; south-facing, tilted at (LAT−10), LAT, (LAT+10), (LAT+20), and 90° from the horizontal. The day-long total irradiation for fixed surfaces is highest for those that face south, but a deviation in azimuth of 15° to 20° causes only a small reduction.

Solar Energy for Flat-Plate Collectors

The preceding data apply to clear days. The irradiation for average days may be estimated for any specific location by referring

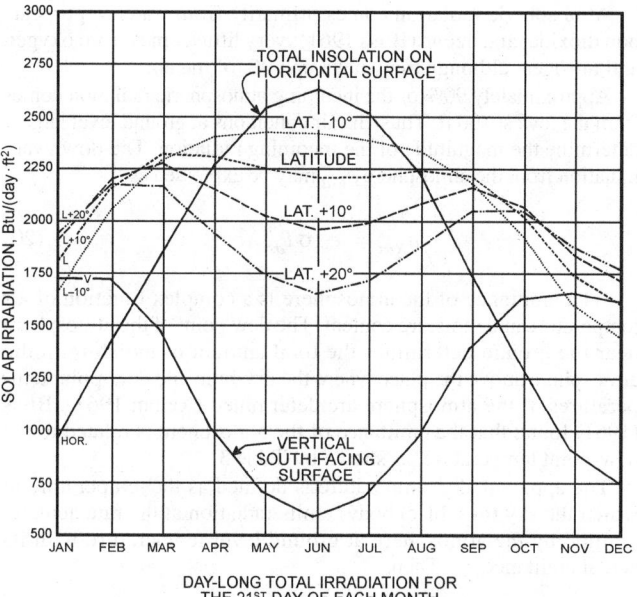

Fig. 6 Total Daily Irradiation for Horizontal, Tilted, and Vertical Surfaces at 40° North Latitude

to publications of the U.S. Weather Service. The *Climatic Atlas of the United States* (U.S. GPO 1968) gives maps of monthly and annual values of percentage of possible sunshine, total hours of sunshine, mean solar radiation, mean sky cover, wind speed, and wind direction.

The total daily horizontal irradiation data reported by the U.S. Weather Bureau for approximately 100 stations prior to 1964 show that the percentage of total clear-day irradiation is approximately a linear function of the percentage of possible sunshine. The irradiation is not zero for days when the percentage of possible sunshine is reported as zero, because substantial amounts of energy reach the earth in the form of diffuse radiation. Instead, the following relationship exists:

$$\frac{\text{Day-long actual } I_{tH}}{\text{Clear day } I_{tH}}\,100 = a + b \qquad (\% \text{ possible sunshine}) \quad (18)$$

where a and b are constants for any specified month at any given location. See also Jordan and Liu (1977) and Duffie and Beckman (1974).

Longwave Atmospheric Radiation

In addition to the shortwave (0.3 to 2.0 μm) radiation it receives from the sun, the earth receives longwave radiation (4 to 100 μm, with maximum intensity near 10 μm) from the atmosphere. In turn, a surface on the earth emits longwave radiation q_{Rs} in accordance with the Stefan-Boltzmann law:

$$q_{Rs} = e_s \sigma T_s^4 \qquad (19)$$

where

e_s = surface emittance
σ = Stefan-Boltzmann constant, 0.1713×10^{-8} Btu/h·ft²·°R⁴
T_s = absolute temperature of the surface, °R

For most nonmetallic surfaces, the longwave hemispheric emittance is high, ranging from 0.84 for glass and dry sand to 0.95 for black built-up roofing. For highly polished metals and certain selective surfaces, e_s may be as low as 0.05 to 0.20.

Atmospheric radiation comes primarily from water vapor, carbon dioxide, and ozone (Bliss 1961); very little comes from oxygen and nitrogen, although they make up 99% of the air.

Approximately 90% of the incoming atmospheric radiation comes from the lowest 300 ft. Thus, the air conditions at ground level largely determine the magnitude of the incoming radiation. The downward radiation from the atmosphere q_{Rat} may be expressed as

$$q_{Rat} = e_{at} \sigma T_{at}^4 \qquad (20)$$

The emittance of the atmosphere is a complex function of air temperature and moisture content. The dew point of the atmosphere near the ground determines the total amount of moisture in the atmosphere above the place where the dry-bulb and dew-point temperatures of the atmosphere are determined (Reitan 1963). Bliss (1961) found that the emittance of the atmosphere is related to the dew-point temperature, as shown by Table 3.

The apparent sky temperature is defined as the temperature at which the sky (as a blackbody) emits radiation at the rate actually emitted by the atmosphere at ground level temperature with its actual emittance e_{at}. Then,

$$\sigma T_{sky}^4 = e_{at} \sigma T_{at}^4 \qquad (21)$$

or

$$T_{sky}^4 = e_{at} T_{at}^4 \qquad (22)$$

Table 3 Sky Emittance and Amount of Precipitable Moisture Versus Dew-Point Temperature

Dew Point, °F	Sky Emittance, e_{at}	Precipitable Water, in.
−20	0.68	0.12
−10	0.71	0.16
0	0.73	0.18
10	0.76	0.22
20	0.77	0.29
30	0.79	0.41
40	0.82	0.57
50	0.84	0.81
60	0.86	1.14
70	0.88	1.61

Example 3. Consider a summer night condition when the ground level temperatures are 65°F dew point and 85°F dry bulb. From Table 3, e_{at} at 65°F dew point is 0.87, and the apparent sky temperature is

$$T_{sky} = 0.87^{0.25}(85 + 459.6) = 526.0°R$$

Thus, $T_{sky} = 526.0 − 459.6 = 66.4°F$, which is 18.6°F below the ground level dry-bulb temperature.

For a winter night in Arizona, when the temperatures at ground level are 60°F dry bulb and 25°F dew point, from Table 3, the emittance of the atmosphere is 0.78, and the apparent sky temperature is 488.3°R or 28.7°F.

A simple relationship, which ignores vapor pressure of the atmosphere, may also be used to estimate the apparent sky temperature:

$$T_{sky} = 0.0552 T_{at}^{1.5} \qquad (23)$$

where T is in degrees Rankine.

If the temperature of the radiating surface is assumed to equal the atmospheric temperature, the loss of heat from a black surface ($e_s = 1.0$) may be found from Figure 7.

Example 4. For the conditions in the previous example for summer, 85°F dry bulb and 65°F dew point, the rate of radiative heat loss is about 23

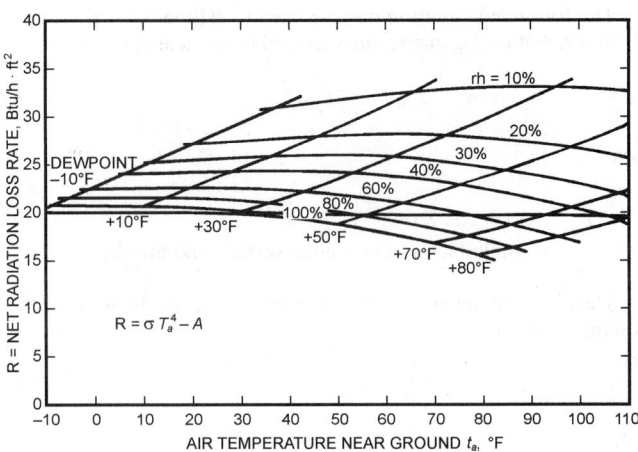

Fig. 7 Radiation Heat Loss to Sky from Horizontal Blackbody

Btu/h·ft². For winter, 60°F dry bulb and 25°F dew point, the heat loss is about 27 Btu/h·ft².

Where a rough, unpainted roof is used as a heat dissipater, the rate of heat loss rises rapidly as the surface temperature goes up. For the summer example, a painted metallic roof, $e_s = 0.96$, at 100°F (559.6°R) will have a heat loss rate of

$$q_{RAD} = 0.96 \times 0.1713 \times 10^{-8}[559.6^4 − 526.0^4]$$
$$= 35.4 \; Btu/h \cdot ft^2$$

This analysis shows that radiation alone is not an effective means of dissipating heat under summer conditions of high dew-point and high ambient temperature. In spring and fall, when both the dew-point and dry-bulb temperatures are relatively low, radiation becomes much more effective.

On overcast nights, when the cloud cover is low, the clouds act much like blackbodies at ground level temperature, and virtually no heat can be lost by radiation. The exchange of longwave radiation between the sky and terrestrial surfaces occurs in the daytime as well as at night, but the much greater magnitude of the solar irradiation masks the longwave effects.

SOLAR ENERGY COLLECTION

Solar energy can be converted to (1) chemical, (2) electrical, and (3) thermal processes. Photosynthesis is a chemical process that produces food and converts CO_2 to O_2. Photovoltaic cells convert solar energy to electricity. The section on Photovoltaic Applications discusses some of the applications for these devices. The thermal conversion process, the primary subject of this chapter, provides thermal energy for space heating and cooling, domestic water heating, power generation, distillation, and process heating.

Solar Heat Collection by Flat-Plate Collectors

The solar irradiation data presented in the foregoing sections may be used to estimate how much energy is likely to be available at any specific location, date, and time of day for collection by either a concentrating device, which uses only the direct rays of the sun, or by a flat-plate collector, which can use both direct and diffuse irradiation. The temperatures needed for space heating and cooling do not exceed 200°F, even for absorption refrigeration, and they can be attained with carefully designed flat-plate collectors. Depending on the load and ambient temperatures, single-effect absorption systems can use energizing temperatures of 110 to 230°F.

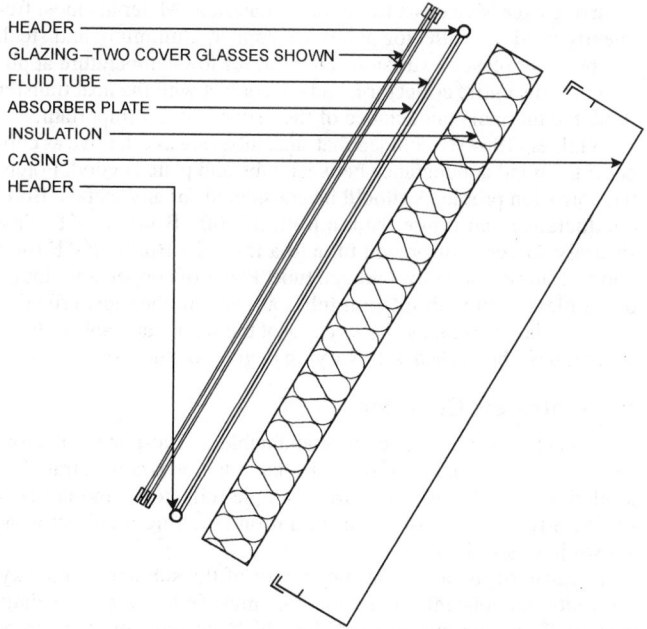

HEADER
GLAZING—TWO COVER GLASSES SHOWN
FLUID TUBE
ABSORBER PLATE
INSULATION
CASING
HEADER

Fig. 8   Exploded Cross Section Through Double-Glazed Solar Water Heater

A flat-plate collector generally consists of the following components (see Figure 8):

- **Glazing.** One or more sheets of glass or other diathermanous (radiation-transmitting) material.
- **Tubes**, **fins**, or **passages.** To conduct or direct the heat transfer fluid from the inlet to the outlet.
- **Absorber plates.** Flat, corrugated, or grooved plates, to which the tubes, fins, or passages are attached. The plate may be integral with the tubes.
- **Headers** or **manifolds.** To admit and discharge the fluid.

- **Insulation.** To minimize heat loss from the back and sides of the collector.
- **Container** or **casing.** To surround the aforementioned components and keep them free from dust, moisture, etc.

Flat-plate collectors have been built in a wide variety of designs from many different materials (Figure 9). They have been used to heat fluids such as water, water plus an antifreeze additive, or air. Their major purpose is to collect as much solar energy as possible at the lowest possible total cost. The collector should also have a long effective life, despite the adverse effects of the sun's ultraviolet radiation; corrosion or clogging because of acidity, alkalinity, or hardness of the heat transfer fluid; freezing or air-binding in the case of water, or deposition of dust or moisture in the case of air; and breakage of the glazing because of thermal expansion, hail, vandalism, or other causes. These problems can be minimized by the use of tempered glass.

Glazing Materials

Glass has been widely used to glaze flat plate solar collectors because it can transmit as much as 90% of the incoming shortwave solar irradiation while transmitting virtually none of the longwave radiation emitted outward by the absorber plate. Glass with low iron content has a relatively high transmittance for solar radiation (approximately 0.85 to 0.90 at normal incidence), but its transmittance is essentially zero for the longwave thermal radiation (5.0 to 50 μm) emitted by sun-heated surfaces.

Plastic films and sheets also possess high shortwave transmittance, but because most usable varieties also have transmission bands in the middle of the thermal radiation spectrum, they may have longwave transmittances as high as 0.40.

Plastics are also generally limited in the temperatures they can sustain without deteriorating or undergoing dimensional changes. Only a few finds of plastics can withstand the sun's ultraviolet radiation for long periods. However, they are not broken by hail and other stones and, in the form of thin films, are completely flexible and have low mass.

The glass generally used in solar collectors may be either single-strength (0.085 to 0.100 in. thick) or double-strength (0.115 to

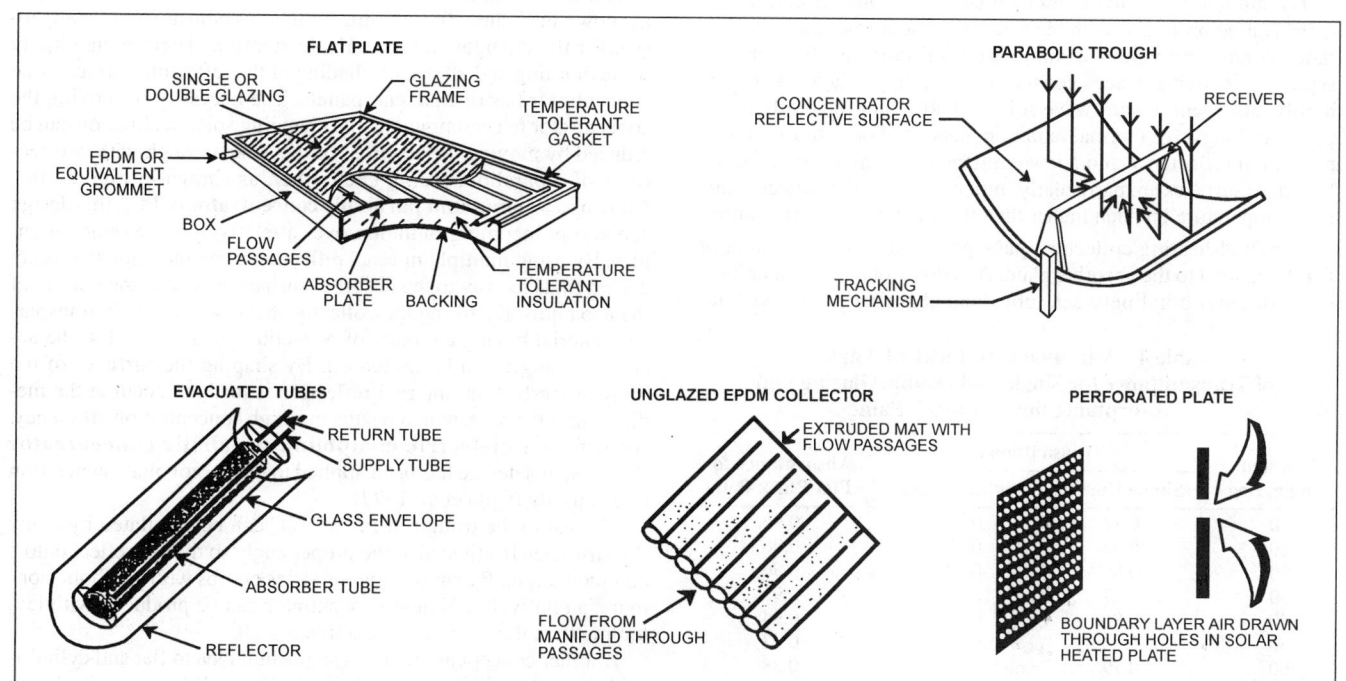

Fig. 9   Various Types of Solar Collectors

0.133 in. thick). The commercially available grades of window and greenhouse glass have normal incidence transmittances of about 0.87 and 0.85, respectively. For direct radiation, the transmittance varies markedly with the angle of incidence, as shown in Table 4, which gives transmittances for single- and double-glazing using double-strength clear window glass.

The 4% reflectance from each glass-air interface is the most important factor in reducing transmission, although a gain of about 3% in transmittance can be obtained by using water-white glass. Antireflective coatings and surface texture can also improve transmission significantly. The effect of dirt and dust on collector glazing may be quite small, and the cleansing effect of an occasional rainfall is usually adequate to maintain the transmittance within 2 to 4% of its maximum.

The glazing should admit as much solar irradiation as possible and reduce the upward loss of heat as much as possible. Although glass is virtually opaque to the longwave radiation emitted by collector plates, absorption of that radiation causes an increase in the glass temperature and a loss of heat to the surrounding atmosphere by radiation and convection. This type of heat loss can be reduced by using an infrared reflective coating on the underside of the glass; however, such coatings are expensive and reduce the effective solar transmittance of the glass by as much as 10%.

In addition to serving as a heat trap by admitting shortwave solar radiation and retaining longwave thermal radiation, the glazing also reduces heat loss by convection. The insulating effect of the glazing is enhanced by the use of several sheets of glass, or glass plus plastic. The loss from the back of the plate rarely exceeds 10% of the upward loss.

Collector Plates

The collector plate absorbs as much of the irradiation as possible through the glazing, while losing as little heat as possible upward to the atmosphere and downward through the back of the casing. The collector plates transfer the retained heat to the transport fluid. The absorptance of the collector surface for shortwave solar radiation depends on the nature and color of the coating and on the incident angle, as shown in Table 4 for a typical flat black paint.

By suitable electrolytic or chemical treatments, selective surfaces can be produced with high values of solar radiation absorptance α and low values of longwave emittance e_s. Essentially, typical selective surfaces consist of a thin upper layer, which is highly absorbent to shortwave solar radiation but relatively transparent to longwave thermal radiation, deposited on a substrate that has a high reflectance and a low emittance for longwave radiation. Selective surfaces are particularly important when the collector surface temperature is much higher than the ambient air temperature.

For fluid-heating collectors, passages must be integral with or firmly bonded to the absorber plate. A major problem is obtaining a good thermal bond between tubes and absorber plates without

Table 4   Variation with Incident Angle of Transmittance for Single and Double Glazing and Absorptance for Flat Black Paint

Incident Angle, Deg	Transmittance		Absorptance for Flat Black Paint
	Single Glazing	Double Glazing	
0	0.87	0.77	0.96
10	0.87	0.77	0.96
20	0.87	0.77	0.96
30	0.87	0.76	0.95
40	0.86	0.75	0.94
50	0.84	0.73	0.92
60	0.79	0.67	0.88
70	0.68	0.53	0.82
80	0.42	0.25	0.67
90	0.00	0.00	0.00

incurring excessive costs for labor or materials. Materials most frequently used for collector plates are copper, aluminum, and steel. UV-resistant plastic extrusions are used for low-temperature application. If the entire collector area is in contact with the heat transfer fluid, the thermal conductance of the material is not important.

Whillier (1964) concluded that steel tubes are as effective as copper if the bond conductance between tube and plate is good. Potential corrosion problems should be considered for any metals. Bond conductance can range from a high of 1000 Btu/h·ft²·°F for a securely soldered or brazed tube to a low of 3 Btu/h·ft²·°F for a poorly clamped or badly soldered tube. Plates of copper, aluminum, or stainless steel with integral tubes are among the most effective types available. Figure 9 shows a few of the solar water and air heaters that have been used with varying degrees of success.

Concentrating Collectors

Temperatures far above those attainable by flat-plate collectors can be reached if a large amount of solar radiation is concentrated on a relatively small collection area. Simple reflectors can markedly increase the amount of direct radiation reaching a collector, as shown in Figure 10A.

Because of the apparent movement of the sun across the sky, conventional concentrating collectors must follow the sun's daily motion. There are two methods by which the sun's motion can be readily tracked. The altazimuth method requires the tracking device to turn in both altitude and azimuth; when performed properly, this method enables the concentrator to follow the sun exactly. Paraboloidal solar furnaces, Figure 10B, generally use this system. The polar, or equatorial, mounting points the axis of rotation at the North Star, tilted upward at the angle of the local latitude. By rotating the collector 15° per hour, it follows the sun perfectly (on March 21 and September 21). If the collector surface or aperture must be kept normal to the solar rays, a second motion is needed to correct for the change in the solar declination. This motion is not essential for most solar collectors.

The maximum variation in the angle of incidence for a collector on a polar mount will be ±23.5° on June 21 and December 21; the incident angle correction would then be cos 23.5° = 0.917.

Horizontal reflective parabolic troughs, oriented east and west, as shown in Figure 10C, require continuous adjustment to compensate for the changes in the sun's declination. There is inevitably some morning and afternoon shading of the reflecting surface if the concentrator has opaque end panels. The necessity of moving the concentrator to accommodate the changing solar declination can be reduced by moving the absorber or by using a trough with two sections of a parabola facing each other, as shown in Figure 10D. Known as a **compound parabolic concentrator** (CPC), this design can accept incoming radiation over a relatively wide range of angles. By using multiple internal reflections, any radiation that is accepted finds its way to the absorber surface located at the bottom of the apparatus. By filling the collector shape with a highly transparent material having an index of refraction greater than 1.4, the acceptance angle can be increased. By shaping the surfaces of the array properly, total internal reflection is made to occur at the medium-air interfaces, which results in a high concentration efficiency. Known as a **dielectric compound parabolic concentrator** (DCPC), this device has been applied to the photovoltaic generation of electricity (Cole et al. 1977).

The parabolic trough of Figure 10C can be simulated by many flat strips, each adjusted at the proper angle so that all reflect onto a common target. By supporting the strips on ribs with parabolic contours, a relatively efficient concentrator can be produced with less tooling than the complete reflective trough.

Another concept applied this segmental idea to flat and cylindrical lenses. A modification is shown in Figure 10F, in which a linear Fresnel lens, curved to shorten its focal distance, can concentrate a relatively large area of radiation onto an elongated receiver. Using

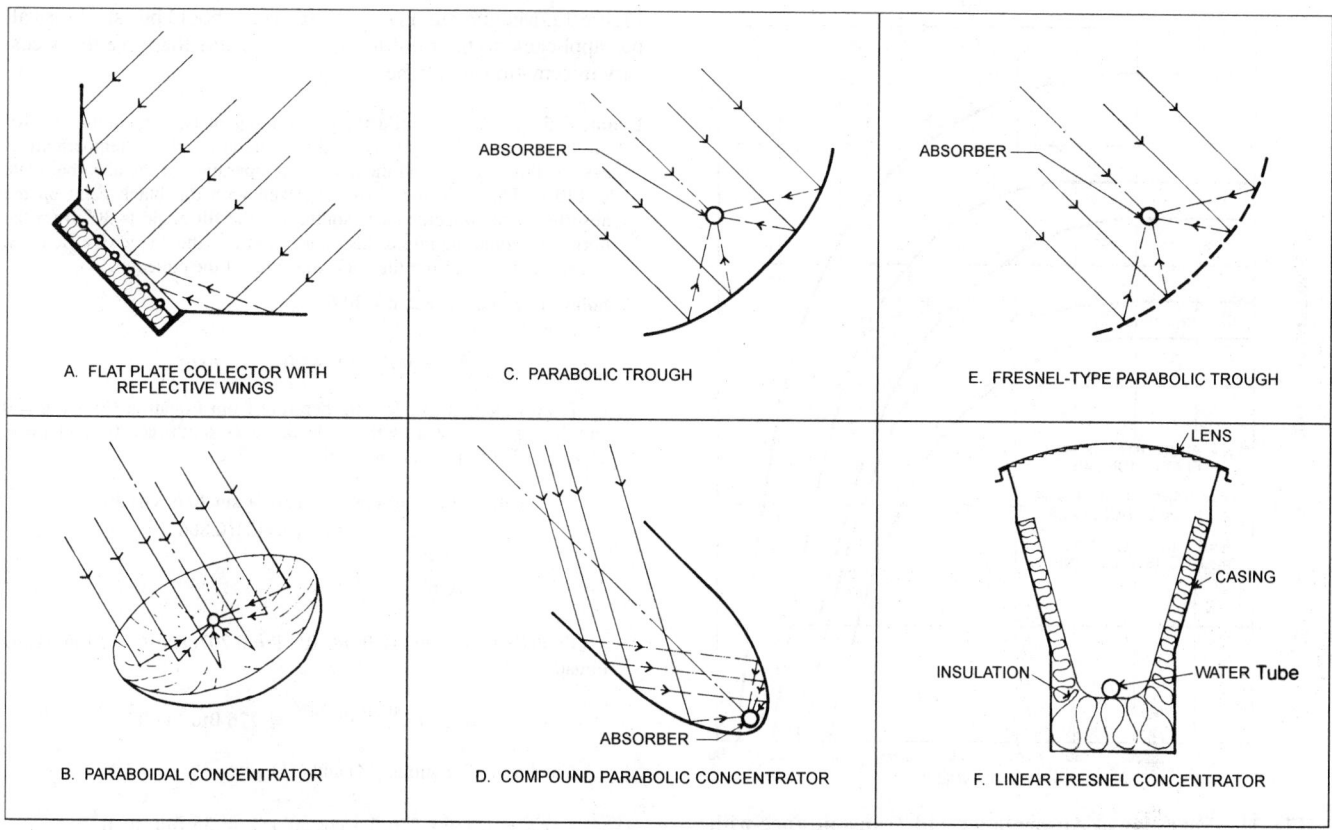

Fig. 10   Types of Concentrating Collectors

the equatorial sun-following mounting, this type of concentrator has been used as a means of attaining temperatures well above those that can be reached with flat-plate collectors.

One disadvantage of concentrating collectors is that, except at low concentration ratios, they can use only the direct component of solar radiation, because the diffuse component cannot be concentrated by most types. However, an advantage of concentrating collectors is that, in summer, when the sun rises and sets well to the north of the east-west line, the sun-follower, with its axis oriented north-south, can begin to accept radiation directly from the sun long before a fixed, south-facing flat plate can receive anything other than diffuse radiation from the portion of the sky that it faces. Thus, at 40° north latitude, for example, the cumulative *direct* radiation available to a sun-follower on a clear day is 3180 Btu/ft^2, while the *total* radiation falling on the flat plate tilted upward at an angle equal to the latitude is only 2220 Btu/ft^2 each day. Thus, in relatively cloudless areas, the concentrating collector may capture more radiation per unit of aperture area than a flat-plate collector.

To get extremely high inputs of radiant energy, many flat mirrors, or **heliostats**, using altazimuth mounts, can be used to reflect their incident direct solar radiation onto a common target. Using slightly concave mirror segments on the heliostats, large amounts of thermal energy can be directed into the cavity of a steam generator to produce steam at high temperature and pressure.

Collector Performance

The performance of collectors may be analyzed by a procedure originated by Hottel and Woertz (1942) and extended by Whillier (ASHRAE 1977). The basic equation is

$$q_u = I_{t\theta}(\tau\alpha)_\theta - U_L(t_p - t_{at}) = \dot{m}\,c_p(t_{fe} - t_{fi})/A_{ap} \qquad (24)$$

Equation (24) also may be adapted for use with concentrating collectors:

$$q_u = I_{DN}(\tau\alpha)_\theta(\rho\Gamma) - U_L(A_{abs}/A_{ap})(t_{abs} - t_a) \qquad (25)$$

where

q_u = useful heat gained by collector per unit of aperture area, Btu/h·ft^2

$I_{t\theta}$ = total irradiation of collector, Btu/h·ft^2

I_{DN} = direct normal irradiation, Btu/h·ft^2

$(\tau\alpha)_\theta$ = transmittance τ of cover times absorptance α of plate at prevailing incident angle θ

U_L = upward heat loss coefficient, Btu/h·ft^2·°F

t_p = temperature of the absorber plate, °F

t_a = temperature of the atmosphere, °F

t_{abs} = temperature of the absorber, °F

$\dot{m}$ = fluid flow rate, lb/h

c_p = specific heat of fluid, Btu/lb·°F

$t_{fe},\,t_{fi}$ = temperatures of the fluid leaving and entering the collector, °F

$\rho\Gamma$ = reflectance of the concentrator surface times fraction of reflected or refracted radiation that reaches the absorber

$A_{abs},\,A_{ap}$ = areas of absorber surface and of aperture that admit or receive radiation, ft^2

The total irradiation and the direct normal irradiation for clear days may be found in ASHRAE *Standard* 93. The transmittance for single and double glazing and the absorptance for flat black paint may be found in Table 4 for incident angles from 0 to 90°. These values, and the products of τ and α, are also shown in Figure 11. Little change occurs in the solar-optical properties of the glazing and absorber plate until θ exceeds 30°, but, since all values reach zero when $\theta = 90°$, they drop off rapidly for values of θ beyond 40°.

For nonselective absorber plates, U_L varies with the temperature of the plate and the ambient air, as shown in Figure 12. For selective surfaces, which effect major reductions in the emittance of the absorber plate, U_L will be much lower than the values shown in

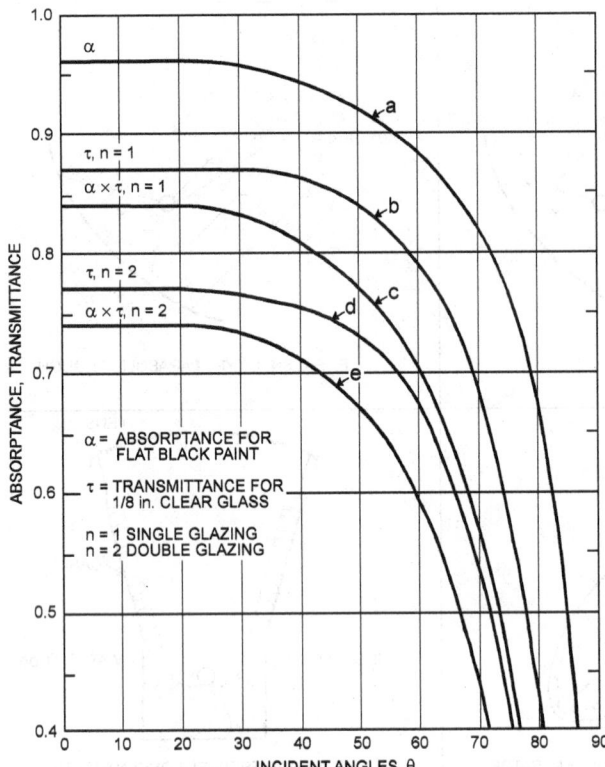

Fig. 11　Variation of Absorptance and Transmittance with Incident Angle

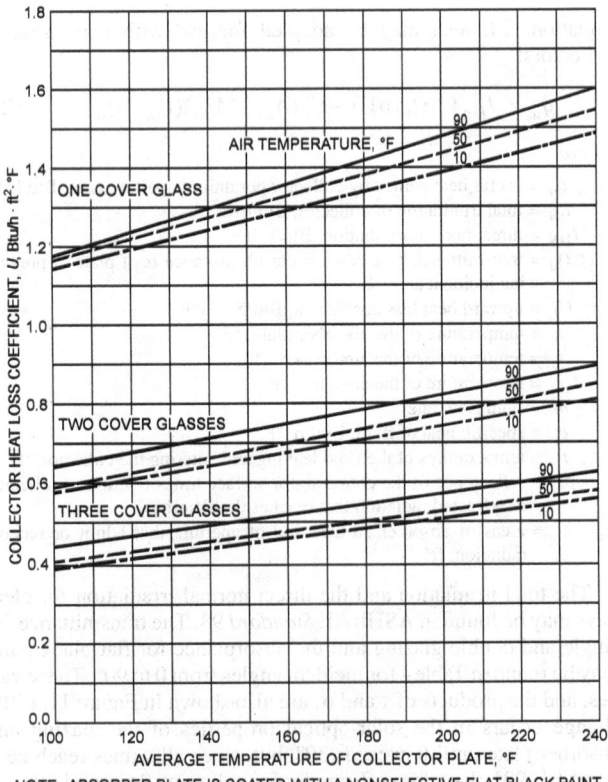

NOTE: ABSORBER PLATE IS COATED WITH A NONSELECTIVE FLAT BLACK PAINT.

Fig. 12　Variation of Upward Heat Loss Coefficient U_L with Collector Plate Temperature and Ambient Air Temperatures for Single-, Double-, and Triple-Glazed Collectors

Figure 12. Manufacturers of such surfaces should be asked for values applicable to their products, or test results that give the necessary information should be consulted.

Example 5. A flat-plate collector is operating in Denver, latitude = 40° north, on July 21 at noon solar time. The atmospheric temperature is assumed to be 85°F, and the average temperature of the absorber plate is 140°F. The collector is single-glazed with flat black paint on the absorber. The collector faces south, and the tilt angle is 30° from the horizontal. Find the rate of heat collection and the collector efficiency. Neglect the losses from the back and sides of the collector.

Solution: From Table 2, $\delta = 20.6°$.
From Equation (2),

$$\beta_N = 90° - 40° + 20.6° = 70.6°$$

From Equation (3), $H = 0$; therefore from Equation (6), $\sin \phi = 0$ and thus, $\phi = 0°$. Because the collector faces south, $\psi = 0°$, and $\gamma = \phi$. Thus $\gamma = 0°$. Then Equation (8) gives

$$\cos \theta = \cos 70.6° \cos 0° \sin 30° + \sin 70.6° \cos 30°$$
$$= (0.332)(1)(0.5) + (0.943)(0.866)$$
$$= 0.983$$
$$\theta = 10.6°$$

From Table 1, $A = 344$ Btu/h·ft^2, $B = 0.207$, and $C = 0.136$. Using Equation (11),

$$I_{DN} = 344 e^{-0.207/\sin 70.6°} = 276 \text{ Btu/h} \cdot \text{ft}^2$$

Combining Equations (14) and (15) gives

$$I_{d\theta} = 0.136 \times 276(1 + \cos 30°)/2 = 35 \text{ Btu/h} \cdot \text{ft}^2$$

Assuming $I_r = 0$, Equation (12) gives a total solar irradiation on the collector of

$$I_{t\theta} = 276 \cos 10.6° + 35(0) = 306 \text{ Btu/h} \cdot \text{ft}^2$$

From Figure 11, for $n = 1$, $\tau = 0.87$ and $\alpha = 0.96$.
From Figure 12, for an absorber plate temperature of 140°F and an air temperature of 85°F, $U_L = 1.3$ Btu/h·ft^2·°F.
Then from Equation (24),

$$q_u = 306(0.87 \times 0.96) - 1.3(140 - 85) = 184 \text{ Btu/h} \cdot \text{ft}^2$$

The collector efficiency η is

$$184/306 = 0.60$$

The general expression for collector efficiency is

$$\eta = (\tau\alpha)_\theta - U_L(t_p - t_{at})/I_{t\theta} \tag{26}$$

For incident angles below about 35°, the product τ times α is essentially constant and Equation (26) is linear with respect to the parameter $(t_p - t_{at})/I_{t\theta}$, as long as U_L remains constant.

ASHRAE (1977) suggested that an additional term, the **collector heat removal factor** F_R, be introduced to permit the use of the fluid inlet temperature in Equations (24) and (26):

$$q_u = F_R[I_{t\theta}(\tau\alpha)_\theta - U_L(t_{fi} - t_{at})] \tag{27}$$

$$\eta = F_R(\tau\alpha)_\theta - F_R U_L(t_{fi} - t_{at})/I_{t\theta} \tag{28}$$

where F_R equals the ratio of the heat actually delivered by the collector to the heat that would be delivered if the absorber were at t_{fi}. F_R is found from the results of a test performed in accordance with ASHRAE *Standard* 93.

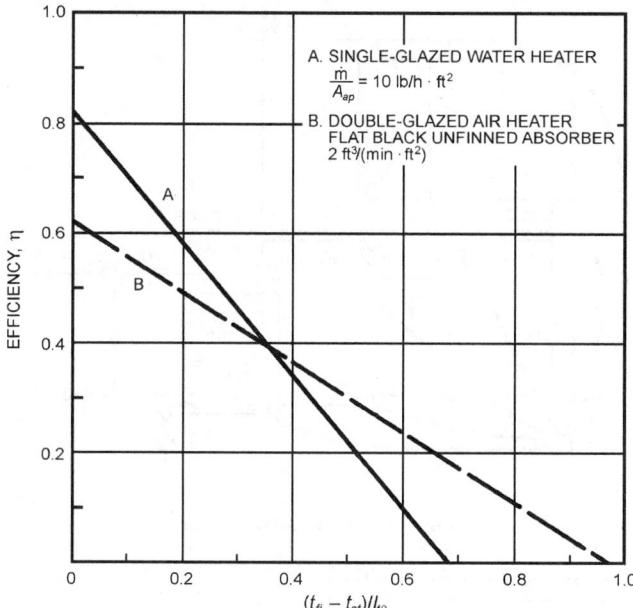

A. SINGLE-GLAZED WATER HEATER
$$\frac{\dot{m}}{A_{ap}} = 10 \text{ lb/h} \cdot \text{ft}^2$$

B. DOUBLE-GLAZED AIR HEATER
FLAT BLACK UNFINNED ABSORBER
$2 \text{ ft}^3/(\text{min} \cdot \text{ft}^2)$

Fig. 13 Efficiency Versus $(t_{fi} - t_{at})/I_{t\theta}$ for Single-Glazed Solar Water Heater and Double-Glazed Solar Air Heater

The results of such a test are plotted in Figure 13. When the parameter is zero, because there is no temperature difference between the fluid entering the collector and the atmosphere, the value of the y-intercept equals $F_R(\tau\alpha)$. The slope of the efficiency line equals the heat loss factor U_L multiplied by F_R. For the single-glazed, nonselective collector with the test results shown in Figure 13, the y-intercept is 0.82, and the x-intercept is $0.69°\text{F} \cdot \text{ft}^2 \cdot \text{h/Btu}$. This collector used high transmittance single glazing, $\tau = 0.91$, and black paint with an absorptance of 0.97, so Equation (28) gives $F_R = 0.82/(0.91 \times 0.97) = 0.93$.

Assuming that the relationship between η and the parameter is actually linear, as shown, then the slope is $-0.82/0.69 = -1.19$; thus $U_L = 1.19/F_R = 1.19/0.93 = 1.28 \text{ Btu/h} \cdot \text{ft}^2 \cdot °\text{F}$. The tests on which Figure 13 is based were run indoors. Factors that affect the measured efficiency are wind speed and fluid velocity.

Figure 13 also shows the efficiency of a double-glazed air heater with an unfinned absorber coated with flat black paint. The y-intercept for the air heater B is considerably less than it is for water heater A because (1) transmittance of the double glazing used in B is lower than the transmittance of the single glazing used in A and (2) F_R is lower for B than for A because of the lower heat transfer coefficient between air and the unfinned metal absorber.

The x-intercept for air heater B is greater than it is for the water heater A because the upward loss coefficient U_L is much lower for the double-glazed air heater than for the single-glazed water heater. The data for both A and B were taken at near-normal incidence with high values of $I_{t\theta}$. For Example 5, using a single-glazed water heater, the value of the parameter would be close to $(140 - 85)/306 = 0.18°\text{F} \cdot \text{ft}^2 \cdot \text{h/Btu}$, and the expected efficiency, 0.60, agrees closely with the test results.

As ASHRAE *Standard* 93 shows, the incident angles encountered with south-facing tilted collectors vary widely throughout the year. Considering a surface located at 40° north latitude with a tilt angle $\Sigma = 40°$, the incident angle θ will depend on the time of day and the declination δ. On December 21, $\delta = 23.456°$; at 4 h before and after solar noon, the incident angle is 62.7°, and it remains close to this value for the same solar time throughout the year. The total irradiation at these conditions varies from a low of 45 Btu/h·ft² on

December 21 to approximately 140 Btu/h·ft² throughout most of the other months.

When the irradiation is below about 100 Btu/h·ft², the losses from the collector may exceed the heat that can be absorbed. This situation varies with the temperature difference between the collector inlet temperature and the ambient air, as suggested by Equation (27).

When the incident angle rises above 30°, the product of the transmittance of the glazing and the absorptance of the collector plate begins to diminish; thus, the heat absorbed also drops. The losses from the collector are generally higher as the time moves farther from solar noon, and consequently the efficiency also drops. Thus, the daylong efficiency is lower than the near-noon performance. During the early afternoon, the efficiency is slightly higher than at the comparable morning time, because the ambient air temperature is lower in the morning than in the afternoon.

ASHRAE *Standard* 93 describes the **incident angle modifier**, which may be found by tests run when the incident angle is set at 30°, 45°, and 60°. Simon (1976) showed that for many flat-plate collectors, the incident angle modifier is a linear function of the quantity $(1/\cos\theta - 1)$. For evacuated tubular collectors, the incident angle modifier may grow with rising values of θ.

ASHRAE *Standard* 93 specifies that the efficiency be reported in terms of the gross collector area A_g rather than the aperture area A_{ap}. The reported efficiency will be lower than the efficiency given by Equation (28), but the total energy collected is not changed by this simplification:

$$\eta_g = \frac{\eta_{ap} A_{ap}}{A_g} \tag{29}$$

HEAT STORAGE

Storage may be part of solar heating, cooling, and power generation. For some applications, such as swimming pool heating, daytime air heating, and irrigation pumping, intermittent operation is acceptable, but most other uses of solar energy require operating at night and when the sun is obscured by clouds. Chapter 33 provides further information on general thermal storage technologies.

WATER HEATING

A solar water heater includes a solar collector that absorbs solar radiation and converts it to heat, which is then absorbed by a heat transfer fluid (water, a nonfreezing liquid, or air) that passes through the collector. The heat transfer fluid's heat is stored or used directly.

Portions of the solar energy system are exposed to the weather, so they must be protected from freezing. The system must also be protected from the overheating caused by high insulation levels during periods of low energy demand.

In solar water heating, water that is heated directly in the collector or indirectly by a heat transfer fluid that is heated in the collector, passes through a heat exchanger and transfers its heat to the domestic or service water. The heat transfer fluid is transported by either natural or forced circulation. Natural circulation occurs by natural convection (thermosiphoning), whereas forced circulation uses pumps or fans. Except for thermosiphon systems, which need no control, solar domestic and service water heaters are controlled by differential thermostats.

Five types of solar energy systems are used to heat domestic and service hot water: thermosiphon, direct circulation, indirect, integral collector storage, and site built. Recirculation and draindown are two methods used to protect direct solar water heaters from freezing.

Thermosiphon Systems

Thermosiphon systems (Figure 14) heat potable water or a heat transfer fluid and rely on natural convection to transport it from the

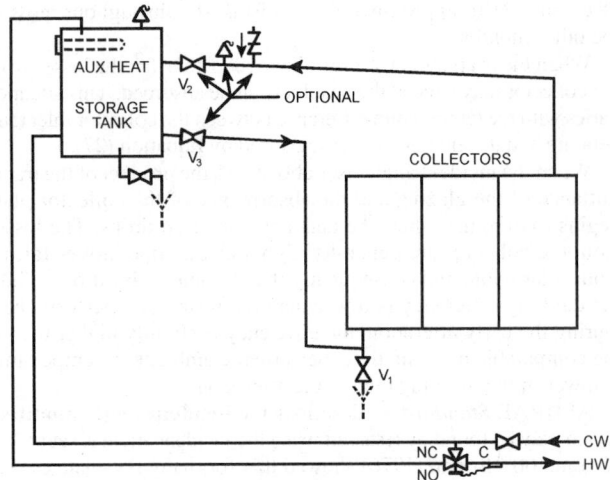

Fig. 14   Thermosiphon System

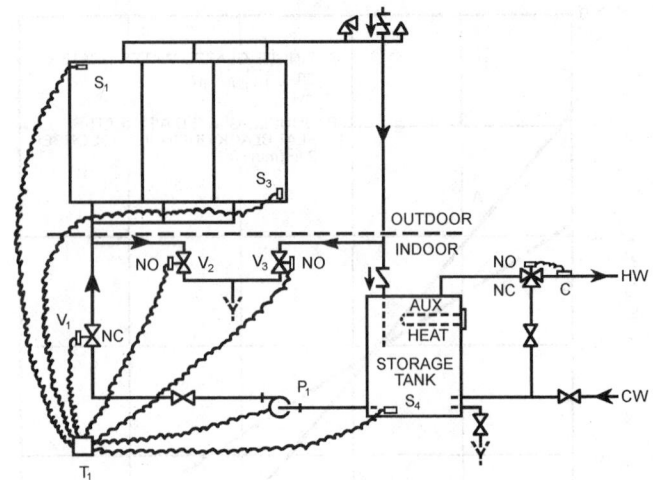

Fig. 16   Draindown System

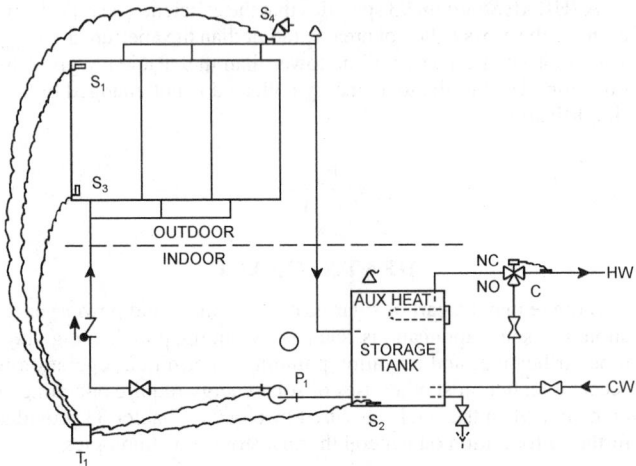

Fig. 15   Direct Circulation System

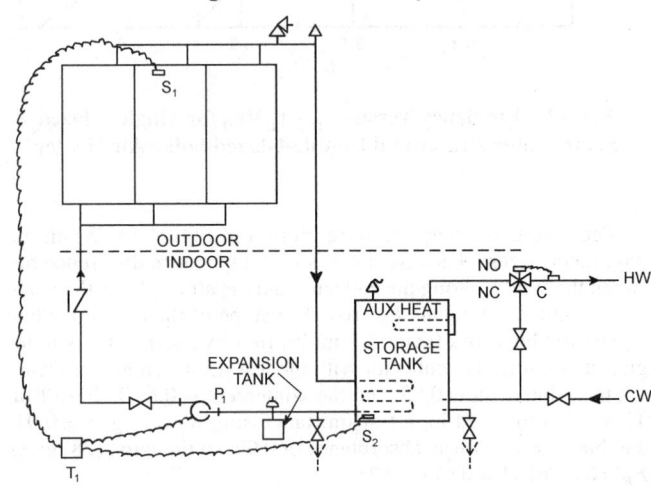

Fig. 17   Indirect Water Heating

collector to storage. For direct systems, pressure-reducing valves are required when the city water pressure is greater than the working pressure of the collectors. In a thermosiphon system, the storage tank must be elevated above the collectors, which sometimes requires designing the upper level floor and ceiling joists to bear this additional load. Extremely hard or acidic water can cause scale deposits that clog or corrode the absorber fluid passages. Thermosiphon flow is induced whenever there is sufficient sunshine, so these systems do not need pumps.

Direct Circulation Systems

A direct circulation system (Figure 15) pump potable water from storage to the collectors when there is enough solar energy available to warm it. They then return the heated water to the storage tank until it is needed. The collectors can be mounted either above or below the storage tank. Direct circulation systems are only feasible in areas where freezing is infrequent. Freeze protection is provided either by recirculating warm water from the storage tank or by flushing the collectors with cold water. Direct water heating systems should not be used in areas where the water is extremely hard or acidic because scale deposits may clog or corrode the absorber fluid passages, rendering the system inoperable.

Direct circulation systems are exposed to city water line pressures and must withstand pressures as required by local codes.

Pressure-reducing valves and pressure relief valves are required when the city water pressure is greater than the working pressure of the collectors. Direct circulation systems often use a single storage tank for both solar energy storage and the auxiliary water heater, but two-tank storage systems can be used.

Draindown systems. Draindown systems (Figure 16) are direct circulation, water heating systems in which potable water is pumped from storage to the collector array where it is heated. Circulation continues until usable solar heat is no longer available. When a freezing condition is anticipated or a power outage occurs, the system drains automatically by isolating the collector array and exterior piping from the city water pressure and using one or more valves for draining. The solar collectors and associated piping must be carefully sloped to drain the collector's exterior piping.

Indirect Water Heating Systems

Indirect water heating systems (Figure 17) circulate a freeze-protected heat transfer fluid through the closed collector loop to a heat exchanger, where its heat is transferred to the potable water. The most commonly used heat transfer fluids are water/ethylene glycol and water/propylene glycol solutions, although other heat transfer fluids such as silicone oils, hydrocarbons, and refrigerants can also be used (ASHRAE 1983). These fluids are nonpotable, sometimes toxic, and normally require double-wall heat exchangers. The double-wall heat exchanger can be located inside the

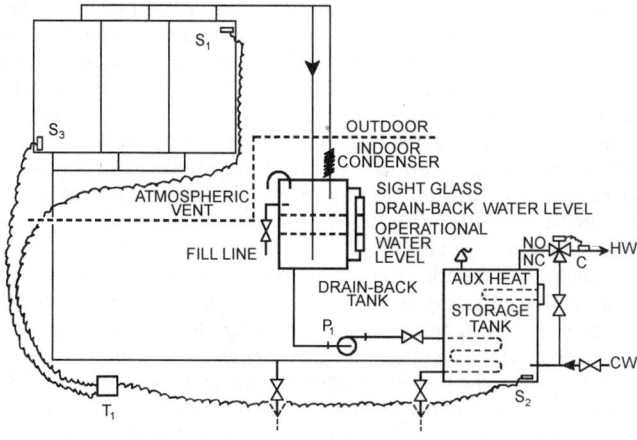

Fig. 18 Drainback System

storage tank, or an external heat exchanger can be used. The collector loop is closed and therefore requires an expansion tank and a pressure relief valve. a one- or two-tank storage can be used. Additional over-temperature protection may be needed to prevent the collector fluid from decomposing or becoming corrosive.

Designers should avoid automatic water makeup in systems using water/antifreeze solutions because a significant leak may raise the freezing temperature of the solution above the ambient temperature, causing the collector array and exterior piping to freeze. Also, antifreeze systems with large collector arrays and long pipe runs may need a time-delayed bypass loop around the heat exchanger to avoid freezing the heat exchanger on startup.

Drainback Systems. Drainback systems are generally indirect water heating systems that circulate treated or untreated water through the closed collector loop to a heat exchanger, where its heat is transferred to the potable water. Circulation continues until usable energy is no longer available. When the pump stops, the collector fluid drains by gravity to a storage or tank. In a pressurized system, the tank also serves as an expansion tank, so it must have a temperature and pressure relief valve to protect against excessive pressure. In an unpressurized system (Figure 18), the tank is open and vented to the atmosphere.

The collector loop is isolated from the potable water, so valves are not needed to actuate draining, and scaling is not a problem. The collector array and exterior piping must be sloped to drain completely, and the pumping pressure must be sufficient to lift water to the top of the collector array.

Integral Collector-Storage Systems

Integral collector storage (ICS) systems use hot water storage as part of the collector. Some types use the surface of a single tank as the absorber, and others use multiple, long, thin tanks placed side-by-side horizontally to form the absorber surface. In this type of ICS, hot water is drawn from the top tank and cold replacement water enters the bottom tank. Because of the greater nighttime heat loss from ICS systems, they are typically less efficient than pumped systems, and selective surfaces are recommended. ICS systems are normally installed as a solar preheater without pumps or controllers. Flow through the ICS system occurs on demand, as hot water flows from the collector to a hot water auxiliary tank in the structure.

Site-Built Systems

Site-built, large volume solar, water, air, or water heating equipment is used in commercial and industrial applications. The site built systems are based on a transpired solar collector for air heating and shallow solar pond technologies.

Transpired Solar Collector. This collector preheats outdoor air by drawing it through small holes in a metal panel. It is typically installed on south facing walls and is designed to heat outdoor air for building ventilation or process applications (Kutscher 1996). The prefabricated panel efficiently heats and captures fresh air by drawing it through a perforated adsorber, eliminating the cost and the reflection losses associated with a glazing. The panel consists of a dark-colored, metal building panel with thousands of small holes. The sun heats the metal panel that in turn, heats a boundary layer of air on its surface. Air is heated as it is drawn through the small holes into a ventilation system for delivery as ventilation air, crop drying, or other process applications.

Shallow Solar Pond. The shallow solar pond (SSP) is a large-scale ICS solar water heater (Figure 19) capable of providing more than 5000 gal of hot water per day for commercial and industrial use. These ponds are built in standard modules and tied together to supply the required load. The SSP module can be ground mounted or installed on a roof. It is typically 16 ft wide and up to 200 ft long. The module contains one or two flat water bags similar to a water bed. The bags rest on a layer of insulation inside concrete or fiberglass curbs. The bag is protected against damage and heat loss by greenhouse glazing. A typical pond filled to a 4 in. depth holds approximately 6000 gal of water.

Pool Heaters

Solar pool heaters do not require a separate storage tank, because the pool itself serves as storage. In most cases, the pool's filtration pump forces the water through the solar panels or plastic pipes. In some retrofit applications, a larger pump may be required to handle the needs of the solar heater, or a small pump may be added to boost the pool water to the solar collectors.

Automatic control may be used to direct the flow of filtered water to the collectors when solar heat is available; this may also be accomplished manually. Normally, solar heaters are designed to drain down into the pool when the pump is turned off; this provides the collectors with freeze protection.

Four primary types of collector designs are used for swimming pool heat: (1) rigid black plastic panels (polypropylene), usually 4 ft by 10 ft or 4 ft by 8 ft; (2) tube-on-sheet panels, which usually have a metal deck (copper or aluminum) with copper water tubes; (3) an EPDM rubber mat, extruded with the water passages running its length; and (4) arrays of black plastic pipe, usually 1.5 in. diameter ABS plastic (Root et al. 1985).

Hot Water Recirculation

Domestic hot water (DHW) recirculation systems (Figures 20 and 21), which continuously circulate domestic hot water throughout a building, are found in motels, hotels, hospitals, dormitories,

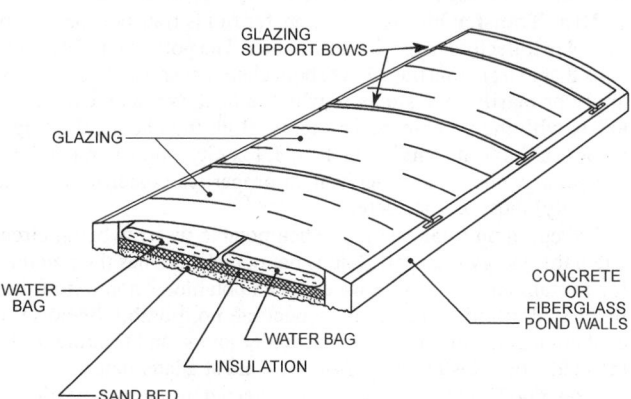

Fig. 19 Shallow Solar Pond

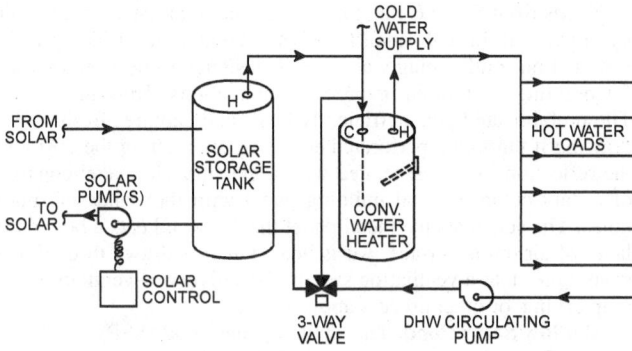

Fig. 20   DHW Recirculation System

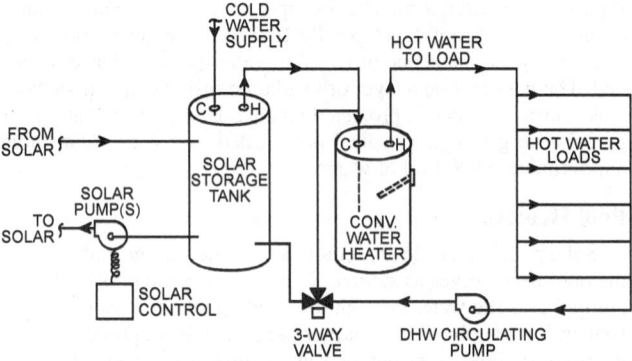

Fig. 21   DHW Recirculation System with Makeup Preheat

office buildings, and other commercial buildings. The recirculation heat losses in these systems are usually a significant part of the total water heating load. A properly integrated solar heater can make up much of this loss.

COMPONENTS

This section describes the major components involved in the collection, storage, transportation, control, and distribution of solar heat for a domestic hot water system.

Collectors. Flat-plate collectors are most commonly used for water heating because of the year-round load requiring temperatures of 80 to 180°F. For discussions of other collectors and applications, see ASHRAE *Standard* 93, Chapter 33 of the 1996 *ASHRAE Handbook—Systems and Equipment*, and previous sections of this chapter. Collectors must withstand extreme weather (such as freezing, stagnation, and high winds), as well as system pressures.

Heat Transfer Fluids. Heat transfer fluids transport heat from the solar collectors to the domestic water. The potential safety problems that exist in this transfer are both chemical and mechanical and apply primarily to systems in which a heat exchanger interface exists with the potable water supply. Both the chemical compositions of the heat transfer fluids (pH, toxicity, and chemical durability), as well as their mechanical properties (specific heat and viscosity) must be considered.

Except in unusual cases, or when potable water is being circulated, the energy transport fluid is nonpotable and has the potential for contaminating potable water. Even potable or nontoxic fluids in closed circuits are likely to become nonpotable because of contamination from metal piping, solder joints, and packing, or by the inadvertent installation of a toxic fluid at a later date.

Thermal Energy Storage. Heat collected by solar domestic and service water heaters is virtually always stored as a liquid in tanks. Storage tanks and bins should be well insulated. In domestic hot

water systems, heat is usually stored in one or two tanks. The hot water outlet is at the top of the tank, and cold water enters the tank through a dip tube that extends down to within 4 to 6 in. of the tank bottom. The outlet on the tank to the collector loop should be approximately 4 in. above the tank bottom to prevent scale deposits from being drawn into the collectors. Water from the collector array returns to the upper portion of the storage tank. This plumbing arrangement may take advantage of thermal stratification, depending on the delivery temperature from the collectors and the flow rate through the storage tank.

Single-tank electric auxiliary systems often incorporate storage and auxiliary heating in the same vessel. Conventional electric water heaters commonly have two heating elements: one near the top and one near the bottom. If a dual element tank is used in a solar energy system, the bottom element should be disconnected and the top left functional to take advantage of fluid stratification. Standard gas- and oil-fired water heaters should not be used in single-tank arrangements. In gas and oil water heaters, heat is added to the bottom of the tanks, which reduces both stratification and collection efficiency in single-tank systems.

Dual-tank systems often use the solar domestic hot water storage tank as a preheat tank. The second tank is normally a conventional domestic hot water tank csystemsontaining the auxiliary heat source. Multiple tanks are sometimes used in large institutions, where they operate similarly to dual-tank heaters. Although the use of two tanks may increase collector efficiency and the solar fraction, it increases tank heat losses. The water inlet is usually a dip tube that extends near the bottom of the tank.

Estimates for sizing storage tanks usually range from 1 to 2.5 gal per square foot of solar collector area. The estimate used most often is 1.8 gal per square foot of collector area, which usually provides enough heat for a sunless period of about a day. Storage volume should be analyzed and sized according to the project water requirements and draw schedule; however, solar applications typically require larger than normal tanks.

Heat Exchangers. Indirect solar water heaters require one or more heat exchangers. The potential exists for contamination in the transfer of heat energy from solar collectors to potable hot water. Heat exchangers influence the effectiveness of energy collected to heat domestic water. They also separate and protect the potable water supply from contamination when nonpotable heat transfer fluids are used. For this reason various codes regulate the need for, and design of, heat exchangers.

Heat exchanger selection should consider thermal performance, cost effectiveness, reliability, safety, and the following:

- Heat exchange effectiveness
- Pressure drop, operating power, and flow rate
- Design pressure, configuration, size, materials, and location
- Cost and availability
- Reliable protection of the potable water supply from contamination by the heat transfer fluid
- Leak detection, inspection, and maintainability
- Material compatibility with other elements such as metals and fluids
- Thermal compatibility with design parameters such as operating temperature, flow rate, and fluid thermal properties

Heat exchanger selection depends on the characteristics of the fluids that pass through the heat exchanger and the properties of the exchanger itself. Fluid characteristics to consider are fluid type, specific heat, mass flow rate, and hot and cold fluid inlet and outlet temperatures. Physical properties of the heat exchanger to consider are the overall heat transfer coefficient of the heat exchanger and the heat transfer surface area.

For most solar domestic hot water designs, only the hot and cold inlet temperatures are known; the other temperatures must be calculated using the physical properties of the heat exchanger. Two quantities that are useful in determining the heat transfer in a heat exchanger and the performance characteristics of a collector when it is combined with a given heat exchanger are: (1) the fluid capacitance rate, which is the product of the mass flow rate and the specific heat of the fluid passing through the heat exchanger and (2) the heat exchanger effectiveness, which relates the capacitance rate of the two fluids to the fluid inlet and outlet temperatures. The effectiveness is equal to the ratio of the actual heat transfer rate to the maximum heat transfer rate theoretically possible. Generally, a heat exchanger effectiveness of 0.4 or greater is desired.

Expansion Tanks. An indirect solar water heater operating in a closed collector loop requires an expansion tank to prevent excessive pressure. Fluid in solar collectors under stagnation conditions can boil, causing excessive pressure to develop in the collector loop, and expansion tanks must be sized for this condition. Expansion tank sizing formulas for closed loop hydronic systems, found in Chapter 12 of the 1996 *ASHRAE Handbook—Systems and Equipment*, may be used for solar heater expansion tank sizing; but the expression for volume change due to temperature increase should be replaced with the total volume of fluid in the solar collectors and of any piping located above the collectors, if significant. This sizing method provides a passive means for eliminating fluid loss due to over temperature or stagnation, common problems in closed loop solar systems. This results in a larger expansion tank than typically found in hydronic systems, but the increase in cost is small compared to the savings in fluid replacement and maintenance costs (Lister and Newell 1989).

Pumps. Pumps circulate heat transfer liquid through collectors and heat exchangers. In solar domestic hot water heaters, the pump is usually a centrifugal circulator driven by a less than 300 W motor. The flow rate for collectors generally ranges from 0.015 to 0.04 gpm/ft^2. Pumps used in drain-back systems must provide pressure to overcome friction and to lift the fluid to the collectors.

Piping. Piping can be plastic, copper, galvanized steel, or stainless steel. The most widely used is nonlead, sweat-soldered L-type copper tubing. M-type copper is also acceptable if permitted by local building codes. If water/glycol is the heat transfer fluid, galvanized pipes or tanks must not be used because unfavorable chemical reactions will occur; copper piping is recommended instead. Also, if glycol solutions or silicone fluids are used, they may leak through joints where water would not. Piping should be compatible with the collector fluid passage material; for example, copper or plastic piping should be used with collectors having copper fluid passages.

Piping that carries potable water can be plastic, copper, galvanized steel, or stainless steel. In indirect systems, corrosion inhibitors must be checked and adjusted routinely, preferably every three months. Inhibitors should also be checked if the system overheats during stagnation conditions. If dissimilar metals are joined, dielectric or nonmetallic couplings should be used. The best protection is sacrificial anodes or getters in the fluid stream. Their location depends on the material to be protected, the anode material, and the electrical conductivity of the heat transfer fluid. Sacrificial anodes consisting of magnesium, zinc, or aluminum are often used to reduce corrosion in storage tanks. Because many possibilities, exist each combination must be evaluated. A copper-aluminum or copper-galvanized steel joint is unacceptable because of severe galvanic corrosion. Aluminum, copper, and iron have a greater potential for corrosion.

Elimination of air, pipe expansion, and piping slope must be considered to avoid possible failures. Collector pipes (particularly manifolds) should be designed to allow expansion from stagnation temperature to extreme cold weather temperature. Expansion control can be achieved with offset elbows in piping, hoses, or expansion couplings. Expansion loops should be avoided unless they are installed horizontally, particularly in systems that must drain for freeze protection. The collector array piping should slope 0.06 in. per foot for drainage (DOE 1978b).

Air can be eliminated by placing air vents at all piping high points and by air purging during filling. Flow control, isolation, and other valves in the collector piping must be chosen carefully so that these components do not restrict drainage significantly or back up water behind them. The collectors must drain completely.

Valves and Gages. Valves in solar domestic hot water systems must be located to ensure system efficiency, satisfactory performance, and the safety of equipment and personnel. Drain valves must be ball-type; gate valves may be used if the stem is installed horizontally. Check valves or other valves used for freeze protection or for reverse thermosiphoning must be reliable to avoid significant damage.

Auxiliary Heat Sources. On sunny days, a typical solar energy system should supply water at a predetermined temperature, and the solar storage tank should be large enough to hold sufficient water for a day or two. Because of the intermittent nature of solar radiation, an auxiliary heater must be installed to handle hot water requirements. If a utility is the source of auxiliary energy, operation of the auxiliary heater can be timed to take advantage of off-peak utility rates. The auxiliary heater should be carefully integrated with the solar energy heater to obtain maximum solar energy use. For example, the auxiliary heater should not destroy any stratification that may exist in the solar-heated storage tank, which would reduce collector efficiency.

Ductwork, particularly in systems with air-type collectors, must be sealed carefully to avoid leakage in duct seams, damper shafts, collectors, and heat exchangers. Ducts should be sized using conventional air duct design methods.

Control. Controls regulate solar energy collection by controlling fluid circulation, activate system protection against freezing and overheating, and initiate auxiliary heating when it is required. The three major control components are sensors, controllers, and actuated devices. Sensors detect conditions or measure quantities, such as temperature. Controllers receive output from the sensors, select a course of action, and signal a component to adjust the condition. Actuators, such as pumps, valves, dampers, and fans, execute controller commands and regulate the system.

Temperature sensors measure the temperature of the absorber plate near the collector outlet and near the bottom of the storage tank. The sensors send signals to a controller, such as a differential temperature thermostat, for interpretation.

The differential thermostat compares the signals from the sensors with adjustable set points for high and low temperature differentials. The controller performs different functions, depending on which set points are met. In liquid systems, when the temperature difference between the collector and storage reaches a high set point, usually 20°F, the pump starts, automatic valves are activated, and circulation begins. When the temperature difference reaches a low set point, usually 4°F, the pump is shut off and the valves are deenergized and returned to their normal positions. To restart the system, the high-temperature set point must again be met. If the system has either freeze or over-temperature protection, the controller opens or closes valves or dampers and starts or stops pumps or fans to protect the system when its sensors detect that either freezing or overheating is about to occur.

Sensors must be selected to withstand high temperature, such as may occur during collector stagnation. Collector loop sensors can be located on the absorber plate, in a pipe above the collector, on a pipe near the collector, or in the collector outlet passage. Although any of these locations may be acceptable, attaching the sensor on the collector absorber plate is recommended. When attached properly, the sensor gives accurate readings, can be installed easily, and is basically unaffected by ambient temperature, as are sensors mounted on exterior piping.

A sensor installed on an absorber plate reads temperatures 3 to 4°F higher than the temperature of the fluid leaving the collector. However, such temperature discrepancies can be compensated for in the differential thermostat settings.

The sensor must be attached to the absorber plate with good thermal contact. If a sensor access cover is provided on the enclosure, it must be gasketed for a watertight fit. Adhesives and adhesive tapes should not be used to attach the sensor to the absorber plate.

The storage temperature sensor should be near the bottom of the storage tank to detect the temperature of the fluid before it is pumped to the collector or heat exchanger. The storage fluid is usually coldest at that location because of thermal stratification and the location of the makeup water supply. The sensor should be either securely attached to the tank and well insulated, or immersed inside the tank near the collector supply.

The freeze protection sensor, if required, should be located so that it will detect the coldest liquid temperature when the collector is shut down. Common locations are the back of the absorber plate at the bottom of the collector, the collector intake or return manifolds, or the center of the absorber plate. The center absorber plate location is recommended because reradiation to the night sky will freeze the collector heat transfer fluid, even though the ambient temperature is above freezing. Some systems, such as the recirculation system, have two sensors for freeze protection, while others, such as the drain-down, use only one.

Control of on-off temperature differentials affect system efficiency. The turn-on temperature differential must be selected properly because, if the differential is too high, the collector starts later than it should, and if it is too low, the collector starts too soon. The turn-on differential for liquid systems usually ranges from 15 to 30°F and is most commonly 20°F. For air systems, the range is usually 25 to 45°F.

The turn-off temperature differential is more difficult to estimate. Selection depends on a comparison between the value of the energy collected and the cost of collecting it. It varies with individual systems, but a value of 4°F is typical.

Water temperature in the collector loop depends on ambient temperature, solar radiation, radiation from the collector to the night sky, and collector loop insulation. Freeze protection sensors should be set to detect 40°F.

Sensors are important but often overlooked control components. They must be selected and installed properly because no control can produce accurate outputs from unreliable sensor inputs. Sensors are used in conjunction with a differential temperature controller and are usually supplied by the controller manufacturer. Sensors must survive the anticipated operating conditions without physical damage or loss of accuracy. Low-voltage sensor circuits must be located away from high voltage lines to avoid electromagnetic interference. Sensors attached to collectors should be able to withstand the stagnation temperature.

Sensor calibration, which is often overlooked by installers and maintenance personnel, is critical to system performance; a routine calibration maintenance schedule is essential.

Another control is for the photovoltaic (PV) that powers a pump. A PV panel converts sunlight into electricity to run a small circulating pump. No additional sensing is required because the PV panel and pump output increase with sunlight intensity and stop when no sunlight (collector energy) is available. Cromer (1984) has shown that with proper matching of pump and PV electrical characteristics, PV panel sizes as low as 5 W per 40 ft^2 of thermal panel may be used successfully. Difficulty with late starting and running too long in the afternoon can be alleviated by tilting the PV panel slightly to the east.

Performance Evaluation Methods

The performance of any solar energy system is directly related to the (1) heating load, (2) amount of solar radiation available, and (3) solar energy system characteristics. Various calculation methods use different procedures and data when considering the available solar radiation. Some simplified methods consider only average annual incident solar radiation, while complex methods may use hourly data.

Solar energy system characteristics, as well as individual component characteristics, are required to evaluate performance. The degree of complexity with which these systems and components are described varies from system to system.

The cost effectiveness of a solar domestic and service hot water heating system depends on the initial cost and energy cost savings. A major task is to determine how much energy is saved. The **annual solar fraction**—the annual solar contribution to the water heating load divided by the total water heating load—can be used to estimate these savings. It is expressed as a decimal fraction or percentage and generally ranges from 0.3 to 0.8 (30 to 80%), although more extreme values are possible.

Water Heating Load

The amount of hot water required must be estimated accurately because it affects component selection. Oversized storage may result in low-temperature water that requires auxiliary heating to reach the desired supply temperature. Undersizing can prevent the collection and use of available solar energy. Chapter 48 gives methods to determine the load.

COOLING BY SOLAR ENERGY

Swartman (1974) emphasizes various absorption systems. Newton (Jordan and Liu 1977) discusses commercially available water vapor/lithium bromide absorption refrigeration systems. Standard absorption chillers are generally designed to give rated capacity for activating fluid temperatures well above 200°F at full load and design condenser water temperature. Few flat-plate collectors can operate efficiently in this range; therefore, a lower hot fluid temperature is used when solar energy provides the heat. Both the temperature of the condenser water and the percentage of design load are determinants of the optimum energizing temperature, which can be quite low, sometimes below 120°F. Proper control can raise the coefficient of performance (COP) at these part-load conditions.

Many large commercial or institutional cooling installations must operate year-round, and Newton (Jordon and Liu 1977) showed that the low-temperature cooling water available in winter enables the LiBr/H$_2$O to function well with a hot fluid inlet temperature below 190°F. Residential chillers in sizes as low as 1.5 tons, with an inlet temperature in the range of 175°F, have been developed.

COOLING BY NOCTURNAL
RADIATION AND EVAPORATION

Radiative cooling is a natural heat loss that causes the formation of dew, frost, and ground fog. Because its effects are the most obvious at night, it is sometimes termed **nocturnal radiation**, although the process continues throughout the day. Thermal infrared radiation, which affects the surface temperature of a building wall or roof, may be estimated by using the sol-air temperature concept. Radiative cooling of window and skylight surfaces can be significant, especially under winter conditions when the dew-point temperature is low.

The most useful parameter for characterizing the radiative heat transfer between horizontal nonspectral emitting surfaces and the sky is the **sky temperature** T_{sky}. If S designates the total down-coming radiant heat flux emitted by the atmosphere, then T_{sky} is defined as

$$T_{sky}^4 = S/\sigma \tag{30}$$

where $\sigma = 0.1713 \times 10^{-8}$ Btu/h·ft^2·°R^4.

The sky radiance is treated as if it originates from a blackbody emitter of temperature T_{sky}. The **net radiative cooling rate** R_{net} of

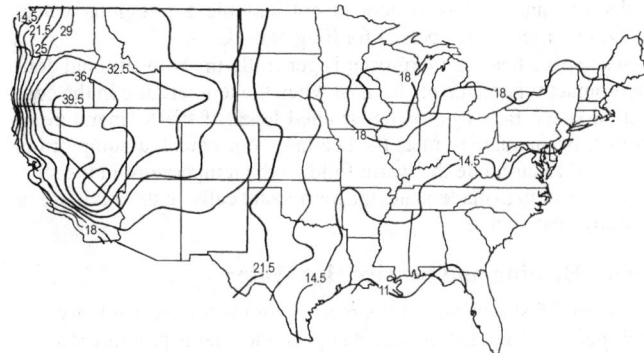

Fig. 22 Average Monthly Sky Temperature Depression
$(T_{air} - T_{sky})$ **for July, °F**
(Adapted from Martin and Berdahl 1984)

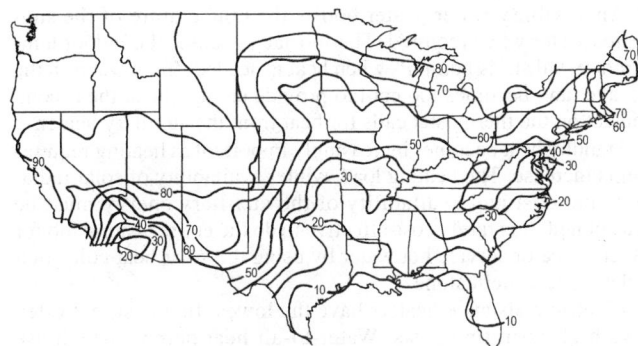

Fig. 23 Percentage of Monthly Hours when
Sky Temperature Falls below 61°F
(Adapted from Martin and Berdahl 1984)

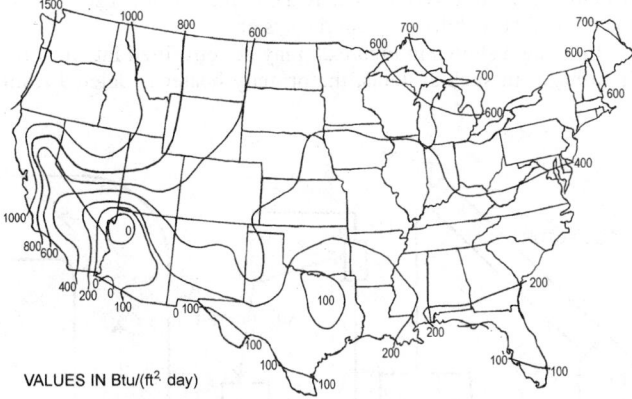

VALUES IN Btu/(ft². day)

Fig. 24 July Nocturnal Net Radiative Cooling Rate from
Horizontal Dry Surface at 76°F
(Adapted from Clark 1981)

a horizontal surface with absolute temperature T_{rad} and a nonspectral emittance ε is then

$$R_{net} = \varepsilon\sigma(T_{rad}^4 - T_{sky}^4) \qquad (31)$$

Values of ε for most nonmetallic construction materials are about 0.9.

Radiative building cooling has not been fully developed. Design methods and performance data compiled by Hay and Yellott (1969) and Marlatt et al. (1984) are available for residential roof ponds that use a sealed volume of water covered by sliding insulation panels as the combined rooftop radiator and thermal storage. Other conceptual radiative cooling designs have been proposed, but more developmental work is required (Givoni 1981; Mitchell and Biggs 1979).

The sky temperature is a function of atmospheric water vapor, the amount of cloud cover, and air temperature; the lowest sky temperatures occur under an arid, cloudless sky. The monthly average sky temperature depression, which is the average of the difference between the ambient air temperature and the sky temperature, typically lies between 9 and 43°F throughout the continental United States. Martin and Berdahl (1984) have calculated this quantity using hourly weather data from 193 sites, as shown in the contour map for the month of July (Figure 22).

The sky temperature may be too high at night to effectively cool the structure. Martin and Berdahl (1984) suggest that the sky temperature should be less than 61°F to achieve reasonable cooling in July (Figure 23). In regions where sky temperatures fall below 61°F 40% or more of the month, all nighttime hours are effectively available for radiative cooling.

Clark (1981) modeled a horizontal radiator at various surface temperatures in convective contact with outdoor air for 77 U.S. locations. The average monthly cooling rates for a surface temperature of 76°F are plotted in Figure 24. If effective steps are taken to reduce the surface convection coefficient by modifying the radiator geometry or using an infrared-transparent glazing, it may be possible to improve performance beyond these values.

SOLAR HEATING AND COOLING SYSTEMS

The components and subsystems discussed earlier may be combined to create a wide variety of solar heating and cooling systems. These systems fall into two principal categories: passive and active.

Passive solar systems require little, if any, nonrenewable energy to make them function (Yellott 1977; Yellott et al. 1976). Every building is passive in the sense that the sun tends to warm it by day, and it loses heat at night. Passive systems incorporate solar collection, storage, and distribution into the architectural design of the building and make minimal or no use of fans to deliver the collected energy to the structure. Passive solar heating, cooling, and lighting design must consider the building envelope and its orientation, the thermal storage mass, and window configuration and design. DOE (1980, 1982), LBL (1981), Mazria (1979), and ASHRAE (1984) give estimates of energy savings resulting from the application of passive solar design concepts.

Active solar systems use either liquid or air as the collector fluid. Active systems must have a continuous availability of nonrenewable energy, generally in the form of electricity, to operate pumps and fans. A complete system includes solar collectors, energy storage devices, and pumps or fans for transferring energy to storage or to the load. The load can be space cooling, heating, or hot water. Although it is technically possible to construct a solar heating and cooling system to supply 100% of the design load, such a system would be uneconomical and oversized. The size of the solar system, and thus its ability to meet the load, is determined by life-cycle cost analysis that weighs the cost of energy saved against the amortized solar cost.

Active solar energy systems have been combined with heat pumps for water and/or space heating. The most economical arrangement in residential heating is a solar system in parallel with a heat pump, which supplies auxiliary energy when the solar source is not available. For domestic water systems requiring high water temperatures, a heat pump placed in series with the solar storage tank may be advantageous. Freeman et al. (1979) and Morehouse and Hughes (1979) present information on performance and estimated energy savings for solar-heat pumps.

Hybrid systems combine elements of both active and passive systems. Hybrid systems require some nonrenewable energy, but the amount is so small that they can maintain a coefficient of performance of about 50.

Passive Systems

Passive systems may be divided into several categories. The first residence to which the name **solar house** was applied used a large expanse of south-facing glass to admit solar radiation; this is known as a **direct gain** passive system.

Indirect gain solar houses use the south-facing wall surface or the roof of the structure to absorb solar radiation, which causes a rise in temperature that, in turn, conveys heat into the building in several ways. This principle was applied to the pueblos and cliff dwellings of the southwestern United States. Glass has led to modern adaptations of the indirect gain principle (Trombe et al. 1977; Balcomb et al. 1977).

By glazing a large south-facing, massive masonry wall, solar energy can be absorbed during the day, and conduction of heat to the inner surface provides radiant heating at night. The mass of the wall and its relatively low thermal diffusivity delays the arrival of the heat at the indoor surface until it is needed. The glazing reduces the loss of heat from the wall back to the atmosphere and increases the collection efficiency of the system.

Openings in the wall, which are near the floor and ceiling, allow convection to transfer heat to the room. The air in the space between the glass and the wall warms as soon as the sun heats the outer surface of the wall. The heated air rises and enters the building through the upper openings. Cool air flows through the lower openings, and convective heat gain can be established as long as the sun is shining.

In another indirect gain passive system, a metal roof-ceiling supports transparent plastic bags filled with water (Hay and Yellott 1969). Movable insulation above these water-filled bags is rolled away during the winter day to allow the sun to warm the stored water. The water then transmits heat indoors by convection and radiation. The insulation remains over the water bags at night or during overcast days. During the summer, the water bags are exposed at night for cooling by (1) convection, (2) radiation, and (3) evaporation of water on the water bags. The insulation covers the water bags during the day to protect them from unwanted irradiation. Pittenger et al. (1978) tested a building for which water rather than insulation was moved to provide summer cooling and winter heating.

Attached greenhouses (sunspaces) can be used as solar attachments when the orientation and other local conditions are suitable. The greenhouse can provide a buffer between the exterior wall of the building and the outdoors. During daylight, warm air from the greenhouse can be introduced into the house by natural convection or a small fan.

In most passive systems, control is accomplished by moving a component that regulates the amount of solar radiation admitted into the structure. Manually operated window shades or venetian blinds are the most widely used and simplest controls.

Passive heating and cooling systems have been effective in field demonstrations (Howard and Pollock 1982; Howard and Saunders 1989).

Active Systems

Active systems absorb solar radiation with collectors and convey it to storage using a suitable fluid. As heat is needed, it is obtained from storage via heated air or water. Control is exercised by several types of thermostats, the first being a differential device that starts the flow of fluid through the collectors when they have been sufficiently warmed by the sun. It also stops the fluid flow when the collectors no longer gain heat. In locations where freezing occurs only rarely, a low-temperature sensor on the collector controls a circulating pump when freezing is impending. This process wastes some stored heat, but it prevents costly damage to the

collector panels. This system is not suitable for regions where freezing temperatures persist for long periods.

The space heating thermostat is generally the conventional double-contact type that calls for heat when the temperature in the controlled space falls to a predetermined level. If the temperature in storage is adequate to meet the heating requirement, a pump or fan is started to circulate the warm fluid. If the temperature in the storage subsystem is inadequate, the thermostat calls on the auxiliary or standby heat source.

Space Heating and Service Hot Water

Figure 25 shows one of the many systems for service hot water and space heating. In this case, a large, atmospheric pressure storage tank is used, from which water is pumped to the collectors by pump P_1 in response to the differential thermostat T_1. Drainback is used to prevent freezing, because the amount of antifreeze required would be prohibitively expensive. Service hot water is obtained by placing a heat exchanger coil in the tank near the top, where, even if stratification occurs, the hottest water will be found.

An auxiliary water heater boosts the temperature of the sun-heated water when required. Thermostat T_2 senses the indoor temperature and starts pump P_2 when heat is needed. If the water in the storage tank becomes too cool to provide enough heat, the second contact on the thermostat calls for heat from the auxiliary heater.

Standby heat becomes increasingly important as heating requirements increase. The heating load, winter availability of solar radiation, and cost and availability of the auxiliary energy must be determined. It is rarely cost-effective to do the entire heating job for either space or service hot water by using the solar heat collection and storage system alone.

Electric resistance heaters have the lowest first cost, but often have high operating costs. Water-to-air heat pumps, which use sun-heated water from the storage tank as the evaporator energy source, are an alternative auxiliary heat source. The heat pump's COP is 10 to 14 Btu of heat for each watt-hour of energy supplied to the compressor. When summer cooling as well as winter heating are needed, the heat pump becomes a logical solution, particularly in large systems where a cooling tower is used to dissipate the heat withdrawn from the system.

The system shown in Figure 25 may be retrofitted into a warm air furnace. In such systems, the primary heater is deleted from

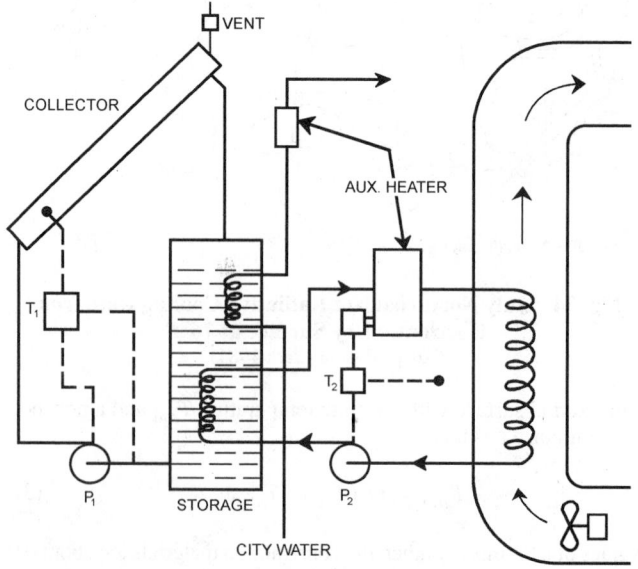

Fig. 25   Solar Collection, Storage, and Distribution System for Domestic Hot Water and Space Heating

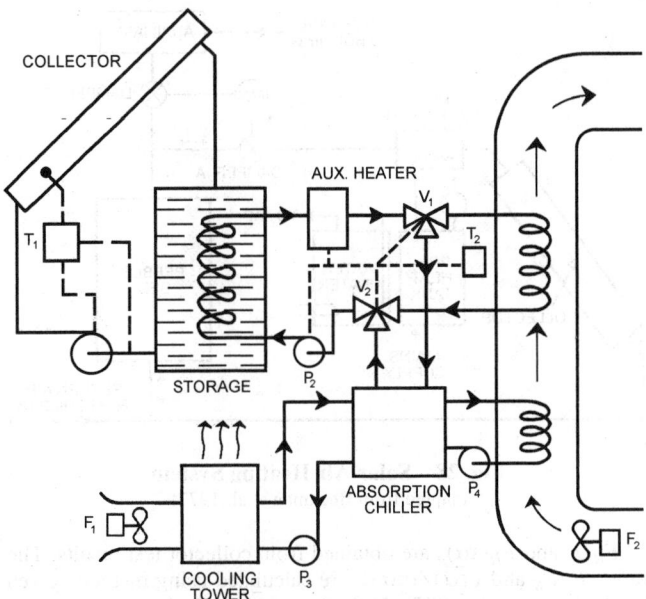

Fig. 26   Space Heating and Cooling System Using Lithium Bromide-Water Absorption Chiller

the space heating circuit, and the coil is located in the return duct of the existing furnace. Full backup is thus obtained, and the auxiliary heater provides only the heat not available at the storage temperature.

Solar Cooling with Absorption Refrigeration

When solar energy is used for cooling as well as for heating, the absorption system shown in Figure 26, or one of its many modifications, may be used. The collector and storage must operate at a temperature approaching 200°F on hot summer days when the water from the cooling tower exceeds 80°F, but considerably lower operating water temperatures may be used when cooler water is available from the tower. The controls for collection, cooling, and distribution are generally separated, with the circulating pump P_1 operating in response to the collector thermostat T_1, which is located within the air-conditioned space. When T_2 calls for heating, valves V_1 and V_2 direct the water flow from the storage tank through the unactivated auxiliary heater to the fan coil in the air distribution system. The fan F_1 in this unit may respond to the thermostat also, or it may have its own control circuit so that it can bring in outdoor air when a suitable temperature condition is present.

When thermostat T_2 calls for cooling, the valves direct the hot water into the absorption unit's generator, and pumps P_3 and P_4 are activated to pump the cooling tower water through the absorber and condenser circuits and the chilled water through the cooling coil in the air distribution system. A relatively large hot water storage tank allows the unit to operate when no sunshine is available. A chilled water storage tank (not shown) may be added so that the absorption unit can operate during the day whenever water is available at a sufficiently high temperature to make the unit function properly. The COP of a typical lithium bromide-water absorption unit may be as high as 0.75 under favorable conditions, but frequent on-off cycling of the unit to meet a high variable cooling load may cause significant loss in performance since the unit must be heated to operating temperature after each shutdown. Modulating systems are analyzed differently than on-off systems.

Water-cooled condensers are required with the absorption cycles, because the lithium bromide-water cycle operates with a relatively delicate balance among the temperatures of the three fluid circuits—cooling tower water, chilled water, and activating water.

The steam-operated absorption systems, from which solar cooling systems are derived, customarily operate at energizing temperatures of 230 to 240°F, but these are above the capability of most flat-plate collectors. The solar cooling units are designed to operate at considerably lower temperature, but unit ratings are also lowered.

Smaller domestic units may operate with natural circulation, or **percolation**, which carries the lithium bromide-water solution from the generator (to which the activating heat is supplied) to the separator and condenser; there, the reconcentrated LiBr is returned to the absorber while the water vapor goes to the condenser before being returned to the evaporator where cooling takes place. Larger units use a centrifugal pump to transfer the fluid.

SIZING SOLAR HEATING AND COOLING SYSTEMS—ENERGY REQUIREMENTS

Methods used to determine solar heating and/or cooling energy requirements for both active and passive/hybrid systems are described by Feldman and Merriam (1979) and Hunn et al. (1987). Descriptions of public and private domain methods are included. An overview of the simulation techniques suitable for active heating and cooling systems analysis, and for passive/hybrid heating, cooling, and lighting analysis follows.

Simplified Analysis Methods

Simplified analysis methods have the advantages of computational speed, low cost, rapid turnaround (especially important during iterative design phases), and ease of use by persons with little technical experience. Disadvantages include limited flexibility for design optimization, lack of control over assumptions, and a limited selection of systems that can be analyzed. Thus, if the application, configuration, or load characteristics under consideration are significantly nonstandard, a detailed computer simulation may be required to achieve accurate results. This section describes the *f*-Chart method for active solar heating and the solar load ratio method for passive solar heating (Dickinson and Cheremisinoff 1980, Lunde 1980, and Klein and Beckman 1979).

Active Heating/Cooling

Beckman et al. (1977) developed the *f*-Chart method using an hourly simulation program (Klein et al. 1976) to evaluate space heating and service water heating in many climates and conditions. The results of these analyses correlate the fraction *f* of the heat load met by solar energy. The correlations give the fraction *f* of the monthly heating load (for space heating and hot water) supplied by solar energy as a function of collector characteristics, heating loads, and weather. The standard error of the differences between detailed simulations in 14 locations in the United States and the *f*-Chart predictions was about 2.5%. Correlations also agree within the accuracy of measurements of long-term performance data. Beckman et al. (1977, 1981), and Duffie and Beckman (1980) discuss the method in detail.

The *f*-Chart method requires the following data:

- Monthly average daily radiation on a horizontal surface
- Monthly average ambient temperatures
- Collector thermal performance curve slope and intercept from standard collector tests, i.e., $F_R U_L$ and $F_R(\tau\alpha)_n$ (see ASHRAE *Standard* 93 and Chapter 33, 1996 *ASHRAE Handbook—Systems and Equipment*)
- Monthly space and water heating loads

Standard Systems

The *f*-Chart assumes several standard systems and applies only to these liquid configurations. The standard **liquid heater** uses water, an antifreeze solution, or air as the heat transfer fluid in the collector loop and water as the storage medium (Figure 27). Energy

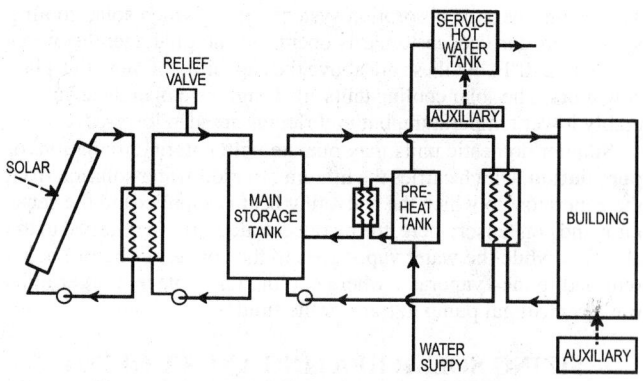

Fig. 27   Liquid-Based Solar Heating System
(Adapted from Beckman et al. 1977)

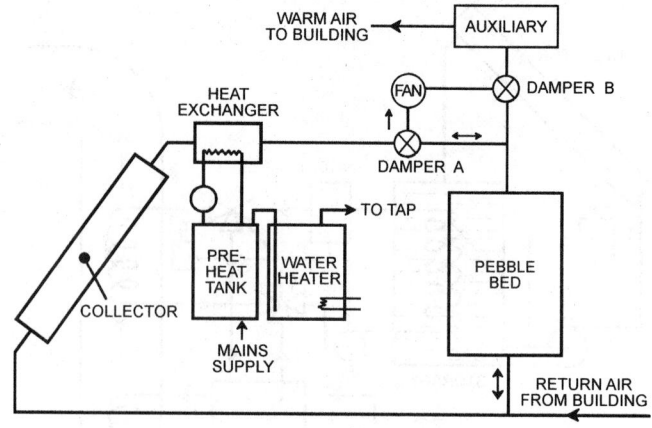

Fig. 28   Solar Air Heating System
(Adapted from Beckman et al. 1977)

is stored in the form of sensible heat in a water tank. A water-to-air heat exchanger transfers heat from the storage tank to the building. A liquid-to-liquid heat exchanger transfers energy from the main storage tank to a domestic hot water preheat tank, which in turn supplies solar heated water to a conventional water heater. A conventional furnace or heat pump is used to meet the space heating load when the energy in the storage tank is depleted.

Figure 28 shows the assumed configuration for a **solar air heater** with a pebble-bed storage unit. Energy for domestic hot water is provided by heat exchange from the air leaving the collector to a domestic water preheat tank as in the liquid system. The hot water is further heated, if necessary, by a conventional water heater. During summer operation, a seasonal, manually operated storage bypass damper is used to avoid heat loss from the hot bed into the building.

The standard **solar domestic water heater** collector heats either air or liquid. Collected energy is transferred by a heat exchanger to a domestic water preheat tank that supplies solar-heated water to a convectional water heater. The water is further heated to the desired temperature by conventional fuel if necessary.

f-Chart Method

Computer simulations correlate dimensionless variables and the long-term performance of the systems. The fraction f of the monthly space and water heating loads supplied by solar energy is empirically related to two dimensionless groups. The first dimensionless group X is collector loss; the second Y is collector gain:

$$X = \frac{F_R U_L A_c \Delta\theta}{L}\left(\frac{F_r}{F_R}\right)(t_{ref} - \bar{t}_a) \qquad (32)$$

$$Y = \frac{F_R(\tau\alpha)_n H_T N A_c}{L}\left(\frac{F_r}{F_R}\right)\left[\frac{(\overline{\tau\alpha})}{(\tau\alpha)_n}\right] \qquad (33)$$

where

A_c = area of solar collector, ft^2
F_r = collector-heat exchanger efficiency factor
F_R = collector efficiency factor
U_L = collector overall energy loss coefficient, Btu/h·ft^2·°F
$\Delta\theta$ = total number of hours in month
$\bar{t}_a$ = monthly average ambient temperature, °F
L = monthly total heating load for space heating and hot water, Btu
H_T = monthly averaged, daily radiation incident on collector surface per unit area, Btu/day·ft^2
N = number of days in month
$(\overline{\tau\alpha})$ = monthly average transmittance-absorptance product
$(\tau\alpha)_n$ = normal transmittance-absorptance product
t_{ref} = reference temperature, 212°F

$F_R U_L$ and $F_R(\tau\alpha)_n$ are obtained from collector test results. The ratios F_r/F_R and $(\overline{\tau\alpha})/(\tau\alpha)_n$ are calculated using methods given by Beckman et al. (1977). The value of $\bar{t}_a$ is obtained from meteorological records for the month and location desired. H_T is calculated from the monthly average, daily radiation on a horizontal surface by the methods in Chapter 33 of the 1996 *ASHRAE Handbook—Systems and Equipment* or in Duffie and Beckman (1980). The monthly load L can be determined by any appropriate load estimating method, including analytical techniques or measurements. Values of the collector area A_c are selected for the calculations. Thus, all the terms in these equations can be determined from available information.

Transmittance of the transparent collector cover τ and the absorptance of the collector plate α depend on the angle at which solar radiation is incident on the collector surface. Collector tests are usually run with the radiation incident on the collector in a nearly perpendicular direction. Thus, the value of $F_R(\tau\alpha)_n$ determined from these tests ordinarily corresponds to the transmittance and absorptance values for radiation at normal incidence. Depending on collector orientation and time of year, the monthly average values of the transmittance and absorptance can be significantly lower. The *f*-Chart method requires a knowledge of the ratio of the monthly average to normal incidence transmittance-absorptance.

The *f*-Chart method for liquid systems is similar to that for air systems. The fraction of the monthly total heating load supplied by the solar air heating system is correlated with the dimensionless groups X and Y, as shown in Figure 29. To determine the fraction of the heating load supplied by solar energy for a month, values of X and Y are calculated for the collector and heating load in question. The value of f is determined at the intersection of X and Y on the *f*-Chart, or from the following equivalent equations.

$$\text{Air system: } f = 1.04\ Y - 0.065\ X - 0.159\ Y^2 + 0.00187\ X^2 - 0.0095\ Y^3 \qquad (34)$$

$$\text{Liquid system: } f = 1.029\ Y - 0.065\ X - 0.245\ Y^2 + 0.0018\ X^2 + 0.025\ Y^3 \qquad (35)$$

This is done for each month of the year. The solar energy contribution for the month is the product of f and the total heating load L for the month. Finally, the fraction F of the annual heating load supplied by solar energy is the sum of the monthly solar energy contributions divided by the annual load:

$$F = \Sigma f L/\Sigma L$$

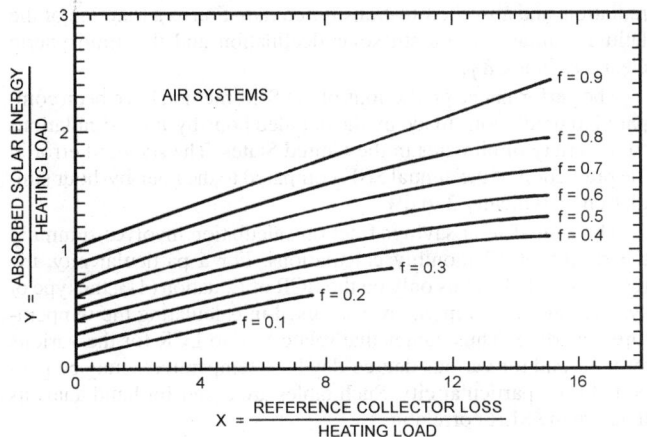

Fig. 29 Chart for Air System
(Adapted from Beckman et al. 1977)

Example 6. Calculating the heating performance of a residence, assume that a solar heating system is to be designed for use in Madison, WI, with two-cover collectors facing south, inclined 58° with respect to the horizontal. The air heating collectors have the characteristics $F_R U_L = 0.50$ Btu/h·ft²·°F and $F_R(\tau\alpha)_n = 0.49$. The $\bar{t}_a$ is 19.4°F, the total space and water heating load for January is calculated to be 34.1×10^6 Btu, and the solar radiation incident on the plane of the collector is calculated to be 1.16×10^3 Btu/day·ft². Determine the fraction of the load supplied by solar energy with a system having a collector area of 538.2 ft².

Solution: For air systems, there is no heat exchanger penalty factor and $F_r/F_R = 1$. The value of $(\overline{\tau\alpha})/(\tau\alpha)_n$ is 0.94 for a two-cover collector in January. Therefore, the values of X and Y are

$$X = (0.50 \text{ Btu/h} \cdot \text{ft}^2 \cdot °\text{F})(1)(212°\text{F} - 19.4°\text{F})(31 \text{ days})$$
$$(24 \text{ h/day})(538.2 \text{ ft}^2)/(34.1 \times 10^6 \text{ Btu}) = 1.13$$

$$Y = (0.49)(1)(0.94)(1.16 \times 10^3 \text{ Btu/day} \cdot \text{ft}^2)$$
$$(31 \text{ days})(538.2 \text{ ft}^2)/(34.1 \times 10^6 \text{ Btu}) = 0.26$$

Then the fraction f of the energy supplied for January is 0.19. The total solar energy supplied by this system in January is

$$fL = 0.19 \times 34.1 \times 10^6 \text{ Btu} = 6.4 \times 10^6 \text{ Btu}$$

The annual system performance is obtained by summing the energy quantities for all months. The result is that 37% of the annual load is supplied by solar energy.

The collector heat removal factor F_R that appears in X and Y is a function of the collector fluid flow rate. Because of the higher cost of power for moving fluid through air collectors than through liquid collectors, the capacitance rate used in air heaters is ordinarily much lower than that in liquid heaters. As a result, air heaters generally have a lower value of F_R. Values of F_R corresponding to the expected airflow in the collector must be used to calculate X and Y.

An increase in airflow rate tends to improve collector performance by increasing F_R, but it tends to decrease performance by reducing the degree of thermal stratification in the pebble bed (or water storage tank). The f-Chart for air systems is based on a collector airflow rate of 2 scfm per square foot of collector area. The performance with different collector airflow rates can be estimated by using the appropriate values of F_R in both X and Y. A further modification to the value of X is required to account for the change in degree of stratification in the pebble bed.

Air system performance is less sensitive to storage capacity than that of liquid systems for two reasons: (1) air systems can operate with air delivered directly to the building in which storage is not used, and (2) pebble beds are highly stratified and additional capacity is effectively added to the cold end of the bed, which is seldom heated and cooled to the same extent as the hot end. The f-Chart for air systems is for a nominal storage capacity. Performance of systems with other storage capacities can be determined by modifying the dimensionless group X as described in Beckman et al. (1977).

With modification, f-Charts can be used to estimate the performance of solar water heating operating in the range of 120 to 160°F. The main water supply temperature and the minimum acceptable hot water temperature (i.e., the desired delivery temperature) both affect the performance of solar water heating. The dimensionless group X, which is related to collector energy loss, can be redefined to include these effects. If monthly values of X are multiplied by a correction factor, the f-Chart for liquid-based solar space and water heating systems can be used to estimate monthly values of f for water heating. Experiments and analysis show that the load profile for a well-designed heater has little effect on long-term performance. Although the f-Chart was originally developed for two-tank systems, it may be applied to single- and double-tank domestic hot water systems with and without collector tank heat exchangers.

For industrial process heating, absorption air conditioning, or other processes for which the delivery temperature is outside the normal f-Chart range, modified f-Charts are applicable (Klein et al. 1976). The concept underlying these charts is that of solar usability, which is the fraction of the total solar energy that is useful in the given process. This fraction depends on the required delivery temperature as well as collector characteristics and solar radiation. The procedure allows the energy delivered to be calculated in a manner similar to that for f-Charts. An example of the application of this method to solar-assisted heat pumps is presented in Svard et al. (1981).

Other Active Collector Methods

The **relative areas method**, based on correlations of the f-Chart method, predicts annual rather than monthly active heating performance (Barley and Winn 1978). An hourly simulation program has been used to develop the **monthly solar-load ratio (SLR) method**, another simplified procedure for residential systems (Dickinson and Cheremisinoff 1980). Based on hour-by-hour simulations, a method was devised to estimate performance based on monthly values of horizontal solar radiation and heating degree-days. This SLR method has also been extended to nonresidential buildings for a range of design water temperatures (Dickinson and Cheremisinoff 1980; Schnurr et al. 1981).

Passive Heating

A widely accepted simplified passive space heating design tool is the solar-load ratio method (DOE 1980, 1982; ASHRAE 1984). It can be applied manually, although like the f-Chart, it is available on microcomputer software. The SLR method for passive systems is based on correlating results of multiple hour-by-hour computer simulations, the algorithms of which have been validated against test cell data for the following generic passive heating types: direct gain, thermal storage wall, and attached sunspace. Monthly and annual performance, as expressed by the auxiliary heating requirement, is predicted by this method. The method applies to single-zone, envelope-dominated buildings. A simplified, annual-basis distillation of SLR results, the **load collector ratio (LCR) method**, and several simple-to-use rules have grown out of the SLR method. Several hand-held calculator and microcomputer programs have been written using the method (Nordham 1981).

The SLR method uses a single dimensionless correlating parameter (SLR), which Balcomb et al. (1982) define as a particular ratio of solar energy gains to building heating load:

$$\text{SLR} = \frac{\text{Solar energy absorbed}}{\text{Building heating load}} \quad (36)$$

A correlation period of 1 month is used; thus the quantities in the SLR are calculated for a 1 month period.

The parameter that is correlated to the SLR, the **solar savings fraction (SSF)**, is defined as

$$SSF = 1 - \frac{\text{Auxiliary heat}}{\text{Net reference load}} \qquad (37)$$

The SSF measures the energy saving expected from the passive solar building, relative to a reference nonpassive solar building.

In Equation (37), the net reference load is equal to the degree-day load DD of the nonsolar elements of the building:

$$\text{Net reference load} = (NLC)(DD) \qquad (38)$$

where NLC is the net load coefficient, which is a modified *UA* coefficient computed by leaving out the solar elements of the building. The nominal units are Btu/°F·day. The term DD is the temperature departure in degree-days computed for an appropriate base temperature. A building energy analysis based on the SLR correlations begins with a calculation of the monthly SSF values. The monthly auxiliary heating requirement is then calculated by

$$\text{Auxiliary heat} = (NLC)(DD)(1 - SSF) \qquad (39)$$

Annual auxiliary heat is calculated by summing the monthly values.

By definition, SSF is the fraction of the heat load of the nonsolar portions of the building met by the solar element. If the solar elements of the building (south-facing walls and window in the northern hemisphere) were replaced by other elements so that the net annual flow of heat through these elements was zero, the annual heat consumption of the building would be the net reference load. The saving achieved by the solar elements would therefore be the net reference load in Equation (39) minus the auxiliary heat in Equation (40), which gives

$$\text{Solar saving} = (NLC)(DD)(SSF) \qquad (40)$$

Although simple, in many situations and climates, Equation (40) is only approximately true because a normal solar-facing wall with a normal complement of opaque walls and windows has a near-zero effect over the entire heating season. In any case, the auxiliary heat estimate is the primary result and does not depend on this assumption.

The hour-by-hour simulations used as the basis for the SLR correlations are done with a detailed model of the building in which all the design parameters are specified. The only parameter that remains a variable is the solar collector area, which can be expressed in terms of the load collector ratio (LCR):

$$LCR = \frac{NLC}{A_p} = \frac{\text{Net load coefficient}}{\text{Projected collector area}} \qquad (41)$$

Performance variations are estimated from the correlations, which allow the user to account directly for thermostat set point, internal heat generation, glazing orientation, and configuration, shading, and other solar radiation modifiers. Major solar system characteristics are accounted for by selecting one of 94 reference designs. Other design parameters, such as thermal storage thickness and conductivity, and the spacing between glazings, are included in a series of sensitivity calculations obtained using hour-by-hour simulations. The results are generally presented in graphic form so that the designer can see the effect of changing a particular parameter.

Solar radiation correlations for the collector area have been determined using hour-by-hour simulation and typical meteorological year (TMY) weather data. These correlations are expressed as ratios of incident-to-horizontal radiation, transmitted-to-incident

radiation, and absorbed-to-transmitted radiation as a function of the latitude minus mid-month solar declination and the atmospheric clearness index K_T.

The performance predictions of the SLR method have been compared to predictions made by the detailed hour-by-hour simulations for a variety of climates in the United States. The standard error in the prediction of the annual SSF, compared to the hour-by-hour simulation, is typically 2 to 4%.

The annual solar savings fraction calculation involves summing the results of 12 monthly calculations. For a particular city, the resulting SSF depends only on the LCR of Equation (41), the type of system, and the temperature base used in calculating the temperature departure. Thus, tables that relate SSF to LCR for the various systems and for various degree-day base temperatures may be generated for a particular city. Such tables are easier for hand analysis than are the SLR correlations.

Annual SSF versus LCR tables have been developed for 209 locations in the United States and 14 cities in southern Canada for 94 reference designs and 12 base temperatures (ASHRAE 1984).

Example 7. Consider a small office building located in Denver, Colorado, with 3000 ft^2 of usable space and a sunspace entry foyer that faces due south; the projected collector area A_p is 420 ft^2. A sketch and preliminary plan are shown in Figure 30. Distribution of solar heat to the offices is primarily by convection through the doorways from the sunspace. The principle thermal mass is in the common wall that separates the sunspace from the offices and in the sunspace floor. This example is abstracted from the detailed version given in Balcomb et al. (1982). Even though lighting and cooling are likely to have the greatest energy costs for this building, heating is a significant energy item and should be addressed by a design that integrates passive solar heating, cooling, and lighting.

Solution: Table 5 shows calculations of the net load coefficient. Then:

$$NLC = 24 \times 525 = 12{,}600 \text{ Btu/°F} \cdot \text{day}$$

Table 5 Calculations for Example 7

	Area A, ft^2	U-Factor, Btu/h·ft^2·°F	UA, Btu/h·°F
Opaque wall	2000	0.04	80
Ceiling	3000	0.03	90
Floor (over crawl space)	3000	0.04	120
Windows (E, W, N)	100	0.55	55
		Subtotal	345
Infiltration			180
		Subtotal	525
Sunspace (treated as unheated space)			151
		Total	676

ASHRAE procedures are approximated with V = volume, ft^3; c = heat capacity of Denver air, Btu/ft^3·°F; ACH = air changes/h; equivalent UA for infiltration = VcACH. In this case, $V = 24{,}000$ ft^3, $c = 0.015$ Btu/ft^3·°F, and ACH = 0.5.

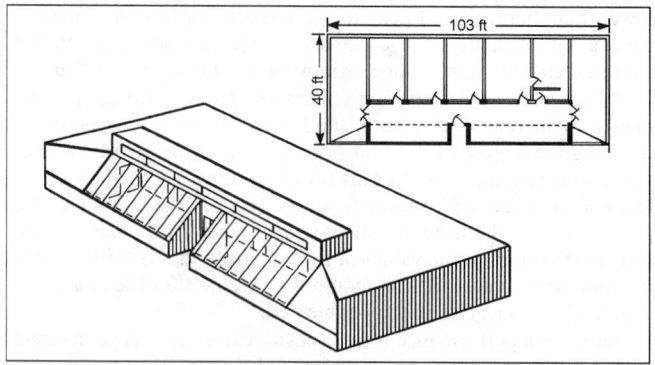

Fig. 30 Commercial Building in Example 7

and the total load coefficient includes the solar aperture:

$$TLC = 24 \times 676 = 16{,}224 \text{ Btu}/°\text{F} \cdot \text{day}$$

From Equation (43), the load collector ratio is

$$LCR = 12{,}600/420 = 30 \text{ Btu/ft}^2 \cdot °\text{F} \cdot \text{day}$$

Suppose the daily internal heat is 130,000 Btu/day, and the average thermostat setting is 68°F. Then,

$$T_{base} = 68 - 130{,}000/16{,}224 = 60°\text{F}$$

(Note that solar gains to the space are not included in the internal gain term as they customarily are for nonsolar buildings.)

In this example, the solar system is type SSD1, defined in ASHRAE (1984) and Balcomb et al. (1982). It is a semiclosed sunspace with a 12-in. masonry common wall between it and the heated space (offices). The aperture is double-glazed, with a 50° tilt and no night insulation. To achieve a projected area of 420 ft², a sloped glazed area of 420/sin 50° = 548 ft² is required.

The SLR correlation for solar system type SSD1 is shown in Figure 31. Values of the absorbed solar energy S and heating degree-days DD to base temperature 60°F are determined monthly. For S, the solar radiation correlation presented in ASHRAE (1984) for tabulated Denver, CO, weather data are used. For January, the horizontal surface incident radiation is 840 Btu/ft²·day. The 60°F base degree-days are 933. Using the tabulated incident-to-absorbed coefficients found in ASHRAE (1984), $S = 43{,}681$ Btu/ft². Thus $S/DD = 46.8$ Btu/ft²·°F·day. From Figure 31, at an LCR = 30 Btu/ft²·°F·day, the SSF = 0.51. Therefore, for January

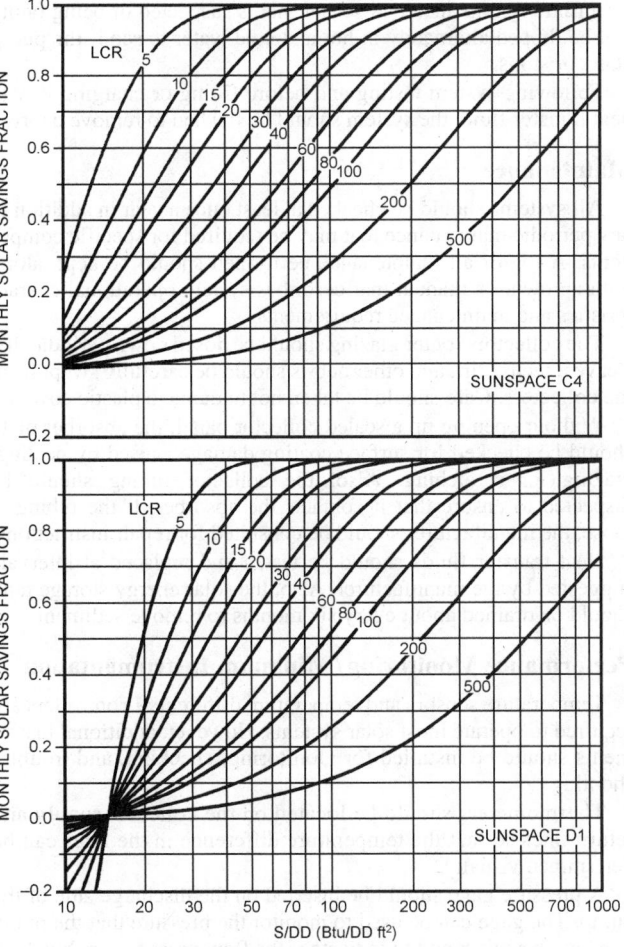

Fig. 31 Monthly SSF Versus Monthly S/DD for Various LCR Values

$$\text{Net reference load} = (933)(12{,}600) = 11.76 \times 10^6 \text{ Btu}$$

$$\text{Solar savings} = (11.76 \times 10^6)(0.51) = 6.00 \times 10^6 \text{ Btu}$$

$$\text{Auxiliary heat} = (11.76 \times 10^6) - (6.00 \times 10^6) = 5.76 \times 10^6 \text{ Btu}$$

Repeating this calculation for each month and adding the results for the year yields an annual auxiliary heat of 19.94×10^6 Btu.

Other Passive Heating Methods

The concept of usability has been applied to passive buildings. In this approach, the energy requirements of zero- and infinite-capacity buildings are calculated. The amount of solar energy that enters the building and exceeds the instantaneous load of the zero-capacity building is then calculated. This excess energy must be dumped in the zero-capacity building, but it can be stored to offset heating loads in a finite-capacity building. Methods are provided to interpolate between the zero- and infinite-capacity limits for finite-capacity buildings. Equations and graphs for direct gain and collector-storage wall systems are given in Monsen et al. (1981, 1982).

INSTALLATION GUIDELINES

Most solar components are the same as those in HVAC and hot water systems (pumps, piping, valves, and controls), and their installation is not much different from a conventional installation. Solar collectors are the most unfamiliar component in a solar heater. They are located outdoors, which requires penetration of the building envelope. They also require a structural element to support them at the proper tilt and orientation toward the sun.

The site must taken into account. Collectors should be (1) located so that shading is minimized and (2) installed so that they are attractive both on and off site. They should also be located to minimize vandalism and to avoid a safety hazard.

Collectors should be placed near the storage tank to reduce piping cost and heat loss. The collector and piping must be installed so that they can be drained without trapping fluid in the system.

For best annual performance, collectors should be installed at a tilt angle above the horizontal that is appropriate for the local latitude. In the northern hemisphere they should be oriented toward true south, not magnetic south. Small variations in tilt (±10°) and orientation (±20°) do not reduce performance significantly.

Collector Mounting

Solar collectors are usually mounted on the ground or on flat or pitched roofs. A roof location necessitates penetration of the building envelope by mounting hardware, piping, and control wiring. Ground or flat roof-mounted collectors are generally rack-mounted.

Pitched roof mounting can be done several ways. Collectors can be mounted on structural **standoffs**, which support them at an angle other than that of the roof to optimize solar tilt. In another pitched roof-mounting technique known as **direct mounting**, collectors are placed on a waterproof membrane on top of the roof sheeting. The finished roof surface together with the necessary collector structural attachments and flashing are then built up around the collector. A weatherproof seal between the collector and the roof must be maintained to prevent leakage, mildew, and rotting.

Integral mounting can be done for new pitched roof construction. The collector is attached to and supported by the structural framing members. The top of the collector then serves as the finished roof surface. Weather tightness is crucial to avoid damage and mildew.

Collectors should support snow loads that occur on the roof area they cover. The collector tilt usually expedites snow sliding with only a small loss in efficiency. The roof structure should be free of objects that could impede snow sliding, and the collectors should be raised high enough to prevent snow buildup over them.

The mounting structure should be built to withstand winds of at least 100 mph, which impose a wind load of 40 lb/ft² on a vertical surface or an average of 25 lb/ft^2 on a tilted roof (HUD 1977). Wind load requirements may be higher, depending on local building codes. Flat-plate collectors mounted flush with the roof surface should be constructed to withstand the same wind loads.

The collector array becomes more vulnerable to wind gusts as the angle of the mount increases. This wind load, in addition to the equivalent roof area wind loads, should be determined according to accepted engineering procedures (ASCE *Standards* 7, 8, and 9).

Expansion and contraction of system components, material compatibility, and the use of dissimilar metals must be considered. Collector arrays and mounting hardware (bolts, screws, washers, and angles) must be well protected from corrosion. Steel-mounting hardware in contact with aluminum, and copper piping in contact with aluminum hardware are both examples of metal combinations that have a high potential for corrosion. Dissimilar metals can be separated by washers made of fluorocarbon polymer, phenolic, or neoprene rubber.

Freeze Protection

Freeze protection is extremely important and is often the determining factor when selecting a system in the United States. Freezing can occur at ambient temperatures as high as 42°F because of radiation to the night sky. Manual freeze protection should not be used for commercial installations.

One simple way of protecting against freezing is to drain the fluid from the collector array and interior piping when potential freezing conditions exist. The drainage may be automatic, as in draindown and drainback systems, or manual, as in direct thermosiphon systems. Automatic systems should be capable of fail-safe drainage operation—even in the event of pump failure or power outage. In some cases water may be designed to drain back through the pump, so the design must allow refilling without causing cavitation.

In areas where freezing is infrequent, recirculating water from storage to the collector array can be used as freeze protection. Freeze protection can also be provided by using fluids that resist freezing. Fluids such as water/glycol solutions, silicone oils, and hydrocarbon oils are circulated by pumps through the collector array and double wall heat exchanger. Draining the collector fluid is not required, because these fluids have freezing points well below the coldest anticipated outdoor temperature.

In mild climates where recirculation freeze protection is used, a second level of freeze protection should be provided by flushing the collector with cold supply water when the collector approaches near-freezing temperatures. This can be accomplished with a temperature-controlled valve that will automatically open a small port at a near-freezing temperature of about 40°F and then close at a slightly higher temperature.

Over-Temperature Protection

During periods of high insolation and low hot water demand, overheating can occur in the collectors or storage tanks. Protection against overheating must be considered for all portions of the solar hot water system. Liquid expansion or excessive pressure can burst piping or storage tanks. Steam or other gases within a system can restrict liquid flow, making the system inoperable.

The most common methods of overheat protection stop circulation in the collection loop until the storage temperature decreases, discharge the overheated water and replace it with cold makeup water, or use a heat exchanger as a means of heat rejection. Some freeze protection methods can also provide overheat protection.

For nonfreezing fluids such as glycol antifreezes, over-temperature protection is needed to limit fluid degradation at high temperatures during collector stagnation.

Safety

Safety precautions required for installing, operating, and servicing a solar domestic hot water heater are essentially the same as those required for a conventional domestic hot water heater. One major exception is that some solar systems use nonpotable heat transfer fluids. Local codes may require a double wall heat exchanger for potable water installations.

Pressure relief must be provided in all parts of the collector array that can be isolated by valves. The outlet of these relief valves should be piped to a container or drain, and not where people could be affected.

Start-Up Procedure

After completing the installation, certain tests must be performed before charging or filling the system. The system must be checked for leakage, and pumps, fans, valves, and sensors must be checked to see that they function. Testing procedures vary with system type.

Closed-loop systems should be hydrostatically tested. The system is filled and pressurized to 1.5 times the operating pressure for one hour and inspected for leaks and any appreciable pressure drop.

Drain-down systems should be tested to be sure that all water drains from the collectors and piping located outdoors. All lines should be checked for proper pitch so that gravity drains them completely. All valves should be verified to be in working order.

Drain-back systems should be tested to ensure that the collector fluid is draining back to the reservoir tank when circulation stops and that the system refills properly.

Air systems should be tested for leaks before insulation is applied by starting the fans and checking the ductwork for leaks.

Pumps and sensors should be inspected to verify that they are in proper working order. Proper cycling of the pumps can be checked by a running time meter. A sensor that is suspected of being faulty can be dipped alternately in hot and cold water to see if the pump starts or stops.

Following system testing and before filling or charging it with heat transfer fluid, the system should be flushed to remove debris.

Maintenance

All systems should be checked at least once a year in addition to any periodic maintenance that may be required for specific components. A log of all maintenance performed should be kept, along with an owner's manual that describes system operational characteristics and maintenance requirements.

The collectors' outer glazing should be hosed down periodically. Leaves, seeds, dirt, and other debris should be carefully swept from the collectors. Care should be taken not to damage plastic covers.

Without opening up a sealed collector panel, the absorber plate should be checked for surface coating damage caused by peeling, crazing, or scratching. Also, the collector tubing should be inspected to ensure that it contacts the absorber. If the tubing is loose, the manufacturer should be consulted for repair instructions.

Heat transfer fluids should be tested and replaced at intervals suggested by the manufacturer. Also, the solar energy storage tank should be drained about every six months to remove sediment.

Performance Monitoring / Minimum Instrumentation

Temperature sensors and temperature differential controllers are required to operate most solar systems. However, additional instruments should be installed for monitoring, checking, and troubleshooting.

Thermometers should be located on the collector supply and return lines so that the temperature difference in the lines can be determined visually.

A pressure gage should be inserted on the discharge side of the pump. The gage can be used to monitor the pressure that the pump must work against and to indicate if the flow passages are blocked.

Running time meters on pumps and fans may be installed to determine if the system is cycling properly.

DESIGN, INSTALLATION, AND OPERATION CHECKLIST

The following checklist is for designers of solar heating and cooling systems. Specific values have not been included because these vary for each application. The designer must decide whether design figures are within acceptable limits for any particular project (see DOE 1978a for further information). The review order listed does not reflect their precedence or importance during design.

Collectors

- Check flow rate for compliance with manufacturer's recommendation.
- Check that collector area matches the application and claimed solar fraction of the load.
- Review collector instantaneous efficiency curve and check match between collector and system requirements.
- Relate collector construction to end use; two cover plates are not required for low-temperature collection in warm climates and may, in fact, be detrimental. Two cover plates are more efficient when the temperature difference between the absorber plate and outdoor air is high, such as in severe winter climates or when collecting at high temperatures for cooling. Radiation loss only become significant at relatively high absorber plate temperatures. Selective surfaces should be used in these cases. Flat black surfaces are acceptable and sometimes more desirable for low collection temperatures.
- Check match between collector tilt angle, latitude, and collector end use.
- Check collector azimuth.
- Check collector location for potential shading and exposure to vandalism or accidental damage.
- Review provisions made for high stagnation temperature. If not used, are liquid collectors drained or left filled in the summer?
- Check for snow hang-up and ice formation. Will casing vents become blocked?
- Review precautions, if any, against outgassing.
- Check access for cleaning covers.
- Check mounting for stability in high winds.
- Check for architectural integration. Do collectors on roof present rainwater drainage or condensation problems? Do roof penetrations present potential leak problems?
- Check collector construction for structural integrity and durability. Will materials deteriorate under operating conditions? Will any pieces fall off?
- Are liquid collector passages organized in such a way as to allow natural fill and drain? Does mounting configuration affect this?
- Does air collector duct connection promote a balanced airflow and an even heat transfer? Are connections potentially leaky?

Hydraulics

- Check that the flow rate through the collector array matches system parameters.
- If antifreeze is used, check that the flow rate has been modified to allow for the viscosity and specific heat.
- Review properties of proposed antifreeze. Some fluids are highly flammable. Check toxicity, vapor pressure, flash point, and boiling and freezing temperatures at atmospheric pressure.
- Check means of makeup into antifreeze system. An automatic water makeup system can result in freezing.
- Check that provisions are made for draining and filling the system. (Air vents at high points, drains at low points, pipes correctly graded in-between, drainback vented to storage or expansion tank.)
- If system uses drain-back freeze protection, check that

  1. Provision is made for drain-back volume and back venting
  2. Pipes are graded for drain back

  3. Solar primary pump is sized for lift head
  4. Pump is self-priming if tank is below pump

- Check that collector pressure drop for drainback is slightly higher than static pressure between the supply and return headers.
- Optimum pipe arrangement is reverse return with collectors in parallel. Series collectors reduce flow rate and increase head. A combination of parallel/series can sometimes be beneficial, but check that equipment has been sized and selected properly.
- Cross-connections under different operating modes sometimes result in pumps operating in opposition or tandem, causing severe hydraulic problems.
- If heat exchangers are used, check that approach temperature differential has been recognized in the calculations.
- Check that adequate provisions are made for water expansion and contraction. Use specific volume/temperature tables for calculation. Each unique circuit must have its own provision for expansion and contraction.
- Three-port valves tend to leak through the closed port. This, together with reversed flows in some modes, can cause potential hydraulic problems. As a general rule, simple circuits and controls are better.

Airflow

- Check that the flow rate through the collector array matches the system design values.
- Check temperature rise across collectors using air mass flow and specific heat.
- Check that duct velocities are within the design limits.
- Check that cold air or water cannot flow from collectors by gravity under "no-sun" conditions.
- Verify duct material and construction methods. Ductwork must be sealed to reduce loss.
- Check duct configuration for balanced flow through collector array.
- Check number of collectors in series. More than two collectors in series can reduce collection efficiency.

Thermal Storage

- Check that thermal storage capacity matches values of collector area, collection temperature, utilization temperature, and load.
- Verify that thermal inertia does not impede effective operation.
- Check provisions for promoting temperature stratification during both collection and use.
- Check that pipe and duct connections to storage are compatible with the control, ensuring that only the coolest air goes to the collectors and connections.
- If liquid storage is used for high temperature (above 200°F), check that tank material and construction can withstand the temperature and pressure.
- Check that storage location does not promote unwanted heat loss or gain and that adequate insulation is provided.
- Verify that liquid storage tanks are treated to resist corrosion. This is particularly important in tanks that are partially filled.
- Check that provision is made to protect liquid tanks from exposure to either an overpressure or vacuum.

Uses

Domestic Hot Water

- Characteristics of domestic hot water loads include short periods of high draw interspersed with long dormant periods. Check that domestic hot water storage matches solar heat input.
- Check that provisions have been made to prevent reverse heating of the solar thermal storage by the domestic hot water backup heater.
- Check that the design allows cold makeup water preheating on days of low solar input.

- Verify that the antiscald valve limits domestic hot water supply to a safe temperature during periods of high solar input.
- Depending on total dissolved solids, city water heated above 150°F may precipitate a calcium carbonate scale. If collectors are used to heat water directly, check provisions for preventing scale formation in absorber plate waterways.
- Check whether the heater is required to have a double wall heat exchanger and that it conforms to appropriate codes if the collector uses nonpotable fluids.

Heating

- Warm air heating systems have the potential of using solar energy directly at moderate temperatures. Check that air volume is sufficient to meet the heating load at low supply temperatures and that the limit thermostat has been reset.
- At times of low solar input, solar heat can still be used to meet part of the load by preheating return air. Check location of solar heating coil in system.
- Baseboard heaters require relatively high supply temperatures for satisfactory operation. Their output varies as the 1.5 power of the log mean temperature difference and falls off drastically at low temperatures. If solar is combined with baseboard heating, check that supply temperature is compatible with heating load.
- Heat exchangers imply an approach temperature difference that must be added to the system operating temperature to derive the minimum collection temperature. Verify calculations.
- Water-to-air heat pumps rely on a constant solar water heat source for operation. When the heat source is depleted, the backup system must be used. Check that storage is adequate.

Cooling

- Solar activated absorption cooling with fossil fuel backup is currently the only commercially available active cooling. Be assured of all design criteria and a large amount of solar participation. Verify calculations.
- Storing both hot water and chilled water may make better use of available storage capacity.

Controls

- Check that control design matches desired modes of operation.
- Verify that collector loop controls recognize solar input, collector temperature, and storage temperature.
- Verify that controls allow both the collector loop and the utilization loop to operate independently.
- Check that control sequences are reversible and will always revert to the most economical mode.
- Check that controls are as simple as possible within the system requirements. Complex controls increase the frequency and possibility of breakdowns.
- Check that all controls are fail-safe.

Performance

- Check building heating, cooling, and domestic hot water loads as applicable. Verify that building thermal characteristics are acceptable.
- Check solar energy collected on a monthly basis. Compare with loads and verify solar participation.

PHOTOVOLTAIC APPLICATIONS

Photovoltaic (PV) devices, or cells, convert light directly into electricity. Chapter 33 of the 1996 *ASHRAE Handbook—Systems and Equipment* discusses the fundamentals of their operation. Photovoltaics are used to power communications, equipment in remote locations, remote monitoring, lighting, water pumping, battery charging, and cathodic protection.

Communications. Photovoltaics can provide reliable power with little maintenance for communications systems, especially those in remote areas (away from the power grid) with extreme weather conditions such as high winds, heavy snows, and ice. Examples include communication relay towers, travelers' information transmitters, cellular telephones, mobile radio systems, emergency call boxes, and military test facilities. These systems range in size from a few watts for call boxes to several kilowatts for microwave repeater stations. For larger systems at remote sites, an engine generator is often combined with the photovoltaic-battery system. These hybrid systems with two or more generators can achieve nearly 100% availability.

Remote Site Electrification. Photovoltaics are used to provide power to rural residences, visitor centers in parks, vacation cabins, island and villages, remote research facilities, and military test areas. The varied loads include lighting, small appliances, water pumps (including circulators on solar water heating systems), and communications equipment. The load demand varies from a few watts to tens of kilowatts.

Remote Monitoring. Photovoltaics provide power at remote sites to sensors, data loggers, and associated transmitters for meteorological monitoring, structural condition measurement, seismic recording, irrigation control, highway/traffic monitoring, and scientific research. Most of these applications require less than 200 W, and many can be powered by a single photovoltaic module. Vandalism may be a problem in some areas, so non-glass-covered modules are sometimes used. Mounting the modules on a tall pole or in an unobtrusive manner may also help avoid damage or theft. The batteries are often located in the same weather-resistant enclosure as the data acquisition/monitoring equipment. This enclosure is sometimes camouflaged or buried for protection. Some data loggers come with their own battery and charge regulator.

Signs and Signals. The most popular application for photovoltaics is warning signs. Typical devices that are powered include navigational beacons, audible signals such as sirens, highway warning signs, railroad signals, and aircraft warning beacons. Because these signals are critical to public safety, they must be operative at all times, and thus the reliability of the photovoltaic system is extremely important. High reliability can be achieved by using large-capacity batteries and no charge control. Many of these systems operate in harsh environments. For maritime applications, special modules are used that are resistant to corrosion from salt water.

Water Pumping and Control. Photovoltaics are used for intermediate-sized water pumping applications—those larger than hand pumps and smaller than large engine-powered pumps. The typical range of sizes for photovoltaic-powered pumps is a few hundred watts to a few kilowatts. Applications include domestic use, water for campgrounds, irrigation, village water supplies, and livestock watering. Photovoltaics for livestock watering can be more economical than maintaining a distribution line to a remote pump on a ranch. Most pumping systems do not use batteries but store the water in holding tanks. The photovoltaic modules are often mounted on tracking frames that maximize energy production by tracking the sun each day.

Rest Room Facilities. Highway rest stops, public beach facilities, outdoor recreation parks, and public campgrounds may have photovoltaics to operate air circulation and ventilation fans, interior and exterior lights, and auxiliary water pumps for sinks and showers. For most of these applications, the initial cost for photovoltaic power is the least expensive option.

Charging Vehicle Batteries. Vehicle batteries self-discharge over time if they are not being used. This is a problem for organizations that maintain a fleet of vehicles such as fire-fighting or snow removal equipment, some of which are infrequently used. Photovoltaic battery chargers can solve this problem by providing a trickle charging current that keeps the battery at a high state of charge. Often the PV module can be placed inside the windshield and plugged into the cigarette lighter, thus using existing wiring and protection circuits and providing a quick disconnect for the module. The modules are installed on the roof or engine hood of larger vehicles. An application under development is the use of PV modules to charge the batteries in electric vehicles.

REFERENCES

ASCE. 1990. Specification for the design of cold-formed stainless steel structural members. ANSI/ASCE *Standard* 8-90. American Society of Civil Engineers, Reston, VA.

ASCE. 1991. Standard practice for construction and inspection of composite slabs. ANSI/ASCE *Standard* 9-91. American Society of Civil Engineers, Reston, VA.

ASCE. 1995. Minimum design loads for buildings and other structures. ANSI/ASCE *Standard* 7-95. American Society of Civil Engineers, Reston, VA.

ASHRAE. 1977. *Applications of solar energy for heating and cooling of buildings.*

ASHRAE. 1983. *Solar domestic and service hot water manual.*

ASHRAE. 1984. *Passive solar heating analysis: A design manual.*

ASHRAE. 1991. Methods of testing to determine the thermal performance of solar collectors. *Standard* 93-1986 (Reaffirmed 1991).

ASHRAE. 1995. Bin and degree-hour weather data for simplified energy calculations.

Balcomb, D. et al. 1977. Thermal storage walls in New Mexico. *Solar Age* 2(8):20.

Balcomb, J.D., R.W. Jones, R.D. McFarland, and W.O. Wray. 1982. Expanding the SLR method. *Passive Solar Journal* 1:2.

Barley, C.D. and C.B. Winn. 1978. Optimal sizing of solar collectors by the method of relative areas. *Solar Energy* 21:4.

Beckman, W.A., S.A. Klein, and J.A. Duffie. 1977. *Solar heating design by the f-Chart method.* John Wiley, New York.

Beckman, W.A., S.A. Klein, and J.A. Duffie. 1981. Performance predictions for solar heating systems. *Solar energy handbook,* J.F. Kreider and F. Kreith, eds. McGraw Hill, New York.

Bliss, R.W. 1961. Atmospheric radiation near the surface of the earth. *Solar Energy* 59(3):103.

Clark, G. 1981. Passive/hybrid comfort cooling by thermal radiation. *Proceedings of the International Passive and Hybrid Cooling Conference.* American Section of the International Solar Energy Society, Miami Beach, FL.

Cole, R.L. et al. 1977. Applications of compound parabolic concentrators to solar energy conversion. *Report* No. AMLw42. Argonne National Laboratory, Chicago.

Cromer, C.J. 1984. Design of a DC-pump, photovoltaic-powered circulation system for a solar domestic hot water system. Florida Solar Energy Center (June).

Dickinson, W.C. and P.N. Cheremisinoff, eds. 1980. *Solar energy technology handbook,* Part B: Application, systems design and economics. Marcel Dekker, Inc., New York.

DOE. 1978a. DOE facilities solar design handbook. DOE/AD-0006/1. U.S. Department of Energy.

DOE. 1978b. SOLCOST—Solar hot water handbook; A simplified design method for sizing and costing residential and commercial solar service hot water systems, 3rd ed. DOE/CS-0042/2. U.S. Department of Energy.

DOE. 1980 and 1982. Passive Solar Design Handbooks. Vols. 2 and 3, Passive solar design analysis. DOE Reports/CS-0127/2 and CS-0127/3. January, July. U.S. Department of Energy.

Duffie, J.A. and W.A. Beckman. 1974. *Solar energy thermal processes.* John Wiley and Sons, New York.

Duffie, J.A. and W.A. Beckman. 1980a. *Solar engineering of thermal processes.* John Wiley and Sons, New York.

Duffie, J.A. and W.A. Beckman. 1980b. *Solar thermal energy processes.* Wiley Interscience, New York.

Edwards, D.K. et al. 1962. Spectral and directional thermal radiation characteristics of selective surfaces. *Solar Energy* 6(1):1.

Feldman, S.J. and R.L. Merriam. 1979. Building energy analysis computer programs with solar heating and cooling system capabilities. Arthur D. Little, Inc. *Report* No. EPRIER-1146 (August) to the Electric Power Research Institute.

Francia, G. 1961. A new collector of solar radiant energy. U.N. Conference on New Sources of Energy (Rome) 4:572.

Freeman, T.L., J.W. Mitchell, and T.E. Audit. 1979. Performance of combined solar-heat pump systems. *Solar Energy* 22:2.

Gates, D.M. 1966. Spectral distribution of solar radiation at the earth's surface. *Science* 151(3710):523.

Givoni, B. 1981. Experimental studies on radiant and evaporative cooling of roofs. *Proceedings of the International Passive and Hybrid Cooling Conference.* American Section of the International Solar Energy Society, Miami Beach, FL.

Hay, H.R. and J.I. Yellott. 1969. Natural air conditioning with roof ponds and movable insulation. *ASHRAE Transactions* 75(1):165-77.

Healey, Henry M. 1988. Site-built large volume solar water heating systems for commercial and industrial facilities. *ASHRAE Transactions* 94(1): 1277-86.

Hottel, H.C. and B.B. Woertz. 1942. The performance of flat-plate solar collectors. *Transactions of ASME* 64:91.

Howard, B.D. and E.O. Pollock. 1982. Comparative report—Performance of passive solar heating systems. Vitro Corp. U.S. DOE National Solar Data Program. U.S. TIC Box 62, Oak Ridge, TN 37829.

Howard, B.D. and D.H. Saunders. 1989. Building performance monitoring—The thermal envelope perspective—Past, present, and future. *Thermal Performance of the Exterior Envelopes of Buildings IV.* ASHRAE.

HUD. 1977. Intermediate minimum property standards supplement for solar heating and domestic hot water systems. SD Cat. No. 0-236-648. U.S. Department of Housing and Urban Development.

Hunn, B.D., N. Carlisle, G. Franta, and W. Kolar. 1987. Engineering principles and concepts for active solar systems. SERI/SP-271-2892. Solar Energy Research Institute, Golden, CO.

Jordan, R.C. and B.Y.H. Liu, eds. 1977. Applications of solar energy for heating and cooling of buildings. ASHRAE Publication GRP 170.

Klein, S.A. and W.A. Beckman. 1979. A general design method for closed-loop solar energy systems. *Solar Energy* 22(3):269-82.

Klein, S.A., W.A. Beckman, J.A. Duffie. 1976. TRNSYS—A transient simulation program. *ASHRAE Transactions* 82(1):623-33.

Kutscher, C.F. 1996. *Proceedings of the 19th World Energy Engineering Congress.* Atlanta, GA.

LBL. 1981. DOE-2 Reference Manual Version 2.1A. Los Alamos Scientific Laboratory, Report LA-7689-M, Version 2.1A. Report LBL-8706 Rev. 2, Lawrence Berkeley Laboratory, May.

Lister, L. and T. Newell. 1989. Expansion tank characteristics of closed loop, active solar energy collection systems; Solar engineering—1989. *American Society of Mechanical Engineers,* New York.

Lunde, P.J. 1980. *Thermal engineering.* John Wiley and Sons, New York.

Macriss, R.A. and R.H. Elkins. 1976. Standing pilot gas consumption. *ASHRAE Journal* 18(6):54-57.

Marlatt, W., C. Murray, and S. Squire. 1984. Roofpond systems energy technology engineering center. Rockwell International, *Report* No. ETEC6, April.

Martin, M. and P. Berdahl. 1984. Characteristics of infrared sky radiation in the United States. *Solar Energy* 33(3/4):321-36.

Mazria, E. 1979. *The passive solar energy book.* Rodale Press, Emmaus, PA.

Mitchell, D. and K.L. Biggs. 1979. Radiative cooling of buildings at night. *Applied Energy* 5:263-75.

Monsen, W.A., S.A. Klein, and W.A. Beckman. 1981. Prediction of direct gain solar heating system performance. *Solar Energy* 27(2):143-47.

Monsen, W.A., S.A. Klein, and W.A. Beckman. 1982. The un-utilizability design method for collector-storage walls. *Solar Energy* 29(5):421-29.

Morehouse, J.H. and P.J. Hughes. 1979. Residential solar-heat pump systems: Thermal and economic performance. Paper 79-WA/SOL-25, ASME Winter Annual Meeting, New York, December.

Nordham, D. 1981. Microcomputer methods for solar design and analysis. Solar Energy Research Institute, SERI-SP-722-1127, February.

Pittenger, A.L., W.R. White, and J.I. Yellott. 1978. A new method of passive solar heating and cooling. *Proceedings of the Second National Passive Systems Conference,* Philadelphia, ISES and DOE.

Reitan, C.H. 1963. Surface dew point and water vapor aloft. *Journal of Applied Meteorology* 2(6):776.

Root, D.E., S. Chandra, C. Cromer, J. Harrison, D. LaHart, T. Merrigan, and J.G. Ventre. 1985. *Solar water and pool heating course manual,* 2 vols. Florida Solar Energy Center, Cape Canaveral, FL.

Schnurr, N.M., B.D. Hunn, and K.D. Williamson. 1981. The solar load ratio method applied to commercial buildings active solar system sizing. *Proceedings of the ASME Solar Energy Division Third Annual Conference on System Simulation, Economic Analysis/Solar Heating and Cooling Operational Results.* Reno, NV, May.

Simon, F.F. 1976. Flat-plate solar collector performance evaluation. *Solar Energy* 18(5):451.

Stephenson, D.G. 1967. Tables of solar altitude and azimuth; Intensity and solar heat gain tables. Technical Paper No. 243, Division of Building Research, National Research Council of Canada, Ottawa.

Svard, C.D., J.W. Mitchell, and W.A. Beckman. 1981. Design procedure and applications of solar-assisted series heat pump systems. *Journal of Solar Energy Engineering* 103(5):135.

Swartman, R.K., Vinh Ha, and A.J. Newton. 1974. Review of solar-powered refrigeration. *Paper* No. 73-WA/SOL-6. American Society of Mechanical Engineers, New York.

Thekaekara, M.P. 1973. Solar energy outside the earth's atmosphere. *Solar Energy* 14(2):109 (January).

Threlkeld, J.L. 1963. Solar irradiation of surfaces on clear days. *ASHRAE Transactions* 69:24.

Threlkeld, J.L. and R.C. Jordan. 1958. Direct radiation available on clear days. *ASHRAE Transactions* 64:45.

Trombe, F. et al. 1977. Concrete walls for heat. *Solar Age* 2(8):13.

U.S. GPO. 1968. *Climatic atlas of the U.S.* U.S. Government Printing Office, Washington, D.C.

U.S. Hydrographic Office. 1958. Tables of computed altitude and azimuth. Hydrographic Office Bulletin No. 214, Vols. 2 and 3. U.S. Superintendent of Documents, Washington, D.C.

Van Straaten, J.F. 1961. Hot water from the sun. Ref. No. D-9, National Building Research Institute of South Africa, Council for Industrial and Scientific Research, Pretoria, South Africa.

Whillier, A. 1964. Thermal resistance of the tube-plate bond in solar heat collectors. *Solar Energy* 8(3):95.

Yellott, J.I. 1977. Passive solar heating and cooling systems. *ASHRAE Transactions* 83(2):429.

Yellott, J.I., D. Aiello, G. Rand, and M.Y. Kung. 1976. Solar-oriented architecture. Arizona State University Architecture Foundation, Tempe, AZ.

BIBLIOGRAPHY

ASHRAE. 1986. Methods of testing to determine the thermal performance of flat-plate solar collectors containing a boiling liquid. *Standard* 109-1986 (Reaffirmed 1996).

ASHRAE. 1996. Methods of testing to determine the thermal performance of solar domestic water heating systems. *Standard* 95-1981 (Reaffirmed 1996).

ASHRAE. 1988. *Active solar heating systems design manual.*

ASHRAE. 1989. Methods of testing to determine the thermal performance of unglazed flat-plate liquid-type solar collectors. *Standard* 96-1980 (Reaffirmed 1989).

ASHRAE. 1991. *Active solar heating systems installation manual.*

Bennett, I. 1965. Monthly maps of daily insolation in the U.S. *Solar Energy* 9(3):145.

Bennett, I. 1967. Frequency of daily insolation in Anglo North America during June and December. *Solar Energy* 11(1):41.

Colorado State University. 1980. Solar heating and cooling of residential buildings: Design of systems. Superintendent of Documents, U.S. Government Printing Office, Washington, D.C.

Colorado State University. 1980. Solar heating and cooling of residential buildings: Sizing, insulation and operation of systems. Superintendent of Documents, U.S. Government Printing Office, Washington, D.C.

Cook, J., ed. 1989. *Passive cooling.* MIT Press, Cambridge, MA.

Diamond, S.C. and J.G. Avery. 1986. Active solar energy system design, installation and maintenance: Technical applications manual. LA-UR-86-4175.

HUD. 1980. Installation guidelines for solar DHW systems in one- and two-family dwellings. U.S. Department of Housing and Urban Development, 2nd ed. (May).

Knapp, C.L., T.L. Stoffel, and S.D. Whitaker. 1980. Insolation data manual. SERI/SP-755-789. Solar Energy Research Institute, Golden, CO.

Kreider, J.F. 1989. *Solar design: Components, systems, economics.* Hemisphere Publishing, New York.

Lameiro, G.F. and P. Bendt. 1978. The GFL method for designing solar energy space heating and domestic hot water systems, 2.1. *Proceedings at the 1978 Annual Meeting of the American Section of the International Solar Energy Society,* Denver, CO.

Lane, G.A. 1986. *Solar heat storage: Latent heat materials,* 2 vols. CRC Press, Boca Raton, FL.

Löf, G.O., J.A. Duffie, and C.D. Smith. 1966. World distribution of solar radiation. *Report* No. 21. Solar Energy Laboratory, University of Wisconsin, Madison, WI.

Mueller Associates, Inc. 1985. Active solar thermal design manual. U.S. Department of Energy, Solar Energy Research Institute and ASHRAE.

Mumma, S.A. 1985. Solar collector tilt and azimuth charts for rotated collectors on sloping roofs. *Proceedings Joint ASME-ASES Solar Energy Conference,* Knoxville, TN.

Parmalee, G.V. and W.W. Aubele. 1952. Radiant energy transmission of the atmosphere. *ASHVE Transactions* 58:85.

Solar Energy Research Institute. 1981. Solar design workbook—Solar federal buildings program. SERI/SP-62-308. U.S. Department of Energy and Los Alamos Scientific Laboratory.

Solar Energy Research Institute. 1981. Solar radiation energy resource atlas of the United States. SERI/SP642-1037. Golden, CO.

Solar Environmental Engineering Co., Inc. 1981. Solar domestic hot water system inspection and performance evaluation handbook SERI/SP-98189-1B. Solar Energy Research Institute, Golden, CO.

THERMAL STORAGE

THERMAL storage systems remove heat from or add heat to a storage medium for use at another time. Thermal storage for HVAC applications can involve storage at various temperatures associated with heating or cooling. High-temperature storage is typically associated with solar energy or high-temperature heating, and cool storage with air-conditioning, refrigeration, or cryogenic-temperature processes. Energy may be charged, stored, and discharged daily, weekly, annually, or in seasonal or rapid batch process cycles. The *Design Guide for Cool Thermal Storage* (Dorgan and Elleson 1993) covers cool storage issues and design parameters in more detail.

Thermal storage may be an economically attractive approach to meeting heating or cooling loads if one or more of the following conditions apply:

- Loads are of short duration
- Loads occur infrequently
- Loads are cyclical in nature
- Loads are not well matched to the availability of the energy source
- Energy costs are time-dependent (e.g., time-of-use energy rates or demand charges for peak energy consumption)
- Utility rebates, tax credits, or other economic incentives are provided for the use of load-shifting equipment
- Energy supply from the utility is limited, thus preventing the use of full-size nonstorage systems

Terminology

Heat storage. As used in this chapter, the storage of thermal energy at temperatures above the nominal temperature of the space or process.

Cool storage. As used in this chapter, the storage of thermal energy at temperatures below the nominal temperature of the space or process.

Mass storage. Storage of energy in building materials in the form of sensible heat.

Sensible energy storage (sensible heat storage). Heat storage or cool storage in which all of the energy stored is in the form of sensible heat associated with a temperature change in the storage medium.

Latent energy storage (latent heat storage). Heat storage or cool storage in which the energy stored is largely as latent heat (usually of fusion) associated with a phase change (usually between solid and liquid states) in the storage medium.

Off-peak air conditioning. An air-conditioning system that uses cool storage during peak periods that was produced during off-peak periods.

Off-peak heating. A heating system that uses heat storage.

The preparation of this chapter is assigned to TC 6.9, Thermal Storage.

Storage Media

A wide range of materials can be used as the storage medium. Desirable characteristics include the following:

- Commonly available
- Low cost
- Environmentally benign
- Nonflammable
- Nonexplosive
- Nontoxic
- Compatible with common HVAC materials
- Noncorrosive
- Inert
- Well-documented physical properties
- High density
- High specific heat (for sensible heat storage)
- High heat of fusion (for latent heat storage)
- High heat transfer characteristics
- Storage at ambient pressure
- Characteristics unchanged over long use

Common storage media for sensible heat storage include water, soil, rock, brick, ceramics, concrete, and various portions of the building structure (or process fluid) being heated or cooled. In HVAC applications such as air conditioning, space heating, and water heating, water is often the chosen thermal storage medium; it provides virtually all of the desirable characteristics when kept between its freezing and boiling points. In lower temperature applications, aqueous secondary coolants (typically glycol solutions) are often used as the heat transfer medium, enabling certain storage media to be used below their freezing or phase-change points. For high-temperature heat storage, the storage medium is often rock, brick, or ceramic materials for residential or small commercial applications and oil, oil-rock combinations, or molten salt for large industrial or solar energy power plant applications. Use of the building structure itself as passive thermal storage offers advantages under some circumstances (Morris et al. 1994).

Common storage media for latent heat storage include water-ice, aqueous brine-ice solutions, and other phase-change materials (PCMs) such as hydrated salts and polymers. Clathrates, carbon dioxide, and paraffin waxes are among the alternative storage media used for latent heat storage at various temperatures. For air-conditioning applications, water-ice is the most common storage medium; it provides virtually all of the previously listed desirable characteristics.

A challenge common to all latent heat storage methods is to find an efficient and economical means of achieving the heat transfer necessary to alternately freeze and thaw the storage medium. Various methods have been developed to limit or deal with the heat transfer approach temperatures associated with freezing and melting; however, leaving fluid temperatures (from storage during melting) must be higher than the freezing point, while entering fluid temperatures (to storage during freezing) must be lower than the freezing point. Ice storage can provide leaving temperatures well below those normally used for comfort and nonstorage air-conditioning applications. However, entering temperatures are also much lower than normal. Certain PCM storage systems can be charged using temperatures near those for comfort cooling, but they produce warmer leaving temperatures.

Benefits of Thermal Storage

The primary reasons for the use of thermal storage are economic. The following are some of the key benefits of storage:

Reduced Equipment Size. If thermal storage is used to meet all or a portion of peak heating or cooling loads, equipment sized to meet the peak load can be downsized to meet an average load.

Capital Cost Savings. Capital savings can result both from equipment downsizing and from certain utility cash incentive programs. Even in the absence of utility cash incentives, the savings from downsizing cooling equipment can offset the cost of the storage. Cool storage integrated with low-temperature air and water distribution systems can also provide an initial cost savings due to the use of smaller chillers, pumps, piping, ducts, and fans. Storage has the potential to provide capital savings for systems having heating or cooling peak loads of extremely short duration.

Energy Cost Savings. The significant reduction of time-dependent energy costs such as electric demand charges and on-peak time-of-use energy charges is a major economic incentive for the use of thermal storage.

Energy Savings. Although thermal storage is generally designed primarily to *shift* energy use rather than to *conserve* energy, storage often reduces energy consumption. Cool storage systems permit chillers to operate more at night when lower condensing temperatures improve equipment efficiency; and storage permits the operation of equipment at full-load, avoiding inefficient part-load performance. Documented examples include chilled water storage installations that reduce annual energy consumption for air conditioning by up to 12% (Fiorino 1994; ITSAC 1992).

Improved HVAC Operation. Storage adds an element of thermal capacitance to a heating or cooling system, allowing the decoupling of the thermal load profile from the operation of the equipment. This decoupling can be used to provide increased flexibility, reliability, or backup capacity for the control and operation of the system.

Other Benefits. Storage can bring about other beneficial synergies. As already noted, cool storage can be integrated with cold air distribution. Thermal storage can be configured to serve a secondary function such as fire protection, as is often the case with chilled water storage. Some cool storage can be configured for recharge via free cooling (Holness 1992; Hussain and Peters 1992; Meckler 1992).

ECONOMICS

Thermal storage is installed for two major reasons: (1) to lower initial cost and (2) to lower operating cost.

Lower Initial Cost. Applications such as churches and sports facilities, where the load is short in duration and there is a long time between load occurrences, generally have the lowest initial cost for thermal storage. These buildings have a relatively large space-conditioning load for fewer than 6 h per day and only a few days per week. The relatively small refrigeration plant for these applications would operate continuously for up to 100 h or more to recharge the thermal storage. The initial cost of such a design could be less than that of a standard cooling plant designed to meet the highest instantaneous cooling load.

Secondary capital costs may also be lower for thermal storage. For example, the electrical supply (kVa) can sometimes be reduced because peak energy demand is lower. For cold air distribution systems, the volume of air is reduced, and smaller air-handling equipment and ducts may generally be used. The smaller air-handling unit sizes may reduce the mechanical room space required, providing more usable space. The smaller duct sizes may allow the floor-to-floor height of the building to be reduced; in some cases, an additional floor may be added without an increase in building height.

An example of reduction in equipment size is shown in Figure 1. The load profiles shown are for a 100,000-ft^2 commercial building. If the 2040 ton-hour load is met by a nonstorage air-conditioning system, as shown in Figure 1A, a 220-ton chiller is required to meet the peak cooling demand.

If a load-leveling partial storage system is used, as shown in Figure 1B, an 85-ton chiller meets the demand. The design-day cooling load in excess of the chiller output (1020 ton-hour) is supplied by the storage. The cost of storage approximates the amount saved by downsizing the chiller, cooling tower, electrical service, etc., so load-leveling partial storage is often competitive with nonstorage systems on an initial cost basis.

If a full storage system is installed, as shown in Figure 1C, the entire peak load is shifted to the storage, and a 120-ton chiller is required. The size of the chiller equipment may be reduced, but the total equipment cost including the storage is usually higher for the full storage system than for nonstorage systems. Although the initial cost is higher than for the load-leveling system (Figure 1B), full storage offers substantially reduced operating costs because the entire chiller demand is shifted to the off-peak period.

In order to prepare a complete cost analysis, the initial cost must be determined. Equipment cost should be obtained from each of the manufacturers under consideration, and an estimate of installation cost should be made.

Lower Operating Cost. The other reason for installing thermal storage is to reduce operating cost. Most electric utilities charge less during the night or weekend off-peak hours than during the time of highest electrical demand, which often occur on hot summer afternoons due to air-conditioning use. Electric rates are normally divided into a demand charge and a consumption charge.

The monthly demand charge is based on the building's highest recorded demand for electricity during the month and is measured over a brief period (usually 1 to 15 min). Ratchet billing, another form of demand charge, is based on the highest annual monthly demand. This charge is assessed each month, even if the maximum demand in the particular month is less than the annual peak.

The consumption charge is based on the total measured use of electricity in kilowatt-hours (kWh) over a longer period and are generally representative of the utility's cost of fuel to operate its generation facilities. In some cases, the consumption charge is lower during off-peak hours because a higher proportion of the electricity is generated by baseload plants that are less expensive to operate. Rates that reflect this difference are known as time-of-use billing structures.

To compare the costs of different systems, the annual operating cost of each system being considered must be estimated, including both electrical demand and consumption costs. To determine demand cost, the monthly peak demand for each system is multiplied by the demand charge and totaled for the year; any necessary adjustments for ratchet billing must be made. The electrical consumption cost is determined by totaling the annual energy use for each system in kilowatt-hours and multiplying it by the cost per kilowatt-hour. For time-of-use billing, energy use must be classified by time-of-use period and multiplied by the corresponding rate.

The first step is to estimate the annual utility cost for nonstorage. For example, the peak demand in Figure 1A, which occurs at 3 P.M. (220 ton × 3.52 kW/ton = 775 kW), is multiplied

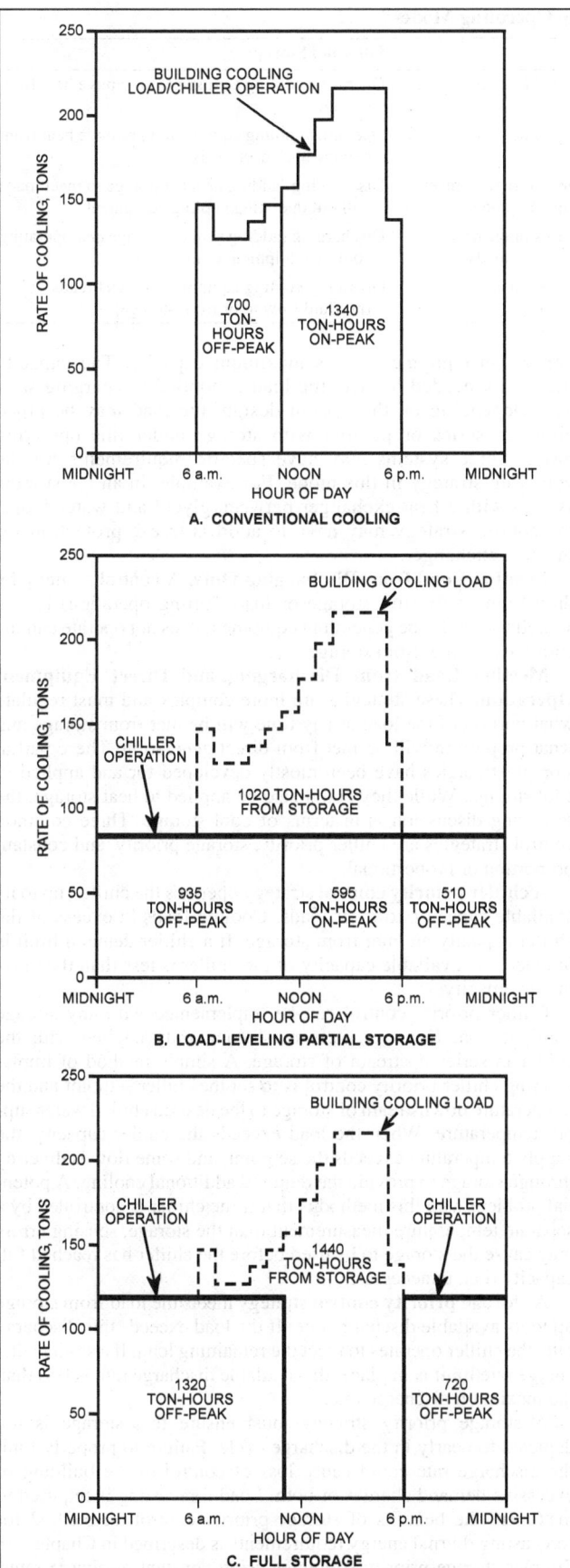

Fig. 1 Hourly Cooling Load Profiles

by the demand charge and, if necessary, adjusted for ratchet billing. The energy used during the off-peak period for the nonstorage system (corresponding to 700 ton-hour of cooling) would be multiplied by the off-peak electrical rate. The energy used during the on-peak period (1340 ton-hour) is multiplied by the on-peak electrical rate. This procedure is repeated (1) for each day of each month to determine the energy charge and (2) for the peak day in each month to determine the demand charge.

Annual cost for a load-leveling partial thermal storage is calculated the same way. The peak demand in Figure 1B (85 ton × 3.52 kW/ton = 300 kW) is multiplied by the demand charge and adjusted for any ratchet billing. This process is repeated for each month. The demand charge for the load-leveling system is lower than for a nonstorage system; the difference represents the demand savings. If time-of-use billing is used, the energy used during the off-peak period (935 + 510 ton-hour) is multiplied by the off-peak electrical rate, and the energy consumed during the on-peak period (595 ton-hour) is multiplied by the on-peak electrical rate. The total energy cost is lower because less energy is used during the on-peak hours.

The full storage system in Figure 1C eliminates all energy use during the on-peak hours, except energy used by the auxiliary equipment that transfers stored energy. The energy cost for the off-peak period (1320 + 720 ton-hour) is multiplied by the off-peak rate. This procedure should be repeated for each month to determine the annual energy cost.

The operating cost and the net capital cost should be analyzed using the life-cycle cost method or other suitable method to determine which design is best for the project. Other items to be considered are space requirements, reliability, and ease of interface to the planned delivery equipment. An optimal application of thermal energy storage balances the savings in utility charges against the initial cost of the installation needed to achieve the savings.

APPLICATIONS

Thermal storage can take many forms to suit a variety of applications. This section addresses several groups of thermal storage applications: off-peak air conditioning, retrofits, industrial/process cooling, off-peak heating, district heating and cooling, and other applications.

Thermal Storage Operation and Control

In general, thermal storage operation and control is more schedule-dependent than that for instantaneous systems. Because thermal storage systems separate the generation of heating or cooling from its use, control of each of these functions must be considered separately. Also, many thermal storage systems offer the ability to provide heating or cooling either directly or from storage. With this flexibility comes the need to define how loads will be met at any time. That is, the methods for providing this level of control are defined in terms of (1) the operating modes available to the system, (2) the control strategies used to implement the operating modes, and (3) the operating strategies that determine when the various operating modes and control strategies are selected.

Operating Modes

A complete thermal storage design includes a detailed description of the intended operating strategy and its associated control strategies and operating modes. This description should address control for the complete storage cycle under full-load and part-load operation, including seasonal variations.

The **operating strategy** defines the overall method of controlling the thermal storage in order to achieve the design intent. The operating strategy determines the logic that governs the selection of operating modes such as charging, meeting the load from chiller(s) only, or meeting the load from discharging storage. The operating

Table 1 Thermal Storage Operating Modes

Operating Mode	For Heat Storage	For Cool Storage
Charging storage	Operating heating equipment to add heat to storage	Operating cooling equipment to remove heat from storage
Charging storage while meeting loads	Operating heating equipment to add heat to storage *and* meet loads	Operating cooling equipment to remove heat from storage *and* meet loads
Meeting loads, from discharging storage only	Discharging (removing heat from) storage to meet loads without operating heating equipment	Discharging (adding heat to) storage to meet loads without operating cooling equipment
Meeting loads, from discharging storage and direct equipment operation	Discharging (removing heat from) storage *and* operating heating equipment to meet loads	Discharging (adding heat to) storage *and* operating cooling equipment to meet loads
Meeting loads, from direct equipment operation only	Operating heating equipment to meet loads (no fluid flow to or from storage)	Operating cooling equipment to meet loads (no fluid flow to or from storage)

strategy also determines which control strategy is implemented within each mode.

A thermal storage **control strategy** defines how the system is controlled when it is in a specific operating mode. The control strategy defines the actions of individual control loops and the values of their setpoints in response to changes in load or other variables.

A thermal storage **operating mode** describes which of several possible functions the system is currently performing. The five basic operating modes are described in Table 1.

The available operating modes differ for individual thermal storage systems. Some systems may include fewer than the five basic modes. For example, the option to meet the load while charging may not be available. In some installations, operation to meet loads may be defined by a single operating mode that includes discharging-only at one end of a continuum and direct equipment-only at the other. In fact, many systems operate with just two modes: daytime and nighttime operation.

Many systems also include other operating modes. Some examples include

• Charging cool storage from free cooling
• Charging cool storage while recovering condenser heat
• Charging heat storage with recovered condenser heat
• Discharging at distinct supply temperatures
• Discharging in conjunction with various combinations of available equipment

In general, the control system selects the current operating mode based on the time of day and the day of the week. Other variables that may also be considered include outdoor temperature, current load, or total facility demand at the billing meter.

In some cases, particularly large cool storage systems with multiple chillers and multiple loads, different operating modes or control strategies may be applied to different parts of a system at one time. For example, one chiller may operate in a charging-only mode, while another chiller operates to meet a load.

Control Strategies

A thermal storage control strategy defines how the system is controlled in a specific operating mode. The control strategy defines what equipment is running and the actions of individual control loops, including the values of their setpoints, in response to changes in load or other variables.

Charging the Storage. Control strategies for charging are generally easily defined. Typically the generation equipment operates at full capacity with a constant supply temperature setpoint and a constant flow through the storage. This operation continues until the storage is fully charged or the period available for charging has ended. Under this basic charging control strategy, the entire capacity of the equipment is applied to charging storage.

Charging Storage while Meeting Load. A control strategy for charging storage while meeting load may also operate the

generation equipment at its maximum capacity. The capacity that is not needed to meet the load is applied to charging storage. Depending on the system design, the load may be piped either in series or parallel with storage under this operating mode. Some systems may have specific requirements for the operating strategy in this mode. For example, in an ice storage system with a heat exchanger between glycol and water loops, the control strategy may have to address freeze protection for the heat exchanger.

Meeting Load from Discharging Only. A control strategy by discharging only (full storage or load shifting operation) is also straightforward. The generating equipment does not operate and the entire load is met from storage.

Meeting Load from Discharging and Direct Equipment Operation. These strategies are more complex and must regulate what portion of the load at any time will be met from storage and what proportion will be met from direct generation. These partial storage strategies have been mostly developed for and applied to cool storage. While they could also be applied to heat storage, the following discussion is in terms of cool storage. Three common control strategies are chiller priority, storage priority, and constant proportion or proportional.

A **chiller priority control** strategy operates the chiller, up to its available capacity, to meet loads. Cooling loads in excess of the chiller capacity are met from storage. If a chiller demand limit is in place, the available capacity of the chiller is less than the maximum capacity.

Chiller priority control can be implemented with any storage configuration. However, it is most commonly applied with the chiller in series upstream of storage. A simple method of implementing chiller priority control is to set the chiller setpoint and the temperature downstream of storage to the desired chilled water supply temperature. When the load exceeds the chiller capacity, the supply temperature exceeds the setpoint, and some flow is diverted through storage to provide the required additional cooling. A potential problem with this method is that if the chiller is controlled by a separate temperature measurement than the storage, sensing errors may cause the storage to be used before the chiller has reached full capacity (i.e., unnecessarily).

A **storage priority control** strategy meets the load from storage up to its available discharge rate. If the load exceeds this discharge rate, the chiller operates to meet the remaining load. If a storage discharge rate limit is in place, the available discharge rate is less than the maximum discharge rate.

A storage priority strategy must ensure that storage is not depleted too early in the discharge cycle. Failure to properly limit the discharge rate could cause loss of control of the building or excessive demand charges or both. Load forecasting is required to maximize the benefits of storage-priority control. A method for forecasting diurnal energy requirements is described in Chapter 40. Simpler storage-priority strategies using constant discharge rates, predetermined discharge rate schedules or pseudo-predictive methods have been used.

A **constant proportion** or **proportional control** strategy divides the load between chiller and storage. The load may be divided equally or in some other proportion. The proportion may change with time in response to changing conditions. A limit on chiller demand or storage discharge may be applied.

Demand-limiting control may be applied to any of the above control strategies. This type of control attempts to limit the facility demand either by setting a maximum capacity above which the chiller is not allowed to operate or by modulating the chiller set-point. Because chiller demand may comprise 30% or more of the total demand, a large demand savings is possible. The demand limit may be a constant value or it may change with time in response to changing conditions. Demand limiting is most effective when the chiller capacity is controlled in response to the facility demand at the billing meter. In such cases, the chiller capacity is controlled to keep the total demand from exceeding a predetermined facility demand limit. A simpler approach, which generally achieves lower demand savings, is simply to limit chiller capacity or the chiller's electric demand without considering the total facility demand.

The storage discharge rate may also be limited, similar to the chiller demand limit. The discharge limit establishes a maximum discharge rate that the storage is allowed to provide. Such a limit is typically used with storage priority control strategies to ensure that sufficient capacity will be available for the entire discharge cycle. The discharge rate limit may change with time in response to changing conditions. The discharge limit may be defined in terms of maximum instantaneous cooling capacity supplied from storage. Alternatively, it may be defined either in terms of the maximum flow through storage, or the minimum mixed temperature leaving the storage and its bypass.

Operating strategies that seek to optimize system operation often recalculate the demand limit and discharge limit on a regular basis during the discharge period.

Nearly any partial storage control strategy can be described by specifying (1) whether it is chiller priority, storage priority, or constant proportion, and (2) by specifying the applicable chiller demand limit and storage discharge rate limit.

Applicable utility rates and system efficiency in various operating modes determine the selection of a control strategy. If on-peak energy cost is significantly higher than off-peak energy cost, the use of stored energy should be maximized and a storage priority strategy is appropriate. If on-peak energy is not significantly more expensive than off-peak energy, a chiller priority strategy is more appropriate. If demand charges are high, some type of demand-limiting control should be implemented.

Operating Strategies

It is important to distinguish between the operating strategy, which defines the higher-level logic by which a system will be operated, and the various control strategies, which implement the operating strategy through specific control actions. The operating strategy defines the overall method of controlling the thermal storage system in order to achieve the design intent. The operating strategy determines the logic that governs when each operating mode is selected, as well as what control strategy is implemented within each mode.

Dorgan and Elleson (1993) use the term **operating strategy** to refer to full storage and partial storage operation. This chapter focuses on design day operation and does not discuss operation under all conditions. For example, a system that is designed for partial storage operation on the design day may operate with a full-storage strategy during many times of the year.

Operating strategies that use sophisticated routines to optimize the use of storage have been investigated (Drees 1994). The cost-saving benefits of such optimal strategies are often small in comparison to well-designed logic that makes full and appropriate use of the principles described previously.

Characterizing Operating and Control Strategies

State diagrams are an effective tool for modeling operating strategy. Rumbaugh et al. (1991) describe the use of state diagrams for modeling systems while Ruchti et al. (1996) describes the application of state diagrams to cool storage control. Figure 2 shows an example of a state diagram of a cool storage system. With structured state diagrams, operating strategies are modeling as a nested hierarchy of superstates and substates. For example, a simple cool storage system may have just three superstates at the highest level in the hierarchy: charging, discharging, and idle. Each of these three superstates may have multiple substates. For example, there may be two strategies of discharging: chiller priority and storage priority. In some cases, particularly large systems with multiple chillers and multiple loads, different operating modes or control strategies may be applied to different parts of a system at one time. For example, one chiller may operate in a charging-only mode, while another chiller operates to meet a load.

Common Mistakes

Poorly designed control logic may induce higher costs than if storage had not been used. The following logic should be avoided.

1. Locking out storage during off-peak periods. Doing so may cause overheating and unnecessary service calls to answer complaints that the building is too hot when the load is higher than the chiller capacity during off-peak periods.
2. Charging an external melt system to full capacity every night. External melt systems should only be charged to the minimum required capacity for the next day.

OFF-PEAK AIR CONDITIONING

Refrigeration Design

Packaged compression or absorption equipment is available for most of the cool storage technologies discussed in this chapter. The maximum cooling load that can be satisfied with a single package and the number of packages that must be installed in parallel to yield the required capacity depend on the specific technology. Individual components such as ice builders, chiller bundles, falling-film chillers, ice makers, compressors, and condensers can be used and interconnected on-site.

For the most part, chillers in chilled water storage systems operate at conditions similar to those for nonstorage applications. However, a greater percentage of the operating hours occur at lower ambient temperatures; special consideration should thus be given to providing a condensing temperature that maintains compressor differential.

The lower suction temperature necessary for making ice imposes a higher compression ratio on the refrigeration equipment. Positive displacement compressors (e.g., reciprocating, screw, and scroll compressors) are usually better suited to these higher compression ratios than centrifugal compressors.

Ice storage systems must operate over a wider range of conditions than nonstorage systems. Care should be taken in design and installation to ensure proper operation at all operating points. The following general categorization of refrigeration equipment applies to ice storage systems:

- Direct-expansion systems feed refrigerant to the heat exchanger through expansion valves. The wide range of operating loads requires careful selection and control of expansion valves.
- Flooded or overfeed systems circulate liquid refrigerant from a large low-pressure receiver to the ice-making heat exchangers. Chapter 1 of the 1998 *ASHRAE Handbook—Refrigeration* covers liquid overfeed systems.
- Secondary coolant systems chill and circulate low-temperature secondary coolant to the ice-producing and storage equipment.

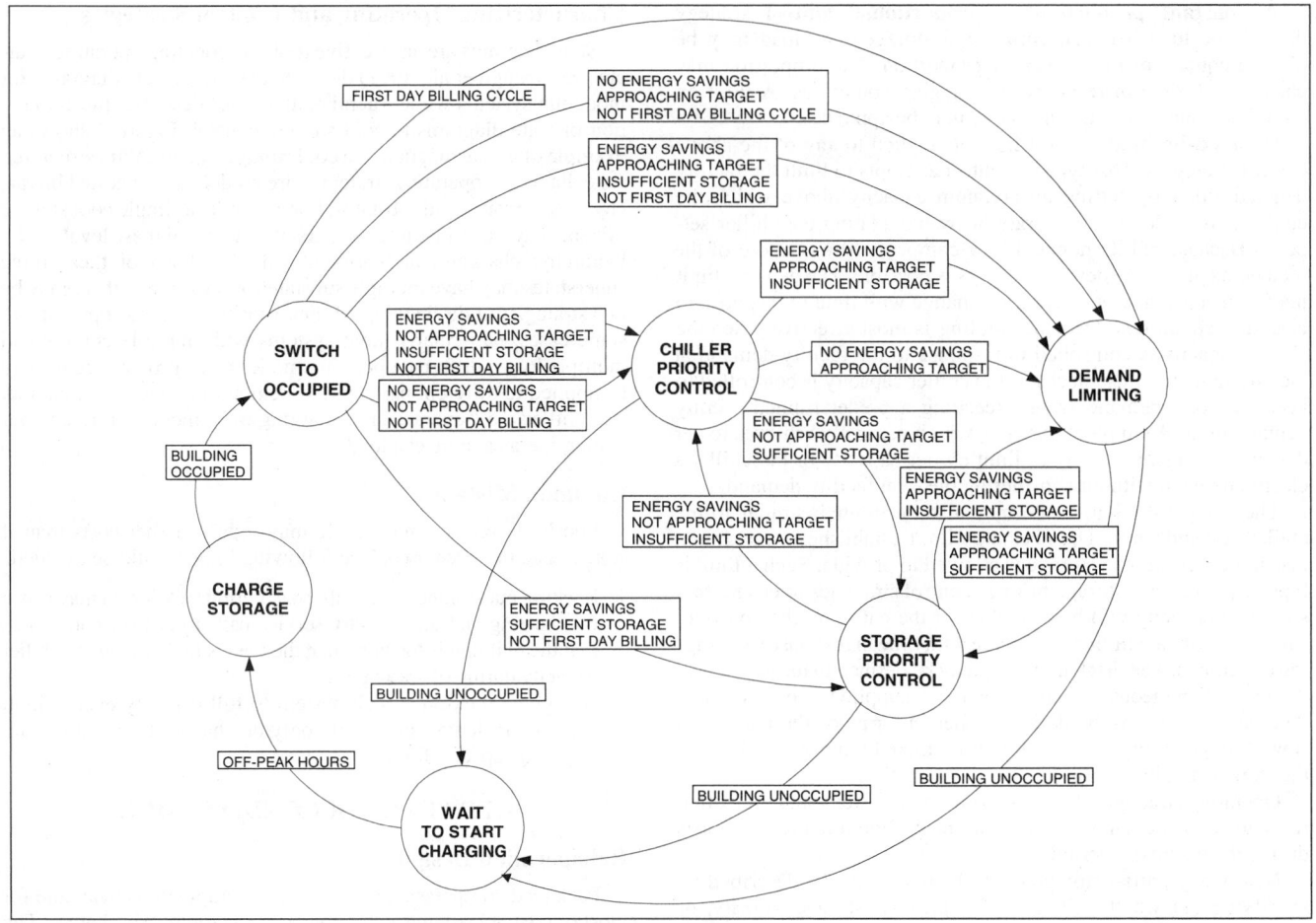

Fig. 2   State Diagram of Cool Storage System

The corrosion properties of the coolant should be considered; a solution containing corrosion inhibitors is normally added. The chiller must also be properly derated both for the lower heat transfer properties of the secondary coolant and for the lower than normal operating temperatures. Secondary coolant systems may use direct-expansion, flooded, or liquid overfeed refrigeration systems.

The following steps should be considered for all ice thermal storage refrigeration:

- **Design for part-load operation.** Refrigerant flow rates, pressure drops, and velocities are reduced during part-load operation. Components and piping must be designed so that control of the system can be maintained and oil can be returned to the compressors at all load conditions.

- **Design for pull-down load.** Because ice-making equipment is designed to operate at water temperatures approaching 32°F, a higher load is imposed on the refrigeration system during the initial start-up, when the inlet water is warmest. The components must be sized to handle this higher load.

- **Plan for chilling versus ice making.** Most ice-making equipment has a much higher instantaneous water/fluid chilling capacity than ice-making capacity. This higher chilling capacity can be used advantageously if the refrigeration equipment and interconnecting piping are properly sized and selected to handle it.

- **Protect compressors from liquid slugging.** Ice builders and ice harvesters tend to contain more refrigerant than chillers of similar capacity used for nonstorage systems. This provides more

opportunity for compressor liquid slugging. Care should be taken to oversize suction accumulators and equip them with high-level compressor cutouts and suction heat exchangers to evaporate any remaining liquid.

- **Oversize the receivers.** Every opportunity should be taken to make the system easy to maintain and service. Maintenance flexibility may be provided in a liquid overfeed system by oversizing the low-pressure receiver and in a direct-expansion system by oversizing the high-pressure receiver.

- **Prevent oil trapping.** Refrigerant lines should be arranged to prevent the trapping of large amounts of oil and to ensure its return to compressors under all operating conditions, especially during periods of low compressor loads. All suction lines should slope toward the suction line accumulators, and all discharge lines should pitch toward the oil separators. Oil tends to collect in the evaporator because that is the location at the lowest temperature and pressure. Because refrigerant accumulators trap oil as well as liquid refrigerant, the larger accumulators needed for ice storage systems require special provisions to ensure adequate oil return to the compressors.

It is very important to fully analyze design trade-offs; **equivalent comparisons** should be made when comparing cool storage to other available cost- and energy-saving opportunities. For example, if energy-efficient motors are assumed for a cool storage design, they should be assumed for the nonstorage system being compared. Current designs, whether based on HCFC, HFC, ammonia, or on any of the proven refrigeration techniques (e.g., direct-expansion, flooded, or liquid overfeed), should be used.

Cold Air Distribution

Reducing the temperature of the distribution air is attractive because smaller air-handling units, ducts, pumps, and piping can be used, resulting in a lower initial cost. In addition, the reduced ceiling space required for ductwork can significantly reduce building height, particularly in high-rise construction. These cost reductions can make thermal storage competitive with nonstorage on an initial cost basis (Landry and Noble 1991).

The optimum supply air temperature should be determined through an analysis of initial and operating cost for the various design options. Depending on the load, the additional latent energy removed at the lower discharge air temperature may be offset by the reduction in fan energy associated with the lower air flow.

The minimum achievable supply air temperature is determined by the chilled water temperature and the temperature rise between the cooling plant and the terminal units. With some ice storage systems, the fluid temperature may rise during discharge; the supply temperatures normally achievable with various types of ice storage plants must therefore be carefully investigated with the equipment supplier.

A heat exchanger, which is sometimes required with storage tanks that operate at atmospheric pressure or between a secondary coolant and chilled water system, adds at least 2°F to the final chilled water supply temperature. The rise in chilled water temperature between the cooling plant discharge and the chilled water coil depends on the length of piping and the amount of insulation; the difference could be up to 0.5°F over long exposed runs.

The difference between the temperature of the chilled water entering the cooling coil and the temperature of the air leaving the coil is generally between 6 and 10°F. A closer approach (smaller temperature differential) can be achieved with more rows or a larger face area on the cooling coil; but extra heat transfer surface to provide a closer approach is often uneconomical. A 3°F temperature rise due to heat gain in the duct between the cooling coil and the terminal units can be assumed for preliminary design analysis. With careful design and adequate insulation, this rise can be reduced to as little as 1°F.

A blow-through configuration provides the lowest supply air temperature and the minimum supply air volume. The lowest temperature rise achievable with a draw-through configuration is 2 to 3°F because heat from the fan is added to the air. A draw-through configuration should be used if space for flow straightening between the fan and coil is limited. A blow-through unit should not be used with a lined duct because air with high relative humidity enters the duct.

Face velocity determines the size of the coil for a given supply air volume. The coil size determines the size of the air-handling unit. A lower face velocity generates a lower supply air temperature, whereas a higher face velocity results in smaller equipment and lower first costs. The face velocity is limited by moisture carryover from the coil. The face velocity for cold air distribution should be 350 to 450 fpm, with an upper limit of 550 fpm.

Cold primary air can be tempered with room air or plenum return air by using fan-powered mixing boxes or induction boxes. The primary air should be tempered before it is supplied to the space. The energy use of fan-powered mixing boxes is significant and negates the savings from downsizing central supply fans (Hittle and Smith 1994; Elleson 1993). Diffusers designed for cold air distribution can provide supply air directly to the space without causing drafts, thereby eliminating the need for fan-powered boxes.

If the supply airflow rate to occupied spaces is expected to be below 0.4 cfm per square foot, fan-powered or induction boxes should be used to boost the air circulation rate. At supply air rates of 0.4 to 0.6 cfm per square foot, a diffuser with a high ratio of induced room air to supply air should be used to ensure adequate dispersion of ventilation air throughout the space. A diffuser that relies on turbulent mixing rather than induction to temper the primary air may not be effective at this flow rate.

Cold air distribution systems normally maintain space humidity between 30 and 45% rh, as opposed to the 50 to 60% rh generally maintained by other systems. At this lower humidity level, equivalent comfort conditions are provided at a higher dry-bulb temperature. The increased dry-bulb set point generally results in decreased energy consumption.

The surfaces of any equipment that may be cooled below the ambient dew point, including air-handling units, ducts, and terminal boxes, should be insulated. All vapor barrier penetrations should be sealed to prevent migration of moisture into the insulation. Prefabricated, insulated round ducts should be insulated externally at joints where internal insulation is not continuous. If ducts are internally insulated, access doors should also be insulated.

Duct leakage is undesirable because it represents cooling capacity that is not delivered to the conditioned space. In cold air distribution, leaking air can cool nearby surfaces to the point that condensation forms. Designers should specify acceptable methods of sealing ducts and air-handling units and establish allowable leakage rates and test procedures. These specifications must be followed up with on-site supervision and inspection during construction.

A thorough commissioning process is important for the optimal operation of any large space-conditioning system, particularly thermal storage and cold air distribution systems. Reductions in initial and operating costs are major selling points for cold air distribution, but the commitment to provide a successful system must not be compromised by the desire to reduce first cost. While a commissioning procedure may appear to involve additional expense, it actually decreases cost by reducing future malfunctions and troubleshooting expense and provides increased value by ensuring optimal operation. The commissioning process is discussed in greater detail in the section on Implementation and Commissioning.

Storage of Heat in Cool Storage Units

Some cool storage installations may be used to provide storage for heating duty and/or heat reclaim. Many commercial buildings have areas that require cooling even during the winter, and the refrigeration plant may be in operation year-round. When cool storage is charged during the off-peak period, the rejected heat may be directed to the heating system or to storage rather than to a cooling tower. Depending on the building loads, the chilled water or ice could be considered a byproduct of heating. One potential use of the heat may be to provide morning warm-up on the following day.

Heating with a cool storage system can be achieved using heat pumps or heat-reclaim equipment such as double bundle condensers, two condensers in series on the refrigeration side, or a heat exchanger in the condenser water circuit. Cooling is withdrawn from storage as needed; when heat is required, the cool storage is recharged and used as the heat source for the heat pump. If needed, additional heat must be obtained from another source because this type of system can supply usable heating energy only when the heat pump or heat-recovery equipment is running. Possible secondary heat sources include a heat storage system, solar collectors, or waste heat recovery from exhaust air. When more cooling than heating is required, excess energy can be rejected through a cooling tower.

Individual storage units may be alternated between heat storage and cool storage service. The stored heat can meet the building's heating requirements, which may include morning warm-up and/or evening heating, and the stored cooling can provide midday cooling. If both the chiller and storage are large enough, all compressor and boiler operation during the on-peak period can be avoided, and the heating or cooling requirements can be satisfied from storage. This method of operation replaces the air-side economizer cycle that involves the use of outside air to cool the building when the enthalpy of the outside air is lower than that of the return air.

The extra cost of equipping a chilled water storage facility for heat storage is small. The necessary plant additions include a partition in the storage tank to convert a portion to warm water storage, some additional controls, and a chiller equipped for heat reclaim. The additional cost may be offset by savings in heating energy. In fact, adding heat storage may increase the economic advantage of cool storage.

Care should be taken to avoid thermal shock in cast-in-place concrete tanks. Tamblyn (1985) showed that the seasonal change from heat storage to cool storage caused sizable leaks to develop if the cool-down period was fewer than five days. Raising the temperature of a concrete tank causes no problems because it generates compressive stresses, which concrete can sustain. Cool-down, on the other hand, causes tensile stresses, under which concrete has low strength.

STORAGE FOR RETROFIT APPLICATIONS

Because latent heat storage (ice and other phase-change materials) has a smaller volume, it is preferable to water storage for retrofit installations where space is limited. Latent heat storage is also preferred for building sites where rock or a high water table make deep excavation for an underground tank expensive. The relatively small size of ice-on-coil units is an advantage in cases where access to the equipment room is limited or where the storage units can be distributed throughout the building and located near the cooling coils.

Existing installations that already contain chilled water piping and chillers are more easily retrofitted to accept chilled water storage than latent heat storage. The requirement for secondary coolant in some ice storage systems may complicate pumping and heat exchange for existing equipment. A secondary coolant temperature colder than the chilled water temperature in the original design typically offsets the reduction in heat transfer.

INDUSTRIAL/PROCESS COOLING

Refrigeration and process cooling typically require lower temperatures than air conditioning. Applications such as vegetable hydrocooling, milk cooling, carcass spray-cooling, and storage room dehumidification can use cold water from an ice storage system. Lower temperatures for freezing food, storing frozen food, and so forth can be obtained by using either a low-temperature hydrated salt phase-change material (PCM) charged and discharged by a secondary coolant or a nonaqueous PCM such as CO_2.

Refrigeration applications have traditionally used heavy-duty industrial refrigeration equipment. The size of the loads has often required custom-engineered, field-erected systems, while the occupancy has often allowed the use of ammonia. Cool storage systems for these applications can be optimized by the choice of refrigerant (halocarbon or ammonia) and the choice of refrigerant feed control (direct-expansion, flooded, mechanically pumped, or gas-pumped).

Some of the advantages of cool storage may be particularly important for refrigeration and process cooling applications. Cool storage can provide a steady supply temperature regardless of large, rapid changes in cooling load or return temperature. Charging equipment and discharging pumps can be placed on a separate electrical service to economically provide cooling during a power blackout or to take advantage of low, interruptible electrical rates. Storage tanks can be oversized to provide water for emergency cooling or fire fighting as well. Cooling can be extracted very quickly from harvested ice storage or chilled water storage to satisfy a very large load and then be recharged slowly using a small charging system. System design may be simplified when the load profile is determined by production scheduling rather than occupancy and outdoor temperatures.

An industrial application for cool storage is the precooling of inlet air for combustion turbines. Oil- or gas-fired combustion turbines typically generate full rated output at an inlet air temperature of 59°F. At higher inlet air temperatures, the mass flow of air is

reduced, as are the shaft power and fuel efficiency. The capacity and efficiency can be increased by precooling the inlet combustion air with chilled water coils or direct contact cooling. The optimum inlet air temperature is typically in the range of 40 to 50°F, depending on the turbine model.

On a 100°F summer day, a combustion turbine driving a generator may lose up to 25% of its rated output. This drop in generator output occurs at the most inopportune time, when an electrical utility or industrial user is most in need of the additional electrical output available from combined cycle or peaking turbines. This loss in capacity can be regained if the inlet air is artificially cooled (MacCracken 1994; Mackie 1994; Ebeling et al. 1994; Andrepont 1994). Storage systems using chilled water or ice are well suited to this application because the cooling is generated and stored during off-peak periods, and maximum net increase in generator capacity is obtained by turning off the refrigeration compressor during peak demand periods. While instantaneous cooling is also used for inlet air precooling, they often use 25% to 50% of the increase in power output to drive the refrigeration compressors. Evaporative coolers are a simple and well established technology for inlet air precooling, but they cannot deliver the low air temperatures available from refrigeration.

OFF-PEAK HEATING

Service Water Heating

The tank-equipped service water heater, which is the standard water heater in North America, is a thermal storage device; some electric utilities provide incentives for off-peak water heating. The heater is equipped with a control system that is activated by a clock or by the electric utility to curtail power use during peak demand. An off-peak water heater generally requires a larger tank than a conventional heater.

An alternate system of off-peak water heating consists of two tanks connected in series, with the hot water outlet of the first tank supplying water to the second tank (ORNL 1985). This arrangement minimizes the mixing of hot and cold water in the second tank. Tests performed on this configuration show that it can supply 80 to 85% of rated capacity at suitable temperatures, compared to 70% for a single-tank configuration. The wiring for the heating elements in the two tanks must be modified to accommodate the dual-tank configuration.

Solar Space and Water Heating

Rock beds, water tanks, or PCMs can be used for solar space and water heating, depending on the type of solar system employed. Various applications are described in Chapter 32.

Space Heating

Thermal storage for space heating can be carried out by radiant floor heating, brick storage heaters (electric thermal storage or ETS heaters), water storage heaters, or PCMs. Radiant floor heating is generally applied to single-story buildings. The choice between the other storage methods depends on the type of heating system. Air heaters use brick, while water heaters may use water or a PCM. Water heaters can be charged either electrically or thermally.

Insulation is more important for heat storage than for cool storage because the difference between storage and ambient temperatures is greater. Methods of determining the cost-effective thickness of insulation are given in Chapter 22 of the 1997 *ASHRAE Handbook—Fundamentals*.

A rational design procedure requires an hourly simulation of the design heat load and discharge capacity of the storage device. The first criterion for satisfactory performance is that the discharge rate be no less than the design heating load at any hour. During the off-peak period, energy is added to storage at the rate determined by the

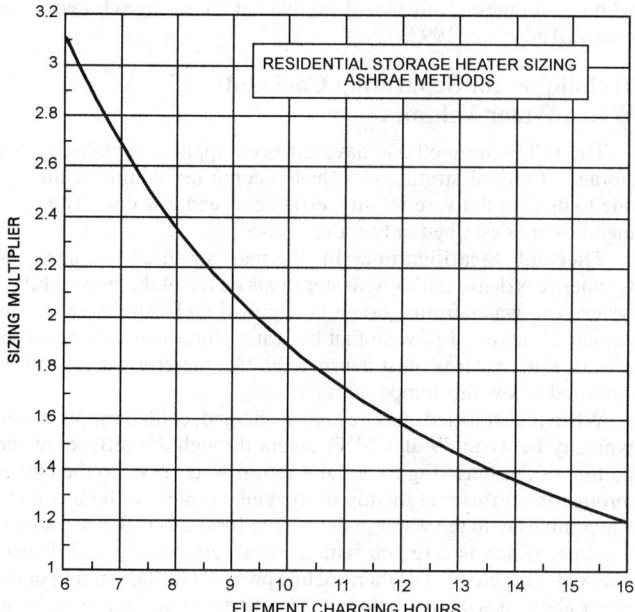

**Fig. 3   Representative Sizing Factor Selection Graph
for Residential Storage Heater**

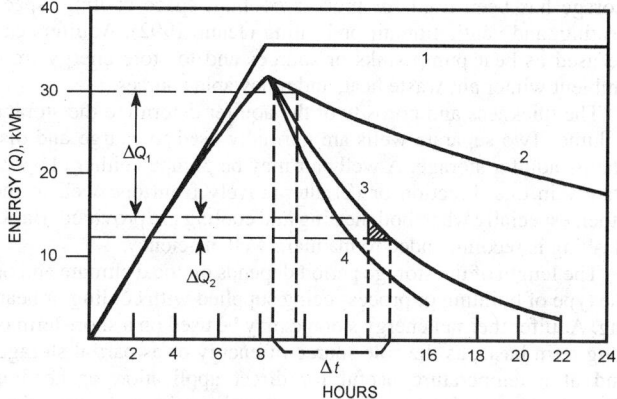

Fig. 4   Storage Heater Performance Characteristics
(Hersh et al. 1982)

connected load of the resistance elements, while the design heating load is subtracted. The second criterion for satisfactory performance, the daily energy balance, is checked at the end of a simulated 24 h period. Assuming that the design day is preceded and followed by similar days, the energy stored at the end of the simulated day should equal that at the start.

A simpler procedure is recommended by Hersh et al. (1982) for typical residential designs. For each zone of the building, the design heat loss is calculated in the usual manner and multiplied by the selected sizing factor. The resulting value (rounded to the next kilowatt) is the required storage heater capacity. Sizing factors in the United States range from 2.0 to 2.5 for an 8 h charge period and from 1.6 to 2.0 for a 10 h charge period. The lower end of the range is marginal for the northeastern United States. The designer should consult the manufacturer for specific sizing information. Figure 3 shows a typical sizing factor selection graph.

Brick Storage (ETS) Heaters. These heaters (commonly called ETS or electric thermal storage heaters) are electrically charged and store heat during off-peak times. In this storage device, air circulates through a hot brick cavity and then discharges into the area in which heat is desired. The brick in these heaters (commonly called ceramic brick) has a very dense magnetite or magnesite composition. The high density of the brick and its ability to store heat at a high temperature give the heater a large thermal storage capacity. Ceramic brick can be heated to approximately 1400°F during off-peak hours by resistance heating elements. Space requirements for brick storage heaters are usually much less than other storage mediums.

Figure 4 shows the operating characteristics of an electrically charged room storage heater. Curve 1 represents theoretical performance. In reality, radiation and convection from the exterior surface of the device continually supplies heat to the room during charging. Curve 2 shows this static discharge. When the thermostat calls for heat, the internal fan starts operating. The resulting faster dynamic discharge corresponds to Curves 3 and 4. Since the heating elements of electrically charged room storage heaters are energized only during off-peak periods, they must store the total daily heating requirement during this period.

The three types of brick storage heaters currently available are room units, heat pump boosters, and central furnaces. The section

on Electrically Charged Heat Storage Devices later in this chapter explains the various types of brick storage heaters.

Water Storage Heaters. Electrically charged pressurized water storage tanks, in most applications, must be able to recharge the full daily heating requirement during the off-peak period, while simultaneously supplying heat during the off-peak period. The heat exchanger design allows a constant discharge and does not decrease near the end of the cycle.

Thermally charged hot water storage tanks are similar in design to the cool storage tanks described in the section on Off-Peak Air Conditioning. Many are also used for cooling. Unlike off-peak cooling, off-peak heating seldom permits a reduction in the size of the heating plant. The size of a storage tank used for both heating and cooling is based more on cooling than on heating for lowest life-cycle cost, except in some residential applications.

District Heating and Cooling. These systems distribute thermal energy (steam, hot and chilled water) from a central energy source (plant) to residential, commercial, and industrial consumers via a distribution system (piping) for use in comfort heating/cooling, service water heating, and process heating/cooling. Thermal energy storage has operated successfully in such district heating and cooling (DHC) applications as off-peak air conditioning, industrial/process cooling, etc. Because most storages tanks are open, attention should be given to system hydraulics when connecting several plants to a common distribution network where storage tank water levels are at different elevations. Chapter 11 of the 1996 *ASHRAE Handbook—Systems and Equipment* has additional information on DHC systems and related topics.

The distribution system in DHC is a large percentage of the total system cost. As a result, the higher temperature differential available from thermal storage media lowers life-cycle costs because is allows the use of smaller pipes, valves, pumps, motors, starters, etc. Furthermore, thermal storage may be used to increase the available cooling capacity of an existing chilled water distribution system by providing cooler supply water. This modification would have little impact to the distribution system itself (i.e., pumping and piping) because the existing system flow rates and pressure drops are not changed.

OTHER APPLICATIONS

Storage in Aquifers

Aquifers can be used to store large quantities of thermal energy. Aquifers are underground, water-yielding geological formations, either unconsolidated (gravel and sand) or consolidated (rocks). In general, the natural aquifer water temperature is slightly warmer than the local mean annual air temperature. Aquifer thermal energy storage has been used for process cooling, space cooling, space heating, and ventilation air preheating (Jenne 1992). Aquifers can be used as heat pump sinks or sources and to store energy from ambient winter air, waste heat, and renewable sources.

The thickness and porosity of the aquifer determine the storage volume. Two separate wells are normally used to charge and discharge aquifer storage. A well pair may be pumped either (1) constantly in one direction or (2) alternatively from one well to the other, especially when both heating and cooling are provided. Backflushing is recommended to maintain well efficiency.

The length of the storage period depends on local climate and on the type of building or process being supplied with cooling or heating. Aquifer thermal energy storage may be used on a short-term or long-term basis, as the sole source of energy or as partial storage, and at a temperature useful for direct application or needing upgrade. It may also be used in combination with a dehumidification system such as desiccant cooling. The cost-effectiveness of aquifer thermal energy storage is based on the avoidance of equipment capital cost and on lower operating cost.

The incorporation of aquifer thermal energy storage into a building system may be accomplished in a variety of ways, depending on the other components present and the intentions of the designer (Hall 1993b; CSA 1993). Control is simplified by separate hot and cold wells that operate on the basis that the last water in is the first water out. This principle ensures that the hottest or coldest water is always available when needed.

Seasonal storage has a large saving potential and began as an environmentally sensitive improvement on the large-scale mining of groundwater (Hall 1993a). The current method reinjects all pumped water to attempt annual thermal balancing (Morofsky 1994; Public Works Canada 1991, 1992; Snijders 1992). Rock caverns have also been used successfully in a manner similar to aquifers to store energy. For example, Oulu, Finland, stores hot water in a cavern for district heating.

STORAGE TECHNOLOGIES

WATER STORAGE

Water is well suited for both hot and cold storage applications, in part because it has the highest specific heat (1 Btu/lb · °F) of all common materials. ASHRAE *Standard* 94.3 covers procedures for measuring the thermal performance of sensible heat storage. Tanks are available in many shapes; however, vertical cylinders are the most common. Tanks can be located above ground, partially buried, or completely buried. They are usually atmospheric, unpressurized tanks and may have clear-span spherical dome roofs or column-supported flat roofs.

Water thermal storage vessels must keep cool and warm water volumes separate. This chapter focuses on chilled water storage because it is the most common storage in use. However, the same techniques apply to hot water.

Temperature Range and Storage Size

The cooling capacity of a chilled water storage vessel is proportional to the volume of water stored and the temperature differential (Δt) between the stored cool water and the returning warm water. For economical storage, the cooling coils should provide as large a Δt as practical. For example, chillers that cool water in storage to 40°F and coils that return water to storage at 60°F provide a Δt of 20°F. The cost of the extra coil surface required to provide this range can be offset by savings in pipe size, insulation, and pumping energy. Storage is likely to be uneconomical if the temperature differential is less than about 10°F because the tank must be so large (Caldwell and Bahnfleth 1997).

The initial cost of chilled water storage benefits from a dramatic economy of scale; that is, large installations can compare favorably

with ice storage as well as with equivalent nonstorage chilled water plants (Andrepont 1992).

Techniques for Separating Cool and Warm Water Volumes

The following methods have all been applied in chilled water storage. Thermal stratification has become the dominant method due to its simplicity, reliability, efficiency, and low cost. The other methods are described only for reference.

Thermal Stratification. In thermal stratified storage, the warmer, less dense returning water floats on top of the stored chilled water. The water from storage is supplied and withdrawn in low velocity, horizontal flow so that buoyancy forces dominate inertial effects. Pure water is most dense at 39.2°F; therefore, it can not be stratified below this temperature.

When the stratified storage tank is charged, chilled supply water, typically between 39 and 44°F, enters through the diffuser at the bottom of the tank (Figure 5), and return water exits to the chiller through the diffuser at the top of the tank. Typically, the incoming water mixes with the water in the tank to form a 1- to 3-ft thick thermocline, which is a region with vertical temperature and density gradients (Figure 6). The thermocline prevents further mixing of the water above it with that below it. The thermocline rises as recharging continues and subsequently falls during discharging. It thickens somewhat during charging and discharging due to heat conduction through the water and heat transfer to and from the walls of the tank. The storage tank may have any cross-section, but the walls are usually vertical. Horizontal cylindrical tanks are generally not good candidates for stratified storage.

Flexible Diaphragm. A flexible diaphragm (normally horizontal) is used in a tank to separate the cold water from the warm water.

Multiple Compartments. Multiple compartments in a single tank or a series of two or more tanks can also be used. Pumping is scheduled so that one compartment is always at least partially empty; water returning from the system during the occupied period and from the chiller during storage regeneration is then received into the empty tank. Water at the different temperatures is thus stored in separate compartments, minimizing blending.

Labyrinth Tank. This tank has both horizontal and vertical traverses. The design commonly takes the form of successive cubicles with high and low ports. Strings of cylindrical tanks and tanks with successive vertical weirs may also be used.

Performance of Chilled Water Storage

A perfect storage tank would deliver water at the same temperature at which it was stored. It would also require that the water returning to storage neither mix nor exchange heat with the stored water or the tank. In practice, however, both types of heat exchange occur.

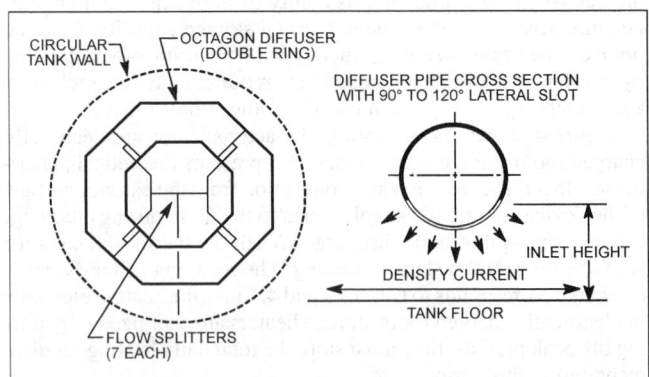

Fig. 5 Octagon Diffuser

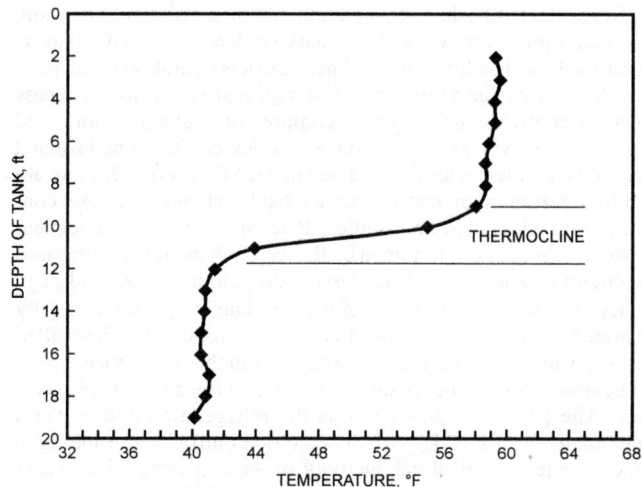

**Fig. 6   Typical Temperature Stratification Profile
in Storage Tank**

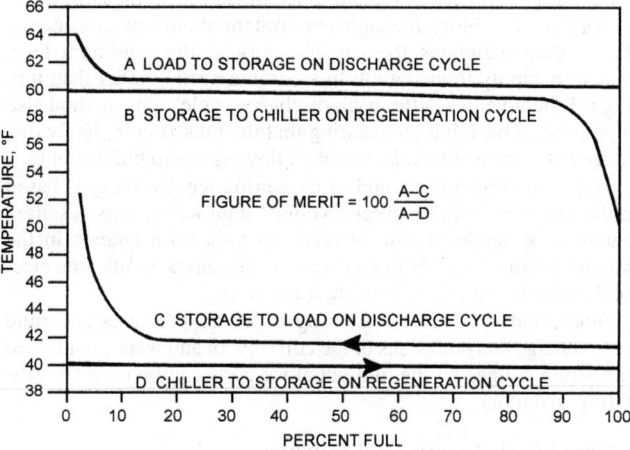

Fig. 7   Chilled Water Storage Profiles

Typical temperature profiles of water entering and leaving a storage tank are shown in Figure 7. Tran et al. (1989) tested several large chilled water storage systems and developed the figure of merit, which is used as a measure of the amount of cooling available from the tank.

$$\text{Figure of Merit (\%)} = \frac{\text{Area between A and C}}{\text{Area between A and D}} \times 100 \quad (1)$$

Well-designed storage tanks have figures of merit of 90% or higher for daily complete charge/discharge cycles and between 80 and 90% for partial charge/discharge cycles.

Design of Stratification Diffusers

Stratification diffusers must be designed and constructed to produce and maintain stratification at the maximum flow through storage. The *Design Guide for Cool Thermal Storage* (Dorgan and Elleson 1993) discusses these design parameters. Wildin (1990) gives more information concerning the design of stratified storage hardware.

Inlet and outlet streams must be kept at sufficiently low velocities, so that buoyancy forces predominate over inertia forces in order to produce a gravity current (density current) across the bottom or top of the tank. Designers typically use a value of 1.0 or less.

Table 2   Chilled Water Density

°F	lb/ft³	°F	lb/ft³	°F	lb/ft³
32	62.419	44	62.424	58	62.378
34	62.424	46	62.421	60	62.368
36	62.426	48	62.417	62	62.357
38	62.427	50	62.411	64	62.344
39	62.428	52	62.404	66	62.331
40	62.427	54	62.396	68	62.316
42	62.426	56	62.387		

However, values up to 2.0 have been successfully applied (Yoo et al. 1986).

The inlet Froude number Fr is defined as

$$\text{Fr} = \frac{Q}{\sqrt{gh^3(\Delta\rho/\rho)}} \quad (2)$$

where

 Q = volume flow rate per unit length of diffuser, ft³/s·ft
 g = gravitational acceleration, ft/s²
 h = inlet opening height, ft
 ρ = inlet water density, lb/ft³
 $\Delta\rho$ = difference in density between stored water and incoming or outflowing water, lb/ft³

The density difference $\Delta\rho$ can be obtained from Table 2. The inlet Reynolds number is defined as follows:

$$\text{Re} = Q/\nu \quad (3)$$

where ν = kinematic viscosity, ft²/s.

Experimental evidence indicates that the intensity of mixing near the inlet diffuser is influenced by the inlet Reynolds number [Equation (3)]. In short tanks, in which the inlet diffuser affects a large portion of the tank volume, the inlet Reynolds number may significantly affect thermal performance. Also, tanks with sloping sidewalls, such as inverted truncated pyramids, may benefit from using lower inlet Reynolds numbers.

Wildin and Truman (1989), observing results from a 15-ft deep, 20-ft diameter vertical cylindrical tank, found that reduction of the inlet Reynolds number from 850 (using a radial disk diffuser) to 240 (using a diffuser comprised of pipes in an octagonal array) reduced mixing to negligible proportions. This is consistent with subsequent results obtained by Wildin (1991) in a 3-ft deep scale model tank, which indicated negligible mixing at Reynolds numbers below approximately 450.

For tall tanks with water depths of 40 ft or more, the inlet Reynolds number is less important. Field measurements indicate that inlet mixing does not significantly affect the thermocline when it is more than about 10 ft away from the inlet diffuser. Bahnfleth and Joyce (1994) and Musser and Bahnfleth (1998) documented successful operation of tanks with water depths greater than 45 ft for design inlet Reynolds numbers as high as 6000. However, improvement in the thermal performance of these systems was observed at lower inlet Reynolds numbers. Some consider values of 1000 to 2000 typical for large tanks.

Storage Tank Insulation

Exposed tank surfaces should be insulated to help maintain the temperature differential in the tank. Insulation is especially important for smaller storage tanks because the ratio of surface area to stored volume is relatively high. Heat transfer between the stored water and the tank contact surfaces (including divider walls) is a primary source of capacity loss. Not only does the stored fluid lose heat to (or gain heat from) the ambient by conduction through the floor and wall, but heat flows vertically along the tank walls from the warmer to the cooler region. Exterior insulation of the tank walls does not inhibit this heat transfer.

Other Factors

The cost of chemicals for water treatment may be significant, especially if the tank is filled more than once during its life. A filter system helps keep the stored water clean. Exposure of the stored water to the atmosphere may require the occasional addition of biocides. While tanks should be designed to prohibit leakage, the designer should understand the potential effect of leakage on the selection of chemical water treatment.

The storage circulating pumps should be installed below the minimum operating water level to assure flooded suction. The required net positive suction pressure (NPSH) must be maintained to avoid subatmospheric conditions at the pumps.

ICE STORAGE AND OTHER PHASE-CHANGE MATERIALS

Thermal energy can be stored in the latent heat of fusion of water (ice) or other materials. Water has the highest latent heat of fusion of all common materials—144 Btu/lb at the melting or freezing point of 32°F. The PCM other than water most commonly used for thermal storage is a hydrated salt with a latent heat of fusion of 41 Btu/lb and a melting or freezing point of 47°F. The volume required to store energy in the form of latent heat is considerably less than that stored as sensible heat, which contributes to lower capital cost.

External Melt Ice-on-Coil Storage

The oldest type of ice storage is the refrigerant-fed ice builder, which consists of refrigerant coils inside a storage tank filled with water (Figure 8). A compressor and evaporative condenser freeze the tank water on the outside of the coils to a thickness of up to 2.5 in. Ice is melted from the outside of the formation (hence the term external melt) by circulating the return water through the tank, whereby it again becomes chilled. Air bubbled through the tank agitates the water to promote uniform ice buildup and melting.

Instead of refrigerant, a secondary coolant (e.g., 25% ethylene glycol and 75% water) can be pumped through the coils inside the storage tank. The coolant has the advantage of greatly decreasing the refrigerant inventory. However, a refrigerant-to-coolant heat exchanger between the refrigerant and the storage tank is required.

Major concerns specific to the control of ice-on-coil storage are (1) to limit ice thickness (and thus excess compressor energy) during the build cycle and (2) to minimize the bridging of ice between individual tubes in the ice bank. Bridging must be avoided because it restricts the free circulation of water during the discharge cycle. While not physically damaging to the tank, this blockage reduces performance, allowing a higher leaving water temperature due to the reduced heat transfer surface.

Regardless of the refrigeration method (direct-expansion, pumped liquid overfeed, or secondary coolant), the compressor is controlled by (1) a time clock, which restricts operation to the periods dictated by the utility rate structure, and (2) an ice-thickness override control, which stops the compressor(s) at a predetermined ice thickness. At least one ice-thickness device should be installed per ice bank; the devices should be connected in series. If there are multiple refrigeration circuits per ice bank, one ice-thickness control per circuit should be installed. Placement of the ice-thickness device(s) should be determined by the ice-bank manufacturer, based on circuit geometry and flow pressure drop, to minimize bridging.

Ice-thickness controls are either mechanically or electrically operated. A mechanical control typically consists of a fluid-filled probe positioned at the desired distance from the coil. As ice builds, it encapsulates the probe, causing the fluid to freeze and apply pressure. The pressure signal controls the refrigeration system via a pneumatic-electric (P/E) switch. Electric controls sense the three times greater electrical conductivity of ice as compared to water. Multiple probes are installed at the desired thickness, and the change in current flow between probes provides a control signal. Consistent water treatment is essential to maintaining constant conductivity and thus accurate control.

Because energy use is related to ice thickness on the coil, a partial-load ice inventory management system should be considered. This system maintains the ice inventory at the minimum level needed to supply immediate future cooling needs, rather than topping off the inventory after each discharge cycle. This method also helps prevent bridging by ensuring that the tank is completely discharged at regular intervals, thereby allowing ice to build evenly.

The most common method of measuring ice inventory is based on the fact that ice has a greater volume than water; thus, a sensed change in water level can be taken to indicate a change in the amount of stored ice. Water level can be measured by either an electrical probe or a pressure-sensing transducer.

Other methods such as metering of all temperatures and fluid flows (energy) on both sides of the coil (glycol and water) combined with strain gauges on the individual ice coils can be used for calibrating inventory.

Internal Melt Ice-on-Coil Storage

In this type of storage device, coils or tubing are placed inside a water tank (Figure 9). The coils occupy approximately 10% of the tank volume; another 10% of the volume is left empty to allow for the expansion of the water upon freezing; and the rest is filled with water. A secondary coolant solution (e.g., 25% ethylene glycol and 75% water) is cooled by a liquid chiller and circulated through the coils to freeze the water in the tank. The thickness of ice on the coils and the percentage of the water in the tank that is frozen depend on the coil configuration and on the type of system. During discharge,

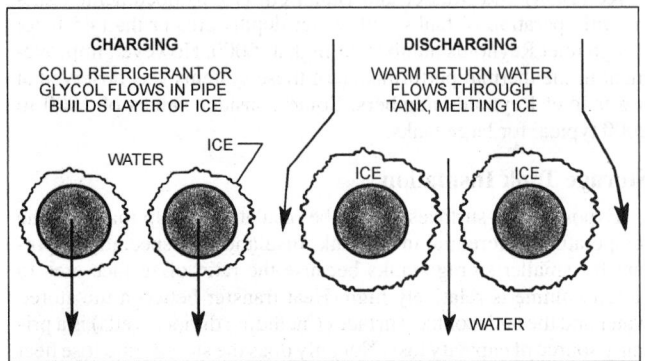

Fig. 8   Charge and Discharge of External Melt Ice Storage

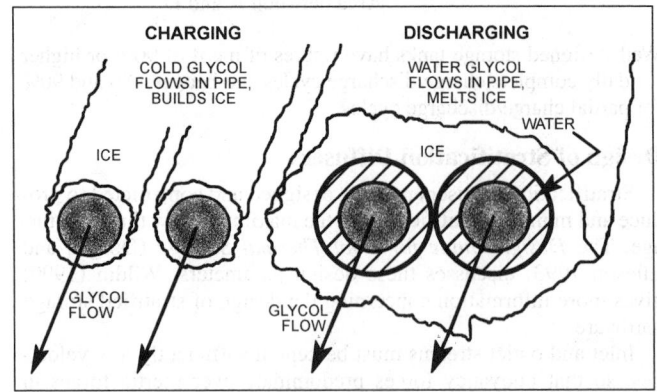

Fig. 9   Charge and Discharge of Internal Melt Ice Storage

the secondary coolant circulates to the system load and returns to the tank to be cooled again by the coils submerged in ice.

A standard chiller can provide the refrigeration for these systems. During the charging cycle for a typical system, the chilled secondary coolant exits from the chiller at a constant 25 to 26°F and returns at 31°F. When the tanks are 90% charged, the chiller inlet and outlet temperatures fall rapidly because there is little water left to freeze. When the chiller exit temperature reaches approximately 22°F, the chiller is shut down and locked off for the night so that it does not short-cycle or recirculate due to convection flow through the pipes. The chiller remains fully loaded through the entire cycle and keeps the system running at its maximum efficiency because the exit-temperature thermostat is set at 22°F. The temperatures for a given system may vary from this example.

Because the same heat transfer surface freezes and melts the water, the coolant may freeze the water completely every night, minimizing loss in efficiency. A temperature-modulating valve at the outlet of the tanks keeps a constant flow of liquid to the load (see Figure 15). Under a full storage control strategy, the chiller is kept off during discharge, and the modulating valve allows some fluid to bypass the tanks to supply the load as needed. If a partial storage control strategy is followed, the chiller thermostat is reset from the 22°F setting up to the design temperature of the cooling coils (e.g., 44°F) during the discharge cycle. If the load on the building is low, the chiller operates to meet the 44°F setting without depleting the storage. If the load is greater than the chiller's capacity, the exiting temperature of the secondary coolant rises, and the temperature-modulating valve automatically opens to maintain the design temperature to the coils (Kirshenbaum 1991).

Because water increases 9% in volume when it turns to ice, the water level varies directly with the amount of ice in the tank as long as all of the ice remains submerged. This water displaced by the ice must not be frozen, or it will trap ice above the original water level. Therefore, no heat exchange surface area can be above the original water level.

The change in water level in the tank due to freezing and thawing is typically 6 in. This change can be measured with either a pressure gage or a standard electrical transducer. On projects with multiple tanks, reverse-return piping ensures uniform flow through all the tanks, so measuring the level on one tank is sufficient to determine the proportion of ice remaining.

Encapsulated Ice

This type of thermal storage relies on plastic containers filled with deionized water and an ice-nucleating agent. Commercially available systems have either spherical containers of approximately 4-in. diameter (Figure 10) or rectangular containers approximately 1-3/8 in. by 12 in. by 30 in. (Figure 14). These primary containers are placed in storage tanks, which may be either steel pressure vessels, open concrete tanks, or suitable fiberglass or polyethylene tanks. In tanks with spherical containers, water flows vertically through the tank, and in tanks with rectangular containers, water flows horizontally. The type, size, and shape of the storage tank is limited only by its ability to achieve even flow of heat transfer fluid between the containers. Refrigeration may be any type of liquid chiller rated for the lower temperatures required.

A secondary coolant (e.g., 25% ethylene glycol or propylene glycol and 75% water) is cooled to 24 to 26°F by a liquid chiller and circulates through the tank and over the outside surface of the plastic containers, causing ice to form inside the containers. As for the internal melt ice-on-coil storage, the temperature at the end of the charge cycle is lower (e.g., 22°F), and the chiller must be capable of operating at this reduced temperature. The plastic containers must be flexible to allow for change of shape during ice formation; the spherical type has preformed dimples in the surface, and the rectangular type is designed for direct flexure of the walls. During discharge, coolant flows either directly to the load or to a heat

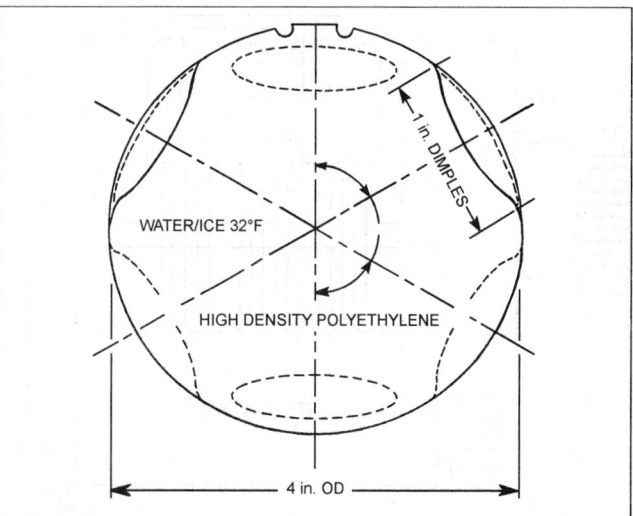

Fig. 10 Encapsulated Ice—Spherical Container

exchanger, thereby removing heat from the load and melting the ice within the plastic containers. As the ice melts, the plastic containers return to their original shape.

Ice inventory is measured and controlled by an inventory/expansion tank normally located at the high point in the assembly and connected directly to the main storage tank. As ice forms, the flexing plastic containers force the surrounding secondary coolant into the inventory tank. The liquid level in the inventory tank may be monitored to account for the ice available at any point during the charge or discharge cycle.

Ice-Harvesting

Ice-harvesting systems separate the ice formation from storage. Ice is typically formed on both sides of a hollow, flat plate or on the outside or inside (or both) of a cylindrical evaporator surface. The evaporators are arranged in vertical banks above the storage tank. Ice is formed to thicknesses between 0.25 and 0.40 in. This ice is then harvested, often by introducing hot refrigerant gas into the evaporator. The gas warms the evaporator, which breaks the bond between the ice and the evaporator surface and allows the ice to drop into the storage tank below. Other types of ice harvesters use a mechanical means of separating the ice from the evaporator surface. Figure 11 shows a an ice-harvesting schematic.

Ice is generated by circulating 32°F water from the storage tank over the evaporators for a 10- to 30-min. build cycle. The defrost time is a function of the amount of energy required to warm the system and break the bond between the ice and the evaporator surface. Depending on the control method, the evaporator configuration, and the discharge conditions of the compressor, defrost can be accomplished in 20 to 90 s. Typically, the evaporators are grouped in sections that are defrosted individually such that the heat of rejection from the active sections provides the energy for defrost. Knebel (1991) showed that the net available capacity from an ice harvester is the compressor gross capacity minus the fraction of time spent in defrost times the total heat of rejection from the compressor.

Tests done by Stovall (1991) indicate that for a four-section plate ice harvester, all of the total heat of rejection is introduced into the evaporator during the defrost cycle. To maximize ice production, harvest time must be kept to a minimum.

In Figure 11 chilled water is pumped from the storage tank to the load and returned to the ice generator. A low-pressure recirculation pump is used to provide minimum flow for wetting the evaporator in the ice-making mode. The system may be applied to load-leveling or load-shifting applications.

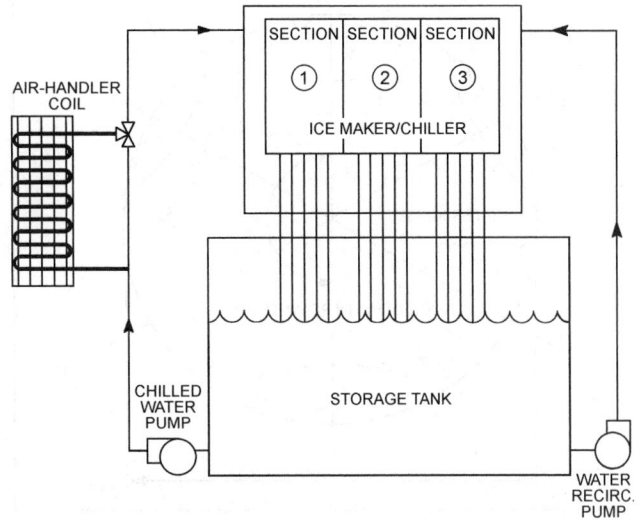

Fig. 11   Ice-Harvesting Schematic

In load-leveling applications, ice is generated and the storage tank charged when there is no building load. When a building load is present, the return chilled water flows directly over the evaporator surface, and the ice generator functions either as a chiller or as both an ice generator and a chiller. Cooling capacity as a chiller is a function of the water velocity on the evaporator surface and the entering water temperature. The defrost cycle must be energized any time the exit water from the evaporator is within a few degrees of freezing. In chiller operation, maximum performance is obtained with minimum water flow and highest entering water temperature. In load-shifting applications, the compressors are turned off during the electric utility on-peak period.

Positive displacement compressors are usually used with ice harvesters, and saturated suction temperatures are usually between 18 and 22°F. The condensing temperature should be kept as low as possible to reduce energy consumption. The minimum allowable condensing temperature depends on the type of refrigeration used and the defrost characteristics of the system. Several systems operating with evaporative cooled condensers have operated with a compressor specific power consumption of 0.9 to 1.0 kW/ton (Knebel 1986, 1988a; Knebel and Houston 1989).

Ice-harvesting systems can melt the stored ice very quickly. Individual ice fragments are characteristically less than 6 in. by 6 in. by 0.25 in. and provide at least 152 ft^2 of surface area per ton-hour of ice stored. When properly wetted, a 24 h charge of ice can be melted in less than 30 min for emergency cooling demands.

During the ice generation mode, the system is energized if the ice is below the high ice level. A partial storage system is energized only when the entering water is at or above a temperature that will permit chilling during the discharge mode; otherwise, the system is off, and the ice tank is discharged during the on-peak period to meet the load. The high ice level sensor can be mechanical, optical, or electronic. The entering water temperature thermostat is usually electronic.

When ice is floating in the tank, the water level will always be constant, so it is impossible to measure ice inventory by measuring water level. The following methods are generally used to determine the ice inventory:

Water Conductivity Method. As water freezes, dissolved solids are forced out of the ice into the liquid water, thus increasing their concentration in the water. Accurate ice inventory information can be maintained by measuring conductivity and recalibrating daily from the high ice level indicator.

Heat Balance Method. The cooling effect obtained from a system may be determined by measuring the power input to, and heat

of rejection from, the compressor. The cooling load on the system is determined by measuring coolant flow and temperature. This may be accomplished by measuring the ice inventory and knowing the time to recharge the storage. The ice inventory is then determined by integrating cooling input minus load and calibrating daily from the high ice level sensor. A variant of the heat balance method is determined by running a heat balance on the compressor only, using performance data from the compressor manufacturer and measured load data (Knebel 1988b).

Optimal performance of ice-harvesting may be achieved by recharging the ice storage tank over a maximum amount of time with minimum compressor capacity. Ice inventory measurement and the known recharge time are used to accomplish this goal. Knebel and Houston (1989) demonstrated that efficiency can be dramatically increased with proper selection of multiple compressors and unloading controls.

Design of the storage tank is important to the operation of the system. The amount of ice stored in a storage tank depends on the shape of the storage, the location of the ice entrance to the tank, the angle of repose of the ice (between 15 and 30°, depending on the shape of the ice fragments), and the water level in the tank. If the water level is high, voids occur under the water due to the buoyancy of the ice. Ice-water slurries have been reported to have a porosity of 0.50 and typical storage densities of 2.92 ft^3 per ton-hour. Gute et al. (1995), and Stewart et al. (1995a, 1995b) describe models for determining the amount of ice that can be stored in rectangular tanks and the discharge characteristics of various tank configurations. Dorgan and Elleson (1993) provide further information.

Ice Slurry Systems

Simply defined, an ice slurry is a suspension of ice crystals in liquid. In general, the working fluid's liquid state consists of a solvent (water) and a solute such as glycol, ethanol, or calcium carbonate. Depending on the specific slurry technology, the initial solute concentration varies from 2% to over 10% by mass. The solute depresses the freezing point of the solvent and buffers the production of ice crystals. Figure 12 illustrates the freezing point for an aqueous solutions with various concentrations of calcium magnesium acetate (CMA).

Slurry generation begins by lowering the working fluid to its initial freezing point. Extracting additional energy from the working fluid initiates the process of solidification. As solidification proceeds, solute is rejected to the solid-liquid interface (only the water is frozen). The solute-rich interface is eventually incorporated into the bulk fluid by convection and diffusion. As freezing progresses, the solute in the bulk field increases and the freezing point needed to sustain ice crystal production decreases.

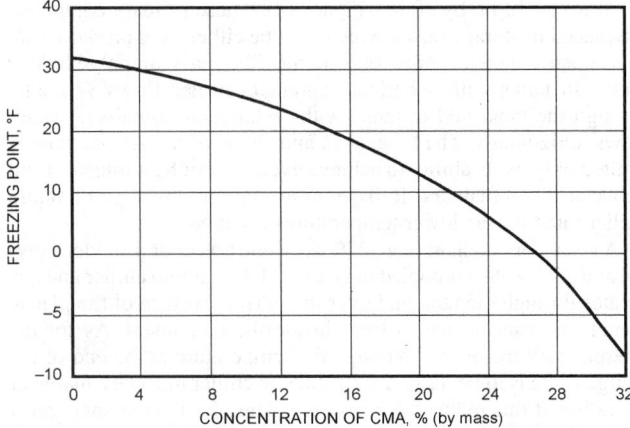

Fig. 12   Freezing Points of Calcium Magnesium Acetate (CMA) Solutions

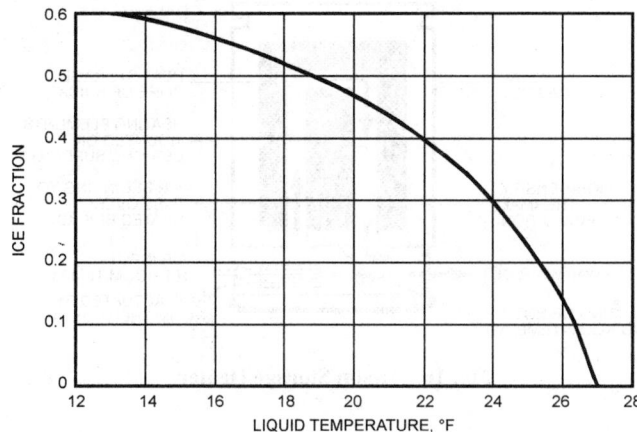

**Fig. 13   Ice Fraction Versus Temperature for
8% CMA Solution**

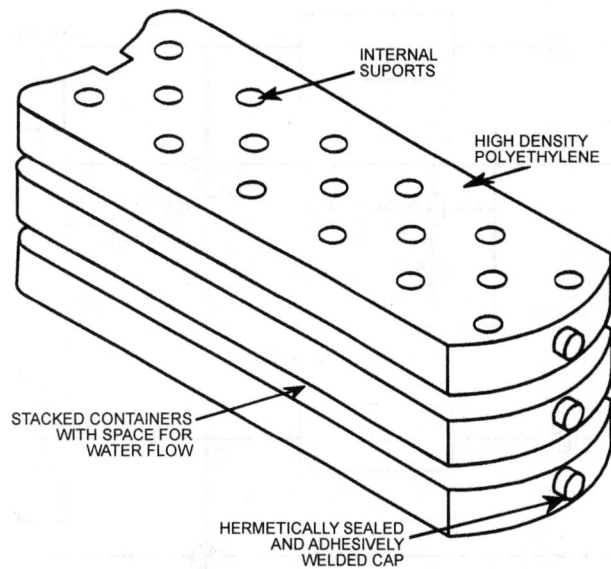

Fig. 14   Container for Phase-Change Material

Assuming that none of the solute is incorporated into the solid, a relationship between the solute concentration and the ice fraction can be established by the following equation:

$$f_{ice} = \frac{C - C_o}{C} \qquad (4)$$

where C is the concentration of the solute in the system, C_o is the initial solute concentration and f_{ice} represents the ice fraction in the system. Thus the ice fraction can be estimated by knowing the initial solute concentration and the solute concentration at any time during slurry production or consumption. The equilibrium solute concentration can be related back to temperature through freezing point curves. A relation between ice fraction and solution temperature is shown in Figure 13, based on an 8% initial concentration of CMA in water.

Thus, by monitoring the liquid temperature during slurry production and slurry consumption, the fraction of ice in a given system can be estimated by knowing the initial solute concentration and the freezing point characteristic of the working fluid. The success of this technique relies on an assumption that all of the solute is being rejected to the liquid phase upon solidification and minimal dilution of the working fluid over time (e.g., due to condensation of moisture from the air in the storage vessel).

Other Phase-Change Materials

Like ice and chilled water storage, hydrated salts have been in use for many decades. PCMs with various phase-change points have been developed. To date, the hydrated salt most commonly used for cool storage applications changes phase at 47°F; it is often encapsulated in plastic containers, as described in the section on Encapsulated Ice. This material is a mixture of inorganic salts, water, and nucleating and stabilizing agents. It has a latent heat of fusion of 41 Btu/lb and a density of 93 lb/ft³. The PCM's latent heat of fusion requires a capacity of about 5.5 ft³/ton-hour for the entire tank assembly, including piping headers and water in the tank.

Another option is an internal melt ice-on-coil system in which the water is replaced by salts or polymers that change phase at 12°F or 28°F. The hydrated salts typically have a latent heat of fusion of 115 to 135 Btu/lb and a density of 65 to 70.5 lb/ft³.

Typically, 40 to 42°F chilled water is used to charge the storage tank of systems using the 47°F PCM. The leaving temperature remains a relatively constant 47°F until the hydrated salts complete the phase-change process, at which point the temperature of the water leaving the tank begins to approach that of the entering water.

It is usually preferable to charge the last 10 to 20% of the tank's capacity while there is a cooling load because of the low (3 to 4°F) temperature difference across the tank.

During discharge, the exit temperature begins at 42°F and rises to 48 to 50°F. The usable storage capacity is determined by the highest discharge temperature that can be used by the system.

Any new or existing centrifugal, screw, or reciprocating chiller can be used to charge the storage because conditions are comparable to those for standard air conditioning. As a result, this technology is particularly appropriate for retrofit applications.

As shown in Figure 14, the PCM is sealed inside self-stacking plastic containers measuring approximately 24 in. by 8 in. by 1-3/4 in. The containers are designed to provide space for water to pass between them in a meandering flow pattern. The containers are placed in an open tank, typically of below-grade concrete or sprayed-on concrete. The containers displace about two-thirds of the volume of the tank, so that one-third of the tank is occupied by the water used as the heat transfer medium. The containers do not float or expand when the PCM changes phase, so the container spacing arrangement is maintained throughout the phase-change cycle. The top of the tank can be landscaped or designed for traffic or use as a parking lot.

Precoolers with partial storage provide some storage and a colder water temperature to the load. The 47°F PCM can be used with the chiller downstream of the storage to supply water at 44°F or lower to the load.

Circuits for Ice Storage

Piping for ice storage can be configured in a variety of ways. Numerous studies have been conducted to determine optimum configurations and control schemes for a variety of designs (Simmonds 1994). The optimum configuration depends on the type of system; the operation, performance, and temperature requirements; and the configuration of the building loop. The chiller may be located upstream or downstream of the building load, and the cooling plant may also be decoupled from the building load.

An example of a basic piping schematic for ice storage providing partial storage is shown in Figure 15. During off-peak periods, the loop supplying the building load is bypassed, and the chiller charges the ice storage unit (Figure 15A). When cooling is required during the on-peak period, the building load may be met by the chiller, the

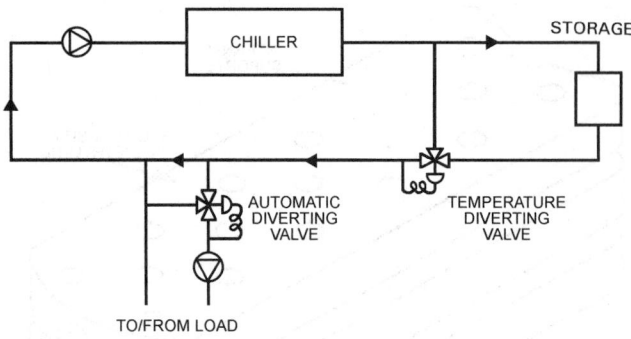

A. CHARGING STORAGE

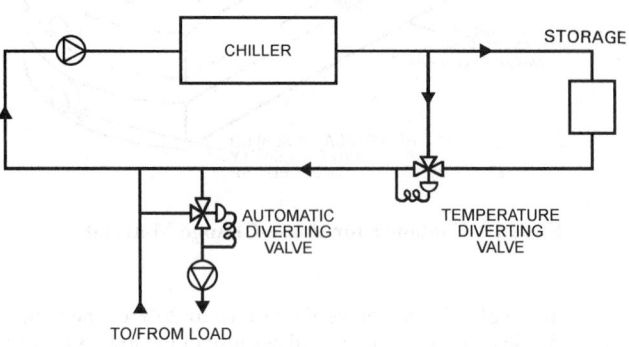

B. DISCHARGING STORAGE

Fig. 15   Thermal Storage with Chiller Upstream

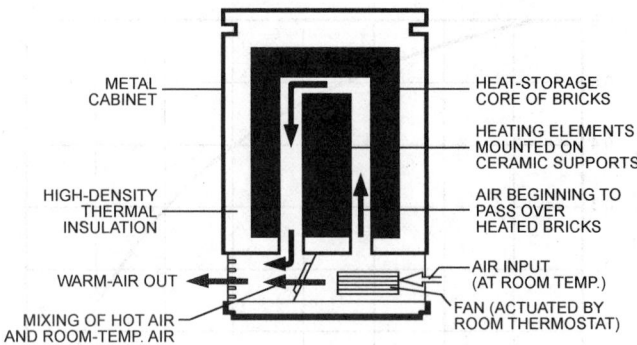

Fig. 16   Room Storage Heater

storage, or a combination of both (Figure 15B). A downstream modulating valve maintains the chilled water supply to the building loop at the desired temperature. If demand limiting is desired, the chiller electrical demand must also be controlled. The remainder of the load is met by the storage.

Storage Tank Insulation

Because of the low temperature associated with ice storage, insulation is a high priority. In retrofit applications, the current insulation must be evaluated to ensure there is no condensation or excessive heat loss. All ice storage tanks located above ground should be insulated to limit standby losses. For external melt ice-on-coil systems and some internal melt ice-on-coil systems, the insulation and vapor barrier are part of the factory-supplied containers; most other storage tanks require that insulation and a vapor barrier by applied in the field. Below-ground tanks used with ice harvesters may not need insulation below the first few feet. Because the tank temperature does not drop below 32°F at any time, there is no danger of freezing and thawing groundwater.

All below-ground tanks using fluids below 32°F during the charge cycle should have a well designed and properly installed insulation and vapor barrier, generally on the exterior. Interior insulation is susceptible to damage from the ice and should be avoided.

Because a hydrated salt solution operates at chilled water temperatures of about 47°F, the same insulation practices apply.

ELECTRICALLY CHARGED HEAT STORAGE DEVICES

Thermal energy can also be stored in electrically-charged, thermally-discharged storage devices. For devices that use a solid mass as the storage medium, equipment size is typically specified by the nominal power rating (to the nearest kilowatt) of the internal heating elements. The nominal storage capacity is taken as the amount of energy supplied during an 8 h charge period. For example, a 5 kW heater would have a nominal storage capacity of 40 kWh. ASHRAE

Standard 94.2 describes methods for testing these devices. If multiple charge/off-peak periods are available during a 24 h period, an alternative method yields a more accurate estimate of equipment size. The method considers not only the nominal power rating, but also fan discharge rate and storage capacity. The equipment manufacturer should have more information on calculating capacity.

Room Storage Heaters (Room Units)

Room storage heaters (commonly called room units) have magnetite or magnesite brick cores encased in shallow metal cabinets (Figure 16). The core can be heated to 1400°F during off-peak hours by resistance heating elements located throughout the cabinet. Room units are generally small heaters that are placed into a particular area or room. These heaters have well-insulated storage cavities, which help retain the heat in the brick cavity. Even though the brick inside the units get very hot, the outside of the heater is relatively cool with surface temperatures generally below 180°F. Storage heaters are discharged by natural convection, radiation, and conduction (static heaters) or by a fan. The air flowing through the core is mixed with room air to limit the outlet air temperature to a comfortable range.

Storage capacities range from 13.5 to 60 kWh. Inputs range from 0.8 to 9.0 kW. In the United States, 120 V, 208 V, 240 V, and 277 V units are commonly available. The 120 V model is useful for heating smaller areas or in geographical areas with moderate heating days. Room storage heaters are for residential, motel, hotel, apartment, and office applications.

Operation is relatively simple. When a room thermostat calls for heat, fans (on dynamic units) located in the lower section of the room unit discharge air through the ceramic brick core and into the room. Depending on the charge level of the brick core, a small amount of radiant heat may also be delivered from the room unit. The amount of heat stored in the brick core of the unit can be regulated either manually or automatically in relation to the outside temperature.

These units fully charge in about 7 h (Figure 17), and they can be fully depleted in as little as 6 h. The equipment retains heat for up to 72 h (3 days) if it has no fan discharge (Figure 18).

Choosing the appropriate size of room unit(s) depends on control strategy of the power company (on-peak versus off-peak hours), outside design climate, and heat loss of the area or space. The manufacturer of the equipment may provide assistance in determining the heat loss for the area requiring heat. Based on the control strategy of the power company, the following two concepts can be used for sizing of the equipment:

Whole House Concept. Under this strategy, room units are placed throughout the home. A room by room heat loss calculation must be performed. This method is used in areas where the power company has long hours of consecutive control (on-peak hours), generally 10 h or more.

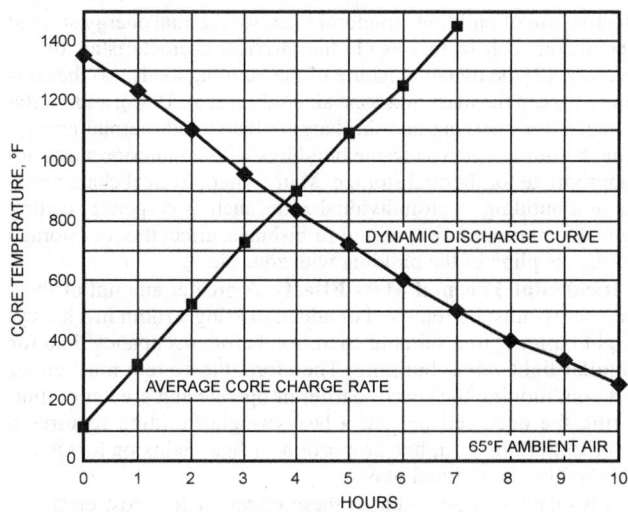

**Fig. 17 Room Storage Heater Dynamic
Discharge and Charge Curves**

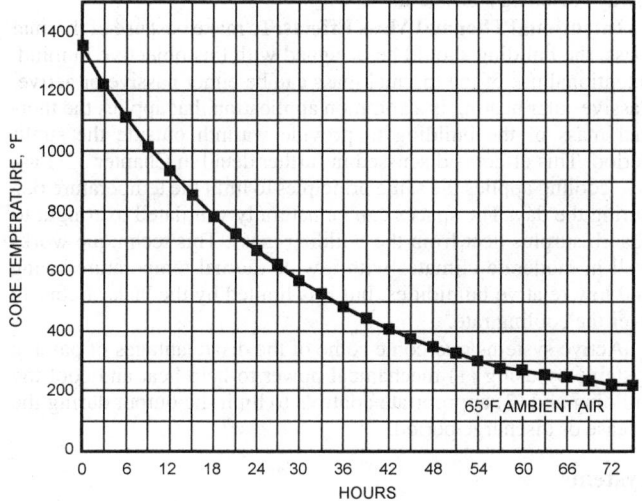

Fig. 18 Static Discharge from Room Storage Heater

Warm Room Concept. Under this strategy, one or two room units are generally used as the primary heating source during the control (on-peak) periods. The units are placed in the area most often occupied (main area). Adjacent areas generally are kept cooler and have no operable heat during the control (on-peak) time; however, some heat migrates from the main area.

When determining the heat loss of the main area, an additional sizing factor of approximately 25% should be added to allow for migration of heat to adjacent areas. Under the warm room concept, sizing factors vary depending upon the control strategy of the power company and the performance of the equipment under those control strategies. The designer should consult the manufacturer of the equipment for specific sizing information.

The warm room concept is the most common method used by power companies for their load management and off-peak marketing programs. It is successful in areas that have a small number of consecutive hours of control (generally 10 hours or less), or have a mid-day block of off-peak time during which the equipment can recharge. The advantage of the warm room concept is that it requires smaller sizes and quantity of equipment than the whole house concept.

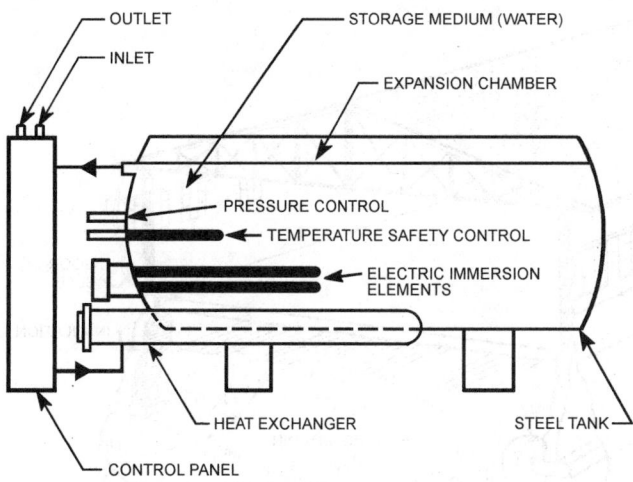

Fig. 19 Pressurized Water Heater

Heat Pump Booster

Air-to-air heat pumps generally perform well when outside temperatures are relatively warm. However, as outside temperature drops, the efficiency, output capacity, and temperature from a heat pump also declines. When the output of the heat pump drops below the heat loss of the structure, supplemental heat (typically from an electric resistance unit) must be added to maintain comfort. To eliminate the use of electric resistance heat during peak periods, storage heaters such as the heat pump booster (HPB) can be used to supplement the output of a heat pump. The HPB can also be used as a booster for stand-alone furnaces or to back up electric or fossil fuel equipment in duel-fuel programs.

Core charging of the HPB is regulated automatically based on outdoor temperature. The brick storage core is well insulated so that radiant or static heat discharge is small. Equipment input power ratings range from 11 to 22 kW. Storage capacities range from 75 to 135 kWh.

Central Furnace

The central storage furnace is a centrally-ducted, heat-storage product for residential and small commercial and industrial applications. These units area available with input ratings ranging from 28.8 to 38.4 kW. Storage capacities range from 180 to 240 kWh.

Pressurized Water Storage Heaters

This storage device consists of an insulated cylindrical steel tank containing immersion electrical resistance elements near the bottom of the tank and a water-to-water heat exchanger near the top (Figure 19). During off-peak periods, the resistance elements are sequentially energized until the storage water reaches a maximum temperature of 280°F, corresponding to 50 psig. The *ASME Boiler Code* considers such vessels unfired pressure vessels, so they are not required to meet the provisions for fired vessels. The heaters are controlled by a pressure sensor, which eliminates problems that could be caused by unequal temperature distributions. A thermal controller gives high-limit temperature protection. Heat is withdrawn from storage by running service water through the heat exchangers and a tempering device that controls the output temperature to a predetermined level. The storage capacity of the device is the sensible heat of water between 10°F above the desired output water temperature and 280°F. The output water can be used for space heating or service hot water. The water in the storage tank is permanently treated and sealed, requires no makeup, and does not interact with the service water.

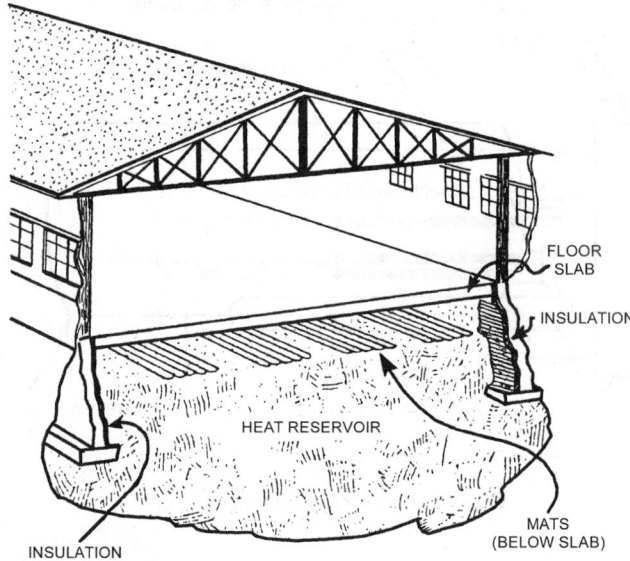

Fig. 20 Underfloor Heat Storage

Underfloor Heat Storage

This storage method typically uses electric resistance cables buried in a bed of sand 1 to 3 ft below the floor of a building. It is suitable for single-story buildings, such as residences, churches, offices, factories, and warehouses. An underfloor storage heater acts as a flywheel; while it is charged only during the nightly off-peak, it maintains the top of the floor slab at a constant temperature slightly higher than the desired space temperature. Because the cables spread heat in all directions, they do not have to cover the entire slab area. For most buildings, a cable location of 18 in. below the floor elevation is optimum. The sand bed should be insulated along its perimeter with 2 in. of rigid, closed-cell foam insulation to a depth of 4 ft (see Figure 20). Even with a well-designed and well-constructed underfloor storage, 10% or more of the input heat may be lost to the ground.

BUILDING MASS

Building Mass Effects

The thermal storage capabilities inherent in building mass can have a significant effect on the temperature within the space as well as on the performance and operation of the HVAC system. Effective use of structural mass for thermal storage reduces building energy consumption and reduces and delays peak heating and cooling loads (Braun 1990). In some cases, it improves comfort (Simmonds 1991; Morris et al. 1994). Perhaps the best-known use of thermal mass to reduce energy consumption is in buildings that include passive solar techniques (Balcomb 1983).

Cooling energy can be reduced by precooling the structure at night using ventilation air. Braun (1990), Ruud et al. (1990), and Andresen and Brandemuehl (1992) suggested that mechanical precooling of a building can reduce and delay peak cooling demand; Simmonds (1991) suggested that the correct building configuration may even eliminate the need for a cooling plant. Mechanical precooling may require more energy use; however, the reduction in electrical demand costs may give lower overall energy costs. Moreover, the installed capacity of air-conditioning equipment may also be reduced, providing lower installation costs.

The effective use of thermal mass can be considered incidental and be allowed for in the heating or cooling design, or it may be considered intentional and form an integral part of the design. The effective use of building structural mass for thermal energy storage depends on such factors as (1) the physical characteristics of the structure, (2) the dynamic nature of the building loads, (3) the coupling between the mass and zone air (Akbari et al. 1986), and (4) the strategies for charging and discharging the stored thermal energy. Some buildings, such as frame buildings with no interior mass, are inappropriate for thermal storage. Many other physical characteristics of a building or an individual zone, such as carpeting, ceiling plenums, interior partitions, and furnishings, affect thermal storage and the coupling of the building with zone air.

Incidental Thermal Mass Effects. A greater amount of thermal energy must be removed or added to bring a room in a heavyweight building to a suitable condition before occupancy than for a similar lightweight building. Therefore, the system must either start conditioning the spaces earlier or operate at a greater output. During the occupied period, a heavyweight building requires a lower output, as a higher proportion of heat gains or losses are absorbed by the thermal mass.

Advantage can be taken of these effects if low-cost electrical energy is available during the night; the air-conditioning system can be operated during this period to precool the building. This can reduce both the peak and total energy required during the following day (Braun 1990; Andresen and Brandemuehl 1992) but may not always be energy-efficient.

Intentional Thermal Mass Effects. To make best use of thermal mass, the building should be designed with this objective in mind. Intentional use of the thermal mass can be either passive or active. Passive solar heating is a common application that applies the thermal mass of the building to provide warmth outside the sunlit period. This effect is discussed in further detail in Chapter 32. Passive cooling applies the same principles to limit the temperature rise during the day. The spaces can be naturally ventilated overnight to absorb surplus heat from the building mass. This technique works well in moderate climates with a wide diurnal temperature swing and low relative humidities, but it is limited by the lack of control over the cooling rate.

Active systems overcome some of the disadvantages of passive systems by using (1) mechanical power to help heat and cool the building and (2) appropriate controls to limit the output during the release or discharge period.

Systems

Both night ventilation and precooling have limitations. The amount of heat stored in a slab equals the product of mass, specific heat and temperature rise. The amount of heat available to the space depends on the rate at which heat can be extracted from the slab, which in simple terms is

$$q_s = \rho c_p V \frac{dt_s}{d\theta} = h_o A(t_s - t) \tag{5}$$

where

q_s = rate of heat flow from slab, Btu/h
ρ = density, lb/ft^3
c_p = specific heat, Btu/lb·°F
V = slab volume, ft^3
θ = time, h
h_o = heat transfer coefficient, Btu/h·ft^2·°F
A = area of slab, ft^2
t_s = temperature of slab, °F
t = temperature of space, °F

Equation (5) also applies to transferring heat to the storage medium; while the potential is equivalent to $c_p V(t_s - t)$, the heat released during the daytime period is related to the transfer coefficients. Building transfer coefficients are quite low; for example, a typical value for room surfaces is 1.4 Btu/h·ft^2·°F, which is the maximum amount of energy that can be released.

Effective Storage Capacity. The total heat capacity (THC) (Ruud et al. 1990) is the maximum amount of thermal energy stored or released due to a uniform change in temperature Δt of the material and is given by

$$THC = \rho c_p V \Delta t \qquad (6)$$

The diurnal heat capacity (DHC) is a measure of the thermal capacity of a building component exposed to periodically varying temperature.

Many factors must be considered when an energy source is time-dependent. The minimum temperature occurs around dawn, which may be at the end of the off-peak tariff; the optimum charge period may run into the working day. Beginning the charge earlier may be less expensive but also less energy-efficient. In addition, the energy stored in the building mass is neither isolated nor insulated, so some energy is lost during charging; and the amount of available free energy varies and must be balanced against the energy cost of mechanical power. As a result, there is a trade-off that varies with time between the amount of free energy that can be stored and the power necessary for charging.

As the cooling capacity is, in effect, embedded in the building thermal mass, conventional techniques of assessing the peak load cannot be used. Detailed weather records that show peaks over 3- to 5-day periods, as well as data on either side of the peaks, should be examined to ensure that (1) the temperature at which the building fabric is assumed to be before the peak period is realistic and (2) the consequences of running with an exhausted storage after peak are considered. This level of analysis can only be carried out effectively using a dynamic simulation program. Experience has shown that these programs should be used with a degree of caution, and the results should be compared with both experience and intuition.

Storage Charging and Discharging

The building mass can be charged (cooled or warmed) either indirectly or directly. Indirect charging is usually accomplished by heating or cooling either the bounded space or an adjacent void. Almost all passive and some active cooling systems are charged by cooling the space overnight (Arnold 1978). Most indirect active systems charge the store by ventilating the void beneath a raised floor (Herman 1980; Crane 1991). Where this is an intermediate floor, cooling can be radiated into the space below and convected from the floor void the following day. By varying the rate of ventilation through the floor void, the rate of discharge can be controlled. Proprietary floor slabs are commonly of the hollow-core type (Anderson et al. 1979; Willis and Wilkins 1993). The cores are continuous, but when used for thermal storage, they are plugged at each end, and holes are drilled to provide the proper airflow. Charging is carried out by circulating cool or warm air through the hollow cores and exhausting it to the room. Discharge can be controlled by a ducted switching unit that directs air through the slab or straight into the space.

A directly charged slab, used commonly for heating and occasionally for cooling, can be constructed with an embedded hydronic coil. The temperature of the slab is only cycled 3 to 5°F to either side of the daily mean temperature of the slab. Consequently the technique can use very low grade free cooling (approximately 66°F) (Meierhans 1993) or low-grade heat rejected from condensers (approximately 82°F). In cooling applications the slab is used as a cool radiant ceiling, and for warming it is usually a heated floor. Little control is necessary due to the small temperature differences and the high heat capacity of the slab.

INSTALLATION, OPERATION, AND MAINTENANCE

The design professional must consider that almost all thermal storage systems require more space than nonstorage systems. Having selected a system, the designer must decide on the physical location; the piping interface to the air-conditioning equipment; and the water treatment, control, and optimization strategies to transfer theoretical benefits into realized benefits. The design must also be documented, the operators trained, and the performance verified (i.e., the system must be properly commissioned). Finally, the system must be properly maintained over its projected service life. For further information on operation and maintenance management, see Chapter 37.

SPECIAL REQUIREMENTS

The location and space required by a thermal storage system are functions of the type of storage and the architecture of the building and site. Building or site constraints often shift the selection from one option to another.

Chilled Water Systems

Chilled water systems are associated with large volume. As a result, many stratified chilled water storage systems are located outdoors (such as in industrial plants or suburban campus locations). A tall tank is desirable for stratification, but a buried tank may be required for architectural or zoning reasons. Tanks are traditionally constructed of steel or prestressed concrete. A supplier who assumes full responsibility for the complete performance often constructs the tank at the site and installs the entire distribution system.

Ice-on-Coil Systems (External and Internal Melt)

Ice-on-coil systems are available in many configurations with differing space and installation requirements. Because of the wide variety available, these often best meet the unique requirements of many types of buildings.

Bare coils are available for installation in concrete cells, which are a part of the building structure. The bare steel coil concept can be used with direct cooling, in which the refrigerant is circulated through the coils, and the water is circulated over the coils to be chilled or frozen. This external melt system has very stringent installation requirements. Coil manufacturers do not normally design or furnish the tank, but they do provide design assistance, which covers distribution and air agitation design as well as side and end clearance requirements. These recommendations must be followed exactly to ensure success.

The bare coil concept can also be used with a secondary coolant to provide the cooling necessary to build the ice. In an internal melt configuration, the ice and water, which remains in the tank and is not circulated to the cooling system, cools the secondary coolant during discharge. This indirect chilling can also be used with an external melt discharge if it is not desirable to circulate the secondary coolant to the cooling load. Indirect chilling can use either steel or plastic tubes in the ice builder.

Coils with factory-furnished containers come in a variety of sizes and shapes. A suitable style can usually be found to fit the available space. Round plastic containers with plastic coils are available in several sizes. These are offered only in an internal melt configuration and can be above ground or partially or completely buried.

Rectangular steel tanks are available with both steel and plastic pipe in a wide variety of sizes and capacities. Steel coil modules have the option of either internal or external melt. These steel tank systems are not normally buried. Each system comes prepackaged; installation requires only placement of the tank and proper piping connections. Any special support or insulation requirements of the manufacturer must be strictly followed.

Encapsulated Ice

Cylindrical steel containers with encapsulated water modules are also available. These offer yet another shape to fit available space. With proper precautions, these containers can be installed below grade. Standards and recommendations for corrosion protection published by the Steel Pipe Institute and the National Association of Corrosion Engineers should be followed, as should the manufacturer's instructions. These systems are not shipped assembled. The containers must be placed in the shell at the job site in a way that channels the secondary coolant through passages where the desired heat transfer will be achieved.

Ice-Harvesting

Field-built concrete ice tanks are generally used with ice harvesting. The ice harvester manufacturer may furnish assistance in tank design and piping distribution in the tank. The tank may be completely or partially buried or installed above ground. Where the ground is dry and free of moving water, tanks have been buried without insulation. In this situation the ground temperature eventually stabilizes, and the heat loss becomes minimal. However, a minimum of 2 in. of closed cell insulation should be applied to the external surface. Because the shifting ice creates strong dynamic forces, internal insulation should not be used except on the underside of the tank cover. In fact, only very rugged components should be placed in the tank; exit water distribution headers should be of stainless steel or rugged plastic suitable for the cold temperatures encountered. PVC is not an acceptable material due to its extreme brittleness at the ice water temperature. An underfloor system that is a part of the concrete structure is preferred.

As with the chilled water and hydrated salt PCM tanks, close attention to the design and construction is critical to prevent leakage. Unlike a system where the manufacturer builds the tank and assumes responsibility for its integrity, an ice-harvesting system needs an on-site engineer familiar with concrete construction requirements to monitor each pour and to check all water stops and pipe seals. Unlined tanks that do not leak can be built. If liners are used, the ice equipment suppliers will provide assistance in determining a suitable type; the liner should be installed only by a qualified installer trained in the proper methods of installation by the liner maker.

The sizing and location of the ice openings is critical; the tank design engineer should check all framed openings against the certified drawings before the concrete is poured.

An ice harvester is generally installed by setting in place a prepackaged unit that includes the ice-making surface, the refrigerant piping, the refrigeration equipment, and, in some cases, the heat rejection equipment and the prewired control. To ensure proper ice harvest, the unit must be properly positioned with respect to the drop opening. As the internal piping is not normally insulated, the drop opening should extend under the piping so that condensate drops into the tank. A grating below the piping is desirable. To prevent air or water leakage, gasketing between the unit frame and caulking must be installed in accordance with the manufacturer's instructions. External piping and power and control wiring complete the installation.

Other PCM Systems

Coolant normally flows horizontally in salt and polymeric systems, so the tanks tend to be shallower than the ideal chilled water storage tank. As in chilled water systems, the chilled water supply to and return from the tank must be designed to distribute water uniformly through the tank without channeling. Tanks are traditionally of concrete. The system supplier normally designs the tank and its distribution system, builds the tank, and installs the salt solution containers.

SYSTEM INTERFACE

Open Systems

Chilled water; salt and polymeric PCMs; external melt ice-on-coil; and ice-harvesting systems are all open chilled water piping systems. Drain-down must be prevented by isolation valves, pressure-sustaining valves, or heat exchangers. Due to the potential for drain-down, the open nature of the system, and the fact that the water being pumped may be saturated with air, the construction contractor must follow the piping details carefully to prevent pumping or piping problems.

Closed Systems

Closed systems normally circulate an aqueous secondary coolant (25 to 30% glycol solution) either directly to the cooling coils or to a heat exchanger interface to the chilled water system. A domestic water makeup system should not be the automatic makeup to the secondary coolant system. An automatic makeup unit that pumps a premixed solution into the system is recommended, along with an alarm signal to the building automation system to indicate makeup operation. The secondary coolant must be an industrial solution (not automotive antifreeze) with inhibitors to protect the steel and copper found in the piping. The water should be deionized; as portable deionizers can be rented, the solution can be mixed on-site. A calculation, backed up by metering the water as it is charged into the piping system for flushing, is needed to determine the specified concentration. Premixed coolant made with deionized water is also available, and tank truck delivery with direct pumping into the system is recommended on large systems. An accurate estimate of volume is required.

INSULATION

Because the chilled water, secondary coolant, or refrigerant temperatures are generally 10 to 20°F below those found in nonstorage systems, special care must be taken to prevent damage. Although fiberglass or other open-cell insulation is theoretically suitable when supplied with an adequate vapor barrier, experience has shown that its success is highly dependent upon workmanship. Therefore, a two-layer closed-cell material with staggered joints and carefully sealed joints is recommended. A thickness of 1.5 to 2 in. is normally adequate to prevent condensation in a normal room. Provisions must be made to ensure that the relative humidity in the equipment room is less than 80%. This can be done with heating or with cooling and dehumidification.

Special attention must be paid to pump and heat exchanger insulation covers. Valve stem, gage, and thermometer penetrations and extensions should be carefully sealed and insulated to prevent condensation. PVC covers over all insulation in the mechanical room improve appearance, provide limited protection, and are easily replaced if damaged. Insulation located outdoors should be protected by an aluminum jacket.

REFRIGERATION EQUIPMENT

The refrigeration system may be packaged chillers, field-built refrigeration, or refrigeration equipment furnished as a part of a package. The refrigeration system must be installed in accordance with the manufacturer's recommendations. Due to the high cost of refrigerant, refrigerant vapor detectors are suggested even for Class A1 refrigerants. Equipment rooms must be designed and installed to meet ASHRAE *Standard* 15. Relief valve lines should be monitored to detect valve weeping; any condensate that collects in the relief lines must be diverted and trapped so that it does not flow to the relief valve and eventually damage the seat.

WATER TREATMENT

Open Systems

Water treatment must be given close scrutiny in open systems. While the evaporation and concentration of solids associated with cooling towers does not occur, the water may be saturated with air, so the corrosion potential is greater than in a closed system. Treatment against algae, scale, and corrosion must be provided. No matter what type of treatment is chosen (i.e., traditional chemical or nontraditional treatment), some type of filtration that is effective at least down to 24 μm should be provided. To prevent damage, water treatment must be operational immediately following the completion of the cleaning procedure. Corrosion coupon assemblies should be included to monitor the effectiveness of the treatment. Water testing and service should be performed at least once a month by the water treatment supplier.

Closed Systems

The secondary coolant should be pretreated by the supplier. A complete analysis should be done annually. Monthly checks on the solution concentration should be made using a refractive indicator. Automotive-type testers are not suitable. For normal use, the solution should be good for many years without needing new inhibitors. However, provision should be made for the injection of new inhibitors through a shot feeder if recommended by the manufacturer. The need for filtering, whether it be the inclusion of a filtering system or filtering the water or solution before it enters the system, should be carefully considered. Combination filter feeders and corrosion coupon assemblies may be needed for monitoring the effect of the solution on copper and steel.

CONTROLS

A direct digital control to monitor and control all of the equipment associated with the central plant is preferred. Monitoring of electrical use by all primary plant components, individually if possible but at least as a group, is strongly recommended. Monitoring of the refrigeration capacity produced by the refrigeration equipment, by direct measurement where possible or by manufacturer's capacity ratings related to suction and condensing pressures, should be incorporated. This ensures that a performance rating can be calculated for use in the commissioning process and reevaluated on an ongoing basis as a management tool to gage performance.

Optimization Software

Optimization software should be installed to obtain the best performance from the system. This software must be able to predict, monitor, and adjust to meet the load, as well as adapt to daily or weekly storage, full or partial storage, chiller or ice priority, and a wide variety of rate schedules.

IMPLEMENTATION AND COMMISSIONING

Elleson (1996) identified the following key steps to designing a cool storage system:

- Calculate an accurate load profile.
- Use an hourly operating profile to size and select equipment.
- Develop a detailed description of the control strategy.
- Produce a schematic diagram.
- Produce a statement of design intent.
- Use safety factors with care.
- Plan for performance monitoring.
- Produce complete design documents.
- Retain an experienced cool storage engineer to review design.

Chapter 37 and Chapter 41 and ASHRAE *Guidelines* 1 and 4 provide information regarding design documentation and operator training.

Performance Verification

The commissioning authority should verify performance and document all operating parameters. This information should be used to establish a database for future reference to normal conditions based on a constant design condensing temperature. Some of the performance data for various systems are as follows:

External Melt Ice-on-Coil Storage

- Evaporator and suction temperatures at start of ice build
- Evaporator and suction temperatures at end of ice build
- Ice thickness at end of ice build
- Time to build ice
- Efficiency at start versus theoretical efficiency
- Efficiency at end versus theoretical efficiency
- Refrigeration capacity based on published ratings (deviation can indicate refrigerant loss or surface fouling)

Internal Melt Ice-on-Coil Storage

- Secondary coolant temperature and suction temperature at start
- Secondary coolant temperature and suction temperature at end
- Secondary coolant flow
- Tank water level at start
- Tank water level at end
- Time to build ice
- Efficiency at start versus theoretical efficiency
- Efficiency at end versus theoretical efficiency
- Capacity based on measured flow, heat balance, and published rating

Ice-Harvesting

- Suction temperature at start
- Suction temperature at harvest
- Harvest time/condensing temperature
- Time from start to bin full signal
- Efficiency
- Tank water level at start
- Tank water level at bin full signal
- Capacity based on published rating

While tank water level cannot be used as an indicator of the amount of ice in storage in a dynamic system, the water level at the end of the discharge cycle is a good indicator of conditions in the system. In systems with no gain or loss of water, the shutdown level should be consistent, and it can be used as a backup to determine when the bin is full for shutdown requirements. Conversely, a change in level at shutdown can indicate a water gain or loss.

Maintenance Requirements

Following the manufacturer's maintenance recommendations is essential to satisfactory long-term operation. These recommendations vary, but their objective is to maintain the refrigeration equipment, the refrigeration charge, the coolant circulation equipment, the ice builder surface, the water distribution equipment, water treatment, and controls so that they continue to perform at the same level as when the system was commissioned. Monitoring ongoing performance against kilowatt-hours per ton of ice built gives a continuing report of system performance.

REFERENCES

Akbari, H. et al. 1986. The effect of variations in convection coefficients on thermal energy storage in buildings: Part 1—Interior partition walls. Lawrence Berkeley Laboratory, Berkeley, CA.

Anderson, L.O., K.G. Bernander, E. Isfalt, and A.H. Rosenfeld. 1979. Storage of heat and cooling in hollow-core concrete slabs. Swedish Experience and Application to Large, American Style Building. 2nd International Conference on Energy Use Management, Los Angeles.

Andrepont, J.S. 1992. Central chilled water plant expansions and the CFC refrigerant issue—Case studies of chilled water storage. Proceedings of the Association of Higher Education Facilities Officers 79th Annual Meeting, Indianapolis, IN.

Andrepont, J.S. 1994. Performance and economics of CT inlet air cooling using chilled water storage. *ASHRAE Transactions* 100(1):587.

Andresen, I. and M.J. Brandemuehl. 1992. Heat storage in building thermal mass: A parametric study. *ASHRAE Transactions* 98(1):910-18.

Arnold, D. 1978. Comfort air conditioning and the need for refrigeration. *ASHRAE Transactions* 84(2):293-303.

ASHRAE. 1981. Methods of testing thermal storage devices with electrical input and thermal output based on thermal performance. *Standard* 94.2-1981 (RA 96).

ASHRAE. 1986. Methods of testing active sensible thermal energy storage devices based on thermal performance. *Standard* 94.3-1986 (RA 96).

ASHRAE. 1996. Commissioning of HVAC systems. *Guideline* 1-1996.

ASHRAE. 1994. Safety code for mechanical refrigeration. *Standard* 15-1994.

ASHRAE. 1993. Preparation of operating and maintenance documentation for building systems. *Guideline* 4-1993.

ASME. Annual. Boiler and pressure vessel codes. American Society of Mechanical Engineers, New York.

Balcomb, J.D. 1983. Heat storage and distribution inside passive solar buildings. Los Alamos National Laboratory, Los Alamos, NM.

Bahnfleth, W.P. and W.S. Joyce. 1994. Energy use in a district cooling system with stratified chilled water storage. *ASHRAE Transactions* 100(1):1767-78.

Braun, J.E. 1990. Reducing energy costs and peak electrical demand through optimal control of building thermal storage. *ASHRAE Transactions* 96(2):876-88.

Caldwell, J. and W. Bahnfleth. 1997. Chilled water storage feasibility without electric rate incentives or rebates. *ASCE Journal of Architectural Engineering* 3(3):133-140.

Crane, J.M. 1991. The Consumer Research Association new office/laboratory complex, Milton Keynes: Strategy for environmental services. CIBSE National Conference (pp. 2-29), Canterbury, England.

CSA. 1993. Design and construction of earth energy heat pump systems for commercial and institutional buildings. *Standard* C447-93. Canadian Standards Association, Rexdale, Ontario.

Dorgan, C.E. and J.S. Elleson. 1993. *Design guide for cool thermal storage.* ASHRAE.

Drees, K.H. 1994. Modeling and control of area constrained ice storage systems. M.S. Thesis, Purdue University, W. Lafayette, IN.

Ebeling, J.A., L. Beaty, and S.K. Blanchard. 1994. Combustion turbine inlet air cooling using ammonia-based refrigeration for capacity enhancement. *ASHRAE Transactions* 100(1):583-6.

Elleson, J.S. 1996. *Successful cool storage projects: From planning to operation.* ASHRAE, Atlanta.

Elleson, J.S. 1993. Energy use of fan-powered mixing boxes with cold air distribution. *ASHRAE Transactions* 99(1):1349-58.

Fiorino, D.P. 1994. Energy conservation with thermally stratified chilled-water storage. *ASHRAE Transactions* 100(1):1754-66.

Gute, G.D., W.E. Stewart, Jr., and J. Chandrasekharan. 1995. Modelling the ice-filling process of rectangular thermal energy storage tanks with multiple ice makers. *ASHRAE Transactions* 101(1).

Hall, S.H. 1993a. Environmental risk assessment for aquifer thermal energy storage. PNL-8365. Pacific Northwest Laboratory, Richland, WA.

Hall, S.H. 1993b. Feasibility studies for aquifer thermal energy storage. PNL-8364. Pacific Northwest Laboratory, Richland, WA.

Herman, A.F.E. 1980. Underfloor, structural storage air-conditioning systems. Proc. Paper 7. FRIGAIR 80, Pretoria, South Africa.

Hersh, H., G. Mirchandani, and R. Rowe. 1982. Evaluation and assessment of thermal energy storage for residential heating. ANL SPG-23. Argonne National Laboratory, Argonne, IL.

Hittle, D.C. and T.R. Smith. 1994. Control strategies and energy consumption for ice storage systems using heat recovery and cold air distribution. *ASHRAE Transactions* 100(1):1221-29.

Holness, G.V.R. 1992. Case study of combined chilled-water thermal energy storage and fire protection storage. *ASHRAE Transactions* 98(1):1119-22.

Hussain, M.A. and D.C.J. Peters. 1992. Retrofit integration of fire protection storage as chilled-water storage—A case study. *ASHRAE Transactions* 98(1):1123-32.

ITSAC. 1992. *Advisory Newsletter* (March). International Thermal Storage Advisory Council, San Diego.

Jenne, E.A. 1992. Aquifer thermal energy (heat and chill) storage. Papers presented at the 1992 Intersociety Energy Conversion Engineering Conference, PNL-8381. Pacific Northwest Laboratory, Richland, WA.

Kirshenbaum, M.S. 1991. Chilled-water production in ice-based thermal storage systems. *ASHRAE Transactions* 97(2):422-27.

Knebel, D.E. 1986. Thermal storage—A showcase on cost savings. *ASHRAE Journal* 28(5):28-31.

Knebel, D.E. 1988a. Economics of harvesting thermal storage systems: A case study of a merchandise distribution center. *ASHRAE Transactions* 94(1):1894-1904. Reprinted in ASHRAE Technical Data Bulletin 5(3):35-39, 1989.

Knebel, D.E. 1988b. Optimal control of harvesting ice thermal storage systems. AICE Proceedings, 1990, pp. 209-214.

Knebel, D.E. and S. Houston. 1989. Case study on thermal energy storage—The Worthington Hotel. *ASHRAE Journal* 31(5):34-42.

Knebel, D.E. 1991. Optimal design and control of ice harvesting thermal energy storage systems. ASME 91-HT-28. American Society of Mechanical Engineers, New York.

Landry, C.M. and C.D. Noble. 1991. Case study of cost-effective low-temperature air distribution, ice thermal storage. *ASHRAE Transactions* 97(1):854-62.

MacCracken, C.D. 1994. An overview of the progress and the potential of thermal storage in off-peak turbine inlet cooling. *ASHRAE Transactions* 100(1):569-71.

Mackie, E.I. 1994. Inlet air cooling for a combustion turbine using thermal storage. *ASHRAE Transactions* 100(1):572-82.

Meckler, M. 1992. Design of integrated fire sprinkler piping and thermal storage systems: Benefits and challenges. *ASHRAE Transactions* 98(1):1140-48.

Meierhans, R.A. 1993. Slab cooling and earth coupling. *ASHRAE Transactions* 99(2):511-18.

Morofsky, E. 1994. Procedures for the environmental impact assessment of aquifer thermal energy storage. PWC/RDD/106E. Environment Canada and Public Works and Government Services Canada, Ottawa K1A OM2.

Morris, F.B., J.E. Braun, and S.J. Treado. 1994. Experimental and simulated performance of optimal control of building thermal storage. *ASHRAE Transactions* 100(1):402-14.

Musser, A. and W. Bahnfleth. 1998. Evolution of temperature distributions in a full-scale stratified chilled water storage tank. *ASHRAE Transactions* 104(1).

ORNL. 1985. Field performance of residential thermal storage systems. EPRI EM 4041. Oak Ridge National Laboratories.

Public Works Canada. 1991. Workshop on generic configurations of seasonal cold storage applications. International Energy Agency, Energy Conservation Through Energy Storage Implementing Agreement (Annex 7). PWC/RDD/89E (September). Public Works Canada, Ottawa K1A OM2.

Public Works Canada. 1992. Innovative and cost-effective seasonal cold storage applications: Summary of national state-of-the-art reviews. International Energy Agency, Energy Conservation Through Energy Storage Implementing Agreement (Annex 7). PWC/RDD/96E (June). Public Works Canada, Ottawa K1A OM2.

Ruchti, T.L., K.H. Drees, and G.M. Decious. 1996. Near optimal ice storage system controller. EPRI International Conference on Sustainable Thermal Energy Storage, pp. 92-98.

Rumbaugh, J., M. Blaha, W. Premerlani, F. Eddy, and W. Lorensen. 1991. *Object-oriented modeling and design.* Prentice Hall, Englewood Cliffs, NJ.

Ruud, M.D., J.W. Mitchell, and S.A. Klein. 1990. Use of building thermal mass to offset cooling loads. *ASHRAE Transactions* 96(2):820-30.

Simmonds, P. 1991. The utilization and optimization of a building's thermal inertia in minimizing the overall energy use. *ASHRAE Transactions* 97(2):1031-42.

Simmonds, P. 1994. A comparison of energy consumption for storage priority and chiller priority for ice-based thermal storage systems. *ASHRAE Transactions* 100(1):1746-53.

Snijders, A.L. 1992. Aquifer seasonal cold storage for space conditioning: Some cost-effective applications. *ASHRAE Transactions* 98(1):1015-22.

Stewart, W.E., G.D. Gute, J. Chandrasekharan, and C.K. Saunders. 1995a. Modelling of the melting process of ice stores in rectangular thermal energy storage tanks with multiple ice openings. *ASHRAE Transactions* 101(1).

Stewart, W.E., G.D. Gute, and C.K. Saunders. 1995b. Ice melting and melt water discharge temperature characteristics of packed ice beds for rectangular storage tanks. *ASHRAE Transactions* 101(1).

Stovall, T.K. 1991. Turbo Refrigerating Company ice storage test report. ORNL/TM-11657. Oak Ridge National Laboratory, Oak Ridge, TN.

Tamblyn, R.T. 1985. College Park thermal storage experience. *ASHRAE Transactions* 91(1B):947-51.

Tran, N., J.F. Kreider, and P. Brothers. 1989. Field measurement of chilled water storage thermal performance. *ASHRAE Transactions* 95(1):1106-12.

Wildin, M.W. 1990. Diffuser design for naturally stratified thermal storage. *ASHRAE Transactions* 96(1):1094-1102.

Wildin, M.W. 1991. Flow near the inlet and design parameters for stratified chilled water storage. ASME 91-HT-27. American Society of Mechanical Engineers, New York.

Wildin, M.W. and C.R. Truman. 1989. Performance of stratified vertical cylindrical thermal storage tanks—Part I: Scale model tank. *ASHRAE Transactions* 95(1):1086-95.

Willis, S. and J. Wilkins. 1993. Mass appeal. *Building Services Journal* 15(1):25-27.

Yoo, J., M. Wildin, and C.W. Truman. 1986. Initial formation of a thermocline in stratified thermal storage tanks. *ASHRAE Transactions* 92(2A):280-90.

Sources of certain figures are as follows: Figure 10 courtesy Cryogel, Figure 14 courtesy Transphase Systems, and Figure 16 courtesy Control Electric Corp.

BIBLIOGRAPHY

Aizawa, J., S. Hamaoka, N. Ino, T. Kuriyama, and M. Ohtsu. 1997. Development of cooling system using slurry ice transportation (No. 1) Dynamic ice making and storage system. Proceedings of International Symposium Air Conditioning in High Rise Buildings '97. I:472-477. Tongji University Press, Shanghai.

ASHRAE. 1996. The HVAC commissioning process. *Guideline* 1-1996.

AWWA. 1984. Standard for welded steel tanks for water storage. *Standard* D100-84. American Water Works Association, Denver, CO.

AWWA. 1984. Standard for wire-wound circular prestressed-concrete water tanks. *Standard* D110-86. American Water Works Association, Denver, CO.

Ames, D.A. 1990. Thermal storage forum: Eutectic cool storage—Current developments. *ASHRAE Journal* 32(4):46-53.

Anderson, E.D., N.W. Hanson, and D.M. Schultz. 1987. Concrete storage tank construction techniques and quality control. EPRI Seminar Proceedings: Commercial Cool Storage, State of the Art, EM-5454-SR (October).

Andersson, O. 1993. Scaling and corrosion, International Energy Agency, Energy Conservation Through Energy Storage Implementing Agreement (Annex 6—Environmental and chemical aspects of thermal energy storage in aquifers and research and development of water treatment methods), Report 38-E.

Arnold, D. 1991. Laboratory performance of an encapsulated-ice store. *ASHRAE Transactions* 97(2):1170-78.

Arnold, D. 1994. Dynamic simulation of encapsulated ice stores—Part II: Model development and validation. *ASHRAE Transactions* 100(1):1245-54.

Athienitis, A.K. and T.Y. Chen. 1993. Experimental and theoretical investigation of floor heating with thermal storage. *ASHRAE Transactions* 99(1):1049-60.

Bellecci, C. and M. Conti. 1993. Transient behaviour analysis of a latent heat thermal storage module. *International Journal of Heat and Mass Transfer* 36(15):3851.

Berlund, L.G. 1991. Comfort benefits for summer air conditioning with ice storage. *ASHRAE Transactions* 97(1):843-47.

Bhansali, A. and D.C. Hittle. 1990. Estimated energy consumption and operating cost for ice storage systems with cold air distribution. *ASHRAE Transactions* 96(1):418-27.

Brady, T.W. 1994. Achieving energy conservation with ice-based thermal storage. *ASHRAE Transactions* 100(1):1735-45.

Cai, L., W.E. Stewart, Jr., and C.W. Sohn. 1993. Turbulent buoyant flows into a two-dimensional storage tank. *International Journal of Heat and Mass Transfer* 36(17):4247-56.

Cao, Y. and A. Faghri. 1992. A study of thermal energy storage systems with conjugate turbulent forced convection. *Journal of Heat Transfer* 114(4): 1019.

CBI. 1991. CBI Strata-Therm: Thermally stratified water storage systems. CBI *Brochure* ST 0191. Chicago Bridge and Iron Company, Chicago.

Conniff, J.P. 1991. Strategies for reducing peak air-conditioning loads by using heat storage in the building structure. *ASHRAE Transactions* 97(1):704-9.

Crane, R.J. and M.J.M. Krane. 1992. The optimum design of stratified thermal energy storage systems—Part II: Completion of the analytical model, presentation and interpretation of the results. *Journal of Energy Resources Technology* 114(3):204.

Dorgan, C.E. and C.B. Dorgan. 1995. *Cold air distribution design guide.* TR-105604. EPRI, Palo Alto, CA.

Dorgan, C.E. and C.B. Dorgan. 1995. Case study of an ice storage system with cold air distribution and heat recovery. TR-105858. EPRI, Palo Alto, CA.Dorgan, C.E. and J.S. Elleson. 1987. Low temperature air distribution: Economics, field evaluation, design. EM-5454-SR (October). EPRI Seminar Proceedings: Commercial Cool Storage, State of the Art.

Dorgan, C.E., C.B. Dorgan, and S. Leight. 1995. *Cool storage total building construction cost benefits—An owner's and architect's guide.* TR-104521. EPRI, Palo Alto, CA.

Dorgan, C.E. and J.S. Elleson. 1989. Design of cold air distribution systems with ice storage. *ASHRAE Transactions* 95(1):1317-22. Reprinted in ASHRAE Technical Data Bulletin 5(4):23-28.

Dorgan, C.E. and J.S. Elleson. 1994. ASHRAE design guide for cool thermal storage. *ASHRAE Transactions* 100(1):33-48.

Dorgan, C.E., J.S. Elleson, S. Dingle, S. Leight and C.B. Dorgan. 1995. Field evaluation of a eutectic salt storage system. TR-104942. EPRI, Palo Alto, CA.

Ebeling, J.A. and R.R. Balsbaugh. 1993. Combustion turbine inlet air cooling with thermal energy storage. *Turbomachinery International* 34(1):30.

Elleson, J.S. 1996. *Successful cool storage projects: From planning to operation.* ASHRAE.

Etaion, Y. and E. Erell. 1991. Thermal storage mass in radiative cooling systems. *Building and Environment* 26(4):389.

Fiorino, D.P. 1991. Case study of a large, naturally stratified, chilled water thermal energy storage system. *ASHRAE Transactions* 97(2):1161-69.

Fiorino, D.P. 1992. Thermal energy storage program for the 1990's. *Energy Engineering* 89(4):23-33.

Gallagher, M.W. 1991. Integrated thermal storage/Life safety systems in tall buildings. *ASHRAE Transactions* 97(1):833-38.

Gatley, D.P. 1992. Cool storage ethylene glycol design guide. EPRI TR-100945. Electric Power Research Institute, Palo Alto, CA.

Goncalves, L.C.C. and S.D. Probert. 1993. Thermal energy storage: Dynamic performance characteristics of cans each containing a phase-change material, assembled as a packed bed. *Applied Energy* 45(2):117.

Gretarsson, S.P., C.O. Pedersen, and R.K. Strand. 1994. Development of a fundamentally based stratified thermal storage tank model for energy analysis calculations. *ASHRAE Transactions* 100(1):1213-20.

Guven, H. and J. Flynn. 1992. Commissioning TES systems. *Heating/Piping/Air Conditioning* (January):82-84.

Harmon, J.J. and H.C. Yu. 1989. Design considerations for low-temperature air distribution systems. *ASHRAE Transactions* 95(1):1295-99.

Harmon, J.J. and H.C. Yu. 1991. Centrifugal chillers and glycol ice thermal storage units. *ASHRAE Journal* 33(12):25-31.

Hensel, E.C. Jr., N.L. Robinson, J. Buntain, J.W. Glover, B.D. Birdsell, and C.W. Sohn. 1991. Chilled-water thermal storage system performance monitoring. *ASHRAE Transactions* 97(2):1151-60.

Holness, G.V.R. 1987. Thermal storage forum: Industrial plants offer opportunities. *ASHRAE Journal* 29(5):24-25.

Holness, G.V.R. 1988. Thermal storage retrofit restores dual-temperature system. *ASHRAE Transactions* 94(1):1866-78. Reprinted in ASHRAE Technical Data Bulletin 5(3):51-59.

Hussain, M.A. and M.W. Wildin. 1991. Studies of mixing on the inlet side of the thermocline in diurnal stratified storage. Proceedings THERMASTOCK 1991, Fifth International Conference on Thermal Energy Storage, Scheveningen, Netherlands, May 13-16.

Jekel, T.B., J.W. Mitchell, and S.A. Klein. 1993. Modeling of ice-storage tanks. *ASHRAE Transactions* 99(1):1016-24.

Joyce, W.S. and W.P. Bahnfleth. 1992. Cornell thermal storage project saves money and electricity. *District Heating and Cooling* (2):22-29.

Kamel, A.A., M.V. Swami, S. Chandra, and C.W. Sohn. 1991. An experimental study of building-integrated off-peak cooling using thermal and moisture ("enthalpy") storage systems. *ASHRAE Transactions* 97(2): 240-44.

Kamimura, T., S. Hamaoka, N. Ino, T. Kuriyama, and H. Morikawa. 1997. Development of cooling system using slurry ice transportation (No. 2) Dynamic ice making and storage system. Proceedings of International Symposium Air Conditioning in High Rise Buildings '97. I:478-483. Tongji University Press, Shanghai.

Kuriyama, T. and Y. Sawahata. 1997. Development of cooling system using slurry ice transportation (No. 1) Dynamic ice making and storage system. Proceedings of International Symposium Air Conditioning in High Rise Buildings '97. I:484-489. Tongji University Press, Shanghai.

Kirkpatrick, A.T. and J.S. Elleson. 1997. *Cold air distribution system design guide*. ASHRAE, Atlanta.

Kleinbach, E., W. Beckman, and S. Klein. 1993. Performance study of one-dimensional models for stratified thermal storage tanks. *Solar Energy* 50(2):155.

Landry, C.M. and C.D. Noble. 1991. Making ice thermal storage first-cost competitive. *ASHRAE Journal* 33(5):19-22.

Laybourn, D.R. 1988. Thermal energy storage with encapsulated ice. *ASHRAE Transactions* 94(1):1971-88. Reprinted in ASHRAE Technical Data Bulletin 5(4):95-102.

Laybourn, D.R. 1990. Encapsulated ice thermal energy storage. AICE Proceedings, pp. 215-20.

MacCracken, C.D. 1985. Control of brine-type ice storage systems. *ASHRAE Transactions* 91(1B):32-43. Reprinted in ASHRAE Technical Data Bulletin: *Thermal Storage* 1985(January):26-36.

Mackie, E.I. and G. Reeves. 1988. *Stratified chilled-water storage design guide*. EPRI EM-4852s. Electric Power Research Institute, Palo Alto, CA.

Mackie, E.I. and W.V. Richards. 1992. Design of off-peak cooling systems. ASHRAE Professional Development Seminar, 1992.

McQuiston, F.C. and J.D. Spitler. 1992. *Cooling and heating load calculation manual*, 2nd ed. ASHRAE.

Midkiff, K.C., Y.K. Song, and C.E. Brett. 1991. Thermal performance and challenges for a seasonal chill energy storage based air-conditioning system. ASME 91-HT-29. American Society of Mechanical Engineers, New York.

Mirth, D.R., S. Ramadhyani, and D.C. Hittle. 1993. Thermal performance of chilled-water cooling coils operating at low water velocities. *ASHRAE Transactions* 99(1):43-53.

Mirza, C. 1992. Proceedings of Workshop on Design and Construction of Water Wells for ATES, International Energy Agency, Energy Conservation Through Energy Storage Implementing Agreement (Annex 6—Environmental and chemical aspects of thermal energy storage in aquifers and research and development of water treatment methods), Report 41-E (August). The Netherlands.

Molson, J.W., E.O. Frind, and C.D. Palmer. 1992. Thermal energy storage in an unconfined aquifer—II: Model development, validation, and application. *Water Resources Research* 28(10):2857.

Najafi, M. and W.J. Schaetzle. 1991. Cooling and heating with clathrate thermal energy storage system. *ASHRAE Transactions* 97(1):177-83.

Nussbaum, O.J. 1990. Using glycol in a closed circuit system. *Heating/Piping/Air Conditioning* (January):75-85.

Ott, V.J. and D. Limaye. 1986. Thermal storage air conditioning with clathrates and direct contact heat transfer. EPRI Proceedings: International Load Management Conference, June 1986, Section 47.

Paris, J., M. Falareau, and C. Villeneuve. 1993. Thermal storage by latent heat: A viable option for energy conservation in buildings. *Energy Sources* 15(1):85.

Paul, J. 1996. Storage of cold energy with binary ice. Proceedings of the EPRI International Conference on Sustainable Thermal Energy Storage, pp. 61-64. Thermal Storage Applications Research Center, Madison, WI.

Peters, D., W. Chadwick, and J. Esformes. 1986. Equipment sizing concepts for ice storage systems. EPRI Proceedings: International Load Management Conference, June 1986, Section 46.

Prusa, J., G.M. Maxwell, and K.J. Timmer. 1991. A Mathematical model for a phase-change, thermal energy storage system utilizing rectangular containers. *ASHRAE Transactions* 97(2):245-61.

Rabl, V.A. 1987. Load management: Issues and opportunities. EM-5454-SR (October). EPRI Seminar Proceedings: Commercial Cool Storage, State of the Art. Electric Power Research Institute, Palo Alto, CA.

Rogers, E.C. and B.A. Stefl. 1993. Ethylene glycol: Its use in thermal storage and its impact on the environment. *ASHRAE Transactions* 99(1): 941-49.

Rudd, A.F. 1993. Phase-change material wallboard for distributed thermal storage in buildings. *ASHRAE Transactions* 99(2):339-46.

Ryu, H.W., S.A. Hong, and B.C. Shin. 1991. Heat transfer characteristics of cool-thermal storage systems. *Energy* 16(4):727.

Slabodkin, A.L. 1992. Integrating an off-peak cooling system with a fire suppression system in an existing high-rise office building. *ASHRAE Transactions* 98(1):1133-39.

Sodha, M.S., J. Kaur, and R.L. Sawhney. 1992. Effect of storage on thermal performance of a building. *International Journal of Energy Research* 16(8):697.

Sohn, C.W., G.L. Cler, and R.J. Kedl. 1990a. Performance of an ice-in-tank diurnal ice storage cooling system at Fort Stewart, GA. U.S.A. CERL *Technical Report* E-90/10. Construction Engineering Research Lab.

Sohn, C.W., G.L. Cler, and R.J. Kedl. 1990b. Ice-on-coil diurnal ice storage cooling system for a barracks/office/dining hall facility at Yuma Proving Ground, AZ. CERL *Technical Report* E-90/13. Construction Engineering Research Lab.

Sohn, C.W. 1991. Field performance of an ice harvester storage cooling system. *ASHRAE Transactions* 97(2):1187-96.

Somasundaram, S., D. Brown, and K. Drost. 1992. Cost evaluation of diurnal thermal energy storage for cogeneration applications. *Energy Engineering* 89(4):8.

Sozen, M., K. Vafai, and L.A. Kennedy. 1992. Thermal charging and discharging of sensible and latent heat storage packed beds. *Journal of Thermophysics and Heat Transfer* 5(4):623.

Spethmann, D.H. 1989. Optimal control for cool storage. *ASHRAE Transactions* 95(1):1189-93. Reprinted in ASHRAE Technical Data Bulletin 5(4):57-61.

Spethmann, D.H. 1993. Application considerations in optimal control of cool storage. *ASHRAE Transactions* 99(1).

Stewart, R.E. 1990. Ice formation rate for a thermal storage system. *ASHRAE Transactions* 96(1):400-5.

Stewart, W.E., R.L. Kaupang, C.G. Tharp, R.D. Wendland, and L.A. Stickler. 1993. An approximate numerical model of falling-film iced crystal growth for cool thermal storage. *ASHRAE Transactions* 99(2):347-55.

Stovall, T.K. 1991a. Baltimore Aircoil Company (BAC) ice storage test report. ORNL/TM-11342. Oak Ridge National Laboratory, Oak Ridge, TN.

Stovall, T.K. 1991b. CALMAC ice storage test report. ORNL/TM-11582. Oak Ridge National Laboratory, Oak Ridge, TN.

Stovall, T.K. and J.J. Tomlinson. 1991. Laboratory performance of a dynamic ice storage system. *ASHRAE Transactions* 97(2):1179-86.

Strand, R.K., C.O. Pedersen, and G.N. Coleman. 1994. Development of direct and indirect ice-storage models for energy analysis calculations. *ASHRAE Transactions* 100(1):1230-44.

Tamblyn, R.T. 1985c. Control concepts for thermal storage. *ASHRAE Transactions* 91(1B):5-11. Reprinted in ASHRAE Technical Data Bulletin: *Thermal Storage* 1985(January):1-6. Also reprinted in *ASHRAE Journal* 27(5):31-34.

Tamblyn, R.T. 1990. Optimizing storage savings. *Heating/Piping/Air Conditioning* (August):43-46.

Wildin, M.W., E.I. Mackie, and W.E. Harrison. 1990. Thermal storage forum—Stratified thermal storage: A new/old technology. *ASHRAE Journal* 32(4):29-39.

Wildin, M.W. and C.R. Truman. 1985a. A summary of experience with stratified chilled water tanks. *ASHRAE Transactions* 91(1B):956-76. Reprinted in ASHRAE Technical Data Bulletin: *Thermal Storage* 1985 (January):104-123.

Wildin, M.W. and C.R. Truman. 1985b. Evaluation of stratified chilled-water storage techniques. EPRI EM-4352. Electric Power Research Institute, Palo Alto, CA.

Yoo, H. and E.-T. Pak. 1993. A theoretical model of the charging process for stratified thermal storage tanks. *Solar Energy* 51(6):513.

ENERGY MANAGEMENT

ENERGY conservation is the more efficient or effective use of energy. As fuel costs rise and environmental concerns grow, more-efficient energy conversion and utilization technologies become cost-effective. However, technology alone cannot produce sufficient results without a continuing management effort. Energy management begins with the commitment and support of an organization's management team.

Suggestions for developing an energy management program are shown in Figure 1. First, a team with the right skills is selected to manage and execute the program. This team establishes objectives, priorities, and a time frame, sometimes with the help of outside consultants. A historic database can help to evaluate future energy conservation opportunities (ECOs). A monitoring system can be set up to obtain detailed use data (see Chapter 39, Building Energy Monitoring). Then, a detailed energy audit, which is discussed later in this chapter, should be performed. After the ECOs and the estimated savings have been determined, the team issues a report and adjusts or fine-tunes the objectives and priorities. A few lower-cost conservation measures that result in substantial savings can be implemented first to show progress. At this stage, it is critical to monitor and collect data, because energy management programs with written proof of progress are often the most successful. Metering can be installed to monitor energy consumption for each major piece of equipment and for each consumption area. Charts showing the progress of everyone in the facility increase awareness of energy conservation. The program requires regular reports and readjustments of the objectives and priorities, and the implemented ECOs need to be maintained.

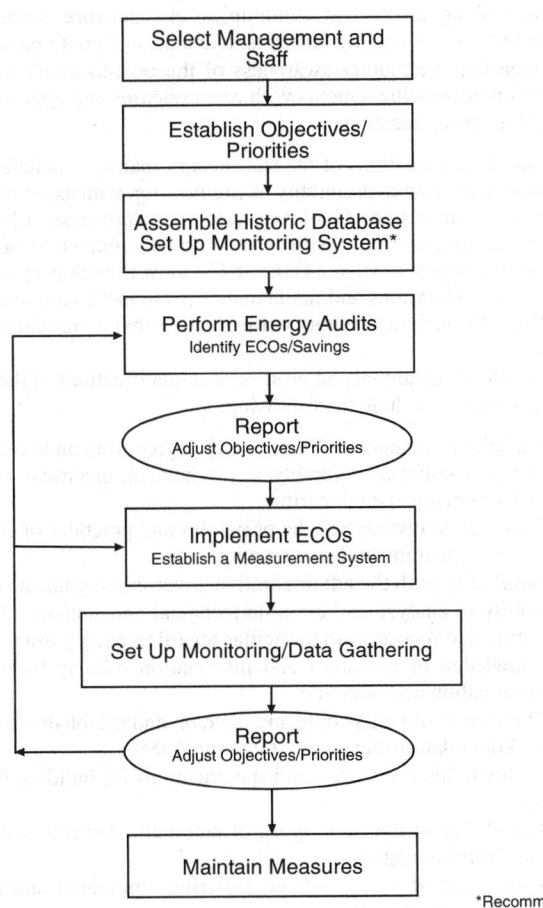

Fig. 1   Energy Management Program

ORGANIZATION

Because energy management is performed in existing facilities, most of this chapter is devoted to these facilities. Information on energy conservation in new design can be found in all volumes of the *ASHRAE Handbook* and in ASHRAE *Standards* 90.1 and 90.2. The area most likely to be overlooked in new design is the ability to measure and monitor energy consumption and trends for each energy use category given in Chapter 39.

To be effective, energy management must be given the same emphasis as management of any other cost/profit center. In this regard, the functions of top management are as follows:

- Establish the energy cost/profit center.
- Assign management responsibility for the program.
- Hire or assign an energy manager.
- Allocate resources.
- Ensure that the energy management program is clearly communicated to all departments to provide necessary support for achieving effective results.
- Monitor the cost-effectiveness of the program.
- Clearly set the program goals.
- Encourage ownership of the program at the lowest possible level in the organization.
- Set up an ongoing reporting and analysis procedure to monitor the energy management program.

An effective energy management program requires that the manager (supported by a suitable budget) act and be held accountable for those actions. It is common for a facility to allocate 3 to 10% of the annual energy cost for the administration of an energy management program. The budget should include funds for additional personnel as needed and for continuing education of the energy manager and staff.

If it is not possible to add a full-time, first-line manager to the staff, an existing employee, preferably with a technical background, should be considered for either a full- or part-time position. This person must be trained to organize an energy management program. Energy management should not be an alternate or collateral duty of an employee who is already fully occupied.

The preparation of this chapter is assigned to TC 9.6, System Energy Utilization.

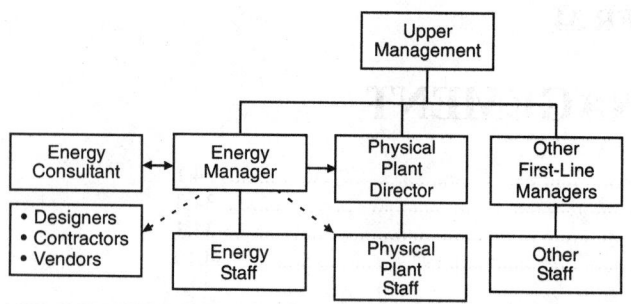

Fig. 2 Organizational Relationships of Effective Energy Management

Another option is to hire a professional energy management consultant to design, implement, and maintain energy efficiency improvements. Some energy services companies (ESCOs) and other firms provide energy management services as part of a contract, with payments based on realized savings.

Figure 2 shows the organizational relationship of effective energy management. The solid lines indicate normal reporting relationships within the organization; the dotted lines indicate new relationships created by energy management. The arrows indicate the primary directions in which initiatives related to energy conservation normally flow. This chart shows the ideal organization. In actuality, the functions shown may overlap, especially in smaller organizations. For example, the assistant principal of a high school may also serve as energy manager and physical plant director. However, it is important that all functions be covered.

Energy Manager

The general functions of an energy manager fall into four categories: technical, policy-related, planning and purchasing, and public relations. The energy manager will need assistance with some of the tasks in the following lists. A list of the specific tasks and plans for their implementation must be clearly documented.

Technical functions include the following:

1. Conducting energy audits and identifying energy conservation opportunities (ECOs)
2. Establishing a baseline from which energy-saving improvements can be measured
3. Acting as an in-house technical consultant on new energy technologies, alternative fuel sources, and energy-efficient practices
4. Evaluating the energy efficiency of proposed new construction, building expansion, remodeling, and new equipment purchases
5. Setting performance standards for efficient operation and maintenance of machinery and facilities
6. Reviewing state-of-the-art energy management hardware
7. Selecting the most appropriate technology
8. Reviewing operation and maintenance
9. Implementing ECOs
10. Establishing an energy accounting program for continuing analysis of energy usage and the results of ECOs
11. Maintaining the effectiveness of ECOs
12. Measuring energy use in the field whenever possible to verify design and operation conditions

Policy-related functions include the following:

1. Fulfilling the energy policy established by upper management
2. Monitoring federal and state legislation and regulatory activities, and recommending policy/response on such issues
3. Representing the organization in energy associations
4. Administering government-mandated reporting programs

Planning and purchasing functions include the following:

1. Monitoring energy supplies and costs to take advantage of fuel-switching and load management opportunities
2. Ensuring that systems and equipment are purchased based on economics (not simply on the lowest initial cost), their energy requirements, and their ability to perform the required functions
3. Maintaining a current understanding of energy conservation, grant programs, and demand side management (DSM) programs offered by utilities and agencies
4. Negotiating or advising on major utility contracts
5. Developing contingency plans for supply interruptions or shortages
6. Forecasting the organization's short- and long-term energy requirements and costs
7. Developing short- and long-term energy conservation plans and budgets
8. Reporting periodically to upper management

Public relations functions include the following:

1. Educating fellow employees on the benefits of efficient energy use
2. Establishing a system to elicit and evaluate energy conservation suggestions from employees
3. Recognizing successful energy conservation projects with awards to plants or employees
4. Setting up a formal system for reporting to upper management
5. Establishing an energy communications network within the organization, including bulletins, manuals, and conferences
6. Increasing community awareness of the organization's energy conservation achievements with press releases and appearances at civic group meetings

General qualifications of the staff energy manager include (1) a technical background, preferably in engineering, with experience in energy-efficient design of building systems and processes; (2) practical, hands-on experience with systems and equipment; (3) a goal-oriented management style; (4) the ability to work with people at all levels, from operations and maintenance personnel to top management; and (5) technical report writing and verbal communications skills.

Desirable educational and professional qualifications of the staff energy manager include the following:

1. Bachelor of Science or Engineering degree from an accredited four-year college, preferably in mechanical, electrical, industrial, or chemical engineering
2. Thorough knowledge of the principles and practices of energy resource planning and conservation
3. Familiarity with the administrative governing organization
4. Ability to analyze and compile technical and statistical information and reports with particular regard to energy usage
5. Knowledge of resources and information relating to energy conservation and planning
6. Ability to motivate people and develop and establish effective working relationships with other employees
7. Ability to interpret plans and specifications for building facilities
8. Knowledge of the basic types of automatic controls and systems instrumentation
9. Knowledge of energy-related metering equipment and practices (see Chapter 39)
10. Knowledge of the organization's manufacturing processes
11. Knowledge of building systems design and operation and/or maintenance
12. Interest in and enthusiasm for efficient energy use
13. Ability to present ideas to all levels in the organization

Energy Consultant

In many instances, an energy manager needs outside assistance to conduct the energy management program. An energy consultant may be called on to assist with any of the energy manager's functions and, in addition, may be responsible for training the energy management, operations, and maintenance staffs. In some cases, the energy consultant may be responsible for implementing and monitoring the energy management program.

The basic qualifications of an energy consultant should be similar to those of an energy manager. The consultant should be an objective party with no connections to the sale of equipment or systems.

Energy service companies (ESCOs) and other contracting firms provide energy management services. Often these companies are compensated by a percentage of realized savings based on measured energy usage before and after the improvements. It is vital that the compensation methods be clearly specified and include provisions to adjust for changes in building use, production levels, and other factors.

Optimum energy conservation in a comprehensive and integrated energy management program can be accomplished only if system interaction is thoroughly understood and accounted for by the energy consultant and manager. For example, a lighting retrofit could reduce lighting energy while simultaneously causing the heating system that interacts with the lighting system to increase energy consumption.

Motivation

The success of an energy management program depends on the interests and motivation of the people implementing it (Turner 1993). Participation and communication are key points. Employees can be motivated to support an energy management program through awareness of the following:

- Amount of energy they use
- Cost of the energy
- Critical role of energy in the continued viability of their jobs
- Meaning of energy saving in their operations
- Relationship between production rate and energy consumption
- Benefits of participation, such as greater comfort and improved air quality

Energy management activities can be made a part of each supervisor's performance or job standards. If the supervisor knows that top management is solidly behind the energy management program, and the overall performance rating depends, to some extent, on the energy savings that the department or group achieves, the supervisor will motivate employee interest and cooperation.

FINANCING A PROGRAM

Several financing options exist for energy management programs.

Internal financing. The building owner, manager, or occupant makes a decision to finance the modifications internally, taking advantage of all the savings and assuming all the risk.

Shared savings contracts. Typically, independent contractors install energy-related modifications at their expense. The building owner then pays a prearranged fee, which is usually related to the savings achieved. The term of these contracts runs from several years to as many as 20 years. The terms and conditions of the contract must be well documented in advance. In particular, methods of determining energy savings and of accounting for other influences on energy consumption must be carefully and precisely described.

Guaranteed savings contracts. These are similar to shared savings contracts, except that the costs of the modifications are specified up front. The owner can pay for the retrofit at the beginning of the contract, finance it over a period of years, or have the contractor finance it with repayment out of the savings. The owner should enter into these agreements carefully, with an understanding of the

financial liability if the savings do not materialize and the contractor is unable to make good on the guarantee. Energy savings are guaranteed, usually on an annual basis. If greater savings are achieved, some contracts require that a bonus be paid to the contractor. If less than the guaranteed savings is realized, the owner is reimbursed for the difference. Insurance policies should be examined carefully. With this type of contract, the performance record of the contractor and insurance company and policy should be checked.

Demand side management. Demand side management (DSM) programs are offered by utilities and most often subsidize the installation of energy modifications. These programs usually take the form of rebates to contractors or building owners for the installation of specified equipment, and they are often carried out by subcontractors or unregulated subsidiaries of utilities. The purpose of these programs is to reduce demand or consumption of the form of energy supplied by the utility.

Utility consultants. Utility consultants review utility bills for building owners, checking for billing errors and proper rates and riders. Usually, no consideration is given to the quantity of energy consumed. Contracts are on a contingent fee basis, frequently with payment being half the savings for five years.

Government programs. There are some federal, state, and local government programs that provide subsidies or tax credits for certain energy modifications.

Tariff analysis. Most utilities offer a variety of rates. The customer selects the rates and riders, provided they meet their criteria. Because rates change frequently, it is often advantageous to examine the applicability of other utility rates that may lower costs.

Tax exemption. Many states exempt certain uses of energy from sales tax. As a general rule, utilities charge sales tax to all customers, unless they are advised that the customer is exempt. Typical exemptions include those for nonprofit organizations, research and development, and some types of manufacturing.

IMPLEMENTING A PROGRAM

There are six basic stages in implementing an energy management program:

1. Develop a thorough understanding of how energy is used.
2. Conduct a planned, comprehensive search to identify all potential opportunities for energy conservation. Contact utility companies to find out about applicable rates and to learn about any DSM program offerings.
3. Determine and clearly specify the energy and cost-savings analysis methods to be used. See Chapter 35, Owning and Operating Costs; MacDonald and Wasserman (1989); and PNL (1990).
4. Identify, acquire, allocate, and prioritize the resources necessary to implement energy conservation opportunities (ECOs).
5. Implement the ECOs in a rational order. This is usually a series of independent activities that take place over several years.
6. Monitor and maintain energy conservation measures taken. Reevaluate them as building functions change over time.

For existing buildings, ASHRAE *Standard* 100 was developed to provide criteria that will result in conservation of nonrenewable energy resources. These standards are directed toward upgrading existing building thermal performance; increasing system energy efficiency; and providing procedures and programs essential to the operation, maintenance, and monitoring of energy conservation measures.

Energy efficiency and/or energy conservation efforts should not be equated with discomfort, nor should they interfere with the primary function of the organization or facility. Energy conservation activities that disrupt or impede the normal functions of workers and/or processes and adversely affect productivity constitute false economies.

Databases

The compilation of a database of past energy usage and cost is important in developing an energy management program. Any reliable utility data that is applicable should be examined. Utilities can usually provide demand metered data on computer disks, often with measurement intervals as short as 15 min. This is preferable to monthly data because anomalies are more apparent. High consumption at certain times may reveal opportunities for conservation (Haberl and Komar 1990). If monthly data are used, they should be analyzed over several years. A base year should be established as a reference point for future energy conservation and energy cost avoidance activities. In tabulating such data, the actual dates of meter readings should be recorded so that energy use can be normalized for differences in the number of days in a period. Any periods during which consumption was estimated rather than measured should be noted.

If energy is available for more than one building and/or department under the authority of the energy manager, each of these should be tabulated separately. Initial tabulations should include both energy and cost per unit area. (In an industrial facility, this may be energy and cost per unit of goods produced.) Available information on variables that may have affected past energy use should also be tabulated. These might include heating or cooling degree-days, percent occupancy for a hotel, quantity of goods produced in a production facility, or average daily weather conditions, which can be obtained from the National Oceanic and Atmospheric Administration (NOAA), National Climatic Data Center Federal Building, Asheville, NC. Because such variables may not be directly proportional to energy use, it is best to plot information separately or to superimpose one plot over another, rather than develop such units as Btu per square foot per degree day. As such data are tabulated, ongoing energy accounting procedures should be developed for regular data collection and future uses.

Comparing a building's energy use with that of many different buildings is a valuable way to check its relative efficiency. Data on buildings in all sectors are summarized in DOE/EIA-0246 and DOE/EIA-0318 for nonresidential buildings and in DOE/EIA-0321/1 for households. The following tables present the DOE/EIA data in a combined format. Table 1 lists physical characteristics of the buildings surveyed. Table 2 lists measured demand and energy consumption. Table 3 lists various residential end uses. The EIA also collects data on household energy consumption, which is summarized in Table 4.

When an energy management program for a new building is established, the energy use database may consist solely of typical energy use data for similar buildings, as illustrated in Table 2. This may be supplemented by energy simulation data for the specific building if such data were developed during the design of the building. In addition, a new building and its systems should be properly commissioned upon completion of construction to ensure proper operation of all systems, including any energy conservation features. Refer to ASHRAE *Guideline* 1, The HVAC Commissioning Process.

All the data presented in these tables are derived from detailed reports of consumption patterns in buildings. Before using the data, however, it is important to understand how it was derived. For example, all the household energy consumption data presented in Table 4 are average data, and they may not reflect variations in appliances or fuel situations for different buildings. Therefore, when using the data, verify the correct use of it with the original EIA documents. Because these surveys are performed regularly, there may be newer data.

ASHRAE *Standard* 105, Standard Methods of Measuring and Expressing Building Energy Performance, contains information that allows uniform, consistent expressions of energy consumption in both proposed and existing buildings. Its use is recommended.

However, the data collected by EIA and presented here are not in accordance with this standard.

Mazzucchi (1992) lists data elements useful for normalizing and comparing utility billing information. Metered energy consumption and cost data are also gathered and published by trade associations that represent building owners. Some trade associations include the Building Owners and Managers Association International (BOMA), the National Restaurant Association (NRA), and the American Hotel and Motel Association (AH&MA).

The quality of published energy consumption data for buildings varies because the data are collected for different purposes by people with different levels of technical knowledge of buildings. The data presented here are primarily national data. In some cases, local energy consumption data may be available from local utility companies or state or provincial energy offices.

At this point in the development of an energy management program, it is useful to compile a list of previously accomplished energy conservation measures and the actual energy and/or cost savings of such measures. Energy and cost savings analysis methods (MacDonald and Wasserman 1989, Section 2) can be used to calculate savings. These measures and savings should be studied during subsequent energy audits to determine their effectiveness and the effort(s) necessary to maintain and/or improve them.

Since most energy management activities are dictated by economics, the energy manager must understand the utility rates that apply to each facility. Special rates are commonly applied for such variables as time of day, interruptible service, on peak/off peak, summer/winter, and peak demand. There are more than 1000 electric rate variations in the United States. The energy manager should work with local utility companies to develop the most cost-effective methods of metering and billing and enable energy cost avoidance to be calculated effectively.

It is common for electric utilities to meter both electric consumption and demand. **Demand** is the peak rate of consumption, typically integrated over a 15 or 30 min period. Electric utilities may also establish a ratchet billing procedure for demand. In a simplified version of ratchet demand billing, the billing demand is established as either the *actual demand* for the month in question or a percentage of the *highest demand* during the previous 11 months, whichever is greatest. Figure 3 illustrates a gas-heated, electrically cooled building with the highest electric demands

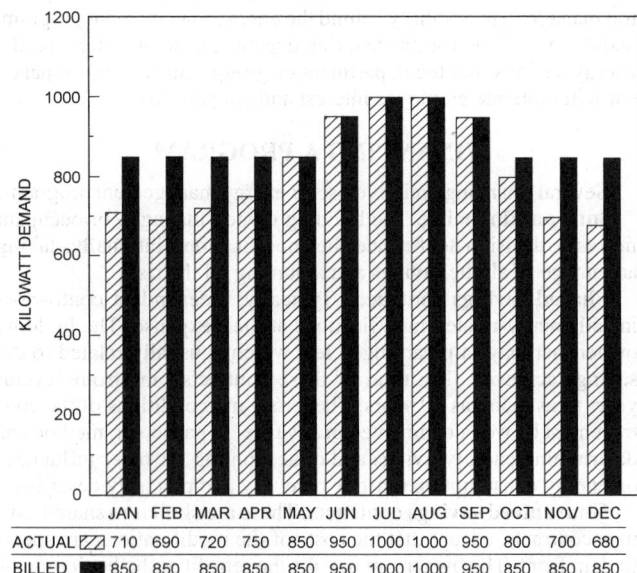

	JAN	FEB	MAR	APR	MAY	JUN	JUL	AUG	SEP	OCT	NOV	DEC
ACTUAL	710	690	720	750	850	950	1000	1000	950	800	700	680
BILLED	850	850	850	850	850	950	1000	1000	950	850	850	850

Fig. 3   Actual Demand Versus Billing Demand
(85%, 11 Month Ratchet)

Table 1 1995 Commercial Building Characteristics
Source: DOE/EIA 0246(95) 1997

Source Table Number	1	1	1	1	1	2	2	2	2
			Mean			Median			
Building Characteristics	Total Number of Buildings, Thousands	Total Floor Space, 10^6 ft^2	Floor Area per Bldg, 1000 ft^2	Floor Area per Person, ft^2	Hours Worked per Week	Floor Area per Bldg, 1000 ft^2	Floor Area per Person, ft^2	Hours Worked per Week	Age of Building, Years
All Buildings	4,579	58,772	12.8	706	62	5.0	938	50	30.5
Building Floor Space									
1001 to 5000 ft^2	2,399	6,338	2.6	599	59	2.5	750	48	30.5
5001 to 10000 ft^2	1,035	7,530	7.3	993	58	7.0	1,250	50	35.5
10001 to 25000 ft^2	745	11,617	15.6	896	67	15.0	1,667	53	27.5
25001 to 50000 ft^2	213	7,676	36.1	752	72	35.0	1,125	56	26.5
50001 to 100000 ft^2	115	7,968	69.3	819	80	65.0	1,316	60	25.5
100001 to 200000 ft^2	48	6,776	140.9	796	87	140.0	1,500	75	26.5
200001 to 500000 ft^2	19	5,553	294.9	731	102	275.0	1,190	80	25.5
Over 500000 ft^2	6	5,313	806.4	553	108	700.0	813	84	24.5
Principal Building Activity									
Education	309	7,740	25.1	767	51	8.5	1,000	45	33.5
Food sales	137	642	4.7	984	112	2.5	1,001	99	25.5
Food service	285	1,353	4.8	578	85	3.0	667	81	22.5
Health care	105	2,333	22.2	520	79	4.5	650	52	23.5
Lodging	158	3,618	22.8	1,317	156	9.0	2,267	168	30.5
Mercantile and service	1,289	12,728	9.9	945	61	4.0	1,083	52	35.5
Office	705	10,478	14.9	387	52	4.0	464	45	23.5
Public assembly	326	3,948	12.1	1,317	53	6.0	1,500	50	31.5
Public order and safety	87	1,271	14.6	746	72	5.0	875	20	32.5
Religious worship	269	2,792	10.4		41	8.0	3,125	20	31.5
Warehouse and storage	580	8,481	14.6	1,730	59	5.5	2,000	48	18.5
Other	67	1,004	14.9	544	79	5.0	1,000	50	26.5
Vacant	261	2,384	9.1	3,735	19	4.0	1,000	0	39.5
Year Constructed									
1919 or before	353	3,673	10.4	1,004	56	5.5	1,250	48	93.5
1920 to 1945	562	6,710	11.9	913	55	4.8	1,000	48	62.5
1946 to 1959	867	9,298	10.7	910	54	4.3	1,000	48	42.5
1960 to 1969	718	10,858	15.1	755	62	5.5	893	50	31.5
1970 to 1979	813	11,333	13.9	754	68	5.0	875	51	20.5
1980 to 1989	846	12,252	14.5	607	67	5.0	882	50	10.5
1990 to 1992	218	2,590	11.9	554	61	3.5	667	50	4.5
1993 to 1995	202	2,059	10.2	1,025	71	3.5	1,250	50	1.5
Floors									
One	3,018	24,552	8.1	966	59	3.8	929	48	25.5
Two	1,002	14,122	14.1	785	66	7.0	882	50	35.5
Three	399	7,335	18.4	867	62	9.5	1,250	52	57.5
Four to nine	148	8,789	59.4	603	84	25.0	1,000	60	42.5
Ten or more	12	3,975	328.9	383	96	200.0	429	68	28.5
Census Region and Division									
Northeast	725	11,883	16.4	784	67	5.0	1,000	52	38.5
Midwest	1,139	14,322	12.6	844	59	4.5	1,250	48	36.5
South	1,750	20,830	11.9	786	60	4.8	893	50	23.5
West	964	11,736	12.2	647	63	5.5	833	50	28.5
Climate Zones: 45 Year Average									
<2000 CDD and >7000 HDD	493	5,098	10.3	855	60	4.0	938	50	29.5
5500 to 7000 HDD	975	14,597	15.0	864	64	5.0	1,100	50	39.5
4000 to 5499 HDD	1,070	15,155	14.2	719	64	5.5	1,000	50	32.5
Fewer than 4000 HDD	1,103	13,491	12.2	656	61	5.0	800	50	25.5
>2000 CDD and <4000 HDD	937	10,430	11.1	849	59	4.8	950	48	25.5
Workers (Main Shift)									
Fewer than 5	2,505	13,885	5.5	2,992	58	3.0	1,583	48	31.5
5 to 9	798	6,291	7.9	1,220	61	4.8	688	50	35.5
10 to 19	625	7,102	11.4	907	63	7.5	600	50	25.5
20 to 49	400	9,132	22.8	792	72	16.3	542	56	26.5
50 to 99	138	6,931	50.3	777	72	37.5	563	55	24.5
100 to 249	71	5,988	84.4	608	85	55.0	400	65	23.5
250 or more	43	9,443	220.1	327	84	120.0	283	63	19.5
Weekly Operating Hours									
39 or fewer	899	6,134	6.8	1,064	14	4.0	1,500	8	32.5
40 to 48	1,257	13,233	10.5	799	43	4.8	833	44	32.5
49 to 60	969	12,242	12.6	689	54	5.5	833	53	29.5
61 to 84	567	10,052	17.7	761	72	6.0	1,000	72	31.5
85 to 167	420	6,202	14.8	891	105	4.3	833	102	24.5
Open continuously	466	10,908	23.4	661	168	6.0	1,250	168	23.5
Ownership and Occupancy									
Nongovernment owned	4,025	46,696	11.6	772	62	4.8	938	50	29.5
Owner occupied	3,158	35,573	11.3	759	63	4.5	1,000	50	29.5
Nonowner occupied	698	9,697	13.9	725	65	5.5	833	53	25.5
Unoccupied	170	1,426	8.4	6,481		3.8	750		39.5

A blank space indicates data are not available, or less that 20 buildings, or error is >50%. A * indicates more than one may apply.

Table 1 1995 Commercial Building Characteristics (*Continued*)

Source: DOE/EIA 0246(95) 1997

Source Table Number	1	1	1	1	1	2	2	2	2
	Total Number of Buildings, Thousands	Total Floor Space, 10^6 ft^2	Mean Floor Area per Bldg, 1000 ft^2	Mean Floor Area per Person, ft^2	Mean Hours Worked per Week	Median Floor Area per Bldg, 1000 ft^2	Median Floor Area per Person, ft^2	Median Hours Worked per Week	Age of Building, Years
Building Characteristics									
Government owned	553	12,076	21.8	742	61	7.0	950	45	35.5
Energy Sources*									
Electricity	4,343	57,076	13.1	746	63	5.0	929	50	30.5
Natural gas	2,478	38,145	15.4	760	64	6.0	917	50	35.5
Fuel oil	607	14,421	23.7	628	64	4.8	1,000	50	35.5
District heat	110	5,658	51.5	543	91	12.5	944	60	36.5
District chilled water	53	2,521	47.7	568	88	12.5	1,250	65	30.5
Propane	589	5,344	9.1	693	66	4.0	1,000	50	20.5
Wood	126	699	5.6	1,278	53	3.3	1,500	50	32.5
Coal		397	22.9	1,997	61	7.0	7,000	53	20.5
Other	71	1,154	16.2	670	72	4.0	750	60	38.5
Space Heating Energy Sources*									
Electricity	1,467	22,156	15.1	651	66	5.5	792	50	21.5
Natural gas	2,211	31,535	14.3	784	61	5.5	917	50	35.5
Fuel oil	504	6,606	13.1	804	61	4.0	1,000	50	38.5
District heat	109	5,606	51.4	540	91	12.5	944	60	37.5
Propane	301	2,025	6.7		64	3.3	1,500	48	18.5
Wood	103	509	5.0	1,271	53	3.3	1,500	50	35.5
Other	25	318	12.9	735	57	3.0	750	48	36.5
Primary Space Heating Energy Source									
Electricity	1,007	13,500	13.4	655	68	5.0	750	50	20.5
Natural gas	2,106	28,808	13.7	791	61	5.5	917	50	35.5
Fuel oil	439	4,207	9.6	1,037	59	4.0	1,000	50	39.5
District heat	107	5,289	49.3	536	91	12.5	944	60	36.5
Propane	260	1,545	5.9		62	3.0	1,500	48	15.5
Other	61	514				11.3	1,607	51	37.5
Cooling Energy Source*									
Electricity	3,293	47,761	14.5	705	65	5.0	833	50	28.5
Natural gas	65	1,314	20.1	638	67	9.5	855	50	40.5
District chilled water	53	2,521	47.4	568	88	12.5	1,250	65	30.5
Water Heating Energy Sources*									
Electricity	1,684	23,056	13.7	681	63	5.0	800	50	22.5
Natural gas	1,577	24,859	15.8	764	68	6.0	893	54	35.5
Fuel oil	120	2,151	17.9	931	59	4.3	875	50	39.5
District heat	54	3,949	73.7	553	97	27.5	1,300	82	30.5
Propane	110	1,020	9.2	875	94	3.5	1,167	88	27.5
Cooking Energy Sources*									
Electricity	487	12,249	25.2	647	82	5.5	750	80	23.5
Natural gas	448	13,195	29.4	658	81	8.5	800	75	33.5
Propane	123	1,480	12.0	816	84	3.3	500	84	19.5
Energy End Uses*									
Buildings with space heating	4,024	54,347	13.5	725	63	5.0	893	50	30.5
Buildings with cooling	3,381	49,935	14.8	702	66	5.0	833	50	28.5
Buildings with water heating	3,486	51,560	14.8	709	66	5.0	859	50	30.5
Buildings with cooking	828	20,713	25.0	650	79	7.0	800	72	28.5
Buildings with manufacturing	204	3,893	19.1	787	53	8.0	1,100	48	29.5
Buildings with electric gen.	247	13,366	54.2	585	90	12.5	833	67	30.5
Percent of Floorspace Heated									
Not heated	554	4,425	8.0	2,394	51	3.8	1,250	36	25.5
1 to 50	555	6,227	11.2	1,466	49	5.5	1,625	45	30.5
51 to 99	633	8,868	14.0	770	69	6.0	875	55	35.5
100	2,836	39,252	13.8	664	64	5.0	833	50	28.5
Percent of Floorspace Cooled									
Not cooled	1,198	8,837	7.4	1,572	49	3.8	1,500	41	34.5
1 to 50	930	15,027	16.2	1,211	57	7.0	1,250	50	35.5
51 to 99	635	12,549	19.8	651	73	5.5	750	56	35.5
100	1,816	22,359	12.3	567	68	4.8	750	50	23.5
Percent Lit When Open									
Zero	36	189	5.2		57	1.8	938	50	16.5
1 to 50	666	6,008	9.0	2,250	62	4.8	2,000	50	35.5
51 to 99	745	9,692	13.0	804	61	5.0	917	49	34.5
100	2,814	40,514	14.4	656	65	5.0	833	50	26.5
Electricity not used	318	2,369	7.5	3,096	33	4.0	4,250	0	35.5
Percent Lit When Closed									
Zero	1,644	13,101	8.0	1,033	42	4.0	1,188	44	30.5
1 to 50	2,109	30,711	14.6	684	59	5.5	761	50	30.5
51 to 100	87	1,914	22.0	744	74	6.0	818	60	30.5
Never closed	421	10,677	25.4	652	168	6.0	1,250	168	25.5
Electricity not used	318	2,369	7.5	9,344	33	4.0	4,250	0	35.5

A blank space indicates data are not available, or less that 20 buildings, or error is >50%. A * indicates more than one may apply.

Table 1   1995 Commercial Building Characteristics (*Continued*)

Source: DOE/EIA 0246(95) 1997

Source Table Number	1	1	1	1	1	2	2	2	2
			Mean			Median			
Building Characteristics	Total Number of Buildings, Thousands	Total Floor Space, 10^6 ft^2	Floor Area per Bldg, 1000 ft^2	Floor Area per Person, ft^2	Hours Worked per Week	Floor Area per Bldg, 1000 ft^2	Floor Area per Person, ft^2	Hours Worked per Week	Age of Building, Years
Heating Equipment*									
Heat pumps	394	5,843	14.8	638	63	5.0	667	48	19.5
Furnaces	1,676	14,923	8.9	853	60	4.8	938	50	34.5
Individual space heaters	1,188	16,809	14.1	796	57	5.0	1,125	49	30.5
District heat	115	5,911	51.5	553	91	12.5	944	65	35.5
Boilers	610	16,754	27.5	682	73	9.0	909	53	39.5
Packaged heating units	1,031	16,893	16.4	720	68	5.5	750	50	21.5
Other	161	6,249	38.8	563	61	10.0	1,300	52	29.5
Cooling Equipment*									
Residential central A/C	878	9,238	10.5	669	65	5.0	750	50	30.5
Heat pumps	457	6,931	15.2	646	65	5.0	700	48	19.5
Individual A/C	862	12,494	14.5	881	68	4.5	1,125	50	40.5
District chilled water	53	2,521	47.7	568	87	12.5	1,250	65	30.5
Central chillers	109	11,065	101.4	526	87	45.0	1,000	65	28.5
Packaged A/C units	1,431	26,628	18.6	697	69	6.5	800	52	23.5
Evaporative coolers	186	2,451	13.2	876	66	4.8	792	56	30.5
Other	18	949	51.9	601	74	18.8	900	60	39.5
Lighting Equipment Types*									
Incandescent	2,479	35,715	14.4	725	65	5.0	917	50	32.5
Standard fluorescent	3,885	53,984	13.9	721	65	5.0	889	50	29.5
Compact fluorescent	364	14,273	39.2	557	85	11.3	750	66	23.5
High intensity discharge	393	16,259	41.4	742	72	12.5	1,067	53	27.5
Halogen	303	9,665	32.0	587	78	8.0	915	60	28.5
Other	30	554	18.7	458	44	2.0	1,500	45	32.5
Water Heating Equipment*									
Central system	2,671	31,656	11.9	752	65	4.8	833	50	31.5
Distributed system	742	16,495	22.2	642	71	9.5	875	55	25.5
Combination central and distr.	73	3,409	46.4	694	73	10.0	1,000	50	26.5
Personal Computers/Terminals									
None	2,039	12,571	6.2	1,568	54	3.5	1,500	45	31.5
1 to 4	1,408	11,401	8.1	983	68	4.3	900	53	30.5
5 to 9	437	5,372	12.3	823	66	6.0	600	50	23.5
10 to 19	344	5,947	17.3	798	61	8.5	563	50	25.5
20 to 49	198	7,048	35.6	815	69	20.0	694	50	25.5
50 to 99	81	4,938	61.2	663	71	40.0	625	54	29.5
100 to 249	46	5,189	112.9	616	84	65.0	600	60	24.5
250 or more	26	6,307	240.9	338	92	140.0	360	70	18.5
Energy Related Functions*									
Comml food preparation	828	20,713	25.0	650	79	7.0	800	72	28.5
Computer room	234	12,890	55.0	502	72	16.3	654	50	19.5
Large hot water activities	243	6,753	27.8	689	99	8.0	882	85	31.5
Shell Conservation Features*									
Roof/ceiling insulation	3,380	46,355	13.7	704	64	5.0	875	50	26.5
Wall insulation	2,372	31,694	13.4	654	62	5.0	833	50	21.5
Storm or multiple glazing	1,897	28,876	15.2	673	66	5.5	813	50	26.5
Tinted, refl., shaded glass	1,202	24,245	20.2	580	66	6.0	750	50	21.5
Shading or awnings	2,271	37,208	16.4	668	65	6.0	813	50	28.5
HVAC Conservation Features*									
Variable volume system	327	13,473	41.2	553	76	12.5	750	50	23.5
Economizer cycle	461	16,550	35.9	573	75	10.0	800	57	21.5
HVAC maintenance	2,403	43,134	18.0	669	68	6.0	833	51	28.5
Other efficient equipment	196	6,453	32.5	568	66	7.0	688	55	23.5
Lighting Conservation Features*									
Specular reflectors	749	17,868	23.9	637	69	6.0	1,000	50	28.5
Energy efficient ballasts	1,363	28,375	20.8	614	67	6.0	750	50	25.5
Natural lighting sensors	237	6,431	27.2	622	83	8.0	885	63	28.5
Occupancy sensors	131	5,958	45.6	515	66	12.5	1,111	50	38.5
Time clock	467	13,262	28.4	593	73	8.0	724	55	27.5
Manual dimmer switches	501	13,056	26.1	655	69	9.5	1,100	55	26.5
Other	79	2,836	35.8	630	71	10.0	688	60	20.5
Energy Conservation Features*									
Any conservation feature	4,075	55,288	13.6	732	63	5.0	900	50	29.5
Building shell	3,906	53,190	13.6	717	63	5.0	875	50	28.5
HVAC	2,529	44,657	17.7	677	68	6.0	850	50	28.5
Lighting	2,084	38,537	18.5	660	67	6.0	875	51	28.5
Off Hour Equipment Reduction*									
Heating	3,211	38,326	11.9	738	52	4.8	875	48	30.5
Cooling	2,707	35,605	13.2	711	54	5.0	833	50	28.5
Lighting	3,753	44,937	12.0	753	52	5.0	889	48	30.5

blank space indicates data are not available, or less that 20 buildings, or error is >50%. A * indicates more than one may apply.

Table 2A 1995 Commercial Building Energy Consumption

Consumption shown is on an annual basis per square foot of building area. *Source*: DOE/EIA 0318(95) 1998

Source Table Number	1	1	3	3	10	10	19	19	21	21	27	31	33	33	33	33	33	33
	Total # Bldgs., Thousands	Total Floor Space, 10⁶ ft²	Major Fuels 1000 Btu/ft²	Major Fuels $/ft²	kWh/ft²	Med. kWh/ft²	Med. Peak W/ft²	Med. Load Fact.	Nat. Gas CF/ft²	Nat. Gas Med. CF/ft²	Fuel Oil Gal/ft²	Distr. Heat lb/ft²	Total	Space Heat	Cool	Ventilation	Water Heat	Lighting
Building Characteristics																		
All Buildings	4,579	58,772	90.5	1.19	13.4	7.2	5.4	0.253	49.7	39.7	0.12	94.14	90.5	29.0	6.0	2.8	13.8	20.4
Building Floor Space																		
1001 to 5000 ft²	2,399	6,338	111.7	1.83	18.7	8.5	8.0	0.239	87.2	51.0	0.34		111.7	39.5	7.0	2.9	9.7	22.7
5001 to 10000 ft²	1,035	7,530	82.8	1.07	9.9	6.2	5.2	0.248	58.8	29.1	0.28		82.8	38.5	4.4	1.7	11.1	13.6
10001 to 25000 ft²	745	11,617	70.9	0.96	10.0	5.6	3.3	0.246	45.9	29.3	0.23	103.56	70.9	27.4	4.8	1.7	9.1	14.7
25001 to 50000 ft²	213	7,676	82.0	1.13	12.1	7.3	3.6	0.295	42.8	27.2	0.17	84.62	82.0	28.2	6.7	2.1	11.6	18.5
50001 to 100000 ft²	115	7,968	87.6	1.11	13.5	8.9	3.2	0.337	42.3	23.4	0.11	81.02	87.6	27.0	7.0	3.2	12.9	21.3
100001 to 200000 ft²	48	6,776	101.4	1.16	15.0	12.4	3.5	0.393	51.3	28.4	0.07	75.38	101.4	26.6	6.2	3.3	19.6	25.0
200001 to 500000 ft²	19	5,553	114.6	1.31	16.2	10.5	3.1	0.458	52.1	25.9	0.06	77.72	114.6	24.0	6.7	4.5	25.2	27.4
Over 500000 ft²	6	5,313	96.8	1.23	16.3	13.2	3.2	0.521	32.9	8.8	0.03	68.71	96.8	18.5	6.0	3.9	18.0	28.6
Principal Building Activity																		
Education	309	7,740	79.3	0.92	8.4	6.1	4.3	0.210	41.1	38.6	0.17	84.51	79.3	32.8	4.8	1.6	17.4	15.8
Food sales	137	642	213.5	4.11	54.1	55.6	14.7	0.463	42.6	31.7			213.5	27.5	13.4	4.4	9.1	33.9
Food service	285	1,353	245.5	3.56	36.0	25.5	12.7	0.333	153.5	135.2			245.5	30.9	19.5	5.3	27.5	37.0
Health care	105	2,333	240.4	2.26	26.5	15.7	5.9	0.295	143.0	66.3	0.10	109.51	240.4	55.2	9.9	7.2	63.0	39.3
Lodging	158	3,618	127.3	1.41	15.2	11.7	4.9	0.364	73.2	57.6		92.63	127.3	22.7	8.1	1.7	51.4	23.2
Mercantile and service	1,289	12,728	76.4	1.10	11.8	6.9	4.9	0.249	45.2	40.4	0.14		76.4	30.6	5.8	2.5	5.1	23.4
Office	705	10,478	97.2	1.51	18.9	12.2	6.0	0.285	35.7	33.2	0.06	49.28	97.2	24.3	9.1	5.2	8.7	28.1
Public assembly	326	3,948	113.7	1.26	12.7	5.8	5.5	0.197	51.9	45.6	0.09		113.7	53.6	6.3	3.5	17.5	21.9
Public order and safety	87	1,271	97.2	1.22	11.3	3.9	5.0	0.280	43.6	43.6	0.22		97.2	27.8	6.1	2.3	23.4	16.4
Religious worship	269	2,792	37.4	0.48	3.4	2.9	4.2	0.092	28.0	27.4	0.21		37.4	23.7	1.9	0.9	3.2	5.0
Warehouse and storage	580	8,481	38.3	0.56	6.4	3.2	2.2	0.265	22.4	21.2	0.09		38.3	15.7	0.9	0.3	2.0	9.8
Other	67	1,004	172.2	1.86	22.1	11.3	7.3	0.227	82.4	35.4			172.2	59.6	9.3	8.3	15.3	26.7
Vacant	261	2,384	21.5	0.27	3.9	2.4	2.4	0.189	38.8	21.0	0.16		21.5	11.9	0.6	0.3	2.4	3.6
Year Constructed																		
1919 or before	353	3,673	79.4	0.90	8.3	4.8	4.0	0.231	49.8	47.6	0.17	55.73	79.4	34.2	2.6	1.6	10.0	14.9
1920 to 1945	562	6,710	75.7	0.84	8.2	5.1	4.7	0.240	44.9	40.4	0.23	98.36	75.7	37.0	3.4	1.6	10.7	12.3
1946 to 1959	867	9,298	88.9	1.06	10.4	6.5	4.2	0.231	58.8	44.1	0.20	60.24	88.9	37.2	4.4	2.1	14.1	15.5
1960 to 1969	718	10,858	94.3	1.21	13.0	7.2	5.9	0.266	50.9	47.5	0.13	88.27	94.3	30.2	5.7	2.7	16.8	20.4
1970 to 1979	813	11,333	99.3	1.36	16.0	8.8	6.2	0.259	51.8	32.7	0.07	92.42	99.3	26.0	7.2	3.6	15.8	25.6
1980 to 1989	846	12,252	86.5	1.30	15.9	10.0	6.0	0.269	39.0	28.4	0.05		86.5	19.8	7.8	3.2	11.5	23.9
1990 to 1992	218	2,590	114.6	1.55	18.8	9.4	8.0	0.292	58.9	29.1	0.03		114.6	26.6	8.4	3.5	17.2	28.7
1993 to 1995	202	2,059	92.2	1.32	17.5	8.4	4.9	0.295	48.2	42.5	0.10		92.2	24.3	7.9	3.2	11.7	22.7
Floors																		
One	3,018	24,552	75.2	1.10									75.2	26.0	5.6	2.1	7.9	17.0
Two	1,002	14,122	79.4	1.09									79.4	28.2	5.7	2.1	10.9	18.3
Three	399	7,335	92.0	1.09									92.0	34.8	5.1	2.3	15.0	18.6
Four to nine	148	8,789	139.8	1.51									139.8	36.5	7.5	4.8	30.2	31.0
Ten or more	12	3,975	113.4	1.54									113.4	23.1	7.3	5.6	21.8	29.6
Census Region and Division																		
Northeast	725	11,883	87.1	1.39	11.2	5.2	3.7	0.260	40.6	29.1	0.22	76.31	87.1	32.4	4.0	2.0	14.2	17.7
New England	204	3,140	87.3	1.28	9.4	5.8	5.3	0.295	50.0	40.8	0.28	100.91	87.3	37.7	3.3	1.6	15.2	16.0
Middle Atlantic	521	8,743	87.1	1.43	11.8	5.2	3.3	0.246	38.3	28.6	0.19	72.71	87.1	30.4	4.3	2.1	13.9	18.4
Midwest	1,139	14,322	104.5	1.05	11.8	6.8	5.1	0.238	66.9	59.8	0.04	91.17	104.5	46.7	4.3	2.5	15.6	18.8
East North Central	739	9,655	102.2	1.05	11.1	5.6	4.5	0.238	65.2	62.1	0.05	93.69	102.2	45.5	4.3	2.2	16.0	17.4
West North Central	401	4,668	109.3	1.06	13.3	8.1	6.8	0.228	71.0	54.3	0.03	86.73	109.3	49.1	4.3	3.0	14.8	21.8
South	1,750	20,830	80.8	1.07	14.9	8.5	5.8	0.261	41.9	29.3	0.08	80.25	80.8	18.0	8.4	3.2	10.5	21.3
South Atlantic	676	9,475	81.5	1.15	15.4	8.5	6.7	0.294	39.9	34.9	0.10	139.92	81.5	17.3	8.8	3.3	9.3	22.2
East South Central	477	4,917	84.8	0.99	14.9	9.8	6.6	0.253	50.5	38.3	0.06		84.8	24.4	7.5	2.7	11.7	21.2
West South Central	597	6,438	76.7	1.00	14.3	7.8	5.2	0.249	37.7	22.0			76.7	14.2	8.7	3.3	11.4	20.1
West	964	11,736	94.2	1.38	14.8	7.3	6.7	0.244	46.1	30.4	0.02		94.2	23.4	5.5	3.1	17.0	23.6
Mountain	319	3,855	111.3	1.15	13.9	6.5	6.8	0.244	55.5	42.2			111.3	40.8	5.9	3.3	21.4	21.7
Pacific	646	7,881	85.9	1.49	15.3	8.8	6.7	0.244	41.3	26.8	0.02	70.65	85.9	14.9	5.4	3.1	14.8	24.5
Climate Zones: 45 Year Average																		
<2000 CDD and >7000 HDD	493	5,098	97.8	0.98	10.6	5.2	5.8	0.263	68.7	50.2	0.20	110.09	97.8	47.3	3.1	1.8	14.2	17.7
5500 to 7000 HDD	975	14,597	109.0	1.22	11.7	5.6	5.0	0.238	62.6	51.0	0.13	134.98	109.0	48.4	4.1	2.4	18.2	18.6
4000 to 5499 HDD	1,070	15,155	92.8	1.24	14.1	6.2	4.0	0.251	48.4	40.0	0.14	67.32	92.8	29.5	5.4	2.9	13.3	22.2
Fewer than 4000 HDD	1,103	13,491	79.9	1.25	14.3	9.3	6.7	0.257	37.7	28.8		93.65	79.9	15.9	6.8	2.9	12.3	21.6
>2000 CDD and <4000 HDD	937	10,430	71.6	1.10	15.0	8.7	5.4	0.253	35.0	22.9			71.6	9.1	9.7	3.3	10.2	20.3
Workers (Main Shift)																		
Fewer than 5	2,505	13,885	56.8	0.79	7.8	5.3	5.3	0.218	44.1	41.0	0.25		56.8	26.9	2.5	1.3	8.1	9.5
5 to 9	798	6,291	80.8	1.10	10.5	8.5	5.3	0.268	52.9	42.6	0.20		80.8	36.1	4.6	1.8	8.8	15.3
10 to 19	625	7,102	86.5	1.18	12.1	9.5	6.1	0.253	55.9	30.1	0.27		86.5	31.9	5.9	2.0	9.9	17.9
20 to 49	400	9,132	95.1	1.29	13.6	12.9	5.5	0.314	50.4	39.8	0.19	106.84	95.1	30.5	6.8	2.2	12.9	20.6
50 to 99	138	6,931	90.9	1.17	13.2	10.6	4.2	0.330	44.6	29.3	0.12	103.56	90.9	27.6	6.8	2.3	16.2	22.3
100 to 249	71	5,988	108.3	1.35	16.4	15.1	4.9	0.366	50.4	32.3	0.09	74.51	108.3	27.4	8.5	3.6	20.2	26.7
250 or more	43	9,443	133.7	1.65	21.7	16.4	5.1	0.454	51.0	27.2	0.03	74.62	133.7	25.7	8.9	6.3	23.5	36.5
Weekly Operating Hours																		
39 or fewer	899	6,134	29.3	0.39	3.6	2.7	4.9	0.142	33.8	29.1	0.21		29.3	16.8	1.6	0.7	3.0	3.6
40 to 48	1,257	13,233	66.4	0.89	9.0	6.8	4.5	0.217	41.6	33.9	0.15		66.4	31.6	4.3	1.9	5.3	12.4
49 to 60	969	12,242	76.6	1.08	12.0	6.9	4.6	0.240	36.6	30.9	0.15	83.27	76.6	29.4	5.1	2.9	6.7	18.3
61 to 84	567	10,052	79.2	1.16	12.7	10.2	5.1	0.276	38.6	45.7	0.10	70.74	79.2	24.7	6.3	2.7	8.6	21.8
85 to 167	420	6,202	134.0	1.87	20.7	31.4	10.2	0.362	59.6	72.6	0.11		134.0	34.0	8.9	3.6	20.5	32.0
Open continuously	466	10,908	155.7	1.76	21.3	12.7	5.8	0.374	81.7	55.4	0.08	87.69	155.7	33.2	9.5	4.4	39.1	34.3
Ownership and Occupancy																		

Blank = data not available, or less that 20 buildings, or error is >50%. * = more than one may apply; ** = for demand metered buildings. Med. = Median

Table 2B 1995 Commercial Building Energy Consumption

Consumption shown is on an annual basis per square foot of building area. *Source*: DOE/EIA 0318(95) 1998

Source Table Number	33	33	33	33	35	35	35	35	35	35	35	35	35	35	37	37	37	37	37
	Sum Major Fuels, 1000 Btu				Energy End Use: Electricity, 1000 Btu/ft²·yr										Natural Gas, 1000 Btu/ft²·yr				
Building Characteristics	Cooking	Refr.	Office Equip	Other	Total	Space Heat	Cool	Ventilation	Water Heat	Lighting	Cooking	Refr.	Office Equip	Other	Total	Space Heat	Water Heat	Cooking	Other
All Buildings	3.7	3.1	5.7	6.1	45.7	2.0	6.0	2.8	0.8	21.1	0.3	3.2	5.8	3.7	51.0	28.6	13.7	5.2	3.5
Building Floor Space																			
1001 to 5000 ft²	8.9	10.4	5.4	5.1	63.9	5.0	7.4	3.1	2.1	24.2	1.1	11.1	5.7	4.1	89.6	56.9	13.3	17.0	
5001 to 10000 ft²	4.3	2.5	3.8	2.9	33.7	2.6	4.6	1.8	1.0	14.5	0.2	2.7	4.0	2.2	60.4	42.9	9.8	6.9	
10001 to 25000 ft²	2.6	2.5	4.3	3.7	34.0	1.9	4.9	1.8	0.6	15.1	0.2	2.6	4.5	2.4	47.1	31.6	9.9	3.8	1.9
25001 to 50000 ft²	2.1	2.5	5.0	5.2	41.4	2.0	6.6	2.1	0.8	18.6	0.2	2.6	5.0	3.3	44.0	26.5	12.3	2.8	
50001 to 100000 ft²	2.0	2.1	6.1	6.0	46.0	1.0	6.9	3.2	0.6	21.5	0.2	2.1	6.2	4.2	43.4	26.0	12.5	2.5	2.4
100001 to 200000 ft²	3.1	1.4	7.2	8.9	51.0	1.3	6.2	3.4	0.4	25.6	0.2	1.5	7.4	4.9	52.6	23.0	20.2	4.2	5.3
200001 to 500000 ft²	4.6	1.6	8.5	11.9	55.3	1.1	6.1	4.5	0.5	27.4	0.3	1.6	8.5	5.1	53.5	16.5	21.6	6.1	9.2
Over 500000 ft²	3.5	2.2	7.0	9.1	55.6	0.8	5.5	4.1	0.5	30.0	0.2	2.3	7.4	4.6	33.8	9.9	13.2	4.4	6.3
Principal Building Activity																			
Education	1.4	1.0	1.5	2.9	28.7	1.8	4.7	1.6	0.8	15.9	0.2	1.0	1.5	1.0	42.3	25.5	12.5	1.6	2.7
Food sales	5.6	110.9	1.3	7.4	184.7		13.4	4.4	2.5	33.9	0.9	110.9	1.3	6.1	43.7	24.3	9.9	7.5	1.9
Food service	77.5	31.6	2.6	13.7	122.8	3.6	19.4	5.3	3.6	37.0	6.2	31.6	2.6	13.5	157.7	33.4	27.7	96.2	
Health care	11.2	4.7	15.5	34.4	90.4	1.4	9.2	7.2	0.9	39.3	0.3	4.7	15.5	12.0	146.9	45.1	59.6	14.4	27.7
Lodging	6.6	2.3	3.8	7.5	52.0	3.2	8.0	1.7	3.4	23.3	0.5	2.3	3.8	5.7	75.2	13.5	51.4	7.9	2.4
Mercantile and service	1.5	0.9	2.9	3.7	40.2	2.0	5.8	2.5	0.5	23.6	0.2	0.9	2.9	1.9	46.4	36.9	5.2	2.0	2.3
Office	1.1	0.4	15.1	5.2	64.5	2.0	8.8	5.2	0.6	28.1	0.1	0.4	15.2	4.2	36.7	24.5	9.0	1.6	1.6
Public assembly	2.8	1.8	2.4	3.8	43.3	2.7	6.1	3.5	0.9	22.0	0.4	1.8	2.5	3.3	53.3	39.7	9.4	3.5	
Public order and safety		0.2	5.8	12.7	38.5	0.2	6.1	2.3		16.4		0.2	5.8	7.2	44.7	20.1	15.4		
Religious worship	0.5	0.6	0.4	1.1	11.7	1.3	1.9	0.9	0.4	5.0	0.2	0.6	0.4	1.0	28.7	24.8	3.4	0.5	
Warehouse and storage		1.7	4.4	3.4	22.0	0.8	1.0	0.3	0.2	10.4		1.8	4.7	2.7	23.0	20.3	1.7	0.1	1.0
Other		0.7	15.2	35.9	75.5	2.2	8.6	8.4	0.2	26.8		0.7	15.3	13.3	84.6	37.7	11.7		33.5
Vacant		0.2	0.5	1.9	13.2	1.0	1.0	0.4	0.1	6.3		0.4	0.9	3.0	39.8	31.9	7.0		
Year Constructed																			
1919 or before	4.0	1.3	3.2	7.5	28.2	0.7	2.7	1.7	0.6	15.5	0.3	1.4	3.3	2.1	51.2	30.9	8.4	5.2	
1920 to 1945	1.8	1.6	3.3	4.1	28.0	1.2	3.6	1.7	0.4	13.3	0.2	1.8	3.6	2.3	46.1	32.8	8.5	2.5	2.4
1946 to 1959	3.0	2.7	4.6	5.2	35.6	1.9	4.4	2.1	0.8	15.8	0.2	2.7	4.7	2.9	60.4	39.4	13.6	4.0	3.3
1960 to 1969	4.0	3.0	5.3	6.1	44.3	1.8	5.7	2.8	0.9	20.8	0.2	3.0	5.4	3.6	52.2	26.2	17.2	5.8	3.2
1970 to 1979	3.2	3.7	6.7	7.5	54.7	2.2	6.9	3.7	0.9	25.8	0.4	3.7	6.7	4.3	53.2	29.0	14.9	4.3	5.0
1980 to 1989	4.2	3.0	7.6	5.9	54.4	2.3	8.0	3.3	0.9	24.2	0.4	3.1	7.8	4.4	40.1	19.1	12.4	6.4	2.2
1990 to 1992	9.3	5.6	7.9	7.4	64.1	2.2	8.2	3.5	0.8	29.2	0.8	5.7	8.1	5.6	60.5	24.0	20.1	13.3	
1993 to 1995	3.3	7.4	4.9	6.8	59.6		8.1	3.4	1.0	24.5	0.6	8.0	5.3	4.1	49.5	24.9	15.3	5.1	4.1
Floors																			
One	4.3	4.6	4.1	3.7	42.0	2.6	5.9	2.2	0.9	17.9	0.4	4.8	4.3	3.0	46.5	30.4	8.2	6.7	1.1
Two	2.4	2.7	4.6	4.6	39.4	1.7	5.7	2.1	0.9	18.5	0.2	2.7	4.6	2.9	48.9	31.0	12.7	3.2	2.0
Three	2.8	1.4	5.2	6.7	39.2	1.6	5.1	2.3	0.9	18.9	0.2	1.4	5.3	3.5	51.2	31.3	12.3	3.4	4.2
Four to nine	4.7	1.8	10.0	13.2	63.3	1.1	7.0	4.9	0.7	31.2	0.3	1.8	10.1	6.1	67.0	24.1	26.8	6.3	9.8
Ten or more	4.1	1.4	10.8	9.6	63.0	1.7	6.5	5.7	0.7	30.3	0.3	1.5	11.1	5.6	45.6	14.8	18.7	6.0	6.1
Census Region and Division																			
Northeast	2.7	3.0	4.5	6.4	38.1	1.6	3.8	2.1	0.7	18.4	0.3	3.2	4.7	3.2	41.7	22.9	10.2	4.0	4.6
New England	1.9	1.9	4.1	5.5	32.1	0.8	3.1	1.6	1.0	16.4	0.3	1.9	4.2	2.7	51.3	28.7	14.8	3.5	4.3
Middle Atlantic	3.0	3.4	4.6	6.7	40.3	1.9	4.1	2.2	0.7	19.2		3.6	4.8	3.4	39.3	21.5	9.1	4.1	4.6
Midwest	3.5	2.4	5.1	5.6	40.2	1.6	4.3	2.6	0.8	19.4	0.3	2.4	5.3	3.4	68.8	47.6	14.1	4.2	2.9
East North Central	4.4	2.5	4.6	5.2	37.8	1.5	4.3	2.3	0.8	17.8	0.4	2.6	4.7	3.3	66.9	45.0	14.4	5.1	2.4
West North Central	1.8	2.1	6.1	6.3	45.2	1.8	4.5	3.1	0.7	22.8	0.2	2.2	6.4	3.6	72.9	53.4	13.4	2.2	
South	4.0	3.4	5.9	6.0	50.9	2.1	8.5	3.3	1.0	22.0	0.3	3.5	6.1	4.1	43.0	20.3	13.2	6.2	3.2
South Atlantic	4.6	3.0	6.6	6.0	52.4	2.2	8.6	3.4	1.1	22.6	0.3	3.1	6.7	4.3	41.0	17.0	11.8	8.5	3.7
East South Central	2.2	3.7	5.3	6.2	50.9	3.2	7.8	2.8	1.0	22.4	0.4	3.9	5.6	3.8	51.9	30.1	15.4	2.8	3.6
West South Central	4.5	3.8	5.2	5.4	48.8	1.2	9.0	3.4	0.7	20.9	0.3	4.0	5.4	3.9	38.7	16.8	13.2	6.3	2.3
West	4.3	3.4	7.2	6.5	50.6	2.4	5.5	3.2	0.7	23.9	0.3	3.5	7.3	3.8	47.3	20.5	16.9	6.0	4.0
Mountain	2.8	3.2	6.8	5.4	47.6	1.9	5.9	3.3	0.7	21.9	0.3	3.2	6.8	3.5	57.0	35.4	15.1	3.7	2.9
Pacific	5.1	3.6	7.5	7.1	52.2	2.6	5.4	3.1	0.7	24.9	0.4	3.6	7.6	4.0	42.4	12.9	17.8	7.1	4.6
Climate Zones: 45 Year Average																			
<2000 CDD and >7000 HDD	2.2	2.3	4.6	4.6	36.2	1.8	3.2	1.9	0.7	18.3	0.2	2.4	4.7	3.0	70.5	51.4	14.5	2.8	
5500 to 7000 HDD	3.4	3.0	5.2	5.8	39.8	1.8	3.9	2.5	0.9	18.9	0.3	3.0	5.3	3.5	64.3	43.1	13.8	4.1	3.2
4000 to 5499 HDD	3.3	2.6	6.2	7.4	48.1	2.3	5.3	3.0	0.6	23.1	0.3	2.7	6.4	3.8	49.7	27.2	12.1	4.9	5.5
Fewer than 4000 HDD	4.4	3.7	6.2	6.1	48.8	1.9	6.8	2.9	0.8	22.0	0.4	3.8	6.3	3.9	38.7	15.7	14.3	5.7	3.1
>2000 CDD and <4000 HDD	4.6	3.6	5.5	5.3	51.3	1.8	10.1	3.5	1.1	21.3	0.3	3.8	5.8	3.8	36.0	10.5	14.4	8.5	2.6
Workers (Main Shift)																			
Fewer than 5	0.9	3.4	1.9	2.3	26.2	2.4	2.8	1.5	0.9	10.6	0.1	3.8	2.2	2.0	45.3	33.6	9.1	1.7	
5 to 9	2.8	3.9	3.5	4.1	35.7	2.5	4.5	1.8	1.1	15.3	0.4	3.9	3.5	2.6	54.4	40.7	8.2	3.4	2.1
10 to 19	7.0	3.5	4.7	3.4	41.2	2.4	5.9	2.0	1.1	17.9	0.6	3.5	4.7	3.0	57.4	36.7	10.4	9.7	0.6
20 to 49	6.1	4.3	5.4	6.2	46.3	2.1	6.8	2.2	0.8	20.6	0.4	4.3	5.5	3.6	51.8	28.4	12.5	7.8	3.0
50 to 99	2.9	2.3	5.3	5.2	45.2	1.4	6.8	2.4	0.5	22.5	0.3	2.3	5.4	3.7	45.8	23.3	17.1	3.8	1.6
100 to 249	3.0	1.5	8.1	9.2	55.8	1.4	8.2	3.6	0.8	26.8	0.2	1.5	8.1	5.1	51.7	24.9	17.6	3.9	5.3
250 or more	4.6	2.2	12.3	13.7	74.2	1.3	8.2	6.3	0.6	36.6	0.4	2.2	12.3	6.3	52.4	16.4	20.1	5.9	10.0
Weekly Operating Hours																			
39 or fewer	0.4	0.5	0.9	1.5	12.4	1.3	2.0		0.3	4.4		0.7	1.2	1.5	34.7	29.2	4.2	1.0	
40 to 48	1.1	0.9	6.1	2.9	30.7	2.1	4.3	1.9	0.5	12.5	0.1	0.9	6.1	2.4	42.8	34.6	5.9	1.6	0.7
49 to 60	0.8	1.1	7.5	4.6	41.0	2.0	5.1	2.9	0.5	18.5	0.1	1.1	7.6	3.1	37.6	28.4	6.2	1.1	1.9
61 to 84	4.1	2.6	4.1	4.3	43.4	2.0	6.1	2.7	0.7	21.9	0.3	2.6	4.1	3.0	39.6	24.6	7.8	5.4	1.9
85 to 167	11.8	11.8	4.6	6.8	70.6	1.9	8.9	3.6	1.2	32.2	1.3	11.9	4.7	5.0	61.2	28.4	14.1	16.4	2.3
Open continuously	6.9	4.9	7.9	15.5	72.7	2.0	9.1	4.5	1.7	35.0	0.4	5.0	8.0	6.8	83.9	26.0	37.8	8.8	11.3
Ownership and Occupancy																			

Blank = data not available, or less that 20 buildings, or error is >50%. * = more than one may apply; ** = for demand metered buildings. Med. = Median

Table 2A   1995 Commercial Building Energy Consumption (*Continued*)

Consumption shown is on an annual basis per square foot of building area. *Source*: DOE/EIA 0318(95) 1998

Source Table Number	1	1	3	3	10	10	19	19	21	21	27	31	33	33	33	33	33	33
	Total # Bldgs., Thousands	Total Floor Space, 10^6 ft^2	Major Fuels 1000 Btu/ft^2	\$/ft^2	kWh/ft^2	Med. kWh/ft^2	Med. Peak W/ft^2	Med. Load Fact.	Nat. Gas CF/ft^2	Med. CF/ft^2	Fuel Oil Gal/ft^2	Distr. Heat lb/ft^2	Total	Space Heat	Cool	Ventilation	Water Heat	Lighting
Nongovernment owned	4,025	46,696	84.6	1.17	13.1	7.2	5.6	0.255	47.4	38.8	0.11	105.65	84.6	25.8	5.9	2.6	12.2	19.4
Owner occupied	3,158	35,573	92.4	1.21	13.4	6.9	5.6	0.253	50.6	39.7	0.11	114.17	92.4	28.7	6.1	2.7	14.5	20.2
Nonowner occupied	698	9,697	66.7	1.15	12.4	9.8	5.8	0.267	35.0	34.0	0.10		66.7	18.1	5.9	2.3	5.7	19.2
Unoccupied	170	1,426	11.0	0.16	2.8	0.6	1.6	0.184		51.0			11.0	5.2		0.1		
Government owned	553	12,076	113.6	1.28	14.6	7.7	4.3	0.246	58.5	46.0	0.13	82.95	113.6	41.2	6.2	3.5	19.8	24.3
Federal	76	1,752	151.8	1.73									151.8	43.9	7.4	6.9	19.5	41.2
State	99	2,851	153.6	1.68									153.6	47.3	8.5	4.9	32.6	33.7
Local	379	7,473	89.4	1.02									89.4	38.2	5.1	2.2	15.0	16.8
Space Vacant for at Least 3 Months																		
Yes	787	15,844	70.7	0.96	11.9	5.1	3.2	0.144	40.9	32.4	0.05	53.43	70.7	20.5	5.3	2.6	9.0	18.3
No	3,791	42,928	97.9	1.27	13.9	7.6	5.8	0.171	52.7	40.0	0.15	109.80	97.9	32.1	6.2	2.8	15.6	21.2
Energy Sources*																		
Electricity	4,343	57,076	93.1	1.22	13.4	7.2	5.4	0.164	49.6	39.7	0.12	94.22	93.1	29.7	6.1	2.8	14.2	21.1
Natural gas	2,478	38,145	103.0	1.26	13.1	7.4	5.1	0.166	49.7	39.7	0.07	83.01	103.0	34.5	6.2	2.9	16.5	21.4
Fuel oil	607	14,421	120.1	1.40	15.9	5.0	4.3	0.178	58.4	28.6	0.12	74.76	120.1	33.9	6.8	4.1	23.1	26.3
District heat	110	5,658	185.8	1.87	18.9	10.2	5.0	0.223	60.6	11.8	0.03	94.14	185.8	64.4	5.0	5.6	41.6	33.4
District chilled water	53	2,521	214.8	2.01	21.9	10.2	4.0	0.228	76.7	24.0	0.02	116.82	214.8	68.0	1.6	7.9	47.4	40.3
Propane	589	5,344	73.4	1.15	12.3	5.9	5.8	0.177	55.8	42.1	0.28		73.4	20.0	5.7	2.4	19.0	17.5
Other	213	2,336	110.7	1.16	10.9	5.1	4.3	0.145	42.0	36.8	0.12		110.7	45.5	4.8	2.3		18.0
Energy End Uses*																		
Buildings with space heating	4,024	54,347	96.5	1.25	13.8	7.7	5.6	0.169	49.7	39.8	0.12	94.38						
Buildings with cooling	3,381	49,935	98.6	1.30	14.6	9.0	5.8	0.180	49.4	38.3	0.10	95.11						
Buildings with water heating	3,486	51,560	98.7	1.29	14.2	8.7	5.9	0.176	50.4	40.0	0.11	91.35						
Buildings with cooking	828	20,713	121.0	1.53	17.7	13.5	8.1	0.200	59.5	58.2	0.09	89.33						
Buildings w/manufacturing	204	3,893	78.8	1.03	11.8	6.3	3.8	0.155	42.4	47.7	0.15							
Buildings w/elec. generation	247	13,366	127.6	1.56	18.9	12.2	5.0	0.223	54.4	40.8	0.06	77.10						
Space Heating Energy Sources*																		
Electricity	1,467	22,156	86.1	1.28			6.9	0.179					86.1	19.5	8.0	3.1	11.7	23.4
Natural gas	2,211	31,535	98.2	1.20					52.0	39.9			98.2	36.8	5.9	2.7	14.5	20.4
Fuel oil	504	6,606	109.3	1.22							0.24		109.3	43.4	4.7	2.7	20.7	20.3
District heat	109	5,606	184.8	1.87								94.55	184.8	64.9	5.0	5.6	41.1	33.4
Propane		2,025	63.8	1.21									63.8	9.7	5.6	2.3	8.6	18.2
Other	135	1,050	72.9	0.91									72.9	19.3	4.3	2.1	10.8	15.2
Electricity					16.2	10.9	6.89	0.266										
Electricity main					17.6	12.9	8.00	0.264										
Electricity secondary					14.0	7.0	5.10	0.268										
Other excluding electricity					12.1	6.7	4.80	0.248										
Buildings without space heating					6.4	1.8	2.22	0.230										
Primary Space Heating Energy Source																		
Electricity	1,007	13,500	74.5	1.28	17.6	12.9	8.0	0.162		27.7	0.03		74.5	8.5	9.5	3.2	8.7	
Natural gas	2,106	28,808	98.5	1.19	12.2	7.3	4.9	0.160	54.5	41.2	0.02		98.5	38.0	5.7	2.6	14.3	20.3
Fuel oil	439	4,207	72.6	0.93	6.5	4.1	3.9	0.173	11.8	6.5	0.33		72.6	35.9	2.8	1.2	13.2	10.9
District heat	107	5,289	184.7	1.89	19.1	10.2	5.0	0.223	56.4	7.8	0.03	95.98	184.7	64.1	5.0	5.6	40.8	33.8
Propane	260	1,545	45.9	1.14	12.4	5.7	6.7	0.197		40.4			45.9		4.9	2.1	2.2	16.9
Other	61	514	31.5	0.54	6.9	1.8	3.1	0.248		44.5			31.5	5.6		1.3	4.0	10.3
Cooling Energy Source*																		
Electricity	3,293	47,761	94.9	1.29	14.4	8.9	5.8	0.256					94.9	28.2	7.2	3.0	14.1	22.2
Natural gas	65	1,314	167.3	1.77					86.0	63.9			167.3	51.4	8.1	5.3	25.1	36.1
District chilled water	53	2,521	214.8	2.01									214.8	68.0	1.6	7.9	47.4	40.3
Water Heating Energy Sources*																		
Electricity	1,684	23,056	71.9	1.16	14.5	9.3	5.9	0.267					71.9	20.9	6.8	2.9	3.2	21.7
Natural gas	1,577	24,859	111.4	1.31					59.3	45.4			111.4	36.2	6.7	2.9	21.7	22.2
Fuel oil	120	2,151	94.2	1.08							0.37		94.2	32.6	4.5	1.3	30.9	12.9
District heat	54	3,949	192.9	1.91								102.89	192.9	58.7	5.3	5.6	51.7	32.9
Propane	110	1,020	73.6	1.55									73.6	11.4	8.6	2.8	4.6	23.0
Cooking Energy Sources*																		
Electricity	487	12,249	122.1	1.58	19.9	15.7	9.6	0.336					122.1	26.4	8.9	4.1	20.7	31.6
Natural gas	448	13,195	128.7	1.53					59.1	66.2			128.7	28.4	8.1	3.5	26.6	26.4
Propane	123	1,480	84.3	1.66									84.3	16.8	9.7	3.1	9.7	23.4
Percent of Floorspace Heated																		
Not heated	554	4,425	16.8	0.41	6.4	1.8	2.2	0.230		7.7			16.8	0.0	1.7	0.6		7.0
1 to 50	555	6,227	39.7	0.65	6.9	4.7	2.8	0.205	22.6	25.0	0.10		39.7	13.8	2.6	1.1	2.4	10.1
51 to 99	633	8,868	90.7	1.32	14.6	7.3	5.3	0.269	44.0	38.3	0.11	54.84	90.7	26.5	6.8	3.1	12.4	23.3
100	2,836	39,252	106.9	1.33	14.7	8.8	6.1	0.260	54.6	42.2	0.12	101.13	106.9	35.2	6.8	3.2	17.4	22.9
Percent of Floorspace Cooled																		
Not cooled	1,198	8,837	45.1	0.55	5.4	2.7	2.5	0.218					45.1	24.0	0.0	0.5	6.8	7.9
1 to 50	930	15,027	69.5	0.79	7.0	5.2	3.3	0.225					69.5	35.1	2.8	1.0	9.3	11.8
51 to 99	635	12,549	108.4	1.47	17.0	9.8	6.0	0.272					108.4	26.8	8.1	4.0	17.9	27.7
100	1,816	22,359	112.6	1.56	18.2	11.9	7.0	0.272					112.6	28.0	9.2	4.2	17.3	27.2
Percent Lit When Open																		
Zero	36	189					1.9	1.3	0.158									
1 to 50	666	6,008	51.2	0.71	6.1	4.9	3.6	0.213					51.2	24.9	2.5	1.1	6.4	7.2
51 to 99	745	9,692	91.2	1.21	12.4	7.5	5.8	0.263					91.2	29.6	5.5	2.7	14.9	19.2
100	2,814	40,514	101.3	1.33	14.9	8.5	5.8	0.261					101.3	30.8	6.9	3.2	15.4	24.0
Electricity not used	318	2,369	10.8	0.10		0.3	0.7	0.152					10.8	7.5		0.1	1.3	0.2
Percent Lit When Closed																		

Blank = data not available, or less that 20 buildings, or error is >50%. * = more than one may apply; ** = for demand metered buildings. Med. = Median

Table 2B 1995 Commercial Building Energy Consumption (*Continued*)

Consumption shown is on an annual basis per square foot of building area. *Source*: DOE/EIA 0318(95) 1998

Source Table Number	33	33	33	33	35	35	35	35	35	35	35	35	35	35	37	37	37	37	37
	Sum Major Fuels, 1000 Btu				Energy End Use: Electricity, 1000 Btu/ft²·yr										Natural Gas, 1000 Btu/ft²·yr				
Building Characteristics	Cooking	Refr.	Office Equip	Other	Total	Space Heat	Cool	Venti-lation	Water Heat	Light-ing	Cooking	Refr.	Office Equip	Other	Total	Space Heat	Water Heat	Cooking	Other
Nongovernment owned	4.0	3.4	5.5	5.7	44.6	2.0	5.9	2.6	0.8	20.1	0.3	3.5	5.7	3.6	48.6	26.8	13.1	5.7	3.0
Owner occupied	4.5	3.8	5.6	6.2	45.8	2.0	6.0	2.8	0.9	20.5	0.4	3.9	5.6	3.7	52.0	27.7	14.7	6.2	3.4
Nonowner occupied	2.9	2.5	5.9	4.4	42.3	2.1	5.8	2.3	0.7	19.6	0.2	2.5	6.0	3.1	35.9	23.3	6.6	4.3	
Unoccupied		0.1	0.2	1.1	9.7		0.9	0.3		4.6		0.3	0.5	2.5					
Government owned	2.3	1.8	6.5	7.8	49.8	1.7	6.1	3.6	0.8	24.8	0.3	1.8	6.6	4.1	60.1	35.7	15.8	3.1	5.4
Federal	1.7	1.7	14.9	14.5	85.3	1.2	7.5	7.2	0.7	43.1		1.8	15.6	7.6	59.6	26.1	14.5	2.6	
State	3.3	2.0	8.6	12.2	68.1		8.1	5.0	1.0	34.2	0.3	2.0	8.8	5.4	71.1	30.6	23.6	4.9	12.0
Local	2.1	1.8	3.7	4.6	34.8	1.3	5.1	2.2	0.7	17.1	0.2	1.8	3.7	2.8	56.7	38.6	13.6	2.6	2.0
Space Vacant for at Least 3 Months																			
Yes	2.3	1.2	5.4	6.0	40.6	1.3	5.4	2.8	0.5	19.8	0.2	1.3	5.9	3.5	42.0	23.6	10.5	3.5	4.4
No	4.2	3.8	5.8	6.1	47.4	2.2	6.2	2.8	0.9	21.5	0.4	3.8	5.8	3.8	54.1	30.4	14.7	5.8	3.2
Energy Sources*																			
Electricity	3.8	3.2	5.8	6.3	45.7	2.0	6.0	2.8	0.8	21.1	0.3	3.2	5.8	3.7	51.0	28.6	13.6	5.2	3.5
Natural gas	5.5	3.1	5.6	7.4	44.8	1.1	6.0	2.9	0.6	20.5	0.3	3.1	5.7	3.8	51.0	28.6	13.7	5.2	3.5
Fuel oil	4.1	2.0	8.1	11.7	54.3	1.1	6.4	4.2	0.7	26.5	0.3	2.0	8.1	5.1	60.0	23.5	22.0	5.9	8.6
District heat	3.6	2.1	10.9	19.1	64.5	0.3	4.5	5.6	0.9	33.5	0.3	2.1	10.9	6.3	62.2		17.1	7.9	
District chilled water	4.8	2.5	13.3	29.1	74.6	0.3	1.5	7.9	0.8	40.3	0.4	2.5	13.3	7.6	78.8	11.6	18.5	8.6	40.1
Propane	1.9	4.6	3.8	6.7	41.9	2.1	5.7	2.4	1.2	17.5	0.8	4.6	3.8	3.8	57.3	31.1	16.2	3.7	6.3
Other	2.9	1.6	4.2		37.2	0.9	4.9	2.4	0.8	18.8	0.3	1.7	4.4	2.9	43.1	20.2	11.2	4.0	
Energy End Uses*																			
Buildings with space heating																			
Buildings with cooling																			
Buildings with water heating																			
Buildings with cooking																			
Buildings with manufacturing																			
Buildings with elec. generation																			
Space Heating Energy Sources*																			
Electricity	4.3	3.5	6.6	6.1															
Natural gas	4.2	3.0	5.5	5.1															
Fuel oil	2.6	1.6	5.6	7.6															
District heat	3.5	2.1	10.9	18.1															
Propane	1.5	7.7	3.8	6.5															
Other	2.6	2.2	3.3																
Electricity																			
Electricity main																			
Electricity secondary																			
Other excluding electricity																			
Buildings without space heating																			
Primary Space Heating Energy Source																			
Electricity	4.8	4.0	6.9	5.4															
Natural gas	4.2	3.0	5.4	5.0															
Fuel oil	1.2	1.2	2.6	3.7															
District heat	3.3	2.1	11.1	18.5															
Propane		8.6	3.2	5.6															
Other	0.8	1.8	1.4	2.5															
Cooling Energy Source*																			
Electricity	4.3	3.4	6.2	6.2															
Natural gas	6.3	2.7	9.2	23.0															
District chilled water	4.8	2.5	13.3	29.1															
Water Heating Energy Sources*																			
Electricity	2.3	3.0	6.3	4.7															
Natural gas	6.1	3.8	5.5	6.2															
Fuel oil	1.5	1.7	3.0	5.7															
District heat	4.4	2.3	10.7	21.1															
Propane			2.7	7.9															
Cooking Energy Sources*																			
Electricity	8.8	6.5	6.4	8.7															
Natural gas	15.6	5.6	5.1	9.4															
Propane	2.5	9.8	2.6	6.6															
Percent of Floorspace Heated																			
Not heated		1.8	1.6	2.3	21.7	0.0	2.5	0.8	0.3	10.5	0.1	2.7	2.3	2.4					
1 to 50	1.4	2.3	3.2	2.8	23.5	1.6	2.5	1.1	0.4	10.3	0.1	2.3	3.2	2.1	23.2	17.1	2.9	2.3	
51 to 99	3.9	3.5	6.0	5.1	49.7	2.5	6.6	3.1	0.9	23.4	0.3	3.5	6.0	3.4	45.2	26.2	11.5	5.1	2.3
100	4.4	3.3	6.5	7.3	50.1	2.0	6.6	3.2	0.9	23.0	0.4	3.3	6.5	4.1	56.0	30.9	15.5	5.5	4.1
Percent of Floorspace Cooled																			
Not cooled	0.6	1.1	1.9	2.3	18.5	1.9	0.0	0.6	0.7	9.6	0.1	1.4	2.4	1.8	53.6	40.2	9.8	1.6	
1 to 50	1.0	1.5	3.0	3.9	24.0	1.3	2.8	1.0	3.0	11.8	0.1	1.5	3.0	2.0	47.6	36.9	7.5	1.2	1.9
51 to 99	5.0	3.5	7.3	8.0	58.0	1.7	7.8	4.0	0.8	27.7	0.4	3.5	7.3	4.8	49.4	22.4	16.6	6.3	4.2
100	6.0	4.7	8.0	8.0	62.2	2.5	9.0	4.2	1.2	27.2	0.5	4.7	8.0	4.8	54.0	24.2	17.1	8.1	4.6
Percent Lit When Open																			
Zero																			
1 to 50	1.0	2.1	2.5	3.5	20.7	2.2	2.5	1.1	0.4	7.2	0.1	2.1	2.5	2.6	36.5	26.8	7.1	1.4	
51 to 99	3.4	2.6	6.1	7.2	42.5	1.7	5.4	2.7	0.8	19.2	0.2	2.6	6.1	3.7	50.3	28.3	12.9	4.4	4.6
100	4.4	3.5	6.4	6.6	51.0	2.0	6.7	3.2	0.9	24.0	0.4	3.5	6.4	3.9	53.4	28.9	14.9	6.0	3.6
Electricity not used			0.3	0.6				0.2		0.6			0.9	2.0					
Percent Lit When Closed																			

Blank = data not available, or less that 20 buildings, or error is >50%. * = more than one may apply; ** = for demand metered buildings. Med. = Median

Table 2A   1995 Commercial Building Energy Consumption (Continued)

Consumption shown is on an annual basis per square foot of building area. *Source*: DOE/EIA 0318(95) 1998

Source Table Number	1	1	3		10	10	19	19	21	21	27	31	33	33	33	33	33	33
	Total #	Total	Major Fuels		Electricity				Nat. Gas		Fuel Oil	Distr. Heat	Energy End Use:					
	Bldgs., Thousands	Floor Space, 10⁶ ft²	1000 Btu/ft²	$/ft²	kWh/ft²	Med. kWh/ft²	Med. Peak W/ft²	Med. Load Fact.	CF/ft²	Med. CF/ft²	Gal/ft²	lb/ft²	Total	Space Heat	Cool	Venti-lation	Water Heat	Light-ing
Building Characteristics																		
Zero	1,644	13,101	57.4	0.79	7.6	4.7	4.2	0.217					57.4	27.9	3.2	1.6	4.5	10.7
1 to 50	2,109	30,711	85.9	1.21	12.8	9.8	6.0	0.263					85.9	30.1	6.1	2.7	9.7	19.9
51 to 100	87	1,914	108.6	1.61	21.6	8.8	4.0	0.260					108.6	17.5	10.1	4.5	13.5	39.9
Never closed	421	10,677	158.9	1.80	21.3	12.7	5.8	0.374					158.9	33.9	9.7	4.5	39.8	35.0
Electricity not used	318	2,369	10.8	0.10		0.3	0.7	0.152					10.8	7.5		0.1		0.2
Energy Conservation Features																		
Any conservation feature	4,075	55,288	95.1	1.24														
Building shell	3,906	53,190	96.5	1.26														
HVAC	2,529	44,657	103.5	1.34														
Lighting	2,084	38,537	104.1	1.36														
Heating Equipment*																		
Heat pumps													85.6	14.9	9.0	2.9	15.9	23.1
Furnaces													77.1	32.2	4.6	2.0	7.4	16.2
Individual space heaters													87.4	31.0	5.4	2.5	11.8	20.2
District heat													179.9	61.6	4.9	5.5	39.2	33.3
Boilers													112.9	40.2	6.4	3.1	22.1	22.4
Packaged heating units													86.7	20.2	8.1	2.8	10.9	23.2
Other													107.2	21.7	8.3	4.8	15.8	32.3
Cooling Equipment*																		
Residential central A/C													102.4	34.0	7.2	2.8	15.3	20.8
Heat pumps													86.5	16.1	9.0	3.0	14.6	23.4
Individual A/C													96.1	36.0	5.5	1.7	19.5	16.5
District chilled water													214.8	68.0	1.6	7.9	47.4	40.3
Central chillers													133.1	29.0	10.0	5.7	26.2	34.0
Packaged A/C units													97.0	26.0	7.9	3.2	12.9	24.6
Evaporative coolers													102.8	21.6	6.5	3.4	21.4	22.7
Other													111.2	28.3	5.9	3.8	14.2	31.8
Lighting Equipment Types*																		
Incandescent													98.4	30.3	6.3	2.9	17.7	21.6
Standard fluorescent													96.4	30.5	6.4	3.0	14.7	21.9
Compact fluorescent													122.3	29.9	8.4	4.3	22.8	29.7
High intensity discharge													102.9	30.9	6.5	3.2	17.5	25.3
Halogen													113.5	33.4	7.1	3.8	20.5	26.4
Other													86.6	12.4	8.9	4.7	11.1	31.2
Water Heating Equipment*																		
Central system													108.8	35.8	6.5	3.1	18.8	22.2
Distributed system													74.7	20.3	6.6	2.6	7.5	21.4
Combination central and distr.													120.8	32.9	6.8	3.8	27.0	27.8
Personal Computers/Terminals																		
None													53.0	23.9	2.6	1.2	7.3	
1 to 4													80.9	30.0	4.9	1.9	9.4	15.9
5 to 9													97.2	36.3	6.8	2.1	11.8	20.7
10 to 19													94.1	32.7	6.3	2.0	16.9	20.7
20 to 49													91.6	28.5	7.1	2.5	15.2	21.6
50 to 99													98.4	28.0	7.6	2.9	15.9	24.6
100 to 249													112.7	27.9	8.0	4.4	20.9	29.2
250 or more													148.2	29.8	9.3	7.6	24.4	40.4
Commercial Refrigeration*																		
Any equipment													122.0	27.2	8.6	3.9	21.3	28.9
Walk in units													133.7	26.3	9.6	4.3	23.8	32.4
Cases and cabinets													124.0	27.5	8.9	3.9	21.8	29.5
None													71.5	30.0	4.4	2.1	9.2	15.3
Shell Conservation Features*																		
Roof/ceiling insulation													98.8	30.5	6.6	3.0	15.5	22.4
Wall insulation													97.3	26.9	7.0	3.1	16.1	22.7
Storm or multiple glazing													106.5	32.9	6.6	3.2	18.1	23.3
Tinted, refl., shaded glass													106.9	27.1	7.8	4.0	16.9	26.5
Shading or awnings													101.4	29.8	6.8	3.3	16.5	23.2
HVAC Conservation Features*																		
Variable volume system													128.1	28.2	8.5	5.3	24.1	32.4
Economizer cycle													127.5	33.2	8.5	4.6	23.0	29.8
HVAC maintenance													104.6	31.7	7.0	3.3	17.0	23.9
Other efficient equipment													120.2	29.5	8.1	4.5	21.6	28.9
Lighting Conservation Features*																		
Specular reflectors													108.2	31.7	7.1	3.5	17.4	26.1
Energy efficient ballasts													109.9	30.7	7.5	3.7	17.9	26.6
Natural lighting sensors													117.1	34.6	8.2	3.4	20.3	28.0
Occupancy sensors													121.2	29.5	7.4	4.4	21.9	28.3
Time clock													103.0	22.3	8.5	4.1	16.5	29.1
Manual dimmer switches													125.7	33.0	8.4	4.6	22.2	29.4
Other													116.9	30.4	8.4	4.3	18.3	30.6
Off Hour Equipment Reduction*																		
Heating													78.9	29.7	5.3	2.5	8.8	
Cooling													80.1	28.3	6.0	2.7	8.8	18.7
Lighting													79.0	29.2	5.4	2.5	8.4	18.1

Blank = data not available, or less that 20 buildings, or error is >50%. * = more than one may apply; ** = for demand metered buildings. Med. = Median

Table 2B 1995 Commercial Building Energy Consumption (*Continued*)

Consumption shown is on an annual basis per square foot of building area. *Source*: DOE/EIA 0318(95) 1998

Source Table Number	33	33	33	33	35	35	35	35	35	35	35	35	35	35	37	37	37	37	37
	Sum Major Fuels, 1000 Btu				Energy End Use: Electricity, 1000 Btu/ft²·yr										Natural Gas, 1000 Btu/ft²·yr				
Building Characteristics	Cooking	Refr.	Office Equip	Other	Total	Space Heat	Cool	Ventilation	Water Heat	Lighting	Cooking	Refr.	Office Equip	Other	Total	Space Heat	Water Heat	Cooking	Other
Zero	0.8	1.4	3.9	3.5	25.9	2.3	3.1	1.6	0.5	10.7	0.1	1.4	3.9	2.3	41.1	33.6	4.7	1.1	1.7
1 to 50	4.0	3.1	6.1	4.3	43.8	1.8	6.0	2.7	0.7	19.9	0.4	3.1	6.1	3.2	43.1	28.6	7.9	5.3	1.3
51 to 100	5.0	6.5		6.1	73.5	2.2	10.1	4.5	0.7	39.9	0.6	6.5		3.5	40.8	16.8	14.3	6.2	
Never closed	7.0	5.0	8.0	15.8	72.7	2.1	9.1	4.5	1.7	35.0	0.4	5.0	8.0	6.9	83.8	26.0	37.7	8.8	11.3
Electricity not used			0.3	0.6							0.6		0.9	2.0					
Energy Conservation Features																			
Any conservation feature																			
Building shell																			
HVAC																			
Lighting																			
Heating Equipment*																			
Heat pumps	3.1	2.4	7.3	6.9	54.9	3.9	9.0	2.9	1.4	23.2	0.3	2.4	7.4	2.4	47.5	13.0	25.1	5.4	3.9
Furnaces	3.2	4.3	3.7	3.5	36.3	1.6	4.5	2.0	0.7	16.3	0.4	4.3	3.8	2.8	46.0	34.2	7.5	3.5	0.8
Individual space heaters	2.2	2.9	5.7	5.6	44.3	2.9	5.3	2.5	0.8	20.3	0.3	3.0	5.7	3.5	47.8	31.2	11.5	2.6	2.5
District heat	3.9	2.1	10.8	18.3	64.1	0.3	4.4	5.5	0.8	33.4	0.3	2.1	10.8	6.2	60.0	7.8	15.4	8.6	28.2
Boilers	3.6	1.6	6.4	7.1	45.7	1.0	6.1	3.1	2.5	22.6	0.2	1.6	6.4	4.1	65.8	37.2	21.5	3.9	3.2
Packaged heating units	6.1	4.4	5.9	5.1	52.0	2.5	8.0	2.8	0.9	23.3	0.4	4.4	5.9	3.8	42.1	21.1	12.2	7.4	1.4
Other	3.6	2.5	9.3	8.9	66.5	2.9	8.1	4.8	0.8	32.3	0.3	2.5	9.4	5.3	40.8	16.9	14.7	4.9	4.3
Cooling Equipment*																			
Residential central A/C	5.8	4.6	5.4	6.4	46.8	1.2	7.1	2.8	0.8	20.9	0.4	4.6	5.4	3.4	60.0	34.9	14.6	7.1	3.3
Heat pumps	3.2	2.8	7.6	6.6	56.0	3.8	9.0	3.0	1.3	23.4	0.5	2.8	7.6	4.4	44.3	14.2	21.5	5.2	3.4
Individual A/C	3.9	2.4	3.9	6.6	36.3	1.7	5.4	1.7	0.9	16.6	0.4	2.4	4.0	3.2	60.1	33.4	18.2	4.9	3.6
District chilled water	4.8	2.5	13.3	29.1	74.6	0.3	1.5	7.9	0.8	40.3	0.4	2.5	13.3	7.6	78.8	11.6	18.5	8.6	40.1
Central chillers	4.6	1.9	10.5	11.2	70.4	1.6	9.2	5.7	0.8	34.0	0.3	1.9	10.5	6.3	60.4	23.8	24.5	5.6	6.5
Packaged A/C units	5.1	4.1	6.7	6.4	54.0	2.0	7.7	3.2	0.8	24.7	0.4	4.1	6.7	4.1	46.2	24.4	12.7	6.2	2.8
Evaporative coolers	8.6	5.2	6.4	6.9	51.5	1.5	6.5	3.4	0.5	22.8	0.5	5.3	6.4	4.6	61.8	23.1	25.5	10.6	2.7
Other	4.5	4.9	9.5	8.3	62.0	1.0	5.9	3.8	0.6	31.8	0.5	4.9	9.5	4.1	40.5	22.1	9.3		4.1
Lighting Equipment Types*																			
Incandescent	4.7	2.9	5.5	6.5	45.7	1.7	6.1	2.9	0.8	21.6	0.3	2.9	5.5	3.7					
Standard fluorescent	4.0	3.3	6.1	6.5	47.4	2.0	6.2	3.0	0.8	21.9	0.3	3.3	6.1	3.8					
Compact fluorescent	5.7	3.1	8.3	10.3	61.5	1.6	7.9	4.3	0.8	29.7	0.5	3.1	8.3	5.3					
High intensity discharge	3.6	2.4	6.3	7.3	50.0	1.5	6.2	3.2	0.6	25.3	0.4	2.4	6.3	4.2					
Halogen	4.5	2.5	7.0	8.3	54.0	1.7	6.8	3.8	0.7	26.4	0.4	2.5	7.0	4.7					
Other	2.6		5.4	8.3	58.2	1.3	8.6	4.7	0.5	31.2	0.4	2.0	5.4	4.1					
Water Heating Equipment*																			
Central system	4.8	4.1	6.3	7.1	49.6	2.0	6.3	3.2	1.8	22.3	0.4	4.1	6.4	4.1	59.8	33.0	16.7	6.1	4.0
Distributed system	2.9	2.1	5.7	5.6	45.2	2.1	6.5	2.7	1.1	21.4	0.2	2.1	5.7	3.3	34.3	19.6	7.7	4.0	2.9
Combination central and distr.	4.9	2.7	7.0	7.9	55.2	1.8	6.6	3.8	0.6	27.8	0.4	2.7	7.0	4.4	56.2	24.3	22.0	5.8	4.1
Personal Computers/Terminals																			
None	2.5	2.7	1.3	3.0	23.6	1.7	3.0	1.4	0.7	9.7	0.2	3.1	1.5	2.4					
1 to 4	6.1	5.8	2.8	4.2	38.5	2.7	4.9	1.9	1.0	15.9	0.4	5.8	2.8	3.1					
5 to 9	5.1	4.4	4.9	5.1	47.5	2.6	6.7	2.1	1.3	20.8	0.7	4.4	4.9	3.8					
10 to 19	2.5	2.4	6.0	4.7	43.9	2.3	6.3	2.0	0.7	20.8	0.2	2.4	6.1	3.2					
20 to 49	2.6	2.1	5.9	6.0	45.4	1.4	7.1	2.5	0.8	21.8	0.2	2.1	6.0	3.5					
50 to 99	2.6	1.4	7.2	8.0	50.2	1.3	7.3	2.9	0.7	24.7	0.2	1.4	7.3	4.5					
100 to 249	3.1	2.3	8.3	8.6	59.2	1.7	7.6	4.4	0.6	29.2	0.2	2.3	8.3	4.9					
250 or more	4.2	1.7	16.2	14.5	83.1	1.6	8.6	7.6	0.7	40.5	0.4	1.7	16.3	5.9					
Commercial Refrigeration*																			
Any equipment	9.5	7.0	6.0	9.7	63.8	2.3	8.4	3.9	1.1	29.1	0.8	7.0	6.0	5.3					
Walk in units	11.5	8.5	6.5	10.8	71.8	2.3	9.3	4.3	1.2	32.6	1.0	8.5	6.6	6.0					
Cases and cabinets	10.0	7.1	5.7	9.6	64.7	2.4	8.6	3.9	1.2	29.6	0.9	7.2	5.8	5.2					
None	0.2	0.7	5.5	3.9	34.3	1.8	4.4	2.2	0.6	16.0		0.8	5.8	2.7					
Shell Conservation Features*																			
Roof/ceiling insulation	4.1	3.5	6.4	6.7	49.4	2.0	6.5	3.1	0.9	22.7	0.4	3.5	6.4	4.0	53.0	29.3	14.6	5.5	3.6
Wall insulation	4.4	3.5	6.8	6.9	50.8	1.9	6.8	3.1	1.0	22.8	0.4	3.6	6.8	4.3	49.6	25.0	15.2	5.8	3.5
Storm or multiple glazing	5.1	3.2	6.9	7.3	51.2	2.0	6.5	3.2	1.0	23.5	0.4	3.2	6.9	4.3	57.9	30.3	17.1	6.7	3.9
Tinted, refl., shaded glass	5.1	3.1	8.3	8.0	57.6	2.1	7.6	4.0	0.8	26.6	0.4	3.1	8.3	4.6	49.8	22.8	15.7	6.7	4.6
Shading or awnings	4.4	2.6	7.1	7.6	50.2	2.0	6.6	3.3	0.9	23.3	0.3	2.6	7.2	4.1	52.0	26.1	15.2	5.9	4.8
HVAC Conservation Features*																			
Variable volume system	5.8	2.7	10.3	10.9	67.8	1.9	8.1	5.3	0.8	32.5	0.4	2.7	10.4	5.8	59.7	23.3	22.0	7.6	6.7
Economizer cycle	5.9	3.1	9.4	10.1	63.8	1.8	8.1	4.6	0.8	30.0	0.5	3.2	9.4	5.4	56.9	25.1	18.5	7.2	6.0
HVAC maintenance	4.4	3.0	6.9	7.4	51.4	1.8	6.8	3.3	0.9	24.0	0.4	3.0	6.9	4.2	53.1	28.0	15.3	5.7	4.1
Other efficient equipment	4.9	3.1	9.0	10.5	61.1	1.8	7.6	4.5	0.7	28.9	0.3	3.1	9.0	5.3	54.6	23.2	18.5	6.1	6.8
Lighting Conservation Features*																			
Specular reflectors	4.6	2.9	7.1	7.8	53.9	1.6	6.9	3.5	0.8	26.1	0.4	2.9	7.1	4.5					
Energy efficient ballasts	4.4	3.4	7.8	8.0	56.8	2.1	7.2	3.7	0.9	26.6	0.4	3.4	7.8	4.6					
Natural lighting sensors	4.4	3.4	6.2	8.6	56.7	1.9	8.0	3.4	0.9	28.0	0.3	3.4	6.2	4.6					
Occupancy sensors	5.7	3.3	9.7	10.9	60.1	1.2	6.9	4.4	0.7	28.3	0.3	3.3	9.7	5.2					
Time clock	4.2	3.1	7.9	7.3	59.4	1.6	8.2	4.1	0.6	29.1	0.3	3.1	7.9	4.5					
Manual dimmer switches	7.2	2.5	8.8	9.6	62.0	1.8	8.1	4.6	0.8	29.4	0.4	2.5	8.8	5.5					
Other	3.3	3.1	10.0	8.4	64.4	1.7	8.3	4.3	0.6	30.6	0.4	3.1	10.0	5.3					
Off Hour Equipment Reduction*																			
Heating	3.2	2.3	5.2	4.0	38.8	1.9	5.2	2.5	0.6	17.9	0.3	2.3	5.3	2.8	41.4	28.6	7.1	4.2	1.5
Cooling	3.4	2.4	5.6	4.1	41.1	1.7	5.9	2.7	0.6	18.8	0.3	2.4	5.6	2.9	41.0	27.7	7.3	4.5	1.5
Lighting	3.1	2.7	5.5	4.1	40.0	2.0	5.3	2.5	0.6	18.1	0.3	2.7	5.5	3.0	42.4	29.3	7.3	4.3	1.5

Blank = data not available, or less that 20 buildings, or error is >50%. * = more than one may apply; ** = for demand metered buildings. Med. = Median

Table 3 1993 Residential Annual End Use Consumption

Annual Electricity Use End Use/Appliance	Millions of Households	kWh per Household
Total Households	96.6	9965
Central air conditioning	41.0	2667
Room air conditioners	33.1	738
Water heating	37.0	2671
Main space heating	25.0	4541
Secondary space heating	12.1	400
Refrigerator	115.7	1155
Appliances (total of list below)	96.6	4933
Lighting (indoor and outdoor)	96.6	940
Television	198.3	360
Clothes dryer	54.7	875
Freezer	33.4	1204
Range/oven	58.3	458
Microwave oven	81.3	191
Waterbed heater	14.6	960
Dishwasher	43.7	299
Swimming pool pump	4.6	2022
Clothes washer	74.5	99
Dehumidifier	9.1	370
Well pump	13.0	228
Personal computer	22.6	77
Hot tub/spa heater	1.9	482
Residual	96.6	1364

Annual Natural Gas Use End Use/Appliance	Millions of Households	Therms per Household
Total Households	58.7	899
Main Space Heating	51.4	709
Secondary Space Heating	1.2	215
Water Heating	51.4	255
Air-conditioning	0.1	238
Appliances	37.8	77

Source: DOE/EIA 0321(93) 1995 Table 3.1 and Table 3.2.

occurring in the summer, and shows actual demand versus billing demand under an 85% ratchet. The winter demand is approximately 700 kW each month, with chiller operation causing the August demand to peak at 1000 kW. Since an 85% ratchet applies, all months following August with actual demand below 850 kW are billed at 850 kW. Therefore, the following can be concluded about this building:

1. ECOs that reduce winter demand would not reduce the billed cost of demand unless they also reduced summer demand; and if such measures were implemented in September, they would not produce savings in demand billings for the first 11 months.
2. ECOs that reduce peak demand in each of the summer months (for example, 50 kW reduction in chiller peaks) may produce savings in demand billings throughout the year.

A billing method called real time pricing (RTP) is used by some utility companies. With this method, the utility calculates the marginal cost of power per hour for the next day, determines the customer price, and sends this hourly price (for the next day) to customers. The customer can then determine the amount of power to be consumed at different times of the day.

There are many variations of the above billing methods, and it is important to understand the applicable rates. Caution is advised in designing or installing energy management and energy retrofit systems that take advantage of utility rate provisions, because these provisions can change. The structure or provisions of utility rates cannot be guaranteed for the life of the ECOs. Some provisions that

change include on-peak times, declining block rates, and demand ratchets.

Priorities

Having established a database, the energy manager should assign priorities to future work efforts. If there is more than one building or department under the energy manager's care, their energy use and cost databases should be compared overall and on the basis of energy use and cost per unit area, of cost per unit of production, or of some other index that demonstrates an acceptable level of accuracy. Comparisons should also be made with realistic energy targets, if they are known. Using such comparisons, it is often possible to set priorities that allocate the available resources most effectively.

One approach to comparing that is used in England is called Monitoring and Targeting. Monitoring involves collecting short-term metered data and analyzing it with respect to occupancy, weather prediction, and other variables. This permits a better understanding of the factors that influence energy consumption so that improvements can be made. Targeting involves comparisons with other similar buildings. When the building being monitored uses more energy than the norm for similar buildings, it is likely that reductions can be achieved.

At this point, a report should be prepared for top management outlining the data collected, the priorities assigned, plans for continued development of the energy management program, and projected budgetary needs. This should be the beginning of a regular monthly, quarterly, or semiannual reporting procedure.

Energy Audits

Three levels of energy audits or analysis have been defined (Mazzucchi 1992). Depending on the physical and energy use characteristics of a building and the needs and resources of the owner, these steps require different levels of effort. Following a preliminary energy use evaluation, an energy analysis can generally be classified into the following three categories:

Level I—Walk-Through Assessment. This level involves the assessment of a building's energy cost and efficiency through the analysis of energy bills and a brief survey of the building. A Level I energy analysis will identify and provide a savings and cost analysis of low-cost/no-cost measures. It also provides a listing of potential capital improvements that merit further consideration, along with an initial judgment of potential costs and savings. The level of detail depends on the experience of the person performing the audit or on the specifications of the client paying for the audit.

Level II—Energy Survey and Analysis. This level includes a more detailed building survey and energy analysis. A breakdown of energy use within the building is provided. A Level II energy analysis identifies and provides the savings and cost analysis of all practical measures that meet the owner's constraints and economic criteria, along with a discussion of any effect on operation and maintenance procedures. It also provides a listing of potential capital-intensive improvements that require more thorough data collection and analysis, along with an initial judgment of potential costs and savings. This level of analysis will be adequate for most buildings and measures.

Level III—Detailed Analysis of Capital-Intensive Modifications. This level focuses on potential capital-intensive projects identified during Level II and involves more detailed field data gathering and engineering analysis. It provides detailed project cost and savings information with a high level of confidence sufficient for major capital investment decisions.

The levels of energy audits do not have sharp boundaries between them. They are general categories for identifying the type of information that can be expected and an indication of the level of confidence in the results; that is, various measures may be subjected

Table 4 1993 Residential Energy Consumption
Source: DOE/EIA 0321(93) 1995

	Total in Millions — Households	Total in Millions — Buildings	Total Floor Space, 10^9 ft^2	Average Consumption — Per Sq. Foot, 10^3 Btu	Average Consumption — Per House, 10^6 Btu
Total Households	96.6	76.5	181.2	55	103.6
Northeast	19.5	13.8	40.1	60	122.4
New England	5.1	3.7	10.6	59	123.1
Middle Atlantic	14.4	10.1	29.4	60	122.1
Midwest	23.3	19.0	50.6	62	134.3
East North Central	16.4	13.1	35.3	64	138.8
West North Central	6.9	5.9	15.2	56	123.8
South	33.5	28.4	57.1	52	87.9
South Atlantic	17.4	14.2	29.9	45	77.8
East South Central	6.0	5.5	10.8	53	94.9
West South Central	10.1	8.7	16.5	62	101.1
West	20.4	15.4	33.5	46	76.0
Mountain	5.4	4.4	9.3	57	98.1
Pacific	15.0	10.9	24.2	42	68.2
Largest Populated States					
California	11.1	8.0	17.9	41	65.2
Florida	5.6	4.3	9.3	31	52.1
New York	6.8	3.7	12.8	64	121.2
Texas	6.4	5.4	10.8	57	94.7
Urban States					
Urban	75.8	57.2	140.5	56	103.2
Central city	30.6	19.8	47.1	63	97.6
Suburban	45.2	37.4	93.4	52	107.0
Rural	20.8	19.3	40.7	54	104.7
Climate Zone					
<2000 CDD, >7000 HDD	8.7	7.6	19.3	56	124.0
5500 to 7000 HDD	26.5	20.4	55.2	62	129.2
4000 to 5499 HDD	22.5	17.0	44.0	55	108.3
Under 4000 HDD	17.8	13.9	28.5	49	78.5
>2000 CDD, <4000 HDD	21.2	17.6	34.2	49	79.0
Type of Housing Unit					
Single-family	66.8	66.8	152.2	52	118.5
Detached	59.5	59.5	139.1	52	121.2
Attached	7.3	7.3	13.1	53	96.3
Mobile Home	5.6	5.6	5.4	84	81.9
Multifamily	24.2	4.1	23.6	69	67.3
2 to 4 Units	8.0	2.9	9.6	83	99.5
5 or More Units	16.2	1.2	14.0	60	51.5
Heated Floorspace					
Fewer than 1000 ft^2	29.3	15.3	24.2	81	66.7
1000 to 1999 ft^2	40.2	34.6	68.2	59	100.7
2000 to 2999 ft^2	17.8	17.4	48.6	50	136.6
3000 ft^2 or more	9.3	9.1	40.3	39	168.8
Total Number of Rooms (Excluding Bathrooms)					
1 or 2	3.2	0.7	1.7	78	41.3
3 to 5	47.4	31.0	59.6	63	79.6
6 to 8	40.2	39.1	97.7	52	126.3
9 or more	5.8	5.7	22.1	46	175.5
Ownership of Unit					
Owned	63.2	60.8	143.5	52	118.5
Rented	33.4	15.7	37.7	67	75.2
Public housing	3.4	0.8	2.9	69	58.2
Not public housing	30.0	14.9	34.8	67	77.2

	Total in Millions — Households	Total in Millions — Buildings	Total Floor Space, 10^9 ft^2	Average Consumption — Per Sq. Foot, 10^3 Btu	Average Consumption — Per House, 10^6 Btu
Rent subsidy	2.0	0.7	2.2	70	75.8
No rent subsidy	28.0	14.2	32.6	66	77.3
Year of Construction					
1939 or before	20.4	16.0	40.6	65	129.4
1940 to 1949	6.9	5.9	11.6	67	111.8
1950 to 1959	13.1	11.7	24.7	60	114.1
1960 to 1969	15.0	11.3	27.2	57	102.9
1970 to 1979	18.1	13.1	31.7	50	87.9
1980 to 1984	8.5	6.5	14.7	46	80.3
1985 to 1987	5.5	4.3	10.8	44	85.2
1988 to 1990	4.7	4.0	10.0	43	90.4
1991 to 1993	4.5	3.7	10.0	40	88.9
All Utilities Paid By Household					
Yes	82.9	73.1	168.8	53	108.3
No	13.8	3.3	12.4	83	75.2
1993 Family Income					
Less Than $5000	4.1	2.3	4.6	71	79.8
$5000 to $9999	10.6	6.9	12.9	67	81.4
$10000 to 14999	11.1	7.8	16.0	62	89.7
$15000 to 19999	9.6	7.5	15.4	62	99.2
$20000 to 24999	8.7	6.8	14.4	59	96.6
$25000 to $34999	14.1	11.5	26.4	55	103.5
$35000 to $49999	17.5	14.6	37.5	51	108.5
$50000 to 74999	12.6	11.5	29.9	50	119.2
$75000 or more	8.3	7.6	24.3	48	139.9
Below Poverty Line					
100%	14.4	9.3	17.7	70	85.8
125%	19.4	13.1	25.0	68	87.8
150%	24.8	16.9	32.8	67	88.6
Eligible for Fed. Assistance	30.7	21.1	42.5	65	90.7
Age of Householder					
Under 25 years	5.7	3.0	6.4	67	75.4
25 to 34 years	19.9	14.0	31.8	60	95.4
35 to 44 years	21.4	18.0	42.8	53	105.9
45 to 59 years	21.9	18.8	46.5	53	113.5
60 years and over	27.8	22.7	53.7	55	105.6
Education of Householder					
12 years or fewer	51.5				
13 to 16 years	33.6				
17 years or more	11.5				
Race of Householder					
White	80.2	66.2	158.6	53	105.3
Black	10.9	7.0	15.2	77	106.7
Other	5.5	3.3	7.4	55	72.7
Householder of Hispanic Descent					
Yes	7.9	5.2	11.2	57	81.2
No	88.7	71.3	170.1	55	105.5
Household Size					
1 person	23.5	14.8	32.1	56	76.7
2 persons	31.7	25.8	61.9	52	101.3
3 persons	16.6	13.9	32.3	58	112.9
4 persons	14.6	13.0	32.4	56	125.2
5 persons	6.8	6.1	15.0	55	122.3
6 or more persons	3.5	3.0	7.5	62	133.9

CDD = cooling degree-days (65°F base)   HDD = heating degree-days (65°F base)

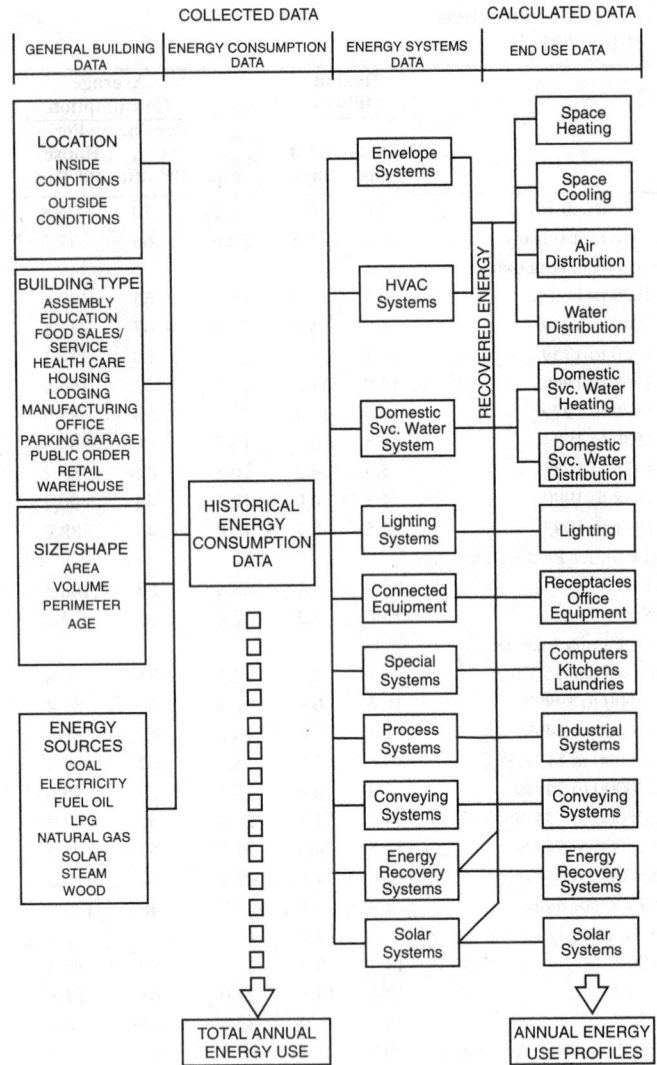

Fig. 4　Energy Audit Input Procedures

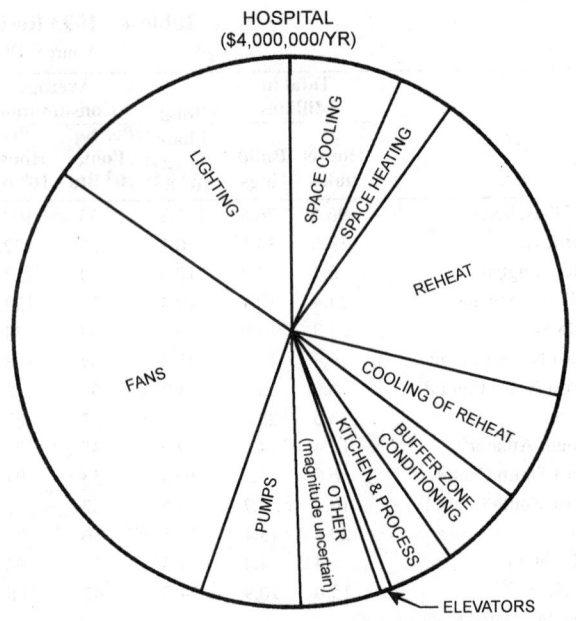

Fig. 5　Energy Cost Distribution

to different levels of analysis during an energy analysis of a particular building.

In the complete development of an energy management program, Level II audits should be performed on all facilities, although Level I audits are useful in establishing the program. Figure 4 illustrates Level II energy audit input procedures in which the following data are collected:

- General building data
- Historic energy consumption data
- Energy systems data

The collected data is used to calculate an energy use profile that includes all end-use categories. From the energy use profiles, it is possible to develop and evaluate energy conservation opportunities.

In conducting an energy audit, a thorough systems approach produces the best results. This approach has been described as starting at the end rather than at the beginning. As an example of this approach, consider a factory with steam boilers in constant operation. An expedient (and often cost-effective) approach would be to measure the combustion efficiency of each boiler and to improve boiler efficiency. Beginning at the end would require observing all or most of the end uses of steam in the plant. It is possible that this would result in the discovery of considerable quantities of steam

being wasted by venting to the atmosphere, venting through defective steam traps, uninsulated lines, and passing through unused heat exchangers. Elimination of such end-use waste could produce greater savings than those easily and quickly developed by improving boiler efficiency. When using this approach, care must be taken to make cost-effective use of the energy auditor's time. It may not be cost-effective to track down every end use.

When conducting an energy audit, it is important to become familiar with operating and maintenance procedures and personnel. The energy manager can then recommend, through the appropriate departmental channels, energy-saving operating and maintenance procedures. The energy manager should determine, through continued personal observation, the effectiveness of the recommendations.

Stewart et al. (1984) tabulated 139 different energy audit input procedures and forms for 10 different building types, each using 62 factors. They discuss features of selected audit forms that can help in developing or obtaining an audit procedure.

To calculate the energy cost avoidance of various energy conservation opportunities, it is helpful to develop an energy cost distribution chart similar to that shown for a hospital in Figure 5. Preliminary information of this nature can be developed from monthly utility data by calculating end-use energy profiles (Spielvogel 1984).

Analysis of electrical operating costs starts with the recording of data from the bills on a form similar to that in Table 5. By dividing the consumption by the days between readings, the average daily consumption can be calculated. This consumption should be plotted to detect errors in meter readings or reading dates and to detect consumption variances (see Figure 6). For this example, 312 kWh/day is chosen as the "base electrical consumption" to cover year-round electrical needs such as lighting, business machines, domestic hot water, terminal reheat, security, and safety lighting. At this point, consumptions or spans that appear to be in error should be reexamined and corrected as necessary. If the reading date for the 10Nov bill in Table 5 was 05Nov, the curve in Figure 6 would be more continuous. On the basis of a 05Nov reading, the minimum daily consumption of 312 kWh on the continuous curve (Figure 6) occurred in the February billing.

To start the analysis, the monthly base consumption is calculated (base daily consumption times billing days) and is subtracted from

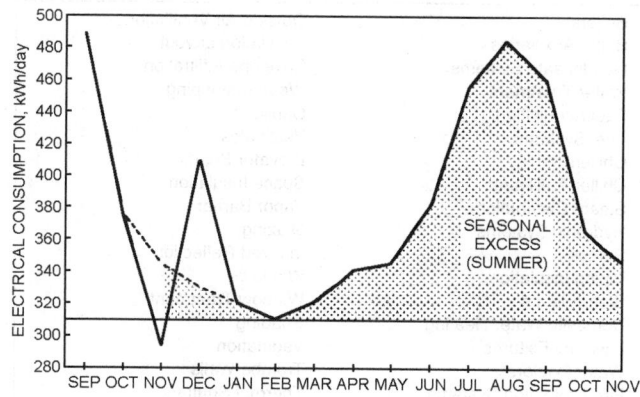

Fig. 6 Average Daily Electrical Consumption

each monthly total to obtain the difference. These differences fall under either summer excess or winter excess, depending on the season. Excess consumption in the summer is primarily due to the air-conditioning load. A similar analysis is made of the actual monthly demand. In Table 5, the base demand is 33.0 kW, and it is usually found in the same or adjacent months as the month with the base consumption. Substantial errors may arise if missing bills are not accounted for.

The base consumption can be further analyzed by calculating the electrical load factor (ELF) associated with this consumption. If the base demand had operated 24 h per day, then base consumption would be

$$33.0 \text{ kW} \times 24 \text{ h/day} = 792 \text{ kWh/day}$$

But if the daily base consumption is 312 kWh/day, then the electrical load factor is

$$\text{ELF} = \frac{\text{Base consumption}}{\text{Base demand} \times 24} = \frac{312}{792} = 0.394 \text{ or } 39.4\%$$

In this example, the electrical load factor (39.4%) is higher than the occupancy factor (29.8%).

$$\text{Occupancy factor} = \frac{\text{Occupied hours}}{24 \text{ h} \times 7 \text{ days}} = \frac{50}{168} = 0.298 = 29.8\%$$

One reason for this difference may be that the lights are left energized beyond the occupied hours.

Because of the air-conditioning demand, summer demand is 46.8 kW. If this additional demand of 13.8 kW had operated 24 h each day, the summer extra would equal 13.8×24 or 331.2 kWh/day. The ratio of each summer month's excess as a percent of 331.2 kWh yields the summer electrical load factor. Summer ELFs higher than the occupancy factor indicate that air conditioning is not shut off as early as possible in the evening. Winter excess demand and consumption are analyzed in the same way to yield winter monthly ELFs.

Base load energy use is the amount of energy consumed independent of weather. When a building has electric cooling and no electric heating, the base load energy use is normally the energy consumed during the winter months. The opposite is true for heating. The annual estimated base load energy consumption can be obtained by establishing the average monthly consumption during the nonheating or noncooling months and multiplying by 12. For many buildings, subtracting the base load energy consumption from the total annual energy consumption yields an accurate estimate of the heating or cooling energy consumption. This approach is not valid when building use differs from summer to winter; when there is cooling in operation 12 months of the year; or when space heating is used during the summer months, as for reheat. In many cases, the base load energy use analysis can be improved by using hourly load data that may be available from the utility company. ELFs and occupancy factors can also be used instead of hourly energy profiles (Haberl and Komor 1990).

Although it is difficult to relate heating and cooling energy used in commercial buildings directly to the severity of the weather, several authors, including Spielvogel (1984) and Fels (1986), suggest that this is possible using a curve-fitting method to calculate the balance point of a building (the balance point of a building is discussed in Chapter 30 of the 1997 *ASHRAE Handbook—Fundamentals*). The pitfalls of such an analysis are (1) estimated rather than actual utility usage data are used, (2) the actual dates of the metered information must be used, together with average billing period weather data, and (3) building use and/or operation are not regular.

A more detailed breakdown of energy usage requires that some metered data be collected on a daily basis (winter days versus summer days, weekdays versus weekends) and that some hourly information be collected to develop profiles for night (unoccupied), morning warmup, day (occupied), and the shutdown period. This discussion presumes that submetering is not installed within the building. Individual metering of the various energy end uses

Table 5 Example of Analytical Method for Analyzing Electrical Operating Costs

Billing Date	Billing Days	Consumption, kWh			Air Conditioning, kWh			Demand, kW		
		Total Actual	Actual per Day	Base[b]	Difference	ELF, %	Excess	Actual	Winter Excess	Summer Excess
12Aug								46.2		0.6
11Sep	30	14,700	490	9,360	5,340	53.7	1,859	46.8		1.2
10Oct	29	10,860	374.5	9,048	1,812	18.9		46.8		1.2
10Nov	31	9,120	294.2	9,672	1,008[a]	11.7		45.6		0.0
06Dec	26	10,680	410.8	8,112	1,008[a]			33.0[c]	0.0	
09Jan	34	10,860	319.4	10,608	252			33.0	0.0	
13Feb	35	10,920	312[b]	10,920	0			33.6	0.6	
11Mar	27	8,700	322.2	8,424	276			33.0	0.0	
10Apr	30	10,140	338	9,360	780			33.6	0.6	
12May	32	11,020	344.4	9,984	1,036	9.8		45.6[d]		0.0
12Jun	31	11,760	379.4	9,672	2,088	20.3		46.8		1.2
13Jul	31	14,160	456.8	9,672	4,488	43.7	893	46.8		1.2
11Aug	29	14,340	494.5	9,048	5,292	55.2	1,937	46.8		1.2
10Sep	30	13,740	458	9,360	4,380	44.1	904	46.2		0.6
14Oct	34	12,120	356.5	10,608	1,512	13.4		45.6		0.0
10Nov	27	9,360	346.7	8,424	936			33.6	0.6	

[a]Estimated from corrected monthly consumption from Figure 6.
[b]Base electrical consumption = 312 kWh/day.

[c]Base winter demand = 33 kW.
[d]Base summer demand = 45.6 kW. Base summer excess demand = (45.6 − 33) = 12.6.

provides the energy manager with the information to apply energy management principles optimally. Ideally, each energy end use in Figure 4 would be metered separately. See also Chapter 39, Building Energy Monitoring.

Energy Conservation Opportunities

It is possible to quantitatively evaluate various energy conservation opportunities (ECOs) from end-use energy profiles. Important considerations in this process are as follows:

- System interaction
- Utility rate structure
- Payback
- Installation requirements
- Life of the measure
- Maintainability
- Impact on building operation and appearance

Accurate energy savings calculations can be made only if system interaction is allowed for and fully understood. Annual simulation models may be necessary to accurately estimate the interactions between various ECOs. The calculated remaining energy use should be verified against a separately calculated zero-based energy target.

Further, the actual energy cost avoidance may not be proportional to the energy saved, depending on the method of billing for energy used. Using average costs per unit of energy in calculating the energy cost avoidance of a particular measure is likely to result in incorrect values.

Figure 7 is a list of potential ECOs. A discussion of 118 ECOs can be found in PNL (1990).

In addition, previously implemented energy conservation measures should be evaluated—first, to ensure that they have remained effective, and second, to consider revising them to reflect changes in technology, building use, and/or energy cost.

Prioritize Resources

Once a list of ECOs is established, it should be evaluated, prioritized, and implemented. In establishing priorities, the capital cost, cost-effectiveness, and resources available must be considered. Factors involved in evaluating the desirability of a particular energy conservation retrofit measure are as follows:

- Rate of return (simple payback, life-cycle cost)
- Total savings (energy, cost avoidance)
- Initial cost (required investment)
- Other benefits (safety, comfort, improved system reliability, and improved productivity)
- Liabilities (increased maintenance costs and potential obsolescence)
- Risk of failure (confidence in predicted savings, rate of increase in energy costs, maintenance complications, and success of others with the same measures)

To reduce the risk of failure, documented performance of ECOs in similar situations should be obtained and evaluated. One common problem is that energy consumption for individual end uses is overestimated, and the predicted savings are not achieved. When doubt exists about the energy consumption, temporary measurements should be made and evaluated. Also, some owners are reluctant to implement ECOs because of past experiences with energy projects. The causes of past failures should be analyzed carefully to minimize the possibility of their reoccurrence.

The resources available to accomplish an energy conservation retrofit opportunity should include the following:

- Management attention, commitment, and follow-through
- Skills
- Manpower
- Investment capital

Boilers	Outside Air Ventilation
Boiler Auxiliaries	Ventilation Layout
Condensate Systems	Envelope Infiltration
Water Treatment	Weatherstripping
Fuel Acquisition	Caulking
Fuel Systems	Vestibules
Chillers	Elevator Shafts
Chiller Auxiliaries	Space Insulation
Steam Distribution	Vapor Barrier
Hydronic Systems	Glazing
Pumps	Infrared Reflection
Piping Insulation	Windows
Steam Traps	Window Treatment
Domestic Water Heating	Shading
Lavatory Fixtures	Vegetation
Water Coolers	Trombe Walls
Fire Protection Systems	Thermal Shutters
Swimming Pools	Surface Color
Cooling Towers	Roof Covering
Condensing Units	Lamps
City Water Cooling	Fixtures
Air Handling Units	Ballasts
Coils	Switch Design
Outside Air Control	Photo Controls
Balancing	Interior Color
Air Volume Control	Demand Limiting
Shutdown	Current Leakage
Air Purging	Power Factor
Minimizing Reheat	Transformers
Air Heat Recovery	Power Distribution
Filters	Cooking Practices
Dampers	Hoods
Humidification	Refrigeration
Duct Resistance	Dishwashing
System Air Leakage	Laundry
Diffusers	Vending Machines
System Interaction	Chiller Heat Recovery
System Reconfiguration	Heat Storage
Space Segregation	Time-of-Day Rates
Equipment Relocation	Computer Controls
Fan-Coil Units	Cogeneration
Heat Pumps	Active Solar Systems
Radiators	Staff Training
System Infiltration	Occupant Indoctrination
Relief Air	Documentation
Space Heaters	Management Structure
Controls	Financial Practices
Thermostats	Building Geometry
Setback	Space Planning
Instrumentation	

Fig. 7 Potential Energy Conservation Opportunities

ECOs may be financed with the following:

- Profit/investment
- Loans
- Rearrangement of budget priorities
- Energy savings
- Shared savings plans with outside firms and investors
- Utility incentives
- Grant programs
- Tax credits, deductions, etc.
- Donations

When all of these considerations are weighed and a prioritized list of recommendations is developed, a report should be prepared for management. Each recommendation should include the following:

- Present condition of the system or equipment to be modified
- Recommended action
- Who should accomplish the action
- Necessary documentation or follow-up required
- Potential interferences to successful completion of the recommendation

- Staff effort required
- Risk of failure
- Interactions with other end uses and ECOs
- Economic analysis (including payback, investment cost, and estimated savings figures) using corporate economic evaluation criteria
- Schedule for implementation

The energy manager must be prepared to sell the plans to upper management. Energy conservation measures must generally be financially justified if they are to be adopted. Every organization has limited funds available and must use these funds in the most effective way. The energy manager is competing with others in the organization for the same funds. A successful plan must be presented in a form that is easily understood by the decision makers. Finally, the energy manager must present nonfinancial benefits, such as improved product quality or the possibility of postponing other expenditures.

Accomplish Measures

Following approval by management, the energy manager directs the completion of energy conservation retrofit measures selected from the above prioritized list. If utility rebates are used, the necessary approvals should be acquired before proceeding with the work. Certain measures require that an architect or engineer prepare plans and specifications for the retrofit work. The package of services required usually includes drawings, specifications, assistance in obtaining competitive bids, evaluation of the bids, selection of the best bid, construction observation, final check-out, and assistance in training personnel in the proper application of the revisions.

Maintain Measures

Once energy conservation measures are underway, procedures need to be established to record, on a frequent and regular basis, energy consumption and costs for each building and/or end-use category in a manner consistent with functional cost accountability. Additional metering may be needed to monitor the energy consumption accurately. Metering can be in the form of devices that automatically read and transmit data to a central location or in the form of less expensive metering devices that require regular readings by building maintenance and/or security personnel. Many energy managers find it beneficial to collect energy consumption information hourly. Data may also be obtained from the utility. However, if the energy manager is not able to evaluate data as frequently as it is collected, it may be more practical to collect data less frequently. The energy manager should review data while it is current and take immediate action if profiles indicate a trend in the wrong direction. Such trends could be caused by uncalibrated controls, changes in operating practices, or mechanical system failure—all problems that should be isolated and corrected as soon as possible.

Energy Accounting

The energy manager continues the meter reading, monitoring, and tabulation of facility energy use and profiles. These tabulations indicate the cost of energy management efforts and the resulting energy cost avoidance. In conjunction with this effort, the energy manager periodically reviews pertinent utility rates, rate structures, and their trends as they affect the facility. Many utilities maintain free mailing lists for changes in their rate tariffs. The energy manager provides periodic reports of the energy management efforts to top management, summarizing the work accomplished, the cost-effectiveness of the work, the plans and suggested budget for future work, and projections of future utility costs. If energy conservation measures are to retain their cost-effectiveness, continued monitoring and periodic reauditing are necessary, because many energy

conservation measures become less effective if they are not carefully monitored and maintained.

BUILDING EMERGENCY ENERGY USE REDUCTION

The need for occasional reductions in energy use during specific periods has become more common due to rising energy costs and supply reductions (which may be voluntary or mandatory) or equipment failures. Emergency periods include short-term shortages of a particular energy source(s) brought about by natural disasters, extreme weather conditions, utility system equipment disruptions, labor strikes, failures in building systems or equipment, self-imposed cut-backs in energy use, world political activities, or other forces beyond the control of the building owner and operator. This section provides information to help building owners and operators maintain near-normal operation of facilities during energy emergencies.

The following terms are applicable to such programs:

Energy Emergency. A period in which energy supply reductions and/or climatic and natural forces or equipment failures preclude the normal operation of a building and necessitate a reduction in building energy use.

Level of Energy Emergency. A measure of the severity of the emergency that calls for the implementation of various energy use reduction measures. Typical levels include the following:

- **Green (Normal)**—All occupant or building functions maintained and systems operating under the normal operating circumstances for the building
- **Blue**—All or most of the occupant or building functions maintained, with reductions in building system output that may result in borderline occupancy comfort
- **Yellow**—A minor reduction of occupant or building functions with reductions in output
- **Red**—A major reduction of occupant or building functions and systems that barely sustains building occupancy capability
- **Black**—The orderly shutdown of systems and occupancy that maintains the minimum building conditions needed to protect the building and its systems

Implementation of Energy Reductions

Each building owner, lessor, and operator should use the energy team approach and identify an individual with the necessary authority to review and fit recommendations into a plan for the particular building. For each class of emergency, the responsible party recommends a specific plan to reduce building energy use that still maintains the best building environment under the given circumstances. Implementation of the particular recommendations should then be coordinated through the building operator with assistance from the responsible party and the building occupants, as necessary. The plan should be tested occasionally.

Depending on the type of building, its use, the energy source(s) for each function, and local conditions such as climate and availability of other similar buildings, the following steps should be taken in developing a building energy plan:

1. Develop a list of measures that are applicable to the building.
2. Estimate the amount and type of energy savings for each measure and appropriate combination of measures (e.g., account for air-conditioning savings resulting from reduced lighting and other internal loads). Tabulate demand and usage savings separately for response to different types of emergencies.
3. For the various levels of energy emergency, develop a plan that would maintain the best building environment under the circumstances. Include both short- and long-term measures in the plan. Operational changes may be implemented quickly and prove adequate for short-term emergencies.

4. Experiment with the plan developed, record energy consumption and demand reduction data, and revise the plan as necessary. Much of the experimentation may be done on weekends to minimize disruptive effects.
5. Meet with the local utility company(s) to review the plan.

In the emergency planning process, some measures can be implemented permanently. Depending on the level of energy emergency and the building priority, the following actions may be considered in developing the plan for emergency energy reduction in the building:

• Change operating hours
• Move personnel into other building areas (consolidation)
• Shut off nonessential equipment

Thermal Envelope

• Use all existing blinds, draperies, and window coverings during summer
• Install interior window insulation
• Caulk and seal around unused exterior doors and windows
• Install solar shading devices in summer
• Seal all unused vents and ducts to outside

Heating, Ventilating, and Air-Conditioning Systems and Equipment

• Modify controls or control set points to raise and lower temperature and humidity as necessary
• Shut off or isolate all nonessential equipment
• Tune up equipment
• Lower thermostat set points in winter
• Raise chilled water temperature
• Lower hot water temperature (*Note*: Keep hot water temperature higher than 145°F if a gas boiler is used.)
• Reduce the level of reheat or eliminate it in winter
• Reduce or eliminate ventilation and exhaust airflow
• Raise thermostat set points in summer
• Reduce the amount of recooling in summer

Lighting Systems

• Remove lamps or reduce lamp wattage
• Use task lighting where appropriate
• Move building functions to exterior or daylight areas
• Turn off electric lights in areas with adequate natural light
• Lower luminaire height where appropriate
• Wash all lamps and luminaires
• Replace fluorescent ballasts with high-efficiency or multilevel ballasts

• Revise building cleaning and security procedures to minimize lighting periods
• Consolidate parking and turn off unused parking security lighting

Special Equipment

• Take transformers off-line during periods of nonuse
• Shut off or regulate the use of vertical transportation systems
• Shut off unused or unnecessary equipment, such as photocopiers, music systems, and computers
• Reduce or turn off hot water supply

Building Operation Demand Reduction

• Sequence or interlock heating or air-conditioning systems
• Disconnect or turn off all nonessential loads
• Turn off some lights
• Preheat or precool prior to the emergency period

REFERENCES

ASHRAE. 1984. Standard methods of measuring and expressing building energy performance. ANSI/ASHRAE *Standard* 105-1984 (RA 90).

ASHRAE. 1996. The HVAC commissioning process. *Guideline* 1-1996.

DOE/EIA. 1995. Household energy consumption and expenditures 1993. Part 1: National Data. DOE/EIA-0321/1(93).

DOE/EIA. 1997. Nonresidential buildings energy consumption survey: Characteristics of commercial buildings 1995. DOE/EIA-0246(95).

DOE/EIA. 1998. Nonresidential buildings energy consumption survey: Commercial buildings consumption and expenditures 1995. DOE/EIA-0318(95).

Fels. M. 1986. *Energy and buildings*, Vol. 9, Nos. 1 and 2, Special issue devoted to the Princeton Scorekeeping Method (PRISM).

Haberl, J.S. and P.S. Komor. 1990a. Improving energy audits—How daily and hourly consumption data can help, Part 1. *ASHRAE Journal* 90(8):26-33.

Haberl, J.S. and P.S. Komor. 1990b. Improving energy audits—How daily and hourly consumption data can help, Part 2. *ASHRAE Journal* 90(9):26-36.

MacDonald, J.M. and D.M. Wasserman. 1989. Investigation of metered data analysis methods for commercial and related buildings. ORNL/CON-279. Oak Ridge National Laboratories, Oak Ridge, TN.

Mazzucchi, R.P. 1992. A guide for analyzing and reporting building characteristics and energy use in commercial buildings. *ASHRAE Transactions* 98(1):1067-80.

PNL (Pacific Northwest Laboratories). 1990. Architect's and engineer's guide to energy conservation in existing buildings. Vol. 2, Chapter 1. DOE/RL/ 01830 P-H4.

Spielvogel, L.G. 1984. One approach to energy use evaluation. *ASHRAE Transactions* 90(1):424-35.

Stewart, R., S. Stewart, and R. Joy. 1984. Energy audit input procedures and forms. *ASHRAE Transactions* 90(1A):350-62.

Turner, W.C. 1993. *Energy management handbook.* John C. Wiley and Sons, New York.

OWNING AND OPERATING COSTS

OWNING and operating cost information for the HVAC system should be part of the investment plan of a facility. This information can be used for preparing annual budgets, managing assets, and selecting design options. Table 1 shows a representative form that summarizes these costs.

A properly engineered system must also be economical. Economics are difficult to assess because of the complexities surrounding the effective management of money and the inherent difficulty of predicting future operating and maintenance expenses. Complex tax structures and the time value of money can affect the final engineering decision. This does not imply the use of either the cheapest or the most expensive system; instead, it demands an intelligent analysis of the financial objectives and requirements of the owner.

Certain tangible and intangible costs or benefits must also be considered when assessing owning and operating costs. Local codes may require highly skilled or certified operators for specific types of equipment. This could be a significant cost over the life of the system. Similarly, such intangible items as aesthetics, acoustics, comfort, safety, security, flexibility, and environmental impact may be important to a particular building or facility.

OWNING COSTS

The following elements must be established to calculate annual owning costs: (1) initial cost, (2) analysis or study period, (3) interest or discount rate, and (4) other periodic costs such as insurance, property taxes, refurbishment, or disposal fees. Once established, these elements are coupled with operating costs to develop an economic analysis, which may be a simple payback evaluation or an in-depth analysis such as outlined in the section on Economic Analysis Techniques.

Initial Cost

Major decisions affecting annual owning and operating costs for the life of the building must generally be made prior to the completion of contract drawings and specifications. To achieve the best performance and economics, comparisons between alternate methods of solving the engineering problems peculiar to each project must be made in the early stages of design. Oversimplified estimates can lead to substantial errors in evaluating the system.

A thorough understanding of the installation costs and accessory requirements must be established. Detailed lists of materials, controls, space and structural requirements, services, installation labor, and so forth can be prepared to increase the accuracy in preliminary cost estimates. A reasonable estimate of the capital cost of components may be derived from cost records of recent installations of comparable design or from quotations submitted by manufacturers and contractors. Table 2 shows a representative checklist for initial costs.

The preparation of this chapter is assigned to TC 1.8, Owning and Operating Costs.

Table 1 Owning and Operating Cost Data and Summary

OWNING COSTS

I. Initial Cost of System _____

II. Periodic Costs

    A. Income taxes _____

    B. Property taxes _____

    C. Insurance _____

    D. Rent _____

    E. Other periodic costs _____

    Total Periodic Costs

III. Replacement Cost _____

IV. Salvage Value _____

    Total Owning Costs _____

OPERATING COSTS

V. Annual Utility, Fuel, Water, etc., Costs

    A. Utilities

        1. Electricity _____

        2. Natural gas _____

        3. Water/Sewer _____

        4. Purchased steam _____

        5. Purchased hot/chilled water _____

    B. Fuels

        1. Propane _____

        2. Fuel oil _____

        3. Diesel _____

        4. Coal _____

    C. On-site generation of electricity _____

    D. Other utility, fuel, water, etc., costs _____

    Total _____

VI. Annual Maintenance Allowances/Costs

    A. In-house labor _____

    B. Contracted maintenance service _____

    C. In-house materials _____

    D. Other maintenance allowances/costs _____

    Total _____

VII. Annual Administration Costs _____

    Total Annual Operating Costs _____

TOTAL ANNUAL OWNING AND OPERATING COSTS _____

Table 2   Initial Cost Checklist

Energy and Fuel Service Costs

Fuel service, storage, handling, piping, and distribution costs
Electrical service entrance and distribution equipment costs
Total energy plant

Heat-Producing Equipment

Boilers and furnaces
Steam-water converters
Heat pumps or resistance heaters
Makeup air heaters
Heat-producing equipment auxiliaries

Refrigeration Equipment

Compressors, chillers, or absorption units
Cooling towers, condensers, well water supplies
Refrigeration equipment auxiliaries

Heat Distribution Equipment

Pumps, reducing valves, piping, piping insulation, etc.
Terminal units or devices

Cooling Distribution Equipment

Pumps, piping, piping insulation, condensate drains, etc.
Terminal units, mixing boxes, diffusers, grilles, etc.

Air Treatment and Distribution Equipment

Air heaters, humidifiers, dehumidifiers, filters, etc.
Fans, ducts, duct insulation, dampers, etc.
Exhaust and return systems

System and Controls Automation

Terminal or zone controls
System program control
Alarms and indicator system

Building Construction and Alteration

Mechanical and electric space
Chimneys and flues
Building insulation
Solar radiation controls
Acoustical and vibration treatment
Distribution shafts, machinery foundations, furring

Analysis Period

The time frame over which an economic analysis is performed greatly affects the results of the analysis. The analysis period is usually determined by specific analysis objectives, such as length of planned ownership or loan repayment period. The chosen analysis period is often unrelated to the equipment depreciation period or service life, although these factors may be important in the analysis.

Table 3 lists representative estimates of the service life of various system components. Service life as used here is the time during which a particular system or component remains in its original service application. Replacement may be for any reason, including, but not limited to, failure, general obsolescence, reduced reliability, excessive maintenance cost, and changed system requirements due to such influences as building characteristics, energy prices, or environmental considerations.

Depreciation periods are usually set by federal, state, or local tax laws, which change periodically. Applicable tax laws should be consulted for more information on depreciation.

Interest or Discount Rate

Most major economic analyses consider the opportunity cost of borrowing money, inflation, and the time value of money. **Opportunity cost** of money reflects the earnings that investing (or loaning) the money can produce. **Inflation** (price escalation) decreases

the purchasing or investing power (value) of future money because it can buy less in the future. **Time value** of money reflects the fact that money received today is more useful than the same amount received a year from now, even with zero inflation, because the money is available earlier for reinvestment.

The cost or value of money must also be considered. When borrowing money, a percentage fee or interest rate must normally be paid. However, the interest rate may not necessarily be the correct cost of money to use in an economic analysis. Another factor, called the **discount rate**, is more commonly used to reflect the true cost of money. Discount rates used for analyses vary depending on individual investment, profit, and other opportunities. Interest rates, in contrast, tend to be more centrally fixed by lending institutions.

To minimize the confusion caused by the vague definition and variable nature of discount rates, the U.S. government has specified particular discount rates that can be used in economic analyses relating to federal expenditures. These discount rates are updated annually (Lippiatt 1994, OMB 1972, NIST) but may not be appropriate for private sector economic analyses.

Periodic Costs

Regularly or periodically recurring costs include insurance, property taxes, income taxes, rent, refurbishment expenses, disposal fees (e.g., refrigerant recycling costs), occasional major repair costs, and decommissioning expenses.

Insurance. Insurance reimburses a property owner for a financial loss so that equipment can be repaired or replaced. Insurance often indemnifies the owner from liability as well. Financial recovery may include replacing income, rents, or profits lost due to property damage.

Some of the principal factors that influence the total annual insurance premium are building size, construction materials, amount and size of mechanical equipment, geographic location, and policy deductibles. Some regulations set minimum required insurance coverages and premiums that may be charged for various forms of insurable property.

Property Taxes. Property taxes differ widely and may be collected by one or more agencies, such as state, county, or local governments or special assessment districts. Furthermore, property taxes may apply to both real (land, buildings) and personal (everything else) property. Property taxes are most often calculated as a percentage of assessed value but are also determined in other ways, such as fixed fees, license fees, registration fees, etc. Moreover, definitions of assessed value vary widely in different geographic areas. Tax experts should be consulted for applicable practices in a given area.

Income Taxes. Taxes are generally imposed in proportion to net income, after allowance for expenses, depreciation, and numerous other factors. Special tax treatment is often granted to encourage certain investments. Income tax experts can provide up-to-date information on income tax treatments.

Additional Periodic Costs. Examples of additional costs include changes in regulations that require unscheduled equipment refurbishment to eliminate use of hazardous substances, and disposal costs for such substances. Moreover, at the end of the equipment's useful life there may be negative salvage value (i.e., removal, disposal, or decommissioning costs).

OPERATING COSTS

Operating costs are those incurred by the actual operation of the system. They include costs of fuel and electricity, wages, supplies, water, material, and maintenance parts and services. Chapter 30 of the 1997 *ASHRAE Handbook—Fundamentals* outlines how fuel and electrical requirements are estimated. Note that total energy consumption cannot generally be multiplied by a per unit energy cost to arrive at annual utility cost.

Table 3 Estimates of Service Lives of Various System Components[a]

Equipment Item	Median Years	Equipment Item	Median Years	Equipment Item	Median Years
Air conditioners		Air terminals		Air-cooled condensers	20
Window unit	10	Diffusers, grilles, and registers	27	Evaporative condensers	20
Residential single or split package	15	Induction and fan-coil units	20	Insulation	
Commercial through-the-wall	15	VAV and double-duct boxes	20	Molded	20
Water-cooled package	15	Air washers	17	Blanket	24
Heat pumps		Ductwork	30	Pumps	
Residential air-to-air	15[b]	Dampers	20	Base-mounted	20
Commercial air-to-air	15	Fans		Pipe-mounted	10
Commercial water-to-air	19	Centrifugal	25	Sump and well	10
Roof-top air conditioners		Axial	20	Condensate	15
Single-zone	15	Propeller	15	Reciprocating engines	20
Multizone	15	Ventilating roof-mounted	20	Steam turbines	30
Boilers, hot water (steam)		Coils		Electric motors	18
Steel water-tube	24 (30)	DX, water, or steam	20	Motor starters	17
Steel fire-tube	25 (25)	Electric	15	Electric transformers	30
Cast iron	35 (30)	Heat exchangers		Controls	
Electric	15	Shell-and-tube	24	Pneumatic	20
Burners	21	Reciprocating compressors	20	Electric	16
Furnaces		Package chillers		Electronic	15
Gas- or oil-fired	18	Reciprocating	20	Valve actuators	
Unit heaters		Centrifugal	23	Hydraulic	15
Gas or electric	13	Absorption	23	Pneumatic	20
Hot water or steam	20	Cooling towers		Self-contained	10
Radiant heaters		Galvanized metal	20		
Electric	10	Wood	20		
Hot water or steam	25	Ceramic	34		

Source: Data obtained from a survey of the United States by ASHRAE Technical Committee TC 1.8 (Akalin 1978).
[a] See Lovvorn and Hiller (1985) and Easton Consultants (1986) for further information.
[b] Data updated by TC 1.8 in 1986.

Electrical Energy

Fundamental changes in the purchase of electrical energy are occurring in the United States, which is opening access to and eventually deregulating the electric energy industry. Individual electric utility rates and regulations may vary widely during this period of deregulation. Consequently, electrical energy providers and brokers or marketers need to be contacted to determine the most competitive supplier. Contract conditions need to be reviewed carefully to be sure that the service will suit the purchaser's requirements.

The total cost of electrical energy is usually a combination of several components: energy consumption charges, fuel adjustment charges, special allowances or other adjustments, and demand charges.

Energy Consumption Charges. Most utility rates have step rate schedules for consumption, and the cost of the last unit of energy consumed may be substantially different from that of the first. The last unit may be cheaper than the first because the fixed costs to the utility may already have been recovered from earlier consumption costs. Alternatively, the last unit of energy may be sold at a higher rate to encourage conservation.

To reflect time-varying operating costs, some utilities charge different rates for consumption according to the time of use and season; typically, costs rise toward the peak period of use. This may justify the cost of shifting the load to off-peak periods.

Fuel Adjustment Charge. Due to substantial variations in fuel prices, electric utilities may apply a fuel adjustment charge to recover costs. This adjustment may not be reflected in the rate schedule. The fuel adjustment is usually a charge per unit of energy and may be positive or negative depending on how much of the actual fuel cost is recovered in the energy consumption rate.

Power plants with multiple generating units that use different fuels typically have the greatest effect on this charge (especially during peak periods, when more expensive units must be brought on-line). Although this fuel adjustment charge can vary monthly, the utility should be able to estimate an average annual or seasonal fuel adjustment for calculations.

Allowances or Adjustments. Special allowances may be available for customers who can receive power at higher voltages or for those who own transformers or similar equipment. Special rates may be available for specific interruptible loads such as domestic water heaters.

Certain facility electrical systems may produce a low power factor, which means that the utility must supply more current on an intermittent basis, thus increasing their costs. These costs may be passed on as an adjustment to the utility bill if the power factor is below a level established by the utility. The power factor is the ratio of active (real) kilowatt power to apparent (reactive) kVA power.

When calculating power bills, utilities should be asked to provide detailed cost estimates for various consumption levels. The final calculation should include any applicable special rates, allowances, taxes, and fuel adjustment charges.

Demand Charges. Electric rates may also have demand charges based on the customer's peak kilowatt demand. While consumption charges typically cover the utility's operating costs, demand charges typically cover the owning costs.

Demand charges may be formulated in a variety of ways:

1. Straight charge—cost per kilowatt per month, charged for the peak demand of the month.
2. Excess charge—cost per kilowatt above a base demand (e.g., 50 kW), which may be established each month.

3. Maximum demand (ratchet)—cost per kilowatt for the maximum annual demand, which may be reset only once a year. This established demand may either benefit or penalize the owner.
4. Combination demand—cost per hour of operation of the demand. In addition to a basic demand charge, utilities may include further demand charges as demand-related consumption charges.

The actual demand represents the peak energy use averaged over a specific period, usually 15, 30, or 60 min. Accordingly, high electrical loads of only a few minutes' duration may never be recorded at the full instantaneous value. Alternatively, peak demand is recorded as the average of several consecutive short periods (i.e., 5 min out of each hour).

The particular method of demand metering and billing is important when load shedding or shifting devices are considered. The portion of the total bill attributed to demand may vary greatly, from 0% to as high as 70%.

Natural Gas

Rates. Conventional natural gas rates are usually a combination of two main components: (1) utility rate for gas consumption and (2) purchased gas adjustment (PGA) charges.

Although gas is usually metered by volume, it is often sold by energy content (therm). The utility rate is the amount the local distribution company charges per unit of energy to deliver the gas to a particular location. This rate may be graduated in steps; the first 100 therms of gas consumed may not be the same price as the last 100 therms. The PGA is an adjustment for the cost of the gas per unit of energy to the local utility. It is similar to the electric fuel adjustment charge. The total cost per therm is then the sum of the appropriate utility rate and the PGA, plus taxes and other adjustments.

Interruptible Gas Rates and Contract/Transport Gas. Large industrial plants usually have the ability to burn alternate fuels at the plant and can qualify for special interruptible gas rates. During peak periods of severe cold weather, these customers may be curtailed by the gas utility and may have to switch to propane, fuel oil, or some other backup fuel. The utility rate and PGA are usually considerably cheaper for these interruptible customers than they are for firm rate (noninterruptible) customers.

Deregulation of the natural gas industry allows end users to negotiate for gas supplies on the open market. The customer actually contracts with a gas producer or broker and pays for the gas at the source. Transport fees must be negotiated with the pipeline companies carrying the gas to the customer's local gas utility. This can be a very complicated administrative process and is usually economically feasible for large gas users only. Some local utilities have special rates for delivering contract gas volumes through their system; others simply charge a standard utility fee (PGA is not applied because the customer has already negotiated with the supplier for the cost of the fuel itself).

When calculating natural gas bills, be sure to determine which utility rate and PGA and/or contract gas price is appropriate for the particular interruptible or firm rate customer. As with electric bills, the final calculation should include any taxes, prompt payment discounts, or other applicable adjustments.

Other Fossil Fuels

Propane, fuel oil, and diesel are examples of other fossil fuels in widespread use. Calculating the cost of these fuels is usually much simpler than calculating typical utility rates. When these fuels are used, other items that can affect owning or operating costs must be considered.

The cost of the fuel itself is usually a simple charge per unit volume or per unit mass. The customer is free to negotiate for the best price. However, trucking or delivery fees must also be included in final calculations. Some customers may have their own transport

trucks, while most shop around for the best delivered price. If storage tanks are not customer-owned, rental fees must be considered. Periodic replacement of diesel-type fuels may be necessary due to storage or shelf-life limitations and must also be considered. The final fuel cost calculation should include any of these costs that are applicable, as well as the appropriates taxes.

MAINTENANCE COSTS

The quality of maintenance and maintenance supervision can be a major factor in the energy cost of a building. Chapter 37 covers the maintenance, maintainability, and reliability of systems. Dohrmann and Alereza (1986) obtained maintenance costs and HVAC system information from 342 buildings located in 35 states in the United States. In 1983 U.S. dollars, data collected showed a mean HVAC system maintenance cost of $0.32 per square foot per year, with a median cost of $0.24/ft^2 per year. The age of the building has a statistically significant but minor effect on HVAC maintenance costs. When analyzed by geographic location, the data revealed that location does not significantly affect maintenance costs. Analysis also indicated that building size is not statistically significant in explaining cost variation.

The type of maintenance program or service agency that the building management contracts for can also have a significant effect on total HVAC maintenance costs. While extensive or thorough routine and preventive maintenance programs cost more to administer, they usually produce benefits such as extended equipment life, improved reliability, and less system downtime.

Estimating Maintenance Costs

Total HVAC maintenance cost for existing buildings with various types of equipment can be estimated using data from Table 4 and the method described below. For equipment types not included in Table 4, or as an alternate estimating method, contact equipment manufacturers or firms that provide maintenance contracts on buildings with similar equipment.

Several important limitations of the data presented here should be noted:

• Only selected data from Table 12 of the report by Dohrmann and Alereza (1986) are presented here.
• Data were collected for office buildings only.

Table 4 Annual HVAC Maintenance Cost Adjustment Factors (in dollars per square foot, 1983 U.S. dollars)

Age Adjustment	$+0.0018n$
Heating Equipment (h)	
Water tube boiler	$+0.0077$
Cast iron boiler	$+0.0094$
Electric boiler	-0.0267
Heat pump	-0.0969
Electric resistance	-0.133
Cooling Equipment (c)	
Reciprocating chiller	-0.04
Absorption chiller (single stage)[a]	$+0.1925$
Water-source heat pump	-0.0472
Distribution System (d)	
Single zone	$+0.0829$
Multizone	-0.0466
Dual duct	-0.0029
Constant volume	$+0.0881$
Two-pipe fan coil	-0.0277
Four-pipe fan coil	$+0.0580$
Induction	$+0.0682$

[a]These results pertain to buildings with older, single-stage absorption chillers. The data from the survey are not sufficient to draw inferences about the costs of HVAC maintenance in buildings equipped with new absorption chillers.

- Data measure total HVAC building maintenance costs, not the costs associated with maintaining particular items or individual pieces of equipment.
- The cost data are not intended for use in selecting new HVAC equipment or systems. This information is for equipment and systems already in place and may not be representative of the maintenance costs expected with newer equipment.

This method assumes that the base HVAC system in the building consists of fire-tube boilers for heating equipment, centrifugal chillers for cooling equipment, and VAV distribution systems. The total annual building HVAC maintenance cost for this system is $0.3338/ft². Adjustment factors from Table 4 are then applied to this base cost to account for building age and variations in type of HVAC equipment as follows:

$$C = \text{Total annual building HVAC maintenance cost (\$/ft}^2)$$
$$= \text{Base system maintenance costs}$$
$$+ \text{(Age adjustment factor)} \times \text{(age in years } n)$$
$$+ \text{Heating system adjustment factor } h$$
$$+ \text{Cooling system adjustment factor } c$$
$$+ \text{Distribution system adjustment factor } d$$
$$C = 0.3338 + 0.0018n + h + c + d$$

Example 1. Estimate the total annual building HVAC maintenance cost per square foot for a building that is 10 years old and has an electric boiler, a reciprocating chiller, and a constant volume distribution system.

$$C = 0.3338 + 0.0018(10) - 0.0267 - 0.04 + 0.0881$$
$$C = \$0.3732/\text{ft}^2 \text{ in 1983 dollars}$$

This estimate can be adjusted to current dollars by multiplying the maintenance cost estimate by the current Consumer Price Index (CPI) divided by the CPI in July 1983. In July 1983, the CPI was 100.1. Monthly CPI statistics are recorded in *Survey of Current Business* (U.S. Department of Commerce). This estimating method is limited to one equipment variable per situation. That is, the method can estimate maintenance costs for a building having either a centrifugal chiller or a reciprocating chiller, but not both. Assessing the effects of combining two or more types of equipment within a single category requires a more complex statistical analysis.

IMPACT OF REFRIGERANT PHASEOUTS

The production phaseout of many commonly used refrigerants has presented building owners with decisions regarding the replacement of existing equipment versus retrofit to alternative refrigerant compatibility. Several factors must be considered, including

- **Initial Cost.** New equipment may have a significantly higher installed cost than the retrofit of existing equipment. For example, the retrofit of an existing centrifugal chiller to operate on R-123 may cost 50% of the cost for a new chiller. The cost of rigging a new unit may significantly raise the installed cost, improving the first cost advantage of conversion.
- **Operating Costs.** The overall efficiency of new equipment is often substantially better than that of existing equipment, depending on age, usage, and level of maintenance performed over the life of the existing unit. In addition, conversion to alternative refrigerants may reduce capacity and/or efficiency.
- **Maintenance Costs.** The maintenance cost for new equipment is generally lower than that for existing equipment. The level of retrofit required to attain compatibility often includes replacement or remanufacture of major unit components, which can bring the maintenance and repair costs in line with those expected of new equipment.
- **Equipment Useful Life.** The impact of a retrofit on equipment useful life is determined by the extent of modification required.

The complete remanufacture of a unit should extend the remaining useful life to a level comparable to that of new equipment.

Decisions regarding the replacement of existing equipment or the conversion to alternative refrigerants offer opportunities to improve overall system efficiency. Reduced capacity requirements and the introduction of new technologies such as variable-speed drives and microprocessor-based controllers can substantially reduce annual operating costs and significantly improve a project's economic benefit.

Information should be gathered to complete Table 1 for each alternative. The techniques described in the section on Economic Analysis Techniques may then be applied to compare the relative values of each option.

OTHER ISSUES

Financing Alternatives

Alternative financing is commonly used in the third-party funding of projects, particularly retrofit projects, and is variously called privatization, third-party financing, energy services, outsourcing, performance contracting, energy savings performance contracting (ESPC), or innovative financing. In these programs, an outside party performs an energy study to identify or quantify attractive energy-saving retrofit projects and then (to varying degrees) designs, builds, and finances the retrofit program on behalf of the owner or host facility. These contracts range in complexity from simple projects such as lighting upgrades to more detailed projects involving all aspects of energy consumption and facility operation.

Alternative financing can be used to accomplish any or all of the following objectives:

- Upgrade capital equipment
- Provide for maintenance of existing facilities
- Speed project implementation
- Conserve or defer capital outlay
- Save energy
- Save money

The benefits of alternative financing are not free. In general terms, these financing agreements transfer the risk of attaining future savings from the owner to the contractor. The contractor will want to be paid for assuming this risk. In addition, these innovative owning and operating cost reduction approaches have important tax consequences that should be investigated on a case-by-case basis.

There are many variations of the basic arrangements and nearly as many terms to define them. Common nomenclature includes guaranteed savings (performance-based), shared savings, paid from savings, guaranteed savings loans, capital leases, municipal leases, and operating leases. For more information, refer to the U.S. Department of Energy's Web page and to the DOE-sponsored document entitled "International Performance Measurement and Verification Protocol."

A few examples of alternative financing techniques follow.

Leasing. Among the most common methods of alternative financing is the lease arrangement. In a true lease or lease-purchase arrangement, outside financing provides capital for the construction of a facility. The institution then leases the facility at a fixed monthly charge and assumes responsibility for fuel and personnel costs associated with its operation.

Leasing is also commonly available for individual pieces of equipment or retrofit systems and often includes all design and installation costs. Equipment suppliers or independent third parties retain ownership of new equipment and lease it to the user.

Outsourcing. For a cogeneration, steam, or chilled water plant, either a lease or an energy output contract can be used. An energy output contract enables a private company to provide all the capital and operating costs, such as personnel and fuel, while the host facility purchases energy from the operating company at a variable monthly charge.

Energy Savings. Retrofit projects that lower energy usage create an income stream that can be used to amortize the investment. In **paid-from-savings** programs, utility payments remain constant over a period of years while the contractor is paid out of savings until the project is amortized. In **shared savings** programs, the institution receives a percentage of savings over a longer period of years until the project becomes its property. In a **guaranteed savings** program, the owner retains all the savings and is guaranteed that a certain level of savings will be attained. A portion of the savings is used to amortize the project. In any type of energy savings project, building operation and utilization can have a major impact on the amount of savings actually realized.

Low-Interest Financing. In this arrangement, the supplier offers the equipment with special financing arrangements at below-market interest rates.

Cost Sharing. Several variations of cost-sharing programs exist. In some instances, two or more groups jointly purchase and share new equipment or facilities, thereby increasing utilization of the equipment and improving the economic benefits for both parties. In other cases, equipment suppliers or independent third parties (such as utilities) who receive an indirect benefit may share part of the equipment or project cost to establish a market foothold for the product.

District Energy Service

District energy service is increasingly available to building owners; district heating and cooling eliminates most on-site heating and cooling equipment. A third party produces treated water or steam and pipes it from a central plant directly to the building. The building owner then pays a metered rate for the energy that is used.

A cost comparison of district energy service versus on-site generation requires careful examination of numerous, often site-specific, factors extending beyond demand and energy charges for fuel. District heating and cooling eliminates or minimizes most costs associated with the installation, maintenance, administration, repair and operation of on-site heating and cooling equipment. Specifically, the costs associated with providing water, water treatment, specialized maintenance services, insurance, staff time, space to house on-site equipment, and structural additions needed to support equipment should be considered. Costs associated with auxiliary equipment, which represent 20 to 30% of the total plant annual operating costs, should also be included.

Any analysis that fails to include all the associated costs does not give a clear picture of the building owner's heating and cooling alternatives. In addition to the tangible costs, there are a number of other factors that should be considered, such as convenience, risk, environmental issues, flexibility, and backup.

On-Site Electric Generation

On-site electric generation covers a broad range of applications, from emergency backup to power for a single piece of equipment to an on-site power plant supplying 100% of the facility's electrical power needs. A variety of system types and fuel sources are available, but the economic principles described in this chapter apply equally to all of them. Other chapters (e.g., Chapter 7, Cogeneration Systems and Engine and Turbine Drives, and Chapter 33, Solar Energy Equipment, of the 1996 *ASHRAE Handbook—Systems and Equipment*) may be helpful in describing system details.

An economic study of on-site electrical power generation should include consideration of all owning, operating, and maintenance costs. Typically, on-site generation is capital intensive (i.e., high first cost) and therefore requires a high utilization rate to produce savings adequate to support the investment. High utilization rates

mean high run time, which requires planned maintenance and careful operation.

Owning costs include any related systems required to adapt the building to on-site power generation. Additional equipment is required if the building will also utilize purchased power from a utility. The costs associated with shared equipment should also be considered. For example, if the power source for the generator is a steam turbine, and a hot water boiler would otherwise be used to meet the HVAC demand, the boiler would need to be a larger, high-pressure steam boiler with a heat exchanger to meet the hot water needs. In addition to higher first cost, operation and maintenance costs for the boiler are increased due to the increased operating hours.

Consideration must also be given to the costs of an initial investment and ongoing inventory of spare parts. Most equipment manufacturers provide a recommended spare parts list as well as recommended maintenance schedules. Typical maintenance schedules consist of daily, weekly, and monthly routine maintenance and periodic major overhauls. Major overhaul frequency depends on equipment utilization and requires taking the equipment off-line. The cost of either lost building utilization or the provision of electricity from an alternate source during the shutdown should be considered.

ECONOMIC ANALYSIS TECHNIQUES

Analysis of overall owning and operating costs and comparisons of alternatives require an understanding of the cost of lost opportunities, inflation, and the time value of money. This process of economic analysis of alternatives falls into two general categories: simple payback analysis and detailed economic analyses (life-cycle cost analyses).

A simple payback analysis reveals options that have short versus long paybacks. Often, however, alternatives are similar and have similar paybacks. For a more accurate comparison, a more comprehensive economic analysis is warranted. Many times it is appropriate to have both a simple payback analysis and a detailed economic analysis. The simple payback analysis shows which options should not be considered further, and the detailed economic analysis determines which of the viable options are the strongest. The strongest options can be accepted or further analyzed if they include competing alternatives.

Simple Payback

In the simple payback technique, a projection of the revenue stream, cost savings, and other factors is estimated and compared to the initial capital outlay. This simple technique ignores the cost of borrowing money (interest) and lost opportunity costs. It also ignores inflation and the time value of money.

Example 2. Equipment item 1 costs $10,000 and will save $2000 per year in operating costs; equipment item 2 costs $12,000 and saves $3000 per year. Which item has the best simple payback?

Item 1 — $10,000/($2000/yr) = 5 year simple payback

Item 2 — $12,000/($3000/yr) = 4 year simple payback

Because the analysis of equipment for the duration of its realistic life can produce a much different result, the simple payback technique should be used with caution.

More Sophisticated Economic Analysis Methods

This section describes a few of the detailed economic techniques available. These techniques are used to examine all costs and incomes to be incurred over the analysis period (1) in terms of their present value (i.e., today's or initial year's value, also called constant value); (2) in terms of equal periodic costs or payments (uniform annualized costs); or (3) in terms of periodic cash flows (e.g., monthly or annual cash flows). Each method provides a slightly different insight. The present value method allows easy comparison of alternatives over the analysis period chosen. The uniform annualized costs method allows comparison of average annual costs of different options. The cash flow method allows comparison of actual cash flows rather than average cash flows; it can identify periods of overall positive and negative cash flow, which is helpful for cash management.

Economic analysis should consider details of both positive and negative costs over the analysis period, such as varying inflation rates, capital and interest costs, salvage costs, replacement costs, interest deductions, depreciation allowances, taxes, tax credits, mortgage payments, and all other costs associated with a particular system. See the section on Symbols for definitions of variables.

Present Value (Present Worth) Analysis. All sophisticated economic analysis methods use the basic principles of present value analysis to account for the time value of money. Therefore, a good understanding of these principles is important.

The total present value (present worth) for any analysis is determined by summing the present worths of all individual items under consideration, both future single payment items and series of equal future payments. The scenario with the highest present value is the preferred alternative.

Single Payment Present Value Analysis. The cost or value of money is a function of the available interest rate and inflation rate. The future value F of a present sum of money P over n periods with compound interest rate i per period is

$$F = P(1 + i)^n \tag{1}$$

Conversely, the present value or present worth P of a future sum of money F is given by

$$P = F/(1 + i)^n \tag{2}$$

or

$$P = F \times \mathrm{PWF}(i,n)_{sgl} \tag{3}$$

where the single payment present worth factor $\mathrm{PWF}(i,n)_{sgl}$ is defined as

$$\mathrm{PWF}(i,n)_{sgl} = 1/(1 + i)^n \tag{4}$$

Example 3. Calculate the value in 10 years at 10% per year interest of a system presently valued at $10,000.

$$F = P(1 + i)^n = \$10,000(1 + 0.1)^{10} = \$25,937.42$$

Example 4. Using the present worth factor for 10% per year interest and an analysis period of 10 years, calculate the present value of a future sum of money valued at $10,000. (Stated another way, determine what sum of money must be invested today at 10% per year interest to yield $10,000 10 years from now.)

$$P = F \times \mathrm{PWF}(i,n)_{sgl}$$

$$P = \$10,000 \times 1/(1 + 0.1)^{10}$$

$$= \$3855.43$$

Series of Equal Payments. The present worth factor for a series of future equal payments (e.g., operating costs) is given by

$$\mathrm{PWF}(i,n)_{ser} = \frac{(1 + i)^n - 1}{i(1 + i)^n} \tag{5}$$

The present value P of those future equal payments (PMT) is then the product of the present worth factor and the payment (i.e., $P = \mathrm{PWF}(i,n)_{sgl} \times \mathrm{PMT}$).

The future equal payments to repay a present value of money is determined by the capital recovery factor (CRF), which is the reciprocal of the present worth factor for a series of equal payments:

$$\mathrm{CRF} = \mathrm{PMT}/P \tag{6}$$

$$\mathrm{CRF}(i,n) = \frac{i(1 + i)^n}{(1 + i)^n - 1} = \frac{i}{1 - (1 + i)^{-n}} \tag{7}$$

The CRF is often used to describe periodic uniform mortgage or loan payments. Table 5 gives abbreviated annual CRF values for several values of analysis period n and annual interest rate i. Some of the values of Table 5 are plotted versus time in Figure 1.

Note that when payment periods other than annual are to be studied, the interest rate must be expressed per appropriate period. For example, if monthly payments or return on investment are being analyzed, then interest must be expressed per month, not per year, and n must be expressed in months, not years.

Example 5. Determine the present value of an annual operating cost of $1000 per year over 10 years, assuming 10% per year interest rate.

$$\mathrm{PWF}(i,n)_{ser} = [(1 + 0.1)^{10} - 1]/[0.1(1 + 0.1)^{10}] = 6.14$$

$$P = \$1000(6.14) = \$6140$$

Table 5 Annual Capital Recovery Factors

Years	Rate of Return or Interest Rate, % per Year								
	3.5	4.5	6	8	10	12	15	20	25
2	0.52640	0.53400	0.54544	0.56077	0.57619	0.59170	0.61512	0.65455	0.69444
4	0.27225	0.27874	0.28859	0.30192	0.31547	0.32923	0.35027	0.38629	0.42344
6	0.18767	0.19388	0.20336	0.21632	0.22961	0.24323	0.26424	0.30071	0.33882
8	0.14548	0.15161	0.16104	0.17401	0.18744	0.20130	0.22285	0.26061	0.30040
10	0.12024	0.12638	0.13587	0.14903	0.16275	0.17698	0.19925	0.23852	0.28007
12	0.10348	0.10967	0.11928	0.13270	0.14676	0.16144	0.18448	0.22526	0.26845
14	0.09157	0.09782	0.10758	0.12130	0.13575	0.15087	0.17469	0.21689	0.26150
16	0.08268	0.08902	0.09895	0.11298	0.12782	0.14339	0.16795	0.21144	0.25724
18	0.07582	0.08224	0.09236	0.10670	0.12193	0.13794	0.16319	0.20781	0.25459
20	0.07036	0.07688	0.08718	0.10185	0.11746	0.13388	0.15976	0.20536	0.25292
25	0.06067	0.06744	0.07823	0.09368	0.11017	0.12750	0.15470	0.20212	0.25095
30	0.05437	0.06139	0.07265	0.08883	0.10608	0.12414	0.15230	0.20085	0.25031
35	0.05000	0.05727	0.06897	0.08580	0.10369	0.12232	0.15113	0.20034	0.25010
40	0.04683	0.05434	0.06646	0.08386	0.10226	0.12130	0.15056	0.20014	0.25006

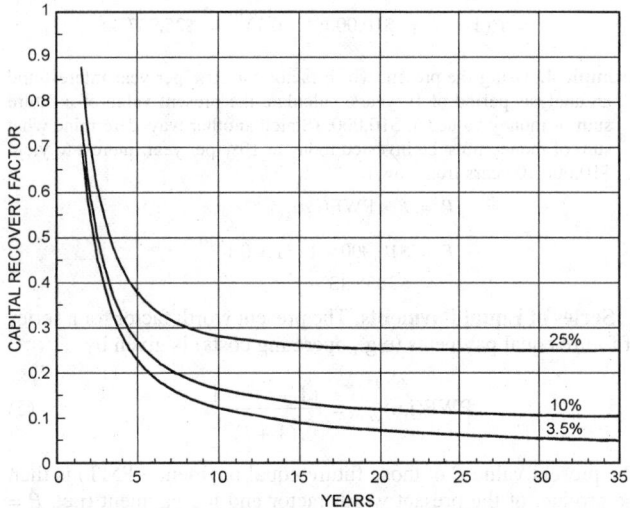

Fig. 1   Capital Recovery Factor Versus Time

Example 6. Determine the uniform monthly mortgage payments for a loan of $100,000 to be repaid over 30 years at 10% per year interest. Since we wish monthly payment periods, the payback duration is 30(12) = 360 monthly periods, and the interest rate per period is 0.1/12 = 0.00833 per month.

$$CRF(i,n) = 0.00833(1 + 0.00833)^{360}/[(1 + 0.00833)^{360} - 1]$$

$$CRF = 0.008773$$

$$PMT = P(CRF)$$

$$PMT = \$100,000(0.008773)$$

$$PMT = \$877.30 \text{ per month}$$

Improved Payback Analysis. This somewhat more sophisticated payback approach is similar to the simple payback method, except that the cost of money (interest rate, discount rate, etc.) is considered. Solving Equation (7) for n yields the following:

$$n = \frac{\ln[CRF/(CRF - i)]}{\ln(1 + i)} \qquad (8)$$

Given known investment amounts and earnings, CRFs can be calculated for the alternative investments. Subsequently, the number of periods until payback has been achieved can be calculated using Equation (8). Alternatively, a period-by-period (e.g., month-by-month or year-by-year) tabular cash flow analysis may be performed, or the necessary period to yield the calculated CRF may be obtained from a plot of CRFs, such as that shown in Figure 1.

Example 7. Compare the years to payback of the same items described in Example 2 if the value of money is 10% per year.

Item 1

cost	=	$10,000
savings	=	$2000/year
CRF	=	$2000/$10,000 = 0.2
n	=	$\ln[0.2/(0.2 - 0.1)]/\ln(1 + 0.1) = 7.3$ years

Item 2

cost	=	$12,000
savings	=	$3000/year
CRF	=	$3000/$12,000 = 0.25
n	=	$\ln[0.25/(0.25 - 0.1)]/\ln(1 + 0.1) = 5.4$ years

Accounting for Inflation. Different economic goods may inflate at different rates. Inflation reflects the rise in the real cost of

a commodity over time and is separate from the time value of money. Inflation must often be accounted for in an economic evaluation. One way to account for inflation is to substitute effective interest rates that account for inflation into the equations given in this chapter.

The effective interest rate i', sometimes called the real rate, accounts for inflation rate j and interest rate i or discount rate i_d; it can be expressed as follows (Kreith and Kreider 1978, Kreider and Kreith 1982):

$$i' = \frac{1 + i}{1 + j} - 1 = \frac{i - j}{1 + j} \qquad (9)$$

Different effective interest rates can be applied to individual components of cost. Projections for future fuel and energy prices are available in the *Annual Supplement to NIST Handbook* 135 (Lippiatt 1994; NIST).

Example 8. Determine the present worth P of an annual operating cost of $1000 over 10 years, given a discount rate of 10% per year and an inflation rate of 5% per year.

$$i' = (0.1 - 0.05)/(1 + 0.05) = 0.0476$$

$$PWF(i',n)_{ser} = \frac{(1 + 0.0476)^{10} - 1}{0.0476(1 + 0.0476)^{10}} = 7.813$$

$$P = \$1000(7.813) = \$7813$$

The following are three commonly used methods of present value analysis that include life-cycle cost factors (life of equipment, analysis period, discount rate, energy escalation rates, maintenance cost, etc., as shown in Table 1). These comparison techniques rely on the same assumptions and economic analysis theories but display the results in different forms. They also use the same definition of each term. All can be displayed as a single calculation or as a cash flow table using a series of calculations for each year of the analysis period.

Savings-to-Investment Ratio. Most large military sponsored work and many other U.S. government entities require a savings-to-investment-ratio (SIR) method. This method produces a ratio that defines the relative economic strength of each option. The higher the ratio, the better the economic strength. If the ratio is less than 1, the measure does not pay for itself within the analysis period. The escalated savings on an annual and a special (nonannual) basis is calculated and discounted. The costs are shown on an annual and special basis for each year over the life of the system or option. The savings and investments are both discounted separately on an annual basis, and then the discounted total cumulative savings is divided by the discounted total cumulative investments (costs). The analysis period is usually the life of the system or equipment being considered.

Life-Cycle Costs. This method of analysis compares the cumulative total of implementation, operating, and maintenance costs. The total costs are discounted over the life of the system or over the loan repayment period. The costs and investments are both discounted and displayed as a total combined life-cycle cost at the end of the analysis period. The options are compared to determine which has the lowest total cost over the anticipated project life.

Internal Rate of Return. The internal rate of return (IRR) method calculates a return on investment over the defined analysis period. The annual savings and costs are not discounted, and a cash flow is established for each year of the analysis period, to be used with an initial cost (or value of the loan). Annual recurring and special (nonannual) savings and costs can be used. The cash flow is then discounted until a calculated discount rate is found that yields a net present value of zero. This method gives a basic comparison of return on investments. This method assumes savings are reinvested at the same calculated rate of return. Therefore, the calculated rates of return can be overstated compared to the actual rates of return.

Another version of this is the modified or adjusted internal rate of return (MIRR or AIRR). In this version, reinvested savings are assumed to have a given rate of return on investment, and the financed moneys a given interest rate. The cash flow is then discounted until a calculated discount rate is found that yields a net present value of zero. This method gives a more realistic indication of expected return on investment, but the difference between alternatives can be small.

Uniform Annualized Costs Method. It is sometimes useful to project a uniform periodic (e.g., average annual) cost over the analysis period. The basic procedure for determining uniform annualized costs is to first determine the present worth of all costs and then apply the capital recovery factor to determine equal payments over the analysis period.

Uniform annualized mechanical system owning, operating, and maintenance costs can be expressed, for example, as

$$
\begin{aligned}
C_y = \ &- \text{capital and interest} + \text{salvage value} - \text{replacements} \\
&- \text{disposals} - \text{operating energy} - \text{property tax} \\
&- \text{maintenance} - \text{insurance} \\
&+ \text{interest tax deduction} + \text{depreciation}
\end{aligned}
$$

where

capital and interest	=	$(C_{s,init} - \text{ITC})\text{CRF}(i',n)$
salvage value	=	$C_{s,salv}\text{PWF}(i',n)\text{CRF}(i',n)(1 - T_{salv})$
replacements or disposals	=	$\sum_{k=1}^{n} [R_k\text{PWF}(i',k)]\text{CRF}(i',n)(1 - T_{inc})$
operating energy	=	$C_e[\text{CRF}(i',n)/\text{CRF}(i'',n)](1 - T_{inc})$
property tax	=	$C_{s,assess}T_{prop}(1 - T_{inc})$
maintenance	=	$M(1 - T_{inc})$
insurance	=	$I(1 - T_{inc})$
interest tax deduction	=	$T_{inc}\sum_{k=1}^{n}[i_mP_{k-1,i}\text{PWF}(i_d,k)]\text{CRF}(i',n)$
depreciation (for commercial systems)	=	$T_{inc}\sum_{k=1}^{n}[D_k\text{PWF}(i_d,k)]\text{CRF}(i',n)$

The outstanding principle P_k during year k at market mortgage rate i_m is given by

$$
P_k = (C_{s,init} - \text{ITC})\left[(1 + i_m)^{k-1} + \frac{(1 + i_m)^{k-1} - 1}{(1 + i_m)^{-n} - 1}\right] \quad (10)
$$

Note: P_k is in current dollars and must, therefore, be discounted by the discount rate i_d, not i'.

Likewise, the summation term for interest deduction can be expressed as

$$
\sum_{k=1}^{n}[i_mP_k/(1 + i_d)^k] = (C_{s,init} - \text{ITC}) \times
$$

$$
\left[\frac{\text{CRF}(i_m,n)}{\text{CRF}(i_d,n)} + \frac{1}{(1 + i_m)}\frac{i_m - \text{CRF}(i_m,n)}{\text{CRF}[(i_d - i_m)/(1 + i_m),n]}\right] \quad (11)
$$

If $i_d = i_m$,

$$
\sum_{k=1}^{n}[i_mP_k/(1 + i_d)^k] = \quad (12)
$$

$$
(C_{s,init} - \text{ITC}) \times \left[1 + \frac{n}{1 + i_m}[i_m - \text{CRF}(i_m,n)]\right]
$$

Depreciation terms commonly used include depreciation calculated by the straight line depreciation method, which is

$$
D_{k,SL} = (C_{s,init} - C_{s,salv})/n \quad (13)
$$

and the sum-of-digits depreciation method:

$$
D_{k,SD} = (C_{s,init} - C_{s,salv})[2(n - k + 1)]/n(n + 1) \quad (14)
$$

Riggs (1977) and Grant et al. (1982) present further information on advanced depreciation methods. Certified accountants may also be consulted for information regarding accelerated methods allowed for tax purposes. The following example illustrates the use of the uniform annualized cost method. Additional examples are presented by Haberl (1993).

Example 9. Calculate the annualized system cost using constant dollars for a $10,000 system considering the following factors: a 5-year life, a salvage value of $1000 at the end of the 5 years, no investment tax credits, a $500 replacement in year 3, a discount rate i_d of 10%, a general inflation rate j of 5%, a fuel inflation rate j_e of 8%, a market mortgage rate i_m of 10%, an annual operating cost for energy of $500, a $100 annual maintenance cost, a $50 annual insurance cost, straight line depreciation, an income tax rate of 50%, a property tax rate of 1% of assessed value, an assessed system value equal to 40% of the initial system value, and a salvage tax rate of 50%.

Effective interest rate i' for all but fuel

$$
i' = (i_d - j)/(1 + j) = (0.10 - 0.05)/(1 + 0.05) = 0.047619
$$

Effective interest rate i'' for fuel

$$
i'' = (i_d - j_e)/(1 + j_e) = (0.10 - 0.08)/(1 + 0.08) = 0.018519
$$

Capital recovery factor $\text{CRF}(i',n)$ for items other than fuel

$$
\begin{aligned}
\text{CRF}(i',n) &= i'/[1 - (1 + i')^{-n}] \\
&= 0.047619/[1 - (1.047619)^{-5}] = 0.229457
\end{aligned}
$$

Capital recovery factor $\text{CRF}(i'',n)$ for fuel

$$
\begin{aligned}
\text{CRF}(i'',n) &= i''/[1 - (1 + i'')^{-n}] \\
&= 0.018519/[1 - (0.018519)^{-5}] = 0.211247
\end{aligned}
$$

Capital recovery factor $\text{CRF}(i_m,n)$ for loan or mortgage

$$
\begin{aligned}
\text{CRF}(i_m,n) &= i_m/[1 - (1 + i_m)^{-n}] \\
&= 0.10/[1 - (1.10)^{-5}] = 0.263797
\end{aligned}
$$

Loan payment = $10,000(0.263797) = $2637.97

Present worth factor $\text{PWF}(i_d$, years 1 to 5)

$$
\text{PWF}(i_d,1) = 1/(1.10)^1 = 0.909091
$$

$$
\text{PWF}(i_d,2) = 1/(1.10)^2 = 0.826446
$$

$$
\text{PWF}(i_d,3) = 1/(1.10)^3 = 0.751315
$$

$$
\text{PWF}(i_d,4) = 1/(1.10)^4 = 0.683013
$$

$$
\text{PWF}(i_d,5) = 1/(1.10)^5 = 0.620921
$$

Present worth factor $\text{PWF}(i'$, years 1 to 5)

$$
\text{PWF}(i',1) = 1/(1.047619)^1 = 0.954545
$$

$$
\text{PWF}(i',2) = 1/(1.047619)^2 = 0.911157
$$

$$
\text{PWF}(i',3) = 1/(1.047619)^3 = 0.869741
$$

$$
\text{PWF}(i',4) = 1/(1.047619)^4 = 0.830207
$$

$$
\text{PWF}(i',5) = 1/(1.047619)^5 = 0.792471
$$

Capital and interest

$$(C_{s,init} - ITC)CRF(i',n) = (\$10,000 - \$0)0.229457 = \$2294.57$$

Salvage value

$$C_{s,salv}PWF(i',n)CRF(i',n)(1 - T_{salv})$$
$$= \$1000 \times 0.792471 \times 0.229457 \times 0.5 = \$90.92$$

Replacements

$$\sum_{k=1}^{n}[R_k PWF(i',k)]CRF(i',n)(1 - T_{inc})$$
$$= \$500 \times 0.869741 \times 0.229457 \times 0.5 = \$49.89$$

Operating energy

$$C_e[CRF(i',n)/CRF(i'',n)](1 - T_{inc})$$
$$= \$500[0.229457/0.211247]0.5 = \$271.55$$

Property tax

$$C_{s,assess}T_{prop}(1 - T_{inc}) = \$10,000 \times 0.40 \times 0.01 \times 0.5 = \$20.00$$

Maintenance

$$M(1 - T_{inc}) = \$100(1 - 0.5) = \$50.00$$

Insurance

$$I(1 - T_{inc}) = \$50(1 - 0.5) = \$25.00$$

Interest deduction

$$T_{inc}\sum_{k=1}^{n}[i_m P_{k-1}PWF(i_d,k)]CRF(i',n) = \dots \text{ see Table 6}$$

Table 6 summarizes the interest and principle payments for this example. Annual payments are the product of the initial system cost

$C_{s,init}$ and the capital recovery factor CRF(i_m,5). Also, Equation (10) can be used to calculate total discounted interest deduction directly.

Next, apply the capital recovery factor CRF(i',5) and tax rate T_{inc} to the total of the discounted interest sum.

$$\$2554.66 \, CRF(i',5)T_{inc} = \$2554.66 \times 0.229457 \times 0.5 = \$293.09$$

Depreciation

$$T_{inc}\sum_{k=1}^{n}[D_{k,SL}PWF(i_d,k)]CRF(i',n)\dots$$

Use the straight line depreciation method to calculate depreciation:

$$D_{k,SL} = (C_{s,init} - C_{s,salv})/n = (\$10,000 - \$1000)/5 = \$1800.00$$

Next, discount the depreciation.

Year	$D_{k,SL}$	PWF(i_d,k)	Discounted Depreciation
1	\$1800.00	0.909091	\$1636.36
2	\$1800.00	0.826446	\$1487.60
3	\$1800.00	0.751315	\$1352.37
4	\$1800.00	0.683013	\$1229.42
5	\$1800.00	0.620921	\$1117.66
		Total	**\$6823.42**

Finally, the capital recovery factor and tax are applied.

$$\$6823.42 \, CRF(i',n)T_{inc} = \$6823.42 \times 0.229457 \times 0.5 = \$782.84$$

U.S. tax code recommends estimating the salvage value prior to depreciating. Then depreciation is claimed as the difference between the initial and salvage value, which is the way depreciation is treated in this example. The more common practice is to initially claim zero salvage value, and at the end of ownership of the item, treat any salvage value as a capital gain.

Table 6 Interest Deduction Summary (for Example 9)

Year	Payment Amount, Current \$	Interest Payment, Current \$	Principal Payment, Current \$	Outstanding Principal, Current \$	PWF(i_d,k)	Discounted Interest, Discounted \$	Discounted Payment, Discounted \$
0	—	—	—	10,000.00	—	—	—
1	2,637.97	1,000.00	1,637.97	8,362.03	0.909091	909.09	2,398.17
2	2,637.97	836.20	1,801.77	6,560.26	0.826446	691.07	2,180.14
3	2,637.97	656.03	1,981.95	4,578.31	0.751315	492.89	1,981.95
4	2,637.97	457.83	2,180.14	2,398.17	0.683013	312.70	1,801.77
5	2,637.97	239.82	2,398.17	0	0.620921	148.91	1,637.97
Total	—	3,189.88	10,000.00	—		2,554.66	10,000.00

Table 7 Summary of Cash Flow (for Example 10)

1	2	3	4	5	6	7	8	9	10	11
							Present Worth of Net Cash Flow			
	Cash Outlay,	Net Income Before Taxes,	Depreciation,	Net Taxable	Income Taxes	Net Cash Flow,[b]	10% Rate		15% Rate	20% Rate
Year	\$	\$	\$	Income,[a] \$	@50%, \$	\$	PWF	P, \$	P, \$	P, \$
0	120,000	0	0	0	0	−120,000	1.000	−120,000	−120,000	−120,000
1	0	20,000	15,000	5,000	2,500	17,500	0.909	15,900	15,200	14,600
2	0	30,000	15,000	15,000	7,500	22,500	0.826	18,600	17,000	15,600
3	0	40,000	15,000	25,000	12,500	27,500	0.751	20,600	18,100	15,900
4	0	50,000	15,000	35,000	17,500	32,500	0.683	22,200	18,600	15,700
5	0	50,000	15,000	35,000	17,500	32,500	0.621	20,200	16,200	13,100
6	0	50,000	15,000	35,000	17,500	32,500	0.564	18,300	14,100	10,900
7	0	50,000	15,000	35,000	17,500	32,500	0.513	16,700	12,200	9,100
8	0	50,000	15,000	35,000	17,500	32,500	0.467	15,200	10,600	7,600
					Total Cash Flow			27,700	2,000	−17,500
					Investment Value			147,500	122,000	102,500

[a]Net taxable income = net income − depreciation.

[b]Net cash flow = net income − taxes.

Summary of terms

Capital and interest	−$2294.57
Salvage value	+$ 90.92
Replacements	−$ 49.89
Operating costs	−$ 271.55
Property tax	−$ 20.00
Maintenance	−$ 50.00
Insurance	−$ 25.00
Interest deduction	+$ 293.09
Depreciation deduction	+$ 782.84
Total annualized cost	−$1544.16

Cash Flow Analysis Method. The cash flow analysis method accounts for costs and revenues on a period-by-period (e.g., year-by-year) basis, both actual and discounted to present value. This method is especially useful for identifying periods when net cash flow will be negative due to intermittent large expenses.

Example 10. An eight-year study for a $120,000 investment with depreciation spread equally over the assigned period. The benefits or incomes are variable. The marginal tax rate is 50%. The rate of return on the investment is required. Table 7 has columns showing year, cash outlays, income, depreciation, net taxable income, taxes and net cash flow.

Solution: To evaluate the effect of interest and time, the net cash flow must be multiplied by the single payment present worth factor. An arbitrary interest rate of 10% has been selected and the PWF_{sgl} is obtained by using Equation (4). Its value is listed in Table 7, column 8. Present worth of the net cash flow is obtained by multiplying columns 7 and 8. Column 9 is then added to obtain the total cash flow. If year 0 is ignored, an investment value is obtained for a 10% required rate of return.

The same procedure is used for 15% interest (column 10, but the PWF is not shown) and for 20% interest (column 11).

Discussion. The interest at which the summation of present worth of net cash flow is zero gives the rate of return. In this example, the investment has a rate of return by interpolation of about 15.4%. If this rate offers an acceptable rate of return to the investor, the proposal should be approved; otherwise, it should be rejected.

Another approach would be to obtain an investment value at a given rate of return. This is accomplished by adding the present worth of the net cash flows, but not including the investment cost. In the example, under the 10% given rate of return, $147,700 is obtained as an investment value. This amount, when using money that costs 10%, would be the acceptable value of the investment.

Computer Analysis

Many computer programs are available that incorporate the economic analysis methods described above. These range from simple macros developed for popular spreadsheet applications to more comprehensive, menu-driven computer programs. Commonly used examples of the latter include Building Life-Cycle Cost (BLCC), Life Cycle Cost in Design (LCCID), and PC-ECONPACK.

BLCC was developed by the National Institute of Standards and Technology (NIST) for the U.S. Department of Energy (DOE). The program follows criteria established by the Federal Energy Management Program (FEMP) and the Office of Management and Budget (OMB). It is intended for the evaluation of energy conservation investments in nonmilitary government buildings; however, it is also appropriate for similar evaluations of commercial facilities.

LCCID is an economic analysis program tailored to the needs of the U.S. Department of Defense (DOD). Developed by the U.S. Army Corps of Engineers and the Construction Engineering Research Laboratory (USA-CERL), LCCID uses economic criteria established by FEMP and OMB.

PC-Econpack, developed by the U.S. Army Corps of Engineers for use by the DOD, uses economic criteria established by the OMB. The program performs standardized life-cycle cost calculations such as net present value, equivalent uniform annual cost, SIR, and discounted payback period.

Table 8   Commonly Used Discount Formulas

Name	Algebrac Form[a,b]
Single compound-amount (SCA) equation	$F = P \cdot [(1+d)^n]$
Single present value (SPW) equation	$P = F \cdot \left[\dfrac{1}{(1+d)^n}\right]$
Uniform sinking-fund (USF) equation	$A = F \cdot \left[\dfrac{d}{(1+d)^n - 1}\right]$
Uniform capital-recovery (UCR) equation	$A = P \cdot \left[\dfrac{d(1+d)^n}{(1+d)^n - 1}\right]$
Uniform compound-account (UCA) equation	$F = A \cdot \left[\dfrac{(1+d)^n - 1}{d}\right]$
Uniform present-value (UPW) equation	$P = A \cdot \left[\dfrac{(1+d)^n - 1}{d(1+d)^n}\right]$
Modified uniform present-value (UPW*) equation	$P = A_0 \cdot \left(\dfrac{1+e}{d-e}\right) \cdot \left[1 - \left(\dfrac{1+e}{1+d}\right)^n\right]$

where

A = end-of-period payment (or receipt) in a uniform series of payments (or receipts) over n periods at d interest or discount rate

A_0 = initial value of a periodic payment (receipt) evaluated at the beginning of the study period

$A_t = A_0 \cdot (1+e)^t$, where $t = 1,\dots, n$

d = interest or discount rate

e = price escalation rate per period

Source: NIST *Handbook* 135 (Ruegg).

[a]Note that the USF, UCR, UCA, and UPW equations yield undefined answers when $d = 0$. The correct algebraic forms for this special case would be as follows: USF formula, $A = F/N$; UCR formula, $A = P/N$; UCA formula, $F = A \cdot n$. The UPW* equation also yields an undefined answer when $e = d$. In this case, $P = A_0 \cdot n$.

[b]The terms by which the known values are multiplied in these equations are the formulas for the factors found in discount factor tables. Using acronyms to represent the factor formulas, the discounting equaitons can also be written as $F = P \cdot SCA$, $P = F \cdot SPW$, $A = F \cdot USF$, $A = P \cdot UCR$, $F = UCA$, $P = A \cdot UPW$, and $P = A_0 \cdot UPW*$.

Macros developed for common spreadsheet programs generally contain preprogrammed functions for the various life-cycle cost calculations. Although typically not as sophisticated as the menu-driven programs, the macros are easy to install and easy to learn.

Reference Equations

Table 8 lists commonly used discount formulas as addressed by NIST. Refer to NIST *Handbook* 135 (Ruegg) and Table 2.3 in that handbook for detailed discussions.

SYMBOLS

c = cooling system adjustment factor

C = total annual building HVAC maintenance cost

C_e = annual operating cost for energy

$C_{s,assess}$ = assessed system value

$C_{s,init}$ = initial system cost

$C_{s,salv}$ = system salvage value at end of study period

C_y = uniform annualized mechanical system owning, operating, and maintenance costs

CRF = capital recovery factor

$CRF(i,n)$ = capital recovery factor for interest rate i and analysis period n

$CRF(i',n)$ = capital recovery factory for interest rate i' for items other than fuel and analysis period n

$CRF(i'',n)$ = capital recovery factor for fuel interest rate i'' and analysis period n

$CRF(i_m,n)$ = capital recovery factor for loan or mortgage rate i_m and analysis period n

d = distribution system adjustment factor

D_k = depreciation during period k

$D_{k,SL}$ = depreciation during period k due to straight line depreciation method

$D_{k,SD}$ = depreciation during period k due to sum-of-digits depreciation method

F = future value of a sum of money

h = heating system adjustment factor

i = compound interest rate per period

i_d = discount rate per period

i_m = market mortgage rate

i' = effective interest rate for all but fuel

i'' = effective interest rate for fuel

I = insurance cost per period

ITC = investment tax credit

j = inflation rate per period

j_e = fuel inflation rate per period

k = end of period(s) during which replacement(s), repair(s), depreciation, or interest are calculated

M = maintenance cost per period

n = number of periods under analysis

P = present value of a sum of money

P_k = outstanding principle on loan at end of period k

PMT = future equal payments

PWF = present worth factor

$PWF(i_d,k)$ = present worth factor for discount rate i_d at end of period k

$PWF(i',k)$ = present worth factor for effective interest rate i' at end of period k

$PWF(i,n)_{sgl}$ = single payment present worth factor

$PWF(i,n)_{ser}$ = present worth factor for a series of future equal payments

R_k = net replacement, repair, or disposal costs at end of period k

T_{inc} = net income tax rate

T_{prop} = property tax rate

T_{salv} = tax rate applicable to salvage value of system

REFERENCES

Akalin, M.T. 1978. Equipment life and maintenance cost survey. *ASHRAE Transactions* 84(2):94-106.

DOE. International performances measurement and verification protocol. *Publication* No. DOE/EE-0157. U.S. Department of Energy.

Dohrmann, D.R. and T. Alereza. 1986. Analysis of survey data on HVAC maintenance costs. *ASHRAE Transactions* 92(2A):550-65.

Easton Consultants. 1986. Survey of residential heat pump service life and maintenance issues. Available from American Gas Association, Arlington, VA (Catalog No. S-77126).

Grant, E., W. Ireson, and R. Leavenworth. 1982. *Principles of engineering economy.* John Wiley and Sons, New York.

Haberl, J. 1993. Economic calculations for ASHRAE Handbook. Energy Systems Laboratory *Report* No. ESL-TR-93/04-07. Texas A&M University, College Station, TX.

Kreider, J. and F. Kreith. 1982. *Solar heating and cooling.* Hemisphere Publishing, Washington, D.C.

Kreith, F. and J. Kreider. 1978. *Principles of solar engineering.* Hemisphere Publishing, Washington, D.C.

Lippiatt, B.L. 1994. Energy prices and discount factors for life-cycle cost analysis 1993. *Annual Supplement to NIST Handbook* 135 and *NBS Special Publication* 709. NISTIR 85-3273.7. National Institute of Standards and Technology, Gaithersburg, MD.

Lovvorn, N.C. and C.C. Hiller. 1985. A study of heat pump service life. *ASHRAE Transactions* 91(2B):573-88.

NIST. *Annual Supplement to NIST Handbook* 135. National Institute of Standards and Technology, Gaithersburg, MD.

NIST and DOE. Building life-cycle cost (BLCC) computer program. Available from National Institute of Standards and Technology, Office of Applied Economics, Gaithersburg, MD.

OMB. 1972. Guidelines and discount rates for benefit-cost analysis of federal programs. *Circular* A-94. Office of Management and Budget, Washington, D.C.

Riggs, J.L. 1977. *Engineering economics.* McGraw-Hill, New York.

Ruegg, R.T. Life-cycle costing manual for the Federal Energy Management Program. NIST *Handbook* 135. National Institute of Standards and Technology, Gaithersburg, MD.

U.S. Department of Commerce, Bureau of Economic Analysis. *Survey of current business.* U.S. Government Printing Office, Washington, D.C.

USA-CERL and USACE. Life cycle cost in design (LCCID) computer program. Available from Building Systems Laboratory, University of Illinois, Urbana.

USACE. PC-Econpack computer program. U.S. Army Corps of Engineers, Huntsville, AL.

BIBLIOGRAPHY

ASTM. 1992. Standard terminology of building economics. *Standard* E 833 Rev A-92. American Society for Testing and Materials, West Conshohoken, PA.

Kurtz, M. 1984. *Handbook of engineering economics: A guide for engineers, technicians, scientists, and managers.* McGraw-Hill, New York.

Quirin, D.G. 1967. *The capital expenditure decision.* Richard D. Win, Inc., Homewood, IL.

TESTING, ADJUSTING, AND BALANCING

THE system that controls the environment in a building is a dynamic entity that changes with time and use, and it must be rebalanced accordingly. The designer must consider initial and supplementary testing and balancing requirements for commissioning. Complete and accurate operating and maintenance instructions that include intent of design and how to test, adjust, and balance the building systems are essential. Building operating personnel must be well trained, or qualified operating service organizations must be employed to ensure optimum comfort, proper process operations, and economy of operation.

This chapter does not suggest which groups or individuals should perform the functions of a complete testing, adjusting, and balancing procedure. However, the procedure must produce repeatable results that meet the intent of the designer and the requirements of the owner. Overall, one source must be responsible for testing, adjusting, and balancing all systems. As part of this responsibility, the testing organization should check all equipment under field conditions to ensure compliance.

Testing and balancing should be repeated as the systems are renovated and changed. The testing of boilers and other pressure vessels for compliance with safety codes is not the primary function of the testing and balancing firm; rather it is to verify and adjust operating conditions in relation to design conditions for flow, temperature, pressure drop, noise, and vibration. ASHRAE *Standard* 111 outlines detailed procedures not covered in this chapter.

TERMINOLOGY

Testing, adjusting, and balancing is the process of checking and adjusting all the environmental systems in a building to produce the design objectives. This process includes (1) balancing air and water distribution systems, (2) adjusting the total system to provide design quantities, (3) electrical measurement, (4) establishing quantitative performance of all equipment, (5) verifying automatic controls, and (6) sound and vibration measurement. These procedures are accomplished by checking installations for conformity to design, measuring and establishing the fluid quantities of the system as required to meet design specifications, and recording and reporting the results.

The following definitions are used in this chapter. Refer to ASHRAE *Terminology of Heating, Ventilation, Air Conditioning, and Refrigeration* (1991) for additional definitions.

Test. Determine quantitative performance of equipment.

Balance. Proportion flows within the distribution system (submains, branches, and terminals) according to specified design quantities.

Adjust. Regulate the specified fluid flow rate and air patterns at the terminal equipment (e.g., reduce fan speed, adjust a damper).

Procedure. An approach to and execution of a sequence of work operations to yield repeatable results.

The preparation of this chapter is assigned to TC 9.7, Testing and Balancing.

Report forms. Test data sheets arranged in logical order for submission and review. The data sheets should also form the permanent record to be used as the basis for any future testing, adjusting, and balancing.

Terminal. A point where the controlled medium (fluid or energy) enters or leaves the distribution system. In air systems, these may be variable air or constant volume boxes, registers, grilles, diffusers, louvers, and hoods. In water systems, these may be heat transfer coils, fan coil units, convectors, or finned-tube radiation or radiant panels.

GENERAL CRITERIA

Effective and efficient testing, adjusting, and balancing require a systematic, thoroughly planned procedure implemented by experienced and qualified staff. All activities, including organization, calibration of instruments, and execution of the actual work, should be scheduled. Air-side must be coordinated with water-side work. Preparatory work includes planning and scheduling all procedures, collecting necessary data (including all change orders), reviewing data, studying the system to be worked on, preparing forms, and making preliminary field inspections.

Leakage can significantly reduce performance; therefore ducts must be designed, constructed, and installed to minimize and control air leakage. During construction, all duct systems should be sealed and tested for air leakage; and water, steam, and pneumatic piping should be tested for leakage.

Design Considerations

Testing, adjusting, and balancing begin as design functions, with most of the devices required for adjustments being integral parts of the design and installation. To ensure that proper balance can be achieved, the engineer should show and specify a sufficient number of dampers, valves, flow measuring locations, and flow balancing devices; these must be properly located in required straight lengths of pipe or duct for accurate measurement. The testing procedure depends on system characteristics and layout. The interaction between individual terminals varies with pressures, flow requirements, and control devices.

The design engineer should specify balancing tolerances. Suggested tolerances are ±10% for individual terminals and branches in noncritical applications and ±5% for main ducts. For critical applications where differential pressures must be maintained, the following tolerances are suggested:

Positive zones
    Supply air   0 to +10%
    Exhaust and return air   0 to −10%
Negative zones
    Supply air   0 to −10%
    Exhaust and return air   0 to +10%

AIR VOLUMETRIC MEASUREMENT METHODS

General

The pitot-tube traverse is the generally accepted method of measuring airflow in ducts. Other methods of measuring airflow at individual terminals are described by the terminal manufacturers. The primary objective is to establish repeatable measurement procedures that correlate with the pitot-tube traverse.

Laboratory tests, data, and techniques prescribed by equipment and air terminal manufacturers must be reviewed and checked for accuracy, applicability, and repeatability of results. Conversion factors that correlate field data with laboratory results must be developed to predict the equipment's actual field performance.

Air Devices

Generally, loss coefficients given by air diffuser manufacturers should be checked for accuracy by field measurement and by comparing actual flow measured by pitot-tube traverse to actual measured velocity. Air diffuser manufacturers usually base their volumetric test measurements on a deflecting vane anemometer. The velocity is multiplied by an empirical effective area to obtain the air diffuser's delivery. Accurate results are obtained by measuring at the vena contracta with the probe of the deflecting vane anemometer.

The methods advocated for measuring the airflow of troffer-type terminals are similar to the methods described for air diffusers. The capture hood is frequently used to measure device airflows, primarily of diffusers and slots. loss coefficients should be established for hood measurements with varying flow and deflection settings. If the air does not fill the measurement grid, the readings will require a correction factor (similar to the loss coefficient).

Rotating vane anemometers are commonly used to measure airflow from sidewall grilles. Effective areas (loss coefficients) should be established with the face dampers fully open and deflection set uniformly on all grilles. Correction factors are required when measuring airflow in open ducts, i.e., damper openings and fume hoods (Sauer and Howell 1990).

All flow measuring instruments should be field verified by running pitot-tube traverses to establish correction and/or density factors.

Duct Flow

Most procedures for testing, adjusting, and balancing air-handling systems rely on measuring volumes in the ducts rather than at the terminals. These measurements are more reliable than those obtained at the terminals, which are based on manufacturer's data. In such procedures, terminal measurements are relied on only for proportionally balancing the distribution within a space or zone.

The preferred method of duct volumetric flow measurement is the pitot-tube traverse average. Care should be taken to obtain the maximum straight run before and after the traverse station. To obtain the best duct velocity profile, measuring points should be located as shown in Chapter 14 of the 1997 *ASHRAE Handbook—Fundamentals* and ASHRAE *Standard* 111. Where factory-fabricated volume measuring stations are used, the measurements should be checked against a pitot-tube traverse.

The power input to a fan's driver should be used only as a guide to indicate its delivery. It may be used to verify performance determined by a reliable method (e.g., pitot-tube traverse of system's main) considering system effects that may be present. The flow rate from some fans is not proportional to the power needed to drive them. In some cases, as with forward-curved blade fans, the same power is required for two or more flow rates. The backward-curved blade centrifugal fan is the only type with suitable characteristics, i.e., flow rate that varies directly with the power input. If

an installation has an inadequate straight length of ductwork or no ductwork to allow a pitot-tube traverse, a procedure prescribed by Sauer and Howell (1990) can be followed. In this procedure a vane anemometer is used to read air velocities at multiple points across the face of a coil to determine a loss coefficient.

Mixture Plenums

Approach conditions are often so unfavorable that the air quantities comprising a mixture (e.g., outdoor air and return air) cannot be determined accurately by volumetric measurements. In such cases, the temperature of the mixture indicates the balance (proportions) between the component airstreams. Temperatures must be measured carefully to account for stratification of the air, and the difference between the outside and return air temperatures must be greater than 20°F. The temperature of the mixture can be calculated from Equation (1) as follows:

$$Q_t t_m = Q_o t_o + Q_r t_r \qquad (1)$$

where

Q_t = total measured air quantity, %
Q_o = outside air quantity, %
Q_r = return air quantity, %
t_m = temperature of outside and return mixture, °F
t_o = outdoor temperature, °F
t_r = return temperature, °F

Pressure Measurement

The air pressures measured include barometric pressure, static pressure, velocity pressure, total pressure, and differential pressure. measurement For field evaluation of air-handling performance, pressure should be measured as recommended in ASHRAE *Standard* 111 and analyzed together with the manufacturers' fan curves and system effect as predicted by applying methods in AMCA *Standard* 210. When measured in the field, pressure readings, air quantity, and power input often do not correlate with the manufacturers' certified performance curves unless proper correction is made.

Pressure drops through equipment such as coils, dampers, or filters should not be used to measure airflow. Pressure is an acceptable means of establishing flow volumes only where it is required by, and performed in accordance with, the manufacturer certifying the equipment.

Stratification

Normal design minimizes conditions causing air turbulence in order to produce the least friction, resistance, and consequent pressure loss. Under certain conditions, however, air turbulence is desirable and necessary. For example, two airstreams of different temperatures can stratify in smooth, uninterrupted flow conditions. In this situation, mixing should be promoted in the design. The return and outside airstreams at the inlet side of the air-handling unit tend to stratify where enlargement of the inlet plenum or casing size decreases the air velocity. Without a deliberate effort to mix the two airstreams (i.e., in cold climates, placing the outdoor air entry at the top of the plenum and the return air at the bottom of the plenum to allow natural mixing), stratification can exist and be carried throughout the system (e.g., filter, coils, eliminators, fans, and ducts). Stratification can cause damage by freezing coils and rupturing tubes. It can also affect temperature control in plenums, spaces, or both.

Stratification can further be reduced by adding vanes to break up and mix the two airstreams. No solution to stratification problems is guaranteed; each condition must be evaluated by field measurements and experimentation.

BALANCING PROCEDURES
FOR AIR DISTRIBUTION

General procedures for testing and balancing are described here, although no one established procedure is applicable to all systems. The bibliography lists sources of additional information.

Instruments for Testing and Balancing

The minimum instruments necessary for air balance are

- Manometer calibrated in 0.005 in. of water divisions
- Combination inclined and vertical manometer (0 to 10 in. of water)
- Pitot tubes in various lengths, as required
- Tachometer (direct contact, self-timing type) or strobe light
- Clamp-on ammeter with voltage scales (rms type)
- Rotating vane anemometer
- Flow hood
- Dial thermometers (2-in. diameter minimum and 1°F graduations minimum) and glass stem thermometers (1°F graduations minimum)
- Sound level meter with octave band filter set, calibrator, and microphone
- Vibration analyzer capable of measuring displacement velocity and acceleration
- Water flowmeters (0 to 50 in. of water and 0 to 400 in. of water ranges)
- Compound gage
- Test gages (100 psi and 300 psi)
- Sling psychrometer
- Etched stem thermometer (30 to 120°F in 0.1°F increments)
- Hygrometers
- Digital thermometers, relative humidity and dew-point instruments

Instruments must be calibrated periodically to verify their accuracy and repeatability prior to use in the field.

Preliminary Procedure for Air Balancing

Before balancing the system, the following steps should be performed:

1. Obtain as-built design drawings and specifications, and become thoroughly acquainted with the design intent.
2. Obtain copies of approved shop drawings of all air-handling equipment, outlets (supply, return, and exhaust), and temperature control diagrams including performance curves. Compare design requirements with shop drawing capacities.
3. Compare design to installed equipment and field installation.
4. Walk the system from the air-handling equipment to terminal units to determine variations of installation from design.
5. Check dampers (both volume and fire) for correct and locked position and temperature control for completeness of installation before starting fans.
6. Prepare report test sheets for both fans and outlets. Obtain manufacturer's outlet factors and recommended test procedure. A summation of required outlet volumes permits a cross-checking with required fan volumes.
7. Determine the best locations in the main and branch ductwork for the most accurate duct traverses.
8. Place all outlet dampers in the full open position.
9. Prepare schematic diagrams of system as-built ductwork and piping layouts to facilitate reporting.
10. Check filters for cleanliness and proper installation (no air bypass). If specifications require, establish procedure to simulate dirty filters.
11. For variable volume air systems, develop a plan to simulate diversity.

Equipment and System Check

1. All fans (supply, return, and exhaust) must be operating before checking the following items:

   (a) Motor amperage and voltage to guard against overload.
   (b) Fan rotation.
   (c) Operability of static pressure limit switch.
   (d) Automatic dampers for proper position.
   (e) Air and water controls operating to deliver required temperatures.
   (f) Air leaks in the casing and in the scarfing around the coils and filter frames must be sealed. Note points where piping enters the casing to ensure that escutcheons are right. Do not rely on pipe insulation to seal these openings because the insulation may shrink. In prefabricated units, check that all panel-fastening holes are filled to prevent whistling.

2. Traverse the main supply ductwork whenever possible. All main branches should also be traversed where duct arrangement permits. Selection of traverse points and method of traverse should be as follows:

   (a) Traverse each main or branch after the longest possible straight run for the duct involved.
   (b) For test hole spacing, refer to Chapter 14 of the 1997 *ASHRAE Handbook—Fundamentals*.
   (c) Traverse using a pitot tube and manometer where velocities are over 600 fpm. Below this velocity, use either a micromanometer and pitot tube or a recently calibrated thermal anemometer.
   (d) Note temperature and barometric pressure to determine if they need to be corrected for standard air quantity. Corrections are normally insignificant below 2000-ft elevation; however, where accurate results are desirable, corrections are justified.
   (e) After establishing the total air being delivered, adjust the fan speed to obtain the design airflow, if necessary. Check power and speed to see that motor power and/or critical fan speed have not been exceeded.
   (f) Proportionally adjust branch dampers until each has the proper air volume.
   (g) With all the dampers and registers in the system open and with the supply, return, and exhaust blowers operating at or near design airflow, set the minimum outdoor and return air ratio. If duct traverse locations are not available, this can be done by measuring the mixture temperature with thermometers in the return air, outdoor air louver, and the filter section. As an approximation, the temperature of the mixture may be calculated from Equation (1).

   The greater the temperature difference between hot and cold air, the easier it is to get accurate damper settings. Take the temperature at many points in a uniform traverse to be sure there is no stratification.

   After the minimum outdoor air damper has been set for the proper percentage of outdoor air, run another traverse of mixture temperatures and install baffling if the variation from the average is more than 5%. Remember that stratified mixed air temperatures vary greatly with the outdoor temperature in cold weather, while return air temperature has only a minor effect.

3. Balance the terminal outlets in each control zone in proportion to each other. The following steps may be followed to balance the terminals:

   (a) Once the preliminary fan quantity is set, proportion the terminal outlet balance from the outlets into the branches to the fan. Concentrate on proportioning the flow rather than the absolute quantity. As changes are made to the fan settings and branch dampers, the outlet terminal quantities remain proportional. Branch dampers should be used for major adjusting and terminal dampers for trim or minor adjustment only. It may be necessary to install additional subbranch dampers to decrease the use of terminal dampers that create objectionable noise.

(b) Normally, several passes through the entire system are necessary to obtain proper outlet values.

(c) The total tested outlet air quantity compared to duct traverse air quantities may be an indicator of duct leakage.

(d) With total design air established in the branches and at the outlets, perform the following: (1) take new fan motor amperage readings, (2) find static pressure across the fan, (3) read and record static pressure across each component (intake, filters, coils, and mixing dampers), and (4) take a final duct traverse.

Dual-Duct Systems

Most constant volume dual-duct systems are designed to handle a portion of the total system's supply through the cold duct and smaller air quantities through the hot duct. Balancing should be accomplished as follows:

1. When adjusting multizone or double-duct constant volume systems, establish the ratio of the design volume through the cooling coil to total fan volume to achieve the desired diversity factor. Keep the proportion of cold to total air constant during the balance. However, check each zone or branch with this component on full cooling. If the design calls for full flow through the cooling coil, the entire system should be set to full flow through the cooling side while making tests. Perform the same procedure for the hot air side.

2. Check the leaving air temperature at the nearest terminal to verify that the hot and cold damper inlet leakage is not greater than the established maximum allowable leakage.

3. Check apparatus and main trunks, as outlined in the section on Equipment and System Check.

4. Determine whether the static pressure at the end of the system (the longest duct run) is at or above the minimum required for mixing box operation. Proceed to the extreme end of the system and check the inlet static pressure with an inclined manometer. The inlet static pressure should exceed the minimum static pressure recommended by the mixing box manufacturer. Additional static pressure is required for the low-pressure distribution system downstream of the box.

5. Proportionately balance the diffusers or grilles on the low-pressure side of the box, as described for low-pressure systems in the previous section.

6. Change the control settings to full heating, and make certain that the controls and dual-duct boxes function properly. Spot-check the airflow at several diffusers. Check for stratification.

7. If the engineer has included a diversity factor in selecting the main apparatus, it will not be possible to get full flow from all boxes simultaneously, as outlined in item 3b in the previous section. Mixing boxes closest to the fan should be set to the opposite hot or cold deck to the more critical season air flow to force the air to the end of the system.

VARIABLE VOLUME SYSTEMS

Many types of variable volume systems have been developed to conserve energy. These systems can be categorized as pressure dependent or pressure independent.

Pressure-dependent systems incorporate air terminal boxes that have a thermostat signal controlling a damper actuator. The air volume to the space varies to maintain the space temperature, while the air temperature supplied to the terminal boxes remains constant.

The balance of this system constantly varies with loading changes; therefore, any balancing procedure will not produce repeatable data unless changes in system load are simulated by using the same configuration of thermostat settings each time the system is tested—that is, the same terminal boxes are fixed in the minimum and maximum positions for the test.

Pressure-independent systems incorporate air terminal boxes that have a thermostat signal used as a master control to open or close the damper actuator and a velocity controller used as a submaster control to maintain the maximum and minimum amounts of air to be supplied to the space. The air volume to the space varies to maintain the space temperature, while the air temperature supplied to the terminal remains constant. Care should be taken to verify the operating range of the damper actuator as it responds to the velocity controller to prevent dead bands or overlap of control in response to other system components (e.g., double-duct VAV, fan-powered boxes, and retrofit systems). Care should also be taken to verify the action of the thermostat with regard to the damper position, as the velocity controller can change the control signal ratio or reverse the control signal.

In a pressure-dependent system, the setting of minimum airflows to the space, other than at no flow, is not suggested, unless the terminal box has a normally closed damper and the manufacturer of the damper actuator provides adjustable mechanical stops. The pressure-independent system requires verification that the velocity controller is operating properly. Inlet duct configuration can adversely affect the operation of the velocity controller (Griggs et al. 1990). The primary difference between the two systems is that the pressure-dependent system supplies a different amount of air to the space as the pressure upstream of the terminal box changes. If the thermostats are not calibrated properly to meet the space load, several zones may overcool or overheat. When the zones overcool and receive greater amounts of supply air than required, they decrease the amount of air that can be supplied to overheated zones. The pressure-independent system is not affected by improper thermostat calibration in the same way that a pressure-dependent system is, because the minimum and maximum airflow limits may be set for each zone.

Static Control

Static control saves energy and prevents overpressurizing the duct system. The following procedures and equipment are some of the means used to control static pressure.

No Fan Volumetric Control. This is sometimes referred to as "riding the fan curve." This type of control should be limited to systems with minimum airflows of 50% of peak design and forward-curved fans with flat pressure curves. Pressure and noise are potential problems, and this control is not energy efficient.

System Bypass Control. As the system pressure increases due to the closing of terminal boxes, a relief damper bypasses air back to the fan inlet. With this type of control, the economy of varied fan output is nonexistent, and the relief damper is usually a major source of duct leakage and noise. The relief damper should be modulated to maintain a minimum duct static pressure.

Discharge Damper. Losses and noise should be considered with this type of damper.

Vortex Damper. Losses due to inlet air conditions are a problem, and the vortex damper does not completely close. The minimum expected airflow should be evaluated.

Variable Inlet Cones. System loss can be a problem because the cone does not typically close completely. The minimum expected airflow should be evaluated.

Varying Fan Speed Mechanically. Slippage, loss of belts, cost of belt replacement, and the initial cost of components are of concern.

Variable Pitch-in-Motion Fans. Maintenance and preventing the fan from running in the stall condition must be evaluated.

Varying Fan Speed Electrically. This method of control, which varies the voltage or frequency to the fan motor, is usually the most efficient. Some versions of motor drives may cause electrical noise and affect other devices.

In controlling VAV fan systems, the location of the static pressure sensors is critical and should be field verified to give the most representative point of operation. After the terminal boxes have been proportioned, the static pressure control can be verified by observing static pressure changes at the fan discharge and the static pressure sensor as the load is simulated from maximum airflow to minimum airflow (i.e., set all terminal boxes to balanced airflow conditions and determine whether any changes in static pressure occur by placing one terminal box at a time to minimum airflow, until all terminals are placed at the minimal airflow setting). Care should be taken to verify that the maximum to minimum air volume changes are within the fan curve performance (speed or total pressure).

Diversity

Diversity may be used on a VAV system, assuming that the total airflow is lower by design and that all terminal boxes will never fully open at the same time. Care should be taken to avoid duct leakage. All ductwork upstream of the terminal box should be considered as medium-pressure ductwork, whether in a low- or medium-pressure system.

A procedure to test the total air on the system should be established by setting terminal boxes to the zero or minimum position nearest the fan. During peak load conditions, care should be taken to verify that an adequate pressure is available upstream of all terminal boxes to achieve design airflow to the spaces.

Outside Air Requirements

Maintaining the space under a slight positive or neutral pressure to atmosphere is difficult with all variable volume systems. In most systems, the exhaust requirement for the space is constant; hence, the outside air used to equal the exhaust air and meet the minimum outside air requirements for the building codes must also remain constant. Due to the location of the outside air intake and the changes in pressure, this does not usually happen. The outside air should enter the fan at a point of constant pressure (i.e., supply fan volume can be controlled by proportional static pressure control, which can control the volume of the return air fan). Makeup air fans can also be used for outside air control.

Return Air Fans

If return air fans are required in series with a supply fan, the type of control and sizing of the fans is most important. Serious over- and underpressurization can occur, especially during the economizer cycle.

Types of VAV Systems

Single-Duct VAV. This system incorporates a pressure-dependent or -independent terminal and usually has reheat at some predetermined minimal setting on the terminal unit or separate heating system.

Bypass. This system incorporates a pressure-dependent damper, which, on demand for heating, closes the damper to the space and opens to the return air plenum. Bypass sometimes incorporates a constant bypass airflow or a reduced amount of airflow bypassed to the return plenum in relation to the amount supplied to the space. No economical value can be obtained by varying the fan speed with this system. A control problem can exist if any return air sensing is done to control a warm-up or cool-down cycle.

VAV Using Single-Duct VAV and Fan-Powered, Pressure-Dependent Terminals. This system has a primary source of air from the fan to the terminal and a secondary powered fan source that pulls air from the return air plenum before the additional heat

source. This system places additional maintenance of terminal filters, motors, and capacitors on the building owner. In certain fan-powered boxes, backdraft dampers are a source of duct leakage when the system calls for the damper to be fully closed. Typical applications include geographic areas where the ratio of heating hours to cooling hours is low.

Double-Duct VAV. This type of terminal incorporates two single-duct variable terminals. It is controlled by velocity controllers that operate in sequence so that both hot and cold ducts can be opened or closed. Some controls have a downstream flow sensor in the terminal unit to maintain either the heating or the cooling. The other flow sensor is in the inlet controlled by the thermostat. As this inlet damper closes, the downstream controller opens the other damper to maintain the set airflow. Often, low pressure in the decks controlled by the thermostat causes unwanted mixing of air, which results in excess energy use or discomfort in the space. On most direct digital controls (DDC) inlet control on both ducts is favored in lieu of the downstream controller.

Balancing the VAV System

The general procedure for balancing a VAV system is

1. Determine the required maximum air volume to be delivered by the supply and return air fans. Diversity of load usually means that the volume will be somewhat less than the outlet total.
2. Obtain fan curves on these units, and request information on surge characteristics from the fan manufacturer.
3. If an inlet vortex damper control is to be used, obtain the fan manufacturer's data pertaining to the deaeration of the fan when used with the damper. If speed control is used, find the maximum and minimum speed that can be used on the project.
4. Obtain from the manufacturer the minimum and maximum operating pressures for terminal or variable volume boxes to be used on the project.
5. Construct a theoretical system curve, including an approximate surge area. The system curve starts at the minimum inlet static pressure of the boxes, plus system loss at minimum flow, and terminates at the design maximum flow. The operating range using an inlet vane damper is between the surge line intersection with the system curve and the maximum design flow. When variable speed control is used, the operating range is between (a) the minimum speed that can produce the necessary minimum box static at minimum flow still in the fan's stable range and (b) the maximum speed necessary to obtain maximum design flow.
6. Position the terminal boxes to the proportion of maximum fan air volume to total installed terminal maximum volume.
7. Set the fan to operate at approximate design speed (increase about 5% for a full open inlet vane damper).
8. Check a representative number of terminal boxes. If a wide variation in static pressure is encountered, or if the airflow at a number of boxes is below minimum at maximum flow, check every box.
9. Run a total air traverse with a pitot tube.
10. Increase the speed if static pressure and/or volume are low. If the volume is correct, but the static is high, reduce the speed. If the static is high or correct, but the volume is low, check for system effect at the fan. If there is no system effect, go over all terminals and adjust them to the proper volume.
11. Run steps (7) through (10) with the return or exhaust fan set at design flow as measured by a pitot-tube traverse and with the system set on minimum outdoor air.
12. Proportion the outlets, and verify the design volume with the VAV box on the maximum flow setting. Verify the minimum flow setting.

13. Set the terminals to minimum, and adjust the inlet vane or speed controller until minimum static pressure and airflow are obtained.

14. The temperature control personnel, the balancing personnel, and the design engineer should agree on the final placement of the sensor for the static pressure controller. This sensor must be placed in a representative location in the supply duct to sense average maximum and minimum static pressures in the system.

15. Check the return air fan speed or its inlet vane damper, which tracks or adjusts to the supply fan airflow, to ensure proper outside air volume.

16. Operate the system on 100% outside air (weather permitting), and check supply and return fans for proper power and static pressure.

Induction Systems

Most induction systems use high-velocity air distribution. Balancing should be accomplished as follows:

1. Perform steps outlined under the basic procedures common to all systems for apparatus and main trunk capacities.
2. Determine the primary airflow at each terminal unit by reading the unit plenum pressure with a manometer and locating the point on the charts (or curves) of air quantity versus static pressure supplied by the unit manufacturer.
3. Normally, about three complete passes around the entire system are required for proper adjustment. Make a final pass without adjustments to record the end result.
4. To provide the quietest possible operation, adjust the fan to run at the slowest speed that provides sufficient nozzle pressure to all units with minimum throttling of all unit and riser dampers.
5. After balancing each induction system with minimum outdoor air, reposition to allow maximum outdoor air and check power and static pressure readings.

Report Information

To be of value to the consulting engineer and owner's maintenance department, the air-handling report should consist of at least the following items:

1. *Design*
   (a) Air quantity to be delivered
   (b) Fan static pressure
   (c) Motor power installed or required
   (d) Percent of outside air under minimum conditions
   (e) Speed of the fan
   (f) Input power required to obtain this air quantity at design static pressure

2. *Installation*
   (a) Equipment manufacturer (indicate model number and serial number)
   (b) Size of unit installed
   (c) Arrangement of the air-handling unit
   (d) Nameplate power, nameplate voltage, phase, cycles, and full-load amperes of the motor installed

3. *Field tests*
   (a) Fan speed
   (b) Power readings (voltage, amperes of all phases at motor terminals)
   (c) Total pressure differential across unit components
   (d) Fan suction and fan discharge static pressure (equals fan total pressure)
   (e) Plot of actual readings on manufacturer's fan performance curve to show the installed fan operating point
   (f) Measured airflow rate

It is important to establish the initial static pressures accurately for the air treatment equipment and the duct system so that the variation in air quantity due to filter loading can be calculated. It enables the designer to ensure that the total air quantity will never be less than the minimum requirements. Because the design air quantity for peak loading of the filters has already been calculated, it also serves as a check of dirt loading in coils.

4. *Terminal Outlets*
   (a) Outlet by room designation and position
   (b) Outlet manufacture and type
   (c) Outlet size (using manufacturer's designation to ensure proper factor)
   (d) Manufacturer's outlet factor (Where no factors are available, or field tests indicate the listed factors are incorrect, a factor must be determined in the field by traverse of a duct leading to a single outlet.)
   (e) Design air quantity and the required velocity to obtain it
   (f) Test velocities and resulting air quantity
   (g) Adjustment pattern for every air terminal

5. *Additional Information (if applicable)*
   (a) Air-handling units
      (1) Belt number and size
      (2) Drive and driven sheave size
      (3) Belt position on adjusted drive sheaves (bottom, middle, and top)
      (4) Motor speed under full load
      (5) Motor heater size
      (6) Filter type and static pressure at initial use and full load; time to replace
      (7) Variations of velocity at various points across the face of the coil
      (8) Existence of vortex or discharge dampers, or both
   (b) Distribution system
      (1) Unusual duct arrangements
      (2) Branch duct static readings in double-duct and induction system
      (3) Ceiling pressure readings where plenum ceiling distribution is being used; tightness of ceiling
      (4) Relationship of building to outdoor pressure under both minimum and maximum outdoor air
      (5) Induction unit manufacturer and size (including required air quantity and plenum pressures for each unit) and a test plenum pressure and resulting primary air delivery from the manufacturer's listed curves
   (c) All equipment nameplates visible and easily readable

Many independent firms have developed detailed procedures suitable to their own operations and the area in which they function. These procedures are often available for information and evaluation upon request.

PRINCIPLES AND PROCEDURES FOR BALANCING HYDRONIC SYSTEMS

Both air- and water-side balance techniques must be performed with sufficient accuracy to ensure that the system operates economically, with minimum energy, and with proper distribution. Air-side balance requires a precise flow measuring technique because air, which is usually the prime heating or cooling transport medium, is more difficult to measure in the field. Reducing the airflow to less than the design requirement directly reduces the heat transfer. In contrast, the heat transfer rate for the water side of the terminal does not vary linearly with the water flow rate. Therefore, the proper balancing valve with the correct control characteristics must be provided at each terminal to proportionally balance the water flow rate according to the design conditions.

Heat Transfer at Reduced Flow Rate

The typical heating-only hydronic terminal gradually reduces its heat output as flow is reduced (Figure 1). Decreasing water flow to 50% of design reduces the heat transfer to 90% of that at full design flow. The control valve must reduce the water flow to 10% to reduce the heat output to 50%. The reason for the relative insensitivity to changing flow rates is that the governing coefficient for heat transfer is the air-side coefficient. A change in internal or water-side coefficient with flow rate does not materially affect the overall heat transfer coefficient. This means that (1) heat transfer for water-to-air terminals is established by the mean air-to-water temperature difference, (2) the heat transfer is measurably changed, and (3) a change in the mean water temperature requires a greater change in the water flow rate.

A secondary concern also applies to heating terminals. Unlike chilled water, hot water can be supplied at a wide range of temperatures. So, in some cases, an inadequate terminal heating capacity caused by insufficient flow can be overcome by raising the supply water temperature. Design below the temperature limit of 250°F (ASME low-pressure boiler code) must be considered.

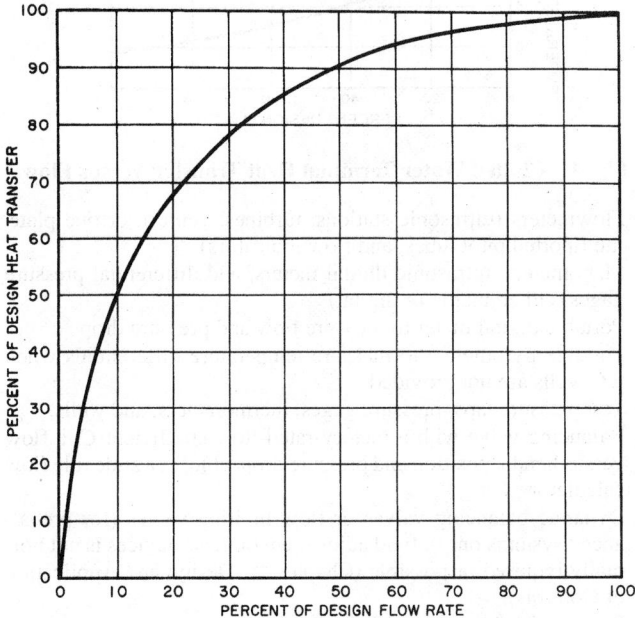

Fig. 1   Effects of Flow Variation on Heat Transfer from a Hydronic Terminal
(Design Δ*t* = 20°F and supply temperature = 200°F)

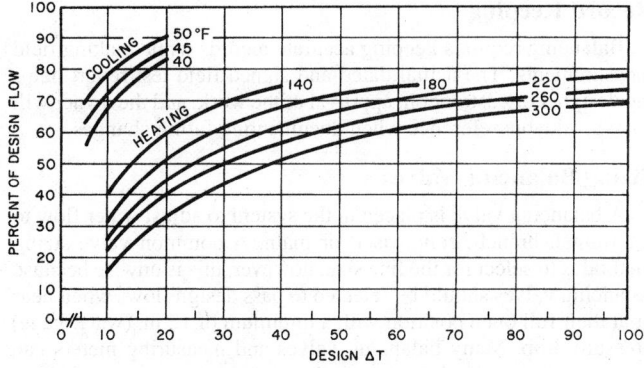

Fig. 2   Percent of Design Flow Versus Design for Various Supply Water Temperatures

The previous comments apply to heating terminals selected for a 20°F temperature drop (Δ*t*) and with a supply water temperature of about 200°F. Figure 2 shows the flow variation when 90% terminal capacity is acceptable. Note that heating tolerance decreases with temperature and flow rates and that chilled water terminals are much less tolerant of flow variation than hot water terminals.

Dual-temperature heating/cooling hydronic systems are sometimes completed and started during the heating season. Adequate heating ability in the terminals may suggest that the system is balanced. Figure 2 shows that 40% of design flow through the terminal provides 90% of design heating with 140°F supply water and a 10°F temperature drop. Increased supply water temperature establishes the same heat transfer at terminal flow rates of less than 40% design.

In some cases, dual-temperature water systems may experience a decreased flow during the cooling season because of the chiller pressure drop; this could cause a flow reduction of 25%. For example, during the cooling season, a terminal that originally heated satisfactorily would only receive 30% of the design flow rate.

While the example of reduced flow rate at Δ*t* = 20°F only affects the heat transfer by 10%, this reduced heat transfer rate may have the following negative effects:

1. The object of the system is to deliver (or remove) heat where required. When the flow is reduced from the design rate, the system must supply heating or cooling for a longer period to maintain room temperature.
2. As the load reaches design conditions, the reduced flow rate is unable to maintain room design conditions.

Terminals with lower water temperature drops have a greater tolerance for unbalanced conditions. However, larger water flows are necessary, requiring larger pipes, pumps, and pumping cost. Also, automatic valve control is more difficult.

System balance becomes more important in terminals with a large temperature difference. Less water flow is required, which reduces the size of pipes, valves, and pumps, as well as pumping costs. A more linear emission curve gives better system control.

Heat Transfer at Excessive Flow

The flow rate should not be increased above design in an effort to increase heat transfer. Figure 3 shows that increasing the flow to 200% of design only increases heat transfer by 6% while increasing the resistance or pressure drop 4 times and the power by the cube of the original power (pump laws).

Generalized Chilled Water Terminal—
Heat Transfer Versus Flow

The heat transfer for a typical chilled water coil in an air duct versus water flow rate is shown in Figure 4. The curves shown are based on ARI rating points: 45°F inlet water at a 10°F rise with entering air at 80°F dry bulb and 67°F wet bulb.

The basic curve applies to catalog ratings for lower dry-bulb temperatures providing a consistent entering air moisture content (e.g., 75°F dry bulb, 65°F wet bulb). Changes in inlet water temperature, temperature rise, air velocity, and dry- and wet-bulb temperatures will cause terminal performance to deviate from the curves. Figure 4 is only a general representation of the total heat transfer change versus flow for a hydronic cooling coil and does not apply to all chilled water terminals. Comparing Figure 4 with Figure 1 indicates the similarity of the nonlinear heat transfer and flow for both the heating and the cooling terminal.

Table 1 shows that if the coil is selected for the load, and the flow is reduced to 90% of the load, three flow variations can satisfy the reduced load at various sensible and latent combinations.

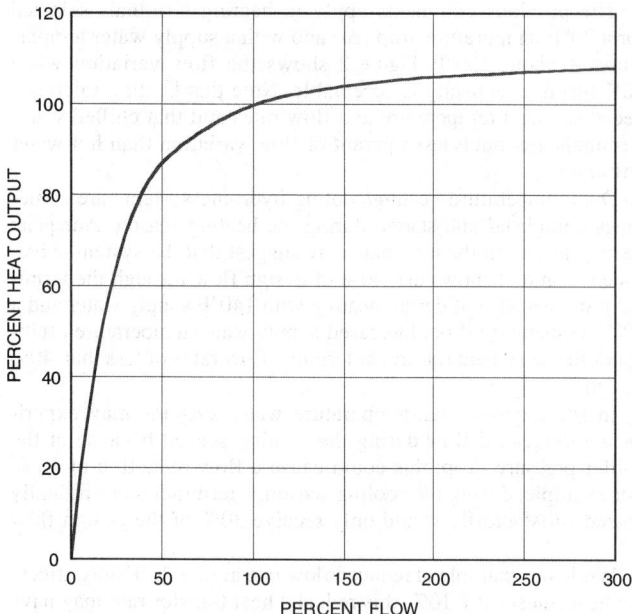

**Fig. 3   Typical Heating Coil Heat Transfer
Versus Water Flow**

Table 1   Load Flow Variation

Load Type	% Design Flow at 90% Load	Other Load, Order of %		
		Sensible	Total	Latent
Sensible	65	90	84	58
Total	75	95	90	65
Latent	90	98	95	90

Note: Dual-temperature systems are designed to chilled flow requirements and often operate on a 10°F temperature drop at full-load heating.

Flow Tolerance and Balance Procedure

The design procedure rests on a design flow rate and an allowable flow tolerance. The designer must define both the terminal's flow rates and feasible flow tolerance, bearing in mind that the cost of balancing rises with tightened flow tolerance. Also, the designer must remember that any overflow increases pumping cost, and any decrease of flow reduces the maximum heating or cooling at design conditions.

WATER-SIDE BALANCING

The water side should be balanced by direct flow measurement. This method is accurate because it avoids the compounding errors introduced by temperature difference procedures. Measuring the flow at each terminal enables proportional balancing and ultimate matching of the pump to the actual system requirements (by trimming the pump impeller or reducing the pump motor power, for example). In many cases, reduction in pump operating cost will pay for the cost of water-side balancing.

Equipment

Proper equipment selection and preplanning are needed to successfully balance hydronic systems. Circumstances sometimes dictate that flow, temperature, and pressure be measured. The designer should specify the water flow balancing devices for installation during construction and testing during the balancing of the hydronic system. The devices may consist of all or some of the following:

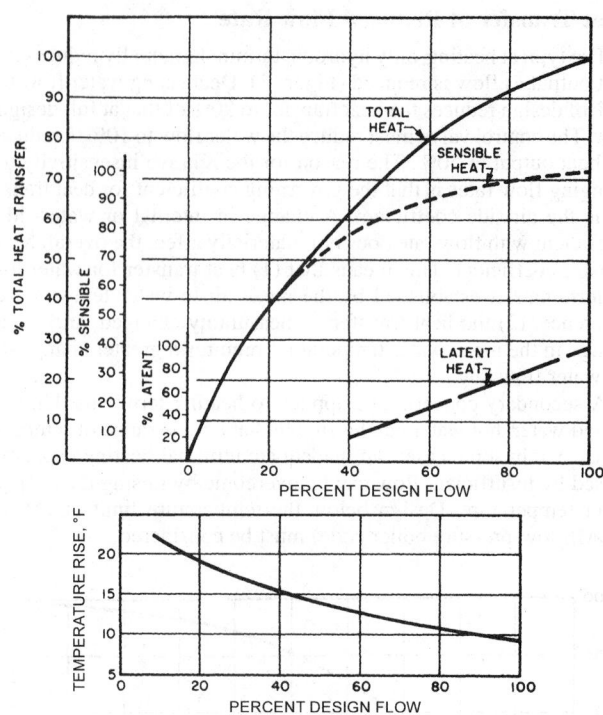

Fig. 4   Chilled Water Terminal Heat Transfer Versus Flow

- Flowmeters (ultrasonic stations, turbines, venturi, orifice plate, multiported pitot tubes, and flow indicators)
- Manometers, ultrasonic digital meters, and differential pressure gages (either analog or digital)
- Portable digital meter to measure flow and pressure drop
- Portable pyrometers to measure temperature differentials when test wells are not provided
- Test pressure taps, pressure gages, thermometers, and wells.
- Balancing valve with a factory-rated flow coefficient C, a flow versus handle position and pressure drop table, or a slide rule flow calculator
- Dynamic balancing valves or flow limiting valves (for prebalanced systems only); field adjustment of these devices is not normally required or possible (Chapter 45, Design and Application of Controls).
- Pumps with factory-certified pump curves
- Components used as flowmeters (terminal coils, chillers, heat exchangers, or control valves if using the manufacturer's factory-certified flow versus pressure drop curves); not recommended as a replacement for metering stations

Record Keeping

Balancing requires keeping accurate records while making field measurements. The actual dated and signed field test report helps the designer or customer in approval of the work, and the owner will have a valuable reference when documenting future changes.

Sizing Balancing Valves

A balancing valve is placed in the system to adjust water flow to a terminal, branch, zone, riser, or main. A common valve-sizing method is to select for the line size; however, this is unwise because balancing valves should be selected to pass design flows when near or at their full open position with a minimum of 12 in. (water gage) pressure drop. Many balancing valves and measuring meters can give an accuracy of ±5% of range down to a pressure drop of 12 in. (water gage) with the balancing valve in the wide open position. Too large a balancing valve pressure drop will affect the performance and flow characteristic of the control valve. Equation (2) may be used to determine the flow coefficient C_v for a balancing valve. The equation is also used to size a control valve.

The flow coefficient C_v is defined as the number of gallons of water per minute (U.S.) that flows through a wide-open valve with a pressure drop of 1 psi at 60°F. This is shown as

$$C_v = Q \sqrt{s_f/\Delta p} \qquad (2)$$

If pressure drop is determined in feet of water, Equation (2) can be shown as

$$C_v = 1.5 Q \sqrt{s_f/\Delta h} \qquad (3)$$

where

C_v = flow coefficient at 1 psi drop
Q = design flow for terminal or valve, gpm
Δp = pressure drop, psi
Δh = pressure drop, ft of water
s_f = specific gravity of fluid

HYDRONIC BALANCING METHODS

A variety of techniques is used to balance hydronic systems. Balance by temperature difference and water balance by proportional method are the most common.

Balance by Temperature Difference

This often-used balancing procedure is based on measurement of the water temperature difference between supply and return at the terminal. The designer selects the cooling and/or heating terminal for a calculated design load at full-load conditions. At less than full load, which is true for most operating hours, the temperature drop is proportionately less. Figure 5 demonstrates this relationship for a heating system at a design Δt of 20°F for outside design of –10°F and room design of 70°F.

For every outside temperature other than design, the balancing technician should construct a similar chart and read off the Δt for balancing. For example, at 50% load, or 30°F outdoor air, the Δt required is 10°F, or 50% of the design drop.

The temperature balance method is a rough approximation and should not be used where great accuracy is required. It is not accurate enough for a cooling system or heating systems with a large temperature drop.

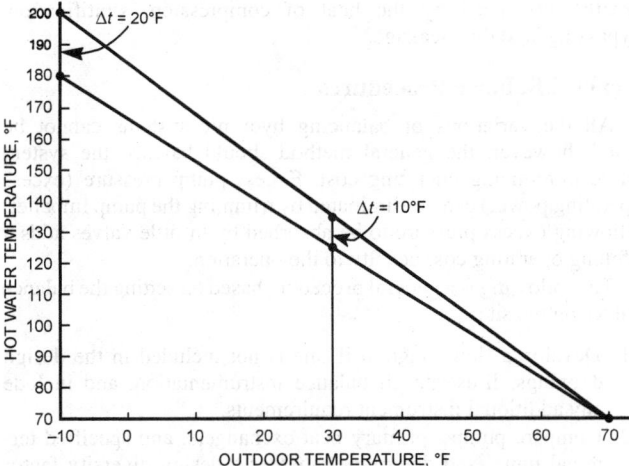

Fig. 5 Water Temperature Versus Outdoor Temperature Showing Approximate Temperature Difference

Water Balance by Proportional Method

The proportional method of water-side balance uses as-built conditions and adapts well to design diversity factors. Circuits in a system are proportionately balanced to each other by a flow quotient, which is the actual flow rate divided by the design flow rate through a circuit:

$$\text{Flow quotient} = \frac{\text{Actual flow rate}}{\text{Design flow rate}} \qquad (4)$$

To balance a branch system proportionally, first open the balancing valves and the control valves in that branch to their wide-open position, and then calculate each balancing valve's quotient based on actual measurements at each terminal. Record these values on the test form, and note the circuit with the lowest flow quotient.

The reference circuit has the lowest quotient and is the circuit with the greatest pressure loss. Adjust all the other balancing valves in that branch until they have the same quotient as the reference circuit (at least one valve in the branch should be fully open).

Note that when a second valve is adjusted, the flow quotient in the reference valve will also change, and continued adjustment is required to make their flow quotients equal. Once they are equal, they will remain equal or in proportional balance to each other while other valves in the branch are adjusted or there is a change in pressure or flow.

When all the balancing valves are adjusted to their branch's respective flow quotient, the total system water flow is adjusted to the design by setting the balancing valve at the pump discharge to a flow quotient of one.

The pressure drop across the balancing valve at the pump discharge is pressure produced by the pump that is not required in providing the design flow. This excess pressure can be removed by trimming the pump impeller or reducing the pump speed. Once the trimming is done, the pump discharge balancing valve must be reopened to its wide-open position to provide the design flow.

As in variable-speed pumping, diversity and flow changes are well accommodated by a system that has been proportionately balanced. Since the balancing valves have been balanced to each other at a particular flow (design), any changes in that flow are proportionately distributed.

Balancing of the water side in a system that uses diversity must be done at full flow. Because the components are selected based on heat transfer at full flow, they must be balanced to this point. To accomplish full-flow proportional balance, shut off part of the system while balancing the remaining sections. When a section has been balanced, shut it off and open the section that was open originally to complete full balance of the system. When balancing, proper care should be taken if the building is occupied or if near full load conditions exist.

Variable-Speed Pumping. To achieve hydronic balance, full flow through the system is required during the balancing procedure. Once balance is achieved, the system can be placed on automatic control and the pump speed allowed to change.

After the full-flow condition is balanced and the system differential pressure set point is established, to control the variable-speed pumps, observe the flow on the circuit with the greatest resistance as the other circuits are closed one at a time. The flow in the observed circuit should remain equal to, or more than, the previously set flow. Note that water flow may become laminar at less than 2 fps, which alters the heat transfer characteristics of the system.

Other Balancing Techniques

Flow Balancing by Rated Differential Procedure. This procedure depends on deriving a performance curve for the test coil that compares water temperature difference Δt_w to entering water temperature t_{ew} minus entering air temperature t_{ea}. One point of the desired curve can be determined from the manufacturer's ratings since these are published as $(t_{ew} - t_{ea})$. A second point is established by observing that the heat transfer from air to water is zero when $(t_{ew} - t_{ea})$ is zero (consequently, $\Delta t_w = 0$). With these two points, an approximate performance curve can be drawn (see Figure 6). Then, for any other $(t_{ew} - t_{ea})$, this curve is used to determine the appropriate Δt_w.

Example 1. From the following manufacturer certified data, determine the required Δt_w:

Capacity = 10,000 Btu/h
t_{ew} = 200°F
t_{ea} = 60°F
Water flow = 1.5 gpm
c_p = 1.0 Btu/lb·°F
ρ = 8.33 lb/gal

Solution:

1. Calculate rated Δt_w.

$$\Delta t_w = \frac{10,000 \text{ Btu/h}}{1 \text{ Btu/lb} \cdot °F \times 8.33 \text{ lb/gal} \times 60 \text{ min/h} \times 1.5 \text{ gpm}}$$
$$= 13.33°F$$

2. Construct a performance curve as illustrated in Figure 6.

3. From test data:

$$t_{ew} = 180°F$$
$$t_{ea} = 70°F$$
$$t_{ew} - t_{ea} = 110°F$$

4. From Figure 6 read $\Delta t_w = 10.5°F$, which is required to balance water flow at 1.5 gpm. The water temperature difference may also be calculated as proportion of the rate value as follows:

$$\frac{(t_{ew} - t_{ea})_{test}}{(t_{ew} - t_{ea})_{rated}}(\Delta t_w)_{rated} = (\Delta t_w)_{required}$$

$$\frac{(180 - 70)}{(200 - 60)}13.3 = 10.5°F$$

This procedure is useful for balancing terminal devices such as finned tube convectors, where flow measuring devices do not exist

and where airflow measurements cannot be made. It may also be used for cooling coils for sensible transfer (dry coil).

Flow Balancing by Total Heat Transfer. This procedure determines water flow by running an energy balance around the coil. From field measurements of airflow, wet- and dry-bulb temperatures both upstream and downstream of the coil, and the difference Δt_w between the entering and leaving water temperatures, water flow can be determined by the following equations:

$$Q_w = q/500 \Delta t_w \tag{5}$$

$$q_{cooling} = 4.5 Q_a(h_1 - h_2) \tag{6}$$

$$q_{heating} = 1.08 Q_a(t_1 - t_2) \tag{7}$$

where

Q_w = water flow rate, gpm
q = load, Btu/h
$q_{cooling}$ = cooling load, Btu/h
$q_{heating}$ = heating load, Btu/h
Q_a = airflow rate, cfm
h = enthalpy, Btu/lb
t = temperature, °F

Example 2. Find the water flow for a cooling system having the following characteristics:

Test data

t_{ewb} = entering wet-bulb temperature = 68.5°F
t_{lwb} = leaving wet-bulb temperature = 53.5°F
Q_a = airflow rate = 22,000 cfm
t_{lw} = leaving water temperature = 59.0°F
t_{ew} = entering water temperature = 47.5°F

From psychrometric chart

h_1 = 32.84 Btu/lb
h_2 = 22.32 Btu/lb

Solution: From Equations (5) and (6),

$$Q_w = \frac{4.5 \times 22,000(32.84 - 22.32)}{500(59.0 - 47.5)} = 181 \text{ gpm}$$

The desired water flow is achieved by successive manual adjustments and recalculations. Note that these temperatures can be greatly influenced by the heat of compression, stratification, bypassing, and duct leakage.

General Balance Procedures

All the variations of balancing hydronic systems cannot be listed; however, the general method should balance the system while minimizing operating cost. Excess pump pressure (excess operating power) can be eliminated by trimming the pump impeller. Allowing excess pressure to be absorbed by throttle valves adds a lifelong operating cost penalty to the operation.

The following is a general procedure based on setting the balance valves on the site:

1. Develop a flow diagram if one is not included in the design drawings. Illustrate all balance instrumentation, and include any additional instrument requirements.
2. Compare pumps, primary heat exchangers, and specified terminal units; and determine whether a design diversity factor can be achieved.
3. Examine the control diagram and determine the control adjustments needed to obtain design flow conditions.

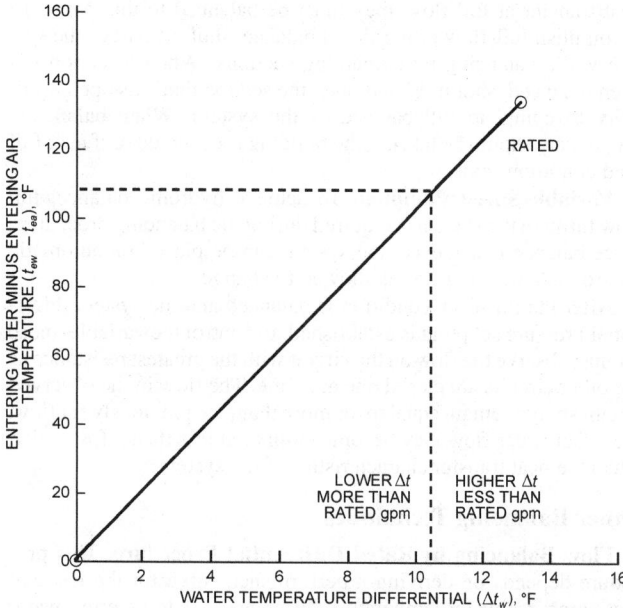

Fig. 6 Coil Performance Curve

Balance Procedure—Primary and Secondary Circuits

1. Inspect the system completely to ensure that (a) it has been flushed out, it is clean, and all air is removed; (b) all manual valves are open or in operating position; (c) all automatic valves are in their proper positions and operative; and (d) the expansion tank is properly charged.
2. Place the controls in position for design flow.
3. Examine the flow diagram and piping for obvious short circuits; check flow and adjust the balance valve.
4. Take pump suction, discharge, and differential pressure readings at both full flow and no flow. For larger pumps, a no flow condition may not be safe. In any event, valves should be closed slowly.
5. Read pump motor amperage and voltage, and determine approximate power.
6. Establish a pump curve, and determine the approximate flow rate.
7. If a total flow station exists determine the flow, and compare this flow with the pump curve flow.
8. If possible, set the total flow about 10% high using the total flow station first and the pump differential pressure second; then maintain pumped flow at a constant value as balance proceeds by adjusting the pump throttle valve.
9. If branch main flow stations exist, these should be tested and set, starting by setting the shortest runs low as balancing proceeds to the longer branch runs.
10. With primary and secondary pumping circuits, a reasonable balance must be obtained in the primary loop before the secondary loop can be considered. The secondary pumps must be running and terminal units must be open to flow when the primary loop is being balanced, unless the secondary loop is decoupled.

FLUID FLOW MEASUREMENT

Flow Measurement Based on Manufacturer's Data

Any component (terminal, control valve, or chiller) that has an accurate, factory-certified flow/pressure drop relationship can be used as a flow-indicating device. The flow and pressure drop may be used to establish an equivalent flow coefficient as shown in Equation (3). According to the Bernoulli equation, pressure drop varies as the square of the velocity or flow rate, assuming density is constant:

$$Q_1^2 / Q_2^2 = \Delta h_1 / \Delta h_2 \qquad (8)$$

For example, a chiller has a certified pressure drop of 25 ft of water at 100 gpm. The calculated flow with a field-measured pressure drop of 30 ft is

$$Q_2 = 100 \sqrt{30/25} = 109.5 \text{ gpm}$$

Flow calculated in this manner is only an estimate. The accuracy of components used as flow indicators depends on (1) the accuracy of cataloged information concerning flow/pressure drop relationships and (2) the accuracy of the pressure differential readings. As a rule, the component should be factory-certified flow tested if it is to be used as a flow indicator.

Pressure Differential Readout by Gage

Gages are used to read differential pressures. Gages are usually used for high differential pressures and manometers for lower differentials. Accurate gage readout is diminished when two gages are used, especially when the gages are permanently mounted and, as such, subject to malfunction.

A single high-quality gage should be used for differential readout (Figure 7). This gage should be alternately valved to the high- and low-pressure side to establish the differential. A single gage needs no static height correction, and errors caused by gage calibration are eliminated.

Differential pressure can also be read from differential gages, thus eliminating the need to subtract outlet from inlet pressures to establish differential pressure. Differential pressure gages are usually dual gages mechanically linked to read differential pressure. The differential pressure gage readout can be stated in terms of psi or in feet of head of 60°F water.

Conversion of Differential Pressure to Head

Pressure gage readings can be restated to fluid head, which is a function of fluid density. The common hydronic system conversion factor is related to water density at about 60°F; 1 psi equals 2.31 ft. Pressure gages can be calibrated to feet of water head using this conversion. Because the calibration only applies to water at 60°F, the readout may require correction when the gage is applied to water at a significantly higher temperature.

Pressure gage conversion and correction factors for various fluid specific gravities (in relation to water at 60°F) are shown in Table 2. The differential gage readout should only be defined in terms of the head of the fluid actually causing the flow pressure differential. When this is done, the resultant fluid head can be applied to the C_v to determine actual flow through any flow device, provided the manufacturer has correctly stated the flow to fluid head relationship.

For example, a manufacturer may test a boiler or control valve with 100°F water. If the test differential pressure is converted to head at 100°F, a C_v independent of test temperature and density may

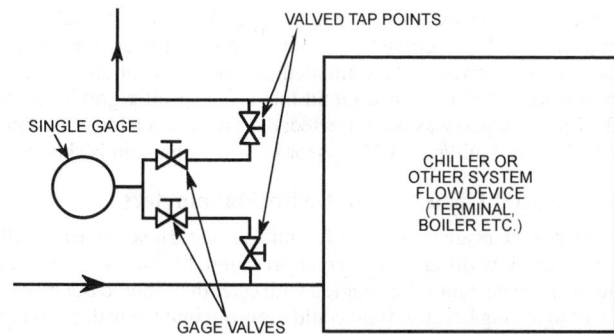

Fig. 7   Single Gage for Reading Differential Pressure

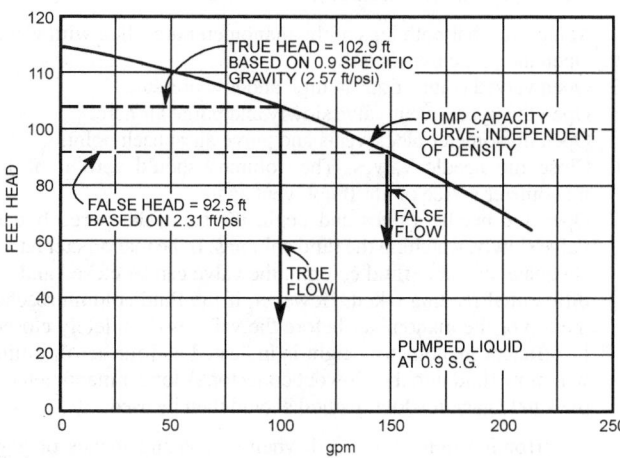

Fig. 8   Fluid Density Correction Chart for Pump Curves

Table 2   Differential Pressure Conversion to Head

Fluid Specific Gravity	Corresponding Water Temperature, °F	Foot Fluid Head Equal to 1 psi[a]	Correction Factor When Gage is Stated to Feet of Water (60°F)[b]
1.5		1.54	
1.4		1.65	
1.3		1.78	
1.2		1.93	
1.1		2.10	
1.0	60	2.31	1.00
0.98	150	2.36	1.02
0.96	200	2.41	1.04
0.94	250	2.46	1.065
0.92	300	2.51	1.09
0.90	340	2.57	1.11
0.80		2.89	
0.70		3.30	
0.60		3.85	
0.50		4.63	

[a]Differential psi readout is multiplied by this number to obtain feet fluid head when gage is calibrated in psi.
[b]Differential feet water head readout is multiplied by this number to obtain feet fluid head when gage calibration is stated to feet head of 60°F water.

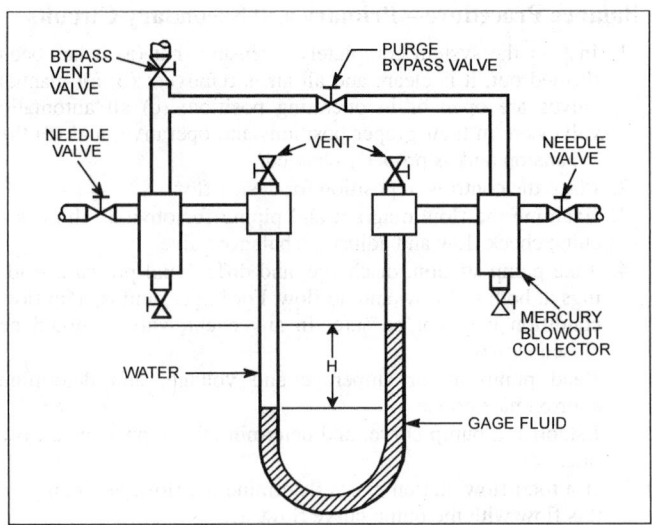

Fig. 9   Fluid Manometer Arrangement for Accurate Reading and Blowout Protection

be calculated. Differential pressures from another test made in the field at 250°F may be converted to head at 250°F. The C_v calculated with this head is also independent of temperature. The manufacturer's data can then be directly correlated with the field test to establish flow rate at 250°F.

A density correction must be made to the gage reading when differential heads are used to estimate pump flows as in Figure 8. This is because of the shape of the pump curve. An incorrect head difference entry into the curve due to an uncorrected gage reading can cause a major error in the estimated pumped flow. In this case, the gage reading for a pumped liquid that has a specific gravity of 0.9 (2.57 ft liquid/psi) was not corrected; the gage conversion is assumed to be 2.31 ft liquid/psi. A 50% error in flow estimation is shown.

Differential Head Readout with Manometers

Manometers are used for differential pressure readout, especially when very low differentials, great precision, or both, are required. But manometers must be handled with care; they should not be used for field testing because fluid could blow out into the water and rapidly deteriorate the components. A proposed manometer arrangement is shown in Figure 9.

Figure 9 and the following instructions provide accurate manometer readings with minimum risk of blowout.

1. Make sure that both legs of the manometer are filled with water.
2. Open the purge bypass valve.
3. Open valved connections to high and low pressure.
4. Open the bypass vent valve slowly and purge air here.
5. Open manometer block vents and purge air at each point.
6. Close the needle valves. The columns should zero in if the manometer is free of air. If not, vent again.
7. Open the needle valves and begin throttling the purge bypass valve slowly, watching the fluid columns. If the manometer has an adequate available fluid column, the valve can be closed and the differential reading taken. However, if the fluid column reaches the top of the manometer before the valve is completely closed, insufficient manometer height is indicated and further throttling will blow fluid into the blowout collector. A longer manometer or the single gage readout method should then be used.

An error is often introduced when converting inches of gage fluid to feet of the test fluid. The conversion factor changes with test fluid temperature, density, or both. Conversion factors shown

in Table 2 are to a water base, and the counterbalancing water height H (Figure 9) is at room temperature.

Orifice Plates, Venturi, and Flow Indicators

Manufacturers provide flow information for several devices used in hydronic system balance. In general, the devices can be classified as (1) orifice flowmeters, (2) venturi flowmeters, (3) velocity impact meters, (4) pitot-tube flowmeters, (5) bypass spring impact flowmeters, (6) calibrated balance valves, (7) turbine flowmeters, and (8) ultrasonic flowmeters.

The **orifice flowmeter** is widely used and is extremely accurate. The meter is calibrated and shows differential pressure versus flow. Accuracy generally increases as the pressure differential across the meter increases. The differential pressure readout instrument may be a manometer, differential gage, or single gage (Figure 7).

The **venturi flowmeter** has lower pressure loss than the orifice plate meter because a carefully formed flow path increases velocity head recovery. The venturi flowmeter is placed in a main flow line where it can be read continuously.

Velocity impact meters have precise construction and calibration. The meters are generally made of specially contoured glass or plastic, which permits observation of a flow float. As flow increases, the flow float rises in the calibrated tube to indicate flow rate. Velocity impact meters generally have high accuracy.

A special version of the velocity impact meter is applied to hydronic systems. This version operates on the velocity head difference between the pipe side wall and the pipe center, which causes fluid to flow through a small flowmeter. Accuracy depends on the location of the impact tube and on a velocity profile that corresponds to theory and the laboratory test calibration base. Generally, the accuracy of this **bypass flow impact** or differential velocity head flowmeter is less than a flow-through meter, which can operate without creating a pressure loss in the hydronic system.

The **pitot-tube flowmeter** is also used for pipe flow measurement. Manometers are generally used to measure velocity head differences because these differences are low.

The **bypass spring impact flowmeter** uses a defined piping pressure drop to cause a correlated bypass side branch flow. The side branch flow pushes against a spring that increases in length with increased side branch flow. Each individual flowmeter is calibrated to relate extended spring length position to main flow. The bypass spring impact flowmeter has, as its principal merit, a direct readout. However, dirt on the spring reduces accuracy. The bypass

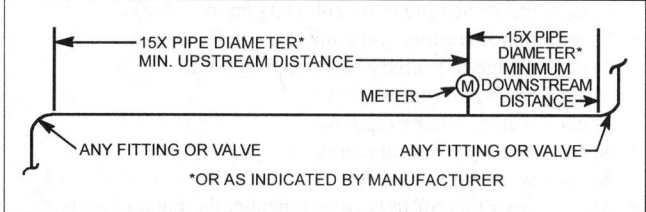

Fig. 10   Minimum Installation Dimensions for Flowmeter

is opened only when a reading is made. Flow readings can be taken at any time.

The **calibrated balance valve** is an adjustable orifice flowmeter. Balance valves can be calibrated so that a flow/pressure drop relationship can be obtained for each incremental setting of the valve. A ball, rotating plug, or butterfly valve may have its setting expressed in percent open or degree open; a globe valve, in percent open or number of turns. The calibrated balance valve must be manufactured with precision and care to ensure that each valve of a particular size has the same calibration characteristics.

The **turbine flowmeter** is a mechanical device. The velocity of the liquid spins a wheel in the meter, which generates a 4 to 20 mA output that may be calibrated in units of flow. The meter must be well maintained, as wear or water impurities on the bearing may slow the wheel, and debris may clog or break the wheel.

The **ultrasonic flowmeter** senses sound signals, which are calibrated in units of flow. The ultrasonic metering station may be installed as part of the piping, or it may be a strap-on meter. In either case, the meter has no moving parts to maintain, nor does it intrude into the pipe and cause a pressure drop. Two distinct types of ultrasonic meter are available: (1) the transit time meter for HVAC or clear water systems and (2) the Doppler meter for systems handling sewage or large amounts of particulate matter.

If any of the above meters are to be useful, the minimum distance of straight pipe upstream and downstream, as recommended by the meter manufacturer and flow measurement handbooks, must be adhered to. Figure 10 presents minimum installation suggestions.

Using a Pump as an Indicator

Although the pump is not a meter, it can be used as an indicator of flow together with the other system components. Differential pressure readings across a pump can be correlated with the pump curve to establish the pump flow rate. Accuracy depends on (1) accuracy of readout, (2) pump curve shape, (3) actual conformance of the pump to its published curve, (4) pump operation without cavitation, (5) air-free operation, and (6) velocity head correction.

When a differential pressure reading must be taken, a single gage with manifold provides the greatest accuracy (Figure 11). The pump suction to discharge differential can be used to establish pump differential pressure and, consequently, pump flow rate. The single gage and manifold may also be used to check for strainer clogging by measuring the pressure differential across the strainer.

If the pump curve is based on fluid head, pressure differential, as obtained from the gage reading, needs to be converted to head, which is pressure divided by the fluid weight per cubic foot. The pump differential head is then used to determine pump flow rate (Figure 12). As long as the differential head used to enter the pump curve is expressed as head of the fluid being pumped, the pump curve shown by the manufacturer should be used as described. The pump curve may state that it was defined by test with 85°F water. This is unimportant, since the same curve applies from 60 to 250°F water, or to any fluid within a broad viscosity range.

Generally, pump-derived flow information, as established by the performance curve, is questionable unless the following precautions are observed:

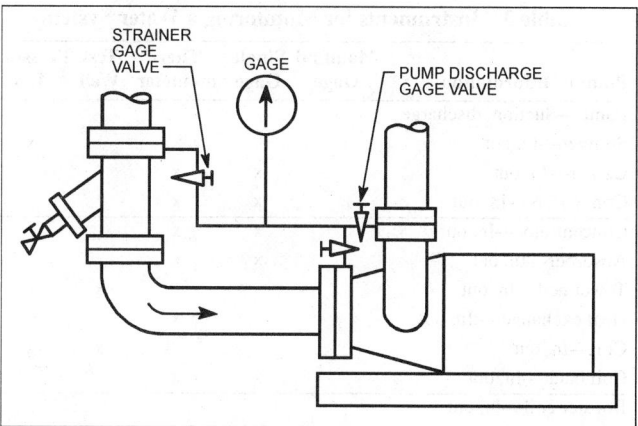

Fig. 11   Single Gage for Differential Readout Across Pump and Strainer

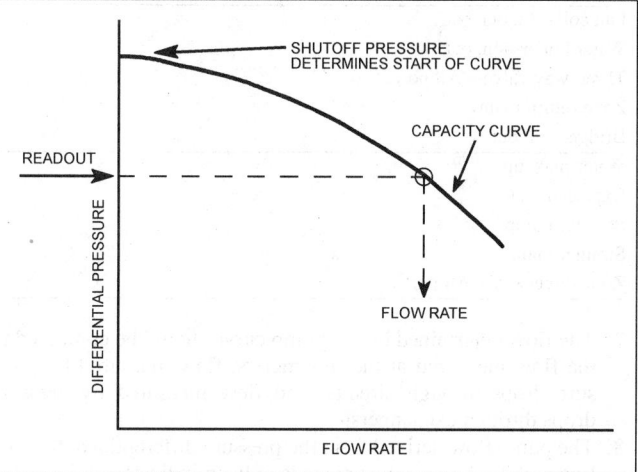

Fig. 12   Differential Pressure Used to Determine Pump Flow

1. The installed pump should be factory calibrated by a test to establish the actual flow-pressure relationship for that particular pump. Production pumps can vary from the cataloged curve because of minor changes in impeller diameter, interior casting tolerances, and machine fits.

2. When a calibration curve is not available for a centrifugal pump being tested, the discharge valve can be closed briefly to establish the no-flow shutoff pressure, which can be compared to the published curve. If the shutoff pressure differs from that published, draw a new curve parallel to the published curve. While not exact, the new curve will usually fit the actual pumping circumstance more accurately. Clearance between the impeller and casing minimize the danger of damage to the pump during a no-flow test, but manufacturer verification is necessary.

3. Differential head should be determined as accurately as possible, especially for pumps with flat flow curves.

4. The pump should be operating air-free and without cavitation. A cavitating pump will not operate to its curve, and differential readings will provide false results.

5. Ensure that the pump is operating above the minimum net positive suction head.

6. Power readings can be used (1) as a check for the operating point when the pump curve is flat or (2) as a reference check when there is suspicion that the pump is cavitating or providing false readings because of air.

Table 3   Instruments for Monitoring a Water System

Point of Information	Manifold Gage	Single Gage	Thermometer	Test Well	Pressure Tap
Pump—Suction, discharge	x				
Strainer—In, out					x
Cooler—In, out		x	x		
Condensers—In, out		x	x		
Concentrator—In, out		x	x		
Absorber—In, out		x	x		
Tower cell—In, out				x	x
Heat exchanger—In, out	x		x		
Coil—In, out				x	x
Coil bank—In, out		x	x		
Booster coil—In, out					x
Cool panel—In, out					x
Heat panel—In, out				x	x
Unit heater—In, out					x
Induction—In, out					x
Fan coil—In, out					x
Water boiler—In, out			x		
Three-way valve—All ports					x
Zone return main			x		
Bridge—In, out			x		
Water makeup		x			
Expansion tank		x			
Strainer pump					x
Strainer main	x				
Zone three-way—All ports				x	x

7. The flow determined by the pump curve should be compared to the flow measured at the flowmeters, flow measured by pressure drops through circuits, and flow measured by pressure drops through exchangers.
8. The pump flow derived from the pressure differential at the suction and discharge connections is only an indicator of the actual flow—it cannot be used to verify the test and balance measurements. If the pump flow is to be used for balancing verification it needs to be determined using the Hydraulic Institute procedure or by measuring the flow through a properly installed metering station 15 to 20 straight pipe diameters downstream from the pump discharge.

The power draw should be measured in watts. Ampere readings cannot be trusted because of voltage and power factor problems. If motor efficiency is known, the wattage drawn can be related to pump brake power (as described on the pump curve) and the operating point determined.

Central Plant Chilled Water Systems

For existing installations, establishing accurate thermal load profiles is of prime importance because it establishes proper primary chilled water supply temperature and flow. In new installations, actual load profiles can be compared with design load profiles to obtain valid operating data.

To perform proper testing and balancing, all interconnecting points between the primary and secondary systems must be designed with sufficient temperature, pressure, and flow connections so that adequate data may be indicated and/or recorded.

Water Flow Instruments

As indicated previously, proper location and use of instruments is vital to accurate balancing. Instruments for testing temperature and pressure at various locations are listed in Table 3. Flow-indicating devices should be placed in water systems as follows:

- At each major heating coil bank (10 gpm or more)
- At each major cooling coil bank (10 gpm or more)
- At each bridge in primary-secondary systems
- At each main pumping station
- At each water chiller evaporator
- At each water chiller condenser
- At each water boiler outlet
- At each floor takeoff to booster reheat coils, fan coil units, induction units, ceiling panels, and radiation (Do not exceed 25 terminals off of any one zone meter probe.)
- At each vertical riser to fan coil units, induction units, and radiation
- At the point of tie-in to existing systems

STEAM DISTRIBUTION

Procedures for Steam Balancing Variable Flow Systems

Steam distribution cannot be balanced by adjustable flow-regulating devices. Instead, fixed restrictions built into the piping in accordance with carefully designed pipe and orifice sizes are used to regulate flow.

It is important to have a balanced distribution of steam to all portions of the steam piping at all loads. This is best accomplished by properly designing the steam distribution piping, which includes carefully considering steam pressure, steam quantities required by each branch circuit, pressure drops, steam velocities, and pipe sizes. Just as other flow systems are balanced, steam distribution systems are balanced by ensuring that the pressure drops are equalized at design flow rates for all portions of the piping. Only marginal balancing can be done by pipe sizing. Therefore, additional steps must be taken to achieve a balanced performance.

Steam flow balance can be improved by using spring-type packless supply valves equipped with precalibrated orifices. The valves should have a tight shutoff between 25 in. of Hg and 60 psig. These valves have a nonrising stem, are available with a lockshield, and have a replaceable disk. Orifice flanges can also be used to regulate and measure steam flow at appropriate locations throughout the system. The orifice sizes are determined by the pressure drop required for a given flow rate at a given location. A schedule should be prepared showing (1) orifice sizes, (2) valve or pipe sizes, (3) required flow rates, and (4) corresponding pressure differentials for each flow rate. It may be useful to calculate pressure differentials for several flow rates for each orifice size. Such a schedule should be maintained for future reference.

After the appropriate regulating orifices are installed in the proper locations, the system should be tested for tightness by sealing all openings in the system and applying a vacuum of 20 in. of Hg, held for 2 hours. Next, the system should be readied for warm-up and pressurizing with steam following the procedures outlined in Section VI of the ASME *Boiler and Pressure Vessel Code*. After the initial warm-up and system pressurization, evaluate system steam flow, and compare it to system requirements. The orifice schedule calculated earlier will now be of value should any of the orifices need to be changed.

Steam Flow Measuring Devices

Many devices are available for measuring flow in steam piping: (1) steam meters, (2) condensate meters, (3) orifice plates, (4) venturi fittings, (5) steam recorders, and (6) manometers for reading differential pressures across orifice plates and venturi fittings. Some of these devices are permanently affixed to the piping system to facilitate taking instantaneous readings that may be necessary for proper operation and control. A surface pyrometer used in conjunction with a pressure gage is a convenient way to determine steam saturation temperature and the degree of superheat at various locations in the system. Such information can be used to evaluate performance characteristics.

COOLING TOWERS

Field testing cooling towers is a demanding and difficult task. ASME *Standard* PTC 23 (1986) and CTI *Standard Specification* ATC-105 (1997) establish procedures for these tests. Certain general guidelines for testing cooling towers are as follows:

Conditions at Time of Test

- Water flow within 15% of design flow
- Heat load within 30% of design heat load and stabilized
- Entering wet bulb within 12°F of entering design wet bulb

Using the above limitations and field readings that are as accurate as possible, a projection to design conditions produces an accuracy of ±5% of tower performance.

Conditions for Performing Test

- Water-circulating system serving the tower should be thoroughly cleaned of all dirt and foreign matter. Samples of water should be clear and indicate clear passage of water through pumps, piping, screens, and strainers.
- Fans serving the cooling tower should operate in proper rotation. Foreign obstructions should be removed. Permanent obstruction should be noted.
- Interior filling of cooling tower should be clean and free of foreign materials such as scale, algae, or tar.
- Water level in the tower basin should be maintained at the proper level. Visually check the basin sump during full flow to determine that the centrifugal action of the water is not causing entrainment of air, which could cause pump cavitation.
- Water-circulating pumps should be tested with full flow through the tower. If flow exceeds design, it should be valved down until design is reached. The flow that is finally set should be maintained throughout the test period. All valves, except necessary balancing valves, should be in the full-open position.
- If makeup and blowdown have facilities to determine flow, set them to design flow at full flow through the tower. If flow cannot be determined, shut off both.

Instruments

Testing and balancing agencies provide instruments to perform the required tests. Mechanical contractors provide and install all components such as orifice plates, venturis, balancing valves, thermometer wells, gage cocks, and corporation cocks. Designers specify measuring point locations.

Instruments used should be recently calibrated as follows:

Temperature

- Thermometer with divisions of 0.2°F in a proper well for water should be used.
- Thermometer with solar shield and 0.2°F divisions or thermocouple with 0.2°F readings having mechanical aspiration for wet-bulb readings should be used.
- Sling psychrometer may be used for rough checks.
- Thermometer with 0.2°F divisions should be used for dry-bulb readings.

Water Flow

- Orifice, venturi, and balancing valve pressure drops can be read using a manometer or a recently calibrated differential pressure gage.
- Where corporation cocks are installed, a pitot tube and manometer traverse can be made by trained technicians.

Test Method

The actual test consists of the following steps:

1. Conduct water flow tests to determine volume of water on the tower, and volume of makeup and blowdown water.
2. Conduct water temperature tests, if possible, in suitable wells as close to the tower as possible. Temperature readings at pumps or the condensing element are not acceptable in tower evaluation. If there are no wells, surface pyrometer readings are acceptable.
3. Take makeup water volume and temperature readings at the point of entry into the tower.
4. Take blowdown volume and temperature readings at the point of discharge from the tower.
5. Take inlet and outlet dry- and wet-bulb temperature readings using the prescribed instruments.
   (a) Use wet-bulb entering and leaving temperatures to determine tower actual performance against design.
   (b) Use wet- and dry-bulb entering and leaving temperatures to determine evaporation involved.
6. If the tower has a ducted inlet or outlet where a reasonable duct traverse can be made, use this air volume as a cross-check of tower performance.
7. Take wet- and dry-bulb temperature readings between 3 and 5 ft from the tower on all inlet sides. These readings shall be taken halfway between the base and the top of the inlet louvers at no more than 5-ft spacing horizontally; then they should be averaged. Note any unusual inlet conditions.
8. Note wind velocity and direction at the time of test.
9. Take test readings continually with a minimum time lapse between readings.
10. If the first test indicates a tower deficiency, perform two additional tests to verify the original readings.

TEMPERATURE CONTROL VERIFICATION

The test and balance technician should work closely with the temperature control installer to ensure that the project is completed correctly. The balancing technician needs to verify proper operation of the control and communicate findings back to the agency responsible for ensuring that the controls have been installed correctly. This is usually the HVAC system designer, although others may be involved. Generally, the balancing technician does not adjust, relocate, or calibrate the controls. However, this is not always the case, and differences do occur with VAV terminal unit controllers. The balancing technician should be familiar with the specifications and design intent of the project so that all responsibilities are understood.

During the design and specification phase of the project, the designer should specify verification procedures for the controls and responsibilities for the contractor who installs the temperature controls. It is important that the designer specify (1) the degree of coordination between the installer of the control and the balancing technician and (2) the testing responsibilities of each.

Verification of control operation starts with the balancing technician reviewing the submitted documents and shop drawings of the control system. In some cases the controls technician should instruct the balancing technician in the operation of certain control elements, such as digital terminal unit controllers. This is followed by schedule coordination between the control and balancing technicians. In addition, the balancing technician and the controls technician need to work together when reviewing the operation of some sections of the HVAC system (not all). This is particularly important with regard to VAV systems and the setting of the flow measurement parameters in digital terminal unit controllers.

Major mechanical systems should be verified after the testing, adjusting, and balancing is completed. The control system should be operated in stages to prove its capability of matching system capacity to varying load conditions. Mechanical subsystem controllers should be verified when balancing data is collected, considering that the entire system may not be completely functional at the time of verification. Testing and verification should account for seasonal variations; tests should be performed under varying outdoor loads to assure operational performance. A retest of a random sampling of terminal units may be desirable to verify the work of the control technician.

Suggested Procedures

Although both electrical and pneumatic controls are used, the following verification procedures may be used with either.

1. Obtain design drawings and documentation, and become well acquainted with the design intent and specified responsibilities.
2. Obtain copies of approved control shop drawings.
3. Compare design to installed field equipment.
4. Obtain recommended operating and test procedures from the manufacturers.
5. Verify with the control contractor that all controllers are calibrated and commissioned.
6. Check the location of transmitters and controllers. Note adverse conditions that would affect control, and suggest relocation as necessary.
7. Note settings on controllers. Note discrepancies between set point for controller and actual measured variable.
8. Verify operation of all limiting controllers, positioners and relays (e.g., high- and low-temperature thermostats, high- and low-differential pressure switches, etc.).
9. Activate controlled devices, checking for free travel and proper operation of stroke for both dampers and valves. Verify normally open (NO) or normally closed (NC) operation.
10. Verify sequence of operation of controlled devices. Note line pressures and controlled device positions. Correlate to air or water flow measurements. Note speed of response to step change.
11. Confirm interaction of electrically operated switch transducers.
12. Confirm interaction of interlock and lockout systems.
13. Coordinate balancing and control technicians' schedules to avoid duplication of work and testing errors.

Pneumatic System Modifications

1. Verify main control supply air pressure and observe compressor and dryer operation.
2. For hybrid systems using electronic transducers for pneumatic actuation, modify procedures accordingly.

Electronic Systems

1. Monitor voltages of power supply and controller output. Determine whether the system operates on a grounded or nongrounded power supply, and check condition. Although electronic controls now have more robust electronic circuits, improper grounding can cause functional variation in the controller and actuator performance from system to system.
2. Note operation of electric actuators using spring return. Generally, actuators should be under control and use springs only upon power failure to return to a fail-safe position.

Direct Digital Controllers

Direct digital control (DDC) offers nontraditional challenges to the balancing technician. Many of the control devices, such as sensors and actuators, are the same as those in electronic and pneumatic systems. Currently DDC is dominated by two types of controllers—one is fully programmable and one is application-specific. Fully programmable controllers offer a group of functions that are linked together in an applications program to control a system such as an air-handling unit. Application-specific controllers are functionally defined with the programming necessary to carry out the functions required for a system, but not all of the adjustments and settings are defined. Both types of controllers and their functions have some variations. One of the functions is adaptive control, which includes

control algorithms that automatically adjust settings of various controller functions.

The balancing technician must understand the controller functions so that they do not interfere with the test and balance functions. The balancing technician does not need to be literate in computer programming, although it does help. When testing the DDC, the following steps should be included:

1. Obtain controller application program. Discuss application of the designer's sequence with the control programmer.
2. Coordinate testing and adjustment of controlled systems with mechanical systems testing. Avoid duplication of efforts between technicians.
3. Coordinate storage (e.g., saving to central DDC database and controller memory) of all required system adjustments with control technician.

In cases where the balancing agency is required to test discrete points in the control system, the following steps are necessary:

1. Establish criteria for test with the designer.
2. Use reference standards that test the end device through the entire controller chain (e.g., device, wiring, controller, communications, and operator monitoring device). An example of this would be using a dry block temperature calibrator (a testing device that allows a temperature to be set, monitored, and maintained in a small chamber) to test a space temperature sensor. The sensor is installed with extra wire so that it may be removed from the wall and placed in the calibrator chamber. After the system is thermally stabilized, the temperature is read at the controller and the central monitor, if installed.
3. Report findings of reference and all points of reading.

FIELD SURVEY FOR ENERGY AUDIT

An energy audit is an organized survey of a specific building to identify and measure all energy uses, determine probable sources of energy losses, and list energy conservation opportunities. This is usually performed as a team effort under the direction of a qualified energy engineer. The field data can be gathered by firms employing technicians trained in testing, adjusting, and balancing.

Instruments

To determine a building's energy use characteristics, an accurate measurement of existing conditions must be made with proper instruments. Accurate measurements point out opportunities to reduce waste and provide a record of the actual conditions in the building before energy conservation measures were taken. They provide a compilation of installed equipment data and a record of equipment performance prior to changes. Judgments will be made based on the information gathered during the field survey; that which is not accurately measured cannot be properly evaluated.

Generally, the instruments required for testing, adjusting, and balancing are sufficient for energy conservation surveying. Possible additional instruments include a power factor meter, a light meter, combustion testing equipment, refrigeration gages, and equipment for recording temperatures, fluid flow rates, and energy use over time. Only high-quality instruments should be used.

Observation of system operation and any information the technician can obtain from the operating personnel pertaining to the operation should be included in the report.

Data Recording

Organized record keeping is extremely important. A camera is also helpful. Photographs of building components and mechanical and electrical equipment can be reviewed later when the data is analyzed.

Data sheets needed for energy conservation field surveys contain different and, in some cases, more comprehensive information than those used for testing, adjusting, and balancing. Generally, the energy engineer determines the degree of fieldwork to be performed; data sheets should be compatible with the instructions received.

Building Systems

The most effective way to reduce building energy waste is to identify, define, and tabulate the energy load by building system. For this purpose, load is defined as the quantity of energy used in a building, or by one of its subsystems, for a given period. By following this procedure, the most effective energy conservation opportunities can be achieved more quickly because high priorities can be assigned to systems that consume the most energy.

A building can be divided into nonenergized systems and energized systems. Nonenergized systems do not require outside energy sources such as electricity and fuel. Energized systems (e.g., mechanical and electrical systems) require outside energy. Energized and nonenergized systems can be divided into subsystems defined by function. Nonenergized subsystems are (1) building site, envelope, and interior; (2) building use; and (3) building operation.

Building Site, Envelope, and Interior. The site, envelope, and interior should be surveyed to determine how they can be modified to reduce the building load that the mechanical and electrical systems must meet without adversely affecting the building's appearance. It is important to compare actual conditions with conditions assumed by the designer, so that the mechanical and electrical systems can be adjusted to balance their capacities to satisfy actual needs.

Building Use. These loads can be classified as people occupancy loads or people operation loads. People occupancy loads are related to schedule, density, and mixing of occupancy types (e.g., process and office). People operation loads are varied, and include (1) operation of manual window shading devices; (2) setting of room thermostats; and (3) conservation-related habits such as turning off lights, closing doors and windows, turning off energized equipment when not in use, and not wasting domestic hot or chilled water.

Building Operation. This subsystem consists of the operation and maintenance of all the building subsystems. The load on the building operation subsystem is affected by factors such as (1) the time at which janitorial services are performed, (2) janitorial crew size and time required to clean, (3) amount of lighting used to perform janitorial functions, (4) quality of the equipment maintenance program, (5) system operational practices, and (6) equipment efficiencies.

Building Energized Systems

The energized subsystems of the building are generally plumbing, heating, ventilating, cooling, space conditioning, control, electrical, and food service. Although these systems are interrelated and often use common components, logical organization of data requires evaluating the energy use of each subsystem as independently as possible. In this way, proper energy conservation measures for each subsystem can be developed.

Process Loads

In addition to building subsystem loads, the process load in most buildings must be evaluated by the energy field auditor. Most tasks not only require energy for performance, but also affect the energy consumption of other building subsystems. For example, if a process releases large amounts of heat to the space, the process consumes energy and also imposes a large load on the cooling system.

Guidelines for Developing a Field Study Form

A brief checklist follows that outlines requirements for a field study form needed to conduct an energy audit.

Inspection and Observation of All Systems. Record physical and mechanical condition of the following:

- Fan blades, fan scroll, drives, belt tightness, and alignment
- Filters, coils, and housing tightness
- Ductwork (equipment room and space, where possible)
- Strainers
- Insulation ducts and piping
- Makeup water treatment and cooling tower

Interview of Physical Plant Supervisor. Record answers to the following survey questions:

- Is the system operating as designed? If not, what changes have been made to ensure its performance?
- Have there been modifications or additions to the system?
- If the system has been a problem, list problems by frequency of occurrence.
- Are any systems cycled? If so, which systems and when, and would building load permit cycling systems?

Recording System Information. Record the following system/equipment identification:

- Type of system—single-zone, multizone, double-duct, low- or high-velocity, reheat, variable volume, or other
- System arrangement—fixed minimum outside air, no relief, gravity or power relief, economizer gravity relief, exhaust return, or other
- Air-handling equipment—fans (supply, return, and exhaust): manufacturer, model, size, type, and class; dampers (vortex, scroll, or discharge); motors: manufacturer, power requirement, full load amperes, voltage, phase, and service factor
- Chilled and hot water coils—area, tubes on face, fin spacing, and number of rows (coil data necessary when shop drawings are not available)
- Terminals—high-pressure mixing box: manufacturer, model, and type (reheat, constant volume, variable volume, induction); grilles, registers, and diffusers: manufacturer, model, style, and loss coefficient to convert field-measured velocity to flow rate
- Main heating and cooling pumps, over 5 hp—manufacturer, pump service and identification, model, size, impeller diameter, speed, flow rate, head at full flow, and head at no flow; motor data: power, speed, voltage, amperes, and service factor
- Refrigeration equipment—chiller manufacturer, type, model, serial number, nominal tons, brake horsepower, total heat rejection, motor (horsepower, amperes, volts), chiller pressure drop, entering and leaving chilled water temperatures, condenser pressure drop, condenser entering and leaving water temperatures, running amperes and volts, no-load running amperes and volts
- Cooling tower—manufacturer, size, type, nominal tons, range, flow rate, and entering wet-bulb temperature
- Heating equipment—boiler (small through medium) manufacturer, fuel, energy input (rated), and heat output (rated)

Recording Test Data. Record the following test data:

- Systems in normal mode of operation (if possible)—fan motor: running amperes and volts and power factor (over 5 hp); fan: speed, total air (pitot-tube traverse where possible), and static pressure (discharge static minus inlet total); static profile drawing (static pressure across filters, heating coil, cooling coil, and dampers); static pressure at ends of runs of the system (identifying locations)
- Cooling coils—entering and leaving dry- and wet-bulb temperatures, entering and leaving water temperatures, coil pressure drop (where pressure taps permit and manufacturer's ratings can be

obtained), flow rate of coil (when other than fan), outdoor wet and dry bulb, time of day, and conditions (sunny or cloudy)

- Heating coils—entering and leaving dry-bulb temperatures, entering and leaving water temperatures, coil pressure drop (where pressure taps permit and manufacturer's ratings can be obtained), and flow rate through the coil (when other than fan)
- Pumps—no-flow head, full-flow discharge pressure, full-flow suction pressure, full-flow differential pressure, motor running amperes and volts, and power factor (over 5 hp)
- Chiller (under cooling load conditions)—chiller pressure drop, entering and leaving chilled water temperatures, condenser pressure drop, entering and leaving condenser water temperatures, running amperes and volts, no-load running amperes and volts, chilled water on and off, and condenser water on and off
- Cooling tower—water flow rate on tower, entering and leaving water temperatures, entering and leaving wet bulb, fan motor [amperes, volts, power factor (over 5 hp), and ambient wet bulb]
- Boiler (full fire)—input energy (if possible), percent CO_2, stack temperature, efficiency, and complete Orsat test on large boilers
- Boiler controls—description of the operation
- Temperature controls—operating and set point temperatures for mixed air controller, leaving air controller, hot deck controller, cold deck controller, outdoor reset, interlock controls, and damper controls; description of complete control system and any malfunctions
- Outside air intake versus exhaust air—total airflow measured by pitot-tube traverses of both outside air intake and exhaust air systems, obtained where possible. Determine whether an imbalance in the exhaust system causes infiltration. Observe exterior walls to determine whether outside air can infiltrate into the return air (record outside air, return air, and return air plenum dry- and wet-bulb temperatures). The greater the differential between outside and return air, the more evident the problem will be.

TESTING FOR SOUND AND VIBRATION

Testing for sound and vibration ensures that equipment is operating satisfactorily and that no objectionable noise and vibration are transmitted to the building structure and occupied space. Although sound and vibration are specialized fields that require expertise not normally developed by the HVAC engineer, the procedures to test HVAC are relatively simple and can be performed with a minimum of equipment by following the steps outlined in this section. Although this section provides useful information for resolving common noise and vibration problems, Chapter 46 should be consulted for information on problem solving or the design of HVAC.

TESTING FOR SOUND

Present technology does not permit tests of whether equipment is operating with desired sound levels. Field tests can only determine sound pressure levels, and equipment ratings are almost always in terms of sound power levels. Until new techniques are developed, the testing engineer can only determine (1) whether sound pressure levels are in desired limits and (2) which equipment, systems, or components are the source of excessive or disturbing transmission.

Sound-Measuring Instruments

Although an experienced listener can often determine whether systems are operating in an acceptably quiet manner, sound-measuring instruments are necessary to determine whether system noise levels are in compliance with specified criteria, and if not, to obtain and report detailed information to evaluate the cause of noncompliance. Instruments normally used in field testing are as follows:

The **precision sound level meter** is used to measure sound pressure level. The most basic sound level meters measure overall sound pressure level and have up to three weighted scales that

provide limited filtering capability. The instrument is useful in assessing outdoor noise levels in certain situations and can provide limited information on the low-frequency content of overall noise levels, but it provides insufficient information for problem diagnosis and solution. Its usefulness in evaluating indoor HVAC sound sources is thus limited.

Proper evaluation of HVAC sound sources requires a sound level meter capable of filtering overall sound levels into frequency increments of one octave or less.

Sound analyzers provide detailed information about the sound pressure levels at various frequencies through filtering networks. The most popular sound analyzers are the octave band and center frequency, which break the sound into the eight octave bands of audible sound. Instruments are also available for 0.33, 0.1, and narrower spectrum analysis; however, these are primarily for laboratory and research applications. Sound analyzers (octave band or center frequency) are required where specifications are based on noise criteria (NC) curves or similar frequency criteria and for problem jobs where a knowledge of frequency is necessary to determine proper corrective action.

Personal computers are a versatile sound-measuring tool. Software used in conjunction with portable computers has all the functional capabilities described previously, plus many that previously required a fully equipped acoustical laboratory. This type of sound-measuring system is many times faster and much more versatile than conventional sound level meters. With suitable accessories, it can also be used to evaluate vibration levels.

A **stethoscope** is an invaluable instrument in measuring sound levels and tracking down problems, as it enables the listener to determine the direction of the sound source.

Regardless of which sound-measuring system is used, it should be calibrated prior to each use. Some systems have built-in calibration, while others use external calibrators. Much information is available on the proper application and use of sound-measuring instruments.

Air noise, caused by air flowing at a velocity of over 1000 fpm or by winds over 12 mph, can cause substantial error in sound measurements due to wind effect on the microphone. For outdoor measurements or in places where air movement is prevalent, either a wind screen for the microphone or a special microphone is required.

Sound Level Criteria

In the absence of specified values, the testing engineer must determine whether sound levels are within acceptable limits (Chapter 46). Note that complete absence of noise is seldom a design criterion, except for certain critical locations such as sound and recording studios. In most locations, a certain amount of noise is desirable to mask other noises and provide speech privacy; it also provides an acoustically pleasing environment, since few people can function effectively in extreme quiet. Table 3 in Chapter 7 of the 1997 *ASHRAE Handbook—Fundamentals* lists typical sound pressure levels. In determining allowable HVAC equipment noise, it is as inappropriate to demand 30 dB for a factory where the normal noise level is 75 dB as it is to specify 60 dB for a private office where normal noise level might be 35 dB.

Most field sound-measuring instruments and techniques yield an accuracy of ±3 dB, the smallest difference in sound pressure level that the average person can discern. A reasonable tolerance for sound criteria is 5 dB; if 35 dBA is considered the maximum allowable noise, the design engineer should specify 30 dBA.

The measured sound level of any location is a combination of all sound sources present, including sound generated by HVAC equipment as well as sound from other sources such as plumbing systems and fixtures, elevators, light ballasts, and outside noises. In testing for sound, all sources from other than HVAC equipment are considered background or ambient noise.

Background sound measurements generally have to be made (1) when the specification requires that the sound levels from HVAC equipment only, as opposed to the sound level in a space, not exceed a certain specified level; (2) when the sound level in the space exceeds a desirable level, in which case the noise contributed by the HVAC system must be determined; and (3) in residential locations where little significant background noise is generated during the evening hours and where generally low allowable noise levels are specified or desired. Because background noise from outside sources such as vehicular traffic can fluctuate widely, sound measurements for residential locations are best made in the normally quiet evening hours.

Sound Testing

Ideally, a building should be completed and ready for occupancy before sound level tests are taken. All spaces in which readings will be taken should be furnished with drapes, carpeting, and furniture, as these affect the room absorption and the subjective quality of the sound. In actual practice, since most tests have to be conducted before the space is completely finished and furnished for final occupancy, the testing engineer must make some allowances. Because furnishings increase the absorption coefficient and reduce to 4 dB the sound pressure level that can be expected between most live and dead spaces, the following guidelines should suffice for measurements made in unfurnished spaces. If the sound pressure level is 5 dB or more over specified or desired criterion, it can be assumed that the criterion will not be met, even with the increased absorption provided by furnishings. If the sound pressure level is 0 to 4 dB greater than specified or desired criterion, recheck when the room is furnished to determine compliance.

Follow this general procedure:

1. Obtain a complete set of accurate, as-built drawings and specifications, including duct and piping details. Review specifications to determine sound and vibration criteria and any special instructions for testing.
2. Visually check for noncompliance with plans and specifications, obvious errors, and poor workmanship. Turn system on for aural check. Listen for noise and vibration, especially duct leaks and loose fittings.
3. Adjust and balance equipment, as described in other sections, so that final acoustical tests are made with the HVAC as it will be operating. It is desirable to perform acoustical tests for both summer and winter operation, but where this is not practical, make tests for the summer operating mode, as it usually has the potential for higher sound levels. Tests must be made for all mechanical equipment and systems, including standby.
4. Check calibration of instruments.
5. Measure sound levels in all areas as required, combining measurements as indicated in item 3 if equipment or systems must be operated separately. Before final measurements are made in any particular area, survey the area using an A-weighted scale reading (dBA) to determine the location of the highest sound pressure level. Indicate this location on a testing form, and use it for test measurements. Restrict the preliminary survey to determine location of test measurements to areas that can be occupied by standing or sitting personnel. For example, measurements would not be made directly in front of a diffuser located in the ceiling, but they would be made as close to the diffuser as standing or sitting personnel might be situated. In the absence of specified sound criteria, the testing engineer should measure sound pressure levels in all occupied spaces to determine compliance with criteria indicated in Chapter 46 and to locate any sources of excessive or disturbing noise.
6. Determine whether background noise measurements must be made.

(a) If specification requires determination of sound level from HVAC equipment only, it will be necessary to take background noise readings by turning HVAC equipment off.
(b) If specification requires compliance with a specific noise level or criterion (e.g., sound levels in office areas shall not exceed 35 dBA), ambient noise measurements must be made only if the noise level in any area exceeds the specified value.
(c) For residential locations and areas requiring very low noise, such as sound recording studios and locations that are used during the normally quieter evening hours, it is usually desirable to take sound measurements in the evening and/or take ambient noise measurements.

7. For outdoor noise measurements to determine noise radiated by outdoor or roof-mounted equipment such as cooling towers and condensing units, the section on Sound Control for Outdoor Equipment in Chapter 46, which presents proper procedure and necessary calculations, should be consulted.

Noise Transmission Problems

Regardless of the precautions taken by the specifying engineer and the installing contractors, situations can occur where the sound level exceeds specified or desired levels, and there will be occasional complaints of noise in completed installations. A thorough understanding of Chapter 46 and the section on Testing for Vibration in this chapter is desirable before attempting to resolve any noise and vibration transmission problems. The following is intended as an overall guide rather than a detailed problem-solving procedure.

All noise transmission problems can be evaluated in terms of the source-path-receiver concept. Objectionable transmission can be resolved by (1) reducing the noise at the source by replacing defective equipment, repairing improper operation, proper balancing and adjusting, and replacing with quieter equipment; (2) attenuating the paths of transmission with silencers, vibration isolators, and wall treatment to increase transmission loss; and (3) reducing or masking objectionable noise at the receiver by increasing room absorption or introducing a nonobjectionable masking sound. The following discussion includes (1) ways to identify actual noise sources using simple instruments or no instruments and (2) possible corrections.

When troubleshooting in the field, the engineer should listen to the offending sound. The best instruments are no substitute for careful listening, as the human ear has the remarkable ability to identify certain familiar sounds such as bearing squeak or duct leaks and is able to discern small changes in frequency or sound character that might not be apparent from meter readings only. The ear is also a good direction and range finder; because noise generally gets louder as one approaches the source, direction can often be determined by turning the head. Hands can also identify noise sources. Air jets from duct leaks can often be felt, and the sound of rattling or vibrating panels or parts often changes or stops when these parts are touched.

In trying to locate noise sources and transmission paths, the engineer should consider the location of the affected area. In areas that are remote from equipment rooms containing significant noise producers but adjacent to shafts, noise is usually the result of structure-borne transmission through pipe and duct supports and anchors. In areas adjoining, above, or below equipment rooms, noise is usually caused by openings (acoustical leaks) in the separating floor or wall or by improper, ineffective, or maladjusted vibration isolation systems.

Unless the noise source or path of transmission is quite obvious, the best way to identify it is by eliminating all sources systematically as follows:

1. Turn off all equipment to make sure that the objectionable noise is caused by the HVAC. If the noise stops, the HVAC components (compressors, fans, and pumps) must be operated

separately to determine which are contributing to the objectionable noise. Where one source of disturbing noise predominates, the test can be performed starting with all equipment in operation and turning off components or systems until the disturbing noise is eliminated. Tests can also be performed starting with all equipment turned off and operating various component equipment singularly, which permits evaluation of noise from each individual component.

Any equipment can be termed a predominant noise source if, when the equipment is shut off, the sound level drops 3 dBA or if, when measurements are taken with equipment operating individually, the sound level is within 3 dBA of the overall objectionable measurement.

When a sound level meter is not used, it is best to start with all equipment operating and shut off components one at a time because the ear can reliably detect differences and changes in noise but not absolute levels.

2. When some part of the HVAC system is established as the source of the objectionable noise, try to further isolate the source. By walking around the room, determine whether the noise is coming through the air outlets or returns, the hung ceiling, or the floors or walls.

3. If the noise is coming through the hung ceiling, check that ducts and pipes are isolated properly and not touching the hung ceiling supports or electrical fixtures, which would provide large noise radiating surfaces. If ducts and pipes are the source of noise and are isolated properly, the possible remedies to reduce the noise are changing flow conditions, installing silencers, and/or wrapping the duct or pipe with an acoustical barrier such as a lead blanket.

4. If noise is coming through the walls, ceiling, or floor, check for any openings to adjoining shafts or equipment rooms, and check vibration isolation systems to ensure that there is no structure-borne transmission from nearby equipment rooms or shafts.

5. Noise traced to air outlets or returns usually requires careful evaluation by an engineer or acoustical consultant to determine the source and proper corrective action (see Chapter 46). In general, air outlets can be selected to meet any acoustical design goal by keeping the velocity sufficiently low. For any given outlet, the sound level increases about 2 dB for each 10% increase in airflow velocity over the vanes, and doubling the velocity increases the sound level by about 16 dB. Also, the sound approach conditions caused by improperly located control dampers or improperly sized diffuser necks can easily increase sound levels by 10 to 20 dB.

A simple yet effective instrument that aids in locating noise sources is a microphone mounted on a pole. It can be used to localize noises in hard-to-reach places, such as hung ceilings and behind heavy furniture.

6. If the noise is traced to an air outlet, measure the A-sound level close to it but with no air blowing against the microphone. Then, remove the inner assembly or core of the air outlet and repeat the reading with the meter and the observer in exactly the same position as before. If the second reading is more than 3 dB below the first, a significant amount of noise is caused by airflow over the vanes of the diffuser or grille. In this case, check whether the system is balanced properly. As little as 10% too much air will increase the sound generated by an air outlet by 2.5 dB. As a last resort, a larger air outlet could be substituted to obtain lower air velocities and hence less turbulence for the same air quality. Before this is considered, however, the air approach to the outlet should be checked.

Noise far exceeding the normal rating of a diffuser or grille is generated when a throttled damper is installed close to it. Air jets impinge on the vanes or cones of the outlet and produce **edge tones** similar to the hiss heard when blowing against the

edge of a ruler. The material of the vanes has no effect on this noise, although loose vanes may cause additional noise from vibration.

When balancing air outlets with integral volume dampers, consider the static pressure drop across the damper, as well as the air quantity. Separate volume dampers should be installed sufficiently upstream from the outlet that there is no jet impingement. Plenum inlets should be brought in from the side, so that the jets do not impinge on the outlet vanes.

7. If the air outlets are eliminated as sources of excessive noise, inspect the fan room. If possible, change the fan speed by about 10%. If resonance is involved, this small change can make a significant difference.

8. Sometimes fans are poorly matched to the system. If a belt-driven fan delivers air at a higher static pressure than is needed to move the design air quantity through the system, reduce the fan speed by changing sheaves. If the fan does not deliver enough air, consider increasing the fan speed only after checking the duct system for unavoidable losses. Turbulence in the air approach to the fan inlet not only increases the fan sound generation, but decreases its air capacity. Other parts that may cause excessive turbulence are dampers, duct bends, and sudden enlargements or contractions of the duct.

When investigating fan noise, seek assistance from the supplier or manufacturer of the fan.

9. If additional acoustical treatment is to be installed in the ductwork, obtain a frequency analysis. This involves the use of an octave-band analyzer and should generally be left to a trained acoustician.

TESTING FOR VIBRATION

Vibration testing is necessary to ensure that (1) equipment is operating within satisfactory vibration levels and (2) objectionable vibration and noise are not transmitted to the building structure. Although these two factors are interrelated, they are not necessarily interdependent. A different solution is required for each, and it is essential to test both the isolation and vibration levels of equipment.

General Procedure

The general order of steps in vibration testing are as follows:

1. Make a visual check of all equipment for obvious errors that must be corrected immediately.

2. Make sure all isolation is free floating and not short-circuited by any obstruction between equipment or equipment base and building structure.

3. Turn on the system for an aural check of any obviously rough operation. Check bearings with a stethoscope. Bearing check is especially important because bearings can become defective in transit and/or if equipment was not properly stored, installed, or maintained. Defective bearings should be replaced immediately to avoid damage to the shaft and other components.

4. Adjust and balance equipment and systems so that final vibration tests are made on equipment as it will actually be operating.

5. Test equipment vibration.

Instruments

Although instruments are not required to test vibration isolation systems, they are essential to test equipment vibration properly.

Sound level meters and **computer-driven sound-measuring systems** are the most useful instruments for measuring and evaluating vibration. Usually, they are fitted with accelerometers or vibration pickups for a full range of vibration measurement and analysis. Other instruments used for testing vibration in the field are described as follows.

Reed vibrometers are relatively inexpensive instruments often used for testing vibration, but relative inaccuracy limits their usefulness.

Vibrometers are moderately priced instruments that measure vibration amplitude by means of a light beam projected on a graduated scale.

Vibration meters are moderately priced electronic instruments that measure vibration amplitude on a meter scale and are simple to use.

Vibrographs are moderately priced mechanical instruments that measure both amplitude and frequency. They are useful for analysis and testing because they provide a chart recording showing amplitude, frequency, and actual wave form of vibration. They can be used for simple yet accurate determination of the natural frequency of shafts, components, and systems by a **bump test**.

Vibration analyzers are relatively expensive electronic instruments that measure amplitude and frequency, usually incorporating a variable filter.

Strobe lights are often used with many of the aforementioned instruments for analyzing and balancing rotating equipment.

Stethoscopes that amplify sound are available as inexpensive mechanic's type (basically, a standard stethoscope with a probe attachment); relatively inexpensive types incorporating a tuneable filter; and moderately priced powered types that electronically amplify sound and provide some type of meter and/or chart recording. Stethoscopes are often used to determine whether bearings are bad.

The choice of instruments depends on the test. A stethoscope should be part of every tester's kit as it is one of the most practical, yet least expensive, instruments and one of the best means of checking bearings. Vibrometers and vibration meters can be used to measure vibration amplitude as an acceptance check. Since they cannot measure frequency, they cannot be used for analysis and primarily function as a go-no-go instrument. The best acceptance criteria consider both amplitude and frequency. However, because vibrometers and vibration meters are moderately priced and easy to use, they are widely used. Anyone seriously concerned with vibration testing should use an instrument that can determine frequency as well as amplitude, such as a vibrograph or vibration analyzer.

Testing Vibration Isolation

The following steps should be taken to ensure that vibration isolators are functioning properly:

1. Ensure that the equipment is **free floating** by applying an unbalanced load, which should cause the equipment to move freely and easily. On floor-mounted equipment, check that there are no obstructions between the base or foundation and the building structure that would cause transmission while still permitting equipment to rock relatively free because of the application of an unbalanced force (Figure 13). On suspended equipment, check that hanger rods are not touching the hanger. Rigid connections such as pipes and ducts can prohibit mounts from functioning properly and from providing a transmission path. Note that the

fact that the equipment is free floating does not mean that the isolators are functioning properly. For example, a 500 rpm fan installed on isolators having a natural frequency of 500 cycles per minute (8.33 Hz) could be free floating but would actually be in resonance, resulting in transmission to the building and excessive movement.

2. Determine whether isolators are adjusted properly and providing desired isolation efficiency. All isolators supporting a piece of equipment should have approximately the same deflection (i.e., they should be compressed the same under the equipment). If not, they have been improperly adjusted, installed, or selected; this should be corrected immediately. Note that isolation efficiency cannot be checked by comparing vibration amplitude on equipment to amplitude on the structure (Figure 14).

The only accurate check of isolation efficiencies is to compare vibration measurements of equipment operating with isolators to measurements of equipment operating without isolators. Because this type of test is usually impractical, it is better to check whether the isolator's deflection is as specified and whether the specified or desired isolation efficiency is being provided. Figure 15 shows natural frequency of isolators as a function of deflection and indicates the theoretical isolation efficiencies for various frequencies at which the equipment operates.

While it is easy to determine the deflection of spring mounts by measuring the difference between the free heights with a ruler (information as shown on submittal drawings or available from a manufacturer), such measurements are difficult with most pad or rubber mounts. Further, most pad and rubber mounts do not lend themselves to accurate determination of natural frequency as a function of deflection. For such mounts, the most practical approach is to check that there is no excessive vibration of the base and no noticeable or objectionable vibration transmission to the building structure.

If isolators are in the 90% efficiency range, and there is transmission to the building structure, either the equipment is operating roughly or there is a flanking path of transmission, such as connecting piping or obstruction, under the base.

Testing Equipment Vibration

Testing equipment vibration is necessary as an acceptance check to determine whether equipment is functioning properly and to ensure that objectionable vibration and noise are not transmitted. Although a person familiar with equipment can determine when it is operating roughly, instruments are usually required to determine accurately whether vibration levels are satisfactory.

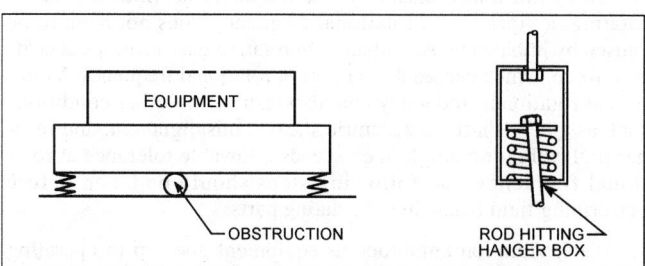

Fig. 13 Obstructed Isolation Systems

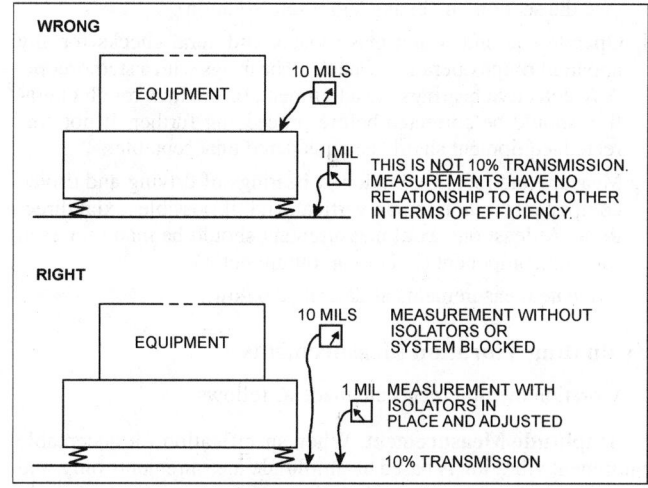

Fig. 14 Testing Isolation Efficiency

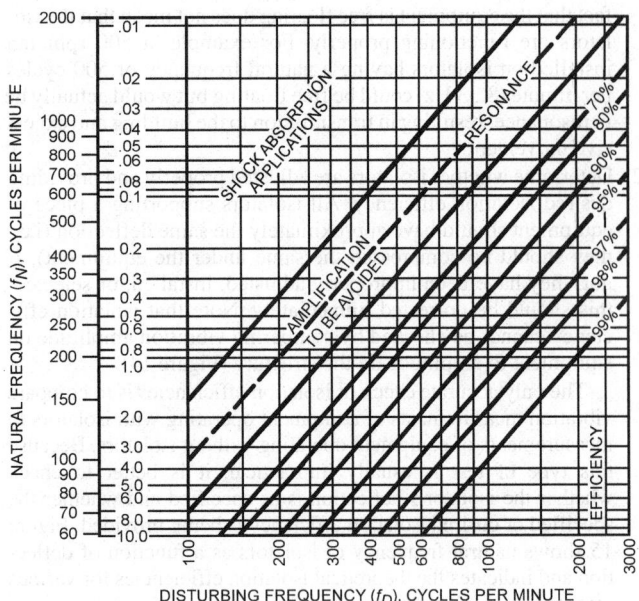

Fig. 15   Isolator Natural Frequencies and Efficiencies

Vibration Tolerances

Vibration tolerance criteria are listed in Table 41 of Chapter 46. These criteria are based on equipment installed on vibration isolators and can be met by any reasonably smoothly running equipment.

Procedure for Testing Equipment Vibration

The following steps should be taken to ensure that equipment vibration is properly tested:

1. Determine operating speeds of equipment from nameplates, drawings, or a speed-measuring device such as a tachometer or strobe, and indicate them on the test form. For any equipment where the driving speed (motor) is different from the driven speed (fan wheel, rotor, impeller) because of belt drive or gear reducers, indicate both driving and driven speeds.
2. Determine acceptance criteria from specifications, and indicate them on the test form. If specifications do not provide criteria, use those shown in Chapter 46.
3. Ensure that the vibration isolation system is functioning properly (see the section on Testing Vibration Isolation).
4. Operate equipment and make visual and aural checks for any apparent rough operation. Check all bearings with a stethoscope. Any defective bearings, misalignment, or obvious rough operation should be corrected before proceeding further. If not corrected, equipment should be considered unacceptable.
5. Measure and record vibration at bearings of driving and driven components in horizontal, vertical, and, if possible, axial directions. At least one axial measurement should be made for each rotating component (fan motor, pump motor).
6. Evaluate measurements as described below.

Evaluating Vibration Measurements

Vibration measurements are made as follows:

Amplitude Measurement. When specification for acceptable equipment vibration is based on amplitude measurements only, and measurements are made with an instrument that measures only amplitude (e.g., a vibration meter or vibrometer):

1. No measurement should exceed specified values or values shown in Tables 43 or 44 of Chapter 46, taking into consideration reduced values for equipment installed on inertia blocks
2. No measurement should exceed values shown in Tables 43 or 44 of Chapter 46 for driving and driven speeds, taking into consideration reduced values for equipment installed on inertia blocks. For example, with a belt-driven fan operating at 800 rpm and having an 1800 rpm driving motor, amplitude measurements at fan bearings must be in accordance with values shown for 800 cpm (13.3 Hz), and measurements at motor bearings must be in accordance with values shown for 1800 cpm (30 Hz). If measurements at motor bearings exceed specified values, take measurements of the motor only with belts removed to determine whether there is feedback vibration from the fan.
3. No axial vibration measurement should exceed maximum radial (vertical or horizontal) vibration at the same location.

Amplitude and Frequency Measurement. When specification for acceptable equipment vibration is based on both amplitude and frequency measurements, and measurements are made with instruments that measure both amplitude and frequency (e.g., a vibrograph or vibration analyzer):

1. Amplitude measurements at driving and driven speeds should not exceed specified values or values shown in Tables 43 or 44 of Chapter 46, taking into consideration reduced values for equipment installed on inertia blocks. Measurements that exceed acceptable amounts may be evaluated as explained in the section on Vibration Analysis.
2. Axial vibration measurements should not exceed maximum radial (vertical or horizontal) vibration at the same location.
3. The presence of any vibration at frequencies other than driving or driven speeds is generally reason to rate operation unacceptable, and such vibration should be analyzed as explained in the section on Vibration Analysis.

Vibration Analysis

The following guide covers most vibration problems that may be encountered.

Axial Vibration Exceeds Radial Vibration. When the amplitude of axial vibration (parallel with shaft) at any bearing exceeds radial vibration (perpendicular to shaft—vertical or horizontal), it usually indicates misalignment. This is most common on direct-driven equipment because flexible couplings accommodate parallel and angular misalignment of shafts. Such misalignment can generate forces that cause axial vibration. Axial vibration can cause premature bearing failure, so misalignment should be checked carefully and corrected promptly. Other possible causes of large-amplitude axial vibration are resonance, defective bearings, insufficient rigidity of bearing supports or equipment, and loose hold-down bolts.

Vibration Amplitude Exceeds Allowable Tolerance at Rotational Speed. The allowable vibration limits established by Table 41 of Chapter 46 are based on vibration caused by rotor imbalance, which results in vibration at rotational frequency. While vibration caused by imbalance must be at the frequency at which the part is rotating, a vibration at rotational frequency does not have to be caused by imbalance. An unbalanced rotating part develops centrifugal force, which causes it to vibrate at rotational frequency. Vibration at rotational frequency can also result from other conditions such as a bent shaft, an eccentric sheave, misalignment, and resonance. If vibration amplitude exceeds allowable tolerance at rotational frequency, the following steps should be taken before performing field balancing of rotating parts.

1. Check vibration amplitude as equipment goes up to operating speed and as it coasts to a stop. Any significant peaks at or near operating speed, as shown in Figure 16, indicate a probable

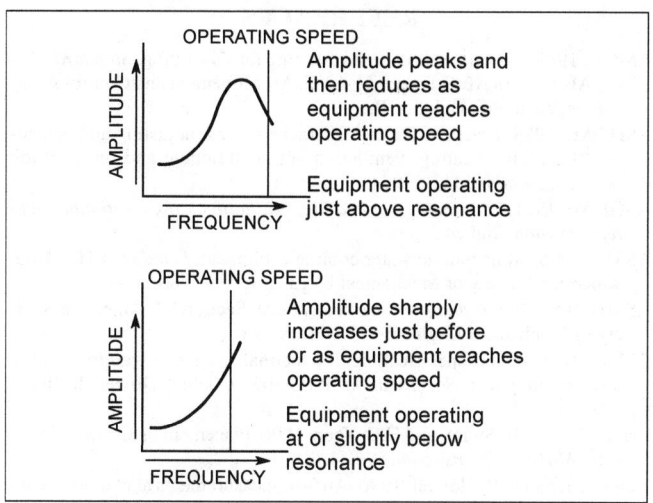

Fig. 16 Vibration from Resonant Condition

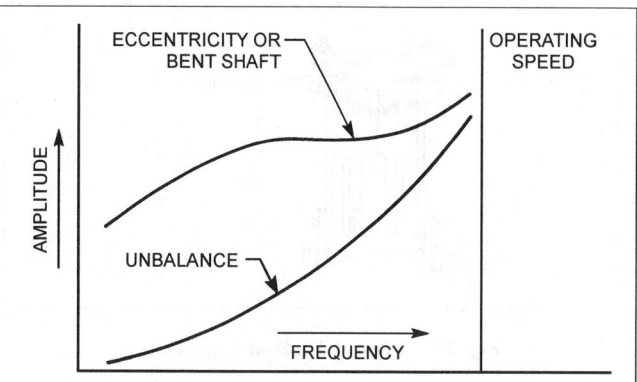

Fig. 17 Vibration Caused by Eccentricity

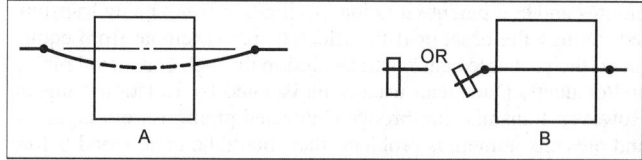

Fig. 18 Bent Shafts

condition of resonance, i.e., some part having a natural frequency close to the operating speed, resulting in greatly amplified levels of vibration.

A bent shaft or eccentricity usually causes imbalance that results in significantly higher vibration amplitude at lower speeds, as shown in Figure 17, whereas vibration caused by imbalance generally increases as speed increases.

If a bent shaft or eccentricity is suspected, check the dial indicator. A bent shaft or eccentricity between bearings as shown in Figure 18A can usually be compensated for by field balancing, although some axial vibration might remain. Field balancing cannot correct vibration caused by a bent shaft on direct-connected equipment, on belt-driven equipment where the shaft is bent at the location of sheave, or if the sheave is eccentric (Figure 18B). This is because the center-to-center distance of the sheaves will fluctuate, each revolution resulting in vibration.

2. For belt- or gear-driven equipment where vibration is at motor driving frequency rather than driven speed, it is best to disconnect the drive to perform tests. If the vibration amplitude of the motor operating by itself does not exceed specified or allowable values, excessive vibration (when the drive is connected) is probably a function of bent shaft, misalignment, eccentricity, resonance, or loose hold-down bolts.

3. Vibration caused by imbalance can be corrected in the field by firms specializing in this service or by testing personnel if they have appropriate equipment and experience.

Vibration at Other than Rotational Frequency. Vibration at frequencies other than driving and driven speeds is generally con-

Table 5 Common Causes of Vibration Other than Unbalance at Rotation Frequency

Frequency	Source
$0.5 \times$ rpm	Vibration at approximately 0.5 rpm can result from improperly loaded sleeve bearings. This vibration will usually disappear suddenly as equipment coasts down from operating speed.
$2 \times$ rpm	Equipment is not tightly secured or bolted down.
$2 \times$ rpm	Misalignment of couplings or shafts usually results in vibration at twice rotational frequency and generally a relatively high axial vibration.
Many $\times$ rpm	Defective antifriction (ball, roller) bearings usually result in low-amplitude, high-frequency, erratic vibration. Because defective bearings usually produce noise rather than any significantly measurable vibration, it is best to check all bearings with a stethoscope or similar listening device.

sidered unacceptable. Table 5 shows some common conditions that can cause vibration at other than rotational frequency.

Resonance. If resonance is suspected, determine which part of the system is in resonance.

Isolation Mounts. The natural frequency of the most commonly used spring mounts is a function of spring deflection, as shown in Figure 11 in Chapter 7 of the 1997 *ASHRAE Handbook—Fundamentals*, and it is relatively easy to calculate by determining the difference between the free and operating height of the mount, as explained in the section on Testing Vibration Isolation. This technique cannot be applied to rubber, pad, or fiberglass mounts, which have a natural frequency in the 300 to 3000 cpm range. Natural frequency for such mounts is determined by a bump test. Any resonance with isolators should be immediately corrected as it results in excessive movement of equipment and more transmission to the building structure than if equipment were attached solidly to the building (installed without isolators).

Components. Resonance can occur with any shaft, structural base, casing, and connected piping. The easiest way to determine natural frequency is to perform a bump test with a vibrograph. This test consists of bumping the part and measuring with an instrument; the part will vibrate at its natural frequency, which is recorded on instrument chart paper. Similar tests, though not as convenient or accurate, can be made with a reed vibrometer or a vibration analyzer. However, most of these instruments are restricted to frequencies above 500 cpm. They therefore cannot be used to determine natural frequencies of most isolation systems, which usually have natural frequencies lower than 500 cpm.

Checking for Vibration Transmission. The source of vibration transmission can be checked by determining frequency with a vibration analyzer and tracing back to equipment operating at this speed. However, the easiest and usually the best method (even if test equipment is being used) is to shut off components one at a time until the source of transmission is located. Most transmission problems cause disturbing noise; listening is the most practical approach to determine a noise source because the ear is usually better than sound-measuring instruments at distinguishing small differences and changes in character and amount of noise. Where disturbing transmission consists solely of vibration, a measuring instrument will probably be helpful, unless vibration is significantly above the

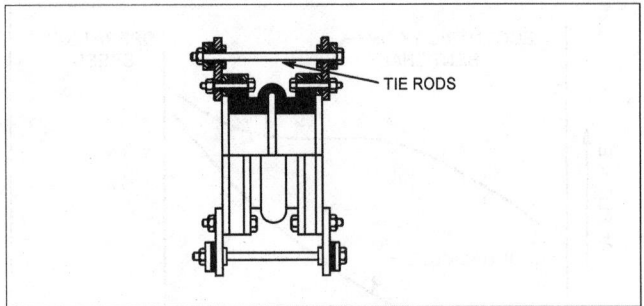

Fig. 19 Typical Tie Rod Assembly

sensory level of perception. Vibration below the sensory level of perception is generally not objectionable.

If equipment is located near the affected area, check isolation mounts and equipment vibration. If vibration is not being transmitted through the base, or if the affected area is remote from equipment, the probable cause is transmission through connected piping and/or ducts. Ducts can usually be isolated by isolation hangers. However, transmission through connected piping is very common and presents numerous problems that should be understood before attempting to correct them as discussed in the following section.

Vibration and Noise Transmission in Piping

Vibration and noise in connected piping can be generated by either equipment (e.g., pump or compressor) or flow (velocity). Mechanical vibration due to equipment can be transmitted through the walls of pipes or by a water column. Flexible pipe connectors, which provide system flexibility to permit isolators to function properly and protect equipment from stress caused by misalignment and thermal expansion, can be useful in attenuating mechanical vibration transmitted through a pipe wall. However, they rarely suppress flow vibration and noise and only slightly attenuate mechanical vibration as transmitted through a water column.

Tie rods are often used with flexible rubber hose and rubber expansion joints (Figure 19). While they accommodate thermal movements, they hinder the isolation of vibration and noise. This is because pressure in the system causes the hose or joint to expand until resilient washers under tie rods are virtually rigid. To isolate noise adequately with a flexible rubber connector, tie rods and anchor piping should not be used. However, this technique generally cannot be used with pumps that are on spring mounts because they would still permit the hose to elongate. Flexible metal hose can be used with spring-isolated pumps since wire braid serves as tie rods; metal hose controls vibration but not noise.

Problems of transmission through connected piping are best resolved by changes in the system to reduce noise (improve flow characteristics, turn down impeller) or by completely isolating piping from the building structure. Note, however, that it is almost impossible to isolate piping completely from the structure, as required resiliency is inconsistent with rigidity requirements of pipe anchors and guides. Chapter 46 contains information on flexible pipe connectors and resilient pipe supports, anchors, and guides, which should help resolve any piping noise transmission problems.

REFERENCES

AMCA. 1985. Laboratory methods of testing fans for rating. *Standard* 210-85. Also ASHRAE *Standard* 51-1985. Air Movement and Control Association, Arlington Heights, IL.

ASHRAE. 1988. Practices for measurement, testing, adjusting and balancing of building heating, ventilation, air conditioning and refrigeration systems. *Standard* 111-1988.

ASHRAE. 1991. *Terminology of heating, ventilation, air conditioning, and refrigeration*, 2nd ed.

ASME. 1986. Atmospheric water cooling equipment. *Standard* PTC 23-86. American Society of Mechanical Engineers, New York.

ASME. 1988. *Boiler and pressure vessel code*, Section VI. American Society of Mechanical Engineers, New York, NY.

CTI. 1997. Standard specifications for thermal testing of wet/dry cooling towers. *Standard Specification* ATC-105. Cooling Tower Institute, Houston, TX.

Griggs, E.I., W.B. Swim, and H.G. Yoon. 1990. Placement of air control sensors. *ASHRAE Transactions* 96(1).

Sauer, H.J. and R.H. Howell. 1990. Airflow measurements at coil faces with vane anemometers: Statistical correction and recommended field measurement procedure. *ASHRAE Transactions* 96(1):502-11.

BIBLIOGRAPHY

AABC. 1989. *National standards for total system balance*, 5th ed. Associated Air Balance Council, Washington, D.C.

AABC. 1997. *Testing and balancing procedures*. Associated Air Balance Council, Washington, D.C.

AMCA. 1987. *Fan application manual*. Air Movement and Control Association, Arlington Heights, IL.

Armstrong Pump. 1986. *Technology of balancing hydronic heating and cooling systems*. Armstrong Pump, North Tonawanda, NY.

ASA. 1983. Specification for sound level meters. *Standard* 1.4-83. Acoustical Society of America, New York.

ASHRAE. 1996. The HVAC commissioning process. *Guideline* 1-1996.

Coad, W.J. 1985. Variable flow in hydronic systems for improved stability, simplicity and energy economics. *ASHRAE Transactions* 91(1B):224-37.

Eads, W.G. 1983. Testing, balancing and adjusting of environmental systems. In *Fan Engineering*, 8th ed. Buffalo Forge Company, Buffalo, NY.

Gladstone, J. 1981. *Air conditioning—Testing and balancing: A field practice manual*. Van Nostrand Reinhold, New York.

Gupton, G. 1989. *HVAC controls, operation and maintenance*. Van Nostrand Reinhold, New York.

Haines, R.W. 1987. *Control systems for heating, ventilating and air conditioning*, 4th ed. Van Nostrand Reinhold, New York.

Hansen, E.G. 1985. *Hydronic system design and operation*. McGraw-Hill, New York.

Miller, R.W. 1983. *Flow measurement engineering handbook*. McGraw-Hill, New York.

NEBB. 1991. *Procedural standards for testing, balancing and adjusting of environmental systems*, 5th ed. National Environmental Balancing Bureau, Vienna, VA.

NEBB. 1986. *Testing, adjusting, balancing manual for technicians*, 1st ed.

SMACNA. 1993. *HVAC systems—Testing, adjusting and balancing*, 2nd ed. Sheet Metal and Air Conditioning Contractors' National Association, Merrifield, VA.

SMACNA. 1995. *HVAC air duct leakage test manual*, 1st ed.

Trane Company. 1988. *Trane air conditioning manual*. The Trane Company, LaCrosse, WI.

OPERATION AND MAINTENANCE MANAGEMENT

ALTHOUGH mechanical maintenance was once the responsibility of trained technical personnel, increasingly sophisticated systems and equipment require overall management programs to handle organization, staffing, planning, and control. These programs should meet present and future personnel and technical requirements. They must also meet system availability and energy use requirements. Additionally, they must upgrade management and operator skills, and increase communication among those who benefit from cost-effective operation and maintenance.

Good maintenance management planning includes proper cost analysis and a process to ensure that occupant comfort, energy planning, and safety and security systems are optimal for all facilities. Appropriate technical expertise, whether in-house or contracted, is also important. This chapter addresses the following issues:

- Cost-effectiveness
- Commissioning
- Management approaches according to the criticality of buildings or systems
- Documentation and record keeping
- Condition monitoring and maintenance
- Operation and maintenance responsibilities of designers, contractors, manufacturers/suppliers, and owners

Operation and maintenance of all HVAC&R systems should be considered during the original design of a building. Any successful operation and maintenance program must include proper documentation of the design intent and criteria. ASHRAE *Guideline* 4 provides a methodology to properly document HVAC systems. Newly installed systems must be commissioned to ensure that they are functioning as designed. ASHRAE *Guideline* 1 provides the methods and procedures for HVAC system commissioning. It is then the responsibility of the management and operational staff to retain the design function throughout the life of the building. Existing systems may need to be recommissioned to accommodate changes.

TERMINOLOGY

System operation defines the parameters under which the building or systems operator can adjust components of the system to satisfy the tenant comfort or process requirements and the strategy for optimum energy use and minimum maintenance.

The **maintenance program** defines maintenance in terms of time and resource allocation. It documents the objectives, establishes evaluation criteria, and commits the maintenance department to basic areas of performance, such as prompt response to mechanical failure and attention to planned functions that protect capital investment and minimize downtime or failure response.

Failure response classifies maintenance department resources expended or reserved to handle interruptions in the operation or function of a system or equipment covered by the maintenance program. This classification includes two response types—repair and service.

The preparation of this chapter is assigned to TC 1.7, Operation and Maintenance Management.

Repair is to make good, or to restore to good or sound condition with the following constraints: (1) operation must be fully restored without embellishment, and (2) failure must trigger the response.

Service provides what is necessary to effect a maintenance program short of repair. It is usually based on manufacturers' recommended procedures.

Planned maintenance classifies maintenance department resources that are invested in prudently selected functions at specified intervals. All functions and resources within this classification must be planned, budgeted, and scheduled. Planned maintenance embodies two concepts—preventive and corrective maintenance.

Preventive maintenance classifies resources allotted to ensure proper operation of a system or equipment under the maintenance program. Durability, reliability, efficiency, and safety are the principal objectives.

Corrective maintenance classifies resources, expended or reserved, for predicting and correcting conditions of impending failure. Corrective action is strictly remedial and always performed before failure occurs. An identical procedure performed in response to failure is classified as a repair. Corrective action may be taken during a shutdown caused by failure, provided the action is optional and unrelated.

Predictive maintenance is a function of corrective maintenance. Statistically supported objective judgment is implied. Nondestructive testing, chemical analysis, vibration and noise monitoring, and routine visual inspection and logging are classified under this function, provided that the item tested or inspected is part of the planned maintenance program.

Durability is the average expected service life of a system or facility. Table 3 in Chapter 35 lists median years of service life of various equipment. Individual manufacturers quantify durability as design life, which is the average number of hours of operation before failure, extrapolated from accelerated life tests and from stressing critical components to economic destruction.

Reliability is the probability that a system or facility will perform its intended function for a specified period of time when used under specific conditions and environment.

QUANTITATIVE MANAGEMENT CONCEPT

The following life-cycle management concept is recommended for operation and maintenance program planning. The concept can be used during the construction cycle as well as for day-to-day operation and maintenance management. Derived from value engineering, this life-cycle concept of management (see Figure 1) involves three interdependent dimensions: effectiveness (value), durability (time), and life-cycle cost (money). Their numerical values are affected by the type and extent of the operation and maintenance programs.

System **effectiveness** is the ability of a system to perform its intended use in terms of performance, availability, and dependability. To be perfectly effective, a facility must provide the required services satisfactorily and must operate dependably without failure

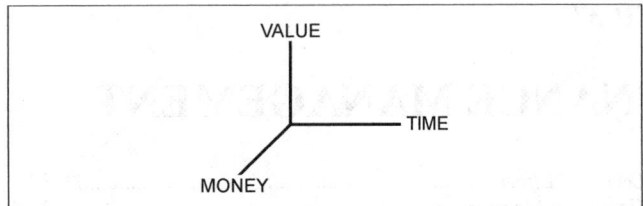

Fig. 1 Life-Cycle Concept of Management

Table 1 Examples of Effectiveness Values

Range of Effectiveness	Effectiveness Category	Example of Service
0.50 to 0.75	Regular	Environmental control
0.76 to 0.85	Essential	Freeze protection
0.86 to 1.00	Critical	Smoke control

for the given period. Mathematically, effectiveness E is the product of capability C and dependability D:

$$E = CD \tag{1}$$

Effectiveness may be assigned a numerical value according to the type of service required to meet a functional need. Table 1 is a guide for assigning effectiveness values. *Note*: The values given in this table are examples only, and values must be developed for each project to create proper relationships between the different systems in each facility.

Capability is the measure of a system's ability to satisfactorily provide required service. It is the probability of meeting functional requirements, provided the system is operated under previously designated operating conditions. An example of capability is the ability of a heating system to cope with the heating load at the design winter temperature. The capability must be verified when the system is first commissioned and whenever the functional requirements change.

Capability can be calculated from the time N the system cannot meet the requirements and the time Y it is expected to operate in one year:

$$C = 1 - (N/Y) \tag{2}$$

Dependability is the measure of a system's condition. Assuming the system was operative at the beginning of its service life, dependability is the probability of its operating at any other given time until the end of its life. For those systems that cannot be repaired during use, dependability is the probability that there will be no failure during use. For systems that can be repaired, dependability is governed by the ease and rapidity in which repairs can be made. This ease is the system maintainability, defined below.

Dependability D can be expressed as a product of reliability R and maintainability M:

$$D = RM \tag{3}$$

Reliability implies that a system will perform its function without failure for a desired portion of a specified time period. Reliability of two or more systems is calculated as follows:

1. If the systems are arranged in **series**, where the output of one is the input of the next (Figure 2), their combined reliability is

$$R = R_1 R_2 \dots R_n \tag{4}$$

2. If the systems are arranged in **parallel**, such as dual pumping system with automatic changeover on failure of one of the two pumps (Figure 3), their combined reliability is calculated from the following equation:

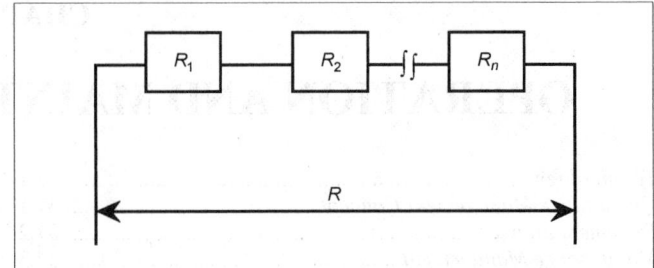

Fig. 2 Series System

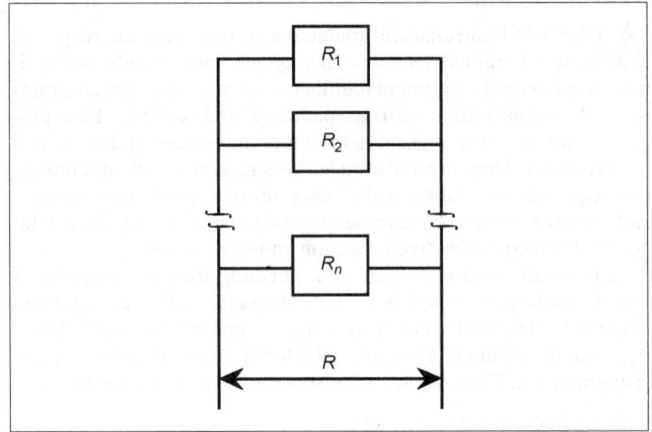

Fig. 3 Parallel System

$$R = 1 - (1 - R_1)(1 - R_2) \dots (1 - R_n) \tag{5}$$

Maintainability is the ease and rapidity with which a system can be repaired. It complements reliability by defining the specific time that a system can operate in a fully restored condition. The value of maintainability is calculated from the time B the system is being repaired in regular occupancy and the time Y the system is expected to operate in one year.

$$M = 1 - (B/Y) \tag{6}$$

DOCUMENTATION

Operation and maintenance documentation should be prepared as outlined in ASHRAE *Guideline* 4. Information should be documented as soon as it becomes available. This supports design and construction activities and the training of operation and maintenance staff in preparation for system commissioning.

A complete operation and maintenance documentation package consists of the following documents:

Operation and maintenance document directory. The operation and maintenance document directory should provide easy access to information in the operation and maintenance documentation, and information should be well organized and clearly identified.

Emergency information. Emergency information should be immediately available during emergency situations and should have emergency and staff and/or agency notification procedures.

Operating manual. The operating manual should contain the following information:

I. General information

   A. Building function
   B. Building description
   C. Operating standards and logs

II. Technical information

    A. System description
    B. Operating routines and procedures
    C. Seasonal start-up and shutdown
    D. Special procedures
    E. Basic troubleshooting

Maintenance manual. The maintenance manual should contain the following information:

I. Equipment data sheets

    A. General data
    B. Warranty

II. Maintenance program information

    A. Installation, operation, and maintenance instructions
    B. Spare parts inventory
    C. Preventive maintenance actions
    D. Schedule of actions
    E. Action descriptions
    F. History

Test reports. The test reports document the observed performance during start-up and commissioning and allows compilation throughout the service life of the facility.

Construction documents. Copies of the construction documents should be included.

All the above documents should be available to the entire facilities department.

MAINTENANCE MANAGEMENT

Maintenance management is the planning, implementation and review of maintenance activities. Myriad levels of maintenance management can be established. These levels are usually determined by their cost-effectiveness and by health and safety concerns. Cost-effectiveness is the balance between system effectiveness [e.g., maintenance levels, equipment (system) availability, reliability, maintainability, and performance] and life-cycle costs.

Three types of maintenance programs that can be adopted are (1) run-to-failure, (2) preventive maintenance, and (3) predictive maintenance. In **run-to-failure**, no money is spent on maintenance until equipment or systems break down. **Preventive maintenance** is maintenance that is scheduled, either by run time or by the calendar. **Predictive maintenance** is based on equipment monitoring and the use of condition and performance indices to maximize repair intervals. Within each of these groups, many levels of effort can be applied, from cursory to detailed. Maintenance programs may incorporate features of all these approaches into a single program. Many arguments can be made about the cost-effectiveness of each of these programs.

HVAC&R maintenance and operating costs represent a significant portion of a facility's operating cost. Therefore, the cost-effectiveness of maintenance management is paramount.

Proper operation and maintenance are important factors in providing good indoor air quality (IAQ) and ensuring that life safety systems operate as designed.

The ultimate success of maintenance management is based on dedicated, trained, and accountable personnel; clearly defined goals and objectives; measurable benefits; management support; and constant examination and reexamination.

KNOWLEDGE AND SKILLS

To be effective and economical, operation and maintenance requires staff with the right combination of technical and managerial skills. Technical skills include not only the skills of operation and maintenance mechanics, but also the engineering skills of the physical plant engineer. Managerial skills include managing the facility in life-cycle terms and on a day-to-day basis. Management on this level may administer contracts with tenants, service contracts, and labor unions. Specialized contractual maintenance companies require yet another level of management.

Physical plant engineers require a variety of skills because they need to coordinate the equipment selected by the designer, the operating and maintenance personnel, and the requirements of the investment plan. Good physical plant engineering solutions are developed when the investment plan is being formulated and continue throughout the life of the facility.

LEVELS OF EFFORT

The criticality and complexity of the system to be operated and maintained must be considered. Building types range from houses to commercial offices, institutional buildings, processing plants, refrigerated storage facilities, and mission critical facilities. All installed systems, no matter how simple, should have a commissioning plan; this gives the building or system owner and operators a means of operating the system in the most economical manner, minimizing energy consumption and maintenance costs while meeting user requirements.

Uncomplicated Systems

Many systems have simple operating procedures such as changing a single-zone thermostat from heating to cooling to control temperature for building occupants. Maintenance procedures for this type of system are usually limited to those recommended by most manufacturers.

Most building managers call on-site maintenance staff to change filters, belts, and motors. However, cleaning of condenser and evaporator coils and assessment of refrigeration and control systems require a more qualified technician. In most small facilities, maintenance contractors provide this service on a maintenance program basis. The frequency of maintenance depends on hours of system use, proximity to major transportation routes (dirt accumulation), and type of operation in the facility.

Smaller facilities may also have complex control systems, especially when zoning is critical for a variety of load conditions. Whenever the operator cannot service and repair the systems or components installed, the owner should ensure that qualified contractors are used.

Dirty evaporator and condenser coils and filters require additional energy to operate due to higher head pressures and inefficient heat transfer. The system must also operate longer to satisfy space conditions. Although these items are variable and difficult to quantify, proper unit and system maintenance improves system operation and equipment life cycles, regardless of system size.

Medium-Complexity Systems

The next level of systems is more complex, with several pieces of mechanical equipment acting together through a control system to provide a variety of comfort zones.

Commissioning is increasingly important with this type of system to ensure optimum comfort at minimal energy cost. Without detailed documentation, the operation staff cannot consider energy budgets when addressing building occupants' comfort complaints.

To manage maintenance programs in medium-sized facilities, the programs must be detailed for the maintenance personnel as well. Maintenance programs should be implemented on a sequential basis to reduce the failure response required. Over time, the maintenance programs can be adjusted for the particular characteristics of the system. Computerized maintenance programs can assist management in overseeing the effectiveness of the program.

For medium-sized systems, some operation and maintenance personnel may be employed by the building owner or lessee. They should have the technical knowledge to operate these systems if they observed a full system commissioning process.

The operating personnel may also be responsible for system maintenance. The maintenance procedures for the system should be detailed in the commissioning documentation (with individual equipment maintenance frequency detailed in manufacturers' literature). The maintenance program should be tailored to each specific building. Again, it may be necessary to contract out certain maintenance program functions for highly technical pieces of equipment should the staff not have the technical expertise.

Complex Systems

Other buildings, including those with central plants, require a management structure to hire, direct, and oversee two staffs—one of operation personnel, and the other of maintenance personnel.

The operations budget should be large enough to support computerized maintenance programs, which detail proper timing of system maintenance procedures so that all systems operate at maximum efficiency while maintaining occupant comfort. Annual and life-cycle cost planning are essential to ensure the most cost-effective operation and maintenance of these systems.

To facilitate proper commissioning, management should be involved with the design process and construction of the facility. Because facilities are long-term investments, any first-cost compromises must consider both life-cycle cost and the ability to satisfy occupant comfort while maintaining reasonable energy budgets.

Logged information can be used with proper database management systems as predictive maintenance to reduce failure response requirements. These systems may be used to indicate the weaknesses in the systems so that management can make appropriate decisions or system changes. Like small and medium-sized systems, large systems may require an outside contractor or manufacturer's support for specific equipment.

CONDITION MONITORING

An important aspect of operation and maintenance engineering systems is the assessment of the condition of the equipment. For complex systems, it may be necessary to monitor several conditions (e.g., temperature, vibration, and load) so that an assessment of the overall condition can be made.

One of the most effective condition-monitoring practices is the routine operating plant inspection conducted by the technician during regularly scheduled plant tours. A technician's knowledge, experience, and familiarity with the plant are invaluable tools in plant diagnostics. Plant familiarity, however, is lost with frequent technician staff changes.

Many physical parameters or conditions can be measured objectively using both special equipment and conventional building management system sensors. One such condition is vibration on rolling element bearings. The vibration data is captured using computers, and special software can analyze the data to determine whether shaft alignment is correct, whether there are excessively unbalanced forces in the rotating mass, the state of lubrication of the bearing, and/or faults with the fixed or moving bearing surfaces or rolling elements. Not only can this technique diagnose failures and repair requirements at an early stage, but it can also be used upon completion of a repair to ensure that the underlying cause of a failure has been removed.

Other condition-monitoring techniques include (1) using thermal infrared images of electrical connections to determine whether mechanical joints are tight, (2) analyzing oil and grease for contamination (e.g., water in fuel oil on diesel engines), (3) analyzing electrical current to diagnose motor winding faults, (4) measuring differential pressure across filter banks and heat exchangers to determine optimum change/cleaning frequency, and (5) measuring

temperature differences for correct control valve response or chiller operation.

In some cases, multiple parameters are required. For example, to determine the degree of contamination of an air filter bank on a variable air volume (VAV) system, it is necessary to measure the differential pressure across the filter and interpret it in terms of the actual flow rate through the filter. This can be done either by forcing the fan on to high speed and then measuring the differential pressure, or by combining a flow rate signal with a differential pressure signal.

CONDITION-BASED MAINTENANCE

Condition-based maintenance is the concept of carrying out maintenance only on the basis of the condition(s) monitored and the interpretation of those conditions. This should result in performing maintenance work only when necessary; therefore, plant reliability will be optimized, and personnel productivity will improve.

Many repairs are a direct result of maintenance-induced failures occurring during scheduled maintenance. Condition-based maintenance prevents unnecessary, repetitive work. There can, however, be added costs involved in supplementary training and instrumentation.

Conditions may be monitored on an on/off basis where comparison can be made against some fixed known values (e.g., tabulated values of acceptable vibration). Generally more useful in engineering systems within buildings is trending over a period of time: either measuring the conditions at a regular interval so that a gradual deterioration can be tracked and remedial work can be planned in advance or, in some circumstances, monitoring on a continuous basis where deterioration can be rapid.

To continue with the example of air filters mentioned in the section on Condition Monitoring, in the case of a VAV system, assuming the main fan is drawing the air through the filter and is controlled to a fixed static pressure, the filter change criterion is a function of the maximum differential pressure the filter can withstand when fully loaded with dust (without bursting) and the energy consumption of the fan. It may be more economical to change the filter before the maximum pressure is reached if the rate of dust loading is slow. Also, monitoring it via the energy management system rather than changing on a fixed time interval means that if something occurs to rapidly accelerate the dust loading (e.g., nearby construction work), the system will alert the building management before the filters are overloaded and could potentially burst.

In a constant volume system (again assuming the main fan is drawing air through the filter), the change criteron is a function of the maximum differential pressure the filter can withstand and the drop in flow rate that can be tolerated by the system users. As the filter blocks, the energy consumption decreases and the filter efficiency improves. A fixed time interval for filter bank changing/cleaning is not optimum. Changing on the basis of a monitored condition should optimize the life of the filters and minimize the cost of labor for building cleaning and this aspect of plant maintenance.

RESPONSIBILITIES

The building owner should be apprised of the design intent of the system. The system design must include proper operation and maintenance information; flowcharts and instrumentation requirements must be indicated. The installing contractor must provide the director of operation and maintenance with an organized and comprehensive turnover of the system that has been installed.

An effective director should be able to organize, staff, train, plan, and control the operation and maintenance of a facility with the cooperation of senior management and all departments. A manager's responsibilities include administering the operation and maintenance budget and protecting the life-cycle objectives. Before selecting a least-cost alternative, a manager should determine the consequential effect on durability and loss prevention (Loveley 1973).

NEW TECHNOLOGY

Operation and maintenance programs are based on the technology available at the time of their preparation. The programs should be adhered to throughout the required service life of the facility or system. In the course of the service life, a new technology may become available that would affect the operation and/or maintenance program. When this occurs, the switch from the existing to the new technology must be assessed in life-cycle terms. The existing technology must be assessed for the degree of loss due to shorter return on investment; the new technology must be assessed for (1) all initial and operation and maintenance cost, (2) the correlation between its service life and the remaining service life of the facility, and (3) the cost of conversion, including the revenue losses due to the associated downtime.

REFERENCES

ASHRAE. 1993. Preparation of operating and maintenance documentation for building systems. *Guideline* 4-1993.

ASHRAE. 1996. Guideline for the HVAC commissioning process. *Guideline* 1-1996.

Loveley, J.D. 1973. Durability, reliability, and serviceability. *ASHRAE Journal* 15(1):67.

BIBLIOGRAPHY

Blanchard, B.S., D. Verma, and E.L. Peterson. 1995. *Maintainability: A key to effective serviceability and maintenance management.* John Wiley and Sons, New York.

Criswell, J.W. 1989. *Planned maintenance for productivity and energy conservation,* 3rd ed. Prentice Hall Press, Englewood Cliffs, NJ.

Fuchs, S.J. 1992. *Complete building equipment maintenance desk book.* Prentice Hall Press, Englewood Cliffs, NJ.

Lawson, C.N. 1989. Commissioning—Construction phase. *ASHRAE Transactions* 95(1):887-94.

Mobely, R.K. 1989. *An introduction to predictive maintenance.* Van Nostrand Reinhold, New York.

Petrocelly, K.L. 1988. *Physical plant operations handbook.* Prentice Hall Press, Englewood Cliffs, NJ.

Trueman, C.S. 1989. Commissioning: An owner's approach for effective operations. *ASHRAE Transactions* 95(1):895-99.

CHAPTER 38

COMPUTER APPLICATIONS

THE use of computers in the heating, refrigerating, and air-conditioning industry has come about because of the variety of engineering analysis programs for the HVAC industry, an even larger number and range of programs for business use, and the low cost of powerful computers on which to run them. The ordinary calculations required in the HVAC industry, such as heating and cooling loads, can be performed easily and inexpensively on a computer. In addition, computers sometimes allow the solution of more complex problems that would otherwise be impractical to solve. The operation and maintenance of buildings has also benefited from computers that monitor, control, and in certain cases diagnose problems in HVAC equipment. This chapter covers computer concepts for both general and specialized application to the HVAC industry.

APPLICATION CONCEPTS

Current technology in the computer industry changes rapidly. By the time a computer is purchased and implemented, new technology may appear on the market that renders it obsolete. This situation, along with the need to share information and applications among various computers, makes a design that can readily adapt to change a necessity. An information system has an architecture in much the same way that a building does. If laid out properly, this architecture can accommodate the introduction of new technologies while allowing the continuing use of existing software.

Before choosing computer products, the business that the information system will serve must be understood. End users can best specify the principles most important to a particular business. The system and applications can then be selected to match the needs of the business, and a technical framework that supports a consistent computing environment can be established.

Standards take several forms. Industry standards, such as ISO, ANSI/IEEE, and SQL, are defined and recognized by industry groups. Proprietary standards are often formed where other standards are not applicable, such as in a company. De facto standards, such as the operating system software for personal computers, are another type of standard.

Hardware

The availability of inexpensive personal computers and workstations at prices affordable to any company has made in-house computing an attractive option for even the smallest firm. The obvious advantages of rapid turnaround and unlimited computing for a fixed investment are compelling.

A **personal computer** operates at low cost to allow interactive usage and storage for the tasks a user generally takes on. Data can be transferred to other computers. This combination can enable a firm to handle a larger number and variety of projects than would be possible with a central system alone.

But owning a computer incurs additional expenses, including maintenance, software, and personnel costs. Software must be

The preparation of this chapter is assigned to TC 1.5, Computer Applications.

either bought or developed, and although it may seem expensive to buy, it is almost always more expensive to develop. A firm should employ at least one person knowledgeable about the systems and the software, or have contractual access to such a person.

Personal and workstation computers offer the small office enormous power at low cost. These machines should be considered in the class of business machines, in that they offer more possibilities than technical analysis (see the section on General Productivity Tools).

Laptop computers are battery-powered, miniature personal computers that are fully functional and more or less portable. These computers are also called palmtops, portables, or notebooks depending on their size.

Any of the hardware options described here may be connected through a **local-area network** (LAN), or a **wide-area network** (WAN). When machines are thus connected, one or more of the computers may be used as a file server. Any computer, once connected locally or via telephone to the server, has access to common data and programs through the network. This method allows data stored in one logical database to be accessed from many separate computers without the necessity of copying the data onto numerous storage devices.

An emerging hardware technology is **parallel processing**. In this computing method, the processors in one or more networked computers are used in parallel, each solving a portion of a problem. A control program determines what processing power is needed and available for a particular computational task. The task is then assigned to a processor. When the computation is complete, the results are passed back. Depending on the system and the control program, several software program steps may be completed at the same time. This method shows great promise in the area of simulations, which tend to be computationally intensive and which may be easily divided into smaller pieces.

Personal digital assistants offer unique capabilities that were previously obtainable only through a combination of equipment (phone, facsimile machine, laptop computer, modem, and more).

Programmable calculators and personal organizers can be programmed to carry out operations of hundreds of steps. They can be outfitted with printers and/or connected to personal computers to exchange data. A range of commercial programs is available for these machines. Many programmable calculators have integral memory or plug-compatible memory modules. These machines can store user-created programs and can read in programs created in machine-readable form by the user or by a vendor.

The primary advantages of the programmable calculator are low cost and portability, although with a printer attached, the calculators are not nearly as portable. For field calculations, they are unexcelled but are being challenged by battery-powered, notebook-sized laptop computers, and even hand-held computers.

Many of these calculators have additional features, such as programmable calendars, alarm clocks, and even language interpreters, which allow them to serve also as personal organizers. In many cases programmable calculators perform the same functions as laptop computers. Likewise, some laptops are becoming nearly as small as programmable calculators.

Software Options

Software can be divided into four major categories: system software, languages, utilities, and application programs.

System software, otherwise known as the operating system, is the environment in which other programs run. It handles input and output (keyboard, video display, and printer) and file transfer between disks and memory; it also supports the operation of other programs. Operating system software can be obtained from the computer manufacturer or from software companies. The operating system is specific to a particular type of computer.

Languages are used to write computer programs. They range from assembly language, which involves coding at the machine instruction level, to high-level languages such as FORTRAN, BASIC, Pascal, C, C++, JAVA, J++, or SQL. Many high-level languages exist to satisfy various programming requirements. FORTRAN and C++ are useful for scientific or mathematical applications. BASIC established microcomputers (personal computers) as viable business machines. New dialects of BASIC are overcoming many of the language's previous limitations. Pascal is a structured language originally designed for teaching programming. C and C++ are emerging as preferred standards for professional programming in many industries, including the HVAC industry, because they compile to very efficient and fast code, and the C source code from one computer can be recompiled with other C compilers to run on many types of machines. The major disadvantage of C is the high level of programming skill required to create programs. JAVA and J++ are favored programming languages for applications that run over computer networks such as the Internet. Structured query language (SQL) allows user-friendly requests to be made for the retrieval of database information.

Utility software programs perform standard organizing and data handling tasks for a specific computer, such as copying files from one disk to another, printing file directories, printing files, and merging files (splicing two or more files together in a specific order). Utility software generally performs one or two specific functions, while applications software is written to accomplish a particular task that can require many function and utility components.

HVAC application programs calculate such items as loads, energy, and piping design. General-purpose applications software, such as accounting and word processing programs, is discussed in the section on General Productivity Tools. Another specialized area of applications software is artificial intelligence, discussed in the section on Advanced Tools.

Purchased Software. PC software can be purchased from the manufacturers of computer equipment, software companies, distributors, and discount houses. Software price, level of support, return policies, and distribution method vary from vendor to vendor. Most software companies support their product in some way, although some companies have a poor reputation for supporting customers. Some vendors allow customers to try software and return it if it is not acceptable; others make demonstration or limited function versions available for a minimal charge.

Some vendors use copy protection schemes that prevent the user from running the software on more than one computer. Copy protection inconveniences the user, so the trend is to move away from this protection method. Upon purchase of PC software, the user is given a license with specific restrictions on how that software may be used, typically that it is to be used on only one computer. A separate, signed license is not common but is occasionally used. Because the distribution of unauthorized copies has been so costly to software companies, many are actively prosecuting illegal software use.

Before purchasing general-purpose software (e.g., spreadsheet, database, or word processor programs), the user should look into how the software runs and how easy it is to use. A full-service computer store may offer advice and demonstrations unavailable from mail-order or discount software sources. The user interface is important; some programs are difficult to operate and understand, while others are very easy to use. Computer magazines often publish comparisons of general-purpose software that can be used for a first appraisal. Once a package is chosen, the user may have difficulty changing to another.

Occasionally suppliers offer the program's source code, which is a set of human-readable instructions in a computer language. Skilled programmers can modify a source code to change the operation of the program. This is not generally recommended, because once the program is changed, support must be supplied internally.

Public Domain Software. Public domain software is available to the public either without charge or for a minimal charge (usually for maintenance and support). These programs are developed through government-supported projects, at universities, and by individuals. The source code for many public domain programs is available, though usually poorly supported. However, some programs are well documented and supported. Some programs offer only executable code to prevent unauthorized use of the source code in proprietary programs.

Public domain software can be obtained from the Internet, computer bulletin boards, other individuals, and companies that distribute it for a small duplication charge. Care should be used when obtaining public domain software because its origin is difficult to trace. Some public domain programs have computer viruses; programs on bulletin boards and the Internet are particularly suspect. A virus is a small program inserted into another program that can destroy stored data, lock up the computer, and otherwise cause problems.

Custom Programming. Three major strategies for obtaining custom programs are followed: (1) contracting to an outside firm, (2) using an outside firm to provide consultation and help, or (3) developing the programs internally.

Contracting to an outside firm for the entire programming effort should be considered if the host organization does not have the personnel or desire to support the software on an ongoing basis. Funds should be budgeted for the outside organization to support any modifications or enhancements that become necessary. Contracting outside is a good approach for an organization that does not want to get involved with programming. The main drawbacks are the expense and the lack of control over the program. Licensing and ownership issues should be carefully spelled out in the contract.

Using an outside firm to provide consulting is practical if internal skill is insufficient for programming, but long-term support and maintenance of the software are to be done internally. Outside firms can provide the expertise to get a project going quickly. A good design specification is critical to the success of the software.

Developing programs internally is viable only if the skills and resources are available. Internal projects are easier to control because the people involved are usually under one roof. Most of the major vendors of software-based HVAC systems develop the software internally with occasional consultation from outside firms.

No matter which approach is chosen, the user must provide a detailed functional design specification. The calculations, human interface, reports, user documents, and testing procedures should be carefully detailed and agreed on by all parties before the development begins. To create a useful software program, a thorough understanding of the subject matter and a solid knowledge of computer programming are required. Software testing should also be specified at the beginning of the project to avoid the common problem of low quality due to hasty and inadequate testing. Design testing should address the human interface, a wide range of input values (including improper inputs), the algorithm, and any outputs (to paper, disk, or other media). Field testing should be done under field conditions with the final users of the software.

With any of the development approaches, good, understandable documentation is required. If for any reason the software cannot be adequately supported, the program will have to be either abandoned, replaced, or recoded, causing a substantial drain on fiscal resources.

Custom programming should be used to create only programs or features unavailable with existing software: no matter how high the cost of existing software, custom software that produces comparable results will cost more in the long-run.

Sources for HVAC Programs. The *ASHRAE Journal* lists sources of HVAC-related software. This resource is available in print (ASHRAE 1999) or through the Internet at www.ashrae.org.

GENERAL PRODUCTIVITY TOOLS

Administrative Tools

Word Processing. Word processing programs range from simple, small programs for creating notes to full-featured, electronic publishing tools. Many word processing programs include spelling and grammar checkers, the ability to create indexes and table of contents, simple databases for merging addresses and other information, drawing and graphing tools, tables, spelling and grammar checkers, simple spreadsheets to perform arithmetic with columns of figures, and the ability to convert documents so they may be read on the Internet.

Specification Writing. The efficiency and flexibility of a computerized master specification in conjunction with a word processor make it a valuable tool for preparing project specifications. Those who use a master specification should standardize language to avoid repeated editing of text. A master specification is most beneficial to firms that do not specialize in unique, one-of-a-kind designs.

If the master can be used with only minor revisions for most specifications, the time spent on editing may be justifiable. In general, the larger the size and the greater the number of documents produced, the greater the savings in time and money compared to traditional preparation and typing. Even without a master specification, however, the writer and typist can save considerable time by copying from previous specifications with minor revisions.

Desktop Publishing. Many high-end word processors have the ability to import graphic images and to print in decorative typefaces, but the full capability to arrange page layouts and incorporate graphics is found only in desktop publishing programs. However, word processing programs continue to include more and more desktop publishing features.

Most desktop publishing programs are limited in their ability to create text and drawings; normally text is imported from a word processing program and images from separate drawing programs. High-resolution graphic displays for personal computers and laser printers have allowed the creation and printing of high-quality graphic images.

Management Planning and Decision Making. Computers are ideal for repetitive operations such as inventory control, accounting, purchasing, and project control, providing managers with up-to-date information. The ability to store and manipulate large amounts of data is a powerful basis for forecasting. With sufficient historical data, statistical methods can be used to forecast sales, inventory needs, production capability, personnel needs, and capital requirements. For example, econometric modeling may be included to help forecast the effects of varying inflation rates.

Much of the data required for such techniques may not be in company files and must be found elsewhere. Technical papers, market trends, industry reports, financial forecasts, codes, standards, regulatory data, and demographic or geographic data are examples of information that can be searched for by computer. In many cases, this information can be obtained on the Internet for a nominal or no user access fee. Management information systems should also have the capability to provide special or supporting information in case it becomes necessary to delve deeper into an operation.

Employee Records and Accounting. All corporate accounting functions, from payroll to aging accounts receivables to taxes, can be automated with a wide choice of engineering-specific accounting software on even the smallest PCs. A firm with unique accounting requirements may often meet them by developing specific applications, called templates or overlays, for common spreadsheet programs. Continuously updated financial reports needed for project evaluation and government use can be generated more readily by computer than by manual methods or off-site accounting services. Labor expenses for various projects or departments can easily be separated and monitored on a regular basis. Personnel records for each employee can also be maintained on a computer.

Project Scheduling and Job Costing. The success of any project, no matter what its size, depends on good scheduling and control to ensure that manpower and materials are where they are supposed to be at the proper time. Whenever portions of a project can be done simultaneously, are dependent on the completion of one or more prior parts, or must share a limited amount of manpower and equipment, a project scheduling technique such as critical path method (CPM) or program evaluation and review technique (PERT) can establish priorities. Current microcomputer programs can also perform personnel leveling to make the best use of available resources, and some may provide graphics suitable for client presentations as well.

Commercially available computerized scheduling techniques quickly provide updated schedules and priorities, even if the critical path changes due to a delay in one part of the job. This capability is valuable for determining whether scheduled completion dates and costs can be met and for finding possible alternatives. Once personnel, payroll, and accounting functions has been established on a computer, a history of project costs, manpower requirements, and material requirements can be maintained. The computer can categorize and tabulate them for future reference, permitting more accurate estimates for future projects.

Security and Integrity. While computers are generally reliable, provision must be made for potential loss of important data in the event of equipment failure, theft, sabotage, natural catastrophe, or a previously undiscovered problem in the software. The most common method of protection is periodic duplication of disk files onto tape, CD-ROM or other digital archival media. Duplication of the most important data files, such as the firm's accounts, should take place with every posting. Duplicate files should be kept in a secure area, such as a fireproof vault on the premises or totally off-site.

Another security concern is unauthorized access to particular data. The implementation of a formal, written data security policy, no matter how minimal it is initially, is a critical step in the computerization of any firm. Careful planning is required to be sure that only authorized personnel have access to confidential files. Most major accounting packages, as well as many popular spreadsheet and database management programs, contain built-in safeguards, such as multilevel password protection, to help keep information secure. Some packages also contain a transactions log to track who has accessed a particular file and what functions were performed.

Design Tools

Spreadsheet Software. Spreadsheets are two- or three-dimensional electronic tables containing thousands of cells into which the user may enter labeling text, numbers, formulas, or even macro-commands. Macro-commands can transfer control to another program, automatically move the user around in the spreadsheet, jump to other spreadsheets or databases, or execute other programs and return with a value. Automatic recalculation of all cells defined by formulas following data entry keeps all areas of the spreadsheet current. Spreadsheets include the following advantages:

- Rapid assembly and testing of a chain of equations
- Preformatted output/input and template structures
- Prepackaged or canned graphics
- Easy file transfer between spreadsheets or between different proprietary spreadsheet programs and ASCII or database files
- Ability to sort lists of information
- Matrix and statistical computations
- Automatic recording of multiple keystrokes into a macro-command
- Additional programs with special features that add on to the existing features of a spreadsheet

Spreadsheets also allow numeric and alphanumeric data to be imported or exported from other non-spreadsheet programs. Graphs from spreadsheets can be imported into sophisticated graphics presentation programs for additional editing and preparation.

Graphics and Imaging Software. Microcomputer software graphics packages readily produce *x-y* and bar graphs, pie charts, and computer-aided design (CAD) drawings, and often have a library of graphic elements that make possible the rapid creation of custom presentations. Many specialty programs allow graphics files to be transferred between different programs and images digitized from photographs or taken using digital cameras to be imported into an existing graphics drawing. Multimedia tools allow video and sound to be combined with traditional text and graphics.

Data Communications Software. Communication between computers takes place through telephone lines or other networks of various configurations. Coaxial cables, twisted pair wiring, superimposed ac carrier frequencies, fiber optics, radio transmissions, and infrared light are among the media used. Features available on such connections include bulletin boards, on-line retrieval services, shared services, web browsers, natural language queries, electronic mail, and file transfer.

Electronic data communication continues to require both appropriate software and a modem, which performs the translation of the digital computer signals into analog signals suitable for current telephone lines. The modems most commonly used at present transfer data at 14,400 to 56,000 bits per second (bps) using one of several transmission standards.

With the simplest of communications programs, the user can type messages at the keyboard to be sent to a remote computer, specify a file in a local storage to be sent to the remote computer, or receive a file from the remote computer and save it locally. Other capabilities include executing error-checking protocols with the remote computer to eliminate transmission errors and allowing the remote computer to take control of the local one, which is called terminal emulation.

Database Software. Microcomputer databases allow the user to record, organize, and store information, and can be viewed as computerized record keeping (Date 1988). Most database software includes a query-based command language, as well as a programming language that can be used to create applications. Some programs allow for query-by-example, windowing, or an icon-driven user interface. With object-oriented design, the query structure is different, and data can be encapsulated and reused in other databases more easily.

Special-Purpose Software. New types of special-purpose software are created almost daily, some of them specific to the HVAC industry. Some programs provide such accessories as notepads, appointment calendars, battery charge (in laptop PCs), name and address files, alarm clocks, telephone dialers, on-line dictionaries, and last command recall. Specialty programs also include file organizers, routine backup programs, file restoration programs, and hardware diagnostics. Many software packages integrate word publishing programs, spreadsheets, communication programs, and graphics programs.

Many special-purpose HVAC application programs exist, including, proprietary equipment selection packages and microcomputer versions of public domain programs such as DOE-2, BLAST, TRNSYS, and other powerful simulation programs previously available only on large mainframe computers. Many non-HVAC advanced analysis programs are also available, including statistical and mathematical packages, specialty artificial intelligence or knowledge-based development shells, and neural network development programs.

Advanced Input and Output Options. Advanced input capabilities include digitizing of images and text, scanners, bar code input, pen-based input, downloading of data directly from portable field instruments, voice input, and facsimile (fax) machines.

While each character of text can be represented internally by a single byte of storage, graphic images are more difficult to store and manipulate; they can be represented in bitmap or in vector format. In the bitmap or raster representation, the image is decomposed into points in a rectangular array (pixels). Storage requires a minimum of one bit per pixel for monochrome images and several bits per pixel for color. In vector representation, only points of line segments are stored (as numbers). While the output of bitmap images is simple on raster devices such as graphic printers, scaling of images does not always produce satisfactory output. Vector representations plot easily on pen plotters and scale to a high degree of accuracy, but must be converted to raster format for output on graphic printers.

Computer-aided design and other drawing programs use vector representation. Paint programs use bitmap images. Scanners can be used to duplicate an image that resides on paper and produce a raster representation (text and images are both in bitmap format). Paint programs can take scanned images directly and facilitate their manipulation by the user. It is possible to scan drawings into raster format, and then use conversion software to change the raster representation into vector form for use in CAD. Similarly, optical character representation (OCR) software can take a raster image of text and convert it to individual text characters in the computer for input into word processing programs.

Fax equipment scans input images into raster format and sends the images bit-by-bit over telephone lines to a receiver, which prints them out as raster images. Computers, can capture incoming fax images, allowing either conversion to internal text form or manipulation of the images.

Just as input capabilities have improved, so have output options, including the porting of graphic routines to video cassette recorder (VCR) tapes for assembly into training tapes, direct 35 mm slide production, and direct transmission of output via fax machine. Newly available hardware and specialty software for laptops allow the user to preassemble and display an animated electronic slide show, digitized images, or even CAD drawings using a special translucent output screen (LCD) that is placed on top of a traditional overhead projector. Certain hardware even allows the user to view a television image in a separate window while working on a document or spreadsheet. Such an application might be the precursor of a video-based energy management and control system (EMCS).

ENGINEERING DESIGN CALCULATIONS

Although computers are now widely used in the design process, most programs perform not design, but simulation. That is, the engineer proposes a design and the computer program calculates the consequences of that design. When alternatives are easily cataloged, a program may aid design by simulating a range of alternatives and then selecting the best according to predetermined criteria. Thus, a program to calculate the cooling load of a building requires specifications for the building and its equipment; it then simulates the performance of that building under certain conditions of weather, occupancy, and scheduling. A duct-design program may actually size ductwork, but usually an engineer must still decide air quantities, duct routing, and so forth.

Since computer programs perform repetitive calculations rapidly, it is possible for the designer to explore a wide range of alternatives and use selection criteria based on annual energy costs or life-cycle costs.

Heating and Cooling Loads

The calculation of design thermal loads in a building is a necessary step in the selection of HVAC equipment for virtually any building project. To ensure that heating equipment can maintain satisfactory building temperature under all conditions, peak heating loads are usually calculated for steady-state conditions without solar or internal heat gains. This relatively simple calculation can be performed with or without a computer.

Peak cooling loads are more transient than heating loads. Radiative heat transfer in a space and thermal storage cause thermal loads to lag behind instantaneous heat gains and losses. Especially with cooling loads this lag can be important, as the peak is both reduced in magnitude and delayed in time compared to the heat gains that cause it. Early methods of calculating peak cooling loads tended to overestimate loads; this resulted in oversized cooling equipment with penalties of a high first cost and part-load operating inefficiencies. Today various calculation methods account for the transient nature of cooling loads. These are described in Chapters 27 and 28 of the 1997 *ASHRAE Handbook—Fundamentals* and in *Load Calculation Principles* (Pederson et al. 1998).

Characteristics of a Loads Program. In general, a loads program requires user input for most or all of the following:

- Full building description, including the construction of the walls, roof, windows, etc., and the geometry of the rooms, zones, and building. Shading geometries may also be included.
- Sensible and latent internal loads due to lights and equipment, and their corresponding operating schedules.
- Sensible and latent internal loads due to people.
- Indoor and outdoor design conditions.
- Geographic data such as latitude and elevation.
- Ventilation requirements and amount of infiltration.
- Number of zones per system and number of systems.

With this input, a load programs calculates both the heating and cooling loads as well as perform a psychrometric analysis. Output typically includes peak room and zone loads, supply air quantities, and total (coil) load.

Selecting a Loads Program. In addition to general characteristics, such as hardware and software requirements, type of interface (icon-based, menu-driven versus command-driven), availability of manuals and support, and cost, some loads program-specific characteristics should be considered when selecting a loads program. The following are among the items to be considered:

- Type of building to be analyzed—residential versus commercial. Residential load programs tend to be simpler to use than the more general-purpose programs meant for commercial and industrial use. However, residential-only programs have limited abilities.
- Method of calculation for the cooling load, as discussed previously in this section.
- Program limits on such items as number of systems, zones, rooms, and surfaces per room.
- Sophistication of modeling techniques, for example, the capability of handling exterior or interior shading devices, tilted walls, daylighting, and skylights.
- Units of input and output.
- Complexity of program. In general, the more sophisticated and flexible programs require more input and are somewhat more difficult to use than the simpler programs.
- Capability of handling the system under investigation.

- Ability to share data with other programs, such as computer-aided design (CAD) and energy analysis.

Duct Design

Two major needs exist in duct design: sizing and flow distribution. Duct sizing and equipment selection are part of any new duct design. Flow distribution is the calculation of flows through the duct sections and terminals for an existing system with known cross sections and fan characteristics.

Duct Sizing. Two major approaches to computerized duct sizing are followed: (1) application of manual procedures, which, although computerized, are still limited in capability, and (2) optimization. Duct design methods are described in Chapter 32 of the 1997 *ASHRAE Handbook—Fundamentals*.

Selecting and Using a Program. Duct design involves laying out the ductwork, selecting the fittings, and sizing the ducts. Computer programs can address many constraints that require recomputation of the duct size. Computer printouts provide detailed documentation. Any calculation requires the preparation of accurate estimates of pressure losses in duct sections and the definition of the interrelations of velocity pressures, static pressures, total pressures, and fitting losses.

The general computer procedure is to designate nodes (the beginning and end of duct sections) by number. Details about each node (e.g., divided flow fitting and terminal) and each section of duct between nodes (maximum velocity, flow rate, length, fitting codes, size limitation, insulation, and acoustic liner) are used as input data (Figure 1).

Characteristics of a duct design program include the following:

- Calculations for supply, return, and exhaust
- Sizing by constant friction, velocity reduction, static regain, and constant velocity methods
- Analysis of existing ducts
- Inclusion of fitting codes for a variety of common fittings
- Identification of the duct run with the highest pressure loss, and tabulation of all individual losses in each run
- Printout of all input data for verification
- Provision for error messages
- Calculation and printout of airflow for each duct section
- Printout of velocity, fitting pressure loss, duct pressure loss, and total static pressure change for each duct section

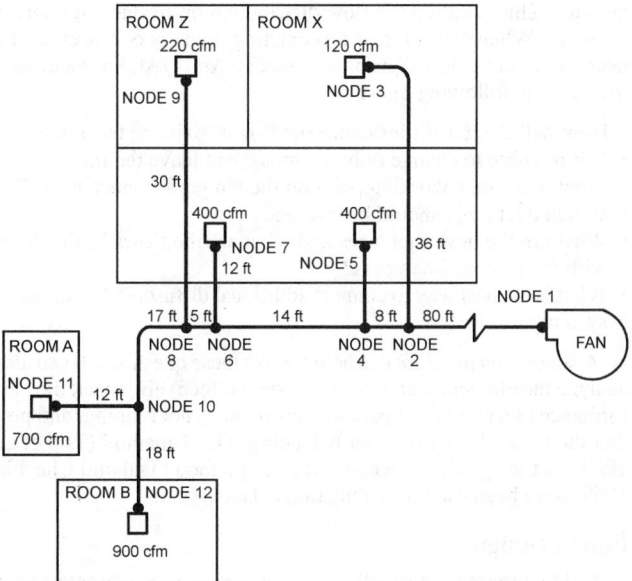

Fig. 1 Example of Duct System Node Designation

- Graphic showing of a schematic or line diagram indicating duct size, shape, flow rate, and temperature in the ducts
- Calculation of heat gain/loss and correction of temperatures and flow rates, including possible resizing
- Specification of maximum velocities, size constraints, and insulation thicknesses
- Consideration of insulated or acoustically lined duct
- Bill of materials for sheet metal, insulation, and acoustic liner
- Acoustic calculations for each section
- File-sharing with other programs, such as spreadsheet and CAD

Because many duct design programs are available, the following factors should be considered in program selection:

- Maximum number of branches that can be calculated
- Maximum number of terminals that can be calculated
- Types of fittings that can be selected
- Number of different types of fittings that can be accommodated in each branch
- Ability of the program to balance pressure losses in branches
- Ability to handle two- and three-dimensional layouts
- Ability to size a double-duct system
- Ability to handle draw-through and blow-through systems
- Ability to prepare cost estimates
- Ability to calculate fan motor power
- Provision for determining acoustical requirements at each terminal
- Ability to update the fitting library

Optimization Techniques for Duct Sizing. For an optimized duct design, fan pressure and duct cross sections are selected by minimizing life-cycle cost, which is an objective function that includes initial cost and energy cost. Many constraints, including constant pressure balancing, acoustic restrictions, and size limitations, must be satisfied. Duct optimization is a mathematical programming problem with a nonlinear objective function and many nonlinear constraints. The solution must be taken from a set of standard diameters and standard equipment. Several numerical methods for duct optimization exist, such as the T-method (Tsal et al. 1988), coordinate descent (Tsal and Chechik 1968), Lagrange multipliers (Stoecker et al. 1971, Kovarik 1971), dynamic programming (Tsal and Chechik 1968), and reduced gradient (Arkin and Shitzer 1979).

Flow Distribution. Another problem is the prediction of airflows in each section of a pre-sized system with known fan characteristics. This is called the flow distribution or air duct simulation problem. Whenever a retrofit to existing ducts is considered, the need to calculate flow distribution occurs. An HVAC engineer may then ask the following questions:

- How will the retrofit influence the flow at existing terminals?
- Is it possible to change only the motor and leave the fan?
- What is the new working point on the fan performance curve?
- Which duct sizes should be changed?
- What are the new duct sizes and what are the flows in the ducts with fully opened dampers?
- What is the best way to connect additional diffusers to an existing system?

A simulation program can help answer these questions. It can also analyze the efficiency of a control system effectively, check the performance of a number of parallel fans if one is not running, and predict the flows during field air balancing. The T-method (Tsal et al. 1988) and the gradient steepest descent method (Tsal and Chechik 1968) have been used for simulating a duct system.

Piping Design

Sizing programs normally size the piping and estimate pump pressure based on velocity and pressure drop limits. Some consider heat gain or loss from piping sections. Several programs produce a

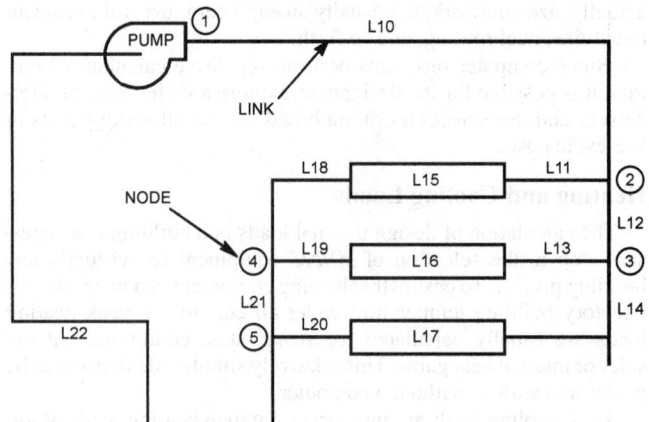

Fig. 2 Example of Nodes for Piping System

bill of materials or cost estimate for the piping. Piping flexibility programs assist in stress and deflection analysis. Many of the piping design programs can account for thermal effects in pipe sizing, as well as deflections, stresses, and moments.

Programs for the analysis of refrigerant piping layouts are also available. These programs aid in installing properly sized pipes or in troubleshooting those already installed. Some programs are generic—they use the physical properties of refrigerants along with system practices to give pipe sizes that may be used. Other programs "mix and match" evaporators and condensers from a particular manufacturer and recommend appropriate pipe sizes.

The general technique for computerizing piping design problems is similar to that for duct design. A typical piping problem in its nodal representation is shown in Figure 2.

Useful piping programs do the following:

- Provide sufficient design information
- Perform calculations for both open and closed systems
- Calculate the flow, pipe size, and pressure drop in each section
- Handle three-dimensional piping systems
- Cover a wide selection of commonly used valves and fittings, including solenoid and pressure-regulating valves
- Consider different piping materials such as steel, copper, and plastic by including generalized friction factor routines
- Accommodate liquids, gases, and steam by providing property information for a multiplicity of fluids
- Calculate pump capacity and pressure required for liquids
- Calculate the available terminal pressure for nonreturn pipes
- Calculate the required expansion tank size
- Estimate heat gain/loss for each portion of the system
- Prepare a cost estimate, including costs of pipe, insulation materials, and associated labor
- Print a bill of materials
- Calculate balance valve requirements
- Perform a pipe flexibility analysis
- Perform a stress analysis for the piping
- Print a graphic display of the pipe network
- Allow customizing of specific design parameters and conditions, such as maximum and minimum velocities, maximum pressure drops, condensing temperature, superheat temperature, and subcooling temperature
- Allow piping to be evaluated for off-design conditions
- Provide links to other programs, such as equipment simulation programs

Limiting factors to consider, in piping program selection include the following:

- Maximum number of terminals program can accommodate

- Maximum number of circuits program can handle
- Maximum number of nodes each circuit can have
- Maximum number of nodes program can handle
- Compressibility effects for gases and steam
- Provision for two-phase fluids

Acoustic Calculations

Chapter 46 summarizes sound generation and attenuation in HVAC applications. Applying this data and methodology often requires a large amount of computation. All sound generation mechanisms and sound transmission paths are potential candidates for analysis. Adding to the computational work load is the need to extend the analysis over, at a minimum, octave bands 1 through 8 (63 Hz through 8 kHz). A computer can save time and reduce the difficulty in analyzing any noise situation, but the designer should be wary of using unfamiliar software.

Caution and critical acceptance of analytical results are mandatory at all frequencies, but particularly at low frequencies. Not all manufacturers of equipment and sound control devices provide data below 125 Hz. Thus, the HVAC designer conducting the analysis and the programmer developing the software must make assumptions based on experience for these critical low-frequency ranges.

The designer/analyst should be well satisfied if predictions are within 5 dB of field-measured results. In the low-frequency rumble regions, results within 10 dB are often as accurate as can be expected, particularly in areas of fan discharge. Conservative analysis and application of the results is necessary, especially if the acoustic environment of the space being served is critical.

Several currently available acoustics programs are generally easy to use, but they are often less detailed than the custom programs developed by acoustic consultants for their own use. Acoustics programs are designed for comparative sound studies and allow the design of comparatively quiet systems. Acoustic analysis should address the following key areas of the HVAC system:

- Sound generation by HVAC equipment
- Sound attenuation and regeneration in duct elements
- Wall and floor sound attenuation
- Ceiling sound attenuation
- Sound break-out or break-in from ducts or casings
- Room absorption effect (relation of sound power criteria to sound pressure experienced)

Algorithm-based programs are preferred because they cover more situations (see Chapter 46). However, assumptions are an essential ingredient of algorithms. These basic algorithms, along with sound data from the acoustics laboratories of equipment manufacturers, are incorporated to various degrees in acoustics programs. The HVAC equipment sound levels in acoustics programs should come from the manufacturer and be based on measured data, because there is a wide variation in the sound generated by similar pieces of equipment. Some generic equipment sound generation data, which may be used as a last resort in the absence of specific measured data, are found in Chapter 46. Whenever possible, equipment sound power data by octave band (including 32 Hz and 63 Hz) should be obtained for the path under study. A good sound prediction program relates all performance data.

Other more specialized acoustics programs are also available. Various manufacturers provide equipment selection programs that not only select the optimum equipment for a specific application, but also provide associated sound power data by octave bands. These programs can help in the design of a specific aspect of a job. Data from these programs should be incorporated in the general acoustic analysis. For example, duct design programs may contain sound predictions for discharge airborne sound based on the discharge sound power of the fans, noise generation/attenuation of duct fittings, attenuation and end reflections of variable air volume (VAV) terminals, attenuation of ceiling tile, and room effect. VAV

terminal selection programs generally contain subprograms that estimate the space NC level near the VAV unit in the occupied space. However, projected space NC levels alone may not be acceptable substitutes for octave-band data. The designer/analyst should be aware of assumptions, such as room effect, made by the manufacturer in the presentation of acoustical data.

Predictive acoustic software allows designers to look at HVAC-generated sound in a realistic, affordable time frame. HVAC oriented acoustic consultants generally assist designers by providing cost-effective sound control ideas for sound-critical applications.

Refrigerant Properties

REFPROP (NIST 1996) is a program that allows the user to examine thermodynamic and transport properties for thirty-eight pure refrigerants and blends. It may be used in an interactive mode. Since the source code is included, it may also be used as part of a program that requires refrigerant properties. A number of other refrigerant property programs are also available from sources such as universities. Refrigerant properties that the user may find useful include, enthalpy, entropy, viscosity, and thermal conductivity.

Ventilation

Several ventilation programs aid the designer in satisfying requirements of ASHRAE *Standard* 62, Ventilation for Acceptable Indoor Air Quality. As with any computer program, the user must qualify the program's technical capabilities, such as

- Ventilation requirements by application
- Calculation of the Multiple Space equation in *Standard* 62
- Use of ventilation effectiveness
- Application for spaces with intermittent or variable occupancy

In addition to technical capabilities the units of input and output and the ability to interface with programs for input or outputs should also be examined.

Equipment Selection and Simulation

Three types of equipment-related computer programs are available—equipment selection, equipment optimization, and equipment simulation programs.

Equipment selection programs are basically computerized catalogs. The program locates an existing equipment model that satisfies the entered criteria. The output is a model number, performance data, and sometimes alternative selections.

Equipment optimization programs display all equipment alternatives and let the user establish ranges of performance data or first cost to narrow the selection. The user continues to narrow the performance ranges until the best selection is found. The performance data used for optimizing selections vary by product family.

Equipment simulation programs calculate the full- and part-load performance of specific equipment over time, generally one year. The calculated performance is matched against an equipment load profile to determine energy requirements. Utility rate structures and related economic data are then used to project equipment operating cost, life-cycle cost, and comparative payback. Before accepting output from an equipment simulation program, the user must understand the assumptions made, especially those assumptions concerning load profile and weather.

Some advantages of equipment programs include the following:

- High speed and accuracy of the selection procedure
- Pertinent data presented in an orderly fashion
- More consistent selections than with manual procedures
- More extensive selection capability
- Multiple or alternate solutions
- Small changes in specifications or operating parameters easily and quickly evaluated
- Data-sharing with other programs, such as spreadsheets and CAD

Simulation programs have the advantage of (1) projecting part-load performance quickly and accurately, (2) establishing minimum part-load performance, and (3) projecting operating costs and pay-back-associated higher performance product options. Programs for nearly every type of HVAC equipment are available, and industry standards apply to the many of them. The more common programs and their optimization parameters include the following:

Air distribution units	Pressure drop, first cost, sound, throw
Air-handling units, rooftop units	Power, first cost, sound, filtration, footprint, heating and cooling capacity
Boilers	First cost, efficiency, stack losses
Cooling towers	First cost, design capacity, power, flow rate, air temperatures
Chillers	Power input, condenser pressure, evaporator pressure, capacity, first cost, compressor size, evaporator size, condenser size
Coils	Capacity, first cost, fluid pressure drop, air pressure drop, rows, fin spacing
Fan coils	Capacity, first cost, sound, power
Fans	Volume flow, power, sound, first cost, minimum volume flow
Heat recovery equipment	Capacity, first cost, air pressure drop, water pressure drop (if used), effectiveness
Pumps	Capacity, pressure, impeller size, first cost, power
Air terminal units (variable and constant volume flow)	Volume flow rate, air pressure drop, sound, first cost

Some selection programs are quite versatile. For example, coil selection programs can select steam, hot water, chilled water, and refrigerant (direct expansion) coils. Generally, they select coils according to procedures in ARI *Standards* 410 and 430.

Chiller and refrigeration equipment selection programs can choose optimal equipment based on such factors as lowest first cost, highest efficiency, best load factor, and best life-cycle performance. In addition, some manufacturers offer modular equipment for customizing their product. This type of equipment is suited for computer selection.

However, equipment selection programs have limitations. The logic of most manufacturers, programs is proprietary and not available to the user. All programs incorporate built-in approximations or assumptions, some of which may not be known to the user. Equipment selection programs should be qualified before use.

SIMULATION PROGRAMS

Energy and System Simulation

Building energy simulation programs in contrast to peak load calculation programs integrate loads over time (usually a year), consider the systems serving the loads, and calculate the energy required by the equipment. Most energy programs simulate the performance of already designed systems, although some programs make selections formerly left to the designer, such as equipment size, air volume, and fan power. Energy programs are necessary for making decisions regarding building energy use and, along with life-cycle cost routines, quantifying the impact of proposed energy conservation measures during the design phase. In new building design, energy programs help determine the appropriate type and size of equipment; they can also be used to explore the effects of design trade-offs and evaluate the benefits of innovative control strategies and the efficiency of new equipment.

Energy programs that track building energy use accurately can help determine whether a building is operating efficiently or wastefully. They can also allocate costs from a central heating/cooling plant among customers of the plant. However, such programs must be adequately calibrated to data measured from the building under consideration.

Characteristics of Building Energy Simulation. Most programs simulate a wide range of buildings, mechanical equipment, and control options. However, computational results differ substantially from program to program. For example, the shading effect from overhangs, side projections, and adjacent buildings is frequently a factor in the energy consumption of a building; however, the diversity of approaches to the load calculation results in a wide range of answers.

The choice of weather data also influences the load calculation. Depending on the requirements of each program, the following types of weather data are used:

- Typical hourly data for one year only, from averaged weather data
- Typical hourly data for one year, as well as design conditions for typical design days
- Reduced data, commonly a typical day or days per month for the year
- Typical reduced data, non-serial or bin format
- Actual hourly data, recorded on-site or nearby, for analysis where the simulation is being compared to actual utility billing data or measured hourly data

Both air-side and energy conversion simulations are required to handle the wide variations among central heating, ventilation, and air conditioning. For proper estimation of energy use, simulations must be performed for each design combination, operating scheme, and control sequence.

Simulation Techniques. Two methods are currently used to simulate energy systems: the fixed schematic technique and the component relation technique.

The **fixed-schematic-with-options technique**, which is the most prevalent program, involves writing a calculation procedure that defines a given set of systems. The schematic is then fixed, with the user's options usually limited to equipment performance characteristics, fuel types, and the choice of certain components.

The **component relation technique** differs from the fixed schematic in that it is organized around components rather than systems. Each component is described mathematically and placed in a library. Input includes the definition of the schematic, as well as equipment characteristics and capacities. Once all components have been identified and a mathematical model for each has been formulated, the components may be connected together and information may be transferred between them. Although this approach has certain inefficiencies because of its more general organization, the component relation technique does offer versatility in defining configurations.

Selecting an Energy Program. In selecting an energy analysis program, factors such as cost, availability, ease of use, technical support, and accuracy are important. However, another major concern is whether the program will do what is expected of it. It should be sensitive to the parameters of concern, and its output should include the necessary data. For other considerations, see Chapter 30 of the 1997 *ASHRAE Handbook—Fundamentals*.

Time can be saved if the initial input file for an energy program can also be used for load calculations. Some programs interface directly with computer-aided design (CAD) files, greatly reducing the time needed to create an energy program input file.

Comparisons of Energy Programs. Many comparisons, verifications, and validations of simulation programs are reported in the literature (see the section on Bibliography). Conclusions from these studies can be summarized as follows:

- The results obtained by using several computer programs on the same building range from good agreement to no agreement at all. The degree of agreement depends on the interpretations of the program user and the ability of the computer programs to model the building.

- Several people using several programs on the same building will probably not agree on the results of an energy analysis.
- The same person using different programs on the same building may or may not find good agreement, depending on the complexity of the building and its systems and on the ability of the computer programs to model the specific conditions in that building.
- Forward computer simulation programs, which calculate the performance of a building given a set of descriptive inputs, weather conditions, and occupancy conditions, are best used for design.
- Calibration of hourly forward computer simulation programs is possible, but can require considerable effort for a moderately complex commercial office building. Detailed information concerning scheduled use, equipment set points, and even certain on-site measurements may be necessary for a closely calibrated model. Special-purpose graphic plots are useful in calibrating a simulation program with data from monthly, daily, or hourly measurements.
- Inverse, empirical, or system parameter identification models may also be useful in determining the characteristics of building energy usage. Such models can determine the relevant building parameters from a given set of actual performance data. This procedure is the inverse of the traditional building modeling approach, hence the name.

Energy Programs to Model Existing Buildings. Many programs are available and they vary widely in cost, degree of complexity, and ease of use. A general procedure should be followed in energy analysis of existing buildings. First, **energy consumption data** must be obtained for a one- to two-year period. For electricity, these data usually consist of metered electrical energy consumption and demand on at least a month-by-month basis. For natural gas, the data are in a similar form and are almost always on a monthly basis. For both electricity and natural gas, it is helpful to record the dates when the meters were read. These dates are important in weather normalization for determining average billing period temperatures. For other types of fuel, such as oil and coal, the only data available may be the delivery amounts and dates.

Unless fuel use is metered or measured daily or monthly, consumption for any specific period shorter than one season or year is difficult to determine. The data should be converted to a per-day usage or adjusted to account for differences in the length of metering periods. The data tell how much energy went into the building on a gross basis. Unless extensive submetering is used, it is nearly impossible to determine when and how that energy was used and what it was used for. It may be necessary to install such meters to determine energy use.

The t**hermal and electricity usage characteristics** of a building and its energy consumption as a time-varying function of ambient conditions and occupancy must also be determined. Most computer programs can use as much detailed information about the building and its mechanical and electrical system as is available. Where the energy implications of these details are significant, it is worth the effort of obtaining them. Testing fans for air quantities, pressures, control set points, and actions can provide valuable information on deviations from design conditions. Test information on pumps can also be useful.

Data on **building occupancy** are among the most difficult to obtain. Since most energy analysis computer programs simulate the building on an hourly basis for a one-year period, it is necessary to know how the building is used for each of those hours. Frequent observation of the building during days, nights, and weekends shows which energy-consuming equipment is being used and to what degree. Measured, submetered hourly data for at least one week is needed to begin to understand weekday/weekend schedule-dependent loads.

Weather data, usually one year of hour-by-hour weather data, are necessary for simulation. The actual weather data for the year in which energy consumption data were recorded significantly improves the simulation of an existing building. Where the energy consuming nature of the building is related more to internal than to external loads, the selection of weather data is less important; however, with residential buildings or buildings with large outside ambient air loads, the selection of weather data can affect results significantly. The purpose of the simulation should also be considered when choosing weather data; i.e., either specific year data, data representative of long-term averages, or data showing temperature extremes may be needed, depending on the goal of the simulation.

Usually the results of the first computer runs do not agree with the actual metered energy consumption data. The following are possible reasons for this discrepancy:

- Insufficient understanding of the energy-consuming systems that create the greatest use
- Inaccurate information on occupancy and time of building use
- Inappropriate design information on air quantities, set points, and control sequences

The input building description must be adjusted and trial runs continued until the results approximate actual energy use. Matching the metered energy consumption precisely is difficult; in any month, results within 10% are considered adequate.

The following techniques for calibrating a simulation program to measured data from a building should be considered:

- Matching submetered loads or 24-hour day profiles of simulated whole-building electric loads to measured data
- Matching x-y scatter plots of simulated daily whole-building thermal loads versus average daily temperature to measured data
- Matching simulated monthly energy use and demand profiles to utility billing data

Having a simulation of the building as it is being used permits subsequent computer runs to evaluate the energy impact of various alternatives or modifications. The evaluation may be accomplished simply by changing the input parameters and running the program again. The impacts of the various alternatives may then be compared and an appropriate one selected.

Computational Fluid Dynamics

Computational fluid dynamics (CFD) is a technique for simulation, study, analysis, and prediction of fluid flow and heat transfer in well defined spaces with boundaries. It is based on equations that govern the physical behavior and the appropriate equation of state. Usually these equations are in the form of nonlinear partial differential equations relating velocities, pressure, temperature, and some scalar variables to space directions, and time. Initial and boundary conditions are specified and numerical methods such as finite difference, finite volume, or finite element analysis methods are used to solve these equations. The domain (typically space) is divided (discretized) into cells or elements, the nodal points are defined, then equations are solved for each of the discrete cells. Upon solving of these equations, the dependent variables (velocities, pressure, temperature, and scalar) are made available at the nodes.

CFD analysis can be successfully applied to a variety of tasks including the location of airflow inlets/outlets in a space, velocities required, flow pattern determination, temperature distribution in space, and smoke movement and buildup in a space with fire.

GRAPHICS APPLICATIONS

Computer-Aided Design

CAD (which refers to computer-aided drafting at times and to computer-aided design at other times) is a subset of CADD (computer-aided design and drafting). CADD encompasses the creation of building data models and the storing and use of attributes of the graphic elements. Drafting, or the creation of construction documents, is becoming a by-product of the design process. The distinction between engineers and drafters in engineering firms is decreasing as design and production work becomes unified. In this section, CADD refers to all aspects of design.

Computer-aided design (CAD) allows the development of a **single-building data model** that integrates complex, three-dimensional (3D) building data into one common location. Engineers can access such building information as room volumes, percentage of glazing, building orientation, and lighting. This information can then be electronically passed on to third-party analysis programs such as load calculation or distribution analysis. Considering the building data model is drawn in true dimensional size, the information is accurate and complete. The architect or engineer can modify the building model at any phase of the design and provide instantaneous updates. Reiterations of analysis can be accomplished at any phase of the building's life cycle.

Computer-generated data models can also help the designer visualize the building and its systems and check for interferences, such as the structure and ductwork occupying the same space. Some software packages now perform interference checking themselves. This layering feature also helps an HVAC designer coordinate interdisciplinary and multidiscipline designs and simplify modifications.

In addition to the concept of the single building data model, the Internet is influencing work flow in the engineering community. Software applications allow users to design projects and share building model data via the Internet on a global scale. These applications can also be used for simple viewing and redlining as well.

Computer-aided design automates (1) material and cost estimating, (2) design and analysis of HVAC, (3) visualization and interference checking, and even (4) manufacturing of HVAC components such as ductwork.

Computer-generated drawings of building parts can be linked to non-graphic characteristics, which can be extracted for reports, schedules, and specifications. For instance, an architectural designer can draw wall partitions and windows, and then have the computer automatically tabulate the number and size of windows, the areas and lengths of walls, and the areas and volumes of rooms and zones. Similarly, the building data model can store information for later use in reports, schedules, design procedures, or drawing notes. For example, the airflow, voltage, weight, manufacturer, model number, cost, and other data about a fan displayed on a drawing can be stored and associated with the fan in the data model. Conversely, the graphics-to-data links allow the designer to enhance graphic items having a particular characteristic by searching for that characteristic in the data files and then making the associated graphics brighter, bolder, or different-colored in the display.

HVAC models and their associated data can be used by building owners and maintenance personnel for ongoing facilities management, strategic planning, and maintenance. For instance, an air-handling unit can be associated with a preventative maintenance schedule in a separate facility management application. Models are useful for computer-aided manufacturing such as duct construction. The three-dimensional building data model can also aid in the development of sections and details for building drawings. CAD programs can also automate the cross-referencing of drawings, drawing notations, and building documents.

Computer Graphics and Modeling

Computer graphics programs may be used to create and manipulate pictorial information between the computer and the user. Combining text with computer graphics enables the design engineer to quickly evaluate design alternatives. Computers can pictorially simulate design problems with three-dimensional color displays that can predict the performance of mechanical systems before they

are constructed. Simulation graphics software is also used to evaluate design conditions that cannot ordinarily be tested with scale models due to high costs or time constraints. Product manufacturability and economic feasibility may thus be determined without the construction of a working prototype.

Integrated CADD and expert systems can help the building designer and planner with construction design drawings and with construction simulation for planning and scheduling complex building construction scenarios (Potter 1987).

A simple yet helpful tool available to the HVAC engineer is the graphic representation of **thermodynamic properties and thermodynamic cycle analysis**. These programs may be as simple as a computer-generated psychrometric chart with cross-hair cursor retrieval of properties and computer display zooming or magnification of chart areas for easier data retrieval. More complex software graphically represent thermodynamic cyclic paths overlaid on two- or three-dimensional thermodynamic property graphs. With a simultaneous graphic display of the calculated results of Carnot efficiency and COP, the user can understand more readily the cycle fundamentals and the practicality of the cycle synthesis.

Typical **piping and duct system** software produces large amounts of data to be analyzed. Pictorially enhanced simulation output of system curves, pump curves, and load curves speeds the sizing and selection process. What-if scenarios with numeric/graphic output increase the designer's understanding of the system and help avoid design problems that may otherwise come to light only after installation. Graphic-assisted fan and duct system design and analysis programs are also available to designers.

Airflow analysis of flow patterns and air streamlines is done by solving the fundamental equations of fluid mechanics. Finite element and finite volume modeling techniques are used to produce two- or three-dimensional pictorial-assisted displays, with the velocity vectors and velocity pressures at individual nodes solved and numerically displayed. An example of pictorial output is shown in Figures 4, 5, and 6 of Chapter 15, Clean Spaces, where calculated airflow streamlines have been overlaid on the computer against a two-dimensional graphic of the clean room. With this display, the clean room designer can see potential problems in the configuration, such as the circular flow pattern in the lower left and right.

While research has shown good correlation between flow modeling by computer and flow modeling done in simple mock-ups, modeling software should not be considered a panacea for design; simulation of flow around complex shapes is still being developed, as are improved simulations of low Reynolds number flow.

MONITORING AND CONTROL

A control system performs two primary functions—monitoring (also referred to as data acquisition or data gathering) and control (also referred to as device control). A control consists of a set of measured (monitored) parameters, a set of controlled parameters, and a control function which translates the measurement data into control signals applied to controlled parameters.

Supervisory control is the total monitoring and overall control of the local subsystems. Overall control includes functions such as manual overrides, optimizing modification or discharge local loop set points, optimizing start-stop of subsystems, and controlled interaction between subsystems. Total system monitoring includes functions such as alarm reporting, energy measurement and calculation, logs, and trend reports.

Supervisory control has been called a **building automation system** (BAS) when the primary focus was on automating as much as possible to save labor. It has been called an **energy monitoring and control system** (EMCS) when the focus was on saving energy by both automatic control and manual control with the aid of energy monitoring. It has been called an **energy management system** (EMS) when the focus was on saving energy by specific automatic control programs. It has been called a **facility management system** (FMS) when the scope of control went beyond HVAC control and/or beyond a single building, such as including fire, security, or manufacturing systems. Building automation system (BAS) has become the most popular term for description of a computerized control in buildings that may provide one or more of the above functions.

Standard and Open Protocols

Standard protocols such as BACnet are developed in committee by professional societies, open protocols are created by manufacturers but available for all to use, and proprietary protocols are developed by manufacturers but not freely distributed. User needs should be determined prior to selecting a particular protocol for a given application.

BACnet, the **ASHRAE Building Automation and Control Networking Protocol**, provides mechanisms by which computerized equipment for a variety of building control functions may exchange information, regardless of the particular building service it performs (ASHRAE *Standard* 135). As a result, the BACnet protocol may be used by head-end computers, general-purpose direct digital controllers, and application specific or unitary controllers.

BACnet is based on a four-layer collapsed architecture that corresponds to the physical, data link, network, and application layers of the ISO/OSI (International Standards Organization/Open Systems Interconnection) model. The application layer and a simple network layer are defined in the BACnet standard.

The **physical layer** provides a means of connecting devices and transmitting the electronic signals that convey the data.

The **data link layer** organizes the data into frames or packets, regulates access to the medium, provides addressing, and handles some error recovery and flow control.

Functions provided by the **network layer** include translation of global addresses to local addresses, routing messages through one or more networks, accommodating differences in network types and in the maximum message size permitted by those networks, sequencing, flow control, error control, and multiplexing. BACnet is designed so that there is only one logical path between devices, thus eliminating the need for optimal path routing algorithms.

The **presentation layer** provides a way for communicating partners to negotiate the transfer syntax that will be used to conduct the communication. This transfer syntax is a translation from the abstract user view of data at the application layer to sequences of octets treated as data at the lower layers.

The **application layer** of the protocol provides the communication services required by the applications to perform their functions, in this case monitoring and control of the HVAC&R and other building functions.

Building automation system networks are local area networks even though some applications must exchange information with devices in a building that is very far away. This long-distance communication is typically done through telephone networks. The use of the Internet is becoming a popular, low-cost alternative to phones. The routing, relaying, and guaranteed delivery issues are handled by the telephone system and can be considered external to the BAS network. BAS devices are static. They do not move from place to place, and the functions that they are asked to perform do not change in the sense that a manufacturing device may make one kind of part today and some very different part tomorrow. These are among the features of BAS networks that can be used to evaluate the appropriateness of the layers in the OSI model.

APPLICATIONS OF ARTIFICIAL INTELLIGENCE

Artificial intelligence (AI) applications range from systems that learn, reason, and recognize speech like human beings to those that can help make decisions based on incomplete, uncertain, and/or complicated information. Artificial intelligence techniques are generally used in three areas—design, control, and diagnosis. The following sections discuss some of the technology of interest to engineers.

Knowledge-Based Systems. Knowledge-based systems (KBSs), often called expert systems, can make decisions similar to those made by human experts. They typically use facts, if-then rules, and/or models to make decisions. Most KBS tools can also link with other programs, access databases, and import graphics. An expert system shell is usually a complete development package that includes rule and database development tools, debugging facilities, and good user interfaces. Some shells take advantage of windowing environments and incorporate object oriented features, hypertext, and graphics. Haberl et al. (1995), Shams et al. (1994a,b) and Nielsen and Walters (1988) discuss KBS in detail.

Knowledge-based systems are used to make decisions, diagnose faults, monitor and control operation, and develop tutorials. Some HVAC applications include KBSs that select HVAC equipment for small office buildings (Shams et al. 1994a,b), analyze building energy consumption (Haberl et al. 1988), recommend energy conservation options (Meadows and Brothers 1989), help with the conceptual design of building energy systems (Doheny and Monaghan 1987, Mayer et al. 1991), incorporate building regulations into design (Cornick et al. 1990), monitor HVAC equipment (Kaler 1990), manage cooling loads (Potter et al. 1991), and diagnose HVAC equipment problems.

Case-Based Reasoning. Case-based reasoning (CBR) techniques use search methods to find a match between entered information and a previous case on file in a case library. This technique is gaining popularity because it requires less programming maintenance than a rule-based system. Generally the user is only required to answer a few questions and the program seeks to match parameters to the cases on file. While not as generic in providing solutions as rule-based systems, CBR tools increase in general applicability each time a new case is added. When a match is found, the solution for the previous case is presented to the user. The most obvious application of CBR is at help desks, where many of the cases handled are similar.

Knowledge Discovery. Modern computing technologies have gathered and stored large amounts of data. Traditional methods of analysis, especially those involving human expertise, often can not be used to find useful knowledge hidden in this data. Knowledge discovery in databases (KDD), also called data mining, describes the tools used to intelligently and automatically discover information in large databases. Some of these tools include classification and clustering techniques, trend and deviation analysis, dependency derivation, and inductive rule learning. More information can be found in Fayyad et al. (1996).

Artificial Neural Networks. Artificial neural networks (ANNs) are collections of small individually interconnected processing units. Information is passed between these units along interconnections. An incoming connection has two values associated with it: an input value and a weight. The output of the unit is a function of the summed values. ANNs, while implemented on computers, are not programmed to perform specific tasks. Instead, they are trained with repeated data sets until they learn the patterns presented to them. Once the net is trained, new patterns may be presented to it for classification.

Numerous techniques and architectures are available for implementing neural nets, which are detailed by Wasserman (1989). Specialized neural computers are currently being developed, but most neural nets can be implemented by software. Some of the areas in which ANNs have been successful include speech recognition, image enhancement, pattern association, global optimization problems, and predicting building energy usage (Haberl et al. 1996).

Fuzzy Logic. This form of logic is based on the concept of the fuzzy set. Membership in fuzzy sets is expressed in probabilities or degrees of truth, i.e., as a continuum of values ranging from 0 to 1. According to fuzzy-logic theorists, classical logic oversimplifies the concept of set membership by flatly including or excluding an individual, whereas fuzzy logic expresses the extent to which an individual pertains to a set. For example, under classical logic, theoretical tree x is a member of the set of tall trees; in contrast, under fuzzy logic, x pertains partly to the set of tall trees and can be described as fairly tall.

Fuzzy logic is used as a form of data processing by advanced computers. In less complex information processors, the possibility that a particular event will occur is expressed as a certainty (either false or true) represented by the binary digits 0 or 1. Fuzzy-logic, in contrast, breaks down the chance of the occurrence into varying degrees of truthfulness or falsehood (e.g., will occur, probably will occur, might occur, might not occur, etc.). This allows the outcome of an event to be expressed as a possibility. Moreover, as additional data is gathered, many fuzzy-logic systems are able to adjust continually the values assigned to different possibilities. Because some fuzzy-logic systems appear able to learn from their mistakes and mimic human thought processes, they are often considered a crude form of artificial intelligence.

INTERNET

The Internet originated in 1969 as an experimental project of the Advanced Research Project Agency (ARPA), and was called ARPANET. From its inception, the network was designed to be a decentralized, self-maintaining series of redundant links between computers and computer networks, capable of rapidly transmitting communications without direct human involvement or control, and with the automatic ability to re-route communications if one or more individual links were damaged or otherwise unavailable. Among other goals, this redundancy of linked computers was designed to allow vital research and communications to continue even if portions of the network were damaged, say, in a war.

Messages between computers on the Internet do not necessarily travel entirely along the same path. The Internet uses packet switching communication protocols that allow individual messages to be subdivided into smaller packets These are then sent independently to the destination and automatically reassembled by the receiving computer. While all packets of a given message often travel along the same path to the destination, if computers along the route become overloaded, packets can be rerouted to less loaded computers.

No single entity—academic, corporate, governmental, or non-profit—administers the Internet. It exists and functions as a result of the fact that hundreds of thousands of separate operators of computers and computer networks independently decided to use common data transfer protocols to exchange communications and information with other computers. No centralized storage location, control point, or communications channel exists for the Internet, and no single entity can control all of the information conveyed on the Internet.

Communication over the Internet

The most common methods of communications on the Internet (as well as within the major on-line services) can be grouped into six categories:

1. One-to-one messaging (such as e-mail)
2. One-to-many messaging (such as listserv)
3. Distributed message databases (such as USENET news groups)
4. Real time communication (such as Internet Relay Chat)
5. Real time remote computer use (such as telnet)
6. Remote information retrieval (such as ftp, gopher, and the World Wide Web)

Most of these methods of communication can be used to transmit text, data, computer programs, sound, visual images (i.e., pictures), and moving video images.

Electronic mail, or **e-mail**, is comparable to sending a letter. The Internet also contains automatic **mailing list services** (such as list-servs) that allow communications about particular subjects of interest to a group of people. Similar in function to listservs—but quite different in how communications are transmitted—are **distributed message databases** such as USENET news groups. Like listservs, newsgroups are open discussions and exchanges on particular topics. Users, however, need not subscribe to the discussion mailing list in advance, but can instead access the database at any time. Some USENET newsgroups are moderated but most are open access. The dissemination of messages to USENET servers around the world is an automated process that does not require direct human intervention or review.

In addition to transmitting messages that can be later read or accessed, individuals on the Internet can **communicate in real time** with other people on the Internet. In its simplest forms, talk allows one-to-one communications and Internet Relay Chat (or IRC) allows two or more to type messages to each other that almost immediately appear on the other computer screens.

Real time remote computer use is another method to access and control remote computers in real time using telnet. For example, a researcher at a university could use the computing power of a supercomputer located at a different university. A student can use telnet to connect to a remote library to access the library's on-line card catalog program.

Information on the Internet is located and retrieved by three primary methods: ftp, gopher, and the World Wide Web.

A **file transfer protocol** (ftp) allows the user to transfer one or more files between the local computer and a remote computer. Another approach uses a program and format named **gopher** to guide a search through the resources available on a remote computer.

World Wide Web

A third approach, and the most well-known on the Internet, is the World Wide Web (WWW). The web serves as the platform for a global, on-line store of knowledge. Though information on the web is contained in individual computer connected to the Internet through WWW protocols. It is currently the most advanced information system developed on the Internet, and embraces most information in previous networked information such as ftp, gopher, and Usenet.

Placing information or publishing on the web simply requires a computer running WWW server software. The computer can be as simple as a small personal computer or as complex as multi-million dollar computers occupying a small building. Many web publishers choose to lease disk storage space from a service that has the necessary computer facilities, eliminating the need for owning any equipment. Web publishers may make their web sites open to all Internet users, or close them, thus making the information accessible only to those authorized.

Searching the Web. Various **search engines** allow users to search for web sites that contain desired information, or to search for key words. For example, a user looking for information on evaporative cooling would type the words "evaporative cooling" into a search engine. The engine then searches the web and presents a list linked to sites that contain this text. The user can browse through the information on each site until the desired material is found.

Common Standards. The web links together disparate information by setting common information storage formats called hypertext markup language (HTML) and a common language for the exchange of web documents, hypertext transfer protocol (HTTP). Although the information itself may be in many different formats, and stored on computers which are not otherwise compatible, the basic web standards provide a basic set of standards which allow communication and exchange of information. The web can display

HTML documents containing text, images, sound, animation, and moving video. Any HTML document can include links to other types of information or resources, even if the information is stored on numerous computers all around the world.

No single authority controls the Internet as a whole. The Internet Society, or ISOC, is a voluntary membership organization whose purpose is to promote global information exchange through Internet technology. The Internet Architecture Board (IAB) develops standards for the ways computers and software talk to each other. The IAB also keeps track of information that must remain unique. For example, each computer on the Internet has a unique 32- bit address; no other computer has the same address. The IAB doe not actually assign the addresses, but it makes the rules about how to assign addresses.

Collaborative Design. The Internet is revolutionizing how project teams collaborate. Sharing or jointly developing drawings or documents can take place efficiently with collaborators around the world. In a similar manner, geographically dispersed teams can now design and develop software. Numerous coordination issues comprise the collaborative design process. Methods of sharing data, such as distributed databases, revision and developmental accounting, and security concerns associated with the transmission of design information must be considered carefully.

ASHRAE DEVELOPED SOFTWARE

ASHRAE has funded the development of several software programs. These applications range from database retrieval for use in HVAC&R calculations to demonstration versions of advanced tools. They may be obtained via the Bookstore on the ASHRAE website at www.ashrae.org.

REFERENCES

Anderson, D., L. Graves, W. Reinert, J.F. Kreider, J. Dow, and H. Wubbena. 1989. A quasi-real-time expert system for commercial building HVAC diagnostics. *ASHRAE Transactions* 95(2):954-60.

ASHRAE. 1999. Focus on software. *ASHRAE Journal* 41(3):69-72.

ASHRAE. 1995. A data communications protocol for building automation and control networks. ASHRAE *Standard* 135-1995.

ARI. 1991. Forced-circulation air-cooling and air-heating coils. ARI *Standard* 410-91. Air-Conditioning and Refrigeration Institute, Arlington, VA.

ARI. 1989. Central station air-handling units. ARI *Standard* 430-89. Air-Conditioning and Refrigeration Institute, Arlington, VA.

Arkin, H. and A. Shitzer. 1979. Computer aided optimal life-cycle design of rectangular air supply duct systems. *ASHRAE Transactions* 85(l):197-213.

Cornick, S.M., D.A. Leishman, and J.R. Thomas. 1990. Incorporating building regulations into design systems: An object-oriented approach. *ASHRAE Transactions* 96(2):542-49.

Date, C. 1988. *Database-A primer.* Addison Wesley Publishing Company, Reading, MA.

Doheny, J.G., and P.F. Monaghan. 1987. IDABES: An expert system for the preliminary stages of conceptual design of building energy systems. *Artificial Intelligence in Engineering* 2(2).

Fayyad, U.M., G. Piatetsky-Shapiro, P. Smyth, and R. Uthurusamy. 1996. *Advances in knowledge discovery and data mining.* AAAI Press/MIT Press, Menlo Park, CA.

Haberl, J.S., L.K. Smith, K.P. Cooney, and F.D. Stern. 1988. An expert system for building energy consumption analysis: Applications at a university campus. *ASHRAE Transactions* 94(1):1037-62.

Haberl, J.S., R. M. Nelson, and C.C. Culp. 1995. The Use of Artificial Intelligence in Building Systems, ASHRAE Special Publication.

Haberl, J.S., S. Thamilseran, and J. Kreider. 1996. Predicting hourly building energy use: The great energy predictor shootout II: Measuring retrofit savings—Overview and discussion of results. *ASHRAE Transactions* 102(2).

Kaler, G.M., Jr. 1990. Embedded expert system development for monitoring packaged HVAC equipment. *ASHRAE Transactions* 96(2):733-42.

Kovarik, M. 1971. Automatic design of optimal duct systems. Use of computers for environmental engineering related to buildings. National Bureau of Standards, *Building Science Series* 39 (October).

Mayer, R., L.O. Degelman, C.J. Su, A. Keen, P. Griffith, J. Huang, D. Brown, and Y.S. Kim. 1991. A knowledge-aided design system for energy-efficient buildings. *ASHRAE Transactions* (97)2:479-99.

Meadows, K.L. and P.W. Brothers. 1989. Decision analysis for prioritizing recommended energy conservation options. *ASHRAE Transactions* 95(2):934-53.

Neilson, N.R. and J.R. Walters. 1988. *Crafting knowledge-based system: Expert systems made realistic.* John Wiley and Sons, New York.

Nesler, C.G. 1986. Automated controller tuning for HVAC applications. *ASHRAE Transactions* 92(2B):189-201.

NIST. 1996. Thermodynamic properties of refrigerants and refrigerant mistures, Version 5.0. *NIST Standard Reference Database* 23. National Institute of Standards and Technology, Gaithersburg, MD. Available from ASHRAE.

Pedersen, C.O., D.E. Fisher, J.D. Spitler, and R.J. Liesen. *Load calculation principles*. ASHRAE.

Potter, C.D. 1987. *CAD in construction*. Penton Publishing, Cleveland, OH.

Potter, R.A., J.F. Kreider, M.J. Brandemuehl, and L.M. Windingland. 1991. Development of a knowledge-based system for cooling load demand management at large installations. *ASHRAE Transactions* 97(2):669-75.

Shams, H., R.M. Nelson, G.M. Maxwell, and C. Leonard. 1994a. Development of a knowledge-based system for the selection of HVAC system types for small buildings: Part I—Knowledge acquisition. *ASHRAE Transactions* 100(1):203-10.

Shams, H., R.M. Nelson, and G.M. Maxwell. 1994b. Development of a knowledge-based system for the selection of HVAC system types for small buildings: Part II-Expert system shell. *ASHRAE Transactions* 100(1):211-17.

Stoecker, W.F., R.C. Winn, and C.O. Pedersen. 1971. Optimization of an air-supply duct system. Use of computers for environmental engineering related to buildings. National Bureau of Standards, *Building Science Series* 39 (October).

Tsal, R.J. and E.I. Chechik. 1968. Use of computers in HVAC systems. Budivelnick Publishing House, Kiev. Available from the Library of Congress, Service - TD153.T77 (Russian).

Tsal, R.J., H.F. Behls, and R. Mangel. 1988. T-method duct design: Part I, optimization theory; Part II, calculation procedure and economic analysis. *ASHRAE Technical Data Bulletin* (June).

Wasserman, P.D. 1989. *Neural computing: Theory and practice.* Van Nostrand Reinhold, New York.

BIBLIOGRAPHY

Brothers, R.W. 1988. Knowledge engineering for HVAC expert systems. *ASHRAE Transactions* 94(1):1063-73.

Brothers, R.W. and K.R. Cooney. 1989. A knowledge-based system for comfort diagnostics. *ASHRAE Journal* 31(9).

Busnaina, A.A. 1987. Modeling of clean rooms on the IBM personal computer. Proceedings of the Institute of Environmental Sciences, 292-97.

Culp, C.C., J.S. Haberl, L.K. Norford, P.W. Brothers, and J.D. Hall. 1990. The impact of AI technology within the HVAC industry. *ASHRAE Journal* 32(12).

Culp, C.H. 1989. Expert systems in preventive maintenance and diagnosis. *ASHRAE Journal* 31(8).

Diamond, S.C., C.C. Cappiello, and B.D. Hunn. 1985. User-effect validation tests of the DOE-2 building energy analysis computer program. *ASHRAE Transactions* 91(2B):712-24.

Haberl, J.S. and D. Claridge. 1985. Retrofit energy studies of a recreation center. *ASHRAE Transactions* 91(2B):1421-33.

Int-Hout, D. 1986. Microprocessor control of zone comfort. *ASHRAE Transactions* 92(1B):528-38.

Judkoff, R., D. Wortman, and B. O'Doherty. 1981. A comparative study of four building energy simulations, phase II: DOE-2. 1, BLAST-3.0, SUNCAT-2.4 and DEROB. Solar Energy Research Institute *Report No.* SERI/TP-721-1326 (July).

Kaplan, M.B., J. McFerran, J. Jansen, and R. Pratt. 1990. Reconciliation of a DOE2.IC model with monitored end-use data for a small office building. *ASHRAE Transactions* 96(1):981-93.

Kosko, B. 1992. *Neural networks and fuzzy systems.* Prentice Hall, Englewood Cliffs, NJ.

Kuehn, T.H. 1988. Computer simulation of airflow and particle transport in cleanrooms. *The Journal of Environmental Sciences,* 31.

Rabl, A. 1988. Parameter estimation in buildings. Methods for dynamic analysis of measured energy use. ASME *Journal of Solar Energy, Engineering,* 110.

Sharimugavelu, I., T.H. Kuehn, and B.Y.H. Liu. 1987. Numerical simulation of flow fields in clean rooms. *Proceedings of the Institute of Environmental Sciences,* 298-303.

Sparks, R., J. Haberl, S. Bhattacharrya, M. Rayaprolu, J. Wang, and S. Vadlamani. 1992. Testing data acquisition systems for HVAC system monitoring. Proceedings of the ASME Solar Energy Conference, New York, pp. 325-28.

Sparks, R., A. Baranowski, K. Weber, and J. Haberl. 1994a. POLLC180 Software, Energy Systems Laboratory, Texas A&M University, College Station, TX.

Sparks, R., C. Sims, and J. Haberl. 1994b. Monitor Software. Energy Systems Laboratory, Texas A&M University, College Station.

BUILDING ENERGY MONITORING

BUILDING energy monitoring provides realistic and empirical information from field data that gives better understanding of actual building energy performance and quantifies any changes in performance over time. Although different building energy monitoring projects can have different objectives and scopes, all have several issues in common that allow methodologies and procedures (monitoring protocols) to be standardized.

This chapter provides guidelines for developing building monitoring projects that provide the necessary measured data at acceptable cost. The intended audience comprises building energy monitoring practitioners and data end users such as energy and energy service suppliers, energy end users, building system designers, public and private research organizations, utility program managers and evaluators, equipment manufacturers, and officials who regulate residential and commercial building energy systems.

Monitoring projects can be **uninstrumented** (that is, no additional instrumentation beyond the utility meter) or **instrumented** (billing data supplemented by additional sources of data, such as an installed instrumentation package, portable data loggers, or a building automation system). Uninstrumented approaches are generally simpler and less costly than instrumented approaches, but they can be subject to more uncertainty in interpretation, especially when the changes made to the building represent a small fraction of total energy use. It is important to (1) determine whether the cost of an instrumented approach is justified by the greater detail and accuracy obtained and (2) decide what accuracy is required and what level of instrumented data gathering and analysis is appropriate.

Instrumented field monitoring projects generally involve data acquisition systems (DASs), which typically comprise various sensors and data-recording devices (e.g., data loggers) or a suitably equipped building automation system. Field monitoring projects may involve a single building or hundreds of buildings and may be carried out over a period ranging from weeks to years. Most monitoring projects involve the following activities:

- Project planning
- Site installation of data acquisition equipment (if required)
- Ongoing data collection, verification, and calibration
- Data analysis and reporting

These activities often require support by various professional disciplines (e.g., engineering, data analysis, and management) and construction trades (electrical installers or plumbers).

Useful building energy performance data cover lighting, HVAC equipment, walls, meter readings, utility load factors, excess capacity, controller actuation, and building and component lifetimes. Current monitoring practices vary considerably. For example, a utility load research project may tend to characterize the average performance of buildings with relatively few data points per building, whereas a test of new technology performance may involve monitoring hundreds of parameters within a single facility. Monitoring projects range from broad research studies to very specific contractually required savings verification

The preparation of this chapter is assigned to TC 9.6, Systems Energy Utilization.

carried out by performance contractors. However, all practitioners should use accepted standards of monitoring practices to communicate results. Key elements in this process are (1) classifying the types of project monitoring and (2) building consensus on the purposes, approaches, and problems associated with each type (Misuriello 1987; Haberl et al. 1990). For example, energy savings from energy savings performance contracts can be specified on either a whole-building or component basis. The monitoring requirements for each approach vary widely and must be carefully matched to the specific project.

REASONS FOR ENERGY MONITORING

Monitoring projects can be broadly categorized by goals, objectives, experimental approach, level of monitoring detail, and uses (Table 1). Other factors such as resources available, data analysis procedures, duration and frequency of data collection, and instrumentation are common to most, if not all, projects.

Energy End Use

Energy end-use projects focus on individual energy systems in particular buildings, typically for large samples. Monitoring usually requires separate meters or data collection channels for each end use, and analysts must account for all factors that may affect energy use. Examples of this approach include detailed utility load research efforts, evaluation of utility incentive programs, and end-use calibration of computer simulations. Depending on the project objectives, the frequency of data collection may range from one-time measurements of full-load operation to continuous time-series measurements.

Specific Technology Assessment

Specific technology assessment projects monitor the field performance of specific equipment or technologies that affect building energy use, such as envelope retrofit measures, major end uses (e.g., lighting), or mechanical equipment.

The typical goal of retrofit performance monitoring projects is to estimate savings resulting from the retrofit despite potentially significant variation in indoor/outdoor conditions, building characteristics, and occupant behavior unrelated to the retrofit. The frequency and complexity of data collection depend on project objectives and site-specific conditions. Projects in this category assess variations in performance between different buildings or for the same building before and after the retrofit.

Field tests of end-use equipment are characterized by detailed monitoring of all critical performance parameters and operational modes. In evaluating energy efficiency improvements, it is preferable to measure in situ performance. Although manufacturers' data and laboratory performance measurements can provide excellent data for sizing and selecting equipment, the installed performance can vary significantly from that at design conditions. The project scope may include reliability, maintenance, design, energy efficiency, sizing, and environmental effects (Phelan et al. 1997a,b).

Table 1 Characteristics of Major Monitoring Project Types

Project Type	Goals and Objectives	General Approach	Level of Detail	Uses
Energy end use	Determine characteristics of specific energy end uses in building.	Often uses large, statistically designed sample. Monitor energy demand or use profile of each end use of interest.	Detailed data on end uses metered. Collect building and operating data that affect end use.	Load forecasting by end use. Identify and confirm energy conservation or demand-side management opportunities. Simulation calculations. Rate design.
Specific technology assessment	Measure field performance of building system technology or retrofit measure in individual buildings.	Characterize individual building or technology, occupant behavior, and operation. Account and correct for variations.	Uses detailed audit, submetering, indoor temperature, on-site weather, and occupant surveys. May use weekly, hourly, or short-term data.	Technology evaluation. Retrofit performance. Validate models and predictions.
Energy savings measurement and verification	Estimate the impact of a retrofit or other building alteration, in order to serve as the basis for payments.	Preretrofit consumption is used to create a baseline model. Postretrofit consumption is measured; the difference between the two is savings.	Varies substantially, including verification of potential to provide savings, retrofit isolation, whole-building, or calibrated simulation.	Focused on specific building component or system. Amount and frequency of data varies widely between projects.
Building operation and diagnostics	Solve problems. Measure physical or operating parameters that affect energy use or that are needed to model building or system performance.	Typically uses one-time and/or short-term measurement with special methods, such as infrared imaging, flue gas analysis, blower door, or coheating.	Focused on specific building component or system. Amount and frequency of data varies widely between projects.	Energy audit. Identify and solve operation and maintenance, indoor air quality, or system problems. Provide input for models. Building commissioning.

Savings Measurement and Verification (M&V)

Accountability is increasingly necessary in energy performance retrofits, whether they are performed as part of an energy savings performance contract (ESPC) or performed directly by the owner. In either case, savings measurement and verification (M&V) is an important part of the project. Measurement and verification is based on comparing the energy use after a retrofit to what it would have been without the retrofit. Because the actual energy savings cannot be measured directly, the appropriate role of energy monitoring methodology is to

- Accurately define baseline conditions and assumptions
- Confirm that the proper equipment and systems were installed and that they have the potential to generate the predicted energy savings
- Take post retrofit measurements
- Estimate the energy savings achieved

See IPMVP (1997) for more information on M&V for performance contracting.

Building Diagnostics

Diagnostic projects measure physical and operating parameters that determine the energy use of buildings and systems. Usually, the project goal is to determine the cause of problems, provide parameters to model energy performance, or isolate the effects of components. Diagnostic tests frequently involve one-time measurements or short-term monitoring. To give insight, the frequency of measurement must be several times faster than the rate of change of the effect being monitored. Some diagnostic tests require intermittent, ongoing data collection.

The residential sector, particularly single-family dwellings, has a large number of diagnostic measurement procedures. Typical measurements for single-family residences include (1) flue gas analysis to determine steady-state furnace combustion efficiency and the efficiency of other end uses, such as air conditioners, refrigerators, and water heaters; (2) fan pressurization tests to measure and locate building envelope air leakage (ASTM *Standard* E 779) and tests to measure airtightness of air distribution systems (Modera 1989; Robison and Lambet 1989); and (3) infrared thermography to locate thermal defects in the building envelope and other methods to determine overall building envelope parameters (Subbarao 1988).

Energy systems in multifamily buildings can be much more complex than those in single-family homes, but the types of diagnostics are similar: combustion equipment diagnostics, air leakage measurements, and infrared thermography to identify thermal defects or moisture porblems (DeCicco 1995). Some techniques are designed to determine the operating efficiency of steam and hot water boilers and to measure the air leakage between apartments.

In research, diagnostic techniques have been designed to measure the overall airtightness of building envelopes and the thermal performance of walls (Persily and Grot 1988). Practicing engineers also employ a host of monitoring techniques to aid in the diagnostics and analysis of equipment energy performance. Portable data loggers are often used to collect time-synchronized distributed data, allowing multiple data sets (such as chiller performance and ambient conditions) to be collected and quickly analyzed. Similar short-term monitoring procedures are used to provide more detailed and complete building system commissioning. Short-term in situ tests have also been developed for pumps, fans, and chillers (Phelan et al. 1997a, 1997b).

Diagnostics are also well suited to support the development and implementation of building energy management programs (see Chapter 34). In this role, diagnostic measurements can be used in conjunction with energy audits to identify existing energy usage patterns, to determine energy performance parameters of building systems and equipment, and to identify and quantify areas of inefficient energy usage (Misuriello 1988). Diagnostic measurement projects can generally be designed using procedures adapted to specific project requirements (see the section on Steps for Project Design and Implementation).

Equipment for diagnostic measurement may be installed temporarily or permanently to aid energy management efforts. Designers should consider providing permanent or portable check metering of major electrical loads in new building designs. The same concept can be extended to fuel and thermal energy use.

PROTOCOLS FOR PERFORMANCE MONITORING

Examples of procedures (**protocols**) for evaluating energy savings for projects involving retrofit of existing building energy systems are presented here. These protocols should also be useful to those interested in more general building energy monitoring.

Building monitoring has been significantly simplified and made more professional in recent years by the development of fairly standardized monitoring protocols. Although there may be no way to define a protocol to encompass all types of monitoring applications, repeatable and understandable methods of measuring and verifying retrofit savings are needed. However, following a protocol does not take the place of adequate project planning and careful assessment of project objectives and constraints.

Residential Retrofit Monitoring

Protocols for residential building retrofit performance can answer specific questions associated with the actual measured performance. For example, Ternes (1986) developed a single-family retrofit monitoring protocol, which consists of a data specification guideline that identifies important parameters to be measured. Both one-time and time-sequential data parameters are covered, and the parameters are defined carefully to ensure consistency and comparability between experiments. Discrepancies between predicted and actual performance, as measured by the energy bill, are common. This protocol improves on billing data methods in two ways: (1) internal temperature is monitored, which eliminates a major unknown variable in data interpretation; and (2) data are taken more frequently than monthly, which potentially shortens the monitoring duration. Utility bill analysis generally requires a full season of pre-retrofit and postretrofit data. The single-family retrofit protocol may require only a single season.

Ternes (1986) identified both a minimum set of data, which must be collected in all field studies that use the protocol, and optional extensions to the minimum data set that can be used to study additional issues. See Table 2 for details. Szydlowski and Diamond (1989) have developed a similar method for multifamily buildings.

The single-family retrofit monitoring protocol recommends a before-after experimental design, and the minimum data set allows performance to be measured on a normalized basis with weekly time-series data. (Some researchers recommend daily.) The protocol also allows hourly recording intervals for time-integrated parameters—an extension of the basic data requirements in the minimum data set. The minimum data set may also be extended through optional data parameter sets for users seeking more information.

The data parameters in this protocol have been grouped into four data sets: basic, occupant behavior, microclimate, and distribution system (Table 2). The minimum data set consists of a weekly option of the basic data parameter set. Time-sequential measurements are monitored continuously throughout the field study period. These are all time-integrated parameters (i.e., the appropriate average value of a parameter over the recording period, rather than the instantaneous values).

This protocol also addresses instrumentation installation, accuracy, and measurement frequency and expected ranges for all time-sequential parameters (Table 3). The minimum data set (weekly option of the basic data) must always be collected. At the user's discretion, hourly data may be collected, which allows two optional parameters to be monitored. Parameters from the optional data sets may be chosen, or other data not described in the protocol added, to arrive at the final data set.

This protocol has standardized the experimental design and data collection specifications, enabling independent researchers to compare project results more readily. Moreover, including both minimum and optional data sets and two recording intervals accommodates projects of varying financial resources.

Commercial Retrofit Monitoring

Several related guidelines have been created for the particular application of retrofit savings (M&V). The International Performance Measurement and Verification Protocol (IPMVP 1997) provides guidance to buyers, sellers, and financiers of energy projects on quantifying energy savings performance of energy retrofits. The Federal Energy Management Program has produced

Table 2 Data Parameters for Residential Retrofit Monitoring

	Recording Period	
	Minimum	Optional
Basic Parameters		
House description		once
Space-conditioning system description		once
Entrance interview information		once
Exit interview information		once
Preretrofit and postretrofit infiltration rates		once
Metered space-conditioning system performance		once
Retrofit installation quality verification		once
Heating and cooling equipment energy consumption	weekly	hourly
Weather station climatic information	weekly	hourly
Indoor temperature	weekly	hourly
House gas or oil consumption	weekly	hourly
House electricity consumption	weekly	hourly
Wood heating use	—	hourly
Domestic hot water energy consumption	weekly	hourly
Optional Parameters		
Occupant behavior		
Additional indoor temperatures	weekly	hourly
Heating thermostat set point	—	hourly
Cooling thermostat set point	—	hourly
Indoor humidity	weekly	—
Microclimate		
Outdoor temperature	weekly	hourly
Solar radiation	weekly	hourly
Outdoor humidity	weekly	hourly
Wind speed	weekly	hourly
Wind direction	weekly	hourly
Shading		once
Shielding		once
Distribution system		
Evaluation of ductwork infiltration		once

Source: Ternes (1986).

guidelines specific to federal projects but having many procedures that could be used for calculating retrofit savings in nonfederal buildings (Schiller 1996).

On a more detailed level, ASHRAE research project RP-827 resulted in separate guidelines for the in situ testing of chillers, fans, and pumps to evaluate installed energy efficiency (Phelan et al. 1997a,b). The guidelines specify the physical characteristics to be measured; the number, range, and accuracy of data points required; methods of artificial loading; and calculation equations with a rigorous uncertainty analysis.

In addition to these specialized protocols for particular monitoring applications, a number of specific laboratory and field measurement standards exist (see Chapter 54), and many monitoring source books are in circulation.

As a final example, a protocol has been developed for use in field monitoring studies of energy improvements (retrofits) for commercial buildings (MacDonald et al. 1989). Similar to the residential protocol, it addresses data requirements for monitoring studies. Commercial buildings are more complex, with a diverse array of potential efficiency improvements. Consequently, the approach to specifying measurement procedures, describing buildings, and determining the range of analysis must differ.

The strategy used for this protocol is to specify data requirements, analysis, performance data with optional extensions, and a building core data set that describes the field performance of efficiency improvements. This protocol requires a description of the approach used for analyzing building energy performance. The necessary performance data, including identification of a minimum data set, are outlined in Table 4.

Table 3 Time-Sequential Parameters for Residential Retrofit Monitoring

Data Parameter	Accuracy[a]	Range	Stored Value per Recording Period	Scan Rate[b] Option 1	Scan Rate[b] Option 2
			Basic Parameters		
Heating and cooling equipment energy consumption	3%		Total consumption	15 s	15 s
Indoor temperature	1.0°F	50 to 95°F	Average temperature	1 h	1 min
House gas or oil consumption	3%		Total consumption	15 s	15 s
House electricity consumption	3%		Total consumption	15 s	15 s
Wood heating use	1.0°F	50 to 800°F	Average surface temperature or total use time		1 min
Domestic hot water	3%		Total consumption	15 s	15 s
			Optional Data Parameter Sets		
Occupant Behavior					
Additional indoor temperatures	1.0°F	50 to 95°F	Average temperature	1 h	1 min
Heating thermostat set point	1.0°F	50 to 95°F	Average set point		1 min
Cooling thermostat set point	1.0°F	50 to 95°F	Average set point		1 min
Indoor humidity	5% rh	10 to 95% rh	Average humidity	1 h	
Microclimate					
Outdoor temperature	1.0°F	−40 to 120°F	Average temperature	1 h	1 min
Solar radiation	10 Btu/h·ft^2	0 to 350 Btu/h·ft^2	Total horizontal radiation	1 min	1 min
Outdoor humidity	5% rh	10 to 95% rh	Average humidity	1 h	1 min
Wind speed	0.5 mph	0 to 20 mph	Average speed	1 min	1 min
Wind direction	5°	0 to 360°	Average direction	1 min	1 min

Source: Ternes (1986).
[a]All accuracies are stated values.
[b]Applicable scan rates if nonintegrating instrumentation is employed.

Table 4 Performance Data Requirements of the Commercial Retrofit Protocol

Projects with Submetering		
	Before Retrofit	**After Retrofit**
Utility billing data (for each fuel)	12 month minimum	3 month minimum (12 months if weather normalization required)
Submetered data (for all recording intervals)	All data for each major end use up to 12 months	All data for each major end use up to 12 months
	Type	**Recording Interval** / **Period Length**
Temperature data (daily maximum and minimum must be provided for any periods without integrated averages)	Maximum and minimum —or— Integrated averages	Daily —or— Same as for submetered data but not longer than daily / Same as billing data length —or— Length of submetering

Projects Without Submetering		
	Before Retrofit	**After Retrofit**
Utility billing data (for each fuel)	12 month minimum	12 month minimum
	Type	**Recording Interval** / **Period Length**
Temperature data	Maximum and minimum —or— Integrated averages	Daily / Same as billing data length

Commercial New Construction Monitoring

New building construction offers the potential for monitoring building subsystem energy consumption at a reasonable cost. The information obtained by monitoring operating hours or direct energy consumption can benefit building owners and managers by

- Alerting them to inefficient or improper operation of equipment
- Providing data that can be useful in determining the benefits of alternative operating strategies or replacement equipment
- Evaluating costs of operation for extending occupancy hours for special conditions or events
- Demonstrating the effects of poor maintenance or identifying when maintenance procedures are not being followed
- Diagnosing power quality problems
- Submetering tenants

In order to provide the data necessary to improve building systems operation, monitoring should be considered for boilers, chillers, heat pumps, air-handling unit fans, major exhaust fans, major pumps, comfort cooling compressors, lighting panels, electric heaters, receptacle panels, substations, motor control centers, major feeders, service water heaters, process loads, and computer rooms.

The construction documents may include provisions for various meters to monitor equipment and system operation. Some equipment can be specified to have factory-installed hour meters that record the actual operating hours of the equipment. Hour meters can also be easily field installed on any electrical motor.

More sophisticated power monitoring systems, with electrical switchgear, substations, switchboards, and motor control centers, can be specified. These systems can monitor energy demand, energy consumption, power factor, neutral current, etc., and can be linked to a personal computer. These same systems can be installed on circuits to existing or retrofit fans, chillers, lighting panels, etc. Some equipment commonly used for improving system efficiency, such as variable-frequency drives, can be provided with capability to monitor kW output, kWh consumed, and other variables.

Use of central energy management control systems (e.g., direct digital control systems) for monitoring is particularly appropriate in new construction. These systems can monitor and record system status, steam flow, gas flow, water flow, energy consumption, demand, and hours of operation, as well as start and stop building systems, control lighting, and print alarms when systems are not operating within specified limits. Initial specification of the new control system should include sensors and trend logging and reporting functions.

COMMON MONITORING ISSUES

The key issues in monitoring are the **accuracy** and **reliability** of the data. Projects have been compromised by inaccurate or missing data. Periodic sensor calibration and ongoing data verification ensure the reliability and quality of the collected data.

Field monitoring projects also require effective management of various professional skills. Project staff must understand the building systems being examined, data management, data acquisition, and sensor technology. In addition to data collection, processing, and analysis, the logistics of field monitoring projects require the coordination of equipment procurement, delivery, and installation.

Planning

Many common problems in monitoring projects can be avoided by effective and comprehensive planning.

Project Goals. Project goals and data requirements should be established before hardware is selected. Unfortunately, projects are often driven by the selection of hardware rather than by project objectives, either because monitoring hardware must be ordered several months before data collection begins or because project initiation procedures are lacking. As a result, the hardware may be inappropriate for the particular monitoring task, or critical data points may be overlooked.

Project Cost and Resources. After goal setting, the feasibility of the anticipated project should be reviewed in light of available resources. Projects to which significant resources can be devoted usually involve different approaches from those for which resources are more limited. This issue should be addressed early on and reviewed throughout the course of the project. Although it is difficult to assess with certainty the cost of an anticipated project at this early stage, rough estimates can be quite helpful.

Data Products. It is important to establish the format and content of the data products before selecting data points. Failure to plan these data products first may lead to failure to answer critical questions.

Data Management. Failure to anticipate the (typically) large amounts of data collected can lead to major difficulties. The computer and personnel resources needed to verify, retrieve, analyze, and archive data can be estimated based on experience with previous projects.

Commitment. Many projects require a long-term commitment of personnel and resources. Project success depends on long-term, daily attention to detail and on staff continuity.

Accuracy Requirements. The accuracy requirements of the final data and the experimental design needed to meet these requirements should be determined early on. After the required accuracy is specified, the sample size (number of buildings, control buildings, or pieces of equipment) must be chosen, and the required measurement precision (including error propagation into the final data products) must be determined. Because trade-offs must usually be made between cost and accuracy, this process is often iterative. It is further complicated by a large number of independent

variables (occupants, operating modes) and the stochastic nature of many variables (weather).

Advice. Expert advice should be sought from others who have experience with the type of monitoring envisioned.

Implementation and Verification

The following steps can facilitate smooth project implementation and data verification:

- Calibrate sensors before installation. Spot-check the calibration on site. During long-term monitoring projects, recalibrate sensors periodically. National Institute of Standards and Technology (NIST) standards should be used in all calibration procedures, which are listed in IPMVP (1997).
- Track sensor performance on a regular basis. Quick detection of sensor failure or calibration problems is essential. Ideally, this should be an automated or a daily task. The value of data is high because it may be difficult or impossible to reconstruct.
- Generate and review data on a timely, periodic basis. Problems often occur in developing final data products. These problems include missing data from failed sensors, data points not installed due to planning oversights, and anomalous data for which there are no explanatory records. If data products are specified as part of general project planning and produced periodically, production problems can be identified and resolved as they occur. Automating the process of checking data reliability and accuracy can be invaluable in keeping the project on track and in preventing sensor failure and data loss.

Data Analysis and Reporting

For most projects, the collected data must be analyzed and put into reports. Because the objective of the project is to translate these data into information and ultimately into knowledge and action, the importance of this step cannot be overemphasized. Clear, convenient, and informative formats should be devised in the planning stages and adhered to throughout the project.

Close attention must be paid to resource allocation to ensure that adequate resources are dedicated to verification, management, and analysis of the data and to ongoing maintenance of the monitoring equipment. As a quality control procedure and to make data analysis more manageable, these activities should be ongoing. The data analysis should be carefully defined before the project begins.

STEPS FOR PROJECT DESIGN AND IMPLEMENTATION

This section describes a methodology for designing effective field monitoring projects. The task components and relationships among the nine activities constituting this methodology are identified in Figure 1. The activities fall into four categories: project management, project development, resolution and feedback, and production quality and data transfer. Field monitoring projects vary in terms of resources, goals and objectives, data product requirements, and other variables, which affects how the methodology should be applied. Nonetheless, the methodology provides a proper framework for advance planning, which minimizes or helps prevent project implementation problems.

An iterative approach to planning activities is best. The scope, accuracy, and techniques can be adjusted based on cost estimates and resource assessments. The initial design should be performed simply and quickly to estimate cost and evaluate resources. An iterative approach to project planning is also necessary when the desired levels of instrumentation exceed the resources available for the project. The planning process should identify and resolve any trade-offs necessary to execute the project within a given budget. In many instances, it is preferable to reduce the scope of the project

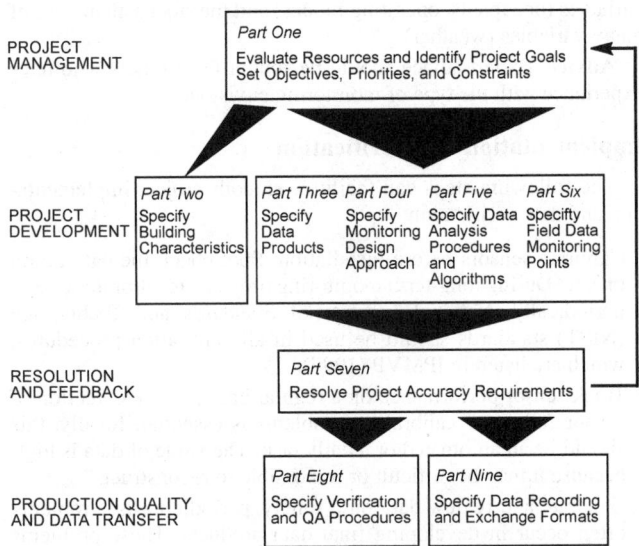

Fig. 1   Methodology for Designing Field Monitoring Projects

rather than to relax instrumentation specifications or accuracy requirements.

One of the most frequent oversights in project planning is failure to reserve sufficient time and resources for later analysis and reporting of data. Unanticipated additional costs associated with data collection and problem resolution should not jeopardize these resources.

Part One: Identify Project Goals and Objectives

Start with a clear understanding of the decision to be made or action to be taken as a result of the project. The goals and objectives statement determines the overall direction and scope of the data collection and analysis effort. The statement should also list the questions to be answered by the empirical data, noting the error or uncertainty associated with the desired result. A realistic assessment of error is needed because requiring too small an uncertainty leads to an overly complex and expensive project. It is important in monitoring projects to resolve that a data acquisition plan will be developed and followed with a clear idea of the **research questions** to be answered.

Even if a project is not research-oriented, it is attempting to obtain some information, and this can be stated in the form of questions. Research questions can have varying scopes and levels of detail, addressing entire systems or specific components. Some examples of research questions follow:

- Measurement and verification: Have the contractors fulfilled their responsibilities in installing equipment capable of producing the agreed-upon energy savings?
- Classes of buildings: To an accuracy of 20%, how much energy has been saved by using a building construction/performance standard mandated in this state?
- Particular buildings: Has a lapse in building maintenance caused energy performance to degrade?
- Particular components: What is the average reduction in demand charges during summer peak periods due to the installation of an ice storage system in this building?

Research questions vary widely in technical complexity, generally taking one of the following three forms:

1. How does the building/component perform?
2. Why does the building/component perform as it does?
3. Which building/component should be targeted to achieve optimal cost-effectiveness?

The first form of question can sometimes be answered generically for a class of typical buildings without detailed monitoring and analysis. The second and third forms usually require detailed monitoring and analysis and, thus, detailed planning.

In general, more detailed and precise goal statements are better. They ensure that the project is constrained in scope and developed to meet specific accuracy and reliability requirements.

Part Two: Specify Building Characteristics

The measured energy data will not be meaningful at a later date to people who were not involved in the project unless the characteristics of the building being monitored and its use have been documented. To meet this need, a data structure (e.g., a characteristic database) can be developed to describe the buildings.

Building characteristics can be collected at many levels of detail, but it is important to provide at least enough detail to document the following:

- General building configuration and envelope (particularly energy-related aspects)
- Building occupant information (number, occupancy schedule, activities)
- Internal loads
- Type and quantity of energy-using systems (including set-point temperatures)
- Any building changes that occur during the monitoring project

The minimum level of detail is known as **summary characteristics** data. **Simulation level characteristics** (detailed information collected for hourly simulation model input) may be desirable for some buildings. Regardless of the level of detail, the data should provide a context for analysts, who may not be familiar with the project, to understand the building and its energy use.

For occupied existing buildings, characteristics information should be collected in four areas:

- Building descriptive information summarizing key building envelope and internal heat gain parameters
- HVAC system descriptive information characterizing key parameters affecting HVAC system performance
- Entrance interview information focusing on the energy-related behavior of building occupants prior to monitoring
- Exit interview information documenting physical or life-style changes at the test site that may affect data analysis

Part Three: Specify Data Products and Project Output

The objective of a monitoring project is typically not to produce data, but to answer a question. However, the data must be of high quality and must be presented to key decision makers and analysts in a convenient and informative format. The specific **data products** (the format and content of data reporting needed to meet the project goals and objectives) must be identified and evaluated for feasibility and usefulness. The final data products must be clearly specified, together with the minimum acceptable data requirements for the project. It is important to clearly define an **analysis path** showing what will be calculated and what data will be necessary to achieve desired results. Clear communication is critical to ensure that project requirements are satisfied and factors contributing to monitoring costs are understood.

Evaluation results can be presented in many forms, often as interim and final reports (possibly by heating and/or cooling season), technical notes, or technical papers. These documents must convey specific results of the field monitoring clearly and concisely. They should also contain estimates of the accuracy of the results.

The composition of data presentations and analysis summaries should be determined early to ensure that no critical parameters are overlooked (Hough et al. 1987). For instance, mock-ups of data tables, charts, and graphs can be used to identify requirements.

Previously reported results can be used to provide examples of useful output. Data products should also be prioritized to accommodate possible cost trade-offs or revisions resulting from error analysis (see Part Seven). A worksheet can be used to determine information needs and to help organize specific research questions and technology issues by project objectives. Next, specific data presentation formats should be determined.

It is important to identify the type and amount of comparative analysis and evaluation needed. For example, any need for data normalization for weather or usage profiles, along with the method to be employed, should be determined in advance.

Although requirements for the minimum acceptable data results can often be specified during planning, data analysis typically reveals further requirements. Thus, budget plans should include allowances and optional data product specifications to handle additional or unique project output requirements uncovered during data analysis.

Longer-term goals and future information needs should be anticipated and explained to project personnel. For example, a project may have short- and long-term data needs (e.g., demonstrating reductions in peak electrical demand versus demonstrating cost-effectiveness or reliability to a target audience). The initial results on demand reduction may not be the ultimate goal, but rather a step toward later presentations on cost reductions achieved. Thus, it is prudent to consider long-term and potential future data needs so that additional supporting information, such as photographs or testimonials, may be identified and obtained.

Part Four: Specify Monitoring Design Approach

A general monitoring design must be developed that defines two interacting factors—the number of buildings admitted to the study and the monitoring approach. A less detailed or precise approach can be considered if the number of buildings is increased, and vice versa. If the goal is related to a specific product, the monitoring design must isolate the effects of that product. Haberl et al. (1990) discuss monitoring designs. For the particular example of retrofit M&V, protocols have been written allowing a range of different monitoring methods, from retrofit verification to retrofit isolation (IPMVP 1997, Schiller 1996).

Specifying the monitoring design approach is particularly important because the total building performance is a complex function of several variables, all of which are subject to changes that are difficult to monitor and to translate into performance. Unless care is taken with measurement organization and accuracy, uncertainties and errors (noise) can make it difficult to detect performance changes of less than 20 to 30% (Fracastoro and Lyberg 1983).

In some cases, judgment may be required in selecting the number of buildings involved in the project. If an owner seeks information about a particular building, the choice is simple—the number of buildings in the experiment is fixed at one. However, for other monitoring applications, such as drawing conclusions regarding effects in a sample population of buildings, some degree of choice is involved. Generally, the error in the derived conclusions decreases as the square root of the number of buildings (Box 1978).

Accuracy requirements depend on the effect to be measured and on the specified experimental design and analysis methods. Without special measurements, energy savings of less than 20% of the total consumption are difficult to detect from an analysis of monthly utility bills. Year-to-year changes in weather conditions alone can cause a significant change in energy use for an average envelope-driven building. A specific project may be directed at

1. Fewer buildings or systems with more detailed measurements
2. Many buildings or systems with less detailed measurements
3. Many buildings or systems with more detailed measurements

For projects of the first type, accuracy requirements are usually resolved initially by determining the expected variations of

measured quantities (dependent variables) about their average values in response to the expected variations of the independent variables. For buildings, a typical concern is the response of heating and cooling loads to changes in temperature or other weather variables. The response of building lighting energy use to daylighting systems using natural light is another example of the relationship between dependent and independent variables. The fluctuations in response are caused by (1) outside influences not quantified by the measured energy use data and (2) limitations and uncertainties associated with measurement equipment and procedures. Thus, accuracy must often be determined using statistical methods to describe mean tendencies of dependent variables.

For projects of the second and third types, the increased number of buildings allows improved confidence in the mean tendencies of the dependent response(s) of interest. Larger sample sizes are also needed for experimental designs with control groups, which are used to adjust for some outside influences. For further information, see Box (1978), Fracastoro and Lyberg (1983), and Hirst and Reed (1991).

Most monitoring procedures use one or more of the following general experimental approaches:

On-Off. If the retrofit or product can be activated or deactivated at will, energy consumption can be measured in a number of repeated on-off cycles. The on-period consumption is then compared to the off-period consumption (Woller 1989; Cohen et al. 1987).

Before-After. Building energy consumption is monitored before and after a new component or retrofit is installed. Changes in the weather and building operation during the two periods must be accounted for, often requiring a model-based analysis (Hirst et al. 1983; Fels 1986; Robison and Lambert 1989; Sharp and MacDonald 1990; Kissock et al. 1992).

Test-Reference. The building energy consumption data of two "identical" buildings, one with the product or retrofit that is being investigated, are compared. Because buildings cannot be identical (e.g., different air leakage distributions, insulation effectiveness, temperature settings, and solar exposure), measurements should be taken prior to installation as well in order to allow calibration. Once the product or retrofit is installed, any deviation from the calibration relationship can be attributed to the product or retrofit (Levins and Karnitz 1986; Fracastoro and Lyberg 1983).

Simulated Occupancy. In some cases, the desire to reduce noise can lead the experimenter to postulate certain standard profiles for temperature set points, internal gains, moisture release, or window manipulation and to introduce this profile into the building by computer-controlled devices. The reference is often given by the test-reference design. In this case, both occupant and weather variations are nearly eliminated in the comparison (Levins and Karnitz 1986).

Nonexperimental Reference. A reference for assessing the performance of a building can be derived nonexperimentally using (1) a normalized, stratified performance database—energy use per unit area classified by building type (MacDonald and Wasserman 1989) or (2) a reasonable standard building, simulated by a calculated hourly or bin-method calibrated building energy performance model subject to the same weather, equipment type, and occupancy as the monitored building.

Engineering Field Test. When the experiment is focused on testing a particular piece of equipment, the total building performance is not of primary interest. The building provides a realistic environment for testing the equipment for reliability, maintenance requirements, and comfort and noise levels, as well as energy usage. The energy consumption of mechanical equipment is significantly affected by the system control strategy. Testing procedures should be designed to incorporate the control strategy of the equipment and its system (Phelan et al. 1997a,b). This type of monitoring and testing can also be used to calibrate computer simulation models of

Table 5 Advantages and Disadvantages of Common Experimental Approaches

Mode	Advantages	Disadvantages
Before-after	No reference building required. Same occupants implies smaller occupant variations. Modeling processes will be mostly identical before/after.	Weather different before/after. More than one heating/cooling season may be needed. Model is required to account for weather and other changes.
Test-reference	One season of data may be adequate. Small climate difference between buildings.	Reference building required. Calibration phase required (may extend testing to two seasons). Occupants in either or both buildings can change behavior.
On-off	No reference building required. One season may be adequate. Modeling processes will be mostly identical before/after. Most occupancy changes will be small.	Requires reversible product. Cycle may be too long if time constants are large. Model is required to account for weather differences in cycles. Dynamic model accounting for transients may be needed.
Simulated occupancy	Noise due to occupancy is eliminated. A variety of standard schedules can be studied.	Not "real" occupants. Expensive apparatus required. Extra cost of keeping building unoccupied.
Nonexperimental reference	Cost of actual reference building eliminated. With simulation, weather variation is eliminated.	Database may be lacking in strata entries. Simulation errors and definition of reference problematic. With database, weather changes usually not possible.
Engineering field test	Information focused on the product of interest. Minimal number of buildings required. Same occupants during the test.	Extensive instrumentation of product processes required. Models required to extrapolate to other buildings and climates. Occupancy effects not determined.

as-built and as-operated buildings, which can then be used to evaluate whole-building energy consumption. The equipment may be extensively instrumented.

Some of the general advantages and disadvantages of these approaches are listed in Table 5 (Fracastoro and Lyberg 1983). Monitoring design choices have been successfully combined (e.g., the before-after and test-reference approaches). Questions to be considered in choosing a monitoring approach include the following:

• Can the building alteration being investigated be turned on and off at will? The on-off design offers considerable advantages.
• Are occupancy and occupant behavior critical? Changes in building tenants, use schedules, internal gains, temperature set points, and natural or forced ventilation practices should be considered because any one of these variables can ruin an experiment if it is not constant or accounted for.
• Are actual baseline energy performance data critical? In before-after designs, time must be allotted to characterize the before case as precisely as the after case. For instances in which heating and cooling systems are being evaluated, data may be required for a complete range of anticipated ambient conditions.
• Is it a test of an individual technology, or are multiple technologies installed as a package being tested? If the impacts of individual technologies are sought, detailed component data and careful model-based analyses are required.
• Does the technology have a single mode or multiple modes of operation? Can the modes be controlled to suit the experiment? If many modes are involved, it will be necessary to test over a variety of conditions and conduct model-based analysis (Phelan et al. 1997a,b).

Part Five: Specify Data Analysis Procedures and Algorithms

Data are useless unless they are distilled into meaningful products that allow conclusions to be drawn. Too often, data are collected and never analyzed. This planning step focuses on specifying the minimum acceptable data analysis procedures and algorithms and detailing how collected data will be processed to produce desired data products. This step determines which parameters will be assumed as constants and which will be calculated by actual continuous data or be monitored continuously as a field data point. Based on this information, monitoring practitioners should do the following:

• Determine the independent variables and analysis constants to be measured in the field (e.g., fan power, lighting and receptacle power, indoor air temperature).
• Develop engineering calculations and equations (algorithms) necessary to convert field data to end products. This may include the use of statistical methods and simulation modeling.
• Specify detailed items, such as the frequency of data collection, the required range of independent variables to be captured in the data set, and the reasons certain data must be obtained at different intervals. For example, 15 min interval demand data is assembled into hourly data streams to match utility billing data.

Determine the proper NIST-traceable calibration standards for each sensor type to be used. For details, see the references cited in IPMVP (1997) for specific types of sensors. However, it is often impractical to implement standards in the field. For example, maintaining the length of straight ductwork required for an airflow sensor is usually difficult in the field, requiring compromise.

Algorithm inputs can be assumed values (such as the energy value of a unit volume of natural gas), one-time measurements (the leakage area of a house), or time-series measurements (fuel consumption and outside and inside temperatures at the site). The algorithms may pertain to (1) utility level aggregates of buildings, (2) particular whole-building performance, or (3) performance of instrumented components.

Chapter 30 of the 1997 *ASHRAE Handbook—Fundamentals* contains a lengthy discussion on inverse modeling procedures. In this chapter, the discussion will focus on analysis methods that can generally be classified as empirical or model-based.

Empirical Methods. Although empirical methods are the simplest, they can have large uncertainty and generate little or no information for small sample sizes. The simplest empirical methods are based on annual consumption values, tracking the annual numbers and looking for degradation. Questions about building performance relative to other buildings are based on comparing certain performance indices between the building and an appropriate reference. ASHRAE/IESNA *Standard* 105 gives an example of how a database could be established for this purpose.

For commercial buildings, the most common index is the **energy use intensity** (EUI), which is the annual consumption, either by fuel type or summed over all fuel types, divided by the conditioned floor area. Comparison is often made only on the basis of general building type, which can ignore potentially large variations in climate, the number of workers in a building, the number and type of computers in a building, and HVAC systems. Such variations can be accommodated to some degree by stratifying the database from which the reference EUI is chosen. Computer simulations are often used to set reasonable standards.

The Commercial Buildings Energy Consumption Survey database, summarized in EIA (1996) and in Chapter 34, has been used to develop an energy use benchmarking method for office buildings in the United States. The initial work in this area covered only electricity use, but the methods have been extended to cover all fuels for office buildings (Sharp 1996). The work showed that the electricity use of office buildings is most significantly explained by the number of workers in the building, the number of personal computers, whether the building is owner-occupied, and the number of operating hours each week. Only a subset of these parameters may be used to determine a benchmark within a specific census division.

Simple empirical methods that are applied to retrofit applications should include at least some periods of data on daily energy use and average daily temperature (recorded locally) to account for variations in occupancy and building schedules. Monthly EUI or billing data provide more information for empirical analysis and can be used for extended analysis of energy impacts of retrofit applications, for example, in **conditional demand analysis** (Hirst and Reed 1991). Monthly data can also be used to detect billing errors, improper equipment operation during unoccupied hours, and seasonal space-conditioning problems (Haberl and Komor 1990a, 1990b). Daily data are often used in these analyses, and raw hourly total building consumption data, when available, provide more detailed information on occupied versus unoccupied performance. Hourly, daily, monthly, and annual EUI across buildings can be directly compared when reduced to average power per unit area (power density). To avoid false correlations, the method of analysis should have statistical significance that can be traced to realistic parameters (Haberl et al. 1996).

Model-Based Methods. These techniques allow a wide range of additional data normalization to potentially improve the accuracy of comparisons and provide estimates of cause-effect relations. The analyst must carefully define the system and postulate a useful form of the governing energy balance/system performance equation or system of equations. Explicit terms are retained for equipment or processes of particular interest. As part of the data analysis, whole-building data (driving forces and thermal or energy response) are used to determine the model's significant parameters. The parameters themselves can provide insight, although parameter interpretation can be difficult, particularly with time-integrated billing data methods. The model can then be used for a number of normalization processes as well as future diagnostic and control applications. Two general classes of models are used in analysis methods: time-integrated methods and dynamic techniques (Burch 1986; Balcomb et al. 1993).

Time-Integrated Methods. Based on algebraic calculation of the building energy balance, time-integrated methods are often used prior to data comparison to correct annual consumption for variations in outdoor temperature, internal gains, and internal temperature (Fels 1986; Busch et al. 1984; Haberl and Claridge 1987; Claridge et al. 1991). This type of correction is essential for most retrofit applications.

Time-integrated methods can be used with whole-building energy consumption data (billing data) or with submetered end-use data. For example, standard time-integrated methods are often used to separately integrate end-use consumption data (heating, cooling, domestic water heating) for comparison and analysis. Time-integrated methods are generally reliable, as long as the following three conditions are accounted for:

- *Appropriate time step.* Generally, the time step should be as long as or longer than the response time of the building or building system for which energy use is being integrated. For example,

the response of daylighting controls to natural illumination levels can be rapid, allowing short time steps for data integration. In contrast, the response of cooling system energy use to changes in cooling load can be comparatively slow. In this instance, either a time step long enough to average over these slow variations or a dynamic model should be used. In general, an appropriate time step should account for the physical behavior of the energy system(s) and the expression of this behavior in model parameters.

- *Linearity of model results.* Generally, time-integrated models should not be applied to data used to estimate nonlinear effects. Air infiltration, for example, is nonlinear when estimated using wind speed and indoor/outdoor temperature difference data in certain models. Estimation errors would result if these parameters were independently time-integrated and then used to calculate air infiltration. Such nonlinear effects should be modeled at each time step (each hour, for example).

- *End-use uniformity within data set.* End-use data sets should be **uniform** (i.e., should not inadvertently contain observations with measurements of end uses other than those intended). During mild weather, for example, HVAC systems may provide both heating and cooling over the course of a day, creating data observations of both heating and cooling measurements. In a time-integrated model of heating energy use, these cooling energy observations would lead to error. Such observations should be identified or otherwise flagged by their true end use.

For whole-building energy consumption data (billing data), reasonable results can be expected from heating analysis models when the building is dominantly responsive to inside-outside temperature differences. The billing data analysis method yields little of interest when internal gains are large compared to skin loads, as in large commercial buildings and industrial applications. Daily, weekly, and monthly whole-building heating season consumption integration steps have been employed (Fels 1986; Ternes 1986; Sharp and MacDonald 1990; Claridge et al. 1991). Cooling analysis results have been less reliable because cooling load is not strictly proportional to variable base cooling degree-days (Kissock et al. 1992; Haberl et al. 1996; Fels 1986). Problems also arise when solar gains are dominant and vary by season.

Dynamic Techniques. Dynamic models, both **macrodynamic** (whole-building model) and **microdynamic** (component-specific), offer great promise for reducing monitoring duration and increasing conclusion accuracy. Furthermore, individual effects from multiple measures and system interactions can be examined explicitly. Dynamic whole-building analysis is generally accompanied by detailed instrumentation of specific technologies.

Dynamic techniques create a dynamic physical model for the building, adjusting the parameters of the model to fit the experimental data (Subbarao 1988; Duffy et al. 1988). In residential applications, computer-controlled electric heaters can be used to maintain a steady interior temperature overnight, extracting from this data an experimental value for the building steady-state load coefficient. A cool-down period can also be used to extract information on internal building mass thermal storage. Daytime data can be used to renormalize the response (computed from a microdynamic model) of the building to solar radiation, which is particularly appropriate for buildings with glazing areas of over 10% of the building floor area (Subbarao et al. 1986). Once the data with electric heaters have been taken, the building can be used as a dynamic calorimeter to assess the performance of auxiliary heating and cooling systems.

Similar techniques have been applied to commercial buildings (Norford et al. 1985; Burch et al. 1990). In these cases, delivered energy from the HVAC system must be monitored directly in lieu of using electric heaters. Because ventilation is a major variable term in the building energy balance, the outside airflow rate should also be monitored directly. Simultaneous heating and cooling, common

Table 6   Whole-Building Analysis Guidelines

| | Class of Method | | |
Project Goal	Empirical (Billing Data)[a]	Time-Integrated Model[a]	Dynamic Model
Building evaluation	Yes, but expect fluctuations in 20 to 30% range.	Yes, extra care needed beyond 15% uncertainty.	Yes, extra care needed beyond 10% uncertainty.
Building retrofit evaluation	Not generally applicable using monthly data, unless large samples are used. Requires daily data and various normalization techniques for reasonable accuracy.	Yes, but difficult beyond 15% uncertainty. Method cannot distinguish multiple retrofit effects.	Yes, can resolve 5% change with short-term tests. Can estimate multiple retrofit effects.
Component evaluation	Not applicable.	Not applicable unless submetering is done to supplement.	Yes, about 5% accuracy, but best with submetering.

Note: Error figures are approximate for total energy use in a single building. All methods improve with selection of more buildings.

[a] Accuracy can be improved by decreasing time step to weekly or daily. These methods are of little use when the *outdoor* temperature approaches the *balance* temperature.

in large buildings, requires a multizone treatment, which has not been adequately tested in any of the dynamic techniques.

Equipment-specific monitoring guidelines using dynamic modeling have been successfully tested in a variety of applications. For fans and pumps, relatively simple regression techniques from short-term monitoring provided accurate estimates of annual energy consumption when combined with an annual equipment load profile. For chillers, a thermodynamic model used with short-term monitoring captured the most important operating parameters for estimating installed annual energy performance. In all cases, the key to accurate model results was capturing a wide enough range of the independent load variable in the monitored data set to reflect annual operating characteristics (Phelan et al. 1997a, 1997b).

Table 6 provides a guide to selecting an analysis method. The error quotations are rough estimates for a single-building scenario.

Part Six: Specify Field Data Monitoring Points

Careful specification of field monitoring points is critical to identifying the variables that need to be monitored or measured in the field to produce the required data.

Because metering projects are often conducted in buildings with changing conditions, special consideration must be given to identifying and controlling significant changes in climate, systems, and operation during the monitoring period. Additional monitoring points may be required to measure variables that are assumed to be constant, insignificant, or related to other measured variables in order to draw sound conclusions from the measurements. Because the necessary data may be obtained in several ways, data analysts, equipment installers, and data acquisition system engineers should work together to develop tactics that best suit the project requirements. It is important to anticipate the need for supplemental measurements in response to project needs that may not become apparent until actual equipment installation occurs.

The cost of data collection is a nonlinear function of the number, accuracy, and duration of measurements that must be considered while planning within budget constraints. If the extent of data applications is unknown, such as in research projects, consideration should be given to the value of other concurrent measurements because the incremental cost of alternative analyses may be small.

For any project involving large amounts of data, data quality verification should be automated (Lopez and Haberl 1992). While this may require adding monitoring points to facilitate energy balances or redundancy checks, the added costs are likely to be offset by savings in data verification for large projects.

If multiple sites are to be monitored, common protocols for the selection and description of all field monitoring points should be established so that data can be more readily verified, normalized, compared, and averaged. Protocols also add consistency in selecting monitoring points. Pilot installations should be conducted to provide data for a test of the system and to ensure that the necessary data points have been properly specified and described.

General considerations in selecting monitoring equipment include the following:

- Evaluate the equipment thoroughly under actual test conditions before committing to large-scale procurement. Particular attention should be paid to any sensitivity to power outages and to protection against power surges and lightning.
- Consider local set up and testing of complex data acquisition systems that are to be installed in the field.
- Avoid data acquisition equipment that has not been proven (Sparks et al. 1992). Untested equipment, even if donated, may not be a good value.
- Consider the costs and benefits of remote data interrogation and programming.
- Evaluate quality and reliability of data loggers and instrumentation; these issues may be more important than cost, particularly when data acquisition sites are distant.
- Verify vendor claims by calling references or obtaining performance guarantees.
- Consider portable battery-powered data loggers in lieu of hard-wired loggers if the monitoring budget is limited and the length of the monitoring period is less than a few months.
- Ensure that the monitoring equipment and installation methods are consistent with prevailing laws, building codes, and standards of good practice.

An examination of the analysis method will determine the data to be measured in the field. The simplest methods require no on-site instrumentation. As the methods become more complex, data channels increase. For engineering field tests conducted with dynamic techniques, up to 100 data channels may be required.

Particular attention must be paid to sensor location. For example, if the method requires an average indoor temperature, examine the potential for internal temperature variation; data from several temperature sensors must often be averaged. Alternatively, temperature sensors adjacent to HVAC thermostats detect the temperature to which the HVAC equipment reacts.

After the channel list is compiled, sensor accuracy and scan rates should be assigned. Some sensors, such as indoor and outdoor temperature sensors, may require low scan rates (once every 5 min). Others, such as total electric sensors, may contain high-frequency transients that require rapid sampling (many times per second). The maximum sampling rate is usually programmed into the logger, and averages are stored at a specified time step (hourly). Some loggers can scan different channels at different rates. The logger's interrupt capability can also be used for rapid, infrequent transients. Interrupt channels signal the data logger to start monitoring an event only once it begins. In some cases, the on-line computation of derived quantities must be considered. For example, if heat flow in an air duct is required, it can be computed from a differential temperature measurement multiplied by an air mass flow rate that was determined from a one-time measurement. However, it should be computed and totaled only when the fan is operating.

Once the required field data monitoring points are specified, these requirements should be clearly communicated to all members of the project team to ensure that the actual monitoring points are accurately described. This can be accomplished by publishing handbooks for measurement plan development and equipment installation and by outlining procedures for diagnostic tests and technology assessments.

Cost savings may be possible through the use of a building energy management system for data acquisition. This should be considered only when its sensors and their accuracy and limitations (scan rate, etc.) are thoroughly understood. When merging data from two data acquisition systems, problems may arise such as differing reliability and low data resolution (e.g., 1 kW resolution of a circuit that draws 10 kW fully loaded). These problems can often be avoided, however, by adding appropriate sensors and setting up custom trend logs with point, memory, and programming capacity.

Scanning and Recording Intervals. The frequency of data measurement and data storage can affect the accuracy of project output. Scanning differs from storage in that data channels may be read (scanned) many times per second, for example, while average data may be recorded and stored every 15 min. Most data loggers maintain temporary storage registers, accumulating an integrated average of channel readings from each scan. The average is then recorded at the specified interval.

The data scan rate must be sufficiently fast to ensure that all significant effects are monitored. The recording frequency required can vary by data channel. As mentioned previously, transient and/or random events are often best monitored using the interrupt capabilities of modern data loggers.

Sensors and Data Acquisition Systems. Sensors should be selected to obtain each measurement on the field data list. Next,

conversion and proportion constants should be specified for each sensor type, and the accuracy, resolution, and repeatability of each sensor should be noted. Sensors should be calibrated before they are installed in the field, preferably with a NIST-traceable calibration procedure. They should be checked periodically for drift, recalibrated, and then postcalibrated at the conclusion of the experiment (Haberl et al. 1995). Calibration of instrumentation is particularly vital for flow and power measurement.

Because hardware needs vary considerably by project, specific selection guidelines are not provided here. However, general characteristics of data acquisition hardware components are shown in Table 7. Some typical concerns for selecting data acquisition hardware are outlined in Table 8. In general, data logger and instrumentation hardware should be standardized, with replacements available in the event of failure. It is also useful to consider redundant measurements for critical data components that are likely to fail include modems, flowmeters, shunt resistors on current transformers, and devices with moving parts (O'Neal et al. 1993). Also, certain measurements are more difficult to obtain accurately than others due to instrumentation limitations, as summarized in Table 9.

Safety must be considered in equipment selection and installation. Installation teams of two or more individuals will reduce risks. When contemplating thermal metering, the presence of asbestos insulation on water piping should be determined. Properly licensed trades personnel, such as an electrician or welder, should be a fundamental part of any team installing electrical monitoring equipment.

In order to prevent inadvertent tampering, any occupants and maintenance personnel should be carefully briefed on what is being done and the purpose of sensors and equipment. Data loggers should have a dedicated (non-occupant-switchable) hard-wired power supply to prevent inadvertent power loss.

Table 7   General Characteristics of Data Acquisition Systems

Type of Data Acquisition System (DAS)	Typical Use	Typical Data Retrieval	Comments
Manual readings	Total energy use	Monthly or daily written logs	Human factors may affect accuracy and reading period. Data must be manually entered for computer analysis.
Pulse counter, cassette tapes (1 to 4 channels)	Total energy use (some end use)	Monthly pickup of cassette tapes	Data loss due to cassette is a common problem. Pulse data must be read and converted before it can be analyzed.
Pulse counter, solid state (1, 4, or 8 channels)	Total energy use (some end use)	Polled by telephone to mainframe or minicomputer	Computer hardware and software is needed for transfer and conversion of pulse data. Can be expensive. Can handle large numbers of sites. User-friendly.
Stick-on battery-powered logger (1 to 8 channels)	Diagnostics, technology assessment, end use	Monthly manual download to PC	Very useful for remote sites. Can record pulse counts, temperature, etc., up to 1500 records.
Plug-in A/D boards for PCs	Diagnostics, technology assessment, control	On-site real-time collection and storage	Usually small quantity, unique applications. PC programming capability needed to set up data software and configure boards.
Simple field DAS (usually 16 to 32 channels)	Technology assessment, residential end use (some diagnostics)	Phone retrieval to host computer for primary storage (usually daily to weekly)	Can use PCs as hosts for data retrieval. Good A/D conversion available. Low cost per channel. Requires programming skills to set up field unit and configure communications for data transfer.
Advanced field DAS (usually > 40 channels/units)	Diagnostics, energy control systems, commercial end use	On-site real-time collection and data storage, or phone retrieval	Usually designed for single buildings. Can be PC-based or stand-alone unit. Can run applications/diagnostic programs. User-friendly.
Building automation system	On-site diagnostics, energy measurement and verification	Proprietary data collection procedures, manual or automated export to spreadsheet	Requires significant coordination with building operation personnel. Sensor accuracy, calibration, and installation require confirmation. Good for projects with limited instrumentation budget.

Table 8   **Practical Concerns for Selecting and Using Data Acquisition Hardware**

Components	Field Application Concerns
Data logger unit and peripherals	• Select equipment for field application. • Equipment should store data in electronic form such as on floppy disks, on magnetic tape, or in memory for easy transfer to the computer that will perform the analysis. • Remote programming capability should be available to minimize on-site software modifications. • Avoid equipment with cooling fans. • Use high-quality, reliable modems. • Make sure logger and modem reset after power outage.
Cabling and interconnection hardware	• Use only signal-grade cable—shielded, twisted-pair with drain wire for analog signals. • Mitigate sources of common mode and normal mode signal noise.
Sensors	• Use rugged, reliable sensors that are rated for field application. • Use a signal splitter if sharing existing sensors or signals with other recorders or energy management control system (EMCS). • Select ranges so sensors operate at 50 to 75% of full scale. • Choose sensors that do not require special signal conditioning. • Precalibrate sensors and recalibrate periodically. • When possible, use redundant channels to cross-check critical channels that can drift.

Table 9   **Instrumentation Accuracy and Reliability**

Instrument	Problems
Hygrometers	Drift, saturation and accuracy over time, need for calibration to remove temperature dependence; aspirated systems need to be cleaned periodically. Chilled mirror systems require frequent maintenance.
Flowmeters	Need for calibration, reliability. Moving parts prone to failure. Pipe size must be verified prior to calibration or installation.
Btu meters	60 Hz noise from surroundings, calibration.
Single-ended voltage	Grounding problems, spurious line voltages, 60 Hz noise.
Outdoor air temperature sensor	Must be properly shielded from solar radiation. Aspiration may reduce solar radiation effects but decrease long-term reliability.
RTD sensors	Signal wire length affects readings.
Power meters	Polarity of current transformers (CTs) often marked incorrectly, problems with shunt resistors and CT output. Devices should be checked prior to installation.

Part Seven: Resolve Data Product Accuracies

Data collected by the monitoring equipment is usually used for the following purposes:

• Direct reporting of the primary measurement data
• Reporting of secondary or deduced quantities (e.g., thermal energy consumed by a building, found by multiplying mass flow rate and temperature difference)
• Subsequent interpretation and analyses (e.g., to develop a statistical model of energy used by a building versus outdoor dry-bulb temperature)

In all three cases, the value of the measurements is dramatically increased if the associated uncertainty can be quantified.

Basic Concepts. The concept of **uncertainty** can be better understood in terms of confidence limits. Confidence limits define the range of values that can be expected to include the true value with a stated probability (ASHRAE *Guideline* 2). Thus, a statement that the 95% confidence limits are 5.1 to 8.2 implies that the true value is contained in the interval bounded by 5.1 and 8.2 in 19 out of 20 predictions or, more loosely, that we are 95% confident that the true value lies between 5.1 and 8.2. For a given set of *n* observations with normal (Gaussian) error distribution, the total variance about the mean predicted value $\bar{X}'$ provides a direct indication of the confidence limits. Thus, the "true" mean value $\bar{X}$ of the random variable is bounded as follows:

$$\bar{X}' \pm t_{\alpha/2,\,n-1}\sqrt{\sigma^2/n} \tag{1}$$

where

α = level of significance
$\bar{X}'$ = mean predicted value of random variable X
$t_{\alpha/2,\,n-1}$ = *t*-statistic with probability of $1 - \alpha/2$ and $n - 1$ degrees of freedom (tabulated in most statistical textbooks)
n = number of observations, with Gaussian error distribution
σ^2 = estimated measurement variance

The terms **accuracy** and **precision** are often used to distinguish between bias errors and random errors. A set of measurements with small bias errors is said to have high accuracy, while a set of measurements with small random errors is said to have high precision. In repeated measurements of a given sample via the same technique (single-sample data), each measurement has the same bias. Bias errors include (1) those that are known and can be calibrated out, (2) those that are negligible and are ignored, and (3) those that are estimated and are included in the uncertainty analysis. It is usually difficult to estimate bias limits, and this effect is often overlooked by most practitioners. However, a proper error analysis should include bias error, which is usually written as a plus-minus error. ASME *Standard* PTC 19.1 has a more complete discussion.

Because bias errors b_m and random errors ε_m are usually uncorrelated, measurement variance σ^2 can be expressed as

$$\sigma^2_{meas}(b_m, \varepsilon_m) = \sigma^2(b_m) + \sigma^2(\varepsilon_m) \tag{2}$$

For further information on uncertainty, see Chapter 14 of the 1997 *ASHRAE Handbook—Fundamentals*.

Primary Measurement Uncertainty. Sensor and measuring equipment manufacturers usually specify measurement variances; frequent recalibration minimizes bias errors. As indicated by Equation (1), increasing the number of measurements *n* reduces the uncertainty bounds.

Uncertainty in Derived Quantities. Once a specific algorithm or equation for obtaining final data from physical measurements has been established, standard techniques can be used that incorporate primary measurement uncertainties into the final data product. For the random errors, the well-known Kline and McClintock (1953) error propagation method, based on a first-order Taylor series expansion, is widely used to determine measurement uncertainties in the derived variables in single-sample experiments. Bias errors are difficult to account for; the usual practice is to calibrate them out and exclude them from the uncertainty analysis.

Uncertainty in Statistical Regression Models. Statistical regression models developed from measured data are usually used for predictive purposes. Measurement errors are much smaller than model errors, which arise because the regression model is imperfect; that is, it is unable to explain the entire variation in the regressor variable (Box 1978). Measurement error is inherently contained in the identified model, so the total prediction variance is simply given by the model prediction uncertainty.

Determination of prediction errors due to regression models is subject to different types of problems. The various sources of error can be classified into three categories (Reddy et al. 1998):

1. Model misspecification errors are due to the fact that the functional form of the regression model is usually an approximation of the true driving function of the response variable.
2. Model prediction errors arise due to the fact that a model is never perfect.
3. Model extrapolation errors arise when a model is used for prediction outside the region covered by the data from which the model was developed. Models developed from short data sets, which do not satisfactorily represent the annual behavior of the system, are subject to this error. This error cannot be quantified in statistical terms alone, but certain experimental conditions are likely to lead to accurate predictive models. This falls under the purview of experimental design (Box 1978).

Sources 1 and 3 are likely to introduce both bias and random error in the predictions. If ordinary least-squares regression is used for parameter estimation, and if the model is subsequently used for prediction, error due to source 2 will be purely random. Thus, models identified from short data sets and used to predict seasonal or annual energy use are affected by error sources 1 and 3.

The least-squares method of calculating linear regression coefficients is incapable of producing unbiased estimators of the slope and intercept if there are errors associated with the measurement of the predictor variable. The uncertainty analysis methodology developed for in situ equipment testing uses standard linear regression practices to find the functional relationship and then estimates the increased uncertainty in the regression prediction due to random and bias errors in both variable measurements (Phelan et al. 1996).

Experimental Design. Errors can also be estimated based on historical experience (e.g., using results from previous similar projects). Alternatively, a pilot study can be launched to obtain an estimate of potential errors in a proposed analysis. Some estimate of potential error must be available to determine whether the project goals and objectives are reasonable.

Estimating data uncertainty is one part of the iterative procedure associated with proper experimental design. If the final data product uncertainty determined using the preceding evaluation procedure is unacceptable, the uncertainty can be reduced in one or more of the following ways:

• Reducing overall measurement uncertainty (improving sensor precision)
• Increasing the duration of the monitoring period to average out stochastic variations
• Increasing the number of buildings tested

On the other hand, if simulations indicate that the expected bias in the final data products is unacceptable, the bias may be reduced by one or more of the following steps:

• Adding sensors to get an unbiased measurement of the quantity
• Using more detailed models and analysis procedures
• Increasing data acquisition frequency, in combination with a more detailed model, to eliminate biases due to sensor or system nonlinearities

Accuracy Versus Cost. The need for accuracy must be carefully balanced against measurement and analysis costs. Accuracy loss can stem from instrumentation or measurement error, normalization or model error, sampling or statistical error, and errors of assumptions. Each of these sources of error can be controlled to varying degrees. However, it is generally true that more accurate methods follow the law of diminishing returns, in which further reductions in error come at progressively greater expense.

Because of this trade-off, the optimal measurement solution is usually found by an iterative approach, where incremental improvements in accuracy are assessed relative to the increase in measurement cost. Such optimization requires that a value be placed on increasing levels of accuracy. One method of evaluating the uncertainty of a proposed method is to calculate results using the highest and lowest values in the confidence interval. The difference between these values can be translated into a monetary amount that is at risk. The question that must be answered is whether further measurement investment is warranted to reduce this risk.

Part Eight: Specify Verification and Quality Assurance Procedures

Establishing and using data quality assurance (QA) procedures can be very important to the success of a field monitoring project. The amount and importance of the data to be collected help determine the extent and formality of the QA procedures. For most projects, the entire data path, from sensor installation to procedures that generate results for the final report, should be considered for verification tests. In addition, the data flow path should be checked routinely for failure of sensors or test equipment, as well as unexpected or unauthorized modifications to equipment.

Quality assurance often requires complex data handling. Building energy monitoring projects collect data from sensors and manipulate that data into results. Data handling in a project with only a few sensors and required readings can consist of a relatively simple data flow on paper. Computers, which are generally used in one or more stages of the process, require a different level of process documentation because much of what occurs has no direct paper trail.

Computers facilitate the collection of large data sets and increase project complexity. To achieve maximum automation, computers require the development of specific software. Often, separate computers are involved in each step, so passing information from one computer to another must be automated in large projects. To move the data as smoothly as possible, an automated **data pipeline** should be developed. This pipeline minimizes the time delay from data collection to results production and maximizes the cost-effectiveness of the entire project.

An automated data verification procedure should be used when possible. Frequent data acquisition (preferably automated) with a quality control review of summarized and plotted data is essential to ensure that reliable data are collected. Verification procedures should be performed at frequent intervals (daily or weekly), depending on the importance of missing data. This minimizes data loss due to equipment failure and/or changes at a building site. It also allows the processed information to be applied quickly.

The following QA actions should take place:

• Calibrate hardware and establish a good control procedure for collection of data. Use NIST-traceable calibration methods.
• Verify data, check for reasonableness, and prepare a summary report to assure the quality of the data after it is collected.
• Perform initial analysis of the data. Significant findings may lead to changes in procedures for checking data quality.
• Thoroughly document and control procedures applied to remedy problems. These procedures may entail changes in hardware or collected data (such as data reconstruction), which can have a fundamental impact on the results reported.
• Maintain an archive of the raw data obtained from the site to ensure project integrity.

Three aspects of a monitoring project that require quality assurance are shown in Table 10: hardware, engineering data, and characteristics data. Three QA reviews are necessary for each of these aspects: (1) initial QA confirms that the project starts correctly; (2) ongoing QA confirms that the information collected by the project continues to satisfy quality requirements; and (3) periodic QA involves additional checks, established at an the beginning of the

Table 10   **Table of Quality Assurance Elements**

Time Frame	Hardware	Engineering Data	Characteristics Data
Initial start-up	Bench calibration (1)	Installation verification (1)	Field verification (1)
	Field calibration (1)	Collection verification (1,2)	Completeness check (1)
	Installation verification (1)	Processing verification (1,2)	Reasonableness check (1,2)
		Result production (1,2)	Result production (1,2)
Ongoing	Functional testing (1)	Quality checking (2)	Problem diagnosis (3)
	Failure mode diagnosis (3)	Reasonableness checking (2)	Data reconstruction (4)
	Repair/maintenance (4)	Failure mode diagnosis (3)	Change control (1)
	Change control (1)	Data reconstruction (4)	
		Change control (1)	
Periodic	Preventive maintenance (1)	Summary report preparation and review (2)	Scheduled updates/resurveys (1)
	Calibration (1)		Summary report preparation and review (2)

(1) Actions to ensure good data.   (2) Actions to check data quality.   (3) Actions to diagnose problems.   (4) Actions to repair problems.

project, to ensure continued performance at an acceptable quality level.

Information about data quality and the quality assurance process should be readily available to data users. Otherwise, significant analytical resources may be expended to determine data quality, or the analyses may never be performed due to uncertainties.

Part Nine: Specify Recording and Data Exchange Formats

This step involves specifying the formats in which the data will be supplied to the end user or other data analysts. Both raw and processed (adjusted for missing data or anomalous readings) data formats should be specified. In addition, if any supplemental analyses are planned, the medium and format to be used (magnetic tape, disk, spreadsheet, ASCII) should be specified. These requirements can be determined by analyzing the software data format specifications. Common formats for raw data are comma-delimited and blank-delimited ASCII, which do not require data conversion.

Data documentation is essential for all monitoring projects, especially when several organizations are involved. Data usability is improved by specifying and adhering to data recording and exchange formats. Most data transfer problems are related to inadequate documentation. Other problems include hardware or software incompatibility, errors in the tape or disk, errors or inconsistencies in the data, and transmittal of the wrong data set. The following precautions can prevent some of these problems:

- Provide documentation to accompany the data transfer (Table 11). Because these guidelines apply to general data, models, programs, and other types of information, the items listed in Table 11 may not all apply to every case.
- Provide documentation of transfer media, including the computer operating system, software used to create the files, media format (ASCII, EBCDIC, binary), and tape or disk characteristics (tracks, density, record length, block, size).
- Provide procedures to check the accuracy and completeness of the data transfer, including statistics or frequency counts for variables and hard copy versions of the file. Test input data and corresponding output results for models on other programs.
- Keep all raw data, including erroneous records.
- Convert and correct data; save routines for later use.
- Limit equipment access to authorized individuals.
- Check incoming data soon after they are collected, using simple time-series and *x-y* inspection plots.
- Automate as many routines as possible to avoid operator error.

Table 11   **Documentation Included with Computer Data to Be Transferred**

1. Title and/or acronym
2. Contact person (name, address, phone number)
3. Description of file (number of records, geographic coverage, spatial resolution, time period covered, temporal resolution, sampling methods, uncertainty/reliability)
4. Definition of data values (variable names, units, codes, missing value representation, location, method of measurement, variable derivation)
5. Original uses of file
6. Size of file (number of records, bytes)
7. Original source (person, agency, citation)
8. Pertinent references (complete citation) on materials providing additional information
9. Appropriate reference citation for the file
10. Credit line (for use in acknowledgments)
11. Restrictions on use of data/program
12. Disclaimer (examples follow):
    - Unverified data; use at your own risk.
    - Draft data; use with caution.
    - Clean data to the best of our knowledge. Please let us know of any possible errors or questionable values.
    - Program under development.
    - Program tested under the following conditions (conditions specified by author).

REFERENCES

ASHRAE. 1984. Standard methods of measuring and expressing building energy performance. ANSI/ASHRAE/IESNA *Standard* 105-1984 (RA 90).

ASHRAE. 1986. Engineering analysis of experimental data. ASHRAE *Guideline* 2-1986 (RA 96).

ASME. 1985. Measurement uncertainty: Instruments and apparatus. ANSI/ASME *Standard* PTC 19.1-85 (R 1990). American Society of Mechanical Engineers, New York.

ASTM. 1987. Test method for determining air leakage rate by fan pressurization. *Standard* E 779-87 (R 1992).

Balcomb, J.D., J.D. Burch, and K. Subbarao. 1993. Short-term energy monitoring of residences. *ASHRAE Transactions* 99(2):935-44.

Box, G.E.P. 1978. Statistics for experimenters: An introduction to design, data analysis and model-building. John Wiley and Sons, New York.

Burch, J.D. 1986. Thermal performance monitoring methods. *Passive Solar Journal* 3.

Burch, J.D., K. Subbarao, A. Lekov, M. Warren, L. Norford, and M. Krarti. 1990. Short-term energy monitoring in a large commercial building. *ASHRAE Transactions* 96(1):1459-77.

Busch, J.F., A.K. Meier, and T.S. Nagpal. 1984. Measured heating performance of new, low-energy homes: Updated results from the BECA-A database. LBL-17883. Lawrence Berkeley Laboratory, Berkeley, CA.

Claridge, D.E., J.S. Haberl, W.D. Turner, D.L. O'Neal, W.M. Heffington, C. Tombari, M. Roberts, and S. Jaeger. 1991. Improving energy conservation retrofits with measured savings. *ASHRAE Journal* 33(10):14-22.

Cohen, R.R., P.W. O'Callaghan, S.D. Probert, N.M. Gibson, D.J. Nevrala, and G.F. Wright. 1987. Energy storage in a central heating system: Spa school field trial. Building Service Engineering Research Technology (Great Britain) 8:79-84.

DeCicco, J., R. Diamond, S.L. Nolden, J. DeBarros, and T. Wilson. 1995. Improving energy efficiency in apartment buildings. ACEEE, Washington, DC. pp. 234-36.

Duffy, J.J., D. Saunders, and J. Spears. 1988. Low-cost method for evaluation of space heating efficiency of existing homes. Proceedings of the 12th Passive Solar Conference, ASES, Boulder, CO.

EIA. 1996. Commercial buildings energy consumption and expenditures: 1995. U.S. Department of Energy, Energy Information Administration. DOE/EIA-0318(95).

Fels, M.F., ed. 1986. Measuring energy savings: The scorekeeping approach. *Energy and Buildings* 9.

Fracastoro, G.V. and M.D. Lyberg. 1983. Guiding principles concerning design of experiments, instruments, instrumentation, and measuring techniques. Swedish Council for Building Research, Stockholm, Sweden (December).

Haberl, J.S. and D.E. Claridge. 1987. An expert system for building energy consumption analysis: Prototype results. *ASHRAE Transactions* 93(1): 979-98.

Haberl, J.S. and P.S. Komor. 1990a. Improving [commercial building] energy audits: How annual and monthly consumption data can help. *ASHRAE Journal* 32(8):26-33.

Haberl, J.S. and P.S. Komor. 1990b. Improving [commercial building] energy audits: How daily and hourly data can help. *ASHRAE Journal* 32(9):26-36.

Haberl, J., D.E. Claridge, and D. Harrje. 1990. The design of field experiments and demonstration. Proceedings of the IEA Field Monitoring for a Purpose Workshop (Gothenburg, Sweden, April), pp. 33-58.

Haberl, J., A. Reddy, D. Claridge, D. Turner, D. O'Neal, and W. Heffington. 1995. Measuring energy-saving retrofits: Experiences from the Texas LoanSTAR Program. ORNL *Report* No. ORNL/Sub/93-SP090-1. Oak Ridge National Laboratory, Oak Ridge, TN.

Hirst, E. and J. Reed, eds. 1991. Handbook of evaluation of utility DSM programs. ORNL/CON-336. Oak Ridge National Laboratory, Oak Ridge, TN.

Hirst, E., D. White, and R. Goeltz. 1983. Comparison of actual electricity savings with audit predictions in the BPA residential weatherization pilot program. ORNL/CON-142. Oak Ridge National Laboratory, Oak Ridge, TN. (See also ORLN/CON-174.)

Hough, R.E., P.J. Hughes, R.J. Hackner, and W.E. Clark. 1987. Results-oriented methodology for monitoring HVAC equipment in the field. *ASHRAE Transactions* 93(1):1569-79.

IPMVP. 1997. International Performance Measurement and Verification Protocol. DOE/EE-0157. (Available as a book from Energy Efficiency and Renewable Energy Clearing House, 1-800-DOE-EREC, or it may be downloaded from www.ipmvp.org.)

Kissock, J.K., D.E. Claridge, J.S. Haberl, and T.A. Reddy. 1992. Measuring retrofit savings for the Texas LoanSTAR program: Preliminary methodology and results. Solar Engineering 1992—Proceedings of the 1992 ASME-JSES-KSES International Solar Engineering Conference (Maui, Hawaii, April 5-8), pp. 299-308.

Kline, S. and F.A. McClintock. 1953. Describing uncertainties in single-sample experiments. *Mech. Eng.* 75:2-8.

Levins, W.P. and M.A. Karnitz. 1986. Cooling-energy measurements of unoccupied single-family houses with attics containing radiant barriers. ORNL/CON-200. Oak Ridge National Laboratory, Oak Ridge, TN. (See also ORNL/CON-2136.)

Lopez, R. and J.S. Haberl. 1992. Data processing routines for monitored building energy data. Solar Engineering 1992—Proceedings of the 1992 ASME-JSES-KSES International Solar Engineering Conference (Maui, Hawaii, April 5-8), pp. 329-36.

MacDonald, J.M. and D.M. Wasserman. 1989. Investigation of metered data analysis methods for commercial and related buildings. ORNL/CON-279. Oak Ridge National Laboratory, Oak Ridge, TN.

MacDonald, J.M., T.R. Sharp, and M.B. Gettings. 1989. A protocol for monitoring energy efficiency improvements in commercial and related buildings. ORNL/CON-291. Oak Ridge National Laboratory, Oak Ridge, TN (September).

Misuriello, H. 1987. A uniform procedure for the development and dissemination of monitoring protocols. *ASHRAE Transactions* 93(1).

Misuriello, H. 1988. Instrumentation applications for commercial building energy audits. *ASHRAE Transactions* 95(2).

Modera, M.P. 1989. Residential duct system leakage: Magnitude, impacts, and potential for reduction. *ASHRAE Transactions* 95(2).

Norford, L.K, A. Rabl, and R.H. Socolow. 1985. Measurement of thermal characteristics of office buildings. ASHRAE Conference on the Thermal Performance of Building Envelopes (Clearwater Beach, FL).

Persily, A.K. and R. Grot. 1988. Diagnostic techniques for evaluating office building envelopes. *ASHRAE Transactions* 94(1).

Phelan, J., M. Brandemuehl, and M. Krarti. 1996. Final Report: Methodology development to measure in-situ chiller, fan, and pump performance. JCEM TR/96/3. Joint Center for Energy Management, University of Colorado, Boulder.

Phelan, J., M. Brandemuehl, and M. Krarti. 1997a. In-Situ performance testing of chillers for energy analysis. *ASHRAE Transactions* 97(1).

Phelan, J., M. Brandemuehl, and M. Krarti. 1997b. In-Situ performance testing of fans and pumps for energy analysis. *ASHRAE Transactions* 97(1).

Reddy, T.A., J.K. Kissock, and D.K. Ruch. 1998. Regression modelling in determination of retrofit savings. *ASME Journal of Solar Energy Engineering* 120:185.

Robison, D.H. and L.A. Lambert. 1989. Field investigation of residential infiltration and heating duct leakage. *ASHRAE Transactions* 95(2):542-50.

Schiller, S.R. 1996. Measurement and verification (M&V) guideline for federal energy projects. DOE/GO-10096-248. U.S. Department of Energy, Washington, D.C. (Available from DOE's Energy Efficiency and Renewable Energy Clearinghouse, 1-800-DOE-EREC.)

Sharp, T. 1996. Energy benchmarking in commercial office buildings. Proceedings of the 1996 ACEEE Summer Study on Energy Efficiency in Buildings, pp. 4.321-4.329.

Sharp, T.R. and J.M. MacDonald. 1990. Effective, low-cost HVAC controls upgrade in a small bank building. *ASHRAE Transactions* 96(1):1011-17.

Sparks, R., J.S. Haberl, S. Bhattacharyya, M. Rayaprolu, J. Wang, and S. Vadlamani. 1992. Use of simplified system models to measure retrofit energy savings. Solar Engineering 1992—Proceedings of the 1992 ASME-JSES-KSES International Solar Engineering Conference (Maui, Hawaii, April 5-8), pp. 325-28.

Subbarao, K. 1988. PSTAR—A unified approach to building energy simulations and short-term monitoring. SERI/TR-254-3175. Solar Energy Research Institute, Golden, CO.

Subbarao, K., J.D. Burch, and H. Jeon. 1986. Building as a dynamic calorimeter: Determination of heating system efficiency. SERI/TR-254-2947. Solar Energy Research Institute, Golden, CO.

Szydlowski, R.F. and R.C. Diamond. 1989. Data specification protocol for multifamily buildings. LBL-27206. Lawrence Berkeley Laboratory, Berkeley, CA.

Ternes, M.P. 1986. Single-family building retrofit performance monitoring protocol: Data specification guideline. ORNL/CON-196. Oak Ridge National Laboratory, Oak Ridge, TN.

Woller, B.E. 1989. Data acquisition and analysis of residential HVAC alternatives. *ASHRAE Transactions* 95(1):679-86.

BIBLIOGRAPHY

ASHRAE. 1986. Energy use in commercial buildings: Measurements and models. *Technical Data Bulletin* 2(4).

ASHRAE. 1987. Energy measurement applications. *Technical Data Bulletin* 3(3).

Baird, G., M.R. Donn, F. Pool, W.D.S. Brander, and C.S. Aun. 1984. *Energy performance of buildings*. CRC Press, Boca Raton, FL.

Doebelin, E. 1990. Measurement systems. McGraw-Hill, New York.

EPRI. 1985. Survey of utility commercial sector activities. EPRI EM-4142 (July). Electric Power Research Institute, Palo Alto, CA.

EPRI. 1985. Survey of residential end-use projects. EPRI EM-4578 (May). Electric Power Research Institute, Palo Alto, CA.

Frey, D.W. and M. Holtz. 1990. The preparation and implementation of a monitoring plan. Proceedings of the IEA Field Monitoring for a Purpose Workshop, (Gothenburg, Sweden, April), pp. 78-91.

Jones, J.R. and S. Boonyatikarn. 1990. Factors influencing overall building efficiency. *ASHRAE Transactions* 96(1):1449-58.

O'Neal, D.L., J. Bryant, C. Boecker, and C. Bohmer. 1993. Instrumenting buildings to determine retrofit savings: Murphy's Law revisited. ESL-PA93/03-02. Proceedings of IETC Conference (College Station, TX).

Palmer, J. 1989. Energy performance assessment: A guide to procedures, Vols. I and II. Energy Technology Support Unit, Harwell, United Kingdom.

Porterfield, J.M. 1988. Alternatives for monitoring multidwelling energy measures. *ASHRAE Transactions* 94(1):1024-33.

Shehadi, M.T. 1984. Automated utility management system. *ASHRAE Transactions* 90(1B):401-10.

Van Hove, J. and L. van Loon. 1990. Long-term cold energy storage in aquifer for air conditioning in buildings. *ASHRAE Transactions* 96(1): 1484-88.

Wexler, A. and S.R. Schiller. 1983. Monitoring methodology handbook for residential HVAC systems. EPRI EM-3003. Electric Power Research Institute, Palo Alto, CA.

CHAPTER 40

SUPERVISORY CONTROL STRATEGIES AND OPTIMIZATION

COMPUTERIZED energy management and control systems provide an excellent means of reducing the utility costs associated with maintaining environmental conditions in commercial buildings. These systems can incorporate advanced control strategies that respond to changing weather and building conditions and minimize operating costs.

HVAC systems are typically controlled using a two-level control structure. Lower level **local-loop control** of a single set point is provided by an actuator. For example, the supply air temperature from a cooling coil is controlled by adjusting the opening of a valve that provides chilled water to the coil. The upper control level, **supervisory control**, specifies the set points and other modes of operation that are time dependent.

The performance of large, commercial HVAC systems can be improved through better local-loop and supervisory control. Proper tuning of local-loop controllers can enhance comfort, reduce energy use, and increase component life. Set points and operating modes for cooling plant equipment can be adjusted by the supervisor to maximize the overall operating efficiency. Dynamic control strategies for ice or chilled water storage systems can significantly reduce on-peak electrical energy and demand costs to minimize total utility costs. Similarly, thermal storage inherent within a building's structure can be dynamically controlled to minimize utility costs. In general, strategies that take advantage of thermal storage work best when forecasts of future energy requirements are available.

This chapter focuses on the opportunities and control strategies associated with using computerized control for centralized cooling systems. The chapter is divided into three major sections. The first section defines the systems and control variables considered and presents background on the impacts and opportunities associated with adjusting control variables. The second section presents a number of supervisory control strategies that can be implemented by computerized control systems and is intended for practitioners. Finally, the third section presents basic methods for optimization of systems both with and without significant thermal energy storage and is intended for researchers and developers of advanced control strategies.

BACKGROUND

SYSTEMS, DEFINITIONS, AND CONTROL VARIABLES

Figure 1 shows a schematic of a typical central cooling system for which control strategies are presented in this chapter. For the most part, the strategies assume that the equipment is electrically

---

The preparation of this chapter is assigned to TC 4.6, Building Operation Dynamics.

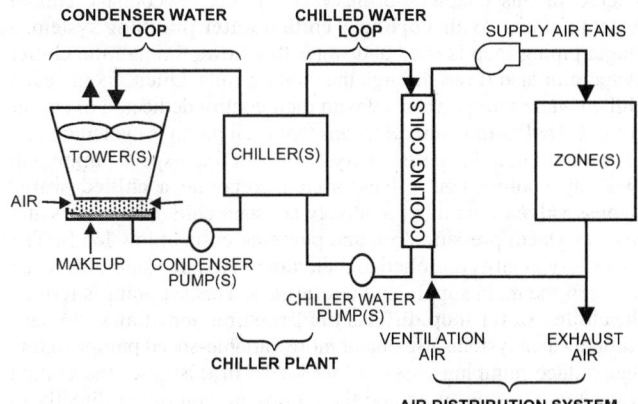

Fig. 1   Schematic of Chilled Water Cooling System

driven and that heat is rejected to the environment by cooling towers. However, some strategies apply to any type of system (e.g., return from night setup). For describing different systems and controls, it is useful to divide the system into the subsystems depicted in Figure 1: air distribution, chilled water loop, chiller plant, condenser water loop.

Subsystems

Air Distribution Systems. An air distribution system includes terminal units (VAV boxes, etc.), air-handling units, ducts, and controls. An air-handling unit (AHU) includes a coiling coil, dampers, fan, and controls (in addition to equipment for heating and humidification). A single air handler may serve several zones and several air handlers may be used in a building. For each AHU, ventilation air is mixed with return air from the zones and fed to the cooling coil. Typically, an **economizer control** is incorporated that selects between minimum and maximum ventilation air depending upon the condition of the outside air.

The air is cooled and dehumidified through the cooling coil. A local-loop controller adjusts the flow of water through the coil using a two- or three-way valve in order to maintain a specified set point temperature for the air leaving the cooling coil. This supervisory control variable is termed the **supply air temperature** (the air temperature supplied to the zones may be different due to fan heating or local reheat). A supply fan and return fan (not shown in Figure 1) provide the necessary airflow to and from the zones.

Two different types of zone controls are considered in this chapter: constant air volume (CAV) and variable air volume (VAV). **CAV systems** have fixed-speed fans and provide no feedback control of airflow to the zones. For this system, zone temperature is controlled to a set point using a feedback controller that regulates the amount of local **reheat** applied to the air entering each zone. With a **VAV system**, a feedback controller regulates the airflow to each zone in order to maintain zone temperature set points. The zone airflows are regulated using dampers located in VAV boxes in each zone. VAV systems also incorporate feedback control of the primary airflow through modulation fans. Typically, the inputs to a fan outlet damper, inlet vanes, blade pitch, or variable speed motor are adjusted in order to maintain a **duct static pressure** set point within the supply duct as described in Chapter 45.

Zone temperature set points are typically fixed values within the comfort zone during the occupied time and zone humidity is allowed to "float" within a range dictated by the system design and choice of the supply air set point temperature. **Night setup** is often used to raise the zone temperature set points during unoccupied times and reduce the cooling requirements.

Chilled Water Loop. The chilled water loop consists of pumps, pipes, valves, and controls. Two different types of systems are considered in this chapter—primary and primary/secondary chilled water pumping. With a **primary chilled water pumping** system, a single piping loop is used and water that flows through the chiller evaporator also flows through the cooling coils. Often, fixed-speed chilled water pumps are used with their control dedicated to chiller control. **Dedicated control** means that each pump is cycled on and off with a chiller that it serves. Systems with fixed-speed pumps and two-way cooling coil valves often incorporate a **chilled water bypass valve** to maintain relatively constant chiller flow rates and reduce system pressure drop and pumping costs at low loads. The valve is typically controlled to maintain a fixed pressure difference between the main supply and return lines. This set point is termed the **chilled water loop differential pressure**. Sometimes, primary chilled water systems use one or more variable-speed pumps to further reduce pumping costs at low loads. In this case, the chilled water bypass is not used and the pumps are controlled directly to maintain a chilled water loop differential pressure set point.

Primary/secondary chilled water systems are designed specifically for variable-speed pumping. In the primary loop, fixed-speed pumps provide a relatively constant flow of water to the chillers. This design ensures good chiller performance and reduces the risk of freezing on evaporator tubes. The secondary loop incorporates one or more variable-speed pumps that are controlled to maintain a chilled water loop differential pressure set point. The primary and secondary loops may be separated by a heat exchanger. However, it is more common to use direct coupling with a common pipe.

Chiller Plant. One or more chillers that are typically arranged in parallel with dedicated pumps provide the primary source of cooling for the system. Individual feedback controllers adjust the cooling capacity of each chiller in order to maintain a specified **chilled water supply temperature**. Additional chiller control variables include the number of chillers operating and the relative loading for each chiller. For a given total cooling requirement, individual chiller cooling loads can be controlled by utilizing different chilled water supply set points for constant individual flow or by adjusting individual flows for identical set points.

Thermal storage can also be used in combination with chillers to provide cooling during occupied periods. During the night, when commercial buildings are unoccupied, the primary cooling equipment is used to cool the thermal storage medium for use the next day. During occupied times, the combination of the primary cooling system and storage must meet the building cooling requirements. The use of thermal storage can significantly reduce the costs of providing cooling in commercial buildings by reducing on-peak demand and energy costs. The term "**charging**" is often used to describe the cooling of storage, while "**discharging**" describes the use of storage to provide cooling. Control of thermal storage is defined by the manner in which the storage medium is charged and discharged over time. This control problem differs from the control of other system variables in that operating costs depend upon a time history of charging and discharging over several hours.

Condenser Water Loop. The condenser water loop consists of cooling towers, pumps, piping, and controls. Cooling towers reject heat to the environment through heat transfer and possibly evaporation (for wet towers) to the ambient air. Typically, large towers incorporate multiple cells sharing a common sump with individual fans having two or more speed settings. Often, a feedback controller adjusts the tower fan speeds in order to maintain a set point for the water temperature leaving the cooling tower, termed the **condenser water supply temperature**. Typically, condenser water pumps are dedicated to individual chillers (i.e., each pump is cycled on and off with a chiller that it serves).

Control and Performance Measures

The control of a VAV cooling system (Figure 1) changes in response to increasing building cooling requirements in the following manner. With no changes in control, the zone temperature rises as energy gains to the zone air increase. However, the zone controller responds to higher temperatures by increasing local airflow by opening a damper. Opening of dampers reduces the static pressure in the primary supply duct, which causes the fan controller to create additional airflow. In the absence of any additional control changes, the supply air temperature of the cooling coils increases due to the greater airflow. However, feedback controllers for the supply air temperature respond to the higher temperatures and increase the water flow by opening the cooling coil valves. This increases the chilled water flow and heat transfer to the chilled water (i.e., cooling load).

For fixed-speed chilled-water pumps, the differential pressure controller closes the chilled-water bypass valve and keeps the overall flow relatively constant. For variable-speed pumping, the differential pressure controller increases the pump speed. The return water temperature and/or flow rate to the chillers increases, which leads to an increase in the chilled water supply temperature. The chiller controller then responds by increasing the chiller cooling capacity in order to maintain the chilled water supply set point (and match the cooling coil loads). The increased energy removed by the chiller increases the heat rejected to the condenser water loop, which increases the temperature of water leaving the condenser. The increased water temperature entering the cooling tower increases the water temperature leaving the tower. However, the tower controller responds to the higher condenser water supply temperature and increases the tower airflow. At some load, the current set of operating chillers is not sufficient to meet the load (i.e., maintain the chilled water supply set points) and an additional chiller is brought online.

For this scenario, several local-loop controllers respond to the load change in order to maintain specified set points. The values of the set points and the modes of operation (e.g., economizer) are established by a supervisory controller. At any given time, the cooling needs can be met with several different combinations of modes of operation and set points. This chapter furnishes several methods for determining supervisory control variables that provide good overall performance.

For all-electric cooling without thermal storage, minimizing power at each point in time is equivalent to minimizing energy costs. Therefore, the "best" performance occurs when supervisory control variables are chosen to maximize the coefficient of performance (COP) at all times while meeting the building load requirements. The **COP** is defined as the ratio of the system power consumption to the total chiller cooling load. In addition to the control variables, the COP depends primarily on the cooling load and the ambient wet-bulb and dry-bulb temperatures. Often, the cooling load is expressed in a dimensionless form as a part-load ratio, **PLR**, which is the cooling load divided by the design cooling capacity.

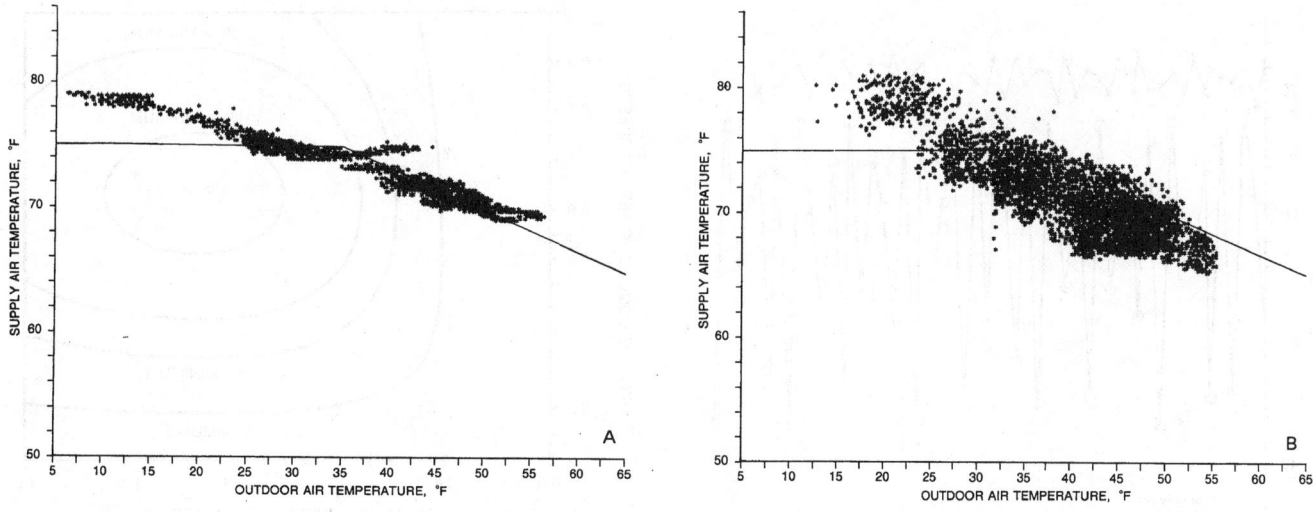

Fig. 2   Outdoor Air Reset History

For systems with thermal storage, performance depends on the time history of charging and discharging. In this case, the "best" performance occurs for controls that minimize operating costs that are integrated over the billing period. In addition, safety features that minimize the risk of prematurely depleting the storage capacity are important.

OPTIMIZING CONTROLS FOR COOLING WITHOUT STORAGE

Local-Loop Control Effects

Proportional-integral (PI) controllers are generally well suited to the control of linear or near-linear processes that occur in HVAC applications. However, the ability of a local-loop controller to maintain a set point depends on the type and quality of the control hardware, the nature of the process being controlled, and the tuning of the controller. Tuning of PI controllers is discussed in the section on commissioning process in Chapter 37, Fundamentals of Control, of the 1997 *ASHRAE Handbook—Fundamentals*. Improper tuning can influence energy consumption and diminish the benefits of optimizing set point choices. Furthermore, there may be interactions between control loops, particularly where a single controller is used to control multiple processes in sequence, such as an air handling system with the heating coil, cooling coil, and return/outside air dampers being modulated by a single controller.

Achieving a desired control response may require considering more than a single process. Reasonably good and poor set point control are shown in Figure 2 for pneumatic control on two air handlers serving an office building (Bushby and Kelly 1988). The control in Figure 2A performs best, maintaining the supply air temperature in a tight temperature band of about 2°F and following the prescribed reset schedule closely for outdoor temperatures ranging from 35 to 55°F. At temperatures below 35°F, the actual maintained set point continues its linear trend instead of maintaining the constant set point supply temperature normally implemented on this particular system.

Of particular interest is the response of the air handler shown in Figure 2B that indicates a similar pattern, except that the data have a much greater spread due to interaction between the outdoor air dampers and the steam preheat valve in this air-handling unit. Analysis of data from this unit shows that the outdoor air dampers tend to oppose the steam preheat valve in the temperature range between 23 and 50°F. In fact, the outdoor air dampers and preheat valve have a tendency to swing open and closed in alternating

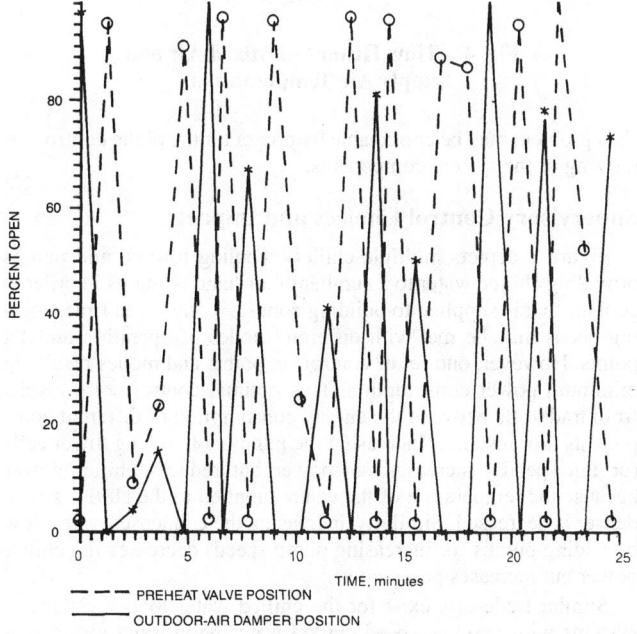

**Fig. 3   Steam Preheat Valve Position and
Outdoor Air Damper Position**

fashion in a very short time. Figure 3 shows the position of the steam preheat valve and outdoor dampers for about 20 min (Bushby and Kelly 1988). The solid line indicates the position of the outdoor air dampers, and the dashed line represents the position of the steam preheat valve.

Both the steam preheat valve and the outdoor air dampers swing open and closed in a cyclic fashion in less than 2 min. The two cycles are roughly 180° out of phase. The effect of the valve and damper cycling on the mixed and supply air temperatures is shown in Figure 4. Both temperatures oscillate in conjunction with the steam valve and the dampers. The maximum peak-to-peak mixed air temperature swing approaches 27°F, while the maximum supply air temperature swing is smaller, approaching 4.5°F. Such control system performance wastes energy by unnecessary use of preheating and has a negative impact on comfort conditions. In addition, it causes needless wear and tear on valves, dampers, and actuators.

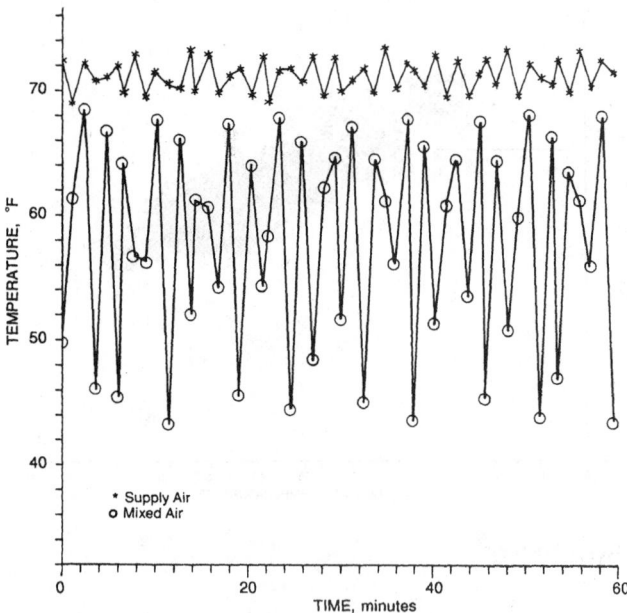

**Fig. 4   Time History of Mixed Air and
Supply Air Temperatures**

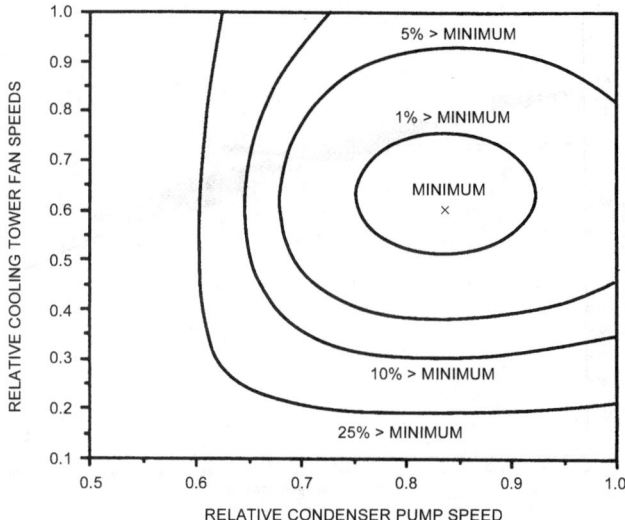

**Fig. 5   Example Power Contours for
Condenser Loop Control Variables**

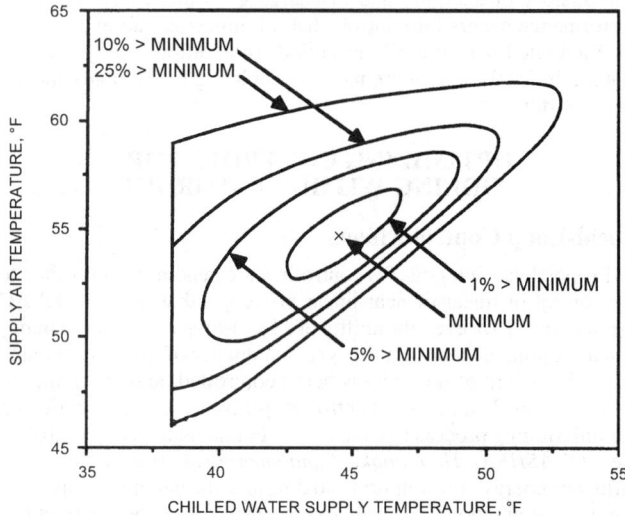

**Fig. 6   Example Power Contours for Chilled Water and
Supply Air Temperatures**

This problem may be correctable by proper tuning of the controls or resizing of the system components.

Supervisory Control Choices and Impacts

Figure 1 depicts multiple chillers, cooling towers, and pumps providing chilled water to a number of air-handling units in order to cool air that is supplied to building zones. At any given time, cooling needs may be met with different modes of operation and set points. However, one set of control set points and modes results in minimum power consumption. This optimal control point results from trade-offs between the energy consumption of different components. For instance, increasing the number of cooling tower cells (or fan speeds) increases fan power but reduces chiller power because the temperature of the water supplied to the chiller's condenser is decreased. Similarly, increasing the condenser water flow by adding pumps (or increasing pump speed) decreases the chiller power but increases pump power.

Similar trade-offs exist for the chilled water loop variables of systems with variable-speed chilled water pumps and air-handler fans. For instance, increasing the chilled water set point reduces chiller power but increases pump power because greater flow is needed to meet the load. Increasing the supply air set point increases fan power, but decreases pump power.

Figure 5 illustrates the sensitivity of the total power consumption to condenser water-loop controls (from Braun et al. 1989a) for a single chiller load, ambient wet-bulb temperature, and chilled water supply temperature. Contours of constant power consumption are plotted versus cooling tower fan and condenser water pump speed for a system with variable-speed fans and pumps. Near the optimum, power consumption is not sensitive to either of these control variables, but increases significantly away from the optimum. The rate of increase in power consumption is particularly large at low condenser pump speeds. A minimum pump speed is necessary to overcome the static pressure associated with the height of the water discharge in the cooling tower above the sump. As the pump speed approaches this value, the condenser flow approaches zero and the chiller power increases dramatically. A pump speed that is too high is generally better than a pump speed that is too low. The broad area near the optimum indicates that the optimal setting does not need to

be accurately determined. However, the optimal settings would change significantly with chiller load.

Figure 6 illustrates the sensitivity of power consumption to chilled-water and supply-air set-point temperatures for a system with variable-speed chilled water pumps and air-handler fans (from Braun et al. 1989a). Within about 3°F of the optimum values, the power consumption is within 1% of the minimum. Outside of this range, the sensitivity to the set points increases significantly. The penalty associated with operation away from the optimum is greater in the direction of smaller differences between the supply air and chilled water set points. As this temperature difference is reduced, the required flow of chilled water to this coil increases and the chilled water pumping power is greater. For a given chilled water or supply air temperature, the temperature difference is limited by the heat transfer characteristics of the coil. As this limit is approached, the required water flow and pumping power would become infinite if the pump speed were not constrained. It is generally better to have too large rather than too small a temperature difference between the supply air and chilled water set points.

For constant chilled-water flow, the trade-offs in energy use with chilled water set point are very different than for variable-flow systems. Increasing the chilled water set point reduces chiller power consumption, but has little effect on chilled-water pumping energy. Therefore, the benefits of chilled-water temperature reset are more significant than for variable-flow systems (although variable-flow systems use less energy). For constant chilled-water flow, the minimum cost strategy is to raise the chilled water set point to the highest value that will keep all discharge air temperatures at their set points and keep zone humidities within acceptable bounds.

For constant-volume air-handling systems (CAV), the trade-offs in energy use with supply air set point are also very different than for variable air volume systems. Increasing the supply-air set point for cooling reduces both the cooling load and reheat required, but does not change fan energy. Again, the benefits of supply air temperature reset for CAV systems are more significant than for VAV systems (although VAV systems use less energy). In general, the set point for a CAV system should be set at the highest value that will keep all zone temperatures at their set points and all humidities within acceptable limits.

In addition to the set points used by local-loop controllers, a number of operational modes can impact performance. For instance, significant energy savings are possible when a system is properly switched over to an economizer cycle. At the onset of economizer operation, return dampers are closed, outside dampers are opened, and the maximum possible outside air is supplied to cooling coils. Two different types of switchover are typically used: (1) dry-bulb and (2) enthalpy. With a dry-bulb economizer, the switchover occurs when the ambient dry-bulb temperature is less than a specified value, typically between 55 and 65°F. With an enthalpy economizer, the switchover happens when the outdoor enthalpy (or wet-bulb temperature) is less than the enthalpy (or wet-bulb temperature) of the return air. Although the enthalpy economizer yields lower overall energy consumption, it requires wet-bulb temperature or dry-bulb and relative humidity measurements.

Another important operation mode is the sequencing of chillers and pumps. Sequencing defines the order and conditions associated with bringing equipment online or offline. Optimal sequencing depends on the individual design and part-load performance characteristics of the equipment. For instance, more efficient chillers should generally be brought online before less efficient ones. Furthermore, the conditions where chillers and pumps should be brought online depend upon their performance characteristics at part-load conditions.

Figure 7 shows the optimal system performance (i.e., optimal set point choices) for different combinations of chillers and fixed-speed pumps in parallel as a function of load relative to the design load for a given ambient wet-bulb temperature. For this system (from Braun et al. 1989a), each component (chillers, chilled water pumps, and condenser water pumps) in each parallel set are identical and sized to meet half of the design requirements. The best performance occurs at about 25% of the design load with one chiller and pump operating. As the load increases, the system COP decreases due to decreasing chiller COP and a nonlinear increase in the power consumption of cooling tower and air handler fans. A second chiller should be brought online at the point where the overall COP of the system is the same with or without the chiller. For this system, this optimal switch point occurs at about 38% of the total design load or about 75% of the individual chiller's capacity. The optimal switch point for bringing a second condenser and chilled water pump online occurs at a much higher relative chilled load (0.62) than the switch point for adding or removing a chiller (0.38). However, pumps are typically sequenced with chillers (i.e., they're brought on-line together). In this case, Figure 7 shows that the optimal switch point for bringing a second chiller online (with pumps) is about

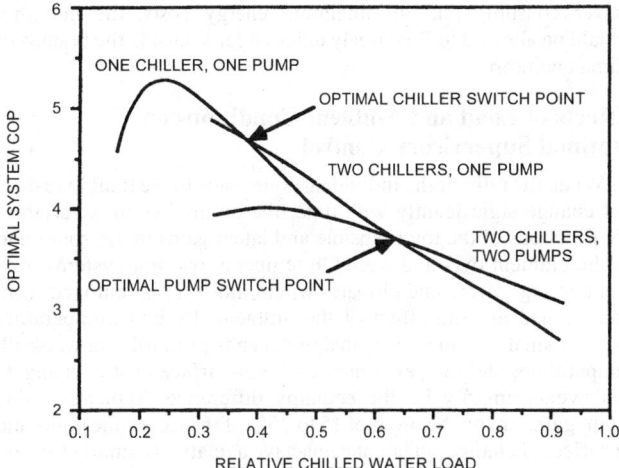

Fig. 7 Effect of Chiller and Pump Sequencing on Optimal Performance

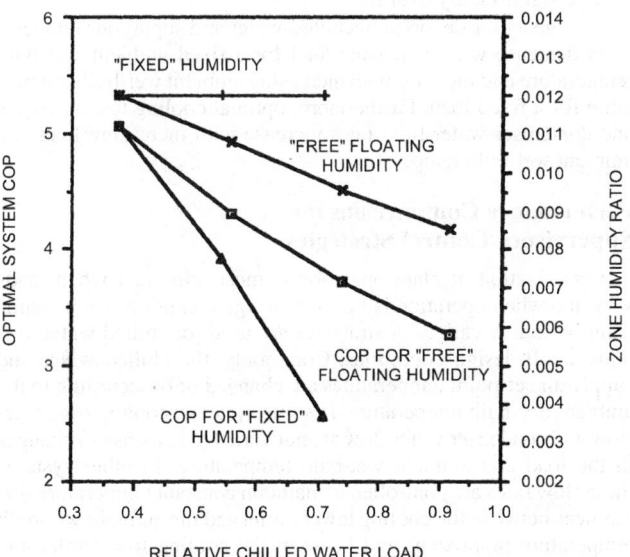

Fig. 8 Comparison of Free Floating and Fixed Humidity

50% of the overall design load or at the design capacity of the individual chiller. This is generally the case for sequencing of chillers with dedicated pumps.

In most cases, zone humidities are allowed to float between upper and lower limits dictated by comfort (see Chapter 8, Thermal Comfort, of the 1997 *ASHRAE Handbook—Fundamentals*). However, variable air volume systems can control the zone humidity and temperature simultaneously. For a zone being cooled, the equipment operating costs are minimized when the zone temperature is at the upper bound of the comfort region. However, operating simultaneously at the upper limit of humidity does not minimize operating costs. Figure 8 shows a comparison of system COP and zone humidity associated with fixed and free-floating zone humidity as a function of the relative load (from Braun et al. 1989a). Over the range of loads, allowing the humidity to float within the comfort zone produces a lower cost and a lower zone humidity than setting the humidity at the highest acceptable value. The largest differences occur at the highest loads. Operation with the zone at the upper humidity bound results in lower latent loads than with a free-floating humidity, but this humidity control constraint requires a higher supply air temperature that, in turn, results in greater air-handler power consumption. For minimum energy costs, the humidity should be allowed to float freely unless it falls outside the bounds of human comfort.

Effects of Load and Ambient Conditions on Optimal Supervisory Control

When the ratio of the individual zone loads to the total load does not change significantly with time, the optimal control variables are functions of the total sensible and latent gains to the zones and of the ambient dry- and wet-bulb temperatures. For systems with wet cooling towers and climates where moisture is removed from conditioned air, the effect of the ambient dry-bulb temperature alone is small because air enthalpy depends primarily on wet-bulb temperature, and the performance of wet-surface heat exchangers is driven primarily by the enthalpy difference. Typically, zone latent gains are on the order of 15 to 25% of the total zone gains and the effect of changes in latent gains have a relatively small effect on performance for a given total load. Consequently, in many cases optimal supervisory control variables depend primarily on ambient wet-bulb temperature and total chilled water load. However, the load distributions between zones may also be important if they change significantly over time.

As a general rule, optimal chilled-water and supply-air temperatures decrease with increasing load for a fixed ambient wet-bulb temperature and increase with increasing ambient wet-bulb temperature for a fixed load. Furthermore, optimal cooling tower airflow and condenser water flow rates increase with increasing load and ambient wet-bulb temperature.

Performance Comparisons for Supervisory Control Strategies

Optimization of plant operation is most important when loads vary and when operation is far from design conditions for a significant period. A variety of strategies are used for chilled water systems at off-design conditions. Commonly, the chilled water and supply air set-point temperatures are changed only according to the ambient dry-bulb temperature. In some systems, cooling tower airflow and condenser water flow are not varied in response to changes in the load and ambient wet-bulb temperature. In other systems these flow rates are controlled to maintain constant temperature differences between the cooling tower outlet and the ambient wet-bulb temperature (approach) and between the cooling tower inlet and outlet (range), regardless of the load and wet-bulb temperature. Although these strategies seem reasonable, they do not generally minimize operating costs.

Figure 9 shows a comparison of the COPs for optimal control and three alternative strategies as a function of load for a fixed ambient wet-bulb temperature. This system (from Braun et al. 1989a) incorporated the use of variable-speed pumps and fans. The three strategies are

1. Fixed chilled water and supply air temperature set points (40 and 52°F, respectively), with optimal condenser loop control
2. Fixed tower approach and range (5 and 12°F, respectively), with optimal chilled water loop control
3. Fixed set points, approach, and range

Since the fixed values were chosen to be optimal at design conditions, the differences in performance for all strategies are minimal at high loads. However, at part-load conditions, Figure 9 shows that the savings associated with the use of optimal control become significant. Optimal control of the chilled water loop results in greater savings than that for the condenser loop for part-load ratios less than about 50%. The overall savings over a cooling season depend on the time variation of the load. If the cooling load is relatively constant and near the design load, fixed values of temperature set points, approach, and range could be chosen to give

near-optimal performance. However, for typical building loads with significant daily and seasonal variations, the penalty for using a fixed set point control strategy is in the range of 5 to 20% of the cooling system energy.

Even greater energy savings are possible with economizer control and discharge air temperature reset with constant volume systems. Kao (1985) investigated the effect of different economizer and supply air reset strategies on both heating and cooling energy use for CAV, VAV, and dual-duct air handling systems for four different buildings. The results were generated using the building energy program BLAST (Hittle 1979) and typical meteorological year (TMY) hourly weather data for six cities in the United States—Lake Charles, LA; Madison, WI; Nashville, TN; Santa Maria, CA; Seattle, WA; and Washington, D.C. For the CAV and dual-duct systems, two different supply air reset strategies were considered: (1) reset based on outdoor air temperature or (2) space demand. In all cases, the air-handlers were operational only during occupied periods in the cooling season, and the conditioned zones used temperature set back at night during the heating season.

Figure 10 and Figure 11 (Kao 1985) show the yearly cooling and heating energy in a large office building. Cases 1 through 6 simulate a CAV, terminal reheat system for the entire building. Adding a dry-bulb economy cycle (Case 2) allows supply air temperature to be maintained by using cooler outside air, consequently saving a large amount of cooling energy but increasing the amount of heating. Changing to an enthalpy economizer (Case 3) results in more cooling savings and greater heating. The increase in heating energy for both economizer cycles results mainly from lowering the supply air temperature by the proportional controls during the period that the economy cycles were in operation. Both the absolute amount and the percentage of cooling savings are greater in low cooling degree-day areas than in high degree-day areas.

Resetting the supply air temperature for the CAV system with reheat benefits both heating and cooling since reheat is not needed if the extra cooling capacity is reduced. Resetting by sensing zone temperature (Case 5) reduces cooling energy approximately twice as much as resetting by using outside air temperature (Case 4). Heating energy reduction is even more dramatic when the reset is performed using the zone air temperature. By adding an enthalpy "economy" cycle to the supply air temperature reset strategy (Case 6), cooling energy is further reduced, while heating energy increases for the reason cited previously.

Cases 7 and 8 use variable air volume (VAV) systems for the entire building. The perimeter zones have reheat coils that operate

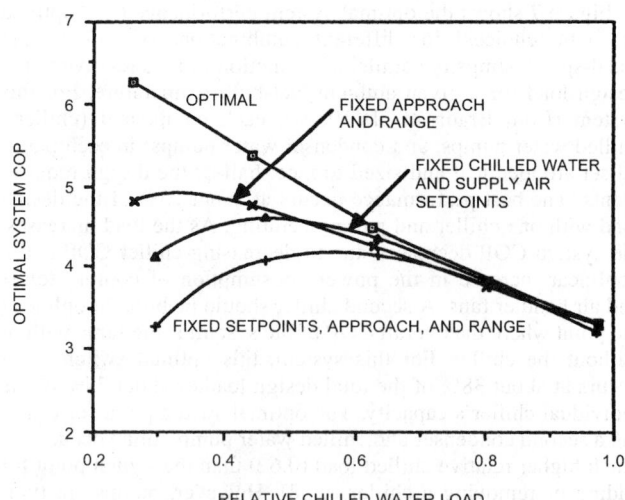

Fig. 9 Comparisons of Optimal Control with Conventional Control Strategies

in sequence with the zone dampers. The dampers allow supply air to be reduced to 20% of design air volume. The interior zones have only damper controls. The base VAV system (Case 7), which has no special strategy, has cooling energy consumption comparable to the best reheat case—Case 6, which uses an enthalpy economy cycle and supply air temperature reset by zone demand. When an enthalpy economy cycle is added to the VAV system, 15% (Lake Charles) to 59% (Seattle) of cooling energy is saved. Contrary to results obtained with the reheat system, adding an enthalpy cycle to the VAV system (Case 8) does not significantly increase the heating energy requirement in most cities. The exceptions were Santa Maria (20% increase) and Lake Charles (7% increase).

Case 9 simulates a base dual-duct system for the perimeter zones and a simple VAV system for interior zones. Case 10 is the result of adding an enthalpy economy cycle to the entire building, with both the hot and the cold air of the dual-duct system reset by sensing the space temperature. It is difficult to pinpoint the effects of the individual strategies and systems for these two cases. The overall results show that much cooling and heating energy is saved as compared to the base VAV and dual-duct systems (Case 9). The perimeter zones are also simulated with a four-pipe fan-coil system having both the cold and hot water available year-round (Case 11). A fixed amount of outside air is introduced directly to the fan-coil units during operating hours. The interior system remains a VAV system with an economy cycle. The energy consumption for both heating and cooling is similar to that in Case 8.

Kao (1985) also analyzed the cooling and heating energy consumption of the combined classroom and office areas of a school building. Reheat, dual-duct, variable air volume, and cooling-type unit ventilator systems were simulated for these areas. Roughly the same relative pattern as for the large office building was observed for the cooling consumption of the first three systems. The cooling results of the various VAV system strategies were close together, the best being the one applying the enthalpy economy cycle and resetting the supply air temperature by zone sensing. The unit ventilator system performed slightly poorer on cooling than do the VAV systems. The school building was unlike the large office building discussed previously, in that applying economy cycles to the reheat systems of the school building did not significantly increase the heating energy. For the retail store, the best strategies and systems were the VAV system with supply air temperature reset by the outside air temperature and the packaged DX system, both with enthalpy economy cycles. A VAV system with enthalpy economy cycle and supply air temperature reset by zone demand sensing was the best choice for the small office building.

Table 1 (Kao 1985) illustrates the energy consumption ratios of some selected cases for Washington, D.C. (1415 cooling degree-days and 4211 heating degree-days). The ratios are defined as the building energy consumption obtained using a particular control strategy, divided by the energy consumption of the reheat base case (Case 1 in Figure 10 and Figure 11). These data indicate that substantial improvements in a building's energy use may be obtained

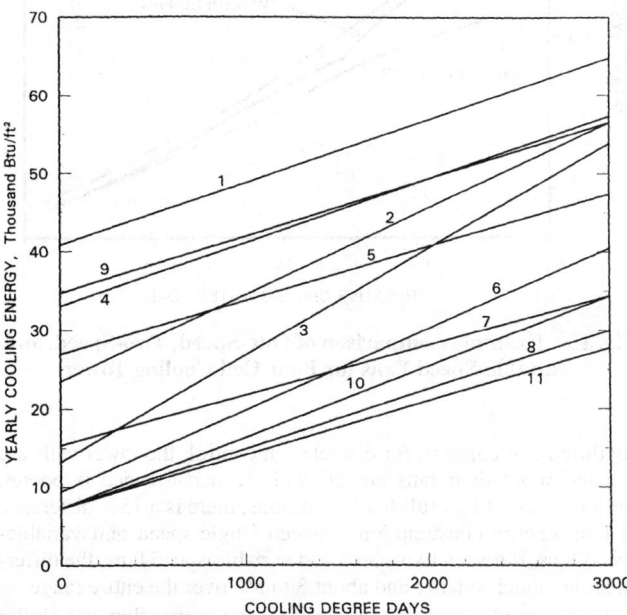

1 Reheat, base case
2 Reheat, dry-bulb economizer
3 Reheat, enthalpy economizer
4 Reheat, supply air reset based on outside air
5 Reheat, supply air reset based on zone
6 Reheat, supply air reset based on zone, enthalpy economizer
7 VAV entire building
8 VAV entire building, enthalpy economizer
9 Int. zones: VAV
   Ext. zones: Dual duct
10 Int. zones: VAV, enthalpy economizer
   Ext. zones: Dual duct, enthalpy economizer, zone reset
11 Int. zones: VAV, enthalpy economizer
   Ext. zones: Dual duct, enthalpy economizer, zone reset, 4-pipe
   zone fan coils

**Fig. 10 Cooling Energy Consumption of
Large Office Building**

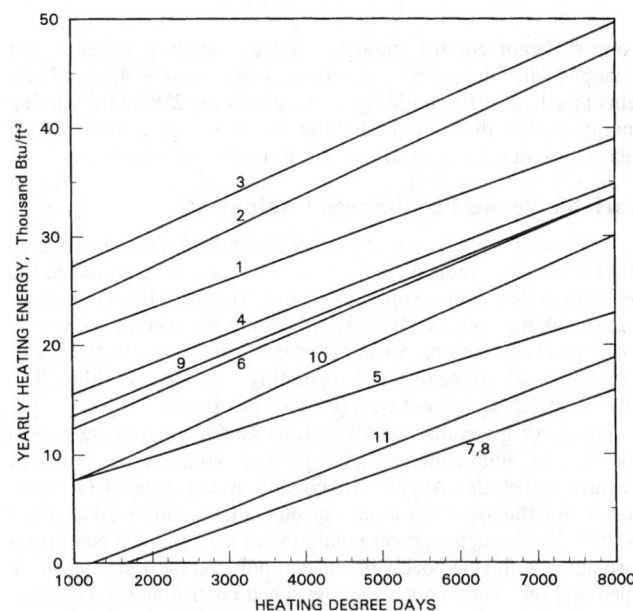

1 Reheat, base case
2 Reheat, dry-bulb economizer
3 Reheat, enthalpy economizer
4 Reheat, supply air reset based on outside air
5 Reheat, supply air reset based on zone
6 Reheat, supply air reset based on zone, enthalpy economizer
7 VAV entire building
8 VAV entire building, enthalpy economizer
9 Int. zones: VAV
   Ext. zones: Dual duct
10 Int. zones: VAV, enthalpy economizer
   Ext. zones: Dual duct, enthalpy economizer, zone reset
11 Int. zones: VAV, enthalpy economizer
   Ext. zones: Dual duct, enthalpy economizer, zone reset, 4-pipe
   zone fan coils

**Fig. 11 Heating Energy Consumption of
Large Office Building**

Table 1 Comparisons of Building Energy Use in Washington, D.C.

Strategies	Small Office Cool	Small Office Heat	Large Office Cool	Large Office Heat	School[a] Cool	School[a] Heat	Retail Store Cool	Retail Store Heat
	Ratios Relative to Base Reheat Cases							
Reheat with enthalpy economy	0.57	1.08	0.58	1.24	0.61	1.03	—	—
Reheat with enthalpy economy and zone reset	—	—	0.43	0.76	0.43	0.73	—	—
Reheat with enthalpy economy and OA reset	—	—	—	—	—	—	0.54	0.67
VAV with enthalpy economy	0.36	0.26	0.25	0.31	0.37	0.45	—	—
VAV with enthalpy economy and OA reset	0.33	0.20	—	—	—	—	0.43	0.17
VAV with enthalpy economy and zone reset	0.33	0.19	—	—	0.33	0.43	—	—
Fan-coil for perimeter	—	—	0.32	0.29	—	—	—	—
Unit ventilator	—	—	—	—	0.42	0.10	—	—

[a]Classroom, office, and library only.

using different control strategies and air-handling systems. For example, when the enthalpy economy cycle is used on the VAV system of the large office building, it consumes only 25% of the cooling energy and less than one-third of the heating energy consumed by a reheat system using a fixed amount of outside air year-round.

Variable Versus Fixed-Speed Equipment

The use of variable-speed motors for chillers, fans, and pumps can significantly reduce energy costs but can also complicate the problem of determining optimal control. The overall savings associated with the use of variable-speed equipment over a cooling season depends on the time variation of the load. Typically, the use of variable-speed drives results in operating costs that are 20 to 50% lower than for equipment with fixed-speed drives.

Figure 12 gives the overall optimal system performance for a cooling plant with either variable speed or variable vane control of a centrifugal chiller. At part-load conditions, the system COP associated with the use of a variable-speed chiller is improved as much as 25%. However, the power requirements are similar at conditions associated with peak loads, because at full load the vanes are wide open and the speed under variable-speed control and fixed-speed operation is the same. The results of Figure 12 are from a single case study of a large chilled water facility (Braun et al. 1989a) that was constructed in the mid 1970s and where the existing chiller was retrofitted with a variable-speed drive. Differences in performance between variable- and fixed-speed chillers may be smaller for current equipment.

The most common design for cooling towers places multiple tower cells in parallel with a common sump. Each tower cell has a fan with one, two, or possibly three operating speeds. Although multiple cells with multiple fan settings offer wide flexibility in control, the use of variable-speed tower fans can provide additional improvements in overall system performance. Figure 13 shows an example comparison of optimal performance for single-speed, two-speed, and variable-speed tower fans as a function of load for a given wet-bulb temperature for a system with four cells (Braun et al. 1989a). The variable-speed option results in higher COP under all

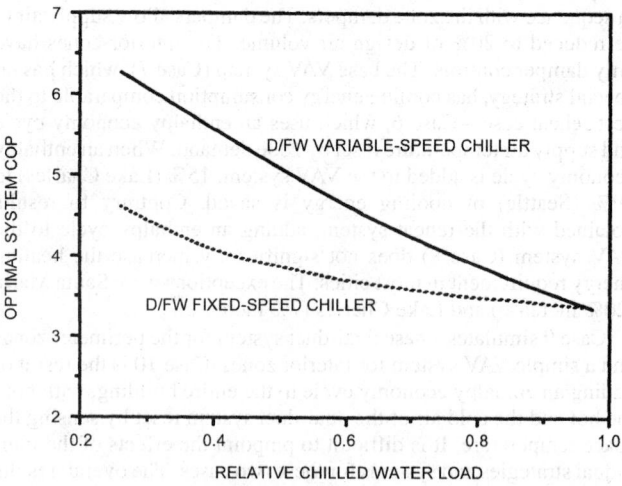

Fig. 12 Example of Optimal Performance for Variable and Fixed-Speed Chillers

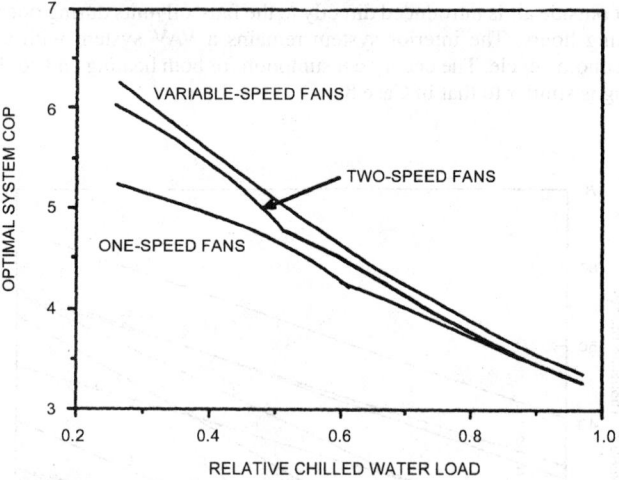

Fig. 13 Example Comparison of One-Speed, Two-Speed, and Variable-Speed Fans for Four Cell Cooling Tower

conditions. In contrast, for discrete fan control, the tower cells are isolated when their fans are off and the performance is poorer. Below about 70% of full-load conditions, there is a 15% difference in total energy consumption between single-speed and variable-speed fans. Between two-speed and variable-speed fans, the differences are much smaller and about 3 to 5% over the entire range.

Fixed-speed pumps that are sized to give proper flow to a chiller at design conditions are oversized for part-load conditions. As a result, the system will have higher operating costs than with a variable-speed pump of the same design capacity. Multiple pumps with different capacities have increased flexibility in control and the use of a smaller fixed-speed pump for low loads can reduce overall power consumption. The optimal performance for variable-speed and fixed-speed pumps applied to both the condenser and chilled water flow loops is shown in Figure 14 (Braun et al. 1989a). Large fixed-speed pumps were sized for design conditions, while the small pumps were sized to have one-half the flow capacity of the large pumps. Below about 60% of full-load conditions, a variable-speed pump showed a significant improvement over the use of a single, large fixed-speed pump. With the addition of a small fixed-speed pump, the improvements with the variable-speed pump were significant at about 40% of the maximum load.

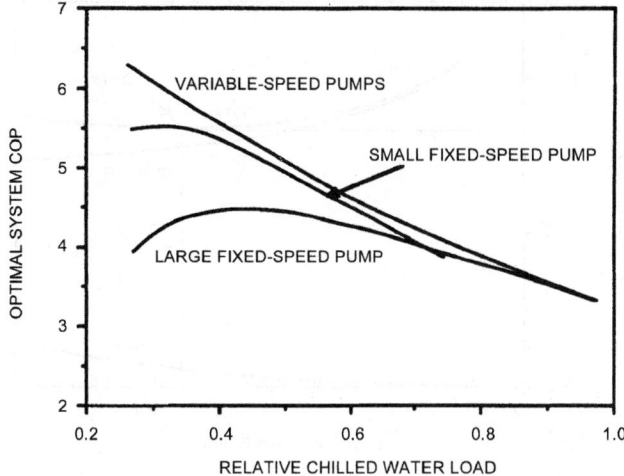

**Fig. 14 Example Comparison of Variable and
Fixed-Speed Pumps**

The fan energy consumed by VAV systems is strongly influenced by the device used to vary the airflow. Brothers and Warren (1986) compared the fan energy consumption for three typical flow modulation devices: (1) dampers on the outlet side of the fan, (2) inlet vanes on the fan, and (3) variable-speed control of the fan motor. The investigation used the DOE 2.1 program to simulate the annual energy consumption of a 10,000 ft² commercial building with a VAV system that used these three flow-control methods. Both centrifugal and vaneaxial fans were considered, even though dampers are not usually used with vaneaxial fans as they may overload the fan motor. The VAV system used an enthalpy economy cycle and provided cooling only during working hours, 7 A.M. to 5 P.M. five days a week. The analysis was performed for five cities: Fresno, CA; Forth Worth, TX; Miami, FL; Phoenix, AZ; and Washington, D.C.

Values for the annual energy consumption are presented in Table 2. In all locations, the centrifugal fan used less energy than the vaneaxial fan. Vaneaxial fans have higher efficiencies at the full-load design point, but centrifugal fans have better off-design characteristics that lead to lower annual energy consumption. For a centrifugal fan, inlet vane control saves about 20% of the energy (over the five locations) compared to damper control. Variable speed control savings, compared to inlet vane control, range from 42% for Miami to 65% for Washington, D.C., with an average savings of 57% over the five locations.

**Table 2 Annual Energy Use with Different
Flow Throttling Techniques**

Type of Fan	Flow Control	Fan Energy Use, MWh				
		Fresno	Fort Worth	Miami	Phoenix	Washington, D.C.
Backward-inclined centrifugal	System damper	12.6	12.9	13.9	17.9	11.5
	Inlet vane	10.2	10.4	10.1	14.4	9.5
	Variable speed	4.1	4.5	5.9	5.8	3.3
Vaneaxial	System damper[a]	16.0	16.2	15.5	22.8	15.0
	Inlet vane	10.8	11.0	10.9	15.4	10.0
	Variable speed	6.8	7.0	7.7	9.6	5.9

[a]Not normally used with vaneaxial fans due to possible overload of fan motor.

SUPERVISORY CONTROL STRATEGIES AND TOOLS

COOLING TOWER FAN CONTROL

Figure 15 shows a schematic of the condenser loop for a typical chilled-water unit consisting of centrifugal chillers, cooling towers, and condenser water pumps. Typically, the condenser water pump control is dedicated to the chiller control in order to provide relatively constant flow for individual chillers. However, the cooling tower cells may be independently controlled so as to maximize the system efficiency.

Typically, cooling tower fans are controlled using a feedback controller that attempts to maintain a temperature set point for the water supplied to the chiller condensers. Often, the condenser water supply temperature set point is held constant. However, a better strategy is to maintain a constant temperature difference between the condenser water supply and the ambient wet bulb (constant approach). Additional savings are possible through optimal control.

With a single feedback controller, the controller output signal must be converted to a specific fan sequence that depends on the number of operating cells and the individual fan speeds. Typically, with the discrete control associated with one- or two-speed tower fans, the set point cannot be realized, resulting in the potential for oscillating tower fan control. Fan cycling can be reduced through the use of deadbands, "sluggish" control parameters, and/or lower limits for on and off periods.

Braun and Diderrich (1990) demonstrated that feedback control for cooling tower fans could be eliminated through the use of an open-loop supervisory control strategy. This strategy requires only the measurement of the chiller loading to specify the control and is inherently stable. The tower fan control is separated into two parts: tower sequencing and optimal airflow. For a given total tower airflow, general rules for optimal tower sequencing are used to specify the number of operating cells and fan speeds that give the minimum power consumption for both the chillers and tower fans. The optimal tower airflow is estimated with an open-loop control equation that uses design information for the cooling tower and chiller. This computational procedure is presented in this section and the control strategy is summarized in a set of steps and sample calculations.

Near-Optimal Tower Fan Sequencing

For variable-speed fans, the minimum power consumption results when all cooling tower cells are operated under all

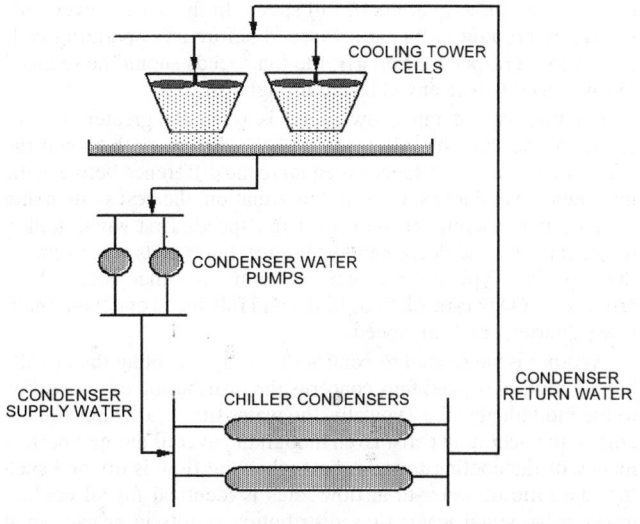

Fig. 15 Condenser Water-Loop Schematic

conditions. Tower airflow varies almost linearly with fan speed, while the fan power consumption varies approximately with the cube of the speed. Thus, for the same total airflow, operating more cells in parallel allows for lower individual fan speeds and lower overall fan power consumption. An additional benefit associated with full-cell operation is lower water pressure drops across the spray nozzles, which results in lower pumping power requirements. However, at very low pressure drops, inadequate spray distribution may adversely affect the thermal performance of the cooling tower.

Most cooling towers use multiple-speed, rather than continuously adjustable variable-speed fans. In this case, it is not optimal to operate all tower cells under all conditions. The optimal number of cells operating and individual fan speeds depend on the system characteristics and ambient conditions. However, simple relationships exist for the best sequencing of cooling tower fans as capacity is added or removed. When additional tower capacity is required, Braun et al. (1989a) showed that in almost all practical cases, the speed of the tower fan operating at the lowest speed (including fans that are off) should be increased first. The rules for bringing cell fans on line are as follows:

Sequencing Rules
- **All Variable-Speed Fans:** Operate all cells with fans at equal speeds.
- **Multi-Speed Fans:** Increment lowest speed fans first when adding tower capacity. Reverse for removing capacity.
- **Variable/Multi-Speed Fans:** Operate all cells with variable-speed fans at equal speeds. Increment lowest speed fans first when adding tower capacity with multi-speed fans. Add multi-speed fan capacity when variable-speed fan speeds match the fan speed associated with the next multi-speed fan increment to be added.

Similarly, for removing tower capacity, the highest fan speeds are the first to be reduced and sequences defined above are reversed.

These guidelines were derived from evaluating the incremental power changes associated with fan sequencing. For two-speed fans, the incremental power increase associated with adding a low-speed fan is less than that for increasing one to high speed if the low speed is less than 79% of the high fan speed. In addition, if the low speed is greater than 50% of the high speed, then the incremental increase in airflow is greater (and therefore better thermal performance) for adding the low-speed fan. Most commonly, the low speed of a two-speed cooling tower fan is between one-half and three-quarters of full speed. In this case, tower cells should be brought online at low speed before any operating cells are set to high speed. Similarly, the fan speeds should be reduced to low speed before any cells are brought offline.

For three-speed fans, low speed is typically greater than or equal to one-third of full speed and the difference between the high and intermediate speeds is equal to the difference between the intermediate and low speeds. In this situation, the best sequencing strategy is to increment the lowest fan speeds first when adding tower capacity and decrement the highest fan speeds when removing capacity. Typical three-speed combinations that satisfy these criteria are (1) one-third, two-thirds, and full speed or (2) one-half, three-quarters, and full speed.

Another issue related to control of multiple cooling tower cells having multiple-speed fans concerns the distribution of water flow to the individual cells. Typically, the water flow is divided equally among the operating cells. Even though the overall thermal performance of the cooling tower is best when the flow is divided such that the ratio of water-to-airflow rates is identical for all cooling tower cells, equal water flow distribution results in near-optimal performance.

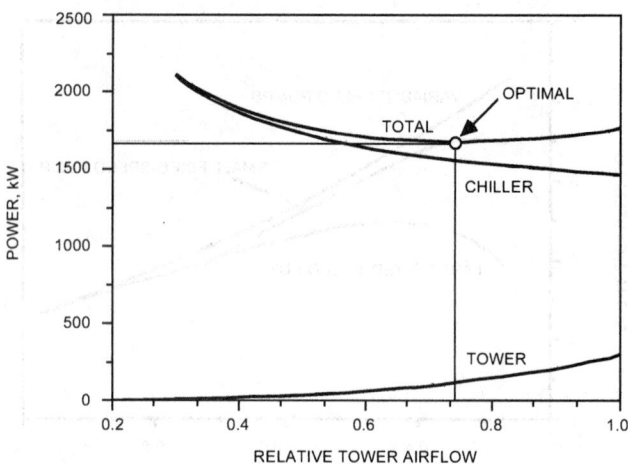

Fig. 16 Trade-offs Between Chiller Power and Fan Power with Tower Airflow

Near-Optimal Tower Airflow

Figure 16 illustrates the trade-off between the chiller and cooling tower fan power associated with increasing tower airflow for variable-speed fans. As the airflow increases, the fan power increases with a cubic relationship. At the same time, there is a reduction in the temperature of the water supplied to the condenser of the chiller resulting in lower chiller power consumption. The minimum total power occurs at a point where the rate of increase in fan power with airflow is equal to the rate of decrease in chiller power. Near the optimum, the total power consumption is not very sensitive to the control. This "flat" optimum indicates extreme accuracy is not needed to determine the optimum control. In general, it is better to have too high rather than too low a fan speed.

Braun et al. (1989a) showed that the tower control that minimizes the instantaneous power consumption of a cooling plant varies as a near linear function of the load over a wide range of conditions. Although the optimal control depends upon the ambient wet-bulb temperature, this dependence is small compared to the load effect. Figure 17 shows an example of how the optimal tower control varies for a specific plant. The tower airflow as a fraction of the design capacity is plotted as a function of load relative to design load for two different wet-bulb temperatures. For a 20°F change in wet-bulb temperature, the optimal control varies only about 5% of the tower capacity. This difference in control results in less than a 1% difference in the plant power consumption. Figure 17 also shows that linear functions work well in correlating the optimal control over a wide range of loads for the two wet-bulb temperatures. Given the insensitivity to wet-bulb temperature and the fact that the load is highly correlated with wet bulb, then a single linear relationship is adequate in correlating the optimal tower control in terms of load.

Figure 18 depicts the general form to determine tower airflow as a function of load. The (unconstrained) relative tower airflow is computed as a linear function of the part-load ratio as

$$G_{twr} = 1 - \beta_{twr}(PLR_{twr,cap} - PLR) \text{ for } 1.0 < PLR < 0.25 \quad (1)$$

where

G_{twr} = tower airflow divided by maximum airflow with all cells operating at high speed

PLR = chilled water load divided by design total chiller plant cooling capacity (part-load ratio)

$PLR_{twr,cap}$ = part-load ratio (value of PLR) at which tower operates at its capacity ($G_{twr} = 1$)

β_{twr} = slope of relative tower airflow (G_{twr}) versus part-load ratio (PLR) function

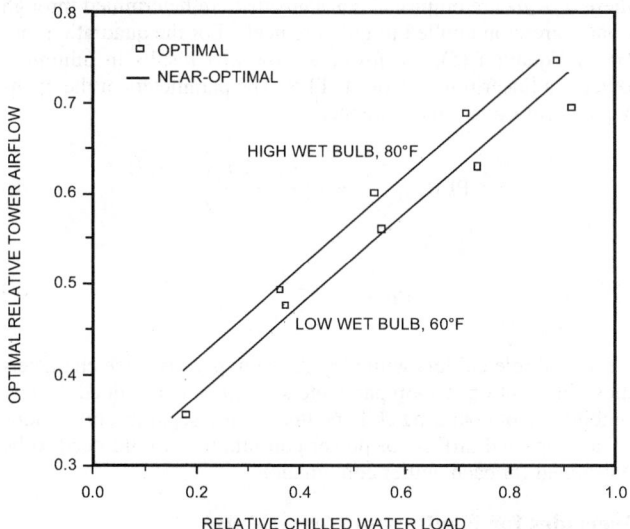

Fig. 17 Example of Optimal Tower Fan Control

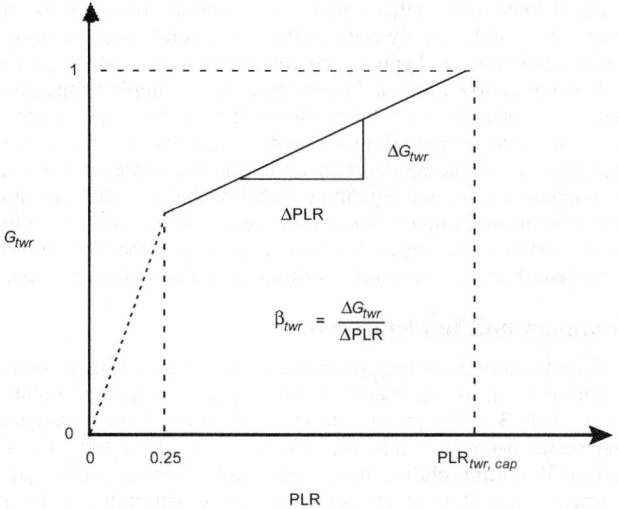

Fig. 18 Fractional Tower Airflow versus Part-Load Ratio

The linear relationship between airflow and load is only valid for loads greater than about 25% of the design load. For many installations, chillers do not operate at these small loads. However, for those situations for which chiller operation is necessary below 25% of full load, the tower airflow should be ramped to zero as the load goes to zero according to

$$G_{twr} = 4\text{PLR}[1 - \beta_{twr}(\text{PLR}_{twr,cap} - 0.25)] \text{ for PLR} < 0.25 \quad (2)$$

The results of either Equation (1) or (2) must be constrained between 0 and 1. This fraction of tower capacity is then converted to a tower control using the sequencing rules of the previous section.

The variables of the open-loop linear control Equation (1) that yield near-optimal control depend on the characteristics of system. Detailed measurements may be taken over a range of conditions and this information can be used to accurately estimate these variables. However, this requires measuring component power consumption along with considerable time and expertise and may not be cost effective, unless performed by on-site plant personnel. Alternatively, simple estimates of these parameters may be obtained using design data.

Table 3 Parameter Estimates for Near-Optimal Tower Control Equation

Parameter	One-Speed Fans	Two-Speed Fans	Variable-Speed Fans
$\text{PLR}_{twr,cap}$	PLO_0	$\sqrt{2}\text{PLO}_0$	$\sqrt{3}\text{PLO}_0$
β_{twr}	$\dfrac{1}{\text{PLR}_{twr,cap}}$	$\dfrac{2}{3\text{PLR}_{twr,cap}}$	$\dfrac{1}{2\text{PLR}_{twr,cap}}$

$$\text{PLR}_0 = \frac{1}{\sqrt{\dfrac{P_{ch,des}}{P_{twr,des}}}S_{cwr,des}(a_{twr,des} + r_{twr,des})}$$

Open-Loop Parameter Estimates Using Design Data. Good estimates of the parameters of Equation (1) may be determined analytically using design information as summarized in Table 3. These estimates were derived by Braun and Diderrich (1990) by applying optimization theory to a simplified mathematical model of the chiller and cooling tower, assuming that the tower fans are sequenced in an near-optimal manner. In general, these estimates are conservative in that they should provide greater rather than less than the optimal tower airflow. The results given in Table 3 for variable-speed fans should also provide adequate estimates for three-speed fans.

The design factors that affect the parameter estimates given in Table 3 are (1) the ratio of chiller power to cooling tower fan power at design conditions $P_{ch,des}/P_{twr,des}$, (2) the sensitivity of the chiller power to changes in condenser water return temperature at design conditions $S_{cwr,des}$, and (3) the sum of the tower approach and range at design conditions $(a_{twr,des} + r_{twr,des})$. The chiller power consumption at design conditions is the total power consumption of all plant chillers operating at their design cooling capacity. Likewise, the design tower fan power is the total power associated with all tower cells operating at high speed. As the ratio of chiller power to tower fan power increases, it becomes more beneficial to operate the tower at higher airflows. This would be reflected in a decrease in the part-load ratio at which the tower reaches its capacity, $\text{PLR}_{twr,cap}$. If the tower airflow were free (i.e., zero fan power), then $\text{PLR}_{twr,cap}$ goes to zero, and the best strategy would be to operate the towers at full capacity independent of the load. A typical value for the ratio of the chiller power to the cooling tower fan power at design conditions is 10.

The chiller sensitivity factor $S_{cwr,des}$ is the incremental increase in chiller power for each degree increase in condenser water temperature as a fraction of the power or

$$S_{cwr,des} = \frac{\text{Change in chiller power}}{\text{Change in cond. water return temp.} \times \text{Chiller power}} \quad (3)$$

If the chiller power increases by 2% for a 1°F increase in condenser water temperature, $S_{cwr,des}$ is equal to 0.02°F^{-1}. A large sensitivity factor means that the chiller power is very sensitive to the cooling tower control favoring operation at higher airflow rates (low $\text{PLR}_{twr,cap}$). The sensitivity factor should be evaluated at design conditions using chiller performance data. Typically, the sensitivity factor is between 0.01°F^{-1} and 0.03°F^{-1}. For multiple chillers with different performance characteristics, the sensitivity factor at design conditions is estimated as

$$S_{cwr,des} = \frac{\displaystyle\sum_{i=1}^{N_{ch}} S_{cwr,des,i}P_{ch,des,i}}{\displaystyle\sum_{i=1}^{N_{ch}} P_{ch,des,i}} \quad (4)$$

where $S_{cwr,des,i}$ is the sensitivity factor and $P_{ch,des,i}$ is the power consumption for the ith chiller at the design conditions and N_{ch} is the total number of chillers.

The design approach to wet bulb $a_{twr,des}$ is the temperature difference between the condenser water supply and the ambient wet bulb for the tower operating at its air and water flow capacity at the plant design conditions. The design range $r_{twr,des}$ is the water temperature difference across the tower at these same conditions (condenser water return minus supply temperature). The sum of $a_{twr,des}$ and $r_{twr,des}$ is the temperature difference between the tower inlet and the ambient wet bulb and represents a measure of the tower's capability to reject heat to ambient relative to the system requirements. A small temperature difference (tower approach plus range) results from a high tower heat transfer effectiveness or high water flow rate and yields lower condenser water temperatures with lower chiller power consumption. Typical values for the design approach and range are 7°F and 10°F.

The part-load ratio associated with the tower operating at full capacity $PLR_{twr,cap}$ may be greater than or less than one. Values less than unity imply that from an "energy point of view" the tower is not sized for optimal operation at design load conditions and that the tower should operate at its capacity for a range of loads less than the design load. Values of $PLR_{twr,cap}$ greater than one imply that the tower is oversized for the design load and that the tower should never operate at its capacity.

For multiple chillers with very different performance characteristics, different open-loop parameters may be used for any combination of operating chillers. The sensitivity factors and chiller design power used to determine the open-loop control parameters in Table 3 should be estimated for each combination of operating chillers and the part-load ratio used in Equation (1) should be determined using the design capacity for the operating chillers (not all chillers). In this case, N_{ch} in Equation (4) represents the number of operating chillers.

Open-Loop Parameter Estimates Using Plant Measurements. Energy consumption can be reduced slightly by determining the open-loop control parameters from plant measurements. However, this results in additional complexity associated with implementation. One method for estimating the open-loop control parameters of Equation (1) from plant measurements involves performing a set of one-time trial-and-error experiments. At a given set of conditions (i.e., cooling load and ambient conditions), the optimal tower control is estimated by varying the fan settings and monitoring the total chiller and fan power consumption. Each tower control setting and load condition must be maintained for a sufficient time for the power consumption to approach steady-state and to hold the chilled-water supply temperature constant. The control setting that produces the minimum total power consumption is deemed optimal. This set of experiments is performed for a number of chilled-water cooling loads and the best fit straight line through the resulting data points is used to estimate the parameters of Equation (1). As initial control settings for each load, Equation (1) may be used with estimates from design data as summarized in the previous section.

Another method for estimating the variables of Equation (1) uses an empirical model for total power consumption that is fit to plant measurements. The control that minimizes the power consumption associated with the model is then determined analytically. The section on Control Optimization Methods describes a general method for determining linear control relations in this manner using a quadratic model. For cooling tower fan control, chiller and fan power consumption are correlated with load and tower airflow for a constant chilled water supply temperature using a quadratic function according to:

$$P = a_0 + a_1 PLR + a_2 PLR^2 + a_3 G_{twr}$$
$$+ a_4 G_{twr}^2 + a_5 PLR \cdot G_{twr} \tag{5}$$

where $a_0 - a_5$ are empirical constants that are determined through linear regression applied to measurements. For the quadratic function of Equation (5), the tower airflow that results in minimum power is a linear function of the PLR. The parameters of the open-loop control Equation (1) are then

$$PLR_{twr,cap} = -\left(\frac{a_3 + 2a_4}{a_5}\right) \tag{6}$$

$$\beta_{twr} = -\left(\frac{a_5}{2a_4}\right) \tag{7}$$

For multiple chillers with very different performance characteristics, different open-loop parameters could be determined for any combination of operating chillers. In this case, separate correlations for near-optimal airflow or power consumption would need to be determined for each chiller combination.

Overrides for Equipment Constraints

The fractional tower airflow as determined by Equation (1) or Equation (2) must be bounded between 0 and 1 according to the physical constraints of the equipment. Additional constraints on the temperature of the supply water to the chiller condensers are necessary to avoid potential chiller maintenance problems. Many (older) chillers have a low limit on the condenser-water supply temperature that is necessary to avoid lubrication migration from the compressor. A high-temperature limit is also necessary to avoid excessively high pressures in the condenser that can lead to compressor surge. If the condenser water temperature falls below the low limit, then it is necessary to override the open-loop tower control and reduce the tower airflow to go above this limit. Similarly, if the high limit is exceeded, then the tower airflow should be increased as required.

Summary and Implementation

Prior to commissioning, the parameters of the open-loop control Equation (1) must be specified. These parameters are estimated using Table 3. After the system is in operation, these parameters may be fined-tuned with measurements as outlined in the previous section. If multiple chillers have significantly different performance characteristics, it may be advantageous to determine different parameters for Equation (1) depending on the combination of operating chillers.

The relative tower airflow must be converted to a specific set of tower fan settings using the sequencing rules defined previously. This involves defining a relationship (i.e., table) for fan settings as a function of tower airflow. The table is constructed by defining the best fan settings for each possible increment of airflow. The conversion process between the continuous output of Equations (1) or (2) and the fan control involves choosing the set of discrete fan settings from the table that produces a tower airflow closest to the desired flow. However, in general, it is better to have greater rather than less than the optimal airflow. A good general rule is to choose the set of discrete fan controls that results in a relative airflow that is closest to, but not more than 10% less than the output of Equations (1) or (2).

With the parameters of Equation (1) specified, the following procedure is applied at each decision interval (e.g, 15 min or less) in order to determine the tower control.

1. If the temperature of the supply water to the chiller condenser is less than the low limit, then reduce the tower airflow by one increment according to the near-optimal sequencing rules and exit the algorithm. Otherwise go to Step 2.
2. If the temperature of the supply water to the chiller condenser is greater than the high limit, then increase the tower airflow by one

increment according to the near-optimal sequencing rules and exit the algorithm. Otherwise go to Step 3.

3. Determine the chilled water load relative to the design load.
4. If the chilled water load has changed by a significant amount (e.g., 10%) since the last control change, then go to Step 5. Otherwise exit the algorithm.
5. If the part-load ratio is greater than 0.25, then compute the near-optimal tower airflow as a fraction of the tower capacity, G_{twr}, with Equation (1). Otherwise, determine G_{twr} with Equation (2).
6. Limit G_{twr} to keep the change from the previous decision interval less than a minimum value (e.g., less than 0.1 change).
7. Restrict the value of G_{twr} between 0 and 1.
8. Convert the value of G_{twr} to a specific set of control functions for each of the tower cell fans according to the near-optimal sequencing rules.

Implementation of this procedure requires some estimate of the chilled water load, along with a measurement of the condenser water supply temperature. However, the accuracy of the load estimates is not extremely critical. In general, near-optimal control determined with load estimates that are accurate to within 5 to 10% results in total power consumption that is within 1% of the minimum. The best method for determining the chilled-water load would be from the product of the measured chilled-water flow rate and the temperature difference between the chilled-water return and supply. For systems that use constant flow pumping to the chillers, the flow rates may be estimated from design data for the pumps and system pressure drop characteristics.

Example 1. Consider an example plant consisting of four 550 ton chillers with four cooling tower cells, each having two-speed fans. Each chiller consumes approximately 330 kW at the design capacity, while each tower fan uses 40 kW at high speed. At design conditions, the chiller power increases approximately 6.6 kW for a 1°F increase in condenser water temperature, giving a sensitivity factor of 6.6/330 or 0.02°F^{-1}. The tower design approach and range from manufacturer's data are 7°F and 10°F.

Solution

The first step in applying the open-loop control algorithm to this problem is the determination of the parameters of Equation (1) from the design data. From Table 3, the part-load ratio at which operation of the tower is at its capacity is estimated for the two-speed fans as

$$\mathrm{PLR}_{twr,cap} = \frac{1}{\sqrt{\dfrac{1}{2}(0.02°F^{-1})\dfrac{4 \times 330\ kW}{4 \times 40\ kW}(7°F + 10°F)}} = 0.84$$

while the slope of the fractional airflow versus part-load ratio is estimated to be

$$\beta_{twr} = \frac{2}{3 \times 0.84} = 0.79$$

Given these parameters and the part-load ratio, the fractional tower airflow is estimated as

IF (PLR > 0.25) THEN

$\quad G_{twr} = 1 - \beta_{twr}(\mathrm{PLR}_{twr,cap} - \mathrm{PLR})$

ELSE

$\quad G_{twr} = 4\mathrm{PLR}[1 - \beta_{twr}(\mathrm{PLR}_{twr,cap} - 0.25)]$

$\quad G_{twr} = \mathrm{MIN}[1, \mathrm{MAX}(0, G_{twr})]$

In order to convert G_{twr} into a specific tower control, the tower sequencing must be defined. The following table gives this information in a form that specifies the relationship between G_{twr} and tower control for this example.

For a specific chilled water load, the fan control should be the sequence of tower fan settings from the table that results in a value of G_{twr} that is closest to, but not more than 10% less than the output of Equations (1) or (2). Note that this example assumes that proper water flow can be maintained over all cooling tower cells.

Cooling Tower Fan Sequencing for Example 1

Sequence No.	G_{twr}	Tower Fan Speeds			
		Cell #1	Cell #2	Cell #3	Cell #4
1	0.125	Low	Off	Off	Off
2	0.250	Low	Low	Off	Off
3	0.375	Low	Low	Low	Off
4	0.500	Low	Low	Low	Low
5	0.625	High	Low	Low	Low
6	0.750	High	High	Low	Low
7	0.875	High	High	High	Low
8	1.000	High	High	High	High

CHILLED WATER RESET WITH FIXED-SPEED PUMPING

Figure 19 shows a common configuration using fixed-speed chilled water pumps with two-way valves at the cooling coils. A two-way bypass valve controlled to maintain a fixed pressure difference between the main supply and return lines is used in order to ensure relatively constant flow through chiller evaporators and reduce pressure drop and pumping costs at low loads. However, additional pump and chiller power savings can be realized by adjusting the chilled water supply temperature in order to keep some cooling coil valves open and thereby minimize the bypass flow.

Ideally, the chilled water temperature should be adjusted in order to maintain all discharge air temperatures with a minimal number of cooling coil control valves in a saturated (fully-open) condition. The procedure described in this section is designed to accomplish this goal in a reliable and stable manner that reacts quickly to changing conditions.

Pump Sequencing

Individual chilled water pumps are commonly physically dedicated to individual chillers. In this case, the sequencing of chilled water pumps is defined by the sequencing of chillers. In some installations the chiller pumps are not dedicated to chillers. Instead they are arranged in parallel, sharing common headers. In this case, the order for bringing pumps online and offline and the conditions for adding or removing chilled water pump capacity must be specified. For pumps of different capacities, the logical order for bringing pumps online is from small to large. For pumps of similar capacity, the most efficient pumps should be brought online first and taken offline last.

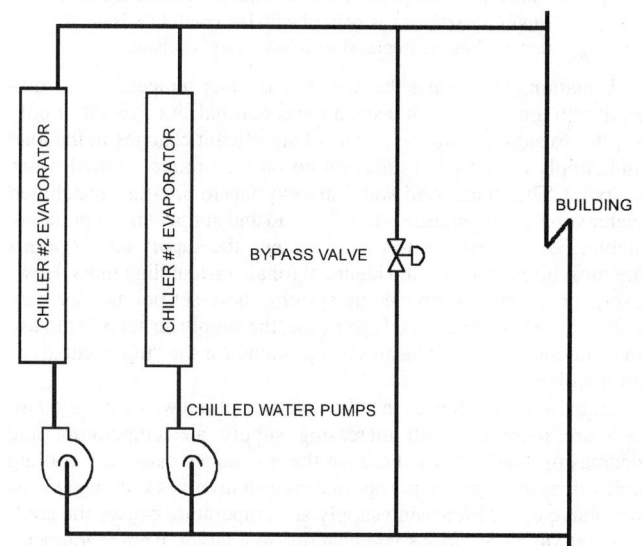

Fig. 19 Typical Chilled Water Distribution for Fixed-Speed Pumping

Optimal Chilled Water Temperature

One method for determining the optimal chilled water temperature is to monitor the water control valve positions of "representative" air handlers and to adjust the set temperature incrementally at fixed decision intervals until a single control valve is fully open. The "representative" air handlers should be chosen to include load diversity at all times and ensure reliable data. One difficulty of the this control approach is that the valve position data is often unreliable. The valve could be stuck open or the saturation indicator could be faulty. This problem can be overcome by also monitoring the discharge air temperatures. The discharge air temperatures allow for a consistency check on the valve position data. If a valve is unsaturated, this implies that the coil has sufficient capacity to maintain the discharge air temperature near the set point. Conversely, if a valve remains saturated at 100% open, the discharge air temperature should ultimately increase above the set point. These considerations lead to the following simple rules for increasing or decreasing the chilled water set point in response to valve position and discharge air temperature data.

1. If all water valves are unsaturated *or* the discharge air temperatures associated with all valves that are saturated are lower than the set point, increase the chilled water set temperature.
2. If more than one valve is saturated at 100% open *and* their corresponding discharge air temperatures are greater than their set points, decrease the chilled water temperature.

In implementing these rules, a fixed increment for increasing or decreasing the chilled water temperature must be chosen. A small increment results in more stable control, but also results in a slow response to sudden changes in load or supply air temperature set points. Using a first-order approximation, the chilled water temperature can be reset in response to sudden changes in load and supply air temperature set point according to

$$t_{chws} = t_{as} - \frac{PLR}{PLR_o}(t_{as,o} - t_{chws,o}) \qquad (8)$$

where

t_{chws} = new chilled water set point temperature
t_{as} = current supply air set point temperature
PLR = current part-load ratio (chiller load divided by total design load for all chillers)
$t_{chws,o}$ = chilled water set point associated with last control decision
$t_{as,o}$ = supply air set point associated with last control decision
PLR_o = part-load ratio associated with last control decision

Equation (8) assumes that the chilled water temperature associated with the last control decision was optimal. As a result, it only applies to anticipating the effects of significant changes in the load and supply-air set-point temperature on the optimal chilled-water set point. The "bump-and-wait" strategy acts to fine-tune the chilled water supply temperature when the load and supply air set point are stable. For a variable-air-volume system, the supply air set points are most often constant and identical for all air-handling units. However, for a constant-air-volume system, these set points may vary with different air handlers. In this case, the supply air set point to use in Equation (8) should be an average value for the "representative" air handlers.

Equation (8) indicates that the optimal chilled water supply temperature increases with increasing supply air temperature and decreasing load. This is because these changes cause the cooling coil valves to close and the optimal control involves keeping at least one valve open. Increasing supply air temperature causes the cooling coil valves to close somewhat due to a larger average temperature difference for heat transfer between the water and air. A lower load requires smaller air-to-water temperature differences, which also leads to control valves closing.

Overrides for Equipment and Comfort Constraints

For a given chiller load, the chilled water temperature has both upper and lower limits. The lower limit is necessary to avoid ice formation on the evaporator tubes of the chiller. This limit depends primarily on the load in relation to the size of the evaporator or, in other words, the temperature difference between the chilled water and refrigerant. At small temperature differences (large area or small load), the evaporator can tolerate a lower chilled water temperature in order to avoid freezing than at large temperature differences. The lower limit on the chilled water set point should be evaluated at the design load, since the overall system performance is improved by increasing chilled water temperature above this limit for loads less than design. This lower limit can range from 38 to 44°F.

An upper limit on the chilled water temperature arises from comfort constraints associated with the zones and the possibility of microbial growth associated with high humidities. For the available flows, the chilled water temperature should be low enough to provide discharge air at a temperature and humidity sufficient to maintain all zones in the comfort region and avoid microbial growth. This upper limit varies with both load and entering air conditions and is accounted for by monitoring the zone conditions to ensure that they are in the comfort zone. If zone temperatures or humidities are not within reasonable bounds then the discharge air temperature set point should be lowered.

Summary and Implementation

At each decision interval (e.g., 5 min), the following algorithm would be applied for determining the optimal chilled water set point temperature.

1. Determine the time-averaged total chilled water load for the previous decision interval.
2. If the chilled water load or supply air set point temperature have changed by a significant amount (e.g., 10%) since the last control change, then estimate a new optimal chilled water set point with Equation (8) and go to Step 6. Otherwise, go to Step 3.
3. Determine the time-averaged position of (or controller output for) the cooling coil water valves and corresponding discharge air temperatures for "representative" air handlers.
4. If more than one valve is saturated at 100% open *and* their corresponding supply air temperatures are greater than set point (e.g., 1°F), then decrease the chilled water temperature by a fixed amount (e.g., 0.5°F) and go to Step 6. Otherwise, go to Step 5.
5. If all water valves are unsaturated *or* the supply air temperatures associated with all valves that are saturated are lower than the set point, then raise the chilled water set temperature by a fixed amount (e.g., 0.5°F). Otherwise, exit the algorithm with the chilled water set point unchanged.
6. Limit the chilled water set point temperature between the upper and lower limits dictated by comfort, humidity, and equipment safety.

Implementation of this algorithm requires some estimate of the chilled water load, along with a measurement of the discharge air temperatures and control valve positions. However, an accurate estimate of the load is not important.

CHILLED WATER RESET WITH VARIABLE-SPEED PUMPING

Figure 20 shows a common configuration for systems utilizing variable-speed chilled water pumps with primary/secondary water loops. The primary pumps are fixed speed and are generally sequenced with chillers to provide a relatively constant flow of water through the chiller evaporators. The secondary chilled water pumps are variable speed and are typically controlled to maintain a

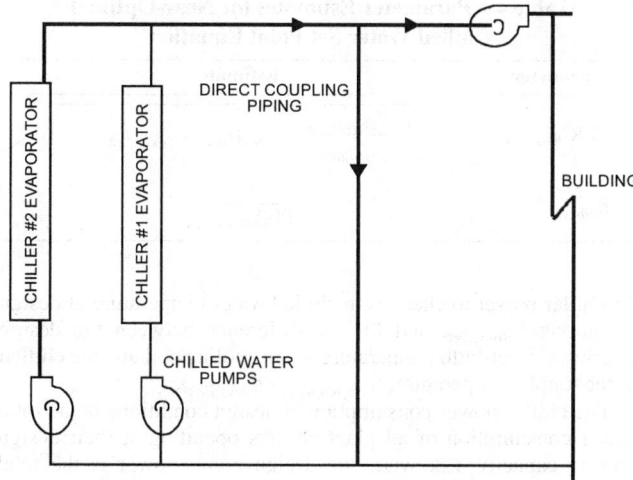

**Fig. 20   Typical Chilled Water Distribution for
Primary/Secondary Pumping**

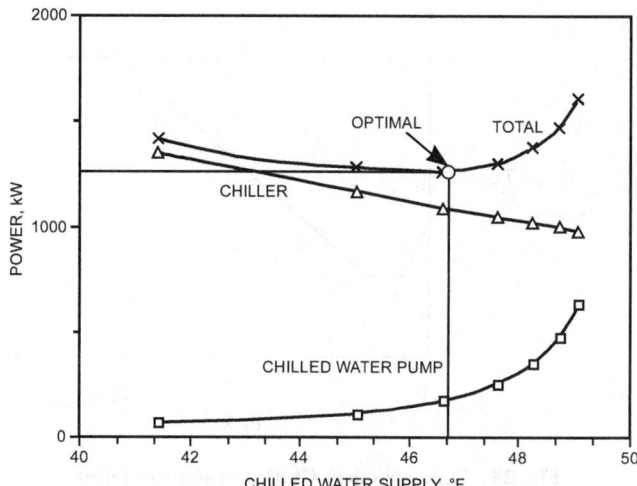

**Fig. 21   Trade-off of Chiller and Pump Power with
Chilled Water Set Point**

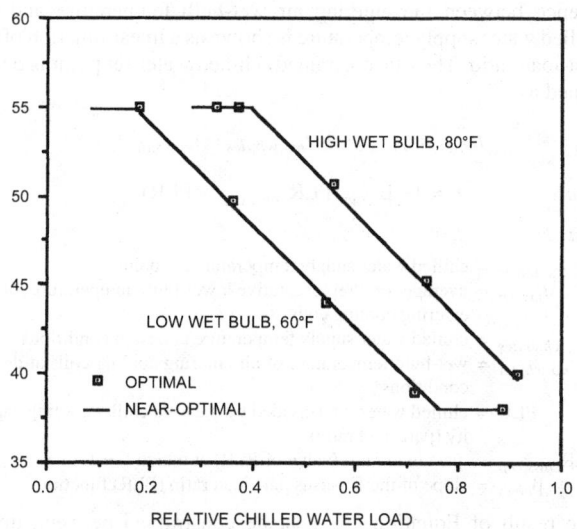

Fig. 22   Comparisons of Optimal Chilled Water Temperature

specified set point for pressure difference between supply and return flows for the cooling coils.

Although variable-speed pumps are usually used with primary/secondary chilled water loops, they may also be applied to systems with a single chilled water loop. In either case, variable-speed pumps offer the potential for a significant operating cost saving when both chilled water and pressure differential set points are optimized in response to changing loads. This section presents an algorithm for determining near-optimal values of these control variables.

Optimal Differential Pressure Set Points

In practically all variable-speed chilled water pumping applications, the pump speed is controlled to maintain a constant pressure differential between the main chilled water supply and return lines. However, this approach is not optimal. In order to maintain a constant pressure differential with changing flow, the control valves for the air-handling units must close as the load (i.e., flow) is reduced, resulting in an increase in the flow resistance. The best strategy for a given chilled water set point would be to reset the differential pressure set point in order to maintain all discharge air temperatures with at least one control valve in a saturated (fully-open) condition. This results in a relatively constant flow resistance and greater pump savings at low loads. With variable differential pressure set points, the optimization of the chilled water loop is described in terms of finding the chilled water temperature that minimizes the sum of the chiller and pumping power with the pump control dependent upon the set point and the load.

Near-Optimal Chilled Water Set Point

The optimal chilled water supply temperature at a given load results from a trade-off between chiller and pumping power as illustrated in Figure 21. As the chilled water temperature increases, the chiller power is reduced due to a reduction in the lift requirements of the chiller. For a higher set temperature, more chilled water flow is necessary to meet the load requirements and the pumping power requirements increase. The minimum total power occurs at a point where the rate of increase in pumping power with chilled water temperature is equal to the rate of decrease in chiller power. This optimal set point moves to lower values as the load increases.

Braun et al. (1989a) demonstrated that the optimal chilled water set point varies as a near linear function of both load and wet-bulb temperature over a wide range of conditions. Figure 22 shows an example of how the optimal set point varies for a specific plant. The set point is plotted as a function of load relative to design load for

two different wet-bulb temperatures. In general, the optimal chilled water temperature decreases with load because the pump power becomes a larger fraction of the total power. A lower set-point limit is set in order to avoid conditions that could form ice on evaporator tubes or too high a chiller "lift", while an upper limit is established to ensure adequate cooling coil dehumidification. For a given load, the chilled water set point increases with wet-bulb temperature because the energy transfer across each cooling coil is proportional to the difference between its entering air wet-bulb temperature and the entering water temperature (the chilled water set point). For a constant load this temperature difference is constant and the chilled water supply temperature increases linearly with entering air wet-bulb temperature.

The results of Figure 22 were obtained for a system where both the chilled water supply and supply air set points to the zones were optimized. For this case, the supply air temperatures varied between 55°F at high loads and 60°F at low loads. More typically, the supply air temperatures are constant at 55°F. In this case, the variation in chilled water supply temperature would be smaller than that shown in Figure 22.

Figure 23 depicts the general form for an algorithm to determine chilled water supply set points as a function of load and the average

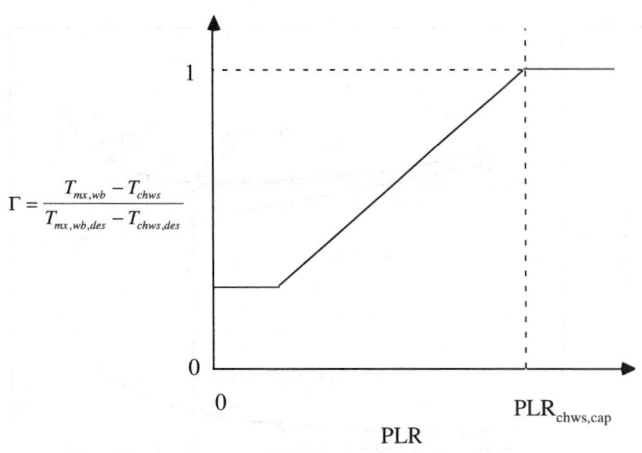

**Fig. 23   Dimensionless Chilled Water Set Point
Versus Part-Load Ratio**

Parameter	Estimate
$PLR_{chws,cap}$	$\sqrt{\dfrac{1}{3}\dfrac{P_{ch,des}}{P_{chwp,des}}S_{chws,des}(t_{mx,wb,des}-t_{chws,des})}$
β_{chws}	$\dfrac{0.5}{PLR_{chws,cap}}$

wet-bulb temperature entering the cooling coils. A normalized difference between the entering air wet-bulb temperature and the chilled water supply temperature is shown as a linear function of the part-load ratio. The (unconstrained) chilled water set point is determined as

$$t_{chws} = t_{mx,wb} - \Gamma(t_{mx,wb,des} - t_{chws,des}) \qquad (9)$$

where $\Gamma = 1 - \beta_{chws}(PLR_{chws,cap} - PLR) \qquad (10)$

and

t_{chws} = chilled water supply temperature set point
$t_{mx,wb}$ = average or "representative" wet-bulb temperature of air entering cooling coils
$t_{chws,des}$ = chilled water supply temperature at design conditions
$t_{mx,wb,des}$ = wet-bulb temperature of air entering cooling coils at design conditions
PLR = chilled water load divided by the total chiller cooling capacity (part-load ratio)
$PLR_{chws,cap}$ = part-load ratio (value of PLR) at which $\Gamma = 1$
β_{chws} = slope of the Γ versus part-load ratio (PLR) function

The result of Equation (9) must be constrained between upper and lower limits dictated by equipment safety (evaporator freezing), the machine operating envelope, and comfort and humidity concerns.

The variables of Equation (10) that yield near-optimal control depend on the characteristics of system. Detailed measurements over a range of conditions may be taken and used to determine estimates of these parameters. However, this requires measurements of component power consumption along with considerable time and expertise and may not be cost effective, unless performed by on-site plant personnel. Alternatively, simple estimates of these parameters may be obtained using design data.

Open-Loop Parameter Estimates Using Design Data. Reasonable estimates of the parameters of Equation (10) may be determined analytically using design information as summarized in Table 4. These estimates were derived by applying optimization theory to a simplified mathematical model of the chiller and secondary-loop water pumps, assuming that a differential pressure reset strategy is used, pump efficiencies are constant, and the supply air temperature is not varied in response to changes in chilled water supply temperature. In general, these parameter estimates are conservative in that they should provide a relatively low estimate of the optimal chilled water set point.

The design factors that affect the parameter estimates given in Table 4 are: (1) the ratio of the chiller power to chilled water pump power at design conditions $P_{ch,des}/P_{chwp,des}$, (2) the sensitivity of the chiller power to changes in chilled water temperature at design conditions $S_{chws,des}$, and (3) the difference between the design entering air wet-bulb temperature to the cooling coil and the chilled water supply temperature, $(t_{mx,wb,des} - t_{chws,des})$.

The chiller power consumption at design conditions is the total power consumption of all plant chillers operating at their design cooling capacity. Likewise, the design pump power is the total power associated with all secondary chilled water supply pumps operating at high speed. As the ratio of chiller power to pump power increases, then it becomes more beneficial to operate the chillers at higher chilled water temperatures and the pumps at higher flows. This is reflected in an increase in $PLR_{chws,cap}$. If the chiller power were free, then $PLR_{chws,cap}$ goes to zero, and the best strategy would be to operate the chillers at the minimum possible set point, resulting in low chilled water flow rates. Typical values for the ratio of the chiller power to the pump power at design conditions are between 10 and 20, depending primarily on whether primary/secondary pumping is used or not.

The chiller sensitivity factor $S_{chws,des}$ is the incremental increase in chiller power for each degree decrease in chilled water temperature as a fraction of the power or:

$$S_{chws,des} = \frac{\text{Increase in chiller power}}{\text{Decrease in chilled water temp.} \times \text{Chiller power}} \qquad (11)$$

If the chiller power increases by 2% for a 1°F decrease in chilled water temperature, then $S_{chws,des}$ is equal to 0.02°F^{-1}. A large sensitivity factor means that the chiller power is very sensitive to the set point control favoring operation at higher set point temperatures and flows (higher $PLR_{chws,cap}$). The sensitivity factor should be evaluated at design conditions using chiller performance data. Typically, the sensitivity factor is between 0.01°F^{-1} and 0.03°F^{-1}. For multiple chillers with different performance characteristics, the sensitivity factor at design conditions is estimated as:

$$S_{chws,des} = \frac{\displaystyle\sum_{i=1}^{N_{ch}} S_{chws,des,i}P_{ch,des,i}}{\displaystyle\sum_{i=1}^{N_{ch}} P_{ch,des,i}} \qquad (12)$$

where $S_{chws,des,i}$ is the sensitivity factor and $P_{ch,des,i}$ is the power consumption for the ith chiller at the design conditions and N_{ch} is the total number of chillers.

The design difference between the coil inlet air wet-bulb temperature and the entering water temperature should be evaluated for a typical air handler operating at its design load and flows. A small temperature difference results from a high coil heat transfer effectiveness or high water flow rate, allowing higher chilled water temperatures with lower chiller power consumption. This is evident from Equation (9), where chilled water set point decreases linearly with $(t_{mx,wb,des} - t_{chws,des})$ for a given Γ, while Γ is inversely related to the square root of $(t_{mx,wb,des} - t_{chws,des})$. Typically, this temperature difference is about 20°F.

Example 2. Consider an example plant with primary/secondary chilled water pumping. There are four 550 ton chillers, each with its own dedicated primary pump. Each chiller consumes approximately 330 kW at the design capacity. At design conditions, the chiller power increases approximately 6.6 kW for a 1°F decrease in chilled water temperature, giving a sensitivity factor of 6.6/330 or 0.02°F^{-1}. The design chilled water set point is 42°F, while the coil entering wet-bulb temperature is 62°F at design conditions. The secondary loop uses three identical 60 hp chilled water pumps: one with a variable-speed and two with fixed-speed motors.

Solution:

The first step in applying the open-loop control algorithm to this problem is the determination of the parameters of Equation (10) from the design data. From Table 4, the part-load ratio at which the chilled water temperature reaches a minimum (with the design entering wet-bulb temperature to the coils) is:

$$PLR_{chws,cap} = \sqrt{\frac{1}{3}\frac{4 \times 330 \text{ kW}}{3 \times 60 \text{ hp} \times 0.75 \text{ kW/hp}}(0.02°F^{-1})(20°F)} = 1.14$$

while the slope of the set point versus part-load ratio is estimated to be:

$$\beta_{chws} = \frac{0.5}{1.14} = 0.44$$

Given these parameters and the part-load ratio, the unconstrained chilled water set point temperature from Equations (9) and (10) is then

$$t_{chws} = t_{mx,wb} - [1 - 0.44(1.14 - PLR)]20$$

Pump Sequencing

Variable-speed pumps are sometimes used in combination with fixed-speed or other variable-speed pumps. The problem of pump sequencing involves determining both the order and point that pumps should be brought online and offline.

Pumps should be brought online in an order that allows a continuous variation in flow rate and maximized operating efficiency of the pumps at each switch point for the specific pressure loss characteristic. For a combination of fixed-speed and variable-speed pumps, at least one variable-speed pump should be brought online prior to any fixed-speed pumps. For single-loop systems (i.e., no secondary-loop) with variable-speed pumps, the pressure drop characteristics change when chillers are added or removed and the optimal sequencing of pumps depends on the sequencing of chillers.

An additional pump should be brought online whenever the current set of pumps are operating at full capacity and can no longer satisfy the differential pressure set point. This situation can be detected by monitoring the differential pressure or the controller output signal. Insufficient pump capacity leads to extended periods with differential pressures that are less than the set point and a controller output that is saturated at 100%. A pump may be taken offline, whenever the remaining pumps have sufficient capacity to maintain the differential pressure set point. This condition can be determined by comparing the current (time-averaged) controller output with the controller output (time-averaged) at the point just after the last pump was brought online. The pump can be brought offline when the current output is less than the switch point value by a specified deadband (e.g., 5%).

Overrides for Equipment and Comfort Constraints

The chilled water temperature is bounded by upper and lower limits dictated by comfort, humidity, and equipment safety concerns. However, within these bounds, the chilled water temperature may not always be low enough to maintain supply air set point temperatures for the cooling coils. This situation might occur at high loads when the chilled water flow is at a maximum and is detectable by monitoring the coil discharge air temperatures. Limits on the pressure differential set point might also be imposed in order to ensure adequate controllability of the cooling coil control valves.

Summary and Implementation

Prior to commissioning, the parameters of the open-loop control for chilled water set point [Equation (10)] need to be estimated using the results of Table 4. After the system is in operation, these parameters may be fine-tuned with measurements as outlined in the section on Control Optimization Methods. With the parameters specified, the control algorithm is separated into two reset strategies: chilled water temperature and pressure differential.

Chilled Water Temperature Reset. The chilled water supply temperature set point is reset at fixed decision intervals (e.g., 15 min) using the following procedure:

1. Determine the time-averaged position of (or controller output for) the cooling coil water valves and corresponding discharge air temperatures for "representative" air handlers over the previous decision interval.
2. If more than one valve is saturated at 100% open *and* their corresponding discharge air temperatures are greater than set point (e.g., 1°F), then decrease the chilled water temperature by a fixed amount (e.g., 0.5°F) and go to Step 5. Otherwise, go to Step 3.
3. Determine the total chilled water flow and load.
4. Estimate an optimal chilled water set point with Equations (9) and (10). Increase or decrease the actual set point in the direction of the near-optimal value by a fixed amount (e.g., 1°F).
5. Limit the new set point between upper and lower constraints dictated by comfort and equipment safety.

Pump Sequencing. Secondary pumps should be brought online or offline at fixed decision intervals (e.g., 15 min) with the following logic:

1. Evaluate the time-averaged pump controller output over the previous decision interval.
2. If the pump controller is saturated at 100%, then bring the next pump online. Otherwise, go to Step 3.
3. If the pump control output is significantly less (e.g., 5%) than the value associated with the first time interval after the last pump was brought online, then bring that pump offline.

Differential Pressure Reset. The set point for differential pressure between supply and return lines should be reset at smaller time intervals than the supply water temperature reset and pump sequencing strategies (e.g., 5 min) using the following procedure:

1. Check the water valve positions (or controller output) for "representative" air handlers and determine the time-averaged values over the last decision interval.
2. If more than one valve has been saturated at 100% open, then increase the differential pressure set point by a fixed value (e.g., 5% of the design value) and go to Step 4. Otherwise, go to Step 3.
3. If none of the valves have been saturated, then decrease the differential pressure set point by a fixed value (e.g., 5% of the design value).
4. Limit the differential pressure set point between upper and lower constraints.

SEQUENCING AND LOADING OF MULTIPLE CHILLERS

Multiple chillers are normally configured in a parallel manner and typically controlled to give identical chilled water supply temperatures. In most cases, controlling for identical set temperatures is the best and simplest strategy. With this approach, the relative loading on operating chillers is controlled by the relative chilled water flow rates. Typically, the distribution of flow rates to heat exchangers for both chilled and condenser water are dictated by chiller pressure drop characteristics and may be adjusted through flow balancing, but are not controlled using a feedback controller. In addition to the distribution of chilled and condenser water flow rates, the chiller sequencing affects energy consumption. Chiller sequencing defines the order that and conditions where chillers are brought online and offline. Simple guidelines may be established for each of these controls that are near optimal.

Near-Optimal Condenser Water Flow Distribution

In general, the condenser water flow to each chiller should be set to give identical leaving condenser water temperatures. This condition approximately corresponds to relative condenser flow rates equal to the relative loads on the chillers, even if the chillers are loaded unevenly. Figure 24 shows results for four sets of two chillers operated in parallel. The curves represent data from chillers at three different installations: (1) a 5500 ton variable-speed chiller at the Dallas-Fort Worth Airport, Texas (Braun et al. 1989a); (2) a 550 ton fixed-speed chiller at an office building in Atlanta (Hackner et al. 1984, 1985); and (3) a 1250 ton fixed-speed chiller at a large office building in Charlotte, North Carolina (Lau et al. 1985). The capacities of the chillers in the two office buildings were scaled up for comparison with the Dallas-Fort Worth airport chiller.

The overall chiller coefficient of performance (COP) is plotted versus the difference between the condenser water return temperatures for equal chiller loading. For identical chillers, and with either variable-speed or fixed-speed motors, the temperature difference between the chillers is almost zero. For situations where chillers do not have identical performance, equal leaving condenser water temperatures result in chiller performance that is close to the optimum. Even for the variable- and fixed-speed chiller combinations that have very different performance characteristics, the penalty associated with the use of identical condenser leaving water temperatures is small. In order to achieve equal condenser leaving water temperatures, it is necessary to properly balance the condenser water flow rates at design operating conditions.

Optimal Chiller Load Distribution

Assuming identical chilled water return and chiller supply temperatures, the relative chilled water load for each parallel chiller (load divided by total load) that is operating could be controlled by its relative chilled water flow rate (flow divided by total flow). In order to change the relative loadings in response to operating conditions, the individual flow rates must be controlled. However, this is typically not done and it is probably sufficient to establish the load

distributions based upon design information and then balance the flow rates to achieve these load distributions. Alternatively, the individual chiller loads can be precisely controlled through variation of individual chiller supply water set points.

Chillers with Similar Performance Characteristics. Braun et al. (1989a) showed that for chillers with identical design COPs and part-load characteristics, a minimum or maximum power consumption occurs when each chiller is loaded according to the ratio of its capacity to the sum total capacity of all operating chillers. This is equivalent to each chiller operating at equal part-load ratios (load divided by cooling capacity at design conditions). For the ith chiller, the optimal chiller loading is then

$$\dot{Q}_{ch,i}^{*} = \frac{\dot{Q}_{load}}{\displaystyle\sum_{i=1}^{N} \dot{Q}_{ch,des,i}} \dot{Q}_{ch,des,i} \tag{13}$$

where $\dot{Q}_{load}$ is the total chiller load, $\dot{Q}_{ch,des,i}$ is the cooling capacity of the ith chiller at design conditions, and N is the number of chillers operating.

The loading determined with Equation (13) could result in either minimum or maximum power consumption. However, this solution gives a minimum when the chillers are operating at loads greater than the point at which the maximum COP occurs (i.e., chiller COP decreases with increased loading). Typically, the maximum COP occurs at loads that are less than the nominal design capacity.

Figure 25 shows the effect of the relative loading on chiller COP for different sets of identical chillers loaded at approximately 70% of their total capacities. Three of the chiller sets have maximum COPs when evenly loaded [matching the criterion of Equation (13)], while the fourth (Dallas-Ft. Worth fixed-speed) obtains a minimum at that point. The part-load characteristic of the Dallas-Ft. Worth fixed-speed chiller is unusual in that the maximum overall COP occurs at its maximum capacity. This chiller was retrofit with a different refrigerant and drive motor, which derated its capacity from 8700 to 5500 tons. As a result, the evaporators and condensers are oversized for its current capacity. Overall, the penalty associated with equally loading the Dallas-Ft. Worth fixed-speed chillers is small as compared with optimal loading and this strategy is probably appropriate. However, a slight reduction in energy consumption is possible if one of the two chillers operates at full capacity. The loading criterion of Equation (13) also works well for many combinations of chillers with different performance characteristics.

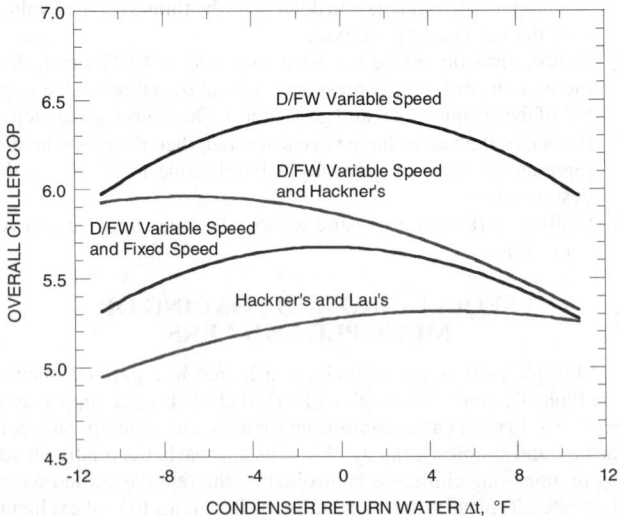

**Fig. 24   Effect of Condenser Water Flow Distribution
for Two Chillers In Parallel**

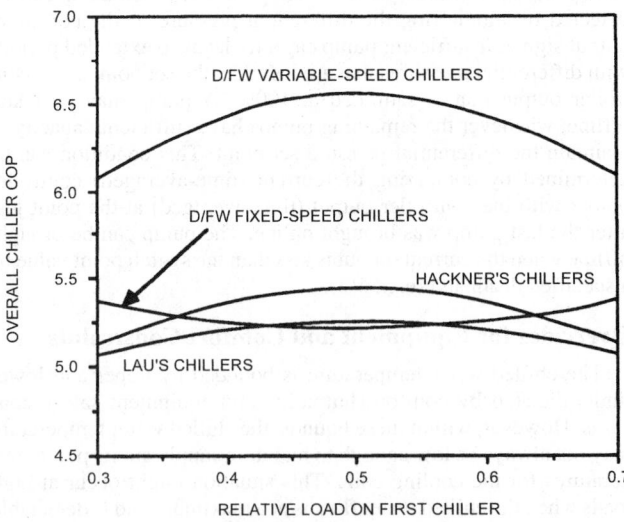

**Fig. 25   Effect of Relative Loading for
Two Identical Parallel Chillers**

In order to achieve specified relative chiller loadings with equal chilled water set points, the chilled water flow rates must be properly balanced. The relative loadings of Equation (13) only depend on design information and the flow balancing can be achieved through proper design and commissioning.

Chillers with Different Performance Characteristics. For the general case of chillers with significantly different part-load characteristics, a point of minimum or maximum overall power occurs where the partial derivatives of the individual chiller's power consumption with respect to their loads are equal, or:

$$\frac{\partial P_{ch,i}}{\partial \dot{Q}_{ch,i}} = \frac{\partial P_{ch,j}}{\partial \dot{Q}_{ch,j}} \quad \text{for all } i \text{ and } j \tag{14}$$

and subject to the constraint that

$$\sum_{i=1}^{N} \dot{Q}_{ch,i} = \dot{Q}_{load} \tag{15}$$

where $\dot{Q}_{ch,i}$ is the cooling load for the ith chiller and $\dot{Q}_{load}$ is the total cooling load.

In general, the power consumption of a chiller can be correlated as a quadratic function of cooling load and difference between the leaving condenser water and chilled water supply temperatures according to:

$$P_{ch,i} = a_{0,i} + a_{1,i}(t_{cwr,i} - t_{chws,i}) + a_{2,i}(t_{cwr,i} - t_{chws,i})^2$$
$$+ a_{3,i}\dot{Q}_{ch,i} + a_{4,i}\dot{Q}_{ch,i}^2 + a_{5,i}(t_{cwr,i} - t_{chws,i})\dot{Q}_{ch,i} \tag{16}$$

where, for the ith chiller, t_{cwr} is the leaving condenser water temperature and t_{chws} is the chilled water supply temperature. The coefficients of Equation (16) ($a_{0,i}$ to $a_{5,i}$) could be determined for each chiller through regression applied to measured or manufacturers' data.

If each chiller has identical leaving condenser and chilled water supply temperatures, the criterion of Equation (14) applied to the correlation of Equation (16) leads to:

$$a_{3,i} + 2a_{4,i}\dot{Q}_{ch,i}^* + a_{5,i}(t_{cwr} - t_{chws})$$
$$= a_{3,j} + 2a_{4,j}\dot{Q}_{ch,j}^* + a_{5,j}(t_{cwr} - t_{chws}) \quad \text{for } i \neq j \tag{17}$$

where $\dot{Q}_{ch,i}^*$ is the optimal load for the ith chiller.

Equations (17) and (15) represent a system of N linear equations in terms of N chiller loads that can be solved to give minimum (or possibly maximum) power consumption. For a given combination of chillers, the solution depends on the operating temperatures and total load. However, the individual chiller loads must be constrained to be less than the maximum chiller capacity at these conditions. If an individual chiller load determined from these equations is greater than its cooling capacity, then this chiller should be fully loaded and Equations (17) and (15) should be resolved for the remaining chillers [Equation (17) should only include unconstrained chillers].

In order to control individual chiller loads with identical chilled water supply temperatures, individual chilled water flow rates need to be controlled with two-way valves, which is not typical. However, the distribution of chiller loads could be changed for a fixed flow distribution by using different chilled water set point temperatures. For a given flow and load distribution, the individual chiller set point for parallel chillers is determined according to:

$$t_{chws,i} = t_{chwr} - \frac{\dot{Q}_{ch,i}}{f_{F,i}\dot{Q}_{load}}(t_{chwr} - t_{chws}^*) \tag{18}$$

where $f_{F,i}$ is the flow for the ith chiller divided by the total flow, t_{chws}^* is the chilled water supply temperature set point for the combination of chillers determined using the previously defined reset strategies, and t_{chwr} is the temperature of the water returned to the chillers from the building.

Substituting Equation (18) into Equation (16) and then applying the criterion of Equation (14) leads to the following:

$$A_i + B_i\dot{Q}_{ch,i}^* = A_j + B_j\dot{Q}_{ch,j}^* \quad \text{for } i \neq j \tag{19}$$

where

$$A_i = a_{3,i} + [a_{1,i} + 2a_{2,i}(t_{cwr} - t_{chwr})]\frac{t_{chwr} - t_{chws}^*}{f_{F,i}\dot{Q}_{load}}$$
$$+ a_{5,i}(t_{cwr} - t_{chws}^*)$$

$$B_i = 2\left[a_{4,i} + a_{5,i}\frac{t_{chwr} - t_{chws}^*}{f_{F,i}\dot{Q}_{load}} + a_{2,i}\left(\frac{t_{chwr} - t_{chws}^*}{f_{F,i}\dot{Q}_{load}}\right)^2\right]$$

The optimal chiller loads are determined by solving the linear system of equations represented by Equations (19) and (15). The individual chiller set points are then evaluated with Equation (18). If any set points are less than the minimum set point or greater than the maximum set point, then the set point should be constrained and Equations (19) and (15) should be resolved for the remaining chillers [Equation (19) should only include unconstrained chillers].

Example 3. The optimal loading for two chillers will be determined using the three methods outlined in this section. Table 5 gives the design cooling capacities and coefficients of the curve-fit of Equation (16) for the two chillers. The chillers are operating with a total cooling load of 1440 tons, condenser water return temperature of 85°F, an overall chilled water supply temperature set point of 45°F, and a chilled water return temperature of 55°F. Figure 26 shows the COPs for the two chillers as a function load relative to their design loads for the given operating temperatures determined using Equation (25). Chiller 1 is more efficient at higher part-ratios and less efficient at lower part-load ratios as compared with Chiller 2.

Solution:

The first case that will be considered is the chillers operating at equal part-load ratios. The ratio of the cooling load to the cooling

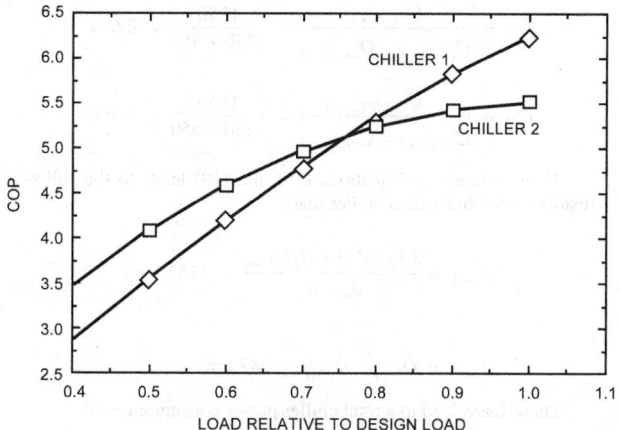

Fig. 26 Chiller COP for Two Chillers

Table 5   Chiller Characteristics for Optimal Loading Example 3

Variable	Units	Chiller 1	Chiller 2
$\dot{Q}_{ch,des,i}$	tons	1250	550
$a_{0,i}$	kW	106.4	119.7
$a_{1,i}$	kW/°F	6.147	0.1875
$a_{2,i}$	kW/°F²	0.1792	0.04789
$a_{3,i}$	kW/ton	−0.0735	−0.3673
$a_{4,i}$	kW/ton²	0.0001324	0.0005324
$a_{5,i}$	kW/ton·°F	−0.001009	0.008526

capacity of the operating chillers is 1440/(1250 + 550) = 0.8. From Equation (13), the individual chiller loads are:

$$\dot{Q}^*_{ch,1} = 0.8(1250 \text{ tons}) = 1000 \text{ tons}$$

$$\dot{Q}^*_{ch,2} = 0.8(550 \text{ tons}) = 440 \text{ tons}$$

The power for each chiller is computed for the specified operating conditions with Equation (16) and the coefficients of Table 5. For the case of equal part-load ratios, the total chiller power consumption is

$$P_{ch} = P_{ch,1} + P_{ch,2} = 657.5 \text{ kW} + 295.3 \text{ kW} = 952.8 \text{ kW}$$

A second solution is determined for optimal chiller loads for the case of equal chilled water temperature set points and controllable flow for each chiller. In this case, algebraic manipulation of Equations (15) and (17) produces the following results for the individual chiller loads:

$$\dot{Q}^*_{ch,1} = \frac{(a_{3,2} - a_{3,1}) + 2a_{4,2}\dot{Q}_{load} + (a_{5,2} - a_{5,1})(t_{cwr} - t_{chws})}{2(a_{4,1} + a_{4,2})}$$

$$= 1219 \text{ tons}$$

$$\dot{Q}_{ch,2} = \dot{Q}_{load} - \dot{Q}_{ch,1} = 221 \text{ tons}$$

The resulting power consumption is then:

$$P_{ch} = P_{ch,1} + P_{ch,2} = 696.9 \text{ kW} + 224 \text{ kW} = 920.9 \text{ kW}$$

Optimal loading of the chillers reduces the overall chiller power consumption by about 4% through heavier loading of Chiller 1 and lighter loading of Chiller 2 (see Figure 26).

Finally, optimal chiller loading is determined for the case where the individual loadings are controlled by using different chilled water temperature set points (individual flow is not controllable). In order to apply Equations (18) and (19), the relative chilled water flow rate for each chiller must be known. For this example, the relative flow for the *i*th chiller is assumed to be equal to the ratio of its design capacity to the design capacity for the operating chillers, so that:

$$f_{F,1} = \frac{\dot{Q}_{ch,des,1}}{\dot{Q}_{ch,des,1} + \dot{Q}_{ch,des,2}} = \frac{1250}{1250 + 550} = 0.694$$

$$f_{F,2} = \frac{\dot{Q}_{ch,des,2}}{\dot{Q}_{ch,des,1} + \dot{Q}_{ch,des,2}} = \frac{1250}{1250 + 550} = 0.306$$

Then, solution of Equations (15) and (19) leads to the following results for the individual chiller loads:

$$\dot{Q}^*_{ch,1} = \frac{(A_2 - A_1) + B_2\dot{Q}_{load}}{B_1 + B_2} = 1153 \text{ tons}$$

$$\dot{Q}_{ch,2} = \dot{Q}_{load} - \dot{Q}_{ch,1} = 287 \text{ tons}$$

These loads lead to a total chiller power consumption of:

$$P_{ch} = P_{ch,1} + P_{ch,2} = 713.8 \text{ kW} + 218.1 \text{ kW} = 931.9 \text{ kW}$$

The individual chilled water set points are determined from Equation (18) and are 43.5°F and 48.5°F for Chillers 1 and 2, respectively. The power consumption has increased slightly from the case of identical chiller set points and variable flow.

Note that either changing flows or chilled water setpoints complicates the overall system control as compared with loading the chillers with fixed part-load ratios and leads to relatively small savings.

Order for Bringing Chillers Online and Offline

For chillers with similar efficiencies, the order in which chillers are brought online and offline may be dictated by their cooling capacities and the desire to provide even runtimes. However, whenever beneficial and possible, chillers should be brought online in an order that minimizes the incremental increase in energy consumption. At a given condition, the power consumption of any chiller can be evaluated using the correlation given by Equation (16), where the coefficients are determined using manufacturers' data or in-situ measurements. Then, the overall power consumption for all operating chillers is

$$P_{ch} = \sum_{i=1}^{N} P_{ch,i} \tag{20}$$

When an additional chiller ($N + 1$) is required (see next section), the best chiller to bring on line should result in the smallest increase (or largest decrease) in overall chiller power consumption as estimated with Equations (20) and (16) with chiller loading determined as outlined in the previous section. Although these estimates could be performed online, it is probably adequate to determine the order for bringing chillers online (and offline) using design operating temperatures and assuming that the current (N) chillers are operating at their design loads.

For chillers with similar design cooling capacities, a simpler approach can be used for determining the order for bringing chillers online and offline. In this case, the chiller with the highest peak COP can be brought online first, followed by the second most efficient chiller, etc. and then brought offline in reverse order. The maximum COP for each chiller can be evaluated using manufacturers' design and part-load data or from curve-fits to in-situ performance.

The chiller load associated with maximum COP for each chiller can be determined by applying a first-order condition for a maximum, using Equation (16) and the definition of COP. For this functional form, the maximum (or possibly minimum) COP occurs for

$$\dot{Q}^*_{ch,i} = \sqrt{\frac{a_{o,i} + a_{1,i}(t_{cwr} - t_{chws}) + a_{2,i}(t_{cwr} - t_{chws})^2}{a_{4,i}}} \tag{21}$$

The load determined from Equation (21) yields a maximum COP whenever it is real and bounded between upper and lower limits. Otherwise, it can be assumed that the maximum COP occurs at full load conditions. Typically, the maximum COP occurs between about 40% and 80% of design load and increases as the temperature difference between the condenser leaving water and chilled water supply decreases. Equation (21) could be applied online to determine the rank ordering of chillers to bring online as a function of operating temperatures. However, it is often sufficient to use Equation (21) at the design temperature difference and establish a chiller sequencing order at the design or commissioning stage.

Example 4. The loads for maximum COP will be determined for two different chillers at a chilled water set point of 45°F and a condenser water return temperature of 80°F. Table 6 gives the design cooling capacities and coefficients of the curve-fit of Equation (16) for the two chillers. Figure 27 shows the COPs of the two chillers determined from the correlations as a function of relative load (PLR) and temperature difference ($t_{cwr} - t_{chws}$). These chillers have identical performance at design

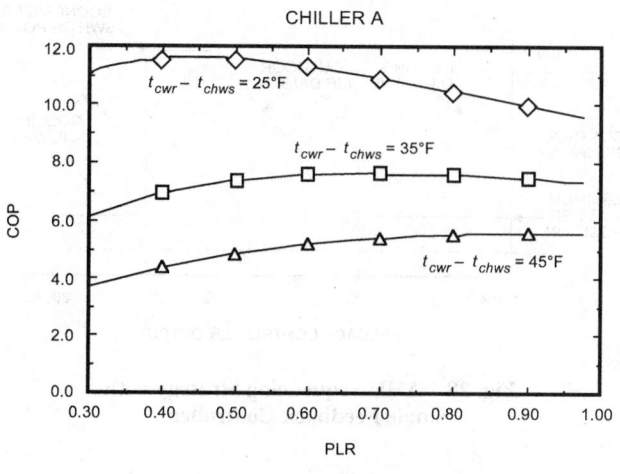
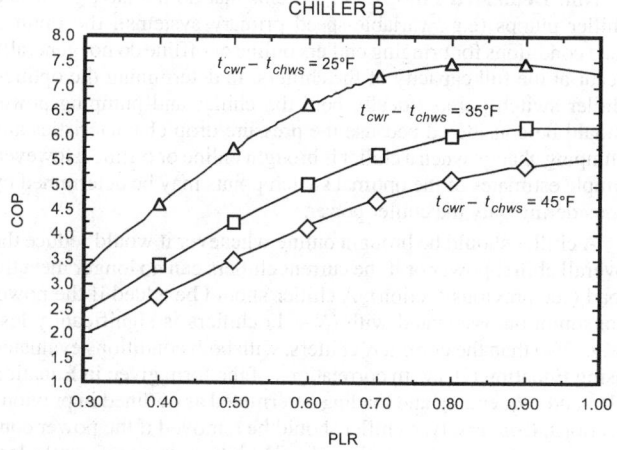

Fig. 27   Chiller A and B Performance Characteristics for Maximum COP Example 4

Table 6   Chiller Characteristics for Maximum COP Example 4

Variable	Unit	Chiller A	Chiller B
$Q_{ch,des,i}$	tons	5421	5421
$a_{0,i}$	kW	262.6	187.2
$a_{1,i}$	kW/°F	−25.36	96.19
$a_{2,i}$	kW/°F²	0.9718	−0.4314
$a_{3,i}$	kW/ton	−0.02568	−0.4314
$a_{4,i}$	kW/ton²	0.00004046	0.0001106
$a_{5,i}$	kW/ton·°F	0.005289	−0.004537

Table 7   Results for Maximum COP Example 4

Variable	Chiller A	Chiller B
$\dot{Q}^*_{ch,i}$	3738 ton	5229 ton
PLR_i	0.690	0.965
$P_{ch,i}$	1727 kW	2964 kW
COP_i	7.61	6.21

conditions, but very different part-load characteristics due to different methods used for capacity control.

Solution:

The loading associated with the maximum COP for each chiller is determined using Equation (21) and the coefficients of Table 6. The power for each chiller is determined using Equation (16) and the COP follows directly. Results of the calculations are given in Table 7. The maximum COP for chiller A is about 20% greater than that for the chiller B at the specified operating temperatures and should be brought on-line first.

Load Conditions for Bringing Chillers Online or Offline

In general, chillers should be brought online at conditions where the total power (including pumps and tower or condenser fans) of operating with the additional chiller would be less than without it. Conversely, a chiller should be taken offline when the total power of operating with that chiller would be less than with it. In practice, the switch point for bringing a chiller online should be greater than that for bringing that same chiller offline (e.g., 10%), in order to ensure a stable control. The optimal sequencing of chillers depends primarily upon their part-load characteristics and the manner with which the chiller pumps are controlled.

Dedicated Pumps. Where individual condenser and chilled water pumps are dedicated to the chiller, Hackner et al. (1985) and Braun et al. (1989a) showed that a chiller should be brought online when the operating chillers reach their capacity. This conclusion is

the result of considering both the chiller and pumping power in determining optimal control. If the pumping power is ignored, the optimal chiller sequencing occurs when the chiller efficiency is maximized at each load. Because the maximum efficiency often occurs at part-load conditions, the optimal point for adding or removing chillers may occur when chillers are operating at less than their capacity. However, the additional pumping power required with bringing additional pumps online with the chiller offset any reductions in overall chiller power consumption associated with part-load operation.

When pumps are dedicated to chillers, situations may arise where the chillers are operating at less than their capacity and yet the chilled water flow to the cooling coils is insufficient to meet the building load. This generally results from an inadequate design or improper maintenance. Under these circumstances, either some zone conditions need to float to reduce the chilled water setpoint (if possible), or an additional chiller needs to be brought online. Monitoring the zone air handler conditions is one method used to detect this situation. If (1) the chilled water set point is at its lower limit *and* (2) any air handler water control valves are saturated at 100% open *and* (3) their corresponding discharge air temperatures are significantly greater (e.g., 2°F) than set point, then the chilled water flow is probably insufficient and an additional chiller/pump combination could be brought online. One advantage of this approach is that it is consistent with the reset strategies for both fixed and variable-speed chilled water systems.

Chillers can be brought on or offline with the following logic:

1. Evaluate the time-averaged values of the chilled water supply temperature and overall cooling load over a fixed time interval (e.g., 5 min).

2. If the chilled water supply temperature is significantly greater than the set point (e.g., 1°F), then bring the next chiller online. Otherwise, go to Step 3.

3. Determine the time-averaged position of the cooling coil water valves and corresponding discharge air temperatures for "representative" air handlers.

4. If the chilled water supply set point is at its lower limit <u>and</u> more than one valve is saturated at 100% open *and* their corresponding discharge air temperatures are significantly greater than set point (e.g., 1°F), then bring another chiller/pump combination online. Otherwise, go to Step 5.

5. If the cooling load is significantly less (e.g., 10%) than the value associated with the first time interval after the last chiller was brought online, then bring that chiller offline.

Non-Dedicated Pumps. For systems that do not have dedicated chiller pumps (e.g., variable-speed primary systems), the optimal load conditions for bringing chillers online or offline do not generally occur at the full capacity of the chillers. In determining the optimal chiller switch points, ideally both the chiller and pumping power should be considered because the pressure drop characteristics and pumping change when a chiller is brought online or offline. However, simple estimates of the optimal switch points may be determined by considering only the chiller power.

A chiller should be brought online whenever it would reduce the overall chiller power or if the current chillers can no longer meet the load (see previous section). A chiller should be added if the power consumption associated with $(N + 1)$ chillers is significantly less (e.g., 5%) than the current N chillers, with both conditions evaluated using Equation (20) with correlations of the form given in Equation (16) and sequencing and loading determined as outlined in previous sections. Conversely, a chiller should be removed if the power consumption associated with the $(N - 1)$ chillers is significantly less (e.g., 5%) than the current N chillers. The decision to add or remove chillers is readily determined using the current load and operating temperatures.

STRATEGIES FOR AIR-HANDLING UNITS (AHUS)

Air Handler Sequencing and Economizer Cooling

Traditional air-handler sequencing strategies use a single PI controller in order to control heating, cooling with outdoor air, mechanical cooling with 100% outside air, and mechanical cooling with minimum outside air. Sequencing between these different modes is accomplished by splitting up the controller output into different regions of operation as shown in Figure 28.

Figure 28 depicts the relationship between the control signal to the valves and dampers and the feedback controller output. The controller adjusts its output in order to maintain the supply air temperature set point. If the output is between 100% and 200%, mechanical cooling is used to cool the air. When the outdoor conditions are suitable, the outdoor air dampers switch from their minimum position (minimum ventilation air) to fully open. For a dry-bulb economizer, this switch point occurs when ambient air is less than a specified value. This switch point should be less than the switch point to return to minimum outside air in order to ensure stable control. The economizer switchover temperature may be significantly lower than the return air temperature (e.g., 10°F lower) in humid climates where latent ventilation loads are significant. However, in dry climates, the switchover temperature may be close to the return temperature (e.g., 75°F). An enthalpy (or wet-bulb) economizer compares the outside and return air enthalpies (or wet-bulb temperatures) in order to initiate or terminate economizer operation. In general, enthalpy economizers yield lower energy costs than dry-bulb economizers, but require a humidity measurement. Humidity sensors require regular maintenance in order to ensure accurate readings. When the controller output is between 0% and 100% (see Figure 28), the cooling coil valve is fully closed and cooling is provided through the use of ambient air only. In this case, the controller output modulates the position of the outside air dampers to maintain the set point. If the controller output signal is between −100% and 0%, the heating coil is used to maintain set point and the outside air dampers are set at their minimum position.

A single feedback controller is difficult to tune to perform well for all four modes of operation associated with an AHU. An alternative to the traditional sequencing strategy is to use three separate feedback controllers as described by Seem et al. (1998). This approach can improve temperature control, reduce actuator usage, and reduce energy costs. Figure 29 shows a state transition diagram for implementing a sequencing strategy that incorporates separate feedback controllers.

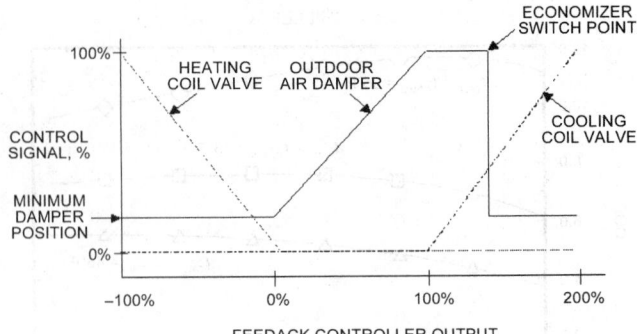

Fig. 28 AHU Sequencing Strategy with Single Feedback Controller

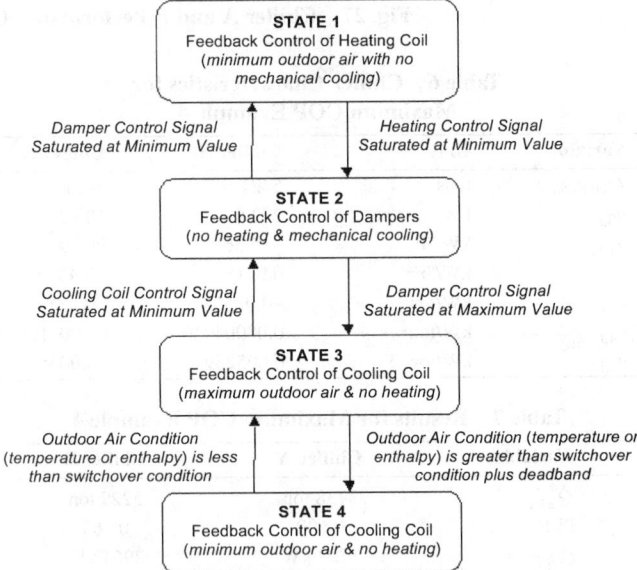

Fig. 29 AHU Sequencing Strategy with Multiple Feedback Controllers

In State 1, a feedback controller adjusts the heating valve to maintain the supply air set point temperature with minimum outside air. The transition to State 2 occurs after the control signal has been saturated at the no-heating position for a period equal to a specified state transition delay (e.g., 3 min). In State 2, a second feedback controller adjusts the outdoor and return air dampers to achieve set point with heating and cooling valves closed. Transition back to State 1 only occurs after the damper control signal is saturated at its minimum value for the state transition delay, whereas transition to State 3 is associated with saturation at the maximum damper position for the state transition delay. In State 3, the damper remains fully open and a third feedback controller is used to adjust the flow of cooling water in order to maintain the supply air temperature at set point. Transition back to State 2 occurs if the controller output is saturated at its minimum value for the state transition delay. For a dry-bulb economizer, transition to State 4 occurs when the ambient dry-bulb temperature is greater than the switchover temperature by a deadband (e.g., 2°F). The feedback controller continues to modulate the cooling coil valve to achieve set point. Transition back to State 3 occurs when the ambient dry-bulb is less than the switchover temperature (e.g., 65°F). For an enthalpy economizer, the ambient enthalpy is compared with return air enthalpy in order to initiate transitions between States 3 and 4.

Supply Air Temperature Reset for Constant Air Volume (CAV)

The benefits of resetting supply air temperature set points for CAV systems are very significant. Increasing the supply air set point for cooling reduces both the cooling load and reheat required, but does not change fan energy. In general, the set point for a CAV system could be set at the highest value that will keep all zone temperatures at their set points and all humidities within acceptable limits. A simple reset strategy based upon this concept follows.

At each the decision interval (e.g., 5 min), the following logic can be applied:

1. Check the controller outputs for "representative" zone reheat units and determine time-averaged values over the last decision interval.
2. If any controller output is less than a threshold value (e.g., 5%), then decrease the supply air set point by a fixed value (e.g., 0.5°F) and go to Step 4. Otherwise, go to Step 3.
3. If all zone humidities are acceptable *and* all controller outputs are greater than a threshold value (e.g., 10%), then increase the discharge air set point by a fixed value (e.g., 0.5°F) and go to Step 4. Otherwise, do not change the set point.
4. Limit the set point between upper and lower limits based on comfort considerations.

Static Pressure Reset for Variable Air Volume (VAV)

Flow may be modulated in a VAV system by using dampers on the outlet side of the fan, inlet vanes on the fan, vane-axial fans with controllable pitch fan blades, or variable-speed control of the fan motor. Typically, the inputs to any of these controlled devices are modulated in order to maintain a duct static pressure set point as described in Chapter 45. In a single duct VAV system, the duct static pressure set point is typically selected by the designer to be a fixed value. The sensor is located at a point in the duct work such that the established set point will ensure proper operation of the zone VAV boxes under varying load (supply airflow) conditions. A shortcoming of this approach is that static pressure is controlled based on a single sensor intended to represent the pressure available to all VAV boxes. Either a poor location or malfunction of this sensor will cause operating problems.

For a fixed static pressure set point, all of the VAV boxes tend to close as the zone loads and flow requirements decrease. Therefore, the flow resistance increases with decreasing load. Significant fan energy savings are possible if the static pressure set point is reset so that at least one of the VAV boxes remains open. With this approach, the flow resistance remains relatively constant. Englander and Norford (1992), Hartman (1993), and Warren and Norford (1993) proposed several different strategies based on this concept. Englander and Norford used simulations to show that either static pressure or fan speed can be controlled directly using a flow error signal from one or more zones and simple rules. Their technique forms the basis of the following reset strategy.

At each the decision interval (e.g., 5 minutes), the following logic can be applied:

1. Check the controller outputs for representative VAV boxes and determine time-averaged values over the last decision interval.
2. If any of the controller outputs are greater than a threshold value (e.g., 98%), then increase the static pressure set point by a fixed value (e.g., 5% of the design range) and go to Step 4. Otherwise, go to Step 3.
3. If *all* of the controller outputs are less than a threshold value (e.g., 90%), then decrease the static pressure set point by a fixed value (e.g., 5% of the design range) and go to Step 4. Otherwise, do not change the set point.
4. Limit the set point between upper and lower limits based on upper and lower flow limits and the duct design.

STRATEGIES FOR BUILDING ZONE TEMPERATURE SET POINTS

Typically, zone temperatures in commercial buildings are maintained at constant set points in the comfort zone during occupied periods. However, during unoccupied times, the set points are set up for cooling and set back for heating in order to reduce energy use. Optimal start algorithms determine times for turning equipment on so that the building zones reach the desired conditions when the building becomes occupied. The goal of these algorithms is to minimize the precool (or preheat) time. In some situations, the thermal mass of a building represents a storage medium that may be used to reduce peak cooling requirements. This section presents some simple strategies for these purposes.

Recovery from Night Setback or Setup

For buildings that are not continuously occupied, a significant savings in operating costs may be realized by raising the building set point temperature for cooling and by lowering the set point for heating during unoccupied times. Bloomfield and Fisk (1977) showed energy savings of 12% for a heavyweight building and 34% for a lightweight building.

An optimal controller for return from night setback or setup returns zone temperatures to the comfort range precisely when the building becomes occupied. Seem et al. (1989) compared seven different algorithms for minimum return time. Each method requires the estimation of parameters from measurements of the actual return times from night setback or setup.

Seem et al. (1989) showed that the optimal return time for cooling was not strongly influenced by the outdoor temperature. The following quadratic function of the initial zone temperature was found to be adequate for estimating the return time:

$$\tau = a_0 + a_1 t_{z,i} + a_2 t_{z,i}^2 \qquad (22)$$

where τ is an estimate of optimal return time, $t_{z,i}$ is initial zone temperature at the beginning of the return period, and a_0, a_1, and a_2 are empirical parameters. The parameters of Equation (22) may be estimated by applying linear least squares techniques to the difference between the actual return time and the estimates. These parameters may be continuously corrected using recursive updating schemes as outlined by Ljung and Söderström (1983).

For heating, Seem et al. found that ambient temperature has a significant effect on the return time and that the following relationship works well in correlating return times.

$$\tau = a_0 + (1-w)(a_1 t_{z,i} + a_2 t_{z,i}^2) + w a_3 t_a \qquad (23)$$

where t_a is the ambient temperature, a_0, a_1, a_2, and a_3 are empirical parameters, and w is a weighting function given by

$$w = 1000^{-(t_{z,i} - t_{unocc})/(t_{occ} - t_{unocc})} \qquad (24)$$

where t_{unocc} and t_{occ} are the zone set points for unoccupied and occupied periods. Within the context of Equation (23), this function weights the outdoor temperature more heavily when the initial zone temperature is close to the set point temperature during the unoccupied time. Again, the parameters of Equation (23) may be estimated by applying linear least squares techniques to the difference between the actual return time and the estimates.

Ideally, separate equations should be used for zones that have significantly different return times. Equipment operation is initiated for the zone with the earliest return time. In a building with a central cooling system, the equipment should be operated above some minimum load limit. With this constraint, some zones need to be returned to their set point earlier than the optimum time.

Optimal start algorithms often use a measure of the building mass temperature rather than the space temperature to determine return time. Although use of space temperature results in lower energy costs (i.e., shorter return time), the mass temperature may result in better comfort conditions at the time of occupancy.

Emergency Strategy to Limit Peak Cooling Requirements

Keeney and Braun (1997) developed a simple control strategy that makes use of building thermal mass in order to reduce peak cooling requirements in the event of a loss of a chiller. This emergency strategy is used only on days where the cooling capacity is not sufficient to keep the building in the comfort range using night setup control. It involves precooling the building during unoccupied times and allowing the temperature to float through the comfort zone during occupancy.

The precooling control strategy is depicted in Figure 30 along with conventional night setup control. Precooling is controlled at a constant temperature set point designated as t_{pre}. The warm-up period is used to reset the zone air temperature set point so that the cooling system turns off without calling for heating. During this time, the zone air warms due to lighting and equipment loads. The occupied set point (t_{occ}) is set at the low end of the comfort region so that the building mass charge is held as long as cooling capacity is available. This set point is maintained until the limit on cooling capacity is reached. After this point, the temperatures in the zones float up and the building thermal mass provides additional cooling. If the precooling and occupied set points have been chosen properly and the cooling capacity is sufficient, the zone conditions will remain in comfortable throughout the occupied period. The peak cooling requirement can be reduced by as much as 25% using this strategy as compared with night setup control. Thus, the loss of one of four identical chillers could be tolerated. This strategy could also be used for spaces such as auditoriums that have a high occupancy density for a short period.

The length of time and temperature for precooling and the occupied temperature set point chosen for this strategy strongly influences the capacity reduction and could impact occupant comfort. A reasonable strategy is to precool at 68°F beginning at midnight, allow a 30 minute warmup period prior to occupancy, and then adjust the occupied set point to 70°F. The zone temperature will then rise above this set point when the chillers are operating at capacity.

Case Study. The control strategy was tested in a 1.4 million square foot office building located near Chicago. The facility has two identical buildings with very similar internal gains and solar radiation loads that are connected by a large separately cooled entrance area. During tests, the east building used the existing building control strategy while the west building used the precooling strategy.

Four 900 ton vapor compression chillers normally provide chilled water to the air handling units. The loss of one chiller results in a 25% reduction of the total capacity. This condition was simulated by limiting the vane position of the two chiller units that cool the west building to 75%. The capacity limitation was imposed directly at the chiller control panels. Set points were provided to local zone controllers from a modern energy management and control system. Chiller cooling loads and zone thermal comfort conditions were monitored throughout the tests.

Consistent with simulation predictions, the precooling control strategy successfully limited the peak load to 75% of the cooling capacity for the west building, while the east building operated at 100% of capacity. Figure 31 shows the total chiller coil load for the east and west buildings for a week of testing in the middle of August 1995. The cooling coil load profile on Monday is the most dramatic example of the load shifting during this test period. The peak cooling load for this facility often occurs on Monday morning. The cooling limit was achieved on Monday during a period in which a heat emergency had been declared in the city. The severe ambient conditions were compounded by a power outage that caused a loss of the west side chiller units for approximately 20 minutes. Under these demanding conditions, the precooling strategy maintained occupant comfort while successfully limiting cooling demand of the west side of the building to less than 75% of that for the east side.

The east side cooling requirement was at or below the 75% chiller capacity target for Tuesday through Friday so the emergency precooling strategy was not necessary. For these off-design days, the emergency strategy is not effective in reducing the on-peak cooling requirements because discharge of the mass is not initiated when the capacity is below the target. The thermal mass remains charged so that peak reduction would occur if the target value on the off design days was reset to a lower value.

Precooling of the top floor of the facility had already been implemented into the conventional control strategy used for the east building. This was necessary to maintain comfort conditions with full cooling capacity on hot days. As a result, even greater peak reduction would have been recorded if the precool strategy had been compared with conventional night setup control. The total electrical use was greater for the precooled west building, however the strategy was designed as an emergency strategy and does not attempt to minimize costs.

This emergency strategy should only be applied on days when the available cooling capacity is not sufficient to maintain comfort

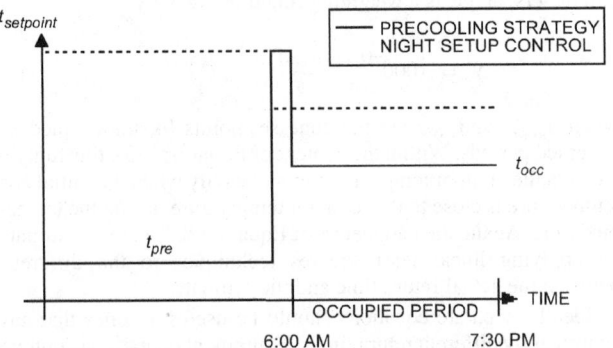

Fig. 30 Zone Air Temperature Set Points

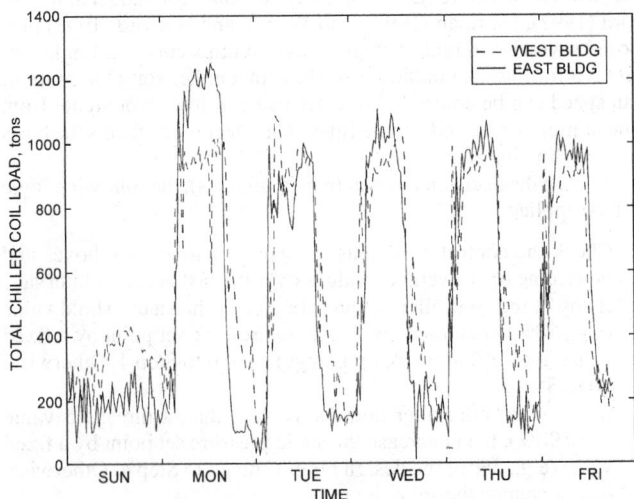

Fig. 31 Total Coil Load for East and West Chiller Units

conditions when using night setup control. Otherwise, the costs associated with providing cooling could increase significantly.

CONTROL OF COOL THERMAL STORAGE

As described in Chapter 33, thermal storage can significantly reduce the costs of providing cooling in commercial buildings by shifting a portion of the daytime cooling requirements to the nighttime hours. The savings result from reductions in both demand and energy charges. This section emphasizes ice storage applications, although much of it is relevant to chilled water storage as well.

Figure 32 shows a schematic of a typical ice storage system. The system consists of one or more chillers, cooling tower cells, condenser water pumps, chilled water/glycol distribution pumps, ice storage tanks, and valves for controlling charging and discharging modes of operation. Ice is made at night and used during the day to provide a portion of a building's cooling requirements. The partial storage terminology comes about because the storage is not sized to handle the full on-peak load requirement on the design day. Typically, in a load-leveling scheme, the storage and chiller capacity are sized such that chiller operates at full capacity during the on-peak period on the design day.

Typical modes of operation for the system shown in Figure 32 are as follows:

1. **Storage Charging Mode.** Typically, the storage only charges (i.e., ice is made) when the building is unoccupied and off-peak electric rates are in effect. In this mode, the load bypass valve (V-2) is fully closed to the building cooling coils, the storage control valve (V-1) is fully open to the ice-storage tank (the total chilled water/gycol flow is through the tank), and the chiller produces low temperatures (e.g., 20°F) sufficient to make ice.

2. **Storage Discharging Mode.** The storage only discharges (i.e., the ice melts) when the building is occupied. In this mode, valve V-2 is open to the building cooling coils and valve V-1 modulates the mixture of flows from the storage tank and chiller in order to maintain a constant supply temperature to the building cooling coils (e.g., 38°F). Individual valves at the cooling coils modulate their chilled water/gycol flow to maintain supply air temperatures to the zones.

3. **Direct Chiller Mode.** The chiller may operate to meet the load directly without the use of storage during the occupied mode (typically when off-peak electric rates are in effect). In this mode, valve V-1 is fully closed with respect to the storage tank.

For a typical partial-storage system, the storage meets only a portion of the on-peak cooling loads on the design day and the chiller operates at capacity during the on-peak period. As a result, the peak power is limited by the capacity of the chiller. For off-design days, many different control strategies would meet the building's cooling requirements. However, each method has a different operating cost.

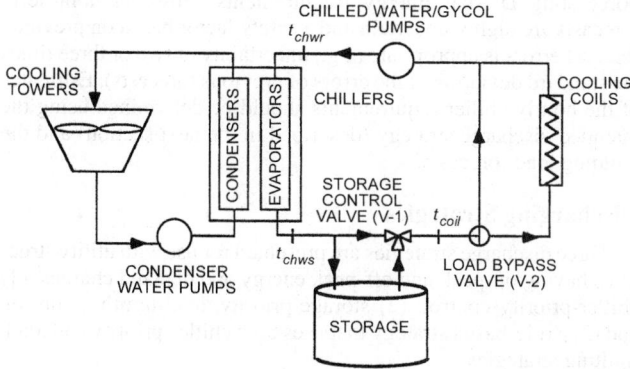

Fig. 32   Schematic of Ice Storage System

The best control strategy for a given day is a complex function of such factors as utility rates, load profile, chiller characteristics, storage characteristics, and weather. For a utility rate structure that includes both time-of-use energy and demand charges, the optimal strategy depends on variables that extend over a month. Consider the following typical discrete cost function associated with electrical use in a building:

$$J = \sum_{k=1}^{N} \{E_k P_k \Delta\tau\} + \text{Max}_{1 \le k \le N}\{D_k P_k\} \qquad (25)$$

where J is the utility cost associated with the billing period (e.g., one month); $\Delta\tau$ is the stage time interval (typically equal to the time over which demand charges are levied, e.g. 0.25 h); N is the number of time stages in a billing period, and for each stage k, P is the average building electrical power (kW); E is the energy cost rate or cost per unit of electrical energy ($/kWh); and D is the demand charge rate or cost per peak power rate over the billing period ($/kW).

The first term in Equation (25) represents the total cost of energy use for a building over a billing period, which is usually a month. Typically, the energy cost rate varies according to time of use with high rates during the daytime on weekdays and low costs at night and on weekends. The second term in Equation (25) is the building demand cost and is the product of the peak power consumption that occurs during the billing period and the demand cost rate for that stage. The demand cost rate can also vary with time of day with higher rates occurring during on-peak periods. In order to determine a control strategy for charging and discharging storage that minimizes utility costs for a given system, it would be necessary to perform a minimization of Equation (25) over the entire billing period because of the influence of the demand charge. An even more complicated cost optimization would result if the utility rate included ratchet clauses whereby the demand charge is the maximum of the monthly peak demand cost and some fraction of the previous monthly peak demand cost during the cooling season. In either case, it is not worthwhile to perform an optimization over periods longer than those for which reliable forecasts of cooling requirements or ambient conditions could be performed (e.g., 1 day). It is therefore important to have simple control strategies for charging and discharging storage over a daily cycle.

The following control strategies for limiting cases of the cost function of Equation (25) provide further insight.

1. In a limiting case where the demand cost rate is zero and the energy cost rate does not vary with time, the minimization of Equation (25) is equivalent to minimizing the total electrical energy use. In general, the cooling plant efficiency is much lower when it makes ice than when it cools the building. Thus, the optimal strategy for minimum energy use would minimize the use of storage. Although this may seem like a simple example, the most common control strategy in use today for partial ice storage attempts to minimize the use of storage and is called **chiller-priority control**.

2. In a situation where the demand cost rate is zero but energy costs are higher during on-peak than off-peak periods, the minimization of Equation (25) involves trade-offs between energy use and energy cost rates. For a relatively small difference between on-peak and off-peak rates of less than about 30%, the energy penalties for ice making typically outweigh the effect of the reduced rates and chiller-priority control is optimal for many cases. However, with differentials higher than about 30% between on-peak and off-peak energy rates or with chillers having smaller charging-mode energy penalties, the optimal strategy might maximize the use of storage. A control strategy that attempts to maximize the load-shifting potential of storage is typically termed **storage-priority control**. With storage-priority control, the chiller operates during the off-peak period to fully charge storage (i.e., make

ice). During the on-peak period, storage is used to cool the building in a manner that minimizes the use of the chiller(s). Partial storage systems that use storage-priority control strategies require forecasts for building cooling requirements in order to avoid the premature depletion of storage.

3. If energy were free and only on-peak demand costs are considered, the optimal control strategy would tend to maximize the use of storage and would control the discharge of storage in a manner that would always minimize the peak building power. A **storage-priority, demand-minimization control** strategy for partial-storage systems would require both cooling load and non-cooling electrical use forecasts.

A number of control strategies based on the three limiting cases described above have been proposed for ice storage (Rawlings 1985, Tamblyn 1985, Spethmann 1989, Grumman and Butkus 1989, Braun 1992, Drees and Braun 1996). Braun (1992) evaluated the performance of chiller-priority and storage-priority control strategies as compared with optimal control. The storage-priority strategy was termed load-limiting control because it attempts to minimize the peak cooling load during the on-peak period. The load-limiting strategy provided near-optimal control in terms of demand costs in all cases and worked well with respect to energy costs when time-of-day energy charges were available. However, the scope of the study was limited.

Krarti et al. (1996) evaluated chiller-priority and storage-priority control strategies as compared with optimal control for a wide range of systems, utility rate structures, and operating conditions. Similar to Braun (1992), they concluded that load-limiting, storage-priority control provides near-optimal performance when there is a significant differential cost between on-peak and off-peak energy and demand charges. However, optimal control provides superior performance in the absence of time-of-day incentives. In general, the monthly utility costs associated with chiller-priority control are significantly higher than optimal and storage-priority control. However, without time-of-use energy charges, chiller-priority control did provide good performance for individual days when the daily peak power was less than the monthly peak.

Drees and Braun (1996) developed a simple rule-based control strategy that combines elements of storage-priority and chiller-priority strategies in a way that results in near-optimal performance under all conditions. The strategy uses rules obtained from both daily and monthly optimization results for several simulated systems.

The choice of control strategy results from a trade-off between performance (i.e., operating cost) and ease of implementation (i.e., initial cost). Chiller-priority control has the lowest implementation cost, but generally leads to the highest operating cost. Storage-priority strategies provide superior performance, but require the use of a forecaster and a measurement of state of charge for storage. The rest of this section presents details of chiller-priority, load-limiting, and a rule-based control. Each of these strategies shares the same procedures for charging storage, but differ in the manner in which storage is discharged. In general, the control strategies presented are appropriate for utility rate structures that include time-of-use energy and demand charges, but would not be appropriate in conjunction with real-time pricing. Additional information on control strategies for cool storage can be found in Chapter 33, Thermal Storage.

Charging Strategies

Ice making should be initiated when the building is both unoccupied and off-peak electrical rates are in effect. During the ice-making period, the chiller should operate at full capacity. Cooling plants for ice storage generally operate most efficiently at full load due to the auxiliaries and the characteristics of ice-making chillers. With feedback control of the chilled water/glycol supply temperature, full capacity control is accomplished by establishing a low enough set point to ensure this condition (e.g., 20°F).

Internal Melt Storage Tanks. The chiller should operate until the tank reaches its maximum state of charge or the charging period (e.g., off-peak, unoccupied period) ends. This strategy ensures that sufficient ice will be available for the next day without the need for a forecaster. Typically, only a small heat transfer penalty is associated with restoring a partially discharged, internal melt storage tank to a full charge. For this type of storage device, the charging cycle always starts with a high transfer effectiveness because water surrounds the tubes regardless of the amount of ice melted. The heat transfer effectiveness drops gradually until the new ice formations intersect with old formations, at which point the tank is fully recharged.

External Melt Storage Tanks. These tanks have a more significant heat transfer penalty associated with recharging after a partial discharge because ice forms on the outside of existing formations during charging. In this case, it is desirable from an efficiency standpoint to fully discharge the tank each day and only recharge as necessary to meet the next day's cooling requirements. To ensure that adequate ice is available, the maximum possible storage capacity needed for the next day must be forecast. The storage requirements for the next day depend on the discharge strategy used and the building load. In general, the state of charge for storage necessary to meet the next day's load can be estimated according to

$$X_{chg} = \sum_{k=1}^{\substack{\text{occupied} \\ \text{period}}} \frac{\hat{Q}_{load,k} - \hat{Q}_{ch,k}}{C_s} \qquad (26)$$

where X_{chg} is the relative state of charge at the end of the charging period, C_s is the maximum change in internal energy of the storage tank that can occur during a normal discharge cycle, and $\hat{Q}_{load,k}$ and $\hat{Q}_{ch,k}$ are forecasts of the building load and chiller cooling requirement for the kth stage (e.g., hour) of the occupied period. The relative state of charge is defined in terms of two reference states: the fully discharged and fully charged conditions that correspond to values of zero and one. These conditions are defined for a given storage based on its particular operating strategy (Elleson 1996, ASHRAE 1997). The fully charged condition is that existing when the control stops the charge cycle as part of its normal sequence. Similarly, the fully discharged condition is the point where no more usable cooling is recovered from the tank. Typically, zero state of charge corresponds to a tank of water at a uniform temperature of 32°F and a complete charge is associated with a tank having maximum ice build at 32°F. (The fully discharged and fully charged conditions are arbitrarily selected reference states.) In abnormal circumstances, a storage tank can be discharged or charged beyond these conditions, resulting in relative states of charge below zero or above one.

Hourly forecasts of the next day's cooling requirement can be determined using the algorithm described in following section on Forecasting Diurnal Energy Requirements. However, long-term forecasts are highly uncertain and a safety factor based on previous forecast errors is appropriate (e.g., uncertainty of two or three times the standard deviation of the errors of previous forecasts). Estimates of the hourly chiller requirements should be determined using the intended discharge strategy (described in the next section) and the building load forecasts.

Discharging Strategies

Three discharge strategies are presented for use with utility structures having on-peak and off-peak energy and demand charges: (1) chiller-priority control, (2) storage-priority, load-limiting control, and (3) a rule-based strategy that uses both chiller-priority and load-limiting strategies.

Chiller-Priority Discharge. During the storage discharge mode, the chiller operates at full cooling capacity (or less if sufficient to meet the load) and storage matches the difference between the building requirement and chiller capacity. For the example system illustrated in Figure 32, the chiller supply temperature set point t_{chws} is set equal to the desired supply temperature for the coils t_{coil}. If the capacity of the chiller is sufficient to maintain this set point at any time, then storage is not used and the system operates in the direct chiller mode. Otherwise, the storage control valve modulates the flow through storage to maintain the supply set point, providing a cooling rate that matches the difference between the building load and the maximum cooling capacity of the chillers.

This strategy is easy to implement and does not require a load forecast. It works well for design conditions, but can result in relatively high demand and energy costs for off-design conditions because the chiller operates at full capacity during the on-peak period.

Load-Limiting, Storage Priority Control. Several storage-priority control approaches ensure that storage is not depleted prematurely. Braun (1992) presented a storage-priority control strategy, termed load-limiting control, which tends to minimize the peak cooling plant power demand. The operation of equipment for load-limiting control during different parts of the occupied period can be described as follows:

1. **Off-Peak, Occupied Period.** During this period, the goal is to minimize the use of storage and the chiller-priority described in the previous section should be applied.
2. **On-Peak, Occupied Period.** During this period, the goal is to operate the chillers at a constant load while discharging the ice storage such that the ice is completely melted when the off-peak period begins. This requires the use of a building cooling load forecaster. At each decision interval (e.g., 15 min), the following steps are applied:
   (a) Forecast the total integrated building cooling requirement until the end of the discharging period.
   (b) Estimate the state of charge of the ice-storage tank from measurements.
   (c) At any time, the chiller loading for load-limiting control is determined as:

$$\dot{Q}_{LLC} = \text{Max}\left\{\frac{\hat{Q}_{load,occ} - (X - X_{min})C_s}{\Delta\tau_{on}}, \dot{Q}_{ch,min}\right\} \quad (27)$$

   where $\hat{Q}_{load,occ}$ is a forecast of the integrated building load for the rest of the on-peak period, $\Delta\tau_{on}$ is the time remaining in the on-peak period, X is the current state of charge defined as the fraction of the maximum storage capacity, X_{min} is a minimum allowable state of charge, C_s is the maximum possible energy that could be added to storage during discharge, and $\dot{Q}_{ch,min}$ is the minimum allowable chiller cooling capacity. If the chiller does not need to be operated during the remainder of the occupied, on-peak period, the minimum allowable cooling capacity could be set to zero. Otherwise, the cooling capacity should be set to the minimum at which the chiller can safely operate.
   (d) Determine the chiller set point temperature necessary to achieve the desired loading as

$$t_{chws} = t_{chwr} - \frac{\dot{Q}_{LLC}}{C_{chw}} \quad (28)$$

   where t_{chwr} is the temperature of water/glycol returned to the chiller and C_{chw} is the capacitance rate (mass flow times specific heat) of the flow stream.

Hourly forecasts of cooling loads can be determined using the algorithm described in the section on Forecasting Diurnal Energy Requirements. The hourly forecasts are then integrated to give a forecast of the total cooling requirement. To ensure sufficient

cooling capacity, a worst-case forecast of cooling requirements could be estimated as the sum of the best forecast and two or three times the standard deviation of the errors of previous forecasts.

Rule-Based Controller. Drees and Braun (1996) presented a rule-based controller that combines elements of chiller-priority and storage-priority strategies, along with a demand-limiting algorithm to achieve near-optimal control. The demand-limiting algorithm requires a measurement of the total building electrical use. A simpler strategy is described here that doesn't require this measurement and yields equivalent performance whenever the peak demand for the billing period is coincident with the peak cooling load.

Figure 33 shows a flowchart for the discharge strategy that is applied during each decision interval (e.g., 15 min.) during the occupied period. Block 1 determines whether the use of storage should be maximized or minimized. Block 2 is used if the use of storage lowers daily energy costs and storage is sufficient to meet the remainder of the load for the occupied period without operating the chillers. Otherwise, the goal of the strategy in Block 3 is to minimize the use of storage while keeping the peak load below a limit. This strategy tends to keep the chiller(s) heavily loaded (and therefore operating efficiently) until they are no longer needed. The logic in each block is as follows.

1. **Block 1: Discharge Strategy Selection.** The discharge of storage will not reduce the energy cost whenever the cost of replenishing the ice is greater than the cost of providing direct cooling by the chiller(s). This situation is always the case during the off-peak, occupied period since the electricity rates are the same as those associated with the charging period and chillers are less efficient in ice making mode than when providing direct cooling. Furthermore, during the on-peak, occupied period the use of storage generally reduces energy costs whenever the following criterion holds.

$$\text{ECR} > \text{COP}_d/\text{COP}_c \quad (29)$$

where ECR is the ratio of on-peak to off-peak energy charges and COP_d and COP_c are the coefficients of performance for the cooling plant (including chiller, pumps, and cooling tower fans) during discharging and charging of the tank. The COPs should be evaluated at the worst-case charging and discharging conditions associated with the design day. Typically, this ratio is between about 1.2 and 1.8 for systems with cooling towers. However, this ratio can be lower for systems with air-cooled condensers in dry climates due to cool nighttime temperatures.

If the criterion of Equation (29) is satisfied, the control will switch from chiller-priority to storage-priority strategy whenever the storage capacity is greater than the remaining integrated load. Therefore, the storage-priority control is enabled whenever

$$(X - X_{min})C_s \geq \hat{Q}_{load,occ} \quad (30)$$

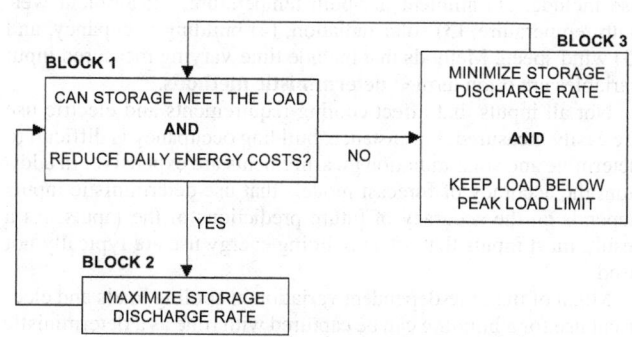

Fig. 33   Flowchart for Rule-Based Controller Discharge Strategy

Hourly forecasts of cooling loads can be determined using the algorithm described in the section on Forecasting and then integrated to give a forecast of the total cooling requirement. To ensure that adequate ice is available, worst-case hourly forecasts could be determined by adding the expected value of the hourly forecasts and the forecast errors associated with a specified confidence interval (e.g., 2 standard deviations for a 95% confidence interval). The worst-case hourly forecasts would then be integrated to give a worst-case integrated forecast.

2. **Block 2: Maximum Use of Storage.** In this mode, the chillers are turned off and storage is used to meet the entire load throughout the remainder of the occupied period. However, a chiller may need to be turned on if the storage discharge rate is not sufficient to meet the building load (i.e., the coil supply temperature set point can not be maintained).

3. **Block 3: Minimize Use of Storage with Peak Load Limiting.** At any time, a target chiller load is determined as

$$\dot{Q}_{ch} = \text{Min}\{\text{Max}[\dot{Q}_{ch,peak}, \dot{Q}_{LCC}], \dot{Q}_{load}\} \qquad (31)$$

where $\dot{Q}_{ch,peak}$ is the peak chiller cooling requirement that has occurred during the on-peak period for the current billing period, $\dot{Q}_{LCC}$ is the chiller load associated with load-limiting control and determined with Equation (27), and $\dot{Q}_{load}$ is the current building load. The chiller set point temperature necessary to achieve the desired loading is determined as:

$$t_{chws} = \text{Max}\left\{t_{chwr} - \frac{\dot{Q}_{ch}}{C_{chw}}, t_{coil}\right\} \qquad (32)$$

On the first day of each billing period, $\dot{Q}_{ch,peak}$ is set to zero. For this first day, application of Equation (31) leads to the load-limiting control strategy described in the previous section. On subsequent days, load-limiting control is used only if the current peak limit would lead to premature depletion of storage. Whenever the current load is less than $\dot{Q}_{ch,peak}$ and $\dot{Q}_{LCC}$, Equations (31) and (32) lead to chiller-priority control.

FORECASTING DIURNAL ENERGY REQUIREMENTS

As discussed previously, forecasts of cooling requirements and electrical use in buildings are often necessary for the control of thermal storage to shift electrical use from on-peak to off-peak periods. In addition, forecasts can help plant operators anticipate major changes in operating modes such as bringing additional chillers online.

In most methods, future predictions are estimated as a function of time-varying input variables that affect the cooling requirements and electrical use. Examples of inputs that affect building energy use include: (1) ambient dry-bulb temperature, (2) ambient wet-bulb temperature, (3) solar radiation, (4) building occupancy, and (5) wind speed. Methods that include time-varying measured input variables are often termed **deterministic methods**.

Not all inputs that affect cooling requirements and electric use are easily measured. For instance, building occupancy is difficult to determine and solar radiation measurements are expensive. In addition, the accuracy of forecast models that use deterministic inputs depends on the accuracy of future predictions of the inputs. As a result, most inputs that affect building energy use are typically not used.

Much of the time-dependent variation in cooling loads and electrical use for a building can be captured with time as a deterministic input. For instance, building occupancy follows a regular schedule that depends on time of the day and time of the year. In addition, variations in ambient conditions follow a regular daily and seasonal

pattern. Many of the forecasting methods use time in place of unmeasured deterministic inputs in a functional form that captures the average time dependence of the variation in energy use.

The use of a deterministic model has a limited accuracy for forecasts due to both unmeasured and unpredictable (random) input variables. Short-term forecasts can be improved significantly by adding previous values of deterministic inputs and previous output measurements (cooling requirements or electrical use) as inputs to the forecasting model. The time history of these inputs provides valuable information about recent trends in the time variation of the forecasted variable and the unmeasured input variables that affect it. Most forecasting methods use variables that reflect past history to predict the future.

Any forecasting method requires that a functional form is defined and the parameters of the model are learned based on measured data. Either offline or online methods can be used to estimate parameters. Offline methods involve estimating parameters from a batch of data that has been collected. Typically, the parameters are determined by minimizing the sum of squares of the forecast errors. The parameters of the process are assumed to be constant over time in the offline methods. Online methods allow the parameters of the forecasting model to vary slowly with time. Again, the sum of squares of the forecast errors is minimized, but this is accomplished in a sequential or recursive manner. Often, a forgetting factor is used in order to give additional weight to the recent data. The ability to track time-varying systems can be important when forecasting cooling requirements or electric use in buildings because of the influence of seasonal variations in weather.

Forrester and Wepfer (1984) presented a forecasting algorithm that uses current and previous ambient temperatures and previous loads to predict future requirements. Trends on an hourly time scale are accounted for with measured inputs for a few hours preceding the current time. Day-to-day trends are considered by the use of the value of the load that occurred 24 h earlier as an input. One of the major limitations of this model is its inability to accurately predict loads when an occupied day (e.g., Monday) follows an unoccupied (e.g., Sunday) or when an unoccupied day follows an occupied day (e.g., Saturday). The cooling load for a particular hour of the day on a Monday depends very little on the requirement 24 h earlier on Sunday. Forrestor and Wepfer (1984) described a number of methods for eliminating this 24 h indicator. MacArthur et al. (1989) also presented a load profile prediction algorithm that uses a 24 h regressor.

Armstrong et al. (1989) presented a very simple method for forecasting either cooling or electrical requirements that does not use the 24 h regressor. Then Seem and Braun (1991) further developed and validated the method. The "average" time-of-day and time-of-week trends are modeled using a lookup table with time of day and type of day (e.g., occupied versus unoccupied) as the deterministic input variables. Entries in the table are updated using an exponentially-weighted, moving-average model. Short-term trends are modeled using previous hourly measurements of cooling requirements in an autoregressive (AR) model. Model parameters adapt to slow changes in the system characteristics. The combination of updating the table and modifying model parameters works well in adapting the forecasting algorithm to changes in season and occupancy schedule.

Kreider and Wang (1991) used artificial neural networks (ANNs) to predict energy consumption of various HVAC equipment in a commercial building. Data inputs to the ANN included (1) previous hour's electrical power consumption, (2) building occupancy, (3) wind speed, (4) ambient relative humidity, (5) ambient dry-bulb temperature, (6) previous hour's ambient dry-bulb temperature, (7) two hours' previous ambient temperature, (8) sine and cosine of the hour number to roughly represent the diurnal change of temperature and solar insolation. The primary purpose in developing these models was to detect changes in equipment and system performance for monitoring purposes. However, the authors suggested that an ANN-based predictor might be valuable when used to predict energy consumption in the future with a network based on recent historical data. Forecasts of all deterministic input variables would be necessary in order to apply this method.

Gibson and Kraft (1993) used an ANN to predict building electrical consumption as part of the operation and control of a thermal energy storage (TES) cooling system. The ANN used the following inputs: (1) electric demand of occupants (lighting and other loads), (2) electric demand of TES cooling tower fans, (3) outside ambient temperature, (4) outside ambient temperature-inside target temperature, (5) outside ambient relative humidity, (6) on-off status for building cooling, (7) cooling system on-off status, (8) Chiller #1 direct-cooling mode on-off status, (9) Chiller #2 direct-cooling mode on-off status, (10) ice storage discharging mode on-off status, (11) ice storage charging mode on-off status, (12) Chiller #1 charging mode on-off status, (13) Chiller #2 charging mode on-off status. In order to use this forecaster, the values of each of these inputs must be predicted. Although the authors suggest that average occupancy demand profile be used as an input, they do not state how the other input variables should be forecast.

A Forecasting Algorithm

This section presents an algorithm for forecasting hourly cooling requirements or electrical use in buildings that is based on the method developed by Seem and Braun (1991). At a given hour n, the forecast value is

$$\hat{E}(n) = \hat{X}(n) + \hat{D}(h,d) \tag{33}$$

where

$\hat{E}(n)$ = forecast cooling load or electrical use for hour n
$\hat{X}(n)$ = stochastic or probabilistic part of forecast for hour n
$\hat{D}(h,d)$ = deterministic part of forecast at hour n associated with hth hour of day and current day type d

The deterministic part of the forecast is simply a lookup table for the forecasted variable in terms of hour of the day h and the type of day d. Seem and Braun recommend the use of three distinct day types: unoccupied days, occupied days following unoccupied days, and occupied days following occupied days. The three day types account for differences between the building response associated with return from night setup and return from weekend setup. The building operator or control engineer must specify the number of day types and a calendar of day types.

Given the hour of the day and the day type, the deterministic part of the forecast is simply the value stored in that location in the table. Table entries are updated when a new measurement becomes available for that hour and day type. Updates are accomplished through the use of an exponentially-weighted, moving-average (EWMA) model as

$$\hat{D}(h,d) = \hat{D}(h,d)_{old} + \lambda[E(n) - \hat{D}(h,d)_{old}] \tag{34}$$

where

$E(n)$ = measured value of cooling load or electrical use for current hour n
λ = exponential smoothing constant, $0 < \lambda < 1$
$\hat{D}(h,d)_{old}$ = previous table entry for $\hat{D}(h,d)$

As λ increases, the more recent observations have more influence on the average. As λ approaches zero, the table entry approaches the average of all data for that hour and day type. When λ equals one, the table entry is updated with the most recent measured value. Seem and Braun recommend using a value of 0.30 for λ in conjunction with three day types and 0.18 with two day types.

The stochastic portion of the forecast is estimated with a third-order autoregressive model, AR(3), of the forecasting errors associated with the deterministic model. With this model, an estimate of

the next hour's error in the deterministic model forecast is given by the following equation:

$$\hat{X}(n + 1) = \phi_1 X(n) + \phi_2 X(n - 1) + \phi_3 X(n - 2) \tag{35}$$

where

$X(n)$ = difference between measurement and deterministic forecast of cooling load or electrical use at any hour n
ϕ_1, ϕ_2, ϕ_3 = parameters of AR(3) model that must be learned

The error in the deterministic forecast at any hour is simply

$$X(n) = E(n) - \hat{D}(h,d) \tag{36}$$

For forecasting more than one hour ahead, conditional expectation is used to estimate the deterministic model forecast errors using the AR(3) model as follows:

$$\hat{X}(n + 2) = \phi_1 \hat{X}(n + 1) + \phi_2 X(n) + \phi_3 X(n - 1)$$

$$\hat{X}(n + 3) = \phi_1 \hat{X}(n + 2) + \phi_2 \hat{X}(n + 1) + \phi_3 X(n)$$

$$\vdots$$

$$\hat{X}(n + k) = \phi_1 \hat{X}(n + k + 1) + \phi_2 \hat{X}(n + k - 2)$$
$$+ \phi_3 \hat{X}(n + k - 3) \qquad \text{for } k > 3 \tag{37}$$

Online estimation of the AR(3) model parameters is accomplished by minimizing the following time-dependent cost function.

$$J(\phi) = \sum_{k=1}^{n} \alpha^{n-k} [X(k) - \hat{X}(k)]^2 \tag{38}$$

where the constant α is called the forgetting factor and has a value between 0 and 1. With this formulation, the residual for the current time step has a weight of one and the residual for k time steps back has a weight of α^k. By choosing a value of α that is positive and less than one, recent data has greater influence on the parameter estimates. In this manner, the model can track changes due to seasonal or other effects. Seem and Braun recommend using a forgetting factor of 0.99. Parameters of the AR(3) model should be updated at each hour when a new measurement becomes available.

Ljung and Söderström (1983) describe online estimation methods for determining coefficients of an AR model. The parameter estimates should be evaluated for stability. If an AR model is not stable, then the forecasts will grow without bound as the time of forecasts increases. Ljung and Söderström discuss methods for checking stability.

Seem and Braun compared forecasts of electrical usage with both simulated and measured data. Figure 34 shows the standard deviation of the 1 h through 24 h errors in electrical use forecasts for annual simulation results. Results are given for the deterministic model alone, deterministic plus AR(2), and deterministic plus AR(3). For the combined models, the standard deviation of the residuals increases as the forecast length increases. For short time steps (i.e., less than six hours), the combined deterministic and stochastic models provide much better forecasts than the purely deterministic model (i.e., lookup table).

Seem and Braun also investigated a method for adjusting the deterministic forecast based upon the use of Weather Service forecasts of maximum daily ambient temperature as an input. For short periods (i.e., less than 4 h), the forecasts for the temperature-dependent model were nearly identical to the forecasts for the temperature

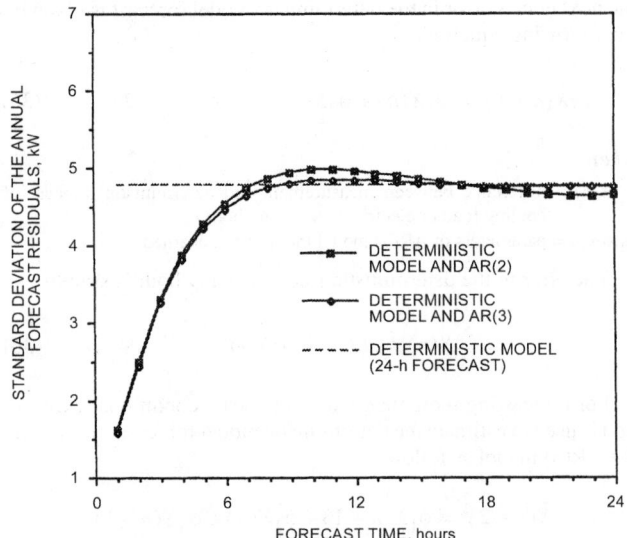

Fig. 34 Standard Deviation of Annual Errors for 1 to 24 h Ahead Forecasts

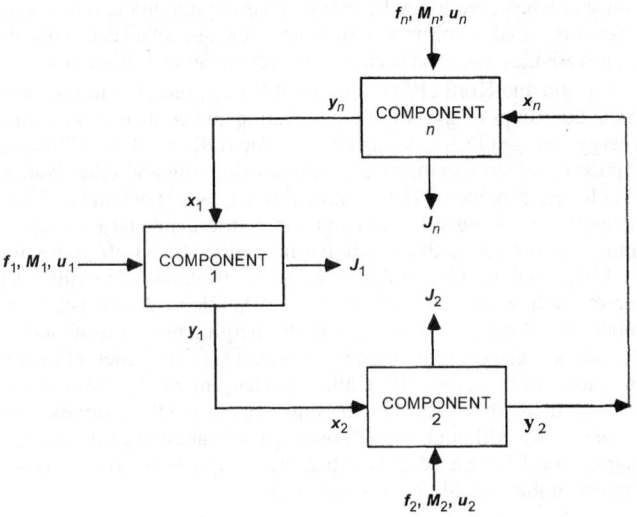

Fig. 35 Schematic of Modular Optimization Problem

independent model. For longer periods, the temperature-dependent model provided better forecasts than the temperature-independent model.

CONTROL OPTIMIZATION METHODS

STATIC OPTIMIZATION

Optimal supervisory control of cooling equipment involves determining the control that minimizes the total operating cost. For an all-electric system without significant storage, the optimization of cost leads to minimization of power at each instant in time. The optimal control does depend on time, through changing cooling requirements and ambient conditions. Static optimization techniques applied to a general simulation can be used to determine the optimal supervisory control variables. The simulation could be based on physical or empirical models. However, for control variable optimization, empirical and semi-empirical models are often used where variables are estimated from measurements. This section presents a framework for determining optimal control and a simplified approach for estimating control laws for cooling plants.

General Static Optimization Problem

Figure 35 depicts the general nature of the static optimization problem for a system of interconnected components. Each component in a system is represented as a separate set of mathematical relationships organized into a computer model. Its output variables and operating cost are functions of parameter, input, output, uncontrolled, and controlled variables. The structure of the complete set of equations to be solved for the entire system is dictated by the manner in which the components are interconnected.

The problem is formally stated as the minimization of the sum of the operating costs of each component J_i with respect to all discrete and continuous controls or

Minimize

$$J(\mathbf{f}, \mathbf{M}, \mathbf{u}) = \sum_{i=1}^{n} J_i(\mathbf{x}_i, \mathbf{y}_i, \mathbf{f}_i, \mathbf{M}_i, \mathbf{u}_i) \qquad (39)$$

with respect to $\mathbf{M}$ and $\mathbf{u}$, subject to equality constraints of the form

$$\mathbf{g}(\mathbf{f}, \mathbf{M}, \mathbf{u}) = \begin{bmatrix} \mathbf{g}_1(\mathbf{f}_1, \mathbf{M}_1, \mathbf{u}_1, \mathbf{x}_1, \mathbf{y}_1) \\ \mathbf{g}_2(\mathbf{f}_2, \mathbf{M}_2, \mathbf{u}_2, \mathbf{x}_2, \mathbf{y}_2) \\ \cdot \\ \mathbf{g}_n(\mathbf{f}_n, \mathbf{M}_n, \mathbf{u}_n, \mathbf{x}_n, \mathbf{y}_n) \end{bmatrix} = 0 \qquad (40)$$

and inequality constraints of the form

$$\mathbf{h}(\mathbf{f}, \mathbf{M}, \mathbf{u}) = \begin{bmatrix} \mathbf{h}_1(\mathbf{f}_1, \mathbf{M}_1, \mathbf{u}_1, \mathbf{x}_1, \mathbf{y}_1) \\ \mathbf{h}_2(\mathbf{f}_2, \mathbf{M}_2, \mathbf{u}_2, \mathbf{x}_2, \mathbf{y}_2) \\ \cdot \\ \mathbf{h}_n(\mathbf{f}_n, \mathbf{M}_n, \mathbf{u}_n, \mathbf{x}_n, \mathbf{y}_n) \end{bmatrix} \geq 0 \qquad (41)$$

where, for any component i,

$\mathbf{x}_i$ = vector of input stream variables
$\mathbf{y}_i$ = vector of output stream variables
$\mathbf{f}_i$ = vector of uncontrolled variables
$\mathbf{M}_i$ = vector of discrete control variables
$\mathbf{u}_i$ = vector of continuous control variables
J_i = operating cost
$\mathbf{g}_i$ = vector of equality constraints
$\mathbf{h}_i$ = vector of inequality constraints

Typical input and output stream variables for thermal systems are temperature and mass flow rate. The uncontrolled variables are measurable quantities that may not be controlled, but that affect the component outputs and/or costs, such as ambient dry-bulb and wet-bulb temperature.

Both equality and inequality constraints arise in the optimization of chilled water systems. One example of an equality constraint that arises when two or more chillers are in operation is that the sum of their loads must equal the total load. The simplest type of inequality constraint is a bound on a control variable. For example, lower and upper limits are necessary for the chilled water set temperature, in order to avoid freezing in the evaporator and to provide adequate dehumidification for the zones. Any equality constraint may be rewritten in the form of Equation (40) such that when it is satisfied, the constraint equation is equal to zero. Similarly, inequality constraints may be expressed as Equation (41), so that the constraint equation is greater than or equal to zero to avoid violation.

Braun (1988) and Braun et al. (1989b) presented a component-based non-linear optimization and simulation tool and used it to investigate optimal performance. Each component is represented as a separate subroutine with its own parameters, controls, inputs, and outputs. The optimization problem is solved in an efficient manner by using second-order representations for costs that arise from curve-fits or Taylor series approximations. Application of the component-based optimization led to many guidelines for control and a simplified system-based optimization methodology. In particular, the results showed that optimal set points could be correlated as a linear function of load and ambient wet-bulb temperature.

Cumali (1988, 1994) presented a method for real-time global optimization of HVAC systems including the central plant and associated piping and duct networks. The method uses a building load model for the zones based on a coupled weighting factor method similar to that used in DOE-2. Variable time steps are used to predict loads over a five to fifteen minute period. The models are based on thermodynamic, heat transfer, and fluid mechanics fundamentals and are calibrated to match the actual performance using data obtained from the building and plant. Pipe and duct networks are represented as incidence and circuit matrices, and both dynamic and static losses are included. The coupling of the fluid energy transfers with the zone loads is done via custom weighting factors calibrated for each zone. The resulting equations are grouped to represent feasible equipment allocations for each range of building loads, and solved using a non-linear solver. The objective function is the cost of delivering or removing energy to meet the loads, and is constrained by the comfort criteria for each zone. The objective function is minimized using the reduced gradient method, subject to constraints on comfort and equipment operation. The optimization starts with the feasible points as determined by a nonlinear equation solver for each combination of equipment allocation. The values of the set points that minimize the objective function are determined; the allocation with the least cost is the desired operation mode. The results obtained from this approach have been applied to high-rise office buildings in San Francisco with central plants, VAV, dual duct, and induction systems. Electrical demand reductions of 8 to 12% and energy savings of 18 to 23% were achieved.

Simplified System-Based Optimization Approach

The component-based optimization method presented by Braun et al. (1989b) was used to develop a simpler method for determining the optimal control. The method involves correlating overall cooling plant power consumption using a quadratic functional form. Minimization of this function leads to linear control laws for control variables in terms of the uncontrolled variables. The technique may be used to tune parameters of the cooling tower and chilled water reset strategies presented in the major section on Supervisory Control Strategies and Tools. It may also be used to define strategies for supply air temperature reset for VAV systems and flow control for variable-speed condenser water pumps.

In the vicinity of any optimal control point, the plant power consumption may be approximated as a quadratic function of the continuous control variables for each of the operating modes (i.e., discrete control mode). A quadratic function also correlates power consumption in terms of the uncontrolled variables (i.e., load, ambient temperature) over a wide range of conditions. This leads to the following general functional relationship between overall cooling plant power and the controlled and uncontrolled variables:

$$J(\mathbf{f}, \mathbf{M}, \mathbf{u}) = \mathbf{u}^T \mathbf{A} \mathbf{u} + \mathbf{b}^T \mathbf{u} + \mathbf{f}^T \mathbf{C} \mathbf{f} + \mathbf{d}^T \mathbf{f} + \mathbf{f}^T \mathbf{E} \mathbf{u} + g \quad (42)$$

where J is the total plant power, $\mathbf{u}$ is a vector of continuous and free control variables, $\mathbf{f}$ is a vector of uncontrolled variables, $\mathbf{M}$ is a vector of discrete control variables, and the superscript T designates the transpose vector. $\mathbf{A}$, $\mathbf{C}$, and $\mathbf{E}$ are coefficient matrices, $\mathbf{b}$ and $\mathbf{d}$ are

coefficient vectors, and g is a scalar. The empirical coefficients of the above function depend on the operating modes so that these constants must be determined for each feasible combination of discrete control modes.

A solution for the optimal control vector that minimizes the power may be determined analytically by applying the first-order condition for a minimum. Equating the Jacobian of Equation (42) with respect to the control vector to zero and solving for the optimal control set points gives

$$\mathbf{u}^* = \mathbf{k} + \mathbf{K}\mathbf{f} \quad (43)$$

where

$$\mathbf{k} = -\mathbf{A}^{-1}\mathbf{b}/2 \quad (44)$$

$$\mathbf{K} = -\mathbf{A}^{-1}\mathbf{E}/2 \quad (45)$$

The cost associated with the unconstrained control of Equation (43) is

$$J^* = \mathbf{f}^T \theta \mathbf{f} + \sigma \mathbf{f} + \tau \quad (46)$$

where

$$\theta = \mathbf{K}^T \mathbf{A}\mathbf{k} + \mathbf{E}\mathbf{K} + \mathbf{C} \quad (47)$$

$$\sigma = 2\mathbf{K}\mathbf{A}\mathbf{k} + \mathbf{K}\mathbf{b} + \mathbf{E}\mathbf{k} + \mathbf{d} \quad (48)$$

$$\tau = \mathbf{K}^T \mathbf{A}\mathbf{k} + \mathbf{b}^T \mathbf{k} + g \quad (49)$$

The control defined by Equation (43) results in a minimum power consumption if $\mathbf{A}$ is positive definite. If this condition holds and if the system power consumption is adequately correlated with Equation (42), then Equation (43) dictates that the optimal continuous, free control variables vary as a nearly linear function of the uncontrolled variables. However, a different linear relationship applies to each feasible combination of discrete control modes. The minimum cost associated with each mode combination must be computed from Equation (46) and compared in order to identify the minimum.

Uncontrolled Variables. As discussed in the background section, optimal control variables primary depend on ambient wet-bulb temperature and total chilled water load. The load affects the heat transfer requirements for all heat exchangers, whereas the wet-bulb temperature impacts chilled and condenser water temperatures necessary to achieve a given heat transfer rate. As discussed in the section on Supervisory Control Strategies and Tools, cooling coil heat transfer depends on the coil entering wet-bulb temperature. However, this reduces to an ambient wet-bulb temperature dependence for a given ventilation mode (e.g., minimum outside air or economizer) and fixed zone conditions. Thus, separate cost functions are necessary for each ventilation mode with load and ambient wet-bulb as uncontrolled variables. Alternatively, for a specified ventilation strategy (e.g., the economizer strategy from the section on Supervisory Control Strategies and Tools), three uncontrolled variables could be used for all ventilation modes: load, ambient wet-bulb temperature, and average cooling coil inlet wet-bulb temperature.

Additional uncontrolled variables that could be important if varied over a wide range are the individual zone latent-to-sensible load ratios and the ratios of individual sensible zone loads to the total sensible loads for all zones. However, these variables are difficult to determine from measurements and are of secondary importance.

Free Control Variables. The number of independent or "free" control variables in the optimization can be reduced significantly by using the material developed in the section on Supervisory Control Strategies and Tools. For instance, the optimal static pressure set point for a VAV system should keep at least one VAV box fully open and should not be considered as a free optimization variable. Similarly, the supply air temperature for a CAV system should be set to minimize reheat. Additional near-optimal guidelines were presented for sequencing of cooling tower fans, sequencing of chillers, loading of chillers, reset of pressure differential set point for variable-speed pumping, and chilled water reset with fixed-speed pumping. Furthermore, Braun et al. (1989a) showed that using identical supply air set points for multiple air handlers gives near optimal results for VAV systems.

For all variable-speed auxiliary equipment (i.e., pumps and fans), the free set point variables to use in Equation (42) could be reduced to the following: (1) supply air set temperature, (2) chilled water set temperature, (3) tower air flow relative to design capacity, and (4) condenser water flow relative to design capacity. All of the other continuous supervisory control variables are dependent upon these variables with the simplified strategies presented in the section on Supervisory Control Strategies and Tools.

Some of the dependent control variables may be discrete control variables. For instance, variable flow pumping may be implemented with multiple fixed and variable-speed pumps where the number of operating pumps is a discrete variable that will change when a variable-speed pump reaches its capacity. These discrete changes could lead to discrete changes in the costs due to changes in overall pump efficiency. However, this has a relatively small effect on the overall power consumption and may be neglected in fitting the overall cost function to changes in the control variables.

Some of the discrete control variables may also be independent variables. In general, different cost functions arise for all operating modes consisting of each possible combination of discrete control variables. With all variable-speed pumps and fans, the only significant discrete control variable is the number of operating chillers. Then, the optimization involves determining optimal values of only four continuous control variables for each of the feasible chiller modes. A chiller mode defines which of the available chillers are to be online. The chiller mode giving the minimum overall power consumption represents the optimum. For a chiller mode to be feasible, the specified chillers must operate safely within their capacity and surge limits. In practice, abrupt changes in the chiller modes should also be avoided. Large chillers should not be cycled on or off except when the savings associated with the change is significant.

Using fixed-speed equipment reduces the number of free continuous control variables. For instance supply air temperature would be removed as a control variable for CAV systems and chilled water temperature would not be included for fixed-speed chilled water pumping. However, for multiple chilled water pumps not dedicated to chillers, the number of operating pumps could become a free discrete control variable. Similarly, for multiple fixed-speed cooling tower fans and condenser water pumps, each of the discrete combinations could be considered as separate modes. However, for multiple cooling tower cells with multiple fan speeds, the number of possible combinations may be large. A simpler approach that works satisfactorily is to treat the relative flows as continuous control variables during the optimization and to select the discrete relative flow that is closest to the optimal value. At least three relative flows (discrete flow modes) are necessary for each chiller mode in order to fit the quadratic cost function. The number of possible sequencing modes for fixed-speed pumps is generally much more limited than that for cooling tower fans, with two or three possibilities (at most) for each chiller mode. In fact, with many current designs, individual pumps are physically coupled with chillers, and it is impossible to operate more or fewer pumps than the number of operating chillers. Thus, it is generally best to treat the control of fixed-speed condenser water pumps with a set of discrete control possibilities rather than using a continuous control approximation.

Training. The coefficients of Equation (42) must be determined empirically and a variety of approaches have been proposed. One approach would be to apply regression techniques directly to measurements of total power consumption. Since the cost function is linear with respect to the empirical coefficients, linear regression techniques may be used. A set of experiments could be performed over the expected range of operating conditions. Large amounts of

data that include the entire range must be taken to account for measurement uncertainty. The regression could possibly be performed online using least-squares recursive parameter updating (Ljung and Söderström 1983).

Rather than fitting empirical coefficients of the system cost function of Equation (42), the coefficients of the optimal control Equation (43) and the minimum cost function of Equation (46) could be estimated directly. At a limited set of conditions, optimal values of the continuous control and free variables could be estimated through trial-and-error variations. Only three independent conditions would be necessary to determine coefficients of the linear control law given by Equation (43) if the load and wet-bulb are the only uncontrolled variables. The coefficients of the minimum cost function could then be determined from system measurements with the linear control law in effect. The disadvantage of this approach is that there is no direct way to handle physical constraints on the controls.

Summary and Constraint Implementation. The methodology for determining the near-optimal control of a chilled water system may be summarized as follows:

1. Change the chiller operating mode if system operation is at the limits of chiller operation (near surge or maximum capacity).
2. For the current set of conditions (load and wet bulb), estimate the feasible modes of operation that would avoid operating the chiller and condenser pump at their limits.
3. For the current operating mode, determine optimal values of the continuous controls using Equation (43).
4. Determine a constrained optimum if controls exceed their bounds.
5. Repeat Steps 3 and 4 for each feasible operating mode.
6. Change the operating mode if the optimal cost associated with the new mode is significantly less than that associated with the current mode.
7. Change the values of the continuous control variables. When treating multiple-speed fan control with a continuous variable, use the discrete control closest to the optimal continuous value.

If the linear optimal control Equation (43) is directly determined from optimal control results, then the constraints on controls may be handled directly. Otherwise, a simple solution is to constrain the individual control variables as necessary and neglect the effects of the constraints on the optimal values of the other controls and the minimum cost function. The variables of primary concern with regard to constraints are the chilled water and supply air set temperatures. These controls must be bounded for proper comfort and safe operation of the equipment. On the other hand, the cooling tower fans and condenser water pumps should be sized so that the system performs efficiently at design loads and constraints on control of this equipment should only occur under extreme conditions.

The optimal value of the chilled water supply temperature is coupled to the optimal value of the supply air temperature, so that decoupling these variables in evaluating constraints is generally not justified. However, optimization studies indicate that when either control is operated at a bound, the optimal value of the other free control is approximately bounded at a value that depends only on the ambient wet-bulb temperature. The optimal value of this free control (either chilled water or supply air set point) may be estimated at the load at which the other control reaches its limit. Coupling between optimal values of the chilled water and condenser water loop controls is not as strong, so that interactions between constraints on these variables may be neglected.

Case Studies. Braun et al. (1987) correlated the power consumption of the Dallas/Fort Worth airport chiller, condenser pumps, and cooling tower fans with the quadratic cost function given by Equation (42) and showed good agreement with data. Because the chilled water loop control was not considered, the chilled water set point was treated as a known uncontrolled variable. The discrete control variables associated with the four tower cells with two-speed fans and the three condenser pumps were treated as continuous control variables. The optimal control determined by the near optimal Equation (43) also agreed well with that determined using a nonlinear optimization applied to a detailed simulation of the system.

In subsequent work, Braun et al. (1989b) considered complete system simulations (cooling plant and air handlers) to evaluate the performance of the quadratic, system-based approach. A number of different system characteristics were considered. Figures 17, 22, 36, and 37 show comparisons between the controls as determined with the component-based and system-based methods for a range of loads, for a relatively low and high ambient wet-bulb temperature (60°F and 80°F).

In Figures 22 and 36, optimal values of the chilled water and supply air temperatures are compared for a system with variable air and water flow. The near-optimal control equation provides a good fit to the optimization results for all conditions considered. The chilled water temperature was constrained between 38°F and 55°F, while the supply air set point was allowed to float freely. Figures 22 and 36 show that for the conditions where the chilled water temperature is constrained, the optimal supply air temperature is also nearly bounded at a value that depends upon the ambient wet bulb.

Optimal relative cooling tower air and condenser water flow rates are compared in Figures 17 and 37 for a system with variable-speed cooling tower fans and condenser water pumps. Although the optimal controls are not exactly linear functions of the load, the linear control equation provides an adequate fit. The differences in these controls result in insignificant differences in overall power consumption, because, as discussed in the background section, the optimum is extremely flat with respect to these variables. The non-linearity of the condenser loop controls is partly due to the constraints imposed upon the chilled water set temperature. However, this effect is not very significant. Figures 17 and 37 also suggest that the optimal condenser loop control is not very sensitive to the ambient wet-bulb temperature.

DYNAMIC OPTIMIZATION

Cool storage is often used in commercial cooling applications as a means of shifting cooling requirements from periods of on-peak or high electrical costs to off-peak periods. As discussed in section 1, cool storage may be implemented using ice, water, or the building structure. In any case, the optimal supervisory control for storage is a complex function of such factors as utility rates, load profile, chiller characteristics, storage characteristics, and weather. For a

utility rate structure that includes both time-of-use energy and demand charges, the optimal strategy can depend on variables that extend over a monthly time scale. The overall problem of minimizing the utility cost over a billing period (e.g., a month) can be mathematically described as follows:

Minimize

$$J = \sum_{k=1}^{N} \{E_k P_k \Delta \tau\} + \text{Max}_{1 \le k \le N}\{D_k P_k\} \qquad (50)$$

with respect to the control variables $(u_1, u_2, \ldots, u_N)$ and subject to the following constraints for each stage k:

$$u_{\min,k} \le u_k \le u_{\max,k} \qquad (51)$$

$$x_k = f(x_{k-1}, u_k, k) \qquad (52)$$

$$x_{\min} \le x_k \le x_{\max} \qquad (53)$$

$$x_N = x_0 \qquad (54)$$

where J is the utility cost associated with the billing period (e.g., a month); $\Delta \tau$ is the stage time interval (typically equal to the time window over which demand charges are levied, e.g. 0.25 h); N is the number of time stages in a billing period, and for each stage k, P is the average building electrical power (kW); E is the energy cost rate or cost per unit of electrical energy ($/kWh); D is the demand charge rate or cost per peak power rate over the billing period ($/kW); u is the control variable that regulates the rate of energy removal from or addition to storage over the stage; u_{max} is the maximum value for u; u_{min} is the minimum value for u; x is the state of storage at the end of the stage; x_{max} is the maximum admissible state of storage; x_{min} is the minimum admissible state of storage; and f is a state equation that relates the state of storage at stage k to the previous state and current control.

The first and second terms in Equation (50) are the total cost of energy use and building demand for the billing period. Both the energy and demand cost rates can vary with time, but typically have two values associated with on-peak and off-peak periods. An even more complex cost optimization results if the utility includes ratchet clauses in which the demand charge is the maximum of the monthly peak demand cost and some fraction of the previous monthly peak

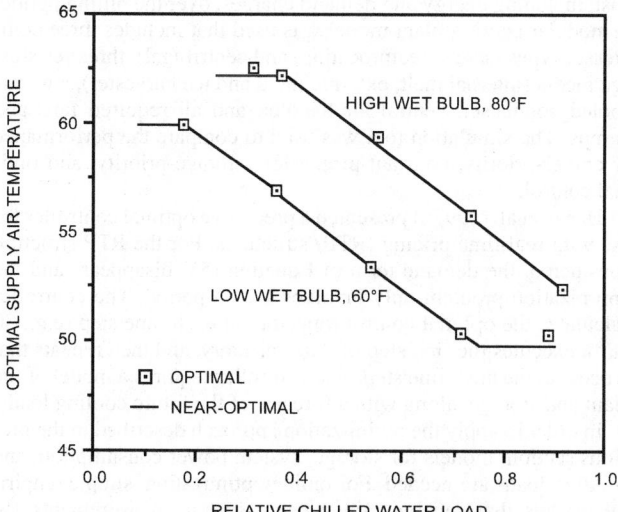

Fig. 36   Comparisons of Optimal Supply Air Temperature

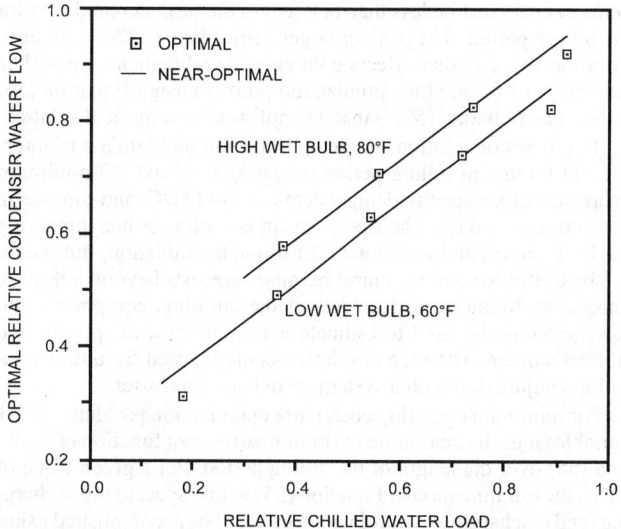

Fig. 37   Comparisons of Optimal Condenser Pump Control

demand cost during the cooling season. With real-time pricing, the demand charge, which is the second term in Equation (50), might not exist and the hourly energy rates would vary over time according to the generation costs.

For ice or chilled water storage systems, the control variable could be the rate at which energy is added or removed from storage. In this case, the constraint given by Equation (51) arises from limits that depend on the chiller and storage heat exchanger and can also depend upon the state of storage. For use of building thermal mass, the control variable could be the zone temperature(s) and the constraint of Equation (51) would be associated with comfort considerations or capacity constraints. Different comfort limits would probably apply for occupied and unoccupied periods.

The equality constraint of Equation (52) is termed the state equation. The state of storage at any stage k is a function of the previous state (x_{k-1}), the control (u_k), and other time-dependent factors (e.g., ambient temperature). For lumped storage systems (e.g., ice), the state of storage can be characterized with a single state variable. However, for a distributed storage (e.g., a building structure), multiple state equations may be necessary to properly characterize the dynamics. The state of storage is also constrained to be between states associated with full discharge and full charge [Equation (53)]. The constraint of Equation (54) forces a steady-periodic solution to the problem. This constraint becomes less important as the length of analysis increases.

In order to determine a control strategy for charging and discharging storage that minimizes utility cost, Equation (50) must be minimized over the entire billing period because of the influence of the demand charge. Alternatively, the optimization problem can be posed as a series of shorter-term (e.g., daily or weekly) optimizations with a constraint on the peak demand charge according to:

Minimize:

$$J = \sum_{k=1}^{N} \{E_k P_k \Delta \tau\} + \text{TDC} \qquad (55)$$

with respect to the control variables ($u_1, u_2, ..., u_N$) and a billing period demand cost target (TDC) and subject to the constraints of Equations (51) through (54) and the following equation:

$$D_k P_k \leq \text{TDC} \qquad (56)$$

The constraint expressed in Equation (56) arises from the form of the cost function chosen for Equation (55). At each stage, the demand cost must be less than or equal to the peak demand cost for the billing period. The peak or target demand cost TDC is an optimization variable that affects both energy and demand costs. The advantage of posing the optimization problem using Equation (55) rather than Equation (50) is that it simplifies the numerical solution.

Two types of solutions to the optimization problem are of interest: (1) minimum billing period operating cost and (2) minimum energy cost for a specified target demand cost (TDC) and short-term horizon (e.g., a day). The first problem is useful for benchmarking the best control and minimum cost through simulation, but would not be useful for online control because forecasts beyond a day are unreliable. Mathematical models of the building, equipment, and storage would be used to estimate load requirements, power, and state of storage. The second solution could be used for online control in conjunction with a system model and forecaster.

For minimum operating costs (first optimization problem), $N + 1$ variables must be determined to minimize the cost function of Equation (55) over the length of the billing period. For a given value of TDC, the minimization of Equation (55) with respect to the N charging (and discharging) control variables could be accomplished using dynamic programming (Bellman 1957) or some other direct search method. The primary advantages of dynamic programming are that

it handles constraints on both state and control variables in a straightforward manner and also guarantees a global minimum. However, the computation becomes excessive if more than one state variable is needed to characterize storage. The N-variable optimization problem would be resolved at each iteration of an outer loop optimization for TDC. Brent's algorithm (1973) is a robust method for solving the one-dimensional optimization for the demand target because it does not require derivative information. This is important because TDC appears as an inequality constraint in the dynamic programming solution and may not always be triggered.

For shorter-term optimizations (second optimization problem), dynamic programming could still be used to minimize Equation (55) with respect to the N charging (and discharging) control variables for a specified TDC. However, an optimal value for TDC cannot be determined when demand charges are imposed. For ice storage, Drees and Braun (1996) found that a simple and near-optimal approach is to set TDC to zero at the beginning of each billing period. Therefore, the optimizer minimizes the demand cost for the first optimization period (e.g., a day) and then uses this demand as the target for the billing period unless it is exceeded. For online optimization, the optimization problem can be resolved at regular intervals (e.g., 1 h) during each day's operation.

Ice Storage Control Optimization

Several researchers have studied optimal supervisory control of ice storage systems. Braun (1992) solved daily optimization problems for two limiting cases: minimum energy (i.e., no demand charge) and minimum demand (no energy charge). Results of the optimizations for different days and utility rates were compared with simple chiller-priority and load-limiting control strategies (see the section on Supervisory Control Strategies and Tools). For the ice-on-pipe system considered, load-limiting control was found to be near optimal for both energy and demand costs with on-peak to off-peak energy cost ratios greater than about 1.4.

Drees and Braun (1996) solved both daily and monthly optimization problems for a range of systems with internal-melt area-constrained ice storage tanks. The optimization results were used to develop rules that became part of a rule-based, near-optimal controller that is presented in the section on Supervisory Control Strategies and Tools. For a range of partial-storage systems, load profiles, and utility rate structures, the monthly electrical costs for the rule-based control strategy were, on average, within about 3% of the optimal costs.

Henze et al. (1997a) developed a simulation environment which determines the optimal control strategy to minimize the operating cost, including energy and demand charges, over the billing period. A modular cooling plant model was used that includes three compressor types (screw, reciprocating, and centrifugal), three ice storage media (internal melt, external melt, and ice harvester), a water-cooled condenser, central air handler, and all required fans and pumps. The simulation tool was used to compare the performance of chiller-priority, constant-proportion, storage-priority, and optimal control.

Henze et al. (1997b) presented a predictive optimal controller for use with real-time pricing (RTP) structures. For the RTP structure considered, the demand term of Equation (55) disappears and the optimization problem only involves a 24-h period. The controller calculates the optimal control trajectory at each time step (e.g., 30 min), executes the first step of that trajectory, and then repeats that process at the next time step. The controller requires a model of the plant and storage, along with a forecast of the future cooling loads.

In order to apply the optimization approach described in the previous section, models for storage, system power consumption, and building loads are needed. For online optimization, simple empirical models that can be trained using system measurements are appropriate. However, physically based models would be best for simulation studies.

The optimization studies that have been performed for ice storage assumed that the state of storage could be represented with a single state variable. Assuming negligible heat gains from the environment, the relative state of charge (i.e., fraction of the maximum available storage capacity) for any stage k is

$$x_k = x_{k-1} + \frac{u_k \Delta t}{C_s} \tag{57}$$

where C_s is the maximum change in internal energy of the storage tank that can occur during a discharge cycle and u_k is the storage charging rate. The state of charge defined in this manner must be between zero and one.

The charging rate for storage depends on the storage heat exchanger area, secondary fluid flow rate and inlet temperature, and the thickness of ice. At any stage, the maximum charging rate can be expressed as

$$u_{k,\,max} = \varepsilon_{c,\,k,\,max} \dot{m}_{f,\,max} c_f (t_s - t_{f,\,i}) \tag{58}$$

where $\varepsilon_{c,k,max}$ is a heat transfer effectiveness for charging at the current state of storage if the secondary fluid flow rate were at its maximum value of $\dot{m}_{f,\,max}$, c_f is the secondary fluid specific heat, $t_{f,i}$ is the temperature of secondary fluid inlet to the tank, and t_s is the temperature at which the storage medium melts or freezes (e.g., 32°F).

The minimum charging rate is actually the negative of the maximum discharging rate and can be given by

$$u_{k,\,min} = \varepsilon_{d,\,k,\,max} \dot{m}_{f,\,max} c_f (t_s - t_{f,\,i}) \tag{59}$$

where $\varepsilon_{d,k,max}$ is the heat transfer effectiveness for discharging at the current state of storage if the secondary fluid flow rate were at its maximum value of $\dot{m}_{f,\,max}$.

In general, the heat transfer effectiveness for charging and discharging at the design flow can be correlated as a function of state of charge using manufacturers' data (e.g., Drees and Braun 1995).

A model for the total building power is also needed. At any time

$$P = P_{noncooling} + P_{plant} + P_{dist} \tag{60}$$

where $P_{noncooling}$ is the building electrical use that is not associated with the cooling system (e.g., lights), P_{plant} is the power needed to operate the cooling plant, and P_{dist} is the power associated with the distribution of secondary fluid and air through the cooling coils. The models used by Henze et al. (1997a) predict cooling plant and distribution system power with a component-based simulation that would be appropriate for simulation studies. Alternatively, for online optimization, plant and distribution system power could be represented with empirical correlations. Drees (1994) used curve fits of plant power consumption in terms of cooling load and ambient wet-bulb temperature. At any time, the cooling requirement for the chiller is the difference between the building load requirement and the storage discharge rate. The chiller supply temperature is then determined from an energy balance on the chiller and used to evaluate the limits on the storage charging and discharging rates in Equations (59) and (60). The chiller cooling rate must be greater than a minimum value for safe operation and less than the chiller capacity. Drees (1994) correlated the maximum cooling capacity as a function of the ambient wet-bulb temperature and the chiller supply temperature. For simulation studies, a building model would be used to estimate building cooling loads. For online optimization, a forecaster would provide estimates of future building cooling loads.

REFERENCES

Armstrong, P.R., T.N. Bechtel, C.E. Hancock, S.E. Jarvis, J.E. Seem, and T.E. Vere. 1989. Environment for structured implementation of general and advanced HVAC controls—Phase II. Small Business Innovative Research Program, Chapter 7. DOE Contract DE-AC02-85ER 80290.

ASHRAE. 1997. Method of testing the performance of cool storage systems. *Standard* 150P. Public review draft.

Bellman, R. 1957. *Dynamic programming.* Princeton University Press, Princeton, N.J.

Bloomfield, D.P. and D.J. Fisk. 1977. The optimization of intermittent heating. *Buildings and Environment* 12:43-55.

Braun, J.E., J.W. Mitchell, S.A. Klein, and W.A. Beckman. 1987. Performance and control characteristics of a large central cooling system. *ASHRAE Transactions* 93(1):1830-52.

Braun, J.E. 1988. *Methodologies for the design and control of central cooling plants.* Ph.D. Dissertation, University of Wisconsin-Madison.

Braun, J.E., S.A. Klein, J.W. Mitchell, and W.A. Beckman. 1989a. Applications of optimal control to chilled water systems without storage. *ASHRAE Transactions* 95(1).

Braun, J.E., S.A. Klein, J.W. Mitchell, and W.A. Beckman. 1989b. Methodologies for optimal control to chilled water systems without storage. *ASHRAE Transactions* 95(1).

Braun, J.E. and G.T. Diderrich. 1990. Near-optimal control of cooling towers for chilled-water systems. *ASHRAE Transactions* 96(2): 806-813.

Braun, J.E. and J.E. Seem. 1991. Adaptive methods for real-time forecasting of building electrical demand. *ASHRAE Transactions* 97(1):710-721.

Braun, J.E. 1992. A comparison of chiller-priority, storage-priority, and optimal control of an ice-storage system. *ASHRAE Transactions* 98(1).

Brent, R.P. 1973. *Algorithms for minimization without derivatives,* Chapter 5. Prentice Hall.

Brothers, P.W. and M.L. Warren. 1986. Fan energy use in variable air volume systems. *ASHRAE Transactions* 92(2).

Bushby, T.B. and G.E. Kelly. 1988. Comparison of digital control and pneumatic control systems in a large office building. NBSIR 88-3739. National Institute of Standards and Technology, Gaithersburg, MD.

Cumali, Z. 1988. Global optimization of HVAC system operations in real time. *ASHRAE Transactions* 94(1).

Cumali, Z. 1994. Application of real-time optimization to building systems. *ASHRAE Transactions* 100(1).

Drees, K.H. 1994. *Modeling and control of area-constrained ice storage systems.* M.S. Thesis, Purdue University, West Lafayette, IN.

Drees, K.H. and J.E. Braun. 1995. Modeling of area-constrained ice storage tanks. *Int. J. of HVAC&R Research* 1(2):143-59.

Drees, K.H. and J.E. Braun. 1996. Development and evaluation of a rule-based control strategy for ice storage systems. *Int. J. of HVAC&R Research* 2(4):312-36.

Elleson, J.S. 1996. *Successful cool storage projects: From planning to operation.* ASHRAE, Atlanta.

Englander, S.L. and L.K. Norford. 1992. Saving fan energy in VAV systems part 2: Supply fan control for static pressure minimization using DDC zone feedback. *ASHRAE Transactions* 98(1):19-32.

Forrester, J.R. and W.J.Wepfer. 1984. Formulation of a load prediction algorithm for a large commercial building. *ASHRAE Transactions* 90(2B):536-51.

Gibson, G.L. and T.T. Kraft. 1993. Electric demand prediction using artificial neural network technology. *ASHRAE Journal* 35(3):60-68.

Grumman, D.L. and A.S. Butkus, Jr. 1988. Ice storage application to an Illinois hospital. *ASHRAE Transactions* 94(1):1879-93.

Hartman, T. 1993. Terminal regulated air volume (TRAV) systems. *ASHRAE Transactions* 99(1):791-800.

Hackner, R.J., J.W. Mitchell, and W.A. Beckman. 1984. HVAC system dynamics and energy use in buildings–Part I. *ASHRAE Transactions* 90(2B):523-35.

Hackner, R.J., J.W. Mitchell, and W.A. Beckman. 1985. HVAC system dynamics and energy use in buildings–Part II. *ASHRAE Transactions* 91(1B):781.

Henze, G.P., M. Krarti, and M.J. Brandemuehl. 1997a. A simulation environment for the analysis of ice storage controls. *Int. J. HVAC&R Research* 3(2):128-48.

Henze, G.P., Dodier, R.H., and M. Krarti. 1997b. Development of a predictive optimal controller for thermal energy storage systems. *Int. J. HVAC&R Research* 3(3):233-64.

Hittle, D.C. 1979. The building loads analysis and systems thermodynamics (BLAST) program, Version 2.0. U.S. Army Construction Engineering Research Laboratory, Users Manual, Vol. 1. (Available from NTIS, Springfield, VA 22151.)

Kao, J.Y. 1985. Control strategies and building energy consumption. *ASHRAE Transactions* 91(2).

Keeney, K.R. and J.E. Braun. 1997. Application of building precooling to reduce peak cooling requirements. *ASHRAE Transactions* 103(1):463-69.

Krarti, M., M.J. Brandemuehl, and G.P. Henze. 1996. Evaluation of optimal control for ice systems. *ASHRAE Research Report* 809-RP.

Kreider, J.F. and X.A. Wang. 1991. Artificial neural networks demonstration for automated generation of energy use predictors for commercial buildings. *ASHRAE Transactions* 97(2):775-79.

Lau, A.S., W.A. Beckman, and J.W. Mitchell. 1985. Development of computer control—Routines for a large chilled water plant. *ASHRAE Transactions* 91(1).

Ljung, L. and T. Söderström. 1983. *Theory and practice of recursive identification.* MIT Press, Cambridge, MA.

MacArthur, J.W, A. Mathur, and J. Zhao. 1989. On-line recursive estimation for load profile prediction. *ASHRAE Transactions* 95(1):621-628.

Rawlings, L.K. 1985. Strategies to optimize ice storage. *ASHRAE Journal* 27(5):39-44.

Seem, J.E., P.R. Armstrong, and C.E. Hancock. 1989. Comparison of seven methods for forecasting the time to return from night setback. *ASHRAE Transactions* 95(2).

Seem, J.E. and J.E. Braun. 1991. Adaptive methods for real-time forecasting of building electrical demand. *ASHRAE Transactions* 97(1).

Seem, J.S., C. Park, and J.M House. 1999. A new sequencing control strategy for air-handling units. *Int. J. of HVAC&R Research* 5(1):35-58.

Spethmann, D.H. 1989. Optimal control for cool storage. *ASHRAE Transactions* 95(1).

Tamblyn, R.T. 1985. Control concepts for thermal storage. *ASHRAE Transactions* 91(1B).

Warren, M. and L. K. Norford. 1993. Integrating VAV zone requirements with supply fan operation. *ASHRAE Journal* 35(4):43-46.

BIBLIOGRAPHY

Avery, G. 1986. VAV—Designing and controlling an outside air economizer cycle. *ASHRAE Journal* 28(12):26-30.

Avery, G. 1989. Updating the VAV outside air economizer cycle. *ASHRAE Journal* 31(4):14-16.

Braun, J.E. 1990. Reducing energy costs and peak electrical demands through optimal control of building thermal storage. *ASHRAE Transactions* 96(1).

Conniff, J.P. 1991. Strategies for reducing peak air-conditioning loads by using heat storage in the building structure. *ASHRAE Transactions* 97(1).

Delp, W.W., R.H. Howell, H.J. Sauer, and B. Subbarao. 1993. Control of outside air and building pressurization in VAV systems. *ASHRAE Transactions* 99(1):565-89.

Hartman, P.E. 1988. Dynamic control: A new approach. *Heating/Piping/Air Conditioning.*

House, J.M., T.F. Smith, and J.S. Arora. 1991. Optimal control of a thermal system. *ASHRAE Transactions* 97(2):991-1001.

Johnson, G.A. 1985. Optimization techniques for a centrifugal chiller plant using a programmable controller. *ASHRAE Transactions* 91(2).

Marseille, T.J. and J.S. Schliesing, 1991. Integration of water loop heat pumps and building structural thermal energy storage. PNL-7850, Pacific Northwest Laboratories.

Miller, D.E. 1980. The impact of HVAC process dynamics on energy use. *ASHRAE Transactions* 86(2):535-53.

Morris, F.B., J.E. Braun, and S. Treado. 1994. Experimental and simulated performance of optimal control of building thermal storage. *ASHRAE Transactions* 100(1).

Nizet, J.L., J. Lecomte, and F.X. Litt. 1984. Optimal control applied to air conditioning in buildings. *ASHRAE Transactions* 90(1B):587-600.

Nugent, D.R., S.A. Klein, and W.A. Beckman. 1988. Investigation of control alternatives for a steam Turbine driven chiller. *ASHRAE Transactions* 94(1).

Nuorkivi, A. 1990. Real-time optimization system of a district heat network operation. *ASHRAE Transactions* 96(1):946-48.

Ruud, M.D., J.W. Mitchell and S.A. Klein. 1990. Use of building thermal mass to offset cooling loads. *ASHRAE Transactions* 96(2).

Seem, J.E. 1987. *Modeling of heat transfer in buildings.* Ph.D. thesis, University of Wisconsin-Madison.

Shapiro, M.M., A.J. Yager, and T.H. Ngan. 1988. Test hut validation of a microcomputer predictive HVAC control. *ASHRAE Transactions* 94(1).

Snder, M.E. and T.A. Newell. 1990. Cooling cost minimization using building mass for thermal storage. *ASHRAE Transactions* 96(2):830-38.

Sud, I. 1984. Control strategies for minimum energy usage. *ASHRAE Transactions* 90(2).

Treichler, W.W. 1985. Variable speed pumps for water chillers, water coils, and other heat transfer equipment. *ASHRAE Transactions* 91(1).

BUILDING COMMISSIONING

COMMISSIONING is a quality assurance process of the installation of the systems in a building. It is a process for achieving, verifying, and documenting the performance of each system to meet the operational needs of the building within the capabilities of the documented design and specified equipment capacities, according to the owner's functional criteria. It is process that ensures the quality of the installation. Successful commissioning includes the preparation of manuals and training of operation and maintenance personnel. The result of commissioning should be fully functional systems that can be properly operated and maintained throughout the useful life of the building. All efforts related to commissioning should be specified in the contract documents.

This definition refers to the building as a total system, which includes the structural elements, building envelope, life safety features, security systems, elevators, escalators, plumbing, electrical, controls, and the HVAC. The commissioning plan needs to consider interface requirements of individual building elements through integrated testing.

ASHRAE *Guideline* 1, *The HVAC Commissioning Process*, provides for documenting and verifying the performance of HVAC systems so that they operate in conformity with the design intent. It is useful building owners, architects, engineers, suppliers, contractors, construction managers, commissioning authorities, operators, and any others involved with the design of the project.

Although testing, adjusting, and balancing (TAB) is an important part of the construction and operation phases of commissioning, building commissioning extends through all phases and systems inherent in a project, from concept through occupancy and the warranty period.

ASHRAE *Guideline* 1 includes the following topics:

- Procedures, methods, and documentation requirements during each phase of the commissioning process, for all types and sizes of HVAC systems, from predesign through final acceptance and post-occupancy; this includes changes in building and occupancy requirements after initial occupancy.
- Documentation
  - Owner's assumptions and requirements
  - Design intent, basis of design, and expected performance
  - Verification and functional performance testing
  - Operation and maintenance criteria
- Specific details to:
  - Conduct verification and functional performance tests necessary to evaluate the HVAC system for acceptance
  - Ensure performance meets design intent after initial occupancy
- Program for training operation and maintenance personnel

Advantages of Building Commissioning

The advantages of the building commissioning process vary from one project to another. The advantages applicable to each

The preparation of this chapter is assigned to TC 9.9, Building Commissioning.

project are important for two reasons. First, an evaluation helps in setting a commissioning scope that is consistent with the owner's desires and with the complexity and criticality (e.g. with respect to life safety and/or environmental impact) of the project. Second, it helps in quantifying both the costs and benefits of commissioning. Building commissioning is needed to

- Open channels of communication
- Create a better understanding of design intent
- Provide early assignment of performance responsibilities
- Set performance goals such as energy, environment, and life cycle
- Improve planning for verification and acceptance using systematic procedures for inspection and testing
- Establish coordination plans
- Improve quality of turnover documents
- Establish continuous monitoring of priorities and schedules
- Improve operation and maintenance programs
- Improve quality of operator personnel training
- Document indoor air quality (IAQ) and comfort control
- Ease building turnover process
- Help ensure on-schedule occupancy
- Reduce callbacks and assist in dealing with warranty claims
- Enable owner to recognize system capabilities and limitations

Owners and design professionals may implement the commissioning process at any stage of a project. However, the commissioning process is most effective when started early.

Commissioning Team

The size and makeup of the commissioning team depends on the size and complexity of the project and the owner's desire to invest in quality assurance. The responsibility of each member of the commissioning team is documented in the commissioning plan.

All participants in the construction project have a commissioning responsibility. Participants include the commissioning authority, owner, design professionals, construction manager, general contractor, subcontractors, operation and maintenance manager, suppliers, and equipment manufacturers. The project building operation, production, and maintenance managers need to be brought into the commissioning process early, preferably during the predesign phase. Their knowledge of occupancy, special lighting, anticipated equipment loads, and other factors should influence the design and set performance objectives. Additionally, the early participation of both the TAB and HVAC controls' contractors is also important because their experience can provide guidance on space requirements, balancing devices, probe and control locations, and accessibility to equipment.

Commissioning Authority

The commissioning authority is a qualified professional, company, or agency that implements the overall commissioning process in cooperation with the commissioning team as specified in the contract documents. The commissioning authority has a significant impact on the project and must be able to take responsibility for acting on or verifying the performance aspects of the project. However, the appointment of a commissioning authority does not alter other professional or contractual obligations. The owner should select a qualified commissioning authority who has proven experience and is registered as a professional in a recognized organization.

The commissioning authority, should understand all building systems being commissioned. The authority should also be familiar with applicable design and building standards, use of test instruments and equipment for functional testing, and interaction of the various systems. Additional qualifications of the commissioning authority may be found in ASHRAE *Guideline* 1.

Responsibility. The primary responsibility of the commissioning authority is to inform the owner of the status, integration, and performance of all tested systems in the facility.

Information. The commissioning authority should function as a catalyst and initiator to disseminate information and assist the design and construction teams in verifying satisfactory completion of construction. The authority verifies system completeness, performance, and adequacy to meet intended performance standards. The authority may provide such services as design review, establishing commissioning specifications, construction observation, spot testing, verification, and functional performance testing. Turnover information services may include providing performance and operating information (i.e., operating and maintenance manuals, training) to the responsible parties.

Quality Assurance. The commissioning authority assists all parties in achieving a high level of installation quality and system performance. Through an effective turnover process, the personnel responsible for the facility should be able to maintain that high level of systems performance.

Observation of Tests. The commissioning authority should observe and coordinate testing as required to ensure the performance meets design intent and contract documents.

Documentation of Tests. The commissioning authority should document or witness the results of the performance testing. The authority should provide or approve test forms to be used by representative parties to ensure a consistent format and to specify the type of information to be recorded.

Resolution of Disputes. The commissioning authority should remain an objective party, present on the project, with specific project knowledge. Should disputes arise, the authority should be available to determine the scope and extent of the problem and educate the involved parties as to the nature and extent of the problem. The owner or their designate should preside over resolution of problems as specified in the contract documents.

Deficiencies. The commissioning authority should verify that deficiencies found during the commissioning process are corrected.

Acceptance. The commissioning authority should determine and advise the owner of the date of acceptance for each component and system. Acceptance may extend to representing the owner with respect to warranty start data, substantial completion, certificate of occupancy, assembling or coordinating turnover documents, and scheduling training.

Turnover. The commissioning authority participates in the successful transfer of control of the facility from the construction team to the owner.

Cost Factors

No reliable data are available to determine the cost or cost-contributing factors of building commissioning. The number of cost-contributing factors depends on the size and complexity of the project and the willingness of the owners to invest in the appropriate commissioning process.

A wide range of professional services, comprehensive documentation, system testing, and operator training are available to be included in the scope of commissioning. One intent of commissioning

is to fully recover its own capital investment over the life of the system through improved management and owner satisfaction.

Commissioning Objectives

The owner should expect the following results from commissioning:

- Improved operator knowledge as to how building systems should operate or be maintained
- Reduced ongoing training requirements
- Performance in accordance with the owner's intent and the contract documents
- Well-structured turnover documents (i.e., manuals, as-builts, submittal data, final sequences of operation) that provide easy reference documents for equipment and system operation and maintenance.
- A method that allows operators to continuously update documents
- Reduced downtime due to better diagnosis of failures
- Improved ability to provide accurate information to occupants regarding maintenance of environmental conditions in the occupied space throughout the year
- Lower operating costs due to optimized performance and improved operational techniques
- Increased comfort and reduced complaints due to poor indoor environmental quality

Designers can expect the following advantages:

- All building systems perform in accordance with the owner's requirements
- Reduced risk exposure because problems are identified earlier
- Improved knowledge for use in future designs and installations
- Benefit of supporting professional input, leading to the most cost-effective design and operation
- Reduced number of interference drawings during construction, due to improved communication and coordination

Contractors can expect the following benefits:

- Clear understanding of the owner's goals, (i.e. program, design intent)
- Improved coordination through implementation of the commissioning plan
- Improved coordination between different trades and reduced likelihood of site interference drawings required throughout the project
- Well-documented successful system tests, which ease turnover and acceptance
- Reduced number of deficiencies at substantial completion
- Reduced number of callbacks
- Reduced number of calls for operation guidance because operations and maintenance personnel participate in training programs

Program and Predesign Commissioning Phase

Objectives. The objectives of the predesign commissioning phase are to

- Document requirements as specified in the owner's program
- Select the commissioning authority
- Identify and assign responsibilities
- Document the initial design intent
- Begin development of the commissioning plan

This phase should begin as close to project inception as possible. The information gathered is used to develop the design of the system and establish a method to evaluate its performance.

Owner's Program. This program outlines the owner's overall vision for the facility and expectations as to how it will be used and operated. The owner's program should contain known performance goals and objectives along with budget constraints, schedules, and other limitations. Typically, the technical aspects of the owner's program includes use of the facility, user needs, occupancy requirements, type of construction, system functions, as well as energy, air quality, power quality, and environmental performance criteria. The program should define the scope of the commissioning process and the preferred organizational structure. It should contain all requirements needed to develop the initial design intent document. This critical document forms the basis from which all other documentation is developed.

Team Selection. The commissioning authority and the design team should be selected at this time. The operations and maintenance (O&M) personnel should be identified and encouraged to lend their knowledge to ensure that important O&M issues are considered in design and during commissioning.

Commissioning Plan. This document or group of documents defines the commissioning process or schedule at the various stages of project development. It continually evolves and is updated as the design and construction of the building progresses.

As the design concepts evolve from the initial design intent document, a preliminary commissioning plan is developed that outlines the process required to commission the building systems. This plan should develop the extent of the commissioning process and communicate it to all project participants. It must include the scope of the process, the time required for completion, and an organizational chart indicating reporting relationships. The scope and potential cost as outlined in the preliminary commissioning plan then should be evaluated and approved.

The roles and responsibilities of each project participant with respect to commissioning are identified in the preliminary commissioning plan. This plan should recognize the owner, commissioning authority, design professional(s), operation and maintenance staff, contractors, vendors, and other specialists as appropriate. The team may be expanded or modified as the project progresses.

Systems Manual. Material for the systems manual should begin to be compiled. The manual provides the information needed to understand, operate, and maintain each system. In this phase, the manual will contain a statement that conveys the required functional operation in all normal and emergency modes of operation. The initial design intent document shall be included in the systems manual, which is updated through all phases of the project.

Design Commissioning Phase

Objectives. The design phase of the commissioning process begins with preparation of schematic design documents. This phase has the following objectives:

- Ensure that clear design intent documents are developed
- Develop or refine commissioning plan and specifications
- Prepare contract documents that clearly identify, describe, and fulfill the design intent
- Review and accept contract documents for compliance with design intent
- Coordinate all building systems with the HVAC equipment and systems

Documentation of Design Intent. Changes to the initial design intent occurring during the design phase should be documented, reviewed, and approved. The revised design intent document should become part of the contract documents and will be represented in the updated systems manual.

Design Intent Documents. The developing of clear design intent documents is a critical aspect of commissioning. The initial design intent document is a detailed explanation of the information developed in the owner's program. This document should describe the facility's functional needs, intended levels and quality of environmental control, environmental needs, and schedules.

When approved, the design intent document should be used as the starting point for the development of the **contract documents** and referred to in case of ambiguities or uncertainties about project design. Changes to the design intent occurring during the course of the project should be documented, reviewed, and approved. The design intent document may be included in the contract documents as a reference. The final design intent document becomes part of the systems manual.

Conceptual designs for HVAC systems and other building systems should be prepared to provide options to the owner that meet the owner's program.

The design intent is updated and defines design assumptions, building energy performance, performance standards of the proposed systems, the most appropriate conceptual approach, space requirements, zoning, and other requirements that affect building use and aesthetics.

Basis of Design. The basis of design should respond to and be consistent with performance criteria specified in the owner's program. The basis of design includes the requirements for each occupancy, activity, and/or physical area of the proposed facility. The basis should reference ASHRAE or other applicable standards, local building codes, and environmental quality objectives.

Because the design intent may be revised and changes may be accepted as design concepts, the basis of design should be reviewed with regard to these changes and updated as needed. The final versions of the design intent document and its reference to the basis of design should be included in the contract documents.

Commissioning Plan. During the design phase, the approved preliminary commissioning plan may expand the planned activities for all participants relative to the commissioning process. Updated information includes work roles and responsibilities, project organization chart, and scheduling of the process. The plan identifies which systems, system components, and functions will be tested on this project. This document also sets the scope for the testing specification.

Observations of the progress of construction are important for updating the commissioning plan, particularly with respect to the scheduling of activities. In addition to construction requirements, observations should be conducted to ensure compliance with manufacturers' installation instructions.

Commissioning Specification. The building systems commissioning specification is part of the project specification. This specification is required to contractually implement the post-design phases of the process. The commissioning specification is a detailed description of the scope and objective of commissioning during the construction, acceptance, and post-acceptance phases of a project. It must specify the scope of work, roles, responsibilities, and requirements of each commissioning team member, and their lines of authority. This specification is needed by contractors so that they can include in their bid the support required by commissioning, which includes meetings, training, testing, etc.

The commissioning specification should detail testing requirements, acceptance phase procedures for verification, functional performance testing acceptance criteria, and any other required acceptance phase procedures. It should include a list of equipment and systems to be evaluated, along with checklist formats and sample test forms.

The scope of work in the specification should identify the required skills and qualifications of the commissioning team, including operation and maintenance personnel. It should include a section for each trade involved in the construction of the building, detailing their scope of work in the commissioning process. Management of turnover documents and training must be specified.

Contract Documents. These documents should clearly reveal, describe, and fulfill the requirements specified in the design intent documents.

Design Intent Compliance. The contract documents should be reviewed to ensure consistency and that they correctly represent the current design intent document. Results are reported to the owner. The contract documents then should be accepted or rejected by the owner or the owner's designee.

System Coordination. Part of the review of drawings includes verifying system coordination. All other specified building systems including other mechanical, electrical, fire-safety, and life-safety service requirements, should be coordinated with the HVAC equipment and systems.

Construction Commissioning Phase

Objectives. During the construction phase, all systems are installed, started, and operated. This phase includes the following objectives:

- Review submittals
- Finalize details of the commissioning plan
- Conduct periodic commissioning team meetings
- Maintain documentation of all tests, observations and issues, verify system installation
- Verify equipment/system start-up and operation
- Verify performance of temperature and building management controls
- Verify testing, adjusting, and balancing (TAB) work
- Coordinate as-built drawings
- Manage assembly, indexing, and turnover of all documents
- Coordinate O&M training

During installation, O&M personnel should observe and monitor the systems and testing. Through this procedure, they should develop a better understanding of the intended operation and performance and become familiar with the location of equipment and devices, including those that may be hidden when construction is complete.

Procedures. During the construction phase, the following activities are part of commissioning:

- Review submittals for performance parameters
- Detail the commissioning plan
- Create project-specific test procedures and checklists needed to supplement the specified tests
- Make necessary observations and inspections
- Manage documentation of all tests/observations including turnover documents
- Report progress and deficiencies to all parties involved
- Coordinate training of operations and maintenance personnel

The updated operation description should be included as part of the systems manual. This description should be updated and combined with equipment data, including performance data, for training of and subsequent use by the operations and maintenance staff.

Regularly scheduled meetings of the entire project team should be held to communicate issues to all concerned, resolve conflicts, report on the progress of the commissioning process, identify urgent work, and identify and resolve deficiencies. This site coordination is critical to ensure a quality commissioning process.

Testing and calibration of controls should begin concurrent with and be completed prior to the completion of the TAB work. Specified TAB work must be performed prior to acceptance procedures. Prior to functional performance testing, operational tests should be conducted on equipment, duct, pipe, and control systems to verify that pressures, flow rates, and control functions meet design requirements. The training program for operations and maintenance personnel should be coordinated with the appropriate participants.

Submittals. Submittals should be reviewed prior to systems installation. Submittals include shop drawings, equipment submittals, and testing and balancing procedures and forms. Submittals should also include test procedures, report forms, data sheets, and checklists that will be used in the functional performance testing.

Equipment submittals should include full and part-load performance data covering the entire operating range for each piece of equipment. The equipment operations and maintenance information (including parts lists, installation instructions, and special tool needs) should be submitted in accordance with specification requirements and reviewed for completeness, clarity, and accessibility. Commissioning submittal review is intended to verify correct information on test documents. It does not supersede or supplant design commissioning review.

Commissioning Plan. The commissioning plan should be updated to define the on-site activities required for implementing the commissioning specifications. This plan must be project specific and reflect the actual equipment that will be installed. The commissioning plan should

- Define each party's role in the inspections, verification, testing and training
- Detail the schedule of inspections during construction
- Develop the schedule for verification and functional performance tests
- Define the process for reporting and correcting any deficiencies identified
- Detail any training sessions for operations and maintenance personnel that are to take place during this phase

Documentation. All commissioning activities that occur during the construction phase should be documented. The equipment and control sequence documentation developed during the design phase should be updated to match the equipment supplied on the project.

Acceptance Commissioning Phase

Objectives. During the acceptance phase, verification, functional performance tests, and other acceptance procedures takes place. The objectives of the acceptance commissioning phase are to

- Verify accuracy of final TAB report
- Verify and document that all systems comply with the contract documents
- Verify functional performance testing of all systems
- Establish an as-delivered performance record for all systems
- Verify completion of the as-built records
- Conduct operations and maintenance personnel training
- Complete the final commissioning report
- Complete the systems manual
- Turnover all documents
- Complete the owner training program

Verification. Verification comprises a full range of checks and tests to determine that all components, equipment, systems, and interfaces between systems operate in accordance with contract documents. All operating modes, all interlocks, all control responses, and all specified responses to abnormal or emergency conditions are tested and verified. Operating modes, etc. can be verified concurrently with a physical point check of controllers and sensors. Verification also validates the TAB report. Each contractor may have some responsibility in verification.

Functional performance testing demonstrates the performance of systems. Documentation should include performance data for all equipment and systems. Operations and maintenance personnel should be made available to observe verification and functional performance tests. Conflicts and deficiencies identified during the acceptance phase should be addressed and resolved to the owner's satisfaction.

At the end of the acceptance procedures, all systems will have been documented as proving operational performance in accordance with the contract documents. This includes both normal operation modes and abnormal or emergency conditions. Before functional performance testing can start, the following should be observed and documented:

- Systems and associated subsystems have been completed, calibrated, and started up and are believed to be operating in accordance with contract documents
- Control systems have been completed and calibrated and are believed to be operating in accordance with contract documents
- Testing, adjusting, and balancing procedures have been completed, and TAB reports have been submitted and reviewed and discrepancies corrected and accepted

A statement should be issued certifying that work has been completed and equipment and systems are operational in accordance with the contract documents.

Deficiencies. Deficiencies that are identified in the building systems or when interfacing with other building systems during verification must be documented and reported to the commissioning team. A decision must be made whether to remedy the situation and if reverification is required.

A **certificate of readiness** should be issued stating that the specified equipment, systems, and controls are complete and ready for functional performance testing. This certificate should be supported by completed prestart/start-up checklists (signed by the responsible parties) and the verification report.

Functional Performance Testing. This testing should progress from testing the individual components within central equipment and systems, to testing the systems that distribute services throughout the building. The extent of these tests, as well as details of the services to be tested, are in accordance with the commissioning specifications.

In each of the individual spaces to be tested, the parties performing the functional performance tests may be required to make temporary modifications to control functions or provide supplementary internal loads to simulate desired load conditions up to design load conditions.

Functional performance testing of the major heat exchange components may be done at less than full-load capacity of the equipment using part-load performance curves provided by the manufacturer for comparison. However, deferred seasonal testing may be required to verify performance at near design conditions.

As each individual check or test is accomplished, physical responses of the system should be observed and compared to the specified requirements in order to verify the test results. Ideally, only actual physical responses of components are observed. Reliance on control signals or other indirect indicators is not adequate unless previously validated. The input and output signals for each control component also need to be observed to confirm that they are correct for each physical condition.

The specific tests required and the order of testing will vary depending on the type and size of system, number of systems, sequence of construction, relationship between building systems, and specific owner/tenant program requirements.

During functional performance testing, a failure in performance of a part of the system or of a component may be revealed. Any performance deficiencies must be evaluated to determine the cause and whether they are part of the contractual obligations. After corrective measures are completed, the functional performance test that failed may be repeated.

Functional Performance Test Procedures. Testing should be performed as defined in the commissioning specification and detailed in the contract documents. If any test cannot be completed because of a deficiency outside the scope of the system, the deficiency should be documented and reported to the owner. Deficiencies should then be resolved and corrected by appropriate parties and the functional performance tests rescheduled.

If the commissioning specification indicates that specific seasonal testing is to be conducted, the appropriate initial performance tests should be completed and documented and the additional tests scheduled. The parties responsible for the seasonal tests should return to the site to complete and document the tests as scheduled. This scheduling is represented in the commissioning plan.

Test Reporting Requirements. All measured data, data sheets, and a comprehensive summary describing the operation of the systems at the time of the test should be submitted to the commissioning authority or the owner's representative. Sample functional performance test reports should be included in the commissioning specifications. Deviations from these contract documents or design intent should be resubmitted and approved, with a description included.

Systems constructed in accordance with contract documents but having performance deficiencies not covered in these documents force a decision as to whether modifications should be implemented to bring the performance up to the design intent criteria or if the test results should be accepted as submitted. If corrective work is performed, a decision must be made as to whether part or all of the functional performance testing should be repeated and a revised report submitted.

After the performance of all systems is evaluated, the results of the evaluation are included in the final commissioning report.

Verification Report. The final tabulated checklist data sheets should be assembled in a verification report and submitted to designated parties on the commissioning team for review. The verification report documents any unresolved deficiencies and may suggest a method of correction. The responsible party should determine if verification is complete and whether building systems are functioning in accordance with the contract documents.

Documentation. Documentation procedures during the acceptance phase include executing verification and functional performance tests forms, completing and assembling the final turnover records, and developing the final commissioning report. The commissioning documentation consists of the following:

- Verification checklist data sheets
- Functional performance test data records
- Final commissioning report, (including a summary of all issues resolved and unresolved)
- As-built drawings and other records
- Final updated operations manuals and maintenance manuals
- Systems manual (system operation description and final design intent)
- Training documents (may include videotapes)

Final Acceptance. For final acceptance, all documents should be turned over to the owner, the final commissioning report should be completed and submitted, and recommendation for acceptance of the facility determined. Some acceptance procedures (such as off-season tests) may not have yet been completed, but this should not impede final acceptance.

The final commissioning report should indicate if the systems have been completed in accordance with the contract documents and if the systems are performing in accordance with the final design intent document. The report should identify and discuss any substitutions, compromises, or variances between the final design intent, contract documents, and as-built conditions. This report should be used to evaluate the system and serve as a future reference document during operation of the systems. It should describe components and performance that exceed design intent and those that do not meet design intent. The report may make recommendations for resolution.

Operations and Maintenance Training Program

The objective of the O&M training program is to provide qualified technicians with the knowledge to operate and maintain building systems in accordance with design intent, manufacturers' recommendations, and procedures contained in the systems manual. The program should be detailed such that it can be repeated for new and replacement personnel. Thorough documentation must be supplied for future training activities. Videotaped training sessions may also be used to assist with future training.

The training program should be implemented as defined in the commissioning plan. Training should be performed by parties with specific expertise in each aspect of the building's systems.

Scope of Training. The training program should furnish a thorough understanding of all equipment, components, systems, and their operation, including appropriate how-to skills. Training should include the following topics:

- Use of the systems manual with an emphasis on
  - Design intent
  - Description, capabilities, and limitations of the systems
  - Operation procedures for all modes of operation
  - Acceptable tolerances for adjustments in all operating modes
  - Procedures for dealing with abnormal conditions and emergency situations for which there is a specified response
  - Use of operation manuals
  - Use of maintenance manuals
- Recommended procedures for collecting and interpreting specific performance data
- Specialized manufacturers' training programs

Goal of Training. The goal of training is to understand system performance and the general theory of operation. This level of understanding should typically include the following:

- Theory of operation on
  - Basic concepts of all pertinent systems
  - Energy efficiency
  - Indoor air quality
  - Occupancy comfort
  - Occupied versus unoccupied or partial occupancy
  - Seasonal modes of operation
  - Emergency conditions and procedures
- System operating procedures
- Operating parameters (i.e., setpoints, performance goals)
- Use of the control system, including
  - Sequence of operation
  - Problem indicators
  - Diagnostics
  - Corrective actions
- Use of reports and logs
- Service, maintenance, diagnostics, and repair

Post-Acceptance Commissioning

Post-acceptance commissioning is the continued adjustment, optimization, and modification of building systems to meet specified requirements. It may be used on existing buildings or even on a facility that has previously been commissioned. It includes updating documentation to reflect minor setpoint adjustment, maintenance and calibration, major modifications, and provision for ongoing training of operations and maintenance personnel.

The objective of post-acceptance commissioning is to maintain the performance throughout the useful life of the facility in accordance with the current design intent. The extent of post-acceptance commissioning is determined by the scope of modifications and occupancy changes that are made to the facility.

Post-acceptance commissioning starts from the base of existing documents. This documentation includes all available original commissioning documents plus any updates from the commissioning of previous modifications.

Post-Acceptance Commissioning Procedures. The three identifiable levels of post-acceptance commissioning are as follows:

- Ongoing commissioning activities
- Minor changes or modifications to systems and changes to occupancy and/or layout
- Major modifications to the facility's layout and/or systems

Major modifications will impact overall system characteristics or performance through significant changes in design or central system capacities. Minor modifications do not anticipate such impacts; they will likely impact individual zones or rooms only.

Minor modifications require, as a minimum, the following:

- Revision of as-built records, such as the final design intent document, systems manual, operations manuals, maintenance manuals, and reports
- Testing, adjusting, and balancing of affected systems or sections
- Training on operations and maintenance procedures that are affected by the modifications

Major modifications require the same efforts, expanded to suit full commissioning of any system or subsystem impacted by modifications. Overall performance must be reestablished. When major modifications are carried out, normal commissioning procedures should be undertaken. For each level of modification, the impact of planned alterations must be reviewed to ensure that a comfortable environment and air quality standards for the occupied portions of the facility are maintained.

Ongoing commissioning activities strive to sustain a specific level of performance and documentation. Proper maintenance programs, training, and familiarization of the systems by the new operating staff are important to the support of the commissioning process. For example, a standard method of recording and responding to complaints should be in place and used consistently.

As equipment and controls are replaced through the maintenance program, calibration and performance should be checked, documents revised, and any changes or new equipment data sheets included in the systems manual.

Systems are periodically retested to measure and document their actual performance. Functional performance test (FPT) checklists used in the acceptance phase and subsequent activities should be a guide for retesting. Discrepancies between predicted performance and actual performance and/or an analysis of the complaints received may indicate a need to re-evaluate the current design intent or to consider returning to the program phase of commissioning.

Ongoing training includes refresher training of existing personnel, training of new personnel, and training of all personnel on newly installed equipment or revised operating procedures. It should be consistent with the training provided by the original commissioning process.

Documentation. Documentation is similar in scope to the requirements outlined for new construction commissioning. Where documents exist they need to be maintained or updated. Regardless of the form, nature, and scope of commissioning, all records (drawings, manuals, etc.) must be catalogued and well referenced.

BUILDING ENVELOPES

THE BUILDING envelope comprises the outer elements of a building, including the foundation, walls, roof, windows, doors, and floors. This chapter describes the nature, functions, and performance of the building envelope. It concentrates on the interaction between the components of the building envelope and on the effects on the interior environment of a building. Understanding this interdependence is essential to executing a reliable building envelope.

The prime functions of the building envelope are to provide

- Shelter for occupants
- Strength and rigidity
- Stability and durability
- Control of heat, air, and moisture vapor flows
- Control of liquid water movement
- Indoor air quality control
- Fire resistance
- Cost-effectiveness
- Acoustical performance
- Aesthetic considerations

In buildings with a large ratio of volume to exterior surface area, the internal heat sources may be larger than gains or losses through the envelope. The effect of the envelope becomes increasingly important as the ratio decreases. In single-family dwellings, heat transfer through the envelope is the dominant factor in determining the overall heating and cooling loads. However, local control (e.g., in business offices) may depend on the ability of glazing to control heat gain or loss.

RESIDENTIAL FOUNDATIONS

Good foundation design entails not only structural soundness and good insulation, but also appropriate moisture, termite, and radon control. In North America, several building energy codes recommend foundation insulation in climates with more than 2500 heating degree-days (base 65°F). An uninsulated, conditioned basement may lose up to 50% of the heat in a tightly sealed house that is well insulated above grade.

Although saving energy is an important incentive, insulating basement foundations also creates a more comfortable, livable space at a relatively low cost. Raising basement temperatures reduces condensation, minimizes problems with mold and mildew, and helps maintain acceptable indoor air quality.

Construction

The three basic types of foundations are basement, crawl space, and slab-on-grade. Each uses specific insulation techniques. The cost of higher insulation levels must be compared with the savings. The comparison can be made in several ways, but a life-cycle cost analysis is the best method (Carmody et al. 1991). It takes into account a number of economic variables, including installation costs, mortgage rates, HVAC efficiencies, and energy costs.

The preparation of this chapter is assigned to TC 4.4, Building Materials and Building Envelope Performance.

Conditioned Basements. A concrete or masonry basement wall may be insulated by (1) insulating the exterior upper half, (2) insulating the entire exterior wall, or (3) insulating the entire interior wall.

Unconditioned Foundation Spaces. Unconditioned basements may be insulated in the same way as conditioned basements, or they may be insulated by placing insulation between or below the floor joists in the ceiling. The latter approach separates the basement thermally from the above-grade space and results in lower basement temperatures in winter, usually making it necessary to insulate exposed ducts and pipes.

A crawl space can be designed as a short basement (with a floor). Because the floor is above grade, it is less subject to moisture hazards than a slab floor. However, if moisture problems occur, they are more likely to be noticed in a basement because it is more accessible than a crawl space.

While most codes require operable louvered vents (1 ft² per 1500 ft²) near each corner of a crawl space, research has shown that a moisture-retardant ground cover is more effective and, in most locations, eliminates the need for vents (ASHRAE 1994).

Crawl Space Insulation. In a vented crawl space, insulation is placed between the floor joists. In an unvented crawl space, either the concrete or masonry walls or the floor above can be insulated. If the walls are insulated, air infiltration paths such as the interface between the crawl space wall and the bottom plate of the floor above must be sealed. Intersections of two hard surfaces are common air infiltration paths. Walls may be insulated by (1) insulating the entire exterior wall, (2) insulating the entire interior wall, or (3) insulating the entire interior wall and the perimeter of the crawl space floor to a depth of 2 ft. When the floor is insulated, an appropriate ground cover must be installed to control moisture in the crawl space; moisture degrades the thermal performance of insulation. In pressure-treated wood construction, insulation is placed in the cavities between the wood studs.

Vented Versus Unvented Crawl Spaces. The disadvantages of a vented crawl space are that (1) pipes and ducts must be insulated; (2) a larger area must be insulated (the floor area is greater than the wall area), which may increase the cost; and (3) in some climates, circulation of warm, humid outside air in the crawl space can cause the wood structure to have an excessive moisture content. Another advantage to designing crawl spaces as unvented, semiconditioned zones is that insulating the foundation at the perimeter instead of the floor above simplifies installation and minimizes condensation. On the other hand, venting can complement other moisture and radon control measures, such as ground covers and proper drainage.

While increased airflow can dilute ground source moisture and radon, it may not eliminate them. Venting crawl spaces may be desirable in areas of high radon hazard, but it should not be considered a reliable radon mitigation strategy. Pressurizing the crawl space is an effective method of minimizing soil gas uptake; however, the crawl space walls and ceiling must be tightly constructed for this approach to be effective. Vented crawl spaces are often provided with operable vents that can be closed to reduce winter heat losses (closing them may, however, increase radon levels). Also, these vents can be closed in summer to keep out moist outside air that may have a dew point above the crawl space temperature. However, unless steps are taken to control air infiltration, closing vents does not ensure that the crawl space will have no air exchange with the outside; any exchange with outside air reduces the effectiveness of any wall insulation.

Slab-on-Grade. For slab-on-grade foundations with concrete or masonry walls, insulation is usually placed on the entire exterior surface of the foundation stem wall, on the interior surface of the foundation wall, or horizontally under the slab perimeter. When the insulation is placed on the interior, it is important to place it in the joint between the slab edge and the foundation wall. The insulation placed in this joint does not need to have an R-value greater than 5 $ft^2 \cdot °F \cdot h/Btu$. Exterior vertical insulation with an R-value of 5 to 10 $ft^2 \cdot °F \cdot h/Btu$ is justified in most climatic zones (Labs et al. 1988).

Another method is to place insulation horizontally on the building exterior extending 2 to 4 ft into the surrounding soil. Shallow footings (less than 2 ft below grade) or slabs with a thickened edge and no foundation wall are common in some regions. In these cases, additional insulation may be placed horizontally along the exterior edge.

Thermal Performance

A key question in foundation design is whether to place insulation inside or outside the basement or crawl space wall. Placing rigid insulation on the **exterior** of a concrete or masonry basement or crawl space wall has some advantages: it provides continuity without thermal bridges, maintains the waterproofing and structural wall at moderate temperatures, minimizes moisture condensation problems, does not reduce interior basement floor area, and includes the foundation in the thermal mass in the building. Exterior insulation at the rim joist leaves joists and sills open for inspection from the interior for termites and decay.

However, exterior insulation can provide a path for termites and can prevent inspection of the wall from the outside. A termite shield may need to be installed through the insulation where the sill plate rests on the foundation wall. Vertical, exterior insulation can extend as deep as the top of the footing and can be supplemented by extending the insulation horizontally from the face of the foundation wall.

Interior insulation is an effective alternative, and it is generally less expensive if the costs of the interior finish materials, including a flame spread cover, are not included. However, interior wall insulation increases the exposure of the wall to thermal stress and freezing and may increase the likelihood of condensation on sill plates, band joists, and joist ends.

Placing insulation in the basement ceiling is an alternative, but it should be used with caution in colder climates where pipes may freeze and structural damage may result from lowering the frost depth in the soil next to the foundation.

In a wood foundation, insulation is placed between studs. Insulation placed around the crawl space floor perimeter can provide additional thermal protection; however, it may also create additional paths for termite entry. A 2 in. air space between the lower end of the insulation in a crawl space or wall cavity and the bottom plate of the foundation wall allows inspection for insect infestation. Batt or tight-fitting rigid foam board insulation is commonly placed inside the rim joist.

Crawl Space Wall Insulation. A common, low-cost approach to insulating crawl space walls is to drape batts with a vapor retarder facing the inside of the wall. (Most building codes require that the vapor retarder be flame resistant.) The batts can be laid loosely on the ground at the perimeter to reduce heat loss through the footing. However, it is difficult to maintain the continuity of the vapor retarder around the joist ends and to seal the vapor retarders where they join the batts. Good installation is difficult because of cramped working conditions, and vapor-proof installation prevents easy inspection for termites. In a foundation system of wood pressure-

treated with preservative, insulation is placed in the stud cavities similarly to above-grade insulation in a wood frame wall. Crawl space walls may also be insulated by spraying an insulating foam.

Insulation may also be incorporated in the concrete or masonry walls. Examples include rigid foam plastic insulation cast within the concrete wall; polystyrene beads or granular insulation poured into the cavities of conventional block walls, or foam insulation field-injected into the cores; concrete blocks with insulating foam inserts; formed interlocking rigid foam units that serve as a permanent, insulating form for cast-in-place concrete; and masonry blocks made with polystyrene beads or a low-density aggregate/regular-density concrete mixture. However, the effectiveness of construction that insulates only a portion of the wall area should be evaluated closely because thermal bridges can reduce the insulation's effectiveness by as much as 80%.

Slab-on-Grade Foundation Insulation. For slab-on-grade systems, the greatest heat loss is through the small area of the foundation wall above grade. Heat is also lost from the slab to the soil. Heat losses to the soil are greatest at the slab edge and diminish rapidly with depth. Heat loss at the edge and through the soil must be considered in designing the insulation system.

Insulation can be placed vertically outside the foundation wall or grade beam. This approach effectively insulates the exposed slab edge above grade and extends downward to reduce heat flow from the floor slab to the ground surface outside the building.

Vertical exterior insulation is the only method of reducing heat loss at the edge of an integral grade beam and slab foundation. A major advantage is that the interior joint between the slab and foundation wall need not be insulated, thus simplifying construction. However, there are several drawbacks to this method. The rigid insulation must be covered above grade with a protective board, coating, or flashing material; and brick facings that contact the covering can create a thermal short that bypasses both the foundation and the above-grade insulation. A further limitation is that the depth of the exterior insulation is controlled by the footing depth.

Thermal insulation also can be placed vertically on the interior of the foundation wall or horizontally under the slab. In both cases, heat loss is reduced and the difficulty of placing and protecting exterior insulation is avoided. Interior vertical insulation is limited to the depth of the footing, but underslab insulation is not limited. Usually, the outer 2 to 4 ft of the slab perimeter is insulated. The joint between the slab and the foundation wall must contain insulation whenever insulation is placed inside the foundation wall or under the slab. Otherwise, a significant amount of heat transfers through the thermal bridge at the slab edge. The insulation is generally limited to a thickness of 1 in. at this point.

Another option for insulating a slab-on-grade foundation is to place insulation above the floor slab. A wood floor deck can be placed on sleepers, leaving cavities that can be filled with rigid board or batt insulation; or a wood floor deck can be placed directly on rigid insulation above the slab. This approach avoids some of the construction detail problems inherent in the more conventional approaches, but it may cause a greater frost depth near the slab edge.

Moisture Control

High moisture levels in a foundation can rapidly degrade the thermal performance of most types of foundation insulation. Properly constructed foundations with good drainage and damp-proofing should keep moisture loads from foundations below 6.3 pints/day; however, Christian (1994) reported on foundations that contribute 40 to 150 pt/day. Elevated moisture levels increase the potential for mold and mildew growth, odors, pathogens, and serious health effects to occupants. Mold and mildew will grow where the relative humidity of a surface is greater than 70%. This level of relative humidity can occur in the foundation of a building that has an interior with 50% rh. Because the surface temperature is colder during winter than the crawl space air temperature, the relative humidity of the crawl space air can be much lower than 70% and still result in 70% rh on the surface. Frequently, methods used to solve the problem in existing buildings are energy-intensive (e.g., year-round operation of dehumidifiers).

Basement. Rain is carried away by gutters, downspouts, grading away from the building, and a cap of low-permeability backfill material. Subgrade drainage by a drain screen (gravel and footer drain connected to daylight, sump, or storm sewer) prevents water from reaching the foundation wall.

Damp-proof or waterproof materials on the exterior of the foundation wall and over the top of the footing control moisture transport. Capillary moisture movement into the slab is controlled by a 4 in. thick granular layer under the slab.

All air leakage openings (i.e., floor slab/wall intersection, rim joist area) are sealed with durable caulking materials, closed with gasket systems, taped, or covered with moisture-vapor-permeable house wrap.

Damp-proofing on the wall and polyethylene under the slab control vapor diffusion. During heating periods, vapor diffusion from the interior of the basement may transport moisture into the rim joist framing, where it accumulates. Rigid insulation installed on the exterior of the rim joist limits periods of potential condensation.

Slab-on-Grade. The ground should be graded to direct water away from the building. A granular layer under the concrete slab provides a capillary break between the soil and the slab. This layer can also be integrated into a subslab ventilation system to reduce radon levels. Extending the vapor retarder over the top of the foundation wall and placing appropriate flashing for any brick facing provide a capillary break to protect the above-grade wall from ground moisture penetration.

A vapor retarder placed under the slab restricts both radon gas entry and water vapor diffusion through the slab. If ducts are placed in the slab, moisture may enter the conditioned space because groundwater and soil gas will likely seep into the ducts.

Air Movement/Radon Control

Construction techniques to minimize radon infiltration through the foundation are appropriate where radon is present. General approaches include sealing joints, cracks, and penetrations in the slab and walls and evacuating soil gas under the foundation.

A tight foundation and a discharge system reduce the chimney or stack effect that draws air from the foundation or basement into the house. If radon levels are excessive, a passive discharge system can be connected. If further relief is needed, an inline duct fan can be installed in the discharge system (Figure 1).

Subslab depressurization effectively reduces radon concentrations to acceptable levels, even in houses with high concentrations (Dudney et al. 1988). This technique lowers the pressure around the foundation envelope, causing the soil gas to be routed into a collection system and discharged to the outdoors.

A continuously operating fan should not be relied on as a discharge method. Ideally, a passive depressurization system should be installed, radon levels tested, and—if necessary—the system activated by adding a fan. The fan should be located in an accessible section of the stack so that any gas leaking from the positive pressure side of the fan does not enter the living space.

Active subslab depressurization also raises some long-term concerns. If the radon barrier techniques are not fully used, indoor air could be drawn into the basement and subsequently discharged, resulting in a larger than expected energy cost. Durability is another concern, particularly with motor-driven components. Also, the system is susceptible to owner interference.

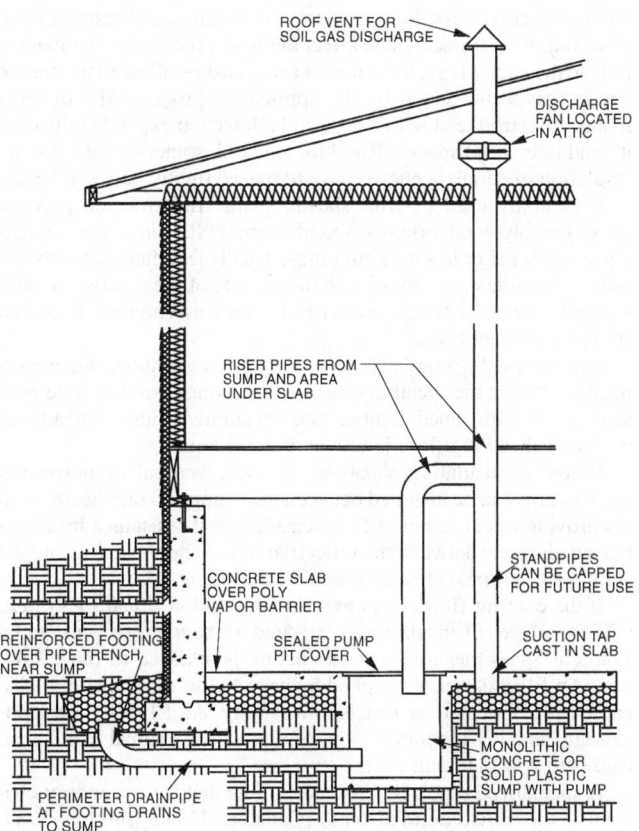

Fig. 1 Soil Gas Collection and Discharge Techniques

RESIDENTIAL WALLS

Lightweight Construction

The most common type of lightweight wall construction is the **wood frame** wall. Most houses built in the United States have exterior walls framed with wood studs, although steel studs offer several advantages; for example, they are more dimensionally stable and are truer than wood studs. Additionally, **steel studs** can be fabricated from recycled material and are themselves recyclable. However, these studs can create a significant thermal bridge across the cavity insulation, reducing its thermal resistance by up to 50%. Exterior foam sheathing insulation can be used to augment the R-value of the wall (Christian and Kosny 1996).

The interior side of the framing is typically sheathed with gypsum wallboard. In wood construction, to enhance the energy efficiency of the wall, foam insulation sheathing can be applied between the gypsum board and the framing. An advantage of this practice is that the low-permeance sheathing can take the place of a vapor retarder. The disadvantage is that it complicates the application of the gypsum board.

The cavities between framing members are filled with thermal insulation including batt, loose fill, stabilized loose fill, site-injected foam, and rigid foam boards. With all products, the level of performance is related to the quality of the installation. Batts should be installed to completely fill the cavities along the back corners and the depth of the cavity. Gaps created by framing members or voids in an insulation caused by material settling or by improper installation can significantly reduce thermal performance. Dry loose fill should be installed at the manufacturer's recommended minimum density to avoid settling or the formation of voids. Wet loose fills should be installed with the proper amount of moisture, as recommended by the manufacturer. Stabilized loose fill insulations are applied with adhesive and water to prevent settling and movement within the cavity. As these products dry, some shrinkage has been observed; this shrinkage can affect the wall's thermal performance. Before the cavity is sealed, enough time must be allowed to remove any excess water added by the application process. Site-injected foams are introduced with a hose and allowed to expand in the cavity and interstitial spaces. Rigid foam board, sometimes used as an insulation material, is cut to size and placed tightly between studs.

Commonly used exterior sheathing materials include gypsum sheathing, plywood, oriented strand board (OSB), and foam insulating sheathing. For low-rise buildings, usually less than three stories, airflow retarders are installed on the outside of the exterior sheathing and covered with an exterior finish such as vinyl or aluminum siding, stucco, or brick.

Airflow can be retarded by constructing a continuous, air-impermeable plane at the sheathing surface (e.g., tongue-and-groove gypsum board with taped joints). The sheathing/window, sheathing/roof, and sheathing/floor junctions are also taped.

When an insulating sheathing is used, vertical or horizontal spacer strips can be installed between the siding and the sheathing to (1) provide a rain screen, (2) increase thermal resistance by maintain an air space between the reflective foil on the sheathing and the siding, and (3) provide a nail base for the siding.

If the exterior finish is an **exterior insulation finishing system** (EIFS), a layer of insulation is attached to the exterior wall, and a synthetic glass-mesh-reinforced coating is attached to the insulation. An EIFS is usually applied by hand to the wall on site; it also can be prefabricated as panels in a factory. An EIFS is non-load-bearing and is vulnerable to wind loads; the critical wind load causes the panel to pull off the structure.

A specialized type of exterior sheathing that is very energy-efficient is the **nonstructural truss** (Canadian Home Builders 1989; Nisson and Dutt 1985; Wilson 1988). These trusses are installed on the exterior of a conventionally framed and sheathed wall. The trusses are easily installed, and they allow high levels of insulation to be installed, although the insulation near the truss must be installed carefully. The additional wall thickness complicates the framing required around wall openings such as windows and doors.

Panelized walls are composed of an insulation layer faced on both sides with a sheathing material. The insulation layer ranges in thickness from 2 to 12 in. Oriented strand board is the most widely used sheathing material, but gypsum sheathing and plywood are also available. Because of evolving techniques for energy-efficient splines, panelized walls can have significantly fewer thermal bridges than wood frame construction.

Interior and exterior finishes for panelized walls are similar to those used in wood frame construction. Some panelized walls may already have suitably finished surfaces. Advantages of using panelized walls include (1) higher energy efficiency than typical wood frame wall systems; (2) little or no thermal bridging in the clear wall area, (3) tighter than average construction due to the reduction in jointing if large panels are used, and (4) integration of the insulation into the panel system, making insulation installation unnecessary. Potential problems with panelized walls are (1) facer delamination in improperly built assemblies, (2) fire safety concerns with the use of cellular plastics (however, this is an issue in any wall with insulating sheathing), and (3) insect control (carpenter ants have been reported to nest in hollows carved in the panels).

Masonry Construction

A common masonry wall is made with concrete masonry units (CMUs), blocks of concrete cast in various shapes and joined with mortar to form a wall. The concrete used to fabricate CMUs is produced with a variety of aggregates and typically varies in density from 85 to 135 lb/ft^3 (Andrews 1989). A CMU wall provides a finished, durable interior and exterior surface. Holes can be drilled through the wall to feed utilities through. Many CMUs are available with patterns cast into the exterior face, thus providing diverse exterior finishes.

Interior or exterior sheathing, and/or an airflow retarder, can be applied to improve energy efficiency or to obtain a desired finish.

If an EIFS is attached to the exterior of the wall, the insulation separates the wall from the fluctuating outdoor temperatures. This method of insulation can be used to advantage in energy-efficient buildings in which the thermal energy stored in the wall mass is used to moderate indoor temperature swings. External placement also allows the wall cavity to be used for utilities that might be susceptible to freezing or heat loss. It is important that the edges of EIFSs be tightly abutted so that thermal bridges are minimized.

Thermal Performance

James and Goss (1993) provide a summary of the heat transmission coefficient measurements and calculations for various wall, roof, ceiling, and floor systems. These data are for clear wall areas only and do not include subsystems or the effects of the joints created at intersections with another portion of the building envelope.

In a series of three-dimensional heat conduction simulations using a finite difference program on a single-family detached house, the average area-weighted R-value for the entire wall was 88% of the R-value of the clear wall. A similar two-dimensional analysis by Tuluca et al. (1997) showed the R-value for the entire wall system to be 40 to 50% of the clear wall R-value. Significant thermal bridging can occur through framing (wood or even more important, steel), metal ties through exterior sheathing, and exposed slab edges.

The thermal bridges created by the webs of CMUs dictate the maximum thermal efficiency that a CMU can attain. To reduce thermal bridging, blocks containing only two webs instead of the usual three have been used, and web thickness has been reduced by up to 40%.

The exterior surface of an EIFS expands and contracts significantly under the influence of changing outdoor temperatures and solar radiation. The thin, highly expansive coating has virtually no thermal storage capacity, but it is reinforced with glass mesh that helps moderate thermal movement. Because the coating is bonded to a layer of plastic insulation, which also is highly thermally expansive, the temperature of the coating fluctuates rapidly. This fluctuation places considerable stress on the coatings themselves and on adjacent sealing joints. This problem is exacerbated when dark colors are used and can become acute because the maximum service temperature of expanded polystyrene insulation, the type most typically used, is about 170°F.

Moisture

Historically, the primary moisture control strategy for walls is to restrict moisture entry. Another approach to envelope design is to provide walls that balance entry with removal and that control accumulation and pressure differential.

Recognizing that lightweight wall systems can get wet or occasionally start out wet, the designer can promote weather protection and drying. A rain screen (Figure 2) can be placed immediately behind the exterior cladding to control rain penetration by providing a pressure-equalized air space; or a house wrap, which is vapor-permeable and water impermeable, can be installed over the exterior sheathing to protect the wall from the weather yet allow it to dry if necessary.

For heating-dominated climates and for mixed climates with both significant heating and significant cooling seasons, a vapor retarder at the interior surface is advisable. Moisture originates from the interior of the building and migrates into the wall until it reaches the interior vapor retarder, which is kept warm because it is inside the thermal envelope. In cooling-dominated climates, exterior sheathing of impermeable rigid insulation can act as a vapor diffusion retarder. Moisture outside the building moves into the wall until it reaches the exterior vapor retarder, which is kept warm because it is outside of the thermal envelope. As a general rule, to keep moisture condensation accumulation to a minimum, this first condensing surface must be kept above the dew point.

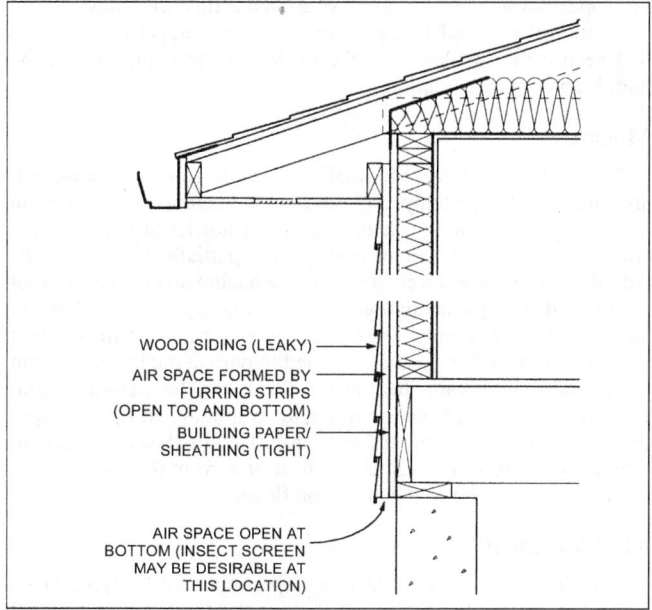

WOOD SIDING (LEAKY)

AIR SPACE FORMED BY
FURRING STRIPS
(OPEN TOP AND BOTTOM)

BUILDING PAPER/
SHEATHING (TIGHT)

AIR SPACE OPEN AT
BOTTOM (INSECT SCREEN
MAY BE DESIRABLE AT
THIS LOCATION)

Fig. 2   Example of Rain Screen

The exterior coating and insulation system of an EIFS is designed to be breathable, allowing moisture to pass through the wall under normal conditions. This design is important, because a vapor retarder could be formed on the wrong side of the wall by the EIFS coatings. Several conditions that can occur should be considered during the design of an EIFS wall system. If a low-vapor-permeance coating is applied on the outside, condensation can occur in the wall. This can happen, for example, if a highly vapor-resistant coating is applied to the outside of the wall for maintenance purposes, or if the wall is refinished with a new EIFS coating. If the ambient temperature and humidity are high in the summer, moisture can flow in and become trapped behind an interior finish that is highly vapor-resistant, such as a heavy vinyl wall covering or polyethylene vapor retarder. This problem is of particular concern in climates with extended hot, humid periods if the space is air-conditioned or refrigerated.

Air Movement

Uncontrolled air movement across walls increases space conditioning costs and may cause moisture to condense. **Air infiltration** is the movement of air from the outside into the conditioned space. **Air exfiltration** is the movement of air from the conditioned space to the outside. **Air intrusion** is the penetration of air into the building envelope without entirely crossing it. For example, outside air can penetrate a wall at sheathing joints, move around or through the insulation and return to the outside. Similarly, inside air can penetrate a wall at electric outlets, move around or through the insulation and return to the conditioned space. **Air leakage** designates any type or combination of air movement in and across the envelope (see the sections on Basic Concepts and Residential Infiltration in Chapter 25 of the 1997 *ASHRAE Handbook—Fundamentals*).

In general, an **airflow retarder** should be installed in walls. The airflow retarder reduces both infiltration and exfiltration. It also prevents intrusion by exterior or interior air, depending on the location of the air seal plane. The airflow retarder reduces energy use and reduces or eliminates moisture condensation.

In heating-dominated climates, the airflow retarder can be positioned on either the exterior or the interior of the wall structure (e.g., framing). If placed to the exterior of stud construction that does *not* have insulating sheathing, the airflow retarder protects the insulation located between framing members from cold air intrusion. To

avoid moisture condensation, such an airflow retarder must be moisture-permeable, because moist interior air can still intrude into the wall even though it cannot easily exfiltrate across the wall. Examples of exterior moisture-permeable airflow retarders are

- Certain thin membranes, such as spun bonded polyolefin and other house wraps
- Gypsum board, particle board, or plywood sheathing taped and sealed at all joints and at the wall/window, wall/roof, and wall/floor junctions
- Pressed paper and cardboard products taped and sealed
- Thin polystyrene accordion-type wraps stapled, taped, and sealed; used only on houses built according to U.S. Housing and Urban Development Department codes

The airflow retarder can also be placed to the exterior of stud construction that has insulating sheathing, such as steel stud walls with R-11 fibrous insulation, gypsum board sheathing, rigid insulation sheathing, and brick finish. If the airflow retarder is located on the winter-cold surface of the gypsum board sheathing, it can be moisture impermeable if the R-value of the rigid insulation is sufficient to maintain the airflow retarder above the dew point during winter conditions. Impermeable airflow retarders are usually bituminous, rubber-based, or plastic membranes. However, the airflow retarder can also be composed of a gypsum board layer with the junctions taped and sealed. Masonry construction usually includes air-impermeable airflow retarders.

If the airflow retarder is placed to the interior of a wall located in a heating-dominated climate, it protects the wall from intrusion by moist interior air. In theory, this type of retarder is more advantageous than an exterior retarder for structures where moisture condensation is likely, such as steel studs without insulating sheathing that define a space with high humidity in winter. In practice, it is difficult to achieve a continuous air seal at the interior surface of a wall, especially in multistory buildings, because of the wall/floor and wall/ceiling junctions. A continuous air seal plane could be achieved by gypsum board with or without a polyethylene vapor retarder if particular attention is paid to details and with special construction supervision.

In hot, humid climates, the airflow retarder for air-conditioned buildings should be vapor-permeable if it is placed nearest the interior of the wall because the outside moist air can condense on cold interior surfaces. The exterior airflow retarder can be either moisture-permeable or moisture-impermeable. If the retarder is moisture-impermeable, the insulated layer near the conditioned space must have a sufficient R-value to keep the membrane above the dew point. This caveat applies to each component of the insulated layer separately.

In climates with significant heating and cooling, the heating-dominated construction is usually indicated, but dew-point calculations need to be performed if layers with low vapor permeance are used. In climates with some heating, significant cooling, and high humidity, it is usually prudent to avoid vapor-impermeable airflow retarders and layers with low vapor permeance.

In general, with the exception of prefabricated wall panels, two layers with low vapor permeance (e.g., an exterior vapor-impermeable airflow retarder and an interior vapor retarder) should be avoided because moisture can be trapped in the wall during construction or operation.

RESIDENTIAL ROOF AND CEILING ASSEMBLIES

Thermal Performance

In pitched roof assemblies, the attic space is typically insulated with mineral fiber batts or blankets and/or mineral fiber or cellulose loose fill insulation. The total thermal resistance of the ceiling plus insulation, attic space, and roof is the sum of the conduction resistance of the ceiling plus insulation and the effective attic resistance. The roof resistance is generally negligible.

The thermal performance of the insulation varies with its mean temperature. In winter conditions, the thermal resistance of the insulation increases; in summer conditions, it decreases. In installing loose fill insulation, care must be taken to avoid overblowing the material to densities below the intended design density for a given R-value. At densities below the design, the material may not deliver the desired R-value. To prevent overblowing, the correct amount of material for a given coverage area and R-value should always be installed. Manufacturers of loose fill insulation are required to provide maximum net coverage information on all product packaging. Some high-density loose fill products may gradually settle to a lower thickness after being installed. The reduced thickness decreases the thermal resistance despite the increased density. For materials subject to settling, initial installed and settled thickness information should be available on package labels or from manufacturers so that the desired settled R-values can be achieved.

Some low-density loose fill materials are porous enough that very low temperatures at the top of the insulation may lead to free convection in the insulation. For example, Wilkes et al. (1991) observed that effective thermal resistance decreased by nearly 50% from nominal values when temperature differences greater than 70°F were imposed. Free convection can be suppressed by placing a layer of batt insulation over the low-density loose fill insulation. Alternately, batts or blankets can be installed first and the low-density loose fill blown in on top of the batts or blankets. Mineral fiber batts or blankets should be installed at the proper width so that there are no air gaps between the batts.

In a roof assembly over a ventilated attic space in a cooling-dominated climate, insulation may be supplemented by a **radiant barrier** of high-reflectance material to reduce radiant heat transfer between the hot roof and the relatively cool ceiling. The radiant barrier can be attached to the roof rafters or fastened directly to the underside of the roof sheathing with the reflective surface facing down. Care should be taken to not disturb and damage the existing insulation when retrofitting an attic with a radiant barrier. Also, dust decreases the effective reflectance of the radiant barrier over time; however, if the operative surface faces down, the reduction in thermal performance should be minimal.

During the cooling season, a radiant barrier reduces temperatures affecting the air, insulated surfaces, and HVAC ducts and equipment located in the attic space. A ventilated radiant barrier combined with an equally divided, continuous soffit and ridge vents provides optimal performance. The most cost-effective use of radiant barriers is in cooling-dominated climates. Use of radiant barriers in other climates may result in a net annual increase in energy use because the beneficial attic heat gain in the noncooling seasons is lost. DOE (1991) compares the present value savings of radiant barriers with those of conventional fibrous insulation for several climates and circumstances.

During the cooling season, a radiant barrier under the roof sheathing increases the efficiency of a duct system located in an attic (Hageman and Modera 1996). DOE (1991) does not consider this case.

With or without a radiant barrier, **powered ventilation** increases the effective attic resistance by reducing the temperature difference between the inside conditioned space and the attic space (see Table 5 in Chapter 24 of the 1997 *ASHRAE Handbook—Fundamentals*). However, this method is rarely cost-effective because of the energy cost of the fan. Additionally, powered ventilation can draw air from the conditioned space into the attic.

In low- or no-slope roofs, the insulation is a layered assembly comprising the deck, insulation, and membrane. Although there are relatively small gaps between the components, the high thermal resistance of the insulation keeps the total heat flux low. However, the heat flow through metal penetrations can be more than 1000 times greater than that through adjacent insulation. Therefore, the area of these thermal bridges should be kept small

in proportion to the roof area. Minimizing this area may require interrupting potential bridges; for example, staggering fasteners that secure the insulation to the deck and those that attach the membrane to the insulation.

Moisture

In addition to moisture normally present in construction materials, moisture can be trapped in the roof and ceiling assembly from rain during construction, leaks that were not fixed before water entered the assembly, or reroofing over partially torn-off roofs. Additional moisture can enter a structure having no insulation vapor retarder. If porous insulation is used, water vapor can diffuse throughout the assembly. The amount of vapor may be insufficient to significantly affect the thermal conductivity of the insulation, but it may add a significant latent heat load. Diurnal temperature variations in the roof can cause water to evaporate, move as vapor through a porous system, and condense (Hedlin 1988). When conditions are right for this effect, the heat flow from this factor alone can easily exceed normal conduction flows.

Air Movement

Whether the roof is pitched, low-slope, or no-slope, a typical roof and ceiling assembly is impervious to significant air movement through the ceiling itself. With pitched roofs, the attic normally is ventilated by airflow from the edges of the assembly, over the insulation, and out through continuous ridge vents or occasional vents in the roof or gable ends of the attic. The ventilation may be by natural or forced convection, but natural convection by a combination of continuous ridge vents and continuous soffit vents seems most effective. A type of low- or no-slope roof called a **ventilated cold deck system** has vents at the eaves and at intervals over the roof to permit the circulation of dry outside air over the insulation. The effectiveness of such vents is uncertain in regions with a cold or a temperate but humid climate (IEA 1994). An alternative is a **nonventilated cold deck system** with a vapor retarder on the ceiling consisting of staggered strips of polyethylene on the bottom and top and exposed fabric between them. Vapor inside the building effectively "sees" a continuous layer of polyethylene because of the staggering. Condensed water wicks through the exposed fabric on the top to the exposed fabric on the bottom and evaporates to the inside (Pedersen et al. 1992).

FENESTRATION

Conduction/Convection and Radiation Effects

Heat transfer through a window resulting from a temperature differential between the inside and outside (i.e., conduction, convection, and radiation) is a complex and interactive phenomenon. Although the glass itself is a poor insulator, it is used effectively in a window to create thin insulating air films on either side of single glazing or dead air/gas spaces between glazing layers. Glass can also decrease the direct transmission of radiant energy coming from the room or ambient sources. Chapter 29 of the 1997 *ASHRAE Handbook—Fundamentals* discusses fenestration in much greater detail.

An examination of the modes of heat transfer in a double-glazed window indicates that approximately 70% of the heat flow is through radiation from one glazing layer to another (Selkowitz 1979; Arasteh et al. 1985). Although the glass blocks the direct transmission of radiant energy, energy is still absorbed and reemitted. Low-emittance coatings that are transparent to the eye significantly reduce the amount of heat transfer through a glazing cavity.

With the radiation mode of heat transfer minimized, the conductive and convective modes dominate. To reduce these modes, gases with low conductivity (e.g., argon or krypton) are used in low-emittance-coated, multiple-glazed windows. Triple or quadruple glazing layers make additional reductions in heat transfer possible.

See Chapter 29 of the 1997 *ASHRAE Handbook—Fundamentals* for design information.

Air Infiltration Effects

Air infiltration is the uncontrolled movement of air from the outside to the inside of a building. It is a function of the pressure differential between the inside and outside environments (which is a function of wind speed and temperature differentials) as well as window sealing characteristics (Weidt and Weidt 1980; Klems 1983). The infiltration rate of a fenestration product is a function of its method of operating (if any), weatherstripping material used, and construction quality See Chapter 25 of the 1997 *ASHRAE Handbook—Fundamentals* for design information.

Solar Gain

Solar gain through windows can play a significant role in the energy balance of a building. Glazings transmit, reflect, or absorb a given wavelength of solar radiation, depending on the glazing characteristics. Transmitted solar radiation contributes heat to a space. Absorbed solar radiation is reemitted and/or conducted either to the inside or to the outside (depending on the window's glazing system configuration). Reflected solar radiation does not contribute heat to a space (Arasteh et al. 1989). Clear glass, the most common glazing material, transmits fairly evenly across the solar spectrum. Tinted or heat-absorbing glass absorbs solar radiation and gives the glass a specific color. Some tints exhibit a significant degree of spectral sensitivity (i.e., they do not transmit, absorb, and reflect evenly across the solar spectrum). These types of glazings offer great flexibility in that they can be tailor-made for specific climates or uses (e.g., provide ample daylighting without overheating an interior space).

Interactions Between Thermal Loss and Solar Gain

In heating-dominated applications, solar gain can provide a significant amount of heat. In many cases, the heat supplied by the window can offset that lost through the window. The amount depends on the site characteristics (i.e., how much solar gain is available, how cold the climate is) and window characteristics (i.e., its U-factor and how much incident solar radiation is transmitted).

Typical **passive** solar applications try to maximize the amount of solar heat gain by installing significant amounts of south-facing glass, which receives the most solar radiation during the winter (in the northern hemisphere). However, high-performance windows facing north in a heating-dominated climate can provide more solar gain to a space than heat loss (Arasteh et al. 1989; Sullivan et al. 1992; Dubrois and Wilson 1992).

WALL/WINDOW INTERFACE

Air infiltration at the wall/window interface can cause serious damage to surrounding building materials and even to remote materials, depending on the leakage path. In cold climates, warm, humid inside air can condense in the wall cavity. The condensate can damage the interior finish, the seals of glazing units, the insulation, the exterior cladding, and possibly the structural elements. Cold air infiltration can affect the health and comfort of the occupants by creating a dry indoor environment, cold drafts, and condensation and mildew on inside window surfaces. Air infiltration can also allow rain to enter the wall. Many materials are installed by many trades at the wall/window interface, making it difficult to inspect and prone to leak.

Control of Air Leakage

The airflow retarder must be continuous across the entire building envelope to provide effective control. To ensure continuity at the window, the components that retard airflow must be unified. Examples include the following:

- In a metal and glass curtainwall, the glazing and sheet metal are the elements that retard airflow in the window and the wall, respectively. Gaskets and caulking make the airtightness plane continuous from the glazing to the metal mullion.
- In a masonry wall, a self-adhesive asphalt membrane mechanically fastened to the masonry makes the wall airtight. A bead of sealant preserves the airtightness between the glazing unit and the inner part of the metal window frame. Thus, the asphalt membrane and the inner part of the metal frame are the two materials joined to withstand large air pressure loads. The membrane is extended and mechanically clamped to the inner part of the frame.

Control of Moisture Diffusion

Vapor diffusion across a small gap at the window/wall junction causes little building damage under average environmental conditions. Moreover, typical interior finishes covering the joint, made of metal or plastic or painted drywall, offer some resistance to moisture diffusion.

Control of Heat Flow

Continuity of the plane of thermal insulation is important primarily to reduce the potential for condensation on interior surfaces of the window and the surrounding interior finish. Insulation can be inserted in the joint between the wall and the window frame to compensate for the expected differential movement between the frame and the wall rough opening.

Control of Surface Condensation

To reduce the potential for condensation on the glazing and the frame of windows, as well as on the surrounding interior finish of the wall, the inside surface temperature can be controlled in the following ways:

- Seal the wall/window interface and between the sash and frame of operable windows to keep air leakage to a minimum.
- Make the area of window frame exposed to the interior larger than the area exposed to the outside. Metal frame extensions on the inside have a higher resistance to condensation but contribute to heat loss.

Control of Rain Entry

Applying the rain screen principle at the wall/window interface requires the same features as applying it to the wall: (1) an airflow retarder, (2) a rain deflector on the outside of the interface, (3) much more venting on the outside face of the interface than on the inside, and (4) a drainage path toward the outside.

The line of airtightness is on the inside of the assembly, so the assembly is protected from water, ultraviolet rays, and extremes of temperature. The rain deflector on the outside acts as a rain deterrent only, not as a watertight/airtight seal. A nonairtight rain deflector does not threaten the weathertightness of the system because the pressure differences across the rain deflector are small. The key is to maintain airtightness on the inside of the joint. With little pressure across the rain deflector and with good detailing for outward drainage of the cavity, rain entry in the wall should be minimal.

Face Seal Approach

The face seal approach calls for perfect sealing of the outer face of the wall/window junction. This sealed surface protects against rain and air infiltration. Such a seal must remain perfect over time; its maintenance schedule is quite demanding. It is also complicated because water leaks can follow indirect paths for entry indoors, and in a large building, it may be difficult to trace these paths back to the entry points.

WALL/ROOF INTERFACE

Control of Air Leakage

The roofing membrane forms a plane of continuous airtightness and therefore can and should be connected to the wall airflow retarder. However, the wall airflow retarder could also be connected to other airtight roof components, such as the roof deck or the roof vapor retarder.

Many building systems provide for differential movement between the wall and roof assemblies; in those systems, a flexible airflow retarder connection is required at the location of the movement control joint. In all systems, the airflow retarder must withstand wind-induced pressures and fully adhere to a rigid, structurally adequate substrate.

Figure 3 shows an assembly that uses the concrete roof deck to achieve a continuous wall/roof airflow retarder. The retarder of the wall assembly is connected to the concrete slab using a sheet airflow retarder material. The connection between the wall and slab edge retarder sheets is designed to accommodate differential movement of the wall and roof.

Figure 4 shows a protected membrane roof assembly in which the roof membrane is an airflow retarder that extends over the edge of the roof perimeter to connect to the wall airflow retarder. A sheet metal backing is used around the roof deck perimeter for attachment and to provide for differential movement at the interface.

In some types of curtain walls, such as the wall in Figure 5, the air seal must be made continuous over the parapet wall. This allows connection of the wall air seal (the metal back pan of the spandrel panel) to the membrane flashing of the roof assembly or to the air/vapor retarder on the parapet, as in asphalt and roofing felt systems.

Control of Moisture Diffusion

The vapor retarder can also act as an airflow retarder, as in the curtain wall shown in Figure 5 in which the metal back pan of the wall spandrel connects to the air/vapor retarder membrane on the parapet upstand wall. Because the parapet is encased in vapor retarder sheets, it must be kept dry before it is enclosed because any moisture left in the parapet cannot disperse. A vapor retarder with wicking properties could provide a better solution.

Control of Condensation

Condensation at the connection between the roof and wall can occur in interstitial spaces because of a lack of continuity in either system. Condensation also occurs if the temperature of the vapor retarder falls below the dew point of the air on the high vapor pressure side. Thus, thermal insulation applied to the exterior (winter-cold surface) of the vapor retarder must be sufficient to ensure that the temperature at the retarder plane remains higher than the dew point of the air. The temperature of the interior space between the exterior wall system and the parapet upstand wall often falls below the temperature of the interior inhabited space because of poor air circulation above the ceiling. Therefore, the insulation applied over the parapet cap and upstand wall may need to have greater thermal resistance than the insulation for the wall or roof.

In exterior insulated wall systems, such as curtain walls and EIFSs, thermal bridging must be minimized to control condensation between the parapet cap and the parapet upstand wall.

THERMAL BRIDGES IN BUILDINGS

A **thermal bridge** is an envelope area with a significantly higher rate of heat transfer than the contiguous enclosure. An example of thermal bridging is a steel truss in an attic with glass fiber insulation placed between trusses. The thermal conductivity of steel is about 1000 times greater than that of glass fiber insulation, so strong thermal bridging results along the trusses. In most construction details, the ratio between the high and low conductivity areas is smaller. An

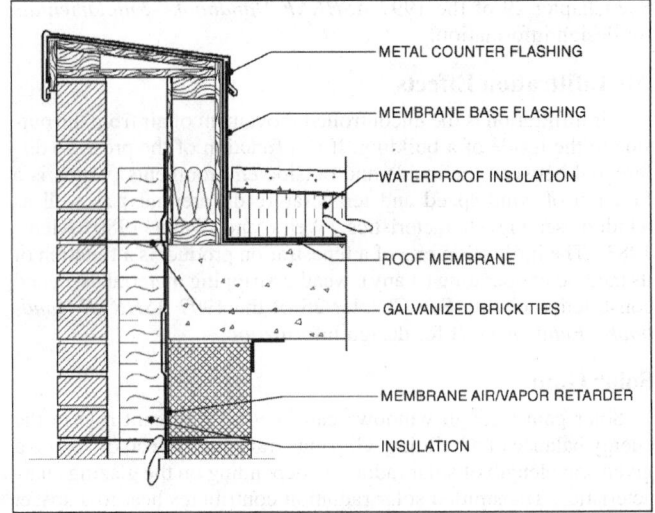

Fig. 3   Detail of Concrete Roof Deck with Continuous Wall/Roof Airflow Retarder

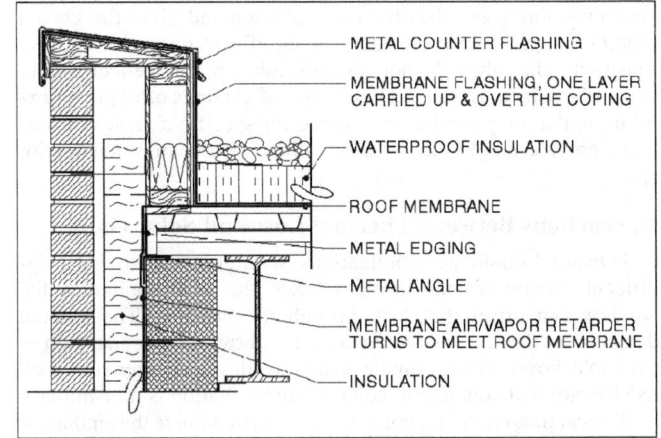

Fig. 4   Protected Membrane Roof Assembly

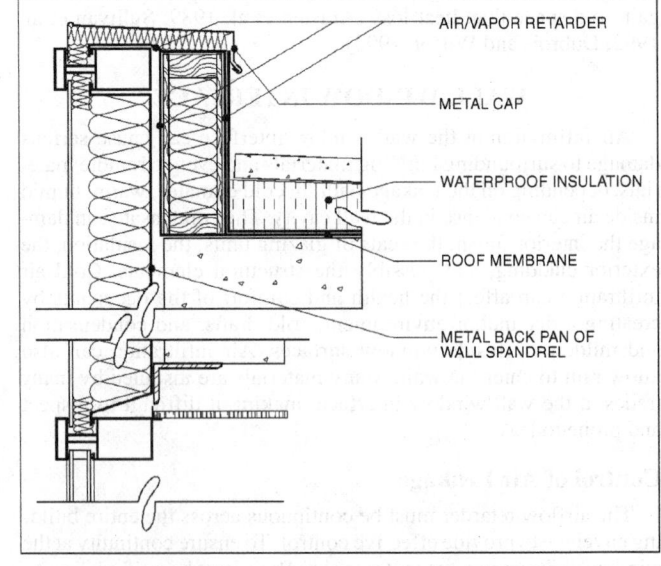

Fig. 5   Continuous Air Seal in Curtain Wall

uninsulated CMU wall with brick facing could have a U-factor of 0.33 Btu/h·ft^2·°F if the CMU is made with sand and gravel aggregate. A concrete column located within that wall has a U-factor about 50% higher. This 50% higher conductivity may justify classifying the column as a thermal bridge, but in typical construction this classification is relevant only if the concrete column proves to be detrimental for reasons other than energy use (e.g., moisture condensation).

Thermal bridges can be created by conduction, as in the previous examples; by radiation; by convection; or by a combination of those three modes of heat transfer (Silvers et al. 1985). Strong radiation effects occur, for instance, in an uninsulated brick wall that is retrofitted with metal cladding. The cladding is placed at some distance from the exterior of the brick wythe. The warmer brick wall radiates to the cold metal sheet. The different temperatures of the two surfaces also engender air convection.

Detrimental Effects of Thermal Bridges

Thermal bridges increase energy use, promote moisture condensation in and on the envelope, and create nonuniform temperatures

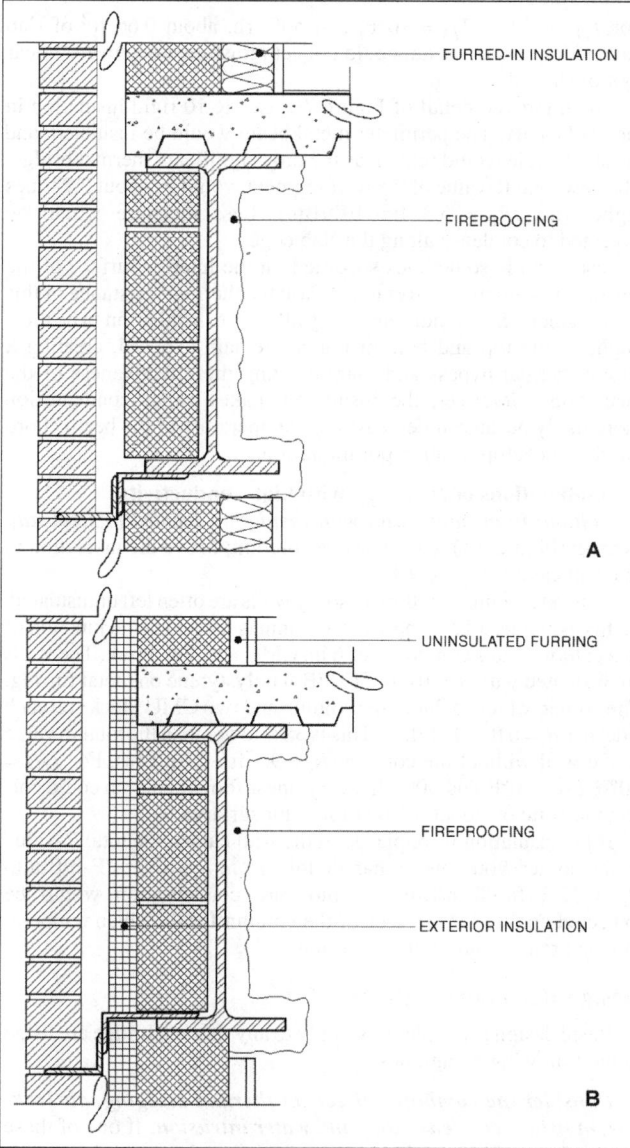

Fig. 6 Detail of Uninsulated (A) and Insulated (B) Slab Edge and Metal Shelf Angle

in conditioned spaces. A comparison between the uninsulated slab edge detail of Figure 6A and the version with slab edge insulation with an R-value of 10 ft^2·°F·h/Btu in Figure 6B illustrates the importance of designing to reduce thermal bridging. [Refer to Steven Winter Associates (1988) for numerical examples summarized in this section.] This example, and the others in this section, use the term **total R-value**, where $R_t = 1/U$; R_t includes the effect of exterior and interior air films.

- **Increased energy use.** The portion of the wall that extends from the top of the slab edge to the shelf angle (Figure 6B) has $R_t = 2.3$ ft^2·°F·h/Btu when the wall insulation is furred in. Moving the insulation into the wall cavity (Figure 6B) increases R_t by about 2.2 times, to 5.3 ft^2·°F·h/Btu.
- **Damage to structural components.** On cold days, the uninsulated wall area collects moisture on the interior surface of the steel beam and on the slab underside. At 20°F outside air temperature, 70°F indoor air temperature, and 50% indoor relative humidity, moisture condenses on about 0.66 ft^2 of beam and slab for every linear foot of wall. This condensation can rust the metal beam and, if it penetrates into the concrete block, can rust the brick ties. The insulated detail (Figure 6B) generates moisture condensation only on the bottom flange of the beam.
- **Damage to interior finishes.** Moisture can drip from the slab underside and from the steel beam, staining the ceiling tiles and resulting in higher maintenance costs and loss of rental value. These effects are specific to the heating season and are magnified when thermal bridging is accompanied by air leakage because air transports and deposits moisture on thermal bridges.

In cold and temperate climates, thermal bridges may slightly decrease the peak cooling load and the cooling energy use. In hot climates, this effect could be reversed. When the climate is hot and humid, moisture condensation can create problems if the building is air conditioned or refrigerated.

Thermal Bridge Mitigation

The primary causes for conduction thermal bridging are (1) high thermal conductivity and (2) geometries that create zones where large exterior surfaces connect to much smaller interior surfaces. Thermal bridges can sometimes be eliminated, but usually they can be mitigated to create systems with significantly lower U-factors. Several examples of conduction thermal bridges and of mitigating details follow.

- **Factory-produced components with high conductivity**
  Sandwich panel with insulation fully encased in concrete (Figure 7A). One possible solution: Connect the exterior and interior wythes of the panel with metal or plastic tie rods (Figure 7B).

  This 6 ft high sandwich panel has a 3 in. extruded polystyrene core completely encased in concrete. The concrete has a conductivity over 50 times higher than that of polystyrene, resulting in $R_t = 5.4$ ft^2·°F·h/Btu. Moisture can collect on the top and bottom concrete edges of the panel. For $T_{OA} = 20$°F, $T_{IA} = 70$°F, and 50% rh, about 20% of the panel area can be covered with moisture.

  When the interior and exterior concrete wythes are attached with metal or plastic tie rods, R_t of the panel increases by a factor of almost 3 to 15.1 ft^2·°F·h/Btu or 16 ft^2·°F·h/Btu, respectively. No moisture condensation on interior surfaces would be likely.

- **Assemblies that contain high-conductance components**
  Steel studs in insulated wall with ceramic finish. One possible solution: Provide insulating sheathing (Figure 8).

  This steel stud wall is used in low-cost multifamily, commercial, and industrial construction. The exterior gypsum board is provided for fire protection. If the wall has R-11 insulation, the studs reduce the wall R_t to 7.75 ft^2·°F·h/Btu. For $T_{OA} = 20$°F, $T_{IA} = 70$°F, and 50% rh, every linear foot of steel stud can generate 0.15 ft^2 of moisture condensation on the interior surface of the gypsum wallboard.

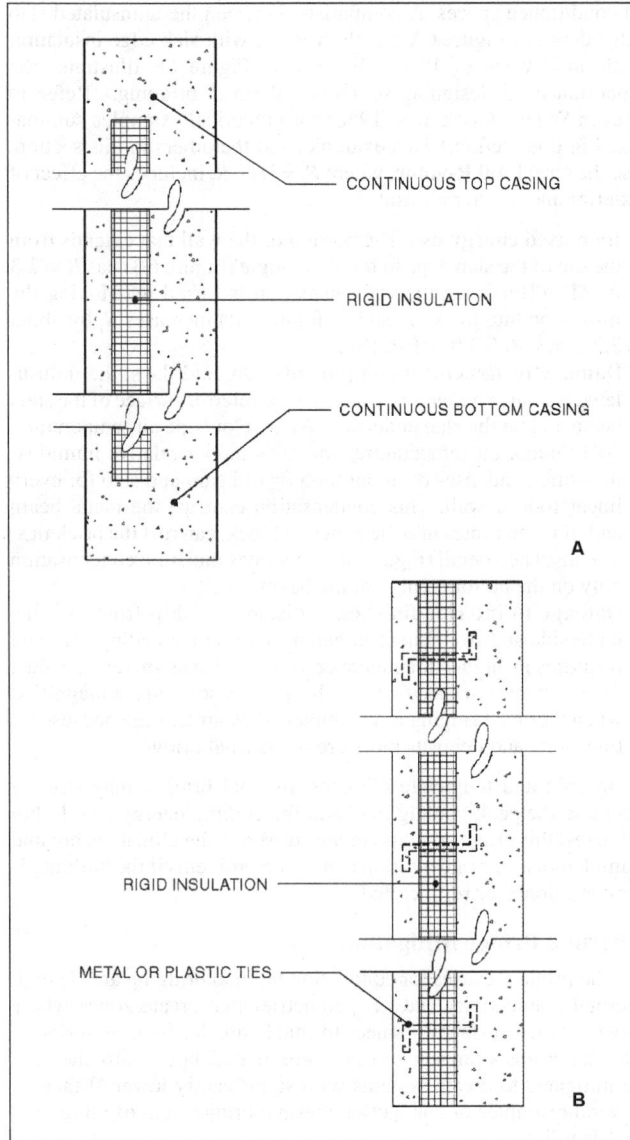

Fig. 7   Sandwich Panel with Insulation Encased in Concrete

Extruded polystyrene sheathing 1 in. thick placed on the winter-cold surface of the gypsum board sheathing would reduce the thermal bridging, resulting in $R_t = 14.5$ ft$^2\cdot$°F·h/Btu. For $T_{OA} = 20$°F, $T_{IA} = 70$°F, and 50% rh, moisture would be less likely to condense on interior surfaces. If polystyrene were placed on the steel studs directly, the increase in R-value would be even higher because the lateral heat flow would be reduced.

If all insulation were placed on the exterior surface of the steel studs, the studs would be kept at a temperature close to that of the conditioned space. Moisture condensation would be practically eliminated. This solution merits special consideration in spaces with high relative humidity such as kitchens, and pool buildings.

- **Junction between envelope systems achieved with high-conductance components**

Uninsulated slab edge and metal shelf angle, at junction between CMU wall and concrete floor (Figure 6A). One possible solution: Place insulation in the wall cavity (Figure 6B).

This detail shows an uninsulated slab edge and perimeter metal beam in a CMU/brick wall. For the wall segment that extends from the top of the slab edge to the shelf angle, $R_t = 2.3$ ft$^2\cdot$°F·h/Btu.

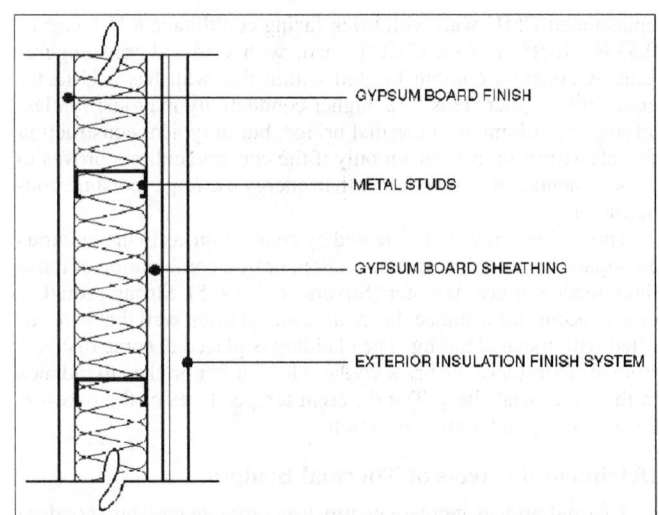

Fig. 8   Steel Studs in Insulated Wall with Ceramic Finish

For $T_{OA} = 20$°F, $T_{IA} = 70$°F, and 50% rh, about 0.66 ft^2 of slab underside and steel beam could collect moisture along every linear foot of slab edge.

The improved detail of Figure 7B uses R-10 rigid insulation in the wall cavity. The perimeter metal beam would be insulated, and the shelf angle would remain as the only significant thermal bridge. The new total R-value of the wall segment would be about 2.2 times higher (i.e., $R_t = 5.3$ ft$^2\cdot$°F·h/Btu). Less moisture would be expected to condense along the slab edge.

Insulation is sometimes specified on the interior surface of the perimeter beam in an effort to increase the thermal resistance of this wall segment. Site conditions rarely allow the installation to be thorough, so the top and bottom flanges remain exposed, creating a strong thermal bypass and marginalizing the effectiveness of the insulation. Moreover, the insulation makes vapor condensation more likely because it decreases the temperature of the beam chord but does not stop water vapor migration.

- **Combinations of geometry with high conductivity**

Column in masonry wall with insulation located in steel stud furring (Figure 9A). One possible solution: Move the insulation in the wall cavity (Figure 9B).

Concrete columns within masonry walls are often left uninsulated. In this example, a 12 in. by 12 in. column is located at the junction of two exterior walls composed of 6 in. CMU and 4 in. brick. The walls are insulated with $R = 10$ ft$^2\cdot$°F·h/Btu polystyrene on metal furring. The corner, which includes the column and two CMU blocks on each side, has $R_t = 6$ ft$^2\cdot$°F·h/Btu. This is 38% lower than the total R-value of the wall without the column ($R_t = 9.7$ ft$^2\cdot$°F·h/Btu). For $T_{OA} = 20$°F, $T_{IA} = 70$°F, and 50% rh, every linear foot of column could collect moisture on about 0.8 ft^2 of interior surfaces.

If the insulation were placed in the wall cavity, the total R-value of the corner would more than double, from $R_t = 6$ ft$^2\cdot$°F·h/Btu to $R_t = 12.3$ ft$^2\cdot$°F·h/Btu. No moisture condensation would be expected on interior surfaces of the column for the given temperature and relative humidity conditions.

Design Recommendations

Basic design principles that apply to any thermal bridge are listed in the following paragraphs.

Consider the combined effects of thermal bridging, air leakage, moisture condensation, and water intrusion. If one of these mechanisms of heat and mass transfer is involved, another one is commonly combined with it in a destructive synergy. In devising a solution for thermal bridging, ensure that moisture condensation

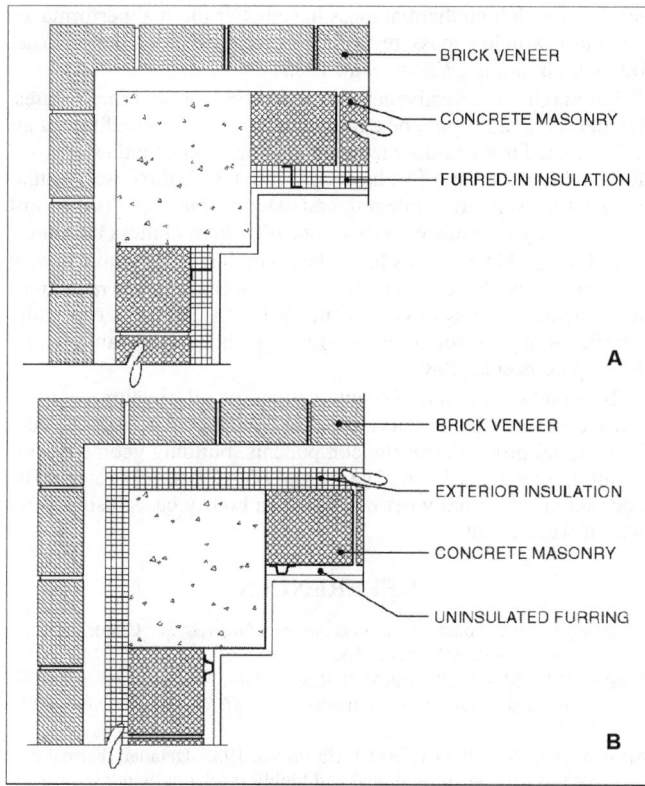

**Fig. 9   Details of Insulation Around Column in
Masonry Wall**

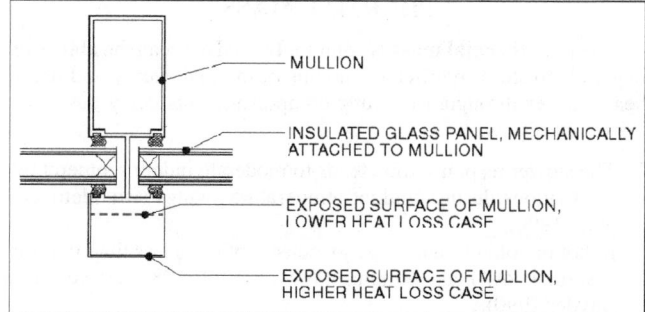

Fig. 10   Mullion with Deep Exterior Projection

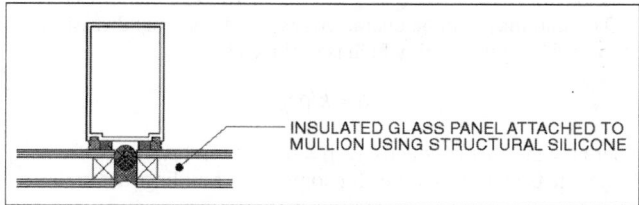

Fig. 11   Mullion Without Exterior Projection

is not increased. In climates with significant heating, this is of special concern for insulation placed on the interior (winter-warm) surface of materials with low vapor permeance, such as CMU or steel beam. In hot, humid climates, the concern relates to insulation placed on the summer-hot surface of materials with low vapor permeance.

In addition, because many thermal bridges are discontinuities in the envelope system, the chance that air will enter the envelope along these paths should be assessed. To cause damage, air does not need to infiltrate into a space from outdoors or to exfiltrate from a space to the outside; it is sufficient that air intrude in the envelope. For example, outside air can enter a stud wall cavity, short-circuiting the insulation and intensifying the thermal bridging effect, after which it exits again to the outside. In another example, indoor air can seep from the conditioned space into a furred-in CMU wall, behind the insulation and vapor retarder, increasing the chance for moisture condensation on the CMU surface. Of all forms of air intrusion, exfiltration of moist air along thermal bridge surfaces can be the most damaging. Note that the entry and exit points of air in the envelope can be distant.

It is particularly important to reduce thermal bridging, air intrusion, and moisture condensation if the following conditions are true:

- The building or spaces in it have uses that are likely to generate a relatively high amount of moisture during winter; examples are residences, hospitals, nursing homes, restaurants, kitchens, natatoriums, and bathrooms.
- The building is located in an area that experiences cold winters or sustained periods of cold during moderate winters. Climates with humid winters with subfreezing periods can also create problems.
- The building is located in a warm but not dry or arid climate and has uses that require low temperatures (e.g., refrigeration), or it is located in a hot, humid area and has uses that require air conditioning.

Consider the potential for thermal bridging and air intrusion during schematic design when decisions about the envelope system are made. Some systems are inherently more difficult to insulate and air-proof, and changes during design development may create budget problems.

Use reasonably accurate methods to determine the effect of thermal bridging. Accurate formulas are available for most steel stud configurations (Kosny and Christian 1995), but they are not sufficiently accurate for many other construction types. Two- and three-dimensional models are available and provide better results.

In mitigating a thermal bridge, several techniques can be used, often in combination. Such techniques include

- Change the thermal bridge material to one with lower conductivity. Plastic ties nearly eliminate thermal bridging in concrete sandwich panels.
- Decouple the thermal bridging elements from the rest of the construction. If the steel stud furring is detached from the CMU wall, the pathway between two materials with high conductivity is broken. The best method to achieve the decoupling is to insert a thin, low-conductivity material, such as foam tape, at each stud. If an air space is created between the insulated steel stud layer and the CMU wythe, air convection can short-circuit the insulation.
- Change the geometry that creates thermal bridging. Figure 10 shows a mullion with a deep exterior projection. This projection acts like a fin during winter, enhancing the heat loss. One method of reducing the thermal bridging effect would be to reduce the exterior projection (Figure 10, dotted line). If this were not acceptable for aesthetic reasons, a better thermal break system should be implemented.
- Insulate the thermal bridge. For example, use insulating sheathing for steel stud walls.
- Change the insulation system. For example, a furred-in masonry wall with all the insulation located in the furring could be changed to a masonry wall with rigid insulation on the exterior (winter-cold) surface of the CMU.
- Change the construction system. In a natatorium, for instance, structural silicone glazing without an exterior projection (Figure 11) might be a solution to avoid moisture condensation on mullions.

THERMAL MASS

The term **thermal mass** is commonly used to mean the ability of materials to store significant amounts of thermal energy and delay heat transfer through a building component. This delay has three important results:

1. The slower response time tends to moderate indoor temperature fluctuations during outdoor temperature swings (Brandemuehl et al. 1990).
2. In hot or cold climates, energy consumption is less than that for a similar low-mass building (Wilcox et al. 1985; Newell and Snyder 1990).
3. Building energy demand can be moved to off-peak periods because energy storage is controlled through correct sizing of the mass and interaction with the HVAC system.

Thermal mass can be characterized by the thermal diffusivity α of the building material, which is defined as

$$\alpha = k/\rho c_p$$

where k = thermal conductivity, ρ = density, and c_p = specific heat. Table 1 lists these properties for some basic building materials.

Table 1 Typical Densities, Thermal Diffusivities, and Specific Heats of Common Building Materials

Description	Density, lb/ft^3	Thermal Diffusivity, ft^2/h	Specific Heat, Btu/lb·°F
Concrete	140	0.027 to 0.054	0.22
Steel	484	0.038	0.12
Wood	22 to 44	0.005 to 0.006	0.40
Insulation	0.6 to 2.0	0.018 to 0.027	0.2 to 0.38

Heat capacity (HC) is defined as the amount of heat necessary to raise the temperature of a given mass by one degree:

$$HC = mc_p$$

where m = component mass. The heat capacity of a building element is the sum of the heat capacities of each of its components.

Heat transfer through a material with high thermal diffusivity is fast, the amount of heat stored in it is relatively small, and the material responds quickly to changes in temperature. The effect of thermal mass on building behavior varies primarily with the climate at the building site and the position of the wall insulation relative to the building mass.

The ideal climate for taking advantage of thermal mass is one that has large daily temperature fluctuations. The mass can be cooled by natural ventilation at night and be allowed to "float" during the warmer day. When outdoor temperatures are at their peak, the inside of the building remains cool because the heat has not yet penetrated the mass. Often, the benefits are greater during spring and fall, when some climates closely approximate this ideal case. In heating-dominated climates, thermal mass can be used effectively to collect and store solar gains or to store heat provided by the mechanical system, allowing the heating system to operate during off-peak hours.

The effectiveness of thermal mass also increases with the allowable temperature swing in the conditioned space. If, for instance, interior temperatures can fluctuate between 68 and 78°F without the intervention of HVAC systems, as is the case in many schools, the mass has the opportunity to charge during warm hours and discharge during cooler periods. Such opportunity is limited in a laboratory that requires temperature control within a 2°F range.

In nonresidential buildings, thermal mass is often more effective in reducing cooling loads than heating loads. In some climates, buildings with high thermal mass have better thermal performance than those with low mass, regardless of the level of insulation in the low-mass building (Wilcox et al. 1985).

For maximum effectiveness in moderating indoor temperatures, the thermal mass should be exposed to the interior, conditioned air and insulated from outdoor temperature variations. Studies to quantify the thermal mass effect have concentrated on three wall insulation strategies: interior, integral, and exterior. The latter is the most effective way to insulate the envelope of a thermal mass building.

Calibrated hot box tests have shown that in walls with identical heat capacities, those with higher R-values cause larger reductions in temperature swings (Van Geem 1986). Insulation is especially significant in very hot climates with large daily temperature variations (Wilcox et al. 1985).

To simulate the complex interactions of all envelope components, computer simulations are necessary. These programs account for material properties of the components, building geometry, orientation, solar gains, internal gains, and HVAC control strategy. The calculations are usually performed on an hourly basis, using a full year of weather data.

REFERENCES

Andrews, S. 1989. *Foam core panels and building systems*. Cutter Information Corporation, Arlington, MA.

Arasteh, D.K., M.S. Reilly, and M.D. Rubin. 1989. A versatile procedure for calculating heat transfer through windows. *ASHRAE Transactions* 95(2): 755-65.

Arasteh, D.K., S. Selkowitz, and J. Hartmann. 1985. Detailed thermal performance data on conventional and highly insulating window systems. *Thermal Performance of the Exterior Envelopes of Buildings* III, pp. 830-45. Available from ASHRAE.

ASHRAE. 1994. Recommended practices for controlling moisture in crawl spaces. *ASHRAE Technical Data Bulletin* 10(3).

Brandemuehl, M.J., J.L. Lepore, and J.F. Kreider. 1990. Modeling and testing the interaction of conditioned air with building thermal mass. *ASHRAE Transactions* 96(2):871-75.

Canadian Home Builders. 1989. *Association builder's manual*. Ottawa, Ontario.

Carmody, J., J. Christian, and K. Labs. 1991. *Builder's foundation handbook*. ORNL/CON-295. Oak Ridge National Laboratory, Oak Ridge, TN.

Christian, J.E. 1994. *Moisture sources*. ASTM *Manual* 18. American Society for Testing and Materials, West Conshohocken, PA.

Christian, J. and J. Kosny 1996. Thermal performance and wall ratings. *ASHRAE Journal* 38(3):56-65.

DOE. 1991. Radiant barrier fact sheet. DOE/CE-0335P. U.S. Department of Energy.

Dubrois, F. and A.G. Wilson. 1992. A simple method for computing energy performance for different locations and orientations. *ASHRAE Transactions* 98(1)841-49.

Dudney, C.S., L.M. Hubbard, T.G. Matthews, R.H. Scolow, A.R. Hawthorne, K.J. Gadsby, D.T. Harrje, D.L. Bohac, and D.L. Wilson. 1988. *Investigation of radon entry and effectiveness of mitigation measures in seven houses in New Jersey*. ORNL-6487 (draft). Oak Ridge National Laboratory, Oak Ridge, TN.

Hageman, R. and M.P. Modera. 1996. Energy savings and HVAC capacity implications of a low-emissivity interior surface for roof sheathing. *ACEEE Summer Study on Energy Efficiency in Buildings*. American Council for an Energy-Efficient Economy, Washington, D.C.

Hedlin, C.P. 1988. Heat flow through a roof insulation having moisture contents between 0 and 1 percent by volume in summer. *ASHRAE Transactions* 94(2):1579-94.

IEA. 1994. *A guidebook for insulated low-slope roof systems*, Annex 19: Low-slope roof systems. International Energy Agency. Prepared by Oak Ridge National Laboratory, Oak Ridge, TN.

James, T.B. and W.P. Goss. 1993. *Heat transmission coefficients for walls, roofs, ceilings, and floors*. ASHRAE *Manual*.

Klems, J. 1983. Methods of estimating air infiltration through windows. *Energy and Buildings* 5:243-252.

Kosny, J. and J. Christian. 1995. Reducing the uncertainties associated with using the ASHRAE-zone method for R-value calculations of metal frame walls. *ASHRAE Transactions* 101(2):779-88.

Labs, K., J. Carmody, R. Sterling, L. Shen, J. Hwang, and D. Parker. 1988. *Building foundation design handbook*. ORNL/Sub/86-72143/1. Oak Ridge National Laboratory, Oak Ridge, TN.

Newell, T.A. and M.E. Snyder. 1990. Cooling cost minimization using building mass for thermal storage. *ASHRAE Transactions* 96(2).

Nisson, J.D. and G. Dutt. 1985. *The superinsulated home book*. John Wiley and Sons, New York.

Pedersen, C. et al. 1992. Moisture effects in low-slope roofs: Drying rates after water addition with various vapor retarders. ORNL/CON-308, Oak Ridge National Laboratory, Oak Ridge, TN.

Selkowitz, S.E. 1979. Thermal performance of insulating window systems. *ASHRAE Transactions* 85(2):669-85.

Silvers, J.P., R.P. Tye, D.L. Brownell, and S.E. Smith. 1985. A survey of building envelope thermal anomalies and assessment of thermal break materials for anomaly correction. ORNL/Sub/83-70376/1. Dynatec R/C Company for Oak Ridge National Laboratory.

Steven Winter Associates. 1988. *Catalog of thermal bridges in commercial and multi-family residential constructions*. Report 88-SA407/1 for Oak Ridge National Laboratory.

Sullivan, R., B. Chin, D. Arasteh, and S. Selkowitz. 1992. A residential fenestration performance design tool. *ASHRAE Transactions* 98(1):832-40.

Tuluca, A., D. Lahiri, and J. Zaidi. 1997. Calculation methods and insulation techniques for steel stud walls in low-rise multifamily housing. *ASHRAE Transactions* 103(1):550-62.

Van Geem, M.G. 1986. Summary of calibrated hot box test results for twenty-one wall assemblies. *ASHRAE Transactions* 92(2B):584-602.

Weidt, J.L. and J. Weidt. 1980. Field air leakage of newly installed residential windows. LBL *Report* 11111. Lawrence Berkeley Laboratory, CA.

Wilcox, B., A. Gumerlock, C. Barnaby, R. Mitchell, and C. Huizenza. 1985. The effects of thermal mass exterior walls on heating and cooling loads in commercial buildings. *Thermal Performance of the Exterior Envelopes of Buildings* III, pp. 1187-1224. Available from ASHRAE.

Wilkes, K.E. et al. 1991. Thermal performance of one loose-fill fiberglass attic insulation. *Insulation Materials: Testing and Applications* 2. ASTM STP 1116:275-91. American Society for Testing and Materials, West Conshohocken, PA.

Wilson, A. 1988. High performance wall systems. *Journal of Light Construction* (April).

BUILDING AIR INTAKE AND EXHAUST DESIGN

A BUILDING outside air intake brings fresh air into a building. Likewise, building exhausts remove air contaminants from a building to the outdoors so that the wind can dilute the emissions. If the intake or exhaust system is not well designed, contaminants from nearby outdoor sources, such as vehicle exhaust, or contaminants emitted from the building itself, such as laboratory fume hood exhaust, can enter the building with insufficient dilution, causing odors, health impacts, and reduced indoor air quality. This chapter discusses the proper design of exhaust stacks and placement of air intakes to avoid air quality impacts. A companion chapter, Chapter 15 of the 1997 *ASHRAE Handbook—Fundamentals,* more fully describes wind and airflow patterns around buildings. Related information can also be found in Chapters 7, 13, 28, 29, and 30 of this volume, Chapters 12 and 13 of the 1997 *ASHRAE Handbook—Fundamentals*, and Chapters 24, 25, and 30 of the 1996 *ASHRAE Handbook—Systems and Equipment.*

EXHAUST STACK AND AIR INTAKE DESIGN STRATEGIES

Stack Design Strategies

The dilution a stack exhaust can provide is limited by the dispersion capability of the atmosphere. Before discharge, exhaust contamination should be reduced by filters, collectors, and scrubbers.

Central exhausts that combine flows from many collecting stations should always be used where safe and practical. By combining several exhaust streams, central systems dilute intermittent bursts of contamination from a single station. Also, the combined flow forms an exhaust plume that rises a greater distance above the emitting building. Additional air volume can be added to the exhaust near the exit with a makeup air unit to increase the initial dilution and

---

The preparation of this chapter is assigned to TC 4.3, Ventilation Requirements and Infiltration.

exhaust plume rise. This added air volume does not need heating or cooling, saving on energy costs.

In some cases, separate exhaust systems are mandatory. The nature of the contaminants to be combined, recommended industrial hygiene practice, and applicable safety codes need to be considered. Separate exhaust stacks should be grouped in a tight cluster to take advantage of the larger plume rise of the resulting combined jet. In addition, a single stack location for a central exhaust system or a tight cluster of stacks allows building air intakes to be positioned as far as possible from the exhaust location. For a tight cluster to be considered as a single stack in dilution calculations, the stacks must be uncapped, have approximately the same exhaust velocities, and all lie within a two-stack diameter radius of the middle of the group. Stacks lined up in a row do not act as a single stack, as shown by Gregoric et al. (1982).

As shown in Figure 1, the effective stack height h_s is the portion of the exhaust stack that extends above local recirculation zones and upwind and downwind obstacles. Wilson and Winkel (1982) demonstrated that stacks terminating below the level of adjacent walls and architectural enclosures do not effectively reduce roof-level exhaust contamination. To take full advantage of their height, stacks should be located on the highest roof of a building.

Architectural screens used to mask rooftop equipment adversely affect exhaust dilution, depending on such variables as porosity, relative height, and distance from the stack. Petersen et al. (1997) found that, in general, the more porous the screen, the better the exhaust dispersion. Stacks should extend above the architectural screen by the height H_c of the flow recirculation zone due to the screen itself, in order to prevent exhaust contamination of equipment within the enclosure. This height H_c depends on screen porosity, and is approximately the value from Equation (1) times (1 − porosity).

Large buildings, structures, and terrain close to the emitting building can have adverse effects on dilution of stack exhaust, because the

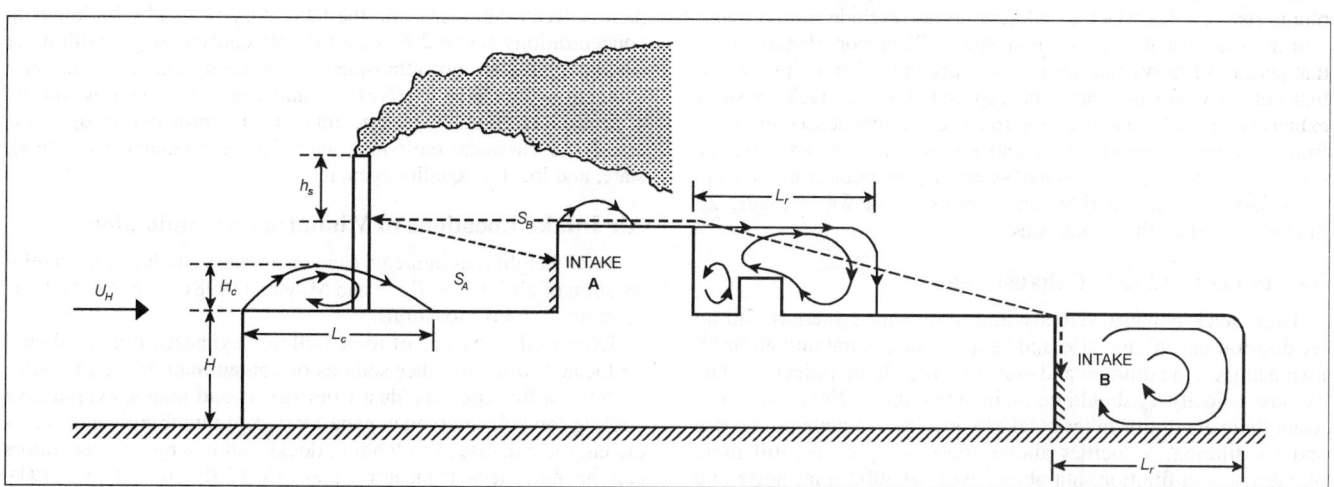

Fig. 1   Flow Recirculation Regions and Exhaust-to-Intake Stretched-String Distances
(Wilson 1982)

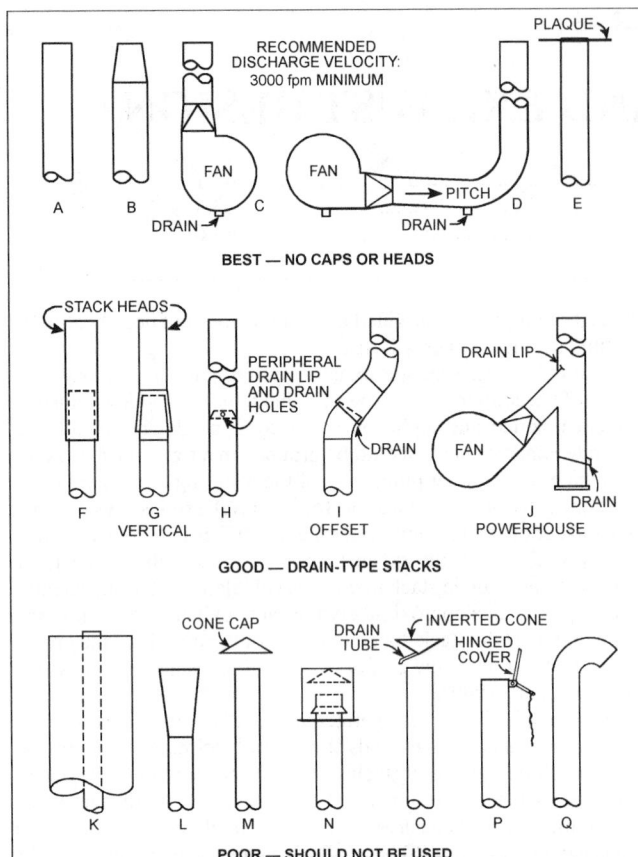

**Fig. 2   Stack Designs Providing Vertical Discharge
and Rain Protection**

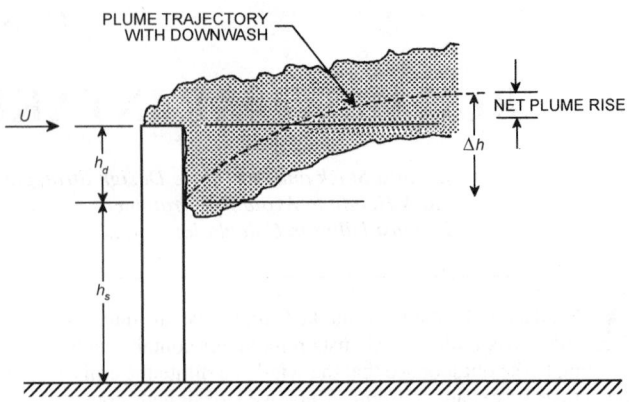

**Fig. 3   Reduction of Effective Stack Height
by Stack Wake Downwash**

An exception to these exhaust velocity recommendations may be required when corrosive condensate droplets are discharged. In this case, a velocity of 1000 fpm in the stack and a condensate drain are recommended to reduce droplet emission. At this low exhaust velocity, an exit nozzle should be used to avoid plume downwash (Figure 2B).

Stack wake downwash occurs where low-velocity exhausts are pulled downward by negative pressures immediately downwind of the stack, as shown in Figure 3. V_e should be 1.5 times the design wind speed U_H at roof level to avoid stack wake downwash. A meteorological station design wind speed U_{met} that is exceeded less than 1% of the time can be used. This value can be obtained from Tables 1A, 2A, or 3A in Chapter 26 of the 1997 *ASHRAE Handbook—Fundamentals*, or it can be estimated by applying Table 2 of Chapter 15 of the 1997 *ASHRAE Handbook—Fundamentals* to the annual average wind speed. Because wind speed increases with height, a correction for roof height should be applied using Equation (4) and Table 1 of Chapter 15 of the 1997 *ASHRAE Handbook—Fundamentals*.

Other Stack Design Standards

Minimum heights for chimneys and other flues are discussed in the Uniform Building Code (ICBO 1994). The American Industrial Hygiene Association (AIHA) *Standard* Z9.5, recommends a minimum stack height of 10 ft above the adjacent roof line; an exhaust velocity V_e of 3000 fpm, or 2000 fpm when internal condensation may occur; and a stack height extending one stack diameter above any architectural screen. The stack must also be situated to avoid reentry (reentrainment) into the laboratory or nearby buildings at concentrations above 20% of allowable concentrations within the laboratory under any atmospheric condition. The National Fire Protection Association (NFPA) *Standard* 45, specifies (in its appendix) a minimum stack height of 10 ft to protect rooftop workers. Toxic chemical emissions may also be regulated by federal, state, and local air quality agencies.

Air Intake Locations to Minimize Contamination

Stack height requirements can sometimes be reduced by careful location of air intakes. Rock and Moylan (1998) reviewed the literature on air intake locations.

Even in the absence of toxic building exhausts, intakes should be located to avoid other sources of contamination such as automobile traffic, kitchens, dust from streets and plants, evaporative cooling towers, emergency generators, and plumbing vents. Vehicle engine exhaust from loading docks and emergency generators can be nuisances (Smeaton et al. 1991, Ratcliff et al. 1994). Kitchen exhaust can be a source of odors and can cause plugging and corrosion of heat exchangers. Exhaust air outlets should also not be located near these sources of outdoor contamination

emitting building can be within the recirculation flow zones downwind of these nearby flow obstacles (Wilson et al. 1998). In addition, an air intake located on a nearby taller building can be contaminated by exhausts from the shorter building. Wherever possible, facilities emitting toxic or highly odorous contaminants should not be located near taller buildings or at the base of steep terrain.

As shown in Figure 2, stacks should be vertically directed and uncapped. Stack caps that deflect the exhaust jet have a detrimental effect on both the initial dilution of the exhaust jet and the exhaust plume rise. Conical stack caps often do not exclude rain, because rain does not usually fall straight down. Changnon (1966) shows that periods of heavy rainfall are often accompanied by high winds that deflect raindrops under the cap and into the stack. A stack exhaust velocity V_e of about 2500 fpm prevents condensed moisture from draining down the stack and keeps rain from entering the stack. For intermittently operated systems, protection from rain and snow should be provided by stack drains, as shown in Figure 2F through 2J, rather than stack caps.

Recommended Stack Exhaust Velocity

High stack exhaust velocity and temperature increase initial jet dilution and plume rise and reduce intake contamination by increasing plume dilution and elevating the plume trajectory. The exhaust velocity V_e should be maintained above 2000 fpm (even when there are drains in the stack) to provide adequate plume rise and jet dilution. Velocities above 2000 fpm provide still more plume rise and dilution, but above 3000 to 4000 fpm, noise and vibration from exhaust fans becomes an important concern. An exit nozzle (Figure 2B) can be used to increase the exhaust velocity and plume rise.

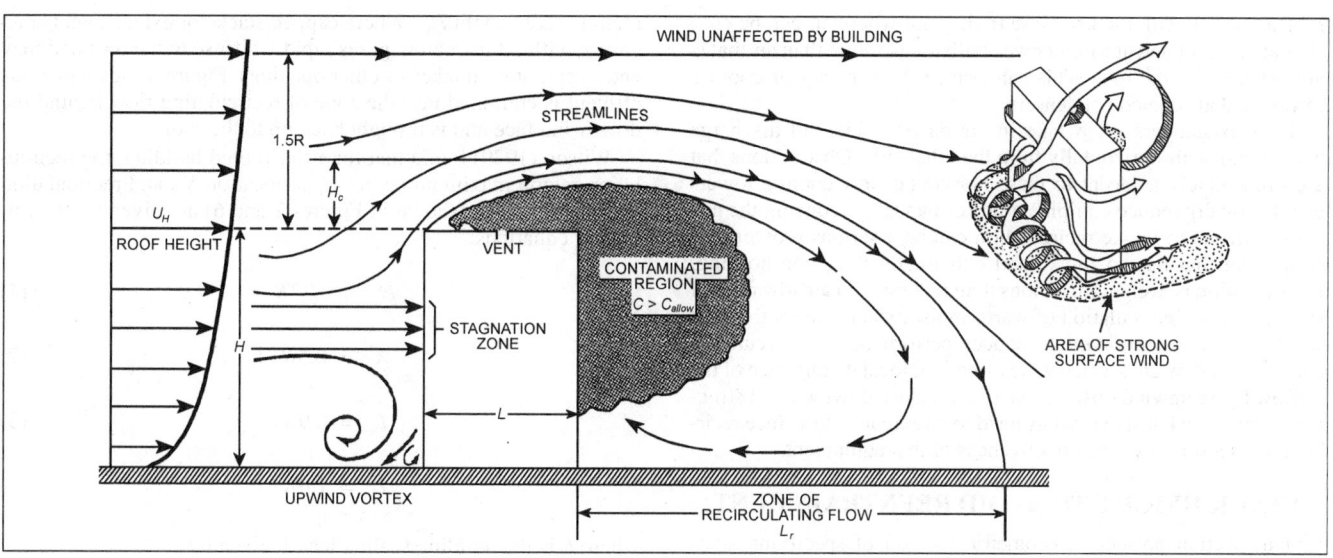

Fig. 4   Flow Patterns Around Rectangular Building

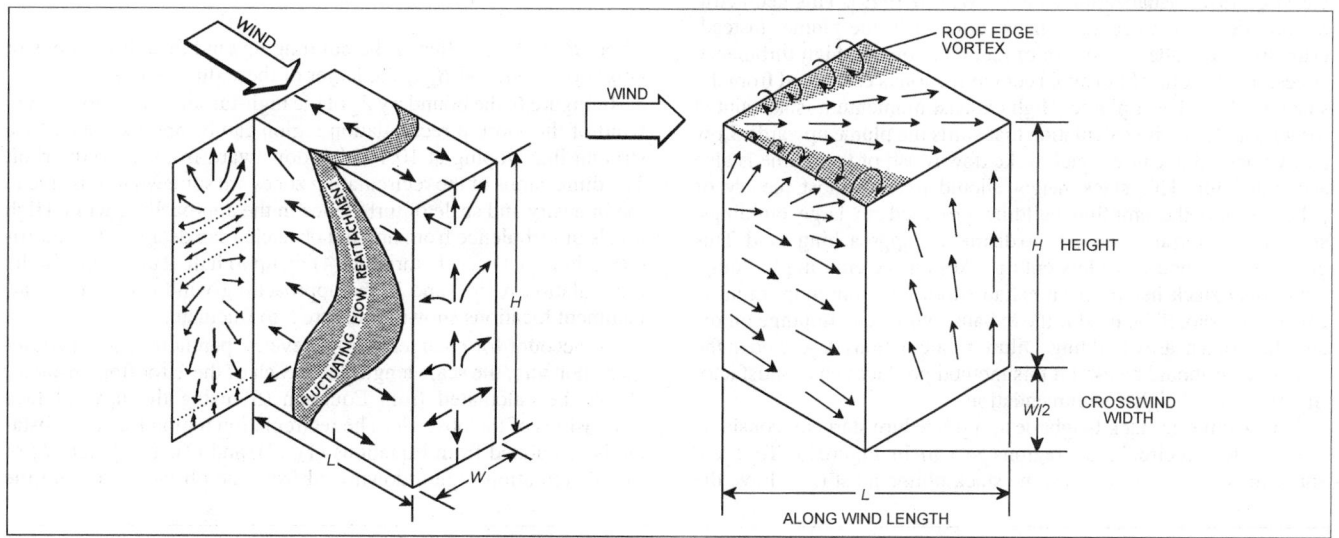

Fig. 5   Surface Flow Patterns and Building Dimensions

because outside air has been sometimes observed to inadvertently enter a building through exhaust air outlets, causing air quality problems (Seem et al. 1998).

Evaporative cooling towers located too closely to air intakes can have several effects: water vapor can increase air-conditioning loads, condensing and freezing water vapor can damage equipment, and ice can block intake grilles and filters. The effect on indoor air quality of escaping droplets of cooling tower water (drift) containing chemicals added to retard scaling and biological contamination is discussed by Vanderhayden and Schuyler (1994).

Intakes should not be located in the same architectural screen enclosure as contaminated exhaust outlets. If exhaust is discharged from several locations on a roof, intakes should be sited to minimize contamination. Where all exhausts of concern are emitted from a single, relatively tall stack or tight cluster of stacks, a possible intake location may be at the base of a tall stack, if this location is not adversely affected by exhaust from nearby buildings. However, leakage of contaminants between the exhaust fans and the stack exit has been observed (Hitchings 1997, Knutson 1997), so air intakes should not be placed next to highly toxic and odorous exhaust stacks.

Intakes near vehicle loading zones should be avoided. Overhead canopies on vehicle docks do not prevent hot vehicle exhaust from rising to intakes above the canopy. When the loading zone is in the flow recirculation region downwind from the building, vehicle exhausts may spread upwind over large sections of the building surface (Ratcliff et al. 1994).

When the wind is perpendicular to the upwind wall, the air flows up and down the wall, dividing at about two-thirds up the wall (Figures 4 and 5). The downward flow creates ground-level swirl (shown in Figure 4) that stirs up dust and debris. To take advantage of the natural separation of wind flow over the upper and lower half of a building, toxic or nuisance exhausts should be located on the roof and intakes located on the lower one-third of the building, but high enough to avoid wind-blown dust, debris, and vehicle exhaust. If ground-level sources such as wind-blown dust and vehicle exhaust are the major sources of contamination, a rooftop intake is desirable. If possible, inlet designs should take into account future sources of contamination, changes to building geometries, and construction of buildings nearby.

Cooling towers and similar heat-rejection devices are very sensitive to airflow around buildings. This equipment is frequently roof-mounted, with intakes close to the roof where air can be considerably hotter and at a higher wet-bulb temperature than air that is not affected by the roof. This can reduce the capacity of cooling towers and air-cooled condensers.

Heat exchangers often take in air on one side and discharge heated, moist air horizontally from the other side. Obstructions that are immediately adjacent to these horizontal-flow cooling towers can drastically reduce equipment performance by reducing the airflow. Exhaust to intake recirculation can be a serious problem for equipment that has an intake and exhaust on the same housing. Recirculation is even more serious than reduction in airflow rate for such devices. Recirculation of warm moist exhaust raises the inlet wet-bulb temperature, which reduces performance. Recirculation can be caused by an adverse wind direction, local disturbance of the airflow by an upwind obstruction, or by a close downwind obstruction. Vertical exhaust ducts may need to be extended to reduce recirculation and improve the effectiveness of this equipment.

STACK HEIGHT TO AVOID REENTRAINMENT

This section presents a geometric method of specifying stack height h_s so that the lower edge of the exhaust plume is above intakes and recirculation zones on the emitting building roof. The method is based on flow visualization studies (Wilson 1979). This geometric method does not calculate exhaust dilution in the plume; instead, estimates are made of the size of recirculation and high turbulence zones; the stack height to avoid contamination is calculated from the shape of the exhaust plume. High exhaust momentum is accounted for with a plume rise calculation that shifts the plume upwards. Low exit velocity that causes stack wake downwash of the plume is also accounted for. This stack height should prevent most reentry of exhausts into the emitting building provided no large buildings, structures, or terrain are nearby to disturb the approaching wind. This geometric method considers only intakes on the emitting building. Additional stack height or an exhaust-to-intake minimum dilution calculation should be used if the exhaust plume can impinge on the air intake of a nearby building. Dilution calculations described in the next section should be used if this method produces an unsatisfactorily high stack height recommendation.

The geometric stack height design procedure starts by considering the flow recirculation regions shown in Figure 6. To avoid entrainment of exhaust gases, the stack plume must rise above the

recirculation height H_c. Where capped stacks or exhaust vents discharge within this region, gases rapidly diffuse to the roof and may enter ventilation intakes or other openings. Figure 4 shows that this effluent is entrained into the zone of recirculating flow behind the downwind face and is brought back up to the roof.

Wilson (1979) found that for a flat-roofed building, the recirculation region maximum height H_c at location X_c, and recirculation lengths L_c and L_r (shown in Figures 1 and 6) are given by the following equations:

$$H_c = 0.22R \tag{1}$$

$$X_c = 0.5R \tag{2}$$

$$L_c = 0.9R \tag{3}$$

$$L_r = 1.0R \tag{4}$$

where R is the building scaling length given by:

$$R = B_s^{0.67} B_L^{0.33} \tag{5}$$

where B_s is the smaller of the building upwind face dimensions of height or width and B_L is the larger of these dimensions.

In Figure 6, the boundary Z_2 of the high-turbulence region downwind of the rooftop recirculation region can be approximated by a straight line sloping at 10:1 (5.7°) downward from H_c to the roof. The dimensions of the recirculating zones are somewhat sensitive to the intensity and scale of turbulence in the approaching wind. High levels of turbulence from upwind obstacles can decrease the coefficients in Equations (1) through (4) by up to half. Turbulence in the recirculation region and in the approaching wind causes the reattachment locations shown in Figure 5 to fluctuate.

To account for changes in roof-level, a penthouse, or an equipment-housing, the scale length R for each of these rooftop obstacles should be calculated from Equation (5) using the upwind face dimensions of the obstacle. The recirculation region for each obstacle is calculated from Equations (1), (2), and (3). The length L_r of the recirculation region downwind from the obstacle, or from the

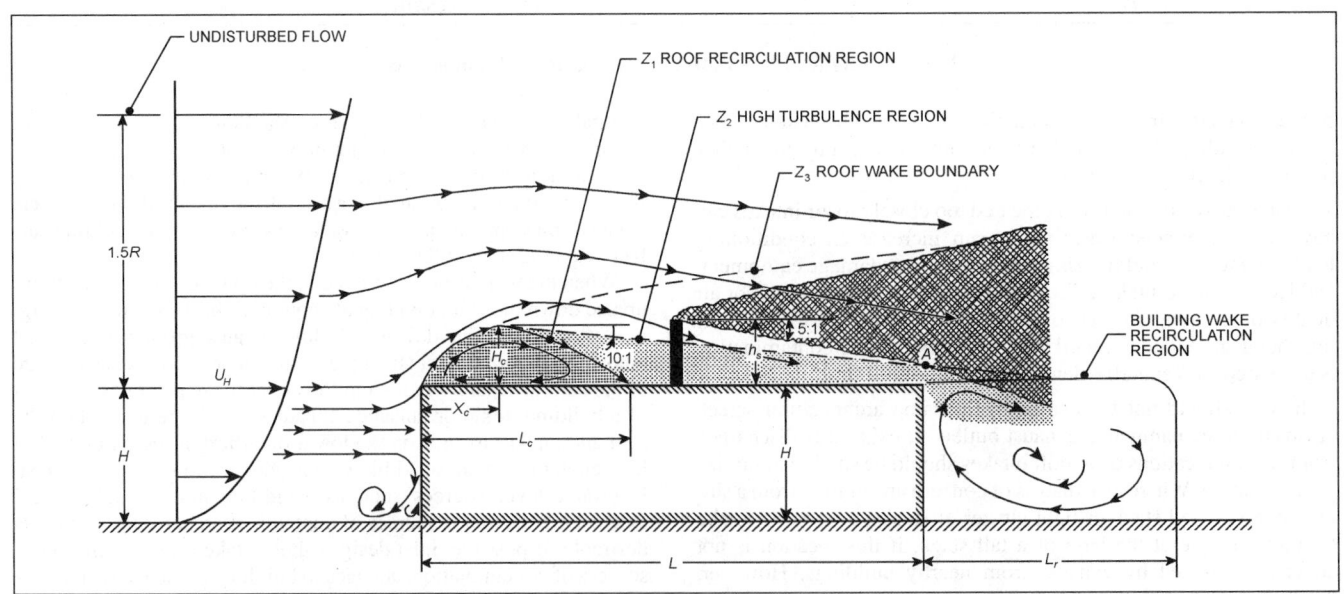

Fig. 6   Design Procedure for Required Stack Height to Avoid Contamination
(Wilson 1979)

entire building, is given by Equation (4), with R based on the dimensions of the downwind face of the obstacle. When an obstacle is close to the upwind edge of a roof or near another obstacle, the flow recirculation zones interact. Wilson (1979) gives methods for dealing with these situations.

Building-generated turbulence is confined to the roof wake region, whose upper boundary Z_3 in Figure 6 is given as

$$Z_3/R = 0.28(X/R)^{0.33} \qquad (6)$$

where X is the distance from the upwind roof edge where the recirculation region forms. Building-generated turbulence decreases with increasing height above roof level. At the edge of the rooftop wake boundary Z_3, the turbulence intensity is close to the background level in the approach wind. The high levels of turbulence below the boundary Z_2 in Figure 6 rapidly diffuse exhaust gases downward to contaminate roof-level intakes.

The next step is to calculate the height h_{sc} of a stack with a rain cap and, therefore, no plume rise. The h_{sc} required to avoid excessive exhaust gas reentry is estimated by assuming that the plume spreads upward and downward from h_{sc} with a 5:1 slope (11.3°), as shown in Figure 6. (This slope represents a downward spread of approximately two standard deviations of a Gaussian plume concentration distribution.) Then, h_{sc} is raised until the lower edge of the plume avoids contact with all recirculation (Zone 1) and high turbulence (Zone 2) boundaries from rooftop obstacles such as air intake housings, architectural screens, or penthouses. The sizes of the recirculation zones are given Equations (1), (2), and (3).

If air intakes are located on the downwind wall, the lower edge of the plume, sloping down at 5:1 must lie above recirculation and high-turbulence zone boundaries (Point A in Figure 6) at the downwind edge of the roof. For a highly toxic contaminant that requires a large dilution factor at a wall intake, the lower edge of the plume should lie above the recirculation zone in the wake downwind of the building. The boundary of the building wake recirculation, shown in Figures 1, 4, and 6, is defined by the horizontal line extending a distance L_r from the downwind edge of the roof. The recirculation length L_r is calculated from Equation (4).

The final step in the geometric stack height design procedure is to reduce the stack height to give credit for plume rise from uncapped stacks, and to increase stack height to account for stack wake downwash caused by low exhaust velocity. Only jet momentum rise is used; buoyancy rise is considered as a safety factor. For an uncapped stack of diameter d, the plume rise h_r due to the vertical momentum of the exhaust is estimated from Briggs (1984) as

$$h_r = 3.0\beta d(V_e/U_H) \qquad (7)$$

where $\qquad d = (4A_e/\pi)^{0.5} \qquad (8)$

For an uncapped stack, the capping factor is $\beta = 1.0$. For a capped stack, $\beta = 0$, so $h_r = 0$, and no credit is given for plume rise. U_H is the maximum design wind speed at roof height for which air intake contamination must be avoided. One possible design wind speed is the wind speed exceeded 1% of the time listed for many cities in Tables 1A, 2A, and 3A of Chapter 26 of the 1997 *ASHRAE Handbook—Fundamentals*. For cities not on this list, set U_H equal to 2.5 times the annual average hourly wind speed as recommended in Table 2 of Chapter 15 of the 1997 *ASHRAE Handbook—Fundamentals*.

Stack wake downwash is described in the section on Recommended Stack Exhaust Velocity. For a vertically directed jet from an uncapped stack ($\beta = 1.0$), Briggs (1973) recommends a stack wake downwash adjustment h_d of

$$h_d = 2.0d(1.5 - \beta V_e/U_H) \qquad (9)$$

for $V_e/U_H < 1.5$. For $V_e/U_H > 1.5$, there is no downwash and $h_d = 0$. Rain caps are frequently used on stacks of gas- and oil-fired furnaces and packaged ventilation units, for which $\beta = 0$ and $h_d = 3.0d$.

The stack height h_s recommended by this geometric method is

$$h_s = h_{sc} - h_r + h_d \qquad (10)$$

The advantage of using an uncapped stack instead of a capped stack is considerable. If the minimum recommended exhaust velocity V_e of $1.5U_H$ is maintained for an uncapped stack ($\beta = 1.0$), plume downwash $h_d = 0$ and $h_r = 4.5d$. For a capped stack ($\beta = 0$), $h_d = 3.0d$ and $h_r = 0$. Using these values in Equation (10), an uncapped stack can be made $7.5d$ shorter than a capped stack.

The largest flow recirculation, high turbulence, and wake regions occur when the wind direction is normal (90°) to the upwind wall of the building. The required stack height is the largest of the heights calculated for the four (or more) wind directions that are normal to an upwind wall.

Example 1. The stack height h_s of the uncapped vertical exhaust on the building in Figure 1 must be specified to avoid excessive contamination of air intakes A and B by stack gases. The stack has a diameter d of 1.64 ft and an exhaust velocity V_e of 1770 fpm. It is located 52.5 ft from the upwind edge of the roof. The penthouse's upwind wall (with intake A) is located 98.4 ft from the upwind edge of the roof, a height of 13.1 ft, and a length of 23.0 ft in the wind direction. The top of intake A is 6.56 ft below the penthouse roof. The building has a height H of 49.2 ft and a length of 203 ft. The top of intake B is 19.7 ft below roof level. The width (measured into the page) of the building is 164 ft, and the penthouse is 29.5 ft wide. What is the required stack height h_s for a design wind speed specified as twice the annual average hourly wind speed of 7.95 mph at a nearby airport with an anemometer height H_{met} of 32.8 ft? The building is located in suburban terrain (Category 2 in Table 1 of Chapter 15 of the 1997 *ASHRAE Handbook—Fundamentals*).

Solution: The first step is to set the height h_{sc} of a capped stack by projecting lines with 5:1 slopes upwind from points of potential plume impact. For intake A, the highest point of impact is the top of the recirculation zone on the roof of the penthouse. To find the height of this recirculation zone, start with Equation (5):

$$R = (13.1)^{0.67}(29.5)^{0.33} = 17.1 \text{ ft}$$

Then use Equations (1) and (2):

$$H_c = 0.22(17.1) = 3.76 \text{ ft}$$

$$X_c = 0.5(17.1) = 8.55 \text{ ft}$$

With the 5:1 slope of the lower plume boundary shown in Figure 6, the capped stack height in Figure 1 must be

$$h_{sc} = 0.2(98.4 - 52.5 + 8.55) + 3.76 = 14.7 \text{ ft}$$

above the penthouse roof to avoid intake A. For intake B on the downwind wall, the plume boundary from the stack in Figure 1 must lie above the end of the roof. To avoid intake B, the capped stack height must be

$$h_{sc} = 0.2(203 - 52.5) - 13.1 = 17.1 \text{ ft}$$

The design stack height is set by the condition of avoiding contamination of intake B, because intake A requires only a 14.7 ft capped stack. Credit for plume rise h_r from the uncapped stack requires calculation of the building wind speed U_H at $H = 49.2$ ft. The design wind speed for $H_{met} = 32.8$ ft at the airport meteorological station is $U_{met} = 2(7.95) = 15.9$ mph $= 1400$ fpm. With the airport in open terrain (Category 3 of Table 1 of Chapter 15 of the 1997 *ASHRAE Handbook—Fundamentals*), and the building in urban terrain (Category 2), the wind speed adjustment parameters are $a_{met} = 0.14$ and $\delta_{met} = 900$ ft at the airport, and $a = 0.22$ and $\delta = 1200$ ft at the building. Using Equation (4) in Chapter 15 of the 1997 *ASHRAE Handbook—Fundamentals*, with the building height $H = 49.2$ ft,

$$U_H = 1400\left(\frac{900}{32.8}\right)^{0.14}\left(\frac{49.2}{1200}\right)^{0.22} = 1102 \text{ fpm}$$

Because $V_e/U_H = 1770/1102 = 1.61$ is greater than 1.5, there is no plume downwash, and $h_d = 0$ from Equation (9). Using Equation (7), the plume rise at the design wind speed is

$$h_r = 3.0(1.64)(1.61) = 7.92 \text{ ft}$$

Deducting this rise from the uncapped height h_{sc} for intake B,

$$h_s = 17.1 - 7.92 = 9.18 \text{ ft}$$

As shown in Figure 1, this stack height is measured above the roof of the nearby penthouse. Adding the penthouse height of 13.1 ft sets the required stack height at 22.2 ft above roof level.

EXHAUST DILUTION CALCULATIONS

This section describes several methods for computing the outdoor dilution of exhausts emitted from a rooftop stack. The resulting dilution can be converted to contaminant concentration for comparison to odor thresholds or health limits. First, definitions will be given, followed by results and equations obtained from wind tunnel studies of simple building shapes.

The dispersion of pollutants from building exhaust depends on the combined effect of atmospheric turbulence in the wind approaching the building and turbulence generated by the building itself. This building-generated turbulence is most intense in and near the flow recirculation zones that occur on the upwind edges of the building (Figures 1 and 5). Because of turbulence and distortion of wind streamlines by the building, the concentration caused by a source near a building cannot be estimated accurately by the design procedures developed for tall isolated stacks. Meroney (1982), Wilson and Britter (1982), Hosker (1984), and Halitsky (1982) review gas diffusion near buildings.

The geometric stack design procedure described in the previous section does not give a quantitative estimate of the worst-case critical dilution factor D_{crit} between the stack and an air intake. If D_{crit} can be specified from knowledge of stack emissions and required health limits odor thresholds, or air quality regulations, the computation of critical dilutions can be an alternative method for specifying stack heights. Smeaton et al. (1991) and Petersen and Ratcliff (1991) discuss the use of emission information and the formulation of dilution requirements in more detail. Exhaust from a single-source dedicated stack may require more atmospheric dilution than a single stack with the same exhausts combined because emissions are diluted in the exhaust manifold.

Dilution and Concentration Definitions

A building exhaust system is used to release a mixture of building air and pollutant gas at concentration C_e (mass of pollutant per volume of air) into the atmosphere through a stack or vent on the building. The exhaust mixes with atmospheric air to produce a pollutant concentration C, which may contaminate an air intake or receptor if the concentration is larger than some specified allowable value C_{allow} (Figure 4). The dilution factor D between source and receptor mass concentrations is defined as

$$D = C_e/C \qquad (11)$$

where

C_e = contaminant mass concentration in exhaust, lb/ft^3
C = contaminant mass concentration at receptor, lb/ft^3

The dilution increases with distance from the source, starting from its initial value of unity. If C is replaced by C_{allow} in Equation (11), the atmospheric dilution D_{req} required to meet the allowable concentration at the intake (receptor) is

$$D_{req} = C_e/C_{allow} \qquad (12)$$

The exhaust (source) concentration is given by

$$C_e = \dot{m}/Q_e = \dot{m}/(A_e V_e) \qquad (13)$$

where

$\dot{m}$ = contaminant mass release rate, lb/s
$Q_e = A_e V_e$ = total exhaust volumetric flow rate, ft^3/s
A_e = exhaust face area, ft^2
V_e = exhaust face velocity, ft/s

The concentration units of mass per mixture volume are appropriate for gaseous pollutants, aerosols, dusts, and vapors. The concentration of gaseous pollutants is usually stated as a volume fraction f (contaminant volume/mixture volume), or as ppm (parts per million) if the volume fraction is multiplied by 10^6. The pollutant volume fraction f_e in the exhaust is

$$f_e = Q/Q_e \qquad (14)$$

where Q is the volumetric release rate of the contaminant gas. Both Q and Q_e are calculated at the exhaust temperature T_e.

The volume concentration dilution factor D_v is

$$D_v = f_e/f \qquad (15)$$

where f is the contaminant volume fraction at the receptor.

If the exhaust gas mixture has a relative molecular mass close to that of air, D_v may be calculated from the mass concentration dilution D by

$$D_v = (T_e/T_a)D \qquad (16)$$

where

T_e = exhaust air absolute temperature, °R
T_a = outdoor ambient air absolute temperature, °R

Many building exhausts are close enough to ambient temperature to assume that volume fraction and mass concentration dilutions D_v and D are equal.

Wind Tunnel Measurements of Exhaust Concentrations

When the exhaust source is in or near a flow recirculation zone, contours of the normalized concentration coefficient K_c, may be used (see Figure 7). For C in contaminant mass per volume,

$$K_c = CU_H HW/\dot{m} \qquad (17)$$

The contours of K_c are developed from wind tunnel model studies on buildings of similar shape. Halitsky (1985) gives details of this technique. Surface concentrations for block buildings with uncapped exhaust vents and short stacks in a uniform nonturbulent wind tunnel airstream are reported by Halitsky (1963), for situations where jet diameters and emission velocities were large enough to project the exhaust plume above the roof and sometimes through the recirculation cavity.

Dilution Prediction Equations

If a representative configuration with published surface concentration coefficients K_c cannot be found to match the existing building, available data may be used to estimate the minimum dilution observed at the same receptor distance from the source.

The most important variables determining minimum dilution are the exhaust-to-intake stretched-string distance S and the effective stack height h_s (Figure 1). Only the stack height extending above large rooftop obstacles such as penthouses and architectural barriers should be used to define the effective stack height.

The stretched-string distance S is defined as the shortest length of string connecting the point on a stack where $h_s = 0$ to the nearest point on an air intake. The point where $h_s = 0$ is defined as the height

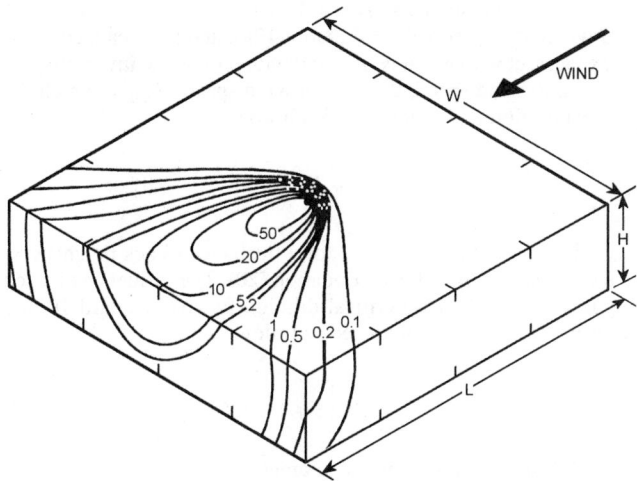

Fig. 7 Normalized Concentration Contours for Central Vent on Roof of a Low-Rise Building
(Wilson 1976)

of obstacles close to the stack, as shown in Figure 1, or the boundary of any roof recirculation region through which the stack passes.

Adjustment for Averaging Time. The following equations for minimum dilution D_{min} were developed for exposures equivalent to 10 min averaging times in the atmosphere, and with nonbuoyant exhaust jets from roof vents. If the exhaust gases are hot, buoyancy increases the rise of the exhaust gas mixture and produces lower concentrations at roof level. By neglecting buoyant plume rise, the D_{min} equations have an inherent safety factor, particularly at low wind speed, where buoyancy is most important.

The averaging time t_a over which exhaust gas concentration exposures are measured is also important in determining minimum dilution. As the averaging time increases, the exhaust gas plume meanders more from side to side, thus reducing the time-averaged concentration at an intake location. The effect of changing the averaging time over a range of about 3 min to 3 h can be estimated by adjusting the 10 min values given in Equations (19) to (26) using the following equation (Wollenweber and Panofsky 1989):

$$D_{min,1} / D_{min,2} = (t_{a1} / t_{a2})^{0.2} \qquad (18)$$

The ratio of minimum dilution at any two averaging times t_{a1} and t_{a2} applies for $S/A_e^{0.5} > 10$ to allow the exhaust jet to begin meandering. If the exhaust and intake are both located in the same flow recirculation region, dilution is less sensitive to averaging time than predicted by Equation (18). In this case, assume the D_{min} values for 3 min averages also apply for averaging times from 3 to 60 min. Use Equation (18) to adjust the 10 min values in Equations (19) to (26) to 3 min values for this flow recirculation situation.

Strong Jets in Flow Recirculation Cavity. For block buildings with surface vents or short stacks that produce significant jet rise because of their large diameter, high-emission velocity, and free discharge, the minimum dilution at a surface intake on the same building or on the jet centerline above the building roof is given by Halitsky (1963) as

$$D_{min} = \left[\alpha + 0.11(1 + 0.2\alpha)S/A_e^{0.5}\right]^2 \qquad (19)$$

where α is a numerical constant related to building shape, emission velocity ratio V_e/U_H, building orientation to the wind, and stack height. Values of α can be found in Halitsky (1963). The smallest D_{min} occurs along the elevated centerline of the jet plume,

where $\alpha = 1.0$ is the appropriate value. Larger values of D_{min} occur on the building surfaces, where α ranges from 2.0 to 20. The larger values are associated with greater stack height and buoyant exhaust.

Strong Jets on Multiwinged Buildings. For a strong vertical exhaust with $V_e/U_H \approx 2$ from surface vents or short stacks on the roof of a multiwinged building with different wing roof heights, Halitsky (1962, 1982) recommends a minimum dilution of

$$D_{min} = N\left[3.16 + 0.1S/A_e^{0.5}\right]^2 \qquad (20)$$

where N is an intake location factor with values of $N = 1.5$ when the intake is on the same roof as the source, $N = 2.0$ when the source and intake are on different wings separated by an air space, and $N = 4.0$ when the intake is substantially lower than the source.

Exhausts with Zero Stack Height. This section presents dilution prediction equations that are an alternative to the strong jet equations of the last two sections. For buildings with exhausts of zero stack height on a flat roof, the minimum dilution $D_{min,o}$ at a roof or wall intake is given by Wilson and Lamb (1994), Wilson and Chui (1985, 1987), and Chui and Wilson (1988) as

$$D_{min,o} = \left[D_o^{0.5} + D_s^{0.5}\right]^2 \qquad (21)$$

where

$$D_o = 1 + 13.0\beta(V_e/U_H) \qquad (22)$$

and

$$D_s = B_1(U_H/V_e)(S^2/A_e) \qquad (23)$$

D_o is the apparent initial dilution at roof level caused by internal turbulence in the exhaust jet. D_s is the distance dilution caused by the combined action of building and atmospheric turbulence. The wind speed U_H at wall height H may be calculated from Equation (4) and Table 1 in Chapter 15 in the 1997 *ASHRAE Handbook—Fundamentals* using either local airport weather data or data in Chapter 26 of the 1997 *ASHRAE Handbook—Fundamentals*. For vertically directed uncapped exhausts, the capping factor $\beta = 1.0$; for capped, louvered, or downward-facing exhausts, $\beta = 0$. The distance dilution parameter B_1 depends on the exhaust jet trajectory and on the intensity of turbulence in the approach wind and generated by the building. This upwind turbulence is represented by σ_θ, which is the standard deviation (in degrees) of wind direction fluctuations averaged over 10 min periods. The full-scale data of Wilson and Lamb (1994) suggest that

$$B_1 = 0.027 + 0.0021\sigma_\theta \qquad (24)$$

Typically, σ_θ varies between 0° and 30° for 10 min averages. For buildings located in urban terrain, the recommended design value is $\sigma_\theta = 15°$, for which $B_1 = 0.059$.

Equations (21), (22), and (23) imply that minimum dilution does not depend on the location of either the exhaust or intake, only on the distance S between them. This is true when exhaust and intake locations are on the same building wall or on the roof. The dilution may increase if the intake and exhaust are located on different faces, as indicated by the N factor in Equation (20). For roof exhausts with wall intakes, the results of Li and Meroney (1983) suggest that the first term in Equation (24) may be increased from 0.027 to 0.10.

For buildings shorter than about 300 ft high and also shorter than twice as high as the surrounding buildings, atmospheric turbulence makes a significant contribution to exhaust gas dilution. Wilson (1976, 1977) gives surface concentration contours (see Figure 7) for flat-roofed buildings in a simulated approach wind that is typical of an urban area. Flush vents with small exhaust velocity make these results suitable for estimates for capped exhaust stacks or louvered exhaust vents.

The effect of atmospheric turbulence is relatively insignificant for isolated highrises that are taller than 300 ft and also twice the average height of buildings for 3000 ft upwind. On buildings where the effects of atmospheric turbulence are small, Wilson and Chui (1987) found that maximum surface concentrations for 10 min exposures were two to ten times higher than on an equivalent low-rise building. For these highrise buildings, use $\sigma_\theta = 0°$ in Equation (24), for which $B_1 = 0.027$.

When exhaust from several collecting stations is combined in a single vent or in a tight cluster of stacks (as recommended in the strategies section), the effective exhaust flow area A_e in Equation (23) increases, which causes the minimum distance dilution in Equation (21) to decrease. However, the exhaust concentration C_e of each contaminant decreases by mixing with other exhaust streams, and the plume rise increases due to the higher momentum in the combined jets. For combined vertical exhaust jets, the roof level intake concentration C is usually significantly lower than the intake concentration caused by separate exhausts. Where possible, exhausts should be combined before release to take advantage of this increase in overall dilution.

Critical Wind Speed and Dilution. At very low wind speed, the exhaust jet from an uncapped stack rises high above roof level, producing a large exhaust dilution D_{min} at a given intake location. Likewise, at high wind speed, the dilution is also large because of longitudinal stretching of the plume by the wind. Between these extremes, a critical wind speed exists at which the smallest amount of dilution occurs for a given exhaust and intake location. This critical, absolute minimum dilution D_{crit} may be used to determine if an exhaust vent will be safe under all wind conditions. The critical wind speed for an uncapped vertical exhaust ($\beta = 1.0$) can be evaluated by finding the absolute minimum in Equations (21), (22), and (23). It is closely approximated by

$$\frac{U_{crit,o}}{V_e} = \frac{3.6}{S}\left(\frac{A_e}{B_1}\right)^{0.5} \tag{25}$$

where $U_{crit,o}$ is the critical wind speed producing the smallest minimum dilution for an uncapped vertical exhaust with negligible stack height. This critical dilution $D_{crit,o}$ is

$$D_{crit,o} = \frac{(1 + 26V_e/U_{crit,o})^2}{1 + 13V_e/U_{crit,o}} \tag{26}$$

To assess the severity of the hazard caused by intake contamination, it is useful to know how often the worst-case D_{crit} is likely to occur. The number of hours per year during which the dilution is no more than twice the critical minimum value may be estimated from weather records by finding the fraction of time that the wind speed is between $0.5U_{crit,o}$ and $3.0U_{crit,o}$ (Wilson 1982, 1983). This fraction is then multiplied by the fraction of time the local wind direction lies in a sector 22.5° on each side of the line joining the exhaust and intake location.

Critical Dilution for Non-Zero Stack Heights. In this section, a method is presented for estimating the worst-case critical dilution D_{crit} for a predetermined stack height, using the critical dilution equations of the last section as a starting point.

An increase in stack height or in exhaust velocity ratio V_e/U_H reduces roof-level contamination by keeping the high concentrations on the plume centerline far enough above the roof so that the intakes see only intermittent concentrations in the fringes of the plume. An increase in stack height or exhaust velocity increases the critical wind speed at which the absolute minimum dilution occurs. This higher critical wind speed often significantly reduces the number of hours per year that high intake contamination (i.e., low dilution) is observed.

Using a Gaussian plume dispersion equation, with a 10-min averaged vertical plume spread standard deviation $\sigma_z = 0.093S$, a crosswind plume spread standard deviation $\sigma_y = (0.093 + 0.0072 \sigma_\theta)S$ where σ_θ is defined for Equation (24), and an uncapped vertical exhaust jet with no buoyancy and with plume rise inversely proportional to wind speed, the critical wind speed U_{crit} at which the smallest minimum dilution D_{crit} is observed is

$$\frac{U_{crit,o}}{U_{crit}} = (Y + 1)^{0.5} - Y^{0.5} \tag{27}$$

$U_{crit,o}$ is the critical wind speed for a flush (zero stack height) vertical exhaust, computed from Equation (20). The influence of stack height on the worst-case critical dilution for the standard 10-min exposure time may be calculated as follows:

$$\frac{D_{crit}}{D_{crit,o}} = \frac{U_{crit}}{U_{crit,o}}\exp\left[Y + Y^{0.5}(Y + 1)^{0.5}\right] \tag{28}$$

where Y is the height-to-spread parameter:

$$Y = \frac{0.25h_s^2}{(\sigma_z + \sigma_o)^2} \tag{29}$$

The initial plume spread σ_o at the exhaust (where $S = 0$) produces the apparent initial dilution D_o caused by the exhaust jet velocity. For a capped stack $\sigma_o = 0$. For an uncapped vertical exhaust jet, the initial dilution at the critical wind speed adds an initial spread $\sigma_o = 0.10S$ to produce D_o. With this value for σ_o and a distance spread of $\sigma_z = 0.093S$ for 10 min averages of atmospheric turbulence, Equation (29) becomes

$$Y = 6.7h_s^2/S^2 \tag{30}$$

$D_{crit,o}$ in Equation (28) is the dilution at critical wind speed for a flush vertical roof exhaust with no stack height, Equation (26). Equations (27) and (28) are reliable only for $Y < 2.0$. Close to the stack, where $Y > 2.0$, use $Y = 2.0$ in Equations (27) and (28). At the upper limit of $Y = 2.0$ where Equations (27) and (28) apply, the maximum critical wind speed is $U_{crit} = 3.15U_{crit,o}$ and $D_{crit} = 270D_{crit,o}$. Because both wind speed and turbulence intensity vary strongly with height above the building roof, the plume rise h_r of the exhaust jet may not be inversely proportional to wind speed so Equations (27) and (28) are only approximations. Because buoyancy is not included, the added rise due to buoyancy provides a safety factor, particularly at low wind speed.

Because Equations (27) and (28) give the effect of a stack relative to a flush exhaust with $h_s = 0$, they are useful for assessing the advantages of increasing stack height as a remedial measure. By comparing two different heights, this calculation allows the relative benefits of a stack to be estimated without knowing any details of the contaminant concentrations or exhaust velocity in the existing stack. For example, the stack height required using the simple geometrical design procedure (see Figure 6) has an h_s/S of at least 0.2. Equations (27) and (28) show that the critical wind speed U_{crit} for this stack height is about 1.6 times as large, and the critical dilution D_{crit} about 3.8 times as large, as the $U_{crit,o}$ and $D_{crit,o}$ for the vertical jet from an uncapped exhaust with zero effective stack height.

Example 2. The uncapped stack from Example 1 (with $h_s = 9.18$ ft above the penthouse roof) is used to exhaust a toxic contaminant. The stack gas concentration must be diluted by 1000:1 to be considered safe for 60 min exposures. (This required dilution is determined from occupational health standards and estimated emission rates.) Calculate the critical minimum dilution factor D_{crit} and the critical wind speed U_{crit} for contamination for intakes A and B. Is the stack height of 9.18 ft sized in Example 1 sufficient to handle this toxic substance?

Solution: At intake A, the stretched-string distance is

$$S_A = \left[(98.4 - 52.5)^2 + (6.56)^2\right]^{0.5} = 46.3 \text{ ft}$$

and the exhaust area of the 1.64 ft diameter stack is $A_e = 2.11$ ft. The first step is to determine the critical wind speed and dilution for an uncapped stack with $h_s = 0$. The normalized exhaust-to-intake distance is

$$S_A/A_e^{0.5} = 46.3/(2.11)^{0.5} = 31.8$$

For dilution calculations, $B_1 = 0.059$ from the text following Equation (24). The critical wind speed for a vent without a stack is calculated as follows from Equation (25), with $V_e = 1770$ fpm:

$$U_{crit,o} = \frac{1770(3.6)}{31.8(0.059)^{0.5}} = 825 \text{ fpm}$$

From Equation (26), the worst-case minimum dilution at intake A for this critical wind speed is

$$D_{crit,o} = \frac{[1 + 26(1770)/825]^2}{1 + 13(1770)/825} = 112$$

The effect of stack height $h_s = 9.18$ ft is calculated from Equations (27) and (28). The height-to-spread parameter Y in Equation (30) is

$$Y = 6.7(9.18/46.3)^2 = 0.260$$

This value of Y is less than the upper limit of 2.0 and can be used in Equation (27),

$$U_{crit} = \frac{825}{(0.260 + 1)^{0.5} - (0.260)^{0.5}} = 1350 \text{ fpm}$$

From Equation (28),

$$D_{crit} = 112\left(\frac{1350}{825}\right)\exp\left[0.260 + (0.260)^{0.5}(0.260 + 1)^{0.5}\right] = 420$$

The stack height has increased U_{crit} to a higher less frequently occurring 1350 fpm (15.3 mph), and increased the worst-case critical dilution by about a factor of 4 from the zero stack height baseline case $D_{crit,o} = 112$ to $D_{crit} = 420$.

The next step is conversion of the dilution from the 10 min exposure averaging time on which all the design equations are based, to the 60 min exposure required for the specified occupational health standard. Using Equation (18),

$$D_{crit} = 420(60/10)^{0.2} = 601$$

Thus, the 9.18 ft stack height produces less than the required 1000:1 hourly-averaged dilution at intake A.

Repeating the calculations for intake B, the stretched-string distance Figure 1 is

$$S_B = (98.4 - 52.5) + 23 + (81.6^2 + 13.1^2)^{0.5} + 19.7 = 172 \text{ ft}$$

and the normalized exhaust-to-intake distance is

$$S_B/A_e^{0.5} = 172/(2.11)^{0.5} = 118$$

For the zero stack height baseline case, $B_1 = 0.059$ and Equation (25) at intake B gives

$$U_{crit,o} = \frac{1770(3.6)}{118(0.059)^{0.5}} = 222 \text{ fpm}$$

From Equation (26), the zero stack height critical dilution for intake B is

$$D_{crit,o} = \frac{[1 + 26(1770)/222]^2}{1 + 13(1770)/222} = 414$$

For the specified stack height $h_s = 9.18$ ft, the height-to-spread parameter Y in Equation (30) is

$$Y = 6.7(9.18/172)^2 = 0.019$$

Then, from Equations (27) and (28),

$$U_{crit} = \frac{222}{(0.019 + 1)^{0.5} - (0.019)^{0.5}} = 254 \text{ fpm}$$

$$D_{crit} = 414(254/222)\exp\left[0.019 + (0.019)^{0.5}(0.019 + 1)^{0.5}\right] = 558$$

For intake B, which is further from the stack, the 9.18 ft stack increases the critical wind speed by a factor of 1.15 (from 222 fpm to 254 fpm), and the critical worst-case dilution increases only by a factor of 1.35 (from 414 to 555). For intake A, which is closer to the stack, the 9.18 ft stack increased the critical wind speed by a factor of 1.6 and increases critical dilution by a factor of 11. This shows that a stack has a diminishing benefit as the exhaust-to-intake distance increases. Using Equation (18), for a 60 min exposure at intake B,

$$D_{crit} = 558(60/10)^{0.2} = 798$$

This does not meet the required 1000:1 minimum hourly averaged dilution specified in the design requirement. Additional stack height must be added using a trial and error solution of Equations (27), (28) and (30).

Scale Model Simulation and Testing

For many routine design applications, exhaust dilution can be estimated using the data and equations presented in the previous sections, which are based on scale modeling of simple building shapes. However, in critical applications, such as where health and safety are of concern, or for complicated building configurations, physical modeling or full-scale field evaluations may be required to obtain more accurate estimates. Measurements on small-scale models in wind tunnels or water channels can provide the required design information. A discussion of scale modeling is presented in Chapter 15 of the *ASHRAE Handbook—Fundamentals*.

SYMBOLS

A_e = stack or exhaust exit face area, ft^2

B_1 = air entrainment parameter in distance dilution D_s, Equation (24)

B_L = larger of the two upwind building face dimensions H and W, ft

B_s = smaller of the two upwind building face dimensions H and W, ft

C = contaminant mass concentration at receptor at ambient air temperature T_e, Equation (11), lb/ft^3

C_{allow} = allowable concentration of contaminant at receptor, Equation (12)

C_e = contaminant mass concentration in exhaust at exhaust temperature T_e, Equation (11), lb/ft^3

d = effective exhaust stack diameter, Equation (8), ft

D = dilution factor between source and receptor mass concentrations, Equation (11)

D_o = apparent initial dilution factor for exhaust jet, Equation (21)

D_{crit} = critical dilution factor at roof level for uncapped vertical exhaust at critical wind speed U_{crit} that produces smallest value of D_{min} for given exhaust-to-intake distance S and stack height h_s, Equation (27)

$D_{crit,o}$ = critical dilution factor D_{crit} at roof level for uncapped vertical exhaust with zero stack height ($h_s = 0$), Equation (26)

D_{min} = minimum dilution factor D at given wind speed for all exhaust locations at same fixed distance S from intake, Equation (18)

$D_{min,o}$ = minimum dilution factor D_{min} at roof level for flush vent with zero stack height ($h_s = 0$), Equation (21)

D_{req} = atmospheric dilution required to meet allowable concentration of contaminant C_{allow}, Equation (12)

D_s = distance dilution factor at fixed wind speed, Equation (23)

D_v = dilution factor between source and receptor volume fraction concentrations, Equation (15)

f = contaminant volume concentration fraction at receptor; ratio of contaminant gas volume to total mixture volume, Equation (15), ppm × 10^{-6}

f_e = contaminant volume concentration fraction in exhaust gas; ratio of contaminant gas volume to total mixture volume, Equation (14), ppm × 10^{-6}

h_d = downwash correction to be subtracted from stack height, Equation (9), ft

h_r = plume rise of uncapped vertical exhaust jet, Equation (7), ft

h_s = effective exhaust stack height above rooftop obstacles and enclosures, Equation (10), ft

h_{sc} = required height of capped exhaust stack to avoid excessive intake contamination, Equation (10), ft

H = wall height above ground on upwind building face, Equation (4), ft

H_c = maximum height above roof level of upwind roof edge flow recirculation zone, Equation (1), ft

K_c = normalized concentration coefficient, Equation (17)

L = length of building in wind direction, Figure 5, ft

L_c = length of upwind roof edge recirculation zone, Equation (3), ft

L_r = length of flow recirculation zone behind rooftop obstacle or building, Equation (25), ft

$\dot{m}$ = contaminant mass release rate, Equation (13), lb/s

N = configuration factor, Equation (20)

Q = contaminant volumetric release rate, Equation (14), ft³/s

Q_e = total exhaust volumetric flow rate, Equation (13), ft³/s

R = scaling length for roof flow patterns, Equation (5), ft

S = stretched-string distance; the shortest distance from exhaust to intake over and along building surface, Equation (9), ft

t_a = time interval over which receptor (intake) concentrations are averaged in computing dilution, Equation (18), s

T_a = outdoor ambient air absolute temperature, Equation (6), °R

T_e = exhaust air mixture absolute temperature, Equation (6), °R

U_{crit} = critical wind speed that produces smallest minimum dilution factor D_{crit} for uncapped vertical exhaust at given S and h_s, Equation (27), ft/s

$U_{crit,o}$ = critical wind speed for smallest minimum dilution factor $D_{crit,o}$ for flush uncapped exhaust with zero stack height ($h_s = 0$), Equation (25), ft/s

U_H = mean wind speed at height H of upwind wall in undisturbed flow approaching building, Equation (7), ft/s

V_e = exhaust face velocity, Equation (13), ft/s

W = width of upwind building face, Equation (17), ft

X = distance from upwind roof edge, Equation (6), ft

X_c = distance from upwind roof edge to H_c, Equation (2), ft

Y = height-to-spread parameter, Equation (29)

Z_1 = height of flow recirculation zone boundary above roof, Figure 6, ft

Z_2 = height of high turbulence zone boundary above roof, Figure 6, ft

Z_3 = height of roof edge wake boundary above the roof, Equation (6) and Figure 6, ft

α = configuration parameter, Equation (19)

β = capping factor; $\beta = 1.0$ for vertical uncapped roof exhaust; $\beta = 0$ for capped, louvered, or downward-facing exhaust, Equation (22)

σ_o = standard deviation of initial plume spread at the exhaust used to account for initial dilution, Equation (29), ft

σ_y = standard deviation of crosswind plume spread, above Equation (27), ft

σ_z = standard deviation of vertical plume spread, Equation (29), ft

σ_θ = standard deviation of wind direction and fluctuations in time t_a, Equation (24) and above Equation (27), degrees

REFERENCES

AIHA. 1992. American national standard for laboratory ventilation. ANSI/AIHA *Standard Z9.5-1992*. American Industrial Hygiene Association, Fairfax, VA.

Briggs. 1973. Diffusion estimates for small emissions. Oak Ridge Atmospheric Turbulence and Diffusion Laboratory, *Draft Report* No. 79.

Briggs. 1984. Plume rise and buoyancy effects. In *Atmospheric Science and Power Production*. D. Randerson, ed. U.S. Department of Energy DOE/TIC-27601 (DE 84005177), Washington, D.C.

Changnon, S.A. 1966. Selected rain-wind relations applicable to stack design. *Heating Piping and Air Conditioning* 38(3):93.

Chui, E.H. and D.J. Wilson. 1988. Effects of varying wind direction on exhaust gas dilution. *Journal of Wind Engineering and Industrial Aerodynamics* 31:87-104.

Gregoric, M., L.R. Davis, and D.J. Bushnell. 1982. An experimental investigation of merging buoyant jets in a crossflow. *Journal of Heat Transfer, Transactions of ASME* 104:236-40.

Halitsky, J. 1962. Diffusion of vented gas around buildings. Journal of the *Air Pollution Control Association* 12:74-80.

Halitsky, J. 1963. Gas diffusion near buildings. *ASHRAE Transactions* 69:464-84.

Halitsky, J. 1966. A method of estimating concentrations in transverse jet plumes. *International Journal of Air and Water Pollution* 10:821-43.

Halitsky, J. 1982. Atmospheric dilution of fume hood exhaust gases. *American Industrial Hygiene Association Journal* 43(3):185-89.

Halitsky, J. 1985. Concentration coefficients in atmospheric dispersion calculations. *ASHRAE Transactions* 91(2B):1722-36.

Hitchings, D.T. 1997. Laboratory fume hood and exhaust fan penthouse exposure risk analysis using the ANSI/ASHRAE 110-1995 and other tracer gas methods. *ASHRAE Transactions* 103(2).

Hosker, R.P. 1984. Flow and diffusion near obstacles. In *Atmospheric Science and Power Production*. D. Randerson, ed. U.S. Department of Energy DOE/TIC-27601 (DE 84005177).

ICBO. 1994. *Uniform building code*. International Conference of Building Officials, Whittier, CA.

Knutson, G.W. 1997. Potential exposure to airborne contamination in fan penthouses. *ASHRAE Transactions* 103(2).

Li, W.W. and R.N. Meroney. 1983. Gas dispersion near a cubical building. *Journal of Wind Engineering and Industrial Aerodynamics* 12:15-33.

Meroney, R.N. 1982. Turbulent diffusion near buildings. In *Engineering Meteorology* 48:525. Elsevier, Amsterdam.

NFPA. 1996. Fire protection for laboratories using chemicals. ANSI/NPPA Standard 45-96. National Fire Protection Association, Quincy, MA.

Petersen, R.L. and M.A. Ratcliff. 1991. An objective approach to laboratory stack design. ASHRAE Transactions 97(2):553-62.

Petersen, R.L., J.J. Carter, and M.A.Ratcliff. 1997. The influence of architectural screens on exhaust dilution. *Final Report* ASHRAE RP 805.

Ratcliff, M.A., R.L. Petersen, and B.C. Cochran. 1994. Wind tunnel modeling of diesel motors for fresh air intake design. *ASHRAE Transactions* 100(2):603-11.

Rock, B.A. and K.A. Moylan. 1999. Placement of ventilation air intakes for improved IAQ. *ASHRAE Transactions* 105(1).

Seem, J.E., J.M. House, and C.J. Klaassen. 1998. Volume matching control: Leave the outdoor air damper wide open. *ASHRAE Journal* 40(2):58-60.

Smeaton, W.H., M.F. Lepage, and G.D. Schuyler. 1991. Using wind tunnel data and other criteria to judge acceptability of exhaust stacks. *ASHRAE Transactions* 97(2):583-88.

Vanderhayden, M.D. and G.D. Schuyler. 1994. Evaluation and quantification of the impact of cooling tower emissions on indoor air quality. *ASHRAE Transactions* 100(2):612-20.

Wilson, D.J. 1976. Contamination of air intakes from roof exhaust vents. *ASHRAE Transactions* 82:1024-38.

Wilson, D.J. 1977. Dilution of exhaust gases from building surface vents. *ASHRAE Transactions* 83(1):168-76.

Wilson, D.J. 1979. Flow patterns over flat roofed buildings and application to exhaust stack design. *ASHRAE Transactions* 85:284-95.

Wilson, D.J. 1982. Critical wind speeds for maximum exhaust gas reentry from flush vents at roof level intakes. *ASHRAE Transactions* 88(1): 503-13.

Wilson, D.J. 1983. A design procedure for estimating air intake contamination from nearby exhaust vents. *ASHRAE Transactions* 89(2):136-52.

Wilson, D.J. and R.E. Britter. 1982. Estimates of building surface concentrations from nearby point sources. *Atmospheric Environment* 16:2631-46.

Wilson, D.J. and E.H. Chui. 1985. Influence of exhaust velocity and wind incidence angle on dilution from roof vents. *ASHRAE Transactions* 91(2b):1693-1706.

Wilson, D.J. and E.H. Chui. 1987. Effect of turbulence from upwind buildings on dilution of exhaust gases. *ASHRAE Transactions* 93(2):2186-97.

Wilson, D.J. and B.K. Lamb. 1994. Dispersion of exhaust gases from roof-level stacks and vents on a laboratory building. *Atmospheric Environment* 28:3099-111.

Wilson, D.J. and G. Winkel. 1982. The effect of varying exhaust stack height on contaminant concentration at roof level. *ASHRAE Transactions* 88(1):513-33.

Wilson, D.J., I. Fabris, and M.Y. Ackerman. 1998. Measuring adjacent building effects on laboratory exhaust stack design. *ASHRAE Transactions* 104(2):1012-28.

Wollenweber, G.C. and H.A. Panofsky. 1989. Dependence of velocity variance on sampling time. *Boundary Layer Meteorology* 47:205-15.

CONTROL OF GASEOUS INDOOR AIR CONTAMINANTS

AMBIENT air contains nearly constant amounts of nitrogen (78% by volume), oxygen (21%), and argon (0.9%), with varying amounts of carbon dioxide (about 0.03%) and water vapor (up to 3.5%). In addition, trace quantities of inert gases (neon, xenon, krypton, helium, etc.) are always present. Gases other than those listed are usually considered contaminants. Their concentrations are almost always small, but they may have serious effects on building occupants, construction materials, or contents. Removal of these gaseous contaminants is often desirable or necessary.

Traditionally, indoor gaseous contaminants are controlled with ventilation air drawn from outdoors, but available outdoor air may contain undesirable gaseous contaminants at unacceptable concentrations. If so, it requires treatment by gaseous contaminant removal equipment before being used for ventilation. In addition, minimizing outdoor airflow by using a high recirculation rate and filtration is an attractive means of energy conservation. However, recirculated air cannot be made equivalent to fresh outdoor air by removing only particulate contaminants. Noxious, odorous, and toxic gaseous contaminants must also be removed by gaseous contaminant control equipment, which is frequently different from particulate filtration equipment.

This chapter covers design procedures for gaseous contaminant control for occupied spaces only. The control of gaseous contaminants from industrial processes and stack gases is covered in Chapter 25 of the 1996 ASHRAE Handbook—Systems and Equipment.

GASEOUS CONTAMINANTS

The only reason to remove a gaseous contaminant from an airstream is that it has harmful or annoying effects on the ventilated space or its occupants. These effects are noticeable at different concentration levels for different contaminants. There are four categories of harmful effects: toxicity, odor, irritation, and material damage. In most cases, contaminants become annoying through irritation or their odors before they reach levels toxic to humans, but this is not always true. For example, the potentially toxic (even deadly) contaminant carbon monoxide has no odor.

Concentration Measures

These effects depend on the contaminant concentrations, which are usually expressed in the following units:

ppmv = parts of contaminant by volume per million (10^6) parts of air by volume

ppbv = parts of contaminant by volume per billion (10^9) parts of air by volume

mg/m³ = milligrams of contaminant per cubic metre of air

μg/m³ = micrograms of contaminant per cubic metre of air

The preparation of this chapter is assigned to TC 2.3, Gaseous Air Contaminants and Gas Contaminant Removal Equipment.

The conversions between ppmv and mg/m³ (based on the ideal gas law) are

$$\text{ppmv} = 0.6699(\text{mg/m}^3)(459.7 + t)/Mp \qquad (1)$$

$$\text{mg/m}^3 = \text{ppmv} \times Mp/[0.6699(459.7 + t)] \qquad (2)$$

where

0.6699 = gas constant in hybrid units
M = molecular mass of contaminant
p = mixture absolute pressure, psia
t = mixture temperature, °F

Concentration data are often needed at a ventilation temperature and pressure of 77°F and 14.7 psia, in which case

$$\text{ppmv} = 24.46(\text{mg/m}^3)/M \qquad (3)$$

Major chemical families of gaseous contaminants and examples of specific compounds are listed in Table 1, which illustrates the wide range of chemicals that may be encountered in gaseous air contaminant control. The groupings are significant because chemically similar contaminants tend to behave similarly in control devices. The *Merck Index* (Budavari 1996), the *Toxic Substances Control Act Chemical Substance Inventory* (EPA 1979), and *Sax's Dangerous Properties of Industrial Materials* (Lewis 1994) are all useful in identifying contaminants, including some known by trade names only. Note that a single chemical compound, especially if it is organic, may have several scientific names.

Toxicity

The harmful effects of gaseous contaminants on a person depend on both short-term peak concentrations and the time-integrated exposure received by the person. Toxic effects are generally considered to be proportional to the exposure dose, though individual response variation can obscure the relationship. The allowable concentration for short exposures is higher than that for long exposures. In the United States, the Occupational Safety and Health Administration (OSHA) has defined three concentration-averaging periods for workplaces and has assigned allowable average concentrations for the three periods for over 490 compounds, mostly gaseous contaminants. The abbreviations for concentrations for the three averaging periods are

AMP = acceptable maximum peak for a short exposure

ACC = acceptable ceiling concentration, not to be exceeded during an 8-h shift, except for periods where AMP applies

TWA8 = time-weighted average, not to be exceeded in any 8-h shift of a 40-h week

Table 1   Major Chemical Families of Gaseous Air Contaminants (with Examples)

Inorganic Contaminants

1. Single-element atoms and molecules
   chlorine
   radon
   mercury
2. Oxidants
   ozone
   nitrogen dioxide
   nitrous oxide
   nitric oxide
3. Reducing agents
   carbon monoxide
4. Acid gases
   sulfur dioxide
   sulfuric acid
   hydrochloric acid
   hydrogen sulfide
   nitric acid
5. Nitrogen compounds
   ammonia
6. Miscellaneous
   arsine

Organic Contaminants

7. *n*-Alkanes
   methane
   n-butane
   n-hexane
   n-octane
   n-hexadecane
8. Branched alkanes
   2-methyl pentane
   2-methyl hexane
9. Alkenes and cyclohexanes
   1-octene
   1-decene
   cyclohexane
10. Chlorofluorocarbons
    R-11 (trichlorofluoromethane)
    R-114 (dichlorotetrafluoroethane)
11. Chlorinated hydrocarbons
    1,1,1-trichloroethane
    carbon tetrachloride
    chloroform

perchloroethane
tetrachloroethylene
12. Halide compounds
    methyl bromide
    methyl iodide
13. Alcohols
    methanol
    ethanol
    2-propanol (isopropanol)
    phenol
    cresol
    diethylene glycol
14. Ethers
    vinyl ether
    methoxyvinyl ether
    n-butoxyethanol
15. Aldehydes
    formaldehyde
    acetaldehyde
    acrolein
    benzaldehyde
16. Ketones
    2-butanone (MEK)
    2-propanone (acetone)
    methyl isobutyl ketone (MIBK)
    chloroacetophenone
17. Esters
    ethyl acetate
    n-butyl acetate
    diethylhexyl phthalate
    dioctyl phthalate (DOP)
    di-*n*-butyl phthalate
    butyl formate
    methyl formate
18. Nitrogen compounds other than amines
    nitromethane
    acetonitrile
    acrylonitrile
    urea
    uric acid
    skatole
    putrescine
    hydrogen cyanide
    peroxyacetal nitrate
    perchloroethane

19. Aromatic hydrocarbons
    benzene
    toluene
    ethyl benzene
    naphthalene
    p-xylene
    benz-α-pyrene
20. Terpenes
    2-pinene
    limonene
21. Heterocylics
    ethylene oxide
    furan
    tetrahydrofuran
    pyrrole
    pyridine
    methyl furfural
    nicotine
    1,4-dioxane
    caffeine
22. Organophosphates
    malathion
    tabun
    sarin
    soman
23. Amines
    methylamine
    diethylamine
    n-nitrosodimethylamine
24. Monomers
    vinyl chloride
    methyl formate
    ethylene
25. Mercaptans and other sulfur compounds
    bis-2-chloroethyl sulfide (mustard gas)
    ethyl mercaptan
    methyl mercaptan
    carbon disulfide
    carbonyl sulfide
26. Organic acids
    formic acid
    acetic acid
    butyric acid
27. Miscellaneous
    phosgene

In non-OSHA literature, ACC is sometimes called STEL (short-term exposure limit), and TWA8 is sometimes called TLV (threshold limit value). The medical community disagrees on what values should be assigned to AMP, ACC, and TWA8 for different contaminants. OSHA values, which change periodically, are published yearly in the *Code of Federal Regulations* (29 CFR 1900, 1000 ff) and intermittently in the *Federal Register.* A similar list is available from the American Conference of Governmental Industrial Hygienists (ACGIH 1997).

The National Institute for Occupational Safety and Health (NIOSH) is charged with researching toxicity problems, and it greatly influences the legally required levels. NIOSH annually publishes the *Registry of Toxic Effects of Chemical Substances* as well as numerous *Criteria for Recommended Standard for Occupational Exposure to (compound)*. Some compounds not in the OSHA list are covered by NIOSH literature, and recommended levels are sometimes lower than the legal requirements set by OSHA. The *NIOSH/OSHA Pocket Guide to Chemical Hazards* (NIOSH 1990) is

a condensation of these references and is convenient for engineering uses. This publication also lists values for the following toxic limit:

IDLH = immediately dangerous to life and health

Although this toxicity limit is rarely a factor in HVAC design, HVAC engineers should consider it when deciding how much recirculation is safe. Ventilation airflow must never be so low that the concentration of any gaseous contaminant could rise to the IDLH level. The levels set by OSHA and NIOSH define acceptable occupational exposures, but cannot be used by themselves as acceptable standards for residential or commercial indoor air concentrations. They do, however, suggest some upper limits for contaminant concentrations for design purposes.

Table 2 presents toxic health effects limits for selected gaseous air contaminants. The chemical family in which the contaminant resides is also given, as are a number of other chemical properties whose use is discussed in later sections.

Table 2 Characteristics of Selected Gaseous Air Contaminants

Contaminants	Allowable Concentration, mg/m³				Odor Threshold, mg/m³	Chemical and Physical Properties			
	IDLH[a]	AMP[a]	ACC[a]	TWA8[a]		Family[b]	BP,[c] °C	M[d]	Retentivity,[e,f] %
Acetaldehyde	18,000			360	1.2	15	21	44	8*
Acetone	4,800		3,200	2,400	47	16	56	58	16
Acetonitrile	7,000	105		70	> 0	18	82	41	1
Acrolein	13		0.75	0.25	0.35	15	52	56	
Acrylonitrile	10			45	50	18	77	53	3
Allyl chloride	810		9	3	1.4	12	44	77	
Ammonia	350		35	38	33	5	−33	17	0
Benzene	10,000		25	5	15	19	80	78	12
Benzyl chloride	50			5	0.2	12	179	127	0.3
2-Butanone (MEK)	8,850			590	30	16	79	72	12
Carbon dioxide	90,000		54,000	9,000	0	4	−78	44	
Carbon monoxide	1,650		220	55	0	3	−192	28	0
Carbon disulfide	1,500	300	90	60	0.6	25	46	76	4*
Carbon tetrachloride	1,800	1,200	150	60	130	11	77	154	8
Chlorine	75		1.5	3	0.007	1	−34	71	2*
Chloroform	4,800		9.6	240	1.5	11	124	119	11
Chloroprene	1,440		3.6	90		12	120	89	
p-Cresol	1,100			22	0.056	13	305	108	5*
Dichlorodifluoromethane	250,000			4,950	5,400	10	−30	121	
Dioxane	720			360	304	21	100	68	13
Ethylene dibromide	3,110	271	233	155		12	131	188	11
Ethylene dichloride	4,100	818	410	205	25	12	84	99	12
Ethylene oxide	1,400		135	90	196	21	10	44	0
Formaldehyde	124	12	6	4	1.2	15	97	30	0.4*
n-Heptane	17,000			2,000	2.4	7	98	100	7*
Hydrogen chloride	140		7	7	12	4	−121	37	0.7*
Hydrogen cyanide	55			11	1	18	26	27	0.4
Hydrogen fluoride	13		5	2	2.7	4	19	20	1*
Hydrogen sulfide	420	70	28	30	0.007	4	−60	34	0.8*
Mercury	28			0.1	0	1	357	201	13
Methane	ASPHY[g]					7	−164	16	0
Methanol	32,500			260	130	13	64	32	6*
Methyl chloride	59,500		1,783	1,189	595	12	74	133	
Methylene chloride	7,500		3,480	1,740	750	12	40	85	9
Nitric acid	250			5		4	84	63	3*
Nitric oxide	120	45		30	> 0	2	−152	30	
Nitrogen dioxide	90		1.8	9	51	2	21	46	2*
Ozone	20			2	0.2	2	−112	48	[h]
Phenol	380		60	19	0.18	13	182	94	5*
Phosgene	8		0.8	0.4	4	27	8	90	1
Propane	36,000				1,800	7	−42	44	5*
Sulfur dioxide	260			13	1.2	4	−10	64	1.7*
Sulfuric acid	80			1	1	4	270	98	1.9*
Tetrachloroethane	1,050			35	24	11	146	108	
Tetrachloroethylene	3,430	2,060	1,372	686	140	11	121	166	
o-Toluidene	440			22	24	23	199	107	
Toluene	7,600	1,900	1,140	760	8	19	111	92	17*
Toluene diisocyanate	70		0.14	0.14	15	18	251	174	
1,1,1-Trichloroethane	2,250			45	1.1	11	113	133	5
Trichloroethylene	5,410	1,620	1,080	541	120	11	87	131	
Vinyl chloride monomer			0.014	0.003	1,400	24	−14	63	0.13
Xylene	43,500		870	435	2	19	137	106	16*

Sources: ASTM *Standard* D 1605 (1988), Balieu et al. (1977), Dole and Klotz (1946), Freedman et al. (1973), Gully et al. (1969), Miller and Reist (1977), Revoir and Jones (1972), Turk (1954).

[a]IDLH, AMP, ACC, and TWA8 are defined in the section on Toxicity.

[b]Chemical family numbers are as given in Table 1.

[c]BP = boiling point at 1 atmosphere pressure

[d]M = relative molecular mass

[e]Retentivities are for typical commercial-grade activated carbon, either measured at TWA8 levels or corrected to TWA8 inlet concentrations using the expression for breakthrough time t_b given in Nelson and Correia (1976):

$$t_{b2} = t_{b1}(C_2/C_1)^{-2/3}$$

Multiplying breakthrough time by inlet concentration gives retentivity, so

$$R_2 = R_1(C_2/C_1)^{1/3}$$

Both concentrations must be in the same units, here mg/m³.

[f]Retentivities marked (*) were calculated from values given in ASTM *Standard* D 1605 (1988) and Turk (1954) assuming that the listed retentivities were measured at 1000 ppm.

[g]ASPHY = Simple asphyxiant; causes breathing problems when concentration reaches about 1/3 atmospheric pressure.

[h]Ozone life is extremely long; activated carbon assists the essentially complete conversion of ozone to normal oxygen by both chemisorption and catalysis.

Personnel exposure to radioactive gases is addressed not by OSHA regulations, but by rules promulgated by the Nuclear Regulatory Commission. Allowable concentrations, in terms of radioactivity, are listed annually in the *Code of Federal Regulations* (10 CFR 50). Radioactive gases are controlled in much the same way as nonradioactive gases, but their high toxicity demands more careful design. Chapter 25, Nuclear Facilities, introduces the procedures unique to nuclear systems.

Another toxic effect that may influence design is the loss of sensory acuity due to gaseous contaminant exposure. Carbon monoxide, for example, affects psychomotor responses and could be a problem in areas such as air traffic control towers. Clearly, waste anesthetic gases in operating suites should not be allowed to reach levels that affect the alertness of any of the personnel. NIOSH recommendations are frequently based on such subtle effects.

Odors

Concentrations as high as those set by OSHA regulations are rarely encountered in commercial or residential spaces. In such spaces, gaseous contaminant problems usually appear as complaints about odors or stuffiness that are the result of concentrations considerably below TWA8 values. Each individual has different sensitivities to odors, and this sensitivity decreases even during a relatively brief exposure. One contaminant may enhance or mask the odor of another. Odors are usually stronger when relative humidity is high. In addition, the human olfactory response S is nonlinear; perceived intensities are approximate power functions of the contaminant concentration C:

$$S = kC^n \qquad (4)$$

where n is about 0.6 and k is specific for each contaminant.

This nonlinear response means that the concentration of some odorants must be reduced substantially before the odor level is perceived to change. For these reasons, determining acceptable concentrations of contaminants on the basis of their odors is as imprecise as attempting to do so from their toxicity. Far fewer data are available on odors, because odors are more of an annoyance than a hazard. In fact, it is desirable for a toxic or explosive material to have an odor threshold well below toxic levels, for this can warn building occupants that the contaminant is present.

At some low concentration of an odorant, an individual ceases to be aware of its presence. Odor studies seek to establish the level at which a percentage of the general population—usually 50%—is no longer aware of the odor of the compound studied. This concentration is the **odor threshold** for that compound. Fazzalari (1978) compiled odor thresholds for a large number of gaseous compounds in commercial and industrial situations. Additional values, reflecting newer measurement techniques but largely limited to compounds found only in industrial workplaces, are listed in AIHA (1989), Moore and Houtala (1983), and Van Gemert and Mettenbreijer (1977). Table 2 shows the wide range of odor threshold values for different contaminants. Chapter 13 of the 1997 *ASHRAE Handbook—Fundamentals* discusses odor perception.

Irritation and Allergic Response

Although some gaseous contaminants may have no discernible continuing health effects, exposure to such contaminants may irritate or cause allergic responses from building occupants. Coughing; sneezing; eye, throat, and skin irritation; nausea; breathlessness; drowsiness; headaches; and depression have all been attributed to gaseous air contaminants. Any of these symptoms could be occur as part of an allergic response, which is differentiated from a toxic response in that the magnitude of the response is not proportional to the exposure dose. A person sensitized to a contaminant can respond strongly at concentrations that are barely detected by other individuals in the space. This makes diagnosis very difficult.

Table 2 does not include a trigger concentration for irritation or allergic effects because none have been determined. In general, it can be said that physical irritation does not occur at odor threshold concentrations. Rask (1988) suggests that when 20% of a single building's occupants suffer such irritations, the structure may be said to suffer from "sick building syndrome" or "tight building syndrome." For the most part, case studies of such occurrences have consisted of analyses of questionnaires submitted to building occupants. Some attempts to relate irritation to gaseous contaminant concentrations are reported (Lamm 1986, Cain et al. 1986, Berglund et al. 1986, Molhave et al. 1982). The correlation of reported complaints with gaseous contaminant concentrations is not strong; many factors affect these nontoxic responses to contamination.

Damage to Materials

Material damage from gaseous contaminants may take such forms as corrosion, embrittlement, and discoloration. Because such effects usually involve chemical reactions that need water, material damage from air contaminants is less in the relatively dry indoor environment than outdoors, even at similar gaseous contaminant concentrations. To maintain this advantage, indoor condensation should be avoided. However, some dry materials can be significantly damaged. These effects are most serious in museums, as any loss of color or texture changes the essence of the object. Libraries and archives are also vulnerable, as are pipe organs and textiles. Ventilation is often a poor method of protecting collections of rare objects because these facilities are usually located in the centers of cities that have relatively polluted ambient outdoor air. Gaseous compounds known to be harmful include ozone, nitrogen oxides, sulfur dioxide, hydrochloric acid, many VOCs, and hydrogen sulfide. Chapter 20 discusses HVAC for museums and libraries.

Lull (1995), Walsh et al. (1977), Chiarenzelli and Joba (1966), NTIS (1982, 1984), Braun and Wilson (1970), Jaffe (1967), Grosjean et al. (1987), AGO (1966), Mathey et al. (1983), Haynie (1978), Graminski et al. (1978), and Thomson (1986) discuss various concerns on material damage by indoor air contaminants.

Chemical Characteristics

Table 2 also includes are several chemical and physical properties that are needed to design control devices. Additional chemical properties can be obtained from handbooks such as the *Merck Index* (Budavari 1996), *Sax's Dangerous Properties of Industrial Materials* (Lewis 1994), *Handbook of Chemistry and Physics* (Lide 1997), *Perry's Chemical Engineers Handbook* (Perry et al. 1997), and chemistry textbooks.

SOURCES

The most commonly cited direct sources of indoor gaseous contaminants are tobacco smoke, cleaning agents and other consumer products, building materials and furnishings, occupants, and outdoor air. In addition, building materials and furnishings adsorb contaminants from indoor air that later desorb, at which time the building materials become secondary sources. This source/sink behavior can be an important contributor to indoor air pollution.

Tobacco Smoke

Tobacco smoke is a prevalent and potent source of indoor air contaminants. Almost all tobacco pollution arises from cigarette smoking. Cigar and pipe smoking produce somewhat different compounds than cigarette smoking, but overall contribute less contamination. Cigarette smoke is frequently categorized as

- Mainstream smoke, which goes directly from the cigarette to the respiratory tract of the smoker
- Sidestream smoke, which is that smoke not directly inhaled by the smoker, including smoke produced between puffs

Table 3 Major Gaseous Contaminants in Typical Cigarette Smoke

Contaminant	Weighted Mean ETS Generation Rate, mg/cigarette	Weighted Standard Error, mg/cigarette	Method
Oxidants			
NO_x	1.801	0.032	Continuous chemi-
NO	1.647	0.032	luminesent analyzer
NO_2	0.198	0.005	
Reducing Agents			
Carbon monoxide	55.101	1.064	Nondispersive IR
Nitrogen Compounds			
Ammonia	4.148	0.107	Cation exchange cartridge
Aldehydes			
Acetaldehyde	2.50	0.054	DNPH cartridge
Formaldehyde	1.33	0.034	
Ketones			
Acetone	1.229	0.040	DNPH cartridge
Nitrogen Compounds			
Acetonitrile	1.145	0.046	Sorbent tube/GC
Aromatic hydrocarbons			
Benzene	0.280	0.005	Sorbent tube/GC
Toluene	0.498	0.011	Sorbent tube/GC
Xylenes	0.297	0.006	Sorbent tube/GC
Styrene	0.094	0.002	Sorbent tube/GC
Sum of five other aromatics	0.140	0.005	Sorbent tube/GC
Alkenes			
Isoprene	6.158	0.123	Sorbent tube/GC
1,3-Butadiene	0.372	0.013	Sorbent tube/GC
Terpenes			
Limonene	0.261	0.005	Sorbent tube/GC
Heterocyclics			
Nicotine	1.585	0.042	Sorbent tube/GC
Pyridine	0.218	0.005	Sorbent tube/GC
Substituted pyridines	0.569	0.013	Sorbent tube/GC
Summary VOC Measurements			
Total HC by FID	27.810	0.483	FID
Ttl sorbed & ID'd VOC	11.270	0.525	Sorbent tube/GC
Total sorbed VOC	19.071		Sorbent tube/GC
Particles			
Respirable	13.674	0.411	Gravimetric

DNPH = 2,4-dinitrophenylhydrazine; GC = gas chromatography.
Source: Martin et al. 1997.

- Environmental tobacco smoke (ETS), which is composed of the aged and diluted combination of exhaled and sidestream smoke. ETS is the smoke category that must be controlled by gaseous air cleaning devices.

Table 3 lists market-share-weighted average concentrations and standard errors for some of the important compounds found in ETS, arranged by chemical family per Table 1. Several points are noteworthy:

- Not all effluents are listed. Carbon dioxide is an example.
- Different methods lead to different results for the VOCs, and different cigarettes have different emission rates.
- Total sorbed VOCs (19 mg/cigarette) was about 2/3 of total hydrocarbon emissions by FID (27.8 mg/cigarette), indicating that approximately 2/3 of the VOCs can be sorbed. The balance are low boiling or otherwise unsuitable for control by adsorption. Of this 2/3 of the total VOCs, about 60% are identified and quantified in Table 3.
- Respirable particulate generated per cigarette is less than the sorbed VOC mass per cigarette.

Both particles and gases should be controlled to control ETS.

Table 4 Generation of Gaseous Contaminants by Building Materials

Contaminant	Caulk	Adhesive	Linoleum	Carpet	Paint	Varnish	Lacquer
	Average Generation Rate, $\mu g/(h \cdot m^2)$						
C-10 Alkane	1200						
n-Butanol	7300						760
n-Decane	6800						
Formaldehyde			44	150			
Limonene		190					
Nonane	250						
Toluene	20	750	110	160	150		310
Ethyl benzene	7300						
Trimethyl benzene		120					
Undecane						280	
Xylene	28						310

Contaminant	GF Insulation	GF Duct Liner	GF Duct Board	UF Insulation	Particleboard	Underlay	Printed Plywood
	Average Generation Rate, $\mu g/(h \cdot m^2)$						
Acetone					40		
Benzene					6		
Benzaldehyde					14		
2-Butanone					2.5		
Formaldehyde	7	2	4	340	250	600	300
Hexanal					21		
2-Propanol					6		

GF = glass fiber; UF = ureaformaldehyde foam
Sources: Matthews et al. (1983, 1985), Nelms et al. (1986), and White et al. (1988).

Cleaning Agents and Other Consumer Products

Commonly used consumer products—liquid detergents and cleaning compounds, waxes, polishes, spot removers, disinfectants, and cosmetics—contain organic solvents that volatilize either slowly or quickly. Mothballs, pesticides, and other pest control agents can also emit organic volatiles that contaminate buildings. Knoeppel and Schauenburg (1989), Black and Bayer (1986), and Tichenor (1989) report data on the release of these volatile organic compounds (VOCs). Field studies have shown that such products contribute significantly to indoor pollution. However, a large variety of compounds are in use, and few studies have been made that allow calculation of typical emission rates.

Manufacturers are reformulating a wide range of products to reduce VOC emissions, but few data are available on the new products, and those that are available are of uncertain applicability.

Building Materials and Furnishings

Particle board is usually made from wood chips bonded with a phenol-formaldehyde or other resin and is a source of indoor gaseous contaminants. It is widely used in current construction, especially for mobile homes, carpet underlay, and case goods. This material, along with ceiling tiles, carpeting, wall coverings, office partitions, adhesives, and paint finishes, emit formaldehyde and other VOCs. Latex paints containing mercury emit mercury vapor. While the emission rates for these materials decline steadily with age, the half-life of emissions is surprisingly long. Black and Bayer (1986), Nelms et al. (1986), and Molhave et al. (1982) report on these sources. Measured emission rates for a few example materials are listed in Table 4. Increasing humidity typically increases emissions.

In addition to being primary sources of emissions, building materials frequently adsorb VOCs and other chemicals during high concentration periods and then re-emit them over an extended period. The complex interaction of primary and secondary emissions is

Table 5 Generation of Gaseous Contaminants by Indoor Combustion Equipment

	Generation Rates, $\mu g/kJ$					Typical Heating Rate, 1000 Btu/h	Typical Use, hour/day	Vented or Unvented	Fuel
	CO_2	NO	NO_2	NO	HCHO				
Convective heater	51,000	83	12	17	1.4	31	4	U	Natural gas
Controlled combustion wood stove		13	0.04	0.07		13	10	V	Oak, pine
Range oven		20	10	22		32	1.0[a]	U	Natural gas
Range-top burner		65	10	17	1.0	9.5/burner	1.7	U	Natural gas

[a]Sterling and Kobayashi (1981) found that gas ranges are used for supplemental heating by about 25% of users in older apartments. This increases the time of use per day to that of unvented convective heaters.

Sources: Wade et al. (1975), Sterling and Kobayashi (1981), Cole (1983), Traynor et al. (1985b), Leaderer et al. (1987), and Moschandreas and Relwani 1989).

extremely difficult to characterize and understand in an actual building. As with consumer products, building materials are also being reformulated to reduce emissions. Emissions measurements are not yet standardized. Currently, the carpet industry has the only labeling program.

Equipment

Commercial and residential spaces contain equipment that produces gaseous contaminants, though generation rates are substantially lower than in the industrial environment. As equipment is rarely hooded, emissions go directly to the occupants. In commercial spaces, the chief sources of gaseous contaminants are office equipment, including electrostatic copiers, laser printers, and fax machines (ozone); diazo printers (ammonia and related compounds); and office supplies, including carbonless copy paper (formaldehyde), correction fluids, inks, and adhesives (various VOCs). Medical and dental activities generate contaminants from the use of disinfectants, escape of anesthetic gases (nitrous oxide and halomethanes), and sterilizers (ethylene oxide). The potential for asphyxiation is always a concern when compressed gases are present, even if the gas is nitrogen.

In residences, the main sources of equipment-derived contaminants are gas ranges, wood stoves, and kerosene heaters. Venting is helpful, but some contaminants escape into the occupied area. The contaminant contribution by gas ranges is somewhat mitigated by the fact that they operate for shorter periods than heaters. The same is true of showers, that can contribute to radon and halocarbon concentrations indoors if the water is contaminated.

Traynor et al. (1985a, 1985b) report on emission rates from indoor combustion devices (Table 5). Emission rates for equipment depend greatly on equipment type and manner of use. Typical values are difficult to obtain (Traynor 1987). Equipment suspected to be a significant source of harmful contaminants should be controlled locally. General ventilation with filtration is not likely to be a safe solution to such problems.

Occupants

Humans and animals emit a wide array of contaminants by breath, sweat, and flatus. Some of these emissions are conversions from solids or liquids within the body. Many volatile organics emitted are, however, re-emissions of contaminants inhaled earlier, with the tracheobronchial system acting like a gas chromatograph or a saturated physical adsorber. Several studies (mostly in relation to spacecraft habitability) have measured contaminant generation by humans (Table 6).

Outdoor Air

The contaminants in outdoor air accompany it into buildings, in some cases to the detriment of the indoor air quality. The contaminants in the outdoor air may be local—such as from restaurants, traffic emissions, sewer vents—or widely distributed in a region—sulfur dioxide, ozone, and nitrogen oxides. The concentrations of these contaminants all vary with time and seasonally, and the air cleaner design must take them into account.

Table 6 Total-Body Emission of Some Gaseous Contaminants by Humans

Contaminant	Typical Emission, $\mu g/h$	Contaminant	Typical Emission, $\mu g/h$
Acetaldehyde	35	Methane	1710
Acetone	475	Methanol	6
Ammonia	15,600	Methylene chloride	88
Benzene	16	Propane	1.3
2-Butanone (MEK)	9700	Tetrachloroethane	1.4
Carbon dioxide	32×10^6	Tetrachloroethylene	1
Carbon monoxide	10,000	Toluene	23
Chloroform	3	1,1,1-Trichloroethane	42
Dioxane	0.4	Vinyl chloride monomer	0.4
Hydrogen sulfide	15	Xylene	0.003

Sources: Anthony and Thibodeau (1980), Brugnone et al. (1989); Cohen et al. (1971), Conkle et al. (1975), Gorban et al. (1964), Hunt and Williams (1977), and Nefedov et al. (1972).

Table 7 Primary Ambient Air Quality Standards for the United States

Contaminant	Long Term		Short Term	
	Concentration, $\mu g/m^3$	Averaging Period	Concentration, $\mu g/m^3$	Averaging Period, h
Sulfur dioxide	80	1 year	365	24
Carbon monoxide			10,000	1
			40,000	8
Nitrogen dioxide	100	1 year		
Ozone[a]			235	1
Hydrocarbons				
Total particulate (PM10)[b]	75	1 year	260	24
Lead particulate	1.5	3 months		

[a]Standard is met when the number of days per year with maximum hour-period concentration above 235 $\mu g/m^3$ is less than one.
[b]PM10 = Particulates below 10 μm diameter.

Outdoor concentrations of a few widely distributed contaminants are available for cities in the United States from Environmental Protection Agency (EPA) summaries of the National Air Monitoring Stations network data and in similar reports for other counties. A summary of this data is available (EPA 1990). In many cases, the concentrations listed in the National Ambient Air Quality Standards, which are the levels the EPA considers acceptable for six contaminants (Table 7), are satisfactory.

Table 8 gives outdoor concentrations for gaseous contaminants at urban sites. These values are typical; however, they may be exceeded if the building under consideration is located near a fossil fuel power plant, refinery, chemical production facility, sewage treatment plant, municipal refuse dump or incinerator, animal feed lot, or other major source of gaseous contaminants. If such sources will have a significant influence on the intake air, a field survey or dispersion model must be run. Many computer programs have been developed to expedite such calculations.

Use of Source Data to Predict Indoor Concentrations

Meckler and Janssen (1988) described a model for calculating the effect of outdoor pollution on indoor air quality, which is outlined in this section. A recirculating air-handling schematic is shown schematically in Figure 1. In this case, mixing is not perfect; the horizontal dashed line represents the boundary of the region close to the ceiling through which air passes directly from the inlet diffuser to the return-air intake. Ventilation effectiveness E_v is the fraction of the total air supplied to the space that mixes with the room air and does not bypass the room along the ceiling. Meckler and Janssen (1988) suggest a value of 0.8 for E_v. Any people in the space are additional sources and sinks for gaseous contaminants. In the ventilated space, the steady-state contaminant concentration

Table 8 Typical Outdoor Concentration of Selected Gaseous Air Contaminants

Contaminants	Typical Concentration, $\mu g/m^3$	Contaminants	Typical Concentration, $\mu g/m^3$
Acetaldehyde	20	Methylene chloride	2.4
Acetone	3	Nitric acid	6
Ammonia	1.2	Nitric oxide	10
Benzene	8	Nitrogen dioxide	51
2-Butanone (MEK)	0.3	Ozone	40
Carbon dioxide	612,000[a]	Phenol	20
Carbon monoxide	3,000	Propane	18
Carbon disulfide	310	Sulfur dioxide	240
Carbon tetrachloride	2	Sulfuric acid	6
Chloroform	1	Tetrachloroethylene	2.5
Ethylene dichloride	10	Toluene	20
Formaldehyde	20	1,1,1-Trichloroethane	4
n-Heptane	29	Trichloroethylene	15
Mercury (vapor)	0.005	Vinyl chloride monomer	0.8
Methane	1,100	Xylene	10
Methyl chloride	9		

[a]Normal concentration of carbon dioxide in air. The concentration in occupied spaces should be maintained at no greater than three times this level (1000 ppm).

Sources: Braman and Shelley (1980), Casserly and O'Hara (1987), Chan et al. (1990), Cohen et al. (1989), Coy (1987), Fung and Wright (1990), Hakov et al. (1987), Hartwell et al. (1985), Hollowell et al. (1982), Lonnemann et al. (1974), McGrath and Stele (1987), Nelson et al. (1987), Sandalls and Penkett (1977), Shah and Singh (1988), Singh et al. (1981), Wallace et al. (1983), and Weschler and Shields (1989).

results from the summation of all contaminants added to the space divided by the total ventilation and amounts removed. The steady-state concentration C_{ss} for a single component can be expressed as

$$C_{ss} = a/b \tag{5}$$

where

$$a = C_x(Q_i + 0.01P\,E_vQ_v/f) + 2119(G_i + NG_O) \tag{6}$$

$$b = Q_e + Q_h + Q_L + k_dA + NQ_O(1 - 0.01P_O)$$
$$+ (E_vQ - Q_v)(1 - 0.01P)/f \tag{7}$$

and

Q = total flow, cfm
Q_e = exhaust airflow, cfm
Q_v = ventilation (makeup) airflow, cfm
Q_h = hood flow, cfm
Q_i = infiltration flow, cfm
Q_L = leakage (exfiltration) flow, cfm
Q_O = average respiratory flow for a single occupant, cfm
P = filter penetration for contaminant, %
P_O = penetration of contaminant through human lung, %
A = surface area inside the ventilated space on which contaminant can be adsorbed, ft^2
k_d = deposition velocity on A for contaminant, fpm
C_x = outdoor concentration of contaminant, mg/m^3
C_{ss} = steady-state indoor concentration of contaminant, mg/m^3
G_O = generation rate for contaminant by an occupant, mg/s
G_i = generation rate for contaminant by nonoccupant sources, mg/s
N = number of occupants
E_v = ventilation effectiveness, fraction
f = $1 - 0.01P(1 - E_v)$

Flow continuity allows the expression for b to be simplified to the following alternate form, which may make it easier to determine flows:

$$b = Q_i + Q_v + k_dA + NQ_O(1 - 0.01P_O) \tag{7a}$$

The parameters for this model must be determined carefully so that nothing significant is ignored. The leakage flow Q_L, for example, may include flow up chimneys or toilet vents.

The steady-state concentration is of interest for design. It may also be helpful to know how rapidly the concentration changes when conditions change suddenly. The dynamic equation for the building in Figure 1 is

$$C_I = C_{ss} + (C_0 - C_{ss})e^{-b\theta/V} \tag{8}$$

where

V = volume of the ventilated space, ft^3
C_0 = concentration in the space at time $\theta = 0$
C_I = concentration in the space θ minutes after a change of conditions.

C_{ss} is given by Equation (5), and b by Equation (7) or (7a), with the parameters for the new condition inserted.

The reduction in air infiltration, leakage, and ventilation air needed to reduce energy consumption raises concerns about indoor contaminant buildup. A low-leakage structure may be simulated by letting $Q_i = Q_L = Q_h = 0$. Then

$$C_{ss} = \frac{0.01PE_vQ_vC_x/f + 2119(G_i + NG_o)}{Q_e + K_dA + NQ_o(1 - 0.01P_O) + (E_vQ - Q_v)(1 - 0.01P)/f} \tag{9}$$

Even if ventilation airflow Q_v is reduced to zero, a low-penetration (high-efficiency) gaseous contaminant filter and a high recirculation rate help lower the internal contaminant concentration. In

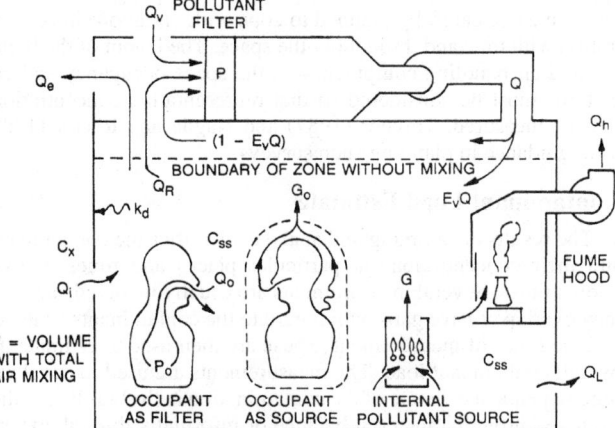

Fig. 1 Recirculatory Air-Handling and Gaseous Contaminant Schematic

commercial structures, infiltration and exfiltration are never zero. The only inhabited spaces operating on 100% recirculated air are space capsules, undersea structures, and those structures with life-support that eliminates carbon dioxide and carbon monoxide and supply oxygen.

Real buildings may have many rooms, with multiple sources of gaseous contaminants and complex room-to-room air changes. In addition, mechanisms other than adsorption may eliminate gaseous contaminants on building interior surfaces. Nazaroff and Cass (1986) provide estimates for contaminant deposition velocity k_d in Equations (5) through (9) that range from 0.0006 to 0.12 fpm for surface adsorption only. A worst-case analysis, yielding the highest estimate of indoor concentration, is obtained by setting $k_d = 0$. Sparks (1988) and Nazaroff and Cass (1986) describe computer programs to handle these calculations.

The assumption of bypass and mixing used in the models presented here approximates the multiple-room case, since gaseous contaminants are readily dispersed by airflows. In addition, a gaseous contaminant diffuses from a zone of high concentration to a zone of low concentration even with low rates of turbulent mixing.

Quantities appropriate for the various flows in Equations (5) through (9) are discussed in the sections on Hooding and Local Exhaust and General Ventilation. Infiltration flow can be determined approximately by the techniques described in Chapter 25, Ventilation and Infiltration, of the 1997 *ASHRAE Handbook— Fundamentals* or, for existing buildings, by tracer or blower-door measurements. ASTM *Standard* E 741 defines procedures for tracer-decay measurements. Tracer and blower-door techniques are discussed by ASTM (1980); DeFrees and Amberger (1987) describe a useful variation on the blower-door technique applicable to large structures.

PROBLEM ASSESSMENT

Ideally, a design for control of gaseous contaminants is based on accurate knowledge of the identity and concentration (as a function of time) of the contaminants to be controlled. This knowledge may be based on close estimates made based on source strength and modeling as described previously or on direct measurements of the contaminants. Unfortunately, a definitive assessment is seldom possible, and then careful observation, experience, and best judgment must supplement data as the basis for the design.

Two general cases exist: (1) new ventilation systems in new buildings for which the contaminant loads must be estimated or measured, and (2) modification of existing ventilation systems to solve particular problems. For the first case, models such as are described previously must be used. Contaminant-generating activities are identified, the building sources estimated and added up, and outside air contaminants identified. Gaps in the contaminant load data must be filled with estimates or measurements. Once the contaminants and loads are identified, the design process can be implemented.

For the second case, measurements may also be required to identify the contaminant. Assessing the problem can become an indoor air quality investigation, including a building inspection, occupant questionnaires, and local sampling and analysis. The EPA Building Air Quality Guide (EPA 1991) is a useful resource for such investigations. Again, once the contaminants and loads are understood, the design process can begin.

Contaminant Measurement and Analysis

Measurements of contaminants have two general purposes: (1) to help determine whether the air quality is acceptable according to published health or odor guidelines (independent of personnel complaints) and (2) to help determine what control strategy and control device should be used in the particular application.

Acceptable carbon dioxide content and odor conditions are defined in ASHRAE *Standard* 62. The alternative indoor air quality procedure outlined in that standard, which applies where outdoor air is reduced to low levels by filtering recirculated air, sets limits for indoor concentrations for several contaminants. The standard also recommends that any unusual contaminants be controlled to one-tenth the TLV levels specified by ACGIH (1997). When outdoor air levels are reduced, the actual gaseous contaminant concentrations must be measured to ensure that the requirements of ASHRAE *Standard* 62 are met.

The concentration of gaseous contaminants at or near levels that are acceptable for indoor air is not as easily measured as temperature or humidity. Relatively costly analytical equipment is needed and it must be calibrated and operated by experienced personnel. Measurement of gaseous contaminant concentrations has two inter-related aspects: sampling and analysis. Sampling and analysis may be (1) short-term (snapshot in time, instantaneous), (2) real-time (a succession of short-term samples with rapid analysis), or (3) integrated over time without the ability to recover detailed time history.

The most useful method depends on the time rate of change of the contaminants, the significance of missing a peak, and the performance characteristics of the control device. For example, if a control device can be saturated and overwhelmed by the peak, it must be redesigned. Also, the application must be understood in order to use the measurements properly. For example, nine hours of a contaminant at zero and one hour at twice a toxic level is not the same as ten hours at 20% of the toxic level, even though the concentration in both cases integrates to the same value. On the other hand, if the problem is odor rather than health effects, short-term excursions may be acceptable. A number of currently available sampling techniques are listed in Table 9, along with their advantages and disadvantages.

Analytical techniques are summarized in Table 10. Some analytical or detection techniques are specific to a single contaminant compound; others are capable of presenting a concentration spectrum for many compounds simultaneously. To be useful for gaseous control, contaminant concentration measurement instruments should be able to detect contaminants of interest at one-tenth TWA8 levels or lower. If odors are of concern, detection sensitivity must be at or below odor threshold levels. As indicated in Table 10, contaminants may be accumulated or concentrated over time so that very low average concentrations can be measured, albeit with the loss of the concentration time history. Procedures for evaluating odor levels are given in Chapter 13 of the 1997 *ASHRAE Handbook—Fundamentals*.

When sampling and analytical procedures appropriate to the application have been selected, a pattern of sampling locations and times must be carefully planned to cope with variations in concentration with time and throughout the space. The layout of the building and air-handling equipment and the space occupancy and use patterns must be considered so that representative concentrations will be measured. Traynor (1987) and Nagda and Rector (1983) offer guidance in planning such surveys.

Contaminant Load Estimates

The results of sampling and analysis identifies the contaminants and their concentrations at particular places and times or over known times. Several measurements are usually involved, and they may overlap or have gaps with regard to the contaminants analyzed and the times of measurement. These are then used to estimate the overall contaminant load. The measurements are used to develop a time-dependent estimate of the contamination in the building, either formally through a material balance or informally through experience with similar buildings and contamination. The degree of formality applied depends on the severity of the potential health or corrosion effects.

Table 9 Gaseous Contaminant Sampling Techniques

Technique[a]	Advantages	Disadvantages
1. Direct flow to detectors	Real-time readout, continuous monitoring possible Several contaminants possible with one sample (when coupled with chromatograph, spectroscope, or multiple detectors)	Average concentration must be determined by integration No preconcentration possible before detector; sensitivity may be inadequate On-site equipment complicated, expensive, intrusive
2. Capture by pumped flow through solid adsorbent; subsequent desorption for concentration measurement	On-site sampling equipment relatively simple and inexpensive Preconcentration and integration over time inherent in method Several contaminants possible with one sample	Sampling media and desorption techniques are compound-specific Interaction between captured compounds and between compounds and sampling media; bias may result Gives only average values over sampling period, no peaks
3. Colorimetric detector tubes	Very simple, relatively inexpensive equipment and materials Immediate readout Integration over time	Rather long sampling period normally required One contaminant per sample Relatively high detection limit Poor precision
4. Collection in evacuated containers	Very simple on-site equipment No pump (silent) Several contaminants possible with one sample	Gives only average over sampling period, no peaks
5. Collection in nonrigid containers (plastic bags) held in an evacuated box	Simple, inexpensive on-site equipment (pumps required) Several contaminants possible with one sample	Cannot hold some contaminants
6. Cryogenic condensation	Wide variety of organic contaminants can be captured Minimal problems with interferences and media interaction Several contaminants possible with one sample	Water vapor interference
7. Passive diffusional samplers	Simple, unobtrusive, inexpensive No pumps; mobile; may be worn by occupants to determine average exposure	Gives only average over sampling period, no peaks
8. Liquid impingers (bubblers)	Integration over time Several contaminants possible with one sample	May be noisy

Sources: NIOSH (1977, 1984), Lodge (1988), Taylor et al. (1977), and ATC (1990).

[a]All techniques except 1 and 3 require laboratory work after completion of field sampling. Only technique 1 is adaptable to continuous monitoring and able to detect short-term excursions.

CONTROL STRATEGIES

Four control strategies may be used to improve the indoor air quality in a building: (1) elimination of sources, (2) local hooding with exhaust or recirculated air cleaning, (3) dilution with increased general ventilation, and (4) general ventilation air cleaning with or without an increase in ventilation rates. In most cases, the first three are favored. By its nature, control by general air cleaning is difficult because it is applied after the contaminants are fully dispersed and at their lowest concentration.

Elimination of Sources

This strategy is the most effective and often the least expensive. For instance, prohibition of smoking in a building, or its isolation to limited areas, greatly reduces indoor pollution, even when rules are poorly enforced (Elliott and Rowe 1975, Lee et al. 1986). Control of radon gas begins with the installation of traps in sewage drains and the sealing and venting of leaky foundations and crawl spaces to prevent entry of the gas into the structure (EPA 1986, 1987). The use of waterborne materials instead of those requiring organic solvents may reduce VOCs, although Girman et al. (1984) show that the reverse is sometimes true. The substitution of carbon dioxide for halocarbons in spray-can propellants is an example of the use of a relatively innocuous contaminant instead of a more troublesome one. The growth of mildew and other organisms that emit odorous

contaminants can be restrained by controlling condensation and applying fungicides and bactericides provided they are registered for the use and carefully chosen to have low off-gassing potential.

Local Source Control

Local source control is more effective than control by general ventilation when discrete sources in a building generate substantial amounts of gaseous contaminants. If these contaminants are toxic, irritating, or strongly odorous, local control and outdoor exhaust is essential. Bathrooms and kitchens are the most common examples. Exhaust rates are sometimes set by local codes. In addition, some office equipment could benefit from direct exhaust. The minimum transport velocity required for the capture of large particles differs from that required for gaseous contaminants; otherwise, the problems of capture are the same for both gases and particles.

Hoods are normally provided with exhaust fans and stacks that vent to the outdoors. Hoods need large quantities of makeup air, which requires a great deal of fan energy for exhaust and makeup air; so hoods waste heating and cooling energy. Makeup for air exhausted by a hood should be supplied so that the general ventilation balance is not upset when a hood exhaust fan is turned on. Back diffusion from an open hood to the general work space can be eliminated by surrounding the work space near the hood with an isolation enclosure. Such a space not only isolates the contaminants, but also helps keep unnecessary personnel out of the area. Glass walls

Table 10 Gaseous Contaminant Concentration Measurement Methods

Method	Description	Typical Application
1. Gas chromatography (Using the following detectors)	Separation of gas mixtures by time of passage down an absorption column	
Flame ionization	Change in flame electrical resistance due to ions of contaminant gas	Volatile, nonpolar organics
Flame photometry	Measures light produced when contaminant is ionized by a flame	Sulfur, phosphorous compounds
Photoionization	Measures ion current for ions created by ultraviolet light	Most organics (except methane)
Electron capture	Radioactively generated electrons attach to contaminant atoms; current measured	Halogenated organics Nitrogenated organics
Mass spectroscopy	Contaminant atoms are charged, passed through electrostatic magnetic fields in a vacuum; path curvature depends on mass of atom, allowing separation and counting of each type	Volatile organics
2. Infrared spectroscopy and Fourier transform infrared spectroscopy (FTIR)	Absorption of infrared light by contaminant gas in a transmission cell; a range of wavelengths is used, allowing identification and measurement of individual pollants	Acid gases Many organics
3. High-performance liquid chromatography (HPLC)	Contaminant is captured in a liquid, which is then passed through a liquid chromatograph (analogous to a gas chromatograph)	Aldehydes, ketones Phosgene Nitrosamines Cresol, phenol
4. Colorimetry	Chemical reaction with contaminant in solution yields a colored product whose light absorption is measured	Ozone Nitrogen oxides Formaldehyde
5. Fluorescence and Pulsed fluorescence	contaminant atoms are stimulated by a monochromatic light beam, often ultraviolet; they emit light at characteristic fluorescent wavelengths, whose intensity is measured	Sulfur dioxide Carbon monoxide
6. Chemiluminescence	Reaction (usually with a specific injected gas) results in photon emission proportional to concentration	Ozone Nitrogen compounds Several organics
7. Electrochemical	Contaminant is bubbled through reagent/water solution, changing its conductivity or generating a voltage	Ozone Hydrogen sulfide Acid gases
8. Titration	Contaminant is absorbed into water	Acid gases
9. Ultraviolet absorption	Absorption of UV light by a cell through which the polluted air passes	Ozone Aromatics Sulfur dioxide Oxides of nitrogen Carbon monoxide
10. Atomic absorption	Contaminant is burned in a hydrogen flame; a light beam with a spectral line specific to the contaminant is passed through the flame; optical absorption of the beam is measured	Mercury vapor

Sources: NIOSH (1977, 1984), Lodge (1988), Taylor et al. (1977), and ATC (1990).

for the enclosure decrease the claustrophobic effect of working in a small space.

Increasingly, codes require filtration of hood exhausts to prevent toxic releases to the outdoors. Hoods should be equipped with controls that decrease their flow when maximum protection is not needed. Hoods are sometimes arranged to exhaust air back into the occupied space, thus saving heating and cooling of that air. This practice must be limited to hoods exhausting the most innocuous contaminants because of the risk of filter failure. The design and operation of effective hoods are described in Chapters 13, 14, 29, and 30 and in *Industrial Ventilation: A Manual of Recommended Practice* (ACGIH 1995).

Dilution Through General Ventilation

In residential and commercial buildings, the chief use of local source control and hooding occurs in kitchens, bathrooms, and occasionally around specific point sources such as diazo printers. Where there is no local control of contaminants, the general ventilation distribution system provides contaminant control through dilution. Such systems must meet both thermal load requirements and contaminant control standards. Complete mixing and a relatively uniform air supply per occupant are desirable for both purposes. The guidelines for air distribution given in Chapters 31 and 32 of the 1997 *ASHRAE Handbook—Fundamentals* are appropriate for contamination control by general ventilation. In addition, the standards set by ASHRAE *Standard 62* must be met.

When local exhaust is combined with general ventilation, a proper supply of makeup air must equal the exhaust flow for any hoods present. Supply fans may be needed to provide enough pressure to maintain flow balance. Clean spaces are designed so that static pressure forces air to flow from clean to less clean spaces, and the effects of doors opening and wind pressure, etc., dictate the need for backdraft dampers. Chapter 15 covers clean spaces in detail.

CONTROL BY AIR CLEANING

If elimination of sources, local hooding, or dilution cannot control contaminants, or are only partially effective, the air must be cleaned. Designing such a system requires an understanding of the capabilities and limitations of the control processes.

Gaseous Contaminant Air Cleaning Terminology

Several methods of measuring the performance of a gaseous control device have been developed, and some are unique to this application. In some cases the meaning of familiar terms from particle filtration are slightly different. In general, gaseous adsorber performance is a function of (1) the specific contaminant, (2) its concentration, (3) the airflow rate, and (4) environmental conditions.

Penetration is the ratio of control device outlet concentration to that at the inlet (the challenge concentration) for a constant challenge. **Efficiency** is a related concept defined as (1 minus penetration). Penetration and efficiency are usually expressed as percentages or decimal fractions. Unlike particulate filters, physical adsorbents and chemisorbents both decline in efficiency as they load. The decline can be very sudden, and it is not likely to be linear with time. Because their efficiency declines with use, the time at which the measurement is made must be given with efficiency or penetration readings.

The sudden decline in efficiency has led to the concept of **breakthrough time**, which is defined as the operating time (at constant operating conditions) before a certain penetration is achieved. For instance, the 10% breakthrough time is the time elapsed between beginning to challenge an adsorber and the time at which air discharged contains 10% of the contaminant feed concentration. Continued operation leads to 50% breakthrough and eventually to 100% breakthrough, at which point a physical adsorbent is **saturated**. For a chemisorber, the media is exhausted. (Some commercial devices are designed to allow a portion of the challenge gas to bypass the sorbent. These devices "breakthrough" immediately, and breakthrough time, as defined above, does not apply.)

At **saturation**, the physical absorbent contains all the contaminant it can hold at the challenge concentration, temperature, and humidity. This point is defined as its **activity**, and is expressed as a percentage of the carbon mass or fraction [i.e., g (contaminant)/g (adsorbent)]. Activity is an equilibrium property, and is not a function of airflow. (In most cases, commercial sorbent bed filters are changed for efficiency reasons well before the sorbent is saturated.) If a saturated adsorbent bed is then exposed to clean air, some of the adsorbed contaminant will desorb. The amount remaining after a saturated bed reaches equilibrium in clean air is the **retentivity** of the adsorbent. Table 2 gives retentivity values for many of the listed contaminants. Those marked with an asterisk were measured or corrected to 1000 ppm challenges. Those not marked with an asterisk were measured or corrected to the TWA8 concentration as the challenge.

Residence time is the period, usually a fraction of a second, that the contaminant molecule is within the boundaries of the media bed and thus exposed to capture (and for chemisorption) chemical change. The efficiency of a particular gaseous contamination adsorbent bed, for a given contaminant and set of environmental conditions, is fixed by the bed residence time. The longer the residence time, the higher the efficiency. For gaseous contaminant control equipment residence time is computed as:

$$\text{Residence time} = \frac{\text{Bed area exposed to airflow} \times \text{Bed depth}}{\text{Airflow rate}} \quad (10)$$

For example, a unitary adsorber containing trays totaling 40 ft^2 of a 1 in. deep bed, challenged at 2000 cfm, has a residence time of 0.1 s. Given this definition, a deeper media bed, lower airflow rate, or adsorbers in series increase the residence time and thus the adsorber performance. Because gaseous contaminant air cleaners have traditionally tended to all have approximately the same granule size and have used the same kind of activated carbon, residence time has become a generally useful indicator of performance. Note that in some engineering disciplines, the volume of the adsorbent media is subtracted from the nominal volume of packed beds when calculating residence time. This definition gives a shorter residence time value and is not normally used for HVAC applications.

Different ways of arranging the carbon, different carbons, or different carbon granule sizes all change the residence time. In addition, the geometry and packaging of some adsorbent technologies makes computation of residence time difficult. For example, the flow pattern in pleated fiber-carbon composite media is difficult to specify, making residence time computation uncertain. Therefore, while residence times can be computed for partial bypass filters, fiber-adsorbent composite filters, or fiber bonded filters, they cannot be compared directly and may serve more as a rating than as an actual residence time. Manufacturers may publish equivalent residence time values that say, in effect, that this adsorber performs the same as a traditional deep bed adsorber. No standard test exists to verify such a rating, however.

Control Processes

Numerous chemical and physical processes remove gases or vapors from air. The gaseous contamination control processes of commercial interest to the HVAC engineer are physical adsorption and chemisorption. Other processes currently have extremely limited application in HVAC work, and they are only briefly discussed.

Physical Adsorption Description. Adsorption is a surface phenomenon similar in many ways to condensation. Adsorbed contaminant gas molecules are those that strike a surface and remain bound to it for an appreciable time. Surfaces of gaseous contamination control adsorption media are expanded in two ways to enhance its adsorptive characteristics. First, the media are provided in granular, pelletized, or fibrous form to increase the gross surface exposed to an airstream. Second, the surface of the adsorbent media is treated to develop pores of microscopic dimensions, greatly increasing the area available for molecular contact. Typical activated alumina has a surface area of 10^6 ft^2 per pound; typical activated carbon has a surface area from 5 to 7.5×10^6 ft^2/lb. Pores of various microscopic sizes and shapes provide minute traps that can fill with condensed contaminant molecules.

The most common granular adsorbents are millimetre-sized, and the granules are used in the form of packed beds. In general, packed beds composed of larger monodisperse particles have lower pressure drops per unit depth of sorbent than those composed of smaller monodisperse particles. On the other hand, more absorbent can be packed in a given volume with smaller particles, and that adsorbent is more accessible to the contaminant, but the pressure drop is higher at a fixed bed depth.

Several steps must occur in the adsorption of a molecule (see Figure 2):

1. The molecule must be transported from the carrier gas stream across the boundary layer surrounding the absorbent granule. This process is random, with molecular movement both to and from the surface; the net flow of molecules is toward the surface when the concentration of the contaminant in the gas flow is greater than at the granule surface. For this reason, adsorption action decreases as contaminant load on the absorbent surface increases. Very low concentrations in the gas flow also result in low adsorption rates.

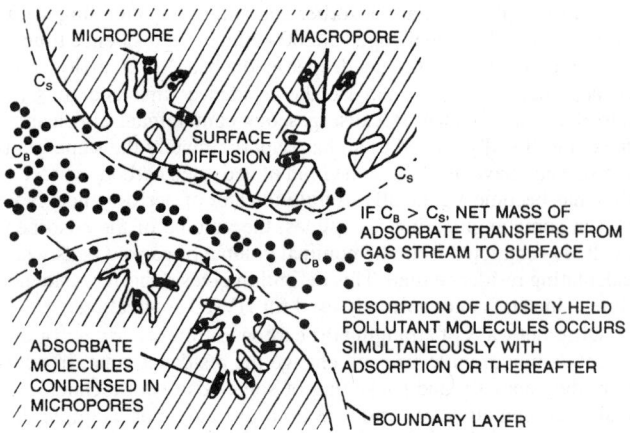

Fig. 2   Steps in Contaminant Adsorption

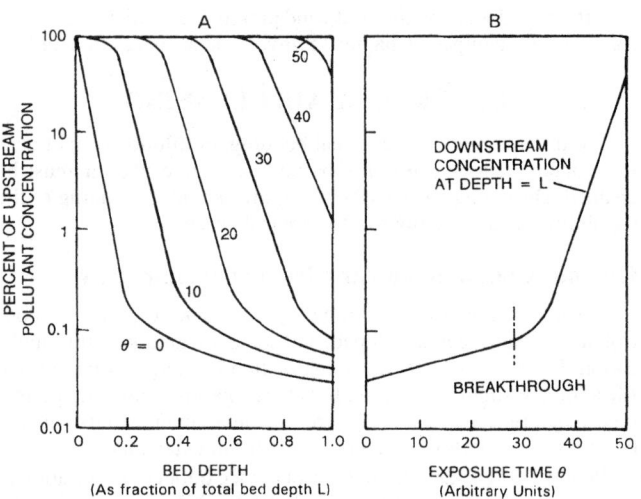

Fig. 3   Dependence of Contaminant Concentration on Bed Depth and Exposure Time

2. The molecules of the contaminant must diffuse into the pores to occupy that portion of the surface. Diffusion distances are lower and adsorption rates higher for smaller particles of absorbent.
3. The contaminant molecules must be bound to the surface. (Sorption is exothermic, releasing energy. At the low concentrations and sorption rates generally encountered in HVAC applications, sorbents operate nearly isothermally.)

Any of these steps may determine the rate at which adsorption occurs. In general, Step 3 is very fast for physical adsorption, but reversible. This means that some adsorbed molecules can be desorbed at a later time, either when cleaner air passes through the adsorber bed or when another contaminant arrives that either binds more tightly to the adsorber surface or is present at a much higher concentration. Complete desorption usually requires that thermal energy be added to the bed.

When a contaminant is fed at constant concentration and constant gas flow rate to an adsorber of sufficient bed depth L, the gas stream concentration varies with time θ and bed depth, as shown in Figure 3A. When bed loading begins ($\theta = 0$), the contaminant concentration decreases logarithmically with bed depth; deeper into the bed, the slope of the concentration-versus-bed depth curve flattens at a very low value. At later times, the entrance portion of the adsorber bed becomes loaded with contaminant, so contaminant concentrations in the gas stream are higher at each bed depth.

For the same constant contaminant feed, the pattern of downstream concentration versus time for an adsorber of bed depth L is shown in Figure 3B. Usually the downstream concentration is very low until time θ_{BT}, when the concentration rises rapidly until the downstream concentration is the same as the upstream. This penetration is called **breakthrough** because it occurs suddenly. However, not all adsorber-contaminant combinations show as sharp a breakthrough as Figure 3B might indicate.

Multiple contaminants produce more complicated penetration patterns than shown in Figure 3B. While individually each contaminant might behave as shown in Figure 3, each has its own time scale. The better adsorbing contaminants are captured in the upstream portion of the bed, and the poorer are adsorbed further downstream. As the challenge continues, the better adsorbing compound progressively displaces the other until the displaced component can leave the adsorber bed at a higher concentration than it entered.

Yoon and Nelson (1988) and Underhill et al. (1988) discuss the effect of relative humidity on physical adsorption. Essentially, water vapor acts as a second contaminant, altering the adsorption parameters by reducing the amount of the first contaminant that can be held by the adsorber and shortening the breakthrough times. For VOCs adsorbed on carbon, the effect of relative humidity is modest up to about 50%, and greater at higher relative humidities.

Chemisorption Description. The three physical adsorption steps explained and then shown in Figure 3 also apply to chemisorption. However, the third step in chemisorption is by chemical reaction with electron exchange between the contaminant molecule and the chemisorber. This action differs in the following ways from physical adsorption:

1. Chemisorption is highly specific; only certain contaminant compounds will react with a particular chemisorbant.
2. Chemisorption improves as temperature increases (reaction rate increases); physical adsorption improves as temperature decreases (vapor pressure drops).
3. Chemisorption does not generate heat, but instead may require heat input.
4. Chemisorption is not generally reversible. Once the adsorbed contaminant has reacted, it is not desorbed. However, one or more reaction products, different from the original contaminant, may be formed in the process, and these reaction products may enter the air as a new contaminant.
5. Water vapor often helps chemisorption or is necessary for it, while it usually hinders physical adsorption.
6. Chemisorption per se is a monomolecular layer phenomenon; the pore-filling effect that takes place in physical adsorption does not occur, except where adsorbed water condensed in the pores forms a reactive liquid.

Most chemisorptive media are formed by coating or impregnating a highly porous, nonreactive substrate (e.g., activated alumina, zeolite, or carbon) with a chemical reactant. The reactant will eventually become exhausted, but the substrate may have physical adsorption ability that remains active when chemisorption ceases.

Other Processes. Physical adsorption and chemisorption are the most commonly used processes; however the following processes are used in some applications.

Liquid absorption devices (scrubbers) and **combustion devices** are generally used to clean exhaust stack gases and process gas effluent. They are not commonly applied to indoor air cleanup. Additional information may be found in Chapter 25 of the 1996 *ASHRAE Handbook—Systems and Equipment.*

Catalysts can clean air by stimulating a chemical reaction on the surface of the media. **Catalytic combustion** or **catalytic oxidation** oxidizes unburned hydrocarbons in moderate concentrations in air. In general, the challenge with catalytic oxidation is to achieve an adequate reaction rate (contaminant destruction rate) at ambient temperature. Similarly, ultraviolet light (photocatalysis) uses ultraviolet energy to initiate reactions (not all oxidation) at temperatures below where they would normally occur. The products of complete combustion of hydrocarbons are CO_2 and water, and at indoor contaminant concentrations these would be acceptable reaction products. Reaction products are a concern, because oxidation of non-hydrocarbons or reactions other than oxidation can produce undesirable by-products. Commercial ultraviolet devices have not been demonstrated at this time.

Cryogenic condensation has been used in nuclear safety systems. **Plants** and **occupants** themselves remove some contaminants from indoor air. **Odor counteractants** and **odor masking** are not truly control methods that may apply only to specific odors and have limited effectiveness.

Equipment

General Considerations. With a few exceptions, adsorption and chemisorption media are supplied in granular or pelletized form, which are held in a retaining structure that allows the air being treated to pass through the media with an acceptable pressure drop at the operating airflow. Granular media has traditionally been a few millimetres in all dimensions. Typical sizes for this media have been on the order of 4×6 or 4×8 U.S. mesh pellets or flakes because these sizes have an acceptable pressure drop at the required operating conditions.

Numerous other sizes are available. Typical configurations of units in which millimetre-sized granular or pelletized media are held between perforated retaining sheets or screens are shown in Figures 4A, 4B, and 4D. The perforated retainers or screens must have holes smaller than the smallest particle of the active media. A margin without perforations must be left around the edges of the retaining sheets or screens to minimize the amount of air that can bypass the active media. Media must be tightly packed in the structure so that open passages through the beds do not develop. Aluminum; stainless, painted, plated, or coated steel; plastics; and kraftboard are all used for retainers.

Adsorptive media may also be retained in fibrous filter media or other porous support structures that can be pleated into large filters as shown in Figures 4C and 4E. The adsorptive media must be bound to the supporting fibers in such a way that media micropores are preserved (i.e., not sealed by binders) and adequate overall adsorptive capacity is maintained. Media supported in this way is generally much smaller in size (down to approximately 80 mesh). As described earlier, small particle adsorbants have higher efficiency than the same adsorbent in larger particles. The smaller particles must be uniformly distributed and supported such that the pressure drop of the composite is acceptable. Because the absorbent is intimately bound to, and its performance affected by the support structure, these adsorbers must be evaluated by tests on the complete composite structure, not the granular adsorbent alone.

Granular media adsorbers are available in two working classes: **total** or **full detention** and **partial detention** units. Full-detention units are those in which the mass transfer zone is initially contained totally within the bed as depicted by the $\theta = 0$ curve of Figure 3A. Other popular commercial designs may allow the challenge gas to bypass the unit or have short residence times. Partial detention units are called partial bypass if large open areas are designed into the bed or device. Partial detention units are useful when 100% control is not needed to achieve design objectives or when high recirculation rates will allow them to effectively remove contaminants from a space.

Generally, total detention units are those in which all the air passes through a media bed that is usually a minimum of 0.5 in. thick with a residence time of 0.02 s or more. They can be much thicker and have longer residence times. At their rated flow, they have an initial efficiency of nearly 100% (full detention) for modest challenge concentrations of the contaminants they control well.

On the other hand, partial detention units operated at capacity have initial efficiencies of 80% or below. They typically have lower pressure drops than full detention units. They may allow 50% or more of the air approaching them to bypass the media bed and remain untreated, as do 50% fill honeycomb panels and some partial bypass unit adsorbers. Similarly, some adsorbent-bonded carbon panels operating at face velocities of approximately 300 fpm or more have very low residence times and are best classed as partial detention units even though all the air passes through the adsorber.

Equipment Configurations. The design engineer has a choice of either **unitary adsorbers** in built-up banks or **side access housings** holding individual media-filled trays, modules, or cells. Built-up banks require space upstream of the filter bank to install the adsorbers and for maintenance. Larger equipment benefits from the maintenance simplicity of built-up front access banks. Side access housings are an integral part of a duct run, and are usually selected if the system is small (about 3000 cfm or less) or if inline space is limited.

A **unitary adsorber** has either a permanent holding frame that is filled with multiple trays, modules, or panels or a single, disposable box-type unit or adsorbent-bonded media cell. Unitary adsorbers incorporate channels to retain trays, modules, or panels in either the vertical or horizontal position. Their size depends on what they are intended to hold.

Media trays (Figure 4A) vary in thickness from 0.625 in. to 2 in. and may be constructed of painted or stainless steel or plastic. Their height and breadth are sufficient to span the dimensions of the housing (Figure 4G) in which they will be placed. **Modules** (Figure 4B) are usually designed to hold media in a V-panel configuration. They may be refillable or completely disposable.

Panel adsorbers usually use a base of fibrous honeycomb material to hold the granular media. The efficiency and pressure drop of these panels can be controlled by the amount of media placed in each honeycomb opening; 50%, 75% and 100% fills are common.

Disposable, box-type units may use multiple media-filled panels arranged in a V configuration and sealed into frames with or without headers. Fiber-adsorbent composite media may be pleated and assembled into disposable box-type unitary filters, which may be of heavy or light-weight construction depending on the capacity of the filter.

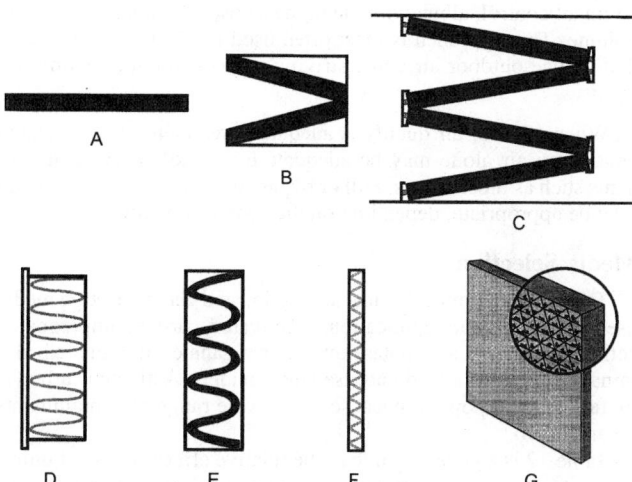

A. Granular media tray, refillable or disposable
B. Granular media module, refillable or disposable
C. Side access adsorber housing with granular media trays or modules
D. Composite media in flanged unitary adsorber, disposable
E. Granular media serpentine cell, refillable
F. Adsorbent-infused panel filter, disposable
G. Honeycomb panel adsorber, disposable

Fig. 4 Sectional and Schematic Views of Typical Adsorber and Chemisorber Configurations

Adsorbent-bonded media is often pleated into a cardboard or metal frame and provide both gaseous and particulate contamination control. These filters usually are only used for extremely light duty peak-shaving applications because the available adsorbent is relatively limited.

A partially-filled granular media holder can have a preferred orientation because the media may have freedom to move. For example, a vertical, 50%-filled honeycomb panel adsorber has about 50% open area and 50% carbon granules in the direction of flow through the panel. The same panel adsorber, oriented horizontally in downflow, becomes a packed bed having a media thickness of about half the panel depth because the media packs against the bottom retaining screen. In upflow, the absorbent is lifted by the airflow and forms a packed or percolating bed against the retaining screen at the top, causing sorbent abrasion. The pressure drop and adsorption performance of each of these orientations is different. Most manufacturers rate their products in only one orientation and do not investigate other applications.

Side access adsorber housings hold the same types of trays, modules, and panels used with unitary adsorbers. The housings have doors on one or both sides and often include channels for particulate prefilters and afterfilters. The adsorber units slide into place on rails or channels that require tight seals to prevent bypass leakage. The housings are often mounted adjacent to the air handler.

AIR CLEANING DESIGN

In an ideal situation, the designer for gaseous contaminant control would have the following information:

1. Exact chemical identity of the contaminants present in significant concentrations
2. Rates at which the contaminants are generated in the space
3. Rates at which the contaminants are brought into the space with outdoor air
4. Time-dependent performance of the proposed air cleaner for the contaminant mixture at the concentration and environmental conditions to be encountered
5. A clear goal concerning what level of air cleaning is needed

This information is usually difficult to obtain. The first three items can be obtained by sampling and analysis, but funding is usually not sufficient to obtain samples except in very simple contamination cases. Designers must often make do with a chemical family name (e.g., aldehydes) and a qualitative description of generation (e.g., from graphics arts department) or perceived concentration (e.g., at odorous levels).

Experimental measurements of air cleaner performance (Item 4) are in most cases not available. However, the performance of an air cleaner can be estimated using the equations presented in the section on Use of Source Data to Predict Indoor Concentrations when the exact chemical identity of a contaminant is known. The chemical and physical properties influencing a contaminant's collection by control devices can usually be obtained from handbooks and technical publications. Contaminant properties of special importance are relative molecular mass, normal boiling point (i.e., at standard pressure), heat of vaporization, polarity, chemical reactivity, and chemisorption velocity.

The performance of air cleaners with mixtures of chemically dissimilar compounds is very difficult to predict. Some gaseous contaminants, including ozone, radon, and sulfur trioxide, have unique properties that require design judgment and experience.

Finally, design goals must be considered. For a museum or archive, the proper design goal is total removal of the target contaminants and no subsequent desorption. For any chemical having or potentially having a health impact, the design goal is to reduce the concentration to below the level of health effects. Again, desorption back into the space must be minimized. For odor control, however,

100% removal may be unnecessary and desorption back into the space at a later time with a lower concentration may be an economical and acceptable mode of operation.

The first step in design is selecting an appropriate adsorption medium. Next, the location of the air cleaner in the HVAC system must be decided. Then the air cleaner must be sized such that sufficient media is used to achieve design efficiency and capacity goals and to estimate media replacement requirements. Finally, the commercial equipment that most economically meets the needs of the particular application can be selected. These steps are not completely independent.

Air Cleaner Location and Other HVAC Concerns

Outside Air Intakes. Proper location of the outside air intake is especially important for applications requiring gaseous contaminant filters because the contaminants load the filters and reduce their operating lifetime. Outside air should not be drawn from areas where point sources of gaseous contaminants are likely. This means outside air intakes should be distant from building exhaust discharge points, roads, loading docks, parking decks and spaces, and similar locations. See Chapter 43 for more information on air inlets.

Outside Air Volume. To further aid in reducing the amount of contaminants from outside air, at least on days of high ambient pollution levels, the quantity of outside air should be minimized.

Air Cleaner Locations. The three principal uses for gaseous contaminant control equipment in an HVAC system are:

Outside Air Treatment. Air cleaning equipment can be located at the outside air intake to treat outside air only. This treatment is used principally when indoor gaseous contaminants are adequately controlled by outdoor air ventilation, but the outdoor air needs to be cleaned to achieve satisfactory air quality.

Bypass or Partial Supply Air Treatment. Bypass can be achieved with a bypass duct and control damper or by installing an air cleaner that allows substantial bypass. Partial supply air treatment may be appropriate where a specific threshold contamination level is targeted, when outside and inside contamination rates are known, and the required level of reduction is small to moderate.

Full Supply Air Treatment. Full treatment achieves the best contaminant control, albeit with the highest cost and largest equipment volume. This approach is most often used in ventilation strategies that reduce outdoor air while striving to maintain good indoor air quality.

When outdoor air quality is adequate, treatment of recirculated ventilation air alone may be adequate to control indoor contaminants such as bioeffluents. Full or bypass treatment of the supply air may be appropriate, depending on the source strength.

Media Selection

Table 11 lists some physical adsorption and chemisorption media used for indoor air applications. Activated carbon, impregnated activated carbon, and potassium permanganate-impregnated alumina are the three adsorbents used most widely. Activated carbon is by far the most popular because of the wide range of contaminants it can adsorb.

Table 12 is a general guide to the relative effectiveness of unimpregnated activated carbon on a range of odors and compounds under typical average air purification duty conditions. The table is widely used in the industry, but it is only a guide. Adsorption capacity for a particular chemical or application may vary from these guidelines with changes in

- **Competitive adsorption.** Multiple contaminants confound performance estimates
- **Temperature.** Activity decreases with a temperature increase
- **Humidity.** Effect of humidity (generally for rh > 50%) depends on the contaminant. Carbon capacity for water-miscible solvents

Table 11 Adsorption and Chemisorption Media Used in HVAC Systems

Material	Impregnant	Typical Vapors or Gases Captured
Physical adsorbers		
Activated carbon	None	Organic vapors, ozone, acid gases
Activated alumina	None	Polar organic compounds[a]
Silica gel	None	Water, polar organic compounds[a]
Molecular sieves (zeolites)	None	Carbon dioxide, iodine
Chemisorbers		
Activated alumina	$KMnO_4$	Hydrogen sulfide, sulfur dioxide
Activated carbon	I_2, Ag, S	Mercury vapor
Activated carbon[b]	I_2, KI_3, amines	Radioactive iodine and organic iodine
Activated carbon	$NaHCO_3$	Nitrogen dioxide
$NaOH + Ca(OH)_2$	None	Acid gases
Activated carbon	KI, I_2	Mercury vapor

[a]Polar organics = alcohols, phenols, aliphatic and aromatic amines, etc.
[b]Mechanism may be isotropic exchange as well as chemisorption.

increases, while capacity for immiscible or partially-miscible solvents decreases

- **Concentration.** Increased contaminant concentration improves activity.

Some chemically reactive gases such as ammonia or formaldehyde are not adsorbed well by standard activated carbon, but specially impregnated carbon can be used successfully. These gases are noted with an asterisk (*) in Table 12. The capacity index values in the table were developed from data obtained at concentrations well above those encountered indoors. The ratings are noted below and remain useful for comparative purposes. With that qualification, the ratings have the following meaning:

4 Activated carbon has a high capacity for these materials, adsorbing 20 to 40% of its mass of the contaminant. (They average 30% at high concentrations.) At indoor concentrations, the capacities are about 8 to 15%. More than 70% of the materials listed fall into this category.

3 Activated carbon has a satisfactory capacity for these materials, adsorbing 10 to 25% of its mass of the contaminant (with an average 17% at high concentrations, lower at indoor concentrations).

2 Activated carbon has a relatively low adsorption capacity for these materials, and its use is borderline.

1 Activated carbon has a very low or negligible capacity for these materials and is generally not recommended.

In general, gaseous contaminants that have the same boiling point as water or greater can be removed by the physical adsorption process using standard activated carbon. Those with a lower boiling point usually require chemisorption for removal. A graphical representation of the media and equipment selection process is shown in Figure 5.

Sizing Gaseous Contaminant Control Equipment

Tables 11 and 12 provide guidance in the selection of sorbent media. Both manufacturers' guidance and absorbent performance Equations (5) through (9) are used to size equipment. A manufacturer's guidance is often followed because the adsorbent performance and contaminant concentration data required to use the equations are not generally available.

When the absorber size is calculated the following approach is applied:

1. Choose an absorbent suited to the contaminant
2. Pick an appropriate efficiency for the adsorber—complete removal or partial bypass—depending on the contaminant.

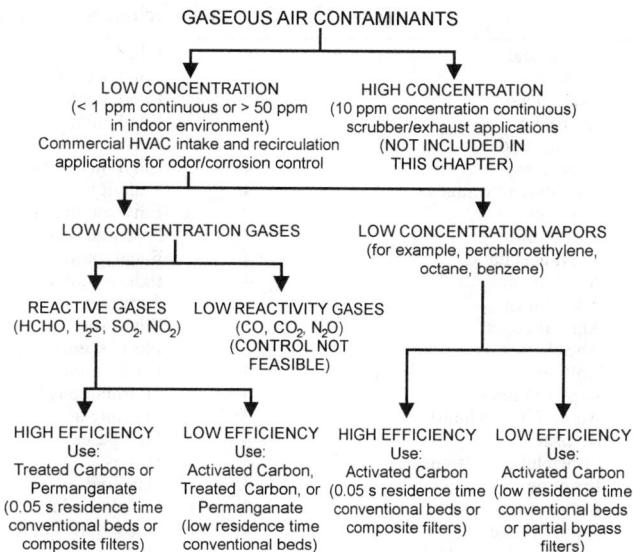

Fig. 5 Media and Equipment Selection Schematic
(Adapted and reprinted with permission. ©1992 Extraction Systems, Inc.)

3. Choose a desired operating adsorber end point of 10%, 50%, or other breakthrough, depending on the application and allowable steady state concentration. A building ventilation performance model, with the adsorber appropriately positioned in the model, allows calculation of the expected indoor concentration at various breakthroughs and efficiencies.
4. Obtain a measurement or estimate of the breakthrough time at adsorber use conditions as developed in step (3).
5. Determine the changeout rate for the adsorbent as set by the breakthrough time.
6. Match the computed design requirements to available air cleaning equipment and specify.

Special Cases

Ozone will reach an equilibrium concentration in a ventilated space without a filtration device. It does so partly because ozone molecules react to form oxygen, but also because it reacts with people, plants, and materials in the space. This oxidation is harmful to all three, and therefore natural ozone decay is not a satisfactory way to control ozone except at low concentrations (< 0.2 mg/m³ or < 0.1 ppmv). Fortunately, activated carbon adsorbs ozone readily, both reacting with it and catalyzing its conversion to oxygen.

Radon is a radioactive gas that decays by alpha-particle emission, eventually yielding individual atoms of polonium, bismuth, and lead. These atoms form extremely fine aerosol particles, called radon daughters or radon progeny, which are also radioactive; they are especially toxic in that they lodge deep in the human lung, where they emit cancer-producing alpha particles. Radon progeny, both attached to larger aerosol particles and unattached, can be captured by particulate air filters. Radon gas itself may be removed with activated carbon, but in HVAC systems this method costs too much for the benefit derived. Control of radon emission at the source and ventilation are the preferred methods of radon control.

Another gaseous contaminant that often appears in particulate form is **sulfur trioxide** (SO_3), which should not be confused with sulfur dioxide (SO_2). It reacts rapidly with water vapor at ambient temperature to form a fine mist of sulfuric acid. If this mist collects on a particulate filter and no means is provided to remove it, the acid will vaporize and reenter the protected space.

Table 12 Activated Carbon Relative Capacity Index for Odors—A General Guide

Acetic acid	4	Ether	3	Mildew odor	3
Acetone	3	Ethyl aceate	4	Mixed odors	4
*Acetylene	1	Ethyl alcohol	4	Naphtha (coal tar or petroleum)	4
*Acrolein	3	Ethyl chloride	3	Naphthalene	4
Acrylic acid	4	Ethyl ether	3	Nicotine	4
Adhesives	4	Ethyl mercaptan	3	*Nitric acid	3
Alcoholic beverages	4	*Ethylene	1	Nitrobenzenes	4
*Amines	2	Ethylene dichloride	4	*Nitrogen dioxide	1
*Ammonia	2	Essential oils	4	Nitroglycerine	4
Amyl acetate	4	Eucalyptole	4	Octane	4
Amyl alcohol	4	Exhaust fumes	3	Odorants	4
Amyl ether	4	Fertilizer	4	Organic chemicals	4
Animal odors	3	Film processing odors	3	Ozone	4
Anesthetics	3	Floral scents	4	Paint and redecorating odors	4
Aniline	4	Food aromas	4	Palmitic acid	4
Asphalt fumes	4	*Formaldehyde	2	Paper deteriorations	4
Automobile exhaust	3	*Formic acid	3	Paradichlorobenzene	4
Benzene	4	Fuel gases	2	Paste and glue	4
*Bleaching solutions	3	Fumes	3	Pentane	3
Bromine	4	Gasoline	4	Perfumes & cosmetics	4
Butane	2	Hospital odors	4	Pet odors	4
Butyl acetate	4	Household smells	4	Phenol	4
Butyl alcohol	4	Hydrogen	1	Phosgene	3
Butyl chloride	4	*Hydrogen bromide	1	Pitch	4
Butyl ether	4	*Hydrogen chloride	1	Plastic	4
Butyric acid	4	*Hydrogen cyanide	1	Propane	2
Carbolic acid	4	*Hydrogen fluoride	1	*Propylene	2
Carbon disulfide	2	*Hydrogen iodide	1	Radiation products	2
*Carbon dioxide	1	*Hydrogen selenide	1	Rancid oils	4
Carbon monoxide	1	*Hydrogen sulfide	1	Resins	4
Carbon tetrachloride	4	Industrial wastes	3	Reodorants	4
Cheese odors	4	Ink odors	3	Ripening fruits	4
*Chlorine	3	Iodine	4	Rubber	4
Chlorobenzene	4	Irritants	4	Sewer odors	4
Chlorobutadiene	4	Isopropyl acetate	4	Smog	4
Chloroform	4	Isopropyl alcohol	4	Smoke	4
Chloropicrin	4	Isopropyl ether	4	Solvents	3
Cigarette smoke odor	4	Kerosene	4	Stoddard solvent	4
Citrus/fruit odors	4	Kitchen odors	4	Styrene monomer	4
Cleaning compounds	4	Lactic acid	4	*Sulfur dioxide	2
Coal smoke odor	3	Liquid fuels	4	*Sulfur trioxide	2
Combustion odor	3	Liquor odors	4	Sulfuric acid	4
Cooking odors	4	Lubricating oils/greases	4	Tar	4
*Corrosive gases	3	Lysol	4	Tetrachloroethane	4
Creosote	4	Masking agents	4	Tetrachloroethylene	4
Decaying substances	4	Medicinal odors	4	Toluene	4
Deodorants	4	Menthol	4	Toluidine	4
Detergents	4	Mercaptans	2	Trichloroethylene	4
Dichloroethane	4	Methane	1	Trichloroethane	4
Dichloroethylene	4	Methyl acetate	3	Turpentine	4
Diesel fumes	4	Methyl acrylate	4	Urea	4
Diethyl ketone	4	Methyl alcohol	3	Uric acid	4
Dimethylsulfide	4	Methyl bromide	3	Valeric acid	4
Disinfectants	4	Methyl butyl ketone	4	Varnish fumes	4
Embalming odors	4	Methyl chloride	3	Vinegar	4
Epoxy	4	Methyl ethyl ketone	4	Vinyl chloride	2
Ethane	1	Methylene chloride	4	Xylene	4

Adapted and reprinted with permission. ©1995 Barneby and Sutcliffe Corporation.
*Requires impregnated carbon.

ENERGY CONCERNS

The pressure drop across the contaminant filter has a direct impact on energy use. Data on the resistance of the filter as a function of airflow and on the resistance of the heating/cooling coils must be provided by the manufacturer. Currently no standard test of the pressure drop across a full-scale gaseous air cleaner is specified, but users could require that the initial pressure drop measurement from the particulate test (ASHRAE *Standard* 52.2) be conducted and reported. In addition to the gaseous contaminant filter itself, the pressure drop through the housing, any added duct elements, and any particulate filters required up and/or downstream of the gaseous contaminant filter must be included in the energy analysis.

The choice between using outside air only and outside air plus filtered recirculated air is complex, but it may be made based on technical or maintenance factors, convenience, economics, or a combination of these. An energy consumption calculation is part of that decision process. Replacing outdoor air with filtered indoor air reduces the amount of air that must be conditioned at an added expense in recirculation pressure drop. Outdoor air or filtered recirculated air may be used in any ratio, provided the air quality level is maintained. Janssen (1989) discusses the logic of these requirements.

Where building habitability can be maintained with ventilation alone, an economizer cycle is feasible under appropriate outdoor conditions. However, the economizer mode may not be feasible at high humidities, because high humidity degrades the performance of carbon adsorbers.

Table 13 Items Included in Economic Comparisons Between Competing Gaseous Contaminant Control Systems

Capital Costs	Operating Costs
Added filtration equipment	Replacement or reactivation of gaseous
Fan	contaminant filter media
Motor	Disposal of spent gaseous contaminant
Controls	filter media
Plenum	Added electric power
Spare media holding units	Maintenance labor
Floor space	

ECONOMIC CONSIDERATIONS

Capital and operating costs for each competing system should be identified. Chapter 35, Owning and Operating Costs, provides general information on performing an economic analysis. Table 13 is a checklist of those filtration items to be considered in an analysis. An important question for an in-line air cleaner is whether the fan can maintain adequate flow with the air cleaner in place. If a larger blower is required, space must be available. Modifying unitary equipment that was not designed to handle the additional pressure drop through air cleaning equipment can be expensive. With built-up designs, the added initial cost of providing air cleaners and their pressure drop can be much less because the increases may be small fraction of the total.

The life of the adsorbent media is very important to any economic analysis. The economic benefits of regenerating spent carbon should be evaluated in light of the cost and generally reduced activity levels of regenerated material. Regeneration of impregnated carbon or any carbon containing hazardous contaminants is never permitted. Spent alumina- or zeolite-based adsorbents cannot be regenerated.

SAFETY

Gaseous contaminant removal equipment generally has a low hazard potential. The contaminant concentrations are low, the temperature moderate, and the equipment is normally not closed in. Alumina- or zeolite-based media do not support combustion, but carbon filter banks have been known to catch fire, usually from an external source such as a welder's torch. Check local codes and fire authorities for their regulations concerning carbon. One such authority requires automatic sprinklers in the duct upstream and downstream of carbon filter banks. As a minimum, a smoke detector should be installed downstream of the filter bank to shut down the fan and sound an alarm in case of fire.

Access for safe maintenance and change out of adsorbent beds must be provided. Adsorbers are much heavier than particulate filters. Suitable lifting equipment must be available during installation and removal to prevent injury.

If adsorbent trays are going to be refilled on-site, safety equipment must be provided to deal with the dust that is generated during the operation. Hooding, dust masks, and gloves are all required to refill adsorber trays from bulk containers.

INSTALLATION, START-UP, AND COMMISSIONING

This section provides general guidance on the installation of gaseous contaminant removal equipment. Most manufacturers can also provide complete details and drawings for design.

Particulate Filters. A minimum 25 to 30% efficiency particulate filter (as per ASHRAE *Standard* 52.1) should be installed ahead of the adsorber bank. A higher efficiency is desirable. Adsorbers and chemisorbers cannot function properly if their surfaces are covered and their pores clogged with dirt. If the air is extremely dirty (from diesel exhaust for example), the filter should be one with a much higher efficiency. One manufacturer requires a 90 to 95% effi-

ciency filter for such applications. Weschler et al. (1994) report that carbon service life for ozone control was lengthened by using improved prefiltration.

Afterfilters are often used in those critical applications where dust from the media at start-up is likely, or where vibration of the adsorber bank may cause the granular media to shed particles. These filters are frequently 25 to 30% efficiency filters, but higher efficiencies may be needed in some applications.

Equipment Weight. Adsorption equipment is much heavier than particulate filtration equipment, so supporting structures and frames must be designed accordingly. A typical 24 in. by 24 in. adsorber consisting of a permanent holding frame and sorbent-loaded trays has an installed weight of approximately 200 lb.

Eliminate Bypass. Adsorbers and ducts in the outside air supply and in the exhaust from hoods must be tightly sealed to prevent bypass of contaminants. Bypass leakage is not critical in most recirculating indoor air systems, but it is good practice to caulk all seams and between individual holding frames. Granular media retainers, such as trays or modules, must be loaded with media per manufacturers' recommendations to eliminate the possibility of bypass through the media bed.

When to Install Media. The decision as to when to install adsorbers in their holding frames depends on building circumstances. If they are installed at the same time as their holding frames and if the HVAC is turned on during the latter phases of construction, the adsorbers will adsorb paint and solvent vapors and other contaminants before the building is ready for beneficial occupancy. In some situations, the adsorption of vapors and gases in the ventilation system prior to official start-up may be what is desired or needed. However, adsorber life will be reduced correspondingly. If adsorber frames are not loaded until the building is ready for occupancy, the unadsorbed contaminants may seriously reduce the initial indoor air quality of the building. Thus the shortened life is an acceptable trade-off for the quality of air at the time of occupancy (NAFA 1997). If the media is not in place during fan testing, the test and balance contractor must be instructed to place blank-offs or restrictions in the frames to simulate adsorber pressure drop. The HVAC designer's job specifications must clearly state when media is to be installed.

Pressure Gages. Provided prefiltration is adequate, adsorber pressure drop will not increase during normal operation. A pressure drop measuring device (gage or manometer) is therefore not required as it is for a particulate filter bank. However, a gage may be useful to detect fouling or unintentional bypass. If the prefilters or afterfilters are installed immediately adjacent to the adsorbers, it may be more feasible to install the gage across the entire assembly.

Provision for Testing. At any time following installation of new media, determining the remaining adsorbent capacity or operating life may be required. (See the section on When to Change Media under Operation and Maintenance.) The installation should provide access ports to the fully mixed air stream both upstream and downstream of the air cleaner. If media samples will be removed to determine remaining life, access must be provided to obtain those samples. No standard method for field evaluation of media life currently exists.

Start-Up and Commissioning

Special procedures are not required during start-up of an air handler incorporating adsorbers. The test and balance contractor will normally be required to take and record a measurement of the resistance of all installed filter banks, including adsorbers, for comparison with design conditions.

The commissioning authority may require an activity test on a sample of media selected at random. This test will determine if the new media has suffered prior exposure that has reduced its life or if it meets specifications. An in-situ air sampling test may also be required on the adsorber filter bank; however, no standard method for this test exists.

OPERATION AND MAINTENANCE

Bypass units and filters with adsorbent-infused media require frequent changing to maintain even low efficiency, but frequent maintenance is not required for complete removal units. Complete removal media adsorbers usually have a replaceable cell that cannot be regenerated or reactivated. This section covers maintenance of complete removal equipment with refillable trays or modules only.

When to Change Media

The changeout point of an adsorbent is difficult to determine. Sometimes the media is changed when breakthrough occurs and occupants complain; but if the application is sensitive, tests for estimated residual activity may be made at periodic intervals. A sample of the media in use is pulled from the adsorber bank or from a pilot cell placed in front of the bank. The sample is sent to the manufacturer or an independent test laboratory for analysis, and the changeout time is estimated knowing the time in service and the life remaining in the sample. In extremely critical installations, "coupons" plated with precious metals are placed in the space being protected by the adsorbent. After some time, usually a month, the coupons are sent to an analytical lab for measurement of corrosion thickness. These measurements indicate the effectiveness of the gaseous contaminant control as well as provide an indication of the system life. Standardized methodology for these tests is not available.

Replacement and Reactivation

The process of replacing media in permanent adsorber trays or modules is not the same as reactivation (regeneration), which is the restoration of spent activated carbon media to its original efficiency (or close to it). Spent carbon is regenerated in special high temperature kilns in the absence of oxygen to drive off the contaminants it has acquired. Chemisorber modules can be replaced (media changed) but chemisorber media, including impregnated carbon, cannot be regenerated.

Building operating personnel may choose to dump and refill trays and modules at the site after replacing those removed with a spare set already loaded with fresh media. They may also choose to dump the trays locally and send the empty trays to a filter service company for refilling; or they may simply exchange their spent trays for fresh ones. Disposing of spent sorbent by dumping must be limited to building air quality applications where no identifiable hazardous chemicals have been collected.

ENVIRONMENTAL INFLUENCES ON AIR CLEANERS

Environmental conditions, particularly temperature and humidity, affect the performance of most gaseous contaminant control equipment. Physical sorbents such as activated carbon are particularly susceptible. The user should confirm performance for any control device at the expected normal environmental conditions as well as at extremes that might be encountered during equipment outages. The following information is an overview.

High relative humidity in the treated airstream lowers the gaseous contaminant removal efficiency of physical adsorbers such as carbon due to competition for sorption sites from the much more numerous water molecules. In many cases performance is relatively stable up to 40 to 50% rh, but some compounds can degrade at higher humidities. The chemical nature of the contaminant(s) and the concentration both affect the degradation of performance as a function of relative humidity. On the other hand, very low relative humidities may make some chemisorption activity impossible. Therefore, the performance of these media must be evaluated over the expected range of operation, and the relative humidity and

temperature of the gaseous contaminant control should be held within design limits.

The impact of relative humidity swings can be better understood by considering a hypothetical adsorber that has a saturation capacity for a contaminant and inlet concentration of 10% at 50% rh and 5% at 70% rh. Over an extended period at its normal operating condition of 50% rh, the sorbent might reach a loading of 2%. At this point a humidity swing to 70% rh would not cause a problem, and the sorbent could load on up to 5% capacity. Should the humidity then swing back to 50%, the sorbent could continue to adsorb up to 10% by weight of the contaminant. However, if the sorbent was loaded to 8% by weight at 50% rh and the humidity rose to 70% rh, the carbon would be above its equilibrium capacity and desorption would occur until equilibrium was reached.

In a similar way, temperature swings and contaminant concentration swings can affect physical sorbent performance. Increasing temperature reduces capacity, while increasing concentration increases capacity. Additionally, changes in the identity of the contaminant can affect overall performance as strongly-sorbed contaminants displace weakly held contaminants.

All adsorption media have a modest ability to capture dust particles and lint, which eventually plug the openings in and between the media granules and cause a rapid rise in the pressure drop across the media or a decrease in airflow. All granular gaseous adsorption beds need to be protected against particulate buildup by installing particulate filters upstream. A prefilter with a minimum ASHRAE *Standard* 52.1 dust spot efficiency of 25 to 30% is recommended.

Vibration breaks up the granules to some degree, depending on the granule hardness. ASTM *Standard* D 3802 describes a test procedure for measuring the resistance of activated carbon to abrasion. Critical systems using activated carbon require hardness above 92%, as described by *Standard* D 3802.

Adsorption and chemisorption media sometimes accelerate the corrosion of metals that they touch. For this reason, media holding cells, trays, and modules should not be constructed of uncoated aluminum or steel. Painted steel or ABS plastic are common and exhibit good material service lives in many applications. Coated or stainless steel components may be required in more aggressive environments.

TESTING OF MEDIA, EQUIPMENT, AND SYSTEMS

Testing may be conducted in the laboratory with small-scale media beds or small pieces of treated fabric or composite material; on full-scale air cleaners in a laboratory test rig capable of generating the test atmosphere; or in the field. Laboratory tests with specific challenge gases are generally intended to evaluate media for developmental, acceptance, or comparative purposes. Full-scale tests using specific challenge contaminants are required to evaluate a complete adsorber as constructed and sold, and ultimately are needed to validate performance claims. Field tests under actual job conditions are used to ensure that the air cleaners were properly installed and to evaluate remaining media life.

Laboratory Testing

Testing of small granular media samples in a laboratory has been done for many years, and most manufacturers have developed their own methods. ASTM *Standard* D 5288 describes a test method. But this method is not entirely applicable to HVAC work because indoor air tends to have a wide range of contaminants at concentrations several orders of magnitude lower than used in testing.

Field Testing

Upstream and downstream measurements are evaluated by converting them to efficiency or fractional penetration and comparing them to measurements made at installation. Efficiency is most directly interpreted if there is a single contaminant and the challenge concentration is relatively constant. For multiple contaminants at multiple concentrations, judgment and experience are needed to interpret downstream measurements. Media samples can be evaluated by the manufacturer for remaining life in fractions of the original capacity.

REFERENCES

ACGIH. 1995. *Industrial ventilation: A manual of recommended practice,* 22nd ed. American Conference of Governmental Industrial Hygienists, Cincinnati, OH.

ACGIH. 1997. *Threshold limit values and biological exposure indices for 1997-98.*

AGO. 1966. Air pollution and organ leathers: Panel discussion. *AGO Quarterly* 11(2):62-73.

AIHA. 1989. *Odor thresholds for chemicals with established occupational health standards.* American Industrial Hygiene Association, Akron, OH.

Anthony, C.P. and G.A. Thibodeau. 1980. *Textbook of anatomy and physiology.* C.V. Mosby, St. Louis, MO.

ASHRAE. 1989. Ventilation for acceptable indoor air quality. ANSI/ASHRAE *Standard* 62-1989.

ASTM. 1980. Building air change rate and infiltration measurements. STP 719. American Society for Testing and Materials, W. Conshohocken, PA.

ASTM. 1990. Standard test method for ball-pan hardness of activated carbon. *Standard* D 3802-79.

ASTM. 1992. Standard test method for the determination of the butane working capacity of activated carbon. *Standard* D 5288-92.

ASTM. 1993. Standard test method for determining air change in a single zone by tracer dilution. *Standard* E 741-93.

ATC. 1990. Technical assistance document for sampling and analysis of toxic organic compounds in ambient air. EPA/600/8-90-005. Environmental Protection Agency, Research Triangle Park, NC.

Balieu, E., T.R. Christiansen, and L. Spindler. 1977. Efficiency of b-filters against hydrogen cyanide. *Staub-Reinhalt. Luft* 37(10):387-90.

Berglund, B., U. Berglund, and T. Lindvall. 1986. Assessment of discomfort and irritation from the indoor air. In *IAQ '86: Managing Indoor Air for Health and Energy Conservation,* 138-49. ASHRAE, Atlanta.

Black, M.S. and C.W. Bayer. 1986. Formaldehyde and other VOC exposures from consumer products. In *IAQ '86.- Managing Indoor Air for Health and Energy Conservation.* ASHRAE, Atlanta.

Braman, R.S. and T.J. Shelley. 1980. Gaseous and particulate ammonia and nitric acid concentrations: Columbus, Ohio area Summer 1980. PB 81-125007. National Technical Information Service, Springfield, VA.

Braun, R.C. and M.J.G. Wilson. 1970. The removal of atmospheric sulphur by building stones. *Atmospheric Environment* 4:371-78.

Brugnone, R., L. Perbellini, R.B. Faccini, G. Pasini, G. Maranelli, L. Romeo, M. Gobbi, and A. Zedde. 1989. Breath and blood levels of benzene, toluene, cumene and styrene in non-occupational exposure. *International Archives of Environmental Health* 61:303-11.

Budavari, S. ed. 1996. *The Merck Index,* 12th ed. Merck and Company, White Station, NJ.

Casserly, D.M. and K.K. O'Hara. 1987. Ambient exposures to benzene and toluene in southwest Louisiana. *Paper* 87-98.1. Air and Waste Management Association, Pittsburgh.

Chan, C.C., L. Vanier, J.W. Martin, and D.T. Williams. 1990. Determination of organic contaminants in residential indoor air using an adsorption-thermal desorption technique. *Journal of Air and Waste Management Association* 40(1):62-67.

Cain, W.S., L.C. See, and T. Tosun. 1986. Irritation and odor from formaldehyde chamber studies, 1986. In *IAQ '86: Managing Indoor Air for Health and Energy Conservation.* ASHRAE, Atlanta.

Chiarenzelli, R.V. and E.L. Joba. 1966. The effects of air pollution on electrical contact materials: A field study. *Journal of the Air Pollution Control Association* 16(3):123-27.

Code of Federal Regulations. 10 CFR 50. U.S. Government Printing Office, Washington, DC. Revised annually.

Code of Federal Regulations. 29 CFR 1900. U.S. Government Printing Office, Washington, DC. Revised annually.

Cohen, S.I., N.M. Perkins, H.K. Ury, and J.R. Goldsmith. 1971. Carbon monoxide uptake in cigarette smoking. *Archives of Environmental Health.* 22(1):55-60.

Cohen, M.A., P.B. Ryan, Y. Yanagisawa, J.D. Spengler, H. Ozkaynak. and P.S. Epstein. 1989. Indoor/outdoor measurements of volatile organic compounds in the Kanawha Valley of West Virginia. *Journal of the Air Pollution Control Association* 39(8).

Cole, J.T 1983. Constituent source emission rate characterization of three gas-fired domestic ranges. APCA *Paper* 83-64.3. Air and Waste Management Association, Pittsburgh.

Conkle, J.P., B.J. Camp, and B.E. Welch. 1975. Trace composition of human respiratory gas. *Archives of Environmental Health* 30(6):290-95.

Coy, C.A. 1987. Regulation and control of air contaminants during hazardous waste site remediation. APCA *Paper* 87-18.1. Air and Waste Management Association, Pittsburgh.

DeFrees, J.A. and R.F. Amberger. 1987. Natural infiltration analysis of a residential high-rise building. In *IAQ '87.- Practical Control of Indoor Air Problems,* 195-210. ASHRAE, Atlanta.

Dole, M. and I.M. Klotz. 1946. Sorption of chloropicrin and phosgene on charcoal from a flowing gas stream. *Industrial Engineering Chemistry* 38(12):1289-97.

Elliott, L.P. and DR. Rowe. 1975. Air quality during public gatherings. *Journal of the Air Pollution Control Association* 25(6):635-36.

EPA. 1979. Toxic substances control act chemical substance inventory, Volumes I-IV.

EPA. 1990. National air quality and emission trends report, 1988. Environmental Protection Agency, Research Triangle Park, NC.

EPA. 1991. Building air quality—A guide for building owners and facility managers. U.S. Environmental Protection Agency. ISBN 0-16-035919-8. December.

Fazzalari, F.A., ed. 1978. Compilation of odor and taste threshold values data. DS 48A. American Society for Testing and Materials, West Conshohocken, PA.

Freedman, R.W., B.I. Ferber, and A.M. Hartstein. 1973. Service lives of respirator cartridges versus several classes of organic vapors. *American Industrial Hygiene Association Journal* (2):55-60.

Fung, K. and B. Wright. 1990. Measurement of formaldehyde and acemldehyde using 2,4-dinitrophenylhydrazine-impregnated cartridges. *Aerosol Science and Technology* 12(1):44-48.

Girman, J.R., A.P. Hodgson. A.F. Newton, and A.R Winks. 1984. Emissions of volatile organic compounds from adhesives for indoor application. *Report* LBL 1 7594. Lawrence Berkeley Laboratory, Berkeley, CA.

Gorban, C.M., I.I. Kondratyeva, and L.Z. Poddubnaya. 1964. Gaseous activity products excreted by man in an airtight chamber. In *Problems of Space Biology.* JPRS/NASA. National Technical Information Service, Springfield, VA.

Graminski, E.L., E.J. Parks, and E.E. Toth. 1978. The effects of temperature and moisture on the accelerated aging of paper. NBSIR 78-1443. National Institute of Standards and Technology, Gaithersburg, MD.

Grosjean, D., P.M. Whitmore, C.P. de Moor, G.R. Cass, and J.R. Druzik. 1987. Fading of alizarinad-related artist's pigments by atmospheric ozone. *Environmental Science and Technology* 21:635-43.

Gully, A.J., R.M. Bethea, R.R. Graham, and M.C. Meador. 1969. Removal of acid gases and oxides of nitrogen from spacecraft cabin atmospheres. NASA-CR-1388. National Aeronautics and Space Administration. National Technical Information Service, Springfield, VA.

Hakov, R., J. Jenks, and C. Ruggeri. 1987. Volatile organic compounds in the air near a regional sewage treatment plant in New Jersey. *Paper* 87-95.1. Air and Waste Management Association, Pittsburgh.

Hartwell, T.D., J.H. Crowder, L.S. Sheldon, and E.D. Pellizzari. 1985. Levels of volatile organics in indoor air. *Paper* 85-30B.3. Air and Waste Management Association, Pittsburgh.

Haynie, F.H. 1978. Theoretical air pollution and climate effects on materials confirmed by zinc corrosion data. In *Durability of building materials and components* (ASTM STP 691). American Society for Testing and Materials, West Conshohocken, PA.

Hollowell, C.D., R.A. Young, J.V. Berk, and S.R. Brown. 1982. Energy-conserving retrofits and indoor air quality in residential housing. *ASHRAE Transactions* 88(1):875-93.

Hunt, R.D. and D.T Williams. 1977. Spectrometric measurement of ammonia in normal human breath. *American Laboratory* (June):10-23.

Jaffe, L.S. 1967. The effects of photochemical oxidants on materials. *Journal of the Air Pollution Control Association* 17(6):375-78.

Janssen, J.H. 1989. Ventilation for acceptable indoor air quality. *ASHRAE Journal* 31(10):40-45.

Knoeppel, H. and H. Schauenburg. 1989. Screening of household products for the emission of volatile organic compounds. *Environment International* 15:413-18.

Lamm, S.H. 1986. Irritancy levels and formaldehyde exposures in U.S. mobile homes. In *Indoor Air Quality in Cold Climates,* 137-47. Air and Waste Management Association, Pittsburgh.

Leaderer, B.P., R.T. Zagranski, M. Berwick, and J.A.J. Stolwijk. 1987. Predicting NO_2 levels in residences based upon sources and source uses: A multi-variate model. *Atmospheric Environment* 21(2):361-68.

Lee, H.K., T.A. McKenna, L.N. Renton, and J. Kirkbride. 1986. Impact of a new smoking policy on office air quality. In *Indoor Air Quality in Cold Climates,* 307-22. Air and Waste Management Association, Pittsburgh.

Lewis, R.J. 1994. *Dangerous properties of industrial materials: 1993 update,* 8th ed. Van Nostrand Reinhold, New York.

Lide, D.R. 1997 *Handbook of chemistry and physics*, 78th Ed. CRC Press, Boca Raton, FL.

Lodge, J.E., ed. 1988. *Methods of air sampling and analysis,* 3rd ed. Lewis Publishers, Chelsea, MI.

Lonnemann, W.A., S.L. Kopczynski, P.E. Darley, and F.D. Sutterfield. 1974. Hydrocarbon composition of urban air pollution. *Environmental Science and Technology* 8(3):229-35.

Lull, W.P. 1995. Conservation environment guidelines for libraries and archives. Canadian Council of Archives.

Martin, P. et al. 1997. Environmental tobacco smoke (ETS): A market cigarette study. *Environment International* 23:(1) 75-89.

Mathey, R.G., T.K. Faison, and S. Silberstein. 1983. Air quality criteria for storage of paper-based archival records. NBSIR 83-2795. National Institute of Standards and Technology, Gaithersburg, MD.

Matthews, T.G., T.J. Reed, B.J. Tromberg, C.R. Daffron, and A.R. Hawthorne. 1983. Formaldehyde emissions from combustion sources and solid formaldehyde resin-containing products. In *Proceedings of Engineering Foundation Conference on Management of Atmosphere in Tightly Enclosed Spaces.* ASHRAE, Atlanta.

Matthews, T.G., W.G. Dreibelbis, C.V. Thompson, and A.R. Hawthorne. 1985. Preliminary evaluation of Formaldehyde mitigation studies in unoccupied research homes. In *Indoor Air Quality in Cold Climates,* 137-47. Air and Waste Management Association, Pittsburgh.

McGrath, T.R. and D.B. Stele. 1987. Characterization of phenolic odors in a residential neighborhood. *Paper* 87-75A.5. Air and Waste Management Association, Pittsburgh.

Meckler, M. and J.E. Janssen. 1988. Use of air cleaners to reduce outdoor air requirements. In *IAQ '88: Engineering Solutions to Indoor Air Problems,* 130-47. ASHRAE, Atlanta.

Miller, G.C. and P.C. Reist. 1977. Respirator cartridge service lives for exposure to vinyl chloride. *American Industrial Hygiene Association Journal* 38:498-502.

Molhave, L., I. Anderson, G.R. Lundquist, and O. Nielson. 1982. Gas emission from building materials. *Report* 137. Danish Building Research Institute, Copenhagen.

Moore, J.E. and E. Houtala. 1983. Odor as an aid to chemical safety: Odor thresholds compared with threshold limit values and volatility for 214 industrial chemicals in air and water dilution. *Journal of Applied Toxicology*, 3:272.

Moschandreas, D.J. and S.M. Relwani. 1989. Field measurement of NO_2 gas-top burner emission rates. *Environment International* 15:499-92.

NAFA. 1997. Installation, operation and maintenance of air filtration systems. National Air Filtration Association, Washington, DC.

Nagda, N.L. and H.E. Rector. 1983. Guidelines for monitoring indoor-air quality. EPA 600/4-83-046. Environmental Protection Agency, Research Triangle Park, NC.

Nazaroff, W.W. and G.R. Cass. 1986. Mathematical modeling of chemically reactive pollutants in indoor air. *Environmental Science and Technology* 20:924-34.

Nefedov, I.G., V.P. Savina, and N.L. Sokolov. 1972. Expired air as a source of spacecraft carbon monoxide. Paper, 23rd International Aeronautical Congress, International Astronautical Federation, Paris.

Nelms, L.H., M.A. Mason, and B.A. Tichenor. 1986. The effects of ventilation rates and product loading on organic emission rates from particleboard. In *IAQ '86: Managing Indoor Air for Health and Energy Conservation,* 469-85. ASHRAE, Atlanta.

Nelson, E.D.P., D. Shikiya, and C.S. Liu. 1987. Multiple air toxics exposure and risk assessment in the south coast air basin. *Paper* 87-97.4. Air and Waste Management Association, Pittsburgh, PA.

Nelson, G.O. and A.N. Correia. 1976. Respirator cartridge efficiency studies: VIII. Summary and Conclusions. *American Industrial Hygiene Association Journal* 37:514-25.

NIOSH. 1977. *NIOSH manual of sampling data sheets.* U.S. Department of Health and Human Services, National Institute for Occupational Safety and Health, Washington, DC.

NIOSH. 1984. *NIOSH manual of analytical methods,* 3rd ed. 2 vols. U.S. Department of Health and Human Services, National Institute for Occupational Safety and Health, Cincinnati.

NIOSH. 1990. *NIOSH/OSHA pocket guide to chemical hazards.* DHHS (NIOSH) *Publication* 90-117. U.S. Department of Labor, OSHA, Washington, DC. Users should read Cohen (1993) before using this reference.

NIOSH. Annual registry of toxic effects of chemical substances. U.S. Department of Health and Human Services, National Institute for Occupational Safety and Health, Washington, DC.

NIOSH. (intermittent). Criteria for recommended standard for occupational exposure to (compound). U.S. Department of Health and Human Services, National Institute for Occupational Safety and Health, Washington, DC.

NTIS. 1982. Air pollution effects on materials. PB 82-809427. National Technical Information Service, Springfield, VA.

NTIS. 1984. Air pollution effects on materials. PB 84-800267.

Perry, Chilton, and Kirkpatrick, 1997. *Perry's chemical engineers' handbook,* 7th ed. McGraw Hill, New York.

Rask, D. 1988. Indoor air quality and the bottom line. *Heating, Piping and Air Conditioning* 60(10).

Revoir, W.H. and J.A. Jones. 1972. Superior adsorbents for removal of mercury vapor from air. AIHA conference paper. American Industrial Hygiene Association, Akron, OH.

Sandalls, E.J. and S.A. Penkett. 1977. Measurements of carbonyl sulfide and carbon disulphide in the atmosphere. *Atmospheric Environment* 11:197-99.

Shah, J.J. and H.B. Singh. 1988. Distribution of volatile organic chemicals in outdoor and indoor air. *Environmental Science and Technology* 22(12): 1391-88.

Singh, H.B., L.J. Salas, A. Smith, R. Stiles, and H. Shigeishi. 1981. Atmospheric measurements of selected hazardous organic chemicals. PB 81-200628, National Technical information Service, Springfield, VA.

Sterling, T.D. and D. Kobayashi. 1981. Use of gas ranges for cooking and heating in urban dwellings. *Journal of the Air Pollution Control Association* 31(2):162-65.

Sparks, L.E. 1988. Indoor air quality model, version 1.0. EPA-60018-88097a. Environmental Protection Agency, Research Triangle Park, NC.

Taylor, D.G., R.E. Kupel, and J.M. Bryant. 1977. *Documentation of the NIOSH validation tests.* U.S. Department of Health and Human Services, National Institute for Occupational Safety and Health, Washington, DC.

Thomson, G. 1986. *The museum environment,* 2nd ed. Butterworth's Publishers, Boston.

Tichenor, B.A. 1989. Measurement of organic compound emissions using small test chambers. *Environment International* 15:389-96.

Traynor, G.W. 1987. Field monitoring design considerations for assessing indoor exposures to combustion pollutants. *Atmospheric Environment* 21(2):377-83.

Traynor, G.W, J.R. Gimm, M.G. Apte, J.E. Dillworth, and P.D. White. 1985a. Indoor air pollution due to emissions from unvented gas-fired space heaters. *J. Air Pollution Control Association* 35:231-37.

Traynor, G.W., I.A. Nitschke, W.A. Clarke, G.P. Adams, and J.E. Rizzuto. 1985b. A detailed study of thirty houses with indoor combustion sources. *Paper* 85-30A.3. Air and Waste Management Association, Pittsburgh.

Turk, A. 1954. Odorous atmospheric gases and vapors: Properties, collection, and analysis. *Annals New York Academy of Sciences* 58:193-214.

Underhill, D.T, G. Mackerel, and M. Javorsky. 1988. Effects of relative humidity on adsorption of contaminants on activated carbon. In *Proceedings of the Symposium on Gaseous and Vaporous Removal Equipment Test Methods* (NBSIR 88-3716). National Institute of Standards and Technology, Gaithersburg, MD.

Van Gemert, L.J. and A.H. Mettenbreijer. 1977. *Compilation of odor threshold values in air and water.* Central Institute for Nutrition and Food Research, TNO Delft, Netherlands or National Institute for Water Supply, Zeist, Voorburg, Netherlands.

Wade, W.A., W.A. Cote, and J.E. Yocum. 1975. A study of indoor air quality. *Journal of the Air Pollution Control Association* 25(9):933-39.

Wallace, L.A., E.D. Pellizari, T.D. Hartwell, C. Sparacino, and H. Zelon. 1983. Personal exposure to volatile organics and other compounds indoors and outdoors—The TEAM study. *APCA Paper*. Air and Waste Management Association, Pittsburgh, PA.

Walsh, M., A. Black, A. Morgan, and G.H. Crawshaw. 1977. Sorption of SO_2 by typical indoor surfaces including wool carpets, wallpaper and paint. *Atmospheric Environment* 11.

Weschler, C.J. and H.C. Shields. 1989. The effects of ventilation, filtration, and outdoor air on the composition of indoor air at a telephone office building. *Environment International* 15:593-604.

Weschler, C.J., H.C. Shields, and D.V. Naik. 1994. Ozone-removal efficiencies of activated carbon filters after more than three years of continuous service. *ASHRAE Transactions* 100(2).

White, J.B., J.C. Reaves, R.C. Reist, and L.S. Mann. 1988. A data base on the sources of indoor air pollution emissions. In *IAQ '88: Engineering Solutions to Indoor Air Problems*. ASHRAE, Atlanta.

Yoon, Y.H. and J.H. Nelson. 1988. A theoretical study of the effect of humidity on respirator cartridge service life. *American Industrial Hygiene Association Journal* 49(7):325-32.

BIBLIOGRAPHY

Cain, W.S., C.R. Shoaf, S.F. Velasquez, S. Selevan, and W. Vickery. 1992. Reference guide to odor thresholds for hazardous air pollutants listed in the Clean Air Act Amendments of 1990. EPA/600/R-92/047. Environmental Protection Agency.

Chung, T-W., T.K. Ghosh, A.L. Hines, and D. Novosel. 1993. Removal of selected pollutants from air during dehumidification by lithium chloride and triethylene glycol solutions. *ASHRAE Transactions* 99(1):834-841.

Cohen, H.J., ed. 1983. AIHA's respiratory protection committee's critique of NIOSH/OSHA pocket guide to chemical hazards. *American Industrial Hygiene Association Journal* 44(9):A6-A7.

Fanger, P.O. 1989. The new comfort equation for indoor air quality. *IAQ '89: The Human Equation: Health and Comfort*, 251-254. ASHRAE.

Foresti, R., Jr. and O. Dennison. 1996. Formaldehyde originating from foam insulation. In *IAQ '86: Managing Indoor Air for Health and Energy Conservation*, 523-37. ASHRAE.

Freedman, R.W., B.I. Ferber, and A.M. Hartstein. 1973. Service lives of respirator cartridges versus several classes of organic vapors. *American Industrial Hygiene Association Journal* (2):55-60.

Hollowell, C.D., R.A. Young, J.V. Berk, and S.R. Brown. 1982. Energy-conserving retrofits and indoor air quality in residential housing. *ASHRAE Transactions* 88(1):875-93.

Jonas, L.A., E.B. Sansone, and T.S. Farris. 1983. Prediction of activated carbon performance for binary vapor mixtures. *American Industrial Hygiene Association Journal* 44:716-19.

Kelly, T.J. and D.H. Kinkead. 1993. Testing of chemically treated adsorbent air purifiers. *ASHRAE Journal* 35(7):14-23.

Mahajan, B.M. 1987. A method of measuring the effectiveness of gaseous contaminant removal devices. NBSIR 87-3666. National Institute of Standards and Technology, Gaithersburg, MD.

Mahajan, B.M. 1989. A method of measuring the effectiveness of gaseous contaminant removal filters. NBSIR 89-4119. National Institute of Standards and Technology, Gaithersburg, MD.

Mehta, M.R, H.E. Ayer, B.E. Saltzman, and R. Romk. 1988. Predicting concentrations for indoor chemical spills. In *IAQ '88: Engineering Solutions to Indoor Air Problems*, 231-50. ASHRAE, Atlanta.

Moschandreas, D.J. and S.M. Relwani. 1989. Field measurement of NO_2 gas-top burner emission rates. *Environment International* 15:499-92.

NAFA. 1996. Guide to air filtration, 2nd ed. National Air Filtration Association, Washington, DC.

NBS. 1959. Maximum permissible body burdens and maximum permissible concentrations of radionuclides in air and water for occupational exposure. NBS *Handbook* 69. U.S. Government Printing Office, Washington, DC.

Nirmalakhandan, N.N. and R.E. Speece. 1993. Prediction of activated carbon adsorption capacities for organic vapors using quantitative structure activity relationship methods. *Environmental Science and Technology* 27:1512-16.

Riggen, R.M., W.T. Wimberly, and N.T. Murphy. 1990. Compendium of methods for determination of toxic organic compounds in ambient air. EPA/600/D-89/186. Environmental Protection Agency, Research Triangle Park, NC. Also second supplement, 1990, EPA/600/ 4-89/017.

Rivers, R.D. 1988. Practical test method for gaseous contaminant removal devices. In *Proceedings of the Symposium on Gaseous and Vaporous Removal Equipment Test Methods* (NBSIR 88-3716). National Institute of Standards and Technology, Gaithersburg, MD.

Shah, J.J. and H.B. Singh. 1988. Distribution of volatile organic chemicals in outdoor and indoor air. *Environmental Science and Technology* 22(12):1391-88.

Singh. H.B., L.J. Salas, A. Smith, R. Stiles, and H. Shigeishi. 1981. Atmospheric measurements of selected hazardous organic chemicals. PB 81-200628. National Technical Information Service, Springfield, VA.

Spicer, C.W., R.W. Coutant, G.F. Ward, A.J. Gaynor, and I.H. Billick. 1986. Removal of nitrogen dioxide from air by residential materials. In *IAQ '86: Managing Indoor Air for Health and Energy Conservation*, 584-90. ASHRAE, Atlanta.

Sterling, T.D., E. Sterling, and H.D. Dimich-Ward. 1983. Air quality in public buildings with health related complaints. *ASHRAE Transactions* 89(2A):198-212.

Surgeon General. 1979. *Smoking and health*. U.S. Department of Health and Human Services, National Institute for Occupational Safety and Health, Washington, DC.

Webb, P. 1964. *Bioastronautics data book*. NASA SP-3006. National Technical Information Service, Springfield, VA.

Weschler, C.J. and H.C. Shields. 1989. The effects of ventilation, filtration, and outdoor air on the composition of indoor air at a telephone office building. *Environment International* 15:593-604.

White, J.B., J.C. Reaves, R.C. Reist, and L.S. Mann. 1988. A data base on the sources of indoor air pollution emissions. In *IAQ '88: Engineering Solutions to Indoor Air Problems*. ASHRAE, Atlanta.

Winberry, W.J. et al. 1990. EPA compendium of methods for the determination of air pollutants in indoor air. EPA/600/S4-90/010. Environmental Protection Agency. (NTIS-PB 90-200288AS).

Woods, J.E., J.E. Janssen, and B.C. Krafthefer. 1996. Rationalization of equivalence between the ventilation rate and air quality procedures in ASHRAE *Standard* 62. In *IAQ '86: Managing Indoor Air for Health and Energy Conservation*. ASHRAE, Atlanta.

Yu, H.H.S. and R.R. Raber. 1992. Air-cleaning strategies for equivalent indoor air quality. *ASHRAE Transactions* 98(1):173-191.

DESIGN AND APPLICATION OF CONTROLS

AUTOMATIC control of HVAC systems and equipment usually includes control of temperature, humidity, pressure, and flow rate. Automatic control sequences equipment operation to meet load requirements and provides safe operation of the equipment, using pneumatic, mechanical, electrical, and electronic control devices.

This chapter covers (1) the control of HVAC elements, (2) control of typical systems, (3) limit control for safe operation, and (4) the design of controls for specific HVAC applications. Chapter 37 of the 1997 *ASHRAE Handbook—Fundamentals* covers the fundamentals of control, types of control components, and commissioning of the control system.

CONTROL OF HVAC ELEMENTS

Boiler

Load affects the rate of heat input to a hydronic system. Rate control is accomplished by cycling and modulating the flame and by turning boilers on and off. Flame cycling and modulation are handled by the boiler control package. The control designer decides under what circumstances to add or drop a boiler and at what temperature to control the boiler supply water.

Hot water distribution control includes temperature control at the hot water boilers or converter, reset of heating water temperature, and control for multiple zones. Other factors that need to be considered include (1) minimum water flow through the boilers, (2) protection of boilers from temperature shock, and (3) coil freeze protection. If multiple or alternate heating sources (such as condenser heat recovery or solar storage) are used, the control strategy must also include a means of sequencing hot water sources or selecting the most economical source.

Figure 1 shows a system for load control of a gas or oil-fired boiler. Boiler safety controls usually include flame-failure, high-temperature, and other cutouts. Intermittent burner firing usually controls capacity, although fuel input modulation is common in larger systems. In most cases, the boiler is controlled to maintain a constant water temperature, although an outdoor air thermostat can reset the temperature if the boiler is not used for domestic water heating. Figure 1 contains a typical reset schedule. To minimize condensation of flue gases and boiler damage, water temperature should not be reset below that recommended by the manufacturer, typically 140°F. Larger systems with sufficiently high pump operating costs can use variable-speed pump drives, pump discharge valves with minimum flow bypass valves, or two-speed drives to reduce secondary pumping capacity to match the load.

Hot water heat exchangers or steam-to-water converters are sometimes used instead of boilers as hot water generators. Converters typically do not include a control package; therefore, the engineer must design the control scheme. The schematic in Figure 2 can be used with either low-pressure steam or boiler water ranging from 200 to 260°F. The supply water thermostat controls a modulating two-way valve in the steam (or hot water) supply line. An outdoor thermostat usually resets the supply water temperature downward

The preparation of this chapter is assigned to TC 1.4, Control Theory and Application.

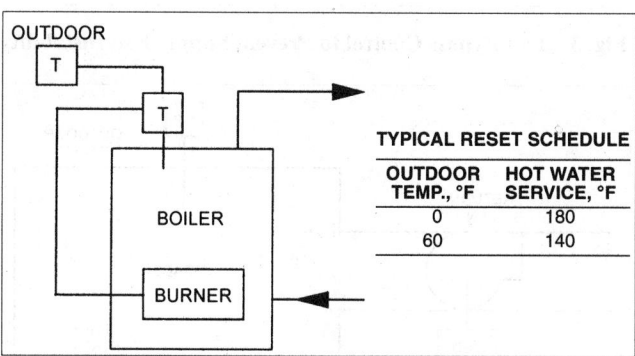

Fig. 1   Boiler Control

TYPICAL RESET SCHEDULE	
OUTDOOR TEMP., °F	HOT WATER SERVICE, °F
0	180
60	140

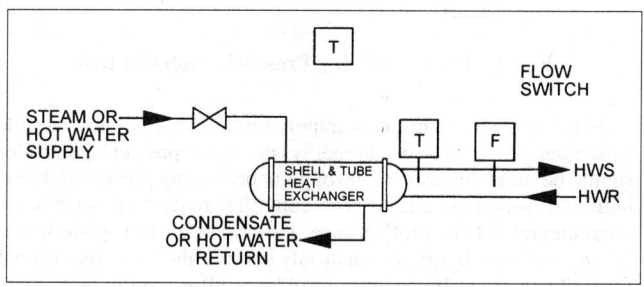

Fig. 2   Steam-to-Water Heat Exchanger Control

as the load decreases in order to improve the controllability of heating valves at low load and to reduce piping losses. A flow switch interlock should close the two-way valve when the hot water pump is not operating. With integrated computer based control, feedback from the zone heating valves can be used to control the starting and stopping of the hot water pumps. On constant flow systems, the feedback can be used to reset the hot water temperature to the lowest temperature that meets the zone requirements.

Fan

The most efficient way to change the output of a fan is to change its speed. Because of their simplicity and high efficiency, variable-frequency drives are widely used. While less efficient, eddy current drives are also an option for electronically controlling the fan speed. Another way of controlling the fan's output is to use various types of dampers including inlet vane dampers, discharge dampers, and scroll dampers. Axial fans can be controlled by varying the pitch of the blade. Also, dampers and ducting can simply bypass some of the air from the supply side of the fan to the return side (Figure 3). While bypassing does not change the output of the fan, it can allow the fan to accommodate flow variations in the distribution system without encountering fan instability. The final selection of a control device is determined by efficiency requirements and available funding.

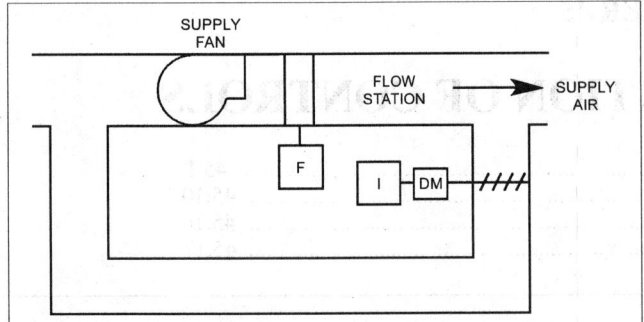

Fig. 3   Fan Bypass Control to Prevent Supply Fan Instability

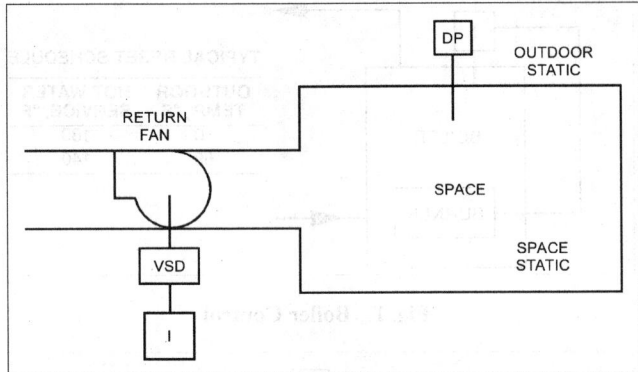

Fig. 4   Direct Building Pressurization Control

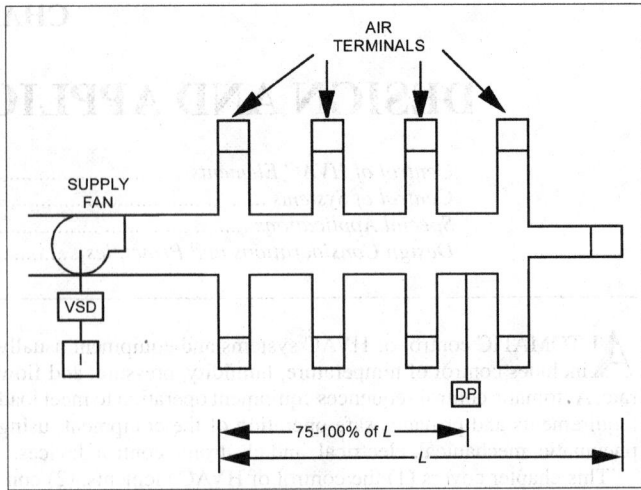

Fig. 5   Duct Static Pressure Control

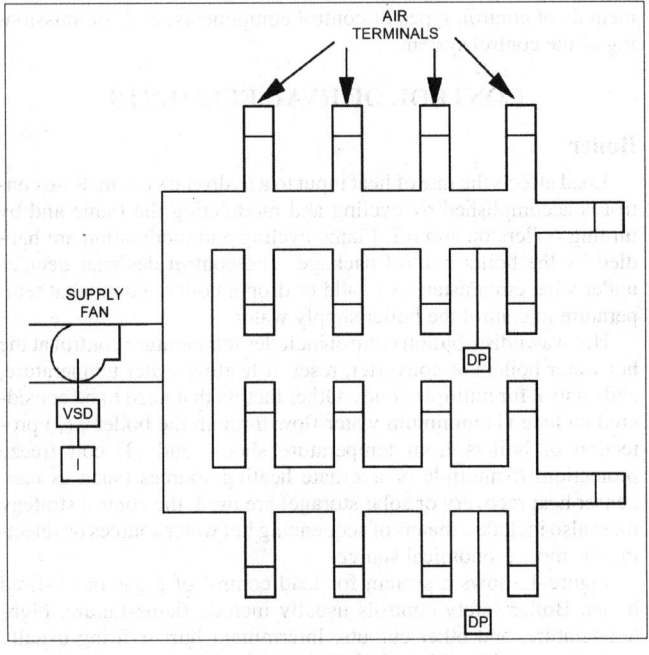

Fig. 6   Multiple Static Sensors

Static pressure control is required in systems having variable flow rates. To conserve fan energy, the static pressure controller should be set at the lowest control point permitting proper air distribution at design conditions. The controller requires proportional-plus-integral (PI) control because it eliminates offset while maintaining stability. In proportional-only control, the low proportional gain required to stabilize fan control loops allows static pressure to offset upward as the load decreases, which causes the supply fan to consume more energy.

Static pressure control is used to **pressurize a building** or space relative to adjacent spaces or the outdoors. Typical applications include clean rooms (positive pressure to prevent infiltration), laboratories (positive or negative, depending on use), and various manufacturing processes, such as spray-painting rooms. The pressure controller usually modulates dampers in the supply duct to maintain the desired pressure as exhaust volumes change. Control of the return fan requires measuring the space and outdoor static pressures (Figure 4). The location for measuring indoor static pressure must be selected carefully—away from doors and openings to the outside, away from elevator lobbies, and, when using a sensor, in a large representative area shielded from drafts. The outdoor location must likewise be selected carefully, typically 10 to 15 ft above the building and oriented to minimize wind effects from all directions. The amount of minimum outdoor air varies with building permeability and exhaust fan operation. Control of building pressurization can affect the amount of outside air entering the building.

Duct static pressure control for variable air volume (VAV) and other terminal systems maintains a static pressure at a measurement point. The most common application for static pressure control is fan output control in VAV systems. The pressure sensor must be properly placed to maintain optimum pressure throughout the supply duct. Experience indicates that performance is satisfactory when the sensor is located at 75 to 100% of the distance from the first to the most remote terminal. If the sensor is located at less than 100% of the distance, the control set point should be adjusted higher to

account for the pressure loss between the sensor and the remote terminal (Figure 5). Care must be taken in selecting the reference sensor location. Controller upset due to opening and closing doors, elevator shafts, and other sources of air turbulence should also be prevented. The pressure selected provides a minimum static pressure to all air terminal units during all supply fan design conditions.

Multiple static sensors (Figure 6) are required when more than one duct runs from the supply fan. The sensor with the highest static requirement controls the fan. Because duct run-outs may vary, a control that uses individual set points for each measurement is preferred.

VAV systems typically incorporate a duct static pressure control loop to control the flow. In a single-duct VAV system, the duct static pressure set point is usually selected by the designer. The sensor should be located in the ductwork where the established set point ensures proper operation of the zone VAV boxes under varying load (supply airflow) conditions. A shortcoming of this approach is that static pressure control is based on the readings of a single sensor that is assumed to represent the pressure available to all VAV boxes. If

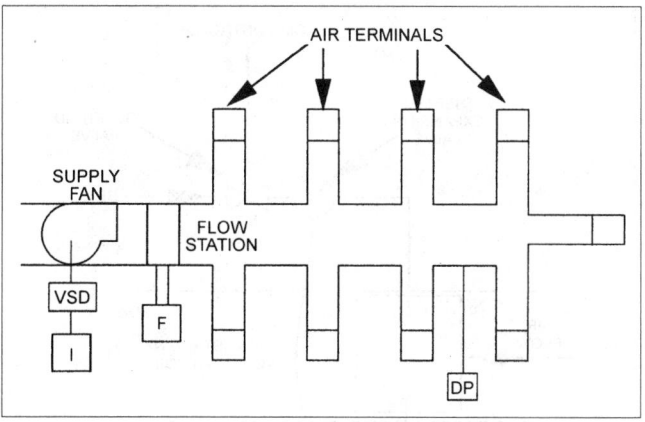

Fig. 7 Supply Fan Warm-Up Control

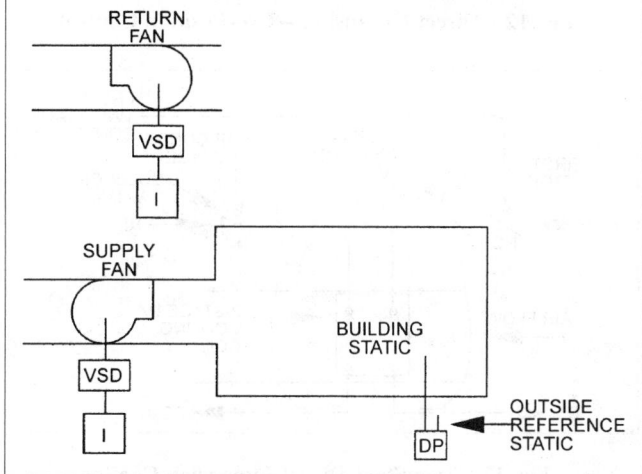

Fig. 8 Duct Static Control of Return Fan

the sensor malfunctions or is placed in a location that is not representative, operating problems will result.

An alternative approach to supply fan control in a VAV system uses flow readings from the direct digital control (DDC) zone terminal boxes to integrate zone VAV requirements with supply fan operation. Englander and Norford (1992) suggest that duct static pressure and fan energy can be reduced without sacrificing occupant comfort or adequate ventilation. They compared modified PI and heuristic control algorithms via simulation and demonstrated that either static pressure or fan speed can be regulated directly using a flow error signal from one or more zones. They noted that component modeling limitations constrain their results primarily to a comparison of the control algorithms. The results show that both PI and heuristic control schemes work, but the authors suggest that a hybrid of the two might be ideal.

Supply fan warm-up control for systems having a return fan must prevent the supply fan from delivering more airflow than the return fan maximum capacity during warm-up mode (Figure 7).

Return fan static control from returns having local (zoned) flow control is identical to supply fan static control (Figure 5). Return fan control for VAV systems provides proper building pressurization and minimum outdoor air. Duct static control of the supply fan is forwarded to the return fan (Figure 8). This open loop (no feedback) control requires similar supply and return fan airflow modulation characteristics. The return fan airflow is adjusted at minimum and maximum airflow conditions. The airflow turndown

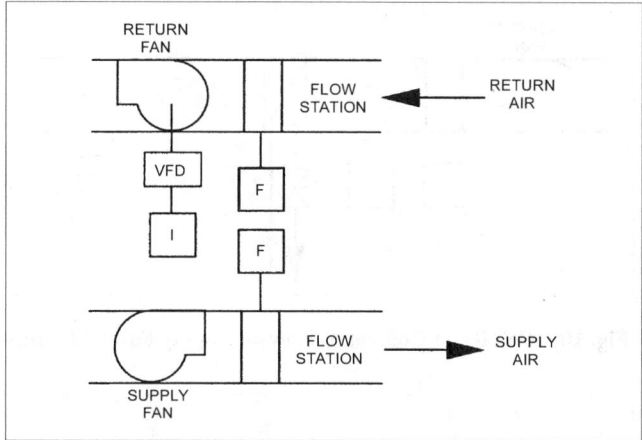

Fig. 9 Airflow Tracking Control

should not be excessive, typically no more than 50%. Provisions for warm-up and exhaust fan switching are impractical.

Airflow tracking uses duct airflow measurements to control the return air fans (Figure 9). Typical sensors, called flow stations, are multiple-point, pitot tube, and averaging. Provisions must be made for exhaust fan switching to maintain pressurization of the building. Warm-up is accomplished by setting the return airflow equal to the supply fan airflow, usually with exhaust fans turned off and limiting supply fan volume to return fan capability. During night cool-down, the return fan operates in the normal mode.

VAV systems that use return or relief fans require control of airflow through the return or relief air duct systems. Return fans are commonly used in VAV systems to help ensure adequate air distribution and acceptable zone pressurization. In a return fan VAV system, there is significant potential for control system instability due to the interaction of control variables (Avery 1986). In a typical system, these variables might include supply fan speed, supply duct static pressure, return fan speed, mixed air temperature, outside and return air damper flow characteristics, and wind pressure effect on the relief louver. The interaction of these variables and the selection of control schemes to minimize or eliminate interaction must be considered carefully. Mixed air damper sizing and selection are particularly important. Zone pressurization, building construction, and outdoor wind velocity must be considered. The resultant design helps ensure proper air distribution, especially through the return air duct. Using the technique described by Dickson, the designer may be able to eliminate the return fan altogether.

Sequencing fans for VAV systems reduces airflow more than other methods and results in greater operating economy and more stable fan operation if airflow reductions are significant. Alternation of fans usually provides greater reliability. Centrifugal fans are controlled to keep system disturbances to a minimum when additional fans are started. The added fan is started and slowly brought to capacity while the capacity of the operating fans is simultaneously reduced. The combined output of all fans then equals the output before fan addition.

Vaneaxial fans usually cannot be sequenced in the same manner as centrifugal fans. To avoid stall, the operating fans must be reduced to some minimum level of airflow. Then, additional fans may be started and all fans modulated to achieve equilibrium.

Unstable fan operation in VAV systems can usually be avoided by proper fan sizing. However, if airflow reduction is large (typically over 60%), fan sequencing is usually required to maintain airflow in the fan's stable range.

Supply air temperature reset can be used to avoid fan instability by resetting the cooling coil discharge temperature higher (Figure 10), so that the building cooling loads require greater airflow.

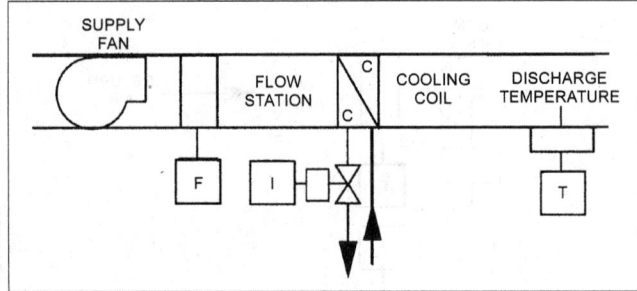

Fig. 10   Coil Reset Control to Prevent Supply Fan Instability

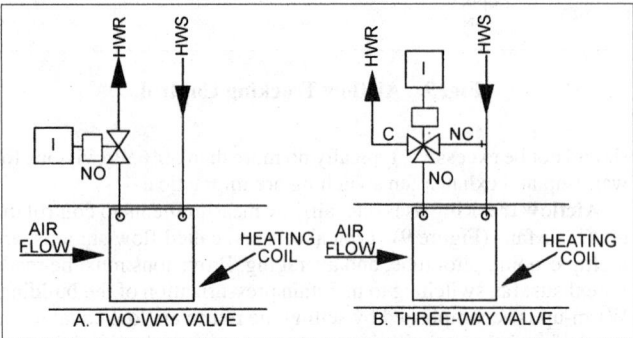

Fig. 11   Control of Hot and Chilled Water Coils

Because of the time lag between temperature reset and demand for more airflow, the value at which reset starts should be selected on the safe side of the fan instability point. When this technique is used, it must be ascertained that dehumidification requirements can be met.

Cooling Coil

Chilled water or **brine cooling coils** are controlled by two- or three-way valves (Figure 11). These valves are similar to those used for heating control, but are usually closed to prevent cooling when the fan is off. The valve typically modulates in response to coil air discharge temperature or space temperature.

Direct-expansion (DX) cooling coils are usually controlled by solenoid valves in the refrigerant liquid line (Figure 12). Face and bypass dampers are not recommended because they permit ice to form on the coil when airflow is reduced. Control can be improved by using two or more stages; the solenoid valves are controlled in sequence, and there is a differential of 1 or 2°F between stages (Figure 13). The first stage should be the first coil row on the entering air side; the following rows form the second and succeeding stages. Side-by-side stages tend to generate icing on the stage in use, which causes reduction of airflow and loss of control. Modulating control is achieved using a variable suction pressure controller (Figure 14). This type of control is uncommon but necessary if accurate control of discharge or space temperature is required.

Cooling Tower

The most common packaged mechanical-draft cooling towers for comfort air-conditioning applications are counterflow induced-draft and forced-draft. They are controlled similarly, depending on the manufacturer's recommendations. On larger towers, two-speed motors or variable-speed drives can reduce fan power consumption at part-load conditions and stabilize condenser water temperature (Figure 15).

In colder areas that require year-round air conditioning, cooling towers may require sump heating and continuous full flow over the tower to prevent ice formation. In that case, the cooling tower sump

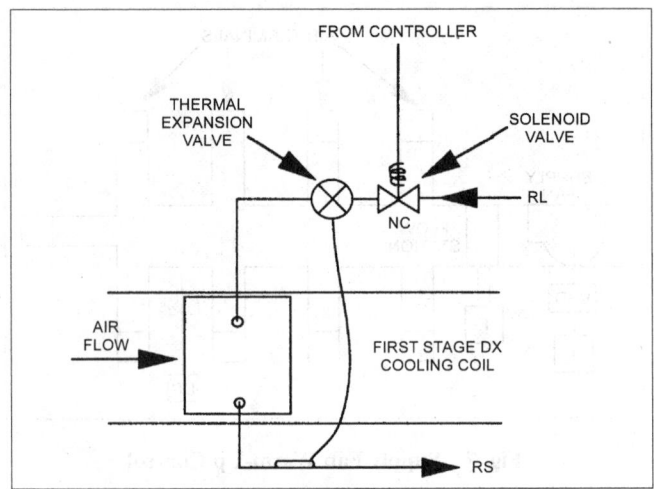

Fig. 12   Direct Expansion—Two-Position Control

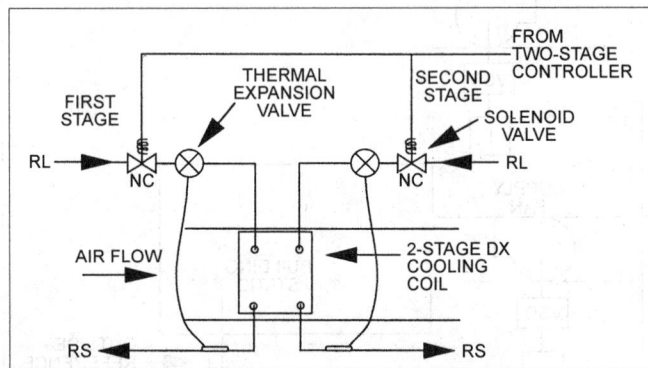

Fig. 13   Two-Stage Direct-Expansion Cooling

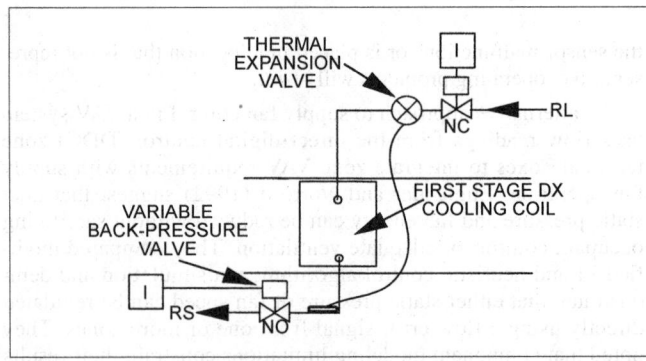

Fig. 14   Modulating Direct-Expansion Cooling

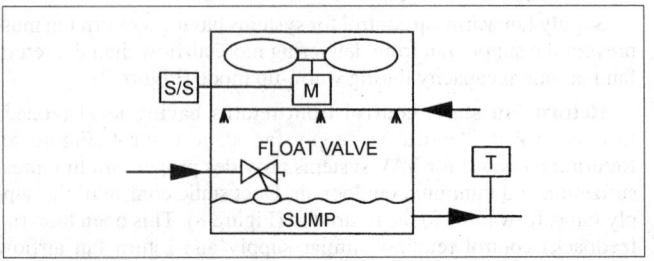

Fig. 15   Cooling Tower

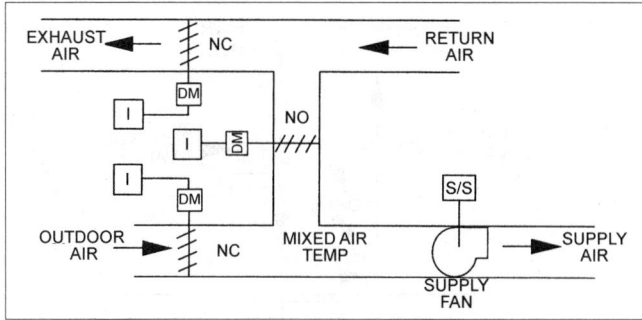

Fig. 16   Economizer Cycle Control

thermostat would control a hot water or steam valve to keep water temperature above freezing.

Economizer Cycle

Economizer cycle control reduces cooling costs when outside conditions are suitable, that is, when the outdoor air is cool enough to be used as a cooling medium. If the outdoor air is below a high-temperature limit, typically 65°F, the return, exhaust, and outdoor air dampers modulate to maintain a ventilation cooling set point, typically 55 to 60°F (Figure 16). The relief dampers are interlocked to close, and the return air dampers to open, when the supply fan is not operating. When the outdoor air temperature exceeds the high-temperature limit set point, the outdoor air damper is closed to a fixed minimum and the exhaust and return air dampers close and open, respectively.

In enthalpy economizer control, the high-temperature limit interlock system of the economizer cycle is replaced in order to further reduce energy costs when latent loads are significant. The interlock function (Figure 16) can be based instead on (1) a fixed enthalpy upper limit, (2) a comparison with return air so as not to exceed return air enthalpy, or (3) a combination of enthalpy and high-temperature limits.

VAV warm-up control during unoccupied periods requires no outdoor air; typically, outdoor and exhaust dampers remain closed. However, in systems with a return fan (Figure 17), the outdoor air damper should be positioned at its minimum position, and supply airflow (volume) should be limited to return air airflow (volume) to minimize positive or negative duct pressurization.

Night cool-down control (night purge) provides 100% outdoor air for cooling during unoccupied periods (Figure 18). The space is cooled to the space set point, typically 9°F above outdoor air temperature. Limit controls prevent operation if outdoor air is above space dry-bulb temperature, if outdoor air dew-point temperature is excessive, or if outdoor air dry-bulb temperature is too cold, typically 50°F or below. The night cool-down cycle is initiated before sunrise, when overnight outside temperatures are usually the coolest. When outside air conditions are acceptable and the space requires cooling, the cool-down cycle is the first phase of the optimum start sequence.

Heating Coil

Heating coils that are not subject to freezing can be controlled by simple two-way or three-way modulating valves (Figure 11). Steam distributing coils are required to ensure proper steam coil control. The valve is controlled by coil discharge air temperature or by space temperature, depending on the HVAC system. Valves are set to open to allow heating if control power fails. In many systems, the outdoor air temperature resets the heating discharge controller.

To provide unoccupied heating or preoccupancy warm-up, a heating coil can be added to the central fan system. During warm-up or unoccupied periods, a constant supply duct heating temperature is maintained and the cooling coil valve is kept closed. Once the

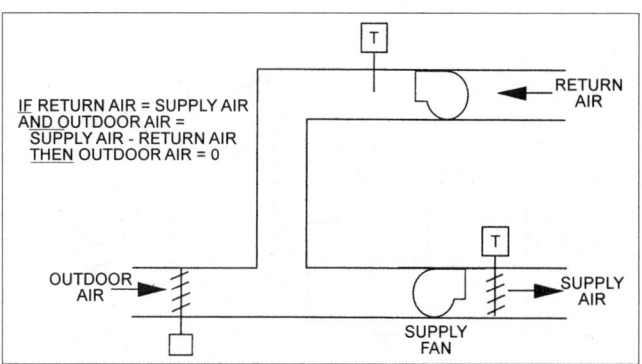

Fig. 17   Warm-Up Control

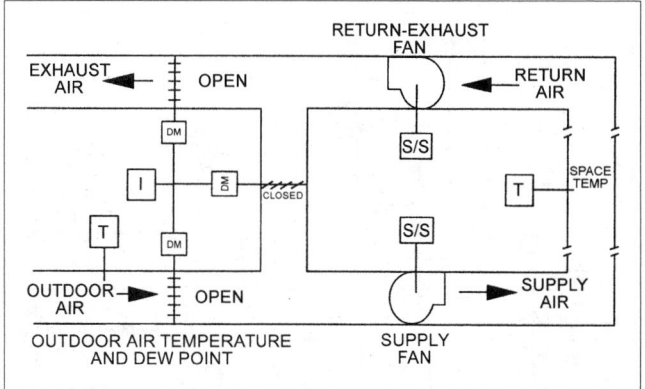

Fig. 18   Night Cool-Down Control

facility has attained the minimum required space temperature, the central air handler will revert back to the occupied mode.

Heating coils in central air-handling units preheat, reheat, or heat, depending on the climate and the amount of minimum outdoor air needed.

Preheating coils using steam or hot water must have protection against freezing, unless (1) the minimum outdoor air quantity is small enough to keep the mixed air temperature above freezing and (2) enough mixing occurs to prevent stratification. That is, even when the average mixed air temperature is above freezing, inadequate mixing may allow freezing air to impinge on the coil.

Steam preheat coils should have two-position valves and vacuum breakers to prevent a buildup of condensate in the coil. The valve should be fully open when outdoor air (or mixed air) temperature is below freezing. This causes unacceptably high coil discharge temperatures at times, necessitating face and bypass dampers for final temperature control (Figure 19). The bypass damper should be sized to provide the same pressure drop at full bypass airflow as the combination of face damper and coil does at full airflow.

Hot water preheat coils must maintain a minimum water velocity in the tubes of 3 fps to prevent freezing. A two-position valve combined with face and bypass dampers can usually be used to control the water velocity. More commonly, a secondary pump control in one of two configurations (Figure 20 and Figure 21) is used. The control valve modulates to maintain the desired coil air discharge temperature, while the pump operates to maintain the minimum tube water velocity when outdoor air is below freezing. The system in Figure 21 uses less pump power, allows variable flow in the hot water supply main, and is preferred for energy conservation. The system in Figure 20 may be required on small systems with only one or two air handlers, or where constant main water flow is needed.

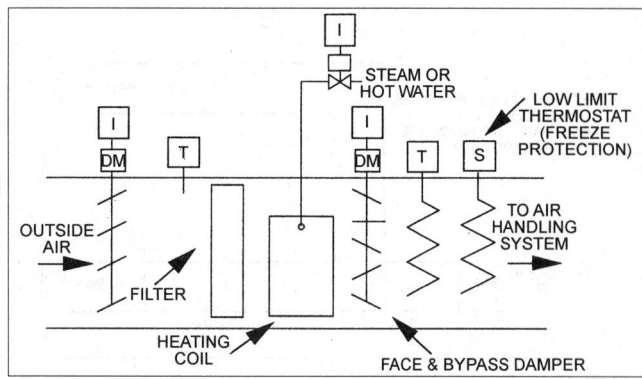

Fig. 19   Preheat with Face and Bypass Dampers

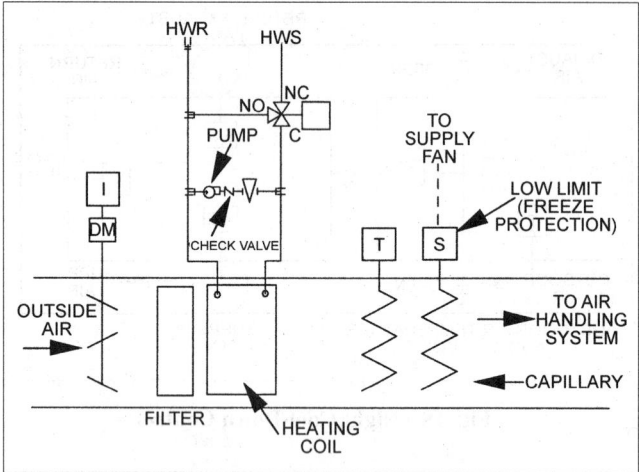

Fig. 20   Preheat with Secondary Pump and Three-Way Valve

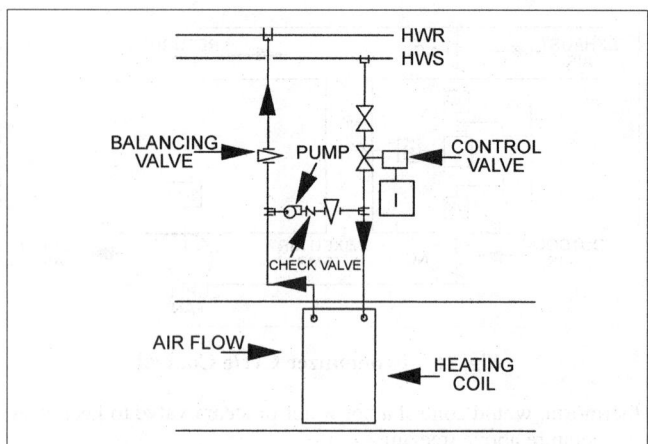

Fig. 21   Preheat with Secondary Pump and Two-Way Valve

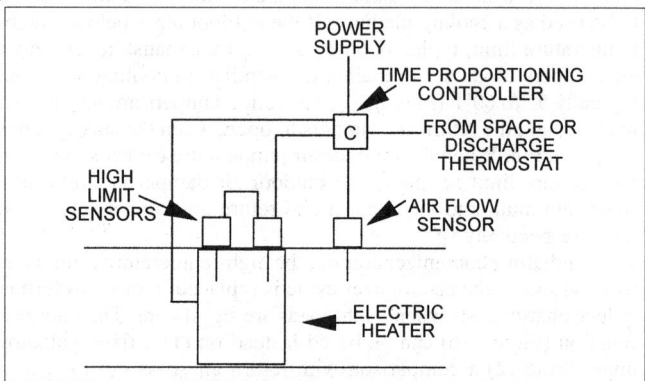

Fig. 22   Electric Heat: Solid-State Controller

Some systems may use a glycol solution in combination with any of these methods.

Electric heating coils (duct heaters) are controlled in either two-position or modulating mode. Two-position operation uses power relays with contacts sized to handle the power required by the heating coil. Timed two-position control requires a timer and contactors. The timer can be electromechanical, but it is usually electronic and provides a time base of 1 to 5 min. Step controllers provide cam-operated sequencing control of up to 10 stages of electric heat. Each stage may require a contactor, depending on the step controller contact rating. Thermostat demand determines the percentage of on-time. Because rapid cycling of mechanical or mercury contactors can cause maintenance problems, solid-state controllers are preferred. These devices make cycling so rapid that the control is proportional; therefore, face and bypass dampers are not used. A control system with a solid-state controller and safety controls is shown in Figure 22.

Heat Pump

A heat pump is a refrigeration device in which the evaporator is used for cooling and heat that is normally rejected from the condenser is used for heating. Many conventional means used to control refrigeration equipment can be used to control heating and cooling cycles. Chapters 8 and 44 of the 1996 *ASHRAE Handbook—Systems and Equipment* include details.

Humidity Control

Humidity control is based on the output of a humidity sensor located either in the space or in the return air duct. Most comfort cooling processes involve some dehumidification. The amount of dehumidification is a function of the effective coil surface temperature and is limited by the freezing point of the coolant. If water condensing out of the airstream freezes on the coil surface, airflow is restricted and, in severe cases, may be shut off. The practical limit is about 40°F dew point on the coil surface. As indicated in Figure 23, this results in a relative humidity of about 30% at a space temperature of 75°F, which is adequate for most commercial applications. When a lower humidity is necessary, a chemical dehumidifier is required.

Dehumidification can be achieved in several ways. One is to override the control of the cooling coil. The temperature of the coil is lowered until sufficient moisture is removed from the supply air to maintain the humidity set point. When maximum relative humidity control is required, a space or return air humidistat is provided in addition to the space thermostat. To limit maximum humidity, a control function selects the higher of the output signals from the two devices and controls the cooling coil valve accordingly. A reheat coil may be required to maintain the space temperature if the moisture removal process results in too low a supply air temperature (Figure 24). If humidification is also provided, this cycle is sometimes called a constant temperature, constant humidity cycle. Although simple cooling by refrigeration maintains an upper limit to space humidity, without additional equipment it does not control humidity.

Sprayed coil dehumidifiers (Figure 25) have been used for dehumidification. Space relative humidity ranging from 35 to 55% at 75°F can be obtained with this equipment; however, the cost of maintenance, reheat, and removal of solid deposits on the coil make the sprayed coil dehumidifier less desirable than other methods.

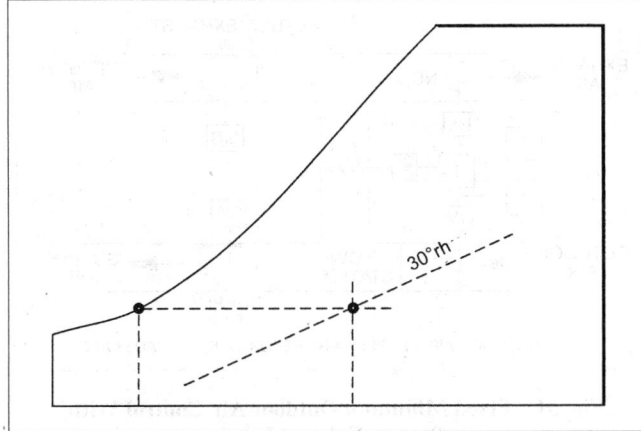

Fig. 23 Cooling and Dehumidifying—Practical Low Limit

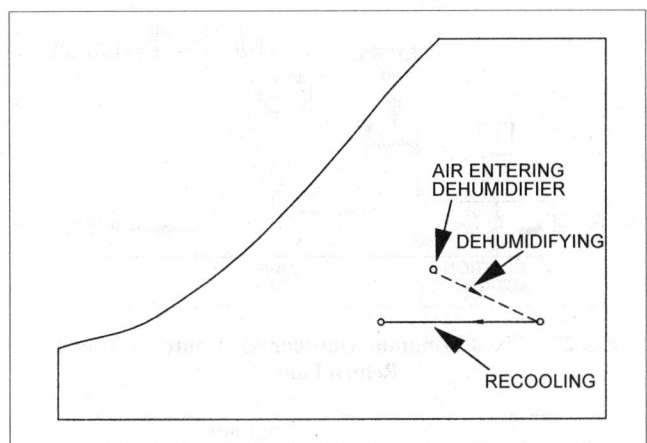

Fig. 26 Psychrometric Chart: Chemical Dehumidification

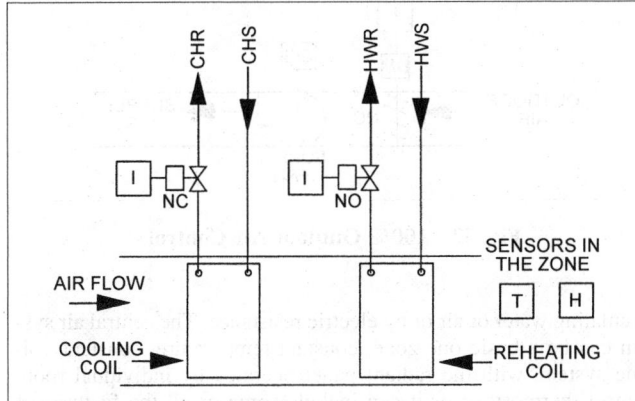

Fig. 24 Cooling and Dehumidifying with Reheat

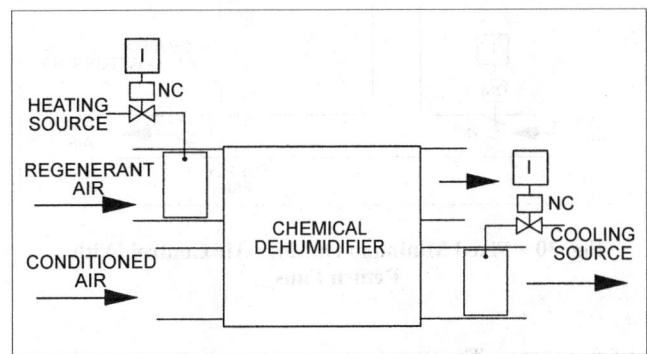

Fig. 27 Chemical Dehumidifier

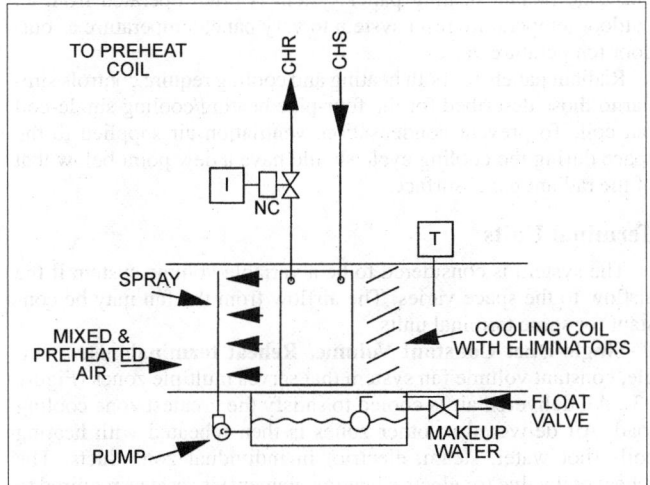

Fig. 25 Sprayed Coil Dehumidifier

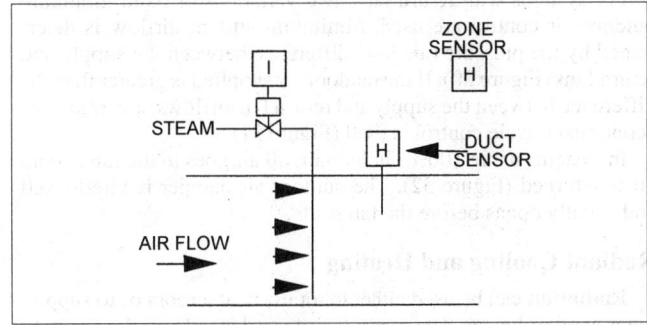

Fig. 28 Steam Jet Humidifier

spray tubes are all used for space humidification. A space or return air humidity sensor provides the necessary signal for the controller. A humidity sensor in the duct should be used to minimize moisture carryover or condensation in the duct (Figure 28). With proper use and control, humidifiers can achieve high space humidity, although they more often maintain design minimum humidity during the heating season.

Outdoor Air Control

Fixed, minimum outdoor air control provides ventilation air, space pressurization (exfiltration), and makeup air for exhaust fans. For systems without return fans, the outdoor air damper is interlocked to remain open only when the supply fan operates (Figure 29). The outdoor air damper should open quickly when the fan turns on to prevent excessive negative duct pressurization. In some applications, the fan on-off switch opens the outdoor air damper before

A **desiccant-based dehumidifier** can lower space humidity below that possible with cooling/dehumidifying coils. This device adsorbs moisture using silica gel or a similar material. For continuous operation, heat is added to regenerate the material. The adsorption process also generates heat (Figure 26). Figure 27 shows a typical control.

Humidification can be achieved by adding moisture to the supply air. Evaporative pans (usually heated), steam jets, and atomizing

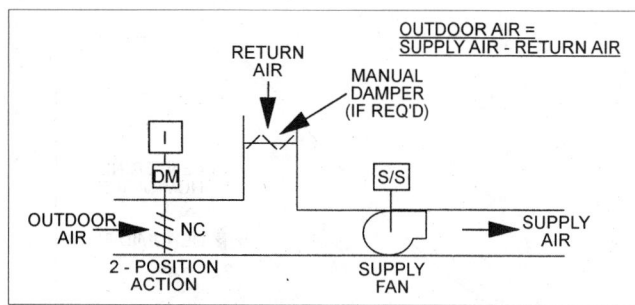

Fig. 29 Fixed Minimum Outdoor Air Control Without Return Fans

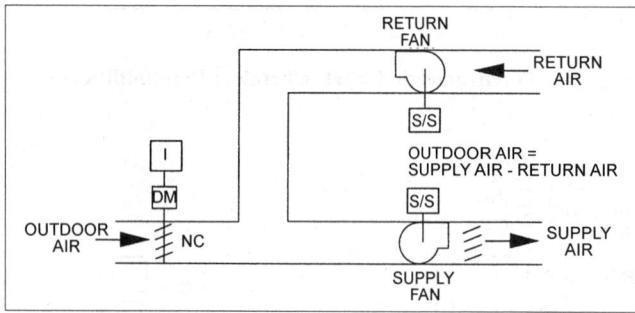

Fig. 30 Fixed Minimum Outdoor Air Control With Return Fans

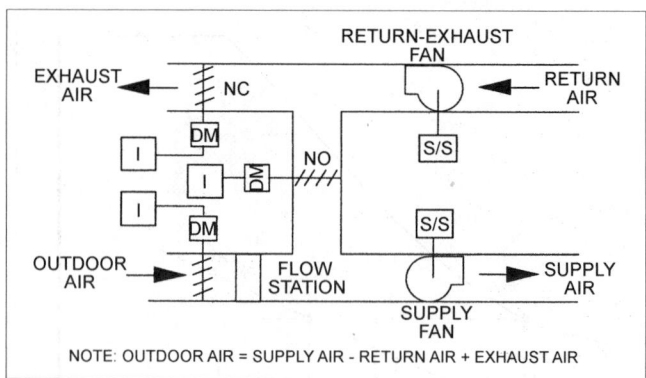

Fig. 31 Fixed Minimum Outdoor Air Control With Return-Exhaust Fans

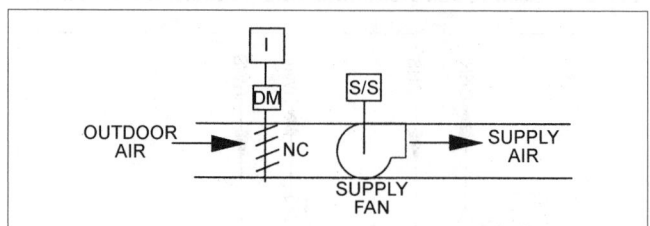

Fig. 32 100% Outdoor Air Control

the fan is started. The rate of outdoor airflow is determined by the opening of the damper and by the pressure difference between the mixed air plenum and the outdoor air plenums.

For systems with return fans, two variations of fixed minimum outdoor air control are used. Minimum outdoor airflow is determined by the pressure (airflow) difference between the supply and return fans (Figure 30). If the outdoor air supplied is greater than the difference between the supply and return fan airflows, a variation of economizer cycle control is used (Figure 31).

In systems using 100% outdoor air, all air goes to the fan and no air is returned (Figure 32). The outdoor air damper is interlocked and usually opens before the fan starts.

Radiant Cooling and Heating

Radiation can be used either to totally heat a room or to supplement another heater. The control strategy depends on the function performed. For a total heating application, rooms are usually controlled individually; each radiator and convector is equipped with an automatic control valve. Depending on room size, one thermostat may control one valve or several valves in unison. The thermostat can be placed in the return air to the unit or on a wall at occupant level. Return air control is generally less accurate and results in wider space temperature fluctuations. When the space is controlled for the comfort of seated occupants, wall-mounted thermostats give the best results.

For supplemental heating applications, where perimeter radiation is used only to offset perimeter heat losses (the zone or space load is handled separately by a zone air system), outdoor reset of the water temperature to the radiation should be considered. Radiation can be zoned by exposure, and the compensating outdoor sensor can be located to sense compensated indoor (outdoor) temperature, solar load, or both.

Radiant panels combine controlled-temperature room surfaces with central air conditioning and ventilation. The radiant panel can be in the floor, walls, or ceiling. Panel temperature is maintained by circulating water or air or by electric resistance. The central air system can be a basic one-zone, constant temperature, constant volume system, with the radiant panel operated by individual room control thermostats; or it can include some or all the features of dual-duct, reheat, multizone, or VAV systems, with the radiant panel operated as a one-zone, constant temperature system. The one-zone radiant heating panel system is often operated from an outdoor temperature reset system to vary panel temperature as outdoor temperature varies.

Radiant panels for both heating and cooling require controls similar to those described for the four-pipe heating/cooling single-coil fan coil. To prevent condensation, ventilation air supplied to the space during the cooling cycle should have a dew point below that of the radiant panel surface.

Terminal Units

The system is considered to be a variable volume system if the airflow to the space varies. The airflow from the fan may be constant for some terminal units.

Single-Duct Constant Volume. Reheat terminals use a single, constant volume fan system that serves multiple zones (Figure 33). All delivered air is cooled to satisfy the greatest zone cooling load. Air delivered to other zones is then reheated with heating coils (hot water, steam, electric) in individual zone ducts. The reheat coil valve (or electric heating element) is reset as required to maintain the space condition. Because these systems consume more energy than VAV systems, they are generally limited to applications with fixed ventilation needs, such as hospitals and special processes or laboratories.

No fan control is required because the design, selection, and adjustment of fan components determine the air volume and duct static pressure. The same temperature air is supplied to all zones. However, the controller can allow the supply temperature to respond to the demand from the greatest cooling load, thus conserving energy.

Single-Duct Variable Volume. A **throttling terminal** has a damper in the inlet that controls the flow of supply air (Figure 34).

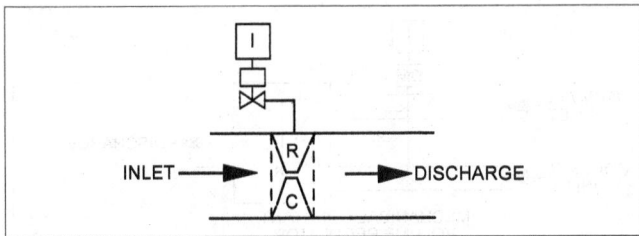

Fig. 33   Constant Volume Single-Duct Zone Reheat

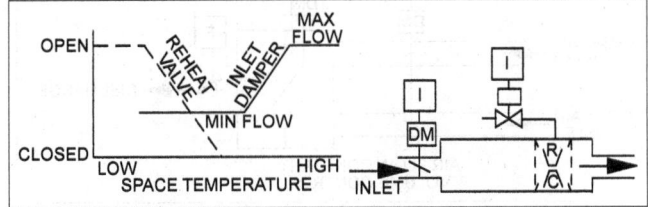

Fig. 34   Throttling VAV Terminal Unit

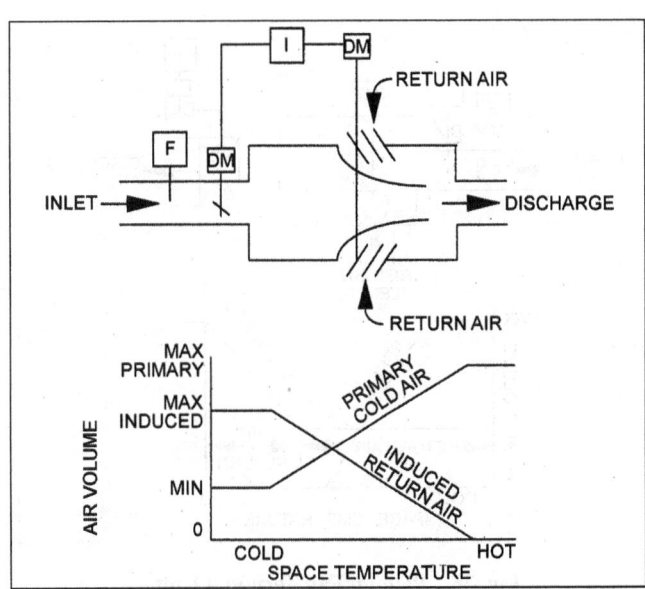

Fig. 35   Induction VAV Terminal Unit

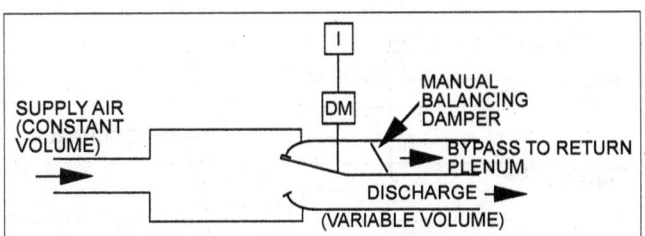

Fig. 36   Bypass VAV Terminal Unit

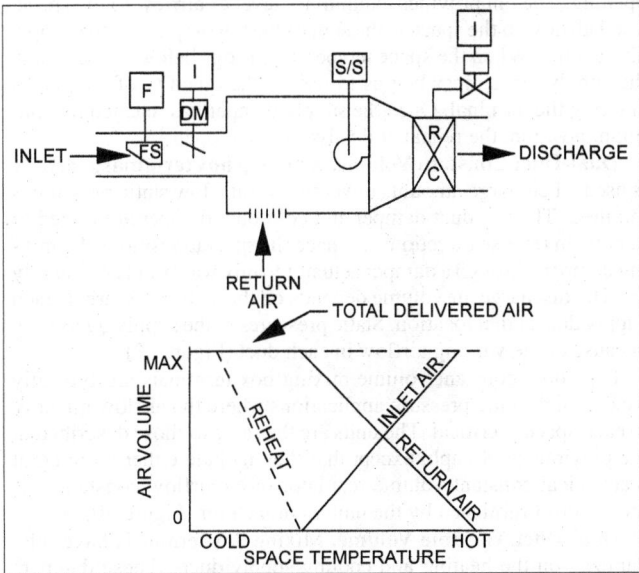

Fig. 37   Fan-Powered VAV Terminal Unit

For spaces requiring heating, a reheat coil can be installed in the discharge. As the temperature in the space drops below the set point, the damper begins to close and reduce the flow of air to the space. When the airflow reaches the minimum limit, the valve on the reheat coil begins to open.

Single-duct VAV systems, which supply warm air to all zones when heating is required and cool air to all zones when cooling is required, have limited application and are used where heating is required only for morning warm-up. They should not be used if some zones require heating at the same time that others require cooling. These systems, like single-duct cooling-only systems, are generally controlled during occupancy.

An **induction terminal** controls the space temperature by reducing the supply airflow to the space and by inducing return air from the plenum space into the airstream for the space (Figure 35). Both dampers are controlled simultaneously, so as the primary air opening decreases, the return air opening increases. When the space temperature drops below the set point, the supply air damper begins to close and the return air damper begins to open.

A **bypass terminal** has a damper that diverts part of the supply air into the return plenum (Figure 36). Control of the diverting damper is based on the output of the space temperature sensor. When the temperature in the space drops below the set point, the bypass damper begins to open, routing some of the supply air to the plenum, which reduces the amount of supply air entering the space. When the bypass is fully open, the control valve for the reheat coil opens as required to maintain the space temperature. A manual balancing damper in the bypass is adjusted to match the resistance in the discharge duct. In this way, the supply of air from the primary system remains at a constant volume. The maximum airflow through the bypass must be restricted in order to maintain the minimum airflow into the space. Although the airflow to the space is reduced, the total airflow of the fan remains constant, so the fan power and associated energy cost are not reduced. These terminals can be added to a single-zone constant volume system to provide zoning without the energy penalty of a conventional reheat system.

A **fan-powered terminal unit** has an integral fan that supplies a constant volume of air to the space (Figure 37). In addition to enhancing air distribution in the space, a reheat coil can be added to maintain a minimum temperature in the space when the primary system is off. When the space is occupied, the fan runs constantly to provide a constant volume of air to the space. The fan can draw air from the return plenum to compensate for the reduced supply air. As the temperature in the space decreases below the set point, the supply air damper begins to close and the fan draws more air from the

return plenum. Units serving the perimeter area of a building can include a reheat coil. Then, when the supply air reaches its minimum level, the valve to the reheat coil begins to open.

A **plenum fan terminal** has a fan that pulls air from the return plenum and mixes it with the supply air (Figure 38). A reheat coil may be placed in the discharge to the space or in the return plenum

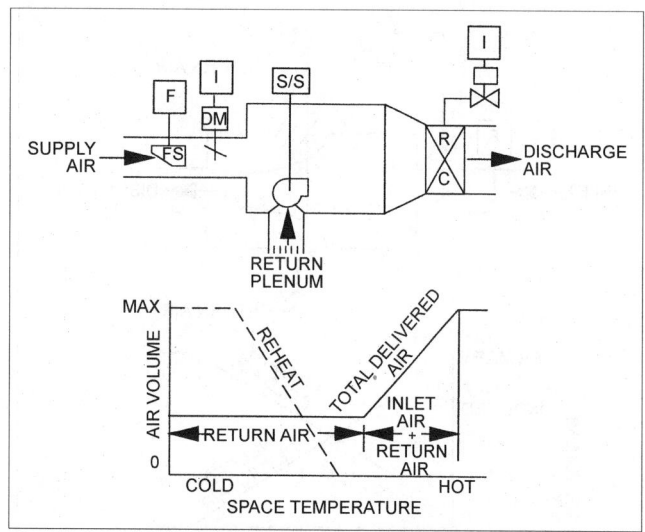

Fig. 38 Plenum Fan Terminal Unit

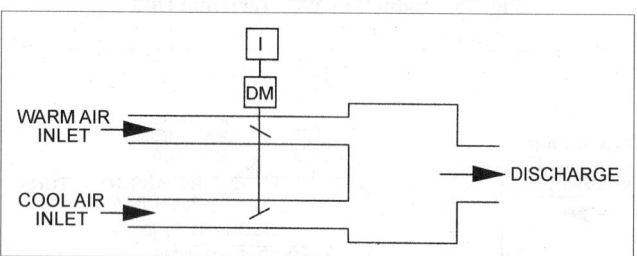

Fig. 39 Dual-Duct Mixing Box Terminal Unit

opening. The fan provides a minimum level of airflow to the space. Total airflow to the space is the sum of the fan output and the supply air quantity. When the space temperature drops below the set point, the supply air damper begins to reduce the quantity of supply air entering the terminal. Once the supply damper has reached its minimum position, the reheat coil valve starts to open.

Dual-Duct Constant Volume. A **mixing box terminal** generally is used where large amounts of ventilation at a low static pressure is required. The hot-duct damper and cold-duct damper are linked to operate in reverse directions. A space thermostat positions the mixing dampers through a damper actuator to mix warm and cool supply air. The discharge air volume depends on the static pressure in each supply duct at that location. Static pressures in the supply ducts vary because of the varying airflow in each duct (Figure 39).

Dual-duct constant volume mixing box terminals are typically used in high static pressure applications where the airflow quantity to each space is critical. The units are the same as those described in the previous paragraph, except that they include either an integral mechanical constant volume regulator or an airflow constant volume control furnished by the unit manufacturer (Figure 40).

Dual Duct Variable Volume. Mixing box terminals have inlet dampers on the heating and cooling supply ducts. These dampers are interlinked to operate in opposite directions, and each requires a single control actuator. Also, a sensor in the discharge monitors total airflow. The space thermostat controls inlet mixing dampers directly, and the airflow controller controls the volume damper. The space thermostat resets the airflow controller from maximum to minimum flow as the thermal load on the conditioned area changes. Figure 41 shows the control schematic and damper operation. Note that in a portion of the control range, heating and cooling supply air mix.

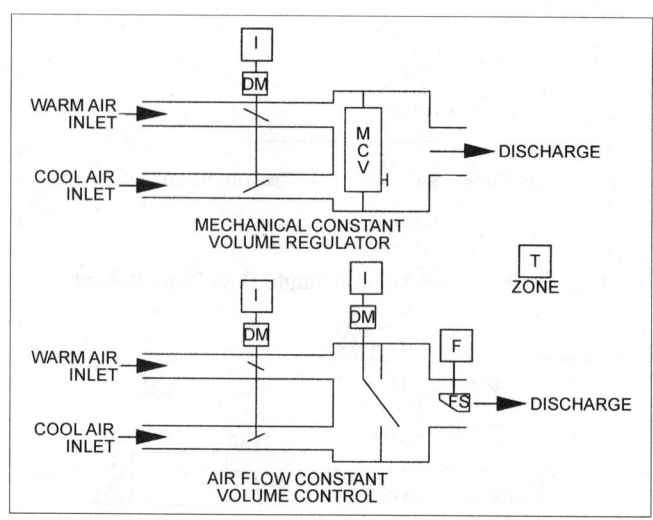

Fig. 40 Dual-Duct Pressure-Independent Constant Volume Mixing Box Terminal Units

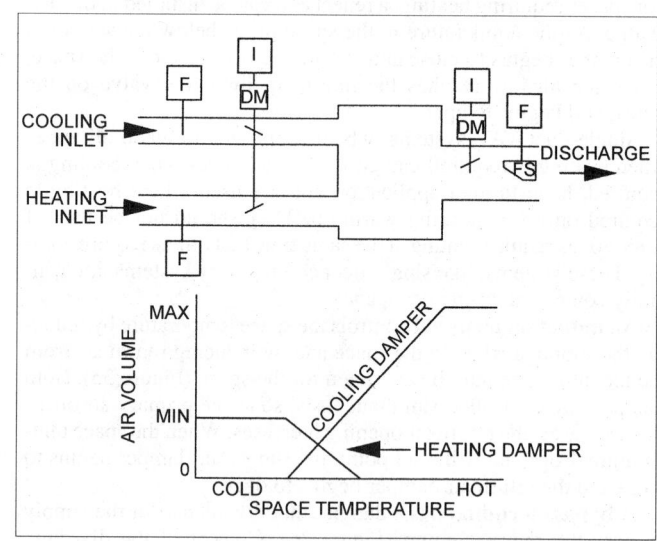

Fig. 41 Pressure-Independent Dual-Duct VAV Terminal Unit

Variable, constant volume (zero energy band) dual-duct terminal units (Figure 42), have inlet dampers (with individual damper actuators and airflow controllers) on the cooling and heating supply ducts and no total airflow volume damper. The zero energy band (ZEB) space thermostat resets the airflow controller set points in sequence as the space load changes. The airflow controllers maintain adjustable minimum flows for ventilation (with no overlap of damper operations) in the ZEB when neither heating nor cooling is required.

CONTROL OF SYSTEMS

Air Handling

Generally, air is distributed by either single-duct or dual-duct systems. Airflow to each zone is controlled by some type of a terminal box. A special case of a dual-duct system is a multizone unit with terminal boxes incorporated into the air handler package. In this case, the hot and cold ducts are referred to as the **hot** and **cold decks.** One special multizone configuration has three decks, which may be used as an alternative in a multizone system (Figure 43). Zone dampers in this unit operate with sequenced dampers to either mix hot supply air with bypass air when the cold deck damper is

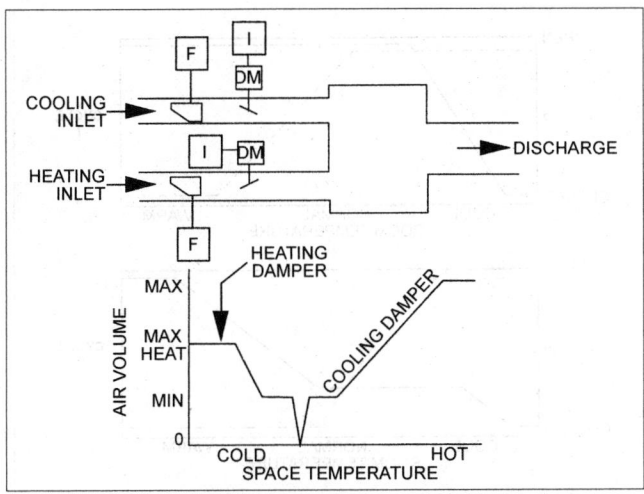

Fig. 42   Variable, Constant Volume (ZEB) Dual-Duct Terminal Unit

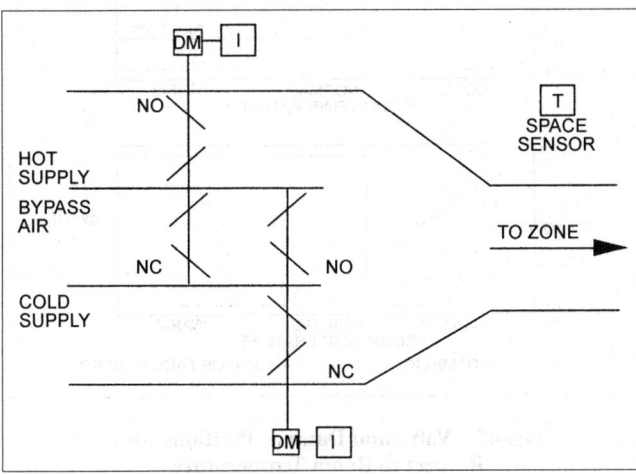

Fig. 43   Zone Mixing Dampers—Three-Deck Multizone System

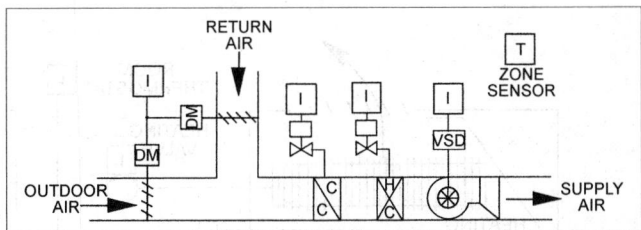

Fig. 44   Single-Zone Fan System

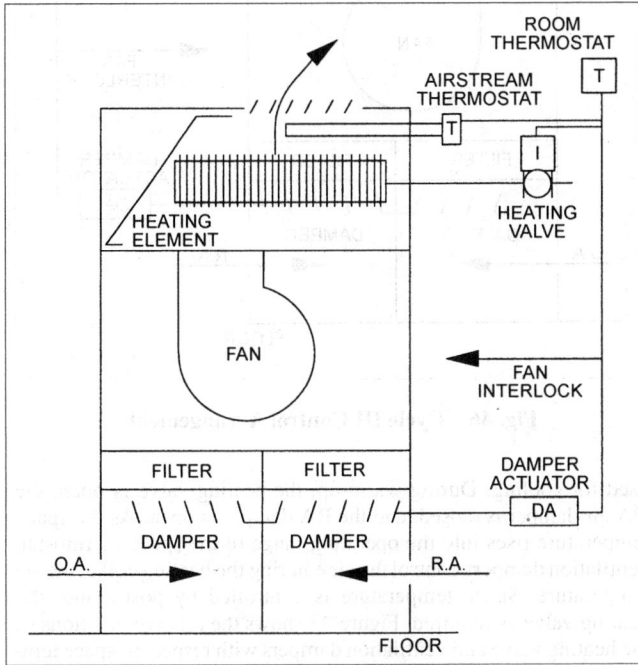

Fig. 45   Cycles I, II, and W Control Arrangements

closed or mix cold supply air with bypass air when the hot deck damper is closed.

A **single-zone system** (Figure 44) uses a constant volume air-handling unit (usually factory-packaged). No fan speed control is required because fan volume and duct static pressure are set by the design and selection of components. Single-zone systems do not require terminal boxes because the zone temperature can be maintained by varying the temperatures of the heating and cooling coils.

During warm-up, as determined by a time clock or manual switch, a constant heating supply air temperature is maintained. Because the terminal unit may be fully open, uncontrolled overheating can occur. It is preferable to allow unit thermostats to maintain complete control of their terminal units by reversing their action to the unit. During warm-up and unoccupied cycles, outdoor air dampers should be closed.

A **unit ventilator** is designed to heat, ventilate, and cool a space by introducing up to 100% outdoor air. Optionally, it can cool and dehumidify with a cooling coil (either chilled water or direct expansion). Heating can be by hot water, steam, or electric resistance. The control of these coils can be by valves or face and bypass dampers. Consequently, controls applied to unit ventilators are many and varied. The three most commonly used control schemes are Cycle I, Cycle II, Cycle III, and Cycle W.

Cycle I Control. Except during the warm-up stage, Cycle I (Figure 45), supplies 100% outdoor air at all times. During warm-up, the heating valve is open, the OA damper is closed, and the RA damper is open. As temperature rises into the operating range of the space thermostat, the OA damper opens fully, and the RA damper closes. The heating valve is positioned to maintain space temperature. The airstream thermostat can override space thermostat action on the heating valve to prevent discharge air from dropping below a minimum temperature. Figure 47 shows the positions of the heating valve and ventilation dampers in relation to space temperature.

Cycle II Control. During the heating stage, Cycle II (Figure 45) supplies a set minimum quantity of outdoor air. Outdoor air is gradually increased as required for cooling. During warm-up, the heating valve is open, the OA damper is closed, and the RA damper is open. As the space temperature rises into the operating range of the space thermostat, ventilation dampers move to their set minimum ventilation positions. The heating valve and ventilation dampers are operated in sequence as required to maintain space temperature. The airstream thermostat can override space thermostat action on the heating valve and ventilation dampers to prevent discharge air from dropping below a minimum temperature. Figure 49 shows the relative positions of the heating valve and ventilation dampers with respect to space temperature.

Cycle III Control. During the heating, ventilating, and cooling stages, Cycle III (Figure 46) supplies a variable amount of outdoor air as required to maintain the air entering the heating coil at fixed temperature (typically 55°F). When heat is not required, this air is

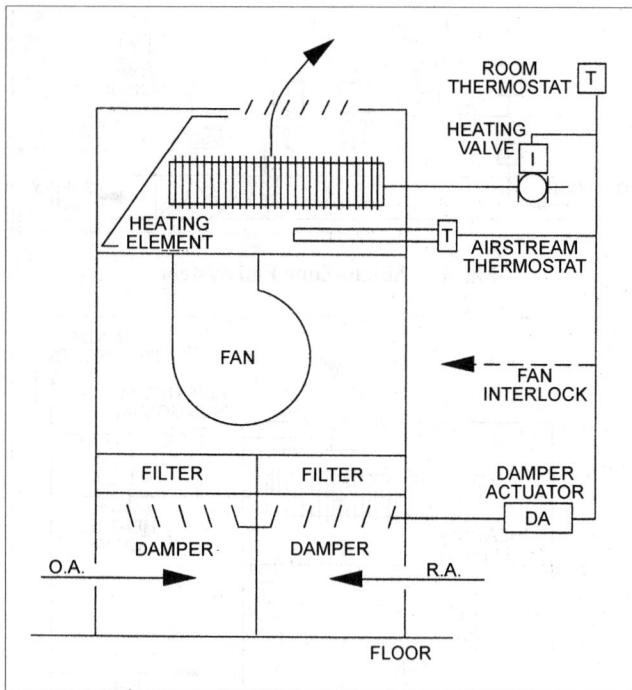

Fig. 46 Cycle III Control Arrangement

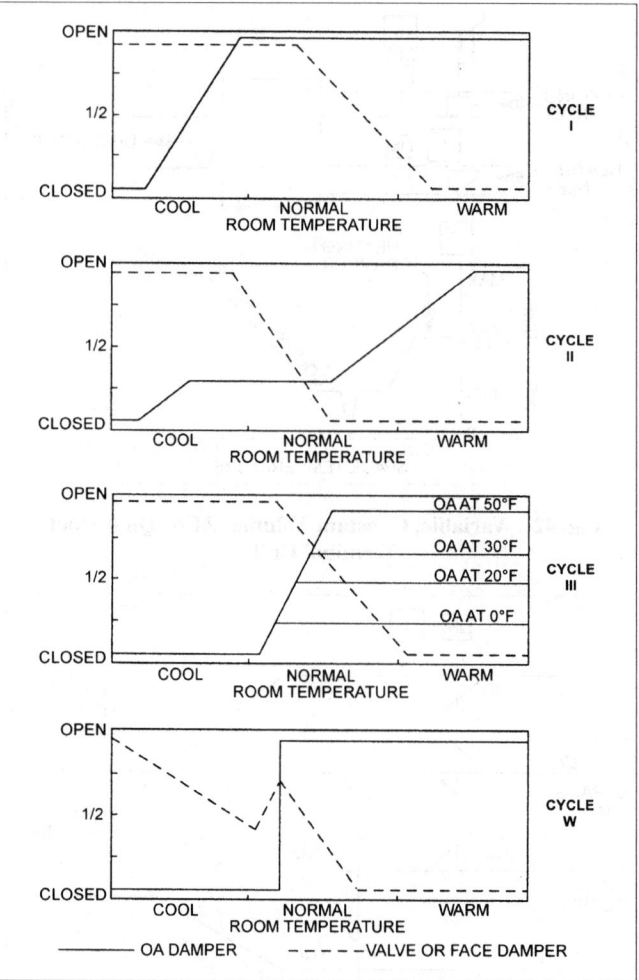

Fig. 47 Valve and Damper Positions with Respect to Room Temperature

used for cooling. During warm-up, the heating valve is open, the OA air damper is closed, and the RA damper is open. As the space temperature rises into the operating range of the space thermostat, ventilation dampers control the air entering the heating coil at the set temperature. Space temperature is controlled by positioning the heating valve as required. Figure 47 shows the relative positions of the heating valve and ventilation dampers with respect to space temperature.

Day/night thermostats are frequently used with any of these control schemes to maintain a lower space temperature during unoccupied periods by cycling the fan with the outdoor air damper closed. Another common option is a freezestat placed next to the heating coil that shuts the unit off when near-freezing temperatures are sensed.

Cycle W Control. Cycle W is similar to Cycle II, except the heating valve is controlled by the room thermostat and the dampers are controlled by the low-limit thermostat (Figures 45 and Figure 47).

Makeup air units (Figure 48) replace air exhausted from the building through exfiltration or by laboratory or industrial processes. Makeup air must be at or near space conditions to minimize uncomfortable air currents. The makeup air fan is usually turned on, either manually or automatically, whenever exhaust fans are turned on. The two-position outdoor air damper remains closed when the makeup fan is not in operation. The outdoor air limit control opens the preheat coil valve when outdoor air temperature drops to the point where the air requires heating to raise it to the desired supply air temperature. The discharge temperature controller then positions the face and bypass dampers to maintain that temperature. A capillary element thermostat located adjacent to the coil shuts the fan down for freeze protection if air temperature approaches freezing at any spot along the sensing element.

Multizone, Single-Duct System

This system (Figure 49) is most effective where all zones are in either heating or cooling mode at the same time. When necessary and where energy regulations allow it, reheat can be used to accommodate systems with simultaneous heating and cooling loads.

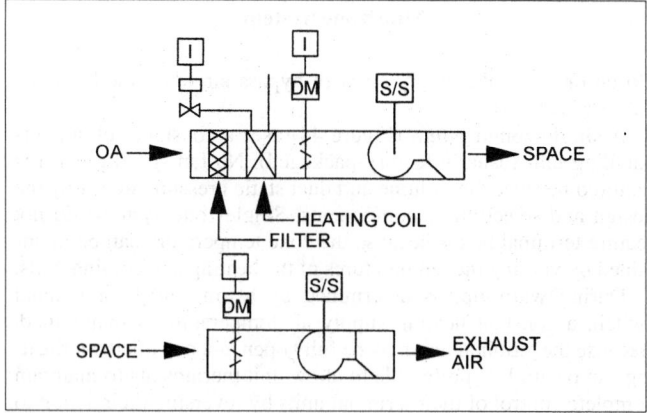

Fig. 48 Makeup Air Unit

Multizone Dual-Duct Systems

A **single supply fan system** uses a single fan to supply separate heating and cooling ducts (Figure 50). Terminal mixing boxes are used to control the zone temperature. Static control is similar to that in VAV single-duct systems, except that static pressure sensors are needed in each supply duct. A controller allows the sensor sensing

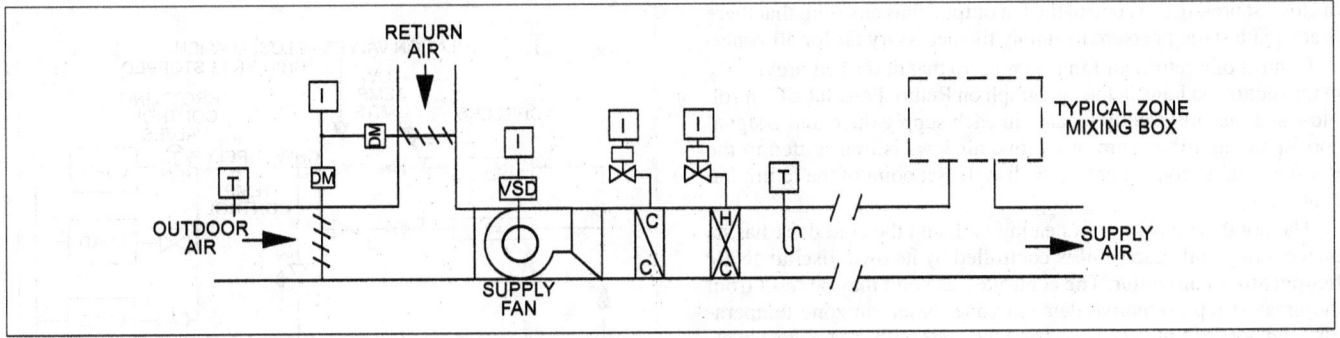

Fig. 49   Multizone Single-Duct System

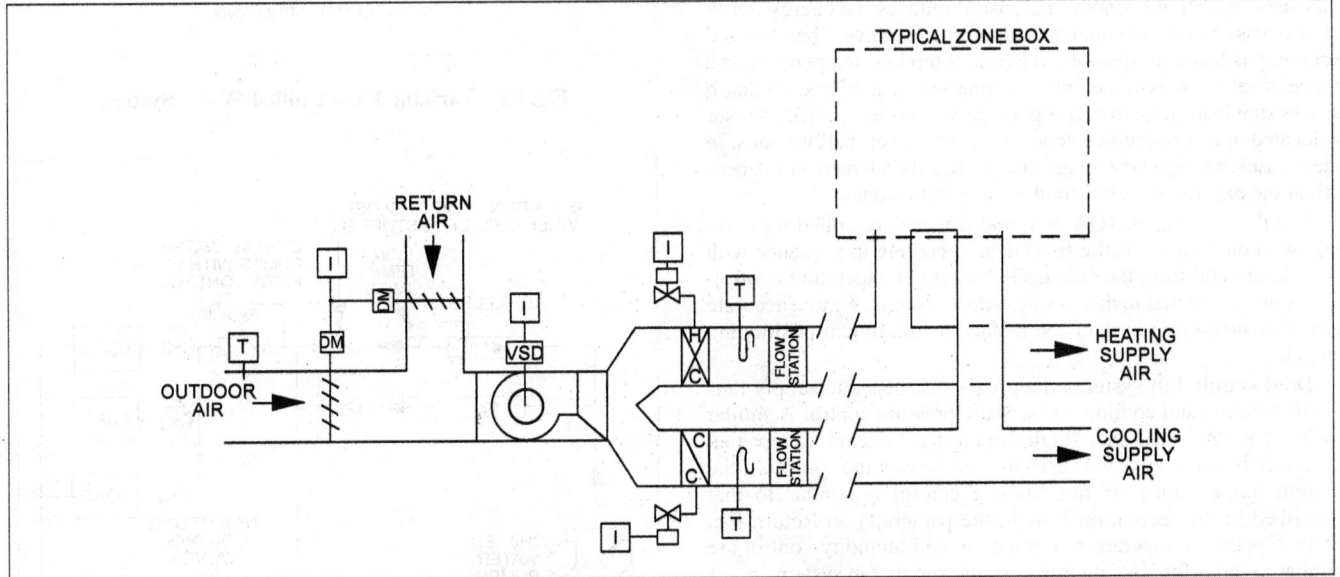

Fig. 50   Dual-Duct Single Supply Fan System

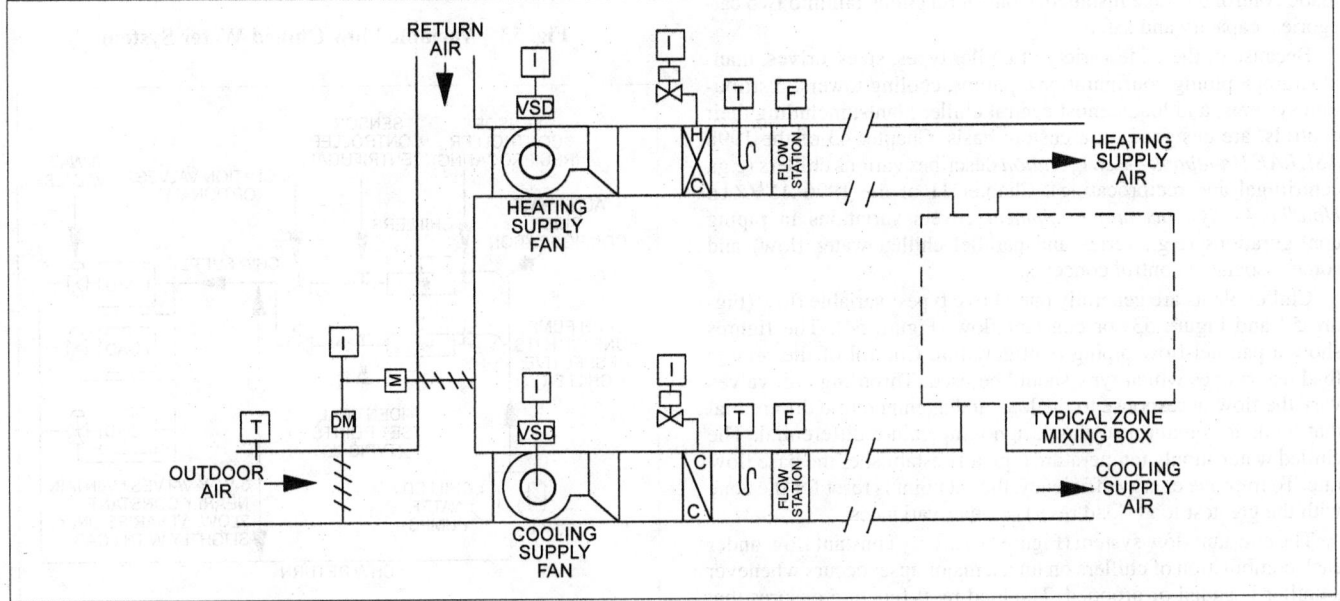

Fig. 51   Dual Supply Fan System

the lowest pressure to control the fan output, thus ensuring that there is adequate static pressure to supply the necessary air for all zones.

Control of a return air fan is similar to that described previously in the section on Fans in the paragraph on Return Fan Static Control. Flow stations are usually located in each supply duct, and a signal corresponding to the sum of the two airflows is transmitted to the RA fan volume controller to establish the set point of the return fan controller.

The hot deck has its own heating coil, and the cold deck has its own cooling coil. Each coil is controlled by its own **discharge air temperature controller**. The controller set point may be reset from the greatest representative demand zone: based on zone temperature, the hot deck may be reset from the zone with the greatest heating demand, and the cold deck from the zone with the greatest cooling demand.

Control based on the zone requiring the most heating or cooling increases operation economy because it reduces the energy delivered at less-than-maximum load conditions. However, the expected economy is lost if air quantity to a zone is too low, temperature in a space is set to an extreme value, a zone sensor is placed so that it senses spot loads (due to coffee pots, the sun, copiers, etc.), a sensor is located in an unoccupied zone, or a zone sensor malfunctions. In these cases, a weighted average of zone signals can recover the benefit at the expense of some comfort in specific zones.

Ventilation dampers (OA, RA, and EA) are controlled for cooling, with outdoor air as the first stage of cooling in sequence with the cooling coil from the cold deck discharge temperature controller. Control is similar to that in single-duct systems. A more accurate OA flow-measuring system can replace the minimum positioning switch.

Dual supply fan systems (Figure 51) use separate supply fans for the heating and cooling ducts. Static pressure control is similar to that for VAV dual-duct single-supply fan systems, except that each supply fan has its own static pressure sensor and control. If the system has a return air fan, volume control is similar to that described in the section on Fans in the paragraph on Return Fan Static Control. Temperature, ventilation, and humidity control are similar to those for VAV dual-duct single supply fan systems.

Chillers

The manufacturer almost always supplies chillers with an automatic control package installed. Control functions fall into two categories: capacity and safety.

Because of the wide variety of chiller types, sizes, drives, manufacturers, piping configurations, pumps, cooling towers, distribution systems, and loads, most central chiller plants, including their controls, are designed on a custom basis. Chapter 43 of the 1998 *ASHRAE Handbook—Refrigeration* describes various chillers (e.g., centrifugal and reciprocating). Chapter 11 of the 1996 *ASHRAE Handbook—Systems and Equipment* covers variations in piping configurations (e.g., series and parallel chilled water flow) and some associated control concepts.

Chiller plants are generally one of two types: variable flow (Figure 52 and Figure 53) or constant flow (Figure 54). The figures show a parallel-flow piping configuration. Control of the remote load determines which type should be used. Throttling coil valves vary the flow in response to the load and a temperature differential that tends to remain near the design temperature differential. The chilled water supply temperature typically establishes the base flow rate. To improve energy efficiency, the set point is reset for the zone with the greatest load (load reset) or other variances.

The constant flow system (Figure 54) is only constant flow under each combination of chillers on line; a major upset occurs whenever a chiller is added or dropped. The load reset function ensures that the zone with the largest load is satisfied, while supply or return water control treats average zone load.

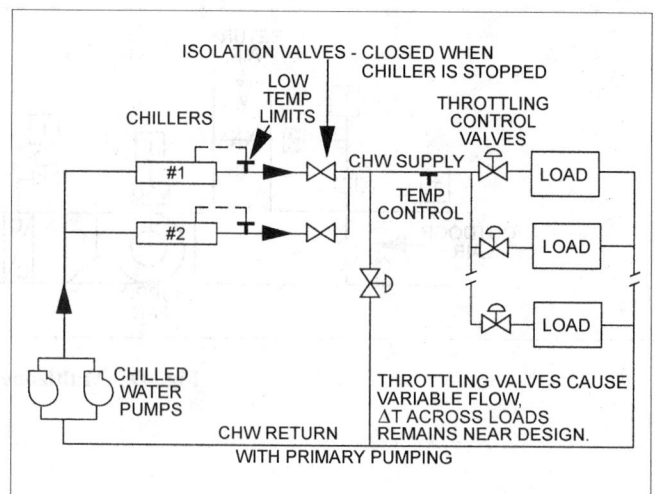

Fig. 52 Variable Flow Chilled Water System

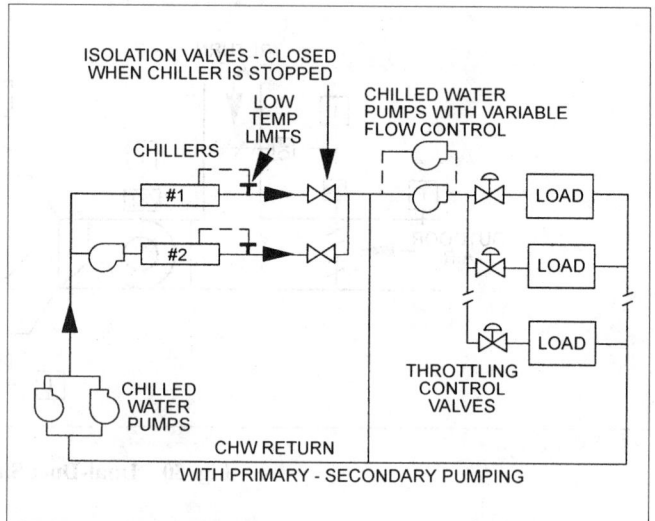

Fig. 53 Variable Flow Chilled Water System

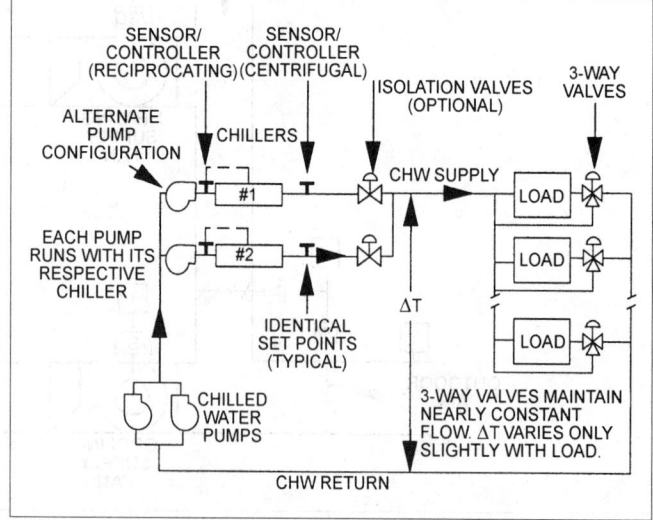

Fig. 54 Constant Flow Chilled Water System

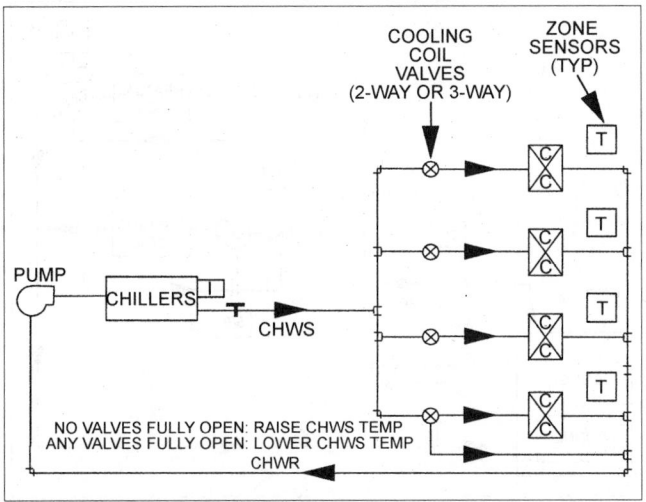

Fig. 55   Chilled Water Load Reset

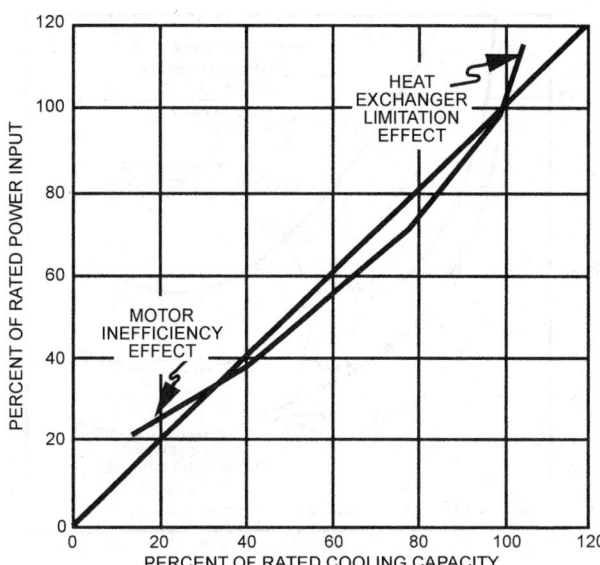

Fig. 56   Chiller Part-Load Characteristics at Design Refrigerant Head

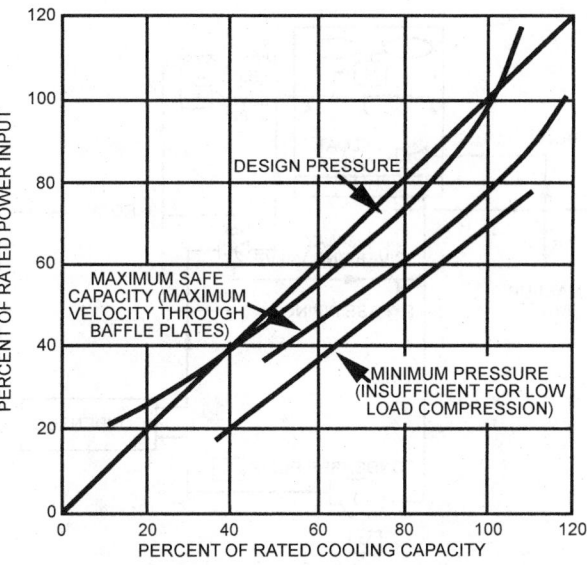

Fig. 57   Chiller Part-Load Characteristics with Variable Head

Refrigerant Pressure Optimization

Chiller efficiency is a function of the percent of full load on the chiller and the difference in refrigerant pressure between the condenser and the evaporator. In practice, the pressure is represented by condenser water exit temperature minus chilled water supply temperature. To reduce the refrigerant pressure, the chilled water supply temperature must be increased and/or the condenser water temperature decreased. An energy saving of 1 to 2% is obtained for each degree 1°F reduction.

The following methods are used to reduce refrigerant pressure:

1. Use chilled water load reset to raise the supply set point as load decreases. Figure 55 shows the basic function of this method. Varying degrees of sophistication are available, including computer control.
2. Lower condenser temperature to the lowest safe temperature (use manufacturer's recommendations) by keeping the cooling tower bypass valve closed, operating at full condenser water pump capacity, and maintaining full airflow in all cells of the cooling tower until water temperature is within about 5°F of the outdoor air wet-bulb temperature. However, the additional pump and fan power as well as the fan power of the VAV air handlers must be considered in calculating net energy savings.

Operation Optimization

Multiple-chiller plants should be operated at the most efficient point on the part-load curve. Figure 56 shows a typical part-load curve for a centrifugal chiller operated at design conditions. Figure 57 shows similar curves at different pressure-limiting conditions. Figure 58 indicates the point at which a chiller should be added or dropped in a two-unit plant. In general, the part-load curves are plotted for all combinations of chillers; then, the break-even point between *n* and *n* + 1 chillers can be determined.

Daily start-up of the chiller plant should be optimized to **minimize run time** based on start-up time of the air-handling units. Chillers are generally started at the same time as the first fan system. Chillers may be started early if the water distribution loop has great thermal mass; they may be started later if outdoor air can provide cooling to fan systems at start-up.

The **condenser water circuit** and control arrangement for the central plant are shown in Figure 59. The control system designer works with liquid chiller control when the equipment is integrated into the central chiller plant. Typically, cooling tower, chiller pump, and condenser pump control must be considered if the overall plant is to be stable and energy-efficient.

With centrifugal chillers, condenser supply water temperature is allowed to float as long as the temperature remains above a low limit. The manufacturer should specify the minimum entering condenser water temperature required for satisfactory performance of the particular chiller. The control schematic in Figure 59 works as follows: for a condenser supply temperature (e.g., above a set point of 75°F), the valve is open to the tower, the bypass valve is closed, and the tower fan or fans are operating. As water temperature decreases (e.g., to 65°F), tower fan speed can be reduced to low-speed operation if a two-speed motor is used. On a further decrease in condenser water supply temperature, the tower fan or fans stop and the bypass valve begins to modulate to maintain the acceptable minimum water temperature.

Water Heating

A basic constant volume hydronic system is shown in Figure 60. A variable speed drive could be added to the pump motor and the

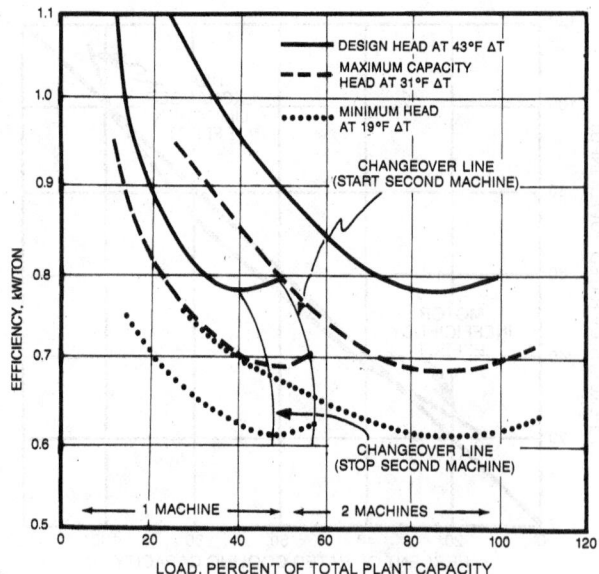

**Fig. 58 Multiple Chiller Operation Changeover Point—
Two Equal-Sized Chillers**

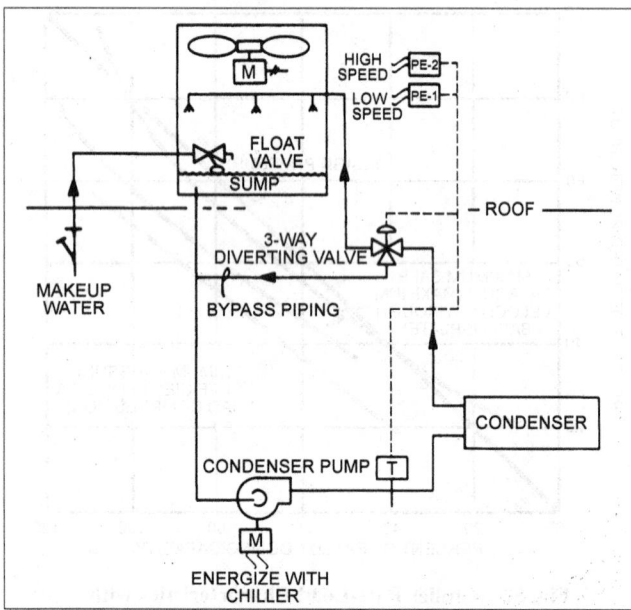

Fig. 59 Condenser Water Temperature Control

three-way valves could be changed to two-way valves to save energy.

SPECIAL APPLICATIONS

Mobile Unit Control

The operating point of any control that relies on pressure to operate a switch or valve varies with atmospheric pressure. Normal variations in atmospheric pressure do not noticeably change the operating point, but a change in altitude affects the control point to an extent governed by the change in absolute pressure. This pressure change is especially important with controls selected for use in land and aerospace vehicles that are subject to wide variations in altitude. The effect can be substantial; for example, barometric pressure decreases by nearly one-third as altitude increases from sea level to 10,000 ft.

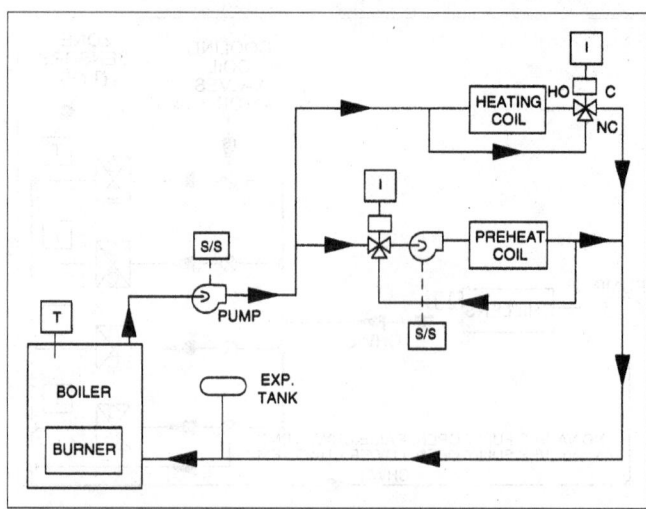

Fig. 60 Load and Zone Control in Simple Hydronic System

In mobile applications, three detrimental factors are always present in varying degrees—vibration, shock, and acceleration forces. Controls selected for service in mobile units must qualify for the specific conditions expected in the installation. In general, devices containing mercury switches, slow-moving or low-force contacts, or mechanically balanced components are unsuitable for mobile applications; electronic solid-state devices are generally less susceptible to the three factors.

Explosive Atmospheres

Sealed-in-glass contacts are not considered explosion-proof; therefore, other means must be provided to eliminate the possibility of a spark in an explosive atmosphere.

When using electric control, the control case and contacts can be surrounded with an explosion-proof case, permitting only the capsule and the capillary tubing to extend into the conditioned space. It is often possible to use a long capillary tube and mount the instrument case in a nonexplosive atmosphere. The latter method can be duplicated with an electronic control by placing an electronic sensor in the conditioned space and feeding its signal to an electronic transducer located in the nonexplosive atmosphere.

Because a pneumatic control uses compressed air, it is safe in otherwise hazardous locations. However, many pneumatic controls interface with electrical components. All electrical components require appropriate explosion-proof protection.

Sections 500 to 503 of the *National Electrical Code* (NFPA *Standard* 70) include detailed information on electrical installation protection requirements for various types of hazardous atmospheres.

Duct Static Pressure Limit Control

In addition to the remote static pressure controllers, a high-limit static pressure controller should be placed at the fan discharge (Figure 61) to turn the fan off or limit discharge static pressure in the event of excessive duct pressure (e.g., when a fire or smoke damper closes between the fan and the remote sensor). Supply-fan static pressure control devices such as inlet guide vanes and variable-speed drives should be interlocked to move to the minimum flow or closed position when the fan is not running; this precaution prevents fan overload or damage to ductwork upon start-up.

When selecting automatic controls, the type needed to control high or low limits (safety) must be considered. This control may be inoperative for days, months, or years, and then it must operate immediately to prevent serious damage to equipment or property. Separate operating and limit controls are always recommended, even for the same functions.

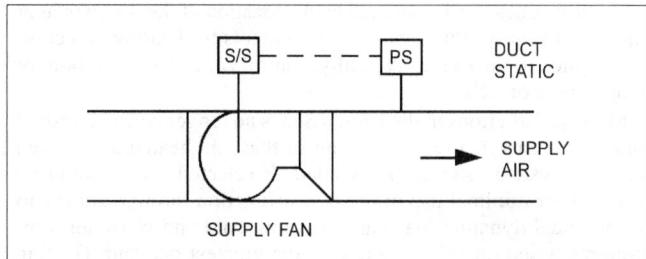

Fig. 61   High-Limit Static Pressure Controller

Duct static limit control prevents excessive duct pressures, usually at the discharge of the supply fan. Two variations are used: (1) the fan shutdown type, which is a safety high-limit control that turns the fans off; and (2) the controlling high-limit type (Figure 28), which is used in systems having zone fire dampers. When the zone fire damper closes, duct pressure drops, causing the duct static control to increase fan modulation; however, the controlling high limit will override.

Steam or hot water exchangers tend to be self-regulating and, in that respect, differ from electrical resistance heat transfer devices. For example, if airflow through a steam or hot water coil stops, coil surfaces approach the temperature of the entering steam or hot water, but cannot exceed it. Convection or radiation losses from the steam or hot water to the surrounding area take place, so the coil is not usually damaged. Electric coils and heaters, on the other hand, can be damaged when air stops flowing around them. Therefore, control and power circuits must interlock with heat transfer devices (pumps and fans) to shut off electrical energy when the device shuts down. Flow or differential pressure switches may be used for this purpose; however, they should be calibrated to energize only when there is airflow. This precaution shuts off power in case a fire damper closes or some duct lining blocks the air passage. Limit thermostats should also be installed to turn off the heaters when temperatures exceed safe operating levels.

Duct Heaters

The current in individual elements of electric duct heaters is normally limited to a maximum safe value established by the *National Electrical Code* or local codes. Two safety devices in addition to the airflow interlock device are usually applied to duct heaters (Figure 62). The automatic reset high-limit thermostat normally turns off the control circuit. If the control circuit has an inherent time delay or uses solid-state switching devices, a separate safety contactor may be desirable. The manual reset backup high-limit safety device is generally set independently to interrupt all current to the heater in case other control devices fail. An electric heater must have a minimum airflow switch and two high-temperature limit sensors; one with manual reset and one with automatic reset.

DESIGN CONSIDERATIONS AND PRINCIPLES

In designing and selecting the HVAC system for the entire building, the type, size, use, and operation of the structure must be considered. Subsystems such as fan and water supply are normally controlled by local automatic control or a local loop control. A local loop control includes the sensors, controllers, and controlled devices used with a single HVAC system and excludes any supervisory or remote functions such as reset and start-stop. However, local control is frequently extended to a central control point to diagnose malfunctions that might result in damage from delay, and to reduce labor and energy costs.

Distributed processing using microprocessors has augmented computer use at many locations other than the central control point. The local loop controller can be a direct digital controller (DDC)

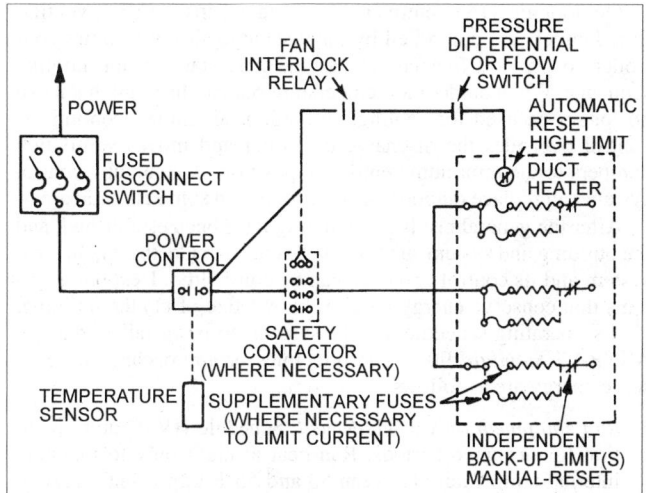

Fig. 62   Duct Heater Control

instead of a pneumatic or electric thermostat, and some energy management functions may be performed by a DDC.

Because HVAC systems are designed to meet maximum design conditions, they nearly always function at partial capacity. Because the system must be adjusted and operated for many years, the simplest control that produces the necessary results is usually the best.

Mechanical and Electrical Coordination

Even a pneumatic control includes wiring, conduit, switchgear, and electrical distribution for many electrical devices. The mechanical designer must inform the electrical designer of the total electrical requirements if the controls are to be wired by the electrical contractor. Requirements include (1) the devices to be furnished and/or connected, (2) electrical load, (3) location of electrical items, and (4) a description of each control function.

Coordination is essential. Proper coordination should produce a control diagram that shows the interface with other control elements to form a complete and usable system. As an option, the control engineer may develop a complete performance specification and require the control contractor to install all wiring related to the specified sequence. The control designer must run the final checks of drawings and specifications. Both mechanical and electrical specifications must be checked for compatibility and uniformity.

Building and System Subdivision

The following factors must be considered in the building and mechanical system subdivision:

- Heating and cooling loads as they vary—the ability to heat or cool the interior or exterior areas of a building at any time
- Occupancy schedules and the flexibility to meet needs without undue initial and/or operating costs
- Fire and smoke control and possibly compartmentation that matches the air-handling layout and operation

Control Principles for Energy Conservation

Temperature and Ventilation Control. VAV systems are typically designed to supply constant temperature air at all times. To conserve central plant energy, the temperature of the supply air can be raised in response to demand from the zone with the greatest load (load analyzer control). However, because more cool air must then be supplied to match a given load, the mechanical cooling energy saved may be offset by an increase in fan energy. Equipment operating efficiency should be studied closely before implementing temperature reset in cooling-only VAV systems.

Outdoor air (OA), return air (RA), and exhaust air (EA) ventilation dampers are controlled by the discharge air temperature controller to provide free cooling as the first stage in the cooling sequence. When outdoor air temperature rises to the point that it can no longer be used for cooling, an outdoor air limit (economizer) control overrides the discharge controller and moves ventilation dampers to the minimum ventilation position. An enthalpy control system can replace outdoor air limit control in some climatic areas.

After the general needs of a building have been established, and the building and system subdivision has been made, the mechanical system and its control approach can be considered. Designing systems that conserve energy requires knowledge of (1) the building, (2) its operating schedule, (3) the systems to be installed, and (4) ASHRAE *Standard* 90.1. The principles or approaches that conserve energy are as follows:

1. *Run equipment only when needed.* Schedule HVAC unit operation for occupied periods. Run heat at night only to maintain internal temperature between 50 and 55°F to prevent freezing. Start morning warm-up as late as possible to achieve design internal temperature by occupancy time, considering residual space temperature, outdoor temperature, and equipment capacity (optimum start control). Under most conditions, equipment can be shut down some time before the end of occupancy, depending on internal and external load and space temperature (optimum stop control). Calculate shutdown time so that space temperature does not drift out of the selected comfort zone before the end of occupancy.
2. *Sequence heating and cooling.* Do not supply heating and cooling simultaneously. Central fan systems should use cool outdoor air in sequence between heating and cooling. Zoning and system selection should eliminate, or at least minimize, simultaneous heating and cooling. Also, humidification and dehumidification should not take place concurrently.
3. *Provide only the heating or cooling actually needed.* Reset the supply temperature of hot and cold air (or water).
4. Supply heating and cooling from the most efficient source. Use free or low-cost energy sources first, then higher cost sources as necessary.
5. *Apply outdoor air control.* When on minimum outdoor air, use no less than that recommended by ASHRAE *Standard* 62. In areas where it is cost-effective, use enthalpy rather than dry-bulb temperature to determine whether outdoor or return air is the most energy-efficient air source for the cooling mode.

System Selection

The mechanical system significantly affects the control of zones and subsystems. The type of system and the number and location of zones influence the amount of simultaneous heating and cooling that occurs. For exterior building sections, heating and cooling should be controlled in sequence to minimize simultaneous heating and cooling. In general, this sequencing must be accomplished by the control system because only a few mechanical systems (e.g., two-pipe systems and single-coil systems) have the ability to prevent simultaneous heating and cooling. Systems that require engineered control systems to minimize simultaneous heating and cooling include the following:

- *VAV cooling with zone reheat.* Reduce cooling energy and/or air volume to a minimum before applying reheat.
- *Four-pipe heating and cooling for unitary equipment.* Sequence heating and cooling.
- *Dual-duct systems.* Condition only one duct (either hot or cold) at a time. The other duct should supply a mixture of outdoor and return air.
- *Single-zone heating/cooling.* Sequence heating and cooling.

Some exceptions exist, such as of dehumidification with reheat.

Control zones are determined by the location of the thermostat or temperature sensor that sets the requirements for heating and cooling supplied to the space. Typically, control zones are for a room or an open area of a floor.

Many jurisdictions in the United States no longer permit constant volume systems that reheat cold air or that mix heated and cooled air. Such systems should be avoided. If selected, they should be designed for minimal use of the reheat function through zoning to match actual dynamic loads and resetting cold and warm air temperatures based on the zone(s) with the greatest demand. Heating and cooling supply zones should be structured to cover areas of similar load. Areas with different exterior exposures should have different supply zones.

Systems that provide changeover switching between heating and cooling prevent simultaneous heating and cooling. Some examples are hot or cold secondary water for fan coils or single-zone fan systems. They usually require small operational zones, which have low load diversity, to permit changeover from warm to cold water without occupant dissatisfaction.

Systems for building interiors usually require year-round cooling and are somewhat simpler to control than exterior systems. These interior areas normally use all-air systems with a constant supply air temperature, with or without VAV control. Proper control techniques and operational understanding can reduce the energy used to treat these areas. Reheat should be avoided. General load characteristics of different parts of a building may lead to selecting different systems for each.

Load Matching

With individual room control, the environment in a space can be controlled more accurately and energy can be conserved if the entire system can be controlled in response to the major factor influencing the load. Thus, water temperature in a water heating system, steam temperature or pressure in a steam heating system, or delivered air temperature in a central fan system can be varied as building load varies. Control on the entire system relieves individual space controls of part of their burden and provides more accurate space control. Also, modifying the basic rate of heating or cooling input in accordance with the entire system load reduces losses in the distribution system.

The system must always satisfy the area or room with the greatest demand. Individual controls handle demand variations in the area the system serves. The more accurate the system zoning, the greater is the control, the smaller are the distribution losses, and the more effectively space conditions are maintained by individual controls.

Buildings or zones with a modular arrangement can be designed for subdivision to meet occupant needs. Before subdivision, operating inefficiencies can occur if a zone has more than one thermostat. In an area where one thermostat activates heating while another activates cooling, the terminals should be controlled from a single thermostat until the area is properly subdivided.

Size of Controlled Area

No individually controlled area should exceed about 5000 ft² because the difficulty of obtaining good distribution and of finding a representative location for the space control increases with zone area. Each individually controlled area must have similar load characteristics throughout. Equitable distribution, provided through competent engineering design, careful equipment sizing, and proper system balancing, is necessary to maintain uniform conditions throughout an area. The control can measure conditions only at its location; it cannot compensate for nonuniform conditions caused by improper distribution or inadequate design. Areas or rooms having dissimilar load characteristics or different conditions to be maintained should be controlled individually. The smaller the controlled area, the better the control and the better the performance and flexibility.

Location of Space Sensors

Space sensors and controllers must be located where they accurately sense the variables they control and where the condition is representative of the area (zone) they serve. In large open areas having more than one zone, thermostats should be located in the middle of their zones to prevent them from sensing conditions in surrounding zones. Typically, space temperature controllers or sensors are placed in the following locations.

- **Wall-mounted thermostats or sensors** are usually placed on inside walls or columns in the space they serve. Avoid outside wall locations. Mount thermostats where they will not be affected by heat from sources such as direct sun rays; wall pipes or ducts; convectors; or direct air currents from diffusers or equipment (e.g., copy machines, coffee makers, or refrigerators). Air circulation should be ample and unimpeded by furniture or other obstructions, and the thermostat should be protected against mechanical injury. Thermostats located in spaces such as corridors, lobbies, or foyers should be used to control those areas only.
- **Return air thermostats** can control floor-mounted unitary conditioners such as induction or fan-coil units and unit ventilators. On induction and fan-coil units, the sensing element is behind the return air grille. On classroom unit ventilators that use up to 100% outdoor air for natural cooling, however, a forced flow sampling chamber should be provided for the sensing element. The sensing element should be located carefully to avoid radiant effect and to ensure adequate air velocity across the element.

If return air sensing is used with a central fan system, locate the sensing element as near as possible to the space being controlled to eliminate any influence from other spaces and the effect of any heat gain or loss in the duct. Where supply/return light fixtures are used to return air to a ceiling plenum, the return air sensing element can be located in the return air opening. Be sure to offset the set point to compensate for the heat from the light fixtures.

- **Diffuser-mounted thermostats** usually have sensing elements mounted on circular or square ceiling supply diffusers and depend on aspiration of room air into the supply airstream. They should be used only on high-aspiration diffusers adjusted for a horizontal air pattern. The diffuser on which the element is mounted should be in the center of the occupied area of the controlled zone.

Lowered Night Temperature

When temperatures during unoccupied periods are lower than those normally maintained during occupied periods, an automatic timer often establishes the proper day and night temperature time cycle. Allow sufficient time in the morning to pick up the conditioning load well before there is any heavy increase. Night setback temperatures are often monitored and controlled more closely with control systems. These computer based systems take into account variables such as outdoor temperature, system capacity, and building mass to determine optimal start-up and shutdown times.

REFERENCES

ASHRAE. 1989. Energy efficient design of new buildings except low-rise residential buildings. ANSI/ASHRAE *Standard* 90.1-1989.

ASHRAE. 1989. Ventilation for acceptable indoor air quality. ANSI/ASHRAE *Standard* 62-1989.

Englander, S.L. and L.K. Norford. 1992. Saving fan energy in VAV systems—Part 2: Supply fan control for static pressure minimization. *ASHRAE Transactions* 98(1):19-32.

NFPA. 1996. *National electrical code.* ANSI/NFPA *Standard* 70-96. National Fire Protection Association, Quincy, MA.

SOUND AND VIBRATION CONTROL

MECHANICAL equipment is one of the major sources of sound in a building. Primary considerations often given to the selection and use of mechanical equipment in buildings have generally been those directly related to the intended use of the equipment, like cooling, heating, and ventilation. However, for environmental considerations in critical listening spaces, like conference rooms and auditoria, and for many other spaces with light-weight building structures and variable-volume air distribution systems, the sound generated by mechanical equipment and its effects on the overall acoustical environment in a building must be considered. Thus, the selection of mechanical equipment and the design of equipment spaces should be undertaken with an emphasis on (1) the intended uses of the equipment and (2) the goal of providing acceptable sound and vibration levels in occupied spaces of the building in which the equipment is located.

The system concept of noise control is used throughout, in that each of the components is related to the source-path-receiver chain. The noise generation is the source; it travels from the source via a path, which can be through the air (airborne) or through the structure (structure-borne) until it reaches the ear of the receiver. When the combination of this chain is complex, it can be referred to it as a **system effect**. So, noise propagates from the sources through the air distribution ducts, through the structure, and through combinations of paths, reaching the occupants. All mechanical components, from dampers to diffusers to junctions, may produce sound by the nature of the airflow through and around them. As a result, almost all components must be considered. Since sound travels effectively in the same or opposite direction of airflow, upstream and down-stream paths are often equally important.

Adequate noise and vibration control in a heating, ventilating, and air-conditioning (HVAC) system is not difficult to achieve during the design phase of the system, providing basic noise and vibration control principles are understood. This chapter discusses basic sound and vibration principles and data needed by HVAC designers. Divided into two main sections, one on sound, the other on vibration, this chapter is organized differently than versions in Handbooks prior to 1995. This chapter includes more information on acoustic design guidelines and system design requirements. Most of the equations associated with sound and vibration control design in HVAC systems have been replaced by related tables and simpler design procedures. The equations that have been removed can be

The preparation of this chapter is assigned to TC 2.6, Sound and Vibration Control.

found in the 1991 and 1992 *ASHRAE Handbooks*. In addition, technical discussions and detailed HVAC component and system design examples can be found in *Algorithms for HVAC Acoustics* (Reynolds and Bledsoe 1991).

Other publications that cover sound and vibration control in HVAC systems include the 1997 *ASHRAE Handbook—Fundamentals*, which covers fundamentals associated with sound and vibration in HVAC; Schaffer (1991), who provides specific guidelines for the acoustic design and related construction phases associated with HVAC systems, troubleshooting sound and vibration problems, and HVAC sound and vibration specifications; Ebbing and Blazier (1998), who interpret and clarify how users can make the best use of HVAC manufacturers' acoustical data and application information; and Reynolds and Bevirt (1994), who cover instrument requirements, instrument and measurement calibration procedures, measurement procedures, and specification and construction installation review procedures associated with sound and vibration measurements relative to HVAC systems.

DATA RELIABILITY

The data in this chapter comes both from consulting experience and research studies. When applying the data, especially to situations that extrapolate from the original data, use caution. While specific uncertainties are not stated for each data set, the sound levels or attenuation data are probably within 2 dB of measured or expected results. However, significantly greater variations may occur, especially in the low frequency ranges and particularly in the 63 Hz octave band. While specific data sets may have a wide uncertainty range, experience has demonstrated the usefulness of combining data sets for estimating the sound level. If done correctly, these estimates usually result in space sound pressure levels within 5 dB of measured levels.

SOUND

ACOUSTICAL DESIGN OF HVAC SYSTEMS

The solution to nearly every HVAC system noise and vibration control problem involves examining the sound sources, the sound transmission paths, and the receivers. For most HVAC systems, the sound sources are associated with the building mechanical and electrical equipment. As indicated in Figure 1, sound travels between a source and receiver through many possible sound and/or vibration

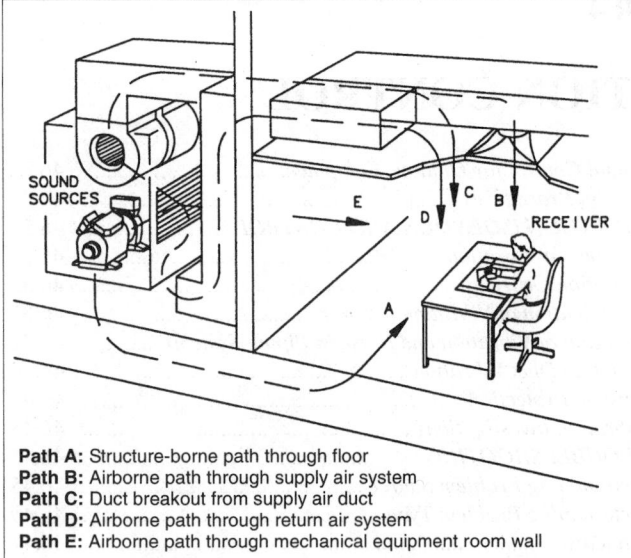

Path A: Structure-borne path through floor
Path B: Airborne path through supply air system
Path C: Duct breakout from supply air duct
Path D: Airborne path through return air system
Path E: Airborne path through mechanical equipment room wall

Fig. 1   Typical Paths in HVAC Systems

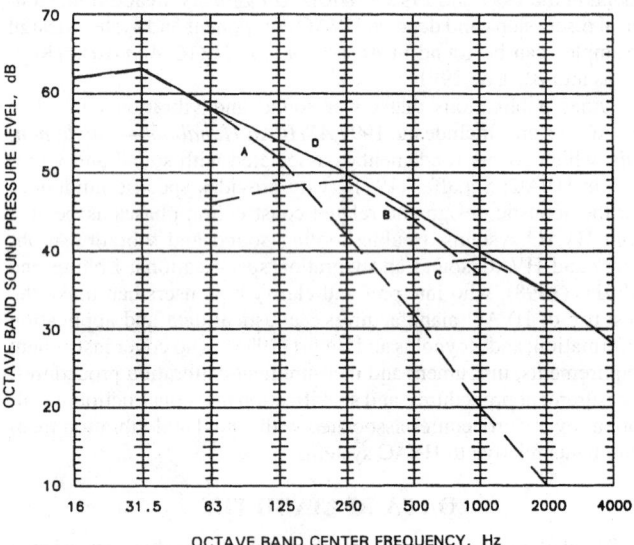

Curve A: Fan sound pressure levels
Curve B: VAV valve sound pressure levels
Curve C: Diffuser sound pressure levels
Curve D: Total sound pressure levels

Fig. 2   Illustration of Well-Balanced HVAC Sound Spectrum
for Occupied Spaces

transmission paths. Sound sources are the components that either generate noise, like electric motors, or produce noise when air passes by them, like dampers or diffusers. Sound receivers are generally the people who occupy a building. For most HVAC systems, designers can not always modify or change the sound source or receiver characteristics. Thus, system designers are most often constrained to modifying the sound transmission paths as a means of achieving desired sound levels in occupied areas of a building.

Different sources produce sounds that have different frequency distributions. For examples, fan noise generally contributes to the sound levels in the 16 Hz through 250 Hz octave frequency bands (Figure 2 as curve A). Variable air volume (VAV) valve noise usually contributes to the sound levels in the 63 Hz through 1000 Hz octave frequency bands (curve B, Figure 2). Diffuser noise is usually in the 250 Hz through 8000 Hz octave frequency bands (curve

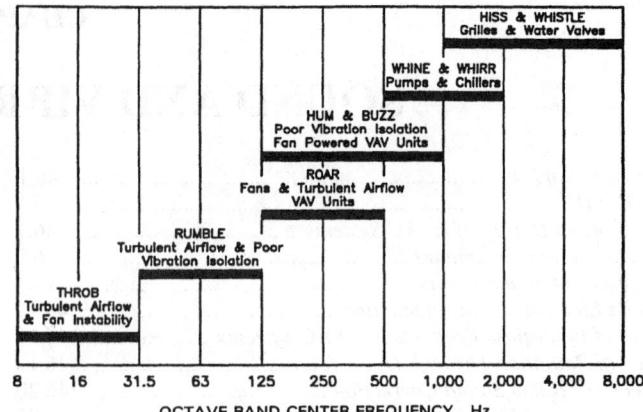

Fig. 3   Frequency Ranges of Likely Sources of
Sound-Related Complaints

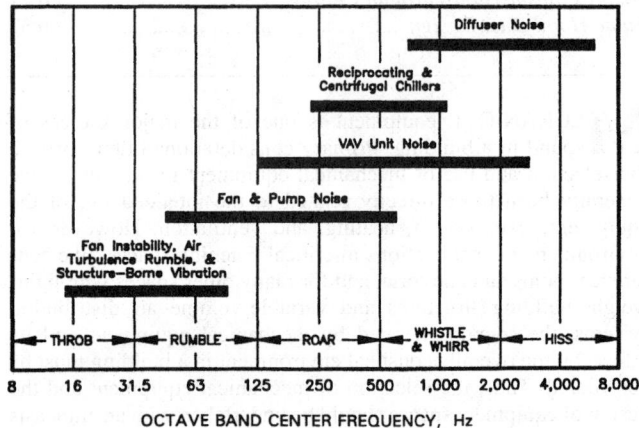

Fig. 4   Frequencies at Which Different Types of Mechanical
Equipment Generally Control Sound Spectra

C, Figure 2). The overall sound pressure level associated with all of these sound sources combined is shown as curve D in Figure 2.

Occupants may complain when an acceptable sound spectrum is not achieved in an occupied space. Figure 3 shows the frequency ranges of the most likely sources of HVAC sound-related complaints. Figure 4 shows the frequencies at which different types of mechanical equipment generally control the sound spectra in a room.

BASIC DESIGN TECHNIQUES

When selecting fans and other related mechanical equipment and when designing air distribution systems to minimize the sound transmitted from different components to the occupied spaces that they serve, consider the following:

1. Design the air distribution system to minimize flow resistance and turbulence. High flow resistance increases the required fan pressure, which results in higher noise being generated by the fan. Turbulence increases the flow noise generated by duct fittings and dampers in the air distribution system, especially at low frequencies.

2. Select a fan to operate as near as possible to its rated peak efficiency when handling the required quantity of air and static pressure. Also, select a fan that generates the lowest possible noise but still meets the required design conditions for which it is selected. Using an oversized or undersized fan that does not operate at or near rated peak efficiency can result in substantially higher noise levels.

3. Design duct connections at both the fan inlet and outlet for uniform and straight air flow. Failure to do this can result in severe turbulence at the fan inlet and outlet and in flow separation at the fan blades. Both of these can significantly increase the noise generated by the fan.

4. Select duct silencers that do not significantly increase the required fan total static pressure. Duct silencers can significantly increase the required fan static pressure if improperly selected. Selecting silencers with static pressure losses of 0.35 in. water gage. or less can minimize silencer airflow regenerated noise.

5. Place fan-powered mixing boxes associated with variable-volume air distribution systems away from noise-sensitive areas.

6. Minimize flow-generated noise by elbows or duct branch takeoffs, whenever possible, by locating them at least four to five duct diameters from each other. For high velocity systems, it may be necessary to increase this distance to up to ten duct diameters in critical noise areas. The use of flow straighteners or honeycomb grids, often called "egg crates", in the necks of short-length takeoffs that lead directly to grilles, registers, and diffusers is preferred to the use of volume extractors that protrude into the main duct airflow.

7. Keep airflow velocity in the duct as low as possible (1500 fpm or less) near critical noise areas by expanding the duct cross-section area. However, do not exceed an included expansion angle of greater than 15°. Flow separation, resulting from expansion angles greater than 15°, may produce rumble noise. Expanding the duct cross-section area will reduce potential flow noise associated with turbulence in these areas.

8. Use turning vanes in large 90° rectangular elbows and branch takeoffs. This provides a smoother transition in which the air can change flow direction, thus reducing turbulence.

9. Place grilles, diffusers and registers into occupied spaces as far as possible from elbows and branch takeoffs.

10. Minimize the use of volume dampers near grills, diffusers and registers in acoustically critical situations.

11. Vibration isolate all vibrating reciprocating and rotating equipment if mechanical equipment is located on upper floors or is roof-mounted. Also, it is usually necessary to vibrate isolate the mechanical equipment that is located in the basement of a building as well as piping supported from the ceiling slab of a basement, directly below tenant space. It may be necessary to use flexible piping connectors and flexible electrical conduit between rotating or reciprocating equipment and pipes and ducts that are connected to the equipment.

12. Vibration isolate ducts and pipes, using spring and/or neoprene hangers for at least the first fifty feet from the vibration-isolated equipment.

13. Use barriers near outdoor equipment when noise associated with the equipment will disturb adjacent properties if barriers are not used. In normal practice, barriers typically produce no more than 15 dB of sound attenuation in the mid frequency range.

Table 1 lists several common sound sources associated with mechanical equipment noise. Anticipated sound transmission paths and recommended noise reduction methods are also listed in the table. Airborne and/or structure-borne sound can follow any or all of the transmission paths associated with a specified sound source. Schaffer (1991) has more detailed information in this area.

Table 1   Sound Sources, Transmission Paths, and Recommended Noise Reduction Methods

Sound Source	Path No.
Circulating fans; grilles; registers; diffusers; unitary equipment in room	1
Induction coil and fan-powered VAV mixing units	1, 2
Unitary equipment located outside of room served; remotely located air-handling equipment, such as fans, blowers, dampers, duct fittings, and air washers	2, 3
Compressors, pumps, and other reciprocating and rotating equipment (excluding air-handling equipment)	4, 5, 6
Cooling towers; air-cooled condensers	4, 5, 6, 7
Exhaust fans; window air conditioners	7, 8
Sound transmission between rooms	9, 10

No.	Transmission Paths	Noise Reduction Methods
1	Direct sound radiated from sound source to ear	Direct sound can be controlled only by selecting quiet equipment.
	Reflected sound from walls, ceiling, and floor	Reflected sound is controlled by adding sound absorption to the room and to equipment location.
2	Air- and structure-borne sound radiated from casings and through walls of ducts and plenums is transmitted through walls and ceiling into room	Design duct and fittings for low turbulence; locate high velocity ducts in noncritical areas; isolate ducts and sound plenums from structure with neoprene or spring hangers.
3	Airborne sound radiated through supply and return air ducts to diffusers in room and then to listener by Path 1	Select fans for minimum sound power; use ducts lined with sound-absorbing material; use duct silencers or sound plenums in supply and return air ducts.
4	Noise transmitted through equipment room walls and floors to adjacent rooms	Locate equipment rooms away from critical areas; use masonry blocks or concrete for equipment room walls and floor.
5	Vibration transmitted via building structure to adjacent walls and ceilings, from which it radiates as noise into room by Path 1	Mount all machines on properly designed vibration isolators; design mechanical equipment room for dynamic loads; balance rotating and reciprocating equipment.
6	Vibration transmission along pipes and duct walls	Isolate pipe and ducts from structure with neoprene or spring hangers; install flexible connectors between pipes, ducts, and vibrating machines.
7	Noise radiated to outside enters room windows	Locate equipment away from critical areas; use barriers and covers to interrupt noise paths; select quiet equipment.
8	Inside noise follows Path 1	Select quiet equipment.
9	Noise transmitted to an air diffuser in a room, into a duct, and out through an air diffuser in another room	Design and install duct attenuation to match transmission loss of wall between rooms.
10	Sound transmission through, over, and around room partition	Extend partition to ceiling slab and tightly seal all around; seal all pipe, conduit, duct, and other partition penetrations.

EQUIPMENT SOUND LEVELS

An accurate acoustical analysis of HVAC systems depends on reliable equipment sound data. These data are often available from equipment manufacturers in the form of sound pressure levels at a specified distance from the equipment or, more preferably, equipment sound power levels. Standards used to determine equipment and component sound data are listed at the end of this chapter.

When reviewing manufacturers' sound data, require certification that the data have been obtained according to one or more of the relevant industry standards. If they have not, the equipment should be rejected in favor of equipment whose data have been obtained according to relevant industry standards.

Fans

Prediction of Fan Sound Power. The sound power generated by a fan performing at a given duty is best obtained from a manufacturer's test data taken under approved test conditions (AMCA *Standard* 300 or ASHRAE *Standard* 68/AMCA *Standard* 330). Applications of air-handling products range from stand-alone fans to systems with various modules and appurtenances. These appurtenances and modules can significantly effect the air handler's sound power levels. In addition, different types of fans that provide similar aerodynamic performance can have significant acoustical differences.

Fan sound levels, even when determined by test, may be quite different once the fan is installed in an air handler, which in effect creates a new acoustic environment. Proper testing to determine the resulting sound power levels once a fan is installed in an air handler is essential. Fan manufacturers are in the best position to supply information on their products, and should be consulted for data when evaluating the acoustic performance of fans for an air handler application. Similarly, air handler manufacturers are in the best position to supply acoustic information on air handlers. For all of these reasons, information obtained directly from manufacturers should be preferred over generic fan-only sound power levels.

For a detailed description of fan operations, see the chapter on fans in this Handbook. Different fan types have different noise characteristics and within a fan type, several factors influence noise.

Point of Fan Operation. Fan selection at the calculated point of maximum efficiency is common practice to ensure minimum power consumption. In general, fan sound is at a minimum near the point of maximum efficiency. Noise increases as the operating point shifts to the right (higher airflow and lower static pressure) and low frequency noise can increase substantially at operating points to the left of maximum efficiency (lower airflow and higher static pressure). These operating points should be avoided.

Figure 5 shows some data taken for a plenum fan. Note that the A-weighted sound power rises to the right as the higher frequency sound increases with airflow. The low frequency (50 Hz, 1/3 octave band) shows a sharp rise left of maximum efficiency.

Blade-Pass Frequency. The blade-pass frequency is represented by the number of times per second a fans impeller passes a stationary item or f_{bp} = (rpm × number of impeller blades)/60. All fans generate a tone at this frequency and its multiples. Whether this tone is objectionable or barely noticeable depends on the type and design of the fan and the point of operation.

Housed Centrifugal Fans. Forward curved (FC) fans are commonly used in many air handlers. The blade pass of FC fans is typically less prominent and at a higher frequency than other fans. The most distinguishing acoustical concern of FC fans is the prevalent occurrence of low-frequency rumble. FC fans are commonly thought to have 16 Hz, 31.5 Hz, and 63 Hz rumble, particularly when operating to the left of the maximum efficiency point.

Backward inclined (BI) fans and airfoil (AF) fans are generally louder at the blade pass-frequency than a given FC fan selected for the same duty; but BI fans are much more energy efficient at higher

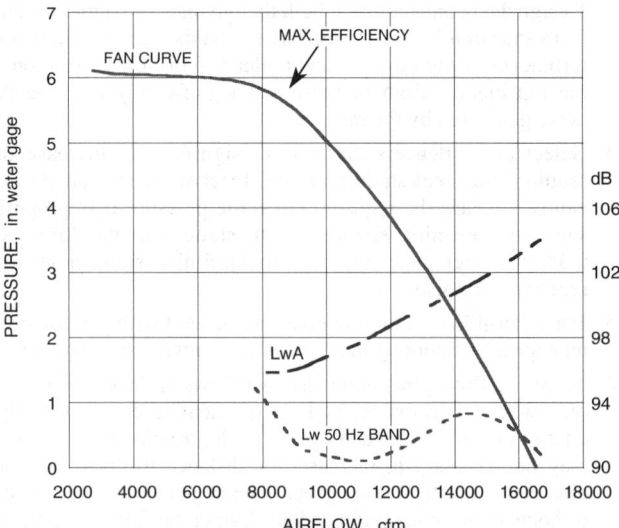

Fig. 5 Plenum Fan Sound Profiles

pressures and airflow. The blade pass tone generally increases in prominence with increasing fan speed and is typically in a frequency range that is difficult to attenuate. Below the blade pass frequency, these fans generally have lower sound amplitude than FC fans and are often quieter at the high frequencies.

Plenum Fans. In this type of fan, air flows into the fan impeller through an inlet bell located in the chamber wall. Airfoil and backward inclined fan impellers are used in a plenum fan configuration. The fan, which has no housing around the fan impeller, discharges directly into the chamber pressurizing the plenum and forcing air through the attached ductwork. These fans produce substantially lower discharge sound power levels if the fan plenum is appropriately sized and acoustically treated with sound absorptive material.

Vaneaxial Fans. Axial fans generally have the lowest amplitudes of sound at low frequency of any fan. For this reason they are often used in applications where the higher frequency noise can be managed with attenuation devices. In the useful operating range, the noise from axial fans is a strong function of the inlet airflow symmetry and blade tip speed.

Propeller Fans. Sound from propeller fans generally has a low-frequency dominated spectrum shape, and the blade passage frequency is typically prominent and occurs in the low-frequency bands due to the small number of blades. Propeller fan blade passage frequency noise is very sensitive to inlet obstructions. For some propeller fan designs, the shape of the fan venturi (inlet) is also a very important variable that affects sound levels. In some applications, the noise of a propeller fan is described as sounding like a helicopter. Propeller fans are most commonly used on condensers and for power exhausts.

Minimizing Fan Sound. To minimize the sound attenuation required in the air distribution system, proper selection and installation of the fan (or fans) are vitally important. The following factors should be considered:

- Design the air distribution system for minimum airflow resistance. High system resistance requires more power, and, consequently, fans generate higher sound power levels.
- Carefully analyze pressure losses. Higher than expected system resistance may result in higher sound power levels than originally estimated.
- Examine the sound power levels of different types of fans. Different fans generate different sound levels and produce different sound spectra. Select a fan (or fans) that will generate the lowest

possible sound power levels, commensurate with other fan selection requirements.

- Many fans generate tones at the blade passage frequency and its harmonics that may require additional acoustical treatment. The amplitude of these tones can be affected by resonance in the duct system, fan design, and inlet flow distortions caused by poor inlet duct design, or by the operation of an inlet volume control damper. When possible, variable speed control for volume control is preferable to volume control dampers to control fan noise.

- Design duct connections at both the fan inlet and outlet for uniform and straight airflow. Avoid unstable, turbulent, and swirling inlet airflow. Deviation from acceptable practice can severely degrade both the aerodynamic and acoustic performance of any fan and invalidate the manufacturer's ratings or other performance predictions.

Variable Air Volume (VAV) Systems

General Design Factors. VAV systems can significantly reduce energy cost due to their ability to modulate air capacity. But they can be the source of fan noise that is very difficult to mitigate. To avoid these potential problems, the designer should carefully design the ductwork and the static pressure control systems and select the fan or air handling unit and its air modulation device.

As in other aspects of HVAC design, the duct system should be designed for the lowest practical static pressure loss, especially in the ductwork closest to the fan or air handling unit. High airflow velocity and convoluted duct routing can cause airflow distortions that result in excessive pressure drop and fan instabilities that are responsible for excessive noise, fan stall, or both.

Many VAV noise complaints have been traced to control problems. While most of the problems are associated with improper installation, many are caused by poor design. The designer should specify high-quality fans or air handling units that will operate in their optimum ranges, not at the edge of their operation ranges where low system tolerances can lead to inaccurate fan flow capacity control. Also, the in-duct static pressure sensors should be placed in duct sections having the lowest possible air turbulence; that is, at least three equivalent duct diameters from any elbow, takeoff, transition, offset, or damper.

VAV noise problems have been traced to improper air balancing. For example, air balance contractors commonly balance an air distribution system by setting all damper positions without considering the possibility of reducing the fan speed. The end result is a duct system in which no damper is completely open and the fan is delivering air at a higher static pressure than would otherwise be necessary. If the duct system is balanced with at least one balancing damper wide open, the fan speed could be reduced with a corresponding reduction in fan noise. Lower sound levels will occur if most balancing dampers are wide open or eliminated.

Fan Selection. For constant-volume systems, fans should be selected to operate at maximum efficiency at the fan design airflow rate. However, VAV systems must be selected to operate with efficiency and stability throughout its range of modulation. For example, a fan selected for peak efficiency at full output may aerodynamically stall at an operating point of 50% of full output resulting in significantly increased low frequency noise. Similarly, a fan selected to operate at the 50% output point may be very inefficient at full output, resulting in substantially increased fan noise at all frequencies. In general, a fan selected for a VAV system should be selected for a peak efficiency at an operating point of around 70 to 80% of the maximum required system capacity. This usually means selecting a fan that is one size smaller than that required for peak efficiency at 100% of maximum required system capacity (Figure 6). When the smaller fan is operated at higher capacities, it will produce up to 5 dB more noise. This occasional increase in sound level is usually more tolerable than the stall-related sound

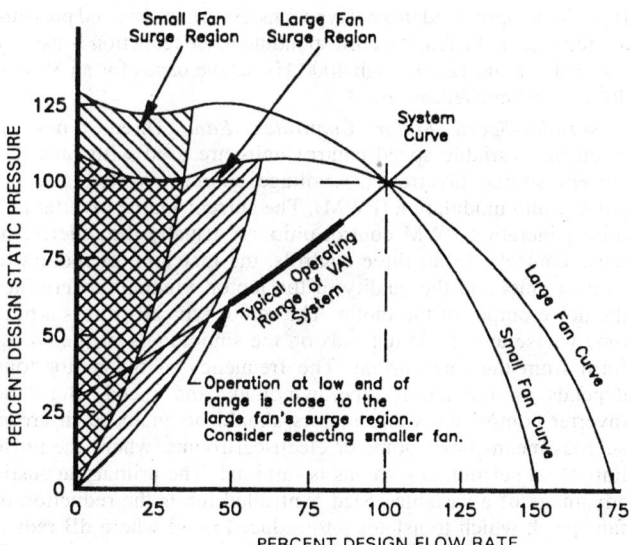

Fig. 6   Basis for Fan Selection in VAV Systems

problems that can occur with a larger fan operating at less than 100% design capacity most of the time.

Air Modulation Devices. Variable capacity control methods can be divided into three general categories: (1) variable inlet vanes (sometimes called inlet guide vanes) or discharge dampers, which yield a new fan system curve at each vane or damper setting; (2) variable pitch fan blades (usually used on in-line axial fans), which adjust the blade angle for optimum efficiency at varying capacity requirements; and (3) variable speed motor drives where the motor speed is varied by modulation of the power line frequency or by mechanical means such as gears or continuous belt adjustment. While inlet vane and discharge damper volume controls can add noise to a fan system at reduced capacities, variable speed motor drives and variable-pitch fan blade systems are quieter at reduced air output than at full air output.

Variable Inlet Vanes and Discharge Dampers. Variable inlet vanes vary airflow capacity by changing the inlet airflow to a fan wheel. This type of air modulation varies the total air volume and pressure at the fan while the fan speed remains constant. While, fan pressure and air volume reductions at the fan result in duct system noise reductions by reduced air velocity and pressures in the duct work, there is an associated increase in fan noise caused by the airflow turbulence and flow distortions at the inlet vanes acting as a fan inlet obstruction. Fan manufacturers' test data have shown that, on airfoil type centrifugal fans, as vanes mounted inside the fan inlet (nested inlet vanes) close, the sound level at the blade passing frequency of the fan increases by 2 to 8 dB, depending on the amount of total air volume restricted. For inlet vanes that are mounted externally the increase is on the order of 2 to 3 dB. Forward curved fan wheels with inlet vanes are about 1 to 2 dB quieter than airfoil fan wheels. In-line axial type fans with inlet vanes generate increased noise levels of 2 to 8 dB in the low frequency octave bands for a 25% to 50% closed vane position.

Discharge dampers are typically located immediately downstream of the supply air fan and reduce airflow and increase pressure drop across the fan while the fan speed remains constant. Because of the air turbulence and flow distortions created by the high-pressure drop across discharge dampers there is a high probability that duct rumble will occur near the damper location. If the dampers are throttled to a very low flow, a stall condition can occur at the fan also resulting in an increase in low-frequency noise.

Variable Pitch Fan Blades for Capacity Control. Variable pitch fan blade controls vary the fan blade angle in order to reduce the overall airflow through the fan. This type of capacity control system is predominantly used in axial type fans. As air volume and pressure are reduced at the fan, the corresponding noise reduction is usually 2 to 5 dB in the 125 through 4000 Hz octave bands for an 80% to 40% air volume reduction.

Variable-Speed Motor Controlled Fan. Three types of electronic variable speed control units are used with fans: (1) current source inverter, (2) voltage source inverter, and (3) pulse-width modulation (PWM). The current source inverter and third generation PWM control units are usually the quietest of these controls. In all three controls, the matching of motors to control units and the quality of the motor windings determines the noise output of the motor. The motor typically emits a pure tone whose amplitude depends on the smoothness of the waveform from the line current. The frequency of the motor tone depends on the motor type, windings, and speed. Both the inverter control units and motors should be enclosed in areas, such as mechanical rooms or electrical rooms, where the noise impact on surrounding rooms is minimal. The primary acoustic advantage of a variable speed controlled fan is the reduction of fan speed, which translates into reduced noise where dB reduction is approximately 50[log (higher speed/lower speed)]. Because this speed reduction generally follows the fan system curve, a fan selected at optimum efficiency initially (lowest noise) does not lose that efficiency as the speed is reduced.

The following guidelines should be observed in the use of variable-speed controllers:

1. Select fan vibration isolators on the basis of the lowest reasonable speed of the fan. For example, the lowest rotational speed might be 600 rpm for a 1000 rpm fan in a commercial system.
2. Select a controller with a feature often called "critical frequency jump band." This feature allows a user to program the controller to avoid certain fan or motor speed settings that might excite vibration isolation system or building structure resonance frequencies.
3. Check the intersection of the fan's various speed curves with the duct system curve, keeping in mind that the system curve does not go to zero static pressure at no flow when selecting a fan that will be controlled by a variable-speed motor controller. (The system curve is asymptotic at the static pressure control setpoint, typically 1 to 1.5 in. water gage) An improperly selected fan may be forced to operate in its stall range at slower fan speeds.

Terminal Units. Fans and pressure reducing valves in VAV units should have manufacturer published sound data that indicate the sound power levels (1) that are discharged from the low pressure end of the unit and (2) that radiate from the exterior shell of the unit. These sound power levels vary as a function of valve position and fan point of operation. Sound data for VAV units should be obtained according to the procedures specified by ARI *Standard* 880.

If the VAV unit is located away from critical areas (such as above a storeroom or corridor), the sound radiated from the shell of the unit may be of no concern. If, however, the unit is located above a critical space and separated from the space by a ceiling with little or no sound transmission loss at low frequencies, the sound radiated from the shell may produce sound in the space below that exceeds the desired noise criterion. In this case, it may be necessary to relocate the unit to a noncritical area or to enclose it with a construction having a high transmission loss. In general, fan-powered VAV units should not be placed above or near any room with a required sound criterion rating of less than RC 40(N).

Systems that use VAV units that will not completely shut off are often helpful. If too many units shut off simultaneously, excessive duct system pressure at low flow can occur. This condition can sometimes cause a fan stall, resulting in accompanying roar, rumble, and surge. Using minimum airflow instead of shutoff VAV units help prevents this condition from occurring.

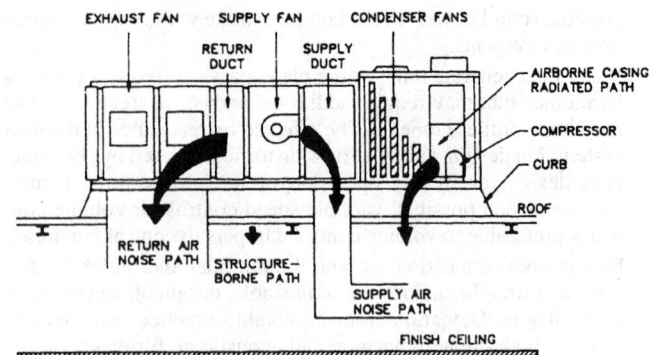

Fig. 7   Sound Paths for Typical Rooftop Installations

Rooftop Mounted Air Handlers

Rooftop air handlers can have unique noise control requirements because these units are often integrated into a light roof construction. Large roof openings are often required for supply and return air duct connections. These ducts run directly from noise-generating rooftop air handlers to the building interior. Generally, the space or distance between the roof-mounted equipment and the closest occupied spaces below the roof is insufficient to apply standard sound control treatments. Rooftop units should be placed above spaces that are not acoustically sensitive and as far as possible from the nearest occupied space. This measure can reduce the amount of sound control treatment necessary.

The four common sound transmission paths associated with rooftop air handlers (Figure 7) are

1. Airborne through the bottom of the rooftop unit to spaces below
2. Structure-borne from vibrating equipment in the rooftop unit to the building structure
3. Duct-borne through the supply air duct from the air handler
4. Duct-borne through return air duct to the air handler

Airborne paths are associated with casing-radiated sound that passes through the air handler enclosure and roof structure to the spaces below. Airborne sound can either be a result of air handler noise or from other equipment in the rooftop unit. When a rooftop unit is placed over a large opening in the roof structure through which the supply and return air ducts pass, the opening should be divided into two openings sized to accommodate only the supply and return air ducts. These openings should be properly sealed after the installation of the ducts. If a large single opening exists under the rooftop unit, it should be structurally and acoustically sealed around the supply and return air ducts with one or more layers of gypsum board or other similar material. Airborne sound transmission to spaces below a rooftop unit can be greatly reduced. One way is by placing a rooftop unit on a structural support extending above the roof structure and running the supply and return air ducts horizontally along the roof for several duct diameters before the ducts turn to penetrate the roof. The roof deck/ceiling system below the unit can be constructed to adequately attenuate the sound radiated from the bottom of the unit.

Proper vibration isolation can minimize structure-borne sound and vibration from vibrating equipment in a rooftop unit. Special curb mounting bases are available to support and isolate rooftop units. For roofs constructed with open web joists, thin long span slabs, wooden construction, and any unusually light construction, evaluate all equipment weighing more than 300 lb to determine the additional deflection of the structure at the mounting points caused by the equipment. Isolator deflection should be a minimum of 15 times the additional deflection. If the required spring isolator deflection exceeds commercially available products, stiffen the supporting structure or change the equipment location.

Table 2   Duct Breakout Insertion Loss—Potential Low-Frequency Improvement over Bare Duct and Elbow

Discharge Duct Configuration, 12 ft of Horizontal Supply Duct	Duct Breakout Insertion Loss at Low Frequencies, dB			Side View	End View
	63 Hz	125 Hz	250 Hz		
Rectangular duct: no turning vanes (reference)	0	0	0		
Rectangular duct: one-dimensional turning vanes	0	1	1		
Rectangular duct: two-dimensional turning vanes	0	1	1		
Rectangular duct: wrapped with foam insulation and two layers of lead	4	3	5		
Rectangular duct: wrapped with glass fiber and one layer 5/8 in. gypsum board	4	7	6		
Rectangular duct: wrapped with glass fiber and two layers 5/8 in. gypsum board	7	9	9		
Rectangular plenum drop (12 ga.): three parallel rectangular supply ducts (22 ga.)	1	2	4		
Rectangular plenum drop (12 ga.): one round supply duct (18 ga.)	8	10	6		
Rectangular plenum drop (12 ga.): three parallel round supply ducts (24 ga.)	11	14	8		
Rectangular (14 ga.) to multiple drop: round mitered elbows with turning vanes, three parallel round supply ducts (24 ga.)	18	12	13		
Rectangular (14 ga.) to multiple drop: round mitered elbows with turning vanes, three parallel round lined double-wall, 22 in. OD supply ducts (24 ga.)	18	13	16		
Round drop: radius elbow (14 ga.), single 37 in. diameter supply duct	15	17	10		

Ductborne transmission of sound through the supply air duct consists of two components: sound transmitted from the air handler through the supply air duct system to occupied areas and sound transmitted via duct breakout through a section or sections of the supply air duct close to the air handler to occupied areas. Experience has indicated that sound transmission below 250 Hz via duct breakout is often a major acoustical limitation for many rooftop installations. Excessive low-frequency noise associated with fan noise and air turbulence in the region of the discharge section of the fan and the first duct elbow results in duct rumble, which is difficult to attenuate. This problem is often made worse by the presence of a duct with a high aspect ratio at the discharge section of the fan.

Rectangular ducts with duct lagging are often ineffective in reducing duct breakout noise. Using either a single- or dual-wall round duct with a radiused elbow coming off the discharge section

of the fan can control duct breakout. If space does not allow for the use of a single duct, the duct can be split into several parallel round ducts. Another method that is effective is the use of an acoustic plenum chamber constructed with a minimum 2 in. thick, dual-wall plenum panel, which is lined with fiberglass and which has a perforated inner liner at the discharge section of the fan. Either round or rectangular ducts can be taken off the plenum as necessary to supply the rest of the air distribution system. Table 2 illustrates twelve possible rooftop discharge duct configurations with their associated low-frequency noise reduction potential (Harold 1986, 1991; Beatty 1987).

Ductborne transmission of sound through the return air duct of a rooftop unit is often a problem. Generally only one short return air duct section runs from the plenum space above a ceiling and the return air section of the air handler. This short run does not adequately attenuate the sound between the fan inlet and the spaces below the air handler. The sound attenuation through the return air duct can be improved by adding at least one (more if possible) branch division where the return air duct is split into two sections that extend several duct diameters before they terminate into the plenum space above the ceiling. The inside surfaces of all the return air ducts should be lined with a minimum of 1 in. thick duct liner. If conditions permit, duct silencers in the duct branches or an acoustic plenum chamber at the air handler inlet gives better sound conditions.

Aerodynamically Generated Sound in Ducts

Although fans are a major source of sound in HVAC systems, they are not the only sound source. Aerodynamic sound is generated at duct elbows, dampers, branch takeoffs, air modulation units, sound attenuators, and other duct elements. Produced by the interaction of moving air with the structure, the sound power levels in each octave frequency band depend on the duct element geometry and the turbulence of the airflow and the airflow velocity in the vicinity of the duct element. Duct-related aerodynamic noise problems can be avoided by

- Sizing ductwork or duct configurations so that air velocity is low (see Tables 3 and 4)
- Avoiding abrupt changes in duct cross-section area
- Providing smooth transitions at duct branches, takeoffs, and bends

Table 3 Maximum Recommended Duct Airflow Velocities Needed to Achieve Specified Acoustic Design Criteria

Main Duct Location	Design RC(N)	Maximum Airflow Velocity, fpm	
		Rectangular Duct	Circular Duct
In shaft or above drywall ceiling	45	3500	5000
	35	2500	3500
	25	1700	2500
Above suspended acoustic ceiling	45	2500	4500
	35	1750	3000
	25	1200	2000
Duct located within occupied space	45	2000	3900
	35	1450	2600
	25	950	1700

Notes:
1. Branch ducts should have airflow velocities of about 80% of the values listed.
2. Velocities in final runouts to outlets should be 50% of the values or less.
3. Elbows and other fittings can increase airflow noise substantially, depending on the type. Thus, duct airflow velocities should be reduced accordingly.

Table 4 Maximum Recommended "Free" Supply Outlet and Return Air Opening Velocities Needed to Achieve Specified Acoustic Design Criteria

Type of Opening	Design RC(N)	"Free" Opening Airflow Velocity, fpm
Supply air outlet	45	625
	40	560
	35	500
	30	425
	25	350
Return air opening	45	750
	40	675
	35	600
	30	500
	25	425

Note: The presence of diffusers or grilles can increase sound levels a little or a lot, depending on how many diffusers or grilles are installed and on their design, construction, installation, etc. Thus, allowable outlet or opening airflow velocities should be reduced accordingly.

- Attenuating sound generated at duct fittings with sufficient sound attenuation elements between a fitting and corresponding air-terminal device

Duct Velocity. The amplitude of aerodynamically generated sound in ducts is generally proportional to between the fifth and sixth power of the duct airflow velocity in the vicinity of a duct fitting. So reducing duct airflow velocity significantly reduces flow generated noise. Table 3 (Schaffer 1991) and Table 4 (Egan 1988) recommend the maximum air velocities in duct sections and duct outlets to avoid problems associated with aerodynamically generated sound in ducts.

Dampers. Depending on its location relative to a duct terminal device, a damper can generate unwanted noise into an occupied area of a building. The noise can be transmitted down the duct to the discharge, or through the ceiling space into the occupied space below.

Volume dampers should not be placed closer than 5 ft from an air outlet for good design. When a volume control damper is installed close to an air outlet, the acoustic performance of the air outlet must be based on the air volume handled and on the pressure drop across the damper. The sound level produced by the damper is accounted for by adding a quantity to the diffuser sound rating. This quantity is proportional to the pressure ratio, which is the throttled pressure drop across the damper divided by the minimum pressure drop across the damper. Table 5 provides quantities to determine the effect of damper location on diffuser sound ratings.

Balancing dampers, equalizers, and other similar devices should not be placed directly upstream of air devices or open-ended ducts in acoustically critical spaces. They should be located 5 to 10 duct diameters from the termination device with acoustically lined duct joining the damper and duct termination device.

Plenums may be used to keep dampers further away from diffusers. Dampers may be installed at the plenum entrance with linear diffusers installed in the distribution plenum. The further a damper is installed from the outlet, the lower the resultant sound level.

Air Devices. Manufacturers' test data should be obtained in accordance with ASHRAE *Standard* 70 or ARI *Standard* 890(P) for room air terminal devices such as grilles, registers, diffusers, air handling light fixtures, and air-handling suspension bars. The room duct termination device should be selected to meet the noise criterion required or specified for the room. However, the manufacturer's sound power rating is obtained with a uniform velocity distribution throughout the diffuser neck or grille collar, which is often not met in practice. If a duct turn precedes the entrance to the diffuser or if a balancing damper is installed immediately before the diffuser, the airflow will be turbulent and the noise generated by the device will be substantially higher than the manufacturer's data by as much as 12 dB. In some cases, placing an equalizer grid in the neck of the diffuser reduces this turbulence substantially. The equalizer grid can help provide a uniform velocity gradient in the neck of the diffuser so the sound power generated in the field is closer to that listed in the manufacturer's catalog.

A flexible duct connection between the diffuser and the air supply duct or VAV unit provides a convenient means to align the diffuser with the ceiling grid. A misalignment in this connection that

Table 5 Decibels to Be Added to Diffuser Sound Rating to Allow for Throttling of Volume Damper

Location of Volume Damper	Damper Pressure Ratio					
	1.5	2	2.5	3	4	6
	dB to Be Added to Diffuser Sound Rating					
In neck of linear diffuser	5	9	12	15	18	24
In inlet of plenum of linear diffuser	2	3	4	5	6	9
In supply duct at least 5 ft from inlet plenum of linear diffuser	0	0	0	2	3	5

exceeds 1/4 the diffuser neck diameter over a length equal to two times the diffuser neck diameter can cause a significant increase in the diffuser sound power levels relative to the levels provided by the manufacturer. If the diffuser offset is less than 1/8 the length of the connection diameter, no appreciable increase in the sound power level will occur. If the offset is equal to or greater than the neck diffuser diameter over a connection length equal to two times the diffuser neck diameter, the sound power levels associated with the diffuser can increase as much as 12 dB.

At present, diffusers are rated in terms of noise criterion (NC) levels, which include a receiver room sound correction of usually 8 to 10 dB. The ratings may be useful for comparison between and among different diffusers, but are not helpful for design. The designer should request from the diffuser manufacturer the component sound power level data in octave bands, and use the sound power to estimate the effects of the diffusers on the sound level in different spaces.

Chillers and Air Cooled Condensers

All chillers and their associated equipment produce significant amounts of both broadband and tonal noise. The broadband noise is due to flows of both refrigerant and water, while the tonal noise is caused by the rotation of compressors, motors, and fans (in fan-cooled equipment). Chiller noise is usually significant in the octave bands from 250 through 1000 Hz.

Compressors. All compressors, except absorption, produce tonal noise. The acoustical differences among compressors generally relate to their tonal content.

- **Centrifugal compressor** tonal noise is due to the rotation of the impeller and the gears in geared machines. The tonal content is typically not very strong, except at reduced capacities. The centrifugal compressor sound pressure level (L_p) typically increases at reduced chiller capacity due to the extra turbulence induced in the refrigerant circuit by the inlet vanes. If, however, capacity is reduced by motor speed control, the resulting compressor L_p values decrease with decreasing capacity.
- **Reciprocating compressor** noise has a drumming quality, caused by the oscillating motion of the pistons. The tonal content is high and the sound level decreases very little with decreasing capacity.
- **Absorption chillers** produce relatively little noise themselves, but the flow of steam in their associated pumps and valves causes significant high-frequency noise. Noise levels increase with decreasing capacity as the valves close. Also, combustion air blowers on direct gas fired units can be noisy.
- **Scroll compressors** have relatively weak tones, and when they are found in capacities less than 75 tons, they may not cause noise-related complaints.
- **Screw compressors** (sometimes called helical rotor or rotary compressors) have very strong tones in the 250 through 2000 Hz octave bands. The rotor-induced tones are amplified by resonances in the oil separation circuit and by efficient sound radiation by the condenser and evaporator shells that are rigidly connected to the compressor via high-pressure piping. Recently, screw compressors have been the sources of chiller noise complaints; therefore, this type of compressor requires the most attention during noise and vibration control design.

Indoor Water-Cooled Chillers. The compressor is the dominant noise source in most water-cooled chillers. Any of the five compressor types listed previously can be used, but most water-cooled chillers use a either centrifugal or screw compressors. Factory sound data for indoor chillers is obtained via ARI *Standard* 575. The standard requires measuring the A-weighted and octave band sound pressure level values at several locations that are 1 m from the chiller and 1.5 m above the floor. Ratings are generally available at operating points of 25%, 50%, and 100%

of a chiller's nominal full capacity. The range of values for typical centrifugal and screw chillers is shown in Figure 8 and Figure 9, respectively.

ARI *Standard* 575 measurements are usually made in very large rooms with large amounts of sound absorption. The measured levels must be adjusted for each chiller installation to account for the size and surface treatment conditions of the mechanical room. For a

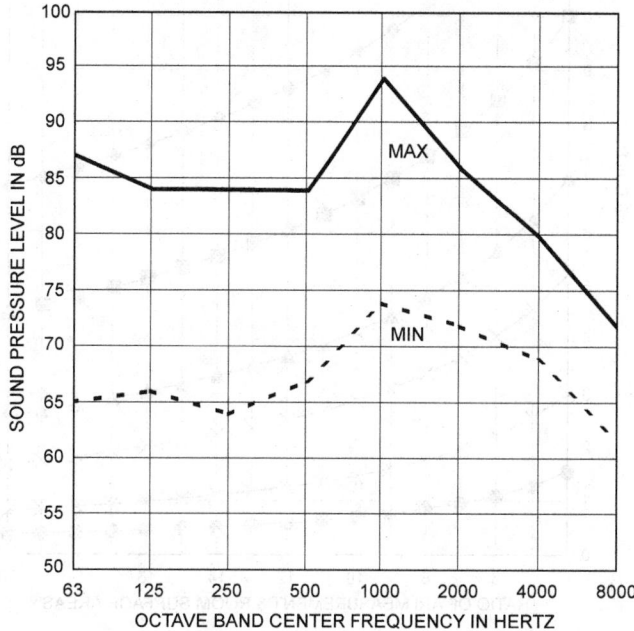

Fig. 8 Typical Minimum and Maximum ARI-575 L_p Values for Centrifugal Chillers—130 to 1300 Tons

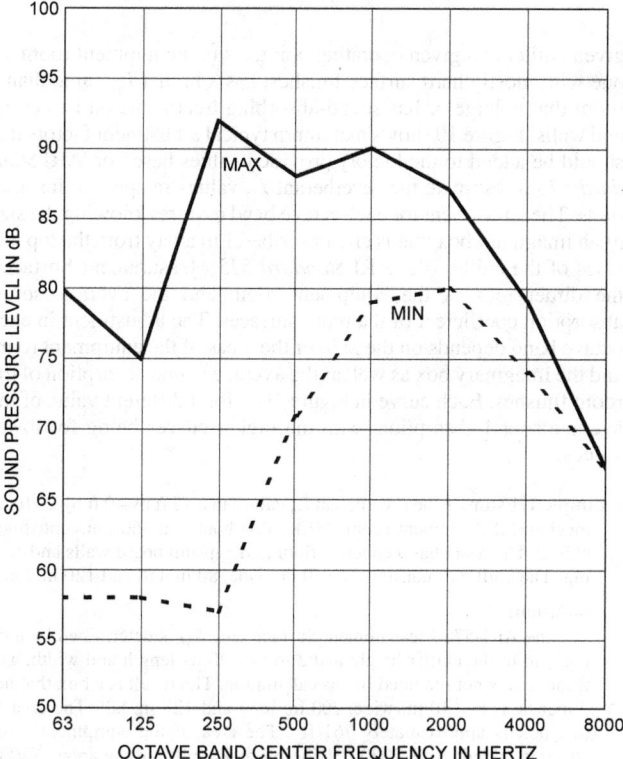

Fig. 9 Typical Minimum and Maximum ARI-575 L_p Values for Screw Chillers—130 to 400 Tons

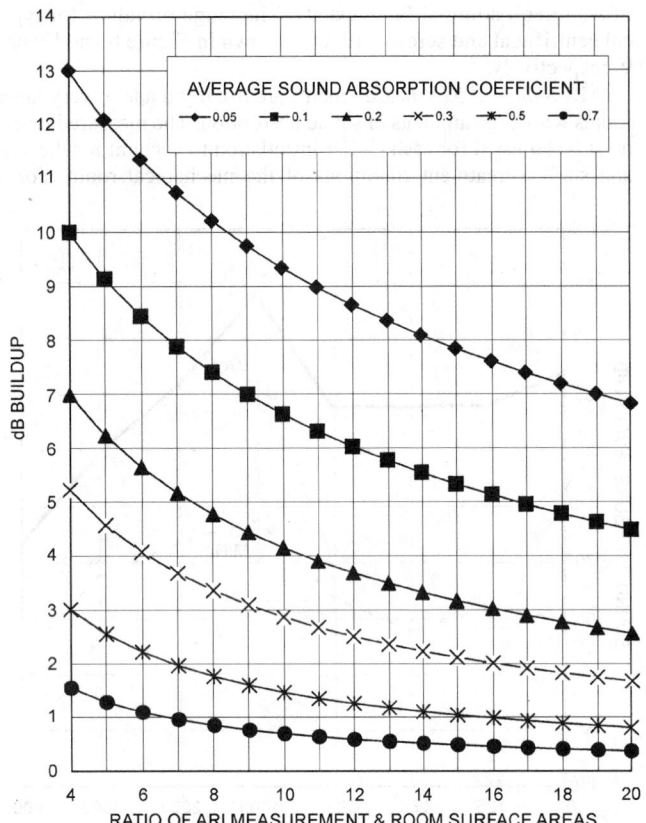

Fig. 10 Estimated dB Buildup in Mechanical Room for ARI-575 Chiller Sound Levels

	63	125	250	500	1000	2000	4000	8000
ARI *Std.* 575 L_p Values	73	74	73	72	74	72	69	63
Adjustment from Fig. 10	7	7	7	7	7	7	7	7
Approximate rev L_p in MER	80	81	80	79	81	79	76	70

The approximate reverberant L_p values in the last line of the above example can be used along with the sound transmission loss information found elsewhere in this chapter to estimate the transmitted L_p values in rooms adjacent to a chiller room.

Indoor chillers are offered with various types of factory noise reduction options ranging from compressor blankets (2 to 6 dBA reduction) to steel panel enclosures with sound absorbing inner surfaces (up to 18 dBA reduction). The amount of reduction is limited because of the structure-borne transmission of compressor vibration into the equipment frame and heat exchanger shells, which act as sounding boards.

Field noise control options include full-size sheet metal housing with specially-treated openings for piping, electrical conduit, and ventilation. Sometimes upgraded building construction is necessary. For more information refer to the section on Mechanical Equipment Room Sound Isolation.

Outdoor Air-Cooled Chillers and Air-Cooled Condensers. Most air-cooled chillers use either reciprocating, scroll, or screw compressors. These chillers are also used as the chiller portion of rooftop packaged units. The dominant noise sources in outdoor air-cooled chillers are the compressors and the condenser fans, which are typically low-cost, high-speed propeller fans. These fans are the only significant noise source.

Factory sound data for outdoor equipment is obtained in accordance with ARI *Standard* 370, which requires that the A-weighted and octave band Sound Power Level (L_w) value of the equipment be determined. The range of ARI *Standard* 370 L_w values for outdoor chillers in the 20 to 380 ton range is shown in Figure 11.

Factory-supplied noise reduction options for outdoor equipment include compressor enclosures, oversized condenser fans, and variable-speed condenser fans. Because air-cooled equipment needs a free-flow of cooling air, full enclosures are not feasible; however, strategically placed barriers can help reduce the noise propagation on a selective basis. For more information see the section on Sound Control for Outdoor Equipment.

Vibration Control. Typical chilled water systems include not only compressors and fans but also water pumps. The associated piping and conduit vibration generated by the compressors and pumps can be transmitted to a building structure via the equipment mounts or the attachments of the piping/conduit. Therefore, vibration isolation of a chiller requires essentially floating the entire chiller water on resilient-attachment floor mounts and hangers. Refer to the section on Vibration Isolation and Control for more information on chiller vibration control.

If a chiller is to be mounted on an on-grade slab, vibration isolators is not usually required unless an adjacent noise sensitive space shares the common slab. In noise-sensitive applications (e.g., rooftop chillers over occupied spaces), high deflection springs or air mounts will likely be required. However, the potential problem of wave resonance (or surge frequencies) common to all steel spring and rubber pad vibration isolators must be considered. This resonance is of particular concern with a screw chiller because the compressor tones are typically in the same frequency range as the isolator's wave resonance. If one of the compressor tones happens to coincide with one of the isolator resonances, the compressor tone can readily be transmitted into the building structure causing excessive structure-borne noise in the occupied space near the unit. It is impossible to predict this occurrence in advance, so if it does happen the

given chiller at a given operating point, a small equipment room, or one with mostly hard surface finishes, has a higher L_p value than a room that is large or has sound-absorbing treatments on its ceiling and walls. Figure 10 shows maximum typical adjustment factors that should be added to the factory provided values based on ARI *Standard* 575 to estimate the reverberant L_p values in specific installations. The adjustment for each octave band requires knowing the size of an imaginary box that is circumscribed 1 m away from the top and sides of the chiller (the ARI *Standard* 575 Measurement Surface), the dimensions of the equipment room and the average sound absorption coefficient of the room surfaces. The adjustment in each octave band depends on the ratio of the areas of the equipment room and the imaginary box as well as the average sound absorption of the room finishes. Each curve in Figure 10 is for a different value of the average sound absorption, with the higher curves being for lower values.

Example 1. Estimate the reverberant L_p values in a 45 ft by 40 ft by 20 ft tall mechanical equipment room (MER) that houses a 360-ton centrifugal chiller. The room has a concrete floor and gypsum board walls and ceiling. The chiller dimensions are 60 in. wide, 80 in. tall and 120 in. long.

Solution:

The ARI-575 Measurement Surface area S_M is determined by adding 1 m to the chiller height and 2 m to both its length and width. The floor area is not included in this calculation. The result is a box that has dimensions of 140 in. wide, 200 in. long and 120 in. tall. The area of this box is approximately 761 ft². The area of the equipment room (floor included) S_R is 7000 ft². Therefore, the ratio of the areas, S_R/S_M, is 7000/761 = 9.2. Assume that the average absorption coefficient value for room is 0.1 for all octave bands; therefore, refer to Figure 10 for the adjustment factor.

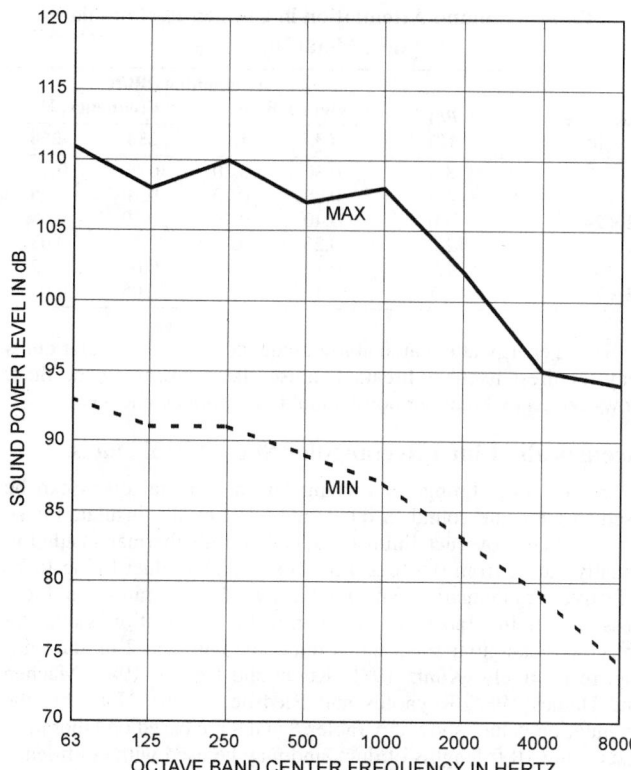

**Fig. 11   Typical ARI-370 L_w Values for
Outdoor Chillers—20 to 380 Tons**

solution is to use either another isolator design (with different surge frequencies) or air mounts, which do not exhibit the surge frequency phenomenon.

DUCT ELEMENT SOUND ATTENUATION

The duct elements covered in this section include sound plenums, unlined rectangular ducts, acoustically lined rectangular ducts, unlined round ducts, acoustically lined round ducts, elbows, acoustically lined round radiused elbows, duct silencers, duct branch power division, duct end reflection loss, and terminal volume regulation units. Simplified tabular procedures for obtaining the sound attenuation associated with these elements are presented in this section.

Plenums

Plenums are often used to smooth the turbulent airflow associated with air as it leaves the outlet section of a fan and before it enters the air distribution ducts. These chambers are usually lined with acoustically absorbent material to reduce fan and other noise. Plenums are usually large rectangular enclosures with an inlet and one or more outlets. The transmission loss associated with a plenum can be expressed as (Beranek 1971, Reynolds and Bledsoe 1991, Reynolds and Bevirt 1994, SMACNA 1990, Wells 1958):

$$TL = -10\log_{10}\left[S_{out}\left(\frac{Q\cos\theta}{4\pi r^2} + \frac{1-\alpha_A}{S\alpha_A}\right)\right] \qquad (1)$$

where (refer to Figure 12)

S_{out} = area of output section of plenum, ft^2
  S = total inside surface area of plenum minus inlet and outlet areas, ft^2
  r = distance between centers of inlet and outlet sections of plenum, ft

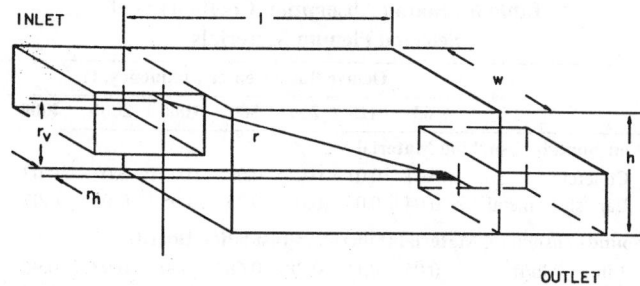

Fig. 12   Schematic of Plenum Chamber

Q = directivity factor, which may be taken as 4
α_A = average absorption coefficient of the plenum lining
θ = angle of vector representing r to long axis l of duct [see Equation (3)]

The average absorption coefficient α_A of plenum lining is

$$\alpha_A = \frac{S_1\alpha_1 + S_2\alpha_2}{S} \qquad (2)$$

where

α_1 = sound absorption coefficient of any bare or unlined inside surfaces of the plenum
S_1 = surface area of any bare or unlined inside surfaces of plenum, ft^2
α_2 = sound absorption coefficient of acoustically lined inside surfaces of plenum
S_2 = surface area of acoustically lined inside surfaces of plenum, ft^2

In many situations, all of the inside surfaces of a plenum chamber are lined with a sound absorbing material. For these situations, $\alpha_A = \alpha_2$. Table 6 gives the values of sound absorption coefficients for selected common plenum materials.

The value of $\cos\theta$ is obtained from

$$\cos\theta = \frac{l}{r} = \frac{l}{\sqrt{l^2 + r_v^2 + r_h^2}} \qquad (3)$$

where (refer to Figure 12)

l = length of plenum, ft
r_v = vertical offset between axes of plenum inlet and outlet, ft
r_h = horizontal offset between axes of plenum inlet and outlet, ft

Equation (1) treats a plenum as if it is a large enclosure. Thus, Equation (1) is valid only for the case where the wavelength of sound is small compared to the characteristic dimensions of the plenum. For frequencies that correspond to plane wave propagation in the duct, the results predicted by Equation (1) are usually not valid. Plane wave propagation in a duct exists at frequencies below

$$f_{co} = \frac{c_o}{2a} \qquad (4)$$

or

$$f_{co} = 0.586\frac{c_o}{d} \qquad (5)$$

where

f_{co} = cutoff frequency, Hz
c_o = speed of sound in air, fps
a = larger cross-section dimension of a rectangular duct, ft
d = diameter of a round duct, ft

The cutoff frequency f_{co} is the frequency above which plane waves no longer propagate in a duct. At these higher frequencies the waves that propagate in the duct are referred to as cross or spinning

Table 6 Sound Absorption Coefficients of Selected Plenum Materials

	Octave Band Center Frequency, Hz						
	63	125	250	500	1000	2000	4000
Non-Sound-Absorbing Material							
Concrete	0.01	0.01	0.01	0.02	0.02	0.02	0.03
Bare sheet metal	0.04	0.04	0.04	0.05	0.05	0.05	0.07
Sound-Absorbing Material (Fiberglass Insulation Board)							
1 in., 3.0 lb/ft^3	0.05	0.11	0.28	0.68	0.90	0.93	0.96
2 in., 3.0 lb/ft^3	0.10	0.17	0.86	1.00	1.00	1.00	1.00
3 in., 3.0 lb/ft^3	0.30	0.53	1.00	1.00	1.00	1.00	1.00
4 in., 3.0 lb/ft^3	0.50	0.84	1.00	1.00	1.00	1.00	0.97

Note: The 63 Hz values are estimated from higher frequency values.

modes. At frequencies below f_{co}, Equation (1) yields conservative results. The actual attenuation usually exceeds the values given by Equation (1) by 5 to 10 dB. Equation (1) usually always applies at frequencies of 1000 Hz and higher.

Example 2. A plenum chamber is 6 ft high, 4 ft wide, and 6 ft long. The configuration of the plenum is similar to that shown in Figure 12. The inlet and outlet are each 36 in. wide by 24 in. high. The horizontal distance between centers of the plenum inlet and outlet is 1 ft. The vertical distance is 4 ft. The plenum is lined with 1 in. thick, 3.0 lb/ft^3 density fiberglass insulation board. All of the inside surfaces of the plenum is lined with fiberglass insulation. Determine the transmission loss associated with this plenum. See Table 6 for the values of the absorption coefficients.

Solution:

The areas of the inlet section, outlet section, and plenum cross-section are

$$S_{in} = 24 \times 36 / 144 = 6 \text{ ft}^2$$

$$S_{out} = 24 \times 36 / 144 = 6 \text{ ft}^2$$

$$S_{pl} = 4 \times 6 = 24 \text{ ft}^2$$

$l = 6$ ft; $r_v = 4$ ft; and $r_h = 1$ ft. The values of r and $\cos\theta$ are

$$r = \sqrt{6^2 + 4^2 + 1^2} = 7.28 \text{ ft}$$

$$\cos\theta = 6/7.28 = 0.824$$

The total inside surface area of the plenum is

$$S = 2(4 \times 6) + 2(4 \times 6) + 2(6 \times 6) - 12 = 156 \text{ ft}^2$$

The value of f_{co} is

$$f_{co} = 1125/(2 \times 3) = 187.5 \text{ Hz}$$

The results are tabulated as follows.

	Octave Band Center Frequency, Hz						
	63	125	250	500	1000	2000	4000
$Q\cos\theta/4\pi r^2$	—	—	0.0050	0.0050	0.0050	0.0050	0.0050
$(1-\alpha_A)/S\alpha_A$	—	—	0.0165	0.0030	0.0007	0.0005	0.0003
TL, dB	—	—	9	13	15	15	15

Unlined Rectangular Sheet Metal Ducts

Straight unlined rectangular sheet metal ducts provide a fairly significant amount of low frequency sound attenuation. Table 7 shows the results of selected unlined rectangular sheet metal ducts (Cummings 1983, Reynolds and Bledsoe 1989b, Ver 1978, Woods 1973). The attenuation values shown in Table 7 apply only to rectangular sheet metal ducts with the lightest gages allowed according to SMACNA duct construction standards.

Table 7 Sound Attenuation in Unlined Rectangular Sheet Metal Ducts

Duct Size, in. × in.	P/A 1/ft	Attenuation, dB/ft Octave Band Center Frequency, Hz			
		63	125	250	>250
6 × 6	8.0	0.30	0.20	0.10	0.10
12 × 12	4.0	0.35	0.20	0.10	0.06
12 × 24	3.0	0.40	0.20	0.10	0.05
24 × 24	2.0	0.25	0.20	0.10	0.03
48 × 48	1.0	0.15	0.10	0.07	0.02
72 × 72	0.7	0.10	0.10	0.05	0.02

Sound energy attenuated at low frequencies in rectangular ducts may manifest itself as breakout noise elsewhere along the duct. Low-frequency breakout noise should therefore be checked.

Acoustically Lined Rectangular Sheet Metal Ducts

Internal duct lining for rectangular sheet metal ducts can be used to attenuate sound in ducts and to thermally insulate ducts. The thickness of duct linings associated with thermal insulation usually varies from 0.5 to 2.0 in. For fiberglass duct lining to be effective for attenuating fan sound, it must have a minimum thickness of 1.0 in. Tables 8 and 9 give the attenuation values of selected rectangular sheet metal ducts for 1 in. and 2 in. duct lining, respectively (Kuntz 1986, Kuntz and Hoover 1987, Machen and Haines 1983, Reynolds and Bledsoe 1989b). Note that the attenuation values shown in Tables 8 and 9 are based on laboratory tests using 10 ft lengths of duct, and may be used with confidence for lined duct lengths at 10 ft. For designs incorporating more than 10 ft of lined rectangular duct, actual dB/ft will be less than shown in Tables 8 and 9 while total attenuated noise will never be below the generated noise level in the duct. The density of the fiberglass lining used in lined rectangular sheet metal ducts usually varies between 1.5 and 3.0 lb/ft^3.

The **insertion loss** values given in Tables 8 and 9 are the difference in the sound pressure level measured in a reverberation room with sound propagating through an unlined section of rectangular duct minus the corresponding sound pressure level that is measured when the unlined section of rectangular duct is replaced with a similar section of acoustically lined rectangular duct. The attenuation of the unlined duct is subtracted out during the process of calculating the insertion loss from measured data.

Insertion loss and attenuation values discussed in this section apply only to rectangular sheet metal ducts made with the lightest gages allowed according to SMACNA duct construction standards.

Unlined Round Sheet Metal Ducts

As with unlined rectangular ducts, unlined round ducts provide some natural sound attenuation that should be considered when designing a duct system. In contrast to rectangular ducts, round ducts are much more rigid and, therefore, do not resonate or absorb as much sound energy. Because of this, round ducts will only provide about 1/10 the sound attenuation at low frequencies as compared to the sound attenuation associated with rectangular ducts. Table 10 list sound attenuation values for unlined round circular ducts (Woods 1973, ASHRAE 1987).

Acoustically Lined Round Sheet Metal Ducts

The literature has little data for the insertion loss of acoustically lined round ducts. The data that are available are usually manufacturer's product data. Tables 11 and 12 give the insertion loss values for dual-wall round sheet metal ducts with 1 in. and 2 in. acoustical lining, respectively (Reynolds and Bledsoe 1989a). The acoustical lining for the ducts is a 0.75 lb/ft^3 density fiberglass blanket, and the fiberglass is covered with an internal liner of perforated galvanized sheet metal that has on open area of

Table 8 Insertion Loss for Rectangular Sheet Metal Ducts with 1 in. Fiberglass Lining

Dimensions, in. × in.	Insertion Loss, dB/ft Octave Band Center Frequency, Hz					
	125	250	500	1000	2000	4000
6 × 6	0.6	1.5	2.7	5.8	7.4	4.3
6 × 10	0.5	1.2	2.4	5.1	6.1	3.7
6 × 12	0.5	1.2	2.3	5.0	5.8	3.6
6 × 18	0.5	1.0	2.2	4.7	5.2	3.3
8 × 8	0.5	1.2	2.3	5.0	5.8	3.6
8 × 12	0.4	1.0	2.1	4.5	4.9	3.2
8 × 16	0.4	0.9	2.0	4.3	4.5	3.0
8 × 24	0.4	0.8	1.9	4.0	4.1	2.8
10 × 10	0.4	1.0	2.1	4.4	4.7	3.1
10 × 16	0.4	0.8	1.9	4.0	4.0	2.7
10 × 20	0.3	0.8	1.8	3.8	3.7	2.6
10 × 30	0.3	0.7	1.7	3.6	3.3	2.4
12 × 12	0.4	0.8	1.9	4.0	4.1	2.8
12 × 18	0.3	0.7	1.7	3.7	3.5	2.5
12 × 24	0.3	0.6	1.7	3.5	3.2	2.3
12 × 36	0.3	0.6	1.6	3.3	2.9	2.2
15 × 15	0.3	0.7	1.7	3.6	3.3	2.4
15 × 22	0.3	0.6	1.6	3.3	2.9	2.2
15 × 30	0.3	0.5	1.5	3.1	2.6	2.0
15 × 45	0.2	0.5	1.4	2.9	2.4	1.9
18 × 18	0.3	0.6	1.6	3.3	2.9	2.2
18 × 28	0.2	0.5	1.4	3.0	2.4	1.9
18 × 36	0.2	0.5	1.4	2.8	2.2	1.8
18 × 54	0.2	0.4	1.3	2.7	2.0	1.7
24 × 24	0.2	0.5	1.4	2.8	2.2	1.8
24 × 36	0.2	0.4	1.2	2.6	1.9	1.6
24 × 48	0.2	0.4	1.2	2.4	1.7	1.5
24 × 72	0.2	0.3	1.1	2.3	1.6	1.4
30 × 30	0.2	0.4	1.2	2.5	1.8	1.6
30 × 45	0.2	0.3	1.1	2.3	1.6	1.4
30 × 60	0.2	0.3	1.1	2.2	1.4	1.3
30 × 90	0.1	0.3	1.0	2.1	1.3	1.2
36 × 36	0.2	0.3	1.1	2.3	1.6	1.4
36 × 54	0.1	0.3	1.0	2.1	1.3	1.2
36 × 72	0.1	0.3	1.0	2.0	1.2	1.2
36 × 108	0.1	0.2	0.9	1.9	1.1	1.1
42 × 42	0.2	0.3	1.0	2.1	1.4	1.3
42 × 64	0.1	0.3	0.9	1.9	1.2	1.1
42 × 84	0.1	0.2	0.9	1.8	1.1	1.1
42 × 126	0.1	0.2	0.9	1.7	1.0	1.0
48 × 48	0.1	0.3	1.0	2.0	1.2	1.2
48 × 72	0.1	0.2	0.9	1.8	1.0	1.0
48 × 96	0.1	0.2	0.8	1.7	1.0	1.0
48 × 144	0.1	0.2	0.8	1.6	0.9	0.9

Table 9 Insertion Loss for Rectangular Sheet Metal Ducts with 2 in. Fiberglass Lining

Dimensions, in. × in.	Insertion Loss, dB/ft Octave Band Center Frequency, Hz					
	125	250	500	1000	2000	4000
6 × 6	0.8	2.9	4.9	7.2	7.4	4.3
6 × 10	0.7	2.4	4.4	6.4	6.1	3.7
6 × 12	0.6	2.3	4.2	6.2	5.8	3.6
6 × 18	0.6	2.1	4.0	5.8	5.2	3.3
8 × 8	0.6	2.3	4.2	6.2	5.8	3.6
8 × 12	0.6	1.9	3.9	5.6	4.9	3.2
8 × 16	0.5	1.8	3.7	5.4	4.5	3.0
8 × 24	0.5	1.6	3.5	5.0	4.1	2.8
10 × 10	0.6	1.9	3.8	5.5	4.7	3.1
10 × 16	0.5	1.6	3.4	5.0	4.0	2.7
10 × 20	0.4	1.5	3.3	4.8	3.7	2.6
10 × 30	0.4	1.3	3.1	4.5	3.3	2.4
12 × 12	0.5	1.6	3.5	5.0	4.1	2.8
12 × 18	0.4	1.4	3.2	4.6	3.5	2.5
12 × 24	0.4	1.3	3.0	4.3	3.2	2.3
12 × 36	0.4	1.2	2.9	4.1	2.9	2.2
15 × 15	0.4	1.3	3.1	4.5	3.3	2.4
15 × 22	0.4	1.2	2.9	4.1	2.9	2.2
15 × 30	0.3	1.1	2.7	3.9	2.6	2.0
15 × 45	0.3	1.0	2.6	3.6	2.4	1.9
18 × 18	0.4	1.2	2.9	4.1	2.9	2.2
18 × 28	0.3	1.0	2.6	3.7	2.4	1.9
18 × 36	0.3	0.9	2.5	3.5	2.2	1.8
18 × 54	0.3	0.8	2.3	3.3	2.0	1.7
24 × 24	0.3	0.9	2.5	3.5	2.2	1.8
24 × 36	0.3	0.8	2.3	3.2	1.9	1.6
24 × 48	0.2	0.7	2.2	3.0	1.7	1.5
24 × 72	0.2	0.7	2.0	2.9	1.6	1.4
30 × 30	0.2	0.8	2.2	3.1	1.8	1.6
30 × 45	0.2	0.7	2.0	2.9	1.6	1.4
30 × 60	0.2	0.6	1.9	2.7	1.4	1.3
30 × 90	0.2	0.5	1.8	2.6	1.3	1.2
36 × 36	0.2	0.7	2.0	2.9	1.6	1.4
36 × 54	0.2	0.6	1.9	2.6	1.3	1.2
36 × 72	0.2	0.5	1.8	2.5	1.2	1.2
36 × 108	0.2	0.5	1.7	2.3	1.1	1.1
42 × 42	0.2	0.6	1.9	2.6	1.4	1.3
42 × 64	0.2	0.5	1.7	2.4	1.2	1.1
42 × 84	0.2	0.5	1.6	2.3	1.1	1.1
42 × 126	0.1	0.4	1.6	2.2	1.0	1.0
48 × 48	0.2	0.5	1.8	2.5	1.2	1.2
48 × 72	0.2	0.4	1.6	2.3	1.0	1.0
48 × 96	0.1	0.4	1.5	2.1	1.0	1.0
48 × 144	0.1	0.4	1.5	2.0	0.9	0.9

25%. Because of structure-borne sound that is transmitted through the duct wall, the total sound attenuation of lined round ducts usually does not exceed 40 dB.

Rectangular Sheet Metal Duct Elbows

Table 13 displays insertion loss values for unlined and lined square elbows without turning vanes (Beranek 1960). For lined square elbows, the duct lining must extend at least two duct widths w beyond the elbow. Table 13 applies only where the duct is lined before and after the elbow. Table 14 gives the insertion loss values associated with radiused elbows. Table 15 gives the insertion loss values for unlined and lined square elbows with turning vanes. The value of $f \times w$ (f_w) in Tables 13 through 15 is the center frequency of the octave frequency band times the width of the elbow (Figure 13) (Beranek 1960; 1984 and 1993 *ASHRAE Handbooks*).

Nonmetallic Insulated Flexible Ducts

Nonmetallic insulated flexible ducts can significantly reduce airborne noise. Insertion loss values for specified duct diameters

Table 10 Sound Attenuation in Straight Round Ducts

Diameter, in.	Attenuation, dB/ft Octave Band Center Frequency, Hz						
	63	125	250	500	1000	2000	4000
$D \leq 7$	0.03	0.03	0.05	0.05	0.10	0.10	0.10
$7 < D \leq 15$	0.03	0.03	0.03	0.05	0.07	0.07	0.07
$15 < D \leq 30$	0.02	0.02	0.02	0.03	0.05	0.05	0.05
$30 < D \leq 60$	0.01	0.01	0.01	0.02	0.02	0.02	0.02

and lengths are given in Table 16 (ARI *Standard* 885). Recommended duct lengths are normally from 3 to 6 ft. Care should be taken to keep flexible ducts straight; bends should have as long a radius as possible. While an abrupt bend may provide some additional insertion loss, the airflow-generated noise associated with the airflow in the bend may be unacceptably high. Because of potentially high breakout sound levels associated with flexible ducts, care should be exercised when using flexible ducts above sound-sensitive spaces.

Table 11   Insertion Loss for Acoustically Lined Round Ducts with 1 in. Lining

Diameter, in.	Insertion Loss, dB/ft Octave Band Center Frequency, Hz							
	63	125	250	500	1000	2000	4000	8000
6	0.38	0.59	0.93	1.53	2.17	2.31	2.04	1.26
8	0.32	0.54	0.89	1.50	2.19	2.17	1.83	1.18
10	0.27	0.50	0.85	1.48	2.20	2.04	1.64	1.12
12	0.23	0.46	0.81	1.45	2.18	1.91	1.48	1.05
14	0.19	0.42	0.77	1.43	2.14	1.79	1.34	1.00
16	0.16	0.38	0.73	1.40	2.08	1.67	1.21	0.95
18	0.13	0.35	0.69	1.37	2.01	1.56	1.10	0.90
20	0.11	0.31	0.65	1.34	1.92	1.45	1.00	0.87
22	0.08	0.28	0.61	1.31	1.82	1.34	0.92	0.83
24	0.07	0.25	0.57	1.28	1.71	1.24	0.85	0.80
26	0.05	0.22	0.53	1.24	1.59	1.14	0.79	0.77
28	0.03	0.19	0.49	1.20	1.46	1.04	0.74	0.74
30	0.02	0.16	0.45	1.16	1.33	0.95	0.69	0.71
32	0.01	0.14	0.42	1.12	1.20	0.87	0.66	0.69
34	0	0.11	0.38	1.07	1.07	0.79	0.63	0.66
36	0	0.08	0.35	1.02	0.93	0.71	0.60	0.64
38	0	0.06	0.31	0.96	0.80	0.64	0.58	0.61
40	0	0.03	0.28	0.91	0.68	0.57	0.55	0.58
42	0	0.01	0.25	0.84	0.56	0.50	0.53	0.55
44	0	0	0.23	0.78	0.45	0.44	0.51	0.52
46	0	0	0.20	0.71	0.35	0.39	0.48	0.48
48	0	0	0.18	0.63	0.26	0.34	0.45	0.44
50	0	0	0.15	0.55	0.19	0.29	0.41	0.40
52	0	0	0.14	0.46	0.13	0.25	0.37	0.34
54	0	0	0.12	0.37	0.09	0.22	0.31	0.29
56	0	0	0.10	0.28	0.08	0.18	0.25	0.22
58	0	0	0.09	0.17	0.08	0.16	0.18	0.15
60	0	0	0.08	0.06	0.10	0.14	0.09	0.07

Table 12   Insertion Loss for Acoustically Lined Round Ducts with 2 in. Lining

Diameter, in.	Insertion Loss, dB/ft Octave Band Center Frequency, Hz							
	63	125	250	500	1000	2000	4000	8000
6	0.56	0.80	1.37	2.25	2.17	2.31	2.04	1.26
8	0.51	0.75	1.33	2.23	2.19	2.17	1.83	1.18
10	0.46	0.71	1.29	2.20	2.20	2.04	1.64	1.12
12	0.42	0.67	1.25	2.18	2.18	1.91	1.48	1.05
14	0.38	0.63	1.21	2.15	2.14	1.79	1.34	1.00
16	0.35	0.59	1.17	2.12	2.08	1.67	1.21	0.95
18	0.32	0.56	1.13	2.10	2.01	1.56	1.10	0.90
20	0.29	0.52	1.09	2.07	1.92	1.45	1.00	0.87
22	0.27	0.49	1.05	2.03	1.82	1.34	0.92	0.83
24	0.25	0.46	1.01	2.00	1.71	1.24	0.85	0.80
26	0.24	0.43	0.97	1.96	1.59	1.14	0.79	0.77
28	0.22	0.40	0.93	1.93	1.46	1.04	0.74	0.74
30	0.21	0.37	0.90	1.88	1.33	0.95	0.69	0.71
32	0.20	0.34	0.86	1.84	1.20	0.87	0.66	0.69
34	0.19	0.32	0.82	1.79	1.07	0.79	0.63	0.66
36	0.18	0.29	0.79	1.74	0.93	0.71	0.60	0.64
38	0.17	0.27	0.76	1.69	0.80	0.64	0.58	0.61
40	0.16	0.24	0.73	1.63	0.68	0.57	0.55	0.58
42	0.15	0.22	0.70	1.57	0.56	0.50	0.53	0.55
44	0.13	0.20	0.67	1.50	0.45	0.44	0.51	0.52
46	0.12	0.17	0.64	1.43	0.35	0.39	0.48	0.48
48	0.11	0.15	0.62	1.36	0.26	0.34	0.45	0.44
50	0.09	0.12	0.60	1.28	0.19	0.29	0.41	0.40
52	0.07	0.10	0.58	1.19	0.13	0.25	0.37	0.34
54	0.05	0.08	0.56	1.10	0.09	0.22	0.31	0.29
56	0.02	0.05	0.55	1.00	0.08	0.18	0.25	0.22
58	0	0.03	0.53	0.90	0.08	0.16	0.18	0.15
60	0	0	0.53	0.79	0.10	0.14	0.09	0.07

Table 13   Insertion Loss of Unlined and Lined Square Elbows Without Turning Vanes

	Insertion Loss, dB	
	Unlined Elbows	Lined Elbows
$fw < 1.9$	0	0
$1.9 \leq fw < 3.8$	1	1
$3.8 \leq fw < 7.5$	5	6
$7.5 \leq fw < 15$	8	11
$15 \leq fw < 30$	4	10
$fw > 30$	3	10

Note: $fw = f \times w$ where f = center frequency, kHz, and w = width, in.

Table 14   Insertion Loss of Round Elbows

	Insertion Loss, dB
$fw < 1.9$	0
$1.9 \leq fw < 3.8$	1
$3.8 \leq fw < 7.5$	2
$fw > 7.5$	3

Note: $fw = f \times w$ where f = center frequency, kHz, and w = width, in.

Table 15   Insertion Loss of Unlined and Lined Square Elbows with Turning Vanes

	Insertion Loss, dB	
	Unlined Elbows	Lined Elbows
$fw < 1.9$	0	0
$1.9 \leq fw < 3.8$	1	1
$3.8 \leq fw < 7.5$	4	4
$7.5 \leq fw < 15$	6	7
$fw > 15$	4	7

Note: $fw = f \times w$ where f = center frequency, kHz, and w = width, in.

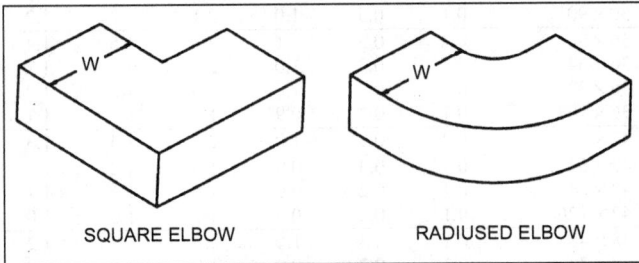

Fig. 13   Rectangular Duct Elbows

Duct Silencers

Duct silencers are used to attenuate sound that is transmitted through HVAC systems; particularly duct systems. They can add pressure and energy losses. Therefore, when selecting silencers, the following parameters should be considered:

- **Airflow pressure drop**, including system effects for less than ideal flow conditions
- **Insertion loss**, which is the reduction in sound power level at a given location due solely to the placement of a sound-attenuating device in the transmission path between the sound source and the given location

Airflow generated noise is created as the air flows into, through, and out of the silencer. In the majority of installations the airflow generated noise is much less than, and does not contribute to, the silenced noise level on the quiet side of the silencer. In general, airflow generated noise should be evaluated if static pressure drops exceed 0.35 in. water gage, the noise criteria is below RC-35, or the silencer is located very close to or in the occupied space.

Table 16 Lined Flexible Duct Insertion Loss

Diameter, in.	Length, ft	Insertion Loss, dB Octave Band Center Frequency, Hz						
		63	125	250	500	1000	2000	4000
4	12	6	11	12	31	37	42	27
	9	5	8	9	23	28	32	20
	6	3	6	6	16	19	21	14
	3	2	3	3	8	9	11	7
5	12	7	12	14	32	38	41	26
	9	5	9	11	24	29	31	20
	6	4	6	7	16	19	21	13
	3	2	3	4	8	10	10	7
6	12	8	12	17	33	38	40	26
	9	6	9	13	25	29	30	20
	6	4	6	9	17	19	20	13
	3	2	3	4	8	10	10	7
7	12	9	12	19	33	37	38	25
	9	6	9	14	25	28	29	19
	6	4	6	10	17	19	19	13
	3	2	3	5	8	9	10	6
8	12	8	11	21	33	37	37	24
	9	6	8	16	25	28	28	18
	6	4	6	11	17	19	19	12
	3	2	3	5	8	9	9	6
9	12	8	11	22	33	37	36	22
	9	6	8	17	25	28	27	17
	6	4	6	11	17	19	18	11
	3	2	3	6	8	9	9	6
10	12	8	10	22	32	36	34	21
	9	6	8	17	24	27	26	16
	6	4	5	11	16	18	17	11
	3	2	3	6	8	9	9	5
12	12	7	9	20	30	34	31	18
	9	5	7	15	23	26	23	14
	6	3	5	10	15	17	16	9
	3	2	2	5	8	9	8	5
14	12	5	7	16	27	31	27	14
	9	4	5	12	20	23	20	11
	6	3	4	8	14	16	14	7
	3	1	2	4	7	8	7	4
16	12	2	4	9	23	28	23	9
	9	2	3	7	17	21	17	7
	6	1	2	5	12	14	12	5
	3	1	1	2	6	7	6	2

Note: The 63 Hz insertion loss values are estimated from higher frequency insertion loss values.

There are three types of HVAC duct silencers: dissipative (with acoustic media), reactive (no media), and active silencers.

- **Dissipative silencers** (Figure 14) generally use perforated metal surfaces covering acoustic grade fiberglass to attenuate sound over a broad range of frequencies. Airflow does not significantly affect the insertion loss if pressure drops are under 0.35 in. water gage.
- **Reactive silencers** use tuned perforated metal facings covering tuned chambers void of any fibrous material. The outside physical appearance of reactive silencers is similar to its dissipative counterpart (Figure 14). Because of tuning, broadband insertion loss is more difficult to achieve than with dissipative silencers. Longer lengths may be required to achieve similar insertion loss performance as dissipative silencers. Airflow generally increases the insertion loss of reactive silencers.
- **Active duct silencers** (Figure 15) reduce noise at lower frequencies by producing inverse sound waves that cancel the unwanted noise. An input microphone measures the noise in the duct and converts it to electrical signals. These signals are processed by a digital computer where exact opposite, mirror-image sound

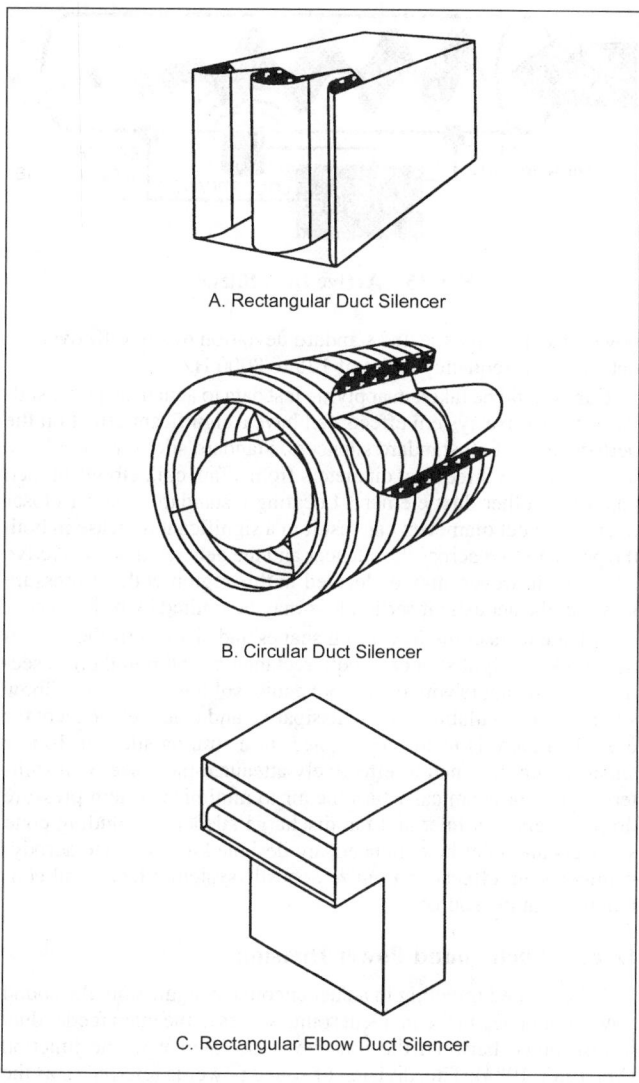

A. Rectangular Duct Silencer

B. Circular Duct Silencer

C. Rectangular Elbow Duct Silencer

Fig. 14 Dissipative Duct Silencers

waves of equal amplitude are generated. This secondary noise source destructively interferes with the noise and cancels a significant portion of the unwanted sound. An error microphone measures the residual sound beyond the silencer and provides feedback to adjust the computer model to increase performance. Because the components are mounted outside the airflow, there is no pressure loss or generated noise. Performance is limited, however, by the presence of excessive turbulence in the airflow detected by the microphones. Manufacturers recommend using active silencers where duct velocity is less than 1500 fpm and where the duct configurations are conducive to smooth evenly distributed airflow.

Data for dissipative and reactive silencers should be obtained from tests in a manner consistent with the procedures outlined in the most current version of ASTM *Standard* E477. (This standard has not been verified for determining performance of active silencers.) Since insertion loss measurements use a substitution technique, reasonable (±3dB) insertion loss values can be achieved down to 63 Hz. Airflow generated noise, however, cannot be measured accurately below 100 Hz due to the lack of a standardized means of qualifying the reverberant rooms for sound power measurements at lower frequencies (a function of room size). A recent series round robin tests has shown that airflow generated sound

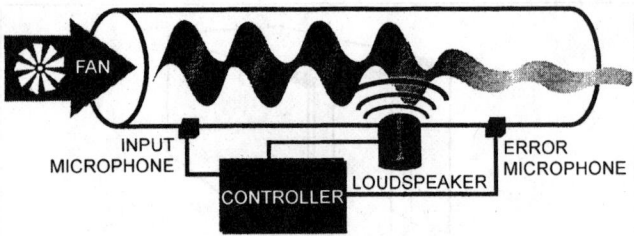

Fig. 15   Active Duct Silencer

power data has an expected standard deviation of 3 to 6dB over the octave band frequency range of 125 to 8000 Hz.

Care should be taken in applying test data to actual project installations. Adverse system effects can have a significant effect on the performance of all standard silencers. Standard silencers should be located at least three duct diameters from a fan, coil, elbow, branch takeoff, or other duct element. Locating a standard silencer closer than three duct diameters can result in a significant increase in both the pressure loss across the silencer and the generated noise. Active silencers, however, may be located in this region without pressure loss, but the acoustic insertion loss may be limited by turbulence.

All silencers come in varying shapes and sizes to fit the project ductwork. Straight silencers, both rectangular and round cross sections, are available with or without center splitters and pods. Elbow silencers are available as both dissipative and reactive silencers for use when there is insufficient space for a straight silencer. Elbow silencers are designed to effectively attenuate the noise with splitters that aerodynamically turn the air to minimize system pressure drop. Special fan inlet and fan discharge silencers including cone silencers and inlet box silencers are designed to minimize aerodynamic system effects, maximize acoustic system effects, and contain noise at the source.

Duct Branch Sound Power Division

When sound traveling in a duct encounters a junction, the sound power contained in the incident sound waves in the main feeder duct is distributed between the branches associated with the junction (Ver 1982, 1984). This division of sound power is referred to as the branch sound power division. The corresponding attenuation of sound power that is transmitted down each branch of the junction is comprised of two components. The first is associated with the reflection of the incident sound wave if the sum of the cross-sectional areas of the individual branches ΣS_{Bi} differs from the cross-sectional area S_M of the main feeder duct. The second component is associated with the energy division according to the ratio of the cross-sectional area ΣS_{Bi} of an individual branch divided by the sum of the cross-sectional areas of the individual branches ΣS_{Bi}. The second component is the dominant component. Values for the attenuation of sound power ΔL_{Bi} at a junction that are related to the sound power transmitted down an individual branch of the junction are given in Table 17.

Table 17   Duct Branch Sound Power Division

$S_i \Sigma S_{Bi}$	ΔL_{Bi}	$S_i \Sigma S_{Bi}$	ΔL_{Bi}
1.00	0	0.10	10
0.80	1	0.08	11
0.63	2	0.063	12
0.50	3	0.050	13
0.40	4	0.040	14
0.32	5	0.032	15
0.25	6	0.025	16
0.20	7	0.020	17
0.16	8	0.016	18
0.12	9	0.012	19

Duct End Reflection Loss

When low frequency plane sound waves interact with openings that discharge into a large room, a significant amount of the sound energy incident on this interface is reflected back into the duct. The sound attenuation values ΔL associated with duct end reflection losses for ducts terminated in free space are given in Table 18 and for ducts terminated flush with a wall are given in Table 19 (Sandbakken et al. 1981, AMCA *Standard* 300). Diffusers that terminate in a suspended lay-in acoustic ceiling can be treated as terminating in free space. To use Table 18, if the duct terminating into a diffuser is rectangular, the duct diameter D is given by

$$D = \sqrt{4A/\pi} \qquad (6)$$

where A = area (in^2) of the rectangular duct.

Tables 18 and 19 have some limitations. The tests on which these equations are based were conducted with straight sections of round ducts. These ducts directly terminated into a reverberation chamber with no restriction on the end of the duct or with a round orifice placed over the end of the duct. Diffusers can be either round or rectangular. They usually have a restriction associated with them that may either be a damper, guide vanes to direct airflow, a perforated metal facing, or a combination of these elements. Currently, there is no data that indicate the effects of these elements. It is not known whether or not these elements react similar to the orifices used in the above described tests. As a result, the effects of an orifice placed over the end of a duct are not included in Tables 18 and 19. Equation (6) does yield reasonable results with diffusers that have low aspect ratios (length/width). However, many types of diffusers (particularly

**Table 18   Duct End Reflection Loss—
Duct Terminated in Free Space**

Duct Diameter, in.	End Reflection Loss, dB Octave Band Center Frequency, Hz					
	63	125	250	500	1000	2000
6	20	14	9	5	2	1
8	18	12	7	3	1	0
10	16	11	6	2	1	0
12	14	9	5	2	1	0
16	12	7	3	1	0	0
20	10	6	2	1	0	0
24	9	5	2	1	0	0
28	8	4	1	0	0	0
32	7	3	1	0	0	0
36	6	3	1	0	0	0
48	5	2	1	0	0	0
72	3	1	0	0	0	0

**Table 19   Duct End Reflection Loss—
Duct Terminated Flush with Wall**

Duct Diameter, in.	End Reflection Loss, dB Octave Band Center Frequency, Hz				
	63	125	250	500	1000
6	18	13	8	4	1
8	16	11	6	2	1
10	14	9	5	2	1
12	13	8	4	1	0
16	10	6	2	1	0
20	9	5	2	1	0
24	8	4	1	0	0
28	7	3	1	0	0
32	6	2	1	0	0
36	5	2	1	0	0
48	4	1	0	0	0
72	2	1	0	0	0

slot diffusers) have high aspect ratios. It is not known whether Tables 18 or 19 can be accurately used with these diffusers. Finally, many diffusers do not have long straight sections (greater than three duct diameters) before they terminate into a room. Many duct sections between a main feed branch and a diffuser are curved or have short, stubby sections. The effects of these configurations on the duct end reflection loss are not known. However, Tables 18 and 19 can be used with reasonable accuracy for many diffuser configurations. However, caution should be exercised and sound from the diffuser may need to be checked if its configuration differs from the test conditions used to derive these tables.

USE OF FIBERGLASS PRODUCTS IN HVAC SYSTEMS

Fiberglass duct liner continues to be a cost-effective solution to noise control in most HVAC air duct systems. However, fiberglass has been banned from or severely limited for use in some institutional, educational, and medical projects. The decisions are based on concerns that the fibers may be carcinogenic and that the products may promote microbial growth. ASHRAE Technical Committee TC 2.6 (Sound and Vibration) conducted an informal review of the available research and studies and concluded the following:

- A review published in ASHRAE (1993) found no clear evidence that any form of manufactured mineral fibers are carcinogenic. The International Agency for Research on Cancer (IARC) has performed research in regard to the carcinogenicity of fiberglass materials and found inadequate evidence to link inhalation of fiberglass wool with cancer in humans (WHO 1986). IARC does indicate there is a possible linkage to cancer from glass wool based on heavy-dosage direct-injection into animals. As a result of this, the U.S. Department of Health & Human Services (USDHHS) has added glasswool to its list of substances "reasonably anticipated" to be a carcinogen. Environment Canada (1994) indicates that studies have failed to show any evidence that fiberglass is carcinogenic and classified the material as "unlikely to be carcinogenic to humans." In addition, the report indicated fiberglass is not entering the environment in quantities or conditions that may constitute a danger to human life or health.
- Morey and Williams (1991) found that both moisture and dirt are required for microbial growth and that microbial growth can grow on any surface including sheet metal, sound traps, filters, and turning vanes, in addition to duct lining. Microbial growth is not related to one component. Thus, removal of fiberglass duct lining by itself does not eliminate microbial growth. Control of moisture and dirt in duct systems is the best method of reducing potential microbial growth. Duct design for access, proper filtration,

humidification, and condensate systems; methods to prevent moisture and dirt accumulation during installation and commissioning; and regular maintenance procedures can reduce the occurrence of microbial growth (ASHRAE *Standard* 62-1989).

Owners and designers should be warned against removing fiberglass duct lining from any building and must understand the acoustic impact of that decision. Compensating for the removal of fiberglass duct liner may require larger and longer duct runs, larger fans and fan plenums, and the use of alternative noise control devices such as removable, lined duct sections, foil-coated liner, no-fill sound traps, and active noise control, some of which may be less effective and more expensive.

Certain spaces in health care facility such as operating rooms and critical care spaces should follow duct liner guidelines issued by the American Institute of Architects Committee on Architecture (AIA 1992-93). The complete removal of fiberglass duct liner from duct systems in spaces such as theaters, concert halls, recording and TV studios, and educational facilities could result in poor acoustic qualities and may render the spaces unusable.

SOUND RADIATION THROUGH DUCT WALLS

Airflow Generated Duct Rumble

An HVAC fan and its connected ductwork can act as a semi-closed, compressible fluid pumping system where both acoustic and aerodynamic air pressure fluctuations at the fan are transmitted to other locations in the duct system. Air pressure fluctuations can

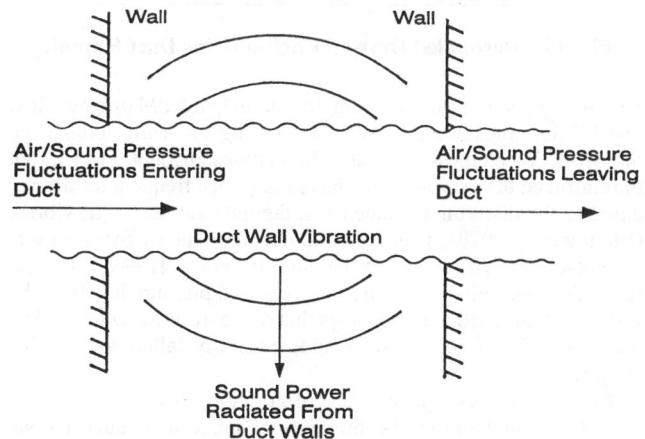

Fig. 16   Airflow-Generated Duct Rumble

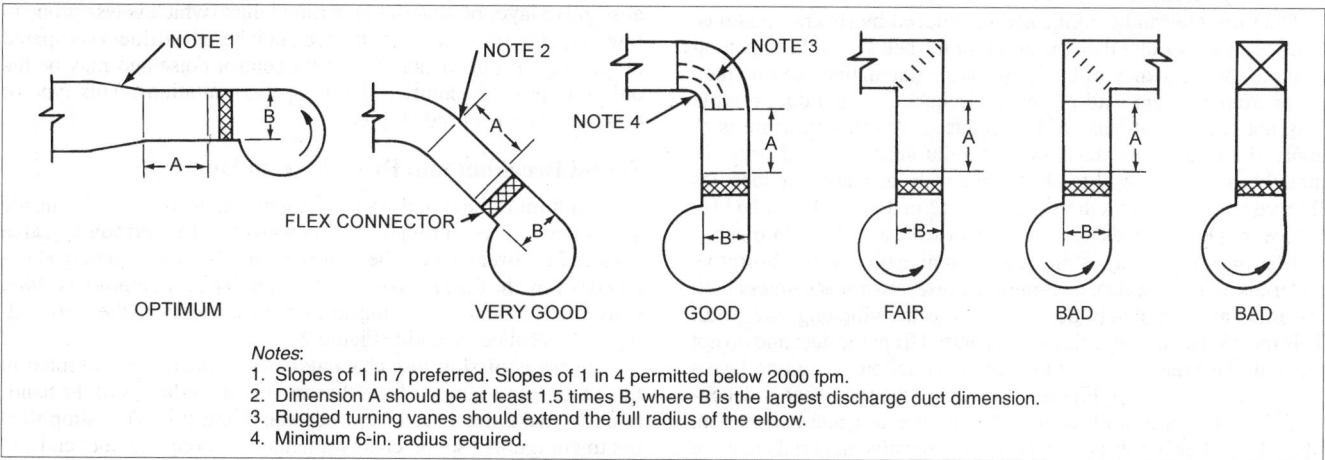

OPTIMUM   VERY GOOD   GOOD   FAIR   BAD   BAD

FLEX CONNECTOR

Notes:
1. Slopes of 1 in 7 preferred. Slopes of 1 in 4 permitted below 2000 fpm.
2. Dimension A should be at least 1.5 times B, where B is the largest discharge duct dimension.
3. Rugged turning vanes should extend the full radius of the elbow.
4. Minimum 6-in. radius required.

Fig. 17   Various Outlet Configurations for Centrifugal Fans and Their Possible Rumble Conditions

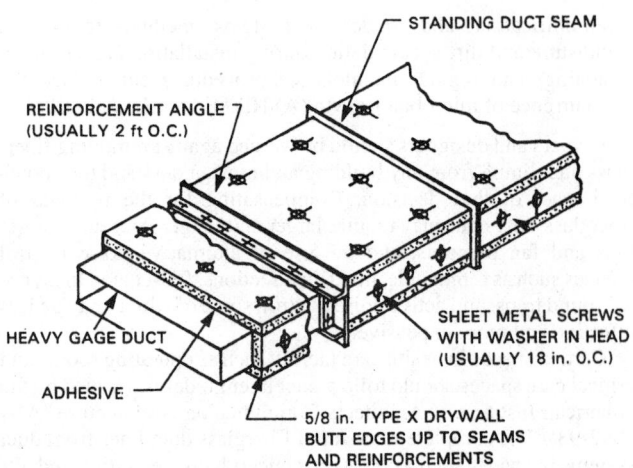

Fig. 18   Drywall Lagging on Duct for Duct Rumble

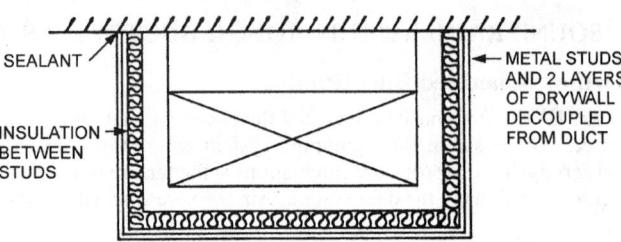

Fig. 19   Decoupled Drywall Enclosure for Duct Rumble

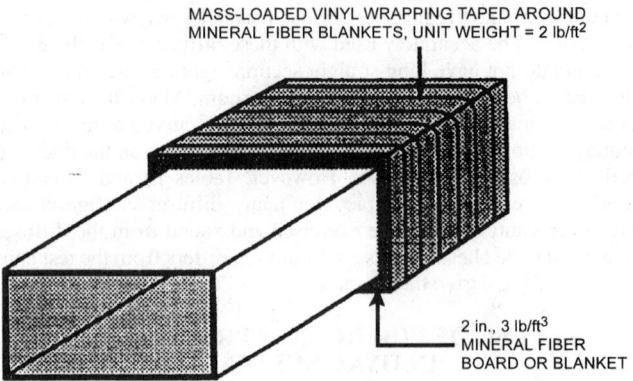

Fig. 20   Rectangular Duct with External Lagging

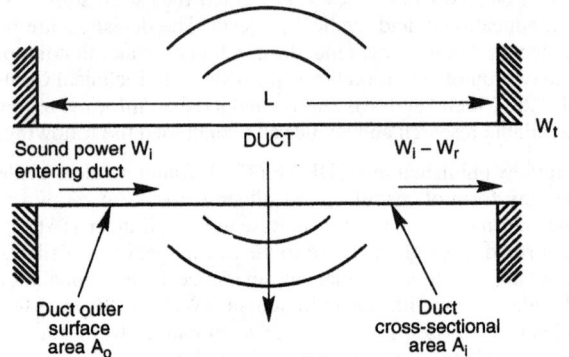

Fig. 21   Duct Breakout

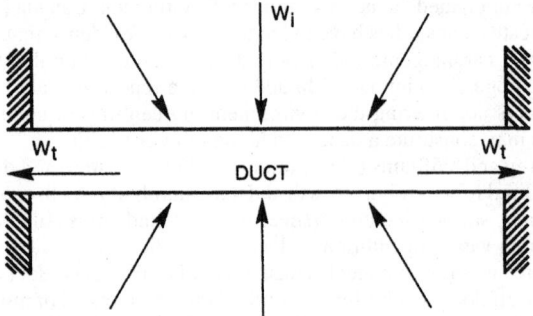

Fig. 22   Duct Breakin

be caused by variations in fan, motor or fan belt RPM or by airflow instabilities transmitted to the fan housing or nearby connected ductwork. When the air pressure fluctuations encounter large, flat, unreinforced duct surfaces that have resonance frequencies near or equal to the disturbing frequencies, the duct surfaces will vibrate (Ebbing et al. 1978). This vibration occurs in the stiffness- or resonance-controlled regions of the duct. In typical HVAC duct systems, duct wall vibration can produce sound pressure levels of the order of 65 to 95 dB at frequencies that range from 10 Hz to 100 Hz. This type of duct-generated sound is generally called duct rumble (Figure 16).

Figure 17 shows typical duct configurations that can exist near a centrifugal fan. Fair to bad configurations can result in duct rumble. Good to optimum designs of fan inlet and discharge transitions minimize the potential for duct rumble; however, these designs may not completely eliminate the potential for duct rumble.

Duct rumble can be eliminated or reduced by several methods. One method is to alter the fan, motor, or fan belt speed. This method changes the frequency of the air pressure fluctuations so that they differ from the duct wall resonance frequencies, and duct rumble may not occur or at least will be reduced. Another measure is to apply rigid materials, such as duct reinforcements and drywall, directly to the duct wall to change the wall resonance frequencies (Figure 18). Noise reductions of 5 to 11 dB in the 31.5 Hz and 63 Hz octave frequency bands have been recorded using this treatment.

Mass-loaded materials applied in combination with absorptive materials do not alleviate duct rumble noise into a space unless both materials are completely decoupled from the vibrating duct wall. This means they are significantly separated from the duct and do not touch it. An example of this type of construction, using two layers of drywall, is shown in Figure 19. Because the treatment is decoupled from the duct wall, it provides the greatest noise reduction. Mass-loaded materials combined with absorptive materials that are directly attached to the duct wall are not effective in reducing duct rumble (Figure 20). Mass-loaded flexible material wrapped over an

absorptive layer of material on a round duct (which is less prone to low frequency rumble due to the greater inherent stiffness compared to rectangular ducts) can effectively control noise and may be the only one possible solution due to space limitations. This type of treatment is often used on piping.

Sound Breakout and Breakin from Ducts

Breakout is the sound associated with fan or airflow noise inside a duct that radiates through the duct walls into the surrounding area (Figure 21). Breakout can be a problem if it is not adequately attenuated before the duct runs over an occupied space (Cummings 1983, Lilly 1987). Sound that is transmitted into a duct from the surrounding area is called **breakin** (Figure 22).

The **transmission loss** characteristics of ducts are presented in this volume in a simpler form than in previous editions of the handbook to make them easier to understand and use. This simplified treatment ignores sound energy attenuation along the duct and can be applied to ducts where the critical length of duct for noise radiation is the first 20 to 30 ft.

Table 20 TL_{out} Versus Frequency for Rectangular Ducts

Duct Size, in. × in.	Gage	TL_{out}, dB Octave Band Center Frequency, Hz							
		63	125	250	500	1000	2000	4000	8000
12 × 12	24	21	24	27	30	33	36	41	45
12 × 24	24	19	22	25	28	31	35	41	45
12 × 48	22	19	22	25	28	31	37	43	45
24 × 24	22	20	23	26	29	32	37	43	45
24 × 48	20	20	23	26	29	31	39	45	45
48 × 48	18	21	24	27	30	35	41	45	45
48 × 96	18	19	22	25	29	35	41	45	45

Note: The data are for duct lengths of 20 ft, but the values may be used for the cross-section shown regardless of length.

Table 21 Experimentally Measured TL_{out} Versus Frequency for Round Ducts

Diameter, in.	Length, ft	Gage	TL_{out}, dB Octave Band Center Frequency, Hz						
			63	125	250	500	1000	2000	4000
Long Seam Ducts									
8	15	26	>45	(53)	55	52	44	35	34
14	15	24	>50	60	54	36	34	31	25
22	15	22	>47	53	37	33	33	27	25
32	15	22	(51)	46	26	26	24	22	38
Spiral Wound Ducts									
8	10	26	>48	>64	>75	72	56	56	46
14	10	26	>43	>53	55	33	34	35	25
26	10	24	>45	50	26	26	25	22	36
26	10	16	>48	53	36	32	32	28	41
32	10	22	>43	42	28	25	26	24	40

Note: In cases where background sound swamped the sound radiated from the duct walls, a lower limit on TL_{out} is indicated by a > sign. Parentheses indicate measurements in which background sound has produced a greater uncertainty than usual.

Table 22 TL_{out} Versus Frequency for Flat Oval Ducts

Duct Size, in. × in.	Gage	TL_{out}, dB Octave Band Center Frequency, Hz						
		63	125	250	500	1000	2000	4000
12 × 6	24	31	34	37	40	43	—	—
24 × 6	24	24	27	30	33	36	—	—
24 × 12	24	28	31	34	37	—	—	—
48 × 12	22	23	26	29	32	—	—	—
48 × 24	22	27	30	33	—	—	—	—
96 × 24	20	22	25	28	—	—	—	—
96 × 48	18	28	31	—	—	—	—	—

Note: The data are for duct lengths of 20 ft, but the values may be used for the cross-section shown regardless of length.

Breakout Sound Transmission from Ducts. The sound power level associated with sound transmitted through a duct wall and then radiated from the exterior surface of the duct wall is given by

$$L_{w(out)} = L_{w(in)} + 10 \log(S/A) - TL_{out} \qquad (7)$$

where

$L_{w(out)}$ = sound power level of sound radiated from outside surface of duct walls, dB

$L_{w(in)}$ = sound power level of sound inside duct, dB

S = surface area of outside sound-radiating surface of duct, in^2

A = cross-sectional area of inside of duct, in^2

TL_{out} = normalized duct breakout transmission loss (independent of S and A), dB

Values for TL_{out} for rectangular ducts are given in Table 20, for round ducts are given in Table 21, and for flat-oval ducts are given in Table 22 (Cummings 1983, 1985). The equations for S and A for rectangular ducts are

$$S = 2 \times 12L(a + b) \qquad (8)$$

$$A = ab \qquad (9)$$

where

a = larger duct cross-section dimension, in.

b = smaller duct cross-section dimension, in.

L = length of the duct sound radiating surface, ft

The equations for S and A for round ducts are

$$S = 12L\pi d \qquad (10)$$

$$A = \pi d^2/4 \qquad (11)$$

where

d = duct diameter, in.

L = length of duct sound radiating surface, ft

The equations for S and A for flat-oval ducts are

$$12L[2(a - b) + \pi b] \qquad (12)$$

$$A = b(a - b) + \frac{\pi b^2}{4} \qquad (13)$$

where

a = length of larger major axis, in.

b = length of minor duct axis, in.

L = length of duct sound radiating surface, ft

Equation (7) assumes no interior sound attenuation along the length of the duct sound radiating surface. Thus, it is valid only for unlined ducts. It is generally valid for duct lengths up to 30 ft. $L_{w(out)}$ must always be equal to or less than $L_{w(in)}$.

For most applications, the sound pressure level in an occupied space as a result of duct sound breakout can be obtained from

$$L_p = L_{w(out)} - 10\log[\pi rL] + 10 \qquad (14)$$

where

L_p = sound pressure level at specified point in the space, dB

$L_{w(out)}$ = sound power level of sound radiated from outside surface of duct walls given by Equation (7), dB

r = distance between duct and position at which L_p is being calculated, ft

L = length of duct sound radiating surface, ft

Example 3. A 12 in. by 48 in. by 15 ft long rectangular supply duct above a mineral fiber lay-in tile ceiling is constructed of 22 gage sheet metal. Given the sound power levels in the duct, what are the sound pressure levels at a listener 5 ft from the duct?

Solution:

	Octave Band Center Frequency, Hz						
	63	125	250	500	1000	2000	4000
$L_{w(in)}$	90	85	80	75	70	65	60
$- TL_{out}$ (Table 20)	−19	−22	−25	−28	−31	−37	−43
10 log (S/A)	16	16	16	16	16	16	16
$L_{w(out)}$	87	79	71	63	55	44	33
− Ceiling tile (Table 23 − 10)	−3	−6	−8	−10	−16	−21	−26
− 10 log (πrL) + 10	−14	−14	−14	−14	−14	−14	−14
L_p, dB	70	59	49	39	23	9	−7

The RC level associated with the above L_p values is RC 24 (LFVb).

Example 4. Repeat Example 3 using a round duct of equivalent airflow area: 27 in. in diameter, 22 gage, 15 ft long.

Solution:

	Octave Band Center Frequency, Hz						
	63	**125**	**250**	**500**	**1000**	**2000**	**4000**
$L_{w(in)}$	90	85	80	75	70	65	60
$- TL_{out}$ (Table 21)	−49	−50	−32	−30	−29	−25	−31
10 log (S/A)	14	14	14	14	14	14	14
$L_{w(out)}$	55	49	62	59	55	54	43
− Ceiling tile (Table 23 − 10)	−3	−6	−8	−10	−16	−21	−26
− 10 log $(\pi r L)$ + 10	−14	−14	−14	−14	−14	−14	−14
L_p, dB	38	29	40	35	25	19	3

The RC level associated with the above L_p values is RC 26 (MF). The use of the round duct results in a slight increase in the RC level and

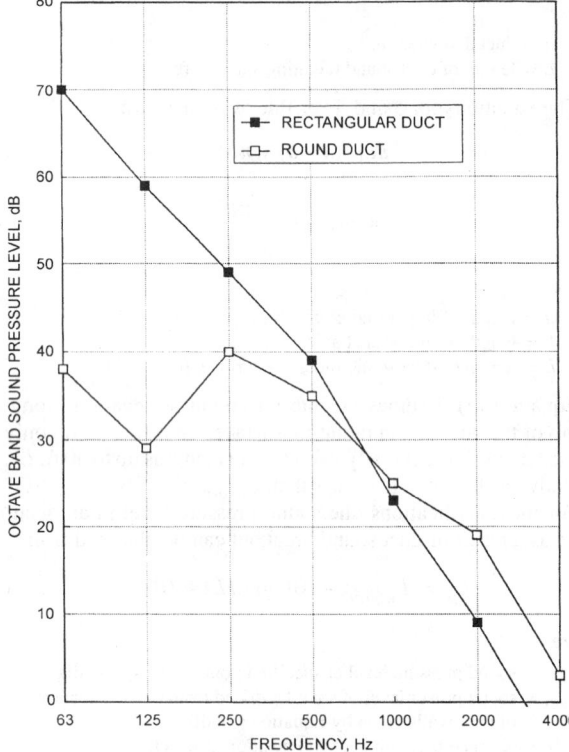

Fig. 23 Duct Breakout Sound Pressure Levels from Examples 3 and 4

eliminates the low-frequency rumble present in Example 3, but it introduces some mid-frequency roar that can be dealt with by another means. The two spectra are shown in Figure 23.

Examples 3 and 4 demonstrate that a round duct is much more effective in reducing the low-frequency sound transmitted through the duct wall (breakout) into an adjacent space than a rectangular duct.

However, when sound is not transmitted through the duct wall of a round duct, it is transmitted down the duct and may become a problem at some other point. Flexible and rigid fiberglass ducts often come in round configurations and may be referred to as round ducts. These types of round ducts do not have high transmission loss properties because they do not have the mass or stiffness associated with round sheet metal ducts. Whenever duct sound breakout is a concern, fiberglass or flexible round duct should not be used.

Breakin Sound Transmission into Ducts. The sound power level associated with sound that is transmitted into (breakin) a duct through the duct walls from the space outside the duct is given by

$$L_{w(in)} = L_{w(out)} - TL_{in} - 3 \qquad (15)$$

where

$L_{w(in)}$ = sound power level of sound transmitted into duct and then transmitted upstream or downstream of point of entry, dB

$L_{w(out)}$ = sound power level of sound incident on outside of duct walls, dB

TL_{in} = duct breakin transmission loss, dB

Values for TL_{in} are given for rectangular ducts in Table 24, for round ducts in Table 25, and for flat-oval ducts in Table 26 (Cummings 1983, 1985).

RECEIVER ROOM SOUND CORRECTION

The sound pressure level at a given location in a room due to a particular sound source is a function of the sound power level and sound radiation characteristics of the sound source, the acoustic properties of the room (surface treatments, furnishings, etc.), the room volume, and the distance between the sound source and the point of observation. Two types of sound sources are typically encountered in HVAC applications: point source and line source. Typical point sources are sound radiated from grilles, registers and diffusers; air-valve and fan-powered air terminal units and fan-coil units located in ceiling plenums; and return-air openings. Line sources are usually associated with sound breakout from air ducts and long slot diffusers.

Table 23 Ceiling/Plenum/Room Attenuations for Generic Ceilings with T-Bar Suspensions (Warnock 1998)

Tile Type	Approximate Density, lb/ft²	Tile Thickness, in.	Octave Band Frequency, Hz						
			63	**125**	**250**	**500**	**1000**	**2000**	**4000**
Mineral fiber	1.0	5/8	13	16	18	20	26	31	36
Mineral fiber	0.5	5/8	13	15	17	19	25	30	33
Glass fiber	0.1	5/8	13	16	15	17	17	18	19
Glass fiber	0.6	2	14	17	18	21	25	29	35
Glass fiber with TL backing	0.6	2	14	17	18	22	27	32	39
Gypsum board tiles	1.8	1/2	14	16	18	18	21	22	22
Solid gypsum board ceiling	1.8	1/2	18	21	25	25	27	27	28
Solid gypsum board ceiling	2.3	5/8	20	23	27	27	29	29	30
Double layer of gypsum board	3.7	1	24	27	31	31	33	33	34
Double layer of gypsum board	4.5	1-1/4	26	29	33	33	35	35	36
Mineral fiber tiles, concealed spline mount	0.5 to 1	5/8	20	23	21	24	29	33	34

Note: To obtain approximate values for the ceiling/plenum attenuation only, subtract 10 from the values given here.

Table 24 TL_{in} Versus Frequency for Rectangular Ducts

Duct Size, in. × in.	Gage	TL_{out}, dB Octave Band Center Frequency, Hz							
		63	125	250	500	1000	2000	4000	8000
12 × 12	24	16	16	16	25	30	33	38	42
12 × 24	24	15	15	17	25	28	32	38	42
12 × 48	22	14	14	22	25	28	34	40	42
24 × 24	22	13	13	21	26	29	34	40	42
24 × 48	20	12	15	23	26	28	36	42	42
48 × 48	18	10	19	24	27	32	38	42	42
48 × 96	18	11	19	22	26	32	38	42	42

Note: The data are for duct lengths of 20 ft, but the values may be used for the cross-section shown regardless of length.

Table 25 Experimentally Measured TL_{in} Versus Frequency for Circular Ducts

Diameter, in.	Length, ft	Gage	TL_{in}, dB Octave Band Center Frequency, Hz						
			63	125	250	500	1000	2000	4000
Long Seam Ducts									
8	15	26	>17	(31)	39	42	41	32	31
14	15	24	>27	43	43	31	31	28	22
22	15	22	>28	40	30	30	30	24	22
32	15	22	(35)	36	23	23	21	19	35
Spiral Wound Ducts									
8	10	26	>20	>42	>59	>62	53	43	26
14	10	26	>20	>36	44	28	31	32	22
26	10	24	>27	38	20	23	22	19	33
26	10	16	>30	>41	30	29	29	25	38
32	10	22	>27	32	25	22	23	21	37

Note: In cases where background sound swamped the sound radiated from the duct walls, a lower limit on TL_{in} is indicated by a > sign. Parentheses indicate measurements in which background sound has produced a greater uncertainty than usual.

Table 26 TL_{in} Versus Frequency for Flat Oval Ducts

Duct Size, in. × in.	Gage	TL_{in}, dB Octave Band Center Frequency, Hz						
		63	125	250	500	1000	2000	4000
12 × 6	24	18	18	22	31	40	—	—
24 × 6	24	17	17	18	30	33	—	—
24 × 12	24	15	16	25	34	—	—	—
48 × 12	22	14	14	26	29	—	—	—
48 × 24	22	12	21	30	—	—	—	—
96 × 24	20	11	22	25	—	—	—	—
96 × 48	18	19	28	—	—	—	—	—

Note: The data are for duct lengths of 20 ft, but the values may be used for the cross-section shown regardless of length.

For a point source in an enclosed space, classical diffuse-field theory predicts that as the distance between the source and point of observation is increased, the sound pressure level initially decreases at the rate of 6 dB per doubling of distance. At some point, the reverberant sound field begins to dominate and the sound pressure level remains at a constant level.

Schultz (1985) and Thompson (1981) found that diffuse-field theory does not apply in real-world rooms with furniture or other sound-scattering objects. Instead, the sound pressure levels decrease at the rate of around 3 dB per every doubling of distance between the sound source and the point of observation. Generally, a reverberant sound field does not exist in small rooms (room volumes less than 15,000 ft^3). In large rooms (room volumes greater than 15,000 ft^3), reverberant fields usually exist, but usually at distances from the sound sources that are significantly greater than those predicted by diffuse-field theory.

Table 27 Values for A in Equation (16)

Room Volume, ft^3	Value for A, dB Octave Band Center Frequency, Hz						
	63	125	250	500	1000	2000	4000
1,500	4	3	2	1	0	−1	−2
2,500	3	2	1	0	−1	−2	−3
4,000	2	1	0	−1	−2	−3	−4
6,000	1	0	−1	−2	−3	−4	−5
10,000	0	−1	−2	−3	−4	−5	−6
15,000	−1	−2	−3	−4	−5	−6	−7

Table 28 Values for B in Equation (16)

Distance from Sound Source, ft	Value for B, dB
3	5
4	6
5	7
6	8
8	9
10	10
13	11
16	12
20	13

Table 29 Values for C in Equation (17)

Distance from Sound Source, ft	Value for C, dB Octave Band Center Frequency, Hz						
	63	125	250	500	1000	2000	4000
3	5	5	6	6	6	7	10
4	6	7	7	7	8	9	12
5	7	8	8	8	9	11	14
6	8	9	9	9	10	12	16
8	9	10	10	11	12	14	18
10	10	11	12	12	13	16	20
13	11	12	13	13	15	18	22
16	12	13	14	15	16	19	24
20	13	15	15	16	17	20	26
25	14	16	16	17	19	22	28
32	15	17	17	18	20	23	30

Point Sound Sources

Most normally furnished rooms with regular proportions have acoustic characteristics that range from average to medium dead. These usually include carpeted rooms that have sound absorptive ceilings. If a normally furnished room has a room volume less than 15,000 ft^3 and the sound source is a single point source, the sound pressure levels associated with the sound source can be obtained from

$$L_p = L_w + A - B \tag{16}$$

where

L_p = sound pressure level at specified distance from sound source, dB
L_w = sound power level of sound source, dB

Values for A and B are given in Tables 27 and 28. If a normally furnished room has a room volume greater than 15,000 ft^3 and the sound source is a single point source, the sound pressure levels associated with the sound source can be obtained from

$$L_p = L_w - C - 5 \tag{17}$$

Values for C are given in Table 29. Equation (17) can be used for room volumes of up to 150,000 ft^3. The accuracy of Equations (16) and (17) is typically within 2 to 5 dB.

Distributed Array of Ceiling Sound Sources

In many office buildings, air supply outlets are located flush with the ceiling of the conditioned space and constitute an array of distributed sound sources in the ceiling. The geometric pattern depends on the floor area served by each outlet, the ceiling height, and the thermal load distribution. In the interior zones of a building where the thermal load requirements are essentially uniform, the air delivery per outlet is usually the same throughout the space; thus, the sound sources tend to have nominally equal sound power levels. One way to calculate the sound pressure levels in a room associated with such a distributed array is first to use Equation (16) or (17) to calculate the sound pressure levels associated with each individual air outlet at specified locations in the room. Then logarithmically add the sound pressure levels associated with each diffuser at each observation point. This calculation procedure can be very tedious and time consuming for a large number of ceiling air outlets.

For a distributed array of ceiling sound sources (air outlets) of nominally equal sound power, the room sound pressure levels tend to be uniform in the horizontal plane parallel to the ceiling. Although the sound pressure levels will decrease with distance from the ceiling along a vertical axis, the sound pressure levels along any selected horizontal plane are nominally constant. The calculation for a distributed ceiling array can be greatly simplified by using Equation (18) instead of Equation (16) or (17). For this case, it is desirable to use a reference plane of 5 ft above the floor (the average distance between seated and standing head height). Thus, $L_{p(5)}$ is obtained from the following equation.

$$L_{p(5)} = L_{w(s)} - D \qquad (18)$$

where

$L_{p(5)}$ = sound pressure level 5 ft above floor, dB
$L_{w(s)}$ = sound power level of single diffuser in array, dB

Values for D are given in Table 30.

Nonstandard Rooms

The previous equations assume the acoustic characteristics of a room range from *average* to *medium dead*, which is generally true of most rooms. However, some rooms may be acoustically *medium live* to *live* (they have little sound absorption). These rooms may be sports or athletic areas, concert halls, or other rooms that are designed to be live; or they may be rooms that are improperly designed from an acoustic standpoint. The previous equations should not be used for acoustically live rooms because they can overestimate the decrease in the sound pressure levels associated with the room sound correction by as much as 10 to 15 dB. When these or other types of nonstandard rooms are encountered, the calculation procedures referred to in Reynolds and Bledsoe (1991) and Reynolds and Bevirt (1994) should be used.

Table 30   Values for *D* in Equation (18)

Floor Area per Diffuser, ft²	Value for *D*, dB Octave Band Center Frequency, Hz						
	63	125	250	500	1000	2000	4000
Ceiling height 8 to 9 ft							
100 to 150	2	3	4	5	6	7	8
200 to 250	3	4	5	6	7	8	9
Ceiling height 10 to 12 ft							
150 to 200	4	5	6	7	8	9	10
250 to 300	5	6	7	8	9	10	11
Ceiling height 14 to 16 ft							
250 to 300	7	8	9	10	11	12	13
350 to 400	8	9	10	11	12	13	14

INDOOR SOUND CRITERIA

HVAC related sound is often the major source of background noise in indoor spaces. Whether an occupant considers the background noise to be acceptable or not generally depends on two factors. First is the **perceived loudness** of the noise relative to that of normal activities; if it is clearly noticeable then it is likely to be distracting and cause complaint. Second is the **quality** of the background noise; noise perceived as a rumble, roar, or hiss may result in complaints of annoyance and stress. The spectrum is then said to be unbalanced.

The acoustical design must ensure that HVAC noise is of sufficiently low level and unobtrusive quality so as not to interfere with the occupancy requirements of the space use. If the background noise reduces speech intelligibility, for example, complaints of lost productivity may result. Accordingly, methods of rating HVAC related noise should assess both the **relative loudness** and **quality** of the background noise.

Sound Rating Methods

Currently, several methods are used to rate indoor sound. They include the traditional A-weighted sound pressure level (dBA) and tangent noise criteria (NC), the more recent room criteria (RC) and balanced noise criteria (NCB), and the new RC Mark II. Each sound rating method was developed based on data for specific applications; hence not all are suitable for the rating of HVAC related noise in the variety of applications encountered. The preferred sound rating methods generally comprise two distinct parts: a family of criterion curves (specifying sound levels by octave bands), and a companion procedure for rating the calculated or measured sound data relative to the criterion curves.

Ideally, HVAC-related background noise should have the following characteristics:

- Balanced contributions from all parts of sound spectrum with no predominant bands of noise
- No audible tones such as a hum or whine
- No fluctuations in level such as a throbbing or pulsing

Unfortunately, no acceptable process easily characterizes the effects of audible tones and level fluctuations. Table 31 summarizes the essential differences, advantages, and disadvantages of the rating methods that characterize HVAC-related background noise.

NC   Noise Criteria Method

The NC method for rating noise, described in Chapter 7 of the 1997 *ASHRAE Handbook—Fundamentals*, is widely used and understood. It is a single-number rating that is somewhat sensitive to the relative loudness and speech interference properties of a given noise spectrum. The method consists of a family of criterion curves extending from 63 to 8000 Hz, and a **tangency rating procedure**. The criterion curves define the limits of octave band spectra that must not be exceeded to meet occupant acceptance in certain spaces. The rating is expressed as NC followed by a number (e.g., NC 40).

The NC method is sensitive to level and has the disadvantage that the tangency method used to determine the rating does not require that the noise spectrum approximate the shape of the NC curves. Thus, many different-sounding noises can have the same numeric rating, but rank differently on the basis of subjective sound quality. With the advent of VAV systems, excessive low frequency noise below the 63 Hz octave band became a serious problem, and the NC rating method does not address this (Ebbing 1992). In many HVAC systems that do not produce excessive low frequency noise, the NC rating correlates relatively well with occupant satisfaction, *if* sound quality is not a significant concern.

RC   Room Criteria Method

The RC method consists of a family of criterion curves and a rating procedure (Blazier 1981a,b; ANSI *Standard* S12.2, ASHRAE

Table 31 Comparison of Sound Rating Methods

Meth.	Overview	Considers Speech Interference	Evaluates Sound Quality	Components Currently Rated by Method
NC	Can rate components No quality assessment Does not evaluate low-frequency rumble	Yes	No	Air terminals Diffusers
RC	Used to evaluate systems Should not be used to evaluate components Can be used to evaluate sound quality Provides some diagnostic capability	Yes	Yes	
dBA	Can be determined using sound level meter No quality assessment Frequently used for outdoor noise ordinances	Yes	No	Cooling towers Water chillers Condensing units
NCB	Can rate components Some quality assessment	Yes	Yes	
RC Mark II	Evaluates sound quality Provides improved diagnostics capability	Yes	Yes	

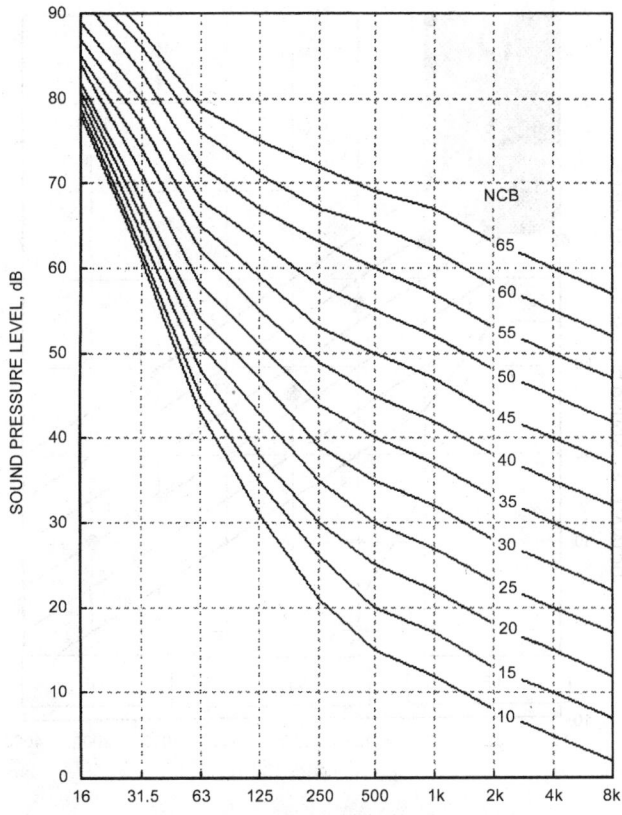

Fig. 24 NCB Noise Criterion Curves Drawn from ANSI *Standard* S 12.2

1995). The shape of these curves differs from the NC curves so as to achieve a well-balanced bland-sounding spectrum, and two additional octave bands (16 and 31 Hz) are added to include possible excessive low-frequency noise. This rating procedure assesses background noise in spaces, both on the basis of its effect on speech communication and on subjective sound quality. The rating is expressed as RC followed by a number to show the level of the noise and a letter to indicate the quality, for example, RC 35(N) where N denotes neutral.

dBA A-Weighted Sound Level

The A-weighted sound level (described in Chapter 7 of the 1997 *ASHRAE Handbook—Fundamentals*) is a single-number measure of the relative loudness of noise that is used extensively in outdoor environmental noise standards. The rating is expressed as a number followed by dBA, for example, 40 dBA.

A-weighted sound levels can be measured with simple sound level meters. The ratings correlate well with human judgments of relative loudness but take no account of spectral balance or sound quality. Thus many different-sounding spectra can result in the same numeric value, but have quite different subjective qualities.

NCB Balanced Noise Criteria Method

The NCB method (Beranek 1989) is used to specify or evaluate room noise and includes noise due to occupant activities. The NCB criterion curves (Figure 24) are intended as replacements for the NC curves, and include two additional low frequency octave bands (16 and 31 Hz), and lower permissible noise levels at high frequencies (4000 and 8000 Hz). The NCB rating procedure is based on a speech interference level (SIL) (SIL = average of four sound pressure levels at octave band mid-frequencies of 500, 1000, 2000, and 4000 Hz), and on additional tests for rumble and hiss compliance. The rating is expressed as NCB followed by a number, for example, NCB 40.

The NCB method is better than the NC method in determining whether a noise spectrum has an unbalanced shape sufficient to demand corrective action, and it addresses the issue of low-frequency noise. The rating procedure is, however, more complicated than the familiar tangency method.

RC Mark II Room Criteria Method

Based on experience and the findings from ASHRAE sponsored research (Broner 1994), the RC method was revised to the RC Mark

II method (Blazier 1997). Like its predecessor, the RC Mark II method is intended for rating the sound performance of an HVAC system as a whole. The method can also be used as a diagnostic tool for analyzing noise problems in the field. The RC Mark II method is more complicated to use than the RC method, but spreadsheet macros are available to do the calculations and graphical analysis.

The RC Mark II method has three parts: (1) a family of criterion curves such as shown in Figure 25, (2) a procedure for determining the RC numerical rating and the noise spectral balance (quality), and (3) a procedure for estimation of occupant satisfaction when the spectrum does not have the shape of an RC curve (Quality Assessment Index) (Blazier 1995).

The rating is expressed as RC followed by a number and a letter, for example, RC 35(N). The number is the arithmetic average rounded to the nearest integer of the sound pressure levels in the 500, 1000, and 2000 Hz octave bands (the principal speech frequency region). The letter identifies the perceived character of the sound: (N) for neutral, (LF) for low frequency rumble, (MF) for mid frequency roar, and (HF) for high frequency hiss. There are also two subcategories of the low frequency descriptor: (LFB), denoting a moderate but perceptible degree of sound induced ceiling/wall vibration, and (LFA), denoting a noticeable degree of sound induced vibration.

Each reference curve in Figure 25 identifies the shape of a neutral, bland-sounding spectrum, indexed to a curve number corresponding to the sound level in the 1000 Hz octave-band. The shape of these curves is based on work by Blazier (1981a,b), modified at 16 Hz following research by Broner (1994). Regions A and B denote levels at which sound can induce vibration in light wall and ceiling construction that can potentially cause rattles in light fixtures, furniture, etc. Curve T is the octave-band threshold of hearing as defined by ANSI *Standard* S12.2.

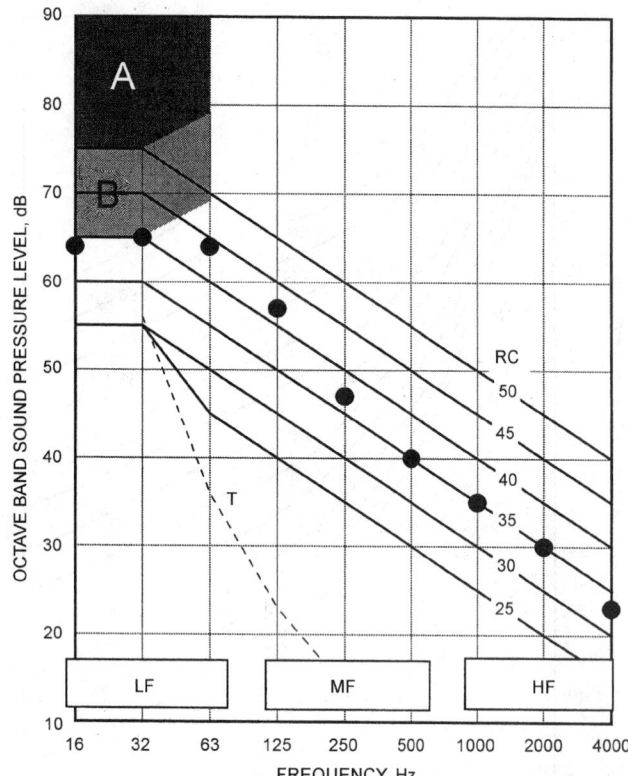

Noise levels in region B are likely to generate perceptible vibration in lightweight walls and ceilings. Rattles are a slight possibility in light fixtures, doors, windows, etc. Noise levels in region A have a high probability of generating easily perceptible noise induced vibration in lightweight walls and ceilings. Audible rattling in light fixtures, doors, windows, etc. is likely. Regions LF, MF, and HF are explained in text. The solid dots are sound pressure levels for Example 5.

Fig. 25 Room Criterion Curves, Mark II

Procedure for Determining RC Mark II Rating

Step 1. Determine the appropriate RC reference curve by obtaining the arithmetic average of the sound levels in the 500, 1000, and 2000 Hz octave bands. (This curve is not to be confused with the speech-interference level (SIL), which is a four-band average obtained by including the 4000 Hz octave-band.). The RC reference curve is chosen to be that which has the same value at 1000 Hz as the calculated average value.

Step 2. Assign a subjective quality by calculating the quality assessment index (QAI) (Blazier 1995). QAI is a measure of the deviation of the shape of the spectrum under evaluation to the shape of the RC reference curve. The procedure requires calculation of **energy-averaged** spectral deviations from the RC reference curve in each of three frequency groups: low frequency, LF (16–63 Hz), medium frequency, MF (125–500 Hz), and high frequency, HF (1000–4000 Hz). (A simple arithmetic average of these deviations is usually adequate for evaluating HVAC sounds.) The procedure for the LF region is given by Equation (19), and is repeated in the MF and HF regions by substituting the corresponding values at each frequency.

$$LF = 10\log[(10^{0.1\Delta L_{16}} + 10^{0.1\Delta L_{31.5}} + 10^{0.1\Delta L_{63}})/3] \quad (19)$$

where the ΔL terms are the differences between the spectrum being evaluated and the RC reference curve in each frequency band. In this way, three specific spectral deviation factors, expressed in dB with either positive or negative values, are associated with the spectrum being rated. QAI is the range in dB between the highest and lowest values of the spectral deviation factors.

If QAI ≤ 5 dB, the spectrum is assigned a neutral (N) rating. If QAI exceeds 5 dB, the sound quality descriptor of the RC rating is the letter designation of the frequency region of the deviation factor having the highest positive value.

Example 5. Figure 25 shows a sound spectrum plotted on an RC Mark II curve plot. The data was processed and is summarized in Table 32. Note that the arithmetic average of the sound levels in the 500, 1000, and 2000 Hz octave bands is 35 dB, so the RC 35 curve is selected as the reference for spectrum quality evaluation.

Table 32 Example 5 Calculation of RC Mart II Rating

	Frequency, Hz								
	16	31	63	125	250	500	1000	2000	4000
Spectrum levels	64	65	64	57	47	40	35	30	23
Average of 500 to 2000 Hz levels							35		
RC 35 contour	60	60	55	50	45	40	35	30	25
Levels - RC contour	4	5	9	7	2	0	0	0	−2
		LF			MF			HF	
Spectral deviations		6.6			4.0			−0.6	
Quality Assessment Index (QAI)					7.2				
RC Mark II rating					RC 35(LF)				

The spectral deviation factors in the LF, MF, and HF regions are 6.6, 4.0 and −0.6, giving a QAI of 7.2. The maximum *positive* deviation

**Table 33 Definition of Sound-Quality Descriptor and Quality-Assessment Index (QAI)
To Aid in Interpreting RC Mark II Ratings of HVAC-Related Sound**

Sound-Quality Descriptor	Description of Subjective Perception	Magnitude of QAI	Probable Occupant Evaluation, Assuming Level of Specified Criterion is Not Exceeded
(N) Neutral (Bland)	Balanced sound spectrum, no single frequency range dominant	QAI ≤ 5 dB, L_{16}, L_{31} ≤ 65 QAI ≤ 5 dB, L_{16}, L_{31} > 65	Acceptable Marginal
(LF) Rumble	Low-frequency range dominant (16 to 63 Hz)	5 dB < QAI ≤ 10 dB QAI > 10 dB	Marginal Objectionable
(LFVB) Rumble, with moderately perceptible room surface vibration	Low-frequency range dominant (16 to 63 Hz)	QAI ≤ 5 dB, 65< L_{16}, L_{31} <75 5 dB < QAI ≤ 10 dB QAI > 10 dB	Marginal Marginal Objectionable
(LFVA) Rumble, with clearly perceptible room surface vibration	Low-frequency range dominant (16 to 63 Hz)	QAI ≤ 5 dB, L_{16}, L_{31} > 75 5 dB < QAI ≤ 10 dB QAI > 10 dB	Marginal Marginal Objectionable
(MF) Roar	Mid-frequency range dominant (125 to 500 Hz)	5 dB < QAI ≤10 dB QAI > 10 dB	Marginal Objectionable
(HF) Hiss	High-frequency range dominant (1000 to 4000 Hz)	5 dB < QAI ≤ 10 dB QAI > 10 dB	Marginal Objectionable

factor occurs in the LF region, QAI exceeds 5, therefore, the rating of the spectrum is RC 35(LF). An average room occupant should perceive this spectrum as marginally rumbly in character.

Estimating Occupant Satisfaction Using QAI. The quality assessment index is useful in estimating the probable reaction of an occupant when the system does not produce optimum sound quality. The basis for estimating occupant satisfaction is that changes in sound level of less than 5 dB do not cause subjects to change their ranking of sounds of similar spectral content. However, level changes greater than 5 dB do significantly affect subjective judgments. As noted already, a QAI of 5 dB or less corresponds to a generally acceptable condition, provided that the perceived level of the sound is in a range consistent with the given type of space occupancy as recommended in Table 34. (An exception to this rule occurs when the sound pressure levels in the 16 Hz or 31 Hz octavebands exceed 65 dB. In such cases, the potential for acoustically-induced vibration in typical lightweight office construction should be considered. If the levels in these bands exceed 75 dB, a significant problem with induced vibration is indicated.)

A QAI that exceeds 5 dB, but is less than or equal to 10 dB, represents a marginal situation in which occupant acceptance is questionable. However, a QAI greater than 10 dB will likely be objectionable to the average occupant. Table 33 lists sound quality descriptors and QAI values and relates them to probable occupant reaction to the noise.

Undoubtedly situations occur in the assessment of HVAC related noise where the numerical part of the RC rating is less than the specified maximum for the space use, but the sound quality descriptor is other than the desirable (N). For example, a maximum of RC 40(N) is specified, but the actual noise environment turns out to be RC 35(MF). Not enough is know in this area to decide which spectrum is preferable.

Even at moderate levels, if the dominant portion of the background noise occurs in the very low-frequency region, Perrson-Waye (1997) observed that some people experience a sense of oppressiveness or depression in the environment. In such situations, the basis for complaint may result from an exposure to that environment for several hours, and may not be noticeable during a short exposure period.

Guideline Criteria for HVAC-Related Background Sound in Rooms

Table 34 lists design guidelines for HVAC-related background sound appropriate to various occupancies. Perceived loudness and task interference are factored into the numerical part of the RC rating. sound quality assumes the preferred design target is a neutral-sounding (N) spectrum, although some spectrum imbalance is likely tolerable in certain limits.

If these levels are used as a basis for contractual requirements, the following additional information must be provided.

1. What sound levels will be measured (specify L_{eq} or L_{max} levels, etc. in each octave frequency band)?
2. Where and how will the sound levels be measured (specify space average over a defined area or specific points, etc.)?
3. What instruments will be used to make the sound measurements (specify ANSI Type 1 or Type 2 sound level meters with octave band filters)?
4. How will the instruments used for the sound measurements be calibrated?
5. How will the sound measurement results be interpreted?

Unless the above five points are clearly stipulated, the specified sound criteria may be unenforceable.

When applying the levels specified in Table 34 as a basis for design, sound from non-HVAC sound sources, such as traffic and office equipment, may be the lower limit for sound levels in a space.

Table 34 Design Guidelines for HVAC-Related Background Sound in Rooms

Room Types	RC(N); QAI ≤ 5dB Criterion [a,b]
Residences, Apartments, Condominiums	25-35
Hotels/Motels	
Individual rooms or suites	25-35
Meeting/banquet rooms	25-35
Corridors, lobbies	35-45
Service/support areas	35-45
Office Buildings	
Executive and private offices	25-35
Conference rooms	25-35
Teleconference rooms	25 (max)
Open-plan offices	30-40
Corridors and lobbies	40-45
Hospitals and Clinics	
Private rooms	25-35
Wards	30-40
Operating rooms	25-35
Corridors and public areas	30-40
Performing Arts Spaces	
Drama theaters	25 (max)
Concert and recital halls [c]	
Music teaching studios	25 (max)
Music practice rooms	35 (max)
Laboratories (with fume hoods)	
Testing/research, minimal speech communication	45-55
Research, extensive telephone use, speech communication	40-50
Group teaching	35-45
Church, Mosque, Synagogue	
General assembly	25-35
With critical music programs [c]	
Schools [d]	
Classrooms up to 750 ft^2	40 (max)
Classrooms over 750 ft^2	35 (max)
Large lecture rooms, without speech amplification	35 (max)
Libraries	30-40
Courtrooms	
Unamplified speech	25-35
Amplified speech	30-40
Indoor Stadiums, Gymnasiums	
Gymnasiums and natatoriums [e]	40-50
Large seating-capacity spaces with speech amplification [e]	45-55

[a]The values and ranges are based on judgment and experience, not on quantitative evaluations of human reactions. They represent general limits of acceptability for typical building occupancies. Higher or lower values may be appropriate and should be based on a careful analysis of economics, space use, and user needs.

[b]When quality of sound in the space is important, specify criteria in terms of RC(N). If the quality of the sound in the space is of secondary concern, the criteria may be specified in terms of NC or NCB levels of similar magnitude.

[c]An experienced acoustical consultant should be retained for guidance on acoustically critical spaces (below RC 30) and for all performing arts spaces.

[d]HVAC-related sound criteria for schools, such as those listed in this table, may be too high and impede learning by children in primary grades whose vocabulary is limited. Some educators and others believe that the HVAC-related background sound should not exceed RC 25 (N).

[e]RC or NC criteria for these spaces need only be selected for the desired speech and hearing conditions.

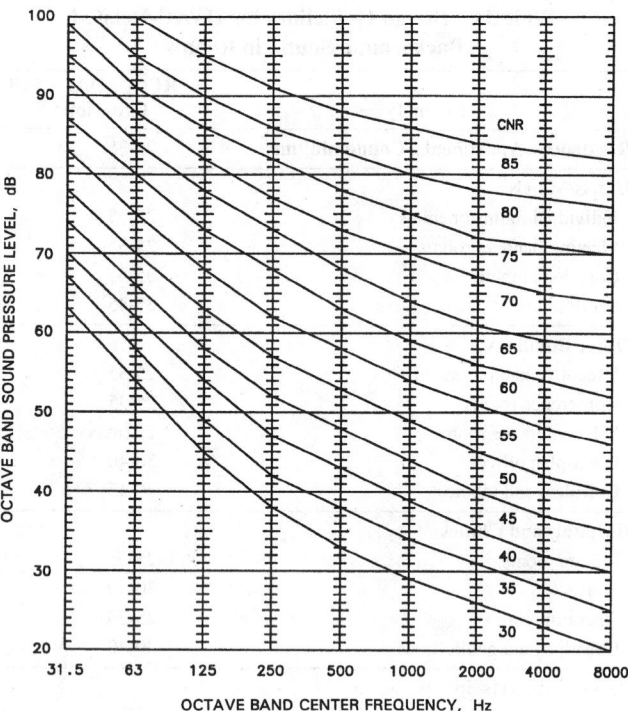

Fig. 26 Non-Normalized Composite Noise Rating (CNR$_{nn}$) Curves

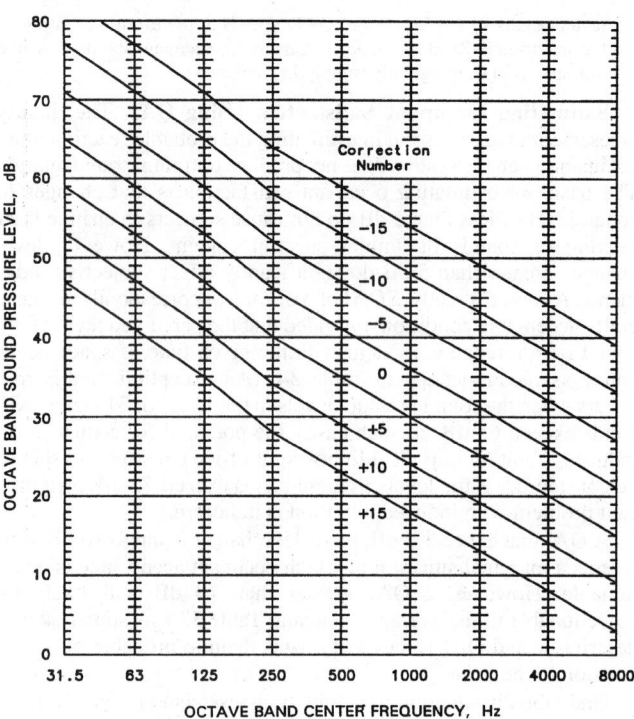

Fig. 27 Correction for Composite Noise Rating (CNR) Associated with Ambient Noise

OUTDOOR SOUND CRITERIA

Acceptable outdoor sound levels are generally specified by local noise ordinances. In the absence of an ordinance, composite noise rating (CNR) procedures can be used to evaluate acceptability of HVAC and other types of mechanical and electrical equipment noise that intrude into communities (Stevens et al. 1955). Figure 26 shows the set of non-normalized composite noise rating (CNR) curves. The non-normalized composite noise rating is determined by plotting the octave band sound pressure levels associated with an intruding noise on Figure 26. The octave band sound pressure levels should be measured at several representative times at each location of interest in the community. The measurements should span a period long enough to give confidence that the average octave band sound pressure levels of the noise are truly representative. If the daytime and nighttime noise signals are different in contents and levels, separate sound measurements should be taken for the two periods.

The non-normalized composite noise rating associated with a noise equals the highest penetration of any of the octave band sound pressure levels into the curves. If the highest penetration falls between two curves, the non-normalized CNR is the interpolated value between the two curves. The non-normalized CNR must then be normalized or corrected for background noise conditions that exist in the absence of the intruding noise and for the time-of-day, seasonal, noise intermittency, noise characteristics, and previous community-exposure-to-similar-noise factors. The correction for background noise that exists in the absence of the intruding noise can be accomplished in one of two ways. If it is possible to measure the octave band sound pressure levels associated with the ambient or background noise in the absence of any intruding noise source, the levels should be measured and plotted on Figure 27.

The zone into which the major portion of the octave band spectrum falls designates the correction to be applied for the background noise. The correction associated with the curve that has a point of tangency that is closest to the octave band ambient sound pressure level curve should be used. It is not necessary to interpolate between curves. Daytime ambient noise levels should be recorded for daytime intruding

noise, and nighttime ambient levels should be used for nighttime intruding noise. If the octave band ambient sound pressure levels cannot be measured, the background sound level corrections given in Table 35 can be used to estimate the correction for background or ambient sound levels. The corrections in Table 35 are based on the general type of community area and nearby traffic activity. The normalized CNR, corrected for background noise level, is obtained by adding the number (keeping track of the sign in front of number) obtained from either Figure 27 or Table 35 to the non-normalized composite noise rating obtained from Figure 26.

The final correction is associated with time-of-day, seasonal, noise intermittency, noise characteristics, and previous community exposure to similar noise factors. These correction factors are obtained from Table 36. The total correction for these factors is the sum of the corrections associated with each individual factor.

The normalized CNR is calculated by taking the non-normalized CNR obtained from Figure 26 and adding to it the correction number for the background noise obtained from either Figure 27 or Table 35 and the total correction number associated with the time-of-day, seasonal, noise intermittency, noise characteristics, and previous community exposure to similar noise factors obtained from Table 36. Once the normalized composite CNR has been calculated, the anticipated community reaction to the intruding noise is obtained from Figure 28.

The composite noise rating procedure is generally a reliable method of determining community reaction to outdoor noise from mechanical and electrical equipment. However, it may not be reliable when dealing with certain types of equipment that generate strong pure tones (e.g., high-pressure blowers, diesel generators, gas turbines, etc.). It is strongly advised that an acoustical consultant be consulted when dealing with these types of sound sources.

Example 6. The octave band sound pressure levels associated with a cooling tower are listed below:

	Octave Band Center Frequency, Hz							
	63	125	250	500	1000	2000	4000	8000
L_p, dB	64	64	62	60	56	53	51	43

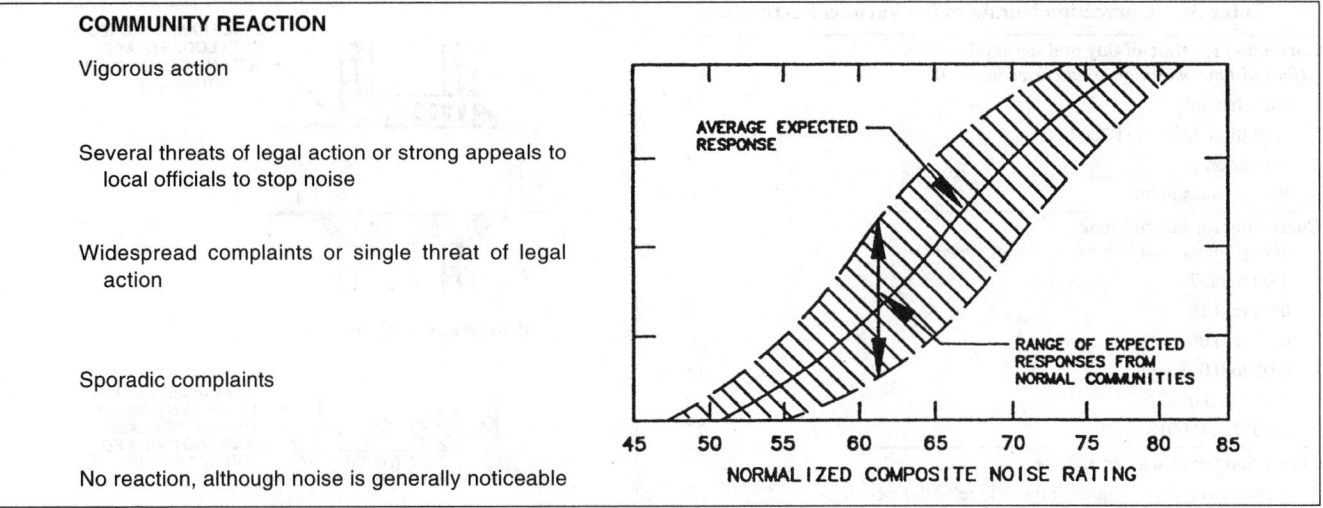

Fig. 28   Estimated Community Reaction to Intruding Noise Versus Normalized Composite Noise Rating (CNR$_n$)

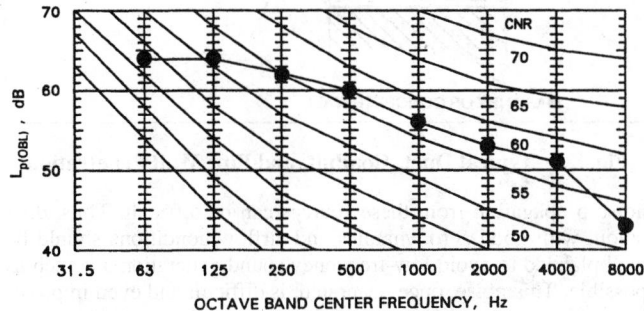

Fig. 29   CNR$_{nn}$ Value for Example 6

The cooling tower runs 24 hours a day. The location at which the sound pressure levels were measured is a business area. Assume there is previous exposure to similar noise and that there are good community relations. Determine the composite noise rating associated with the cooling tower noise, and make some statement relative to the anticipated community reaction to the noise.

Solution:

The non-normalized composite noise rating (CNR$_{nn}$) is obtained by plotting the above octave band sound pressure levels on Figure 26. The resulting plot is shown in Figure 29. An examination of the plot indicates that the CNR$_{nn}$ is CNR$_{nn}$ 58.

Because the cooling tower runs 24 hours a day, the normalized composite noise rating (CNR$_n$) for both daytime and nighttime use must be determined. The background noise correction numbers are obtained from Table 35. The numbers for a business area are

Daytime: −5
Nighttime: 0

The correction numbers from Table 36 are:

Time-of-day: 0
Intermittency: 0
Character of noise: +5
Previous exposure: 0

Thus, the normalized CNR$_n$ values are

Daytime: CNR$_n$ = 58 − 5 + 0 + 5 + 0 = 58
Nighttime: CNR$_n$ = 58 + 0 + 0 + 5 + 0 = 63.

An examination of Figure 28 indicates there will be no complaints during the daytime hours; and there will be some sporadic complaints during the nighttime hours.

Table 35   Background Noise Correction Numbers

Condition	Background Correction Number
Nighttime, rural; no nearby traffic of concern	+15
Daytime, rural; no nearby traffic of concern	+10
Nighttime, suburban; no nearby traffic of concern	+10
Daytime, suburban; no nearby traffic of concern	+5
Nighttime, urban; no nearby traffic of concern	+5
Daytime, urban; no nearby traffic of concern	0
Nighttime, business or commercial area	0
Daytime, business or commercial area	−5
Nighttime, industrial or manufacturing area	−5
Daytime, industrial or manufacturing area	−10
Within 300 ft of intermittent light traffic	0
Within 300 ft of continuous light traffic	−5
Within 300 ft of continuous medium-density traffic	−10
Within 300 ft of continuous heavy-density traffic	−15
300 to 1000 ft from intermittent light traffic	+5
300 to 1000 ft from continuous light traffic	0
300 to 1000 ft from continuous medium-density traffic	−5
300 to 1000 ft from continuous heavy-density traffic	−10
1000 to 2000 ft from intermittent light traffic	+10
1000 to 2000 ft from continuous light traffic	+5
1000 to 2000 ft from continuous heavy-density traffic	−5
2000 to 4000 ft from intermittent light traffic	+15
2000 to 4000 ft from continuous light traffic	+10
2000 to 4000 ft from continuous medium-density traffic	+5
2000 to 4000 ft from continuous heavy-density traffic	0

MECHANICAL EQUIPMENT ROOM SOUND ISOLATION

Mechanical equipment is inherently noisy. Pumps, chillers, cooling towers, boilers, and fans create noise that is difficult to contain. Noise-sensitive spaces must be isolated from these sources. One of the best precautions is to locate mechanical spaces away from acoustically critical spaces. Space planning early in the design stage is critical for successful new projects. Buffer zones (storage rooms, corridors, or less noise-sensitive spaces) can be placed between mechanical equipment rooms and rooms requiring relatively quiet conditions. Sound transmission through roofs and exterior walls is usually less of a problem, making corner rooms and top-floor spaces reasonably good for housing mechanical equipment.

Table 36 Correction Numbers for Various Factors

Correction for time-of-day and seasonal factors	
(for full-time operation, total correction is 0)	
Daytime only	–5
Nighttime (2200 to 0700 hrs)	0
Winter only	–5
Winter and summer	0
Correction for intermittency	
(ratio of source "on" time to reference time period)	
1.00 to 0.57	0
0.56 to 0.18	–5
0.17 to 0.06	–10
0.05 to 0.018	–15
0.017 to 0.0057	–20
0.0057 to 0.0018	–25
Correction for character of noise	
Noise is very low frequency (peak level at 1/1 octave center frequency of 125 Hz or lower)	+5
Noise contains tonal components	+5
Impulsive sound	+5
Correction for previous exposure and community attitude	
No prior exposure	+5
Some previous exposure but poor community relations	+5
Some previous exposure and good community relations	0
Considerable previous exposure and good community relations	–5

Construction enclosing a mechanical equipment room should be poured concrete or masonry units with enough surface weight to provide adequate sound transmission loss and inertia to resist low frequency noise often inherent with mechanical equipment. Heavy-duty drywall construction can also be used successfully sometimes. When it can be used, it saves weight on upper floors, but it typically has a lower frequency sound transmission loss than masonry. Walls must be airtight and caulked at the edges to prevent sound leaks. Floors and ceilings should be concrete slabs, except for the ceiling or roof deck above a top-floor mechanical space.

Penetrations of the mechanical equipment room enclosure create potential paths for sound to escape into adjacent spaces. Therefore, wherever ducts, pipes, conduits, and the like penetrate the walls, floor, or ceiling of a mechanical equipment room, the opening must be acoustically treated for adequate noise control. A 1/2 to 5/8 in. clear space should be left all around the penetrating element and filled with fibrous material for the full depth of the penetration. Both sides of the penetration should be sealed airtight with a nonhardening resilient sealant (Figure 30).

Doors into mechanical equipment rooms are frequently the weak link in the enclosure. Where noise control is important, they should be as heavy as possible, gasketed around the perimeter, have no grilles or other openings, and be self closing. If such doors lead to sensitive spaces, two doors separated by a 3 to 10 ft corridor may be necessary. Mechanical room doors should open out, not in, so that the negative pressure in the room keeps the door sealed against the jamb instead of holding it against the latch.

Penetration of Walls by Ducts

Ducts passing through the mechanical equipment room enclosure pose an additional problem. Sound can be transmitted to either side of the wall via the duct walls. Airborne sound in the mechanical room can be transmitted into the duct (breakin) and enter an adjacent space by reradiating (breakout) from the duct walls, even if the duct contains no grilles, registers, diffusers, or other openings.

Sound levels in ducts close to fans are usually high. Sound can come not only from the fan but also from pulsating duct walls, excessive air turbulence, and air buffeting caused by tight or

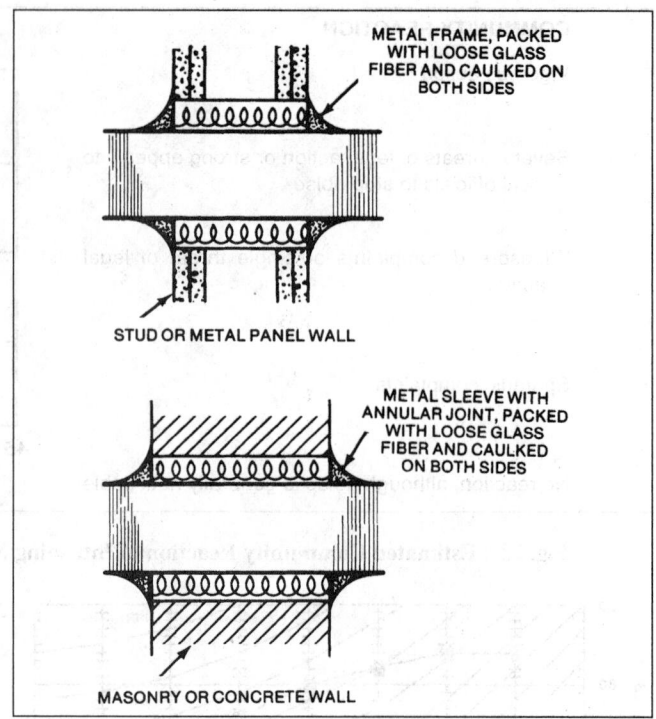

Fig. 30 Typical Duct, Conduit, and Pipe Wall Penetration

noise propagation from these sources can be difficult. Thus, duct layout with good aerodynamics and airflow conditions should be well planned to avoid low-frequency sound generation as much as possible. This noise, once generated, is difficult and even impossible to remove especially when they are near noise-sensitive areas.

Mechanical Chases

Mechanical chases and shafts should be acoustically treated the same way as mechanical equipment rooms, especially if they contain noise-producing ductwork, pipes, and equipment such as fans and pumps. The shaft should be closed at the mechanical equipment room, and shaft wall construction must provide sufficient reduction of mechanical noise from the shaft to noise-sensitive areas to obtain acceptable noise levels. Chases should not be allowed to become "speaking tubes" between spaces requiring different acoustical environments. Thus crosstalk through the shaft must be prevented. Any vibrating pipes, ducts, conduits, or equipment should be vibration isolated so that mechanical vibration and structure-borne noise is not transmitted to the shaft walls and into the building.

If mechanical equipment rooms are used as supply or return plenums, all openings into the equipment room may require noise control treatment, especially if any sound critical space is immediately adjacent. This is especially true if the ceiling space just outside the equipment room is used as a return air plenum and the ceiling is acoustical tile. Most acoustical tile ceilings are almost acoustically transparent at low frequencies.

Often supply ducts are run inside a chase that is also used for the return air. It is best to attenuate the supply duct path at the fan rather than let duct breakout noise force the use of additional noise control at the return air inlets to the chase. Care should be used to prevent turbulent noise generation in the supply duct through proper supply duct design. Also the control of return path noise at the fan rather than at the inlet to the chase is best.

Special Wall Construction

When mechanical equipment rooms must be placed adjacent to offices, conference rooms, or other noise-sensitive areas, the

Table 37   Sound Transmission Class, STC, and Transmission Loss Values of Gypsumboard Wall Configurations, dB

(All exposed joints taped and finished.)

	STC	63	125	250	500	1000	2000	4000
			Octave Band Frequency, Hz					

1. 2-1/2 in. metal studs; one layer 5/8 in. gypsum wallboard screwed to each side of studs

| | 35 | 17 | 12 | 29 | 44 | 54 | 39 | 46 |

2. 3-5/8 in. metal studs; 5/8 in. gypsum wallboard screwed to each side of studs

| | 38 | 12 | 13 | 32 | 46 | 55 | 41 | 46 |

3. Same as (2) with 3.5 in. fiberglass batts between studs

| | 47 | 15 | 17 | 39 | 57 | 62 | 49 | 51 |

4. 3-5/8 in. metal studs; two layers 5/8 in. gypsum wallboard screwed to each side of studs

| | 48 | 17 | 24 | 39 | 50 | 54 | 47 | 51 |

5. Same as (4) with 3.5 in. fiberglass batts between studs

| | 56 | 19 | 25 | 47 | 62 | 66 | 57 | 60 |

6. 2-1/2 in. metal studs; two layers 5/8 in. gypsum wallboard screwed to each side of studs

| | 49 | 17 | 24 | 41 | 50 | 55 | 48 | 53 |

7. Same as (6) with 2.5 in fiberglass batts between studs

| | 55 | 18 | 26 | 47 | 60 | 64 | 55 | 57 |

8. 3-5/8 in. metal studs; three layers 5/8 in. gypsum wallboard screwed to each side of studs; 3 in. rockwool between studs

| | 61 | 20 | 37 | 54 | 61 | 64 | 59 | 64 |

9. 3-5/8 in. metal studs; two layers 5/8 in. gypsum wallboard screwed to horizontal resilient channels on one side; three layers 5/8 in. gypsum wallboard screwed to studs on other side; 3 in.rockwool between studs

| | 62 | 27 | 40 | 54 | 61 | 65 | 63 | 67 |

10. Double wood or steel stud wall with 8 in. cavity and two layers of 3.5 in. fiberglass batts between studs on each side

| | 69 | 30 | 43 | 59 | 70 | 81 | 77 | 85 |

building construction around the equipment space must reduce the noise enough to satisfy the acoustical requirements for the nearby spaces. Depending on the requirements, walls with high sound transmission loss may need to be considered. The low frequency performance should be carefully considered. Tables 37 through 39 give the transmission loss values for selected wall configurations. Note that the solid concrete data also apply to floors. Also note that the data in the tables are for laboratory tests. In buildings, special care must be taken to deal with all possible flanking paths for the transmission of sound. Sound propagation along floors, ceilings, and wall surfaces perpendicular to the common wall can greatly reduce the sound isolation between rooms. When high sound isolation is sought and simple wall construction is inadequate, then double walls, isolated from each other, should be considered. An acoustical consultant can provide information on any special wall sound isolation requirements. .

Floating Floors

Floating floor construction can be used to further reduce sound transmission between the mechanical room and noise-sensitive areas above or below it. Floating floors are composed of two reinforced concrete slabs: the floating or isolated wearing slab and the structural floor slab. The floating slab is designed with a minimum thickness of 3 (nominal 4) in. and rests on a stable, long lasting resilient load-bearing material with a 1 to 4 in.separation between the structural and the floating slabs. A 2 in. separation is most common. The edges of the floating slab are isolated from the structure with a resilient material, and the joint between the floating slab and the adjacent curb is resiliently caulked.

Table 38   Sound Transmission Class and Transmission Loss Values of Concrete Masonry Walls and Concrete Slabs, dB

STC	63	125	250	500	1000	2000	4000
		Octave Band Frequency, Hz					

Concrete Masonry Units

1. 4 in., 50% solid, 24 lb/block

| 44 | 31 | 27 | 31 | 35 | 39 | 44 | 49 |

2. 6 in., 75% solid, 40 lb/block

| 47 | 34 | 31 | 31 | 37 | 46 | 52 | 60 |

3. 8 in., 50% solid, 38 lb/block

| 50 | 35 | 30 | 36 | 40 | 45 | 52 | 58 |

4. 10 in., 50% solid, 37 lb/block

| 44 | 34 | 27 | 30 | 35 | 39 | 46 | 52 |

5. 12 in., 50% solid, 45 lb/block

| 51 | 35 | 30 | 37 | 41 | 47 | 52 | 55 |

6. 10 in., 50% solid, 46 lb/block

| 48 | 35 | 27 | 31 | 38 | 46 | 52 | 57 |

7. 6 in., 100% solid, 49 lb/block

| 50 | 34 | 31 | 32 | 42 | 50 | 57 | 63 |

8. 12 in., 50% solid, 55 lb/block

| 49 | 38 | 29 | 36 | 40 | 46 | 52 | 56 |

9. 12 in., 50% solid, 64 lb/block

| 53 | 40 | 33 | 36 | 43 | 51 | 56 | 59 |

Solid Concrete, About 160 lb/ft³

10. 2 in. thick

| 43 | 30 | 28 | 32 | 31 | 37 | 43 | 30 |

11. 3 in. thick

| 47 | 33 | 28 | 30 | 38 | 46 | 51 | 56 |

12. 4 in. thick

| 54 | 36 | 35 | 40 | 44 | 48 | 55 | 62 |

13. 6 in. thick

| 53 | 38 | 35 | 36 | 44 | 51 | 60 | 70 |

14. 8 in. thick

| 56 | 35 | 34 | 40 | 48 | 52 | 55 | 58 |

15. 10 in. thick

| 59 | 40 | 36 | 44 | 51 | 56 | 61 | 62 |

The isolation material supporting the floating slab must be resilient, have permanent dynamic and static properties, and safely support both the floating slab and the imposed live load. The performance characteristics of the floating slab should be as follows:

- Have resilient isolation media material with a natural frequency in the range of 7 to 12 Hz with up to 15 Hz being generally acceptable under the load conditions that exist in the building both during and after construction,

- Be tested for load versus deflection and for natural frequency with known aging properties and a history of applications in similar floating floor installations,

- Be tested in service for impact insulation class (IIC) with a recommended minimum value of 68 and sound transmission class (STC) with a recommended minimum value of 73 by an independent testing laboratory, and

- Be designed for the loads imposed on the isolator in service and during construction.

Airborne and structure-borne sound flanking occurs to some degree in all buildings. Thus IIC and STC values should be such that with reasonable flanking, the noise levels produced in the occupied space adjacent to the mechanical room are suitable for the intended use.

Table 39   Transmission Loss Values of Painted Masonry Block Walls and Painted Block Walls with Resiliently Mounted Gypsum Wallboard

	STC	63	125	250	500	1000	2000	4000
					Octave Band Frequency, Hz			

8 in. concrete masonry walls with 5/8 in. gypsum board on each side

1. Gypsum board on 1.5 in. wood furring

	STC	63	125	250	500	1000	2000	4000
1	54	28	20	40	49	58	56	60

2. Gypsum board on 1.5 in. wood furring with fiberglass filling furring cavities

| 2 | 59 | 25 | 21 | 48 | 50 | 58 | 55 | 59 |

3. Gypsum board on 1/2 in. resilient metal channels

| 3 | 49 | 34 | 26 | 29 | 51 | 60 | 63 | 67 |

4. Gypsum board on 1/2 in. resilient metal channels with fiberglass in cavities

| 4 | 49 | 30 | 20 | 43 | 58 | 62 | 64 | 68 |

5. Gypsum board on 2 in. deep resilient metal supports

| 5 | 52 | 26 | 20 | 34 | 57 | 66 | 64 | 65 |

6. Gypsum board on 2 in. deep resilient metal supports with fiberglass in cavities

| 6 | 64 | 23 | 31 | 53 | 65 | 70 | 65 | 68 |

7. Gypsum board on 2.5 in. steel studs

| 7 | 57 | 24 | 25 | 46 | 67 | 70 | 66 | 69 |

8. Gypsum board on 2.5 in. steel studs with fiberglass in cavities

| 8 | 72 | 27 | 39 | 60 | 69 | 69 | 68 | 74 |

Two common methods for constructing floating floors are

1. Individual pads, which are typically of neoprene rubber or pre-compressed fiberglass, are spaced on 1 to 2 ft centers each way and covered with plywood or sheet metal. Low-density sound absorption material can be placed between the pads to increase air damping effect. Tests using this method of construction show an STC value of around 73. Flanking and variations in construction produce either lesser or greater STC values. Reinforced concrete is placed directly on the waterproofed panels and cured; then the equipment is set in place (Figure 31A). Typically 1 to 2 in. deep air spaces are used. Instead of individual pads, sheets of resilient material designed for this purpose may be used.

2. Isolation mounts, which are typically neoprene rubber or pre-compressed fiberglass pads or even steel coil springs housed in protective canisters, are cast-in-place on up to 52 in. centers each way on the structural floor. Concrete is poured over the canisters and the reinforcement steel bars or heavy steel screen mesh; after curing, the entire slab is raised into operating position, the canister access holes are grouted, and the equipment is set in place (Figure 31B). A concrete bond breaking material such as heavy plastic film is used between slabs. Cast-in-place systems have achieved typical STC values of 73 without absorption materials. Again note that flanking and variations in construction will produce either lesser or greater STC values. Typically 1 to 4 in. deep air spaces are used with lower speed equipment requiring the deeper air space.

With either system, mechanical equipment with vibration isolators can be placed on the floating slab unless the equipment operates at shaft rotation speeds of 0.7 to 1.4 times the resonance frequency of the floating slab. The equipment should be supported on isolators with a maximum resonant frequency of 2 Hz. In case the rotating speed is coincident with the floor resonant frequency, the equipment should be supported with isolators on structural slab extensions that penetrate the floating slab. This type of installation requires careful attention to detail to avoid acoustical flanking associated with improper direct contact between the floating slab and structural penetrations. Floating floors primarily control

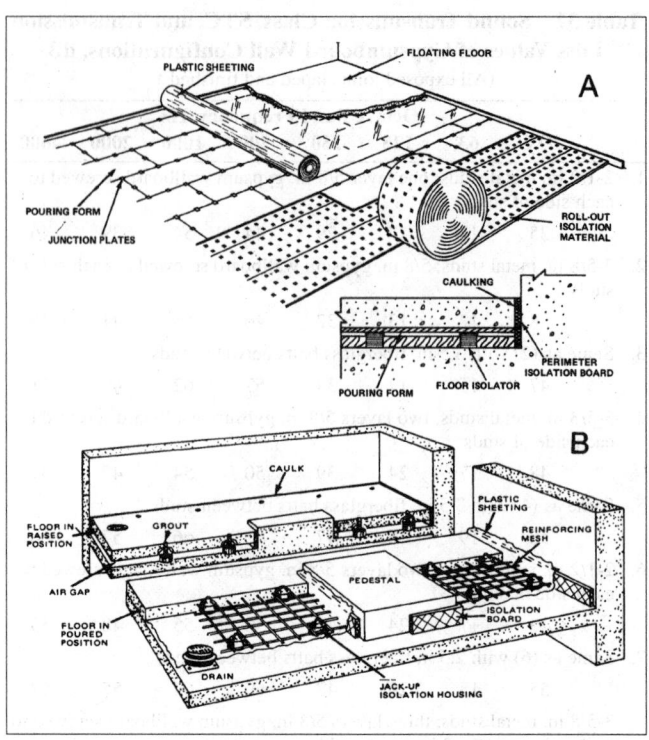

Fig. 31   Two Typical Floating Floor Constructions

airborne sound transmission; they are not intended to be used in place of vibration isolators and inertia bases. The actual resonant frequency of the floated slab is determined by both the stiffness of the resilient elements used to support the floated slab and the stiffness of the airspace between the structural and floated slabs.

The floating slab must be designed to operate within the safe bending limits of the concrete. While floating floor tests in laboratories have shown transmission loss ratings exceeding STC of 75; these can only be approached (realized) in field installations where all flanking sound transmission paths or short circuits have been minimized. Table 39 shows improvement of sound transmission loss by floating floor slabs, with or without flanking noise control. Often flanking can be controlled by installing the mechanical equipment room wall on the edge of the floating floor slab. Note that sound flanking can occur in high rise curtain wall construction where the exterior wall is lightweight and continuous from floor to floor past a noisy mechanical equipment room. Use of sound isolated walls inside the curtain wall may be required to eliminate or at least minimize the sound flanking path at the curtain wall.

In seismic zones it is important to contain the floating floor within the structure in the event of an earthquake. Horizontal containment should be provided by a material such as 0.75 in. thick Neoprene pads or other suitable resilient material that would allow a maximum motion of 0.2 in. of compression in any direction based on the maximum acceleration force used in the building code in that particular area. The resilient pads must bear on suitable building structure or curbs built to the proper capacity.

Buckling of the floor can be of concern, particularly where snubbers are attached to the floating floor around high center of gravity machinery, such as chillers. In those areas additional snubbers such as double acting vertical floor snubbers should be located near the machinery snubber locations or just use of additional concrete reinforcement may be necessary to prevent floor buckling.

Enclosed Air Cavity

If acoustically critical spaces are located over a mechanical equipment room, it may be necessary to

1. Install a floating floor in this space above it, or
2. Form a totally enclosed air cavity between the spaces by installing a dense plaster or gypsum board ceiling, resiliently suspended and acoustically isolated, in the equipment room. Usually sound absorptive batts are placed in the enclosed air cavity to greatly enhanced the acoustical performance.

Ideally, such a ceiling can be positioned between building beams, leaving the beam bottoms available for hanging equipment, piping, ducts, and so forth without penetrating the ceiling with hanger rods or other devices. In the case of bar joist construction, it may be easier to resiliently attach the ceiling to the underside of the joists and then support all equipment, ducts, and pipes from the mechanical room floor, wall, column or combination of all surfaces with supports that are properly vibration isolated. This type of ceiling can be difficult to implement especially if the hanging piping and ductwork are already installed. Thus the design, construction schedule, and actual installation must be carefully coordinated.

The same type of sound isolation construction can be used for acoustically critical spaces located below a mechanical equipment room. This room is often encountered with the use of builtup air handling units, chillers, pumps, and cooling towers located on the building penthouse floor level. An acoustically isolated ceiling can be installed above the typical decorative acoustical tile ceiling, and it is often found to be very effective with lighter weight than a floating floor. Such a ceiling can approximate the sound transmission loss performance of a floating floor in the space above. Some extreme sound isolation conditions require both a floating floor and acoustically isolated ceiling in series. If used, the sound isolation ceiling must be designed and installed to minimize flanking, which reduces the sound isolation.

SOUND TRANSMISSION IN RETURN-AIR SYSTEMS

The fan return air system provides a sound path (through ducts or through an unducted ceiling plenum) between a fan and occupied rooms. Often a direct opening goes into the mechanical equipment room through the ceiling plenum. This condition can result in high sound levels in adjacent spaces. The high sound levels are caused by the close proximity of the fan and other sound sources in the mechanical equipment room and the low system attenuation between the mechanical equipment room and the adjacent spaces.

Sound in ducted return-air systems is controlled by the fan intake sound power levels. Unducted plenum return air is impacted by the sound power levels of the fan intake and casing-radiated noise components. In certain installations, sound from other equipment in the mechanical equipment room may also radiate through the wall opening and into adjacent spaces. Good design yields room return- air sound levels that are approximately 5 dB below the corresponding supply-air system sound levels.

When sound levels in spaces adjacent to mechanical equipment rooms are too high, noise control measures must be provided. Before specifying modifications, the controlling sound paths between the mechanical equipment room and adjacent spaces must be identified. Ducted return-air systems can be modified using methods applicable to ducted supply systems. Good planning is required with unducted ceiling plenums so that the mechanical equipment room is located away from noise-sensitive spaces.

Unducted plenums may still require several modifications to satisfy design goals. Prefabricated silencers can be effective when installed at the mechanical equipment room wall opening or at the suction side of the fan. Improvements to the ceiling transmission loss are often limited by typical ceiling penetrations and lighting fixtures. Modifications to the mechanical equipment room wall can also be effective for some construction. Adding acoustical absorption in the mechanical equipment room reduces the buildup of reverberant sound energy in this space; however, this typically reduces noise only slightly in the low frequency range in nearby areas.

SOUND TRANSMISSION THROUGH CEILINGS

When terminal units, fan-coil units, air-handling units, ducts, or return air openings to mechanical equipment rooms are located in a ceiling plenum above an occupied room, sound transmission through the ceiling system can be high enough to cause excessive noise levels in that room. There are no standard test procedures associated with measuring the direct transmission of sound through ceilings from sources close to the ceiling. As a result, ceiling product manufacturers rarely publish data that can be used in calculations. This problem seems further complicated by the presence of light fixtures, diffusers, grilles, speakers, and so forth, which might be expected to reduce the transmission loss of the ceiling. Experiments have shown, however, that for ceiling panels supported in a T-bar grid system, the leakage between the panels and the grid is the major transmission path; differences among panel types are small and light fixtures, diffusers and the like, have only a localized effect.

To estimate the sound levels in a room associated with sound transmission through the ceiling, the sound power levels in the ceiling plenum must be attenuated by factors that account for the transmission loss of the ceiling and plenum. As well, an adjustment to the measured data must be made to account for the reverberant nature of the room. The procedures given here are based on results found as the result of an ASHRAE research project (Warnock 1998).

Proceed as follows:

1. Obtain the octave band radiated sound power levels of the device.
2. Subtract the environmental correction in Table 40.
3. Calculate the surface area of the bottom panel of the source closest to the ceiling tiles.
4. From Table 41 find the adjustment to be subtracted from the sound power values at the three frequencies given there.
5. Select the Ceiling/Plenum attenuations from Table 23 according to the ceiling type in use.
6. Subtract the three sets of values, taking account of sign where necessary, from the sound power values. The result is the average sound pressure level in the room.
7. For practical purposes, the sound field in the room may be assumed to uniform up to distances of about 16 ft from the source.

Table 40 Environmental Correction To Be Subtracted From Device Sound Power

Octave Band Frequency, Hz						
63	125	250	500	1000	2000	4000
4	2	1	0	0	0	0

Table 41 Compensation Factors for Source Area Effect

Area Range, ft²			
63 Hz	125 Hz	250 Hz	Adjustment, dB
Less than 2.6	Less than 2.2		−3
2.8 to 4.9	2.4 to 4.6	Less than 2.3	−2
5.1 to 7.2	4.9 to 7.1	2.7 to 6.3	−1
7.4 to 9.4	7.3 to 9.5	6.7 to 10.3	0
9.7 to 11.7	9.8 to 12.0	10.7 to 14.3	1
11.9 to 14.0	12.2 to 14.4	14.7 to 18.3	2
14.2 to 16.3	14.6 to 16.8	18.7 to 22.3	3
16.5 to 18.5	17.1 to 19.3		4
18.8 to 20.8	19.5 to 21.7		5
21.0 to 23.1			6

Note: Find correct area in each frequency column and read adjustment from last column on right.

Example 7. A terminal unit having an area of 14 ft^2 is to be used above a standard 5/8 in. thick mineral fiber ceiling system. What room sound pressure levels can be expected?

Solution:

		Octave Band Frequency, Hz						
		63	125	250	500	1000	2000	4000
Step 1	Sound power	71	71	65	55	54	53	45
Step 2	Environment	−4	−2	−1	0	0	0	0
Step 4	Area adjustment	−2	−2	−1	0	0	0	0
Step 5	Ceiling/plenum	−13	−15	−17	−19	−25	−30	−33
Step 6	Room sound pressure levels, dB	52	52	46	36	29	23	12

FUME HOOD DUCT DESIGN

Fume hood exhaust systems are often the major sound source in a laboratory and, as such, require noise control. The exhaust system may consist of individual exhaust fans ducted to separate fume hoods, or a central exhaust fan connected through a collection duct system to a large number of hoods. In either case, the sound levels produced in the laboratory space can be estimated using procedures described in this section. Recommended noise level design criteria for laboratory spaces using fume hoods are given in Table 34.

To minimize static pressure loss and blower power consumption in a duct system, fume hood ducts should be of a sufficient size to permit the rated flow of air through the duct at an airflow velocity no greater than 2000 fpm or which are consistent with regulatory requirements. a duct velocity in excess of 2000 fpm should be avoided for acoustical reasons and to conserve energy, unless the resulting increases in sound levels, static pressure, and blower power requirements are deemed acceptable for the spaces served.

Noise control measures that have been successfully applied to a variety of fume hood systems consist of the following:

1. Use backward inclined or forward curved, rather than radial blade, fans where conditions permit.
2. Select the fan to operate at a low tip speed and maximum efficiency.
3. Use prefabricated duct silencers or sections of lined ducts where conditions permit.

All potential noise control measures should be evaluated for compliance with applicable codes, safety requirements, and corrosion resistance requirements of the specific fume hood system. In addition, vibration isolation for fume hood exhaust fans is generally required. However, for some laboratory facilities, particularly those having highly vibration sensitive instruments such as electron microscopes, vibration control can be critical, and a vibration specialist should be consulted.

SOUND CONTROL FOR OUTDOOR EQUIPMENT

Outdoor mechanical equipment should be carefully selected, installed, and maintained to prevent sound radiated by the equipment from annoying people outdoors or in nearby building and from violating local noise codes. The difference in sound levels at the listening location with the equipment operating and the ambient conditions with the equipment not operating determines if the noise from the equipment is annoying. Equipment with strong tonal components will more likely generate complaints than equipment with a broadband noise spectrum.

Sound Propagation Outdoors

If the equipment sound power level spectrum and ambient sound pressure level spectrum are known, the contribution of the equipment to the sound level at any location can be estimated by

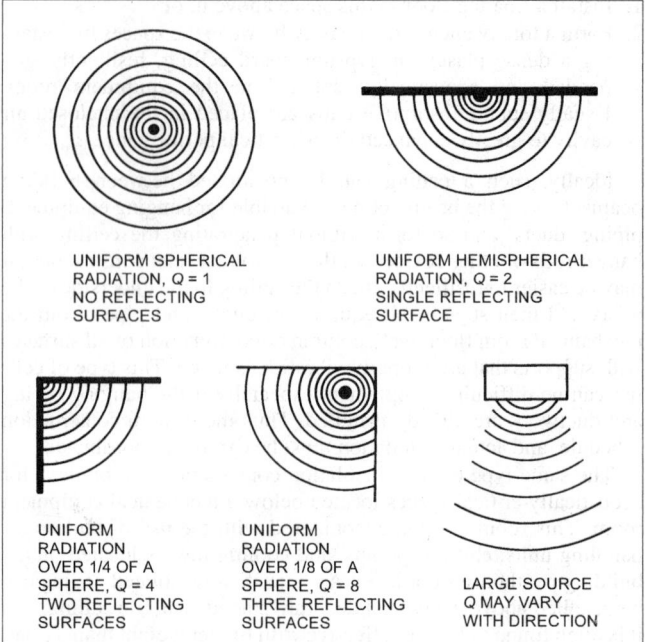

Fig. 32 Directivity Factors for Various Radiation Patterns

analyzing the sound transmission paths involved. When there are no intervening barriers, the principal factors outdoors are reflections from buildings near the equipment and the distance to the specific location.

The following equation may be used to estimate the sound pressure level of the equipment at a distance from it and at any frequency when given the sound power level:

$$L_p = L_w + 10 \log Q - 20 \log d + 10 \tag{20}$$

where

L_p = sound pressure level at distance d (ft) from sound source, dB
L_w = sound power level of sound source, dB
Q = directivity factor associated with the way sound radiates from sound source (refer to Figure 32)

Equation (20) does not apply where d is less than twice the maximum dimension of the sound source. L_p may be low by up to 5 dB where d is between two and five times the maximum sound source dimension. Also, if the distance is greater than about 500 ft downwind, thermal gradients and atmospheric sound absorption need to be considered.

Sound Barriers

A sound barrier is a solid structure that intercepts the direct sound path from a sound source to a receiver. It reduces the sound pressure level within its shadow zone. Figure 33 illustrates the geometrical aspects of an outdoor barrier where no extraneous surfaces reflect sound into the protected area. Here the barrier is treated as an intentionally constructed noise control structure. If a sound barrier is placed between a sound source and receiver location, the sound pressure level L_p in Equation (20) is reduce by the insertion loss IL associated with the barrier.

Table 42 gives the insertion loss of an outdoor ideal solid barrier when

1. No surfaces reflect sound into the shadow zone, and
2. The sound transmission loss of the barrier wall or structure is at least 10 dB greater at all frequencies than the insertion loss expected of the barrier.

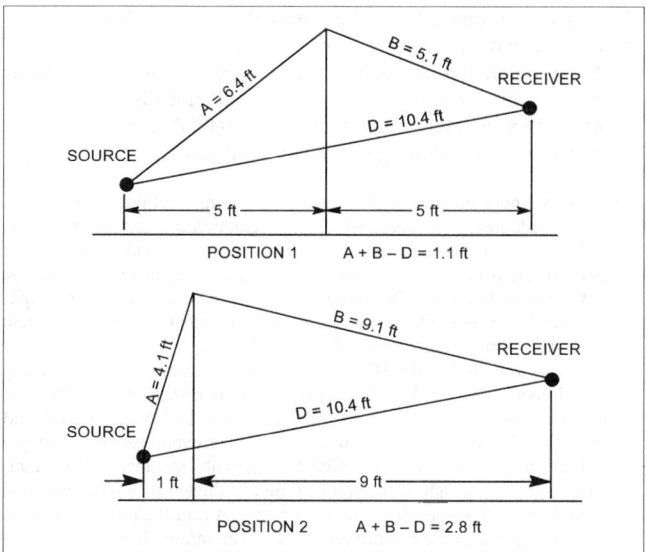

Fig. 33   Noise Barrier

Table 42   Insertion Loss Values of an Ideal Solid Barrier

Path-Length Difference, ft	Insertion Loss, dB Octave Band Center Frequency, Hz							
	31	63	125	250	500	1000	2000	4000
0.01	5	5	5	5	5	6	7	8
0.02	5	5	5	5	5	6	8	9
0.05	5	5	5	5	6	7	9	10
0.1	5	5	5	6	7	9	11	13
0.2	5	5	6	8	9	11	13	16
0.5	6	7	9	10	12	15	18	20
1	7	8	10	12	14	17	20	22
2	8	10	12	14	17	20	22	23
5	10	12	14	17	20	22	23	24
10	12	15	17	20	22	23	24	24
20	15	18	20	22	23	24	24	24
50	18	20	23	24	24	24	24	24

The path-length difference referred to in Table 42 is given by

$$\text{Path-length difference} = A + B - D \qquad (21)$$

where A, B, and D are as specified in Figure 33.

The limiting value of about 24 dB is caused by scattering and refraction of sound into the shadow zone formed by the barrier. Practical constructions, size and space restrictions often limit sound barrier performance to 10 to 15 dBA. For large distances outdoors, this scattering and bending of sound waves into the shadow zone reduces the effectiveness of the barrier. At large distances atmospheric conditions can significantly affect the sound path losses by amounts even greater than provided by the barrier with typical differences of 10 dBA observed. For a conservative estimate, the height of the sound source location should be taken as the topmost part of the sound source, and the height of the receiver should be taken as the topmost location of a sound receiver, such as the top of the second-floor windows in a two-floor house or at a height of 5 ft for a standing person.

Reflecting Surfaces. No other surfaces should be located where they can reflect sound around the ends or over the top of the barrier into the barrier shadow zone. Figure 34 shows examples of reflecting surfaces that should be avoided because they can reduce the effectiveness of a barrier wall.

Width of Barrier. Each end of the barrier should extend horizontally beyond the line of sight from the outer edge of the source to the outer edge of the receiver position by at least three times the path-length distance. Near the end of the barrier, the barrier effectiveness is reduced because some sound is diffracted over the top of the barrier, some sound is diffracted around the end of the barrier, and some sound is reflected or scattered from various nonflat surfaces along the ground near the end of the barrier. In critical situations, the barrier should completely enclose the sound source to eliminate or reduce the effects of reflecting surfaces.

Reflection from a Barrier. A large, flat reflecting surface, such as a barrier wall, may reflect more sound in the opposite direction than if no wall was present. If the wall produces no focusing effect, it may produce only a 2 or 3 dB higher noise level (at most) in the direction of the reflected sound. Acoustical absorption on the barrier surface should be considered to reduce the noise.

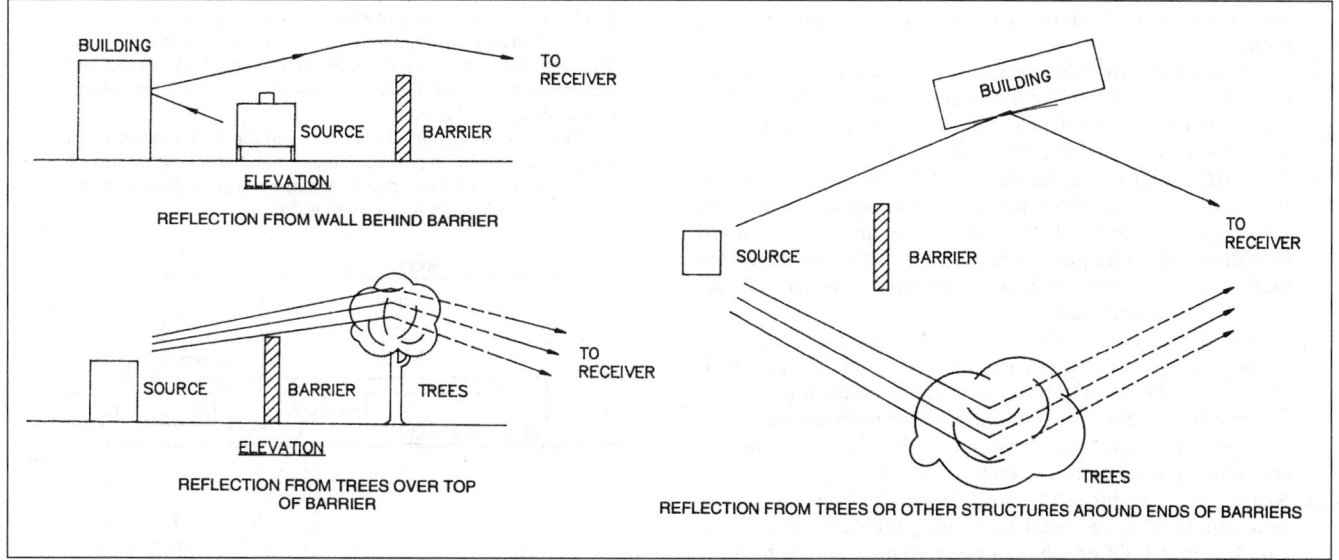

Fig. 34   Examples of Surfaces That Can Reflect Sound Around or Over a Barrier Wall

DESIGN PROCEDURES

The following design procedures are suggested for managing each of the different sound sources and related sound transmission paths associated with an HVAC system.

1. Determine the design goal for HVAC system noise for each critical area according to its use and construction. Choose desirable RC criterion from Table 34. A balanced sound spectrum is as important as the overall sound level.

2. Relative to equipment such as air inlet and outlet grilles, registers, diffusers, and air terminal and fan coil units that radiate sound directly into a room, select equipment that is quiet enough to meet the desired design goal.

3. If ducted central or roof-mounted mechanical equipment such as air handling units are to be used, complete an initial design and layout of the HVAC system using acoustical treatment such as lined ductwork and duct silencers where appropriate. Consider the return air, exhaust air, and supply paths.

4. Starting at the fan, appropriately add the sound attenuation and sound power levels associated with the central fan(s), fan-powered terminal units (if used), and duct elements between the central fan(s) and the room of interest. Then convert to the corresponding sound pressure levels in the room. For a more complete estimate of resultant sound levels, consider regenerated and self noise from duct silencers and air inlets and outlets due to the airflow itself. Investigate both the supply and return air paths in similar ways. Investigate and control possible duct sound breakout when fans are adjacent to the room of interest or roof-mounted fans are above the room of interest. Be sure to combine the sound contribution from all paths into the occupied space of concern. The following example shows the calculation procedure for supply and return air paths along with duct breakout noise contributions.

5. If the mechanical equipment room is adjacent to the room of interest, determine the sound pressure levels in the room of interest that are associated with sound transmitted through the mechanical equipment room wall. Typical equipment to consider include air handling units, ventilation and exhaust fans, chillers, pumps, electrical transformers, and instrument air compressors. Also consider the vibration isolation requirements for all the equipment along with piping and ductwork.

6. Combine on an energy basis (see the example for sample calculation procedures) the sound pressure levels in the room of interest that are associated with all sound paths between the mechanical equipment room or roof-mounted unit and the room.

7. Determine the corresponding RC level associated with the calculated total sound pressure levels in the room of interest. Take special note of the sound quality indicators for possible rumble, roar, hiss, tones, and perceivable vibration.

8. If the RC level exceeds the design goal, determine the octave frequency bands in which the corresponding sound pressure levels are exceeded and the sound paths that are associated with these octave frequency bands. If resultant noise levels are high enough to cause perceivable vibration, consider both airborne and structure-borne noise.

9. Redesign the system, adding additional sound attenuation to the paths that contribute to the excessive sound pressure levels in the room of interest. If resultant noise levels are high enough to cause perceivable vibration, then major redesign and possibly use of supplemental vibration isolation for the equipment and building systems will often be required.

10. Repeat Steps 4 through 9 until the desired design goal is achieved. Involve the complete design team where major problems are found. Often simple design changes to the building architectural and equipment systems can eliminate potential problems once the problems are identified.

11. Steps 3 through 10 must be repeated for every room that is to be analyzed.

12. Make sure that noise radiated by outdoor equipment such as air cooled chillers and cooling towers will not disturb adjacent properties or interfere with criteria established in Step (1) or any applicable building or zoning ordinances.

Example 8. Individual examples in the preceding sections demonstrate how to calculate equipment and airflow-generated sound power levels and sound attenuation values associated with the elements of HVAC air distribution systems. This example shows how the information can be combined to determine the sound pressure levels associated with a specific HVAC system. Only a summary of the results is shown rather than showing complete calculations for each element.

Air is supplied to the HVAC system in this example by the rooftop unit shown in Figure 35. The receiver room is directly below the unit. The room has the following dimensions: length = 20 ft, width = 20 ft; and height = 9 ft. This example assumes the roof penetrations for the supply and return air ducts are well sealed and there are no other roof penetrations. The supply side of the rooftop unit is ducted to a VAV terminal control unit that serves the room in question. A return air grille conducts air to a common ceiling return air plenum. The return air is then directed to the rooftop unit through a short rectangular return air duct.

The following three sound paths are examined:

Path 1. Fan airborne supply air sound that enters the room from the supply air system through the ceiling diffuser
Path 2. Fan airborne supply air sound that breaks out through the wall of the main supply air duct into the plenum space above the room
Path 3. Fan airborne return air sound that enters the room from the inlet of the return air duct

The sound power levels associated with the supply air and return air sides of the fan in the rooftop unit are specified by the manufacturer as follows:

	Octave Band Center Frequency, Hz						
	63	125	250	500	1000	2000	4000
Rooftop supply air = 7000 cfm at 2.5 in. w.g.	92	86	80	78	78	74	71
Rooftop return air = 7000 cfm at 2.5 in. w.g.	82	79	73	69	69	67	59

Solution:

Paths 1 and 2 are associated with the supply air side of the system. Figure 36 shows a layout of the part of the supply air system that is associated with the receiver room. The main duct is a 22 in. diameter, 26 gage, unlined, round sheet metal duct. The flow volume in the main duct is 7000 cfm. The silencer after the radiused elbow is a 22 in. diameter by 44 in. long, high pressure, circular silencer.

The branch junction that occurs 8 ft from the silencer is a 45° wye. The branch duct between the main duct and the VAV control unit is a 10 in. diameter, unlined, round sheet metal duct. The flow volume in the branch duct is 800 cfm.

The straight section of duct between the VAV control unit and the diffuser is a 10 in. diameter, unlined round sheet metal duct. The diffuser is 15 in. by 15 in. square. Assume a typical distance between the diffuser and a listener in the room is 5 ft.

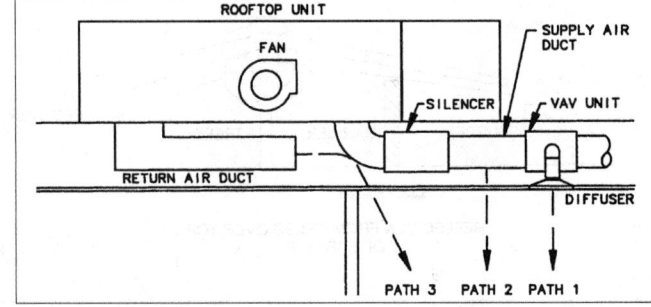

Fig. 35   Sound Paths Layout for Example 8

With regard to the duct breakout sound associated with the main duct, the length of the duct that runs over the room is 20 ft. The ceiling of the room is comprised of 2 ft by 4 ft by 5/8 in. lay-in ceiling tiles that have a surface weight of 0.6 to 0.7 lb/ft^2. The ceiling has integrated lighting and diffusers.

Path 3 is associated with the return air side of the system. Figure 37 shows a layout of the part of the return air system that is associated with the receiver room. The rectangular return air duct is lined with 2 in. thick, 3 lb/ft^3 density fiberglass duct liner. For the return air path, assume the typical distance between the inlet of the return air duct and a listener is 10 ft.

The calculations associated with this example were conducted with a computer program such as provided in *Algorithms for HVAC Acoustics* (ASHRAE 1991) or other commercially available ones. The analysis associated with each path begins at the rooftop unit (fan) and proceeds progressively through the different elements to the receiver room. The element numbers in the tables correspond to the element numbers contained in brackets in Figures 36 and 37.

The first table is associated with Path 1. The first entry in the table is the manufacturer's values for supply air fan sound power levels (1). The second entry is the sound attenuation associated with the 22 in. diameter unlined radius elbow (3). Because the next entry is associated with the regenerated sound power levels associated with the elbow (4), the results associated with the elbow attenuation must be tabulated to determine the sound power levels at the exit of the elbow.

These sound power levels and the elbow regenerated sound power levels are then added logarithmically, which is denoted by (Log Sum) in the tables where the value in decibels equals $10\log(10^{A/10} + 10^{B/10})$ with A and B being the values in decibels. In a like fashion, the dynamic insertion loss values of the duct silencer (5) and the silencer regenerated sound power levels (6) are included in the table and tabulated.

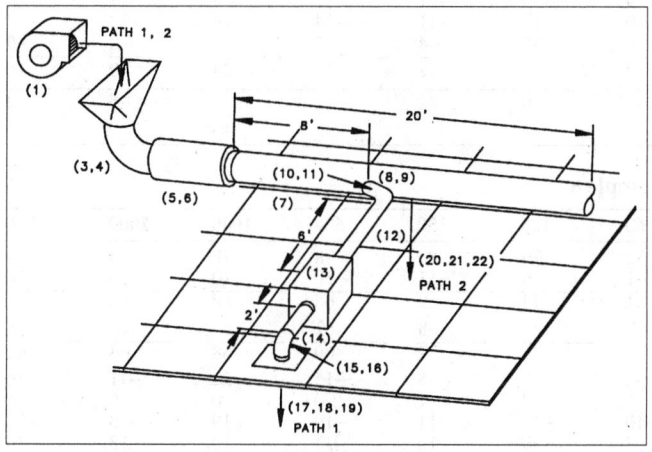

Fig. 36 Supply Air Portion Layout for Example 7

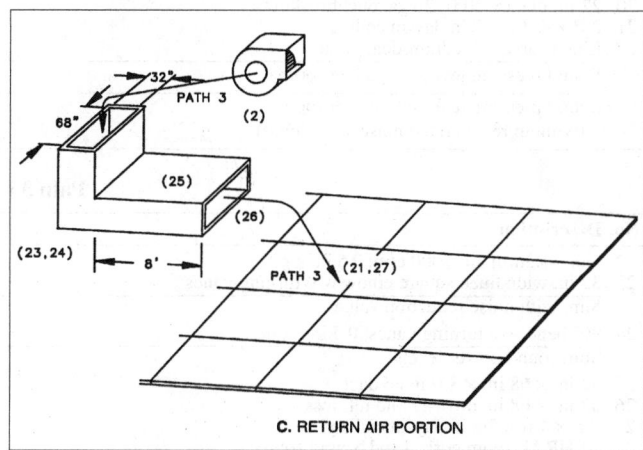

Fig. 37 Return Air Portion Layout for Example 7

Path 1 in Example 8

No. Description	Octave Band Center Frequency, Hz						
	63	125	250	500	1000	2000	4000
1 Fan—Supply air, 7000 cfm, 2.5 in. s.p.	92	86	80	78	78	74	71
3 22 in. wide (dia.) unlined radius elbow	0	−1	−2	−3	−3	−3	−3
Sum with noise reduction values	92	85	78	75	75	71	68
4 90° bend without turning vanes, 12 in. radius	56	54	51	47	42	37	29
Sum sound power levels	92	85	78	75	75	71	68
5 22 in. dia. by 44 in. high pressure silencer	−4	−7	−19	−31	−38	−38	−27
Sum with noise reduction values	88	78	59	44	37	33	41
6 Regenerated noise from above silencer	68	79	69	60	59	59	55
Sum sound power levels	88	82	69	60	59	59	55
7 22 in. dia. by 8 ft unlined circular duct	0	0	0	0	0	0	0
10 Branch pwr. div., M-22 in. dia., B-10 in. dia.	−8	−8	−8	−8	−8	−8	−8
Sum with noise reduction values	80	74	61	52	51	51	47
11 Duct 90° branch takeoff, 2 in. radius	56	53	50	47	43	37	31
Sum sound power levels	80	74	61	53	52	51	47
12 10 in. dia. by 6 ft unlined circular duct	0	0	0	0	0	0	0
13 Terminal volume register unit (gen. attn.)	0	−5	−10	−15	−15	−15	−15
14 10 in. dia. by 2 ft unlined circular duct	0	0	0	0	0	0	0
15 10 in. wide (dia.) unlined radius elbow	0	0	−1	−2	−3	−3	−3
Sum with noise reduction values	80	69	50	36	34	33	29
16 90° bend without turning vanes, 2 in. radius	49	45	41	37	31	24	16
Sum sound power levels	80	69	51	40	36	34	29
17 10 in. dia. diffuser end ref. loss	−16	−10	−6	−2	−1	0	0
Sum with noise reduction values	64	59	45	38	35	34	29
18 15 in. by 15 in. rectangular diffuser	31	36	39	40	39	36	30
Sum sound power levels	64	59	46	42	40	38	33
19 ASHRAE room correction, 1 ind. sound source	−5	−6	−7	−8	−9	−10	−11
Sound pressure levels—receiver room	59	53	39	34	31	28	22
Sound pressure levels—receiver room (without regenerated noise considered)	59	49	27	9	1	−3	4

Path 2 in Example 8

No.	Description	63	125	250	500	1000	2000	4000
1	Fan—Supply air, 7000 cfm, 2.5 in. s.p.	92	86	80	78	78	74	71
3	22 in. wide (dia.) unlined radius elbow	0	−1	−2	−3	−3	−3	−3
	Sum with noise reduction values	92	85	78	75	75	71	68
4	90° bend without turning vanes, 12 in. radius	56	54	51	47	42	37	29
	Sum sound power levels	92	85	78	75	75	71	68
5	22 in. dia. by 44 in. high pressure silencer	−4	−7	−19	−31	−38	−38	−27
	Sum with noise reduction values	88	78	59	44	37	33	41
6	Regenerated noise from above silencer	68	79	69	60	59	59	55
	Sum sound power levels	88	82	69	60	59	59	55
7	22 in. dia. by 8 ft unlined circular duct	0	0	0	0	0	0	0
8	Branch pwr. div., M-22 in. dia., B-22 in. dia.	−1	−1	−1	−1	−1	−1	−1
	Sum with noise reduction values	87	81	68	59	58	58	54
9	Duct 90° branch takeoff, 2 in. radius	63	60	57	54	50	44	34
	Sum sound power levels	87	81	68	60	59	58	54
20	22 in. dia. by 20 ft, 26 ga. duct breakout	−29	−29	−21	−11	−9	−7	−5
21	2 ft × 4 ft × 5/8 in. lay-in ceiling	−10	−13	−11	−14	−19	−23	−24
22	Line source—Medium-dead room	−6	−5	−4	−6	−7	−8	−9
	Sound pressure levels—receiver room	42	34	32	29	24	20	16
	Sound pressure levels—receiver room (without regenerated noise considered)	42	30	22	12	1	−6	2

Path 3 in Example 8

No.	Description	63	125	250	500	1000	2000	4000
2	Fan—Return air, 7000 cfm, 2.5 in. s.p.	82	79	80	78	78	74	71
23	32 in. wide lined square elbow w/o turning vanes	−1	−6	−11	−10	−10	−10	−10
	Sum with noise reduction values	81	73	69	68	68	64	61
24	90° bend w/o turning vanes; 0.5 in radius	77	73	68	62	55	48	38
	Sum sound power levels	82	76	72	69	68	64	61
25	32 in. × 68 in. × 8 ft lined duct	−2	−2	−5	−15	−22	−11	−10
26	32 in. × 68 in. diffuser end ref. loss	−5	−2	−1	0	0	0	0
21	2 ft × 4 ft × 5/8 in. lay-in ceiling	−10	−13	−11	−19	−19	−23	−24
27	ASHRAE room corr., 1 ind. sound source	−8	−9	−10	−11	−12	−13	−14
	Sound pressure levels—receiver room	57	50	45	29	15	17	13
	Sound pressure levels—receiver room (without regenerated noise considered)	56	47	42	28	15	17	13

Total Sound Pressure Levels from All Paths in Example 8

Description	63	125	250	500	1000	2000	4000
Sound pressure levels Path 1	59	53	39	34	31	28	22
Sound pressure levels Path 2	42	34	32	29	24	20	16
Sound pressure levels Path 3	57	50	45	29	15	17	13
Total sound pressure levels—All paths	61	55	46	36	32	29	23
Sound pressure levels—receiver room (without regenerated noise considered)	61	51	42	28	15	17	14

Next, the attenuation associated with the 8 ft section of 22 in. diameter duct (7) and the branch power division (10) associated with sound propagation in the 10 in. diameter branch duct are included in the table. After element 10, the sound power levels that exist in the branch duct after the branch takeoff are calculated so that the regenerated sound power levels (11) in the branch duct associated with the branch takeoff can be logarithmically added to the results.

Next, the sound attenuation values associated with the 6 ft section of 10 in. diameter unlined duct (12), the terminal volume regulation unit (13), the 2 ft section of 10 in. diameter unlined duct (14), and 10 in. diameter radius elbow (15) are included in the table. The sound power levels that exist at the exit of the elbow are then calculated so that the regenerated sound power levels (16) associated with the elbow can be logarithmically added to the results. The diffuser end reflection loss (17) and the diffuser regenerated sound power levels (18) are appropriately included in the table. The sound power levels that are tabulated after element 18 are the sound power levels that exist at the diffuser in the receiver room. Note that the end reflection from a duct in free space and flush with a suspended acoustical ceiling are assumed to be the same.

The final entry in the table is the "room correction" that converts the sound power levels at the diffuser to their corresponding sound pressure levels at the point of interest in the receiver room.

Elements 1 through 7 in Path 2 are the same as Path 1. Elements 8 and 9 are associated with the branch power division (8) and the corresponding regenerated sound power levels (9) associated with sound that propagates down the main duct beyond the duct branch. The next three entries in the table are the sound transmission loss associated with the duct breakout sound (20), the sound transmission loss associated with the ceiling (21), which considers the integrated lighting and diffuser including the return air openings, and the room correction (22), converting the sound power levels at the ceiling to corresponding sound pressure levels in the room. While not specifically considered in this example, noise radiated by a VAV terminal unit can be significant. Consult with the manufacturer for both radiated and discharge sound data.

The first element in Path 3 is the manufacturer's values for return air fan sound power levels (2). The next two elements are the sound attenuation associated with a 32 in. wide, lined square elbow without turning vanes (23) and the regenerated sound power levels associated with the square elbow (24). The final four elements are the insertion loss associated with a 32 in. by 68 in. by 8 ft long rectangular sheet metal duct lined with 2 in. thick, 3 lb/ft³ fiberglass duct lining (26), the diffuser end reflection loss (27), the transmission loss through the ceiling (21), which considers the integrated lighting and diffuser system including the return air openings, and the room correction (27) converting the sound power levels at the ceiling to corresponding sound pressure levels in the room.

The total sound pressure levels in the receiver room from the three paths are obtained by logarithmically adding the individual sound pressure levels associated with each path. From the total sound pressure levels for all three paths, the NC value in the room is NC 42, and the RC value is RC 34 (R-H), which is a combination of lower frequency rumble and higher frequency hiss.

If the regenerated noise due to airflow through the ductwork, silencer, and diffuser are not considered, the NC value in the room is NC 42, and the RC value is RC 26 (R-H). While the calculation procedure is simplified, the typically higher-frequency regenerated noise is not accounted for in the overall ratings especially in the RC value, whose numeric magnitude is often set by the higher frequency noise contribution. At a minimum, the self-noise or regenerated noise of the silencers and outlet or inlet devices such as grilles, registers, and diffusers should be considered along with the attenuation provided by the duct elements and dynamic insertion loss of the silencers.

VIBRATION ISOLATION AND CONTROL

Mechanical vibration and vibration-induced noise are often major sources of occupant complaints in modern buildings. Lighter weight construction in new buildings has made these buildings more susceptible to vibration and vibration-related problems. Increased interest in energy conservation in buildings has resulted in many new buildings being designed with variable air volume systems. This often results in mechanical equipment being located in penthouses on the roof, in the use of roof-mounted HVAC units, and in mechanical equipment rooms located on intermediate level floors. These trends have resulted in an increase in the number of pieces of mechanical equipment located in a building, and they often have resulted in mechanical equipment being located adjacent to or above occupied areas.

Occupant complaints associated with building vibration typically take one of three forms:

1. The level of vibration perceived by building occupants is of sufficient magnitude to cause concern or alarm.
2. Vibration energy from mechanical equipment, which is transmitted to the building structure, is transmitted to various parts of the building and then is radiated as structure-borne noise.
3. Vibration present in a building may interfere with proper operation of sensitive equipment or instrumentation.

The following sections present basic information to properly select and specify vibration isolators and to analyze and correct field vibration problems. Chapter 7 in the 1997 *ASHRAE Handbook—Fundamentals* and Reynolds and Bevirt (1994) provide more detailed information.

EQUIPMENT VIBRATION

Vibration can be isolated or reduced to a fraction of the original force with resilient mounts between the equipment and the supporting structure. To determine the excessive forces that must be isolated or that adversely affect the performance or life of the equipment, criteria should be established for equipment vibration. Figures 38 and 39 show the relation between equipment vibration levels and vibration isolators that have a fixed vibration isolation efficiency. In this case, the magnitude of transmission to the building is a function of the magnitude of the vibration force.

VIBRATION CRITERIA

Vibration criteria can be specified relative to three areas: (1) human response to vibration, (2) vibration levels associated with potential damage to sensitive equipment in a building, and (3) vibration severity of a vibrating machine. Figure 40 and

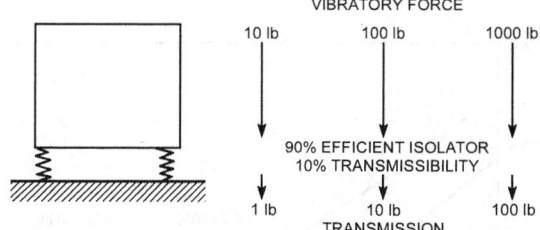

Fig. 38   Transmission to Structure Varies as Function of Magnitude of Vibration Force

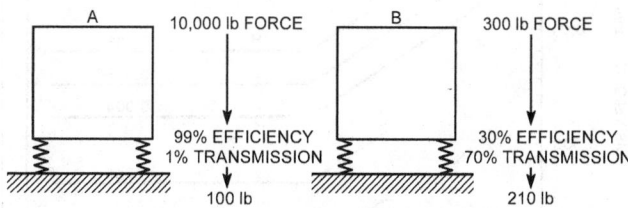

Fig. 39   Interrelationship of Equipment Vibration, Isolation Efficiency, and Transmission

Table 43 present recommended acceptable vibration criteria for vibration that can exist in a building structure (Ungar et al. 1990). Vibration values associated with Figure 40 are measured by vibration transducers (usually accelerometers) that are placed on the building structure in the vicinity of vibrating equipment or in areas of the building that contain building occupants or sensitive equipment. The occupant vibration criteria are based on guidelines specified by ANSI *Standard* S3.29, and ISO *Standard* 2631-2.

The manufacturer's vibration criteria should be followed for sensitive equipment. If acceptable vibration values are not available from manufacturers, the values specified in Figure 41 can be used. Figure 41 gives recommended equipment vibration severity ratings based on measured RMS velocity values (IRD 1988). The vibration values associated with Figure 41 are measured by vibration transducers (usually accelerometers) mounted directly on equipment, equipment structures, or bearing caps. Vibration levels measured on equipment and equipment components can be affected by unbalance, misalignment of components, and resonance interaction between a vibrating piece of equipment and the structural floor on which it is placed. If a piece of equipment is balanced within acceptable tolerances and excessive vibration levels still exist, the equipment and its installation should be checked for possible resonant conditions. Table 44 gives maximum allowable RMS velocity levels for selected pieces of equipment.

With regard to maintenance and preventive maintenance requirements, the vibration levels measured on equipment structures should be in the "Good" region or below in Figure 41. Machine vibration levels in the "Fair" or "Slightly Rough" regions may indicate potential problems. Machines with vibration levels in these regions should be monitored to ensure problems do not arise. Machine vibration levels in the "Rough" and "Very Rough" regions indicate a potentially serious problem exists, and immediate action should be taken to identify and correct the problem.

SPECIFICATION OF VIBRATION ISOLATORS

Vibration isolators must be selected to compensate for floor stiffness. Longer spans also allow the structure to be more flexible, permitting the building to be more easily set into vibration. Building

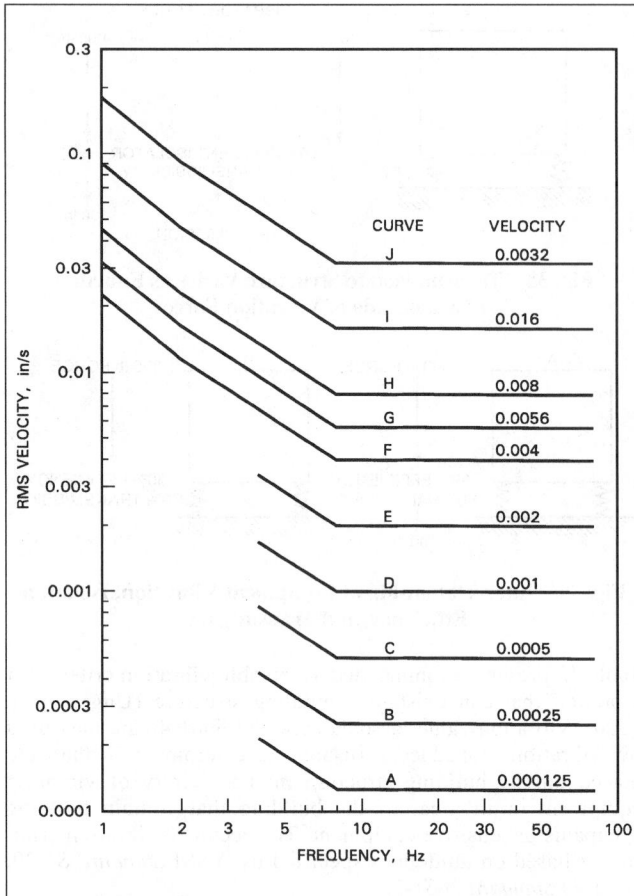

Fig. 40 Building Vibration Criteria for Vibration Measured on Building Structure

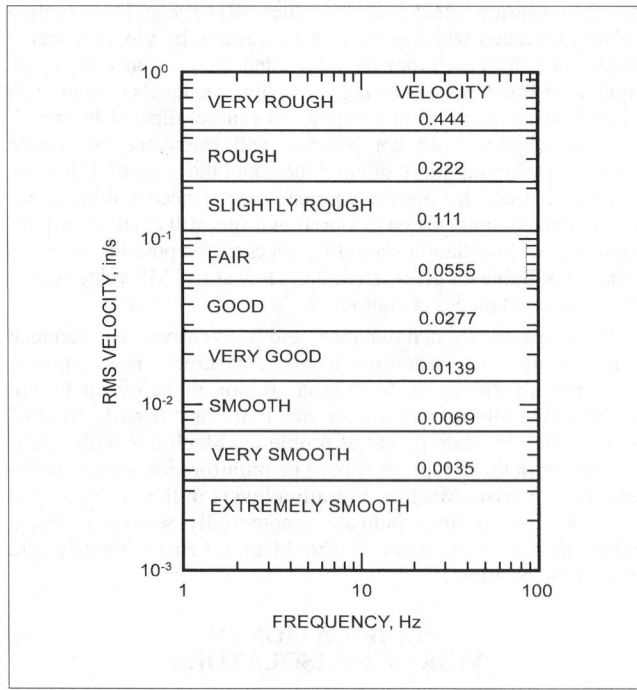

Fig. 41 Equipment Vibration Severity Rating for Vibration Measured on Equipment Structure or Bearing Caps

Table 43 Equipment Vibration Criteria

Human Occupancy	Time of Day	Curve[a]
Workshops	All	J
Office areas	All[b]	I
Residential (good environmental standards)	0700-2200[b]	H-I
	2200-0700[b]	G
Hospital operating rooms and critical work areas	All	F

Equipment Requirements	Curve[a]
Computer areas	H
Bench microscopes up to 100× magnification; laboratory robots	F
Bench microscopes up to 400× magnification; optical and other precision balances; coordinate measuring machines; metrology laboratories; optical comparators; microelectronics manufacturing equipment—Class A[c]	E
Microsurgery, eye surgery, neurosurgery; bench microscope at magnification greater than 400×; optical equipment on isolation tables; microelectronic manufacturing equipment—Class B[c]	D
Electron microscopes up to 30,000× magnification; microtomes; magnetic resonance imagers; microelectronics manufacturing equipment—Class C[c]	C
Electron microscopes at magnification greater than 30,000×; mass spectrometers; cell implant equipment; microelectronics manufacturing equipment—Class D[c]	B
Unisolated laser and optical research systems; microelectronics manufacturing equipment—Class E[c]	A

[a]See Figure 40 for corresponding curves.
[b]In areas where individuals are sensitive to vibration, use curve H.
[c]Classes of microelectronics manufacturing equipment:

Class A: Inspection, probe test, and other manufacturing support equipment.
Class B: Aligners, steppers, and other critical equipment for photolithography with line widths of 3 μm or more.
Class C: Aligners, steppers, and other critical equipment for photolithography with line widths of 1 μm.
Class D: Aligners, steppers, and other critical equipment for photolithography with line widths of 0.5 μm; includes electron-beam systems.
Class E: Aligners, steppers, and other critical equipment for photolithography with line widths of 0.25 μm; includes electron-beam systems.

Table 44 Equipment Vibration Criteria

Equipment	Allowable rms Velocity, in/s
Pumps	0.13
Centrifugal compressors	0.13
Fans (vent sets, centrifugal, axial)	0.09

spans, equipment operating speeds, equipment power, damping, and other factors have been considered in the vibration isolator selection guide in Table 45.

By specifying isolator deflection rather than isolation efficiency, transmissibility, or other theoretical parameters, a designer can compensate for floor stiffness and building resonances by selecting isolators that provide minimum vibration transmission and that have more deflection than the supporting floor. To apply the information from Table 45, base type, isolator type, and minimum deflection columns are added to the equipment schedule. These isolator specifications are then incorporated into mechanical specifications for the project.

The minimum deflections listed in Table 45 are based on the experience of acoustical and mechanical consultants and vibration control manufacturers. Recommended isolator type, base type, and minimum static deflection are reasonable and safe recommendations for 70% to 80% of HVAC equipment installations. The selections are based on concrete equipment room floors 4 to 12 in. thick with typical floor stiffness. The type of equipment, proximity to noise-sensitive areas, and the type of building construction may alter these choices.

Table 45　Selection Guide for Vibration Isolation

Equipment Type	Horsepower and Other	Rpm	Slab on Grade Base Type	Iso- lator Type	Min. Defl., in.	Up to 20 ft Floor Span Base Type	Iso- lator Type	Min. Defl., in.	20 to 30 ft Floor Span Base Type	Iso- lator Type	Min. Defl., in.	30 to 40 ft Floor Span Base Type	Iso- lator Type	Min. Defl., in.	Reference Notes
Refrigeration Machines and Chillers															
Bare compressors	All	All	A	2	0.25	C	3	0.75	C	3	1.75	C	4	2.50	2,3,12
Reciprocating	All	All	A	2	0.25	A	4	0.75	A	3	1.75	A	4	2.50	2,3,12
Centrifugal	All	All	A	1	0.25	A	4	0.75	A	3	1.75	A	3	1.75	2,3,4,12
Open centrifugal	All	All	C	1	0.25	C	4	0.75	C	3	1.75	C	3	1.75	2,3,12
Absorption	All	All	A	1	0.25	A	4	0.75	A	3	1.75	A	3	1.75	
Air Compressors and Vacuum Pumps															
Tank-mounted	Up to 10	All	A	3	0.75	A	3	0.75	A	3	1.75	A	3	1.75	3,13,15
	15 and over	All	C	3	0.75	C	3	0.75	C	3	1.75	C	3	1.75	3,13,15
Base-mounted	All	All	C	3	0.75	C	3	0.75	C	3	1.75	C	3	1.75	3,13,14,15
Large reciprocating	All	All	C	3	0.75	C	3	0.75	C	3	1.75	C	3	1.75	3,13,14,15
Pumps															
Closed coupled	Up to 7.5	All	B	2	0.25	C	3	0.75	C	3	0.75	C	3	0.75	16
	10 and over	All	C	3	0.75	C	3	0.75	C	3	1.75	C	3	1.75	16
Large inline	5 to 25	All	A	3	0.75	A	3	1.75	A	3	1.75	A	3	1.75	
	30 and over	All	A	3	1.75	A	3	1.75	A	3	1.75	A	3	2.50	
End suction and split case	Up to 40	All	C	3	0.75	C	3	0.75	C	3	1.75	C	3	1.75	16
	50 to 125	All	C	3	0.75	C	3	0.75	C	3	1.75	C	3	2.50	10,16
	150 and over	All	C	3	0.75	C	3	1.75	C	3	1.75	C	3	2.50	10,16
Cooling Towers	All	Up to 300	A	1	0.25	A	4	3.50	A	4	3.50	A	4	3.50	5,8,18
		301 to 500	A	1	0.25	A	4	2.50	A	4	2.50	A	4	2.50	5,18
		500 and over	A	1	0.25	A	4	0.75	A	4	0.75	A	4	1.75	5,18
Boilers—Fire-tube	All	All	A	1	0.25	B	4	0.75	B	4	1.75	B	4	2.50	4
Axial Fans, Fan Heads, Cabinet Fans, and Fan Sections															
Up to 22 in. dia.	All	All	A	2	0.25	A	3	0.75	A	3	0.75	C	3	0.75	4,9
24 in. dia. and over	Up to 2 in. s.p.	Up to 300	B	3	2.50	C	3	3.50	C	3	3.50	C	3	3.50	9
		300 to 500	B	3	0.75	B	3	1.75	C	3	2.50	C	3	2.50	9
		501 and over	B	3	0.75	B	3	1.75	B	3	1.75	B	3	1.75	9
	2.1 in. s.p. and over	Up to 300	C	3	2.50	C	3	3.50	C	3	3.50	C	3	3.50	3,9
		300 to 500	C	3	1.75	C	3	1.75	C	3	2.50	C	3	2.50	3,8,9
		501 and over	C	3	0.75	C	3	1.75	C	3	1.75	C	3	2.50	3,8,9
Centrifugal Fans															
Up to 22 in. dia.	All	All	B	2	0.25	B	3	0.75	B	3	0.75	C	3	1.75	9,19
24 in. dia. and over	Up to 40	Up to 300	B	3	2.50	B	3	3.50	B	3	3.50	B	3	3.50	8,19
		300 to 500	B	3	1.75	B	3	1.75	B	3	2.50	B	3	2.50	8,19
		501 and over	B	3	0.75	B	3	0.75	B	3	0.75	B	3	1.75	8,19
	50 and over	Up to 300	C	3	2.50	C	3	3.50	C	3	3.50	C	3	3.50	2,3,8,9,19
		300 to 500	C	3	1.75	C	3	1.75	C	3	2.50	C	3	2.50	2,3,8,9,19
		501 and over	C	3	1.00	C	3	1.75	C	3	1.75	C	3	2.50	2,3,8,9,19
Propeller Fans															
Wall-mounted	All	All	A	1	0.25	A	1	0.25	A	1	0.25	A	1	0.25	
Roof-mounted	All	All	A	1	0.25	A	1	0.25	B	4	1.75	D	4	1.75	
Heat Pumps	All	All	A	3	0.75	A	3	0.75	A	3	0.75	A/D	3	1.75	
Condensing Units	All	All	A	1	0.25	A	4	0.75	A	4	1.75	A/D	4	1.75	
Packaged AH, AC, H and V Units															
All	Up to 10	All	A	3	0.75	A	3	0.75	A	3	0.75	A	3	0.75	19
	15 and over, up to 4 in. s.p.	Up to 300	A	3	0.75	A	3	3.50	A	3	3.50	C	3	3.50	2,4,8,19
		301 to 500	A	3	0.75	A	3	2.50	A	3	2.50	A	3	2.50	4,19
		501 and over	A	3	0.75	A	3	1.75	A	3	1.75	A	3	1.75	4,19
	15 and over, 4 in. s.p. and over	Up to 300	B	3	0.75	C	3	3.50	C	3	3.50	C	3	3.50	2,3,4,8,9
		301 to 500	B	3	0.75	C	3	1.75	C	3	2.50	C	3	2.50	2,3,4,9
		501 and over	B	3	0.75	C	3	1.75	C	3	1.75	C	3	2.50	2,3,4,9
Packaged Rooftop Equipment	All	All	A/D	1	0.25	D	3	0.75		——— See Note 17 ———						5,6,8,17
Ducted Rotating Equipment															
Small fans, fan-powered boxes	Up to 600 cfm	All	A	3	0.50	A	3	0.50	A	3	0.50	A	3	0.50	7
	601 cfm and over	All	A	3	0.75	A	3	0.75	A	3	0.75	A	3	0.75	7
Engine-Driven Generators	All	All	A	3	0.75	C	3	1.75	C	3	2.50	C	3	3.50	2,3,4

Base Types:
A. No base, isolators attached directly to equipment (Note 27)
B. Structural steel rails or base (Notes 28 and 29)
C. Concrete inertia base (Note 30)
D. Curb-mounted base (Note 31)

Isolator Types:
1. Pad, rubber, or glass fiber (Notes 20 and 21)
2. Rubber floor isolator or hanger (Notes 20 and 25)
3. Spring floor isolator or hanger (Notes 22, 23, and 25)
4. Restrained spring isolator (Notes 22 and 24)
5. Thrust restraint (Note 26)

NOTES FOR VIBRATION ISOLATOR SELECTION GUIDE (TABLE 45)

The notes in this section are keyed to the numbers listed in the column titled "Reference Notes" and to other reference numbers throughout the table. While the guide is conservative, cases may arise where vibration transmission to the building is still excessive. If the problem persists after all short circuits have been eliminated, it can almost always be corrected by increasing isolator deflection, using low-frequency air springs, changing operating speed, reducing vibratory output by additional balancing or, as a last resort, changing floor frequency by stiffening or adding more mass.

Note 1. Isolator deflections shown are based on a floor stiffness that can be reasonably expected for each floor span and class of equipment.

Note 2. For large equipment capable of generating substantial vibratory forces and structure-borne noise, increase isolator deflection, if necessary, so isolator stiffness is at least 0.10 times the floor stiffness.

Note 3. For noisy equipment adjoining or near noise-sensitive areas, see the text section on Mechanical Equipment Room Sound Isolation.

Note 4. Certain designs cannot be installed directly on individual isolators (Type A), and the equipment manufacturer or a vibration specialist should be consulted on the need for supplemental support (Base Type).

Note 5. Wind load conditions must be considered. Restraint can be achieved with restrained spring isolators (Type 4), supplemental bracing, or limit stops.

Note 6. Certain types of equipment require a curb-mounted base (Type D). Airborne noise must be considered.

Note 7. See the text section on Resilient Pipe Hangers and Supports for hanger locations adjoining equipment and in equipment rooms.

Note 8. To avoid isolator resonance problems, select isolator deflection so that resonance frequency is 40% or less of the lowest operating speed of equipment.

Note 9. To limit undesirable movement, thrust restraints (Type 5) are required for all ceiling-suspended and floor-mounted units operating at 2 in. and more total static pressure.

Note 10. Pumps over 75 hp may require extra mass and restraining devices.

Isolation for Specific Equipment

Note 12. Refrigeration Machines: Large centrifugal, hermetic, and reciprocating refrigeration machines generate very high noise levels, and special attention is required when such equipment is installed in upper stories or near noise-sensitive areas. If such equipment is to be located near extremely noise-sensitive areas, confer with an acoustical consultant.

Note 13. Compressors: The two basic reciprocating compressors are (1) single- and double-cylinder vertical, horizontal or L-head, which are usually air compressors; and (2) Y, W, and multihead or multicylinder air and refrigeration compressors. Single- and double-cylinder compressors generate high vibratory forces requiring large inertia bases (Type C) and are generally not suitable for upper-story locations. If such equipment must be installed in upper stories or on grade locations near noise-sensitive areas, unbalanced forces should be obtained from the equipment manufacturer, and a vibration specialist should be consulted for design of the isolation system.

Note 14. Compressors: When using Y, W, and multihead and multicylinder compressors, obtain the magnitude of unbalanced forces from the equipment manufacturer so that the necessity for an inertia base can be evaluated.

Note 15. Compressors: Base-mounted compressors through 5 hp and horizontal tank-type air compressors through 10 hp can be installed directly on spring isolators (Type 3) with structural bases (Type B) if required, and compressors 15 to 100 hp on spring isolators (Type 3) with inertia bases (Type C) weighing one to two times the compressor weight.

Note 16. Pumps: Concrete inertia bases (Type C) are preferred for all flexible-coupled pumps and are desirable for most close-coupled pumps, although steel bases (Type B) can be used. Close-coupled pumps should not be installed directly on individual isolators (Type A)

because the impeller usually overhangs the motor support base, causing the rear mounting to be in tension. The primary requirements for Type C bases are strength and shape to accommodate base elbow supports. Mass is not usually a factor, except for pumps over 75 hp where extra mass helps limit excess movement due to starting torque and forces. Concrete bases (Type C) should be designed for a thickness of one-tenth the longest dimension with minimum thickness as follows: (1) for up to 30 hp, 6 in.; (2) for 40 to 75 hp, 8 in.; and (3) for 100 hp and higher, 12 in.

Pumps over 75 hp and multistage pumps may exhibit excessive motion at start-up; supplemental restraining devices can be installed if necessary. Pumps over 125 hp may generate high starting forces, so a vibration specialist should be consulted for installation recommendations.

Note 17. Packaged Rooftop Air-Conditioning Equipment: This equipment is usually on light structures that are susceptible to sound and vibration transmission. The noise problem is further compounded by curb-mounted equipment, which requires large roof openings for supply and return air.

The table shows Type D vibration isolator selections for all spans up to 20 ft, but extreme care must be taken for equipment located on spans of over 20 ft, especially if construction is open web joists or thin lightweight slabs. The recommended procedure is to determine the additional deflection caused by equipment in the roof. If additional roof deflection is 0.25 in. or under, the isolator can be selected for 15 times the additional roof deflection. If additional roof deflection is over 0.25 in., supplemental stiffening should be installed or the unit should be relocated.

For units, especially large units, capable of generating high noise levels, consider (1) mounting the unit on a platform above the roof deck to provide an air gap (buffer zone) and (2) locating the unit away from the roof penetration, thus permitting acoustical treatment of ducts before they enter the building.

Some rooftop equipment has compressors, fans, and other equipment isolated internally. This isolation is not always reliable because of internal short circuiting, inadequate static deflection, or panel resonances. It is recommended that rooftop equipment be isolated externally, as if internal isolation were not used.

Note 18. Cooling Towers: These are normally isolated with restrained spring isolators (Type 4) directly under the tower or tower dunnage. Occasionally, high deflection isolators are proposed for use directly under the motor-fan assembly, but this arrangement must be used with extreme caution.

Note 19. Fans and Air-Handling Equipment: The following should be considered in selecting isolation systems for fans and air-handling equipment:

Fans with wheel diameters of 22 in. and under and all fans operating at speeds to 300 rpm do not generate large vibratory forces. For fans operating under 300 rpm, select isolator deflection so that the isolator natural frequency is 40% or less of the fan speed. For example, for a fan operating at 275 rpm, an isolator natural frequency over 110 rpm (1.8 Hz) or lower is required ($0.4 \times 275 = 110$ rpm). A 3-in. deflection isolator (Type 3) can provide this isolation.

Flexible duct connectors should be installed at the intake and discharge of all fans and air-handling equipment to reduce vibration transmission to air ducts.

Inertia bases (Type C) are recommended for all Class 2 and 3 fans and air-handling equipment because extra mass permits the use of stiffer springs, which limit movement.

Thrust restraints (Type 5) that incorporate the same deflection as isolators should be used for all fan heads, all suspended fans, and all base-mounted and suspended air-handling equipment operating at 2 in. and over total static pressure.

Vibration Isolators: Materials, Types, and Configurations

Notes 20 through 31 are useful for evaluating commercially available isolators for HVAC equipment. The isolator selected for a particular application depends on the required deflection, but life, cost, and suitability must also be considered.

RUBBER PADS (Type 1)

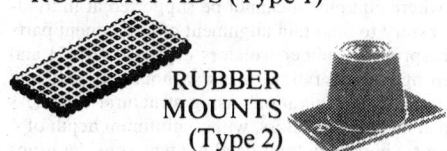

RUBBER MOUNTS (Type 2)

Note 20. Rubber isolators are available in pad (Type 1) and molded (Type 2) configurations. Pads are used in single or multiple layers. Molded isolators come in a range of 30 to 70 durometer (a measure of stiffness). Material in excess of 70 durometer is usually ineffective as an isolator. Isolators are designed for up to 0.5 in. deflection, but are used where 0.3 in. or less deflection is required. Solid rubber and composite fabric and rubber pads are also available. They provide high load capacities with small deflection and are used as noise barriers under columns and for pipe supports. These pad types work well only when they are properly loaded and the weight load is evenly distributed over the entire pad surface. Metal loading plates can be used for this purpose.

GLASS FIBER PADS (Type 1)

Note 21. Precompressed glass fiber isolation pads (Type 1) constitute inorganic inert material and are available in various sizes in thicknesses of 1 to 4 in., and in capacities of up to 500 psi. Their manufacturing process assures long life and a constant natural frequency of 7 to 15 Hz over the entire recommended load range. Pads are covered with an elastomeric coating to increase damping and to protect the glass fiber. Glass fiber pads are most often used for the isolation of concrete foundations and floating floor construction.

SPRING ISOLATOR (Type 3)

Note 22. Steel springs are the most popular and versatile isolators for HVAC applications because they are available for almost any deflection and have a virtually unlimited life. All spring isolators should have a rubber acoustical barrier to reduce transmission of high-frequency vibration and noise that can migrate down the steel spring coil. They should be corrosion-protected if installed outdoors or in a corrosive environment. The basic types include

1. **Note 23.** Open spring isolators (Type 3) consist of a top and bottom load plate with an adjustment bolt for leveling. Springs should be designed with a horizontal stiffness at least 100% of the vertical stiffness to assure stability, 50% travel beyond rated load and safe solid stresses.

RESTRAINED SPRING ISOLATOR (Type 4)

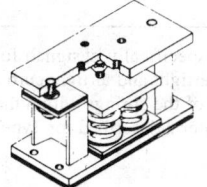

2. **Note 24.** Restrained spring isolators (Type 4) have hold-down bolts to limit vertical movement. They are used with (a) equipment with large variations in mass (boilers, refrigeration machines) to restrict movement and prevent strain on piping when water is removed, and (b) outdoor equipment, such as cooling towers, to prevent excessive movement because of wind load. Spring criteria should be the same as for open spring isolators, and restraints should have adequate clearance so that they are activated only when a temporary restraint is needed.

3. Housed spring isolators consist of two telescoping housings separated by a resilient material. Depending on design and installation, housed spring isolators can bind and short circuit. Their use should be avoided.

AIR SPRINGS

ROLLING LOBE   BELLOWS

Air springs can be designed for any frequency but are economical only in applications with natural frequencies of 1.33 Hz or less (6 in. or greater deflection). Their use is advantageous in that they do not transmit high-frequency noise and are often used to replace high deflection springs on problem jobs. Constant air supply is required, and there should be an air dryer in the air supply.

RUBBER HANGER (Type 2)
SPRING HANGER (Type 3)

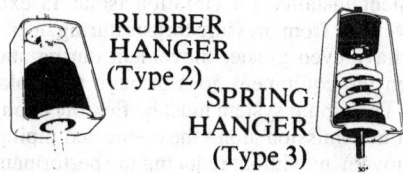

Note 25. Isolation hangers (Types 2 and 3) are used for suspended pipe and equipment and have rubber, springs, or a combination of spring and rubber elements. Criteria should be the same as for open spring isolators. To avoid short circuiting, hangers should be designed for 20 to 35° angular hanger rod misalignment. Swivel or traveler arrangements may be necessary for connections to piping systems subject to large thermal movements.

THRUST RESTRAINT (Type 5)

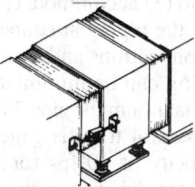

Note 26. Thrust restraints (Type 5) are similar to spring hangers or isolators and are installed in pairs to resist the thrust caused by air pressure.

DIRECT ISOLATION (Type A)

Note 27. Direct isolation (Type A) is used when equipment is unitary and rigid and does not require additional support. Direct isolation can be used with large chillers, packaged air-handling units, and air-cooled condensers. If there is any doubt that the equipment can be supported directly on isolators, use structural bases (Type B) or inertia bases (Type C), or consult the equipment manufacturer.

STRUCTURAL BASES (Type B)

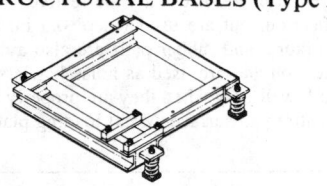

Note 28. Structural bases (Type B) are used where equipment cannot be supported at individual locations and/or where some means is necessary to maintain alignment of component parts in equipment. These bases can be used with spring or rubber isolators (Types 2 and 3) and should have enough rigidity to resist all starting and operating forces without supplemental hold-down devices. Bases are made in rectangular configurations using structural members with a depth equal to one-tenth the longest span between isolators, with a minimum depth of 4 in. Maximum depth is limited to 12 in., except where structural or alignment considerations dictate otherwise.

STRUCTURAL RAILS (Type B)

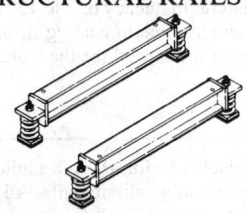

Note 29. Structural rails (Type B) are used to support equipment that does not require a unitary base or where the isolators are outside the equipment and the rails act as a cradle. Structural rails can be used with spring or rubber isolators and should be rigid enough to support the equipment without flexing. Usual industry practice is to use structural members with a depth one-tenth of the longest span between isolators with a minimum depth of 4 in. Maximum depth is limited to 12 in., except where structural considerations dictate otherwise.

CONCRETE BASES (Type C)

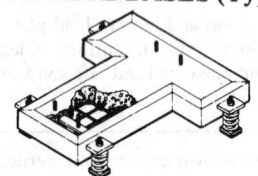

Note 30. Concrete bases (Type C) consist of a steel pouring form usually with welded-in reinforcing bars, provision for equipment hold-down, and isolator brackets. Like structural bases, concrete bases should be rectangular or T-shaped and, for rigidity, have a depth equal to one-tenth the longest span between isolators, with a minimum of 6 in. Base depth need not exceed 12 in. unless it is specifically required for mass, rigidity, or component alignment.

CURB ISOLATION (Type D)

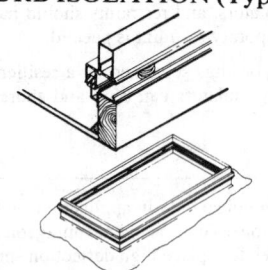

Note 31. Curb isolation systems (Type D) are specifically designed for curb-supported rooftop equipment and have spring isolation with a watertight and airtight curb assembly. The roof curbs are narrow to accommodate the small diameter of the springs within the rails, with static deflection in the 1 to 3 in. range to meet the design criteria described for Type 3.

The following approach is suggested to develop isolator selections for specific applications:

1. Use Table 45 for floors specifically designed to accommodate mechanical equipment.
2. Use recommendations for the 20 ft span column for equipment on ground-supported slabs adjacent to noise-sensitive areas.
3. For roofs and floors constructed with open web joists, thin long span slabs, wooden construction, and any unusual light construction, evaluate all equipment weighing more than 300 lb to determine the additional deflection of the structure caused by the equipment. Isolator deflection should be 15 times the additional deflection or the deflection shown in Table 45, whichever is greater. If the required spring isolator deflection exceeds commercially available products, consider air springs, stiffen the supporting structure, or change the equipment location.
4. When mechanical equipment is adjacent to noise-sensitive areas, isolate mechanical equipment room noise.

ISOLATION OF VIBRATION AND NOISE IN PIPING SYSTEMS

All piping has mechanical vibration generated by the equipment and impeller-generated and flow-induced vibration and noise, which is transmitted by the pipe wall and the water column. In addition, equipment installed on vibration isolators exhibits some motion or movement from pressure thrusts during operation. Vibration isolators have even greater movement during start-up and shutdown, when the equipment goes through the isolators' resonant frequency. The piping system must be flexible enough to (1) reduce vibration transmission along the connected piping, (2) permit equipment movement without reducing the performance of vibration isolators, and (3) accommodate equipment movement or thermal movement of the piping at connections without imposing undue strain on the connections and equipment.

Flow noise in piping can be minimized by sizing pipe so that the velocity is 4 fps maximum for pipe 2 in. and smaller and using a pressure drop limitation of 4 ft water gage per 100 ft of pipe length with a maximum velocity of 10 fps for larger pipe sizes. Flow noise and vibration can be reintroduced by turbulence, sharp pressure drops, and entrained air. Care should be taken to avoid these conditions.

Resilient Pipe Hangers and Supports

Resilient pipe hangers and supports are necessary to prevent vibration and noise transmission from the piping to the building structure and to provide flexibility in the piping.

Suspended Piping. Isolation hangers described in the vibration isolation section should be used for all piping in equipment rooms or for 50 ft from vibrating equipment, whichever is greater. To avoid reducing the effectiveness of equipment isolators, at least the first three hangers from the equipment should provide the same deflection as the equipment isolators, with a maximum limitation of 2 in. deflection; the remaining hangers should be spring or combination spring and rubber with 0.75 in. deflection.

Good practice requires the first two hangers adjacent to the equipment to be the positioning or precompressed type, to prevent load transfer to the equipment flanges when the piping is filled. The positioning hanger aids in installing large pipe, and many engineers specify this type for all isolated pipe hangers for piping 8 in. and over.

While isolation hangers are not often specified for branch piping or piping beyond the equipment room for economic reasons, they should be used for all piping over 2 in. in diameter and for any piping suspended below or near noise-sensitive areas. Hangers adjacent to noise-sensitive areas should be the spring and rubber combination Type 3.

Floor Supported Piping. Floor supports for piping in equipment rooms and adjacent to isolated equipment should use vibration isolators as described in the vibration isolation section. They should be selected according to the guidelines for hangers. The first two adjacent floor supports should be the restrained spring type, with a blocking feature that prevents load transfer to equipment flanges as the piping is filled or drained. Where pipe is subjected to large thermal movement, a slide plate (PTFE, graphite, or steel) should be installed on top of the isolator, and a thermal barrier should be used when rubber products are installed directly beneath steam or hot water lines.

Riser Supports, Anchors, and Guides. Many piping systems have anchors and guides, especially in the risers, to permit expansion joints, bends, or pipe loops to function properly. Anchors and guides eliminate or limit (guide) pipe movement, but must be rigidly attached to the structure; this is inconsistent with the resiliency required for effective isolation. The engineer should try to locate the pipe shafts, anchors, and guides in noncritical areas, such as next to elevator shafts, stairwells, and toilets, rather than adjoining noise-sensitive areas. Where concern about vibration transmission exists, some type of vibration isolation support or acoustical support is required for the pipe support, anchors, and guides.

Because anchors or guides must be rigidly attached to the structure, the isolator cannot deflect in the sense previously discussed, and the primary interest is to create an acoustical barrier. Such acoustical barriers can be provided by heavy-duty rubber and duck and rubber pads that can accommodate large loads with minimal deflection. Figure 42 shows some arrangements for resilient anchors and guides. Similar resilient-type supports can be used for the pipe.

Resilient supports for pipe, anchors, and guides can attenuate noise transmission, but they do not provide the resiliency required to isolate vibration. Vibration must be controlled in an anchor guide by designing flexible pipe connectors and resilient isolation hangers or supports.

Completely spring-isolated risers that eliminate the anchors and guides have been used successfully in many instances and give effective vibration and acoustical isolation. In this type of isolation, the springs are sized to accommodate thermal growth as well as to guide and support the pipe. Such systems require careful engineering to accommodate the movements encountered not only in the riser but also in the branch takeoff to avoid overstressing the piping.

Piping Penetrations. Most HVAC systems have many points at which piping must penetrate floors, walls, and ceilings. If such penetrations are not properly treated, they provide a path for airborne noise, which can destroy the acoustical integrity of the occupied space. Seal the openings in the pipe sleeves between noisy areas,

such as equipment rooms, and occupied spaces with an acoustical barrier such as fibrous material and caulking or with engineered pipe penetration seals as shown in Figure 43.

Flexible Pipe Connectors

Flexible pipe connectors (1) provide piping flexibility to permit isolators to function properly, (2) protect equipment from strain from misalignment and expansion or contraction of piping, and (3)

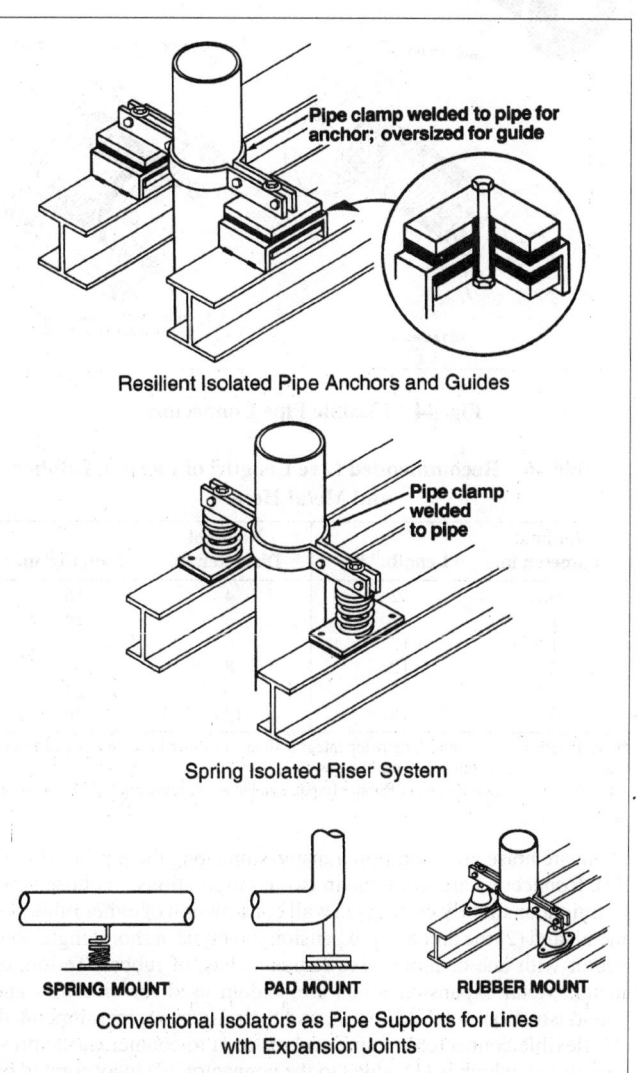

Resilient Isolated Pipe Anchors and Guides

Spring Isolated Riser System

Conventional Isolators as Pipe Supports for Lines with Expansion Joints

Fig. 42   Resilient Anchors and Guides for Pipes

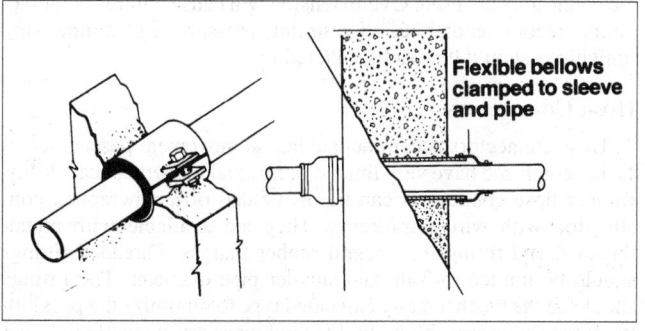

Fig. 43   Acoustical Pipe Penetration Seals

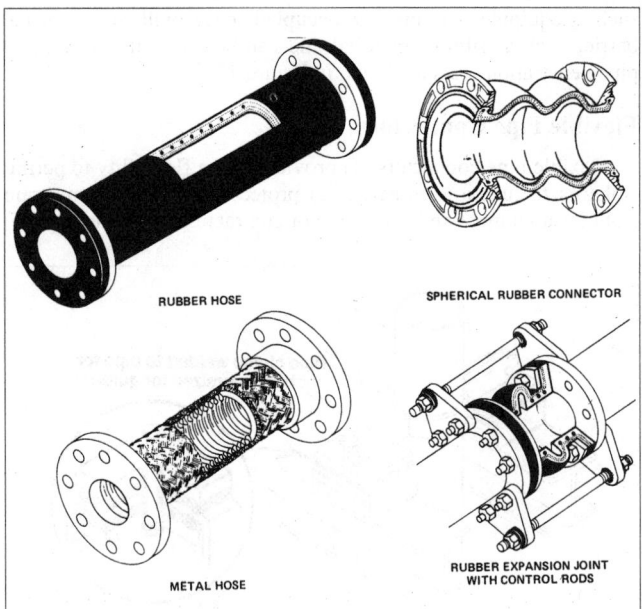

Fig. 44   Flexible Pipe Connectors

Table 46   Recommended Live Length[a] of Flexible Rubber and Metal Hose

Nominal Diameter, in.	Length,[b] in.	Nominal Diameter, in.	Length,[b] in.
0.75	12	4	18
1	12	5	24
1.5	12	6	24
2	12	8	24
2.5	12	10	24
3	18	12	36

[a]Live length is end-to-end length for integral flanged rubber hose and is end-to-end less total fitting length for all other types.

[b]Based on recommendations of Rubber Expansion Division, Fluid Sealing Association.

attenuate noise and vibration transmission along the piping (Figure 44). Connectors are available in two configurations: (1) hose type, a straight or slightly corrugated wall construction of either rubber or metal; and (2) the arched or expansion joint type, a short length connector with one or more large radius arches, of rubber, Teflon, or metal. Metal expansion joints are seldom used for vibration and sound isolation in HVAC work, and their use is not recommended. All flexible connectors require end restraint to counteract the pressure thrust, which is (1) added to the connector, (2) incorporated by its design, (3) added to the piping (anchoring), or (4) built in by the stiffness of the system. Connector extension caused by pressure thrust on isolated equipment should also be considered when flexible connectors are used. Overextension will cause failure. Manufacturers' recommendations on restraint, pressure, and temperature limitations should be strictly adhered to.

Hose Connectors

Hose connectors accommodate lateral movement perpendicular to the length and have very limited or no axial movement capability. Rubber hose connectors can be of molded or handwrapped construction with wire reinforcing. They are available with metal-threaded end fittings or integral rubber flanges. Threaded fittings should be limited to 3 in. and smaller pipe diameter. The fittings should be the mechanically expanded type to minimize the possibility of pressure thrust blowout. Flanged types are available in larger pipe sizes. Table 46 lists recommended lengths.

Metal hose is constructed with a corrugated inner core and a braided cover, which helps attain a pressure rating and provides end restraints that eliminate the need for supplemental control assemblies. Short lengths of metal hose or corrugated metal bellows, or pump connectors, are available without braid and have built-in control assemblies. Metal hose is used to control misalignment and vibration rather than noise and is used primarily where temperature or the pressure of flow media precludes the use of other material. Table 46 provides recommended lengths.

Expansion Joint or Arched-Type Connectors

Expansion joint or arched-type connectors have one or more convolutions or arches and can accommodate all modes of axial, lateral, and angular movement and misalignment. These connectors are available in flanged rubber and PTFE (Teflon) construction. PTFE expansion joints and couplings are similar in construction to rubber expansion joints with reinforcing metal rings. When made of rubber, they are commonly called expansion joints, spool joints, or spherical connectors, and in PTFE, as couplings or expansion joints.

Rubber expansion joints or spool joints are available in two basic types: (1) handwrapped with wire and fabric reinforcing, and (2) molded with fabric and wire or with high-strength fabric only (instead of metal) for reinforcing. The handmade joint is available in a variety of materials and lengths for special applications. Rubber spherical connectors are molded with high-strength fabric or tire cord reinforcing instead of metal. Their distinguishing characteristic is a large radius arch. The shape and construction of some designs permit use without control assemblies in systems operating to 150 psi. Where thrust restraints are not built in, they must be used as described for rubber hose joints.

In evaluating these devices, temperature, pressure, and service conditions must be considered as well as the ability of each device to attenuate vibration and noise. Metal hose connections can accommodate misalignment and attenuate mechanical vibration transmitted through the pipe wall but do little to attenuate noise. This type of connector has superior resistance to long-term temperature effects.

Rubber hose, expansion joints, and spherical connectors attenuate vibration and impeller-generated noise transmitted through the pipe wall. Because the rubber expansion joint and spherical connector walls are flexible, they have the ability to grow volumetrically and attenuate noise and vibration at blade passage frequencies. This feature is particularly desirable for uninsulated piping, such as condenser water and domestic water, which may run adjacent to noise-sensitive areas. However, high pressure has a detrimental effect on the ability of the connector to attenuate vibration and noise.

Because none of the flexible pipe connectors control flow or velocity noise or completely isolate vibration and noise transmission to the piping, resilient pipe hangers and supports should be used; these are shown in Note 25, Table 45 and are described in the Resilient Pipe Hangers and Supports section.

ISOLATING DUCT VIBRATION

Flexible canvas and rubber duct connections should be used at fan intake and discharge. However, they are not completely effective since they become rigid under pressure and allow the vibrating fan to pull on the duct wall. To maintain a slack position of the flexible duct connections, thrust restraints (see note 26, Table 45) should be used on all equipment as indicated in Table 45 and in the Vibration Isolator Selection section.

While vibration transmission from ducts isolated by flexible connectors is not a common problem, flow pulsations in the duct can cause vibration in the duct walls, which can be transmitted through rigid hangers. Spring or combination spring and rubber hangers should be used on ducts suspended below or near a noise-sensitive area. These hangers are especially recommended for large ducts with a velocity above 1500 fpm and for all size ducts when duct static pressure is 2 in. of water gage and over.

SEISMIC PROTECTION

Seismic restraint requirements are specified by applicable building codes that define the design forces to be resisted by the mechanical system, depending on the building location and occupancy, location of the system in the building, and whether it is used for life safety. Where required, seismic protection of resiliently mounted equipment poses a unique problem, because resiliently mounted systems are much more susceptible to earthquake damage due to overturning forces and to resonances inherent in vibration isolators.

As a deficiency in seismic restraint design or anchorage would not become apparent until an earthquake occurs, with possible catastrophic consequences, the adequacy of the restraints and anchorage to resist design forces must be verified before the event. This verification should be either by equipment tests, calculations, or dynamic analysis, depending on the item; with calculations or dynamic analysis performed under the direction of a professional engineer. These items are often supplied as a package by the vibration isolation vendor.

The restraints for floor-mounted equipment should have adequate clearances so that they are not engaged during normal operation of the equipment. Contact surfaces should be protected with resilient pads to limit shock during an earthquake, and restraints should be sufficiently strong to resist forces in any direction. The integrity of these devices can be verified by a comprehensive analysis but is more frequently verified by laboratory tests.

Calculations or dynamic analysis should have an engineer's seal to verify that input forces are obtained in accordance with code or specification requirements. The anchorage calculations should also be made by a professional engineer in accordance with accepted standards. Chapter 53, Seismic and Wind Restraint Design, has more information on this topic.

VIBRATION INVESTIGATIONS

Theoretically, a vibration isolation system can be selected to isolate vibration forces of extreme magnitude. However, isolators should not be used to mask a condition that should be corrected before it damages the equipment and its operation. High transmitted vibration levels can indicate a faulty equipment operating condition in need of correction or they can be a symptom of a resonance interaction between a vibrating piece of equipment and the structural floor on which it is placed.

Vibration investigations should include

- Measurement of the imbalance of reciprocating or rotating equipment components.
- Measurement of the vibration levels on vibrating equipment. Refer to Figure 41 for recommended vibration severity ratings of vibrating equipment.
- Measurement of vibration levels in building structures on which vibrating equipment is placed. Refer to Figure 40 and Table 43 for recommended building vibration criteria.
- Examination of equipment vibration generated by components, such as bearings, drives, etc.
- Examination of equipment installation factors, such as equipment alignment, vibration isolator placement, etc. Refer to Table 45.

TROUBLESHOOTING

In spite of the efforts taken by specifying engineers, consultants, and installing contractors, some situations arise that have disturbing noise and vibration. Fortunately, many problems can be readily identified and corrected by

- Determining which equipment or system is the problem source
- Determining if the problem is one of airborne sound, vibration and structure-borne noise, or a combination of both
- Applying appropriate solutions

Troubleshooting is time-consuming, expensive, and often difficult. In addition, once a transmission problem exists, the occupants become more sensitive and require greater reduction of the sound and vibration levels than would initially have been satisfactory. Therefore, the need for troubleshooting should be avoided by carefully designing, installing, and testing the system as soon as it is operational and before the building is occupied.

DETERMINING PROBLEM SOURCE

The system or equipment that is the source of the problem can often be determined without instrumentation. Vibration and noise levels are usually well above the sensory level of perception and are readily felt or heard. A simple and accurate method of determining the problem source is to turn individual pieces of equipment on and off until the vibration or noise is eliminated. Since the source of the problem is often more than one piece of equipment or the interaction of two or more systems, it is good practice to double check by shutting off the system and operating the equipment individually. Reynolds and Bevirt (1994) provides information relative to the measurement and assessment of sound and vibration in buildings.

DETERMINING PROBLEM TYPE

The next step is to determine if the problem is one of noise or vibration.

1. If vibration is perceptible, vibration transmission is usually the major cause of the problem. The possibility that lightweight wall or ceiling panels are excited by airborne noise should be considered. If vibration is not perceptible, the problem may still be one of vibration transmission causing structure-borne noise, which can be checked by following the procedure below.
2. If a sound level meter is available, check C-weighted and overall readings. If the difference is greater than 6 dB, or if the slope of the curve is greater than 5 to 6 dB per octave in the low frequencies, vibration is probably the cause.
3. If the affected area is remote from the source equipment, no problem is apparent in intermediary spaces, and noise does not appear to be coming from the duct system or diffusers, structure-borne noise is the probable cause.

Noise Problems

Noise problems are more complex than vibration problems and usually require the services of an acoustical engineer or consultant. If the affected area adjoins the room where the source equipment is located, structure-borne noise must be considered as part of the problem, and the vibration isolation should be checked. A simple but reasonably effective test is to have one person listen in the affected area while another shouts loudly in the equipment room. If the voice cannot be heard, the problem is likely one of structure-borne noise. If the voice can be heard, check for openings in the wall or floor separating the areas. If no such openings exist, the structure separating the areas does not provide adequate transmission loss. In such situations, refer to the Mechanical Equipment Room Sound Isolation section of this chapter for possible solutions.

If ductborne sound, i.e. noise from grilles or diffusers or duct breakout noise, is the problem, measure the sound pressure levels and compare them with the design goal RC curves. Where the measured curve differs from the design goal RC curve, the potential noise source(s) can be identified. Once the noise sources have been identified, the engineer can determine whether sufficient attenuation has been provided by analyzing each sound source using the procedures presented in this chapter.

If the sound source is a fan, pump, or similar rotating equipment, determine if it is operating at the most efficient part of its operating curve. Excessive vibration and noise can occur if a fan or pump is trying to move too little or too much air or water. In this respect, check that vanes, dampers, and valves are in the correct operating position and that the system has been properly balanced.

Vibration Problems

Vibration and structure-borne noise problems can occur from

- Equipment operating with excessive levels of vibration, usually caused by unbalance
- Lack of vibration isolators
- Improperly selected or installed vibration isolators that do not provide the required isolator deflection
- Flanking transmission paths such as rigid pipe connections or obstructions under the base of vibration-isolated equipment
- Floor flexibility
- Resonances in equipment, the vibration isolation, or the building structure

Most field-encountered problems are the result of improperly selected or installed isolators and flanking paths of transmission, which can be simply evaluated and corrected. Floor flexibility and resonance problems are sometimes encountered and usually require analysis by experts. However, the information provided below will identify such problems. If the equipment lacks vibration isolators, isolators recommended in Table 45 can be added by using structural brackets without altering connected ducts or piping.

Testing Vibration Isolation. Improperly functioning vibration isolation is the cause of most field-encountered problems and can be evaluated and corrected by the following procedures.

1. Ensure that the system is free-floating by bouncing the base, which should cause the equipment to move up and down freely and easily. On floor-mounted equipment, check that there are no obstructions between the base and the floor that would short circuit the isolation system. This check is best accomplished by passing a rod under the equipment. A small obstruction might permit the base to rock, giving the impression that it is free-floating when it is not. On suspended equipment, make sure that rods are not touching the hanger box. Rigid connections such as pipes and ducts can prevent equipment from floating freely, prohibit isolators from functioning properly, and provide flanking paths for the transmission of vibration.
2. Determine if the isolator deflection is as specified or required, changing it if necessary, as recommended in Table 45. A common problem is inadequate deflection caused by underloaded isolators. Overloaded isolators are not generally a problem as long as the system is free floating and there is space between the spring coils.

With the most spring isolators, determine the spring deflection by measuring the operating height and comparing it to the free height information available from the manufacturer. Once the actual isolator deflection is known, determine its adequacy by comparing it with the recommended deflection in Table 45.

If the natural frequency of the isolator is 25% or less than the disturbing frequency (usually considered the operating speed of the equipment), the isolators should be amply efficient except for heavy equipment installed on extremely long span floors or very flexible floors. If a transmission problem exists, it may be caused by (1) excessively rough equipment operation, (2) the system not being free floating or flanking path transmission, or (3) a resonance or floor stiffness problem, as described below.

While it is easy to determine the natural frequency of spring isolators by height measurements, such measurements are difficult with pad and rubber isolators and are not accurate in determining their natural frequencies. Although such isolators can theoretically provide natural frequencies as low as 4 Hz, they actually provide higher natural frequencies and generally do not provide the desired isolation efficiencies for upper floor equipment locations.

Generally, vibration isolation efficiency can not be determined in field installations by field vibration measurements. However, vibration measurements can be made on vibrating equipment, on equipment supports, on floors supporting vibration-isolated equipment, and on floors in adjacent areas to determine if vibration criteria specified in Table 43 or in Figures 40 and 41 have been achieved in field installations.

Floor Flexibility Problems. Floor flexibility is not a problem with most equipment and structures; however, such problems can occur with heavy equipment installed on long span floors or on thin slabs and with rooftop equipment installed on light structures with open web joist construction. If floor flexibility is suspected, the isolators should be one-tenth or less as stiff as the floor to eliminate the problem. Floor stiffness can be determined by calculating the additional deflection in the floor caused by a specific piece of equipment.

For example, if a 10,000 lb piece of equipment causes floor deflection of an additional 0.1 in., floor stiffness is 100,000 lb/in, and an isolator combined stiffness of 10,000 lb/in or less must be used. Note that the floor stiffness or spring rate, not the total floor deflection, is determined. In this example, the total floor deflection might be 1 in., but if the problem equipment causes 0.1 in. of that deflection, 0.1 in. is the important figure, and floor stiffness k is 100,000 lb/in.

Resonance Problems. These problems occur when the operating speed of the equipment is the same as or close to the resonance frequency of (1) an equipment component such as a fan shaft or bearing support pedestal, (2) the vibration isolation, or (3) the resonance frequency of the floor or other building component, such as a wall. Vibration resonances can cause excessive equipment vibration levels, as well as objectionable and possibly destructive vibration transmission in a building. These conditions must always be identified and corrected.

When vibrating mechanical equipment is mounted on vibration isolators on a flexible floor, there are two resonance frequencies that must be considered. The lower frequency is associated with and primarily controlled by the stiffness (and consequently the static deflection) of the vibration isolators. This frequency is generally significantly less than the operating speed (or frequency) of the mechanical equipment and is generally not a problem. The higher resonance frequency is associated with and primarily controlled by the stiffness of the floor. This resonance frequency is usually not affected by increasing or decreasing the static deflection of the mechanical equipment vibration isolators. Sometimes when the floor on which mechanical equipment is located is flexible (occurs with some long-span floors and with roofs supporting rooftop packaged units) the operating speed of the mechanical equipment can coincide with the higher resonance frequency. When this occurs, changing the static deflection of the vibration isolators will not solve the problem.

Vibration Isolation Resonance. Always characterized by excessive equipment vibration, vibration isolation resonance usually results in objectionable transmission to the structure. However, transmission might not occur if the equipment is on-grade or on a stiff floor. Vibration isolation resonance can be measured with instrumentation or, more simply, by determining the isolator natural frequency as described in the section Testing Vibration Isolation and comparing this figure to the operating speed of the equipment.

When vibration isolation resonance exists, the isolator natural frequency must be changed using the following guidelines:

1. If the equipment is installed on pad or rubber isolators, isolators with the deflection recommended in Table 45 should be installed.
2. If the equipment is installed on spring isolators and there is objectionable vibration or noise transmission to the structure, determine if the isolator is providing maximum deflection. For

example, an improperly selected or installed nominal 2 in. deflection isolator could be providing only 1/8 in. deflection, which would be in resonance with equipment operating at 500 rpm. If this is the case, the isolators should be replaced with ones having enough capacity to provide 2 in. deflection. Since there was no transmission problem with the resonant isolators, it is not necessary to use greater deflection isolators than can be conveniently installed.

3. If the equipment is installed on spring isolators and there is objectionable noise or vibration transmission, replace the isolators with spring isolators with the deflection recommended in Table 45.

Building Resonances. These problems occur when some part of the structure has a resonance frequency the same as the disturbing frequency or the operating speed of some of the equipment. These problems can exist even if the isolator deflections recommended in Table 45 are used. The resulting objectionable noise or vibration should be evaluated and corrected. Often, the resonant problem is in the floor on which the equipment is installed, but it can also occur in a remotely located floor, wall, or other building component. If a noise or vibration problem has a remote source which cannot be associated with piping or ducts, resonance must be suspected.

Building resonance problems can be resolved by the following:

1. Reduce the vibration force by balancing the equipment. This is not a practical solution for a true resonant problem; however, it is viable when the disturbing frequency equals the floor natural frequency, as evidenced by the equal displacement of the floor and the equipment, especially when the equipment is operating with excessive vibration.

2. Change the isolator resonance frequency by increasing or decreasing the static deflection of the isolator. Only small changes are necessary to "detune" the system. Generally, increasing the deflections is preferred. If the initial deflection is 1 in., a 2 or 3 in. deflection isolator should be installed. However, if the initial isolator deflection is 4 in., it may be more practical and economical to replace it with a 3 or 2 in. deflection isolator.

3. Change the structure stiffness or the structure resonance frequency. A change in structure stiffness changes the structure resonance frequency. The greater the stiffness, the higher the resonance frequency. However, the structure resonance frequency can also be changed by increasing or decreasing the floor deflection without changing the floor stiffness. While this approach is not recommended, it may be the only solution in certain cases.

4. Change the disturbing frequency by changing the equipment operating speed. This is practical only for belt-driven equipment, or equipment driven by variable frequency drives.

STANDARDS

AMCA 300. Reverberant Room Method for Sound Testing of Fans.

ANSI S3.29. 1983 (Reviewed 1990). Guide to Evaluation of Human Exposure to Vibration in Buildings.

ANSI S12.2. 1995. Criteria for Evaluating Room Noise.

ANSI S12.31. 1990. Precision Methods for the Determination of Sound Power Levels of Broad-Band Noise Sources in Reverberation Rooms.

ANSI S12.32. 1990. Precision Methods for the Determination of Sound Power Levels of Discrete-Frequency and Narrow-Band Noise Sources in Reverberation Rooms.

ANSI S12.34. 1988. Engineering Methods for the Determination of Sound Power Levels of Noise Sources for Essentially Free-Field Conditions over a Reflecting Plane.

ARI 270. 1995. Sound Rating of Outdoor Unitary Equipment.

ARI 275. 1997. Application of Sound Rating Levels of Outdoor Unitary Equipment.

ARI 300.1988. Rating the Sound Level and Transmission Loss of Packaged Terminal Equipment.

ARI 350. 1986. Sound Rating of Non-Ducted Indoor Air-Conditioning Equipment.

ARI 370. 1986. Sound Rating of Large Outdoor Refrigerating and Air-Conditioning Equipment.

ARI 530. 1995. Method of Rating Sound and Vibration of Refrigerant Compressors.

ARI 575. 1994. Method of Measuring Machinery Sound within an Equipment Space.

ARI 880. 1994. Air Terminals.

ARI 885. 1990. Procedure for Estimating Occupied Space Sound Levels in the Application of Air Terminals and Air Outlets.

ARI 890. 1994. Rating of Air Diffusers and Air Diffuser Assemblies.

ASHRAE 68R/AMCA 330. 1986. Laboratory Methods of Testing In-Duct Sound Power Measurement Procedure for Fans.

ASHRAE 70. 1991. Method of Testing for Rating the Performance of Air Outlets and Inlets.

ASTM E 477. 1996. Standard Test Method for Measuring Acoustical and Airflow Performance of Duct Liner Materials and Prefabricated Silencers.

ISO 2631-2. Continuous and Shock-Induced Vibration in Buildings.

REFERENCES

AIA. 1992-93. *Guidelines for construction and equipment of hospital and Medical facilities.* AIA Press, Washington, DC.

ASHRAE. 1987 *ASHRAE handbook*, Chapter 52.

ASHRAE. 1993 *ASHRAE handbook*, Chapter 37.

ASHRAE. 1995 *ASHRAE Handbook—HVAC Applications*, Chapter 43.

Beatty, J. 1987. Discharge duct configurations to control rooftop sound. *Heating/Piping/Air Conditioning* (July).

Beranek, L.L. 1960. *Noise Reduction.* McGraw-Hill, New York.

Beranek, L.L. 1971. *Noise and vibration control.* McGraw-Hill, New York.

Beranek, L.L. 1989. Balanced noise criterion (NCB) curves. *J. Acous. Soc. Am.* (86):650-54.

Blazier, W.E., Jr. 1981a. Revised noise criteria for design and rating of HVAC systems. *ASHRAE Transactions* 87(1).

Blazier, W.E., Jr. 1981b. Revised noise criteria for application in the acoustical design and rating of HVAC systems. *Noise Control Eng.* 16(2):64-73.

Blazier, W.E., Jr. 1995. Sound quality considerations in rating noise from heating, ventilating and air-conditioning (HVAC) systems in buildings. *Noise Control Eng. J.* 43(3).

Blazier, W.E., Jr. 1997. RC Mark II; a refined procedure for rating the noise of heating, ventilating and air-conditioning (HVAC) systems in buildings. *Noise Control Eng. J.* 45(6).

Broner, N. 1994. Determination of the relationship between low-frequency HVAC noise and comfort in occupied spaces—Objective Phase. ASHRAE 714-RP.

Cummings, A. 1983. Acoustic noise transmission through the walls of air-conditioning ducts. *Final Report.* Department of Mechanical and Aerospace Engineering, University of Missouri-Rolla.

Cummings, A. 1985. Acoustic noise transmission through duct walls. *ASHRAE Transactions* 91(2A).

Ebbing, C.E., D. Fragnito, and S. Inglis. 1978. Control of low frequency duct-generated noise in building air distribution systems. *ASHRAE Transactions* 84(2).

Ebbing, C.E. and W.E. Blazier, Jr. 1992. HVAC low frequency noise in buildings. *Proc.* INTER-NOISE 92(2):767-70.

Ebbing, C.E. and W.E. Blazier, Jr. 1998. *Application of manufacturers' sound data.* ASHRAE.

Egan, M.D. 1988. *Architectural acoustics.* McGraw-Hill, New York.

Environment Canada. 1994. Mineral fibres: Priority substances list assessment report. Canadian Environmental Protection Act, Ottawa.

Harold, R.G. 1986. Round duct can stop rumble noise in air-handling installations. *ASHRAE Transactions* 92(2).

Harold, R.G. 1991. Rooftop installation sound and vibration considerations. *ASHRAE Transactions* 97(1).

IRD. 1988. *Vibration technology*-1. IRD Mechanalysis, Columbus, OH.

Kuntz, H.L. 1986. The determination of the interrelationship between the Physical and acoustical properties of fibrous duct liner materials and lined duct sound attenuation. *Report No.* 1068. Hoover Keith and Bruce, Houston, TX.

Kuntz, H.L. and R.M Hoover. 1987. The interrelationships between the physical properties of fibrous duct lining materials and lined duct sound attenuation. *ASHRAE Transactions* 93(2).

Lilly, J. 1987. Break-out in HVAC duct systems. *Sound & Vibration* (October).

Machen, J. and J.C. Haines. 1983. Sound insertion loss properties of linacoustic and competitive duct liners. *Report No.* 436-T-1778. Johns-Manville Research and Development Center, Denver, CO.

Morey, P.R. and C.M. Williams. 1991. Is porous insulation inside an HVAC system compatible with healthy building? *ASHRAE IAQ Symposium.*

Persson-Waye, K., et al. 1997. Effects on performance and work quality due to low-frequency ventilation noise. *Journal of Sound and Vibration* 205(4):467-74.

Reynolds, D.D. and J.M. Bledsoe. 1989a. Sound attenuation of acoustically lined circular ducts and radiused elbows. *ASHRAE Transactions* 95(1).

Reynolds, D.D. and J.M. Bledsoe. 1989b. Sound attenuation of unlined and acoustically lined rectangular ducts. *ASHRAE Transactions* 95(1).

Reynolds, D.D. and J.M. Bledsoe. 1991. *Algorithms for HVAC acoustics.* ASHRAE, Atlanta.

Reynolds, D.D. and W.D. Bevirt. 1994. *Procedural standards for the measurement and assessment of sound and vibration.* National Environmental Balancing Bureau, Rockville, MD.

Reynolds, D.D. and W.D. Bevirt. 1989. *Sound and vibration design and analysis.* National Environmental Balancing Bureau, Rockville, MD.

Sandbakken, M., L. Pande, and M.J. Crocker. 1981. Investigation of end reflection of coefficient accuracy problems with AMCA *Standard* 300-67. HL 81-16. Ray W. Herrick Laboratories, Purdue University, West Lafayette, IN.

Schaffer, M.E. 1991. A *practical guide to noise and vibration control for HVAC systems.* ASHRAE, Atlanta.

Schultz, T.J. 1985. Relationship between sound power level and sound pressure level in dwellings and offices. *ASHRAE Transactions* 91(1).

SMACNA. 1990. *HVAC systems duct design,* 3rd ed. Sheet Metal and Air Conditioning Contractors' National Association, Vienna, VA.

Stevens, K.N., W.A. Rosenblith, and R.H. Bolt. 1955. A Community's reaction to noise: Can it be forecast? *Noise Control* (January).

Thompson, J.K. 1981. The room acoustics equation: Its limitation and potential. *ASHRAE Transactions* 87(2).

Ungar, E.E., D.H. Sturz, and C.H. Amick. 1990. Vibration control design of high technology facilities. *Sound and Vibration* (July).

Ver, I.L. 1978. A review of the attenuation of sound in straight lined and unlined ductwork of rectangular cross section. *ASHRAE Transactions* 84(1).

Ver, I.L. 1982. A study to determine the noise generation and noise attenuation of lined and unlined duct fittings. *Report No.* 5092. Bolt, Beranek and Newman, Boston.

Ver, I.L. 1984a. Noise generation and noise attenuation of duct fittings–A review: Part II. *ASHRAE Transactions* 90(2A).

Ver, I.L. 1984b. Prediction of sound transmission through duct walls: Break-out and pickup. *ASHRAE Transactions* 90(2A).

Warnock, A.C.C. 1998. Transmission of sound from air terminal devices through ceiling systems. *ASHRAE Transactions* 194(1A):650-57.

Wells, R.J. 1958. Acoustical plenum chambers. *Noise Control* (July).

Woods Fan Division. 1973. *Design for sound.* The English Electric Company.

WHO. 1986. *International Symposium on Man-Made Mineral Fibers in the Working Environment.* World Health Organization, International Agency for Research on Cancer, Copenhagen.

BIBLIOGRAPHY

Cummings, A. 1979. The effects of external lagging on low frequency sound transmission through the walls of rectangular ducts. *Journal of Sound Vibration* 67(2):187-201.

Departments of the Army, the Air Force, and the Navy. 1983. *Noise and Vibration Control for Mechanical Equipment.* Army TM 5-805-4, Air Force AFM 88-37, Navy NAVFAC DM-3, 10.

Fry, A., ed. 1988. *Noise control in building services.* Pergamon Press, Oxford, UK.

Goodfriend, L.S. 1980. Indoor sound rating criteria. *ASHRAE Transactions* 86(2).

Kahn, Greenberg, and Essert. 1987. Break-out noise from lined air-conditioning ducts. *Noise-Con 87.*

Office of Noise Control. 1981. Catalog of STC and IIC ratings for wall and floor/ceiling assemblies. California Department of Health Services, Sacramento.

Owens-Corning Fiberglass Corp. 1981. *Noise Control Manual,* 4th ed.

Reynolds, D.D. and J.M. Bledsoe. 1989. Sound transmission through mechanical equipment room walls, floor, or ceiling. *ASHRAE Transactions* 95(1).

Reynolds, D.D. and W.P. Zeng. 1994. New relationship between sound power level and sound pressure level in rooms. *Report No.* Urp-93001-1. Ventilation & Acoustic Systems Technology Laboratory. University of Nevada, Las Vegas.

WATER TREATMENT

THIS chapter covers the fundamentals of water treatment and some of the common problems associated with water in heating and air-conditioning equipment.

WATER CHARACTERISTICS

Chemical Characteristics

When rain falls, it dissolves carbon dioxide and oxygen in the atmosphere. The carbon dioxide mixes with the water to form carbonic acid (H_2CO_3). When carbonic acid contacts soil that contains limestone ($CaCO_3$), it dissolves the calcium to form calcium carbonate. Calcium carbonate in water used in heating or air-conditioning applications can eventually become scale, which can increase energy costs, maintenance time, equipment shutdowns, and could eventually lead to equipment replacement.

The following paragraphs discuss typical chemical and physical properties of water used for HVAC applications.

Alkalinity is a measure of the capacity of a water to neutralize strong acids. In natural waters, the alkalinity almost always consists of bicarbonate, although some carbonate may also be present. Borate, hydroxide, phosphate, and other constituents, if present, are included in the alkalinity measurement in treated waters. Alkalinity also contributes to scale formation.

Alkalinity is measured using two different end-point indicators. The **phenolphthalein alkalinity** (P alkalinity) measures the strong alkali present; the **methyl orange alkalinity** (M alkalinity), or **total alkalinity**, measures all of the alkalinity present in the water. Note that the total alkalinity includes the phenolphthalein alkalinity. For most natural waters, in which the concentration of phosphates, borates, and other noncarbonated alkaline materials is small, the actual chemical species present can be estimated from the two alkalinity measurements (Table 1).

Alkalinity or acidity is often confused with pH. Such confusion may be avoided by keeping in mind that the pH is a measure of hydrogen ion concentration expressed as the logarithm of its reciprocal.

Chlorides have no effect on scale formation but do contribute to corrosion because of their conductivity and because the small size of the chloride ion permits the continuous flow of corrosion current when surface films are porous. The amount of chlorides in the water is a useful measuring tool in evaporative systems. Virtually all other constituents in the water increase or decrease when common treatment chemicals are added or because of chemical changes that take place in normal operation. With few exceptions, only evaporation affects chloride concentration, so the ratio of chlorides in a water sample from an operating system to those of the makeup water provides a measure of how much the water has been concentrated. (*Note*: Chloride levels will change if the system is continuously chlorinated.)

Dissolved solids consist of salts and other materials that combine with water as a solution. They can affect the formation of corrosion and scale. Low-solids waters are generally corrosive because they have less tendency to deposit protective scale. If a high-solids water is nonscaling, it tends to produce more intensive corrosion because of its high conductivity. Dissolved solids are often referred to as total dissolved solids (TDS).

Conductivity or **specific conductance** measures the ability of a water to conduct electricity. Conductivity increases with the total dissolved solids. Specific conductance can be used to estimate total dissolved solids.

Silica can form particularly hard-to-remove deposits if allowed to concentrate. Fortunately, silicate deposition is less likely than other deposits.

Soluble iron in water can originate from metal corrosion in water systems or as a contaminant in the makeup water supply. The iron can form heat-insulating deposits by precipitation as iron hydroxide or iron phosphate (if a phosphate-based water treatment product is used or if phosphate is present in the makeup water).

Sulfates also contribute to scale formation in high-calcium waters. Calcium sulfate scale, however, forms only at much higher concentrations than the more common calcium carbonate scale. High sulfates also contribute to increased corrosion because of their high conductivity.

Suspended solids include both organic and inorganic solids suspended in water (particularly unpurified water from surface sources or those that have been circulating in open equipment). Organic matter in surface supplies may be colloidal. Naturally occurring compounds such as lignins and tannins are often colloidal. At high velocities, hard suspended particles can abrade equipment. Settled suspended matter of all types can contribute to concentration cell corrosion.

Turbidity can be interpreted as a lack of clearness or brilliance in a water. It should not be confused with color. A water may be dark in color but still clear and not turbid. Turbidity is due to suspended matter in a finely divided state. Clay, silt, organic matter, microscopic organisms, and similar materials are contributing causes of turbidity. Although suspended matter and turbidity are closely related they are not synonymous. Suspended matter is the quantity of material in a water that can be removed by filtration. The turbidity of water used in HVAC systems should be as low as possible. This is particularly true of boiler feedwater. The turbidity can concentrate in the boiler and may settle out as sludge or mud and lead to deposition. It can also cause increased boiler blowdown, plugging, overheating, priming, and foaming.

Biological Characteristics

Bacteria, algae, and fungi can be present in water systems, and their growth can cause operating, maintenance, and health problems.

Table 1 Alkalinity Interpretation for Waters[a]

If	Then, Carbonate	Bicarbonate	Free CO_2
P Alk = 0	0	M Alk	Present
P Alk < 0.5M Alk	2P Alk	M Alk − 2P Alk	0
P Alk = 0.5M Alk	2P Alk = M Alk	0	0
P Alk > 0.5M Alk[b]	2(M Alk − P Alk)	0	0

[a] P Alk = Phenolphthalein alkalinity. M Alk = Methyl orange (total) alkalinity.
[b] Treated waters only. Hydroxide also present.

The preparation of this chapter is assigned to TC 3.6, Corrosion and Water Treatment.

Microorganism growth is affected by temperature, food sources, and water availability. Biological growth can occur in most water systems below 150°F. Problems caused by biological materials range from green algae growth in cooling towers to slime formation from bacteria in secluded and dark areas. The result can be plugging of equipment where flow is essential. Dead algae are often characterized by foul odors and mud-like deposits, which are often high in silica from the cell walls of the diatoms. These problems can be eliminated, or at least reduced to reasonable levels, by mechanical or chemical treatment.

CORROSION CONTROL

Corrosion is the destruction of a metal or alloy by chemical or electrochemical reaction with its environment. In most instances, this reaction is electrochemical in nature, much like that in an electric battery. For corrosion to occur, a corrosion cell consisting of an anode, a cathode, an electrolyte, and an electrical connection must exist. Metal ions dissolve into the electrolyte (water) at the anode. Electrically charged particles (electrons) are left behind. These electrons flow through the metal to other points (cathodes) where electron-consuming reactions occur. The result of this activity is the loss of metal and often the formation of a deposit.

Types of Corrosion

Corrosion can often be characterized as general, localized or pitting, galvanic, caustic cracking, corrosion fatigue, erosion corrosion, and microbiological corrosion.

General corrosion is uniformly distributed over the metal surface. The considerable amount of iron oxide produced by generalized corrosion contributes to fouling.

Localized or pitting corrosion exists when only small areas of the metal corrode. Pitting is the most serious form of corrosion because the action is concentrated in a small area. Pitting may perforate the metal in a short time.

Galvanic corrosion can occur when two different metals are in contact. The more active (less noble) metal corrodes rapidly. Common examples of galvanic corrosion in water systems are steel and brass, aluminum and steel, zinc and steel, and zinc and brass. If galvanic corrosion occurs, the metal named first corrodes.

Corrosion fatigue may occur by one of two mechanisms. In the first mechanism, cyclic stresses (for example, those created by rapid heating and cooling) are concentrated at points where corrosion has roughened or pitted the metal surface. In the second type, cracks often originate where metal surfaces are covered by a dense protective oxide film, and cracking occurs from the action of applied cyclic stresses.

Caustic cracking occurs when metal is stressed, water has a high caustic content, a trace of silica is present, and a mechanism of concentration occurs.

Stress corrosion cracking can occur in alloys exposed to a characteristic corrosive environment and subject to tensile stress. The stress may be either applied, residual (from processing), or a combination. The resulting crack formation is generally intergranular.

Erosion corrosion occurs as a result of flow or impact that physically removes the protective metal oxide surface. This type of corrosion usually takes place due to altered flow patterns or a flow rate that is above design.

Microbiologically influenced corrosion occurs when bacteria secrete corrosive acids on the metal surface. Often bacteria commingle with dirt and silt that protect the bacteria from biocides.

Factors That Contribute to Corrosion

Moisture. Corrosion does not occur in dry environments. However, some moisture is present as water vapor in most environments. In pure oxygen, almost no iron corrosion occurs at relative humidities up to 99%. However, when contaminants such as sulfur dioxide

or solid particles of charcoal are present, corrosion can proceed at relative humidities of 50% or more. The corrosion reaction proceeds on surfaces of exposed metals such as iron and unalloyed steel as long as the metal remains wet. Many alloys develop protective corrosion-product films or oxide coatings and are thus unaffected by moisture.

Oxygen. In electrolytes consisting of water solutions of salts or acids, the presence of dissolved oxygen accelerates the corrosion rate of ferrous metals by depolarizing the cathodic areas through reaction with hydrogen generated at the cathode. In many systems, such as boilers or water heaters, most of the dissolved oxygen and other dissolved gases are removed from the water by deaeration to reduce the potential for corrosion.

Solutes. In ferrous materials such as iron and steel, mineral acids accelerate the corrosion rate, whereas alkalies decrease it. The likelihood of cathodic polarization, which slows down corrosion, decreases as the concentration of hydrogen ions (i.e., the acidity of the environment) increases. Relative acidity or alkalinity of a solution is defined as pH; a neutral solution has a pH of 7. The corrosivity of most salt solutions depend on their pH. Because alkaline solutions are generally less corrosive to ferrous systems, it is practical in many closed water systems to minimize corrosion by adding alkali or alkaline salt to raise the pH to 9 or higher.

Differential Solute Concentration. For the corrosion reaction to proceed, a potential difference between anode and cathode areas is required. This potential difference can be established between different locations on a metal surface because of differences in solute concentration in the environment at these locations. Corrosion caused by such conditions is called **concentration cell corrosion**. These cells can be metal ions or oxygen concentration cells.

In the metal ion cell, the metal surface in contact with the higher concentration of dissolved metal ion becomes the cathodic area, and the surface in contact with the lower concentration becomes the anode. The metal ions involved may be a constituent of the environment or may come from the corroding surface itself. The concentration differences may be caused by the sweeping away of the dissolved metal ions at one location and not at another.

In an oxygen concentration cell, the surface area in contact with the surface environment of higher oxygen concentration becomes the cathodic area, and the surface in contact with the surface environment of lower oxygen concentration becomes the anode. Crevices or foreign deposits on the metal surface can create conditions that contribute to corrosion. The anodic area, where corrosion proceeds, is in the crevice or under the deposit. Although they are manifestations of concentration cell corrosion, **crevice corrosion** and **deposit attack** are sometimes referred to as separate forms of corrosion.

Galvanic or Dissimilar Metal Corrosion. Another factor that can accelerate the corrosion process is the difference in potential of dissimilar metals coupled together and immersed in an electrolyte. The following factors control the severity of corrosion resulting from such dissimilar metal coupling:

- Relative differences in position (potential) in the galvanic series, with reference to a standard electrode. The greater the difference, the greater the force of the reaction. The galvanic series for metals in flowing aerated seawater is shown in Table 2.

- Relative area relationship between anode and cathode areas. Because the amount of current flow and, therefore, total metal loss is determined by the potential difference and resistance of the circuits, a small anodic area corrodes more rapidly; it is penetrated at a greater rate than a large anodic area.

- Polarization of either the cathodic or anodic area. Polarization can reduce the potential difference and thus reduce the rate of attack of the anode.

- The mineral content of water. As mineral content increases, the resulting higher electrical conductivity increases the rate of galvanic corrosion.

Table 2   Galvanic Series of Metals and Alloys in Flowing Aerated Seawater at 40 to 80°F

Corroded End (Anodic or Least Noble)

Magnesium alloys
Zinc
Beryllium
Aluminum alloys
Cadmium
Mild steel, wrought iron
Cast iron, flake or ductile
Low-alloy high-strength steel
Ni-Resist, Types 1 & 2
Naval Brass (CA464), yellow brass (CA268), Al brass (CA687),
   red brass (CA230), admiralty brass (CA443), Mn Bronze
Tin
Copper (CA102, 110), Si Bronze (CA655)
Lead-tin solder
Tin bronze (G & M)
Stainless steel, 12 to 14% Cr (AISI types 410,416)
Nickel silver (CA 732, 735, 745, 752, 764, 770, 794)
90/10 Copper-nickel (CA 706)
80/20 Copper-nickel (CA 710)
Stainless steel, 16 to 18% Cr (AISI Types 430)
Lead
70/30 Copper-nickel (CA 715)
Nickel-aluminum bronze
Silver braze alloys
Nickel 200
Silver
Stainless steel, 18% Cr, 8% Ni (AISI Types 302, 304, 321, 347)
Stainless steel, 18% Cr, 12% Ni-Mo (AISI Types 316, 317)
Titanium
Graphite, graphitized cast iron

Protected End (Cathodic or Most Noble)

Stress. Stresses in metallic structures rarely have significant effects on the uniform corrosion resistance of metals and alloys. Stresses in specific metals and alloys can cause corrosion cracking when the metals are exposed to specific corrosive environments. The cracking can have catastrophic effects on the usefulness of the metal.

Almost all metals and alloys exhibit susceptibility to stress corrosion cracking in at least one environment. Common examples are steels in hot caustic solutions, high zinc content brasses in ammonia, and stainless steels in hot chlorides. Metal manufacturers have technical details on specific materials and their resistance to stress corrosion.

Temperature. According to studies of chemical reaction rates, corrosion rates double for every 18°F rise in temperature. However, such a ratio is not necessarily valid for nonlaboratory corrosion reactions. The effect of temperature on a particular system is difficult to predict without specific knowledge of the characteristics of the metals involved and the environmental conditions.

An increase in temperature may increase the corrosion rate, but only to a point. Oxygen solubility decreases as temperature increases and, in an open system, may approach zero as water boils. Beyond a critical temperature level, the corrosion rate may decrease due to a decrease in oxygen solubility. However, in a closed system, where oxygen cannot escape, the corrosion rate may continue to increase with an increase in temperature.

For those alloys, such as stainless steel, that depend on oxygen in the environment for maintaining a protective oxide film, the reduction in oxygen content due to an increase in temperature can accelerate the corrosion rate by preventing oxide film formation.

Temperature can affect corrosion potential by causing a salt dissolved in the environment to precipitate on the metal surface as a protective layer of scale. One example is calcium carbonate scale in hard waters. Temperature can also affect the nature of the corrosion product, which may be relatively stable and protective in certain temperature ranges and unstable and non protective in others. An example of this is zinc in distilled water; the corrosion product is non protective from 140 to 190°F but reasonably protective at other temperatures.

Pressure. Where dissolved gases such as oxygen and carbon dioxide affect the corrosion rate, pressure on the system may increase their solubility and thus increase corrosion. Similarly, a vacuum on the system reduces the solubility of the dissolved gas, thus reducing corrosion. In a heated system, pressure may rise with temperature. It is difficult and impractical to control system corrosion by pressure control alone.

Flow Velocity. The effect of flow velocity on the corrosion rate of systems depends on several factors, including

- Amount of oxygen in the water
- Type of metal (iron and steel are most susceptible)
- Flow rate

In metal systems where corrosion products retard corrosion by acting as a physical barrier, high flow velocities may cause the removal of those protective barriers and increase the potential for corrosion. A turbulent environment may cause uneven attack, from both erosion and corrosion. This corrosion is called **erosion corrosion**. It is commonly found in piping with sharp bends where the flow velocity is high. Copper and softer metals are more susceptible to this type of attack.

Preventive and Protective Measures

Materials Selection. Any piece of heating or air-conditioning equipment can be made of metals that are virtually corrosion-proof under normal and typical operating conditions. However, economics usually dictate material choices. When selecting construction materials the following factors should be considered:

- Corrosion resistance of the metal in the operating environment
- Corrosion products that may be formed and their effects on equipment operation
- Ease of construction using a particular material
- Design and fabrication limitations on corrosion potential
- Economics of construction, operation, and maintenance during the projected life of the equipment; i.e., expenses may be minimized in the long run by paying more for a corrosion-resistant material and avoiding regular maintenance.
- Use of dissimilar metals should be avoided. Where dissimilar materials are used, insulating gaskets and/or organic coatings must be used to prevent galvanic corrosion.
- Compatibility of chemical additives with materials in the system

Protective Coatings. The operating environment has a significant role in the selection of protective coatings. Even with a coating suited for that environment, the protective material depends on the adhesion of the coating to the base material, which itself depends on the surface preparation and application technique.

Maintenance. Defects in a coating are difficult to prevent. These defects can be either flaws introduced into the coating during application or mechanical damage sustained after application. In order to maintain corrosion protection, defects must be repaired.

Cycles of Concentration. Some corrosion control may be achieved by optimizing the cycles of concentration (the degree to which soluble mineral solids in the makeup water have increased in the circulating water due to evaporation). Generally, adjustment of the blowdown rate and pH to produce a slightly scale-forming condition (see section on Scale Control) will result in an optimum condition between excess corrosion and excess scale.

Chemical Methods. Chemical protective film-forming chemical inhibitors reduce or stop corrosion by interfering with the corrosion mechanism. Inhibitors usually affect either the anode or the cathode.

Table 3   Typical Corrosion Inhibitors

Anodic Corrosion Inhibitors
  Molybdate
  Nitrite
  Orthophosphate
  Silicate
Mainly Cathodic Corrosion Inhibitors
  Bicarbonate
  Polyphosphate
  Phosphonate
  Zinc
  Polysilicate
General
  Soluble oils
  Other organics, such as azole or carboxylate

Anodic corrosion inhibitors establish a protective film on the anode. Though these inhibitors can be effective, they can be dangerous—if insufficient anodic inhibitor is present, the entire corrosion potential occurs at the unprotected anode sites. This causes severe localized (or pitting) attack.

Cathodic corrosion inhibitors form a protective film on the cathode. These inhibitors reduce the corrosion rate in direct proportion to the reduction of the cathodic area.

General corrosion inhibitors protect by forming a film on all metal surfaces whether anodic or cathodic.

Table 3 lists typical corrosion inhibitors. The most important factor in an effective corrosion inhibition program is the consistent control of both the corrosion inhibition chemicals and the key water characteristics. No program will work without controlling these factors.

Cathodic Protection. Sacrificial anodes reduce galvanic attack by providing a metal (usually zinc, but sometimes magnesium) that is higher on the galvanic series than either of the two metals that are coupled together. The sacrificial anode thereby becomes anodic to both metals and supplies electrons to these cathodic surfaces. Proper design and placement of these anodes are important. When properly used, they can reduce loss of steel from the tube sheet of exchangers using copper tubes. Sacrificial anodes have helped supplement chemical programs in many cooling water and process water systems.

Impressed-current protection is a similar corrosion control technique that reverses the corrosion cell's normal current flow by impressing a stronger current of opposite polarity. Direct current is applied to an anode—inert (platinum, graphite) or expendable (aluminum, cast iron)—reversing the galvanic flow and converting the steel from a corroding anode to a protective cathode. The method is very effective in protecting essential equipment such as elevated water storage tanks, steel tanks, or softeners.

White Rust on Galvanized Steel Cooling Towers

White rust is a zinc corrosion product that forms on galvanized surfaces. It appears as a white, waxy or fluffy deposit composed of loosely adhering zinc carbonate. The loose crystal structure allows continued access of the corrosive water to exposed zinc. Unusually rapid corrosion of galvanized steel, as evidenced by white rust, can affect galvanized steel cooling towers under certain conditions.

Before chromates in cooling tower water were banned, the common treatment system consisted of chromates for corrosion control and sulfuric acid for scale control. This control method generally has been replaced by alkaline treatment involving scale inhibitors at a higher pH. Alkaline water chemistry is naturally less corrosive to steel and copper, but create an environment where white rust on galvanized steel can occur. Also, some scale prevention programs soften the water to reduce hardness, rather than use acid to reduce alkalinity. The resulting soft water is corrosive to galvanized steel.

Prevention. White rust can be prevented by promoting the formation of a nonporous surface layer of basic zinc carbonate. This barrier layer is formed during a process called **passivation** and normally protects the galvanized steel for many years. Passivation is best accomplished by controlling pH during initial operation of the cooling tower. Control of the cooling water pH in the range of 7 to 8 for 45 to 60 days usually allows passivation of galvanized surfaces to occur. In addition to pH control, operation with moderate hardness levels of 100 to 300 ppm as $CaCO_3$ and alkalinity levels of 100 to 300 ppm as $CaCO_3$ will promote passivation. Where pH control is not possible, certain phosphate-based inhibitors may help protect galvanized steel. A water treatment company should be consulted for specific formulations.

SCALE CONTROL

Scale is a dense coating of predominantly inorganic material formed from the precipitation of water-soluble constituents. Some common scales are:

- Calcium carbonate
- Calcium phosphate
- Magnesium salts
- Silica

The following principal factors determine whether or not a water is scale forming:

- Temperature
- Alkalinity or acidity (pH)
- Amount of scale-forming material present
- Influence of other dissolved materials, which may or may not be scale-forming

As any of these factors changes, scaling tendencies also change. Most salts become more soluble as temperature increases. However, some salts, such as calcium carbonate, become less soluble as temperature increases. Therefore, they often cause deposits at higher temperatures.

A change in pH or alkalinity can greatly affect scale formation. For example, as pH or alkalinity increases, calcium carbonate—the most common scale constituent in cooling systems—decreases in solubility and deposits on surfaces. Some materials, such as silica (SiO_2), are less soluble at lower alkalinities. When the amount of scale-forming material dissolved in water exceeds its saturation point, scale may result. In addition, other dissolved solids may influence scale-forming tendencies. In general, a higher level of scale-forming dissolved solids results in a greater chance for scale formation. Indices such as the Langelier Saturation Index (Langelier 1936) and the Ryznar Stability Index (Ryznar 1944) can be useful tools to predict the calcium carbonate scaling tendency of water. These indices are calculated using the pH, alkalinity, calcium hardness, temperature, and total dissolved solids of the water, and indicate whether the water will favor precipitating or dissolving of calcium carbonate.

Methods used to control scale formation include

- Limit the concentration of scale-forming minerals by controlling cycles of concentration or by removing the minerals before they enter the system (see the section on External Treatments later in this section). **Cycles of concentration** is the ratio of makeup rate to the sum of blowdown and drift rates. The cycles of concentration can be monitored by calculating the ratio of chloride ion, which is highly soluble, in the system water to that in the makeup water.
- Make mechanical changes in the system to reduce the chances for scale formation. Increased water flow and exchangers with larger surface areas are examples.
- Feed acid to keep the common scale forming minerals (such as calcium carbonate) dissolved.

- Treat with chemicals designed to prevent scale. Chemical scale inhibitors work by the following mechanisms:
  1. **Threshold inhibition chemicals** prevent scale formation by keeping the scale forming minerals in solution and not allowing a deposit to form. Threshold inhibitors include organic phosphates, polyphosphates, and polymeric compounds.
  2. **Scale conditioners** modify the crystal structure of scale, creating a bulky, transportable sludge instead of a hard deposit. Scale conditioners include lignins, tannins, and polymeric compounds.

Nonchemical Methods

Equipment based on magnetic, electromagnetic, or electrostatic technology has been used for scale control in boiler water, cooling water, and other process applications.

Magnetic systems are designed to cause scale-forming minerals to precipitate in a low-temperature area away from heat exchanger surfaces, thus producing nonadherent particles (e.g., aragonite form of calcium carbonate versus the hard, adherent calcite form). The precipitated particles can then be removed by blowdown, mechanical means, or physical flushing. The effectiveness of a magnetic system can be diminished by a low ratio of dissolved calcium to silica, by the presence of excessive iron in the water, or if it is installed in close proximity to high-voltage power lines.

The objective of **electrostatics** is to prevent scale-forming reactions by imposing a surface charge on dissolved ions that causes them to repel.

Results of side-by-side comparative tests with conventional water treatment have been mixed. A Federal Technology Alert report regarding these technologies (DOE 1998) stresses that success of the application depends largely on the experience of the installer. The report includes a discussion of the potential benefits achieved and the necessary precautions to consider when applying these systems.

External Treatments

Minerals may also be removed by various external pretreatment methods such as reverse osmosis and ion exchange. Zeolite softening, demineralization, and dealkalization are examples of ion exchange processes.

BIOLOGICAL GROWTH CONTROL

Biological growth (algae, bacteria, and fungi) can interfere with a cooling operation due to fouling or corrosion, and may present a health hazard if present in aerosols produced by the equipment. Heating equipment operates above normal biological limits and therefore has fewer microbial problems. When considering biological growth in a cooling system, it is important to distinguish between free-living planktonic organisms and sessile (attached) organisms. Sessile organisms cause the majority of the problems, though they may have entered and multiplied as planktonic organisms.

Biological fouling can be caused by a wide variety of organisms that produce biofilm and slime masses. Slimes can be formed by bacteria, algae, yeasts, or molds and frequently consist of a mixture of these organisms combined with organic and inorganic debris. Organisms such as barnacles and mussels may cause fouling when river, estuarine, or sea water is used. Biological fouling can significantly reduce the efficiency of cooling by reducing heat transfer, increasing back pressure on recirculation pumps, disrupting flow patterns over cooling media, plugging heat exchangers, and blocking distribution systems. In extreme cases the additional mass of slime has caused the cooling media to collapse.

Microorganisms can dramatically enhance, accelerate or, in some cases, initiate localized corrosion (pitting). Microorganisms can influence localized corrosion directly by their metabolism or indirectly by the deposits they form. Indirect influence may not be mediated by simply killing the microorganisms; deposit removal is usually necessary, while direct influence can be substantially mediated by inhibiting microorganism metabolism.

Algae use energy from the sun to convert bicarbonate or carbon dioxide into biomass. Masses of algae can block piping, distribution holes, and nozzles. A distribution deck cover, which drastically reduces the sunlight reaching the algae, is one of the most cost-effective control devices for a cooling tower. Biocides are also used to assist in the control of algae.

Algae can also provide nutrients for other microorganisms in the cooling system, increasing the biomass in the water. Bacteria can grow in systems even when nutrient levels are relatively low. Yeasts and fungi are much slower growing than bacteria and find it difficult to compete in bulk waters for the available food. Fungi do thrive in partially wetted and high humidity areas such as the cooling media. Wood-destroying species of fungi can be a major concern for wooden cooling towers as the fungi consume the cellulose and/or lignin in the wood, reducing its structural integrity.

Most waters contain organisms capable of producing biological slime, but optimal conditions for growth are poorly understood. Equipment near nutrient sources or that has process leaks acting as a food source is particularly susceptible to slime formation. Even a thin layer of biofilm significantly reduces heat transfer rates in exchangers.

Control Measures

Eliminating sunlight from wetted surfaces such as distribution troughs, cooling media, and sumps significantly reduces algae growth. Eliminating deadlegs and low flow areas in the piping and the cooling loop reduces biological growth in those areas. Careful selection of materials of construction can remove nutrient sources and environmental niches for growth. Maintaining a high quality makeup water supply with low bacteria counts also helps minimize biological growth. Equipment should also be designed with adequate access for inspection, sampling, and manual cleaning.

Sometimes the effective control of slime and algae requires a combination of mechanical and chemical treatments. For example, when a system already contains a considerable accumulation of slime, a preliminary mechanical cleaning makes the subsequent application of a biocidal chemical more effective in killing the growth and more effective in preventing further growth. A build-up of scale deposits, corrosion product, and sediment in a cooling system also reduces the effectiveness of chemical biocides. Routine manual cleaning of cooling towers, including the use of high level chlorination and a biodispersant (surfactant), helps control Legionella bacteria as well as other microorganisms.

Microbiocides. Chemical biocides used to control biological growth in cooling systems fall into two broad categories: oxidizing and nonoxidizing biocides.

Oxidizing biocides (chlorine, chlorine-yielding compounds, bromine, bromochlorodimethylhydantoin (BCDMH, or BCD), ozone, iodine, and chlorine dioxide) are among the most effective microbiocidal chemicals. However, they are not always appropriate for control in cooling systems with a high organic loading. In air washers, the odor may become offensive; in wooden cooling towers, excessive concentrations of oxidizing biocides can cause delignification; and overdosing of oxidizing biocides may cause corrosion of metallic components. In systems large enough to justify the cost of equipment to control feeding of oxidizing biocides accurately, the application may be safe and economical. The most effective use of oxidizing biocides is to maintain a constant low level residual in the system. However, if halogen-based oxidizing biocides are fed intermittently (slug dosed), a pH near 7 is advantageous because, at this neutral pH, halogens are present as the hypohalous acid (HOR, where R represents the halogen) form over the hypohalous ion (OR^-) form. The effectiveness of this shock feeding is enhanced due to the faster killing action of hypohalous acid over that of hypohalous ion. The residual biocide concentration should be tested, using a field test kit, on a routine basis. Most halogenation programs can benefit from the use of dispersants or surfactants (chlorine helpers) to break up microbiological masses.

Chlorine has been the oxidizing biocide of choice for many years, either as chlorine gas or in the liquid form as sodium hypochlorite. Other forms of chlorine, such as powders or pellets, are also available. Use of chlorine gas is declining due to the health and safety concerns involved in handling this material and in part due to environmental pressures concerning the formation of chloramines and trihalomethane.

Bromine is produced either by the reaction of sodium hypochlorite with sodium bromide on site, or by release from pellets. Bromine has certain advantages over chlorine: it is less volatile, and bromamines break down more rapidly than chloramines in the environment. Also, when slug feeding biocide in high pH systems, hypobromous acid may have an advantage because its dissociation constant is lower than that of chlorine. This effect is less important when biocides are fed continuously.

Ozone has several advantages compared to chlorine: it does not produce chloramines or trihalomethane, it breaks down to nontoxic compounds rapidly in the environment, it controls biofilm better, and it requires significantly less chemical handling. The use of ozone-generating equipment in an enclosed space however, requires care be taken to protect operators from the toxic gas. Also, research by ASHRAE has shown that ozone is only marginally effective as a scale and corrosion inhibitor (Gan et al. 1996, Nasrazadani and Chao 1996).

Water conditions should be reviewed to determine the need for scale and corrosion inhibitors and then, as with all oxidizing biocides, inhibitor chemicals should be carefully selected to ensure compatibility. To maximize the biocidal performance of the ozone, the injection equipment should be designed to provide adequate contact of the ozone with the circulating water. In larger systems, care should be taken to ensure that the ozone is not depleted before the water has circulated through the entire system.

Iodine is provided in pelletized form, often from a rechargeable cartridge. Iodine is a relatively expensive chemical for use on cooling towers and is probably only suitable for use on smaller systems.

Nonoxidizing Biocides. When selecting a nonoxidizing microbiocide, the pH of the circulating water and the chemical compatibility with the corrosion and/or scale inhibitor product must be considered. The following list, while not exhaustive, identifies some of these products:

- Quaternary ammonium compounds
- Methylene bis(thiocyanate) (MBT)
- Isothiazolones
- Thiadiazine thione
- Dithiocarbamates
- Decyl thioethanamine (DTEA)
- Glutaraldehyde
- Dodecylguanidine
- Benzotriazole
- Tetrakis(hydroxymethyl)phosphonium sulfate (THPS)
- Dibromo-nitrilopropionamide (DBNPA)
- Bromo-nitropropane-diol
- Bromo-nitrostyrene (BNS)
- Proprietary blends

The manner in which nonoxidizing biocides are fed is important. Sometimes the continuous feeding of low dosages is neither effective nor economical. Slug feeding large concentrations to achieve a toxic level of the chemical in the water for a sufficient time to kill the organisms present can show better results. Water blowdown rate and biocide hydrolysis (chemical degradation) rate affect the required dosage. The hydrolysis rate of the biocide is affected by the type of biocide, along with the temperature and pH of the system

water. Dosage rates are proportional to system volume; dosage concentrations should be sufficient to ensure that the contact time of the biocide is long enough to obtain a high kill rate of microorganisms before the minimum inhibitory concentration of the biocide is reached. The period between nonoxidizing biocide additions should be based on the system half life, with sequential additions timed to prevent regrowth of bacteria in the water.

Handling Microbiocides. All microbiocides must be handled with care to ensure personal safety. In the United States, cooling water microbiocides are approved and regulated through the EPA and, by law, must be handled in accordance with labeled instructions. Maintenance staff handling the biocides should read the material safety data sheets and be provided with all the appropriate safety equipment to handle the substance. Automatic feed systems should be used that minimize and eliminate the handling of biocides by maintenance personnel.

Other Biocides. Ultraviolet irradiation deactivates the microorganisms as the water passes through a quartz tube. The intensity of the light and thorough contact with the water are critical in obtaining a satisfactory kill of microorganisms. Suspended solids in the water or deposits on the quartz tube significantly reduce the effectiveness of this treatment method. Therefore, a filter is often installed upstream of the lamp to minimize these problems. Because the ultraviolet light leaves no residual material in the water, sessile organisms and organisms that do not pass the light source are not affected by the ultraviolet treatment. Ultraviolet irradiation may be effective on humidifiers and air washers where the application of biocidal chemicals is unacceptable and where 100% of the recirculating water passes the lamp. Ultraviolet irradiation is less effective where all the microorganisms cannot be exposed to the treatment, such as in cooling towers. Ultraviolet lamps require replacement after approximately every 8000 h of operation.

Metallic ions, namely copper and silver, effectively control microbial populations under very specific circumstances. Either singularly or in combination, copper and silver ions are released into the water via electrochemical means to generate 1 to 2 ppm of copper and/or 0.5 to 1.0 ppm of silver. The ions assist in the control of bacterial populations in the presence of a free chlorine residual of at least 0.2 ppm. Copper, in particular, effectively controls algae.

Liu et. al. (1994) reported control of *Legionella pneumophila* bacteria in a hospital hot water supply using copper-silver ionization. In this case, *Legionella* colonization decreased significantly when copper and silver concentrations exceeded 0.4 and 0.04 ppm, respectively. Also, residual disinfection prevented *Legionella* colonization for two months after the copper-silver unit was inactivated.

Significant limitations exist in the use of copper and silver ion for cooling systems. Many states are restricting the discharge of these ions to surface waters, and if the pH of the system water rises above 7.8, the efficacy of the treatment is significantly reduced. Systems that have steel or aluminum heat exchangers should not be treated by this method, as the potential for the deposition of the copper ion and subsequent galvanic corrosion is significant.

Legionnaires' Disease

Like other living things, *Legionella pneumophila*, the bacterium that causes Legionnaires' disease (legionellosis), requires moisture for survival. *Legionella* bacteria are widely distributed in natural water systems and are present in many drinking water supplies. Potable hot water systems between 80 and 120°F, cooling towers, certain types of humidifiers, evaporative condensers, whirlpools and spas, and the various components of air conditioners are considered to be amplifiers. These bacteria are killed in a matter of minutes when exposed to temperatures above 140°F.

Legionellosis can be acquired by inhalation of *Legionella* organisms in aerosols. Aerosols can be produced by cooling towers, evaporative condensers, decorative fountains, showers, and misters. It has been reported that the aerosol from cooling towers can be transmitted over a distance of up to 2 miles. If air inlet ducts of nearby air conditioners draw the aerosol from contaminated cooling towers into the building, the air distribution system itself can transmit the disease. When an outbreak of Legionnaires' disease occurs, cooling towers are often the suspected source. However, other water systems may produce an aerosol and should not be neglected. Amplification of *Legionella* within protozoans has been demonstrated, and *Legionella* bacteria are thought to be protected from biocides while growing intracellularly. Amplification of *Legionella* bacteria in biofilm and slime masses has been shown by a number of researchers. Microbial control programs should consider the effectiveness of the products against slimes as part of the *Legionella* control program.

Humidifiers. Units that generate a water aerosol for humidity control can become amplifiers of bacteria if their reservoirs are poorly maintained. Manufacturers' recommendations on cleaning and maintenance should be followed closely. Steam humidifiers should not pose a bacterial problem because the steam does not contain bacteria.

Whirlpools and Spas. These units pose a potential hazard to users if not properly maintained. The complexity of some units makes cleaning difficult, so a firm that specializes in cleaning these devices may need to be hired. Manufacturers' instructions should be followed carefully.

Decorative Fountains. If these systems become contaminated, *Legionella* may multiply and pose a hazard to those nearby. The recirculating water should be kept clean and clear with proper filtration. Continuous chlorination or use of another biocide with EPA approval for decorative fountains is recommended. Indoor decorative fountains with ponds containing fish and other aquatic life cannot be chlorinated or treated; steps should be taken to ensure that water does not become airborne via sprays or cascading waterfalls.

Roof Ponds. These are used to lower the roof temperature and thus the demand on the air conditioning. If contaminated, such ponds can pose a risk to the staff and general public. Roof ponds should be monitored and treated in the same manner as recirculating cooling water systems (see the section on Selection of Water Treatment).

Safety Showers. Safety showers are mandated for safety, but they are used infrequently. Standing reservoirs of nonsterile water at room temperature should be avoided. A preventative maintenance program that includes flushing is suggested.

Vegetable Misters or Sprays. Although an older-style ultrasonic mist machine has been implicated in at least one case of this disease, once-through cool-water sprayers, (directly connected to a cold, potable water line with no reservoir) do not appear to be a problem.

Machine Shop Cooling. Water used in machining operations may be mixed with oil or other ingredients to improve cooling and cutting efficiency. It is typically stored in holding tanks that are subject to bacterial contamination, including by *Legionella*. Bacterial activity should be monitored and controlled.

Ice-Making Machines. While *Legionella* bacteria are not likely to amplify in this cold environment, bacterial amplification can occur prior to freezing the water. Graman et al. (1997) described a case in which *Legionella* were presumed to have been transmitted by aspiration of ice or ice water from an ice machine.

Prevention and Control. The *Legionella* count required to cause illness has not been firmly established because many factors are involved, including (1) virulence and number of *Legionella* in the air, (2) rate at which the aerosol dries, (3) wind direction, and (4) susceptibility to the disease of the person breathing the air. The organism is often found in sites not associated with an outbreak of the disease. It has been shown that it is feasible to operate cooling systems with *Legionella* bacteria below the limit of detection and that the only method to prove that a system is operating at these levels is to specifically test for *Legionella* bacteria, rather than to infer from total bacteria count measurements.

Periodic monitoring of circulating water for total bacteria count and *Legionella* count can be accomplished using culture methods. However, routine culturing of samples from building water systems may not yield an accurate prediction of the risk of *Legionella* transmission. Monitoring system cleanliness and using a microbial control agent that has proven efficacy or is generally regarded as effective in controlling *Legionella* populations are also important. Other measures to decrease risk include optimizing cooling tower design to minimize drift, eliminating deadlegs or low flow areas, selecting materials that do not promote the growth of *Legionella*, and locating the tower so that drift is not injected into the air handlers. The *Legionellosis Position Statement*, an ASHRAE Position Paper (1998) has further information on this topic.

SUSPENDED SOLIDS AND DEPOSITATION CONTROL

Mechanical Filtration

Strainers, filters, and separators may be used to reduce suspended solids to an acceptable low level. Generally, if the screen is 200 mesh, equivalent to about 0.003 in., it is called a strainer; if it is finer than 200 mesh, it is called a filter.

Strainers. A strainer is a closed vessel with a cleanable screen designed to remove and retain foreign particles down to 0.001 in. diameter from various flowing fluids. Strainers extract material that is not wanted in the fluid, and allow saving the extracted product if it is valuable. Strainers are available as single-basket or duplex units, manual or automatic cleaning units, and may be made of cast iron, bronze, stainless steel, copper-nickel alloys, or plastic. Magnetic inserts are available where microscopic iron or steel particles are present in the fluid.

Cartridge Filters. These are typically used as final filters to remove nearly all suspended particles from about 0.004 in. down to 0.00004 in. or less. Cartridge filters are typically disposable (i.e., once plugged, they must be replaced). The frequency of replacement, and thus the economical feasibility of their use, depends on the concentration of suspended solids in the fluid, the size of the smallest particles to be removed, and the removal efficiency of the cartridge filter selected.

In general, cartridge filters are favored in systems where contamination levels are less than 0.01% by mass (< 100 ppm). They are available in many different materials of construction and configurations. Filter media materials include yarns, felts, papers, nonwoven materials, resin-bonded fabric, woven wire cloths, sintered metal, and ceramic structures. The standard configuration is a cylinder with an overall length of approximately 10 in., an outside diameter of approximately 2.5 to 2.75 in., and an inside diameter of about 1 to 1.5 in., where the filtered fluid collects in the perforated internal core. Overall lengths from 4 to 40 in. are readily available.

Cartridges made of yarns, resin-bonded, or melt-blown fibers normally have a structure that increases in density towards the center. These depth-type filters capture particles throughout the total media thickness. Thin media, such as pleated paper (membrane types), have a narrow pore size distribution design to capture particles at or near the surface of the filter. Surface-type filters can normally handle higher flow rates and provide higher removal efficiency than equivalent depth filters. Cartridge filters are rated according to manufacturers' guidelines. Surface-type filters have an absolute rating, while depth-type filters have a nominal rating that reflects their general classification function. Higher efficiency melt-blown depth filters are available with absolute ratings as needed.

Sand Filters. A downflow filter is used to remove suspended solids from a water stream. The degree of suspended solids removal depends on the combinations and grades of the medium being used in the vessel. During the filtration mode, water enters the top of the filter vessel. After passing through a flow impingement plate, it enters the quiescent (calm) freeboard area above the medium.

In multimedia downflow vessels, various grain sizes and types of media are used to filter the water. This design increases the suspended solids holding capacity of the system, which in turn increases the backwashing interval. Multimedia vessels might also be used for low suspended solids applications, where chemical additives are required. In the multimedia vessel, the fluid enters the top layer of anthracite media, which has an effective size of 0.04 in. This relatively coarse layer removes the larger suspended particles, a substantial portion of the smaller particles, and small quantities of free oil. Flow continues down through the next layer of fine garnet material, which has an effective size of 0.012 in. A more finely divided range of suspended solids is removed in this polishing layer. The fluid continues into the final layer, a coarse garnet material that has an effective size of 0.08 in. Contained in this layer is the header/lateral assembly that collects the filtered water.

When the vessel has retained enough suspended solids to develop a substantial pressure drop, the unit must be backwashed either manually or automatically by reversing the direction of flow. This operation removes the accumulated solids out through the top of the vessel.

Centrifugal-Gravity Separators. In this type of separator, liquids/solids enter the unit tangentially, which sets up a circular flow. Liquids/solids are drawn through tangential slots and accelerated into the separation chamber. Centrifugal action tosses the particles heavier than the liquid to the perimeter of the separation chamber. Solids gently drop along the perimeter and into the separator's quiescent collection chamber. Solids-free liquid is drawn into the separator's vortex (low-pressure area) and up through the separator's outlet. Solids are either purged periodically or continuously bled from the separator by either a manual or automatic valve system.

Bag Type Filters. These filters are composed of a bag of mesh or felt supported by a removable perforated metal basket, placed in a closed housing with an inlet and outlet. The housing is a welded, tubular pressure vessel with a hinged cover on top for access to the bag and basket. Housings are made of carbon or stainless steel. The inlet can be in the cover, in the side (above the bag), or in the bottom (and internally piped to the bag). The side inlet is the simplest type. In any case, the liquid enters the top of the bag. The outlet is located at the bottom of the side (below the bag). Pipe connections can be threaded or flanged. Single-basket housings can handle up to 220 gpm, multibaskets up to 3500 gpm.

The support basket is usually of 304 stainless steel perforated with 1/8-in. holes. (Heavy wire mesh baskets also exist.) The baskets can be lined with fine wire mesh and used by themselves as strainers, without adding a filter bag. Some manufacturers offer a second, inner basket (and bag) that fits inside the primary basket. This provides for two-stage filtering: first a coarse filtering stage, then a finer one. The benefits are longer service time and possible elimination of a second housing to accomplish the same function.

The filter bags are made of many materials (cotton, nylon, polypropylene, and polyester) with a range of ratings from 0.00004 to 0.033 in. Most common are felted materials because of their depth-filtering quality, which provides high dirt-loading capability, and their fine pores. Mesh bags are generally coarser, but are reusable and, therefore, less costly. The bags have a metal ring sewn into their opening; this holds the bag open and seats it on top of the basket rim.

In operation, the liquid enters the bag from above, flows out through the basket, and exits the housing cleaned of particulate down to the desired size. The contaminant is trapped inside the bag, making it easy to remove without spilling any downstream.

Special Methods. Localized areas frequently can be protected by special methods. Thus, pump-packing glands or mechanical shaft seals can be protected by fresh water makeup or by circulating water from the pump casing through a cyclone separator or filter, then into the lubricating chamber.

In smaller equipment, a good dirt-control measure is to install backflush connections and shutoff valves on all condensers and heat exchangers so that accumulated settled dirt can be removed by backflushing with makeup water or detergent solutions. These connections can also be used for acid cleaning to remove calcium carbonate scale.

In specifying filtration systems, third-party testing by a qualified university or private test agency should be requested. The test report documentation should include a description of methods, piping diagrams, performance data, and certification.

Filters described in this section may also be used where industrial process cooling water is involved. For this type of service, consultation with the filtration equipment manufacturer is essential to ensure proper application.

START-UP AND SHUTDOWN OF COOLING TOWER SYSTEMS

The following guidelines are for start-up (or recommissioning) and shutdown of cooling tower systems:

Start-Up and Recommissioning for Drained Systems

- Clean all debris, such as leaves and dirt from the cooling tower.
- Close building air intakes in the area of the cooling tower to prevent entrainment of biocide and biological aerosols in the building air handling systems.
- Fill the system with water. While operating the condensing water pump(s) and *prior to operating the cooling tower fans*, execute one of the following two biocidal treatment programs:

  1. Resume treatment with the biocide that had been used prior to shutdown. Use the services of the water treatment supplier. Maintain the maximum recommended biocide residual (for the specific biocide) for a period sufficient to bring the system under good biological control (residual and time varies with the biocide).

  2. Treat the system with sodium hypochlorite at a level of 4 to 5 ppm free chlorine residual at a pH of 7.0 to 7.6. The residual level of free chlorine must be held at 4 to 5 ppm for 6 h. Commercially available test kits can be used to measure the residual of free chlorine.

- Once one of the two biocidal treatment programs has been successfully completed, turn on the fan and the put the system in service. The standard water treatment program (including biocidal treatment) should be resumed at this time.

Start-Up and Recommissioning for Undrained (Stagnant) Systems

- Remove accessible solid debris from bulk water storage vessel.
- Close building air intakes in the area of the cooling tower to prevent entrainment of biocide and biological aerosols in the building air handlers.
- Perform one of the two biocide pretreatment procedures (described in the section on Start-Up for Drained Systems) directly to the bulk water storage vessel (cooling tower sump, drain-down tank, etc.). *Do not circulate stagnant bulk cooling water over cooling tower fill or operate cooling tower fans during pretreatment.*
- Stagnant cooling water may be circulated with condenser water pumps if tower fill is bypassed. Otherwise, add approved biocide directly to the bulk water source and mix with manual or sidestream flow methods. Take care to prevent the creation of aerosol spray from the stagnant cooling water from any point in the cooling water system.
- When one of the two biocidal pretreatments has been successfully completed, the cooling water should be circulated over the tower fill. If biocide residual is maintained at a satisfactory level for at least 6 h, the cooling tower fans may then be operated safely.

Shutdown

When the system is to be shut down for an extended period, the entire system (cooling tower, system piping, heat exchangers, etc.) should be flushed and drained using the following procedure:

- Add a dispersant and biocide to the system and recirculate for 12 to 24 h. Confer with a water treatment consultant for suitable chemicals and dosage levels.
- Shut down pumps and completely drain all water distribution piping and headers, as well as the cooling loop. Remove water and debris from dead heads and low areas in the piping, which may not have completely drained.
- Rinse silt and debris from the sump. Pay special attention to corners and crevices. Add a mild solution of detergent and disinfectant to the sump and rinse. If the sump does not completely drain, pump out the remaining water and residue.
- If the equipment cannot be completely drained and is exposed to cold temperatures, freeze protection may be required.

SELECTION OF WATER TREATMENT

As discussed in the previous sections, many methods are available to prevent or correct water-caused problems. The selection of the proper water treatment method, and the chemicals and equipment necessary to apply that method, depends on many factors. The chemical characteristics of the water, which change with the operation of the equipment, are important. Other factors contributing to the selection of proper water treatment are

- Economics
- Chemistry control mechanisms
- Dynamics of the operating system
- Design of major components (e.g., the cooling tower or boiler)
- Number of operators available
- Training and qualifications of personnel
- Preventive maintenance program

Once-Through Systems

Economics is an overriding concern in treating water for once-through systems (in which a very large volume of water passes through the system only once). Protection can be obtained with relatively little treatment per unit mass of water because the water does not change significantly in composition while passing through equipment. However, the quantity of water to be treated is usually so large that any treatment other than simple filtration or the addition of a few parts per million of a polyphosphate, silicate, or other inexpensive chemical may not be practical or affordable. Intermittent treatment with polyelectrolytes can help maintain clean conditions when the cooling water is sediment-laden. In such systems, it is generally less expensive to invest more in corrosion-resistant construction materials than to attempt to treat the water.

Open Recirculating Systems

In an open recirculating system with chemical treatment, more chemical must be present because the water composition changes significantly by evaporation. Corrosive and scaling constituents are concentrated. However, treatment chemicals also concentrate by evaporation; therefore, after the initial dosage, only moderate dosages maintain the higher level of treatment needed. The selection of a water treatment program for an open recirculating system depends on the following major factors:

- Economics
- Water quality
- Performance criteria (e.g., corrosion rate, bacteria count, etc.)
- System metallurgy
- Available staffing
- Automation capabilities
- Environmental requirements

- Water treatment supplier (some technologies are superior to others in terms of economics, ease of use, safety, and impact on the environment)

An open recirculating system is typically treated with a scale inhibitor, corrosion inhibitor, oxidizing biocide, nonoxidizing biocide, and possibly a dispersant. The exact treatment program depends on the previously mentioned conditions.

A water treatment control scheme for a cooling tower might include

- Chemistry and cycles of concentration control using a conductivity controller
- Alkalinity control using automatic injection of sulfuric acid based on pH
- Scale control using contacting water meters, proportional feed, or traced control technology
- Oxidizing biocide control using an ORP (oxidation-reduction potential) controller
- Nonoxidizing biocide control using timers and pump systems

Air Washers and Sprayed Coil Units

A water treatment program for an air washer or a sprayed coil unit is usually complex and depends on the purpose and function of the system. Some systems, such as sprayed coils in office buildings, are used primarily to control the temperature and humidity. Other systems are intended to remove dust, oil vapor, and other airborne contaminants from an airstream. Unless the water is properly treated, the fouling characteristics of the contaminants removed from the air can cause operating problems.

Scale control is important in air washers or sprayed coils providing humidification. The minerals in the water may become concentrated (by evaporation) enough to cause problems. Inhibitor/dispersant treatment commonly used in cooling towers are often used in air washers to control scale formation and corrosion.

Suitable dispersants and surfactants are often needed to control oil and dust removed from the airstream. The type of dispersant depends on the nature of the contaminant and the degree of contamination. For maximum operating efficiency, dispersants should produce minimal amounts of foam.

Control of slime and bacterial growth is also necessary in the treatment of air washers and sprayed coils. The potential for biological growth is enhanced, especially if the water contains contaminants that are nutrients for the microorganisms. Due to variations in conditions and applications of air-washing installations and the possibility of toxicity problems, individual treatment options should be discussed with water treatment experts before a program is chosen. All microbiocides applied in air washers must have specific regulatory approval.

Ice Machines

Lime scale formation, cloudy or "milky" ice, objectionable taste and odor, and sediment are the most frequently encountered water problems in ice machines. Lime scale formation is probably the most serious problem because it interferes with the harvest cycle by forming on the freezing surfaces, preventing the smooth release of ice from the surface to the harvest bin.

Scale is caused by dissolved minerals in the water. Water freezes in a pure state and dissolved minerals concentrate in the unfrozen water, and some eventually deposit on the machine's freezing surfaces as scale. As shown in Table 4, the probability of scale formation in an ice machine varies directly with the concentrations of carbonate and bicarbonate in the water circulating in the machine.

Dirty or scaled-up ice makers should be thoroughly cleaned before the water treatment program is started. Water distributor holes should be cleared, and all loose sediment and other material should be flushed from the system. Existing scale can be removed by circulating an acid solution through the system.

Table 4 Amount of Scale in an Ice Machine

Total Alkalinity as Bicarbonate, ppm	Hardness as Calcium Carbonate, ppm			
	0 to 49	50 to 99	100 to 199	200 and up
0 to 49	None	Very light	Very light	Very light
50 to 99	Very light	Moderate	Moderate	Moderate
100 to 199	Very light	Troublesome	Troublesome	Heavy
200 and up	Very light	Troublesome	Heavy	Very heavy

Slowly soluble food-grade polyphosphates inhibit scale formation on freezing surfaces during normal operation by keeping hardness in solution. Although polyphosphates can inhibit lime scale in the ice-making section and help prevent sludge deposits of loose particles of lime scale in the sump, they do not eliminate the soft, milky, or white ice caused by a high concentration of dissolved minerals in the water.

Even with proper chemical treatment, recirculating water having a mineral content above 500 to 1000 ppm cannot produce clear ice. Increasing bleed off or reducing the thickness of the ice slab or the size of the cubes may mitigate the problem. However, demineralizing or distillation equipment is usually needed to prevent white ice production.

Another problem frequently encountered with ice machines is objectionable taste or odor. When water containing a material having an offensive taste or odor is used in an ice machine, the taste or odor is trapped in the ice. An activated carbon filter on the makeup water line can remove the objectionable material from the water. Carbon filters must be serviced or replaced regularly to avoid organic buildup in the carbon bed.

Occasionally, slime growth causes an odor problem in an ice machine. This problem can be controlled by regularly cleaning the machine with a food-grade acid. If slime deposits persist, sterilization of the ice machine may be helpful.

Feedwater often contains suspended solids such as mud, rust, silt, and dirt. To remove these contaminants, a sediment filter of appropriate size can be installed in the feed lines.

Closed Recirculating Systems

In a closed recirculating system, water composition remains fairly constant with very little loss of either water or treatment chemical. Closed systems are often defined as those requiring less than 5% makeup per year. The need for water treatment in such systems (i.e., water heating, chilled water, combined cooling and heating, and closed loop condenser water systems) is often ignored based on the rationalization that the total amount of scale from the water initially filling the system would be insufficient to interfere significantly with heat transfer, and that corrosion would not be serious. However, leakage losses are common, and corrosion products can accumulate sufficiently to foul heat transfer surfaces. Therefore, all systems should be adequately treated to control corrosion. Systems with high makeup rates should be treated to control scale as well.

The selection of a treatment program for closed systems should consider the following factors:

• Economics
• System metallurgy
• Operating conditions
• Makeup rate
• System size

Possible treatment technologies include

• Buffered nitrite
• Molybdate
• Silicates

• Polyphosphates
• Oxygen scavengers
• Organic blends

Before new systems are treated, they must be cleaned and flushed. Grease, oil, construction dust, dirt, and mill scale are always present in varying degrees and must be removed from the metallic surfaces to ensure adequate heat transfer and to reduce the opportunity for localized corrosion. Detergent cleaners with organic dispersants are available for proper cleaning and preparation of new closed systems.

Water Heating Systems

Secondary and Low-Temperature. Closed, chilled-water systems that are converted to secondary water heating during winter and primary low-temperature water heating, both of which usually operate in the range of 140 to 250°F, require sufficient inhibitors to control corrosion to less than 0.005 in. per year. Ethylene glycol or propylene glycol may be used as antifreeze in secondary hot water systems. Such glycols are available commercially, with inhibitors such as sodium nitrite, potassium phosphate, and organic inhibitors for nonferrous metals added by the manufacturer. These require no further treatment, but softened water should be used for all filling and makeup requirements. Samples should be checked periodically to ensure that the inhibitor has not been depleted. Analytical services are available from the glycol manufacturers and others for this purpose.

Environment- and High-Temperature. Environment-temperature water heating systems (250 to 350°F) and high-temperature, high-pressure hot water systems (above 350°F) require careful consideration of treatment for corrosion and deposit control. Makeup water for such systems should be demineralized or softened to prevent scale deposits. For corrosion control, oxygen scavengers such as sodium sulfite can be added to remove dissolved oxygen.

Electrode boilers are sometimes used to supply low- or high-temperature hot water. Such systems use heat generated due to the electrical resistance of the water between electrodes. The conductivity of the recirculating water must be in a specific range depending on the voltage used. Treatment of this type of system for corrosion and deposit control varies. In some cases oil-based corrosion inhibitors that do not contribute to the conductivity of the recirculating water are used.

Brine Systems

Systems containing brine, a strong solution of sodium chloride or calcium chloride, must be treated to control corrosion and deposits. Sodium nitrite at a minimum 3000 ppm in calcium brines or 4000 ppm in sodium brines, and a pH between 7.0 and 8.5 should provide adequate protection. Organic inhibitors are available that may provide adequate protection where nitrites cannot be used. Molybdates should not be used with calcium brines because insoluble calcium molybdate will precipitate.

Boiler Systems

Many treatment methods are available for steam-producing boilers; the method selected depends on

• Makeup water quality
• Makeup water quantity (or percentage condensate return)
• Pretreatment equipment
• Boiler operating conditions
• Steam purity requirements
• Economics

Pretreatment for makeup water could consist of

• Water softeners (for removal of calcium and magnesium hardness)

- Deaerators (for removal of dissolved gases; especially oxygen and carbon dioxide)
- Dealkalizers (to remove alkalinity in systems with high makeup rates and high alkalinity makeup water)
- Demineralizers (to remove almost all the hardness, alkalinity, and solids, depending on the application)
- Reverse osmosis (can remove up to 99% of minerals and dissolved solids from the water leaving almost pure H_2O)

Contaminants that are not removed mechanically by pretreatment equipment must be treated chemically. Once in the feedwater, dissolved gases, hardness, and dissolved minerals must be treated to prevent deposition and corrosion. ASME (1994) and ABMA (1982) have published recommended water chemistry limits for boiler water.

Treatment of the boiler water is often determined by the end use of the steam. Treatment regimens may include reduction of alkalinity; hardness removal; silica removal; oxygen reduction; and the feed of scale and corrosion inhibitors, oxygen scavengers, condensate treatment chemicals, and antifoams. These regimens are described in the following paragraphs.

Prevention of Scale. After the feedwater is pretreated, scale is controlled with phosphates, acrylates, polymers, chelates, and coagulation programs. Chelates, polymers, and acrylates work by binding the hardness, thereby preventing precipitation and scale formation. Phosphates and coagulation programs work in combination with sludge conditioners (tannins, lignins, starches, and synthetic polymers) to produce a softened precipitate that is removed by blowdown of the boiler.

Prevention of Corrosion and Oxygen Pitting. While boilers can corrode as the result of low boiler water pH or misuse of certain chemicals, corrosion is primarily caused by oxygen. After mechanical deaeration, boiler feedwater must be treated chemically. Effective deaeration removes most of the dissolved oxygen. (Most properly operating deaerators can remove oxygen down to 0.007 ppm and deaerating heaters can remove oxygen down to 0.04 ppm.) Oxygen scavengers such as catalyzed sodium sulfite and proprietary organic oxygen scavengers should then be fed to react with the residual oxygen in feedwater after deaeration. Oxygen scavengers will not only provide added protection to the boiler but to the steam and condensate system as well. Oxygen at levels as low as 0.005 ppm can cause oxygen pitting in the steam and condensate system if not chemically reduced by oxygen scavengers.

Most of the corrosion damage to boilers and associated equipment occurs during idle periods. The corrosion is caused by the exposure of wet metal to oxygen in the air or water. For this reason, special precautions must be taken to prevent corrosion while boilers are out of service.

Wet Boiler Lay-Up. This is a method of storing boilers full of water so that they can be returned to service. It involves adding extra chemicals (usually something to increase alkalinity, an oxygen scavenger, and a dispersant) to the boiler water. The water level is raised in the idle boiler to eliminate air spaces, and the boiler is kept completely full of treated water. Superheaters require special protection. Nitrogen gas can also be used on airtight boilers to maintain a positive pressure on the boiler, thereby preventing oxygen in-leakage.

Dry Boiler Lay-Up. This method of lay-up is usually for longer boiler outages. It involves draining, cleaning, and drying the boiler. A material that absorbs moisture, such as hydrated lime or silica gel, is placed in trays inside the boiler. The boiler is then sealed carefully to keep out air. Periodic inspection and replacement of the drying chemical are required during long storage periods.

Steam and Condensate Systems

Two problems associated with steam and condensate systems include general corrosion and pitting corrosion. In order to prevent condensate corrosion, the systems must be protected from acidic conditions, which lead to general corrosion, and oxygen, which leads to pitting corrosion. Protection can be by mechanical or chemical means or a combination of both. The following methods are commonly used for condensate system protection.

Protection from General Corrosion

Mechanical Protection. Reduce alkalinity from boiler feedwater to minimize the amount of carbon dioxide in the system. Carbon dioxide reacts with water (condensate) and forms carbonic acid that will corrode condensate system metal. Alkalinity can be reduced by dealkalization, demineralization, and reverse osmosis. Note: In many cases mechanical reduction of alkalinity is not needed due to low alkalinity makeup water and/or feedwater.

Chemical Protection. Use volatile amines, such as morpholine, diethylaminoethanol (DEAE), and cyclohexylamine, to neutralize carbonic acid and keep the condensate pH between 8.0 and 9.0.

Protection from Oxygen Corrosion

Mechanical Protection. Reduce oxygen from all boiler feedwater to prevent oxygen carryover to the steam and condensate system (via mechanical methods and chemical oxygen scavengers).

Chemical Protection. Feed filming amines, such as octadecylamine, to the steam to form a thin, hydrophobic film on the condensate system surfaces.

Chemical Protection. Feed a volatile oxygen scavenger to the steam to scavenge the oxygen.

The need for chemical treatment can be reduced by designing and maintaining tight return systems so that the condensate is returned to the boiler and less makeup is required in the boiler feedwater. The greater the amount of makeup, the more the system requires increased chemical treatment.

TERMINOLOGY

The following terms are commonly used in the water treatment industry as they pertain to corrosion, scale formation and fouling.

Alkalinity. The sum of bicarbonate, carbonate, and hydroxide ions in water. Other ions, such as borate, phosphate, or silicate, can also contribute to alkalinity.

Anion. A negatively charged ion of an electrolyte that migrates toward the anode influenced by an electric potential gradient.

Anode. The electrode of an electrolytic cell at which oxidation occurs.

Biological deposits. Water-formed deposits of biological organisms or the products of their life processes, such as barnacles, algae, or slimes.

Cathode. The electrode of an electrolytic cell at which reduction occurs.

Cation. A positively charged ion of an electrolyte that migrates toward the cathode influenced by an electric potential gradient.

Corrosion. The deterioration of a material, usually a metal, by reaction with its environment.

Corrosivity. The capacity of an environment or environmental factor to bring about destruction of a specific metal by the process of corrosion.

Electrolyte. A solution through which an electric current can flow.

Filtration. Process of passing a liquid through a porous material in such a manner as to remove suspended matter from the liquid.

Galvanic corrosion. Corrosion resulting from the contact of two dissimilar metals in an electrolyte or from the contact of two similar metals in an electrolyte of nonuniform concentration.

Hardness. The sum of the calcium and magnesium ions in water; usually expressed in ppm as $CaCO_3$.

Inhibitor. A chemical substance that reduces the rate of corrosion, scale formation, fouling, or slime production.

Ion. An electrically charged atom or group of atoms.

Passivity. The tendency of a metal to become inactive in a given environment.

pH. The logarithm of the reciprocal of the hydrogen ion concentration of a solution. pH values below 7 are increasingly acidic; those above 7 are increasingly alkaline.

Polarization. The deviation from the open circuit potential of an electrode resulting from the passage of current.

ppm. Parts per million by mass. In water, ppm are essentially the same as milligrams per liter (mg/L); 10,000 ppm (mg/L) = 1%.

Scale. 1. The formation at high temperature of thick corrosion product layers on a metal surface. 2. The precipitation of water-insoluble constituents on a surface.

Sludge. A sedimentary water-formed deposit, either of biological origin or suspended particles from the air.

Tuberculation. The formation over a surface of scattered, knob-like mounds of localized corrosion products.

Water-formed deposit. Any accumulation of insoluble material derived from water or formed by the reaction with water on surfaces in contact with it.

REFERENCES

ABMA. 1982. Boiler water limits and steam purity recommendations for water tube boilers, 3rd edition. American Boiler Manufacturers Association, Arlington, VA.

ASHRAE. 1998. Legionellosis position statement and Legionellosis position paper.

ASME. 1994. Consensus on operating practices for the control of feedwater and boiler water chemistry in modern industrial boilers. Research Committee on Water in Thermal Power Systems, Industrial Boiler Subcommittee. American Society of Mechanical Engineers, New York.

DOE. 1998. Non-chemical technologies for scale and hardness control. *Federal Technology Alert* DOE/EE-0162. U.S. Department of Energy. Web site: http://www.pnl.gov/fta/.

Gan, F., D.-T. Chin, and A. Meitz. 1996. Laboratory evaluation of ozone as a corrosion inhibitor for carbon steel, copper, and galvanized steel in cooling water. *ASHRAE Transactions* 102(1).

Graman, P.S., G.A. Quinlan, and J.A. Rank. 1997. Nosocomial legionellosis traced to a contaminated ice machine. *Infection Control and Hospital Epidemiology* 18(9):637-40.

Langelier, W.F. 1936. The analytical control of anticorrosion water treatment. *Journal of the American Water Works Association* 28:1500.

Liu, Z., J.E. Stout, L. Tedesco, M. Boldin, C. Hwang, W.F. Diven, and V.L. Yu. 1994. Controlled evaluation of copper-silver ionization in eradicating *Legionella pneumophila* from a hospital water distribution system. *The Journal of Infectious Diseases* 169:919-22.

Nasrazadani, S. and T.J. Chao. 1996. Laboratory evaluations of ozone as a scale inhibitor for use in open recirculating cooling systems. *ASHRAE Transactions* 102(2).

Ryznar, J.W. 1944. A new index for determining amount of calcium carbonate scale formed by a water. *Journal of the American Water Works Association* 36:472.

SERVICE WATER HEATING

A SERVICE water heating system has (1) a heat energy source, (2) heat transfer equipment, (3) a distribution system, and (4) terminal hot water usage devices.

Heat energy sources may be (1) fuel combustion, (2) electrical conversion, (3) solar energy, and/or (4) recovered waste heat from such sources as flue gases, ventilation and air-conditioning systems, refrigeration cycles, and process waste discharge.

Heat transfer equipment is either of the direct or indirect type. For direct equipment, heat is derived from combustion of fuel or direct conversion of electrical energy into heat and is applied within the water heating equipment. For indirect heat transfer equipment, heat energy is developed from remote heat sources, such as boilers, solar energy collection, cogeneration, refrigeration, or waste heat, and is then transferred to the water in a separate piece of equipment. Storage tanks may be part of or associated with either type of heat transfer equipment.

Distribution systems transport the hot water produced by the water heating equipment to terminal hot water usage devices. The water consumed must be replenished from the building water service main. For locations where constant supply temperatures are desired, circulation piping or a means of heat maintenance must be provided.

Terminal hot water usage devices are plumbing fixtures and equipment requiring hot water that may have periods of irregular flow, constant flow, and no flow. These patterns and their related water usage vary with different buildings, process applications, and personal preference.

In this chapter, it is assumed that an adequate supply of service water is available. If this is not the case, alternate strategies such as water accumulation, pressure control, and flow restoration should be considered.

SYSTEM PLANNING

Flow rate and temperature are the primary factors to be determined in the hydraulic and thermal design of a water heating and piping system. Operating pressures, time of delivery, and water quality are also factors to consider. Separate procedures are used to select water heating equipment and to design the piping system.

Water heating equipment, storage facilities, and piping should (1) have sufficient capacity to provide the required hot water while minimizing the waste of energy or water and (2) allow economical system installation, maintenance, and operation.

Water heating equipment types and designs are based on (1) the energy source, (2) the application of the developed energy to heating the water, and (3) the control method used to deliver the necessary hot water at the required temperature under varying water

The preparation of this chapter is assigned to TC 6.6, Service Water Heating.

demand conditions. Application of water heating equipment within the overall design of the hot water system is based on (1) location of the equipment within the system, (2) related temperature requirements, and (3) the volume of water to be used.

Energy Sources

The choice among available energy sources is interrelated with the choices among equipment types and locations. These decisions should be made only after evaluating purchase, installation, operating, and maintenance costs. A life-cycle analysis is highly recommended.

In making energy conservation choices, current editions of the following energy conservation guides should be consulted: the ANSI/ASHRAE/IES Standard 90 series or the sections on Service Water Heating of ANSI/ASHRAE/IESNA Standard 100 (see also the section on Design Considerations in this chapter).

WATER HEATING EQUIPMENT

Gas-Fired or Oil-Fired

Residential water heating equipment is usually the automatic storage type. For industrial and commercial applications, commonly used types of heaters are (1) automatic storage, (2) circulating tank, (3) instantaneous, and (4) hot water supply boilers.

Installation guidelines for gas-fired water heaters can be found in the National Fuel Gas Code, NFPA Standard 54 (ANSI Z223.1). This code also covers the sizing and installation of venting equipment and controls. Installation guidelines for oil-fired water heaters can be found in NFPA Standard 31, Installation of Oil-Burning Equipment (ANSI Z95.1).

Automatic storage water heaters incorporate the burner(s), storage tank, outer jacket, insulation, and controls in a single unit.

Circulating tank water heaters are classified in two types: (1) automatic, in which the thermostat is located in the water heater, and (2) nonautomatic, in which the thermostat is located within an associated storage tank.

Automatic instantaneous water heaters are produced in two distinctly different types. Tank-type instantaneous heaters have an input-to-storage capacity ratio of 4000 Btu/h per gallon or more and a thermostat to control energy input to the heater. Water-tube type instantaneous heaters have minimal water storage capacity. They usually have a flow switch that controls the burner. They may have a modulating fuel valve that varies fuel flow as water flow changes.

Hot water supply boilers are capable of providing service hot water. They are typically installed with separate storage tanks and applied as an alternative to circulating tank water heaters.

Direct vent models are available in nearly all types of water heating equipment. They are to be install indoors, but are not vented via a conventional chimney or gas vent, nor do they use ambient air for combustion. They must be installed with the means for venting (typically horizontal) and the means for supplying combustion air from outside the building that is supplied or specified by the equipment manufacturer.

Electric

Electric water heaters are generally of the automatic storage type, consisting of a tank with one or more immersion heating elements. The heating elements consist of resistance wire embedded in refractories having good heat conduction properties and electrical insulating values. Heating elements are fitted into a threaded or flanged mounting for insertion into a tank. Thermostats controlling heating elements may be of the immersion or surface-mounted type.

Residential storage tank water heaters range up to 120 gal with input up to 12 kW. They have a primary resistance heating element near the bottom and often a secondary element located in the upper portion of the tank. Each element is controlled by its own thermostat. In dual element heaters, the thermostats are usually interlocked so that the lower heating element cannot operate if the top element is operating. Thus, only one heating element operates at a time to limit the current draw.

Commercial storage tank water heaters are available in many combinations of element quantity, wattage, voltage, and storage capacity. Storage tanks may be horizontal or vertical. Compact, low-volume models are used in point-of-use applications to reduce hot water piping length. Location of the water heater near the point of use makes recirculation loops unnecessary.

Instantaneous electric water heaters are sometimes used in lavatory, hot tub, whirlpool bath, and swimming pool applications. The smaller sizes are commonly used as boosters for dishwasher final rinse applications.

Heat pump water heaters (HPWHs) use a vapor-compression refrigeration cycle to extract energy from an air or water source to heat water. Most HPWHs are air-to-water units. As the HPWH collects heat, it provides a potentially useful cooling effect and dehumidifies the air. HPWHs typically have a maximum output temperature of 140°F. Where a higher delivery temperature is required, a conventional storage-type or booster water heater downstream of the heat pump storage tank should be used. HPWH function most efficiently where the inlet water temperature is low and the entering air is warm and humid. Systems should be sized to allow high HPWH run time. The effect of HPWH cooling output on the building's energy balance should be considered. Cooling output should be directed to provide occupant comfort and avoid interfering with temperature-sensitive equipment (EPRI 1990).

Demand controlled water heating can significantly reduce the cost of heating water electrically. Demand controllers operate on the principle that a building's peak electrical demand exists for a short period, during which heated water can be supplied from storage rather than hot water recovery. Shifting the use of electricity for service water heating from peak demand periods allows water heating at the lowest electric energy cost in many electric rate schedules. The building electrical load must be detected and compared with peak demand data. When the load is below peak demand, the control device allows the water heater to operate. Some controllers can program deferred loads in steps as capacity is available. The priority sequence may involve each of several banks of elements in (1) a water heater, (2) multiple water heaters, or (3) water heating and other equipment having a deferrable load, such as pool heating and snow melting. When load controllers are used, hot water storage must be used.

Instantaneous and hot water supply boilers as described in the section on Gas-Fired or Oil-Fired water heating equipment are also available with electric heating elements.

Electric off-peak storage water heating is a water heating equipment load management strategy whereby electrical demand to a water heating system is time-controlled, primarily in relation to the building or utility electrical load profile. This approach may require an increase in tank storage capacity and/or stored water temperature to accommodate water use during peak periods.

Sizing recommendations in this chapter apply only to water heating without demand or off-peak control. When demand control devices are used, the storage and recovery rate may need to be increased to supply all the hot water needed during the peak period and during the ensuing recovery period. Manian and Chackeris (1974) include a detailed discussion on load-limited storage heating system design.

Indirect

In indirect water heating, the heating medium is steam, hot water, or another fluid that has been heated in a separate generator or boiler. The water heater extracts heat through an external or internal heat exchanger.

When the heating medium is at a higher pressure than the service water, the service water may be contaminated by leakage of the heating medium through a damaged heat transfer surface. In the United States, some national, state, and local codes require double-wall, vented tubing in indirect water heaters to reduce the possibility of cross-contamination. When the heating medium is at a lower pressure than the service water, other jurisdictions allow single-wall tubing heaters because any leak would be into the heating medium.

If the heating medium is steam, high rates of condensation occur, particularly when a sudden demand causes an inflow of cold water. The steam pipe and condensate return pipes should be of ample size. Condensate should drain by gravity, without lifts, to a vented condensate receiver located below the level of the heater. Otherwise, water hammer, reduced capacity, or heater damage may result. The condensate may be cooled by preheating the cold water supply to the heater.

Corrosion is minimized on the heating medium side of the heat exchanger because no makeup water, and hence no oxygen, is brought into that system. The metal temperature of the service water side of the heat exchanger is usually less than that in direct-fired water heaters. This minimizes scale formation from hard water.

Storage water heaters are designed for service conditions where hot water requirements are not constant, i.e., where a large volume of heated water is held in storage for periods of peak load. The amount of storage required depends on the nature of the load and the recovery capacity of the water heater. An individual tank or several tanks joined by a manifold may be used to provide the required storage.

External storage water heaters are designed for connection to a separate tank (Figure 1). The boiler water circulates through the heater shell, while service water from the storage tank circulates through the tubes and back to the tank. Circulating pumps are usually installed in both the boiler water piping circuit and the circuits

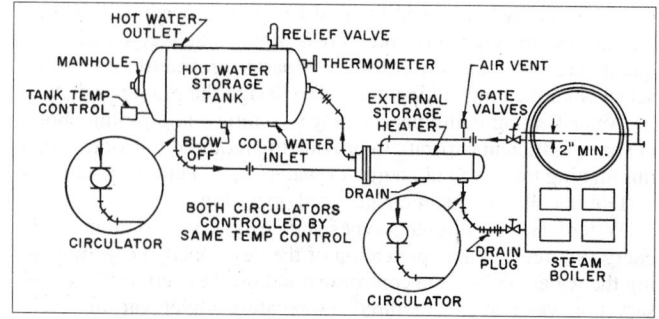

Fig. 1   Indirect, External Storage Water Heater

between the heat exchanger and the storage tank. Steam can also be used as the heating medium in a similar scheme.

Instantaneous indirect water heaters (tankless coils) are best used for a steady, continuous supply of hot water. In these units, the water is heated as it flows through the tubes. Because the heating medium flows through a shell, the ratio of hot water volume to heating medium volume is small. As a result, variable flow of the service water causes uncertain temperature control unless a thermostatic mixing valve is used to maintain the hot water supply to the plumbing fixtures at a more uniform temperature.

Some indirect instantaneous water heaters are located inside a boiler. The boiler is provided with a special opening through which the coil can be inserted. While the coil can be placed in the steam space above the water line of a steam boiler, it is usually placed below the water line. The water heater transfers heat from the boiler water to the service water. The gross output of the boiler must be sufficient to serve all loads.

Semi-Instantaneous

These water heaters have limited storage to meet the average momentary surges of hot water demand. They usually consist of a heating element and control assembly devised for close control of the temperature of the leaving hot water.

Circulating Tank

These water heaters are instantaneous or semi-instantaneous types used with a separate storage tank and a circulating pump. The storage acts as a flywheel to accommodate variations in the demand for hot water.

Blending Injection

These water heaters inject steam or hot water directly into the process or volume of water to be heated. This is often associated with point-of-use applications, e.g., certain types of commercial laundry, food, and process equipment. *Caution*: Cross-contamination of potable water is possible.

Solar

The availability of solar energy at the building site, the efficiency and cost of solar collectors, and the availability and cost of other fuels determine whether solar energy collection units should be used as a primary heat energy source. Solar energy equipment can also be included to supplement other energy sources and conserve fuel or electrical energy.

The basic elements of a solar water heater include solar collectors, a storage tank, piping, controls, and a transfer medium. The system may use natural convection or forced circulation. Auxiliary heat energy sources may be added, if needed.

Collector design must allow operation in below-freezing conditions, where applicable. Antifreeze solutions in a separate collector piping circuit arrangement are often used, as are systems that allow water to drain back to heated areas when low temperatures occur. Uniform flow distribution in the collector or bank of collectors and stratification in the storage tank are important for good system performance.

The application of solar water heaters depends on (1) auxiliary energy requirements; (2) collector orientation; (3) temperature of the cold water; (4) general site, climatic, and solar conditions; (5) installation requirements; (6) area of collectors; and (7) amount of storage. Chapter 32, Solar Energy Use, has more detailed design information.

Waste Heat Use

Waste heat recovery can reduce energy cost and the energy requirement of the building heating and service water heating equipment. Waste heat can be recovered from equipment or processes by using appropriate heat exchangers in the hot gaseous or liquid streams. Heat recovered is frequently used to preheat the water entering the service water heater. Refrigeration heat reclaim uses refrigerant-to-water heat exchangers connected to the refrigeration circuit between the compressor and condenser of a host refrigeration or air-conditioning system to extract heat. Water is heated only when the host is operating at the same time there is a call for water heating. A conventional water heater is typically required to augment the output of the heat reclaim and to provide hot water during periods when the host system is not in operation.

Refrigeration Heat Reclaim

These systems heat water with heat that would otherwise be rejected through a refrigeration condenser. Because many simple systems reclaim only superheat energy from the refrigerant, they are often called desuperheaters. However, some units are also designed to provide partial or full condensing. The refrigeration heat reclaim heat exchanger is generally of vented, double-wall construction to isolate potable water from refrigerant. Some heat reclaim devices are designed for use with multiple refrigerant circuits. Controls are required to limit high water temperature, prevent low condenser pressure, and provide for freeze protection. Refrigeration systems with higher run time and lower efficiency provide more heat reclaim potential. Most systems are designed with a preheat water storage tank connected in series with a conventional water heater (EPRI 1992). In all installations, care must be taken to prevent the inappropriate venting of refrigerants.

Combination Heating

A **combo system** is an integrated appliance that provides both space heating and domestic hot water. A space heating coil and a space cooling coil are often included with the air handler to provide year-round comfort. Combo systems also can use other types of heat exchangers for space heating, such as baseboard convectors or floor heating coils. A method of testing combo systems is contained in the ASHRAE *Standard* 124. The test procedures allow the calculation of a Combined Annual Efficiency (CAE), as well as space-heating and water-heating efficiency factors. Subherwal (1986), Pietsch and Talbert (1989), Talbert et al. (1992), Kweller (1992), and Pietsch et al. (1994) provide additional design information on these heaters.

DISTRIBUTION

Piping Material

Traditional piping materials have included galvanized steel used with galvanized cast iron or galvanized malleable iron screwed fittings. Copper piping and copper tube types K, L, or M have been used with brass, bronze, or wrought copper solder fittings. Legislation or plumbing code changes have banned the use of lead in solders or pipe-jointing compounds in potable water piping because of possible lead contamination of the water supply. See the current ASHRAE *Standard* 90 series for pipe insulation requirements.

Today, most potable water supplies require treatment before distribution; this may cause the water to become more corrosive. Therefore, depending on the water supply, traditional galvanized steel piping or copper tube may no longer be satisfactory, due to accelerated corrosion. Galvanized steel piping is particularly susceptible to corrosion (1) when hot water is between 140 and 180°F and (2) where repairs have been made using copper tube without a nonmetallic coupling.

Before selecting any water piping material or system, consult the local code authority. The local water supply authority should also be consulted about any history of water aggressiveness causing failures of any particular material.

Alternate piping materials that may be considered are (1) stainless steel tube and (2) various plastic piping and tubes. Particular

Table 1   Hot Water Demand in Fixture Units (140°F Water)

	Apartments	Club	Gymnasium	Hospital	Hotels and Dormitories	Industrial Plant	Office Building	School	YMCA
Basin, private lavatory	0.75	0.75	0.75	0.75	0.75	0.75	0.75	0.75	0.75
Basin, public lavatory	—	1	1	1	1	1	1	1	1
Bathtub	1.5	1.5	—	1.5	1.5	—	—	—	—
Dishwasher	1.5	Five fixture units per 250 seating capacity							
Therapeutic bath	—	—	—	5	—	—	—	—	—
Kitchen sink	0.75	1.5	—	3	1.5	3	—	0.75	3
Pantry sink	—	2.5	—	2.5	2.5	—	—	2.5	2.5
Service sink	1.5	2.5	—	2.5	2.5	2.5	2.5	2.5	2.5
Shower[a]	1.5	1.5	1.5	1.5	1.5	3.5	—	1.5	1.5
Circular wash fountain	—	2.5	2.5	2.5	—	4	—	2.5	2.5
Semicircular wash fountain	—	1.5	1.5	1.5	—	3	—	1.5	1.5

[a]In applications where the principal use is showers, as in gymnasiums or at end of shift in industrial plants, use conversion factor of 1.00 to obtain design water flow rate in gpm.

care must be taken to make sure that the application meets the design limitations set by the manufacturer and that the correct materials and methods of joining are used. These precautions are easily taken with new projects but become more difficult to control during repairs of existing work. The use of incompatible piping, fittings, and jointing methods or materials must be avoided, as they can cause severe problems.

Pipe Sizing

Sizing of hot water supply pipes involves the same principles as sizing of cold water supply pipes (see Chapter 33 of the 1997 *ASHRAE Handbook—Fundamentals*). The water distribution system must be correctly sized for the total hot water system to function properly. Hot water demand varies with the type of establishment, usage, occupancy, and time of day. The piping system should be capable of meeting peak demand at an acceptable pressure loss.

Supply Piping

Table 1, Figures 23 and 24, and manufacturers' specifications for fixtures and appliances can be used to determine hot water demands. These demands, together with procedures given in Chapter 33 of the 1997 *ASHRAE Handbook—Fundamentals*, are used to size the mains, branches, and risers.

Allowance for pressure drop through the heater should not be overlooked when sizing hot water distribution systems, particularly where instantaneous water heaters are used and where the available pressure is low.

Pressure Differential

Sizing of both cold and hot water piping requires that the pressure differential at the point of use of blended hot and cold water be kept to a minimum. This required minimum differential is particularly important for tubs and showers, since sudden changes in flow at fixtures cause discomfort and a possible scalding hazard. Pressure-compensating devices are available.

Return Piping

Return piping is commonly provided for hot water supply supplies in which it is desirable to have hot water available continuously at the fixtures. This includes cases where the hot water piping exceeds 100 ft.

The water circulation pump may be controlled by a thermostat (in the return line) set to start and stop the pump over an acceptable temperature range. This thermostat can significantly reduce both heat loss and pumping energy in some applications. An automatic time switch or other control should turn the water circulation off when hot water is not required. Because hot water is corrosive, circulating pumps should be made of corrosion-resistant material.

For small installations, a simplified pump sizing method is to allow 1 gpm for every 20 fixture units in the system, or to allow 0.5

gpm for each 3/4- or 1-in. riser; 1 gpm for each 1-1/4- or 1-1/2-in. riser; and 2 gpm for each riser 2 in. or larger.

Werden and Spielvogel (1969) and Dunn et al. (1959) cover heat loss calculations for large systems. For larger installations, piping heat losses become significant. A quick method to size the pump and return for larger systems is as follows:

1. Determine the total length of all hot water supply and return piping.
2. Choose an appropriate value for piping heat loss from Table 2 or other engineering data (usually supplied by insulation companies, etc.). Multiply this value by the total length of piping involved.

Table 2   Heat Loss of Pipe at 140°F Inlet, 70°F Ambient

Nominal Pipe Size, in.	Bare Copper Tubing, Btu/h·ft	1/2-in. Glass Fiber Insulated Copper Tubing, Btu/h·ft
3/4	30	17.7
1	38	20.3
1-1/4	45	23.4
1-1/2	53	25.4
2	66	29.6
2-1/2	80	33.8
3	94	39.5
4	120	48.4

A rough estimation can be made by multiplying the total length of covered pipe by 30 Btu/h·ft or uninsulated pipe by 60 Btu/h·ft. Table 2 gives actual heat losses in pipes at a service water temperature of 140°F and ambient temperature of 70°F. The values of 30 or 60 Btu/h·ft are only recommended for ease in calculation.

3. Determine pump capacity as follows:

$$Q_p = \frac{q}{60 \rho c_p \Delta t} \tag{1}$$

where

Q_p = pump capacity, gpm
q = heat loss, Btu/h
ρ = density of water = 8.25 lb/gal (120°F)
c_p = specific heat of water = 1 Btu/lb·°F
Δt = allowable temperature drop, °F

For a 20°F allowable temperature drop,

$$Q_p(\text{gpm}) = \frac{q}{60 \times 8.25 \times 1 \times 20} = \frac{q}{9900} \tag{2}$$

Caution: This calculation assumes that a 20°F temperature drop is acceptable at the last fixture.

4. Select a pump to provide the required flow rate, and obtain from the pump curves the pressure created at this flow.

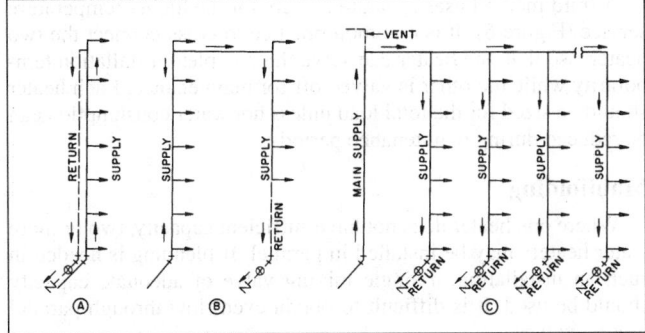

Fig. 2 Arrangements of Hot Water Circulation Lines

5. Multiply the head by 100 and divide by the total length of hot water return piping to determine the allowable friction loss per 100 ft of pipe.

6. Determine the required flow in each circulating loop, and size the hot water return pipe based on this flow and the allowable friction loss from Step 5.

Where multiple risers or horizontal loops are used, balancing valves with means of testing are recommended in the return lines. A swing-type check valve should be placed in each return to prevent entry of cold water or reversal of flow, particularly during periods of high hot water demand.

Three common methods of arranging circulation lines are shown in Figure 2. Although the diagrams apply to multistory buildings, arrangements (A) and (B) are also used in residential designs. In circulation systems, air venting, pressure drops through the heaters and storage tanks, balancing, and line losses should be considered. In Figures 2A and 2B, air is vented by connecting the circulating line below the top fixture supply. With this arrangement, air is eliminated from the system each time the top fixture is opened. Generally, for small installations, a NPS 1/2- or 3/4-in. hot water return is ample.

All storage tanks and piping on recirculating systems should be insulated as recommended by the ASHRAE *Standard* 90 series and *Standard* 100.

Heat Traced, Nonreturn Piping

In this system, the fixtures can be as remote as in the return piping. The hot water supply piping is heat traced with electric resistance heating cable preinstalled under the pipe insulation. Electrical energy input is self-regulated by the cable's construction to maintain the required water temperature at the fixtures. No return piping system or circulation pump is required.

Special Piping—Commercial Dishwashers

An adequate flow rate and pressure must be maintained for automatic dishwashers in commercial kitchens. To reduce operating difficulties, piping for automatic dishwashers should be installed according to the following recommendations:

1. The cold water feed line to the water heater should be no smaller than NPS 1.
2. The supply line that carries 180°F water from the water heater to the dishwasher should not be smaller than NPS 3/4.
3. No auxiliary feed lines should connect to the 180°F supply line.
4. A return line should be installed if the source of 180°F water is more than 5 ft from the dishwasher.
5. Forced circulation by a pump should be used if the water heater is installed on the same level as the dishwasher, if the length of return piping is more than 60 ft, or if the water lines are trapped.

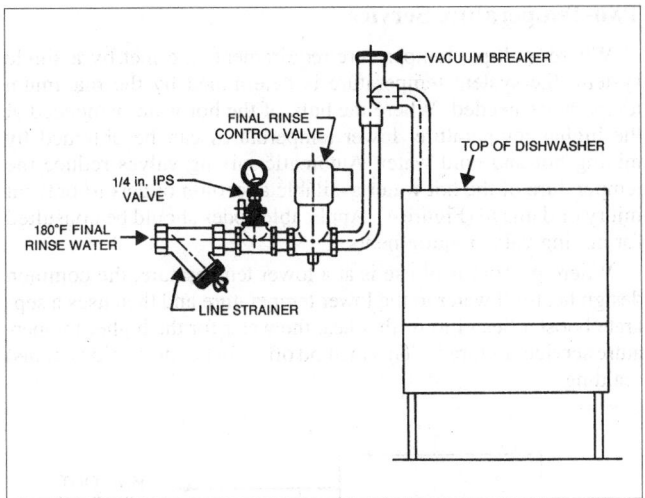

Fig. 3 National Sanitation Foundation (NSF) Plumbing Requirements for Commercial Dishwasher

6. If a circulating pump is used, it is generally installed in the return line. It may be controlled by (a) the dishwasher wash switch, (b) a manual switch located near the dishwasher, or (c) an immersion or strap-on thermostat located in the return line.
7. A pressure-reducing valve should be installed in the low-temperature supply line to a booster water heater, but external to a recirculating loop. It should be adjusted, with the water flowing, to the value stated by the washer manufacturer (typically 20 psi).
8. A check valve should be installed in the return circulating line.
9. If a check valve type of water meter or a backflow prevention device is installed in the cold water line ahead of the heater, it is necessary to install a properly sized diaphragm-type expansion tank between the water meter or prevention device and the heater.
10. National Sanitation Foundation (NSF) standards require the installation of a NPS 1/4 IPS connection for a pressure gage mounted adjacent to the supply side of the control valve. They also require a water line strainer ahead of any electrically operated control valve (Figure 3).
11. NSF standards do not allow copper water lines that are not under constant pressure, except for the line downstream of the solenoid valve on the rinse line to the cabinet.

Water Pressure—Commercial Kitchens

Proper flow pressure must be maintained to achieve efficient dishwashing. NSF standards for dishwasher water flow pressure are 15 psig minimum, 25 psig maximum, and 20 psig ideal. Flow pressure is the line pressure measured when water is flowing through the rinse arms of the dishwasher.

Low flow pressure can be caused by undersized water piping, stoppage in piping, or excess pressure drop through heaters. Low water pressure causes an inadequate rinse, resulting in poor drying and sanitizing of the dishes. If flow pressure in the supply line to the dishwasher is below 15 psig, a booster pump or other means should be installed to provide supply water at 20 psig.

A flow pressure in excess of 25 psig causes atomization of the 180°F rinse water, resulting in an excessive temperature drop. The temperature drop between the rinse nozzle and the dishes can be as much as 15°F. A pressure regulator should be installed in the supply water line adjacent to the dishwasher and external to the return circulating loop (if used). The regulator should be set to maintain a pressure of 20 psig.

Two-Temperature Service

Where multiple temperature requirements are met by a single system, the system temperature is determined by the maximum temperature needed. Where the bulk of the hot water is needed at the higher temperature, lower temperatures can be obtained by mixing hot and cold water. Automatic mixing valves reduce the temperature of the hot water available at certain outlets to prevent injury or damage (Figure 4). Applicable codes should be consulted for mixing valve requirements.

Where predominant use is at a lower temperature, the common design heats all water to the lower temperature and then uses a separate booster heater to further heat the water for the higher temperature service (Figure 5). This method offers better protection against scalding.

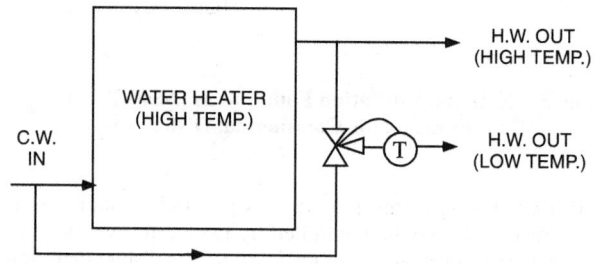

Fig. 4   Two-Temperature Service with Mixing Valve

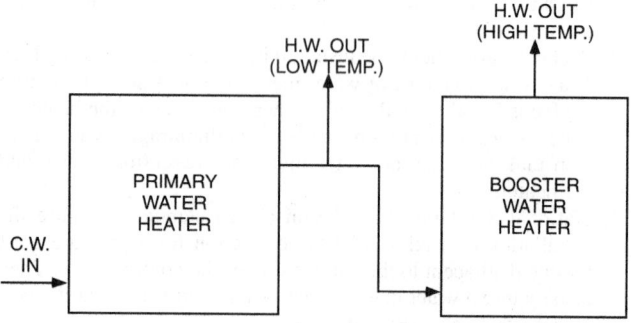

Fig. 5   Two-Temperature Service with Primary Heater and Booster Heater in Series

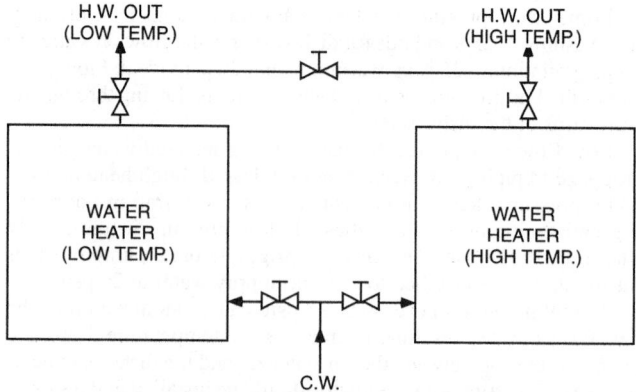

Fig. 6   Two-Temperature Service with Separate Heater for Each Service

A third method uses separate heaters for the higher temperature service (Figure 6). It is common practice to cross-connect the two heaters, so that one heater can serve the complete installation temporarily while the other is valved off for maintenance. Each heater should be sized for the total load unless hot water consumption can be reduced during maintenance periods.

Manifolding

Where one heater does not have sufficient capacity, two or more water heaters may be installed in parallel. If blending is needed in such an installation, a single mixing valve of adequate capacity should be used. It is difficult to obtain even flow through parallel mixing valves.

Heaters installed in parallel should have similar specifications, i.e., have the same input and storage capacity, with inlet and outlet piping arranged so that an equal flow is received from each heater under all demand conditions.

An easy way to get balanced, parallel flow is to use reverse/return piping (Figure 7). The unit having its inlet closest to the cold water supply is piped so that its outlet will be farthest from the hot water supply line. Quite often this results in a hot water supply line that reverses direction (see dotted line, Figure 7) to bring it back to the first unit in line; hence the name reverse/return.

TERMINAL HOT WATER USAGE DEVICES

Details on the vast number of devices using service hot water are beyond the scope of this chapter. Nonetheless, they are important to a successful overall design. Consult the manufacturer's literature for information on required flow rates, temperature limits, and/or other operating factors for specific items.

WATER HEATING TERMINOLOGY

The following terms apply to water heaters and water heating:

Recovery efficiency. Heat absorbed by the water divided by the heat input to the heating unit during the period that water temperature is raised from inlet temperature to final temperature.

Recovery rate. The amount of hot water that a residential water heater can continually produce, usually reported as flow rate in gallons per hour that can be maintained for a specified temperature rise through the water heater.

Thermal efficiency. Heat in the water flowing from the heater outlet divided by the heat input of the heating unit over a specific period of steady-state conditions.

Energy factor. Is the delivered efficiency of a water heater when operated as specified in DOE test procedures (10 CFR Part 430).

First hour rating. An indicator of the amount of hot water a residential water heater can supply. This rating is used by the Federal Trade Commission (FTC) for comparative purposes.

Standby loss. As applied to a tank type water heater (under test conditions with no water flow), the average hourly energy consumption divided by the average hourly heat energy contained in the stored water, expressed as a percent per hour. This can be converted to the average Btu/h energy consumption required to maintain any water-air temperature difference by multiplying the percent by the temperature difference, 8.25 Btu/gal·°F (a nominal specific heat for water), the tank capacity, and then dividing by 100.

Hot water distribution efficiency. Heat contained in the water at points of use divided by the heat delivered at the heater outlet at a given flow.

Heater/system efficiency. Heat contained in the water at points of use divided by the heat input to the heating unit at a given flow rate (thermal efficiency times distribution efficiency).

Overall system efficiency. Heat energy in the water delivered at points of use divided by the total energy supplied to the heater for any selected period.

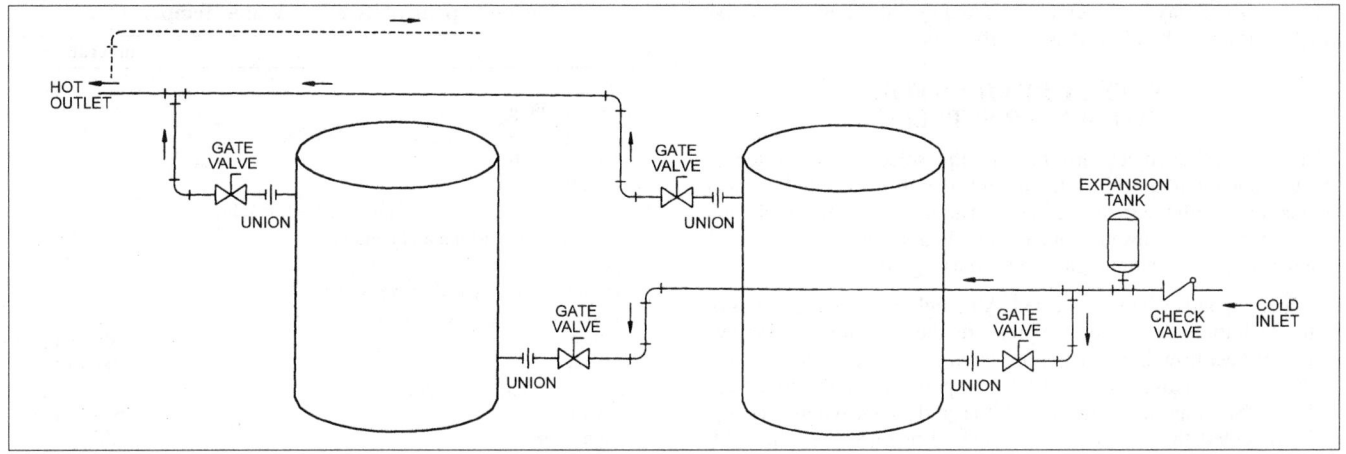

Fig. 7   Reverse/Return Manifold Systems

System standby loss. The amount of heat lost from the water heating system and the auxiliary power consumed during periods of nonuse of service hot water.

DESIGN CONSIDERATIONS

Hot water system design should consider the following:

- Water heaters of different sizes and insulation may have different standby losses, recovery efficiency, thermal efficiency, or energy factors.
- A distribution system should be properly designed, sized, and insulated to deliver adequate water quantities at temperatures satisfactory for the uses served. This reduces standby loss and improves hot water distribution efficiency.
- Heat traps between recirculation mains and infrequently used branch lines reduce convection losses to these lines and improve heater/system efficiency.
- Controlling circulating pumps to operate only as needed to maintain proper temperature at the end of the main reduces losses on return lines.
- Provision for shutdown of circulators during building vacancy reduces standby losses.

WATER QUALITY, SCALE, AND CORROSION

A complete water analysis and an understanding of system requirements are needed to protect water heating systems from scale and corrosion. Analysis shows whether water is hard or soft. Hard water, unless treated, will cause scaling or liming of heat transfer and water storage surfaces; soft water may aggravate corrosion problems and anode consumption (Talbert 1986).

Scale formation is also affected by system requirements and equipment. As shown in Figure 8, the rate of scaling increases with temperature and usage; this is because calcium carbonate and other scaling compounds lose solubility at higher temperatures. In water tube-type equipment, scaling problems can be offset by increasing the water velocity over the heat transfer surfaces, which reduces the tube surface temperature. Also, the turbulence of the flow, if high enough, works to keep any scale that does precipitate off the surface. When water hardness is over 8 gr/gal, water softening or other water treatment is often recommended.

Corrosion problems increase with temperature because corrosive oxygen and carbon dioxide gases are released from the water. Electrical conductivity also increases with temperature, enhancing electrochemical reactions such as rusting (Toaborek et al. 1972). A deposit of scale provides some protection from corrosion; however, this deposit also reduces the heat transfer

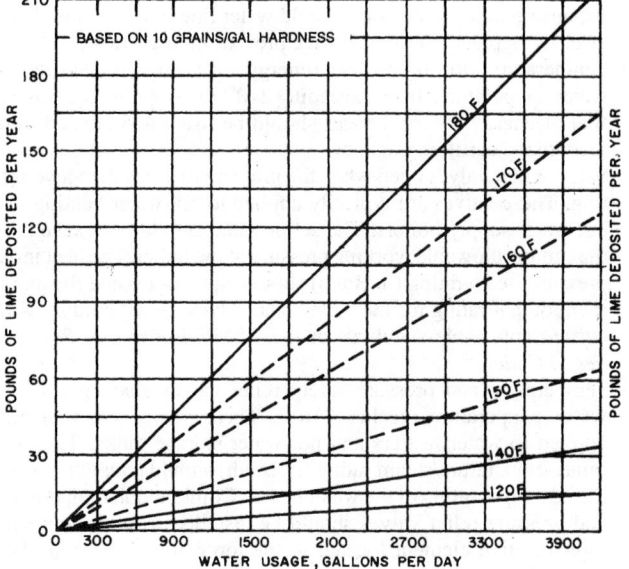

Fig. 8   Lime Deposited Versus Temperature and Water Use
(Purdue University Bulletin No. 74)

rate, and it is not under the control of the system designer (Talbert et al. 1986).

Steel vessels can be protected to varying degrees by galvanizing them or by lining them with copper, glass, cement, electroless nickel-phosphorus, or other corrosion-resistant material. Glass-lined vessels are almost always supplied with electrochemical protection. Typically, one or more anode rods of magnesium, aluminum, or zinc alloy are installed in the vessel by the manufacturer. This electrochemically active material sacrifices itself to reduce or prevent corrosion of the tank (the cathode). A higher temperature, softened water, and high water use may lead to rapid anode consumption. Manufacturers recommend periodic replacement of the anode rod(s) to prolong the life of the vessel. Some waters have very little electrochemical activity. In this instance, a standard anode shows little or no activity, and the vessel is not adequately protected. If this condition is suspected, the equipment manufacturer should be consulted on the possible need for a high potential anode, or the use of vessels made of nonferrous material should be considered.

Water heaters and hot water storage tanks constructed of stainless steel, copper, or other nonferrous alloys are protected against oxygen corrosion. However, care must still be taken, as some stainless steel may be adversely affected by chlorides, and copper may be attacked by ammonia or carbon dioxide.

SAFETY DEVICES FOR HOT WATER SUPPLIES

Regulatory agencies differ as to the selection of protective devices and methods of installation. It is therefore essential to check and comply with the manufacturer's instructions and the applicable local codes. In the absence of such instructions and codes, the following recommendations may be used as a guide:

- Water expands when it is heated. Although the water heating system is initially under service pressure, the pressure will rise rapidly if backflow is prevented by devices such as a check valve, pressure-reducing valve, or backflow preventer in the cold water line or by temporarily shutting off the cold water. When backflow is prevented, the pressure rise during heating may cause the safety relief valve to weep to relieve the pressure. However, if the safety relief valve is inadequate, inoperative, or missing, the pressure rise may rupture the tank or cause other damage. Systems having this potential problem must be protected by a properly sized expansion tank located on the cold water line downstream of and as close as practical to the device preventing back flow.
- Temperature limiting devices (energy cutoff/high limit) prevent water temperatures from exceeding 210°F by stopping the flow of fuel or energy. These devices should be listed and labeled by a recognized certifying agency.
- Safety relief valves open when the pressure exceeds the valve setting. These valves are typically applied to hot water heating and hot water supply boilers. The set pressure should not exceed the maximum allowable working pressure of the boiler. The heat input pressure steam rating (in Btu/h) should equal or exceed the maximum output rating for the boiler. The valves should comply with current applicable standards or the ASME *Boiler and Pressure Vessel Code*.
- Temperature and pressure safety relief valves also open if the water temperature reaches 210°F. These valves are typically applied to water heaters and hot water storage tanks. The heat input temperature/steam rating (in Btu/h) should equal or exceed the heat input rating of the water heater. Combination temperature and pressure relief valves should be installed with the temperature-sensitive element located in the top 6 in. of the tank, i.e., where the water is hottest.
- To reduce scald hazards, discharge temperature at fixtures accessible to the occupant should not exceed 120°F. Thermostatically controlled mixing valves can be used to blend hot and cold water in order to maintain safe service hot water temperatures.
- A relief valve should be installed in any part of the system containing a heat input device that can be isolated by valves. The heat input device may be solar water heating panels, desuperheater water heaters, heat recovery devices, or similar equipment.

SPECIAL CONCERNS

Legionella pneumophila (Legionnaires' Disease)

Legionnaires' disease (a form of severe pneumonia) is caused by inhaling the bacteria *Legionella pneumophila*. It has been discovered in the service water systems of various buildings throughout the world. Infection has often been traced to *Legionella pneumophila* colonies in shower heads. Ciesielki et al. (1984) determined that *Legionella pneumophila* can colonize in hot water maintained at 115°F or lower. Segments of service water systems in which the water stagnates (e.g., shower heads, faucet aerators, and certain sections of storage-type water heaters) provide ideal breeding locations.

Service water temperature in the 140°F range is recommended to limit the potential for *Legionella pneumophila* growth. This high

Table 3 Representative Hot Water Temperatures

Use	Temperature, °F
Lavatory	
Hand washing	105
Shaving	115
Showers and tubs	110
Therapeutic baths	95
Commercial or institutional laundry, based on fabric	up to 180
Residential dish washing and laundry	140
Surgical scrubbing	110
Commercial spray-type dish washing[a]	
Single- or multiple-tank hood or rack type	
Wash	150 minimum
Final rinse	180 to 195
Single-tank conveyor type	
Wash	160 minimum
Final rinse	180 to 195
Single-tank rack or door type	
Single-temperature wash and rinse	165 minimum
Chemical sanitizing types[b]	140
Multiple-tank conveyor type	
Wash	150 minimum
Pumped rinse	160 minimum
Final rinse	180 to 195
Chemical sanitizing glass washer	
Wash	140
Rinse	75 minimum

[a]As required by NSF.
[b]See manufacturer for actual temperature required.

temperature increases the potential for scalding, so care must be taken such as installing an anti scald or mixing valve. Supervised periodic flushing of fixture heads with 170°F water is recommended in hospitals and health care facilities because the already weakened patients are generally more susceptible to infection.

Temperature Requirement

Typical temperature requirements for some services are shown in Table 3. A 140°F water temperature minimizes flue gas condensation in the equipment.

Hot Water from Tanks and Storage Systems

With storage systems, 60 to 80% of the hot water in a tank is assumed to be usable before dilution by cold water lowers the temperature below an acceptable level. However, better designs can exceed 90%. Thus, the hot water available from a self-contained storage heater is usually considered to be

$$V_t = Rd + MS_t \qquad (3)$$

where

    V_t = available hot water, gal
    R = recovery rate at the required temperature, gph
    d = duration of peak hot water demand, h
    M = ratio of usable water to storage tank capacity
    S_t = storage capacity of the heater tank, gal

Usable hot water from an unfired tank is

$$V_a = MS_a \qquad (4)$$

where

    V_a = usable water available from an unfired tank, gal
    S_a = capacity of unfired tank, gal

Note: Assumes tank water at required temperature.

Hot water obtained from a water heater using a storage heater with an auxiliary storage tank is

$$V_z = V_t + V_a = Rd + M(S_t + S_a) \qquad (5)$$

where V_z = total hot water available during one peak in gallons.

Placement of Water Heaters

Many types of water heaters may be expected to leak at the end of their useful life. They should be placed where such leakage will not cause damage. Alternatively, suitable drain pans piped to drains must be provided.

Water heaters not requiring combustion air may generally be placed in any suitable location, as long as relief valve discharge pipes open to a safe location.

Water heaters requiring ambient combustion air must be located in areas with air openings large enough to admit the required combustion/dilution air (see NFPA *Standard* 54/ANSI Z223.1).

For water heaters located in areas where flammable vapors are likely to be present, precautions should be taken to eliminate the probable ignition of flammable vapors. For water heaters installed in residential garages, additional precautions should be taken. Consult local codes for additional requirements or see sections 5.1.9 through 5.1.12 of NFPA *Standard* 54/ANSI Z223.1.

Outdoor models with a weather-proofed jacket and that are to be installed outdoors are available. Direct-vent gas and oil-fired models are also available. They are to be installed indoors, but are not vented via a conventional chimney or gas vent. They use ambient air for combustion. They must be installed with the means for venting (typically horizontal) and the means for supplying air for combustion from outside the building that is supplied or specified by the equipment manufacturer.

HOT WATER REQUIREMENTS AND STORAGE EQUIPMENT SIZING

Methods for sizing storage water heaters vary. Those using recovery versus storage curves are based on extensive research. All methods provide adequate hot water if the designer allows for unusual conditions.

RESIDENTIAL

Table 4 shows typical hot water usage in a residence. In its Minimum Property Standards for One- and Two-Family Living Units, No. 4900.1-1982, HUD-FHA established minimum permissible water heater sizes (Table 5). Storage water heaters may vary from the sizes shown in the table if combinations of recovery and storage are used that produce the required 1-h draw.

The first hour rating (FHR) is the amount of hot water that the water heater can supply in one hour of operation under specific test conditions (DOE 1993). The FHR is under review regarding its use for system sizing. The linear regression lines shown in Figure 9 represent the FHR for 1556 electric heaters and 2901 gas heaters (GAMA 1997, Hiller 1998). Regression lines are not included for

Table 4 Typical Residential Use of Hot Water

Use	High Flow, Gallons/Task	Low Flow (Water Savers Used), Gallons/Task
Food preparation	5	3
Hand dish washing	4	4
Automatic dishwasher	15	15
Clothes washer	32	21
Shower or bath	20	15
Face and hand washing	4	2

Table 5 HUD-FHA Minimum Water Heater Capacities for One- and Two-Family Living Units

Number of Baths	1 to 1.5			2 to 2.5				3 to 3.5			
Number of Bedrooms	1	2	3	2	3	4	5	3	4	5	6
GAS[a]											
Storage, gal	20	30	30	30	40	40	50	40	50	50	50
1000 Btu/h input	27	36	36	36	36	38	47	38	38	47	50
1-h draw, gal	43	60	60	60	70	72	90	72	82	90	92
Recovery, gph	23	30	30	30	30	32	40	32	32	40	42
ELECTRIC[a]											
Storage, gal	20	30	40	40	50	50	66	50	66	66	80
kW input	2.5	3.5	4.5	4.5	5.5	5.5	5.5	5.5	5.5	5.5	5.5
1-h draw, gal	30	44	58	58	72	72	88	72	88	88	102
Recovery, gph	10	14	18	18	22	22	22	22	22	22	22
OIL[a]											
Storage, gal	30	30	30	30	30	30	30	30	30	30	30
1000 Btu/h input	70	70	70	70	70	70	70	70	70	70	70
1-h draw, gal	89	89	89	89	89	89	89	89	89	89	89
Recovery, gph	59	59	59	59	59	59	59	59	59	59	59
TANK-TYPE INDIRECT[b,c]											
I-W-H-rated draw, gal in 3 h, 100°F rise		40	40		66	66[e]	66	66	66	66	66
Manufacturer-rated draw, gal in 3 h, 100°F rise		49	49		75	75[e]	75	75	75	75	75
Tank capacity, gal		66	66		66	66[e]	82	66	82	82	82
TANKLESS-TYPE INDIRECT[c,d]											
I-W-H-rated draw, gpm, 100°F rise		2.75	2.75		3.25	3.25[e]	3.75	3.25	3.75	3.75	3.75
Manufacturer-rated draw, gal in 5 min, 100°F rise		15	15		25	25[e]	35	25	35	35	35

[a]Storage capacity, input, and recovery requirements indicated in the table are typical and may vary with each individual manufacturer. Any combination of these requirements to produce the stated 1-h draw will be satisfactory.
[b]Boiler-connected water heater capacities (180°F boiler water, internal, or external connection).

[c]Heater capacities and inputs are minimum allowable. Variations in tank size are permitted when recovery is based on 4 gph/kW at 100°F rise for electrical, AGA recovery ratings for gas, and IBR ratings for steam and hot water heaters.
[d]Boiler-connected heater capacities (200°F boiler water, internal, or external connection).
[e]Also for 1 to 1.5 baths and 4 bedrooms for indirect water heaters.

oil-fired and heat-pump water heaters because of limited data. The FHR represents water heater performance characteristics that are similar to those represented by the one-hour draw values listed in Table 5.

Another factor to consider when sizing water heaters is the set point temperature. At lower storage tank water temperatures, the tank volume and/or energy input rate may need to be increased to meet a given hot water demand. Currently, manufacturers are shipping residential water heaters with a recommendation that the initial set point be approximately 120°F to minimize the potential for scalding. Reduced set points generally lower standby losses and increase the water heater's efficiency and recovery capacity.

Over the last decade, the structure and life-style of a typical family has altered the household's hot water consumption. Due to variations in family size, age of family members, presence and age of children, hot water use volume and temperature, and other factors, demand patterns fluctuate widely in both magnitude and time distribution.

Perlman and Mills (1985) developed the overall and peak average hot water use volumes shown in Table 6. Average hourly patterns and 95% confidence level profiles are illustrated in Figures 10 and 11. Samples of results from the analysis of similarities in hot water use are given in Figures 12 and 13.

COMMERCIAL AND INSTITUTIONAL

Most commercial and institutional establishments use hot or warm water. The specific requirements vary in total volume, flow rate, duration of peak load period, and temperature. Water heaters and systems should be selected based on these requirements.

This section covers sizing recommendations for central storage water heating systems. Hot water usage data and sizing curves for dormitories, motels, nursing homes, office buildings, food service establishments, apartments, and schools are based on EEI-sponsored research (Werden and Spielvogel 1969). Caution must be taken in applying these data to small buildings. Also, within any given category there may be significant variation. For example, the motel category encompasses standard, luxury, resort, and convention motels.

Table 6 Overall (OVL) and Peak Average Hot Water Use

	Average Hot Water Use, gal							
	Hourly		**Daily**		**Weekly**		**Monthly**	
Group	OVL	Peak	OVL	Peak	OVL	Peak	OVL	Peak
All families	2.6	4.6	62.4	67.1	436	495	1897	2034
"Typical" families	2.6	5.8	63.1	66.6	442	528	1921	2078

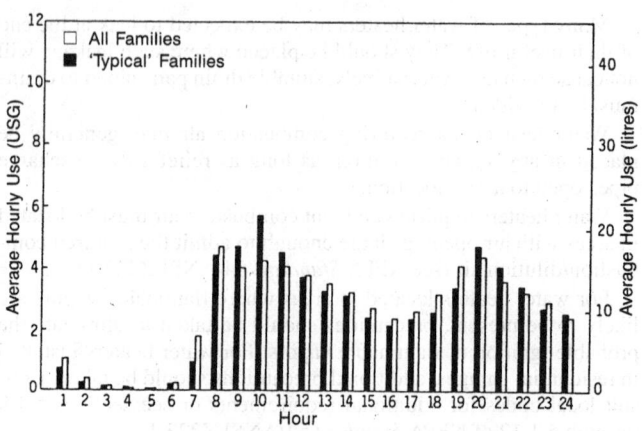

Fig. 10   Residential Average Hourly Hot Water Use

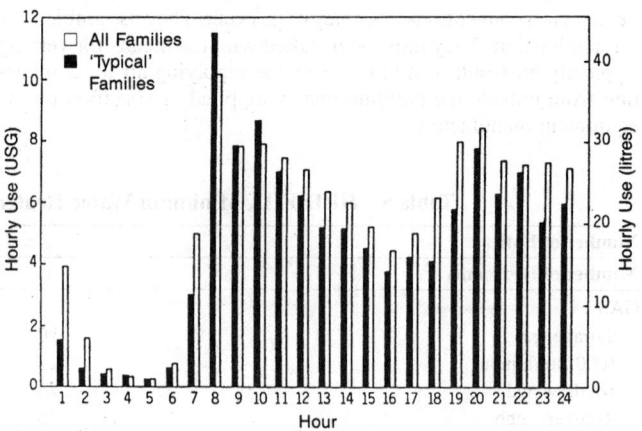

**Fig. 11   Residential Hourly Hot Water Use—
95% Confidence Level**

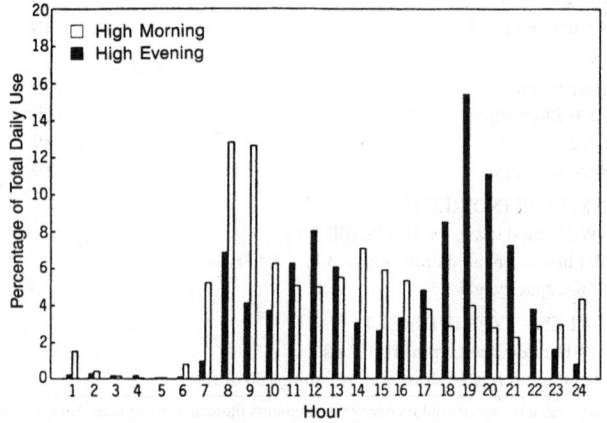

**Fig. 12   Residential Hourly Water Use Pattern for
Selected High Morning and High Evening Users**

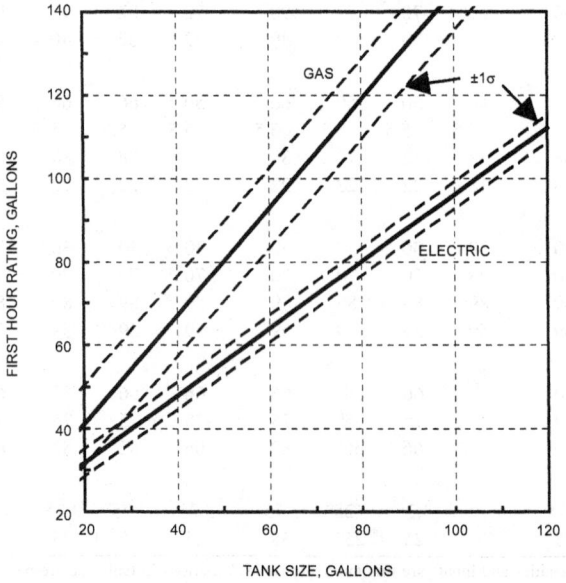

**Fig. 9   First Hour Rating Relationships for
Residential Water Heaters**

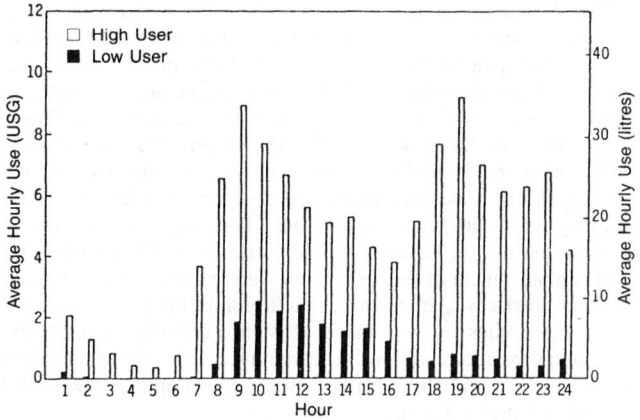

Fig. 13 Residential Average Hourly Hot Water Use for Low and High Users

When additional hot water requirements exist, a designer should increase the recovery and/or storage capacity accordingly. For example, if there is food service in an office building, the recovery and storage capacities required for each additional hot water use should be added when sizing a single central water heating system.

Peak hourly and daily demands for various categories of commercial and institutional buildings are shown in Table 7. These demands for central storage hot water represent the maximum flows metered in this 129-building study, excluding extremely high and very infrequent peaks. Table 7 also shows average hot water consumption figures for these buildings. Averages for schools and food service establishments are based on actual days of operation, while all others are based on total days. These averages can be used to estimate monthly consumption of hot water.

Research conducted for ASHRAE (Becker et al. 1991, Thrasher and DeWerth 1994) and others (Goldner 1993, 1994a,b) included a compilation and review of service hot water use information in commercial and multifamily structures along with new monitoring data. Some of this work found consumption comparable to those shown in Table 7; however, many of the studies showed higher consumption.

Dormitories

Hot water requirements for college dormitories generally include showers, lavatories, service sinks, and clothes washers. Peak demand usually results from the use of showers. Load profiles and hourly consumption data indicate that peaks may last 1 or 2 h and then taper off substantially. Peaks occur predominantly in the evening, mainly around midnight. The figures do not include hot water used for food service.

Military Barracks

Design criteria for military barracks are available from the engineering departments of the U.S. Department of Defense. Some measured data exist for hot water use in these facilities. For published data, contact the U.S. Army Corps of Engineers or Naval Facilities Engineering Command.

Motels

Domestic hot water requirements are for tubs and showers, lavatories, and general cleaning purposes. Recommendations are based on tests at low- and high-rise motels located in urban, suburban, rural, highway, and resort areas. Peak demand, usually from shower use, may last 1 or 2 h and then drop off sharply. Food service, laundry, and swimming pool requirements are not included.

Nursing Homes

Hot water is required for tubs and showers, wash basins, service sinks, kitchen equipment, and general cleaning. These figures include hot water for kitchen use. When other equipment, such as that for heavy laundry and hydrotherapy purposes, is to be used, its hot water requirement should be added.

Office Buildings

Hot water requirements are primarily for cleaning and lavatory use by occupants and visitors. Hot water use for food service within office buildings is not included.

Food Service Establishments

Hot water requirements are primarily for dish washing. Other uses include food preparation, cleaning pots and pans and floors, and hand washing for employees and customers. The

Table 7 Hot Water Demands and Use for Various Types of Buildings

Type of Building	Maximum Hourly	Maximum Daily	Average Daily
Men's dormitories	3.8 gal/student	22.0 gal/student	13.1 gal/student
Women's dormitories	5.0 gal/student	26.5 gal/student	12.3 gal/student
Motels: Number of units[a]			
20 or less	6.0 gal/unit	35.0 gal/unit	20.0 gal/unit
60	5.0 gal/unit	25.0 gal/unit	14.0 gal/unit
100 or more	4.0 gal/unit	15.0 gal/unit	10.0 gal/unit
Nursing homes	4.5 gal/bed	30.0 gal/bed	18.4 gal/bed
Office buildings	0.4 gal/person	2.0 gal/person	1.0 gal/person
Food service establishments			
Type A—full meal restaurants and cafeterias	1.5 gal/max meals/h	11.0 gal/max meals/day	2.4 gal/average meals/day[b]
Type B—drive-ins, grilles, luncheonettes, sandwich, and snack shops	0.7 gal/max meals/h	6.0 gal/max meals/day	0.7 gal/average meals/day[b]
Apartment houses: Number of apartments			
20 or less	12.0 gal/apartment	80.0 gal/apartment	42.0 gal/apartment
50	10.0 gal/apartment	73.0 gal/apartment	40.0 gal/apartment
75	8.5 gal/apartment	66.0 gal/apartment	38.0 gal/apartment
100	7.0 gal/apartment	60.0 gal/apartment	37.0 gal/apartment
200 or more	5.0 gal/apartment	50.0 gal/apartment	35.0 gal/apartment
Elementary schools	0.6 gal/student	1.5 gal/student	0.6 gal/student[b]
Junior and senior high schools	1.0 gal/student	3.6 gal/student	1.8 gal/student[b]

[a]Interpolate for intermediate values.
[b]Per day of operation.

recommendations are for establishments serving food at tables, counters, booths, and parked cars. Establishments that use disposable service exclusively are not covered in Table 7.

Dish washing, as metered in these tests, is based on the normal practice of dish washing after meals, not on indiscriminate or continuous use of machines irrespective of the flow of soiled dishes. The recommendations include hot water supplied to dishwasher booster heaters.

Apartments

Hot water requirements for both garden-type and high-rise apartments are for one- and two-bath apartments, showers, lavatories, kitchen sinks, dishwashers, clothes washers, and general cleaning purposes. Clothes washers can be either in individual apartments or centrally located. These data apply to central water heating systems only.

Elementary Schools

Hot water requirements are for lavatories, cafeteria and kitchen use, and general cleaning purposes. When showers are used, their additional hot water requirements should be added. The recommendations include hot water for dishwashers but not for extended school operation such as evening classes.

High Schools

Senior high schools, grades 9 or 10 through 12, require hot water for showers, lavatories, dishwashers, kitchens, and general cleaning. Junior high schools, grades 7 through 8 or 9, have requirements similar to those of the senior high schools. Junior high schools having no showers follow the recommendations for elementary schools.

Requirements for high schools are based on daytime use. Recommendations do not take into account hot water usage for additional activities such as night school. In such cases, the maximum hourly demand remains the same, but the maximum daily and the average daily usage increases, usually by the number of additional people using showers and, to a lesser extent, eating and washing facilities.

Additional Data

Fast Food Restaurants. Hot water is used for food preparation, cleanup, and rest rooms. Dish washing is usually not a significant load. In most facilities, peak usage occurs during the cleanup period, typically soon after opening and immediately prior to closing. Hot water consumption varies significantly among individual facilities. Fast food restaurants typically consume 250 to 500 gallons per day (EPRI 1994).

Supermarkets. The trend in supermarket design is to incorporate food preparation and food service functions, substantially increasing the usage of hot water. Peak usage is usually associated with cleanup periods, often at night, with a total consumption of 300 to 1000 gallons per day (EPRI 1994).

Apartments. Table 8 differs from Table 7 in that it represents low-medium-high (LMH) guidelines rather than specific singular volumes. The values presented in Table 8 are based on monitored domestic hot water consumption found in the studies by Becker et al. (1991), Thrasher and DeWerth (1994), Goldner (1993, 1994a,b), and Goldner and Price (1996).

Table 8   Hot Water Demand and Use Guidelines for Apartment Buildings (Gallons per Person)

Guideline	Maximum Hourly	Peak 15 Minutes	Maximum Daily	Average Daily
Low	3.0	1.0	20.0	14.0
Medium	5.0	2.0	49.0	30.0
High	9.0	3.0	90.0	54.0

"Med" is the overall average. "Low" is the lowest peak value. Such values are generally associated with apartment buildings having such occupant demographics as (1) all occupants working, (2) seniors, (3) couples with children, (4) middle income, or (5) higher population density. "High" is the maximum recurring value. These values are generally associated with (1) high percentage of children, (2) low income, (3) public assistance, or (4) no occupants working.

In applying these guidelines, the designer should note that a building may outlast its current use. This may be a reason to increase the design capacity for domestic hot water or allow for future enhancement of the service hot water system. Building management practices, such as the explicit prohibition (in the lease) of apartment clothes washers or the existence of bath/kitchen hookups, should be factored into the design process. A diversity factor that lowers the probability of coincident consumption should also be used in larger buildings.

SIZING EXAMPLES

Figures 14 through 21 show the relationships between recovery and storage capacity for the various building categories. Any combination of storage and recovery rates that falls on the proper curve will satisfy the building requirements. Using the minimum recovery rate and the maximum storage capacity on the curves yields the smallest hot water capacity capable of satisfying the building requirement. The higher the recovery rate, the greater the 24-h heating capacity and the smaller the storage capacity required.

These curves can be used to select recovery and storage requirements to accommodate water heaters that have fixed storage or recovery rates. Where hot water demands are not coincident with peak electric, steam, or gas demands, greater heater inputs can be selected if they do not create additional energy system demands, and the corresponding storage tank size can be selected from the curves.

Ratings of gas-fired water heating equipment are based on sea level operation and apply for operation at elevations up to 2000 ft. For operation at elevations above 2000 ft, and in the absence of specific recommendations from the local authority, equipment ratings should be reduced at the rate of 4% for each 1000 ft above sea level before selecting appropriately sized equipment.

Recovery rates in Figures 14 through 21 represent the actual hot water required without considering system heat losses. Heat losses from storage tanks and recirculating hot water piping should be calculated and added to the recovery rates shown. Storage tanks and hot water piping must be insulated.

The storage capacities shown are net usable requirements. Assuming that 60 to 80% of the hot water in a storage tank is usable, the actual storage tank size should be increased by 25 to 66% to compensate for unusable hot water.

Examples

In the following examples, results are often rounded to the accuracy appropriate for the assumptions made.

Example 1. Determine the required water heater size for a 300-student women's dormitory for the following criteria:

    a. Storage with minimum recovery rate.

    b. Storage with recovery rate of 2.5 gph per student.

    c. With the additional requirement for a cafeteria to serve a maximum of 300 meals per hour for minimum recovery rate, combined with item *a*; and for a recovery rate of 1.0 gph per maximum meals per hour, combined with item *b*.

Solution:

    a. The minimum recovery rate from Figure 15 for women's dormitories is 1.1 gph per student or 330 gph total. At this rate, storage required is 12 gal per student or 3600 gal total. On a 70% net usable basis, the necessary tank size is 3600/0.7 = 5150 gal.

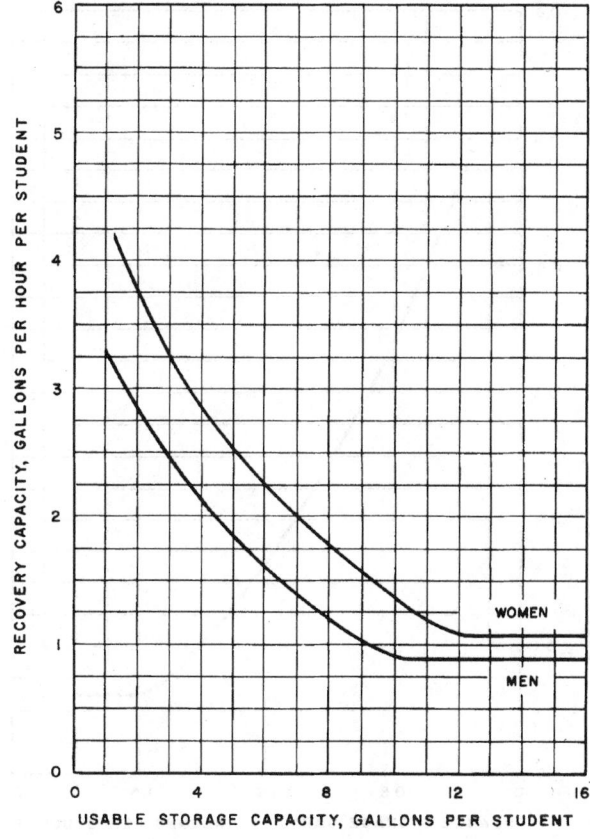

Fig. 14   Dormitories

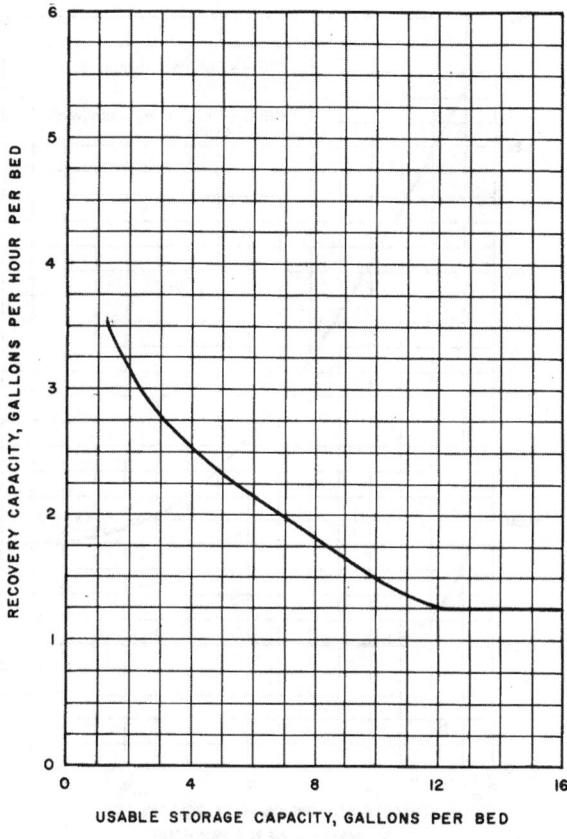

Fig. 16   Nursing Homes

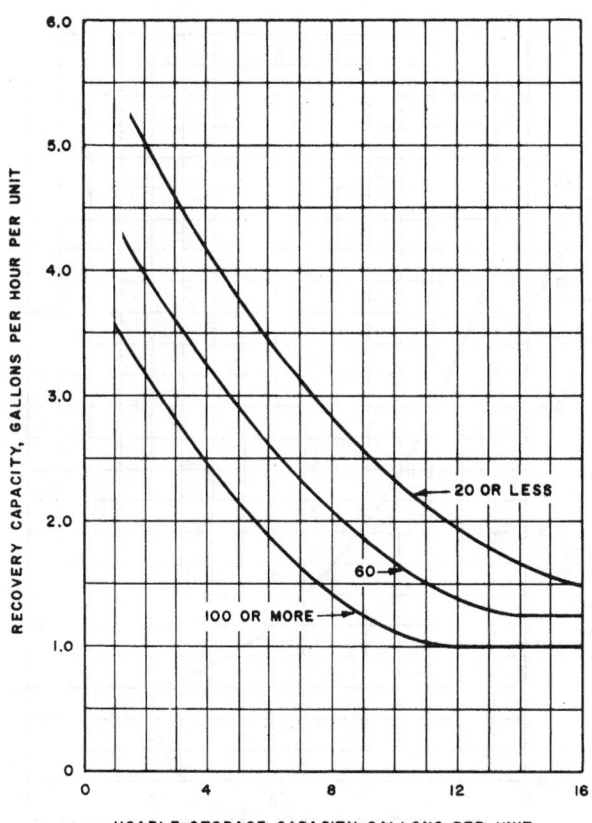

Fig. 15   Motels

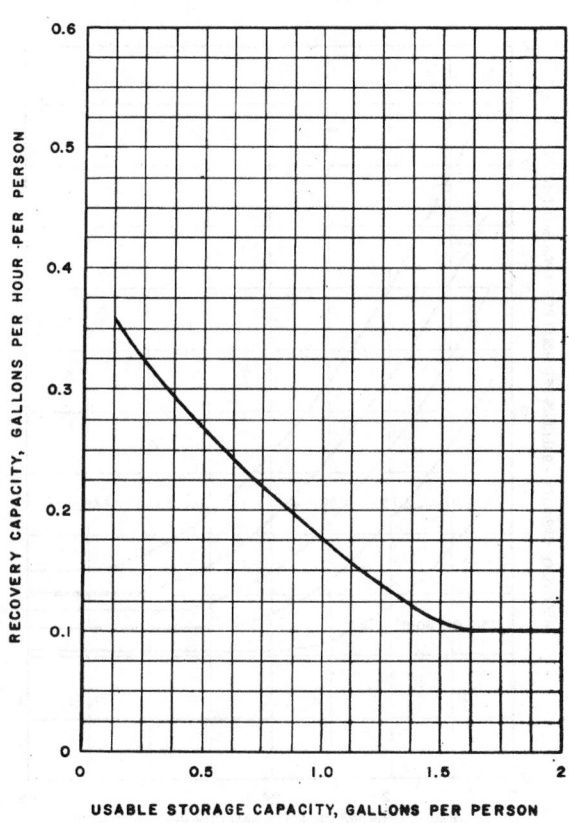

Fig. 17   Office Buildings

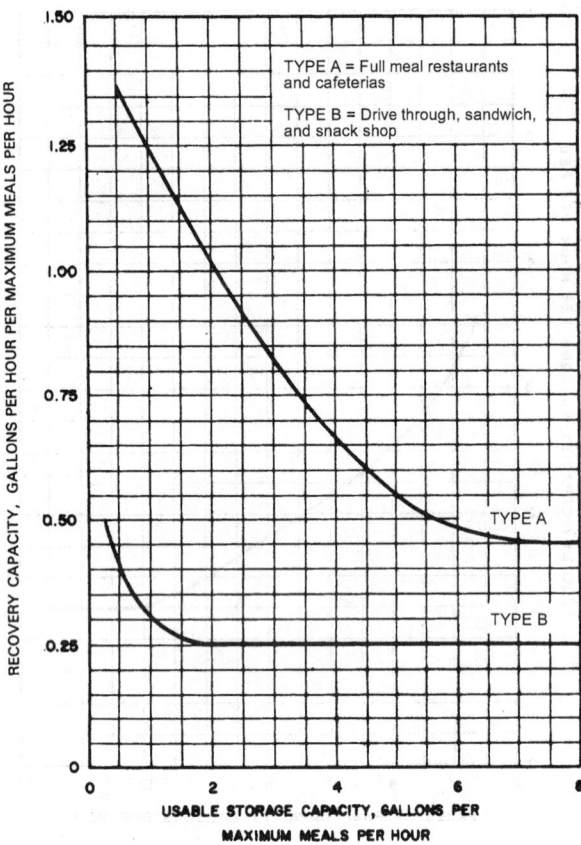

Fig. 18   Food Service

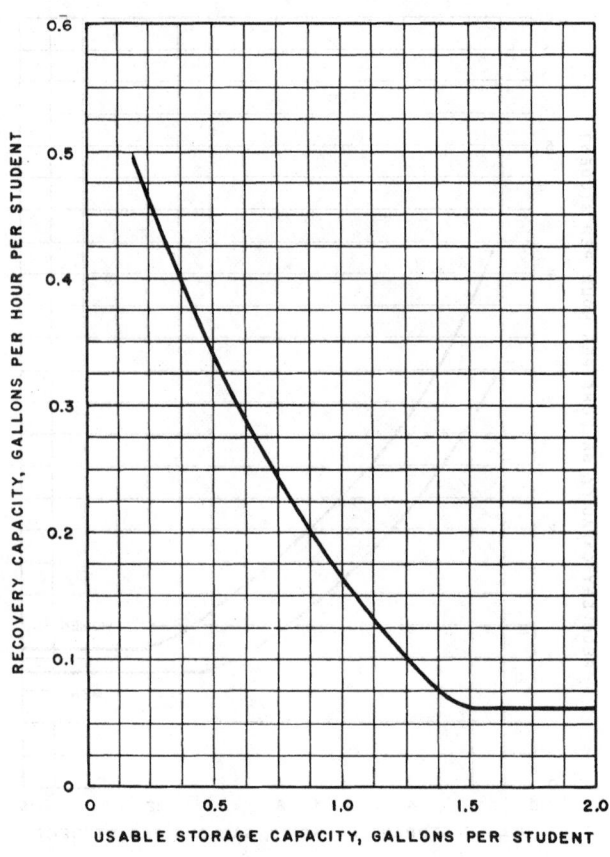

Fig. 20   Elementary Schools

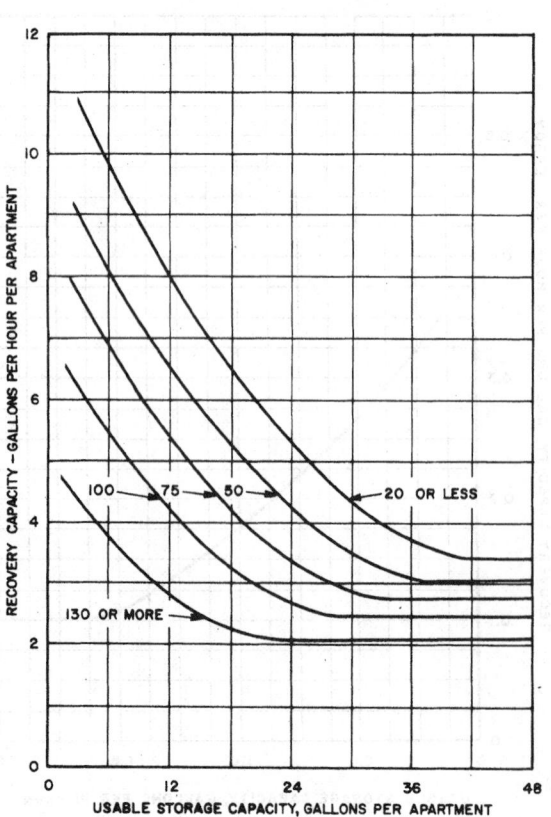

Fig. 19   Apartments

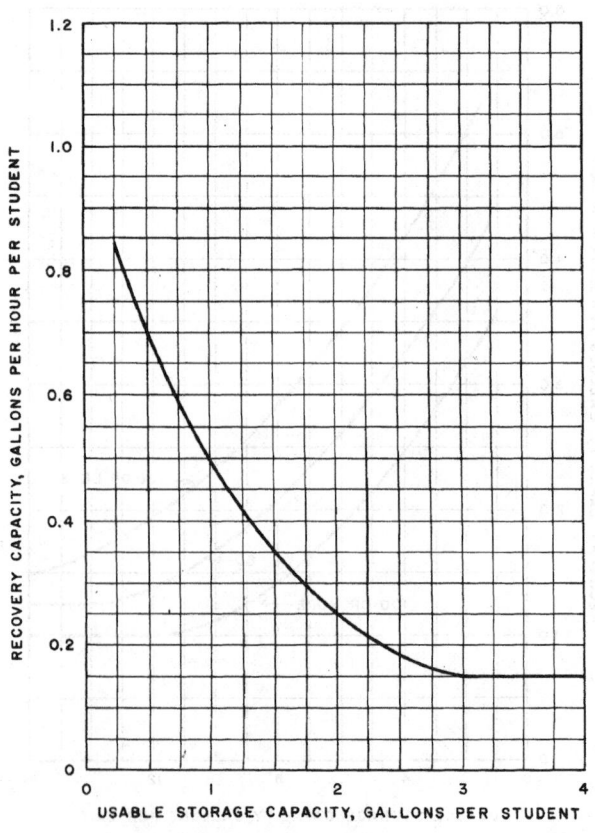

Fig. 21   High Schools

b. The same curve shows 5 gal storage per student at 2.5 gph recovery, or 300 × 5 = 1500 gal storage with recovery of 300 × 2.5 = 750 gph. The tank size will be 1500/0.7 = 2150 gal.

c. The requirements for a cafeteria can be determined from Figure 19 and added to those for the dormitory. For the case of minimum recovery rate, the cafeteria (Type A) requires 300 × 0.45 = 135 gph recovery rate and 300 × 7/0.7 = 3000 gal of additional storage. The entire building then requires 330 + 135 = 465 gph recovery and 5150 + 3000 = 8150 gal of storage.

With 1 gph recovery at the maximum hourly meal output, the recovery required is 300 gph, with 300 × 2.0/0.7 = 860 gal of additional storage. Combining this with item *b*, the entire building requires 750 + 300 = 1050 gph recovery and 2150 + 860 = 3010 gal of storage.

Note: Recovery capacities shown are for heating water only. Additional capacity must be added to offset the system heat losses.

Example 2. Determine the water heater size and monthly hot water consumption for an office building to be occupied by 300 people:
>   *a.* Storage with minimum recovery rate.
>   *b.* Storage with 1.0 gal per person storage.
>   *c.* Additional minimum recovery rate requirement for a luncheonette open 5 days a week, serving a maximum of 100 meals per hour and an average of 200 meals per day.
>   *d.* Monthly hot water consumption.

Solution:
a. With minimum recovery rate of 0.1 gph per person from Figure 18, 30 gph recovery is required, while the storage is 1.6 gal per person, or 300 × 1.6 = 480 gal. If 70% of the hot water is usable, the tank size will be 480/0.7 = 690 gal.

b. The curve also shows 1.0 gal storage per person at 0.175 gph per person recovery, or 300 × 0.175 = 52.5 gph. The tank size will be 300/0.7 = 430 gal.

c. The hot water requirements for a luncheonette (Type B) are contained in Figure 19. With a minimum recovery capacity of 0.25 gph per maximum meals per hour, 100 meals per hour would require 25 gph recovery, while the storage would be 2.0 gal per maximum meals per hour, or 100 × 2.0/0.7 = 290 gal storage. The combined requirements with item *a* would then be 55 gph recovery and 980 gal storage.

Combined with item *b*, the requirement is 77.5 gph recovery and 720 gal storage.

d. Average day values are found in Table 7. The office building will consume an average of 1.0 gal per person per day × 30 days per month × 300 people = 9000 gal per month, while the luncheonette will consume 0.7 gal per meal × 200 meals per day × 22 days per month = 3100 gal per month, for a total of 12,100 gal per month.

Note: Recovery capacities shown are for heating water only. Additional capacity must be added to offset the system heat losses.

Example 3. Determine the water heater size for a 200-unit apartment house:
>   *a.* Storage with minimum recovery rate.
>   *b.* Storage with 4 gph per apartment recovery rate.
>   *c.* Storage for each of two 100-unit wings.
>   1. Minimum recovery rate.
>   2. Recovery rate of 4 gph per apartment.

Solution:
a. The minimum recovery rate, from Figure 20, for apartment buildings with 200 apartments is 2.1 gph per apartment, or a total of 420 gph. The storage required is 24 gal per apartment, or 4800 gal. If 70% of this hot water is usable, the necessary tank size is 4800/0.7 = 6900 gal.

b. The same curve shows 5 gal storage per apartment at a recovery rate of 4 gph per apartment, or 200 × 4 = 800 gph. The tank size will be 200 × 5/0.7 = 1400 gal.

c. Solution for a 200-unit apartment house having two wings, each with its own hot water system.

1. With minimum recovery rate of 2.5 gph per apartment (see Figure 20), a 250 gph recovery is required, while the necessary storage is 28 gal per apartment, or 100 × 28 = 2800 gal. The required tank size is 2800/0.7 = 4000 gal for each wing.

2. The curve shows that for a recovery rate of 4 gph per apartment the storage would be 14 gal per apartment, or 100 × 14 = 1400 gal, with

recovery of 100 × 4 = 400 gph. The necessary tank size is 1400/0.7 = 2000 gal in each wing.

Note: Recovery capacities shown are for heating water only. Additional capacity must be added to offset the system heat loss.

Example 4. Determine the water heater size and monthly hot water consumption for a 2000-student high school.
>   *a.* Storage with minimum recovery rate.
>   *b.* Storage with 4000-gal maximum storage capacity.
>   *c.* Monthly hot water consumption.

Solution:
a. With the minimum recovery rate of 0.15 gph per student (from Figure 22) for high schools, 300 gph recovery is required. The storage required is 3.0 gal per student, or 2000 × 3.0 = 6000 gal. If 70% of the hot water is usable, the tank size is 6000/0.7 = 8600 gal.

b. The net storage capacity will be 0.7 × 4000 = 2800 gal, or 1.4 gal per student. From the curve, a recovery capacity of 0.37 gph per student or 2000 × 0.37 = 740 gph is required.

c. From Table 7, monthly hot water consumption is 2000 students × 1.8 gal per student per day × 22 days = 79,000 gal.

Note: Recovery capacities shown are for heating water only. Additional capacity must be added to offset the system heat loss.

Table 9 can be used to determine the size of water heating equipment from the number of fixtures. To obtain the probable maximum demand, multiply the total quantity for the fixtures by the demand factor in line 19. The heater or coil should have a water heating capacity equal to this probable maximum demand. The storage tank should have a capacity equal to the probable maximum demand multiplied by the storage capacity factor in line 20.

Example 5. Determine heater and storage tank size for an apartment building from a number of fixtures.

Solution:

60 lavatories	×	2 gph	=	120 gph
30 bathtubs	×	20 gph	=	600 gph
30 showers	×	30 gph	=	900 gph
60 kitchen sinks	×	10 gph	=	600 gph
15 laundry tubs	×	20 gph	=	300 gph
Possible maximum demand			=	2520 gph
Probable maximum demand		=2520 × 0.30= 756 gph		
Heater or coil capacity		=	756 gph	
Storage tank capacity		=756 × 1.25=945 gal		

Showers

In many housing installations such as motels, hotels, and dormitories, the peak hot water load is usually due to the use of showers. Tables 1 and 9 indicate the probable hourly hot water demand and the recommended demand and storage capacity factors for various types of buildings. Hotels could have a 3- to 4-h peak shower load. Motels require similar volumes of hot water, but the peak demand may last for only a 2-h period. In some types of housing, such as barracks, fraternity houses, and dormitories, all occupants may take showers within a very short period. In this case, it is best to find the peak load by determining the number of shower heads and the rate of flow per head; then estimate the length of time the shower will be on. It is estimated that the average shower time per individual is 1 to 3 min.

The flow rate from a shower head varies depending on type, size, and water pressure. At 40 psi water pressure, available shower heads have nominal flow rates of blended hot and cold water from about 2.5 to 6 gpm. In multiple shower installations, flow control valves on shower heads are recommended because they reduce the flow rate and maintain it regardless of fluctuations in water pressure. Flow can usually be reduced to 50% of the manufacturer's maximum flow rating without adversely affecting the spray pattern of the shower head. Flow control valves are commonly available with capacities from 1.5 to 4.0 gpm.

Table 9   Hot Water Demand per Fixture for Various Types of Buildings
(Gallons of water per hour per fixture, calculated at a final temperature of 140°F)

	Apartment House	Club	Gymnasium	Hospital	Hotel	Industrial Plant	Office Building	Private Residence	School	YMCA
1. Basin, private lavatory	2	2	2	2	2	2	2	2	2	2
2. Basin, public lavatory	4	6	8	6	8	12	6	—	15	8
3. Bathtub[c]	20	20	30	20	20	—	—	20	—	30
4. Dishwasher[a]	15	50-150	—	50-150	50-200	20-100	—	15	20-100	20-100
5. Foot basin	3	3	12	3	3	12	—	3	3	12
6. Kitchen sink	10	20	—	20	30	20	20	10	20	20
7. Laundry, stationary tub	20	28	—	28	28	—	—	20	—	28
8. Pantry sink	5	10	—	10	10	—	10	5	10	10
9. Shower	30	150	225	75	75	225	30	30	225	225
10. Service sink	20	20	—	20	30	20	20	15	20	20
11. Hydrotherapeutic shower				400						
12. Hubbard bath				600						
13. Leg bath				100						
14. Arm bath				35						
15. Sitz bath				30						
16. Continuous-flow bath				165						
17. Circular wash sink				20	20	30	20		30	
18. Semicircular wash sink				10	10	15	10		15	
19. DEMAND FACTOR	0.30	0.30	0.40	0.25	0.25	0.40	0.30	0.30	0.40	0.40
20. STORAGE CAPACITY FACTOR[b]	1.25	0.90	1.00	0.60	0.80	1.00	2.00	0.70	1.00	1.00

[a]Dishwasher requirements should be taken from this table or from manufacturers' data for the model to be used, if this is known.
[b]Ratio of storage tank capacity to probable maximum demand/h. Storage capacity may be reduced where an unlimited supply of steam is available from a central street steam system or large boiler plant.
[c]Whirlpool baths require specific consideration based on their capacity. They are not included in the bathtub category.

If the manufacturer's flow rate for a shower head is not available, and no flow control valve is used, the following average flow rates may serve as a guide for sizing the water heater:

Small shower head	2.5 gpm
Medium shower head	4.5 gpm
Large shower head	6 gpm

Food Service

In a restaurant, bacteria are usually killed by rinsing the washed dishes with 180 to 195°F water for several seconds. In addition, an ample supply of general-purpose hot water, usually 140 to 150°F, is required for the wash cycle of dishwashers. Although a water temperature of 140°F is reasonable for dish washing in private dwellings, in public places, the NSF or local health departments require 180 to 195°F water in the rinsing cycle. However, the NSF allows a lower temperature when certain types of machines and chemicals are used. The two-temperature hot water requirements of food service establishments present special problems. The lower temperature water is distributed for general use, but the 180°F water should be confined to the equipment requiring it and should be obtained by boosting the temperature. It would be dangerous to distribute 180°F water for general use. NSF *Standard* 26 covers the design of dishwashers and water heaters used by restaurants. The American Gas Association (Dunn et al. 1959) has published a recommended procedure for sizing water heaters for restaurants that consists of determining the following:

1. Types and sizes of dishwashers used (manufacturers' data should be consulted to determine the initial fill requirements of the wash tanks)
2. Required quantity of general-purpose hot water
3. Duration of peak hot water demand period
4. Inlet water temperature
5. Type and capacity of existing water heating system
6. Type of water heating system desired

After the quantity of hot water withdrawn from the storage tank each hour has been taken into account, the following equation may be used to size the required heater(s). The general-purpose and 180 to 195°F water requirements are determined from Tables 10 and 11.

$$q_i = Q_h c_p \rho \Delta t / \eta \qquad (6)$$

where

q_i = heater input, Btu/h
Q_h = flow rate, gph
c_p = specific heat of water = 1.00 Btu/lb·°F
ρ = density of water = 8.33 lb/gal
Δt = temperature rise, °F
η = heater efficiency

To determine the quantity of usable hot water from storage, the duration of consecutive peak demand must be estimated. This peak usually coincides with the dishwashing period during and after the main meal and may last from 1 to 4 h. Any hour in which the dishwasher is used at 70% or more of capacity should be considered a peak hour. If the peak demand lasts for 4 h or more, the value of a storage tank is reduced, unless especially large tanks are used. Some storage capacity is desirable to meet momentary high draws.

NSF *Standard* 5 recommendations for hot water rinse demand are based on 100% operating capacity of the machines, as are the data provided in Table 10. NSF *Standard* 5 states that 70% of operating rinse capacity is all that is normally attained, except for rackless-type conveyor machines.

Examples 6, 7, and 8 demonstrate the use of Equation (6) in conjunction with Tables 10 and 11.

Example 6. Determine the hot water demand for water heating in a cafeteria kitchen with one vegetable sink, five lavatories, one prescrapper, one utensil washer, and one two-tank conveyor dishwasher (dishes inclined) with makeup device. The initial fill requirement for the tank of the utensil washer is 85 gph at 140°F. The initial fill requirement for the dishwasher is 20 gph for each tank, or a total of 40 gph, at 140°F. The maximum period of consecutive operation of the dishwasher at or above 70% capacity is assumed to be 2 h. The supply water temperature is 60°F.

Solution: The required quantities of general purpose (140°F) and rinse (180°F) water for the equipment, from Tables 10 and 11 and given values, are shown in the following tabulation:

Item	Quantity Required at 140°F, gph	Quantity Required at 180°F, gph
Vegetable sink	45	—
Lavatories (5)	25	—
Prescrapper	180	—
Dishwasher	—	277
Initial tank fill	40	—
Makeup water	—	139
Utensil washer	—	75
Initial tank fill	85	—
Total requirements	375	491

The total consumption of 140°F water is 375 gph. The total consumption at 180°F depends on the type of heater to be used. For a heater that has enough internal storage capacity to meet the flow demand, the total consumption of 180°F water is (277 + 139 + 75) = 491 gph. Based on the requirements taken from Table 10, or approximately 350 gph (0.70 × 491 = 344 gph). For an instantaneous heater without internal storage capacity, the total quantity of 180°F water consumed must be based on the flow demand. From Table 10, the quantity required for the dishwasher is 277 gph; for the makeup, 139 gph; and for the utensil washer, 480 gph. The total consumption of 180°F water is 277 + 139 + 480 = 896 gph, or approximately 900 gph.

Table 10 NSF Final Rinse Water Requirement for Dish Washing Machines[a]

Type and Size of Dishwasher	Flow Rate, gpm	180 to 195°F Hot Water Requirements	
		Heaters Without Internal Storage,[b] gph	Heaters with Internal Storage to Meet Flow Demand,[c] gph
Door type:			
16 × 16 in.	6.94	416	69
18 × 18 in.	8.67	520	87
20 × 20 in.	10.4	624	104
Undercounter	5	300	70
Conveyor type:			
Single tank	6.94	416	416
Multiple tank (dishes flat)	5.78	347	347
Multiple tank (dishes inclined)	4.62	277	277
Silver washers	7	420	45
Utensil washers	8	480	75
Makeup water requirements	2.31	139	139

Note: Values are extracted from a previous version of NSF *Standard* 3. The current version of NSF *Standard* 3-1996 is a performance based standard that no longer lists minimum flow rates.
[a]Flow pressure at dishwashers is assumed to be 20 psig.
[b]Based on the flow rate in gpm.
[c]Based on dishwasher operation at 100% of mechanical capacity.

Table 11 General-Purpose Hot Water (140°F) Requirement for Various Kitchens Uses[a,b]

Equipment	gph
Vegetable sink	45
Single pot sink	30
Double pot sink	60
Triple pot sink	90
Prescrapper (open type)	180
Preflush (hand-operated)	45
Preflush (closed type)	240
Recirculating preflush	40
Bar sink	30
Lavatories (each)	5

Source: Dunn et al. (1959).
[a]Supply water pressure at equipment is assumed to be 20 psig.
[b]Dishwasher operation at 100% of mechanical capacity.

Example 7. Determine fuel input requirements (assume 75% heater efficiency) for heating water in the cafeteria kitchen described in Example 6, by the following systems, which are among many possible solutions:
a. Separate, self-contained, storage-type heaters.
b. Single instantaneous-type heater having no internal storage, to supply both 180°F and 140°F water through a mixing valve.
c. Separate instantaneous-type heaters having no internal storage.

Solution:
a. The temperature rise for 140°F water is 140 − 60 = 80°F. From Equation (6), the fuel input required to produce 375 gph of 140°F water with an 80°F temperature rise at 75% efficiency is about 333,000 Btu/h. One or more heaters may be selected to meet this total requirement.
From Equation (6), the fuel input required to produce 491 gph of 180°F water with a temperature rise of 180 − 60 = 120°F at 75% efficiency is 654,000 Btu/h. One or more heaters with this total requirement may be selected from manufacturers' catalogs.
b. The correct sizing of instantaneous-type heaters depends on the flow rate of the 180°F rinse water. From Example 6, the consumption of 180°F water based on the flow rate is 900 gph; consumption of 140°F water is 375 gph.
Fuel input required to produce 900 gph (277 + 480 + 139) of 180°F water with a 120°F temperature rise is 1,200,000 Btu/h. Fuel input to produce 375 gph of 140°F water with a temperature rise of 80°F is 333,000 Btu/h. Total heater requirement is 1,200,000 + 333,000 = 1,533,000 Btu/h. One or more heaters meeting this total input requirement can be selected from manufacturers' catalogs.
c. Fuel input required to produce 140°F water is the same as for Solution *b*, 333,000 Btu/h. One or more heaters meeting this total requirement can be selected.
Fuel input required to produce 180°F water is also the same as in Solution *b*, 1,200,000 Btu/h. One or more heaters meeting this total requirement can be selected.

Example 8. A luncheonette has purchased a door-type dishwasher that will handle 16 × 16-in. racks. The existing hot water system is capable of supplying the necessary 140°F water to meet all requirements for general-purpose use and for the booster heater that is to be installed. Determine the size of the following booster heaters operating at 75% thermal efficiency required to heat 140°F water to provide sufficient 180°F rinse water for the dishwasher:
a. Booster heater with no storage capacity.
b. Booster heater with enough storage capacity to meet flow demand.

Solution:
a. Because the heater is the instantaneous type, it must be sized to meet the 180°F water demand at a rated flow. From Table 10, this rated flow is 6.94 gpm, or 416 gph. From Equation (6), the required fuel input with a 40°F temperature rise is 185,000 Btu/h. A heater meeting this input requirement can be selected from manufacturers' catalogs.
b. In designing a system with a booster heater having storage capacity, the dishwasher's hourly flow demand can be used instead of the flow demand used in Solution *a*. The flow demand from Table 10 is 69 gph when the dishwasher is operating at 100% mechanical capacity. From Equation (6), with a 40°F temperature rise, the fuel input required is 30,700 Btu/h. A booster heater with this input can be selected from manufacturers' catalogs.

Estimating Procedure. Hot water requirements for kitchens are sometimes estimated on the basis of the number of meals served (assuming eight dishes per meal). Demand for 180°F water for a dishwasher is

$$D_1 = C_1 N/\theta \qquad (7)$$

where

D_1 = water for dishwasher, gph
N = number of meals served
θ = hours of service
C_1 = 0.8 for single-tank dishwasher
   = 0.5 for two-tank dishwasher

Demand for water for a sink with gas burners is

$$D_2 = C_2 V \qquad (8)$$

where

 D_2 = water for sink, gph
 C_2 = 3
 V = sink capacity (15 in. depth), gal

Demand for general-purpose hot water at 140°F is

$$D_3 = C_3 N/(\theta + 2) \qquad (9)$$

where

 D_3 = general-purpose water, gph
 C_3 = 1.2

Total demand is

$$D = D_1 + D_2 + D_3$$

For soda fountains and luncheonettes, use 75% of the total demand. For hotel meals or other elaborate meals, use 125%.

Schools

Service water heating in schools is needed for janitorial work, lavatories, cafeterias, shower rooms, and sometimes swimming pools.

Hot water used in cafeterias is about 70% of that usually required in a commercial restaurant serving adults and can be estimated by the method used for restaurants. Where NSF sizing is required, follow *Standard 5*.

Shower and food service loads are not ordinarily concurrent. Each should be determined separately, and the larger load should determine the size of the water heater(s) and the tank. Provision must be made to supply 180°F sanitizing rinse. The booster must be sized according to the temperature of the supply water. If feasible, the same water can be used for both needs. If the distance between the two points of need is great, a separate water should be used.

A separate heater system for swimming pools can be sized as outlined in the section on Swimming Pools/Health Clubs.

Domestic Coin-Operated Laundries

Small domestic machines in coin laundries or apartment house laundry rooms have a wide range of draw rates and cycle times. Domestic machines provide a wash water temperature (normal) as low as 120°F. Some manufacturers recommend a temperature of 160°F; however, the average appears to be 140°F. Hot water sizing calculations must assure a supply to both the instantaneous draw requirements of a number of machines filling at one time and the average hourly requirements.

The number of machines that will be drawing at any one time varies widely; the percentage is usually higher in smaller installations. One or two customers starting several machines at about the same time has a much sharper effect in a laundry with 15 or 20 machines than in one with 40 machines. Simultaneous draw may be estimated as follows:

1 to 11 machines	100% of possible draw
12 to 24 machines	80% of possible draw
25 to 35 machines	60% of possible draw
36 to 45 machines	50% of possible draw

Possible peak draw can be calculated from

$$F = NPV_f/T \qquad (10)$$

where

 F = peak draw, gpm
 N = number of washers installed
 P = number of washers drawing hot water divided by N
 V_f = quantity of hot water supplied to machine during hot wash fill, gal
 T = wash fill period, min

Recovery rate can be calculated from

$$R = 60NPV_f/(\theta + 10) \qquad (11)$$

where

 R = total hot water (machines adjusted to hottest water setting), gph
 θ = actual machine cycle time, min

Note: (θ + 10) is the cycle time plus 10 min for loading and unloading.

Commercial Laundries

Commercial laundries generally use a storage water heater. The water may be softened to reduce soap use and improve quality. The trend is toward installation of high-capacity washer-extractor wash wheels, resulting in high peak demand.

Sizing Data. Laundries can normally be divided into five categories. The required hot water is determined by the weight of the material processed. Average hot water requirements at 180°F are

Institutional	2 gal/lb·h
Commercial	2 gal/lb·h
Linen supply	2.5 gal/lb·h
Industrial	2.5 gal/lb·h
Diaper	2.5 gal/lb·h

Total weight of the material times these values give the average hourly hot water requirements. The designer must consider peak requirements; for example, a 600-lb machine may have a 20 gpm average requirement, but the peak requirement could be 350 gpm.

In a multiple-machine operation, it is not reasonable to fill all machines at the momentary peak rate. Diversity factors can be estimated by using 1.0 of the largest machine plus the following balance:

	Total number of machines				
	2	3 to 5	6 to 8	9 to 11	12 and over
1.0 +	0.6	0.45	0.4	0.35	0.3

For example, four machines would have a diversity factor of 1.0 + 0.45 = 1.45.

Types of Systems. Service water heating systems for laundries are pressurized or vented. The pressurized system uses city water pressure, and the full peak flow rates are received by the softeners, reclaimer, condensate cooler, water heater, and lines to the wash wheels. The flow surges and stops at each operation in the cycle. A pressurized system depends on an adequate water service.

The vented system uses pumps from a vented (open) hot water heater or tank to supply hot water. The tank's water level fluctuates from about 6 in. above the heating element to a point 12 in. from the top of the tank; this fluctuation defines the working volume. The level drops for each machine fill, while makeup water runs continuously at the average flow rate under water service pressure during the complete washing cycle. The tank is sized to have full working volume at the beginning of each cycle. Lines and softeners may be sized for the average flow rate from the water service to the tank, not the peak machine fill rate as with a closed, pressurized system.

The waste heat exchangers have a continuous flow across the heating surface at a low flow rate, with continuous heat reclamation from the wastewater and flash steam. Automatic flow-regulating valves on the inlet water manifold control this low flow rate. Rapid fill of machines increases production; i.e., more batches can be processed.

Heat Recovery. Commercial laundries are ideally suited for heat recovery because 135°F wastewater is discharged to the sewer. Fresh water can be conservatively preheated to within 15°F of the wastewater temperature for the next operation in the wash cycle. Regions with an annual average temperature of 55°F can increase to 120°F the initial temperature of fresh water going into the hot water heater. For each 1000 gph or 8330 lb per hour of water preheated 65°F (55 to 120°F), heat reclamation and associated energy savings is 540,000 Btu/h.

Flash steam from a condensate receiving tank is often wasted to the atmosphere. The heat in this flash steam can be reclaimed with a suitable heat exchanger. Makeup water to the heater can be preheated 10 to 20°F above the existing makeup temperature with the flash steam.

Swimming Pools/Health Clubs

The desirable temperature for swimming pools is 80°F. Most manufacturers of water heaters and boilers offer specialized models for pool heating; these include a pool temperature controller and a water bypass to prevent condensation. The water heating system is usually installed prior to the return of treated water to the pool. A circulation rate to generate a change of water every 8 h for residential pools and 6 h for commercial pools is acceptable. An indirect heater, in which piping is embedded in the walls or floor of the pool, has the advantage of reduced corrosion, scaling, and condensation because pool water does not flow through the pipes. The disadvantage of this type of system is the high initial installation cost.

The installation should have a pool temperature control and a water pressure or flow safety switch. The temperature control should be installed at the inlet to the heater, while the pressure or flow switch can be installed at either the inlet or outlet, depending on the manufacturer's instructions. It affords protection against inadequate water flow.

Sizing should be based on four considerations:

1. Conduction through the pool walls
2. Convection from the pool surface
3. Radiation from the pool surface
4. Evaporation from the pool surface

Except in aboveground pools and in rare cases where cold groundwater flows past the pool walls, conduction losses are small and can be ignored. Because convection losses depend on temperature differentials and wind speed, these losses can be greatly reduced by the installation of windbreaks such as hedges, solid fences, or buildings.

Radiation losses occur when the pool surface is subjected to temperature differentials; these frequently occur at night, when the sky temperature may be as much as 80°F below ambient air temperature. This usually occurs on clear, cool nights. During the daytime, however, an unshaded pool will receive a large amount of radiant energy, often as much as 100,000 Btu/h. These losses and gains may offset each other. An easy method of controlling nighttime radiation losses is to use a floating pool cover; this also substantially reduces evaporative losses.

Evaporative losses constitute the greatest heat loss from the pool—50 to 60% in most cases. If it is possible to cut evaporative losses drastically, the heating requirement of the pool may be cut by as much as 50%. The floating pool cover can accomplish this reduction.

A pool heater with an input great enough to provide a heat-up time of 24 h would be the ideal solution. However, it may not be the most economical system for pools that are in continuous use during an extended swimming season. In this instance, a less expensive unit providing an extended heat-up period of as much as 48 h can be used. Pool water may be heated by several methods. Fuel-fired water heaters and boilers, electric boilers, tankless electric circulation water heaters, air-source heat pumps, and solar heaters have all been used successfully. Air-source heat pumps and solar heating systems would probably be used to extend a swimming season rather than to allow intermittent use with rapid pickup.

The following equations provide some assistance in determining the area and volume of pools:

Elliptical

Area = 3.14 AB
A = short radius
B = long radius
Volume = 7.5 gal/ft^3 × Area × Average Depth

Kidney Shape

Area = 0.45 $L(A+B)$ (approximately)
L = length
A = width at one end
B = width at other end
Volume = 7.5 gal/ft^3 × Area × Average Depth

Oval (for circular, set $L = 0$)

Area = 3.14 $R^2 + LW$
L = length of straight sides
W = width or $2R$
R = radius of ends
Volume = 7.5 gal/ft^3 × Area × Average Depth

Rectangular

Area = LW
L = length
W = width
Volume = 7.5 gal/ft^3 × Area × Average Depth

The following is an effective method for heating outdoor pools. Additional equations can be found in Chapter 4.

1. Obtain pool water capacity, in gallons, from the above.
2. Determine the desired heat pickup time in hours.
3. Determine the desired pool temperature—if not known, use 80°F.
4. Determine the average temperature of the coldest month of use.

The required heater output q_t can now be determined by the following equations:

$$q_1 = \rho c_p V(t_f - t_i)/\theta \tag{12}$$

where

q_T = pool heat-up rate, Btu/h
ρ = density of water = 8.33 lb/gal
c_p = specific heat of water = 1.00 Btu/lb·°F
V = pool volume, gal
t_f = desired temperature (usually 80°F)
t_i = initial temperature of pool, °F
θ = pool heat-up time, h

$$q_2 = UA(t_p - t_a) \tag{13}$$

where

q_2 = heat loss from pool surface, Btu/h
U = surface heat transfer coefficient = 10.5 Btu/h·ft^2·°F
A = pool surface area, ft^2
t_p = pool temperature, °F
t_a = ambient temperature, °F

$$q_t = q_1 + q_2 \tag{14}$$

Notes: These heat loss equations assume a wind velocity of 3 to 5 mph. For pools sheltered by nearby fences, dense shrubbery, or buildings, an average wind velocity of less than 3.5 mph can be assumed. In this case, use 75% of the values calculated by Equation (13). For a velocity of 5 mph, multiply by 1.25; for 10 mph, multiply by 2.0.

Because Equation (13) applies to the coldest monthly temperatures, the calculated results may not be economical. Therefore, a value of one-half the surface loss plus the heat-up value yields a more viable heater output figure. The heater input then equals the output divided by the efficiency of the fuel source.

Whirlpools and Spas

Hot water requirements for whirlpool baths and spas depend on temperature, fill rate, and total volume. Water may be stored separately at the desired temperature or, more commonly, regulated at the point of entry by blending. If rapid filling is desired, provide storage at least equal to the volume needed; fill rate can then be varied at will. An alternative is to establish a maximum fill rate and provide an instantaneous water heater that will handle the flow.

Industrial Plants

Hot water (potable) is used in industrial plants for cafeterias, showers, lavatories, gravity sprinkler tanks, and industrial processes. Employee cleanup load is usually heaviest and not concurrent with other uses. The other loads should be checked before sizing, however, to be certain that this is true.

The employee cleanup load consists of one or more of the following: (1) wash troughs or standard lavatories, (2) multiple wash sinks, and (3) showers. Hot water requirements for employees using standard wash fixtures can be estimated at 1 gal of hot water for each clerical and light-industrial employee per work shift and 2 gal for each heavy-industrial worker.

For sizing purposes, the number of workers using multiple wash fountains is disregarded. Hot water demand is based on full flow for the entire cleanup period. This usage over a 10-min period is indicated in Table 12. The shower load depends on the flow rate of the shower heads and their length of use. Table 12 may be used to estimate flow based on a 15-min period.

Water heaters used to prevent freezing in gravity sprinkler tanks or water storage tanks should be part of a separate system. The load depends on tank heat loss, tank capacity, and winter design temperature.

Process hot water load must be determined separately. Volume and temperature vary with the specific process. If the process load occurs at the same time as the shower or cafeteria load, the system must be sized to reflect this total demand. Separate systems can also be used, depending on the size of the various loads and the distance between them.

Ready-Mix Concrete

In cold weather, ready-mix concrete plants need hot water to mix the concrete so that it will not be ruined by freezing before it sets. Operators prefer to keep the mix at about 70°F by adding hot water to the cold aggregate. Usually, water at about 150°F is considered proper for cold weather. If the water temperature is too high, some of the concrete will flash set.

Generally, 30 gal of hot water per cubic yard of concrete mix is used for sizing. To obtain the total hot water load, this number is multiplied by the number of trucks loaded each hour and the capacity of the trucks. The hot water is dumped into the mix as quickly as possible at each loading, so ample hot water storage or large heat exchangers must be used. Table 13 shows a method of sizing water heaters for concrete plants.

Table 12 Hot Water Usage for Industrial Wash Fountains and Showers

	Multiple Wash Fountains	Showers	
Type	Gal of 140°F Water Required for 10-min Period[a]	Flow Rate, gpm	Gal of 140°F Water Required for 15-min Period[b]
36 in. Circular	40	3	29.0
36 in. Semicircular	22	4	39.0
54 in. Circular	66	5	48.7
54 in. Semicircular	40	6	58.0

[a]Based on 110°F wash water and 40°F cold water at average flow rates.
[b]Based on 105°F shower water and 40°F cold water.

Table 13 Water Heater Sizing for Ready-Mix Concrete Plant
(Input and Storage Tank Capacity to Supply 150°F Water at 40°F Inlet Temperature)

Truck Capacity, yd³	Water Heater Storage Tank Volume, gal	Time Interval Between Trucks, min[a]					
		50	35	25	10	5	0
		Water Heater Capacity, 1000 Btu/h					
6	430	458	612	785	1375	1830	2760
7.5	490	527	700	900	1580	2100	3150
9	560	596	792	1020	1790	2380	3580
11	640	687	915	1175	2060	2740	4120

[a]This table assumes that there is 10-min loading time for each truck. Thus, for a 50-min interval between trucks, it is assumed that 1 truck/h is served. For 0 min between trucks, it is assumed that one truck loads immediately after the truck ahead has pulled away. Thus, 6 trucks/h are served.

It is also assumed that each truck carries a 120-gal storage tank of hot water for washing down at the end of dumping the load. This hot water is drawn from the storage tank and must be added to the total hot water demands. This has been included in the table.

Part of the heat may be obtained by heating the aggregate bin by circulating hot water through pipe coils in the walls or sides of the bin. If the aggregate is warmed in this manner, the temperature of the mixing water may be lower, and the aggregate will flow easily from the bins. When aggregate is not heated, it often freezes into chunks, which must be thawed before they will pass through the dump gates. If hot water is used for thawing, too much water accumulates in the aggregate, and control of the final product may vary beyond allowable limits. Therefore, jets of steam supplied by a small boiler and directed on the large chunks are often used for thawing.

SIZING INSTANTANEOUS AND SEMI-INSTANTANEOUS WATER HEATERS

The methods for sizing storage water heating equipment should not be used for instantaneous and semi-instantaneous heaters. The following is based on the Hunter method for sizing hot and cold water piping, with diversity factors applied for hot water and various building types.

Fixture units (Table 1) are assigned to each fixture using hot water and totalled. Maximum hot water demand is obtained from Figure 22 or Figure 23 (Hunter 1941) by matching total fixture units to the curve for the type of building. Hot water for fixtures and outlets requiring constant flows should be added to the demand.

The heater can then be selected for the total demand and the total temperature rise required. For critical applications such as hospitals, multiple heaters with 100% standby are recommended. Consider multiple heaters for buildings in which continuity of service is important. The minimum recommended size for semi-instantaneous

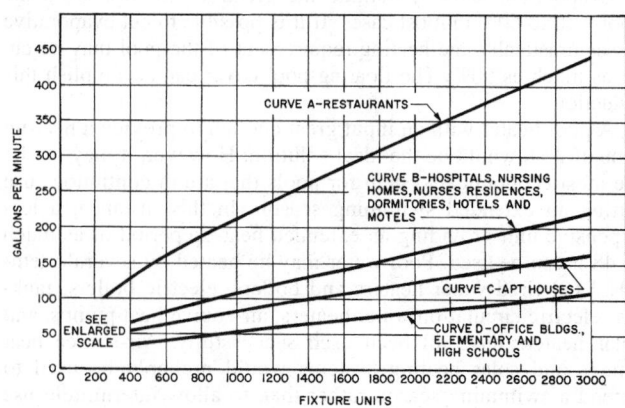

Fig. 22 Modified Hunter Curve for Calculating Hot Water Flow Rate

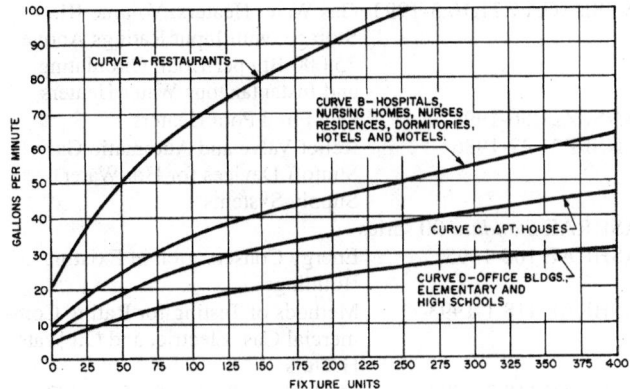

Fig. 23 Enlarged Section of Figure 22
(Modified Hunter Curve)

Table 14 Preliminary Hot Water Demand Estimate

Type of Building	Fixture Units
Hospital or nursing home	2.50 per bed
Hotel or motel	2.50 per room
Office building	0.15 per person
Elementary school	0.30 per student[a]
Junior and senior high school	0.30 per student[a]
Apartment house	3.00 per apartment

[a]Plus shower load.

heaters is 10 gpm, except for restaurants, in which it is 15 gpm. When the flow for a system is not easily determined, the heater may be sized for the full flow of the piping system. Heaters with low flows must be sized carefully, and care should be taken in the estimation of diversity factors. Unusual hot water requirements should be analyzed to determine whether additional capacity is required. One example is a dormitory in a military school, where all showers and lavatories are used simultaneously when students return from a drill. In this case, the heater and piping should be sized for the full flow of the system.

While the fixture count method bases heater size of the diversified system on hot water flow, hot water piping should be sized for the full flow to the fixtures. Recirculating hot water systems are adaptable to instantaneous heaters. When recirculating systems are installed, the heater capacity should be checked and increased if necessary to offset heat losses of the recirculating system.

To make preliminary estimates of hot water demand when the fixture count is not known, use Table 14 with Figure 22 or Figure 23. The result will usually be higher than the demand determined from the actual fixture count. Actual heater size should be determined from Table 1. Hot water consumption over time can be assumed the same as that in the section on Hot Water Requirements and Storage Equipment Sizing.

Example 9. A 600-student elementary school has the following fixture count: 60 public lavatories, 6 service sinks, 4 kitchen sinks, 6 showers, and 1 dishwasher at 8 gpm. Determine the hot water flow rate for sizing a semi-instantaneous heater based on the following:

a. Estimating the number of fixture units.
b. Actual fixture count.

Solution:

a. Use Table 14 to find the estimated fixture count: 600 students × 0.3 fixture units per student = 180 fixture units. As showers are not included, Table 1 shows 1.5 fixture units per shower × 6 showers = 9 additional fixture units. The basic flow is determined from curve D of Figure 24, which shows that the total flow for 189 fixture units is 23 gpm.

b. To size the unit based on actual fixture count and Table 1, the calculation is as follows:

60	public lavatories	×	1.0	FU	=	60 FU
6	service sinks	×	2.5	FU	=	15 FU
4	kitchen sinks	×	0.75	FU	=	3 FU
6	showers	×	1.5	FU	=	9 FU
	Subtotal					87 FU

At 87 fixture units, curve D of Figure 21 shows 16 gpm, to which must be added the dishwasher requirement of 8 gpm. Thus, the total flow is 24 gpm.

Comparing the flow based on actual fixture count to that obtained from the preliminary estimate shows the preliminary estimate to be slightly lower in this case. It is possible that the preliminary estimate could have been as much as twice the final fixture count. To prevent such oversizing of equipment, use the actual fixture count method to select the unit.

SIZING REFRIGERANT-BASED WATER HEATERS

Refrigerant-based water heaters such as heat pump water heaters and refrigeration heat reclaim systems cannot be sized like conventional systems to meet peak loads. The seasonal and instantaneous efficiency and output of these systems vary greatly with operating conditions. Computer software that performs detailed performance simulations taking these factors into account should be used for sizing and analysis. The capacities of these systems and any related supplemental water heating equipment should be selected to achieve a high average daily run time and the lowest combination of operating and equipment cost. For heat pump water heaters, the adequacy of the heat source and the potential effect of the cooling output must be addressed.

BOILERS FOR INDIRECT WATER HEATING

When service water is heated indirectly by a space heating boiler, Figure 24 may be used to determine the additional boiler capacity required to meet the recovery demands of the domestic water heating load. Indirect heaters include immersion coils in boilers as well as heat exchangers with space heating media.

Because the boiler capacity must meet not only the water supply requirement but also the space heating loads, Figure 24 indicates the reduction of additional heat supply for water heating if the ratio of water heating load to space heating load is low. This reduction is possible because

1. Maximum space heating requirements do not occur at the time of day when the maximum peak hot water demands occur.
2. Space heating requirements are based on the lowest outdoor design temperature, which may occur for only a few days of the total heating season.
3. An additional heat supply or boiler capacity to compensate for pickup and radiation losses is usual. The pickup load cannot occur at the same time as the peak hot water demand because the building must be brought to a comfortable temperature before the occupants will be using hot water.

The factor obtained from Figure 24 is multiplied by the peak water heating load to obtain the additional boiler output capacity required.

For reduced standby losses in summer and improved efficiency in winter, step-fired modular boilers may be used. Units not in operation cool down and reduce or eliminate jacket losses. The heated boiler water should not pass through an idle boiler. Figure 25 shows a typical modular boiler combination space and water heating arrangement.

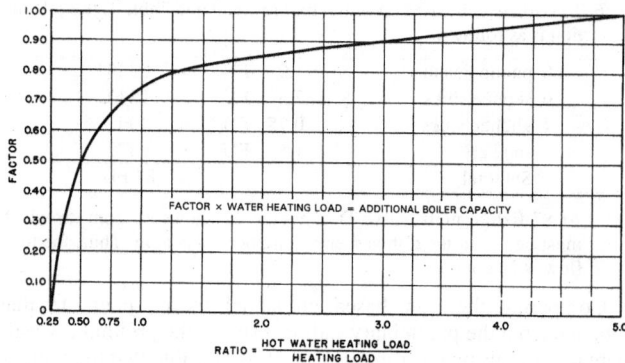

Fig. 24 Sizing Factor for Combination Heating and Water Heating Boilers

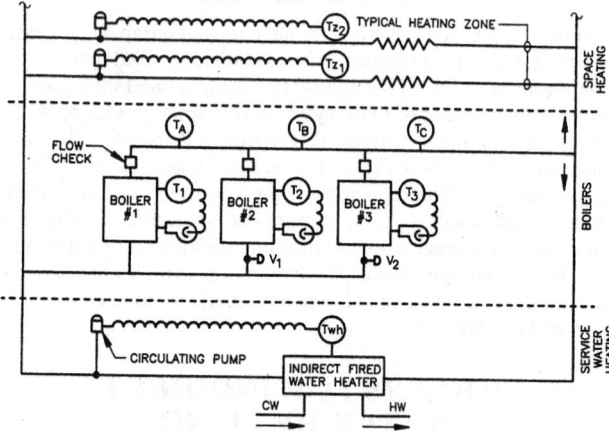

Fig. 25 Typical Modular Boiler for Combined Space and Water Heating

Typical Control Sequence

1. Any control zone or indirectly fired water heater thermostat starts its circulating pump and supplies power to boiler No. 1 control circuit.
2. If T1 is not satisfied, burner is turned on, boiler cycles as long as any circulating pump is on.
3. If after 5 min TA is not satisfied, V1 opens and boiler No. 2 comes on line.
4. If after 5 min TB is not satisfied, V2 opens and boiler No. 3 comes on line.
5. If TC is satisfied and two boilers or fewer are firing for a minimum of 10 min, V2 closes.
6. If TB is satisfied and only one boiler is firing for a minimum of 10 min, V1 closes.
7. If all circulating pumps are off, boiler No. 1 shuts down.

The ASHRAE/IES *Standard* 90 series discusses combination service water heating/space heating boilers and establishes restrictions on their use. The ASHRAE/IES *Standard* 100 section on Service Water Heating also has information on this subject.

CODES AND STANDARDS

ANSI/AGA Z21.10.1-1993 Gas Water Heaters, Volume I: Storage Water Heaters with Input Ratings of 75,000 Btu per Hour or Less

ANSI/AGA Z21.10.3-1993 Gas Water Heaters, Volume III: Storage, with Input Ratings Above 75,000 Btu per Hour, Circulating, and Instantaneous Water Heaters

ANSI Z21.56-1994 Gas-Fired Pool Heaters
ANSI Z21.22-1986 Relief Valve and Automatic Gas Shutoff Devices for Hot Water Supply Systems

ASHRAE *Standard* 90 series
ASHRAE 100-1995 Energy Conservation in Existing Buildings
ASHRAE 118.1-1993 Methods of Testing for Rating Commercial Gas, Electric, and Oil Water Heaters
ASHRAE 118.2-1993 Methods of Testing for Rating Residential Water Heaters

ASME Boiler and Pressure Vessel Code Section IV-98: Rules for Construction of Heating Boilers
Section VIII D1-98: Rules for Construction of Pressure Vessels

HUD-FHA 4900.1 Minimum Property Standards for One- and Two-Family Living Units

NFPA 54-96 National Fuel Gas Code
  (ANSI Z223.1-1996)
NFPA 31-92 Installation of Oil-Burning Equipment

NSF 5-92 Water Heaters, Hot Water Supply Boilers, and Heat Recovery Equipment

NSF 26-80 Pot, Pan, and Utensil Washers
UL 174-1996 Household Electric Storage Tank Water Heaters
UL 732-1997 Oil-Fired Unit Heaters
UL 1261-1992 Electric Water Heaters for Pools and Tubs
UL 1453-1994 Electric Booster and Commercial Storage Tank Water Heaters

REFERENCES

AGA. Sizing and equipment data for specifying swimming pool heaters. Catalog No. R-00995. American Gas Association, Cleveland, OH.

Becker, B.R., W.H. Thrasher, and D.W. DeWerth. 1991. Comparison of collected and compiled existing data on service hot water use patterns in residential and commercial establishments. *ASHRAE Transactions* 97(2): 231-39.

Ciesielki, C.A. et al. 1984. Role of stagnation and obstruction of water flow in isolation of *Legionella pneumophila* from hospital plumbing. *Applied and Environmental Microbiology* (November):984-87.

DOE. 1990. Final rule regarding test procedures and energy conservation standards for water heaters. 10 CFR Part 430, *Federal Register* 55(201). U.S. Department of Energy.

DOE. 1998. Uniform test method for measuring the energy consumption of water heaters. 10 CFR Part 430, Subpart B, Appendix E. U.S. Department of Energy.

Dunn, T.Z., R.N. Spear, B.E. Twigg, and D. Williams. 1959. Water heating for commercial kitchens. *Air Conditioning, Heating and Ventilating* (May):70. Also published as a bulletin, *Enough hot water—hot enough.* American Gas Association (1959).

EPRI. 1990. *Commercial heat pump water heaters applications handbook.* CU-6666. Electric Power Research Institute, Palo Alto, CA.

EPRI. 1992. *Commercial water heating applications handbook.* TR-100212. Electric Power Research Institute, Palo Alto, CA.

EPRI. 1994. High-efficiency electric technology fact sheet: Commercial heat pump water heaters. BR-103415. Electric Power Research Institute, Palo Alto, CA.

GAMA. 1997. Consumers' directory of certified efficiency rating for residential heating and water heating equipment. Gas Appliance Manufacturers Association, Arlington, VA. April issue.

Goldner, F.S. 1994a. Energy use and domestic hot water consumption: Final Report—Phase 1. Report No. 94-19. New York State Energy Research and Development Authority, Albany, NY.

Goldner, F.S. 1994b. DHW system sizing criteria for multifamily buildings. *ASHRAE Transactions* 100(1):963-77.

Goldner, F.S. and D.C. Price. 1996. DHW modeling: System sizing and selection criteria, Phase 2. Interim Project Research Report #1. New York State Energy Research and Development Authority, Albany.

Hiller, C.C. 1998. New hot water consumption analysis and water-heating system sizing methodology. *ASHRAE Transactions* 104(1B):1864-77.

Hunter, R.B. 1941. Water distributing systems for buildings. National Bureau of Standards *Report* BMS 79.

Kweller, E.R. 1992. Derivation of the combined annual efficiency of space/water heaters in ASHRAE 124-1991. *ASHRAE Transactions* 98(1):665-75.

Manian, V.S. and W. Chackeris. 1974. Off peak domestic hot water systems for large apartment buildings. *ASHRAE Transactions* 80(1):147-65.

NFPA. 1996. National fuel gas code. *Standard* 54, also ANSI Z223.1. National Fire Protection Association, Quincy, MA.

NSF. 1992. Water heaters, hot water supply boilers, and heat recovery equipment. *Standard* 5-92. National Sanitation Foundation, Ann Arbor, MI.

Perlman, M. and B. Mills. 1985. Development of residential hot water use patterns. *ASHRAE Transactions* 91(2A):657-79.

Pietsch, J.A. and S.G. Talbert. 1989. Equipment sizing procedures for combination space-heating/water-heating systems. *ASHRAE Transactions* 95(2):250-58.

Pietsch, J.A., S.G. Talbert, and S.H. Stanbouly. 1994. Annual cycling characteristics of components in gas-fired combination space/water systems. *ASHRAE Transactions* 100(1):923-34.

Subherwal, B.R. 1986. Combination water-heating/space-heating appliance performance. *ASHRAE Transactions* 92(2B):415-32.

Talbert, S.G., G.H. Stickford, D.C. Newman, and W.N. Stiegelmeyer. 1986. The effect of hard water scale buildup and water treatment on residential water heater performance. *ASHRAE Transactions* 92(2B):433-47.

Talbert, S.G., J.G. Murray, R.A. Borgeson, V.P. Kam, and J.A. Pietsch. 1992. Operating characteristics and annual efficiencies of combination space/water-heating systems. *ASHRAE Transactions* 98(1):655-64.

Toaborek, J. et al. 1972. Fouling—The major unresolved problem in heat transfer. *Chemical Engineering Progress* (February):59.

Thrasher, W.H. and D.W. DeWerth. 1994. New hot-water use data for five commercial buildings (RP-600). *ASHRAE Transactions* 100(1):935-47.

Werden, R.G. and L.G. Spielvogel. 1969. Sizing of service water heating equipment in commercial and institutional buildings, Part I. *ASHRAE Transactions* 75(I):81.

Werden, R.G. and L.G. Spielvogel. 1969. Sizing of service water heating equipment in commercial and institutional buildings, Part II. *ASHRAE Transactions* 75(II):181.

BIBLIOGRAPHY

AGA. Comprehensive on commercial and industrial water heating. Catalog No. R-00980. American Gas Association, Cleveland, OH.

AGA. 1962. Water heating application in coin operated laundries. Catalog No. C-10540. American Gas Association, Cleveland, OH.

AGA. 1965. *Gas engineers handbook*. American Gas Association, Cleveland, OH.

Brooks, F.A. Use of solar energy for heating water. Smithsonian Institution, Washington, D.C.

Carpenter, S.C. and J.P. Kokko. 1988. Estimating hot water use in existing commercial buildings. *ASHRAE Transactions* 94(2):3-12

Coleman, J.J. 1974. Waste water heat reclamation. *ASHRAE Transactions* 80(2):370.

EPRI. 1992. WATSIM 1.0: Detailed water heating simulation model user's manual. TR-101702. Electric Power Research Institute, Palo Alto, CA.

EPRI. 1993. HOTCALC 2.0: Commercial water heating performance simulation tool, Version 2.0. SW-100210-R1. Electric Power Research Institute, Palo Alto, CA.

GRI. 1993. TANK computer program user's manual with diskettes. GRI-93/0186 *Topical Report* available only to licensees. Gas Research Institute.

Hebrank, E.F. 1956. Investigation of the performance of automatic storage-type gas and electric domestic water heaters. *Engineering Experiment Bulletin* No. 436. University of Illinois.

Jones, P.G. 1982. The consumption of hot water in commercial building. *Building Services Engineering, Research and Technology* 3:95-109.

Olivares, T.C. 1987. Hot water system design for multi-residential buildings. Report No. 87-239-K. Ontario Hydro Research Division.

Schultz, W.W. and V.W. Goldschmidt. 1978. Effect of distribution lines on stand-by loss of service water heater. *ASHRAE Transactions* 84(1):256-65.

Smith, F.T. 1965. Sizing guide for gas water heaters for in-ground swimming pools. Catalog No. R-00999. American Gas Association, Cleveland, OH.

Vine, E., R. Diamond, and R. Szydlowski. 1987. Domestic hot water consumption in four low income apartment buildings. *Energy* 12(6).

Wetherington, T.I., Jr. 1975. Heat recovery water heating. *Building Systems Design* (December/January).

SNOW MELTING

THE practicality of melting snow by supplying heat to the snow-covered surface has been demonstrated in a large number of installations, including sidewalks, roadways, ramps, and runways. Melting eliminates the need for snow removal, provides greater safety for pedestrians and vehicles, and reduces the labor of slush removal.

This chapter covers three types of snow-melting systems:

1. Hot fluid circulated in embedded pipes (hydronic)
2. Embedded electric heater cables or wire (electric)
3. Overhead high-intensity infrared radiant heating (infrared)

Components of the system design include (1) heating requirement, (2) pavement design, (3) control, and (4) hydronic or electric system design.

HEATING REQUIREMENT (HYDRONIC AND ELECTRIC)

The heat required for snow melting depends on five atmospheric factors: (1) rate of snowfall, (2) air dry-bulb temperature, (3) humidity, (4) wind speed, and (5) apparent sky temperature. The dimensions of the snow-melting slab affect the heat and mass transfer rates at the surface. Other factors such as back and edge heat losses must be considered in the complete design.

The processes that establish the heating requirement at the snow-melting surface can be described by inspecting the terms in the following equation, which is the steady-state energy balance for the required total heat flux (power per unit surface area) q_o at the upper surface of a snow-melting slab during snowfall. The general discussion of the heat balance will be followed by a detailed description of how each of the terms is evaluated.

$$q_o = q_s + q_m + A_r(q_h + q_e) \qquad (1)$$

where

q_o = total required heat flux, Btu/h·ft^2
q_s = sensible heat flux, Btu/h·ft^2
A_r = snow-free area ratio, dimensionless
q_m = latent heat flux, Btu/h·ft^2
q_h = convective and radiative heat flux from snow-free surface, Btu/h·ft^2
q_e = heat flux needed for evaporation, Btu/h·ft^2

Heat Balance

Sensible and Latent Heat Requirements. The sensible heat flux q_s is the heat flux required to raise the temperature of the snow falling on the slab to the melting temperature plus, after the snow has been melted, to raise the temperature of the liquid to the assigned temperature t_f of the liquid film. The snow is assumed to fall at atmospheric temperature t_a. The latent heat flux q_m is the heat flux required to melt the snow. Under steady-state conditions, both q_s and q_m are directly proportional to the snowfall rate s.

The preparation of this chapter is assigned to TC 6.1, Hydronic and Steam Equipment and Systems.

Free Area Ratio. The heating loads due to sensible and latent (melting) heat flux are imposed on the entire slab during snowfall. On the other hand, the rates of heat and mass transfer from the surface depend on whether there is a snow layer on the surface of the slab. Any snow accumulation on the slab acts to partially insulate the surface from heat losses and evaporation. The insulating effect of partial snow cover can be large. Because snow may cover a portion of the slab area, it is convenient to think of the insulating effect in terms of an effective or equivalent snow-covered area A_s, which is perfectly insulated and from which no evaporation occurs. The balance is then considered to be the equivalent snow-free area A_f. This area is assumed to be completely covered with a thin liquid film; therefore, both heat and mass transfer occur at the maximum rates for the existing environmental conditions. It is convenient to define a dimensionless **snow-free area ratio** A_r:

$$A_r = \frac{A_f}{A_t} \qquad (2)$$

where

A_f = equivalent snow-free area, ft^2
A_s = equivalent snow-covered area, ft^2
$A_t = A_f + A_s$ = total area, ft^2

Therefore,

$$0 \le A_r \le 1$$

For $A_r = 1$, the system must melt snow rapidly enough that no accumulation occurs. For $A_r = 0$, the surface must be covered with snow of sufficient thickness to prevent heat and evaporation losses. Practical snow-melting systems operate somewhere between these limits. Earlier studies indicate that sufficient snow-melting system design information is obtained by considering three values of the free area ratio: 0, 0.5, and 1.0 (Chapman 1952).

Heat Losses due to Surface Convection, Radiation, and Evaporation. Using the concept of the snow-free area ratio, the appropriate heat and mass transfer relations can then be written for the snow-free fraction of the slab, A_r. These appear as the third term on the right-hand side of Equation (1). On the snow-free surface, maintained at film temperature t_f, there is heat transfer to the surroundings and evaporation from the liquid film. The heat flux q_h includes the convective losses to the ambient air at temperature t_a and radiative losses to the surroundings, which are at a mean radiant temperature T_{MR}. The convection heat transfer coefficient is a function of the wind speed and a characteristic dimension of the snow-melting surface. This heat transfer coefficient is also a function of the thermodynamic properties of the air, which vary slightly over the temperature range for various snowfall events. The mean radiant temperature depends on air temperature, relative humidity, cloudiness, cloud height, and whether precipitation is falling.

The heat flux q_e needed for the evaporation is equal to the evaporation rate multiplied by the heat of vaporization. The evaporation rate is driven by the difference in vapor pressure between the wet surface of the snow-melting slab and the ambient air. The mass transfer coefficient is a function of the wind speed, a characteristic dimension of the slab, and the thermodynamic properties of the ambient air.

Heat Flux Equations

Sensible Heat Flux. The sensible heat flux q_s is given by the following equation. The snow is assumed to fall at temperature t_a.

$$q_s = \rho_{water} s [c_{p, ice}(t_s - t_a) + c_{p, water}(t_f - t_s)] / c_1 \qquad (3)$$

where

$c_{p, ice}$ = specific heat of ice, Btu/lb·°F
$c_{p, water}$ = specific heat of water, Btu/lb·°F
s = snowfall rate, in. of liquid water equivalent per hour
t_a = ambient temperature, °F
t_f = liquid film temperature, °F
t_s = melting temperature, °F
ρ_{water} = density of water, lb/ft^3
c_1 = 12 in/ft

The density of water, specific heat of ice, and specific heat of water are approximately constant over the temperature range of interest and are evaluated at 32°F. The ambient temperature and snowfall rate are available from the weather data. The liquid film temperature is usually taken as 33°F.

Melting Heat Flux. The heat flux q_m required to melt the snow is given by the following equation:

$$q_m = \rho_{water} s h_{if} / c_1 \qquad (4)$$

where h_{if} = heat of fusion of snow, Btu/lb.

Convective and Radiative Heat Flux from a Snow-Free Surface. The corresponding heat flux q_h is given by the following equation:

$$q_h = h_c(t_f - t_a) + \sigma \varepsilon_s (T_f^4 - T_{MR}^4) \qquad (5)$$

where

h_c = convection heat transfer coefficient for turbulent flow, Btu/h·ft^2·°F
T_f = liquid film temperature, °R
T_{MR} = mean radiant temperature of surroundings, °R
σ = Stefan-Boltzmann constant = 0.1714 × 10^{-8} Btu/h·ft^2·°R^4
ε_s = emittance of surface

The convection heat transfer coefficient over the slab is given by the following equation (Incropera and DeWitt 1996):

$$h_c = 0.037 \left(\frac{k_{air}}{L} \right) \text{Re}_L^{0.8} \text{Pr}^{1/3} \qquad (6)$$

where

k_{air} = thermal conductivity of air at t_a, Btu·ft/h·ft^2·°F
L = characteristic length of slab measured in the direction of the wind, ft
Pr = Prandtl number of the air, taken as Pr = 0.7
Re_L = Reynolds number based on characteristic length L

and

$$\text{Re}_L = \frac{VL}{v_{air}} c_2 \qquad (7)$$

where

V = design wind speed, mph
v_{air} = kinematic viscosity of air, ft^2/h
c_2 = 5280 ft/mile

From Equations (6) and (7), it can be seen that the turbulent convective heat transfer coefficient is a function of $L^{-0.2}$. Because of this relationship, shorter snow-melting slabs have higher convective heat transfer coefficients than longer snow-melting slabs. For design, the shortest dimension should be used (e.g., for a long narrow driveway or sidewalk, use the shorter width). A snow-melting slab length L = 20 ft is used in the heat transfer calculations which resulted in Tables 1, 2, and 3.

The **mean radiant temperature** T_{MR}, which appears in Equation (5), is the equivalent blackbody temperature of the surroundings of the snow-melting slab. Under snowfall conditions, the entire surroundings are approximately at the ambient air temperature (i.e., $T_{MR} = T_a$). When there is no snow precipitation (e.g., during idling and after snowfall operations for $A_r < 1$), the mean radiant temperature is approximated by the following equation:

$$T_{MR} = [T_{cloud}^4 F_{sc} + T_{sky\ clear}^4 (1 - F_{sc})]^{1/4} \qquad (8)$$

where

F_{sc} = fraction of the radiation exchange that takes place between the slab and clouds
T_{cloud} = temperature of clouds, °R
$T_{sky\ clear}$ = temperature of clear sky, °R

The equivalent blackbody temperature of a clear sky is primarily a function of the ambient air temperature and the water content of the atmosphere. An approximation for the clear sky temperature is given by the following equation, which is a curve fit of data in Ramsey et al. (1982).

$$T_{sky\ clear} = T_a - (1.99036 \times 10^3 - 7.562 T_a$$
$$+ 7.407 \times 10^{-3} T_a^2 - 56.325 \phi + 26.25 \phi^2) \qquad (9)$$

where

T_a = ambient temperature, °R
ϕ = relative humidity of the air at the elevation for which typical weather measurements are made, decimal

The cloud-covered portion of the sky is assumed to be at T_{cloud}. The height of the clouds may be assumed to be 10,000 ft. The temperature of the clouds at 10,000 ft is calculated by subtracting the product of the average lapse rate (rate of decrease of atmospheric temperature with height), and the altitude from the atmospheric temperature T_a. The average lapse rate, determined from the tables of U.S. Standard Atmospheres (1962), is 3.5°F per 1000 ft of elevation (Ramsey et al. 1982). Therefore, for clouds at 10,000 ft,

$$T_{cloud} = T_a - 35 \qquad (10)$$

Under most conditions, this method of approximating the temperature of the clouds provides an acceptable estimate. However, when the atmosphere contains a very high water content, the temperature calculated for a clear sky using Equation (9) may be warmer than the temperature estimated for the clouds using Equation (10). When that condition exists, the temperature T_{cloud} of the clouds is set equal to the calculated clear sky temperature $T_{sky\ clear}$.

Evaporation Heat Flux. The heat flux q_e required to evaporate water from a wet surface is given by the following equation.

$$q_e = \rho_{dry\ air} h_m (W_f - W_a) h_{fg} \qquad (11)$$

where

h_m = mass transfer coefficient, ft/h
W_a = humidity ratio of ambient air, lb$_{vapor}$/lb$_{air}$
W_f = humidity ratio of saturated air at the film surface temperature, lb$_{vapor}$/lb$_{air}$

h_{fg} = heat of vaporization (enthalpy difference between saturated water vapor and saturated liquid water), Btu/lb

$\rho_{dry\,air}$ = density of dry air, lb/ft^3

The determination of the mass transfer coefficient is based on the analogy between heat transfer and mass transfer. A detailed discussion of the analogy is given in Chapter 5 of the 1997 *ASHRAE Handbook—Fundamentals*. For external flow where mass transfer occurs at the convective surface and the water vapor component is dilute, the following equation relates the mass transfer coefficient h_m to the heat transfer coefficient h_c [Equation (6)]:

$$h_m = \left(\frac{Pr}{Sc}\right)^{2/3} \frac{h_c}{\rho_{dry\,air}c_{p,\,air}} \qquad (12)$$

where Sc = Schmidt number. In applying Equation (11), the values Pr = 0.7 and Sc = 0.6 are used to generate the values in Tables 1 through 4.

The humidity ratios both in the atmosphere and at the surface of the water film are calculated using the standard psychrometric relation given in the following equation (from Chapter 6 of the 1997 *ASHRAE Handbook—Fundamentals*).

$$W = 0.622\left(\frac{p_v}{p - p_v}\right) \qquad (13)$$

where

p = atmospheric pressure
p_v = partial pressure of water vapor

The atmospheric pressure in Equation (13) is corrected for altitude using the following equation (Kuehn et al. 1998):

$$p = p_{std}\left(1 - \frac{Az}{T_o}\right)^{5.265} \qquad (14)$$

where

p_{std} = standard atmospheric pressure
A = 0.00356 °R/ft
z = altitude of the location above sea level, ft
T_o = 518.7°R

The vapor pressure p_v for the calculation of W_a is equal to the saturation vapor pressure p_s at the dew-point temperature of the air. Saturated conditions exist at the water film surface. Therefore, the vapor pressure used in calculating W_f is the saturation pressure at the film temperature t_f. The saturation partial pressures of water vapor for temperatures above and below freezing can be found in tables of the thermodynamic properties of water at saturation or can be calculated using appropriate equations. Both are presented in Chapter 6 of the 1997 *ASHRAE Handbook—Fundamentals*.

Heat Load Calculations. Equations (1) through (14) can be used to determine the required heat loads of a snow-melting system. However, calculations must be made for coincident values of the climatic factors, snowfall rate, wind speed, ambient temperature, and dew-point temperature (or another measure of humidity). By computing the load for each snowfall hour over a period of several years, a frequency distribution of hourly loads can be developed. Annual averages or maximums for climatic factors should never be used in sizing a system because they are unlikely to occur simultaneously. Finally, it is critical for the designer to note that the above analysis only describes what is happening at the upper surface of the snow-melting surface. Edge losses and back losses have not been taken into account.

Example 1. During the snowfall that occurred during the 8 P.M. hour on December 26, 1985, in the Detroit metropolitan area, the following simultaneous conditions existed: air dry-bulb temperature = 17°F,

dew-point temperature = 14°F, wind speed = 19.7 mph, and snowfall rate = 0.10 in. of liquid water equivalent per hour. Assuming L = 20 ft, Pr = 0.7, and Sc = 0.6, calculate the heat flux (load) q_o for a snow-free area ratio of A_r = 1.0. The thermodynamic and transport properties used in the calculation are taken from the Chapters 6 and 36 of the 1997 *ASHRAE Handbook—Fundamentals*.

Solution:

By Equation (3),

$$q_s = 62.4 \times \frac{0.10}{12}[0.49(32 - 17) + 1.0(33 - 32)] = 4.3 \text{ Btu/h} \cdot \text{ft}^2$$

By Equation (4),

$$q_m = 62.4 \times \frac{0.10}{12} \times 143.3 = 74.5 \text{ Btu/h} \cdot \text{ft}^2$$

By Equation (7),

$$Re_L = \frac{19.7 \times 20 \times 5280}{0.49} = 4.24 \times 10^6$$

By Equation (6),

$$h_c = 0.037\left(\frac{0.0135}{20}\right)(4.24 \times 10^6)^{0.8}(0.7)^{1/3} = 4.44 \text{ Btu/h} \cdot \text{ft}^2 \cdot °F$$

By Equation (5),

$$q_h = 4.44(33 - 17) + (0.1714 \times 10^{-8})(1.0)(493^4 - 477^4)$$

$$= 83.6 \text{ Btu/h} \cdot \text{ft}^2$$

By Equation (12),

$$h_m = \left(\frac{0.7}{0.6}\right)^{2/3}\frac{4.44}{0.083 \times 0.24} = 247 \text{ ft/h}$$

Obtain the values of the saturation vapor pressures at the dew-point temperature 14°F and the film temperature 33°F from Table 3 in Chapter 6 of the 1997 *ASHRAE Handbook—Fundamentals*. Then, use Equation (13) to obtain W_a = 0.00160 lb$_{vapor}$/lb$_{air}$ and W_f = 0.00393 lb$_{vapor}$/lb$_{air}$. By Equation (11),

$$q_o = 0.083 \times 247(0.00393 - 0.00160) \times 1075 = 51.3 \text{ Btu/h} \cdot \text{ft}^2$$

By Equation (1),

$$q_o = 4.3 + 74.5 + 1.0(83.6 + 51.3) = 214 \text{ Btu/h} \cdot \text{ft}^2$$

It should be emphasized that this is the heat flux needed at the snow-melting surface of the slab. Back and edge losses must be added as discussed in the section on Back and Edge Heat Losses.

Weather Data and Load Calculation Results

Table 1 shows frequencies of snow-melting loads for 46 cities in the United States (Ramsey et al. 1999). For the calculations, the temperature of the surface of the snow-melting slab was taken to be 33°F. Any time the ambient temperature was below 32°F and it was not snowing, it was assumed that the system was idling (i.e., that heat was supplied to the slab so that melting would start immediately when snow began to fall).

Weather data were taken for the years 1982 through 1993. These years were selected because of their completeness of data. The weather data included hourly values of the precipitation amount in equivalent depth of liquid water, precipitation type, ambient dry-bulb and dew-point temperatures, wind speed, and sky cover. All weather elements for the years 1982 to 1990 were obtained from the *Solar and Meteorological Surface Observation Network 1961 to 1990 (SAMSON), Version 1.0* (NCDC 1993). For the years 1991 to 1993, all weather elements except precipitation were taken from *DATSAV2* data obtained from the National Climatic Data Center as described in Colliver et al. (1998). The precipitation data for these years were taken from NCDC's *Hourly Cooperative Dataset* (NCDC 1990).

Table 1 Frequencies of Snow-Melting Loads [a]

Location	Snowfall Hours per Year	Snow-Free Area Ratio	Loads Not Exceeded During Indicated Percentage of Snowfall Hours from 1982 Through 1993, Btu/h·ft²					
			75%	90%	95%	98%	99%	100%
Albany, NY	156	1	89	125	149	187	212	321
		0.5	60	86	110	138	170	276
		0	37	62	83	119	146	276
Albuquerque, NM	44	1	70	118	168	191	242	393
		0.5	51	81	96	117	156	229
		0	30	46	61	89	92	194
Amarillo, TX	64	1	113	150	168	212	228	318
		0.5	71	88	108	124	142	305
		0	24	46	62	89	115	292
Billings, MT	225	1	112	164	187	212	237	340
		0.5	64	89	102	116	128	179
		0	22	33	45	60	68	113
Bismarck, ND	158	1	151	199	231	275	307	477
		0.5	83	107	124	148	165	243
		0	16	30	39	60	73	180
Boise, ID	85	1	58	79	100	126	146	203
		0.5	38	52	66	80	89	164
		0	22	31	40	53	62	164
Boston, MA	112	1	96	137	165	202	229	365
		0.5	65	95	112	149	190	365
		0	37	75	93	121	172	365
Buffalo, NY	292	1	115	166	210	277	330	570
		0.5	68	97	127	164	188	389
		0	23	39	55	93	112	248
Burlington, VT	204	1	91	130	154	184	200	343
		0.5	58	78	92	113	128	343
		0	23	40	55	78	94	343
Cheyenne, WY	224	1	119	172	201	229	261	354
		0.5	70	97	111	132	149	288
		0	16	37	52	77	100	285
Chicago, IL, O'Hare Int'l AP	124	1	96	126	153	186	235	521
		0.5	58	77	94	113	137	265
		0	23	38	53	75	83	150
Cleveland, OH	188	1	85	124	157	195	230	432
		0.5	52	73	92	118	147	235
		0	23	37	47	69	92	225
Colorado Springs, CO	159	1	89	135	167	202	219	327
		0.5	57	82	99	124	140	218
		0	23	45	61	87	112	165
Columbus, OH, Int'l AP	92	1	71	101	123	149	175	328
		0.5	45	60	71	87	95	184
		0	15	30	45	60	62	135
Des Moines, IA	127	1	120	174	208	255	289	414
		0.5	74	102	120	149	180	310
		0	24	46	69	94	108	231
Detroit, MI, Metro	153	1	92	130	156	192	212	360
		0.5	57	77	94	118	134	227
		0	23	38	47	75	89	194
Duluth, MN	238	1	123	171	201	238	250	370
		0.5	71	97	114	131	142	213
		0	22	32	46	68	77	196
Ely, NV	153	1	67	97	116	134	162	242
		0.5	44	66	83	111	129	241
		0	23	45	67	97	112	240
Eugene, OR	18	1	59	110	139	165	171	224
		0.5	47	77	93	119	122	164
		0	30	53	70	102	120	164
Fairbanks, AK	288	1	91	121	144	174	202	391
		0.5	52	68	78	94	108	200
		0	15	23	31	40	48	87
Baltimore, MD, BWI Airport	56	1	87	139	172	235	282	431
		0.5	69	108	147	200	238	369
		0	46	84	119	181	214	306
Great Falls, MT	233	1	123	171	193	233	276	392
		0.5	71	93	107	129	144	210
		0	17	31	45	60	75	143
Indianapolis, IN	96	1	95	134	158	194	215	284
		0.5	58	80	96	116	124	209
		0	23	38	52	83	99	209

[a] Loads are losses from top surface only. See text for calculation of back and edge losses.

Table 1 Frequencies of Snow-Melting Loads [a] (*Continued*)

Location	Snowfall Hours per Year	Snow-Free Area Ratio	Loads Not Exceeded During Indicated Percentage of Snowfall Hours from 1982 Through 1993, Btu/h·ft²					
			75%	90%	95%	98%	99%	100%
Lexington, KY	50	1	81	108	123	150	170	233
		0.5	49	65	74	85	95	197
		0	16	30	39	46	55	162
Madison, WI	161	1	99	138	164	206	241	449
		0.5	61	82	98	129	163	245
		0	23	39	60	91	113	194
Memphis, TN	13	1	106	141	172	200	206	213
		0.5	75	96	115	118	130	157
		0	40	75	76	90	97	123
Milwaukee, WI	161	1	101	135	164	196	207	431
		0.5	62	83	101	128	147	246
		0	23	46	68	98	120	239
Minneapolis-St. Paul, MN	199	1	119	169	193	229	254	332
		0.5	73	99	114	138	154	287
		0	23	45	61	91	113	245
New York, NY, JFK Airport	61	1	91	134	164	207	222	333
		0.5	63	93	118	145	164	325
		0	38	68	86	113	133	316
Oklahoma City, OK	35	1	117	168	215	248	260	280
		0.5	72	101	123	133	144	208
		0	24	46	68	78	113	190
Omaha, NE	94	1	108	148	189	222	259	363
		0.5	65	89	105	128	135	186
		0	23	38	60	90	100	136
Peoria, IL	91	1	95	139	166	201	227	436
		0.5	58	83	99	119	130	250
		0	23	38	53	76	92	228
Philadelphia, PA, Int'l AP	56	1	94	129	154	208	246	329
		0.5	65	90	112	162	185	267
		0	38	63	79	111	150	225
Pittsburgh, PA, Int'l AP	168	1	83	125	159	194	219	423
		0.5	51	75	94	111	129	216
		0	16	31	46	68	77	136
Portland, ME	157	1	120	168	195	234	266	428
		0.5	76	108	132	168	199	376
		0	39	67	90	130	152	324
Portland, OR	15	1	50	78	102	177	239	296
		0.5	39	55	81	114	130	199
		0	23	45	60	78	102	128
Rapid City, SD	177	1	139	203	252	312	351	482
		0.5	78	111	132	164	183	245
		0	16	30	38	53	65	179
Reno, NV	63	1	50	72	89	116	137	191
		0.5	36	55	75	105	115	172
		0	23	45	68	91	113	159
Salt Lake City, UT	142	1	52	77	89	110	120	171
		0.5	39	62	76	96	104	171
		0	30	60	75	89	104	171
Sault Ste. Marie, MI	425	1	112	153	183	216	249	439
		0.5	66	88	104	125	142	239
		0	23	37	47	68	83	188
Seattle, WA	27	1	56	107	138	171	205	210
		0.5	45	72	97	122	133	175
		0	37	52	75	96	123	151
Spokane, WA	144	1	67	98	116	141	159	227
		0.5	45	61	73	84	95	145
		0	23	37	45	54	67	112
Springfield, MO	58	1	110	155	179	215	224	292
		0.5	70	95	117	142	171	240
		0	32	54	76	115	129	227
St. Louis, MO, Int'l AP	62	1	97	147	170	193	227	344
		0.5	66	90	105	126	144	269
		0	31	53	68	97	104	194
Topeka, KS	61	1	102	153	192	234	245	291
		0.5	64	92	110	132	139	185
		0	23	39	52	68	84	167
Wichita, KS	60	1	115	163	209	248	285	326
		0.5	71	96	116	137	153	168
		0	24	45	57	75	83	158

[a] Loads are losses from top surface only. See text for calculation of back and edge losses.

All wind speeds used in the calculations were taken directly from the weather data. Wind speed is usually measured at an elevation of approximately 33 ft and will be referred to as V_{met}. As indicated in the section on Heat Balance, the heat and mass transfer coefficients are functions of a characteristic dimension of the snow-melting slab. The dimension used in generating the values of Table 1 was 20 ft. Sensitivity of the load to both the wind speed and the characteristic dimension is included in Table 2. During snowfall conditions, the sky temperature was taken as equal to the ambient temperature.

The first data column in Table 1 presents the average number of snowfall hours per year for each location. All loads were computed for snow-free area ratios of 1, 0.5, and 0, and the frequency of the snow-melting loads are presented. This frequency indicates the percentage of time that the required snow-melting load does not exceed the value in the table for the particular area ratio.

Figures 1 and 2 are maps showing the distribution of snow-melting loads for snow-free area ratios of 1.0 and 0, respectively. The values presented are those needed to satisfy the loads 99% of the time (as listed in the 99% column of Table 1). These maps show how the loads vary throughout the United States. If local data are unavailable, values can be approximated by interpolating between values given on the figure. However, care must be taken because special local climatological conditions exist for some areas (e.g., lake effect snows).

Example for Heat Load Calculation Using Table 1

Example 2. Consider the design of a system for Albany, New York, which has an installed heat flux capacity at the top surface of approximately 149 Btu/h·ft². Based on the data in Table 1, this system will keep the surface completely free of snow 95% of the time. Because there are 156 snowfall hours in an average year, this design would have some accumulation of snow during approximately 8 h per year (the remaining 5% of the 156 h). This design will also meet the load for more than 98% of the time (i.e., more than 153 of the 156 snowfall hours) at an area ratio of 0.5 and more than 99% of the time for an area ratio of 0. $A_r = 0.5$ means that there is a thin layer of snow over all or part of the slab such that it acts as though half the slab is insulated by a snow layer; $A_r = 0$ means that a snow layer exists which is sufficient to insulate the surface from heat and evaporation losses, but that snow is melting at the base of this layer at the same rate that it is falling on the top of the layer.

Therefore, the results for this system can be interpreted to mean the following (all times are rounded to the nearest hour):

(a) For all but 8 h of the year, the slab will be snow-free.
(b) For the less than 5 h between the 95% and 98% nonexceedance values, there will be a thin build-up of snow on part of the slab.
(c) For the less than 2 h between the 98% and 99% nonexceedance values, snow will accumulate on the slab to a thickness at which the snow blanket insulates the slab, but the thickness will not increase beyond that level.
(d) For less than 2 h, the system can not keep up with the snowfall.

An examination of the 100% column shows that to keep up with the snowfall the last 1% of the time, in this case less than 2 h for an average year, would require a system capacity of approximately 276 Btu/h·ft²; to attempt to keep the slab completely snow-free the entire season requires a capacity of 321 Btu/h·ft². Based on this interpretation of the data in Table 1, the designer and customer must decide what are acceptable operating conditions. Note that the heat flux values in this example do not include back or edge losses, which must be added in sizing energy and supply systems.

Sensitivity of Design Load to Wind Speed and Surface Size

Some snow-melting systems are in sheltered areas, while others may exist in locations where surroundings create a wind tunnel effect. Similarly, systems vary in size from the baseline characteristic length of 20 ft. For example, sidewalks exposed to a crosswind may have a characteristic length on the order of 5 ft. In such cases,

Table 2 Mean Sensitivity of Snow-Melting Loads to Wind Speed and Slab Length

For loads not exceeded during 99% of snowfall hours, 1982 through 1993

Snow-Free Area Ratio, A_r	Ratio of Load at Stated Condition to Load at $L = 20$ ft and $V = V_{met}$				
	$L = 20$ ft		$L = 5$ ft		
	$V = 0.5V_{met}$	$V = 2V_{met}$	$V = V_{met}$	$V = 0.5V_{met}$	$V = 2V_{met}$
1	0.7	1.6	1.2	0.8	2.0
0.5	0.8	1.4	1.2	0.9	1.7
0	1.0	1.0	1.0	1.0	1.0

Note: Based on data from U.S. locations.
   L = characteristic length
   V_{met} = meteorological wind speed from NCDC

the wind speed will be either less than or greater than the meteorological value V_{met} used in establishing Table 1. To establish the sensitivity of the load to wind speed and characteristic length, calculations were performed at combinations of wind speeds $0.5V_{met}$, V_{met}, and $2V_{met}$ and L-values of 5 ft and 20 ft for area ratios A_r of 1.0 and 0.5. Wind speed and system size do not affect the load values for $A_r = 0$ because the calculations assume that no heat or mass transfer from the surface occurs at this condition. Table 2 presents a set of mean values of multipliers that can be applied to the loads presented in Table 1; these multipliers were established by looking at the effect on the 99% nonexceedance values. The designer is cautioned that these are to be used only as guidelines on the impact of wind speed and size variations.

Back and Edge Heat Losses

The heat loads presented in Table 1 do not account for heat losses from the back and edges of the slab. Adlam (1950) demonstrated that these back and edge losses may vary from 4 to 50%, depending on factors such as pavement construction, operating temperature, ground temperature, and back and edge insulation and exposure. With the construction shown in Figure 3 or Figure 5 and ground temperature of 40°F at a depth of 24 in., back losses are on the order of 20%. Higher losses would be expected with (1) colder ground, (2) more cover over the pipe or cable, or (3) exposed back, such as on bridges or parking decks.

Annual Operating Data

Annual operating data for the cities included in Table 1 are presented in Table 3. The number of melting and idling hours are summarized in this table along with the energy per unit area needed to operate the system during an average year based on calculations performed for the years 1982 to 1993. Back and edge losses are not included in the energy values in Table 3. Data are presented for snow-free area ratios of 1.0, 0.5, and 0.

The melting loads are based on systems designed to satisfy the loads 99% of the time (i.e., at the levels indicated in the 99% column in Table 1) for each of the three area ratios. The energy use for each melting hour is taken as either (1) the actual output required to maintain the surface at 33°F or (2) the design output, whichever is less. The design output differs, of course, depending on whether the design is for $A_r = 1.0$, 0.5, or 0; therefore, the annual melting energy differs as well.

The idling hours include all non-snowfall hours when the ambient temperature is below 32°F. The surface is dry; therefore, Equation (5) approximates the surface heat loss. The mean radiant temperature that appears in Equation (5) is evaluated using Equations (8), (9), and (10). The fraction F_{sc} of radiation between the surface and the clouds is equal to the cloud cover fraction in the meteorological data. The energy consumption for each idling hour is taken as either (1) the actual output required to maintain the surface at 32°F or (2) the design output, whichever is less.

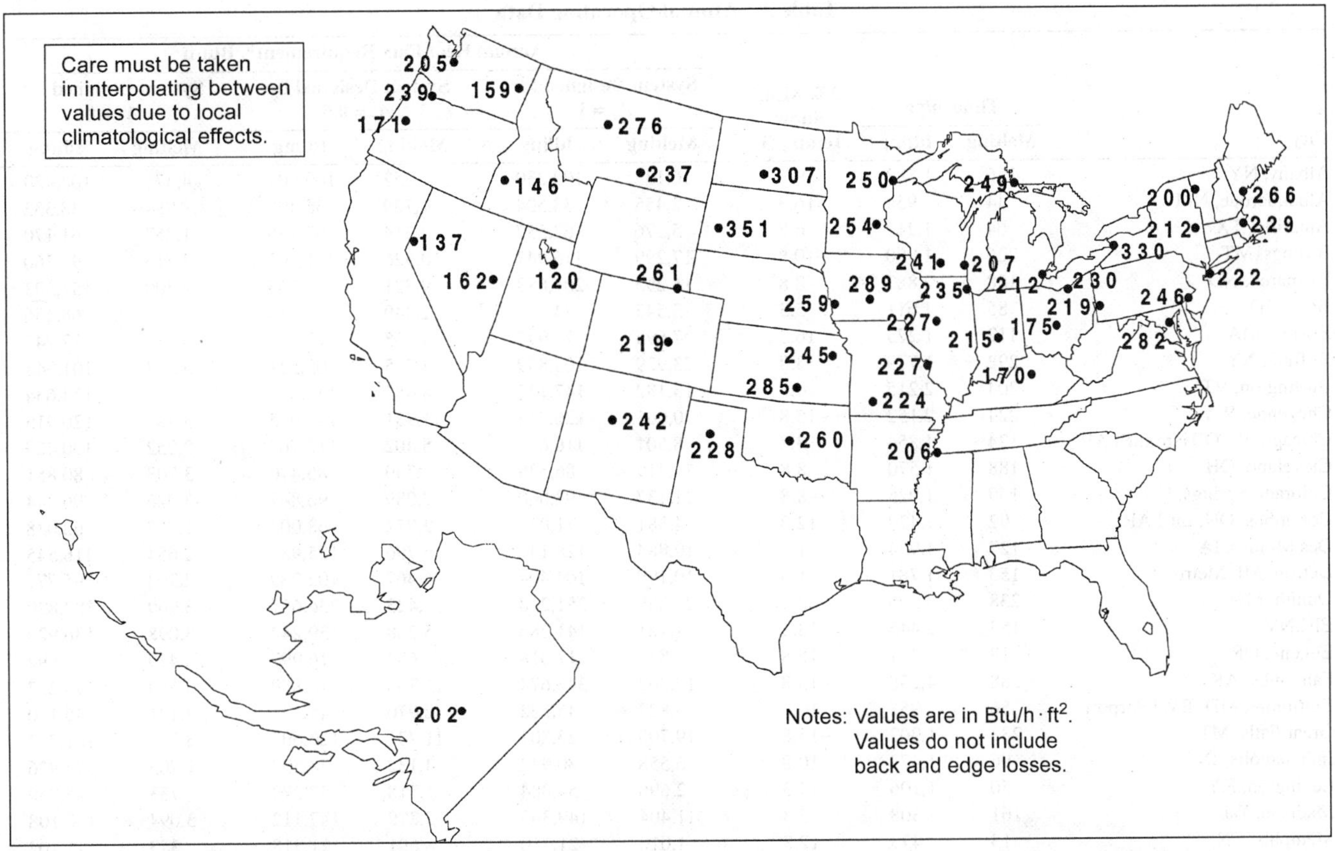

Care must be taken
in interpolating between
values due to local
climatological effects.

205
239 159
171

276
146 237 307 250 249 200 266
137 351 254 212 229
162 120 261 241 207 330
289 235 212 230 222
259 227 219 246
215 175
219 245 227 282
285 224 170
242 206
228 260

202

Notes: Values are in Btu/h · ft².
Values do not include
back and edge losses.

Fig. 1   Snow-Melting Loads Required to Provide a Snow-Free Area Ratio of 1.0 for 99% of the Time

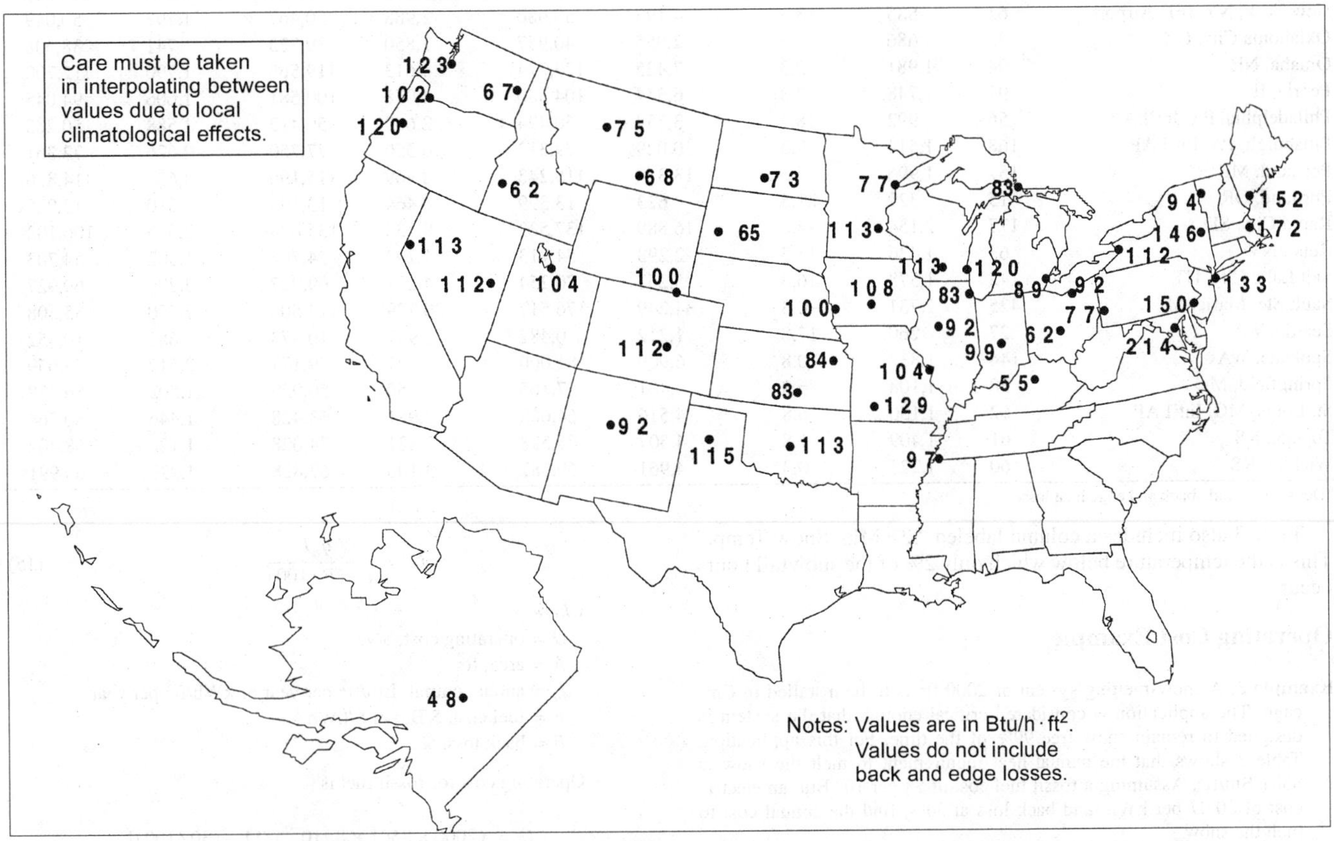

Care must be taken
in interpolating between
values due to local
climatological effects.

123
102 67
120

75
62 68 73 77 83 94 152
65 113 146 172
113 112 133
112 104 100 113 120
108 83 89 92 150
100 92 77 214
84 92 99 62
104 55
83 129
92 115 113 97

48

Notes: Values are in Btu/h · ft².
Values do not include
back and edge losses.

Fig. 2   Snow-Melting Loads Required to Provide a Snow-Free Area Ratio of 0 for 99% of the Time

Table 3 Annual Operating Data

City	Time, h/yr Melting	Time, h/yr Idling	2% Min. Snow Temp., °F	System Designed for $A_r = 1$ Melting	System Designed for $A_r = 1$ Idling	System Designed for $A_r = 0.5$ Melting	System Designed for $A_r = 0.5$ Idling	System Designed for $A_r = 0$ Melting	System Designed for $A_r = 0$ Idling
Albany, NY	156	1,883	9.3	10,132	109,230	7,252	109,004	4,371	108,420
Albuquerque, NM	44	954	16.3	2,455	38,504	1,729	38,495	984	38,332
Amarillo, TX	64	1,212	6.8	5,276	62,557	3,314	62,136	1,357	61,170
Billings, MT	225	1,800	−10.8	17,299	116,947	10,526	111,803	3,716	91,360
Bismarck, ND	158	2,887	−8.8	16,295	207,888	9,321	201,565	2,300	157,503
Boise, ID	85	1,611	5.3	3,543	74,724	2,449	73,015	1,345	68,456
Boston, MA	112	1,273	16.3	7,694	77,992	5,455	77,907	3,218	77,747
Buffalo, NY	292	1,779	3.8	23,929	105,839	14,735	105,521	5,563	101,945
Burlington, VT	204	2,215	4.3	13,182	147,122	8,485	143,824	3,783	134,634
Cheyenne, WY	224	2,152	−15.8	20,061	126,714	11,931	125,635	3,782	120,915
Chicago, IL, O'Hare Int'l AP	124	1,854	3.8	8,501	116,663	5,402	112,763	2,252	100,427
Cleveland, OH	188	1,570	8.8	11,419	86,539	7,359	85,470	3,208	80,851
Colorado Springs, CO	159	1,925	−8.8	11,137	97,060	7,089	96,847	3,026	96,244
Columbus, OH, Int'l AP	92	1,429	12.8	4,581	71,037	2,972	68,002	1,367	62,038
Des Moines, IA	127	1,954	−1.8	10,884	128,140	6,796	125,931	2,654	116,545
Detroit, MI, Metro	153	1,781	11.3	10,199	104,404	6,467	102,289	2,704	95,777
Duluth, MN	238	3,206	0.3	20,838	251,218	12,423	236,657	3,969	187,820
Ely, NV	153	2,445	13.3	7,421	141,288	5,268	139,242	3,098	136,920
Eugene, OR	18	481	15.8	841	17,018	634	16,997	429	16,992
Fairbanks, AK	288	4,258	−15.8	19,803	343,674	11,700	318,880	3,559	194,237
Baltimore, MD, BWI Airport	56	957	16.3	3,827	45,132	2,970	45,132	2,121	45,130
Great Falls, MT	233	1,907	−15.8	19,703	123,801	11,731	120,603	3,736	101,712
Indianapolis, IN	96	1,473	10.8	6,558	80,942	4,132	78,532	1,705	75,926
Lexington, KY	50	1,106	13.3	2,696	54,084	1,718	52,278	733	45,859
Madison, WI	161	2,308	5.3	11,404	149,363	7,279	147,112	3,094	140,108
Memphis, TN	13	473	12.8	1,010	21,756	691	21,518	373	21,102
Milwaukee, WI	161	1,960	7.3	11,678	127,230	7,564	123,960	3,431	119,945
Minneapolis-St. Paul, MN	199	2,513	0.3	16,532	183,980	10,325	178,495	4,097	166,921
New York, NY, JFK Airport	61	885	18.3	4,193	50,680	2,988	50,467	1,797	50,049
Oklahoma City, OK	35	686	6.8	2,955	40,957	1,850	39,725	741	38,308
Omaha, NE	94	1,981	−2.3	7,425	124,274	4,613	119,565	1,790	112,700
Peoria, IL	91	1,748	2.3	6,544	104,380	4,078	100,581	1,606	94,045
Philadelphia, PA, Int'l AP	56	992	18.3	3,758	50,494	2,669	50,412	1,588	50,203
Pittsburgh, PA, Int'l AP	168	1,514	9.3	10,029	79,312	6,350	77,750	2,626	72,361
Portland, ME	157	1,996	7.3	13,318	115,248	8,969	115,196	4,630	114,836
Portland, OR	15	329	21.8	623	13,399	464	13,194	310	12,918
Rapid City, SD	177	2,154	−4.8	16,889	137,523	9,738	135,024	2,535	106,102
Reno, NV	63	1,436	16.3	2,293	54,713	1,792	54,706	1,302	54,703
Salt Lake City, UT	142	1,578	16.3	5,263	70,254	4,271	69,927	3,286	69,927
Sault Ste. Marie, MI	425	2,731	−0.3	34,249	176,517	20,779	174,506	7,250	155,508
Seattle, WA	27	260	17.8	1,212	10,482	943	10,473	682	10,452
Spokane, WA	144	1,832	10.8	6,909	81,000	4,721	79,177	2,512	75,659
Springfield, MO	58	1,108	6.8	4,401	57,165	2,950	56,929	1,503	56,238
St. Louis, MO, Int'l AP	62	1,150	6.8	4,516	64,668	2,981	63,428	1,446	60,764
Topeka, KS	61	1,409	−1.8	4,507	75,598	2,821	74,028	1,126	68,402
Wichita, KS	60	1,223	0.3	4,961	69,187	3,106	67,828	1,229	60,991

[a] Does not include back and edge heat losses.

Table 3 also includes a column labeled "2% Min. Snow Temp." This is the temperature below which only 2% of the snowfall hours occur.

Operating Cost Example

Example 3. A snow-melting system of 2000 ft^2 is to be installed in Chicago. The application is considered critical enough that the system is designed to remain snow-free 99% of the time. For this application, Table 3 shows that the annual heat requirement to melt the snow is 8501 Btu/ft^2. Assuming a fossil fuel cost of $8 per 10^6 Btu, an electric cost of $0.07 per kWh, and back loss at 30%, find the annual cost to melt the snow.

Solution: Operating cost O may be expressed as follows:

$$O = \frac{A q_a F}{1 - (B/100)} \quad (15)$$

where

O = operating cost, $/yr
A = area, ft^2
q_a = annual output, Btu/ft^2 per year or kWh/ft^2 per year
F = fuel cost, $/Btu or $/kWh
B = back loss, %

Operating cost for fossil fuel is

$$O = (2000 \times 8501 \times 8 \times 10^{-6})/[1 - (30/100)]$$
$$= \$194/yr$$

Operating cost for electric heat is

$$O = [2000(8501 \times 0.293 \times 10^{-3})(0.07)]/[1-(30/100)]$$
$$= \$498/yr$$

PAVEMENT DESIGN
(HYDRONIC AND ELECTRIC)

Either concrete or asphalt pavement may be used for snow-melting systems. The thermal conductivity of asphalt is less than that of concrete; pipe or cable spacing and required fluid temperatures are thus different. Hot asphalt may damage plastic or electric snow-melting systems unless adequate precautions are taken. For specific recommendations, refer to the sections on Hydronic System Design and Electric System Design.

Concrete slabs containing hydronic or electric snow-melting apparatus must be designed and constructed with a subbase, expansion-contraction joints, reinforcement, and drainage to prevent slab cracking; otherwise, crack-induced shearing or tensile forces would break the pipe or cable. The pipe or cable must not run through expansion-contraction joints, keyed construction joints, or control joints (dummy grooves); however, the pipe or cable may be run under 0.12 in. score marks (block and other patterns). Control joints must be placed wherever the slab changes size or incline. The maximum distance between control joints for ground-supported slabs should be less than 15 ft, and the length should be no greater than twice the width, except for ribbon driveways or sidewalks. In ground-supported slabs, most cracking occurs during the early cure. Depending on the amount of water used in the concrete mix, shrinkage during cure may be up to 0.75 in. per 100 ft. If the slab is more than 15 ft long, the concrete does not have sufficient strength to overcome friction between it and the ground while shrinking during the cure period.

If the slabs are poured in two separate layers, the top layer, which contains the snow-melting apparatus, usually does not contribute toward total slab strength; therefore, the lower layer must be designed to provide the total strength.

The concrete mix of the top layer should give maximum weatherability. The compressive strength should be 4000 to 5000 psi; recommended slump is 3 in. maximum, 2 in. minimum. Aggregate size and air content should be as follows:

Maximum Size Crushed Rock Aggregate, in.	Air Content, %
2.5	5 ± 1
1	6 ± 1
0.5	7.5 ± 1

Note: Do not use river gravel or slag.

The pipe or cable may be placed in contact with an existing sound pavement (either concrete or asphalt) and then covered as described in the sections on Hydronic System Design and Electric System Design. If there are signs of cracking or heaving, the pavement should be replaced. Pipe or cable should not be placed over existing expansion-contraction, control, or construction joints. The finest grade of asphalt is best for the top course; stone diameter should not exceed 0.38 in.

A moisture barrier should be placed between any insulation and the fill. The joints in the barrier should be sealed and the fill made smooth enough to eliminate holes or gaps for moisture transfer. Also, the edges of the barrier should be flashed to the surface of the pavement to seal the ends.

Snow-melting systems should have good surface drainage. When the ambient air temperature is 32°F or below, runoff from melting snow freezes immediately on leaving the heated area. Any water that is able to get under the pavement also freezes when the system is deenergized, causing extreme frost heaving. Runoff should be piped away in drains that are heated or below the frost line.

The area to be protected by the snow-melting system must first be measured and planned. For total snow removal, hydronic or electric heat must cover the entire area. In larger installations, it may be desirable to melt snow and ice from only the most frequently used areas, such as walkways and wheel tracks for trucks and autos. Planning for separate circuits should be considered so that areas within the system can be heated individually, as required.

Where snow-melting apparatus must be run around obstacles (e.g., a storm sewer grate), the pipe or cable spacing should be uniformly reduced. Because some drifting will occur adjacent to walls or vertical surfaces, extra heating capacity should be provided in these areas, if possible also in the vertical surface. Drainage flowing through the area expected to be drifted tends to wash away some snow.

CONTROL (HYDRONIC AND ELECTRIC)

Snow-melting systems can be controlled either manually or automatically.

Manual Control

Manual operation is strictly by on-off control; an operator must activate and deactivate the system when snow falls. If the system is not turned off after snowfall, operating cost increases.

Automatic Control

If the snow-melting system is not turned on until snow starts falling, it may not melt snow effectively for several hours, giving additional snowfall a chance to accumulate and increasing the time needed to melt the area. Automatic controls provide satisfactory operation because they turn on the system when light snow starts, allowing adequate warm-up before heavy snowfall develops. Automatic turn-off reduces operating costs.

Snow Detectors. Snow detectors monitor precipitation and temperature. They allow operation only when snow is present and may incorporate a delay-off timer. Snow detectors located in the heated area activate the snow-melting system when precipitation (snow) occurs at a temperature below the preset pavement temperature (usually 40°F). Another type of snow detector is mounted above ground, adjacent to the heated area, without cutting into the existing system; however, it does not detect tracked or drifting snow. Both types of sensors should be located so that they are not affected by overhangs, trees, blown snow, or other local conditions.

Pavement Temperature Sensor. To limit energy waste during normal and light snow conditions, it is common to include a remote temperature sensor installed midway between two pipes or cables in the pavement; the set point is adjusted between 40 and 60°F. Thus, during mild weather snow conditions, the system is automatically modulated or cycled on and off to keep the pavement temperature at the sensor at set point.

Outdoor Thermostat. The control system may include an outdoor thermostat that turns the system off when the outdoor ambient temperature rises above 35 to 40°F as automatic protection against accidental operation in summer or mild weather.

Control Selection

For optimum operating convenience and minimum operating cost, all of the aforementioned controls should be incorporated in the snow-melting system.

Operating Cost

To evaluate operating cost during idling or melting, use the annual output data from Table 3. Idling and melting data are based on pavement surface temperature control at 32°F during idling, which requires a pavement temperature sensor. Without a pavement temperature sensor, operating costs will be substantially higher.

HYDRONIC SYSTEM DESIGN

Hydronic system design includes selection of the following components: (1) heat transfer fluid, (2) piping, (3) fluid heater, (4) pump(s) to circulate the fluid, and (5) controls. With concrete pavement, thermal stress is also a design consideration.

Heat Transfer Fluid

A variety of fluids, including brine, oils, and glycol-water, is suitable for transferring the heat from the fluid heater to the pavement. Freeze protection is essential because most systems will not be operated continuously in subfreezing weather. Without freeze protection, power loss or pump failure could cause freeze damage to the piping and pavement.

Brine is the least costly heat transfer fluid, but it has a lower specific heat than glycol. The use of brine may be discouraged because of the cost of heating equipment that can resist its corrosive potential.

Although **heat transfer oils** are not corrosive, they are more expensive than and have a lower specific heat and higher viscosity than brine or glycol. Petroleum distillates used as fluids in snow-melting systems are classified as nonflammable but have fire points between 300 and 350°F. When oils are used as heat transfer fluids, any oil dripping from the seals on the pump should be collected. It is good practice to place a barrier between the oil lines and the boiler so that a flashback from the boiler cannot ignite a possible oil leak. Other nonflammable fluids, such as those used in some transformers, can be used as antifreeze.

Glycols (ethylene and propylene) are used the most often in snow-melting systems because of their moderate cost, high specific heat, and low viscosity; ease of corrosion control is another advantage. Automotive glycols containing silicates are not recommended because they can cause fouling, pump seal wear, fluid gelation, and reduced heat transfer. The piping should be designed for periodic addition of inhibitor. Glycols should be tested annually to determine any change in reserve alkalinity and freeze protection. Only inhibitors obtained from the manufacturer of the glycol should be added. Heat exchanger surfaces should be kept below 285°F, which corresponds to about 40 psig steam. Temperatures above 300°F accelerate the deterioration of the inhibitors.

Because ethylene glycol and petroleum distillates are toxic, no permanent connection should be installed between the snow-melting system and the drinking water supply. Gordon (1950) discusses precautions concerning internal corrosion, flammability, toxicity, cleaning, joints, and hook-up that should be taken during the installation of hydronic piping. The properties of brine and glycol are discussed in Chapter 20 of the 1997 *ASHRAE Handbook—Fundamentals*. The effect of glycol on system performance is detailed in Chapter 12 of the 1996 *ASHRAE Handbook—Systems and Equipment*.

Piping

Piping may be metal, plastic, or ethylene-propylene terpolymer (EPDM). Steel, iron, and copper pipes have long been used. Steel and iron may corrode rapidly if the pipe is not shielded by a coating and/or cathodic protection. Both the use of salts for deicing and the elevated temperature accelerate corrosion of metallic components. NACE (1978) states that the corrosion rate roughly doubles for each 18°F rise in temperature.

Chapman (1952) derived the equation for the fluid temperature required to provide an output q_o. For construction similar to that shown in Figure 3, the equation is

$$t_m = 0.5q_o + t_f \qquad (16)$$

where t_m = mean fluid (antifreeze solution) temperature, °F. Equation (16) applies to 1 in. as well as 3/4 in. IPS pipe (Figure 3).

For specific conditions or for cities other than those given in Table 1, Equations (1) and (16) are used. Table 4 gives solutions to these equations at a relative humidity of 80%.

It is satisfactory to use 3/4 in. pipe or tube on 12 in. centers as a standard coil. If pumping loads require reduced friction, the pipe size can be increased to 1 in., but the pavement depth must be increased accordingly. The piping should be supported by a minimum of 2 in. of concrete above and below. This requires 5 in. pavement for 3/4 in. pipe and 5.4 in. pavement for 1 in. pipe.

Plastic Pipe. Plastic pipe [polyethylene (PE), polybutylene (PB), cross-linked polyethylene (PEX)] is popular due to its lower material cost, lower installation cost, and corrosion resistance. Considerations when using plastic pipe include stress crack resistance, temperature limitations, and thermal conductivity. Heat transfer oils should not be used with plastic pipe. Plastic pipe is furnished in coils. The smaller sized pipe can be bent to form a variety of heating panel designs without elbows or joints. Mechanical compression connections can be used to connect the heating panel pipe to the larger supply and return piping leading to the pump and fluid heater. PE and PB plastic pipe may be fused using the appropriate fittings and fusion equipment. Fusion joining eliminates metallic components and thus the possibility of corrosion in the piping; however, it requires considerable installation training.

When plastic pipe is used, the system must be designed so that the fluid temperature required will not damage the pipe. If a design requires a temperature above the tolerance of plastic pipe, the heat output will never meet design requirements. A logical solution is to decrease the pipe spacing. Adlam (1950) addresses the parameter of pipe size and the effect of pipe spacing on heat output. A typical solution is summarized in Table 5, which shows a way of designing pipe spacing according to heating requirements. This table also shows adjustments for the effect of more than 2 in. of concrete or paver over the pipe.

Pipe Installation. It is good design practice to avoid passing any embedded piping through a concrete expansion joint; otherwise, the pipe may be stressed and possibly ruptured. Figure 4 shows a method of protecting piping that must pass through a concrete expansion joint from stress under normal conditions.

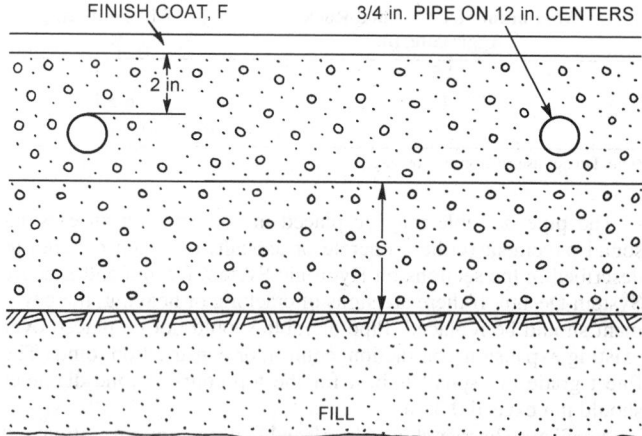

F = depth of finish coat—assumed to be 0.5 in. of concrete. Finish coat may be asphalt, but then cover slab should be reduced from 3 in. Depth of slab should always keep thermal resistance equal to 3 in. of concrete.

S = depth required by structural design (should be at least 2 in. of concrete).

Fig. 3   Detail of Hydronic Snow-Melting System

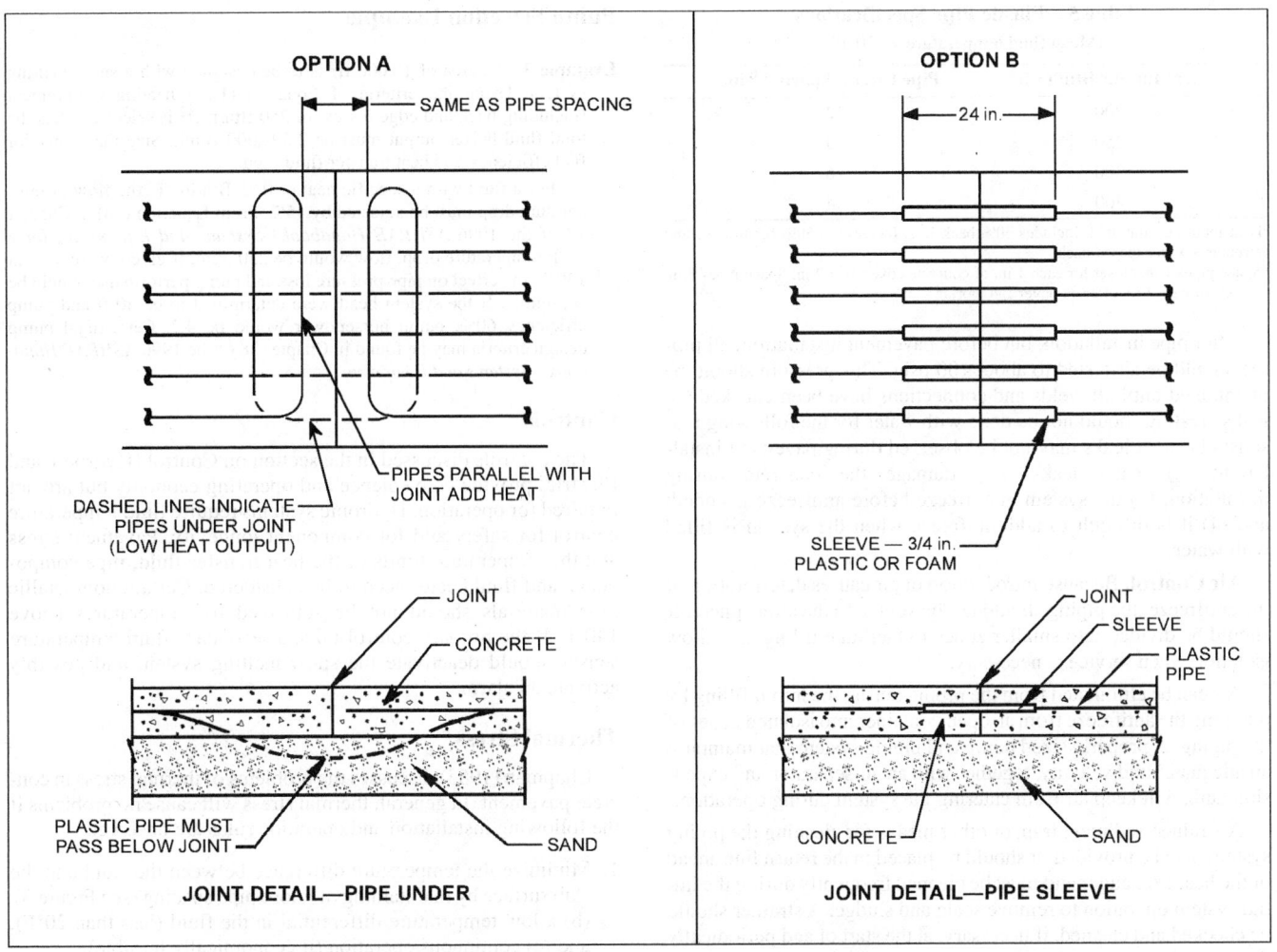

Fig. 4   Piping Details for Concrete Construction Jobs

Table 4   Heat Ouput and Mean Fluid Temperature for Hydronic Snow-Melting System
(Mean fluid temperature based on 12 in. tube spacing)

s, Rate of Snowfall in/h	A_r		$t_a = 0°F$ Wind Speed V, mph			$t_a = 10°F$ Wind Speed V, mph			$t_a = 20°F$ Wind Speed V, mph			$t_a = 30°F$ Wind Speed V, mph		
			5	10	15	5	10	15	5	10	15	5	10	15
0.08	1.0	q_o	166	222	272	138	180	217	108	135	159	76	86	94
		t_m	116	144	169	102	123	142	87	100	112	71	76	80
	0.0	q_o	67	67	67	65	65	65	63	63	63	61	61	61
		t_m	66	66	66	65	65	65	64	64	64	63	63	63
0.16	1.0	q_o	233	289	339	203	244	282	171	197	221	136	146	155
		t_m	149	177	202	134	155	174	118	132	144	101	106	110
	0.0	q_o	133	133	133	129	129	129	125	125	125	121	121	121
		t_m	100	100	100	98	98	98	96	96	96	93	93	93
0.25	1.0	q_o	308	363	414	275	317	354	241	268	292	204	214	223
		t_m	187	215	240	171	192	210	154	167	179	135	140	144
	0.0	q_o	208	208	208	202	202	202	195	195	195	189	189	189
		t_m	137	137	137	134	134	134	131	131	131	127	127	127

Note: Table based on a characteristic pavement length of 20 ft, standard atmospheric pressure, a water film temperature of 33°F, and relative humidity of 80%.

A_r = free area ratio
q_o = slab heating load, Btu/h·ft^2

t_a = atmospheric dry-bulb temperature, °F
t_m = mean fluid temperature based on construction shown in Figure 3, °F

Table 5 Plastic Pipe Specifications

(Mean fluid temperature = 130°F)

Heat Input,[a] Btu/h · ft^2	Pipe Circuit Spacing,[b] in.
200	12
250	9
300	6
400	4

[a]Heat input per unit area. Includes 30% back loss. Increase to 50% for bridges and structures with exposed back.

[b]Space pipes 1 in. closer for each 1 in. of concrete cover over 2 in. Space pipes 2 in. closer for each 1 in. of brick paver and mortar.

After pipe installation, but before pavement installation, all piping should be air-tested to about 100 psig. This pressure should be maintained until all welds and connections have been checked for leaks. Testing should not be done with water for the following reasons: (1) small leaks may not be observed during pavement installation; (2) water leaks may damage the concrete during installation; (3) the system may freeze before antifreeze is added; and (4) it is difficult to add antifreeze when the system is filled with water.

Air Control. Because introduction of air causes deterioration of the antifreeze, the piping should not be vented to the atmosphere. It should be divided into smaller zones to facilitate filling and allow isolation when service is necessary.

Air can be eliminated from the piping during the initial filling by pumping the antifreeze from an open container into isolated zones of the piping. A properly sized pump and piping system that maintains an adequate fluid velocity, together with an air separator and expansion tank, will keep air from entering the system during operation.

A strainer, sediment trap, or other means for cleaning the piping system may be provided. It should be placed in the return line ahead of the heat exchanger and must be cleaned frequently during the initial system operation to remove scale and sludge. A strainer should be checked and cleaned, if necessary, at the start of and periodically during each snow-melting season.

An ASME safety relief valve of adequate capacity should be installed on a closed system.

Fluid Heater

The heat transfer fluid can be heated using any of a variety of energy sources, depending on availability at the location of the snow-melting system. A fluid heater can use steam, hot water, gas, oil, or electricity. In some applications, heat may be available from secondary sources, such as engine generators, condensate, and other waste heat sources.

The design capacity of the fluid heater can be established by evaluating the data in the section on Heating Requirement (Hydronic and Electric); it is usually 200 to 300 Btu/h·ft^2, which includes back and edge losses.

Design of the fluid heater should follow standard practice, with adjustments for the film coefficient. Consideration should be given to flue gas condensation and thermal shock in boilers due to low fluid temperatures. Bypass flow controls may be necessary to maintain recommended boiler temperatures. Boilers should be derated for high-altitude applications.

Pump Selection

The proper pump is selected based on (1) the fluid flow rate; (2) the energy requirements of the piping system; (3) the specific heat of the fluid; and (4) the viscosity of the fluid, particularly during a cold start-up.

Pump Selection Example

Example 3. An area of 10,000 ft^2 is to be designed with a snow-melting system. Using the criteria of Equation (1), a heating requirement (including back and edge losses) of 250 Btu/h·ft^2 is selected. Thus the total fluid heater output must be 2,500,000 Btu/h. Size the heater for fuel efficiency and heat transfer fluid used.

For a fluid with a specific heat of 0.85 Btu/lb·°F, the flow or temperature drop must be adjusted by 15%. From Equation (18) in Chapter 12 of the 1996 *ASHRAE Handbook—Systems and Equipment*, for a 23°F temperature drop, flow would be 250 gpm. If glycol were used at 130°F, the effect on pipe pressure loss and pump performance would be negligible. If the system head were determined to be 40 ft and pump efficiency 60%, pump horsepower would be 4.2. Centrifugal pump design criteria may be found in Chapter 38 of the 1996 *ASHRAE Handbook—Systems and Equipment.*

Controls

The controls discussed in the section on Control (Hydronic and Electric) provide convenience and operating economy but are not required for operation. Hydronic systems require fluid temperature control for safety and for component longevity. Pavement stress and the temperature limits of the heat transfer fluid, pipe components, and fluid heater need to be considered. Certain nonmetallic pipe materials should not be subjected to temperatures above 140°F. If the primary control fails, a secondary fluid temperature sensor should deactivate the snow-melting system and possibly activate an alarm.

Thermal Stress

Chapman (1955) discusses the problems of thermal stress in concrete pavement. In general, thermal stress will cause no problems if the following installation and operation rules are observed:

1. Minimize the temperature difference between the fluid and the slab surface by maintaining (a) close pipe spacing (see Figure 3), (b) a low temperature differential in the fluid (less than 20°F), and (c) continuous operation (if economically feasible).
2. Install pipe within about 2 in. of the surface.
3. Use reinforcing steel designed for thermal stress if high structural loads are expected (such as on highways).

Thermal shock to the pavement may occur if heated fluid is introduced from a large source of residual heat such as a storage tank, a large piping system, or another snow-melting area. The pavement should be brought up to temperature by maintaining the fluid temperature differential at less than 20°F.

ELECTRIC SYSTEM DESIGN

Snow-melting systems using electricity as an energy source have heating elements in the form of (1) mineral insulated (MI) cable, (2) self-regulating cable, (3) constant wattage cable, or (4) high-intensity infrared heaters.

Heat Flux

The basic load calculations for electric systems are the same as presented in the section on Heating Requirement (Hydronic and Electric). However, because electric system output is determined by the resistance installed and the voltage impressed, it cannot be altered by fluid flow rates or temperatures. Consequently, neither safety factors nor marginal capacity systems are design considerations.

Heat flux within a slab can be varied by altering the cable spacing to compensate for anticipated drift areas or other high heat loss areas. Power density should not exceed 120 W/ft^2 (NFPA *Standard* 70).

Electrical Equipment

The installation and design of electric snow-melting systems is governed by Article 426 of the *National Electrical Code* (NFPA *Standard* 70). The *NEC* requires that each electric snow-melting circuit be provided with a ground fault protection device. An equipment protection device (EPD) with a trip level of 30 mA should be used to reduce the likelihood of nuisance tripping.

Double-pole, single-throw switches or tandem circuit breakers should be used to open both sides of the line. The switchgear may be in any protected, convenient location. It is also advisable to include a pilot lamp on the load side of each switch so that there is a visual indication when the system is energized.

Junction boxes located at grade level are susceptible to water ingress. Weatherproof junction boxes installed above grade should be used for terminations.

The power supply conduit is run underground, outside the slab, or in a prepared base. With concrete pavement, this conduit should be installed before the reinforcing mesh.

Mineral Insulated Cable

Mineral insulated (MI) heating cable is a magnesium oxide (MgO)-filled, die-drawn cable with one or two copper or copper alloy conductors and a seamless copper or stainless steel alloy sheath. The metal outer sheath is protected from salts and other chemicals by a high-density PE jacket that is important whenever MI cable is embedded in a medium. Although it is heavy-duty cable, MI cable is practical in any snow-melting installation.

Cable Layout. To determine the characteristics of the MI heating cable needed for a specific area, the following must be known:

- Heated area size
- Power density required
- Voltage(s) available
- Approximate cable length needed

To find the approximate MI cable length, estimate 2 linear ft of cable per square foot of concrete. This corresponds to 6 in. on-center spacing. Actual cable spacing will vary between 3 and 9 in. to provide the proper power density.

Cable spacing is dictated primarily by the heat-conducting ability of the material in which the cable is embedded. Concrete has a higher heat transmission coefficient than asphalt, permitting wider cable spacing. The following is a procedure to select the proper MI heating cable:

1. Determine total power required for each heated slab.

$$W = Aw \qquad (17)$$

2. Determine total resistance.

$$R = E^2/W \qquad (18)$$

3. Calculate cable resistance per foot.

$$r_1 = R/L_1 \qquad (19)$$

where

W = total power needed, W
A = heated area of each heated slab, ft^2
w = required power density input, W/ft^2
R = total resistance of cable, Ω
E = voltage available, V
r_1 = calculated cable resistance, Ω per foot of cable
L_1 = estimated cable length, ft
L = actual cable length needed, ft
r = actual cable resistance, Ω/ft
S = cable on-center spacing, in.
I = total current per MI cable, A

Commercially available mineral insulated heating cables have actual resistance values (if there are two conductors, the value is the total of the two resistances) ranging from 0.0016 to 0.6 Ω/ft. Manufacturing tolerances are ±10% on these values. MI cables are die-drawn, with the internal conductor drawn to size indirectly via pressures transmitted through the mineral insulation. Special cables are not economical unless the quantity needed is 100,000 ft or more.

4. From manufacturers' literature, choose a cable with a resistance r closest to the calculated r_1. Note that r is generally listed at ambient room temperature. At the specific temperature, r may drift from the listed value. It may be necessary to make a correction as described in Chapter 6 of the 1996 *ASHRAE Handbook— Systems and Equipment.*

5. Determine the actual cable length needed to give the wattage desired.

$$L = R/r \qquad (20)$$

6. Determine cable spacing within the heated area.

$$S = 12A/L \qquad (21)$$

For optimum performance, heating cable spacing should be within the following limits: in concrete, 3 to 9 in.; in asphalt, 3 to 6 in.

Because the manufacturing tolerance on cable length is ±1%, and installation tolerances on cable spacing must be compatible with field conditions, it is usually necessary to adjust the installed cable as the end of the heating cable is rolled out. Cable spacing in the last several passes may have to be altered to give uniform heat distribution.

The installed cable within the heated areas follows a serpentine path originating from a corner of the heated area (Figure 5). As heat is conducted evenly from all sides of the heating cable, cables in a concrete slab can be run within half the spacing dimension of the perimeter of the heated area.

7. Determine the current required for the cable.

$$I = E/R, \text{ or } I = W/E \qquad (22)$$

8. Choose cold lead cable as dictated by typical design guidelines and local electrical codes (see Table 6).

Cold Lead Cable. Every MI heating cable is factory-fabricated with a non-heat-generating cold lead cable attached. The cold lead cable must be long enough to reach a dry location for termination and of sufficient wire gage to comply with local and *NEC* standards. The *NEC* requires a minimum cold lead length of 6 in. within the

**Table 6 Mineral Insulated Cold Lead Cables
(Maximum Voltage—600 V)**

Single-Conductor Cable		Two-Conductor Cable	
Current Capacity, A	American Wire Gage	Current Capacity, A	American Wire Gage
35	14	25	14/2
40	12	30	12/2
55	10	40	10/2
80	8	55	8/2
105	6	75	6/2
140	4	95	4/2
165	3		
190	2		
220	1		

Source: National Electrical Code (NFPA *Standard* 70).

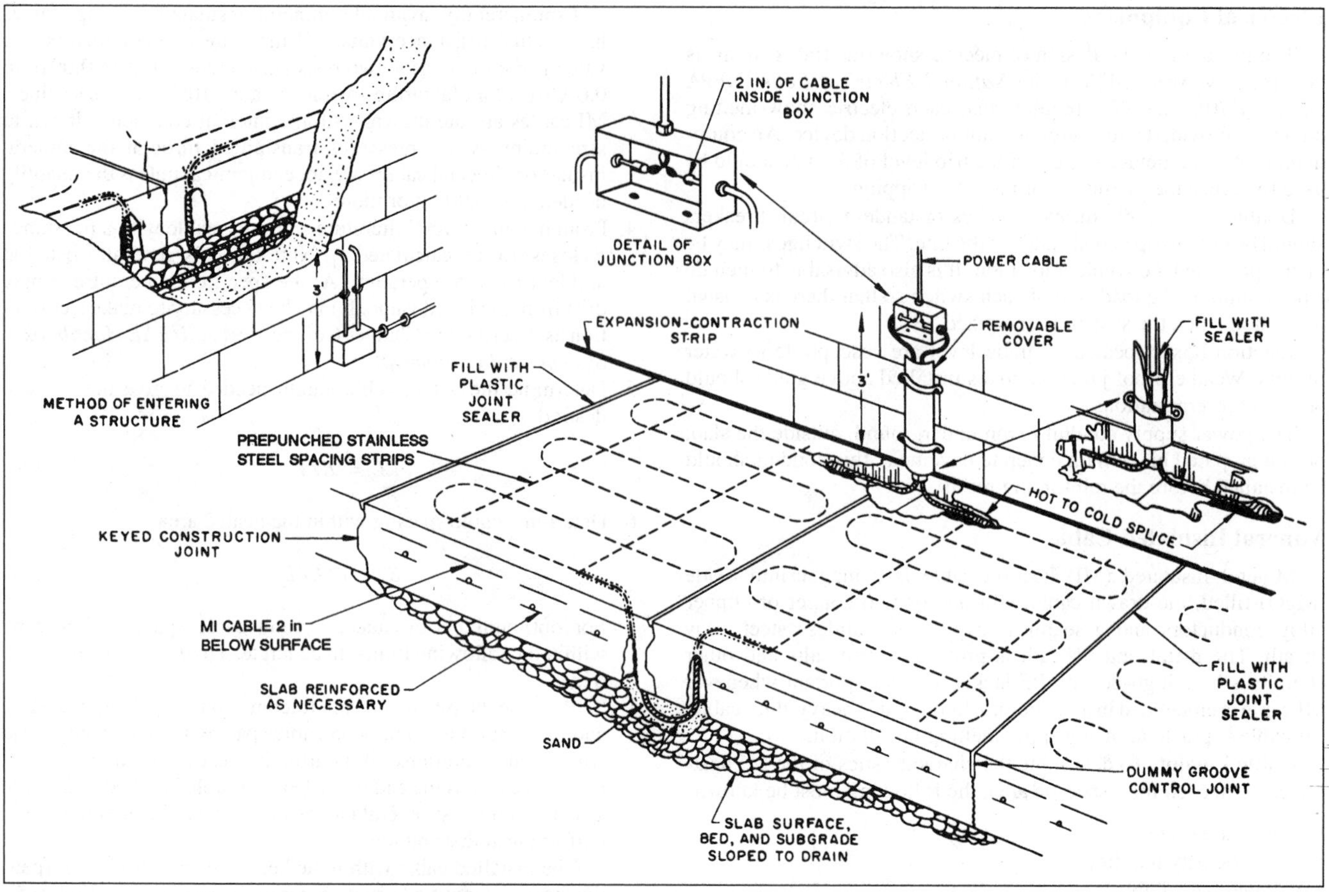

Fig. 5 Typical Mineral Insulated Heating Cable Installation in Concrete Slab

junction box. Mineral insulated cable junction boxes must be located such that the box remains dry and at least 3 ft of cold lead cable is available at the end for any future service (Figure 5). Preferred junction box locations are indoors; on the side of a building, utility pole, or wall; or inside a manhole on the wall. Boxes should have a hole in the bottom to drain condensation. Outdoor boxes should be completely watertight except for the condensation drain hole. Where junction boxes are mounted below grade, the cable end seals must be coated with an epoxy to prevent moisture entry. Cable end seals should extend into the junction box far enough to allow the end seal to be removed if necessary.

Although MgO, the insulation in MI cable, is hygroscopic, the only vulnerable part of the cable is the end seal. However, should moisture penetrate the seal, it can easily be detected with a megohmmeter and driven out by applying a torch 2 to 3 ft from the end and working the flame toward the end.

Installation. When mineral insulated electric heating cable is installed in a concrete slab, the slab may be poured in one or two layers. In single-pour application, the cable is hooked on top of the reinforcing mesh before the pour is started. In two-layer application, the cable is laid on top of the bottom structural slab and embedded in the finish layer. For a proper bond between the layers, the finish slab should be poured within 24 h of the bottom slab, and a bonding grout should be applied. The finish slab should be at least 2 in. thick. Cable should not run through expansion, control, or dummy joints (score or groove). If the cable must cross such a joint, the cable should exit the bottom of the slab at least 4 in. from one side of the joint and reenter the slab through the bottom at least 4 in. from the joint on the opposite side.

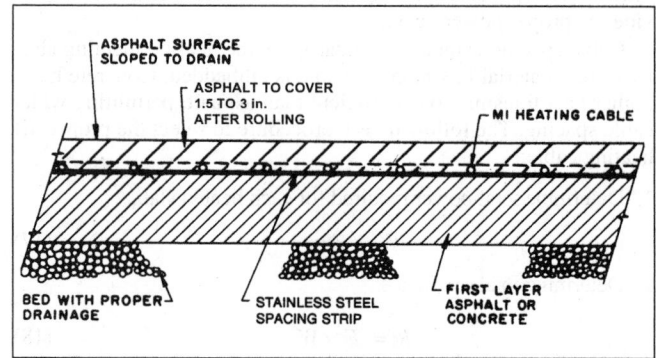

**Fig. 6 Typical Section, Mineral Insulated
Heating Cable in Asphalt**
(Potter 1967)

The cable is uncoiled from reels and laid as described in the section on Cable Layout. Prepunched copper or stainless steel spacing strips are often nailed to the lower slab for uniform spacing.

A polyvinyl chloride (PVC) jacket is extruded over the cable to protect the cold lead from chemical damage and to protect the cable from physical damage without adding excessive thermal insulation.

Calcium chloride or other chloride additives should not be added to a concrete mix in winter because chlorides are destructive to copper. Cinder or slag fill under snow-melting panels should also be avoided. Where the cold lead cable exits the slab, it should be

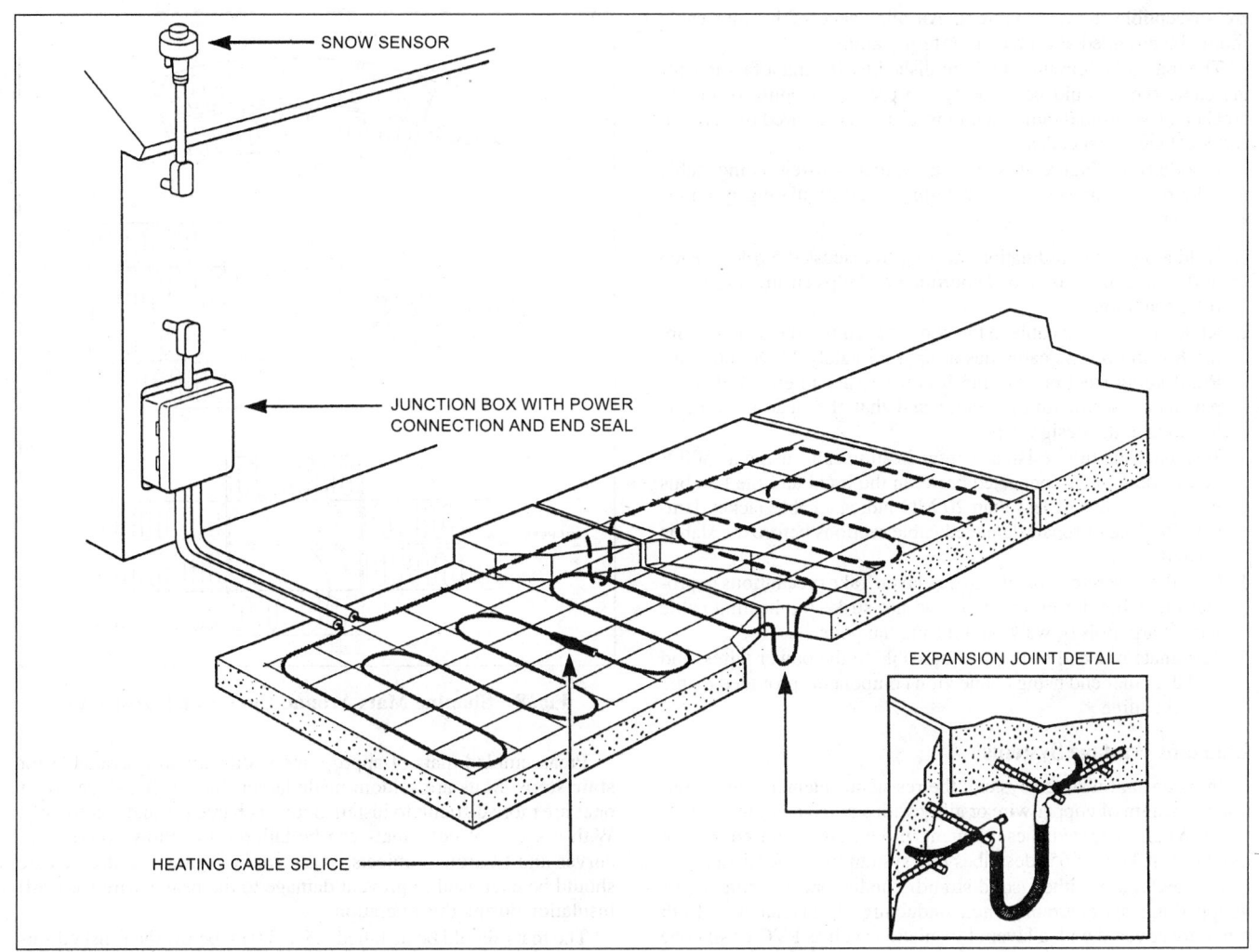

Fig. 7 Typical Self-Regulating Cable Installation

wrapped with PVC or PE tape to protect it from fertilizer corrosion and other ground attack. Within 2 ft of the heating section, only PE should be used because heat may break down PVC. Underground, the leads should be installed in suitable conduits to protect them from physical damage.

In asphalt slabs, the MI cable is fixed in place on top of the base pour with prepunched stainless steel strips or 6 in. by 6 in. wire mesh. A coat of bituminous binder is applied over the base and the cable to prevent them from floating when the top layer is applied. The layer of asphalt over the cable should be 1.5 to 3 in. thick (Figure 6).

Testing. Mineral insulated heating cables should be thoroughly tested before, during, and after installation to ensure they have not been damaged either in transit or during installation.

Because of the hygroscopic nature of the MgO insulation, damage to the cable sheath is easily detectable with a 500 V field megohmmeter. Cable insulation resistance should be measured on arrival of the cable. Cable with insulation resistance of less than 20 MΩ should not be used. Cable that shows a marked loss of insulation resistance after installation should be investigated for damage. Cable should also be checked for electrical continuity.

Self-Regulating Cable

Self-regulating heating cables consist of two parallel conductors embedded in a heating core made of conductive polymer. These cables automatically adjust their power output to compensate for local temperature changes. Heat is generated as electric current passes through the core between the conductors. As the slab temperature drops, the number of electrical paths increases, and more heat is produced. Conversely, as the slab temperature rises, the core has fewer electrical paths, and less heat is produced.

Power output of self-regulating cables may be specified as watts per unit length at a particular temperature or in terms of snow-melting performance at a given cable spacing. In typical slab-on-grade applications, adequate performance may be achieved with cables spaced up to 12 in. apart. Narrower cable spacings may be required to achieve the desired snow-melting performance. The parallel construction of the self-regulating cable allows it to be cut to length in the field without affecting the rated power output.

Layout. For uniform heating, the heating cable should be arranged in a serpentine pattern that covers the area with 12 in. on-center spacing (or alternate spacing determined for the design). The heating cable should not be routed closer than 4 in. to the edge of the pavement, drains, anchors, or other material in the concrete.

Crossing expansion, control, or other pavement joints should be avoided. Self-regulating heating cables may be crossed or overlapped as necessary. Because the cables limit power output locally, they will not burn out.

Both ends of the cable should terminate in an above-ground weatherproof junction box. Junction boxes installed at grade level are susceptible to water ingress. An allowance of heating cable should be provided at each end for termination.

The maximum circuit length published by the manufacturer for the cable type should be respected to prevent tripping of circuit breakers. Use ground fault circuit protection as required by national and local electrical codes.

Installation. Figure 7 shows a typical self-regulating cable installation. The procedure for installing a self-regulating system is as follows:

1. Hold a project coordination meeting to discuss the role of each trade and contractor. Good coordination helps ensure a successful installation.
2. Attach the heating cable to the concrete reinforcing steel or wire mesh using plastic cable ties at approximately 12 in. intervals. Reinforcing steel or wire mesh is necessary to ensure that the pavement is structurally sound and that the heating cable is installed at the design depth.
3. Test the insulation resistance of the heating cable using a 2500 V dc megohmmeter connected between the braid and the two bus wires. Readings of less than 20 MΩ indicate cable jacket damage. Replace or repair damaged cable sections before the slab is poured.
4. Pour the concrete, typically in one layer. Take precautions to protect the cable during the pour. Do not strike the heating cable with sharp tools or walk on it during the pour.
5. Terminate one end of the heating cable to the power wires, and seal the other end using connection components provided by the manufacturer.

Constant Wattage Systems

In a constant wattage system, the resistance elements may consist of a length of copper wire or alloy with a given amount of resistance. When energized, these elements produce the required amount of heat. Witsken (1965) describes this system in further detail.

Elements are either solid-strand conductors or conductors wrapped in a spiral around a nonconducting fibrous material. Both types are covered with a layer of insulation such as PVC or silicone rubber.

The heat-generating portion of an element is the conductive core. The resistance is specified in ohms per linear foot of core. Alternately, a manufacturer may specify the wire in terms of watts per foot of core, where the power is a function of the resistance of the core, the applied voltage, and the total length of core. As with MI cable, the power output of constant wattage cable does not change with temperature.

Considerations in the selection of insulating materials for heating elements are power density, chemical inertness, application, and end use. Polyvinyl chloride is the least expensive insulation and is widely used because it is inert to oils, hydrocarbons, and alkalies. An outer covering of nylon is often added to increase its physical strength and to protect it from abrasion. The heat output of embedded PVC is limited to 5 W per linear foot. Silicone rubber is not inert to oils or hydrocarbons. It requires an additional covering—metal braid, conduit, or fiberglass braid—for protection. This material can dissipate up to 10 W/ft.

Lead can be used to encase resistance elements insulated with glass fiber. The lead sheath is then covered with a vinyl material. Output is limited to approximately 10 W/ft by the PVC jacket.

Teflon has good physical and electrical properties and can be used at temperatures up to 500°F.

Low watt density (less than 10 W/ft) resistance wires may be attached to plastic or fiber mesh to form a mat unit. Prefabricated factory-assembled mats are available in a variety of watt densities for embedding in specified paving materials to match desired snow-melting capacities. Mats of lengths up to 60 ft are available for installation in asphalt sidewalks and driveways.

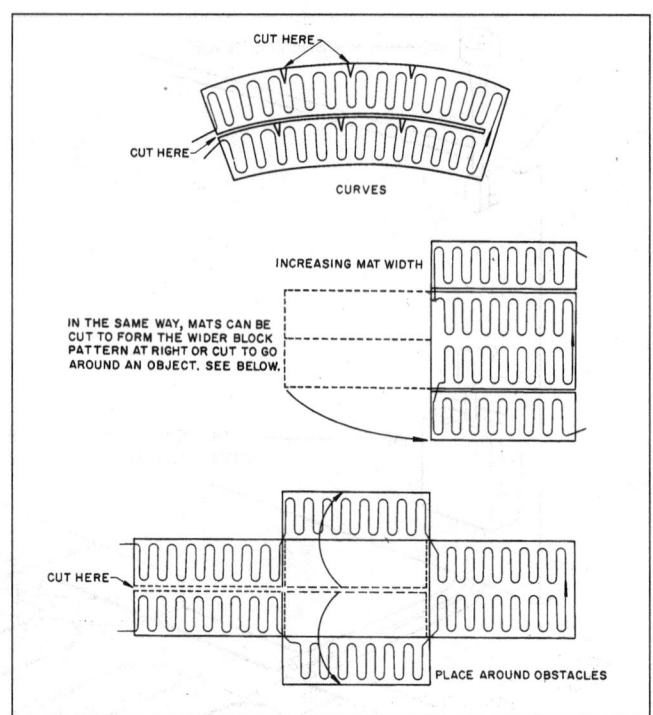

Fig. 8 Shaping Mats Around Curves and Obstacles

Preassembled mats of appropriate widths are also available for **stair steps**. Mats are seldom made larger than 60 ft², since larger ones are more difficult to install, both mechanically and electrically. With a series of cuts, mats can be tailored to follow contours of curves and fit around objects, as shown in Figure 8. Extreme care should be exercised to prevent damage to the heater wire (or lead) insulation during this operation.

The mats should be installed 1.5 to 3.0 in. below the finished surface of asphalt or concrete. Installing the mats deeper decreases the snow-melting efficiency. Only mats that can withstand hot asphalt compaction should be used for asphalt paving.

Layout. Heating wires should be long enough to fit between the concrete slab dummy groove control or construction joints. Because concrete forms may be inaccurate, 2 to 4 in. of clearance should be allowed between the edge of the concrete and the heating wire. Approximately 4 in. should be allowed between adjacent heating wires at the control or construction joints.

For asphalt, the longest wire or largest heating mat that can be used on straight runs should be selected. The mats must be placed at least 12 in. in from the pavement edge. Adjacent mats must not overlap. Junction boxes should be located so that each accommodates the maximum number of mats. Wiring must conform to requirements of the *NEC* (NFPA *Standard* 70). It is best to position junction boxes adjacent to or above the slab.

Installation

General

1. Check the wire or mats with an ohmmeter before, during, and after installation.
2. Temporarily lay the mats in position and install conduit feeders and junction boxes. Leave enough slack in the lead wires to permit temporary removal of the mats during the first pour. Carefully ground all leads using the grounding braids provided.
3. Secure all splices with approved crimped connectors or set screw clamps. Tape all of the power splices with plastic tape to make them waterproof. All junction boxes, fittings, and snug bushings

must be approved for this class of application. The entire installation must be completely waterproof to ensure trouble-free operation.

In Concrete

1. Pour and finish each slab area between the expansion joints individually. Pour the base slab and rough level to within 1.5 to 2 in. of the desired finish level. Place the mats in position and check for damage.
2. Pour the top slab over the mats while the rough slab is still wet, and cover the mats to a depth of at least 1.5 in., but not more than 2 in.
3. Do not walk on the mats or strike them with shovels or other tools.
4. Except for brief testing, do not energize the mats until the concrete is completely cured.

In Asphalt

1. Pour and level the base course. If units are to be installed on an existing asphalt surface, clean it thoroughly.
2. Apply a bituminous binder course to the lower base, install the mats, and apply a second binder coating over the mats. The finish topping over the mats should be applied in a continuous pour to a depth of 1.25 to 1.5 in. *Note*: Do not dump a large mass of hot asphalt on the mats because the heat could damage the insulation.
3. Check all circuits with an ohmmeter to be sure that no damage occurred during the installation.
4. Do not energize the system until the asphalt has completely hardened.

Infrared Snow-Melting Systems

While overhead infrared systems can be designed specifically for snow-melting and pavement drying, they are usually installed for the additional features they offer. Infrared systems provide comfort heating, which can be particularly useful at the entrances of plants, office buildings, and hospitals or on loading docks. Infrared lamps can improve the security, safety, and appearance of a facility. These additional benefits may justify the somewhat higher cost of infrared systems.

Infrared fixtures can be installed under entrance canopies, along building facades, and on freestanding poles. Approved equipment is available for recess, surface, and pendant mounting.

Infrared Fixture Layout. The same infrared fixtures used for comfort heating installations (as described in Chapter 15 of the 1996 *ASHRAE Handbook—Systems and Equipment*) can be used for snow-melting systems. The major differences in fixture selection result from the difference in the orientation of the target area. Whereas in comfort applications the *vertical* surfaces of the human body constitute the target of irradiation, in snow-melting applications it is a *horizontal* surface that is targeted. When snow melting is the primary design concern, fixtures with narrow beam patterns confine the radiant energy within the target area for more efficient operation. Asymmetric reflector fixtures, which aim the thermal radiation primarily to one side of the fixture centerline, are often used near the periphery of the target area.

Infrared fixtures usually have a longer energy pattern parallel to the long dimension of the fixture than at right angles to it (Frier 1965). Therefore, fixtures should be mounted in a row parallel to the longest dimension of the area. If the target area is 8 ft or more in width, it is best to locate the fixtures in two or more parallel rows. This arrangement also provides better comfort heating because radiation is directed across the target area from both sides at a more favorable incident angle.

Radiation Spill. In theory, the most desirable energy distribution would be uniform throughout the snow-melting target area at a density equal to the design requirement. The design of heating fixture reflectors determines the percentage of the total fixture radiant output scattered outside the target area design pattern.

Even the best controlled beam fixtures do not produce a completely sharp cutoff at the beam edges. Therefore, if uniform distribution is maintained for the full width of the area, a considerable amount of radiant energy falls outside the target area. For this reason, infrared snow-melting systems are designed so that the intensity on the pavement begins to decrease near the edge of the area (Frier 1964). This design procedure minimizes stray radiant energy losses.

Figure 9 shows the power density values obtained in a sample snow-melting problem (Frier 1965). The sample design average is 45 W/ft^2. It is apparent that the incident power density is above the design average value at the center of the target area and below average at the periphery. Figure 9 shows how the power density and distribution in the snow-melting area depend on the number, wattage, beam pattern, and mounting height of the heaters, and on their position relative to the pavement (Frier 1964).

With distributions similar to the one in Figure 9, snow begins to collect at the edges of the area as the energy requirements for snow melting approach or exceed system capacity. As the snowfall lessens, the snow at the edges of the area and possibly beyond is then melted if the system continues to operate.

Target Area Power Density. Theoretical target area power densities for snow melting with infrared systems are the same as those for commercial applications of constant wattage systems; however, it should be emphasized that theoretical density values are for radiation incident on the pavement surface, not that emitted from the lamps. Merely multiplying the recommended snow-melting power density by the pavement area to obtain the total power input for the system does not result in good performance. Experience has shown that multiplying this product by a correction factor of 1.6 gives a more realistic figure for the total required power input. The resulting wattage compensates not only for the radiant inefficiency involved, but also for the radiation falling outside the target area. For small areas, or when the fixture mounting height exceeds 16 ft, the multiplier can be as large as 2.0; large areas with sides of approximately equal length can have a multiplier of about 1.4.

The point-by-point method is the best way to calculate the fixture requirements for an installation. This method involves dividing the target area into 1 ft squares and adding the radiant energy from each infrared fixture incident on each square (Figure 9). The radiant energy distribution of a given infrared fixture can be obtained from the equipment manufacturer and should be followed for that fixture size and placement.

INTENSITY ON PAVEMENT FROM FOUR INFRARED FIXTURES, W/ft^2														
14.7	16.0	19.75	23.7	25.5	27.5	28.2	28.0	28.2	27.5	25.5	23.7	19.75	16.0	14.7
23.7	24.8	28.7	31.7	35.7	38.2	38.5	39.4	38.5	38.2	35.7	31.7	28.7	24.8	23.7
25.7	31.7	37.5	42.7	46.4	49.4	52.2	53.0	52.2	49.4	46.4	42.7	37.5	31.7	25.7
28.2	34.3	42.8	46.7	51.2	55.7	58.5	63.0	58.5	55.7	51.2	46.7	42.8	34.3	28.2
28.2	34.3	42.8	46.7	51.2	55.7	58.5	63.0	58.5	55.7	51.2	46.7	42.8	34.3	28.2
25.7	31.7	37.5	42.7	46.4	49.4	52.2	53.0	52.2	49.4	46.4	42.7	37.5	31.7	25.7
23.7	24.8	28.7	31.7	35.7	38.2	38.5	39.4	38.5	38.2	35.7	31.7	28.7	24.8	23.7
14.7	16.0	19.75	23.7	25.5	27.5	28.2	28.0	28.2	27.5	25.5	23.7	19.75	16.0	14.7

Fig. 9 Typical Power Density Distribution for Infrared Snow-Melting System

(Potter 1967)

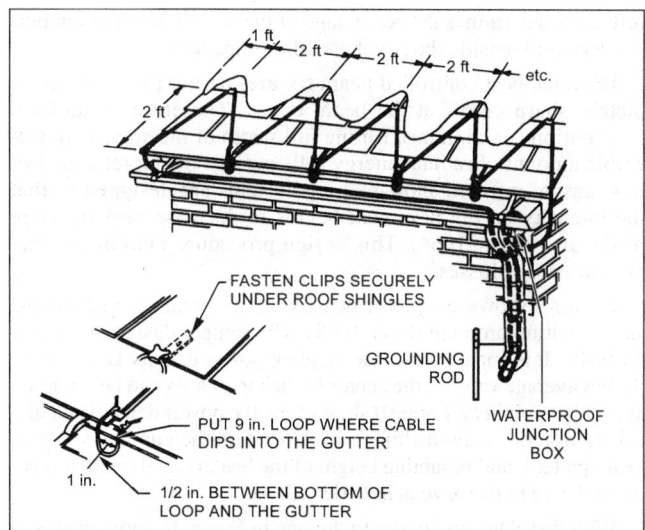

Fig. 10 Typical Insulated Wire Layout to Protect Roof Edge and Downspout

System Operation. With infrared energy, the target area can be preheated to snow-melting temperatures in 20 to 30 min, unless the air temperature is well below 20°F or wind velocity is high (Frier 1965). This short warm-up time makes it unnecessary to turn on the system before snow begins to fall. The equipment can be turned on either manually or with a snow detector. A timer is sometimes used to turn the system off 4 to 6 h after snow stops falling, allowing time for the pavement to dry completely.

If the snow is allowed to accumulate before the infrared system is turned on, there will be a delay in clearing the pavement, as is the case with embedded hydronic or electric systems. Because the infrared energy is absorbed in the top layer of snow rather than by the pavement surface, the length of time needed depends on the snow depth and on atmospheric conditions. Generally, a system that maintains a clear pavement by melting 1 in. of snow per hour as it falls requires 1 h to clear 1 in. of accumulated snow under the same conditions.

To ensure maximum efficiency, fixtures should be cleaned at least once a year, preferably at the beginning of the winter season. Other maintenance requirements are minimal.

Snow Melting in Gutters and Downspouts

Electrical heating cables are used to prevent heavy snow and ice accumulation on roof overhangs and to prevent ice dams from forming in gutters and downspouts (Lawrie 1966). Figure 10 shows a typical cable layout for protecting a roof edge and downspout. Cable for this purpose is generally rated at approximately 6 to 16 W/ft, and about 2.5 ft of wire is installed per linear foot of roof edge. One foot of heated wire per linear foot of gutter or downspout is usually adequate.

If the roof edge or gutters (or both) are heated, downspouts that carry away melted snow and ice must also be heated. A heated length of cable (weighted, if necessary) is dropped inside the downspout to the bottom, even if it is underground.

Lead wires should be spliced or plugged into the main power line in a waterproof junction box, and a ground wire should be installed

from the downspout or gutter. Ground fault circuit protection is required per the *NEC* (NFPA *Standard* 70).

Manual switch control is generally used, although a protective thermostat that senses outdoor temperature should be used to prevent system operation at ambient temperatures above 40°F.

REFERENCES

Adlam, T.N. 1950. *Snow melting.* The Industrial Press, New York; University Microfilms, Ann Arbor, MI.

Chapman, W.P. 1952. Design of snow melting systems. *Heating and Ventilating* (April):95 and (November):88.

Chapman, W.P. 1955. Are thermal stresses a problem in snow melting systems? *Heating, Piping and Air Conditioning* (June):92 and (August):104.

Colliver, D.G., R.S. Gates, H. Zhang, T. Burks, and K.T. Priddy. 1998. Updating the tables of design weather conditions in the *ASHRAE Handbook—Fundamentals.* ASHRAE RP-890 *Final Report.*

Frier, J.P. 1964. Design requirements for infrared snow melting systems. *Illuminating Engineering* (October):686. Also discussion, December.

Frier, J.P. 1965. Snow melting with infrared lamps. *Plant Engineering* (October):150.

Gordon, P.B. 1950. Antifreeze protection for snow melting systems. *Heating, Piping and Air Conditioning Contractors National Association Official Bulletin* (February):21.

Incropera, F.P. and D.P. DeWitt. 1996. *Introduction to heat transfer,* pp. 332-34. John Wiley and Sons, New York.

Kuehn, T.H., J.W. Ramsey, and J.L. Threlkeld. 1998. *Thermal environmental engineering,* 3rd ed., p. 179. Prentice Hall, Upper Saddle River, NJ.

Lawrie, R.J. 1966. Electric snow melting systems. *Electrical Construction and Maintenance* (March):110.

NCDC. 1990. Precipitation—Hourly cooperative TD-3240 documentation manual. National Climatic Data Center, Asheville, NC.

NCDC. 1993. Solar and meteorological surface observation network 1961-1990 (SAMSON), Version 1.0.

NACE. 1978. *Basic corrosion course text* (October). National Association of Corrosion Engineers, Houston, TX.

NFPA. 1996. *National electrical code.* ANSI/NFPA *Standard* 70-96. National Fire Protection Association, Quincy, MA.

Potter, W.G. Electric snow melting systems. *ASHRAE Journal* 9(10):35-44.

Ramsey, J.W., H.D. Chiang, and R.J. Goldstein. 1982. A study of the incoming long-wave atmospheric radiation from a clear sky. *Journal of Applied Meteorology* 21:566-78.

Ramsey, J.W., M.J. Hewett, T.H. Kuehn, and S.D. Petersen. 1999. Updated design guidelines for snow melting system. *ASHRAE Transactions* 105(1).

Witsken, C.H. 1965. Snow melting with electric wire. *Plant Engineering.* (September):129.

BIBLIOGRAPHY

Chapman, W.P. 1955. Snow melting system hydraulics. *Air Conditioning, Heating and Ventilating* (November).

Chapman, W.P. 1957. Calculating the heat requirements of a snow melting system. *Air Conditioning, Heating and Ventilating* (September through August).

Chapman, W.P. and S. Katunich. 1956. Heat requirements of snow melting systems. *ASHAE Transactions* 62:359.

Hydronics Institute. 1994. *Snow melting calculation and installation guide.* Berkeley Heights, NJ.

Kilkis, i.B. 1994. Design of embedded snow-melting systems: Part 1, Heat requirements—An overall assessment and recommendations. *ASHRAE Transactions* 100(1):423-33.

Kilkis, i.B. 1994. Design of embedded snow-melting systems: Part 2, Heat transfer in the slab—A simplified model. *ASHRAE Transactions* 100(1):434-41.

EVAPORATIVE COOLING APPLICATIONS

EVAPORATIVE cooling is energy-efficient, environmentally benign, and cost-effective in many applications. Applications range from comfort cooling in residential, agricultural, commercial and institutional buildings, to industrial applications for spot cooling in mills, foundries, power plants, and other hot environments. Several types of apparatus cool by evaporating water directly in the airstream, including (1) direct evaporative coolers, (2) spray-filled and wetted surface air washers, (3) sprayed coil units, and (4) humidifiers. Indirect evaporative cooling equipment combine the evaporative cooling effect in a secondary air stream with a heat exchanger to produce cooling without adding moisture to the primary airstream.

Direct evaporative cooling reduces the dry bulb temperature and increases the relative humidity of the air. It is most commonly applied to dry climates or to applications requiring high air exchange rates. Innovative schemes combining evaporative cooling with other equipment have resulted in energy efficient designs.

When temperature and/or humidity must be controlled within narrow limits, heat and mechanical refrigeration can be combined with evaporative cooling in stages. Evaporative cooling equipment, including unitary equipment and air washers, is covered in Chapter 19 of the 1996 *ASHRAE Handbook—Systems and Equipment*.

GENERAL APPLICATIONS

Cooling

Evaporative cooling is used in almost all climates. The wet-bulb temperature of the entering airstream limits direct evaporative cooling. The wet bulb temperature of the secondary airstream limits indirect evaporative cooling.

Design wet-bulb temperatures are rarely higher than 78°F, in which case direct evaporative cooling is economical for spot cooling, kitchens, laundries, agricultural, and industrial applications. At lower wet-bulb temperatures, evaporative cooling can be effectively used for comfort cooling, although some climates may require mechanical refrigeration for part of the year.

Indirect applications lower the air wet-bulb temperature and can produce leaving dry-bulb temperatures that approach the wet-bulb temperature of the secondary airstream. Using room exhaust as secondary air or incorporating precooled air in the secondary airstream lowers the wet-bulb temperature of the secondary air and further enhances the cooling capability of the indirect evaporative cooler.

The direct evaporative cooling process is an adiabatic exchange of heat. Heat must be added to evaporate water. The air into which water is evaporated supplies the heat. The dry-bulb temperature is lowered and sensible cooling results. The amount of heat removed from the air equals the amount of heat absorbed by the water evaporated as heat of vaporization. If water is recirculated in the direct evaporative cooling apparatus, the water temperature in the reservoir approaches the wet-bulb temperature of the air entering the process. By definition, no heat is added to, or extracted from, an adiabatic process. The initial and final conditions of an adiabatic

process fall on a line of constant total heat (enthalpy), which nearly coincides with a line of constant wet-bulb temperatures.

The maximum reduction in dry-bulb temperature is the difference between the entering air dry- and wet-bulb temperatures. If the air is cooled to the wet-bulb temperature, it becomes saturated and the process would be 100% effective. Effectiveness is the depression of the dry-bulb temperature of the air leaving the apparatus divided by the difference between the dry- and wet-bulb temperatures of the entering air. Theoretically, adiabatic direct evaporative cooling is less than 100% effective, although evaporative coolers are 85 to 95% or even more effective.

When a direct evaporative cooling unit alone cannot provide the desired conditions, several alternatives can satisfy application requirements and still be energy effective and economical to operate. The recirculating water supplying the direct evaporative cooling unit can be increased in volume and chilled by mechanical refrigeration to provide lower leaving wet-and dry-bulb temperatures and lower humidity. Compared to the cost of using mechanical refrigeration only, this arrangement reduces operating costs by as much as 25 to 40%. Indirect evaporative cooling applied as a first stage, upstream from a second, direct evaporative stage, reduces both the entering dry- and wet-bulb temperatures before the air enters the direct evaporative cooler. Indirect evaporative cooling may save as much as 60 to 75% or more of the total cost of operating mechanical refrigeration to produce the same cooling effect. Systems may combine indirect evaporative cooling, direct evaporative cooling, heaters, and mechanical refrigeration, or any combination of these processes.

The psychrometric chart in Figure 1 illustrates what happens when air is passed through a direct evaporative cooler. In the example shown, assume an entering condition of 95°F db and 75°F wb. The initial difference is 95 − 75 = 20°F. If the effectiveness is 80%, the depression is 0.80 × 20 = 16°F db. The dry-bulb temperature leaving the direct evaporative cooler is 95 − 16 = 79°F. In the adiabatic evaporative cooler, only a portion of the water recirculated is assumed to evaporate and the water supply is recirculated. The recirculated water will reach an equilibrium temperature that is approximately the same as the wet bulb temperature of the entering air.

The performance of an indirect evaporative cooler can also be shown on a psychrometric chart (Figure 1). Many manufacturers of indirect evaporative cooling equipment use a similar definition of effectiveness as is used for a direct evaporative cooler. In indirect evaporative cooling, the cooling process in the primary airstream follows a line of constant moisture content (constant dew point). Indirect evaporative cooling effectiveness is the dry-bulb depression in the primary airstream divided by the difference between the entering dry-bulb temperature of the primary airstream and the entering wet-bulb temperature of the secondary air. Depending on the heat exchanger design and relative air quantities of primary and secondary air, effectiveness ratings may be as high as 85%.

Assuming an effectiveness of 60%, and assuming both primary air and secondary air enter the apparatus at the outdoor condition of 95°F db and 75°F wb, the dry-bulb depression is 0.60 (95 − 75) = 12°F. The dry-bulb temperature leaving the indirect evapora-

---

The preparation of this chapter is assigned to TC 5.7, Evaporative Cooling.

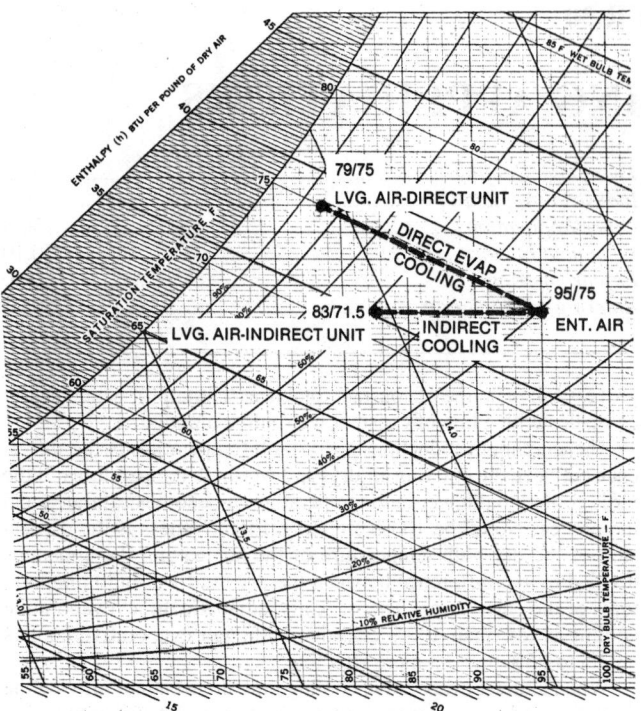

Fig. 1 Psychrometrics of Evaporative Cooling

cooling process is 95 − 12 = 83°F. Because the process cools without adding moisture, the wet-bulb temperature is also reduced. Plotting on the psychrometric chart shows that the final wet-bulb temperature is 71.5°F. Because both the wet- and the dry-bulb temperatures in the indirect evaporative cooling process are reduced, indirect evaporative cooling can be used as a substitute for a portion of the refrigeration load in many applications.

Humidification

Air can be humidified with a direct evaporative cooler by three methods: (1) Using recirculated water without prior treatment of the air, (2) preheating the air and treating it with recirculated water, or (3) heating recirculated water. The air leaving an evaporative cooler that is being used as either a humidifier or a dehumidifier will be substantially saturated when in operation. Usually, the spread between leaving dry- and wet-bulb temperatures is less than 1°F. The temperature difference between leaving air and leaving water depends on the difference between entering dry- and wet-bulb temperatures and on certain physical features, such as the length and height of a spray chamber, the cross-sectional area and depth of the media being used, quantity and velocity of air, quantity of water, and the spray pattern. In any direct evaporative humidifier installation, the air should not enter with a dry-bulb temperature of less than 39°F; otherwise, the water may freeze.

Recirculated Water. Except for the small amount of energy added by shaft work from the recirculating pump and the small amount of heat leakage into the apparatus through the unit enclosure, the evaporative humidification process is strictly adiabatic. As evaporation occurs from the recirculated liquid its temperature approaches the thermodynamic wet-bulb temperature of the entering air.

The airstream can not be brought to complete saturation, but its state point changes adiabatically along a line of constant enthalpy. Typical saturation or humidifying effectiveness of various air washer spray arrangements is between 50 and 98%. The degree of saturation depends on the extent of the contact between air and

water. Other conditions being equal, a low-velocity airflow is conducive to higher humidifying effectiveness.

Preheated Air. Preheating the air increases both the dry- and wet-bulb temperatures and lowers the relative humidity; it does not, however, alter the humidity ratio (i.e., the mass ratio of water vapor to dry air) or the dew-point temperature of the air. At a higher wet-bulb temperature, but with the same humidity ratio, more water can be absorbed per unit mass of dry air in passing through the direct evaporative humidifier. The analysis of the process that occurs in the direct evaporative humidifier is the same as that for recirculated water. The desired conditions are achieved by heating to the desired wet-bulb temperature and evaporatively cooling at constant wet-bulb temperature to the desired dry-bulb temperature and relative humidity. Relative humidity of the leaving air may be controlled by (1) bypassing air around the direct evaporative humidifier or (2) reducing the number of operating spray nozzles or the area of media wetted.

Heated Recirculated Water. Heating of humidifier water increases the effectiveness of a direct evaporative humidifier. When heat is added to the recirculated water, the mixing in the direct evaporative humidifier may still be modeled as an adiabatic process. The state point of the mixture should move toward the specific enthalpy of the heated water. By elevating the water temperature, the air temperature (both dry and wet-bulb) may be raised above the dry-bulb temperature of the entering air. The relative humidity of the leaving air may be controlled by methods similar to those used with preheated air.

Dehumidification and Cooling

Direct evaporative coolers may also be used to cool and dehumidify air. If the entering water temperature is cooled below the entering wet-bulb temperature, both the dry- and wet-bulb temperatures of the leaving air are lowered. Dehumidification results if the leaving water temperature is maintained below the entering air dew point. Moreover, the final water temperature is determined by the sensible and latent heat absorbed from the air and the amount of water circulated. However, the final water temperature cannot exceed the final required dew-point temperature, with 1 to 2°F below dew-point being common.

The air leaving a direct evaporative cooler that is being used as a dehumidifier is substantially saturated. Usually, the spread between dry- and wet-bulb temperatures is less than 1°F. The temperature difference between leaving air and leaving water depends on the difference between entering dry- and wet-bulb temperatures and on certain design features, such as the cross-sectional area and depth of the media or spray chamber, quantity and velocity of air, quantity of water, and the water distribution.

Air Cleaning

Direct evaporative coolers of all types perform some air cleaning. Rigid media direct evaporative coolers are effective at removing particulate down to about 1 μm in size. Air washers are effective down to about 10 μm.

The dust removal efficiency of direct evaporative coolers depends largely on the size, density, wettability, and solubility of the dust particles. Larger, more wettable particles are the easiest to remove. Separation is largely a result of the impingement of particles on the wetted surface of the eliminator plates or on the surface of the media. Because the force of impact increases with the size of the solid, the impact (together with the adhesive quality of the wetted surface) determines the cooler's usefulness as a dust remover. The standard low-pressure spray is relatively ineffective in removing most atmospheric dusts. Direct evaporative coolers are of little use in removing soot particles because their greasy surface will not adhere to the wet plates or media. Direct evaporative coolers are also ineffective in removing smoke, because the small particles (less than 1 μm) do not impinge with sufficient impact to pierce the water film and be held on the media. Instead, the particles follow the air path between the media surfaces.

Control of Gaseous Contaminants. When used in a makeup air system comprised of a mixture of outside air and recirculated air, direct evaporative coolers function as scrubbers and reduce some of the gaseous contaminants found in the atmosphere. These contaminants may concentrate in the recirculating water, so some water needs to be bled off. For more information regarding the control of gaseous contaminants, see Chapter 44.

INDIRECT EVAPORATIVE COOLING

Outdoor Air Systems

Because indirect evaporative cooling does not increase the absolute humidity in the primary airstream, it is well suited for precooling the air entering a refrigerated coil. This precooling reduces the sensible load on the refrigerated coil and compressor. As a result, the size of refrigeration equipment may be reduced, which reduces energy use and operating cost. By contrast, direct evaporative cooling equipment, when used with a refrigerated coil, exchanges latent heat for sensible heat, increasing the latent load on the coil in proportion to the sensible cooling achieved. The enthalpy of the air entering the coil is not changed. The power input per ton of cooling effect is substantially lower with indirect evaporative cooling than with conventional refrigerated equipment.

Peterson and Runn (1991) observed in a study in Dallas that (1) the seasonal energy efficiency ratio (SEER) of an indirect evaporative cooler can be 70% higher than that of a conventional air conditioner and (2) nearly 12% of the air-conditioning capacity can be displaced by the indirect evaporative cooler. However, the additional static pressure created by the indirect equipment increases the motor power of the primary air fan and must be considered, even when continuous cooling is not required. The additional static pressure loss may be as low as 0.2 in. of water, which increases the supply fan power very little. Also, the equipment may require additional energy for water pumping and for moving secondary air across the evaporative surfaces.

The cooling configuration is shown in Figure 2. The primary air side of the indirect unit is positioned at the intake to the refrigerated cooling coil. The secondary air to the unit can come from outdoor ambient air or from room exhaust air. Exhaust air from the space may have a lower wet-bulb temperature than outdoor ambient, depending on climate, time of year, and space latent load. Latent cooling may be possible in the primary airstream using room exhaust air as secondary air. This may occur if the dew-point temperature of the primary air is above the exhaust (secondary air) wet-bulb temperature. If this is possible, provision to drain the water condensed from the primary airstream may be necessary. In many areas, an indirect precooler can satisfy more than one-half of the annual cooling load. For example, Supple (1982) showed that 30% of the annual cooling load for Chicago could be accomplished by indirect evaporative precooling. Indirect evaporative precooling using exhaust air for secondary air may be as effective in warm, humid climates as in drier areas.

Mixed Air Systems

Indirect evaporative cooling can also save energy in systems that use a mixture of return and outside air. The configuration is similar to that shown in Figure 2, except that a mixing section with outside- and return-air dampers is added upstream of the indirect precooling section. Outdoor air would be used as secondary air for the indirect evaporative cooler.

A typical indirect evaporative precooling stage can reduce the dry-bulb temperature by as much as 60 to 80% of the difference between the entering dry-bulb temperature and the wet-bulb temperature of the secondary air. When the dry-bulb temperature of the

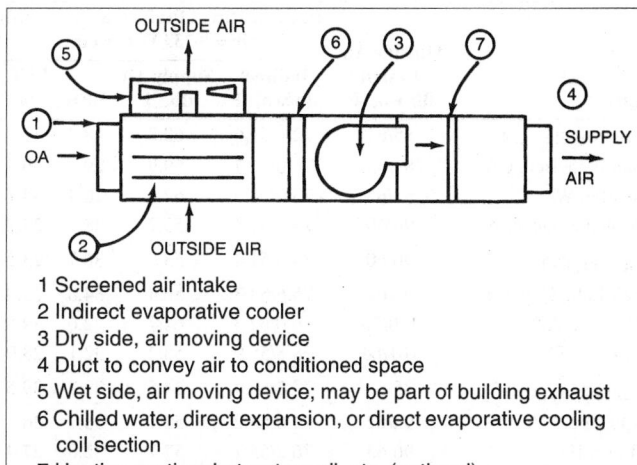

1 Screened air intake
2 Indirect evaporative cooler
3 Dry side, air moving device
4 Duct to convey air to conditioned space
5 Wet side, air moving device; may be part of building exhaust
6 Chilled water, direct expansion, or direct evaporative cooling coil section
7 Heating section, hot water coil, etc. (optional)

Fig. 2 Indirect Evaporative Cooling Configuration

mixed air is more than a few degrees above the wet-bulb temperature of the secondary airstream, indirect evaporative cooling of the mixed air-stream may reduce the amount of refrigerated cooling required. The cooling contribution depends on the differential between the mixed air dry-bulb temperature and secondary air wet-bulb temperature. As mixture temperature increases, or as secondary air wet-bulb decreases, the precooling contribution becomes more significant.

In variable-air-volume (VAV) systems, a decrease in supply air volume (during periods of reduced load) results in lower air velocity through the indirect evaporative cooler; this increases equipment effectiveness. Lower static pressure loss reduces the energy consumed by the supply fan motor.

Indirect Evaporative Cooling With Heat Recovery

In indirect evaporative cooling, outside supply air passes through an air-to-air heat exchanger and is cooled by evaporatively cooled air exhausted from the building or application. The two air streams never mix or come into contact, so no moisture is added to the supply air stream. Cooling the building's exhaust air results in a larger overall temperature difference across the heat exchanger and a greater cooling of the supply air. Indirect evaporative cooling requires only fan and water pumping power, so the coefficient of performance tends to be high. The principle of indirect evaporative cooling is effective in most air-conditioned buildings, because the evaporative cooling process is applied to the exhaust air rather than the outside air.

Indirect evaporative cooling has been applied in a number of heat recovery applications (Mathur 1993a). For example, plate type heat exchangers (Scofield and DesChamps 1984, Wu and Yellot 1987), heat pipe exchangers (Mathur 1991, 1998; Scofield 1986), rotary regenerative heat exchangers (Woolridge et al. 1976), and two phase thermosiphon loop heat exchangers (Mathur 1992, 1990). The principle of indirect evaporative cooling has also been applied to residential air conditioning where the outside condensing unit is evaporatively cooled to enhance performance (Mathur 1997, Mathur and Goswami 1995, Mathur et al. 1993b). Indirect evaporative cooling with heat recovery is covered in detail in Chapter 42, Air-To-Air Energy Recovery, of the 1996 *ASHRAE Handbook—Systems and Equipment.*

BOOSTER REFRIGERATION

Staged evaporative coolers can totally cool office buildings, schools, gymnasiums, sports facilities, department stores, restaurants, factory space, and other buildings. These coolers can control

City	Outside Air Design db/wb, °F	Indirect/Direct Performance (Supply Air = 0.733 W per cfm)			
		Indirect db/wb, °F	Supply Air db, °F	EER	EUC, %
Los Angeles, CA	85/64	72.4/59.6	60.8	28.2	31.9
San Francisco, CA	83/63	71.0/58.7	59.9	29.6	30.4
Seattle, WA	85/65	73.0/60.9	62.1	26.4	34.1
Albuquerque, NM	96/60	74.4/51.4	53.7	38.7	23.2
Denver, CO	96/60	74.4/51.4	53.7	38.7	23.2
Salt Lake City, UT	96/62	75.6/54.3	56.4	34.8	25.9
Phoenix, AZ	110/70	86.0/62.3	64.7	22.6	39.8
El Paso, TX	101/64	78.8/55.9	58.2	32.1	28.0
Santa Rosa, CA	85/67	74.2/63.4	64.5	22.8	39.5
Spokane, WA	92/62	74.0/55.2	57.1	33.7	26.7
Boise, ID	96/63	76.2/55.7	57.7	32.8	27.4
Billings, MT	93/63	75.0/56.4	58.3	32.0	28.1
Portland, OR	90/67	76.2/62.4	63.8	23.9	37.7
Sacramento, CA	100/69	81.4/63.0	64.8	22.3	40.3
Fresno, CA	103/71	83.8/65.1	66.9	19.3	46.7
Austin, TX	98/74	83.8/69.9	71.3	12.9	69.9

INDIRECT/DIRECT SYSTEM PERFORMANCE

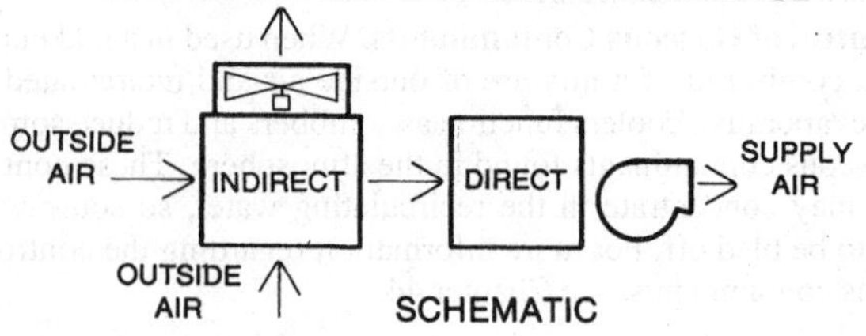

SCHEMATIC

Outdoor air design condition: 0.4% dry bulb/mean coincident wet bulb (1997 *ASHRAE Handbook—Fundamentals*, Chapter 26).

EER = Energy Efficiency Ratio = /h cooling output per watt of electrical input. Comparison base to conventional refrigeration with 60°F supply air and 20°F temperature drop.

EUC = Energy Use Comparison to conventional refrigeration with EER = 9.

I/D effectiveness: Indirect = 60% or 0.6 (dry bulb − wet bulb);
Direct = 90% or 0.9 (dry bulb − wet bulb).

Note: Sea level psychrometric chart used. At 5000 ft elevation, supply air temperature increases 3 to 4%.

Fig. 3 Indirect/Direct Two-Stage System Performance

room dry-bulb temperature and relative humidity, even though one stage is a direct evaporative-cooling stage. In many cases, booster refrigeration is not required. Supple (1982) showed that even in higher humidity areas with a 1% mean wet-bulb design temperature of 75°F, 42% of the annual cooling load can be satisfied by two-stage evaporative cooling. Refrigerated cooling need supply only 58% of the load.

Figure 3 shows indirect/direct two-stage performance for 16 cities in the United States. Performance is based on 60% effectiveness of the indirect stage and 90% for the direct stage. Supply air temperatures (leaving the direct stage) at the 0.4% design dry-bulb mean coincident wet-bulb condition range from 52.7 to 71°F. Energy use ranges from 23.2 to 69.9%, compared to conventional refrigerated equipment.

Booster mechanical refrigeration provides indoor design comfort conditions regardless of the outdoor wet-bulb temperature without having to size the mechanical refrigeration equipment for the total cooling load. If the indoor humidity level becomes uncomfortable, the quantity of moisture introduced into the airstream must be limited in order to control room humidity. Where the upper relative humidity design level is critical, a life cycle cost analysis would favor a design composed of an indirect cooling stage and a mechanical refrigeration stage.

RESIDENTIAL OR COMMERCIAL COOLING

In dry climates, evaporative cooling is effective, with lower air velocities than those required in humid climates. Packaged direct evaporative coolers are used for residential and commercial application. Cooler capacity may be determined from standard heat gain calculations (see Chapters 27 and 28 of the 1997 *ASHRAE Handbook—Fundamentals*).

Detailed calculation of heat load, however, is usually not economically justified. Instead, one of several estimates give satisfactory results. In one method, the difference between dry-bulb design temperature and coincident wet-bulb temperature divided by 10 is equal to the number of minutes needed for each air change. This or any other arbitrary method for equating cooling capacity with airflow depends on a direct evaporative cooler effectiveness of 70 to 80%. Obviously, the method must be modified for unusual conditions such as large unshaded glass areas, uninsulated roof exposure,

or high internal heat gain. Also, such empirical methods make no attempt to predict air temperature at specific points; they merely establish an air quantity for use in sizing equipment.

Example 1. An indirect evaporative cooler is to be installed in a 50 by 80 ft one-story office building with a 10 ft ceiling and a flat roof. Outdoor design conditions are assumed to be 95°F dry bulb and 65°F wet bulb. The following heat gains are to be used in the design:

	Gains, Btu/h
All walls, doors, and roof	78,500
Glass area	5,960
Occupants (sensible load)	17,000
Lighting	62,700
Total sensible heat load	164,160
Total latent load (occupants)	21,250
Total heat load	185,410

Find the required air quantity, the temperature and humidity ratio of the air leaving the cooler (entering the office), and the temperature and humidity ratio of the air leaving the office.

Solution: A temperature rise of 10°F in the cooling air is assumed. The airflow rate that must be supplied by the indirect evaporative cooler may be found from the following equation:

$$Q_{ra} = \frac{q_s}{60\rho c_p(t_1 - t_s)} = \frac{164,160}{60 \times 0.018 \times 10} = 15,200 \text{ cfm} \quad (1)$$

where:

Q_{ra} = required airflow, cfm

q_s = instantaneous sensible heat load, Btu/h

t_1 = indoor air dry-bulb temperature, °F

t_s = room supply air dry-bulb temperature, °F

ρc_p = density times specific heat of air ≈ 0.018 Btu/ft^3·°F

This air volume represents a 2.6 min [50 × 80 × 10/15,200] air change for a building of this size. The indirect evaporative air cooler is assumed to have a saturation effectiveness of 80%. This is the ratio of the reduction of the dry-bulb temperature to the wet-bulb depression of the entering air. The dry-bulb temperature of the air leaving the indirect evaporative cooler is found from the following equation:

$$t_2 = t_1 - \frac{e_h}{100}(t_1 - t') = 95 - \frac{80}{100}(95 - 65) = 71°F \qquad (2)$$

where:

t_2 = dry-bulb temperature of leaving air, °F

t_1 = dry-bulb temperature of entering air, °F

e_h = humidifying or saturating effectiveness, %

t' = thermodynamic wet-bulb temperature of entering air, °F

From the psychrometric chart, the humidity ratio W_2 of the cooler discharge air is 0.01185 lb/lb dry air. The humidity ratio W_3 of the air leaving the space being cooled is found from the following equation:

$$W_3 = \frac{q_e}{4840\, Q_{ra}} + W_2 \qquad (3)$$

$$W_3 = \frac{21250}{4840 \times 15200} + 0.01185 = 0.01214 \text{ lb/lb (dry air)}$$

where q_e = latent heat load in Btu/h.

The remaining values of wet-bulb temperature and relative humidity for the problem may be found from the psychrometric chart. Figure 4 illustrates the various relationships of outdoor air, supply air to the space. and discharge air.

The wet-bulb depression (WBD) method to estimate airflow gives the following result:

$$\text{WBD}/10 = (95 - 65)/10 = 3.0 \text{ min per air change}$$

$$Q_{ra} = \frac{\text{volume}}{\text{air change rate}} = \frac{80 \times 50 \times 10}{3.0} = 13,300 \text{ cfm}$$

While not exactly alike, these two air volume calculations are close enough to select cooler equipment of the same size.

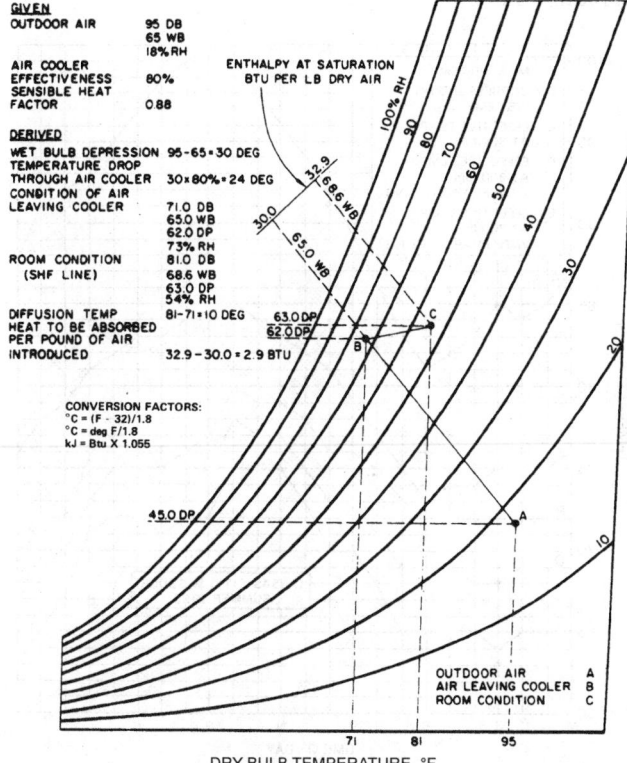

Fig. 4   Psychrometric Diagram for Example 1

EXHAUST REQUIRED

If air is not exhausted freely, the increased static pressure will reduce airflow through the evaporative cooler. The result is a marked increase in the moisture and the heat absorbed per unit mass of air leaving the evaporative cooler. This reduced airflow also reduces the air velocity in the room. The combination of these effects reduces the comfort level. Properly designed systems should have a minimum of 2 ft² of exhaust area for every 1000 cfm. If the exhaust area is not sufficient, a powered exhaust should be used. The amount of power depends on the total airflow and the amount of free or gravity exhaust. Some applications require that the powered exhaust capacity equal the cooler output.

TWO-STAGE COOLING

Two-stage coolers for commercial applications can extend the range of atmospheric conditions under which comfort requirements can be met, as well as reduce the energy cost. (For the same design conditions, two-stage cooling provides lower cool air temperatures, which reduces the required airflow.)

INDUSTRIAL APPLICATIONS

In factories with large internal heat loads, it is difficult to simply ventilate and approach outdoor conditions during the summer without using extremely large quantities of outdoor air. Both direct and indirect evaporative cooling may be used to reduce heat stress with less outdoor air. Evaporative cooling normally results in lower effective temperatures than ventilation alone, regardless of the ambient relative humidity.

Effective Temperature. Comfort cooling in air-conditioned spaces is usually based on providing space temperature and relative humidity conditions designed to provide human comfort without a draft. The effective temperature relates the cooling effects of air motion and air relative humidity to the effect of air-conditioned (refrigerated) air. Figure 5 shows an effective temperature chart for air velocities ranging from 20 to 700 fpm. Although the maximum velocity shown on the chart is 700 fpm, workers exposed to high heat-producing operations may request air movement up to 4000 fpm to offset the radiant heat effect of the equipment. Because the normal working range of the chart is approximately midway between the vertical dry- and wet-bulb scales, changes in either dry-bulb or wet-bulb temperatures have similar impacts on worker comfort. A reduction in either one will decrease the effective temperature by about one-half of the reduction. Lines ED and CD on the chart graphically illustrate this.

A condition of 95°F dry bulb and 75°F wet bulb was chosen as the original state, because this condition is usually considered the summer design criteria in most areas. Reducing the temperature 15°F by evaporating water adiabatically provides an effective temperature reduction of 5.5°F for air moving at 20 fpm and a reduction of 9.5°F for air moving at 700 fpm—an improvement of 4°F.

The reduction in dry-bulb temperature through water evaporation increases the effectiveness of the cooling power of moving air in this example by 137%. On Line ED, the effective temperature varies from 83°F at 20 fpm to 79.5°F at 700 fpm with unconditioned air, whereas Line CD indicates an effective temperature of 77.5°F at 20 fpm and 70°F at 700 fpm with air cooled by a simple direct evaporative process. In the unconditioned case, increasing the air velocity from 20 to 700 fpm resulted in only a 3.5°F decrease in effective temperature. This contrasts with a 7.5°F decrease in effective temperature for the same range of air movement when the dry-bulb temperature was lowered by water evaporation. This demonstrates that direct evaporative cooling can provide a more comfortable environment regardless of geographical location.

Two methods are demonstrated to illustrate the environmental improvement that may be achieved with evaporative coolers. In one

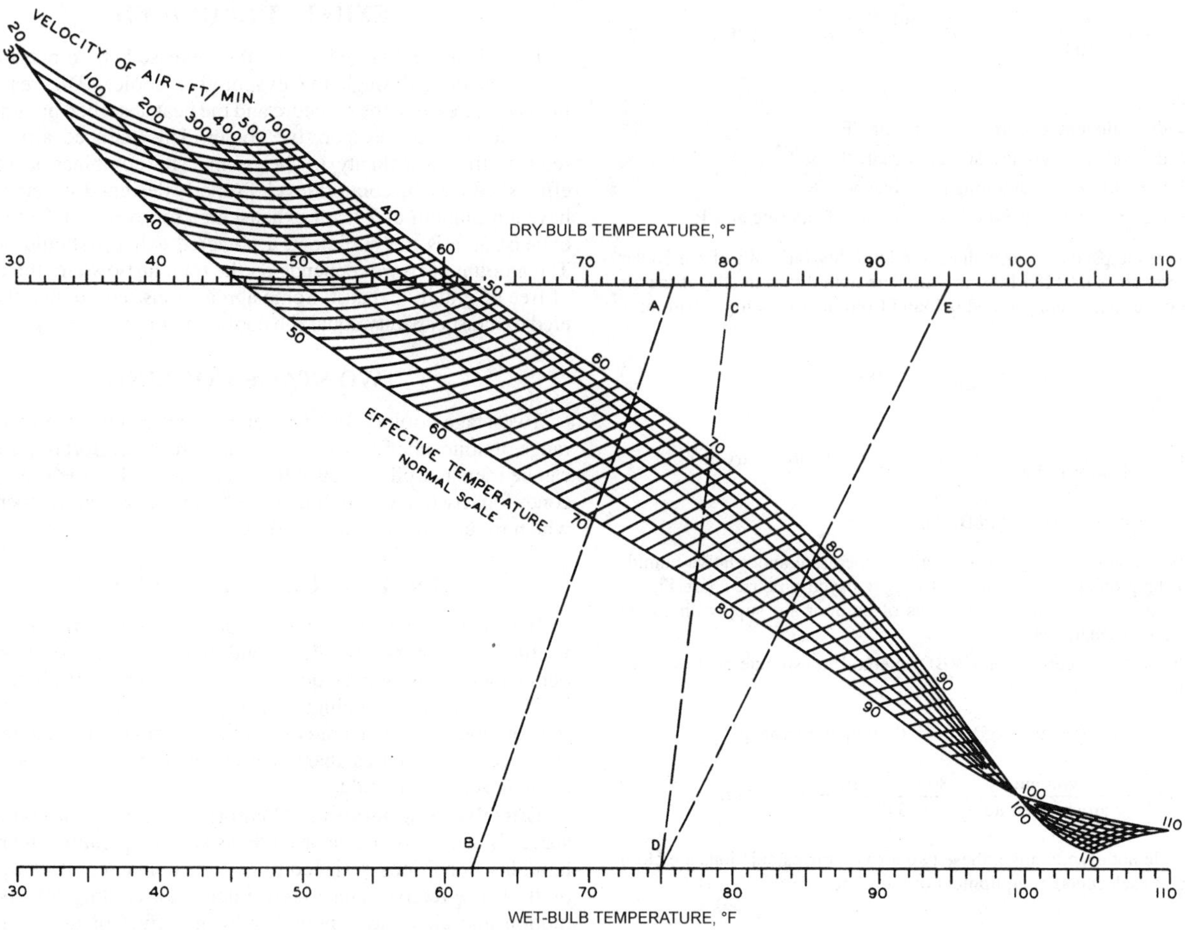

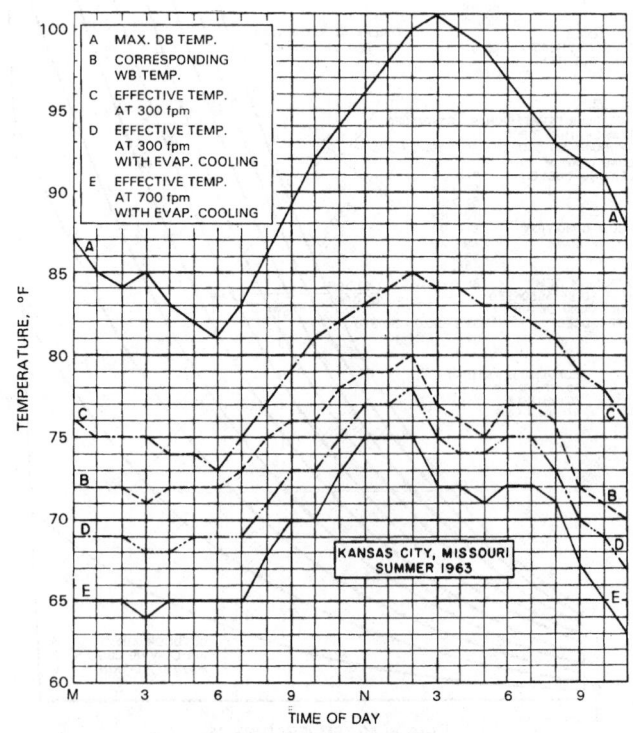

Fig. 5 Effective Temperature Chart

method, shown in Figure 6, temperature is plotted against time of day to illustrate effective temperature depression over time. Curve A shows ambient maximum dry-bulb temperature recordings. Curve B shows the corresponding wet-bulb temperatures. Curve C depicts the effective temperature when unconditioned air is moved over a person at 300 fpm. Curve D illustrates air conditioned in an 80% effective direct evaporative cooler before being projected over the person at 300 fpm. Curve E shows the additional decrease in effective temperature with air velocities of 700 fpm. While a maximum suggested effective temperature of 80°F is briefly exceeded with unconditioned air at 300 fpm (Curve C), both the differential and the total hours are substantially reduced from still air conditions. Curves D and E illustrate that, in spite of the high wet-bulb temperatures, the in-plant environment can be continuously maintained below the suggested upper limit of 80°F effective temperature. This demonstration assumes that the combination of air velocity, duct length, and insulation between the evaporative cooler and the duct outlet is such that there is little heat transfer between air in the ducts and warmer air under the roof.

Figure 7 illustrates another method of demonstrating the effect of using direct evaporative coolers by plotting effective comfort zones using ambient wet- and dry-bulb temperatures on an ASHRAE Psychrometric Chart (Crow 1972). The dashed lines show the improvement to expect when using an 80% effective direct evaporative cooler.

Area Cooling

Both direct and indirect evaporative cooling may be used for area or spot cooling of industrial buildings. Both can be controlled either

Fig. 6 Effective Temperature for Summer Day in Kansas City, Missouri (Worst Case Basis)

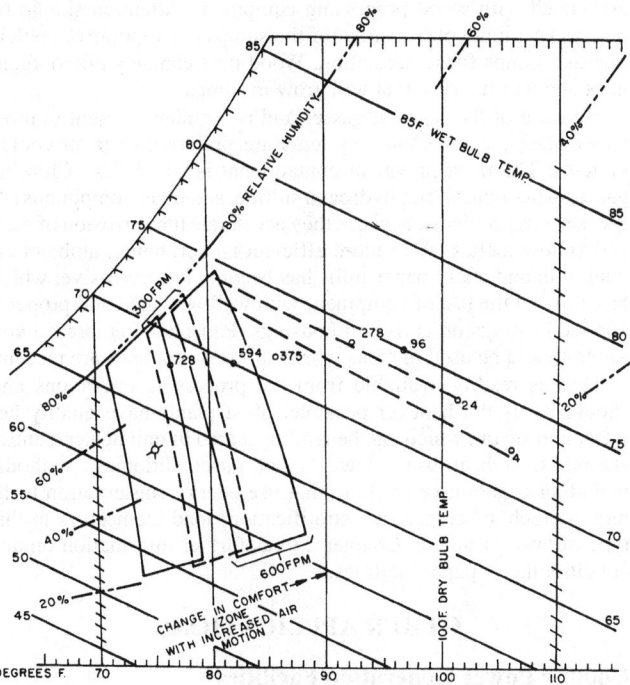

**Fig. 7   Change in Human Comfort Zone as
Air Movement Increases**

automatically or manually. In addition, evaporative coolers can supply tempered air during fall, winter, and spring. Gravity or power ventilators exhaust the air. Area cooling works well in buildings where personnel move about and workers are not subjected to concentrated, radiant heat sources. Area cooling may be used in either high- or low-bay industrial buildings, but may provide significant advantages in high-bay construction where cooling loads associated with roofs, lighting, and heat from equipment may be effectively eliminated by taking advantage of stratification. When cooling an area, the ductwork should be designed to distribute air to the lower 10 ft of the space to ensure that air is supplied to the workers.

Cooling requirements change from day to day and season to season, so, if discharge grilles are used, they should be adjustable to prevent drafts. The horizontal blades of an adjustable grille can be adjusted so that the air is discharged above the workers' heads rather than directly on them. In some cases, the air volume can be adjusted, either at each outlet or the entire system, in which case, the exhaust volume may need to be varied accordingly.

Spot Cooling

Spot cooling yields more efficient use of the equipment when personnel work in one spot. The cool air is brought to the spot at levels below 10 ft, and may even be delivered from floor outlets. Selection of the duct height may depend on the location of other equipment in the area. For best results, air velocity should be kept low. Controls may be automatic or manual, with the fan often operating throughout the year. Workers are especially appreciative of spot cooling in hot environments, such as in chemical plants and die casting shops, and near glass-forming machines, billet furnaces, and pig and ingot casting.

When spot cooling a worker, the air volume depends on the throw of the air jet, the activity of the worker, and the amount of heat that must be overcome. Air volumes can vary from 200 to 5000 cfm per worker, with target velocities ranging between 200 to 4000 fpm. The outlets should be between 4 to 10 ft from the workstations to avoid entrainment of warm air and to effectively blanket workers with cooler air. Provisions should be made for workers to control the

direction of air discharge, because air motion that is appropriate for hot weather may be too great for cool weather or even cool mornings. Volume controls may be required to prevent overcooling of the building and to minimize excessive grille blade adjustment.

Spot cooling is useful in rooms with elevated temperatures, regardless of climatic or geographical location. When the dry-bulb temperature of the air is below skin temperature, convection rather than evaporation cools workers. An 80°F airstream feels so good that its relative humidity is inconsequential.

Cooling Large Motors

Electrical generators and motors are generally rated for a maximum ambient temperature of 104°F. When this temperature is exceeded, excessive temperatures develop in the electrical windings unless the load on the motor or generator is reduced. By providing evaporatively cooled air to the windings, this equipment may be safely operated without reducing the load. Likewise, transformer capacity can be increased using evaporative cooling.

The heat emitted by high-capacity electrical equipment may also be sufficient to raise the ambient condition to an uncomfortable level. With mill drive motors, an additional problem is often encountered with the commutator. If the air used to ventilate the motor is dry, the temperature rise through the motor results in a still lower relative humidity. At a lower relative humidity, the brush film can be destroyed, with unusual brush and commutator wear as well as the occurrence of dusting.

As a rule, a motor having a temperature rise of 25°F requires approximately 120 cfm of ventilating air per kilowatt hour of loss. If the inlet air to the motor is 95°F, the air leaving the motor would be 120°F. This average motor temperature rise of over 107°F is 3°F higher than it should be for the normal 104°F ambient. The same quantity of 95°F inlet air, having a 75°F wet-bulb can be cooled with

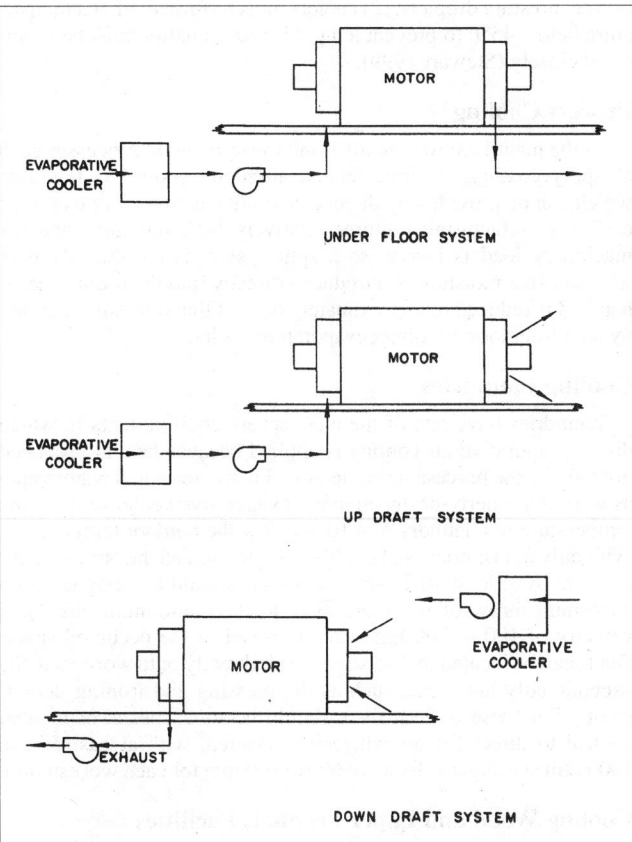

Fig. 8   Arrangements for Cooling Large Motors

a direct evaporative cooler having a 97% saturation effectiveness. The resulting 88°F average motor temperature would eliminate the need for special high-temperature insulation and improve the ability of the motor to absorb temporary overloads. By comparison, an air quantity of 185 cfm would be required if supplied by a cooler with 80% saturation effectiveness. Figure 8 shows three basic arrangements for motor cooling.

The air from the evaporative cooler may be directed on the motor windings or into the room. Doing so requires an increased air volume to compensate for the building heat load. Operation of a direct evaporative cooler should be keyed to the motor operation to ensure that (1) saturated or nearly saturated air is never introduced into a motor until it has had time to warm up, and (2) if more than one motor is served by a single system, air circulation through idle motors should be prevented.

Cooling Gas Turbine Engines and Generators

Combustion turbines used for electric power production are normally rated at 59°F. Their performance is greatly influenced by the compressor inlet air temperature because temperature effects air density and therefore mass flow. As the ambient temperature increases, the demand on electric utilities also increases. At the same time, the capacity of the combustion turbine decreases. Recovery of capacity due to inlet air cooling is approximately 0.4%/°F (cooling). Direct and indirect evaporative cooling is beneficial to gas turbine performance in almost all climates because when the air is the hottest, it generally has the lowest relative humidity. Expected increases in output using direct evaporative cooling ranges from 5.8% in Albany, New York, to 14% in Yuma, Arizona. In addition to increasing gas turbine output, direct evaporative cooling also improves heat rate and reduces NO_x emissions.

For an installation of this type, the following precautions must be taken: (1) mist eliminators must be provided to stop the entrainment of free moisture droplets, (2) coolers must be turned off at a temperature below 45°F to prevent icing. (3) water quality must be monitored closely (Stewart 1999).

Process Cooling

In the manufacture of textiles and tobacco and in processes such as spray coating, accurate relative humidity control is required, which can be provided by direct evaporative coolers. For example, textile manufacturing requires relatively high humidity and the machinery load is heavy; so a split system is customarily used whereby free moisture is introduced directly into the room. The air handled is reduced to approximately 60% of that normally required by an all-outdoor air, direct evaporative cooler.

Cooling Laundries

Laundries have one of the most severe environments in which direct evaporative air cooling is applied because heat is produced not only by the processing equipment, but by steam and water vapor as well. A properly-designed direct evaporative cooler reduces the temperature in a laundry 5 to 10°F below the outdoor temperature. With only fan ventilation, laundries usually exceed the outdoor temperature by at least 10°F. Air distribution should be designed for a maximum throw of not more than 30 ft. A minimum circulated velocity of 100 to 200 fpm should prevail in the occupied space. Ducts can be located to discharge the air directly onto workers in the exceptionally hot areas, such as the pressing and ironing departments. For these outlets, there should be some means of manual control to direct the air where it is desired, with at east 500 to 1000 cfm at a target velocity of 600 to 900 fpm for each workstation.

Cooling Wood and Paper Products Facilities

Wood processing plants and paper mills are good applications for evaporative cooling because of the high temperatures and gasses associated with wood processing equipment. Attention should be paid to keeping wood dust out of the sumps of evaporative coolers that use sumps for recirculation. Wood dust contains microorganisms and worm larvae that will grow in sumps.

Because of the types of gasses and particulates present in most paper plants, water-cooled systems are preferred over air-cooled systems. The most prevalent contaminant is wood dust. Chlorine gas, caustic soda, sulfur, hydrogen sulfide, and other compounds are also serious problems, because they accelerate the corrosion of steel and yellow metals. With more efficient air scrubbing, ambient air quality in and about paper mills has become less corrosive, which has allowed the use of equipment with well-analyzed and properly applied coatings on coils and housings. Phosphor-free brazed coil joints should be used in areas where sulfur compounds are present.

Heat is readily available from the processing operations and should be used whenever possible. Most plants have quality hot water and steam, which can be readily geared to unit heater, central station, or reheat use. Newer plant air-conditioning methods, including evaporative cooling, that use energy conservation techniques (such as temperature stratification) lend themselves to this type of large structure. Chapter 23 has further information on air-conditioning of paper facilities.

OTHER APPLICATIONS

Cooling Power Generation Facilities

An appropriate air cooling system can be selected once the preliminary heating and cooling loads are determined and the criteria are established for temperature, humidity, pressure, and airflow control. The same considerations for selection apply to power generation facilities and industrial facilities.

Cooling Mines

Chapter 26 describes various evaporative cooling methods developed for cooling mines.

Cooling Animals

The design criteria for farm animal environments and the need for cooling animal shelters are discussed in Chapter 21. Direct evaporative cooling is ideally suited to farm animal shelters because 100% outside air is used. The fresh air removes odors and reduces the harmful effects of ammonia fumes. At night and in the spring and fall, direct evaporative cooling can also be used for ventilation.

The equipment should be sized to change the air in the shelter in 1 to 2 min, assuming the ceiling height does not exceed 10 ft. This flow rate will usually keep the shelter at 80°F or lower. In addition, conditions can be improved with portable or packaged spot coolers.

For poultry housing most applications require an air change every 0.75 to 1.5 min., with the majority at 1 min. Placement of the fans at the ends or the center of the house, with the direct evaporative cooler located at the opposite end, creates a tunnel ventilation system with an air velocity of 300 to 500 fpm. The fans are generally selected for a total pressure drop of 0.125 in. of water, which means that the direct evaporative cooling media cannot have a pressure drop in excess of 0.075 in. of water. Thus, to guard against an inadequate volume of air being pulled through the poultry house, the designer must carefully size media being selected.

Using direct evaporative cooling for poultry broiler houses decreases bird mortality, improves feed conversion ratio, and increases the growth rate. Poultry breeder houses are evaporatively cooled to improve egg production and fertility during warm weather. Evaporative cooling of egg layers improves feed conversion, shell quality, and egg size. When the ambient outside temperature exceeds 100°F, evaporative cooling is often the only way to keep a flock alive. Direct evaporative cooling is also used to cool swine farrowing and gestation houses to improve production.

Produce Storage Cooling

Potatoes. Direct evaporative cooling for bulk potato storage should pass air directly through the pile. The ventilation and cooling system should provide 1.0 to 1.5 cfm/100 lb of potatoes. Average potato density is 45 lb/ft^3 in the pile. Pile depth range from 12 to 20 ft deep, which creates a static pressure of 0.15 to 0.25 in. of water. Ventilation consists of fresh air inlets, return air openings, exhaust air openings, main air ducts, and lateral ducts with holes or slots to distribute air uniformly through the pile. Distribution ducts should be placed no further apart than 80% of the potato pile depth, and should extend to within 18 in. of the walls of the storage. Ducts, the direct evaporative cooling media, and any refrigeration coils cause a static pressure ranging from 0.5 to 1.0 in. of water. Typically the static pressure ranges from 0.75 to 1.25 in. of water depending on the equipment. The velocity through each of the openings in the ventilation/cooling system should be as listed in Table 1:

Table 1   Air Velocities for Potato Storage Evaporative Cooler

Opening	Min. Velocity, fpm	Max. Velocity, fpm	Desired Velocity, fpm
Fresh air inlet	1000	1400	1200
Return air opening	1000	1400	1200
Exhaust opening	1000	1200	1100
Main duct	500	900	700
Lateral duct	750	1100	900
Slot	900	1300	1050

Direct evaporative cooling media should be 90 to 95% effective depending on the climate. In arid regions, 95% effective media is recommended. In the more humid climates, such as in the midwestern and eastern United States, 90% effective media is commonly used. The velocity through the media should be 500 to 550 fpm to assure high pad efficiency with low static pressure penalty.

Apples. Direct evaporative cooling for apple storage without refrigeration should distribute cool air to all parts of the storage. The evaporative cooler may be floor-mounted or located near the ceiling in a fan room. Air should be discharged horizontally at ceiling level. Because the prevailing wet-bulb temperature limits the degree of cooling, a cooler with the maximum reasonable size should be installed to reduce the storage temperature rapidly and as close to the wet-bulb temperature as possible. Generally, a cooler designed to exchange air at a rate of 3 min. per hour is the largest that can be installed. This capacity results in 1 to 1.5 min per hour air change when the storage is loaded.

For further information on apple storage, see Chapter 21 of the 1998 *ASHRAE Handbook—Refrigeration*.

Citrus. The chief purpose of evaporative cooling as it is applied to fruits and vegetables is to provide an effective, yet inexpensive, means of improving storage. However, it also serves a special function in the case of oranges, grapefruit, and lemons. Although mature and ready for harvest, citrus fruits are often still green. Color change (degreening) is achieved through a sweating process in rooms equipped with direct evaporative cooling. Air with a high relative humidity and a moderate temperature is circulated continuously during the operation. Ethylene gas, the concentration depending on the variety and intensity of green pigment in the rind, is discharged into the rooms. Ethylene destroys the chlorophyll in the rind, allowing the yellow or orange color to become evident. During the degreening operation, a temperature of 70°F and a relative humidity of 88 to 90% are maintained in the sweat room. (In the Gulf States, 82 to 85°F with 90 to 92% rh is used.) The evaporative cooler is designed to deliver 4 cfm per field box of fruit (0.11 cfm/lb).

Direct and indirect evaporative cooling is also used as a supplement to refrigeration in the storage of citrus fruit. Citrus storage requires refrigeration in the summer, but the required conditions can often be obtained using evaporative cooling during the fall, winter, and spring when the outdoor wet-bulb temperature is low. For further information, see Chapter 22 of the 1998 *ASHRAE Handbook-Refrigeration*.

Cooling Greenhouses

Proper regulation of greenhouse temperatures during the summer is essential for developing high-quality crops. The principal load on a greenhouse is solar radiation, which at sea level at about noon in the temperate zone is approximately 200 Btu/h·ft^2. Smoke, dust, or heavy clouds reduce the radiation load. Table 2 gives solar radiation loads for representative cities in the United States. Note that the values cited are average solar heat gains, not peak loads. Temporary rises in temperature inside a greenhouse can be tolerated; an occasional rise above design conditions is not likely to cause damage.

Not all the solar radiation that reaches the inside of the greenhouse becomes a cooling load. About 2% of the total solar radiation is used in photosynthesis. Transpiration of moisture varies from crop to crop, but typically uses about 48% of the solar radiation. This leaves 50% to be removed by the cooler. Example 2 shows a method for calculating the size of a greenhouse evaporative cooling system.

Example 2. A direct evaporative cooler is to be installed in a 50 ft by 100 ft greenhouse. Design conditions are assumed to be 92°F db and 73°F wb, and the average solar radiation is 138 Btu/h·ft^2. An indoor temperature of 90°F db must not be exceeded at design conditions.

Solution: The direct evaporative air cooler is assumed to have a saturation effectiveness of 80%. Equation (2) may be used to determine the dry-bulb temperature of the air leaving the direct evaporative cooler:

$$t_2 = 92 - \frac{80}{100}(92 - 73) = 77°F$$

The following equation may be used to calculate the airflow rate that must be supplied by the direct evaporative cooler:

$$Q_{ra} = \frac{0.5 A I_t}{\rho c_p (t_1 - t_2)} \qquad (4)$$

where

A = greenhouse floor area, ft^2
I_t = total incident solar radiation, Btu/h·ft^2 of receiving surface
$60\rho c_p$ = density times specific heat of air times 60 min/h ≈ 1.0 Btu/ft^3·°F at design conditions

For this problem

$$Q_{ra} = \frac{0.5 \times 50 \times 100 \times 138}{1.0(2)(90 - 77)} = 26{,}500 \text{ cfm}$$

Horizontal illumination from the direct rays of a noonday summer sun with a clear sky can be as much as 10,000 footcandles (fc); under clear glass, this is approximately 8500 fc. Crops such as chrysanthemums and carnations grow best in full sun, but many foliage plants, such as gloxinias and orchids, do not need more than 1500 to 2000 fc. Solar radiation is nearly proportional to light intensity. Thus, the greater the amount of shade, the smaller the cooling capacity required. A value of 100 fc per hour is approximately equivalent to 3 Btu/h·ft^2. Although atmospheric conditions, such as clouds and haze, affect the relationship, this is a safe conversion factor. This relationship should be used instead of the information given in Table 2 where illumination can be determined by design or by measurement.

Direct evaporative cooling for greenhouses may be under either positive or negative pressure. Regardless of the type of system used, the length of air travel should not exceed 160 ft. The rise in temperature of the cool air limits the throw to this value. Air movement must be kept low due to possible mechanical damage to the plants,

Table 2 Three-Year Average Solar Radiation for Horizontal Surface During Peak Summer Month

City	Btu/h·ft²	City	Btu/h·ft²
Albuquerque, NM	198	Lemont, IL	142
Apalachicola, FL	170	Lexington, KY	170
Astoria, OR	132	Lincoln, NE	150
Atlanta, GA	158	Little Rock, AR	148
Bismarck, ND	140	Los Angeles, CA	162
Blue Hill, MA	128	Madison, WI	138
Boise, ID	155	Medford, OR	170
Boston, MA	125	Miami, FL	153
Brownsville, TX	175	Midland, TX	177
Caribou, ME	115	Nashville, TN	154
Charleston, SC	152	Newport, RI	138
Cleveland, OH	152	New York, NY	140
Columbia, MO	153	Oak Ridge, TN	148
Columbus, OH	127	Oklahoma City, OK	165
Davis, CA	184	Phoenix, AZ	200
Dodge City, KS	184	Portland, ME	133
East Lansing, MI	132	Prosser, WA	176
East Wareham, MA	132	Rapid City, SD	152
El Paso, TX	195	Richland, WA	137
Ely, NV	175	Riverside, CA	176
Fort Worth, TX	176	St. Cloud, MN	132
Fresno, CA	188	San Antonio, TX	176
Gainesville, FL	156	Santa Maria, CA	188
Glasgow, MT	152	Sault Ste. Marie, MI	138
Grandby, CO	149	Sayville, NY	148
Grand Junction, CO	173	Schenectady, NY	117
Great Falls, MT	150	Seabrook, NJ	135
Greensboro, NC	155	Seattle, WA	117
Griffin, GA	164	Spokane, WA	139
Hatteras, NC	177	State College, PA	141
Indianapolis, IN	140	Stillwater, OK	167
Inyokern, CA	218	Tallahassee, FL	134
Ithaca, NY	145	Tampa, FL	167
Lake Charles, LA	160	Upton, NY	148
Lander, WY	177	Washington, D.C.	142
Las Vegas, NV	195		

but it should generally not be less than 100 fpm in areas occupied by workers.

ECONOMIC FACTORS

Design of direct and indirect evaporative cooling systems and sizing of equipment is based on the load requirements of the application and on the local dry- and wet-bulb design conditions, which may be found in Chapter 26 of the 1997 *ASHRAE Handbook—Fundamentals*. Total energy use for a specific application during a set period may be forecasted by using annual weather data. Dry-bulb and mean coincident wet-bulb temperatures, with the hours of occurrence, can be summarized and used in a modified bin procedure. The calculations must reflect the hours of use, conditions of load, and occupancy. Because of annual variations in dry and wet-bulb temperatures, and the affect of increasing cooling capacity with decreasing wet-bulb temperatures, bin calculations using mean coincident wet-bulb temperatures generally produce conservative results. When comparing the various cooling systems, the cost analysis should include the annual energy reduction at the applicable electrical rate, plus the anticipated energy cost escalation over the expected life.

Many areas have time-of-day electrical metering as an incentive to use energy during off-peak hours when rates are lowest. Reducing air-conditioning kilowatt demand is especially important in

areas with ratcheted demand rates (Scofield and DesChamps, 1980). Thermal storage using ice banks or chilled water storage may be used as part of a multistage evaporative-refrigerated cooler to combine the energy-saving advantages of evaporative cooling and off-peak savings of thermal storage (Eskra 1980).

Direct Evaporation Energy Saving

Direct evaporative cooling may be used in all climates to save cooling and humidification energy. In humid climates, the benefits of direct evaporation is realized during periods when outside air is warm and dry, but cooling savings are unlikely to be realized during peak design conditions. In more arid areas, direct evaporative cooling may partially, or fully, offset mechanical cooling at peak load conditions. Humidification energy savings may be realized during the heating season when outside air is used to provide cooling and humidification. If properly controlled, direct evaporative cooling can use waste heat otherwise rejected from buildings when outside air is used for cooling.

Indirect Evaporation Energy Saving

Indirect evaporative cooling may be used in all climates to save cooling and, in some applications, heating energy. In humid climates, indirect evaporative cooling may be used throughout the cooling cycle to precool outside air. Indirect evaporative cooling can be used to extend the range of 100% outside air ventilation to both higher and lower temperatures, and it can be used to increase the percentage of outdoor air a system can support at any given temperature through heat recovery. In high-humidity areas, indirect evaporative cooling may be used (1) to partially offset mechanical cooling requirements at peak load conditions and (2) to provide better control over low-load humidity conditions by permitting the use of smaller refrigeration equipment to provide ventilation over a wider range of outside air conditions. The cost of heating may be reduced when operating below temperatures when minimum outside air quantities exceed the rates of ventilation required for free cooling by using heat recovered from building exhausts.

Water Cost for Evaporative Cooling

Typically, tap water is used for evaporative cooling to avoid excessive scaling and associated problems with poor water quality. In designing evaporative coolers, the cost of water treatment is included in the overall project cost. However, water cost is typically ignored for evaporative coolers because it is usually an insignificant part of the operational cost. Depending on the ambient dry-bulb temperature and wet-bulb depression for a specific location, the cost of water could become a significant part of the operational cost (Mathur 1997, 1998).

PSYCHROMETRICS

Figure 9 shows the two-stage (indirect/direct) process applied to nine western cities in the United States. While the examples indicated are primarily shown for arid areas, the principles also apply to moderately humid and humid areas when weather conditions permit. For each city indicated, the entering conditions to the first-stage indirect unit are at or near the 0.4% design dry- and wet-bulb temperatures in Chapter 26 of the 1997 *ASHRAE Handbook—Fundamentals*. Although higher effectiveness can be achieved for both the indirect and direct evaporative processes modeled, the effectiveness ratings are 60% for the first (indirect) stage and 90% for the second (direct) stage. The leaving air temperatures range from 52 to 70°F with leaving conditions approaching saturation.

Figure 9 projects space conditions in each city at 78°F dry-bulb for these second-stage supply temperatures based on a 95% room sensible heat factor, i.e., room sensible heat/room total heat. With the exception of Wichita, Los Angeles, and Seattle, room conditions can be maintained in the comfort zone without a refrigerated third

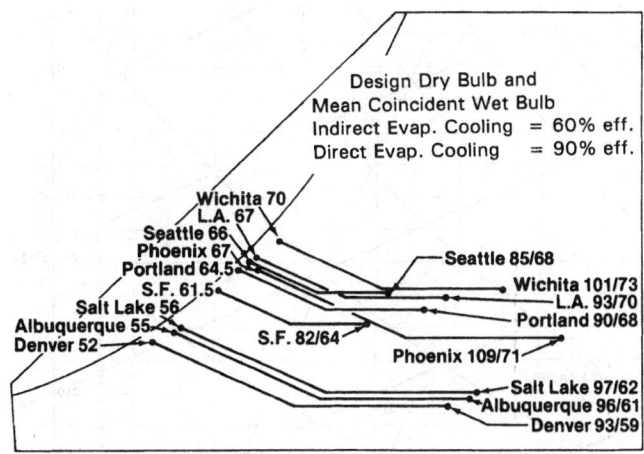

Fig. 9   Two-Stage Evaporative Cooling at 0.4% Design Condition in Various Cities in the Western United States

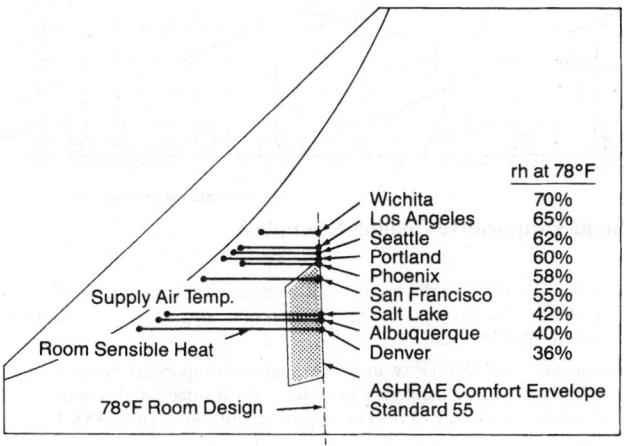

Fig. 10   Final Room Design Conditions After Two-Stage Evaporative Cooling

stage. But even in these cities, third-stage refrigeration requirements are sharply reduced as compared to conventional mechanical cooling. However, Figures 9 and 10 indicate the need to consider the following factors when deciding whether or not to include a third stage of cooling:

1. As the room sensible heat factor decreases, the required supply air temperature must decrease to maintain a given room condition.
2. As the supply air temperature increases the supply air quantity must increase to maintain space temperature, which results in higher air-side initial cost and increased supply air fan power.
3. A decrease in the required room dry-bulb temperature requires an increase in the supply air quantity. For a given room sensible heat factor, a decrease in room dry-bulb temperature may cause the relative humidity to exceed the comfort zone.
4. The suggested 0.4% entering design (dry-bulb/mean wet-bulb) conditions are only one concern. Partial load conditions must also be considered, along with the effect (extent and duration) of spike wet-bulb temperatures. Mean wet-bulb temperatures can be used to determine energy use of the indirect/direct system. However, the higher wet-bulb temperature spikes should be considered to determine their effect on room temperatures.

An ideal condition for maximum use with minimum energy consumption of a two- and three-stage indirect/direct system is a room

sensible heat factor of 90% and higher, a supply air temperature of 60°F, and a dry-bulb room design temperature of 78°F. In many cases, third-stage refrigeration is required to ensure satisfactory dry-bulb temperature and relative humidity. Example 3 shows a method for determining the refrigeration capacity for three-stage cooling. Figure 11 is a psychrometric diagram of the process.

Example 3. Assume the following:
1. Supply air quantity = 24,000 cfm; supply air temperature = 60°F
2. Design condition = 99°F db and 68°F wb
3. Effectiveness of indirect unit = 60%;
4. Effectiveness of direct unit = 90%

Using Equation (2), indirect unit performance is

$$99 - 0.60(99 - 68) = 80.4°F \text{ leaving dry-bulb (61.8°F wb)}$$

Using Equation (2), direct unit performance is

$$80.4 - 0.90(80.4 - 61.8) = 63.7°F \text{ db supply air temperature}$$

Calculate booster refrigeration capacity to drop the supply air temperature from 63.7°F to the required 60°F.

If the refrigerating coil is located ahead of the direct unit,

$$\text{Btu/h cooling} = \frac{60(h_1 - h_2)(\text{supply air, cfm})}{\text{Specific volume dry air at leaving air condition}}$$

With numeric values of enthalpies h_1 and h_2 (in Btu/lb) and the specific volume of air (in ft³/lb dry air) taken from ASHRAE Psychrometric Chart No. 1, the cooling load is calculated as follows:

$$60(27.6 - 25.5)24,000/13.78 = 219,400 \text{ Btu/h} = 18.3 \text{ tons}$$

The load for a coil located in the leaving air of the direct unit is

$$60(27.6 - 25.5)24,000/13.43 = 225,000 \text{ Btu/h} = 18.8 \text{ tons}$$

Depending on the location of the booster coil, the above calculations can be used to determine third-stage refrigeration capacity and to select a cooling coil.

Using this example, refrigeration sizing can be compared to conventional refrigeration without staged evaporative cooling. Assuming mixed air conditions to the coil of 81°F db and 66.5°F wb, and the same 60°F db supply air as shown in Figure 11, the refrigerated capacity is

$$60(31.1 - 25.7)24,000/13.31 = 584,200 \text{ Btu/h} = 48.7 \text{ tons}$$

This represents an increase of 30.4 tons. The staged evaporative effect reduces the required refrigeration by 62.4%.

ENTERING AIR CONSIDERATIONS

The effectiveness of direct and indirect evaporative cooling depends on the condition of the entering air. Where outdoor air is used in a direct evaporative cooler, the design is affected by the prevailing outdoor dry- and wet-bulb temperatures as well as by the application. Where conditioned exhaust air is used as the secondary air for indirect evaporative cooling, the design is less affected by local weather conditions, which makes evaporative cooling viable in hot and humid environments.

For example, in arid areas like Reno, Nevada, a simple, direct evaporative cooler with as effectiveness of 80% provides a leaving air temperature of 68°F when dry-bulb and wet-bulb temperatures of the entering air are 96 and 61°F, respectively. In the same location, the addition of an indirect evaporative precooling stage, with an effectiveness of 80%, produces a leaving air condition of 53.6°F.

In a geographic location such as Atlanta, Georgia, with design temperatures of 94 and 74°F, the same direct evaporative cooler could supply only 78°F. This could be reduced to 71.1°F by the addition of an indirect evaporative precooling stage that is 80% effective (Supple 1982). If exhaust air from the building served is provided at a stable 75°F dry-bulb and 62.5°F wet-bulb, an indirect evaporative precooler could deliver air at 68.8°F, substantially reducing outside air cooling loads. Under these conditions, indirect evaporative precoolers can provide limited dehumidification capabilities.

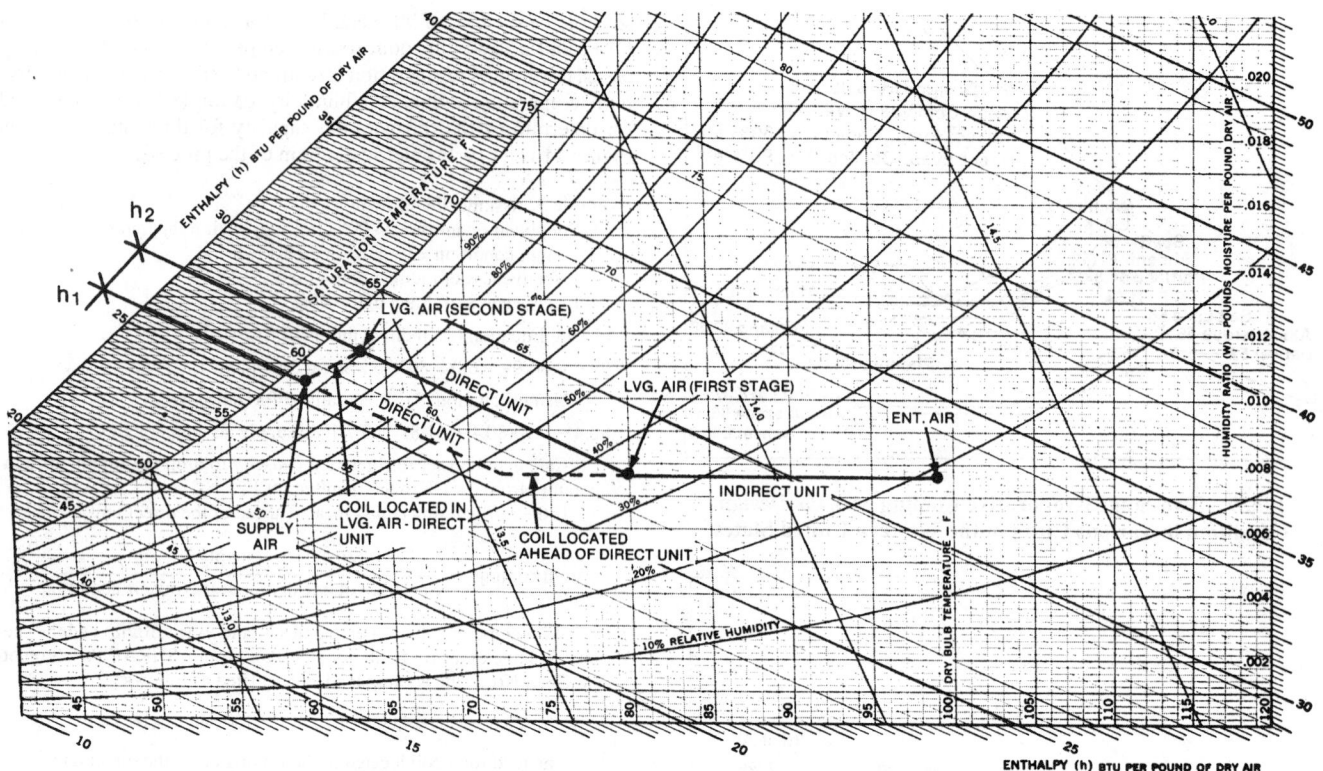

Fig. 11   Psychrometric Diagram of Three-Stage Evaporative Cooling Example 3

Long-term benefits to owners of direct evaporative cooling systems include a 20 to 40% reduction of utility costs compared to mechanical refrigeration (Watt 1988). When used to provide humidity control, the reduction in cooling and humidification energy use ranges from 35 to 90% (Lentz 1991). While direct evaporative cooling does not reduce peak cooling loads in other than arid areas, direct evaporative cooling can reduce both total cooling energy and humidification energy requirements in a wide range of environments, including environments that are considered hot and humid.

Indirect evaporative cooling lowers the temperature (both dry and wet-bulb) of the air entering a direct evaporative cooling stage and, consequently, lowers the supply air temperature. When used with mechanical cooling on 100% outside air systems, where the secondary air is taken from the conditioned space, the precooling affect may reduce peak cooling loads between 50 and 70%. Total cooling requirements may be reduced between 40 and 85% annually depending on location, system configuration, and load characteristics. Indirect evaporative coolers may also function as heat recovery systems, which expands the range of conditions over which the process is used. Indirect evaporative cooling, when used with building exhaust air, is especially effective in hot and humid climates.

REFERENCES

ASHRAE. 1992. Thermal environmental conditions for human occupancy. *Standard* 55-1992.

Crow, L.W. 1972. Weather data related to evaporative cooling. Research Report No. 2223. *ASHRAE Transactions* 78(1):153-64.

Eskra, N. 1980. Indirect/direct evaporative cooling systems. *ASHRAE Journal* 22(5):22.

Lentz, M.S. 1991. Adiabatic saturation and variable-air-volume: a prescription for economy in close environmental control. *ASHRAE Transactions* 97(1):477-85.

Mathur, G.D. 1998. Predicting yearly energy savings using bin weather data with heat pipe exchangers with indirect evaporative cooling. *Intersociety Energy Conversation Engineering Conference*, Paper #98-IECEC-049.

Mathur, G.D. 1997. Performance enhancement of existing air conditioning systems. *Intersociety Energy Conversion Engineering Conference, AIChE* 3:1618-1623

Mathur, G.D. and D.Y. Goswami. 1995. Indirect evaporative cooling retrofit as a demand side management strategy for residential air conditioning. *Intersociety Energy Conversion Engineering Conference, ASME* 2:317-22.

Mathur, G.D., D.Y. Goswami, and S.M. Kulkarni. 1993. Experimental investigation of a residential air conditioning system with an evaporatively cooled condenser. *J. Solar Energy Engineering* 115:206-11.

Mathur, G.D. 1990. Indirect evaporative cooling with two-phase thermosiphon coil loop heat exchangers. *ASHRAE Transactions* 96(1):1241-49.

Peterson, J.L. and B.D. Hunn. 1992. Experimental performance of an indirect evaporative cooler. *ASHRAE Transactions* 98(2):15-23.

Scofield, M. and N. DesChamps. 1980. EBTR compliance and comfort too. *ASHRAE Journal* 22(6):61.

Scofield, C.M. and DesChamps, N.H. 1984. Indirect evaporative cooling using plate-type heat exchangers. *ASHRAE Transactons* 90(1B):148-53.

Scofield, C.M. 1986. The heat pipe used for dry evaporative cooling. *ASHRAE Transactions* 92(1B):371-81.

Stewart, W.E., Jr. 1999. *Design guide: Combustion turbine inlet air cooling systems.* ASHRAE.

Supple, R.G. 1982. Evaporative cooling for comfort. *ASHRAE Journal* 24(8):42.

Watt, J.R. 1988. Power cost comparisons: Evaporative vs. refrigerative cooling. *ASHRAE Transactions* 94(2):1108-15.

Wu, H. and J.L. Yellot. 1987. Investigation of a plate-type indirect evaporative cooling system for residences in hot and arid climates. *ASHRAE Transactions* 93(1):1252-60.

BIBLIOGRAPHY

Stewart, W.E., Jr. and L.A. Stickler. 1999. Designing for combustion turbine inlet air cooling. *ASHRAE Transactions* 105(1)

Watt, J.R. 1997. *Evaporative air conditioning handbook*, 3rd ed. Chapman & Hall, New York.

FIRE AND SMOKE MANAGEMENT

IN building fires, smoke often flows to locations remote from the fire, threatening life and damaging property. Stairwells and elevators frequently fill with smoke, thereby blocking or inhibiting evacuation. Smoke causes the most deaths in fires. **Smoke** is defined as the airborne solid and liquid particulates and gases produced when a material undergoes pyrolysis or combustion, together with the air that is entrained or otherwise mixed into the mass.

The idea of using pressurization to prevent smoke infiltration of stairwells began to attract attention in the late 1960s. This concept was followed by the idea of the pressure sandwich (i.e., venting or exhausting the fire floor and pressurizing the surrounding floors). Frequently, a building's ventilation system is used for this purpose. **Smoke control** systems use fans to pressurize appropriate areas to limit smoke movement in fire situations. **Smoke management** systems include pressurization and all other methods that can be used singly or in combination to modify smoke movement.

This chapter discusses fire protection and smoke control systems in buildings as they relate to heating, ventilating, and air-conditioning (HVAC). For a more complete discussion, refer to *Design of Smoke Management Systems* (Klote and Milke 1992). National Fire Protection Association (NFPA) *Standard* 204, Guide for Smoke and Heat Venting, provides information about venting of large industrial and storage buildings. For further information, refer to NFPA *Standard* 92A, Recommended Practice for Smoke Control Systems, and NFPA *Standard* 92B, Guide for Smoke Management Systems in Malls, Atria, and Large Areas.

The objective of fire safety is to provide some degree of protection for a building's occupants, the building and the property inside it, and neighboring buildings. Various forms of analysis have been used to help quantify protection. Specific life safety objectives differ with occupancy; for example, nursing home requirements are different from those for office buildings.

Two basic approaches to fire protection are (1) to prevent fire ignition and (2) to manage fire impact. Figure 1 shows a decision tree for fire protection. The building occupants and managers have the primary role in preventing fire ignition. The building design team may incorporate features into the building to assist the occupants and managers in this effort. Because it is impossible to prevent fire ignition completely, managing fire impact has become significant in fire protection design. Examples of fire impact management include compartmentation, suppression, control of construction materials, exit systems, and smoke management. The *Fire Protection Handbook* (NFPA 1997a) and *Smoke Movement and Control in High-Rise Buildings* (Tamura 1994) contain detailed fire safety information.

Historically, fire safety professionals have considered the HVAC system a potentially dangerous penetration of natural building membranes (walls, floors, and so forth) that can readily transport smoke and fire. For this reason, the HVAC has traditionally been shut down when fire is discovered. Although shutting down

The preparation of this chapter is assigned to TC 5.6, Control of Fire and Smoke.

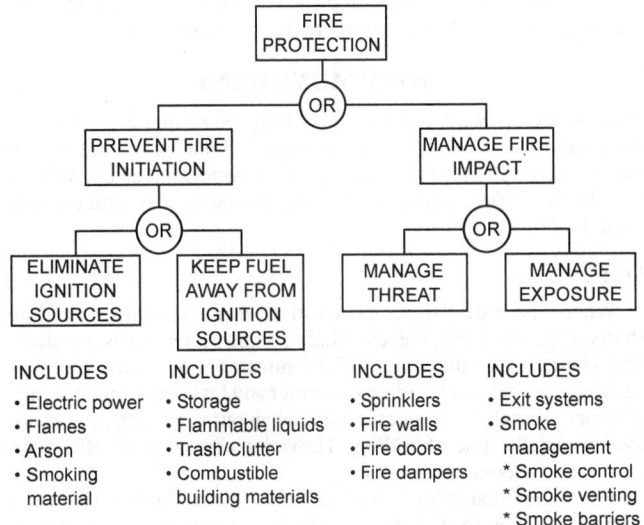

Fig. 1 Simplified Fire Protection Decision Tree

the HVAC prevents fans from forcing smoke flow, it does not prevent smoke movement through ducts due to smoke buoyancy, stack effect, or wind. To solve the problem of smoke movement, methods of smoke control have been developed; smoke control should be viewed as only one part of the overall building fire protection system.

FIRE MANAGEMENT

Although most of this chapter discusses smoke management, fire management at HVAC penetrations is an additional concern for the HVAC engineer. The most efficient way to limit the damage from a fire is through compartmentation. Fire-rated assemblies, such as the floor or the walls, keep the fire in a given area for a specific period. However, fire can easily pass through openings for plumbing, HVAC ductwork, communication cables, or other services. Therefore, fire stop systems are installed to maintain the rating of the fire-rated assembly. The rating of a fire stop system depends on the type of penetration, the number of penetrations, the size of penetration, and the construction assembly in which it is installed.

The performance of the entire fire stop system, which includes the construction assembly with its penetrations, is tested under real fire conditions by recognized independent testing laboratories. ASTM *Standard* E 814, UL *Standard* 1479, which is based on ASTM E 814, and CAN4-S115 describe methods used to determine the performance of **through-penetration fire stopping (TPFS)**.

TPFS is required by building codes under certain circumstances for specific construction types and occupancies. In the United States, the model building codes require that most penetrations meet the ASTM E 814 test standard. TPFS classifications are published by the testing laboratories. Each classification is proprietary, and each applies to use with a specific set of conditions, so numerous types are usually required on any given project.

The construction manager and general contractor, not the architects and engineers, make work assignments. Sometimes they assign fire stopping to the discipline making the penetration; other times, they assign it to a specialty fire stopping subcontractor. The Construction Specifications Institute (CSI) assigns fire stopping specifications to Division 7. Such placement accomplishes three goals:

- It encourages continuity of fire stopping products on the project by consolidating their requirements (e.g., TPFS, expansion joint fire stopping, floor-to-wall joint fire stopping, etc.).
- It maintains flexibility of work assignments for the general contractor and and the construction manager.
- It encourages prebid discussions between the contractor and subcontractors regarding appropriate work assignments.

SMOKE MOVEMENT

A smoke control system must be designed so that it is not overpowered by the driving forces that cause smoke movement, which include stack effect, buoyancy, expansion, wind, and the HVAC system. During a fire, smoke is generally moved by a combination of these forces.

Stack Effect

When it is cold outside, air tends to move upward within building shafts (e.g., stairwells, elevator shafts, dumbwaiter shafts, mechanical shafts, or mail chutes). This **normal stack effect** occurs because the air in the building is warmer and less dense than the outside air. Normal stack effect is large when outside temperatures are low, especially in tall buildings. However, normal stack effect can exist even in a one-story building.

When the outside air is warmer than the building air, there is a natural tendency for downward airflow, or **reverse stack effect**, in shafts. At standard atmospheric pressure, the pressure difference due to either normal or reverse stack effect is expressed as

$$\Delta p = 7.64 \left(\frac{1}{T_o} - \frac{1}{T_i} \right) h \qquad (1)$$

where

Δp = pressure difference, in. of water
T_o = absolute temperature of outside air, °R
T_i = absolute temperature of air inside shaft, °R
h = distance above neutral plane, ft

For a building 200 ft tall with a neutral plane at midheight, an outside temperature of 0°F (460°R), and an inside temperature of 70°F (530°R), the maximum pressure difference due to stack effect would be 0.22 in. of water. This means that at the top of the building, a shaft would have a pressure 0.22 in. of water greater than the outside pressure. At the base of the building, the shaft would have a pressure 0.22 in. of water lower than the outside pressure. Figure 2 diagrams the pressure difference between a building shaft and the outside. A positive pressure difference indicates that the shaft pressure is higher than the outside pressure, and a negative pressure difference indicates the opposite. Figure 3 illustrates the air movement in buildings caused by both normal and reverse stack effect.

Figure 4 can be used to determine the pressure difference due to stack effect. For normal stack effect, $\Delta p/h$ is positive, and the pressure difference is positive above the neutral plane and negative below it. For reverse stack effect, $\Delta p/h$ is negative, and the pressure difference is negative above the neutral plane and positive below it.

In unusually tight buildings with exterior stairwells, Klote (1980) observed reverse stack effect even with low outside air temperatures. In this situation, the exterior stairwell temperature is

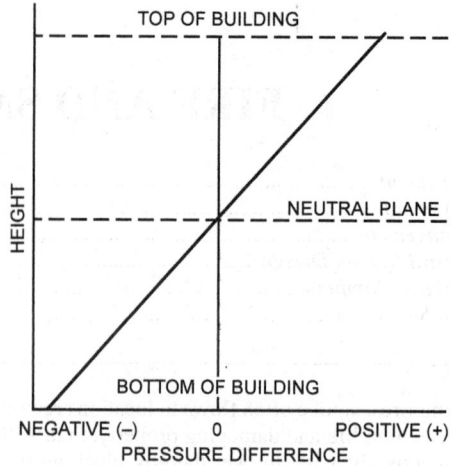

Fig. 2 Pressure Difference Between a Building Shaft and the Outside due to Normal Stack Effect

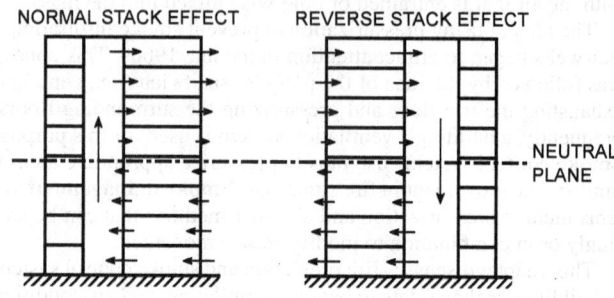

Note: Arrows indicate direction of air movement

Fig. 3 Air Movement due to Normal and Reverse Stack Effect

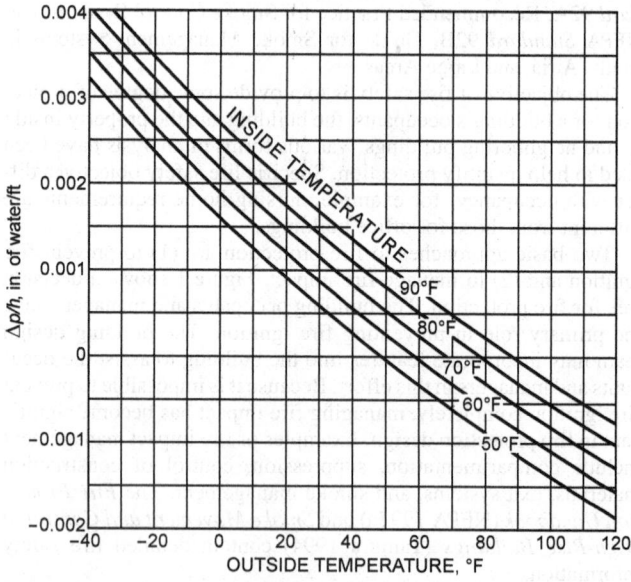

Fig. 4 Pressure Difference due to Stack Effect

considerably lower than the building temperature. The stairwell represents the cold column of air, and other shafts within the building represent the warm columns of air.

If the leakage paths are uniform with height, the neutral plane is near the midheight of the building. However, when the leakage paths are not uniform, the location of the neutral plane can vary considerably, as in the case of vented shafts. McGuire and Tamura (1975) provide methods for calculating the location of the neutral plane for some vented conditions.

Smoke movement from a building fire can be dominated by stack effect. In a building with normal stack effect, the existing air currents (as shown in Figure 3) can move smoke considerable distances from the fire origin. If the fire is below the neutral plane, smoke moves with the building air into and up the shafts. This upward smoke flow is enhanced by buoyancy forces due to the temperature of the smoke. Once above the neutral plane, the smoke flows from the shafts into the upper floors of the building. If the leakage between floors is negligible, the floors below the neutral plane (except the fire floor) remain relatively smoke-free until the quantity of smoke produced is greater than can be handled by stack effect flows.

Smoke from a fire located above the neutral plane is carried by the building airflow to the outside through exterior openings in the building. If the leakage between floors is negligible, all floors other than the fire floor remain relatively smoke-free until the quantity of smoke produced is greater than can be handled by stack effect flows. When the leakage between floors is considerable, the smoke flows to the floor above the fire floor.

The air currents caused by reverse stack effect (Figure 3) tend to move relatively cool smoke down. In the case of hot smoke, buoyancy forces can cause smoke to flow upward, even during reverse stack effect conditions.

Buoyancy

High-temperature smoke from a fire has a buoyancy force due to its reduced density. The pressure difference between a fire compartment and its surroundings can be expressed as follows:

$$\Delta p = 7.64\left(\frac{1}{T_s} - \frac{1}{T_f}\right)h \tag{2}$$

where

Δp = pressure difference, in. of water
T_s = absolute temperature of surroundings, °R
T_f = average absolute temperature of fire compartment, °R
h = distance above neutral plane, ft

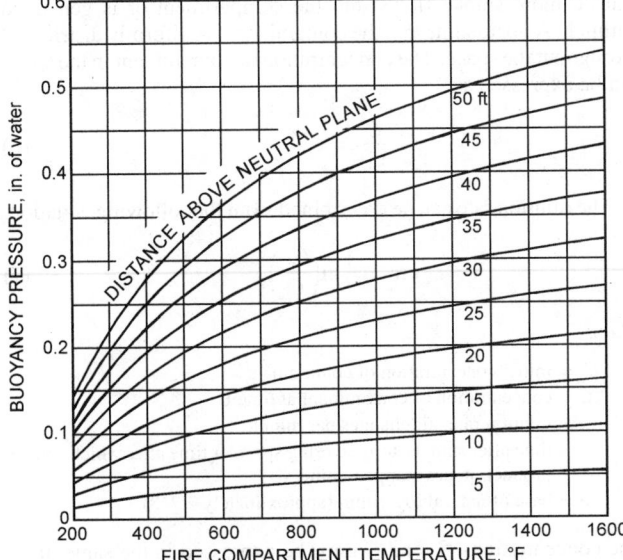

Fig. 5 Pressure Difference due to Buoyancy

The pressure difference due to buoyancy can be obtained from Figure 5 for the surroundings at 68°F (328°R). The neutral plane is the plane of equal hydrostatic pressure between the fire compartment and its surroundings. For a fire with a fire compartment temperature at 1470°F (1930°R), the pressure difference 5 ft above the neutral plane is 0.052 in. of water. Fang (1980) studied pressures caused by room fires during a series of full-scale fire tests. During these tests, the maximum pressure difference reached was 0.064 in. of water across the burn room wall at the ceiling.

Much larger pressure differences are possible for tall fire compartments where the distance h from the neutral plane can be larger. If the fire compartment temperature is 1290°F (1750°R), the pressure difference 35 ft above the neutral plane is 0.35 in. of water. This is a large fire, and the pressures it produces are beyond present smoke control methods. However, the example illustrates the extent to which Equation (2) can be applied.

In sprinkler-controlled fires, the temperature in the fire room remains at that of the surroundings except for a short time before sprinkler activation. Sprinklers are activated by a thin (2 to 4 in.) layer of hot gas under the ceiling. This layer is called the **ceiling jet**. The maximum temperature of the ceiling jet depends on the location of the fire, the activation temperature of the sprinkler, and the thermal lag of the sprinkler heat-responsive element. For most residential and commercial applications, the ceiling jet is between 180 and 300°F. In Equation (2), T_f is the average temperature of the fire compartment. For a sprinkler-controlled fire,

$$T_f = \frac{T_s(H - H_j) + T_j H_j}{H} \tag{3}$$

where

H = floor to ceiling height, ft
H_j = thickness of ceiling jet, ft
T_j = absolute temperature of ceiling jet, °R

For example, for H = 96 in., H_j = 4 in., T_s = 68 + 460 = 528 °R, and T_j = 300 + 460 = 760°R,

$$T_f = [528(96 - 4) + 760 \times 4]/96 = 538°R \text{ or } 78°F$$

In Equation (2), this results in a pressure difference of 0.002 in. of water, which is insignificant for smoke control applications.

Expansion

The energy released by a fire can also move smoke by expansion. In a fire compartment with only one opening to the building, building air will flow in, and hot smoke will flow out. Neglecting the added mass of the fuel, which is small compared to the airflow, the ratio of volumetric flows can be expressed as a ratio of absolute temperatures:

$$\frac{Q_{out}}{Q_{in}} = \frac{T_{out}}{T_{in}} \tag{4}$$

where

Q_{out} = volumetric flow rate of smoke out of fire compartment, cfm
Q_{in} = volumetric flow rate of air into fire compartment, cfm
T_{out} = absolute temperature of smoke leaving fire compartment, °R
T_{in} = absolute temperature of air into fire compartment, °R

For a smoke temperature of 1290°F (1750°R) and an entering air temperature of 67°F (527°R), the ratio of volumetric flows is 3.32. Note that absolute temperatures are used in the calculation. In such a case, if the air flowing into the fire compartment is 3000 cfm, the smoke flowing out of the fire compartment would be 9960 cfm, with the gas expanding to more than three times its original volume.

For a fire compartment with open doors or windows, the pressure difference across these openings due to expansion is negligible. However, for a tightly sealed fire compartment, the pressure differences due to expansion may be important.

Wind

In many instances, wind can have a pronounced effect on smoke movement within a building. The pressure the wind exerts on a surface can be expressed as

$$p_w = 0.00643 C_w \rho_o V^2 \tag{5}$$

where

p_w = pressure exerted by wind, in. of water
C_w = dimensionless pressure coefficient
ρ_o = outside air density, lb_m/ft^3
V = wind velocity, mph

The pressure coefficients C_w are in the range of −0.8 to 0.8, with positive values for windward walls and negative values for leeward walls. The pressure coefficient depends on building geometry and varies locally over the wall surface. In general, wind velocity increases with height from the surface of the earth. Sachs (1972), Houghton and Carruther (1976), Simiu and Scanlan (1978), and MacDonald (1975) give detailed information concerning wind velocity variations and pressure coefficients. Shaw and Tamura (1977) have developed specific information about wind data with respect to air infiltration in buildings.

With a pressure coefficient of 0.8 and air density of 0.075 lb_m/ft^3, a 35 mph wind produces a pressure on a structure of 0.47 in. of water. The effect of wind on air movement within tightly constructed buildings with all exterior doors and windows closed is slight. However, the effects of wind can be important for loosely constructed buildings or for buildings with open doors or windows. Usually, the resulting airflows are complicated, and computer analysis is required.

Frequently in fire situations, a window breaks in the fire compartment. If the window is on the leeward side of the building, the negative pressure caused by the wind vents the smoke from the fire compartment. This reduces smoke movement throughout the building. However, if the broken window is on the windward side, the wind forces the smoke throughout the fire floor and to other floors, which endangers the lives of building occupants and hampers fire fighting. Pressure induced by the wind in this situation can be large and can dominate air movement throughout the building.

HVAC Systems

Before methods of smoke control were developed, HVAC systems were shut down when fires were discovered because the systems frequently transported smoke during fires.

In the early stages of a fire, the HVAC system can aid in fire detection. When a fire starts in an unoccupied portion of a building, the system can transport the smoke to a space where people can smell it and be alerted to the fire. However, as the fire progresses, the system transports smoke to every area it serves, thus endangering life in all those spaces. The system also supplies air to the fire space, which aids combustion. Although shutting the system down prevents it from supplying air to the fire, it does not prevent smoke movement through the supply and return air ducts, air shafts, and other building openings due to stack effect, buoyancy, or wind.

SMOKE MANAGEMENT

In this chapter, smoke management includes all methods that can be used singly or in combination to modify smoke movement for the benefit of occupants or fire fighters or for the reduction of property damage. Barriers, smoke vents, and smoke shafts are traditional methods of smoke management. The effectiveness of barriers is limited by the extent to which they are free of leakage paths. Smoke vents and smoke shafts are limited by the fact that smoke must be sufficiently buoyant to overcome any other driving forces that could be present. In the last few decades, fans have been used with the intent of overcoming the limitations of traditional approaches. The mechanisms of compartmentation, dilution, pressurization, airflow, and buoyancy are used by themselves or in combination to manage smoke conditions in fire situations. These mechanisms are discussed in the following sections.

Compartmentation

Barriers with sufficient fire endurance to remain effective throughout a fire exposure have a long history of providing protection against fire spread. In such fire compartmentation, walls, partitions, floors, doors, and other barriers provide some level of smoke protection to spaces remote from the fire. This section refers to passive compartmentation. The use of compartmentation with pressurization is discussed in the section on Pressurization (Smoke Control). Many codes, such as NFPA *Standard* 101, provide specific criteria for the construction of smoke barriers (including doors) and smoke dampers in these barriers. The extent to which smoke leaks through such barriers depends on the size and shape of the leakage paths in the barriers and the pressure difference across the paths.

Dilution Remote from Fire

Dilution of smoke is sometimes referred to as **smoke purging**, **smoke removal**, **smoke exhaust**, or **smoke extraction**. Dilution can be used to maintain acceptable gas and particulate concentrations in a compartment subject to smoke infiltration from an adjacent space. It can be effective if the rate of smoke leakage is small compared to either the total volume of the safeguarded space or the rate of purging air supplied to and removed from the space. Also, dilution can be beneficial to the fire service for removing smoke after a fire has been extinguished. Sometimes, when doors are opened, smoke flows into areas intended to be protected. Ideally, the doors are only open for short periods during evacuation. Smoke that has entered spaces remote from the fire can be purged by supplying outside air to dilute the smoke.

The following is a simple analysis of smoke dilution for spaces in which there is no fire. Assume that at time zero ($\theta = 0$), a compartment is contaminated with some concentration of smoke and that no more smoke flows into the compartment or is generated within it. Also, assume that the contaminant is uniformly distributed throughout the space. The concentration of contaminant in the space can be expressed as

$$\frac{C}{C_o} = e^{-at} \tag{6}$$

The dilution rate can be determined from the following equation:

$$a = \frac{1}{t} \ln\left(\frac{C_o}{C}\right) \tag{7}$$

where

C_o = initial concentration of contaminant
C = concentration of contaminant at time θ
a = dilution rate, air changes per minute
t = time after smoke stops entering space or time after which smoke production has stopped, min
e = base of natural logarithm (approximately 2.718)

The concentrations C_o and C must be expressed in the same units, and they can be any units appropriate for the particular contaminant being considered.

McGuire et al. (1970) evaluated the maximum levels of smoke obscuration from a number of fire tests and a number of proposed criteria for tolerable levels of smoke obscuration. Based on this evaluation, they state that the maximum levels of smoke obscuration are greater by a factor of 100 than those relating to the limit of tolerance. Thus, they indicate that a space can be considered "reasonably safe" with respect to smoke obscuration if the concentration of contaminants in the space is less than about 1% of the concentration in the immediate fire area. This level of dilution increases visibility by about a factor of 100 (for example from 0.5 ft to 50 ft) and reduces the concentrations of toxic smoke components. Toxicity is a more complex problem, and no parallel statement has been made regarding dilution needed to obtain a safe atmosphere with respect to toxic gases.

In reality, it is impossible to ensure that the concentration of the contaminant is uniform throughout the compartment. Because of buoyancy, it is likely that higher concentrations are near the ceiling. Therefore, exhausting smoke near the ceiling and supplying air near the floor probably dilutes smoke even faster than indicated by Equation (6). The supply and exhaust points should be placed to prevent the supply air from blowing into the exhaust inlet, thereby short-circuiting the dilution.

Example 1. A space is isolated from a fire by smoke barriers and self-closing doors, so that no smoke enters the compartment when the doors are closed. When a door is opened, smoke flows through the open doorway into the space. If the door is closed when the contaminant in the space is 20% of the burn room concentration, what dilution rate is required to reduce the concentration to 1% of that in the burn room in 6 min?

The time $t = 6$ min and $C_o/C = 20$. From Equation (7), the dilution rate is about 0.5 air changes per minute or 30 air changes per hour.

Caution about Dilution near Fire

Many people have unrealistic expectations about what dilution can accomplish in the fire space. Neither theoretical nor experimental evidence indicates that using a building's HVAC system for smoke dilution will significantly improve tenable conditions in a fire space. The exception is an unusual space where the fuel is such that fire size can not grow above a specific limit; this occurs in some tunnels and underground transit situations. Because HVAC systems promote a considerable degree of air mixing within the spaces they serve and because very large quantities of smoke can be produced by building fires, it is generally believed that dilution of smoke by an HVAC system in the fire space will not improve the tenable conditions in that space. Thus, any attempt to improve hazard conditions within the fire space, or in spaces connected to the fire space by large openings, with smoke purging will be ineffective.

Pressurization (Smoke Control)

Systems that pressurize an area using mechanical fans are referred to as smoke control in this chapter and in NFPA *Standard 92A*. A pressure difference across a barrier can control smoke movement, as illustrated in Figure 6. Within the barrier is a door. The high-pressure side of the door can be either a refuge area or an egress route. The low-pressure side is exposed to smoke from a fire. Airflow through the gaps around the door and through construction cracks prevents smoke infiltration to the high-pressure side.

For smoke control analysis, the orifice equation can be used to estimate the flow through building flow paths:

$$Q = 776CA\sqrt{2\Delta p/\rho} \qquad (8)$$

where

Q = volumetric airflow rate, cfm
C = flow coefficient
A = flow area (leakage area), ft^2
Δp = pressure difference across flow path, in. of water
ρ = density of air entering flow path, lb$_m$/ft^3

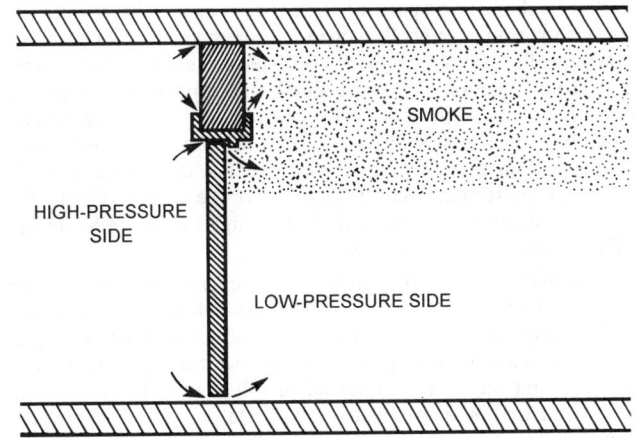

Fig. 6   Smoke Control System Preventing Smoke Infiltration to High-Pressure Side of Barrier

The flow coefficient depends on the geometry of the flow path, as well as on turbulence and friction. In the present context, the flow coefficient is generally in the range of 0.6 to 0.7. For $\rho = 0.075$ lb$_m$/ft^3 and $C = 0.65$, the flow equation above can be expressed as

$$Q = 2610A\sqrt{\Delta p} \qquad (9)$$

The flow area is frequently the same as the cross-sectional area of the flow path. A closed door with a crack area of 0.11 ft^2 and a pressure difference of 0.01 in. of water has an air leakage rate of approximately 29 cfm. If the pressure difference across the door is increased to 0.30 in. of water, the flow is 157 cfm.

Frequently, in field tests of smoke control systems, pressure differences across partitions or closed doors have fluctuated by as much as 0.02 in. of water. These fluctuations have generally been attributed to wind, although they could have been due to the HVAC system or some other source. To control smoke movement, the pressure difference produced by a smoke control system must be sufficiently large to overcome pressure fluctuations, stack effect, smoke buoyancy, and wind pressure. However, the pressure difference should not be so large that the door is difficult to open.

Airflow

Airflow has been used extensively to manage smoke from fires in subway, railroad, and highway tunnels (see Chapter 12). Large airflow rates are needed to control smoke flow, and these flow rates can supply additional oxygen to the fire. Because of the need for complex controls, airflow is not used as extensively in buildings. The control problem consists of having very small flows when a door is closed and then significantly increased flows when that door is open. Furthermore, it is a major concern that the airflow supplies oxygen to the fire. This section presents the basics of smoke control by airflow and demonstrates why this technique is rarely recommended.

Thomas (1970) determined that in a corridor in which there is a fire, airflow can almost totally prevent smoke from flowing upstream of the fire. Molecular diffusion is believed to transfer trace amounts of smoke, which are not hazardous but which are detectable as the smell of smoke upstream. Based on work by Thomas, the critical air velocity for most applications can be approximated as

$$V_k = 5.68\left(\frac{q_c}{W}\right)^{1/3} \qquad (10)$$

where

V_k = critical air velocity to prevent smoke backflow, fpm
q_c = heat release rate into corridor, Btu/h
W = corridor width, ft

This relation can be used when the fire is located in the corridor or when the smoke enters the corridor through an open doorway, air transfer grille, or other opening. While the critical velocities calculated from Equation (10) are general and approximate, they are indicative of the kind of air velocities required to prevent smoke backflow from fires of different sizes. For specific applications, other equations may be more appropriate. For the critical velocity for tunnel applications, see Chapter 12. For the critical velocity for smoke management in atriums and other large spaces see Klote and Milke (1992) and NFPA *Standard* 92B.

Although Equation (10) can be used to estimate the airflow rate necessary to prevent smoke backflow through an open door, the oxygen supplied is a concern. Huggett (1980) evaluated the oxygen consumed in the combustion of numerous natural and synthetic solids. He found that for most materials involved in building fires, the energy released is approximately 5630 Btu per pound of oxygen. Air is 23.3% oxygen by mass. Thus, if all the oxygen in a pound of air is consumed, 1300 Btu is liberated. If all the oxygen in 1 cfm of air with a density of 0.075 lb_m/ft^3 is consumed by fire, 5850 Btu/h will be liberated.

Examples 2 and 3 demonstrate that the air needed to prevent smoke backflow can support an extremely large fire. In most commercial and residential buildings, sufficient fuel (paper, cardboard, furniture, etc.) is present to support very large fires. Even when the amount of fuel is normally very small, short-term fuel loads (during building renovation, material delivery, etc.) can be significant. Therefore, the use of airflow for smoke control is not recommended, except when the fire is suppressed or in the rare cases when fuel can be restricted with confidence.

Example 2. What airflow at a doorway is needed to stop smoke backflow from a room fully involved in fire, and how large a fire can this airflow support?

A room fully involved in fire can have an energy release rate on the order of 8×10^6 Btu/h. Assume the door is 3 ft wide and 7 ft high. From Equation (10), $V_k = 5.68(8 \times 10^6/3)^{1/3} = 790$ fpm. A flow through the doorway of $790 \times 3 \times 7 = 16,600$ cfm is needed to prevent smoke from backflowing into the area.

If all the oxygen in this airflow is consumed in the fire, the heat liberated is 16,600 cfm × 5850 Btu·cfm = 9.7×10^7 Btu/h. This is over 10 times more than the heat generated by the fully involved room fire and indicates why airflow is generally not recommended for smoke control in buildings.

Example 3. What airflow is needed to stop smoke backflow from a wastebasket fire, and how large a fire can this airflow support?

A wastebasket fire can have an energy release rate on the order of 5×10^5 Btu/h. As in Example 2, $V_k = 5.68(5 \times 10^5/3)^{1/3} = 310$ fpm. A flow through the doorway of $310 \times 3 \times 7 = 6500$ cfm is needed to prevent smoke backflow.

If all the oxygen in this airflow is consumed in the fire, the heat liberated is 6500 cfm × 5850 Btu/h·cfm = 3.8×10^7 Btu/h. This is still many times greater than the fully involved room fire and further indicates why airflow is generally not recommended for smoke control in buildings.

Buoyancy

The buoyancy of hot combustion gases is employed in both fan-powered and non-fan-powered venting systems. Fan-powered venting for large spaces is commonly employed for atriums and covered shopping malls, and non-fan-powered venting is commonly used for large industrial and storage buildings. There is a concern that the sprinkler flow will cool the smoke, reducing buoyancy and thus the system effectiveness. Research is needed in this area. Refer to Klote and Milke (1992) and NFPA *Standards* 92B and 204 for detailed design information about these systems.

SMOKE CONTROL SYSTEM DESIGN

Door-Opening Forces

The door-opening forces resulting from the pressure differences produced by a smoke control system must be considered. Unreasonably high door-opening forces can make it difficult or impossible for occupants to open doors to refuge areas or escape routes.

The force required to open a door is the sum of the forces to overcome the pressure difference across the door and to overcome the door closer. This can be expressed as

$$F = F_{dc} + \frac{5.20 WA \Delta p}{2(W - d)} \qquad (11)$$

where

F = total door-opening force, lb_f
F_{dc} = force to overcome door closer, lb_f
W = door width, ft
A = door area, ft^2
Δp = pressure difference across door, in. of water
d = distance from doorknob to edge of knob side of door, ft

This relation assumes that the door-opening force is applied at the knob. Door-opening force F_p due to pressure difference can be determined from Figure 7 for a value of $d = 3$ in. The force to overcome the door closer is usually greater than 3 lb_f and, in some cases, can be as great as 20 lb_f. For a door that is 7 ft high and 3 ft wide and subject to a pressure difference of 0.30 in. of water, the total door-opening force is 30 lb_f, if the force to overcome the door closer is 12 lb_f.

Flow Areas

In designing smoke control systems, airflow paths must be identified and evaluated. Some leakage paths are obvious, such as cracks around closed doors, open doors, elevator doors, windows, and air transfer grilles. While construction cracks in building walls are less obvious, they are equally important.

The flow area of most large openings, such as open windows, can be calculated easily. However, flow areas of cracks are more difficult to evaluate. The area of these leakage paths depends on such features as workmanship, door fit, and weatherstripping. A

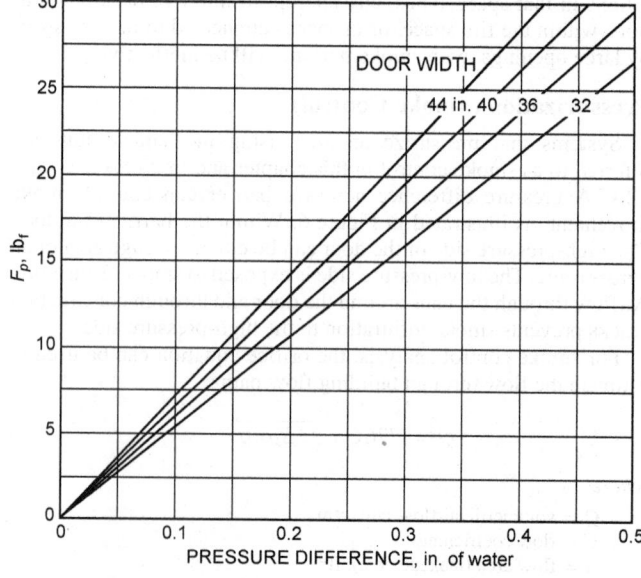

Fig. 7 Door-Opening Force due to Pressure Difference

Table 1 Typical Leakage Areas for Walls and Floors of Commercial Buildings

Construction Element	Wall Tightness	Area Ratio
		A/A_w
Exterior building walls[a]	Tight	0.70×10^{-4}
(includes construction cracks and	Average	0.21×10^{-3}
cracks around windows and doors)	Loose	0.42×10^{-3}
	Very Loose	0.13×10^{-2}
Stairwell walls[a]	Tight	0.14×10^{-4}
(includes construction cracks but not	Average	0.11×10^{-3}
cracks around windows or doors)	Loose	0.35×10^{-3}
Elevator shaft walls[a]	Tight	0.18×10^{-3}
(includes construction cracks but	Average	0.84×10^{-3}
not cracks around doors)	Loose	0.18×10^{-2}
		A/A_f
Floors[b]	Average	0.52×10^{-4}
(includes construction cracks and		
areas around penetrations)		

A = leakage area; A_w = wall area; A_f = floor area
[a]Flow areas evaluated at 0.30 in. of water.
[b]Flow areas evaluated at 0.10 in. of water.

door that is 3 ft by 7 ft with an average crack width of 1/8 in. has a leakage area of 0.21 ft². However, if this door is installed with a 3/4 in. undercut, the leakage area is 0.36 ft²—a significant difference. The leakage area of elevator doors is in the range of 0.55 to 0.70 ft² per door.

For open stairwell doorways, Cresci (1973) found that complex flow patterns exist and that the resulting flow through open doorways was considerably below the flow calculated using the geometric area of the doorway as the flow area in Equation (8). Based on this research, it is recommended that the design flow area of an open stairwell doorway be *half* the geometric area (door height times width) of the doorway. An alternate approach for open stairwell doorways is to use the geometric area as the flow area and use a reduced flow coefficient. Because it does not allow the direct use of Equation (8), this alternate approach is not used here.

Typical leakage areas for walls and floors of commercial buildings are tabulated as area ratios in Table 1. These data are based on a relatively small number of tests performed by the National Research Council of Canada (Tamura and Shaw 1976a, 1976b, 1978; Tamura and Wilson 1966). Actual leakage areas depend primarily on workmanship rather than on construction materials, and in some cases, the flow areas in particular buildings may vary from the values listed. Data concerning air leakage through building components are also provided in Chapter 25, Ventilation and Infiltration, of the 1997 *ASHRAE Handbook—Fundamentals*.

Because a vent surface is usually covered by a louver and screen, the flow area of a vent is less than its area (vent height times width). Calculation of the flow area is further complicated by the fact that the slats in louvers are frequently slanted. Manufacturer's data should be used for specific information.

Effective Flow Areas

The concept of effective flow areas is useful for analyzing smoke control systems. The paths in the system can be in parallel with one another, in series, or a combination of parallel and series. The effective area of a system of flow areas is the area that gives the same flow as the system when it is subjected to the same pressure difference over the total system of flow paths. This is similar to the effective resistance of a system of electrical resistances. The effective flow area A_e for **parallel** leakage areas is the sum of the individual leakage paths:

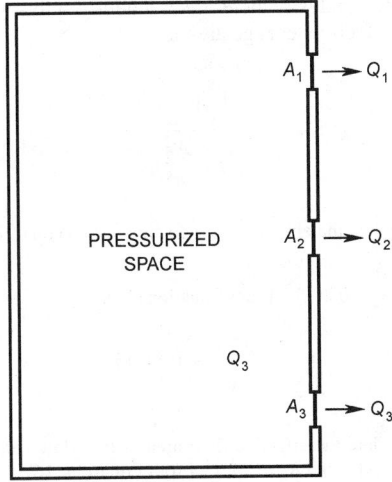

Fig. 8 Leakage Paths in Parallel

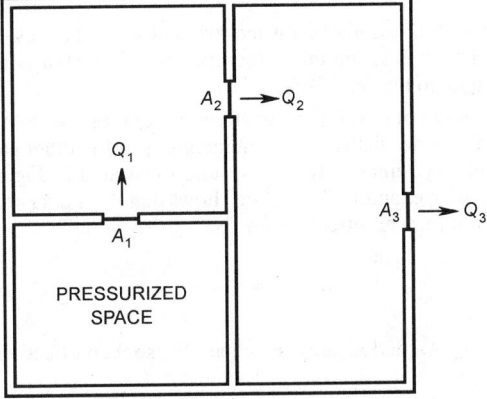

Fig. 9 Leakage Paths in Series

$$A_e = \sum_{i=1}^{n} A_i \qquad (12)$$

where n is the number of flow areas A_i in parallel.

For example, the effective area A_e for the three parallel leakage areas in Figure 8 is

$$A_e = A_1 + A_2 + A_3 \qquad (13)$$

If A_1 is 1.0 ft² and A_2 and A_3 are each 0.5 ft², then the effective flow area A_e is 2.0 ft².

The general rule for any number of leakage areas in series is

$$A_e = \left[\sum_{i=1}^{n} \frac{1}{A_i^2} \right]^{-0.5} \qquad (14)$$

where n is the number of leakage areas A_i in series.

Three leakage areas in series from a pressurized space are illustrated in Figure 9. The effective flow area of these paths is

$$A_e = \left(\frac{1}{A_1^2} + \frac{1}{A_2^2} + \frac{1}{A_3^2} \right)^{-0.5} \qquad (15)$$

In smoke control analysis, there are frequently only two paths in series, and the effective leakage area is

$$A_e = \frac{A_1 A_2}{\sqrt{A_1^2 + A_2^2}} \qquad (16)$$

Example 4. Calculate the effective leakage area of two equal flow paths in series.

Let $A = A_1 = A_2 = 0.20$ ft^2. From Equation (16),

$$A_e = \frac{A^2}{\sqrt{2A^2}} = 0.14 \text{ ft}^2$$

Example 5. Calculate the effective flow area of two flow paths in series, where $A_1 = 0.20$ ft^2 and $A_2 = 2.0$ ft^2. From Equation (16),

$$A_e = 0.199 \text{ ft}^2$$

Example 5 illustrates that when two paths are in series, and one is much larger than the other, the effective flow area is approximately equal to the smaller area.

The method of developing an effective area for a system of both parallel and series paths is to combine groups of parallel paths and series paths systematically. The system illustrated in Figure 10 is analyzed as an example. The figure shows that A_2 and A_3 are in parallel; therefore, their effective area is

$$(A_{23})_e = A_2 + A_3$$

Areas A_4, A_5, and A_6 are also in parallel, so their effective area is

$$(A_{456})_e = A_4 + A_5 + A_6$$

These two effective areas are in series with A_1. Therefore, the effective flow area of the system is given by

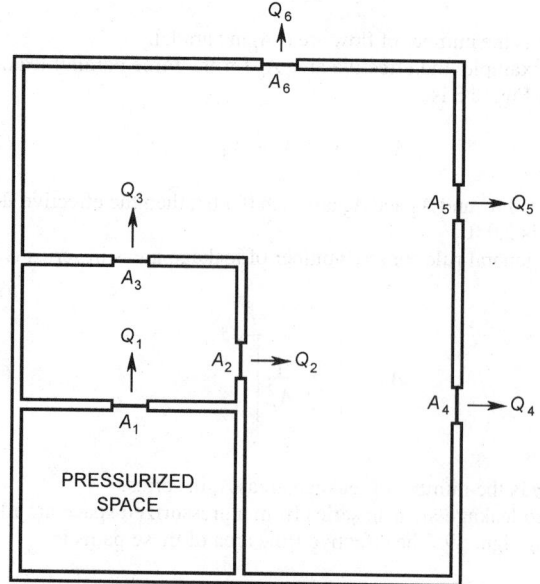

Fig. 10 Combination of Leakage Paths in Parallel and Series

$$A_e = \left[\frac{1}{A_1^2} + \frac{1}{(A_{23})_e^2} + \frac{1}{(A_{456})_e^2} \right]^{-0.5}$$

Example 6. Calculate the effective area of the system in Figure 10, if the leakage areas are $A_1 = A_2 = A_3 = 0.2$ ft^2 and $A_4 = A_5 = A_6 = 0.1$ ft^2.

$$(A_{23})_e = 0.4 \text{ ft}^2$$
$$(A_{456})_e = 0.3 \text{ ft}^2$$
$$A_e = 0.15 \text{ ft}^2$$

Symmetry

The concept of symmetry is useful in simplifying problems. Figure 11 illustrates the floor plan of a multistory building that can be divided in half by a plane of symmetry. Flow areas on one side of the plane of symmetry are equal to corresponding flow areas on the other side. For a building to be treated in this manner, every floor of the building must be such that it can be divided in the same manner by the plane of symmetry. If wind effects are not considered in the analysis, or if the wind direction is parallel to the plane of symmetry, the airflow is only one-half the total for the building analyzed. It is not necessary that the building be geometrically symmetric, as shown in Figure 11; it must be symmetric only with respect to flow.

Design Weather Data

Little weather data has been developed specifically for the design of smoke control systems. A designer may use the design temperatures for heating and cooling found in Chapter 26 of the 1997 *ASHRAE Handbook—Fundamentals*. Extreme temperatures can be considerably lower than the winter design temperatures. For example, the 99% design temperature for Tallahassee, Florida, is 28°F, but the lowest temperature observed there was −2°F (NOAA 1979).

Temperatures are generally below the design values for short periods, and because of the thermal lag of building materials, these short intervals of low temperature usually do not cause problems with heating. However, there is no time lag for a smoke control system; it is therefore subjected to all the extreme forces of stack effect that exist the moment it is operated. If the outside temperature is below the winter design temperature for which the smoke control system was designed, problems from stack effect may result. A similar situation can occur with respect to summer design temperatures and reverse stack effect.

Extreme wind data for smoke management design are listed in Chapter 26 of the 1997 *ASHRAE Handbook—Fundamentals*.

Design Pressure Differences

Both the maximum and minimum allowable pressure differences across the boundaries of smoke control should be considered. The maximum allowable pressure difference should not cause excessive door-opening forces.

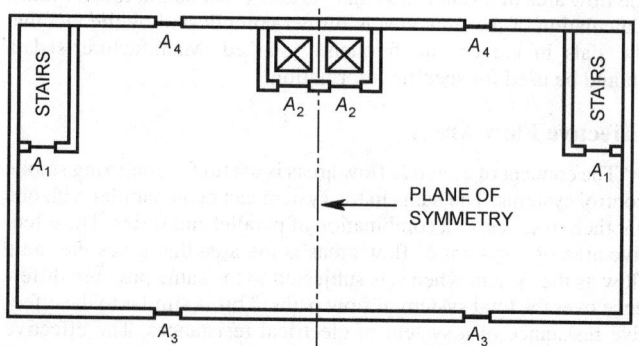

Fig. 11 Building Floor Plan Illustrating Symmetry Concept

The minimum allowable pressure difference across a boundary of a smoke control system might be the difference such that no smoke leakage occurs during building evacuation. In this case, the smoke control system must produce sufficient pressure differences to overcome forces of wind, stack effect, or buoyancy of hot smoke. The pressure differences due to wind and stack effect can be large in the event of a broken window in the fire compartment. Evaluation of these pressure differences depends on evacuation time, rate of fire growth, building configuration, and the presence of a fire suppression system. NFPA *Standard* 92A suggests values of minimum and maximum design pressure difference.

Open Doors

Another design concern is the number of doors that could be opened simultaneously when the smoke control system is operating. A design that allows all doors to be open simultaneously may ensure that the system always works, but it probably adds to the cost of the system.

The number of doors that may be open simultaneously depends largely on building occupancy. For example, in a densely populated building, it is likely that all doors will be open during evacuation. However, if a staged evacuation plan or refuge area concept is incorporated in the building fire emergency plan, or if the building is sparsely occupied, only a few of the doors may be open during a fire.

FIRE AND SMOKE DAMPERS

Opening for ducts in walls and floors with fire resistance ratings should be protected by fire dampers and ceiling dampers, as required by local codes. Air transfer openings should also be protected. These dampers should be classified and labeled in accordance with Underwriters Laboratories (UL) *Standard* 555. Figure 12 shows recommended damper positions for smoke control.

A smoke damper can be used for either traditional smoke management (smoke containment) or smoke control. In **smoke management**, a smoke damper inhibits the passage of smoke under the forces of buoyancy, stack effect, and wind. However, smoke dampers are only one of many elements (partitions, floors, doors) intended to inhibit smoke flow. In smoke management applications, the leakage characteristics of smoke dampers should be selected to be appropriate with the leakage of the other system elements.

In a **smoke control system**, a smoke damper inhibits the passage of air that may or may not contain smoke. A damper does not need

low leakage characteristics when outside (fresh) air is on the high-pressure side of the damper, as is the case for dampers that shut off supply air from a smoke zone or that shut off exhaust air from a non-smoke zone. In these cases, moderate leakage of smoke-free air through the damper does not adversely affect the control of smoke movement. It is best to design smoke control systems so that only smoke-free air is on the high-pressure side of a closed smoke damper.

Smoke dampers should be classified and listed in accordance with UL *Standard* 555S for temperature, leakage, and operating velocity. The velocity rating of a smoke damper is the velocity at which the actuator will open and close the damper.

At locations requiring both smoke and fire dampers, combination dampers meeting the requirements of both UL *Standard* 555 and UL *Standard* 555S can be used. The combination fire/smoke dampers must close when they reach their UL *Standard* 555S temperature rating to maintain the integrity of the firewall.

Fire, ceiling, and smoke dampers should be installed in accordance with the manufacturers' instructions. NFPA *Standard* 90A gives general guidelines regarding locations requiring these dampers.

The supply and return/smoke dampers should be a minimum of Class II leakage at 250°F. The return air damper should be a minimum of Class I leakage at 250°F to prevent recirculation of smoke exhaust. The operating velocity of the dampers should be evaluated when the dampers are in a smoke control mode. To minimize the velocity buildup, only zones adjacent to the fire need to be pressurized.

The exhaust ductwork and fan must be designed to handle the temperature of the exhaust smoke. The temperature of the exhaust smoke can be lowered by making the smoke control zones large or by pressurizing only the zones adjacent to the fire zone and leaving all the other zones operating normally.

PRESSURIZED STAIRWELLS

Many pressurized stairwells have been designed and built to provide a smoke-free escape route in the event of a building fire. They also provide a smoke-free staging area for fire fighters. On the fire floor, a pressurized stairwell must maintain a positive pressure difference across a closed stairwell door to prevent smoke infiltration.

During building fire situations, some stairwell doors are opened intermittently during evacuation and fire fighting, and some doors may even be blocked open. Ideally, when the stairwell door is opened on the fire floor, airflow through the door should be sufficient to prevent smoke backflow. Designing a system to achieve this goal is difficult because of the many combinations of open stairwell doors and weather conditions affecting airflow.

Stairwell pressurization systems may be single- or multiple-injection systems. A **single-injection system** supplies pressurized air to the stairwell at one location—usually at the top. Associated with this system is the potential for smoke to enter the stairwell through the pressurization fan intake. Therefore, automatic shutdown during such an event should be considered.

For tall stairwells, single-injection systems can fail when a few doors are open near the air supply injection point. Such a failure is especially likely in bottom injection systems when a ground-level stairwell door is open.

For tall stairwells, supply air can be supplied at a number of locations over the height of the stairwell. Figures 13 and 14 show two examples of **multiple-injection systems** that can be used to overcome the limitations of single-injection systems. In these figures, the supply duct is shown in a separate shaft. However, systems have been built that have eliminated the expense of a separate duct shaft by locating the supply duct in the stairwell itself. In such a case, care must be taken that the duct does not become an obstruction to orderly building evacuation.

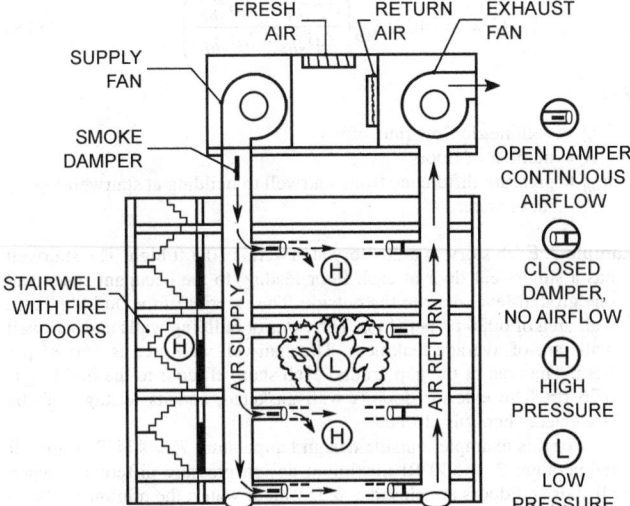

Note: If fans are off, all dampers should be closed.

Fig. 12   Smoke Control System Damper Recommendation

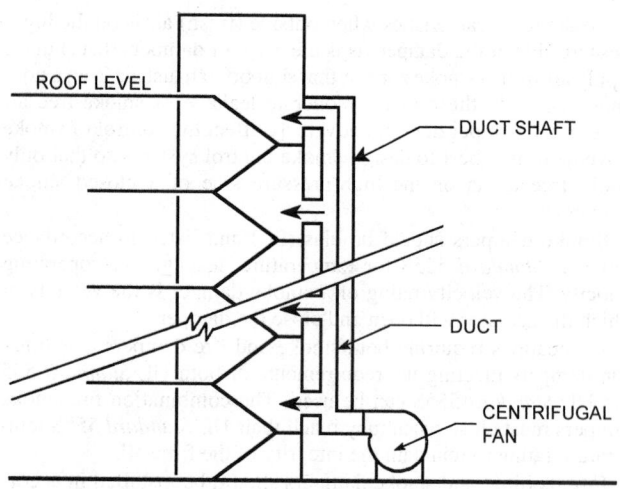

Fig. 13 Stairwell Pressurization by Multiple Injection with Fan Located at Ground Level

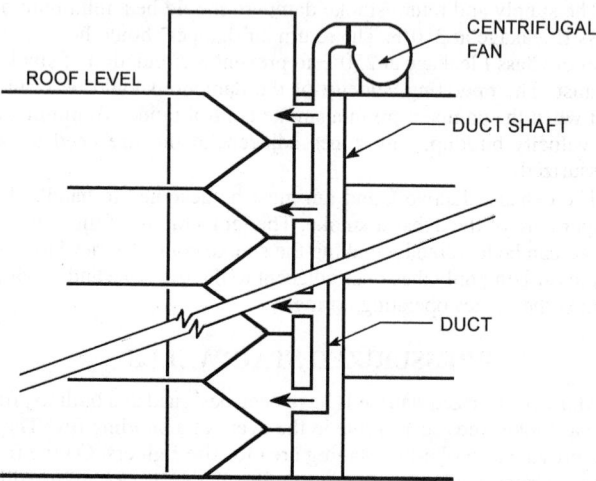

Fig. 14 Stairwell Pressurization by Multiple Injection with Roof-Mounted Fan

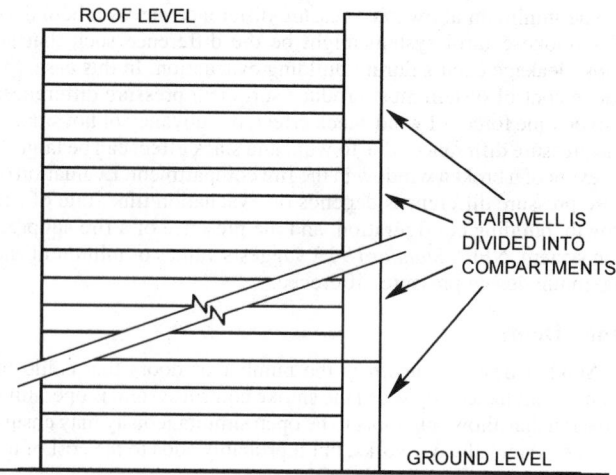

Note: Each four-floor compartment has at least one supply air injection point.

Fig. 15 Compartmentation of Pressurized Stairwell

Stairwell Compartmentation

Compartmentation of the stairwell into a number of sections is one alternative to multiple injection (Figure 15). When the doors between compartments are open, the effect of compartmentation is lost. For this reason, compartmentation is inappropriate for densely populated buildings where total building evacuation by the stairwell is planned in the event of fire. However, when a staged evacuation plan is used and when the system is designed to operate successfully with the maximum number of doors between compartments open, compartmentation can provide stairwell pressurization for tall stairwells effectively.

Stairwell Analysis

This section presents an analysis for a pressurized stairwell in a building without vertical leakage. This method closely approximates the performance of pressurized stairwells in buildings without elevators. It is also useful for buildings with vertical leakage because it yields conservative results. Only one stairwell is considered in the building; however, the analysis can be extended to any number of stairwells by the concept of symmetry. For evaluation of vertical leakage through the building or with open stairwell doors, computer analysis is recommended. The analysis is for buildings where the leakage areas are the same for each floor of the building

and where the only significant driving forces are the stairwell pressurization system and the indoor-outdoor temperature difference.

The pressure difference Δp_{sb} between the stairwell and the building can be expressed as

$$\Delta p_{sb} = \Delta p_{sbb} + \frac{By}{1 + (A_{sb}/A_{bo})^2} \tag{17}$$

where

Δp_{sbb} = pressure difference between stairwell and building at stairwell bottom, in. of water
B = $7.64(1/T_o - 1/T_s)$
y = distance above stairwell bottom, ft
A_{sb} = flow area between stairwell and building (per floor), ft^2
A_{bo} = flow area between building and outside (per floor), ft^2
T_o = temperature of outside air, °R
T_s = temperature of stairwell air, °R

For a stairwell with no leakage directly to the outside, the flow rate of pressurization air is

$$Q = 1740NA_{sb}\left(\frac{\Delta p_{sbt}^{3/2} - \Delta p_{sbb}^{3/2}}{\Delta p_{sbt} - \Delta p_{sbb}}\right) \tag{18}$$

where

Q = volumetric flow rate, cfm
N = number of floors
Δp_{sbt} = pressure difference from stairwell to building at stairwell top, in. of water

Example 7. Each story of a 20-story stairwell is 10.8 ft high. The stairwell has a single-leaf door at each floor leading to the occupant space and one ground-level door to the outside. The exterior of the building has a wall area of 6030 ft^2 per floor. The exterior building walls and stairwell walls are of average leakiness. The stairwell wall area is 560 ft^2 per floor. The area of the gap around each stairwell door to the building is 0.26 ft^2. The exterior door is well gasketed, and its leakage can be neglected when it is closed.

For this example, outside design temperature T_o = 474°R; stairwell temperature T_s = 530°R; maximum design pressure differences when all stairwell doors are closed is 0.551 in. of water; the minimum allowable pressure difference is 0.052 in. of water.

Using the leakage ratio for an exterior building wall of average tightness from Table 1, A_{bo} = 6030(0.21 × 10^{-3}) = 1.27 ft^2. Using the leakage ratio for a stairwell wall of average tightness from Table 1, the leakage area of the stairwell wall is $560(0.11 \times 10^{-3}) = 0.06 \text{ ft}^2$. The value of A_{sb} equals the leakage area of the stairwell wall plus the gaps around the closed doors: $A_{sb} = 0.06 + 0.26 = 0.32 \text{ ft}^2$. The temperature factor B is calculated at 0.00170 in. of water/ft. The pressure difference at the stairwell bottom is selected as $\Delta p_{sbb} = 0.080$ in. of water to provide an extra degree of protection above the minimum allowable value of 0.052 in. of water. The pressure difference Δp_{sbt} is calculated from Equation (17) at 0.425 in. of water, using $y = 20(10.8) = 216$ ft. Thus, Δp_{sbt} does not exceed the maximum allowable pressure. The flow rate of pressurization air is calculated from Equation (18) at 8200 cfm.

The flow rate depends strongly on the leakage area around the closed doors and on the leakage area in the stairwell walls. In practice, these areas are difficult to evaluate and even more difficult to control. If the flow area A_{sb} in Example 7 were 0.54 ft^2 rather than 0.32 ft^2, a flow rate of pressurization air of 13,800 cfm would be calculated from Equation (18). A fan with a sheave allows adjustment of supply air to offset for variations in actual leakage from the values used in design calculations.

Stairwell Pressurization and Open Doors

The simple pressurization system discussed in the previous section has two limitations regarding open doors. First, when a stairwell door to the outside and building doors are open, the simple system cannot provide sufficient airflow through building doorways to prevent smoke backflow. Second, when stairwell doors are open, the pressure difference across the closed doors can drop to low levels. Two systems used to overcome these problems are overpressure relief (Tamura 1990) and supply fan bypass.

Overpressure Relief. The total airflow rate is selected to provide the minimum air velocity when a specific number of doors is open. When all the doors are closed, part of this air is relieved through a vent to prevent excessive pressure buildup, which could cause excessive door-opening forces. This excess air can be vented either to the building or to the outside. Because exterior vents can be subject to adverse wind effects, wind shields are recommended.

Barometric dampers that close when the pressure drops below a specified value can minimize air loss through the vent when doors are open. Figure 16 illustrates a pressurized stairwell with overpressure relief vents to the building at each floor. In systems with vents between the stairwell and the building, the vents typically have a fire damper in series with the barometric damper. As an energy conservation feature, these fire dampers are normally closed, but they open when the pressurization system is activated. This arrangement also reduces the possibility of the annoying damper chatter that frequently occurs with barometric dampers.

An exhaust duct can provide overpressure relief in a pressurized stairwell. The system is designed so that the normal resistance of a nonpowered exhaust duct maintains pressure differences within the design limits.

Exhaust fans can also relieve excess pressure when all stairwell doors are closed. An exhaust fan should be controlled by a differential pressure sensor, so that it will not operate when the pressure difference between the stairwell and the building falls below a specified level. This control should prevent the fan from pulling smoke into the stairwell when a number of open doors have reduced stairwell pressurization. Such an exhaust fan should be specifically sized so that the pressurization system will perform within design limits. A wind shield is recommended because an exhaust fan can be adversely affected by the wind.

An alternate method of venting a stairwell is through an automatically opening stairwell door to the outside at ground level. Under normal conditions, this door would be closed and, in most cases, locked for security reasons. Provisions need to be made to prevent this lock from conflicting with the automatic operation of the system. Possible adverse wind effects are also a concern with a system that uses an open outside door as a vent. Occasionally, high local

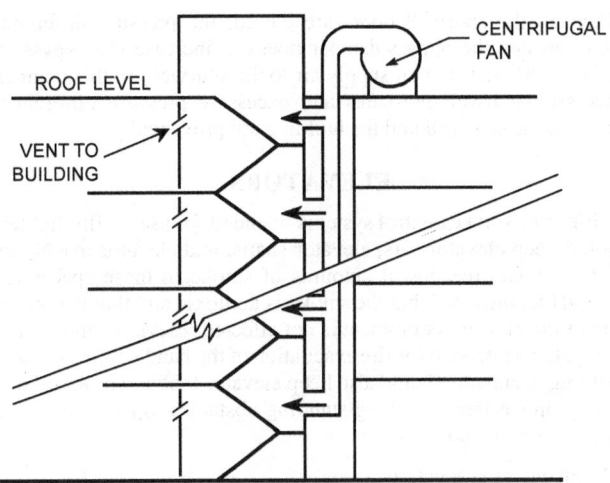

Notes:

1. Vents to the building have a barometric damper and a fire damper in series.
2. A roof-mounted supply fan is shown; however, the fan may be located at any level.
3. A manually operated damper may be located at the stairwell top for smoke purging by the fire department.

Fig. 16   Stairwell Pressurization with Vents to the Building at Each Floor

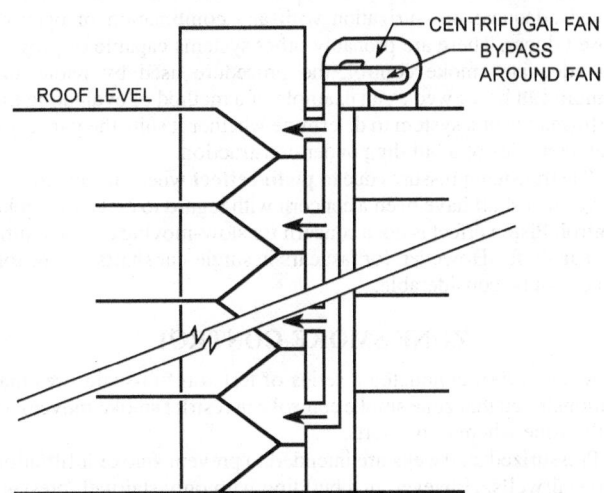

Notes:

1. Fan bypass controlled by one or more static pressure sensors located between the stairwell and the building.
2. A roof-mounted supply fan is shown; however, the fan may be located at any level.
3. A manually operated damper may be located at the stairwell top for smoke purging by the fire department.

Fig. 17   Stairwell Pressurization with Bypass Around Supply Fan

wind velocities develop near the exterior stairwell door; such winds are difficult to estimate without expensive modeling. Nearby obstructions can act as wind breaks or wind shields.

Supply Fan Bypass. In this system, the supply fan is sized to provide at least the minimum air velocity when the design number of doors are open. Figure 17 illustrates such a system. The flow rate of air into the stairwell is varied by modulating bypass dampers, which are controlled by one or more static pressure sensors that sense the pressure difference between the stairwell and the building.

When all the stairwell doors are closed, the pressure difference increases and the bypass damper opens to increase the bypass air and decrease the flow of supply air to the stairwell. In this manner, excessive stairwell pressures and excessive pressure differences between the stairwell and the building are prevented.

ELEVATORS

Elevator smoke control systems intended for use by fire fighters should keep elevator cars, elevator shafts, and elevator machinery rooms smoke-free. Small amounts of smoke in these spaces are acceptable, provided that the smoke is nontoxic and that the operation of the elevator equipment is not affected. Elevator smoke control systems intended for fire evacuation of the handicapped or other building occupants should also keep elevator lobbies smoke-free or nearly smoke-free. The long-standing obstacles to fire evacuation by elevators include

- Logistics of evacuation
- Reliability of electrical power
- Jamming of elevator doors
- Fire and smoke protection

All these obstacles, except smoke protection, can be addressed by existing technology (Klote 1984).

Klote and Tamura (1986) studied conceptual elevator smoke control systems for the evacuation of handicapped individuals. The major problem was maintaining pressurization with open building doors, especially doors on the ground floor. Of the systems evaluated, only one with a supply fan bypass with feedback control maintained adequate pressurization with any combination of open or closed doors. There are probably other systems capable of providing adequate smoke control; the procedure used by Klote and Tamura can be viewed as an example of a method of evaluating the performance of a system to determine whether it suits the particular characteristics of a building under construction.

The transient pressures due to **piston effect** when an elevator car moves in a shaft have been a concern with regard to elevator smoke control. Piston effect is not a concern for slow-moving cars in multiple-car shafts. However, for fast cars in single-car shafts, the piston effect can be considerable.

ZONE SMOKE CONTROL

Klote (1990) conducted a series of tests on full-scale fires that demonstrated that zone smoke control can restrict smoke movement to the zone where a fire starts.

Pressurized stairwells are intended to prevent smoke infiltration into stairwells. However, in a building with only stairwell pressurization, smoke can flow through cracks in floors and partitions and through shafts to damage property and threaten life at locations remote from the fire. Zone smoke control is intended to limit such smoke movement.

A building is divided into a number of smoke control zones, each separated from the others by partitions, floors, and doors that can be closed to inhibit smoke movement. In the event of a fire, pressure differences and airflows produced by mechanical fans limit the spread of smoke from the zone in which the fire started. The concentration of smoke in this zone goes unchecked; thus, in zone smoke control systems, the occupants should evacuate the smoke zone as soon as possible after fire detection.

A smoke control zone can consist of one floor, more than one floor, or part of a floor. Sprinkler zones and smoke control zones should be coordinated so that sprinkler water flow activates the zone's smoke control system. Some arrangements of smoke control zones are illustrated in Figure 18. All the nonsmoke zones in the building may be pressurized. The term **pressure sandwich** describes cases where only zones adjacent to the smoke zone are pressurized, as in Figures 18B and 18D.

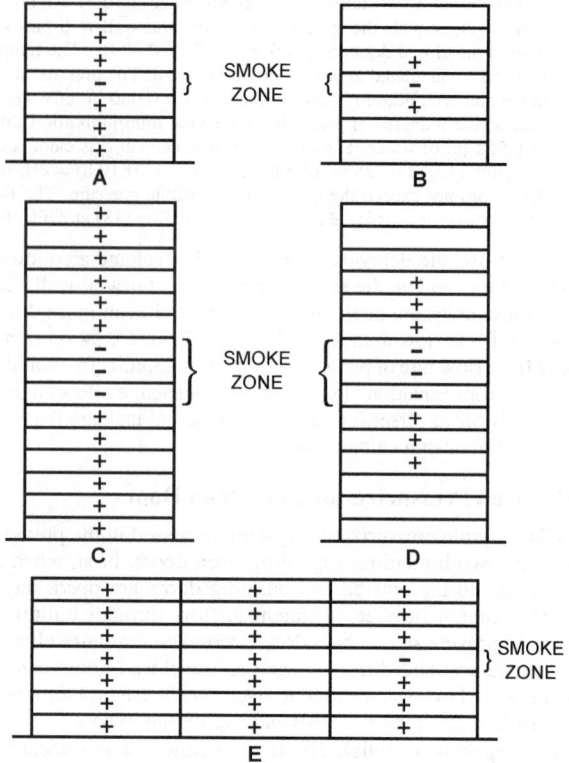

Note:
In the above figures, the smoke zone is indicated by a minus sign, and pressurized spaces are indicated by a plus sign. Each floor can be a smoke control zone as in A and B, or a smoke zone can consist of more than one floor as in C and D. All the nonsmoke zones adjacent to the smoke zone may be pressurized as in A and C or only nonsmoke zones adjacent to the smoke zone may be pressurized as in B and D. A smoke zone can also be limited to a part of a floor as in E.

Fig. 18   Some Arrangements of Smoke Control Zones

Zone smoke control is intended to limit smoke movement to the smoke zone by the use of pressurization. Pressure differences in the desired direction across the barriers of a smoke zone can be achieved by supplying outside (fresh) air to nonsmoke zones, by venting the smoke zone, or by a combination of these methods.

Venting smoke from a smoke zone prevents significant overpressure due to the thermal expansion of gases caused by the fire. This venting can be accomplished by exterior wall vents, smoke shafts, and mechanical venting (exhausting). However, venting only slightly reduces smoke concentration in the smoke zone.

COMPUTER ANALYSIS FOR PRESSURIZATION SYSTEMS

Because of the complexity of airflow in buildings, network computer programs were developed to model the airflow with pressurization systems. These models represent rooms and shafts by nodes; airflow is from nodes of high pressure to nodes of lower pressure. Some programs calculate steady-state airflow and pressures throughout a building (Sander 1974, Sander and Tamura 1973). Other programs go beyond this to calculate the smoke concentrations that would be produced throughout a building in the event of a fire (Yoshida et al. 1979, Evers and Waterhouse 1978, Wakamatsu 1977, and Rilling 1978).

The ASCOS program was developed specifically for analysis of pressurization smoke control systems (Klote 1982). ASCOS was the most widely used program for smoke control analysis

(Said 1988), and it has been validated against field data from flow experiments at an eight-story tower in Champs sur Marne, France (Klote and Bodart 1985). ASCOS and the other network models have been used extensively for design and for parametric analysis of the performance of smoke control systems. However, ASCOS was originally intended as a research tool for application to 10- and 20-story buildings. It is not surprising that convergence failures have been encountered with applications to much larger buildings.

Wray and Yuill (1993) evaluated several flow algorithms to find the most appropriate one for analysis of smoke control systems. They selected the AIRNET flow routine developed by Walton (1989) as the best algorithm based on computational speed and use of computer memory. None of the algorithms from this study takes advantage of the repetitive nature of building flow networks, so data entry is difficult. However, Walton (1994) developed CONTAM96, a public domain program with an improved version of the AIRNET flow routine and an easier method of input.

These models are appropriate for analyzing systems that use pressurization to control smoke flow. For systems that rely on buoyancy of hot smoke (such as atrium smoke exhaust), **zone fire models** are appropriate. The concepts behind zone fire modeling are discussed by Bukowski (1991), Jones (1983), and Mitler (1985). Some frequently used zone models are ASET (Cooper 1985), CCFM (Cooper and Forney 1987), and CFAST (Peacock et al. 1993). Milke and Mowrer (1994) have enhanced the CCFM model for atrium applications.

SMOKE MANAGEMENT IN LARGE SPACES

In recent years, atrium buildings have become commonplace. Other large, open spaces include enclosed shopping malls, arcades, sports arenas, exhibition halls, and airplane hangars. For simplicity, the term **atrium** is used in this chapter in a generic sense to mean any of these large spaces. Traditional fire protection by compartmentation is not applicable to these large-volume spaces.

Most atrium smoke management systems are designed to prevent exposure of occupants to smoke during evacuation. An alternate goal is to avoid subjecting occupants to untenable conditions. This approach is not commonly accepted, possibly because engineers are reluctant to design systems that expose occupants to any amount of smoke, even if the exposure is not lethal. This chapter deals only with atrium systems designed to prevent occupant exposure to smoke.

The following are approaches that can be used to manage smoke in atriums:

1. **Smoke filling.** This approach consists of allowing smoke to fill the atrium space while occupants evacuate the atrium. It applies only to spaces where the smoke filling time is sufficient for both decision making and evacuation. Nelson and MacLennan (1995) and Pauls (1995) have information on people movement during evacuation. The filling time can be estimated either by zone fire models or by filling equations [Equation (21) or (22)].
2. **Unsteady clear height with upper layer exhaust.** This approach consists of exhausting smoke from the top of the atrium at a rate such that occupants have sufficient time for decision making and evacuation. This approach requires an analysis of people movement and a fire model analysis of smoke filling.
3. **Steady clear height with upper layer exhaust.** This approach consists of exhausting smoke from the top of the atrium in order to achieve a steady clear height for a steady fire (Figure 19). A calculation method is presented in the section on Steady Clear Height with Upper Layer Exhaust.

Design Fires

The design fire has a major impact on the atrium smoke management system. Fire size is expressed in terms of rate of heat release.

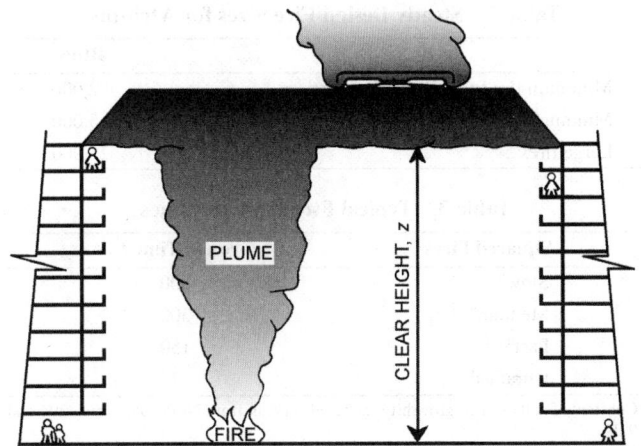

Fig. 19   Smoke Exhaust to Maintain Steady Clear Height

Fire growth is the rate of change of the heat release rate and is sometimes expressed as a growth constant that identifies the time required for the fire to attain a particular rate of heat release. Designs may be based on either steady fires or unsteady fires.

It is the nature of fires to be unsteady, but the steady fire is a very useful idealization. **Steady fires** have a constant heat release rate. In many applications, use of a steady design fire leads to straightforward and conservative design.

Morgan (1979) suggests 44 Btu/s·ft² as a typical rate of heat release per unit floor area for mercantile occupancies. Fang and Breese (1980) determined about the same rate of heat release for residential occupancies. Morgan and Hansell (1987) and Law (1982) suggest a heat release rate per unit floor area for office buildings of 20 Btu/s·ft².

In many atriums, the fuel loading is severely restricted with the intent of restricting fire size. Such atriums are characterized by interior finishes of metal, brick, stone, or gypsum board and furnished with objects made of similar materials plus plants. Even in such a **fuel-restricted atrium**, many combustible objects are present for short periods. Packing materials, holiday decorations, displays, construction materials, and furniture being moved into another part of the building are a few examples of **transient fuels**.

In this chapter, a heat release rate per floor area of 20 Btu/s·ft² will be used for a fuel-restricted atrium, and 44 Btu/s·ft² will be used for atriums containing furniture, wood, or other combustible materials.

Transient fuels must not be overlooked when selecting a design fire. Klote and Milke (1992) suggest incorporating transient fuels in a design fire by considering the fire occurring over 100 ft² of floor space. This results in a design fire of 2000 Btu/s for fuel-restricted atriums. In an atrium with combustibles, the design fire would be 4400 Btu/s. However, the area involved in fire may be much greater; flame spread considerations must be taken into account (NFPA *Standard* 92B; Klote and Milke 1992). A large atrium fire of 25,000 Btu/s would involve an area of 568 ft² at 44 Btu/s·ft². Table 2 lists some steady design fires.

Unsteady fires are often characterized by the following equation:

$$q = 1000\left(\frac{t}{t_g}\right)^2 \tag{19}$$

where

q = heat release rate of fire, Btu/s
t = time, s
t_g = growth time, s

These unsteady fires are called ***t*-squared fires**; typical growth times are listed in Table 3.

Table 2 Steady Design Fire Sizes for Atriums

	Btu/s
Minimum fire for fuel-restricted atrium	2,000
Minimum fire for atrium with combustibles	5,000
Large fires	25,000

Table 3 Typical Fire Growth Times

t-Squared Fires	Growth Time t_g, s
Slow[a]	600
Medium[a]	300
Fast[a]	150
Ultrafast[b]	75

[a]Constants for these fire growth types based on data from NFPA *Standards* 204 and 92B.

[b]Constants for ultrafast fire based on data from Nelson (1987).

Zone Fire Models

Atrium smoke management design is based on the zone fire model concept. This concept has been applied to several computer models, which can be used for atrium smoke management design analysis. These computer models include the Harvard Code (Mitler and Emmons 1981), ASET (Cooper 1985), the BRI Model (Tanaka 1983), CCFM (Cooper and Forney 1987), and CFAST (Peacock et al. 1993). The University of Maryland has modified CCFM specifically for atrium smoke management design (Milke and Mowrer 1994). Although each of these models has unique features, they all share the same basic two-zone model concept.

For more information about zone models, see Mitler and Rockett (1986), Mitler (1984), and Quintiere (1989). The ASET-B model (Walton 1985) is a good starting point for learning about zone models.

Zone models were originally developed for room fires. In a room fire, hot gases rise above the fire, forming a **plume**. As the plume rises, it entrains air from the room so that the diameter and mass flow rate of the plume increase with elevation. Accordingly, the plume temperature decreases with elevation. The fire gases from the plume flow up to the ceiling and form a hot stratified layer under the ceiling. The hot gases can flow through openings in walls to other spaces; such flow is referred to as a **door jet**. A door jet is similar to a plume, except that it flows through an opening in a wall.

Figure 20A is a sketch of a room fire. Zone modeling is an idealization of the room fire conditions, as illustrated in Figure 20B. For this idealization, the temperature of the hot upper layer of the room is uniform, and the temperature of the lower layer is also uniform. The height of the discontinuity between these layers is the same everywhere. The dynamic effects on pressure are considered negligible, so pressures are treated as hydrostatic. Other properties are considered uniform for each layer. Algebraic equations are used to calculate the mass flows due to plumes and door jets.

Many computer zone models allow exhaust from the upper layer, which is essential for simulation of atrium smoke exhaust systems. Many of the computer models estimate heat transfer by methods ranging from a simple allowance as a fraction of the heat released by the fire to a complicated simulation including the effects of conduction, convection, and radiation.

Atrium Smoke Filling by a Steady Fire

The following experimental correlation of the accumulation of smoke in a space due to a steady fire is the **steady filling equation**:

$$\frac{z}{H} = 0.67 - 0.28 \ln\left(\frac{tq^{1/3}H^{-4/3}}{A/H^2}\right) \qquad (20)$$

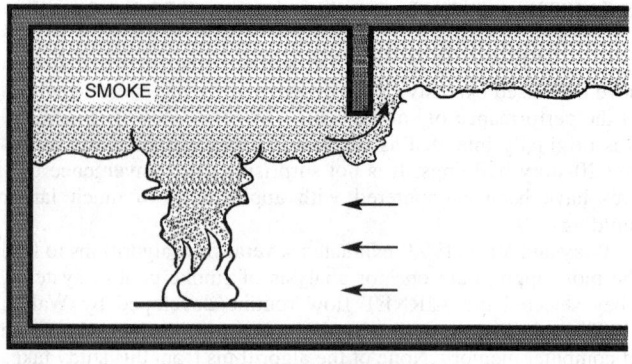

A. SKETCH OF ROOM FIRE

B. ZONE MODEL IDEALIZATION OF ROOM FIRE

Fig. 20 Room Fire and Zone Fire Model Idealization

where

z = height of first indication of smoke above fire, ft
H = ceiling height above fire, ft
t = time, s
q = heat release rate from steady fire, Btu/s
A = cross-sectional area of atrium, ft²

Equation (20) is conservative in that it estimates the height of the first indication of smoke above the fire rather than the smoke interface, as illustrated in Figure 21. In the idealized zone model, the smoke interface is considered to be a height where there is smoke above and none below. In actual fires, there is a gradual transition zone between the lower cool layer and upper hot layer. The first indication of smoke can be thought of as the bottom of the transition zone. Another factor making Equation (20) conservative is that it is based on a plume that has no contact with the walls, which would reduce entrainment of air.

Equation (20) is for a constant cross-sectional area with respect to height. For other atrium shapes, physical modeling or computational fluid dynamics can be used. Alternatively, a sensitivity analysis can be made using Equation (20) to set bounds on the filling time for an atrium of complex shape. The equation is appropriate for A/H^2 from 0.9 to 14 and for values of z greater than or equal to 20% of H. A value of z/H greater than 1 means that the smoke layer under the ceiling has not yet begun to descend. These conditions can be expressed as

$$A = \text{Constant with respect to } H$$

$$0.2 \le \frac{z}{H} < 1.0$$

$$0.9 \le \frac{A}{H^2} \le 14$$

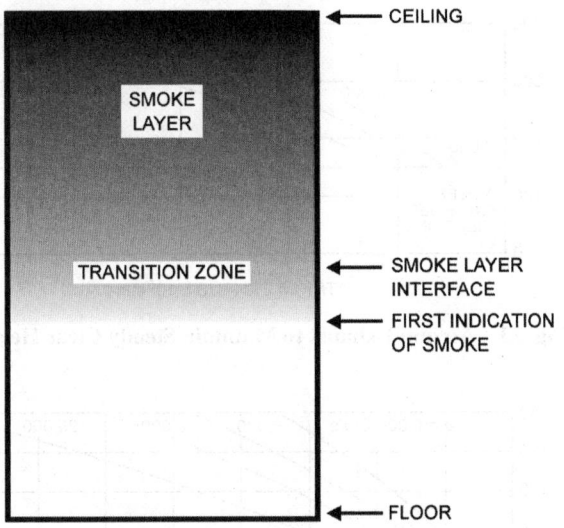

Fig. 21 Smoke Layer Interface

When Equation (20) is solved for z/H, z/H is often outside the acceptable range. Equation (20) can be solved for time.

$$t = \frac{A}{H^2}\frac{H^{4/3}}{q^{1/3}}\exp\left[\frac{1}{0.28}\left(0.67 - \frac{z}{H}\right)\right] \quad (21)$$

Atrium Smoke Filling by an Unsteady Fire

For a t-squared fire, the location of the smoke layer interface can be estimated by the **unsteady filling equation**:

$$\frac{z}{H} = 0.23\left[\frac{t}{t_g^{2/5}H^{4/5}}\left(\frac{A}{H^2}\right)^{-3/5}\right]^{-1.45} \quad (22)$$

This equation is based on experimental data and is conservative in that it estimates the height of the first indication of smoke and is for a plume that has no wall contact. The conditions can be expressed as

$$A = \text{Constant with respect to } H$$

$$0.2 \le \frac{z}{H} < 1.0$$

$$1 \le \frac{A}{H^2} \le 23$$

Values of t_g for various characteristic fire growths are listed in Table 2. Like the steady filling equation, the unsteady filling equation can be solved for time:

$$t = 0.363t_g^{2/5}H^{4/5}\left(\frac{A}{H^2}\right)^{3/5}\left(\frac{z}{H}\right)^{-0.69} \quad (23)$$

Steady Clear Height with Upper Layer Exhaust

Figure 18 illustrates smoke exhaust from the hot smoke layer at the top of an atrium to maintain a steady clear height. The smoke flow into the upper layer from the fire plume depends on the heat release rate of the fire, clear height, fuel type, and fuel orientation. The following is a generalized plume approximation that does not take into account the specifics of the material being burned.

$$\dot{m} = 0.022q_c^{1/3}z^{5/3} + 0.0042q_c \quad (24)$$

where

$\dot{m}$ = mass flow of plume, lb_m/s
q_c = convective heat release rate of fire, Btu/s
z = clear height above top of fuel, ft

The clear height z is the distance from the top of the fuel to the interface between the "clear" space and the smoke layer. Because a smoke management system generally must protect against a fire at any location, it is suggested that the top of the fuel be considered at the floor level.

Equation (27) is not applicable when the mean flame height is greater than the clear height. An approximate relationship for the mean flame height is

$$z_f = 0.533q_c^{2/5} \quad (25)$$

where z_f = mean flame height, ft.

The convective portion q_c of the heat release rate can be expressed as

$$q_c = \xi q \quad (26)$$

where ξ is the convective fraction of heat release. The convective fraction depends on the material being burned, heat conduction through the fuel, and the radiative heat transfer of the flames, but a value of 0.7 is often used.

The temperature of the smoke entering the upper smoke layer is

$$T_p = T_a + \frac{q_c}{\dot{m}c_p} \quad (27)$$

where

T_p = plume temperature at clear height, °R
T_a = ambient temperature, °R
$\dot{m}$ = mass flow of plume, lb_m/s
q_c = convective heat release rate of fire, Btu/s
c_p = specific heat of plume gases, Btu/lb·°F

Figure 22 shows plume temperature as a function of height above the fuel as calculated from Equations (24) and (27). Smoke plumes consist primarily of air mixed with the products of combustion, and the specific heat of plume gases is generally taken to be the same as that of air (c_p = 0.24 Btu/lb·°F). Equation (24) was developed for strongly buoyant plumes. For small temperature differences between the plume and ambient, errors due to low buoyancy could be significant. This topic needs study, and, in the absence of better data, it is recommended that the plume equations not be used when this temperature difference is small (less than 4°F).

The density of smoke gases can be calculated from the perfect gas law:

$$\rho = \frac{p}{RT} \quad (28)$$

where

ρ = density, lb_m/ft^3
p = absolute pressure, lb_f/ft^2
R = gas constant, ft·lb_f/lb_m·°R
T = absolute temperature of smoke gases, °R

Volumetric flow is expressed as

$$Q = \frac{60\dot{m}}{\rho} \quad (29)$$

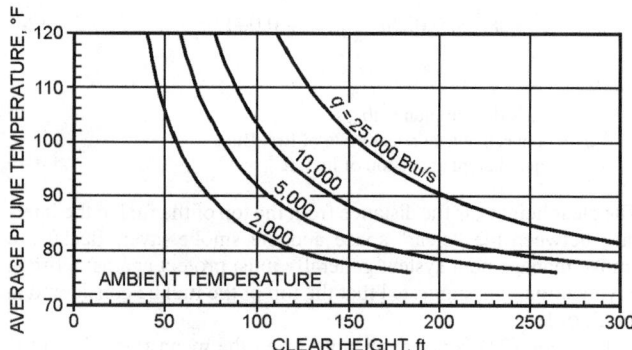

Plume equations should not be used when the plume temperature is less than 4°F above ambient.

Fig. 22   Average Plume Temperature

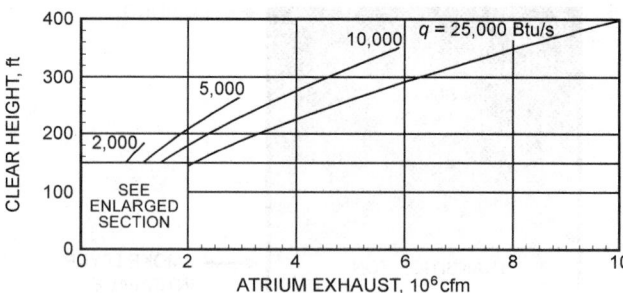

Fig. 23   Atrium Exhaust to Maintain Steady Clear Height

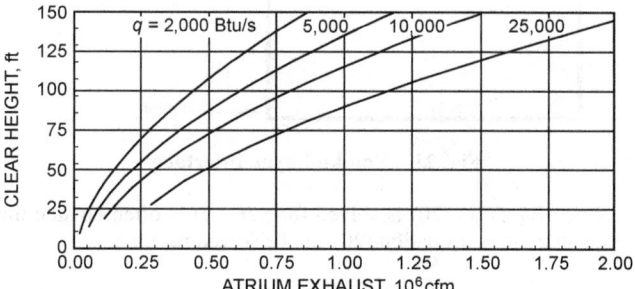

Fig. 24   Enlarged Scale for Figure 23

where

$\dot{m}$ = mass flow of plume or exhaust air, lb_m/s
Q = volumetric flow of exhaust gases, cfm
ρ = density of plume or exhaust gases, lb_m/ft^3

The atrium exhaust should equal the mass flow of the plume plus any leakage flow into the atrium above the clear height.

For an atrium with negligible heat loss from the smoke layer and negligible air leakage into the smoke layer from the outside, the exhaust equals the mass flow rate of the plume from Equation (24) at the same temperature as the plume from Equation (27). Figures 23 and 24 show the exhaust rate needed to maintain a constant clear height for an atrium with negligible heat loss from the smoke layer and negligible air leakage into the smoke layer from the outside.

The major assumptions of the analysis plotted in Figures 23 and 24 are as follows:

1. The plume has space to flow to the top of the atrium without obstructions.
2. The heat release rate of the fire is constant.
3. The clear height is greater than the mean flame height.
4. The smoke layer is adiabatic.
5. The plume flow and the exhaust are the only significant mass flows into or out of the smoke layer (i.e., outside airflow, either as leakage or as makeup air, into the smoke layer is insignificant).

Minimum Smoke Layer Depth

When the smoke layer depth below an exhaust inlet is relatively shallow, a high exhaust rate can lead to entrainment of cold air from the clear layer. This phenomenon is called **plugholing**. Accordingly, more than one exhaust point may be needed. The maximum volumetric flow rate that can be extracted through an exhaust line is

$$Q_{max} = 0.537 \beta d^{5/2} \sqrt{T_o(T_s - T_o)} \qquad (30)$$

where

Q_{max} = maximum volumetric flow rate at T_s, cfm
T_s = absolute temperature of smoke layer, °R
T_o = absolute ambient temperature, °R
d = depth of smoke layer below exhaust inlet, ft
β = exhaust location factor, dimensionless

Based on limited information, suggested values of β are 2.0 for a ceiling exhaust inlet near a wall, 2.0 for a wall exhaust inlet near the ceiling, and 2.8 for a ceiling exhaust inlet far from any walls. The inlet velocity should not exceed 2000 fpm and d/D should not exceed 2, where D is the diameter of the inlet. For exhaust inlets, use

$D = 2\ ab/(a + b)$, where a and b are the length and width of the inlet. Equation (30) was adapted from CIBSE (1995), and predictions using this equation are consistent with the experimental results of Lougheed and Hadjisophocleous (1997).

Prestratification and Detection

A hot layer of air often forms under the ceiling of an atrium due to solar radiation on the atrium roof. Although no studies have been made of this **prestratification layer**, building designers indicate that the temperature of such a layer can exceed 120°F. Temperatures below this layer are controlled by the building's heating and cooling system; the temperature can be considered to increase significantly over a small increase in elevation, as shown in Figure 25. The analysis of smoke stratification given in NFPA *Standard* 92B is not appropriate for the temperature profile addressed in this section because it is for a constant temperature increase per unit elevation.

When the average temperature of the plume is lower than that of the prestratification layer, the smoke will form a stratified layer beneath the prestratification layer, as shown in Figure 26. Average plume temperatures can be calculated from Equations (24) and (27); they are plotted in Figure 22, which shows that the average plume temperature is usually less than expected temperatures of the hot air layer. Thus, when there is a hot prestratified air layer, smoke cannot be expected to reach the ceiling of the atrium, and smoke detectors mounted on that ceiling cannot be expected to go into alarm.

Beam smoke detectors oriented horizontally to detect smoke in the plume can overcome the limitations of ceiling-mounted detectors in atriums. Beam detectors are often mounted on balconies, where they are generally easier to reach for maintenance than ceiling-mounted detectors. The light beams must be below the prestratified hot layer, and the space between beams must be small enough that plumes will be detected regardless of the location of the fire. The visible plume diameter is approximately half the clear height, with an estimated uncertainty in the range of +5% to −40%. Therefore, it is suggested that the spacing between beams be

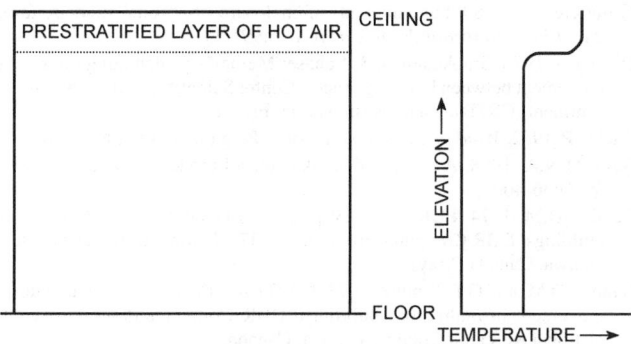

Note: Temperature below the hot layer is controlled by the building's heating and cooling system.

Fig. 25 Prestratified Layer of Hot Air under Atrium Ceiling and Resulting Temperature Profile

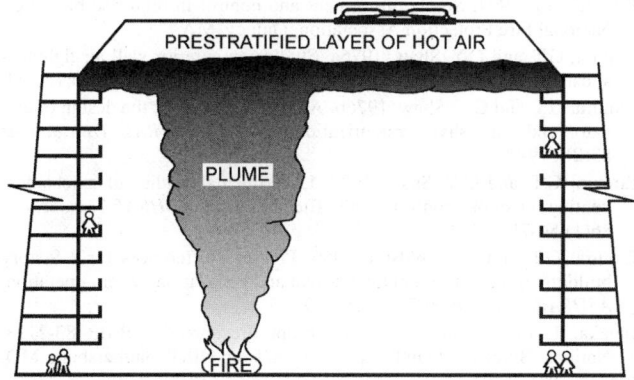

Fig. 26 Smoke Filling a Prestratified Atrium

$$x = \frac{H_b}{4} \qquad (31)$$

where

x = minimum spacing between light beams, ft
H_b = height of beams above floor, ft

The height H_b of beams above the floor must be well below the prestratification layer so that the plume will reach the level of the beams and be detected. Figure 27 is an example arrangement of beam detectors in an atrium.

ACCEPTANCE TESTING

Regardless of the care, skill, and attention to detail with which a smoke control system is designed, an acceptance test is needed as assurance that the system, as built, operates as intended.

An acceptance test should be composed of two levels of testing. The first is of a functional nature, that is, an initial check of the system components. The importance of the initial check has become apparent because of the problems encountered during tests of smoke control systems. These problems include fans operating backward, fans to which no electrical power was supplied, and controls that did not work properly.

The second level of testing is of a performance nature to determine whether the system performs adequately under all required modes of operation. This can consist of measuring pressure differences across barriers under various modes of smoke control system operation. If airflows through open doors are important, these

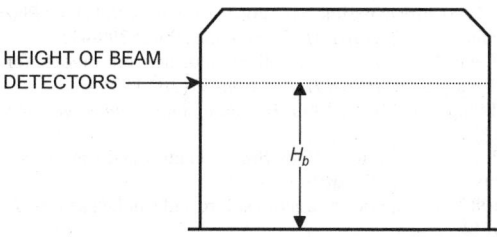

Note: Suggested spacing of beam detectors is $x = H_b/4$.

Fig. 27 Beam Detectors Used for Activation of Atrium Smoke Management System

should be measured. Chemical smoke from smoke candles (sometimes called smoke bombs) is not recommended for performance testing because it normally lacks the buoyancy of hot smoke from a real building fire. Smoke near a flaming fire has a temperature in the range of 1000 to 2000°F. Heating chemical smoke to such temperatures to emulate smoke from a real fire is not recommended unless precautions are taken to protect life and property. The same comments about buoyancy apply to tracer gases. Thus, pressure difference testing is the most practical performance test. However, chemical smoke can be used to aid flow visualization.

ASHRAE *Guideline* 5 covers the commissioning of smoke management systems.

REFERENCES

ASHRAE. 1994. Commissioning of smoke management systems. *Guideline* 5.

ASTM. 1994. Test method for fire tests of through-penetration fire stops. *Standard* E814 Rev. B. American Society for Testing and Materials, West Conshohocken, PA.

Bukowski, R.W. 1991. Fire models, the future is now! *NFPA Journal* 85(2): 60-69.

CIBSE. 1995. Relationships for smoke control calculations. TM19:1995. Chartered Institute of Building Service Engineers, London.

Cooper, L.Y. 1985. ASET—A computer program for calculating available safe egress time. *Fire Safety Journal* 9:29-45.

Cooper, L.Y. and G.P. Forney. 1987. Fire in a room with a hole: A prototype application of the consolidated compartment fire model (CCFM) computer code. Presented at the 1987 Combined Meetings of Eastern Section of Combustion Institute and NBS Annual Conference on Fire Research.

Cresci, R.J. 1973. Smoke and fire control in high-rise office buildings—Part II, Analysis of stair pressurization systems. Symposium on Experience and Applications on Smoke and Fire Control, ASHRAE Annual Meeting, June.

Evers, E. and A. Waterhouse. 1978. A computer model for analyzing smoke movement in buildings. Building Research Est., Fire Research Station, Borehamwood, Herts, UK.

Fang, J.B. 1980. Static pressures produced by room fires. NBSIR 80-1984. National Bureau of Standards. Available from NIST, Gaithersburg, MD.

Fang, J.B. and J.N. Breese. 1980. Fire development in residential basement rooms. NBSIR 80-2120. National Bureau of Standards. Available from NIST, Gaithersburg, MD.

Houghton, E.L. and N.B. Carruther. 1976. *Wind forces on buildings and structures.* John Wiley and Sons, New York.

Huggett, C. 1980. Estimation of heat release by means of oxygen consumption measurements. *Fire and Materials* 4(2).

Jones, W.W. 1983. A review of compartment fire models. NBSIR 83-2684. National Bureau of Standards. Available from NIST, Gaithersburg, MD.

Klote, J.H. 1980. Stairwell pressurization. *ASHRAE Transactions* 86(1): 604-73.

Klote, J.H. 1982. A computer program for analysis of smoke control systems. NBSIR 82-2512. National Bureau of Standards. Available from NIST, Gaithersburg, MD.

Klote, J.H. 1984. Smoke control for elevators. *ASHRAE Journal* 26(4):23-33.

Klote, J.H. 1990. Fire experiments of zoned smoke control at the Plaza Hotel in Washington, DC. *ASHRAE Transactions* 96(2):399-416.

Klote, J.H. and X. Bodart. 1985. Validation of network models for smoke control analysis. *ASHRAE Transactions* 91(2B):1134-45.

Klote, J.H. and J.A. Milke. 1992. *Design of smoke management systems.* ASHRAE.

Klote, J.H. and G.T. Tamura. 1986. Smoke control and fire evacuation by elevators. *ASHRAE Transactions* 92(1).

Law, M. 1982. Air-supported structures: Fire and smoke hazards. *Fire Prevention* 148:24-28.

Lougheed, G.D. and G.V. Hadjisophocleous. 1997. Investigation of atrium smoke exhaust effectiveness. *ASHRAE Transactions* 103(2):519-33.

MacDonald, A.J. 1975. *Wind loading on buildings.* John Wiley and Sons, New York.

McGuire, J.H. and G.T. Tamura. 1975. Simple analysis of smoke flow problems in high buildings. *Fire Technology* 11(1):15-22.

McGuire, J.H., G.T. Tamura, and A.G. Wilson. 1970. Factors in controlling smoke in high buildings. Symposium on Fire Hazards in Buildings. ASHRAE Winter Meeting.

Milke, J.A. and F.W. Mowrer. 1994. Computer-aided design for smoke management in atria and covered malls. *ASHRAE Transactions* 100(2):448-56.

Mitler, H.E. 1984. Zone modeling of forced ventilation fires. *Combustion Science and Technology* 39:83-106.

Mitler, H.E. 1985. Comparison of several compartment fire models: An interim report. NBSIR 85-3233. National Bureau of Standards. Available from NIST, Gaithersburg, MD.

Mitler, H.E. and H.W. Emmons. 1981. Documentation for CFC V, the fifth Harvard computer code. Home Fire Project *Technical Report #45.* Harvard University, Cambridge, MA.

Mitler, H.E. and J.A. Rockett. 1986. How accurate is mathematical fire modeling? NBSIR 86-3459. National Bureau of Standards. Available from NIST, Gaithersburg, MD.

Morgan, H.P. 1979. Smoke control methods in enclosed shopping complexes of one or more stories: A design summary. Fire Research Station, Borehamwood, Herts, UK.

Morgan, H.P. and G.O. Hansell. 1987. Atrium buildings: Calculating smoke flows in atria for smoke control design. *Fire Safety Journal* 12:9-12.

Nelson, H.E. 1987. An engineering analysis of the early stages of fire development—The fire at the Dupont Plaza Hotel and Casino—Dec. 31, 1986. NISTIR 87-3560. National Institute of Standards and Technology.

Nelson, H.E. and H.A. MacLennan. 1995. Emergency movement. *SFPE Handbook of fire protection engineering*, Society of Fire Protection Engineers, Boston.

NFPA. 1995. Guide for smoke management systems in malls, atria, and large areas. ANSI/NFPA *Standard* 92B-95. National Fire Protection Association, Quincy, MA.

NFPA. 1996. Recommended practice for smoke-control systems. ANSI/NFPA *Standard* 92A-96.

NFPA. 1996. Installation of air conditioning and ventilating systems. ANSI/NFPA *Standard* 90A-96.

NFPA. 1997a. *Fire protection handbook*, 18th ed.

NFPA. 1997b. *Life safety code.* ANSI/NFPA *Standard* 101-94.

NFPA 1998. Guide for smoke and heat venting. ANSI/NFPA *Standard* 204.

NOAA. 1979. Temperature extremes in the United States. National Oceanic and Atmospheric Administration (U.S.), National Climatic Center, Asheville, NC.

Pauls, J. 1995. People movement, *SFPE Handbook of fire protection engineering.* Society of Fire Protection Engineers, Boston.

Peacock, R.D., G.P. Forney, P. Reneke, R. Portier, and W.W. Jones. 1993. CFAST, the consolidated model of fire growth and smoke transport. NIST *Technical Note* 1299. National Institute of Standards and Technology, Gaithersburg, MD.

Quintiere, J.G. 1989. Fundamentals of enclosure fire "zone" models. *Journal of Fire Protection Engineering* 1(3):99-119.

Rilling, J. 1978. Smoke study, 3rd phase: Method of calculating the smoke movement between building spaces. Centre Scientifique et Technique du Batiment (CSTB), Champs sur Marne, France.

Sachs, P. 1972. *Wind forces in engineering.* Pergamon Press, New York.

Said, M.N.A. 1988. A review of smoke control models. *ASHRAE Journal* 30(4):36-40.

Sander, D.M. 1974. FORTRAN IV program to calculate air infiltration in buildings. DBR Computer Program No. 37. National Research Council, Ottawa, Canada (May).

Sander, D.M. and G.T. Tamura. 1973. FORTRAN IV program to simulate air movement in multi-story buildings. DBR Computer Program No. 35. National Research Council, Ottawa, Canada.

Shaw, C.Y. and G.T. Tamura. 1977. The calculation of air infiltration rates caused by wind and stack action for tall buildings. *ASHRAE Transactions* 83(2):145-58.

Simiu, E. and R.H. Scanlan. 1978. *Wind effects on structures: An introduction to wind engineering.* John Wiley and Sons, New York.

Tamura, G.T. 1990. Field tests of stair pressurization systems with overpressure relief. *ASHRAE Transactions* 96(1):951-58.

Tamura, G.T. 1994. Smoke movement and control in high-rise buildings. National Fire Protection Association, Quincy, MA.

Tamura, G.T. and C.Y. Shaw. 1976a. Studies on exterior wall air tightness and air infiltration of tall buildings. *ASHRAE Transactions* 83(1):122-34.

Tamura, G.T. and C.Y. Shaw. 1976b. Air leakage data for the design of elevator and stair shaft pressurization systems. *ASHRAE Transactions* 83(2):179-90.

Tamura, G.T. and C.Y. Shaw. 1978. Experimental studies of mechanical venting for smoke control in tall office buildings. *ASHRAE Transactions* 86(1):54-71.

Tamura, G.T. and A.G. Wilson. 1966. Pressure differences for a 9-story building as a result of chimney effect and ventilation system operation. *ASHRAE Transactions* 72(1):180-89.

Tanaka, T. 1983. A model of multiroom fire spread. NBSIR 83-2718. National Bureau of Standards. Available from NIST, Gaithersburg, MD.

Thomas, P.H. 1970. Movement of smoke in horizontal corridors against an airflow. *Institution of Fire Engineers Quarterly* 30(77):45-53.

UL. 1994. Fire tests of through-penetration firestops. ANSI/UL *Standard* 1479-94. Underwriters Laboratories, Northbrook, IL.

UL. 1995. Fire dampers, 5th ed. *Standard* 555-95.

UL. 1996. Leakage rated dampers for use in smoke control systems, 3rd ed. *Standard* 555S-96.

Wakamatsu, T. 1977. Calculation methods for predicting smoke movement in building fires and designing smoke control systems. Fire *Standards and Safety*, ASTM STP 614, pp. 168-93. American Society for Testing and Materials, West Conshohocken, PA.

Walton, G.N. 1989. AIRNET—A computer program for building airflow network modeling. National Institute of Standards and Technology, Gaithersburg, MD.

Walton, G.N. 1994. CONTAM93 user manual. NISTIR 5385. National Institute of Standards and Technology, Gaithersburg, MD.

Walton, W.D. 1985. ASET-B: A room fire program for personal computers. NBSIR 85-3144-1. National Bureau of Standards. Available from NIST, Gaithersburg, MD.

Wray, C.P. and G.K. Yuill. 1993. An evaluation of algorithms for analyzing smoke control systems. *ASHRAE Transactions* 99(1):160-74.

Yoshida, H., C.Y. Shaw, and G.T. Tamura. 1979. A FORTRAN IV program to calculate smoke concentrations in a multi-story building. DBR Computer Program No. 45. National Research Council, Ottawa, Canada.

RADIANT HEATING AND COOLING

RADIANT heating and cooling applications are classified as panel heating or cooling if the surface temperature is below 300°F and are classified as low-, medium-, or high-intensity if the surface or source temperature range exceeds 300°F. Radiant energy is transmitted by electromagnetic waves that travel in straight lines, can be reflected, and heat solid objects but do not heat the air through which the energy is transmitted. Because of these characteristics, radiant systems are effective for both spot heating and space heating or cooling requirements for an entire building.

LOW-, MEDIUM-, AND HIGH-INTENSITY INFRARED HEATING

Low-, medium-, and high-intensity infrared heaters are compact, self-contained direct-heating devices used in hangars, warehouses, factories, greenhouses, and gymnasiums, as well as in areas such as loading docks, racetrack stands, outdoor restaurants, animal breeding areas, swimming pool lounge areas, and areas under marquees. Infrared heating is also used for snow melting (e.g., on stairs and ramps) and process heating (e.g., paint baking and drying). An infrared heater may be electric, gas-fired, or oil-fired and is classified by the source temperature as follows:

- Low intensity for source temperatures to 1200°F
- Medium intensity for source temperatures to 1800°F
- High intensity for source temperatures to 5000°F

The source temperature is determined by such factors as the source of energy, the configuration, and the size. Reflectors can be used to direct the distribution of radiation in specific patterns. Chapter 15 of the 1996 *ASHRAE Handbook—Systems and Equipment* covers radiant equipment in detail.

PANEL HEATING AND COOLING

Panel heating and cooling systems provide a comfortable environment by controlling surface temperatures and minimizing air motion within a space. They include the following designs:

- Metal ceiling panels
- Embedded hydronic tubing or attached piping in ceilings, walls, or floors
- Air-heated floors or ceilings
- Electric ceiling or wall panels
- Electric heating cable in ceilings or floors
- Deep heat, a modified storage system using electric heating cable or embedded hydronic tubing in ceilings or floors

In these systems, generally more than 50% of the heat transfer from the controlled temperature surface to other surfaces is by radiation. Panel heating and cooling systems are used in residences, office buildings (for perimeter heating), classrooms, hospital patient rooms, swimming pool areas, repair garages, and in industrial and warehouse

The preparation of this chapter is assigned to TC 6.5, Radiant Space Heating and Cooling.

applications. Additional information is available in Chapter 6 of the 1996 *ASHRAE Handbook—Systems and Equipment*.

Some radiant panel systems combine controlled heating and cooling with central air conditioning and are referred to as hybrid conditioning systems. They are used more for cooling than for heating (Wilkins and Kosonen 1992). The controlled temperature surfaces may be in the floor, walls, or ceiling, with the temperature maintained by electric resistance or circulation of water or air. The central station can be a basic, one-zone, constant-temperature, or constant-volume system, or it can incorporate some or all the features of dual-duct, reheat, multizone, or variable-volume systems. When used in combination with other water/air systems, radiant panels provide zone control of temperature and humidity.

Metal ceiling panels may be integrated into the central heating and cooling system to provide individual room or zone heating and cooling. These panels can be designed as small units to fit the building module, or they can be arranged as large continuous areas for economy. Room thermal conditions are maintained primarily by direct transfer of radiant energy, normally using four-pipe hot and chilled water. These metal ceiling panel systems have generally been used in hospital patient rooms.

The application of metal ceiling panel systems is discussed in Chapter 6 of the 1996 *ASHRAE Handbook—Systems and Equipment*.

ELEMENTARY DESIGN RELATIONSHIPS

When considering radiant heating or cooling for human comfort, the following terms describe the temperature and energy characteristics of the total radiant environment:

- **Mean radiant temperature** (MRT) $\bar{t}_r$ is the temperature of an imaginary isothermal black enclosure in which an occupant would exchange the same amount of heat by radiation as in the actual nonuniform environment.
- **Ambient temperature** t_a is the temperature of the air surrounding the occupant.
- **Operative temperature** t_o is the temperature of a uniform isothermal black enclosure in which the occupant would exchange the same amount of heat by radiation and convection as in the actual nonuniform environment.

  For air velocities less than 80 fpm and mean radiant temperatures less than 120°F, the operative temperature is approximately equal to the adjusted dry-bulb temperature, which is the average of the air and mean radiant temperatures.
- **Adjusted dry-bulb temperature** is the average of the air temperature and the mean radiant temperature at a given location. The adjusted dry-bulb temperature is approximately equivalent to the operative temperature for air motions less than 80 fpm and mean radiant temperatures less than 120°F.
- **Effective radiant flux** (ERF) is defined as the net radiant heat exchanged at the ambient temperature t_a between an occupant, whose surface is hypothetical, and all enclosing surfaces and directional heat sources and sinks. Thus, ERF is the net radiant energy received by the occupant from all surfaces and sources

whose temperatures *differ* from t_a. ERF is particularly useful in high-intensity radiant heating applications.

The relationship between these terms can be shown for an occupant at surface temperature t_{sf}, exchanging sensible heat H_m in a room with ambient air temperature t_a and mean radiant temperature $\bar{t}_r$. The linear radiative and convective heat transfer coefficients are h_r and h_c, respectively; the latter coefficient is a function of the air movement V. The heat balance equation is

$$H_m = h_r(t_{sf} - \bar{t}_r) + h_c(t_{sf} - t_a) \qquad (1)$$

During thermal equilibrium, H_m is equal to metabolic heat minus work and evaporative cooling by sweating. By definition of operative temperature,

$$H_m = (h_r + h_c)(t_{sf} - t_o) \qquad (2)$$

Using Equations (1) and (2) to solve for t_o yields

$$t_o = \frac{h_r \bar{t}_r + h_c t_a}{h_r + h_c} \qquad (3)$$

Thus, t_o is an average of $\bar{t}_r$ and t_a, weighted by their respective heat transfer coefficients; it represents how people sense the thermal level of their total environment as a single temperature.

The combined heat transfer coefficient is h, where $h = h_r + h_c$. Rearranging Equation (1) and substituting $h - h_r$ for h_c,

$$H_m + h_r(\bar{t}_r - t_a) = h(t_{sf} - t_a) \qquad (4)$$

where $h_r(\bar{t}_r - t_a)$ is, by definition, the effective radiant flux (ERF) and represents the radiant energy absorbed by the occupant from all sources whose temperatures differ from t_a.

The principal relationships between $\bar{t}_r$, t_a, t_o, and ERF are as follows:

$$\text{ERF} = h_r(\bar{t}_r - t_a) \qquad (5)$$

$$\text{ERF} = h(t_o - t_a) \qquad (6)$$

$$\bar{t}_r = t_a + \text{ERF}/h_r \qquad (7)$$

$$t_o = t_a + \text{ERF}/h \qquad (8)$$

$$\bar{t}_r = t_a + (h/h_r)(t_o - t_a) \qquad (9)$$

$$t_o = t_a + (h_r/h)(\bar{t}_r - t_a) \qquad (10)$$

In Equations (1) through (10), the radiant environment is treated as a blackbody with temperature $\bar{t}_r$. The effect of the emittance of the source, radiating at absolute temperature in degrees Rankin, and the absorptance of the skin and clothed surfaces is reflected in the effective values of $\bar{t}_r$ or ERF and not in the coefficient h_r, which is generally given by

$$h_r = 4\sigma f_{eff}[(\bar{t}_r + t_a)/2 + T]^3 \qquad (11)$$

where

$\quad h_r$ = linear radiative heat transfer coefficient, Btu/h·ft²·°F
$\quad f_{eff}$ = ratio of radiating surface of the human body to its total DuBois
$\qquad$ surface area $A_D = 0.71$
$\quad \sigma$ = Stefan-Boltzmann constant = 0.1713×10^{-8} Btu/h·ft²·°R⁴
$\quad T$ = 460 for temperatures in °F

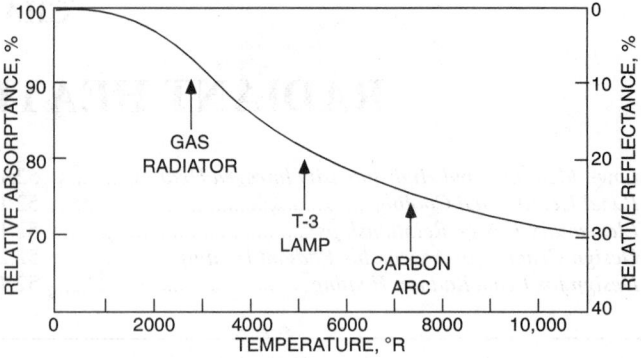

Fig. 1   Relative Absorptance and Reflectance of Skin and Typical Clothing Surfaces at Various Color Temperatures

The convective heat transfer coefficient for an occupant is

$$h_c = 0.107 \, V^{0.5} \qquad (12)$$

where h_c is in Btu/h·ft²·°F and V is air movement in fpm.

When $\bar{t}_r > t_a$, ERF adds heat to the body system; when $t_a > \bar{t}_r$, heat is lost from the body due to radiant cooling. ERF is independent of the surface temperature of the occupant and can be measured directly by a black globe thermometer or any blackbody radiometer or flux meter using the ambient air t_a as its heat sink.

In the above definitions and for radiators below 1700°F (2160°R), the body clothing and skin surface are treated as blackbodies, exchanging radiation with an imaginary blackbody surface at temperature $\bar{t}_r$. The effectiveness of a radiating source on human occupants is governed by the absorptance α of the skin and clothing surface for the color temperature (in °R) of that radiating source. The relationship between α and temperature is illustrated in Figure 1. The values for α are those expected relative to the matte black surface normally found on globe thermometers or radiometers measuring radiant energy. A gas radiator usually operates at 1700°F (2160°R); a quartz lamp, for example, radiates at 4000°F (4460°R) with 240 V; and the sun's radiating temperature is 10,000°F (10,460°R). The use of α in estimating the ERF and t_o caused by sources radiating at temperatures above 1700°F (2160°R) is discussed in the section on Testing Instruments for Radiant Heating.

DESIGN CRITERIA FOR ACCEPTABLE RADIANT HEATING

Perceptions of comfort, temperature, and thermal acceptability are related to activity, the transfer of body heat from the skin to the environment, and the resulting physiological adjustments and body temperature. Heat transfer is affected by the ambient air temperature, thermal radiation, air movement, humidity, and clothing worn. Thermal sensation is described as feelings of hot, warm, slightly warm, neutral, slightly cool, cool, and cold. An acceptable environment is defined as one in which at least 80% of the occupants perceive a thermal sensation between "slightly cool" and "slightly warm." Comfort is associated with a neutral thermal sensation during which the human body regulates its internal temperature with minimal physiological effort for the activity concerned. In contrast, warm discomfort is primarily related to the physiological strain necessary to maintain the body's thermal equilibrium rather than to the temperature sensation experienced. For a full discussion on the interrelation of physical, psychological, and physiological factors, refer to Chapter 8 of the 1997 *ASHRAE Handbook—Fundamentals*.

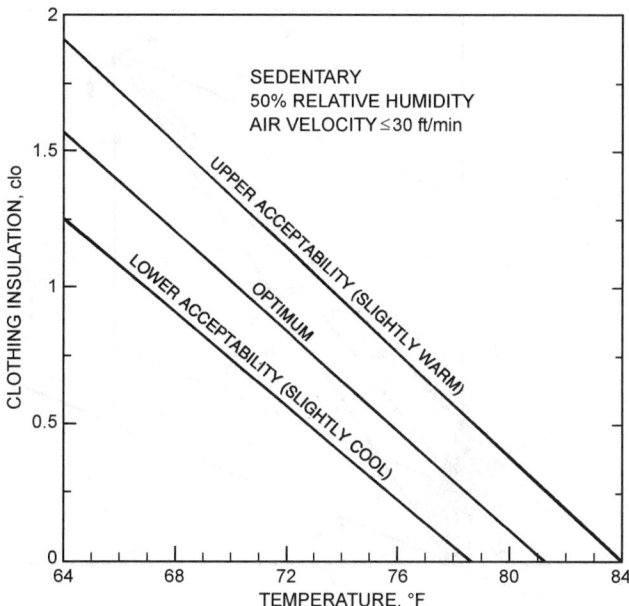

**Fig. 2 Range of Thermal Acceptability for Various
Clothing Insulations and Operative Temperatures**

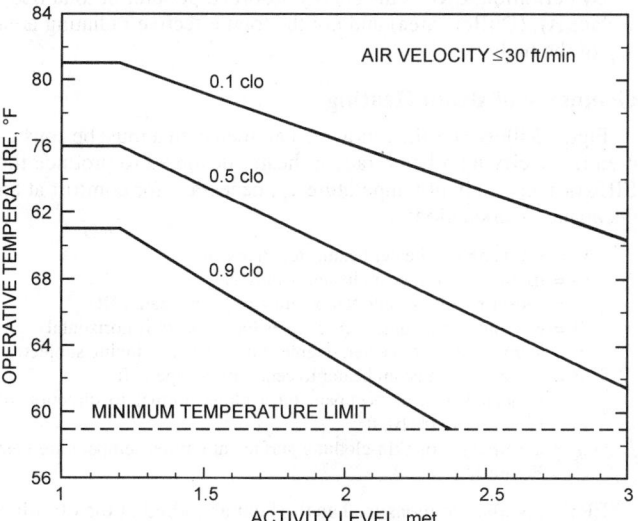

**Fig. 3 Optimum Operative Temperatures for Active
People in Low Air Movement Environments**

ASHRAE *Standard* 55 shows a linear relationship between the clothing insulation worn and the operative temperature t_o for comfort (Figure 2). Figure 3, which is adapted from the standard, shows the effect of both activity and clothing on the t_o for comfort. Figure 4 shows the slight effect humidity has on the comfort of a sedentary person wearing average clothing.

A comfortable t_o at 50% rh is perceived as slightly warmer as the humidity increases or is perceived as slightly cooler as the humidity decreases. Changes in humidity have a much greater effect on "warm" and "hot" discomfort. In contrast, "cold" discomfort is only slightly affected by humidity and is very closely related to a "cold" thermal sensation.

Determining the specifications for a radiant heating installation designed for human occupancy and acceptability involves the following steps:

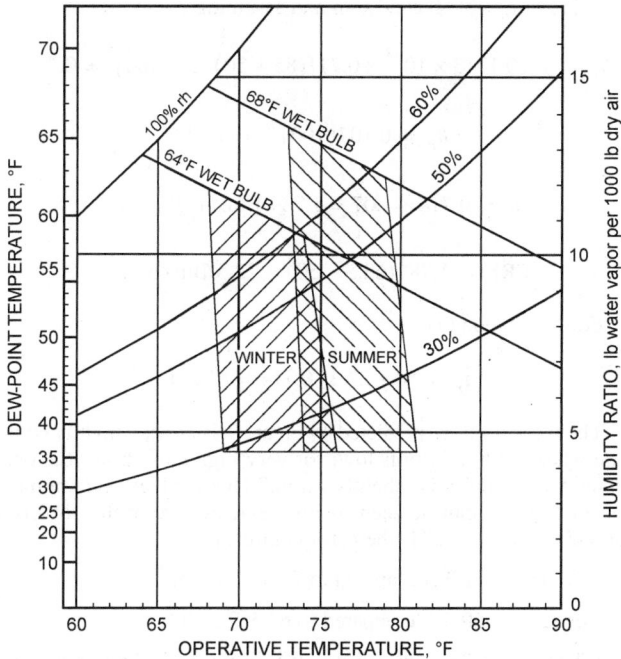

Fig. 4 ASHRAE Comfort Zones for Sedentary Individuals

1. Define the probable activity (metabolism) level of, and clothing worn by, the occupant and the air movement in the occupied space. The following are two examples:

Case 1: Sedentary (1.1 met)

  Clothing insulation = 0.6 clo; air movement = 30 ft/min

Case 2: Light work (2 met)

  Clothing insulation = 0.9 clo; air movement = 100 ft/min

2. From Figure 2 or 3, determine the optimum t_o for comfort and acceptability:

$$Case\ 1:\ t_o\ =\ 74.5°F;\quad Case\ 2:\ t_o\ =\ 62.5°F$$

3. For the ambient air temperature t_a, calculate the mean radiant temperature $\bar{t}_r$ and/or ERF necessary for comfort and thermal acceptability.

Case 1: For $t_a = 60°F$ and assuming $\bar{t}_r = 90°F$,

  Solve for h_r from Equation (11):

$$h_r\ =\ 4 \times 0.1713 \times 10^{-8} \times 0.71[(90 + 60)/2 + 460]^3\ =\ 0.745$$

Solve for h_c from Equation (12):

$$h_c\ =\ 0.107(30)^{0.5}\ =\ 0.586$$

Then,

$$h\ =\ h_r + h_c\ =\ 0.745 + 0.586\ =\ 1.331\ Btu/h \cdot ft^2 \cdot °F$$

From Equation (6), for comfort,

$$ERF\ =\ 1.331(74.5 - 60)\ =\ 19.3\ Btu/h \cdot ft^2$$

From Equation (9),

$$\bar{t}_r\ =\ 60 + (1.331/0.745)(74.5 - 60)\ =\ 85.9°F$$

Case 2: For $t_a = 50°F$ at 50% rh and assuming $\bar{t}_r = 85°F$.

$$h_r = 4 \times 0.1713 \times 10^{-8} \times 0.71[(85 + 50)/2 + 460]^3 = 0.714$$

$$h_c = 0.107(100)^{0.5} = 1.07$$

$$h = 0.714 + 1.07 = 1.784 \text{ Btu/h} \cdot \text{ft}^2 \cdot °F$$

$$\text{ERF} = 1.784(62.5 - 50) = 22.3 \text{ Btu/h} \cdot \text{ft}^2$$

From Equation (7),

$$\bar{t}_r = 50 + 22.3/0.714 = 81.2°F$$

The t_o for comfort, predicted by Figure 2, is on the "slightly cool" side when the humidity is low; for very high humidities, the predicted t_o for comfort is "slightly warm." This small effect of humidity on comfort can be seen in Figure 4. For example, for high humidity at $t_{dp} = 55.5°F$, the t_o for comfort is

Case 1: $t_o = 72°F$, compared to 74.5°F at 50% rh

Case 2: $t_o = 60.5°F$ compared to 62.5°F at 50% rh

When thermal acceptability is the primary consideration in an installation, humidity can sometimes be ignored in preliminary design specifications. However, for conditions where radiant heating and the work level cause sweating and high heat stress, humidity is a major consideration.

Equations (3) through (12) can also be used to determine the ambient air temperature t_a required when the mean radiant temperature MRT is maintained by a specified radiant system.

When calculating heat loss, t_a must be determined. For a radiant system that is to maintain a MRT of $\bar{t}_r$, the operative temperature t_o can be determined from Figure 4. Then, t_a can be calculated by recalling that t_o is approximately equal to the average of t_a and t_r. For example, for a t_o of 73°F and a radiant system designed to maintain an MRT of 78°F, the t_a would be 68°F.

When the surface temperature of outside walls, particularly those with large areas of glass, deviates too much from the room air temperature and from the temperature of other surfaces, simplified calculations for the load and the operative temperature may lead to errors in sizing and locating the panels. In such cases, more detailed radiant exchange calculations may be required, with separate estimation of heat exchange between the panels and each surface. A large window area may lead to significantly lower mean radiant temperatures than expected. For example, Athienitis and Dale (1987) reported an MRT 5.4°F lower than room air temperature for a room with a glass area equivalent to 22% of its floor area.

DESIGN FOR BEAM RADIANT HEATING

Spot beam radiant heat can improve comfort at a specific location in a large, poorly heated work area. The design problem is specifying the type, capacity, and orientation of the beam heater.

Using the same reasoning as in Equations (1) through (10), the effective radiant flux ΔERF that must be added to an unheated work space with an operative temperature t_{uo} to result in a t_o for comfort (as given by Figure 2 or 3) is

$$\Delta\text{ERF} = h(t_o - t_{uo}) \tag{13}$$

or

$$t_o = t_{uo} + \Delta\text{ERF}/h \tag{14}$$

This equation is unaffected by air movement. The heat transfer coefficient h for the occupant in Equation (13) is given by Equations (11) and (12), with $h = h_r + h_c$.

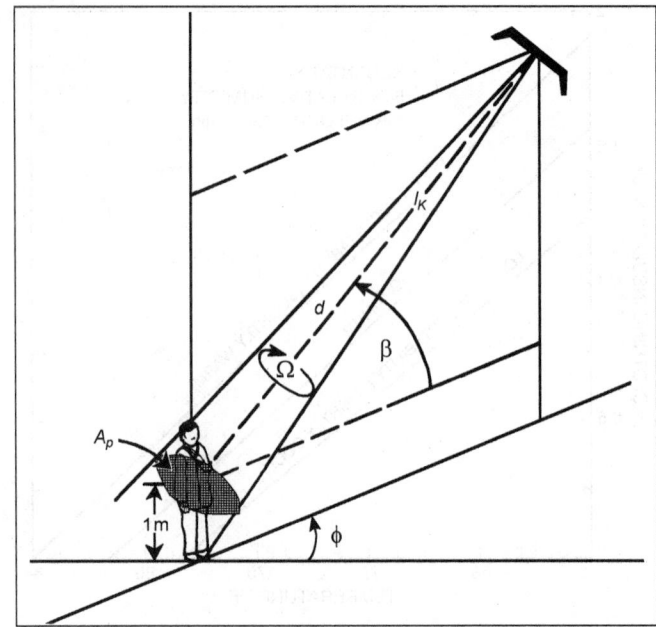

Fig. 5 Geometry and Symbols for Describing Beam Heaters

By definition, ERF is the energy absorbed per unit of total body surface A_D (DuBois area) and *not* the total effective radiating area A_{eff} of the body.

Geometry of Beam Heating

Figure 5 illustrates the following parameters that must be considered in specifying a beam radiant heater designed to produce the ERF, or mean radiant temperature $\bar{t}_r$, necessary for comfort at an occupant's workstation:

Ω = solid angle of heater beam, steradians (sr)
I_K = irradiance from beam heater, Btu/h $\cdot$ sr
K = subscript for absolute temperature of beam heater, °R
β = elevation angle of heater, degrees (at 0°, beam is horizontal)
ϕ = azimuth angle of heater, degrees (at 0°, beam is facing subject)
d = distance from beam heater to center of occupant, ft
A_p = projected area of occupant on a plane normal to direction of heater beam (ϕ, β), ft^2
α_K = absorptance of skin-clothing surface at emitter temperature (see Figure 1)

ERF may also be measured as the heat absorbed at the clothing and skin surface of the occupant from a beam heater at absolute temperature:

$$\text{ERF} = \frac{\alpha_K I_K A_p}{d^2 A_D} \tag{15}$$

where ERF is in Btu/h·ft^2 and (A_p/d^2) is the solid angle subtended by the projected area of the occupant from the radiating beam heater I_K, which is treated here as a point source. A_D is the DuBois area:

$$A_D = 0.0621 W^{0.425} H^{0.725}$$

where

W = occupant weight, lb
H = occupant, height, ft

For additional information on radiant flux distribution patterns and sample calculations of radiation intensity I_K and ERF, refer to Chapter 15 of the 1996 *ASHRAE Handbook—Systems and Equipment.*

Floor Reradiation

In most low-, medium-, and high-intensity radiant heater installations, local floor areas are strongly irradiated. The floor absorbs most of this energy and warms to an equilibrium temperature t_f, which is higher than that of the ambient air temperature t_a and the unheated room enclosure surfaces. Part of the energy directly absorbed by the floor is transmitted by conduction to the cooler underside (or, for slabs-on-grade, to the ground), part is transferred by natural convection to room air, and the remainder is reradiated. The warmer floor will raise ERF or $\bar{t}_r$ over that caused by the heater alone.

For a person standing on a large, flat floor that has a temperature raised by direct radiation t_f, the linearized $\bar{t}_r$ due to the floor and unheated walls is

$$\bar{t}_{rf} = F_{p-f} t_f + (1 - F_{p-f}) t_a \qquad (16)$$

where the unheated walls, ceiling, and ambient air are assumed to be at t_a, and F_{p-f} is the angle factor governing the radiation exchange between the heated floor and the person.

The ERF_f from the floor affecting the occupant, which is due to the $(t_f - t_a)$ difference, is

$$\begin{aligned} ERF_f &= h_r(\bar{t}_{rf} - t_a) \\ &= h_r F_{p-f}(t_f - t_a) \end{aligned} \qquad (17)$$

where h_r is the linear radiative heat transfer coefficient for a person as given by Equation (11). For a standing or sitting subject when the walls are farther than 16 ft away, F_{p-f} is 0.44 (Fanger 1973). For an average-sized 16 ft by 16 ft room, a value of 0.35 for F_{p-f} is suggested. For detailed information on floor reradiation, see Chapter 15 in the 1996 *ASHRAE Handbook—Systems and Equipment*.

In summary, when radiant heaters warm occupants in a selected area of a poorly heated space, the radiation heat necessary for comfort consists of two additive components: (1) ERF directly caused by the heater and (2) reradiation ERF_f from the floor. The effectiveness of floor reradiation can be improved by choosing flooring with a low specific conductivity. Flooring with high thermal inertia may be desirable during radiant transients, which may occur as the heaters are cycled by a thermostat set to the desired operative temperature t_o.

Asymmetric Radiant Fields

In the past, comfort heating has required flux distribution in occupied areas to be uniform, which is not possible with beam radiant heaters. Asymmetric radiation fields, such as those experienced when lying in the sun on a cool day or when standing in front of a warm fire, can be pleasant. Therefore, a limited amount of asymmetry, which is allowable for comfort heating, is referred to as "reasonable uniform radiation distribution" and is used as a design requirement.

To develop criteria for judging the degree of asymmetry allowable for comfort heating, Fanger et al. (1980) proposed defining radiant temperature asymmetry as the difference in the plane radiant temperature between two opposing surfaces. Plane radiant temperature is the equivalent $\bar{t}_{r1}$ caused by radiation on one side of the subject, compared with the equivalent $\bar{t}_{r2}$ caused by radiation on the opposite side. Gagge et al. (1967) conducted a study of subjects (eight clothed and eight unclothed) seated in a chair and heated by two lamps. Unclothed subjects found a $(\bar{t}_r - t_a)$ asymmetry as high as 20°F to be comfortable, but clothed subjects were comfortable with an asymmetry as high as 31°F.

For an unclothed subject lying on an insulated bed under a horizontal bank of lamps, neutral temperature sensation occurred for a t_o of 72°F, which corresponds to a $(t_o - t_a)$ asymmetry of 20°F or a

$(\bar{t}_r - t_a)$ asymmetry of 27°F, both averaged for eight subjects (Stevens et al. 1969). In studies of heated ceilings, 80% of eight male and eight female clothed subjects voted conditions as comfortable and acceptable for asymmetries as high as 20°F. The study compared the floor and heated ceilings. The asymmetry in the MRTs for direct radiation from three lamps and for floor reradiation is about 1°F, which is negligible.

In general, the human body has a great ability to sum sensations from many hot and cold sources. For example, Australian aborigines sleep unclothed next to open fires in the desert at night, where t_a is 43°F. The $\bar{t}_r$ caused directly by three fires alone is 171°F, and the cold sky $\bar{t}_r$ is 30°F; the resulting t_o is 82°F, which is acceptable for human comfort (Scholander 1958).

According to the limited field and laboratory data available, an allowable design radiant asymmetry of 22 ± 5°F should cause little discomfort over the comfortable t_o range used by ANSI/ASHRAE *Standard* 55 and in Figures 2 and 3. Increased clothing insulation allows increases in the acceptable asymmetry, but increased air movement reduces it. Increased activity also reduces human sensitivity to changing $\bar{t}_r$ or t_o and, consequently, increases the allowable asymmetry. The design engineer should use caution with an asymmetry greater than 27°F, as measured by a direct beam radiometer or estimated by calculation.

RADIATION PATTERNS

Figure 6 indicates the basic radiation patterns commonly used in design for radiation from point or line sources (Boyd 1962). A point source radiates over an area that is proportional to the square of the distance from the source. The area for a (short) line source also varies substantially as the square of the distance, with about the same area as the circle actually radiated at that distance. For line sources, the width of the pattern is determined by the reflector shape and position of the element within the reflector. The rectangular area used for installation purposes as the pattern of radiation from a line source assumes a length equal to the width plus the fixture length. This assumed length is satisfactory for design, but is often two or three times the pattern width.

Electric infrared fixtures are often identified by their beam pattern (Rapp and Gagge 1967), which is the radiation distribution normal to the line source element. The beam of a high-intensity infrared fixture may be defined as that area in which the intensity is at least 80% of the maximum intensity encountered anywhere within the beam. This intensity is measured in the plane in which maximum control of energy distribution is exercised.

The beam size is usually designated in angular degrees and may be symmetrical or asymmetrical in shape. For adaptation to their design specifications, some manufacturers indicate beam characteristics based on 50% maximum intensity.

The control used for an electric system affects the desirable maximum end-to-end fixture spacing. Actual pattern length is about three times the design pattern length, so control in three equal stages is achieved by placing every third fixture on the same circuit. If all fixtures are controlled by input controllers or variable voltage to electric units, end-to-end fixture spacing can be nearly three times the design pattern length. Side-to-side minimum spacing is determined by the distribution pattern of the fixture and is not influenced by the method of control.

Low-intensity equipment typically consists of a steel tube hung near the ceiling and parallel to the outside wall. Circulation of combustion products inside the tube elevates the tube temperature and radiant energy is emitted. The tube is normally provided with a reflector to direct the radiant energy down into the space to be conditioned.

Radiant ceiling panels for heating only are installed in a narrow band around the perimeter of an occupied space and are usually the primary heat source for the space. The radiant source is (long)

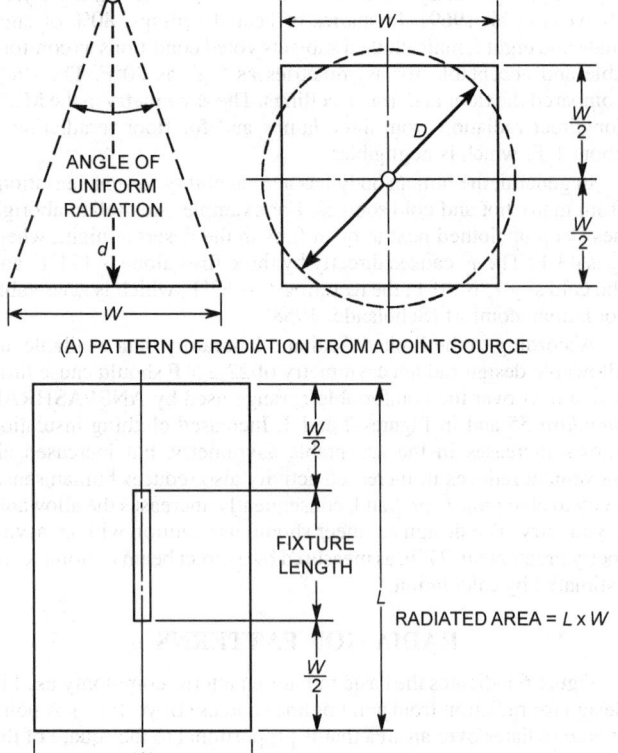

(A) PATTERN OF RADIATION FROM A POINT SOURCE

(B) PATTERN OF RADIATION FROM A LINE SOURCE

Note: The projected area W^2 normal to a beam that is Ω steradians wide at distance d is $\Omega\ d^2$. The floor area irradiated by a beam heater at an angle elevation β is $W^2/\sin\beta$. Fixture length L increases the area irradiated by the factor $(1 + L/W)$.

Fig. 6   Basic Radiation Patterns for System Design
(Boyd 1962)

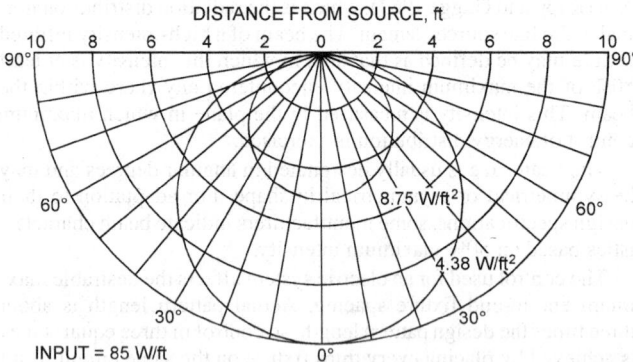

Fig. 7   Lines of Constant Radiant Flux for a Line Source

linear. The flux density is inversely proportional to the distance from the source.

The rate of radiation exchange between a panel and a particular object depends on their temperatures, emissivities, and geometrical orientation (i.e., shape factor). It also depends on the temperatures and configurations of all the other objects and walls within the space.

The energy flux for an ideal line source is shown in Figure 7. All objects (except the radiant source) are at the same temperature, and the radiant source is suspended symmetrically in the room. The flux density is inversely proportional to the distance from the source.

DESIGN FOR TOTAL SPACE HEATING

Radiant heating differs from conventional heating by a moderately elevated ERF, $\bar{t}_r$, or t_o over the ambient temperature t_a. Standard methods of design are normally used, although informal studies indicate that radiant heating requires a lower heating capacity than convection heating (Zmeureanu et al. 1987). Buckley and Seel (1987) demonstrated that a combination of elevated floor temperature, higher mean radiant temperature, and reduced ambient temperature results in lower thermostat settings, reduced temperature differential across the building envelope, and thus, lower heat loss and heating load for the structure. In addition, the peak load may be decreased due to heat (cool) stored in the structure (Kilkis 1990, 1992).

Most gas radiation systems for full-building heating concentrate the bulk of capacity at mounting heights of 10 to 16 ft at the perimeter, directed at the floor near the walls. Units can be mounted considerably higher. Successful application depends on supplying the proper amount of heat in the occupied area. Heaters should be located to take maximum advantage of the pattern of radiation produced. Exceptions to perimeter placement include walls with high transmission losses and extreme height, as well as large roof areas where roof heat losses exceed perimeter and other heat losses.

Electric infrared systems installed indoors for complete building heating have used layouts that uniformly distribute the radiation throughout the area used by people, as well as layouts that emphasize perimeter placement, such as in ice hockey rinks. Some electric radiant heaters emit a significant amount of visible radiation and provide both heating and illumination.

The orientation of equipment and people is less important for general area heating (large areas within larger areas) than it is for spot heating. With reasonably uniform radiation distribution in work or living areas, the exact orientation of the units is not important. Higher intensities of radiation may be desirable near walls with outside exposure. Radiation shields (reflective to infrared) fastened a few inches from the wall to allow free air circulation between the wall and shield are effective for frequently occupied work locations close to outside walls.

In full-building heating, units should be placed where their radiant and convective output best compensates for the structure's heat loss. The objective of a complete heating system is to provide a warm floor with low conductance to the heat sink beneath the floor. This thermal storage may permit cycling of units with standard controls.

TESTING INSTRUMENTS FOR RADIANT HEATING

In designing a radiant heating system, the calculation of radiant heat exchange may involve some untested assumptions. During field installation of the radiant heating equipment, the designer must test and adjust the equipment to ensure that it provides acceptable comfort conditions. The black globe thermometer and the directional radiometer can be used to evaluate the installation.

Black Globe Thermometer

The classic (Bedford) globe thermometer is a thin-walled, matte-black, hollow sphere with a thermocouple, thermistor, or thermometer placed at the center. It can directly measure $\bar{t}_r$, ERF, and t_o. When a black globe is in thermal equilibrium with its environment, the gain in radiant heat from various sources is balanced by the convective loss to ambient air. Thus, in terms of the globe's linear radiative and convective heat transfer coefficients, h_{rg} and h_{cg}, respectively, the heat balance at equilibrium is

$$h_{rg}(\bar{t}_{rg} - t_g) = h_{cg}(t_g - t_a) \qquad (18)$$

where $\bar{t}_{rg}$ is the mean radiant temperature measured by the globe and t_g is the temperature in the globe.

In general, the $\bar{t}_{rg}$ of Equation (18) equals the $\bar{t}_r$ affecting a person when the globe is placed at the center of the occupied space and when the radiant sources are distant from the globe.

The effective radiant flux measured by a black globe is

$$ERF_g = h_{rg}(\bar{t}_{rg} - t_a) \tag{19}$$

which is analogous to Equation (5) for occupants. From Equations (18) and (19), it follows that

$$ERF_g = (h_{rg} + h_{cg})(t_g - t_a) \tag{20}$$

If the ERF_g of Equation (20) is modified by the skin-clothing absorptance α_K and the shape f_{eff} of an occupant relative to the black globe, the corresponding ERF affecting the occupant is

$$ERF \text{ (for a person)} = f_{eff}\alpha_K ERF_g \tag{21}$$

where α_K is defined in the section on Geometry of Beam Heating and f_{eff}, which is defined after Equation (11), is approximately 0.71 and equals the ratio h_r/h_{rg}. The t_o affecting a person, in terms of t_g and t_a, is given by

$$t_o = Kt_g + (1 - K)t_a \tag{22}$$

where the coefficient K is

$$K = \alpha_K f_{eff}(h_{rg} + h_{cg})/(h_r + h_c) \tag{23}$$

Ideally, when K is unity, the t_g of the globe would equal the t_o affecting a person.

For an average comfortable equilibrium temperature of 77°F and noting that f_{eff} for the globe is unity, Equation (11) yields

$$h_{rg} = 1.06 \text{ Btu/h} \cdot \text{ft}^2 \cdot °F \tag{24}$$

and Equation (12) yields

$$h_{cg} = 0.345D^{-0.4}V^{0.5} \tag{25}$$

where

   D = globe diameter, in.
   V = air velocity, fpm

Equation (25) is Bedford's convective heat transfer coefficient for a 6 in.globe's convective loss, modified for D. For any radiating source below 1700°F, the ideal diameter of a sphere that makes $K = 1$ and that is independent of air movement is 8 in. (see Table 1). Table 1 shows the value of K for various values of globe diameter D and ambient air movement V. The table shows that the uncorrected temperature of the traditional 6 in. globe would overestimate the true $(t_o - t_a)$ difference by 6% for velocities up to 200 fpm, and the probable error of overestimating t_o by t_g uncorrected would be less than 0.9°F. Globe diameters between 6 and 8 in. are optimum for using the uncorrected t_g measurement for t_o. The exact value for K may be used for the smaller-sized globes when estimating t_o from t_g and t_a measurements. The value of $\bar{t}_r$ may be found by substituting Equations (24) and (25) in Equation (18), because $\bar{t}_r$ (person) is equal to $\bar{t}_{rg}$. The smaller the globe, the greater the variation in K caused by air movement. Globes with D greater than 8 in. will overestimate the importance of radiation gain versus convection loss.

For sources radiating at high temperatures (1300 to 10,000°F), the ratio α_m/α_g may be set near unity by using a pink-colored globe surface, whose absorptance for the sun is 0.7, a value similar to that of human skin and normal clothing (Madsen 1976).

In summary, the black globe thermometer is simple and inexpensive and may be used to determine $\bar{t}_r$ [Equation (18)] and the ERF

Table 1   Value of K for Various Air Velocities and Globe Diameters ($\alpha_g = 1$)

Air Velocity, fpm	Approximate Globe Diameter, in.			
	2	4	6	8
50	1.35	1.15	1.05	0.99
100	1.43	1.18	1.06	0.99
200	1.49	1.21	1.07	1.00
400	1.54	1.23	1.08	1.00
800	1.59	1.26	1.09	1.00

[Figure 1 and Equations (20) and (21)]. When the radiant heater temperature is less than 1700°F, the uncorrected t_g of a 6 to 8 in. black globe is a good estimate of the t_o affecting the occupants. A pink globe extends its usefulness to sun temperatures (10,000°F). A globe with a low mass and low thermal capacity is more useful because it reaches thermal equilibrium in less time.

Using the heat exchange principles described, many instruments of various shapes, heated and unheated, have been designed to measure acceptability in terms of t_o, $\bar{t}_r$, and ERF, as sensed by their own geometric shapes. Madsen (1976) developed an instrument that can determine the predicted mean vote (PMV) from the $(t_g - t_a)$ difference, as well as correct for clothing insulation, air movement, and activity (ISO 1984).

Directional Radiometer

The angle of acceptance (in steradians) in commercial radiometers allows the engineer to point the radiometer directly at a wall, floor, or high-temperature source and read the average temperature of that surface. Directional radiometers are calibrated to measure either the radiant flux accepted by the radiometer or the equivalent blackbody radiation temperature of the emitting surface. Many are collimated to sense small areas of body, clothing, wall, or floor surfaces. A directional radiometer allows rapid surveys and analyses of important radiant heating factors such as the temperature of skin, clothing surfaces, and walls and floors, as well as the radiation intensity I_K of heaters on the occupants. One radiometer for direct measurement of the equivalent radiant temperature has an angle of acceptance of 2.8° or 0.098 sr, so that at 1 ft, it measures the average temperature over a projected circle about 3/8 in. in diameter.

APPLICATIONS

When installing radiant heaters in specific applications, consider the following factors:

- Gas and electric high-temperature infrared heaters must not be placed where they could ignite flammable dust or vapors, or decompose vapors into toxic gases.
- Fixtures must be located with recommended clearances to ensure proper heat distribution. Stored materials must be kept far enough from the fixtures to avoid hot spots. Manufacturers' recommendations must be followed.
- Unvented gas heaters inside tight, poorly insulated buildings may cause excessive humidity with condensation on cold surfaces. Proper insulation, vapor barriers, and ventilation prevent these problems.
- Combustion-type heaters in tight buildings may require makeup air to ensure proper venting of combustion gases. Some infrared heaters are equipped with induced draft fans to relieve this problem.
- Some transparent materials may break due to uneven application of high-intensity infrared. Infrared energy is transmitted without loss from the radiator to the absorbing surfaces. The system must produce the proper temperature distribution at the absorbing surfaces. Problems are rarely encountered with glass 0.25 in. or less in thickness.

- Comfort heating with infrared heaters requires a reasonably uniform flux distribution in the occupied area. While thermal discomfort can be relieved in warm areas with high air velocity, such as on loading docks, the full effectiveness of a radiant heater installation is reduced by the presence of high air velocity.
- Radiant spot heating and zoning in large undivided areas with variable occupancy patterns provides localized heating just where and when people are working, which reduces the heating cost.

Low-, Medium-, and High-Intensity Infrared Applications

Low-, medium-, and high-intensity infrared equipment is used extensively in industrial, commercial, and military applications. This equipment is particularly effective in large areas with high ceilings, such as in buildings with large air volumes and in areas with high infiltration rates, large access doors, or large ventilation requirements.

Factories. Low-intensity radiant equipment suspended near the ceiling around the perimeter of facilities with high ceilings enhances the comfort of employees because it warms floors and equipment in the work area. For older uninsulated buildings, the energy cost for low-intensity radiant equipment is less than that of other heating systems. High-intensity infrared for spot heating and low-intensity infrared for zone temperature control effectively heat large unheated facilities.

Warehouses. Low- and high-intensity infrared are used for heating warehouses, which usually have a large volume of air, are often poorly insulated, and have high infiltration. Low-intensity infrared equipment is installed near the ceiling around the perimeter of the building. High-level mounting near the ceiling leaves floor space available for product storage. Both low- and high-intensity infrared are arranged to control radiant intensity and provide uniform heating at the working level and frost protection areas, which is essential for perishable goods storage.

Garages. Low-intensity infrared provides comfort for mechanics working near or on the floor. With elevated MRT in the work area, comfort is provided at a lower ambient temperature.

In winter, opening the large overhead doors to admit equipment for service causes a substantial entry of cold outdoor air. On closing the doors, the combination of reradiation from the warm floor and radiant heat warming the occupants (not the air) provides rapid recovery of comfort. Radiant energy rapidly warms the cold (perhaps snow-covered) vehicles. Radiant floor panel heating systems are also effective in garages.

Low-intensity equipment is suspended near the ceiling around the perimeter, often with greater concentration near overhead doors. High-intensity equipment is also used to provide additional heat near doors.

Aircraft Hangars. Equipment suspended near the roofs of hangars, which have high ceilings and large access doors, provides uniform radiant intensity throughout the working area. A heated floor is particularly effective in restoring comfort after an aircraft has been admitted. As in garages, the combination of reradiation from the warm floor and radiation from the radiant heating system provides rapid regain of comfort. Radiant energy also heats aircraft moved into the work area.

Greenhouses. In greenhouse applications, a uniform flux density must be maintained throughout the facility to provide acceptable growing conditions. In a typical application, low-intensity units are suspended near, and run parallel to, the peak of the greenhouse.

Outdoor Applications. Applications include loading docks, racetrack stands, outdoor restaurants, and under marquees. Low-, medium-, and high-intensity infrared are used in these facilities, depending on their layout and requirements.

Other Applications. Radiant heat may be used in a variety of large facilities with high ceilings, including churches, gymnasiums, swimming pools, enclosed stadiums, and facilities that are open to the outdoors. Radiant energy is also used to control condensation on surfaces such as large glass exposures. One example is at the Chicago/O'Hare Airport.

Low-, medium-, and high-intensity infrared are also used for other industrial applications, including process heating for component or paint drying ovens, humidity control for corrosive metal storage, and snow control for parking or loading areas.

Panel Heating and Cooling

Residences. Embedded pipe coil systems, electric resistance panels, and forced warm-air panel systems have all been used in residences. The embedded pipe coil system is most common, using plastic or rubber tubing in the floor slab or copper tubing in older plaster ceilings. These systems are suitable for conventionally constructed residences with normal amounts of glass. Lightweight hydronic metal panel ceiling systems have also been applied to residences, and prefabricated electric panels are advantageous, particularly in rooms that have been added on.

Office Buildings. A panel system is usually applied as a perimeter heating system. Panels are typically piped to provide exposure control with one riser on each exposure and all horizontal piping incorporated in the panel piping. In these applications, the air system provides individual room control. Perimeter radiant panel systems have also been installed with individual zone controls. However, this type of installation is usually more expensive and, at best, provides minimal energy savings and limited additional occupant comfort. Radiant panels can be used for cooling as well as heating. Cooling installations are generally limited to retrofit or renovation jobs where ceiling space is insufficient for the required duct sizes. In these installations, the central air supply system provides ventilation air, dehumidification, and some sensible cooling. Water distribution systems using the two- and four-pipe concept may be used. Hot water supply temperatures are commonly reset by outside temperature, with additional offset or flow control to compensate for solar load. Panel systems are readily adaptable to accommodate most changes in partitioning. Electric panels in lay-in ceilings have been used for full perimeter heating.

Schools. In all areas except gymnasiums and auditoriums, panels are usually selected for heating only, and may be used with any type of approved ventilation system. The panel system is usually sized to offset the transmission loads plus any reheating of the air. If the school is air conditioned by a central air system and has perimeter heating panels, single-zone piping may be used to control the panel heating output, and the room thermostat modulates the supply air temperature or volume. Heating and cooling panel applications are similar to those in office buildings. Panel heating and cooling for classroom areas has no mechanical equipment noise to interfere with instructional activities.

Hospitals. The principal application of heating and cooling radiant panels has been for hospital patient rooms. Perimeter radiant heating panels are typically applied in other areas of hospitals. Compared to conventional systems, radiant heating and cooling systems are well suited to hospital patient rooms because they (1) provide a draft-free, thermally stable environment, (2) have no mechanical equipment or bacteria and virus collectors, and (3) do not take up space in the room. Individual room control is usually achieved by throttling the water flow through the panel. The supply air system is often 100% outdoor air; minimum air quantities delivered to the room are those required for ventilation and exhaust of the toilet room and soiled linen closet. The piping system is typically a four-pipe design. Water control valves should be installed in corridors so that they can be adjusted or serviced without entering the patient rooms. All piping connections above the ceiling should be soldered or welded and thoroughly tested. If cubicle tracks are applied to the ceiling surface, track installation should be coordinated with the radiant ceiling. Security panel ceilings are often used in areas occupied by mentally disturbed patients so that equipment cannot be damaged by a patient or used to inflict injury.

Swimming Pools. A partially clothed person emerging from a pool is very sensitive to the thermal environment. Panel heating systems are well suited to swimming pool areas. Floor panel temperatures must be controlled so they do not cause foot discomfort. Ceiling panels are generally located around the perimeter of the pool, not directly over the water. Panel surface temperatures are higher to compensate for the increased ceiling height and to produce a greater radiant effect on partially clothed bodies. Ceiling panels may also be placed over windows to reduce condensation.

Apartment Buildings. For heating, pipe coils are embedded in the masonry slab. The coils must be carefully positioned so as not to overheat one apartment while maintaining the desired temperature in another. The slow response of embedded pipe coils in buildings with large glass areas may be unsatisfactory. Installations for heating and cooling have been made with pipes embedded in hung plaster ceilings. A separate minimum-volume dehumidified air system provides the necessary dehumidification and ventilation for each apartment. The application of electric resistance elements embedded in floors or behind a skim coat of plaster at the ceiling has increased. Electric panels are easy to install and simplify individual room control.

Industrial Applications. Panel systems are widely used for general space conditioning of industrial buildings in Europe. For example, the walls and ceilings of an internal combustion engine test cell are cooled with chilled water. Although the ambient air temperature in the space reaches up to 95°F, the occupants work in relative comfort when 55°F water is circulated through the ceiling and wall panels.

Other Buildings. Metal panel ceiling systems can be operated as heating systems at elevated water temperatures and have been used in airport terminals, convention halls, lobbies, and museums, especially those with large glass areas. Cooling may also be applied. Because radiant energy travels through the air without warming it, ceilings can be installed at any height and remain effective. One particularly high ceiling installed for a comfort application is 50 ft above the floor, with a panel surface temperature of approximately 285°F for heating. The ceiling panels offset the heat loss from a single-glazed, all-glass wall.

The high lighting levels in television studios make them well suited to panels that are installed for cooling only and are placed above lighting to absorb the radiation and convection heat from the lights and normal heat gains from the space. The panel ceiling also improves the acoustical properties of the studio.

Metal panel ceiling systems are also installed in minimum and medium security jail cells and in facilities where disturbed occupants are housed. The ceiling is strengthened by increasing the gage of the ceiling panels, and security clips are installed so that the ceiling panels cannot be removed. Part of the perforated metal ceiling can be used for air distribution.

New Techniques. The introduction of thermoplastic and rubber tubing and new design techniques have improved radiant panel heating and cooling equipment. The systems are energy-efficient and use low water temperatures available from solar collectors and heat pumps (Kilkis 1993). Metal radiant panels can be integrated into the ceiling design to provide a narrow band of radiant heating around the perimeter of the building. These new radiant systems are more attractive, provide more comfortable conditions, operate more efficiently, and have a longer life than some baseboard or overhead air systems.

SYMBOLS

A_D = total DuBois surface area of person, ft^2
A_{eff} = effective radiating area of person, ft^2
A_p = projected area of occupant normal to the beam, ft^2
clo = unit of clothing insulation equal to 0.880 $\text{ft}^2 \cdot °\text{F} \cdot \text{h/Btu}$
D = diameter of globe thermometer, in.
d = distance of beam heater from occupant, ft
ERF = effective radiant flux (person), $\text{Btu/h} \cdot \text{ft}^2$
ERF_f = radiant flux caused by heated floor on occupant, $\text{Btu/h} \cdot \text{ft}^2$

ERF_g = effective radiant flux (globe), $\text{Btu/h} \cdot \text{ft}^2$
$F_{p\text{-}f}$ = angle factor between occupant and heater floor
f_{eff} = ratio of radiating surface (person) to its total area (DuBois)
H_m = net metabolic heat loss from body surface, $\text{Btu/h} \cdot \text{ft}^2$
h = combined heat transfer coefficient (person), $\text{Btu/h} \cdot \text{ft}^2 \cdot °\text{F}$
h_c = convective heat transfer coefficient for person, $\text{Btu/h} \cdot \text{ft}^2 \cdot °\text{F}$
h_{cg} = convective heat transfer coefficient for globe, $\text{Btu/h} \cdot \text{ft}^2 \cdot °\text{F}$
h_r = linear radiative heat transfer coefficient (person), $\text{Btu/h} \cdot \text{ft}^2 \cdot °\text{F}$
h_{rg} = linear radiative heat transfer coefficient for globe, $\text{Btu/h} \cdot \text{ft}^2 \cdot °\text{F}$
I_K = irradiance from beam heater, $\text{Btu/h} \cdot \text{sr}$
K = coefficient that relates t_a and t_g to t_o [Equation (22)]
K = subscript indicating absolute irradiating temperature of beam heater, °R
L = fixture length, ft
met = unit of metabolic energy equal to 18.4 $\text{Btu/h} \cdot \text{ft}^2$
t_a = ambient air temperature near occupant, °F
t_f = floor temperature, °F
t_g = temperature in globe, °F
t_o = operative temperature, °F
$\bar{t}_r$ = mean radiant temperature affecting occupant, °F
$\bar{t}_{rf}$ = linearized $\bar{t}_r$ caused by floor and unheated walls on occupant, °F
t_{sf} = exposed surface temperature of occupant, °F
t_{uo} = operative temperature of unheated workspace, °F
V = air velocity, fpm
W = width of a square equivalent to the projected area of a beam of angle Ω steradians at a distance d, ft
α = relative absorptance of skin-clothing surface to that of matte black surface
α_g = absorptance of globe
α_K = absorptance of skin-clothing surface at emitter temperature
α_m = absorptance of skin-clothing surface at emitter temperatures above 1700°F
β = elevation angle of beam heater, degrees
Ω = radiant beam width, sr
ϕ = azimuth angle of heater, degrees
σ = Stefan-Boltzmann constant = 0.1713×10^{-8} $\text{Btu/h} \cdot \text{ft}^2 \text{t} \cdot °\text{R}^4$

REFERENCES

ASHRAE. 1992. Thermal environmental conditions for human occupancy. ANSI/ASHRAE *Standard* 55-1992.

Athienitis, A.K. and J.D. Dale. 1987. A study of the effects of window night insulation and low emissivity coating on heating load and comfort. *ASHRAE Transactions* 93(1A):279-94.

Boyd, R.L. 1962. Application and selection of electric infrared comfort heaters. *ASHRAE Journal* 4(10):57.

Buckley, N.A. and T.P. Seel. 1987. Engineering principles support an adjustment factor when sizing gas-fired low-intensity infrared equipment. *ASHRAE Transactions* 93(1):1179-91.

Fanger, P.O. 1973. *Thermal comfort*. McGraw-Hill, New York.

Fanger, P.O., L. Banhidi, B.W. Olesen, and G. Langkilde. 1980. Comfort limits for heated ceiling. *ASHRAE Transactions* 86(2):141-56.

Gagge, A.P., G.M. Rapp, and J.D. Hardy. 1967. The effective radiant field and operative temperature necessary for comfort with radiant heating. *ASHRAE Transactions* 73(1):I.2.1-9; and *ASHRAE Journal* 9(5):63-66.

ISO. 1994. Moderate thermal environments—Determination of the PMV and PPD indices and specification of the conditions for thermal comfort. *Standard* 7730-1984. International Standard Organization, Geneva.

Kilkis, B. 1990. Panel cooling and heating of buildings using solar energy. ASME Winter Meeting: Solar Energy in the 1990s. Vol. 10:1-7.

Kilkis, B. 1992. Enhancement of heat pump performance using radiant floor heating systems. ASME Winter Meeting: Advanced Energy Systems, Recent Research in Heat Pump Design, Analysis, and Application. Vol. 28:119-27.

Kilkis, B. 1993. Radiant ceiling cooling with solar energy: Fundamentals, modeling, and a case design. *ASHRAE Transactions* 99(2):521-33.

Madsen, T.L. 1976. Thermal comfort measurements. *ASHRAE Transactions* 82(1):60-70.

Rapp, G.M. and A.P. Gagge. 1967. Configuration factors and comfort design in radiant beam heating of man by high temperature infrared sources. *ASHRAE Transactions* 73(3):1.1-1.8.

Scholander, P.E. 1958. Cold adaptation in the Australian aborigines. *Journal of Applied Physiology* 13:211-18.

Stevens, J.C., L.E. Marks, and A.P. Gagge. 1969. The quantitative assessment of thermal comfort. *Environmental Research* 2:149-65.

Wilkins, C.K. and R. Kosonen. 1992. Cool ceiling system: A European air-conditioning alternative. *ASHRAE Journal* 34(8):41-5.

Zmeureanu, R., P.P. Fazio, and F. Haghighat. Thermal performance of radiant heating panels. *ASHRAE Transactions* 94(2):13-27.

CHAPTER 53

SEISMIC AND WIND RESTRAINT DESIGN

ALMOST all inhabited areas of the world are susceptible to the damaging effects of either earthquakes or wind. Restraints that are designed to resist one may not be adequate to resist the other. Consequently, when exposure to either earthquake or wind loading is a possibility, strength of equipment and attachments should be evaluated for both conditions.

Earthquake damage to inadequately restrained HVAC&R equipment can be extensive. Mechanical equipment that is blown off the support structure can become a projectile, threatening life and property. The cost of properly restraining the equipment is small compared to the high costs of replacing or repairing damaged equipment, or compared to the cost of building down-time due to damaged facilities.

Design and installation of seismic and wind restraints has the following primary objectives:

- Life safety to reduce the threat to life
- Reduce long-term costs due to equipment damage and the resultant down time

This chapter covers the design of restraints to limit the movement of equipment and to keep the equipment captive during an earthquake or during extreme wind loading. Seismic restraints and seismic isolators do not reduce the forces transmitted to the equipment to be restrained. Instead, properly designed and installed seismic restraints and seismic isolators have the necessary strength to withstand the imposed forces. However, equipment that is to be restrained must also have the necessary strength to remain attached to the restraint. Equipment manufacturers should review structural aspects of the design in the areas of attachment to ensure the equipment will remain attached to the restraint.

For mechanical systems, analysis of seismic and wind loading conditions is typically a static analysis, and conservative safety factors are applied to reduce the complexity of earthquake and wind loading response analysis and evaluation. Three aspects are considered in a properly designed restraint system.

1. *Attachment of equipment to restraint.* The equipment must be positively attached to the restraint, and must have sufficient strength to withstand the imposed forces, and to transfer the forces to the restraint.

2. *Restraint design.* Strength of the restraint must also be sufficient to withstand the imposed forces. This should be determined by the manufacturer by tests and/or analyses.

3. *Attachment of restraint to substructure.* Attachment may be by means of bolts, welds or concrete anchors. The sub structure must be capable of surviving the imposed forces.

The preparation of this chapter is assigned to TC 2.7, Seismic Restraint Design.

SEISMIC RESTRAINT DESIGN

Most seismic requirements adopted by local jurisdictions in North America are based on model codes developed by the International Conference of Building Officials (ICBO), Building Officials and Code Administrators International (BOCA), the Southern Building Code Conference, Inc. (SBCCI), and the National Building Code of Canada (NBCC); or on the requirements of the National Earthquake Hazards Reduction Program (NEHRP). The model code bodies are working through the International Code Council (ICC) to unify their model codes into the International Building Code (IBC) by the year 2000. Local building officials must be contacted for specific requirements that may be more stringent than those presented in this chapter and to determine if the unified specification has been invoked.

Other sources of seismic restraint information include

- *Seismic Restraint Manual: Guidelines for Mechanical Systems*, published by SMACNA (1998), includes seismic restraint information for mechanical equipment subjected to seismic forces of up to 0.48g.
- The National Fire Protection Association (NFPA) has developed standards on restraint design for fire protection systems.
- U.S. Department of Energy DOE 6430.1A and ASME AG-1 cover restraint design for nuclear facilities.
- *Technical Manual* TM 5-809-10, published by the United States Army, Navy, and Air Force (1992), also provides guidance for seismic restraint design.

In seismically active areas where governmental agencies regulate the earthquake-resistive design of buildings (e.g., California), the HVAC engineer usually does not prepare the code-required seismic restraint calculations. The HVAC engineer selects all the heating and cooling equipment and, with the assistance of the acoustical engineer (if the project has one), selects the required vibration isolation devices. The HVAC engineer specifies these devices and calls for shop drawing submittals from the contractors, but the manufacturer employs a registered engineer to design and detail the installation. The HVAC engineer reviews the shop design and details the installation, reviews the shop drawings and calculations, and obtains the approval of the architect and structural engineer before issuance to the contractors for installation.

Anchors for tanks, brackets, and other equipment supports that do not require vibration isolation are designed by the building's structural engineer, or by the supplier of the seismic restraints, based on layout drawings prepared by the HVAC engineer. The building officials maintain the code-required quality control over the design by requiring that all building design professionals are registered (licensed) engineers. Upon completion of installation, the supplier of the seismic restraints, or a qualified representative, should inspect the installation and verify that all restraints are installed properly and in compliance with specifications.

TERMINOLOGY

Base plate thickness. Thickness of the equipment bracket fastened to the floor.

Effective shear force (V_{eff}). Maximum shear force of one seismic restraint or tie-down bolt.

Effective tension force (T_{eff}). Maximum tension force or pullout force on one seismic restraint or tie-down bolt.

Equipment. Any HVAC&R component that must be restrained from movement during an earthquake.

Fragility level. Maximum lateral acceleration force that the equipment is able to withstand. This data may be available from the equipment manufacturer and is generally on the order of four times the acceleration of gravity ($4g$) for mechanical equipment.

Resilient support. An active seismic device (such as a spring with a bumper) to prevent equipment from moving more than a specified amount.

Response spectra. Relationship between the acceleration response of the ground and the peak acceleration of the earthquake in a damped single degree of freedom at various frequencies. The ground motion response spectrum varies with soil conditions.

Rigid support. Passive seismic device used to restrict any movement.

Shear force (V). Force generated at the plane of the seismic restraints, acting to cut the restraint at the base.

Seismic restraint. Device designed to withstand an earthquake.

Seismic zones. The geographical location of a facility determines its seismic zone, as given in the Uniform Building Code or International Building Code.

Snubber. Device made of steel-housed resilient bushings arranged to prevent equipment from moving beyond an established gap.

Tension force (T). Force generated by overturning moments at the plane of the seismic restraints, acting to pull out the bolt.

CALCULATIONS

The calculations presented here assume that the equipment support is an integrated resilient support and restraint device. When the two functions of resilient support and motion restraint are separate or act separately, additional spring loads may need to be added to the anchor load calculation for the restraint device. Internal loads within integrated devices are not addressed in this chapter. Such devices must be designed to withstand the full anchorage loads plus any internal spring loads.

Both static and dynamic analyses reduce the force generated by an earthquake to an equivalent static force, which acts in a horizontal direction at the component's center of gravity. The resulting overturning moment is resisted by shear and tension (pullout) forces on the tie-down bolts. Static analysis is used for both rigid-mounted and resilient-mounted equipment.

Table 1   Coefficients for Mechanical Components

Mechanical and Electrical Component or Element	a_p	R_p
General Mechanical		
Boilers and furnaces	1.0	2.5
Piping		
High deformability elements and attachments	1.0	3.5
Limited deformability elements and attachments	1.0	2.5
Low deformability elements or attachments	1.0	1.25
HVAC Equipment		
Vibration isolated	2.5	2.5
Non-vibration isolated	1.0	2.5
Mounted in line with ductwork	1.0	2.5

Dynamic Analysis

A dynamic analysis is based on site-specific ground motions developed by a geotechnical or soils engineer. A common approach assumes an elastic response spectrum. The results of the dynamic analysis are then scaled up or down as a percentage of the total lateral force obtained from the static analysis performed on the building. The scaling coefficient is established by the UBC or by the governing building official. The scaled acceleration calculated by the structural engineer at any level in the structure can be determined and compared to the force calculated in Equation (1). The greater of the two should be used in the anchor design. The horizontal force factor C_p should be multiplied by a factor of two in either case, as shown in Table 5.

Static Analysis as Defined in the International Building Code

The final draft of the International Building Code (ICC 1998) specifies a design lateral force F_p for nonstructural components as

$$F_p = (0.4a_p S_{DS} W_p)\frac{I_p}{R_p}\left(1 + 2\frac{z}{h}\right) \qquad (1)$$

but F_p need not be greater than $F_p = 1.6 S_{DS} I_p W_p$ $\qquad (2)$

nor less than $F_p = 0.35 S_{DS} I_p W_p$ $\qquad (3)$

A vertical force is specified as

$$F_{pv} = 0.2 S_{DS} W_p \qquad (4)$$

where

a_p = component amplification factor in accordance with Table 1

S_{DS} = design spectral response acceleration at short periods as determined by $S_{DS} = 2F_a S_s/3$. S_s is the mapped spectral accelerations from Figure 1 and $0.8 \le F_a \le 2.5$. F_a is a function of the site soil characteristics and must be determined in consultation with either the project geotechnical (soils) or structural engineer. Values for F_a are given in Table 2.

R_p = component response modification factor in accordance with Table 1. *Note*: If expansive anchors, chemical anchors, or shallow embedded cast-in-place anchors are used, then $R_p = 1.25$.

I_p = component importance factor in accordance with Table 3.

$1 + 2z/h$ = height amplification factor where z is the height of attachment in the structure and h is the average height of the roof above grade. The value of z should not be taken as less than 0.

W_p = mass of equipment, which includes all items attached or contained in the equipment

Table 2   Values of Site Coefficient F_a as Function of Site Class and Mapped Spectral Response Acceleration at 1 s Period (S_s)

Site Class	Soil Profile Name	Mapped Spectral Response Acceleration At Short Periods[a]				
		$S_s \le 0.25$	$S_s = 0.50$	$S_s = 0.75$	$S_s = 1.00$	$S_s \ge 1.25$
A	Hard rock	0.8	0.8	0.8	0.8	0.8
B	Rock	1.0	1.0	1.0	1.0	1.0
C	Very dense soil and soft rock	1.2	1.2	1.1	1.0	1.0
D	Stiff soil profile	1.6	1.4	1.2	1.1	1.0
E	Soft soil profile	2.5	1.7	1.2	0.9	a
F		See IBC Table 1615.1.1 and Note b				

[a] Use straight line interpolation for intermediate values of mapped spectral acceleration at short period S_s.

[b] Site specific geotechnical investigation and dynamic site response analyses shall be performed to determine appropriate values.

Table 3   Seismic Group and Occupancy Importance Factor

Seismic Group	Occupancy Importance Factor	Nature of Occupancy
I	1.0	All occupancies except those listed below
II	1.25	1. Assembly Group A in which 300 or more people congregate in one area 2. Educational Group E with an occupant load greater than 250 3. Day care centers in Educational Group E with a capacity greater than 150 4. Institutional Group I-2 (medical, etc.) with an occupant load greater than 50, not otherwise designated a Seismic Group III structure 5. Institutional Group I-3 inhabited by more than five persons who are under restraint or security 6. Any other occupancy with an occupant load greater than 5000 7. Power generating stations and other public utility facilities not included in Seismic Group III and required for continued operation 8. Water treatment facilities required for primary treatment and disinfection for potable water 9. Waste water treatment facilities required for primary treatment.
III	1.5	1. Fire, rescue, and police stations 2. Institutional Group I-2 facilities that are hospitals 3. Designated Institutional Group I-2 facilities having surgery or emergency treatment facilities 4. Designated emergency preparedness centers 5. Designated emergency operation centers 6. Designated emergency shelters 7. Power generating stations or other utilities required as emergency backup facilities for seismic group III facilities 8. Emergency vehicle garages and emergency aircraft hangars 9. Designated communication centers 10. Aviation control towers and air traffic control centers 11. Structures containing highly toxic materials 12. Water treatment facilities required to maintain water pressure for fire suppression

Table 4   Seismic Zone Factor Z

Zone	Z
1	0.075
2A	0.15
2B	0.20
3	0.30
4	0.40

Table 5   Horizontal Force Factor C_p

Equipment or Nonstructural Components	1
Mechanical equipment, plumbing, and electrical equipment and associated piping rigidly mounted	0.75
All equipment resiliently mounted (maximum 2.0)	2

Note: Stacks and tanks should be evaluated for compliance with the applicable codes by a qualified engineer.

Static Analysis as Defined in 1994 Uniform Building Code

The total design lateral seismic force is given as

$$F_p = ZI_pC_pW_p \tag{5}$$

where

F_p = total design lateral seismic force
Z = seismic zone factor
I_p = importance factor (set equal to 1.5 for equipment)
C_p = horizontal force factor
W_p = weight of equipment

Figure 2 and Table 6 may be used to determine the seismic zone. The seismic zone factor Z can then be determined from Table 4. *Seismic Design for Buildings* (Army, Navy, and Air Force 1992) can also be used to determine the seismic zone.

The importance factor I_p from the UBC (ICBO 1994) ranges from 1 to 1.5, depending on the building occupancy and hazard level. For equipment, I_p should be conservatively set at 1.5.

The horizontal force factor C_p is determined from Table 5 based on the type of equipment, the tie-down configuration, and the type of base.

The weight W_p of the equipment should include all the items attached or contained in the equipment.

APPLYING STATIC ANALYSIS USING 1994 UBC

The forces acting on the equipment are the lateral and vertical forces resulting from the earthquake, the force of gravity, and the forces of the restraint holding the equipment in place. The analysis assumes the equipment does not move during an earthquake, thus, the sum of the forces and moments must be zero. When calculating the overturning moment, the vertical component F_{pV} at the center of gravity is given as

$$F_{pv} = F_p/3 \tag{6}$$

The forces of the restraint holding the equipment in position include shear and tension forces. It is important to determine the number of bolts that are affected by the earthquake forces. The direction of the lateral force should be evaluated in both horizontal directions as shown in Figure 3. All bolts or as few as a single bolt may be affected.

Figure 3 shows a typical rigid floor mount installation of a piece of equipment. To calculate the shear force, the sum of the forces in the horizontal plane is

$$0 = F_p - V \tag{7}$$

The effective shear force V_{eff} is

$$V_{eff} = F_p/N_{bolt} \tag{8}$$

where N_{bolt} = the number of bolts in shear.

Fig. 1   Maximum Considered Earthquake Ground Motion for the United States
0.2 s spectral response acceleration (% *g*) (5% of critical damping) Site Class B

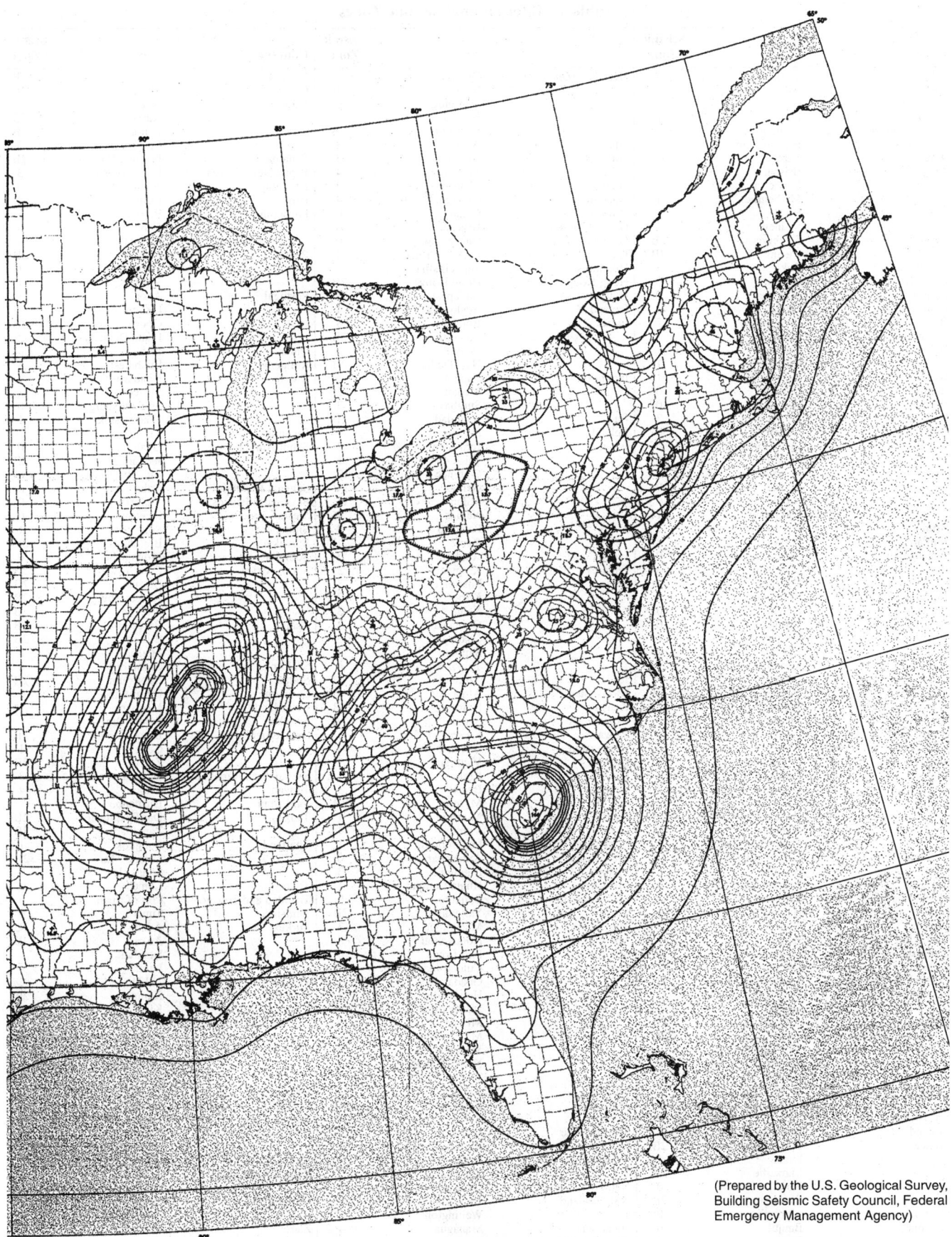

(Prepared by the U.S. Geological Survey,
Building Seismic Safety Council, Federal
Emergency Management Agency)

Fig. 1   Maximum Considered Earthquake Ground Motion for the United States (*Continued*)

0.2 s spectral response acceleration (% *g*) (5% of critical damping) Site Class B

Table 6 International Seismic Zones

Country	City	Seismic Zone	Country	City	Seismic Zone	Country	City	Seismic Zone
Albania	Tirana	3	Germany (*Continued*)	Dusseldorf	1	Norway	Oslo	2B
Algeria	Algiers	4		Frankfurt	1	Oman	Muscat	2B
	Oran	4		Hamburg	0	Pakistan	Islamabad	4
Angola	Luanda	0		Munich	1		Karachi	2B
Antigua and Barbuda	St. Johns	3		Stuttgart	2B		Lahore	2B
Argentina	Buenos Aires	0	Ghana	Accra	3		Peshawar	3
Armenia	Yerevan	3	Greece	Athens	3	Panama	Panama City	2B
Australia	Brisbane	1		Thessaloniki	4	Papua New Guinea	Port Moresby	3
	Canberra	1	Grenada	St. George's	3	Paraguay	Asuncion	0
	Melbourne	1	Guatemala	Guatemala City	4	Peru	Lima	4
	Perth	1	Guinea	Conakry	0	Philippines	Baguio	3
	Sydney	1	Guinea-Bissau	Bissau	0		Cebu	4
Austria	Salzburg	2B	Guyana	Georgetown	0		Manila	4
	Vienna	2B	Haiti	Port-au-Prince	3	Poland	Krakow	2B
Azerbaijan	Baku	3	Honduras	Tegucigalpa	3		Poznan	1
Bahamas	Nassau	0	Hong Kong	Hong Kong	2B		Warsaw	1
Bahrain	Manama	0	Hungary	Budapest	2B	Portugal	Azor	3
Bangladesh	Dhaka	3	Iceland	Reykjavik	4		Lisbon	3
Barbados	Bridgetown	3	India	Bombay	3		Oporto	2B
Belarus	Minsk	1		Calcutta	2B		Ponta Delgada	3
Belgium	Antwerp	1		Madras	1	Qatar	Doha	0
	Brussels	1		New Delhi	2B	Romania	Bucharest	3
Belize	Belize City	1	Indonesia	Jakarta	3	Russia	Khabarovsk	1
Benin	Cotonou	0		Medan	3		Moscow	1
Bermuda	Hamilton	0		Surabaya	3		St. Petersburg	0
Bolivia	La Paz	3	Iraq	Baghdad	2B		Vladivostok	1
Botswana	Gaborone	0	Ireland	Dublin	0	Rwanda	Kigali	3
Brazil	Belo Horizonte	0	Israel	Jerusalem	3	Saudi Arabia	Dhahran	0
	Brasilia	0		Tel Aviv	1		Jeddah	2B
	Porto Alegre	0	Italy	Florence	3		Riyadh	0
	Recife	0		Genoa	2B	Senegal Republic	Dakar	0
	Rio de Janeiro	0		Milan	2B	Seychelles Islands	Victoria	0
	São Paulo	1		Naples	2B	Sierra Leone	Freetown	0
Brunei	Bandar Seri Begawan	1		Palermo	4	Singapore	Singapore	1
Bulgaria	Sofia	3		Rome	2B	Slovakia	Bratislava	2B
Burkina Faso	Ouagadougou	0	Ivory Coast	Abidjan	3	Somalia	Mogadishu	0
Burma	Mandalay	3	Jamaica	Kingston	3	South Africa	Cape Town	2B
	Rangoon	2B	Japan	Fukuoka	3		Durban	2B
Burundi	Bujumbura	3		Kobe	3		Johannesburg	2B
Cameroon	Douala	0		Naha	3		Pretoria	2B
	Yaounde	0		Okinawa	3	Spain	Barcelona	2B
Canada	Calgary	1		Osaka	3		Bilbao	2B
	Halifax	1		Sapporo	3		Madrid	0
	Montreal	2A		Tokyo	4	Sri Lanka	Colombo	0
	Ottawa	2A	Jordan	Amman	3	Sudan	Khartoum	2B
	Quebec	3	Kazakhstan	Alma-Ata	4	Suriname	Paramaribo	0
	Toronto	1	Kenya	Nairobi	2B	Swaziland	Mbabane	2B
	Vancouver	3	Korea	Seoul	2A	Sweden	Stockholm	0
Cape Verde	Praia	0	Kuwait	Kuwait	1	Switzerland	Bern	2B
Central African Republic	Bangui	0	Kyrgyzstan	Bishkek	4		Geneva	1
Chad Republic	N'Djamena	0	Laos	Vientiane	1		Zurich	2B
Chile	Santiago	4	Latvia	Riga	1	Syrian Arab Republic	Damascus	3
China	Beijing (Peking)	3	Lebanon	Beirut	3	Taiwan	Taipei	4
	Chengdu	3	Lesotho	Maseru	2B	Tajikistan	Dushanbe	4
	Guangzhou (Canton)	2B	Liberia	Monrovia	1	Tanzania	Dar Es Salaam	2B
	Shanghai	2B	Lithuania	Vilnius	1		Zanzibar	2B
	Shenyang (Mukden)	4	Luxembourg	Luxembourg	1	Thailand	Bangkok	1
Colombia	Barranquilla	2B	Madagascar	Antananarivo	0		Chiang Mai	2B
	Bogota	3	Malawi	Lilongwe	3		Songkhla	0
Congo	Brazzaville	0	Malaysia	Kuala Lumpur	1	Togo	Lome	1
Costa Rica	San Jose	3	Mali Republic	Bamako	0	Trinidad and Tobago	Port of Spain	3
Cuba	Havana	1	Malta	Valletta	2B	Tunisia	Tunis	3
Cyprus	Nicosia	3	Martinique	Martinique	3	Turkey	Adana	2B
Czech Republic	Prague	1	Mauritania	Nouakchott	0		Ankara	2B
Denmark	Copenhagen	1	Mauritius	Port Louis	0		Istanbul	4
Djibouti	Djibouti	3	Mexico	Cuidad Juarez	2B		Izmir	4
Dominican Republic	Santo Domingo	3		Guadalajara	3	Turkmenistan	Ashkhabad	4
Ecuador	Guayaquil	3		Hermosillo	3	Uganda	Kampala	2B
	Quito	3		Matamoros	0	Ukraine	Kiev	1
Egypt	Alexandria	2B		Merida	0	United Arab Emirates	Abu Dhabi	0
	Cairo	2B		Mexico City	3		Dubai	0
El Salvador	San Salvador	4		Monterrey	0	United Kingdom	Belfast	0
Equatorial Guinea	Malabo	0		Nuevo Laredo	0		Edinburgh	1
Estonia	Tallinn	2A		Tijuana	3		London	1
Ethiopia	Addis Ababa	3	Moldova	Kishinev	2B	Uruguay	Montevideo	0
	Asmara	3	Morocco	Casablanca	2B	Uzbekistan	Tashkent	3
Fiji Islands	Suva	3		Rabat	2B	Vatican City	Vatican City	2B
Finland	Helsinki	1	Mozambique	Maputo	2B	Venezuela	Caracas	3
France	Bordeaux	2B	Nepal	Kathmandu	3		Maracaibo	2B
	Lyon	1	Netherlands	Amsterdam	0	Vietnam	Ho Chi Minh City	0
	Marseille	2B		The Hague	0	Yemen Arab Republic	Aden City	3
	Paris	0	Netherlands Antilles	Curacao	3		Sana	3
	Strasbourg	2B	New Zealand	Auckland	2B	Yugoslavia	Belgrade	2A
Gabon	Libreville	0		Wellington	4		Zagreb	3
Gambia	Banjul	0	Nicaragua	Managua	4	Zaire	Kinshasa	0
Georgia	Tbilisi	3	Niger Republic	Niamey	0		Lubumbashi	2B
Germany	Berlin	0	Nigeria	Kaduna	0	Zambia	Lusaka	2B
	Bonn	1		Lagos	0	Zimbabwe	Harare	2B

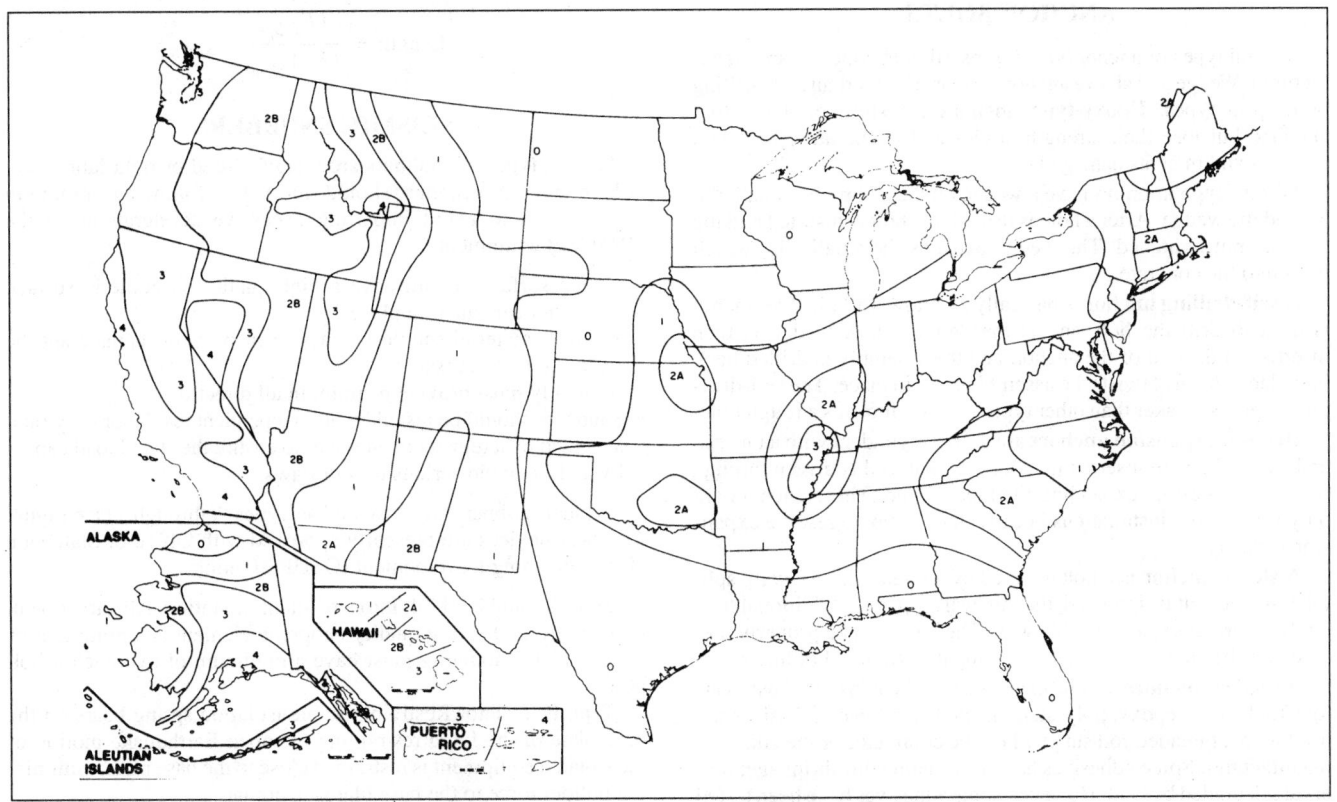

Fig. 2 Seismic Zone Map of the United States

(Reproduced from the 1994 edition of the *Uniform Building Code* with permission of the publisher, the International Conference of Building Officials.)

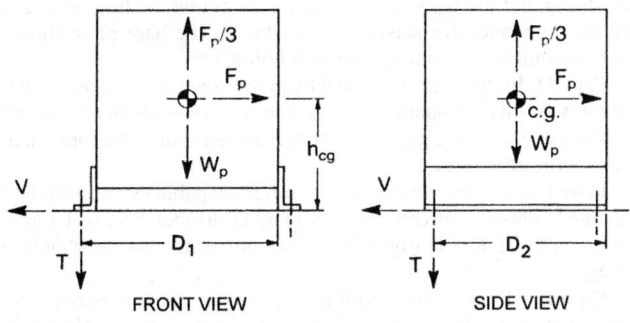

Fig. 3 Equipment with Rigidly Mounted Structural Bases

The restraints shown in Figure 3 have two bolts on each side, so that four bolts are in shear. To calculate the tension force, the sum of the moments for overturning are as follows:

$$F_p h_{cg} - (W_p - F_p/3)(D_1/2) - TD_1 = 0$$

Thus,

$$T = \frac{1}{D_1}\left[F_p h_{cg} - \left(W_p - \frac{F_p}{3}\right)\left(\frac{D_1}{2}\right)\right] \qquad (9)$$

For the example shown in Figure 3, two bolts are in tension. The effective tension force T_{eff}, where overturning affects only one side, is

$$T_{eff} = T/N_{bolt} \qquad (10)$$

The shear and tension forces (V and T) should be calculated independently for both axes as shown in the front and side views. The worst case governs seismic restraint design; however, the direction of seismic loading that governs the design in not always obvious.

For example, if three bolts were installed on each side, the lateral force applied as shown in the side view affect six bolts in shear and a minimum of two in tension. The lateral force applied as shown on the front view results in six bolts affected in shear and three in tension. Also, D_1 and D_2 are different for each axis.

Equations (6), (7), and (8) may be applied to ceiling-mounted equipment. Equation (9) must be modified to Equation (11) because the mass of the equipment adds to the overturning moment. Summing the moments determines the effective tension force as

$$T = \frac{1}{D}\left[F_p h_{cg} + \left(W_p + \frac{F_p}{3}\right)\left(\frac{D}{2}\right)\right] \qquad (11)$$

Interaction Formula. To evaluate the combined effective tension and shear forces that act simultaneously on the bolt, the following equation applies:

$$\left(\frac{T_{eff}}{T_{allow}}\right)^{5/3} + \left(\frac{V_{eff}}{V_{allow}}\right)^{5/3} \leq 1.0 \qquad (12)$$

The allowable forces T_{allow} and V_{allow} are the generic allowable capacities given in Table 7 for wedge-type anchor bolts.

Table 7 Typical Allowable Loads for Wedge-Type Anchors

Diameter, in.	T_{allow}, lb	V_{allow}, lb
0.5	600	1200
0.625	900	2200
0.75	1350	3000

Notes:
1. The allowable tensile forces are for installations without special inspection (torque test) and may be doubled if the installation is inspected.
2. Additional tension and shear values may be obtained from published ICBO reports.

ANCHOR BOLTS

Several types of anchor bolts for insertion in concrete are manufactured. Wedge and sleeve anchors perform better than self-drilling or drop-in types. **Epoxy-type anchors** are stronger than other anchors, but lose their strength at elevated temperatures (i.e., on rooftops and in areas damaged by fire).

Wedge-type anchors have a wedge on the end with a small clip around the wedge. After a hole is drilled, the bolt is inserted and the external nut tightened. The wedge expands the small clip, which bites into the concrete.

A **self-drilling anchor** is basically a hollow drill bit. The anchor is used to drill the hole and is then removed. A wedge is then inserted on the end of the anchor, and the assembly is drilled back into place; the drill twists the assembly fully in place. The self-drilling anchor is weaker than other types because it forms a rough hole.

Drop-in expansion anchors are hollow cylinders with a tapered end. After they are inserted in a hole, a small rod is driven through the hollow portion, expanding the tapered end. These anchors are only for shallow installations because they have no reserve expansion capacity.

A **sleeve anchor** is a bolt covered by a threaded, thin-wall, split tube. As the bolt is tightened, the thin wall expands. Additional load tends to further expand the thin wall. The bolt must be properly preloaded or friction force will not develop the required holding force.

Adhesive anchors may be in glass capsules or installed with various tools. Pure epoxy, polyester, or vinyl ester resin adhesives are used with a threaded rod supplied by the contractor or the adhesive manufacturer. Some adhesives have a problem with shrinkage; others are degraded by heat. However, some adhesives have been tested without protection to 1100°F before they fail (all mechanical anchors will fail at this temperature). Where required, or if there is a concern, anchors should be protected with fire retardants similar to those applied to steel decks in high-rise buildings.

The manufacturer's instructions for installing the anchor bolts should be followed. Performance test data published by manufacturers should include shock, fatigue, and seismic resistance. IBCO reports have further information on allowable forces for design. Add a safety factor of two if the installation has not been inspected by a qualified firm or individual.

WELD CAPACITIES

Weld capacities may be calculated to determine the size of welds needed to attach equipment to a steel plate or to evaluate raised support legs and attachments. A static analysis provides the effective tension and shear forces. The capacity of a weld is given per unit length of weld based on the shear strength of the weld material. For steel welds, the allowable shear strength capacity if 16,000 psi on the throat section of the weld. The section length is 0.707 times the specified weld size.

For a 1/16 in. weld, the length of shear in the weld is $0.707 \times 1/16 = 0.0442$ in. The allowable weld force $(F_w)_{allow}$ for a 1/16 in. weld is

$$(F_w)_{allow} = 0.0442 \times 16,000 = 700 \text{ lb per inch of weld}$$

For a 1/8 in. weld, the capacity is 1400 lb/in.

The effective weld force is the sum of the vectors calculated in Equations (8) and (10). Because the vectors are perpendicular, they are added by the method of the square root of the sum of the squares (SRSS), or:

$$(F_w)_{eff} = \sqrt{(T_{eff})^2 + (V_{eff})^2}$$

The length of weld required is given by the following equation:

$$\text{Length} = \frac{(F_w)_{eff}}{(F_w)_{allow}} \tag{13}$$

SEISMIC SNUBBERS

Several types of snubbers are manufactured or field fabricated. All snubber assemblies should meet the following minimum requirements to avoid imparting excessive accelerations to the HVAC&R equipment:

- Impact surface should have a high-quality elastomeric surface that is not cemented in place
- Resilient material should be easy to inspect for damage and be replaceable if necessary
- Assembly must provide restraint in all directions
- Snubbers should be tested by an independent test laboratory (and analyzed by a registered engineer to ensure the stated load capacity and) to avoid serious design flaws.

Typical snubbers are classified as Types A through J (see Figure 4). Many devices are presently approved with Office of Statewide Health Planning Development (OSHPD) ratings.

Type A. Snubber built into a resilient mounting. All-directional, molded bridge-bearing quality neoprene element is a minimum of 1/8 in. thick. Mounting must have a minimum of two anchor bolt holes.

Type B. Isolator/Restraint. Stable isolation spring bears on the base plate of the fixed restraining member. Earthquake motion of the isolated equipment is restrained close to the base plate, minimizing pullout force to the base plate anchorage.

Type C. Spring isolator with built-in all directional restraints. Restraints have molded neoprene elements with a minimum thickness of 1/8 in. A neoprene sound pad should be installed between the spring and the base plate. Sound pads below the base plate are not recommended for seismic installations. The base plate should have a minimum of four anchor bolt holes.

Type D. Integral all directional snubber/restrained spring isolator with AASHTO quality replaceable neoprene element. The all welded housing has a minimum of two anchor bolt holes for attachment to the structure.

Type E. Fully bonded neoprene mount capable of withstanding seismic loads in all directions with no metal-to-metal contact. Outer hosing must be ductile iron and have a minimum of two anchor bolt holes.

Type F. All-directional with molded, replaceable neoprene element. Neoprene element of bridge-bearing quality is a minimum of 3/16 in. Snubber must have a minimum of two anchor bolt holes.

Type G. All-directional lateral snubber. Reinforced AASHTO quality neoprene element is a minimum of 1/4 in. thick. Upper bracket is welded to the equipment and the base plate has a minimum of two anchor bolt holes.

Type H. Restraint for floor mounted equipment consisting of interlocking steel assemblies lined with resilient elastomer. Bolt to equipment and anchored to structure through slotted holes to allow field adjustment of restraint for 1/4 in. clearance in the horizontal and vertical directions. After final adjustment, weld anchor to floor bracket and angle clip to equipment to assure no slip can occur. Alternately, fill slots with epoxy grout to prevent slip. The restraint assembly rating is certified by independent laboratory test.

Type I. Single axis, single direction lateral snubber, ribbed AASHTO quality neoprene element is a minimum of 1/4 in. thick. Minimum floor mounting is with two anchor bolts. Must be used in sets of 4 or more.

Type J. Prestretched aircraft wire rope with galvanized end connections that avoid bending the wire rope across sharp edges. This type of snubber is mainly used with suspended pipe duct and equipment.

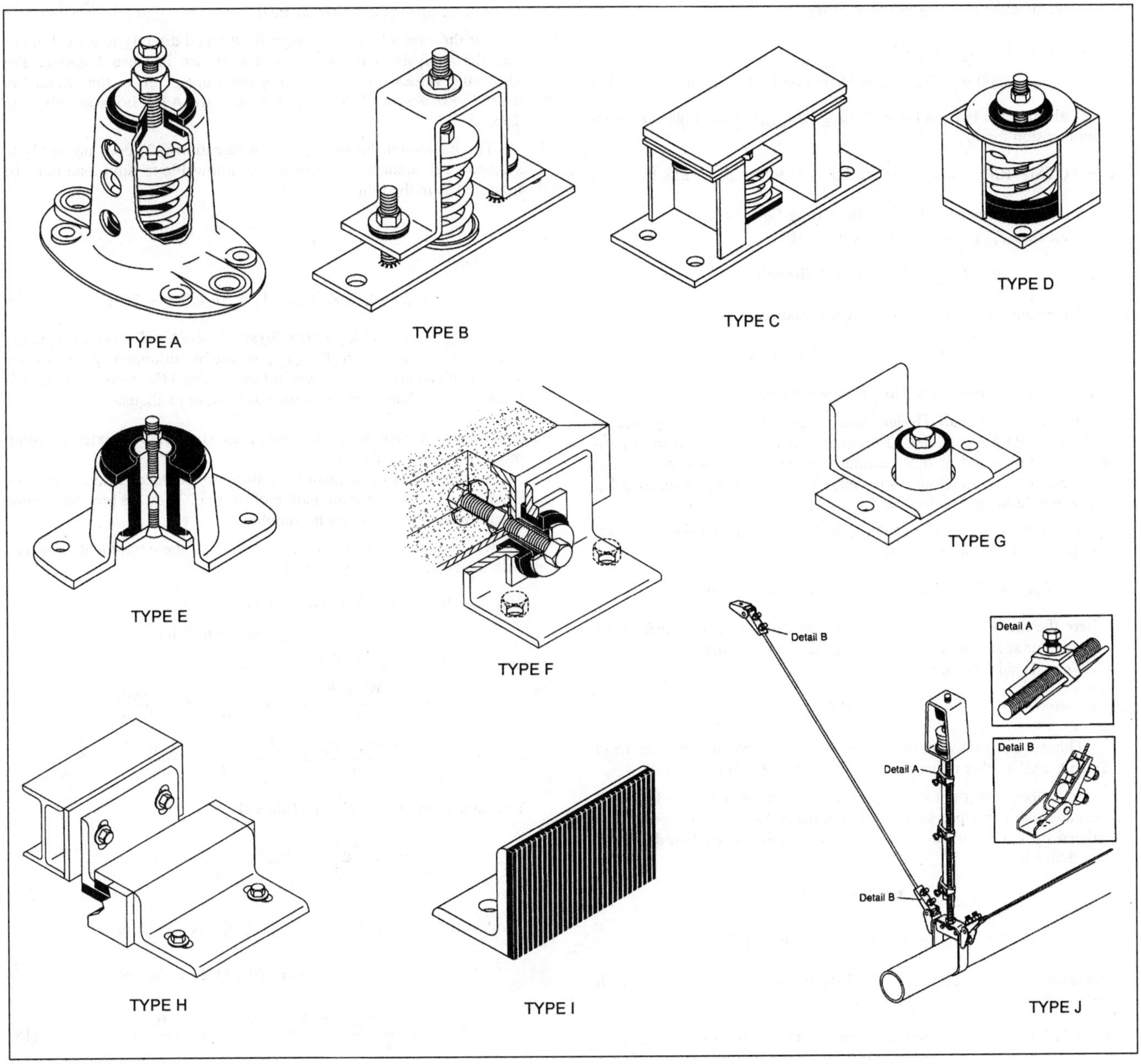

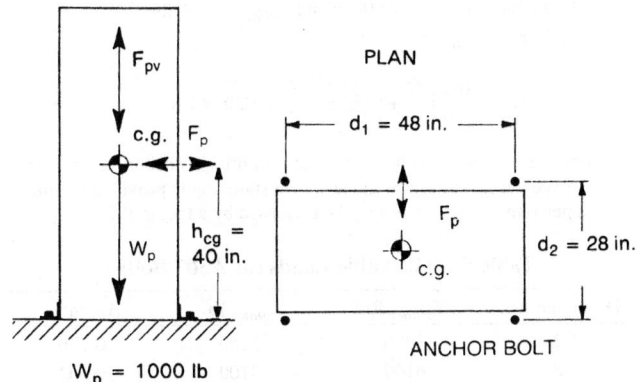

Fig. 4 Seismic Snubbers

EXAMPLES

The following examples are provided to assist in the design of equipment anchorage to resist seismic forces. Assume seismic zone 4 for all examples.

Example 1. Anchorage design for equipment rigidly mounted to the structure (see Figure 5).

From Equations (5) and (6), calculate the lateral seismic force and its vertical component:

$$F_p = 0.4 \times 1.5 \times 0.75 \times 1000 = 450 \text{ lb}$$

$$F_{pv} = F_p/3 = 450/3 = 150 \text{ lb}$$

Calculate the overturning moment (OTM):

$$OTM = F_p h_{cg} = (450 \times 40) = 18,000 \text{ in} \cdot \text{lb} \qquad (14)$$

Fig. 5 Equipment Rigidly Mounted to Structure (Example 1)

Calculate the resisting moment (RM):

$$RM = (W_p \pm F_{pv})d_{min}/2$$
$$= (1000 \pm 150)28/2 = 16,000 \text{ or } 11,900 \text{ in} \cdot \text{lb} \qquad (15)$$

Calculate the tension force T, using RM_{min} to determine the maximum tension force:

$$T = (OTM - RM_{min})/d_{min} = (18,000 - 11,900)/28 = 218 \text{ lb} \qquad (16)$$

This force is the same as that obtained using Equation (9).

Calculate T_{eff} per bolt from Equation (10):

$$T_{eff} = 218/2 = 109 \text{ lb/bolt}$$

Calculate shear force per bolt from Equation (8):

$$V_{eff} = 450/4 = 112.5 \text{ lb/bolt}$$

Case 1. *Equipment attached to a timber structure*

From the *National Design Specification for Wood Construction* (NDS) (AFPA 1997). Selected fasteners must be secured to solid lumber, not to plywood or other similar material. The following calculations are made to determine whether a 1/2 in. diameter, 4 in. long lag screw will hold the required load.

From Table 8.1A in the NDS, for Group IV (redwood), $G = 0.37$, and from Table 8.6A in the NDS,

$$T_{allow} = (241 \text{ lb/in} \times 3.5 \text{ in. penetration})2/3 = 562 \text{ lb}$$

where the factor 2/3 accounts for the fact that about one-third of the length of a lag screw or bolt has no threads on the shank.

From Table 8.6C in the NDS,

$$V_{allow} = 180 \text{ lb}$$

Other types of wood may be used with appropriate factors from Table 8 and/or other reductions as specified in Part II of the NDS.

In timber construction, the interaction formula given in Equation (12) does not apply per Section 8.6.8 of the NDS. The ratios of the calculated shear and tension values to the allowable values should each be less than 1.0:

$$T/T_{allow} = 109/562 = 0.19 < 1.0.$$

$$V/V_{allow} = 112.5/180 = 0.63 < 1.0$$

Therefore, a 1/2 in. diameter, 4 in. long lag screw can be used at each corner of the equipment.

Case 2. *Equipment attached to concrete with post installed anchors.*

It is good design practice to specify a minimum of 1/2 in. diameter bolts to attach roof or floor-mounted equipment to the structure. Determine whether 1/2 in. wedge anchors without special inspection provisions will hold the required load.

From Table 7, $T_{allow} = 600 \text{ lb}$ and $V_{allow} = 1200 \text{ lb}$

From Equation (12),

$$\left(\frac{109}{600}\right)^{5/3} + \left(\frac{112.5}{1200}\right)^{5/3} = 0.08 < 1.0$$

Therefore, 1/2 in. diameter, 4 in. long post drill-in anchors can be used.

If special inspection of the anchor installation is provided by qualified personnel, T_{allow} only may be increased by a factor of 2.

Table 8   Allowable Loads for A307 Bolts

Diameter, in.	T_{allow}, lb	V_{allow}, lb	A_b, in^2
1/2	3900	1950	0.196
5/8	6100	3100	0.307
3/4	8800	4400	0.442
1	15700	7900	0.785

Case 3. *Equipment attached to steel*

For the case where equipment is attached directly to a steel member, the analysis is the same as that shown in Case 1 above. The allowable values for the attaching bolts are given in the *Manual of Steel Construction* (AISC 1989). Values for A307 bolts are given in Table 8.

The interaction formula given in Equation (12) does not apply to steel-to-steel connections. Instead, the allowable tension load must be modified as in the following equation:

$$(T_{allow})_{mod} = F_t A_b$$

where

$$F_t = 26 - 1.8(V/N_{bolt})A_b \le 20(4/3) = 26.67$$

V/N_{bolt} is in kips (1 kip = 1000 lb) and F_t is in kips/in^2 (ksi). If F_t is less than 20 ksi, the calculated $(T_{allow})_{mod}$ should be multiplied by 1000 to give values equivalent to those shown in Table 8. The 33% stress increase (4/3) is allowed for short-term loads such as wind or earthquakes.

Example 2. Anchorage design for equipment supported by external spring mounts (see Figure 6).

A mechanical or acoustical consultant should choose the type of isolator or snubber or combination of the two. Then the product vendor should select the actual spring snubber.

Assume that the center of gravity (cg) of the equipment coincides with the center of gravity of the isolator group.

If T = maximum tension on isolator,

     C = maximum compression on isolator, and

     $F_{pv} = F_p/3$, then

$$T = \frac{-W_p + F_{pv}}{4} + F_p h_{cg}\frac{\cos\theta}{2b} + F_p h_{cg}\frac{\sin\theta}{2a}$$

$$= \frac{-W_p + F_{pv}}{4} + \frac{F_p h_{cg}}{2}\left(\frac{\cos\theta}{b} + \frac{\sin\theta}{a}\right)$$

To find maximum T or C, set $dT/dq = 0$:

$$\frac{dT}{d\theta} = \frac{F_p h_{cg}}{2}\left(-\frac{\sin\theta}{b} + \frac{\cos\theta}{a}\right) = 0$$

$$\theta_{max} = \tan^{-1}(b/a) = \tan^{-1}(28/48) = 30.26°$$

$$\theta_{max} = \tan^{-1}(b/a) = \tan^{-1}(0.7/1.2) = 30.26° \qquad (17)$$

$$T = \frac{-W_p + F_{pv}}{4} + \frac{F_p h_{cg}}{2}\left(\frac{\cos\theta_{max}}{b} + \frac{\sin\theta_{max}}{a}\right) \qquad (18)$$

$$C = \frac{-W_p - F_{pv}}{4} - \frac{F_p h_{cg}}{2}\left(\frac{\cos\theta_{max}}{b} + \frac{\sin\theta_{max}}{a}\right) \qquad (19)$$

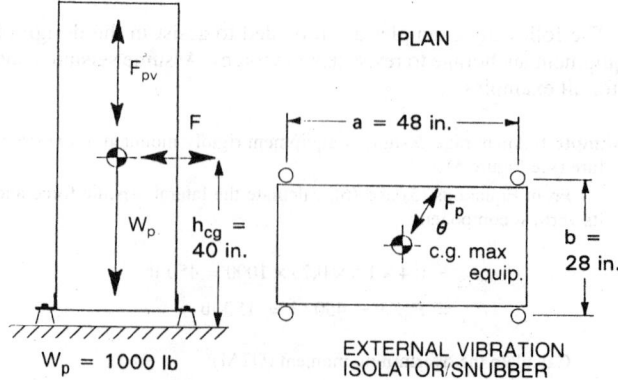

Fig. 6   Equipment Supported by External Spring Mounts

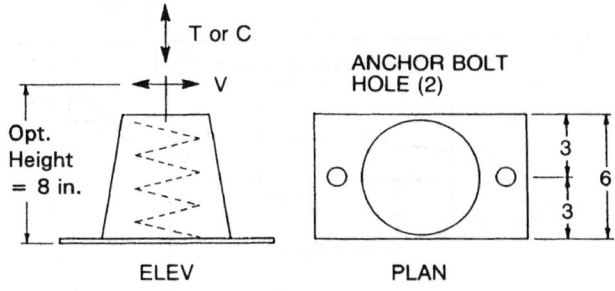

Fig. 7   Spring Mount Detail (Example 2)

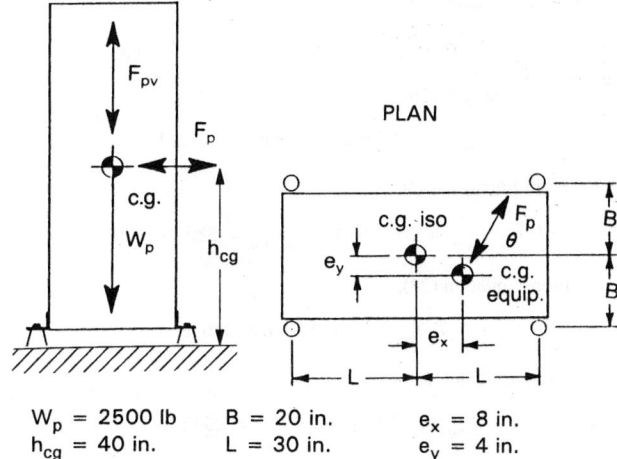

W_p = 2500 lb   B = 20 in. e_x = 8 in.
h_{cg} = 40 in.   L = 30 in. e_y = 4 in.

Fig. 8   Equipment with Different Center of Gravity than Isolator Group (in Plan View)

From Equations (5) and (6),

$$F_p = 0.4 \times 1.5 \times 2 \times 0.75 \times 2500 = 2250 \text{ lb}$$

$$F_{pv} = F_p/3 = 2250/3 = 750 \text{ lb}$$

From Equations (18) and (19),

$$T = -438 + 1860 = 1422 \text{ lb}$$

$$C = -812 - 1860 = -2672 \text{ lb}$$

Calculate the shear force per isolator:

$$V = (F_p/N_{iso}) = 2250/4 = 563 \text{ lb} \qquad (20)$$

This shear force is applied at the operating height of the isolator. Uplift tension T on the vibration isolator is the worst condition for the design of the anchor bolts. The compression force C must be evaluated to check the adequacy of the structure to resist the loads (Figure 7).

$$(T_1)_{eff} \text{ per bolt} = T/2 = 1422/2 = 711 \text{ lb}$$

The value of $(T_2)_{eff}$ per bolt due to overturning on the isolator is

$$(T_2)_{eff} = \frac{V \times \text{Optimum ht.}}{0.85 d N_{bolt}}$$

where d = distance from edge of isolator base plate to center of bolt hole.

$$(T_2)_{eff} = \frac{563 \times 8}{0.85 \times 3 \times 2} = 883 \text{ lb}$$

$$(T_{max})_{eff} = (T_1)_{eff} + (T_2)_{eff} = 711 + 883 = 1595 \text{ lb}$$

$$V_{eff} = 563/2 = 282 \text{ lb}$$

Determine whether 5/8 in. post drill-in anchors *with special inspection* will handle this load. From Equation (12) and Table 7,

$$\left(\frac{1595}{2 \times 900}\right)^{5/3} + \left(\frac{282}{2200}\right)^{5/3} = 0.85 < 1.0$$

Therefore, 5/8 in. post drill-in anchors will carry the load.

Example 3. Anchorage design for equipment with a center of gravity different from that of the isolator group (see Figure 8).

Anchor properties: $I_x = 4B^2$; $I_y = 4L^2$

Angles:

$$\theta = \tan^{-1}(B/L) \qquad (21)$$

$$\alpha = \tan^{-1}(e_x/e_y) \qquad (22)$$

$$\beta = 180 - |\alpha - \theta| \qquad (23)$$

$$\phi = \tan^{-1}(LI_x/BI_y) \qquad (24)$$

Vertical reactions:

$$(W_n)_{max/min} = W_p \pm F_{pv} \qquad (25)$$

Vertical reaction due to overturning moment:

$$T_m = F_p h_{cg}\left(\frac{B}{I_x}\cos\phi + \frac{L}{I_y}\sin\phi\right) \qquad (26)$$

Vertical reaction due to eccentricity:

$$(T_e)_{max/min} = (W_n)_{max/min}\left(\frac{Be_y}{I_x} + \frac{Le_x}{I_y}\right) \qquad (27)$$

Vertical reaction due to W_p:

$$(T_w)_{max/min} = (W_n)_{max/min}/4 \qquad (28)$$

$$T_{max} = T_m + (T_e)_{max} + (T_w)_{max} \qquad (29)$$

$$T_{min} = T_m + (T_e)_{min} + (T_w)_{min} \text{ (tension if positive)} \qquad (30)$$

Horizontal reactions:
Horizontal reaction due to rotation:

$$V_{rot} = F_p\left[\frac{e_x^2 + e_y^2}{16(B^2 + L^2)}\right]^{0.5} \qquad (31)$$

$$V_{dir} = F_p/4 \qquad (32)$$

$$V_{max} = (V_{rot}^2 + V_{dir}^2 - 2V_{rot}V_{dir}\cos\beta)^{0.5} \qquad (33)$$

From Equations (5) and (6),

$$F_p = 0.4 \times 1.5 \times 2 \times 0.75 \times 2500 = 2250 \text{ lb}$$

$$F_{pv} = 2250/3 = 750 \text{ lb}$$

$$I_x = 4(20)^2 = 1600; \quad I_y = 4(30)^2 = 3600$$

From Equations (21) through (24),

$$\theta = 33.69° \qquad \alpha = 63.43°$$

$$\beta = 150.26° \qquad \phi = 33.69°$$

From Equation (25),

$$(W_n)_{max/min} = 3250 \text{ lb or } 1750 \text{ lb}$$

From Equation (26),

$$T_m = 90,000(0.01 + 0.005) = 1352 \text{ lb}$$

From Equation (27),

$$(T_e)_{max/min} = 0.1167(W_n)_{max/min} = 380 \text{ lb or } 205 \text{ lb}$$

From Equation (28),

$$(T_w)_{max/min} = 810 \text{ lb or } 440 \text{ lb}$$

From Equation (29),

$$T_{max} = 1352 + 380 + 810 = 2542 \text{ lb}$$

From Equation (30),

$$T_{min} = 1352 + 205 + 440 = 1997 \text{ lb (tension)}$$

From Equations (31), (32), and (33),

$$V_{rot} = 2250 \times 0.062 = 140 \text{ lb}$$

$$V_{dir} = 2250/4 = 563 \text{ lb}$$

$$V_{max} = 687 \text{ lb}$$

The values of T_{min} and V_{max} are used to design the anchorage of the isolators and/or snubbers, and T_{max} is used to verify the adequacy of the structure to resist the vertical loads.

Example 4. Anchorage design for equipment with supports and bracing for suspended equipment (see Figure 9). Equipment weight $W_p = 500$ lb.

Because post drill-in anchors may not withstand published allowable static loads when subjected to vibratory loads, vibration isolators should be used between the equipment and the structure to dampen vibrations generated by the equipment.

From Equations (5) and (6),

$$F_p = 0.4 \times 1.5 \times 2 \times 0.75 \times 500 = 450 \text{ lb}$$
$$F_{pv} = 450/3 = 150 \text{ lb}$$

From Equation (14),

$$OTM = 450 \times 12 = 5400 \text{ in} \cdot \text{lb}$$

From Equation (10),

$$RM = (500 \pm 150)36/2 = 11,700 \text{ or } 6300 \text{ in} \cdot \text{lb}$$

Because RM is greater than OTM, overturning is not critical.

Force to the hanger rods:

$$T_{eff} = (W_p + F_{pv})/4 = (500 + 150)/4 = 163 \text{ lb}$$

Force in the splay brace = $\sqrt{2} \, F_p = 636$ lb at a 1:1 slope

Due to the force being applied at the critical angle, as in Example 2, only one splay brace is effective in resisting the lateral load F_p. If eccentricities occur, as in Example 3, a similar method of analysis must be done to obtain the design forces.

Design of hanger rod/vibration isolator and connection to structure

When installing post drill-in anchors in the underside of a concrete beam or slab, the allowable tension loads on the anchors must be reduced to account for the cracking of the concrete. A general rule is to use half the allowable load.

Determine whether a 1/2 in. wedge anchor with special inspection provisions will hold the required load.

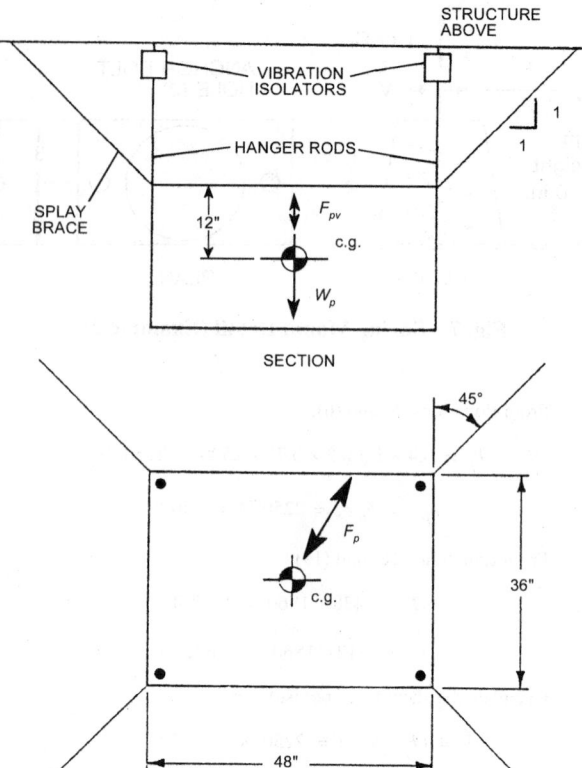

Note: The splay braces are prestretched aircraft cables with enough slack so that the isolators can fully function vertically.

Fig. 9 Supports and Bracing for Suspended Equipment

$$T_{allow} = 600 \times 0.5 = 300 \text{ lb} > T_{eff} = 163 \text{ lb}$$

Therefore, a 1/2 in. rod and post drill-in anchor should be used at each corner of the unit.

For anchors installed without special inspection,

$$T_{allow} = 300 \times 0.5 = 150 \text{ lb} < T_{eff} = 163 \text{ lb}$$

Therefore, a larger anchor would have to be chosen.

Design of splay brace and connection to structure

Force in the slack cable = 636 lb

Force in the connection to the structure:

$$V_{max} = 636/\sqrt{2} = 450 \text{ lb} \qquad T_{max} = F_p = 450 \text{ lb}$$

Determine whether a 3/4 in. wedge-type anchor will hold the required load. From Table 7:

$$T_{allow} = 1350/2 = 675 \text{ lb} \qquad V_{allow} = 3000 \text{ lb}$$

From Equation (12),

$$\left(\frac{450}{1350}\right)^{5/3} + \left(\frac{450}{3000}\right)^{5/3} = 0.2 < 1.0$$

Therefore, it is permissible to use a 3/4 in. anchor or multiple anchors of a smaller size bolted through a clip and to the structure.

Because the cable forces are relatively small, a 3/8 in. aircraft cable attached to clips with cable clamps should be used. The clips, in turn, may be attached to either the structure or the equipment.

INSTALLATION PROBLEMS

The following should be considered when installing seismic restraints.

- Anchor location affects the required strengths. Concrete anchors should be located away from edges, stress joints, or existing fractures. ASTM *Standard* E488 should be followed as a guide for edge distances and center-to-center spacing.

- Concrete anchors should not be too close together. Epoxy-type anchors can be closer together than expansion-type anchors. Expansion-type anchors (self-drilling and drop-in) can crush the concrete where they expand and impose internal stresses in the concrete. Spacing of all anchor bolts should be carefully reviewed. (See manufacturer's recommendations.)

- Supplementary steel bases and frames, concrete bases, or equipment modifications may void some manufacturer's warranties. Snubbers, for example, should be properly attached to a subbase. Bumpers may be used with springs.

- Static analysis does not account for the effects of resonant conditions within a piece of equipment or its components. Because all equipment has different resonant frequencies during operation and nonoperation, the equipment itself might fail even if the restraints do not. Equipment mounted inside a housing should be seismically restrained to meet the same criteria as the exterior restraints.

- Snubbers used with spring mounts should withstand motion in all directions. Some snubbers are only designed for restraint in one direction; sets of snubbers or snubbers designed for multidirectional purposes should be used.

- Equipment must be strong enough to withstand the high deceleration forces developed by resilient restraints.

- Flexible connections should be provided between equipment that is braced and piping and ductwork that need not be braced.

- Flexible connections should be provided between isolated equipment and braced piping and ductwork.

- Bumpers installed to limit horizontal motion should be outfitted with resilient neoprene pads to soften the potential impact loads of the equipment.

- Anchor installations should be inspected; in many cases, damage occurs because bolts were not properly installed. To develop the rated restraint, bolts should be installed according to manufacturer's recommendations.

- Brackets in structural steel attachments should be matched to reduce bending and internal stresses at the joint. Rigid seismic restraints should not have slotted holes.

WIND RESTRAINT DESIGN

Damage done to HVAC&R equipment by both sustained and gusting wind forces has increased concern about the adequacy of the equipment protection that is defined in design documents. The following calculative procedure generates the same type of total design lateral force that is used in the static analysis of the seismic restraint. This means that the value that is determined for the design wind force F_w can be substituted for the total design lateral seismic force F_p when evaluating and choosing restraint devices.

ASCE *Standard* 7-93, *Minimum Design Loads for Buildings and Other Structures*, includes design guidelines for wind, snow, rain and earthquake loads. The equations, guidelines, and data presented here are from an earlier version of this standard and only cover nonstructural components. The current standard includes more comprehensive and rigorous procedures for evaluating wind forces and wind restraint.

TERMINOLOGY

Classification. Buildings and other structures are classified for wind load design according to Table 9.

Basic wind speed. The fastest mile-per-hour wind speed at 33 ft (10 m) above the ground of Terrain Exposure C (see Table 10) having an annual probability of occurrence of 0.02. Data in ASCE *Standard* 7 or regional climatic data may be used to determine basic wind speeds. ASCE data does not include all special wind regions (such as mountainous terrains, gorges, and ocean promontories) where records or experience indicate that the wind speeds are higher than what is shown in appropriate wind data tables. For these circumstances, regional climatic data may be used provided that both acceptable extreme-value statistical-analysis procedures were used in reducing the data and due regard was given to the

**Table 9   Classification of Buildings and
Other Structures for Wind Loads**

Reprinted with permission from ASCE *Standard* 7-93.

Nature of Occupancy	Category
All buildings and structures except those listed below	I
Buildings and structures where the primary occupancy is one in which more than 300 people congregate in one area	II
Buildings and structures designated as essential facilities, including, but not limited to: - Hospital and other medical facilities having surgery or emergency treatment areas - Fire or rescue and police stations - Structures and equipment in government - Communication centers and other facilities required for emergency response - Power stations and other utilities required in an emergency - Structures having critical national defense capabilities - Designated shelters for hurricanes	III
Buildings and structures that represent a low hazard to human life in the event of failure, such as agricultural buildings, certain temporary facilities, and minor storage facilities	IV

Table 10   Definition of Exposure Categories

Reprinted with permission from ASCE *Standard* 7-93.

Exposure A. Large city centers with at least 50% of the buildings having a height in excess of 70 ft. Use of this exposure category is limited to those areas for which terrain representative of Exposure A prevails in the upwind direction for a distance of at least 0.5 mile or 10 times the height of the building or structure, whichever is greater.
Possible channeling effects or increased velocity pressures due to the building or structure being located in the wake of adjacent buildings needs to be considered.

Exposure B. Urban and suburban areas, wooded areas, or other terrain with numerous closely spaced obstructions having the size of single-family dwellings or larger. Use of this exposure category shall be limited to those areas for which terrain representative of Exposure B prevails in the upwind direction for a distance of at least 1500 ft or 10 times the height of the building or structure, whichever is greater.

Exposure C. Open terrain with scattered obstructions having heights generally less than 30 ft. This category includes flat open country and grasslands.

Exposure D. Flat, unobstructed areas exposed to wind flowing over large bodies of water. This exposure shall apply only to those buildings and other structures exposed to the wind coming from over the water.

Notes:
1. HVAC components for buildings with a mean roof height of 60 ft or less are designed on the basis of Exposure C.
2. HVAC components on buildings with a mean roof height greater than 60 ft and other structures are designed on the basis of the exposure categories defined in this table except assume Exposure B for buildings and other structures sited in terrain representative of Exposure A.

Table 11   Importance Factor (Wind Loads)

Reprinted with permission from ASCE *Standard* 7-93.

	Importance Factor I_w	
Category	100 miles from hurricane ocean line and in other areas	At hurricane ocean line
I	1.00	1.05
II	1.07	1.11
III	1.07	1.11
IV	0.95	1.00

Notes:
1. Table 9 lists the building and structure classification categories.
2. Determine I_w by linear interpolation for regions between the hurricane ocean line and 100 miles inland.
3. Typical hurricane ocean lines are the Atlantic and Gulf of Mexico coastal areas.

length of record, averaging time, anemometer height, data quality, and terrain exposure. One final exclusion is that tornadoes have not been considered in developing the basic wind-speed distributions.

Design wind force. Equivalent static force that is assumed to act on a component in a direction parallel to the wind and not necessarily normal to the surface area of the component. This force varies with respect to height above ground level.

Importance factor (I_w). A factor that accounts for the degree of hazard to human life and damage to HVAC components (see Table 11). For hurricanes, the value of the importance factor can be linearly interpolated between the ocean line and 100 miles inland because wind effects are assumed negligible at this distance inland.

Gust response factor (G_z). A factor that accounts for the fluctuating nature of wind and the corresponding additional loading effects on HVAC components.

Minimum design wind load. The wind load may not be less than 10 lb/ft^2 multiplied by the area of the HVAC component projected on a vertical plane that is normal to the wind direction.

CALCULATIONS

Two procedures are used to determine the design wind load on HVAC components. The **analytical procedure**, which is described here, is the most common analysis method for standard component shapes. The second method, the **wind-tunnel procedure**, is incorporated in the analysis of complex and unusual shaped components or equipment that are located on sites that produce wind channeling or buffeting due to upwind obstructions. The analytical procedure produces design wind forces that are expected to act on HVAC components for durations of 1 to 10 s. The various factors, pressure, and force coefficients incorporated in this procedure are based on a mean wind speed that corresponds to the fastest mile-per-hour wind speed.

Analytical Procedure

The design wind force is determined by the following equation:

$$F_w = Q_z G_z C_f A_f \qquad (34)$$

where

F_w = design wind force, lb
Q_z = velocity pressure evaluated at height z above ground level, lb/ft^2
G_z = gust response factor for HVAC components evaluated at height z above ground level
C_f = force coefficient (Table 12)
A_f = area of HVAC component projected on a plane normal to wind direction. ft^2

Certain of the above factors must be calculated from equations that incorporate site-specific conditions that are defined as follows:

Velocity Pressure. The design wind speed must be converted to a velocity pressure that is acting on an HVAC component at a height z above the ground. The equation is

Table 12   Force Coefficients for HVAC Components, Tanks, and Similar Structures

Reprinted with permission from ASCE *Standard* 7-93.

Shape	Type of Surface	C_f for h/D Values of:		
		1	7	25
Square (wind normal to face)	All	1.3	1.4	2.0
Square (wind along diagonal)	All	1.0	1.1	1.5
Hexagonal or octagonal ($D\sqrt{Q_z} > 2.5$)	All	1.0	1.2	1.4
Round ($D\sqrt{Q_z} > 2.5$)	Moderately smooth	0.5	0.6	0.7
	Rough ($D'/D \approx 0.02$)	0.7	0.8	0.9
	Very rough ($D'/D \approx 0.08$)	0.8	1.0	1.2
Round ($D\sqrt{Q_z} \le 2.5$)	All	0.7	0.8	1.2

Notes:
1. The design wind force is calculated based on the area of the structure projected on a plane normal to the wind direction. The force is assumed to act parallel to the wind direction.
2. Linear interpolation may be used for h/D values other than shown.
3. Nomenclature:
   D = diameter or least horizontal dimension, ft
   D' = depth of protruding elements such as ribs and spoilers, ft
   h = height of structure, ft
   Q_z = velocity pressure evaluated at height z above ground level, lb/ft^2

Table 13   Exposure Category Constants

Reprinted with permission from ASCE *Standard* 7-93.

Exposure Category	α	z_g	D_o
A	3.0	1500	0.025
B	4.5	1200	0.010
C	7.0	900	0.005
D	10.0	700	0.003

Note: See Table 10 for definitions of Exposure Categories.

$$Q_z = \rho K_z (I_w V)^2 / 2g = 0.00256 K_z (I_w V)^2 \qquad (35)$$

where

ρ = air mass density. At 59°F and 29.92 in.Hg the density of dry air is 0.0765 lb/ft^3.

K_z = velocity pressure exposure coefficient evaluated at height z above ground level. The value of K_z is determined by the following equation that uses data from Table 13.

$$K_z = 2.58 (z/z_g)^{2/\alpha} \qquad (36)$$

Values for K_z from $0 \le z \le 500$ ft are given in Table 14. The definitions of Exposure A through D categories are shown in Table 10.

I_w = importance factor. Values are given in Table 11. Definitions for categories I, II, III, and IV are found in Table 9.

V = basic wind speed determined from data in ASCE *Standard* 7 or appropriate regional data. The basic wind speed used should be at least 70 mph.

Gust response factor. The values of G_z are determined by the following equations. Table 15 lists values for G_z for $0 \le z \le 500$ ft.

$$G_z = 0.65 + 3.65 T_z \qquad (37)$$

where

$$T_z = 2.35 \frac{(D_o)^{1/2}}{(z/30)^{1/\alpha}} \qquad (38)$$

Table 14 Velocity Pressure Exposure Coefficient K_z

Reprinted with permission from ASCE *Standard* 7-93.

Ht. Above Ground, ft	K_z Exposure A	Exposure B	Exposure C	Exposure D
0–15	0.12	0.37	0.80	1.20
20	0.15	0.42	0.87	1.27
25	0.17	0.46	0.93	1.32
30	0.19	0.50	0.98	1.37
40	0.23	0.57	1.06	1.46
50	0.27	0.63	1.13	1.52
60	0.30	0.68	1.19	1.58
70	0.33	0.73	1.24	1.63
80	0.37	0.77	1.29	1.67
90	0.40	0.82	1.34	1.71
100	0.42	0.86	1.38	1.75
120	0.48	0.93	1.45	1.81
140	0.53	0.99	1.52	1.87
160	0.58	1.05	1.58	1.92
180	0.63	1.11	1.63	1.97
200	0.67	1.16	1.68	2.01
250	0.78	1.28	1.79	2.10
300	0.88	1.39	1.88	2.18
350	0.98	1.49	1.97	2.25
400	1.07	1.58	2.05	2.31
450	1.16	1.67	2.12	2.36
500	1.24	1.75	2.18	2.41

Notes: 1. Values of K_z are calculated with Equation (36) and data from Table 13.
2. Linear interpolation of table values is acceptable for intermediate values of height z.

Table 15   Gust Response Factor

Reprinted with permission from ASCE *Standard* 7-93.

Height Above Ground, ft	Gust Response Factor G_z Exposure A	Exposure B	Exposure C	Exposure D
0–15	2.36	1.65	1.32	1.15
20	2.20	1.59	1.29	1.14
25	2.09	1.54	1.27	1.13
30	2.01	1.51	1.26	1.12
40	1.88	1.45	1.23	1.11
50	1.79	1.42	1.21	1.10
60	1.73	1.39	1.20	1.09
70	1.67	1.36	1.19	1.08
80	1.63	1.34	1.18	1.08
90	1.59	1.32	1.17	1.07
100	1.56	1.31	1.16	1.07
120	1.50	1.28	1.15	1.06
140	1.46	1.26	1.14	1.05
160	1.43	1.24	1.13	1.05
180	1.40	1.23	1.12	1.04
200	1.37	1.21	1.11	1.04
250	1.32	1.19	1.10	1.03
300	1.28	1.16	1.09	1.02
350	1.25	1.15	1.08	1.02
400	1.22	1.13	1.07	1.01
450	1.20	1.12	1.06	1.01
500	1.18	1.11	1.06	1.00

Notes:
1. Values of G_z are calculated with Equations (37) and (38) and data from Table 13.
2. Linear interpolation of table values is acceptable for intermediate values of height z.
3. The value of the G_z may not be less than 1.0.

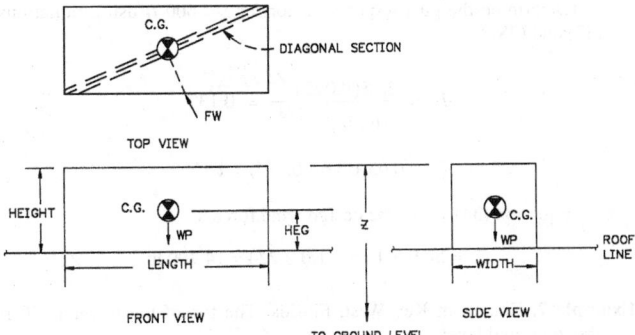

Fig. 10   Equipment Dimensions and Force Locations for Wind Examples 5, 6, and 7

Example Calculations: Analytical Procedure

The following example calculations are for a 400 ton cooling tower with dimensions shown in Figure 10:

Tower height = h = 10 ft

Tower width = D = 10 ft

Tower length = l = 20 ft

Tower operating weight = W_p = 19,080 lb

Tower diagonal dimension = $(10^2 + 20^2)^{0.5}$ = 22.4 ft

Area normal to wind direction = A_f = 10 × 22.4 = 224 ft^2

From Table 12; C_f = 1.0 for wind acting along diagonal with h/D = 10/10 = 1.

Example 5. Suburban hospital in Omaha, Nebraska. The top of the cooling tower is 100 ft. above ground level.

Solution:

From appropriate wind data tables, the design wind speed is found to be 90 mph.

From Table 9, use Category III

From Table 10, use Exposure B

From Table 11, I_w = 1.07

From Table 14 or Equation (36), K_z = 0.86

Substitution into Equation (35) yields:

$$Q_z = 0.00256 \times 0.86(1.07 \times 90)^2 = 20.42 \text{ lb/ft}^2$$

From Table 15 or Equations (37) and (38); G_z = 1.31. Substitution into Equation (34) yields the design wind force as

$$F_w = 20.42 \times 1.31 \times 1.0 \times 224 = 5990 \text{ lb}$$

Example 6. Office building in New York City. Top of tower is 600 ft above ground level.

Solution:

From wind data tables, the design wind speed is 120 mph.

From Table 9, use Category I.

From Table 10, use Exposure A

From Table 11, I_w = 1.05

Because z > 500 ft, use Equations (36), (37), (38), and Table 15 to determine K_z and G_z.

From Table 13, α = 3.0, z_g = 1500, and D_o = 0.025

From Equation (36):

$$K_z = 2.58(600/1500)^{2/3} = 1.40$$

From Equation (35)

$$Q_z = 0.00256 \times 1.40(1.05 \times 120)^2 = 56.9 \text{ lb/ft}^2$$

Determine the gust response factor for $z = 600$ ft using Equations (37) and (38).

$$T_z = \frac{2.35(0.025)^{1/2}}{(600/30)^{1/3}} = 0.137$$

$$G_z = 0.65 + 3.65(0.137) = 1.15$$

Equation (34) yields the design wind force as

$$F_w = 56.9 \times 1.15 \times 1.0 \times 224 = 14{,}700 \text{ lb}$$

Example 7. Church in Key West, Florida. The top of the tower is 50 ft above ground level.

Solution:

From Table 9, use Category II

From Table 10, use Exposure D

From Table 11, $I_z = 1.11$

From Table 14, $K_z = 1.52$.

From appropriate wind data, the design wind speed is found to be 150 mph.

From Equation (35):

$$Q_z = 0.00256(1.52)(1.11 \times 150)^2 = 107.9 \text{ lb/ft}^2$$

From Table 15, $G_z = 1.10$. Again using Equation (34), the design wind force is

$$F_w = 107.9 \times 1.10 \times 1.0 \times 224 = 26{,}600 \text{ lb}$$

REFERENCES

AFPA. 1997. *National design specification (NDS) for wood construction.* American Forest and Paper Association, Washington, DC.

AISC. 1989. *Manual of steel construction—Allowable stress design*, 9th ed. American Institute of Steel Construction, Chicago.

Army, Navy, and Air Force. 1992. *Seismic design for buildings.* TMS-809-10, NAVFAC P-355, AFN 88-3, Chapter 13.

ASCE. 1993. Minimum design loads for buildings and other structures. *Standard* ASCE 7. American Society of Civil Engineers, Reston, VA.

BOCA. 1996. *The BOCA National building code*, 13th ed. Building Officials & Code Administrators International, Inc., Country Club Hills, IL.

ICC. 1998. *International building code* 2000, Final draft. International Code Council, Falls Church, VA.

ICBO. 1997. *Uniform building code.* International Conference of Building Officials, Whittier, CA.

SBCCI. 1994. *Standard building code* 1996. Southern Building Code Congress International, Inc., Birmingham, AL.

SMACNA. 1998. *Seismic restraint manual: Guidelines for mechanical systems*, 2nd ed. Sheet Metal and Air Conditioning Contractors' National Association, Chantilly, VA.

BIBLIOGRAPHY

ACI. 1995. Building code requirements for structural concrete. *Standard* 318-95 and *commentary* 318R-95. American Concrete Institute, Farmington Hills, MI.

AISC. 1995. *Manual of steel construction—Load and resistance factor design*, 2nd ed. American Institute of Steel Construction, Chicago.

Associate Committee on the National Building Code. 1985. *National Building Code of Canada* 1985, 9th ed. National Research Council of Canada, Ottawa.

Associate Committee on the National Building Code. 1986. *Supplement to the National Building Code of Canada* 1985, 2nd ed. National Research Council of Canada, Ottawa. First errata, January.

AWS. 1996. Structural welding code. AWS D1.1-96. Steel American Welding Society, Miami.

Ayres, J.M. and R.J. Phillips. 1998. Water damage in hospitals resulting from the Northridge earthquake. *ASHRAE Transactions* 104(1B):1286-96.

Batts, M.E., M.R. Cordes, L.R Russell, J.R. Shaver, and E. Simiu. 1980. Hurricane wind speeds in the United States. NBS BSS 124. National Institute of Standards and Technology, Gaithersburg, MD.

Bolt, B.A. 1988. *Earthquakes.* W.H. Freeman, New York.

DOE. 1989. General design criteria. DOE Order 6430.1A. U.S. Department of Energy, Washington, D.C.

FEMA 302 & 303. NEHRP recommended provisions for seismic regulations for new buildings and other structures. Part 1, Provisions; Part 2, Commentary. Building Seismic Safety Council, Washington, DC.

Jones, R.S. 1984. *Noise and vibration control in buildings.* McGraw-Hill, New York.

Kennedy, R.P., S.A. Short, J.R. McDonald, M.W. McCann, and R.C. Murray. 1989. Design and evaluation guidelines for the Department of Energy facilities subjected to natural phenomena hazards.

Lama, P.J. 1998. Seismic codes, HVAC pipe systems and practical solutions. *ASHRAE Transactions* 104(1B):1297-1304.

Maley, R., A. Acosta, F. Ellis, E. Etheredge, L. Foote, D. Johnson, R. Porcella, M. Salsman, and J. Switzer. 1989. Department of the Interior, U.S. geological survey. U.S. geological survey strong-motion records from the Northern California (Loma Prieta) earthquake of October 17, 1989. Open-file *report* 89-568.

Naeim, F. 1989. *The seismic design handbook.* Van Nostrand Reinhold International Company Ltd., London, England.

NFPA. 1991. *Installation of sprinkler systems.* National Fire Protection Association, Quincy, MA.

Peterka, J.A., and J.E. Cermak. 1974. Wind pressures on buildings—Probability densities. *J. Structural Div.*, ASCE 101(6):1255-67.

Simiu, E., M.J. Changery, and J.J. Filliben. 1979. Extreme wind speeds at 129 stations in the contiguous United States. U.S. NBS BSS 118. National Institute of Standards and Technology, Gaithersburg, MD.

SMACNA. 1995. *HVAC duct construction standard—metal and flexible*, 2nd ed. Sheet Metal and Air Conditioning Contractors' National Association, Chantilly, VA.

Wasilewski, R.J. 1998. Seismic restraints for piping systems. *ASHRAE Transactions* 104(1B):1273-95.

Weigels, R.L. 1970. *Earthquake engineering*, 10th ed. Prentice-Hall, Englewood Cliffs, New Jersey.

CHAPTER 54

CODES AND STANDARDS

THE Codes and Standards listed here represent practices, methods, or standards published by the organizations indicated. They are useful guides for the practicing engineer in determining test methods, ratings, performance requirements, and limits applying to the equipment used in heating, refrigerating, ventilating, and air conditioning. *Copies of the publications can usually be obtained from the organizations listed in the Publisher column.* Addresses of the organizations are given at the end of the chapter. A comprehensive database with over 250,000 industry, government, and international standards is on the Internet at **www.nssn.org**.

Codes and Standards Published by Various Societies and Associations

Subject	Title	Publisher	Reference
Air Conditioners	Commercial Applications Systems and Equipment (1993)	ACCA	ACCA Manual CS
	Residential Equipment and Selection (1995)	ACCA	ACCA Manual S
	Methods of Testing for Rating Ducted Air Terminal Units	ASHRAE	ANSI/ASHRAE 130-1996
	Ducted Air-Conditioners and Air-to-Air Heat Pumps—Testing and Rating for Performance	ISO	ISO 13253:1995
	Non-Ducted Air Conditioners and Heat Pumps—Testing and Rating for Performance	ISO	ISO 5151:1994
	Heating and Cooling Equipment (1995)	UL	UL 1995
		CSA	CAN/CSA-C22.2 No. 236-95
Central	Performance Standard for Split-System Central Air-Conditioners and Heat Pumps	CSA	CAN/CSA-C273.3-M91
	Performance Standard for Single Package Central Air-Conditioners and Heat Pumps	CSA	CAN/CSA-C656-M92
	Performance Standard for Rating Large Air Conditioners and Heat Pumps	CSA	CAN/CSA-C746-93
	Heating and Cooling Equipment (1995)	UL	UL 1995
Gas-Fired	Gas-Fired Absorption Summer Air Conditioning Appliances (with 1982 addenda)	AGA	ANSI Z21.40.1-1994
	Requirements for Gas-Fired, Engine-Driven Air Conditioning Appliances	AGA	4-89
	Requirements for Gas-Fired Desiccant Type Dehumidifiers and Air Conditioners	AGA	9-90
Packaged Terminal	Packaged Terminal Air Conditioners and Packaged Terminal Heat Pumps	ARI	ANSI/ARI 310/380-93
	Standards for Packaged Terminal Air-Conditioners and Heat Pumps	CSA	C744-93
Room	Room Air Conditioners	AHAM	ANSI/AHAM RA C-1-1992
	Method of Testing for Rating Room Air Conditioners and Packaged Terminal Air Conditioners	ASHRAE	ANSI/ASHRAE 16-1983 (R 88)
	Method of Testing for Rating Room Air Conditioner and Packaged Terminal Air Conditioner Heating Capacity	ASHRAE	ANSI/ASHRAE 58-1986 (R 90)
	Methods of Testing for Rating Room Fan-Coil Air Conditioners	ASHRAE	ANSI/ASHRAE 79-1984 (R 91)
	Performance Standard for Room Air Conditioners	CSA	CAN/CSA-C368.1-M90
	Room Air Conditioners	CSA	C22.2 No. 117-1970 (R 1992)
	Room Air Conditioners (1993)	UL	ANSI/UL 484
Unitary	Application of Sound Rating Levels - Outdoor Unitary Equipment	ARI	ARI 275-97
	Commercial and Industrial Unitary Air-Conditioning and Heat Pump Equipment	ARI	ANSI/ARI 340/360-93
	Sound Rating of Outdoor Unitary Equipment	ARI	ARI 270-95
	Unitary Air-Conditioning and Air-Source Heat Pump Equipment	ARI	ANSI/ARI 210/240-94
	Method of Testing for Rating Computer and Data Processing Room Unitary Air Conditioners	ASHRAE	ANSI/ASHRAE 127-1988
	Method of Rating Unitary Spot Air Conditioners	ASHRAE	ANSI/ASHRAE 128-1989
	Methods of Testing for Rating Heat-Operated Unitary Air-Conditioning Equipment for Cooling	ASHRAE	ANSI/ASHRAE 40-1980 (R 92)
	Methods of Testing for Rating Unitary Air-Conditioning and Heat Pump Equipment	ASHRAE	ANSI/ASHRAE 37-1988
	Methods of Testing for Rating Seasonal Efficiency of Unitary Air Conditioners and Heat Pumps	ASHRAE	ANSI/ASHRAE 116-1995
Air Conditioning	Comfort, Air Quality and Efficiency by Design (1997)	ACCA	ACCA Manual RS
	Commercial Applications Systems and Equipment (1993)	ACCA	ACCA Manual CS
	Load Calculation for Commercial Summer and Winter Air Conditioning (1988)	ACCA	ACCA Manual N
	Residential Load Calculation	ACCA	ACCA Manual J, 1986
	Environmental System Technology (1984)	NEBB	NEBB
	Installation of Air Conditioning and Ventilating Systems	NFPA	ANSI/NFPA 90A-1996
	Standard of Purity for Use in Mobile Air-Conditioning Systems	SAE	SAE J 1991-1989
	HVAC Systems Applications, 1st ed.	SMACNA	SMACNA
	HVAC Systems Duct Design, 3rd ed.	SMACNA	SMACNA
	Heating and Cooling Equipment (1995)	UL	UL 1995
		CSA	CAN/CSA-22.2 No. 36-95
Aircraft	Air Conditioning of Aircraft Cargo	SAE	SAE AIR 806B-1997
	Air Conditioning Systems for Subsonic Airplanes	SAE	ANSI/SAE ARP 85E-1991 (R 96)
	Aircraft Fuel Weight Penalty Due to Air Conditioning	SAE	SAE AIR 1168/8-1989
	Aircraft Ground Air Conditioning Service Connection	SAE	SAE AS 4262A-1997

Codes and Standards Published by Various Societies and Associations (*Continued*)

Subject	Title	Publisher	Reference
Aircraft (*Continued*)	Air Cycle Air Conditioning Systems for Military Air Vehicles	SAE	ANSI/SAE ARP 4073-1993
	Control of Excess Humidity in Avionics Cooling	SAE	SAE ARP 1987A-1997
	Engine Bleed Air Systems for Aircraft	SAE	ANSI/SAE ARP 1796-1987
	Guide for Qualification Testing of Aircraft Air Valves	SAE	ANSI/SAE ARP 986C-1997
	Nomenclature, Aircraft Air-Conditioning Equipment	SAE	SAE ARP 147C-1978 (R 1992)
	Testing of Commercial Airplane Environmental Control Systems	SAE	SAE ARP 217C-1997
Automotive	Automotive Air-Conditioning Hose	SAE	ANSI/SAE J 51-1989
	Design Guidelines for Air Conditioning Systems for Off-Road Operator Enclosures	SAE	SAE J 169-1985
	Extraction and Recycle Equipment for Mobile Automotive Air-Conditioning Systems	SAE	SAE J 1990-1992
	Guide to the Application and Use of Passenger Car Air-Conditioning Compressor Face Seals	SAE	SAE J 1954-1995
	Information Relating to Duty Cycles and Average Power Requirements of Truck and Bus Engine Accessories	SAE	SAE J 1343-1981
	Rating Air Conditioner Evaporator Air Delivery and Cooling Capacities	SAE	ANSI/SAE J 1487-1985
	Service Hose for Automotive Air Conditioning	SAE	SAE J 2196-1992 (R 1997)
	Test Method for Measuring Power Consumption of Air Conditioning and Brake Compressors for Trucks and Buses	SAE	ANSI/SAE J 1340-1981 (R 1996)
Ships	Mechanical Refrigeration and Air-Conditioning Installations Aboard Ship	ASHRAE	ANSI/ASHRAE 26-1996
Air Curtains	Air Curtains for Entranceways in Food and Food Service Establishments	NSF	ANSI/NSF 37-1992
	Test Methods for Air Curtain Units	AMCA	AMCA 220-91
	Air Terminals	ARI	ARI 880-94
	Method of Testing for Rating the Performance of Air Outlets and Inlets	ASHRAE	ANSI/ASHRAE 70-1991
	Standard Methods for Laboratory Airflow Measurement	ASHRAE	ANSI/ASHRAE 41.2-1987 (R 92)
	Rating the Performance of Residential Mechanical Ventilating Equipment	CSA	CAN/CSA-C260-M90
	Direct Gas-Fired Door Heaters	AGA	ANSI Z83.17-1990, Z83.17A-1991, Z83.17B-1992
Air Diffusion	Air Distribution Basics for Residential and Small Commercial Buildings	ACCA	ACCA Manual T, 1989
	Test Code for Grilles, Registers and Diffusers	ADC	ADC 1062:GRD-84
	Method of Testing for Rating the Performance of Air Outlets and Inlets	ASHRAE	ANSI/ASHRAE 70-1991
	Method of Testing for Room Air Diffusion	ASHRAE	ANSI/ASHRAE 113-1990
Air Filters	Industrial Ventilation: A Manual of Recommended Practice, 23rd ed. (1998), Selection of Air Filtration Equipment, p. 4.9	AGCIH	AGCIH
	Method for Measuring Performance of Portable Household Electrical Cord Connected Room Air Cleaners	AHAM	ANSI/AHAM AC-1-1988
	Commercial and Industrial Air Filter Equipment	ARI	ARI 850-93
	Residential Air Filter Equipment	ARI	ARI 680-93
	Agricultural Cabs-Environmental Air Quality. Part 1: Definitions, Test Methods, and Safety Practices	ASAE	ASAE S525-1-1-1997
	Part 2: Pesticide Vapor Filters-Procedure and Performance Criteria	ASAE	ASAE S525-2-1997
	Gravimetric and Dust-Spot Procedures for Testing Air-Cleaning Devices Used in General Ventilation for Removing Particulate Matter	ASHRAE	ANSI/ASHRAE 52.1-1992
	Method for Sodium Flame Test for Air Filters	BSI	BS 3928
	Particulate Air Filters for General Ventilation—Requirements, Testing Marking	BSI	BS EN 779:1993
	Electrostatic Air Cleaners (1995)	UL	ANSI/UL 867-1997
	High-Efficiency, Particulate, Air Filter Units (1996)	UL	UL 586
	Test Performance of Air Filter Units (1994)	UL	ANSI/UL 900-1995
Air-Handling Units	Commercial Applications Systems and Equipment (1993)	ACCA	ACCA Manual CS
	Central Station Air-Handling Units	ARI	ANSI/ARI 430-89
	Direct Gas-Fired Make-Up Air Heaters	AGA	ANSI Z83.4-1991, Z83.4A-1992
Air Leakage	Air Leakage Performance for Detached Single-Family Residential Buildings	ASHRAE	ANSI/ASHRAE 119-1988 (R 94)
	Method of Determining Air Change Rates in Detached Dwellings	ASHRAE	ANSI/ASHRAE 136-1993
	Standard Practices for Air Leakage Site Detection in Building Envelopes and Air Retarder Systems	ASTM	ASTM E 1186-87 (R 1998)
	Test Method for Determining Air Change in a Single Zone by Means of a Tracer Gas Dilution	ASTM	ASTM E 741-95
	Test Method for Determining Air Leakage Rate by Fan Pressurization	ASTM	ASTM E 779-87 (R 1992)
	Test Method for Determining the Rate of Air Leakage Through Exterior Windows, Curtain Walls, and Doors Under Specified Pressure and Temperature Differences Across the Specimen	ASTM	ASTM E 1424-91
	Test Method for Determining the Rate of Air Leakage Through Exterior Windows, Curtain Walls, and Doors Under Specified Pressure Differences Across the Specimen	ASTM	ASTM E 283-91
	Test Method for Field Measurement of Air Leakage Through Installed Exterior Window and Doors	ASTM	ASTM E 783-93
Boilers	A Guide to Clean and Efficient Operation of Coal Stoker-Fired Boilers	ABMA	ABMA
	Boiler Water Limits and Steam Purity Recommendations for Watertube Boilers	ABMA	ABMA
	Boiler Water Requirements and Associated Steam Purity—Commercial Boilers	ABMA	ABMA

Codes and Standards Published by Various Societies and Associations (*Continued*)

Subject	Title	Publisher	Reference
Boilers	Fluidized Bed Combustion Guidelines	ABMA	ABMA
(*Continued*)	Guidelines for Industrial Boiler Performance Improvement	ABMA	ABMA
	Matrix of Recommended Quality Control Requirements	ABMA	ABMA
	Operation and Maintenance Safety Manual	ABMA	ABMA
	Recommended Design Guidelines for Stoker Firing of Bituminous Coals	ABMA	ABMA
	(Selected) Summary of Codes and Standards of the Boiler Industry	ABMA	ABMA
	Thermal Shock Damage to Hot Water Boilers as a Result of Energy Conservation Measures	ABMA	ABMA
	Commercial Applications Systems and Equipment (1993)	ACCA	ACCA Manual CS
	Methods of Testing for Annual Fuel Utilization Efficiency of Residential Central Furnaces and Boilers	ASHRAE	ANSI/ASHRAE 103-1993
	Boiler and Pressure Vessel Code (11 sections) (1998)	ASME	ASME
	Boiler, Pressure Vessel, and Pressure Piping Code	CSA	CSA B51-95
	Testing and Rating Standard for Heating Boilers (1989)	HYDI	IBR
	Prevention of Furnace Explosions/Implosions in Multiple Burner Boilers	NFPA	ANSI /NFPA 8502-1995
	Heating, Water Supply, and Power Boilers—Electric (1995)	UL	ANSI/UL 834-1998
Gas or Oil	Gas-Fired Low-Pressure Steam and Hot Water Boilers	AGA	ANSI Z21.13-1991, Z21.13A-1993, Z21.13B-1994
	Gas Utilization Equipment in Large Boilers	AGA	ANSI Z83.3-1971, Z83.3A-1972, Z83.3B-1976 (R 1989)
	Requirements for High Pressure Steam Boilers	AGA	AGA 3-89
	Control and Safety Devices for Automatically Fired Boilers	ASME	ANSI/ASME CSD-1-1998
	Industrial and Commercial Gas-Fired Package Boilers	CGA	CAN1-3.1-77 (R 1985)
	Oil-Fired Steam and Hot-Water Boilers for Residential Use	CSA	B140.7.1-1976 (R 1991)
	Oil-Fired Steam and Hot-Water Boilers for Commercial and Industrial Use	CSA	B140.7.2-1967 (R 1991)
	Prevention of Furnace Explosions/Implosions in Multiple Burner Boilers	NFPA	ANSI/NFPA 8502-1995
	Single Burner Boiler Operations	NFPA	ANSI/NFPA 8501-1997
	Commercial-Industrial Gas Heating Equipment (1994)	UL	UL 795
	Oil-Fired Boiler Assemblies (1995)	UL	UL 726
	Standards and Typical Specifications for Deaerators, 6th ed. (1998)	HEI	HEI
Building Codes	ASTM Standards Used in Building Codes	ASTM	ASTM
	BOCA National Building Code, 14th ed. (1999)	BOCA	BOCA
	ICC International Property Maintenance Code, 1st ed. (1998)	ICC	ICC
	National Building Code of Canada (1995)	NRCC	NRCC
	International One- and Two-Family Dwelling Code (1998)	ICC	ICC
	International Energy Conservation Code (1998)	ICC	BOCA/ICBO/SBCCI
	Uniform Building Code, three volumes (1997)	ICBO	ICBO
	Directory of Building Codes and Regulations, State and City Volumes (annual)	NCSBCS	NCSBCS
	Standard Building Code (1997 with 1998 revisions)	SBCCI	SBCCI
Mechanical	Safety Code for Elevators and Escalators (plus two yearly supplements)	ASME	ANSI/ASME A 17.1-1996
	Natural Gas Installation Code	CGA	CAN/CGA-B149.1-M95
	Propane Installation Code	CGA	CAN/CGA-B149.2-M91
	Safety Code for Elevators	CSA	CAN/CSA-B44-94
	Uniform Mechanical Code (1997) (with Uniform Mechanical Code Standards)	ICBO	ICBO
		IAPMO	IAPMO
	Standard Gas Code (1997)	SBCCI	SBCCI
	International Mechanical Code (1998)	ICC	ICC
	International Fuel Gas Code (1997)	ICC	ICC
	Standard Mechanical Code (1997)	SBCCI	SBCCI
Burners	Guidelines for Burner Adjustments of Commercial Oil-Fired Boilers	ABMA	ABMA
	Domestic Gas Conversion Burners	AGA	ANSI Z21.17-1991, Z21.17A-1993, Z21.17B-1994
	Installation of Domestic Gas Conversion Burners	AGA	ANSI Z21.8-1994
	General Requirements for Oil Burning Equipment	CSA	CAN/CSA-B140.0-M87 (R 1991)
	Installation Code for Oil Burning Equipment	CSA	CAN/CSA-B139-M91
	Oil Burners: Atomizing-Type	CSA	CAN/CSA-B140.2.1-M90
	Pressure Atomizing Oil Burner Nozzles	CSA	B140.2.2-1971 (R 1991)
	Replacement Burners and Replacement Combustion Heads for Residential Oil Burners	CSA	B140.2.3-M1981 (R 1991)
	Vapourizing-Type Oil Burners	CSA	B140.1-1966 (R 1991)
	Commercial/Industrial Gas and/or Oil-Burning Assemblies with Emission Reduction Equipment (1993)	UL	ANSI/UL 2096-1995
	Commercial-Industrial Gas Heating Equipment (1994)	UL	UL 795-1994
	Oil Burners (1994)	UL	ANSI/UL 296-1995
	Waste Oil-Burning Air-Heating Appliances (1995)	UL	ANSI/UL 296A-1997
Chillers	Methods of Testing Liquid-Chilling Packages	ASHRAE	ASHRAE 30-1995
	Commercial Applications Systems and Equipment (1993)	ACCA	ACCA Manual CS
	Absorption Water-Chilling and Water Heating Packages	ARI	ARI 560-92
	Centrifugal and Rotary Screw Water-Chilling Packages	ARI	ANSI/ARI 550-92

Codes and Standards Published by Various Societies and Associations (*Continued*)

Subject	Title	Publisher	Reference
Chillers (*Continued*)	Positive Displacement Compressor Water-Chilling Packages	ARI	ANSI/ARI 590-92
	Performance Standard for Rating Packaged Water Chillers	CSA	C743-93
Chimneys	Design and Construction of Masonry Chimneys and Fireplaces	CSA	CAN/CSA-A405-M87
	Chimneys, Fireplaces, Vents, and Solid Fuel-Burning Appliances	NFPA	ANSI/NFPA 211-1996
	Medium Heat Appliance Factory-Built Chimneys (1995)	UL	ANSI/UL 959-1992
	Factory-Built Chimneys for Residential Type and Building Heating Appliance (1994)	UL	ANSI/UL 103-1995
Clean Rooms	Procedural Standards for Certified Testing of Cleanrooms, 2nd ed. (1996)	NEBB	NEBB
	Standard Practice for Continuous Sizing and Counting of Airborne Particles in Dust-Controlled Areas and Clean Rooms Using Instruments Capable of Detecting Single Sub-Micrometre and Larger Particles	ASTM	ASTM F 50-92 (R 1996)
Coils	Forced-Circulation Air-Cooling and Air-Heating Coils	ARI	ANSI/ARI 410-91
	Methods of Testing Forced Circulation Air Cooling and Air Heating Coils	ASHRAE	ASHRAE 33-1978
Comfort Conditions	Threshold Limit Values for Physical Agents (updated annually)	AGCIH	ACGIH
	Thermal Environmental Conditions for Human Occupancy	ASHRAE	ANSI/ASHRAE 55-1992
	Ergonomics—Determination of Metabolic Heat Production	ISO	ISO 8996:1990
	Comfort, Air Quality and Efficiency by Design (1997)	ACCA	ACCA Manual RS
	Ergonomics of the Thermal Environment—Estimation of the Thermal Insulation and Evaporative Resistance of a Clothing Ensemble	ISO	ISO 9920:1995
	Hot Environments—Estimation of the Heat Stress on Working Man, Based on the WBGT Index (Wet Bulb Globe Temperature)	ISO	ISO 7243:1989
	Moderate Thermal Environments—Determination of the PMV and PPD Indices and Specification of the Conditions for Thermal Comfort	ISO	ISO 7730:1994
Compressors	Compressors and Exhausters (reaffirmed 1986)	ASME	ANSI/ASME PTC 10-1997
	Displacement Compressors, Vacuum Pumps and Blowers	ASME	ANSI/ASME PTC 9-1970
	Safety Standard for Air Compressor Systems	ASME	ANSI/ASME B19.1-1995
	Safety Standard for Compressors for Process Industries	ASME	ANSI/ASME B19.3-1991
	Compressed Air and Gas Handbook, 5th ed. (1988)	CAGI	CAGI
Refrigerant	Ammonia Compressor Units	ARI	ANSI/ARI 510-93
	Method for Presentation of Compressor Performance Data	ARI	ARI 540-91
	Positive Displacement Condensing Units	ARI	ANSI/ARI 520-97
	Methods of Testing for Rating Positive Displacement Refrigerant Compressors and Condensing Units	ASHRAE	ANSI/ASHRAE 23-1993
	Safety Code for Mechanical Refrigeration	ASHRAE	ANSI/ASHRAE 15-1994
	Testing of Refrigerant Compressors	ISO	ISO 917:1989
	Refrigerant Compressors—Presentation of Performance Data	ISO	ISO 9309:1989
	Hermetic Refrigerant Motor-Compressors (1996)	UL	UL 984
		CSA	CAN/CSA-C22.2 No.140.2-M91
Computers	Method of Rating Computer and Data Processing Room Unitary Air Conditioners	ASHRAE	ANSI/ASHRAE 127-1988
	Protection of Electronic Computer/Data Processing Equipment	NFPA	ANSI/NFPA 75-1995
Condensers	Commercial Applications Systems and Equipment (1993)	ACCA	ACCA Manual CS
	Remote Mechanical-Draft Air-Cooled Refrigerant Condensers	ARI	ARI 460-94
	Remote Mechanical Draft Evaporative Refrigerant Condensers	ARI	ANSI/ARI 490-89
	Water-Cooled Refrigerant Condensers, Remote Type	ARI	ARI 450-93
	Methods of Testing for Rating Remote Mechanical-Draft Air-Cooled Refrigerant Condensers	ASHRAE	ASHRAE 20-1970
	Methods of Testing Remote Mechanical-Draft Evaporative Refrigerant Condensers	ASHRAE	ANSI/ASHRAE 64-1995
	Methods of Testing for Rating Water-Cooled Refrigerant Condensers	ASHRAE	ANSI/ASHRAE 22-1992
	Safety Code for Mechanical Refrigeration	ASHRAE	ANSI/ASHRAE 15-1994
	Steam Condensing Apparatus	ASME	ANSI/ASME PTC 12.2-1998
	Standards for Steam Surface Condensers, 9th ed. (1995)	HEI	HEI
	Standards for Direct Contact Barometric and Low Level Condensers, 6th ed. (1995)	HEI	HEI
Condensing Units	Commercial Applications Systems and Equipment (1993)	ACCA	ACCA Manual CS
	Commercial and Industrial Unitary Air-Conditioning Condensing Units	ARI	ARI 365-94
	Methods of Testing for Rating Positive Displacement Refrigerant Compressors and Condensing Units	ASHRAE	ANSI/ASHRAE 23-1993
	Heating and Cooling Equipment (1995)	UL	UL 1995
		CSA	CAN/CSA-C22.2 No. 236-M95
Containers	Series I Freight Containers—Classifications, Dimensions, and Ratings	ISO	ISO 668:1995
	Series I Freight Containers—Specifications and Testing—Part 2: Thermal Containers	ISO	ISO 1496-2:1996
	Animal Environment in Cargo Compartments	SAE	SAE AIR 1600A-1997

Codes and Standards Published by Various Societies and Associations (*Continued*)

Subject	Title	Publisher	Reference
Controls	Control Systems (1989)	AABC	National Standards, Ch 25
	Quick-Disconnect Devices for Use with Gas Fuel	AGA	ANSI Z21.41-1989, Z21.41A-1990, Z21.41B-1992
	Energy Management Control Systems Instrumentation	ASHRAE	ANSI/ASHRAE 114-1986
	BACnet—A Data Communication Protocol for Building Automation and Control Networks	ASHRAE	ANSI/ASHRAE 135-1995
	Performance Requirements for Electric Heating Line-Voltage Wall Thermostats	CSA	C273.4-M1978 (R 1992)
	Temperature-Indicating and Regulating Equipment	CSA	C22.2 No. 24-93
	Control Centers for Changing Message Type Electric Signals (1996)	UL	UL 1433
	Limit Controls (1994)	UL	ANSI/UL 353-1995
	Primary Safety Controls for Gas- and Oil-Fired Appliances (1994)	UL	ANSI/UL 372-1994
	Solid State Controls for Appliances (1994)	UL	ANSI/UL 244A-1995
	Temperature-Indicating and -Regulating Equipment (1994)	UL	UL 873
	Tests for Safety-Related Controls Employing Solid-State Devices (1995	UL	UL 991
	Process Control Equipment (1998)	UL	UL 3121-1
	Electrical Controls for Household and Similar Use, Part 1: General Requirements (1993)	UL	UL 8730-1
	Part 2: Particular Requirements for Thermal Motor Protectors for Motor-Compressors of Hermetic and Semi-Hermetic Type (1995)	UL	UL 8730-2-4
Commercial and Industrial	Industrial Control and Systems General Requirements	NEMA	ANSI/NEMA ICS 1-1993
	Industrial Control and Systems, Controllers, Contactors, and Overload Relays Rated Not More than 2000 Volts AC or 750 Volts DC	NEMA	ANSI/NEMA ICS 2-1993
	Instructions for the Handling, Installation, Operation and Maintenance of Motor Control Centers	NEMA	NEMA ICS 2.3-1995
	Preventive Maintenance of Industrial Control and Systems Equipment	NEMA	NEMA ICS 1.3-1986 (R 1991)
	Industrial Control Equipment (1993)	UL	UL 508
Residential	Automatic Gas Ignition Systems and Components	AGA	ANSI Z21.20-1993, Z21.20A-1994
	Gas Appliance Pressure Regulators	AGA	ANSI Z21.18-1993, Z21.18A-1994
	Gas Appliance Thermostats	AGA	ANSI Z21.23-1993, Z21.23A-1994
	Manually Operated Gas Valves for Appliances, Appliance Connector Valves and Hose End Valves	AGA	ANSI Z21.15-1992
	Manually-Operated Piezo Electric Spark Gas Ignition Systems and Components	AGA	ANSI Z21.77-1989, Z21.77A-1993
	Hot-Water Immersion Controls	NEMA	NEMA DC-12-1985 (R 1991)
	Line-Voltage Integrally Mounted Thermostats for Electric Heaters	NEMA	NEMA DC 13-1991 (1997)
	Residential Controls—Electrical Wall-Mounted Room Thermostats	NEMA	NEMA DC 3-1989
	Residential Controls—Surface Type Controls for Electric Storage Water Heaters	NEMA	NEMA DC 5-1989
	Residential Controls—Temperature Limit Controls for Electric Baseboard Heaters	NEMA	NEMA DC 10-1983 (R 1989)
	Residential Controls—Class 2 Transformers	NEMA	NEMA DC 20-1992
	Safety Guidelines for the Application, Installation, and Maintenance of Solid State Controls	NEMA	NEMA ICS 1.1-1984 (R 1988)
	Electrical Quick-Connect Terminals (1995)	UL	ANSI/UL 310-1996
Coolers	Refrigeration Equipment	CSA	CAN/CSA-C22.2 No. 120-M91
	Unit Coolers for Refrigeration	ARI	ARI 420-94
Air	Methods of Testing Forced Convection and Natural Convection Air Coolers for Refrigeration	ASHRAE	ANSI/ASHRAE 25-1990
	Commercial Bulk Milk Dispensing Equipment	NSF	ANSI/NSF 20-1998
Drinking Water	Self-Contained, Mechanically-Refrigerated Drinking-Water Coolers	ARI	ARI 1010-94
	Methods of Testing for Rating Drinking-Water Coolers with Self-Contained Mechanical Refrigeration Systems	ASHRAE	ANSI/ASHRAE 18-1987 (R 97)
	Drinking-Water Coolers (1993)	UL	ANSI/UL 399-1992
Food and Beverage	Methods of Testing for Rating Bottled and Canned Beverage Vending Machines	ASHRAE	ANSI/ASHRAE 32.1-1997
	Methods of Testing for Rating Pre-Mix and Post-Mix Soft Drink Vending and Dispensing Equipment	ASHRAE	ANSI/ASHRAE 32.2-1997
	Refrigerated Vending Machines (1995)	UL	UL 541
	Manual Food and Beverage Dispensing Equipment	NSF	ANSI/NSF 18-1996
Liquid	Refrigerant-Cooled Liquid Coolers, Remote Type	ARI	ARI 480-95
	Methods of Testing for Rating Liquid Coolers	ASHRAE	ANSI/ASHRAE 24-1989
	Liquid Cooling Systems	SAE	SAE AIR 1811A-1997
Cooling Towers	Special Systems: Cooling Tower Performance Tests (1989)	AABC	National Standards, Ch 22
	Commercial Applications Systems and Equipment (1993)	ACCA	ACCA Manual CS
	Bioaerosols: Assessment and Control (1999)	AGCIH	AGCIH
	Atmospheric Water Cooling Equipment	ASME	ANSI/ASME PTC 23-1986 (R 97)
	Water-Cooling Towers	NFPA	ANSI/NFPA 214-1996
	Acceptance Test Code for Spray Cooling Systems (1985)	CTI	CTI ATC-133
	Acceptance Test Code for Water Cooling Towers: Mechanical Draft, Natural Draft Fan Assisted Types, Evaluation of Results, and Thermal Testing of Wet/Dry Cooling Towers (1990)	CTI	CTI ATC-105
	Certification Standard for Commercial Water Cooling Towers (1991)	CTI	CTI STD-201
	Code for Measurement of Sound from Water Cooling Towers (1981)	CTI	CTI ATC-128
	Fiberglass-Reinforced Plastic Panels for Application on Industrial Water-Cooling Towers (1986)	CTI	CTI STD-131

Codes and Standards Published by Various Societies and Associations (*Continued*)

Subject	Title	Publisher	Reference
Cooling Towers (*Continued*)	Nomenclature for Industrial Water-Cooling Towers (1997)	CTI	CTI NCL-109
	Recommended Practice for Airflow Testing of Cooling Towers (1994)	CTI	CTI PFM-143
Crop Drying	Density, Specific Gravity, and Mass-Moisture Relationships of Grain for Storage	ASAE	ANSI/ASAE D241.4-1993
	Moisture Measurement—Forages	ASAE	ASAE S358.2-1993
	Moisture Measurement—Unground Grain and Seeds	ASAE	ASAE S352.2-1997
	Moisture Relationships of Plant-Based Agricultural Products	ASAE	ASAE D245.5-1995
	Resistance to Airflow of Grains, Seeds, Other Agricultural Products, and Perforated Metal Sheets	ASAE	ASAE D272.3-1996
Dehumidifiers	Commercial Applications Systems and Equipment (1993)	ACCA	ACCA Manual CS
	Bioaerosols: Assessment and Control (1999)	AGCIH	AGCIH
	Dehumidifiers	AHAM	ANSI/AHAM DH-1-1992
	Dehumidifiers	CSA	C22.2 No. 92-1971 (R 1992)
	Dehumidifiers (1993)	UL	ANSI/UL 474-1992
Desiccants	Method of Testing Desiccants for Refrigerant Drying	ASHRAE	ANSI/ASHRAE 35-1992
Dryers	Method of Testing Liquid Line Refrigerant Driers	ASHRAE	ANSI/ASHRAE 63.1-1995
	Liquid-Line Driers	ARI	ANSI/ARI 710-86
Ducts and Fittings	Fibrous Glass Duct Liner Standards (1994)	NAIMA	NAIMA AH 124
	Hose, Air Duct, Flexible Nonmetallic, Aircraft	SAE	SAE AS 1501C-1994
	Ducted Electric Heat Guide for Air Handling Systems, 1st ed.	SMACNA	SMACNA 1971
	Factory-Made Air Ducts and Air Connectors (1996)	UL	UL 181
	Marine Rigid and Flexible Air Ducting (1986)	UL	ANSI/UL 1136-1986
Construction	Industrial Ventilation: A Manual of Recommended Practice, 23rd ed. (1998) Construction Guidelines for Local Exhaust Systems, p. 5.19	ACGIH	ACGIH
	Preferred Metric Sizes for Flat Metal Products	ASME	ANSI/ASME B32.3M-1984 (R 94)
	Sheet Metal Welding Code	AWS	ANSI/AWS D9.1-90
	Pipes, Ducts and Fittings for Residential Type Air Conditioning Systems	CSA	B228.1-1968
	Fibrous Glass Duct Construction Standards	NAIMA	NAIMA AH 116
	Fibrous Glass Duct Construction with 1-1/2" Duct Boards (1997)	NAIMA	NAIMA AH 120
	Fibrous Glass Residential Duct Construction Standards (1993)	NAIMA	NAIMA AH 119
	Fibrous Glass Duct Construction Standards, 6th ed.	SMACNA	SMACNA 1992
	HVAC Duct Construction Standards Metal and Flexible, 2nd ed.	SMACNA	SMACNA 1995
	Rectangular Industrial Duct Construction Standards, 1st ed.	SMACNA	SMACNA 1980
	Round Industrial Duct Construction Standards, 1st ed.	SMACNA	SMACNA 1977
	Thermoplastic Duct (PVC) Construction Manual, 1st ed.	SMACNA	SMACNA 1974
Installation	Flexible Duct Performance and Installation Standards, 3rd ed. (1996)	ADC	ADC-91
	Installation of Air Conditioning and Ventilating Systems	NFPA	ANSI/NFPA 90A-1996
	Installation of Warm Air Heating and Air-Conditioning Systems	NFPA	ANSI/NFPA 90B-1996
Materials Specifications	Specification for General Requirements for Flat-Rolled Stainless and Heat-Resisting Steel Plate, Sheet and Strip	ASTM	ASTM A 480/A 480M-98A
	Specification for General Requirements for Steel, Sheet, Carbon, and High-Strength, Low-Alloy, Hot-Rolled and Cold-Rolled	ASTM	ASTM A 568/A 568M-98
	Specification for General Requirements for Steel Sheet, Metallic-Coated by the Hot-Dip Process	ASTM	ASTM A 924/A 924M-97A
	Specification for Steel, Carbon (0.15 Maximum, Percent), Hot-Rolled Sheet and Strip Commercial	ASTM	ASTM A 569/A 569M-98
	Specification for Commercial Steel (CS) Sheet, Carbon (0.15 Maximum, Percent), Cold-Rolled	ASTM	ASTM A 366/A 366-97
	Specification for Steel Sheet, Zinc-Coated (Galvanized) or Zinc-Iron Alloy-Coated (Galvannealed) by the Hot-Dipped Process	ASTM	ASTM A 653/A 653-98
System Design	Commercial Low Pressure, Low Velocity Duct System Design (1990)	ACCA	ACCA Manual Q
	Installation Techniques for Perimeter Heating and Cooling, 11th ed.	ACCA	ACCA Manual 4
	Residential Duct Systems (1995)	ACCA	ACCA Manual D
	Closure Systems for Use with Rigid Air Ducts and Air Connectors (1994)	UL	UL 181A
	Closure Systems for Use with Flexible Air Ducts and Air Connectors (1995)	UL	UL 181B
Testing	Special Systems: Duct Testing (1989)	AABC	National Standards, Ch 23
	Flexible Air Duct Test Code	ADC	ADC FD-72 (R 1979)
	Test Method for Measuring Acoustical and Airflow Performance of Duct Liner Materials and Prefabricated Silencers	ASTM	ASTM E 477-96
	HVAC Air Duct Leakage Test Manual, 1st ed.	SMACNA	SMACNA 1985
	HVAC Duct Systems Inspection Guide, 1st ed.	SMACNA	SMACNA 1989
Electrical	Voltage Ratings for Electrical Power Systems and Equipment	ANSI	ANSI C84.1-1989
	Test Method for Bond Strength of Electrical Insulating Varnishes by the Helical Coil Test	ASTM	ASTM D 2519-96
	Definite Purpose and Limited Duty Definite Purpose Magnetic Contactors	ARI	ANSI/ARI 780/790-97
	Canadian Electrical Code, Part I (17th ed.)	CSA	C22.1-1994
	Canadian Electrical Code, Part II—General Requirements	CSA	CAN/CSA-C22.2 No. 0-M91
	Application Guide for Ground Fault Circuit Interrupters	NEMA	NEMA 280-1990
	Application Guide for Ground Fault Protective Devices for Equipment	NEMA	ANSI/NEMA PB 2.2-1988 (R 94)

Codes and Standards Published by Various Societies and Associations (*Continued*)

Subject	Title	Publisher	Reference
Electrical (*Continued*)	Enclosures for Electrical Equipment (1000 Volts Maximum)	NEMA	ANSI/NEMA 250-1997
	Industrial Control and Systems Enclosures	NEMA	ANSI/NEMA ICS 6-1993
	General Requirements for Wiring Devices	NEMA	NEMA WD 1-1983 (R 1989)
	Low Voltage Cartridge Fuses	NEMA	ANSI/NEMA FU 1-1986
	Molded Case Circuit Breakers and Molded Case Switches	NEMA	NEMA AB 1-1993
	Industrial Control and Systems Terminal Blocks	NEMA	NEMA ICS 4-1993
	National Electrical Code	NFPA	ANSI/NFPA 70-1999
	National Fire Alarm Code	NFPA	ANSI/NFPA 72-1996
	Compatibility of Electrical Connectors and Wiring	SAE	ANSI/SAE AIR 1329A-1988
Energy	Air Conditioning and Refrigerating Equipment Nameplate Voltages	ARI	ARI 110-97
	Comfort, Air Quality and Efficiency by Design (1997)	ACCA	ACCA Manual RS
	Energy Efficient Design of New Buildings Except Low-Rise Residential Buildings	ASHRAE	ANSI/ASHRAE/IESNA 90.1-1989
	Energy-Efficient Design of New Low-Rise Residential Buildings	ASHRAE	ASHRAE 90.2-1993
	Energy Conservation in Existing Buildings	ASHRAE	ASHRAE 100-1995
	Standard Methods of Measuring and Expressing Building Energy Performance	ASHRAE	ANSI/ASHRAE 105-1984 (R 90)
	International Energy Conservation Code (1998)	ICC	ICC
	Uniform Solar Energy Code (1997)	IAPMO	IAPMO
	Model Energy Code, Thermal Envelope Compliance Guide	NAIMA	NAIMA BI407
	Energy Management Guide for Selection and Use of Polyphase Motors	NEMA	NEMA MG 10-1994
	Energy Management Guide for Selection and Use of Single-Phase Motors	NEMA	NEMA MG 11-1977 (R 1997)
	Energy Conservation Guidelines, 1st ed.	SMACNA	SMACNA 1984
	Energy Recovery Equipment and Systems, 2nd ed.	SMACNA	SMACNA 1991
	HVAC Systems Commissioning Manual, 1st ed.	SMACNA	SMACNA 1994
	Retrofit of Building Energy Systems and Processes	SMACNA	SMACNA 1982
	Energy Management Equipment (1994)	UL	ANSI/UL 916-1993
Exhaust Systems	Return and Exhaust Air Systems (1989)	AABC	National Standards, Ch 20
	Commercial Applications Systems and Equipment (1993)	ACCA	ACCA Manual CS
	Industrial Ventilation: A Manual of Recommended Practice, 23rd ed. (1998)	ACGIH	ACGIH
	Recirculation of Air from Industrial Process Exhaust Systems	AIHA	ANSI/AIHA Z9.7-1998
	Fundamentals Governing the Design and Operation of Local Exhaust Systems	ANSI	ANSI/AIHA Z9.2-1979 (R 1991)
	Laboratory Ventilation	AIHA	ANSI/AIHA Z9.5-1992
	Open-Surface Tanks—Ventilation and Operation	ANSI	ANSI/AIHA Z9.1-1991
	Safety Code for Design, Construction, and Ventilation of Spray Finishing Operations	AIHA	ANSI/AIHA Z9.3-1994
	Abrasive Blasting Operations—Ventilation and Safe Practices	AIHA	ANSI/AIHA Z9.4-1997
	Method of Testing Performance of Laboratory Fume Hoods	ASHRAE	ANSI/ASHRAE 110-1995
	Compressors and Exhausters	ASME	ANSI/ASME PTC 10-1997
	Mechanical Flue-Gas Exhausters	CSA	CAN 3-B255-M81
	Exhaust Systems for Air Conveying of Materials	NFPA	ANSI/NFPA 91-1995
	Draft Equipment (1993)	UL	UL 378
Expansion Valves	Thermostatic Refrigerant Expansion Valves	ARI	ARI 750-94
	Method of Testing for Capacity Rating of Thermostatic Refrigerant Expansion Valves	ASHRAE	ANSI/ASHRAE 17-1986 (R 90)
Fan-Coil Units	Industrial Ventilation: A Manual of Recommended Practice, 23rd ed. (1998) Installing Fan Coil Units, P. 8.4.11	ACGIH	ACGIH
	Room Fan-Coils and Unit Ventilators	ARI	ARI 440-97
	Methods of Testing for Rating Room Fan-Coil Air Conditioners	ASHRAE	ANSI/ASHRAE 79-1984 (R 91)
	Heating and Cooling Equipment (1995)	UL	UL 1995
Fans	Commercial Low Pressure, Low Velocity Duct System Design (1990)	ACCA	ACCA Manual Q
	Residential Duct Systems (1995)	ACCA	ACCA Manual D
	Industrial Ventilation: A Manual of Recommended Practice, 23rd ed. (1998)	ACGIH	ACGIH
	Designation of Rotation and Discharge of Centrifugal Fans	AMCA	AMCA 99-2406-98
	Drive Arrangements for Centrifugal Fans	AMCA	AMCA 99-2404-98
	Drive Arrangements for Tubular Centrifugal Fans	AMCA	AMCA 99-2410-98
	Recommended Safety Practices for Users and Installers of Industrial and Commercial Fans	AMCA	AMCA 410-90
	Inlet Box Positions for Centrifugal Fans	AMCA	AMCA 99-2405-98
	Motor Positions for Belt or Chain Drive Centrifugal Fans	AMCA	AMCA 99-2407-66-98
	Industrial Process/Power Generation Fans: Site Performance Test Standard	AMCA	AMCA 803-96
	Standards Handbook	AMCA	AMCA 99-86
	Fans and Blowers	ARI	ARI 670-96
	Methods for the Measurement of Noise Emitted by Small Air-Moving Devices	ASA	ANSI S12.11-1987 (R 1997)
	Laboratory Methods of Testing Fans for Rating AMCA Standard 210-85	ASHRAE AMCA	ANSI/ASHRAE 51-1985 ANSI/AMCA 210-85
	Laboratory Method of Testing In-Duct Sound Power Measurement Procedure for Fans	ASHRAE AMCA	ANSI/ASHRAE 68-1986 ANSI/AMCA 330-86
	Methods of Testing Fan Vibration—Blade Vibrations and Critical Speeds	ASHRAE	ANSI/ASHRAE 87.1-1992
	Fans	ASME	ANSI/ASME PTC 11-1984 (R 95)
	Fans and Ventilators	CSA	C22.2 No. 113-M1984 (R 93)

Codes and Standards Published by Various Societies and Associations (*Continued*)

Subject	Title	Publisher	Reference
Fans (*Continued*)	Rating the Performance of Residential Mechanical Ventilating Equipment	CSA	CAN/CSA-C260-M90
	Acoustics—Method for the Measurement of Airborne Noise Emitted by Small Air-Moving Devices	ISO	ISO 10302:1996
	Electric Fans (1994)	UL	UL 507
Fenestration	Specification for Classification of the Durability of Sealed Insulating Glass Units	ASTM	ASTM E 774-97
	Standard Practice for Calculation of Photometric Transmittance and Reflectance of Materials to Solar Radiation	ASTM	ASTM 971-88 (R 1996)
	Practice for Determining the Load Resistance of Glazing	ASTM	ASTM E 1300-97
	Standard Tables for Terrestrial Direct Normal Solar Spectral Irradiance for Air Mass 1.5	ASTM	ASTM E 891-87 (R 1992)
	Standard Tables for Terrestrial Solar Spectral Irradiance at Air Mass 1.5 for a 37° Tilted Surface	ASTM	ASTM E 892-87 (R 1992)
	Test Method for Accelerated Weathering of Sealed Insulating Glass Units	ASTM	ASTM E 773-97
	Test Method for Solar Absorptance, Reflectance and Transmittance of Materials Using Integrating Spheres	ASTM	ASTM E 903-96A
	Test Method for Solar Photometric Transmittance of Sheet Materials Using Sunlight	ASTM	ASTM E 972-96
	Test Method for Solar Transmittance (Terrestrial) of Sheet Materials Using Sunlight	ASTM	ASTM E 1084-86 (R 1996)
	Energy Performance Evaluation of Swinging Doors	CSA	CSA-A453-95
	Energy Performance Evaluation of Windows and Sliding Glass Doors	CSA	CAN/CSA-A440.2-93
	Windows	CSA	CAN/CSA-A440-M90
Filters	Comfort, Air Quality and Efficiency by Design (1997)	ACCA	ACCA Manual RS
	Industrial Ventilation: A Manual of Recommended Practice, 23rd ed. (1998)	ACGIH	ACGIH
	Flow-Capacity Rating and Application of Suction-Line Filters and Filter Driers	ARI	ANSI/ARI 730-86
	Specification for Octave-Band and Fractional-Octave-Band Analog and Digital Filters	ASA	ANSI S1.11-1986 (R 1998)
	Method of Testing Flow Capacity of Suction Line Filters and Filter Driers	ASHRAE	ANSI/ASHRAE 78-1985 (R 97)
	Method of Testing Liquid Line Filter-Drier Filtration Capability	ASHRAE	ANSI/ASHRAE 63.2-1996
	Exhaust Hoods for Commercial Cooking Equipment (1995)	UL	UL 710
	Grease Filters for Exhaust Ducts (1979)	UL	UL 1046
Fireplaces	Factory-Built Fireplaces (1996)	UL	UL 127
	Fireplace Stoves (1996)	UL	UL 737
Fire Protection	Standard Test Methods for Fire Teste of Building Construction and Materials	ASTM	ASTM E 119-98
	Standard Test Method for Surface Burning Characteristics of Building Materials	ASTM	ASTM E 84-98
		NFPA	NFPA 255-1996
	BOCA National Fire Prevention Code, 11th ed. (1999)	BOCA	BOCA
	Uniform Fire Code	IFCI	IFCI 1997
	Fire-Resistance Tests—Elements of Building Construction Part 1: General Requirements	ISO	ISO 834:1975
	Fire-Resistance Tests—Methods of Test of Fire Doors and Shutters	ISO	ISO 3008:1976
	Reaction to Fire Tests—Ignitability of Building Products Using a Radiant Heat Source	ISO	ISO 5657:1997
	Fire-Resistance Tests—Service Installations In Buildings—Fire-Resisting Ducts	ISO	ISO 6944:1985
	Interconnection Circuitry of Noncoded Remote-Station Protective Signalling Systems	NEMA	NEMA SB 3-1969 (R 1989)
	Fire Doors and Fire Windows	NFPA	ANSI/NFPA 80-1995
	Fire Hazard Properties of Flammable Liquids, Gases, and Volatile Solids	NFPA	ANSI/NFPA 325-1994
	Fire Prevention Code	NFPA	ANSI/NFPA 1-1997
	Fire Protection for Laboratories Using Chemicals	NFPA	ANSI/NFPA 45-1996
	Fire Protection Handbook, 18th ed. (1996)	NFPA	NFPA
	Flammable and Combustible Liquids Code	NFPA	ANSI/NFPA 30-1996
	Health Care Facilities	NFPA	ANSI/NFPA 99-1996
	Installation of Sprinkler Systems	NFPA	NFPA 13-1996
	Life Safety Code	NFPA	ANSI/NFPA 101-1997
	National Fire Alarm Code	NFPA	ANSI/NFPA 72-1996
	National Fire Codes (issued annually)	NFPA	NFPA
	Methods of Fire Tests of Door Assemblies	NFPA	ANSI/NFPA 252-1995
	Standard Fire Prevention Code (1997)	SBCCI	SBCCI
	Fire, Smoke and Radiation Damper Installation Guide for HVAC Systems, 4th ed.	SMACNA	SMACNA 1992
	Fire Dampers (1995)	UL	UL 555
	Fire Tests of Building Construction and Materials (1997)	UL	ANSI/UL 263-1996
	Fire Tests of Door Assemblies (1997)	UL	UL 10B
	Fire Tests of Through-Penetration Firestops (1994)	UL	ANSI/UL 1479-1995
	Heat Responsive Links for Fire-Protection Service (1993)	UL	ANSI/UL 33-1995
Smoke Management	Commissioning Smoke Management Systems	ASHRAE	ASHRAE Guideline 5-1994
	Recommended Practice for Smoke Control Systems	NFPA	ANSI/NFPA 92A-1996
	Guide for Smoke Management Systems in Malls, Atria, and Large Areas	NFPA	ANSI/NFPA 92B-1995
	Leakage Rated Dampers for Use in Smoke Control Systems (1996)	UL	UL 555S

Codes and Standards Published by Various Societies and Associations (*Continued*)

Subject	Title	Publisher	Reference
Freezers	Capacity Measurement and Energy Consumption Test Methods for Refrigerators, Combination Refrigerator-Freezers, and Freezers	CSA	CAN/CSA-C300-M91
	Refrigeration Equipment	CSA	CAN/CSA-C22.2 No. 120-M91
Commercial	Dispensing Freezers	NSF	ANSI/NSF 6-1996
	Ice Cream Makers (1993)	UL	ANSI/UL 621-1992
	Commercial Refrigerators and Freezers (1995)	UL	ANSI/UL 471-1996
	Ice Makers (1995)	UL	ANSI/UL 563-1995
Household	Household Refrigerators, Combination Refrigerator-Freezers and Household Freezers	AHAM	ANSI/AHAM HRF-1-1988
	Household Refrigerators and Freezers (1993)	UL	ANSI/UL 250-1997
		CSA	C22.2 No. 63-93
Fuels	Threshold Limit Values for Chemical Substances (updated annually)	AGCIH	AGCIH
	Standard Classification of Coals by Rank	ASTM	ANSI/ASTM D388-98
	Specification for Diesel Fuel Oils	ASTM	ANSI/ASTM D 975-97
	Specification for Fuel Oils	ASTM	ANSI/ASTM D 396-97
	Specification for Gas Turbine Fuel Oils	ASTM	ANSI/ASTM D 2880-98
	Reporting of Fuel Properties when Testing Diesel Engines with Alternative Fuels Derived from Biological Materials	ASAE	ASAE EP552-1996
Furnaces	Commercial Applications Systems and Equipment (1993)	ACCA	ACCA Manual CS
	Residential Equipment Selection (1995)	ACCA	ACCA Manual S
	Methods of Testing for Annual Fuel Utilization Efficiency of Residential Central Furnaces and Boilers	ASHRAE	ANSI/ASHRAE 103-1993
	Prevention of Furnace Explosions/Implosions in Multiple Burner Boilers	NFPA	ANSI /NFPA 8502-1995
	Standard Mechanical Code (1997)	SBCCI	SBCCI
	Heating and Cooling Equipment (1995)	UL	UL 1995
		CSA	CAN/CSA-C22.2 No. 236-95
	Residential Gas Detectors (1994)	UL	ANSI/UL 1484-1994
	Single and Multiple Station Carbon Monoxide Alarms (1996)	UL	UL 2034
Gas	Direct Vent Central Furnaces	AGA	ANSI Z21.64-1990
	Gas-Fired Central Furnaces	AGA	ANSI Z21.47-1993
		CGA	CAN/CGA-2.3-M93
	Gas-Fired Duct Furnaces	AGA	ANSI Z83.9-1990, Z83.9A-1992
	Gas-Fired Gravity and Fan Type Direct Vent Wall Furnaces	AGA	ANSI Z21.44-1995
	Gas-Fired Gravity and Fan Type Floor Furnaces	AGA	Z21.48-1992
	Gas-Fired Gravity and Fan Type Vented Wall Furnaces	AGA	Z21.49-1992
	Gas-Fired Duct Furnaces	CGA	CAN/CGA-2.8-M86
	Gas-Fired Gravity and Fan Type Direct Vent Wall Furnaces	CGA	CAN1-2.19-M81
	Gas-Fired Gravity and Fan Type Vented Wall Furnaces	CGA	CAN/CGA-2.5-M86
	Industrial and Commercial Gas-Fired Package Furnaces	CGA	CGA 3.2-1976
	International Fuel Gas Code (1997)	ICC	ICC
	Standard Gas Code(1997)	SBCCI	SBCCI
	Commercial-Industrial Gas Heating Equipment (1994)	UL	UL 795
Oil	Standard Specification for Fuel Oils	ASTM	ANSI/ASTM D 396-97
	Test Method for Smoke Density in Flue Gases from Burning Distillate Fuels	ASTM	ANSI/ASTM D 2156-94
	Oil Burning Stoves and Water Heaters	CSA	B140.3-1962 (R 1991)
	Oil-Fired Warm Air Furnaces	CSA	B140.4-1974 (R 1991)
	Installation of Oil-Burning Equipment	NFPA	ANSI/NFPA 31-1997
	Oil-Fired Central Furnaces (1994)	UL	UL 727
	Oil-Fired Floor Furnaces (1994)	UL	ANSI/UL 729-1995
	Oil-Fired Wall Furnaces (1994)	UL	ANSI/UL 730-1995
Solid Fuel	Standard Classification of Coals by Rank	ASTM	ANSI/ASTM D975-98
	Installation Code for Solid-Fuel-Burning Appliances and Equipment	CSA	CAN/CSA-B365-M91
	Solid-Fuel-Fired Central Heating Appliances	CSA	CAN/CSA-B366.1-M91
	Solid-Fuel and Combination-Fuel Central and Supplementary Furnaces (1995)	UL	ANSI/UL 391-1997
Heaters	Gas-Fired Infrared Heaters	AGA	ANSI Z83.6-1990, Z83.6A-1992, Z83.6B-1993
	Requirements for Gas-Fired Infrared Patio	AGA	5-90
	Threshold Limit Values for Chemical Substances (updated annually)	AGCIH	AGCIH
	Industrial Ventilation: A Manual of Recommended Practice, 23rd ed. (1998) Replacement Air Heating Equipment, p. 7.11	AGCIH	AGCIH
	Requirements for Residential Radiant Tube Heaters	AGA	7-89
	Thermal Performance Testing of Solar Ambient Air Heaters	ASAE	ASAE S423-1993
	Air Heaters	ASME	ANSI/ASME PTC 4.3-1968 (R 91)
	Direct Gas-Fired Non-Recirculating Make-up Air Heaters	CGA	CAN1-3.7-77
	Gas-Fired Infra-Red Heaters	CGA	CAN1-2.16-M81
	Electric Air Heaters	CSA	C22.2 No.46-M1988
	Electric Duct Heaters	CSA	C22.2 No. 155-M1986 (R 1992)
	Portable Kerosine-Fired Heaters	CSA	CAN 3-B140.9.3 M86

Codes and Standards Published by Various Societies and Associations (*Continued*)

Subject	Title	Publisher	Reference
Heaters	Standards for Closed Feedwater Heaters, 6th ed. (1998)	HEI	HEI
(*Continued*)	Electric Dry Bath Heaters (1994)	UL	UL 875
	Electric Heating Appliances (1997)	UL	UL 499
	Electric Oil Heaters (1996)	UL	UL 574
	Oil-Burning Stoves (1993)	UL	ANSI/UL 896-1997
	Oil-Fired Air Heaters and Direct-Fired Heaters (1993)	UL	UL 733
Combination	Requirements for Gas-Fired Combination Space Heating/Water Heating Appliances	AGA	11-90
Engine	Electric Engine Preheaters and Battery Warmers for Diesel Engines	SAE	SAE J 1310-1993
	Fuel Warmer—Diesel Engines	SAE	ANSI/SAE J 1422-1996
	Selection and Application Guidelines for Diesel, Gasoline, and Propane Fired Liquid Cooled Engine Pre-Heaters	SAE	SAE J 1350-1981
Nonresidential	Direct Gas-Fired Industrial Air Heaters	AGA	ANSI Z83.18-1990, Z83.18A-1991, Z83.18B-1992
	Gas-Fired Construction Heaters	AGA	ANSI Z83.7-1990, Z83.7A-1991, Z83.7B-1993
	Gas-Fired Unvented Commercial and Industrial Heaters	AGA	ANSI Z83.16-1982, Z83.16A-1984, Z83.16B-1989
	Requirements for Direct Gas-Fired Circulating Heaters for Agricultural Buildings	AGA	5-88
	Requirements for High Pressure LP Infrared Poultry and Livestock Heating Systems	AGA	4-87
	Portable Industrial Oil-Fired Heaters	CSA	B140.8-1967 (R 1991)
	Fuel-Fired Heaters—Air Heating—for Construction and Industrial Machinery	SAE	ANSI/SAE J 1024-1989
	Commercial-Industrial Gas Heating Equipment (1994)	UL	UL 795
	Electric Heaters for Use in Hazardous (Classified) Locations (1995)	UL	ANSI/UL 823-1996
Pool	Gas-Fired Pool Heaters	AGA	ANSI Z21.56-1994
	Method of Testing and Rating Pool Heaters	ASHRAE	ANSI/ASHRAE 146-1998
	Oil-Fired Service Water Heaters and Swimming Pool Heaters	CSA	B140.12-1976 (R 1991)
Room	Gas-Fired Room Heaters, Vol. I, Vented Room Heaters	AGA	ANSI Z21.11.1-1991, Z21.11.1A-1993, Z21.11.1B-1995
	Gas-Fired Room Heaters, Vol. II, Unvented Room Heaters	AGA	ANSI Z21.11.2-1992 Z21.11.2A-1993, Z21.11.2B-1995
	Gas-Fired Unvented Catalytic Room Heaters for Use with Liquefied Petroleum (LP) Gases	AGA	ANSI Z21.76-1994
	Requirements for Gas-Fired Vented Catalytic Type Room Heaters	AGA	1-81
	Requirements for Unvented Room Heaters Equipped with Oxygen Depletion Safety Shutoff Systems	AGA	2-79
	Fixed and Location-Dedicated Electric Room Heaters (1997)	UL	UL 2021
	Movable and Wall- or Ceiling-Hung Electric Room Heaters (1994)	UL	UL 1278
	Room Heaters, Solid Fuel-Type (1996)	UL	ANSI/UL 1482-1998
	Unvented Kerosene-Fired Room Heaters and Portable Heaters (1993)	UL	UL 647
Transport	Heater, Aircraft Internal Combustion Heat Exchanger Type	SAE	SAE AS 8040A-1996
	Heater, Airplane, Engine Exhaust Gas to Air Heat Exchanger Type	SAE	SAE ARP 86A-1952 (R 1992)
	Installation, Heaters, Airplane, Internal Combustion Heater Exchange Type	SAE	SAE ARP 266-1952 (R 1992)
	Motor Vehicle Heater Test Procedure	SAE	SAE J 638-1982 (R 1993)
Unit	Gas Unit Heaters	AGA	ANSI Z83.8-1990, Z83.8A-1990, Z83.8B-1992
	Gas Unit Heaters	CGA	CAN/CGA-2.6-M86
	Oil-Fired Unit Heaters (1995)	UL	ANSI/UL 731-1995
Heat Exchangers	Remote Mechanical-Draft Evaporative Refrigerant Condensers	ARI	ANSI/ARI 490-89
	Method of Testing Air-to-Air Heat Exchangers	ASHRAE	ANSI/ASHRAE 84-1991
	Standard Methods of Test for Rating the Performance of Heat-Recovery Ventilators	CSA	CAN/CSA-C439-88
	Standards for Power Plant Heat Exchangers, 3rd ed. (1998)	HEI	HEI
	Standards of Tubular Exchanger Manufacturers Association, 7th ed. (1988)	TEMA	TEMA
Heating	Commercial Applications Systems and Equipment (1993)	ACCA	ACCA Manual CS
	Comfort, Air Quality and Efficiency by Design (1997)	ACCA	ACCA Manual RS
	Residential Equipment Selection (1995)	ACCA	ACCA Manual S
	Determining the Required Capacity of Residential Space Heating and Cooling Appliances	CSA	CAN/CSA-F280-M90
	Automatic Flue-Pipe Dampers for Use with Oil-Fired Appliances	CSA	B140.14-M1979 (R 1991)
	Heater Elements	CSA	C22.2 No.72-M1984 (R 1992)
	Advanced Installation Guide for Hydronic Heating Systems (1991)	HYDI	IBR 250
	Heat Loss Calculation Guide	HYDI	IBR H-21 (1984), IBR H-22 (1998)
	Installation Guide for Residential Hydronic Heating Systems, 6th ed. (1988)	HYDI	IBR 200
	Radiant Floor Heating (1993)	HYDI	IBR 400
	Environmental System Technology (1984)	NEBB	NEBB
	Pulverized Fuel Systems	NFPA	ANSI/NFPA 8503-1993
	Aircraft Electrical Heating Systems	SAE	ANSI/SAE AIR 860-1965 (R 1992)
	Performance Test for Air-Conditioned, Heated, and Ventilated Off-Road Self-Propelled Work Machines	SAE	SAE J 1503-1995
	Heating Value of Fuels	SAE	SAE J 1498-1990
	HVAC Systems Applications, 1st ed.	SMACNA	SMACNA 1987

Codes and Standards Published by Various Societies and Associations (*Continued*)

Subject	Title	Publisher	Reference
Heating (*Continued*)	Electric Baseboard Heating Equipment (1994)	UL	ANSI/UL 1042-1995
	Electric Duct Heaters (1996)	UL	UL 1996
	Heating and Cooling Equipment (1995)	UL	UL 1995
		CSA	CAN/CSA-C22.2 No. 236-95
Heat Pumps	Commercial Applications Systems and Equipment (1993)	ACCA	ACCA Manual CS
	Geothermal Heat Pump Training Certification Program (1996)	ACCA	ACCA Training Manual
	Heat Pumps Systems, Principles and Applications, 2nd ed. (1984)	ACCA	ACCA Manual H
	Residential Equipment Selection (1995)	ACCA	ACCA Manual S
	Industrial Ventilation: A Manual of Recommended Practice, 23rd ed. (1998) HVAC Components and System Types, p. 8.3	ACGIH	ACGIH
	Commercial and Industrial Unitary Air-Conditioning and Heat Pump Equipment	ARI	ARI 340/360-93
	Ground Source Closed-Loop Heat Pumps	ARI	ANSI/ARI 330-93
	Ground Water-Source Heat Pumps	ARI	ANSI/ARI 325-93
	Water-Source Heat Pumps	ARI	ANSI/ARI 320-93
	Methods of Testing for Rating Unitary Air-Conditioning and Heat Pump Equipment	ASHRAE	ANSI/ASHRAE 37-1988
	Methods of Testing for Rating Seasonal Efficiency of Unitary Air-Conditioners and Heat Pumps	ASHRAE	ANSI/ASHRAE 116-1995
	Installation Requirements for Air-to-Air Heat Pumps	CSA	C273.5-1980 (R 1991)
	Performance Standard for Split-System Central Air-Conditioners and Heat Pumps	CSA	CAN/CSA-C273.3-M91
	Heating and Cooling Equipment (1995)	UL	UL 1995
		CSA	CAN/CSA C22.2 No. 236-95
Gas-Fired	Requirements for Gas-Fired, Absorption and Adsorption Heat Pumps	AGA	10-90
Heat Recovery	Gas Turbine Heat Recovery Steam Generators	ASME	ANSI/ASME PTC 4.4-1981 (R 1992)
	Energy Recovery Equipment and Systems, 2nd ed.	SMACNA	SMACNA 1991
	Requirements for Heat Reclaimer Devices for Use with Gas-Fired Appliances	AGA	ANSI Z21.40.1-1994
Humidifiers	Method for Measuring Performance of Appliance Humidifiers	AHAM	ANSI/AHAM HU-1-1987
	Comfort, Air Quality and Efficiency by Design (1997)	ACCA	ACCA Manual RS
	Commercial Applications Systems and Equipment (1993)	ACCA	ACCA Manual CS
	Bioaerosols: Assessment and Control (1999)	AGCIH	AGCIH
	Central System Humidifiers for Residential Applications	ARI	ANSI/ARI 610-96
	Commercial and Industrial Humidifiers	ARI	ANSI/ARI 640-96
	Self-Contained Humidifiers for Residential Applications	ARI	ANSI/ARI 620-96
	Humidifiers (1993)	UL	UL 998
		CSA	C22.2 No. 104-93
Ice Makers	Automatic Commercial Ice Makers	ARI	ARI 810-95
	Ice Storage Bins	ARI	ARI 820-95
	Methods of Testing Automatic Ice Makers	ASHRAE	ANSI/ASHRAE 29-1988
	Refrigeration Equipment	CSA	CAN/CSA-C22.2 No. 120-M91
	Standards for Safety, Ice Makers	CSA	CSA 133-1964
	Performance of Automatic Ice-Makers and Ice Storage Bins	CSA	CAN/CSA-C742-94
	Automatic Ice Making Equipment	NSF	ANSI/NSF 12-1992
	Ice Makers (1995)	UL	ANSI/UL 563-1997
Incinerators	Large Incinerators	ASME	ANSI/ASME PTC 33-1978 (R 1991)
	Incinerators and Waste and Linen Handling Systems and Equipment	NFPA	ANSI/NFPA 82-1994
	Residential Incinerators (1993)	UL	UL 791
Indoor Air Quality	Ventilation for Acceptable Indoor Air Quality	ASHRAE	ANSI/ASHRAE 62-1989
	Bioaerosols: Assessment and Control (1999)	AGCIH	AGCIH
	Standard Practice for Continuous Sizing and Counting of Airborne Particles in Dust-Controlled Areas and Clean Rooms Using Instruments Capable of Detecting Single Sub-Micrometre and Larger Particles	ASTM	ASTM F 50-92 (R 1996)
	Practice for Referencing Suprathreshold Odor Intensity	ASTM	ASTM E 544-75 (R 1997)
	Ambient Air—Determination of Mass Concentration of Nitrogen Dioxide— Modified Griess-Saltzman Method	ISO	ISO 6768:1985
	Air Quality- Presentation of Ambient Air Quality Data in Alphanumerical Form	ISO	ISO 7168:1985
	Comfort, Air Quality and Efficiency by Design (1997)	ACCA	ACCA Manual RS
Induction Units	Room Air-Induction Units	ARI	ARI 445-87 (RA 93)
	Frame Assignments for Alternating Current Integral-Horsepower Induction Motors	NEMA	NEMA MG1-1993
Industrial Duct	Rectangular Industrial Duct Construction Standards, 1st ed.	SMACNA	SMACNA 1980
	Round Industrial Duct Construction Standards, 1st ed.	SMACNA	SMACNA 1977
Insulation	Classification for Rating Sound Insulation	ASTM	ASTM E 413-87 (R 1994)
	Classification of Potential Health and Safety Concerns Associated with Thermal Insulation Materials and Accessories	ASTM	ASTM C 930-92
	Practice for Prefabrication and Field Fabrication of Thermal Insulating Fitting Covers for NPS Piping, Vessel Lagging, and Dished Head Segments	ASTM	ASTM C450-94

Codes and Standards Published by Various Societies and Associations (*Continued*)

Subject	Title	Publisher	Reference
Insulation (*Continued*)	Specification for Adhesives for Duct Thermal Insulation	ASTM	ASTM C 916-85 (R 1996)
	Specification for Fibrous Glass Duct Lining Insulation (Thermal and Sound Absorbing Material)	ASTM	ASTM C 1071-98
	Specification for Preformed Flexible Elastometric Cellular Thermal Insulation in Sheet and Tabular Form	ASTM	ASTM C534-94
	Specification for Cellular Glass Thermal Insulation	ASTM	ASTM C552-91
	Specification for Rigid, Cellular Polystyrene Thermal Insulation	ASTM	ASTM C578-95
	Specification for Unfaced Preformed Rigid Cellular Polyisocyanurate Thermal Insulation	ASTM	ASTM C591-94
	Specification for Faced or Unfaced Rigid Cellular Phenolic Thermal Insulation	ASTM	ASTM C1126-98
	Standard Practice for Determination of Heat Gain or Loss and the Surface Temperature of Insulated Pipe and Equipment Systems by the Use of a Computer Program	ASTM	ASTM C 680-89 (R 1995)
	Practice for Inner and Outer Diameters of Rigid Thermal Insulation for Nominal Sizes of Pipe and Tubing (NPS System))	ASTM	ASTM C 585-90 (R 1998)
	Practice for Thermographic Inspection of Insulation Installations in Envelope Cavities of Frame Buildings	ASTM	ASTM C 1060-90 (R 1997)
	Terminology Relating to Thermal Insulating Materials	ASTM	ASTM C 168-97
	Test Method for Steady-State Heat Flux Measurements and Thermal Transmission Properties by Means of the Guarded-Hot-Plate Apparatus	ASTM	ASTM C 177-97
	Test Method for Steady-State Heat Flux Measurements and Thermal Transmission Properties by Means of the Heat Flow Meter Apparatus	ASTM	ASTM C 518-98
	Test Method for Steady-State Heat Transfer Properties of Horizontal Pipe Insulations	ASTM	ASTM C 335-95
	Test Method for Steady-State and Thermal Performance of Building Assemblies by Means of a Guarded Hot Box	ASTM	ASTM C 236-89 (R 1993)
	Guidelines for Use of Thermal Insulation in Agricultural Buildings	ASAE	ASAE S401.2-1993
	Thermal Insulation, Mineral Fibre, for Buildings	CSA	A101-M1983
	Thermal Insulation—Definition of Terms	ISO	ISO 9229:1991
	National Commercial and Industrial Insulation Standards, 5th ed.	MICA	MICA 1999
Louvers	Test Methods for Louvers, Dampers and Shutters	AMCA	AMCA 500-89
	Laboratory Methods for Testing Dampers for Rating	AMCA	AMCA 500D-97
Lubricants	Methods of Testing the Floc Point of Refrigeration Grade Oils	ASHRAE	ANSI/ASHRAE 86-1994
	Classification of Industrial Fluid Lubricants by Viscosity System	ASTM	ANSI/ASTM D 2422-97
	Test Method for Mean Molecular Weight of Mineral Insulating Oils by the Cryoscopic Method	ASTM	ASTM D 2224-78 (R 1983)
	Test Method for Molecular Weight (Relative Molecular Mass) of Hydrocarbons by Thermoelectric Measurement of Vapor Pressure	ASTM	ASTM D 2503-92 (R 1997)
	Test Method for Pour Point of Petroleum Products	ASTM	ASTM D 97-96A
	Petroleum Industry—Corrosiveness to Copper—Copper Strip Test	ISO	ISO 2160:1985
	Semiconductor Graphite	NEMA	NEMA CB 4-1989 (R 1995)
Measurements	Industrial Ventilation: A Manual of Recommended Practice, 23rd ed. (1998)	AGCIH	AGCIH
	Engineering Analysis of Experimental Data	ASHRAE	ASHRAE Guideline 2-1986(R 96)
	Method of Measuring Solar-Optical Properties of Materials	ASHRAE	ANSI/ASHRAE 74-1988
	Standard Method for Measurement of Proportion of Lubricant in Liquid Refrigerant	ASHRAE	ASHRAE 41.4-1996
	Standard Method for Measurement of Moist Air Properties	ASHRAE	ANSI/ASHRAE 41.6-1994
	Standard Methods of Measuring and Expressing Building Energy Performance	ASHRAE	ANSI/ASHRAE 105-1984 (R 90)
	Measurement of Industrial Sound	ASME	ANSI/ASME PTC 36-1985
	Measurement Uncertainty	ASME	ANSI/ASME PTC 19.1-1998
	Method for Establishing Installation Effects on Flowmeters	ASME	ANSI/ASME MFC-10M-1994
	Procedure for Bench Calibration of Tank Level Gaging Tapes and Sounding Rules	ASME	ANSI/ASME MC88.2-1974 (R 95)
	Specification for Temperature-Electromotive Force (EMF) Tables for Standardized Thermocouples	ASTM	ASTM E 230-96
	Standard Practice for Continuous Sizing and Counting of Airborne Particles in Dust-Controlled Areas and Clean Rooms Using Instruments Capable of Detecting Single Sub-Micrometre and Larger Particles	ASTM	ASTM F 50-92 (R 1996)
	Standard for Use of the International System of Units (SI): The Modern Metric System	IEEE/ ASTM	ANSI/IEEE/ASTM SI 10-97
	Test Methods for Water Vapor Transmission of Materials	ASTM	ASTM E 96-95
	Ergonomics—Determination of Metabolic Heat Production	ISO	ISO 8996:1990
	Ergonomics of the Thermal Environment—Estimation of the Thermal Insulation and Evaporative Resistance of a Clothing Ensemble	ISO	ISO 9920:1995
	Thermal Environments—Instruments and Methods for Measuring Physical Quantities	ISO	ISO 7726:1985
Fluid Flow	Standard Methods of Measurement of Flow of Liquids in Pipes Using Orifice Flowmeters	ASHRAE	ANSI/ASHRAE 41.8-1989
	Standard Calorimeter Test Method for Flow Measurement of a Volatile Refrigerant	ASHRAE	ANSI/ASHRAE 41.9-1988
	Application of Fluid Meters	ASME	ASME 19.5-1972
	Fluid Flow in Closed Conduits—Connections for Pressure Signal Transmissions Between Primary and Secondary Devices	ASME	ANSI/ASME MFC-8M-1988

Codes and Standards Published by Various Societies and Associations (*Continued*)

Subject	Title	Publisher	Reference
Fluid Flow (*Continued*)	Glossary of Terms Used in the Measurement of Fluid Flow in Pipes	ASME	ANSI/ASME MFC-1M-1991
	Measurement of Fluid Flow by Means of Coriolis Mass Flowmeters	ASME	ANSI/ASME MFC-11M-1989 (R 94)
	Measurement of Fluid Flow in Pipes Using Orifice, Nozzle, and Venturi	ASME	ASME MFC-3M-1989 (R 1995)
	Measurement of Fluid Flow in Pipes Using Vortex Flow Meters	ASME	ASME/ANSI MFC-6M-1998
	Measurement of Fluid Flow Using Small Bore Precision Orifice Meters	ASME	ANSI/ASME MFC-14M-1995
	Measurement of Liquid Flow in Closed Conduits by Weighting Method	ASME	ANSI/ASME MFC-9M-1988
	Measurement of Liquid Flow in Closed Conduits Using Transit-Time Ultrasonic Flowmeters	ASME	ANSI/ASME MFC-5M-1985 (R 1994)
	Measurement Uncertainty for Fluid Flow in Closed Conduits	ASME	ANSI/ASME MFC-2M-1983 (R 88)
	Measurement of Fluid Flow in Closed Conduits—Velocity Area Method Using Pitot Static Tubes	ISO	ISO 3966:1977
Gas Flow	Standard Methods for Laboratory Airflow Measurement	ASHRAE	ANSI/ASHRAE 41.2-1987 (R 92)
	Standard Method for Measurement of Flow of Gas	ASHRAE	ANSI/ASHRAE 41.7-1984 (R 91)
	Measurement of Gas Flow by Means of Critical Flow Venturi Nozzles	ASME	ASME/ANSI MFC-7M-1987 (R 92)
	Measurement of Gas Flow by Turbine Meters	ASME	ANSI/ASME MFC-4M-1986 (R 97)
Pressure	Standard Method for Pressure Measurement	ASHRAE	ANSI/ASHRAE 41.3-1989
	Gauges—Pressure Indicating Dial Type—Elastic Element	ASME	ANSI/ASME B40.1-1991
	Guide for Dynamic Calibration of Pressure Transducers	ASME	ANSI MC88.1-1972 (R 1995)
	Pressure Measurement	ASME	ANSI/ASME PTC 19.2-1987 (R 98)
Temperature	Standard Method for Temperature Measurement	ASHRAE	ANSI/ASHRAE 41.1-1986 (R 91)
	Temperature Measurement	ASME	ANSI/ASME PTC 19.3-1974 (R 98)
	Total Temperature Measuring Instruments (Turbine Powered Subsonic Aircraft)	SAE	SAE AS 793-1966 (R 1991)
Thermal	Method of Testing Thermal Energy Meters for Liquid Streams in HVAC Systems	ASHRAE	ANSI/ASHRAE 125-1992
	Standard Practice for Determining Thermal Resistance of Building Envelope Components from In-Situ Data	ASTM	ASTM C 1155-95
	Standard Practice for In-Situ Measurement of Heat Flux and Temperature on Building Envelope Components	ASTM	ASTM C 1046-95
	Test Method for Steady-State Heat Flux Measurements and Thermal Transmission Properties by Means of the Guarded-Hot-Plate Apparatus	ASTM	ASTM C 177-97
	Test Method for Steady-State Heat Flux Measurements and Thermal Transmission Properties by Means of the Heat Flow Meter Apparatus	ASTM	ASTM C 518-98
	Test Method for Thermal Performance of Building Assemblies by Means of a Calibrated Hot Box	ASTM	ASTM C 976-90 (R 1996)
Mobile Homes and Recreational Vehicles	Residential Load Calculation (1986)	ACCA	ACCA Manual J
	Recreational Vehicle Cooking Gas Appliances	AGA	ANSI Z21.57-1993
	Mobile Homes	CSA	CAN/CSA-Z240 MH Series-92
	Mobile Home Parks	CSA	Z240.7.1-1972
	Oil-Fired Warm Air Heating Appliances for Mobile Housing and Recreational Vehicles	CSA	B140.10-1974 (R 1991)
	Park Model Trailers	CSA	CAN/CSA-Z41 Series-92
	Recreational Vehicle Parks	CSA	Z240.7.2-1972
	Recreational Vehicles	CSA	CAN/CSA-Z240 RV Series-M86 (R 1992)
	Gas Supply Connectors for Manufactured Homes	IAPMO	IAPMO TSC 9-1997
	Manufactured Home Installations	NCSBCS	ANSI/NCSBCS A225.1-1994
	Recreational Vehicles	NFPA	NFPA 501C-1996 (ANSI A119.2)
	Plumbing System Components for Manufactured Homes and Recreational Vehicles	NSF	ANSI/NSF 24-1988 (R 1996)
	Gas Burning Heating Appliances for Manufactured Homes and Recreational Vehicles (1995)	UL	UL 307B
	Gas-Fired Cooking Appliances for Recreational Vehicles (1993)	UL	UL 1075
	Liquid Fuel-Burning Heating Appliances for Manufactured Homes and Recreational Vehicles (1995)	UL	ANSI/UL 307A-1997
	Low Voltage Lighting Fixtures for Use in Recreational Vehicles (1994)	UL	ANSI/UL 234-1995
Motors and Generators	Steam Generating Units	ASME	ANSI/ASME PTC 4.1-1964 (R 1991)
	Testing of Nuclear Air-Treatment Systems	ASME	ANSI/ASME N510-1989 (R 1995)
	Nuclear Power Plant Air Cleaning Units and Components	ASME	ANSI/ASME N509-1989 (R 1996)
	Test Methods for Film-Insulated Magnet Wire	ASTM	ANSI/ASTM D 1676-95
	Energy Efficiency Test Methods for Three-Phase Induction Motors (Efficiency Quoting Method and Permissible Efficiency Tolerance)	CSA	C390-93
	Motors and Generators	CSA	C22.2 No. 100-95
	Energy Management Guide for Selection and Use of Polyphase Motors	NEMA	NEMA MG 10-1994
	Energy Management Guide for Selection and Use of Single-Phase Motors	NEMA	NEMA MG 11-1977 (R 1997)
	Motion/Position Control Motors, Controls, and Feedback Devices	NEMA	NEMA MG 7-1993
	Motors and Generators	NEMA	NEMA MG 1-1993
	Electric Motors (1994)	UL	UL 1004
	Electric Motors and Generators for Use in Division 1 Hazardous (Classified) Locations (1994)	UL	ANSI/UL 674-1993
	Overheating Protection for Motors (1987)	UL	UL 2111
	Standard Test Procedure for Polyphase Induction Motors and Generators	IEEE	IEEE 112-1996

Codes and Standards Published by Various Societies and Associations (*Continued*)

Subject	Title	Publisher	Reference
Operation and Maintenance	Preparation of Operating and Maintenance Documentation for Building Systems	ASHRAE	ASHRAE Guideline 4-1993
Pipe, Tubing, and Fittings	Process Piping	ANSI	ANSI/ASME B31.3-1996
	Building Services Piping	ASME	ANSI/ASME B31.9-1996
	Pipe Threads, General Purpose (Inch)	ASME	ANSI/ASME B1.20.1-1983 (R 92)
	Power Piping	ASME	ANSI/ASME B31.1-1995
	Refrigeration Piping	ASME	ASME/ANSI B31.5-1992
	Scheme for the Identification of Piping Systems	ASME	ANSI/ASME A13.1-1996
	Standard Practice for Obtaining Hydrostatic or Pressure Design Basis for "Fiberglass" (Glass-Fiber-Reinforced Thermosetting-Resin) Pipe and Fittings	ASTM	ANSI/ASTM D 2992-96
	Standards of the Expansion Joint Manufacturers Association, Inc., 7th ed. (1998)	EJMA	EJMA
	Guideline for Quality Piping Installation	MCAA	MCAA
	Pipe Hangers and Supports—Materials, Design and Manufacture	MSS	MSS SP-58-93
	Pipe Hangers and Supports—Selection and Application	MSS	MSS SP-69-96
	Welding Procedure Specifications	NCPWB	NCPWB
	Electrical Nonmetallic Tubing (ENT)	NEMA	NEMA TC 13-1993
	Filament-Wound Reinforced Thermosetting Resin Conduit and Fittings	NEMA	NEMA TC 14-1984 (R 1986)
	National Fuel Gas Code	NFPA	ANSI/NFPA 54-1996
		AGA	ANSI Z223.1-1992, Z223.1A-1994
	Refrigeration Tube Fittings	SAE	SAE J 513-1997
	Seismic Restraint Manual Guidelines for Mechanical Systems, 2nd ed.	SMACNA	SMACNA 1998
	Tube Fittings for Flammable and Combustible Fluids, Refrigeration Service, and Marine Use (1997)	UL	UL 109
Plastic	Specification for Acrylonitrile-Butadiene-Styrene (ABS) Plastic Pipe, Schedules 40 and 80	ASTM	ASTM D 1527-96A
	Specification for Chlorinated Polyvinyl Chloride (CPVC) Plastic Pipe, Schedules 40 and 80	ASTM	ASTM F 441/F 441M-97
	Specification for Plastic Insert Fittings for Polybutylene (PB) Tubing	ASTM	ASTM F 845-96
	Specification for Polybutylene (PB) Plastic Hot-Water Distribution Systems	ASTM	ASTM D 3309-96A
	Specification for Polybutylene (PB) Plastic Pipe (SDR-PR) Based on Controlled Inside Diameter	ASTM	ASTM D 26-96A
	Specification for Polybutylene (PB) Plastic Pipe (SDR-PR) Based on Outside Diameter	ASTM	ASTM D 3000-95A
	Specification for Polybutylene (PB) Plastic Tubing	ASTM	ASTM D 2666-96A
	Specification for Polyethylene (PE) Plastic Pipe, Schedule 40	ASTM	ASTM D 2104-96
	Specification for Polyvinyl Chloride (PVC) Plastic Pipe, Schedules 40, 80, and 120	ASTM	ASTM D 1785-96B
	Test Method for Obtaining Hydrostatic Design Basis for Thermoplastic Pipe Materials	ASTM	ASTM D 2837-98
	Electrical Plastic Tubing (EPT) and Conduit Schedule EPC-40 and EPC-80	NEMA	NEMA TC 2-1990
	Extra-Strength PVC Plastic Utilities Duct for Underground Installation	NEMA	NEMA TC 8-1990
	Fittings for ABS and PVC Plastic Utilities Duct for Underground Installation	NEMA	NEMA TC 9-1990
	PVC and ABS Plastic Utilities Duct for Underground Installation	NEMA	NEMA TC 6-1990
	Smooth-Wall Coilable Polyethylene Electrical Plastic Duct	NEMA	ANSI/NEMA TC 7-1990
	Plastics Piping Components and Related Materials	NSF	ANSI/NSF 14-1998
	Rubber Gasketed Fittings for Fire-Protection Service (1993)	UL	UL 213
Metal	Welded and Seamless Wrought Steel Pipe	ASME	ASME/ANSI B36.10M-1996
	Specification for Pipe, Steel, Black and Hot-Dipped, Zinc-Coated, Welded and Seamless	ASTM	ASTM A 53/53M-98A
	Specification for Seamless Carbon Steel Pipe for High-Temperature Service	ASTM	ASTMA 106-97A
	Specification for Seamless Copper Pipe, Standard Sizes	ASTM	ASTM B 42-98
	Specification for Seamless Copper Tube	ASTM	ASTM B 75-97
	Specification for Hand-Drawn Copper Capillary Tube for Restrictor Applications	ASTM	ASTM B 360-95
	Specification for Seamless Copper Tube for Air Conditioning and Refrigeration Field Service	ASTM	ASTM B 280-98
	Specification for Seamless Copper Water Tube	ASTM	ASTM B 88-96
	Specification for Welded Copper and Copper Alloy Tube for Air Conditioning and Refrigeration Service	ASTM	ASTM B 640-93
	Thickness Design of Ductile-Iron Pipe	AWWA	ANSI/AWWA C150/A21.50-96
	Fittings, Cast Metal Boxes, and Conduit Bodies for Conduit and Cable Assemblies	NEMA	NEMA FB 1-1993
	Polyvinyl-Chloride (PVC) Externally Coated Galvanized Rigid Steel Conduit and Intermediate Metal Conduit	NEMA	NEMA RN 1-1989
Plumbing	Uniform Plumbing Code (1997) (with IAPMO Installation Standards)	IAPMO	IAPMO
	International Plumbing Code (1997 with 1998 supplement)	ICC	ICC
	International Private Sewage Disposal Code (1997 with 1998 supplement)	ICC	ICC
	National Standard Plumbing Code (NSPC)	PHCC	NSPC 1996
	Standard Plumbing Code (1997)	SBCCI	SBCCI
Pumps	Centrifugal Pumps	ASME	ASME PTC 8.2-1990
	Displacement Compressors, Vacuum Pumps and Blowers	ASME	ANSI/ASME PTC 9-1970 (R 1992)

Codes and Standards Published by Various Societies and Associations (*Continued*)

Subject	Title	Publisher	Reference
Pumps (*Continued*)	Liquid Pumps	CSA	CAN/CSA C.22.2 No. 108-M89
	Performance Standard for Liquid Ring Vacuum Pumps, 1st ed. (1987) (R 1994)	HEI	HEI
	Centrifugal Pumps	HI	ANSI/HI 1.1-1.5 (1994)
	Centrifugal Pumps - Horizontal Baseplate Design	HI	ANSI/HI 1.3.4 (1997)
	Vertical Pumps	HI	ANSI/HI 2.1-2.5 (1994)
	Rotary Pumps	HI	ANSI/HI 3.1-3.5 (1994)
	Sealless Rotary Pumps	HI	ANSI/HI 4.1-4.6 (1994)
	Sealless Centrifugal Pumps	HI	ANSI/HI 5.1-5.6 (1994)
	Reciprocating Power Pumps	HI	ANSI/HI 6.1-6.5 (1994)
	Controlled Volume Pumps	HI	ANSI/HI 7.1-7.5 (1994)
	Direct Acting (Steam) Pumps	HI	ANSI/HI 8.1-8.5 (1994)
	Pumps—General Guideline	HI	ANSI/HI 9.1-9.6 (1994)
	Pumps—Polymer Material Selections	HI	ANSI/HI 9.3.3 (1997)
	Centrifugal and Vertical Pumps—NPSH Margin	HI	ANSI/HI 9.6.1 (1998)
	Centrifugal and Vertical Pumps—Allowable Operating Region	HI	ANSI/HI 9.6.3 (1997)
	Pump Intake Design	HI	ANSI/HI 9.8 (1998)
	Engineering Data Book, 2nd ed. (1990)	HI	HI
	Circulation System Components and Related Materials for Swimming Pools, Spas/Hot Tubs	NSF	ANSI/NSF 50-1996
	Swimming Pool Pumps, Filters and Chlorinators (1997)	UL	UL 1081
	Motor-Operated Water Pumps (1996)	UL	UL 778
	Pumps for Oil-Burning Appliances (1997)	UL	ANSI/UL 343-1998
Radiators	Testing and Rating Standard for Baseboard Radiation, 6th ed. (1990)	HYDI	IBR
	Testing and Rating Standard for Finned-Tube (Commercial) Radiation (1990)	HYDI	IBR
Receivers	Refrigerant Liquid Receivers	ARI	ARI 495-93
Refrigerant-Containing Components	Refrigerant-Containing Components for Use in Electrical Equipment	CSA	C22.2 No. 140.3-M1987 (R 1993)
	Refrigerant-Containing Components and Accessories, Non-Electrical (1993)	UL	ANSI/UL 207-1994
Refrigerants	Threshold Limit Values for Chemical Substances (updated annually)	AGCIH	AGCIH
	Refrigerant Recovery/Recycling Equipment	ARI	ARI 740-95
	Specifications for Fluorocarbon and Other Refrigerants	ARI	ARI 700-95
	Format for Information on Refrigerants	ASHRAE	ASHRAE Guideline 6-1996
	Method of Testing Flow Capacity of Refrigerant Capillary Tubes	ASHRAE	ANSI/ASHRAE 28-1996
	Number Designation and Safety Classification of Refrigerants	ASHRAE	ANSI/ASHRAE 34-1997
	Methods of Testing Discharge Line Refrigerant-Oil Separators	ASHRAE	ANSI/ASHRAE 69-1990
	Reducing Emission of Fully Halogenated Refrigerants in Refrigeration and Air-Conditioning Equipment and Systems	ASHRAE	ASHRAE Guideline 3-1996
	Refrigeration Oil Description	ASHRAE	ANSI/ASHRAE 99-1981 (R 87)
	Sealed Glass Tube Method to Test the Chemical Stability of Material for Use Within Refrigerant Systems	ASHRAE	ANSI/ASHRAE 97-1983 (R 89)
	Test Method for Acid Number of Petroleum Products by Potentiometric Titration	ASTM	ASTM D 664-95
	Test Method for Concentration Limits of Flammability of Chemical (Vapors and Gases)s	ASTM	ASTM E 681-98
	Refrigerants-Number Designation	ISO	ISO 817:1974
	Recommended Service Procedure for the Containment of HFC-134a	SAE	ANSI/SAE J 2211-1991
	HFC-134a Recycling Equipment for Mobile Air-Conditioning Systems	SAE	ANSI/SAE J 2210-1991
	CFC-12 (R-12) Extraction Equipment for Mobile Automotive Air-Conditioning Systems	SAE	ANSI/SAE J 2209-1992
	HFC-134a (R-134a) Service Hose Fittings for Automotive Air-Conditioning Service Equipment	SAE	SAE J 2197-1992 (R 1997)
	Standard of Purity for Recycled HFC-134a for Use in Mobile Air-Conditioning Systems	SAE	ANSI/SAE J 2099-1991
	Recommended Service Procedure for the Containment of R-12, Highway Vehicle	SAE	SAE J 1989-1989
	Procedure for Retrofitting CFC-12 (R12) Mobile Air Conditioning Systems to HFC-134a (R134a)	SAE	ANSI/SAE J 1661-1993
	Field Conversion/Retrofit of Products to Change to an Alternate Refrigerant—Construction and Operation (1993)	UL	ANSI/UL 2170-1995
	Field Conversion/Retrofit of Products to Change to an Alternate Refrigerant—Insulating Material and Refrigerant Compatibility (1993)	UL	ANSI/UL 2171-1995
	Field Conversion/Retrofit of Products to Change to an Alternate Refrigerant—Procedures and Methods (1993)	UL	ANSI/UL 2172-1995
	Refrigerant Recovery/Recycling Equipment (1995)	UL	UL 1963
	Refrigerants (1994)	UL	UL 2182
Refrigeration	Safety Code for Mechanical Refrigeration	ASHRAE	ANSI/ASHRAE 15-1994
	Mechanical Refrigeration Code	CSA	B52-95
	Refrigeration Equipment	CSA	CAN/CSA-C22.2 No. 120-M91
	Equipment, Design and Installation of Ammonia Mechanical Refrigerating Systems	IIAR	ANSI/IIAR 2-1992
	Refrigerated Medical Equipment (1993)	UL	ANSI/UL 416-1996

Codes and Standards Published by Various Societies and Associations (*Continued*)

Subject	Title	Publisher	Reference
Refrigeration Systems	Ejectors	ASME	ASME PTC 24-1976 (R 1982)
	Testing of Refrigerating Systems	ISO	ISO 916-1968
	Standards for Steam Jet Vacuum Systems, 4th ed. (1988) (R 1995)	HEI	HEI
Transport	Mechanical Transport Refrigeration Units	ARI	ARI 1110-92
	Mechanical Refrigeration and Air-Conditioning Installations Aboard Ship	ASHRAE	ASHRAE 26-1996
	General Requirements for Application of Vapor Cycle Refrigeration Systems for Aircraft	SAE	SAE ARP 731B-1997
	Safety and Containment of Refrigerant for Mechanical Vapor Compression Systems Used for Mobile Air-Conditioning Systems	SAE	ANSI/SAE J 639-1994
Refrigerators	Method of Testing Open Refrigerators for Food Stores	ASHRAE	ANSI/ASHRAE 72-1983
	Methods of Testing Closed Refrigerators	ASHRAE	ANSI/ASHRAE 117-1992
Commercial	Energy Performance Standard for Commercial Refrigerated Display Cabinets and Merchandise	CSA	CAN/CSA-C657-95
	Mobile Food Carts	NSF	ANSI/NSF 59-1997
	Food Equipment	NSF	ANSI/NSF 2-1996
	Commercial Refrigerators and Storage Freezers	NSF	ANSI/NSF 7-1997
	Commercial Refrigerators and Freezers (1995)	UL	ANSI/UL 471-1996
	Refrigerating Units (1994)	UL	ANSI/UL 427-1996
	Refrigeration Unit Coolers (1993)	UL	ANSI/UL 412-1992
Household	Refrigerators Using Gas Fuel	AGA	ANSI Z21.19-1990, Z21.19A-1992
	Household Refrigerators, Combination Refrigerator-Freezers and Household Freezers	AHAM	ANSI/AHAM HRF-1-1988
	Capacity Measurement and Energy Consumption Test Methods for Refrigerators, Combination Refrigerator-Freezers, and Freezers	CSA	CAN/CSA C300-M91
	Household Refrigerators and Freezers (1993)	UL	ANSI/UL 250-1997
		CSA	CAN/CSA C22.2 No. 63-93
Retrofitting			
Building	Retrofit of Building Energy Systems and Processes, 1st ed.	SMACNA	SMACNA 1982
Refrigerant	Procedure for Retrofitting CFC-12 (R12) Mobile Air Conditioning Systems to HFC-134a (R134a)	SAE	ANSI/SAE J 1661-1993
	Field Conversion/Retrofit of Products to Change to an Alternate Refrigerant—Construction and Operation (1993)	UL	ANSI/UL 2170-1995
	Field Conversion/Retrofit of Products to Change to an Alternate Refrigerant—Insulating Material and Refrigerant Compatibility (1993)	UL	ANSI/UL 2171-1995
	Field Conversion/Retrofit of Products to Change to an Alternate Refrigerant—Procedures and Methods (1993)	UL	ANSI/UL 2172-1995
Roof Ventilators	Commercial Low Pressure, Low Velocity Duct System Design (1990)	ACCA	ACCA Manual Q
	Power Ventilators (1994)	UL	ANSI/UL 705-1994
Safety Devices	Safety Devices for Protection Against Excessive Pressure		
	Part 2: Bursting Disc Safety Devices	ISO	ISO 4126-2:1981
	Part 3: Safety Valves and Bursting Disc Safety Devices in Combination	ISO	ISO 4126-3: 1981
Solar Equipment	Method of Measuring Solar-Optical Properties of Materials	ASHRAE	ANSI/ASHRAE 74-1988
	Methods of Testing to Determine the Thermal Performance of Flat-Plate Solar Collectors Containing a Boiling Liquid	ASHRAE	ANSI/ASHRAE 109-1986 (R 96)
	Methods of Testing to Determine the Thermal Performance of Solar Collectors	ASHRAE	ANSI/ASHRAE 93-1986 (R 91)
	Methods of Testing to Determine the Thermal Performance of Solar Domestic Water Heating Systems	ASHRAE	ASHRAE 95-1981 (R 87)
	Methods of Testing to Determine the Thermal Performance of Unglazed Flat-Plate Liquid-Type Solar Collectors	ASHRAE	ANSI/ASHRAE 96-1980 (R 89)
	Reference Solar Spectral Irradiance at the Ground at Different Receiving Conditions—Part 1: Direct Normal and Hemispherical Solar Irradiance for Air Mass 1.5	ISO	ISO 9845-1:1992
	Solar Energy—Calibration of a Pyranometer Using a Pyrheliometer	ISO	ISO 9846:1993
	Solar Heating—Domestic Water Heating Systems—Part 2: Outdoor Test Methods for System Performance Characterization and Yearly Performance Prediction of Solar-Only Systems	ISO	ISO 9459-2:1995
	Solar Heating—Swimming Pool Heating Systems—Dimensions, Design and Installation Guidelines	ISO	ISO 12596:1995
	Solar Water Heaters—Elastomeric Materials for Absorbers, Connecting Pipes and Fittings—Method of Assessment	ISO	ISO 9808:1990
	Test Methods for Solar Collectors—Part 2: Qualification Test Procedures	ISO	ISO 9806-2:1995
	Test Methods for Solar Collectors—Part 3: Thermal Performance of Unglazed Liquid Heating Collectors (Sensible Heat Transfer Only) Including Pressure Drop	ISO	ISO 9806-3:1995
Solenoid Valves	Solenoid Valves for Use with Volatile Refrigerants	ARI	ARI 760-94
Sound Measurement	Threshold Limit Values for Physical Agents (updated annually)	AGCIH	AGCIH
	Method for the Calibration of Microphones (reaffirmed 1997)	ASA	ANSI S1.10-1966 (R 1997)
	Specification for Sound Level Meters	ASA	ANSI S1.4-1983, ANSI S1.4A-1985 (R 1997)

Codes and Standards Published by Various Societies and Associations (*Continued*)

Subject	Title	Publisher	Reference
Sound Measurement (*Continued*)	Test Method for Laboratory Measurement of Airborne Sound Transmission Loss of Building Partitions and Elements	ASTM	ASTM E 90-97
	Test Method for Measuring Acoustical and Airflow Performance of Duct Liner Materials and Prefabricated Silencers	ASTM	ASTM E 477-96
	Sound and Vibration Design and Analysis (1994)	NEBB	NEBB
Fans	Methods for Calculating Fan Sound Ratings from Laboratory Test Data	AMCA	AMCA 301-90
	Reverberant Room Method for Sound Testing of Fan	AMCA	AMCA 300-96
	Balance Quality and Vibration Levels for Fans	AMCA	AMCA 204-96
	Laboratory Methods of testing Air Circulator Fans for Rating	AMCA	AMCA 230-96
	Methods for the Measurement of Noise Emitted by Small Air-Moving Devices	ASA	ANSI S12.11-1987 (R 1997)
	Laboratory Method of Testing In-Duct Sound Power Measurement Procedure for Fans	ASHRAE	ANSI/ASHRAE 68-1986
		AMCA	ANSI/AMCA 330-86
	Acoustics—Method for the Measurement of Airborne Noise Emitted by Small Air-Moving Devices	ISO	ISO 10302:1996
Other Equipment	Application of Sound Rating Levels of Outdoor Unitary Equipment	ARI	ARI 275-97
	Method of Measuring Machinery Sound Within Equipment Space	ARI	ARI 575-94
	Method of Measuring Sound and Vibration of Refrigerant Compressors	ARI	ARI 530-89
	Rating the Sound Levels and Sound Transmission Loss of Packaged Terminal Equipment	ARI	ANSI/ARI 300-88
	Sound Rating of Large Outdoor Refrigerating and Air-Conditioning Equipment	ARI	ARI 370-86
	Sound Rating of Non-Ducted Indoor Air-Conditioning Equipment	ARI	ARI 350-86
	Sound Rating of Outdoor Unitary Equipment	ARI	ARI 270-95
	Statistical Methods for Determining and Verifying Stated Noise Emission Values of Machinery and Equipment	ASA	ANSI S12.3-85 (R 1996)
	Sound Level Prediction for Installed Rotating Electrical Machines	NEMA	NEMA MG 3-1974 (R 1990)
Techniques	Methods for Measurement of Sound Emitted by Machinery and Equipment at Workstations and Other Specified Positions	ANSI	ANSI S12.43-1997
	Methods for Calculation of Sound Emitted by Machinery and Equipment at Workstations and Other Specified Positions from Sound Power Level	ANSI	ANSI S12.44-1997
	Criteria for Evaluating Room Noise	ASA	ANSI S12.2-1995
	Engineering Method for the Determination of Sound Power Levels of Noise Sources Using Sound Intensity	ASA	ANSI S12.12-1992 (R 1997)
	Engineering Methods for the Determination of Sound Power Levels of Noise Sources for Essentially Free-Field Conditions over a Reflecting Plane	ASA	ANSI S12.34-1988 (R 1997)
	Guidelines for the Use of Sound Power Standards and for the Preparation of Noise Test Codes	ASA	ANSI S12.30-1990 (R 1997)
	Measurement of Sound Pressure Levels in Air	ASA	ANSI S1.13-1995
	Methods for Determination of Insertion Loss of Outdoor Noise Barriers	ASA	ANSI S12.8-1998
	Methods for the Determination of Sound Power Levels of Noise Sources in a Special Reverberation Test Room	ASA	ANSI S12.33-1990 (R 1997)
	Precision Methods for the Determination of Sound Power Levels of Broad-Band Noise Sources in Reverberation Rooms	ASA	ANSI S12.31-1990 (R 1996)
	Precision Methods for the Determination of Sound Power Levels of Discrete-Frequency and Narrow-Band Noise Sources in Reverberation Rooms	ASA	ANSI S12.32-1990 (R 1996)
	Precision Methods for the Determination of Sound Power Levels of Noise Sources in Anechoic and Hemi-Anechoic Rooms	ASA	ANSI S12.35-1990 (R 1996)
	Preferred Frequencies, Frequency Levels, and Band Numbers for Acoustical Measurements	ASA	ANSI S1.6-1984 (R 1997)
	Procedure for the Computation of Loudness of Noise	ASA	ANSI S3.4-1980 (R 1997)
	Procedures for Outdoor Measurement of Sound Pressure Level	ASA	ANSI S12.18-1994
	Reference Quantities for Acoustical Levels	ASA	ANSI S1.8-1989 (R 1997)
	Survey Methods for the Determination of Sound Power Levels of Noise Sources	ASA	ANSI S12.36-1990 (R 1997)
	Measurement of Industrial Sound	ASME	ASME/ANSI PTC 36-1985
	Test Method for Evaluating Masking Sound in Open Offices Using A-Weighted and One-Third Octave Band Sound Pressure Levels	ASTM	ASTM E 1573-93 (R 1998)
	Test Method for Measurement of Sound in Residential Spaces	ASTM	ASTM E 1574-98
	Acoustics–Measurement of Sound Insulation in Buildings and of Building Elements	ISO	ISO 140-1: 1997
	Part 1: Requirements for Laboratory Test Facilities with Suppressed Flanking Transmission		
	Part 4: Field Measurements of Airborne Sound Insulation Between Rooms	ISO	ISO 140-4: 1978
	Part 5: Field Measurements of Airborne Sound Insulation of Facade Elements and Facade	ISO	ISO 140-5: 1990
	Part 6: Laboratory Measurements of Impact Sound Insulation of Floors	ISO	ISO 140-6: 1978
	Part 7: Field Measurements of Impact Sound Insulation of Floors	ISO	ISO 140-7: 1990
	Part 8: Laboratory Measurements of the Reduction of Transmitted Impact Noise by Floor Coverings on a Solid Standard Floor	ISO	ISO 140-8: 1978
	Acoustics—Determination of Sound Power Levels of Noise Sources Using Sound Intensity—Part 1: Measurement at Discrete Points	ISO	ISO 9614-1:1993
	Acoustics—Determination of Sound Power Levels of Noise Sources Using Sound Intensity—Part 2: Measurement by Scanning	ISO	ISO 9614-2:1996
	Acoustics—Method for Calculating Loudness Level	ISO	ISO 532:1975

Codes and Standards Published by Various Societies and Associations (*Continued*)

Subject	Title	Publisher	Reference
Sound Measurement (*Continued*)	Procedural Standards for the Measurement and Assessment of Sound and Vibration (1994)	NEBB	NEBB
Terminology	Acoustical Terminology	ASA	ANSI S1.1-1994
	Terminology Relating to Environmental Acoustics	ASTM	ASTM C 634-98
Space Heaters	Methods of Testing for Rating Combination Space-Heating and Water-Heating Appliances	ASHRAE	ANSI/ASHRAE 124-1991
	Electric Air Heaters	CSA	C22.2 No. 46-M1988
	Fixed and Location-Dedicated Electric Room Heaters (1997)	UL	UL 2021
	Movable and Wall- or Ceiling-Hung Electric Room Heaters (1994)	UL	UL 1278
	Gas-Fired Room Heaters, Vol. I, Vented Room Heaters	AGA	ANSI Z21.11.1-1991, Z21.11.1A-1993, Z21.11.1B-1995
	Gas-Fired Room Heaters, Vol. II, Unvented Room Heaters	AGA	ANSI Z21.11.2-1992, Z21.11.2A-1993, Z21.11.2B-1995
Symbols	Graphic Electrical/Electronic Symbols for Air-Conditioning and Refrigeration Equipment	ARI	ARI 130-88
	Graphic Symbols for Heating, Ventilating, and Air Conditioning	ASME	ANSI/ASME Y32.2.4-1949 (R 1998)
	Graphic Symbols for Pipe Fittings, Valves and Piping	ASME	ANSI/ASME Y32.2.3-1949 (R 1994)
	Graphic Symbols for Plumbing Fixtures for Diagrams used in Architecture and Building Construction	ASME	ANSI/ASME Y32.4-1977 (R 1994)
	Symbols for Mechanical and Acoustical Elements as used in Schematic Diagrams	ASME	ANSI/ASME Y32.18-1972 (R 1998)
	Graphic Symbols for Electrical and Electronic Diagrams	IEEE	ANSI/IEEE 315-1975 (R 1994)
	Recommended Practice for the Preparation and Use of Symbols	IEEE	IEEE 267-1966
	Standard Letter Symbols for Quantities Used in Electrical Science and Electrical Engineering	IEEE	IEEE 280-1985 (R 1997)
	Standard for Logic Circuit Diagrams	IEEE	IEEE 991-1986 (R 1994)
	Standard for Use of the International System of Units (SI): The Modern Metric System	IEEE/ ASTM	ANSI/IEEE/ASTM SI 10-1997
	Abbreviations for Use on Drawings and in Text	ASME	ANSI/ASME Y1.1-1989
	Safety Color Code	NEMA	ANSI/NEMA Z535.1-1997
Terminals, Wiring	Electrical Quick-Connect Terminals (1995)	UL	ANSI/UL 310-1996
	Equipment Wiring Terminals for Use with Aluminum and/or Copper Conductors (1994)	UL	ANSI/UL 486E-1994
	Splicing Wire Connectors (1997)	UL	ANSI/UL 486C-1998
	Wire Connectors and Soldering Lugs for Use with Copper Conductors (1997)	UL	ANSI/UL 486A-1998
	Wire Connectors for Use with Aluminum Conductors (1997)	UL	UL 486B
Testing and Balancing	Industrial Process/Power Generating Fans: Site Performance Test Standard	AMCA	AMCA 803-96
	Guideline for the HVAC Commissioning Process	ASHRAE	ASHRAE Guideline 1-1996
	Practices for Measurement, Testing, Adjusting, and Balancing of Building Heating, Ventilation, Air-Conditioning, and Refrigeration Systems	ASHRAE	ANSI/ASHRAE 111-1988
	Centrifugal Pump Test	HI	ANSI/HI 1.6-1994
	Vertical Pump Tests	HI	ANSI/HI 2.6-1994
	Rotary Pump Tests	HI	ANSI/HI 3.6-1994
	Reciprocating Pump Tests	HI	ANSI/HI 6.6-1994
	Pumps—General Guidelines (Including "Measurement of Airborne Sound")	HI	HI 9.1-9.6-1994
	Procedural Standards for Certified Testing of Cleanrooms, 2nd ed. (1996)	NEBB	NEBB
	Procedural Standards for Testing, Adjusting, Balancing of Environmental Systems, 6th ed. (1998)	NEBB	NEBB
	HVAC Systems Testing, Adjusting and Balancing, 2nd ed.	SMACNA	SMACNA 1993
Thermal Storage	Method of Testing Active Sensible Thermal Energy Storage Devices Based on Thermal Performance	ASHRAE	ANSI/ASHRAE 94.3-1986 (R 96)
	Method of Testing Active Latent-Heat Storage Devices Based on Thermal Performance	ASHRAE	ANSI/ASHRAE 94.1-1985 (R 91)
	Methods of Testing Thermal Storage Devices with Electrical Input and Thermal Output Based on Thermal Performance	ASHRAE	ANSI/ASHRAE 94.2-1981 (R 96)
	Practices for Measurement, Testing, Adjusting, and Balancing of Building Heating, Ventilation, Air-Conditioning, and Refrigeration Systems	ASHRAE	ANSI/ASHRAE 111-1988
Turbines	Steam Turbines	ASME	ANSI/ASME PTC 6-1996
	Wind Turbines	ASME	ANSI/ASME PTC 42-1988 (R 98)
	Specification for Gas Turbine Fuel Oils	ASTM	ANSI/ASTM D 2880-98
	Land Based Steam Turbine Generator Sets	NEMA	NEMA SM 24-1991
	Steam Turbines for Mechanical Drive Service	NEMA	ANSI/NEMA SM 23-1991
Valves	Methods of Testing Nonelectric, Nonpneumatic Thermostatic Radiator Valves	ASHRAE	ANSI/ASHRAE 102-1983 (R 89)
	Face-to-Face and End-to-End Dimensions of Valves	ASME	ANSI/ASME B16.10-1992
	Pressure Relief Devices	ASME	ANSI/ASME PTC 25-1994
	Valves—Flanged Threaded, and Welding End	ASME	ANSI/ASME B16.34-1996
	Control Valve Capacity Test Procedure	ISA	ANSI/ISA-S75.02-1996
	Flow Equations for Sizing Control Valves	ISA	ANSI/ISA-S75.01-1985 (R 1995)

Codes and Standards Published by Various Societies and Associations (*Continued*)

Subject	Title	Publisher	Reference
Valves	Industrial Valves—Part-Turn Valve Actuator Attachments—Part 1	ISO	ISO 5211-1:1977
(*Continued*)	Industrial Valves—Part-Turn Valve Actuator Attachments—Part 2	ISO	ISO 5211-2:1979
	Industrial Valves—Part-Turn Valve Actuator Attachments—Part 3	ISO	ISO 5211-3:1982
	Metal Valves for Use in Flanged Pipe Systems—Face-to-Face Dimensions	ISO	ISO 5752:1982
	Safety Valves, Part 1: General Requirements	ISO	ISO 4126-1:1991
	Oxygen System Fill/Check Valve	SAE	SAE AS 1225A-1997
	Electrically Operated Valves (1994)	UL	UL 429
	Pressure Regulating Valves for LP-Gas (1994)	UL	UL 144
	Safety Relief Valves for Anhydrous Ammonia and LP-Gas (1997)	UL	UL 132
	Valves for Anhydrous Ammonia and LP-Gas (Other than Safety Relief) (1997)	UL	UL 125
	Valves for Flammable Fluids (1997)	UL	UL 842
Gas	Automatic Gas Valves for Gas Appliances	AGA	ANSI Z21.21-1993
	Combination Gas Controls for Gas Appliances	AGA	ANSI Z21.78-1992, Z21.78A-1993, Z21.78B-1994
	Manually Operated Gas Valves for Appliances, Appliance Connection Valves, and Hose End Valves	AGA	ANSI Z21.15-1992
	Relief Valves and Automatic Gas Shutoff Devices for Hot Water Supply Systems	AGA	ANSI Z21.22-1986, Z21.22A-1990
	Requirements for Automatic Non-Shutoff Modulating Gas Valves	AGA	1-92
	Requirements for Gas Operated Valves for High Pressure Natural Gas	AGA	3-93
	Requirements for Manually Operated Gas Valves for Use in House Piping Systems	AGA	3-88
	Requirements for Manually Operated Valves for High Pressure Natural Gas	AGA	2-93
	Large Metallic Valves for Gas Distribution (Manually Operated, NPS-2 1/2 to 12, 125 psig Maximum)	ASME	ANSI/ASME B16.38-1985 (R 1994)
	Manually Operated Metallic Gas Valves for Use in Gas Piping Systems up to 125 psig (Sizes 1/2 through 2)	ASME	ANSI/ASME B16.33-1990
	Manually Operated Thermoplastic Gas Shutoffs and Valves in Gas Distribution Systems	ASME	ANSI/ASME B16.40-1985 (R 1994)
Refrigerant	Refrigerant Access Valves and Hose Connectors	ARI	ARI 720-97
	Refrigerant Pressure Regulating Valves	ARI	ARI 770-94
	Solenoid Valves for Use with Volatile Refrigerants	ARI	ANSI/ARI 760-94
	Thermostatic Refrigerant Expansion Valves	ARI	ANSI/ARI 750-94
Vapor Retarders	Practice for Selection of Vapor Retarders for Thermal Insulation	ASTM	ASTM C 755-97
	Practice for Determining the Properties of Jacketing Materials for Thermal Insulation	ASTM	ASTM C 921-89 (R 1996)
	Specification for Flexible, Low Permeance Vapor Retarders for Thermal Insulation	ASTM	ASTM C 1136-95
	Test Method for Water Vapor Transmission Rate of Flexible Barrier Materials Using an Infrared Detection Technique	ASTM	ASTM F 372-94
Vending Machines	Methods of Testing for Rating Bottled and Canned Beverage Vending Machines	ASHRAE	ANSI/ASHRAE 32.1-1997
	Methods of Testing for Rating Pre-Mix and Post-Mix Soft Drink Vending and Dispensing Equipment	ASHRAE	ANSI/ASHRAE 32.2-1997
	Vending Machines	CSA	CAN/CSA-C22.2 No.128-95
	Vending Machines for Food and Beverages	NSF	ANSI/NSF 25-1997
	Vending Machines (1995)	UL	ANSI/UL 751-1997
	Refrigerated Vending Machines (1995)	UL	UL 541
Vent Dampers	Automatic Vent Damper Devices for Use with Gas-Fired Appliances	AGA	ANSI Z21.66-1994
	Vent or Chimney Connector Dampers for Oil-Fired Appliances (1994)	UL	UL 17-1995
Ventilation	Commercial Low Pressure, Low Velocity Duct System Design (1990)	ACCA	ACCA Manual Q
	Comfort, Air Quality and Efficiency by Design (1997)	ACCA	ACCA Manual RS
	Commercial Applications Systems and Equipment (1993)	ACCA	ACCA Manual CS
	Guide for Testing Ventilation Systems	ACGIH	ACGIH
	Industrial Ventilation: A Manual of Recommended Practice, 23rd ed. (1998)	ACGIH	ACGIH
	Ventilation for Acceptable Indoor Air Quality	ASHRAE	ANSI/ASHRAE 62-1989
	Method of Testing for Room Air Diffusion	ASHRAE	ANSI/ASHRAE 113-1990
	Measuring Air Change Effectiveness	ASHRAE	ASHRAE 129-1997
	Method of Determining Air Change Rates in Detached Dwellings	ASHRAE	ANSI/ASHRAE 136-1993
	Design of Ventilation Systems for Poultry and Livestock Shelters	ASAE	ASAE D270.5-1994
	Design Values for Emergency Ventilation and Care of Livestock and Poultry	ASAE	ASAE EP282.2-1993
	Energy Ratings for Selection of Energy Efficient Agricultural Ventilation Fans	ASAE	ASAE S566-1997
	Heating, Ventilating and Cooling Greenhouses	ASAE	ASAE EP406.3-1997
	Residential Mechanical Ventilation Systems	CSA	CAN/CSA F326-M91
	Installation of Air Conditioning and Ventilating Systems	NFPA	ANSI/NFPA 90A-1996
	Parking Structures	NFPA	ANSI/NFPA 88A-1995
	Repair Garages	NFPA	ANSI/NFPA 88B-1997
	Ventilation Control and Fire Protection of Commercial Cooking Operations	NFPA	ANSI/NFPA 96-1994
	Food Equipment	NSF	ANSI/NSF 2-1996
	Class II (Laminar Flow) Biohazard Cabinetry	NSF	NSF 49-1992
	(R) Test Procedure for Battery Flame Retardant Venting Systems	SAE	ANSI/SAE J 1495-1992
	Heater, Airplane, Engine Exhaust Gas to Air Heat Exchanger Type	SAE	SAE ARP 86A-1952 (R 1992)
	Aerothermodynamic Systems Engineering and Design	SAE	SAE AIR 1168/3-1990

Codes and Standards Published by Various Societies and Associations (*Continued*)

Subject	Title	Publisher	Reference
Venting	Commercial Applications Systems and Equipment (1993)	ACCA	ACCA Manual CS
	Draft Hoods	AGA	ANSI Z21.12-1990, Z21.12A-1993, Z21.12B-1994
	National Fuel Gas Code	AGA	ANSI Z223.1-1992, Z223.1A-1994
		NFPA	ANSI/NFPA 54-1996
	Requirements for Electrically Operated Automatic Combustion and Ventilation Air Control Devices for Use with Gas-Fired Appliances	AGA	1-88
	Requirements for Mechanical Venting System	AGA	6-9
	Balance Quality and Vibration Levels for Fans	AMCA	AMCA 204-96
	Chimneys, Fireplaces, Vents and Solid Fuel-Burning Appliances	NFPA	ANSI/NFPA 211-1996
	Explosion Prevention Systems	NFPA	ANSI/NFPA 69-1997
	Smoke and Heat Venting	NFPA	ANSI/NFPA 204M
	Guide for Steel Stack Construction, 2nd ed.	SMACNA	SMACNA 1996
	Draft Equipment (1993)	UL	UL 378
	Gas Vents (1996)	UL	UL 441
	Low-Temperature Venting Systems, Type L (1995)	UL	ANSI/UL 641-1995
Vibration	Mechanical Vibration of Rotating and Reciprocating Machinery—Requirements for Instruments for Measuring Vibration Severity	ASA	ANSI S2.40-1984 (R 1997)
	Methods for Analysis and Presentation of Shock and Vibration Data	ASA	ANSI S2.10-1971 (R 1997)
	Selection of Calibrations and Tests for Electrical Transducers Used for Measuring Shock and Vibration	ASA	ANSI S2.11-1969 (R 1997)
	Techniques of Machinery Vibration Measurement	ASA	ANSI S2.17-1980 (R 1997)
	Vibrations of Buildings—Guidelines for the Measurement of Vibrations and Evaluation of Their Effects on Buildings	ASA	ANSI S2.47-1990 (R 1997)
	Evaluation of Human Exposure to Whole-Body Vibration—Part 2: Continuous and Shock-Induced Vibrations in Buildings (1 to 80 Hz)	ISO	ISO 2631-2:1989
	Guidelines for the Evaluation of the Response of Occupants of Fixed Structures, Especially Buildings and Off-Shore Structures, to Low-Frequency Horizontal Motion (0.063 to 1 Hz)	ISO	ISO 6897:1984
	Procedural Standards for the Measurement and Assessment of Sound and Vibration (1994)	NEBB	NEBB
	Sound and Vibration Design and Analysis (1994)	NEBB	NEBB
Water Heaters	Gas Water Heaters, Vol. I, Storage Water Heaters with Input Ratings of 75,000 Btu per Hour or Less	AGA	ANSI Z21.10.1-1993, Z21.10.1A-1993, Z21.10.1B-1994
	Gas Water Heaters, Vol. III, Storage, with Input Ratings Above 75,000 Btu per Hour, Circulating and Instantaneous Water Heaters	AGA	ANSI Z21.10.3-1993, Z21-10.3A-1993, Z21-10.3B-1994
	Requirements for Non-Metallic Dip Tubes for Use in Gas-Fired Water Heaters	AGA	1-89
	Requirements for Indirect Water Heaters for Use with External Heat Source	AGA	1-91
	Desuperheater/Water Heaters	ARI	ARI 470-95
	Method of Testing for Rating Commercial Gas, Electric, and Oil Water Heaters	ASHRAE	ANSI/ASHRAE 118.1-1993
	Method of Testing for Rating Residential Water Heaters	ASHRAE	ANSI/ASHRAE 118.2-1993
	Methods of Testing for Efficiency of Space-Conditioning/Water-Heating Appliances That Include a Desuperheater Water Heater	ASHRAE	ANSI/ASHRAE 137-1995
	Methods of Testing to Determine the Thermal Performance of Solar Domestic Water Heating Systems	ASHRAE	ASHRAE 95-1981 (R 87)
	Methods of Testing for Rating Combination Space-Heating and Water-Heating Appliances	ASHRAE	ANSI/ASHRAE 124-1991
	Construction and Test of Electric Storage-Tank Water Heaters	CSA	CAN/CSA-C22.2 No. 110-M90
	Performance of Electric Storage Tank Water Heaters	CSA	CAN/CSA-C191 Series-M90
	Oil Burning Stoves and Water Heaters	CSA	B140.3-1962 (R 1991)
	Oil-Fired Service Water Heaters and Swimming Pool Heaters	CSA	B140.12-1976 (R 1991)
	Water Heaters, Hot Water Supply Boilers, and Heat Recovery Equipment	NSF	NSF 5-1992
	Commercial-Industrial Gas Heating Equipment (1994)	UL	UL 795
	Electric Booster and Commercial Storage Tank Water Heaters (1995)	UL	UL 1453
	Household Electric Storage Tank Water Heaters (1996)	UL	ANSI/UL 174-1996
	Oil-Fired Storage Tank Water Heaters (1995)	UL	ANSI/UL 732-1997
Wood-Burning Appliances	Threshold Limit Values for Chemical Substances (updated annually)	AGCIH	AGCIH
	Installation Code for Solid Fuel Burning Appliances and Equipment	CSA	CAN/CSA-B365-M91
	Solid-Fuel-Fired Central Heating Appliances	CSA	CAN/CSA-B366.1-M91
	Chimneys, Fireplaces, Vents, and Solid Fuel-Burning Appliances	NFPA	ANSI/NFPA 211-1996
	Commercial Cooking, Rethermalization and Powered Hot Food Holding and Transport Equipment	NSF	ANSI/NSF 4-1997

ABBREVIATIONS AND ADDRESSES

AABC	Associated Air Balance Council, 1518 K Street NW, Washington, D.C. 20005
ABMA	American Boiler Manufacturers Association, 950 North Glebe Road, Suite 160, Arlington, VA 22203-1824
ACCA	Air Conditioning Contractors of America, 1712 New Hampshire Avenue, NW, Washington, D.C. 20009
ACGIH	American Conference of Governmental Industrial Hygienists, 1330 Kemper Meadow Drive, Cincinnati, OH 45240
ADC	Air Diffusion Council, 104 South Michigan Avenue, Suite 1500, Chicago, IL 60603
AGA	American Gas Association, 1515 Wilson Boulevard, Arlington, VA 22209
	Also available through International Approval Services U.S., Inc., 8501 East Pleasant Valley Road, Cleveland, OH 44131
AHAM	Association of Home Appliance Manufacturers, 20 North Wacker Drive, Suite 1600, Chicago, IL 60606
AIHA	American Industrial Hygiene Association, 2700 Prosperity Avenue, Suite 250, Fairfax, VA 22031
AMCA	Air Movement and Control Association International, Inc., 30 West University Drive, Arlington Heights, IL 60004-1893
ANSI	American National Standards Institute, 11 West 42nd Street, 13th floor, New York, NY 10036-8002
ARI	Air-Conditioning and Refrigeration Institute, 4301 North Fairfax Drive, Suite 425, Arlington, VA 22203
ASA	Acoustical Society of America, Standards Secretariat, 120 Wall Street, 32nd floor, New York, NY 10005-3993
	For ordering publications: Standards and Publications Fulfillment Center, P.O. Box 1020, Sewickley, PA 15143-9998
ASAE	American Society of Agricultural Engineers, 2950 Niles Road, St. Joseph, MI 49085-9659
ASHRAE	American Society of Heating, Refrigerating and Air-Conditioning Engineers, Inc., 1791 Tullie Circle, NE, Atlanta, GA 30329
ASME Intl.	The American Society of Mechanical Engineers, 3 Park Avenue, New York, NY 10016-5990
	For ordering publications: ASME Information Central, P.O. Box 2900, Fairfield, NJ 07007-2900
ASTM	American Society for Testing and Materials, 100 Barr Harbor Drive, West Conshohocken, PA 19428-2959
AWS	American Welding Society, Inc., 550 N.W. LeJeune Road, Miami, FL 33126
AWWA	American Water Works Association, 6666 W. Quincy Avenue, Denver, CO 80235
BOCA	Building Officials and Code Administrators International, Inc., 4051 West Flossmoor Road, Country Club Hills, IL 60478-5795
BSI	British Standards Institution, 389 Chiswick High Road, London W4 4AL, England
CAGI	Compressed Air and Gas Institute, 1300 Sumner Avenue, Cleveland, OH 44115-2851
CGA	Canadian Gas Association, 243 Consumers Road, Suite 1200, North York, ON M2J 5E3
CSA	Canadian Standards Association, 178 Rexdale Boulevard, Etobicoke (Toronto), ON M9W 1R3
CTI	Cooling Tower Institute, P.O. Box 73383, Houston, TX 77273
EJMA	Expansion Joint Manufacturers Association, Inc., 25 North Broadway, Tarrytown, NY 10591-3201
HEI	Heat Exchange Institute, 1300 Sumner Avenue, Cleveland, OH 44115-2851
HI	Hydraulic Institute, 9 Sylvan Way, Parsippany, NJ 07054-3802
HYDI	Hydronics Institute, 35 Russo Place, P.O. Box 218, Berkeley Heights, NJ 07922
IAPMO	International Association of Plumbing and Mechanical Officials, 20001 Walnut Drive South, Walnut, CA 91789-2825
ICBO	International Conference of Building Officials, 5360 Workman Mill Road, Whittier, CA 90601
ICC	International Code Council, 5203 Leesburg Pike, Suite 708, Falls Church, VA 22041
IEEE	Institute of Electrical and Electronics Engineers, 445 Hoes Lane, P.O. Box 1331 Piscataway, NJ 08855-1331
IESNA	Illuminating Engineering Society of North America, 120 Wall Street, 17th floor, New York, NY 10005-4001
IFCI	International Fire Code Institute, 5360 Workman Mill Road, Whittier, CA 90601-2298
IIAR	International Institute of Ammonia Refrigeration, 1200 19th Street, NW, Suite 300, Washington, DC 20036-2412
ISA	Instrument Society of America, 67 Alexander Drive, P.O. Box 12777, Research Triangle Park, NC 27709
ISO	International Organization for Standardization, 1, rue de Varembé, Case postale 56, -1211 Genève 20, Switzerland
	Publications available in the U.S. from ANSI, 11 West 42nd Street, 13th floor, New York, NY 10036-8002
MCAA	Mechanical Contractors Association of America, 1385 Piccard Drive, Rockville, MD 20850-4329
MICA	Midwest Insulation Contractors Association, 2017 South 139th Circle, Omaha, NE 68144
	Manufacturers Standardization Society of the Valve and Fittings Industry, Inc., 127 Park Street, N.E., Vienna, VA 22180-4602
NAIMA	North American Insulation Manufacturers Association, 44 Canal Center Plaza, Suite 310, Alexandria, VA 22314
PHCC	Plumbing-Heating-Cooling Contractors National Association, P.O. Box 6808, 180 South Washington, Falls Church, VA 22046-1148
NCPWB	National Certified Pipe Welding Bureau, 1385 Piccard Drive, Rockville, MD 20850
NCSBCS	National Conference of States on Building Codes and Standards, 505 Huntmar Park Drive, Suite 210, Herndon, VA 22070
NEBB	National Environmental Balancing Bureau, 8575 Grovemont Circle, Gaithersburg, MD 20877-4121
NEMA	National Electrical Manufacturers Association, 1300 North 17th Street, Suite 1847, Rosslyn, VA 22209
NFPA	National Fire Protection Association, 1 Batterymarch Park, P.O. Box 9101, Quincy, MA 02269-9101
NRCC	National Research Council of Canada, Client Services, M-20, 1200 Montreal Road, Ottawa, ON K1A 0R6, Canada
NSF	NSF International, P.O. Box 130140, Ann Arbor, MI 48113-0140
SAE Intl.	Society of Automotive Engineers International, 400 Commonwealth Drive, Warrendale, PA 15096-0001
SBCCI	Southern Building Code Congress International, Inc., 900 Montclair Road, Birmingham, AL 35213-1206
SMACNA	Sheet Metal and Air Conditioning Contractors' National Association, 4201 Lafayette Center Drive, Chantilly, VA 22021-1209
TEMA	Tubular Exchanger Manufacturers Association, Inc., 25 North Broadway, Tarrytown, NY 10591-3201
UL	Underwriters Laboratories Inc., 333 Pfingsten Road, Northbrook, IL 60062-2096

ADDITIONS AND CORRECTIONS

This section includes additional information and notes technical errors found between June 15, 1996 and April 1, 1999 in the inch-pound (I-P) edition of the *ASHRAE Handbooks*. No additions and corrections were published in the 1997 *ASHRAE Handbook—Fundamentals*. Changes marked with a * were published in the Additions and Corrections section of the 1998 *ASHRAE Handbook—Refrigeration*. Occasional typographical errors and nonstandard symbol labels will be corrected in future volumes.

The authors and editor encourage you to notify them if you find other technical errors. Please send corrections to: Handbook Editor, ASHRAE, 1791 Tullie Circle NE, Atlanta, GA 30329, or e-mail bparsons@ashrae.org.

1996 HVAC SYSTEMS AND EQUIPMENT

*p. 6.14. Figure 20 shows earth banked up to the top of the foundation wall. Most building codes require some clearance between the earth and any wood framing.

*p. 12.4, Equation (13). A brace in the equation is misplaced. The equation should read

$$V_t = 2V_s\left[\left(\frac{v_2}{v_1} - 1\right) - 3\alpha\Delta t\right] \tag{13}$$

*p. 16.3, Table 4A, 2nd column. The nominal thickness for 22 gage galvanized sheet is 0.0336 in., not 0.0396 in.

p. 18.2, Table 1. The last impeller design feature for the tubeaxial fan should read

- Hub is usually less than half the fan tip diameter

*p. 20.3, Table 1. The limiting relative humidity for double glazing at 0°F should be 30%, not 40%.

*p. 28.2, 1st column. The operating efficiencies of the listed furnace categories should be switched. Category I furnaces operate at a steady-state efficiency less than 83%, Category II at greater than 83%, Category III at less than 83%, and Category IV at greater than 83%.

*p. 30.8, Equation (14). The right side of the equation should be multiplied by ρ_m, which is the mean gas density calculated from Equation (4).

1997 FUNDAMENTALS

*p. 1.8, Equation (39). The equation for h should read $h_3 = h_4$ instead of $h_1 = h_2$.

*p. 6.7. Change saturated solid/liquid enthalpy at –43°F to read –178.29 instead of –178.79.

*p. 6.8. Change saturated solid/liquid enthalpy at –6°F to read –161.75 instead of –162.75.

*p. 6.8. In the 9°F row, delete the negative sign from the values for saturated solid enthalpy, evaporation enthalpy, saturated vapor enthalpy, and saturated vapor entropy.

*p. 6.9. Change comma in saturated vapor entropy value at 56°F to a decimal so that it reads 2.1064.

*p. 6.9. Change saturated vapor entropy at 58°F to read 2.1002 instead of 2.0002.

*p. 6.9. Change saturated vapor specific volume at 77°F to read 699.80 instead of 794.40.

*p. 6.10. Change saturated vapor entropy at 256°F to read 1.6691 instead of 1.6916.

p. 6.14, Equation (38). The second constant in the equation should read 26.142, not 26.412.

p. 6.14. Equation (40) shoud read

$$\text{Enthalpy} \qquad h = h_a + \mu h_{as} \tag{40}$$

p. 8.7, Table 5. Oxygen consumed for Light work is < 1 ft³/h.

*p. 10.8, 2nd column. In the section on Heat Production, 7th line, add a minus sign to 40°F.

*p. 15.11, 1st column, 4th line up. The AIHA standard number should read Z9.5.

p. 18.8, Table 8, last line. Change the suction temperature to read 40°F instead of 47°F.

p. 21.4, Figure 6. Delete Column 4 in the table with the heading, Total Volume of Capillaries.

p 24.12, 1st column. In Line 2 of the section on Windows and Doors delete "in Chapter 29."

p 24.14, Figure 7. Dimensions on curves should read 3-1/2, 4, and 6 in., respectively.

*p. 25.12, Equation (35). Insert a multiplier of 60 in the denominator to convert minutes to hours. Also, the specific heat (c_p) of air should read 0.24 Btu/lb·°F.

*p. 26.3, 1st column. The extreme value distribution is a **double** exponential distribution, so the equation for F at the bottom of the column should read

$$F = -\frac{\sqrt{6}}{\pi}\left[0.5772 + \ln\ln\left(\frac{n}{n-1}\right)\right]$$

*pp. 26.6–52, Tables 1A, 2A, and 3A. In the MWS/MWD to DB columns, the columns headed PWD should read MWD.

*pp. 26.6–52, Tables 1A, 2A, and 3A. The humidity ratio (HR) units are grains of water vapor per pound of dry air (7000 gr = 1 lb).

*p. 26.9. DP/MDB and HR data for Hartford, Brainard Field, Connecticut, should be corrected to read as follows:

DP/MDB AND HR								
0.4%			1%			2%		
DP	HR	MDB	DP	HR	MDB	DP	HR	MDB
73	121	81	71	116	79	70	110	78

*p. 26.26. The latitude and longitude for the Salta Airport weather station in Argentina are S and W, respectively.

*p. 26.52. The latitude and longitude for the Treinta y Tres weather station in Uruguay are S and W, respectively.

*p. 27.4, Figure 1. To reflect the revised weather data in Chapter 26, the caption on the x-axis of the graph should read, 1% DESIGN HUMIDITY RATIO (*W*).

*p. 27.12, Table 16. The values in the first and third degree-day column headings have been switched. The first value should read 2950, and the value in the last column should read 7433.

p. 28.19. The R-value for B-4 is 10.0, not 1.19.

*pp. 28.26–27, Tables 18 and 19. The heading in the first column of these tables should read Wall Group.

p. 28.36. Lines 4 and 8 list values for the North (not East) exposure. Lines 5 and 9 list values for the South exposure and lines 6 and 10 for the East exposure.

*p. 28.38, 1st column, 2nd equation. Add "Btu/h" to the right side of the equation.

*p. 28.47. The headings for the tables should read:

**Table 33B   Wall Types, Mass Evenly Distributed,
for Use with Table 32**

p. 28.48. Table 33 C. The table is incorrect. The correct table is on page 26.48 in the 1993 *ASHRAE Handbook—Fundamentals* or it may be obtained from the Handbook Editor.

p. 28.54, 1st column, line 2 up. Equation should read

$$C_2 = \frac{(94 + 74)}{2} - 85 = \text{daily average temperature correction}$$

*p. 28.57, 1st column. In Equation (49), the letter l should be a λ.

*p. 28.57, 2nd column. Under "Appliances," change energy input to power input in the definition of q_{input}.

p. 29.13, Equation (6). Insert a multiplier of 100 to the right side of the equation.

*p. 29.28, 2nd column. The first sentence in the third paragraph up should refer to Chapter 26, not Chapter 24.

p. 29.37, Equation (49). Change α_j (alpha sub j) to read a_j.

p. 29.37, Table 23. Correct the spelling of **Absorptance** in caption.

*p. 30.5, Figure 1. The dimensions for the office floor plan are as follows:

Outside length	= 147.6 ft (45 m)
Outside width	= 98 ft (30 m)
Interior space length	= 124 ft (37.8 m)
Exterior space width	= 11.8 ft (3.6 m)
Example zone width	= 11.8 ft (3.6 m)

*p. 30.12, 1st column, equation nomenclature. Change definition of c to read, c = fluid capacity, lb/h.

p. 30.19, Equation (42). Change θ to φ in two exponents.

p. 30.19, Table 2. Change σ to φ in the heading of the fifth column.

*p. 32.9, Figure 9. The chart should include a shaded area to show the suggested range of air velocity and friction loss for design. Below about 17,000 cfm, the suggested limits for friction loss are 0.08 and 0.6 in. water gage per 100 ft of duct. Above 17,000 cfm the limiting factor is air velocity, which ranges from 1800 and 4000 fpm. The Addition and Corrections section in 1998 *ASHRAE Handbook—Refrigeration* includes the figure with the shaded area.

*p. 35.2, last line of Table 2. Conversion of Fahrenheit to °R should read x + 459.67.

*p. 39.1. The page numbers in the table of contents should read 39.#, not 38.#.

1998 REFRIGERATION

p. 9.2, 1st line after Equation (4). Change t to θ.

p. 9.3, Table 2. The variable on the left side of the fifth equation (below $0.1 \leq Bi \leq 100$) should read v, not w.

p. 9.5, Equation (15). ω is in radians.

p. 16.11, 1st column. Change the third sentence in the second paragraph in the section on Chilling and Freezing Variety Meats to read:

Quick Chilling. A better and more widely used method consists of quick ...

p. 18.4, 1st column. Change heading to read "Irradiation of Freah Seafood."

p. 20.2, 1st column, 2nd para. Change second sentence to read, "Most of the shell's organic matter is protein."

p. 23.2, 2nd column. Second heading should read, "Packaging, Loading, and Handling."

p. 23.3, Table 3. Beans, green should be in the third column. (Store at 50°F and 80 to 85% rh). Green beans suffer chilling injury if stored at 32to 41°F.

p. 32.1, 1st para. Change the paragraph to read:

This chapter is a guide to specifying insulation systems for refrigeration piping, fittings, and vessels. It does not deal with HVAC systems or applications such as chilled water systems. Refer to Chapters 22, 23, 24, and 39 in the 1997 *ASHRAE Handbook—Fundamentals* for information about insulation and vapor barriers for these systems.

p. 32.3, Table 2. Change footnote a to read:

a Tested in accordance with ASTM E 84 for 1 in. thick insulation.

p. 34.2, 1st column, 7th line up. Change sentence to read,

The amount of control over each load source is indicated as an approximate percentage of the maximum reduction possible through effective design and operation.

p. 45.27, 2nd column. In line 2 of the second paragraph in the section on Short Tube Restrictors, the length-to-diameter ratio should read $3 < L/D < 20$.

p. 47.12, Figure 17. Relabel valve on upper left of heat reclaim coil to read, 3-WAY SOLENOID VALVE. Relabel valve on upper right of heat reclaim coil to read, CHECK VALVE.

COMPOSITE INDEX

ASHRAE HANDBOOK SERIES

This index covers the current Handbook series published by ASHRAE. The four volumes in the series are identified as follows:

S = 1996 Systems and Equipment

F = 1997 Fundamentals

R = 1998 Refrigeration

A = 1999 Applications

Alphabetization of the index is letter by letter; for example, **Heaters** precedes **Heat exchangers**, and **Floors** precedes **Floor slabs**.

The page reference for an index entry includes the book letter and the chapter number, which may be followed by a decimal point and the beginning page within the chapter. For example, the page number A31.4 means the information may be found in the 1999 Applications volume, Chapter 31, beginning on page 4.

Each Handbook is revised and updated on a four-year cycle. Because technology and the interests of ASHRAE members change, some topics are not included in the current Handbook series but may be found in the earlier Handbook editions cited in the index.

Duct design *(cont.)*
friction losses *(cont.)*
rectangular ducts, F32.8
round equivalent dimensions (table), F32.10
roughness factors, F32.7
table, F32.7
head, F32.2
static head, F32.2
velocity head, F32.2
industrial exhaust systems, F32.26
example, F32.27
insulation, F32.16
leakage, F32.17
duct seal levels (table), F32.18
leakage class, F32.17
table, F32.18
methods, F32.20
equal friction, F32.20
static regain, F32.20
T-method optimization, F32.20
T-method simulation, F32.21
noise, F32.19
pressure, F32.2
static pressure, F32.2
total pressure, F32.2
velocity pressure, F32.2
smoke management, F32.16
testing, adjusting, and balancing, F32.20
Ductility
metals and alloys (figure), R39.6
Ducts
acoustical lining, A46.12
fiberglass, A46.17
in hospitals, A7.11
air diffusers, F31.12
airflow measurement in, A36.2
central systems, S1.6
classification, S16.1
commercial, S16.1
industrial, S16.1
residential, S16.1
cleaning, S16.2
construction, S16
codes, S16.1
commercial, S16.2
acoustical treatment, S16.4
fibrous glass ducts, S16.3
flat-oval ducts, S16.3
flexible ducts, S16.4
hangers, S16.4
materials, S16.2
plenums/apparatus casings, S16.4
rectangular ducts, S16.2
round ducts, S16.2
grease-/moisture-laden vapors, S16.5
industrial, S16.4
construction details, S16.5
hangers, S16.5
materials, S16.4
rectangular ducts, S16.5
round ducts, S16.4
master specifications, S16.6
outside, S16.6
residential, S16.2
seismic qualification, S16.6
sheet metal welding, S16.6
standards, S16.1
commercial, S16.2
industrial, S16.4
residential, S16.2
thermal insulation, S16.6
underground, S16.5
dehumidification, S22.7

equivalent diameter, F2.10
equivalent lengths, S9.5
flat oval ducts, F32.8
round equivalent dimensions (table), F32.12
fluid flow, F2
friction chart, F32.8
heaters for
control, A45.17
industrial exhaust systems, A29.16
insulation, F23.18; F32.16
leakage, S16.2; F32.17
duct seal levels (table), F32.18
leakage class, F32.17
table, F32.18
noise in, A46.8
noncircular ducts, F32.8
plastic, S16.5
rectangular ducts, F32.8
round equivalent dimensions (table), F32.10
road tunnels, A12.6
roughness factors, F32.7
table, F32.7
ships, A10.5
merchant (table), A10.5
naval (table), A10.7
silencers, A46.14
small forced-air systems, S9.1, 5
sound attenuation, A46.11
sound control, F7.11
sound transmission, A46.17
velocity measurement in, F14.17
vibration control, A46.44
Dump boxes
air distribution, S17.7
Dynamic test chambers, R37
Dynamometers, A14.1, 3, 4
Earth
stabilization, R35.3, 4
Earthquakes
seismic restraint design, A53.1
U.S. maximum ground motion (figure), A53.4
U.S. seismic zones (figure), A53.7
world seismic zones (table), A53.6
Economic analysis techniques, A35.6
cash flow, A35.11
computer analysis, A35.11
present value (worth), A35.7
equal payment series, A35.7
improved payback, A35.8
single payment, A35.7
simple payback, A35.6
uniform annualized costs, A35.9
Economics *(see also* **Costs**)
cogeneration, S7.42
district heating and cooling, S11.1
energy management program, A34.3, 18
energy recovery equipment, S42.1
evaporative cooling, A50.10, 12
indoor gaseous contaminant control, A44.17
insulation thickness, F22.22
laboratory systems, A13.17
owning and operating costs, A35
pipe insulation thickness, S11.19
steam turbines, S7.18
thermal storage, A33.2
Economizers
airside, S5.7
compressors
single-screw, S34.13, 14
twin-screw, S34.20
control, A40.22
cycle control, A45.5

kitchen ventilation, A30.12
nonresidential ventilation, F25.24
waterside, S5.7
Eddy diffusivity, F5.7
Educational facilities
air conditioning, A6
design considerations, A6.2
design temperatures (table), A6.2
sound design levels (table), A6.2
energy conservation, A6.1
equipment selection, A6.4
system selection, A6.4
hot water demand/use (tables), 4, 11, 16
service water heating, A48.18
EER. *See* **Energy efficiency ratio (EER)**
Effectiveness
heat transfer, F3.4
Efficiency
air conditioners, room, S43.2
best efficiency point (BEP)
pumps, centrifugal, S38.6
boilers, S27.5
combustion, F17.12
compressors
centrifugal, S34.30
orbital (scroll), S34.24
positive-displacement, S34.2
reciprocating, S34.4
rotary, S34.9
single-screw, S34.15
twin-screw, S34.21
dehumidifiers, room, S43.6
fins, F3.18
furnaces, S28.5, 7, 9
heating equipment, F30.18
industrial gas cleaning, S25.3, 10
infrared heaters, S15.4
motors, S39.2
pumps, centrifugal, S38.6
unitary systems, S44.5
Eggs, R20
egg products, R20.8
dehydrated, R20.11
frozen, R20.11
processing, R20.9
refrigerated, R20.10
processing plant sanitation, R20.12
shell eggs, R20.1
processing, R20.5
refrigeration, R20.5
spoilage prevention, R20.4
storage, R20.7
transportation, R20.8
storage requirements (table), R10.3
thermal properties, R8
Electric
boilers, S27.4
coils
air-heating, S23.2
furnaces, residential, S28.4, 6, 8
heaters (in-space), S29.4
baseboard, S29.4
control, S29.4
panel, S29.4
portable, S29.4
radiant, S29.4
wall/floor/toe space/ceiling, S29.4
infrared heaters, S15.2
metal sheath, S15.3
quartz tube, S15.3
reflector lamp, S15.3
tubular quartz lamp, S15.3
makeup air units, S31.9
panel systems, S6.15
